CHEMISTRY

THE CENTRAL SCIENCE

BROWN · LEMAY · BURSTEN · MURPHY
WOODWARD · LANGFORD · SAGATYS · GEORGE

3RD EDITION

Copyright © Pearson Australia (a division of Pearson Australia Group Pty Ltd) 2014

Pearson Australia
Unit 4, Level 3
14 Aquatic Drive
Frenchs Forest NSW 2086

www.pearson.com.au

Authorised adaptation from the United States edition entitled *Chemistry: The Central Science,* 12th edition, ISBN 01321696727 by Brown, Theodore L.; LeMay, H. Eugene Jr; Bursten, Bruce E.; Murphy, Catherine J.; Woodward, Patrick M., published by Pearson Education, Inc., copyright © 2012.

Third adaptation edition published by Pearson Australia Group Pty Ltd, Copyright © 2014.

Senior Acquisitions Editor: Mandy Sheppard
Manager—Product Development: Michael Stone
Senior Project Editor: Katie Millar
Development Editors: Catherine du Peloux Menage, David Chelton
Media Content Developer, MasteringChemistry: Adam Catarius
Editorial Coordinator: Camille Layt
Production Coordinator: Julie McArthur
Copy Editor: Jennifer Coombs
Proofreader: Judi Walters
Copyright and Pictures Editor: Lisa Woodland
Indexer: Mary Coe
Cover design by Natalie Bowra
Typeset by Midland Typesetters, Australia

Printed in China (CTPS/01)

1 2 3 4 5 18 17 16 15 14

National Library of Australia
Cataloguing-in-Publication Data

Title:	Chemistry: the central science/Theodore L. Brown ... [et al.].
Edition:	3rd ed.
ISBN:	9781442554603 (pbk)
Notes:	Includes index.
Subjects:	Chemistry—Textbooks.
Other Authors/Contributors:	Brown, Theodore L.
Dewey Number:	540

Every effort has been made to trace and acknowledge copyright. However, should any infringement have occurred, the publishers tender their apologies and invite copyright owners to contact them.

ALWAYS LEARNING

PEARSON

CHEMISTRY

3RD EDITION

To our families for their love, support and understanding and to our students, whose enthusiasm and curiosity have inspired us to undertake this project.

brief contents

detailed contents

chapter 1

Introduction: Matter and measurement 2

chapter 2

Atoms, molecules and ions 28

chapter 5
Nuclear chemistry: Changes within the core of an atom   142

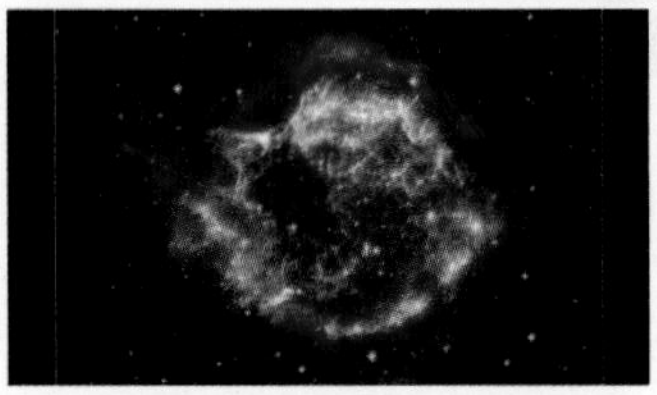

chapter 6
Electronic structure of atoms   178

chapter 7

Periodic properties of the elements 220

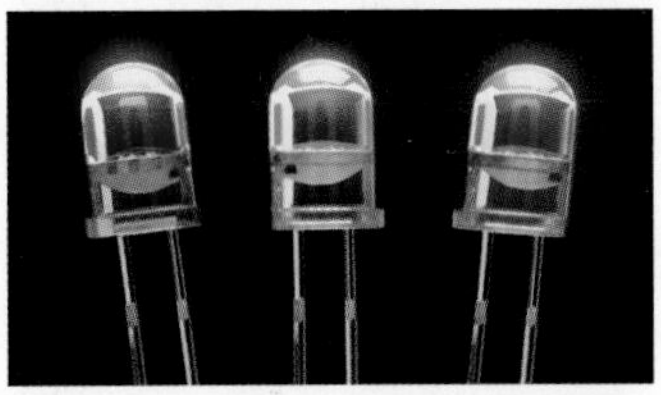

chapter 8

Basic concepts of chemical bonding 250

chapter 9
Molecular geometry and bonding theories   292

chapter 10
Intermolecular forces: Gases   336

chapter 17
Acid–base equilibria 654

chapter 20
Chemistry of the non-metals 802

preface

Philosophy

This is the third Australian edition of a text that has enjoyed significant global success over a number of decades. Our original aim in adapting *Chemistry: The Central Science* for a wider market was to ensure that the text remained a central, indispensable learning tool for the student of chemistry. In this book we aim to provide a comprehensive coverage of all aspects of chemistry that may be used at introductory university level. It will provide students with the depth of knowledge they require in their first-year undergraduate curriculum and to arm chemistry academics across Australia with a broad and balanced view of chemistry with which to set their curricula. Throughout the text we have maintained a conversational style of explaining information rather than simply stating facts. We have found that this style has been highly appreciated by the student.

Organisation and Contents

In this edition the first four chapters give a largely macroscopic overview of chemistry. The basic concepts presented—such as atomic structure, the nature of chemical reactions, stoichiometry, measurement and quantification and the main types of reaction in aqueous solution—provide a necessary background for many of the laboratory experiments usually performed in first year general chemistry.

The following five chapters (Chapters 5–9) focus on the atomic scale, starting with the transformations that occur within the nucleus of an atom before moving on to deal with the electronic structure of the atom, the consequent effects on the properties of the elements and the basic theories of chemical bonding and molecular geometry. Chapters 10 and 11 consider the macroscopic properties of the three states of matter—gas, liquid and solid—and the forces which influence their behaviour. A more detailed look at solubility and solutions and our interaction with the atmosphere and oceans are examined in the next two chapters (Chapters 12 and 13). All chemical reactions involve energy changes and Chapter 14 comprehensively covers the thermodynamic processes operating in all chemical reactions.

The next several chapters examine the factors that determine the speed and extent of chemical reactions: kinetics (Chapter 15) and equilibria (Chapters 16–18). These are followed by electrochemistry, which discusses the use of chemical reactions to produce electrical energy and *vice versa* (Chapter 19). Chapters 20 and 21 introduce the chemistry of non-metals, metals and coordination compounds. Throughout Chapters 1–21 there are extensive areas of modern chemistry that are dealt with broadly. For example, we introduce students to descriptive inorganic chemistry by integrating examples throughout the text. You will find pertinent and relevant examples of organic and inorganic chemistry woven into all chapters as a means of illustrating principles and applications and the relationship between all areas of chemistry. Some chapters, of course, more directly address the properties of elements and their compounds, especially Chapters 7, 20 and 21.

Organic chemistry is central to all living things and Chapters 22–30 lead us on a journey from elementary hydrocarbons to elaborate bioorganic molecules. Much of what we discuss is treated from a fundamental level so your transition to tertiary studies in organic chemistry is smooth and rapid. We place emphasis on the core reactions observed in organic chemistry and treated many cases mechanistically. This fosters a deep understanding of why organic molecules react in the way they do thereby giving you an opportunity to understand much more chemistry than is discussed. Chapter 22 provides a foundation to our examination of organic chemistry by using hydrocarbons to illustrate how we represent and name organic molecules. It goes on to provide an overview of the functional groups—the reactive parts of the molecule—on which we build our understanding of organic chemistry. The shape of a molecule may be pivotal in determining its reactivity, particularly in a biological context, and Chapter 23 leads to an in-depth discussion of stereochemistry.

The next six chapters cover the fundamental reactions encountered in organic chemistry, at each step building to the application of these reaction in a modern world (for example, polymerisation in Chapters 24 and 27) and their essential role in the chemistry of life (for example, carbohydrates in Chapter 26, fats in Chapter 27, proteins and nucleic acids in Chapter 29). Chapter 28 investigates aromatic compounds as a separate class. Here it is important for the student to note the differences in reactivity to the alkenes studied in Chapter 24.

Finally, Chapter 30 stands alone as a reference guide to mass spectrometry, NMR spectroscopy and IR spectroscopy. Whether these topics are taught with much emphasis on the technology is up to the instructor. What we believe is most important is students' development at complex problem-solving, bringing two or more concepts together to draw a logical conclusion. The approach to solving molecular structure also confirms their knowledge of the basic principles of organic chemistry, bonding, functional groups and drawing structural formulae. Our coverage of organic chemistry gives students a unique perspective and challenges the very 'standard format' often seen in a first-year text.

Our topic sequence provides a logical progression through chemistry, but we recognise that not everyone teaches all the topics in exactly the order we have chosen. We have therefore made sure that instructors can make common changes in teaching sequence with no loss in student comprehension. In particular, many instructors prefer to introduce gases (Chapter 10) after stoichiometry or after thermochemistry rather than with states of matter. The chapter on gases has been written to permit this change with no disruption in the flow of material. It is also possible to treat the balancing of redox equations (Sections 19.1 and 19.2) earlier, after the introduction of redox reactions in Section 4.4. Finally, some instructors like to cover organic chemistry (Chapters 22 to 30) earlier than its position in this text. Throughout the text we have introduced linkages (indicated by the symbol ∞) to sections in other parts of the book. This allows the reader to quickly find relevant material and highlights the integrated nature of chemistry. A glossary of terms provides succinct definitions for quick reference and a comprehensive index ensures the extensive information contained in this book is easily accessible.

▲ **Steven Langford**   School of Chemistry
Monash University, Clayton VIC 3800
steven.langford@sci.monash.edu.au

▲ **Dalius Sagatys**   Science and Engineering Faculty
Queensland University of Technology, Brisbane QLD 4001
d.sagatys@qut.edu.au

▲ **Adrian George**   School of Chemistry
University of Sydney, NSW 2006
adrian.george@sydney.edu.au

guided tour for students

MAKING CONNECTIONS

The 3rd edition of *Chemistry: The Central Science* includes several key features to help you see the bigger picture: to move beyond memorisation and have a deeper understanding of the relationships between concepts in chemistry.

Making connections across different topics

New enhanced blue links ∞ are featured in the margins and include voice balloons which direct you to other relevant sections that will enrich your understanding of the current topic.

∞ Review this on page xx

Making connections between chemistry and the real world

My World of Chemistry

Chemistry occurs all around us, throughout every day. Recognising the importance of chemistry in your daily life can improve your understanding of chemical concepts. *My World of Chemistry* showcases chemistry's connection to world events, scientific discoveries, and medical breakthroughs throughout the text.

MY WORLD OF CHEMISTRY

GLUCOSE MONITORING

Over 1 million Australians (although estimates vary) have diabetes, and globally the number approaches 172 million. Diabetes is a metabolic disorder in which the body either cannot produce or cannot properly use the hormone insulin. One signal that a person is diabetic is that the concentration of glucose in the blood is higher than normal. Therefore, people who are diabetic need to measure their blood glucose concentrations regularly. Untreated dia-

H^+ Cl^- Mg $Mg(s)$ is oxidised (loses electrons) $H^+(aq)$ is reduced (gains electrons) Mg^{2+}

Reactants
$2\ HCl(aq) + Mg(s)$
Oxidation +1 −1   0
number

Products
$H_2(g) + MgCl_2(aq)$
   0   +2 −1

Making connections visually

Micro to Macro Art

These illustrations offer three parts: a macroscopic image (what you can see with your eyes); a molecular image (what the molecules are doing); and a symbolic representation (how chemists represent the process with symbols and equations).

A new intermediate step has been added, showing where chemistry occurs in the problem-solving process.

NEW Figure It Out questions and Voice Balloons

Figure It Out questions encourage you to stop and analyse the artwork in the text, for conceptual understanding. 'Voice Balloons' in selected figures help you break down and understand the components of the image.

▲ FIGURE IT OUT

Which species is reduced in this reaction? Which specie

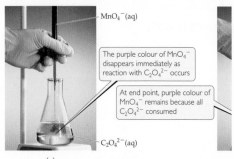

$MnO_4^-(aq)$

The purple colour of MnO_4^- disappears immediately as reaction with $C_2O_4^{2-}$ occurs

At end point, purple colour of MnO_4^- remains because all $C_2O_4^{2-}$ consumed

$C_2O_4^{2-}(aq)$

(a)

SAMPLE EXERCISE 3.9 Calculating mo

What is the molar mass of glucose, $C_6H_{12}O_6$?

SOLUTION

Analyse We are given a molecular formula wh
their number in the molecule.

Plan The molar mass of any substance is nur
which will have units of u, whereas the molar
proceed as in Sample Exercise 3.5.

Solve Our first step is to determine the formula

$$6\text{ C atoms} = 6(12.0\text{ u}$$
$$12\text{ H atoms} = 12(1.0\text{ u}$$
$$6\text{ O atoms} = 6(16.0\text{ u}$$

Because glucose has a formula mass of 180.0 u,
of 180.0 g. In other words, $C_6H_{12}O_6$ has a molar

Check The magnitude of our answer seems rea
ate unit for the molar mass.

Making connections to problem-solving and critical thinking skills

Analyse/Plan/Solve/Check

This four-step problem-solving method helps you understand what you are being asked to solve, to plan how you will solve each problem, to work your way through the solution, and to check your answers. This method is introduced in Chapter 3 and reinforced throughout the book.

Dual-Column Problem-Solving Strategies

Found in Selected Sample Exercises, these strategies explain the thought process involved in each step of a mathematical calculation using a unique layout for clarity. They help you develop a conceptual understanding of those calculations.

Solve The number of moles of Na_2SO_4 is obtained by dividing its mass by its molar mass:	$\text{Moles Na}_2\text{SO}_4 = \dfrac{23.4\text{ g}}{142\text{ g mol}^{-1}} = 0.165\text{ mol}$
Converting the volume of the solution to litres:	$\text{Litres solution} = \left(\dfrac{125}{1000}\right)\text{dm}^3 = 0.125\text{ dm}^3$
Thus the molarity is	$\text{Molarity} = \dfrac{0.165\text{ mol}}{0.125\text{ dm}^3} = 1.32\,\dfrac{\text{mol}}{\text{dm}^3} = 1.32\text{ M}$

Strategies in Chemistry

Strategies in Chemistry teach ways to analyse information and organise thoughts, helping to improve your problem-solving and critical-thinking abilities.

STRATEGIES IN CHEMISTRY

PROBLEM SOLVING

Practice is the key to success in solving problems. As you practise, you can improve your skills by following these steps.

Step 1: Analyse the problem. Read the problem carefully. What is it asking you to do? What information does it provide you with? List both the data you are given and the quantity you need to obtain (the unknown).

Step 2: Develop a plan for solving the problem. Consider a possible path between the given information and the unknown. This is usually a formula, an equation or some principle you

learnt earlier. Recognise that some data may not be given explicitly in the problem; you may be expected to know certain quantities (such as Avogadro's number) or look them up in tables (such as atomic masses). Recognise also that your plan may involve either a single step or a series of steps with intermediate answers.

Step 3: Solve the problem. Use the known information and suitable equations or relationships to solve for the unknown. Be careful with significant figures, signs and units.

Step 4: Check the solution. Read the problem again to make sure you have found all the solutions asked for in the problem. Does your answer make sense? That is, is the answer outrageously large or small or is it in the ballpark? Finally, are the units and significant figures correct?

MasteringChemistry

for

Chemistry: The Central Science, 3rd Edition

A Guided Tour for Students and Educators

Reading Quizzes: The Item Library in MasteringChemistry includes Reading Quizzes that educators can assign to ensure students have completed their readings and are prepared for class discussion and activities.

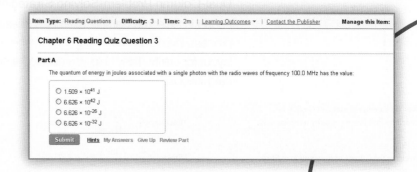

Gradebook: MasteringChemistry is the only system to capture the step-by-step work of each student in class, including wrong answers submitted, hints requested, and time taken on every step. This data powers an unprecedented gradebook.

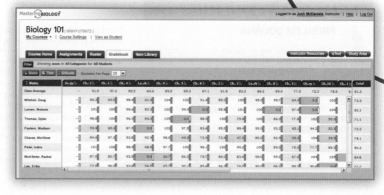

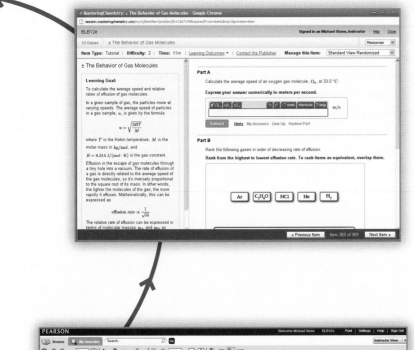

Personalised coaching and feedback:

MasteringChemistry is the only system to provide instantaneous feedback specific to the most common wrong answers. Students can submit an answer and receive immediate error-specific feedback. Simpler sub-problems— hints—are provided upon request.

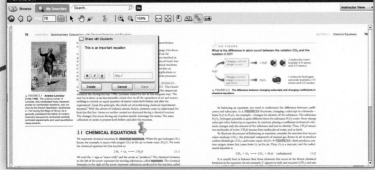

Online and tablet eText: The eText gives

students and educators access to the text whenever and wherever they can access the internet. The eText pages look exactly like the printed text and include powerful interactive and customisation features.

Students and educators can:

- create notes (educators can share these with a whole class)
- highlight text in different colours
- create book marks
- click hyperlinked words and phrases to view definitions
- view in single-page or two-page format
- perform a full-text search and save or export notes.

guided tour for educators

Learning and Teaching Tools

Instructors Solution Manual

Organised by chapter, this manual offers detailed lecture outlines and complete descriptions of all available lecture demonstrations, the interactive media assets, common student misconceptions, and more. It also offers solutions to all end-of-chapter exercises in the textbook.

Computerised TestBank

The test bank allows educators to customise the bank of questions to meet specific needs and add/revise questions as needed. It consists of more than 2000 true–false, multiple choice, short-answer, essay and matching questions complete with solutions. Using Pearson's TestGen software, lecturers can create professional-looking exams in just minutes by building tests from the existing database of questions, editing questions, or adding your own. TestGen also supports the creation of printed, network or online testsk.

Digital Image Library

The digital image library provides all images and artwork from the book.

acknowledgements

TECHNICAL EDITORS

Dr Simon Bedford, University of Wollongong

Chris Fellows, University of New England

EDITORIAL REVIEW BOARD

Dr Simon Bedford, University of Wollongong

Penny Commons, University of Melbourne

Professor Bice Martincigh, University of KwaZulu-Natal

Professor Joe Shapter, Flinders University

Dr David Wilson, La Trobe University

PROPOSAL REVIEWERS

Associate Professor Chris L. Brown, Griffith University

Penny Commons, University of Melbourne

Associate Professor Greg Dicinoski, University of Tasmania

Dr Greg Doran, Charles Sturt University

Dr Damian Laird, Murdoch University

Dr Gwendolyn Laurie, The University of Queensland

Professor Joe Shapter, Flinders University

Associate Professor Kieran F. Lim, Deakin University

Dr Evan Robertson, La Trobe University

Dr Andrew J. Seen, University of Tasmania

Ms Rosemary Ward, University of Technology, Sydney

Dr Magdalena Wajrak, Edith Cowan University

Dr Danny K. Y. Wong, Macquarie University

We would also like to express our gratitude to our many team members at Pearson Australia whose hard work, imagination, and commitment have contributed so greatly to the final form of this edition: Mandy Sheppard, our Chemistry Editor, for many fresh ideas and her unflagging enthusiasm, continuous encouragement, and support; Catherine du Peloux Menage, our Development Editor, who very effectively coordinated the scheduling and tracked the multidimensional deadlines that come with a project of this magnitude; Michael Stone, Manager—Product Development, whose diligence and careful attention to detail were invaluable to this revision, especially in keeping us on task in terms of consistency and student understanding; Katie Millar, our Senior Project Editor, who managed the complex responsibilities of bringing the design, photos, artwork, and writing together with efficiency and good cheer; and Lisa Woodland, our Copyright & Pictures Editor, who researched and secured rights for stunning photographs to bring the concepts to life.

Finally, to Theodore Brown, Eugene LeMay, Bruce Bursten, Catherine Murphy and Patrick Woodward we thank you sincerely for allowing us to use your textbook as the foundation to a broad perspective.

STEVEN J. LANGFORD received his BSc (Hons I) and PhD from The University of Sydney. After postdoctoral work in the UK under the auspices of a Ramsay Memorial Fellowship, and at the University of UNSW as an ARC Postdoctoral Fellow, he joined the School of Chemistry at Monash University in 1998. He was appointed Professor of Organic Chemistry in 2006 and is currently Deputy Dean and Associate Dean (Research) of the Faculty of Science. He teaches all aspects of organic and supramolecular chemistry in Monash's undergraduate program and is known for his entertaining and enthusiastic teaching style. In 2005 Professor Langford was awarded the inaugural Faculty of Science Dean's Excellence in Science Teaching Award and in 2006 was one of only a handful of scientists to receive a Carrick Citation For Outstanding Contributions to Student Learning in Australian university teaching. He was also awarded the Centenary of Federation teaching award from the Royal Australian Chemical Institute—its premier teaching award—in that same year. His research interests focus on concept transfers from nature, particularly in the areas of photosynthesis and genetic encoding. He has published over 100 research articles and was awarded the 2006 Young Investigator Award by the Society of Porphyrins and Phthalocyanines.

DALIUS S. SAGATYS received his BSc(Hons) degree in Chemistry from The University of Queensland (Brisbane) and his PhD from the Illinois Institute of Technology (Chicago) in 1970. After three years as Joliot Curie Fellow of the Commissariat á L'Energie Atomique, Université de Paris VII (Paris), he worked at the International Patents Institute in Rijswijk, Holland, and from there returned to Brisbane where he joined the then Queensland Institute of Technology in 1982. From the beginning he became interested in the design and implementation of chemistry courses for very different student requirements, such as those in the fields of nursing, engineering and the built environment, as well as developing a chemistry bridging course for students with no chemistry background at all. His research interests have been centred on the synthesis and structure determination of complexes of the Group 15 elements, specifically arsenic, antimony and bismuth. He is currently a Visiting Academic at Queensland University of Technology.

ADRIAN V. GEORGE received his BSc(Hons) and PhD degrees from The University of Reading in England and joined the staff there as a lecturer in 1984. After a short spell as a guest scientist at The University of California, Berkeley he moved to The University of Sydney in 1988. His research has ranged from organic synthesis at extremely high pressures and the development of new organometallic materials to the use of isotope ratio mass spectrometry in the detection of doping in competitive sports and chemistry education. He has conducted research in Japan and taught University level chemistry in Sweden. He has always had a passion for teaching and obtained a graduate  certificate of education in 2000. He has been awarded a University of Sydney Excellence in Teaching award (1999), Vice Chancellor's award for Support of the Student Experience twice (2007, 2011), the inaugural Royal Australian Chemical Institute Centenary of Federation Teaching Award (2001), Australian College of Education Teaching Award (2001) and was part of a team that received the Carrick Institute Award for Programs that Enhance Learning (2007). He has been Director of First Year Studies in the School of Chemistry and the Associate Dean (Teaching and Learning) in the Faculty of Science at The University of Sydney. He currently divides his time between academic pursuits at The University of Sydney and rain forest regeneration in northern New South Wales.

THEODORE L. BROWN received his PhD from Michigan State University in 1956. Since then, he has been a member of the faculty of the University of Illinois, Urbana-Champaign, where he is now Professor of Chemistry, Emeritus. He served as Vice Chancellor for Research, and Dean of The Graduate College, from 1980 to 1986, and as Founding Director of the Arnold and Mabel Beckman Institute for Advanced Science and Technology from 1987 to 1993. Professor Brown has been an Alfred P Sloan Foundation Research Fellow and has been awarded a Guggenheim Fellowship. In 1972 he was awarded the American Chemical Society Award for Research in Inorganic Chemistry and received the American Chemical Society Award for Distinguished Service in the Advancement of Inorganic Chemistry in 1993. He has been elected a Fellow of the American Association for the Advancement of Science, the American Academy of Arts and Sciences, and the American Chemical Society.

H. EUGENE LEMAY, JR. received his BS degree in Chemistry from Pacific Lutheran University (Washington) and his PhD in Chemistry in 1966 from the University of Illinois, Urbana-Champaign.He then joined the faculty of the University of Nevada, Reno, where he is currently Professor of Chemistry, Emeritus. He has enjoyed Visiting Professorships at The University of North Carolina at Chapel Hill, at The University College of Wales in Great Britain, and at The University of California, Los Angeles. Professor LeMay is a popular and effective teacher, who has taught thousands of students during more than 40 years of university teaching. Known for the clarity of his lectures and his sense of humour, he has received several teaching awards, including the University Distinguished Teacher of the Year Award (1991) and the first Regents' Teaching Award given by the State of Nevada Board of Regents (1997).

BRUCE E. BURSTEN received his PhD in Chemistry from the University of Wisconsin in 1978. After two years as a National Science Foundation Postdoctoral Fellow at Texas A&M University, he joined the faculty of The Ohio State University, where he rose to the rank of Distinguished University Professor. In 2005, he moved to The University of Tennessee, Knoxville, as Distinguished Professor of Chemistry and Dean of the College of Arts and Sciences. Professor Bursten has been a Camille and Henry Dreyfus Foundation Teacher-Scholar and an Alfred P Sloan Foundation Research Fellow, and he is a Fellow of both the American Association for the Advancement of Science and the American Chemical Society. At Ohio State he has received the University Distinguished Teaching Award in 1982 and 1996, the Arts and Sciences Student Council Outstanding Teaching Award in 1984, and the University Distinguished Scholar Award in 1990. He received the Spiers Memorial Prize and Medal of the Royal Society of Chemistry in 2003, and the Morley Medal of the Cleveland Section of the American Chemical Society in 2005. He was President of the American Chemical Society for 2008. In addition to his teaching and service activities, Professor Bursten's research program focuses on compounds of the transition-metal and actinide elements.

CATHERINE J. MURPHY received two BS degrees, one in Chemistry and one in Biochemistry, from The University of Illinois, Urbana-Champaign, in 1986. She received her PhD in Chemistry from The University of Wisconsin in 1990. She was a National Science Foundation and National Institutes of Health Postdoctoral Fellow at the California Institute of Technology from 1990 to 1993. In 1993, she joined the faculty of The University of South Carolina, Columbia, becoming the Guy F Lipscomb Professor of Chemistry in 2003. In 2009 she moved to The University of Illinois, Urbana-Champaign, as the Peter C and Gretchen Miller Markunas Professor of Chemistry. Professor Murphy has been honoured for both research and teaching as a Camille Dreyfus Teacher-Scholar, an Alfred P Sloan Foundation Research Fellow, a Cottrell Scholar of the Research Corporation, a National Science Foundation CAREER Award winner, and a subsequent NSF Award for Special Creativity. She has also received a USC Mortar Board Excellence in Teaching Award, the USC Golden Key Faculty Award for Creative Integration of Research and Undergraduate Teaching, the USC Michael J Mungo Undergraduate Teaching Award, and the USC Outstanding Undergraduate Research Mentor Award. Since 2006, Professor Murphy has served as a Senior Editor for the Journal of Physical Chemistry. In 2008 she was elected a Fellow of the American Association for the Advancement of Science. Professor Murphy's research program focuses on the synthesis and optical properties of inorganic nanomaterials, and on the local structure and dynamics of the DNA double helix.

PATRICK M. WOODWARD received BS degrees in both Chemistry and Engineering from Idaho State University in 1991. He received a MS degree in Materials Science and a PhD in Chemistry from Oregon State University in 1996. He spent two years as a postdoctoral researcher in the Department of Physics at Brookhaven National Laboratory. In 1998, he joined the faculty of the Chemistry Department at The Ohio State University where he currently holds the rank of Professor. He has enjoyed visiting professorships at the University of Bordeaux in France and the University of Sydney in Australia. Professor Woodward has been an Alfred P Sloan Foundation Research Fellow and a National Science Foundation CAREER Award winner. He currently serves as an Associate Editor to the Journal of Solid State Chemistry and as the director of the Ohio REEL program, an NSF-funded centre that works to bring authentic research experiments into the laboratories of first- and second-year chemistry classes in 15 colleges and universities across the state of Ohio. Professor Woodward's research program focuses on understanding the links between bonding, structure, and properties of solidstate inorganic functional materials

INTRODUCTION:
MATTER AND MEASUREMENT

The giant elliptical galaxy Centaurus A, 10 to 16 million light-years from Earth, as seen by the Hubble Space Telescope, shows the blue glow of young star clusters among clouds of dust and gas. Compression of hydrogen gas clouds causes new star formation, seen as red patches in the image.

KEY CONCEPTS

1.1 THE STUDY OF CHEMISTRY
We begin with a brief description of what chemistry is and why it is useful to learn chemistry.

1.2 CLASSIFICATIONS OF MATTER
We discuss some fundamental ways of classifying matter, distinguishing between *pure substances* and *mixtures* and between *elements* and *compounds*.

1.3 PROPERTIES OF MATTER
We describe the different characteristics or *properties* of matter, used to characterise, identify and separate substances.

1.4 UNITS OF MEASUREMENT
We observe that many properties of matter rely on quantitative measurements involving numbers and units that are based on the *metric system*.

1.5 UNCERTAINTY IN MEASUREMENT
We observe that all measured quantities have an inherent uncertainty that is expressed by the number of *significant figures* used to report the quantity. Significant figures are also used to express the uncertainty associated with calculations involving measured quantities.

The universe is full of mysteries that we will probably never comprehend. And even on Earth some of the simple things that we see and experience can be quite mysterious. How do we obtain electricity from a battery? How does a plant grow? How does a modern LED television screen work? There are innumerable questions, which while seemingly unanswerable, can actually be answered by the study of chemistry.

Chemistry is the study of the properties of matter and the changes that matter undergoes.

This first chapter lays a foundation for our studies by providing an overview of what chemistry is about and what chemists do. The preceding Key Concepts list indicates the chapter organisation and some of the ideas that we will consider.

1.1 | THE STUDY OF CHEMISTRY

The Atomic and Molecular Perspective of Chemistry

Chemistry involves studying the properties and behaviour of matter. **Matter** is the *physical* material of the universe: it is anything that has mass and occupies space. A **property** is any characteristic that allows us to recognise a particular type of matter and to distinguish it from other types. This book, your body, the clothes you are wearing, the water you drink and the air you are breathing are all examples of matter. It has long been known that all matter is composed of infinitesimally small building blocks called **atoms**. Despite the tremendous variety of matter in the universe, there are only about 100 different types of atoms that occur in nature and these, combined in various combinations and proportions, constitute all of the matter of the universe. We will see that the properties of matter relate not only to the kinds of atoms it contains (*composition*), but also to the arrangements of these atoms (*structure*).

Atoms can combine to form **molecules** in which two or more atoms are joined in specific shapes. Throughout this text you will see molecules represented using coloured spheres to show how their component atoms connect to each other (▼ FIGURE 1.1). The colour merely provides a convenient way to distinguish between different kinds of atoms. As examples, compare the molecules of ethanol and ethylene glycol, depicted in Figure 1.1. Notice that these molecules differ somewhat in composition. Ethanol contains one red sphere, which represents one oxygen atom, whereas ethylene glycol contains two.

Even apparently minor differences in the composition or structure of molecules can cause profound differences in their properties. Ethanol, also called grain alcohol, is the alcohol in beverages such as beer and wine. Ethylene glycol, however, is a viscous liquid used as coolant in car radiators.

Every change in the observable world—from boiling water to the changes that occur as our bodies combat invading viruses—has its basis in the world of

---

▲ **FIGURE IT OUT**

How many carbon atoms are in one aspirin molecule?

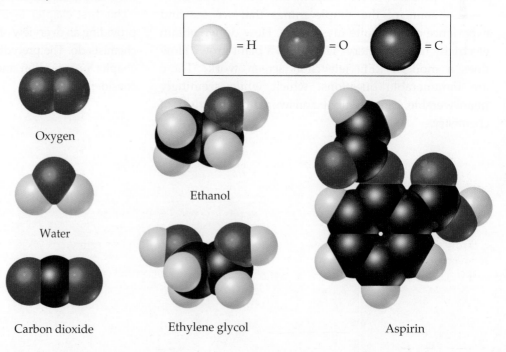

= H   = O   = C

Oxygen

Water

Carbon dioxide

Ethanol

Ethylene glycol

Aspirin

▶ **FIGURE 1.1 Molecular models.** The white, black and red spheres represent atoms of hydrogen, carbon and oxygen, respectively.

atoms and molecules. Thus as we proceed with our study of chemistry, we will find ourselves thinking in two realms: the *macroscopic* realm of ordinary-sized objects (*macro* = large) and the *submicroscopic* realm of atoms and molecules. We make our observations in the macroscopic world—in the laboratory and in our everyday surroundings. In order to understand that world, however, we must visualise how atoms and molecules behave at the submicroscopic level. Chemistry is the science that seeks to understand the properties and behaviour of matter by studying the properties and behaviour of atoms and molecules.

 CONCEPT CHECK 1

   a. In round numbers, about how many elements are there?
   b. What submicroscopic particles are the building blocks of matter?

Why Study Chemistry?

You will note, when studying any scientific discipline, whether it be biology, engineering, medicine, agriculture, geology and so forth, that chemistry is an integral part of your curriculum. This is because chemistry, by its very nature, is the *central science*, central to a fundamental understanding of other sciences and technologies. Chemistry provides an important understanding of our world and how it works. It is an extremely practical science that greatly impacts on our daily living. Indeed, chemistry lies near the heart of many matters of public concern: improvement of health care, conservation of natural resources, protection of the environment and provision of our everyday needs for food, clothing and shelter.

    Using chemistry, we have discovered pharmaceutical chemicals that enhance our health and prolong our lives. We have increased food production through the development of fertilisers and pesticides. We have developed plastics and other materials that are used in almost every facet of our lives. Unfortunately, some chemicals also have the potential to harm our health or the environment. It is in our best interests as educated citizens and consumers to understand the profound effects, both positive and negative, that chemicals have on our lives and to strike an informed balance about their uses.

1.2 | CLASSIFICATIONS OF MATTER

Let's begin our study of chemistry by examining some fundamental ways in which matter is classified and described. Two principal ways of classifying matter are according to its physical state (gas, liquid or solid) and according to its composition (element, compound or mixture) as explained below.

States of Matter

A sample of matter can be a gas, a liquid or a solid. These three forms of matter are called the **states of matter**. The states of matter differ in some of their simple observable properties. A **gas** (also known as *vapour*) has no fixed volume or shape; rather, it conforms to the volume and shape of its container. A gas can be compressed to occupy a smaller volume, or it can expand to occupy a larger one. A **liquid** has a distinct volume independent of its container but has no specific shape: it assumes the shape of the portion of the container that it occupies. A **solid** has both a definite shape and a definite volume. Neither liquids nor solids can be compressed to any appreciable extent.

    The properties of the states can be understood on the molecular level (▼ FIGURE 1.2). In a gas the molecules are far apart and are moving at high speeds, colliding repeatedly with each other and with the walls of the container. In a liquid the molecules are packed more closely together, but still move rapidly, allowing them to slide over each other; thus liquids pour easily. In a solid the molecules are held tightly together, usually in definite arrangements, so

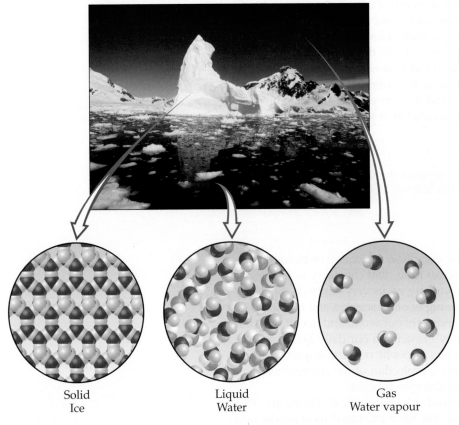

Solid
Ice

Liquid
Water

Gas
Water vapour

▲ **FIGURE 1.2** **The three physical states of water: water vapour, liquid water and ice.** Here we see both the liquid and solid states of water. We cannot see water vapour. What we see when we look at steam or clouds is tiny droplets of liquid water dispersed in the atmosphere. The molecular views show that the molecules in the solid are arranged in a more orderly way than in the liquid. The molecules in the gas are much further apart than those in the liquid or the solid.

the molecules can wiggle only slightly in their otherwise fixed positions. Changes in temperature and/or pressure can lead to a conversion from one state of matter to another, illustrated by such familiar processes as ice melting or water evaporating.

Composition of Matter

When we discuss matter in daily language we often use the word substance as in, 'This is a peculiar substance!' In fact, the word substance is used in everyday language as a substitute for matter which may be one kind of matter or a mixture of more than one kind of matter. In chemistry, however, the word substance means *matter of uniform composition throughout a sample*, as well as having distinct properties. To emphasise this, we usually use the term **pure substance**. However, even when we use the word substance by itself, it is understood to refer to a pure form of matter. For example, oxygen, water, table sugar (sucrose), table salt (sodium chloride) should be referred to as pure substances but more usually are referred to simply as substances.

All substances are either elements or compounds. **Elements** cannot be decomposed into simpler substances; they may be atoms, or molecules composed of only one kind of atom (▼ FIGURE 1.3(a), (b)). **Compounds** are substances composed of two or more different elements, so they contain two or more kinds of atoms (Figure 1.3(c)). Water, for example, is a compound composed of two elements, hydrogen and oxygen. Figure 1.3(d) shows a mixture of substances. **Mixtures** are combinations of two or more substances in which each

🔺 **FIGURE IT OUT**

How do the molecules of a compound differ from the molecules of an element?

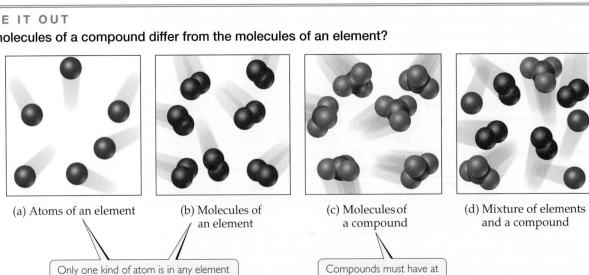

(a) Atoms of an element

(b) Molecules of an element

(c) Molecules of a compound

(d) Mixture of elements and a compound

Only one kind of atom is in any element

Compounds must have at least two kinds of atoms

▲ **FIGURE 1.3**   Molecular comparison of elements, compounds and mixtures.

substance retains its own chemical identity and which can be separated into the individual pure substances by various means.

Some of the more common elements are listed in ▼ TABLE 1.1, along with the chemical abbreviations of their names—**chemical symbols**—used to denote them. All the known elements and their symbols are listed on the inside front cover of this text. The table in which the symbol for each element is enclosed in a box is called the periodic table which is discussed later (∞ Section 2.5, 'The Periodic Table').

The symbol for each element consists of one or two letters, with the first letter capitalised. These symbols are often derived from the English name for the element, but sometimes they are derived from a foreign name (usually Latin) instead (last column in Table 1.1). You will need to know these symbols and to learn others as we encounter them in the text.

The observation that the elemental composition of a pure compound is always the same is known as the **law of constant composition** (or the **law of definite proportions**). It was first put forth by the French chemist Joseph Louis Proust (1754–1826) in about 1800. Although this law has been known for 200 years, the general belief persists among some people that a fundamental difference exists between compounds prepared in the laboratory and the corresponding compounds found in nature. This is not true: a pure compound has the same composition and properties regardless of its source. Both chemists and nature must use the same elements and operate under the same natural laws to form compounds.

∞ Find out more on page 39

(a)

> ### CONCEPT CHECK 2
> Hydrogen, oxygen and water are all composed of molecules. What is it about the molecules of water that makes water a compound?

Most of the matter we encounter consists of mixtures of different substances. Each substance in a mixture retains its own chemical identity and hence its own properties. Whereas pure substances have fixed compositions, the compositions of mixtures can vary. A cup of sweetened coffee, for example, can contain either a little sugar or a lot. The substances making up a mixture (such as sugar and water) are called *components* of the mixture.

Some mixtures do not have the same composition, properties and appearance throughout. Both rocks and wood, for example, vary in texture and appearance throughout any typical sample. Such mixtures are *heterogeneous* (▶ FIGURE 1.4(a)). Mixtures that are uniform throughout are *homogeneous*. Air is a homogeneous mixture of the gaseous substances nitrogen, oxygen and smaller amounts of other substances. The nitrogen in air has all the properties that pure nitrogen does because both the pure substance and the mixture contain the same nitrogen molecules. Salt, sugar and many other substances dissolve in water to form homogeneous mixtures (Figure 1.4(b)). Homogeneous mixtures are also called **solutions**.

(b)

▲ FIGURE 1.4   **Mixtures.** (a) Many common materials, including rocks, are heterogeneous. This granite shows a heterogeneous mixture of silicon dioxide and other metal oxides. (b) Homogeneous mixtures are called solutions. Many substances, including the blue solid shown in this photo (copper sulfate), dissolve in water to form solutions.

TABLE 1.1 • Some common elements and their symbols					
Carbon	C	Aluminium	Al	Copper	Cu (from *cuprum*)
Fluorine	F	Bromine	Br	Iron	Fe (from *ferrum*)
Hydrogen	H	Calcium	Ca	Lead	Pb (from *plumbum*)
Iodine	I	Chlorine	Cl	Mercury	Hg (from *hydrargyrum*)
Nitrogen	N	Helium	He	Potassium	K (from *kalium*)
Oxygen	O	Lithium	Li	Silver	Ag (from *argentum*)
Phosphorus	P	Magnesium	Mg	Sodium	Na (from *natrium*)
Sulfur	S	Silicon	Si	Tin	Sn (from *stannum*)

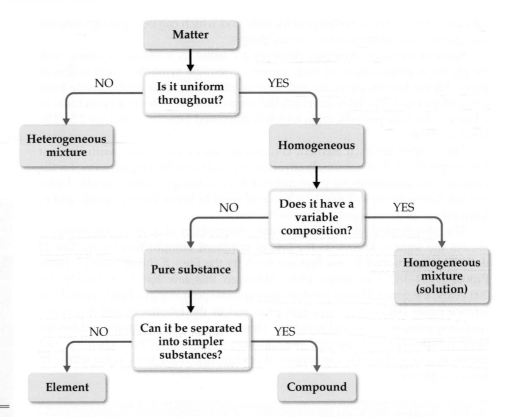

▶ **FIGURE 1.5** **Classification of matter.** All matter is classified ultimately as either an element or a compound.

▲ **FIGURE 1.5** summarises the classification of matter into elements, compounds and mixtures.

SAMPLE EXERCISE 1.1   **Distinguishing between elements, compounds and mixtures**

'White gold', used in jewellery, contains two elements: gold and a 'white' metal such as palladium. Two different samples of white gold differ in the relative amounts of gold and palladium that they contain. Both samples are uniform in composition throughout. Without knowing any more about the materials, use Figure 1.5 to characterise and classify white gold.

SOLUTION

Because the material is uniform throughout, it is homogeneous. Because its composition differs for the two samples, it cannot be a compound. Instead, it must be a homogeneous mixture. Gold and palladium can be said to form a solid solution with one another.

PRACTICE EXERCISE

All aspirin is composed of 60.0% carbon, 4.5% hydrogen and 35.5% oxygen by mass, regardless of its source. Use Figure 1.5 to characterise and classify aspirin.

Answer: It is a compound because it has constant composition and can be separated into several elements.

(See also Exercises 1.6, 1.7, 1.31.)

1.3 | PROPERTIES OF MATTER

Every substance has a unique set of properties. The properties of matter can be categorised as physical or chemical. **Physical properties** can be measured without changing the identity and composition of the substance. These properties include colour, odour, density, melting point, boiling point and hardness.

Chemical properties describe the way a substance may change, or *react*, to form other substances. A common chemical property is flammability, the ability of a substance to burn in the presence of oxygen.

Some properties—such as temperature, melting point and density—do not depend on the amount of the sample being examined. These properties, called **intensive properties**, are particularly useful in chemistry because many can be used to *identify* substances. **Extensive properties** of substances depend on the quantity of the sample, with two examples being mass and volume. Extensive properties relate to the *amount* of substance present.

Physical and Chemical Changes

As with the properties of a substance, the changes that substances undergo can be classified as either physical or chemical. During **physical changes** a substance changes its physical appearance, but not its composition. The evaporation of water is a physical change. When water evaporates, it changes from the liquid state to the gas state, but it is still composed of water molecules, as depicted earlier in Figure 1.2. All **changes of state** (for example, from liquid to gas or from liquid to solid) are physical changes.

In **chemical changes** (also called **chemical reactions**) a substance is transformed into a chemically different substance. When hydrogen burns in air, for example, it undergoes a chemical change because it combines with oxygen to form water. The molecular-level view of this process is depicted in ▼ **FIGURE 1.6**.

CONCEPT CHECK 3

Which of the following is a physical change, and which is a chemical change? Explain.

a. Plants use carbon dioxide and water to make sugar.
b. Water vapour in the air on a cold day forms frost.

Separation of Mixtures

Because each component of a mixture retains its own properties, we can separate a mixture into its components by taking advantage of the differences in their properties. For example, a heterogeneous mixture of iron filings and gold filings could be sorted individually by colour into iron and gold. A less tedious approach would be to use a magnet to attract the iron filings, leaving the gold ones behind. We can also take advantage of an important chemical difference

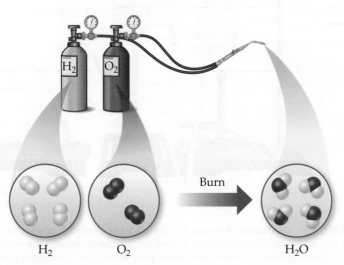

H_2 O_2 Burn H_2O

▲ **FIGURE 1.6**   **A chemical reaction.**

(a)

(b)

▲ **FIGURE 1.7 Separation by filtration.**
(a) A mixture of a solid and a liquid is poured through a porous medium, in this case filter paper. (b) The liquid has passed through the paper but the solid remains on the paper.

between these two metals: many acids dissolve iron but not gold. Thus if we put our mixture into an appropriate acid, the iron would dissolve and the gold would be left behind. The two could then be separated by *filtration*, a procedure illustrated in ◀ **FIGURE 1.7**. We would have to use other chemical reactions, which we will learn about later, to transform the dissolved iron back into metal.

An important method of separating the components of a homogeneous mixture is *distillation*, a process that depends on the different abilities of substances to form gases. For example, if we boil a solution of salt and water, the water evaporates, forming a gas, and the salt is left behind. The gaseous water can be converted back to a liquid on the walls of a condenser, as shown in the apparatus depicted in ▼ **FIGURE 1.8**.

1.4 | UNITS OF MEASUREMENT

Science depends on making accurate measurements of the phenomena being observed. A measurement consists of a number and a scale, which is referred to as a unit. When a number represents a measured quantity, the units of that quantity must be specified. To say that the length of a pencil is 17.5 is meaningless. Expressing the number with its units, 17.5 centimetres (cm), properly specifies the length. The units used for scientific measurements are those of the **metric system**, which was first developed in France in the eighteenth century. Most countries in the world also use this system in their daily life.

SI Units

In 1960 an international agreement was reached specifying a particular choice of metric units for use in scientific measurements. These preferred units are called **SI units**, after the French *Système International d'Unités*. This system has seven *base units* from which all other units are derived. ▶ **TABLE 1.2** lists these base units and their symbols. In this chapter we consider the base units for length, mass and temperature.

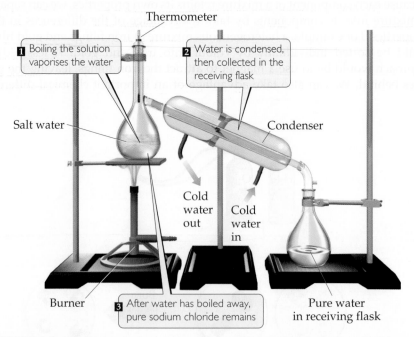

▲ **FIGURE 1.8 Distillation.** A simple apparatus for the separation of a sodium chloride solution (salt water) into its components. Boiling the solution evaporates the water, which is condensed then collected in the receiving flask. After all the water has boiled away, pure sodium chloride remains in the boiling flask.

A CLOSER LOOK

THE SCIENTIFIC METHOD

Although two scientists rarely approach the same problem in exactly the same way, there are guidelines for the practice of science that have come to be known as the scientific method. These guidelines are outlined in ▶ FIGURE 1.9. We begin by collecting information, or *data*, by observation and experiment. The collection of information, however, is not the ultimate goal. The goal is to find a pattern or sense of order in our observations and to understand the origin of this order.

As we perform our experiments, we may begin to see patterns that lead us to a *tentative explanation*, or hypothesis, that guides us in planning further experiments. A key feature of a good hypothesis is that it proposes a mechanism and can be used to make predictions about new experiments. If a hypothesis is sufficiently general and is continually effective in predicting facts yet to be observed, it is called a theory. A theory is *an explanation of the general causes of certain phenomena, with considerable evidence or facts to support it.* For example, Einstein's theory of relativity was a revolutionary new way of thinking about space and time. It was more than just a simple hypothesis, however, because it could be used to make predictions that could be tested experimentally. The results of these experiments were generally in agreement with Einstein's predictions and were not explainable by earlier theories. Despite the landmark achievements of Einstein's theory, scientists can never say the theory is proven. A theory that has excellent predictive power today may not work as well in the future as more data are collected and improved scientific equipment is developed. Thus science is always a work in progress.

Eventually, we may be able to tie together a great number of observations in a single statement or equation called a scientific law. A scientific law is *a concise verbal statement or a mathematical equation that summarises a broad variety of observations and experiences.* We tend to think of the laws of nature as the basic rules under which nature operates. However, it is not so much that matter obeys the laws of nature, but rather that the laws of nature describe the behaviour of matter.

As we proceed through this text, we will rarely have the opportunity to discuss the doubts, conflicts, clashes of personalities and revolutions of perception that have led to our present ideas. We need to be aware that just because we can spell out the results of science so concisely and neatly in textbooks does not mean that scientific progress is smooth, certain and predictable. Some of the ideas we present in this text took centuries to develop and involved large numbers of scientists. We gain our view of the natural world by standing on the shoulders of the scientists who came before us. Take advantage of this view. As you study, exercise your imagination. Don't be afraid to ask daring questions when they occur to you. You may be fascinated by what you discover!

RELATED EXERCISE: 1.32

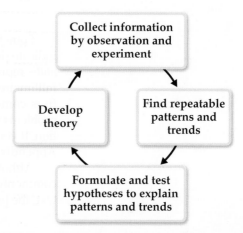

▲ **FIGURE 1.9** **The scientific method.** The scientific method is a general approach to problems that involves making observations, confirming that they are reproducible, seeking patterns in the observations, formulating hypotheses to explain the observations and testing these hypotheses by further experiments. Those hypotheses that withstand such tests and prove themselves useful in explaining and predicting behaviour become known as theories.

TABLE 1.2 ● SI base units		
Physical quantity	**Name of unit**	**Abbreviation**
Mass	Kilogram	kg
Length	Metre	m
Time	Second	s
Temperature	Kelvin	K
Amount of substance	Mole	mol
Electric current	Ampere	A
Luminous intensity	Candela	cd

TABLE 1.3 • Selected prefixes used in the metric system			
Prefix	Abbreviation	Meaning	Example
Giga	G	10^9	1 gigametre (Gm) = 1×10^9 m
Mega	M	10^6	1 megametre (Mm) = 1×10^6 m
Kilo	k	10^3	1 kilometre (km) = 1×10^3 m
Deci	d	10^{-1}	1 decimetre (dm) = 1×10^{-1} m
Centi	c	10^{-2}	1 centimetre (cm) = 1×10^{-2} m
Milli	m	10^{-3}	1 millimetre (mm) = 1×10^{-3} m
Micro	μ^a	10^{-6}	1 micrometre (μm) = 1×10^{-6} m
Nano	n	10^{-9}	1 nanometre (nm) = 1×10^{-9} m
Pico	p	10^{-12}	1 picometre (pm) = 1×10^{-12} m
Femto	f	10^{-15}	1 femtometre (fm) = 1×10^{-15} m
Atto	a	10^{-18}	1 attometre (am) = 1×10^{-18} m
Zepto	z	10^{-21}	1 zeptometre (zm) = 1×10^{-21} m

[a] This is the Greek letter mu (pronounced 'mew').

Sometimes the base units are not convenient and so prefixes are used to indicate decimal fractions or multiples of various units. For example, the prefix *milli*- represents a 10^{-3} fraction of a unit: a milligram (mg) is 10^{-3} gram (g), a millimetre (mm) is 10^{-3} metre (m), and so forth. ▲ TABLE 1.3 presents the prefixes commonly encountered in chemistry. In using SI units and in working problems throughout this text, you must be comfortable using exponential notation. If you are unfamiliar with exponential notation or want to review it, refer to Appendix A.

Although non-SI units are being phased out, there are still some that are commonly used by scientists. Whenever we first encounter a non-SI unit in the text, the proper SI unit will also be given.

CONCEPT CHECK 4

Which of the following quantities is the smallest: 1 mg, 1 μg or 1 pg?

Length and Mass

The SI base unit of *length* is the **metre** (m).

Mass* (*m*) is a measure of the amount of material in an object. The SI base unit of mass is the **kilogram** (kg). This base unit is unusual because it uses a prefix, *kilo*-, instead of the word *gram* alone. We obtain other units for mass by adding prefixes to the word *gram*.

SAMPLE EXERCISE 1.2 Using metric prefixes

What is the name given to the unit that equals (a) 10^{-9} gram, (b) 10^{-6} second, (c) 10^{-3} metre?

SOLUTION

In each case we can refer to Table 1.3, finding the prefix related to each of the decimal fractions: (a) nanogram, ng, (b) microsecond, μs, (c) millimetre, mm.

---

* Mass and weight are not interchangeable terms and are often incorrectly thought to be the same. The weight of an object is the force that its mass exerts due to gravity. In space, where gravitational forces are very weak, astronauts can be weightless, but they cannot be massless. In fact, an astronaut's mass in space is the same as it is on Earth.

PRACTICE EXERCISE

(a) What decimal fraction of a second is a picosecond, ps? **(b)** Express the measurement 6.0×10^3 m using a prefix to replace the power of ten. **(c)** Use exponential notation to express 3.76 mg in grams.

Answers: **(a)** 10^{-12} second, **(b)** 6.0 km, **(c)** 3.76×10^{-3} g

(See also Exercises 1.17, 1.18.)

Temperature

Temperature is a measure of the hotness or coldness of an object. Indeed, temperature is a physical property that determines the direction of heat flow. Heat always flows spontaneously from a substance at higher temperature to one at lower temperature. Thus we feel the influx of heat when we touch a hot object, and we know that the object is at a higher temperature than our hand.

The temperature scales commonly employed in scientific studies are the Celsius and Kelvin scales. The **Celsius scale** is also the everyday scale of temperature in most countries (▼ **FIGURE 1.10**). It was originally based on the assignment of 0 °C to the freezing point of water and 100 °C to its boiling point at sea level (Figure 1.10).

The **Kelvin scale** is the SI temperature scale, and the SI unit of temperature is the kelvin (K). Historically, the Kelvin scale was based on the properties of gases; its origins are considered in Chapter 10. Zero on this scale is the lowest attainable temperature, –273.15 °C, a temperature referred to as *absolute zero*. Both the Celsius and Kelvin scales have equal-sized units—that is, a kelvin is the same size as a degree Celsius. Thus the Kelvin and Celsius scales are related as follows:

$$K = {}^\circ C + 273.15 \qquad\qquad [1.1]$$

The freezing point of water, 0 °C, is 273.15 K (Figure 1.10). Notice that we do not use a degree sign (°) with temperatures on the Kelvin scale.

FIGURE IT OUT

True or false: The 'size' of a degree on the Celsius scale is the same as the 'size' of a degree on the Kelvin scale.

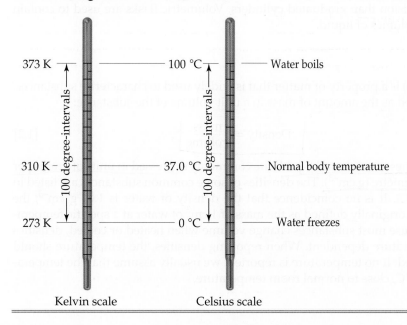

▲ **FIGURE 1.10** **Comparison of the Kelvin and Celsius temperature scales.**

◢ FIGURE IT OUT

How many 1 dm³ bottles are required to contain 1 m³ of liquid?

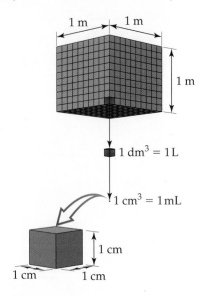

▲ **FIGURE 1.11** **Volume relationships.** The volume occupied by a cube 1 m on each edge is one cubic metre, 1 m³. Each cubic metre contains 1000 dm³. One litre is the same volume as one cubic decimetre, 1 L = 1 dm³. Each cubic decimetre contains 1000 cubic centimetres, 1 dm³ = 1000 cm³. One cubic centimetre equals one millilitre, 1 cm³ = 1 mL.

TABLE 1.4 • Densities of some selected substances at 25 °C	
Substance	**Density ($g\ cm^{-3}$)**
Air	0.001
Balsa wood	0.16
Ethanol	0.79
Water	1.00
Ethylene glycol	1.09
Table sugar	1.59
Table salt	2.16
Iron	7.9
Gold	19.32

SAMPLE EXERCISE 1.3 Converting units of temperature

If a weather forecaster predicts that the temperature for the day will reach 31 °C, what is the predicted temperature in K?

SOLUTION

Referring to Equation 1.1, we have K = 31 + 273 = 304 K.

PRACTICE EXERCISE

Ethylene glycol, the major ingredient in antifreeze, freezes at –11.5 °C. What is the freezing point in K?

Answer: 261.7 K

Derived SI Units

The SI base units in Table 1.2 are used to derive the units of other quantities. To do so, we use the defining equation for the quantity, substituting the appropriate base units. For example, speed is defined as the ratio of distance travelled to elapsed time. Thus the SI unit for speed is the SI unit for distance (length) divided by the SI unit for time, $m\ s^{-1}$, which we read as 'metres per second'. We will encounter many derived units, such as those for force, pressure and energy, later in this text. In this chapter we examine the derived units for volume and density.

Volume

The **volume** (V) of a cube is given by its length cubed (length)³. Thus the SI unit of volume is the SI unit of length raised to the third power. The cubic metre, or m³, is the volume of a cube that is 1 m on each edge. Smaller units are usually used in chemistry. These are the cubic decimetre (dm³), often referred to as the litre (L), and the cubic centimetre (cm³), frequently referred to as the millilitre (mL). We will frequently use the terms litre and millilitre in this text because of common usage, but when the abbreviations are written, they will appear as dm³ and cm³ (◀ **FIGURE 1.11**). The relationship between the cubic decimetre and the cubic centimetre is given by:

$$1\ dm^3 = (1\ dm) \times (1\ dm) \times (1\ dm) = (10\ cm) \times (10\ cm) \times (10\ cm) = 1000\ cm^3$$

The devices used most frequently in chemistry to measure volume are illustrated in ▶ **FIGURE 1.12**. Syringes, burettes and pipettes deliver liquids with more precision than graduated cylinders. Volumetric flasks are used to contain specific volumes of liquid.

Density

Density (ρ) is a property of matter that is widely used to characterise substances. It is defined as the amount of mass in a unit volume of the substance:

$$\text{Density} = \frac{\text{mass}}{\text{volume}} \qquad [1.2]$$

The densities of solids and liquids are commonly expressed in units of grams per cubic centimetre ($g\ cm^{-3}$). The densities of some common substances are listed in ◀ TABLE 1.4. It is no coincidence that the density of water is 1.00 $g\ cm^{-3}$; the gram was originally defined as the mass of 1 cm³ of water at a specific temperature. Because most substances change volume when heated or cooled, densities are temperature dependent. When reporting densities, the temperature should be specified. If no temperature is reported, we usually assume that the temperature is 25 °C, close to normal room temperature.

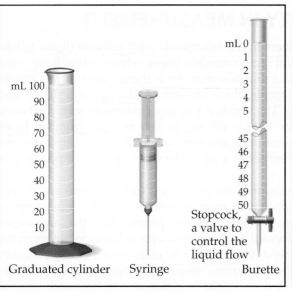

These deliver variable volumes   **Pipette delivers a specific volume**   **Volumetric flask contains a specific volume**

▲ **FIGURE 1.12** **Common volumetric glassware.**

SAMPLE EXERCISE 1.4 **Determining density and using density to determine volume or mass**

(a) Calculate the density of mercury if 1.00×10^2 g occupies a volume of 7.36 cm³.

(b) Calculate the volume of 65.0 g of the liquid methanol (wood alcohol) if its density is 0.791 g cm⁻³.

(c) What is the mass in grams of a cube of gold (density = 19.32 g cm⁻³) if the length of the cube is 2.00 cm?

SOLUTION

(a) We are given mass and volume, so Equation 1.2 yields

$$\text{Density} = \frac{\text{mass}}{\text{volume}} = \frac{1.00 \times 10^2 \text{ g}}{7.36 \text{ cm}^3} = 13.6 \text{ g cm}^{-3}$$

(b) Solving Equation 1.2 for volume and then using the given mass and density gives

$$\text{Volume} = \frac{\text{mass}}{\text{density}} = \frac{65.0 \text{ g}}{0.791 \text{ g cm}^{-3}} = 82.2 \text{ cm}^3$$

(c) We can calculate the mass from the volume of the cube and its density. The volume of a cube is given by its length cubed

$$\text{Volume} = (2.00 \text{ cm})^3 = (2.00)^3 \text{ cm}^3 = 8.00 \text{ cm}^3$$

Solving Equation 1.2 for mass and substituting the volume and density of the cube, we have

$$\text{Mass} = \text{volume} \times \text{density} = (8.00 \text{ cm}^3)(19.32 \text{ g cm}^{-3}) = 155 \text{ g}$$

PRACTICE EXERCISE

(a) Calculate the density of a 374.5 g sample of copper if it has a volume of 41.8 cm³.

(b) A student needs 15.0 g of ethanol for an experiment. If the density of ethanol is 0.789 g cm⁻³, how many millilitres of ethanol are needed?

(c) What is the mass, in grams, of 25.0 cm³ of mercury?

Answers: (a) 8.96 g/cm³, (b) 19.0 cm³, (c) 340 g

(See also Exercises 1.19–1.21, 1.34, 1.35.)

1.5 | UNCERTAINTY IN MEASUREMENT

There are two kinds of numbers in scientific work: *exact numbers* (those whose values are known exactly) and *inexact numbers* (those whose values have some uncertainty). Most of the exact numbers that we will encounter in this course have defined values. For example, there are exactly 12 eggs in a dozen and exactly 1000 g in a kilogram. The number 1 in any conversion factor between units, as in 1 m = 100 cm or 1 kg = 10^6 mg, is also an exact number. Exact numbers can also result from counting numbers of objects. For example, we can count the exact number of marbles in a jar or the exact number of people in a classroom.

Numbers obtained by measurement are always *inexact*. There are always inherent limitations in the equipment used to measure quantities (equipment errors), and there are differences in how different people make the same measurement (human errors). Suppose that 10 students with 10 balances are given the same coin and told to determine its mass. The 10 measurements will probably vary slightly from one another for various reasons. The balances might be calibrated slightly differently, and there might be differences in how each student reads the mass from the balance. Remember: *uncertainties always exist in measured quantities*. Counting very large numbers of objects usually has some associated error as well. Consider, for example, how difficult it is to obtain accurate census information for a city or vote counts for an election.

CONCEPT CHECK 5

Which of the following is an inexact quantity?
a. The number of people in your chemistry class.
b. The mass of a coin.
c. The number of grams in a kilogram.

MY WORLD OF CHEMISTRY

CHEMISTRY IN THE NEWS

Chemistry is a very lively, active field of science. Because it is so central to our lives, there are reports on matters of chemical significance in the news nearly every day. Some tell of recent breakthroughs in the development of new pharmaceuticals, materials and processes. Others deal with environmental and public safety issues. As you study chemistry, we hope you will develop the skills to understand the impact of chemistry on your life more effectively. You need these skills to take part in public discussions and debates about matters related to chemistry that affect your community, the nation and the world. By way of examples, here are summaries of a few recent stories in which chemistry plays a role.

Biofuels are coming

The term biofuel describes fuels that are derived from biomass; the two most common biofuels are bioethanol and biodiesel. Currently only a very small proportion of the world's transport fuels is derived from biomass, being estimated at approximately 3%. However, since biofuels are produced from crops such as corn, soybean and sugar cane, it is rapidly being realised that using food crops to produce transport fuel can actually cause food shortages as farmers sell their crops for a better price to the fuel industry. In Australia, the government

has legislated that ethanol produced mainly from sugar cane but also from many other sugar-containing plants can be added to petrol with a maximum of 10%. This is the common E10 petrol now sold at most service stations (▼ FIGURE 1.13). Biodiesel is made from vegetable/animal fats and recycled oils

◀ FIGURE 1.13
E10. A petrol pump dispensing E10 unleaded petrol.

and greases. It is usually used as a diesel additive but is increasingly being used by itself (B100).

Research into biofuel production is being conducted on an increasing scale often using complicated biomatter such as algae and agricultural wastes in an attempt to minimise the use of food crops for biofuel production.

Super batteries

Electric cars are becoming more common but are still held back by the lack of suitable energy sources. A promising recent development is the lithium iron phosphate ($LiFePO_4$, LFP) battery, invented in 1996, which is used in the General Motors 'Volt' hybrid car. It offers a high current rate, a long cycle life of up to 2000 charge cycles and good energy density of 95–140 Wh kg^{-1}. Recharge time is only 2.5–3 hours but work is in progress to reduce this to 1 hour. The strong advantage of this battery over its lithium cobalt oxide or lithium magnesium oxide predecessors is that the materials are cheap and non-toxic. The batteries are very stable and virtually incombustible during charging and discharging since the P–O bonds are very strong so that oxygen is not readily released as can happen with the lithium cobalt oxide batteries. The *Killacycle* electric motorcycle, specifically built for drag racing, uses a 374 volt pack of $LiFePO_4$ batteries weighing 79.4 kg. It achieved a top speed of 274 km h^{-1} in November 2007.

New solar cell design

A new flexible and lightweight solar cell that achieves a high conversion rate of solar energy to electrical energy has been developed in the United States. The cells use micrometre-sized rods of silicon instead of the conventional silicon wafers. Light entering the cells bounces back and forth many times between the rods guided by aluminium nano-particle reflectors until it is absorbed. The claim is that 85% of usable sunlight is absorbed compared with 17% for current commercial panels. In addition the cells are much better at absorbing light in the near-infrared spectrum. Because of their flexibility, they are even suitable for inserting solar-powered devices in clothes. In addition to these inorganic photovoltaic cells, organic photovoltaic cells, which use highly conjugated organic molecules, are being rapidly developed. They are also known as polymer cells depending on the size of the organic molecules that are used to convert solar energy to electrical energy. Although their possible uses are very interesting they do have very low efficiencies, that is, less than 3% compared with the inorganic devices.

New lighting

The city of Sydney has embarked on a project involving the installation of LED (light emitting diode) street lighting across the Sydney CBD (▶ FIGURE 1.14). Recently developed LED light bulbs emit as much light as incandescent or fluorescent light bulbs but use a small fraction of the electrical energy. Instead of emitting light from a vacuum as in an incandescent light bulb or a gas as in a compact fluorescent light bulb (CFL), an LED emits light from a solid, that is, a semiconductor, which is made of a positively and a negatively charged component. When an electric charge is applied to the semiconductor it activates the flow of electrons from the negative to the positive layer. These excited electrons emit light of a certain wavelength. The diode material can be varied but is commonly aluminium gallium arsenide (AlGaAs). A new material gallium nitride promises to deliver light bulbs that will be operational for

▲ FIGURE 1.14 **LED lighting in Sydney.**

100 000 hours. Although the initial costs of the bulbs are much higher than those of conventional lighting, the savings involved in replacement costs and power usage seem to be very significant.

Important antibiotic modified to combat bacterial resistance

Vancomycin is an antibiotic of last resort. It is used only when other antibacterial agents are ineffective. Some bacteria have now developed a resistance to vancomyscin, causing researchers to modify the molecular structure of the substance to make it more effective in killing bacteria. This approach was based on the knowledge that vancomycin works by binding to a protein that is essential to forming bacterial cell walls. Researchers have synthesised a vancomycin analogue in which a CO group has been converted to a CH_2 group (▼ FIGURE 1.15). This modification increases the compound's binding affinity in the cell walls of vancomycin-resistant bacteria, making the analogue 100 times more active than vancomycin itself.

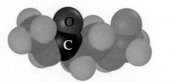 

▲ FIGURE 1.15 **Comparing CO and CH$_2$ groups.** The molecule on the left contains a CO group and the one on the right contains a CH_2 group. This subtle difference is similar to how the much more complex vancomycin molecule was modified.

▲ FIGURE IT OUT

How would the darts be positioned on the target for the case of 'good accuracy, poor precision'?

Good accuracy
Good precision

Poor accuracy
Good precision

Poor accuracy
Poor precision

▲ FIGURE 1.16   **Precision and accuracy.** The distribution of darts in a target illustrates the difference between accuracy and precision.

Precision and Accuracy

The terms 'precision' and 'accuracy' are often used in discussing the uncertainties of measured values. **Precision** is a measure of how closely individual measurements agree with one another. **Accuracy** refers to how closely individual measurements agree with the correct, or 'true', value. The analogy of darts stuck in a dartboard pictured in ◄ FIGURE 1.16 illustrates the difference between these two concepts.

In the laboratory we often perform several different 'trials' of the same experiment. We gain confidence in the accuracy of our measurements if we obtain nearly the same value each time. Figure 1.16 should remind us, however, that precise measurements could be inaccurate. For example, if a very sensitive balance is poorly calibrated, the masses we measure will be consistently either high or low. They will be inaccurate even if they are precise.

Significant Figures

Suppose you determine the mass of a 5 cent piece on a balance capable of measuring to the nearest 0.0001 g. You could report the mass as 2.2405 ± 0.0001 g. The ± notation (read as 'plus or minus') expresses the magnitude of the uncertainty of your measurement. In much scientific work we drop the ± notation with the understanding that there is always some uncertainty in the last digit of the measured quantity. That is, *measured quantities are generally reported in such a way that only the last digit is uncertain*.

▼ FIGURE 1.17 shows a thermometer with its liquid column between the scale marks. We can read the certain digits from the scale and estimate the uncertain one. From the scale marks on the thermometer, we see that the liquid is between the 25 °C and 30 °C marks. We might estimate the temperature to be 27 °C, being somewhat uncertain of the second digit of our measurement.

All digits of a measured quantity, including the uncertain one, are called **significant figures**. A measured mass reported as 2.2 g has two significant figures, whereas one reported as 2.2405 g has five significant figures. The greater the number of significant figures, the greater is the certainty implied for the measurement. When multiple measurements are made of a quantity, the results can be averaged, and the number of significant figures estimated by using statistical methods.

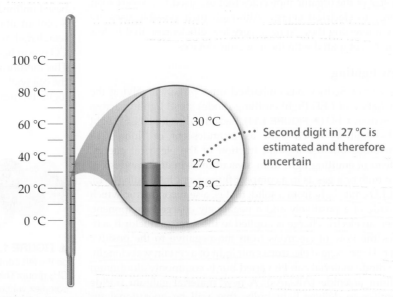

▲ FIGURE 1.17   **Significant figures in measurements.**

SAMPLE EXERCISE 1.5 | Relating significant figures to the uncertainty of a measurement

What difference exists between the measured values 4.0 g and 4.00 g?

SOLUTION

Many people would say there is no difference, but a scientist would note the difference in the number of significant figures in the two measurements. The value 4.0 has two significant figures, but 4.00 has three. This difference implies that the first measurement has more uncertainty. A mass of 4.0 g indicates that the uncertainty is in the first decimal place of the measurement. Thus the mass might be anything between 3.9 and 4.1 g, which we can represent as 4.0 ± 0.1 g. A measurement of 4.00 g implies that the uncertainty is in the second decimal place. Thus the mass might be anything between 3.99 and 4.01 g, which we can represent as 4.00 ± 0.01 g. Without further information, we cannot be sure whether the difference in uncertainties of the two measurements reflects the precision or accuracy of the measurement.

PRACTICE EXERCISE

A balance has a precision of ± 0.001 g. A sample that has a mass of about 25 g is placed on this balance. How many significant figures should be reported for this measurement?

Answer: Five, as in the measurement 24.995 g

To determine the number of significant figures in a reported measurement, read the number from left to right, counting the digits starting with the first digit that is not zero. *In any measurement that is properly reported, all non-zero digits are significant.* Zeros, however, can be used either as part of the measured value or merely to locate the decimal point. Thus zeros may or may not be significant, depending on how they appear in the number. The following guidelines describe the different situations involving zeros.

1. Zeros *between* non-zero digits are always significant—1005 kg (four significant figures); 1.03 cm (three significant figures).
2. Zeros *at the beginning* of a number are never significant; they merely indicate the position of the decimal point—0.02 g (one significant figure); 0.0026 cm (two significant figures).
3. Zeros *at the end* of a number are significant if the number contains a decimal point—0.0200 g (three significant figures); 3.0 cm (two significant figures).

A problem arises when a number ends with zeros but contains no decimal point. In such cases, it is normally assumed that the zeros are not significant. Exponential notation (Appendix A) can be used to indicate clearly whether zeros at the end of a number are significant. For example, a mass of 10 300 g can be written in exponential notation showing three, four or five significant figures depending on how the measurement is obtained:

1.03×10^4 g   (three significant figures)
1.030×10^4 g   (four significant figures)
1.0300×10^4 g   (five significant figures)

In these numbers all the zeros to the right of the decimal point are significant (rules 1 and 3). (The exponential term does not add to the number of significant figures.)

SAMPLE EXERCISE 1.6 | Determining the number of significant figures in a measurement

How many significant figures are in each of the following numbers (assume that each number is a measured quantity): **(a)** 4.003, **(b)** 6.023×10^{23}, **(c)** 5000?

SOLUTION

(a) Four; the zeros are significant figures. **(b)** Four; the exponential term does not add to the number of significant figures. **(c)** One; we assume that the zeros are not significant when there is no decimal point shown. If the number has more significant figures, it should be written in exponential notation. Thus 5000×10^4 has four significant figures, whereas 5.00×10^3 has three.

PRACTICE EXERCISE

How many significant figures are in each of the following measurements: (a) 3.549 g, **(b)** 2.3×10^4 cm, **(c)** 0.00134 m³?

Answers: **(a)** Four, **(b)** two, **(c)** three

(See also Exercises 1.24, 1.25.)

Significant Figures in Calculations

When carrying measured quantities through calculations, *the least certain measurement limits the certainty of the calculated quantity and thereby determines the number of significant figures in the final answer*. The final answer should be reported with only one uncertain digit. To keep track of significant figures in calculations, we will make frequent use of two rules, one for multiplication and division and another for addition and subtraction.

1. *For multiplication and division*, the result contains the same number of significant figures as the measurement with the fewest significant figures. When the result contains more than the correct number of significant figures, it must be rounded off. For example, the area of a rectangle whose measured edge lengths are 6.221 cm and 5.2 cm should be reported as 32 cm² even though a calculator shows the product of 6.221 and 5.2 to have more digits:

$$\text{Area} = (6.221 \text{ cm})(5.2 \text{ cm}) = 32.3492 \text{ cm}^2 \Rightarrow \text{round off to } 32 \text{ cm}^2$$

We round off to two significant figures because the least precise number—5.2 cm—has only two significant figures.

2. *For addition and subtraction*, the result has the same number of decimal places as the measurement with the fewest decimal places. Consider the following example in which the uncertain digits appear in colour.

This number limits	20.42	← two decimal places
the number of significant	1.322	← three decimal places
figures in the result →	83.1	← one decimal place
	104.842	← round off to 104.8 (one decimal place)

We report the result as 104.8 because 83.1 has only one decimal place.

Notice that for multiplication and division, significant figures are counted. For addition and subtraction, decimal places are counted. In determining the final answer for a calculated quantity, exact numbers can be treated as if they have an infinite number of significant figures. This rule applies to many definitions between units. Thus when we say, 'There are 100 centimetres in 1 metre', the number 100 is exact, and we need not worry about the number of significant figures in it.

- In *rounding off* numbers, numbers ending in 0–4 are rounded down and those ending in 5–9 are rounded up. For example, rounding 2.780 to three significant figures would give 2.78 whereas 2.785 would result in 2.79. Similarly, rounding off 2.780 to two significant figures gives 2.8 whereas 4.645 would give 4.6.

SAMPLE EXERCISE 1.7 | **Determining the number of significant figures in a calculated quantity**

The width, length and height of a small box are 15.5 cm, 27.3 cm and 5.4 cm, respectively. Calculate the volume of the box, using the correct number of significant figures in your answer.

SOLUTION

The volume of a box is determined by the product of its width, length and height. In reporting the product, we can show only as many significant figures as given in the dimension with the fewest significant figures, that for the height (two significant figures):

$$\text{Volume} = \text{width} \times \text{length} \times \text{height}$$
$$= (15.5 \text{ cm})(27.3 \text{ cm})(5.4 \text{ cm}) = 2285.01 \text{ cm}^3 \Rightarrow 2.3 \times 10^3 \text{ cm}^3$$

When we use a calculator to do this calculation, the display shows 2285.01, which we must round off to two significant figures. Because the resulting number is 2300, it is best reported in exponential notation, 2.3×10^3, to indicate clearly two significant figures.

PRACTICE EXERCISE

It takes 10.5 s for a sprinter to run 100.00 m. Calculate the average speed of the sprinter in metres per second, and express the result to the correct number of significant figures.

Answer: 9.52 m s^{-1} (three significant figures)

(See also Exercise 1.28.)

STRATEGIES IN CHEMISTRY

THE IMPORTANCE OF PRACTICE and ESTIMATING ANSWERS

If you listen to lectures and think to yourself, 'Yes, I understand that', be assured, that if you have done nothing else to reinforce what you heard, when questioned about the material the following day, you would not be able to explain it to someone. This is because listening to lectures or reading a textbook gives you *passive* knowledge. To become proficient in chemistry you must apply what you have learned into *active* knowledge by doing the sample exercises, the practice exercises and the problems at the end of the chapter. The more you practise doing problems, the more active your knowledge and the better your understanding of the principles of chemistry involved becomes. Remember, however, if you are stuck on a problem ask for help from your instructor, a tutor or a fellow student. Spending an inordinate amount of time on a single exercise that you cannot solve is rarely effective and leads to frustration.

When you are performing calculations to solve a problem, remember that a calculator is an important instrument that provides an answer quickly. However, the accuracy of the answer depends on the accuracy of the input. Very simply, if you were told to multiply 3.17×4.282, you should obtain the answer 13.6 (three significant figures). If you happened to obtain 135.74, you should immediately realise that the answer is wrong by a magnitude of 10 due to an incorrect input of data, because simply by rounding off the above numbers to 3 and 4 you would realise that the answer should be of the order of magnitude of 12. So by making a rough calculation using numbers that are rounded off in such a way that the arithmetic can be done without a calculator you will be able to check whether the answers to your calculations are reasonable. This requires some practice, but learning chemistry is all about practice as mentioned above.

When a calculation involves two or more steps and you write down answers for intermediate steps, retain at least one additional digit—past the number of significant figures—for the intermediate answers. This procedure ensures that small errors from rounding at each step do not combine to affect the final result. When using a calculator, you may enter the numbers one after another, rounding only the final answer. Accumulated rounding-off errors may account for small differences between results you obtain and answers given in the text for numerical problems.

SAMPLE EXERCISE 1.8 | **Determining the number of significant figures in a calculated quantity**

A gas at 25 °C fills a container whose volume is 1.05×10^3 cm^3. The container plus gas have a mass of 837.6 g. The container, when emptied of all gas, has a mass of 836.2 g. What is the density of the gas at 25 °C?

SOLUTION

To calculate the density, we must know both the mass and the volume of the gas. The mass of the gas is just the difference in the masses of the full and empty container:

$$(837.6 \text{ g} - 836.2) \text{ g} = 1.4 \text{ g}$$

In subtracting numbers, we determine the number of significant figures in our result by counting decimal places in each quantity. In this case each quantity has one decimal place. Thus the mass of the gas, 1.4 g, has one decimal place.

Using the volume given in the question, 1.05×10^3 cm^3, and the definition of density, we have

$$\text{Density} = \frac{\text{mass}}{\text{volume}} = \frac{1.4 \text{ g}}{1.05 \times 10^3 \text{ cm}^3}$$

$$= 1.3 \times 10^{-3} \text{ g cm}^{-3} = 0.0013 \text{ g cm}^{-3}$$

In dividing numbers, we determine the number of significant figures in our result by counting the number of significant figures in each quantity. There are two significant figures in our answer, corresponding to the smaller number of significant figures in the two numbers that form the ratio.

PRACTICE EXERCISE

To how many significant figures should the mass of the container be measured (with and without the gas) in Sample Exercise 1.8 in order for the density to be calculated to three significant figures?

Answer: Five; in order for the difference in the two masses to have three significant figures, there must be two decimal places in the masses of the filled and empty containers.

STRATEGIES IN CHEMISTRY

THE FEATURES OF THIS BOOK

To help you understand chemistry, this book includes features that help you organise your thoughts. At the beginning of each chapter, Key Concepts, which outline the chapter by section, will prepare you for the material in the chapter. At the end of each chapter, the Summary of Key Concepts, Key Skills, and Key Equations will help you remember what you have learned and prepare you for quizzes and exams.

The Concept Check features are placed in the text to test your understanding of what you have just read and the Figure It Out features are associated with artwork and ask you to interpret a concept visually. Sample Exercises, with worked-out solutions and answers, and Practice Exercises, which provide only the answer, test your problem-solving skills in chemistry.

At the end of each chapter is a series of exercises to allow you to practise your problem-solving skills further. The first few exercises, called Visualising Concepts, are meant to test how well you understand a concept without plugging a lot of numbers into a formula. The other exercises are divided into sections that reflect the order of the material in the chapter, with answers provided online. Additional Exercises appear after the regular exercises; the chapter sections that they cover are identified. Integrative Exercises, which start appearing in Chapter 3, are problems that require skills learned in previous chapters.

Throughout the book boxed essays highlight the importance of chemistry to our everyday lives. The My World of Chemistry boxes focus on biological and environmental and industrial aspects of chemistry. Strategies in Chemistry boxes, like this one, are meant to help you think about the material you are learning. Finally, boxes entitled A Closer Look provide in-depth coverage of a key chemical concept.

Many chemical databases are available, usually through your university or school. The *CRC Handbook of Chemistry and Physics* is the standard reference for many types of data and is available in libraries. The *Merck Index* is a standard reference for the properties of many small organic compounds, especially ones of biological interest. WebElements (www.webelements.com) is a good website for looking up the properties of the elements.

CHAPTER SUMMARY AND KEY TERMS

SECTION 1.1 Chemistry is the study of the composition, structure, properties and changes of **matter**. The composition of matter relates to the kind of elements it contains. The structure of matter relates to the ways the **atoms** of these elements are arranged. A **property** is any characteristic that gives a sample of matter its unique identity. A **molecule** is an entity composed of two or more atoms with the atoms attached to one another in a specific way.

SECTION 1.2 Matter exists in three physical states, **gas**, **liquid** and **solid**, which are known as the **states of matter**. There are two kinds of **pure substances: elements** and **compounds**. Each element has a single kind of atom and is represented by a **chemical symbol** consisting of one or two letters, with the first letter capitalised. Compounds are composed of two or more kinds of atoms joined in some specific fashion. The **law of constant composition**, also called the **law of definite proportion**, states that the elemental composition of a pure compound is always the same. Most matter consists of mixtures of pure substances. **Mixtures** can have variable composition and can be either homogeneous or heterogeneous. A homogeneous mixture is called a **solution**.

SECTION 1.3 Each form of matter (substance) has a unique set of **physical** and **chemical properties** that can be used to distinguish it from other forms of matter. **Physical changes**, such as **changes of state**, do not alter the composition of matter and are reversible, whereas **chemical changes** (**chemical reactions**) transform matter into chemically different substances. **Intensive properties** are independent of the amount of matter being examined and are used to identify substances. **Extensive properties** relate to the amount of substance present.

SECTION 1.4 Measurements in chemistry are made using the **metric system** using an internationally accepted system of units called **SI units**. SI units are based on the **metre** (m) and the **kilogram** (kg) as the basic units of length and mass. Although the SI temperature scale is the **Kelvin** (K) **scale**, the **Celsius** (°C) **scale** is most commonly used in chemistry. The SI unit for **volume** (V) is the cubic metre (m^3), but the cubic centimetre (cm^3) is commonly used in chemistry. **Density** (ρ) is an important quantity and is reported and equals mass divided by volume.

SECTION 1.5 All measured quantities are inexact to some extent. The **precision** of a measurement indicates how closely different measurements of a quantity agree with each other whereas **accuracy** of a measurement indicates how well a measurement agrees with the accepted or 'true' value. The number of **significant figures** (digits) in a measured quantity include one estimated digit, the last digit of the measurement. The significant figures indicate the degree of uncertainty in the measurement and must be reported according to a set of rules.

KEY SKILLS

- Distinguish between elements, compounds and mixtures. (Section 1.2)

- Memorise symbols of common elements and common prefixes for units. (Section 1.2)

- Memorise common SI units and metric prefixes. (Section 1.4)

- Use significant figures, scientific notation and SI units. (Section 1.5)

KEY EQUATIONS

- Interconversion between Celsius (°C) and Kelvin (K) temperature scale

$$K = °C + 273.15$$ [1.1]

- Definition of density

$$\text{Density} = \frac{\text{mass}}{\text{volume}}; \rho = \frac{m}{V}$$ [1.2]

EXERCISES

VISUALISING CONCEPTS

The exercises in this section are intended to probe your understanding of key concepts rather than your ability to utilise formulae and perform calculations.

1.1 Which of the following figures represents **(a)** a pure element, **(b)** a mixture of two elements, **(c)** a pure compound, **(d)** a mixture of an element and a compound? (More than one picture might fit each description.) [Section 1.2]

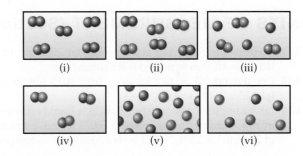
(i) (ii) (iii)
(iv) (v) (vi)

1.2 Does the following diagram represent a chemical or physical change? How do you know? [Section 1.3]

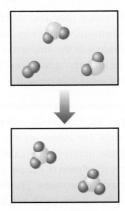

1.3 Identify each of the following as measurements of length, area, volume, mass, density, time or temperature: **(a)** 5 ns, **(b)** 5.5 kg m^{-3}, **(c)** 0.88 pm, **(d)** 540 km^2, **(e)** 173 K, **(f)** 2 mm^3, **(g)** 23 °C. [Section 1.4]

1.4 The following dartboards illustrate the types of errors often seen when one measurement is repeated several times. The bull's-eye represents the 'true value' and the darts represent the experimental measurements. Which board best represents each of the following scenarios: **(a)** measurements both accurate and precise, **(b)** measurements precise but inaccurate, **(c)** measurements imprecise but yield an accurate average? [Section 1.5]

(i)   (ii)   (iii)

1.5 What is the length of the pencil in the following figure if the scale reads in centimetres? How many significant figures are there in this measurement? [Section 1.5]

CLASSIFICATION AND PROPERTIES OF MATTER (Sections 1.2 and 1.3)

The following exercises are divided into sections that deal with specific topics in this chapter.

1.6 Classify each of the following as a pure substance or a mixture; if a mixture, indicate whether it is homogeneous or heterogeneous: **(a)** rice pudding, **(b)** seawater, **(c)** magnesium, **(d)** petrol.

1.7 Classify each of the following as a pure substance or a mixture; if a mixture, indicate whether it is homogeneous or heterogeneous: **(a)** air, **(b)** tomato juice, **(c)** iodine crystals, **(d)** sand.

1.8 Give the chemical symbols for the following elements: **(a)** sulfur, **(b)** potassium, **(c)** chlorine, **(d)** copper, **(e)** silicon, **(f)** nitrogen, **(g)** calcium, **(h)** helium.

1.9 Give the chemical symbol for each of the following elements: **(a)** carbon, **(b)** sodium, **(c)** fluorine, **(d)** iron, **(e)** phosphorus, **(f)** argon, **(g)** nickel, **(h)** silver.

1.10 Name the chemical elements represented by the following symbols: **(a)** Li, **(b)** Al, **(c)** Pb, **(d)** S, **(e)** Br, **(f)** Sn, **(g)** Cr, **(h)** Zn.

1.11 Name each of the following elements: **(a)** Co, **(b)** I, **(c)** Kr, **(d)** Hg, **(e)** As, **(f)** Ti, **(g)** K, **(h)** Ge.

1.12 A solid white substance A is heated strongly in the absence of air. It decomposes to form a new white substance B and a gas C. The gas has exactly the same properties as the product obtained when carbon is burned in an excess of oxygen. Based on these observations, can we determine whether solids A and B and the gas C are elements or compounds? Explain your conclusions for each substance.

1.13 In the process of attempting to characterise a substance, a chemist makes the following observations. The substance is a silvery white, lustrous metal. It melts at 649° C and boils at 1105 °C. Its density at 20 °C is 1.738 g cm^{-3}. The substance burns in air, producing an intense white light. It reacts with chlorine to give a brittle white solid. The substance can be pounded into thin sheets or drawn into wires. It is a good conductor of electricity. Which of these characteristics are physical properties and which are chemical properties?

1.14 Label each of the following as either a physical process or a chemical process: **(a)** corrosion of aluminium metal, **(b)** melting of ice, **(c)** pulverising an aspirin, **(d)** digesting a chocolate bar, **(e)** explosion of nitroglycerin.

1.15 Suggest a method of separating each of the following mixtures into two components: **(a)** sugar and sand, **(b)** iron and sulfur.

1.16 A beaker contains a clear, colourless liquid. If it is water, how could you determine whether it contained dissolved table salt? Do *not* taste it!

UNITS OF MEASUREMENT (Section 1.4)

1.17 Perform the following conversions: **(a)** 25.5 mg to g, **(b)** 4.0×10^{-10} m to nm, **(c)** 0.575 mm to μm.

1.18 Convert **(a)** 9.5×10^{-2} kg to g, **(b)** 0.0023 μm to nm, **(c)** 7.25×10^{-4} s to ms.

1.19 **(a)** A sample of carbon tetrachloride, a liquid once used in dry cleaning, has a mass of 39.73 g and a volume of 25.0 cm^3 at 25 °C. What is its density at this temperature? Will carbon tetrachloride float on water? (Materials that are less dense than water will float.) **(b)** The density of platinum is 21.45 g cm^{-3} at 20 °C. Calculate the mass of 75.00 cm^3 of platinum at this temperature. **(c)** The density of magnesium is 1.738 g cm^{-3} at 20 °C. What is the volume of 87.50 g of this metal at this temperature?

1.20 **(a)** A cube of osmium metal 1.500 cm on a side has a mass of 76.31 g at 25 °C. What is its density in g cm^{-3} at this temperature? **(b)** The density of titanium metal is 4.51 g cm^{-3} at 25 °C. What mass of titanium displaces 65.8 cm^3 of water at 25 °C? **(c)** The density of benzene at 15° C is 0.8787 g cm^{-3}. Calculate the mass of 0.1500 dm^3 of benzene at this temperature.

1.21 **(a)** To identify a liquid substance, a student determined its density. Using a graduated cylinder, she measured out a 45 cm^3 sample of the substance. She then measured the mass of the sample, finding that it weighed 38.5 g.

She knew that the substance had to be either isopropyl alcohol (density 0.785 g cm^{-3}) or toluene (density 0.866 g cm^{-3}). What are the calculated density and the probable identity of the substance? **(b)** An experiment requires 45.0 g of ethylene glycol, a liquid whose density is 1.114 g cm^{-3}. Rather than weigh the sample on a balance, a chemist chooses to dispense the liquid using a graduated cylinder. What volume of the liquid should he use? **(c)** A cubic piece of metal measures 5.00 cm on each edge. If the metal is nickel, which has a density of 8.90 g cm^{-3}, what is the mass of the cube?

UNCERTAINTY IN MEASUREMENT (Section 1.5)

1.22 Indicate which of the following are exact numbers: **(a)** the mass of a paper clip, **(b)** the surface area of a coin, **(c)** the number of microseconds in a week, **(d)** the number of pages in this book.

1.23 Indicate which of the following are exact numbers: **(a)** the number of students in your chemistry class, **(b)** the temperature of the surface of the sun, **(c)** the mass of a postage stamp, **(d)** the number of millilitres in a cubic metre of water, **(e)** the average height of students in your class.

1.24 What is the number of significant figures in each of the following measured quantities? **(a)** 358 kg, **(b)** 0.054 s, **(c)** 6.3050 cm, **(d)** 0.0105 dm^3, **(e)** 7.0500 × 10^{-3} m^3.

1.25 Indicate the number of significant figures in each of the following measured quantities: **(a)** 3.7745 km, **(b)** 205 m^2, **(c)** 1.700 cm, **(d)** 350.0 K, **(e)** 307.080 g.

1.26 Round each of the following numbers to four significant figures, and express the result in standard exponential notation: **(a)** 102.53070, **(b)** 656980, **(c)** 0.008543210, **(d)** 0.000257870, **(e)** − 0.0357202.

1.27 **(a)** The diameter of Earth at the equator is 12 784.49 km. Round this number to three significant figures and express it in standard exponential notation. **(b)** The circumference of Earth through the poles is 40 008 km. Round this number to four significant figures and express it in standard exponential notation.

1.28 Compute the following and express the answers with the appropriate number of significant figures:
(a) 12.0550 + 9.05
(b) 257.2 − 19.789
(c) (6.21 × 10^3) (0.1050)
(d) 0.0577/0.753.

1.29 Compute the following and express the answer with the appropriate number of significant figures:
(a) 320.55 − (6104.5/2.3)
(b) [(285.3 × 10^5) − (1.200 × 10^3)] × 2.8954
(c) (0.0045 × 20 000.0) + (2813 × 12)
(d) 863 × [1255 − (3.45 × 108)].

ADDITIONAL EXERCISES

The exercises in this section are not divided by category, although they are roughly in the order of the topics in the chapter.

1.30 What is meant by the terms 'composition' and 'structure' when referring to matter?

1.31 **(a)** Classify each of the following as a pure substance, a solution or a heterogeneous mixture: a gold coin, a cup of coffee, a wood plank. **(b)** What ambiguities are there in answering part **(a)** from the descriptions given?

1.32 **(a)** What is the difference between a hypothesis and a theory? **(b)** Explain the difference between a theory and a scientific law. Which addresses how matter behaves, and which addresses why it behaves that way?

1.33 The liquid substances mercury (density = 13.5 g cm^{-3}), water (1.00 g cm^{-3}) and cyclohexane (0.778 g cm^{-3}) do not form a solution when mixed, but separate in distinct layers. Sketch how the liquids would position themselves in a test tube.

1.34 **(a)** You are given a bottle that contains 4.59 cm^3 of a metallic solid. The total mass of the bottle and solid is 35.66 g. The empty bottle weighs 14.23 g. What is the

density of the solid? **(b)** Mercury is traded by the 'flask', a unit that has a mass of 34.5 kg. What is the volume of a flask of mercury if the density of mercury is 13.5 g cm^{-3}? **(c)** A thief plans to steal a gold sphere with a radius of 28.9 cm from a museum. If the gold has a density of 19.3 g cm^{-3} what is the mass of the sphere? (The volume of a sphere is $V = (4/3)\pi r^3$.) Is he likely to be able to walk off with it unassisted?

1.35 Car batteries contain sulfuric acid, which is commonly referred to as 'battery acid'. Calculate the number of grams of sulfuric acid in 0.500 dm^3 of battery acid if the solution has a density of 1.28 g cm^{-3} and is 38.1% sulfuric acid by mass.

1.36 Gold is alloyed (mixed) with other metals to increase its hardness in making jewellery. **(a)** Consider a piece of gold jewellery that weighs 9.85 g and has a volume of 0.0675 cm^3. The jewellery contains only gold and silver, which have densities of 19.3 g cm^{-3} and 10.5 g cm^{-3}, respectively. Assuming that the total volume of the jewellery is the sum of the volumes of the gold and silver that it contains, calculate the percentage of gold (by mass) in the jewellery. **(b)** The relative amount of gold in an alloy is commonly expressed in units of carats. Pure gold is 24 carat, and the percentage of gold in an alloy is given as a percentage of this value. For example, an alloy that is 50% gold is 12 carat. State the purity of the gold jewellery in carats.

1.37 Suppose you are given a sample of a homogeneous liquid. What would you do to determine whether it is a solution or a pure substance?

MasteringChemistry (www.pearson.com.au/masteringchemistry)

Make learning part of the grade. Access:

- tutorials with personalised coaching
- study area
- Pearson eText

PHOTO/ART CREDITS

2 NASA, ESA and the Hubble Heritage (STScI/AURA)-ESA/Hubble Collaboration. R. O'Connell (University of Virginia) and the WFC3 Scientific Oversight Committee; **7 (a)** © Sergei Petrakov/iStockphoto.com, **(b)** © Richard Megna/Fundamental Photographs, NYC; **10 (a)** and **(b)** Donald Clegg and Roxy Wilson/Pearson Education/PH College Author series; **17** Courtesy of City of Sydney.

2

ATOMS, MOLECULES AND IONS

A helicopter engine is composed of many smaller parts, just as any substance on Earth is composed of countless atoms and molecules, to give it its unique characteristics.

ATOMS, MOLECULES AND IONS

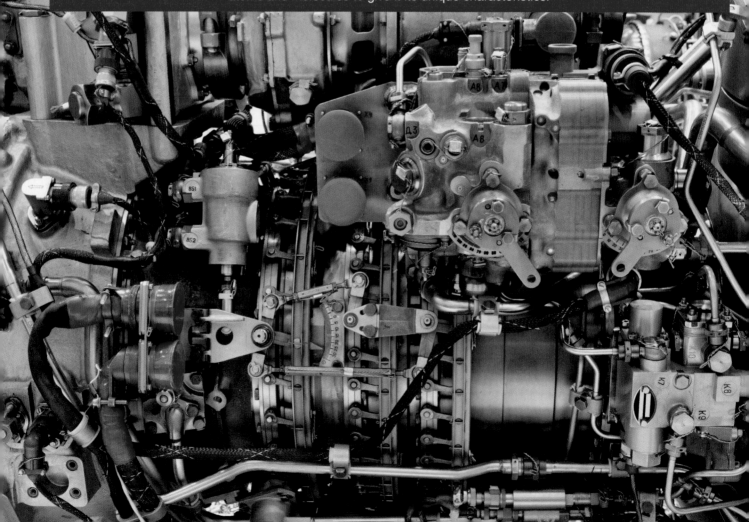

A helicopter engine is composed of many smaller parts, just as any substance on Earth is composed of countless atoms and molecules to give it its unique characteristics.

KEY CONCEPTS

Look around at the great variety of colours, textures and other properties in the materials that surround you—the colours in a garden, the texture of the fabric in your clothes, the solubility of sugar in a cup of coffee or the transparency and beauty of a diamond. The materials in our world exhibit a striking and seemingly infinite variety of properties, but how do we understand and explain them? What makes diamonds transparent and hard, whereas table salt is brittle and dissolves in water? Why does paper burn, and why does water quench fires? The structure and behaviour of atoms are the key to understanding both the physical and chemical properties of matter.

Although the materials in our world vary greatly in their properties, everything is formed from only about 100 elements and therefore from only about 100 chemically different kinds of atoms. In a sense, the atoms are like the 26 letters of the English alphabet that join in different combinations to form the immense number of words in our language. But what rules govern the ways in which atoms combine? How do the properties of a substance relate to the kinds of atoms it contains? Indeed, what is an atom like, and what makes the atoms of one element different from those of another?

In this chapter we examine the basic structure of atoms and discuss the formation of molecules and ions, thereby providing a foundation for exploring chemistry more deeply in later chapters.

2.1 | ATOMIC THEORY OF MATTER

Philosophers from the earliest times have speculated about the nature of the fundamental 'stuff' from which the world is made. Democritus (460–370 BC) and other early Greek philosophers thought that the material world must be made up of tiny indivisible particles that they called *atomos*, meaning 'uncuttable' or indivisible. Later Plato and Aristotle formulated the notion that there can be no ultimately indivisible particles and the atomic view of matter faded for many centuries.

The notion of atoms re-emerged in Europe during the seventeenth century, when scientists tried to explain the properties of gases. Air is composed of something invisible and in constant motion; we can feel the motion of the wind against us, for example. It is natural to think of tiny invisible particles as giving rise to these familiar effects.

As chemists learned to measure the amounts of elements that reacted with one another to form new substances, the ground was laid for an atomic theory that linked the idea of elements with the idea of atoms. That theory came into being during the period 1803–1807 due to the work of an English schoolteacher, John Dalton. Dalton's atomic theory was based on the four postulates given in (▼ FIGURE 2.1).

∞ Review this on page 7

Dalton's theory explains several laws of chemical combination that were known during his time, including the *law of constant composition* (∞ Section 1.2, 'Classifications of Matter'), based on postulate 4:

> In a given compound, the relative numbers and kinds of atoms are constant.

It also explains the *law of conservation of mass,* based on postulate 3:

> The total mass of materials present after a chemical reaction is the same as the total mass present before the reaction.

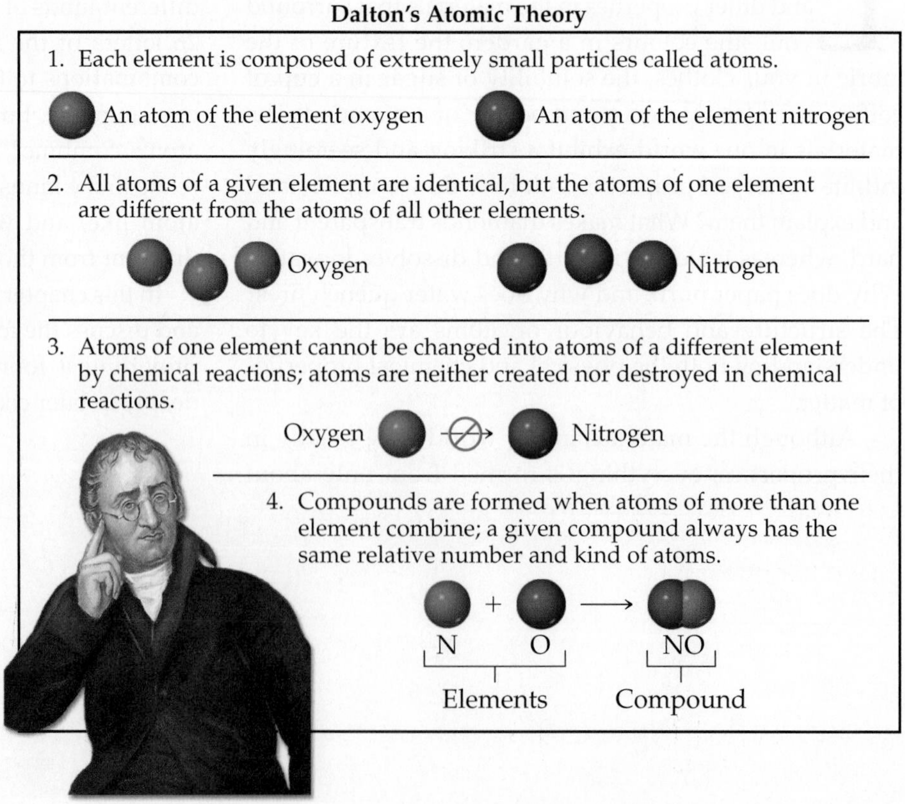

▶ FIGURE 2.1 **Dalton's atomic theory.** John Dalton (1766–1844), the son of a poor English weaver, began teaching at age 12. He spent most of his years in Manchester, where he taught both grammar school and college. His lifelong interest in meteorology led him to study gases, then chemistry, and eventually atomic theory. Despite his humble beginnings, Dalton gained a strong scientific reputation during his lifetime.

Dalton's Atomic Theory

1. Each element is composed of extremely small particles called atoms.

   ⬤ An atom of the element oxygen   ⬤ An atom of the element nitrogen

2. All atoms of a given element are identical, but the atoms of one element are different from the atoms of all other elements.

   ⬤⬤⬤ Oxygen   ⬤⬤⬤ Nitrogen

3. Atoms of one element cannot be changed into atoms of a different element by chemical reactions; atoms are neither created nor destroyed in chemical reactions.

   Oxygen ⬤ –⊘→ ⬤ Nitrogen

4. Compounds are formed when atoms of more than one element combine; a given compound always has the same relative number and kind of atoms.

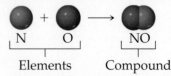

   N   O   NO
   └─ Elements ─┘   └─ Compound ─┘

A good theory explains known facts and predicts new ones. Dalton used his theory to deduce the *law of multiple proportions*:

> If two elements A and B combine to form more than one compound, the masses of B that can combine with a given mass of A are in the ratio of small whole numbers.

We can illustrate this law by considering water and hydrogen peroxide, both of which are composed of the elements hydrogen and oxygen. In forming water, 8.0 g of oxygen combine with 1.0 g of hydrogen. In forming hydrogen peroxide, 16.0 g of oxygen combine with 1.0 g of hydrogen. Thus the ratio of the mass of oxygen per gram of hydrogen in the two compounds is 2 : 1. Using Dalton's atomic theory, we conclude that hydrogen peroxide contains twice as many atoms of oxygen per hydrogen atom as does water.

CONCEPT CHECK 1

One compound of carbon and oxygen contains 1.333 g of oxygen per gram of carbon, whereas a second compound contains 2.666 g of oxygen per gram of carbon.
a. What chemical law do these data illustrate?
b. If the first compound has an equal number of oxygen and carbon atoms, what can we conclude about the composition of the second compound?

2.2 | THE DISCOVERY OF ATOMIC STRUCTURE

Dalton reached his conclusion about atoms on the basis of chemical observations in the macroscopic world of the laboratory. Neither he nor those who followed him during the century after his work was published had direct evidence for the existence of atoms. Today, however, we can use powerful instruments to measure the properties of individual atoms and even provide images of them (▶ FIGURE 2.2).

As scientists began to develop methods for more detailed probing of the nature of matter, the atom, which was supposed to be indivisible, began to show signs of a more complex structure. We now know that the atom is composed of still smaller **subatomic particles**. Before we summarise the current model of atomic structure, we briefly consider a few of the landmark discoveries that led to that model. We'll see that the atom is composed in part of electrically charged particles, some with a positive (+) charge and some with a negative (−) charge. As we discuss the development of our current model of the atom, keep in mind a simple statement of the behaviour of charged particles: *particles with the same charge repel one another, whereas particles with unlike charges are attracted to one another.*

▲ **FIGURE 2.2   An image of the surface of silicon.** The image was obtained by a technique called scanning tunnelling microscopy. The colour was added to the image by computer to help distinguish its features. Each purple sphere is a silicon atom.

Cathode Rays and Electrons

In the mid-1800s numerous studies of electrical discharges through partially evacuated tubes, such as those shown in ▼ FIGURE 2.3, revealed that the discharges occurred at the cathode (negative electrode) and streamed to the anode, hence the name **cathode rays**. In 1897 the British physicist J. J. Thomson noted that the rays were identical irrespective of the cathode material and that they were deflected by electric and magnetic fields. He concluded that this was consistent with the rays being composed of negatively charged particles which later came to be known as *electrons*. By quantitatively measuring the deflection of a stream of electrons by both electric and magnetic fields, Thomson calculated that the ratio of the electron's electrical charge to its mass was $1.78 \times 10^8 \text{ C g}^{-1}$, where C represents a *coulomb*, the SI unit for electric charge.

Once the charge-to-mass ratio of the electron was known, measuring either the charge or the mass of an electron would also yield the value of the other quantity. In 1909 Robert Millikan of the University of Chicago succeeded in

⚠ **FIGURE IT OUT**

How do we know that the cathode rays travel from cathode to anode?

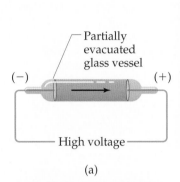

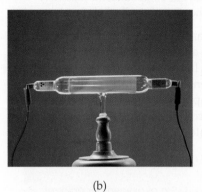

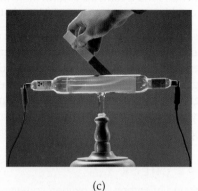

(a)

In a cathode-ray tube, electrons move from the negative electrode (cathode) to the positive electrode (anode).

(b)

A photo of a cathode-ray tube containing a fluorescent screen to show the path of the cathode rays.

(c)

The path of the cathode rays is deflected by the presence of a magnet.

▲ **FIGURE 2.3 Cathode-ray tube.**

measuring the charge of an electron by performing a series of experiments described in ▼ **FIGURE 2.4**. He then calculated the mass of the electron by using his experimental value for the charge, 1.60×10^{-19} C, and Thomson's charge-to-mass ratio, 1.76×10^8 C g^{-1}:

$$\text{Electron mass} = \frac{1.60 \times 10^{-19}\,\text{C}}{1.76 \times 10^8\,\text{C g}^{-1}} = 9.10 \times 10^{-28}\,\text{g}$$

This result agrees well with the presently accepted value for the mass of the electron, 9.10938×10^{-28} g. This mass is about 2000 times smaller than that of hydrogen, the lightest atom.

Radioactivity

In 1896 the French scientist Henri Becquerel was studying a uranium compound when he discovered that it spontaneously emitted high-energy radiation. This spontaneous emission of radiation is called **radioactivity**. At Becquerel's suggestion Marie Curie and her husband, Pierre, began experiments to isolate the radioactive components of the compound.

Further study of the nature of radioactivity, principally by the New Zealand scientist Ernest Rutherford (▶ **FIGURE 2.5**), revealed three types of radiation: alpha (α), beta (β) and gamma (γ) radiation. Each type differs in its response to an electric field, as shown in ▶ **FIGURE 2.6**. The paths of both α and β radiation are bent by the electric field, although in opposite directions, whereas γ radiation is unaffected.

Rutherford showed that both α and β rays consist of fast-moving particles, which were called α and β particles. In fact, β particles turned out to be high-speed electrons and can be considered the high-energy equivalent of cathode rays. They are therefore attracted to a positively charged plate. The α particles have a positive charge and are therefore attracted towards a negative plate. In

▲ **FIGURE 2.4 Millikan's oil-drop experiment used to measure the charge of the electron.** Small drops of oil were allowed to fall between electrically charged plates. The drops picked up extra electrons as a result of irradiation by X-rays and so became negatively charged. Millikan measured how varying the voltage between the plates affected the rate of fall. From these data he calculated the negative charge on the drops. Because the charge on any drop was always some integral multiple of 1.602×10^{-19} C, Millikan deduced this value to be the charge of a single electron.

units of the charge of the electron, β particles have a charge of 1− and α particles a charge of 2+. Each α particle is considerably heavier than an electron, having a mass about 7300 times that of an electron. Gamma radiation is high-energy radiation similar to X-rays; it does not consist of particles and carries no charge.

The Nuclear Atom

With the growing realisation that the atom is composed of smaller particles, attempts were made to rationalise how the particles fitted together. It was the work of Rutherford that paved the way for the modern concept of the atom. In 1910 Rutherford and his co-workers were studying the angles at which α particles were deflected or *scattered* as they passed through a thin gold foil a few thousand atoms thick. Although almost all of the α particles passed through the foil, one of his undergraduate students, Eric Marsden (also a New Zealander), noted that about 1 in 8000 α particles were scattered at quite large angles and in fact some particles simply bounced back in the direction from which they had come.

By 1911 Rutherford was able to explain these observations. He postulated that most of the mass of each gold atom in his foil and all of its positive charge reside in a very small, extremely dense region, which he called the **nucleus**. He postulated further that most of the total volume of an atom is empty space in which electrons move around the nucleus. In the α scattering experiment, most α particles passed directly through the foil because they did not encounter the minute nucleus of any gold atom; they merely passed through the empty space making up the greatest part of all the atoms in the foil. Occasionally, however, an α particle came close to a gold nucleus. The repulsion between the highly charged gold nucleus and the α particle was strong enough to deflect the less massive α particle, as shown in ▼ **FIGURE 2.7**.

Subsequent experimental studies led to the discovery of both positive particles (*protons*) and neutral particles (*neutrons*) in the nucleus. Protons were discovered in 1919 by Rutherford. Neutrons were discovered in 1932 by the British scientist James Chadwick (a student of Rutherford).

▲ **FIGURE 2.5   Ernest Rutherford (1871–1937).** Rutherford, whom Einstein called 'the second Newton', was born and educated in New Zealand. In 1895 he was the first overseas student ever to be awarded a position at the Cavendish Laboratory at Cambridge University in England, where he worked with J. J. Thomson. In 1898 he joined the faculty of McGill University in Montreal. While at McGill, Rutherford did his research on radioactivity that led to his being awarded the 1908 Nobel Prize in Chemistry. In 1907 Rutherford moved back to England to be a faculty member at Manchester University, where in 1910 he performed his famous α-particle scattering experiments that led to the nuclear model of the atom. In 1992 his native New Zealand honoured Rutherford by putting his likeness, along with his Nobel Prize medal, on its $100 currency note.

 CONCEPT CHECK 2

What happens to most of the α particles that strike the gold foil in Rutherford's experiment, and why do they behave that way?

🔺 **FIGURE IT OUT**

Which of the three kinds of radiation shown consists of electrons? Why are these rays deflected to a greater extent than the others?

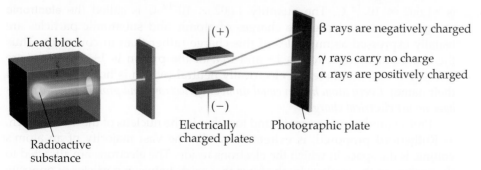

Lead block

Radioactive substance

Electrically charged plates

Photographic plate

β rays are negatively charged

γ rays carry no charge

α rays are positively charged

▲ **FIGURE 2.6   Behaviour of alpha (α), beta (β) and gamma (γ) rays in an electric field.**

What is the charge on the particles that form the beam?

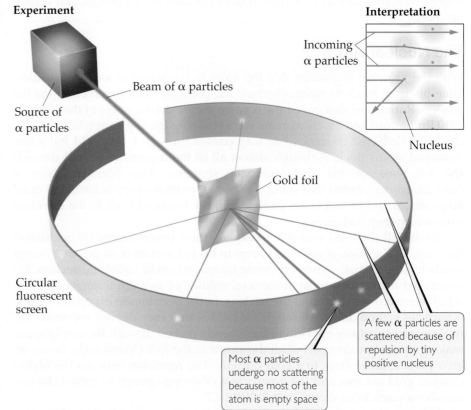

Experiment

Interpretation

Source of
α particles

Beam of α particles

Incoming
α particles

Nucleus

Gold foil

Circular
fluorescent
screen

A few α particles are
scattered because of
repulsion by tiny
positive nucleus

Most α particles
undergo no scattering
because most of the
atom is empty space

▶ **FIGURE 2.7 Rutherford's
α-scattering experiment.** When α particles
pass through a gold foil, most pass through
undeflected but some are scattered, a few at
very large angles. The nuclear model of the
atom explains why a few α particles are
deflected at large angles. For clarity, the
nuclear atom is shown here as a coloured
sphere, but most of the space around the
nucleus is empty except for the tiny electrons
moving around.

2.3 | THE MODERN VIEW OF ATOMIC STRUCTURE

Since the time of Rutherford, physicists have learned much about the detailed
composition of atomic nuclei. In the course of these discoveries, the list of subatomic particles has grown long and continues to increase. As chemists,
however, we can take a very simple view of the atom because only three subatomic particles—the **proton**, **neutron** and **electron**—have a bearing on chemical
behaviour.

The charge of an electron is -1.602×10^{-19} C, and that of a proton
is $+1.602 \times 10^{-19}$ C. The quantity 1.602×10^{-19} C is called the **electronic
charge**. For convenience, the charges of atomic and subatomic particles are
usually expressed as multiples of this charge rather than in coulombs. Thus
the charge of the electron is 1– and that of the proton is 1+. Neutrons are
uncharged and are therefore electrically neutral (which is how they received
their name). *Every atom has an equal number of electrons and protons, and so atoms
have no net electrical charge.*

Protons and neutrons are bound together in the nucleus of the atom, which,
as Rutherford proposed, is extremely small. The vast majority of an atom's
volume is the space in which the electrons reside. The electrons are attracted to
the protons in the nucleus by the force that exists between particles of opposite
electrical charge. In later chapters we will see that the strength of the attractive
forces between electrons and nuclei can be used to explain many of the differences between different elements.

TABLE 2.1 • Comparison of the proton, neutron and electron

Particle	Charge	Mass (u)
Proton	Positive (1+)	1.0073
Neutron	None (neutral)	1.0087
Electron	Negative (1−)	5.486×10^{-4}

CONCEPT CHECK 3

a. If an atom has 15 protons, how many electrons does it have?
b. Where do the protons reside in an atom?

Atoms have extremely small masses. The mass of the heaviest known atom, for example, is in the order of 4×10^{-22} g. Because it would be cumbersome to express such small masses in grams, we use instead the **unified atomic mass unit (u)**. One u equals 1.66054×10^{-24} g. The masses of the proton and neutron are very nearly equal, and both are much greater than that of the electron: a proton has a mass of 1.0073 u, a neutron 1.0087 u and an electron 5.486×10^{-4} u. Because it would take 1836 electrons to equal the mass of 1 proton, the nucleus contains most of the mass of an atom. ▲ TABLE 2.1 summarises the charges and masses of the subatomic particles. We have more to say about atomic masses in Section 2.4.

Atoms are also extremely small. Most atoms have diameters between 1×10^{-10} m and 5×10^{-10} m. The SI unit used to express atomic dimensions is the **picometre (pm)** ($1 \text{ pm} = 1 \times 10^{-12}$ m). The angstrom (Å), where one angstrom equals 10^{-10} m, is still commonly used but we shall employ the picometre exclusively as the unit of atomic dimensions. Thus atoms have diameters in the order of 100–500 pm. The diameter of a chlorine atom, for example, is 200 pm.

SAMPLE EXERCISE 2.1 **Illustrating the size of an atom**

The diameter of an Australian 5 cent coin is 20 mm. The diameter of a silver atom, by comparison, is only 288 pm. How many silver atoms could be arranged side by side in a straight line across the diameter of the coin?

SOLUTION
We can start with the diameter of the coin, first converting this distance into picometres and then using the diameter of the Ag atom to convert distance to the number of Ag atoms:

$$20 \text{ mm} = 20 \times 10^{9} \text{ pm}$$

$$\text{Ag atoms} = \frac{20 \times 10^{9} \text{ pm}}{288 \text{ pm}} = 6.9 \times 10^{7} \text{ Ag atoms}$$

That is, 69 million silver atoms could sit side by side across a 5 cent coin!

PRACTICE EXERCISE
The diameter of a carbon atom is 154 pm. How many carbon atoms could be aligned side by side in a straight line across the width of a pencil line that is 0.20 mm wide?

Answers: 1.3×10^{6} C atoms
(See also Exercises 2.14, 2.15, 2.61.)

The diameters of atomic nuclei are in the order of 10^{-2} pm, only a small fraction of the diameter of the atom as a whole. You can appreciate the relative sizes of the atom and its nucleus by imagining that if the atom were as large as a football stadium, the nucleus would be the size of a small marble. Because the tiny nucleus carries most of the mass of the atom in such a small volume, it has an incredible density—in the order of 10^{13} to 10^{14} g cm^{-3}. A matchbox full of material of such density would weigh over 2.5 billion tonnes! Astrophysicists have suggested that the interior of a collapsed star may approach this density.

A CLOSER LOOK

BASIC FORCES

There are four basic forces, or interactions, known in nature: gravitational, electromagnetic, strong nuclear and weak nuclear. *Gravitational forces* are attractive forces that act between all objects in proportion to their masses. Gravitational forces between atoms or between subatomic particles are so small that they are of no chemical significance.

Electromagnetic forces are attractive or repulsive forces that act between either electrically charged or magnetic objects. Electric and magnetic forces are intimately related. Electric forces are of fundamental importance in understanding the chemical behaviour of atoms. The magnitude of the electric force between two charged particles is given by *Coulomb's law* (∞ Section 14.1, 'The Nature of Energy'): $F = kq_1q_2/r^2$ where q_1 and q_2 are the magnitudes of the charges on the two particles,

r is the distance between their centres and k is a constant determined by the units for q and r. A negative value for the force indicates attraction, whereas a positive value indicates repulsion.

All nuclei except those of hydrogen atoms contain two or more protons. Because like charges repel, electrical repulsion would cause the protons to fly apart if a stronger attractive force did not keep them together. This force is called the *strong nuclear force*. It acts between subatomic particles, as in the nucleus. At this distance, the strong nuclear force is stronger than the electric force, and as a result, the nucleus holds together. The *weak nuclear force* is weaker than the electric force but stronger than the gravitational force. We are aware of its existence only because it shows itself in certain types of radioactivity.

RELATED EXERCISE: 2.64

Atomic Numbers, Mass Numbers and Isotopes

What makes an atom of one element different from an atom of another element? For example, how does an atom of carbon differ from an atom of oxygen? The significant difference is in their subatomic compositions: the atoms of each element have a characteristic number of protons. Indeed, the number of protons in the nucleus of an atom of any particular element is called that element's **atomic number** (Z). Because an atom has no net electrical charge, the number of electrons it contains must equal the number of protons. All atoms of carbon, for example, have six protons and six electrons, whereas all atoms of oxygen have eight protons and eight electrons. Thus carbon has atomic number 6, whereas oxygen has atomic number 8. The atomic number of each element is listed with the name and symbol of the element on the inside front cover of this text.

Atoms of a given element can differ in the number of neutrons they contain and consequently in mass. For example, most atoms of carbon have six neutrons, although some have more and some have less. The symbol $^{12}_{6}C$, or carbon-12 (read as 'carbon twelve') represents the carbon atom containing six protons and six neutrons. The atomic number is shown by the subscript, and the superscript, called the **mass number** (A), is the total number of protons plus neutrons in the atom.

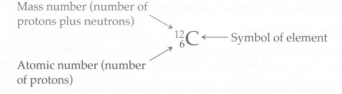

Because all atoms of a given element have the same atomic number, the subscript is redundant and is often omitted. Thus the symbol for carbon-12 can be represented simply as ^{12}C. As one more example of this notation, atoms that contain six protons and eight neutrons have a mass number of 14 and are represented as $^{14}_{6}C$ or ^{14}C and referred to as carbon-14.

Atoms with identical atomic numbers but different mass numbers (that is, same number of protons but different numbers of neutrons) are called **isotopes** of one another. Several isotopes of carbon are listed in ▼ TABLE 2.2. We will generally use the notation with superscripts only when referring to a particular isotope of an element.

TABLE 2.2 • Some isotopes of carbon*			
Symbol	Number of protons	Number of electrons	Number of neutrons
^{11}C	6	6	5
^{12}C	6	6	6
^{13}C	6	6	7
^{14}C	6	6	8

*Almost 99% of the carbon found in nature is ^{12}C.

SAMPLE EXERCISE 2.2   Determining the number of subatomic particles in atoms

How many protons, neutrons and electrons are in **(a)** an atom of ^{197}Au; **(b)** an atom of strontium-90?

SOLUTION

(a) The superscript 197 is the mass number, the sum of the number of protons plus the number of neutrons. According to the list of elements given inside the front cover, gold has an atomic number of 79. Consequently, an atom of ^{197}Au has 79 protons, 79 electrons and $197 - 79 = 118$ neutrons. **(b)** The atomic number of strontium (listed inside the front cover) is 38. Thus all atoms of this element have 38 protons and 38 electrons. The strontium-90 isotope has $90 - 38 = 52$ neutrons.

PRACTICE EXERCISE

How many protons, neutrons and electrons are in **(a)** a ^{138}Ba atom, **(b)** an atom of phosphorus-31?

Answers: **(a)** 56 protons, 56 electrons and 82 neutrons; **(b)** 15 protons, 15 electrons and 16 neutrons
(See also Exercises 2.18–2.20, 2.60, 2.63.)

SAMPLE EXERCISE 2.3   Writing symbols for atoms

Magnesium has three isotopes, with mass numbers 24, 25 and 26. **(a)** Write the complete chemical symbol (superscript and subscript) for each of them. **(b)** How many neutrons are in an atom of each isotope?

SOLUTION

(a) Magnesium has atomic number 12, and so all atoms of magnesium contain 12 protons and 12 electrons. The three isotopes are therefore represented by $^{24}_{12}Mg$, $^{25}_{12}Mg$ and $^{26}_{12}Mg$. **(b)** The number of neutrons in each isotope is the mass number minus the number of protons. The numbers of neutrons in an atom of each isotope are therefore 12, 13 and 14, respectively.

PRACTICE EXERCISE

Give the complete chemical symbol for the atom that contains 82 protons, 82 electrons and 126 neutrons.

Answer: $^{208}_{82}Pb$
(See also Exercises 2.16, 2.17, 2.21, 2.63(a).)

2.4 | ATOMIC MASS

Atoms are small pieces of matter and so they have mass. In this section we discuss the mass scale used for atoms and introduce the concept of *atomic mass*.

The Atomic Mass Scale

Although scientists of the nineteenth century knew nothing about subatomic particles, they were aware that atoms of different elements have different masses. They found, for example, that each 100.0 g of water contains 11.1 g of hydrogen and 88.9 g of oxygen. Thus water contains $88.9/11.1 = 8$ times as much oxygen, by mass, as hydrogen. Once scientists understood that water contains two hydrogen atoms for each oxygen atom, they concluded that an oxygen atom must have $2 \times 8 = 16$ times as much mass as a hydrogen atom. Hydrogen, the lightest atom, was arbitrarily assigned a relative mass of 1 (no units), and atomic masses of other elements were at first determined relative to this value. Thus oxygen was assigned an atomic mass of 16.

Today we can determine the masses of individual atoms with a high degree of accuracy. For example, we know that the ^{1}H atom has a mass of 1.6735×10^{-24} g and the ^{16}O atom has a mass of 2.6560×10^{-23} g. As we noted in Section 2.3, it is convenient to use the *unified atomic mass unit* (u) when dealing with these extremely small masses:

$$1\,u = 1.66054 \times 10^{-24}\,g \text{ and } 1\,g = 6.02214 \times 10^{23}\,u$$

The unified atomic mass unit is presently defined by assigning a mass of exactly 12 u to an atom of the ^{12}C isotope of carbon. In these units, a ^{1}H atom has a mass of 1.0078 u and an ^{16}O atom has a mass of 15.9949 u.

Average Atomic Masses

Most elements occur in nature as mixtures of isotopes. We can determine the **average atomic mass** of an element by using the masses of its various isotopes and their relative abundances. Naturally occurring carbon, for example, is composed of 98.93% ^{12}C and 1.07% ^{13}C. The masses of these isotopes are 12 u (exactly) and 13.00335 u, respectively. We calculate the average atomic mass of carbon from the fractional abundance of each isotope and the mass of that isotope:

$$\text{Atomic mass of carbon} = \left[\left(\frac{98.93}{100} \times 12\,u\right) + \left(\frac{1.07}{100} \times 12.00335\,u\right)\right]$$
$$= [(0.9893)(12\,u) + (0.0107)(13.00335\,u)]$$
$$= 12.01\,u$$

Although the term *average atomic mass* is more correct, it is usually abbreviated to simply *atomic mass*. The atomic masses of the elements are listed in both the periodic table and the table of elements on the inside front cover of this text.

 CONCEPT CHECK 4

A particular atom of chromium has a mass of 52.94 u, whereas the atomic mass of chromium is 51.99 u. Explain the difference in the two masses.

SAMPLE EXERCISE 2.4   **Calculating the atomic mass of an element from isotopic abundances**

Naturally occurring chlorine is 75.78% ^{35}Cl, which has an atomic mass of 34.969 u, and 24.22% ^{37}Cl, which has an atomic mass of 36.966 u. Calculate the average atomic mass of chlorine.

SOLUTION

The average atomic mass is found by multiplying the fractional abundance of each isotope by its atomic mass and summing these products. Because 75.78% = 0.7578 and 24.22% = 0.2422 in terms of fractional abundances, we have

$$\text{Average atomic mass} = (0.7578)(34.969\,u) + (0.2422)(36.966\,u)$$
$$= 26.50\,u + 8.953\,u$$
$$= 35.45\,u$$

This answer makes sense. The average atomic mass of Cl is between the masses of the two isotopes and is closer to the value of ^{35}Cl, which is the more abundant isotope.

PRACTICE EXERCISE

Three isotopes of silicon occur in nature: ^{28}Si (92.23%), which has an atomic mass of 27.97693 u; ^{29}Si (4.68%), which has an atomic mass of 28.97649 u; and ^{30}Si (3.09%), which has an atomic mass of 29.97377 u. Calculate the atomic mass of silicon.

Answer: 28.09 u

(See also Exercises 2.25, 2.65.)

A CLOSER LOOK

THE MASS SPECTROMETER

The most direct and accurate means for determining atomic and molecular masses is provided by the **mass spectrometer** (▼ FIGURE 2.8). A gaseous sample is introduced at *A* and bombarded by a stream of high-energy electrons at *B*. Collisions between the electrons and the atoms or molecules of the gas produce positively charged particles, mostly with a 1+ charge. These charged particles are accelerated towards a negatively charged wire grid (*C*). After they pass through the grid, the particles encounter two slits that allow only a narrow beam of particles to pass. This beam then passes between the poles of a magnet, which deflects the particles into a curved path, much as electrons are deflected by a magnetic field. For charged particles with the same charge, the extent of deflection depends on mass—the more massive the particle, the less the deflection. The particles are thereby separated according to their masses. By changing the strength of the magnetic field or the accelerating voltage on the negatively charged grid, charged particles of various masses can be selected to enter the detector at the end of the instrument.

A graph of the intensity of the detector signal versus particle atomic mass is called a *mass spectrum*. The mass spectrum of chlorine atoms, shown in ▼ FIGURE 2.9, reveals the presence of two isotopes. Analysis of a mass spectrum gives both the masses of the charged particles reaching the detector and their relative abundances. The abundances are obtained from the signal intensities. Knowing the atomic mass and the abundance of each isotope allows us to calculate the atomic mass of an element, as shown in Sample Exercise 2.4.

Mass spectrometers are used extensively today to identify chemical compounds and analyse mixtures of substances. Any molecule that loses electrons falls apart, forming an array of positively charged fragments. The mass spectrometer measures the masses of these fragments, producing a chemical 'fingerprint' of the molecule and providing clues about how the atoms were connected in the original molecule. Thus a chemist might use this technique to determine the molecular structure of a newly synthesised compound or to identify a pollutant in the environment.

RELATED EXERCISE: 2.62

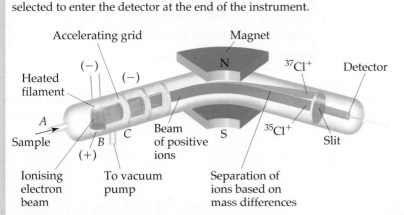

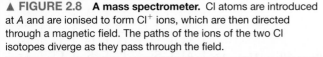

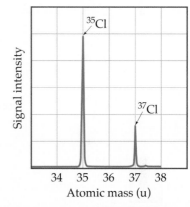

▲ **FIGURE 2.8** **A mass spectrometer.** Cl atoms are introduced at *A* and are ionised to form Cl⁺ ions, which are then directed through a magnetic field. The paths of the ions of the two Cl isotopes diverge as they pass through the field.

▲ **FIGURE 2.9** **Mass spectrum of atomic chlorine.** The fractional abundances of the isotopes ^{35}Cl and ^{37}Cl are indicated by the relative signal intensities of the beams reaching the detector of the mass spectrometer.

2.5 | THE PERIODIC TABLE

Dalton's atomic theory set the stage for a vigorous growth in chemical experimentation during the early 1800s. As the body of chemical observations grew and the list of known elements expanded, attempts were made to find regular patterns in chemical behaviour. These efforts culminated in the development of the periodic table in 1869. You will quickly learn that *the periodic table is the most significant tool that chemists use for organising and remembering chemical facts.*

Many elements show very strong similarities to one another. The elements lithium (Li), sodium (Na) and potassium (K) are all soft, very reactive metals, for example; and the elements helium (He), neon (Ne) and argon (Ar) are all very non-reactive gases. If the elements are arranged in order of increasing atomic number, their chemical and physical properties are found to show a repeating, or periodic, pattern. For example, each of the soft, reactive metals—lithium, sodium and potassium—comes immediately after one of the non-reactive gases—helium, neon and argon—as shown in ▼ FIGURE 2.10.

FIGURE IT OUT

If F is a reactive non-metal, which other elements shown here do you expect to also be reactive non-metals?

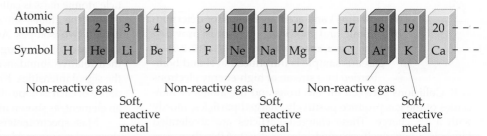

▶ **FIGURE 2.10** **Arranging elements by atomic number reveals a periodic pattern of properties.** This pattern is the basis of the periodic table.

The arrangement of elements in order of increasing atomic number, with elements having similar properties placed in vertical columns, is known as the **periodic table**. The periodic table is shown in ▼ **FIGURE 2.11** and is also given inside the front cover of this text. For each element in the table, the atomic number and atomic symbol are given, and the atomic mass is often given as well, as in the following typical entry for potassium:

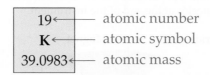

19 ← atomic number
K ← atomic symbol
39.0983 ← atomic mass

The horizontal rows of the periodic table are called **periods**. The first period consists of only two elements, hydrogen (H) and helium (He). The second and

1																	18
1 H	2											13	14	15	16	17	2 He
3 Li	4 Be											5 B	6 C	7 N	8 O	9 F	10 Ne
11 Na	12 Mg	3	4	5	6	7	8	9	10	11	12	13 Al	14 Si	15 P	16 S	17 Cl	18 Ar
19 K	20 Ca	21 Sc	22 Ti	23 V	24 Cr	25 Mn	26 Fe	27 Co	28 Ni	29 Cu	30 Zn	31 Ga	32 Ge	33 As	34 Se	35 Br	36 Kr
37 Rb	38 Sr	39 Y	40 Zr	41 Nb	42 Mo	43 Tc	44 Ru	45 Rh	46 Pd	47 Ag	48 Cd	49 In	50 Sn	51 Sb	52 Te	53 I	54 Xe
55 Cs	56 Ba	71 Lu	72 Hf	73 Ta	74 W	75 Re	76 Os	77 Ir	78 Pt	79 Au	80 Hg	81 Tl	82 Pb	83 Bi	84 Po	85 At	86 Rn
87 Fr	88 Ra	103 Lr	104 Rf	105 Db	106 Sg	107 Bh	108 Hs	109 Mt	110 Ds	111 Rg	112 Cn	113 Uut	114 Fl	115 Uup	116 Lv		

Metals

57 La	58 Ce	59 Pr	60 Nd	61 Pm	62 Sm	63 Eu	64 Gd	65 Tb	66 Dy	67 Ho	68 Er	69 Tm	70 Yb
89 Ac	90 Th	91 Pa	92 U	93 Np	94 Pu	95 Am	96 Cm	97 Bk	98 Cf	99 Es	100 Fm	101 Md	102 No

Metalloids

Non-metals

▲ **FIGURE 2.11** **Periodic table of the elements.** Different colours are used to show the division of the elements into metals, metalloids and non-metals.

TABLE 2.3 • Names of some groups in the periodic table		
Group	Name	Elements
1	Alkali metals	Li, Na, K, Rb, Cs, Fr
2	Alkaline earth metals	Be, Mg, Ca, Sr, Ba, Ra
16	Chalcogens	O, S, Se, Te, Po
17	Halogens	F, Cl, Br, I, At
18	Noble gases (or rare gases)	He, Ne, Ar, Kr, Xe, Rn

third periods, which begin with lithium (Li) and sodium (Na), respectively, consist of eight elements each. The fourth and fifth periods contain 18 elements. The sixth period has 32 elements, but in order for it to fit on a page, 14 of these elements (those with atomic numbers 57–70) appear at the bottom of the table. The seventh and last period is incomplete, but it also has 14 of its elements placed in a row at the bottom of the table.

The vertical columns of the periodic table are called **groups**. The groups are labelled 1–18 according to the system proposed by the International Union of Pure and Applied Chemistry (IUPAC).

Elements that belong to the same group often exhibit similarities in physical and chemical properties. For example, the 'coinage metals'—copper (Cu), silver (Ag) and gold (Au)—all belong to group 11. As their name suggests, the coinage metals are used throughout the world to make coins. Many other groups in the periodic table also have names, as listed in ▲ TABLE 2.3.

Except for hydrogen, all the elements on the left side and in the middle of the periodic table are **metallic elements**, or **metals**. The majority of elements are metallic, and they all share many characteristic properties, such as lustre and high electrical and heat conductivity. All metals, with the exception of mercury (Hg), are solids at room temperature. The metals are separated from the **non-metallic elements**, or **non-metals**, by a diagonal steplike line that runs from boron (B) to astatine (At), as shown in Figure 2.11. Hydrogen, although on the left side of the periodic table, is a non-metal. At room temperature some of the non-metals are gaseous, some are solid and one is liquid. Non-metals generally differ from the metals in appearance (▶ FIGURE 2.12) and in other physical properties. Many of the elements that lie along the line that separates metals from non-metals, such as antimony (Sb), have properties that fall between those of metals and those of non-metals. These elements are often referred to as **metalloids**.

▲ FIGURE 2.12 Some familiar examples of metals and non-metals.

 CONCEPT CHECK 5

Chlorine is a halogen. Locate this element in the periodic table.
a. What is its symbol?
b. In what period and in what group is the element located?
c. What is its atomic number?
d. Is chlorine a metal or non-metal?

SAMPLE EXERCISE 2.5   Using the periodic table

Which two of the following elements would you expect to show the greatest similarity in chemical and physical properties: B, Ca, F, He, Mg, P?

SOLUTION

Elements that are in the same group of the periodic table are most likely to exhibit similar chemical and physical properties. We therefore expect that Ca and Mg should be most alike because they are in the same group (2, the alkaline earth metals).

2.6 | MOLECULES AND MOLECULAR COMPOUNDS

The atom is the smallest representative sample of an element, but only the noble gas elements are normally found in nature as isolated atoms. Most matter is composed of molecules or ions, both of which are formed from atoms. We examine molecules here and ions in Section 2.7.

A **molecule** is an assembly of two or more atoms tightly bound together. The resultant 'package' of atoms behaves in many ways as a single, distinct object, just as a television set composed of many parts can be recognised as a single object.

Molecules and Chemical Formulae

Many elements are found in nature in molecular form; that is, two or more of the same type of atom are bound together. For example, the oxygen normally found in air consists of molecules that contain two oxygen atoms. We represent this molecular form of oxygen by the **chemical formula** O_2 (read as 'oh two'). The subscript in the formula tells us that two oxygen atoms are present in each molecule. A molecule that is made up of two atoms is called a **diatomic molecule**. Oxygen also exists in another molecular form known as *ozone*. Molecules of ozone consist of three oxygen atoms, making the chemical formula for this substance O_3. Molecules that contain more than two atoms, such as ozone and water (H_2O), are known as **polyatomic molecules**. Even though 'normal' oxygen (O_2) and ozone are both composed only of oxygen atoms, they exhibit very different chemical and physical properties. For example, O_2 is essential for life, but O_3 is toxic; O_2 is odourless, whereas O_3 has a sharp, pungent smell.

The elements that normally occur as diatomic molecules are hydrogen, nitrogen, oxygen and the halogens. Their locations in the periodic table are shown in ▼ **FIGURE 2.13**. When we speak of the substance hydrogen, we mean H_2 unless we explicitly indicate otherwise. Likewise, when we speak of oxygen, nitrogen or any of the halogens, we are referring to N_2, O_2, F_2, Cl_2, Br_2 or I_2.

Compounds that are composed of molecules are called **molecular compounds** and may contain more than one type of atom. A molecule of water, for example, consists of two hydrogen atoms and one oxygen atom. It is therefore represented by the chemical formula H_2O. Lack of a subscript on the O indicates one atom of O per water molecule. Another compound composed of these same elements (in different relative proportions) is hydrogen peroxide, H_2O_2. The properties of hydrogen peroxide are very different from the properties of water.

Several common molecules are shown in ▶ **FIGURE 2.14**. Notice how the composition of each compound is given by its chemical formula. Notice also that these substances are composed only of non-metallic elements. *Most molecular substances that we encounter contain only non-metals.*

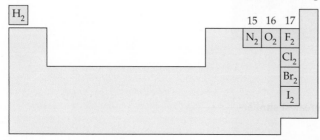

▲ **FIGURE 2.13   Diatomic molecules.**
Seven common elements exist as diatomic molecules at room temperature.

Molecular and Empirical Formulae

Chemical formulae that indicate the actual numbers and types of atoms in a molecule are called **molecular formulae**. (The formulae in Figure 2.14 are molecular formulae.) Chemical formulae that give only the relative number of atoms of each type in a molecule are called **empirical formulae** or **simplest formulae**. The subscripts in an empirical formula are always the smallest possible whole-number ratios. The molecular formula for hydrogen peroxide is H_2O_2, for example, whereas its empirical formula is HO. The molecular formula for ethylene is C_2H_4, and its empirical formula is CH_2. For many substances, the molecular formula and the empirical formula are identical, as in the case for water, H_2O.

Molecular formulae provide more information about molecules than do empirical formulae. Whenever we know the molecular formula of a compound, we can determine its empirical formula. The converse is not true, however; if we know the empirical formula of a substance, we can't determine its molecular formula unless we have more information. So why do chemists bother with empirical formulae? Once the empirical formula is known, additional experiments can give the information needed to convert the empirical formula to the molecular one. In addition, there are substances, such as the most common forms of elemental carbon, that don't exist as isolated molecules. For these substances, we must rely on empirical formulae. Thus all the common forms of elemental carbon are represented by the element's chemical symbol C, which is the empirical formula for all the forms.

Hydrogen, H_2 Oxygen, O_2

Water, H_2O Hydrogen peroxide, H_2O_2

Carbon monoxide, CO   Carbon dioxide, CO_2

SAMPLE EXERCISE 2.6   Relating empirical and molecular formulae

Write the empirical formulae for the following molecules: **(a)** glucose, a substance also known as either blood sugar or dextrose, whose molecular formula is $C_6H_{12}O_6$; **(b)** nitrous oxide, a substance used as an anaesthetic and commonly called laughing gas, whose molecular formula is N_2O.

SOLUTION

(a) The subscripts of an empirical formula are the smallest whole-number ratios. The smallest ratios are obtained by dividing each subscript by the largest common factor, in this case 6. The resultant empirical formula for glucose is CH_2O. **(b)** Because the subscripts in N_2O are already the lowest integral numbers, the empirical formula for nitrous oxide is the same as its molecular formula, N_2O.

PRACTICE EXERCISE

Give the empirical formula for the substance called diborane, whose molecular formula is B_2H_6.

Answer: BH_3

(See also Exercises 2.28–2.30.)

Methane, CH_4 Ethene, C_2H_4

▲ **FIGURE 2.14 Molecular models.** Notice how the chemical formulae of these simple molecules correspond to their compositions.

Picturing Molecules

The molecular formula of a substance summarises the composition of the substance but does not show how the atoms come together to form the molecule. The **structural formula** of a substance shows which atoms are attached to which within the molecule. For example, the structural formulae for water, hydrogen peroxide and methane (CH_4) can be written as follows:

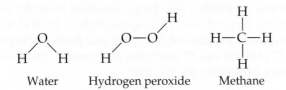

Water   Hydrogen peroxide   Methane

The atoms are represented by their chemical symbols, and lines are used to represent the bonds that hold the atoms together.

A structural formula usually does not depict the actual geometry of the molecule, that is, the actual angles between the lines joining the nuclei. A structural formula can be written as a *perspective drawing*, however, to give some sense of three-dimensional shape, as shown in ▼ **FIGURE 2.15**.

▲ FIGURE IT OUT

What advantage does a ball-and-stick model have over a space-filling model?

CH_4

Molecular formula

Structural formula

Dashed wedge is bond behind page

Solid line is bond in plane of page

Wedge is bond out of page

Perspective drawing

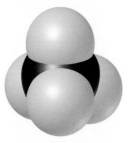

Ball-and-stick model

Space-filling model

▲ **FIGURE 2.15 Different representations of the methane (CH_4) molecule.** Structural formulae, perspective drawings, ball-and-stick models and space-filling models correspond to the molecular formula, and each helps us visualise the ways atoms are attached to each other.

Scientists also rely on various models to help visualise molecules. *Ball-and-stick models* show atoms as spheres and bonds as sticks. This type of model has the advantage of accurately representing the angles at which the atoms are attached to one another within the molecule (Figure 2.15). All atoms may be represented by balls of the same size, or the relative sizes of the balls may reflect the relative sizes of the atoms. Sometimes the chemical symbols of the elements are superimposed on the balls, but often the atoms are identified simply by colour.

A *space-filling model* depicts what the molecule would look like if the atoms were scaled up in size (Figure 2.15). These models show the relative sizes of the atoms, but the angles between atoms, which help define their molecular geometry, are often more difficult to see than in ball-and-stick models. As in ball-and-stick models, the identities of the atoms are indicated by their colours, but they may also be labelled with the element's symbol.

▲ **CONCEPT CHECK 6**

The structural formula for the substance ethane is shown here:

$$H-\overset{\displaystyle H}{\underset{\displaystyle H}{C}}-\overset{\displaystyle H}{\underset{\displaystyle H}{C}}-H$$

a. What is the molecular formula for ethane?
b. What is its empirical formula?
c. Which kind of molecular model would most clearly show the angles between atoms?

2.7 | IONS AND IONIC COMPOUNDS

The nucleus of an atom is unchanged by chemical processes, but atoms can readily gain or lose electrons. If electrons are removed from or added to a neutral atom, a charged particle called an **ion** is formed. An ion with a positive charge is called a **cation** (pronounced CAT-ion); a negatively charged ion is called an **anion** (AN-ion).

To see how ions form, consider the sodium atom, which has 11 protons and 11 electrons. This atom may lose one electron. The resulting cation has 11 protons and 10 electrons, which means it has a net charge of 1+

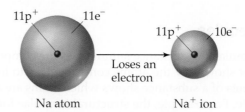

The net charge on an ion is represented by a superscript; the superscripts +, 2+ and 3+ for instance, mean a net charge resulting from the loss of one, two and three electrons, respectively. The superscripts −, 2− and 3− represent net charges resulting from the gain of one, two and three electrons, respectively. Chlorine, with 17 protons and 17 electrons, for example, can gain an electron in chemical reactions, producing the Cl^- ion:

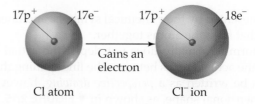

In general, metal atoms tend to lose electrons to form cations, whereas non-metal atoms tend to gain electrons to form anions.

SAMPLE EXERCISE 2.7   Writing chemical symbols for ions

Give the chemical symbol, including mass number, for each of the following ions: **(a)** the ion with 22 protons, 26 neutrons and 19 electrons; **(b)** the ion of sulfur that has 16 neutrons and 18 electrons.

SOLUTION

(a) The number of protons (22) is the atomic number of the element, which means this element is titanium (Ti). The mass number of this isotope is $22 + 26 = 48$ (the sum of the protons and neutrons). Because the ion has three more protons than electrons, it has a net charge of 3+. Thus the symbol for the ion is $^{48}Ti^{3+}$.

(b) By referring to a periodic table or a table of elements, we see that sulfur (S) has an atomic number of 16. Thus, each atom or ion of sulfur must contain 16 protons. We are told that the ion also has 16 neutrons, meaning the mass number of the ion is $16 + 16 = 32$. Because the ion has 16 protons and 18 electrons, its net charge is 2−. Thus the symbol for the ion is $^{32}S^{2-}$.

In general, we will focus on the net charges of ions and ignore their mass numbers.

PRACTICE EXERCISE

How many protons and electrons does the Se^{2-} ion possess?

Answer: 34 protons and 36 electrons

(See also Exercises 2.36, 2.37.)

In addition to simple ions, such as Na^+ and Cl^- there are **polyatomic ions**, such as NH_4^+ (ammonium ion) and SO_4^{2-} (sulfate ion). These latter ions consist of atoms joined as in a molecule, but they have a net positive or negative charge. We consider further examples of polyatomic ions in Section 2.8.

It is important to realise that the chemical properties of ions are very different from the chemical properties of the atoms from which the ions are derived. The difference is like the change from Dr Jekyll to Mr Hyde: although a given atom and its ion may be essentially the same (plus or minus a few electrons), the behaviour of the ion is very different from that of the atom.

Predicting Ionic Charges

Many atoms gain or lose electrons so as to end up with the same number of electrons as the noble gas closest to them in the periodic table. This is called the octet rule (⚬⚬ Section 8.1, 'Chemical Bonds, Lewis Symbols and the Octet Rule'). The members of the noble gas family are chemically very non-reactive and form very few compounds. We might deduce that this is because their electron arrangements are very stable. Nearby elements can obtain these same stable arrangements by losing or gaining electrons. For example, loss of one electron from an atom of sodium leaves it with the same number of electrons as the neutral neon atom (atomic number 10). Similarly, when chlorine gains an electron, it ends up with 18, the same number of electrons as in argon (atomic number 18). We will use this simple observation to explain the formation of ions until Chapter 8, where we discuss chemical bonding.

⚬⚬ Find out more on page 253

SAMPLE EXERCISE 2.8   Predicting the charges of ions

Predict the charge expected for the most stable ion of barium and for the most stable ion of oxygen.

SOLUTION

We will assume that these elements form ions that have the same number of electrons as the nearest noble gas atom. From the periodic table, we see that barium has atomic number 56. The nearest noble gas is xenon, atomic number 54. Barium can attain a stable arrangement of 54 electrons by losing two of its electrons, forming the Ba^{2+} cation.

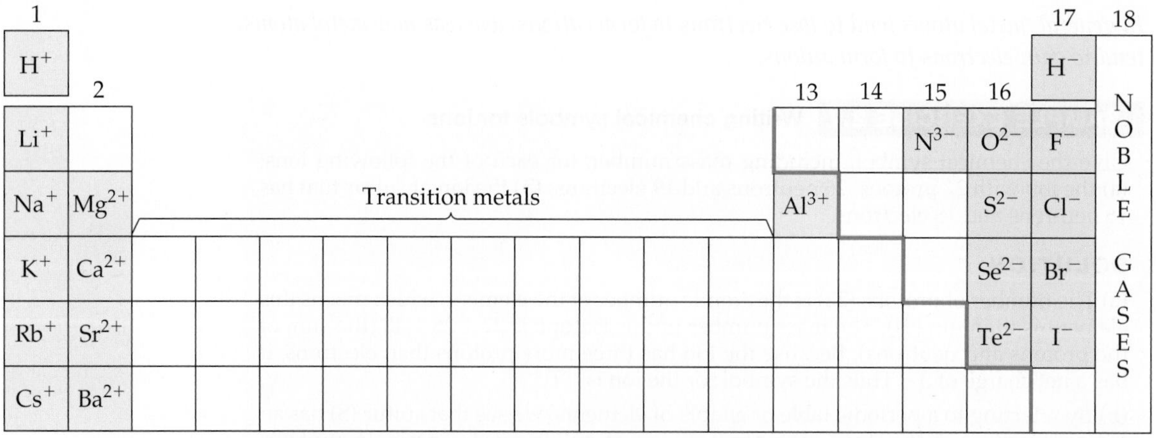

1													13	14	15	16	17	18
H^+	2																H^-	N O B L E G A S E S
Li^+															N^{3-}	O^{2-}	F^-	
Na^+	Mg^{2+}				Transition metals								Al^{3+}			S^{2-}	Cl^-	
K^+	Ca^{2+}															Se^{2-}	Br^-	
Rb^+	Sr^{2+}															Te^{2-}	I^-	
Cs^+	Ba^{2+}																	

► **FIGURE 2.16 Charges of some common ions.** Notice that the steplike line that divides metals from non-metals also separates cations from anions.

Oxygen has atomic number 8. The nearest noble gas is neon, atomic number 10. Oxygen can attain this stable electron arrangement by gaining two electrons, thereby forming the O^{2-} anion.

PRACTICE EXERCISE
Predict the charge expected for the most stable ion of **(a)** aluminium and **(b)** fluorine.
Answer: **(a)** 3+, and **(b)** 1−
(See also Exercises 2.38, 2.39.)

The periodic table is very useful for remembering the charges of ions, especially those of the elements on the left and right sides of the table. As ▲ FIGURE 2.16 shows, the charges of these ions relate in a simple way to their positions in the table. On the left side of the table, for example, the group 1 elements (the alkali metals) form 1+ ions and the group 2 elements (the alkaline earth metals) form 2+ ions. On the other side of the table, the group 17 elements (the halogens) form 1− ions (17 − 18 = 1−) and the group 16 elements form 2− ions (16 − 18 = 2−). As we will see later in the text, many of the other groups do not lend themselves to such simple rules.

Ionic Compounds

A great deal of chemical activity involves the transfer of electrons from one substance to another, and, as we just saw, ions form when one or more electrons transfer from one neutral atom to another. ▼ FIGURE 2.17 shows that when ele-

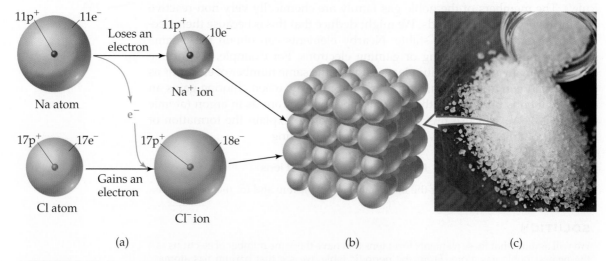

▲ **FIGURE 2.17 Formation of an ionic compound.** (a) The transfer of an electron from a Na atom to a Cl atom leads to the formation of a Na^+ ion and a Cl^- ion. (b) Arrangement of these ions in solid sodium chloride, NaCl. (c) A sample of sodium chloride crystals.

mental sodium is allowed to react with elemental chlorine, an electron transfers from a neutral sodium atom to a neutral chlorine atom. We are left with a Na^+ ion and a Cl^- ion. Because objects of opposite charge attract, the Na^+ and the Cl^- ions bind together to form the compound sodium chloride (NaCl). Sodium chloride, which we know better as common table salt, is an example of an **ionic compound**, a compound that contains both positively and negatively charged ions.

We can often tell whether a compound is ionic (consisting of ions) or molecular (consisting of molecules) from its composition. In general, cations are metal ions, whereas anions are non-metal ions. Consequently, *ionic compounds are generally combinations of metals and non-metals*, as in NaCl. In contrast, *molecular compounds are generally composed of non-metals only*, as in H_2O.

SAMPLE EXERCISE 2.9   **Identifying ionic and molecular compounds**

Which of the following compounds would you expect to be ionic: N_2O, Na_2O, $CaCl_2$, SF_4?

SOLUTION

We would predict that Na_2O and $CaCl_2$ are ionic compounds because they are composed of a metal combined with a non-metal. The other two compounds, composed entirely of non-metals, are predicted (correctly) to be molecular compounds.

PRACTICE EXERCISE

Which of the following compounds are molecular: CBr_4, FeS, P_4O_6, PbF_2?

Answer: CBr_4 and P_4O_6

The ions in ionic compounds are arranged in three-dimensional structures. The arrangement of Na^+ and Cl^- ions in NaCl is shown in Figure 2.17. Because there is no discrete molecule of NaCl, we are able to write only an empirical formula for this substance. In fact, only empirical formulae can be written for ionic compounds.

We can readily write the empirical formula for an ionic compound if we know the charges of the ions of which the compound is composed. Chemical compounds are always electrically neutral. Consequently, the ions in an ionic compound always occur in such a ratio that the total positive charge equals the total negative charge. Thus there is one Na^+ to one Cl^- (giving NaCl), one Ba^{2+} to two Cl^- (giving $BaCl_2$) and so forth.

As you consider these and other examples, you will see that if the charges on the cation and anion are equal, the subscript on each ion will be 1. If the charges are not equal, the charge on one ion (without its sign) will become the subscript on the other ion. For example, the ionic compound formed from Mg (which forms Mg^{2+} ions) and N (which forms N^{3-} ions) is Mg_3N_2:

$$Mg^{2+} \quad N^{3-} \longrightarrow Mg_3N_2$$

⚠ **CONCEPT CHECK 7**

Why don't we write the formula for the compound formed by Ca^{2+} and O^{2-} as Ca_2O_2?

SAMPLE EXERCISE 2.10   **Using ionic charge to write empirical formulae for ionic compounds**

What are the empirical formulae of the compounds formed by **(a)** Al^{3+} and Cl^- ions, **(b)** Al^{3+} and O^{2-} ions, **(c)** Mg^{2+} and NO_3^- ions?

SOLUTION

(a) Three Cl^- ions are required to balance the charge of one Al^{3+} ion. Thus the formula is $AlCl_3$. **(b)** Two Al^{3+} ions are required to balance the charge of three O^{2-} ions (that is, the total positive charge is 6+ and the total negative charge is 6−). Thus the formula is Al_2O_3. **(c)** Two NO_3^- ions are needed to balance the charge of one Mg^{2+}. Thus the formula is $Mg(NO_3)_2$. In this case the formula for the entire polyatomic ion NO_3^- must be enclosed in parentheses so that it is clear that the subscript 2 applies to all the atoms of that ion.

PRACTICE EXERCISE

Write the empirical formulae for the compounds formed by the following ions:
(a) Na^+ and PO_4^{3-}, **(b)** Zn^{2+} and SO_4^{2-}, **(c)** Fe^{3+} and CO_3^{2-}

Answers: **(a)** Na_3PO_4, **(b)** $ZnSO_4$, **(c)** $Fe_2(CO_3)_3$
(See also Exercises 2.40, 2.41.)

MY WORLD OF CHEMISTRY

ELEMENTS REQUIRED BY LIVING ORGANISMS

The coloured regions of ▶ **FIGURE 2.18** shows the elements essential to life. More than 97% of the mass of most organisms is made up of just six of these elements—oxygen, carbon, hydrogen, nitrogen, phosphorus and sulfur. Water is the most common compound in living organisms, accounting for at least 70% of the mass of most cells. In the solid components of cells, carbon is the most prevalent element by mass. Carbon atoms are found in a vast variety of organic molecules, bonded either to other carbon atoms or to atoms of other elements. All proteins, for example, contain the group

$$
\begin{array}{c}
\quad\ \ O \\
\quad\ \ \| \\
-N-C- \\
\ | \\
\ R
\end{array}
$$

which occurs repeatedly in the molecules. (R is either an H atom or a combination of atoms, such as CH_3.)

In addition, 23 more elements have been found in various living organisms. Five are ions required by all organisms: Ca^{2+}, Cl^-, Mg^{2+}, K^+, and Na^+. Calcium ions, for example, are necessary for the formation of bone and transmission of nervous system signals. Many other elements are needed in only very small quantities and consequently are called *trace* elements. For example, trace quantities of copper are required in the diet of humans to aid in the synthesis of haemoglobin.

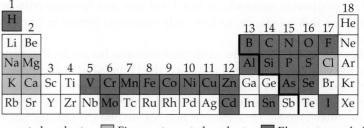

| | Six most abundant essential elements | | Five next most abundant essential elements | | Elements needed only in trace quantities |

▲ **FIGURE 2.18   Elements essential to life.**

2.8 | NAMING INORGANIC COMPOUNDS

To obtain information about a particular substance, you must know its name and chemical formula. The names and formulae of compounds are essential vocabulary in chemistry. The system used in naming substances is called **chemical nomenclature**, from the Latin words *nomen* (name) and *calare* (to call).

There are now more than 50 million known chemical substances. Naming them all would be a hopelessly complicated task if each had a special name independent of all others. Many important substances that have been known for a long time, such as water (H_2O) and ammonia (NH_3), do have individual, traditional names (so-called 'common' names). For most substances, however, we rely on a systematic set of rules that leads to an informative and unique name for each substance, a name based on the composition of the substance.

The rules for chemical nomenclature are based on the division of substances into categories. The major division is between organic and inorganic compounds. *Organic compounds* contain carbon, usually in combination with hydrogen, oxygen, nitrogen or sulfur. All others are *inorganic compounds*. Early chemists associated organic compounds with plants and animals, and they associated inorganic compounds with the non-living portion of our world. Although this distinction between living and non-living matter is no longer pertinent, the classification between organic and inorganic compounds continues to be useful. Among inorganic compounds we consider three categories: ionic compounds, molecular compounds and acids.

Names and Formulae of Ionic Compounds

Recall from Section 2.7 that ionic compounds usually consist of metal ions combined with non-metal ions. The metals form positive ions, and the non-metals form negative ions. Let's examine the naming of positive ions, then the naming of negative ones. After that, we will consider how to put the names of the ions together to identify the complete ionic compound.

1. *Positive ions (cations)*

  (a) *Cations formed from metal atoms have the same name as the metal:*

  Na^+ sodium ion Zn^{2+} zinc ion Al^{3+} aluminium ion

  Ions formed from a single atom are called *monatomic ions*.

  (b) *If a metal can form cations with different charges, the positive charge is indicated by a Roman numeral in parentheses following the name of the metal:*

Fe^{2+} iron(II) ion	Cu^+ copper(I) ion
Fe^{3+} iron(III) ion	Cu^{2+} copper(II) ion

  Ions of the same element that have different charges exhibit different properties, such as different colours (▼ **FIGURE 2.19**).

  Most of the metals that can form cations with different charges are *transition metals*, elements that occur in the middle block of elements, from group 3 to group 12 in the periodic table. The charges of these ions are indicated by Roman numerals as mentioned above. The metals that form only one cation are those of group 1 (Li^+, Na^+, K^+, Rb^+ and Cs^+) and group 2 (Mg^{2+}, Ca^{2+}, Sr^{2+} and Ba^{2+}), as well as Al^{3+} (group 13) and two transition metal ions: Ag^+ (group 11) and Zn^{2+} (group 12). Charges are not expressed explicitly when naming these ions. However, if there is any doubt in your mind whether a metal forms more than one cation, use a Roman numeral to indicate the charge. It is never wrong to do so, even though it may be unnecessary.

  An older method still widely used for distinguishing between two differently charged ions of a metal is to apply the ending *-ous* or *-ic*. These endings represent the lower and higher charged ions, respectively. They are added to the root of the element's Latin name:

Fe^{2+} ferrous ion	Cu^+ cuprous ion
Fe^{3+} ferric ion	Cu^{2+} cupric ion

  Although we will avoid using these older names in this text, you might encounter them elsewhere.

  (c) *Cations formed from non-metal atoms have names that end in* -ium:

  NH_4^+ ammonium ion H_3O^+ hydronium ion

▲ **FIGURE 2.19 Different ions of the same element have different properties.** Both substances shown are compounds of iron. The substance on the left is Fe_3O_4, which contains Fe^{2+} and Fe^{3+} ions. The substance on the right is Fe_2O_3, which contains only Fe^{3+} ions.

TABLE 2.4 • Common cations*

Charge	Formula	Name	Formula	Name
1+	**H⁺**	**hydrogen ion**	**NH₄⁺**	**ammonium ion**
	Li⁺	lithium ion	Cu⁺	copper(I) or cuprous ion
	Na⁺	**sodium ion**		
	K⁺	**potassium ion**		
	Cs⁺	caesium ion		
	Ag⁺	**silver ion**		
2+	**Mg²⁺**	**magnesium ion**	Co²⁺	cobalt(II) or cobaltous ion
	Ca²⁺	**calcium ion**	**Cu²⁺**	**copper(II)** or cupric ion
	Sr²⁺	strontium ion	**Fe²⁺**	**iron(II)** or ferrous ion
	Ba²⁺	barium ion	Mn²⁺	manganese(II) or manganous ion
	Zn²⁺	**zinc ion**	Hg₂²⁺	mercury(I) or mercurous ion
	Cd²⁺	cadmium ion	**Hg²⁺**	**mercury(II)** or mercuric ion
			Ni²⁺	nickel(II) or nickelous ion
			Pb²⁺	**lead(II)** or plumbous ion
			Sn²⁺	tin(II) or stannous ion
3+	**Al³⁺**	**aluminium ion**	Cr³⁺	chromium(III) or chromic ion
			Fe³⁺	**iron(III)** or ferric ion

*The ions we use most often in this course are in bold. Learn them first.

These two ions are the only ions of this kind that we encounter frequently in the text. They are both polyatomic. The vast majority of cations you will encounter in this text are monatomic metal ions.

The names and formulae of some common cations are shown in ▲ TABLE 2.4 and are also included in a table of common ions at the end of this text. The ions listed on the left in Table 2.4 are the monatomic ions that do not have variable charges. Those listed on the right are either polyatomic cations or cations with variable charges. The Hg_2^{2+} ion is unusual because this metal ion is not monatomic. It is called the mercury(I) ion because it can be thought of as two Hg^+ ions fused together. The cations that you will encounter most frequently are shown in bold. These are the ones you should learn first.

 CONCEPT CHECK 8

Why is CrO named using a Roman numeral, chromium(II) oxide, whereas CaO is named without a Roman numeral in the name, calcium oxide?

2. *Negative ions (anions)*

   (a) *The names of monatomic anions are formed by replacing the ending of the name of the element with* -ide:

H^- hydride ion	O^{2-} oxide ion	N^{3-} nitride ion

   A few simple polyatomic anions also have names ending in *-ide*:

OH^- hydroxide ion	CN^- cyanide ion	O_2^{2-} peroxide ion

   (b) *Polyatomic anions containing oxygen have names ending in* -ate *or* -ite. These anions are called **oxyanions**. The ending *-ate* is used for the most

FIGURE IT OUT

Name the anion obtained by removing one oxygen atom from the perbromate ion, BrO_4^-.

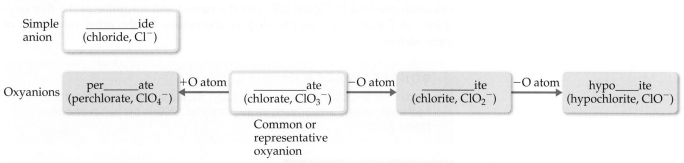

▲ **FIGURE 2.20   Summary of the procedure for naming anions.** The first part of the name, such as 'chlor' for chlorine or 'sulf' for sulfur, goes in the blank.

common oxyanion of an element. The ending *-ite* is used for an oxyanion that has the same charge but one O atom fewer:

NO_3^-	nitrate ion	SO_4^{2-}	sulfate ion
NO_2^-	nitrite ion	SO_3^{2-}	sulfite ion

Prefixes are used when the series of oxyanions of an element extends to four members, as with the halogens. The prefix *per-* usually indicates one more O atom than the oxyanion ending in *-ate*; the prefix *hypo-* normally indicates one O atom fewer than the oxyanion ending in *-ite*:

ClO_4^-	perchlorate ion (one more O atom than chlorate)
ClO_3^-	**chlorate ion**
ClO_2^-	chlorite ion (one O atom fewer than chlorate)
ClO^-	hypochlorite ion (one O atom fewer than chlorite)

These rules are summarised in ▲ **FIGURE 2.20**.

 CONCEPT CHECK 9

What information is conveyed by the endings *-ide*, *-ate* and *-ite* in the name of an anion?

Students often have a hard time remembering the number of oxygen atoms in the various oxyanions and the charges of these ions. ▼ **FIGURE 2.21** lists the oxyanions of C, N, P, S and Cl that contain the maximum number of O atoms. There is a periodic pattern to these formulae that can help you remember them. Notice that C and N, which are in the second period of the periodic table, have only three O atoms each, whereas P, S and Cl, which are in the third period, have four O atoms each. If we begin at the lower right side of the figure, with

	14	15	16	17
2	CO_3^{2-} **Carbonate ion**	NO_3^- **Nitrate ion**		
3		PO_4^{3-} **Phosphate ion**	SO_4^{2-} **Sulfate ion**	ClO_4^- **Perchlorate ion**

◀ **FIGURE 2.21   Common oxyanions.** The composition and charge of common oxyanions are related to their location in the periodic table.

Cl, we see that the charges increase from right to left, from 1− for Cl (ClO_4^-) to 3− for P (PO_4^{3-}). In the second period the charges also increase from right to left, from 1− for N (NO_3^-) to 2− for C (CO_3^{2-}). Each anion shown in Figure 2.21 has a name ending in -*ate*. The ClO_4^- ion also has a *per-* prefix. If you know the rules summarised in Figure 2.20 and the names and formulae of the five oxyanions in Figure 2.21, you can deduce the names for the other oxyanions of these elements.

⚠ **CONCEPT CHECK 10**

Predict the formulae for the borate ion and silicate ion, assuming that they contain a single B and Si atom, respectively, and follow the trends shown in Figure 2.21.

SAMPLE EXERCISE 2.11 **Determining the formula of an oxyanion from its name**

Based on the formula for the sulfate ion, predict the formula for **(a)** the selenate ion and **(b)** the selenite ion. (Sulfur and selenium are both members of group 16 and form analogous oxyanions.)

SOLUTION

(a) The sulfate ion is SO_4^{2-}. The analogous selenate ion is therefore SeO_4^{2-}. **(b)** The ending -*ite* indicates an oxyanion with the same charge but one O atom fewer than the corresponding oxyanion that ends in -*ate*. Thus the formula for the selenite ion is SeO_3^{2-}.

PRACTICE EXERCISE

The formula for the bromate ion is analogous to that for the chlorate ion. Write the formula for the hypobromite and perbromate ions.

Answer: BrO^- and BrO_4^-

(See also Exercises 2.42, 2.43.)

(c) *Anions derived by adding H^+ to an oxyanion are named by adding as a prefix the word hydrogen or dihydrogen, as appropriate:*

CO_3^{2-}	carbonate ion	PO_4^{3-}	phosphate ion
HCO_3^-	hydrogen carbonate ion	$H_2PO_4^-$	dihydrogen phosphate ion

Notice that each H^+ reduces the negative charge of the parent anion by one. An older method for naming some of these ions is to use the prefix *bi-*. Thus the HCO_3^- ion is commonly called the bicarbonate ion, and HSO_4^- is sometimes called the bisulfate ion.

The names and formulae of the common anions are listed in ▶ TABLE 2.5 and on the back inside cover of this text. Those anions whose names end in -*ide* are listed on the left of Table 2.5, and those whose names end in -*ate* are listed on the right. The most common of these ions are shown in bold. These are the ones you should learn first. The formulae of the ions whose names end with -*ite* can be derived from those ending in -*ate* by removing an O atom. Notice the location of the monatomic ions in the periodic table. Those of group 17 always have a 1− charge (F^-, Cl^-, Br^- and I^-) and those of group 16 have a 2− charge (O^{2-} and S^{2-}).

1. *Ionic compounds*

   Names of ionic compounds consist of the cation name followed by the anion name:

$CaCl_2$	calcium chloride
$Al(NO_3)_3$	aluminium nitrate
$Cu(ClO_4)_2$	copper(II) perchlorate (or cupric perchlorate)

TABLE 2.5 • Common anions[*]				
Charge	**Formula**	**Name**	**Formula**	**Name**
1−	H⁻	hydride ion	CH_3COO^-	**acetate ion**
			(or $C_2H_3O_2^-$)	
	F⁻	**fluoride ion**	ClO_3^-	chlorate ion
	Cl⁻	**chloride ion**	ClO_4^-	**perchlorate ion**
	Br⁻	**bromide ion**	NO_3^-	**nitrate ion**
	I⁻	**iodide ion**	MnO_4^-	permanganate ion
	CN⁻	cyanide ion		
	OH⁻	**hydroxide ion**		
2−	**O²⁻**	**oxide ion**	CO_3^{2-}	**carbonate ion**
	O_2^{2-}	peroxide ion	CrO_4^{2-}	chromate ion
	S²⁻	**sulfide ion**	$Cr_2O_7^{2-}$	dichromate ion
			SO_4^{2-}	**sulfate ion**
3−	N³⁻	nitride ion	PO_4^{3-}	**phosphate ion**

*The ions we use most often are in bold. Learn them first.

In the chemical formulae for aluminium nitrate and copper(II) perchlorate, parentheses followed by the appropriate subscript are used because the compounds contain two or more polyatomic ions.

SAMPLE EXERCISE 2.12 Determining the names of ionic compounds from their formulae

Name the following compounds: **(a)** K_2SO_4, **(b)** $Ba(OH)_2$, **(c)** $FeCl_3$.

SOLUTION

Each compound is ionic and is named using the guidelines we have already discussed. In naming ionic compounds, it is important to recognise polyatomic ions and to determine the charge of cations with variable charge. **(a)** The cation in this compound is K^+, and the anion is SO_4^{2-}. (If you thought the compound contained S^{2-} and O^{2-} ions, you failed to recognise the polyatomic sulfate ion.) Putting together the names of the ions, we have the name of the compound, potassium sulfate. **(b)** In this case the compound is composed of Ba^{2+} and OH^- ions. Ba^{2+} is the barium ion and OH^- is the hydroxide ion. Thus the compound is called barium hydroxide. **(c)** You must determine the charge of Fe in this compound because an iron atom can form more than one cation. Because the compound contains three Cl^- ions, the cation must be Fe^{3+}, which is the iron(III), or ferric, ion. The Cl^- ion is the chloride ion. Thus the compound is iron(III) chloride or ferric chloride.

PRACTICE EXERCISE

Name the following compounds: **(a)** NH_4Br, **(b)** Cr_2O_3, **(c)** $Co(NO_3)_2$.

Answers: **(a)** Ammonium bromide, **(b)** chromium(III) oxide, **(c)** cobalt(II) nitrate

(See also Exercises 2.44, 2.45.)

SAMPLE EXERCISE 2.13 Determining the formulae of ionic compounds from their names

Write the chemical formulae for the following compounds: **(a)** potassium sulfide, **(b)** calcium hydrogen carbonate, **(c)** nickel(II) perchlorate.

SOLUTION

In going from the name of an ionic compound to its chemical formula, you must know the charges of the ions to determine the subscripts. **(a)** The potassium ion is K^+ and the sulfide ion is S^{2-}. Because ionic compounds are electrically neutral, two K^+ ions are required to balance the charge of one S^{2-} ion, giving the empirical formula of the compound, K_2S. **(b)** The calcium ion is Ca^{2+}. The carbonate ion is CO_3^{2-}, so the hydrogen carbonate ion is HCO_3^-. Two HCO_3^- ions are needed to balance the positive charge of Ca^{2+}, giving $Ca(HCO_3)_2$. **(c)** The nickel(II) ion is Ni^{2+}. The perchlorate ion is ClO_4^-. Two ClO_4^- ions are required to balance the charge on one Ni^{2+} ion, giving $Ni(ClO_4)_2$.

PRACTICE EXERCISE
Give the chemical formula for **(a)** magnesium sulfate, **(b)** silver sulfide, **(c)** lead(II) nitrate.

Answers: **(a)** $MgSO_4$, **(b)** Ag_2S, **(c)** $Pb(NO_3)_2$
(See also Exercises 2.46, 2.47.)

Names and Formulae of Acids

Acids are an important class of hydrogen-containing compounds and are named in a special way. For our present purposes, an *acid* is a substance whose molecules yield hydrogen ions (H^+) when dissolved in water. When we encounter the chemical formula for an acid at this stage of the course, it will be written with H as the first element, as in HCl and H_2SO_4.

An acid is composed of an anion connected to enough H^+ ions to neutralise, or balance, the anion's charge. Thus the SO_4^{2-} ion requires two H^+ ions, forming H_2SO_4. The name of an acid is related to the name of its anion, as summarised in ▼ **FIGURE 2.22**.

1. *Acids containing anions whose names end in* -ide *are named by changing the* -ide *ending to* -ic, *adding the prefix* hydro- *to this anion name, and then following with the word acid, as in the following examples:*

Anion	Corresponding acid
Cl^- (chloride)	HCl (hydrochloric acid)
S^{2-} (sulfide)	H_2S (hydrosulfuric acid)

2. *Acids containing anions whose names end in* -ate *or* -ite *are named by changing* -ate *to* -ic *and* -ite *to* -ous, *and then adding the word* acid. Prefixes in the anion name are retained in the name of the acid. These rules are illustrated by the oxyacids of chlorine:

Anion	Corresponding acid
ClO_4^- (perchlorate)	$HClO_4$ (perchloric acid)
ClO_3^- (chlorate)	$HClO_3$ (chloric acid)
ClO_2^- (chlorite)	$HClO_2$ (chlorous acid)
ClO^- (hypochlorite)	HClO (hypochlorous acid)

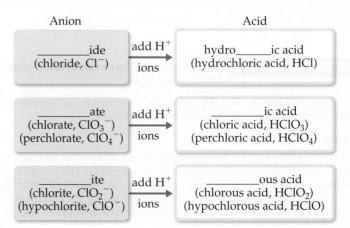

▶ **FIGURE 2.22 How anion names and acid names relate.** The prefixes *per-* and *hypo-* are retained in going from the anion to the acid.

SAMPLE EXERCISE 2.14   **Relating the names and formulae of acids**

Name the following acids: **(a)** HCN, **(b)** HNO_3, **(c)** H_2SO_4, **(d)** H_2SO_3.

SOLUTION

(a) The anion from which this acid is derived is CN^-, the cyanide ion. Because this ion has an *-ide* ending, the acid is given a *hydro-* prefix and an *-ic* ending: hydrocyanic acid. Only water solutions of HCN are referred to as hydrocyanic acid: the pure compound, which is a gas under normal conditions, is called hydrogen cyanide. Both hydrocyanic acid and hydrogen cyanide are *extremely* toxic. **(b)** Because NO_3^- is the nitrate ion, HNO_3 is called nitric acid (the *-ate* ending of the anion is replaced with an *-ic* ending in naming the acid). **(c)** Because SO_4^{2-} is the sulfate ion, H_2SO_4 is called sulfuric acid. **(d)** Because SO_3^{2-} is the sulfite ion, H_2SO_3 is sulfurous acid (the *-ite* ending of the anion is replaced with an *-ous* ending).

PRACTICE EXERCISE

Give the chemical formulae for **(a)** hydrobromic acid, **(b)** carbonic acid.

Answers: **(a)** HBr, **(b)** H_2CO_3

(See also Exercises 2.48, 2.49.)

Names and Formulae of Binary Molecular Compounds

The procedures used for naming *binary* (two-element) molecular compounds are similar to those used for naming ionic compounds.

1. *The name of the element further to the left in the periodic table is usually written first.* An exception to this rule occurs in the case of compounds that contain oxygen. Oxygen is always written last except when combined with fluorine.

2. *If both elements are in the same group in the periodic table, the one having the higher atomic number is named first.*

3. *The name of the second element is given an* -ide *ending.*

4. *Greek prefixes (▶ TABLE 2.6) are used to indicate the number of atoms of each element.* The prefix *mono-* is never used with the first element. When the prefix ends in *a* or *o* and the name of the second element begins with a vowel (such as *oxide*), the *a* or *o* of the prefix is often dropped.
    The following examples illustrate these rules:

Cl_2O	dichlorine monoxide	NF_3	nitrogen trifluoride
N_2O_4	dinitrogen tetroxide	P_4S_{10}	tetraphosphorus decasulfide

It is important to realise that you cannot predict the formulae of most molecular substances in the same way that you predict the formulae of ionic compounds. That is why we name molecular compounds using prefixes that explicitly indicate their composition. However, molecular compounds that contain hydrogen and one other element are an important exception. These compounds can be treated as if they were neutral substances containing H^+ ions and anions. Thus you can predict that the substance whose name is hydrogen chloride has the formula HCl, containing one H^+ to balance the charge of one Cl^-. (The name hydrogen chloride is used only for the pure compound; water solutions of HCl are called hydrochloric acid.) Similarly, the formula for hydrogen sulfide is H_2S because two H^+ are needed to balance the charge on S^{2-}.

TABLE 2.6 • Prefixes used in naming binary compounds formed between non-metals

Prefix	Meaning
Mono-	1
Di-	2
Tri-	3
Tetra-	4
Penta-	5
Hexa-	6
Hepta-	7
Octa-	8
Nona-	9
Deca-	10

SAMPLE EXERCISE 2.15 Relating the names and formulae of binary molecular compounds

Name the following compounds: **(a)** SO_2, **(b)** PCl_5, **(c)** N_2O_3.

SOLUTION

The compounds consist entirely of non-metals, so they are probably molecular rather than ionic. Using the prefixes in Table 2.6, we have **(a)** sulfur dioxide, **(b)** phosphorus pentachloride and **(c)** dinitrogen trioxide.

PRACTICE EXERCISE

Give the chemical formula for **(a)** silicon tetrabromide, **(b)** disulfur dichloride.

Answers: **(a)** $SiBr_4$, **(b)** S_2Cl_2

(See also Exercises 2.50, 2.51.)

2.9 | SOME SIMPLE ORGANIC COMPOUNDS

The study of compounds of carbon is called **organic chemistry** and, as noted earlier, compounds that contain carbon and hydrogen, often in combination with oxygen, nitrogen or other elements, are called *organic compounds*. We will examine organic compounds in detail in Chapter 22, but here we present a brief introduction to some of the simplest organic compounds.

Alkanes

Compounds that contain only carbon and hydrogen are called **hydrocarbons**. In the simplest class of hydrocarbons, **alkanes**, each carbon is bonded to four other atoms. The three smallest alkanes are methane (CH_4), ethane (C_2H_6) and propane (C_3H_8). The structural formulae of these three alkanes are as follows:

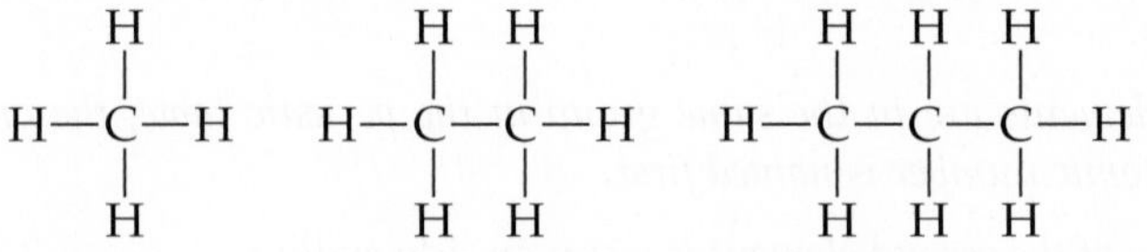

Methane   Ethane   Propane

Although hydrocarbons are binary molecular compounds, they are not named like the binary inorganic compounds discussed in Section 2.8. Instead, each alkane has a name that ends in *-ane*. The alkane with four carbons is called *butane*. For alkanes with five or more carbons, the names are derived from prefixes like those in Table 2.6. An alkane with eight carbon atoms, for example, is *octane* (C_8H_{18}), where the *octa-* prefix for eight is combined with the *-ane* ending for an alkane.

Some Derivatives of Alkanes

Other classes of organic compounds are obtained when one or more hydrogen atoms in an alkane are replaced with *functional groups*, which are specific groups of atoms. An **alcohol**, for example, is obtained by replacing a H atom of an alkane with an −OH group. The name of the alcohol is derived from that of the alkane by adding an *-ol* ending:

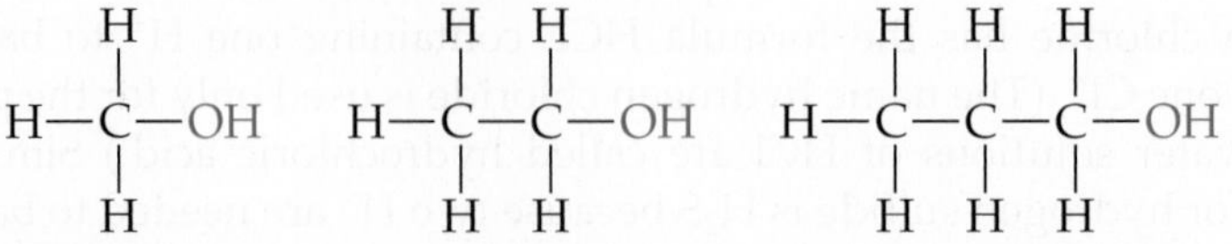

Methanol   Ethanol   1-Propanol

Alcohols have properties that are very different from the properties of the alkanes from which the alcohols are obtained. For example, methane, ethane and propane are all colourless gases under normal conditions, whereas methanol, ethanol and propanol are colourless liquids. We will discuss the reasons for these differences in Chapter 11.

The prefix '1' in the name 1-propanol indicates that the replacement of H with OH has occurred at one of the 'outer' carbon atoms rather than the 'middle' carbon atom. A different compound, called 2-propanol, is obtained when the OH functional group is attached to the middle carbon atom (▶ FIGURE 2.23).

Compounds with the same molecular formula but different arrangements of atoms are called **isomers**. There are many different kinds of isomers, as we will discover later in this book. What we have here with 1-propanol and 2-propanol are *structural isomers,* compounds having the same molecular formula but different structural formulae.

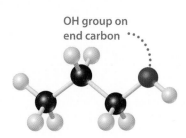

1-Propanol

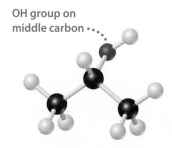

2-Propanol

▲ FIGURE 2.23 **The two forms (isomers) of propanol.**

CONCEPT CHECK 11

Draw the structural formulae of the two isomers of butane, C_4H_{10}.

Much of the richness of organic chemistry is possible because organic compounds can form long chains of carbon–carbon bonds. The series of alkanes that begins with methane, ethane and propane and the series of alcohols that begins with methanol, ethanol and propanol can both be extended for as long as we desire, in principle. The properties of alkanes and alcohols change as the chains get longer. Octanes, which are alkanes with eight carbon atoms, are liquids under normal conditions. If the alkane series is extended to tens of thousands of carbon atoms, we obtain *polyethene,* a solid substance that is used to make thousands of plastic products, such as plastic bags, food containers and laboratory equipment.

SAMPLE EXERCISE 2.16 **Writing structural and molecular formulae for hydrocarbons**

Assuming the carbon atoms in *pentane* are in a linear chain, write **(a)** the structural formula and **(b)** the molecular formula for this alkane.

SOLUTION

(a) Alkanes contain only carbon and hydrogen, and each carbon is attached to four other atoms. The name pentane contains the prefix *penta-* for five (Table 2.6), and we are told that the carbons are in a linear chain. If we then add enough hydrogen atoms to make four bonds to each carbon, we obtain the structural formula:

$$\begin{array}{ccccc} H & H & H & H & H \\ | & | & | & | & | \\ H-C-C-C-C-C-H \\ | & | & | & | & | \\ H & H & H & H & H \end{array}$$

This form of pentane is often called *n*-pentane, where the *n*- stands for 'normal' because all five carbon atoms are in one line in the structural formula.

(b) Once the structural formula is written, we determine the molecular formula by counting the atoms present. Thus *n*-pentane has the molecular formula C_5H_{12}.

PRACTICE EXERCISE

(a) What is the molecular formula of butane, the alkane with four carbons? **(b)** What are the name and molecular formula of an alcohol derived from butane?

Answers: **(a)** C_4H_{10}, **(b)** butanol, $C_4H_{10}O$ or C_4H_9OH

(See also Exercise 2.55(c).)

SAMPLE INTEGRATIVE EXERCISE | **Complex ions and subatomic particles**

How many protons, neutrons and electrons are there in the **(a)** phosphate ion and **(b)** hypochlorite ion?

SOLUTION

(a) The phosphate ion has the formula PO_4^{3-} (Table 2.5). There is one phosphorus atom and four oxygen atoms each having the number of protons, neutrons and electrons that can be obtained from the periodic table. In addition there are three extra electrons provided by the 3− charge. The atomic number of phosphorus is 15 (Figure 2.11) meaning there are 15 protons in the nucleus of the atom and 15 electrons surrounding the nucleus. The mass number (front page) is 31 (to a first approximation) so the number of neutrons is given by mass number − atomic number = 31 − 15 = 16. For each oxygen atom, the atomic number is 8, so there are eight protons and eight electrons and the mass number is 16, so the number of neutrons is 16 − 8 = 8. Since there are four oxygen atoms the contribution from the oxygen atoms is 4 × 8 = 32 protons, 4 × 8 = 32 electrons and 4 × 8 = 32 neutrons. In addition there are three extra electrons from the 3− charge. In total the number of protons = 15 + 32 = 47, number of electrons is 15 + 32 + 3 = 50 and number of neutrons is 16 + 32 = 48.

(b) The hypochlorite ion (Figure 2.20) has the formula ClO^-. The atomic number of chlorine is 17 so there are 17 protons and 17 electrons in the atom. The mass number (to a first approximation) is 35 so the number of neutrons is 18. As seen in (a) the oxygen atom has eight protons, eight electrons and eight neutrons. In addition there is one extra electron from the 1− charge. So in total there are 17 + 8 = 25 protons, 17 + 8 + 1 = 26 electrons and 18 + 8 = 26 neutrons.

CHAPTER SUMMARY AND KEY TERMS

SECTIONS 2.1 and 2.2 **Atoms** are the smallest units of an element that can combine with another element. Atoms themselves are composed of even smaller particles called **subatomic particles**. Some of the subatomic particles are charged and follow the usual behaviour of charged particles: particles with the same charge repel one another, whereas particles with unlike charges are attracted to one another. We considered some of the experiments that led to the discovery and characterisation of subatomic particles. Thomson's experiments on the behaviour of **cathode rays** in magnetic and electric fields led to the discovery of the electron and allowed its charge-to-mass ratio to be measured. Millikan's oil-drop experiment determined the charge of the electron. Becquerel's discovery of **radioactivity**, the spontaneous emission of radiation by atoms, gave further evidence that the atom has a substructure. Rutherford's studies of how thin metal foils scatter α particles showed that the atom has a dense positively charged **nucleus**.

SECTION 2.3 The structure of the atom consists of a nucleus containing **protons** and **neutrons**, with electrons moving in the space around the nucleus. The magnitude of the charge of the electron, 1.602×10^{-19} C, is called the **electronic charge**. The charges of particles are usually represented as multiples of this charge; thus an electron has a 1− charge and a proton has a 1+ charge. The masses of atoms (**atomic mass**) are expressed in **unified atomic mass units (u)** where $1 \, u = 1.66054 \times 10^{-24}$ g. The dimensions of atoms are expressed in **picometres (pm)** where $1 \, pm = 1 \times 10^{-12}$ m. Elements can be classified by **atomic number**, the number of protons in the nucleus of an atom. All atoms of a given element have the same atomic number. The **mass number** of an atom is the sum of the number of protons and neutrons in the nucleus. Atoms of the same element that differ in mass number are known as **isotopes**.

SECTION 2.4 The atomic mass scale is defined by assigning a mass of exactly 12 u to the ^{12}C atom. The **average atomic mass** of an element can be calculated from the relative abundances and masses of that element's isotopes. The **mass spectrometer** provides the most direct and accurate means of experimentally measuring atomic and molecular masses.

SECTION 2.5 The **periodic table** is an arrangement of the elements in order of increasing atomic number such that elements with similar properties are placed in vertical rows called **groups**. The horizontal rows are called **periods**. The **metallic elements (metals)**, which comprise the majority of the elements, dominate the left side and middle of the periodic table, whereas the **non-metallic elements (non-metals)** are located on the upper right side. The **metalloids** are a line of elements with properties intermediate to those of the metals and non-metals.

SECTION 2.6 Atoms can combine to form **molecules**. Compounds composed of molecules are known as **molecular compounds** and usually contain only non-metallic elements. A molecule that contains only two atoms is called a **diatomic molecule** whereas one containing more than two atoms is a **polyatomic molecule**. The composition of a substance is given by its **chemical formula**. A molecular substance can be represented by its **empirical formula**, which gives the relative numbers of atoms of each kind. However, it is usually represented by its molecular formula, which gives the actual number of each type of atom in the molecule. **Structural formulae** show the order in which the atoms in a molecule are connected. Ball-and-stick models and space-filling models are often used to represent molecules.

SECTION 2.7 The loss or gain of electrons by atoms leads to the formation of **ions**, which are charged particles. Metals tend to lose electrons to form positively charged ions (**cations**) and non-metals tend to gain electrons to form negatively charged particles (**anions**). Because **ionic compounds** result from a combination of cations and anions such that the amount of negative charge exactly equals the amount of positive charge, they are electrically neutral, and usually contain metallic and non-metallic elements. Atoms that are joined together, as in a molecule, but carry a net charge are called **polyatomic ions**. The chemical formulae used for ionic compounds are empirical formulae that can be written readily if the charges of the ions are known.

SECTION 2.8 The set of rules for naming chemical compounds is called **chemical nomenclature**. These sets of rules are introduced for naming three classes of inorganic substances: ionic compounds, acids and binary molecular compounds. In naming an ionic compound, the cation is named first and then the anion. Cations formed from metal atoms have the same name as the atom. If the metal can form cations of different charges, the charge is given using Roman numerals. Monatomic anions have names ending in *-ide*. Polyatomic anions containing oxygen and another element (**oxyanions**) have names ending in *-ate* or *-ite*.

SECTION 2.9 Organic chemistry is the study of compounds that contain carbon. The simplest class of organic molecules is the **hydrocarbons**, which contain only carbon and hydrogen. Hydrocarbons in which each carbon atom is attached to four other atoms are called **alkanes**. Alkanes have names that end in *-ane*, such as methane and ethane. Other organic compounds are formed when an H atom of a hydrocarbon is replaced with a functional group. An **alcohol**, for example, is a compound in which an H atom of a hydrocarbon is replaced by an OH functional group. Alcohols have names that end in *-ol*, such as methanol and ethanol. Compounds with the same molecular formula but a different bonding arrangement of their constituent atoms are called **isomers**.

KEY SKILLS

- Describe the basic postulates of Dalton's atomic theory. (Section 2.1)
- Describe the key experiments that led to the discovery of electrons and to the nuclear model of the atom. (Section 2.2)
- Describe the structure of the atom in terms of protons, neutrons and electrons. (Section 2.3)
- Describe the electrical charge and relative masses of protons, neutrons and electrons. (Section 2.3)
- Use chemical symbols together with atomic numbers and mass numbers to express the subatomic composition of isotopes. (Section 2.3)
- Understand how atomic masses relate to the masses of individual atoms and to their natural abundances. (Section 2.4)
- Describe how elements are organised in the periodic table by atomic number and by similarities in chemical behaviour, giving rise to periods and groups. (Section 2.5)
- Describe the locations of metals and non-metals in the periodic table. (Section 2.5)
- Distinguish between molecular substances and ionic substances in terms of their composition. (Sections 2.6 and 2.7)
- Distinguish between empirical formulae and molecular formulae. (Section 2.6)
- Describe how molecular formulae and structural formulae are used to represent the composition of molecules. (Section 2.6)
- Explain how ions are formed by gaining or losing electrons and be able to use the periodic table to predict the charges of common ions. (Section 2.7)
- Write the empirical formulae of ionic compounds, given the charges of their component ions. (Section 2.7)
- Write the name of an ionic compound given its chemical formula, or write the chemical formula given its name. (Section 2.8)
- Name or write chemical formulae for binary inorganic compounds and for acids. (Section 2.8)
- Identify organic compounds and name simple alkanes and alcohols. (Section 2.9)

EXERCISES

VISUALISING CONCEPTS

2.1 A charged particle is caused to move between two electrically charged plates, as shown below.

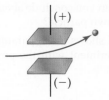

(a) Why does the path of the charged particle bend? **(b)** What is the sign of the electrical charge on the particle? **(c)** As the charge on the plates is increased, would you expect the bending to increase, decrease or stay the same? **(d)** As the mass of the particle is increased while the speed of the particles remains the same, would you expect the bending to increase, decrease or stay the same? [Section 2.2]

2.2 Four of the boxes in the following periodic table are coloured. Which of these are metals and which are non-metals? Which one is an alkaline earth metal? Which one is a noble gas? [Section 2.5]

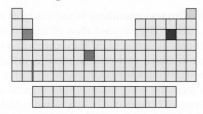

2.3 Does the following drawing represent a neutral atom or an ion? Write its complete chemical symbol including mass number, atomic number and net charge (if any). [Sections 2.3 and 2.7]

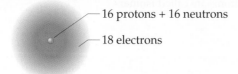

— 16 protons + 16 neutrons

— 18 electrons

2.4 Which of the following diagrams is most likely to represent an ionic compound and which a molecular one? Explain your choice. [Sections 2.6 and 2.7]

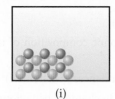 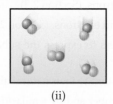

(i)   (ii)

2.5 Write the chemical formula for the following compound. Is the compound ionic or molecular? Name the compound. [Sections 2.6 and 2.8]

2.6 The following diagram represents an ionic compound in which the cations are indicated by the red spheres and the anions are indicated by the blue spheres. Which of the following formulae is consistent with the drawing: KBr, K_2SO_4, $Ca(NO_3)_2$, $Fe_2(SO_4)_3$? Name the compound. [Sections 2.7 and 2.8]

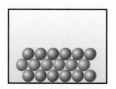

ATOMIC THEORY AND THE DISCOVERY OF ATOMIC STRUCTURE (Sections 2.1 and 2.2)

2.7 How does Dalton's atomic theory account for the fact that when 1.000 g of water is decomposed into its elements, 0.111 g of hydrogen and 0.889 g of oxygen are obtained, regardless of the source of the water?

2.8 Hydrogen sulfide is composed of two elements: hydrogen and sulfur. In an experiment, 6.500 g of hydrogen sulfide is fully decomposed into its elements. **(a)** If 0.384 g of hydrogen is obtained in this experiment, how many grams of sulfur must be obtained? **(b)** What fundamental law does this experiment demonstrate? **(c)** How is this law explained by Dalton's atomic theory?

2.9 A chemist finds that 30.82 g of nitrogen will react with 17.60 g, 35.20 g, 70.40 g or 88.00 g of oxygen to form four different compounds. **(a)** Calculate the mass of oxygen per gram of nitrogen in each compound. **(b)** How do the numbers in part (a) support Dalton's atomic theory?

2.10 In a series of experiments, a chemist prepared three different compounds that contain only iodine and fluorine and determined the mass of each element in each compound.

Compound	Mass of iodine (g)	Mass of fluorine (g)
1	4.75	3.56
2	7.64	3.43
3	9.41	9.86

(a) Calculate the mass of fluorine per gram of iodine in each compound. **(b)** How do the numbers in part (a) support the atomic theory?

2.11 Summarise the evidence used by J. J. Thomson to argue that cathode rays consist of negatively charged particles.

2.12 An unknown particle is caused to move between two electrically charged plates, as illustrated in Figure 2.6. Its path is deflected in the opposite direction from that of a beta particle, and it is deflected by a smaller magnitude. What can you conclude about the charge and mass of this unknown particle?

2.13 **(a)** Figure 2.4 shows the apparatus used in the Millikan oil-drop experiment with the positively charged plate above the negatively charged plate. What do you think would be the effect on the rate of oil drops descending if the charges on the plates were reversed (negative above positive)? **(b)** In his original series of experiments, Millikan measured the charge on 58 separate oil drops. Why do you suppose he chose so many drops before reaching his final conclusions?

MODERN VIEW OF ATOMIC STRUCTURE; ATOMIC MASS (Sections 2.3 and 2.4)

2.14 The radius of an atom of krypton (Kr) is about 190 pm. **(a)** Express this distance in nanometres (nm). **(b)** How many krypton atoms would have to be lined up to span 1.0 mm? **(c)** If the atom is assumed to be a sphere, what is the volume in cm^3 of a single Kr atom?

2.15 An atom of tin (Sn) has a diameter of about 2.8×10^{-8} cm. **(a)** What is the radius of a tin atom in picometres (pm) and in metres (m)? **(b)** How many Sn atoms would have to be placed side by side to span a distance of 6.0 μm? **(c)** If the atom is assumed to be a sphere, what is the volume in m^3 of a single Sn atom?

2.16 **(a)** Define atomic number and mass number. **(b)** Which of these can vary without changing the identity of the element?

2.17 **(a)** Which two of the following are isotopes of the same element: $^{31}_{16}X$, $^{31}_{15}X$, $^{32}_{16}X$? **(b)** What is the identity of the element whose isotopes you have selected?

2.18 How many protons, neutrons and electrons are in the following atoms: **(a)** ^{40}Ar, **(b)** ^{65}Zn, **(c)** ^{70}Ga, **(d)** ^{80}Br, **(e)** ^{184}W, **(f)** ^{243}Am.

2.19 Each of the following isotopes is used in medicine. Indicate the number of protons and neutrons in each isotope: **(a)** phosphorus-32, **(b)** chromium-51, **(c)** cobalt-60, **(d)** technetium-99, **(e)** iodine-131, **(f)** thallium-201.

2.20 Fill in the gaps in the following table, assuming each column represents a neutral atom.

Symbol	^{52}Cr				
Protons		25			82
Neutrons		30	64		
Electrons			48	86	
Mass no				222	207

2.21 Write the correct symbol, with both superscript and subscript, for each of the following. Use the list of elements on the inside front cover of this text as needed: **(a)** the isotope of platinum that contains 118 neutrons, **(b)** the isotope of krypton with mass number 84, **(c)** the isotope of arsenic with mass number 75, **(d)** the isotope of magnesium that has an equal number of protons and neutrons.

2.22 One way in which Earth's evolution as a planet can be understood is by measuring the amounts of certain isotopes in rocks. One quantity recently measured is the ratio of ^{129}Xe to ^{130}Xe in some minerals. In what way do these two isotopes differ from one another, and in what respects are they the same?

2.23 **(a)** What isotope is used as the standard in establishing the atomic mass scale? **(b)** The atomic mass of boron is reported as 10.81, yet no atom of boron has the mass of 10.81 u. Explain.

2.24 **(a)** What is the mass in u of a carbon-12 atom? **(b)** Why is the atomic mass of carbon reported as 12.011 in the table of elements and the periodic table on the inside front cover of this text?

2.25 The element lead (Pb) consists of four naturally occurring isotopes with atomic masses 203.97302, 205.97444, 206.97587 and 207.97663 u. The relative abundances of these four isotopes are 1.4, 24.1, 22.1 and 52.4%, respectively. From these data, calculate the average atomic mass of lead.

THE PERIODIC TABLE; MOLECULES AND IONS (Sections 2.5 to 2.7)

2.26 For each of the following elements, write its chemical symbol, locate it in the periodic table and indicate whether it is a metal, metalloid or non-metal: **(a)** chromium, **(b)** helium, **(c)** phosphorus, **(d)** zinc, **(e)** magnesium, **(f)** bromine, **(g)** arsenic.

2.27 Locate each of the following elements in the periodic table; indicate whether it is a metal, metalloid or non-metal; and give the name of the element: **(a)** Na, **(b)** Ti, **(c)** Ga, **(d)** U, **(e)** Pd, **(f)** Se, **(g)** Kr.

2.28 What can we tell about a compound when we know the empirical formula? What additional information is conveyed by the molecular formula? By the structural formula? Explain in each case.

2.29 Two compounds have the same empirical formula. One substance is a gas, the other is a viscous liquid. How is it possible for two substances with the same empirical formula to have markedly different properties?

2.30 Write the empirical formula corresponding to each of the following molecular formulae: **(a)** Al_2Br_6, **(b)** C_8H_{10}, **(c)** $C_4H_8O_2$, **(d)** P_4O_{10}, **(e)** $C_6H_4Cl_2$, **(f)** $B_3N_3H_6$.

2.31 Determine the molecular and empirical formulae of the following: **(a)** the organic solvent *benzene*, which has six carbon atoms and six hydrogen atoms; **(b)** the compound *silicon tetrachloride*, which has a silicon atom and four chlorine atoms and is used in the manufacture of computer chips; **(c)** the reactive substance *diborane*, which has two boron atoms and six hydrogen atoms; **(d)** the sugar called *glucose*, which has six carbon atoms, 12 hydrogen atoms and six oxygen atoms.

2.32 How many hydrogen atoms are in each of the following: **(a)** C_2H_5OH, **(b)** $Ca(CH_3COO)_2$, **(c)** $(NH_4)_3PO_4$?

2.33 How many of the indicated atoms are represented by each chemical formula: **(a)** carbon atoms in $C_2H_5COOCH_3$, **(b)** oxygen atoms in $Ca(ClO_3)_2$, **(c)** hydrogen atoms in $(NH_4)_2HPO_4$?

2.34 Write the molecular and structural formulae for the compounds represented by the following molecular models:

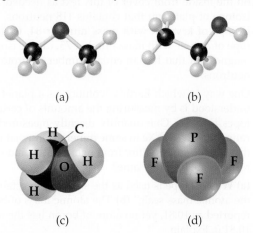

(a)   (b)

(c)   (d)

2.35 Write the molecular and structural formulae for the compounds represented by the following models:

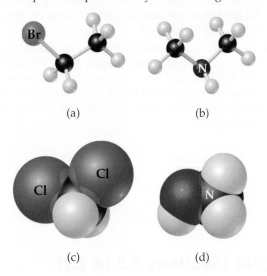

(a)   (b)

(c)   (d)

2.36 Fill in the gaps in the following table:

Symbol	$^{59}Co^{3+}$			
Protons		34	76	80
Neutrons		46	116	120
Electrons		36		78
Net charge			2+	

2.37 Fill in the gaps in the following table:

Symbol	$^{75}As^{3-}$			
Protons		34	53	
Neutrons		31	74	118
Electrons		26		78
Net charge			1−	3+

2.38 Each of the following elements is capable of forming an ion in chemical reactions. By referring to the periodic table, predict the charge of the most stable ion of each: **(a)** Mg, **(b)** Al, **(c)** K, **(d)** S, **(e)** F.

2.39 Using the periodic table, predict the charges of the ions of the following elements: **(a)** Sr, **(b)** Sc, **(c)** P, **(d)** I, **(e)** Se.

2.40 Using the periodic table to guide you, predict the chemical formula and name of the compound formed by the following elements: **(a)** Ga and F, **(b)** Li and H, **(c)** Al and I, **(d)** K and S.

2.41 The most common charge associated with silver in its compounds is 1+. Indicate the chemical formulae you would expect for compounds formed between Ag and **(a)** iodine, **(b)** sulfur, **(c)** fluorine.

NAMING INORGANIC COMPOUNDS; ORGANIC MOLECULES (Sections 2.8 and 2.9)

2.42 Give the chemical formula for **(a)** chlorite ion, **(b)** chloride ion, **(c)** chlorate ion, **(d)** perchlorate ion, **(e)** hypochlorite ion.

2.43 Selenium, an element required nutritionally in trace quantities, forms compounds analogous to sulfur. Name the following ions: **(a)** SeO_4^{2-}, **(b)** Se^{2-}, **(c)** HSe^-, **(d)** $HSeO_3^-$.

2.44 Name the following ionic compounds: **(a)** MgO, **(b)** $AlCl_3$, **(c)** Li_3PO_4, **(d)** $Ba(ClO_4)_2$, **(e)** $Cu(NO_3)_2$,

(f) $Fe(OH)_2$, **(g)** $Ca(CH_3COO)_2$, **(h)** $Cr_2(CO_3)_3$, **(i)** K_2CrO_4, **(j)** $(NH_4)_2SO_4$.

2.45 Name the following ionic compounds: **(a)** Li_2O, **(b)** NaClO, **(c)** $Sr(CN)_2$, **(d)** $Cr(OH)_3$, **(e)** $Fe_2(CO_3)_3$, **(f)** $Co(NO_3)_2$, **(g)** $(NH_4)_2SO_3$, **(h)** NaH_2PO_4, **(i)** $KMnO_4$, **(j)** $Ag_2Cr_2O_7$.

2.46 Write the chemical formulae for the following compounds: **(a)** aluminium hydroxide, **(b)** potassium sulfate, **(c)** copper(I) oxide, **(d)** zinc nitrate,

(e) mercury(II) bromide, **(f)** iron(III) carbonate, **(g)** sodium hypobromite.

2.47 Give the chemical formula for each of the following ionic compounds: **(a)** sodium phosphate, **(b)** cobalt(II) nitrate, **(c)** barium bromate, **(d)** copper(II) perchlorate, **(e)** magnesium hydrogen carbonate, **(f)** chromium(III) acetate, **(g)** potassium dichromate.

2.48 Give the name or chemical formula, as appropriate, for each of the following acids: **(a)** $HBrO_3$, **(b)** HBr, **(c)** H_3PO_4, **(d)** hypochlorous acid, **(e)** iodic acid, **(f)** sulfurous acid.

2.49 Provide the name or chemical formula, as appropriate, for each of the following acids: **(a)** hydrobromic acid, **(b)** hydrosulfuric acid, **(c)** nitrous acid, **(d)** H_2CO_3, **(e)** $HClO_3$, **(f)** CH_3COOH.

2.50 Give the name or chemical formula, as appropriate, for each of the following binary molecular substances: **(a)** SF_6, **(b)** IF_5, **(c)** XeO_3, **(d)** dinitrogen tetroxide, **(e)** hydrogen cyanide, **(f)** tetraphosphorus hexasulfide.

2.51 The oxides of nitrogen are very important ingredients in determining urban air pollution. Name each of the following compounds: **(a)** N_2O, **(b)** NO, **(c)** NO_2, **(d)** N_2O_5, **(e)** N_2O_4.

2.52 Write the chemical formula for each substance mentioned in the following word descriptions (use the inside front cover to find the symbols for the elements you don't know). **(a)** Zinc carbonate can be heated to form zinc oxide and carbon dioxide. **(b)** On treatment with hydrofluoric acid, silicon dioxide forms silicon tetrafluoride and water. **(c)** Sulfur dioxide reacts with water to form sulfurous acid. **(d)** The substance phosphorus trihydride, commonly called phosphine, is a toxic gas. **(e)** Perchloric acid reacts with cadmium to form cadmium(II) perchlorate. **(f)** Vanadium(III) bromide is a coloured solid.

2.53 **(a)** What is a hydrocarbon? **(b)** Butane is the alkane with a chain of four carbon atoms. Write a structural formula for this compound and determine its molecular and empirical formulae.

2.54 **(a)** What ending is used for the names of alkanes? **(b)** n-Hexane is an alkane whose structural formula has all its carbon atoms in a straight chain. Draw the structural formula for this compound and determine its molecular and empirical formulae. (*Hint:* You might need to refer to Table 2.6.)

2.55 **(a)** What is a functional group? **(b)** What functional group characterises an alcohol? **(c)** With reference to Exercise 2.53, write a structural formula for 1-butanol, the alcohol derived from butane, by making a substitution on one of the carbon atoms.

2.56 **(a)** What do ethane and ethanol have in common? **(b)** How does 1-propanol differ from propane?

2.57 Chloropropane is a compound derived from propane by substituting Cl for H on one of the carbon atoms. **(a)** Draw the structural formulae for the two isomers of chloropropane. **(b)** Suggest names for these two compounds.

2.58 Draw the structural formulae for three isomers of pentane, C_5H_{12}.

ADDITIONAL EXERCISES

The exercises in this section are not divided by category, although they are roughly in the order of the topics in the chapter.

2.59 How did Rutherford interpret the following observations made during his α-particle scattering experiments? **(a)** Most α particles were not appreciably deflected as they passed through the gold foil. **(b)** A few α particles were deflected at very large angles. **(c)** What differences would you expect if beryllium foil were used instead of gold foil in the α-particle scattering experiment?

2.60 The natural abundance of ^3He is 0.000137%. **(a)** How many protons, neutrons and electrons are in an atom of ^3He? **(b)** Based on the sum of the masses of their subatomic particles, which is expected to be more massive, an atom of ^3He or an atom of ^3H (which is also called *tritium*)? **(c)** Based on your answer for part (b), what would need to be the precision of a mass spectrometer that is able to differentiate between signals that are due to $^3He^+$ and $^3H^+$?

2.61 A cube of gold that is 1.00 cm on a side has a mass of 19.3 g. A single gold atom has a mass of 197.0 u. How many gold atoms are in the cube?

2.62 **(a)** The mass spectrometer in Figure 2.8 has a magnet as one of the components. What is the purpose of the magnet? **(b)** The atomic mass of Cl is 35.5 u. However the mass spectrum of Cl (Figure 2.9) does not show a peak at this mass. Explain. **(c)** A mass spectrum of phosphorus (P) atoms shows only a single peak at a mass of 31 u. What can you conclude from this observation?

2.63 The element oxygen has three naturally occurring isotopes, with 8, 9 and 10 neutrons in the nucleus, respectively. **(a)** Write the full chemical symbols for these three isotopes. **(b)** Describe the similarities and differences between the three kinds of atoms of oxygen.

2.64 Use Coulomb's law, $F = \dfrac{kq_1q_2}{r^2}$, to calculate the electric force on an electron ($Q = -1.6 \times 10^{-19}$ C) exerted by a single proton if the particles are 0.53×10^{-10} m apart. The constant k in Coulomb's law is 9.0×10^9 $Nm^2 C^{-2}$. (The unit abbreviated N is the newton, the SI unit of force.)

2.65 Gallium (Ga) consists of two naturally occurring isotopes with masses of 68.926 and 70.925 u. **(a)** How many protons and neutrons are in the nucleus of each isotope? Write the complete atomic symbol for each, showing the atomic number and mass number. **(b)** The average atomic mass of Ga is 69.72 u. Calculate the abundance of each isotope.

2.66 *Bronze* is a metallic alloy often used in decorative applications and in sculpture. A typical bronze consists of copper, tin and zinc, with lesser amounts of phosphorus and lead. Locate each of these five elements in the periodic table, write their symbols and identify the group of the periodic table to which they belong.

2.67 From the following list of elements—Ar, H, Ga, Al, Ca, Br, Ge, K, O—pick the one that best fits each description; use each element only once: **(a)** an alkali metal, **(b)** an alkaline earth metal, **(c)** a noble gas, **(d)** a halogen, **(e)** a metalloid, **(f)** a non-metal listed in group 1, **(g)** a metal that forms a 3+ ion, **(h)** a non-metal that forms a 2− ion, **(i)** an element that resembles aluminium.

2.68 From the molecular structures shown here, identify the one that corresponds to each of the following species: **(a)** chlorine gas, **(b)** propane, **(c)** nitrate ion, **(d)** sulfur trioxide, **(e)** chloromethane, CH_3Cl.

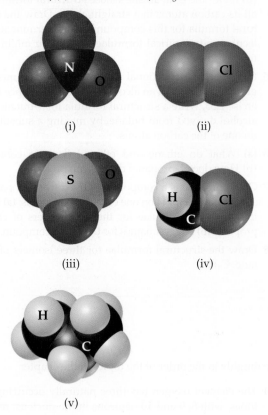

(i)

(ii)

(iii)

(iv)

(v)

2.69 Name each of the following oxides. Assuming that the compounds are ionic, what charge is associated with the metallic element in each case? **(a)** NiO, **(b)** MnO_2, **(c)** Cr_2O_3, **(d)** MoO_3.

2.70 Iodic acid has the molecular formula HIO_3. Write the formulae for the following: **(a)** the iodate anion, **(b)** the periodate anion, **(c)** the hypoiodite anion, **(d)** hypoiodous acid, **(e)** periodic acid.

2.71 Elements in the same group of the periodic table often form oxyanions with the same general formula. The anions are also named in a similar fashion. Based on these observations, suggest a chemical formula or name, as appropriate, for each of the following ions: **(a)** BrO_4^-, **(b)** SeO_3^{2-}, **(c)** arsenate ion, **(d)** hydrogen tellurate ion.

2.72 Give the chemical names of each of the following familiar compounds: **(a)** NaCl (table salt), **(b)** $NaHCO_3$ (baking soda), **(c)** NaOCl (in many bleaches), **(d)** NaOH (caustic soda), **(e)** $(NH_4)_2CO_3$ (smelling salts), **(f)** $CaSO_4$ (plaster of Paris).

MasteringChemistry (www.pearson.com.au/masteringchemistry)

Make learning part of the grade. Access:

- tutorials with peronalised coaching
- study area
- Pearson eText

PHOTO/ART CREDITS

28 © Igor Dolgov/Dreamstime.com; **30** Corbis Australia Pty Ltd; **32** Richard Megna, Fundamental Photographs, NYC; **33** George Grantham Bain Collection (Library of Congress). Wikipedia Creative Commons Licenses/CC-BY-2.5; **46** iStockphoto.

3

STOICHIOMETRY: CALCULATIONS WITH CHEMICAL FORMULAE AND EQUATIONS

A burning match. The head and flame are visible evidence of a chemical reaction. Combustion reactions were among the first systematically studied chemical reactions.

KEY CONCEPTS

3.1 CHEMICAL EQUATIONS
We introduce the writing of *chemical equations*, which represent (or describe) chemical reactions, by the use of chemical formulae.

3.2 SOME SIMPLE PATTERNS OF CHEMICAL REACTIVITY
We examine some simple chemical reactions: combination reactions, decomposition reactions and combustion reactions.

3.3 FORMULA MASS
We see how to obtain quantitative information from chemical formulae by using formula mass.

3.4 AVOGADRO'S NUMBER AND THE MOLE
We use chemical formulae to relate the masses of substances to the numbers of atoms, molecules or ions contained in the substances, a relationship that leads to the crucially important concept of the *mole*, defined as 6.022×10^{23} objects (e.g. atoms, molecules, ions).

3.5 EMPIRICAL FORMULAE FROM ANALYSES
We use the mole concept to determine chemical formulae from the masses of each element in a given quantity of a compound.

3.6 QUANTITATIVE INFORMATION FROM BALANCED CHEMICAL EQUATIONS
We use the quantitative information that is inherently present in chemical formulae and equations, together with the mole concept, to predict the amounts of substances consumed or produced in chemical reactions.

3.7 LIMITING REACTANTS
We recognise that one reactant may be used up before others in a chemical reaction. This is the *limiting reactant*. The reaction stops, leaving some excess starting material.

You pour vinegar into a glass of water containing baking powder, and bubbles form. You strike a match and use the flame to light a candle. You heat sugar in a pan and it turns brown. The bubble, flame and colour change are visual evidence that something is happening.

To an experienced eye, it indicates a chemical change, a kind of change that we also call a chemical reaction. The study of chemical changes is at the heart of chemistry. Some chemical changes are simple and some are complex. Some are dramatic and some are very subtle. Even as you sit reading this chapter, there are chemical changes occurring in you. The changes that occur in your eyes and brain, for example, are what allows you to see these words and think about them. Although such chemical changes are not as obvious as the reaction shown in the chapter-opening photograph, they are nevertheless remarkable because of how they allow us to function.

In this chapter we begin to explore some important aspects of chemical changes. Our focus will be on the use of chemical formulae to represent reactions and on the quantitative information we can obtain about the amounts of substances involved in reactions. The area of study that examines the quantities of substances consumed and produced in chemical reactions is known as stoichiometry (pronounced stoy-key-OM-uh-tree), a name derived from the Greek *stoicheion* ('element') and *metron* ('measure'). This study provides an essential set of tools that are widely used in chemistry. Such diverse problems as measuring the concentration of ozone in the atmosphere, determining the potential yield of gold from an ore and assessing different processes for converting coal into gaseous fuels all use aspects of stoichiometry.

Stoichiometry is built on an understanding of atomic masses, chemical formulae and the law of conservation of mass (∞ Section 2.1, 'Atomic theory of matter').

The French nobleman and scientist Antoine Lavoisier (1743–1794) (▼ FIGURE 3.1) discovered this important chemical law in the late 1700s. In a chemistry text published in 1789, Lavoisier stated the law in this eloquent way, 'We may lay it down as an incontestable axiom that, in all the operations of art and nature, nothing is created; an equal quantity of matter exists both before and after the experiment. Upon this principle, the whole art of performing chemical experiments depends.' With the advent of Dalton's atomic theory, chemists came to understand the basis for this law: *Atoms are neither created nor destroyed during any chemical reaction.* The changes that occur during any reaction merely rearrange the atoms. The same collection of atoms is present both before and after the reaction.

▲ **FIGURE 3.1 Antoine Lavoisier (1734–1794).** The scientific career of Lavoisier, who conducted many important studies on combustion reactions, was cut short by the French Revolution. Guillotined in 1794 during the Reign of Terror, he is generally considered the father of modern chemistry because he conducted carefully controlled experiments and measured the weights of reactants and products rather than only observing changes.

3.1 | CHEMICAL EQUATIONS

Chemical reactions are represented in a concise way by **chemical equations**. When hydrogen (H_2) burns, for example, it reacts with oxygen (O_2) in the air to form water (H_2O). We write the chemical equation for this reaction as follows:

$$2\,H_2 + O_2 \longrightarrow 2\,H_2O \qquad\qquad [3.1]$$

We read the + sign as 'reacts with' and the arrow as 'produces'. The chemical formulae to the left of the arrow represent the starting substances, called **reactants**. The chemical formulae to the right of the arrow represent substances produced in the reaction, called **products**. The numbers in front of the formulae are *coefficients*. (As in algebraic equations, the numeral 1 is usually not written.) The coefficients indicate the relative numbers of molecules of each kind involved in the reaction.

Because atoms are neither created nor destroyed in any reaction, a chemical equation must have an equal number of atoms of each element on each side of the arrow. When this condition is met, the equation is said to be *balanced*. On the right side of Equation 3.1, for example, there are two molecules of H_2O, each composed of two atoms of hydrogen and one atom of oxygen. Thus 2 H_2O (read 'two molecules of water') contains $2 \times 2 = 4$ H atoms and $2 \times 1 = 2$ O atoms. Notice that *the number of atoms is obtained by multiplying the coefficient and the subscripts in the chemical formula*. Because there are four H atoms and two O atoms on each side of the equation, the equation is balanced. We can represent the balanced equation by the following molecular models, which illustrate that the number of atoms of each kind is the same on both sides of the arrow (◀ **FIGURE 3.2**).

Reactants   Products

$$2\,H_2 + O_2 \longrightarrow 2\,H_2O$$

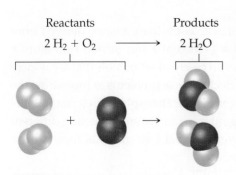

▲ **FIGURE 3.2 A balanced chemical equation.**

> ⚠ **CONCEPT CHECK 1**
> How many atoms of Mg, O and H are represented by 3 $Mg(OH)_2$?

Balancing Equations

Once we know the formulae of the reactants and products in a reaction, we can write the unbalanced equation. We then balance the equation by determining the coefficients that provide equal numbers of each type of atom on each side of the equation. For most purposes, a balanced equation should contain the smallest possible whole number coefficients.

In balancing equations, it is important to understand the difference between a coefficient in front of a formula and a subscript in a formula. Refer to ▼ **FIGURE 3.3**. Notice that changing a subscript in a formula—from H_2O to

> ▲ **FIGURE IT OUT**
> **What is the difference in atom count between the notation CO_2 and the notation 2 CO?**

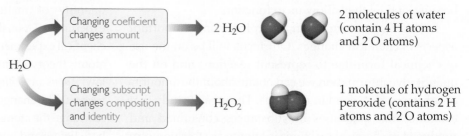

▲ **FIGURE 3.3 The difference between changing subscripts and changing coefficients in chemical equations.**

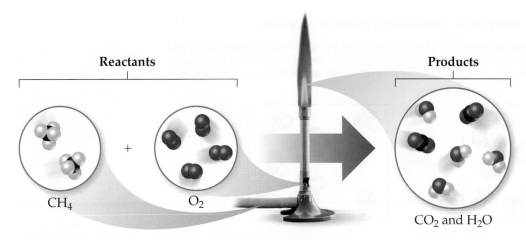

▲ **FIGURE 3.4   Methane reacts with oxygen in a Bunsen burner.**

H_2O_2 for example—changes the identity of the substance. The substance H_2O_2, hydrogen peroxide, is quite different from the substance H_2O, water. *Subscripts must never be changed when balancing an equation.* In contrast, placing a coefficient in front of a formula changes only the *amount* of the substance and not its *identity*. Thus $2 H_2O$ means two molecules of water, $3 H_2O$ means three molecules of water and so forth.

To illustrate the process of balancing equations, consider the reaction that occurs when methane (CH_4), the principal component of natural gas, burns in air to produce carbon dioxide gas (CO_2) and water vapour (H_2O) (▲ **FIGURE 3.4**). Both of these products contain oxygen atoms that come from O_2 in the air. Thus O_2 is a reactant, and the unbalanced equation is:

$$CH_4 + O_2 \longrightarrow CO_2 + H_2O \qquad \text{(unbalanced)} \qquad [3.2]$$

It is usually best to balance first those elements that occur in the fewest chemical formulae on each side of the equation. In our example both C and H appear in only one reactant and, separately, in one product each, so we begin by focusing on CH_4. Let's consider first carbon and then hydrogen.

One molecule of the reactant CH_4 contains the same number of C atoms (one) as one molecule of the product CO_2. The coefficients for these substances *must* be the same in the balanced equation. Therefore we start by choosing the coefficient 1 (unwritten) for both CH_4 and CO_2. Next we focus on H. Because CH_4 contains four H atoms and H_2O contains two H atoms, we balance the H atoms by placing the coefficient 2 in front of H_2O. There are then four H atoms on each side of the equation:

$$CH_4 + O_2 \longrightarrow CO_2 + 2 H_2O \qquad \text{(unbalanced)} \qquad [3.3]$$

Finally, a coefficient 2 in front of O_2 balances the equation by giving four O atoms on each side (2×2 left, $2 + 2 \times 1$ right):

$$CH_4 + 2 O_2 \longrightarrow CO_2 + 2 H_2O \qquad \text{(balanced)} \qquad [3.4]$$

The molecular view of the balanced equation is shown in ▶ **FIGURE 3.5**. We see that there are one C, four H and four O atoms on both sides of the arrow, indicating the equation is balanced.

The approach we have taken in arriving at balanced Equation 3.4 is largely trial and error. We balance each kind of atom in succession, adjusting coefficients as necessary. This approach works for most chemical equations.

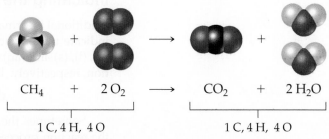

$$CH_4 \quad + \quad 2 O_2 \quad \longrightarrow \quad CO_2 \quad + \quad 2 H_2O$$

$$\underbrace{\text{1 C, 4 H, 4 O}} \qquad\qquad \underbrace{\text{1 C, 4 H, 4 O}}$$

▲ **FIGURE 3.5   Balanced chemical equation for the combustion of CH_4.**

SAMPLE EXERCISE 3.1 Interpreting and balancing chemical equations

The following diagram represents a chemical reaction in which the red spheres are oxygen atoms and the blue spheres are nitrogen atoms. **(a)** Write the chemical formulae for the reactants and products. **(b)** Write a balanced equation for the reaction. **(c)** Is the diagram consistent with the law of conservation of mass?

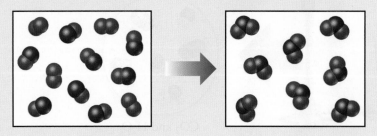

SOLUTION

(a) The left box, which represents the reactants, contains two kinds of molecules, those composed of two oxygen atoms (O_2) and those composed of one nitrogen atom and one oxygen atom (NO). The right box, which represents the products, contains only molecules composed of one nitrogen atom and two oxygen atoms (NO_2).

(b) The unbalanced chemical equation is

$$O_2 + NO \longrightarrow NO_2 \quad \text{(unbalanced)}$$

In this equation there are three O atoms on the left side of the arrow and two O atoms on the right side. We can increase the number of O atoms by placing a coefficient 2 on the product side:

$$O_2 + NO \longrightarrow 2\,NO_2 \quad \text{(unbalanced)}$$

Now there are two N atoms and four O atoms on the right. Placing a coefficient 2 in front of NO brings both the N atoms and O atoms into balance:

$$O_2 + 2\,NO \longrightarrow 2\,NO_2 \quad \text{(balanced)}$$

(c) The left box (reactants) contains four O_2 molecules and eight NO molecules.

Thus the molecular ratio is one O_2 for each two NO as required by the balanced equation. The right box (products) contains eight NO_2 molecules. The number of NO_2 molecules on the right equals the number of NO molecules on the left as the balanced equation requires. Counting the atoms, we find eight N atoms in the eight NO molecules in the box on the left. There are also $4 \times 2 = 8$ O atoms in the O_2 molecules and eight O atoms in the NO molecules, giving a total of 16 O atoms. In the box on the right, we find eight N atoms and $8 \times 2 = 16$ O atoms in the eight NO_2 molecules. Because there are equal numbers of both N and O atoms in the two boxes, the drawing is consistent with the law of conservation of mass.

PRACTICE EXERCISE

In order to be consistent with the law of conservation of mass, how many NH_3 molecules should be shown in the right box of the following diagram?

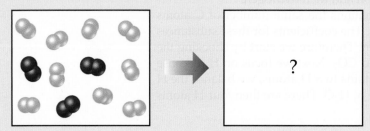

Answer: Six NH_3 molecules
(See also Exercises 3.1, 3.2.)

Indicating the States of Reactants and Products

Additional information is often added to the formulae in balanced equations to indicate the physical state of each reactant and product. We use the symbols (g), (l), (s) and (aq) for gas, liquid, solid and aqueous (dissolved in water) solution, respectively. Thus Equation 3.4 can be written

$$CH_4(g) + 2\,O_2(g) \longrightarrow CO_2(g) + 2\,H_2O(g) \quad [3.5]$$

Sometimes the conditions (such as temperature or pressure) under which the reaction proceeds appear above or below the reaction arrow. The symbol Δ (the Greek upper-case letter delta) is often placed above the arrow to indicate the addition of heat.

SAMPLE EXERCISE 3.2   **Balancing chemical equations**

Balance this equation:

$$Na(s) + H_2O(l) \longrightarrow NaOH(aq) + H_2(g)$$

SOLUTION

We begin by counting the atoms of each kind on both sides of the arrow. The Na and O atoms are balanced (one Na and one O on each side), but there are two H atoms on the left and three H atoms on the right. Thus we need to increase the number of H atoms on the left. As a trial beginning in our effort to balance H, let's place a coefficient 2 in front of H_2O:

$$Na(s) + 2\,H_2O(l) \longrightarrow NaOH(aq) + H_2(g)$$

Beginning this way doesn't balance H, but introducing the coefficient 2 does increase the number of H atoms among the reactants, which we need to do. The fact that it causes O to be unbalanced is something we will take care of after we balance H. Now that we have 2 H_2O on the left, we can balance H by putting a coefficient 2 in front of NaOH on the right:

$$Na(s) + 2\,H_2O(l) \longrightarrow 2\,NaOH(aq) + H_2(g)$$

Balancing H in this way fortuitously brings O into balance, but notice that Na is now unbalanced, with one on the left but two on the right. To rebalance Na, we put a coefficient 2 in front of the reactant:

$$2\,Na(s) + 2\,H_2O(l) \longrightarrow 2\,NaOH(aq) + H_2(g)$$

Finally, we check the number of atoms of each element and find that we have two Na atoms, four H atoms and two O atoms on each side of the equation. The equation is balanced.

Comment Notice that, in balancing this equation, we moved back and forth placing a coefficient in front of H_2O, then NaOH, and finally Na. In balancing equations, we often find ourselves following this pattern of moving back and forth from one side of the arrow to the other, placing coefficients first in front of a formula on one side and then in front of a formula on the other side until the equation is balanced.

PRACTICE EXERCISE

Balance the following equations by providing the missing coefficients:

(a) __ Fe(s) + __ O_2(g) $\longrightarrow$ __ Fe_2O_3(s)

(b) __ C_2H_4(g) + __ O_2(g) $\longrightarrow$ __ CO_2(g) + __ H_2O(g)

(c) __ Al(s) + __ HCl(aq) $\longrightarrow$ __ $AlCl_3$(aq) + __ H_2(g)

Answers: **(a)** 4, 3, 2; **(b)** 1, 3, 2, 2; **(c)** 2, 6, 2, 3

(See also Exercises 3.8–3.10.)

3.2 | SOME SIMPLE PATTERNS OF CHEMICAL REACTIVITY

In this section we examine three simple kinds of reactions that we will see frequently throughout this chapter. Our first reason for examining these reactions is merely to become better acquainted with chemical reactions and their balanced equations. Our second reason is to consider how we might predict the products of some of these reactions knowing only their reactants. The key to predicting the products formed by a given combination of reactants is recognising general patterns of chemical reactivity. Recognising a pattern of reactivity for a class of substances gives you a broader understanding than merely memorising a large number of unrelated reactions.

Combination and Decomposition Reactions

▼ TABLE 3.1 summarises two simple types of reactions, combination and decomposition reactions. In **combination reactions** two or more substances react to form one product. There are many examples of such reactions, especially those in which elements combine to form compounds. For example,

TABLE 3.1 • Combination and decomposition reactions	
Combination reactions	
$A + B \longrightarrow C$ $C(s) + O_2(g) \longrightarrow CO_2(g)$ $N_2(g) + 3\,H_2(g) \longrightarrow 2\,NH_3(g)$ $CaO(s) + H_2O(l) \longrightarrow Ca(OH)_2(s)$	Two reactants combine to form a single product. Many elements react with one another in this fashion to form compounds.
Decomposition reactions	
$C \longrightarrow A + B$ $2\,KClO_3(s) \longrightarrow 2\,KCl(s) + 3\,O_2(g)$ $PbCO_3(s) \longrightarrow PbO(s) + CO_2(g)$ $Cu(OH)_2(s) \longrightarrow CuO(s) + H_2O(l)$	A single reactant breaks apart to form two or more substances. Many compounds react this way when heated.

magnesium metal burns in air with a dazzling brilliance to produce magnesium oxide, as shown in ▼ **FIGURE 3.6**:

$$2\,Mg(s) + O_2(g) \longrightarrow 2\,MgO(s) \qquad \qquad [3.6]$$

This reaction is used to produce the bright flame generated by flares.

When a combination reaction occurs between a metal and a non-metal, as in Equation 3.6, the product is usually an ionic solid. Recall that the formula of an ionic compound can be determined from the charges of the ions involved (∞ Section 2.7, 'Ions and ionic compounds').

∞ Review this on page 46

When magnesium reacts with oxygen, for example, the magnesium loses electrons and forms the magnesium ion, Mg^{2+}. The oxygen gains electrons and

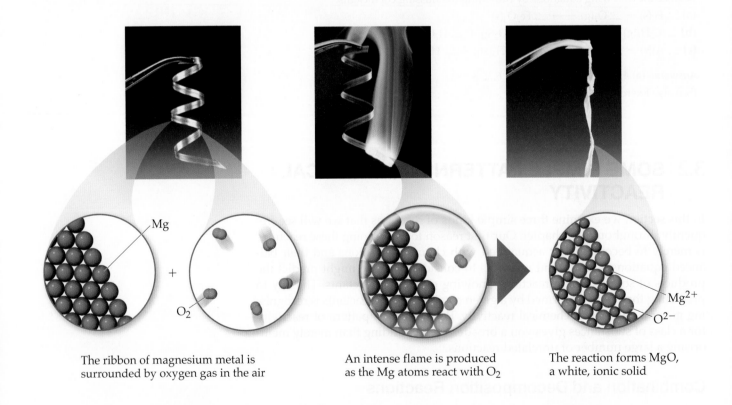

The ribbon of magnesium metal is surrounded by oxygen gas in the air

An intense flame is produced as the Mg atoms react with O_2

The reaction forms MgO, a white, ionic solid

Reactants

$2\,Mg(s) + O_2(g)$ ⟶ *Products* $2\,MgO(s)$

▲ FIGURE 3.6 **Combustion of magnesium metal in air, a combination reaction.**

forms the oxide ion, O^{2-}. Thus the reaction product is MgO. You should be able to recognise when a reaction is a combination reaction and to predict the products of a combination reaction in which the reactants are a metal and a non-metal.

CONCEPT CHECK 2

When Na and S combine in a combination reaction, what is the chemical formula of the product?

In a **decomposition reaction** one substance undergoes a reaction to produce two or more other substances. Many compounds undergo decomposition reactions when heated. For example, many metal carbonates decompose to form metal oxides and carbon dioxide when heated:

$$CaCO_3(s) \longrightarrow CaO(s) + CO_2(g) \qquad [3.7]$$

The decomposition of $CaCO_3$ is an important commercial process. Limestone or seashells, which are both primarily $CaCO_3$, are heated to prepare CaO, which is known as lime or quicklime. About 1.5×10^9 kg (1.5 million tonnes) of CaO is produced in Australia each year, principally in making glass, in obtaining iron from its ores and in making mortar to bind bricks.

The decomposition of sodium azide (NaN_3) rapidly releases $N_2(g)$, so this reaction is used to inflate safety air bags in vehicles (▶ **FIGURE 3.7**):

$$2\,NaN_3(s) \longrightarrow 2\,Na(s) + 3\,N_2(g) \qquad [3.8]$$

The system is designed so that an impact ignites a detonator cap, which in turn causes NaN_3 to decompose explosively. A small quantity of NaN_3 (about 100 g) forms a large quantity of gas (about 50 dm^3). The volume of gases produced in chemical reactions will be considered in Chapter 10.

▲ **FIGURE 3.7   Decomposition of sodium azide, NaN_3, is used to inflate car air bags.**

   **Writing balanced equations for combination and decomposition reactions**

Write balanced equations for the following reactions: **(a)** the combination reaction that occurs when lithium metal and fluorine gas react and **(b)** the decomposition reaction that occurs when solid barium carbonate is heated. (Two products form: a solid and a gas.)

SOLUTION

(a) The symbol for lithium is Li. With the exception of mercury, all metals are solids at room temperature. Fluorine occurs as a diatomic molecule (see Figure 2.13). Thus the reactants are Li(s) and $F_2(g)$. The product will consist of a metal and a non-metal, so we expect it to be an ionic solid. Lithium ions have a 1+ charge, Li^+, whereas fluoride ions have a 1− charge, F^-. Thus the chemical formula for the product is LiF. The balanced chemical equation is

$$2\,Li(s) + F_2(g) \longrightarrow 2\,LiF(s)$$

(b) The chemical formula for barium carbonate is $BaCO_3$. As noted in the text, many metal carbonates decompose to form metal oxides and carbon dioxide when heated. In Equation 3.7, for example, $CaCO_3$ decomposes to form CaO and CO_2. Thus we would expect that $BaCO_3$ decomposes to form BaO and CO_2. Barium and calcium are both in group 2 in the periodic table, moreover, which further suggests they would react in the same way:

$$BaCO_3(s) \longrightarrow BaO(s) + CO_2(g)$$

PRACTICE EXERCISE

Write balanced chemical equations for the following reactions: **(a)** solid mercury(II) sulfide decomposes into its component elements when heated, **(b)** the surface of aluminium metal undergoes a combination reaction with oxygen in the air.

Answers: **(a)** $HgS(s) \longrightarrow Hg(l) + S(s)$; **(b)** $4\,Al(s) + 3\,O_2(g) \longrightarrow 2\,Al_2O_3(s)$

(See also Exercise 3.12.)

Combustion Reactions

Combustion reactions are rapid reactions that produce a flame. Most of the combustion reactions we observe involve O_2 from air as a reactant. Equation 3.5 illustrates a general class of reactions involving the burning, or combustion, of hydrocarbon compounds (compounds that contain only carbon and hydrogen, such as CH_4 and C_2H_6) (∞ Section 2.9, 'Some Simple Organic Compounds').

∞ Review this on page 56

When hydrocarbons are combusted in air, they react with O_2 to form CO_2 and H_2O.* The number of molecules of O_2 required in the reaction and the number of molecules of CO_2 and H_2O formed depend on the composition of the hydrocarbon, which acts as the fuel in the reaction. For example, the combustion of propane (C_3H_8), a gas used for cooking and home heating, is described by the following equation:

$$C_3H_8(g) + 5\,O_2(g) \longrightarrow 3\,CO_2(g) + 4\,H_2O(g) \qquad [3.9]$$

The state of the water, $H_2O(g)$ or $H_2O(l)$, depends on the conditions of the reaction. Water vapour, $H_2O(g)$, is formed at high temperature in an open container. The blue flame produced when propane burns is shown in ◄ FIGURE 3.8.

Combustion of oxygen-containing derivatives of hydrocarbons, such as CH_3OH, also produces CO_2 and H_2O. The simple rule that hydrocarbons and related oxygen-containing derivatives of hydrocarbons form CO_2 and H_2O when they burn in air summarises the behaviour of about 3 million compounds. Many substances that our bodies use as energy sources, such as the sugar glucose ($C_6H_{12}O_6$), similarly react in our bodies with O_2 to form CO_2 and H_2O. In our bodies, however, the reactions take place in a series of steps that occur at body temperature. The reactions are then described as *oxidation reactions* rather than combustion reactions.

▲ FIGURE IT OUT

In what ways are the reactions depicted in Figures 3.4 and 3.8 alike?

▲ FIGURE 3.8 Propane burning in air. The liquid propane, C_3H_8, vaporises and mixes with air as it escapes through the nozzle. The combustion reaction of C_3H_8 and O_2 produces a blue flame.

SAMPLE EXERCISE 3.4 **Writing balanced equations for combustion reactions**

Write the balanced equation for the reaction that occurs when methanol, $CH_3OH(l)$, is burned in air.

SOLUTION

When any compound containing C, H and O is combusted, it reacts with the $O_2(g)$ in air to produce $CO_2(g)$ and $H_2O(g)$. Thus the unbalanced equation is:

$$CH_3OH(l) + O_2(g) \longrightarrow CO_2(g) + H_2O(g)$$

The C atoms are balanced, with one on each side of the arrow. Because CH_3OH has four H atoms, we place a coefficient 2 in front of H_2O to balance the H atoms:

$$CH_3OH(l) + O_2(g) \longrightarrow CO_2(g) + 2\,H_2O(g)$$

This balances H but gives four O atoms in the products. Because there are only three O atoms in the reactants (one in CH_3OH and two in O_2), we are not finished yet. We can place the fractional coefficient $\frac{3}{2}$ in front of O_2 to give a total of four O atoms in the reactants (there are $\frac{3}{2} \times 2 = 3$ O atoms in $\frac{3}{2}O_2$):

$$CH_3OH(l) + \frac{3}{2}O_2(g) \longrightarrow CO_2(g) + 2\,H_2O(g)$$

Although the equation is now balanced, it is not in its most conventional form because it contains a fractional coefficient. If we multiply each side of the equation by 2, we will remove the fraction and achieve the following balanced equation:

$$2\,CH_3OH(l) + 3\,O_2(g) \rightarrow 2\,CO_2(g) + 4\,H_2O(g)$$

---

*When there is an insufficient quantity of O_2 present, carbon monoxide (CO) will be produced along with the CO_2; this is called *incomplete* combustion. If the amount of O_2 is severely restricted, fine particles of carbon that we call soot will be produced. *Complete* combustion produces only CO_2 and H_2O. Unless specifically stated to the contrary, we will always take *combustion* to mean *complete combustion*.

3.3 | FORMULA MASS

Chemical formulae and chemical equations both have a *quantitative* significance; the subscripts in formulae and the coefficients in equations represent precise quantities. The formula H_2O indicates that a molecule of this substance contains exactly two atoms of hydrogen and one atom of oxygen. Similarly, the coefficients in a balanced chemical equation indicate the relative quantities of reactants and products. But how do we relate the numbers of atoms or molecules to the amounts we measure out in the laboratory? Although we cannot directly count atoms or molecules, we can indirectly determine their numbers if we know their masses. Therefore, before we can pursue the quantitative aspects of chemical formulae or equations, we must examine the masses of atoms and molecules, which we do in this section and the next.

Formula and Molecular Masses

The **formula mass** of a substance is the sum of the atomic masses of each atom in its chemical formula. Using atomic masses from a periodic table, we find, for example, that the formula mass of sulfuric acid (H_2SO_4) is 98.1 u:

$$\text{Formula mass} = 2\,(\text{atomic mass of H}) + (\text{atomic mass of S}) + 4\,(\text{atomic mass of O})$$
$$= 2(1.0\ \text{u}) + 32.1\ \text{u} + 4(16.0\ \text{u})$$
$$= 98.1\ \text{u}$$

For convenience, we have rounded off all the atomic masses to one place beyond the decimal point. We will round off the atomic masses in this way for most problems.

If the chemical formula is merely the chemical symbol of an element, such as Na, then the formula mass equals the **atomic mass** of the element. If the chemical formula is that of a molecule, then the formula mass is also called the **molecular mass**. The molecular mass of glucose ($C_6H_{12}O_6$), for example, is:

$$\text{Molecular mass of } C_6H_{12}O_6 = 6(12.0\ \text{u}) + 12(1.0\ \text{u}) + 6(16.0\ \text{u}) = 180.0\ \text{u}$$

Because ionic substances, such as NaCl, exist as three-dimensional arrays of ions (∞∞ Section 2.7, 'Ions and Ionic Compounds'), it is inappropriate to speak of molecules of NaCl. Instead, we speak of *formula units*, represented by the chemical formula of the substance. The formula unit of NaCl consists of one Na^+ ion and one Cl^- ion. Thus the formula mass of NaCl is the mass of one formula unit:

$$\text{Formula mass of NaCl} = 23.0\ \text{u} + 35.5\ \text{u} = 58.5\ \text{u}$$

∞∞ Review this on page 46

SAMPLE EXERCISE 3.5   **Calculating formula mass**

Calculate the formula mass of (**a**) sucrose, $C_{12}H_{22}O_{11}$ (table sugar) and (**b**) calcium nitrate, $Ca(NO_3)_2$.

SOLUTION

(**a**) By adding the atomic masses of the atoms in sucrose, we find it to have a formula mass of 342.0 u:

$$12\ \text{C atoms} = 12(12.0\ \text{u}) = 144.0\ \text{u}$$
$$22\ \text{H atoms} = 22(1.0\ \text{u}) = 22.0\ \text{u}$$
$$11\ \text{O atoms} = 11(16.0\ \text{u}) = \underline{176.0\ \text{u}}$$
$$342.0\ \text{u}$$

(b) If a chemical formula has parentheses, the subscript outside the parentheses is a multiplier for all atoms inside. Thus for $Ca(NO_3)_2$, we have

$$
\begin{aligned}
1 \text{ Ca atom} &= 1(40.1 \text{ u}) = 40.1 \text{ u} \\
2 \text{ N atoms} &= 2(14.0 \text{ u}) = 28.0 \text{ u} \\
6 \text{ O atoms} &= 6(16.0 \text{ u}) = \underline{96.0 \text{ u}} \\
& \hspace{3.2cm} 164.1 \text{ u}
\end{aligned}
$$

PRACTICE EXERCISE

Calculate the formula mass of **(a)** $Al(OH)_3$ and **(b)** CH_3OH.

Answers: **(a)** 78.0 u, **(b)** 32.0 u

(See also Exercises 3.17, 3.18.)

Percentage Composition from Formulae

Occasionally we must calculate the *percentage composition* of a compound (that is, the percentage by mass contributed by each element in the substance). For example, in order to verify the purity of a compound, we may wish to compare the calculated percentage composition of the substance with that found experimentally. Calculating percentage composition is a straightforward matter if the chemical formula is known.

$$
\% \text{ Element} = \frac{\left(\begin{array}{c} \text{number of atoms} \\ \text{of that element} \end{array} \right) \times \left(\begin{array}{c} \text{atomic mass} \\ \text{of the element} \end{array} \right)}{\text{formula mass of the compound}} \times 100\% \qquad [3.10]
$$

STRATEGIES IN CHEMISTRY

PROBLEM SOLVING

Practice is the key to success in solving problems. As you practise, you can improve your skills by following these steps.

Step 1: Analyse the problem. Read the problem carefully. What is it asking you to do? What information does it provide you with? List both the data you are given and the quantity you need to obtain (the unknown).

Step 2: Develop a plan for solving the problem. Consider a possible path between the given information and the unknown. This is usually a formula, an equation or some principle you

learnt earlier. Recognise that some data may not be given explicitly in the problem; you may be expected to know certain quantities (such as Avogadro's number) or look them up in tables (such as atomic masses). Recognise also that your plan may involve either a single step or a series of steps with intermediate answers.

Step 3: Solve the problem. Use the known information and suitable equations or relationships to solve for the unknown. Be careful with significant figures, signs and units.

Step 4: Check the solution. Read the problem again to make sure you have found all the solutions asked for in the problem. Does your answer make sense? That is, is the answer outrageously large or small or is it in the ballpark? Finally, are the units and significant figures correct?

SAMPLE EXERCISE 3.6 Calculating percentage composition

Calculate the percentage of carbon, hydrogen and oxygen (by mass) in $C_{12}H_{22}O_{11}$.

SOLUTION

Analyse We are given a chemical formula which shows the elements and the number of atoms of each element in the molecule.

Plan We see that Equation 3.10 relates % composition of an element in a molecule to the number of atoms of each element, the atomic mass of each element and the formula mass of the molecule. The formula mass can be calculated as in Sample Exercise 3.5 and the atomic masses can be obtained from the periodic table.

Solve Using Equation 3.10 and the periodic table to obtain atomic masses, we have

$$
\%C = \frac{(12)(12.0 \text{ u})}{342.0 \text{ u}} \times 100\% = 42.1\%
$$

$$\%H = \frac{(22)(1.0\ u)}{342.0\ u} \times 100\% = 6.4\%$$

$$\%O = \frac{(11)(16.0\ u)}{342.0\ u} \times 100\% = 51.5\%$$

Check The percentages of the individual elements must add up to 100%, which they do in this case. We could have used more significant figures for our atomic masses, giving more significant figures for our percentage composition, but we have adhered to our suggested guideline of rounding atomic masses to one digit beyond the decimal point.

PRACTICE EXERCISE

Calculate the percentage of nitrogen, by mass, in $Ca(NO_3)_2$.

Answer: 17.1%

(See also Exercises 3.19, 3.20.)

3.4 | AVOGADRO'S NUMBER AND THE MOLE

Even the smallest samples that we deal with in the laboratory contain enormous numbers of atoms, ions or molecules. For example, a teaspoon of water (about 5 cm^3) contains about 2×10^{23} water molecules, a number so large that it almost defies comprehension. Chemists, therefore, have devised a special counting unit for describing such large numbers of atoms or molecules.

In everyday life we use counting units like a dozen (12 objects) and a gross (144 objects) to deal with modestly large quantities. In chemistry the unit for dealing with the number of atoms, ions or molecules in a common-sized sample is the **mole**, abbreviated mol.[*] One mole is the amount of matter that contains as many objects (atoms, molecules or whatever objects we are considering) as the number of atoms in exactly 12 g of isotopically pure ^{12}C. From experiments, scientists have determined this number to be 6.0221421×10^{23}. Scientists call this number **Avogadro's number**, in honour of Amedeo Avogadro (1776–1856), an Italian scientist. For most purposes we will use 6.02×10^{23} or 6.022×10^{23} for Avogadro's number throughout the text. Avogadro's number is represented by N_A.

A mole of atoms, a mole of molecules or a mole of anything else all contain Avogadro's number of these objects:

$$N_A = 6.022 \times 10^{23}$$
$$1\ mol\ ^{12}C\ atoms = 6.02 \times 10^{23}\ ^{12}C\ atoms$$
$$1\ mol\ H_2O\ molecules = 6.02 \times 10^{23}\ H_2O\ molecules$$
$$1\ mol\ NO_3^-\ ions = 6.02 \times 10^{23}\ NO_3^-\ ions$$

Avogadro's number is so large that it is difficult to imagine. Spreading 6.02×10^{23} marbles over the entire surface of the Earth would produce a layer about 5 km thick.

SAMPLE EXERCISE 3.7 Estimating numbers of atoms

Without using a calculator, arrange the following samples in order of increasing numbers of carbon atoms: 12 g ^{12}C, 1 mol C_2H_2, 9×10^{23} molecules of CO_2.

SOLUTION

Analyse We are given amounts of three different substances expressed in grams, moles and number of molecules and are asked to rank the samples in increasing number of C atoms.

Plan Each quantity must be converted to numbers of C atoms. To do this it is most convenient to go via the number of moles in each quantity where necessary.

Solve A mole is defined as the amount of matter that contains as many units of the matter as there are C atoms in exactly 12 g of ^{12}C. Thus 12 g of ^{12}C contains 1 mol of C atoms (that is, 6.02×10^{23} C atoms). In 1 mol C_2H_2, there are 6×10^{23} C_2H_2 molecules. Because there are two C atoms in each C_2H_2 molecule, this sample contains 12×10^{23} C atoms. Because each CO_2 molecule contains one C atom, the sample of CO_2 contains 9×10^{23} C atoms. Hence, the

---

[*]The term *mole* comes from the Latin word *moles*, meaning 'a mass'. The term *molecule* is the diminutive form of this word and means 'a small mass'.

order is 12 g ^{12}C (6×10^{23} C atoms) $< 9 \times 10^{23}$ CO$_2$ molecules (9×10^{23} C atoms) < 1 mol C$_2$H$_2$ (12×10^{23} C atoms).

Check We can check our results by comparing the number of moles of C atoms in each sample because the number of

moles is proportional to the number of atoms. Thus 12 g of ^{12}C is 1 mol C, 1 mol of C$_2$H$_2$ contains 2 mol C and 9×10^{23} molecules of CO$_2$ contain 1.5 mol C, giving the same order as above: 12 g ^{12}C (1 mol C) $< 9 \times 10^{23}$ CO$_2$ molecules (1.5 mol C) < 1 mol C$_2$H$_2$ (2 mol C).

PRACTICE EXERCISE

Without using a calculator, arrange the following samples in order of increasing number of O atoms: 1 mol H$_2$O, 1 mol CO$_2$, 3×10^{23} molecules O$_3$.

Answer: 1 mol H$_2$O (6×10^{23} O atoms) $< 3 \times 10^{23}$ molecules O$_3$ (9×10^{23} O atoms) < 1 mol CO$_2$ (12×10^{23} O atoms)

(See also Exercises 3.23, 3.24.)

SAMPLE EXERCISE 3.8   Converting moles to number of atoms

Calculate the number of H atoms in 0.350 mol of C$_6$H$_{12}$O$_6$.

SOLUTION

Analyse We are given the amount in mol of C$_6$H$_{12}$O$_6$ and must find the number of H atoms in the sample.

Plan First we must convert the number of moles to the number of molecules and then multiply this by the number of C atoms in each molecule.

Solve   Moles $\times 6.02 \times 10^{23} \longrightarrow$ number of molecules

Molecules C$_6$H$_{12}$O$_6$ = 0.350 mol $\times 6.02 \times 10^{23}$ molecules mol^{-1}

$= 2.11 \times 10^{23}$ molecules

There are 12 H atoms in each molecule.

So H atoms $= 2.11 \times 10^{23}$ molecules $\times 12$ atoms molecule^{-1}

$= 2.53 \times 10^{24}$ atoms

Check The magnitude of our answer is reasonable: it is a large number about the magnitude of Avogadro's number. We can also make the following rough calculation: multiplying $0.35 \times 6 \times 10^{23}$ gives about 2×10^{23} molecules. Multiplying this result by 12 gives $24 \times 10^{23} = 2.4 \times 10^{24}$ H atoms, which agrees with the previous, more detailed calculation. Because we were asked for the number of H atoms, the units of our answer are correct. The given data had three significant figures, so our answer has three significant figures.

PRACTICE EXERCISE

How many oxygen atoms are in **(a)** 0.25 mol Ca(NO$_3$)$_2$ and **(b)** 1.50 mol of sodium carbonate?

Answers: **(a)** 9.0×10^{23}, **(b)** 2.71×10^{24}

(See also Exercises 3.25, 3.26.)

Molar Mass

A dozen is the same number (12) whether we have a dozen eggs or a dozen elephants. Clearly, however, a dozen eggs does not have the same mass as a dozen elephants. Similarly, a mole is always the *same number* (6.02×10^{23}), but 1 mole samples of different substances will have *different masses*. Compare, for example, 1 mol of ^{12}C with 1 mol of ^{24}Mg. A single ^{12}C atom has a mass of 12 u, whereas a single ^{24}Mg atom is twice as massive, 24 u (to two significant figures). Because a mole always has the same number of particles, a mole of ^{24}Mg must be twice as massive as a mole of ^{12}C. Because a mole of ^{12}C has a mass of 12 g (by definition), then a mole of ^{24}Mg must have a mass of 24 g. This example illustrates a general rule relating the mass of an atom to the mass of Avogadro's number (1 mol) of these atoms: *The mass of a single atom of an element (in u) is numerically equal to the mass (in grams) of 1 mol of that element.* This statement is true regardless of the element.

1 atom of ^{12}C has a mass of 12 u $\Rightarrow$ 1 mol ^{12}C has a mass of 12 g

1 atom of Cl has an atomic mass of 35.5 u $\Rightarrow$ 1 mol Cl has a mass of 35.5 g

1 atom of Au has an atomic mass of 197 u $\Rightarrow$ 1 mol Au has a mass of 197 g

Notice that when we are dealing with a particular isotope of an element, we use the mass of that isotope; otherwise, we use the atomic mass (the average atomic mass) of the element.

For other kinds of substances, the same numerical relationship exists between the formula mass (in u) and the mass (in grams) of one mole of that substance:

1 H_2O molecule has a mass of 18.0 u $\Rightarrow$ 1 mol H_2O has a mass of 18.0 g

1 NO_3^- ion has a mass of 62.0 u $\Rightarrow$ 1 mol NO_3^- has a mass of 62.0 g

1 NaCl unit has a mass of 58.5 u $\Rightarrow$ 1 mol NaCl has a mass of 58.5 g

▶ **FIGURE 3.9** illustrates the relationship between the mass of a single molecule of H_2O and that of a mole of H_2O.

The mass in grams of one mole of a substance (that is, the mass in grams per mol) is called the **molar mass** (M) of the substance. *The molar mass (in g mol^{-1}) of any substance is always numerically equal to its formula mass (in u).* The substance NaCl, for example, has a formula mass of 58.5 u and a molar mass of 58.5 g mol^{-1}. Further examples of mole relationships are shown in ▼ **TABLE 3.2.** ▼ **FIGURE 3.10** shows 1 mole quantities of several common substances.

FIGURE IT OUT

How many H_2O molecules are in a 9.00 g sample of water?

Single molecule

1 molecule H_2O
(18.0 u)

Avogadro's number of molecules
(6.02×10^{23})

Laboratory-size sample

1 mol H_2O
(18.0 g)

▲ **FIGURE 3.9 Comparing the mass of 1 molecule and 1 mol of H_2O.** Both masses have the same number but different units (atomic mass units and grams). Expressing both masses in grams indicates their huge difference: 1 molecule H_2O has a mass of 2.99×10^{-23} g whereas 1 mol H_2O has a mass of 18.0 g.

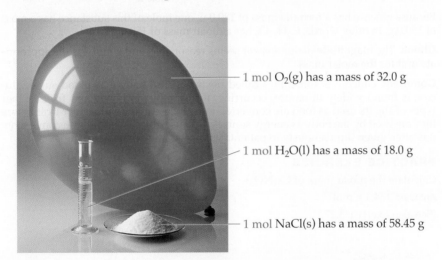

1 mol $O_2(g)$ has a mass of 32.0 g

1 mol $H_2O(l)$ has a mass of 18.0 g

1 mol NaCl(s) has a mass of 58.45 g

▲ **FIGURE 3.10   One mole each of a solid (NaCl), a liquid (H_2O) and a gas (O_2).** In each case, the mass in grams of 1 mol—that is, the molar mass—is numerically equal to the formula weight in atomic mass units. Each of these samples contains 6.02×10^{23} formula units.

TABLE 3.2 • Mole relationships

Name of substance	Formula	Formula weight (u)	Molar mass (g mol^{-1})	Number and kind of particles in one mole
Atomic nitrogen	N	14.0	14.0	6.02×10^{23} N atoms
Molecular nitrogen	N_2	28.0	28.0	$\begin{cases} 6.02 \times 10^{23}\ N_2\ \text{molecules} \\ 2(6.02 \times 10^{23})\ \text{N atoms} \end{cases}$
Silver	Ag	107.9	107.9	6.02×10^{23} Ag atoms
Silver ions	Ag$^+$	107.9*	107.9	6.02×10^{23} Ag$^+$ ions
Barium chloride	$BaCl_2$	208.2	208.2	$\begin{cases} 6.02 \times 10^{23}\ BaCl_2\ \text{formula units} \\ 6.02 \times 10^{23}\ Ba^{2+}\ \text{ions} \\ 2(6.02 \times 10^{23})\ Cl^-\ \text{ions} \end{cases}$

*Recall that the electron has negligible mass; thus ions and atoms have essentially the same mass.

The entries in Table 3.2 for N and N_2 point out the importance of stating the chemical form of a substance exactly when we use the mole concept. Suppose you read that 1 mol of nitrogen is produced in a particular reaction. You might interpret this statement to mean 1 mol of nitrogen atoms (14.0 g). Unless otherwise stated, however, what is probably meant is 1 mol of nitrogen molecules, N_2 (28.0 g), because N_2 is the usual chemical form of the element. To avoid ambiguity, it is important to state explicitly the chemical form being discussed. Using the chemical formula N_2 avoids ambiguity.

SAMPLE EXERCISE 3.9 Calculating molar mass

What is the molar mass of glucose, $C_6H_{12}O_6$?

SOLUTION

Analyse We are given a molecular formula which gives us the types of atoms and their number in the molecule.

Plan The molar mass of any substance is numerically equal to its formula mass which will have units of u, whereas the molar mass will have units of g mol^{-1}. We proceed as in Sample Exercise 3.5.

Solve Our first step is to determine the formula mass of glucose.

$$6\,C\text{ atoms} = 6(12.0\text{ u}) = 72.0\text{ u}$$
$$12\,H\text{ atoms} = 12(1.0\text{ u}) = 12.0\text{ u}$$
$$6\,O\text{ atoms} = 6(16.0\text{ u}) = \underline{96.0\text{ u}}$$
$$180.0\text{ u}$$

Because glucose has a formula mass of 180.0 u, one mole of this substance has a mass of 180.0 g. In other words, $C_6H_{12}O_6$ has a molar mass of 180.0 g mol^{-1}.

Check The magnitude of our answer seems reasonable, and g mol^{-1} is the appropriate unit for the molar mass.

Comment Glucose is sometimes called dextrose. Also known as blood sugar, glucose is found widely in nature, occurring, for example, in honey and fruits. Other types of sugars used as food are converted into glucose in the stomach or liver before they are used by the body as energy sources. Because glucose requires no conversion, it is often given intravenously to patients who need immediate nourishment.

PRACTICE EXERCISE

Calculate the molar mass of $Ca(NO_3)_2$.

Answer: 164.1 g mol^{-1}

(See also Exercise 3.27.)

MY WORLD OF CHEMISTRY

GLUCOSE MONITORING

Over 1 million Australians (although estimates vary) have diabetes, and globally the number approaches 172 million. Diabetes is a metabolic disorder in which the body either cannot produce or cannot properly use the hormone insulin. One signal that a person is diabetic is that the concentration of glucose in the blood is higher than normal. Therefore, people who are diabetic need to measure their blood glucose concentrations regularly. Untreated diabetes can cause severe complications such as blindness and loss of limbs.

The body converts most of the food we eat into glucose. After digestion, glucose is delivered to cells via the blood. Cells need glucose to live, and insulin must be present in order for glucose to enter the cells. Normally, the body adjusts the concentration of insulin automatically, in concert with the glucose concentration after eating. However, in a diabetic person, either little or no insulin is produced (Type 1 diabetes) or insulin is produced but the cells cannot take it up properly (Type 2 diabetes). The result is that the blood glucose concentration is too high. People normally have a range of 70–120 mg glucose per litre of blood. A person who has not eaten for 8 hours or more is diagnosed as diabetic if his or her glucose level is 1260 mg dm^{-3} or higher.

Glucose meters work by the introduction of blood from a person, usually by a prick of the finger, onto a small strip of paper that contains chemicals that react with glucose. Insertion of the strip into a small battery-operated reader gives the glucose concentration. The mechanism of the readout varies from one monitor to another—it may be a measurement of a small electrical current or measurement of light produced in a chemical reaction. Depending on the reading on any given day, a diabetic person may need to receive an injection of insulin or simply stop eating sweets for a while.

Interconverting Masses and Moles

Conversions of mass to moles and of moles to mass are frequently encountered in calculations using the mole concept. These calculations are simplified using dimensional analysis, as shown in Sample Exercises 3.10 and 3.11.

There is a simple relationship between moles, mass and molar mass. This is given by Equation 3.11.

$$\text{Moles} = \frac{\text{mass in grams}}{\text{molar mass in g mol}^{-1}} \qquad [3.11]$$

$$\text{or, } n = \frac{m}{M}$$

where $n = $ moles
       $m = $ mass in grams
       $M = $ molar mass in g mol^{-1}

SAMPLE EXERCISE 3.10   **Converting grams to moles**

Calculate the number of moles of glucose ($C_6H_{12}O_6$) in 5.380 g of $C_6H_{12}O_6$.

SOLUTION

Analyse We are given a molecular formula and the mass of the compound.

Plan We remember the relationship between moles, mass and molar mass given in Equation 3.11.

Solve
Molar mass $C_6H_{12}O_6 = 180.0$ g mol^{-1} (Sample Exercise 3.9).
From Equation 3.11

$$\text{Since } n = \frac{m}{M}$$

$$\text{Moles } C_6H_{12}O_6 = \frac{5.380 \text{ g}}{180.0 \text{ g mol}^{-1}}$$

$$= 0.02989 \text{ mol}$$

PRACTICE EXERCISE

How many moles of sodium bicarbonate ($NaHCO_3$) are there in 508 g of $NaHCO_3$?

Answer: 6.05 mol $NaHCO_3$

(See also Exercises 3.25, 3.26.)

SAMPLE EXERCISE 3.11   **Converting moles to grams**

Calculate the mass, in grams, of 0.433 mol of calcium nitrate.

SOLUTION

Analyse We are given the name of a compound and the number of moles and are asked to calculate the mass corresponding to that number of moles.

Plan Again we can apply Equation 3.11, but to do so we need to find the molar mass of calcium nitrate. To do this we need the molecular formula.

Solve Because the calcium ion is Ca^{2+} and the nitrate ion is NO_3^-, calcium nitrate is $Ca(NO_3)_2$. Adding the atomic masses of the elements in the compound gives a formula mass of 164.1 u or 164.1 g mol^{-1} for 1 mole of $Ca(NO_3)_2$. From Equation 3.11 we have

$$n = \frac{m}{M}$$

therefore, by rearranging the equation,

$$m = n \times M$$

$$\text{Grams } Ca(NO_3)_2 = 0.433 \text{ mol} \times 164.1 \text{ g mol}^{-1}$$

$$= 71.1 \text{ g}$$

PRACTICE EXERCISE

What is the mass, in grams, of **(a)** 6.33 mol of $NaHCO_3$ and **(b)** 3.0×10^{-5} mol of sulfuric acid?

Answers: (a) 532 g, **(b)** 2.9×10^{-3} g

(See also Exercises 3.25, 3.26.)

Interconverting Masses and Numbers of Particles

The mole concept provides the bridge between masses and numbers of particles. To illustrate how we can interconvert masses and numbers of particles, let's calculate the number of copper atoms in an old copper coin. Such a coin weighs about 3 g, and we'll assume that it is 100% copper:

$$\text{Number of atoms} = \text{moles} \times 6.02 \times 10^{23} \text{ atoms mol}^{-1}$$

$$\text{Cu atoms} = \frac{3 \text{ g}}{63.5 \text{ g mol}^{-1}} \times 6.02 \times 10^{23} \text{ atoms mol}^{-1}$$

$$= 3 \times 10^{22} \text{ atoms}$$

The molar mass and Avogadro's number are used as conversion factors to convert grams $\rightarrow$ moles $\rightarrow$ atoms. Notice also that our answer is a very large number. Any time you calculate the number of atoms, molecules or ions in an ordinary sample of matter, you can expect the answer to be very large. In contrast, the number of moles in a sample will usually be much smaller, often less than 1. The general procedure for interconverting mass and number of formula units (atoms, molecules, ions or whatever is represented by the chemical formula) of a substance is summarised in ▼ **FIGURE 3.11**.

SAMPLE EXERCISE 3.12	**Calculating the number of molecules and number of atoms from mass**

(a) How many glucose molecules are in 5.23 g of $C_6H_{12}O_6$? **(b)** How many oxygen atoms are in this sample?

SOLUTION

Analyse We are given the number of grams and the molecular formula and are asked to calculate (a) the number of molecules and (b) the number of O atoms in the sample.

Plan A summary of the strategy involved is given in Figure 3.11. Grams are converted to moles and moles to numbers of molecules of glucose. Then knowing the number of O atoms in each molecule we can calculate the total number of O atoms in each sample.

⚠ **FIGURE IT OUT**

What number would you use to convert (a) moles of CH_4 to grams of CH_4 and (b) number of molecules of CH_4 to moles of CH_4?

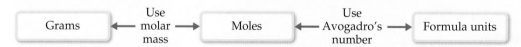

▲ **FIGURE 3.11 Procedure for interconverting mass and number of formula units.**
The number of moles of the substance is central to the calculation. Thus the mole concept can be thought of as the bridge between the mass of a sample in grams and the number of formula units contained in the sample.

Solve

(a) First convert the grams of $C_6H_{12}O_6$ to moles.

$$n = \frac{m}{M}$$

$$\text{Moles } C_6H_{12}O_6 = \frac{5.23 \text{ g}}{180.0 \text{ g mol}^{-1}}$$

$$= 0.0291 \text{ mol}$$

$$\text{Molecules } C_6H_{12}O_6 = \text{moles} \times N_A$$

$$= 0.0291 \text{ mol} \times 6.022 \times 10^{23} \text{ molecules mol}^{-1}$$

$$= 1.75 \times 10^{22} \text{ molecules}$$

(b) There are six oxygen atoms in each molecule of $C_6H_{12}O_6$.

$$\text{Atoms O} = 1.75 \times 10^{22} \text{ molecules} \times 6$$

$$= 1.05 \times 10^{23} \text{ atoms}$$

Check The answer is simply six times as large as the answer to part (a). The number of significant figures (three) and the units (atoms O) are correct.

PRACTICE EXERCISE

(a) How many nitric acid molecules are in 4.20 g of HNO_3? **(b)** How many O atoms are in this sample?

Answers: **(a)** 4.01×10^{22} molecules HNO_3, **(b)** 1.20×10^{23} atoms O

(See also Exercise 3.31.)

3.5 | EMPIRICAL FORMULAE FROM ANALYSES

As we learned in Section 2.6, the empirical formula for a substance tells us the relative number of atoms of each element it contains. Thus the empirical formula H_2O indicates that water contains two H atoms for each O atom. This ratio also applies on the molar level: 1 mol of H_2O contains 2 mol of H atoms and 1 mol of O atoms. Conversely, the ratio of the number of moles of each element in a compound gives the subscripts in a compound's empirical formula. Thus the mole concept provides a way of calculating the empirical formulae of chemical substances, as shown in the following examples.

Mercury and chlorine combine to form a compound that is 73.9% mercury and 26.1% chlorine by mass. This means that if we had a 100.0 g sample of the solid, it would contain 73.9 g of mercury (Hg) and 26.1 g of chlorine (Cl). (Any size sample can be used in problems of this type, but we will generally use 100.0 g to simplify the calculation of mass from percentage.) Using the atomic masses of the elements to give us molar masses, we can calculate the number of moles of each element in the sample:

$$n = \frac{m}{M}$$

$$\text{mol Hg} = \frac{73.9 \text{ g}}{200.6 \text{ g mol}^{-1}} = 0.368 \text{ mol Hg}$$

$$\text{mol Cl} = \frac{26.1 \text{ g}}{35.5 \text{ g mol}^{-1}} = 0.735 \text{ mol Cl}$$

We then divide the larger number of moles (0.735) by the smaller (0.368) to obtain a Cl:Hg mole ratio of 1.99:1:

$$\frac{\text{Moles of Cl}}{\text{Moles of Hg}} = \frac{0.735 \text{ mol Cl}}{0.368 \text{ mol Hg}} = \frac{1.99 \text{ mol Cl}}{1 \text{ mol Hg}}$$

Because of experimental errors, the results generally will not lead to exact numbers for the ratios of moles. The number 1.99 is very close to 2, so we can

FIGURE IT OUT

How do you calculate the mole ratio of each element in any compound?

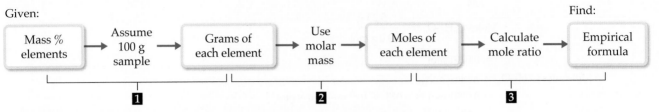

▲ **FIGURE 3.12   Procedure for calculating an empirical formula from percentage composition.** The key step in the calculation is step 2, determining the number of moles of each element in the compound.

∞ Review this on page 42

confidently conclude that the empirical formula for the compound is $HgCl_2$. This is the empirical formula because its subscripts are the smallest integers that express the *ratios* of atoms present in the compound (∞ Section 2.6, 'Molecules and Molecular Compounds').

The general procedure for determining empirical formulae is outlined in ▲ **FIGURE 3.12.**

SAMPLE EXERCISE 3.13   Calculating an empirical formula

The koala dines exclusively on eucalyptus leaves. Its digestive system detoxifies the eucalyptus oil which is a poison to other animals. The chief constituent of eucalyptus oil is a substance called eucalyptol, which contains 77.87% C, 11.76% H and the remainder O. What is the empirical formula of this compound?

SOLUTION

Analyse We are given the percentage composition of a compound and asked to calculate the empirical formula.

Plan In such problems we have to assume a mass of the compound. Any mass would do, but for simplicity we usually assume a mass of 100 g. We also obtain the % O by subtracting the sum of the other percentages from 100.

Solve *First*, we assume that we have exactly 100 g of material (although any mass can be used). So in 100 g of eucalyptol, we have

77.87 g C, 11.76 g H and 10.37 g O (from the percentages given)

Second, we calculate the number of moles of each element:

$$\text{Moles C} = \frac{77.87 \text{ g}}{12.01 \text{ g mol}^{-1}} = 6.484 \text{ mol C}$$

$$\text{Moles H} = \frac{11.76 \text{ g}}{1.008 \text{ g mol}^{-1}} = 11.67 \text{ mol H}$$

$$\text{Moles O} = \frac{10.37 \text{ g}}{16.00 \text{ g mol}^{-1}} = 0.6481 \text{ mol O}$$

Third, we determine the simplest whole-number ratio of moles by dividing each number of moles by the smallest number of moles, 0.6481:

$$\text{C: } \frac{6.484}{0.6481} = 10.00 \quad \text{H: } \frac{11.67}{0.6481} = 18.00 \quad \text{O: } \frac{0.6481}{0.6481} = 1.000$$

$$C : H : O = 10 : 18 : 1$$

The whole-number mole ratio gives us the subscripts for the empirical formula:

$$C_{10}H_{18}O$$

Check It is reassuring that the subscripts are moderately sized whole numbers. Otherwise, we have little by which to judge the reasonableness of our answer.

PRACTICE EXERCISE

A 5.325 g sample of methyl benzoate, a compound used in the manufacture of perfumes, is found to contain 3.758 g of carbon, 0.316 g of hydrogen and 1.251 g of oxygen. What is the empirical formula of this substance?

Answer: C_4H_4O

(See also Exercises 3.32–3.35, 3.38.)

Molecular Formulae from Empirical Formulae

For any compound, the formula obtained from percentage compositions is always the empirical formula. We can obtain the molecular formula from the empirical formula if we are given the molecular mass or molar mass of the compound. *The subscripts in the molecular formula of a substance are always a whole-number multiple of the corresponding subscripts in its empirical formula.* (∞ Section 2.6, 'Molecules and Molecular Compounds').This multiple can be found by comparing the empirical formula mass with the molecular mass:

∞ Review this on page 42

$$\text{Whole-number multiple} = \frac{\text{molecular mass}}{\text{empirical formula mass}} \qquad [3.12]$$

The empirical formula of ascorbic acid (vitamin C) was determined to be $C_3H_4O_3$, giving an empirical formula mass of $3(12.0\text{ u}) + 4(1.0\text{ u}) + 3(16.0\text{ u}) = 88.0$ u. The experimentally determined molecular mass is 176 u. Thus the molecular mass is twice the empirical formula mass $(176/88.0 = 2.00)$, and the molecular formula must therefore have twice as many of each kind of atom as the empirical formula. Consequently, we multiply the subscripts in the empirical formula by 2 to obtain the molecular formula: $C_6H_8O_6$.

SAMPLE EXERCISE 3.14   **Determining a molecular formula**

Eucalyptol, the constituent of eucalyptus oil (Sample Exercise 3.13), has an empirical formula of $C_{10}H_{18}O$. The experimentally determined molecular mass of this substance is 152 u. What is the molecular formula of eucalyptol?

SOLUTION

Analyse We are given an empirical formula and a molecular mass and asked to determine a molecular formula.

Plan The subscripts in a molecular formula are whole-number multiples of the subscripts in the compound's empirical formula. We can find the appropriate multiple by using Equation 3.12.

Solve First, we calculate the formula mass of the empirical formula, $C_{10}H_{18}O$:

$$10(12.0\text{ u}) + 18(1.0\text{ u}) + 1(16.0\text{ u}) = 154\text{ u}$$

Next, we divide the molecular mass by the empirical formula mass to obtain the multiple used to multiply the subscripts in $C_{10}H_{18}O$:

$$\frac{\text{Molecular mass}}{\text{Empirical formula mass}} = \frac{154}{152} = 1.01$$

Only whole-number ratios make physical sense because we must be dealing with whole atoms. The 1.01 in this case results from a small experimental error in the molecular mass. In this case the molecular mass and empirical formula mass are the same and so the molecular formula must be the same as the empirical formula, $C_{10}H_{18}O$.

PRACTICE EXERCISE

Ethylene glycol, the substance used in car antifreeze, is composed of 38.7% C, 9.7% H and 51.6% O by mass. Its molar mass is 62.1 g mol⁻¹. **(a)** What is the empirical formula of ethylene glycol? **(b)** What is its molecular formula?

Answers: **(a)** CH_3O, **(b)** $C_2H_6O_2$

(See also Exercises 3.38, 3.39.)

Combustion Analysis

The empirical formula of a compound is based on experiments that give the number of moles of each element in a sample of the compound. That is why we use the word 'empirical', which means 'based on observation and experiment'. Chemists have devised a number of experimental techniques to determine

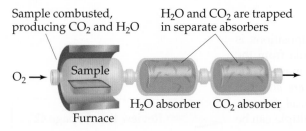

Sample combusted, producing CO_2 and H_2O

H_2O and CO_2 are trapped in separate absorbers

$O_2 \rightarrow$ Sample

H_2O absorber CO_2 absorber

Furnace

Mass gained by each absorber corresponds to mass of CO_2 or H_2O produced

▲ **FIGURE 3.13** **Apparatus for combustion analysis.**

empirical formulae. One of these is *combustion analysis*, which is commonly used for compounds containing principally carbon and hydrogen as their component elements.

When a compound containing carbon and hydrogen is completely combusted in an apparatus such as that shown in ◄ **FIGURE 3.13**, the carbon in the compound is converted to CO_2 and the hydrogen is converted to H_2O (Section 3.2). The amounts of CO_2 and H_2O produced are determined by measuring the mass increase in the CO_2 and H_2O absorbers. From the masses of CO_2 and H_2O we can calculate the number of moles of C and H in the original compound and thereby the empirical formula. If a third element is present in the compound, its mass can be determined by subtracting the masses of C and H from the compound's original mass. Sample Exercise 3.15 shows how to determine the empirical formula of a compound containing C, H and O.

SAMPLE EXERCISE 3.15 **Determining empirical formulae by combustion analysis**

Isopropyl alcohol, a substance sold as rubbing alcohol, is composed of C, H and O. Combustion of 0.255 of isopropyl alcohol produces 0.561 g of CO_2 and 0.306 g of H_2O. Determine the empirical formula of isopropyl alcohol.

SOLUTION

Analyse We are given the number of grams of CO_2 and H_2O produced when 0.255 g of isopropy alcohol is combusted and are asked to determine the empirical formula.

Plan We can use the mole concept to calculate the number of grams of C present in the CO_2 and the number of grams of H present in the H_2O. These are the quantities of C and H present in the isopropyl alcohol before combustion. The number of grams of O in the compound equals the mass of the isopropyl alcohol minus the sum of the C and H masses. Once we have the number of grams of C, H and O in the sample, we can then proceed as in Sample Exercise 3.13 to calculate the number of moles of each element and determine the mole ratio, which gives the subscripts in the empirical formula.

Solve To calculate the number of grams of C, we first use the molar mass of CO_2 (44.0 g mol^{-1}) to convert grams of CO_2 to moles of CO_2. Because there is only 1 C atom in each CO_2 molecule, there is 1 mol of C atoms per mole of CO_2 molecules. This fact allows us to convert the moles of CO_2 to moles of C. Finally, we use the molar mass of C (12.0 g mol^{-1}) to convert moles of C to grams of C. The calculation of the number of grams of H from the grams of H_2O is similar, although we must remember that there are 2 mol of H atoms per 1 mol of H_2O molecules.

$$\text{Moles } CO_2 = \frac{0.561 \text{ g}}{44.0 \text{ g mol}^{-1}} = 0.0128 \text{ mol}$$

$$\text{Moles C} = 0.0128 \text{ mol}$$

$$\text{Moles } H_2O = \frac{0.306 \text{ g}}{18.0 \text{ g mol}^{-1}} = 0.0170 \text{ mol}$$

$$\text{Moles H} = 2 \times 0.017 \text{ mol} = 0.0340 \text{ mol}$$

Therefore,

$$\text{Grams C} = 0.0128 \text{ mol} \times 12.0 \text{ g mol}^{-1} = 0.154 \text{ g C}$$

$$\text{Grams H} = 0.0340 \text{ mol} \times 1.01 \text{ g mol}^{-1} = 0.0343 \text{ g H}$$

The total mass of the sample, 0.255 g, is the sum of the masses of C, H and O. Thus we can calculate the mass of O as follows:

$$\text{Mass of O} = \text{mass of sample} - (\text{mass of C} + \text{mass of H})$$
$$= 0.255 \text{ g} - (0.154 \text{ g} + 0.0343 \text{ g}) = 0.067 \text{ g O}$$

We then calculate the number of moles of C, H and O in the sample:

$$\text{Moles C} = \frac{0.154 \text{ g}}{12.0 \text{ g mol}^{-1}} = 0.0128 \text{ mol C}$$

$$\text{Moles H} = \frac{0.0343 \text{ g}}{1.01 \text{ g mol}^{-1}} = 0.0340 \text{ mol H}$$

$$\text{Moles O} = \frac{0.068 \text{ g}}{16.0 \text{ g mol}^{-1}} = 0.0043 \text{ mol O}$$

To find the empirical formula, we must compare the relative number of moles of each element in the sample. The relative number of moles of each element is found by dividing each number by the smallest number, 0.0043. The mole ratio of C : H : O so obtained is 2.98 : 7.91 : 1.00. The first two numbers are very close to the whole numbers 3 and 8, giving the empirical formula C_3H_8O.

Check The subscripts work out to be moderately sized whole numbers, as expected.

▲ **CONCEPT CHECK 3**

In Sample Exercise 3.15, how do you explain the fact that the ratios C : H : O are 2.98 : 7.91 : 1.00, rather than exact integers 3 : 8 : 1?

3.6 | QUANTITATIVE INFORMATION FROM BALANCED CHEMICAL EQUATIONS

The coefficients in a chemical equation represent the relative numbers of molecules involved in a reaction. The mole concept allows us to convert this information to the masses of the substances. Consider the following balanced equation:

$$2 H_2(g) + O_2(g) \longrightarrow 2 H_2O(l) \qquad \text{[3.13]}$$

The coefficients indicate that two molecules of H_2 react with each molecule of O_2 to form two molecules of H_2O. It follows that the relative numbers of moles are identical to the relative numbers of molecules:

$2 H_2(g)$	$+$	$O_2(g)$	$\longrightarrow$	$2 H_2O(l)$
2 molecules		1 molecule		2 molecules
$2(6.02 \times 10^{23}$ molecules)		$1(6.02 \times 10^{23}$ molecules)		$2(6.02 \times 10^{23}$ molecules)
2 mol		1 mol		2 mol

We can generalise this observation for all balanced chemical equations: *The coefficients in a balanced chemical equation indicate both the relative numbers of molecules (or formula units) involved in the reaction and the relative numbers of moles.*
▼ **FIGURE 3.14** further summarises this result and also shows how it corresponds to the law of conservation of mass. Notice that the total mass of the reactants (4.0 g + 32.0 g) equals the total mass of the products (36.0 g).

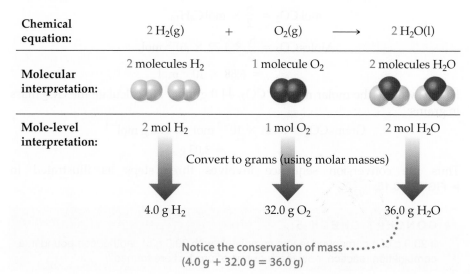

Chemical equation:	$2 H_2(g)$	$+$	$O_2(g)$	$\longrightarrow$	$2 H_2O(l)$
Molecular interpretation:	2 molecules H_2		1 molecule O_2		2 molecules H_2O
Mole-level interpretation:	2 mol H_2		1 mol O_2		2 mol H_2O

Convert to grams (using molar masses)

4.0 g H_2 32.0 g O_2 36.0 g H_2O

Notice the conservation of mass
(4.0 g + 32.0 g = 36.0 g)

◄ **FIGURE 3.14 Interpreting a balanced chemical equation quantitatively.**

The quantities 2 mol H_2, 1 mol O_2 and 2 mol H_2O, which are given by the coefficients in Equation 3.13, may be related by the ratios of the coefficients.

$$\text{So,} \quad \frac{\text{moles } H_2}{\text{moles } O_2} = \frac{2}{1}; \frac{\text{moles } H_2}{\text{moles } H_2O} = \frac{2}{2}; \frac{\text{moles } O_2}{\text{moles } H_2O} = \frac{1}{2}$$

In other words, Equation 3.13 shows 2 mol of H_2 and 1 mol of O_2 forming 2 mol of H_2O. These stoichiometric relations can be used to convert between quantities of reactants and products in a chemical reaction. For example, the number of moles of H_2O produced from 1.57 mol of O_2 can be calculated as follows:

$$\frac{\text{Moles } H_2O}{\text{Moles } O_2} = \frac{2}{1}$$

So, moles $H_2O = \dfrac{2}{1} \times$ mol $O_2 = \dfrac{2}{1} \times 1.57$ mol

$$= 3.14 \text{ mol}$$

 CONCEPT CHECK 4

When 1.57 mol O_2 reacts with H_2 to form H_2O, how many moles of H_2 are consumed in the process?

As an additional example, consider the combustion of butane (C_4H_{10}), the fuel in disposable cigarette lighters:

$$2 C_4H_{10}(l) + 13 O_2(g) \longrightarrow 8 CO_2(g) + 10 H_2O(g) \qquad [3.14]$$

Let's calculate the mass of CO_2 produced when 1.00 g of C_4H_{10} is burned. The coefficients in Equation 3.14 tell how the amount of C_4H_{10} consumed is related to the amount of CO_2 produced: $\dfrac{\text{mol } C_4H_{10}}{\text{mol } CO_2} = \dfrac{2}{8}$. In order to use this relationship, however, we must use the molar mass of C_4H_{10} (58.0 g mol^{-1}) to convert grams of C_4H_{10} to moles of C_4H_{10}.

$$\text{Moles } C_4H_{10} = \frac{1.0 \text{ g}}{58.0 \text{ g mol}^{-1}}$$

$$= 1.72 \times 10^{-2} \text{ mol}$$

We can then use the stoichiometric ratio from the balanced equation to calculate moles of CO_2:

$$\text{mol } CO_2 = \frac{8}{2} \times \text{mol } C_4H_{10}$$

$$\text{Moles } CO_2 = \frac{8}{2} \times 1.72 \times 10^{-2} \text{ mol}$$

$$= 6.88 \times 10^{-2} \text{ mol}$$

Finally, we use the molar mass of CO_2, 44.0 g mol^{-1}, to calculate the CO_2 mass in grams:

$$\text{Grams } CO_2 = 6.88 \times 10^{-2} \text{ mol} \times 44.0 \text{ g mol}^{-1}$$

$$= 3.03 \text{ g}$$

Thus the conversion sequence involves three steps as illustrated in ▶ FIGURE 3.15.

CONCEPT CHECK 5

If 20.0 g of a compound reacts completely with 30.0 g of another compound in a combination reaction, how many grams of product are formed?

Given:

Find:

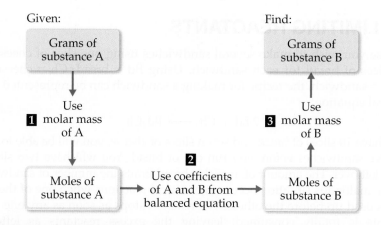

◄ **FIGURE 3.15 Procedure for calculating amounts of reactants consumed or products formed in a reaction.** The number of grams of a reactant consumed or product formed can be calculated in three steps, starting with the number of grams of any reactant or product. Notice how molar masses and the coefficients in the balanced equation are used.

SAMPLE EXERCISE 3.16 Calculating amounts of reactants and products

How many grams of water are produced in the oxidation of 1.00 g of glucose, $C_6H_{12}O_6$?

$$C_6H_{12}O_6(s) + 6\,O_2(g) \longrightarrow 6\,CO_2(g) + 6\,H_2O(l)$$

SOLUTION

Analyse We are given the mass of a reactant and must determine the mass of product in the given reaction.

Plan The general strategy, as outlined in Figure 3.15, requires three steps.

1. Convert grams of $C_6H_{12}O_6$ to moles using the molar mass of $C_6H_{12}O_6$.
2. Convert moles of $C_6H_{12}O_6$ to moles of H_2O using the stoichiometric relationship that 1 mol $C_6H_{12}O_6$ gives 6 mol H_2O.
3. Convert moles of H_2O to grams using the molar mass of H_2O.

Solve

$$\text{Moles } C_6H_{12}O_6 = \frac{1.00\ \text{g}}{180.0\ \text{g mol}^{-1}} = 0.00556\ \text{mol}$$

$$\text{Since } \frac{\text{mol } C_6H_{12}O_6}{\text{mol } H_2O} = \frac{1}{6} \text{ or } \frac{\text{mol } H_2O}{\text{mol } C_6H_{12}O_6} = \frac{6}{1}$$

$$\text{Therefore, Moles } H_2O = \frac{6}{1} \times 0.00556\ \text{mol}$$
$$= 0.0333\ \text{mol}$$

$$\text{And Grams } H_2O = 0.0333\ \text{mol} \times 18.0\ \text{g mol}^{-1}$$
$$= 0.600\ \text{g } H_2O$$

Check We can check how reasonable our result is by doing a ballpark estimate of the mass of H_2O. Because the molar mass of glucose is 180 g mol^{-1}, 1 gram of glucose equals 1/180 mol. Because one mole of glucose yields 6 mol H_2O, we would have 6/180 = 1/30 mol H_2O. The molar mass of water is 18 g mol^{-1}, so we have 1/30 × 18 = 6/10 = 0.6 g of H_2O, which agrees with the full calculation. The units, grams H_2O, are correct. The initial data had three significant figures, so three significant figures for the answer is correct.

Comment An average person ingests 2 dm^3 of water daily and eliminates 2.4 dm^3. The difference between 2 dm^3 and 2.4 dm^3 is produced in the metabolism of foodstuffs, such as in the oxidation of glucose. (*Metabolism* is a general term used to describe all the chemical processes of a living animal or plant.) The koala, however, apparently never drinks water. It survives on its metabolic water.

PRACTICE EXERCISE

The decomposition of $KClO_3$ is commonly used to prepare small amounts of O_2 in the laboratory: $2\,KClO_3(s) \longrightarrow 2\,KCl(s) + 3\,O_2(g)$. How many grams of O_2 can be prepared from 4.50 g of $KClO_3$?

Answer: 1.77 g

(See also Exercises 3.44–3.49.)

If O_2 had been the limiting reactant, how many moles of H_2O would have been formed?

Before reaction

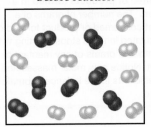

10 H_2 and 7 O_2

After reaction

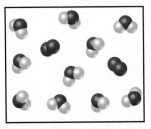

10 H_2O and 2 O_2 (no H_2 molecules)

▲ **FIGURE 3.16 Limiting reactant.**
Because H_2 is completely consumed, it is the limiting reactant. Because some O_2 is left over after the reaction is complete, O_2 is the excess reactant. The amount of H_2O formed depends on the amount of limiting reactant, H_2.

3.7 | LIMITING REACTANTS

Suppose you wish to make several sandwiches using one slice of cheese and two slices of bread for each sandwich. Using Bd = bread, Ch = cheese and Bd_2Ch = sandwich, the recipe for making a sandwich can be represented like a chemical equation:

$$2\,Bd + Ch \longrightarrow Bd_2Ch$$

If you have 10 slices of bread and seven slices of cheese, you will be able to make only five sandwiches before you run out of bread. You will have two slices of cheese left over. The amount of available bread limits the number of sandwiches.

An analogous situation occurs in chemical reactions when one of the reactants is used up before the others. The reaction stops as soon as any one of the reactants is totally consumed, leaving the excess reactants as leftovers. Suppose, for example, that we have a mixture of 10 mol H_2 and 7 mol O_2, which react to form water:

$$2\,H_2(g) + O_2(g) \longrightarrow 2\,H_2O(g)$$

Because $\dfrac{mol\ O_2}{mol\ H_2} = \dfrac{1}{2}$, the number of moles of O_2 needed to react with all the H_2 is

$$Moles\,O_2 = \frac{1}{2} \times mol\,H_2 = \frac{1}{2} \times 10\,mol = 5\,mol$$

Because 7 mol O_2 was available at the start of the reaction, 7 mol $O_2 - 5$ mol $O_2 = 2$ mol O_2 will still be present when all the H_2 is consumed. The example we have considered is depicted on a molecular level in ◄ **FIGURE 3.16**.

The reactant that is completely consumed in a reaction is called either the **limiting reactant** or *limiting reagent* because it determines, or limits, the amount of product formed. The other reactants are sometimes called either *excess reactants* or *excess reagents*. In our example, H_2 is the limiting reactant, which means that once all the H_2 has been consumed, the reaction stops. O_2 is the excess reactant, and some is left over when the reaction stops.

There are no restrictions on the starting amounts of the reactants in any reaction. Indeed, many reactions are conducted using an excess of one reagent. The quantities of reactants consumed and the quantities of products formed, however, are restricted by the quantity of the limiting reactant. When a combustion reaction takes place in the open air, oxygen is plentiful and is therefore the excess reactant. You may have had the unfortunate experience of running out of petrol while driving. The car stops because you've run out of the limiting reactant in the combustion reaction, the fuel.

Before we leave our present example, let's summarise the data in a tabular form.

	2 H_2(g)	+	O_2(g)	$\longrightarrow$	2 H_2O(g)
Initial quantities:	10 mol		7 mol		0 mol
Change (reaction):	−10 mol		−5 mol		+10 mol
Final quantities:	0 mol		2 mol		10 mol

The initial amounts of the reactants are what we started with (10 mol H_2 and 7 mol O_2). The second line in the table (Change) summarises the amounts of the reactants consumed and the amount of the product formed in the reaction. These quantities are restricted by the quantity of the limiting reactant and depend on the coefficients in the balanced equation. The mole ratio $H_2 : O_2 : H_2O = 10 : 5 : 10$ conforms to the ratio of the coefficients in the balanced equation, $2 : 1 : 2$. The changes are negative for the reactants because they are consumed during the reaction and positive for the product because it is formed during the reaction. The final quantities in the third line of the table depend on the initial quantities and their changes, and these entries are found by adding the entries for the initial quantity and change for each column. There is none of the limiting reactant (H_2) left at the end of the reaction. All that remains is 2 mol O_2 and 10 mol H_2O.

SAMPLE EXERCISE 3.17 **Calculating the amount of product formed from a limiting reactant**

The most important commercial process for converting N_2 from the air into nitrogen-containing compounds is based on the reaction of N_2 and H_2 to form ammonia (NH_3):

$$N_2(g) + 3\,H_2(g) \longrightarrow 2\,NH_3(g)$$

How many moles of NH_3 can be formed from 3.0 mol of N_2 and 6.0 mol of H_2?

SOLUTION

Analyse We are asked to calculate the number of moles of product, NH_3, given the quantities of each reactant N_2 and H_2, available in a reaction. This is a limiting reactant problem.

Plan If we assume one reactant is completely consumed, we can calculate how much of the second reactant is needed. By comparing the calculated quantity of the second reactant with the amount available, we can determine which reactant is limiting. We then proceed with the calculation, using the quantity of the limiting reactant.

Solve
From the stoichiometry of the equation, the required ratio of H_2 to N_2 is 3:1.

$$\text{mol } H_2 = \text{mol } N_2 \times \frac{3}{1}$$

$$= 3.0 \text{ mol} \times \frac{3}{1} = 9.0 \text{ mol}$$

So for three moles of N_2 to react completely we would need 9.0 mol of H_2. However, we have only 6.0 mol of H_2 so H_2 is the limiting reactant and so the quantity of it available (6.0 mol) will be used to calculate the yield of NH_3.

$$\text{Since } \frac{\text{mol } NH_3}{\text{mol } H_2} = \frac{2}{3}$$

$$\text{Then mol } NH_3 = \frac{2}{3} \times \text{mol } H_2 = \frac{2}{3} \times 6.0 \text{ mol}$$

$$= 4.0 \text{ mol}$$

PRACTICE EXERCISE

Consider the reaction $2\,Al(s) + 3\,Cl_2(g) \longrightarrow 2\,AlCl_3(s)$. A mixture of 1.50 mol of Al and 3.00 mol of Cl_2 is allowed to react. **(a)** Which is the limiting reactant? **(b)** How many moles of $AlCl_3$ are formed? **(c)** How many moles of the excess reactant remain at the end of the reaction?

Answers: **(a)** Al, **(b)** 1.50 mol, **(c)** 0.75 mol Cl_2

(See also Exercises 3.52, 3.54, 3.56.)

SAMPLE EXERCISE 3.18 **Calculating the amount of product formed from a limiting reactant**

Consider the following reaction:

$$2\,Na_3PO_4(aq) + 3\,Ba(NO_3)_2(aq) \longrightarrow Ba_3(PO_4)_2(s) + 6\,NaNO_3(aq)$$

Suppose a solution containing 3.50 g of Na_3PO_4 is mixed with a solution containing 6.40 g of $Ba(NO_3)_2$. How many grams of $Ba_3(PO_4)_2$ can be formed?

SOLUTION

Analyse We are given the number of grams of each reactant and are asked to calculate the number of grams of products.

Plan We must first identify the limiting reagent. To do so, we can calculate the number of moles of each reactant and compare their ratio with that required by the balanced equation. We then use the quantity of the limiting reagent to calculate the mass of $Ba_3(PO_4)_2$ that forms.

Solve From the balanced equation, we have the following stoichiometric relations:

$$\frac{\text{Moles } Na_3PO_4}{\text{Moles } Ba(NO_3)_2} = \frac{2}{3} = \frac{1}{1.5}$$

Using the molar mass of each substance, we can calculate the number of moles of each reactant:

$$\text{Moles } Na_3PO_4 = \frac{3.50 \text{ g}}{164 \text{ g mol}^{-1}} = 0.0213 \text{ mol}$$

$$\text{Moles } Ba(NO_3)_2 = \frac{6.40 \text{ g}}{261 \text{ g mol}^{-1}} = 0.0245 \text{ mol}$$

and, experimentally,

$$\frac{\text{mol Na}_3\text{PO}_4}{\text{mol Ba(NO}_3)_2} = \frac{0.0213}{0.0245} = \frac{1}{1.15}$$

Comparing the two ratios, we see that there is insufficient $\text{Ba(NO}_3)_2$ to completely consume the Na_3PO_4. That means that $\text{Ba(NO}_3)_2$ is the limiting reagent. We therefore use the quantity of $\text{Ba(NO}_3)_2$ to calculate the quantity of product formed. To calculate the grams of $\text{Ba}_3(\text{PO}_4)_2$ formed, we again use our ratios.

$$\frac{\text{mol Ba}_3(\text{PO}_4)_2}{\text{mol Ba(NO}_3)_2} = \frac{1}{3}$$

$$\text{Moles Ba}_3(\text{PO}_4)_2 = \frac{1}{3} \times 0.0245 \text{ mol}$$

$$\text{Grams Ba}_3(\text{PO}_4)_2 = \frac{1}{3} \times 0.0245 \text{ mol} \times 602 \text{ g mol}^{-1}$$
$$= 4.92 \text{ g}$$

Check The magnitude of the answer seems reasonable: starting with the numbers in the two conversion factors on the right, we have $600/3 = 200$; $200 \times 0.025 = 5$. The units are correct, and the number of significant figures (three) corresponds to the number in the quantity of $\text{Ba(NO}_3)_2$.

Comment The quantity of the limiting reagent, $\text{Ba(NO}_3)_2$, can also be used to determine the quantity of NaNO_3 formed (4.16 g) and the quantity of Na_3PO_4 used (2.67 g). The number of grams of the excess reagent, Na_3PO_4, remaining at the end of the reaction equals the starting amount minus the amount consumed in the reaction, $3.50 \text{ g} - 2.67 \text{ g} = 0.83 \text{ g}$.

PRACTICE EXERCISE

A strip of zinc metal having a mass of 2.00 g is placed in an aqueous solution containing 2.50 g of silver nitrate, causing the following reaction to occur:

$$\text{Zn(s)} + 2\,\text{AgNO}_3(\text{aq}) \longrightarrow 2\,\text{Ag(s)} + \text{Zn(NO}_3)_2(\text{aq})$$

(a) Which reactant is limiting? **(b)** How many grams of Ag will form? **(c)** How many grams of $\text{Zn(NO}_3)_2$ will form? **(d)** How many grams of the excess reactant will be left at the end of the reaction?

Answers: **(a)** AgNO_3, **(b)** 1.59 g, **(c)** 1.39 g, **(d)** 1.52 g Zn

(See also Exercise 3.57.)

Theoretical Yields

The quantity of product that is calculated to form when all of the limiting reactant reacts is called the **theoretical yield**. The amount of product actually obtained in a reaction is called the *actual yield*. The actual yield is almost always less than (and can never be greater than) the theoretical yield. There are many reasons for this difference. Part of the reactants may not react, for example, or they may react in a way different from that desired (side reactions). In addition, it is not always possible to recover all of the product from the reaction mixture. The **percent yield** of a reaction relates the actual yield to the theoretical (calculated) yield:

$$\text{Percent yield} = \frac{\text{actual yield}}{\text{theoretical yield}} \times 100\% \qquad [3.15]$$

In the experiment described in Sample Exercise 3.18, for example, we calculated that 4.92 g of $\text{Ba}_3(\text{PO}_4)_2$ should form when 3.50 g of Na_3PO_4 is mixed with 6.40 g of $\text{Ba(NO}_3)_2$. The 4.92 g is the theoretical yield of $\text{Ba}_3(\text{PO}_4)_2$ in the reaction. If the actual yield turned out to be 4.70 g, the percent yield would be

$$\frac{4.70 \text{ g}}{4.92 \text{ g}} \times 100\% = 95.5\%$$

SAMPLE EXERCISE 3.19 **Calculating the theoretical yield and percent yield for a reaction**

Adipic acid, $C_6H_{10}O_4$, is used to produce nylon. The acid is made commercially by a controlled reaction between cyclohexane (C_6H_{12}) and O_2:

$$2\,C_6H_{12}(l) + 5\,O_2(g) \longrightarrow 2\,C_6H_{10}O_4(l) + 2\,H_2O(g)$$

(a) Assume that you conduct this reaction starting with 25.0 g of cyclohexane and that cyclohexane is the limiting reactant. What is the theoretical yield of adipic acid?

(b) If you obtain 33.5 g of adipic acid from your reaction, what is the percent yield of adipic acid?

SOLUTION

Analyse We are given a chemical equation and the quantity of the limiting reactant (25.0 g of C_6H_{12}). We are asked to calculate the theoretical yield of a product $C_6H_{10}O_4$ and the percent yield if only 33.5 g of product is obtained.

Plan (a) The theoretical yield, which is the calculated quantity of adipic acid formed, can be calculated using the sequence of conversions shown in Figure 3.15.

(b) The percent yield is calculated by using Equation 3.15 to compare the given actual yield (33.5 g) with the theoretical yield.

Solve (a) The theoretical yield can be calculated using the following sequence of conversions:

$$\text{g } C_6H_{12} \longrightarrow \text{mol } C_6H_{12} \longrightarrow \text{mol } C_6H_{10}O_4 \longrightarrow \text{g } C_6H_{10}O_4$$

$$\text{Grams } C_6H_{10}O_4 = \frac{25.0\,\text{g}}{84.0\,\text{g mol}^{-1}} \times \frac{2}{2} \times 146.0\,\text{g mol}^{-1}$$

$$= 43.5\,\text{g } C_6H_{10}O_4$$

(b) $\text{Percent yield} = \dfrac{\text{actual yield}}{\text{theoretical yield}} \times 100\% = \dfrac{33.5\,\text{g}}{43.5\,\text{g}} \times 100\% = 77.0\%$

Check Our answer in (a) has the appropriate magnitude, units and significant figures. In (b) the answer is less than 100%, as necessary.

PRACTICE EXERCISE

Imagine that you are working on ways to improve the process by which iron ore containing Fe_2O_3 is converted into iron. In your tests you conduct the following reaction on a small scale:

$$Fe_2O_3(s) + 3\,CO(g) \longrightarrow 2\,Fe(s) + 3\,CO_2(g)$$

(a) If you start with 150 g of Fe_2O_3 as the limiting reagent, what is the theoretical yield of Fe? **(b)** If the actual yield of Fe in your test was 87.9 g, what was the percent yield?

Answers: **(a)** 105 g Fe, **(b)** 83.7%

(See also Exercise 3.58.)

SAMPLE INTEGRATIVE EXERCISE

Copper is an excellent electrical conductor widely used in making electric circuits. In producing a printed circuit board for the electronics industry, a layer of copper is laminated on a plastic board. A circuit pattern is then printed on the copper using a chemically resistant polymer. The board is then exposed to a chemical bath that reacts with the exposed copper, leaving the desired copper circuit, which has been protected by the polymer. One reaction used to remove the exposed copper from the circuit board is

$$Cu(s) + Cu(NH_3)_4Cl_2(aq) + 4\,NH_3(aq) \rightarrow 2\,Cu(NH_3)_4Cl(aq)$$

Finally, a solvent removes the polymer.

A plant needs to produce 500 circuit boards, each with a surface area measuring 5.00 cm × 7.50 cm. The boards are covered with a 0.650 mm layer of copper. In subsequent processing, 85.0% of the copper is removed by the reaction above. Copper has a density of 8.96 g cm^{-3}. Calculate the masses of $Cu(NH_3)_4Cl_2$ and NH_3 needed to produce the circuit boards, assuming that the reaction used gives a 97% yield.

SOLUTION

First we need to know how much copper is deposited to cover the board to a thickness of 0.65 mm, and then how much is removed by the reaction.

$$\text{Volume } (V) \text{ of Cu deposited} = 5.00 \text{ cm} \times 7.50 \text{ cm} \times (0.650 \times 10^{-1}) \text{ cm}$$

$$= 2.4375 \text{ cm}^3 = 2.44 \text{ cm}^3$$

$$\text{Mass of Cu deposited on each board} = \rho \times V = 8.96 \text{ g cm}^{-3} \times 2.44 \text{ cm}^3$$

$$= 21.86 \text{ g} = 21.9 \text{ g}$$

$$\text{Mass of Cu on 500 boards} = 21.9 \text{ g} \times 500$$

$$= 1.0931 \times 10^4 \text{ g} = 1.09 \times 10^4 \text{ g}$$

Since 85.0% of the copper is removed, then mass of Cu removed

$$= 1.0931 \times 10^4 \text{ g} \times 85.0\%$$

$$= 9.28 \times 10^3 \text{ g}$$

However, the process is only 97% efficient, so to remove all of the 9.28×10^4 g, we need to calculate our reactant quantities based on the removal of a theoretical mass slightly higher.

$$\text{Calculated mass of Cu to be removed} = 9.28 \times 10^3 \text{ g} \times \frac{100}{97}$$

$$= 9.56 \times 10^3 \text{ g}$$

From the above chemical equation, we see that for every mole of Cu, we need 1 mole of $Cu(NH_3)_4Cl_2$ and 4 mol NH_3.

$$\text{Mol of Cu to be removed} = \frac{9.56 \times 10^3 \text{ g}}{63.546 \text{ g mol}^{-1}} = 1.51 \times 10^2 \text{ mol}$$

$$\text{Therefore mol } Cu(NH_3)_4Cl_2 \text{ required} = 1.51 \times 10^2 \text{ mol}$$

$$\text{Mol } NH_3 \text{ required} = 4 \times 1.51 \times 10^2 \text{ mol} = 6.04 \times 10^2 \text{ mol}$$

$$\text{Mass } Cu(NH_3)_4Cl_2 \text{ required} = 1.51 \times 10^2 \text{ mol} \times 202.6 \text{ g mol}^{-1}$$

$$= 3.06 \times 10^4 \text{ g} = 3.06 \times 10^1 \text{ kg} = 31 \text{ kg}$$

$$\text{Mass } NH_3 \text{ required} = 6.04 \times 10^2 \text{ mol} \times 17.03 \text{ g mol}^{-1}$$

$$= 1.02 \times 10^4 \text{ g} = 10 \text{ kg}$$

(Note that in the calculation involving the efficiency of the Cu removal reaction, 97% has only two significant figures.)

CHAPTER SUMMARY AND KEY TERMS

SECTION 3.1 The study of the quantitative relationships between chemical formulae and chemical equations is known as **stoichiometry**. One of the important concepts of stoichiometry is the law of conservation of mass, which states that the total mass of the products of a chemical reaction is the same as the total mass of the reactants. The same numbers of atoms of each element are present before and after a chemical reaction. A balanced **chemical equation** shows equal numbers of atoms of each element on each side of the equation. Equations are balanced by placing coefficients in front of the chemical formulae for the **reactants** and **products** of a reaction, *not* by changing the subscripts in chemical formulae.

SECTION 3.2 Among the reaction types described in the chapter are (1) **combination reactions**, in which two reactants combine to form one product; (2) **decomposition reactions**, in which a single reactant forms two or more products; and (3) **combustion reactions** in oxygen, in which a hydrocarbon or related compound reacts with O_2 to form CO_2 and H_2O.

SECTION 3.3 Much quantitative information can be determined from chemical formulae and balanced chemical equations by using **atomic masses**. The **formula mass** of a compound equals the sum of the atomic masses of the atoms in the formula. If the formula is a molecular formula, the formula mass is also called the **molecular mass**. Atomic masses and formula masses can be used to determine the elemental composition of a compound.

SECTION 3.4 A **mole** of any substance is **Avogadro's number** (approximately 6.02×10^{23}) of formula units of that substance. The mass of a mole of atoms, molecules or ions (the **molar mass**) equals the formula mass of that substance expressed in grams. The mass of one molecule of H_2O is 18 g. That is, the molar mass of H_2O is 18 g mol^{-1}.

SECTION 3.5 The empirical formula of any substance can be determined from its percentage composition by calculating the relative number of moles of each atom in 100 g of the substance. If the substance is molecular in nature, the molecular formula can be determined from the empirical formula if the molecular mass is also known.

SECTION 3.6 The mole concept can be used to calculate the relative quantities of reactants and products involved in chemical reactions. The coefficients in a balanced chemical equation give the relative numbers of the reactants and products. Therefore to calculate the number of grams of a product from the number of grams of a reactant, first convert grams of reactant to moles of reactant. We then use the coefficients in the balanced equation to convert the moles of reactant to moles of product. Finally we convert moles of product to grams of product.

SECTION 3.7 A **limiting reactant** is completely consumed in a reaction. When it is used up, the reaction stops, thus limiting the quantities of products formed. The **theoretical yield** of a reaction is the quantity of product calculated to form when all of the limiting reagent reacts. The actual yield of a reaction is always less than the theoretical yield. The **percent yield** compares the actual and theoretical yields.

KEY SKILLS

- Balance chemical equations. (Section 3.1)
- Predict the products of simple combination, decomposition and combustion reactions. (Section 3.2)
- Calculate formula masses. (Section 3.3)
- Convert grams to moles and moles to grams using molar masses. (Section 3.4)
- Convert number of molecules to moles and moles to number of molecules using Avogadro's number. (Section 3.4)
- Calculate the empirical and molecular formula of a compound from percentage composition and molecular mass. (Section 3.5)
- Calculate amounts, in grams or moles, of reactants and products for a reaction. (Section 3.6)
- Determine the limiting reactants and calculate the percent yield for a reaction. (Section 3.7)

KEY EQUATIONS

- This is the formula to calculate the mass percentage of each element in a compound.

$$\text{Percentage element} = \frac{(\text{number of atoms of that element}) \times (\text{atomic mass of the element})}{\text{formula mass of the compound}} \times 100\% \qquad [3.10]$$

- This is the formula to calculate the number of moles of any substance knowing its mass and molar mass.

$$\text{Moles} = \frac{\text{mass in grams of a substance}}{\text{molar mass of the substance in g mol}^{-1}}; n = \frac{m}{M} \qquad [3.11]$$

- This is the formula to calculate the percent yield in a reaction. The percent yield can never be more than 100%.

$$\text{Percent yield} = \frac{\text{actual yield}}{\text{theoretical yield}} \times 100\% \qquad [3.15]$$

EXERCISES

VISUALISING CONCEPTS

3.1 The reaction between reactant A (blue spheres) and reactant B (red spheres) is shown in the following diagram:

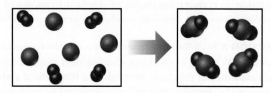

Based on this diagram, which equation best describes the reaction? [Section 3.1]
(a) $A_2 + B \longrightarrow A_2B$
(b) $A_2 + 4B \longrightarrow 2AB_2$
(c) $2A + B_4 \longrightarrow 2AB_2$
(d) $A + B_2 \longrightarrow AB_2$

3.2 The following diagram represents the collection of elements formed by a decomposition reaction. (a) If the blue spheres represent N atoms and the red ones represent O atoms, what was the empirical formula of the original compound? (b) Could you draw a diagram representing the molecules of the compound that was decomposed? Why or why not? [Section 3.2]

3.3 The following diagram represents the collection of CO_2 and H_2O molecules formed by complete combustion of a hydrocarbon. What is the empirical formula of the hydrocarbon? [Section 3.2]

3.4 Glycine, an amino acid used by organisms to make proteins, is represented by the molecular model below. (a) Write its molecular formula. (b) Determine its molecular mass. (c) Calculate the percent nitrogen by mass in glycine. [Sections 3.3 and 3.5]

3.5 The following diagram represents a high-temperature reaction between CH_4 and H_2O. Based on this reaction, how many moles of each product can be obtained starting with 4.0 mol CH_4? [Section 3.6]

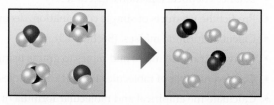

3.6 Nitrogen (N_2) and hydrogen (H_2) react to form ammonia (NH_3). Consider the mixture of N_2 and H_2 shown in the accompanying diagram. The blue spheres represent N, and the white ones represent H. Draw a representation of the product mixture, assuming that the reaction goes to completion. How did you arrive at your representation? What is the limiting reactant in this case? [Section 3.7]

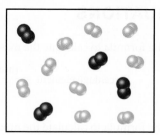

BALANCING CHEMICAL EQUATIONS (Section 3.1)

3.7 (a) What scientific principle or law is used in the process of balancing chemical equations? (b) In balancing equations, why shouldn't subscripts in chemical formulae be changed? (c) What are the symbols used to represent gases, liquids, solids and aqueous solutions in chemical equations?

3.8 (a) What is the difference between adding a subscript 2 to the end of the formula for CO to give CO_2 and adding a coefficient in front of the formula to give 2 CO? (b) Is

the following chemical equation, as written, consistent with the law of conservation of mass?

$3 Mg(OH)_2(s) + 2 H_3PO_4(aq) \longrightarrow$
$$Mg_3(PO_4)_2(s) + 6 H_2O(l)$$

Why or why not?

3.9 Balance the following equations:
(a) $CO(g) + O_2(g) \longrightarrow CO_2(g)$
(b) $N_2O_5(g) + H_2O(l) \longrightarrow HNO_3(aq)$

(c) $CH_4(g) + Cl_2(g) \longrightarrow CCl_4(l) + HCl(g)$
(d) $Al_4C_3(s) + H_2O(l) \longrightarrow Al(OH)_3(s) + CH_4(g)$
(e) $C_5H_{10}O_2(l) + O_2(g) \longrightarrow CO_2(g) + H_2O(g)$
(f) $Fe(OH)_3(s) + H_2SO_4(aq) \longrightarrow Fe_2(SO_4)_3(aq) + H_2O(l)$
(g) $Mg_3N_2(s) + H_2SO_4(aq) \longrightarrow$
$$MgSO_4(aq) + (NH_4)_2\,SO_4(aq)$$

3.10 Write balanced chemical equations to correspond to each of the following descriptions. **(a)** Solid calcium carbide, CaC_2, reacts with water to form an aqueous solution of calcium hydroxide and acetylene gas, C_2H_2. **(b)** When solid potassium chlorate is heated it decomposes to form solid potassium chloride and oxygen gas. **(c)** Solid zinc metal reacts with sulfuric acid to form hydrogen gas and an aqueous solution of zinc sulfate. **(d)** When liquid phosphorus trichloride is added to water, it reacts to form aqueous phosphorous acid, $H_3PO_3(aq)$, and aqueous hydrochloric acid. **(e)** When hydrogen sulfide gas is passed over solid hot iron(III) hydroxide, the resultant reaction produces solid iron(III) sulfide and gaseous water.

PATTERNS OF CHEMICAL REACTIVITY (Section 3.2)

3.11 **(a)** When the metallic element sodium combines with the non-metallic element bromine, $Br_2(l)$, how can you determine the chemical formula of the product? How do you know whether the product is solid, liquid or gas at room temperature? Write the balanced chemical equation for the reaction. **(b)** When a hydrocarbon burns in air, what reactant besides the hydrocarbon is involved in the reaction? What products are formed? Write a balanced chemical equation for the combustion of benzene, $C_6H_6(l)$, in air.

3.12 **(a)** Determine the chemical formula of the product formed when the metallic element calcium combines with the non-metallic element oxygen, O_2. Write the balanced chemical equation for the reaction. **(b)** What products form when a compound containing C, H and O is completely combusted in air? Write a balanced chemical equation for the combustion of acetone, $C_3H_6O(l)$, in air.

3.13 Write a balanced chemical equation for the reaction that occurs when **(a)** $Mg(s)$ reacts with $Cl_2(g)$; **(b)** barium carbonate decomposes into barium oxide and carbon dioxide gas when heated; **(c)** the hydrocarbon styrene, $C_8H_8(l)$, is combusted in air; **(d)** dimethylether, $CH_3OCH_3(g)$, is combusted in air.

3.14 Write a balanced chemical equation for the reaction that occurs when **(a)** aluminium metal undergoes a combination reaction with $O_2(g)$; **(b)** copper(II) hydroxide decomposes into copper(II) oxide and water when heated; **(c)** heptane, $C_7H_{16}(l)$, burns in air; **(d)** the petrol additive MTBE (methyl tert-butyl ether), $C_5H_{12}O(l)$, burns in air.

3.15 Balance the following equations and indicate whether they are combination, decomposition or combustion reactions:
(a) $Al(s) + Cl_2(g) \longrightarrow AlCl_3(s)$
(b) $C_2H_4(g) + O_2(g) \longrightarrow CO_2(g) + H_2O(g)$
(c) $Li(s) + N_2(g) \longrightarrow Li_3N(s)$
(d) $PbCO_3(s) \longrightarrow PbO(s) + CO_2(g)$
(e) $C_7H_8O_2(l) + O_2(g) \longrightarrow CO_2(g) + H_2O(g)$

3.16 Balance the following equations and indicate whether they are combination, decomposition or combustion reactions:
(a) $C_3H_6(g) + O_2(g) \longrightarrow CO_2(g) + H_2O(g)$
(b) $NH_4NO_3(s) \longrightarrow N_2O(g) + H_2O(g)$
(c) $C_5H_6O(l) + O_2(g) \longrightarrow CO_2(g) + H_2O(g)$
(d) $N_2(g) + H_2(g) \longrightarrow NH_3(g)$
(e) $K_2O(s) + H_2O(l) \longrightarrow KOH(aq)$

FORMULA MASS (Section 3.3)

3.17 Determine the formula masses of each of the following compounds: **(a)** N_2O_5, **(b)** $CuSO_4$, **(c)** $(NH_4)_3PO_4$, **(d)** $Ca(HCO_3)_2$, **(e)** aluminium sulfide, **(f)** iron(III) sulfate, **(g)** disilicon hexabromide.

3.18 Determine the formula masses of each of the following compounds: **(a)** nitrous oxide, N_2O, known as laughing gas and used as an anaesthetic in dentistry; **(b)** benzoic acid, $C_7H_6O_2$, a substance used as a food preservative; **(c)** $Mg(OH)_2$, the active ingredient in milk of magnesia; **(d)** urea, $(NH_2)_2CO$, a compound used as a nitrogen fertiliser; **(e)** isopentyl acetate, $CH_3CO_2C_5H_{11}$, responsible for the odour of bananas.

3.19 Calculate the percentage by mass of the indicated element in the following compounds: **(a)** carbon in acetylene, C_2H_2, a gas used in welding; **(b)** hydrogen in ascorbic acid, $C_6H_8O_6$, also known as vitamin C; **(c)** hydrogen in ammonium sulfate, $(NH_4)_2\,SO_4$, a substance used as a nitrogen fertiliser; **(d)** platinum in $PtCl_2(NH_3)_2$, a chemotherapy agent called cisplatin; **(e)** oxygen in the female sex hormone estradiol, $C_{18}H_{24}O_2$; **(f)** carbon in capsaicin, $C_{18}H_{27}NO_3$, the compound that gives the hot taste to chilli peppers.

3.20 Based on the following structural formulae, calculate the percentage of carbon by mass present in each compound:

(a) Benzaldehyde (almond fragrance)

(b) HO—C ... C=C H₃CO / H ... C=C ... O ... C—C—H Vanillin (vanilla flavour)

(c) $H_3C-C-C-C-O-C-CH_3$ with H H H and H₃C H H and O

Isopentyl acetate
(banana flavour)

AVOGADRO'S NUMBER AND THE MOLE (Section 3.4)

3.21 **(a)** What is Avogadro's number, and how is it related to the mole? **(b)** What is the relationship between the formula mass of a substance and its molar mass?

3.22 **(a)** What is the mass, in grams, of a mole of ^{12}C? **(b)** How many carbon atoms are present in a mole of ^{12}C?

3.23 Without doing any detailed calculations (but using a periodic table to give atomic masses), rank the following samples in order of increasing number of atoms: 0.50 mol H_2O, 23 g Na, 6.0×10^{23} N_2 molecules.

3.24 Without doing any detailed calculations (but using a periodic table to give atomic masses), rank the following samples in order of increasing number of atoms: 3.0×10^{23} molecules of H_2O_2, 2.0 mol CH_4, 32 g O_2.

3.25 Calculate the following quantities:
(a) mass, in grams, of 0.773 mol CaH_2
(b) moles of $Mg(NO_3)_2$ in 5.35 g of this substance
(c) number of molecules in 0.0305 mol CH_3OH
(d) number of C atoms in 0.585 mol C_4H_{10}

3.26 Calculate the following quantities:
(a) mass, in grams, of 1.906×10^{-2} mol BaI_2
(b) number of moles of NH_4Cl in 48.3 g of this substance
(c) number of molecules in 0.05752 mol CH_2O_2
(d) number of O atoms in 4.88×10^{-3} mol $Al(NO_3)_3$

3.27 The molecular formula of allicin, the compound responsible for the characteristic smell of garlic, is $C_6H_{10}OS_2$.

(a) What is the molar mass of allicin? **(b)** How many moles of allicin are present in 5.00 mg of this substance? **(c)** How many molecules of allicin are in 5.00 mg of this substance? **(d)** How many S atoms are present in 5.00 mg of allicin?

3.28 A sample of glucose, $C_6H_{12}O_6$, contains 1.250×10^{21} atoms of carbon. **(a)** How many atoms of hydrogen does it contain? **(b)** How many molecules of glucose does it contain? **(c)** How many moles of glucose does it contain? **(d)** What is the mass of this sample in grams?

3.29 A sample of the male sex hormone testosterone, $C_{19}H_{28}O_2$, contains 7.08×10^{20} atoms of hydrogen. **(a)** How many atoms of carbon does it contain? **(b)** How many molecules of testosterone does it contain? **(c)** How many moles of testosterone does it contain? **(d)** What is the mass of this sample in grams?

3.30 The allowable concentration level of vinyl chloride, C_2H_3Cl, in the atmosphere in a chemical plant is 2.0×10^{-6} g dm^{-3}. How many moles of vinyl chloride in each litre does this represent? How many molecules per litre?

3.31 At least 25 μg of tetrahydrocannabinol (THC), the active ingredient in marijuana, is required to produce intoxication. The molecular formula of THC is $C_{21}H_{30}O_2$. How many moles of THC does this 25 μg represent? How many molecules?

EMPIRICAL FORMULAE (Section 3.5)

3.32 Give the empirical formula of each of the following compounds if a sample contains **(a)** 0.0130 mol C, 0.0390 mol H and 0.0065 mol O; **(b)** 11.66 g iron and 5.01 g oxygen; **(c)** 40.0% C, 6.7% H and 53.3% O by mass.

3.33 Determine the empirical formula of each of the following compounds if a sample contains **(a)** 0.104 mol K, 0.052 mol C and 0.156 mol O; **(b)** 5.28 g Sn and 3.37 g F; **(c)** 87.5% N and 12.5% H by mass.

3.34 Determine the empirical formulae of the compounds with the following compositions by mass:
(a) 10.4% C, 27.8% S and 61.8% Cl
(b) 21.7% C, 9.6% O and 68.7% F
(c) 32.79% Na, 13.02% Al and 54.19% F

3.35 Determine the empirical formulae of the compounds with the following compositions by mass:
(a) 55.3% K, 14.6% P and 30.1% O
(b) 24.5% Na, 14.9% Si and 60.6% F
(c) 62.1% C, 5.20% H, 12.1% N and 20.6% O

3.36 What is the molecular formula of each of the following compounds?
(a) Empirical formula CH_2, molar mass = 84 g mol^{-1}
(b) Empirical formula NH_2Cl, molar mass = 51.5 g mol^{-1}

3.37 What is the molecular formula of each of the following compounds?
(a) Empirical formula HCO_2, molar mass = 90.0 g mol^{-1}
(b) Empirical formula C_2H_4O, molar mass = 88 g mol^{-1}

3.38 **(a)** Combustion analysis of toluene, a common organic solvent, gives 5.86 mg of CO_2 and 1.37 mg of H_2O. If the compound contains only carbon and hydrogen, what is its empirical formula? **(b)** Menthol, the substance we can smell in mentholated cough drops, is composed of C, H and O. A 0.1005 g sample of menthol is combusted, producing 0.2829 g of CO_2 and 0.1159 g of H_2O. What is the empirical formula for menthol? If the compound has a molar mass of 156 g mol^{-1}, what is its molecular formula?

3.39 **(a)** The characteristic odour of pineapple is due to ethyl butyrate, a compound containing carbon, hydrogen and oxygen. Combustion of 2.78 mg of ethyl butyrate produces 6.32 mg of CO_2 and 2.58 mg of H_2O. What is the empirical formula of the compound? **(b)** Nicotine, a component of tobacco, is composed of C, H and N. A 5.250 mg sample of nicotine was combusted, producing 14.242 mg of CO_2 and 4.083 mg of H_2O. What is the empirical formula for nicotine? If nicotine has a molar mass of $160 \pm 5\,g\,mol^{-1}$, what is its molecular formula?

3.40 Washing soda, a compound used to soften hard water for washing laundry, is a hydrate, which means that a certain number of water molecules are included in the solid structure. Its formula can be written as $Na_2CO_3 \cdot xH_2O$, where x is the number of moles of H_2O per mole of Na_2CO_3. When a 2.558 g sample of washing soda is heated at 25 °C, all the water of hydration is lost, leaving 0.948 g of Na_2CO_3. What is the value of x?

3.41 Epsom salts, a strong laxative used in veterinary medicine, is a hydrate, which means that a certain number of water molecules are included in the solid structure. The formula for Epsom salts can be written as $MgSO_4 \cdot xH_2O$, where x indicates the number of moles of H_2O per mole of $MgSO_4$. When 5.061 g of this hydrate is heated to 250 °C, all the water of hydration is lost, leaving 2.472 g of $MgSO_4$. What is the value of x?

CALCULATIONS BASED ON CHEMICAL EQUATIONS (Section 3.6)

3.42 Why is it essential to use balanced chemical equations when determining the quantity of a product formed from a given quantity of a reactant?

3.43 What parts of balanced chemical equations give information about the relative numbers of moles of reactants and products involved in a reaction?

3.44 Hydrofluoric acid, HF(aq), cannot be stored in glass bottles because compounds called silicates in the glass are attacked by the HF(aq). Sodium silicate (Na_2SiO_3), for example, reacts as follows:

$$Na_2SiO_3(s) + 8\,HF(aq) \longrightarrow$$
$$H_2SiF_6(aq) + 2\,NaF(aq) + 3\,H_2O(l)$$

(a) How many moles of HF are needed to react with 0.300 mol of Na_2SiO_3?
(b) How many grams of NaF form when 0.500 mol of HF reacts with excess Na_2SiO_3?
(c) How many grams of Na_2SiO_3 can react with 0.800 g of HF?

3.45 The fermentation of glucose ($C_6H_{12}O_6$) produces ethyl alcohol (C_2H_5OH) and CO_2:

$$C_6H_{12}O_6(aq) \longrightarrow 2\,C_2H_5OH(aq) + 2\,CO_2(g)$$

(a) How many moles of CO_2 are produced when 0.400 mol of $C_6H_{12}O_6$ reacts in this fashion?
(b) How many grams of $C_6H_{12}O_6$ are needed to form 7.50 g of C_2H_5OH?
(c) How many grams of CO_2 form when 7.50 g of C_2H_5OH are produced?

3.46 Several brands of antacids use $Al(OH)_3$ to react with stomach acid, which contains primarily HCl:

$$Al(OH)_3(s) + HCl(aq) \longrightarrow AlCl_3(aq) + H_2O(l)$$

(a) Balance this equation.
(b) Calculate the number of grams of HCl that can react with 0.500 g of $Al(OH)_3$.
(c) Calculate the number of grams of $AlCl_3$ and the number of grams of H_2O formed when 0.500 g of $Al(OH)_3$ reacts.

(d) Show that your calculations in parts (b) and (c) are consistent with the law of conservation of mass.

3.47 An iron ore sample contains Fe_2O_3 together with other substances. Reaction of the ore with CO produces iron metal:

$$Fe_2O_3(s) + CO(g) \longrightarrow Fe(s) + CO_2(g)$$

(a) Balance this equation.
(b) Calculate the number of grams of CO that can react with 0.150 kg of Fe_2O_3.
(c) Calculate the number of grams of Fe and the number of grams of CO_2 formed when 0.150 kg of Fe_2O_3 reacts.
(d) Show that your calculations in parts (b) and (c) are consistent with the law of conservation of mass.

3.48 Car air bags inflate when sodium azide, NaN_3, rapidly decomposes to its component elements:

$$2\,NaN_3(s) \longrightarrow 2\,Na(s) + 3\,N_2(g)$$

(a) How many moles of N_2 are produced by the decomposition of 1.50 mol of NaN_3?
(b) How many grams of NaN_3 are required to form 10.0 g of nitrogen gas?
(c) How many grams of NaN_3 are required to produce 280 dm^3 of nitrogen gas if the gas has a density of 1.25 $g\,dm^{-3}$?

3.49 The complete combustion of octane, C_8H_{18}, a component of petrol, proceeds as follows:

$$2\,C_8H_{18}(l) + 25\,O_2(g) \longrightarrow 16\,CO_2(g) + 18\,H_2O(g)$$

(a) How many moles of O_2 are needed to burn 1.25 mol of C_8H_{18}?
(b) How many grams of O_2 are needed to burn 10.0 g of C_8H_{18}?
(c) Octane has a density of 0.692 $g\,cm^{-3}$ at 20 °C. How many grams of O_2 are required to burn 4.54 dm^3 of C_8H_{18}?

LIMITING REACTANTS; THEORETICAL YIELDS (Section 3.7)

3.50 **(a)** Define the terms *limiting reactant* and *excess reactant*. **(b)** Why are the amounts of products formed in a reaction determined only by the amount of the limiting reactant?

3.51 **(a)** Define the terms *theoretical yield, actual yield* and *percent yield*. **(b)** Why is the actual yield in a reaction almost always less than the theoretical yield?

3.52 Sodium hydroxide reacts with carbon dioxide as follows:

$$2\,NaOH(s) + CO_2(g) \longrightarrow Na_2CO_3(s) + H_2O(l)$$

Which reagent is the limiting reactant when 1.85 mol NaOH and 1.00 mol CO_2 are allowed to react? How many moles of Na_2CO_3 can be produced? How many moles of the excess reactant remain after the completion of the reaction?

3.53 Aluminium hydroxide reacts with sulfuric acid as follows:

$$2\,Al(OH)_3(s) + 3\,H_2SO_4(aq) \longrightarrow$$
$$Al_2(SO_4)_3\,(aq) + 6\,H_2O(l)$$

Which reagent is the limiting reactant when 0.500 mol $Al(OH)_3$ and 0.500 mol H_2SO_4 are allowed to react? How many moles of $Al_2(SO_4)_3$ can form under these conditions? How many moles of the excess reactant remain after the completion of the reaction?

3.54 The fizz produced when an Alka-Seltzer® tablet is dissolved in water is due to the reaction between sodium bicarbonate ($NaHCO_3$) and citric acid ($C_6H_8O_7$):

$$3\,NaHCO_3(aq) + C_6H_8O_7(aq) \longrightarrow$$
$$3\,CO_2(g) + 3\,H_2O(l) + Na_3C_6H_5O_7(aq)$$

In a certain experiment 1.00 g of sodium bicarbonate and 1.00 g of citric acid are allowed to react. **(a)** Which is the limiting reactant? **(b)** How many grams of carbon dioxide form? **(c)** How many grams of the excess reactant remain after the limiting reactant is completely consumed?

3.55 One of the steps in the commercial process for converting ammonia to nitric acid is the conversion of NH_3 to NO:

$$4\,NH_3(g) + 5\,O_2(g) \longrightarrow 4\,NO(g) + 6\,H_2O(g)$$

In a certain experiment, 1.50 g of NH_3 reacts with 2.75 g of O_2. **(a)** Which is the limiting reactant? **(b)** How many grams of NO and of H_2O form? **(c)** How many grams of the excess reactant remain after the limiting reactant is completely consumed? **(d)** Show that your calculations in parts (b) and (c) are consistent with the law of conservation of mass.

3.56 Solutions of sodium carbonate and silver nitrate react to form solid silver carbonate and a solution of sodium nitrate. A solution containing 3.50 g of sodium carbonate is mixed with one containing 5.00 g of silver nitrate. How many grams of sodium carbonate, silver nitrate, silver carbonate and sodium nitrate are present after the reaction is complete?

3.57 Solutions of sulfuric acid and lead(II) acetate react to form solid lead(II) sulfate and a solution of acetic acid. If 7.50 g of sulfuric acid and 7.50 g of lead(II) acetate are mixed, calculate the number of grams of sulfuric acid, lead(II) acetate, lead(II) sulfate and acetic acid present in the mixture after the reaction is complete.

3.58 When benzene (C_6H_6) reacts with bromine (Br_2), bromobenzene (C_6H_5Br) is obtained:

$$C_6H_6 + Br_2 \longrightarrow C_6H_5Br + HBr$$

(a) What is the theoretical yield of bromobenzene in this reaction when 30.0 g of benzene reacts with 65.0 g of bromine? **(b)** If the actual yield of bromobenzene was 56.7 g, what was the percentage yield?

ADDITIONAL EXERCISES

3.59 Write the balanced chemical equation for **(a)** the complete combustion of butyric acid, $C_4H_8O_2(l)$, a compound produced when butter becomes rancid; **(b)** the decomposition of solid nickel(II) hydroxide into solid nickel(II) oxide and water vapour; **(c)** the combination reaction between zinc metal and chlorine gas.

3.60 **(a)** Diamond is a natural form of pure carbon. How many moles of carbon are in a 1.25 carat diamond (1 carat = 0.200 g)? How many atoms are in this diamond? **(b)** The molecular formula of acetylsalicylic acid (aspirin), one of the most common pain relievers, is $C_9H_8O_4$. How many moles of $C_9H_8O_4$ are in a 0.500 g tablet of aspirin? How many molecules of $C_9H_8O_4$ are in this tablet?

3.61 **(a)** One molecule of the antibiotic known as penicillin G has a mass of 5.342×10^{-21} g. What is the molar mass of penicillin G? **(b)** Haemoglobin, the oxygen-carrying protein in red blood cells, has four iron atoms per molecule and contains 0.340% iron by mass. Calculate the molar mass of haemoglobin.

3.62 Very small crystals composed of 1000 to 100 000 atoms, called quantum dots, are being investigated for use in electronic devices.
   (a) Calculate the mass in grams of a quantum dot consisting of 10 000 atoms of silicon.
   (b) Assuming that the silicon in the dot has a density of 2.3 g cm^{-3}, calculate its volume.
   (c) Assuming that the dot has the shape of a cube, calculate the length of each edge of the cube.

3.63 Serotonin is a compound that conducts nerve impulses in the brain. It contains 68.18 mass percent C, 6.84 mass percent H, 15.9 mass percent N and 9.08 mass percent O. Its molar mass is 176 g mol^{-1}. Determine its molecular formula.

3.64 An organic compound was found to contain only C, H and Cl. When a 1.50 g sample of the compound was completely combusted in air, 3.52 g of CO_2 were formed. In a separate experiment the chlorine in a 1.00 g sample of the compound was converted to 1.27 g of AgCl. Determine the empirical formula of the compound.

3.65 An element X forms an iodide (XI_3) and a chloride (XCl_3). The iodide is quantitatively converted to the chloride when it is heated in a stream of chlorine:

$$2\,XI_3 + 3\,Cl_2 \longrightarrow 2\,XCl_3 + 3\,I_2$$

If 0.5000 g of XI_3 is treated, 0.2360 g of XCl_3 is obtained.
(a) Calculate the atomic weight of the element X.
(b) Identify the element X.

3.66 If 1.5 mol of each of the following compounds is completely combusted in oxygen, which one will produce the largest number of moles of H_2O? Which will produce the least? Explain. C_2H_5OH, C_3H_8, $CH_3CH_2COCH_3$.

3.67 A chemical plant uses electrical energy to decompose aqueous solutions of NaCl to give Cl_2, H_2 and NaOH:

$$2\,NaCl(aq) + 2\,H_2O(l) \longrightarrow$$
$$2\,NaOH(aq) + H_2(g) + Cl_2(g)$$

If the plant produces 1.5×10^6 kg (1500 tonnes) of Cl_2 daily, estimate the quantities of H_2 and NaOH produced.

3.68 When a mixture of 10.0 g of acetylene (C_2H_2) and 10.0 g of oxygen (O_2) is ignited, the resultant combustion reaction produces CO_2 and H_2O. **(a)** Write the balanced chemical equation for this reaction. **(b)** Which is the limiting reactant? **(c)** How many grams of C_2H_2, O_2, CO_2 and H_2O are present after the reaction is complete?

INTEGRATIVE EXERCISES

These exercises require skills from Chapters 1 and 2 as well as skills from the present chapter.

3.69 Consider a sample of calcium carbonate in the form of a cube measuring 5.093 cm on each edge. If the sample has a density of $2.71\ \text{g cm}^{-3}$, how many oxygen atoms does it contain?

3.70 **(a)** You are given a cube of silver metal that measures 1.000 cm on each edge. The density of silver is $10.49\ \text{g cm}^{-3}$. How many atoms are in this cube? **(b)** Because atoms are spherical, they cannot occupy all the space of the cube. The silver atoms pack in the solid in such a way that 74% of the volume of the solid is actually filled with the silver atoms. Calculate the volume of a single silver atom. **(c)** Using the volume of a silver atom and the formula for the volume of a sphere, calculate the radius of a silver atom in picometres.

3.71 In 1865 a chemist reported that he had treated a weighed amount of pure silver with nitric acid and had recovered all the silver as pure silver nitrate. The mass ratio of silver to silver nitrate was found to be 0.634985. Using only this ratio and the presently accepted values for the atomic masses of silver and oxygen, calculate the atomic mass of nitrogen. Compare this calculated atomic mass with the currently accepted value.

3.72 A particular coal contains 2.5% sulfur by mass. When this coal is burned, the sulfur is converted into sulfur dioxide gas. The sulfur dioxide reacts with calcium oxide to form solid calcium sulfite. **(a)** Write the balanced chemical equation for the reaction. **(b)** If the coal is burned in a power plant that uses 2000 tonnes of coal per day, what is the daily production of calcium sulfite?

MasteringChemistry (www.pearson.com.au/masteringchemistry)

Make learning part of the grade. Access:

- tutorials with peronalised coaching
- study area
- Pearson eText

REACTIONS IN AQUEOUS SOLUTIONS

The waters of an ocean, such as the Southern Ocean, are simply aqueous solutions containing numerous dissolved substances.

KEY CONCEPTS

4.1 GENERAL PROPERTIES OF AQUEOUS SOLUTIONS
Substances dissolved in water can exist as ions, molecules or a mixture of the two.

4.2 PRECIPITATION REACTIONS
We identify reactions in which *soluble* reactants yield an *insoluble* product.

4.3 ACIDS, BASES AND NEUTRALISATION REACTIONS
We examine reactions in which *protons* (H^+ ions) are transferred from one reactant to another.

4.4 OXIDATION–REDUCTION REACTIONS
We look at reactions in which electrons are transferred from one reactant to another.

4.5 CONCENTRATIONS OF SOLUTIONS
We learn how the amount of a compound dissolved in a given volume of a solution can be expressed as a *concentration*. Concentration can be defined in a number of ways, the most commonly used being moles of compound per litre of solution (*molarity*).

4.6 SOLUTION STOICHIOMETRY AND CHEMICAL ANALYSIS
We see how the concepts of stoichiometry and concentration can be used to calculate amounts or concentrations of substances in solution through a process called *titration*.

The waters of the Southern Ocean are part of the oceans that cover almost two-thirds of our planet. Life itself almost certainly originated in water, and the need for water by all forms of life has helped determine diverse biological structures; your own body is about 60% water by mass. We see repeatedly throughout this text that water possesses many unusual properties essential to supporting life on Earth.

Water has an exceptional ability to dissolve a wide variety of substances. Water in nature—whether it is drinking water from a tap, water from a clear mountain stream or seawater—invariably contains a variety of dissolved substances. Solutions in which water is the dissolving medium are called **aqueous solutions**. Seawater differs from what we call 'fresh water' in having a much higher total concentration of dissolved ionic substances.

Water is the medium for most of the chemical reactions that take place within us and around us. Nutrients dissolved in blood are carried to our cells, where they enter into reactions that help keep us alive. Car parts rust when they come into frequent contact with aqueous solutions that contain various dissolved substances. Spectacular limestone caves (▼ FIGURE 4.1) are formed by the dissolving action of underground water containing carbon dioxide, $CO_2(aq)$:

$$CaCO_3(s) + H_2O(l) + CO_2(aq) \rightarrow Ca(HCO_3)_2(aq) \qquad [4.1]$$

We saw in Chapter 3 a few simple types of chemical reactions and how they are described. In this chapter we continue to examine chemical reactions by focusing on aqueous solutions. A great deal of important chemistry occurs in aqueous solutions, and we need to learn the vocabulary and concepts used to describe and understand this chemistry. In addition, we will extend the concepts of stoichiometry that we learned in Chapter 3 by considering how solution concentrations are expressed and used.

◀ **FIGURE 4.1 Limestone cave.** When CO_2 dissolves in water the resulting solution is slightly acidic. Limestone caves are formed by the dissolving action of this acidic solution acting on $CaCO_3$ in the limestone.

4.1 | GENERAL PROPERTIES OF AQUEOUS SOLUTIONS

∞ Review this on page 7

A *solution* is a homogeneous mixture of two or more substances (∞ Section 1.2, 'Classifications of Matter').

The substance present in greatest quantity is usually called the **solvent**. The other substances in the solution are known as the **solutes**; they are said to be dissolved in the solvent. When a small amount of sodium chloride (NaCl) is dissolved in a large quantity of water, for example, the water is the solvent and the sodium chloride is the solute.

Electrolytic Properties

At a young age we learn not to bring electrical devices into the bathtub so as not to electrocute ourselves. That's a useful lesson because most of the water you encounter in daily life is electrically conducting. *Pure* water, however, is a very poor conductor of electricity. The conductivity of bathwater originates from the substances dissolved in the water, not from the water itself.

Not all substances that dissolve in water make the resulting solution conducting. Imagine preparing two aqueous solutions—one by dissolving a teaspoon of table salt (sodium chloride) in a glass of water and the other by dissolving a teaspoon of table sugar (sucrose) in a glass of water (▼ FIGURE 4.2).

In order for the bulb in the device of Figure 4.2 to light up, there must be a current (that is, a *flow* of electrically charged particles) between two electrodes immersed in the solution. The conductivity of pure water is not sufficient to complete the electrical circuit and light the bulb. The situation changes when ions are present in solution because the ions carry electrical charge from one electrode to the other, completing the circuit. Thus the conductivity of NaCl solutions indicates the presence of ions. The lack of conductivity of sucrose solutions indicates the absence of ions. When NaCl dissolves in water, the solution contains Na^+ and Cl^- ions, each surrounded by water molecules. When sucrose ($C_{12}H_{22}O_{11}$) dissolves in water, the solution contains only neutral sucrose molecules surrounded by water molecules.

A substance (such as NaCl) whose aqueous solutions contain ions is called an **electrolyte**. A substance (such as $C_{12}H_{22}O_{11}$) that does not form ions in solution is called a **non-electrolyte**. The different behaviour of NaCl and $C_{12}H_{22}O_{11}$ arises because NaCl is ionic, whereas $C_{12}H_{22}O_{11}$ is molecular.

Pure water,
$H_2O(l)$
does not conduct electricity

Sucrose solution,
$C_{12}H_{22}O_{11}(aq)$
non-electrolyte
does not conduct electricity

Sodium chloride solution,
NaCl(aq)
electrolyte
conducts electricity

▲ **FIGURE 4.2 Electrical conductivities of water and two aqueous solutions.** One way to differentiate two aqueous solutions is to employ a device that measures their electrical conductivities. The ability of a solution to conduct electricity depends on the number of ions it contains. An electrolyte solution contains ions that serve as charge carriers, causing the bulb to light.

Ionic Compounds in Water

Recall from Section 2.7 and Figure 2.17 that solid NaCl consists of an orderly arrangement of Na^+ and Cl^- ions. When NaCl dissolves in water, each ion separates from the solid structure and disperses throughout the solution, as shown in ▼ FIGURE 4.3(a). The ionic solid *dissociates* into its component ions as it dissolves.

Water is a very effective solvent for ionic compounds. Although water is an electrically neutral molecule, one end of the molecule (the O atom) is rich in electrons and thus possesses a partial negative charge, denoted by δ–. The other end (the H atoms) has a partial positive charge, denoted by δ+. Note that the amounts of positive and negative charge cancel each other and the molecule is neutral overall. Positive ions (cations) are attracted by the negative end of H_2O and negative ions (anions) are attracted by the positive end.

As an ionic compound dissolves, the ions become surrounded by H_2O molecules, as shown in Figure 4.3(a). The ions are said to be solvated. The **solvation** process helps stabilise the ions in solution and prevents cations and anions from recombining. Furthermore, because the ions and their shells of surrounding water molecules are free to move about, the ions become dispersed uniformly throughout the solution.

We can usually predict the nature of the ions present in a solution of an ionic compound from the chemical name of the substance. Sodium sulfate (Na_2SO_4), for example, dissociates into sodium ions (Na^+) and sulfate ions (SO_4^{2-}). You must remember the formulae and charges of common ions (Tables 2.4 and 2.5) to understand the forms in which ionic compounds exist in aqueous solution.

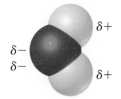

δ+

δ–

δ–

δ+

▲ **FIGURE IT OUT**

Which solution, NaCl(aq) or CH₃OH(aq), conducts electricity?

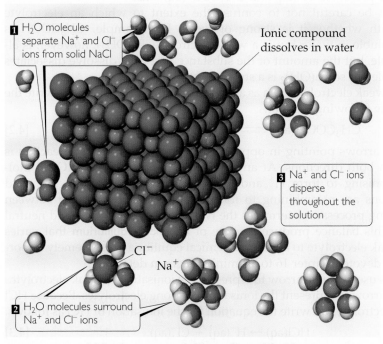

1 H_2O molecules separate Na^+ and Cl^- ions from solid NaCl

Ionic compound dissolves in water

3 Na^+ and Cl^- ions disperse throughout the solution

Cl^-

Na^+

2 H_2O molecules surround Na^+ and Cl^- ions

(a) Ionic compounds like sodium chloride, NaCl, form ions when they dissolve

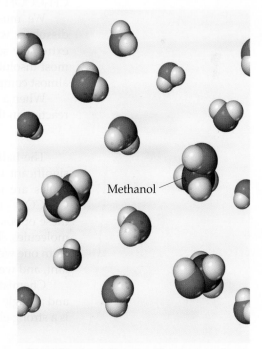

Methanol

(b) Molecular substances like methanol, CH_3OH, dissolve without forming ions

▲ **FIGURE 4.3   Dissolution in water.** (a) When an ionic compound, such as sodium chloride, NaCl, dissolves in water, H_2O molecules separate, surround and uniformly disperse the ions into the liquid. (b) Molecular substances that dissolve in water, such as methanol, CH_3OH, usually do so without forming ions. We can think of this as a simple mixing of two molecular species. In both (a) and (b) the water molecules have been moved apart so that the solute particles can be seen clearly.

⚠ **CONCEPT CHECK 1**

What dissolved species are present in a solution of
a. KCN?
b. $NaClO_4$?

Molecular Compounds in Water

When a molecular compound dissolves in water, the solution usually consists of intact molecules dispersed throughout the solution. Consequently, most molecular compounds are non-electrolytes. As we have seen, table sugar (sucrose) is an example of a non-electrolyte. As another example, a solution of methanol (CH_3OH) in water consists entirely of CH_3OH molecules dispersed throughout the water (Figure 4.3(b)).

There are, however, a few molecular substances whose aqueous solutions contain ions. The most important of these are acids. For example, when HCl(g) dissolves in water to form hydrochloric acid, HCl(aq), it *ionises;* that is, it dissociates into H^+(aq) and Cl^-(aq) ions.

Strong and Weak Electrolytes

There are two categories of electrolytes, strong and weak, which differ in the extent to which they conduct electricity. **Strong electrolytes** are those solutes that exist in solution completely or nearly completely as ions. Essentially, all soluble ionic compounds (such as NaCl) and a few molecular compounds (such as HCl) are strong electrolytes. **Weak electrolytes** are those solutes that exist in solution mostly in the form of molecules with only a small fraction in the form of ions. For example, in a solution of acetic acid (CH_3COOH) most of the solute is present as CH_3COOH molecules. Only a small fraction (about 1%) of the CH_3COOH is present as H^+(aq) and CH_3COO^-(aq) ions.

We must be careful not to confuse the extent to which an electrolyte dissolves with whether it is strong or weak. For example, CH_3COOH is extremely soluble in water but is a weak electrolyte. $Ba(OH)_2$, however, is almost insoluble, but the amount of the substance that does dissolve dissociates almost completely, so $Ba(OH)_2$ is a strong electrolyte.

When a weak electrolyte such as acetic acid ionises in solution, we write the reaction in the following manner:

$$CH_3COOH(aq) \rightleftharpoons CH_3COO^-(aq) + H^+(aq) \qquad [4.2]$$

The half arrows pointing in opposite directions mean that the reaction is significant in both directions. At any given moment, some CH_3COOH molecules are ionising to form H^+ and CH_3COO^-. At the same time, H^+ and CH_3COO^- ions are recombining to form CH_3COOH. The balance between these opposing processes determines the relative numbers of ions and neutral molecules. This balance produces a state of **chemical equilibrium** that varies from one weak electrolyte to another. Chemical equilibria are extremely important, and we devote Chapter 16 to examining them in detail.

Chemists use a double arrow to represent the ionisation of weak electrolytes and a single arrow to represent the ionisation of strong electrolytes. Because HCl is a strong electrolyte, we write the equation for the ionisation of HCl as follows:

$$HCl(aq) \rightarrow H^+(aq) + Cl^-(aq) \qquad [4.3]$$

The absence of a reverse arrow indicates that the H^+ and Cl^- ions have no tendency to recombine in water to form HCl molecules.

In the sections ahead we begin to look more closely at how we can use the composition of a compound to predict whether it is a strong electrolyte, weak electrolyte or non-electrolyte. For the moment, it is important only to remember that *water-soluble ionic compounds are strong electrolytes.* We identify ionic compounds as those where a metal is combined with non-metals (such as NaCl,

$FeSO_4$ and $Al(NO_3)_3$) or compounds containing the ammonium ion, NH_4^+ (such as NH_4Br and $(NH_4)_2CO_3$).

CONCEPT CHECK 2

Which solute will cause the light bulb in the experiment shown in Figure 4.2 to glow more brightly: CH_3OH, $NaOH$ or CH_3COOH?

4.2 | PRECIPITATION REACTIONS

▼ **FIGURE 4.4** shows two clear solutions being mixed, one containing lead nitrate ($Pb(NO_3)_2$) and the other containing potassium iodide (KI). The reaction between these two solutes produces an insoluble yellow product. Reactions that result in the formation of an insoluble product are known as **precipitation reactions**. A **precipitate** is an insoluble solid formed by a reaction in solution. In Figure 4.4 the precipitate is lead iodide (PbI_2), a compound that has a very low solubility in water:

$$Pb(NO_3)_2(aq) + 2\ KI(aq) \rightarrow PbI_2(s) + 2\ KNO_3(aq) \qquad [4.4]$$

The other product of this reaction, potassium nitrate, remains in solution.

Precipitation reactions occur when certain pairs of oppositely charged ions attract each other so strongly that they form an insoluble ionic solid. To predict whether certain combinations of ions form insoluble compounds, we must consider some guidelines concerning the solubilities of common ionic compounds.

FIGURE IT OUT

Which ions remain in solution after PbI_2 precipitation is complete?

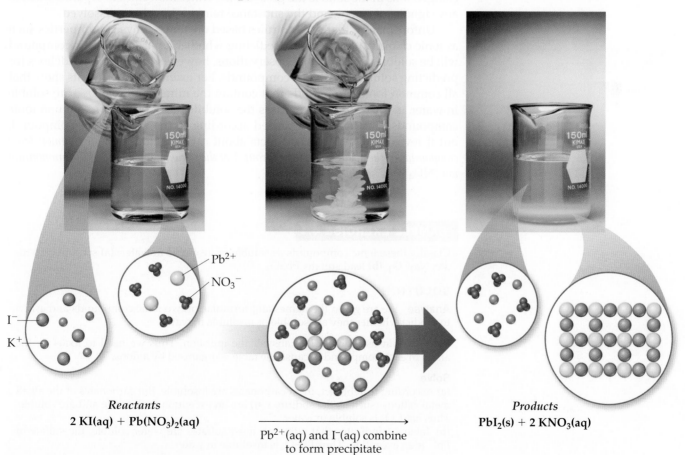

Reactants
$2\ KI(aq) + Pb(NO_3)_2(aq)$

$Pb^{2+}(aq)$ and $I^-(aq)$ combine
to form precipitate

Products
$PbI_2(s) + 2\ KNO_3(aq)$

▲ **FIGURE 4.4   A precipitation reaction.**

TABLE 4.1 • Solubility guidelines for common ionic compounds in water		
Soluble ionic compounds		**Important exceptions**
Compounds containing	NO_3^-	None
	CH_3COO^-	None
	Cl^-	Compounds of Ag^+, Hg_2^{2+} and Pb^{2+}
	Br^-	Compounds of Ag^+, Hg_2^{2+} and Pb^{2+}
	I^-	Compounds of Ag^+, Hg_2^{2+} and Pb^{2+}
	SO_4^{2-}	Compounds of Sr^{2+}, Ba^{2+}, Hg_2^{2+} and Pb^{2+}
Insoluble ionic compounds		**Important exceptions**
Compounds containing	S^{2-}	Compounds of NH_4^+, the alkali metal cations, Ca^{2+}, Sr^{2+} and Ba^{2+}
	CO_3^{2-}	Compounds of NH_4^+ and the alkali metal cations
	PO_4^{3-}	Compounds of NH_4^+ and the alkali metal cations
	OH^-	Compounds of NH_4^+, the alkali metal cations, Ca^{2+}, Sr^{2+} and Ba^{2+}

Solubility Guidelines for Ionic Compounds

The **solubility** of a substance at a given temperature is the amount of that substance that can be dissolved in a given quantity of solvent at that temperature. For instance, only 1.2×10^{-3} mol of PbI_2 dissolves in a litre of water at 25 °C. In our discussions any substance with a solubility less than 0.01 mol dm^{-3} will be referred to as *insoluble*. In those cases the attraction between the oppositely charged ions in the solid is too great for the water molecules to separate them to any significant extent, and the substance remains largely undissolved.

Unfortunately, there are no rules based on simple physical properties such as ionic charge to guide us in predicting whether a particular ionic compound will be soluble. Experimental observations, however, have led to guidelines for predicting solubility for ionic compounds. For example, experiments show that all common ionic compounds that contain the nitrate anion, NO_3^-, are soluble in water. ▲ TABLE 4.1 summarises the solubility guidelines for common ionic compounds. The table is organised according to the anion in the compound, but it reveals many important facts about cations. Note that *all common ionic compounds of the alkali metal ions (group 1 of the periodic table) and of the ammonium ion (NH_4^+) are soluble in water.*

SAMPLE EXERCISE 4.1

Classify these ionic compounds as soluble or insoluble in water: **(a)** sodium carbonate, Na_2CO_3, **(b)** lead sulfate, $PbSO_4$.

SOLUTION

Analyse We are given the names and formulae of two ionic compounds and asked to predict whether they are soluble or insoluble in water.

Plan We can use Table 4.1 to answer the question. Thus we need to focus on the anion in each compound because the table is organised by anions.

Solve
(a) According to Table 4.1, most carbonates are insoluble. But carbonates of the alkali metal cations (such as the sodium ion) are an exception to this rule and are soluble. Thus Na_2CO_3 is soluble in water.
(b) Table 4.1 indicates that although most sulfates are water soluble, the sulfate of Pb^{2+} is an exception. Thus $PbSO_4$ is insoluble in water.

PRACTICE EXERCISE

Classify the following compounds as soluble or insoluble in water: **(a)** cobalt(II) hydroxide, **(b)** barium nitrate, **(c)** ammonium phosphate.

Answers: (a) insoluble, **(b)** soluble, **(c)** soluble

(See also Exercises 4.18, 4.19.)

To predict whether a precipitate forms when we mix aqueous solutions of two strong electrolytes, we must (1) note the ions present in the reactants, (2) consider the possible combinations of the cations and anions, and (3) use Table 4.1 to determine whether any of these combinations are insoluble. For example, will a precipitate form when solutions of $Mg(NO_3)_2$ and $NaOH$ are mixed? Both $Mg(NO_3)_2$ and $NaOH$ are soluble ionic compounds and they are both strong electrolytes. Mixing $Mg(NO_3)_2$(aq) and $NaOH$(aq) first produces a solution containing Mg^{2+}, NO_3^-, Na^+ and OH^- ions. Will either of the cations interact with either of the anions to form an insoluble compound? In addition to the reactants, the other possible interactions are Mg^{2+} with OH^- and Na^+ with NO_3^-. From Table 4.1 we see that hydroxides are generally insoluble. Because Mg^{2+} is not an exception, $Mg(OH)_2$ is insoluble and will form a precipitate. $NaNO_3$, however, is soluble, so Na^+ and NO_3^- will remain in solution. The balanced equation for the precipitation reaction is:

$$Mg(NO_3)_2(aq) + 2\,NaOH(aq) \rightarrow Mg(OH)_2(s) + 2\,NaNO_3(aq) \qquad [4.5]$$

Exchange (Metathesis) Reactions

Notice that in Equation 4.5 the cations in the two reactants exchange anions: Mg^{2+} ends up with OH^-, and Na^+ ends up with NO_3^-. The chemical formulae of the products are based on the charges of the ions: two OH^- ions are needed to give a neutral compound with Mg^{2+}, and one NO_3^- ion is needed to give a neutral compound with Na^+ (∞ Section 2.7, 'Ions and Ionic Compounds').

∞ Review this on page 47

It is only after the chemical formulae of the products have been determined that the equation can be balanced.

Reactions in which positive ions and negative ions appear to exchange partners conform to the following general equation:

$$AX + BY \rightarrow AY + BX \qquad [4.6]$$

Example: $AgNO_3(aq) + KCl(aq) \rightarrow AgCl(s) + KNO_3(aq)$

Such reactions are known as **exchange reactions**, or **metathesis reactions** (meh-TATH-eh-sis, which is the Greek word for 'to place differently'). Precipitation reactions conform to this pattern, as do many acid–base reactions, as we see in Section 4.3.

To complete and balance a metathesis equation, we follow these steps.

1. Use the chemical formulae of the reactants to determine the ions that are present.

2. Write the chemical formulae of the products by combining the cation from one reactant with the anion of the other. (Use the charges of the ions to determine the subscripts in the chemical formulae.)

3. Check the water solubilities of the products. For a precipitation reaction to occur, at least one product must be insoluble in water.

4. Finally, balance the equation.

SAMPLE EXERCISE 4.2 **Predicting a metathesis reaction**

(a) Predict the identity of the precipitate that forms when solutions of $BaCl_2$ and K_2SO_4 are mixed. **(b)** Write the balanced chemical equation for the reaction.

SOLUTION

Analyse We are given two ionic reactants and asked to predict the insoluble product that they form.

Plan We need to write down the ions present in the reactants and exchange the anions between the two cations. Once we have written the chemical formulae for these products, we can use Table 4.1 to determine which is insoluble in water. Knowing the products also allows us to write the equation for the reaction.

Solve

(a) The reactants contain Ba^{2+}, Cl^-, K^+ and SO_4^{2-} ions. If we exchange the anions, we will have $BaSO_4$ and KCl. According to Table 4.1, most compounds of SO_4^{2-} are soluble, but those of Ba^{2+} are not. Thus $BaSO_4$ is insoluble and will precipitate from solution. KCl, however, is soluble.

(b) From part (a) we know the chemical formulae of the products, $BaSO_4$ and KCl. The balanced equation with phase labels shown is

$$BaCl_2(aq) + K_2SO_4(aq) \rightarrow BaSO_4\,(s) + 2\,KCl(aq)$$

PRACTICE EXERCISE

(a) What compound precipitates when solutions of $Fe_2(SO_4)_3$ and $LiOH$ are mixed? **(b)** Write a balanced equation for the reaction. **(c)** Will a precipitate form when solutions of $Ba(NO_3)_2$ and KOH are mixed?

Answers: **(a)** $Fe(OH)_3$; **(b)** $Fe_2(SO_4)_3(aq) + 6\,LiOH(aq) \rightarrow 2\,Fe(OH)_3\,(s) + 3\,Li_2SO_4(aq)$; **(c)** no (both possible products are water soluble)

(See also Exercise 4.21.)

Ionic Equations

In writing chemical equations for reactions in aqueous solution, it is often useful to indicate explicitly whether the dissolved substances are present predominantly as ions or as molecules. Let's reconsider the precipitation reaction between $Pb(NO_3)_2$ and $2\,KI$, shown previously in Figure 4.4:

$$Pb(NO_3)_2(aq) + 2\,KI(aq) \rightarrow PbI_2\,(s) + 2\,KNO_3(aq)$$

An equation written in this fashion, showing the complete chemical formulae of the reactants and products, is called a **molecular equation** because it shows the chemical formulae of the reactants and products without indicating their ionic character. Because $Pb(NO_3)_2$, KI and KNO_3 are all soluble ionic compounds and therefore strong electrolytes, we can write the chemical equation to indicate explicitly the ions that are in the solution:

$$Pb^{2+}(aq) + 2\,NO_3^-(aq) + 2\,K^+(aq) + 2\,I^-(aq) \longrightarrow$$
$$PbI_2(s) + 2\,K^+(aq) + 2\,NO_3(aq) \qquad [4.7]$$

An equation written in this form, with all soluble strong electrolytes shown as ions, is known as a **complete ionic equation**.

Notice that $K^+(aq)$ and $NO_3^-(aq)$ appear on both sides of Equation 4.7. Ions that appear in identical forms among both the reactants and products of a complete ionic equation are called **spectator ions**. They are present but play no direct role in the reaction. When spectator ions are omitted from the equation (they cancel out like algebraic quantities), we are left with the **net ionic equation**:

$$Pb^{2+}(aq) + 2\,I^-(aq) \rightarrow PbI_2(s) \qquad [4.8]$$

A net ionic equation includes only the ions and molecules directly involved in the reaction. Charge is conserved in reactions, so the sum of the charges of the ions must be the same on both sides of a balanced net ionic equation. In this case the 2+ charge of the cation and the two 1− charges of the anions add to give zero, the charge of the electrically neutral product. *If every ion in a complete ionic equation is a spectator, then no reaction occurs.*

Which ions, if any, are spectator ions in the reaction
$AgNO_3(aq) + NaCl(aq) \longrightarrow AgCl(s) + NaNO_3(aq)$?

Net ionic equations are widely used to illustrate the similarities between large numbers of reactions involving electrolytes. For example, Equation 4.8 expresses the essential feature of the precipitation reaction between any strong electrolyte containing Pb^{2+} and any strong electrolyte containing I^-: the $Pb^{2+}(aq)$ and $I^-(aq)$ ions combine to form a precipitate of PbI_2. Thus a net ionic equation demonstrates that more than one set of reactants can lead to the same net reaction. The complete equation, however, identifies the actual reactants that participate in a reaction.

The following steps summarise the procedure for writing net ionic equations.

1. Write a balanced molecular equation for the reaction.

2. Rewrite the equation to show the ions that form in solution when each soluble strong electrolyte dissociates into its component ions. *Only strong electrolytes dissolved in* aqueous *solution are written in ionic form.*

3. Identify and cancel spectator ions.

SAMPLE EXERCISE 4.3 Writing a net ionic equation

Write the net ionic equation for the precipitation reaction that occurs when solutions of calcium chloride and sodium carbonate are mixed.

SOLUTION

Analyse We are required to write a net ionic equation for a precipitation reaction, given the names of the reactants present in solution.

Plan We write the chemical formulae of the reactants and products and then determine which product is insoluble. We then write and balance the molecular equation. Next, we write each soluble strong electrolyte as separated ions to obtain the complete ionic equation. Finally, we eliminate the spectator ions to obtain the net ionic equation.

Solve Calcium chloride is composed of calcium ions, Ca^{2+}, and chloride ions, Cl^-; hence an aqueous solution of the substance is $CaCl_2(aq)$. Sodium carbonate is composed of Na^+ ions and CO_3^{2-} ions; so an aqueous solution of the compound is $Na_2CO_3(aq)$. In the molecular equations for precipitation reactions, the anions and cations appear to exchange partners. Thus we put Ca^{2+} and CO_3^{2-} together to give $CaCO_3$ and Na^+ and Cl^- together to give $NaCl$. According to the solubility guidelines in Table 4.1, $CaCO_3$ is insoluble and $NaCl$ is soluble. The balanced molecular equation is

$$CaCl_2(aq) + Na_2CO_3(aq) \rightarrow CaCO_3(s) + 2\,NaCl(aq)$$

In a complete ionic equation, only dissolved strong electrolytes (such as soluble ionic compounds) are written as separate ions. The (aq) designations remind us that $CaCl_2$, Na_2CO_3 and $NaCl$ are all dissolved in the solution. Furthermore, they are all strong electrolytes. $CaCO_3$ is an ionic compound, but it is not soluble. We do not write the formula of any insoluble compound as its component ions. Thus the complete ionic equation is

$$Ca^{2+}(aq) + 2\,Cl^-(aq) + 2\,Na^+(aq) + CO_3^{2-}(aq) \longrightarrow$$
$$CaCO_3(s) + 2\,Na^+(aq) + 2\,Cl^-(aq)$$

Cl^- and Na^+ are spectator ions. Cancelling them gives the following net ionic equation:

$$Ca^{2+}(aq) + CO_3^{2-}(aq) \rightarrow CaCO_3(s)$$

Check We can check our result by confirming that both the elements and the electric charge are balanced. Each side has one Ca, one C and three O, and the net charge on each side equals 0.

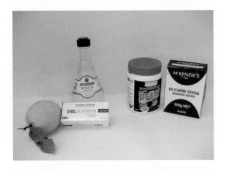

▲ **FIGURE 4.5** Some common household acids (left) and bases (right).

Hydrochloric acid, HCl

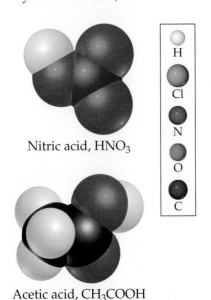

Nitric acid, HNO_3

Acetic acid, CH_3COOH

▲ **FIGURE 4.6** Molecular models of three common acids.

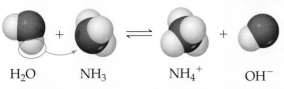

H_2O NH_3 NH_4^+ OH^-

▲ **FIGURE 4.7** **Hydrogen ion transfer.** An H_2O molecule acts as a proton donor (acid), and NH_3 acts as a proton acceptor (base). Only a fraction of the NH_3 molecules react with H_2O. Consequently, NH_3 is a weak electrolyte.

Comment If none of the ions in an ionic equation is removed from solution or changed in some way, then they are all spectator ions and a reaction does not occur.

PRACTICE EXERCISE

Write the net ionic equation for the precipitation reaction that occurs when aqueous solutions of silver nitrate and potassium phosphate are mixed.

Answer: $3\,Ag^+(aq) + PO_4^{3-}(aq) \rightarrow Ag_3PO_4\,(s)$
(See also Exercise 4.23.)

4.3 | ACIDS, BASES AND NEUTRALISATION REACTIONS

Many acids and bases are industrial and household substances (◄ FIGURE 4.5), and some are important components of biological fluids. Hydrochloric acid, for example, is not only an important industrial chemical but also the main constituent of gastric juice in your stomach. Acids and bases also happen to be common electrolytes.

Acids

Acids are substances that ionise in aqueous solutions to form hydrogen ions, thereby increasing the concentration of $H^+(aq)$ ions. Because a hydrogen atom consists of a proton and an electron, H^+ is simply a proton. Thus acids are often called *proton donors*. Molecular models of three common acids, HCl, HNO_3 and CH_3COOH, are shown in ◄ FIGURE 4.6.

Just as cations are surrounded and bound by water molecules (see Figure 4.3(a)), the proton is also solvated by water molecules. The nature of the proton in water is discussed in detail in Section 17.2. In writing chemical equations involving the proton in water, we usually represent it simply as $H^+(aq)$.

Molecules of different acids can ionise to form different numbers of H^+ ions. Both HCl and HNO_3 are *monoprotic* acids, which yield one H^+ per molecule of acid. Sulfuric acid, H_2SO_4, is a *diprotic* acid, one that yields two H^+ per molecule of acid. The ionisation of H_2SO_4 and other diprotic acids occurs in two steps:

$$H_2SO_4(aq) \rightarrow H^+(aq) + HSO_4^-(aq) \qquad [4.9]$$

$$HSO_4^-(aq) \rightleftharpoons H^+(aq) + SO_4^{2-}(aq) \qquad [4.10]$$

Although H_2SO_4 is a strong electrolyte, only the first ionisation is complete. Thus aqueous solutions of sulfuric acid contain a mixture of $H^+(aq)$, $HSO_4^-(aq)$ and $SO_4^{2-}(aq)$.

Bases

Bases are substances that accept (react with) H^+ ions. Bases produce hydroxide ions (OH^-) when they dissolve in water. Ionic hydroxide compounds such as NaOH, KOH and $Ca(OH)_2$ are among the most common bases. When dissolved in water, they dissociate into their component ions, introducing OH^- ions into the solution.

Compounds that do not contain OH^- ions can also be bases. For example, ammonia (NH_3) is a common base. When added to water, it accepts an H^+ ion from the water molecule and thereby produces an OH^- ion (◄ FIGURE 4.7):

$$NH_3(aq) + H_2O(l) \rightleftharpoons NH_4^+(aq) + OH^-(aq) \qquad [4.11]$$

Ammonia is a weak electrolyte because only a small fraction of the NH_3 (about 1%) forms NH_4^+ and OH^- ions.

TABLE 4.2 • **Common strong acids and bases**	
Strong acids	**Strong bases**
Hydrochloric, HCl	Group 1 metal hydroxides (LiOH, NaOH, KOH, RbOH, CsOH)
Hydrobromic, HBr	Heavy group 2 metal hydroxides (Ca(OH)$_2$, Sr(OH)$_2$, Ba(OH)$_2$)
Hydroiodic, HI	
Chloric, HClO$_3$	
Perchloric, HClO$_4$	
Nitric, HNO$_3$	
Sulfuric, H$_2$SO$_4$	

Strong and Weak Acids and Bases

Acids and bases that are strong electrolytes (completely ionised in solution) are called **strong acids** and **strong bases**. Those that are weak electrolytes (partly ionised) are called **weak acids** and **weak bases**. Strong acids are more reactive than weak acids when the reactivity depends only on the concentration of H$^+$(aq). The reactivity of an acid, however, can depend on the anion as well as on H$^+$(aq). For example, hydrofluoric acid (HF) is a weak acid (only partly ionised in aqueous solution), but it is very reactive and vigorously attacks many substances, including glass. This reactivity is due to the combined action of H$^+$(aq) and F$^-$(aq).

▲ TABLE 4.2 lists the common strong acids and bases. You should commit these to memory. As you examine this table, notice that some of the most common acids, such as HCl, HNO$_3$ and H$_2$SO$_4$, are strong. Three of the strong acids are the hydrogen compounds of the halogen family. (HF, however, is a weak acid.) The list of strong acids is very short. Most acids are weak. The only common strong bases are the hydroxides of Li$^+$, Na$^+$, K$^+$, Rb$^+$ and Cs$^+$ (the alkali metals, group 1) and the sparingly soluble hydroxides of Ca^{2+}, Sr^{2+} and Ba^{2+} (the heavy alkaline earths, group 2). These are the common soluble metal hydroxides. Most other metal hydroxides are insoluble in water. The most common weak base is NH$_3$, which reacts with water to form OH$^-$ ions (Equation 4.11).

 CONCEPT CHECK 4

Which of the following is a strong acid: H$_2$SO$_3$, HBr, CH$_3$COOH?

SAMPLE EXERCISE 4.4 **Comparing acid strengths**

The following diagrams represent aqueous solutions of three acids (HX, HY and HZ) with water molecules omitted for clarity. Rank them from strongest to weakest.

HX   HY   HZ

 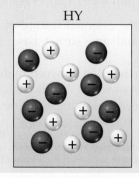 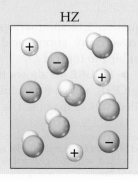

SOLUTION

Analyse We are asked to rank the three acids from strongest to weakest, based on schematic drawings of their solutions.

Plan We can determine the relative numbers of uncharged molecular species in the diagrams. The strongest acid is the one with the most H^+ ions and fewest undissociated acid molecules in solution. The weakest is the one with the largest number of undissociated molecules.

Solve The order is HY > HZ > HX. HY is a strong acid because it is totally ionised (no HY molecules in solution), whereas both HX and HZ are weak acids, whose solutions consist of a mixture of molecules and ions. Because HZ contains more H^+ ions and fewer molecules than HX, it is a stronger acid.

PRACTICE EXERCISE

Imagine a diagram showing 10 Na^+ ions and 10 OH^- ions. If this solution were mixed with the one pictured above for HY, what would the diagram look like if it were to represent the solution after any possible reaction? (H^+ ions will react with OH^- ions to form H_2O.)

Answer: The final diagram would show 10 Na^+ ions, two OH^- ions, eight Y^- ions and eight H_2O molecules.

Neutralisation Reactions and Salts

The properties of acidic solutions are quite different from those of basic solutions. Acids have a sour taste, whereas bases have a bitter taste.* Acids can change the colours of certain dyes in a way that differs from the way bases affect the same dyes (◀ FIGURE 4.8). This is the principle behind the dye known as litmus paper. In addition, acidic and basic solutions differ in chemical properties in several important ways that we explore in this chapter and in later chapters.

When a solution of an acid and a solution of a base are mixed, a **neutralisation reaction** occurs. The products of the reaction have none of the characteristic properties of either the acidic solution or the basic solution. For example, when hydrochloric acid is mixed with a solution of sodium hydroxide, the following reaction occurs:

$$HCl(aq) + NaOH(aq) \longrightarrow H_2O(l) + NaCl(aq) \qquad [4.12]$$
$$\text{(acid)} \qquad \text{(base)} \qquad \qquad \text{(water)} \qquad \text{(salt)}$$

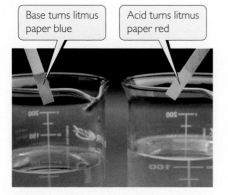

▲ **FIGURE 4.8 Litmus paper.** Litmus paper is coated with dyes that change colour in response to exposure to either acids or bases.

Water and table salt, NaCl, are the products of the reaction. By analogy to this reaction, the term **salt** has come to mean any ionic compound whose cation comes from a base (for example, Na^+ from NaOH) and whose anion comes from an acid (for example, Cl^- from HCl). In general, *a neutralisation reaction between an acid and a metal hydroxide produces water and a salt*.

Because HCl, NaOH and NaCl are all soluble strong electrolytes, the complete ionic equation associated with Equation 4.12 is

$$H^+(aq) + Cl^-(aq) + Na^+(aq) + OH^-(aq) \longrightarrow H_2O(l) + Na^+(aq) + Cl^-(aq) \qquad [4.13]$$

Therefore, the net ionic equation is

$$H^+(aq) + OH^-(aq) \rightarrow H_2O(l) \qquad [4.14]$$

Equation 4.14 summarises the essential feature of the neutralisation reaction between any strong acid and any strong base: $H^+(aq)$ and $OH^-(aq)$ ions combine to form H_2O.

▶ FIGURE 4.9 shows the reaction between hydrochloric acid and the base $Mg(OH)_2$, which is insoluble in water. A milky white suspension of $Mg(OH)_2$, called milk of magnesia, is seen dissolving as the neutralisation reaction occurs:

In the figure:
Base turns litmus paper blue
Acid turns litmus paper red

---

*Tasting chemical solutions is not a good practice. However, we have all had acids such as ascorbic acid (vitamin C), acetylsalicylic acid (aspirin) and citric acid (in citrus fruits) in our mouths and we are familiar with their characteristic sour taste. Soaps, which are basic, have the characteristic bitter taste of bases.

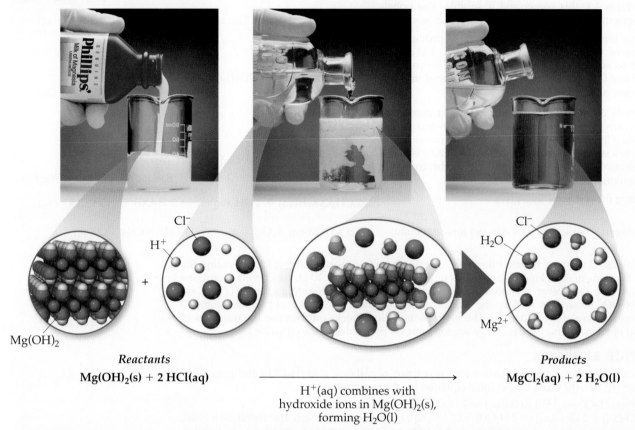

FIGURE IT OUT

Adding just a few drops of hydrochloric acid would not be sufficient to dissolve all the $Mg(OH)_2(s)$. Why not?

Reactants   *Products*

$Mg(OH)_2(s) + 2 HCl(aq)$ ———————→ $MgCl_2(aq) + 2 H_2O(l)$

$H^+(aq)$ combines with
hydroxide ions in $Mg(OH)_2(s)$,
forming $H_2O(l)$

▲ **FIGURE 4.9   Neutralisation reaction between $Mg(OH)_2(s)$ and hydrochloric acid.** Milk of magnesia is a suspension of water-insoluble magnesium hydroxide, $Mg(OH)_2(s)$, in water. When sufficient hydrochloric acid, HCl(aq), is added a reaction ensues that leads to an aqueous solution containing $Mg^{2+}(aq)$ and $Cl^-(aq)$ ions.

Molecular equation:

$$Mg(OH)_2 \text{ (s)} + 2 HCl(aq) \rightarrow MgCl_2(aq) + 2 H_2O \qquad [4.15]$$

Net ionic equation:

$$Mg(OH)_2 \text{ (s)} + 2 H^+(aq) \rightarrow Mg^{2+}(aq) + 2 H_2O \qquad [4.16]$$

Notice that the OH^- ions and H^+ ions combine to form H_2O. Because the ions exchange partners, neutralisation reactions between acids and metal hydroxides are also metathesis reactions.

SAMPLE EXERCISE 4.5   Writing chemical equations for a neutralisation reaction

(a) Write a balanced molecular equation for the reaction between aqueous solutions of acetic acid (CH_3COOH) and barium hydroxide ($Ba(OH)_2$). **(b)** Write the net ionic equation for this reaction.

SOLUTION

Analyse We are given the chemical formulae for an acid and a base and asked to write a balanced molecular equation and a net ionic equation for the neutralisation reaction.

Plan As Equation 4.12 and the italicised statement that follows it indicate, neutralisation reactions form two products, H_2O and a salt. We examine the cation of the base and the anion of the acid to determine the composition of the salt.

Solve

(a) The salt will contain the cation of the base (Ba^{2+}) and the anion of the acid (CH_3COO^-). Thus the formula of the salt is $Ba(CH_3COO)_2$. According to the solubility guidelines in Table 4.1, this compound is soluble. The unbalanced molecular equation for the neutralisation reaction is

$$CH_3COOH(aq) + Ba(OH)_2(aq) \longrightarrow H_2O(l) + Ba(CH_3COO)_2(aq)$$

To balance this molecular equation, we must provide two molecules of CH_3COOH to furnish the two CH_3COO^- ions and to supply the two H^+ ions needed to combine with the two OH^- ions of the base. The balanced molecular equation is

$$2\,CH_3COOH(aq) + Ba(OH)_2(aq) \longrightarrow$$
$$2\,H_2O(l) + Ba(CH_3COO)_2(aq)$$

(b) To write the net ionic equation, we must determine whether each compound in aqueous solution is a strong electrolyte. CH_3COOH is a weak electrolyte (weak acid), $Ba(OH)_2$ is a strong electrolyte and $Ba(CH_3COO)_2$ is also a strong electrolyte (ionic compound). Thus the complete ionic equation is

$$2\,CH_3COOH(aq) + Ba^{2+}(aq) + 2\,OH^-(aq) \longrightarrow$$
$$2\,H_2O(l) + Ba^{2+}(aq) + 2\,CH_3COO^-(aq)$$

Eliminating the spectator ions gives

$$2\,CH_3COOH(aq) + 2\,OH^-(aq) \longrightarrow 2\,H_2O(l) + 2\,CH_3COO^-(aq)$$

Simplifying the coefficients gives the net ionic equation:

$$CH_3COOH(aq) + OH^-(aq) \longrightarrow H_2O(l) + CH_3COO^-(aq)$$

Check We can determine whether the molecular equation is correctly balanced by counting the number of atoms of each kind on both sides of the arrow. (There are 10 H, six O, four C and one Ba on each side.) However, it is often easier to check equations by counting groups. (There are two CH_3COO^- groups, as well as one Ba, and four additional H atoms and two additional O atoms on each side of the equation.) The net ionic equation checks out because the numbers of each kind of element and the net charge are the same on both sides of the equation.

PRACTICE EXERCISE

(a) Write a balanced molecular equation for the reaction of carbonic acid (H_2CO_3) and potassium hydroxide (KOH). **(b)** Write the net ionic equation for this reaction.

Answers: **(a)** $H_2CO_3(aq) + 2\,KOH(aq) \rightarrow 2\,H_2O\,(l) + K_2CO_3(aq)$;
(b) $H_2CO_3(aq) + 2\,OH^-(aq) \rightarrow 2\,H_2O(l) + CO_3{}^{2-}(aq)$; H_2CO_3 is a weak acid and therefore a weak electrolyte, whereas KOH, a strong base, and K_2CO_3, an ionic compound, are strong electrolytes.
(See also Exercises 4.31, 4.32.)

Neutralisation Reactions with Gas Formation

There are many bases besides OH^- that react with H^+ to form molecular compounds. Two of these that you might encounter in the laboratory are the sulfide ion and the carbonate ion. Both of these anions react with acids to form gases that have low solubilities in water. Hydrogen sulfide, H_2S, the substance that gives rotten eggs their foul odour, forms when an acid such as HCl(aq) reacts with a metal sulfide such as Na_2S:

Molecular equation:

$$2\,HCl(aq) + Na_2S(aq) \rightarrow H_2S\,(g) + 2\,NaCl(aq) \qquad [4.17]$$

Net ionic equation:

$$2\,H^+(aq) + S^{2-}(aq) \rightarrow H_2S(g) \qquad [4.18]$$

Carbonates and bicarbonates react with acids to form CO_2 gas. Reaction of $CO_3{}^{2-}$ or $HCO_3{}^-$ with an acid first gives carbonic acid (H_2CO_3). For example, when hydrochloric acid is added to sodium bicarbonate, the following reaction occurs

$$HCl(aq) + NaHCO_3(aq) \rightarrow NaCl(aq) + H_2CO_3(aq) \qquad [4.19]$$

Carbonic acid is unstable; if present in solution in sufficient concentrations, it decomposes to form CO_2, which escapes from the solution as a gas

$$H_2CO_3(aq) \rightarrow H_2O(l) + CO_2(g) \qquad [4.20]$$

The decomposition of H_2CO_3 produces bubbles of CO_2 gas, as shown in ▶ FIGURE 4.10. The overall reaction is summarised by the following equations:

Molecular equation:

$$HCl(aq) + NaHCO_3(aq) \rightarrow NaCl(aq) + H_2O(l) + CO_2(g) \qquad [4.21]$$

Net ionic equation:

$$H^+(aq) + HCO_3^-(aq) \rightarrow H_2O(l) + CO_2(g) \qquad [4.22]$$

Both $NaHCO_3$ and Na_2CO_3 are used as acid neutralisers in acid spills. The bicarbonate or carbonate salt is added until the fizzing due to the formation of $CO_2(g)$ stops. Sometimes, sodium bicarbonate is used as an antacid to soothe an upset stomach. In that case, the HCO_3^- reacts with stomach acid to form $CO_2(g)$ The fizz when Alka-Seltzer® tablets are added to water arises from the reaction of sodium bicarbonate and citric acid.

CONCEPT CHECK 5

By analogy to examples already given in the text, predict what gas forms when $Na_2SO_3(s)$ is treated with $HCl(aq)$.

▲ **FIGURE 4.10 Carbonates react with acids to form carbon dioxide gas.** Here $NaHCO_3$ (white solid) reacts with hydrochloric acid; the bubbles contain CO_2.

MY WORLD OF CHEMISTRY

ANTACIDS

The stomach secretes acids to help digest foods. These acids, which include hydrochloric acid, contain about 0.1 mol of H^+ per litre of solution. The stomach and digestive tract are normally protected from the corrosive effects of stomach acid by a mucosal lining. Holes can develop in this lining, however, allowing the acid to attack the underlying tissue, causing painful damage. These holes, known as ulcers, are mainly caused by the bacterium *Helicobacter pylori* (▼ FIGURE 4.11). Two Australian scientists, Barry J. Marshall and J. Robin

Warren, received the Nobel Prize in Physiology or Medicine in 2005 for its discovery. Many Australians suffer from ulcers at some point in their lives, and many others experience occasional indigestion or heartburn that is due to digestive acids entering the oesophagus.

We can address the problem of excess stomach acid in two simple ways: (1) removing the excess acid, or (2) decreasing the production of acid. Those substances that remove excess acid are called antacids, whereas those that decrease the production of acid are called acid inhibitors.

Antacids are simple bases that neutralise digestive acids. Their ability to neutralise acids is due to the hydroxide, carbonate or bicarbonate ions they contain. ▼ TABLE 4.3 lists the active ingredients in some antacids.

The newer generation of anti-ulcer drugs, such as Tagamet® and Zantac®, are acid inhibitors. They act on acid-producing cells in the lining of the stomach. Formulations that control acid in this way are now available as over-the-counter drugs.

RELATED EXERCISE: 4.69

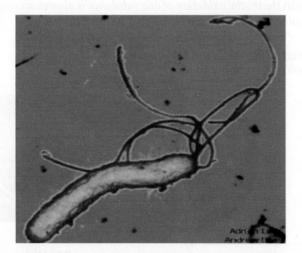

▲ **FIGURE 4.11 Bacterium *Helicobacter pylori*, which is responsible for most stomach ulcers.**

TABLE 4.3 • Some common antacids

Commercial name	Acid-neutralising agents
Alka-Seltzer®	$NaHCO_3$
Amphojel®	$Al(OH)_3$
Milk of Magnesia	$Mg(OH)_2$
Mylanta®	$Mg(OH)_2$ and $Al(OH)_3$
Rolaids®	$NaAl(OH)_2CO_3$
Tums®	$CaCO_3$

▲ **FIGURE 4.12 Corrosion of iron** is caused by chemical attack of oxygen and water on exposed metal surfaces. The surface of this 20 m sculpture, the 'Angel of the North', has oxidised to form a patina that mellows with age to a rich red-brown colour.

4.4 | OXIDATION–REDUCTION REACTIONS

In precipitation reactions cations and anions come together to form an insoluble ionic compound. In neutralisation reactions H^+ ions and OH^- ions come together to form H_2O molecules. We will now consider a third important kind of reaction, one in which electrons are transferred between reactants. Such reactions are called **oxidation–reduction**, or **redox**, **reactions**.

Oxidation and Reduction

One of the most familiar redox reactions is the *corrosion* of a metal. Corrosion of iron (rusting) and of other metals, such as the corrosion of the terminals of a car battery, are familiar processes. What we call corrosion is the conversion of a metal into a metal compound by a reaction between the metal and a substance in its environment. Rusting for example (◄ FIGURE 4.12), involves the reaction of oxygen with iron in the presence of water.

When a metal corrodes, each metal atom loses electrons and forms a cation, which can combine with an anion to form an ionic compound.

When an atom, ion or molecule has become more positively charged (that is, when it has lost electrons), we say that it has been oxidised. *Loss of electrons by a substance is called* **oxidation**.

The term 'oxidation' is used because the first reactions of this sort to be studied thoroughly were reactions with oxygen. Many metals react directly with O_2 in air to form metal oxides. In these reactions the metal loses electrons to oxygen, forming an ionic compound of the metal ion and oxide ion. For example, when calcium metal is exposed to air, the bright metallic surface of the metal tarnishes as CaO forms:

$$2\,Ca(s) + O_2(g) \longrightarrow 2\,CaO(g) \qquad [4.23]$$

As Ca is oxidised in Equation 4.23, oxygen is transformed from neutral O_2 to two O^{2-} ions (▼ FIGURE 4.13). When an atom, ion or molecule has become more negatively charged (gained electrons), we say that it is reduced. *Gain of electrons by a substance is called* **reduction**. When one reactant loses electrons, another reactant must gain them: *the oxidation of one substance is always accompanied by the reduction of another as electrons are transferred between them.*

Ca(s) is oxidised O_2(g) is reduced Ca^{2+} and O^{2-} ions
(loses electrons)   (gains electrons)   combine to form CaO(s)

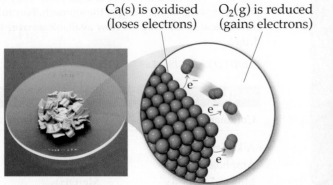

Reactants
$2\,Ca(s) + O_2(g)$

Products
$2\,CaO(s)$

▲ **FIGURE 4.13 Oxidation of calcium metal by molecular oxygen.** The oxidation involves transfer of electrons from the calcium metal to the O_2, leading to the formation of CaO.

Oxidation Numbers

Before we can properly identify an oxidation–reduction reaction, we must have a kind of bookkeeping system, a way of keeping track of the electrons gained by the substance reduced and those lost by the substance oxidised. The concept of oxidation numbers (also called *oxidation states*) was devised as a way of doing this. Each atom in a neutral molecule or charged species is assigned an **oxidation number**. For monatomic ions, the oxidation number is the same as the charge on the ion. For neutral molecules and polyatomic ions, the oxidation number of a given atom is a *hypothetical charge* assigned to the atom, by artificially dividing up the electrons among the atoms in the molecule or ion. This is done by assuming that the electrons in a bond are *completely* held by one atom or the other.

We use the following rules for assigning oxidation numbers.

1. *For an atom in its **elemental form**, the oxidation number is always zero.* Thus each H atom in the H_2 molecule has an oxidation number of 0, and each P atom in the P_4 molecule has an oxidation number of 0.

2. *For any **monatomic ion** the oxidation number equals the charge on the ion.* Thus K^+ has an oxidation number of +1, S^{2-} has an oxidation number of −2, and so forth. The alkali metal ions (group 1) always have a 1+ charge, and therefore the alkali metals always have an oxidation number of +1 in their compounds. Similarly, the alkaline earth metals (group 2) are always +2, and aluminium (group 3) is always +3 in ionic compounds. (In writing oxidation numbers, we will write the sign before the number to distinguish them from the actual electronic charges, which we write with the number first.)

3. ***Non-metals*** usually have negative oxidation numbers, although they can sometimes be positive.

   (a) *The oxidation number of **oxygen** is usually* −2 in both ionic and molecular compounds. The major exception is in compounds called peroxides, which contain the O_2^{2-} ion, giving each oxygen an oxidation number of −1.

   (b) *The oxidation number of **hydrogen** is* +1 *when bonded to non-metals and* −1 *when bonded to metals.*

   (c) *The oxidation number of **fluorine** is −1 in all compounds. The other halogens have an oxidation number of −1 in most binary compounds. When combined with oxygen, as in oxyanions, however, they have positive oxidation numbers.*

4. ***The sum of the oxidation numbers*** *of all atoms in a neutral compound is zero.* In hydrogen peroxide, H_2O_2, the oxidation number of each hydrogen is +1 and that of each oxygen is −1. The sum of the oxidation numbers is $2(+1) + 2(-1) = 0$.

5. ***The sum of the oxidation numbers*** *of all atoms in a polyatomic ion equals the charge of the ion.* For example, in the hydronium ion, H_3O^+, the oxidation number of each hydrogen is +1 and that of oxygen is +2. Thus the sum of the oxidation numbers is $3(+1) + (-2) = +1$, which equals the net charge of the ion.

The oxidation numbers of certain atoms change in an oxidation–reduction reaction. The oxidation number increases for any atom that is oxidised and decreases for any atom that is reduced.

Rules 4 and 5 are very useful in obtaining the oxidation number of one atom in a compound or ion if you know the oxidation numbers of the other atoms, as illustrated in Sample Exercise 4.6.

CONCEPT CHECK 6

   a. What noble gas element has the same number of electrons as the fluoride ion?
   b. What is the oxidation number of that species?

SAMPLE EXERCISE 4.6 | Determining oxidation numbers

Determine the oxidation number of sulfur in each of the following: **(a)** H_2S, **(b)** S_8, **(c)** SCl_2, **(d)** Na_2SO_3, **(e)** SO_4^{2-}.

SOLUTION

Analyse We are asked to determine the oxidation number of sulfur in two molecular species, in the elemental form, and in two substances containing ions.

Plan In each species the sum of oxidation numbers of all the atoms must equal the charge on the species. We use the rules outlined above to assign oxidation numbers.

Solve

(a) When bonded to a non-metal, hydrogen has an oxidation number of +1 (rule 3b). Because the H_2S molecule is neutral, the sum of the oxidation numbers must equal zero (rule 4). Letting x equal the oxidation number of S, we have $2(+1) + x = 0$. Thus S has an oxidation number of −2.

(b) Because this is an elemental form of sulfur, the oxidation number of S is 0 (rule 1).

(c) Because this is a binary compound, we expect chlorine to have an oxidation number of −1 (rule 3c). The sum of the oxidation numbers must equal zero (rule 4). Letting x equal the oxidation number of S, we have $x + 2(-1) = 0$. Consequently, the oxidation number of S must be +2.

(d) Sodium, an alkali metal, always has an oxidation number of +1 in its compounds (rule 2). Oxygen has a common oxidation state of −2 (rule 3a). Letting x equal the oxidation number of S, we have $2(+1) + x + 3(-2) = 0$. Therefore, the oxidation number of S in this compound is +4.

(e) The oxidation state of O is −2 (rule 3a). The sum of the oxidation numbers equals −2 the net charge of the SO_4^{2-} ion (rule 4). Thus, we have $x + 4(-2) = -2$. From this relation we conclude that the oxidation number of S in this ion is +6.

Comment These examples illustrate that the oxidation number of a given element depends on the compound in which it occurs. The oxidation numbers of sulfur, as seen in these examples, range from −2 to +6.

PRACTICE EXERCISE

What is the oxidation number of the element in bold in each of the following: **(a)** P_2O_5, **(b)** Na**H**, **(c)** $Cr_2O_7^{2-}$, **(d)** $SnBr_4$, **(e)** BaO_2?

Answers: **(a)** +5, **(b)** −1, **(c)** +6, **(d)** +4, **(e)** −1

(See also Exercises 4.37, 4.38.)

Oxidation of Metals by Acids and Salts

The reaction of a metal with either an acid or a metal salt conforms to the following general pattern:

$$A + BX \longrightarrow AX + B \qquad [4.24]$$

Examples:

$$Zn(s) + 2\,HBr(aq) \longrightarrow ZnBr_2(aq) + H_2(g)$$

$$Mn(s) + Pb(NO_3)_2(aq) \longrightarrow Mn(NO_3)_2(aq) + Pb(s)$$

These reactions are called **displacement reactions** because the ion in solution is displaced (replaced) through oxidation of an element.

Many metals undergo displacement reactions with acids, producing salts and hydrogen gas. For example, magnesium metal reacts with hydrochloric acid to form magnesium chloride and hydrogen gas (▶ **FIGURE 4.14**). To show that oxidation and reduction have occurred, the oxidation number for each atom is shown below the chemical equation for this reaction:

$$Mg(s) + 2\,HCl(aq) \longrightarrow MgCl_2(aq) + H_2(g) \qquad [4.25]$$

$$\begin{array}{ccccc} 0 & +1 & -1 & +2 & -1 & 0 \end{array}$$

Notice that the oxidation number of Mg changes from 0 to +2. The increase in the oxidation number indicates that the atom has lost electrons and has

$$\underset{\substack{\text{Oxidation} \\ \text{number}}}{\underbrace{\text{2 HCl(aq)}}_{+1 \ -1} + \underset{0}{\text{Mg(s)}}} \quad\longrightarrow\quad \underset{0}{\text{H}_2\text{(g)}} + \underset{+2 \ -1}{\text{MgCl}_2\text{(aq)}}$$

Reactants — *Products*

▲ **FIGURE 4.14 Reaction of magnesium metal with hydrochloric acid.** The metal is readily oxidised by the acid, producing hydrogen gas, H_2(g), and $MgCl_2$(aq).

therefore been oxidised. The H^+ ion of the acid decreases in oxidation number from +1 to 0, indicating that this ion has gained electrons and has therefore been reduced. The oxidation number of the Cl^- ion remains −1, so it is a spectator ion in the reaction. The net ionic equation is as follows:

$$\text{Mg(s)} + 2\,\text{H}^+\text{(aq)} \longrightarrow \text{Mg}^{2+}\text{(aq)} + \text{H}_2\text{(g)} \qquad [4.26]$$

Metals can also be oxidised by aqueous solutions of various salts. Iron metal, for example, is oxidised to Fe^{2+} by aqueous solutions of Ni^{2+} (such as $Ni(NO_3)_2$(aq)):

Molecular equation: $\quad \text{Fe(s)} + \text{Ni(NO}_3)_2\text{(aq)} \longrightarrow \text{Fe(NO}_3)_2\text{(aq)} + \text{Ni(s)} \quad [4.27]$

Net ionic equation: $\quad \text{Fe(s)} + \text{Ni}^{2+}\text{(aq)} \longrightarrow \text{Fe}^{2+}\text{(aq)} + \text{Ni(s)} \quad [4.28]$

The oxidation of Fe to form Fe^{2+} in this reaction is accompanied by the reduction of Ni^{2+} to Ni. Remember: *whenever one substance is oxidised, some other substance must be reduced.*

SAMPLE EXERCISE 4.7 | **Writing equations for oxidation–reduction reactions**

Write the balanced molecular and net ionic equations for the reaction of aluminium with hydrobromic acid.

SOLUTION

Analyse We must write two equations—molecular and net ionic—for the redox reaction between a metal and an acid.

Plan Metals react with acids to form salts and H_2 gas. To write the balanced equations, we must write the chemical formulae for the two reactants and then determine the formula of the salt. The salt is composed of the cation formed by the metal and the anion of the acid.

Solve The formulae of the given reactants are Al and HBr. The cation formed by Al is Al^{3+} and the anion from hydrobromic acid is Br^-. Thus the salt formed in the reaction is $AlBr_3$. Writing the reactants and products and then balancing the equation gives this molecular equation:

$$2\,Al(s) + 6\,HBr(aq) \longrightarrow 2\,AlBr_3(aq) + 3\,H_2(g)$$

Both HBr and $AlBr_3$ are soluble strong electrolytes. Thus the complete ionic equation is

$$2\,Al(s) + 6\,H^+(aq) + 6\,Br^-(aq) \longrightarrow 2\,Al^{3+}(aq) + 6\,Br^-(aq) + 3\,H_2(g)$$

Because Br^- is a spectator ion, the net ionic equation is

$$2\,Al(s) + 6\,H^+(aq) \longrightarrow 2\,Al^{3+}(aq) + 3\,H_2(g)$$

Comment The substance oxidised is the aluminium metal because its oxidation state changes from 0 in the metal to +3 in the cation, thereby increasing in oxidation number. The H^+ is reduced because its oxidation state changes from +1 in the acid to 0 in H_2.

PRACTICE EXERCISE

(a) Write the balanced molecular and net ionic equations for the reaction between magnesium and cobalt (II) sulfate. **(b)** What is oxidised and what is reduced in the reaction?

Answers: **(a)** $Mg(s) + CoSO_4(aq) \rightarrow MgSO_4(aq) + Co(s)$,
$Mg(s) + Co^{2+}(aq) \rightarrow Mg^{2+}(aq) + Co(s)$; **(b)** Mg is oxidised and Co^{2+} is reduced.
(See also Exercise 4.40.)

The Activity Series

Can we predict whether a certain metal will be oxidised either by an acid or by a particular salt? This question is of practical importance as well as chemical interest. According to Equation 4.27, for example, it would be unwise to store a solution of nickel nitrate in an iron container because the solution would dissolve the container. When a metal is oxidised, it appears to be eaten away as it reacts to form various compounds. Extensive oxidation can lead to the failure of metal machinery parts or the deterioration of metal structures.

Different metals vary in the ease with which they are oxidised. Zn is oxidised by aqueous solutions of Cu^{2+}, for example, but Ag is not. Zn, therefore, loses electrons more readily than Ag; that is, Zn is easier to oxidise than Ag.

A list of metals arranged in order of decreasing ease of oxidation is called an **activity series**. ▶ TABLE 4.4 gives the activity series in aqueous solution for many of the most common metals. Hydrogen is also included in the table. The metals at the top of the table, such as the alkali metals and the alkaline earth metals, are most easily oxidised; that is, they react most readily to form compounds. They are called the *active metals*. The metals at the bottom of the activity series, such as the transition elements from groups 8–11, are very stable and form compounds less readily. The metals of periods 5 and 6 of these groups, which are used to make coins and jewellery, are called *noble metals* because of their low reactivity.

The activity series can be used to predict the outcome of reactions between metals and either metal salts or acids. *Any metal on the list can be oxidised by the ions of elements below it.* For example, copper is above silver in the series. Thus copper metal will be oxidised by silver ions.

$$Cu(s) + 2\,Ag^+(aq) \longrightarrow Cu^{2+}(aq) + 2\,Ag(s) \qquad [4.29]$$

The oxidation of copper to copper ions is accompanied by the reduction of silver ions to silver metal.

TABLE 4.4 • **Activity series of metals in aqueous solution**

Metal	Oxidation reaction
Lithium	$Li(s) \longrightarrow Li^+(aq) + e^-$
Potassium	$K(s) \longrightarrow K^+(aq) + e^-$
Barium	$Ba(s) \longrightarrow Ba^{2+}(aq) + 2e^-$
Calcium	$Ca(s) \longrightarrow Ca^{2+}(aq) + 2e^-$
Sodium	$Na(s) \longrightarrow Na^+(aq) + e^-$
Magnesium	$Mg(s) \longrightarrow Mg^{2+}(aq) + 2e^-$
Aluminium	$Al(s) \longrightarrow Al^{3+}(aq) + 3e^-$
Manganese	$Mn(s) \longrightarrow Mn^{2+}(aq) + 2e^-$
Zinc	$Zn(s) \longrightarrow Zn^{2+}(aq) + 2e^-$
Chromium	$Cr(s) \longrightarrow Cr^{3+}(aq) + 3e^-$
Iron	$Fe(s) \longrightarrow Fe^{2+}(aq) + 2e^-$
Cobalt	$Co(s) \longrightarrow Co^{2+}(aq) + 2e^-$
Nickel	$Ni(s) \longrightarrow Ni^{2+}(aq) + 2e^-$
Tin	$Sn(s) \longrightarrow Sn^{2+}(aq) + 2e^-$
Lead	$Pb(s) \longrightarrow Pb^{2+}(aq) + 2e^-$
Hydrogen	$H_2(g) \longrightarrow 2H^+(aq) + 2e^-$
Copper	$Cu(s) \longrightarrow Cu^{2+}(aq) + 2e^-$
Silver	$Ag(s) \longrightarrow Ag^+(aq) + e^-$
Mercury	$Hg(l) \longrightarrow Hg^{2+}(aq) + 2e^-$
Platinum	$Pt(s) \longrightarrow Pt^{2+}(aq) + 2e^-$
Gold	$Au(s) \longrightarrow Au^{3+}(aq) + 3e^-$

Ease of oxidation increases ↑

CONCEPT CHECK 7

Does a reaction occur
a. when an aqueous solution of $NiCl_2(aq)$ is added to a test tube containing strips of metallic zinc?
b. when $NiCl_2(aq)$ is added to a test tube containing $Zn(NO_3)_2(aq)$?

Only those metals above hydrogen in the activity series are able to react with acids to form H_2. For example, Ni reacts with HCl(aq) to form H_2:

$$Ni(s) + 2HCl(aq) \longrightarrow NiCl_2(aq) + H_2(g) \qquad [4.30]$$

Because elements below hydrogen in the activity series are not oxidised by H^+, Cu does not react with HCl(aq). Note that copper does react with nitric acid. This reaction, however, is not a simple oxidation of Cu by the H^+ ions of the acid. Instead, the metal is oxidised to Cu^{2+} by the nitrate ion of the acid, accompanied by the formation of brown nitrogen dioxide, $NO_2(g)$:

$$Cu(s) + 4HNO_3(aq) \longrightarrow Cu(NO_3)_2(aq) + 2H_2O(l) + 2NO_2(g) \quad [4.31]$$

As the copper is oxidised in this reaction, the NO_3^- where the oxidation number of nitrogen is +5, is reduced to NO_2, where the oxidation number is +4.

We will examine reactions of this type in Chapter 19.

SAMPLE EXERCISE 4.8 | **Determining when an oxidation–reduction reaction can occur**

Will an aqueous solution of iron(II) chloride oxidise magnesium metal? If so, write the balanced molecular and net ionic equations for the reaction.

SOLUTION

Analyse We are given two substances—an aqueous salt, $FeCl_2$, and a metal, Mg—and asked if they react with each other.

Plan A reaction occurs if the reactant that is a metal in its elemental form (Mg) is located above the reactant that is an ion in its oxidised form (Fe^{2+}) in Table 4.4. If the reaction occurs, the Fe^{2+} ion in $FeCl_2$ is reduced to Fe, and the Mg is oxidised to Mg^{2+}.

Solve Because Mg is above Fe in Table 4.4, the reaction will occur. To write the formula for the salt that is produced in the reaction, we must remember the charges on common ions. Magnesium is always present in compounds as Mg^{2+}; the chloride ion is Cl^-. The magnesium salt formed in the reaction is $MgCl_2$, meaning the balanced molecular equation is

$$Mg(s) + FeCl_2(aq) \longrightarrow MgCl_2(aq) + Fe(s)$$

Both $FeCl_2$ and $MgCl_2$ are soluble strong electrolytes and can be written in ionic form. Cl^-, then, is a spectator ion in the reaction. The net ionic equation is

$$Mg(s) + Fe^{2+}(aq) \longrightarrow Mg^{2+}(aq) + Fe(s)$$

The net ionic equation shows that Mg is oxidised and Fe^{2+} is reduced in this reaction.

Check Note that the net ionic equation is balanced with respect to both charge and mass.

PRACTICE EXERCISE

Which of the following metals will be oxidised by $Pb(NO_3)_2$: Zn, Cu, Fe?

Answer: Zn and Fe

(See also Exercises 4.41, 4.42.)

A CLOSER LOOK

THE AURA OF GOLD

Gold has been known since the earliest records of human existence. Throughout history people have cherished gold, fought for it and died for it.

The physical and chemical properties of gold serve to make it a special metal. First, its intrinsic beauty and rarity make it precious (▶ FIGURE 4.15). Second, gold is soft and can be easily formed into artistic objects, jewellery and coins. Third, gold is one of the least active metals (Table 4.4). It is not oxidised in air and does not react with water. It is unreactive towards basic solutions and nearly all acidic solutions. As a result, gold can be found in nature as a pure element rather than combined with oxygen or other elements, which accounts for its early discovery.

Many of the early studies of the reactions of gold arose from the practice of alchemy, in which people attempted to turn cheap metals, such as lead, into gold. Alchemists discovered that gold can be dissolved in a 3:1 mixture of concentrated hydrochloric and nitric acids, known as aqua regia ('royal water'). The action of nitric acid on gold is similar to that on copper (Equation 4.32) in that the nitrate ion, rather than H^+, oxidises the metal to Au^{3+}. The Cl^- ions interact with Au^{3+} to form highly stable $AuCl_4^-$ ions. The net ionic equation for the reaction of gold with aqua regia is

$$Au(s) + NO_3^-(aq) + 4\,H^+(aq) + 4\,Cl^-(aq) \longrightarrow$$
$$AuCl_4^-(aq) + 2\,H_2O(l) + NO(g) \qquad [4.32]$$

All the gold ever mined would easily fit into a cube 21 m on a side, weighing about 1.5×10^8 kg (150 000 tonnes). Worldwide production of gold amounts to about 2.5×10^6 kg (2500 tonnes). By contrast, about 3.0×10^{10} kg (30 million tonnes) of

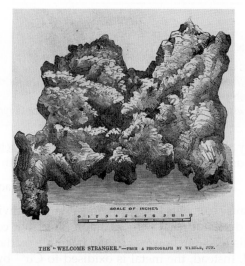

▲ **FIGURE 4.15** **A 69 kg gold nugget, the 'Welcome Stranger', found at Moliagul, Victoria in 1869.**

aluminium are produced annually. Gold is used mainly in jewellery, coins and electronics. Its use in electronics relies on its excellent conductivity and its corrosion resistance. Gold is used, for example, to plate contacts in electrical switches, relays and connections. Gold is also used in computers and other microelectronic devices where fine gold wire is used to link components.

Because of its resistance to corrosion by acids and other substances found in saliva, gold is an ideal metal for dental crowns and caps. The pure metal is too soft to use in dentistry, so it is combined with other metals to form alloys.

4.5 | CONCENTRATIONS OF SOLUTIONS

The behaviour of solutions often depends not only on the nature of the solutes but also on their concentrations. Scientists use the term **concentration** to designate the amount of solute dissolved in a given quantity of solvent or quantity of solution. The concept of concentration is intuitive: the greater the amount of solute dissolved in a certain amount of solvent, the more concentrated the resulting solution. In chemistry we often need to express the concentrations of solutions quantitatively.

Molarity

Molarity (M) expresses the concentration of a solution as the number of moles of solute in a litre of solution:

$$\text{Molarity (M)} = \frac{\text{moles of solute}}{\text{volume of solution in litres}} = \frac{n}{V} \qquad [4.33]$$

A 1.00 molar solution (written 1.00 M) contains 1.00 mol of solute in every litre of solution. ▼ **FIGURE 4.16** shows the preparation of 250.0 cm³ of a 1.00 M solution of $CuSO_4$ by using a volumetric flask that is calibrated to hold exactly 250.0 cm³. First, 0.250 mol of $CuSO_4$ (39.9 g) is weighed out and placed in the volumetric flask. Water is added to dissolve the salt, and the resultant solution is diluted to a total volume of 250.0 cm³. The molarity of the solution is

$$\frac{(0.250 \text{ mol CuSO}_4)}{(0.250 \text{ dm}^3 \text{ soln})} = 1.00 \text{ M}$$

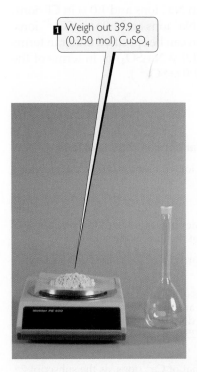 **1** Weigh out 39.9 g (0.250 mol) $CuSO_4$

 **2** Put $CuSO_4$ (solute) into 250 cm³ volumetric flask; add water and swirl to dissolve solute

 **3** Add water until solution just reaches calibration mark on neck of flask

▲ **FIGURE 4.16** Preparing 0.250 dm³ of a 1.00 M solution of $CuSO_4$.

SAMPLE EXERCISE 4.9   **Calculating molarity**

Calculate the molarity of a solution made by dissolving 23.4 g of sodium sulfate (Na_2SO_4) in enough water to form 125 cm^3 of solution.

SOLUTION

Analyse We are given the number of grams of solute (23.4 g), its chemical formula (Na_2SO_4) and the volume of the solution (125 cm^3) and asked to calculate the molarity of the solution.

Plan We can calculate molarity using Equation 4.33. To do so, we must convert the grams of solute to moles and the volume of the solution from millilitres to litres.

Solve The number of moles of Na_2SO_4 is obtained by dividing its mass by its molar mass:

$$\text{Moles } Na_2SO_4 = \frac{23.4 \text{ g}}{142 \text{ g mol}^{-1}} = 0.165 \text{ mol}$$

Converting the volume of the solution to litres:

$$\text{Litres solution} = \left(\frac{125}{1000}\right) dm^3 = 0.125 \text{ dm}^3$$

Thus the molarity is

$$\text{Molarity} = \frac{0.165 \text{ mol}}{0.125 \text{ dm}^3} = 1.32 \frac{\text{mol}}{\text{dm}^3} = 1.32 \text{ M}$$

PRACTICE EXERCISE

Calculate the molarity of a solution made by dissolving 5.00 g of glucose ($C_6H_{12}O_6$) in sufficient water to form exactly 100 cm^3 of solution.

Answer: 0.278 M

(See also Exercises 4.45, 4.46.)

Expressing the Concentration of an Electrolyte

When an ionic compound dissolves the relative concentrations of the ions introduced into the solution depend on the chemical formula of the compound. For example, a 1.0 M solution of NaCl is 1.0 M in Na^+ ions and 1.0 M in Cl^- ions. Similarly, a 1.0 M solution of Na_2SO_4 is 2.0 M in Na^+ ions and 1.0 M in SO_4^{2-} ions. Thus the concentration of an electrolyte solution can be specified either in terms of the compound used to make the solution (1.0 M Na_2SO_4) or in terms of the ions that the solution contains (2.0 M Na^+ and 1.0 M SO_4^{2-}).

SAMPLE EXERCISE 4.10   **Calculating molar concentrations of ions**

What are the molar concentrations of each of the ions present in a 0.025 M aqueous solution of calcium nitrate?

SOLUTION

Analyse We are given the concentration of the ionic compound used to make the solution and asked to determine the concentrations of the ions in the solution.

Plan We can use the subscripts in the chemical formula of the compound to determine the relative ion concentrations.

Solve Calcium nitrate is composed of calcium ions (Ca^{2+}) and nitrate ions (NO_3^-), so its chemical formula is $Ca(NO_3)_2$. Because there are two NO_3^- ions for each Ca^{2+} ion in the compound, each mole of $Ca(NO_3)_2$ that dissolves dissociates into 1 mol of Ca^{2+} and 2 mol of NO_3^-. Thus a solution that is 0.025 M in $Ca(NO_3)_2$ is 0.025 M in Ca^{2+} and 2 × 0.025 M = 0.050 M in NO_3^-.

Check The concentration of NO_3^- ions is twice that of Ca^{2+} ions, as the subscript 2 after the NO_3^- in the chemical formula $Ca(NO_3)_2$ suggests it should be.

PRACTICE EXERCISE

What is the molar concentration of K^+ ions in a 0.015 M solution of potassium carbonate?

Answer: 0.030 M K^+

(See also Exercises 4.50, 4.51.)

Interconverting Molarity, Moles and Volume

The definition of molarity (Equation 4.33) contains three quantities—molarity, moles of solute and litres of solution. If we know any two of these, we can calculate the third. For example, if we know the molarity of a solution, we can calculate the number of moles of solute in a given volume. Molarity, therefore, is a conversion factor between volume of solution and moles of solute. Calculation of the number of moles of HNO_3 in 2.0 dm^3 of 0.200 M HNO_3 solution illustrates the conversion of volume to moles:

$$\text{M} = \frac{n}{V}$$

$$n = \text{M}V$$

$$\text{Moles HNO}_3 = 0.200 \text{ mol dm}^{-3} \times 2.0 \text{ dm}^3$$

$$= 0.40 \text{ mol}$$

To illustrate the conversion of moles to volume, let's calculate the volume of 0.30 M HNO_3 solution required to supply 2.0 mol of HNO_3:

$$\text{M} = \frac{n}{V}$$

$$V = \frac{n}{\text{M}}$$

$$\text{Volume of solution (dm}^3) = \frac{2.0 \text{ mol}}{0.30 \text{ mol dm}^{-3}} = 6.7 \text{ dm}^3$$

SAMPLE EXERCISE 4.11 Using molarity to calculate grams of solute

How many grams of Na_2SO_4 are required to make 0.350 dm^3 of 0.500 M Na_2SO_4?

SOLUTION

Analyse We are given the volume of the solution (0.350 dm^3), its concentration (0.500 M), and the identity of the solute Na_2SO_4 and asked to calculate the number of grams of the solute in the solution.

Plan We can use the definition of molarity (Equation 4.33) to determine the number of moles of solute, and then convert moles to grams using the molar mass of the solute.

$$\text{M}_{Na_2SO_4} = \frac{\text{moles Na}_2\text{SO}_4}{V_{Na_2SO_4}}$$

Solve Calculating the moles of Na_2SO_4 gives:

$$\text{Moles Na}_2\text{SO}_4 = \text{M}_{Na_2SO_4} \times V_{Na_2SO_4}$$

$$= 0.500 \text{ mol dm}^{-3} \times 0.350 \text{ dm}^3$$

$$= 0.175 \text{ mol}$$

Because each mole of Na_2SO_4 weighs 142 g, and $m = n \times$ molar mass

$$\text{Grams Na}_2\text{SO}_4 = 0.175 \text{ mol} \times 142 \text{ g mol}^{-1} = 24.9 \text{ g Na}_2\text{SO}_4$$

Check The magnitude of the answer, the units and the number of significant figures are all appropriate.

PRACTICE EXERCISE

(a) How many grams of Na_2SO_4 are there in 15 cm^3 of 0.50 M Na_2SO_4? **(b)** How many millilitres of 0.50 M Na_2SO_4 solution are needed to provide 0.038 mol of this salt?

Answers: **(a)** 1.1 g, **(b)** 76 cm^3

(See also Exercise 4.49.)

Dilution

Solutions that are used routinely in the laboratory are often purchased or prepared in concentrated form (called *stock solutions*). Hydrochloric acid, for example, is purchased as a 12 M solution (concentrated HCl). Solutions of lower concentrations can then be obtained by adding water, a process called **dilution**.[*]

To illustrate the preparation of a dilute solution from a concentrated one, suppose we wanted to prepare 250.0 cm^3 (that is, 0.250 dm^3) of 0.100 M CuSO$_4$ solution by diluting a stock solution containing 1.00 M CuSO$_4$. When solvent is added to dilute a solution, the number of moles of solute remains unchanged.

$$\text{Moles solute before dilution} = \text{moles solute after dilution} \qquad [4.34]$$

MY WORLD OF CHEMISTRY

DRINKING TOO MUCH WATER CAN KILL YOU

For a long time it was thought that dehydration was a potential danger for people engaged in extended vigorous activity. Athletes were encouraged to drink lots of water while engaged in active sport. The trend towards extensive hydration has spread throughout society; many people carry water bottles everywhere and dutifully keep well hydrated.

It turns out, though, that in some circumstances drinking too much water is a greater danger than not drinking enough. Excess water consumption can lead to hyponatraemia, a condition in which the concentration of sodium ion in the blood is too low. In the past decade at least four marathon runners have died from hyponatraemia-related trauma, and dozens more have become seriously ill. For example, in 2003 a first-time marathon runner collapsed near kilometre 35 and died the next day. Death was attributed to hyponatraemia-induced brain swelling, the result of drinking too much water before and during the race.

The normal blood sodium level is 135–145 mM. When that level drops to as low as 125 mM, dizziness and confusion set in. A concentration below 120 mM can be critical. Low sodium level in the blood causes brain tissue to swell. Dangerously low levels can occur in a marathon runner or other active athlete who is sweating out salt at the same time that excessive salt-free water is being drunk to compensate for water loss. The condition affects women more than men because of their generally different body composition and patterns of metabolism. Drinking a sport drink that contains some electrolytes helps to prevent hyponatraemia (▶ FIGURE 4.17).

Contrary to popular belief, dehydration is not as likely to present a life-threatening situation as overhydration, though it can contribute to heat stroke when the temperature is high.

Athletes frequently lose several kilograms in the course of extreme workouts, all in the form of water loss, with no lasting adverse effects. Runners typically lose up to 6% of their body weight during a marathon. Weight losses of this magnitude are typical of elite marathon runners, who produce tremendous amounts of heat and sweat and can't afford to slow down for much drinking.

RELATED EXERCISES: 4.47, 4.48

▲ FIGURE 4.17   **Water stations.** To help prevent overhydration the number of water stations has been reduced in many marathon events.

---

[*]In diluting a concentrated acid or base, the acid or base should be added to water and then further diluted by adding more water. Adding water directly to concentrated acid or base can cause spattering because of the intense heat generated.

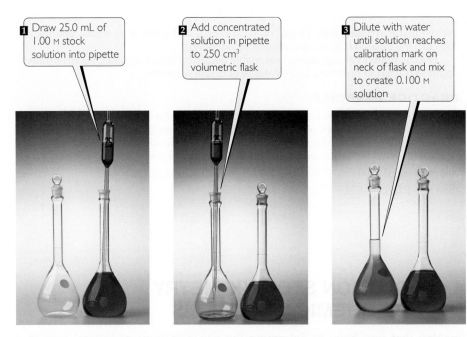

1 Draw 25.0 mL of 1.00 M stock solution into pipette

2 Add concentrated solution in pipette to 250 cm³ volumetric flask

3 Dilute with water until solution reaches calibration mark on neck of flask and mix to create 0.100 M solution

▲ **FIGURE 4.18** **Procedure for preparing 250 dm³ of 0.100 M CuSO₄ by dilution of 1.00 M CuSO₄.** (1) Draw 25.0 cm³ of the 1.00 M solution into a pipette. (2) Add this to a 250 cm³ volumetric flask. (3) Add water to dilute the solution to a total volume of 250 cm³.

But we know that $n = MV$, so

$$M_{conc} \times V_{conc} = M_{dil} \times V_{dil} \qquad [4.35]$$

$$V_{conc} = \frac{M_{dil} \times V_{dil}}{M_{conc}}$$

$$= \frac{0.100 \text{ mol dm}^{-3} \times 0.250 \text{ dm}^3}{1.00 \text{ mol dm}^{-3}}$$

$$= 0.0250 \text{ dm}^3$$

So this dilution is achieved by withdrawing 0.0250 dm³ (that is, 25.0 cm³) of the 1.00 M solution using a pipette, adding it to a 250 cm³ volumetric flask and then diluting it to a final volume of 250.0 cm³, as shown in ▲ **FIGURE 4.18**.

 CONCEPT CHECK 8

How is the molarity of a 0.50 M KBr solution changed when water is added to double its volume?

SAMPLE EXERCISE 4.12   **Preparing a solution by dilution**

How many millilitres of 3.0 M H_2SO_4 are needed to make 450 cm³ of 0.10 M H_2SO_4?

SOLUTION

Analyse We need to dilute a concentrated solution. We are given the molarity of a more concentrated solution (3.0 M) and the volume and molarity of a more dilute one containing the same solute (450 cm³ of 0.10 M solution). We must calculate the volume of the concentrated solution needed to prepare the dilute solution.

Plan We can directly apply Equation 4.35.

Solve

$$(3.0 \text{ M})(V_{conc}) = (0.10 \text{ M})(450 \text{ cm}^3)$$

$$V_{conc} = \frac{(0.10 \text{ M})(450 \text{ cm}^3)}{3.0 \text{ M}} = 15 \text{ cm}^3$$

We see that if we start with 15 cm³ of 3.0 M H_2SO_4 and dilute it to a total volume of 450 cm³, the desired 0.10 M solution will be obtained.

The calculated volume seems reasonable because a small volume of concentrated solution is used to prepare a large volume of dilute solution.

PRACTICE EXERCISE

(a) What volume of 2.50 M lead(II) nitrate solution contains 0.0500 mol of Pb^{2+}?
(b) How many millilitres of 5.0 M $K_2Cr_2O_7$ solution must be diluted to prepare 250 cm³ of 0.10 M solution? **(c)** If 10.0 cm³ of a 10.0 M stock solution of NaOH is diluted to 250 cm³, what is the concentration of the resulting stock solution?

Answers: **(a)** 0.0200 dm³ = 20.0 cm³ **(b)** 5.0 cm³, **(c)** 0.40 M

(See also Exercise 4.54.)

4.6 | SOLUTION STOICHIOMETRY AND CHEMICAL ANALYSIS

Imagine that you have to determine the concentrations of several ions in a sample of lake water. In Chapter 3 we learned that if you know the chemical equation and the amount of one reactant consumed in the reaction, you can calculate the quantities of other reactants and products. In this section we briefly explore such analyses of solutions. Although many instrumental methods have been developed for such analyses, chemical reactions such as those discussed in this chapter continue to be used.

Recall that the coefficients in a balanced equation give the relative number of moles of reactants and products (∞ Section 3.6, 'Quantitative Information from Balanced Chemical Equations').

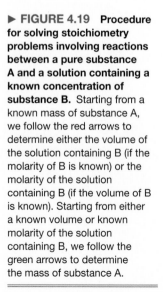

∞ Review this on page 88

To use this information, we must convert the quantities of substances involved in a reaction into moles. When we are dealing with grams of substances, as we were in Chapter 3, we use the molar mass to achieve this conversion. When we are working with solutions of known molarity, however, we use molarity and volume to determine the number of moles (moles solute = M × V). ▼ FIGURE 4.19 summarises this approach to using stoichiometry.

▶ **FIGURE 4.19** **Procedure for solving stoichiometry problems involving reactions between a pure substance A and a solution containing a known concentration of substance B.** Starting from a known mass of substance A, we follow the red arrows to determine either the volume of the solution containing B (if the molarity of B is known) or the molarity of the solution containing B (if the volume of B is known). Starting from either a known volume or known molarity of the solution containing B, we follow the green arrows to determine the mass of substance A.

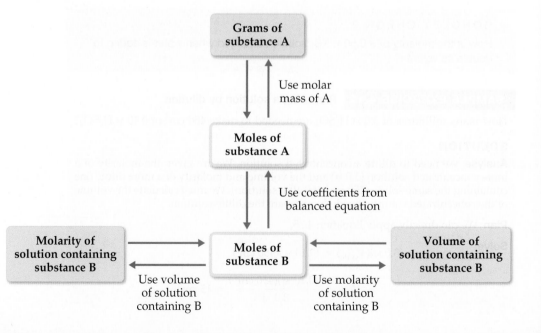

SAMPLE EXERCISE 4.13   **Using mass relations in a neutralisation reaction**

How many grams of $Ca(OH)_2$ are needed to neutralise 25.0 cm^3 of 0.100 M HNO_3?

SOLUTION

Analyse The reactants are an acid, HNO_3, and a base, $Ca(OH)_2$. The volume and molarity of HNO_3 are given, and we are asked how many grams of $Ca(OH)_2$ are needed to neutralise this quantity of HNO_3.

Plan We can use the molarity and volume of the HNO_3 solution to calculate the number of moles of HNO_3. We then use the balanced equation for the neutralisation reaction to relate the moles of HNO_3 to moles of $Ca(OH)_2$ Finally, we can convert moles of $Ca(OH)_2$ to grams.

Solve Because this is an acid–base neutralisation reaction, HNO_3 and $Ca(OH)_2$ react to form H_2O and the salt containing Ca^{2+} and NO_3^-:

$$2\ HNO_3(aq) + Ca(OH)_2(s) \rightarrow 2\ H_2O(l) + Ca(NO_3)_2(aq)$$

$$\frac{mol\ Ca(OH)_2}{mol\ HNO_3} = \frac{1}{2}$$

$$mol\ Ca(OH)_2 = \frac{1}{2} \times mol\ HNO_3$$

$$= \frac{1}{2}(0.100\ mol\ dm^{-3}) \times (0.0250\ dm^3) = 0.00125\ mol$$

$$Grams\ Ca(OH)_2 = 0.00125\ mol \times 74.1\ g\ mol^{-1}$$
$$= 0.0926\ g$$

Check The answer is reasonable because a small volume of dilute acid requires only a small amount of base to neutralise it.

PRACTICE EXERCISE

(a) How many grams of NaOH are needed to neutralise 20.0 cm^3 of 0.150 M H_2SO_4 solution? **(b)** How many litres of 0.500 M HCl(aq) are needed to react completely with 0.100 mol of $Pb(NO_3)_2(aq)$, forming a precipitate of $PbCl_2(s)$?

Answers: **(a)** 0.240 g, **(b)** 0.400 dm^3

(See also Exercises 4.57, 4.58.)

Titrations

To determine the concentration of a particular solute in a solution, chemists often carry out a **titration**, which involves combining a sample of the solution with a reagent solution of known concentration, called a **standard solution**. Titrations can be conducted using acid–base, precipitation or oxidation–reduction reactions. Suppose we have an HCl solution of unknown concentration and an NaOH solution we know to be 0.100 M. To determine the concentration of the HCl solution, we take a specific volume of that solution, say 20.00 cm^3. We then slowly add the standard NaOH solution to it until the neutralisation reaction between the HCl and NaOH is complete. The point at which this happens is known as the **equivalence point** of the titration.

  CONCEPT CHECK 9

25.00 cm^3 of a 0.100 M HBr solution is titrated with a 0.200 M NaOH solution. How many cm^3 of the NaOH solution are required to reach the equivalence point?

In order to titrate an unknown with a standard solution, there must be some way to determine when the equivalence point of the titration has been reached. In acid–base titrations, dyes known as acid–base **indicators** are used for this purpose. For example, the dye known as phenolphthalein is colourless in acidic solution but pink in basic solution. If we add phenolphthalein to an unknown

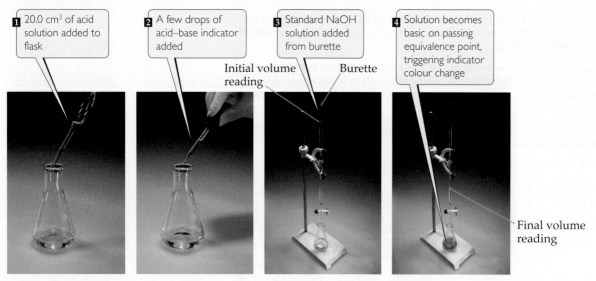

How would the volume of standard solution added change if that solution
were Ba(OH)$_2$(aq) instead of NaOH(aq) at the same concentration?

1 20.0 cm^3 of acid solution added to flask

2 A few drops of acid–base indicator added

3 Standard NaOH solution added from burette

4 Solution becomes basic on passing equivalence point, triggering indicator colour change

Initial volume reading Burette

Final volume reading

▲ **FIGURE 4.20** **Procedure for titrating an acid against a standard solution of NaOH.**
The acid–base indicator, phenolphthalein, is colourless in acidic solution but takes on a pink colour
in basic solution.

solution of acid, the solution will be colourless, as seen in ▲ FIGURE 4.20(2). We
can then add standard base from a burette until the solution barely turns from
colourless to pink, as seen in Figure 4.20(4). This colour change indicates that
the acid has been neutralised and the drop of base that caused the solution to
become coloured has no acid to react with. The solution therefore becomes
basic, and the dye turns pink. The colour change signals the *end point* of the
titration, which usually coincides very nearly with the equivalence point.
Care must be taken to choose indicators whose end points correspond to the
equivalence point of the titration. The titration procedure is summarised in
▼ FIGURE 4.21.

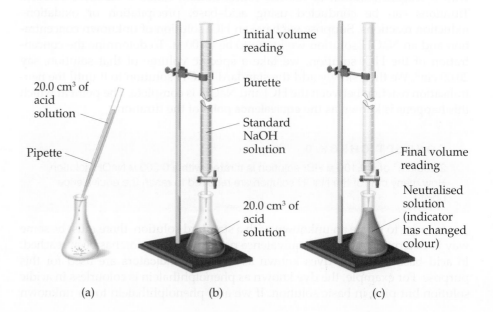

Initial volume reading

20.0 cm^3 of acid solution

Pipette

Burette

Standard NaOH solution

Final volume reading

20.0 cm^3 of acid solution

Neutralised solution (indicator has changed colour)

► **FIGURE 4.21** **Procedure for titrating an acid against a standardised solution of NaOH.** (a) A known quantity of acid is added to a flask. (b) An acid–base indicator is added, and standardised NaOH is added from a burette. (c) The equivalence point is signalled by a colour change in the indicator.

(a)   (b)   (c)

SAMPLE EXERCISE 4.14 **Determining solution concentration via an acid–base titration**

One commercial method used to peel potatoes is to soak them in a solution of NaOH for a short time, remove them from the NaOH and spray off the peel. The concentration of NaOH is normally in the range of 3 to 6 M. The NaOH is analysed periodically. In one such analysis, 45.7 cm^3 of 0.500 M H$_2$SO$_4$ is required to neutralise a 20.0 cm^3 sample of NaOH solution. What is the concentration of the NaOH solution?

SOLUTION

Analyse We are given that 45.7 cm^3 of 0.500 M H$_2$SO$_4$ is required to neutralise a 20.0 cm^3 sample of NaOH and are asked to calculate the concentration of the NaOH solution.

Plan We can use the volume and molarity of the H$_2$SO$_4$ to calculate the number of moles of this substance. Then, we can use this quantity and the balanced equation for the reaction to calculate the number of moles of NaOH. Finally, we can use the moles of NaOH and the volume of this solution to calculate molarity.

Solve The number of moles of H$_2$SO$_4$ is given by the product of the molarity and the volume of this solution:

$$\text{Moles H}_2\text{SO}_4 = 0.500 \text{ mol dm}^{-3} \times \left(\frac{45.7}{1000}\right)\text{dm}^3$$
$$= 2.28 \times 10^{-2} \text{ mol}$$

Acids react with metal hydroxides to form water and a salt. Thus the balanced equation for the neutralisation reaction is

$$\text{H}_2\text{SO}_4(\text{aq}) + 2\,\text{NaOH}(\text{aq}) \longrightarrow 2\,\text{H}_2\text{O(l)} + \text{Na}_2\text{SO}_4(\text{aq})$$

According to the balanced equation

$$\frac{\text{mol NaOH}}{\text{mol H}_2\text{SO}_4} = \frac{2}{1}$$
$$\text{Moles NaOH} = \frac{2}{1} \times 2.28 \times 10^{-2} \text{ mol}$$
$$= 4.56 \times 10^{-2} \text{ mol}$$

Knowing the number of moles of NaOH present in 20.0 cm^3 of solution allows us to calculate the molarity of this solution:

$$\text{Molarity NaOH} = \frac{n \text{ NaOH}}{V \text{ NaOH}} = \frac{4.56 \times 10^{-2} \text{ mol}}{(20/1000) \text{ dm}^3}$$

$$= 2.28 \text{ mol dm}^{-3} = 2.28 \text{ M}$$

PRACTICE EXERCISE

What is the molarity of an NaOH solution if 48.0 cm^3 is needed to neutralise 35.0 cm^3 of 0.144 M H$_2$SO$_4$?

Answer: 0.210 M

(See also Exercises 4.59, 4.60.)

SAMPLE EXERCISE 4.15 **Determining the quantity of solute by titration**

The quantity of Cl$^-$ in a municipal water supply is determined by titrating the sample with Ag$^+$. The reaction taking place during the titration is

$$\text{Ag}^+(\text{aq}) + \text{Cl}^-(\text{aq}) \longrightarrow \text{AgCl(s)}$$

The end point in this type of titration is marked by a change in colour of a special type of indicator. **(a)** How many grams of chloride ion are in a sample of the water if 20.2 cm^3 of 0.100 M Ag$^+$ is needed to react with all the chloride in the sample? **(b)** If the sample has a mass of 10.0 g, what percentage of Cl$^-$ does it contain?

SOLUTION

Analyse (a) We are given the volume and concentration of Ag$^+$ solution required to react with the chloride ions in water and are asked to determine the mass of chloride ions present. **(b)** Given the mass of the sample containing the chloride ions, we are asked to calculate the percentage of chloride in the sample.

Plan (a) We begin by using the volume and molarity of Ag^+ to calculate the number of moles of Ag^+ used in the titration. We can then use the balanced equation to determine the moles of Cl^- in the sample and from that, the grams of Cl^-:

$$\text{Moles } Ag^+ = 0.100 \text{ mol dm}^{-3} \times \left(\frac{20.2}{1000}\right) \text{dm}^3$$

$$= 2.02 \times 10^{-3} \text{ mol}$$

From the balanced equation we see that $\frac{\text{mol } Cl^-}{\text{mol } Ag^+} = \frac{1}{1}$. Using this information and the molar mass of Cl, we have

$$\text{Grams } Cl^- = \frac{1}{1} \times 2.02 \times 10^{-3} \text{ mol} \times 35.5 \text{ g mol}^{-1}$$

$$= 7.17 \times 10^{-2} \text{ g}$$

(b) To calculate the percentage of Cl^- in the sample, we compare the number of grams of Cl^- in the sample, 7.17×10^{-2} g, with the original mass of the sample, 10.0 g.

$$\text{Percentage of } Cl^- = \frac{7.17 \times 10^{-2} \text{ g}}{10.0 \text{ g}} \times 100\% = 0.717\% \text{ } Cl^-$$

PRACTICE EXERCISE

A sample of an iron ore is dissolved in acid and the iron is converted to Fe^{2+}. The sample is then titrated with 47.20 cm^3 of 0.02240 M MnO_4^- solution. The oxidation–reduction reaction that occurs during titration is:

$$MnO_4^-(aq) + 5 Fe^{2+}(aq) + 8 H^+(aq) \longrightarrow Mn^{2+}(aq) + 5 Fe^{3+}(aq) + 4 H_2O(l)$$

(a) How many moles of MnO_4^- were added to the solution? **(b)** How many moles of Fe^{2+} were in the sample? **(c)** How many grams of iron were in the sample? **(d)** If the sample had a mass of 0.8890 g, what is the percentage of iron in the sample?

Answers: **(a)** 1.057×10^{-3} mol MnO_4^-, **(b)** 5.286×10^{-3} mol Fe^{2+}, **(c)** 0.2952 g, **(d)** 33.21%

(See also Exercises 4.66, 4.67.)

SAMPLE INTEGRATIVE EXERCISE **Putting concepts together**

Note: *Integrative exercises require skills from earlier chapters as well as ones from the present chapter.*

A sample of 70.5 mg of potassium phosphate is added to 15.0 cm^3 of 0.050 M silver nitrate, resulting in the formation of a precipitate. **(a)** Write the molecular equation for the reaction. **(b)** What is the limiting reactant in the reaction? **(c)** Calculate the theoretical yield, in grams, of the precipitate that forms.

SOLUTION

(a) Potassium phosphate and silver nitrate are both ionic compounds. Potassium phosphate contains K^+ and PO_4^{3-} ions, so its chemical formula is K_3PO_4. Silver nitrate contains Ag^+ and NO_3^- ions, so its chemical formula is $AgNO_3$. Because both reactants are strong electrolytes, the solution contains K^+, PO_4^{3-}, Ag^+ and NO_3^- ions before the reaction occurs. According to the solubility guidelines in Table 4.1, Ag^+ and PO_4^{3-} form an insoluble compound, so Ag_3PO_4 will precipitate from the solution. In contrast, K^+ will remain in solution because KNO_3 is water soluble. Thus the balanced molecular equation for the reaction is

$$K_3PO_4(aq) + 3 AgNO_3(aq) \longrightarrow Ag_3PO_4(s) + 3 KNO_3(aq)$$

(b) To determine the limiting reactant, we must examine the number of moles of each reactant (∞ Section 3.7, 'Limiting Reactants').

The number of moles of K_3PO_4 is calculated from the mass of the sample using the molar mass as a conversion factor.

The molar mass of K_3PO_4 is $3(39.1) + 31.0 + 4(16.0) = 212.3$ g mol^{-1}.

Converting milligrams of K_3PO_4 to grams and then to moles, we have

$$\frac{70.5 \text{ mg}}{1000} \times \frac{1}{212.3 \text{ g mol}^{-1}} = 3.32 \times 10^{-4} \text{ mol } K_3PO_4$$

∞ Review this on page 90

We determine the number of moles of $AgNO_3$ from the volume and molarity of the solution (∞ Section 4.5, 'Concentrations of Solutions').

∞ Review this on page 125

Converting cm^3 to dm^3 and then to moles, we have

$$mol\ AgNO_3 = 0.050\ mol\ dm^{-3} \times 0.15\ dm^3 = 7.5 \times 10^{-4}\ mol$$

Comparing the molar amounts of the two reactants, we find that there are $(7.5 \times 10^{-4})/(3.32 \times 10^{-4}) = 2.3$ times as many moles of $AgNO_3$ as there are moles of K_3PO_4. According to the balanced equation, however, 1 mol K_3PO_4 requires 3 mol $AgNO_3$. Thus there is insufficient $AgNO_3$ to consume the K_3PO_4, and $AgNO_3$ is the limiting reactant.

(c) The precipitate is Ag_3PO_4, whose molar mass is $3(107.9) + 31.0 + 4(16.0) = 418.7\ g\ mol^{-1}$. To calculate the number of grams of Ag_3PO_4 that could be produced in this reaction (the theoretical yield), we use the number of moles of the limiting reactant, converting mol $AgNO_3 \Rightarrow$ mol $Ag_3PO_4 \Rightarrow$ g Ag_3PO_4. We use the coefficients in the balanced equation to convert moles of $AgNO_3$ to moles Ag_3PO_4, and we use the molar mass of Ag_3PO_4 to convert the number of moles of this substance to grams:

$$mol\ Ag_3PO_4 = \frac{1}{3} \times mol\ AgNO_3 = \frac{1}{3} \times 7.5 \times 10^{-4}\ mol = 2.5 \times 10^{-4}\ mol$$

$$Mass\ Ag_3PO_4 = 2.5 \times 10^{-4}\ mol \times 418.7\ g\ mol^{-1}$$
$$= 0.10\ g$$

The answer has only two significant figures because the quantity of $AgNO_3$ is given to only two significant figures.

CHAPTER SUMMARY AND KEY TERMS

SECTION 4.1 Solutions in which water is the dissolving medium are called **aqueous solutions**. The component of the solution that is present in the greatest quantity is the **solvent**. The other components are **solutes**.

Any substance whose aqueous solution contains ions is called an **electrolyte**. Any substance that forms a solution containing no ions is a **non-electrolyte**. Electrolytes that are present in solution entirely as ions are **strong electrolytes**, whereas those that are present partly as ions and partly as molecules are **weak electrolytes**. Ionic compounds dissociate into ions when they dissolve, and they are strong electrolytes. The solubility of ionic substances is made possible by **solvation**, the interaction of ions with polar solvent molecules. Most molecular compounds are non-electrolytes, although some are weak electrolytes, and a few are strong electrolytes. When representing the ionisation of a weak electrolyte in solution, half-arrows in both directions are used, indicating that the forward and reverse reactions can achieve a chemical balance called a **chemical equilibrium**.

SECTION 4.2 **Precipitation reactions** are those in which an insoluble product, called a **precipitate**, forms. Solubility guidelines help determine whether or not an ionic compound will be soluble in water. (The **solubility** of a substance is the amount that dissolves in a given quantity of solvent.) Reactions such as precipitation reactions, in which cations and anions appear to exchange partners, are called **exchange reactions**, or **metathesis reactions**.

Chemical equations can be written to show whether dissolved substances are present in solution predominantly as ions or molecules. When the complete chemical formulae of all reactants and products are used, the equation is called a **molecular equation**. A **complete ionic equation** shows all dissolved strong electrolytes as their component ions. In a **net ionic equation**, those ions that go through the reaction unchanged (**spectator ions**) are omitted.

SECTION 4.3 Acids and bases are important electrolytes. **Acids** are proton donors; they increase the concentration of $H^+(aq)$ in aqueous solutions to which they are added. **Bases** are proton acceptors; they increase the concentration of $OH^-(aq)$ in aqueous solutions. Those acids and bases that are strong electrolytes are called **strong acids** and **strong bases**, respectively. Those that are weak electrolytes are **weak acids** and **weak bases**. When solutions of acids and bases are mixed, a neutralisation reaction occurs. The **neutralisation reaction** between an acid and a metal hydroxide produces water and a **salt**. Gases can also be formed as a result of neutralisation reactions. The reaction of a sulfide with an acid forms $H_2S(g)$; the reaction between a carbonate and an acid forms $CO_2(g)$.

SECTION 4.4 **Oxidation** is the loss of electrons by a substance, whereas **reduction** is the gain of electrons by a substance. **Oxidation numbers** keep track of electrons during chemical reactions and are assigned to atoms using specific rules. The oxidation of an element results in an increase in its oxidation number, whereas reduction is accompanied by a decrease in oxidation number. Oxidation is always accompanied by reduction, giving **oxidation–reduction**, or **redox**, **reactions**.

Many metals are oxidised by O_2, acids and salts. The redox reactions between metals and acids as well as those between metals and salts are called **displacement reactions**. The products of these displacement reactions are always an element (H_2 or a metal) and a salt. Comparing such reactions allows us to rank metals according to their ease of oxidation. A list of metals arranged in order of decreasing ease of oxidation is called an **activity series**. Any metal on the list can be oxidised by ions of metals (or H^+) below it in the series.

SECTION 4.5 The **concentration** of a solution expresses the amount of a solute dissolved in the solution. One of the common ways to express the concentration of a solute is in terms of molarity. The **molarity (M)** of a solution is the number of moles of solute per litre of solution. Molarity makes it possible to interconvert solution volume and number of moles of solute. Solutions of known molarity can be formed either by weighing out the solute and diluting it to a known volume or by the dilution of a more concentrated solution of known concentration (a stock solution). Adding solvent to the solution (the process of dilution) decreases the concentration of the solute without changing the number of moles of solute in the solution ($M_{conc} \times V_{conc} = M_{dil} \times V_{dil}$).

SECTION 4.6 In the process called **titration**, we combine a solution of known concentration (a **standard solution**) with a solution of unknown concentration to determine the unknown concentration or the quantity of solute in the unknown. The point in the titration at which stoichiometrically equivalent quantities of reactants are brought together is called the **equivalence point**. An **indicator** can be used to show the end point of the titration, which coincides closely with the equivalence point.

KEY SKILLS

- Recognise compounds as acids or bases and as strong electrolytes, weak electrolytes or non-electrolytes. (Sections 4.1 and 4.3)

- Recognise reactions by type and be able to predict the products of simple acid–base, precipitation metathesis or redox reactions. (Sections 4.2–4.4)

- Be able to calculate molarity and use it to convert between moles of a substance in solution and volume of the solution. (Section 4.5)

- Understand how to conduct a dilution to achieve a desired solution concentration. (Section 4.5)

- Understand how to perform and interpret the results of a titration. (Section 4.6)

KEY EQUATIONS

- Molarity is the most commonly used unit of concentration in chemistry.

$$\text{Molarity} = \frac{\text{moles solute}}{\text{volume of solution in litres}}; M = \frac{n}{V(\text{dm}^3)} \qquad [4.33]$$

- When adding solvent to a concentrated solution to make a dilute solution, molarities and volumes of both dilute and concentrated solutions can be calculated if three of the quantities are known.

$$M_{conc} \times V_{conc} = M_{dil} \times V_{dil} \qquad [4.35]$$

EXERCISES

VISUALISING CONCEPTS

4.1 Which of the following schematic drawings best describes a solution of Li_2SO_4 in water (water molecules not shown for simplicity)? [Section 4.1]

     (a)   (b)   (c)

4.2 Methanol, CH_3OH, and hydrogen chloride, HCl, are both molecular substances, yet an aqueous solution of methanol does not conduct an electrical current, whereas a solution of HCl does conduct. Account for this difference. [Section 4.1]

4.3 Aqueous solutions of three different substances, AX, AY and AZ, are represented by the three diagrams below. Identify each substance as a strong electrolyte, weak electrolyte or non-electrolyte. [Section 4.1]

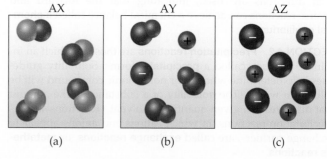

     AX   AY   AZ

     (a)   (b)   (c)

4.4 A 0.1 M solution of acetic acid, CH_3COOH, causes the light bulb in the apparatus of Figure 4.2 to glow about as brightly as a 0.001 M solution of HBr. How do you account for this fact? [Section 4.1]

4.5 You are presented with three white solids, A, B and C, which are glucose (a sugar substance), NaOH and AgBr. Solid A dissolves in water to form a conducting solution. B is not soluble in water. C dissolves in water to form a non-conducting solution. Identify A, B and C. [Section 4.2]

4.6 We have seen that ions in aqueous solution are stabilised by the attractions between the ions and the water molecules. Why then do some pairs of ions in solution form precipitates? [Section 4.2]

4.7 Which of the following ions will *always* be a spectator ion in a precipitation reaction? **(a)** Cl^-, **(b)** NO_3^-, **(c)** NH_4^+, **(d)** S^{2-}, **(e)** SO_4^{2-}. Explain briefly. [Section 4.2]

4.8 The labels have fallen off two bottles, one containing $Mg(NO_3)_2$ and the other containing $Pb(NO_3)_2$. You have a bottle of dilute H_2SO_4. How could you use it to test a portion of each solution to identify which solution is which? [Section 4.2]

4.9 Which of the following chemical equations is represented schematically in the drawing below? [Section 4.2]

(a) $BaCl_2(aq) + Na_2SO_4(aq) \longrightarrow$

$$BaSO_4(s) + 2\,NaCl(aq)$$

(b) $SrCO_3(s) + 2\,HBr(aq) \longrightarrow$

$$SrBr_2(aq) + H_2O(l) + CO_2(g)$$

(c) $2\,NaOH(aq) + CdCl_2(aq) \longrightarrow$

$$Cd(OH)_2(s) + 2\,NaCl(aq)$$

GENERAL PROPERTIES OF AQUEOUS SOLUTIONS (Section 4.1)

4.10 When asked what causes electrolyte solutions to conduct electricity, a student responds that it is due to the movement of electrons through the solution. Is the student correct? If not, what is the correct response?

4.11 When methanol, CH_3OH, is dissolved in water, a non-conducting solution results. When acetic acid, CH_3COOH, dissolves in water, the solution is weakly conducting and acidic in nature. Describe what happens upon dissolution in the two cases, and account for the different results.

4.12 We have learned in this chapter that many ionic solids dissolve in water as strong electrolytes, that is, as separated ions in solution. What properties of water facilitate this process?

4.13 What does it mean to say that ions are hydrated when an ionic substance dissolves in water?

4.14 Specify what ions are present in solution upon dissolving each of the following substances in water: **(a)** $ZnCl_2$, **(b)** HNO_3, **(c)** $(NH_4)_2SO_4$, **(d)** $Ca(OH)_2$.

4.15 Specify what ions are present upon dissolving each of the following substances in water: **(a)** MgI_2, **(b)** $Al(NO_3)_3$, **(c)** $HClO_4$, **(d)** CH_3COOK.

4.16 Formic acid, $HCOOH$, is a weak electrolyte. What solute particles are present in an aqueous solution of this compound? Write the chemical equation for the ionisation of $HCOOH$.

4.17 Acetone, CH_3COCH_3, is a non-electrolyte; hypochlorous acid, $HClO$, is a weak electrolyte; and ammonium chloride, NH_4Cl, is a strong electrolyte. **(a)** What are the solute particles present in aqueous solutions of each compound? **(b)** If 0.1 mol of each compound is dissolved in solution, which one contains 0.2 mol of solute particles, which contains 0.1 mol of solute particles and which contains somewhere between 0.1 and 0.2 mol of solute particles?

PRECIPITATION REACTIONS AND NET IONIC EQUATIONS (Section 4.2)

4.18 Using solubility guidelines, predict whether each of the following compounds is soluble or insoluble in water: **(a)** $NiCl_2$, **(b)** Ag_2S, **(c)** Cs_3PO_4, **(d)** $SrCO_3$, **(e)** $PbSO_4$.

4.19 Predict whether each of the following compounds is soluble in water: **(a)** $Ni(OH)_2$, **(b)** $PbBr_2$, **(c)** $Ba(NO_3)_2$, **(d)** $AlPO_4$, **(e)** $AgC_2H_3O_2$.

4.20 Will precipitation occur when the following solutions are mixed? If so, write a balanced chemical equation for the reaction. **(a)** Na_2CO_3 and $AgNO_3$, **(b)** $NaNO_3$ and $NiSO_4$, **(c)** $FeSO_4$ and $Pb(NO_3)_2$.

4.21 Identify the precipitate (if any) that forms when the following solutions are mixed, and write a balanced equation for each reaction. **(a)** $Ni(NO_3)_2$ and $NaOH$, **(b)** $NaOH$ and K_2SO_4, **(c)** Na_2S and $Cu(CH_3COO)_2$.

4.22 Name the spectator ions in any reactions that may be involved when each of the following pairs of solutions are mixed.

(a) $Na_2CO_3(aq)$ and $MgSO_4(aq)$

(b) $Pb(NO_3)_2(aq)$ and $Na_2S(aq)$

(c) $(NH_4)_3PO_4(aq)$ and $CaCl_2(aq)$

4.23 Write balanced net ionic equations for the reactions that occur in each of the following cases. Identify the spectator ion or ions in each reaction.

(a) $Cr_2(SO_4)_3(aq) + (NH_4)_2CO_3(aq) \rightarrow$

(b) $Ba(NO_3)_2(aq) + K_2SO_4(aq) \rightarrow$

(c) $Fe(NO_3)_2(aq) + KOH(aq) \rightarrow$

4.24 You know that an unlabelled bottle contains a solution of one of the following: $AgNO_3$, $CaCl_2$ or $Al_2(SO_4)_3$. A friend suggests that you test a portion of the solution with $Ba(NO_3)_2$ and then with $NaCl$ solutions. Explain how these two tests together would be sufficient to determine which salt is present in the solution.

ACIDS, BASES AND NEUTRALISATION REACTIONS (Section 4.3)

4.25 Which of the following solutions has the largest concentration of solvated protons: **(a)** 0.1 M LiOH, **(b)** 0.1 M HI, **(c)** 0.5 M methanol (CH_3OH)? Explain.

4.26 Which of the following solutions is the most basic: **(a)** 0.5 M NH_3, **(b)** 0.1 M KOH, **(c)** 0.1 M $Ca(OH)_2$? Explain.

4.27 What is the difference between **(a)** a monoprotic acid and a diprotic acid, **(b)** a weak acid and a strong acid, **(c)** an acid and a base?

4.28 Explain the following observations. **(a)** NH_3 contains no OH^- ions and yet its aqueous solutions are basic. **(b)** HF is called a weak acid and yet it is very reactive. **(c)** Although sulfuric acid is a strong electrolyte, an aqueous solution of H_2SO_4 contains more HSO_4^- ions than SO_4^{2-} ions.

4.29 Label each of the following substances as an acid, base, salt or none of these. Indicate whether the substance exists in aqueous solution entirely in molecular form, entirely as ions or as a mixture of molecules and ions. **(a)** HF, **(b)** acetonitrile, CH_3CN, **(c)** $NaClO_4$, **(d)** $Ba(OH)_2$.

4.30 An aqueous solution of an unknown solute is tested with litmus paper and found to be acidic. The solution is weakly conducting compared with a solution of NaCl of the same concentration. Which of the following substances could the unknown be: KOH, NH_3, HNO_3 $KClO_2$, H_3PO_3, CH_3COCH_3 (acetone)?

4.31 Write the balanced molecular and net ionic equations for each of the following neutralisation reactions.

(a) Aqueous acetic acid is neutralised by aqueous potassium hydroxide.

(b) Solid chromium(III) hydroxide reacts with nitric acid.

(c) Aqueous hypochlorous acid and aqueous calcium hydroxide react.

4.32 Write balanced molecular and net ionic equations for the following reactions, and identify the gas formed in each: **(a)** solid cadmium sulfide reacts with an aqueous solution of sulfuric acid; **(b)** solid magnesium carbonate reacts with an aqueous solution of perchloric acid.

OXIDATION–REDUCTION REACTIONS (Section 4.4)

4.33 Define oxidation and reduction in terms of **(a)** electron transfer and **(b)** oxidation numbers.

4.34 Can oxidation occur without an accompanying reduction? Explain.

4.35 Which circled region of the periodic table shown here contains the most readily oxidised elements? Which contains the least readily oxidised?

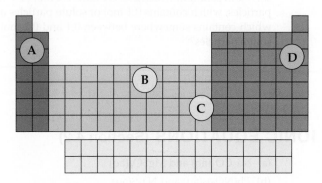

4.36 From the elements listed in Table 4.4, select an element that lies in region A of the periodic table shown above and an element that lies in region C. Write a balanced oxidation–reduction equation that shows the oxidation of one metal and reduction of an ion of the other. You will need to decide which element is oxidised and which is reduced.

4.37 Determine the oxidation number for the indicated element in each of the following substances: **(a)** S in SO_2, **(b)** C in $COCl_2$, **(c)** Mn in MnO_4^-, **(d)** Br in HBrO, **(e)** As in As_4, **(f)** O in K_2O_2.

4.38 Determine the oxidation number for the indicated element in each of the following compounds: **(a)** Ti in TiO_2, **(b)** Sn in $SnCl_3^-$, **(c)** C in $C_2O_4^{2-}$, **(d)** N in N_2H_4, **(e)** N in HNO_2, **(f)** Cr in $Cr_2O_7^{2-}$.

4.39 Which element is oxidised and which is reduced in the following reactions?

(a) $Ni(s) + Cl_2(g) \rightarrow NiCl_2(s)$

(b) $3\,Fe(NO_3)_2(aq) + 2\,Al(s) \rightarrow 3\,Fe(s) + 2\,Al(NO_3)_3(aq)$

(c) $Cl_2(aq) + 2\,NaI(aq) \rightarrow I_2(aq) + 2\,NaCl(aq)$

(d) $PbS(s) + 4\,H_2O_2(aq) \rightarrow PbSO_4(s) + 4\,H_2O(l)$

4.40 Which of the following are redox reactions? For those that are, indicate which element is oxidised and which is reduced. For those that are not, indicate whether they are precipitation or acid–base reactions.

(a) $Cu(OH)_2(s) + 2\,HNO_3(aq) \rightarrow$
$$Cu(NO_3)_2(aq) + 2\,H_2O(l)$$

(b) $Fe_2O_3(s) + 3\,CO(g) \rightarrow 2\,Fe(s) + 3\,CO_2(g)$

(c) $Sr(NO_3)_2(aq) + H_2SO_4(aq) \rightarrow SrSO_4(s) + 2\,HNO_3(aq)$

(d) $4\,Zn(s) + 10\,H^+(aq) + 2\,NO_3^-(aq) \rightarrow$
$$4\,Zn^{2+}(aq) + N_2O(g) + 5\,H_2O(l)$$

4.41 Using the activity series (Table 4.4), write balanced chemical equations for the following reactions. If no reaction occurs, simply write NR. **(a)** Iron metal is added to a solution of copper(II) nitrate; **(b)** zinc metal is added to a solution of magnesium sulfate; **(c)** hydrobromic acid is added to tin metal; **(d)** hydrogen gas is bubbled through an aqueous solution of nickel(II) chloride; **(e)** aluminium metal is added to a solution of cobalt(II) sulfate.

4.42 Based on the activity series (Table 4.4), what is the outcome (if any) of each of the following reactions?

(a) $Mn(s) + NiCl_2(aq) \rightarrow$

(b) $Cu(s) + Cr(C_2H_3O_2)_3(aq) \rightarrow$

(c) $Cr(s) + NiSO_4(aq) \rightarrow$

(d) $Pt(s) + HBr(aq) \rightarrow$

(e) $H_2(g) + CuCl_2(aq) \rightarrow$

4.43 The metal cadmium tends to form Cd^{2+} ions. The following observations are made: (i) When a strip of zinc metal is placed in $CdCl_2(aq)$ cadmium metal is deposited on the strip. (ii) When a strip of cadmium metal is placed in $Ni(NO_3)_2$, nickel metal is deposited on the strip. **(a)** Write net ionic equations to explain each of the observations made above. **(b)** What can you conclude about the position of cadmium in the activity series? **(c)** What experiments would you need to perform to locate more precisely the position of cadmium in the activity series?

4.44 (a) Use the following reactions to prepare an activity series for the halogens:

$$Br_2(aq) + 2\,NaI(aq) \rightarrow 2\,NaBr(aq) + I_2(aq)$$
$$Cl_2(aq) + 2\,NaBr(aq) \rightarrow 2\,NaCl(aq) + Br_2(aq)$$

(b) Relate the positions of the halogens in the periodic table with their locations in this activity series. **(c)** Predict whether a reaction occurs when the following reagents are mixed: $Cl_2(aq)$ and $KI(aq)$; $Br_2(aq)$ and $LiCl(aq)$.

CONCENTRATIONS OF SOLUTIONS (Section 4.5)

4.45 (a) Calculate the molarity of a solution that contains 0.0345 mol NH_4Cl in exactly 400 cm^3 of solution. **(b)** How many moles of HNO_3 are present in 35.0 cm^3 of a 2.20 M solution of nitric acid? **(c)** How many millilitres of 1.50 M KOH solution are needed to provide 0.125 mol of KOH?

4.46 (a) Calculate the molarity of a solution made by dissolving 0.145 mol Na_2SO_4 in enough water to form exactly 750 cm^3 of solution. **(b)** How many moles of $KMnO_4$ are present in 125 cm^3 of a 0.0850 M solution? **(c)** How many millilitres of 11.6 M HCl solution are needed to obtain 0.255 mol of HCl?

4.47 The average adult human male has a total blood volume of 5.0 dm^3. If the concentration of sodium ion in this average individual is 0.135 M, what is the mass of sodium ion circulating in the blood?

4.48 A person suffering from hyponatraemia has a sodium ion concentration in the blood of 0.118 M and a total blood volume of 4.6 dm^3. What mass of sodium chloride would need to be added to the blood to bring the sodium ion concentration up to 0.138 M, assuming no change in blood volume?

4.49 Calculate **(a)** the number of grams of solute in 0.250 dm^3 of 0.150 M KBr, **(b)** the molar concentration of a solution containing 4.75 g of $Ca(NO_3)_2$ in 0.200 dm^3, **(c)** the volume of 1.50 M Na_3PO_4 in millilitres that contains 5.00 g of solute.

4.50 (a) Which will have the highest concentration of potassium ion: 0.20 M KCl, 0.15 M K_2CrO_4 or 0.080 M K_3PO_4? **(b)** Which will contain the greater number of moles of potassium ion: 30.0 cm^3 of 0.15 M K_2CrO_4 or 25.0 cm^3 of 0.080 M K_3PO_4?

4.51 In each of the following pairs, indicate which has the higher concentration of Cl^- ion: **(a)** 0.10 M $CaCl_2$ or 0.15 M KCl solution, **(b)** 100 cm^3 of 0.10 M KCl solution or 400 cm^3 of 0.080 M LiCl solution, **(c)** 0.050 M HCl solution or 0.020 M $CdCl_2$ solution.

4.52 Indicate the concentration of each ion present in the solution formed by mixing **(a)** 16.0 cm^3 of 0.130 M HCl and 12.0 cm^3 of 0.600 M HCl, **(b)** 18.0 cm^3 of 0.200 M Na_2SO_4 and 15.0 cm^3 of 0.150 M KCl, **(c)** 2.38 g of NaCl in 50.0 cm^3 of 0.400 M $CaCl_2$ solution. Assume the volumes are additive.

4.53 (a) You have a stock solution of 14.8 M NH_3. How many millilitres of this solution should you dilute to make 100.0 cm^3 of 0.250 M NH_3? **(b)** If you take a 10.0 cm^3 portion of the stock solution and dilute it to a total volume of 0.250 dm^3, what will be the concentration of the final solution?

4.54 (a) How many millilitres of a stock solution of 10.0 M HNO_3 would you have to use to prepare 0.350 dm^3 of 0.400 M HNO_3? **(b)** If you dilute 25.0 cm^3 of the stock solution to a final volume of 0.500 dm^3, what will be the concentration of the diluted solution?

4.55 Pure acetic acid, known as glacial acetic acid, is a liquid with a density of 1.049 g cm^{-3} at 25 °C. Calculate the molarity of a solution of acetic acid made by dissolving 20.00 cm^3 of glacial acetic acid at 25 °C in enough water to make 250.0 cm^3 of solution.

4.56 Glycerol, $C_3H_8O_3$, is a substance used extensively in the manufacture of cosmetics, foodstuffs, antifreeze and plastics. Glycerol is a water-soluble liquid with a density of 1.266 g cm^{-3} at 15 °C. Calculate the molarity of a solution of glycerol made by dissolving 50.00 cm^3 of glycerol at 15 °C in enough water to make 250.0 cm^3 of solution.

SOLUTION STOICHIOMETRY AND CHEMICAL ANALYSIS (Section 4.6)

4.57 What mass of NaCl is needed to precipitate the silver ions from 20.0 cm^3 of 0.100 M $AgNO_3$ solution?

4.58 What mass of NaOH is needed to precipitate the Cd^{2+} ions from 25.0 cm^3 of 0.500 M $Cd(NO_3)_2$ solution?

4.59 (a) What volume of 0.115 M $HClO_4$ solution is needed to neutralise 50.00 cm^3 of 0.0875 M NaOH? **(b)** What volume of 0.128 M HCl is needed to neutralise 2.87 g of $Mg(OH)_2$? **(c)** If 25.8 cm^3 of $AgNO_3$ solution is needed to precipitate all the Cl^- ions in a 785 mg sample of KCl (forming AgCl), what is the molarity of the $AgNO_3$ solution? **(d)** If 45.3 cm^3 of 0.108 M HCl solution is needed to neutralise a solution of KOH, how many grams of KOH must be present in the solution?

4.60 (a) How many millilitres of 0.120 M HCl are needed to completely neutralise 50.0 cm^3 of 0.101 M $Ba(OH)_2$ solution? **(b)** How many millilitres of 0.125 M H_2SO_4 are needed to neutralise 0.200 g of NaOH? **(c)** If 55.8 cm^3 of $BaCl_2$ solution is needed to precipitate all the sulfate ion in a 752 mg sample of Na_2SO_4, what is the molarity of the solution? **(d)** If 42.7 cm³ of 0.208 M HCl solution is needed to neutralise a solution of $Ca(OH)_2$, how many grams of $Ca(OH)_2$ must be in the solution?

4.61 Some sulfuric acid is spilled on a lab bench. It can be neutralised by sprinkling sodium bicarbonate on it and then mopping up the resultant solution. The sodium bicarbonate reacts with sulfuric acid as follows:

$$2\,NaHCO_3(s) + H_2SO_4(aq) \rightarrow$$
$$Na_2SO_4(aq) + 2\,H_2O(l) + 2\,CO_2(g)$$

Sodium bicarbonate is added until the fizzing due to the formation of $CO_2(g)$ stops. If 27 cm³ of 6.0 M H_2SO_4 was spilled, what is the minimum mass of $NaHCO_3$ that must be added to the spill to neutralise the acid?

4.62 A sample of solid $Ca(OH)_2$ is stirred in water at 30 °C until the solution contains as much dissolved $Ca(OH)_2$ as it can hold. A 100 cm³ sample of this solution is withdrawn and titrated with 5.00×10^{-2} M HBr. It requires 48.8 cm³ of the acid solution for neutralisation. What is the molarity of the $Ca(OH)_2$ solution? What is the solubility of $Ca(OH)_2$ in water at 30 °C in grams of $Ca(OH)_2$ per 100 cm³ of solution?

4.63 In the laboratory 6.82 g of $Sr(NO_3)_2$ is dissolved in enough water to form 0.500 dm³. A 0.100 dm³ sample is withdrawn from this stock solution and titrated with a 0.0335 M solution of Na_2CrO_4. What volume of Na_2CrO_4 solution is needed to precipitate all the Sr^{2+}(aq) as $SrCrO_4$?

4.64 A solution of 100.0 cm³ of 0.200 M KOH is mixed with a solution of 200.0 cm³ of 0.150 M $NiSO_4$. **(a)** Write the balanced chemical equation for the reaction that occurs. **(b)** What precipitate forms? **(c)** What is the limiting reactant? **(d)** How many grams of this precipitate form? **(e)** What is the concentration of each ion that remains in solution?

4.65 A solution is made by mixing 12.0 g of NaOH and 75.0 cm³ of 0.200 M HNO_3. **(a)** Write a balanced equation for the reaction that occurs between the solutes. **(b)** Calculate the concentration of each ion remaining in solution. **(c)** Is the resultant solution acidic or basic?

4.66 A 0.5895 g sample of impure magnesium hydroxide is dissolved in 100.0 cm³ of 0.2050 M HCl solution. The excess acid then needs 19.85 cm³ of 0.1020 M NaOH for neutralisation. Calculate the percentage by mass of magnesium hydroxide in the sample, assuming that it is the only substance reacting with the HCl solution.

4.67 A 1.248 g sample of limestone rock is pulverised and then treated with 30.00 cm³ of 1.035 M HCl solution. The excess acid then requires 11.56 cm³ of 1.010 M NaOH for neutralisation. Calculate the percentage by mass of calcium carbonate in the rock, assuming that it is the only substance reacting with the HCl solution.

ADDITIONAL EXERCISES

4.68 The accompanying photo shows the reaction between a solution of $Cd(NO_3)_2$ and one of Na_2S. What is the identity of the precipitate? What ions remain in solution? Write the net ionic equation for the reaction.

4.69 Antacids are often used to relieve pain and promote healing in the treatment of mild stomach ulcers. Write balanced net ionic equations for the reactions between the HCl(aq) in the stomach and each of the following substances used in various antacids: **(a)** $Al(OH)_3$(s), **(b)** $Mg(OH)_2$(s), **(c)** $MgCO_3$(s), **(d)** $NaAl(CO_3)(OH)_2$(s), **(e)** $CaCO_3$(s).

4.70 The commercial production of nitric acid involves the following chemical reactions:

$$4\,NH_3(g) + 5\,O_2(g) \rightarrow 4\,NO(g) + 6\,H_2O(g)$$
$$2\,NO(g) + O_2(g) \rightarrow 2\,NO_2(g)$$
$$3\,NO_2(g) + H_2O(l) \rightarrow 2\,HNO_3(aq) + NO(g)$$

(a) Which of these reactions are redox reactions? **(b)** In each redox reaction identify the element undergoing oxidation and the element undergoing reduction.

4.71 Use Table 4.4 to predict which of the following ions can be reduced to their metal forms by reacting with zinc: **(a)** Na^+(aq), **(b)** Pb^{2+}(aq), **(c)** Mg^{2+}(aq), **(d)** Fe^{2+}(aq), **(e)** Cu^{2+}(aq), **(f)** Al^{3+}(aq). Write the balanced net ionic equation for each reaction that occurs.

4.72 Lanthanum metal forms cations with a charge of 3+. Consider the following observations about the chemistry of lanthanum. When lanthanum metal is exposed to air, a white solid (compound A) is formed that contains lanthanum and one other element. When lanthanum metal is added to water, gas bubbles are observed and a different white solid (compound B) is formed. Both A and B dissolve in hydrochloric acid to give a clear solution. When either of these solutions is evaporated, a soluble white solid (compound C) remains. If compound C is dissolved in water and sulfuric acid is added, a white precipitate (compound D) forms. **(a)** Propose identities for the substances A, B, C and D. **(b)** Write net ionic equations for all the reactions described. **(c)** Based on the preceding observations, what can be said about the position of lanthanum in the activity series (Table 4.4)?

4.73 A 35.0 cm³ sample of 1.00 M KBr and a 60.0 cm³ sample of 0.600 M KBr are mixed. The solution is then heated to evaporate water until the total volume is 50.0 cm³. What is the molarity of the KBr in the final solution?

4.74 Calculate the molarity of the solution produced by mixing **(a)** 40.0 cm³ of 0.160 M NaCl and 65.0 cm³ of 0.150 M NaCl, **(b)** 32.5 cm³ of 0.750 M NaOH and 26.8 cm³ of 0.750 M NaOH. Assume that the volumes are additive.

4.75 Using modern analytical techniques, it is possible to detect sodium ions in concentrations as low as 50 pg cm⁻³ (pg = picogram). What is this detection limit expressed in **(a)** molarity of Na^+, **(b)** Na^+ ions per cubic centimetre?

4.76 Hard water contains Ca^{2+}, Mg^{2+} and Fe^{2+}, which interfere with the action of soap and leave an insoluble coating on the insides of containers and pipes when heated. Water softeners replace these ions with Na^+. If 1.0×10^3 dm^3 of hard water contains 0.010 M Ca^{2+} and 0.0050 M Mg^{2+}, how many moles of Na^+ are needed to replace these ions?

4.77 Tartaric acid, $C_4H_6O_6$, has two acidic hydrogens. The acid is often present in wines and precipitates from solution as the wine ages. A solution containing an unknown concentration of the acid is titrated with NaOH. It requires 22.62 cm^3 of 0.2000 M NaOH solution to titrate both acidic protons in 40.00 cm^3 of the tartaric acid solution. Write a balanced net ionic equation for the neutralisation reaction and calculate the molarity of the tartaric acid solution.

4.78 The concentration of hydrogen peroxide in a solution is determined by titrating a 10.0 cm^3 sample of the solution with permanganate ion.

$$2\ MnO_4^-(aq) + 5\ H_2O_2(aq) + 6\ H^+(aq) \rightarrow$$
$$2\ Mn^{2+}(aq) + 5\ O_2(g) + 8\ H_2O(l)$$

If it takes 16.8 cm^3 of 0.124 M MnO_4^- solution to reach the equivalence point, what is the molarity of the hydrogen peroxide solution?

INTEGRATIVE EXERCISES

4.79 A 3.455 g sample of a mixture is analysed for barium ions by adding a small excess of sulfuric acid to an aqueous solution of the sample. The resultant reaction produces a precipitate of barium sulfate, which is collected by filtration, washed, dried and weighed. If 0.2815 g of barium sulfate is obtained, what is the mass percentage of barium in the sample?

4.80 A sample of 5.53 g of $Mg(OH)_2$ is added to 25.0 cm^3 of 0.200 M HNO_3. **(a)** Write the chemical equation for the reaction that occurs. **(b)** Which is the limiting reactant in the reaction? **(c)** How many moles of $Mg(OH)_2$, HNO_3 and $Mg(NO_3)_2$ are present after the reaction is complete?

4.81 A sample of 1.50 g of lead(II) nitrate is mixed with 125 cm^3 of 0.100 M sodium sulfate solution. **(a)** Write the chemical equation for the reaction that occurs. **(b)** Which is the limiting reactant in the reaction? **(c)** What are the concentrations of all ions that remain in solution after the reaction is complete?

4.82 A mixture contains 76.5% NaCl, 6.5% $MgCl_2$ and 17.0% Na_2SO_4 by mass. What is the molarity of Cl^- ions in a solution formed by dissolving 7.50 g of the mixture in enough water to form 500.0 cm^3 of solution?

4.83 The average concentration of bromide ion in seawater is 65 mg of bromide ion per kg of seawater. What is the molarity of the bromide ion if the density of the seawater is 1.025 g cm^{-3}?

4.84 The mass percentage of chloride ion in a 25.00 cm^3 sample of seawater was determined by titrating the sample with silver nitrate, precipitating silver chloride. It took 42.58 cm^3 of 0.2997 M silver nitrate solution to reach the equivalence point in the titration. What is the mass percentage of chloride ion in the seawater if its density is 1.025 g cm^{-3}?

4.85 The standard for arsenate in drinking water requires that public water supplies must contain no greater than 10 parts per billion (ppb) arsenic. Assuming that this arsenic is present as arsenate, AsO_4^{3-}, what mass of sodium arsenate would be present in a 1.00 dm^3 sample of drinking water that just meets the standard?

4.86 WorkSafe Australia sets an upper limit of 25 ppm of NH_3 in the air on a work environment (that is, 25 molecules of $NH_3(g)$ for every million molecules in the air). Air from a manufacturing operation was drawn through a solution containing 1.00×10^2 cm^3 of 0.0105 M HCl. The NH_3 reacts with HCl as follows:

$$NH_3(aq) + HCl(aq) \rightarrow NH_4Cl(aq)$$

After drawing air through the acid solution for 10.0 min at a rate of 10.0 dm^3 min^{-1}, the acid was titrated. The remaining acid needed 13.1 cm^3 of 0.0588 M NaOH to reach the equivalence point. **(a)** How many grams of NH_3 were drawn into the acid solution? **(b)** How many ppm of NH_3 were in the air? (Air has a density of 1.20 g dm^{-3} and an average molar mass of 29.0 g mol^{-1} under the conditions of the experiment.) **(c)** Is this manufacturer in compliance with the regulations?

MasteringChemistry (www.pearson.com.au/masteringchemistry)

Make learning part of the grade. Access:

- tutorials with peronalised coaching
- study area
- Pearson eText

5

NUCLEAR CHEMISTRY: CHANGES WITHIN THE CORE OF AN ATOM

The remnant of the supernova cassiopeia as viewed from the Chandra X-ray Observatory.

KEY CONCEPTS

5.1 RADIOACTIVITY
In this chapter we learn how to describe nuclear reactions by equations analogous to chemical equations, in which the nuclear charges and masses of reactants and products are in balance. Radioactive nuclei most commonly decay by emission of *alpha*, *beta* or *gamma* radiation.

5.2 PATTERNS OF NUCLEAR STABILITY
We recognise that nuclear stability is determined largely by the *neutron-to-proton ratio*. For stable nuclei, this ratio increases with increasing atomic number. All nuclei with 84 or more protons are radioactive. Heavy nuclei gain stability by a series of nuclear disintegrations leading to stable nuclei.

5.3 NUCLEAR TRANSMUTATIONS
We study *nuclear transmutations*, which are nuclear reactions induced by bombardment of a nucleus by a neutron or an accelerated charged particle.

5.4 RATES OF RADIOACTIVE DECAY
We learn that radioisotope decays are first-order kinetic processes with characteristic half-lives. Decay rates can be used to determine the age of ancient artifacts and geological formations.

5.5 DETECTION OF RADIOACTIVITY
We see that the radiation emitted by a radioactive substance can be detected by dosimeters, Geiger counters and scintillation counters.

5.6 ENERGY CHANGES IN NUCLEAR REACTIONS
We recognise that energy changes in nuclear reactions are related to mass changes via Einstein's equation, $E = mc^2$. The *nuclear binding energy* of a nucleus is the difference between the mass of the nucleus and the sum of the masses of its nucleons.

5.7 NUCLEAR POWER: FISSION
We learn that in *nuclear fission* a heavy nucleus splits to form two or more product nuclei. This type of nuclear reaction is the energy source for nuclear power plants, and we look at the operating principles of these plants.

5.8 NUCLEAR POWER: FUSION
We learn that in *nuclear fusion* two light nuclei are fused together to form a more stable, heavier nucleus.

5.9 RADIATION IN THE ENVIRONMENT AND LIVING SYSTEMS
We discover that naturally occurring radioisotopes bathe our planet—and us—with low levels of radiation. The radiation emitted in nuclear reactions can damage living cells but also has diagnostic and therapeutic applications.

All energy that fuels life on Earth comes ultimately from sunlight. Life on Earth could not exist without energy from the sun, but where does the sun get its energy? Stars, including our sun, use **nuclear reactions** that involve changes in atomic nuclei to generate their energy. For example, the sun produces energy by fusing hydrogen atoms to form helium, releasing vast amounts of energy in the process.

The fusion of hydrogen to form helium is the dominant nuclear reaction for most of a star's lifetime. Towards the end of its life, the hydrogen in the star's core is exhausted and the helium atoms fuse to form progressively heavier elements. A select few stars end their lives in dramatic supernova explosions. The nuclear reactions that occur when a star goes supernova are responsible for the existence of all naturally occurring elements heavier than nickel.

Nuclear chemistry is the study of nuclear reactions, with an emphasis on their uses in chemistry and their effects on biological systems. Nuclear chemistry affects our lives in many ways, particularly in energy and medical applications. In radiation therapy, for example, gamma rays from a radioactive substance such as cobalt-60 are directed to cancerous tumours to destroy them. Positron emission tomography (PET) is one example of a medical diagnostic tool that relies on decay of a radioactive element injected into the body.

The emanation of gamma rays from cobalt-60 is an example of a *nuclear reaction*, in which a change in matter originates in the nucleus of an atom. When nuclei change spontaneously, emitting radiation, they are said to be **radioactive**.

Radioactivity is also used to help determine the mechanisms of chemical reactions, to trace the movement of atoms in biological systems and the environment and to date historical artifacts.

Nuclear reactions are also used to generate electricity. Roughly 15% of the electricity generated worldwide comes from nuclear power plants, though the percentage varies from one country to the next.

The use of nuclear energy for power generation is a controversial social and political issue because of the public's conceptions about the safety of nuclear reactors and, more importantly, the difficulty of disposing of nuclear reactor waste.

5.1 | RADIOACTIVITY

∞ Review this on page 34

The nucleus of an atom contains two types of subatomic particles, protons and neutrons, which are referred to as **nucleons** (∞ Section 2.3, 'The Modern View of Atomic Structure'). All atoms of a given element have the same number of protons (*atomic number*). They can have different numbers of neutrons, however, resulting in different *mass numbers*. The mass number is the total number of nucleons in the nucleus. Atoms with the same atomic number but different mass numbers are known as *isotopes* of the element. For example, the three naturally occurring isotopes of uranium are labelled $^{234}_{92}$U, $^{235}_{92}$U and $^{238}_{92}$U or may be written uranium-234, uranium-235 and uranium-238.

The nucleus of an element with a specified number of neutrons and protons is a **nuclide** (essentially synonymous with the word isotope). Some nuclides are stable and some are not, depending on the number of protons and neutrons in the nucleus. An atom whose nucleus is unstable is said to be radioactive and is called a **radioisotope**, and its nucleus is called a **radionuclide**.

Nuclear Equations

The vast majority of nuclei found in nature are stable and remain intact indefinitely. Radionuclides, however, are unstable and spontaneously emit particles and electromagnetic radiation (energy). Emission of radiation is one of the ways in which an unstable nucleus is transformed into a more stable one with less energy. Uranium-238, for example, is radioactive, undergoing a nuclear reaction in which helium-4 nuclei are spontaneously emitted. The helium-4 particles are known as **alpha (α) particles**, and a stream of these particles is called **alpha radiation**. When a uranium-238 nucleus loses an alpha particle, the remaining fragment has an atomic number of 90 and a mass number of 234. It is therefore a thorium-234 nucleus. We represent this reaction by the following *nuclear equation*:

$$^{238}_{92}\text{U} \longrightarrow {}^{234}_{90}\text{Th} + {}^{4}_{2}\text{He} \qquad \text{[5.1]}$$

When a nucleus spontaneously decomposes in this way, it is said to have decayed, or to have undergone *radioactive decay*. Because an alpha particle is involved in this reaction, scientists also describe the process as **alpha decay**.

 CONCEPT CHECK 1

What change in the mass number of a nucleus occurs when the nucleus emits an alpha particle?

In Equation 5.1 the sum of the mass numbers is the same on both sides of the equation (238 = 234 + 4). Likewise, the sum of the atomic numbers on both sides of the equation is equal (92 = 90 + 2). Mass numbers and atomic numbers must be balanced in all nuclear equations.

The radioactive properties of the nucleus are essentially independent of the state of chemical combination of the atom. In writing nuclear equations, therefore, we are not concerned with the chemical form of the atom in which the nucleus resides. It makes no difference whether we are dealing with the atom in the form of an element or of one of its compounds.

SAMPLE EXERCISE 5.1   Predicting the product of a nuclear reaction

What product is formed when radium-226 undergoes alpha decay?

SOLUTION

Analyse We are asked to determine the nucleus that results when radium-226 loses an alpha particle.

Plan We can best do this by writing a balanced nuclear equation for the process.

Solve

The periodic table shows that radium has an atomic number of 88. The complete chemical symbol for radium-226 is therefore $^{226}_{88}\text{Ra}$. An alpha particle is a helium-4 nucleus, and so its symbol is $^{4}_{2}\text{He}$ (sometimes written as $^{4}_{2}\alpha$). The alpha particle is a product of the nuclear reaction, and so the equation is of the form

$$^{226}_{88}\text{Ra} \longrightarrow {}^{A}_{Z}\text{X} + {}^{4}_{2}\text{He}$$

where A is the mass number of the product nucleus and Z is its atomic number. Mass numbers and atomic numbers must balance, so

$$226 = A + 4$$

and

$$88 = Z + 2$$

Hence,

$$A = 222 \text{ and } Z = 86$$

Again, from the periodic table, the element with $Z = 86$ is radon (Rn). The product, therefore, is $^{222}_{86}\text{Rn}$ and the nuclear equation is

$$^{226}_{88}\text{Ra} \longrightarrow {}^{222}_{86}\text{Rn} + {}^{4}_{2}\text{He}$$

PRACTICE EXERCISE

Which element undergoes alpha decay to form lead-208?

Answer: $^{212}_{84}\text{Po}$

(See also Exercises 5.11(d), 5.13(d).)

Types of Radioactive Decay

The three most common kinds of radioactive decay are alpha (α), beta (β) and gamma (γ) decay (Section 2.2, 'The Discovery of Atomic Structure').
▼ **TABLE 5.1** summarises some of the important properties of these kinds of radiation. As we have just discussed, alpha radiation consists of a stream of helium-4 nuclei known as alpha particles, which we denote as $^{4}_{2}\text{He}$ or $^{4}_{2}\alpha$.

Review this on page 32

Beta radiation consists of streams of **beta (β) particles**, which are high-speed electrons emitted by an unstable nucleus. Beta particles are represented in nuclear equations by the symbol $^{0}_{-1}\text{e}$ or sometimes $^{0}_{-1}\beta$. The superscript zero indicates that the mass of the electron is exceedingly small compared with the mass of a nucleon. The subscript -1 represents the negative charge of the particle, which is opposite that of the proton. Iodine-131 is an isotope that undergoes decay by beta emission:

$$^{131}_{53}\text{I} \longrightarrow {}^{131}_{54}\text{Xe} + {}^{0}_{-1}\text{e} \qquad [5.2]$$

In Equation 5.2 beta decay causes the atomic number to increase from 53 to 54. Beta emission is equivalent to the conversion of a neutron ($^{1}_{0}\text{n}$) to a proton ($^{1}_{1}\text{p}$ or $^{1}_{1}\text{H}$), thereby increasing the atomic number by 1:

$$^{1}_{0}\text{n} \longrightarrow {}^{1}_{1}\text{p} + {}^{0}_{-1}\text{e} \qquad [5.3]$$

TABLE 5.1 • Properties of alpha, beta and gamma radiation

Property	Type of radiation		
	α	β	γ
Charge	2+	1−	0
Mass	6.64×10^{-24} kg	9.11×10^{-28} kg	0
Relative penetrating power	1	100	10 000
Nature of radiation	$^{4}_{2}\text{He}$ nuclei	High-energy electrons	High-energy photons

It must be stressed that the emitted electron is created in the nucleus when a nuclear reaction occurs. It is not part of the normal electron structure of the isotope.

Gamma radiation (or gamma rays) consists of high-energy photons (that is, electromagnetic radiation of very short wavelength). Gamma radiation changes neither the atomic number nor the mass number of a nucleus and is represented as $^0_0\gamma$, or merely γ. It almost always accompanies other radioactive emission because it represents the energy lost when the remaining nucleons reorganise into more stable arrangements. Generally, the gamma rays are not shown when writing nuclear equations.

Two other types of radioactive decay are **positron emission** and **electron capture**. A *positron* is a particle that has the same mass as an electron, but an opposite charge.* The positron is represented as ^0_1e or sometimes $^0_1\beta$. The isotope carbon-11 decays by positron emission:

$$^{11}_6C \longrightarrow {}^{11}_5B + {}^0_1e \qquad [5.4]$$

Positron emission causes the atomic number to decrease from 6 to 5. The emission of a positron has the effect of converting a proton to a neutron, thereby decreasing the atomic number of the nucleus by 1:

$$^1_1p \longrightarrow {}^1_0n + {}^0_1e \qquad [5.5]$$

Electron capture is the capture by the nucleus of an electron from the electron cloud surrounding the nucleus. Rubidium-81 undergoes decay in this fashion, as shown in Equation 5.6:

$$^{81}_{37}Rb + {}^0_{-1}e \text{ (orbital electron)} \longrightarrow {}^{81}_{36}Kr \qquad [5.6]$$

Because the electron is consumed rather than formed in the process, it is shown on the reactant side of the equation. Electron capture, like positron emission, has the effect of converting a proton to a neutron:

$$^1_1p + {}^0_{-1}e \longrightarrow {}^1_0n \qquad [5.7]$$

◀ TABLE 5.2 summarises the symbols used to represent the various elementary particles commonly encountered in nuclear reactions.

TABLE 5.2 • Common particles in radioactive decay and nuclear transformations

Particle	Symbol
Neutron	^1_0n
Proton	^1_1H or ^1_1p
Electron	$^0_{-1}e$
Alpha particle	^4_2He or $^4_2\alpha$
Beta particle	$^0_{-1}e$ or $^0_{-1}\beta$
Positron	^0_1e or $^0_1\beta$

▲ **CONCEPT CHECK 2**

Which of the particles listed in Table 5.2 result in no change in nuclear charge when emitted in nuclear decay?

SAMPLE EXERCISE 5.2   **Writing nuclear equations**

Write nuclear equations for the following processes: **(a)** mercury-201 undergoes electron capture; **(b)** thorium-231 decays to form protactinium-231.

SOLUTION

Analyse We must write balanced nuclear equations in which the masses and charges of reactants and products are equal.

Plan We can begin by writing the complete chemical symbols for the nuclei and decay particles that are given in the problem.

Solve
(a) The information given in the question can be summarised as

$$^{201}_{80}Hg + {}^0_{-1}e \longrightarrow {}^A_ZX$$

The mass numbers must have the same sum on both sides of the equation:

$$201 + 0 = A$$

---

* The positron has a very short life because it is annihilated when it collides with an electron, producing gamma rays: $^0_1e + {}^0_{-1}e \longrightarrow 2{}^0_0\gamma$.

Thus the product nucleus must have a mass number of 201. Similarly, balancing the atomic numbers gives

$$80 - 1 = Z$$

Thus, the atomic number of the product nucleus must be 79, which identifies it as gold (Au):

$$^{201}_{80}\text{Hg} + ^{0}_{-1}\text{e} \longrightarrow ^{201}_{79}\text{Au}$$

(b) In this case we must determine what type of particle is emitted in the course of the radioactive decay:

$$^{231}_{90}\text{Th} \longrightarrow ^{231}_{91}\text{Pa} + ^{A}_{Z}\text{X}$$

From $231 = 231 + A$ and $90 = 91 + Z$, we deduce $A = 0$ and $Z = -1$. According to Table 5.2, the particle with these characteristics is the beta particle (electron). We therefore write

$$^{231}_{90}\text{Th} \longrightarrow ^{231}_{91}\text{Pa} + ^{0}_{-1}\text{e}$$

PRACTICE EXERCISE

Write a balanced nuclear equation for the reaction in which oxygen-15 undergoes positron emission.

Answer: $^{15}_{8}\text{O} \longrightarrow ^{15}_{7}\text{N} + ^{0}_{1}\text{e}$

(See also Exercises 5.11(c), 5.12(d).)

5.2 | PATTERNS OF NUCLEAR STABILITY

The stability of a particular nucleus depends on a variety of factors, and no single rule allows us to predict whether a particular nucleus is radioactive or how it might decay. There are, however, several empirical observations that will help you predict the stability of a nucleus.

Neutron-to-Proton Ratio

Because like charges repel each other, it may seem surprising that a large number of protons can reside within the small volume of the nucleus. At close distances, however, a strong force of attraction, called the *strong nuclear force*, exists between nucleons. This force is sufficient to bind neutrons and protons to generate the stable nuclei that we know. All nuclei other than $^{1}_{1}\text{H}$ contain neutrons. As the number of protons in the nucleus increases, there is an ever greater need for neutrons to counteract the effect of proton–proton repulsions. Stable nuclei with low atomic numbers (up to about 20) have approximately equal numbers of neutrons and protons. For nuclei with higher atomic numbers, the number of neutrons exceeds the number of protons. Indeed, the number of neutrons necessary to create a stable nucleus increases more rapidly than the number of protons, as shown in ▼ FIGURE 5.1. Thus the neutron-to-proton ratios of stable nuclei increase with increasing atomic number as illustrated by the most common isotopes of carbon, $^{12}_{6}\text{C}$ (n/p = 1), manganese, $^{55}_{25}\text{Mn}$ (n/p = 1.20) and gold, $^{197}_{79}\text{Au}$ (n/p = 1.49).

The dark blue dots in Figure 5.1 represent stable (non-radioactive) elements. The region of the graph covered by these dark blue dots is known as the *belt of stability*. The belt of stability ends at element 83 (bismuth). *All nuclei with 84 or more protons (atomic number ≥ 84) are radioactive.* For example, all isotopes of uranium, atomic number 92, are radioactive.

CONCEPT CHECK 3

Fluorine, sodium, aluminium and phosphorus each have only one stable isotope. What can you say about the number of neutrons in these isotopes?

◢ FIGURE IT OUT

Estimate the optimal number of neutrons for a nucleus containing 70 protons.

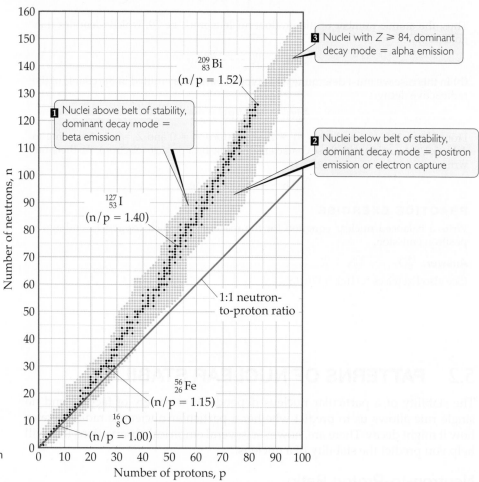

▶ **FIGURE 5.1 Stable and radioactive isotopes as a function of numbers of neutrons and protons in a nucleus.** The stable nuclei (dark blue dots) define a region known as the belt of stability.

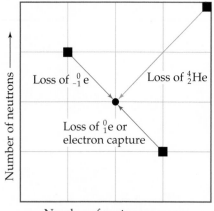

▲ **FIGURE 5.2 Proton and neutron changes in nuclear processes.** The graph shows the results of alpha emission ($_2^4$He), beta emission ($_{-1}^0$e), positron emission ($_1^0$e) and electron capture on the number of protons and neutrons in a nucleus. Moving from left to right or from bottom to top, each square represents an additional proton or neutron, respectively. Moving in the reverse direction indicates the loss of a proton or neutron.

The type of radioactive decay that a particular radionuclide undergoes depends to a large extent on how its neutron-to-proton ratio compares with those of nearby nuclei within the belt of stability. We can envisage three general situations:

1. Nuclei above the belt of stability (high neutron-to-proton ratios). These neutron-rich nuclei can lower their ratio and move towards the belt of stability by emitting a beta particle. Beta emission decreases the number of neutrons and increases the number of protons in a nucleus, as shown in Equation 5.3.

2. Nuclei below the belt of stability (low neutron-to-proton ratios). These proton-rich nuclei can increase their ratio by either positron emission or electron capture. Both kinds of decay increase the number of neutrons and decrease the number of protons, as shown in Equations 5.5 and 5.7. Positron emission is more common than electron capture among the lighter nuclei; however, electron capture becomes increasingly common as nuclear charge increases.

3. Nuclei with atomic numbers ⩾ 84. These heavy nuclei, which lie beyond the upper right edge of the band of stability, tend to undergo alpha emission. Emission of an alpha particle decreases both the number of neutrons and the number of protons by 2, moving the nucleus diagonally towards the belt of stability.

These three situations are summarised in ◀ **FIGURE 5.2.**

| SAMPLE EXERCISE 5.3 | Predicting modes of nuclear decay |

Predict the mode of decay of **(a)** carbon-14, **(b)** xenon-118.

SOLUTION

Analyse We are asked to predict the modes of decay of two nuclei.

Plan To do this, we must calculate the neutron-to-proton ratios and compare the values with those for nuclei that lie within the belt of stability shown in Figure 5.1.

Solve
(a) Carbon has an atomic number of 6. Thus carbon-14 has 6 protons and $14 - 6 = 8$ neutrons, giving it a neutron-to-proton ratio of $\frac{8}{6} = 1.3$. Elements with low atomic numbers normally have stable nuclei with approximately equal numbers of neutrons and protons. Thus carbon-14 has a high neutron-to-proton ratio and we expect that it will decay by emitting a beta particle:

$$^{14}_{6}\text{C} \longrightarrow {}^{0}_{-1}\text{e} + {}^{14}_{7}\text{N}$$

This is indeed the mode of decay observed for carbon-14.

(b) Xenon has an atomic number of 54. Thus xenon-118 has 54 protons and $118 - 54 = 64$ neutrons, giving it a neutron-to-proton ratio of $\frac{64}{54} = 1.2$. According to Figure 5.1, stable nuclei in this region of the belt of stability have higher neutron-to-proton ratios than xenon-118. The nucleus can increase this ratio by either positron emission or electron capture:

$$^{118}_{54}\text{Xe} \longrightarrow {}^{0}_{1}\text{e} + {}^{118}_{53}\text{I}$$

$$^{118}_{54}\text{Xe} + {}^{0}_{-1}\text{e} \longrightarrow {}^{118}_{53}\text{I}$$

In this case, both modes of decay are observed.

Comment Keep in mind that our guidelines don't always work. For example, thorium-233, $^{233}_{90}\text{Th}$, which we might expect to undergo alpha decay, actually undergoes beta decay. Furthermore, a few radioactive nuclei actually lie within the belt of stability. Both $^{146}_{60}\text{Nd}$ and $^{148}_{60}\text{Nd}$, for example, are stable and lie in the belt of stability. $^{147}_{60}\text{Nd}$, however, which lies between them, is radioactive.

PRACTICE EXERCISE

Predict the mode of decay of **(a)** plutonium-239, **(b)** indium-120.

Answers: **(a)** α decay, **(b)** β decay

(See also Exercises 5.16, 5.17.)

Radioactive Series

Some nuclei, like uranium-238, cannot gain stability by a single nuclear reaction, so progress through a series of successive emissions before a stable nucleus is achieved. As shown in ▼ **FIGURE 5.3**, uranium-238 decays to thorium-234, which is radioactive and decays to protactinium-234. This nucleus is also unstable and subsequently decays. Such successive reactions continue until a stable nucleus, lead-206, is formed. A series of nuclear reactions that begins with an unstable nucleus and terminates with a stable one is known as a **radioactive series**, or a **nuclear disintegration series**. Three such series occur in nature. In addition to the series that begins with uranium-238 and terminates with lead-206, there is one that begins with uranium-235 and ends with lead-207, and one that begins with thorium-232 and ends with lead-208.

Further Observations

Two further observations can help you predict nuclear stability:

- Nuclei with 2, 8, 20, 28, 50 or 82 protons or 2, 8, 20, 28, 50, 82 or 126 neutrons are generally more stable than nuclei that do not contain these numbers of nucleons. These numbers of protons and neutrons are called **magic numbers**.

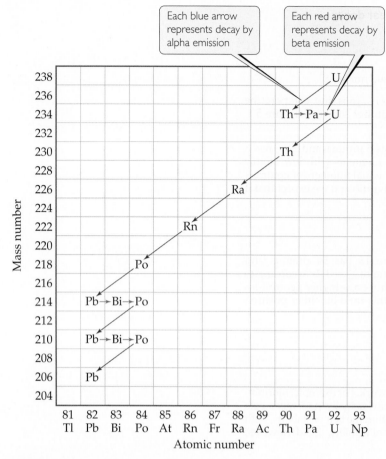

Each blue arrow represents decay by alpha emission

Each red arrow represents decay by beta emission

▲ FIGURE 5.3 **Nuclear disintegration series for uranium-238.** The $^{238}_{92}$U nucleus decays to $^{234}_{90}$Th. Subsequent decay processes eventually form the stable $^{206}_{82}$Pb nucleus.

• Nuclei with even numbers of protons, neutrons, or both, are generally more stable than those with odd numbers of nucleons, as shown in ▼ TABLE 5.3.

⚠ CONCEPT CHECK 4

For a nucleus containing 20 protons, how many neutrons would be likely to produce the most stable nucleus: 30, 24 or 25?

These observations can be understood in terms of the *shell model of the nucleus*, in which nucleons are described as residing in shells analogous to the shell structure for electrons in atoms. Just as certain numbers of electrons (2, 8, 18, 36, 54 and 86) correspond to stable closed-shell electron configurations, so also the magic numbers of nucleons represent closed shells in nuclei. As an example of the stability of nuclei with magic numbers of nucleons, note that the radioactive series depicted in Figure 5.3 ends with formation of the stable $^{206}_{82}$Pb nucleus, which has a magic number of protons (82).

Evidence also suggests that pairs of protons and pairs of neutrons have a special stability, analogous to the pairs of electrons in molecules. Thus stable nuclei with an even number of protons and an even number of neutrons are far more numerous than those with odd numbers (Table 5.3).

TABLE 5.3 • **The number of stable isotopes with even and odd numbers of protons and neutrons**

Number of stable isotopes	Protons	Neutrons
157	Even	Even
53	Even	Odd
50	Odd	Even
5	Odd	Odd

SAMPLE EXERCISE 5.4 Predicting nuclear stability

Which of the following nuclei are especially stable: $^{4}_{2}$He, $^{40}_{20}$Ca, $^{98}_{43}$Tc?

SOLUTION

Analyse We are asked to identify especially stable nuclei, given their mass numbers and atomic numbers.

Plan We look to see whether the numbers of protons and neutrons correspond to magic numbers.

Solve The $^{4}_{2}$He nucleus (the alpha particle) has a magic number of both protons (2) and neutrons (2) and is very stable. The $^{40}_{20}$Ca nucleus also has a magic number of both protons (20) and neutrons (20) and is especially stable.

The $^{98}_{43}$Tc nucleus does not have a magic number of either protons or neutrons. In fact, it has an odd number of both protons (43) and neutrons (55). There are very few stable nuclei with odd numbers of both protons and neutrons. Indeed, technetium-98 is radioactive.

PRACTICE EXERCISE

Which of the following nuclei would you expect to exhibit especially high stability: $^{118}_{50}$Sn, $^{210}_{85}$At, $^{208}_{82}$Pb?

Answer: $^{118}_{50}$Sn, $^{208}_{82}$Pb

(See also Exercises 5.18, 5.19.)

5.3 | NUCLEAR TRANSMUTATIONS

So far we have examined nuclear reactions in which a nucleus spontaneously decays. A nucleus can also change identity if it is struck by a neutron or by another nucleus. Nuclear reactions that are induced in this way are known as **nuclear transmutations**.

The first conversion of one nucleus into another was performed in 1919 by Ernest Rutherford. He succeeded in converting nitrogen-14 into oxygen-17, plus a proton, using the high-velocity alpha particles emitted by radium. The reaction is

$$^{14}_{7}\text{N} + ^{4}_{2}\text{He} \longrightarrow ^{17}_{8}\text{O} + ^{1}_{1}\text{H} \qquad [5.8]$$

This reaction demonstrated that nuclear reactions can be induced by striking nuclei with particles such as alpha particles. Such reactions made it possible to synthesise hundreds of radioisotopes in the laboratory.

Nuclear transmutations are sometimes represented by listing, in order, the target nucleus, the bombarding particle, the ejected particle and the product nucleus. Using this condensed notation, Equation 5.8 becomes:

Target nucleus   Product nucleus

$$^{14}_{7}\text{N} \ (\alpha, \ \text{p})^{17}_{8}\text{O}$$

Bombarding particle   Ejected particle

SAMPLE EXERCISE 5.5   **Writing a balanced nuclear equation**

Write the balanced nuclear equation for the process summarised as $^{27}_{13}\text{Al}(\text{n}, \alpha)^{24}_{11}\text{Na}$.

SOLUTION

Analyse We must go from the condensed descriptive form of the reaction to the balanced nuclear equation.

Plan We arrive at the balanced equation by writing n and α, each with its associated subscripts and superscripts.

Solve The n is the abbreviation for a neutron ($^{1}_{0}\text{n}$) and α represents an alpha particle ($^{4}_{2}\text{He}$). The neutron is the bombarding particle and the alpha particle is a product. Therefore, the nuclear equation is

$$^{27}_{13}\text{Al} + ^{1}_{0}\text{n} \longrightarrow ^{24}_{11}\text{Na} + ^{4}_{2}\text{He}$$

PRACTICE EXERCISE

Using a shorthand notation, write the nuclear reaction

$$^{16}_{8}\text{O} + ^{1}_{1}\text{H} \longrightarrow ^{13}_{7}\text{N} + ^{4}_{2}\text{He}$$

Answer: $^{16}_{8}\text{O}(\text{p}, \alpha)^{13}_{7}\text{N}$

(See also Exercises 5.26, 5.27.)

Accelerating Charged Particles

Charged particles, such as alpha particles, must be moving very fast in order to overcome the electrostatic repulsion between them and the target nucleus. The higher the nuclear charge on either the projectile or the target, the faster the projectile must be moving to bring about a nuclear reaction. Many methods have been devised to accelerate charged particles, using strong magnetic and electrostatic fields. These **particle accelerators**, popularly called 'atom smashers', bear such names as *cyclotron* and *synchrotron*. The cyclotron is illustrated in ▶ FIGURE 5.4. The hollow D-shaped electrodes are called 'dees'. The projectile particles are introduced into a vacuum chamber within the cyclotron. The particles are then accelerated by making the dees alternately positively and negatively charged. Magnets placed above and below the dees keep the particles

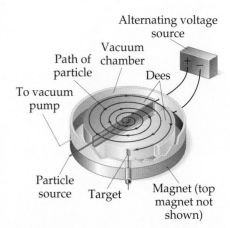

▲ **FIGURE 5.4   Schematic drawing of a cyclotron.** Charged particles are accelerated around the ring by applying alternating voltage to the dees.

∞ Review this on page 191

moving in a spiral path until they are finally deflected out of the cyclotron and emerge to strike a target substance. Particle accelerators have been used mainly to synthesise heavy elements and to investigate the fundamental structure of matter (∞ Section 6.4, 'The Wave Behaviour of Matter').

Reactions Involving Neutrons

Most synthetic isotopes used in medicine and scientific research are made using neutrons as projectiles. Because neutrons are neutral, they are not repelled by the nucleus. Consequently, they do not need to be accelerated, as do charged particles, in order to cause nuclear reactions. (Indeed, they cannot be accelerated.) The necessary neutrons are produced by the reactions that occur in nuclear reactors. Cobalt-60, for example, used in radiation therapy for cancer, is produced by neutron capture. Iron-58 is placed in a nuclear reactor, where it is bombarded by neutrons. The following sequence of reactions takes place:

$$\ce{^{58}_{26}Fe} + \ce{^{1}_{0}n} \longrightarrow \ce{^{59}_{26}Fe} \tag{5.9}$$

$$\ce{^{59}_{26}Fe} \longrightarrow \ce{^{59}_{27}Co} + \ce{^{0}_{-1}e} \tag{5.10}$$

$$\ce{^{59}_{27}Co} + \ce{^{1}_{0}n} \longrightarrow \ce{^{60}_{27}Co} \tag{5.11}$$

Transuranium Elements

Nuclear transmutations have collectively been used to produce the elements with atomic number above 92. These are collectively known as the **transuranium elements** because they occur immediately following uranium in the periodic table. Elements 93 (neptunium, Np) and 94 (plutonium, Pu) were first discovered in 1940. They were produced by bombarding uranium-238 with neutrons:

$$\ce{^{238}_{92}U} + \ce{^{1}_{0}n} \longrightarrow \ce{^{239}_{92}U} \longrightarrow \ce{^{239}_{93}Np} + \ce{^{0}_{-1}e} \tag{5.12}$$

$$\ce{^{239}_{93}Np} \longrightarrow \ce{^{239}_{94}Pu} + \ce{^{0}_{-1}e} \tag{5.13}$$

Elements with still larger atomic numbers are normally formed in small quantities in particle accelerators. Curium-242, for example, is formed when a plutonium-239 target is struck with accelerated alpha particles:

$$\ce{^{239}_{94}Pu} + \ce{^{4}_{2}He} \longrightarrow \ce{^{242}_{96}Cm} + \ce{^{1}_{0}n} \tag{5.14}$$

In 1994 a team of European scientists synthesised roentgenium (element 111) by bombarding a bismuth target for several days with a beam of nickel atoms:

$$\ce{^{209}_{83}Bi} + \ce{^{64}_{28}Ni} \longrightarrow \ce{^{272}_{111}Rg} + \ce{^{1}_{0}n}$$

Amazingly, their discovery was based on the detection of only three atoms of the new element. The nuclei are very short-lived and undergo alpha decay within milliseconds of their synthesis. More recently, elements 112–118 (with the exception of 117) have also been synthesised.

5.4 | RATES OF RADIOACTIVE DECAY

Some radioisotopes, such as uranium-238, are found in nature even though they are not stable. Other radioisotopes do not exist in nature but can be synthesised in nuclear reactions. To understand this distinction, we must realise that different nuclei undergo radioactive decay at different rates. Many radioisotopes decay essentially completely in a matter of seconds, so we do not find them in nature. Uranium-238, in contrast, decays very slowly. Therefore, despite its instability, we can still observe what remains from its formation in the early history of the universe.

Radioactive decay is a first-order kinetic process. A first-order process has a characteristic **half-life**, which is the time required for half of any given quantity of a substance to react (∞ Section 15.4, 'The Change of Concentration with Time (Integrated Rate Equations)'). Nuclear decay rates are commonly expressed in terms of half-lives. Each isotope has its own characteristic half-life.

For example, the half-life of strontium-90 is 28.8 years. If we started with 10.0 g of strontium-90, only 5.0 g of that isotope would remain after 28.8 years, 2.5 g would remain after another 28.8 years, and so on. Strontium-90 decays to yttrium-90:

$$\ce{^{90}_{38}Sr} \longrightarrow \ce{^{90}_{39}Y} + \ce{^{0}_{-1}e} \qquad [5.15]$$

The loss of strontium-90 as a function of time is shown in ▶ FIGURE 5.5.

Half-lives as short as millionths of a second and as long as billions of years are known. The half-lives of some radioisotopes are listed in ▼ TABLE 5.4. One important feature of half-lives for nuclear decay is that they are unaffected by external conditions such as temperature, pressure or state of chemical combination. Unlike toxic chemicals, therefore, radioactive atoms cannot be reacted to make them harmless, only contained or diluted to reduce the impact of their emissions.

∞ Find out more on page 578

FIGURE IT OUT

If we start with a 50.0 g sample, how much of it remains after three half-lives have passed?

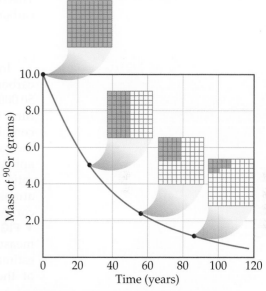

▲ **FIGURE 5.5   Decay of a 10.0 g sample of strontium-90 ($t_{1/2}$ = 28.8 years).** The 10 × 10 grids show how much of the radioactive isotope remains after various amounts of time.

TABLE 5.4 • The half-lives and type of decay for several radioisotopes

	Isotope	Half-life (years)	Type of decay
Natural radioisotopes	$\ce{^{238}_{92}U}$	4.5×10^9	Alpha
	$\ce{^{235}_{92}U}$	7.0×10^8	Alpha
	$\ce{^{232}_{90}Th}$	1.4×10^{10}	Alpha
	$\ce{^{40}_{19}K}$	1.3×10^9	Beta
	$\ce{^{14}_{6}C}$	5715	Beta
Synthetic radioisotopes	$\ce{^{239}_{94}Pu}$	24 000	Alpha
	$\ce{^{137}_{55}Cs}$	30	Beta
	$\ce{^{90}_{38}Sr}$	28.8	Beta
	$\ce{^{131}_{53}I}$	0.022	Beta

SAMPLE EXERCISE 5.6   Calculation involving half-life

The half-life of cobalt-60 is 5.3 years. How much of a 1.000 mg sample of cobalt-60 is left after a 15.9 years period?

SOLUTION

Analyse We are given the half-life for cobalt-60 and asked to calculate the amount of cobalt-60 remaining from an initial 1.000 mg sample after 15.9 years.

Plan We will use the fact that the amount of radioactive substance decreases by 50% for every half-life that passes.

Solve A period of 15.9 years is three half-lives for cobalt-60. At the end of one half-life, 0.500 mg of cobalt-60 remains, 0.250 mg at the end of two half-lives and 0.125 mg at the end of three half-lives.

PRACTICE EXERCISE

Carbon-11, used in medical imaging, has a half-life of 20.4 min. The carbon-11 nuclides are formed and the carbon atoms are then incorporated into an appropriate compound. The resulting sample is injected into a patient and the medical image obtained. If the entire process takes five half-lives, what percentage of the original carbon-11 remains at this time?

Answer: 3.12%

(See also Exercises 5.30–5.32.)

Radiometric Dating

Because the half-life of any particular nuclide is constant, the half-life can serve as a nuclear clock to determine the ages of different objects. The method of dating objects based on their isotopes and isotope abundances is known as **radiometric dating**. Carbon-14, for example, has been used to determine the age of organic materials. This is known as **radiocarbon dating**. The procedure is based on the formation of carbon-14 by capture of solar neutrons in the upper atmosphere:

$$\ce{^{14}_{7}N + ^{1}_{0}n \longrightarrow ^{14}_{6}C + ^{1}_{1}p} \qquad [5.16]$$

This reaction provides a small but reasonably constant source of carbon-14. The carbon-14 is radioactive, undergoing beta decay with a half-life of 5715 years:

$$\ce{^{14}_{6}C \longrightarrow ^{14}_{7}N + ^{0}_{-1}e} \qquad [5.17]$$

In using radiocarbon dating, we generally assume that the ratio of carbon-14 to carbon-12 in the atmosphere has been constant for the past 50 000 years. The carbon-14 is incorporated into carbon dioxide, which is in turn incorporated, through photosynthesis, into more complex carbon-containing molecules within plants. When the plants are eaten by animals, the carbon-14 becomes incorporated within them. Because a living plant or animal has a constant intake of carbon compounds, it is able to maintain a ratio of carbon-14 to carbon-12 that is nearly identical with that of the atmosphere. Once the organism dies, however, it no longer ingests carbon compounds to replenish the carbon-14 that is lost through radioactive decay ▼ **FIGURE 5.6**. The ratio of carbon-14 to carbon-12 therefore decreases. By measuring this ratio and contrasting it with that of the atmosphere, we can estimate the age of an object. For example, if the ratio diminishes to half that of the atmosphere, we can conclude that the object is one half-life, or

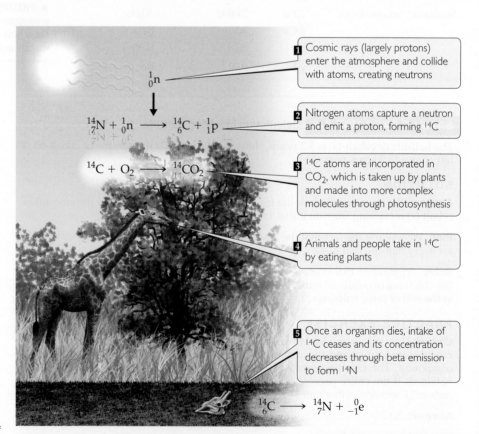

1 Cosmic rays (largely protons) enter the atmosphere and collide with atoms, creating neutrons

2 Nitrogen atoms capture a neutron and emit a proton, forming $\ce{^{14}C}$

3 $\ce{^{14}C}$ atoms are incorporated in CO_2, which is taken up by plants and made into more complex molecules through photosynthesis

4 Animals and people take in $\ce{^{14}C}$ by eating plants

5 Once an organism dies, intake of $\ce{^{14}C}$ ceases and its concentration decreases through beta emission to form $\ce{^{14}N}$

▶ **FIGURE 5.6 Creation and distribution of carbon-14.** The ratio of carbon-14 to carbon-12 in a dead animal or plant is related to the time since death occurred.

5715 years, old. This method cannot be used to date objects older than about 50 000 years. After this length of time the radioactivity is too low to be measured accurately.

The radiocarbon dating technique has been checked by comparing the ages of trees determined by counting their rings and by radiocarbon analysis. As a tree grows, it adds a ring each year. In the old growth the carbon-14 decays, but the concentration of carbon-12 remains constant. The two dating methods agree to within about 10%.

Calculations Based on Half-Life

We now consider the topic of half-lives from a more quantitative point of view. This approach enables us to answer questions of the following types: How do we determine the half-life of uranium-238? Similarly, how do we quantitatively determine the age of an object?

Radioactive decay is a first-order kinetic process. Its rate is therefore proportional to the number of radioactive nuclei, N, in the sample:

$$\text{Rate} = kN \tag{5.18}$$

The first-order rate constant, k, is called the *decay constant*. The rate at which a sample decays is called its **activity** (A); it is often expressed as the number of disintegrations observed per unit time. The **becquerel** (Bq) is the SI unit for expressing the activity of a particular radiation source (that is, the rate at which nuclear disintegrations are occurring). A becquerel is defined as one nuclear disintegration per second. An older, but still widely used, unit of activity especially in medicine is the **curie** (Ci), defined as 3.7×10^{10} disintegrations per second, which is the rate of decay of 1 g of radium. Thus 1 Ci = 3.7×10^{10} Bq.

As we will see in Section 15.4, a first-order rate law can be transformed into the following equation (⬥ Section 15.4, 'The Change of Concentration with Time (Integrated Rate Equations)'):

⬥ Find out more on page 578

$$\ln \frac{N_t}{N_0} = -kt \tag{5.19}$$

In this equation t is the time interval of decay, k is the decay constant, N_0 is the initial number of nuclei (at time zero) and N_t is the number remaining after the time interval. Both the mass of a particular radioisotope and its activity are proportional to the number of radioactive nuclei. Thus either the ratio of the mass at any time t to the mass at time $t = 0$ or the ratio of the activities at time t and $t = 0$ can be substituted for N_t/N_0 in Equation 5.19.

From Equation 5.19 we can obtain the relationship between the decay constant, k, and half-life, $t_{1/2}$ (⬥ Section 15.4, 'The Change of Concentration with Time (Integrated Rate Equations)').

⬥ Find out more on page 578

$$k = \frac{0.693}{t_{1/2}} \tag{5.20}$$

Thus if we know the value of either the decay constant or the half-life, we can calculate the value of the other.

CONCEPT CHECK 5

a. Would doubling the mass of a radioactive sample change the amount of radioactivity the sample shows?

b. Would doubling the mass change the half-life for the radioactive decay?

SAMPLE EXERCISE 5.7 Calculating the age of a mineral

A rock contains 0.257 mg of lead-206 for every milligram of uranium-238. The half-life for the decay of uranium-238 to lead-206 is 4.5×10^9 years. How old is the rock?

SOLUTION

Analyse We are told that a rock sample has a certain amount of lead-206 for every unit mass of uranium-238 and asked to estimate the age of the rock.

Plan We must assume the lead-206 is due entirely to radioactive decay of uranium-228 to form lead-206, with a known half-life. To apply first-order kinetics expressions (Equations 5.19 and 5.20) to calculate the time elapsed since the rock was formed, we need first to calculate how much initial uranium-238 there was for every 1 milligram that remains today.

Solve Let's assume that the rock contains 1.000 mg of uranium-238 at present. The amount of uranium-238 in the rock when it was first formed, therefore, equals 1.000 mg plus the quantity that decayed to lead-206. We obtain the latter quantity by multiplying the present mass of lead-206 by the ratio of the mass number of uranium to that of lead, into which it has decayed. The total original $^{238}_{92}U$ was thus

$$\text{Original } ^{238}_{92}U = 1.000 \text{ mg} + \frac{238}{206}(0.257 \text{ mg})$$

$$= 1.297 \text{ mg}$$

Using Equation 5.20, we can calculate the decay constant for the process from its half-life:

$$k = \frac{0.693}{4.5 \times 10^9 \text{ yr}} = 1.5 \times 10^{-10} \text{ yr}^{-1}$$

Rearranging Equation 5.19 to solve for time, t, and substituting known quantities gives

$$t = -\frac{1}{k}\ln\frac{N_t}{N_0} = -\frac{1}{1.5 \times 10^{-10} \text{ yr}^{-1}}\ln\frac{1.000 \text{ mg}}{1.297 \text{ mg}} = 1.7 \times 10^9 \text{ yr}$$

PRACTICE EXERCISE

A wooden object from an archaeological site is subjected to radiocarbon dating. The activity of the sample that is due to ^{14}C is measured to be 11.6 disintegrations per second. The activity of a carbon sample of equal mass from fresh wood is 15.2 disintegrations per second. The half-life of ^{14}C is 5715 years. What is the age of the archaeological sample?

Answer: 2230 years

(See also Exercises 5.35, 5.36.)

SAMPLE EXERCISE 5.8 Calculations involving radioactive decay

If we start with 1.000 g of strontium-90, 0.953 g will remain after 2.00 years. **(a)** What is the half-life of strontium-90? **(b)** How much strontium-90 will remain after 5.00 years? **(c)** What is the initial activity of the sample in Bq?

SOLUTION

(a)

Analyse We are asked to calculate a half-life, $t_{1/2}$, based on data that tell us how much of a radioactive nucleus has decayed in a time interval $t = 2.00$ years and the information $N_0 = 1.000$ g, $N_t = 0.935$ g.

Plan We first calculate the rate constant for the decay k, and then use that to calculate $t_{1/2}$.

Solve Equation 5.19 is solved for the decay constant, k, and then Equation 5.20 is used to calculate half-life, $t_{1/2}$:

$$k = -\frac{1}{t}\ln\frac{N_t}{N_0} = -\frac{1}{2.00 \text{ yr}}\ln\frac{0.953 \text{ g}}{1.000 \text{ g}}$$

$$= -\frac{1}{2.00 \text{ yr}}(-0.0481) = 0.0241 \text{ yr}^{-1}$$

$$t_{1/2} = \frac{0.693}{k} = \frac{0.693}{0.0241 \text{ yr}^{-1}} = 28.8 \text{ yr}$$

(b)

Analyse We are asked to calculate the amount of a radionuclide remaining after a given period of time.

Plan We need to calculate N_t, the amount of strontium at time t, using the initial quantity, N_0, and the rate constant for decay, k, calculated in part (a).

Solve Again using Equation 5.19, with $k = 0.0241 \text{ yr}^{-1}$, we have

$$\ln \frac{N_t}{N_0} = -kt = -(0.0241 \text{ yr}^{-1})(5.00 \text{ yr}) = -0.120$$

$$\frac{N_t}{N_0} = e^{-0.120} = 0.887$$

Because $N_0 = 1.000$ g, we have

$$N_t = (0.887)N_0 = (0.887)(1.000 \text{ g}) = 0.887 \text{ g}$$

(c)

Analyse We are asked to calculate the activity of the sample in Bq.

Plan We must calculate the number of disintegrations per second, then multiply by the number of atoms in the sample.

Solve The number of disintegrations per atom per second is given by the rate constant, k.

$$k = 0.0241 \text{ disintegrations yr}^{-1} = -\frac{0.0241 \text{ disintegrations}}{(365 \times 24 \times 3600) \text{ s}}$$

$$= 7.64 \times 10^{-10} \text{ disintegrations s}^{-1} \text{ atom}^{-1}$$

To obtain the total number of disintegrations per second, we calculate the number of atoms in the sample. We multiply this quantity by k, where we express k as the number of disintegrations per atom per second, to obtain the number of disintegrations per second:

$$\frac{1.00 \text{ g }^{90}\text{Sr}}{90 \text{ g mol}^{-1} \, {}^{90}\text{Sr}} \times 6.022 \times 10^{23} \text{ atoms }^{90}\text{Sr mol}^{-1} = 6.7 \times 10^{21} \text{ atoms }^{90}\text{Sr}$$

Total disintegrations s^{-1} atom^{-1} =

$$(7.64 \times 10^{-10} \text{ disintegrations s}^{-1} \text{ atom}^{-1})(6.7 \times 10^{21} \text{ atoms})$$

$$= 5.1 \times 10^{12} \text{ disintegrations s}^{-1} = 5.1 \times 10^{12} \text{ Bq}$$

We have used only two significant figures in products of these calculations because we don't know the atomic weight of ^{90}Sr to more than two significant figures without looking it up in a special source. Using a more accurate mass of ^{90}Sr of 89.9 g mol^{-1} would give an answer of 5.12×10^{12} Bq.

PRACTICE EXERCISE

A sample to be used for medical imaging is labelled with ^{18}F, which has a half-life of 110 min. What percentage of the original activity in the sample remains after 300 min?

Answer: 15.1%

(See also Exercise 5.32.)

5.5 | DETECTION OF RADIOACTIVITY

A variety of methods have been devised to detect emissions from radioactive substances. Henri Becquerel discovered radioactivity because of the effect of radiation on photographic plates. Photographic plates and film have long been used to detect radioactivity. The radiation affects photographic film in much the same way as X-rays do. With care, film can be used to give a quantitative measure of activity. The greater the extent of exposure to radiation, the darker the area of the developed negative. People who work with radioactive substances can carry a radiation monitor or film badges to record the extent of their exposure to radiation (▼ FIGURE 5.7(a), (b)).

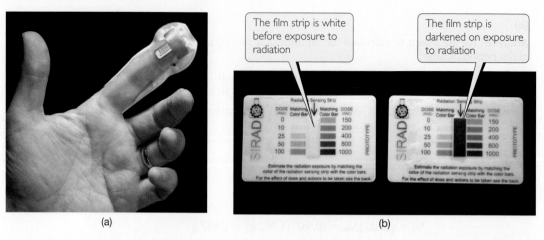

(a)   (b)

▲ **FIGURE 5.7** **(a) Personal radiation monitor on a man's finger.** This is an extremity TLD (thermoluminescent dosimeter). It is worn by people who are exposed to radiation regularly during their work. A piece of plastic within the dosimeter becomes pitted when hit by radiation. The number of pits is proportional to the amount of radiation, allowing a person's daily exposure to be calculated. **(b) Badge dosimeters monitor the extent to which the individual has been exposed to high-energy radiation.** The radiation dose is determined from the extent of darkening of the film in the dosimeter.

Radioactivity can also be detected and measured using a device known as a **Geiger counter** (▼ **FIGURE 5.8**), which detects high-energy (ionising) radiation. It consists of a metal tube with a thin plastic or glass window at one end that can be penetrated by alpha, beta and gamma rays. The tube is filled with a gas at low pressure and has a wire through the centre of the tube connected to the anode of a high-voltage direct current source via an amplifier and counter. The tube itself is connected to the cathode of the direct current source. Entering radiation ionises the gas molecules, producing positive ions that move to the negatively charged metal tube (cathode) and electrons that are attracted to the positively charged wire (anode). The resulting small electric current (pulse) is amplified and used to cause a clicking sound or a flashing light. The number of clicks or flashes per given time depends on the intensity of the radiation.

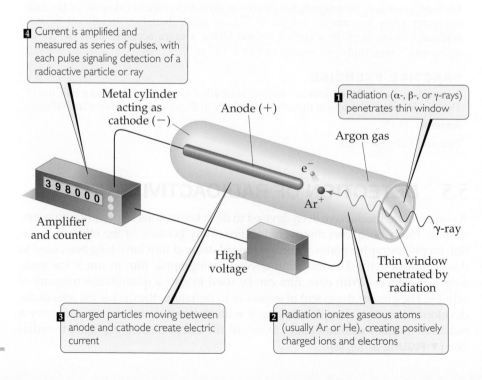

▶ **FIGURE 5.8** **Schematic drawing of a Geiger counter.**

CONCEPT CHECK 6
Will alpha, beta and gamma rays pass through the window of a Geiger counter detection tube with equal efficiency?

Certain substances that are electronically excited by radiation can also be used to detect and measure radiation. For example, some substances excited by radiation give off light as electrons return to their lower-energy states. These substances are called *phosphors*. Different substances respond to different particles. Zinc sulfide, for example, responds to alpha particles. An instrument called a **scintillation counter** is used to detect and measure radiation, based on the tiny flashes of light produced when radiation strikes a suitable phosphor. The flashes are magnified electronically and counted to measure the amount of radiation.

Radiotracers

Because radioisotopes can be detected so readily, they can be used to follow an element through its chemical reactions. The incorporation of carbon atoms from CO_2 into glucose in photosynthesis, for example, has been studied using CO_2 containing carbon-14:

$$6\,^{14}CO_2 + 6\,H_2O \xrightarrow[\text{Chlorophyll}]{\text{Sunlight}} {}^{14}C_6H_{12}O_6 + 6\,O_2 \qquad [5.21]$$

The CO_2 containing carbon-14 is said to be 'labelled' with carbon-14. Detection devices such as scintillation counters follow the carbon-14 as it moves from the CO_2 through the various intermediate compounds to glucose.

The use of radioisotopes is possible because all isotopes of an element have essentially identical chemical properties. When a small quantity of a radioisotope is mixed with the naturally occurring stable isotopes of the same element, all the isotopes go through the same reactions together. The element's path is revealed by the radioactivity of the radioisotope. Because the radioisotope can be used to trace the path of the element through a trail of chemical reactions, it is called a **radiotracer**.

5.6 | ENERGY CHANGES IN NUCLEAR REACTIONS

The energy associated with nuclear reactions can be considered with the aid of Einstein's famous equation relating mass and energy:

$$E = mc^2 \qquad [5.22]$$

In this equation E stands for energy, m for mass and c for the speed of light, 3.00×10^8 m s^{-1}. This equation states that the mass and energy of an object are proportional. If a system loses mass, it loses energy (exothermic); if it gains mass, it gains energy (endothermic) (Section 14.2, 'The First Law of Thermodynamics'). Because the proportionality constant in the equation, c^2, is such a large number, even small changes in mass are accompanied by large changes in energy.

Find out more on page 505

The mass changes in chemical reactions are too small to detect. Because the mass change is so small, it is possible to treat chemical reactions as though mass is conserved.

The mass changes and the associated energy changes in nuclear reactions are much greater than those in chemical reactions. The mass change accompanying the radioactive decay of a mol of uranium-238, for example, is

50 000 times greater than that for the combustion of 1 mol of CH_4. Let's examine the energy change for the nuclear reaction:

$$^{238}_{92}U \longrightarrow {}^{234}_{90}Th + {}^{4}_{2}He$$

The nuclei in this reaction have the following masses: $^{238}_{98}U$, 238.0003 u; $^{234}_{90}Th$, 233.9942 u; and $^{4}_{2}He$, 4.0015 u. The mass change, Δm, is the total mass of the products minus the total mass of the reactants. The mass change for the decay of a *mole* of uranium-238 can then be expressed in grams:

$$233.9942 \text{ g} + 4.0015 \text{ g} - 238.0003 \text{ g} = -0.0046 \text{ g}$$

MY WORLD OF CHEMISTRY

MEDICAL APPLICATIONS OF RADIOTRACERS

Radiotracers are widely used as diagnostic tools in medicine. ▼ TABLE 5.5 lists some radioisotopes and their uses. These radioisotopes are incorporated into a compound that is administered to the patient, usually intravenously. The diagnostic use of these isotopes is based on the ability of the radioactive compound to localise and concentrate in the organ or tissue under investigation. Iodine-131, for example, has been used to test the activity of the thyroid gland. This gland is the only important user of iodine in the body. The patient is injected with a solution of NaI containing iodine-131. Only a very small amount is used so that the patient does not receive a harmful dose of radioactivity. A Geiger counter placed close to the thyroid, in the neck region, determines the ability of the thyroid to take up the iodine. A normal thyroid will absorb about 12% of the iodine within a few hours.

The medical applications of radiotracers are further illustrated by positron emission tomography (PET). PET is used for clinical diagnosis of many diseases. In this method, compounds containing radionuclides that decay by positron emission are injected into a patient. These compounds are chosen to enable researchers to monitor blood flow, oxygen and glucose metabolic rates and other biological functions. Some of the most interesting work involves the study of the brain, which depends on glucose for most of its energy. Changes in how this sugar is metabolised or used by the brain may signal a disease such as cancer, epilepsy, Parkinson's disease or schizophrenia.

The compound to be detected in the patient must be labelled with a radionuclide that is a positron emitter. The most widely used nuclides are carbon-11 (half-life 20.4 min), fluorine-18 (half-life 110 min), oxygen-15 (half-life 2 min) and nitrogen-13 (half-life 10 min). Glucose, for example, can be labelled with ^{11}C. Because the half-lives of positron emitters are so short, the chemist must quickly incorporate the radionuclide into the sugar (or other appropriate) molecule and inject the compound immediately. The patient is placed in an elaborate instrument (▼ FIGURE 5.9(a)) that measures the positron emission and constructs a computer-based image of the organ in which the emitting compound is localised. The nature of this image (Figure 5.9(b)) provides clues to the presence of disease or other abnormality and helps medical researchers understand how a particular disease affects the functioning of the brain.

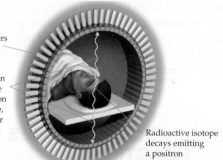

Scintillation counters detect gamma rays

Gamma rays moving in opposite directions are created when a positron and an electron collide, annihilating each other

Radioactive isotope decays emitting a positron

(a)

TABLE 5.5 • Some radionuclides used as radiotracers		
Nuclide	**Half-life**	**Area of the body studied**
Iodine-131	8.04 days	Thyroid
Iron-59	44.5 days	Red blood cells
Phosphorus-32	14.3 days	Eyes, liver, tumours
Technetium-99	6.0 hours	Heart, bones, liver and lungs
Thallium-201	73 hours	Heart, arteries
Sodium-24	14.8 hours	Circulatory system

(b)

Normal   Mild cognitive impairment   Alzheimer's disease

▲ **FIGURE 5.9** **(a) Schematic representation of positron emission tomography (PET) scanner. (b) PET scans showing metabolism levels in the brain.** Yellow colours show higher levels of metabolism.

The fact that the system has lost mass indicates that the process is exothermic. All spontaneous nuclear reactions are exothermic.

The energy change per mole associated with this reaction can be calculated using Einstein's equation:

$$\Delta E = \Delta(mc^2) = c^2\,\Delta m$$

$$= (2.9979 \times 10^8\ \text{m s}^{-1})^2 \times \left(\frac{-0.0046}{1000}\right)\text{kg}$$

$$= -4.1 \times 10^{11}\ \text{kg m}^2\,\text{s}^{-2} = -4.1 \times 10^{11}\ \text{J}$$

Notice that Δm is converted to kilograms, the SI unit of mass, to obtain ΔE in joules, J, the SI unit for energy.

SAMPLE EXERCISE 5.9   Calculating mass change in a nuclear reaction

How much energy is lost or gained when a mole of cobalt-60 undergoes beta decay: $^{60}_{27}\text{Co} \longrightarrow ^{60}_{28}\text{Ni} + ^{0}_{-1}\text{e}$? The mass of the $^{60}_{27}\text{Co}$ atom is 59.933819 u, and that of a $^{60}_{28}\text{Ni}$ atom is 59.930788 u.

SOLUTION

Analyse We are asked to calculate the energy change in a nuclear reaction.

Plan We must first calculate the mass change in the process. We are given atomic masses, but we need the masses of the nuclei in the reaction. We calculate these by taking account of the masses of the electrons that contribute to the atomic masses.

Solve A $^{60}_{27}\text{Co}$ atom has 27 electrons. The mass of an electron is 5.4858×10^{-4} u. (See the list of fundamental constants at the back of the text.) We subtract the mass of the 27 electrons from the mass of the $^{60}_{27}\text{Co}$ *atom* to find the mass of the $^{60}_{27}\text{Co}$ *nucleus*:

$$59.933819\ \text{u} - (27)(5.4858 \times 10^{-4}\ \text{u}) = 59.919007\ \text{u}\ (\text{or }59.919007\ \text{g mol}^{-1})$$

Likewise, for $^{60}_{28}\text{Ni}$, the mass of the nucleus is

$$59.930788\ \text{u} - (28)(5.4858 \times 10^{-4}\ \text{u}) = 59.915428\ \text{u}\ (\text{or }59.915428\ \text{g mol}^{-1})$$

The mass change in the nuclear reaction is the total mass of the products minus the mass of the reactant:

$$\Delta m = \text{mass of electron} + \text{mass }^{60}_{28}\text{Ni nucleus} - \text{mass of }^{60}_{27}\text{Co nucleus}$$

$$= 0.00054858\ \text{u} + 59.915428\ \text{u} - 59.919007\ \text{u}$$

$$= -0.003031\ \text{u}$$

Thus when a mole of cobalt-60 decays,

$$\Delta m = -0.003031\ \text{g}$$

Because the mass decreases ($\Delta m < 0$), energy is released ($\Delta E < 0$). The quantity of energy released *per mole* of cobalt-60 is calculated using Equation 5.22:

$$\Delta E = c^2\,\Delta m$$

$$= (2.9979 \times 10^8\ \text{m s}^{-1})^2\left(\frac{-0.003031}{1000}\right)\text{kg}$$

$$= -2.724 \times 10^{11}\ \text{kg m}^2\,\text{s}^{-2} = 2.724 \times 10^{11}\ \text{J}$$

PRACTICE EXERCISE

Positron emission from ^{11}C, $^{11}_{6}\text{C} \longrightarrow ^{11}_{5}\text{B} + ^{0}_{1}\text{e}$ occurs with release of 2.87×10^{11} J per mole of ^{11}C. What is the mass change per mole of ^{11}C in this nuclear reaction?

Answer: -3.19×10^{-3} g

(See also Exercise 5.43.)

Nuclear Binding Energies

Scientists discovered in the 1930s that the masses of nuclei are always less than the masses of the individual nucleons of which they are composed. For example, the helium-4 nucleus has a mass of 4.00150 u. The mass of a proton is 1.00728 u and that of a neutron is 1.00866 u. Consequently, two protons and two neutrons have a total mass of 4.03188 u:

$$\text{Mass of two protons} = 2(1.00728 \text{ u}) = 2.01456 \text{ u}$$

$$\text{Mass of two neutrons} = 2(1.00866 \text{ u}) = \underline{2.01732 \text{ u}}$$

$$\text{Total mass} = 4.03188 \text{ u}$$

The mass of the individual nucleons is 0.03038 u greater than that of the helium-4 nucleus:

$$\text{Mass of two protons and two neutrons} = 4.03188 \text{ u}$$

$$\text{Mass of } {}_{2}^{4}\text{He nucleus} = \underline{4.00150 \text{ u}}$$

$$\text{Mass difference} = 0.03038 \text{ u}$$

The mass difference between a nucleus and its constituent nucleons is called the **mass defect**. The origin of the mass defect is readily understood if we consider that energy must be added to a nucleus in order to break it into separated protons and neutrons:

$$\text{Energy} + {}_{2}^{4}\text{He} \longrightarrow 2\,{}_{1}^{1}\text{p} + 2\,{}_{0}^{1}\text{n} \qquad [5.23]$$

The addition of energy to a system must be accompanied by a proportional increase in mass. The mass change for the conversion of helium-4 into separated nucleons is $\Delta m = 0.03038$ u, as shown in these calculations. The energy required for this process is

$$\Delta E = c^2 \, \Delta m$$

$$= (2.9979 \times 10^8 \text{ m s}^{-1})^2 (0.03038 \text{ u}) \left(\frac{1 \text{ g}}{6.022 \times 10^{23} \text{ u}} \right) \left(\frac{1 \text{ kg}}{1000 \text{ g}} \right)$$

$$= 4.534 \times 10^{-12} \text{ J}$$

The energy required to separate a nucleus into its individual nucleons is called the **nuclear binding energy**. The larger the binding energy, the more stable is the nucleus towards decomposition. The nuclear binding energies of helium-4 and two other nuclei (iron-56 and uranium-238) are compared in TABLE 5.6 ▼. The binding energies per nucleon (that is, the binding energy of each nucleus divided by the total number of nucleons in that nucleus) are also compared in the table.

The binding energies per nucleon can be used to compare the stabilities of different combinations of nucleons (such as 2 protons and 2 neutrons arranged either as ${}_{2}^{4}\text{He}$ or $2\,{}_{1}^{2}\text{H}$.) ▶ FIGURE 5.10 shows the average binding energy per nucleon plotted against mass number. The binding energy per nucleon at first increases in magnitude as the mass number increases, reaching about 1.4×10^{-12} J for nuclei

TABLE 5.6 • **Mass defects and binding energies for three nuclei**					
Nucleus	Mass of nucleus (u)	Mass of individual nucleons (u)	Mass defect (u)	Binding energy (J)	Binding energy per nucleon (J)
${}_{2}^{4}\text{He}$	4.00150	4.03188	0.03038	4.53×10^{-12}	1.13×10^{-12}
${}_{26}^{56}\text{Fe}$	55.92068	56.44914	0.52846	7.90×10^{-11}	1.41×10^{-12}
${}_{92}^{238}\text{U}$	238.00031	239.93451	1.93420	2.89×10^{-10}	1.21×10^{-12}

whose mass numbers are in the vicinity of iron-56. It then decreases slowly to about 1.2×10^{-12} J for very heavy nuclei. This trend indicates that nuclei of intermediate mass numbers are more tightly bound (and therefore more stable) than those with either smaller or larger mass numbers. This trend has two significant consequences. First, heavy nuclei gain stability and therefore give off energy if they are fragmented into two mid-sized nuclei. This process, known as **fission**, is used to generate energy in nuclear power plants. Second, even greater amounts of energy are released if very light nuclei are combined or fused together to give more massive nuclei. This **fusion** process is the essential energy-producing process in the sun. We look more closely at fission and fusion in Sections 5.7 and 5.8.

CONCEPT CHECK 7
Could fusing two stable nuclei that have mass numbers in the vicinity of 100 cause an energy-releasing process?

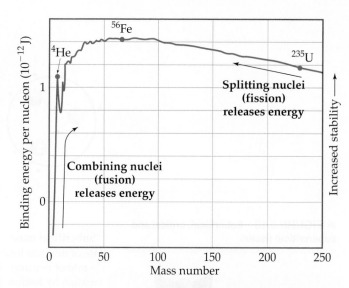

▲ **FIGURE 5.10   Nuclear binding energies.** The average binding energy per nucleon increases initially as the mass number increases and then decreases slowly. Because of these trends, fusion of light nuclei and fission of heavy nuclei are exothermic processes.

5.7 | NUCLEAR POWER: FISSION

According to our discussion of nuclear binding energies (Section 5.6), both the splitting of heavy nuclei (fission) and the union of light nuclei (fusion) are exothermic processes. Commercial nuclear power plants and the most common forms of nuclear weaponry depend on the process of nuclear fission for their operation. The first nuclear fission to be discovered was that of uranium-235. This nucleus, as well as those of uranium-233 and plutonium-239, undergoes fission when struck by a slow-moving neutron.* The induced fission process is illustrated in **FIGURE 5.11 ▶**. A heavy nucleus can split in many different ways. Two ways that the uranium-235 nucleus splits are shown in Equations 5.24 and 5.25:

$$_0^1 n + \, _{92}^{235}U \quad \Big\langle \begin{array}{l} _{52}^{137}Te \; + \; _{40}^{97}Zr \; + \; 2 \, _0^1 n \qquad [5.24] \\[2mm] _{56}^{142}Ba \; + \; _{36}^{91}Kr \; + \; 3 \, _0^1 n \qquad [5.25] \end{array}$$

More than 200 isotopes of 35 elements have been found among the fission products of uranium-235. Most of these isotopes are radioactive.

On the average, 2.4 neutrons are produced by every fission of a uranium-235 nucleus. If one fission produces two neutrons, these two neutrons can cause two additional fissions. The four neutrons thereby released can produce four fissions, and so on, as shown in ▶ **FIGURE 5.12**. The number of fissions and the energy released quickly escalate and, if the process is unchecked, the result is a violent explosion. Reactions that multiply in this fashion are called **branching chain reactions**.

In order for a fission chain reaction to occur, the sample of fissionable material must have a certain minimum mass. Otherwise, neutrons escape from the sample before they have the opportunity to strike other nuclei and cause additional fission. The chain stops if too many neutrons are lost. The amount of fissionable material large enough to maintain the chain reaction with a

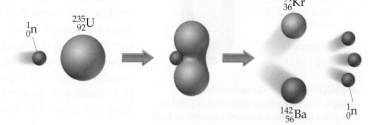

▲ **FIGURE 5.11   Uranium-235 fission.** The diagram shows just one of many fission patterns. In the process shown, 3.5×10^{-11} J of energy is produced per uranium-235 nucleus.

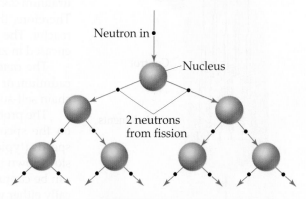

▲ **FIGURE 5.12   Chain fission reaction.** Assuming that each fission produces two neutrons, the process leads to an accelerating rate of fission, with the number of fissions potentially doubling at each stage.

---

* Other heavy nuclei can be induced to undergo fission. However, these three are the only ones of practical importance.

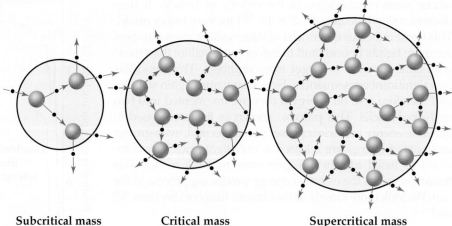

Subcritical mass	Critical mass	Supercritical mass
Rate of neutron loss > rate of neutron creation by fission	Rate of neutron loss = rate of neutron creation by fission	Rate of neutron loss < rate of neutron creation by fission

▶ **FIGURE 5.13   Subcritical, critical and supercritical fission.**

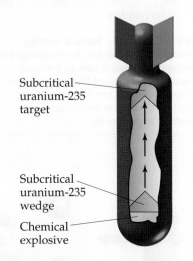

Subcritical uranium-235 target

Subcritical uranium-235 wedge

Chemical explosive

▲ **FIGURE 5.14   An atomic bomb design.** A conventional explosive is used to bring two subcritical masses together to form a supercritical mass.

constant rate of fission is called the **critical mass**. When a critical mass of material is present, one neutron on average from each fission is subsequently effective in producing another fission. The critical mass of uranium-235 is about 1 kg. If more than a critical mass of fissionable material is present, very few neutrons escape. The chain reaction thus multiplies the number of fissions, which can lead to a nuclear explosion. A mass in excess of a critical mass is referred to as a **supercritical mass**. The effect of mass on a fission reaction is illustrated in ▲ **FIGURE 5.13**.

◀ **FIGURE 5.14** shows a schematic diagram of the first atomic bomb used in warfare, the bomb that was dropped on Hiroshima, Japan, on 6 August 1945. To trigger a fission reaction, two subcritical masses of uranium-235 are slammed together using chemical explosives. The combined masses of the uranium form a supercritical mass, which leads to a rapid, uncontrolled chain reaction and, ultimately, a nuclear explosion. The energy released by the bomb dropped on Hiroshima was equivalent to that of 20 000 tonnes of 1,3,5-trinitrotoluene (TNT) (thus it is called a *20-kiloton* bomb).

Nuclear Reactors

Nuclear power plants use nuclear fission to generate energy. The core of a typical nuclear reactor consists of four principal components: fuel elements, control rods, a moderator and a primary coolant (◀ **FIGURE 5.15**). The fuel is a fissionable substance, such as uranium-235. The natural isotopic abundance of uranium-235 is only 0.7%, too low to sustain a chain reaction in most reactors. Therefore, the ^{235}U content of the fuel must be enriched to 3–5% for use in a reactor. The *fuel elements* contain enriched uranium in the form of UO_2 pellets encased in zirconium or stainless steel tubes.

The *control rods* are composed of materials that absorb neutrons, such as cadmium or boron. These rods regulate the flux of neutrons to keep the reaction chain self-sustaining and also prevent the reactor core from overheating.*

The probability that a neutron will trigger fission of a ^{235}U nucleus depends on the speed of the neutron. The neutrons produced by fission have high speeds (typically in excess of 10 000 km s^{-1}). The function of the *moderator* is to slow down the neutrons (to speeds of a few kilometres per second) so that they can be captured more readily by the fissionable nuclei. The moderator is typically either water or graphite.

The *primary coolant* is a substance that transports the heat generated by the

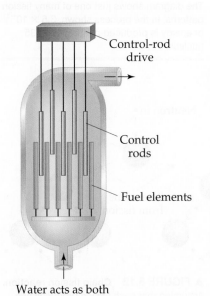

Control-rod drive

Control rods

Fuel elements

Water acts as both moderator and coolant

▲ **FIGURE 5.15   Diagram of a pressurised water reactor core.**

*The reactor core cannot reach supercritical levels and explode with the violence of an atomic bomb because the concentration of uranium-235 is too low. However, if the core overheats, sufficient damage can lead to the release of radioactive materials into the environment.

nuclear chain reaction away from the reactor core. In a *pressurised water reactor*, which is the most common commercial reactor design, water acts as both the moderator and the primary coolant.

The design of a nuclear power plant is basically the same as that of a power plant that burns fossil fuel (except that the burner is replaced by a reactor core). The nuclear power plant design shown in ▼ **FIGURE 5.16**, a pressurised water reactor, is currently the most widely used design. The primary coolant passes through the core in a closed system, which lessens the chance that radioactive products could escape the core. As an added safety precaution, the reactor is surrounded by a reinforced concrete *containment shell* to shield personnel and nearby residents from radiation and to protect the reactor from external forces. After passing through the reactor core, the very hot primary coolant passes through a heat exchanger where much of its heat is transferred to a *secondary coolant*, converting the latter to high-pressure steam that is used to drive a turbine. The secondary coolant is then condensed by transferring heat to an external source of water, such as a river or lake.

Approximately two-thirds of all commercial reactors are pressurised water reactors, but there are several variations on this basic design, each with advantages and disadvantages. A *boiling water reactor* generates steam by boiling the primary coolant; thus no secondary coolant is needed. Pressurised water reactors and boiling water reactors are collectively referred to as *light-water reactors* because they use H_2O as moderator and primary coolant. A *heavy-water reactor* uses D_2O (D = deuterium, ^2H) as moderator and primary coolant, and a *gas-cooled reactor* uses a gas, typically CO_2, as primary coolant and graphite as the moderator. Use of either D_2O or graphite as the moderator has the advantage that both substances absorb fewer neutrons than H_2O. Consequently, the uranium fuel does not need to be enriched (although the reactor can also be run with enriched fuel).

▲ **FIGURE IT OUT**

Why are nuclear power plants usually located near a large body of water?

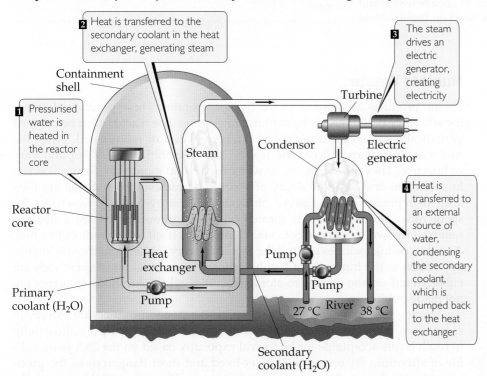

Containment shell

2 Heat is transferred to the secondary coolant in the heat exchanger, generating steam

3 The steam drives an electric generator, creating electricity

Turbine

1 Pressurised water is heated in the reactor core

Steam

Condensor

Electric generator

Reactor core

Heat exchanger

Pump

Condensor

Pump

Pump

4 Heat is transferred to an external source of water, condensing the secondary coolant, which is pumped back to the heat exchanger

Primary coolant (H_2O)

Pump

27 °C River 38 °C

Secondary coolant (H_2O)

▲ **FIGURE 5.16** Basic design of a pressurised water reactor nuclear power plant.

A CLOSER LOOK

THE DAWNING OF THE NUCLEAR AGE

The fission of uranium-235 was first achieved in the late 1930s by Enrico Fermi and his colleagues in Rome, and shortly thereafter by Otto Hahn and his co-workers in Berlin. Both groups were trying to produce transuranium elements. In 1938 Hahn identified barium among his reaction products. He was puzzled by this observation and questioned the identification because the presence of barium was so unexpected. He sent a detailed letter describing his experiments to Lise Meitner, a former co-worker. Meitner had been forced to leave Germany because of the anti-Semitism of the Third Reich and had settled in Sweden. She surmised that Hahn's experiment indicated that a new nuclear process was occurring in which the uranium-235 split. She called this process *nuclear fission*.

Meitner passed word of this discovery to her nephew, Otto Frisch, a physicist working at Niels Bohr's institute in Copenhagen. He repeated the experiment, verifying Hahn's observations and finding that tremendous energies were involved. In January 1939 Meitner and Frisch published a short article describing this new reaction. In March 1939 Leo Szilard and Walter Zinn at Columbia University discovered that more neutrons are produced than are used in each fission. As we have seen, this allows a chain reaction process to occur.

News of these discoveries and an awareness of their potential use in explosive devices spread rapidly within the scientific community. Several scientists finally persuaded Albert Einstein, the most famous physicist of the time, to write a letter to President Roosevelt, explaining the implications of these discoveries. Einstein's letter, written in August 1939, outlined the possible military applications of nuclear fission and emphasised the danger that weapons based on fission would pose if they were to be developed by the Nazis. Roosevelt judged it imperative that the United States investigate the possibility of such weapons. Late in 1941 the decision was made to build a bomb based on the fission reaction. An enormous research project, known as the Manhattan Project, began.

On 2 December 1942, the first artificial self-sustaining nuclear fission chain reaction was achieved by Enrico Fermi and co-workers in an abandoned squash court at the University of Chicago (▼ FIGURE 5.17). This accomplishment led to the development of the first atomic bomb, at Los Alamos National Laboratory in New Mexico in July 1945. In August 1945 the United States dropped atomic bombs on two Japanese cities, Hiroshima and Nagasaki. The nuclear age had arrived.

▲ FIGURE 5.17 **The first self-sustaining nuclear fission reactor.** The reactor was built in a squash court at the University of Chicago. The painting depicts the scene in which the scientists witnessed the reactor as it became self-sustaining on 2 December 1942.

Nuclear Waste

Fission products accumulate as the reactor operates. These products decrease the efficiency of the reactor by capturing neutrons. The reactor must be stopped periodically so that the nuclear fuel can be replaced or reprocessed. When the fuel rods are removed from the reactor, they are initially very hot and very radioactive. They are stored under water for several months at the reaction site to cool them and to allow decay of short-lived radionuclides. Some are then sent for reprocessing to remove plutonium-239, a by-product present in the spent rods. Plutonium-239 is formed when uranium-238 absorbs a neutron followed by two successive beta emissions and is itself a very radioactive fuel. As reprocessing facilities around the world are very limited and there is intense opposition to the transport of highly radioactive material, most spent rods are either stored on site or sent to storage facilities whose locations are generally not made public.

Storage poses a major problem because the fission products are extremely radioactive. It is estimated that 20 half-lives are required for their radioactivity to reach levels acceptable for biological exposure. Based on the 28.8 years half-life of strontium-90, one of the longer-lived and most dangerous of the products, the wastes must be stored for 600 years. Plutonium-239 is one of the by-products present in the expended fuel rods. (Remember that most of the uranium in the fuel rods is ^{238}U.) If the fuel rods are processed, the ^{239}Pu is largely recovered because it can be used as a nuclear fuel. However, if the plutonium is not removed, fuel rod storage must be for a very long period because plutonium-239 has a half-life of 24 000 years.

A considerable amount of research is being devoted to disposal of radioactive wastes. At present, the most attractive possibilities appear to be formation of glass, ceramic or synthetic rock from the wastes, as a means of immobilising them. These solid materials would then be placed in containers of high corrosion resistance and durability and buried deep underground. Central Australia is receiving a lot of attention from the nuclear industry as a possible site. Because the radioactivity will persist for a long time, there must be assurances that the solids and their containers will not crack from the heat generated by nuclear decay, allowing radioactivity to find its way into underground water supplies.

5.8 | NUCLEAR POWER: FUSION

Recall from Section 5.6 that energy is produced when light nuclei are fused into heavier ones. Reactions of this type are responsible for the energy produced by the sun. Spectroscopic studies indicate that the sun is composed of 73% H, 26% He and only 1% of all other elements, by mass. Among the several fusion processes that are believed to occur are the following:

$$^{1}_{1}\text{H} + ^{1}_{1}\text{H} \longrightarrow ^{2}_{1}\text{H} + ^{0}_{1}\text{e} \qquad [5.26]$$

$$^{1}_{1}\text{H} + ^{2}_{1}\text{H} \longrightarrow ^{3}_{2}\text{He} \qquad [5.27]$$

$$^{3}_{2}\text{He} + ^{3}_{2}\text{He} \longrightarrow ^{4}_{2}\text{He} + 2\,^{1}_{1}\text{H} \qquad [5.28]$$

$$^{3}_{2}\text{He} + ^{1}_{1}\text{H} \longrightarrow ^{4}_{2}\text{He} + ^{0}_{1}\text{e} \qquad [5.29]$$

Theories have been proposed for the generation of the other elements through fusion processes.

Fusion is appealing as an energy source because of the availability of light isotopes and because fusion products are generally not radioactive. Despite this fact, fusion is not presently used to generate energy. The problem is that high energies are needed to overcome the repulsion between nuclei. The required energies are achieved by high temperatures. Fusion reactions are therefore also known as **thermonuclear reactions**. The lowest temperature required for any fusion is that needed to fuse deuterium ($^{2}_{1}\text{H}$) and tritium ($^{3}_{1}\text{H}$), shown in Equation 5.30. This reaction requires a temperature of about 40 000 000 K:

$$^{2}_{1}\text{H} + ^{3}_{1}\text{H} \longrightarrow ^{4}_{2}\text{He} + ^{1}_{0}\text{n} \qquad [5.30]$$

Such high temperatures have been achieved by using an atomic bomb to initiate the fusion process. This is done in the thermonuclear, or hydrogen, bomb. This approach is unacceptable, however, for controlled power generation.

Numerous problems must be overcome before fusion becomes a practical energy source. In addition to the high temperatures necessary to initiate the reaction, there is the problem of confining the reaction. No known structural material is able to withstand the enormous temperatures necessary for fusion. Research has centred on the use of an apparatus called a *tokamak**, which uses strong magnetic fields to contain and to heat the reaction (▼ **FIGURE 5.18**). Temperatures of nearly 3 000 000 K have been achieved in a tokamak, but this is not yet hot enough to initiate continuous fusion. Much research has also been directed to the use of powerful lasers to generate the necessary temperatures.

---

* This word derives from a Russian acronym describing an apparatus that generates a toroidal magnetic field.

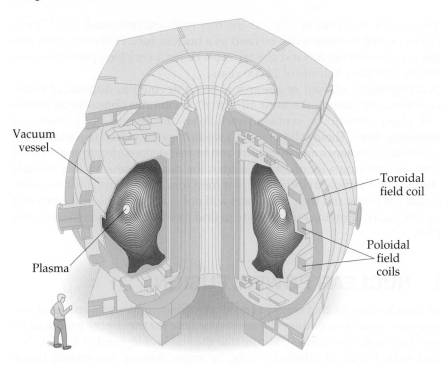

▶ FIGURE 5.18 A tokamak fusion test reactor. A tokamak is essentially a magnetic 'bottle' for confining and heating nuclei in an effort to cause them to fuse.

Vacuum vessel

Toroidal field coil

Poloidal field coils

Plasma

5.9 | RADIATION IN THE ENVIRONMENT AND LIVING SYSTEMS

We are continually bombarded by radiation from both natural and artificial sources. We are exposed to infrared, ultraviolet and visible radiation from the sun, for example, in addition to radio waves from radio and television stations, microwaves from microwave ovens and X-rays from various medical procedures. We are also exposed to radioactivity from the soil and other natural materials. Understanding the different energies of these various kinds of radiation is necessary to understand their different effects on matter.

When matter absorbs radiation, the energy of the radiation can cause either excitation or ionisation of the matter. Excitation occurs when the absorbed radiation excites electrons to higher energy states or increases the motion of molecules, causing them to move, vibrate or rotate. Ionisation occurs when the radiation removes an electron from an atom or molecule. In general, radiation that causes ionisation, called **ionising radiation**, is far more harmful to biological systems than radiation that does not cause ionisation. The latter, called **non-ionising radiation**, is generally of lower energy, such as radiofrequency electromagnetic radiation or slow-moving neutrons.

Most living tissue contains at least 70% water by mass. When living tissue is irradiated, most of the energy of the radiation is absorbed by water molecules. Thus it is common to define ionising radiation as radiation that can ionise water, a process requiring a minimum energy of 1216 kJ mol^{-1}. Alpha, beta and gamma rays (as well as X-rays and higher-energy ultraviolet radiation) possess energies in excess of this quantity and are therefore forms of ionising radiation.

When ionising radiation passes through living tissue, electrons are removed from water molecules, forming highly reactive H_2O^+ ions. An H_2O^+ ion can react with another water molecule to form an H_3O^+ ion and a neutral $\cdot OH$ molecule:

$$H_2O^+ + H_2O \longrightarrow H_3O^+ + \cdot OH \qquad [5.31]$$

The unstable and highly reactive $\cdot OH$ molecule is a **free radical**. In cells and tissues such particles can attack a host of surrounding biomolecules to produce new free radicals, which, in turn, attack yet other compounds. Thus the formation of a single free radical can initiate a large number of chemical reactions that are ultimately able to disrupt the normal operations of cells.

The damage produced by radiation depends on the activity and energy of the radiation, the length of exposure and whether the source is inside or outside the body. Gamma rays are particularly harmful outside the body because they penetrate human tissue very effectively, just as X-rays do. Consequently, their damage is not limited to the skin. In contrast, most alpha rays are stopped by skin, and beta rays are able to penetrate only about 1 cm beyond the surface of the skin (▶ FIGURE 5.19). Neither is as dangerous as gamma rays, therefore, *unless* the radiation source somehow enters the body. Within the body, alpha rays are particularly dangerous because they transfer their energy efficiently to the surrounding tissue, initiating considerable damage.

In general, the tissues that show the greatest damage from radiation are those that reproduce rapidly, such as bone marrow and lymph nodes. The principal effect of extended exposure to low doses of radiation is to cause cancer. Cancer is caused by damage to the growth-regulation mechanism of cells, inducing cells to reproduce in an uncontrolled manner. Leukaemia, which is characterised by excessive growth of white blood cells, is probably the major type of cancer caused by radiation.

In light of the biological effects of radiation, it is important to determine whether any levels of exposure are safe. Unfortunately, we are hampered in our attempts to set realistic standards because we don't fully understand the effects of long-term exposure to radiation. Scientists concerned with setting health standards have used the hypothesis that the effects of radiation are proportional to exposure, even down to low doses. Any amount of radiation is assumed to cause some finite risk of injury, and the effects of high dosage rates are extrapolated to those of lower ones. Other scientists believe, however, that there is a threshold below which there are no radiation risks. Until scientific evidence enables us to settle the matter with some confidence, it is safer to assume that even low levels of radiation present some danger.

Radiation Doses

Two units commonly used to measure the amount of exposure to radiation are the *gray* and the *rad*. The **gray (Gy)**, which is the SI unit of absorbed dose, corresponds to the absorption of 1 J of energy per kilogram of tissue. The **rad** is commonly used in medicine instead of the gray, where 1 rad = 10^{-2} Gy.

Not all forms of radiation harm biological materials with the same efficiency. A gray of alpha radiation, for example, can produce more damage than a gray of beta radiation. To correct for these differences, the radiation dose is multiplied by a factor that measures the relative biological damage caused by the radiation. This multiplication factor is known as the *relative biological effectiveness* of the radiation, abbreviated RBE. The RBE is approximately 1 for gamma and beta radiation and 10 for alpha radiation.

The exact value of the RBE varies with dose rate, total dose and the type of tissue affected. The product of the radiation dose in grays and the RBE of the radiation gives the effective dosage in units of sieverts where the **sievert (Sv)** is the SI unit for effective dosage. In medicine, the **rem** is the common unit of effective dosage, where 1 rem = 10^{-2} Sv.

$$\text{Number of sieverts} = (\text{number of Gy})(\text{RBE}) \qquad [5.32]$$

The effects of short-term exposures to radiation appear in ▶ TABLE 5.7. An exposure of 6 Sv is fatal to most humans. To put this number in perspective, a typical dental X-ray entails an exposure of about 5×10^{-6} Sv (5×10^{-3} mSv). Globally, the average exposure for a person in 1 year due to all natural sources of ionising radiation (called *background radiation*) is about 3.6 mSv.

▲ FIGURE IT OUT

Why are alpha rays much more dangerous when the source of radiation is located inside the body?

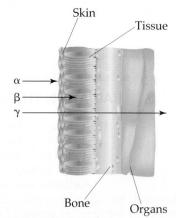

▲ **FIGURE 5.19** Relative penetrating abilities of alpha, beta and gamma radiation.

TABLE 5.7 ∘ Effects of short-term exposures to radiation

Dose (Sv)	Effect
0–0.25	No detectable clinical effects
0.25–0.50	Slight, temporary decrease in white blood cell counts
1.0–2.0	Nausea; marked decrease in white blood cells
5.0	Death of half the exposed population within 30 days after exposure

Radon

The radioactive noble gas radon has been much publicised in recent years as a potential risk to health. Radon-222 is a product of the nuclear disintegration series of uranium-238 (Figure 5.3) and is continually generated as uranium in rocks and soil decays. Globally, radon is estimated to account for about 50% of the average annual exposure to background ionising radiation. However, expo-

MY WORLD OF CHEMISTRY

RADIATION THERAPY

High-energy radiation poses a health hazard because of the damage it does to cells. Healthy cells are either destroyed or damaged by radiation, leading to physiological disorders. Radiation can also destroy *unhealthy* cells, however, including cancerous cells. All cancers are characterised by the runaway growth of abnormal cells. This growth can produce masses of abnormal tissue, called *malignant tumours*. Malignant tumours can be caused by the exposure of healthy cells to high-energy radiation. Somewhat paradoxically, however, malignant tumours can be destroyed by exposing them to the same radiation because rapidly reproducing cells are very susceptible to radiation damage. Thus cancerous cells are more susceptible to destruction by radiation than healthy ones, allowing radiation to be used effectively in the treatment of cancer. As early as 1904, physicians attempted to use the radiation emitted by radioactive substances to treat tumours by destroying the mass of unhealthy tissue. The treatment of disease by high-energy radiation is called *radiation therapy*.

Many different radionuclides are currently used in radiation therapy. Some of the more commonly used ones are listed in ▼ TABLE 5.8, along with their half-lives. Most of the half-lives are quite short, meaning that these radioisotopes emit a great deal of radiation in a short period of time (▶ FIGURE 5.20).

The radiation source used in radiation therapy may be inside or outside the body. In almost all cases radiation therapy is designed to use the high-energy gamma radiation emitted by radioisotopes. Alpha and beta radiation, which are not as penetrating as gamma radiation, can be blocked by appropriate packaging. For example, ^{192}Ir is often administered as 'seeds' consisting of a core of radioactive isotope coated with 0.1 mm of platinum metal. The platinum coating stops the alpha and beta rays, but the gamma rays penetrate it readily. The radioactive seeds can be surgically implanted in a tumour. In other cases

human physiology allows the radioisotope to be ingested. For example, most of the iodine in the human body ends up in the thyroid gland (see My World of Chemistry box in Section 5.6), so thyroid cancer can be treated by using large doses of ^{131}I. Radiation therapy on deep organs, where a surgical implant is impractical, often uses a ^{60}Co 'gun' outside the body to shoot a beam of gamma rays at the tumour. Particle accelerators are also used as an external source of high-energy radiation for radiation therapy.

Because gamma radiation is so strongly penetrating, it is nearly impossible to avoid damaging healthy cells during radiation therapy. Most cancer patients undergoing radiation treatments experience unpleasant and dangerous side effects such as fatigue, nausea, hair loss, a weakened immune system and even death. In many cases, therefore, radiation therapy is used only if other cancer treatments, such as *chemotherapy* (the treatment of cancer with powerful drugs), are unsuccessful. Nevertheless, radiation therapy is one of the major weapons we have in the fight against cancer.

▲ FIGURE 5.20   **Storage of radioactive caesium.** The vials contain a salt of caesium-137, a beta emitter, which is used in radiation therapy. The blue glow is from the radioactivity of the caesium.

Table 5.8 • Some radioisotopes used in radiation therapy			
Isotope	**Half-life**	**Isotope**	**Half-life**
^{32}P	14.3 days	^{137}Cs	30 years
^{60}Co	5.26 years	^{192}Ir	74.2 days
^{90}Sr	28.8 years	^{198}Au	2.7 days
^{125}I	60.25 days	^{222}Rn	3.82 days
^{131}I	8.04 days	^{226}Ra	1600 years

sure varies strongly from country to country and it is estimated by the Australian Radiation Protection and Nuclear Safety Agency (ARPANSA) that the value in Australia is much less and forms only a small part of the total annual exposure to background ionising radiation. This is approximately 1.5 mSv and is based on radiation from sources such as rocks and soil, cosmic rays, food and drinking water and medical X-rays and other nuclear diagnostic tools.

The interplay between the chemical and nuclear properties of radon makes it a health hazard. Because radon is a noble gas, it is extremely unreactive and is therefore free to escape from the ground without chemically reacting along the way. It is readily inhaled and exhaled with no direct chemical effects. The half-life of ^{222}Rn, however, is only 3.82 days. It decays by losing an alpha particle to form a radioisotope of polonium:

$$^{222}_{86}\text{Rn} \longrightarrow {}^{218}_{84}\text{Po} + {}^{4}_{2}\text{He} \qquad\qquad [5.33]$$

Because radon has such a short half-life and alpha particles have a high RBE, inhaled radon is considered a probable cause of lung cancer. Even worse, however, is the decay product, polonium-218, which is an alpha-emitting chemically active element that has an even shorter half-life (3.11 min) than radon-222:

$$^{218}_{84}\text{Po} \longrightarrow {}^{214}_{82}\text{Pb} + {}^{4}_{2}\text{He} \qquad\qquad [5.34]$$

The atoms of polonium-218 can become trapped in the lungs, where they continually bathe the delicate tissue with harmful alpha radiation. The resulting damage is estimated to result in as many as 10–14% of all lung cancer deaths worldwide.

SAMPLE INTEGRATIVE EXERCISE | Putting concepts together

Potassium ion is present in foods and is an essential nutrient in the human body. One of the naturally occurring isotopes of potassium, potassium-40, is radioactive. Potassium-40 has a natural abundance of 0.0117% and a half-life of $t_{1/2} = 1.28 \times 10^9$ years. It undergoes radioactive decay in three ways: 98.2% is by electron capture, 1.35% is by beta emission and 0.49% is by positron emission. **(a)** Why should we expect ^{40}K to be radioactive? **(b)** Write the nuclear equations for the three modes by which ^{40}K decays. **(c)** How many ^{40}K$^+$ ions are present in 1.00 g of KCl? **(d)** How long does it take for 1.00% of the ^{40}K in a sample to undergo radioactive decay?

SOLUTION

(a) The ^{40}K nucleus contains 19 protons and 21 neutrons. There are very few stable nuclei with odd numbers of both protons and neutrons (∞ Section 5.2, 'Patterns of Nuclear Stability').

∞ Review this on page 150

(b) Electron capture is capture of an inner-shell electron ($^{0}_{-1}$e) by the nucleus:

$$^{40}_{19}\text{K} + {}^{0}_{-1}\text{e} \longrightarrow {}^{40}_{18}\text{Ar}$$

Beta emission is loss of a beta particle by the nucleus:

$$^{40}_{19}\text{K} \longrightarrow {}^{40}_{20}\text{Ca} + {}^{0}_{-1}\text{e}$$

Positron emission is loss of a positron by the nucleus:

$$^{40}_{19}\text{K} \longrightarrow {}^{40}_{18}\text{Ar} + {}^{0}_{1}\text{e}$$

(c) Mol KCl $= \dfrac{1.00 \text{ g}}{74.55 \text{ g mol}^{-1}} = 1.34 \times 10^{-2}$ mol $=$ mol K$^+$ ions

Number of K$^+$ ions in the sample $= 1.34 \times 10^{-2}$ mol $\times 6.022 \times 10^{23}$ ions mol^{-1}
$$= 8.08 \times 10^{21} \text{ K}^+ \text{ ions}$$

Of these, 0.0117% are $^{40}_{19}$K ions:

$$(8.08 \times 10^{21} \text{ K}^+ \text{ ions})\left(\frac{0.0117 \text{ }^{40}\text{K}^+ \text{ ions}}{100 \text{ K}^+ \text{ ions}}\right) = 9.45 \times 10^{17} \text{ }^{40}\text{K}^+ \text{ ions}$$

(d) The decay constant (the rate constant) for the radioactive decay can be calculated from the half-life, using Equation 5.20:

$$k = \frac{0.693}{t_{1/2}} = \frac{0.693}{1.28 \times 10^9 \text{ yr}} = (5.41 \times 10^{-10}) \text{ yr}^{-1}$$

The rate equation, Equation 5.19, then allows us to calculate the time required:

$$\ln \frac{N_t}{N_0} = -kt$$

$$\ln \frac{99}{100} = -5.41 \times 10^{-10} \text{ yr}^{-1} t$$

$$-0.01005 = -5.41 \times 10^{-10} \text{ yr}^{-1} t$$

$$t = \frac{-0.01005}{(-5.41 \times 10^{-10}) \text{ yr}^{-1}} = 1.86 \times 10^7 \text{yr}$$

That is, it would take 18.6 million years for just 1.00% of the ^{40}K in a sample to decay.

CHAPTER SUMMARY AND KEY TERMS

SECTION 5.1 The nucleus of an atom contains protons and neutrons, both of which are called **nucleons**: the nucleus itself containing a specified number of protons and neutrons is referred to as **nuclide**. Reactions that involve changes in atomic nuclei are called **nuclear reactions**. Nuclei that spontaneously change by emitting radiation are said to be **radioactive**. Radioactive nuclei are called **radionuclides**, and the atoms containing them are called **radioisotopes**. Radionuclides spontaneously change through a process called radioactive decay. The three most important types of radiation given off as a result of radioactive decay are **alpha (α) particles** ($^4_2$He), **beta (β) particles** ($^0_{-1}$e) and **gamma (γ) radiation** ($^0_0$γ). Streams of alpha particles and beta particles are known as **alpha radiation** and **beta radiation**, respectively. **Positrons** ($^0_1$e), which are particles with the same mass as an electron but the opposite charge, can also be produced when a radioisotope decays.

In nuclear equations, reactant and product nuclei are represented by giving their mass numbers and atomic numbers, as well as their chemical symbol. The totals of the mass numbers on both sides of the equation are equal; the totals of the atomic numbers on both sides are also equal. There are four common modes of radioactive decay: **alpha decay**, which reduces the atomic number by 2 and the mass number by 4, **beta emission**, which increases the atomic number by 1 and leaves the mass number unchanged, **positron emission** and **electron capture**, both of which reduce the atomic number by 1 and leave the mass number unchanged.

SECTION 5.2 The neutron-to-proton ratio is an important factor determining nuclear stability. By comparing a nuclide's neutron-to-proton ratio with those in the band of stability, we can predict the mode of radioactive decay. In general, neutron-rich nuclei tend to emit beta particles; proton-rich nuclei tend to

either emit positrons or undergo electron capture; and heavy nuclei tend to emit alpha particles. The presence of **magic numbers** of nucleons and an even number of protons and neutrons also help determine the stability of a nucleus. A nuclide may undergo a series of decay steps before a stable nuclide forms. This series of steps is called a **radioactive series** or a **nuclear disintegration series**.

SECTION 5.3 **Nuclear transmutations**, induced conversions of one nucleus into another, can be brought about by bombarding nuclei with either charged particles or neutrons. **Particle accelerators** increase the kinetic energies of positively charged particles, allowing these particles to overcome their electrostatic repulsion by the nucleus. Nuclear transmutations are used to produce the **transuranium elements**, those elements with atomic numbers greater than that of uranium.

SECTIONS 5.4 and 5.5 The SI unit for the activity of a radioactive source is the **becquerel (Bq)**, defined as one nuclear disintegration per second. A related unit, the **curie (Ci)**, corresponds to 3.7×10^{10} disintegrations per second. Nuclear decay is a first-order process. The decay rate (**activity**) is therefore proportional to the number of radioactive nuclei. The **half-life** of a radionuclide, which is a constant, is the time needed for one-half of the nuclei to decay. Some radioisotopes can be used to date objects. The process is known as **radiometric dating** and if carbon-14 is used it is referred to as **radiocarbon dating**. **Geiger counters** and **scintillation counters** count the emissions from radioactive samples. The ease of detection of radioisotopes also permits their use as **radiotracers** to follow elements through reactions.

SECTION 5.6 The energy produced in nuclear reactions is accompanied by measurable changes of mass in accordance with Einstein's relationship, $\Delta E = c^2 \Delta m$. The difference in mass between nuclei and the nucleons of which they are composed is known as the **mass defect**. The mass defect of a nuclide makes it possible to calculate its **nuclear binding energy**, the energy required to separate the nucleus into individual nucleons. Energy is produced when heavy nuclei split (**fission**) and when light nuclei fuse (**fusion**).

SECTIONS 5.7 and 5.8 Uranium-235, uranium-233 and plutonium-239 undergo fission when they capture a neutron, splitting into lighter nuclei and releasing more neutrons. The neutrons produced in one fission can cause further fission reactions, which can lead to a nuclear **chain reaction**. A reaction that maintains a constant rate is said to be critical, and the mass necessary to maintain this constant rate is called a **critical mass**. A mass in excess of the critical mass is termed a **supercritical mass**.

In nuclear reactors the fission rate is controlled to generate a constant power. The reactor core consists of fuel elements containing fissionable nuclei, control rods, a moderator and a primary coolant. A nuclear power plant resembles a conventional power plant except that the reactor core replaces the fuel burner. There is concern about the disposal of highly radioactive nuclear wastes that are generated in nuclear power plants.

Nuclear fusion requires high temperatures because nuclei must have large kinetic energies to overcome their mutual repulsions. Fusion reactions are therefore called **thermonuclear reactions**. It is not yet possible to generate power on Earth through a controlled fusion process.

SECTION 5.9 **Ionising radiation** is energetic enough to remove an electron from a water molecule; radiation with less energy is called **non-ionising radiation**. Ionising radiation generates **free radicals**, reactive substances with one or more unpaired electrons. The effects of long-term exposure to low levels of radiation are not completely understood, but it is usually assumed that the extent of biological damage varies in direct proportion to the level of exposure.

The amount of energy deposited in biological tissue by radiation is called the radiation dose and is measured in units of gray (SI unit) or rad (used in medicine). One **gray (Gy)** corresponds to a dose of 1 J kg^{-1} of tissue. The **rad** is a smaller unit; 100 rad = 1 Gy. The effective dose, which measures the biological damage created by the deposited energy, is measured in units of rem or **sievert (Sv)**. The **rem** is obtained by multiplying the number of rad by the relative biological effectiveness (RBE); 100 rem = 1 Sv.

KEY SKILLS

- Write balanced nuclear equations. (Section 5.1)

- Predict nuclear stability and expected type of nuclear decay from the neutron-to-proton ratio of an isotope. (Section 5.2)

- Write balanced nuclear equations for nuclear transmutations. (Section 5.3)

- Calculate ages of objects and/or the amount of a radionuclide remaining after a given period of time using the half-life of the radionuclide in question. (Section 5.4)

- Calculate mass and energy changes for nuclear reactions. (Section 5.6)

- Calculate the binding energies for nuclei. (Section 5.6)

- Describe the difference between fission and fusion. (Sections 5.7 and 5.8)

- Understand the basic workings of a nuclear power plant. (Section 5.7)

- Understand the meaning of radiation dosage terms. (Section 5.9)

- Understand the biological effects of different kinds of radiation. (Section 5.9)

KEY EQUATIONS

- First-order rate law for nuclear decay

$$\ln \frac{N_t}{N_0} = -kt \qquad [5.19]$$

- Relationship between nuclear decay constant and half-life

$$k = \frac{0.693}{t_{1/2}} \qquad [5.20]$$

- Einstein's equation relating mass and energy

$$E = mc^2 \qquad [5.22]$$

EXERCISES

VISUALISING CONCEPTS

5.1 Indicate whether each of the following nuclides lies within the belt of stability in Figure 5.1: **(a)** neon-24, **(b)** chlorine-32, **(c)** tin-108, **(d)** polonium-216. For any that do not, describe a nuclear decay process that would alter the neutron-to-proton ratio in the direction of increased stability. [Section 5.2]

5.2 Write the balanced nuclear equation for the reaction represented by the diagram shown here. [Section 5.2]

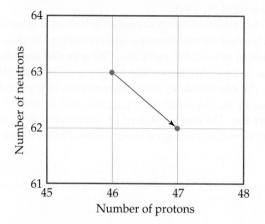

5.3 Draw a diagram similar to that shown in Exercise 5.2 that illustrates the nuclear reaction $^{211}_{83}Bi \longrightarrow {}^{4}_{2}He + {}^{207}_{81}Tl$. [Section 5.2]

5.4 The graph shown opposite illustrates the decay of $^{88}_{42}Mo$, which decays via positron emission. **(a)** What is the half-life of the decay? **(b)** What is the rate constant for the decay? **(c)** What fraction of the original sample of $^{88}_{42}Mo$ remains after 12 minutes? **(d)** What is the product of the decay process? [Section 5.4]

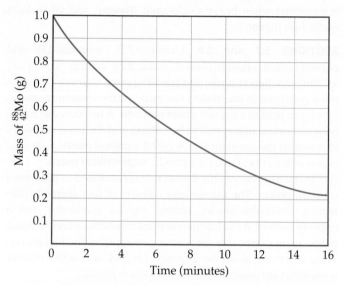

5.5 A $^{100}_{44}Ru$ atom has a mass of 99.90422 u. **(a)** Why is the mass of this nuclide not much closer to 100? **(b)** Explain the observed mass using Figure 5.9. **(c)** Calculate the binding energy per nucleon for $^{100}_{44}Ru$. [Section 5.6]

5.6 The diagram below illustrates a fission process. **(a)** What is the second nuclear product of the fission? **(b)** Use Figure 5.1 to predict whether the nuclear products of this fission reaction are stable. [Section 5.7]

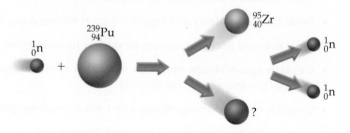

RADIOACTIVITY (Section 5.1)

5.7 Indicate the number of protons and neutrons in the following nuclei: **(a)** $^{55}_{25}Mn$, **(b)** ^{201}Hg, **(c)** potassium-39.

5.8 Indicate the number of protons and neutrons in the following nuclei: **(a)** $^{126}_{55}Cs$, **(b)** ^{119}Sn, **(c)** barium-141.

5.9 Give the symbol for **(a)** a proton, **(b)** a positron, **(c)** a beta particle.

5.10 Give the symbol for **(a)** a neutron, **(b)** an electron, **(c)** an alpha particle.

5.11 Write balanced nuclear equations for the following processes: **(a)** rubidium-90 undergoes beta emission; **(b)** selenium-72 undergoes electron capture; **(c)** krypton-76 undergoes positron emission; **(d)** radium-226 emits alpha radiation.

5.12 Write balanced nuclear equations for the following transformations: **(a)** neodymium-141 undergoes electron capture; **(b)** gold-201 decays to a mercury isotope;

(c) selenium-81 undergoes beta decay; **(d)** strontium-83 decays by positron emission.

5.13 Decay of which nucleus will lead to the following products: **(a)** bismuth-211 by beta decay; **(b)** chromium-50 by positron emission; **(c)** tantalum-179 by electron capture; **(d)** radium-226 by alpha decay?

5.14 The naturally occurring radioactive decay series that begins with $^{235}_{92}U$ stops with formation of the stable $^{207}_{82}Pb$ nucleus. The decays proceed through a series of alpha-particle and beta-particle emissions. How many of each type of emission are involved in this series?

5.15 A radioactive decay series that begins with $^{232}_{90}Th$ ends with formation of the stable nuclide How many alpha-particle emissions and how many beta-particle emissions are involved in the sequence of radioactive decays?

NUCLEAR STABILITY (Section 5.2)

5.16 Predict the type of radioactive decay process for the following radionuclides: **(a)** ^8_5B, **(b)** $^{68}_{29}Cu$, **(c)** neptunium-241, **(d)** chlorine-39.

5.17 Each of the following nuclei undergoes either beta or positron emission. Predict the type of emission for each: **(a)** $^{66}_{32}Ge$, **(b)** $^{105}_{45}Rh$, **(c)** iodine-137, **(d)** cerium-133.

5.18 One of the nuclides in each of the following pairs is radioactive. Predict which is radioactive and which is stable: **(a)** $^{39}_{19}K$ and $^{40}_{19}K$, **(b)** ^{209}Bi and ^{208}Bi, **(c)** magnesium-25 and neon-24. Explain.

5.19 In each of the following pairs, which nuclide would you expect to be the more abundant in nature:

(a) $^{115}_{48}Cd$ or $^{112}_{48}Cd$, **(b)** $^{30}_{13}Al$ or $^{27}_{13}Al$, **(c)** palladium-106 or palladium-113, **(d)** xenon-128 or caesium-128? Justify your choices.

5.20 Which of the following nuclides have magic numbers of both protons and neutrons: **(a)** helium-4, **(b)** carbon-12, **(c)** calcium-40, **(d)** nickel-58, **(e)** lead-208?

5.21 Tin-112 is a stable nuclide but indium-112 is radioactive, with a half-life of only 14 min. How can we explain this difference in nuclear stability?

5.22 Which of the following nuclides of group 16 elements would you expect to be radioactive: $^{14}_8O$, $^{32}_{16}S$, $^{78}_{34}Se$, $^{115}_{52}Te$ or $^{208}_{84}Po$? Justify your choices.

NUCLEAR TRANSMUTATIONS (Section 5.3)

5.23 Why are nuclear transmutations involving neutrons generally easier to accomplish than those involving protons or alpha particles?

5.24 Complete and balance the following nuclear equations by supplying the missing particle:

(a) $^{32}_{16}S + ^1_0n \longrightarrow ^1_1p + ?$

(b) $^7_4Be + ^0_{-1}e$ (orbital electron) $\longrightarrow ?$

(c) $? \longrightarrow ^{187}_{76}Os + ^0_{-1}e$

(d) $^{98}_{42}Mo + ^2_1H \longrightarrow ^1_0n + ?$

(e) $^{235}_{92}U + ^1_0n \longrightarrow ^{135}_{54}Xe + 2^1_0n + ?$

5.25 Complete and balance the following nuclear equations by supplying the missing particle:

(a) $^{252}_{98}Cf + ^{10}_5B \longrightarrow 3^1_0n + ?$

(b) $^2_1H + ^3_2He \longrightarrow ^4_2He + ?$

(c) $^1_1H + ^{11}_5B \longrightarrow 3?$

(d) $^{122}_{53}I \longrightarrow ^{122}_{54}Xe + ?$

(e) $^{59}_{26}Fe \longrightarrow ^0_{-1}e + ?$

5.26 Write balanced equations for each of the following nuclear reactions: **(a)** $^{238}_{92}U(n, \gamma)^{239}_{92}U$, **(b)** $^{14}_7N(p, \alpha)^{11}_6C$, **(c)** $^{18}_8O(n, \beta)^{19}_9F$.

5.27 Write balanced equations for **(a)** $^{238}_{92}U(\alpha, n)^{241}_{94}Pu$, **(b)** $^{14}_7N(\alpha, p)^{17}_8O$, **(c)** $^{56}_{26}Fe(\alpha, \beta)^{60}_{29}Cu$.

RATES OF RADIOACTIVE DECAY (Section 5.4)

5.28 Harmful chemicals are often destroyed by chemical treatment. An acid, for example, can be neutralised by a base. Why can't chemical treatment be used to destroy the radioactive products produced in a nuclear reactor?

5.29 It has been suggested that strontium-90 (generated by nuclear testing) deposited in the hot desert will undergo radioactive decay more rapidly because it will be exposed to much higher average temperatures. Is this a reasonable suggestion?

5.30 The half-life of tritium (hydrogen-3) is 12.3 years. If 48.0 mg of tritium is released from a nuclear power plant during the course of an accident, what mass of this nuclide will remain after 12.3 years? After 49.2 years?

5.31 It takes 5.2 minutes for a 1.000 g sample of ^{210}Fr to decay to 0.250 g. What is the half-life of ^{210}Fr?

5.32 Cobalt-60 has a half-life of 5.26 years. The cobalt-60 in a radiotherapy unit must be replaced when its radioactivity falls to 75% of the original sample. If the original sample was purchased in August 2005, when will it be necessary to replace the cobalt-60?

5.33 Radium-226, which undergoes alpha decay, has a half-life of 1600 years. **(a)** How many alpha particles are emitted in 1.0 min by a 5.0 mg sample of ^{226}Ra? **(b)** What is the activity of the sample in mCi?

5.34 Cobalt-60, which undergoes beta decay, has a half-life of 5.26 years. **(a)** How many beta particles are emitted in 45.5 s by a 2.44 mg sample of ^{60}Co? **(b)** What is the activity of the sample in Bq?

5.35 A wooden artifact from a Chinese temple has a ^{14}C activity of 24.9 counts per minute compared with an activity of 32.5 counts per minute for a standard of zero age. From the half-life for ^{14}C decay, 5715 years, determine the age of the artifact.

5.36 The cloth shroud from around a mummy is found to have a ^{14}C activity of 8.9 disintegrations per minute per gram of carbon compared with living organisms that undergo 15.2 disintegrations per minute per gram of carbon. From the half-life for ^{14}C decay, 5715 years, calculate the age of the shroud.

ENERGY CHANGES (Section 5.6)

5.37 The combustion of one mole of graphite releases 393.5 kJ of energy. What is the mass change that accompanies the loss of this energy?

5.38 An analytical laboratory balance typically measures mass to the nearest 0.1 mg. What energy change would accompany the loss of 0.1 mg in mass?

5.39 How much energy must be supplied to break a single sodium-23 nucleus into separated protons and neutrons if the nucleus has a mass of 22.983733 u? How much energy is required per mole of this nucleus?

5.40 How much energy must be supplied to break a single ^{21}Ne nucleus into separated protons and neutrons if the nucleus has a mass of 20.98846 u? What is the nuclear binding energy for 1 mol of ^{21}Ne?

5.41 Calculate the binding energy per nucleon for the following nuclei: **(a)** $^{12}_{6}$C (nuclear mass = 11.966708 u); **(b)** ^{37}Cl (nuclear mass, 36.956576); **(c)** barium-137 (atomic mass, 136.905812 u).

5.42 Calculate the binding energy per nucleon for the following nuclei: **(a)** $^{14}_{7}$N (nuclear mass, 13.999234 u); **(b)** ^{48}Ti (nuclear mass, 47.935878 u); **(c)** mercury-201 (atomic mass, 200.970277 u).

5.43 The solar radiation falling on the Earth amounts to 1.07×10^{16} kJ min^{-1}. **(a)** What is the mass equivalence of the solar energy falling on the Earth in a 24-hr period? **(b)** If the energy released in the reaction

$$^{235}U + {}^{1}_{0}n \longrightarrow {}^{141}_{56}Ba + {}^{92}_{36}Kr + 3\,{}^{1}_{0}n$$

(^{235}U nuclear mass, 234.9935 u; ^{141}Ba nuclear mass, 140.8833 u; ^{92}Kr nuclear mass, 91.9021 u) is taken as typical of that occurring in a nuclear reactor, what mass of uranium-235 is required to equal 0.10% of the solar energy that falls on the Earth in 1.0 day?

5.44 Based on the following atomic mass values—^{1}H, 1.00782 u; ^{2}H, 2.01410 u; ^{3}H, 3.01605 u; ^{3}He, 3.01603 u; ^{4}He, 4.00260 u—and the mass of the neutron given in the text, calculate the energy released per mole in each of the following nuclear reactions, all of which are possibilities for a controlled fusion process:

(a) $^{2}_{1}H + {}^{3}_{1}H \longrightarrow {}^{4}_{2}He + {}^{1}_{0}n$

(b) $^{2}_{1}H + {}^{2}_{1}H \longrightarrow {}^{3}_{2}He + {}^{1}_{0}n$

(c) $^{2}_{1}H + {}^{3}_{2}He \longrightarrow {}^{4}_{2}He + {}^{1}_{1}H$

5.45 Which of the following nuclei is likely to have the largest mass defect per nucleon: **(a)** ^{59}Co, **(b)** ^{11}B, **(c)** ^{118}Sn, **(d)** ^{243}Cm? Explain your answer.

5.46 Based on Figure 5.10, explain why energy is released in the course of the fission of heavy nuclei.

EFFECTS AND USES OF RADIOISOTOPES (Sections 5.7 to 5.9)

5.47 Explain how you might use radioactive ^{59}Fe (a beta emitter with $t_{1/2} = 44.5$ days) to determine the extent to which rabbits are able to convert a particular iron compound in their diet into blood haemoglobin, which contains iron atoms.

5.48 Chlorine-36 is a convenient radiotracer. It is a weak beta emitter, with $t_{1/2} = 3 \times 10^{5}$ yr. Describe how you would use this radiotracer to conduct each of the following experiments. **(a)** Determine whether trichloroacetic acid, CCl_3COOH, undergoes any ionisation of its chlorines as chloride ion in aqueous solution. **(b)** Demonstrate that the equilibrium between dissolved and solid $BaCl_2$ in a saturated solution is a dynamic process. **(c)** Determine the effects of soil pH on the uptake of chloride ion from the soil by soybeans.

5.49 Explain the function of the following components of a nuclear reactor: **(a)** control rods, **(b)** moderator.

5.50 Explain the following terms that apply to fission reactions: **(a)** chain reaction, **(b)** critical mass.

5.51 Complete and balance the nuclear equations for the following fission reactions:

(a) $^{235}_{92}U + {}^{1}_{0}n \longrightarrow {}^{160}_{62}Sm + {}^{72}_{30}Zn + \underline{\quad}\,{}^{1}_{0}n$

(b) $^{239}_{94}Pu + {}^{1}_{0}n \longrightarrow {}^{144}_{58}Ce + \underline{\quad} + 2\,{}^{1}_{0}n$

5.52 Complete and balance the nuclear equations for the following fission or fusion reactions:

(a) $^{2}_{1}H + {}^{2}_{1}H \longrightarrow {}^{3}_{2}He + \underline{\quad}$

(b) $^{233}_{92}U + {}^{1}_{0}n \longrightarrow {}^{133}_{51}Sb + {}^{98}_{41}Nb + \underline{\quad}\,{}^{1}_{0}n$

5.53 A portion of the sun's energy comes from the reaction

$$4\,{}^{1}_{1}H \longrightarrow {}^{4}_{2}He + 2\,{}^{0}_{1}e$$

This reaction requires a temperature of about 10^{6} to 10^{7} K. Why is such a high temperature required?

5.54 The spent fuel rods from a fission reactor are much more intensely radioactive than the original fuel rods. **(a)** What does this tell you about the products of the fission process in relationship to the belt of stability, Figure 5.1? **(b)** Given that only two or three neutrons are released per fission event, and knowing that the nucleus undergoing fission has a neutron-to-proton ratio characteristic of a heavy nucleus, what sorts of decay would you expect to be dominant among the fission products?

5.55 Why is $\cdot OH$ more dangerous to an organism than OH^-?

5.56 A laboratory rat is exposed to an alpha-radiation source whose activity is 3.2×10^{8} Bq. **(a)** What is the activity of the radiation in disintegrations per second? **(b)** The rat has a mass of 250 g and is exposed to the radiation for 2.0 s, absorbing 65% of the emitted alpha particles, each having an energy of 9.12×10^{-13} J. Calculate the absorbed dose in grays. **(c)** If the RBE of the radiation is 9.5, calculate the effective absorbed dose in Sv.

ADDITIONAL EXERCISES

5.57 Radon-222 decays to a stable nucleus by a series of three alpha emissions and two beta emissions. What is the stable nucleus that is formed?

5.58 A free neutron is unstable and decays into a proton with a half-life of 10.4 min. **(a)** What other particle forms? **(b)** Why don't neutrons in atomic nuclei decay at the same rate?

5.59 The 13 known nuclides of zinc range from ^{60}Zn to ^{72}Zn. The naturally occurring nuclides have mass numbers 64, 66, 67, 68 and 70. What mode or modes of decay would you expect for the least massive radioactive nuclides of zinc? What mode for the most massive nuclides?

5.60 Chlorine has two stable nuclides, ^{35}Cl and ^{37}Cl. In contrast, ^{36}Cl is a radioactive nuclide that decays by beta emission. **(a)** What is the product of decay of ^{36}Cl? **(b)** Based on the empirical rules about nuclear stability, explain why the nucleus of ^{36}Cl is less stable than either ^{35}Cl or ^{37}Cl.

5.61 The synthetic radioisotope technetium-99, which decays by beta emission, is the most widely used isotope in nuclear medicine. The following data were collected on a sample of ^{99}Tc:

Disintegrations per minute	Time (hours)
180	0
130	2.5
104	5.0
77	7.5
59	10.0
46	12.5
24	17.5

Make a graph of these data similar to Figure 5.5 and determine the half-life. (You may want to make a graph of the natural log of the disintegration rate versus time; a little rearranging of Equation 5.19 will produce an equation for a linear relation between $\ln N_t$ and t; from the slope, you can obtain k.)

5.62 According to current regulations, the maximum permissible dose of strontium-90 in the body of an adult is 3.7×10^4 Bq. Using the relationship rate $= kN$, calculate the number of atoms of strontium-90 to which this dose corresponds. To what mass of strontium-90 does this correspond ($t_{1/2}$ for strontium-90 is 28.8 years)?

5.63 Suppose you had a detection device that could count every decay from a radioactive sample of plutonium-239 ($t_{1/2}$ is 24 000 years). How many counts per second would you obtain from a sample containing 0.173 g of plutonium-239? [*Hint*: Look at Equations 5.19 and 5.20.]

5.64 Methyl acetate (CH_3COOCH_3) is formed by the reaction of acetic acid with methanol. If the methanol is labelled with oxygen-18, the oxygen-18 ends up in the methyl acetate:

$$\underset{\substack{\| \\ }}{CH_3\overset{O}{C}OH} + H^{18}OCH_3 \longrightarrow CH_3\overset{O}{C}^{18}OCH_3 + H_2O$$

Do the C–OH bond of the acid and the O–H bond of the alcohol break in the reaction, or do the O–H bond of the acid and the C–OH bond of the alcohol break? Explain.

5.65 The nuclear masses of ^7Be, ^9Be and ^{10}Be are 7.0147, 9.0100 and 10.0113 u, respectively. Which of these nuclei has the largest binding energy per nucleon?

5.66 The sun radiates energy into space at the rate of 3.9×10^{26} J s^{-1}. **(a)** Calculate the rate of mass loss from the sun in kg s^{-1}. **(b)** How does this mass loss arise?

5.67 The average energy released in the fission of a single uranium-235 nucleus is about 3×10^{-11} J. If the conversion of this energy to electricity in a nuclear power plant is 40% efficient, what mass of uranium-235 undergoes fission in a year in a plant that produces 1000 MW (megawatts)? Recall that a watt is 1 J s^{-1}.

INTEGRATIVE EXERCISES

5.68 A 36.9 mg sample of sodium perchlorate contains radioactive chlorine-36 (whose atomic mass is 36.0 u). If 25.6% of the chlorine atoms in the sample are chlorine-36 and the remainder are naturally occurring non-radioactive chlorine atoms, how many disintegrations per second are produced by this sample? The half-life of chlorine-36 is 3.0×10^5 yr.

5.69 Charcoal samples from Stonehenge in England were burned in O_2 and the resultant CO_2 gas bubbled into a solution of $Ca(OH)_2$ (limewater), resulting in the precipitation of $CaCO_3$. The $CaCO_3$ was removed by filtration and dried. A 788 mg sample of the $CaCO_3$ had a radioactivity of 1.5×10^{-2} Bq because of carbon-14. By comparison, living organisms undergo 15.3 disintegrations per minute per gram of carbon. Using the half-life of carbon-14, 5715 years, calculate the age of the charcoal sample.

MasteringChemistry (www.pearson.com.au/masteringchemistry)

Make learning part of the grade. Access:

- tutorials with peronalised coaching
- study area
- Pearson eText

PHOTO/ART CREDITS

142 Courtesy NASA/JPL-Caltech, O. Krause (Steward Observatory); **158 (Figure 5.7(a))** © Health Protection Agency/Science Photo Library, **(b)** Getty Images Australia; **160 (b)** Drs Suzanne Baker, William Jagust and Susan Landau; **166** Gary Sheahan, 'Birth of the Atomic Age', Chicago (Illinois), Chicago Historical Society; **170** Earl Roberge/Photo Researchers Inc.

ELECTRONIC STRUCTURE OF ATOMS

The glass tubes of neon lights contain various gases that can be excited by electricity. Light is produced when electrically excited atoms return to their lowest-energy states.

KEY CONCEPTS

6.1 THE WAVE NATURE OF LIGHT
We learn that light (radiant energy, or *electromagnetic radiation*) has wave-like properties and is characterised by *wavelength*, *frequency* and *speed*.

6.2 QUANTISED ENERGY AND PHOTONS
We recognise that in addition to the wave-like properties, electromagnetic radiation also has particle-like properties and can be described in terms of *photons*, 'particles' of light.

6.3 LINE SPECTRA AND THE BOHR MODEL
We observe that when atoms are given an appropriate amount of energy, they give off light (*line* spectra). Line spectra indicate that electrons exist only at certain energy levels around a nucleus and that energy is involved when an electron moves from one level to another. The Bohr model of the atom pictures the electrons moving only in certain allowed orbits around the nucleus.

6.4 THE WAVE BEHAVIOUR OF MATTER
We recognise that matter also has wave-like properties. As a result, it is impossible to determine simultaneously the exact position and the exact motion of an electron in an atom (*Heisenberg's uncertainty principle*).

6.5 QUANTUM MECHANICS AND ATOMIC ORBITALS
We can describe how an electron exists in atoms by treating it as a standing wave. The *wave functions* that mathematically describe the electron's position and energy in an atom are called *atomic orbitals*. The orbitals can be described in a shorthand notation using *quantum numbers*.

6.6 REPRESENTATIONS OF ORBITALS
We describe the three-dimensional shapes of orbitals and how they can be represented by graphs of electron density.

6.7 MANY-ELECTRON ATOMS
We recognise that the energy levels for an atom with one electron are altered when the atom contains multiple electrons. Each electron has a quantum-mechanical property called *spin*. The *Pauli exclusion principle* states that no two electrons in an atom can have the same four quantum numbers (three for the orbital and one for the spin). Therefore an orbital can hold a maximum of two electrons.

6.8 ELECTRON CONFIGURATIONS
We learn that knowing orbital energies as well as some fundamental characteristics of electrons described by *Hund's rule* allows us to determine how electrons are distributed in an atom.

6.9 ELECTRON CONFIGURATIONS AND THE PERIODIC TABLE
We observe that the electron configuration of an atom is related to the location of the element in the periodic table.

What happens when we turn on a neon light? The electrons in the neon atoms, excited to a higher energy by electricity, emit light when they return to a lower energy. The pleasing glow that results is explained by one of the most revolutionary discoveries of the twentieth century, namely the *quantum theory*. This theory explains much of the behaviour of electrons in atoms—and we discover that the behaviour of subatomic electrons is quite unlike anything we see in our macroscopic world.

In this chapter we explore the quantum theory and its importance in chemistry. We begin by looking more closely at the nature of light and how our description of light was changed by the quantum theory. We explore some of the tools used in *quantum mechanics*, the 'new' physics that was developed to describe atoms correctly. We then use the quantum theory to describe the arrangements of electrons in atoms, or the *electronic structure* of atoms. The electronic structure of an atom refers not only to the number of electrons in an atom but also to their distribution around the nucleus and to their energies. We find that the quantum description of the electronic structure of atoms helps us to understand the beautiful arrangement of the elements in the periodic table—why, for example, helium and neon are both unreactive gases, whereas sodium and potassium are both soft, reactive metals. In the chapters that follow we see how the concepts of quantum theory are used to explain trends in the periodic table and the formation of bonds between atoms.

▲ **FIGURE 6.1   Water waves.** The dropping of a stone in the water forms waves. The regular variation of the peaks and troughs enables us to sense the motion, or *propagation*, of the waves.

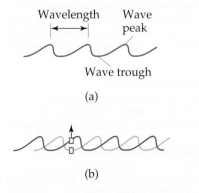

(a)

(b)

▲ **FIGURE 6.2   Characteristics of water waves.** (a) The distance between corresponding points on each wave is called the *wavelength*. In this drawing, the two corresponding points are two peaks, but they could be any other two corresponding points, such as two adjacent troughs. (b) The number of times per second that the cork bobs up and down is called the *frequency* of the wave.

6.1 | THE WAVE NATURE OF LIGHT

Much of our present understanding of the electronic structure of atoms has come from analysis of the light either emitted or absorbed by substances. The light that we see with our eyes, *visible light*, is an example of **electromagnetic radiation**. There are many types of electromagnetic radiation in addition to visible light. These different forms—such as the radio waves that carry music to our radios, the infrared radiation (heat) from a glowing fireplace and the X-rays used by a dentist—all share certain fundamental characteristics.

All types of electromagnetic radiation move through a vacuum at a speed of 3.00×10^8 m s^{-1}, the *speed of light*. Furthermore, all have wave-like characteristics similar to those of waves that move through water. Water waves are the result of energy imparted to the water, perhaps by the dropping of a stone or the movement of a boat on the water surface (◄ **FIGURE 6.1**). This energy is expressed as the up and down movements of the water.

A cross-section of a water wave (◄ **FIGURE 6.2**) shows that it is periodic, which means that the pattern of peaks and troughs repeats itself at regular intervals. The distance between two adjacent peaks (or between two adjacent troughs) is called the **wavelength**. The number of complete wavelengths, or *cycles*, that passes a given point each second is the **frequency** of the wave. We can measure the frequency of a water wave by counting the number of times per second that a cork bobbing on the water moves through a complete cycle of upward and downward motion.

> △ **CONCEPT CHECK 1**
>
> What is the difference between visible light and electromagnetic radiation?

Just as with water waves, we can assign a frequency and wavelength to electromagnetic waves, as illustrated in ▼ **FIGURE 6.3**. These and all other wave characteristics of electromagnetic radiation are due to the periodic oscillations in the intensities of the electric and magnetic fields associated with the radiation.

All electromagnetic radiation moves at the same speed, namely the speed of light. As a result, the wavelength and frequency of electromagnetic radiation are always related in a straightforward way. If the wavelength is long, there

---

△ **FIGURE IT OUT**

If wave (a) has a wavelength of 1.0 m and a frequency of 3.0×10^8 cycles s^{-1}, what are the wavelengths and frequencies of waves (b) and (c)?

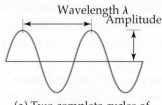

(a) Two complete cycles of wavelength λ

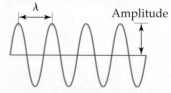

(b) Wavelength half of that in (a); frequency twice as great as in (a)

(c) Same frequency as (b), smaller amplitude

▲ **FIGURE 6.3   Characteristics of electromagnetic waves.** Radiant energy has wave characteristics; it consists of electromagnetic waves. All electromagnetic radiation travels at the same speed, represented as a horizontal line in this figure. Notice that the shorter the wavelength, λ, the higher the frequency, ν. The wavelength in (b) is half as long as that in (a), and the frequency of the wave in (b) is therefore twice as great as the frequency in (a). In these diagrams amplitude is measured as the vertical distance from the midline of the wave to its peak. The waves in (a) and (b) have the same amplitude. The wave in (c) has the same frequency as that in (b) but its amplitude is lower.

◢ FIGURE IT OUT

How do the wavelength and frequency of an X-ray compare with those of the red light from a neon sign?

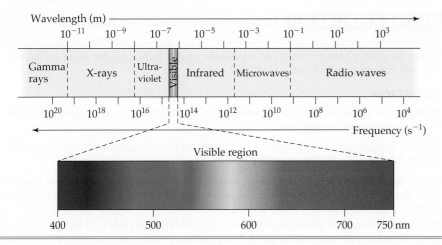

◀ **FIGURE 6.4   The electromagnetic spectrum.** Wavelengths in the spectrum range from very short gamma rays to very long radio waves.

will be fewer cycles of the wave passing a point per second; thus the frequency will be low. Conversely, for a wave to have a high frequency, the distance between the peaks of the wave must be small (short wavelength). This inverse relationship between the frequency and the wavelength of electromagnetic radiation can be expressed by the equation

$$\nu = c/\lambda \qquad\qquad [6.1]$$

where ν (nu) is the frequency, λ (lambda) is the wavelength and c is the speed of light. The **amplitude** of the wave relates to the intensity of the radiation. It is the maximum extent of the oscillation of the wave.

Why do different forms of electromagnetic radiation have different properties? Their differences are due to their different wavelengths, which are expressed in units of length. ▲ **FIGURE 6.4** shows the various types of electromagnetic radiation arranged in order of increasing wavelength, a display called the *electromagnetic spectrum*. Notice that the wavelengths span an enormous range. The wavelengths of gamma rays are similar to the diameters of atomic nuclei, whereas the wavelengths of radio waves can be longer than a football field. Notice also that visible light, which corresponds to wavelengths of about 4×10^{-7} m to 7×10^{-7} m (400 nm to 700 nm) is an extremely small portion of the electromagnetic spectrum. We can see visible light because of the chemical reactions it triggers in our eyes. The unit of length normally chosen to express wavelength depends on the type of radiation, as shown in ▼ **TABLE 6.1**.

Frequency is expressed in cycles per second, a unit also called a *hertz* (Hz). Because it is understood that cycles are involved, the units of frequency are

TABLE 6.1 • Common wavelength units for electromagnetic radiation			
Unit	Symbol	Length (m)	Type of radiation
Picometre	pm	10^{-12}	Gamma ray, X-ray
Nanometre	nm	10^{-9}	Ultraviolet, visible
Micrometre	μm	10^{-6}	Infrared
Millimetre	mm	10^{-3}	Microwave
Centimetre	cm	10^{-2}	Microwave
Metre	m	1	Television, radio
Kilometre	km	1000	Radio

normally given simply as 'per second', which is denoted by s^{-1}. For example, a frequency of 820 kilohertz (kHz), a typical frequency for an AM radio station, could be written as $820\ 000\ s^{-1}$.

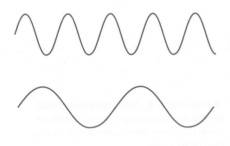

SAMPLE EXERCISE 6.1 Concepts of wavelength and frequency

Two electromagnetic waves are represented in the margin. **(a)** Which wave has the higher frequency? **(b)** If one wave represents visible light and the other represents infrared radiation, which wave is which?

SOLUTION

Analyse We are asked about the frequency of two waves and their relative position in the electromagnetic spectrum.

Plan Frequency and wavelength are inversely related to one another ($\nu = c/\lambda$): the longer the wavelength, the lower the frequency.

Solve **(a)** The lower wave has a longer wavelength (greater distance between peaks). Thus the lower wave has the lower frequency, and the upper one has the higher frequency.

(b) The electromagnetic spectrum (Figure 6.4) indicates that infrared radiation has a longer wavelength than visible light. Thus the lower wave would represent the infrared radiation.

PRACTICE EXERCISE

If one of the waves in the margin represents blue light and the other red light, which is which?

Answer: The expanded visible-light portion of Figure 6.4 tells you that red light has a longer wavelength than blue light. The lower wave has the longer wavelength (lower frequency) and would represent the red light.

(See also Exercises 6.1, 6.2.)

SAMPLE EXERCISE 6.2 Calculating frequency from wavelength

The yellow light given off by a sodium vapour lamp used for public lighting has a wavelength of 589 nm. What is the frequency of this radiation?

SOLUTION

Analyse Here we need to convert a wavelength to a frequency.

Plan The relationship between the wavelength (which is given) and the frequency (which is the unknown) is given by Equation 6.1. We can solve this equation for ν and then use the values of λ and c to obtain a numerical answer. (The speed of light, c, is a fundamental constant whose value is given in the text and in the table of fundamental constants on the inside back cover.)

Solve Equation 6.1 is $\nu = c/\lambda$. When we insert the values for c and λ, we note that the units of length in these two quantities are different. We can convert the wavelength from nanometres to metres, so the units cancel:

$$1\ \text{nm} = 1 \times 10^{-9}\ \text{m}$$

$$\nu = \frac{c}{\lambda} = \frac{(3.00 \times 10^8\ \text{m s}^{-1})}{589 \times 10^{-9}\ \text{m}} = 5.09 \times 10^{14}\ \text{s}^{-1}$$

Check The high frequency is reasonable because of the short wavelength. The units are proper because frequency has units of 'per second', or s^{-1}.

PRACTICE EXERCISE

(a) A laser used in eye surgery to fuse detached retinas produces radiation with a wavelength of 640.0 nm. Calculate the frequency of this radiation. **(b)** An FM radio station broadcasts electromagnetic radiation at a frequency of 103.4 MHz (megahertz: MHz = $10^6\ s^{-1}$). Calculate the wavelength of this radiation.

Answers: **(a)** $4.688 \times 10^{14}\ s^{-1}$, **(b)** 2.901 m

(See also Exercises 6.11, 6.12.)

▲ **FIGURE 6.5   Colour as a function of temperature.** The colour and intensity of the light emitted by a hot object depend on the temperature of the object. The temperature is highest at the centre of this pour of molten steel. As a result, the light emitted from the centre is most intense and has the shortest wavelength.

> ⚠ **CONCEPT CHECK 2**
>
> Human skin is not penetrated by visible light but is penetrated by X-rays. Which travels faster, visible light or X-rays?

6.2 | QUANTISED ENERGY AND PHOTONS

Although the wave model of light explains many aspects of its behaviour, there are several phenomena this model can't explain. Three of these are particularly pertinent to our understanding of how electromagnetic radiation and atoms interact: (1) the emission of light from hot objects (referred to as *blackbody radiation* because the objects studied appear black before heating); (2) the emission of electrons from metal surfaces on which light shines (the *photoelectric effect*); and (3) the emission of light from electronically excited gas atoms (*emission spectra*). We examine the first two here and the third in Section 6.3.

Hot Objects and the Quantisation of Energy

When solids are heated, they emit radiation, as seen in the red glow of an electric stove element and the bright white light of a tungsten light bulb. The wavelength distribution of the radiation depends on temperature, a red-hot object being cooler than a white-hot one (▶ **FIGURE 6.5**).

In 1900 a German physicist, Max Planck (1858–1947), explained the phenomenon by making a daring assumption. He assumed that energy can be either released or absorbed by atoms only in discrete 'chunks' of some minimum size. Planck gave the name **quantum** (meaning 'fixed amount') to the smallest quantity of energy that can be emitted or absorbed as electromagnetic radiation. He proposed that the energy, E, of a single quantum equals a constant times the frequency of the radiation:

$$E = h\nu \qquad [6.2]$$

The constant h is called **Planck's constant** and has a value of 6.626×10^{-34} joule seconds (J s). According to Planck's theory, matter is allowed to emit and absorb energy only in whole-number multiples of $h\nu$, such as $h\nu$, $2h\nu$, $3h\nu$ and so forth. If the quantity of energy emitted by an atom is $3h\nu$, for example, we say that three quanta of energy have been emitted (quanta is the plural of quantum). Because the energy can be released only in specific amounts, we say that the allowed energies are **quantised**—their values are restricted to certain quantities. Planck was awarded the 1918 Nobel Prize in Physics for his work on quantum theory.

If the notion of quantised energies seems strange, it might be helpful to draw an analogy by comparing an escalator with a staircase (▼ **FIGURE 6.6**). As

Potential energy of person walking up steps increases in stepwise, quantised manner

Potential energy of person riding escalator increases in uniform, continuous manner

◀ **FIGURE 6.6   Quantised versus continuous change in energy.**

you travel up an escalator, your potential energy increases in a uniform, continuous manner. When you climb a staircase, you can step only *on* individual stairs, not *between* them, so that your potential energy is restricted to certain values and is therefore quantised.

If Planck's quantum theory is correct, why aren't its effects more obvious in our daily lives? Why do energy changes seem continuous rather than quantised, or 'jagged'? Notice that Planck's constant is an extremely small number. Thus a quantum of energy, $h\nu$, is an extremely small amount. Planck's rules about the gain or loss of energy are always the same, whether we are concerned with objects on the scale of our ordinary experience or with microscopic objects. With everyday *macro*scopic objects, however, the gain or loss of a single quantum of energy goes completely unnoticed as it is small compared to the overall change. In contrast, when dealing with matter at the atomic level, the impact of quantised energies is far more significant.

CONCEPT CHECK 3

The temperature of stars is gauged by their colours. For example, red stars have a lower temperature than blue-white stars. How is this temperature scale consistent with Planck's assumption?

The Photoelectric Effect and Photons

A few years after Planck presented his theory, scientists began to see its applicability to a great many experimental observations. In 1905, Albert Einstein (1879–1955) used Planck's quantum theory to explain the **photoelectric effect**. Experiments had shown that light shining on a clean metal surface causes the surface to emit electrons. For each metal, there is a minimum frequency of light below which no electrons are emitted, irrespective of the intensity of the light. For example, light with a frequency of $4.60 \times 10^{14}\,\text{s}^{-1}$ or greater will cause caesium metal to emit electrons, but light of lower frequency has no effect.

To explain the photoelectric effect, Einstein assumed that the radiant energy striking the metal surface is behaving not like a wave but rather as if it were a stream of tiny energy packets. Each energy packet, called a **photon**, behaves like a tiny particle. Extending Planck's quantum theory, Einstein deduced that each photon must have an energy equal to Planck's constant times the frequency of the light:

$$\text{Energy of photon} = E = h\nu \qquad [6.3]$$

Thus radiant energy itself is quantised.

Under the right conditions, a photon can strike a metal surface and be absorbed. When this happens, the photon can transfer its energy to an electron in the metal. A certain amount of energy—called the *work function*—is required for an electron to overcome the attractive forces that hold it in the metal. If the photons of the radiation impinging on the metal have less energy than the work function, electrons do not acquire sufficient energy to escape from the metal surface, even if the light beam is intense. If the photons of radiation have sufficient energy, electrons are emitted from the metal. If the photons have more than the minimum energy required to free electrons, the excess energy appears as the kinetic energy of the emitted electrons. Einstein won the 1921 Nobel Prize in Physics for his explanation of the photoelectric effect.

To understand more clearly what a photon is, imagine that you have a light source that produces radiation with a single wavelength. Further suppose that you could switch the light on and off faster and faster to provide ever-smaller bursts of energy: Einstein's photon theory tells us that you would eventually come to a smallest energy burst, given by $E = h\nu$. This smallest burst of energy consists of a single photon of light.

SAMPLE EXERCISE 6.3 **Energy of a photon**

Calculate the energy of one photon of yellow light whose wavelength is 589 nm.

SOLUTION

Analyse This question involves the conversion of wavelength to energy.

Plan We can use Equation 6.1 to convert the wavelength to frequency:

$$\nu = c/\lambda$$

We can then use Equation 6.3 to calculate energy:

$$E = h\nu$$

Solve The frequency, ν, is calculated from the given wavelength, as shown in Sample Exercise 6.2:

$$\nu = c/\lambda = \frac{(3.00 \times 10^8 \text{ m s}^{-1})}{(589 \times 10^{-9} \text{ m})} = 5.09 \times 10^{14} \text{ s}^{-1}$$

The value of Planck's constant, h, is given both in the text and in the table of fundamental constants inside the back cover of this text, and so we can easily calculate E:

$$E = (6.626 \times 10^{-34} \text{ J s})(5.09 \times 10^{14} \text{ s}^{-1}) = 3.37 \times 10^{-19} \text{ J}$$

Comment If one photon of radiant energy supplies 3.37×10^{-19} J, then one mole of these photons will supply

$$(6.02 \times 10^{23} \text{ photons mol}^{-1})(3.37 \times 10^{-19} \text{ J photon}^{-1})$$
$$= 2.03 \times 10^5 \text{ J mol}^{-1}$$

This is the magnitude of energy involved in chemical reactions (⚬⚬⚬ Section 14.4, 'Enthalpies of Reaction') so radiation can break chemical bonds, producing what are called *photochemical reactions*.

PRACTICE EXERCISE

(a) A laser emits light with a frequency of 4.69×10^{14} s^{-1}. What is the energy of one photon of the radiation from this laser? **(b)** If the laser emits a pulse of energy containing 5.0×10^{17} photons of this radiation, what is the total energy of that pulse? **(c)** If the laser emits 1.3×10^{-2} J of energy during a pulse, how many photons are emitted during the pulse?

Answers: **(a)** 3.11×10^{-19} J, **(b)** 0.16 J, **(c)** 4.2×10^{16} photons

(See also Exercises 6.15, 6.17.)

The idea that the energy of light depends on its frequency helps us understand the diverse effects that different kinds of electromagnetic radiation have on matter. For example, the high frequency (short wavelength) of X-rays (Figure 6.4) causes X-ray photons to have high energy, sufficient to cause tissue damage and even cancer. Thus signs are normally posted around X-ray equipment warning of high-energy radiation.

Although Einstein's theory of light as a stream of particles rather than a wave explains the photoelectric effect and a great many other observations, it does pose a dilemma. Is light a wave, or does it consist of particles? The only way to resolve this dilemma is to adopt what might seem to be a bizarre position. We must consider that light possesses both wave-like and particle-like characteristics and, depending on the situation, will behave more like a wave or more like particles. We soon see that this dual nature is also characteristic of matter.

 CONCEPT CHECK 4

Suppose that yellow visible light can be used to eject electrons from a certain metal surface. What would happen if ultraviolet light were used instead?

6.3 | LINE SPECTRA AND THE BOHR MODEL

The work of Planck and Einstein paved the way for understanding how electrons are arranged in atoms.

In 1913 the Danish physicist Niels Bohr offered a theoretical explanation of *line spectra*, another phenomenon that had puzzled scientists in the nineteenth century. Let's first examine this phenomenon and then consider how Bohr used the ideas of Planck and Einstein.

▲ **FIGURE 6.7 Monochromatic radiation.** Lasers produce light of one wavelength, which we call *monochromatic light*. Different lasers produce light of different wavelength. The photo shows beams from a variety of lasers that produce visible light of different colours. Other lasers produce light that is not visible, including infrared and ultraviolet light.

Line Spectra

A particular source of radiant energy may emit a single wavelength, as in the light from a laser (◄ **FIGURE 6.7**). Radiation composed of a single wavelength is said to be *monochromatic*. However, most common radiation sources, including light bulbs and stars, produce radiation containing many different wavelengths. When radiation from such sources is separated into its different wavelength components, a **spectrum** is produced. ▼ **FIGURE 6.8** shows how a prism spreads light from a light bulb into its component wavelengths. The spectrum so produced consists of a continuous range of colours: violet merges into blue, blue into green and so forth, with no blank spots. This range of colours, containing light of all wavelengths, is called a **continuous spectrum**. The most familiar example of a continuous spectrum is the rainbow, produced when raindrops or mist act as a prism for sunlight.

Not all radiation sources produce a continuous spectrum. When different gases are placed under reduced pressure in a tube and a high voltage is applied, the gases emit different colours of light (▼ **FIGURE 6.9**). The light emitted by neon gas is the familiar red-orange glow of many 'neon' lights, quite different from the colour emitted from a discharge tube containing hydrogen. When light coming from such tubes is passed through a prism, only a few wavelengths are present in the resultant spectra, as shown in ▶ **FIGURE 6.10**. Each wavelength is represented by a coloured line in one of these spectra. The coloured lines are separated by black regions, which correspond to wavelengths that are absent from the light. A spectrum containing radiation of only specific wavelengths is called a **line spectrum**.

When scientists first detected the line spectrum of hydrogen in the mid-1800s, they were fascinated by its simplicity. At that time, only the four lines in the visible portion of the spectrum were observed, as shown in Figure 6.10. In 1885 a Swiss schoolteacher, Johann Balmer, showed that the wavelengths of these four visible lines of hydrogen fit an intriguingly simple formula. This formula is given by the Rydberg equation, where $n_1 = 2$. Later, additional lines were found to occur in the ultraviolet and infrared regions of the hydrogen spectrum. Soon Balmer's equation was extended to a more general one, called the *Rydberg equation*, which allowed the calculation of the wavelengths of all the spectral lines of hydrogen:

$$\frac{1}{\lambda} = (R_H)\left(\frac{1}{n_1^2} - \frac{1}{n_2^2}\right) \qquad [6.4]$$

In this formula λ is the wavelength of a spectral line, R_H is the *Rydberg constant* (1.096776×10^7 m^{-1}), and n_1 and n_2 are positive integers, with n_2 being larger than n_1.

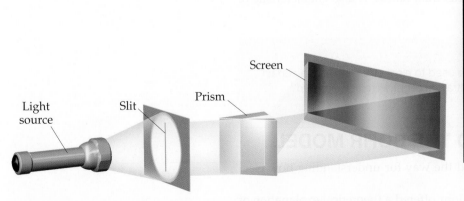

▲ **FIGURE 6.8 Creating a spectrum.** A continuous visible spectrum is produced when a narrow beam of white light is passed through a prism. The white light could be sunlight or light from an incandescent lamp.

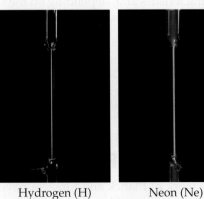

Hydrogen (H)   Neon (Ne)

▲ **FIGURE 6.9 Atomic emission.** Different gases emit light of different characteristic colours upon excitation by an electrical discharge.

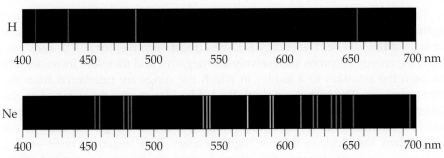

▲ FIGURE 6.10 **Line spectra.** Spectra obtained from the electrical discharge from hydrogen and neon. Light of only a few specific wavelengths are produced, as shown by coloured lines in the spectra.

Bohr's Model

To explain the line spectrum of hydrogen, Bohr assumed that electrons move in circular orbits around the nucleus and that these orbits have certain defined values of energy. That is, the energies of the orbits, and consequently of the electrons in them, are quantised. An electron does not lose energy while in its orbit and thus spiral into the nucleus because it does not obey the laws of classical physics.

Bohr based his model of the atom on three postulates.

1. Only orbits of certain radii, corresponding to certain definite energies, are permitted for the electron in a hydrogen atom.

2. An electron in a permitted orbit has a specific energy and is in an 'allowed' energy state. An electron in an allowed energy state will not radiate energy and thus will not spiral into the nucleus.

3. Energy is emitted or absorbed by the electron only as the electron changes from one allowed energy state to another. This energy is emitted or absorbed as a photon, $E = h\nu$.

 CONCEPT CHECK 5

Before reading further about the details of Bohr's model, speculate as to how it explains the fact that hydrogen gas emits a line spectrum (Figure 6.10) rather than a continuous spectrum.

The Energy States of the Hydrogen Atom

Starting with his three postulates and using classical equations for motion and for interacting electrical charges, Bohr calculated the energies corresponding to each allowed orbit for the electron in the hydrogen atom. These energies fit the formula

$$E = (-hcR_H)\left(\frac{1}{n^2}\right) = (-2.18 \times 10^{-18}\,\text{J})\left(\frac{1}{n^2}\right) \qquad [6.5]$$

In this equation, h, c and R_H are Planck's constant, the speed of light and the Rydberg constant, respectively. The product of these three constants equals $2.18 \times 10^{-18}\,\text{J}$.

The integer n, which can have values from one to infinity, is called the **principal quantum number**. Each orbit corresponds to a different value of n, and the radius of the orbit gets larger as n increases. Thus the first allowed orbit (the one closest to the nucleus) has $n = 1$, the next allowed orbit (the one second closest to the nucleus) has $n = 2$, and so forth. The electron in the hydrogen atom can be in any allowed orbit, and Equation 6.5 tells us the energy that the electron will have, depending on which orbit it is in.

If the transition of an electron from the $n = 3$ state to the $n = 2$ state results in emission of visible light, is the transition from the $n = 2$ state to the $n = 1$ state more likely to result in the emission of infrared or ultraviolet radiation?

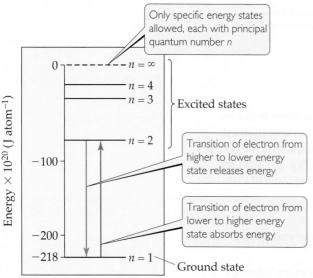

▲ **FIGURE 6.11   Energy states in the hydrogen atom.** Only states for $n = 1$ to $n = 4$ and $n = \infty$ are shown. Energy is released or absorbed when an electron moves from one energy state to another.

The energies of the electron of a hydrogen atom given by Equation 6.5 are negative for all values of n. The lower (more negative) the energy is, the more stable the atom will be. The energy is lowest (most negative) for $n = 1$. As n gets larger, the energy becomes successively less negative and therefore increases. We can liken the situation to a ladder in which the rungs are numbered from the bottom rung up. The higher we climb the ladder (the greater the value of n), the higher the energy. When the electron of the hydrogen atom is in the lowest energy state ($n = 1$, analogous to the bottom rung) it is called the **ground state** of the atom. When the electron is in a higher energy (less negative) orbit—$n = 2$ or higher—the atom is said to be in an **excited state**. ◀ **FIGURE 6.11** shows the energy of the electron in a hydrogen atom for several values of n.

What happens to the orbit radius and the energy as n becomes infinitely large? The radius increases as n^2, so we reach a point at which the electron is completely separated from the nucleus. When $n = \infty$, the energy is zero:

$$E = (-2.18 \times 10^{-18}\,\text{J})\left(\frac{1}{\infty^2}\right) = 0$$

In his third postulate, Bohr assumed that the electron could 'jump' from one allowed energy state to another by either absorbing or emitting photons whose radiant energy corresponds exactly to the energy difference between the two states. Energy must be absorbed for an electron to move to a higher energy state (one with a higher value of n). Conversely, radiant energy is emitted when the electron jumps to a lower energy state (one with a lower value of n). Thus if the electron jumps from an initial state that has energy E_i to a final state of energy E_f, the change in energy is

$$\Delta E = E_f - E_i = E_{\text{photon}} = h\nu \qquad [6.6]$$

Bohr's model of the hydrogen atom states, therefore, that only the specific frequencies of light that satisfy Equation 6.6 can be absorbed or emitted by the atom.

Substituting the energy expression in Equation 6.5 into Equation 6.6 and recalling that $\nu = c/\lambda$, we have

$$\Delta E = h\nu = \frac{hc}{\lambda} = (-2.18 \times 10^{-18}\,\text{J})\left(\frac{1}{n_f^2} - \frac{1}{n_i^2}\right) \qquad [6.7]$$

In this equation n_i and n_f are the principal quantum numbers of the initial and final states of the atom, respectively. If n_f is smaller than n_i, the electron moves closer to the nucleus and ΔE is a negative number, indicating that the atom releases energy. For example, if the electron moves from $n_i = 3$ to $n_f = 1$, we have

$$\Delta E = (-2.18 \times 10^{-18}\,\text{J})\left(\frac{1}{1^2} - \frac{1}{3^2}\right) = (-2.18 \times 10^{-18}\,\text{J})\left(\frac{8}{9}\right) = -1.94 \times 10^{-18}\,\text{J}$$

Knowing the energy for the emitted photon, we can calculate either its frequency or its wavelength. For the wavelength, we have

$$\lambda = \frac{c}{\nu} = \frac{hc}{\Delta E} = \frac{(6.63 \times 10^{-34}\,\text{J s})(3.00 \times 10^8\,\text{m s}^{-1})}{1.94 \times 10^{-18}\,\text{J}} = 1.03 \times 10^{-7}\,\text{m}$$

We have not included the negative sign of the energy in this calculation because wavelength and frequency are always reported as positive quantities. The direction of energy flow is indicated by saying that a photon of wavelength 1.03×10^{-7} has been *emitted*.

If we solve Equation 6.7 for $1/\lambda$, we find that this equation, derived from Bohr's theory, corresponds to the Rydberg equation, Equation 6.4, which was obtained using experimental data:

$$\frac{1}{\lambda} = \frac{-hcR_H}{hc}\left(\frac{1}{n_f^2} - \frac{1}{n_i^2}\right) = R_H\left(\frac{1}{n_i^2} - \frac{1}{n_f^2}\right)$$

Thus the existence of discrete spectral lines can be attributed to the quantised jumps of electrons between energy levels.

CONCEPT CHECK 6

As the electron in a hydrogen atom jumps from the $n = 3$ orbit to the $n = 7$ orbit, does it absorb energy or emit energy?

SAMPLE EXERCISE 6.4   **Electronic transitions in the hydrogen atom**

Using Figure 6.11, predict which of the following electronic transitions produces the spectral line having the longest wavelength: $n = 2$ to $n = 1$, $n = 3$ to $n = 2$ or $n = 4$ to $n = 3$.

SOLUTION

Analyse For a series of spectral lines, the one of longest wavelength corresponds to the lowest energy transition.

Plan The wavelength increases as frequency decreases ($\lambda = c/\nu$). Hence, the longest wavelength will be associated with the lowest frequency. According to Planck's equation, $E = h\nu$, the lowest frequency is associated with the lowest energy.

Solve In Figure 6.11 the shortest vertical line represents the smallest energy change. Thus the $n = 4$ to $n = 3$ transition produces the longest wavelength (lowest frequency) line.

PRACTICE EXERCISE

Indicate whether each of the following electronic transitions emits energy or requires the absorption of energy: **(a)** $n = 3$ to $n = 1$, **(b)** $n = 2$ to $n = 4$.

Answers: **(a)** emits energy, **(b)** requires absorption of energy

(See also Exercises 6.25, 6.26.)

Limitations of the Bohr Model

Although the Bohr model offers an explanation for the line spectrum of the hydrogen atom, it cannot explain the spectra of other atoms, except in a rather crude way. Furthermore, there is a problem with describing an electron merely as a small particle circling the nucleus. As we see in Section 6.4, the electron exhibits wave-like properties, a fact that any acceptable model of electronic structure must accommodate. As it turns out, the Bohr model was only one important step along the way towards the development of a more comprehensive model. What is most significant about Bohr's model is that it introduces two important ideas that are also incorporated into our current model: (1) electrons exist only in certain discrete energy levels, which are described by quantum numbers, and (2) energy is involved in moving an electron from one level to another.

We now start to develop the successor to the Bohr model, which requires that we take a closer look at the behaviour of matter.

6.4 | THE WAVE BEHAVIOUR OF MATTER

In the years following the development of Bohr's model for the hydrogen atom, the dual nature of radiant energy became a familiar concept. Depending on the experimental circumstances, radiation appears to have either a wave-like or a particle-like (photon) character. Louis de Broglie (1892–1987), who was working on his PhD thesis in physics at the Sorbonne in Paris, boldly extended this idea. If radiant energy could, under appropriate conditions, behave as though it were a stream of particles, could matter, under appropriate conditions, possibly show the properties of a wave? Suppose that the electron orbiting the nucleus of a hydrogen atom could be thought of not as a particle but rather as a wave, with a characteristic wavelength. De Broglie suggested that, in its movement about the nucleus, the electron has associated with it a particular wavelength. He went on to propose that the characteristic wavelength of the electron, or of any other particle, depends on its mass, m, and on its velocity, u:

$$\lambda = \frac{h}{mu} \qquad [6.8]$$

where h is Planck's constant. The quantity mu for any object is called its **momentum**. De Broglie used the term **matter waves** to describe the wave characteristics of material particles.

Because de Broglie's hypothesis is applicable to all matter, any object of mass m and velocity u would give rise to a characteristic matter wave. However, Equation 6.8 indicates that the wavelength associated with an object of ordinary size, such as a golf ball, is so tiny as to be completely out of the range of any possible observation. This is not so for an electron because its mass is so small, as we see in Sample Exercise 6.5.

SAMPLE EXERCISE 6.5 | **Matter waves**

What is the wavelength of an electron moving with a speed of 5.97×10^6 m s^{-1}? (The mass of the electron is 9.11×10^{-28} g.)

SOLUTION

Analyse Here we consider the dual nature of an electron by converting its momentum to wavelength.

Plan The wavelength of a moving particle is given by Equation 6.8, so λ is calculated by inserting the known quantities h, m and u. In doing so, however, we must pay attention to units.

Solve Using the value of Planck's constant and recalling that we have the following:

$h = 6.63 \times 10^{-34}$ J s

$1\,\text{J} = 1\,\text{kg m}^2\,\text{s}^{-2}$

$$\lambda = \frac{h}{mu}$$

$$= \frac{(6.63 \times 10^{-34}\,\text{J s})}{(9.11 \times 10^{-28}\,\text{g})(5.97 \times 10^6\,\text{m s}^{-1})}$$

$$= \frac{(6.63 \times 10^{-34}\,\text{kg m}^2\,\text{s}^{-1})}{(9.11 \times 10^{-28} \times 1 \times 10^{-3}\,\text{kg})(5.97 \times 10^6\,\text{m s}^{-1})}$$

$$= 1.22 \times 10^{-10}\,\text{m} = 0.122\,\text{nm}$$

Comment By comparing this value with the wavelengths of electromagnetic radiation shown in Figure 6.4, we see that the wavelength of this electron is about the same as that of X-rays.

PRACTICE EXERCISE

Calculate the velocity of a neutron whose de Broglie wavelength is 500 pm. The mass of a neutron is given in the table at the back of this text.

Answer: 7.92×10^2 m s^{-1}

(See also Exercises 6.30, 6.31.)

Within a few years after de Broglie published his theory, the wave properties of the electron were demonstrated experimentally. Electrons passed through a crystal were diffracted by the crystal, just as X-rays are diffracted. Thus a stream of moving electrons exhibits the same kinds of wave behaviour as electromagnetic radiation.

MY WORLD OF CHEMISTRY

AUSTRALIAN SYNCHROTRON

The Australian Synchrotron, which is an advanced 'third generation' device, was opened on 31 July 2007, after four years of construction, across the road from Monash University Clayton Campus (▶ FIGURE 6.12). The Australian Synchrotron is known as a 'Light Source facility' and is a device (as large as a football field) that produces synchrotron light. This is the electromagnetic radiation emitted when electrons moving at velocities close to the speed of light are forced to change direction under the action of a magnetic field.

In a synchrotron, electrons are generated by thermionic emission from a heated metal cathode (the electron gun (1)) and then pass into the linear accelerator (linac (2)) where they are accelerated to almost the speed of light. They are then transferred to the booster ring (3) where they are increased in energy and then moved to the outer storage ring (4). By this time the electrons have reached a velocity of 99.9997% of the speed of light. In the storage ring the electrons are circulated by a series of magnets separated by straight sections. As they are deflected through the magnetic field created by the magnets, they give off electromagnetic radiation, so that a beam of synchrotron light is produced at each bending magnet. In addition, the beam also passes through a series of 'insertion devices' in the straight sections of the ring which increase the intensity of the light significantly by providing a varying magnetic field (▶ FIGURE 6.13).

There are two classes of these devices. One is the multipole wiggler (MPW) and the other is the undulatory. Together with the use of very sophisticated monochromators, this allows the emitted radiation to be very finely tuned to a specific wavelength. This extremely bright light of the desired wavelength is channelled down beam lines (5) to experimental work stations (6) where it is used for research. Synchrotron light is unique in its intensity and brilliance and it can be generated across a range of wavelengths from infrared to X-rays.

The immense value of synchrotron light can actually be defined by the word *bright*. The brightness of the beam gives the synchrotron immense power to act essentially as a spectromicroscope for all regions of the electromagnetic spectrum. Radiation in the shorter wavelength region (X-rays) is of the order of 10^5 times more intense than that emitted by conventional X-ray tubes. This allows the resolution of very complex

▲ **FIGURE 6.12**   Precise wavelength radiation from speedy electrons: the Australian Synchrotron, Clayton, Victoria.

▲ **FIGURE 6.13**   (1) Electron gun, (2) linac, (3) booster ring (4) outer storage ring. As the electrons are deflected through magnetic fields they create extremely bright light. The light is channelled down beamlines (5) to experimental workstations (6) where it is used for research.

structures such as proteins, which are normally very difficult to resolve.

The uses of synchrotron light are boundless, ranging from medical imaging to the study of engineering materials to forensic science.

The Uncertainty Principle

The discovery of the wave properties of matter raised some new and interesting questions about classical physics. Consider, for example, a ball rolling down a ramp. Using the equations of classical physics, we can calculate its position, direction of motion and speed at any time, with great accuracy. Can we do the same for an electron that exhibits wave properties? A wave extends in space and thus its location is not precisely defined. We might therefore anticipate that it is impossible to determine exactly where an electron is located at a specific time.

The German physicist Werner Heisenberg proposed that the dual nature of matter places a fundamental limitation on how precisely we can know both the location and the momentum of any object. The limitation becomes important only when we deal with matter at the subatomic level (that is, with masses as small as that of an electron). Heisenberg's principle is called the **uncertainty principle**. When applied to the electrons in an atom, this principle states that it is inherently impossible for us to know simultaneously both the exact momentum of the electron and its exact location in space.

Heisenberg mathematically related the uncertainty of the position (Δx) and the uncertainty in momentum $\Delta(mu)$ to a quantity involving Planck's constant:

$$\Delta x \cdot \Delta(mu) \geqslant \frac{h}{4\pi} \qquad [6.9]$$

A brief calculation illustrates the dramatic implications of the uncertainty principle. The electron has a mass of 9.11×10^{-31} kg and moves at an average speed of about 5×10^6 m s^{-1} in a hydrogen atom. Let's assume that we know the speed to an uncertainty of 1% (that is, an uncertainty of $(0.01)(5 \times 10^6$ m s$^{-1}) = 5 \times 10^4$ m s^{-1}) and that this is the only important source of uncertainty in the momentum, so that $\Delta(mu) = m\Delta u$. We can then use Equation 6.9 to calculate the uncertainty in the position of the electron:

$$\Delta x \geqslant \frac{h}{4\pi m \Delta u} = \frac{(6.63 \times 10^{-34}\,\text{J s})}{4\pi(9.11 \times 10^{-31}\,\text{kg})(5 \times 10^4\,\text{m s}^{-1})} = 1 \times 10^{-9}\,\text{m}$$

Because the diameter of a hydrogen atom is only about 1×10^{-10} m, the uncertainty is an order of magnitude greater than the size of the atom. Thus we have essentially no idea of where the electron is located within the atom. In contrast, if we were to repeat the calculation with an object of ordinary mass, such as a tennis ball, the uncertainty would be so small that it would be inconsequential. In that case, m is large and Δx is out of the realm of measurement and therefore of no practical consequence.

De Broglie's hypothesis and Heisenberg's uncertainty principle set the stage for a new and more broadly applicable theory of atomic structure. In this new approach, any attempt to define precisely the instantaneous location and momentum of the electron is abandoned. The wave nature of the electron is recognised, and its behaviour is described in terms appropriate to waves. The result is a model that precisely describes the energy of the electron while describing its location not precisely, but in terms of *probabilities*.

 CONCEPT CHECK 8

What is the main reason why the uncertainty principle seems very important when discussing electrons and other subatomic particles, but rather unimportant in our macroscopic world?

6.5 | QUANTUM MECHANICS AND ATOMIC ORBITALS

In 1926 the Austrian physicist Erwin Schrödinger (1887–1961) proposed an equation, now known as Schrödinger's wave equation, that incorporates both the wave-like behaviour and the particle-like behaviour of the electron. His work opened a new way of dealing with subatomic particles, known as either **quantum mechanics** or *wave mechanics*. We can consider the results he obtained in a qualitative fashion, by examining the structure of the hydrogen atom.

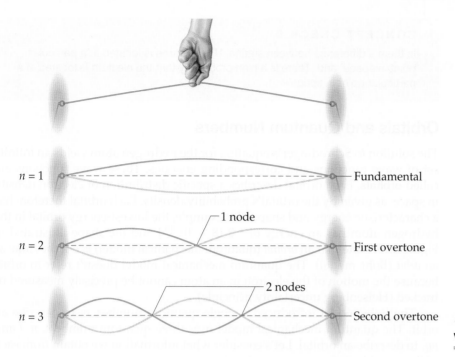

$n = 1$ — Fundamental

1 node

$n = 2$ — First overtone

2 nodes

$n = 3$ — Second overtone

◀ **FIGURE 6.14 Standing waves in a vibrating string.**

Schrödinger treated the electron in a hydrogen atom like the wave on a plucked guitar string (▲ **FIGURE 6.14**). Because such waves do not travel in space, they are called *standing waves*. Just as the plucked guitar string produces a standing wave that has a fundamental frequency and higher overtones (harmonics), the electron exhibits a lowest-energy standing wave and higher-energy ones. Furthermore, just as the overtones of the guitar string have *nodes*, points where the amplitude of the wave is zero, so do the waves characteristic of the electron.

Solving Schrödinger's equation for the hydrogen atom leads to a series of mathematical functions, called **wave functions**, which describe the electron. These wave functions are usually represented by the symbol ψ (the Greek lower-case letter *psi*). Although the wave function itself has no direct physical meaning, the square of the wave function, ψ^2, provides information about an electron's location when the electron is in an allowed energy state.

For the hydrogen atom, the allowed energies are the same as those predicted by the Bohr model. However, the Bohr model assumes that the electron is in a circular orbit of some particular radius about the nucleus. In the quantum mechanical model, the electron's location cannot be described so simply. According to the uncertainty principle, if we know the momentum of the electron with high accuracy, our simultaneous knowledge of its location is very uncertain. Thus we cannot hope to specify the exact location of an individual electron around the nucleus. Rather, we must be content with a kind of statistical knowledge. In the quantum mechanical model, we therefore speak of the *probability* that the electron will be in a certain region of space at a given instant. The square of the wave function, ψ^2, at a given point in space represents the probability that the electron will be found at that location. For this reason, ψ^2 is called either the **probability density** or the **electron density**.

One way of representing the probability of finding the electron in various regions of an atom is shown in ▶ **FIGURE 6.15**. In this figure the density of the dots represents the probability of finding the electron. The regions with a high density of dots correspond to relatively large values for ψ^2, and are therefore regions where there is a high probability of finding the electron. In Section 6.6 we say more about the ways in which we can represent electron density.

▲ **FIGURE IT OUT**

Where in the figure is the region of highest electron density?

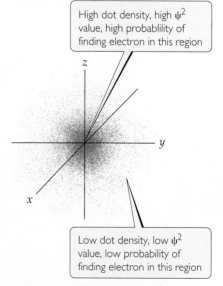

High dot density, high ψ^2 value, high probablility of finding electron in this region

Low dot density, low ψ^2 value, low probability of finding electron in this region

▲ **FIGURE 6.15 Electron-density distribution.** This rendering represents the probability, ψ^2, of finding the electron in a hydrogen atom in its ground state. The origin of the co-ordinate system is at the nucleus.

⚠ **CONCEPT CHECK 9**

Is there a difference between stating, 'The electron is located at a particular point in space' and 'There is a high probability that the electron is located at a particular point in space'?

Orbitals and Quantum Numbers

The solution to Schrödinger's equation for the hydrogen atom yields an infinite set of wave functions and corresponding energies. These wave functions are called **orbitals**. Each orbital describes a specific distribution of electron density in space, as given by the orbital's probability density. Each orbital therefore has a characteristic energy and shape. For example, the lowest-energy orbital in the hydrogen atom has an energy of -2.18×10^{-18} J and the shape illustrated in Figure 6.15. Note that an *orbital* (quantum mechanical model) is not the same as an *orbit* (Bohr model). The quantum mechanical model doesn't refer to orbits because the motion of the electron in an atom cannot be precisely measured or tracked (Heisenberg uncertainty principle).

The Bohr model introduced a single quantum number, n, to describe an orbit. The quantum mechanical model uses three quantum numbers, n, l and m_l, to describe an orbital. Let's consider what information we obtain from each of these and how they are interrelated.

1. The **principal quantum number**, **n**, can have positive integral values of 1, 2, 3 and so forth. This quantum number describes the *size* of the orbital. As n increases, the orbital becomes larger, and the electron spends more time further from the nucleus. An increase in n also means that the electron has a higher energy and is therefore less tightly bound to the nucleus. For the hydrogen atom,

$$E_n = -\left(\frac{1}{n^2}\right)(2.18 \times 10^{-18})\text{J}$$

as in the Bohr model.

2. The second quantum number—the **angular momentum quantum number**, **l**—can have integral values from 0 to $n - 1$ for each value of n. This quantum number defines the *shape* of the orbital. (We consider these shapes in Section 6.6.) So for $n = 2$, l can take the values of 0 or 1 and for $n = 3$, $l = 0, 1$ or 2. The value of l for a particular orbital is generally designated by the letters s, p, d and f,* corresponding to l values of 0, 1, 2 and 3, respectively, as summarised here:

Value of l	0	1	2	3
Letter used	s	p	d	f

3. The **magnetic quantum number**, **m_l**, can have integral values between $-l$ and l including zero. This quantum number describes the *orientation* of the orbital in space, as discussed in Section 6.6. For example, if $l = 1$, then m_l can take the values of $-1, 0$ or $+1$.

Notice that, because the value of n can be any positive integer, there is an infinite number of orbitals for the hydrogen atom. The electron in a hydrogen atom is described by only one of these orbitals at any given time—we say that the electron *occupies* a certain orbital. The remaining orbitals are *unoccupied* for that particular state of the hydrogen atom. We will see that we are mainly interested in the orbitals of the hydrogen atom with small values of n.

---

*The letters s, p, d and f come from the words sharp, principal, diffuse and fundamental, which were used to describe certain features of spectra before quantum mechanics was developed.

TABLE 6.2 • **Relationship between values of *n*, *l*, and *m$_l$* to *n* = 4**

n	Possible values of *l*	Subshell designation	Possible values of *m$_l$*	Number of orbitals in subshell	Total number of orbitals in shell
1	0	1*s*	0	1	1
2	0	2*s*	0	1	
	1	2*p*	1, 0, −1	3	4
3	0	3*s*	0	1	
	1	3*p*	1, 0, −1	3	
	2	3*d*	2, 1, 0, −1, −2	5	9
4	0	4*s*	0	1	
	1	4*p*	1, 0, −1	3	
	2	4*d*	2, 1, 0, −1, −2	5	
	3	4*f*	3, 2, 1, 0, −1, −2, −3	7	16

> **△ CONCEPT CHECK 10**
>
> What is the difference between an *orbit* (Bohr model) and an *orbital* (quantum mechanical model)?

The collection of orbitals with the same value of *n* is called an **electron shell**. For example, all the orbitals that have *n* = 3 are said to be in the third shell. Further, the set of orbitals that has the same *n* and *l* values is called a **subshell**. Each subshell is designated by a number (the value of *n*) and a letter (*s*, *p*, *d* or *f*, corresponding to the value of *l*). For example, the orbitals that have *n* = 3 and *l* = 2 are called 3*d* orbitals and are in the 3*d* subshell.

▲ **TABLE 6.2** summarises the possible values of the quantum numbers *l* and *m$_l$* for values of *n* up to *n* = 4. The restrictions on the possible values of the quantum numbers give rise to the following very important observations.

1. The shell with principal quantum number *n* will consist of exactly *n* subshells. Each subshell corresponds to a different allowed value of *l* from 0 to *n* − 1. Thus the first shell (*n* = 1) consists of only one subshell, the 1*s* (*l* = 0); the second shell (*n* = 2) consists of two subshells, the 2*s* (*l* = 0) and 2*p* (*l* = 1); the third shell consists of three subshells, 3*s*, 3*p* and 3*d* and so forth.

2. Each subshell consists of a specific number of orbitals. Each orbital corresponds to a different allowed value of *m$_l$*. For a given value of *l*, there are 2*l* + 1 allowed values of *m$_l$*, ranging from −*l* to +*l*. Thus each *s* (*l* = 0) subshell consists of one orbital; each *p* (*l* = 1) subshell consists of three orbitals; each *d* (*l* = 2) subshell consists of five orbitals and so forth.

3. The total number of orbitals in a shell is *n*², where *n* is the principal quantum number of the shell. The resulting number of orbitals for the shells—1, 4, 9, 16—is related to a pattern seen in the periodic table: we see that the number of elements in the rows of the periodic table—2, 8, 18 and 32—equal twice these numbers. We discuss this relationship further in Section 6.9.

▶ **FIGURE 6.16** shows the relative energies of the hydrogen atom orbitals to *n* = 3. Each box represents an orbital; orbitals of the same subshell, such as the 2*p*, are grouped together. When the electron occupies the lowest-energy orbital (1*s*), the hydrogen atom is said to be in its *ground state*. When the electron occupies

> **△ FIGURE IT OUT**
>
> If the fourth shell (the *n* = 4 energy level) were shown, how many sub-shells would it contain? How would they be labelled?

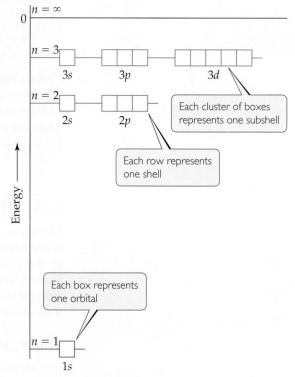

n = 1 shell has one orbital
n = 2 shell has two subshells composed of four orbitals
n = 3 shell has three subshells composed of nine orbitals

▲ **FIGURE 6.16** **Energy levels in the hydrogen atom.**

any other orbital, the atom is in an *excited state*. At ordinary temperatures, essentially all hydrogen atoms are in the ground state. The electron can be excited to a higher-energy orbital by absorption of a photon of appropriate energy. As Figure 6.16 suggests, the energy of the shells becomes closer and closer together as *n* increases. So although there are an infinite number of orbitals obtained from solving Schrödinger's equation, the energies converge on $E = 0$.

 CONCEPT CHECK 11

In Figure 6.16, why is the energy difference between the $n = 1$ and $n = 2$ levels so much greater than the energy difference between the $n = 2$ and $n = 3$ levels?

SAMPLE EXERCISE 6.6 **Subshells of the hydrogen atom**

(a) Without referring to Table 6.2, predict the number of subshells in the fourth shell, that is, for $n = 4$. **(b)** Give the label for each of these subshells. **(c)** How many orbitals are in each of these subshells?

SOLUTION

Analyse The number of subshells in a shell increase with the size of the shell and the number of orbitals in a subshell depends on the type of subshell involved.

Plan The value of *n* indicates the number of the shell and also how many subshells it contains. The value of *l*, which depends on *n*, indicates the type of orbital within each subshell. The number of each orbital within a subshell depends on its type.

Solve **(a)** There are four subshells in the fourth shell, corresponding to the four possible values of *l* (0, 1, 2 and 3).

(b) These subshells are labelled 4*s*, 4*p*, 4*d* and 4*f*. The number given in the designation of a subshell is the principal quantum number, *n*; the letter designates the value of the angular momentum quantum number, *l*: for $l = 0$, *s*; for $l = 1$, *p*; for $l = 2$, *d*; for $l = 3$, *f*.

(c) There is one 4*s* orbital (when $l = 0$, there is only one possible value of m_l: 0). There are three 4*p* orbitals (when $l = 1$, there are three possible values of m_l: 1, 0 and -1). There are five 4*d* orbitals (when $l = 2$, there are five allowed values of m_l: 2, 1, 0, -1, -2). There are seven 4*f* orbitals (when $l = 3$, there are seven permitted values of m_l: 3, 2, 1, 0, -1, -2, -3).

PRACTICE EXERCISE

(a) What is the designation for the subshell with $n = 5$ and $l = 1$? **(b)** How many orbitals are in this subshell? **(c)** Indicate the values of m_l for each of these orbitals.

Answers: **(a)** 5*p*, **(b)** 3, **(c)** 1, 0, –1

(See also Exercises 6.38, 6.40.)

6.6 | REPRESENTATIONS OF ORBITALS

The wave function also provides information about the electron's location in space when it occupies an orbital. First, we look at the three-dimensional shape of the orbital—is it spherical, for example, or does it have directionality? Second, we examine how the probability density changes as we move on a straight line further and further from the nucleus. Finally, we look at the typical three-dimensional sketches that chemists use in describing the orbitals.

The *s* Orbitals

One representation of the lowest-energy orbital of the hydrogen atom, the 1*s*, is shown in Figure 6.15. This type of drawing, which shows the distribution of electron density around the nucleus, is one of the several ways we use to help us visualise orbitals. The electron density for the 1*s* orbital is *spherically symmetric*—in other words, the electron density at a given distance from the nucleus is the same regardless of the direction in which we proceed from the nucleus. All the other *s* orbitals (2*s*, 3*s*, 4*s* and so forth) are spherically symmetric as well.

FIGURE IT OUT

How many maxima would you expect to find in the radial probability function for the 4s orbital of the hydrogen atom? How many nodes would you expect in this function?

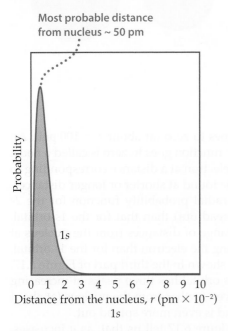

Most probable distance
from nucleus ~ 50 pm

Probability

1s

0 1 2 3 4 5 6 7 8 9 10
Distance from the nucleus, r (pm $\times 10^{-2}$)
1s

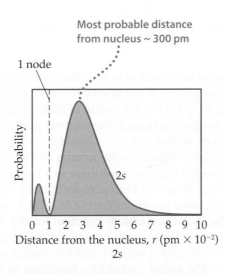

Most probable distance
from nucleus ~ 300 pm

1 node

Probability

2s

0 1 2 3 4 5 6 7 8 9 10
Distance from the nucleus, r (pm $\times 10^{-2}$)
2s

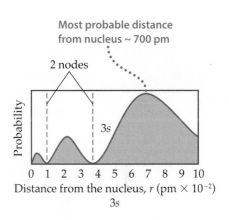

Most probable distance
from nucleus ~ 700 pm

2 nodes

Probability

3s

0 1 2 3 4 5 6 7 8 9 10
Distance from the nucleus, r (pm $\times 10^{-2}$)
3s

▲ **FIGURE 6.17   Radial probability distributions for the 1s, 2s and 3s orbitals of hydrogen.** These graphs of the radial probability function plot probability of finding the electron as a function of distance from the nucleus. As n increases, the most likely distance at which to find the electron (the highest peak) moves further from the nucleus.

So what is different about the *s* orbitals having different *n* quantum numbers? For example, how does the electron-density distribution of the hydrogen atom change when the electron is excited from the 1s orbital to the 2s orbital? To address questions like this, we must take a look at the *radial probability density*, that is, the probability that we will find the electron at a specific distance from the nucleus. In ▲ **FIGURE 6.17** we have plotted the radial probability density for the 1s orbital as a function of r, the distance from the nucleus—the resulting curve is the **radial probability function** for the 1s orbital. (Radial probability functions are described more fully in the 'A Closer Look' box in this section.) We see that the probability of finding the electron rises rapidly as we move away from the nucleus, maximising at a distance of 52.9 pm from the nucleus, and then falls off rapidly. Thus when the electron occupies the 1s orbital, it is *most likely* to be found 52.9 pm from the nucleus[*]—we still use the probabilistic description, consistent with the uncertainty principle. Notice also that the probability of finding the electron at a distance greater than 300 pm from the nucleus is essentially zero.

The second part of Figure 6.17 shows the radial probability function for the 2s orbital of the hydrogen atom. We can see three significant differences between this plot and that for the 1s orbital. (1) There are two separate maxima in the radial probability function for the 2s orbital, namely a small peak at about $r = 50$ pm and a much larger peak at about $r = 300$ pm. (2) Between these two

---

[*]In the quantum mechanical model, the most probable distance at which to find the electron in the orbital—52.9 pm—is identical to the radius of the orbit predicted by Bohr for $n = 1$. The distance 52.9 pm is often called the *Bohr radius*.

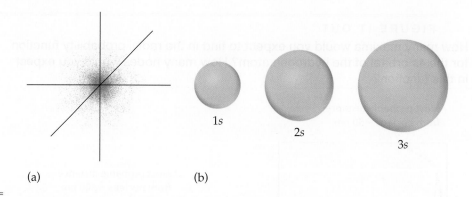

► **FIGURE 6.18 Comparison of the 1s, 2s and 3s orbitals.** (a) Electron-density distribution of a 1s orbital. (b) Contour representations of the 1s, 2s and 3s orbitals. Each sphere is centred on the atom's nucleus and encloses the volume in which there is a 90% probability of finding the electron.

peaks is a point at which the function goes to zero (at about $r = 100$ pm). An intermediate point at which a probability function goes to zero is called a **node**. There is a zero probability of finding the electron at a distance corresponding to a node, even though the electron might be found at shorter or longer distances. Every orbital has $n - 1$ nodes. (3) The radial probability function for the 2s orbital is significantly broader (more spread out) than that for the 1s orbital. Thus for the 2s orbital there is a larger range of distances from the nucleus at which there is a high probability of finding the electron than for the 1s orbital. This trend continues for the 3s orbital, as shown in the third part of Figure 6.17. The radial probability function for the 3s orbital has three peaks of increasing size, with the largest peak maximising even further from the nucleus (at about $r = 700$ pm) at which it has two nodes and is even more spread out.

The radial probability functions in Figure 6.17 tell us that, as n increases, there is also an increase in the most likely distance from the nucleus to find the electron. In other words, the size of the orbital increases with increasing n, just as it did in the Bohr model.

One widely used method of representing orbitals is to display a boundary surface that encloses some substantial portion, say 90%, of the total electron density for the orbital. For the s orbitals, these contour representations are spheres. The contour representations of the 1s, 2s and 3s orbitals are shown in ▲ **FIGURE 6.18**. They all have the same shape, but they differ in size. Although the details of how the electron density varies within the contour representation are lost in these representations, this is not a serious disadvantage. For more qualitative discussions, the most important features of orbitals are their shapes and their relative sizes, which are adequately displayed by contour representations.

⚠ CONCEPT CHECK 12

How many maxima would you expect to find in the radial probability function for the 3s orbital of the hydrogen atom? How many nodes would you expect in the 3s radial probability function?

The p Orbitals

The distribution of electron density for a 2p orbital is shown in ► **FIGURE 6.19(a)**. As we can see from this figure, the electron density is not distributed in a spherically symmetric fashion as in an s orbital. Instead, the electron density is concentrated in two regions on either side of the nucleus, separated by a node at the nucleus. We say that this dumbbell-shaped orbital has two *lobes*. It is useful to recall that we are making no statement of how the electron is moving within the orbital; the only thing Figure 6.19(a) portrays is the *averaged* distribution of the electron density in a 2p orbital.

Beginning with the $n = 2$ shell, each shell has three p orbitals. Thus there are three 2p orbitals, three 3p orbitals and so forth. Each set of p orbitals has the dumbbell shape shown in Figure 6.19(a) for the 2p orbital. For each value of n,

▲ FIGURE IT OUT

(a) Note on the left that the colour is deep pink in the interior of each lobe but fades to pale pink at the edges. What does this change in colour represent? (b) What label is applied to the 2p orbital aligned along the x axis?

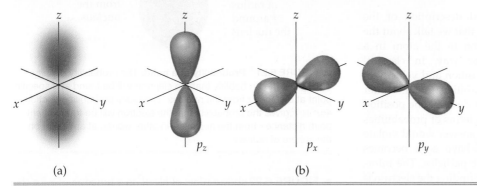

(a)   (b)

◀ **FIGURE 6.19 The p orbitals.**
(a) Electron-density distribution of a 2p orbital.
(b) Contour representations of the three p orbitals. The subscript on the orbital label indicates the axis along which the orbital lies.

the three p orbitals have the same size and shape but differ from one another in spatial orientation. We usually represent p orbitals by drawing the shape and orientation of their wave functions, as shown in Figure 6.19(b). It is convenient to label these the p_x, p_y and p_z orbitals. The letter subscript indicates the Cartesian axis along which the orbital is oriented. Like s orbitals, p orbitals increase in size as we move from 2p to 3p to 4p and so forth.

The d and f Orbitals

When n is 3 or greater, we encounter the d orbitals (for which $l = 2$). There are five 3d orbitals five 4d orbitals and so forth. The different d orbitals in a given shell have different shapes and orientations in space, as shown in ▼ **FIGURE 6.20**. Four of the d-orbital contour representations have a 'four-leaf clover' shape, and each lies primarily in a plane. The d_{xy}, d_{xz} and d_{yz} lie in the xy, xz and yz planes, respectively, with the lobes oriented *between* the axes. The lobes of the $d_{x^2-y^2}$ orbital also lie in the xy plane, but the lobes lie *along* the x and y axes. The d_{z^2} orbital looks very different from the other four: it has two lobes along the z axis and a 'doughnut' in the xy plane. Even though the d_{z^2} orbital looks different from the other d orbitals, it has the same energy as the other four d orbitals. The representations in Figure 6.20 are commonly used for all d orbitals, regardless of principal quantum number.

When n is 4 or greater, there are seven equivalent f orbitals (for which $l = 3$). The shapes of the f orbitals are even more complicated than those of the d orbitals and are not presented here. In the next section, however, you must be aware of f orbitals as we consider the electronic structure of atoms in the lower part of the periodic table.

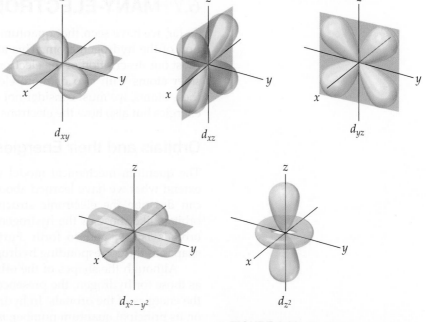

▲ **FIGURE 6.20 Contour representations of the five d orbitals.**

▲ CONCEPT CHECK 13

Note in Figure 6.19(a) that the pink colour fades to zero at the intersection of the axis. What does this change in colour represent?

A CLOSER LOOK

PROBABILITY DENSITY AND RADIAL PROBABILITY FUNCTIONS

The quantum mechanical description of the hydrogen atom requires that we talk about the position of the electron in the atom in a somewhat unfamiliar way. In classical physics, we can pinpoint exactly the position and velocity of an orbiting object, such as a planet orbiting a star. Under quantum mechanics, however, we must describe the position of the electron in the hydrogen atom in terms of probabilities rather than an exact location—an exact answer would violate the uncertainty principle, which we have seen becomes important when considering subatomic particles. The information we need about the probability of finding the electron is contained in the wave functions, ψ, that are obtained when Schrödinger's equation is solved. Remember that there is an infinite number of discrete wave functions (orbitals) for the hydrogen atom, but the electron can occupy only one of them at any given time. Here we discuss briefly how we can use the orbitals to obtain radial probability functions, such as those in Figure 6.17.

In Section 6.5 we stated that the square of the wave function, ψ^2, gives the probability that the electron is at any one given point in space—recall that this quantity is called the *probability density* for the point. For a spherically symmetric *s* orbital, the value of ψ depends only on the distance from the nucleus, r. Let's consider a straight line outward from the nucleus, as shown in ▶ **FIGURE 6.21**. The probability of finding the electron at distance r from the nucleus along that line is $[\psi(r)]^2$ where $\psi(r)$ is the value of ψ at distance r.

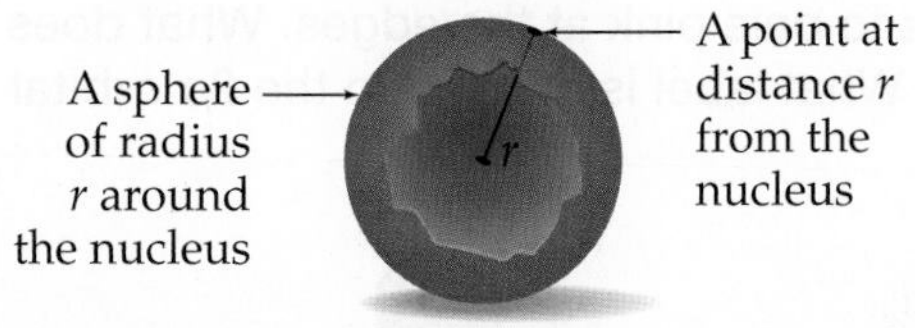

▲ **FIGURE 6.21** **Probability at a point.** The probability density, $\psi(r)^2$, gives the probability that the electron will be found at a specific point at distance r from the nucleus. The radial probability function, $4\pi r^2\psi(r)^2$, gives the probability that the electron will be found at *any* point distance r from the nucleus—in other words, at any point on the sphere of radius r.

▶ **FIGURE 6.22** shows plots of $[\psi(r)]^2$ as a function of r for the 1*s*, 2*s* and 3*s* orbitals of the hydrogen atom.

You will notice that the plots in Figure 6.22 look distinctly different from the radial probability functions plotted in Figure 6.17. These two types of plots for the *s* orbitals are very closely related, but they provide somewhat different information. The probability density, $[\psi(r)]^2$, tells us the probability of finding the electron at *a specific point* in space that is at distance r from the nucleus. The radial probability function, which we will denote $P(r)$, tells us the probability of finding the electron at *any* point that is distance r from the nucleus—in other words, to get $P(r)$ we need to 'add up' the probabilities of finding the electron over all the points at distance r from the nucleus.

As shown in Figure 6.21, the collection of points at distance r from the nucleus is simply a sphere of radius r. The probability density at every point on that sphere is $[\psi(r)]^2$. To

6.7 | MANY-ELECTRON ATOMS

So far, we have seen that quantum mechanics leads to a very elegant description of the hydrogen atom. This atom, however, has only one electron. How must our description of the electronic structure of atoms change when we consider atoms with two or more electrons (a *many-electron* atom)? To describe these atoms, we must consider not only the nature of orbitals and their relative energies but also how the electrons populate the available orbitals.

Orbitals and their Energies

The quantum mechanical model would not be very useful if we could not extend what we have learned about hydrogen to other atoms. Fortunately, we can describe the electronic structure of a many-electron atom in terms of orbitals like those of the hydrogen atom. Thus we can continue to designate orbitals 1*s*, 2*p*$_x$ and so forth. Further, these orbitals have the same general shapes as the corresponding hydrogen orbitals.

Although the shapes of the orbitals for many-electron atoms are the same as those for hydrogen, the presence of more than one electron greatly changes the energies of the orbitals. In hydrogen the energy of an orbital depends only on its principal quantum number, n (Figure 6.15); the 3*s*, 3*p* and 3*d* subshells all have the same energy, for instance. In a many-electron atom, however, the electron–electron repulsions cause the different subshells to be at different

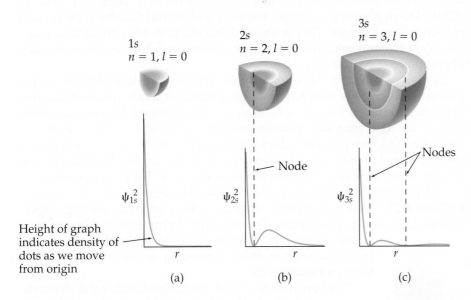

1s
n = 1, l = 0

2s
n = 2, l = 0

3s
n = 3, l = 0

← Node

Nodes

ψ_{1s}^2

ψ_{2s}^2

ψ_{3s}^2

Height of graph
indicates density of
dots as we move
from origin

r

r

r

(a)

(b)

(c)

◀ **FIGURE 6.22   Probability density distribution in 1s, 2s and 3s orbitals.** The lower part of the figure shows how the probability density, $\psi(r)^2$, varies as a function of distance r from the nucleus. The upper part of the figure shows a cutaway of the spherical electron density in each of the s orbitals.

add up all the individual probability densities requires the use of calculus and is beyond the scope of this text (in the language of calculus, 'we integrate the probability density over the surface of the sphere'). The result we obtain is easy to describe, however. The radial probability function at distance r, $P(r)$, is simply the probability density at distance r, $[\psi(r)]^2$, multiplied by the surface area of the sphere, which is given by the formula $4\pi r^2$.

$$P(r) = 4\pi r^2 [\psi(r)]^2$$

Thus the plots of $P(r)$ in Figure 6.17 are equal to the plots of $[\psi(r)]^2$ in Figure 6.22 multiplied by $4\pi r^2$. The fact that $4\pi r^2$

increases rapidly as we move away from the nucleus makes the two sets of plots look very different. For example, the plot of $[\psi(r)]^2$ for the 3s orbital (Figure 6.22) shows that the function generally gets smaller the further we go from the nucleus. But when we multiply by $4\pi r^2$ we see peaks that get larger and larger as we move away from the nucleus (Figure 6.17). We will see that the radial probability functions in Figure 6.17 provide us with the more useful information because they tell us the probability for finding the electron at *all* points distance r from the nucleus, not just one particular point.

RELATED EXERCISES: 6.35, 6.42, 6.43

energies, as shown in ▶ **FIGURE 6.23.** To understand why this is so, we must consider the forces between the electrons and how these forces are affected by the shapes of the orbitals. We forgo this analysis until Chapter 7.

The important idea is this: *In a many-electron atom, for a given value of* n, *the energy of an orbital increases with increasing value of* l. You can see this illustrated in Figure 6.23. Notice, for example, that the n = 3 orbitals (red) increase in energy in the order 3s < 3p < 3d. Figure 6.23 is a *qualitative* energy-level diagram; the exact energies of the orbitals and their spacings differ from one atom to another. Notice that all orbitals of a given subshell (such as the five 3d orbitals) still have the same energy as one another, just as they do in the hydrogen atom. Orbitals with the same energy are said to be **degenerate**.

> **CONCEPT CHECK 14**
>
> For a many-electron atom, can we predict unambiguously whether the 4s orbital is lower in energy or higher in energy than the 3d orbitals?

Electron Spin and the Pauli Exclusion Principle

When scientists examined the line spectra of atoms in great detail, they noticed a very puzzling feature. Lines that were originally

FIGURE IT OUT

Not all of the orbitals in the n = 4 shell are shown in this figure. Which subshells are missing?

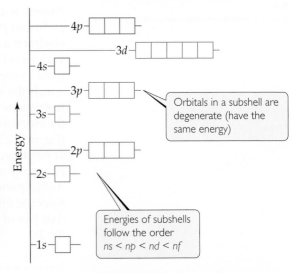

4p

3d

4s

3p

Orbitals in a subshell are degenerate (have the same energy)

3s

Energy →

2p

2s

Energies of subshells follow the order
ns < np < nd < nf

1s

▲ **FIGURE 6.23   General energy ordering of orbitals for a many-electron atom.**

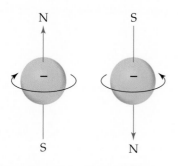

▲ **FIGURE 6.24** **Electron spin.** The electron behaves as if it were spinning about an axis, thereby generating a magnetic field whose direction depends on the direction of spin. The two directions for the magnetic field correspond to the two possible values for the spin quantum number, m_s.

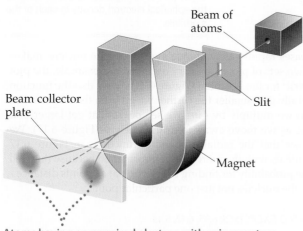

Atoms having an unpaired electron with spin quantum number $m_s = +\frac{1}{2}$ deflect in one direction; those having an unpaired electron with $m_s = -\frac{1}{2}$ deflect in opposite direction

▲ **FIGURE 6.25** **The Stern–Gerlach experiment.** Atoms with an odd number of electrons are passed through a magnetic field.

thought to be single were actually closely spaced pairs. This meant, in essence, that there were twice as many energy levels as there were 'supposed' to be. In 1925 the Dutch physicists George Uhlenbeck and Samuel Goudsmit proposed a solution to this dilemma. They postulated that electrons have an intrinsic property, called **electron spin**, that causes each electron to behave as if it were a tiny sphere spinning on its own axis.

By now it probably does not surprise you to learn that electron spin is quantised. This observation led to the assignment of a new quantum number for the electron, in addition to n, l and m_l already discussed. This new quantum number, the **spin magnetic quantum number**, is denoted m_s (the subscript s stands for *spin*). Two possible values are allowed for m_s, $+\frac{1}{2}$ or $-\frac{1}{2}$, which was first interpreted as indicating the two opposite directions in which the electron can spin. A spinning charge produces a magnetic field. The two opposite directions of spin therefore produce oppositely directed magnetic fields, as shown in ◄ **FIGURE 6.24**.[*] These two opposite magnetic fields lead to the splitting of spectral lines into closely spaced pairs. The effects of electron spin can be observed on the macroscopic scale by using an external magnetic field to split a beam of atoms possessing an odd number of electrons (◄ **FIGURE 6.25**).

Electron spin is crucial for understanding the electronic structures of atoms. In 1925 the Austrian-born physicist Wolfgang Pauli (1900–1958) discovered the principle that governs the arrangements of electrons in many-electron atoms. The **Pauli exclusion principle** states that *no two electrons in an atom can have the same set of four quantum numbers n, l, m_l and m_s*. For a given orbital ($1s$ $2p_z$ and so forth), the values of n, l and m_l are fixed. Thus if we want to put more than one electron in an orbital *and* satisfy the Pauli exclusion principle, our only choice is to assign different m_s values to the electrons. Because there are only two such values, we conclude that *an orbital can hold a maximum of two electrons and they must have opposite spins*. This restriction allows us to index the electrons in an atom, giving their quantum numbers and thereby defining the region in space where each electron is most likely to be found.

6.8 | ELECTRON CONFIGURATIONS

Armed with a knowledge of the relative energies of orbitals and the Pauli exclusion principle, we are now in a position to consider the arrangements of electrons in atoms. The way in which the electrons are distributed among the various orbitals of an atom is called the **electron configuration** of the atom. The most stable electron configuration of an atom—the ground state—is that in which the electrons are in the lowest possible energy states. If there were no restrictions on the possible values for the quantum numbers of the electrons, all the electrons would crowd into the $1s$ orbital because it is the lowest in energy (Figure 6.23). The Pauli exclusion principle tells us, however, that there can be at most two electrons in any single orbital. Thus *the orbitals are filled in order of increasing energy, with no more than two electrons per orbital*. This is called the *Aufbau principle*, from a German word meaning 'to build up'. For example, consider the lithium atom, which has three electrons. The $1s$ orbital can accommodate two of the electrons. The third one goes into the next lowest energy orbital, the $2s$.

---

[*]As discussed earlier, the electron has both wave-like and particle-like properties. Thus the picture of an electron as a spinning charged sphere is, strictly speaking, just a useful pictorial representation that helps us understand the two directions of magnetic field that an electron can possess.

We can represent any electron configuration by writing the symbol for the occupied subshell and adding a superscript to indicate the number of electrons in that subshell. For example, for lithium we write $1s^2 2s^1$ (read as '1s two, 2s one'). We can also show the arrangement of the electrons as

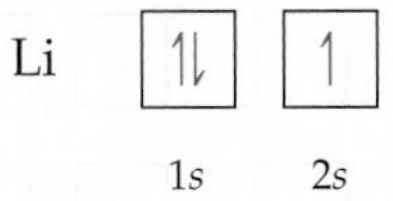

Li   [↑↓] [↑]

    1s   2s

In this kind of representation, which we call an *orbital diagram*, each orbital is denoted by a box and each electron by a half arrow. A half arrow pointing up (↑) represents an electron with a positive spin magnetic quantum number ($m_s = +\frac{1}{2}$), and a half arrow pointing down (↓) represents an electron with a negative spin magnetic quantum number ($m_s = -\frac{1}{2}$). This pictorial representation of electron spin is quite convenient. In fact, chemists and physicists often refer to electrons as 'spin-up' and 'spin-down' rather than specifying the value for m_s.

Electrons with opposite spins are said to be *paired* when they are in the same orbital (↑↓). An *unpaired electron* is one not accompanied by a partner of opposite spin. In the lithium atom the two electrons in the 1s orbital are paired and the electron in the 2s orbital is unpaired.

Hund's Rule

Consider now how the electron configurations of the elements change as we move from element to element across the periodic table. Hydrogen has one electron, which occupies the 1s orbital in its ground state.

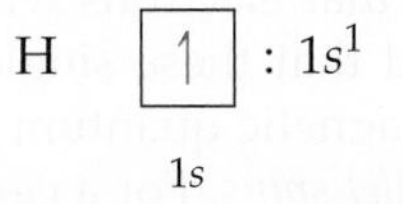

H   [↑]   : $1s^1$

    1s

The choice of a spin-up electron here is arbitrary; we could equally well show the ground state with one spin-down electron in the 1s orbital. It is customary, however, to show unpaired electrons with their spins up.

The next element, helium, has two electrons. Because two electrons with opposite spins can occupy an orbital, both of helium's electrons are in the 1s orbital.

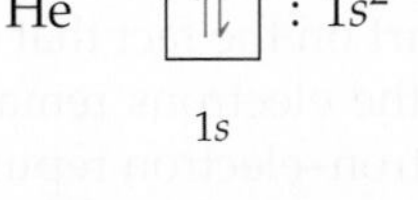

He   [↑↓]   : $1s^2$

     1s

The two electrons present in helium complete the filling of the first shell. This arrangement represents a very stable configuration, as is evidenced by the chemical inertness of helium.

The electron configurations of lithium and several elements that follow it in the periodic table are shown in ▼ TABLE 6.3. For the third electron of lithium, the change in principal quantum number represents a large jump in energy and a corresponding jump in the average distance of the electron from the nucleus. It represents the start of a new shell occupied with electrons. As you can see by examining the periodic table, lithium starts a new row of the table. It is the first member of the alkali metals (group 1).

The element that follows lithium is beryllium; its electron configuration is $1s^2 2s^2$ (Table 6.3). Boron, atomic number 5, has the electron configuration $1s^2 2s^2 2p^1$. The fifth electron must be placed in a 2p orbital because the 2s orbital is filled. Because all the three 2p orbitals are of equal energy, it doesn't matter which 2p orbital is occupied.

TABLE 6.3 • **Electron configurations of several lighter elements**

Element	Total electrons	Orbital diagram				Electron configuration
		1s	2s	2p	3s	
Li	3	[↑↓]	[↑]	[ ][ ][ ]	[ ]	$1s^2 2s^1$
Be	4	[↑↓]	[↑↓]	[ ][ ][ ]	[ ]	$1s^2 2s^2$
B	5	[↑↓]	[↑↓]	[↑][ ][ ]	[ ]	$1s^2 2s^2 2p^1$
C	6	[↑↓]	[↑↓]	[↑][↑][ ]	[ ]	$1s^2 2s^2 2p^2$
N	7	[↑↓]	[↑↓]	[↑][↑][↑]	[ ]	$1s^2 2s^2 2p^3$
Ne	10	[↑↓]	[↑↓]	[↑↓][↑↓][↑↓]	[ ]	$1s^2 2s^2 2p^6$
Na	11	[↑↓]	[↑↓]	[↑↓][↑↓][↑↓]	[↑]	$1s^2 2s^2 2p^6 3s^1$

With the next element, carbon, we encounter a new situation. We know that the sixth electron must go into a 2p orbital. However, does this new electron go into the 2p orbital that already has one electron, or into one of the other two 2p orbitals? This question is answered by **Hund's rule**, which states that *for degenerate orbitals, the lowest energy is attained when the number of electrons with the same spin is maximised*. This means that electrons will occupy orbitals singly to the maximum extent possible and that these single electrons in a given subshell will all have the same spin magnetic quantum number. Electrons arranged in this way are said to have *parallel spins*. For a carbon atom to achieve its lowest energy, therefore, the two 2p electrons will have the same spin. In order for this to happen, the electrons must be in different 2p orbitals, as shown in Table 6.3. Thus a carbon atom in its ground state has two unpaired electrons. Similarly, for nitrogen in its ground state, Hund's rule requires that the three 2p electrons singly occupy each of the three 2p orbitals. This is the only way that all three electrons can have the same spin. For oxygen and fluorine, we place four and five electrons, respectively, in the 2p orbitals. To achieve this, we pair up electrons in the 2p orbitals, as we see in Sample Exercise 6.7.

Hund's rule is based in part on the fact that electrons repel one another. By occupying different orbitals, the electrons remain as far as possible from one another, thus minimising electron–electron repulsions.

SAMPLE EXERCISE 6.7 Orbital diagrams and electron configurations

Draw the orbital diagram for the electron configuration of an oxygen atom, atomic number 8. How many unpaired electrons does an oxygen atom possess?

SOLUTION

Analyse An orbital diagram portrays the electron configuration of an atom.

Plan We first need to arrange the orbitals in terms of increasing energy. Then the electrons are fed into the orbitals starting with the lowest-energy orbital and putting no more than two electrons in each orbital. Orbitals in the same subshell are first filled singly before electrons are paired up.

Solve Because oxygen has an atomic number of 8, each oxygen atom has 8 electrons. Figure 6.23 shows the ordering of orbitals. The electrons (represented as 'half-arrows') are placed in the orbitals (represented as boxes) beginning with the lowest-energy orbital, the $1s$. Each orbital can hold a maximum of two electrons (the Pauli exclusion principle). Because the $2p$ orbitals are degenerate, we place one electron in each of these orbitals (spin-up) before pairing any electrons (Hund's rule).

Two electrons each go into the $1s$ and $2s$ orbitals with their spins paired. This leaves four electrons for the three degenerate $2p$ orbitals. Following Hund's rule, we put one electron into each $2p$ orbital until all three orbitals have one electron each. The fourth electron is then paired up with one of the three electrons already in a $2p$ orbital, so that the representation is

$$\boxed{\uparrow\downarrow} \quad \boxed{\uparrow\downarrow} \quad \boxed{\uparrow\downarrow}\;\boxed{\uparrow}\;\boxed{\uparrow}$$

$$1s \qquad 2s \qquad\qquad 2p$$

The corresponding electron configuration is written $1s^2 2s^2 2p^4$. The atom has two unpaired electrons.

PRACTICE EXERCISE

(a) Write the electron configuration for phosphorus, element 15. **(b)** How many unpaired electrons does a phosphorus atom possess?
Answers: **(a)** $1s^2 2s^2 2p^6 3s^2 3p^3$, **(b)** three
(See also Exercises 6.52–6.54.)

Electron Configurations

The filling of the $2p$ subshell is complete at neon (Table 6.3), which has a stable configuration with eight electrons (an *octet*) in the outermost occupied shell. The next element, sodium, atomic number 11, marks the beginning of a new row of the periodic table. Sodium has a single $3s$ electron beyond the stable configuration of neon. We can therefore abbreviate the electron configuration of sodium as

$$\text{Na:} \qquad [\text{Ne}]3s^1$$

The symbol [Ne] represents the electron configuration of the 10 electrons of neon, $1s^2 2s^2 2p^6$. Writing the electron configuration as $[\text{Ne}]3s^1$ helps focus attention on the outermost electrons of the atom, which are the ones largely responsible for the chemical behaviour of an element.

We can generalise what we have just done for the electron configuration of sodium. In writing the *condensed electron configuration* of an element, the electron configuration of the nearest noble gas element of lower atomic number is represented by its chemical symbol in brackets. For example, we can write the electron configuration of lithium as

$$\text{Li:} \qquad [\text{He}]2s^1$$

We refer to the electrons represented by the symbol for a noble gas as the *noble gas core* of the atom. More usually, these inner-shell electrons are referred to merely as the *core electrons*. The electrons given after the noble gas core are called the *outer-shell electrons*. The outer-shell electrons include the electrons involved in chemical bonding, which are called the *valence electrons* (∞ Section 8.1, 'Chemical Bonds, Lewis Symbols and the Octet Rule').

∞ Find out more on page 252

For lighter elements (those with atomic number of 30 or less), all the outer-shell electrons are valence electrons. As we discuss later, many of the heavier elements have completely filled subshells in their outer-shell electrons that are not involved in bonding and are therefore not considered valence electrons.

By comparing the condensed electron configuration of lithium with that of sodium, we can appreciate why these two elements are so chemically similar: they have the same type of electron configuration in the outermost occupied shell. Indeed, all the members of the alkali metal group (1) have a single s valence electron beyond a noble gas configuration.

Transition Metals

The noble gas element argon marks the end of the row started by sodium. The configuration for argon is $1s^2 2s^2 2p^6 3s^2 3p^6$. The element following argon in the periodic table is potassium (K), atomic number 19. In all its chemical properties, potassium is clearly a member of the alkali metal group. The experimental facts about the properties of potassium leave no doubt that the outermost electron of this element occupies an s orbital. But this means that the electron that has the highest energy has *not* gone into a $3d$ orbital, which we might have expected it to do. Here the ordering of energy levels is such that the $4s$ orbital is lower in energy than the $3d$ (Figure 6.23). Hence, the condensed electron configuration of potassium is

$$\text{K:} \qquad [\text{Ar}]4s^1$$

Following complete filling of the $4s$ orbital (this occurs in the calcium atom), the next set of orbitals to be filled is the $3d$. (You will find it helpful as we go along to refer often to the periodic table on the inside front cover.) Beginning with scandium and extending through to zinc, electrons are added to the five $3d$ orbitals until they are completely filled. Thus the fourth row of the periodic table is 10 elements wider than the two previous rows. These 10 elements are known as either **transition elements** or **transition metals** (Section 21.1, 'Transition Metals'). Note the position of these elements in the periodic table.

Find out more on page 852

In deriving the electron configurations of the transition elements, the orbitals are filled in accordance with Hund's rule—electrons are added to the $3d$ orbitals singly until all five orbitals have one electron each. Additional electrons are then placed in the $3d$ orbitals with spin pairing until the shell is completely filled. The condensed electron configurations and the corresponding orbital diagram representations of two transition elements are as follows:

		$4s$	$3d$
Mn: $[\text{Ar}]4s^2 3d^5$	or $[\text{Ar}]$	$\uparrow\downarrow$	$\uparrow$ $\uparrow$ $\uparrow$ $\uparrow$ $\uparrow$
Zn: $[\text{Ar}]4s^2 3d^{10}$	or $[\text{Ar}]$	$\uparrow\downarrow$	$\uparrow\downarrow$ $\uparrow\downarrow$ $\uparrow\downarrow$ $\uparrow\downarrow$ $\uparrow\downarrow$

Once all the $3d$ orbitals have been filled with two electrons each, the $4p$ orbitals begin to be occupied until the completed octet of outer electrons ($4s^2 4p^6$) is reached with krypton (Kr), atomic number 36, another of the noble gases. Rubidium (Rb) marks the beginning of the fifth row. Refer again to the periodic table on the inside front cover. Notice that this row is in every respect like the preceding one, except that the value for n is greater by 1.

CONCEPT CHECK 15

Based on the structure of the periodic table, which becomes occupied first, the $6s$ orbital or the $5d$ orbitals?

The Lanthanides and Actinides

The sixth row of the periodic table begins similarly to the preceding one: one electron in the $6s$ orbital of caesium (Cs) and two electrons in the $6s$ orbital of barium (Ba). Notice, however, that the periodic table then has a break and the subsequent set of elements (elements 57–70) is placed below the main portion of the table. It is at this place that we begin to encounter a new set of orbitals, the $4f$.

There are seven degenerate $4f$ orbitals, corresponding to the seven allowed values of m_l, ranging from 3 to -3. Thus it takes 14 electrons to fill the $4f$ orbitals completely. The 14 elements corresponding to the filling of the $4f$ orbitals are known as either the **lanthanide elements** or the **rare earth elements**. They are set below the other elements to avoid making the periodic table unduly wide. The properties of the lanthanide elements are all quite similar, and these elements often occur together in nature. For many years it was virtually impossible to separate them from one another.

Because the energies of the $4f$ and $5d$ orbitals are very close to each other, the electron configurations of some of the lanthanides involve $5d$ electrons. For example, the elements lanthanum (La), cerium (Ce) and praseodymium (Pr) have the following electron configurations:

$$[Xe]6s^2 5d^1 \qquad [Xe]6s^2 5d^1 4f^1 \qquad [Xe]6s^2 4f^3$$
$$\text{Lanthanum} \qquad \text{Cerium} \qquad \text{Praseodymium}$$

Because La has a single $5d$ electron, it is sometimes placed below yttrium (Y) as the first member of the third series of transition elements, and Ce is then placed as the first member of the lanthanides. Based on their chemistry, however, La can be considered the first element in the lanthanide series. Arranged this way, there are fewer apparent exceptions to the regular filling of the $4f$ orbitals among the subsequent members of the series.

After the lanthanide series, the third transition element series is completed by the filling of the $5d$ orbitals, followed by the filling of the $6p$ orbitals. This brings us to radon (Rn), heaviest of the known noble gas elements.

The final row of the periodic table begins by filling the $7s$ orbitals. The **actinide elements**, of which uranium (U, element 92) and plutonium (Pu, element 94) are the best known, are then built up by completing the $5f$ orbitals. The actinide elements are radioactive and most of them are not found in nature.

6.9 | ELECTRON CONFIGURATIONS AND THE PERIODIC TABLE

The periodic table is structured so that elements with the same pattern of outer-shell (valence) electron configuration are arranged in columns. For example, the electron configurations for the elements in groups 2 and 13 are given in ▶ TABLE 6.4. We see that the group 2 elements all have ns^2 outer configurations, while the group 13 elements all have ns^2np^1 configurations.

Earlier, in Table 6.2, we saw that the total number of orbitals in each shell is equal to n^2: 1, 4, 9 or 16. Because each orbital can hold two electrons, each shell can accommodate up to $2n^2$ electrons: 2, 8, 18 or 32. The structure of the periodic table reflects this orbital structure. The first row has two elements, the second and third rows have eight elements, the fourth and fifth rows have 18 elements and the sixth row has 32 elements (including the lanthanide metals). Some of the numbers repeat because we reach the end of a row of the periodic table before a shell completely fills. For example, the third row has eight elements, which corresponds to filling the $3s$ and $3p$ orbitals. The remaining orbitals of the third shell, the $3d$ orbitals, do not begin to fill until the fourth row of the periodic table (and after the $4s$ orbital is filled). Likewise, the $4d$ orbitals don't begin to fill until the fifth row of the table, and the $4f$ orbitals don't begin filling until the sixth row.

All these observations are evident in the structure of the periodic table. For this reason, we emphasise that *the periodic table is your best guide to the order in which orbitals are filled*. You can easily write the electron configuration of an element based on its location in the periodic table. The pattern is summarised in ▼ FIGURE 6.26. Notice that the elements can be grouped by the *type* of orbital into which the electrons are placed. On the left are *two* columns of elements, depicted in blue. These elements, known as the alkali metals (group 1) and alkaline earth metals (group 2), are those in which the valence s orbitals are being filled. On the right is a pink block of *six* columns. These are the elements in which the valence p orbitals are being filled. The s block and the p block of

TABLE 6.4 •
Electron configurations of group 2 and 13 elements

Group 2	
Be	$[He]2s^2$
Mg	$[Ne]3s^2$
Ca	$[Ar]4s^2$
Sr	$[Kr]5s^2$
Ba	$[Xe]6s^2$
Ra	$[Rn]7s^2$

Group 13	
B	$[He]2s^2 2p^1$
Al	$[Ne]3s^2 3p^1$
Ga	$[Ar]3d^{10} 4s^2 4p^1$
In	$[Kr]4d^{10} 5s^2 5p^1$
Tl	$[Xe]4f^{14} 5d^{10} 6s^2 6p^1$

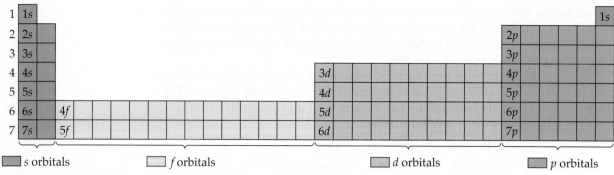

▲ **FIGURE 6.26** **Regions of the periodic table.** The order in which electrons are added to orbitals is read left to right beginning in the top left corner.

∞ Find out more on page 804

the periodic table together are the **main-group elements** (∞ Section 20.1, 'Periodic Trends and Chemical Reactions'), which are sometimes called the *representative elements*.

In the middle of Figure 6.26 is a gold block of 10 columns containing the transition metals. These are the elements in which the valence *d* orbitals are being filled. There are also two tan rows containing 14 columns, usually placed under the main part of the periodic table to save space. These elements are often referred to as the *f-block metals* because they are the ones in which the valence *f* orbitals are being filled. Recall that the numbers 2, 6, 10 and 14 are precisely the number of electrons that can fill the *s*, *p*, *d* and *f* subshells, respectively. Recall also that the 1*s* subshell is the first *s* subshell, the 2*p* is the first *p* subshell, the 3*d* is the first *d* subshell and the 4*f* is the first *f* subshell.

SAMPLE EXERCISE 6.8   Electron configurations for a group

What is the characteristic valence electron configuration of the group 17 elements, the halogens?

SOLUTION

Analyse Elements in the same group have the same number of valence electrons. We need to determine this number for group 17.

Plan We first locate the halogens in the periodic table, write the electron configurations for the first two elements and then determine the general similarity between them.

Solve The first member of the halogen group is fluorine, atomic number 9. The condensed electron configuration for fluorine is

$$\text{F:}\quad [\text{He}]2s^2\, 2p^5$$

Similarly, that for chlorine, the second halogen, is

$$\text{Cl:}\quad [\text{Ne}]3s^2 3p^5$$

From these two examples, we see that the characteristic valence electron configuration of a halogen is ns^2np^5, where n ranges from 2 in the case of fluorine to 6 in the case of astatine.

PRACTICE EXERCISE

Which family of elements is characterised by an ns^2np^2 electron configuration in the outermost occupied shell?

Answer: Group 14

(See also Exercises 6.54, 6.55.)

SAMPLE EXERCISE 6.9   Electron configurations from the periodic table

(a) Write the electron configuration for bismuth, element number 83. **(b)** Write the condensed electron configuration for this element. **(c)** How many unpaired electrons does each atom of bismuth possess?

SOLUTION

Analyse The electron configuration of an element relates to its position in the periodic table.

Plan Locate bismuth in the periodic table. The period number indicates the number of shells containing electrons and the group number is related to the number of electrons in the outermost shell.

Solve (a) We write the electron configuration by moving across the periodic table one row at a time and writing the occupancies of the orbital corresponding to each row (refer to Figure 6.26).

First row	$1s^2$
Second row	$2s^2 2p^6$
Third row	$3s^2 3p^6$
Fourth row	$4s^2 3d^{10} 4p^6$
Fifth row	$5s^2 4d^{10} 5p^6$
Sixth row	$6s^2 4f^{14} 5d^{10} 6p^3$
Total:	$1s^2 2s^2 2p^6 3s^2 3p^6 3d^{10} 4s^2 4p^6 4d^{10} 4f^{14} 5s^2 5p^6 5d^{10} 6s^2 6p^3$

Note that 3 is the lowest possible value that n may have for a d orbital and that 4 is the lowest possible value of n for an f orbital.

The total of the superscripted numbers should equal the atomic number of bismuth, 83. The electrons may be listed, as shown above in the 'Total' row, in the order of increasing principal quantum number. However, it is equally correct to list the orbitals in the order in which they are read from Figure 6.26:

$$1s^2 2s^2 2p^6 3s^2 3p^6 4s^2 3d^{10} 4p^6 5s^2 4d^{10} 5p^6 6s^2 4f^{14} 5d^{10} 6p^3$$

(b) We write the condensed electron configuration by locating bismuth on the periodic table and then moving *backward* to the nearest noble gas, which is Xe, element 54. Thus the noble gas core is [Xe]. The outer electrons are then read from the periodic table as before. Moving from Xe to Cs, element 55, we find ourselves in the sixth row. Moving across this row to Bi gives us the outer electrons. Thus the abbreviated electron configuration is $[Xe]6s^2 4f^{14} 5d^{10} 6p^3$ or $[Xe]4f^{14} 5d^{10} 6s^2 6p^3$.

(c) We can see from the abbreviated electron configuration that the only partially occupied subshell is the $6p$. The orbital diagram representation for this subshell is

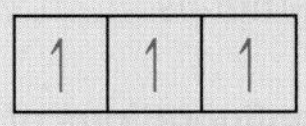

In accordance with Hund's rule, the three $6p$ electrons occupy the three $6p$ orbitals singly, with their spins parallel. Thus there are three unpaired electrons in each atom of bismuth.

PRACTICE EXERCISE

Use the periodic table to write the condensed electron configurations for **(a)** Co (atomic number 27), **(b)** Te (atomic number 52).

Answers: (a) $[Ar]4s^2 3d^7$ or $[Ar]3d^7 4s^2$, **(b)** $[Kr]5s^2 4d^{10} 5p^4$ or $[Kr]4d^{10} 5s^2 5p^4$

(See also Exercises 6.70, 6.71.)

▼ **FIGURE 6.27** gives the valence ground-state electron configurations for all the elements. You can use this figure to check your answers as you practise writing electron configurations. We have written these configurations with orbitals listed in order of increasing principal quantum number. As in Sample Exercise 6.9, the orbitals can also be listed in order of filling, as they would be read off the periodic table.

The electron configurations in Figure 6.27 allow us to re-examine the concept of *valence electrons*. Notice, for example, that as we proceed from Cl ($[Ne]3s^2 3p^5$) to B ($[Ar]3d^{10} 4s^2 4p^5$) we have added a complete subshell of $3d$ electrons to the outer-shell electrons beyond the noble gas core of Ar. Although the $3d$ electrons are outer-shell electrons, they are not involved in chemical bonding and are therefore not considered valence electrons. Thus we consider

FIGURE 6.27 Valence electron configurations of the elements.

Periodic table with valence electron configurations (Core noble-gas cores shown at left: [He], [Ne], [Ar], [Kr], [Xe], [Rn]):

Period 1
- 1 H $1s^1$
- 2 He $1s^2$

Period 2 — Core [He]
- 3 Li $2s^1$ | 4 Be $2s^2$
- 5 B $2s^2 2p^1$ | 6 C $2s^2 2p^2$ | 7 N $2s^2 2p^3$ | 8 O $2s^2 2p^4$ | 9 F $2s^2 2p^5$ | 10 Ne $2s^2 2p^6$

Period 3 — Core [Ne]
- 11 Na $3s^1$ | 12 Mg $3s^2$
- 13 Al $3s^2 3p^1$ | 14 Si $3s^2 3p^2$ | 15 P $3s^2 3p^3$ | 16 S $3s^2 3p^4$ | 17 Cl $3s^2 3p^5$ | 18 Ar $3s^2 3p^6$

Period 4 — Core [Ar]
- 19 K $4s^1$ | 20 Ca $4s^2$
- 21 Sc $3d^1 4s^2$ | 22 Ti $3d^2 4s^2$ | 23 V $3d^3 4s^2$ | 24 Cr $3d^5 4s^1$ | 25 Mn $3d^5 4s^2$ | 26 Fe $3d^6 4s^2$ | 27 Co $3d^7 4s^2$ | 28 Ni $3d^8 4s^2$ | 29 Cu $3d^{10} 4s^1$ | 30 Zn $3d^{10} 4s^2$
- 31 Ga $3d^{10} 4s^2 4p^1$ | 32 Ge $3d^{10} 4s^2 4p^2$ | 33 As $3d^{10} 4s^2 4p^3$ | 34 Se $3d^{10} 4s^2 4p^4$ | 35 Br $3d^{10} 4s^2 4p^5$ | 36 Kr $3d^{10} 4s^2 4p^6$

Period 5 — Core [Kr]
- 37 Rb $5s^1$ | 38 Sr $5s^2$
- 39 Y $4d^1 5s^2$ | 40 Zr $4d^2 5s^2$ | 41 Nb $4d^4 5s^1$ | 42 Mo $4d^5 5s^1$ | 43 Tc $4d^5 5s^2$ | 44 Ru $4d^7 5s^1$ | 45 Rh $4d^8 5s^1$ | 46 Pd $4d^{10}$ | 47 Ag $4d^{10} 5s^1$ | 48 Cd $4d^{10} 5s^2$
- 49 In $4d^{10} 5s^2 5p^1$ | 50 Sn $4d^{10} 5s^2 5p^2$ | 51 Sb $4d^{10} 5s^2 5p^3$ | 52 Te $4d^{10} 5s^2 5p^4$ | 53 I $4d^{10} 5s^2 5p^5$ | 54 Xe $4d^{10} 5s^2 5p^6$

Period 6 — Core [Xe]
- 55 Cs $6s^1$ | 56 Ba $6s^2$
- 71 Lu $4f^{14} 5d^1 6s^2$ | 72 Hf $4f^{14} 5d^2 6s^2$ | 73 Ta $4f^{14} 5d^3 6s^2$ | 74 W $4f^{14} 5d^4 6s^2$ | 75 Re $4f^{14} 5d^5 6s^2$ | 76 Os $4f^{14} 5d^6 6s^2$ | 77 Ir $4f^{14} 5d^7 6s^2$ | 78 Pt $4f^{14} 5d^9 6s^1$ | 79 Au $4f^{14} 5d^{10} 6s^1$ | 80 Hg $4f^{14} 5d^{10} 6s^2$
- 81 Tl $4f^{14} 5d^{10} 6s^2 6p^1$ | 82 Pb $6s^2 6p^2$ | 83 Bi $6s^2 6p^3$ | 84 Po $6s^2 6p^4$ | 85 At $4f^{14} 5d^{10} 6s^2 6p^5$ | 86 Rn $4f^{14} 5d^{10} 6s^2 6p^6$

Period 7 — Core [Rn]
- 87 Fr $7s^1$ | 88 Ra $7s^2$
- 103 Lr $5f^{14} 6d^1 7s^2$ | 104 Rf $5f^{14} 6d^2 7s^2$ | 105 Db $5f^{14} 6d^3 7s^2$ | 106 Sg $5f^{14} 6d^4 7s^2$ | 107 Bh $5f^{14} 6d^5 7s^2$ | 108 Hs $5f^{14} 6d^6 7s^2$ | 109 Mt $5f^{14} 6d^7 7s^2$ | 110 Ds | 111 Rg | 112 Cn
- 113 | 114 Fl | 115 | 116 Lv | 117 | 118

Lanthanide series — [Xe]
- 57 La $5d^1 6s^2$ | 58 Ce $4f^1 5d^1 6s^2$ | 59 Pr $4f^3 6s^2$ | 60 Nd $4f^4 6s^2$ | 61 Pm $4f^5 6s^2$ | 62 Sm $4f^6 6s^2$ | 63 Eu $4f^7 6s^2$ | 64 Gd $4f^7 5d^1 6s^2$ | 65 Tb $4f^9 6s^2$ | 66 Dy $4f^{10} 6s^2$ | 67 Ho $4f^{11} 6s^2$ | 68 Er $4f^{12} 6s^2$ | 69 Tm $4f^{13} 6s^2$ | 70 Yb $4f^{14} 6s^2$

Actinide series — [Rn]
- 89 Ac $6d^1 7s^2$ | 90 Th $6d^2 7s^2$ | 91 Pa $5f^2 6d^1 7s^2$ | 92 U $5f^3 6d^1 7s^2$ | 93 Np $5f^4 6d^1 7s^2$ | 94 Pu $5f^6 7s^2$ | 95 Am $5f^7 7s^2$ | 96 Cm $5f^7 6d^1 7s^2$ | 97 Bk $5f^9 7s^2$ | 98 Cf $5f^{10} 7s^2$ | 99 Es $5f^{11} 7s^2$ | 100 Fm $5f^{12} 7s^2$ | 101 Md $5f^{13} 7s^2$ | 102 No $5f^{14} 7s^2$

Legend: ☐ Metals ☐ Metalloids ☐ Non-metals

---

only the 4s and 4p electrons of Br to be valence electrons. Similarly, if we compare the electron configuration of Ag and Au, Au has a completely full $4f^{14}$ subshell beyond its noble gas core, but those 4f electrons are not involved in bonding. In general, *for the main group elements we do not consider completely full d or f subshells to be among the valence electrons*, and *for transition elements we likewise do not consider a completely full f subshell to be among the valence electrons*.

Anomalous Electron Configurations

If you inspect Figure 6.27 closely you will see that the electron configurations of certain elements appear to violate the rules we have just discussed. For example, the electron configuration of chromium is [Ar]$3d^5 4s^1$ rather than the [Ar]$3d^4 4s^2$ configuration we might have expected. Similarly, the configuration of copper is [Ar]$3d^{10} 4s^1$ instead of [Ar]$3d^9 4s^2$. This anomalous behaviour is largely a consequence of the closeness of the 3d and 4s orbital energies. It frequently occurs when there are enough electrons to lead to precisely half-filled sets of degenerate orbitals (as in chromium) or to completely fill a d subshell (as in copper). There are a few similar cases among the heavier transition metals (those with partially filled 4d or 5d orbitals) and among the f-block metals.

CONCEPT CHECK 16

The elements Ni, Pd and Pt are all in the same group. By examining the electron configurations for these elements in Figure 6.27, what can you conclude about the relative energies of the nd and $(n + 1)s$ orbitals for this group?

SAMPLE INTEGRATIVE EXERCISE   Putting concepts together

An English physicist, H. Moseley, established the concept of atomic number by studying X-rays emitted by the elements. The X-rays emitted by some of the elements have the following wavelengths

Element	Wavelength (pm)
Ne	1461.0
Ca	335.8
Zn	143.5
Zr	78.6
Sn	49.1

(a) Calculate the frequency, ν, of the X-rays emitted by each of the elements, in Hz. **(b)** Using graph paper (or suitable computer software), plot the square root of ν, versus the atomic number of the element. What do you observe about the plot? **(c)** Explain how the plot in part (b) allowed Moseley to predict the existence of undiscovered elements. **(d)** Use the result from part (b) to predict the X-ray wavelength emitted by iron. **(e)** A particular element emits X-rays with a wavelength of 98.0 pm. What element do you think it is?

SOLUTION

Analyse We first convert a wavelength to frequency and use these data to plot a graph. Analysis of the graph allows us to estimate the X-ray wavelength emitted by iron and also to determine the atomic number of an element that emits X-rays of known wavelength.

Solve (a) Solving Equation 6.1 for frequency gives $\nu = c/\lambda$. When we insert the values of c and λ, we note that the units of length are different in the two quantities. We need to convert the wavelength from picometres to metres, so the units cancel:

$$1\ pm = 1 \times 10^{-12}\ m$$

Using the X-rays emitted by neon as an example:

$$\nu = c/\lambda = 3.00 \times 10^8\ m\ s^{-1}/1461.0 \times 10^{-12}\ m = 2.05 \times 10^{17}\ s^{-1}$$

The frequency of the X-rays emitted by the given selection of elements is summarised in the following table

Element	Wavelength (pm)	Frequency (Hz)	Square root of frequency (Hz$^{\frac{1}{2}}$)	Atomic number
Ne	1461.0	2.05×10^{17}	4.53×10^8	10
Ca	335.8	8.93×10^{17}	9.45×10^8	20
Zn	143.5	2.09×10^{18}	1.45×10^9	30
Zr	78.6	3.82×10^{18}	1.95×10^9	40
Sn	49.1	6.11×10^{18}	2.47×10^9	50

Note. Hertz (Hz) is frequently used instead of s^{-1} as the unit of frequency.

(b) A plot of the square root of v versus the atomic number of the element is a straight line.

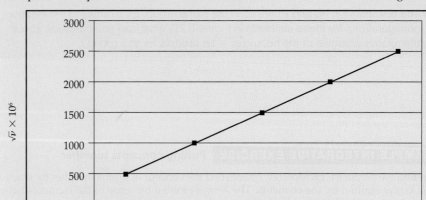

(c) The linear nature of the plot enabled it to be used to identify the atomic number, and wavelength of X-ray emission of elements that had yet to be discovered by 'filling in the gaps' in the known data.

(d) Iron has the atomic number of 26. From the plot the square root of the frequency is 1.3×10^9 and the frequency is 1.69×10^{18} s^{-1}. Converting this to wavelength using $\lambda = c/v$ we get a wavelength of 180 pm.

(e) A wavelength of 98.0 pm corresponds to a frequency of 3.06×10^{18} Hz and the square root of this (1.75×10^9) corresponds to an atomic number of 35 from the plot. This is the element bromine.

CHAPTER SUMMARY AND KEY TERMS

SECTIONS 6.1 and 6.2   Electromagnetic radiation (or radiant energy) has wave-like properties and so can be characterised in terms of its **wavelength** (λ) and **frequency** (v). In addition, the **amplitude** relates to intensity of the wave. Matter may absorb energy (such as heat) but only in discrete 'chunks' of some minimum size. The minimum amount of energy that an atom can absorb (or lose) is called a **quantum** of energy and is related to the frequency of the absorbed or emitted radiation by the relationship $E = hv$, where h is **Planck's constant**. When we say that the energy absorbed or emitted by an atom is **quantised**, we mean it can only have certain allowed values which are integral multiples of hv. This concept is used to explain the **photoelectric effect** by assuming that light behaves as though it consists of quantised energy packets which have particle-like properties, called **photons**.

SECTION 6.3   Radiation can be dispersed into its component wavelengths, producing a **spectrum**, which can be a **continuous spectrum** (it contains all the wavelengths in a region) or a **line spectrum** (it contains only specific wavelengths). The line spectrum of hydrogen was explained by Bohr as indicating that electrons move around the nucleus in circular orbits and that these orbits have certain defined energies; that is, they are quantised. We consider the model of the hydrogen atom proposed by Bohr in which the possible energy levels of the electron are represented by a **quantum number** n, where $n = 1, 2, 3, \ldots, \infty$. Each value of n corresponds to a different specific value of energy E_n.

The lowest energy of the electron in a hydrogen atom is when $n = 1$ (the **ground state**) and all larger values of n correspond to **excited states** of the atom.

SECTION 6.4   De Broglie postulated that an object in motion has wave-like properties with a characteristic wavelength depending on its **momentum** (mu). These **matter waves** are expressed by the relationship, $\lambda = \frac{h}{mu}$. This leads to **Heisenberg's uncertainty principle**, which states that the dual nature of matter places a fundamental limitation on the precision with which we can know both the location and momentum of any object.

SECTION 6.5   The arrangement of electrons in atoms can be described by **quantum mechanics** by assigning a mathematical function called the **wave function** (ψ) to each electron. Each allowed value of ψ has a precisely known energy but the location of the electron at any point in space can only be described by the **probability density** (ψ^2) or **electron density**. The allowed wave functions of the hydrogen atom are called **orbitals** and are described in energy and shape by three sets of quantum numbers. These are the **principle quantum number** (n), **the angular momentum quantum number** (l) and **the magnetic quantum number** (m_l). We define an **electron shell** as the set of orbitals with the same value of n, for example 3s, 3p, 3d, and an **electron subshell** is the set of one or more orbitals with the same n and l values; for example, 3s, 3p and 3d are each subshells of the $n = 3$ shell. Remember that there is one orbital in an s subshell, three in a p subshell, five in a d subshell and seven in an f subshell.

SECTION 6.6 We show how contour representations can be used to visualise the shapes of the orbitals and we use the **radial probability function** to indicate the probability that the electron will be found at a certain distance from the nucleus.

SECTION 6.7 In the case of many-electron atoms, different subshells of the same shell have different energies. For a given value of n the energy of a subshell increases as the value of l increases ($ns < np < nd < nf$). Orbitals within the same subshell have the same energy and are said to be **degenerate**. Electrons have an intrinsic property called **electron spin**, which is quantised. The **spin magnetic quantum number**, m_s, can have two possible values, $+\frac{1}{2}$ and $-\frac{1}{2}$, which can be envisaged as the two directions of an electron spinning about an axis. The **Pauli exclusion principle** states that no two electrons in an atom can have the same values of n, l, m_l and m_s. This principle places a limit of two on the number of electrons that can occupy any

one atomic orbital. These two electrons differ in their value of m_s.

SECTIONS 6.8 and 6.9 The **electron configuration** of an atom describes how the electrons are distributed among the orbitals of the atom. The ground-state electron configuration may be obtained by placing the electrons successively in the atomic orbitals of lowest possible energy according to the Pauli exclusion principle and **Hund's rule**. We show that elements in the same group of the periodic table have similar properties because they have the same type of electron arrangements in their outermost shells. We note that the periodic table partitions elements into different types depending on whether the outermost subshell is an s or p subshell (**main-group elements**) or a d subshell (**transition elements** or **transition metals**). The elements in which the $4f$ subshell is being filled are called the **lanthanides** or **rare earth elements** and the **actinides** are those in which the $5f$ subshell is being filled.

KEY SKILLS

- Calculate the frequency of electromagnetic radiation given its frequency or calculate its frequency given its wavelength. (Section 6.1)
- Explain the concept of photons and be able to calculate their energies given either their frequency or their wavelength. (Section 6.2)
- Explain how the line spectra of the elements relate to the idea of quantised energy states of electrons in atoms. (Section 6.3)
- Understand the concept of the wave-like properties of matter. (Section 6.4)
- Explain how the uncertainty principle limits how precisely we can specify the position and momentum of subatomic particles. (Section 6.4)
- Relate the quantum numbers to the number and types of orbitals and recognise the shapes of different orbitals. (Section 6.5)
- Interpret the probability function graphs of orbitals. (Section 6.6)
- Differentiate the energy-level diagrams for one electron and many electron atoms. (Section 6.7)
- Be able to populate the orbitals in the ground state of an atom using the Pauli exclusion principle and Hund's rule. (Section 6.8)
- Use the periodic table to write abbreviated electron configurations and determine the number of unpaired electrons in an atom. (Section 6.9)

KEY EQUATIONS

- Light as a wave; relationship between the speed of light ($m\ s^{-1}$) and wavelength (m) and frequency (s^{-1})

$$v = \frac{c}{\lambda}$$ [6.1]

- Light as a particle (photon); relationship between energy of light (J) and its frequency (s^{-1}) where h is Planck's constant

$$E = h\nu$$ [6.2]

- Matter as a wave; relationship between mass (kg) velocity ($m\ s^{-1}$) and wavelength

$$\lambda = \frac{h}{mu}$$ [6.8]

- Heisenberg's uncertainty principle

$$\Delta x \cdot \Delta(mu) \geqslant \frac{h}{4\pi}$$ [6.9]

EXERCISES

VISUALISING CONCEPTS

6.1 A popular kitchen appliance produces electromagnetic radiation with a wavelength of 1 cm. With reference to Figure 6.4, answer the following: **(a)** Would the radiation produced by the appliance be visible to the human eye? **(b)** If the radiation is not visible, do photons of this radiation have more or less energy than photons of

visible light? **(c)** Propose an identity for the kitchen appliance. [Section 6.1]

6.2 The familiar phenomenon of a rainbow results from the diffraction of sunlight through raindrops. **(a)** Does the wavelength of light increase or decrease as we proceed outward from the innermost band of the rainbow?

(b) Does the frequency of light increase or decrease as we proceed outward? **(c)** Suppose that, instead of sunlight, the visible light from a hydrogen discharge tube (Figure 6.9) was used as the light source. What do you think the resulting 'hydrogen discharge rainbow' would look like? [Section 6.3]

6.3 A certain quantum mechanical system has the energy levels shown in the diagram below. The energy levels are indexed by a single quantum number, n, which is an integer. **(a)** As drawn, which quantum numbers are involved in the transition that requires the most energy? **(b)** Which quantum numbers are involved in the transition that requires the least energy? **(c)** Based on the drawing, put the following in order of increasing wavelength of the light absorbed or emitted during the transition: (i) $n = 1$ to $n = 2$; (ii) $n = 3$ to $n = 2$; (iii) $n = 2$ to $n = 4$; (iv) $n = 3$ to $n = 1$. [Section 6.3]

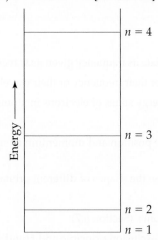

6.4 Consider a fictitious one-dimensional system with one electron. The wave function for the electron, drawn below, is $\psi(x) = \sin x$ from $x = 0$ to $x = 2\pi$. **(a)** Sketch the probability density, $\psi^2(x)$, from $x = 0$ to $x = 2\pi$. **(b)** At what value or values of x will there be the greatest probability of finding the electron? **(c)** What is the probability that the electron will be found at $x = \pi$? What is such a point in a wave function called? [Section 6.5]

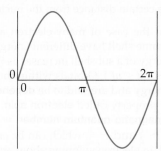

6.5 The contour representation of one of the orbitals for the $n = 3$ shell of a hydrogen atom is shown below. **(a)** What is the quantum number l for this orbital? **(b)** How do we label this orbital? **(c)** How would you modify this sketch to show the analogous orbital for the $n = 4$ shell? [Section 6.6]

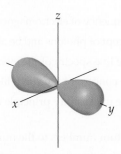

6.6 The drawing below shows part of the orbital diagram for an element. **(a)** As drawn, the drawing is *incorrect*. Why? **(b)** How would you correct the drawing without changing the number of electrons? **(c)** To which group in the periodic table does the element belong? [Section 6.8]

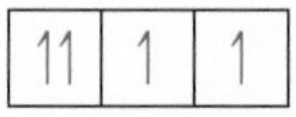

| 11 | 1 | 1 |

THE WAVE NATURE OF LIGHT (Section 6.1)

6.7 What are the basic SI units for **(a)** the wavelength of light, **(b)** the frequency of light, **(c)** the speed of light?

6.8 **(a)** What is the relationship between the wavelength and the frequency of radiant energy? **(b)** Ozone in the upper atmosphere absorbs energy in the 210–230 nm range of the spectrum. In what region of the electromagnetic spectrum does this radiation occur?

6.9 Label each of the following statements as true or false. For those that are false, correct the statement. **(a)** Visible light is a form of electromagnetic radiation. **(b)** The frequency of radiation increases as the wavelength increases. **(c)** Ultraviolet light has longer wavelengths than visible light. **(d)** Electromagnetic radiation and sound waves travel at the same speed.

6.10 Determine which of the following statements are false, and correct them. **(a)** Electromagnetic radiation is incapable of passing through water. **(b)** Electromagnetic radiation travels through a vacuum at a constant speed, regardless of wavelength. **(c)** Infrared light has higher frequencies than visible light. **(d)** The glow from a fireplace, the energy within a microwave oven and a foghorn blast are all forms of electromagnetic radiation.

6.11 **(a)** What is the frequency of radiation that has a wavelength of 955 μm? **(b)** What is the wavelength of radiation that has a frequency of $5.50 \times 10^{14} \text{ s}^{-1}$? **(c)** Would the radiations in part (a) or part (b) be visible to the human eye? **(d)** What distance does electromagnetic radiation travel in 50.0 μs?

6.12 **(a)** What is the frequency of radiation whose wavelength is 1000 pm? **(b)** What is the wavelength of radiation that has a frequency of $7.6 \times 10^{10} \text{ s}^{-1}$? **(c)** Would the radiations in part (a) or part (b) be detected by an X-ray detector?

QUANTISED ENERGY AND PHOTONS (Section 6.2)

6.13 **(a)** What does it mean when we say energy is quantised? **(b)** Why don't we notice the quantisation of energy in everyday activities?

6.14 Einstein's 1905 paper on the photoelectric effect was the first important application of Planck's quantum hypothesis. Describe Planck's original hypothesis and explain how Einstein made use of it in his theory of the photoelectric effect.

6.15 **(a)** Calculate the smallest increment of energy (a quantum) that can be emitted or absorbed at a wavelength of 438 nm. **(b)** Calculate the energy of a photon of frequency $6.75 \times 10^{12}\,\text{s}^{-1}$. **(c)** What wavelength of radiation has photons of energy $2.87 \times 10^{-18}\,\text{J}$? In what portion of the electromagnetic spectrum would this radiation be found?

6.16 **(a)** Calculate the smallest increment of energy that can be emitted or absorbed at a wavelength of 10.8 mm. **(b)** Calculate the energy of a photon from an FM radio station at a frequency of 101.1 MHz. **(c)** For what frequency of radiation will a mole of photons have energy 24.7 kJ? In what region of the electromagnetic spectrum would this radiation be found?

6.17 One type of sunburn occurs on exposure to UV light of wavelength in the vicinity of 325 nm. **(a)** What is the energy of a photon of this wavelength? **(b)** What is the energy of a mole of these photons? **(c)** How many photons are in a 1.00 mJ burst of this radiation?

6.18 The energy from radiation can be used to cause the rupture of chemical bonds. A minimum energy of $941\,\text{kJ mol}^{-1}$ is required to break the nitrogen–nitrogen bond in N_2. What is the longest wavelength of radiation that possesses the necessary energy to break the bond? What type of electromagnetic radiation is this?

6.19 A diode laser emits at a wavelength of 987 nm. **(a)** In what portion of the electromagnetic spectrum is this radiation found? **(b)** All of its output energy is absorbed in a detector that measures a total energy of 0.52 J over a period of 32 s. How many photons per second are being emitted by the laser?

6.20 A stellar object emits radiation at 3.55 mm. **(a)** What type of electromagnetic spectrum is this radiation? **(b)** If the detector is capturing 3.2×10^8 photons per second at this wavelength, what is the total energy of the photons detected in one hour?

6.21 Molybdenum metal must absorb radiation with a minimum frequency of $1.09 \times 10^{15}\,\text{s}^{-1}$ before it can emit an electron from its surface via the photoelectric effect. **(a)** What is the minimum energy needed to produce this effect? **(b)** What wavelength radiation will provide a photon of this energy? **(c)** If molybdenum is irradiated with light of wavelength 120 nm, what is the maximum possible kinetic energy of the emitted electrons?

6.22 It requires a photon with a minimum energy of $4.41 \times 10^{-19}\,\text{J}$ to emit electrons from sodium metal. **(a)** What is the minimum frequency of light necessary to emit electrons from sodium via the photoelectric effect? **(b)** What is the wavelength of this light? **(c)** If sodium is irradiated with light of 439 nm, what is the maximum possible kinetic energy of the emitted electrons? **(d)** What is the maximum number of electrons that can be freed by a burst of light whose total energy is 1.00 μJ?

BOHR'S MODEL; MATTER WAVES (Sections 6.3 and 6.4)

6.23 Explain how the existence of line spectra is consistent with Bohr's theory of quantised energies for the electron in the hydrogen atom.

6.24 **(a)** In terms of Bohr's theory of the hydrogen atom, what process is occurring when excited hydrogen atoms emit radiant energy of certain wavelengths and only those wavelengths? **(b)** Does a hydrogen atom 'expand' or 'contract' as it moves from its ground state to an excited state?

6.25 According to Bohr's model, is energy emitted or absorbed when the following electronic transitions occur in hydrogen? **(a)** From $n = 4$ to $n = 2$. **(b)** From an orbit of radius 212 pm to one of radius 846 pm. **(c)** An electron adds to the H^+ ion and ends up in the $n = 3$ shell.

6.26 Using Equation 6.5, calculate the energy of an electron in the hydrogen atom when $n = 2$ and when $n = 6$. Calculate the wavelength of the radiation released when an electron moves from $n = 6$ to $n = 2$. Is this line in the visible region of the electromagnetic spectrum? If so, what colour is it?

6.27 For each of the following electronic transitions in the hydrogen atom, calculate the energy, frequency and wavelength of the associated radiation, and determine whether the radiation is emitted or absorbed during the transition: **(a)** from $n = 4$ to $n = 1$, **(b)** from $n = 5$ to $n = 2$, **(c)** from $n = 3$ to $n = 6$. Do any of these transitions emit or absorb visible light?

6.28 One of the emission lines of the hydrogen atom has a wavelength of 93.8 nm. **(a)** In what region of the electromagnetic spectrum is this emission found? **(b)** Determine the initial and final values of n associated with this emission.

6.29 The hydrogen atom can absorb light of wavelength 2626 nm. **(a)** In what region of the electromagnetic spectrum is this absorption found? **(b)** Determine the initial and final values of n associated with this absorption.

6.30 Use the de Broglie relationship to determine the wavelengths of the following objects: **(a)** an 85 kg person skiing at 50 km hr^{-1}, **(b)** a 10.0 g bullet fired at 250 m s^{-1}, **(c)** a lithium atom moving at 2.5×10^5 m s^{-1}.

6.31 Among the elementary subatomic particles of physics is the muon, which decays within a few nanoseconds after formation. The muon has a rest mass 206.8 times that of an electron. Calculate the de Broglie wavelength associated with a muon travelling at a velocity of 8.85×10^5 cm s^{-1}.

6.32 Neutron diffraction is an important technique for determining the structures of molecules. Calculate the velocity of a neutron that has a characteristic wavelength of 95.5 pm. (Refer to the back of the text for the mass of the neutron.)

6.33 The electron microscope has been widely used to obtain highly magnified images of biological and other types of materials. When an electron is accelerated through a particular potential field, it attains a speed of 9.38×10^6 m s^{-1}. What is the characteristic wavelength of this electron? Is the wavelength comparable with the size of atoms?

QUANTUM MECHANICS AND ATOMIC ORBITALS (Sections 6.5 and 6.6)

6.34 (a) Why does the Bohr model of the hydrogen atom violate the uncertainty principle? (b) In what way is the description of the electron using a wave function consistent with de Broglie's hypothesis? (c) What is meant by the term *probability density*? Given the wave function, how do we find the probability density at a certain point in space?

6.35 (a) According to the Bohr model, an electron in the ground state of a hydrogen atom orbits the nucleus at a specific radius of 53 pm. In the quantum mechanical description of the hydrogen atom, the most probable distance of the electron from the nucleus is 53 pm. Why are these two statements different? (b) Why is the use of Schrödinger's wave equation to describe the location of a particle very different from the description obtained from classical physics? (c) In the quantum mechanical description of an electron, what is the physical significance of the square of the wave function, ψ^2?

6.36 (a) For $n = 4$, what are the possible values of l? (b) For $l = 2$, what are the possible values of m_l?

6.37 How many possible values for l and m_l are there when (a) $n = 3$, (b) $n = 5$?

6.38 Give the numerical values of n and l corresponding to each of the following designations: (a) $3p$, (b) $2s$, (c) $4f$, (d) $5d$.

6.39 Give the values for n, l and m_l for (a) each orbital in the $2p$ subshell, (b) each orbital in the $5d$ subshell.

6.40 Which of the following represent impossible combinations of n and l: (a) $1p$, (b) $4s$, (c) $5f$, (d) $2d$?

6.41 Which of the following are permissible sets of quantum numbers for an electron in a hydrogen atom: (a) $n = 2$, $l = 1$, $m_l = 1$; (b) $n = 1$, $l = 0$, $m_l = -1$; (c) $n = 4$, $l = 2$, $m_l = -2$; (d) $n = 3$, $l = 3$, $m_l = 0$? For those combinations that are permissible, write the appropriate designation for the subshell to which the orbital belongs (that is, $1s$ and so on).

6.42 (a) What are the similarities and differences between the $1s$ and $2s$ orbitals of the hydrogen atom? (b) In what sense does a $2p$ orbital have directional character? Compare the 'directional' characteristics of the p_x and $d_{x^2-y^2}$ orbitals (that is, in what direction or region of space is the electron density concentrated)? (c) What can you say about the average distance from the nucleus of an electron in a $2s$ orbital compared with a $3s$ orbital? (d) For the hydrogen atom, list the following orbitals in order of increasing energy (that is, most stable ones first): $4f$, $6s$, $3d$, $1s$, $2p$.

6.43 (a) With reference to Figure 6.17, what is the relationship between the number of nodes in an s orbital and the value of the principal quantum number? (b) Identify the number of nodes; that is, identify places where the electron density is zero, in the $2p_x$ orbital and in the $3s$ orbital. (c) What information is obtained from the radial probability functions in Figure 6.17? (d) For the hydrogen atom, list the following orbitals in order of increasing energy: $3s$, $2s$, $2p$, $5s$, $4d$.

MANY-ELECTRON ATOMS AND ELECTRON CONFIGURATIONS (Sections 6.7 to 6.9)

6.44 For a given value of the principal quantum number, n, how do the energies of the s, p, d and f subshells vary for (a) hydrogen, (b) a many-electron atom?

6.45 (a) The average distance from the nucleus of a $3s$ electron in a chlorine atom is smaller than that for a $3p$ electron. In light of this fact, which orbital is higher in energy? (b) Would you expect it to require more or less energy to remove a $3s$ electron from the chlorine atom than a $2p$ electron? Explain.

6.46 (a) What are the possible values of the electron spin quantum number? (b) What piece of experimental equipment can be used to distinguish electrons that have different values of the electron spin quantum number? (c) Two electrons in an atom both occupy the $1s$ orbital. What quantity must be different for the two electrons? What name do we apply to this requirement?

6.47 (a) State the Pauli exclusion principle in your own words. (b) The Pauli exclusion principle is, in an important sense, the key to understanding the periodic table. Explain why.

6.48 What is the maximum number of electrons that can occupy each of the following subshells: (a) $3p$, (b) $5d$, (c) $2s$, (d) $4f$?

6.49 What is the maximum number of electrons in an atom that can have the following quantum numbers: (a) $n = 2$, $m_s = -\frac{1}{2}$; (b) $n = 5$, $l = 3$; (c) $n = 4$, $l = 3$, $m_l = -3$; (d) $n = 4$, $l = 1$, $m_l = 1$.

6.50 (a) What does each box in an orbital diagram represent? (b) What quantity is represented by the direction (either up or down) of the half arrows in an orbital diagram? (c) Is Hund's rule needed to write the electron configuration of beryllium? Explain.

6.51 (a) What are 'valence electrons'? (b) What are 'unpaired electrons'? (c) How many valence electrons does a P atom possess? How many of these are unpaired?

6.52 Write the condensed electron configurations for the following atoms, using the appropriate noble gas core abbreviations: (a) Cs, (b) Ni, (c) Se, (d) Cd, (e) Ac, (f) Pb.

6.53 Write the condensed electron configurations for the following atoms, and indicate how many unpaired electrons each has: (a) Ga, (b) Ca, (c) V, (d) I, (e) Y, (f) Pt, (g) Lu.

6.54 Identify the specific element that corresponds to each of the following electron configurations: (a) $1s^2 2s^2 2p^6 3s^2$, (b) [Ne]$3s^2 3p^1$, (c) [Ar]$4s^1 3d^5$, (d) [Kr]$5s^2 4d^{10} 5p^4$.

6.55 Identify the group of elements that corresponds to each of the following generalised electron configurations:

(a) [noble gas] $ns^2 np^5$

(b) [noble gas] $ns^2(n-1)d^2$

(c) [noble gas] $ns^2(n-1)d^{10}np^1$

(d) [noble gas] $ns^2(n-2)f^6$

6.56 What is wrong with the following electron configurations for atoms in their ground states?

(a) $1s^2 2s^2 3s^1$

(b) [Ne]$2s^2 2p^3$

(c) [Ne]$3s^2 3d^5$

6.57 The following electron configurations represent excited states. Identify the element, and write its ground-state condensed electron configuration.

(a) $1s^2 2s^2 3p^2 4p^1$

(b) [Ar]$3d^{10} 4s^1 4p^4 5s^1$

(c) [Kr]$4d^6 5s^2 5p^1$

ADDITIONAL EXERCISES

6.58 Certain elements emit light of a specific wavelength when they are burned. Historically, chemists used such emission wavelengths to determine whether specific elements were present in a sample. Some characteristic wavelengths for some of the elements are

Ag	328.1 nm	Fe	372.0 nm
Au	267.6 nm	K	404.7 nm
Ba	455.4 nm	Mg	285.2 nm
Ca	422.7 nm	Na	589.6 nm
Cu	324.8 nm	Ni	341.5 nm

(a) Determine which elements emit radiation in the visible part of the spectrum. (b) Which element emits photons of highest energy and of lowest energy? (c) When burned, a sample of an unknown substance is found to emit light of frequency 6.59×10^{14} s^{-1}. Which of these elements is probably in the sample?

6.59 The rays of the sun that cause tanning and burning are in the ultraviolet portion of the electromagnetic spectrum. These rays are categorised by wavelength: so-called UV-A radiation has wavelengths in the range of 320–380 nm, whereas UV-B radiation has wavelengths in the range of 290–320 nm. (a) Calculate the frequency of light that has a wavelength of 320 nm. (b) Calculate the energy of a mole of 320 nm photons. (c) Which are more energetic: photons of UV-A radiation or photons of UV-B radiation? (d) The UV-B radiation from the sun is considered a greater cause of sunburn in humans than UV-A radiation. Is this observation consistent with your answer to part (c)?

6.60 A photocell is a device used to measure the intensity of light. In a certain experiment, when light of wavelength 630 nm is directed onto the photocell, electrons are emitted at the rate of 2.6×10^{-12} C s^{-1}. Assume that each photon that impinges on the photocell emits one electron. How many photons per second are striking the photocell? How much energy per second is the photocell absorbing?

6.61 In an experiment to study the photoelectric effect, a scientist measures the kinetic energy of ejected electrons as a function of the frequency of radiation hitting a metal surface. She obtains the following plot.

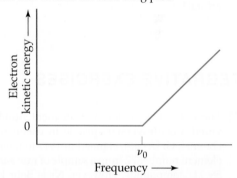

The point labelled ν_0 corresponds to light with a wavelength of 680 nm. (a) What is the value of ν_0 in s^{-1}? (b) What is the value of the work function of the metal in units of kJ mol^{-1} of ejected electrons? (c) What happens when the metal is irradiated with light of frequency less than ν_0? (d) Note that when the frequency of the light is greater than ν_0, the plot shows a straight line with a non-zero slope. Why is this the case? (e) Can you determine the slope of the line segment discussed in part (d)? Explain.

6.62 The series of emission lines of the hydrogen atom for which $n_f = 3$ is called the *Paschen series*. (a) Determine the region of the electromagnetic spectrum in which the lines of the Paschen series are observed. (b) Calculate the wavelengths of the first three lines in the Paschen series—those for which $n_i = 4$, 5 and 6.

6.63 Bohr's model can be used for hydrogen-like ions—ions that have only one electron, such as He$^+$ and Li^{2+}. (a) Why is the Bohr model applicable to He$^+$ ions but not to neutral He atoms? (b) The ground-state energies of H, He$^+$ and Li^{2+} are tabulated as follows

Atom or ion	H	He$^+$	Li^{2+}
Ground-state energy	-2.18×10^{-18} J	-8.72×10^{-18} J	-1.96×10^{-17} J

By examining these numbers, propose a relationship between the ground-state energy of hydrogen-like systems and the nuclear charge, Z. **(c)** Use the relationship you derive in part (b) to predict the ground-state energy of the C^{5+} ion.

6.64 Under appropriate conditions, molybdenum emits X-rays that have a characteristic wavelength of 71.1 pm. These X-rays are used in diffraction experiments to determine the structures of molecules. How fast would an electron have to be moving in order to have the same wavelength as these X-rays?

6.65 An electron is accelerated through an electric potential to a kinetic energy of 18.6 keV. What is its characteristic wavelength? [*Hint:* Recall that the kinetic energy of a moving object is $E = \frac{1}{2}mu^2$, where m is the mass of the object and u is the speed of the object.]

6.66 What is the difference between an *orbit* (Bohr model of the hydrogen atom) and an *orbital* (quantum mechanical model of the hydrogen atom)?

6.67 Which of the quantum numbers governs **(a)** the shape of an orbital, **(b)** the energy of an orbital, **(c)** the spin properties of the electron, **(d)** the spatial orientation of the orbital?

6.68 For non-spherically symmetric orbitals, the contour representations (as in Figures 6.18 and 6.21) suggest where nodal planes exist (that is, where the electron density is zero). For example, the p_x orbital has a node wherever $x = 0$; this equation is satisfied by all points on the yz plane, so this plane is called a nodal plane of the p_x orbital. **(a)** Determine the nodal plane of the p_z orbital. **(b)** What are the two nodal planes of the d_{xy} orbital? **(c)** What are the two nodal planes of the $d_{x^2-y^2}$ orbital?

6.69 Suppose that the spin quantum number, m_s, could have *three* allowed values instead of two. How would this affect the number of elements in the first four rows of the periodic table?

6.70 Using only a periodic table as a guide, write the condensed electron configurations for the following atoms: **(a)** Se, **(b)** Rh, **(c)** Si, **(d)** Hg, **(e)** Hf.

6.71 Scientists have speculated that element 126 might have a moderate stability, allowing it to be synthesised and characterised. Predict what the condensed electron configuration of this element might be.

INTEGRATIVE EXERCISES

6.72 The discovery of hafnium, element number 72, provided a controversial episode in chemistry. A French chemist, G. Urbain, claimed in 1911 to have isolated an element number 72 from a sample of rare earth (elements 58–71) compounds. However, Niels Bohr believed that hafnium was more likely to be found along with zirconium than with the rare earths. D. Coster and G. von Hevesy, working in Bohr's laboratory in Copenhagen, showed in 1922 that element 72 was present in a sample of Norwegian zircon, an ore of zirconium. (The name 'hafnium' comes from the Latin name for Copenhagen, *Hafnia*). **(a)** How would you use electron configuration arguments to justify Bohr's prediction? **(b)** Zirconium, hafnium's neighbour in group 4, can be produced as a metal by reduction of solid $ZrCl_4$ with molten sodium metal. Write a balanced chemical equation for the reaction. Is this an oxidation–reduction reaction? If yes, what is reduced and what is oxidised? **(c)** Solid zirconium dioxide, ZrO_2, is reacted with chlorine gas in the presence of carbon. The products of the reaction are $ZrCl_4$ and two gases, CO_2 and CO in the ratio 1:2. Write a balanced chemical equation for the reaction. Starting with a 55.4 g sample of ZrO_2, calculate the mass of $ZrCl_4$ formed, assuming that ZrO_2 is the limiting reagent and assuming 100% yield. **(d)** Using their electron configurations, account for the fact that Zr and Hf form chlorides MCl_4 and oxides MO_2.

6.73 The first 25 years of the twentieth century were momentous for the rapid pace of change in scientists' understanding of the nature of matter. **(a)** How did Rutherford's experiments on the scattering of α particles by a gold foil set the stage for Bohr's theory of the hydrogen atom? **(b)** In what ways is de Broglie's hypothesis, as it applies to electrons, consistent with J. J. Thomson's conclusion that the electron has mass? In what sense is it consistent with proposals that preceded Thomson's work, that the cathode rays are a wave phenomenon?

6.74 The two most common isotopes of uranium are ^{235}U and ^{238}U. **(a)** Compare the number of protons, the number of electrons and the number of neutrons in atoms of these two isotopes. **(b)** Using the periodic table on the inside front cover, write the electron configuration for a U atom. **(c)** Compare your answer to part (b) with the electron configuration given in Figure 6.27. How can you explain any differences between these two electron configurations? **(d)** ^{238}U undergoes radioactive decay to ^{234}Th. How many protons, electrons and neutrons are gained or lost by the ^{238}U atom during this process? **(e)** Examine the electron configuration for Th in Figure 6.27. Are you surprised by what you find? Explain.

MasteringChemistry (www.pearson.com.au/masteringchemistry)

Make learning part of the grade. Access:

- tutorials with peronalised coaching
- study area
- Pearson eText

PHOTO/ART CREDITS

PERIODIC PROPERTIES OF THE ELEMENTS

Light-emitting diodes, LEDs.

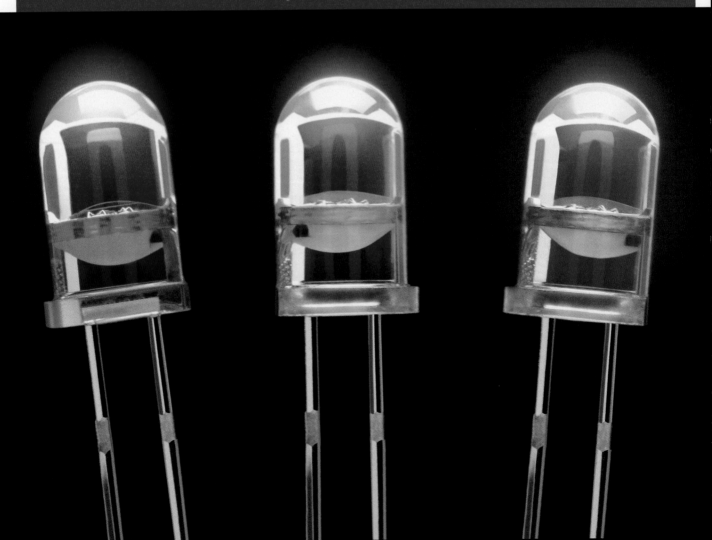

KEY CONCEPTS

7.1 DEVELOPMENT OF THE PERIODIC TABLE
We begin our discussion with a brief history of the periodic table.

7.2 EFFECTIVE NUCLEAR CHARGE
We explore the many properties of atoms that depend on the net attraction of the outer electrons to the nucleus and on the average distance of those electrons from the nucleus. The net positive charge of the nucleus experienced by the outer electrons is called the *effective nuclear charge*.

7.3 SIZES OF ATOMS AND IONS
We explore the relative sizes of atoms and ions, both of which follow trends that are related to their placement in the periodic table.

7.4 IONISATION ENERGY
We discover *ionisation energy*, which is the energy required to remove one or more electrons from an atom. The periodic trends in ionisation energy depend on variations in effective nuclear charge and atomic radii.

7.5 ELECTRON AFFINITIES
We examine periodic trends in the energy released when an electron is added to an atom.

7.6 METALS, NON-METALS AND METALLOIDS
We learn that the physical and chemical properties of metals are different from those of non-metals. These properties arise from the fundamental characteristics of atoms, particularly ionisation energy. Metalloids display properties that are intermediate between those of metals and those of non-metals.

The brilliant colours of light-emitting diodes (LEDs) arise from the composition of the materials from which they are made. The LEDs shown here are compounds of gallium and aluminium mixed with nitrogen, phosphorus and arsenic. GaN, GaP and GaAs can make solid solutions with each other and with aluminium nitride, aluminium phosphide and aluminium arsenide; the composition of each solid solution dictates the wavelength of light emitted by a given LED.

Today the periodic table is still the most significant tool chemists have for organising and remembering chemical facts. As we saw in Chapter 6, the periodic nature of the table arises from the repeating patterns in the electron configurations of the elements. Elements in the same group (column) of the table contain the same number of electrons in their **valence orbitals**, the occupied orbitals that hold the electrons involved in bonding. For example, O ([He]$2s^2 2p^4$) and S ([Ne]$3s^2 3p^4$) are both members of group 16; the similarity of the electron distribution in their valence s and p orbitals leads to similarities in the properties of these two elements.

When we compare O and S, however, it is apparent that they exhibit differences as well, not the least of which is that oxygen is a colourless gas at room temperature, whereas sulfur is a yellow solid (▶ FIGURE 7.1).

In this chapter we explore how some of the important properties of elements change as we move across a period or down a group of the periodic table. In many cases the trends within a period or group allow us to make predictions about the physical and chemical properties of the elements.

◀ **FIGURE 7.1**
Discovering the elements. Because they are both group 16 elements, oxygen and sulfur have many chemical similarities. They also have many differences, however, including the forms they take at room temperature. Oxygen consists of O_2 molecules that appear as a colourless gas (shown here enclosed in a glass container on the right). In contrast, sulfur consists of S_8 molecules that form a yellow solid.

7.1 | DEVELOPMENT OF THE PERIODIC TABLE

The discovery of the chemical elements has been an ongoing process since ancient times (▼ FIGURE 7.2). Certain elements, such as gold, appear in nature in elemental form and were thus discovered thousands of years ago. In contrast, some elements are radioactive and intrinsically unstable. We know about them only because of technology developed in the twentieth century.

The majority of the elements, although stable, are dispersed widely in nature and are incorporated into numerous compounds. For centuries, therefore, scientists were unaware of their existence. In the early nineteenth century, advances in chemistry made it easier to isolate elements from their compounds. As a result, the number of known elements more than doubled from 31 in 1800 to 63 by 1865.

As the number of known elements increased, scientists began to investigate the possibilities of classifying them in useful ways. In 1869 Dmitri Mendeleev (1834–1907) in Russia and Lothar Meyer (1830–1895) in Germany published nearly identical classification schemes. Both scientists noted that similar chemical and physical properties recur periodically when the elements are arranged in order of increasing atomic mass. Scientists at that time had no knowledge of atomic numbers. Atomic masses, however, generally increase with increasing atomic number, so both Mendeleev and Meyer fortuitously arranged the elements in the proper sequence. The tables of elements advanced by Mendeleev and Meyer were the forerunners of the modern periodic table.

Mendeleev's insistence that elements with similar characteristics be listed in the same family forced him to leave several blank spaces in his table. For example, both gallium (Ga) and germanium (Ge) were unknown at that time. Mendeleev boldly predicted their existence and properties, referring to them as *eka-aluminium* ('under' aluminium) and *eka-silicon* ('under' silicon), respectively, after the elements under which they appear in the periodic table. When these elements were discovered, their properties closely matched those predicted by Mendeleev, as shown in ▶ TABLE 7.1.

In 1913, two years after Rutherford proposed the nuclear model of the atom (∞ Section 2.2, 'The Discovery of Atomic Structure'), English physicist Henry Moseley (1887–1915) developed the concept of atomic numbers. Bombarding different elements with high-energy electrons, Moseley found that each

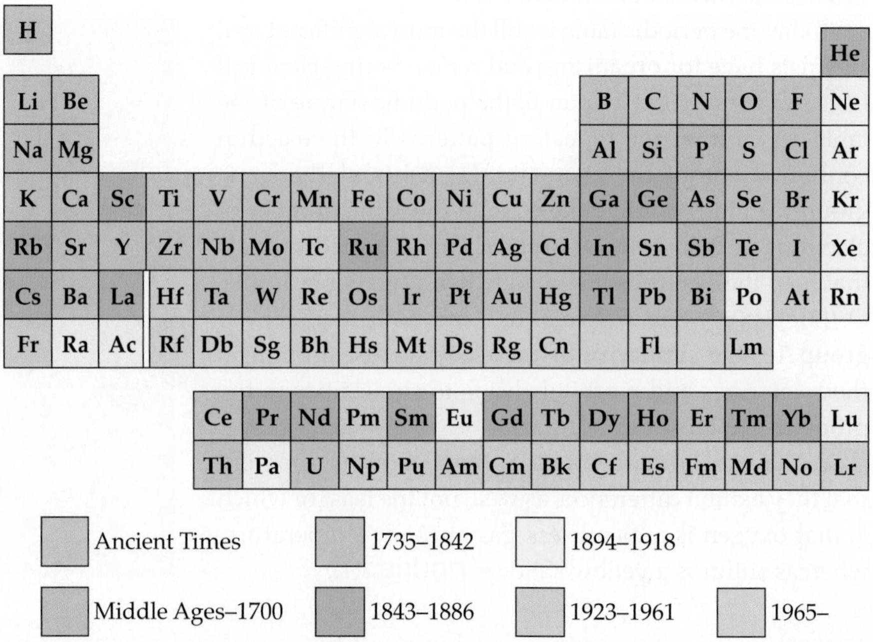

∞ Review this on page 33

▶ **FIGURE 7.2 Discovering the elements.** Periodic table showing the dates of discovery of the elements.

| | | Ancient Times | | 1735–1842 | | 1894–1918 | |
| Middle Ages–1700 | | 1843–1886 | | 1923–1961 | | 1965– |

TABLE 7.1 • Comparison of the properties of eka-silicon predicted by Mendeleev with the observed properties of germanium

Property	Mendeleev's predictions for eka-silicon (made in 1871)	Observed properties of germanium (discovered in 1886)
Atomic mass	72	72.59
Density (g cm^{-3})	5.5	5.35
Specific heat (J g^{-1} K^{-1})	0.305	0.309
Melting point (°C)	High	947
Colour	Dark grey	Greyish white
Formula of oxide	XO$_2$	GeO$_2$
Density of oxide (g cm^{-3})	4.7	4.70
Formula of chloride	XCl$_4$	GeCl$_4$
Boiling point of chloride (°C)	A little under 100	84

element produced X-rays of a unique frequency and that the frequency generally increased as the atomic mass increased. He arranged the X-ray frequencies in order by assigning a unique whole number, called an *atomic number,* to each element. Moseley correctly identified the atomic number as the number of protons in the nucleus of the atom (Section 2.3, 'The Modern View of Atomic Structure').

 Review this on page 36

The concept of atomic number clarified some problems in the early version of the periodic table, which was based on atomic masses. For example, the atomic mass of Ar (atomic number 18) is greater than that of K (atomic number 19). However, when the elements are arranged in order of increasing atomic number, rather than increasing atomic mass, Ar and K appear in their correct places in the table. Moseley's studies also made it possible to identify 'holes' in the periodic table, which led to the discovery of other previously unknown elements.

CONCEPT CHECK 1

Arranging the elements by atomic mass leads to a slightly different order than arranging them by atomic number. How can this happen?

7.2 | EFFECTIVE NUCLEAR CHARGE

Because electrons are negatively charged, they are attracted to nuclei that are positively charged. The force of attraction between an electron and the nucleus depends on the magnitude of the net nuclear charge acting on the electron and on the average distance between the nucleus and the electron, according to Coulomb's law (Equation 14.3). The force of attraction increases as the nuclear charge increases and decreases as the electron moves further from the nucleus.

In a many-electron atom, each electron is simultaneously attracted to the nucleus and repelled by the other electrons. In considering the effect of the attraction of the nucleus to the valence electrons, we note that the inner (core) electrons are particularly effective at partially cancelling the attraction of the nucleus to the valence electrons and that electrons in the same valence shell do not have a significant effect in diminishing the nuclear attraction to a particular electron. The **actual nuclear charge**, Z, is the number of protons in the nucleus, but because the core electrons *shield* or *screen* the valence electrons from the full charge, a valence electron will only see an **effective nuclear charge**, Z_{eff}, given by the relationship

$$Z_{eff} = Z - S \qquad [7.1]$$

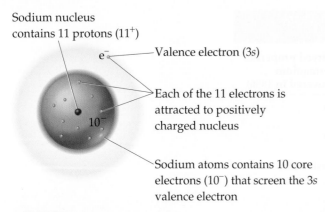

Sodium nucleus contains 11 protons (11⁺)

Valence electron (3s)

Each of the 11 electrons is attracted to positively charged nucleus

Sodium atoms contains 10 core electrons (10⁻) that screen the 3s valence electron

▲ **FIGURE 7.3 Effective nuclear charge.** The effective nuclear charge experienced by the valence electron in a sodium atom depends mostly on the 11+ charge of the nucleus and the 10− charge of the core electrons.

where S is a positive number called the *screening constant* and whose value is usually close to the number of core electrons around the nucleus. In essence, we are treating an electron as though it were moving in the *net electric field created by the nucleus and the electron density of the other electrons.*

Let's take a look at a Na atom to see what we would expect for the magnitude of Z_{eff}. Sodium (atomic number 11) has a condensed electron configuration of $[Ne]3s^1$. The nuclear charge of the atom is 11+, and the Ne inner core consists of 10 electrons ($1s^2 2s^2 2p^6$). Very roughly then, we would expect the 3s valence electron of the Na atom to experience an effective nuclear charge of about $11 - 10 = 1+$, as pictured in a simplified way in ◄ **FIGURE 7.3.**

The notion of effective nuclear charge also explains an important effect we noted in Section 6.7, namely, that for a many-electron atom the energies of orbitals with the same n value increase with increasing l value. For example, consider a carbon atom, for which the electron configuration is $1s^2 2s^2 2p^2$. The energy of the 2p orbital ($l = 1$) is slightly higher than that of the 2s orbital ($l = 0$) even though both of these orbitals are in the $n = 2$ shell (Figure 6.23). The fact that these orbitals have different energy in a many-electron atom can be explained by considering orbital shapes. A 2s orbital has a high probability function close to the nucleus, whereas a 2p orbital has a node at the nucleus. In essence, an electron in a 2s orbital spends more time near the nucleus and so is less screened by the core electrons than an electron in the 2p orbital. So, for a 2s orbital, Z_{eff} is greater than for a 2p orbital. This reasoning applies to d and f orbitals, such that the trend for Z_{eff} is $ns > np > nd$. Conversely, the orbital energies are in the order $ns < np < nd$ in many-electron atoms.

Finally, let's examine the trends in Z_{eff} for valence electrons as we move from one element to another in the periodic table. The effective nuclear charge increases as we move from left to right across a period of the table. Although the number of core electrons stays the same as we move across the period, the actual nuclear charge increases. The valence electrons added to counterbalance the increasing nuclear charge shield one another very ineffectively. Thus the effective nuclear charge increases steadily. For example, the $1s^2$ core electrons of lithium ($1s^2 2s^1$) shield the 2s valence electron from the 3+ nucleus fairly efficiently. Consequently, the outer electron experiences an effective nuclear charge of roughly $3 - 2 = 1+$. For beryllium ($1s^2 2s^2$), the effective nuclear charge experienced by each 2s valence electron is larger; in this case, the inner $1s^2$ electrons are shielding a 4+ nucleus, and each 2s electron only partially shields the other from the nucleus. Consequently, the effective nuclear charge experienced by each 2s electron is about $4 - 2 = 2+$.

Going down a group, the effective nuclear charge experienced by valence electrons changes far less than it does across a period. For example, we would expect the effective nuclear charge for the outer electrons in lithium and sodium to be about the same, roughly $3 - 2 = 1+$ for lithium and $11 - 10 = 1+$ for sodium. In fact, however, the effective nuclear charge increases slightly as we go down a group because larger electron cores are less able to screen the outer electrons from the nuclear charge. In effect, the value for sodium is 2.5+. Nevertheless, the small change in effective nuclear charge that occurs moving down a group is generally of less importance than the increase that occurs when moving across a period.

CONCEPT CHECK 2

Which would you expect to experience a greater effective nuclear charge: a 2p electron of an Ne atom or a 3s electron of an Na atom?

A CLOSER LOOK

EFFECTIVE NUCLEAR CHARGE

To get a sense of how effective nuclear charge varies as both nuclear charge and number of electrons increase, consider ▼ **FIGURE 7.4**. Although the details of how the Z_{eff} values in the graph were calculated are beyond the scope of our discussion, the trends are instructive.

The effective nuclear charge felt by the outermost electrons is smaller than that felt by inner electrons because of screening by the inner electrons. In addition, the effective nuclear charge felt by the outermost electrons does not increase as steeply with increasing atomic number because the valence electrons make a small but non-negligible contribution to the screening constant S. The most striking feature associated with the Z_{eff} value for the outermost electrons is the sharp drop between the last period 2 element (Ne) and the first period 3 element (Na). This drop reflects the fact that the core electrons are much more effective than the valence electrons at screening the nuclear charge.

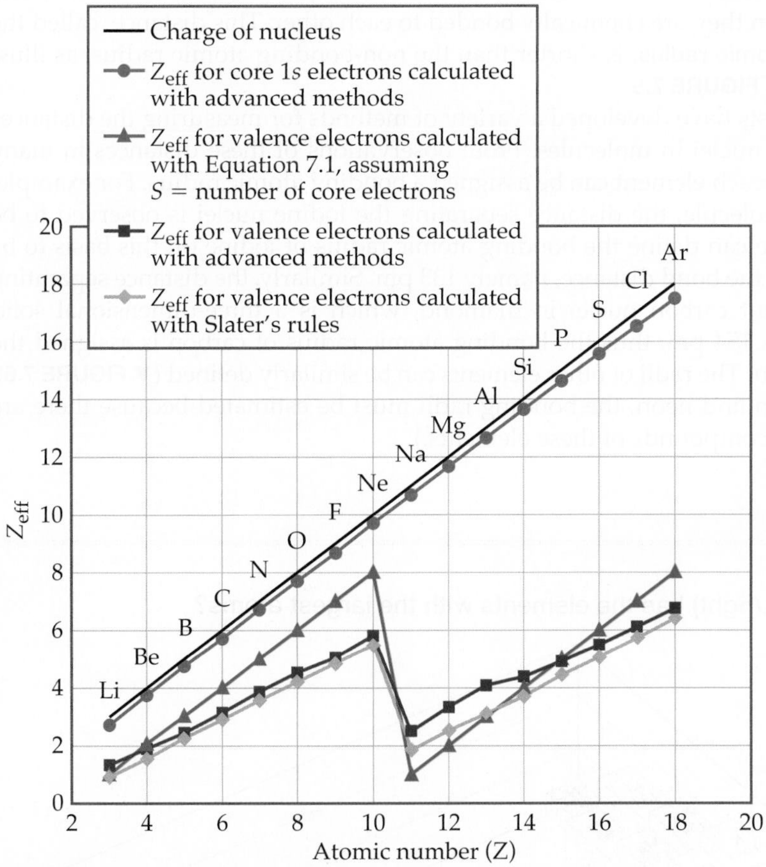

Because Z_{eff} can be used to understand many physically measurable quantities, it is desirable to have a simple method for estimating it. The value of Z in Equation 7.1 is known exactly, so the challenge boils down to estimating the value of S. In the text, we estimated S by assuming that each core electron contributes 1.00 to S and the outer electrons contribute nothing. A more accurate approach was developed by John Slater, however, and we can use his approach if we limit ourselves to elements that do not have electrons in d or f subshells.

Electrons for which the principal quantum number, n, is larger than the value of n for the electron of interest contribute 0 to the value of S. Electrons with the same value of n as the electron of interest contribute 0.35 to the value of S. Electrons for which n is 1 less than n for the electron of interest contribute 0.85, while those with even smaller values of n contribute 1.00. For example, consider fluorine, which has the ground-state electron configuration $1s^2 2s^2 2p^5$. For a valence electron in fluorine, Slater's rules tell us that $S = (0.35 \times 6) + (0.85 \times 2) = 3.8$. (Slater's rules ignore the contribution of an electron to itself in screening; therefore, we consider only six $n = 2$ electrons, not all seven). Thus $Z_{eff} = Z - S = 9 - 3.8 = 5.2+$.

Values of Z_{eff} estimated using the simple method outlined in the text, as well as those estimated with Slater's rules, are plotted in Figure 7.4. Although neither of these methods exactly replicate the values of Z_{eff} obtained from more sophisticated calculations, both methods effectively capture the periodic variation in Z_{eff}. Hence, Slater's approach is more accurate, but the method outlined in the text does a reasonably good job of estimating Z_{eff} despite its simplicity. For our purposes, therefore, we can assume that the screening constant S in Equation 7.1 is roughly equal to the number of core electrons.

RELATED EXERCISES: 7.7–7.10

▲ **FIGURE 7.4 Variations in effective nuclear charge for period 2 and period 3 elements.** Moving from one element to the next in the periodic table, the increase in Z_{eff} felt by the innermost (1s) electrons (red circles) closely tracks the increase in nuclear charge Z (black line) because these electrons are not screened. The results of several methods to calculate Z_{eff} for valence electrons are shown in other colours.

7.3 | SIZES OF ATOMS AND IONS

One of the important properties of an atom or an ion is its size. We often think of atoms and ions as hard, spherical objects. According to the quantum mechanical model, however, atoms and ions do not have sharply defined boundaries at which the electron distribution becomes zero (Section 7.5).

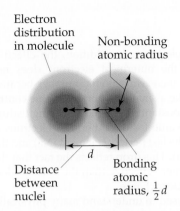

Electron distribution in molecule

Non-bonding atomic radius

Distance between nuclei

d

Bonding atomic radius, $\frac{1}{2}d$

▲ **FIGURE 7.5** **Distinction between non-bonding and bonding atomic radii.** The non-bonding atomic radius is the effective radius of an atom when it is not involved in bonding to another atom. Values of bonding atomic radii are obtained from measurements of interatomic distances in chemical compounds.

Thus we can define atomic size in several ways, based on the distances between atoms in various situations.

Imagine a collection of argon atoms in the gas phase. When two atoms collide with each other in the course of their motions, they ricochet apart—somewhat like billiard balls. This happens because the electron clouds of the colliding atoms cannot penetrate each other to any significant extent. The closest distances separating the nuclei during such collisions determine the *apparent* radii of the argon atoms. We might call this radius the *non-bonding atomic radius* of an atom.

When two atoms are chemically bonded to each other, as in the Cl_2 molecule, there is an attractive interaction between the two atoms, leading to a chemical bond. We discuss the nature of such bonding in Chapter 8. For now, the important point to note is that this attractive interaction brings the two atoms closer together than they would be in a non-bonding collision. We can define an atomic radius as half the distance between the nuclei of two identical atoms when they are chemically bonded to each other. This distance, called the **bonding atomic radius**, is shorter than the non-bonding atomic radius, as illustrated in ◀ **FIGURE 7.5**.

Scientists have developed a variety of methods for measuring the distances separating nuclei in molecules. From observations of these distances in many molecules, each element can be assigned a bonding atomic radius. For example, in the I_2 molecule, the distance separating the iodine nuclei is observed to be 266 pm. We can define the bonding atomic radius of iodine on this basis to be one-half of the bond distance, namely 133 pm. Similarly, the distance separating two adjacent carbon nuclei in diamond, which is a three-dimensional solid network, is 154 pm; thus the bonding atomic radius of carbon is assigned the value 77 pm. The radii of other elements can be similarly defined (▼ **FIGURE 7.6**). (For helium and neon, the bonding radii must be estimated because there are no known compounds of these elements.)

FIGURE IT OUT

Which part of the periodic table (top/bottom, left/right) has the elements with the largest atoms?

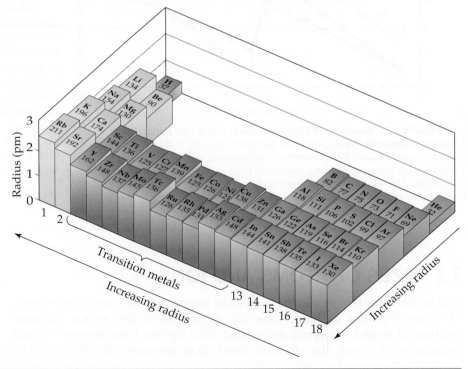

▶ **FIGURE 7.6** **Trends in atomic radii.** Bonding atomic radii for the first 54 elements of the periodic table. The height of the bar for each element is proportional to its radius, giving a 'relief map' view of the radii.

Knowing atomic radii allows us to estimate the bond lengths between different elements in molecules. For example, the Cl–Cl bond length in Cl_2 is 199 pm, so a radius of 99 pm is assigned to Cl. In the compound CCl_4 the measured length of the C–Cl bond is 177 pm, very close to the sum (77 pm + 99 pm) of the atomic radii of C and Cl.

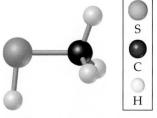

Methyl mercaptan

SAMPLE EXERCISE 7.1 **Bond lengths in a molecule**

Natural gas used in home heating and cooking is odourless. Because natural gas leaks pose the danger of explosion or suffocation, various smelly substances are added to the gas to allow detection of a leak. One such substance is methyl mercaptan, CH_3SH, whose structure is shown in the margin. Use Figure 7.6 to predict the lengths of the C–S, C–H and S–H bonds in this molecule.

SOLUTION

Analyse We are asked to calculate bond lengths from atomic radii.

Plan We are given three bonds and the list of bonding atomic radii. We assume that each bond length is the sum of the radii of the two atoms involved.

Solve Using radii for C, S and H from Figure 7.6, we predict

$$C–S \text{ bond length } = \text{ radius of C } + \text{ radius of S}$$
$$= 77 \text{ pm } + 102 \text{ pm } = 179 \text{ pm}$$
$$C–H \text{ bond length } = 77 \text{ pm } + 37 \text{ pm } = 114 \text{ pm}$$
$$S–H \text{ bond length } = 102 \text{ pm } + 37 \text{ pm } = 139 \text{ pm}$$

Check The experimentally determined bond lengths in methyl mercaptan are C–S = 182 pm, C–H = 110 pm and S–H = 133 pm. (In general, the lengths of bonds involving hydrogen show larger deviations from the values predicted by the sum of the atomic radii than do those bonds involving larger atoms.)

Comment Notice that the estimated bond lengths using bonding atomic radii are close to, but not exact matches of, the experimental bond lengths. Atomic radii must be used with some caution in estimating bond lengths. In Chapter 8 we examine some of the average lengths of common types of bonds.

PRACTICE EXERCISE

Using Figure 7.6, predict which will be greater, the P–Br bond length in PBr_3 or the As–Cl bond length in $AsCl_3$.

Answer: P–Br

(See also Exercise 7.13, 7.14.)

Periodic Trends in Atomic Radii

If we examine the 'relief map' of atomic radii shown in Figure 7.6, we observe two interesting trends in the data.

1. Within each group, atomic radius tends to increase from top to bottom. This trend results primarily from the increase in the principal quantum number (n) of the outer electrons. As we go down a group, the outer electrons have a greater probability of being further from the nucleus, causing the atom to increase in size.

2. Within each period, atomic radius tends to decrease from left to right. The major factor influencing this trend is the increase in the effective nuclear charge (Z_{eff}) as we move across a period. The increasing effective nuclear charge steadily draws the valence electrons closer to the nucleus, causing the atomic radius to decrease.

> ⚠ **CONCEPT CHECK 3**
>
> As we proceed across a period of the periodic table, atomic mass increases but atomic radius decreases. Are these trends a contradiction?

SAMPLE EXERCISE 7.2 | **Atomic radii**

Referring to the periodic table on the inside front cover of this text, arrange (as much as possible) the following atoms in order of increasing size: $_{15}P$, $_{16}S$, $_{33}As$, $_{34}Se$. (Atomic numbers are given for the elements to help you locate them quickly in the periodic table.)

SOLUTION

Analyse Here we use the regular trend in atomic radii across a period and down a group to predict the relative size of atoms of four elements.

Plan Recall that radii decrease as we move from left to right across the periodic table and increase as we move down a group.

Solve Notice that P and S are in the same period of the periodic table, with S to the right of P. Therefore we expect the radius of S to be smaller than that of P. Likewise, the radius of Se is expected to be smaller than that of As. We also notice that As is directly below P and that Se is directly below S. We expect, therefore, that the radius of As is greater than that of P and the radius of Se is greater than that of S. From these observations, we predict S < P, P < As, S < Se and Se < As. We can therefore conclude that S has the smallest radius of the four elements and that As has the largest radius.

Using just the two trends described above, we cannot determine whether P or Se has the larger radius; to go from P to Se in the periodic table, we must move down (radius tends to increase) and to the right (radius tends to decrease). In Figure 7.6 we see that the radius of Se (116 pm) is greater than that of P (106 pm). If you examine the figure carefully, you will discover that for the main-group elements the increase in radius moving down a group tends to be the greater effect. There are exceptions, however.

Check From Figure 7.6, we have S (102 pm) < P (106 pm) < Se (116 pm) < As (119 pm).

Comment Note that the trends we have just discussed are for the main-group elements. You will see in Figure 7.6 that the transition elements do not show a regular decrease from left to right in a period.

PRACTICE EXERCISE

Arrange the following atoms in order of increasing atomic radius: Na, Be, Mg.

Answer: Be < Mg < Na

(See also Exercises 7.15–7.18.)

Periodic Trends in Ionic Radii

The radii of ions are based on the distances between ions in ionic compounds. Like the size of an atom, the size of an ion depends on its nuclear charge, on the number of electrons it possesses and on the orbitals in which the valence electrons reside. The formation of a cation vacates the most spatially extended occupied orbitals in an atom and also decreases the number of electron–electron repulsions. As a consequence, *cations are smaller than their parent atoms*, as illustrated in ▶ **FIGURE 7.7**. The opposite is true of anions. When electrons are added to a neutral atom to form an anion, the increased electron–electron repulsions cause the electrons to spread out more in space. Thus *anions are larger than their parent atoms*.

For ions carrying the same charge, size increases as we go down a group in the periodic table. This trend is also seen in Figure 7.7. As the principal quantum number of the outermost occupied orbital of an ion increases, the radius of the ion increases.

FIGURE IT OUT

How do cations of the same charge change in radius as you move down a group in the periodic table?

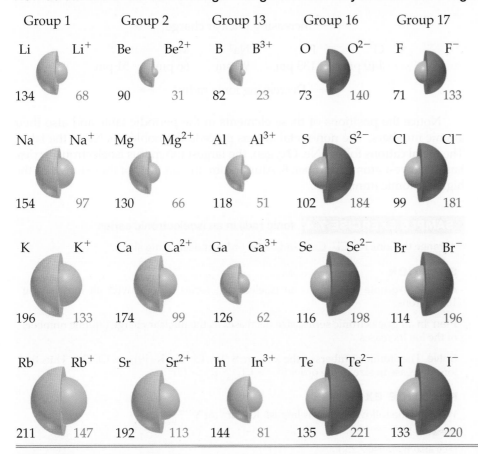

◄ **FIGURE 7.7** **Cation and anion size.** Comparisons of the radii, in pm, of neutral atoms and ions for several of the groups of main-group elements. Neutral atoms are shown in grey, cations in pink and anions in blue.

SAMPLE EXERCISE 7.3   **Atomic and ionic radii**

Arrange these atoms and ions in order of decreasing size: Mg^{2+}, Ca^{2+} and Ca.

SOLUTION

Analyse We are asked to compare the size of two ions in the same group and the size of an atom and an ion of the same element.

Plan Ions from the same group that have the same charge increase in size down the group. Comparing ions and atoms of the same element, cations are always smaller and anions always larger than their parent atoms.

Solve Cations are smaller than their parent atoms, and so the Ca^{2+} ion is smaller than the Ca atom. Because Ca is below Mg in group 2 of the periodic table, Ca^{2+} is larger than Mg^{2+}. Consequently, Ca $>$ Ca^{2+} $>$ Mg^{2+}.

PRACTICE EXERCISE

Which of the following atoms and ions is largest: S^{2-}, S, O^{2-}?

Answer: S^{2-}

(See also Exercises 7.20, 7.22.)

An **isoelectronic series** is a group of ions all containing the same number of electrons. For example, each ion in the isoelectronic series O^{2-}, F^-, Na^+, Mg^{2+}, Al^{3+} has 10 electrons. In any isoelectronic series we can list the members in order of increasing atomic number, and therefore nuclear charge increases as we move through the series. (Recall that the charge on the nucleus of an atom or monatomic ion is given by the atomic number of the element.) Because the number of electrons remains constant, the radius of the ion decreases with increasing nuclear charge, as the electrons are more strongly attracted to the nucleus:

$$\longrightarrow \text{Increasing nuclear charge} \longrightarrow$$

O^{2-}	F^-	Na^+	Mg^{2+}	Al^{3+}
140 pm	133 pm	97 pm	66 pm	51 pm

$$\longrightarrow \text{Decreasing ionic radius} \longrightarrow$$

Notice the positions of these elements in the periodic table and also their atomic numbers. The non-metal anions precede the noble gas Ne in the table. The metal cations follow Ne. Oxygen, the largest ion in this isoelectronic series, has the lowest atomic number, 8. Aluminium, the smallest of these ions, has the highest atomic number, 13.

SAMPLE EXERCISE 7.4 | **Ionic radii in an isoelectronic series**

Arrange the ions K^+, Cl^-, Ca^{2+} and S^{2-} in order of decreasing size.

SOLUTION

Analyse We note that this is an isoelectronic series of ions, with all ions having 18 electrons.

Plan In an isoelectronic series, size decreases as the nuclear charge (atomic number) of the ion increases.

Solve The atomic numbers of the ions are S (16), Cl (17), K (19) and Ca (20). Thus the ions decrease in size in the order $S^{2-} > Cl^- > K^+ > Ca^{2+}$.

PRACTICE EXERCISE

Which of the following ions is largest: Rb^+ Sr^{2+} or Y^{3+}?

Answer: Rb^+

(See also Exercise 7.25.)

7.4 | IONISATION ENERGY

The ease with which electrons can be removed from an atom or ion has a major impact on chemical behaviour. The **ionisation energy** of an atom or ion is the minimum energy required to remove an electron from the ground state of the isolated gaseous atom or ion. The *first ionisation energy*, I_1, is the energy needed to remove the first electron from a neutral atom. For example, the first ionisation energy for the sodium atom is the energy required for the process

$$Na(g) \longrightarrow Na^+(g) + e^- \qquad [7.2]$$

The *second ionisation energy*, I_2, is the energy needed to remove the second electron, and so forth, for successive removals of additional electrons. Thus I_2 for the sodium atom is the energy associated with the process

$$Na^+(g) \longrightarrow Na^{2+}(g) + e^- \qquad [7.3]$$

The greater the ionisation energy, the more difficult it is to remove an electron.

⚠ **CONCEPT CHECK 4**

Light can be used to ionise atoms and ions, as in Equations 7.2 and 7.3. What physical process introduced in Chapter 6 is related to the ionisation of atoms and molecules?

Element	I_1	I_2	I_3	I_4	I_5	I_6	I_7
Na	495	4562			(inner-shell electrons)		
Mg	738	1451	7733				
Al	578	1817	2745	11 577			
Si	786	1577	3232	4 356	16 091		
P	1012	1907	2914	4 964	6 274	21 267	
S	1000	2252	3357	4 556	7 004	8 496	27 107
Cl	1251	2298	3822	5 159	6 542	9 362	11 018
Ar	1521	2666	3931	5 771	7 238	8 781	11 995

TABLE 7.2 • Successive values of ionisation energies, I, for the elements sodium to argon ($kJ\ mol^{-1}$)

Variations in Successive Ionisation Energies

Ionisation energies for the elements sodium to argon are listed in
▲ **TABLE 7.2**. Notice that the values for a given element increase as successive
electrons are removed: $I_1 < I_2 < I_3$ and so forth. This trend exists because,
with each successive removal, an electron is being pulled away from an increasingly more positive ion, requiring increasingly more energy.

A second important feature shown in Table 7.2 is the sharp increase in ionisation energy that occurs when an inner-shell electron is removed. For
example, consider silicon, whose electron configuration is $1s^2 2s^2 2p^6 3s^2 3p^2$ or
$[Ne]3s^2 3p^2$. The ionisation energies increase steadily from $786\ kJ\ mol^{-1}$ to
$4360\ kJ\ mol^{-1}$ for the loss of the four electrons in the outer $3s$ and $3p$ subshells.
Removal of the fifth electron, which comes from the $2p$ subshell, requires a
great deal more energy: $16\ 100\ kJ\ mol^{-1}$. The large increase occurs because the
$(n = 2)$ $2p$ electron is much more likely to be found close to the nucleus than are
the four $n = 3$ electrons, and therefore the $2p$ electron experiences a much
greater effective nuclear charge than do the $3s$ and $3p$ electrons.

 CONCEPT CHECK 5

Which would you expect to be greater, I_1 for a boron atom or I_2 for a carbon
atom?

Every element exhibits a large increase in ionisation energy when electrons
are removed from its noble gas core. This observation supports the idea that
only the outermost electrons, the valence electrons, are involved in the sharing
and transfer of electrons that give rise to chemical bonding and reactions. The
inner electrons are too tightly bound to the nucleus to be lost from the atom or
even shared with another atom.

SAMPLE EXERCISE 7.5 Trends in ionisation energy

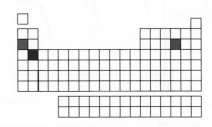

Three elements are indicated in the periodic table in the margin. Based on their locations, predict the one with the largest second ionisation energy.

SOLUTION

Analyse The locations of the elements in the periodic table allow us to predict the
electron configurations. The greatest ionisation energies involve removal of core
electrons.

Plan We should look first for an element with only one electron in the outermost
occupied shell. Removing a second electron from this element would break into
the core electrons.

Solve The element in group 1 (Na), indicated by the red box, has only one valence electron. The second ionisation energy of this element is associated, therefore, with the removal of a core electron. The other elements indicated, S (green box) and Ca (blue box), have two or more valence electrons. Thus Na should have the largest second ionisation energy.

Check If we consult a chemistry handbook, we find the following values for the second ionisation energies (I_2) of the respective elements: Ca (1145 kJ mol^{-1}) < S (2252 kJ mol^{-1}) < Na (4562 kJ mol^{-1}).

PRACTICE EXERCISE

Which will have the greater third ionisation energy, Ca or S?

Answer: Ca

(See also Exercises 7.31, 7.32.)

Periodic Trends in First Ionisation Energies

We have seen that the ionisation energy for a given element increases as we remove successive electrons. What trends do we observe in ionisation energy as we move from one element to another in the periodic table? ▼ FIGURE 7.8 shows a graph of I_1 versus atomic number for the first 54 elements. The important trends are as follows.

1. Within each period of the table, I_1 generally increases with increasing atomic number. The alkali metals show the lowest ionisation energy in each period, and the noble gases the highest. There are slight irregularities in this trend that we discuss shortly.

2. Within each group of the table, the ionisation energy generally decreases with increasing atomic number. For example, the ionisation energies of the noble gases follow the order He > Ne > Ar > Kr > Xe.

3. The main-group elements show a larger range of values of I_1 than do the transition metal elements. Generally, the ionisation energies of the transition metals increase slowly as we proceed from left to right in a period. The f-block metals, which are not shown in Figure 7.8, also show only a small variation in the values of I_1.

The periodic trends in the first ionisation energies of the main-group elements are further illustrated in ▶ FIGURE 7.9.

In general, smaller atoms have higher ionisation energies. The same factors that influence atomic size also influence ionisation energies. The energy needed to remove an electron from the outermost occupied shell depends on both the effective nuclear charge and the average distance of the electron from the nucleus. Either increasing the effective nuclear charge or decreasing the distance from the nucleus increases the attraction between the electron and the nucleus. As this attraction increases, it becomes harder to remove the electron and thus the ionisation energy increases. As we move across a period, there is both an increase in effective nuclear charge

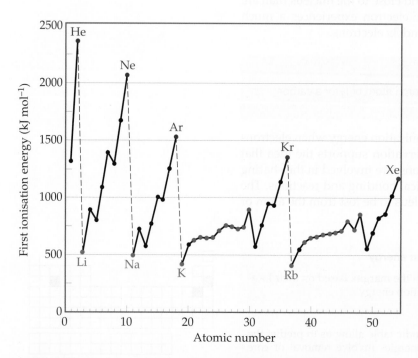

▲ FIGURE 7.8 **First ionisation energy versus atomic number.** The red dots mark the beginning of a period (alkali metals), the blue dots mark the end of a period (noble gases) and the black dots indicate other main-group elements. Green dots are used for the transition metals.

▲ FIGURE IT OUT

Which has a larger first ionisation energy, Ar or As? Why?

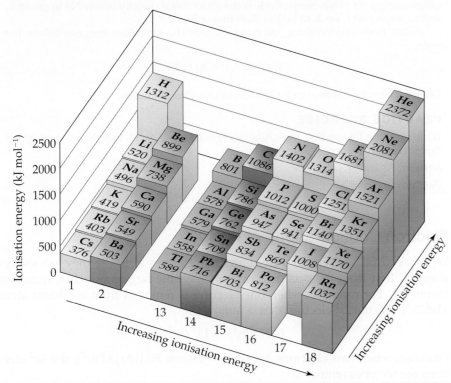

◄ **FIGURE 7.9   Trends in first ionisation energy.** First ionisation energies for the main-group elements in the first six periods. The ionisation energy generally increases from left to right and decreases from top to bottom. The ionisation energy of the highly unstable group 17 element astatine is unknown.

and a decrease in atomic radius, causing the ionisation energy to increase. As we move down a group, however, the atomic radius increases, although the effective nuclear charge changes little. Thus the attraction between the nucleus and the electron decreases, causing the ionisation energy to decrease.

The irregularities within a given period are somewhat more subtle but still readily explained. For example, the decrease in ionisation energy from beryllium ($[He]2s^2$) to boron ($[He]2s^22p^1$), shown in Figures 7.8 and 7.9, occurs because the third valence electron of B must occupy the $2p$ subshell, which is empty for Be. Recall that the $2p$ subshell is at a higher energy than the $2s$ (Figure 6.23). The decrease in ionisation energy on going from nitrogen ($[He]2s^22p^3$) to oxygen ($[He]2s^22p^4$) is because of repulsion of paired electrons in the p^4 configuration. Remember that, according to Hund's rule, each electron in the p^3 configuration resides in a different p orbital, which minimises the electron–electron repulsion between the three $2p$ electrons but that adding a fourth electron results in forming an electron pair (∞ Section 6.8, 'Electron Configurations').

∞ Review this on page 203

SAMPLE EXERCISE 7.6   **Periodic trends in ionisation energy**

Referring to the periodic table, arrange the following atoms in order of increasing first ionisation energy: Ne, Na, P, Ar, K.

SOLUTION

Analyse In order to rank them according to increasing first ionisation energy, we need to locate each element in the periodic table. We can then use their relative positions and the trends in first ionisation energies to predict their order.

Plan First locate the elements in the periodic table. Ionisation energy increases as we move left to right across a period. It decreases as we move from the top of a group to the bottom.

Solve Because Na, P and Ar are in the same period of the periodic table, we expect I_1 to vary in the order Na < P < Ar.

Because Ne is above Ar in group 18, we expect Ne to have the greater first ionisation energy: Ar < Ne. Similarly, K is the alkali metal directly below Na in group 1, and so we expect I_1 for K to be less than that of Na: K < Na.

From these observations, we conclude that the ionisation energies follow the order

$$K < Na < P < Ar < Ne$$

Check The values shown in Figure 7.9 confirm this prediction.

PRACTICE EXERCISE

Which has the lowest first ionisation energy: B, Al, C or Si? Which has the highest first ionisation energy?

Answer: Al lowest, C highest

(See also Exercise 7.33.)

Electron Configurations of Ions

When electrons are removed from an atom to form a cation, they are always removed first from the occupied orbitals with the largest principal quantum number, n. For example, when one electron is removed from a lithium atom $(1s^2\,2s^1)$, it is the $2s^1$ electron that is removed:

$$Li\ (1s^2 2s^1) \Rightarrow Li^+\ (1s^2)$$

Likewise, when two electrons are removed from Fe ([Ar]$3d^6 4s^2$), the $4s^2$ electrons are the ones removed:

$$Fe\ ([Ar]3d^6 4s^2) \Rightarrow Fe^{2+}\ ([Ar]3d^6)$$

If an additional electron is removed, forming Fe^{3+}, it now comes from a $3d$ orbital because all the orbitals with $n = 4$ are empty:

$$Fe^{2+}\ ([Ar]3d^6) \Rightarrow Fe^{3+}\ ([Ar]3d^5)$$

When electrons are added to an atom to form an anion, they are added to the empty or partially filled orbital with the lowest value of n. For example, when an electron is added to a fluorine atom to form the F$^-$ ion, the electron goes into the one remaining vacancy in the $2p$ subshell:

$$F\ (1s^2 2s^2 2p^5) \Rightarrow F^-\ (1s^2 2s^2 2p^6)$$

 CONCEPT CHECK 6

Would Cr^{3+} and V^{2+} have the same or different electron configurations?

SAMPLE EXERCISE 7.7 Electron configurations of ions

Write the electron configuration for **(a)** Ca^{2+}, **(b)** Co^{3+} and **(c)** S^{2-}.

SOLUTION

Analyse We are asked to determine the electron configuration of a selection of ions.

Plan We first write the electron configuration of the parent atom. We then remove electrons to form cations or add electrons to form anions. Electrons are first removed from the orbitals with the highest value of n. They are added to the empty or partially filled orbitals with the lowest value of n.

Solve (a) Calcium (atomic number 20) has the electron configuration

$$Ca:\ [Ar]4s^2$$

To form a 2+ ion, the two outer electrons must be removed, giving an ion that is iso-electronic with Ar:

$$Ca^{2+}: [Ar]$$

(b) Cobalt (atomic number 27) has the electron configuration

$$Co: [Ar]3d^7 4s^2$$

To form a 3+ ion, three electrons must be removed. As discussed in the text preceding this Sample Exercise, the 4s electrons are removed before the 3d electrons. Consequently, the electron configuration for Co^{3+} is

$$Co^{3+}: [Ar]3d^6$$

(c) Sulfur (atomic number 16) has the electron configuration

$$S: [Ne]3s^2 3p^4$$

To form a 2− ion, two electrons must be added. There is room for two additional electrons in the 3p orbitals. Thus the S^{2-} electron configuration is

$$S^{2-}: [Ne]3s^2 3p^6 = [Ar]$$

Comment Remember that many of the common ions of the main-group elements, such as Ca^{2+} and S^{2-}, have the same number of electrons as the closest noble gas.

PRACTICE EXERCISE

Write the electron configuration for **(a)** Ga^{3+}, **(b)** Cr^{3+} and **(c)** Br^-.

Answers: **(a)** $[Ar]3d^{10}$, **(b)** $[Ar]3d^3$, **(c)** $[Ar]3d^{10}4s^2 4p^6 = [Kr]$

(See also Exercises 7.35, 7.36.)

7.5 | ELECTRON AFFINITIES

We have seen that the first ionisation energy of an atom is a measure of the energy change associated with removing an electron from the atom to form a positively charged ion. For example, the first ionisation energy of Cl(g), 1251 kJ mol^{-1}, is the energy change associated with the process:

Ionisation energy: $Cl(g) \longrightarrow Cl^+(g) + e^-$ $\Delta E = 1251$ kJ mol^{-1} [7.4]
$\quad\quad\quad\quad\quad$ [Ne]3s²3p⁵ [Ne]3s²3p⁴

The positive value of the ionisation energy means that energy must be put into the atom in order to remove the electron.

In addition, most atoms can gain electrons to form negatively charged ions. The energy change that occurs when an electron is added to a gaseous atom or ion is called the **electron affinity** because it measures the attraction, or *affinity*, of the atom for the added electron. For most atoms, energy is released when an electron is added. For example, the addition of an electron to a chlorine atom is accompanied by an energy change of −349 kJ mol^{-1}, the negative sign indicating that energy is released during the process. We therefore say that the electron affinity of Cl is −349 kJ mol^{-1}:

Electron affinity: $Cl(g) + e^- \longrightarrow Cl^-(g)$ $\Delta E = -349$ kJ mol^{-1} [7.5]
$\quad\quad\quad\quad$ [Ne]3s²3p⁵ $\quad\quad$ [Ne]3s²3p⁶

It is important to understand the difference between ionisation energy and electron affinity: ionisation energy measures the ease with which an atom *loses* an electron, whereas electron affinity measures the ease with which an atom *gains* an electron.

Which of the groups shown here has the most negative electron affinities? Why does this make sense?

1							18
H −73	2	13	14	15	16	17	**He** >0
Li −60	**Be** >0	**B** −27	**C** −122	**N** >0	**O** −141	**F** −328	**Ne** >0
Na −53	**Mg** ≈0	**Al** −43	**Si** −134	**P** −72	**S** −200	**Cl** −349	**Ar** >0
K −48	**Ca** −2	**Ga** −30	**Ge** −119	**As** −78	**Se** −195	**Br** −325	**Kr** >0
Rb −47	**Sr** −5	**In** −30	**Sn** −107	**Sb** −103	**Te** −190	**I** −295	**Xe** >0

▲ **FIGURE 7.10 Electron affinity.**
Electron affinities in kJ mol^{-1} for the main-group elements in the first five periods of the periodic table.

The greater the attraction between a given atom and an added electron, the more negative the atom's electron affinity. For some elements, such as the noble gases, the electron affinity has a positive value, meaning that the anion is higher in energy than the separated atom and electron:

$$Ar(g) + e^- \longrightarrow Ar^-(g) \quad \Delta E > 0 \qquad [7.6]$$
$$[Ne]3s^23p^6 \qquad\qquad [Ne]3s^23p^64s^1$$

The fact that the electron affinity is a positive number means that an electron will not attach itself to an Ar atom; the Ar$^-$ ion is unstable and cannot be formed. However, positive electron affinities are difficult to measure and obtaining precise values is not possible.

◀ **FIGURE 7.10** shows the electron affinities for the main group elements in the first five periods of the periodic table. Notice that the trends in electron affinity as we proceed through the periodic table are not as evident as they were for ionisation energy. The halogens, which are one electron short of a filled p subshell, have the most-negative electron affinities. By gaining an electron, a halogen atom forms a stable negative ion that has a noble gas configuration (Equation 7.5). The addition of an electron to a noble gas, however, would require that the electron reside in a higher-energy subshell that is empty in the neutral atom (Equation 7.6). Because occupying a higher-energy subshell is energetically very unfavourable, the electron affinity is highly positive. The electron affinities of Be and Mg are positive for the same reason; the added electron would reside in a previously empty p subshell that is higher in energy.

The electron affinities of the group 15 elements (N, P, As, Sb) are also interesting. Because these elements have half-filled p subshells, the added electron must be put in an orbital that is already occupied, resulting in larger electron–electron repulsions. Consequently, these elements have electron affinities that are either positive (N) or less negative than their neighbours to the left (P, As, Sb).

Electron affinities do not change greatly as we move down a group. For example, consider the electron affinities of the halogens (Figure 7.10). For F, the added electron goes into a $2p$ orbital, for Cl a $3p$ orbital, for Br a $4p$ orbital and so forth. As we proceed from F to I, therefore, the average distance between the added electron and the nucleus steadily increases, causing the electron–nucleus attraction to decrease. The orbital that holds the outermost electron is increasingly spread out, however, as we proceed from F to I, thereby reducing the electron–electron repulsions. A lower electron–nucleus attraction is thus counterbalanced by lower electron–electron repulsions.

CONCEPT CHECK 7

Suppose you were asked for a value for the first ionisation energy of a Cl$^-$(g) ion. What is the relationship between this quantity and the electron affinity of Cl(g)?

7.6 | METALS, NON-METALS AND METALLOIDS

Atomic radii, ionisation energies and electron affinities are properties of individual atoms. With the exception of the noble gases, however, none of the elements exists in nature as an individual atom. To get a broader understanding of the properties of elements, we must also examine periodic trends in properties that involve large collections of atoms.

The elements can be broadly grouped into the categories of metals, non-metals and metalloids (Section 2.5, 'The Periodic Table'). This classification is shown in ▶ **FIGURE 7.11**. Roughly three-quarters of the elements are metals, situated in the left and middle portions of the table. The non-metals are located at the top right corner, and the metalloids lie between the metals and non-metals. Hydrogen, which is located at the top left corner, is a non-metal. This

∞ Review this on page 40

◤ FIGURE IT OUT

Notice that germanium, Ge, is a metalloid but tin, Sn, is a metal. What changes in atomic properties do you think are important in explaining this difference?

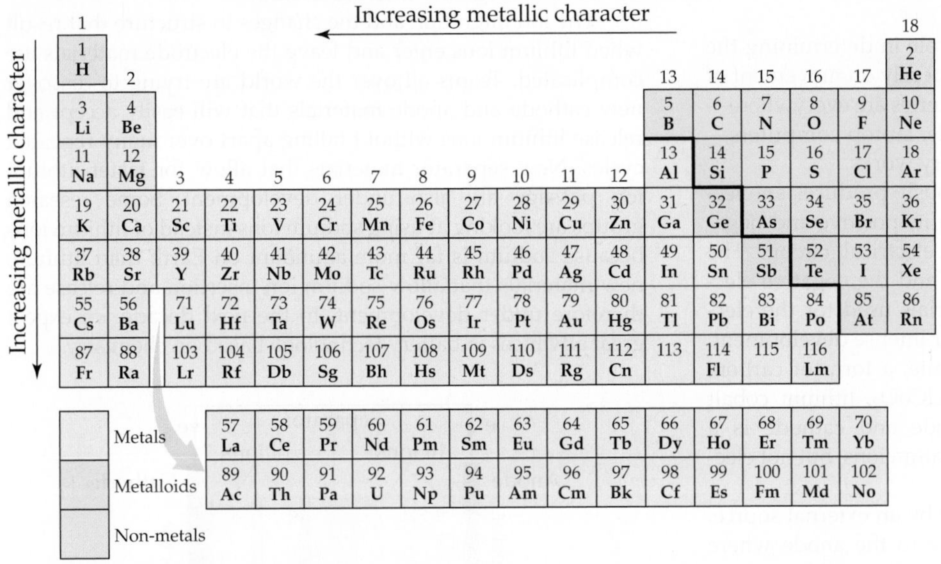

▲ FIGURE 7.11   Metals, metalloids and non-metals. The majority of elements are metals. Metallic character increases from right to left across a period and also increases from top to bottom in a group.

is why we off set hydrogen from the remaining group 1 elements in Figure 7.11 by inserting a space between the H box and the Li box. Some of the distinguishing properties of metals and non-metals are summarised in ▼ TABLE 7.3.

The more an element exhibits the physical and chemical properties of metals, the greater its metallic character. As indicated in Figure 7.11, metallic character generally increases as we proceed down a group of the periodic table and increases as we proceed from right to left in a period. Let's now examine the close relationships that exist between electron configurations and the properties of metals, non-metals and metalloids.

Metals

Most metallic elements exhibit the shiny lustre that we associate with metals (▼ FIGURE 7.13). Metals conduct heat and electricity. They are malleable (can be pounded into thin sheets) and ductile (can be drawn into wires). All are solids at room temperature except mercury (melting point $= -39\ °C$), which is a liquid. Two melt at slightly above room temperature, cesium at 28.4 °C and gallium at 29.8 °C. (Though it is difficult to obtain enough material to make measurements, the melting point of francium is believed to be in the range 22–27 °C.) At the other extreme, many metals melt at very high temperatures. For example, chromium melts at 1900 °C.

Table 7.3 • Characteristic properties of metals and non-metals	
Metals	**Non-metals**
Have a shiny lustre; various colours, although most are silvery	Do not have a lustre; various colours
Solids are malleable and ductile	Solids are usually brittle; some are hard, some are soft
Good conductors of heat and electricity	Poor conductors of heat and electricity
Most metal oxides are ionic solids that are basic	Most non-metal oxides are molecular substances that form acidic solutions
Tend to form cations in aqueous solution	Tend to form anions or oxyanions in aqueous solution

MY WORLD OF CHEMISTRY

ION MOVEMENT POWERS ELECTRONICS

Ionic size plays a major role in determining the properties of devices that rely on movement of ions. 'Lithium ion' batteries are everywhere—mobile phones, iPods, laptop computers—and so let's see how a lithium ion battery works.

A fully charged battery spontaneously produces an electric current and therefore power when its positive and negative electrodes are connected in an electrical circuit. The positive electrode is called the anode, and the negative electrode is called the cathode. The materials used for the electrodes in lithium ion batteries are under intense development. Currently the anode material is graphite, a form of carbon, and the cathode is most frequently $LiCoO_2$, lithium cobalt oxide (▶ FIGURE 7.12). Between anode and cathode is a *separator*, a solid material that allows lithium ions, but not electrons, to pass through.

When the battery is being charged by an external source, lithium ions migrate from the cathode to the anode where they insert between the layers of carbon atoms. Lithium ions are smaller and lighter than most other elements, which means that many can fit between the layers. When the battery discharges and its electrodes are properly connected, it is energetically favourable for the lithium ions to move from anode to cathode. In order to maintain charge balance, electrons simultaneously migrate from anode to cathode through an external circuit, thereby producing electricity.

At the cathode, lithium ions then insert in the oxide material. Again, the small size of lithium ions is an advantage. For every lithium ion that inserts into the lithium cobalt oxide

cathode, a Co^{4+} ion is reduced to a Co^{3+} by an electron that has travelled through the external circuit.

The ion migration and the changes in structure that result when lithium ions enter and leave the electrode materials are complicated. Teams all over the world are trying to discover new cathode and anode materials that will easily accept and release lithium ions without falling apart over many repeated cycles. New separator materials that allow for faster lithium ion passage are also under development. Some research groups are looking at using sodium ions instead of lithium ions because sodium is far more abundant on Earth than lithium; new materials that allow sodium ion insertion and release are therefore under development. In the next decade we expect great advances in battery technology based on chemistry.

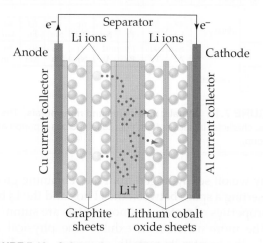

▲ **FIGURE 7.12** **Schematic of a lithium ion battery.**

∞ Review this on page 118

▲ **FIGURE 7.13** **Metals are shiny and malleable.**

Metals tend to have low ionisation energies (often < 800 kJ mol^{-1}) and therefore form positive ions more easily than non-metals. As a result, metals are oxidised (lose electrons) when they undergo chemical reactions. The relative ease of oxidation of common metals is discussed in Chapter 4 (∞ Section 4.4, 'Oxidation–Reduction Reactions'). As we noted, many metals are oxidised by a variety of common substances, including O_2 and acids.

▶ FIGURE 7.14 shows the charges of some common ions of both metals and non-metals. As noted in Section 2.7, the charge on any alkali metal ion is always 1+ and that on any alkaline earth metal is always 2+ in their compounds. In atoms in both these groups, the outer *s* electrons are lost, yielding a noble-gas electron configuration. The charge on transition metal ions does not follow an obvious pattern. Many transition metal ions carry a charge of 2+ but charges of 1+ and 3+ are also encountered. One of the characteristic features of the transition metals is their ability to form more than one positive ion. For example, iron may be 2+ in some compounds and 3+ in others.

The red stepped line divides metals from non-metals. How are common oxidation states divided by this line?

⚠ CONCEPT CHECK 8

Based on periodic trends discussed in this chapter, can you see a general relationship between the trends in metallic character and those for ionisation energy?

FIGURE IT OUT

The red stepped line divides metals from non-metals. How are common oxidation states divided by this line?

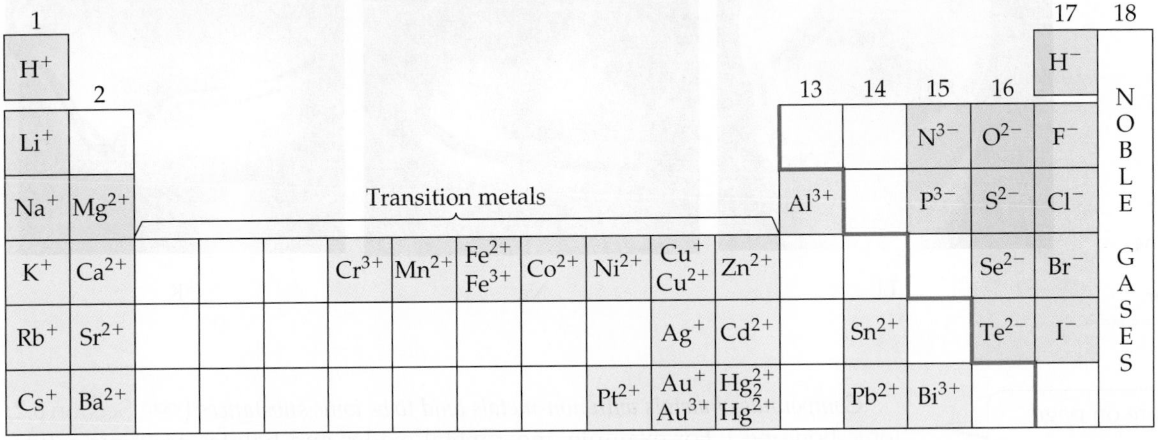

1												13	14	15	16	17	18
H^+																H^-	N O B L E G A S E S
Li^+														N^{3-}	O^{2-}	F^-	
Na^+	Mg^{2+}		Transition metals									Al^{3+}		P^{3-}	S^{2-}	Cl^-	
K^+	Ca^{2+}				Cr^{3+}	Mn^{2+}	Fe^{2+} Fe^{3+}	Co^{2+}	Ni^{2+}	Cu^+ Cu^{2+}	Zn^{2+}				Se^{2-}	Br^-	
Rb^+	Sr^{2+}									Ag^+	Cd^{2+}		Sn^{2+}		Te^{2-}	I^-	
Cs^+	Ba^{2+}								Pt^{2+}	Au^+ Au^{3+}	Hg_2^{2+} Hg^{2+}		Pb^{2+}	Bi^{3+}			

▲ **FIGURE 7.14 Common ions.** Charges of some common ions found in ionic compounds. Notice that the red line that divides metals from non-metals also separates cations from anions.

Note that some metals with low first ionisation energies, notably the alkali (group 1) and alkaline earth (group 2) metals, shown in ▼ **TABLE 7.4**, react with water. The reaction of the alkali metals with water is vigorous and exothermic, producing hydrogen gas and a metal hydroxide. The reaction becomes increasingly more vigorous (▼ **FIGURE 7.15**) until, in the case of rubidium and cesium whose hold on their respective electrons is quite weak, the reaction becomes explosive.

$$2\,M(s) + 2\,H_2O(l) \longrightarrow 2\,MOH(aq) + H_2(g) \qquad [7.7]$$

The symbol M in the above equation represents any one of the alkali metals.

In comparison, the alkaline earth metals react with water quite slowly. In fact, beryllium does not react with water (or steam) and magnesium reacts only with steam. The other three, calcium, strontium and barium, react quite sedately again to produce the metal hydroxide and hydrogen gas (▼ **FIGURE 7.16**).

$$M'(s) + 2\,H_2O(l) \longrightarrow M'(OH)_2(aq) + H_2(g) \qquad [7.8]$$

The symbol M' represents Ca, Sr or Ba.

Table 7.4 ● Some properties of the alkali and alkaline earth metals

Element	Electron configuration	Melting point (°C)	Density (g cm^{-3})	Atomic radius (pm)	I_1 (kJ mol^{-1})
Alkali metals					
Lithium	$[He]2s^1$	181	0.53	134	520
Sodium	$[Ne]3s^1$	98	0.97	154	496
Potassium	$[Ar]4s^1$	63	0.86	196	419
Rubidium	$[Kr]5s^1$	39	1.53	211	403
Cesium	$[Xe]6s^1$	28	1.88	225	376
Alkaline earth metals					
Beryllium	$[He]2s^2$	1287	1.85	90.0	899
Magnesium	$[Ne]3s^2$	650	1.74	130	738
Calcium	$[Ar]4s^2$	842	1.55	174	590
Strontium	$[Kr]5s^2$	777	2.63	192	549
Barium	$[Xe]6s^2$	727	3.51	198	503

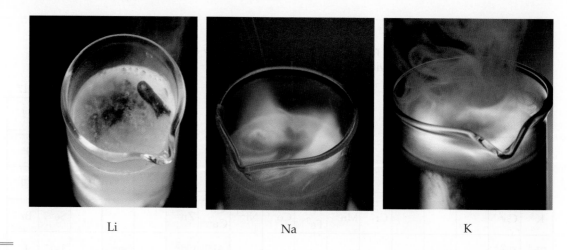

▶ **FIGURE 7.15 The alkali metals react vigorously with water.**

Li   Na   K

∞ Find out more on page 253

Compounds of metals with non-metals tend to be ionic substances (∞ Section 8.2, 'Ionic Bonding'). For example, most metal oxides and halides are ionic solids. To illustrate, the reaction between nickel metal and oxygen produces nickel oxide, an ionic solid containing Ni^{2+} and O^{2-} ions:

$$2\,Ni(s) + O_2(g) \longrightarrow 2\,NiO(s) \qquad [7.9]$$

The oxides are particularly important because of the great abundance of oxygen in our environment.

Most metal oxides are basic. Those that dissolve in water react to form metal hydroxides, as in the following examples:

▲ **FIGURE 7.16 Elemental calcium solution.** Calcium metal reacts with water to form hydrogen gas and aqueous calcium hydroxide, $Ca(OH)_2(aq)$.

$$\text{Metal oxide} + \text{water} \longrightarrow \text{metal hydroxide}$$

$$Na_2O(s) + H_2O(l) \longrightarrow 2\,NaOH(aq) \qquad [7.10]$$

$$CaO(s) + H_2O(l) \longrightarrow Ca(OH)_2(aq) \qquad [7.11]$$

The basicity of metal oxides is due to the oxide ion, which reacts with water according to the net ionic equation

$$O^{2-}(aq) + H_2O(l) \longrightarrow 2\,OH^-(aq) \qquad [7.12]$$

Metal oxides also demonstrate their basicity by reacting with acids to form a salt plus water, as illustrated in ▼ **FIGURE 7.17.**

$$\text{Metal oxide} + \text{acid} \longrightarrow \text{salt} + \text{water}$$

$$NiO(s) + 2\,HCl(aq) \longrightarrow NiCl_2(aq) + H_2O(l) \qquad [7.13]$$

In contrast, we see below that non-metal oxides are acidic, dissolving in water to form acidic solutions and reacting with bases to form salts.

▶ **FIGURE 7.17 Metal oxides react with acids.** (a) Nickel oxide (NiO), nitric acid (HNO_3) and water. (b) NiO is insoluble in water, but reacts with HNO_3 to give a green solution of the salt $Ni(NO_3)_2$.

(a)

(b)

SAMPLE EXERCISE 7.8   **Metal oxides**

(a) Would you expect aluminium oxide to be a solid, liquid or gas at room temperature? **(b)** Write the balanced chemical equation for the reaction of aluminium oxide with nitric acid.

SOLUTION

Analyse We are asked to predict a physical property of a compound and the chemical change that occurs when it reacts with acid.

Plan Locating aluminium and oxygen in the periodic table is the starting point. A combination of a metal with a non-metal results in an ionic compound which is likely to be solid, while a combination of two non-metals will give a covalent compound which may be a solid, liquid or gas. Metal oxides are generally basic in character whereas non-metal oxides are acidic.

Solve (a) Because aluminium oxide is the oxide of a metal, we would expect it to be an ionic solid. Indeed it is, with the very high melting point of 2072 °C.

(b) In its compounds, aluminium has a 3+ charge, Al^{3+}; the oxide ion is O^{2-}. Consequently, the formula of aluminium oxide is Al_2O_3. Metal oxides tend to be basic and therefore to react with acids to form a salt plus water. In this case the salt is aluminium nitrate, $Al(NO_3)_3$. The balanced chemical equation is

$$Al_2O_3(s) + 6\,HNO_3(aq) \longrightarrow 2\,Al(NO_3)_3(aq) + 3\,H_2O(l)$$

PRACTICE EXERCISE

Write the balanced chemical equation for the reaction between copper(II) oxide and sulfuric acid.

Answer: $CuO(s) + H_2SO_4(aq) \longrightarrow CuSO_4(aq) + H_2O(l)$

(See also Exercise 7.48.)

Non-metals

Non-metals vary greatly in appearance. They can be solid, liquid or gas. They are not lustrous and are generally poor conductors of heat and electricity. Their melting points are usually lower than those of metals (although diamond, a form of carbon, melts at 3570 °C). Under ordinary conditions, seven non-metals exist as diatomic molecules. Five of these are gases (H_2, N_2, O_2, F_2 and Cl_2), one is a liquid (Br_2) and one is a volatile solid (I_2). The remaining non-metals are solids that can be either hard, such as diamond, or soft, such as sulfur (▶ **FIGURE 7.18**).

▲ **FIGURE 7.18** **Sulfur, known to the medieval world as 'brimstone', is a non-metal.**

Because of their electron affinities, non-metals tend to gain electrons when they react with metals. For example, the reaction of aluminium with bromine produces aluminium bromide, an ionic compound containing the aluminium ion, Al^{3+}, and the bromide ion, Br^-:

$$2\,Al(s) + 3\,Br_2(l) \longrightarrow 2\,AlBr_3(s) \qquad [7.14]$$

A non-metal will typically gain enough electrons to fill its outermost occupied *p* subshell, giving a noble gas electron configuration. For example, the bromine atom gains one electron to fill its 4*p* subshell:

$$Br\,([Ar]4s^2 3d^{10} 4p^5) \Rightarrow Br^-\,([Ar]4s^2 3d^{10} 4p^6)$$

Compounds composed entirely of non-metals are molecular substances (co Section 8.3, 'Covalent Bonding'). For example, the oxides, halides and hydrides of the non-metals are molecular substances that tend to be gases, liquids or low-melting solids at room temperature.

Most non-metal oxides are acidic; those that dissolve in water react to form acids, as in the following examples:

co Find out more on page 258

$$Non\text{-}metal\ oxide + water \longrightarrow acid$$

$$CO_2(g) + H_2O(l) \longrightarrow H_2CO_3(aq) \qquad [7.15]$$

$$P_4O_{10}(s) + 6\,H_2O(l) \longrightarrow 4\,H_3PO_4(aq) \qquad [7.16]$$

The reaction of carbon dioxide with water accounts for the acidity of carbonated water and, to some extent, rainwater. Because sulfur is present in oil and coal, combustion of these common fuels produces sulfur dioxide and sulfur trioxide. These substances dissolve in water to produce *acid rain*, a major pollution problem in many parts of the world.

$$S(s) + O_2(g) \longrightarrow SO_2(g) \qquad [7.17]$$

$$2\,SO_2(g) + O_2(g) \longrightarrow 2\,SO_3(g) \qquad [7.18]$$

$$SO_3(g) + H_2O(l) \longrightarrow H_2SO_4(aq) \qquad [7.19]$$

Like acids, most non-metal oxides dissolve in basic solutions to form a salt plus water:

$$\text{Non-metal oxide} + \text{base} \longrightarrow \text{salt} + \text{water}$$

$$CO_2(g) + 2\,NaOH(aq) \longrightarrow Na_2CO_3(aq) + H_2O(l) \qquad [7.20]$$

 CONCEPT CHECK 9

A compound ACl_3 (A is an element) has a melting point of $-112\ °C$. Would you expect the compound to be a molecular or ionic substance? Is element A more likely to be scandium (Sc) or phosphorus (P)?

SAMPLE EXERCISE 7.9 **Non-metal oxides**

Write the balanced chemical equations for the reactions of solid selenium dioxide with **(a)** water, **(b)** aqueous sodium hydroxide.

SOLUTION

Analyse We are asked to write the chemical equation for the reaction of a non-metal oxide with water and with a basic solution.

Plan Non-metal oxides are acidic, reacting with water to form an acid and with bases to form a salt and water.

Solve (a) The formula of selenium dioxide is SeO_2. Its reaction with water is like that of carbon dioxide (Equation 7.15):

$$SeO_2(s) + H_2O(l) \longrightarrow H_2SeO_3(aq)$$

(It doesn't matter that SeO_2 is a solid and CO_2 is a gas; the point is that both are water-soluble non-metal oxides.)

(b) The reaction with sodium hydroxide is similar to the reaction summarised by Equation 7.20:

$$SeO_2(s) + 2\,NaOH(aq) \longrightarrow Na_2SeO_3(aq) + H_2O(l)$$

PRACTICE EXERCISE

Write the balanced chemical equation for the reaction of solid tetraphosphorus hexoxide with water.

Answer: $P_4O_6(s) + 6\,H_2O(l) \longrightarrow 4\,H_3PO_3(aq)$

(See also Exercises 7.48(c)(d), 7.49(b)(d).)

Metalloids

Metalloids have properties intermediate between those of metals and those of non-metals. They may have *some* characteristic metallic properties but lack others. For example, silicon *looks* like a metal, but it is brittle rather than malleable and is a much poorer conductor of heat and electricity than metals. Compounds of metalloids can have characteristics of the compounds of metals or non-metals, depending on the specific compound.

Several of the metalloids, most notably silicon, are electrical semiconductors and are the principal elements used in the manufacture of integrated circuits and computer chips.

SAMPLE INTEGRATIVE EXERCISE   Putting concepts together

One way to measure ionisation energies is photoelectron spectroscopy (PES), a technique based on the photoelectric effect (∞ Section 6.2, 'Quantised Energy and Photons'). In PES, monochromatic light is directed onto a sample, causing electrons to be emitted. The kinetic energy of the emitted electrons is measured. The difference between the energy of the photons and the kinetic energy of the electrons corresponds to the energy needed to remove the electrons (that is, the ionisation energy). Suppose that a PES experiment is performed in which mercury vapour is irradiated with ultraviolet light of wavelength 58.4 nm. **(a)** What is the energy of this light, in electron volts ($1\ eV = 1.602 \times 10^{-19}$)? **(b)** Write an equation that shows the process corresponding to the first ionisation energy of Hg. **(c)** The kinetic energy of the emitted electrons is measured to be 10.75 eV. What is the first ionisation energy of Hg, in kJ mol^{-1}? **(d)** With reference to Figure 7.9, determine which of the halogen elements has a first ionisation energy closest to that of mercury.

∞ Review this on page 183

Analyse First we determine the energy of a photon of UV light and convert it to eV. This photon is able to cause ionisation of a mercury atom. The difference in energy between the photon and the ejected electron is the ionisation energy of mercury.

Solve **(a)** Combining Equation 6.1 ($\nu = c/\lambda$) and Equation 6.2 ($E = h\nu$) we can determine the energy of a photon of known wavelength. We need to convert the wavelength from nanometres to metres, so the units cancel:

$1\ nm = 1 \times 10^{-9}\ m$

$E = hc/\lambda = (6.626 \times 10^{-34}\ J\ s)(3.00 \times 10^8\ m\ s^{-1})/58.4 \times 10^{-9}\ m = 3.40 \times 10^{-18}\ J$

$1\ eV = 1.602 \times 10^{-19}\ J$, hence the energy of the photon is

$$E = \frac{3.40 \times 10^{-18}\ J}{1.602 \times 10^{-19}\ J\ eV^{-1}} = 21.2\ eV$$

(b) First ionisation energy: $Hg\ (g) \longrightarrow Hg^+\ (g) + e^-$

(c) The first ionisation energy of mercury is the difference in the kinetic energy of the ejected electron and the energy of the incident photon:

$$I_1 = (21.2 - 10.75)\ eV = 10.45\ eV$$

Converting this energy to J we get the energy needed to ionise one electron to be

$$I_1 = (10.45\ eV)(1.602 \times 10^{-19}\ J\ eV^{-1}) = 1.67 \times 10^{-18}\ J$$

And for a mole of electrons

$$I_1 = (1.67 \times 10^{-18}\ J)(6.022 \times 10^{23}\ mol^{-1}) = 1.01 \times 10^6\ J\ mol^{-1}$$

or 1010 kJ mol^{-1} to three significant figures.

(d) Referring to Figure 7.9 we see that iodine has a first ionisation energy of 1008 kJ mol^{-1}.

CHAPTER SUMMARY AND KEY TERMS

SECTION 7.1 The periodic table was first developed by Mendeleev and Meyer on the basis of the similarity in chemical and physical properties exhibited by certain elements. Moseley established that each element has a unique atomic number, which added more order to the periodic table. We now recognise that the periodic table lists elements in order of their atomic number and that elements in the same group have the same number of electrons in their **valence orbitals**. As a consequence, elements in the same group have similar properties. However, differences in the properties arise as we move down the group because the valence orbitals are in different shells.

SECTION 7.2 The number of protons in the nucleus of an atom is called the **actual nuclear charge**. However, the attraction of the outer shell electrons to the nucleus is diminished by the shielding or screening effect of the core electrons, resulting in an **effective nuclear charge** (Z_{eff}). Many properties of atoms are determined by the magnitude of this charge. The core electrons are very effective in screening the outer electrons from the full charge on the nucleus although electrons in the same shell do not screen each other very effectively.

SECTION 7.3 Atomic size is based on the **bonding atomic radius** (or simply **atomic radius**) and the atomic radii increase as we go down a group of the periodic table and decrease as we move left to right across a period.

Cations are smaller than their parent atom whereas anions are larger than their parent atom. An **isoelectronic series** is a series of ions that has the same number of electrons, whether the ions are positively or negatively charged. For isoelectronic ions, size decreases with increasing nuclear charge.

SECTIONS 7.4 and 7.5 The first **ionisation energy** of an atom is the minimum energy needed to remove an electron from the atom in the gas phase, forming a cation. The second ionisation energy is the energy needed to remove a second electron and so forth. First ionisation energy decreases as we go down a group in the periodic table and increase as we move left to right across a period. **Electron affinity** is the addition of one or more electrons to an atom and is generally an exothermic process. We see that the higher the value of ΔE, the more stable is the resultant anion relative to the neutral atom and the separated electron.

Electron configurations for ions can be written by first writing the electron configuration of the neutral atom and then removing or adding the appropriate number of electrons. Electrons are removed first from the orbitals with the largest value of n. If there are two valence orbitals with the same value of n (such as $4s$ and $4p$), then the electrons are lost first from the orbital with a higher value of l (in this case, $4p$). Electrons are added to orbitals in the reverse order.

SECTION 7.6 The elements can be characterised as metals, metalloids and non-metals with the majority of the elements being metals. They occupy the left side and middle of the periodic table. Non-metals are found at the upper right side of the periodic table and metalloids occupy a narrow band between the metals and non-metals. The metallic character of the elements increases down a group of the periodic table and decreases as we move left to right across it.

Metals have a characteristic luster, and they are good conductors of heat and electricity. When metals react with non-metals, the metal atoms are oxidised to cations and ionic substances are generally formed. Most metal oxides are basic; they react with acids to form salts and water.

Non-metals lack metallic luster and are generally poor conductors of heat and electricity. Several are gases at room temperature. Compounds composed entirely of non-metals are generally molecular. Non-metals usually form anions in their reactions with metals. Non-metal oxides are acidic; they react with bases to form salts and water. Metalloids have properties that are intermediate between those of metals and non-metals. We see that metals can react with non-metals during which the metals are oxidised to cations and ionic substances are generally formed. Metals are good conductors of heat and electricity whereas non-metals are poor conductors of heat and electricity.

KEY SKILLS

- Understand the concept of effective nuclear charge and its effect on valence electrons. (Section 7.2)

- Use the periodic table to predict the trends in atomic radii, ionic radii, ionisation energies and electron affinities. (Sections 7.2–7.5)

- Understand how the ionisation energy changes as we remove successive electrons and recognise the large increase in ionisation energy when a core electron is removed. (Section 7.4)

- Be able to write the electron configurations for atoms and ions. (Section 7.4)

- Be able to explain the irregularities in the periodic table for electron affinities. (Section 7.5)

- Be able to write the chemical equations for the reactions of metals with water and oxygen and of their oxides with acids. (Section 7.6)

- Be able to write the chemical equations for the reactions of non-metals with water and of their oxides with bases. (Section 7.6)

KEY EQUATION

- Estimating effective nuclear charge $Z_{eff} = Z - S$ [7.1]

EXERCISES

VISUALISING CONCEPTS

7.1 Neon has atomic number 10. What are the similarities and differences in describing the radius of a neon atom and the radius of the billiard ball illustrated here? Could you use billiard balls to illustrate the concept of bonding atomic radius? Explain. [Section 7.3]

7.2 Consider the A_2X_4 molecule depicted below, where A and X are elements. The A–A bond length in this molecule is d_1, and the four A–X bond lengths are each d_2. **(a)** In terms of d_1 and d_2, how could you define the bonding atomic radii of atoms A and X? **(b)** In terms of d_1 and d_2, what would you predict for the X–X bond length of an X_2 molecule? [Section 7.3]

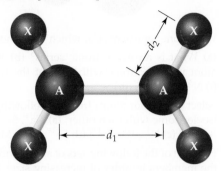

7.3 In the chemical process called *electron transfer*, an electron is transferred from one atom or molecule to another. A simple electron transfer reaction is

$$A(g) + A(g) \longrightarrow A^+(g) + A^-(g)$$

In terms of the ionisation energy and electron affinity of atom A, what is the energy change for this reaction? [Sections 7.4 and 7.5]

7.4 An element X reacts with $F_2(g)$ to form the molecular product shown below. **(a)** Write a balanced equation for this reaction (do not worry about the phases for X and the product). **(b)** Do you think that X is a metal or non-metal? Explain. **(c)** If X is a non-metal, what are some of the physical properties you would expect it to possess? **(d)** Would you expect the product to be reactive with water? [Section 7.6]

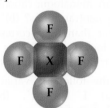

PERIODIC TABLE; EFFECTIVE NUCLEAR CHARGE (Sections 7.1 and 7.2)

7.5 Why did Mendeleev leave blanks in his early version of the periodic table? How did he predict the properties of the elements that belonged in those blanks?

7.6 **(a)** In the period from about 1800 to about 1865, the atomic masses of many elements were accurately measured. Why was this important to Mendeleev's formulation of the periodic table? **(b)** What property of the atom did Moseley associate with the wavelength of X-rays emitted from an element in his experiments? In what ways did this affect the meaning of the periodic table?

7.7 **(a)** What is meant by the term effective nuclear charge? **(b)** How does the effective nuclear charge experienced by the valence electrons of an atom vary going from left to right across a period of the periodic table?

7.8 **(a)** How is the concept of effective nuclear charge used to simplify the numerous electron–electron repulsions in a many-electron atom? **(b)** Which experiences a greater effective nuclear charge in a Be atom, the $1s$ electrons or the $2s$ electrons? Explain.

7.9 **(a)** If each core electron were totally effective in screening the valence electrons from the full charge of the nucleus and the valence electrons provided no screening for each other, what would be the values of the screening constant, S, and the effective nuclear charge, Z_{eff}, for the $4s$ electron in a potassium atom? **(b)** Detailed calculations show that the value of Z_{eff} for a K atom is 3.49. Explain the difference between this number and the one you obtained in part (a).

7.10 **(a)** If the core electrons were totally effective at shielding the valence electrons from the full charge of the nucleus and the valence electrons provided no shielding for each other, what would be the values of S and Z_{eff} for a $3p$ electron in a sulfur atom? **(b)** Detailed calculations indicate that the value of S for a $3p$ electron in a sulfur atom is 10.52. Can you explain any difference between this number and the one you obtained in part (a)?

ATOMIC AND IONIC RADII (Section 7.3)

7.11 Because an exact outer boundary cannot be measured or even calculated for an atom, how are atomic radii determined? What is the difference between a bonding radius and a non-bonding radius?

7.12 **(a)** Why does the quantum mechanical description of many-electron atoms make it difficult to define a precise atomic radius? **(b)** When non-bonded atoms come up against one another, what determines how closely the nuclear centres can approach?

7.13 The distance between W atoms in tungsten metal is 274 pm. What is the atomic radius of a tungsten atom in

this environment? (This radius is called the metallic radius.)

7.14 Based on the radii presented in Figure 7.6, predict the distance between Ge atoms in solid germanium.

7.15 How do the sizes of atoms change as we move **(a)** from left to right across a period in the periodic table, **(b)** from top to bottom in a group in the periodic table? **(c)** Arrange the following atoms in order of increasing atomic radius: F, P, S, As.

7.16 **(a)** Among the non-metallic elements, the change in atomic radius in moving one place left or right in a period is smaller than the change in moving one period up or down. Explain these observations. **(b)** Arrange the following atoms in order of increasing atomic radius: Si, S, Ge, Se.

7.17 Using only the periodic table, arrange each set of atoms in order of increasing radius: **(a)** Ca, Mg, Be; **(b)** Ga, Br, Ge; **(c)** Al, Tl, Si.

7.18 Using only the periodic table, arrange each set of atoms in order of increasing radius: **(a)** Cs, K, Rb; **(b)** In, Te, Sn; **(c)** P, Cl, Sr.

7.19 **(a)** Why are monatomic cations smaller than their corresponding neutral atoms? **(b)** Why are monatomic anions larger than their corresponding neutral atoms? **(c)** Why does the size of ions increase as one proceeds down a group in the periodic table?

7.20 Explain the following variations in atomic or ionic radii: **(a)** $I^- > I > I^+$; **(b)** $Ca^{2+} > Mg^{2+} > Be^{2+}$; **(c)** $Fe > Fe^{2+} > Fe^{3+}$.

7.21 Consider a reaction represented by the following spheres:

Reactants   Products

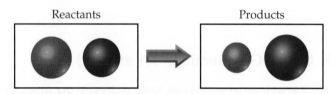

Which sphere represents a metal and which a non-metal? Explain.

7.22 Consider the following spheres:

Which one represents Ca, which Ca^{2+} and which Mg^{2+}?

7.23 **(a)** What is an isoelectronic series? **(b)** Which neutral atom is isoelectronic with each of the following ions: (i) N^{3-}, (ii) Ba^{2+}, (iii) Se^{2-}, (iv) Bi^{3+}?

7.24 Select the ions or atoms from the following sets that are isoelectronic with each other: **(a)** Na^+, Sr^{2+}, Br^-; **(b)** Y^{3+}, Br^-, Kr; **(c)** N^{3-}, P^{3-}, Ti^{4+}; **(d)** Fe^{3+}, Co^{3+}, Mn^{2+}.

7.25 For each of the following sets of atoms and ions, arrange the members in order of increasing size: **(a)** Se^{2-}, Te^{2-}, Se; **(b)** Co^{3+}, Fe^{2+}, Fe^{3+}; **(c)** Ca, Ti^{4+}, Sc^{3+}; **(d)** Be^{2+}, Na^+, Ne.

7.26 For each of the following statements, provide an explanation: **(a)** Cl^- is larger than Cl; **(b)** S^{2-} is larger than O^{2-}; **(c)** K^+ is larger than Ca^{2+}.

IONISATION ENERGIES; ELECTRON AFFINITIES (Sections 7.4 and 7.5)

7.27 Write equations that show the processes that describe the first, second and third ionisation energies of a boron atom.

7.28 Write equations that show the process for **(a)** the first two ionisation energies of tin and **(b)** the fourth ionisation energy of titanium.

7.29 **(a)** Why are ionisation energies always positive quantities? **(b)** Why does F have a larger first ionisation energy than O? **(c)** Why is the second ionisation energy of an atom always greater than its first ionisation energy?

7.30 **(a)** Why does Li have a larger first ionisation energy than Na? **(b)** The difference between the third and fourth ionisation energies of scandium is much larger than the difference between the third and fourth ionisation energies of titanium. Why? **(c)** Why does Li have a much larger second ionisation energy than Be?

7.31 **(a)** What is the general relationship between the size of an atom and its first ionisation energy? **(b)** Which element in the periodic table has the largest ionisation energy? Which has the smallest?

7.32 **(a)** What is the trend in first ionisation energies as one proceeds down the group 17 elements? Explain how this trend relates to the variation in atomic radii. **(b)** What is the trend in first ionisation energies as one moves across the fourth period from K to Kr? How does this trend compare with the trend in atomic sizes?

7.33 Based on their positions in the periodic table, predict which atom of the following pairs will have the larger first ionisation energy: **(a)** Cl, Ar; **(b)** Be, Ca; **(c)** K, Co; **(d)** S, Ge; **(e)** Sn, Te.

7.34 For each of the following pairs, indicate which element has the larger first ionisation energy: **(a)** Rb, Mo; **(b)** N, P; **(c)** Ga, Cl; **(d)** Pb, Rn. (In each case use electron configuration and effective nuclear charge to explain your answer.)

7.35 Write the electron configurations for the following ions: **(a)** Si^{2+}, **(b)** Bi^{3+}, **(c)** Te^{2-}, **(d)** V^{3+}, **(e)** Hg^{2+}, **(f)** Ni^{2+}.

7.36 Write electron configurations for the following ions and determine which have noble gas configurations: **(a)** Mn^{3+}, **(b)** Se^{2-}, **(c)** Sc^{3+}, **(d)** Ru^{2+}, **(e)** Tl^+, **(f)** Au^+.

7.37 Write equations, including electron configurations beneath the species involved, that explain the difference between the first ionisation energy of Se(g) and the electron affinity of Se(g).

7.38 The first ionisation energy of Ar and the electron affinity of Ar are both positive values. What is the significance of the positive value in each case?

7.39 The electron affinity of lithium is a negative value, whereas the electron affinity of beryllium is a positive value. Use electron configurations to account for this observation.

7.40 The electron affinity of bromine is a negative quantity, but it is positive for Kr. Use the electron configurations of the two elements to explain the difference.

7.41 What is the relationship between the ionisation energy of an anion with a 1− charge such as F^- and the electron affinity of the neutral atom, F?

7.42 Write an equation for the process that corresponds to the electron affinity of the Mg^+ ion. Also write the electron configurations of the species involved. What process does this electron affinity equation correspond to? What is the magnitude of the energy change in the process? [*Hint:* The answer is in Table 7.2.]

METALS, NON-METALS AND METALLOIDS (Section 7.6)

7.43 How are metallic character and first ionisation energy related?

7.44 Arrange the following pure solid elements in order of increasing electrical conductivity: Ge, Ca, S and Si. Explain the reasoning you used.

7.45 Predict whether each of the following oxides is ionic or molecular: SO_2, MgO, Li_2O, P_2O_5, Y_2O_3, N_2O and XeO_3. Explain the reasons for your choices.

7.46 **(a)** What is meant by the terms acidic oxide and basic oxide? **(b)** How can we predict whether an oxide will be acidic or basic, based on its composition?

7.47 An element X reacts with oxygen to form XO_2 and with chlorine to form XCl_4. XO_2 is a white solid that melts at high temperatures (above 1000 °C). Under usual conditions, XCl_4 is a colourless liquid with a boiling point of 58 °C. **(a)** XCl_4 reacts with water to form XO_2 and another product. What is the likely identity of the other product? **(b)** Do you think that element X is a metal, non-metal or metalloid? Explain. **(c)** By using a source-book such as the CRC *Handbook of Chemistry and Physics,* try to determine the identity of element X.

7.48 Write balanced equations for the following reactions: **(a)** barium oxide with water, **(b)** iron(II) oxide with perchloric acid, **(c)** sulfur trioxide with water, **(d)** carbon dioxide with aqueous sodium hydroxide.

7.49 Write balanced equations for the following reactions: **(a)** potassium oxide with water, **(b)** diphosphorus trioxide with water, **(c)** chromium(III) oxide with dilute hydrochloric acid, **(d)** selenium dioxide with aqueous potassium hydroxide.

7.50 Compare the elements sodium and magnesium with respect to the following properties: **(a)** electron configuration, **(b)** most common ionic charge, **(c)** first ionisation energy, **(d)** reactivity towards water, **(e)** atomic radius. Account for the differences between the two elements.

7.51 **(a)** Compare the electron configurations and atomic radii (see Figure 7.6) of rubidium and silver. In what respects are their electronic configurations similar? Account for the difference in radii of the two elements. **(b)** As with rubidium, silver is most commonly found as the 1+ ion, Ag^+. However, silver is far less reactive. Explain these observations.

7.52 **(a)** Why is calcium generally more reactive than magnesium? **(b)** Why is calcium generally less reactive than potassium?

7.53 Write a balanced equation for the reaction that occurs in each of the following cases. **(a)** Potassium metal burns in an atmosphere of chlorine gas. **(b)** Strontium oxide is added to water. **(c)** A fresh surface of lithium metal is exposed to oxygen gas. **(d)** Sodium metal is reacted with molten sulfur.

7.54 Write a balanced equation for the reaction that occurs in each of the following cases. **(a)** Potassium is added to water. **(b)** Barium is added to water. **(c)** Lithium is heated in nitrogen, forming lithium nitride. **(d)** Magnesium burns in oxygen.

7.55 Use electron configurations to explain why hydrogen exhibits properties similar to those of both Li and F.

7.56 Compare the elements fluorine and chlorine with respect to the following properties: **(a)** electron configuration, **(b)** most common ionic charge, **(c)** first ionisation energy, **(d)** reactivity towards water **(e)** electron affinity, **(f)** atomic radius. Account for the differences between the two elements.

7.57 Little is known about the properties of astatine, At, because of its rarity and high radioactivity. Nevertheless, it is possible for us to make many predictions about its properties. **(a)** Do you expect the element to be a gas, liquid or solid at room temperature? Explain. **(b)** What is the chemical formula of the compound it forms with Na?

ADDITIONAL EXERCISES

7.58 **(a)** Which will have the lower energy, a $4s$ or a $4p$ electron in an As atom? **(b)** How can we use the concept of effective nuclear charge to explain your answer to part (a)?

7.59 **(a)** If the core electrons were totally effective at shielding the valence electrons and the valence electrons provided no shielding for each other, what would be the effective nuclear charge acting on the valence electron in P? **(b)** Detailed calculations indicate that the effective nuclear charge is 5.6+ for the $3s$ electrons and 4.9+ for the $3p$ electrons. Why are the values for the $3s$ and $3p$ electrons different? **(c)** If you remove a single electron from a P atom, which orbital will it come from? Explain.

7.60 Nearly all the mass of an atom is in the nucleus, which has a very small radius. When atoms bond together (for example, two fluorine atoms in F_2), why is the distance separating the nuclei so much larger than the radii of the nuclei?

7.61 Consider the change in effective nuclear charge experienced by a $2p$ electron as we proceed from C to N. **(a)** Based on a simple model in which core electrons screen the valence electrons completely and valence electrons do not screen other valence electrons, what do you predict for the change in Z_{eff} from C to N? **(b)** The actual calculated change in Z_{eff} from C to N is 0.70+. How can we explain the difference between this number and the one obtained in part (a)? **(c)** The calculated change in Z_{eff} from N to O is smaller than that from C to N. Can you provide an explanation for this observation?

7.62 As we move across a period of the periodic table, why do the sizes of the transition elements change more gradually than those of the main group elements?

7.63 The ionic substance strontium oxide, SrO, forms from the direct reaction of strontium metal with molecular oxygen. The arrangement of the ions in solid SrO is analogous to that in solid NaCl. This is illustrated below. **(a)** Write a balanced equation for the formation of SrO(s) from the elements. **(b)** In the figure, do the large spheres represent Sr^{2+} ions or O^{2-} ions? Explain. **(c)** Based on the ionic radii in Figure 7.7, predict the length of the side of the cube in the figure. **(d)** The experimental density of SrO is 5.10 g cm^{-3}. Given your answer to part (c), what is the number of formula units of SrO that are contained in the cube in the figure?

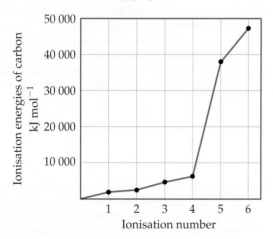

7.64 Explain the variation in ionisation energies of carbon, as displayed in the following graph.

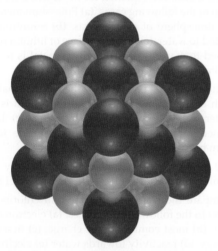

7.65 Listed here are the atomic and ionic (2+) radii for calcium and zinc.

Radii (pm)			
Ca	174	Ca^{2+}	99
Zn	131	Zn^{2+}	74

(a) Explain why the ionic radius in each case is smaller than the atomic radius. **(b)** Why is the atomic radius of calcium larger than that of zinc? **(c)** Suggest a reason why the difference in the ionic radii is much less than the difference in the atomic radii.

7.66 Do you agree with the following statement? 'A negative value for the electron affinity of an atom occurs when the outermost electrons only incompletely shield one another from the nucleus.' If not, change it to make it more nearly correct in your view. Apply either the statement as given or your revised statement to explain why the electron affinity of bromine is −325 kJ mol^{-1} and that for its neighbour Kr is > 0.

7.67 Use orbital diagrams to illustrate what happens when an oxygen atom gains two electrons. Why is it extremely difficult to add a third electron to the atom?

7.68 Use electron configurations to explain the following observations. **(a)** The first ionisation energy of phosphorus is greater than that of sulfur. **(b)** The electron affinity of nitrogen is lower (less negative) than those of both carbon and oxygen. **(c)** The second ionisation energy of oxygen is greater than that of fluorine. **(d)** The third ionisation energy of manganese is greater than those of both chromium and iron.

7.69 The following table gives the electron affinities, in kJ mol^{-1}, for the group 11 and group 12 metals.

Cu	Zn
−119	> 0
Ag	Cd
−126	> 0
Au	Hg
−223	> 0

(a) Why are the electron affinities of the group 12 elements approximately zero? **(b)** Why do the electron affinities of the group 11 elements become more negative as we move down the group? [*Hint*: Examine the trends in the electron affinity of other groups as we proceed down the periodic table.]

7.70 Hydrogen is an unusual element because it behaves in some ways like the alkali metal elements and in other ways like a non-metal. Its properties can be explained in part by its electron configuration and by the values for its ionisation energy and electron affinity. **(a)** Explain why the electron affinity of hydrogen is much closer to the values for the alkali elements than for the halogens. **(b)** Is the following statement true? 'Hydrogen has the smallest bonding atomic radius of any element that forms chemical compounds.' If not, correct it. If it is, explain in terms of electron configurations. **(c)** Explain why the ionisation energy of hydrogen is closer to the values for the halogens than for the alkali metals.

7.71 The first ionisation energy of the oxygen molecule is the energy required for the following process:

$$O_2(g) \longrightarrow O_2^+(g) + e^-$$

The energy needed for this process is 1175 kJ mol^{-1}, very similar to the first ionisation energy of Xe. Would you expect O_2 to react with F_2? If so, suggest a product or products of this reaction.

7.72 There are certain similarities in properties that exist between the first member of any periodic family and the element located below it and to the right in the periodic table. For example, in some ways Li resembles Mg, Be resembles Al and so forth. This observation is called the diagonal relationship. Using what we have learned in this chapter, offer a possible explanation for this relationship.

INTEGRATIVE EXERCISES

7.73 **(a)** Write the electron configuration for Li and estimate the effective nuclear charge experienced by the valence electron. **(b)** The energy of an electron in a one-electron atom or ion equals $(-2.18 \times 10^{-18} \text{ J}) \left(\dfrac{Z^2}{n^2} \right)$, where Z is the nuclear charge and n is the principal quantum number of the electron. Estimate the first ionisation energy of Li. **(c)** Compare the result of your calculation with the value reported in Table 7.4, and explain the difference. **(d)** What value of the effective nuclear charge gives the proper value for the ionisation energy? Does this agree with your explanation in part (c)?

7.74 **(a)** The experimental Bi–Br bond length in bismuth tribromide, $BiBr_3$, is 263 pm. Based on this value and the data in Figure 7.6, predict the atomic radius of Bi. **(b)** Bismuth tribromide is soluble in acidic solution. It is formed by treating solid bismuth(III) oxide with aqueous hydrobromic acid. Write a balanced chemical equation for this reaction. **(c)** Bismuth(III) oxide is soluble in acidic solutions, but it is insoluble in basic solutions such as NaOH(aq). On the basis of these properties, is bismuth characterised as a metallic, metalloid or non-metallic element? **(d)** Treating bismuth with fluorine gas forms BiF_5. Use the electron configuration of Bi to explain the formation of a compound with this formulation. **(e)** Although it is possible to form BiF_5 in the manner just described, pentahalides of bismuth are not known for the other halogens. Explain why the pentahalide might form with fluorine, but not with the other halogens. How does the behaviour of bismuth relate to the fact that xenon reacts with fluorine to form compounds, but not with the other halogens?

7.75 Potassium superoxide, KO_2, is often used in oxygen masks (such as those used by firefighters) because KO_2 reacts with CO_2 to release molecular oxygen. Experiments indicate that 2 moles of KO_2(s) react with each mole of CO_2(g). **(a)** The products of the reaction are K_2CO_3(s) and O_2(g). Write a balanced equation for the reaction between KO_2(s) and CO_2(g). **(b)** Indicate the oxidation number for each atom involved in the reaction in part (a). What elements are being oxidised and reduced? **(c)** What mass of KO_2(s) is needed to consume 18.0 g CO_2(g)? What mass of O_2(g) is produced during this reaction?

MasteringChemistry (www.pearson.com.au/masteringchemistry)

Make learning part of the grade. Access:

- tutorials with peronalised coaching
- study area
- Pearson eText

PHOTO/ART CREDITS

220 Steven Puetzer/Getty Images; **240 (Figure 7.15)** Richard Megna, Fundamental Photographs, NYC, **(Figure 7.16)** Tom Pantages; **241** Richard Treptow/Photo Researchers, Inc; **244** Mishakov/Shutterstock.

BASIC CONCEPTS OF
CHEMICAL BONDING

A crocolite crystal found in the Zeehan District of Tasmania. Its shape is determined not only by the atoms it contains (Pb, Cr, O) but also by the types of bonds between the atoms.

KEY CONCEPTS

8.1 CHEMICAL BONDS, LEWIS SYMBOLS AND THE OCTET RULE

There are three main types of chemical bonds: *ionic, covalent* and *metallic*. In evaluating bonding, *Lewis symbols* provide a useful shorthand for keeping track of valence electrons.

8.2 IONIC BONDING

In ionic substances the atoms are held together by the electrostatic attractions between ions of opposite charge. We discuss the energetics of forming ionic substances and describe the *lattice energy* of these substances.

8.3 COVALENT BONDING

We then examine the bonding in molecular substances in which atoms bond by sharing one or more electron pairs. In general, the electrons are shared in such a way that each atom attains an *octet* of electrons

8.4 BOND POLARITY AND ELECTRONEGATIVITY

Electronegativity is defined as the ability of an atom in a compound to attract electrons to itself. In general, electron pairs are shared unequally between atoms with different electronegativities, leading to *polar covalent bonds*.

8.5 DRAWING LEWIS STRUCTURES

Lewis structures are a simple yet powerful way of predicting covalent bonding patterns in molecules. In addition to the octet rule, we see that the concept of *formal charge* can be used to identify the dominant Lewis structure.

8.6 RESONANCE STRUCTURES

In some cases more than one equivalent Lewis structure can be drawn for a molecule or polyatomic ion. The bonding description in such cases is a blend of two or more *resonance structures*.

8.7 EXCEPTIONS TO THE OCTET RULE

We recognise that the octet rule is more of a guideline than an absolute rule. Exceptions to the rule include molecules with an odd number of electrons, molecules where large differences in electronegativity prevent an atom from completing its octet, and molecules where an element from period 3 or below in the periodic table attains more than an octet of electrons.

8.8 STRENGTHS OF COVALENT BONDS

Bond strengths vary with the number of shared electron pairs as well as other factors. We use *average bond enthalpy* values to estimate the enthalpies of reactions in cases where thermodynamic data are unavailable.

The bonds that form between atoms and ions are not as complex as one would be led to believe by the varied shapes and sizes of the substances found on Earth. In fact there are essentially only three types of chemical bonds that hold atoms or ions strongly together. These are ionic, covalent and metallic bonds. The properties of substances depend strongly on what types of bonds hold their atoms together. Table salt, which consists of Na^+ ions and Cl^- ions, is held together by the attractions between oppositely charged ions which we call *ionic bonds*. It has totally different physical and chemical properties from that of granulated sugar, which does not contain ions but consists of molecules of sucrose, $C_{12}H_{22}O_{11}$, in which the atoms are held together by attractive forces called *covalent bonds*. The atoms in metals are held together by *metallic bonds* in which the atoms all share the valence electrons. Other substances such as

crocolite, a lead chromite ($PbCrO_4$) mineral, shown in the opening photograph, are composed of both ionic and covalent bonds, which give it its particular shape.

The properties of substances are determined in large part by the *chemical bonds* that hold their atoms together. What determines the type of bonding in each substance, and just how do the characteristics of these bonds give rise to different physical and chemical properties? The keys to answering the first question are found in the electronic structures of the atoms involved, discussed in Chapters 6 and 7. In this chapter and the next, we examine the relationships between electronic structure, chemical bonding forces and chemical bond type. We also see how the properties of ionic and covalent substances arise from the distributions of electronic charge within atoms, ions and molecules.

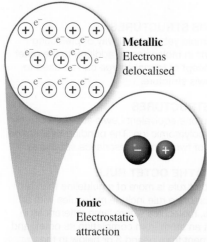

▲ **FIGURE 8.1   Ionic, covalent and metallic bonds.** Different types of interactions between atoms lead to different types of chemical bonds.

∞ Review this on page 205

8.1 | CHEMICAL BONDS, LEWIS SYMBOLS AND THE OCTET RULE

Whenever two atoms or ions are strongly attached to each other, we say there is a **chemical bond** between them. There are three general types of chemical bonds: ionic, covalent and metallic. We can imagine these three types of bonds by thinking about the simple act of using a stainless-steel spoon (metallic bonds) to add table salt (ionic bonds) to a glass of water (covalent bonds) (◄ **FIGURE 8.1**).

The term **ionic bond** refers to electrostatic forces that exist between ions of opposite charge. Ions may be formed from atoms by the transfer of one or more electrons from one atom to another. Ionic substances generally result from the interaction of metals on the far left side of the periodic table with non-metals on the far right side (excluding the noble gases, group 18). Ionic bonding is discussed in Section 8.2.

A **covalent bond** results from the sharing of electrons between two atoms. The most familiar examples of covalent bonding are seen in the interactions of non-metallic elements with one another. We devote much of this chapter and the next to describing and understanding covalent bonds.

Metallic bonds are found in metals, such as copper, iron and aluminium. Each atom in a metal is bonded to several neighbouring atoms. The bonding electrons are relatively free to move throughout the three-dimensional structure of the metal. Metallic bonds give rise to such typical metallic properties as high electrical conductivity and lustre.

Lewis Symbols

The electrons involved in chemical bonding are the *valence electrons*, which, for most atoms, are those residing in the outermost occupied shell of an atom (∞ Section 6.8, 'Electron Configurations'). The American chemist Gilbert N. Lewis (1875–1946) suggested a simple way of showing the valence electrons in an atom and tracking them in the course of bond formation, using what are now known as *Lewis electron-dot symbols* or merely *Lewis symbols*.

The **Lewis symbol** for an element consists of the chemical symbol for the element plus a dot for each valence electron. Sulfur, for example, has the electron configuration $[Ne]3s^23p^4$; its Lewis symbol therefore shows six valence electrons:

·Ṡ·

The dots are placed singly on the four sides of the atomic symbol: the top, the bottom and the left and right sides. After four single dots have been placed around the chemical symbol, extra dots (electrons) are paired with those already present. All four sides of the symbol are equivalent, which means that the choice of which side to place two electrons versus one electron is arbitrary.

The electron configurations and Lewis symbols for the main group elements of the second and third periods of the periodic table are shown in ▶ **TABLE 8.1**. Notice that the number of valence electrons in any main-group element (groups 1, 2, 13–17) is the *same as the group number of the element* or, for those in groups 13–17, it is *the group number minus 10*. For example, the Lewis symbols for oxygen and sulfur, members of group 16, both show six dots.

△ **CONCEPT CHECK 1**

Which of the following three possible Lewis symbols for Cl is correct?

:C̈l·   :C̈l: :C̈l·

Table 8.1 • Lewis symbols

Element	Electron configuration	Lewis symbol	Element	Electron configuration	Lewis symbol
Li	$[\text{He}]2s^1$	Li·	Na	$[\text{Ne}]3s^1$	Na·
Be	$[\text{He}]2s^2$	·Be·	Mg	$[\text{Ne}]3s^2$	·Mg·
B	$[\text{He}]2s^2 2p^1$	·Ḃ·	Al	$[\text{Ne}]3s^2 3p^1$	·Àl·
C	$[\text{He}]2s^2 2p^2$	·Ċ·	Si	$[\text{Ne}]3s^2 3p^2$	·Ṡi·
N	$[\text{He}]2s^2 2p^3$	·N̈:	P	$[\text{Ne}]3s^2 3p^3$	·P̈:
O	$[\text{He}]2s^2 2p^4$	:Ö:	S	$[\text{Ne}]3s^2 3p^4$	:S̈:
F	$[\text{He}]2s^2 2p^5$	·F̈:	Cl	$[\text{Ne}]3s^2 3p^5$	·C̈l:
Ne	$[\text{He}]2s^2 2p^6$	:N̈e:	Ar	$[\text{Ne}]3s^2 3p^6$	:Är:

The Octet Rule

Atoms often gain, lose or share electrons to achieve the same number of electrons as the noble gas closest to them in the periodic table. The noble gases have very stable electron arrangements, as shown by their high ionisation energies, low affinity for additional electrons and general lack of chemical reactivity. Because all noble gases (except He) have eight valence electrons, many atoms undergoing reactions also end up with eight valence electrons. This observation has led to a guideline known as the **octet rule:** *Atoms tend to gain, lose or share electrons until they are surrounded by eight valence electrons.*

An *octet of electrons* consists of *full s and p subshells* in an atom. In terms of Lewis symbols, an octet can be thought of as four pairs of valence electrons arranged around the atom, as in the Lewis symbol for Ne in Table 8.1. There are many exceptions to the octet rule, but it provides a useful framework for introducing many important concepts of bonding.

8.2 | IONIC BONDING

When sodium metal, Na(s), is brought into contact with chlorine gas, $Cl_2(g)$, a violent reaction ensues (▼ **FIGURE 8.2**). The product of this very exothermic reaction is sodium chloride, NaCl(s):

$$\text{Na(s)} + \frac{1}{2}\text{Cl}_2(\text{g}) \longrightarrow \text{NaCl(s)} \quad \Delta_f H^\circ = -410.9 \text{ kJ} \quad [8.1]$$

where $\Delta_f H^\circ$ is the heat of formation of NaCl from its elements (∞ Section 14.7, 'Enthalpies of Formation'). Using Lewis symbols for the atoms, we can represent this reaction as an *electron transfer* from a sodium atom—a metal of low ionisation energy—to a chlorine atom—a non-metal of high electron affinity (Equation 8.2). This results in an ionic compound in which the Na^+ and Cl^- ions are arranged in a regular three-dimensional array, as shown in ▼ **FIGURE 8.3**.

∞ Find out more on page 523

$$\text{Na·} + \text{·C̈l:} \longrightarrow \text{Na}^+ + [\text{:C̈l:}]^- \quad [8.2]$$

Each ion has an octet of electrons, the octet on Na^+ being the $2s^2 2p^6$ electrons that lie below the single $3s$ valence electron of the Na atom. We've put a bracket around the chloride ion to emphasise that all eight electrons are located exclusively on the Cl^- ion.

CONCEPT CHECK 2

Describe the electron transfers that occur in the formation of magnesium fluoride from elemental magnesium and fluorine.

FIGURE IT OUT

Do you expect a similar reaction between potassium metal and elemental bromine?

▲ **FIGURE 8.2** **Reaction of sodium metal with chlorine gas to form the ionic compound sodium chloride.**

FIGURE IT OUT

If there was no indication of which colour sphere represented Na⁺ and which represented Cl⁻, is there a way for you to guess which is which?

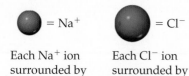

Each Na⁺ ion surrounded by six Cl⁻ ions

Each Cl⁻ ion surrounded by six Na⁺ ions

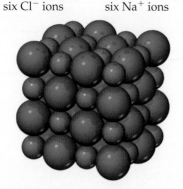

▲ **FIGURE 8.3** **The crystal structure of sodium chloride.**

∞ Review this on page 230

∞ Review this on page 235

Energetics of Ionic Bond Formation

As seen in Figure 8.2, and given by Equation 8.1, the reaction of sodium with chlorine is *very* exothermic; that is, energy is released as the reaction proceeds. In fact, Equation 8.1 is the reaction for the formation of NaCl(s) from its elements, so that the enthalpy change for the reaction is $\Delta_f H°$ for NaCl(s). In Appendix C we see that the heat of formation of other ionic substances is also quite negative. What factors make the formation of ionic compounds so exothermic?

In Equation 8.2 we represent the formation of NaCl as the transfer of an electron from Na to Cl. Recall from our discussion of ionisation energies, however, that the loss of electrons from an atom is always an endothermic process, which means energy is required for the reaction to proceed (∞ Section 7.4, 'Ionisation Energy').

Removing an electron from Na(g) to form Na⁺(g), for instance, requires 496 kJ mol⁻¹.

Conversely, when a non-metal gains an electron, the process is generally exothermic, as seen from the negative electron affinities of the elements (∞ Section 7.5, 'Electron Affinities').

Adding an electron to Cl(g), for example, releases 349 kJ mol⁻¹. If the transfer of an electron from one atom to another were the only factor in forming an ionic bond, the overall process would rarely be exothermic. So, removing an electron from Na(g) and adding it to Cl(g) is an endothermic process that requires $496 - 349 = 147$ kJ mol⁻¹. This endothermic process corresponds to the formation of sodium and chloride ions that are infinitely far apart—in other words, the positive energy change assumes that the ions are not interacting with one another, which is quite different from the situation in ionic solids.

The principal reason that ionic compounds are stable is the attraction between ions of opposite charge. This attraction draws the ions together, releasing energy and causing the ions to form a solid array, or lattice, such as that shown for NaCl in Figure 8.3. A measure of just how much stabilisation results from arranging oppositely charged ions in an ionic solid is given by the

lattice energy, which is *the energy required to separate completely a mole of a solid ionic compound into its gaseous ions.*

To get a picture of this process for NaCl, imagine that the structure shown in Figure 8.3 expands from within, so that the distances between the ions increase until the ions are very far apart. This process requires 788 kJ mol⁻¹, which is the value of the lattice energy:

$$NaCl(s) \longrightarrow Na^+(g) + Cl^-(g) \quad \Delta_{lattice} H = +788 \text{ kJ mol}^{-1} \quad \text{[8.3]}$$

Notice that this process is highly endothermic (⚭⚭ Section 14.2, 'The First Law of Thermodynamics').

⚭⚭ Find out more on page 505

The reverse process—the coming together of Na(g)⁺ and Cl(g)⁻ to form NaCl(s)—is therefore highly exothermic ($\Delta H = -788$ kJ mol⁻¹).

▼ TABLE 8.2 lists the lattice energies of NaCl and other ionic compounds. All are large positive values, indicating that the ions are strongly attracted to one another in these solids. The energy released by the attraction between ions of unlike charge more than makes up for the endothermic nature of ionisation energies, making the formation of ionic compounds an exothermic process. The strong attractions also cause most ionic materials to be hard and brittle, with high melting points—for example, NaCl melts at 801 °C.

The magnitude of the lattice energy of a solid depends on the charges of the ions, their sizes and their arrangement in the solid. We will see in Chapter 14 that the potential energy of two interacting charged particles is given by

$$E_{el} = \frac{kq_1q_2}{r^2} \quad \text{[8.4]}$$

In this equation q_1 and q_2 are the charges on the particles, r is the distance between their centres and k is a constant, 8.99×10^9 J m C⁻¹ (⚭⚭ Section 14.1, 'The Nature of Energy'). Equation 8.4 indicates that the attractive interaction between two oppositely charged ions increases as the magnitudes of their charges increase and as the distance between their centres decreases. Thus *for a given arrangement of ions, the lattice energy increases as the charges on the ions increase and as their radii decrease.* The magnitude of lattice energies depends primarily on the ionic charges because ionic radii do not vary over a very wide range. We will take a closer look at the structure of ionic solids in Chapter 11 (⚭⚭ Section 11.8, 'Structures of Solids').

⚭⚭ Find out more on page 502

⚭⚭ Find out more on page 405

TABLE 8.2 ● Lattice energies for some ionic compounds			
Compound	Lattice energy (kJ mol⁻¹)	Compound	Lattice energy (kJ mol⁻¹)
LiF	1030	MgCl₂	2326
LiCl	834	SrCl₂	2127
LiI	730		
NaF	910	MgO	3795
NaCl	788	CaO	3414
NaBr	732	SrO	3217
NaI	682		
KF	808	ScN	7547
KCl	701		
KBr	671		
CsCl	657		
CsI	600		

SAMPLE EXERCISE 8.1 **Magnitudes of lattice energies**

Without consulting Table 8.2, arrange the following ionic compounds in order of increasing lattice energy: NaF, CsI and CaO.

SOLUTION

Analyse We need to estimate the relative strengths of attractive forces between cations and anions. This will depend on the magnitude of the charges and sizes of the ions involved.

Plan We need to determine the charges and relative sizes of the ions in the compounds. We can then use Equation 8.4 qualitatively to determine the relative energies, knowing that the larger the ionic charges, the greater the energy and the further apart the ions are, the lower the energy.

Solve NaF consists of Na^+ and F^- ions, CsI of Cs^+ and I^- ions and CaO of Ca^{2+} and O^{2-} ions. Because the product of the charges, $q_1 q_2$, appears in the numerator of Equation 8.4, the lattice energy will increase dramatically when the charges of the ions increase. Thus we expect the lattice energy of CaO, which has 2+ and 2− ions, to be the greatest of the three.

The ionic charges in NaF and CsI are the same. As a result, the difference in their lattice energies will depend on the difference in the distance between the centres of the ions in their lattice. Because ionic size increases as we go down a group in the periodic table (Section 7.3), we know that Cs^+ is larger than Na^+ and I^- is larger than F^-. Therefore, the distance between the Na^+ and F^- ions in NaF will be less than the distance between the Cs^+ and I^- ions in CsI. As a result, the lattice energy of NaF should be greater than that of CsI. In order of increasing energy, therefore, we have CsI < NaF < CaO.

Check Table 8.2 confirms that our predicted order is correct.

PRACTICE EXERCISE

Which substance would you expect to have the greatest lattice energy, AgCl, CuO or CrN?

Answer: CrN. It has the highest charge on the ions.

(See also Exercises 8.18–8.20.)

Electron Configurations of Ions of the Main-Group Elements

We began considering the electron configurations of ions in Section 7.4. In light of our examination of ionic bonding, we continue with that discussion here. The energetics of ionic bond formation helps explain why many ions tend to have noble gas electron configurations. For example, when a sodium atom ionises it forms Na^+, which has the same electron configuration as Ne:

$$Na \qquad 1s^2 2s^2 2p^6 3s^1 = [Ne]3s^1$$

$$Na^+ \qquad 1s^2 2s^2 2p^6 \quad [Ne]$$

Even though lattice energy increases with increasing ionic charge, we never find ionic compounds that contain Na^{2+} ions. The second electron removed would have to come from an inner shell of the sodium atom, and removing electrons from an inner shell requires a very large amount of energy (∞ Section 7.4, 'Ionisation Energy').

∞ Review this on page 231

The increase in lattice energy is not enough to compensate for the energy needed to remove an inner-shell electron. Thus sodium and the other group 1 metals are found in ionic substances only as 1+ ions.

Similarly, the addition of electrons to non-metals is either exothermic or only slightly endothermic as long as the electrons are being added to the valence shell. Thus a Cl atom easily adds an electron to form Cl^-, which has the same electron configuration as Ar:

$$Cl \qquad 1s^2 2s^2 2p^6 3s^2 3p^5 = [Ne]3s^2 3p^5$$

$$Cl^- \qquad 1s^2 2s^2 2p^6 3s^2 3p^6 = [Ne]3s^2 3p^6 = [Ar]$$

In order to form a Cl^{2-} ion, the second electron would have to be added to the next higher shell of the Cl atom, which is energetically very unfavourable. Therefore, we never observe Cl^{2-} ions in ionic compounds.

Based on these concepts, we expect that ionic compounds of the main-group metals from groups 1, 2 and 13 will contain cations with charges of 1+, 2+ and 3+, respectively. Likewise, ionic compounds of the main-group non-metals of groups 15, 16 and 17 usually contain anions of charge 3–, 2– and 1–, respectively. Although we rarely find ionic compounds of the non-metals from group 14 (C, Si and Ge), the heaviest elements in group 14 (Sn and Pb) are metals and are usually found as 2+ cations in ionic compounds: Sn^{2+} and Pb^{2+}. This behaviour is consistent with the increasing metallic character found as we proceed down a group in the periodic table (∞ Section 7.6, 'Metals, Non-metals and Metalloids').

∞ Review this on page 236

SAMPLE EXERCISE 8.2 | **Charges on ions**

Predict the ion generally formed by **(a)** Sr, **(b)** S, **(c)** Al.

SOLUTION

Analyse The magnitude and sign of the charge on an ion is characteristic of the atom from which it has formed.

Plan We use the element's position in the periodic table to predict whether it will form an anion or a cation and then use its electron configuration to determine the charge on the ion that is likely to be formed.

Solve (a) Strontium is a metal in group 2 and will therefore form a cation. Its electron configuration is $[Kr]5s^2$, and so we expect that the two valence electrons can be lost easily to give an Sr^{2+} ion. **(b)** Sulfur is a non-metal in group 16 and will thus tend to be found as an anion. Its electron configuration ($[Ne]3s^23p^4$) is two electrons short of a noble gas configuration. Thus we expect that sulfur tends to form S^{2-} ions. **(c)** Aluminium is a metal in group 13. We therefore expect it to form Al^{3+} ions.

PRACTICE EXERCISE

Predict the charges on the ions formed when magnesium reacts with nitrogen.

Answer: Mg^{2+} and N^{3-}

(See also Exercises 8.14, 8.16, 8.17.)

Transition Metal Ions

Because ionisation energies increase rapidly for each successive electron removed, the lattice energies of ionic compounds are generally large enough to compensate for the loss of up to only three electrons from each atom. Thus we find cations with charges of 1+, 2+ or 3+ in ionic compounds. Most transition metals, however, have more than three electrons beyond a noble gas core. Silver, for example, has a $[Kr]4d^{10}5s^1$ electron configuration. Metals of group 11 (Cu, Ag, Au) often occur as 1+ ions (as in CuBr and AgCl). In forming Ag^+, the 5s electron is lost, leaving a completely filled 4d subshell. As in this example, transition metals generally do not form ions that have a noble gas configuration. The octet rule, although useful, is clearly limited in scope.

Recall from our discussion in Section 7.4 that, when a positive ion is formed from an atom, electrons are always lost first from the subshell with the largest value of n. Thus *in forming ions, transition metals lose the valence-shell s electrons first, then as many d electrons as are required to reach the charge of the ion.* Let's consider Fe, which has the electron configuration $[Ar]3d^64s^2$. In forming the Fe^{2+} ion, the two 4s electrons are lost, leading to an $[Ar]3d^6$ configuration. Removal of an additional electron gives the Fe^{3+} ion, whose electron configuration is $[Ar]3d^5$.

⚠ CONCEPT CHECK 3

Which element forms a 3+ ion that has the electron configuration $[Kr]4d^6$?

Polyatomic Ions

Several cations and many common anions are polyatomic. Examples include the ammonium ion, NH_4^+, and the carbonate ion, CO_3^{2-}. In polyatomic ions, two or more atoms are bound together by predominantly covalent bonds. They form a stable grouping that carries a charge, either positive or negative. We examine the covalent bonding forces in these ions in Chapter 9. For now, the important point to understand about any polyatomic ion is that the group of atoms as a whole acts as a charged species when the ion forms an ionic compound with an ion of opposite charge.

8.3 | COVALENT BONDING

The vast majority of chemical substances do not have the characteristics of ionic materials. Most of the substances with which we come into daily contact—such as water—tend to be gases and liquids, or solids with low melting points. Many, such as gasoline, vaporise readily. Many are pliable in their solid forms—for example, plastic bags and paraffin.

For the very large class of substances that do not behave like ionic substances, we need a different model for the bonding between atoms. G. N. Lewis reasoned that atoms might acquire a noble gas electron configuration by sharing electrons with other atoms. A chemical bond formed by sharing a pair of electrons is a *covalent bond*.

The hydrogen molecule, H_2, provides the simplest example of a covalent bond. As we bring two isolated hydrogen atoms towards one another, an attraction between the electron of one atom towards the nucleus of the other atom occurs. This attraction will increase as the atoms get closer together until it is balanced by repulsion between the two positively charged nuclei and repulsion between the two negatively charged electrons, as shown in ◀ **FIGURE 8.4**. Because the H_2 molecule exists as a stable entity, the attractive forces must exceed the repulsive ones. Why is this so?

By using quantum mechanical methods analogous to those employed for atoms, it is possible to calculate the distribution of electron density in molecules. Such a calculation for H_2 shows that the attractions between the nuclei and the electrons cause electron density to concentrate between the nuclei, as shown in Figure 8.4(b). As a result, the overall electrostatic interactions are attractive. Thus the atoms in H_2 are held together principally because the two nuclei are electrostatically attracted to the concentration of negative charge between them. In essence, the shared pair of electrons in any covalent bond acts as a kind of 'glue' to bind atoms together.

⚠ **CONCEPT CHECK 4**

Because it is less stable than two separated He atoms, the He_2 molecule has never been observed. What are the attractive forces in He_2? What are the repulsive forces in He_2? Which are greater, the attractive or the repulsive forces?

Lewis Structures

The formation of covalent bonds can be represented using Lewis symbols for the constituent atoms. The formation of the H_2 molecule from two H atoms, for example, can be represented as

$$H\cdot + \cdot H \longrightarrow \boxed{H\!:\!H}$$

In this way, each hydrogen atom acquires a second electron, achieving the stable, two-electron, noble gas electron configuration of helium.

The formation of a bond between two Cl atoms to give a Cl_2 molecule can be represented in a similar way:

$$:\ddot{C}l\cdot + \cdot\ddot{C}l: \longrightarrow :\ddot{C}l\!:\!\ddot{C}l:$$

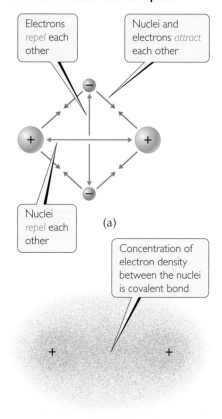

FIGURE IT OUT

What would happen to the magnitudes of the attractions and repulsions represented in (a) if the nuclei were further apart?

Electrons *repel* each other

Nuclei and electrons *attract* each other

Nuclei *repel* each other

(a)

Concentration of electron density between the nuclei is covalent bond

(b)

▲ **FIGURE 8.4 The covalent bond in H_2.**
(a) The attractions and repulsions between electrons and nuclei in the hydrogen molecule. (b) Electron distribution in the H_2 molecule. The concentration of electron density between the nuclei leads to a net attractive force that constitutes the covalent bond holding the molecule together.

By sharing the bonding electron pair, each chlorine atom has eight electrons (an octet) in its valence shell. It thus achieves the noble gas electron configuration of argon. Although many other combinations of chlorine atoms may be possible, they would not possess an octet of electrons around every atom and hence would not be predicted to be stable.

The structures shown here for H_2 and Cl_2 are called **Lewis structures** (or Lewis electron-dot structures). In writing Lewis structures, we usually show each electron pair shared between atoms as a line and the unshared electron pairs as dots. Written this way, the Lewis structures for H_2 and Cl_2 are:

$$H—H \qquad :\ddot{Cl}—\ddot{Cl}:$$

For the non-metals, the number of valence electrons in a neutral atom is the same as the group number minus 10. Therefore, one might predict that group 17 elements, such as F, would form one covalent bond to achieve an octet; group 16 elements, such as O, would form two covalent bonds; group 15 elements, such as N, would form three covalent bonds; and group 14 elements, such as C, would form four covalent bonds. These predictions are borne out in many compounds. For example, consider the simple hydrogen compounds of the non-metals of the second period of the periodic table:

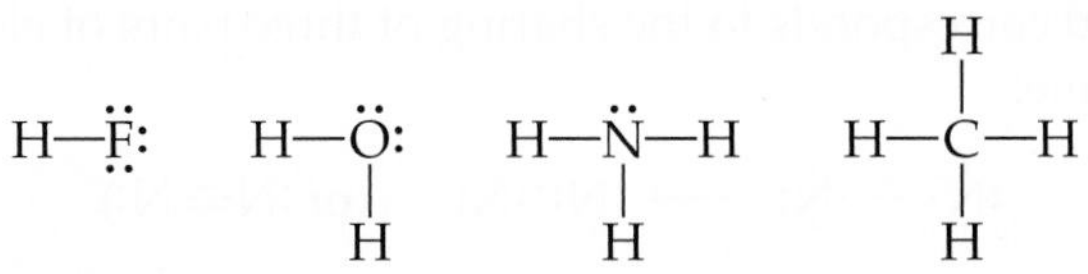

$$H—\ddot{\underset{..}{F}}: \qquad H—\overset{..}{\underset{|}{O}}: \qquad H—\overset{..}{\underset{|}{N}}—H \qquad H—\overset{\overset{\textstyle H}{|}}{\underset{\underset{\textstyle H}{|}}{C}}—H$$
$$\qquad\qquad\quad H \qquad\quad H$$

Thus the Lewis model succeeds in accounting for the compositions of many compounds of non-metals, in which covalent bonding predominates.

SAMPLE EXERCISE 8.3 Lewis structure of a compound

Given the Lewis symbols for the elements nitrogen and fluorine shown in Table 8.1, predict the formula of the stable binary compound (a compound composed of two elements) formed when nitrogen reacts with fluorine, and draw its Lewis structure.

SOLUTION

Analyse Here we are asked to identify a stable bonding arrangement of nitrogen with fluorine based on the electron configuration of the two atoms.

Plan Nitrogen has five valence electrons and fluorine has seven, so we need to find a combination of the two elements that results in an octet of electrons around each atom in the compound.

Solve Nitrogen requires three additional electrons to complete its octet, whereas fluorine requires only one. Sharing a pair of electrons between one N atom and one F atom will result in an octet of electrons for fluorine but not for nitrogen. We therefore need to figure out a way to get two more electrons for the N atom.

Nitrogen must share a pair of electrons with three fluorine atoms to complete its octet. Thus the Lewis structure for the resulting compound, NF_3, is

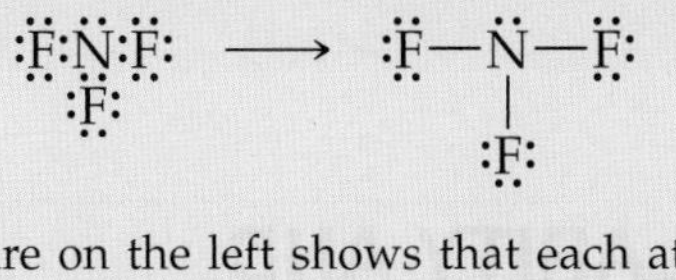

$$:\ddot{F}:\overset{..}{\underset{..}{N}}:\ddot{F}: \quad\longrightarrow\quad :\ddot{F}—\overset{..}{\underset{|}{N}}—\ddot{F}:$$
$$:\ddot{F}: \qquad\qquad\qquad :\ddot{F}:$$

Check The Lewis structure on the left shows that each atom is surrounded by an octet of electrons. Once you are accustomed to thinking of each line in a Lewis structure as representing *two* electrons, you can just as easily use the structure on the right to check for octets.

PRACTICE EXERCISE

Compare the Lewis symbol for neon with the Lewis structure for methane, CH_4. In what important way are the electron arrangements about neon and carbon alike? In what important respect are they different?

Answer: Both atoms have an octet of electrons about them. However, the electrons about neon are unshared electron pairs, whereas those about carbon are shared with four hydrogen atoms.

(See also Exercises 8.26, 8.27.)

Multiple Bonds

The sharing of a pair of electrons constitutes a single covalent bond, generally referred to simply as a **single bond**. In many molecules, atoms attain complete octets by sharing more than one pair of electrons. When two electron pairs are shared, two lines are drawn, representing a **double bond**. In carbon dioxide, for example, bonding occurs between carbon, with four valence electrons, and oxygen, with six:

$$:\dot{\underset{\cdot}{O}}: + \cdot\dot{\underset{\cdot}{C}}\cdot + :\dot{\underset{\cdot}{O}}: \longrightarrow \ddot{\underset{\cdot\cdot}{O}}::C::\ddot{\underset{\cdot\cdot}{O}} \qquad (\text{or } \ddot{\underset{\cdot\cdot}{O}}{=}C{=}\ddot{\underset{\cdot\cdot}{O}})$$

As the diagram shows, each oxygen acquires an octet of electrons by sharing two electron pairs with carbon. Carbon, on the other hand, acquires an octet of electrons by sharing two pairs with two oxygen atoms.

A **triple bond** corresponds to the sharing of three pairs of electrons, such as in the N_2 molecule:

$$:\dot{\underset{\cdot}{N}}\cdot + \cdot\dot{\underset{\cdot}{N}}: \longrightarrow :N:::N: \qquad (\text{or } :N{\equiv}N:)$$

Because each nitrogen atom possesses five electrons in its valence shell, three electron pairs must be shared to achieve the octet configuration.

The properties of N_2 are in complete accord with its Lewis structure. Nitrogen is a diatomic gas with exceptionally low reactivity that results from the very stable nitrogen–nitrogen bond. Study of the structure of N_2 reveals that the nitrogen atoms are separated by only 110 pm. The short N–N bond distance is a result of the triple bond between the atoms. From studies of the structure of many different substances in which nitrogen atoms share one or two electron pairs, we have learned that the average distance between bonded nitrogen atoms varies with the number of shared electron pairs:

N—N	N=N	N≡N
147 pm	124 pm	110 pm

As a general rule, the distance between bonded atoms decreases as the number of shared electron pairs increases. The distance between the nuclei of the atoms involved in a bond is called the **bond length** for the bond.

CONCEPT CHECK 5

The C–O bond length in carbon monoxide, CO, is 113 pm, whereas the C–O bond length in CO_2 is 124 pm. Without drawing a Lewis structure, do you think that carbon monoxide has a single, double or triple bond between the C and O atoms?

8.4 BOND POLARITY AND ELECTRONEGATIVITY

When two identical atoms bond, as in Cl_2 or N_2, the electron pairs must be shared equally. In ionic compounds, however, such as NaCl, there is essentially no sharing of electrons, which means that NaCl is best described as composed of Na^+ and Cl^- ions. The 3*s* electron of the Na atom is, in effect, transferred completely to chlorine. The bonds occurring in most covalent substances fall somewhere between these extremes.

The concept of **bond polarity** helps describe the sharing of electrons between atoms. A **non-polar covalent bond** is one in which the electrons are shared equally between two atoms, as in the Cl_2 and N_2 examples just cited. In a **polar covalent bond**, one of the atoms exerts a greater attraction for the bonding electrons than the other. If the difference in relative ability to attract electrons is large enough, an ionic bond is formed.

Electronegativity

We use a quantity called electronegativity to estimate whether a given bond will be non-polar covalent, polar covalent or ionic. **Electronegativity** is defined as the ability of an atom *in a molecule* to attract electrons to itself. The greater an atom's electronegativity, the greater its ability to attract electrons to itself. The electronegativity of an atom in a molecule is related to its ionisation energy and electron affinity, which are properties of isolated atoms.

The *ionisation energy* measures how strongly an atom holds on to its electrons (∞ Section 7.4, 'Ionisation Energy'). Likewise, the *electron affinity* is a measure of how strongly an atom attracts additional electrons. (∞ Section 7.5, 'Electron Affinities'). An atom with a very high electron affinity and high ionisation energy will attract electrons from other atoms and resist having its electrons attracted away; it will be highly electronegative.

Numerical estimates of electronegativity can be based on a variety of properties, not just ionisation energy and electron affinity. The first and most widely used electronegativity scale was developed by the American chemist Linus Pauling (1901–1994), who based his scale on thermochemical data. ▼ FIGURE 8.5 shows Pauling's electronegativity values for many of the elements. The values are unitless. Fluorine, the most electronegative element, has an electronegativity of 4.0. The least electronegative element, cesium, has an electronegativity of 0.7. The values for all other elements lie between these two extremes.

Within each period there is generally a steady increase in electronegativity from left to right; that is, from the most metallic to the most non-metallic elements. With some exceptions (especially within the transition metals), electronegativity decreases with increasing atomic number in any one group. This is what we might expect because we know that ionisation energies tend to decrease with increasing atomic number in a group and electron affinities

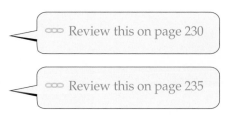

∞ Review this on page 230

∞ Review this on page 235

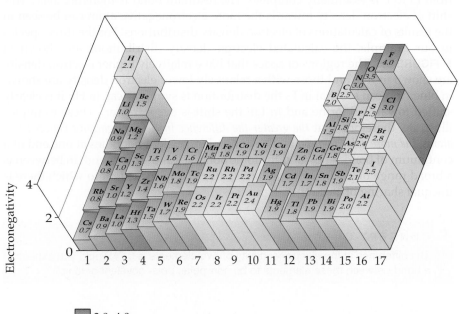

3.0–4.0
2.0–2.9
1.5–1.9
<1.5

◄ **FIGURE 8.5 Electronegativities of the elements.** Electronegativity generally increases from left to right across a period and decreases from top to bottom down a group.

don't change very much. You do not need to memorise numerical values for electronegativity. Instead, you should know the periodic trends so that you can predict which of two elements is more electronegative.

CONCEPT CHECK 6
How does the *electronegativity* of an element differ from its *electron affinity*?

Electronegativity and Bond Polarity

We can use the difference in electronegativity between two atoms to gauge the polarity of the bonding between them. Consider these three fluorine-containing compounds:

Compound	F_2	HF	LiF
Electronegativity difference	$4.0 - 4.0 = 0$	$4.0 - 2.1 = 1.9$	$4.0 - 1.0 = 3.0$
Type of bond	Non-polar covalent	Polar covalent	Ionic

In F_2, the electrons are shared equally between the fluorine atoms, and thus the covalent bond is *non-polar*.

In HF the fluorine atom has a greater electronegativity than the hydrogen atom, with the result that the sharing of electrons is unequal—the bond is polar. In HF the more electronegative fluorine atom attracts electron density away from the less electronegative hydrogen atom, leaving a partial positive charge on the hydrogen atom and a partial negative charge on the fluorine atom. We can represent this charge distribution as

$$\overset{\delta+}{H} - \overset{\delta-}{F}$$

The $\delta+$ and $\delta-$ (read 'delta plus' and 'delta minus') symbolise the partial positive and negative charges, respectively.

In LiF the electronegativity difference is very large, meaning that the electron density is shifted far towards F. In the three-dimensional structure of LiF, analogous to that shown for NaCl in Figure 8.3, the transfer of electronic charge from Li to F is essentially complete. The resultant bond is therefore *ionic*. This shift of electron density towards the more electronegative atom can be seen in the results of calculations of electron density distributions. For the three species in our example, the calculated electron density distributions are shown in ▶ FIGURE 8.6. The regions of space that have relatively higher electron density are shown in red, and those with a relatively lower electron density are shown in blue. You can see that in F_2 the distribution is symmetrical, in HF it is clearly shifted towards fluorine and in LiF the shift is even greater.[*] These examples illustrate, therefore, that *the greater the difference in electronegativity between two atoms, the more polar their bond*. The non-polar covalent bond lies at one end of a continuum of bond types, and the ionic bond lies at the other end. In between is a broad range of polar covalent bonds, differing in the extent to which there is unequal sharing of electrons.

CONCEPT CHECK 7
The difference in the electronegativity of two elements is 0.7. Would you expect a bond between these elements to be non-polar, polar covalent or ionic?

---

[*] The calculated electron density distribution for LiF is for an isolated LiF 'molecule', not the ionic solid. Although the bond in this isolated diatomic system is very polar, it is not 100% ionic, as is the bonding in solid lithium fluoride. The solid state promotes a more complete electron transfer from Li to F because of the stabilising effects of the ionic lattice.

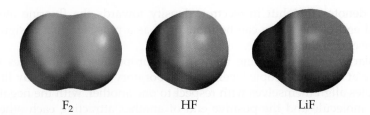

F_2 HF   LiF

▲ **FIGURE 8.6** **Electron density distribution.** This computer-generated rendering shows the calculated electron density distribution on the surface of the F_2, HF and LiF molecules. The regions of relatively low electron density (net positive charge) appear blue, those of relatively high electron density (net negative charge) appear red and regions that are close to electrically neutral appear green.

SAMPLE EXERCISE 8.4 | **Bond polarity**

In each case, which bond is more polar: **(a)** B–Cl, or C–Cl, **(b)** P–F or P–Cl? Indicate in each case which atom has the partial negative charge.

SOLUTION

Analyse We need to identify which one of a pair of bonds is the more polar and indicate the relatively negative end of the bond.

Plan A polar bond exists whenever two atoms of different electronegativities are bonded together. As we are not asked for quantitative answers, we can use the periodic table and our knowledge of electronegativity trends to answer the question. Alternatively, we can get the electronegativity values for all the atoms involved from Figure 8.5.

Solve **(a)** Using the periodic table—because boron is to the left of carbon in the periodic table, we predict that boron has the lower electronegativity. Chlorine, being on the right side of the table, has a higher electronegativity. The more polar bond will be the one between the atoms having the lowest electronegativity (boron) and the highest electronegativity (chlorine).

Using Figure 8.5—the difference in the electronegativities of chlorine and boron is $3.0 - 2.0 = 1.0$; the difference between chlorine and carbon is $3.0 - 2.5 = 0.5$. Consequently, the B–Cl bond is more polar; the chlorine atom carries the partial negative charge because it has a higher electronegativity.

(b) The electronegativities are P = 2.1, F = 4.0, Cl = 3.0. Consequently, the P–F bond will be more polar than the P–Cl bond. The fluorine atom carries the partial negative charge. We reach the same conclusion by noting that fluorine is above chlorine in the periodic table, and so fluorine should be more electronegative and will form the more polar bond with P.

PRACTICE EXERCISE

Which of the following bonds is most polar: S–Cl, S–Br, Se–Cl, Se–Br?

Answer: Se–Cl

(See also Exercises 8.32, 8.33.)

Dipole Moments

The difference in electronegativity between H and F leads to a polar covalent bond in the HF molecule. As a consequence, there is a concentration of negative charge on the more electronegative F atom, leaving the less electronegative H atom at the positive end of the molecule. A molecule such as HF, in which the centres of positive and negative charge do not coincide, is said to be a **polar molecule**. Thus we describe not only bonds as being polar and non-polar but also entire molecules.

We can indicate the polarity of the HF molecule in two ways:

$$\overset{\delta+}{H} - \overset{\delta-}{F} \quad \text{or} \quad \overset{\longmapsto}{H - F}$$

Recall from the preceding subsection that $\delta+$ and $\delta-$ indicate the partial positive and negative charges on the H and F atoms. In the notation on the right, the

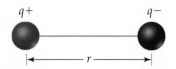

▲ **FIGURE 8.7   Dipole and dipole moment.** When charges of equal magnitude and opposite sign $q+$ and $q-$ are separated by a distance r, a dipole is produced.

arrow denotes the shift in electron density towards the fluorine atom. The crossed end of the arrow can be thought of as a plus sign that designates the positive end of the molecule.

Polarity helps determine many of the properties of substances that we observe at the macroscopic level, in the laboratory and in everyday life. Polar molecules align themselves with respect to one another, with the negative end of one molecule and the positive end of another attracting each other. Polar molecules are likewise attracted to ions. The negative end of a polar molecule is attracted to a positive ion, and the positive end is attracted to a negative ion. These interactions account for many properties of liquids, solids and solutions.

How can we quantify the polarity of a molecule? Whenever two electrical charges of equal magnitude but opposite sign are separated by a distance, a **dipole** is established. The quantitative measure of the magnitude of a dipole is called its **dipole moment**, denoted μ. If two equal and opposite charges, $q+$ and $q-$, are separated by a distance, r, the magnitude of the dipole moment is the product of q and r (◄ **FIGURE 8.7**):

$$\mu = qr \qquad [8.5]$$

The dipole moment increases as the magnitude of charge that is separated increases and as the distance between the charges increases. For a non-polar molecule, such as F_2, the dipole moment is zero because there is no charge separation.

⚠ **CONCEPT CHECK 8**

The molecules chlorine monofluoride, ClF, and iodine monofluoride, IF, are examples of *interhalogen* compounds—compounds that contain bonds between different halogen elements. Which of these molecules will have the larger dipole moment?

Dipole moments are usually reported in *debyes* (D), a unit that equals 3.34×10^{-30} coulomb metres (C m). For molecules, we usually measure charge in units of the electronic charge e, 1.60×10^{-19} C, and distance in units of metres. Suppose that two charges, $1+$ and $1-$ (in units of e), are separated by a distance of exactly 100 pm (100×10^{-12} m). The dipole moment produced is

$$\mu = qr = (1.60 \times 10^{-19} \text{ C})(100 \times 10^{-12} \text{ m}) = 1.60 \times 10^{-29} \text{ C m}$$

Since
$$1 \text{ D} = 3.34 \times 10^{-30} \text{ C m}$$

Then
$$\mu = \frac{1.60 \times 10^{-29} \text{ C m}}{3.34 \times 10^{-30} \text{ C m D}^{-1}} = 4.479 \text{ D}$$

Measurement of the dipole moments can provide us with valuable information about the charge distributions in molecules, as illustrated in Sample Exercise 8.5.

SAMPLE EXERCISE 8.5   **Dipole moments of diatomic molecules**

The bond length in the HCl molecule is 127 pm. **(a)** Calculate the dipole moment, in debyes, that would result if the charges on the H and Cl atoms were $1+$ and $1-$, respectively. **(b)** The experimentally measured dipole moment of HCl(g) is 1.08 D. What magnitude of charge, in units of e, on the H and Cl atoms would lead to this dipole moment?

SOLUTION

Analyse We calculate the dipole moment of HCl assuming a full charge on each of the two atoms then compare this with the experimentally determined dipole moment and determine the actual charge on the atoms.

Plan In both cases we use Equation 8.5.

Solve (a) The charge on each atom is the electronic charge, $e = 1.60 \times 10^{-19}$ C. The separation is 127 pm. The dipole moment is therefore

$$\mu = qr = (1.60 \times 10^{-19} \text{ C})(127 \times 10^{-12} \text{ m}) = 2.03 \times 10^{-29} \text{ C m}$$

since
$$1 \text{ C m} = \frac{1 \text{ D}}{3.34 \times 10^{-30}}$$

$$\mu = \frac{2.03 \times 10^{-29} \text{ D}}{3.34 \times 10^{-30}} = 6.08 \text{ D}$$

(b) We know the value of μ, 1.08 D, and the value of r, 127 pm, and we want to calculate the value of q. So, rearranging Equation 8.5:

$$q = \frac{\mu}{r} = \frac{1.08 \text{ D}}{127 \times 10^{-12} \text{ m}} = \frac{1.08 \times 3.34 \times 10^{-30} \text{ C m}}{127 \times 10^{-12} \text{ m}}$$

$$= 2.84 \times 10^{-20} \text{ C}$$

We can readily convert this charge to units of e:

$$\text{Charge in } e = \frac{2.84 \times 10^{-20} \text{ C}}{1.60 \times 10^{-19} \text{ C } e^{-1}} = 0.178 \, e$$

Thus the experimental dipole moment indicates that the charge separation in the HCl molecule is

$$\overset{0.178+}{\text{H}} — \overset{0.178-}{\text{Cl}}$$

Because the experimental dipole moment is less than that calculated in part (a), the charges on the atoms are less than a full electronic charge. We could have anticipated this because the H–Cl bond is polar covalent rather than ionic.

PRACTICE EXERCISE

The dipole moment of chlorine monofluoride, ClF, is 0.88 D. The bond length of the molecule is 163 pm. **(a)** Which atom is expected to have the partial negative charge? **(b)** What is the charge on that atom, in units of e?

Answers: (a) F, **(b)** 0.11 e

(See also Exercises 8.34, 8.35.)

▼ **TABLE 8.3** presents the bond lengths and dipole moments of the hydrogen halides. Notice that, as we proceed from HF to HI, the electronegativity difference decreases and the bond length increases. The first effect decreases the amount of charge separated and causes the dipole moment to decrease from HF to HI, even though the bond length is increasing. We can 'observe' the varying degree of electronic charge shift in these substances from computer-generated renderings based on calculations of electron distribution, as shown in ▼ **FIGURE 8.8**. For these molecules, the change in the electronegativity difference has a greater effect on the dipole moment than does the change in bond length.

 CONCEPT CHECK 9

How do you interpret the fact that there is no red in the HBr and HI representations in Figure 8.8?

Differentiating Ionic and Covalent Bonding

To understand the interactions responsible for chemical bonding, it is advantageous to treat ionic and covalent bonding separately. That is the approach taken in this chapter, as well as in most other undergraduate-level chemistry texts. In reality, however, there is a continuum between the extremes of ionic and covalent bonding. This lack of a well-defined separation between the two types of bonding may seem unsettling or confusing at first.

TABLE 8.3 • **Bond lengths, electronegativity differences and dipole moments of the hydrogen halides**

Compound	Bond length (pm)	Electronegativity difference	Dipole moment (D)
HF	92.0	1.9	1.82
HCl	127	0.9	1.08
HBr	141	0.7	0.82
HI	161	0.4	0.44

▲ FIGURE IT OUT

Does the change in electronegativity difference or the change in bond strength have the greater effect on bond polarisation in the series?

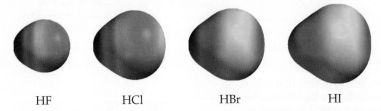

HF   HCl   HBr   HI

▲ **FIGURE 8.8 Charge separation in the hydrogen halides.** In HF, the strongly electronegative F pulls much of the electron density away from H. In HI, the I, being much less electronegative than F, does not attract the shared electrons as strongly and consequently there is far less polarisation of the bond.

The simple models of ionic and covalent bonding presented in this chapter go a long way towards understanding and predicting the structures and properties of chemical compounds. When covalent bonding is dominant, more often than not we expect compounds to exist as molecules,* having all the properties we associate with molecular substances, such as relatively low melting and boiling points and non-electrolyte behaviour when dissolved in water. When ionic bonding is dominant, we expect the compounds to be brittle, high-melting solids with extended lattice structures and exhibit strong electrolyte behaviour when dissolved in water.

There are, of course, exceptions to these general characterisations, some of which we examine later in the text. Nonetheless, the ability to quickly categorise the predominant bonding interactions in a substance as covalent or ionic imparts considerable insight into the properties of that substance. The question then becomes the best way to recognise which type of bonding dominates.

The simplest approach is to assume that the interaction between a metal and a non-metal is ionic and that between two non-metals is covalent. Although this classification scheme is reasonably predictive, there are far too many exceptions to use it blindly. For example, tin is a metal and chlorine is a non-metal, but $SnCl_4$ is a molecular substance that exists as a colourless liquid at room temperature. It freezes at $-33\,°C$ and boils at $114\,°C$. Clearly this substance does not have the characteristics of an ionic substance. A more sophisticated approach is to use the difference in electronegativity as the main criterion for determining whether ionic or covalent bonding will be dominant. This approach correctly predicts the bonding in $SnCl_4$ to be polar covalent based on an electronegativity difference of 1.2 and at the same time correctly predicts the bonding in NaCl to be predominantly ionic based on an electronegativity difference of 2.1.

Evaluating bonding based on electronegativity difference is a useful system, but it has one shortcoming. The electronegativity values given in Figure 8.5 do not take into account changes in bonding that accompany changes in the oxidation state of the metal. For example, Figure 8.5 gives the electronegativity difference between manganese and oxygen as $3.5 - 1.5 = 2.0$, which falls in the range where the bonding is normally considered ionic (the electronegativity difference for NaCl is $3.0 - 0.9 = 2.1$). Therefore, it is not surprising to learn that manganese(II) oxide, MnO, is a green solid that melts at $1842\,°C$ and has the same crystal structure as NaCl.

---

* There are some exceptions to this statement, such as network solids, including diamond, silicon, and germanium, where an extended structure is formed even though the bonding is clearly covalent.

However, the bonding between manganese and oxygen is not always ionic. Manganese(VII) oxide, Mn_2O_7, is a green liquid that freezes at 5.9 °C, which indicates that covalent rather than ionic bonding dominates. The change in the oxidation state of manganese is responsible for the change in bonding. In general, as the oxidation state of a metal increases, so does the degree of covalent bonding. When the oxidation state of the metal is highly positive (roughly speaking, +4 or larger), we should expect significant covalency in the bonds it forms with non-metals. Thus with metals in high oxidation states we find molecular substances, such as Mn_2O_7, or polyatomic ions, such as MnO_4^- and CrO_4^{2-}, rather than ionic compounds.

 CONCEPT CHECK 10

You have a yellow solid that melts at 41 °C and boils at 131 °C and a green solid that melts at 2320 °C. If you are told that one of them is Cr_2O_3 and the other is OsO_4, which one do you expect to be the yellow solid?

8.5 | DRAWING LEWIS STRUCTURES

Lewis structures can help us understand the bonding in many compounds and are frequently used when discussing the properties of molecules. For this reason, drawing Lewis structures is an important skill that you should practise. To do so, you should follow a regular procedure. First, we'll outline the procedure, and then we'll go through several examples.

1. **Sum the valence electrons from all atoms**. (Use the periodic table as necessary to help you determine the number of valence electrons in each atom.) For an anion, add one electron to the total for each negative charge. For a cation, subtract one electron from the total for each positive charge. Don't worry about keeping track of which electrons come from which atoms. Only the total number is important.

2. **(a) Select a central atom. (b) Arrange the other atoms around the central atom and connect them with a single bond** (a dash, representing *two* electrons). Chemical formulae are often written in the order in which the atoms are connected in the molecule or ion; the formula HCN, for example, tells you that the carbon atom is bonded to the H and to the N. When a central atom has a group of other atoms bonded to it, the central atom is usually written first, as in CO_3^{2-} and SF_4. It also helps to remember that the central atom is generally less electronegative than the atoms surrounding it. In other cases you may need more information before you can draw the Lewis structure.

3. **Complete the octets around all the atoms bonded to the central atom**. (Remember, however, that you use only a single pair of electrons around hydrogen.)

4. **Place any leftover electrons on the central atom**, even if doing so results in more than an octet of electrons around the atom. In Section 8.7 we discuss molecules that don't adhere to the octet rule.

5. **If there are not enough electrons to give the central atom an octet, try multiple bonds**. Use one or more of the unshared pairs of electrons on the atoms bonded to the central atom to form double or triple bonds.

SAMPLE EXERCISE 8.6 | **Drawing Lewis structures**

Draw the Lewis structure for phosphorus trichloride, PCl_3.

SOLUTION

Analyse Drawing the Lewis structure of PCl_3 will enable us to determine a stable bonding arrangement for this molecule.

Plan Determine the total number of valence electrons, bond the atoms together and arrange the remaining electrons to achieve an octet of electrons on each atom.

Solve

First, we sum the valence electrons. Phosphorus (group 15) has five valence electrons and each chlorine (group 17) has seven. The total number of valence electrons is therefore

$$5 + (3 \times 7) = 26$$

Second, we arrange the atoms to show which atom is connected to which, and draw a single bond between them. There are various ways the atoms might be arranged. In binary (two-element) compounds, however, the first element listed in the chemical formula is generally surrounded by the remaining atoms. Thus we begin with a skeleton structure that shows a single bond between the phosphorus atom and each chlorine atom:

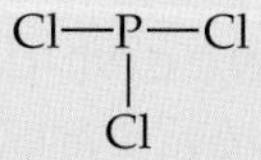

$$\text{Cl—P—Cl} \atop \;\;\;\;\;|\;\;\;\;\; \atop \;\;\;\text{Cl}\;\;\;$$

(It is not crucial to place the atoms in exactly this arrangement.)

Third, we complete the octets on the atoms bonded to the central atom. Placing octets around each Cl atom accounts for 24 electrons (remember, each line in our structure represents *two* electrons):

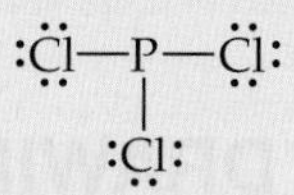

Fourth, we place the remaining two electrons on the central atom, completing the octet around it:

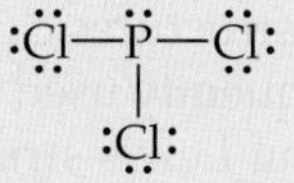

This structure gives each atom an octet, so we stop at this point. (Remember that in achieving an octet, the bonding electrons are counted for both atoms.)

PRACTICE EXERCISE

(a) How many valence electrons should appear in the Lewis structure for CH_2Cl_2?
(b) Draw the Lewis structure.

Answers: **(a)** 20, **(b)**

$$:\!\ddot{\text{Cl}}\!\!-\!\!\text{C}\!\!-\!\!\ddot{\text{Cl}}\!:$$

(See also Exercises 8.38, 8.39.)

SAMPLE EXERCISE 8.7   **Lewis structures with multiple bonds**

Draw the Lewis structure for HCN.

SOLUTION

Analyse Drawing the Lewis structure of HCN will enable us to determine a stable bonding arrangement for this molecule.

Plan Determine the total number of valence electrons, bond the atoms together and arrange the remaining electrons to achieve an octet of electrons around carbon and nitrogen.

Solve Hydrogen has one valence electron, carbon (group 14) has four and nitrogen (group 15) has five. The total number of valence electrons is therefore $1 + 4 + 5 = 10$. In principle, there are different ways in which we might choose to arrange the atoms. Because hydrogen can accommodate only one electron pair, it always has only one single bond associated with it in any compound. Therefore, C–H–N is an impossible arrangement. The remaining two possibilities are H–C–N and H–N–C The first is the arrangement found experimentally. You might have guessed this to be the atomic arrangement because the formula is written with the atoms in this order. Thus, we begin with a skeleton structure that shows single bonds between hydrogen, carbon and nitrogen:

$$\text{H—C—N}$$

These two bonds account for four electrons. If we then place the remaining six electrons around N to give it an octet, we do not achieve an octet on C:

$$H—C—\ddot{\underset{..}{N}}:$$

We therefore try a double bond between C and N, using one of the unshared pairs of electrons we placed on N. Again, there are fewer than eight electrons on C, and so we next try a triple bond. This structure gives an octet around both C and N:

$$H—C\overset{\frown}{\underset{\smile}{+}}\ddot{N}: \longrightarrow H—C\equiv N:$$

We see that the octet rule is satisfied for the C and N atoms, and the H atom has two electrons around it, so this appears to be a correct Lewis structure.

PRACTICE EXERCISE

Draw the Lewis structure for **(a)** NO^+ ion, **(b)** C_2H_4.

Answers: **(a)** $[:N\equiv O:]^+$, **(b)** 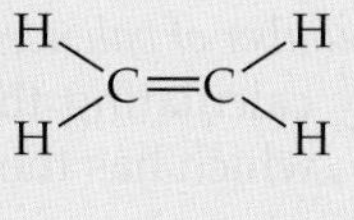
$$\begin{array}{c} H \\ \diagdown \\ H \end{array} C=C \begin{array}{c} H \\ \diagup \\ H \end{array}$$

(See also Exercise 8.42.)

SAMPLE EXERCISE 8.8   **Lewis structure for a polyatomic ion**

Draw the Lewis structure for the BrO_3^- ion.

SOLUTION

Analyse Drawing the Lewis structure of BrO_3^- will enable us to determine a stable bonding arrangement for this polyatomic ion.

Plan Determine the total number of valence electrons, allowing for the overall charge on the ion, bond the atoms together and arrange the remaining electrons to achieve an octet of electrons on each atom.

Solve Bromine (group 17) has seven valence electrons, and oxygen (group 16) has six. We must now add one more electron to our sum to account for the $1-$ charge of the ion. The total number of valence electrons is therefore $7 + (3 \times 6) + 1 = 26$. For oxyanions—$BrO_3^-$, SO_4^{2-}, NO_3^-, CO_3^{2-} and so forth—the oxygen atoms surround the central non-metal atoms. After following this format and then putting in the single bonds and distributing the unshared electron pairs, we have

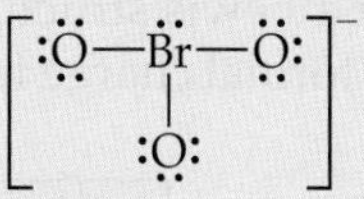

$$\left[:\ddot{\underset{..}{O}}—\ddot{Br}—\ddot{\underset{..}{O}}: \\ \quad \mid \\ \quad :\ddot{\underset{..}{O}}: \right]^-$$

Notice here and elsewhere that the Lewis structure for an ion is written in brackets with the charge shown outside the brackets at the upper right.

PRACTICE EXERCISE

Draw the Lewis structure for **(a)** ClO_2^- ion, **(b)** PO_4^{3-} ion.

Answers: **(a)** $\left[:\ddot{\underset{..}{O}}—\ddot{\underset{..}{Cl}}—\ddot{\underset{..}{O}}: \right]^-$ **(b)** $\left[\begin{array}{c} :\ddot{O}: \\ \mid \\ :\ddot{\underset{..}{O}}—P—\ddot{\underset{..}{O}}: \\ \mid \\ :\ddot{\underset{..}{O}}: \end{array} \right]^{3-}$

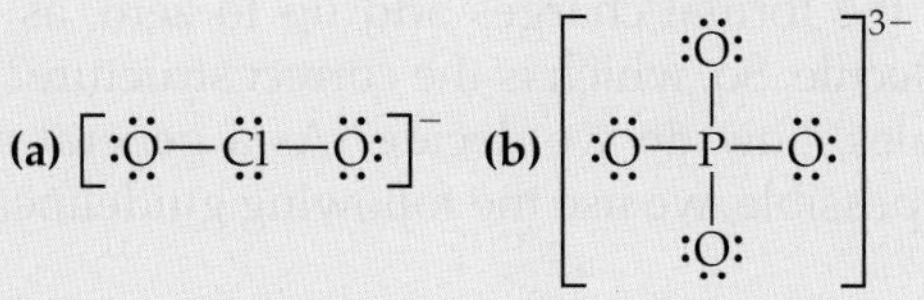

(See also Exercises 8.43–8.45.)

Formal Charge

When we draw a Lewis structure we are describing how the electrons are distributed in a molecule (or polyatomic ion). In some instances we can draw several different Lewis structures that all obey the octet rule. How do we decide which one is the most reasonable? One approach is to do some 'bookkeeping' of the valence electrons to determine the formal charge of each atom in each

Lewis structure. The **formal charge** of any atom in a molecule is the charge the atom would have if all the atoms in the molecule had the same electronegativity (that is, if each bonding electron pair in the molecule were shared equally between its two atoms). The Lewis structure that most closely represents the actual structure usually corresponds to the one that gives the smallest formal charges.

To calculate the formal charge on any atom in a Lewis structure, we assign the electrons to the atom as follows.

1. *All* unshared (non-bonding) electrons are assigned to the atom on which they are found.

2. For any bond—single, double or triple—*half* the bonding electrons are assigned to each atom in the bond.

The formal charge of each atom is then calculated *by subtracting the number of electrons assigned to the atom from the number of valence electrons in the isolated atom.*

Let's illustrate this procedure by calculating the formal charges on the C and N atoms in the cyanide ion, CN^-, which has the Lewis structure

$$[:C\equiv N:]^-$$

For the C atom, there are two non-bonding electrons and three electrons from the six in the triple bond ($\frac{1}{2} \times 6 = 3$), giving a total of five. The number of valence electrons on a neutral C atom is four. Thus the formal charge on C is $4 - 5 = -1$. For N, there are two non-bonding electrons and three electrons from the triple bond. Because the number of valence electrons on a neutral N atom is five, its formal charge is $5 - 5 = 0$. Thus the formal charges on the atoms in the Lewis structure of CN^- are

$$[:\overset{-1}{C}\equiv\overset{0}{N}:]^-$$

Notice that the sum of the formal charges equals the overall charge on the ion, $1-$. In general, the formal charges on a neutral molecule add up to zero, whereas those on an ion add up to give the overall charge on the ion.

The concept of formal charge can help us choose between alternative Lewis structures. We show how this is done by considering the CO_2 molecule. As shown in Section 8.3, CO_2 is represented as having two double bonds. The octet rule is also obeyed, however, in a Lewis structure having one single bond and one triple bond. Calculating the formal charge for each atom in these structures, we have

	$\ddot{O}=C=\ddot{O}$			$:\ddot{O}-C\equiv O:$		
Valence electrons:	6	4	6	6	4	6
−(Electrons assigned to atom):	6	4	6	7	4	5
Formal charge:	0	0	0	−1	0	+1

Note that in both cases the formal charges add up to zero, as they must because CO_2 is a neutral molecule. So, which is the correct structure? With both choices following all our rules, how do we decide? As a general rule, when several Lewis structures are possible, we use the following guidelines to choose the most correct one.

1. *We generally choose the Lewis structure in which the atoms bear formal charges closest to zero.*

2. *We generally choose the Lewis structure in which any negative charges reside on the more electronegative atoms.*

3. *We generally choose the Lewis structure in which any positive charges reside on the less electronegative atoms.*

Thus the first Lewis structure of CO_2 is preferred because the atoms carry no formal charges and so satisfy the first guideline.

Although the concept of formal charge helps us choose between alternative Lewis structures, it is very important that you remember that *formal charges do not represent real charges on atoms*. These charges are just a bookkeeping convention. The actual charge distributions in molecules and ions are determined not by formal charges but by the electronegativity differences between atoms.

CONCEPT CHECK 11

Suppose that a Lewis structure for a neutral fluorine-containing molecule results in a formal charge on the fluorine atom of +1. What conclusion would you draw?

SAMPLE EXERCISE 8.9   **Lewis structures and formal charges**

The following are three possible Lewis structures for the thiocyanate ion, NCS^-:

$$[:\ddot{N}—C≡S:]^- \qquad [\ddot{N}=C=\ddot{S}]^- \qquad [:N≡C—\ddot{S}:]^-$$

(a) Determine the formal charges of the atoms in each structure. (b) Which Lewis structure is the preferred one?

SOLUTION

Analyse We need to assign a formal charge to each atom of each structure and use them to predict which of the three valid Lewis structures is most likely to represent the actual structure.

Plan The formal charge is the difference between the number of valence electrons of an atom and the number of atoms assigned to it in the structure. The preferred Lewis structure is the one in which the formal charges are closest to zero and any negative charge is on the most electronegative atom.

Solve (a) Neutral N, C and S atoms have five, four and six valence electrons, respectively. We can determine the following formal charges in the three structures by using the rules discussed above:

$$\begin{array}{ccc} -2 \quad 0 \quad +1 & -1 \quad 0 \quad 0 & 0 \quad 0 \quad -1 \\ [:\ddot{N}—C≡S:]^- & [\ddot{N}=C=\ddot{S}]^- & [:N≡C—\ddot{S}:]^- \end{array}$$

As they must, the formal charges in all three structures sum to $1-$, the overall charge of the ion.

(b) We now use the guidelines for the best Lewis structure to determine which of the three structures is the most correct. As discussed in Section 8.4, N is more electronegative than C or S. Therefore, we expect that any negative formal charge will reside on the N atom (guideline 2). Further, we usually choose the Lewis structure that produces the formal charges of smallest magnitude (guideline 1). For these two reasons, the middle structure is the preferred Lewis structure of the NCS^- ion.

PRACTICE EXERCISE

The cyanate ion (NCO^-), like the thiocyanate ion, has three possible Lewis structures. (a) Draw these three Lewis structures, and assign formal charges to the atoms in each structure. (b) Which Lewis structure is the preferred one?

$$\begin{array}{ccc} -2 \quad 0 \quad +1 & -1 \quad 0 \quad 0 & 0 \quad 0 \quad -1 \\ \textbf{\textit{Answers: (a)}} \ [:\ddot{N}—C≡O:]^- & [\ddot{N}=C=\ddot{O}]^- & [:N≡C—\ddot{O}:]^- \\ \text{(i)} & \text{(ii)} & \text{(iii)} \end{array}$$

(b) Structure (iii), which places a negative charge on oxygen, the most electronegative of the three elements, is the preferred Lewis structure

(See also Exercises 8.40, 8.41.)

A CLOSER LOOK

OXIDATION NUMBERS, FORMAL CHARGES AND ACTUAL PARTIAL CHARGES

In Chapter 4 we introduced the rules for assigning *oxidation numbers* to atoms. The oxidation number of an atom is the charge it would have if its bonds were completely ionic. That is, in determining the oxidation number, all shared electrons are counted with the more electronegative atom. For example, consider the Lewis structure of HCl shown in ▼ FIGURE 8.9(a). To assign oxidation numbers, the pair of electrons in the covalent bond between the atoms is assigned to the more electronegative Cl atom. This procedure gives Cl eight valence-shell electrons, one more than the neutral atom. Thus it is assigned an oxidation number of −1. Hydrogen has no valence electrons when they are counted this way, giving it an oxidation number of +1.

In this section we have just considered another way of counting electrons that gives rise to *formal charges*. The formal charge is assigned by completely ignoring electronegativity and assigning the electrons in bonds equally between the bonded atoms. Consider again the HCl molecule, but this time divide the bonding pair of electrons equally between H and Cl, as shown in Figure 8.9(b). In this case Cl has seven assigned electrons, the same as that of the neutral Cl atom. Thus, the formal charge of Cl in this compound is 0. Likewise, the formal charge of H is also 0.

Neither the oxidation number nor the formal charge gives an accurate depiction of the actual charges on atoms. Oxidation numbers overstate the role of electronegativity, and formal charges ignore it completely. It seems reasonable that electrons in covalent bonds should be apportioned according to the relative electronegativities of the bonded atoms. From Figure 8.5 we see that Cl has an electronegativity of 3.0 whereas that of H is 2.1. The more electronegative Cl atom might therefore be expected to have roughly $3.0/(3.0 + 2.1) = 0.59$ of the electrical charge in the bonding pair, whereas the H atom has $2.1/(3.0 + 2.1) = 0.41$ of the charge. Because the bond consists of two electrons, the Cl atom's share is $0.59 \times 2\,e = 1.18\,e$, or $0.18e$ more than the neutral Cl atom. This gives rise to a partial charge of 0.18− on Cl and 0.18+ on H (notice again that we place the + and − signs before the magnitude when speaking about oxidation numbers and formal charges but after the magnitude when talking about actual charges).

The dipole moment of HCl gives an experimental measure of the partial charges on each atom. In Sample Exercise 8.5 we saw that the dipole moment of HCl indicates a charge separation with a partial charge of 0.178+ on H and 0.178− on Cl, in remarkably good agreement with our simple approximation based on electronegativities. Although that type of calculation provides rough numbers for the magnitude of charge on atoms, the relationship between electronegativities and charge separation is generally more complicated. As we have already seen, computer programs employing quantum mechanical principles have been developed to calculate the partial charges on atoms, even in complex molecules. Figure 8.9(c) shows a graphical representation of the charge distribution in HCl.

RELATED EXERCISES: 8.40–8.43

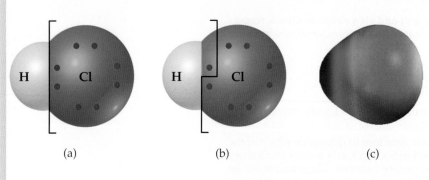

(a) (b) (c)

◀ **FIGURE 8.9 Oxidation number and formal charge.** (a) The oxidation number for any atom in a molecule is determined by assigning all shared electrons to the more electronegative atom (in this case, Cl). (b) Formal charges are derived by dividing all shared electron pairs equally between the bonded atoms. (c) The calculated distribution of electron density on an HCl molecule. Regions of relatively more negative charge are red; those of more positive charge are blue. Negative charge is clearly localised on the chlorine atom.

8.6 | RESONANCE STRUCTURES

We sometimes encounter molecules and ions in which the experimentally determined arrangement of atoms is not adequately described by a single Lewis structure. Consider a molecule of ozone, O_3, which is a bent molecule with two equal O–O bond lengths (▶ FIGURE 8.10). Because each oxygen atom contributes six valence electrons, the ozone molecule has 18 valence electrons. In writing the Lewis structure, we find that we must have one O–O single bond and one O–O double bond to attain an octet of electrons about each atom:

$$:\ddot{\text{O}} \diagdown \overset{\displaystyle \ddot{\text{O}}}{} \diagup \ddot{\text{O}}:$$

However, this structure cannot by itself be correct because it requires one O–O bond to be different from the other, contrary to the observed structure—we would expect the O–O double bond to be shorter than the O–O single bond (Section 8.3). In drawing the Lewis structure, however, we could just as easily have put the O–O double bond on the left:

The placement of the atoms in these two alternative but completely equivalent Lewis structures for ozone is the same, but the placement of the electrons is different. Lewis structures of this sort are called **resonance structures**. To describe the structure of ozone properly, we write both Lewis structures and use a double-headed arrow to indicate that the real molecule is described by an average of the two resonance structures:

To understand why certain molecules require more than one resonance structure, we can draw an analogy with the mixing of paint (▶ FIGURE 8.11). Blue and yellow are both primary colours of paint pigment. An equal blend of blue and yellow pigments produces green pigment. We can't describe green paint in terms of a single primary colour, yet it still has its own identity. Green paint does not oscillate between its two primary colours. It is not blue part of the time and yellow the rest of the time. Similarly, molecules such as ozone cannot be described by a single Lewis structure in which the electrons are 'locked into' a particular arrangement.

The true arrangement of the electrons in molecules such as O_3 must be considered as a blend of two (or more) Lewis structures. By analogy with the green paint, the molecule has its own identity separate from the individual resonance structures. For example, the ozone molecule always has two equivalent O–O bonds whose lengths are intermediate between the lengths of an oxygen–oxygen single bond and an oxygen–oxygen double bond. Another way of looking at it is to say that the rules for drawing Lewis structures don't allow us to have a single structure that adequately represents the ozone molecule. For example, there are no rules for drawing half-bonds. We can get around this limitation by drawing two equivalent Lewis structures that, when averaged, amount to something very much like what is observed experimentally. Although Lewis structures have their limitations, they do predict the most likely bonding arrangement of a group of atoms most of the time.

 CONCEPT CHECK 12

The O–O bonds in ozone are often described as 'one-and-a-half' bonds. Is this description consistent with the idea of resonance?

As an additional example of resonance structures, consider the nitrate ion, $NO_3{}^-$, for which three equivalent Lewis structures can be drawn:

Notice that the arrangement of atoms is the same in each structure; only the placement of electrons differs. In writing resonance structures, the same atoms must be bonded to each other in all structures, so that the only differences are in the arrangements of electrons. All three Lewis structures taken together adequately describe the nitrate ion, in which all three N–O bond lengths are the same.

 FIGURE IT OUT

What feature of this structure suggests that the two outer O atoms are in some way equivalent to each other?

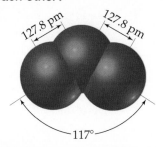

▲ **FIGURE 8.10**   Molecular structure of ozone.

 FIGURE IT OUT

Is the electron density consistent with equal weights for the two resonance structures for O_3? Explain.

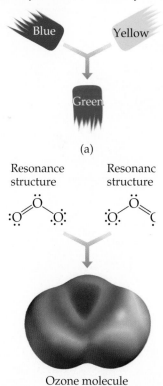

Ozone molecule

(b)

▲ **FIGURE 8.11   Resonance.** Describing a molecule as a blend of different resonance structures is similar to describing a paint colour as a blend of primary colours. (a) Green paint is a blend of blue and yellow. We cannot describe green as a single primary colour. (b) The ozone molecule is a blend of two resonance structures. We cannot describe the ozone molecule in terms of a single Lewis structure.

CONCEPT CHECK 13

In the same sense that we describe the O–O bonds in O_3 as 'one-and-a-half' bonds, how would you describe the N–O bonds in NO_3^-?

In some instances, all the possible Lewis structures for a species may not be equivalent to one another; instead, one or more may represent a more stable arrangement than other possibilities. We encounter examples of this as we proceed.

SAMPLE EXERCISE 8.10 Resonance structures

Which is predicted to have the shorter sulfur–oxygen bonds, SO_3 or SO_3^{2-}?

SOLUTION

Analyse These two similar looking species have different numbers of valence electrons. This may change the bonding present and hence the length of the sulfur–oxygen bonds.

Plan Draw the Lewis structure, including any resonance structures, of both species and determine the average number of bonds between the sulfur and oxygen atoms.

Solve The sulfur atom has six valence electrons, as does oxygen. Thus SO_3 contains 24 valence electrons. In writing the Lewis structure, we see that three equivalent resonance structures can be drawn:

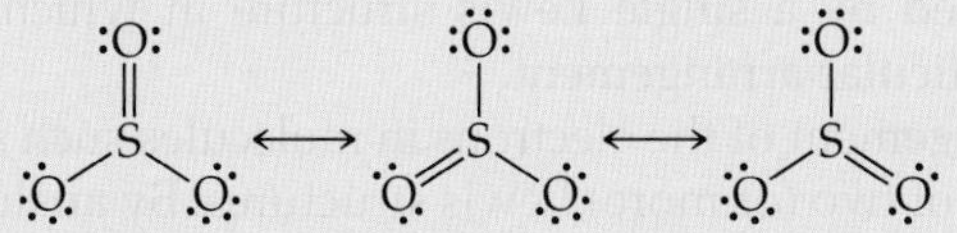

As was the case for NO_3^-, the actual structure of SO_3 is an equal blend of all three. Thus each S–O bond distance should be about one-third of the way between that of a single and that of a double bond (see Concept Check 13). That is, they should be shorter than single bonds but not as short as double bonds.

The SO_3^{2-} ion has 26 electrons, which leads to a Lewis structure in which all the S–O bonds are single bonds:

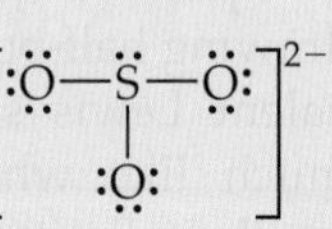

There are no other reasonable Lewis structures for this ion—it can be described quite well by a single Lewis structure rather than by multiple resonance structures.

Our analysis of the Lewis structures leads us to conclude that SO_3 should have the shorter S–O bonds and SO_3^{2-} the longer ones. This conclusion is correct: the experimentally measured S–O bond lengths are 142 pm in SO_3 and 151 pm in SO_3^{2-}.

PRACTICE EXERCISE

Draw two equivalent resonance structures for the formate ion, HCO_2^-.

Answer:

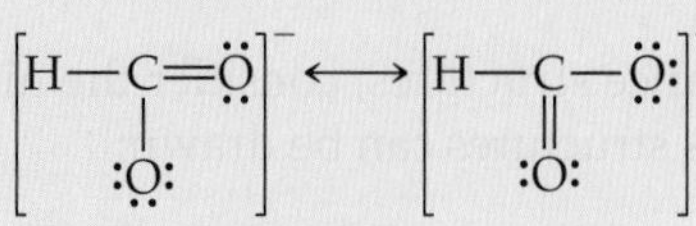

(See also Exercises 8.44, 8.45.)

8.7 EXCEPTIONS TO THE OCTET RULE

The octet rule is so simple and useful in introducing the basic concepts of bonding that you might assume it is always obeyed. In Section 8.2, however, we noted its limitation in dealing with ionic compounds of the transition metals. The octet rule also fails in situations involving covalent bonding. These exceptions to the octet rule are of three main types.

1. Molecules and polyatomic ions containing an odd number of electrons.
2. Molecules and polyatomic ions in which an atom has less than an octet of valence electrons.
3. Molecules and polyatomic ions in which an atom has more than an octet of valence electrons.

Odd Number of Electrons

In the vast majority of molecules and polyatomic ions, the total number of valence electrons is even, and complete pairing of electrons occurs. In a few molecules and polyatomic ions, however, such as ClO_2, NO, NO_2 and $O_2{}^-$, the number of valence electrons is odd. Complete pairing of these electrons is impossible, and an octet around each atom cannot be achieved. For example, NO contains $5 + 6 = 11$ valence electrons.

The two most important Lewis structures for this molecule are

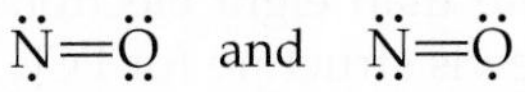

$$\ddot{N}=\ddot{O} \quad \text{and} \quad \dot{N}=\ddot{O}$$

Note on the grounds of formal charge, the left-hand structure is preferred.

Less than an Octet of Valence Electrons

A second type of exception occurs when there are less than eight valence electrons around an atom in a molecule or polyatomic ion. This situation is also relatively rare, most often encountered in compounds of boron and beryllium. As an example, let's consider boron trifluoride, BF_3. If we follow the first four steps of the procedure at the beginning of Section 8.5 for drawing Lewis structures, we obtain the structure

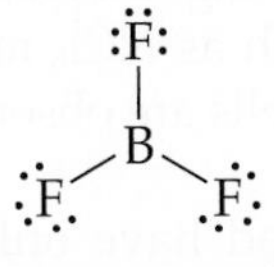

There are only six electrons around the boron atom. In this Lewis structure the formal charges on both the B and the F atoms are zero. We could complete the octet around boron by forming a double bond (step 5). In so doing, we see that there are three equivalent resonance structures (the formal charges on each atom are shown in red):

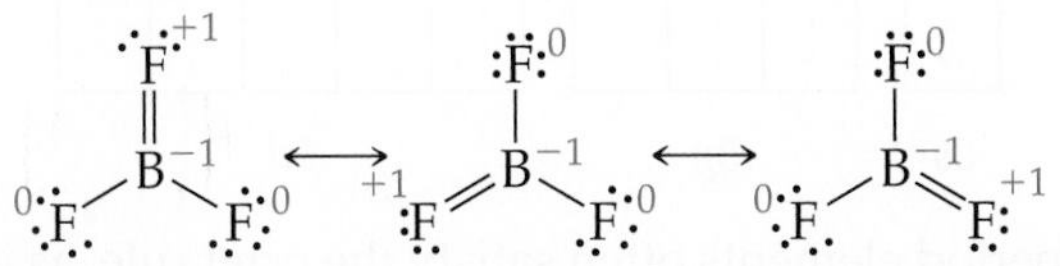

These Lewis structures force a fluorine atom to share additional electrons with the boron atom, which is inconsistent with the high electronegativity of fluorine. In fact, the formal charges tell us that this is an unfavourable situation: in each of the Lewis structures, the F atom involved in the B–F double bond has a formal charge of +1, whereas the less electronegative B atom has a formal charge of −1. Thus the Lewis structures in which there is a B–F double bond are less important than the one in which there are fewer than an octet of valence electrons around boron:

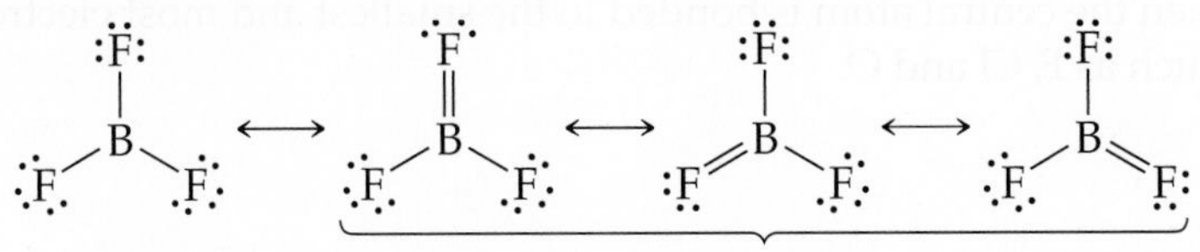

Most important   Less important

We usually represent BF_3 solely by the leftmost resonance structure, in which there are only six valence electrons around boron. The chemical behaviour of

BF_3 is consistent with this representation. In particular, BF_3 reacts very energetically with molecules having an unshared pair of electrons that can be used to form a bond with boron. For example, it reacts with ammonia, NH_3, to form the compound NH_3BF_3:

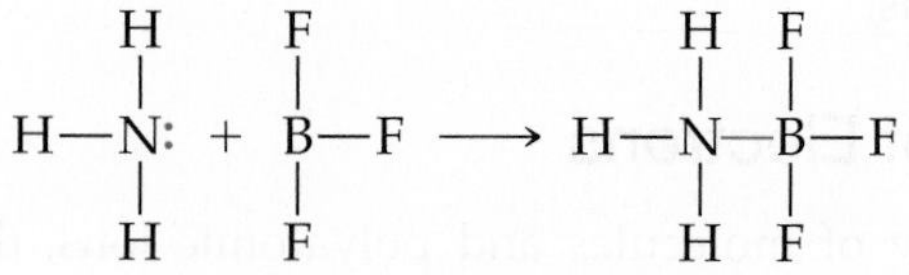

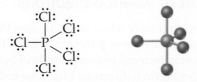

In this stable compound, boron has an octet of valence electrons.

More than an Octet of Valence Electrons

The third and largest class of exceptions consists of molecules or polyatomic ions in which there are more than eight electrons in the valence shell of an atom. When we draw the Lewis structure for PCl_5, for example, we are forced to 'expand' the valence shell and place 10 electrons around the central phosphorus atom:

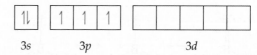

Other examples of molecules and ions with expanded valence shells are SF_4, AsF_6 and ICl_4^-. The corresponding molecules with a second-period atom bonded to the halogen atom, such as NCl_5 and OF_4, do *not* exist. Let's take a look at why expanded valence shells are observed only for elements in period 3 and beyond in the periodic table.

Elements of the second period have only the 2s and 2p valence orbitals available for bonding. Because these orbitals can hold a maximum of eight electrons, we never find more than an octet of electrons around elements from the second period. Elements from the third period and beyond, however, have *ns*, *np* and unfilled *nd* orbitals that can be used in bonding. For example, the orbital diagram for the valence shell of a phosphorus atom is

↑↓	↑	↑	↑					

    3s   3p   3d

Although third-period elements often satisfy the octet rule, as in PCl_3, they also may exceed an octet by seeming to use their empty *d* orbitals to accommodate additional electrons.*

Size also plays an important role in determining whether an atom in a molecule or polyatomic ion can accommodate more than eight electrons in its valence shell. The larger the central atom, the larger the number of atoms that can surround it. The number of molecules and ions with expanded valence shells therefore increases with increasing size of the central atom. The size of the surrounding atoms is also important. Expanded valence shells occur most often when the central atom is bonded to the smallest and most electronegative atoms, such as F, Cl and O.

---

* On the basis of theoretical calculations, some chemists have questioned whether valence *d* orbitals are actually used in the bonding of molecules and ions with expanded valence shells. Nevertheless, the presence of valence *d* orbitals in period 3 and beyond provides the simplest explanation of this phenomenon, especially within the scope of a general chemistry textbook.

SAMPLE EXERCISE 8.11   **Lewis structure for an ion with an expanded valence shell**

Draw the Lewis structure for ICl_4^-.

SOLUTION

Analyse Drawing the Lewis structure of ICl_4^- will enable us to determine a stable bonding arrangement for this polyatomic ion.

Plan Determine the total number of valence electrons, allowing for the overall charge on the ion, bond the atoms together and arrange the remaining electrons appropriately.

Solve Iodine (group 17) has seven valence electrons; each chlorine (group 17) also has seven; an extra electron is added to account for the 1− charge of the ion. Therefore, the total number of valence electrons is

$$7 + 4(7) + 1 = 36$$

The I atom is the central atom in the ion. Putting eight electrons around each Cl atom (including a pair of electrons between I and each Cl to represent the single bond between these atoms) requires $8 \times 4 = 32$ electrons.

We are thus left with $36 - 32 = 4$ electrons to be placed on the larger iodine:

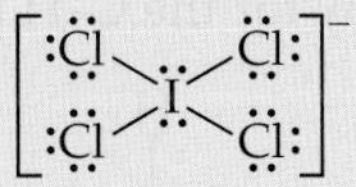

Iodine has 12 valence electrons around it, four more than needed for an octet.

PRACTICE EXERCISE

(a) Predict which of the following atoms is never found with more than an octet of valence electrons around it: S, C, P, Br. **(b)** Draw the Lewis structure for XeF_2.

Answers: **(a)** C, **(b)** $:\!\ddot{F}\!-\!\ddot{X}\dot{e}\!-\!\ddot{F}\!:$

(See also Exercises 8.50–8.53.)

At times you may see Lewis structures written with an expanded valence shell even though structures can be written with an octet. For example, consider these Lewis structures for the phosphate ion, PO_4^{3-}:

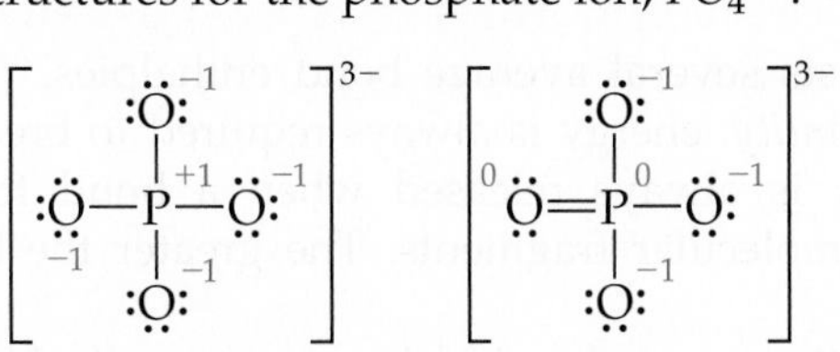

The formal charges on the atoms are shown in red. On the left, the P atom has an octet; on the right, the P atom has an expanded valence shell of five electron pairs. The structure on the right is often used for PO_4^{3-} because it has smaller formal charges on the atoms. The best representation of PO_4^{3-} is a series of such Lewis structures in resonance with one another. However, theoretical calculations based on quantum mechanics suggest that the structure on the left is the best single Lewis structure for the phosphate ion. In general, when choosing between alternative Lewis structures, you should choose one that satisfies the octet rule if it is possible to do so.

8.8 | STRENGTHS OF COVALENT BONDS

The stability of a molecule is related to the strengths of the covalent bonds it contains. The strength of a covalent bond between two atoms is determined by the energy required to break that bond. It is easiest to relate bond strength to

Find out more on page 513

the enthalpy change in reactions in which bonds are broken (Section 14.4, 'Enthalpies of Reaction').

The **bond enthalpy** is the enthalpy change, ΔH, for the breaking of a particular bond in one mole of a gaseous substance. For example, the bond enthalpy for the bond between chlorine atoms in the Cl_2 molecule is the enthalpy change when 1 mol of Cl_2 is dissociated into chlorine atoms:

$$:\ddot{C}l{-}\ddot{C}l:(g) \longrightarrow 2:\ddot{C}l\cdot(g)$$

We use the designation D (bond type) to represent bond enthalpies.

It is relatively simple to assign bond enthalpies to bonds that are found in diatomic molecules, such as the Cl–Cl bond in Cl_2 or the H–Br bond in HBr. The bond enthalpy is just the energy required to break the diatomic molecule into its component atoms. Many important bonds, such as the C–H bond, exist only in polyatomic molecules. For these types of bonds, we usually utilise *average* bond enthalpies. For example, the enthalpy change for the following process in which a methane molecule is decomposed to its five atoms (a process called *atomisation*) can be used to define an average bond enthalpy for the C–H bond:

$$\underset{\underset{H}{|}}{\overset{\overset{H}{|}}{H{-}C{-}H}}(g) \longrightarrow \cdot\dot{C}\cdot(g) \ + \ 4\,H\cdot(g) \qquad \Delta H = 1660 \text{ kJ}$$

Because there are four equivalent C–H bonds in methane, the heat of atomisation is equal to the sum of the bond enthalpies of the four C–H bonds. Therefore, the average C–H bond enthalpy for CH_4 is $D(C{-}H) = (1660/4)$ kJ mol^{-1} = 415 kJ mol^{-1}.

The bond enthalpy for a given set of atoms, say C–H, depends on the rest of the molecule of which the atom pair is a part. However, the variation from one molecule to another is generally small, which supports the idea that bonding electron pairs are localised between atoms. If we consider C–H bond enthalpies in many different compounds, we find that the average bond enthalpy is 413 kJ mol^{-1}, which compares closely with the 415 kJ mol^{-1} value calculated from CH_4.

▶ **TABLE 8.4** lists several average bond enthalpies. *The bond enthalpy is always a positive quantity*: energy is always required to break chemical bonds. Conversely, energy is always released when a bond forms between two gaseous atoms or molecular fragments. The greater the bond enthalpy, the stronger the bond.

A molecule with strong chemical bonds generally has less tendency to undergo chemical change than one with weak bonds. This relationship between strong bonding and chemical stability helps explain the chemical form in which many elements are found in nature. For example, Si–O bonds are among the strongest that silicon forms. It should not be surprising, therefore, that SiO_2 and other substances containing Si–O bonds (silicates) are so common; it is estimated that over 90% of the Earth's crust is composed of SiO_2 and silicates.

Bond Enthalpies and the Enthalpies of Reactions

We can use average bond enthalpies to estimate the enthalpies of reactions in which bonds are broken and new bonds are formed. This procedure allows us to estimate quickly whether a given reaction will be endothermic ($\Delta H > 0$) or exothermic ($\Delta H < 0$) even if we do not know $\Delta_f H°$ for all the chemical species involved.

Our strategy for estimating reaction enthalpies is a straightforward application of Hess's law (Section 14.6, 'Hess's Law').

Find out more on page 520

Table 8.4 • Average bond enthalpies (kJ mol⁻¹)

Single bonds

C—H	413	N—H	391	O—H	463	F—F	155
C—C	348	N—N	163	O—O	146		
C—N	293	N—O	201	O—F	190	Cl—F	253
C—O	358	N—F	272	O—Cl	203	Cl—Cl	242
C—F	485	N—Cl	200	O—I	234		
C—Cl	328	N—Br	243			Br—F	237
C—Br	276					Br—Cl	218
C—I	240			S—H	339	Br—Br	193
C—S	259	H—H	436	S—F	327		
		H—F	567	S—Cl	253		
		H—Cl	431	S—Br	218	I—Cl	208
Si—H	323	H—Br	366	S—S	266	I—Br	175
Si—Si	226	H—I	299			I—I	151
Si—C	301						
Si—O	466						
Si—Cl	464						

Multiple bonds

C=C	614	N=N	418	O=O	495
C≡C	839	N≡N	941		
C=N	615	N=O	607	S=O	523
C≡N	891			S=S	418
C=O	799				
C≡O	1072				

We use the fact that breaking bonds is always an endothermic process and bond formation is always exothermic. We therefore imagine that the reaction occurs in two steps. (1) We supply enough energy to break those bonds in the reactants that are not present in the products. In this step the enthalpy of the system is increased by the sum of the bond enthalpies of the bonds that are broken. (2) We form the bonds in the products that were not present in the reactants. This step releases energy and therefore lowers the enthalpy of the system by the sum of the bond enthalpies of the bonds that are formed. The enthalpy of the reaction, $\Delta_{rxn}H$, is estimated as the sum of the bond enthalpies of the bonds broken minus the sum of the bond enthalpies of the bonds formed:

$$\Delta_{rxn}H = \Sigma(\text{bond enthalpies of bonds broken})$$
$$- \Sigma(\text{bond enthalpies of bonds formed}) \qquad [8.6]$$

Consider, for example, the gas-phase reaction between methane, CH_4, and chlorine to produce methyl chloride, CH_3Cl, and hydrogen chloride, HCl:

$$H—CH_3(g) + Cl—Cl(g) \longrightarrow Cl—CH_3(g) + H—Cl(g) \quad \Delta_{rxn}H = ? \quad [8.7]$$

Our two-step procedure is outlined in ▼ **FIGURE 8.12**. We note that in the course of this reaction the following bonds are broken and made:

> *Bonds broken:*   1 mol C—H, 1 mol Cl—Cl
>
> *Bonds made:*   1 mol C—Cl, 1 mol H—Cl

We first supply enough energy to break the C–H and Cl–Cl bonds, which will raise the enthalpy of the system. We then form the C–Cl and H–C bonds, which

Is this reaction exothermic or endothermic?

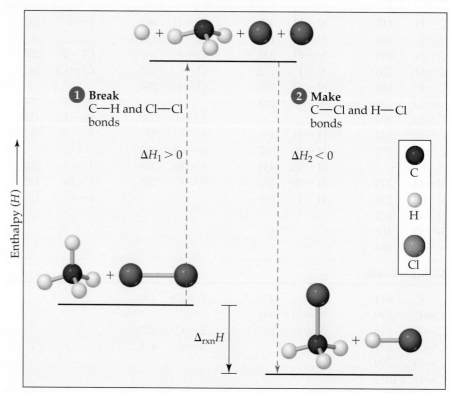

▲ **FIGURE 8.12 Using bond enthalpies to calculate $\Delta_{rxn}H$.** Average bond enthalpies are used to estimate $\Delta_{rxn}H$ for the reaction in Equation 8.7.

will release energy and lower the enthalpy of the system. By using Equation 8.6 and the data in Table 8.4, we estimate the enthalpy of the reaction as

$$\Delta_{rxn}H = [D(C—H) + D(Cl—Cl)] - [D(C—Cl) + D(H—Cl)]$$
$$= (413 \text{ kJ} + 242 \text{ kJ}) - (328 \text{ kJ} + 431 \text{ kJ}) = -104 \text{ kJ}$$

The reaction is exothermic because the bonds in the products (especially the H–Cl bond) are stronger than the bonds in the reactants (especially the Cl–Cl bond).

We usually use bond enthalpies to estimate $\Delta_{rxn}H$ only if we do not have the needed $\Delta_fH°$ values readily at hand. For the above reaction, we cannot calculate $\Delta_{rxn}H$ from $\Delta_fH°$ values and Hess's law because the value of $\Delta_fH°$ for $CH_3Cl(g)$ is not given in Appendix C. If we obtain the value of $\Delta_fH°$ for $CH_3Cl(g)$ from another source (such as the *CRC Handbook of Chemistry and Physics*) and use Equation 14.27, we find that $\Delta_{rxn}H = -99.8$ kJ for the reaction in Equation 8.7. Thus the use of average bond enthalpies provides a reasonably accurate estimate of the actual reaction enthalpy change.

It is important to remember that bond enthalpies are derived for *gaseous* molecules and that they are often *averaged* values. Nonetheless, average bond enthalpies are useful for estimating reaction enthalpies quickly, especially for gas-phase reactions.

MY WORLD OF CHEMISTRY

EXPLOSIVES AND ALFRED NOBEL

Enormous amounts of energy can be stored in chemical bonds. Perhaps the most graphic illustration of this fact is seen in certain molecular substances that are used as explosives. Our discussion of bond enthalpies allows us to examine more closely some of the properties of such explosive substances.

An explosive must have the following characteristics. (1) It must decompose very exothermically; (2) the products of its decomposition must be gaseous, so that a tremendous gas pressure accompanies the decomposition; (3) its decomposition must occur very rapidly; and (4) it must be stable enough so that it can be detonated predictably. The combination of the first three effects leads to the violent evolution of heat and gases.

To give the most exothermic reaction, an explosive should have weak chemical bonds and should decompose into molecules with very strong bonds. Looking at bond enthalpies (Table 8.4), the $N \equiv N$, $C \equiv O$ and $C = O$ bonds are among the strongest. Not surprisingly, explosives are usually designed to produce the gaseous products $N_2(g)$, $CO(g)$ and $CO_2(g)$. Water vapour is nearly always produced as well.

Many common explosives are organic molecules that contain nitro (NO_2) or nitrate (NO_3) groups attached to a carbon skeleton. The structure of the most familiar explosive, nitroglycerin, is shown here.

Nitroglycerin

Nitroglycerin is a pale yellow, oily liquid. It is highly *shock-sensitive*. Merely shaking the liquid can cause its explosive decomposition into nitrogen, carbon dioxide, water and oxygen gases:

$$4\, C_3H_5N_3O_9(l) \longrightarrow 6\, N_2(g) + 12\, CO_2(g) + 10\, H_2O(g) + O_2(g)$$

The large bond enthalpies of the N_2 molecules ($941\ \text{kJ mol}^{-1}$), CO_2 molecules ($2 \times 799\ \text{kJ mol}^{-1}$) and water molecules ($2 \times 463\ \text{kJ mol}^{-1}$) make this reaction enormously exothermic. Nitroglycerin is an exceptionally unstable explosive because it is in nearly perfect *explosive balance*. With the exception of a small amount of $O_2(g)$, the only products are N_2, CO_2 and H_2O. Note also that, unlike combustion reactions (Section 3.2), explosions are entirely self-contained: no other reagent, such as $O_2(g)$, is needed for the explosive decomposition.

Because nitroglycerin is so unstable, it is difficult to use as a controllable explosive. The Swedish inventor Alfred Nobel (▼ **FIGURE 8.13**) found that mixing nitroglycerin with an absorbent solid material such as diatomaceous earth or cellulose gives a solid explosive (*dynamite*) that is much safer than liquid nitroglycerin.

RELATED EXERCISES: 8.72, 8.73

▲ **FIGURE 8.13   Alfred Nobel (1833–1896), the Swedish inventor of dynamite.** By many accounts Nobel's discovery that nitroglycerin could be made more stable by absorbing it onto cellulose was an accident. This discovery made Nobel a very wealthy man. He was also a complex and lonely man, however, who never married, was frequently ill and suffered from chronic depression. He had invented the most powerful military explosive to date, but he strongly supported international peace movements. His will stated that his fortune should be used to establish prizes awarding those who 'have conferred the greatest benefit on mankind', including the promotion of peace and 'fraternity between nations'. The Nobel Prize is probably the most coveted award that a scientist, economist, writer or peace advocate can receive.

SAMPLE EXERCISE 8.12   Using average bond enthalpies

Using Table 8.4, estimate $\Delta_{rxn}H$ for the following reaction (where we explicitly show the bonds involved in the reactants and products):

SOLUTION

Analyse We use bond enthalpies to estimate the enthalpy change in a reaction.

Plan The enthalpy of reaction is the difference between the energy needed to break the bonds in the reactants and the energy gained from forming the new bonds in the products.

Solve Among the reactants, we must break six C–H bonds and a C=C bond in C_2H_6; we also break $\frac{7}{2}$ O=O bonds. Among the products, we form four C–O double bonds (two in each CO_2) and six O–H bonds (two in each H_2O).

Using Equation 8.6 and data from Table 8.4, we have

$$\Delta_{rxn}H = 6D(C\!-\!H) + D(C\!-\!C) + \tfrac{7}{2}D(O\!=\!O) - 4D(C\!=\!O) - 6D(O\!-\!H)$$

$$= 6(413\text{ kJ}) + 348\text{ kJ} + \tfrac{7}{2}(495\text{ kJ}) - 4(799\text{ kJ}) - 6(463\text{ kJ})$$

$$= 4558\text{ kJ} - 5974\text{ kJ}$$

$$= -1416\text{ kJ}$$

Check This estimate can be compared with the value of −1428 kJ calculated from more accurate thermochemical data; the agreement is good.

PRACTICE EXERCISE

Using Table 8.4, estimate $\Delta_{rxn}H$ for the reaction

$$\text{H}\!-\!\underset{\underset{\text{H}}{|}}{\text{N}}\!-\!\underset{\underset{\text{H}}{|}}{\text{N}}\!-\!\text{H}(g) \longrightarrow \text{N}\!\equiv\!\text{N}(g) + 2\,\text{H}\!-\!\text{H}(g)$$

Answer: −86 kJ
(See also Exercises 8.54–8.56.)

Bond Enthalpy and Bond Length

Just as we can define an average bond enthalpy, we can also define an average bond length for a number of common bond types. Some of these are listed in ▼ TABLE 8.5. Of particular interest is the relationship between bond enthalpy, bond length and the number of bonds between the atoms. For example, we can use the data in Tables 8.4 and 8.5 to compare the bond lengths and bond enthalpies of carbon–carbon single, double and triple bonds:

C—C	C=C	C≡C
154 pm	134 pm	120 pm
348 kJ mol^{-1}	614 kJ mol^{-1}	839 kJ mol^{-1}

As the number of bonds between the carbon atoms increases, the bond enthalpy increases and the bond length decreases; that is, the carbon atoms are held more closely and more tightly together. In general, *as the number of bonds between two atoms increases, the bond grows shorter and stronger*.

Table 8.5 • Average bond lengths for some single, double and triple bonds

Bond	Bond length (pm)	Bond	Bond length (pm)
C—C	154	N—N	147
C=C	134	N=N	124
C≡C	120	N≡N	110
C—N	143	N—O	136
C=N	138	N=O	122
C≡N	116		
		O—O	148
C—O	143	O=O	121
C=O	123		
C≡O	113		

SAMPLE INTEGRATIVE EXERCISE   Putting concepts together

Phosgene, a substance used in poisonous gas warfare in World War I, is so named because it was first prepared by the action of sunlight on a mixture of carbon monoxide and chlorine gases. Its name comes from the Greek words *phos* (light) and *genes* (born of). Phosgene has the following elemental composition: 12.14% C, 16.17% O and 71.69% Cl by mass. Its molar mass is 98.9 g mol^{-1}. **(a)** Determine the molecular formula of this compound. **(b)** Draw three Lewis structures for the molecule that satisfy the octet rule for each atom. (The Cl and O atoms bond to C.) **(c)** Using formal charges, determine which Lewis structure is the most important one. **(d)** Using average bond enthalpies, estimate $\Delta_{rxn}H$ for the formation of gaseous phosgene from CO(g) and Cl$_2$(g).

SOLUTION

Analyse We first determine a molecular formula for phosgene before we can draw a Lewis structure. We use formal charges to indicate the most probable Lewis structure and use this structure, together with that of CO and Cl$_2$, to estimate the enthalpy change when it is formed from CO and Cl$_2$.

Plan We need to work out the answer to each part of this integrative exercise in the order given and the answer to part (a) is the starting point to part (b) and so on.

Solve **(a)** The empirical formula of phosgene can be determined from its elemental composition (Section 3.5, 'Empirical Formulae from Analyses'). Assuming 100 g of the compound and calculating the number of moles of C, O and Cl in this sample, we have

Review this on page 83

$$\frac{12.14 \text{ g C}}{12.01 \text{ g mol}^{-1}} = 1.011 \text{ mol C}$$

$$\frac{16.17 \text{ g O}}{16.00 \text{ g mol}^{-1}} = 1.011 \text{ mol O}$$

$$\frac{71.69 \text{ g Cl}}{35.45 \text{ g mol}^{-1}} = 2.022 \text{ mol Cl}$$

The ratio of the number of moles of each element, obtained by dividing each number of moles by the smallest quantity, indicates that there is one C and one O for each two Cl in the empirical formula, COCl$_2$.

The molar mass of the empirical formula is $12.01 + 16.00 + 2(35.45) = 98.91$ g mol^{-1}, the same as the molar mass of the molecule. Thus COCl$_2$ is the molecular formula.

(b) Carbon has four valence electrons, oxygen has six and chlorine has seven, giving $4 + 6 + 2(7) = 24$ electrons for the Lewis structures. Drawing a Lewis structure with all single bonds does not give the central carbon atom an octet. Using multiple bonds, three structures satisfy the octet rule:

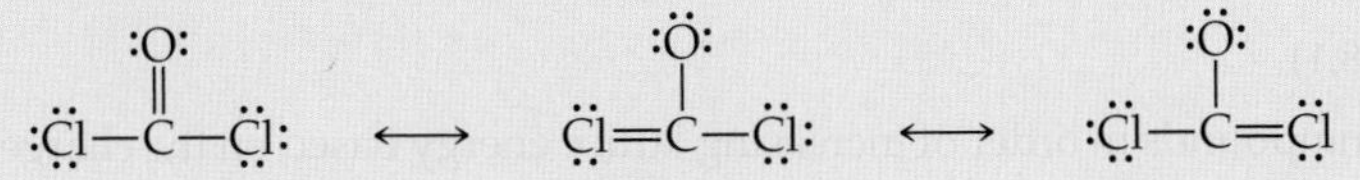

(c) Calculating the formal charges on each atom gives

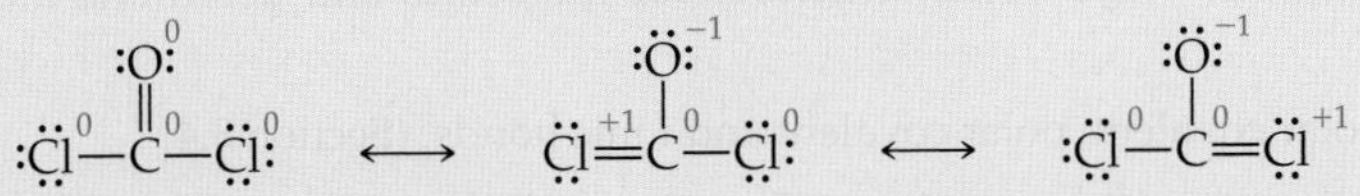

The first structure is expected to be the most important one because it has the lowest formal charges on each atom. Indeed, the molecule is usually represented by this Lewis structure.

(d) Writing the chemical equation in terms of the Lewis structures of the molecules, we have

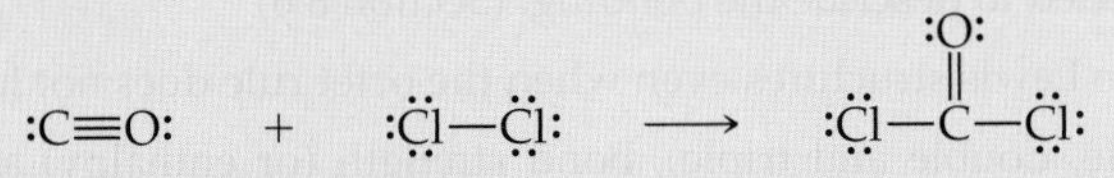

Thus the reaction involves breaking a C–O triple bond and a Cl–Cl bond and forming a C–O double bond and two C–Cl bonds. Using bond enthalpies from Table 8.4, we have

$$\Delta_{rxn}H = D(C{\equiv}O) + D(Cl{-}Cl) - D(C{=}O) - 2D(C{-}Cl)$$

$$= 1072 \text{ kJ} + 242 \text{ kJ} - 799 \text{ kJ} - 2(328 \text{ kJ}) = -141 \text{ kJ}$$

CHAPTER SUMMARY AND KEY TERMS

SECTION 8.1 In this chapter we have seen that **chemical bonds** can be classified into three broad groups: **ionic bonds**, which result from electrostatic forces that exist between ions of opposite charge; **covalent bonds**, which result from sharing of electrons by two atoms; and **metallic bonds**, which result from delocalised sharing of electrons in metals. When we consider bonding, **Lewis symbols** are a convenient method of representing the valence electrons in atoms and ions. In general, atoms gain or lose electrons to end up with eight valence electrons (as in the noble gases, apart from He). This is known as the **octet rule**.

SECTION 8.2 Ionic bonding results from the transfer of electrons from one atom to another, leading to the formation of a three-dimensional lattice of charged particles. The stabilities of ionic substances result from the strong electrostatic attractions between an ion and the surrounding ions of opposite charge. The magnitude of these interactions is measured by the **lattice energy**, which is the energy needed to separate an ionic lattice into gaseous ions. Lattice energy increases with increasing charge on the ions and with decreasing distance between the ions. We see that we can predict the type of ion an element will form from its position in the periodic table.

SECTION 8.3 We observe that atoms in molecular substances can be held together by sharing one or more electron pairs. **Lewis structures** indicate how many valence electrons are involved in forming bonds and how many remain as unshared electron pairs. The octet rule helps determine how many bonds will be formed between two atoms. We see how **single**, **double** and **triple bonds** are formed and how **bond length** decreases as the number of bonds between atoms increases.

SECTION 8.4 In covalent bonds, the electrons may not necessarily be shared equally between two atoms. **Bond polarity** helps describe unequal sharing of electrons in a bond. When the electron pairs of a covalent bond are shared equally between the two bonded atoms, a **non-polar covalent bond** results. This occurs in molecules in which identical atoms are bonded together. More commonly, electrons are not shared equally between two bonded atoms, resulting in **polar covalent bonds**.

Electronegativity is a numerical method of measuring the ability of an atom in a compound to attract electrons to itself. A **polar molecule** has a positive and negative side and this separation of charge produces a **dipole**, which is measured by the **dipole moment**.

SECTIONS 8.5 and 8.6 Lewis structures are a convenient way of predicting the covalent bonding pattern in a molecule. By determining the **formal charge** of each atom, we can identify the most favourable structure of the molecule. If it is found that more than one equivalent Lewis structure can be drawn for a molecule, the actual structure is envisaged as a blend of these structures called a **resonance structure**.

SECTION 8.7 The octet rule is not obeyed in all cases. Exceptions to the rule occur when (a) a molecule has an odd number of electrons, (b) it is not possible to complete an octet around an atom without forcing an unfavourable distribution of electrons, or (c) a large atom is surrounded by so many small electronegative atoms that it has more than an octet around it. Lewis structures with more than an octet of electrons are observed for atoms in the third period and beyond in the periodic table.

SECTION 8.8 Bond strengths can vary for a number of reasons and are measured by **bond enthalpies**. The strengths of covalent bonds increase with the number of electron pairs shared between two atoms. These can be used to estimate the heats of reaction.

KEY SKILLS

- Write Lewis symbols for atoms and ions. (Section 8.1)
- Understand lattice energy and be able to arrange compounds in order of increasing lattice energy based on the charges and sizes of the ions involved. (Section 8.2)
- Use atomic electron configurations and the octet rule to write Lewis structures for molecules and determine their electron distribution. (Section 8.3)
- Use electronegativity differences to identify non-polar covalent, polar covalent and ionic bonds. (Section 8.4)
- Calculate charge separation in diatomic molecules based on the experimentally measured dipole moment and bond distance. (Section 8.4)
- Calculate formal charges from Lewis structures and use those formal charges to identify the most favourable Lewis structures for a molecule or ion. (Section 8.5)
- Recognise molecules where resonance structures are needed to describe the bonding. (Section 8.6)
- Recognise exceptions to the octet rule and draw accurate Lewis structures even when the octet rule does not hold. (Section 8.7)
- Understand the relationship between bond type (single, double and triple), bond strength (or enthalpy) and bond length. (Section 8.8)
- Use bond enthalpies to estimate enthalpy changes for reactions involving gas-phase reactants and products. (Section 8.8)

KEY EQUATIONS

- The potential energy of two interacting charges

$$E_{el} = \frac{kq_1q_2}{r^2}$$ [8.4]

- The dipole moment of two charges of equal magnitude but opposite sign, separated by a distance r

$$\mu = qr$$ [8.5]

- The enthalpy change as a function of bond enthalpies for reactions involving gas-phase molecules

$$\Delta_{rxn}H = \Sigma(\text{bond enthalpies of bonds broken}) - \Sigma(\text{bond enthalpies of bonds formed})$$ [8.6]

EXERCISES

VISUALISING CONCEPTS

8.1 For each of these Lewis symbols, indicate the group in the periodic table to which the element X belongs:
(a) $\cdot\ddot{X}\cdot$ (b) $\cdot X\cdot$ (c) $:\ddot{X}\cdot$ [Section 8.1]

8.2 Illustrated below are four ions—A_1, A_2, Z_1 and Z_2—showing their relative ionic radii. The ions shown in red carry a 1+ charge, and those shown in blue carry a 1− charge. (a) Would you expect to find an ionic compound of formula A_1A_2? Explain. (b) Which combination of ions leads to the ionic compound with the largest lattice energy? (c) Which combination of ions leads to the ionic compound with the smallest lattice energy? [Section 8.2]

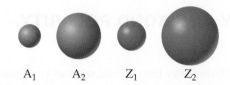

| A_1 | A_2 | Z_1 | Z_2 |

8.3 The orbital diagram below shows the valence electrons for a 2+ ion of an element. (a) What is the element? (b) What is the electron configuration of an atom of this element? [Section 8.2]

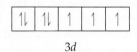

$3d$

8.4 In the Lewis structure shown below, A, D, E, Q, X and Z represent elements in the first two periods of the periodic table. Identify all six elements. [Section 8.3]

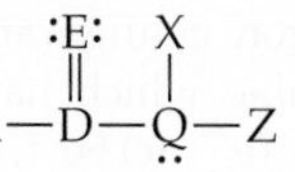

$$:\ddot{E}: \quad X$$
$$:\ddot{A}-D-\underset{..}{\overset{..}{Q}}-Z$$

8.5 The partial Lewis structure below is for a hydrocarbon molecule. In the full Lewis structure, each carbon atom satisfies the octet rule and there are no unshared electron pairs in the molecule. The carbon–carbon bonds are labelled 1, 2 and 3. (a) Determine where the hydrogen atoms are in the molecule. (b) Rank the carbon–carbon bonds in order of increasing bond length. (c) Rank the carbon–carbon bonds in order of increasing bond enthalpy. [Sections 8.3 and 8.8]

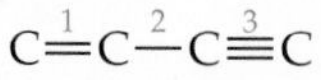

$$C\overset{1}{=}C\overset{2}{-}C\overset{3}{\equiv}C$$

8.6 One possible Lewis structure for the compound xenon trioxide, XeO_3, is shown below. (a) Prior to the 1960s this compound was thought to be impossible. Why? (b) How many other equivalent resonant structures are there for this Lewis structure? (c) Does this Lewis structure satisfy the octet rule? Explain why or why not. (d) Do you think this is the best choice of Lewis structure for XeO_3? [Sections 8.5–8.7]

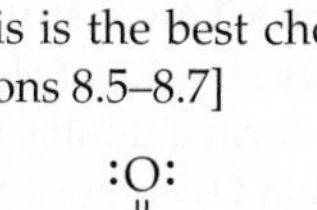

$$:\ddot{O}:$$
$$\ddot{O}=\ddot{X}e-\ddot{O}:$$

LEWIS SYMBOLS (Section 8.1)

8.7 (a) What are valence electrons? (b) How many valence electrons does a nitrogen atom possess? (c) An atom has the electron configuration $1s^22s^22p^63s^23p^2$. How many valence electrons does the atom have?

8.8 (a) What is the octet rule? (b) How many electrons must a sulfur atom gain to achieve an octet in its valence shell? (c) If an atom has the electron configuration $1s^22s^22p^3$, how many electrons must it gain to achieve an octet?

8.9 Write the electron configuration for phosphorus. Identify a valence electron in this configuration and a non-valence electron. From the standpoint of chemical reactivity, what is the important difference between them?

8.10 Write the Lewis symbol for atoms of each of the following elements: (a) Al, (b) Br, (c) Ar, (d) Sr.

8.11 What is the Lewis symbol for each of the following atoms or ions: (a) K, (b) Si, (c) Mg^{2+}, (d) P^{3-}?

IONIC BONDING (Section 8.2)

8.12 Using Lewis symbols, diagram the reaction between magnesium and oxygen atoms to give the ionic substance MgO.

8.13 Use Lewis symbols to represent the reaction that occurs between Mg and Br atoms.

8.14 Predict the chemical formula of the ionic compound formed between the following pairs of elements: **(a)** Al and F, **(b)** K and S, **(c)** Y and O, **(d)** Mg and N.

8.15 Which ionic compound is expected to form from combination of the following pairs of elements: **(a)** barium and oxygen, **(b)** rubidium and iodine, **(c)** lithium and sulfur, **(d)** bromine and magnesium?

8.16 Write the electron configuration for each of the following ions, and determine which ones possess noble gas configurations: **(a)** Sr^{2+}, **(b)** Ti^{2+}, **(c)** Se^{2-}, **(d)** Ni^{2+}, **(e)** Br^-, **(f)** Mn^{3+}.

8.17 Write electron configurations for the following ions, and determine which have noble gas configurations: **(a)** Zn^{2+}, **(b)** Te^{2-}, **(c)** Se^{3+}, **(d)** Ru^{2+}, **(e)** Tl^+, **(f)** Au^+.

8.18 **(a)** Define the term *lattice energy*. **(b)** Which factors govern the magnitude of the lattice energy of an ionic compound?

8.19 **(a)** The lattice energies of NaF and MgO are given in Table 8.2. Account for the difference in these two quantities. **(b)** Account for the difference in the lattice energies of $MgCl_2$ and $SrCl_2$, which are also listed in that table.

8.20 **(a)** Does the lattice energy of an ionic solid increase or decrease (i) as the charges of the ions increase, (ii) as the sizes of the ions increase? **(b)** Using a periodic table, arrange the following substances according to their expected lattice energies, listing them from lowest lattice energy to the highest: LiCl, NaBr, RbBr, MgO. Compare your list with the data in Table 8.2.

8.21 Energy is required to remove two electrons from Ca to form Ca^{2+} and is also required to add two electrons to O to form O^{2-}. Why, then, is CaO stable relative to the free elements?

8.22 Use data from Appendix C, Figure 7.9 and Figure 7.10 to calculate the lattice energy of RbCl. Is this value greater than or less than the lattice energy of NaCl? Explain.

8.23 By using data from Appendix C, Figure 7.9, Figure 7.10 and the value of the second ionisation energy for Ca, 1145 kJ mol^{-1}, calculate the lattice energy of $CaCl_2$. Is this value greater or less than the lattice energy of NaCl? Explain.

COVALENT BONDING, ELECTRONEGATIVITY AND BOND POLARITY (Sections 8.3 and 8.4)

8.24 **(a)** What is meant by the term *covalent bond*? **(b)** Give three examples of covalent bonding. **(c)** A substance XY, formed from two different elements, boils at −33 °C. Is XY likely to be a covalent or an ionic substance? Explain.

8.25 Which of these elements is unlikely to form covalent bonds: S, H, K, Ar, Si? Explain your choices.

8.26 **(a)** Construct a Lewis structure for O_2 in which each atom achieves an octet of electrons. **(b)** Explain why it is necessary to form a double bond in the Lewis structure. **(c)** The bond in O_2 is shorter than the O–O bond in compounds that contain an O–O single bond. Explain this observation.

8.27 The C–S bond lengths in carbon disulfide, CS_2, are shorter than would be expected for C–S single bonds. Use a Lewis structure to rationalise this observation.

8.28 **(a)** What is meant by the term *electronegativity*? **(b)** On the Pauling scale what is the range of electronegativity values for the elements? **(c)** Which element has the greatest electronegativity? **(d)** Which element has the smallest electronegativity?

8.29 **(a)** What is the trend in electronegativity going from left to right in a period of the periodic table? **(b)** How do electronegativity values generally vary going down a group in the periodic table? **(c)** How do periodic trends in electronegativity relate to those for ionisation energy and electron affinity?

8.30 Using only the periodic table as your guide, select the most electronegative atom in each of the following sets: **(a)** As, Se, Br, I; **(b)** Al, B, C, Si; **(c)** Ge, As, P, Sn; **(d)** Li, Rb, Be, Sr.

8.31 By referring only to the periodic table, select **(a)** the most electronegative element in group 16; **(b)** the least electronegative element in the group Al, Si, P; **(c)** the most electronegative element in the group Ga, P, Cl, Na; **(d)** the element in the group K, C, Zn, F that is most likely to form an ionic compound with Ba.

8.32 Which of the following bonds are polar: **(a)** B–F, **(b)** Cl–Cl, **(c)** Se–O, **(d)** H–I? Which is the more electronegative atom in each polar bond?

8.33 Arrange the bonds in each of the following sets in order of increasing polarity: **(a)** C–F, O–F, Be–F; **(b)** O–Cl, S–Br, C–P; **(c)** C–S, B–F, N–O.

8.34 From the data in Table 8.3, calculate the effective charges on the H and F atoms of the HF molecule in units of the electronic charge e.

8.35 The iodine monobromide molecule, IBr, has a bond length of 249 pm and a dipole moment of 1.21 D. **(a)** Which atom of the molecule is expected to have a negative charge? Explain. **(b)** Calculate the effective charges on the I and Br atoms in IBr, in units of the electronic charge e.

8.36 Give the name or chemical formula, as appropriate, for each of the following substances, and in each case predict whether the bonding is better described by the ionic-bonding or covalent-bonding model: **(a)** manganese(IV) oxide, **(b)** phosphorus(III) sulfide, **(c)** cobalt(II) oxide, **(d)** Cu_2S, **(e)** ClF_3, **(f)** VF_5.

8.37 Give the name or chemical formula, as appropriate, for each of the following substances, and in each case predict whether the bonding is better described by the ionic-bonding or covalent-bonding model:**(a)** manganese(III) fluoride, **(b)** chromium(VI) oxide, **(c)** arsenic(V) bromide, **(d)** SF_4, **(e)** $MoCl_4$, **(f)** $ScCl_3$.

LEWIS STRUCTURES; RESONANCE STRUCTURES (Sections 8.5 and 8.6)

8.38 Draw Lewis structures for the following: **(a)** SiH_4, **(b)** CO, **(c)** H_2SO_4 (H is bonded to O), **(d)** ClO_2^-, **(e)** NH_2OH.

8.39 Write Lewis structures for the following: **(a)** H_2CO (both H atoms are bonded to C), **(b)** H_2O_2, **(c)** C_2F_6 (contains a C–C bond), **(d)** AsO_3^{3-}, **(e)** H_2SO_3 (H is bonded to O).

8.40 **(a)** When talking about atoms in a Lewis structure, what is meant by the term *formal charge*? **(b)** Does the formal charge of an atom represent the actual charge on that atom? Explain. **(c)** How does the formal charge of an atom in a Lewis structure differ from the oxidation number of the atom?

8.41 **(a)** Write a Lewis structure for the phosphorus trifluoride molecule, PF_3. Is the octet rule satisfied for all the atoms in your structure? **(b)** Determine the oxidation numbers of the P and F atoms. **(c)** Determine the formal charges of the P and F atoms. **(d)** Is the oxidation number for the P atom the same as its formal charge? Explain why or why not.

8.42 Write Lewis structures that obey the octet rule for each of the following, and assign oxidation numbers and formal charges to each atom: **(a)** NO^+, **(b)** $POCl_3$ (P is bonded to the three Cl atoms and to the O), **(c)** ClO_4^-, **(d)** $HClO_3$ (H is bonded to O).

8.43 For each of the following molecules or ions of sulfur and oxygen, write a single Lewis structure that obeys the octet rule, and calculate the oxidation numbers and formal charges on all the atoms: **(a)** SO_2, **(b)** SO_3, **(c)** SO_3^{2-}, **(d)** SO_4^{2-}.

8.44 **(a)** Write one or more appropriate Lewis structures for the nitrite ion, NO_2^-. **(b)** With what compound of oxygen is it isoelectronic? **(c)** What would you predict for the lengths of the bonds in this species relative to N–O single bonds?

8.45 Consider the nitryl cation, NO_2^+. **(a)** Write one or more appropriate Lewis structures for this species. **(b)** Are resonance structures needed to describe the structure? **(c)** With what familiar species is it isoelectronic?

EXCEPTIONS TO THE OCTET RULE (Section 8.7)

8.46 **(a)** State the octet rule. **(b)** Does the octet rule apply to ionic as well as to covalent compounds? Explain, using examples as appropriate.

8.47 Considering the main-group non-metals, what is the relationship between the group number for an element (carbon, for example, belongs to group 14; see the periodic table on the inside front cover) and the number of single covalent bonds that an element needs to form to conform to the octet rule?

8.48 What is the most common exception to the octet rule? Give two examples.

8.49 For elements in the third period of the periodic table and beyond, the octet rule is often not obeyed. What factors are usually cited to explain this fact?

8.50 Draw the Lewis structures for each of the following ions or molecules. Identify those that do not obey the octet rule, and explain why they do not. **(a)** SO_3^{2-}, **(b)** AlH_3, **(c)** N_3^-, **(d)** CH_2Cl_2, **(e)** SbF_5.

8.51 Draw the Lewis structures for each of the following molecules or ions. Which do not obey the octet rule? **(a)** CO_2, **(b)** IO_3^-, **(c)** BH_3, **(d)** BF_4^-, **(e)** XeF_2.

8.52 In the vapour phase, $BeCl_2$ exists as a discrete molecule. **(a)** Draw the Lewis structure of this molecule, using only single bonds. Does this Lewis structure satisfy the octet rule? **(b)** What other resonance forms are possible that satisfy the octet rule? **(c)** Using formal charges, select the resonance form from among all the Lewis structures that is most important in describing $BeCl_2$.

8.53 **(a)** Describe the molecule chlorine dioxide, ClO_2, using three possible resonance structures. **(b)** Do any of these resonance structures satisfy the octet rule for every atom in the molecule? Why or why not? **(c)** Using formal charges, select the resonance structure(s) that is (are) most important.

BOND ENTHALPIES (Section 8.8)

8.54 Using the bond enthalpies tabulated in Table 8.4, estimate $\Delta_{rxn}H$ for each of the following gas-phase reactions:

(a)

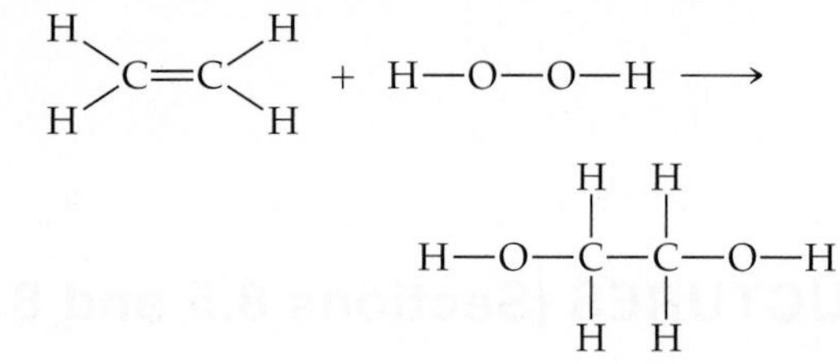

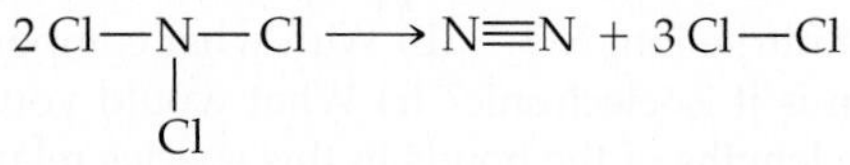

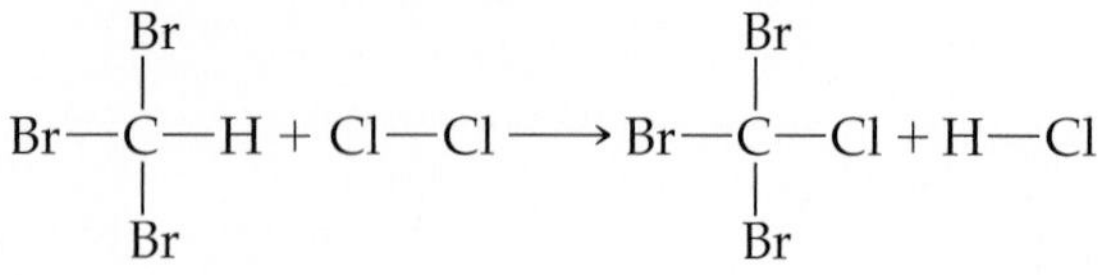

8.55 Using bond enthalpies (Table 8.4), estimate $\Delta_{rxn}H$ for the following gas-phase reactions:

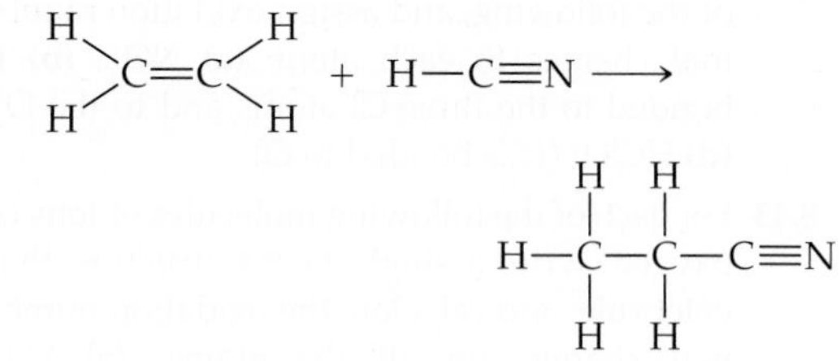

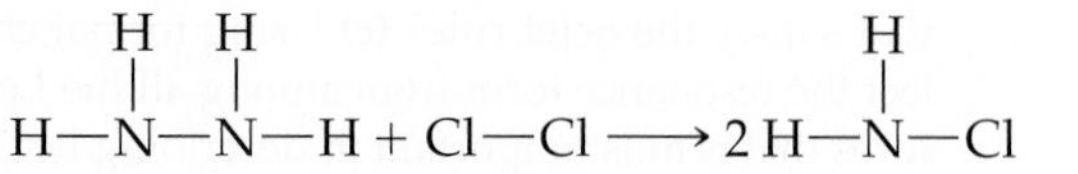

8.56 Using bond enthalpies (Table 8.4), estimate $\Delta_{rxn}H$ for each of the following reactions:

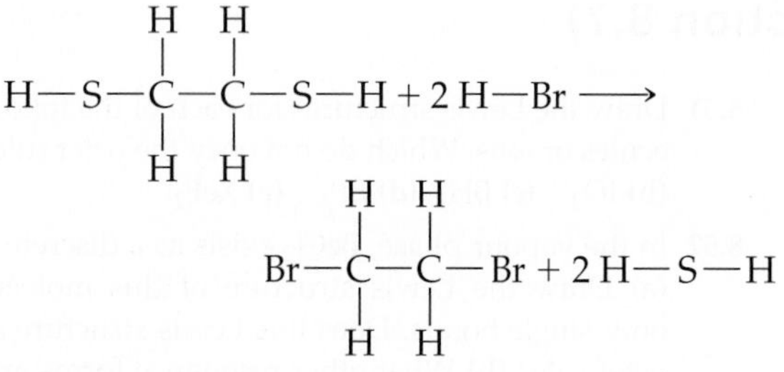

(a) $2 CH_4(g) + O_2(g) \longrightarrow 2 CH_3OH(g)$
(b) $H_2(g) + Br_2(g) \longrightarrow 2 HBr(g)$
(c) $2 H_2O_2(g) \longrightarrow 2 H_2O(g) + O_2(g)$

8.57 Ammonia is produced directly from nitrogen and hydrogen by the following reaction:

$$N_2(g) + 3 H_2(g) \longrightarrow 2 NH_3(g)$$

(a) Use bond enthalpies (Table 8.4) to estimate the enthalpy change for the reaction, and state whether this reaction is exothermic or endothermic. (b) Compare the enthalpy change you calculate in (a) with the true enthalpy change as obtained using $\Delta_f H°$ values.

8.58 Given the following bond-dissociation energies, calculate the average bond enthalpy for the Ti–Cl bond.

	ΔH (kJ mol^{-1})
$TiCl_4(g) \longrightarrow TiCl_3(g) + Cl(g)$	335
$TiCl_3(g) \longrightarrow TiCl_2(g) + Cl(g)$	423
$TiCl_2(g) \longrightarrow TiCl(g) + Cl(g)$	444
$TiCl(g) \longrightarrow Ti(g) + Cl(g)$	519

8.59 (a) Using average bond enthalpies, predict which of the following reactions will be most exothermic:

(i) $C(g) + 2 F_2(g) \longrightarrow CF_4(g)$
(ii) $CO(g) + 3 F_2(g) \longrightarrow CF_4(g) + OF_2(g)$
(iii) $CO_2(g) + 4 F_2(g) \longrightarrow CF_4(g) + 2 OF_2(g)$

(b) Explain the trend, if any, that exists between reaction exothermicity and the extent to which the carbon atom is bonded to oxygen.

ADDITIONAL EXERCISES

8.60 How many elements in the periodic table are represented by a Lewis symbol with a single dot? Are all these elements in the same group? Explain.

8.61 (a) Explain the following trend in lattice energy: BeH$_2$, 3205 kJ mol^{-1}, MgH$_2$, 2791 kJ mol^{-1}; CaH$_2$, 2410 kJ mol^{-1}; SrH$_2$, 2250 kJ mol^{-1}; BaH$_2$, 2121 kJ mol^{-1}. (b) The lattice energy of ZnH$_2$ is 2870 kJ mol^{-1}. Based on the data given in part (a), the radius of the Zn^{2+} ion is expected to be closest to that of which group 2 element?

8.62 Based on data in Table 8.2, estimate the lattice energy for **(a)** LiBr, **(b)** CsBr, **(c)** $CaCl_2$.

8.63 **(a)** How does a polar molecule differ from a non-polar one? **(b)** Atoms X and Y have different electronegativities. Will the diatomic molecule X–Y necessarily be polar? Explain. **(c)** What factors affect the size of the dipole moment of a diatomic molecule?

8.64 Which of the following molecules or ions contain polar bonds: **(a)** P_4, **(b)** H_2S, **(c)** NO_2^-, **(d)** S_2^{2-}?

8.65 For the following collection of non-metallic elements, O, P, Te, I, B, **(a)** which two would form the most polar single bond? **(b)** Which two would form the longest single bond? **(c)** Which two would be likely to form a compound of formula XY_2? **(d)** Which combinations of elements would be likely to yield a compound of empirical formula X_2Y_3? In each case, explain your answer.

8.66 Using the electronegativities of Br and Cl, estimate the partial charges on the atoms in the Br–Cl molecule. Using these partial charges and the atomic radii given in Figure 7.7, estimate the dipole moment of the molecule. The measured dipole moment is 0.57 D.

8.67 Although I_3^- is known, F_3^- is not. Using Lewis structures, explain why F_3^- does not form.

8.68 Calculate the formal charge on the indicated atom in each of the following molecules or ions: **(a)** the central oxygen atom in O_3, **(b)** phosphorus in PF_6^-, **(c)** nitrogen in NO_2, **(d)** iodine in ICl_3, **(e)** chlorine in $HClO_4$ (hydrogen is bonded to O).

8.69 **(a)** Determine the formal charge on the chlorine atom in the hypochlorite ion, ClO^-, and the perchlorate ion, ClO_4^-, if the Cl atom has an octet. **(b)** What are the oxidation numbers of chlorine in ClO^- and in ClO_4^-? **(c)** What are the essential differences in the definitions of formal charge and oxidation number that lead to the differences in your answers to parts (a) and (b)?

8.70 The following three Lewis structures can be drawn for N_2O:

$$:N{\equiv}N{-}\ddot{\ddot{O}}: \longleftrightarrow :\ddot{N}{-}N{\equiv}O: \longleftrightarrow :\ddot{N}{=}N{=}\ddot{O}:$$

(a) Using formal charges, which of these three resonance forms is likely to be the most important? **(b)** The N–N bond length in N_2O is 112 pm, slightly longer than a typical N–N triple bond, and the N–O bond length is 119 pm, slightly shorter than a typical N–O double bond (see Table 8.5.). Rationalise these observations in terms of the resonance structures shown previously and your conclusion for (a).

8.71 An important reaction for the conversion of natural gas to other useful hydrocarbons is the conversion of methane to ethane.

$$2\,CH_4(g) \longrightarrow C_2H_6(g) + H_2(g)$$

In practice, this reaction is carried out in the presence of oxygen, which converts the hydrogen produced to water.

$$2\,CH_4(g) + \tfrac{1}{2}O_2(g) \longrightarrow C_2H_6(g) + H_2O(g)$$

Use bond enthalpies (Table 8.4) to estimate $\Delta_{rxn}H$ for these two reactions. Why is the conversion of methane to ethane more favourable when oxygen is used?

8.72 With reference to the My World of Chemistry box on explosives, use bond enthalpies to estimate the enthalpy change for the explosion of 1.00 g of nitroglycerin.

8.73 The 'plastic' explosive C-4, often used in action movies, contains the molecule *cyclotrimethylenetrinitramine*, which is often called RDX (for Royal Demolition eXplosive):

Cyclotrimethylenetrinitramine (RDX)

(a) Complete the Lewis structure for the molecule by adding unshared electron pairs where they are needed. **(b)** Does the Lewis structure you drew in part (a) have any resonance structures? If so, how many? **(c)** The molecule causes an explosion by decomposing into $CO(g)$, $N_2(g)$ and $H_2O(g)$. Write a balanced equation for the decomposition reaction. **(d)** With reference to Table 8.4, which is the weakest type of bond in the molecule? **(e)** Use average bond enthalpies to estimate the enthalpy change when 5.0 g of RDX decomposes.

8.74 Consider this reaction involving the hypothetical molecule A=A, which contains a double bond:

$$\underset{\substack{A\\ \|\\ A}}{A} + \underset{\substack{A\\ \|\\ A}}{A} \longrightarrow \underset{\substack{A{-}A\\ | \quad |\\ A{-}A}}{}$$

(a) In terms of $D(A{=}A)$ and $D(A{-}A)$, write a general expression for the enthalpy change for this reaction. **(b)** What must be the relationship between $D(A{=}A)$ and $D(A{-}A)$ for the reaction to be exothermic? **(c)** If the reaction is exothermic, what can you say about the strength of the second bond between the A atoms in A=A relative to the strength of the first bond?

8.75 The bond lengths of carbon–carbon, carbon–nitrogen, carbon–oxygen and nitrogen–nitrogen single, double and triple bonds are listed in Table 8.5. Plot bond enthalpy (Table 8.4) versus bond length for these bonds. What do you conclude about the relationship between bond length and bond enthalpy? What do you conclude about the relative strengths of C–C, C–N, C–O and N–N bonds?

8.76 Use the data in Table 8.5 and the following data S–S distance in S_8 = 205 pm; S–O distance in SO_2 = 143 pm— to answer the following questions. **(a)** Predict the bond length in an S–N single bond. **(b)** Predict the bond length in an S–O single bond. **(c)** Why is the S–O bond length in SO_2 considerably shorter than your predicted value for the S–O single bond? **(d)** When elemental sulfur, S_8, is carefully oxidised, a compound S_8O is formed in which one of the sulfur atoms in the S_8 ring is bonded to an oxygen atom. The S–O bond length in this compound is 148 pm. In light of this information, write Lewis structures that can account for the observed S–O bond length. Does the sulfur bearing the oxygen in this compound obey the octet rule?

INTEGRATIVE EXERCISES

8.77 The electron affinity of oxygen is -141 kJ mol^{-1}, corresponding to the reaction

$$O(g) + e^- \longrightarrow O^-(g)$$

The lattice energy of $K_2O(s)$ is 2238 kJ mol^{-1}. Use these data along with data in Appendix C and Figure 7.9 to calculate the 'second electron affinity' of oxygen, corresponding to the reaction

$$O^-(g) + e^- \longrightarrow O^{2-}(g)$$

8.78 The reaction of indium, In, with sulfur leads to three binary compounds, which we will assume to be purely ionic. The three compounds have the following properties:

Compound	Mass % In	Melting point (°C)
A	87.7	653
B	78.2	692
C	70.5	1050

(a) Determine the empirical formulae of compounds A, B and C. **(b)** Give the oxidation state of In in each of the three compounds. **(c)** Write the electron configuration for the In ion in each compound. Do any of these configurations correspond to a noble gas configuration? **(d)** In which compound is the ionic radius of In expected to be smallest? Explain. **(e)** The melting point of ionic compounds often correlates with the lattice energy. Explain the trends in the melting points of compounds A, B and C in these terms.

8.79 One scale for electronegativity is based on the concept that the electronegativity of any atom is proportional to the ionisation energy of the atom minus its electron affinity: electronegativity = $k(\text{IE} - \text{EA})$, where k is a proportionality constant. **(a)** How does this definition explain why the electronegativity of F is greater than that of Cl even though Cl has the greater electron affinity? **(b)** Why are both ionisation energy and electron affinity relevant to the notion of electronegativity? **(c)** By using data in Chapter 7, determine the value of k that would lead to an electronegativity of 4.0 for F under this definition. **(d)** Use your result from part (c) to determine the electronegativities of Cl and O using this scale. Do these values follow the trend shown in Figure 8.5?

8.80 The compound chloral hydrate, known in detective stories as knockout drops, is composed of 14.52% C, 1.83% H, 64.30% Cl and 19.35% O by mass and has a molar mass of 165.4 g mol^{-1}. **(a)** What is the empirical formula of this substance? **(b)** What is the molecular formula of this substance? **(c)** Draw the Lewis structure of the molecule, assuming that the Cl atoms bond to a single C atom and that there is a C–C bond and two C–O bonds in the compound.

8.81 Barium azide is 62.04% Ba and 37.96% N. Each azide ion has a net charge of 1−. **(a)** Determine the chemical formula of the azide ion. **(b)** Write three resonance structures for the azide ion. **(c)** Which structure is most important? **(d)** Predict the bond lengths in the ion.

8.82 Acetylene (C_2H_2) and nitrogen (N_2) both contain a triple bond, but they differ greatly in their chemical properties. **(a)** Write the Lewis structures for the two substances. **(b)** By referring to the index, look up the chemical properties of acetylene and nitrogen and compare their reactivities. **(c)** Write balanced chemical equations for the complete oxidation of N_2 to form $N_2O_5(g)$ and of acetylene to form $CO_2(g)$ and $H_2O(g)$. **(d)** Calculate the enthalpy of oxidation per mole of N_2 and C_2H_2 (the enthalpy of formation of $N_2O_5(g)$ is 11.30 kJ mol^{-1}). How do these comparative values relate to your response to part (b)? Both N_2 and C_2H_2 possess triple bonds with quite high bond enthalpies (Table 8.4). What aspect of chemical bonding in these molecules or in the oxidation products seems to account for the difference in chemical reactivities?

8.83 Under special conditions, sulfur reacts with anhydrous liquid ammonia to form a binary compound of sulfur and nitrogen. The compound is found to consist of 69.6% S and 30.4% N. Measurements of its molecular mass yield a value of 184.3 g mol^{-1}. The compound occasionally detonates on being struck or when heated rapidly. The sulfur and nitrogen atoms of the molecule are joined in a ring. All the bonds in the ring are of the same length. **(a)** Calculate the empirical and molecular formulae for the substance. **(b)** Write Lewis structures for the molecule, based on the information you are given. [*Hint:* You should find a relatively small number of dominant Lewis structures.] **(c)** Predict the bond distances between the atoms in the ring. (*Note:* The S–S distance in the S_8 ring is 205 pm.) **(d)** The enthalpy of formation of the compound is estimated to be 480 kJ mol^{-1}. $\Delta_f H°$ of S(g) is 222.8 kJ mol^{-1}. Estimate the average bond enthalpy in the compound.

8.84 A common form of elemental phosphorus is the tetrahedral P_4 molecule:

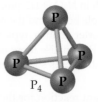

At room temperature phosphorus is a solid. **(a)** Do you think there are any unshared pairs of electrons in the P_4 molecule? **(b)** How many P–P bonds are there in the molecule? **(c)** Use data in Appendix C to determine the enthalpy of atomisation of $P_4(g)$. **(d)** Based on these results, what value would you give for $D(\text{P–P})$? **(e)** Is the P–P bond stronger or weaker than an N–N single bond?

8.85 Average bond enthalpies are generally defined for gas-phase molecules. Many substances are liquids in their standard state. By using appropriate thermochemical data from Appendix C, calculate average bond enthalpies in the liquid state for the following bonds, and compare these values with the gas-phase values given in Table 8.4: **(a)** Br–Br, from $Br_2(l)$; **(b)** C–Cl, from $CCl_4(l)$; **(c)** O–O, from $H_2O_2(l)$ (assume that the O–H bond enthalpy is the same as in the gas phase). **(d)** What can you conclude about the process of breaking bonds in the liquid as compared with the gas phase? Explain the difference in the $\Delta_{rxn}H$ values between the two phases.

MasteringChemistry (www.pearson.com.au/masteringchemistry)

Make learning part of the grade. Access:

- tutorials with peronalised coaching
- study area
- Pearson eText

PHOTO/ART CREDITS

250 © 2009 Robert O. Meyer. Reproduced with permission by Robert Meyer; **281** Courtesy of The Nobel Foundation.

9

MOLECULAR GEOMETRY AND BONDING THEORIES

One attraction of orchid flowers is their diverse and often unusual shape.

KEY CONCEPTS

9.1 MOLECULAR SHAPES
We begin by discussing *molecular shapes* and examining some shapes commonly encountered in molecules.

9.2 THE VSEPR MODEL
We consider how molecular geometries can be predicted using the *valence-shell electron-pair repulsion*, or *VSEPR*, model, which is based on Lewis structures and the repulsion between regions of high electron density.

9.3 MOLECULAR SHAPE AND MOLECULAR POLARITY
Once we know the geometry of a molecule and the types of bonds it contains, we can determine whether the molecule is *polar* or *non-polar*.

9.4 COVALENT BONDING AND ORBITAL OVERLAP
We recognise that electrons are shared between atoms in a covalent bond. In *valence-bond theory*, the bonding electrons are visualised as originating in atomic orbitals on two atoms. A covalent bond is formed when these orbitals overlap.

9.5 HYBRID ORBITALS
To account for molecular shape, we consider how the orbitals of one atom mix with one another, or *hybridise*, to create *hybrid orbitals*.

9.6 MULTIPLE BONDS
Atomic orbitals that contribute to covalent bonding in a molecule can overlap in multiple ways to produce *sigma* and *pi* bonds between atoms. Single bonds generally consist of one sigma bond; multiple bonds involve one sigma and one or more pi bonds. We examine the geometric arrangements of these bonds and how they are exemplified in organic molecules.

9.7 MOLECULAR ORBITALS
We examine a more sophisticated treatment of bonding called *molecular orbital theory*, which introduces the concepts of *bonding* and *antibonding molecular orbitals*.

9.8 SECOND-PERIOD DIATOMIC MOLECULES
We consider how molecular orbital theory is used to construct *energy-level diagrams* for second-row diatomic molecules.

W e saw in chapter 8 that Lewis structures help us understand the compositions of molecules and their covalent bonds. However, Lewis structures do not show one of the most important aspects of molecules—their overall shapes. Molecules have shapes and sizes that are defined by the angles and distances between the nuclei of their component atoms. Indeed, chemists often refer to molecular *architecture* in describing the distinctive shapes and sizes of molecules.

The chapter opening photograph shows the flower of the windswept helmet-orchid, *Nematoceras dienemum*. Orchids are one of the largest families of flowering plants and are distributed over a diverse range of habitats. Most are found in tropical areas, but there are three species of orchid found on Macquarie Island, close to the Antarctic. *Nematoceras* is one of these. The colours and shape of orchid flowers have evolved as part of a complex mechanism to ensure cross-pollination. Here the structure of the flower is closely related to its function. In the world of chemistry the shape and size of a molecule of a particular substance, together with the strength and polarity of its bonds, largely determine the properties of that substance.

Our first goal in this chapter is to learn the relationship between two-dimensional Lewis structures and three-dimensional molecular shapes. Armed with this knowledge, we can then examine more closely the nature of covalent bonds. The lines used to depict bonds in Lewis structures provide important clues about the orbitals that molecules use in bonding. By examining these orbitals, we can gain a greater understanding of the behaviour of molecules. Mastering the material in this chapter will help you in later discussions of the physical and chemical properties of substances.

9.1 | MOLECULAR SHAPES

∞ Review this on page 267

In Chapter 8 we used Lewis structures to account for the formulae of covalent compounds (∞ Section 8.5, 'Drawing Lewis Structures').

Lewis structures, however, do not indicate the shapes of molecules; they simply show the number and types of bonds between atoms. For example, the Lewis structure of CCl_4 tells us only that four Cl atoms are bonded to a central C atom:

$$:\ddot{C}l: \\ | \\ :\ddot{C}l-C-\ddot{C}l: \\ | \\ :\ddot{C}l:$$

The Lewis structure is drawn with the atoms all in the same plane. As shown in ▼ FIGURE 9.1, however, the actual three-dimensional arrangement of the atoms shows the Cl atoms at the corners of a *tetrahedron*, a geometric object with four corners and four faces, each of which is an equilateral triangle. This geometry gives the largest distance between the chlorine atoms. The three-dimensional representation of a molecule is termed the **molecular geometry** of the molecule.

The overall shape of a molecule is determined by its **bond angles**, the angles made by the lines joining the nuclei of the atoms in the molecule. The bond angles of a molecule, together with the **bond lengths** (Section 8.8), accurately define the shape and size of the molecule. In CCl_4, the bond angles are defined as the angles between the C–Cl bonds. You should be able to see that there are six Cl–C–Cl angles in CCl_4, and they all have the same value (109.5°, which is characteristic of a tetrahedron). In addition, all four C–Cl bonds are the same length (178 pm). Thus the shape and size of CCl_4 are completely described by stating that the molecule is tetrahedral with C–Cl bonds of length 178 pm.

In our discussion of the shapes of molecules, we begin with molecules (and ions) that, like CCl_4, have a single central atom bonded to two or more atoms of the same type. Such molecules conform to the general formula AB_n, in which the central atom, A, is bonded to *n* B atoms. Both CO_2 and H_2O are AB_2 molecules, for example, whereas SO_3 and NH_3 are AB_3 molecules and so on.

The possible shapes of AB_n molecules depend on the value of *n*. For a given value of *n*, only a few general shapes are observed. Those commonly found for AB_2 and AB_3 molecules are shown in ▶ FIGURE 9.2. Thus an AB_2 molecule must be either linear (bond angle = 180°) or bent (bond angle ≠ 180°). For example, CO_2 is linear and SO_2 is bent. For AB_3 molecules, the two most common shapes

▲ **FIGURE IT OUT**

In the space-filling model, what determines the relative sizes of the spheres?

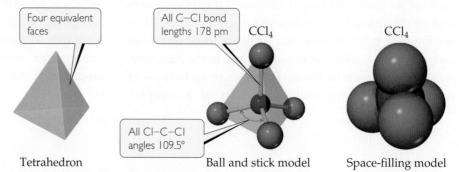

Four equivalent faces

All C–Cl bond lengths 178 pm CCl_4

CCl_4

All Cl–C–Cl angles 109.5°

Tetrahedron Ball and stick model Space-filling model

▲ **FIGURE 9.1** **Tetrahedral shape of CCl_4.**

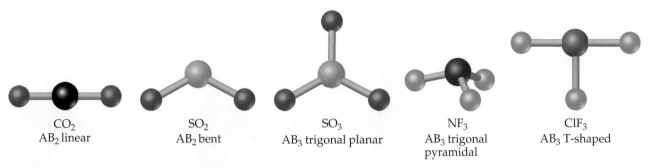

▲ FIGURE 9.2 **Shapes of AB$_2$ and AB$_3$ molecules.**

place the B atoms at the corners of an equilateral triangle. If the A atom lies in the same plane as the B atoms, the shape is called *trigonal planar*. If the A atom lies above the plane of the B atoms, the shape is called *trigonal pyramidal* (a pyramid with an equilateral triangle as its base). For example, SO$_3$ is trigonal planar and NF$_3$ is trigonal pyramidal. Some AB$_3$ molecules such as ClF$_3$, exhibit the more unusual *T shape* shown in Figure 9.2.

The shape of any particular AB$_n$ molecule can usually be derived from one of the five basic geometric structures shown in **▼ FIGURE 9.3**. Starting with a tetrahedron, for example, we can remove atoms successively from the corners as shown in **▼ FIGURE 9.4**. When an atom is removed from one corner of the tetrahedron, the remaining fragment has a trigonal-pyramidal geometry such as that found for NF$_3$. When two atoms are removed, a bent geometry results.

Why do so many AB$_n$ molecules have shapes related to the basic structures in Figure 9.3, and can we predict these shapes? When A is a main-group element (one of the elements from the *s* block or *p* block of the periodic table), we can answer these questions by using the *valence-shell electron-pair repulsion (VSEPR)* model. Although the name is rather imposing, the model is quite simple and useful, as we will see in Section 9.2.

FIGURE IT OUT

Which of these molecular shapes do you expect for the SF$_6$ molecule?

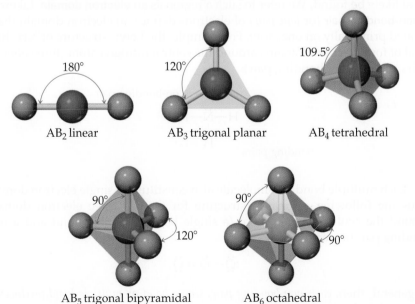

◄ FIGURE 9.3 **Shapes allowing maximum distance between atoms in AB$_n$ molecules.** For molecules whose formula is of the general form AB$_n$, there are five fundamental shapes.

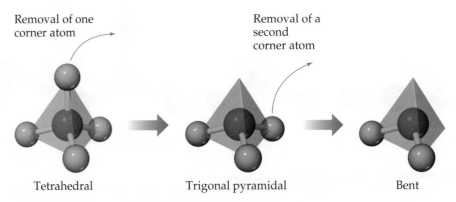

Removal of one corner atom →

Removal of a second corner atom →

Tetrahedral   Trigonal pyramidal   Bent

▲ **FIGURE 9.4** **Derivatives of the tetrahedral molecular shape.**

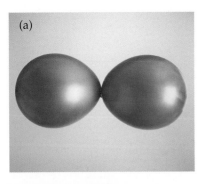

(a)

Two balloons adopt a linear orientation

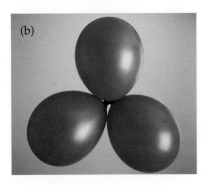

(b)

Three balloons adopt a trigonal-planar orientation

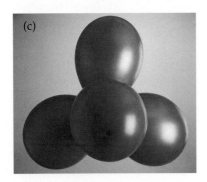

(c)

Four balloons adopt a tetrahedral orientation

▲ **FIGURE 9.5** **A balloon analogy for electron domains.**

∞ Review this on page 258

> ## CONCEPT CHECK 1
>
> One of the common shapes for AB$_4$ molecules is *square planar*: all five atoms lie in the same plane, the atoms B lie at the corners of a square and the atom A is at the centre of the square. Which of the shapes in Figure 9.3 could lead to a square planar geometry upon the removal of one or more atoms?

9.2 | THE VSEPR MODEL

The **valence-shell electron-pair repulsion (VSEPR) model** can be visualised as follows. Imagine tying two identical balloons together at their ends. As shown in ◀ **FIGURE 9.5(a)**, the balloons naturally orient themselves to point away from each other; that is, they try to 'get out of each other's way' as much as possible. If we add a third balloon, the balloons orient themselves towards the vertices of an equilateral triangle, as in Figure 9.5(b). If we add a fourth balloon, they adopt a tetrahedral shape (Figure 9.5(c)). We see that there is an optimum geometry for each number of balloons.

In some ways the electrons in molecules behave like the balloons in Figure 9.5. We have seen that a single covalent bond is formed between two atoms when a pair of electrons occupies the space between the atoms (∞ Section 8.3, 'Covalent Bonding').

A **bonding pair** of electrons thus defines a region in which the electrons will most likely be found. We refer to such a region as an **electron domain**. Likewise, a **non-bonding pair** (or *lone pair*) of electrons defines an electron domain that is located principally on one atom. For example, the Lewis structure of NH$_3$ has a total of four electron domains around the central nitrogen atom (three bonding pairs and one non-bonding pair):

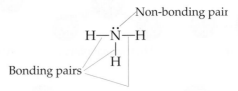

Non-bonding pair

H—N̈—H

Bonding pairs

H

Each multiple bond in a molecule also constitutes a single electron domain. Thus the following resonance structure for O$_3$ has three electron domains around the central oxygen atom (a single bond, a double bond and a non-bonding pair of electrons):

$$\ddot{\text{O}}\!:\!—\!\ddot{\text{O}}\!=\!\ddot{\text{O}}$$

In general, then, *each non-bonding pair, single bond or multiple bond produces an electron domain around the central atom.*

▲ CONCEPT CHECK 2

An AB_3 molecule has the resonance structure

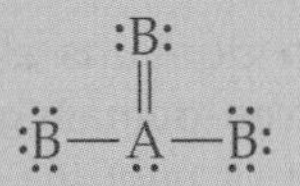

$$
\begin{array}{c}
:\ddot{B}: \\
\parallel \\
:\ddot{B}-A-\ddot{B}:
\end{array}
$$

Does this Lewis structure follow the octet rule? How many electron domains are there around the atom A?

Because electron domains are negatively charged, they repel one another. Therefore, like the balloons in Figure 9.5, electron domains try to stay out of one another's way. *The best arrangement of a given number of electron domains is the one that minimises the repulsions between them.* This simple idea is the basis of the VSEPR model. In fact, the analogy between electron domains and balloons is so close that the same preferred geometries are found in both cases. Thus like the balloons in Figure 9.5, two electron domains are arranged *linearly*, three domains are arranged in a *trigonal-planar* fashion and four are arranged *tetrahedrally*. These arrangements, together with those for five electron domains (*trigonal bipyramidal*) and six electron domains (*octahedral*), are summarised in ▼ TABLE 9.1. If you compare the

TABLE 9.1 • Electron-domain geometries as a function of number of electron domains

Number of electron domains	Arrangement of electron domains	Electron-domain geometry	Predicted bond angles
2	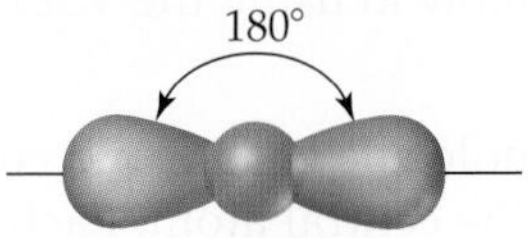 180°	Linear	180°
3	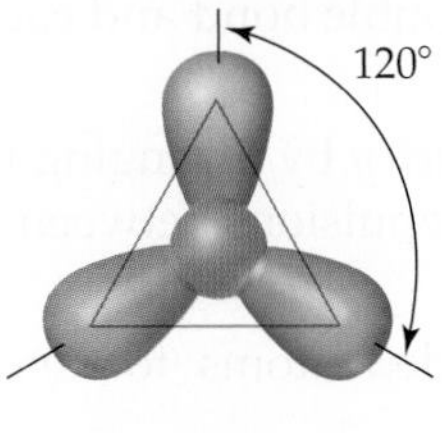 120°	Trigonal planar	120°
4	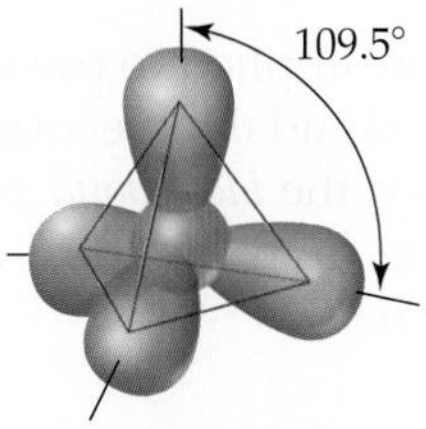 109.5°	Tetrahedral	109.5°
5	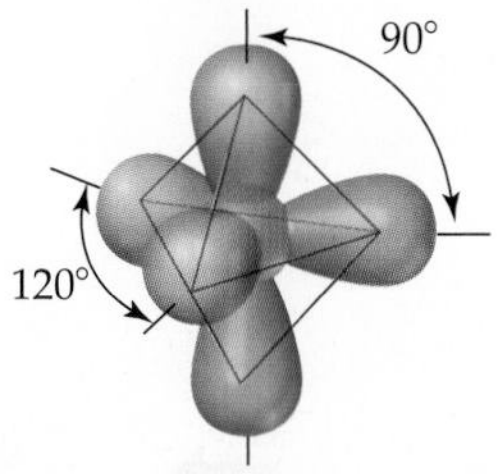 90° 120°	Trigonal bipyramidal	120° 90°
6	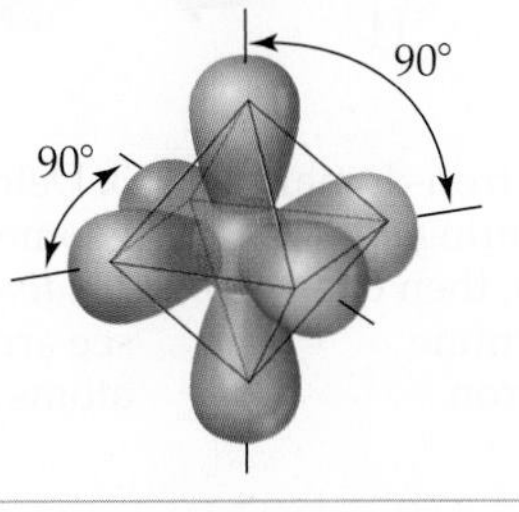 90° 90°	Octahedral	90°

geometries in Table 9.1 with those in Figure 9.3, you will see that they are the same. *The shapes of different AB_n molecules or ions depend on the number of electron domains surrounding the central A atom.*

The arrangement of electron domains about the central atom of an AB_n molecule or ion is called its **electron-domain geometry**. In contrast, the molecular geometry is the arrangement of *only the atoms* in a molecule or ion—any non-bonding pairs are not part of the description of the molecular geometry. In the VSEPR model, we predict the electron-domain geometry and, by knowing how many domains are due to non-bonding pairs, we can then predict the molecular geometry of a molecule or ion from its electron-domain geometry.

When all the electron domains in a molecule arise from bonds, the molecular geometry is identical to the electron-domain geometry. When, however, one or more of the domains involve non-bonding pairs of electrons, we must remember the molecular geometry is derived from, but not identical to, the electron-domain geometry. Consider the NH_3 molecule, for instance, which has four electron domains around the nitrogen atom (▼ **FIGURE 9.6**). We know from Table 9.1 that the repulsions between four electron domains are minimised when the domains point towards the vertices of a tetrahedron—the electron-domain geometry of NH_3 is tetrahedral. We know from the Lewis structure of NH_3 that one of the electron domains is due to a non-bonding pair of electrons, which will occupy one of the four vertices of the tetrahedron. Hence, the molecular geometry of NH_3 is trigonal pyramidal, as shown in Figure 9.6. Notice that it is the tetrahedral arrangement of the four electron domains that leads us to predict the trigonal-pyramidal molecular geometry.

We can generalise the steps we follow in using the VSEPR model to predict the shapes of molecules or ions.

1. Draw the *Lewis structure* of the molecule or ion, and count the total number of electron domains around the central atom. Each non-bonding electron pair, each single bond, each double bond and each triple bond counts as an electron domain.

2. Determine the *electron-domain geometry* by arranging the electron domains about the central atom so that the repulsions between them are minimised, as shown in Table 9.1.

3. Use the arrangement of the bonded atoms to determine the *molecular geometry*.

Figure 9.6 shows how these steps are applied to predict the geometry of the NH_3 molecule. Because the trigonal-pyramidal molecular geometry is based on a tetrahedral electron-domain geometry, the *ideal bond angles* are 109.5°. As we will soon see, bond angles can deviate from the ideal angles unless the surrounding electron domains are identical.

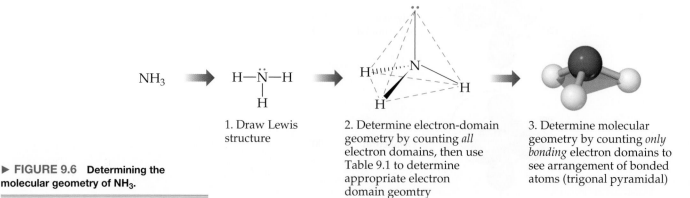

NH_3 ⟶ H—N̈—H (with H below) ⟶ [structure] ⟶ [molecular model]

1. Draw Lewis structure

2. Determine electron-domain geometry by counting *all* electron domains, then use Table 9.1 to determine appropriate electron domain geomtry

3. Determine molecular geometry by counting *only bonding* electron domains to see arrangement of bonded atoms (trigonal pyramidal)

▶ **FIGURE 9.6 Determining the molecular geometry of NH₃.**

Let's apply these steps to determine the shape of the CO_2 molecule. We first draw its Lewis structure, which reveals two electron domains (two double bonds) around the central carbon:

$$\ddot{O}\!=\!C\!=\!\ddot{O}$$

Two electron domains orient in a linear electron-domain geometry (Table 9.1). Because neither of the domains is a non-bonding pair of electrons, the molecular geometry is also linear and the O–C–O bond angle is 180°.

▼ **TABLE 9.2** summarises the possible molecular geometries when an AB_n molecule has four or fewer electron domains about A. These geometries are important because they include all the commonly occurring shapes found for molecules or ions that obey the octet rule.

TABLE 9.2 • Electron-domain and molecular geometries for two, three and four electron domains around a central atom

Number of electron domains	Electron-domain geometry	Bonding domains	Non-bonding domains	Molecular geometry	Example
2	 Linear	2	0	 Linear	$\ddot{O}\!=\!C\!=\!\ddot{O}$
3	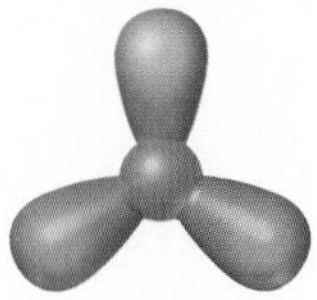 Trigonal planar	3	0	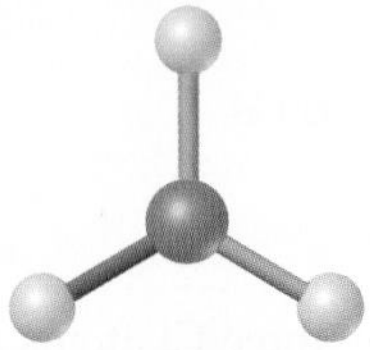 Trigonal planar	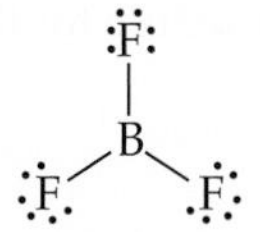
		2	1	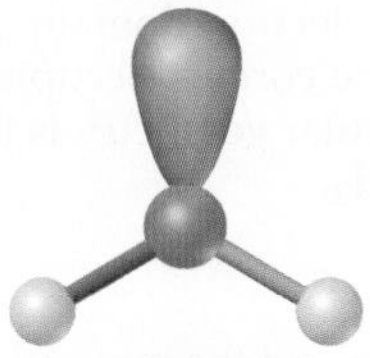 Bent	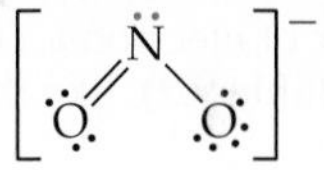
4	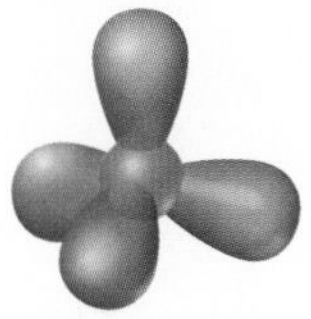 Tetrahedral	4	0	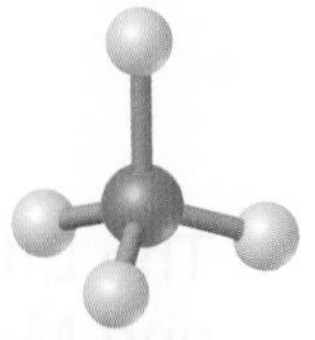 Tetrahedral	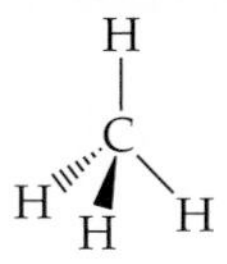
		3	1	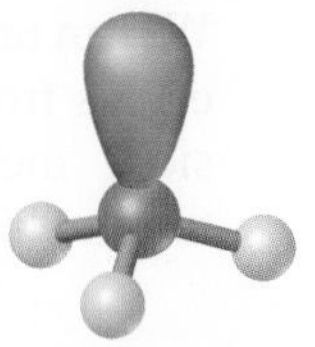 Trigonal pyramidal	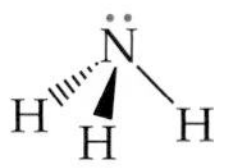
		2	2	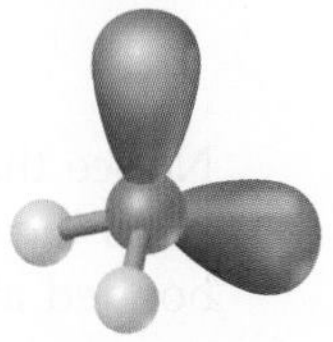 Bent	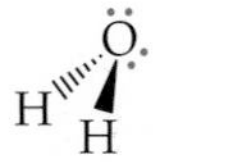

SAMPLE EXERCISE 9.1 | **Using the VSEPR model**

Use the VSEPR model to predict the molecular geometry of **(a)** O_3, **(b)** $SnCl_3^-$.

SOLUTION

Analyse Here we apply the VSEPR model to predict the shape of a molecule and an ion.

Plan To predict the molecular geometries of these species, we first draw their Lewis structures and then count the number of electron domains around the central atom. The number of electron domains gives the electron-domain geometry. We then obtain the molecular geometry from the arrangement of the domains that are due to bonds.

Solve

(a) We can draw two resonance structures for O_3

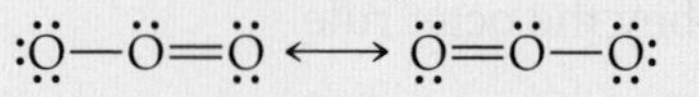

:Ö—Ö=Ö: ⟷ :Ö=Ö—Ö:

Because of resonance, the bonds between the central O atom and the outer O atoms are of equal length. In both resonance structures the central O atom is bonded to the two outer O atoms and has one non-bonding pair. Thus there are three electron domains about the central O atoms. (Remember that a double bond counts as a single electron domain.) The best arrangement of three electron domains is trigonal planar (Table 9.1). Two of the domains are from bonds and one is due to a non-bonding pair, so the molecule has a bent shape with an ideal bond angle of 120° (Table 9.2).

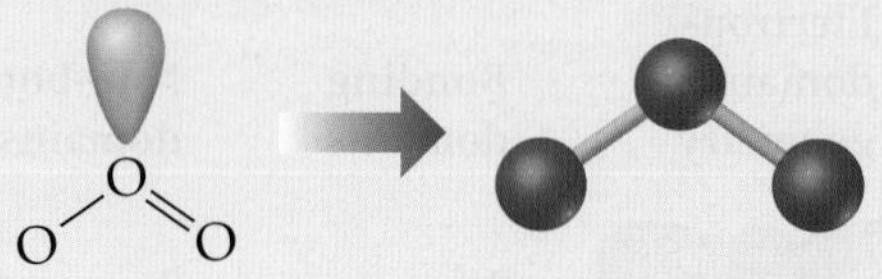

As this example illustrates, when a molecule exhibits resonance, any one of the resonance structures can be used to predict the molecular geometry.

(b) The Lewis structure for the $SnCl_3^-$ ion is

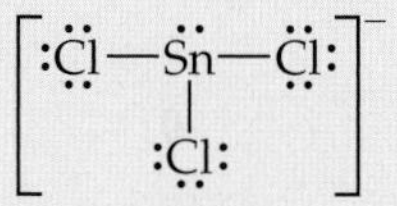

$$\left[\begin{array}{c} :\ddot{Cl}-\ddot{Sn}-\ddot{Cl}: \\ :\ddot{Cl}: \end{array}\right]^-$$

The central Sn atom is bonded to the three Cl atoms and has one non-bonding pair. Therefore, the Sn atom has four electron domains around it. The resulting electron-domain geometry is tetrahedral (Table 9.1) with one of the corners occupied by a non-bonding pair of electrons. The molecular geometry is thus trigonal pyramidal (Table 9.2), like that of NH_3.

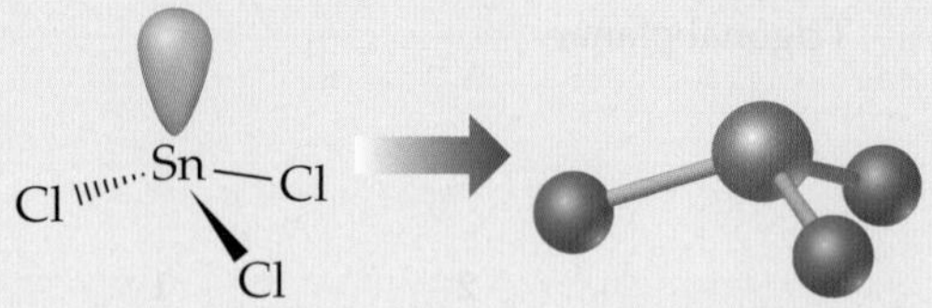

PRACTICE EXERCISE

Predict the electron-domain geometry and the molecular geometry for **(a)** $SeCl_2$, **(b)** CO_3^{2-}.

Answers: (a) tetrahedral, bent; **(b)** trigonal planar, trigonal planar

(See also Exercises 9.15–9.17.)

The Effect of Non-bonding Electrons and Multiple Bonds on Bond Angles

We can refine the VSEPR model to predict and explain slight distortions of molecules from the ideal geometries summarised in Table 9.2. For example, consider methane (CH_4), ammonia (NH_3) and water (H_2O). All three have tetrahedral electron-domain geometries, but their bond angles differ slightly:

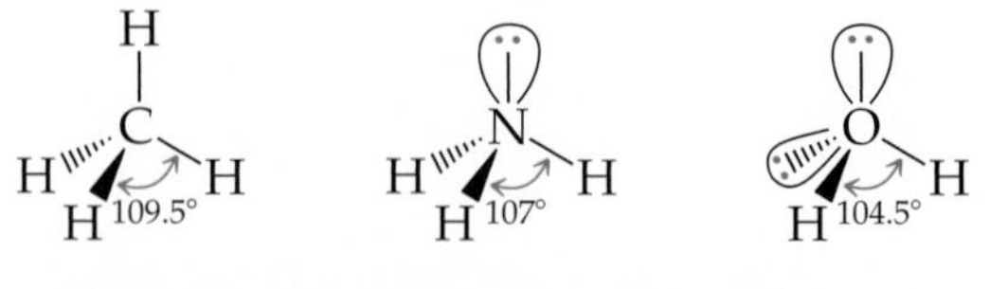

Notice that the bond angles decrease as the number of non-bonding electron pairs increases. A bonding pair of electrons is attracted by both nuclei of the bonded atoms. In contrast, a non-bonding pair is attracted primarily by only one nucleus. Because a non-bonding pair experiences less nuclear attraction, its electron domain is spread out more in space than the electron domain for a bonding pair, as shown in ▶ **FIGURE 9.7**. As a result, *electron domains for non-bonding electron pairs exert greater repulsive forces on adjacent electron domains and thus tend to compress the bond angles.* Using the analogy in Figure 9.5, we can envision the domains for non-bonding electron pairs as represented by balloons that are slightly larger and slightly fatter than those for bonding pairs.

Because multiple bonds contain a higher electronic-charge density than single bonds, multiple bonds also represent larger electron domains ('fatter balloons'). Consider the Lewis structure of *phosgene*, Cl_2CO:

$$:\ddot{C}l$$
$$\overset{|}{C}=\ddot{O}$$
$$:\ddot{C}l$$

Because the central carbon atom is surrounded by three electron domains, we might expect a trigonal-planar geometry with 120° bond angles. The double bond seems to act much like a non-bonding pair of electrons, however, reducing the Cl–C–Cl bond angle from the ideal angle of 120° to an actual angle of 111.4°

$$\overset{Cl}{\underset{Cl}{\overset{\nwarrow 124.3°}{\underset{\swarrow 124.3°}{111.4°\, C=O}}}}$$

In general, electron domains for multiple bonds exert a greater repulsive force on adjacent electron domains than do electron domains for single bonds.

CONCEPT CHECK 3

One of the resonance structures of the nitrate ion, NO_3^-, is

$$\left[\begin{array}{c} :O: \\ \| \\ N \\ \ddot{\underset{..}{O}} \quad \ddot{\underset{..}{O}}: \end{array} \right]^-$$

The bond angles in this ion are exactly 120°. Is this observation consistent with the above discussion of the effect of multiple bonds on bond angles?

Molecules with Expanded Valence Shells

Our discussion of the VSEPR model so far has involved molecules with no more than an octet of electrons around the central atom. Recall, however, that when the central atom of a molecule is from the third period of the periodic table and beyond, that atom may have more than four electron pairs around it (∞ Section 8.7, 'Exceptions to the Octet Rule').

Molecules with five or six electron domains around the central atom display a variety of molecular geometries based on the *trigonal-bipyramidal* (five electron domains) or the *octahedral* (six electron domains) electron-domain geometries, as shown in ▼ **TABLE 9.3**.

The most stable electron-domain geometry for five electron domains is the trigonal bipyramid (two trigonal pyramids sharing a base). Unlike the arrangements we have seen to this point, the electron domains in a trigonal bipyramid can point towards two geometrically distinct types of positions. Two of the five domains point towards what are called *axial positions*, and the remaining three domains point towards *equatorial positions* (▶ **FIGURE 9.8**). Each axial domain makes a 90° angle with any equatorial domain. Each equatorial domain makes a 120° angle with either of the other two equatorial domains and a 90° angle with either axial domain.

Suppose a molecule has five electron domains, one or more of which originates from a non-bonding pair. Will the electron domains from the

FIGURE IT OUT

Why is the volume occupied by the non-bonding electron pair domain larger than the volume occupied by the bonding domain?

Bonding electron pair

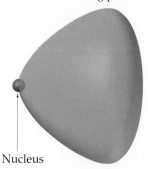

Non-bonding pair

Nuclei

Nucleus

▲ **FIGURE 9.7** Relative volumes occupied by bonding and non-bonding electron domains.

∞ Review this on page 276

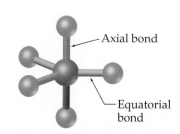

Axial bond

Equatorial bond

▲ **FIGURE 9.8** **Trigonal-bipyramidal geometry.** Five electron domains arrange themselves around a central atom as a trigonal bipyramid. The three *equatorial* electron domains define an equilateral triangle. The two *axial* domains lie above and below the plane of the triangle. If a molecule has non-bonding electron domains, they will occupy the equatorial positions.

TABLE 9.3 • Electron-domain and molecular geometries for five and six electron domains around a central atom

Number of electron domains	Electron-domain geometry	Bonding domains	Non-bonding domains	Molecular geometry	Example
5	 Trigonal bipyramidal	5	0	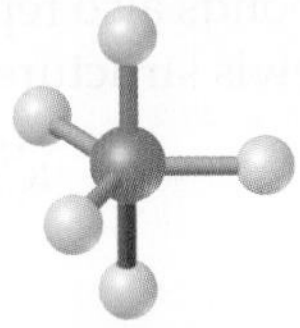 Trigonal bipyramidal	PCl_5
		4	1	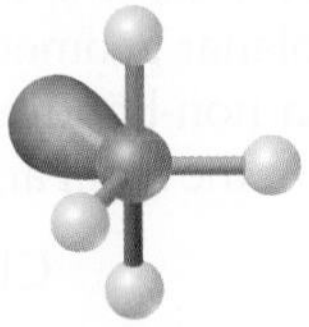 Seesaw	SF_4
		3	2	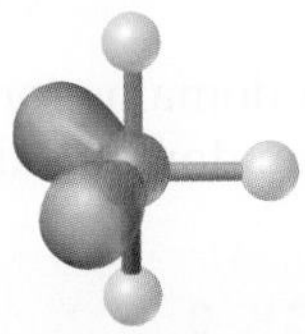 T-shaped	ClF_3
		2	3	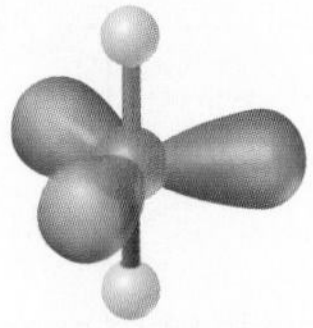 Linear	XeF_2
6	 Octahedral	6	0	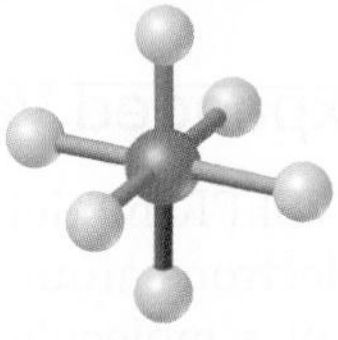 Octahedral	SF_6
		5	1	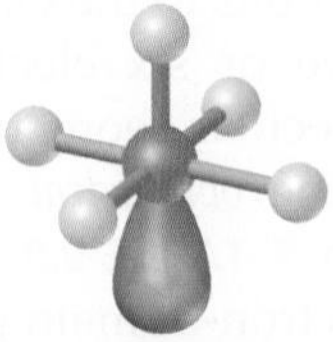 Square pyramidal	BrF_5
		4	2	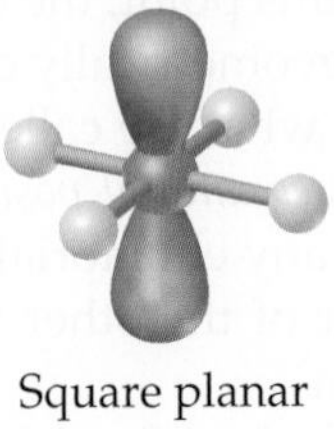 Square planar	XeF_4

non-bonding pairs occupy axial or equatorial positions? In order to answer this question, we must determine which location minimises the repulsions between the electron domains. Repulsions between domains are much greater when they are situated 90° from each other than when they are at 120°. An equatorial domain is 90° from only two other domains (the two axial domains). By contrast, an axial domain is situated 90° from *three* other domains (the three equatorial domains). Hence, an equatorial domain experiences less repulsion than an axial domain. Because the domains from non-bonding pairs exert larger repulsions than those from bonding pairs, they always occupy the equatorial positions in a trigonal bipyramid.

The most stable electron-domain geometry for six electron domains is the *octahedron*. An octahedron is a polyhedron with six vertices and eight faces, each of which is an equilateral triangle (▶ FIGURE 9.9). If an atom has six electron domains around it, that atom can be visualised as being at the centre of the octahedron with the electron domains pointing towards the six vertices. All the bond angles in an octahedron are 90°, and all six vertices are equivalent. Therefore, if an atom has five bonding electron domains and one non-bonding domain, we can point the non-bonding domain at any of the six vertices of the octahedron. The result is always a *square-pyramidal* molecular geometry. When there are two non-bonding electron domains, however, their repulsions are minimised by pointing them towards opposite sides of the octahedron, producing a *square-planar* molecular geometry, as shown in Table 9.3.

▲ FIGURE 9.9 **An octahedron.** The octahedron is an object with eight faces and six vertices. Each face is an equilateral triangle.

SAMPLE EXERCISE 9.2 **Molecular geometries of molecules with expanded valence shells**

Use the VSEPR model to predict the molecular geometry of **(a)** SF_4, **(b)** IF_5.

SOLUTION

Analyse The VSEPR model is used here to predict the molecular geometry of two molecules both of which have more that an octet of electrons on their central atom.

Plan We can predict the structure of SF_4 and IF_5 by first drawing Lewis structures and then using the VSEPR model to determine the electron-domain geometry and molecular geometry.

Solve
(a) The Lewis structure for SF_4 is

The sulfur has five electron domains around it: four from the S–F bonds and one from the non-bonding pair. Each domain points towards a vertex of a trigonal bipyramid. The domain from the non-bonding pair will point towards an equatorial position. The four bonds point towards the remaining four positions, resulting in a molecular geometry that is described as seesaw-shaped:

186°   116°

Comment The experimentally observed structure has the bond angles indicated above, and we can infer that the non-bonding electron domain occupies an equatorial position, as predicted. The axial and equatorial S–F bonds are slightly bent back away from the non-bonding domain, suggesting that the bonding domains are 'pushed' by the non-bonding domain, which is larger and has greater repulsion (Figure 9.7).

(b) The Lewis structure of IF_5 is

The iodine has six electron domains around it, one of which is from a non-bonding pair. The electron-domain geometry is therefore octahedral, with one position occupied by the non-bonding electron pair. The resulting molecular geometry is therefore *square pyramidal* (Table 9.3).

Comment Because the domain for the non-bonding pair is larger than the other domains, the four F atoms in the base of the pyramid are tipped up slightly towards the F atom on top. Experimentally, it is found that the angle between the base and top F atoms is 82°, smaller than the ideal 90° angle of an octahedron.

PRACTICE EXERCISE
Predict the electron-domain geometry and molecular geometry of **(a)** ClF_3, **(b)** ICl_4^-.
Answers: (a) trigonal bipyramidal, T-shaped; **(b)** octahedral, square planar
(See also Exercise 9.18.)

Shapes of Larger Molecules

Although the molecules and ions whose structures we have so far considered contain only a single central atom, the VSEPR model can be extended to more complex molecules. Consider the acetic acid molecule, whose Lewis structure is

$$\begin{array}{c} H \quad :O: \\ | \quad\quad || \\ H-C-C-\ddot{O}-H \\ | \\ H \end{array}$$

Acetic acid has three interior atoms: the leftmost C atom, the central C atom and the rightmost O atom. We can use the VSEPR model to predict the geometry about each of these atoms individually.

	H—C (H, H)	:O:—C	Ö—H
Number of electron domains	4	3	4
Electron-domain geometry	Tetrahedral	Trigonal planar	Tetrahedral
Predicted bond angles	109.5°	120°	109.5°

The leftmost C has four electron domains (all from bonding pairs) and so the geometry around that atom is tetrahedral. The central C has three electron domains (counting the double bond as one domain). Thus the geometry around that atom is trigonal planar. The O atom has four electron domains (two from bonding pairs and two from non-bonding pairs), so its electron-domain geometry is tetrahedral and the molecular geometry around O is bent. The bond angles about the central C atom and the O atom are expected to deviate slightly from the ideal values of 120° and 109.5°, because multiple bonds and non-bonding electron pairs are present. The structure of the acetic acid molecule is shown in ◀ **FIGURE 9.10**.

▲ **FIGURE IT OUT**

Although the electron-domain geometry around the right O is tetrahedral, the C–O–H bond is slightly less than 109.5°. Explain.

▲ FIGURE 9.10 Ball-and-stick (top) and space-filling (bottom) representations of acetic acid, CH_3COOH.

SAMPLE EXERCISE 9.3 Predicting bond angles

Eyedrops for dry eyes usually contain a water-soluble polymer called *poly(vinyl alcohol)*, which is based on the unstable organic molecule called *vinyl alcohol*:

$$\begin{array}{c} H \quad H \\ | \quad\ | \\ H-\ddot{O}-C=C-H \end{array}$$

Predict the approximate values for the H–O–C and O–C–C bond angles in vinyl alcohol.

SOLUTION

Analyse Bond angles can be predicted if the molecular geometry of the atoms of interest is known. In turn, this may be predicted using the VSEPR model.

Plan To predict a particular bond angle, we consider the middle atom of the angle and determine the number of electron domains surrounding that atom. The ideal angle corresponds to the electron-domain geometry around the atom. The angle will be compressed somewhat by non-bonding electrons or multiple bonds.

Solve For the H–O–C bond angle, there are four electron domains around the middle O atom (two bonding and two non-bonding). The electron-domain geometry around O is therefore tetrahedral, which gives an ideal angle of 109.5°. The H–O–C angle will be compressed somewhat by the non-bonding pairs, so we expect this angle to be slightly less than 109.5°.

To predict the O–C–C bond angle, we must examine the leftmost C atom, which is the central atom for this angle. There are three atoms bonded to this C atom and no non-bonding pairs, and so it has three electron domains about it. The predicted electron-domain geometry is trigonal planar, resulting in an ideal bond angle of 120°. Because of the larger size of the C–C double bond domain, however, the O–C–C bond angle should be slightly greater than 120°.

PRACTICE EXERCISE

Predict the H–C–H and C–C–C bond angles in the following molecule, called *propyne*:

$$
\begin{array}{c}
\text{H} \\
| \\
\text{H}-\text{C}-\text{C}\equiv\text{C}-\text{H} \\
| \\
\text{H}
\end{array}
$$

Answer: 109.5°, 180°

(See also Exercises 9.19, 9.20.)

∞ Review this on page 262

9.3 | MOLECULAR SHAPE AND MOLECULAR POLARITY

We now have a sense of the shapes that molecules adopt and why they do so. We spend the rest of this chapter looking more closely at the ways in which electrons are shared to form the bonds between atoms in molecules. We begin by returning to a topic that we first discussed in Section 8.4, namely *bond polarity* and *dipole moments*.

Recall that bond polarity is a measure of how equally the electrons in a bond are shared between the two atoms of the bond: as the difference in electronegativity between the two atoms increases, so does the bond polarity (∞ Section 8.4, 'Bond Polarity and Electronegativity').

We saw that the dipole moment of a diatomic molecule is a quantitative measure of the amount of charge separation in the molecule.

For a molecule that consists of more than two atoms, *the dipole moment depends on both the polarities of the individual bonds and the geometry of the molecule.* For each bond in the molecule, we can consider the **bond dipole**, which is the dipole moment that is due only to the two atoms in that bond. Consider the linear CO_2 molecule, for example. As shown in ▶ **FIGURE 9.11**, each C–O double bond is polar and, because the C–O double bonds are identical, the bond dipoles are equal in magnitude. A plot of the electron density of the CO_2 molecule, shown in the second part of Figure 9.11, clearly shows that the bonds are polar—the regions of high electron density (red) are at the ends of the molecule, on the oxygen atoms—and the regions of low electron density (blue) are in the centre, on the carbon atom. But what can we say about the *overall* dipole moment of the CO_2 molecule?

FIGURE IT OUT

Explain how the directions of the red bond dipole arrows relate to the electron-density picture.

Equal and oppositely directed bond dipoles

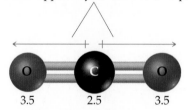

3.5   2.5   3.5

Overall dipole moment = 0

High electron density

Low electron density

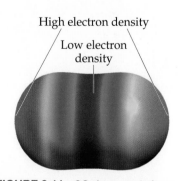

▲ **FIGURE 9.11 CO_2 is a non-polar molecule.** The numbers are electronegativity values for these two atoms.

Bond dipoles

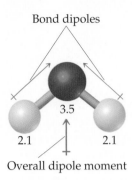

Overall dipole moment

▲ **FIGURE 9.12   H₂O is a polar molecule.** The numbers are electronegativity values.

Bond dipoles and dipole moments are *vector* quantities; that is, they have both a magnitude and a direction. The *overall* dipole moment of a polyatomic molecule is the vector sum of its bond dipoles. Both the magnitudes *and* the directions of the bond dipoles must be considered when summing these vectors. The two bond dipoles in CO_2, although equal in magnitude, are exactly opposite in direction. Adding them together is the same as adding two numbers that are equal in magnitude but opposite in sign, such as $100 + (-100)$: the bond dipoles, like the numbers, 'cancel' each other. Therefore, the overall dipole moment of CO_2 is zero, even though the individual bonds are polar. Thus the geometry of the molecule dictates that the overall dipole moment be zero, making CO_2 a *non-polar* molecule.

Now let's consider H_2O, which is a bent molecule with two polar bonds (◀ **FIGURE 9.12**). Again, the two bonds in the molecule are identical and so the bond dipoles are equal in magnitude. Because the molecule is bent, however, the bond dipoles do not directly oppose each other and therefore do not cancel each other. Hence, the H_2O molecule has an overall non-zero dipole moment ($\mu = 1.85$ D). Because H_2O has a non-zero dipole moment, it is a *polar* molecule. The oxygen atom carries a partial negative charge and the hydrogen atoms each have a partial positive charge, as shown by the electron-density model in the lower part of Figure 9.12.

> **CONCEPT CHECK 4**
>
> The molecule $O\!=\!C\!=\!S$ has a Lewis structure analogous to that of CO_2 and is a linear molecule. Will it necessarily have a zero dipole moment like CO_2?

▼ **FIGURE 9.13** shows examples of polar and non-polar molecules, all of which have polar bonds. The molecules in which the central atom is symmetrically surrounded by identical atoms (BF_3 and CCl_4) are non-polar. For AB_n molecules in which all the B atoms are the same, certain symmetrical shapes—linear (AB_2), trigonal planar (AB_3), tetrahedral and square planar (AB_4), trigonal bipyramidal (AB_5) and octahedral (AB_6)—must lead to non-polar molecules even though the individual bonds might be polar.

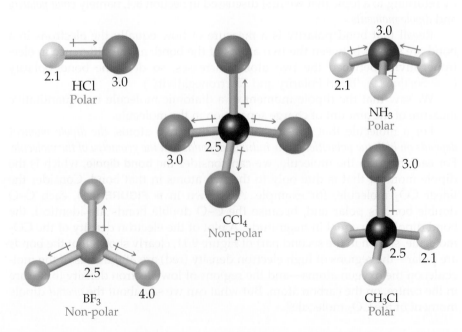

▶ **FIGURE 9.13   Polar and non-polar molecules containing polar bonds.** The numbers are electronegativity values.

SAMPLE EXERCISE 9.4   **Polarity of molecules**

Predict whether the following molecules are polar or non-polar: **(a)** BrCl, **(b)** SO_2, **(c)** SF_6.

SOLUTION

Analyse For a molecule to be polar, it must have atoms of different electronegativities bonded together as well as an asymmetric molecular geometry. We need to determine this for three molecules.

Plan If the molecule contains only two atoms, it will be polar if the atoms differ in electronegativity. If it contains three or more atoms, its polarity depends on both its molecular geometry and the polarity of its bonds. Thus we must draw a Lewis structure for each molecule containing three or more atoms and determine its molecular geometry. We then use the relative electronegativities of the atoms in each bond to determine the direction of the bond dipoles. Finally, we see if the bond dipoles cancel each other to give a non-polar molecule or reinforce each other to give a polar one.

Solve

(a) Chlorine is more electronegative than bromine. All diatomic molecules with polar bonds are polar molecules. Consequently, BrCl will be polar, with chlorine carrying the partial negative charge:

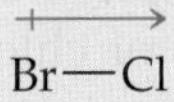

$$Br \longrightarrow Cl$$

The actual dipole moment of BrCl, as determined by experimental measurement, is $\mu = 0.57$ D.

(b) Because oxygen is more electronegative than sulfur, SO_2 has polar bonds. Three resonance forms can be written for SO_2

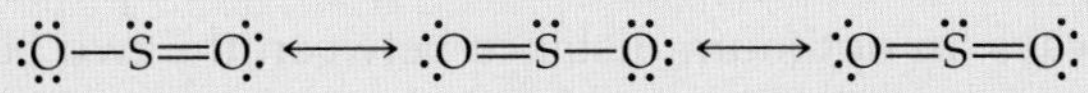

$$\ddot{\underset{..}{O}}-\ddot{S}=\ddot{\underset{..}{O}} \longleftrightarrow \ddot{\underset{..}{O}}=\ddot{S}-\ddot{\underset{..}{O}} \longleftrightarrow \ddot{\underset{..}{O}}=\ddot{S}=\ddot{\underset{..}{O}}$$

For each of these, the VSEPR model predicts a bent geometry. Because the molecule is bent, the bond dipoles do not cancel and the molecule is polar:

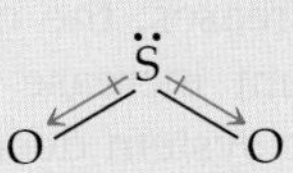

Experimentally, the dipole moment of SO_2 is $\mu = 1.63$ D.

(c) Fluorine is more electronegative than sulfur, so the bond dipoles point towards fluorine. The six S–F bonds are arranged octahedrally around the central sulfur:

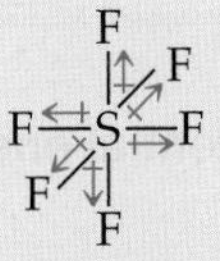

Because the octahedral geometry is symmetrical, the bond dipoles cancel and the molecule is non-polar, meaning that $\mu = 0$.

PRACTICE EXERCISE

Determine whether the following molecules are polar or non-polar: **(a)** NF_3, **(b)** BCl_3.

Answers: **(a)** polar, because polar bonds are arranged in a trigonal-pyramidal geometry, **(b)** non-polar, because polar bonds are arranged in a trigonal-planar geometry.

(See also Exercises 9.27, 9.28.)

9.4 | COVALENT BONDING AND ORBITAL OVERLAP

The VSEPR model provides a simple means for predicting the shapes of molecules. However, it does not explain why bonds exist between atoms. In developing theories of covalent bonding, chemists have approached the problem from another direction, using quantum mechanics. In this approach, the question becomes: How can we use atomic orbitals to explain bonding and

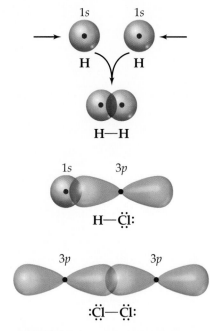

▲ **FIGURE 9.14** **Covalent bonds in H₂, HCl and Cl₂ result from overlap of atomic orbitals.**

to account for the geometries of molecules? The marriage of Lewis's notion of electron-pair bonds and the idea of atomic orbitals leads to a model of chemical bonding called **valence-bond theory**. By extending this approach to include the ways in which atomic orbitals can mix with one another, we obtain a picture that corresponds nicely to the VSEPR model.

In the Lewis theory, covalent bonding occurs when atoms share electrons. Such sharing concentrates electron density between the nuclei. In the valence-bond theory—the buildup of electron density between two nuclei—is visualised as occurring when a valence atomic orbital of one atom merges with that of another atom. The orbitals are then said to share a region of space, or to **overlap**. The overlap of orbitals allows two electrons of opposite spin to share the common space between the nuclei, forming a covalent bond.

The coming together of two H atoms to form H₂ is depicted in the first part of ◀ **FIGURE 9.14**. Each atom has a single electron in a 1*s* orbital. As the orbitals overlap, electron density is concentrated between the nuclei. Because the electrons in the overlap region are simultaneously attracted to both nuclei, they hold the atoms together, forming a covalent bond.

The idea of orbital overlap producing a covalent bond applies equally well to other molecules. In HCl, for example, chlorine has the electron configuration [Ne]3*s*²3*p*⁵. All the valence orbitals of chlorine are full except one 3*p* orbital, which contains a single electron. This electron pairs up with the single electron of H to form a covalent bond. The second part of Figure 9.14 shows the overlap of the 3*p* orbital of Cl with the 1*s* orbital of H. Likewise, we can explain the covalent bond in the Cl₂ molecule in terms of the overlap of the 3*p* orbital of one atom with the 3*p* orbital of another, as shown in the third part of Figure 9.14.

There is always an optimum distance between the two bonded nuclei in any covalent bond. ▼ **FIGURE 9.15** shows how the potential energy of the system changes as two H atoms come together to form an H₂ molecule. As the distance between the atoms decreases, the overlap between their 1*s* orbitals increases. Because of the resultant increase in electron density between the nuclei, the potential energy of the system decreases. That is, the strength of the bond increases, as shown by the decrease in the energy on the curve. However, the curve also shows that as the atoms come very close together, the energy

▲ **FIGURE IT OUT**

On the left part of the curve the potential energy rises above zero. What causes this to happen?

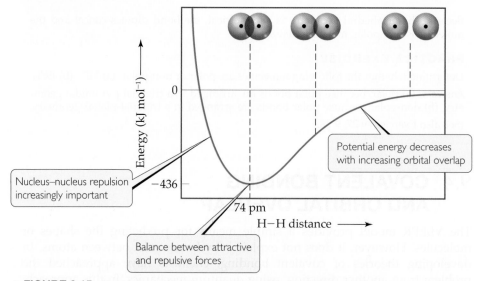

▲ **FIGURE 9.15** **Formation of the H₂ molecule as atomic orbitals overlap.**

increases rapidly. This rapid increase is due mainly to the electrostatic repulsion between the nuclei, which becomes significant at short internuclear distances. The internuclear distance at a minimum of the potential-energy curve corresponds to the observed bond length. Thus the observed bond length is the distance at which the attractive forces between unlike charges (electrons and nuclei) are balanced by the repulsive forces between like charges (electron–electron and nucleus–nucleus).

CONCEPT CHECK 5

Can orbital overlap explain why the bond length in Cl_2 is longer than the bond length in F_2?

9.5 | HYBRID ORBITALS

Although the idea of orbital overlap allows us to understand the formation of covalent bonds, it is not always easy to extend these ideas to polyatomic molecules. When we apply valence-bond theory to polyatomic molecules, we must explain both the formation of electron-pair bonds *and* the observed geometries of the molecules.

To explain geometries, we often assume that the atomic orbitals on an atom mix to form new orbitals called **hybrid orbitals**. The shape of any hybrid orbital is different from the shapes of the original atomic orbitals. The process of mixing atomic orbitals as atoms approach each other to form bonds is called **hybridisation**. The total number of atomic orbitals on an atom remains constant, however, and so the number of hybrid orbitals on an atom equals the number of atomic orbitals mixed.

Let's examine the common types of hybridisation. As we do so, notice the connection between the type of hybridisation and the five basic electron-domain geometries—linear, trigonal planar, tetrahedral, trigonal bipyramidal and octahedral—predicted by the VSEPR model.

sp Hybrid Orbitals

To illustrate the process of hybridisation, consider the BeF_2 molecule, which is generated when solid BeF_2 is heated to high temperatures. The Lewis structure of BeF_2 is

$$:\!\ddot{F}\!-\!Be\!-\!\ddot{F}\!:$$

The VSEPR model correctly predicts that BeF_2 is linear with two identical Be–F bonds, but how can we use valence-bond theory to describe the bonding? The electron configuration of F $(1s^2 2s^2 2p^5)$ indicates there is an unpaired electron in a $2p$ orbital. This $2p$ electron can be paired with an unpaired electron from the Be atom to form a polar covalent bond. Which orbitals on the Be atom, however, overlap with those on the F atoms to form the Be–F bonds?

The orbital diagram for a ground-state Be atom is

Because it has no unpaired electrons, the Be atom in its ground state is incapable of forming bonds with the fluorine atoms. It could form two bonds, however, by 'promoting' one of the $2s$ electrons to a $2p$ orbital:

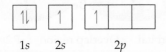

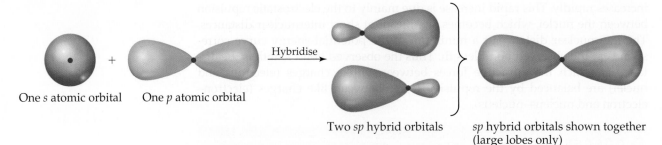

One *s* atomic orbital   One *p* atomic orbital   Two *sp* hybrid orbitals   *sp* hybrid orbitals shown together (large lobes only)

▲ **FIGURE 9.16**   **Formation of *sp* hybrid orbitals.**

The Be atom now has two unpaired electrons and can therefore form two polar covalent bonds with the F atoms. The two bonds would not be identical, however, because a Be 2*s* orbital would be used to form one of the bonds and a 2*p* orbital would be used for the other. Therefore, although the promotion of an electron allows two Be–F bonds to form, we still haven't explained the structure of BeF$_2$.

We can solve this dilemma by 'mixing' the 2*s* orbital and one of the 2*p* orbitals to generate two new orbitals, as shown in ▲ **FIGURE 9.16**. Like *p* orbitals, each of the new orbitals has two lobes. Unlike *p* orbitals, however, one lobe is much larger than the other. The two new orbitals are identical in shape, but their large lobes point in opposite directions. We have created two hybrid orbitals. In this case we have hybridised one *s* and one *p* orbital, so we call each hybrid an *sp* hybrid orbital. *According to the valence-bond model, a linear arrangement of electron domains implies* sp *hybridisation*.

For the Be atom of BeF$_2$, we write the orbital diagram for the formation of two *sp* hybrid orbitals as follows:

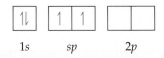

             1*s*   *sp*   2*p*

The electrons in the *sp* hybrid orbitals can form two-electron bonds with the two fluorine atoms (▼ **FIGURE 9.17**). Because the *sp* hybrid orbitals are equivalent but point in opposite directions, BeF$_2$ has two identical bonds and a linear geometry. The remaining two unoccupied 2*p* orbitals remain unhybridised.

Our first step in constructing the *sp* hybrids, namely the promotion of a 2*s* electron to a 2*p* orbital in Be, requires energy. Why, then, do the orbitals hybridise? Hybrid orbitals have one large lobe and can therefore be directed at

▲ **FIGURE IT OUT**

Why is it reasonable to take account of only the large lobes of the Be hybrid orbitals in considering the bonding to F?

Large lobes from two Be *sp* hybrid orbitals

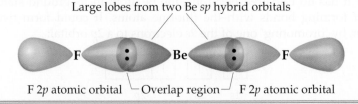

▶ **FIGURE 9.17**   **Formation of two equivalent Be–F bonds in BeF$_2$.**

F 2*p* atomic orbital   Overlap region   F 2*p* atomic orbital

other atoms better than unhybridised atomic orbitals. Hence, they can overlap more strongly with the orbitals of other atoms than atomic orbitals, and stronger bonds result. The energy released by the formation of bonds more than offsets the energy that must be expended to promote electrons.

 CONCEPT CHECK 6

Suppose that two unhybridised $2p$ orbitals were used to make the Be–F bonds in BeF_2. Would the two bonds be equivalent to each other? What would be the expected F–Be–F bond angle?

sp^2 and sp^3 Hybrid Orbitals

Whenever we mix a certain number of atomic orbitals, we get the same number of hybrid orbitals. Each of these hybrid orbitals is equivalent to the others but points in a different direction. Thus mixing one $2s$ and one $2p$ orbital yields two equivalent sp hybrid orbitals that point in opposite directions (Figure 9.16). Other combinations of atomic orbitals can be hybridised to obtain different geometries. In BF_3, for example, a $2s$ electron on the B atom can be promoted to a vacant $2p$ orbital. Mixing the $2s$ and two of the $2p$ orbitals yields three equivalent sp^2 (pronounced 's-p-two') hybrid orbitals:

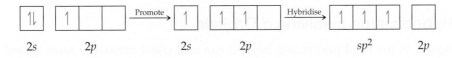

The three sp^2 hybrid orbitals lie in the same plane, 120° apart from one another (▼ **FIGURE 9.18**). They are used to make three equivalent bonds with the three fluorine atoms, leading to the trigonal-planar geometry of BF_3. Notice that an unfilled $2p$ orbital remains unhybridised.

 CONCEPT CHECK 7

In an sp^2 hybridised atom, what is the orientation of the unhybridised p orbital relative to the three sp^2 hybrid orbitals?

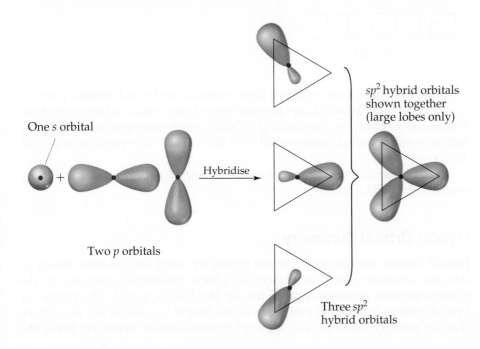

One s orbital

Two p orbitals

sp^2 hybrid orbitals shown together (large lobes only)

Three sp^2 hybrid orbitals

◄ **FIGURE 9.18   Formation of sp^2 hybrid orbitals.**

An s orbital can also mix with all three p orbitals in the same subshell. For example, the carbon atom in CH_4 forms four equivalent bonds with the four hydrogen atoms. We envision this process as resulting from the mixing of the $2s$ and all three $2p$ atomic orbitals of carbon to create four equivalent sp^3 (pronounced 's-p-three') hybrid orbitals:

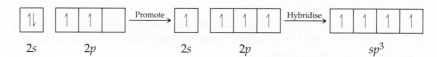

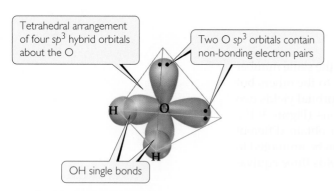

Tetrahedral arrangement of four sp^3 hybrid orbitals about the O

Two O sp^3 orbitals contain non-bonding electron pairs

OH single bonds

▲ **FIGURE 9.19 Hybrid orbital description of H_2O.**

Each sp^3 hybrid orbital has a large lobe that points towards a vertex of a tetrahedron. We reinvestigate sp^3 hybridisation and discuss its use to describe the geometry of organic molecules in Chapter 22. Carbon is not, however, the only atom that can be sp^3 hybridised. In H_2O, for example, the electron-domain geometry around the central O atom is approximately tetrahedral. Thus the four electron pairs can be envisioned as occupying sp^3 hybrid orbitals. Two of the hybrid orbitals contain non-bonding pairs of electrons, and the other two are used to form bonds with hydrogen atoms, as shown in ◄ **FIGURE 9.19.**

Hybridisation Involving d Orbitals

Atoms in the third period and beyond can also use d orbitals to form hybrid orbitals. Mixing one s orbital, three p orbitals and one d orbital leads to five sp^3d hybrid orbitals. These hybrid orbitals are directed towards the vertices of a trigonal bipyramid. The formation of sp^3d hybrids is exemplified by the phosphorus atom in PF_5

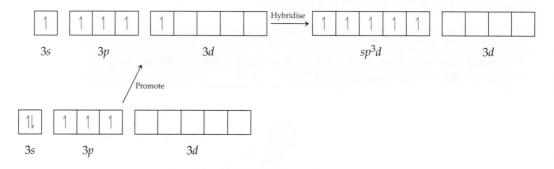

∞ Review this on page 276

Similarly, mixing one s orbital, three p orbitals and two d orbitals gives six sp^3d^2 hybrid orbitals, which are directed towards the vertices of an octahedron. The use of d orbitals in constructing hybrid orbitals corresponds nicely to the notion of an expanded valence shell (∞ Section 8.7, 'Exceptions to the Octet Rule').

The geometrical arrangements characteristic of hybrid orbitals are summarised in ► **TABLE 9.4.**

Hybrid Orbital Summary

Hybrid orbitals provide a convenient model for using valence-bond theory to describe covalent bonds in molecules whose geometries conform to the electron-domain geometries predicted by the VSEPR model. The picture of hybrid orbitals has limited predictive value; that is, we cannot say in advance that the nitrogen atom in NH_3 uses sp^3 hybrid orbitals. When we know the

TABLE 9.4 • Geometric arrangements characteristic of hybrid orbital sets

Atomic orbital set	Hybrid orbital set	Geometry	Examples
s,p	Two sp	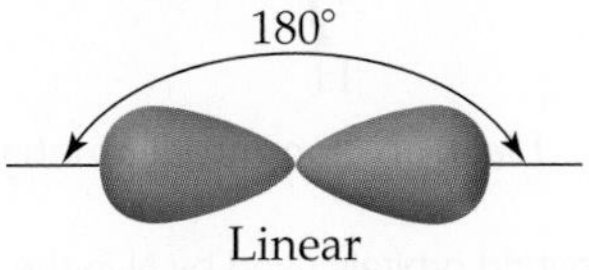 180° Linear	BeF_2, $HgCl_2$
s,p,p	Three sp^2	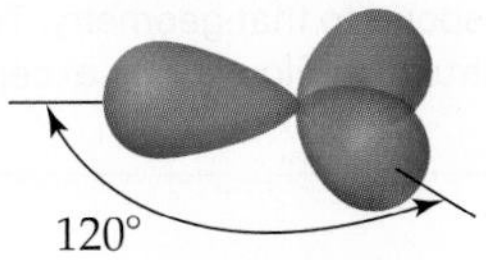 120° Trigonal planar	BF_3, SO_3
s,p,p,p	Four sp^3	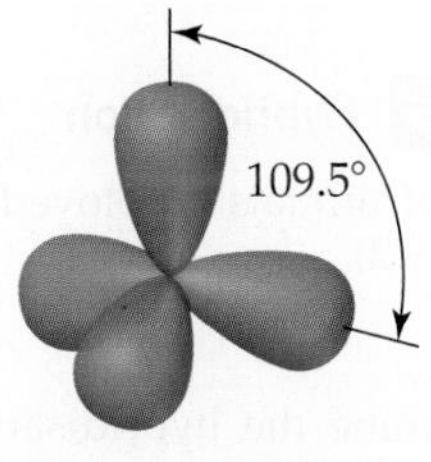 109.5° Tetrahedral	CH_4, NH_3, H_2O, $NH_4{}^+$
s,p,p,p,d	Five sp^3d	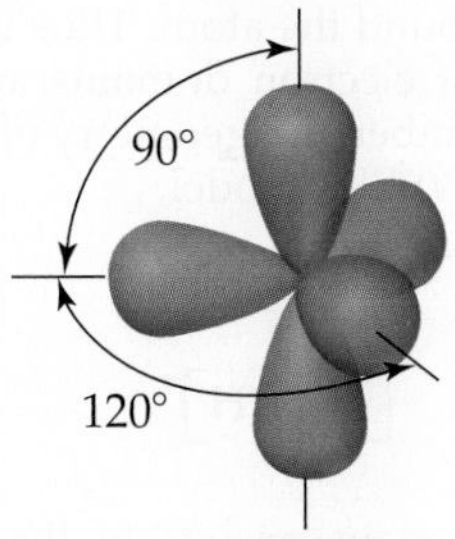 90° 120° Trigonal bipyramidal	PF_5, SF_4, BrF_3
s,p,p,p,d,d	Six sp^3d^2	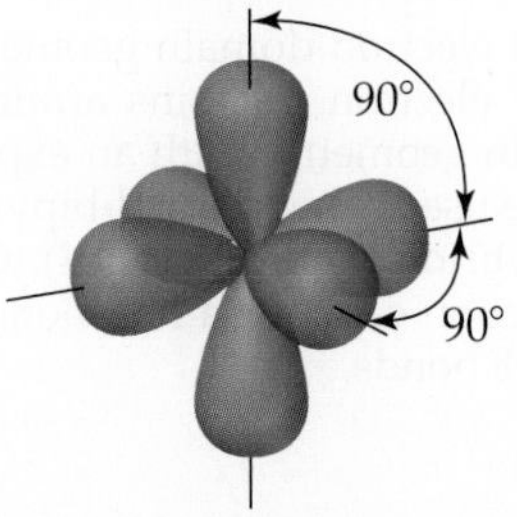 90° 90° Octahedral	SF_6, ClF_5, XeF_4, $PF_6{}^-$

electron-domain geometry, however, we can employ hybridisation to describe the atomic orbitals used by the central atom in bonding.

The following steps allow us to predict the hybrid orbitals used by an atom in bonding.

1. Draw the *Lewis structure* for the molecule or ion.
2. Determine the electron-domain geometry using the *VSEPR model*.
3. Specify the *hybrid orbitals* needed to accommodate the electron pairs based on their geometric arrangement (Table 9.4).

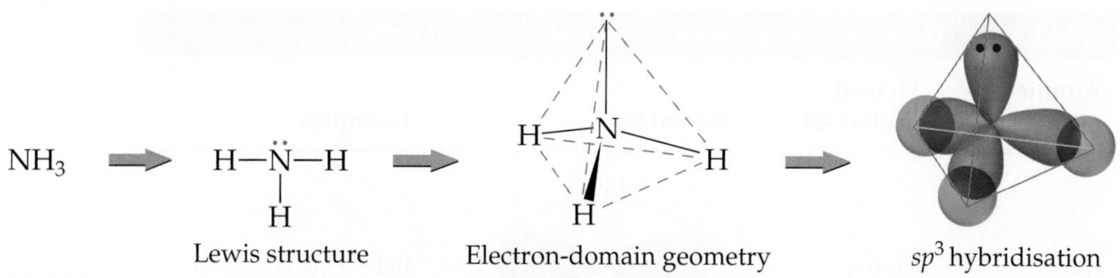

NH$_3$ $\Rightarrow$ $H-\overset{..}{N}-H$ (with H below N) $\Rightarrow$ Electron-domain geometry $\Rightarrow$ sp^3 hybridisation

Lewis structure   Electron-domain geometry sp^3 hybridisation

▲ **FIGURE 9.20 Bonding in NH$_3$.** The hybrid orbitals used by N in the NH$_3$ molecule are predicted by first drawing the Lewis structure, then using the VSEPR model to determine the electron-domain geometry, and then specifying the hybrid orbitals that correspond to that geometry. This is essentially the same procedure as that used to determine molecular structure (Figure 9.6), except we focus on the orbitals used to make bonds and to hold non-bonding pairs.

These steps are illustrated in ▲ **FIGURE 9.20**, which shows how the hybridisation employed by N in NH$_3$ is determined.

SAMPLE EXERCISE 9.5 **Hybridisation**

Indicate the hybridisation of orbitals employed by the central atom in **(a)** NH$_2{}^-$, **(b)** SF$_4$ (see Sample Exercise 9.2).

SOLUTION

Analyse We need to determine the hybridisation of N and S in NH$_2{}^-$ and SF$_4$, respectively.

Plan To determine the hybrid orbitals used by an atom in bonding, we must know the electron-domain geometry around the atom. Thus we first draw the Lewis structure to determine the number of electron domains around the central atom. The hybridisation conforms to the number and geometry of electron domains around the central atom, as predicted by the VSEPR model.

Solve (a) The Lewis structure of NH$_2{}^-$ is

$$\left[H:\overset{..}{\underset{..}{N}}:H \right]^-$$

Because there are four electron domains around N, the electron-domain geometry is tetrahedral. The hybridisation that gives a tetrahedral electron-domain geometry is sp^3 (Table 9.4). Two of the sp^3 hybrid orbitals contain non-bonding pairs of electrons, and the other two are used to make two-electron bonds with the hydrogen atoms.

(b) The Lewis structure and electron-domain geometry of SF$_4$ are shown in Sample Exercise 9.2. There are five electron domains around S, giving rise to a trigonal-bipyramidal electron-domain geometry. With an expanded octet of 10 electrons, a d orbital on the sulfur must be used. The trigonal-bipyramidal electron-domain geometry corresponds to sp^3d hybridisation (Table 9.4). One of the hybrid orbitals that points in an equatorial direction contains a non-bonding pair of electrons; the other four are used to form the S–F bonds.

PRACTICE EXERCISE

Predict the electron-domain geometry and the hybridisation of the central atom in **(a)** SO$_3{}^{2-}$, **(b)** SF$_6$.

Answers: (a) tetrahedral, sp^3; **(b)** octahedral, sp^3d^2

(See also Exercises 9.32, 9.34, 9.35.)

9.6 | MULTIPLE BONDS

Covalent bonds in which the electron density is equally distributed about the *internuclear axis* are called **sigma (σ) bonds**. The formation of these bonds is illustrated in Figures 9.14 and 9.17.

There is a second type of bond that results from the sideways overlap of two p orbitals (▶ FIGURE 9.21), in which the overlap regions lie above and below the internuclear axis. This is called a **pi (π) bond**. Single bonds are σ-bonds and additional bonds between atoms are π-bonds.

H—H

One σ-bond

$\ddot{O}=\ddot{O}$

One σ-bond plus one π-bond

:N≡N:

One σ-bond plus two π-bonds

Multiple bonds are considered in detail in Section 24.1.

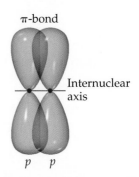

▲ **FIGURE 9.21** **The π-bond.** When two p orbitals overlap in a sideways fashion, the result is a π-bond. Note that the two regions of overlap constitute a single π-bond.

General Conclusions

Together with the discussion of hybrid orbitals in Section 9.5, we can now draw a few helpful conclusions for using the concept of hybrid orbitals to describe molecular structures.

1. Every pair of bonded atoms shares one or more pairs of electrons. In every bond at least one pair of electrons is localised in the space between the atoms in a σ-bond. The appropriate set of hybrid orbitals used to form the σ-bonds between an atom and its neighbours is determined by the observed geometry of the molecule. The correlation between the set of hybrid orbitals and the geometry about an atom is given in Table 9.4.

2. The electrons in σ-bonds are localised in the region between two bonded atoms and do not make a significant contribution to the bonding between any other two atoms.

3. When atoms share more than one pair of electrons, one pair is used to form a σ-bond and the additional pairs form π-bonds. The centres of charge density in a π-bond lie above and below the bond axis.

9.7 | MOLECULAR ORBITALS

Valence-bond theory and hybrid orbitals allow us to move in a straightforward way from Lewis structures to rationalising the observed geometries of molecules in terms of atomic orbitals. For example, we can use this theory to understand why methane has the formula CH_4, how the carbon and hydrogen atomic orbitals are used to form electron-pair bonds, and why the arrangement of the C–H bonds about the central carbon is tetrahedral. This model, however, does not explain all aspects of bonding. It is not successful, for example, in describing the excited states of molecules, which we must understand in order to explain how molecules absorb light, giving them colour.

Some aspects of bonding are better explained by an alternative model called **molecular orbital theory**. In Chapter 6 we saw that electrons in atoms can be described by certain wave functions, which we call atomic orbitals. In a similar way, molecular orbital theory describes the electrons in molecules by using specific wave functions called **molecular orbitals**. Chemists use the abbreviation **MO** for molecular orbital.

MOs have many of the same characteristics as atomic orbitals. For example, a MO can hold a maximum of two electrons (with opposite spins), it has a definite energy and we can visualise its electron-density distribution by using a contour representation, as we did when discussing atomic orbitals. Unlike atomic orbitals, however, MOs are associated with the entire molecule, not with a single atom.

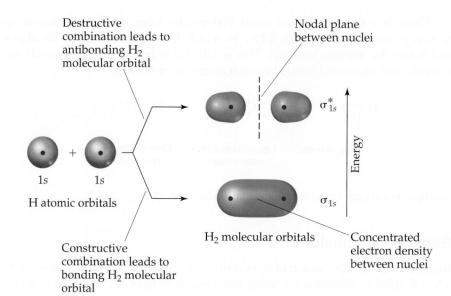

Destructive combination leads to antibonding H_2 molecular orbital

Nodal plane between nuclei

$1s$ $1s$

H atomic orbitals

σ_{1s}^*

Energy

σ_{1s}

H_2 molecular orbitals

Constructive combination leads to bonding H_2 molecular orbital

Concentrated electron density between nuclei

▶ **FIGURE 9.22** **The two molecular orbitals of H_2: one a bonding MO and one an antibonding MO.**

The Hydrogen Molecule

To get a sense of the approach taken in MO theory, we begin with the simplest molecule: the hydrogen molecule, H_2. We use the two $1s$ atomic orbitals (one on each H atom) to 'build' MOs for the H_2 molecule. *Whenever two atomic orbitals overlap, two molecular orbitals form.* Thus the overlap of the $1s$ orbitals of two hydrogen atoms to form H_2 produces two MOs (▲ **FIGURE 9.22**).

The lower-energy MO of H_2 concentrates electron density between the two hydrogen nuclei and is called the **bonding molecular orbital**. This sausage-shaped MO results from summing the two atomic orbitals so that the atomic orbital wave functions enhance each other in the bond region. Because an electron in this MO is strongly attracted to both nuclei, the electron is more stable (in other words, it has lower energy) than it is in the $1s$ atomic orbital of an isolated hydrogen atom. Further, because it concentrates electron density between the nuclei, the bonding MO holds the atoms together in a covalent bond.

The higher-energy MO in Figure 9.22 has very little electron density between the nuclei and is called the **antibonding molecular orbital**. Instead of enhancing each other in the region between the nuclei, the atomic orbitals cancel each other in this region, and the greatest electron density is on opposite sides of the nuclei. This MO excludes electrons from the very region in which a bond must be formed. An electron in this MO is repelled from the bonding region and is therefore less stable (in other words, it has higher energy) than it is in the $1s$ atomic orbital of a hydrogen atom. The electron density in both the bonding MO and the antibonding MO of H_2 is centred about the internuclear axis, an imaginary line passing through the two nuclei. MOs of this type are called *sigma (σ) molecular orbitals.* The bonding sigma MO of H_2 is labelled σ_{1s} the subscript indicating that the MO is formed from two $1s$ orbitals. The antibonding sigma MO of H_2 is labelled σ_{1s}^* (read 'sigma-star-one-s'), the asterisk denoting that the MO is antibonding.

The interaction between two $1s$ atomic orbitals and the MOs that result can be represented by an **energy-level diagram** (also called a **molecular orbital diagram**), like those in ▶ **FIGURE 9.23**. Such diagrams show the interacting atomic orbitals in the left and right columns and the MOs in the middle column. Note that the bonding MO, σ_{1s}, is lower in energy than the atomic $1s$ orbitals, whereas the antibonding orbital, σ_{1s}^* is higher in energy than the $1s$ orbitals. Like atomic orbitals, each MO can accommodate two electrons with their spins paired (Pauli exclusion principle) (⚭ Section 6.7, 'Many–Electron Atoms').

⚭ Review this on page 202

FIGURE IT OUT

By referring to Figure 9.22, determine which molecular orbital in He_2 has a node between the nuclei.

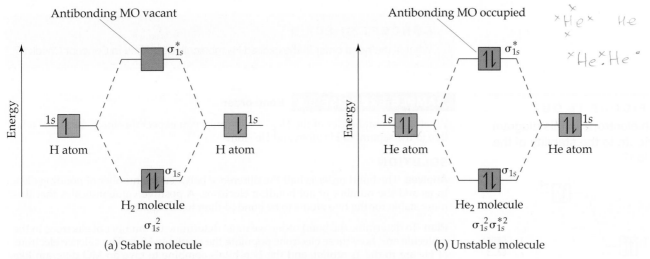

▲ FIGURE 9.23   Energy-level diagrams and electron configurations for H_2 and He_2.

The MO diagram of the H_2 molecule is shown in Figure 9.23(a). Each H atom has one electron, and so there are two electrons in H_2. These two electrons occupy the lower-energy bonding (σ_{1s}) MO, and their spins are paired. Electrons occupying a bonding MO are called *bonding electrons*. Because the σ_{1s} MO is lower in energy than the isolated $1s$ atomic orbitals, the H_2 molecule is more stable than the two separate H atoms.

In contrast, the hypothetical He_2 molecule requires four electrons to fill its MOs, as in Figure 9.23(b). Because only two electrons can be put in the σ_{1s} MO, the other two must be placed in the σ_{1s}^* MO. The energy decrease from the two electrons in the bonding MO is offset by the energy increase from the two electrons in the antibonding MO.[*] Hence He_2 is an unstable molecule. MO theory correctly predicts that hydrogen forms diatomic molecules but helium does not.

CONCEPT CHECK 8

Suppose light is used to excite one of the electrons in the H_2 molecule from the σ_{1s} MO to the σ_{1s}^* MO. Would you expect the H atoms to remain bonded to each other, or would the molecule fall apart?

Bond Order

In MO theory the stability of a covalent bond is related to its **bond order**, defined as half the difference between the number of bonding electrons and the number of antibonding electrons:

$$\text{Bond order} = \tfrac{1}{2}(\text{no. of bonding electrons} - \text{no. of antibonding electrons}) \quad [9.1]$$

We take half the difference because we are used to thinking of bonds as pairs of electrons. A bond order of 1 represents a single bond, a bond order of 2 represents a double bond and a bond order of 3 represents a triple bond. Because MO theory also treats molecules containing an odd number of electrons, bond orders of $\tfrac{1}{2}$, $\tfrac{3}{2}$ and $\tfrac{5}{2}$ are possible.

---

[*] Antibonding MOs are slightly more energetically unstable than bonding MOs are energetically favourable. Thus whenever there are an equal number of electrons in bonding and antibonding orbitals, the energy of the molecule is slightly higher than that for the separated atoms. Consequently no bond is formed.

Because H_2 has two bonding electrons and zero antibonding electrons (Figure 9.23(a)), it has a bond order of 1. Because He_2 has two bonding and two antibonding electrons (Figure 9.23(b)), it has a bond order of 0. A bond order of 0 means that no bond exists.

△ CONCEPT CHECK 9

What is the bond order in the excited H_2 molecule described in Concept Check 8?

△ FIGURE IT OUT

Which electrons in this diagram contribute to the stability of the He_2^+ ion?

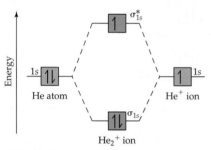

▲ **FIGURE 9.24** **Energy-level diagram for the He_2^+ ion.**

SAMPLE EXERCISE 9.6 | **Bond order**

What is the bond order of the He_2^+ ion? Would you expect this ion to be stable relative to the separated He atom and He^+ ion?

SOLUTION

Analyse The bond order is half the difference between the number of bonding electrons and the number of antibonding electrons. A positive value indicates that it is more stable for the two atoms to be bonded than to be separate.

Plan To determine the bond order, we must determine the number of electrons in the molecule and how these electrons populate the available MOs. The valence electrons of He are in the 1s orbital, and the 1s orbitals combine to give an MO diagram like that for H_2 or He_2 (Figure 9.23). If the bond order is greater than 0, we expect a bond to exist and the ion is stable.

Solve The energy-level diagram for the He_2^+ ion is shown in ◄ **FIGURE 9.24**. This ion has three electrons. Two are placed in the bonding orbital, the third in the antibonding orbital. Thus the bond order is

$$\text{Bond order} = \tfrac{1}{2}(2 - 1) = \tfrac{1}{2}$$

Because the bond order is greater than 0, the He_2^+ ion is predicted to be stable relative to the separated He and He^+. Formation of He_2^+ in the gas phase has been demonstrated in laboratory experiments.

PRACTICE EXERCISE

Determine the bond order of the H_2^- ion.

Answer: $\dfrac{1}{2}$

(See also Exercise 9.43.)

9.8 | SECOND-PERIOD DIATOMIC MOLECULES

Just as we treated the bonding in H_2 by using MO theory, we can consider the MO description of other diatomic molecules. Initially, we restrict our discussion to *homonuclear* diatomic molecules (those composed of two identical atoms) of elements in the second period of the periodic table. The procedure for determining the distribution of electrons in these molecules closely follows the one we used for H_2.

Second-period atoms have valence 2s and 2p orbitals and we need to consider how they interact to form MOs. The following rules summarise some of the guiding principles for the formation of MOs and for how they are populated by electrons.

1. The number of MOs formed equals the number of atomic orbitals combined.

2. Atomic orbitals combine most effectively with other atomic orbitals of similar energy.

3. The effectiveness with which two atomic orbitals combine is proportional to their overlap; that is, as the overlap increases, the energy of the bonding MO is lowered and the energy of the antibonding MO is raised.

4. Each MO can accommodate, at most, two electrons, with their spins paired (Pauli exclusion principle) (∞ Section 6.7, 'Many-Electron Atoms').

5. When MOs of the same energy are populated, one electron enters each orbital (with the same spin) before spin pairing occurs (Hund's rule) (∞ Section 6.8, 'Electron Configurations').

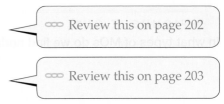

∞ Review this on page 202

∞ Review this on page 203

Molecular Orbitals for Li₂ and Be₂

Lithium, the first element of the second period, has a $1s^2 2s^1$ electron configuration. When lithium metal is heated above its boiling point (1342 °C), Li_2 molecules are found in the vapour phase. The Lewis structure for Li_2 indicates an Li–Li single bond. We now use MOs to describe the bonding in Li_2.

Because the 1s and 2s orbitals of Li are so different in energy, we can assume that the 1s orbital on one Li atom interacts only with the 1s orbital on the other atom (rule 2). Likewise, the 2s orbitals interact only with each other. The resultant energy-level diagram is shown in ▶ **FIGURE 9.25**. Notice that combining four atomic orbitals produces four MOs (rule 1).

The 1s orbitals of Li combine to form σ_{1s} and σ_{1s}^* bonding and antibonding MOs, as they did for H_2. The 2s orbitals interact with one another in exactly the same way, producing bonding (σ_{2s}) and antibonding (σ_{2s}^*) MOs. Because the 2s orbitals of Li extend further from the nucleus than the 1s orbitals, the 2s orbitals overlap more effectively. As a result, the energy separation between the σ_{2s} and σ_{2s}^* orbitals is greater than that for the 1s-based MOs. The 1s orbitals of Li are so much lower in energy than the 2s orbitals, however, that the σ_{1s}^* antibonding MO is still well below the σ_{2s} bonding MO.

Each Li atom has three electrons, so six electrons must be placed in the MOs of Li_2. As shown in Figure 9.25, these occupy the σ_{1s}, σ_{1s}^* and σ_{2s} MOs, each with two electrons. There are four electrons in bonding orbitals and two in antibonding orbitals, so the bond order equals $\frac{1}{2}(4 - 2) = 1$. The molecule has a single bond, in accord with its Lewis structure.

Because both the σ_{1s} and σ_{1s}^* MOs of Li_2 are completely filled, the 1s orbitals contribute almost nothing to the bonding. The single bond in Li_2 is due essentially to the interaction of the valence 2s orbitals on the Li atoms. This example illustrates the general rule that *core electrons usually do not contribute significantly to bonding in molecule formation*. The rule is equivalent to using only the valence electrons when drawing Lewis structures. Thus we need not consider further the 1s orbitals while discussing the other second-period diatomic molecules.

The MO description of Be_2 follows readily from the energy-level diagram for Li_2. Each Be atom has four electrons ($1s^2 2s^2$), so we must place eight electrons in MOs. Thus we completely fill the σ_{1s}, σ_{1s}^*, σ_{2s} and σ_{2s}^* MOs. We have an equal number of bonding and antibonding electrons, so the bond order equals 0. Consistent with this analysis, Be_2 does not exist.

▲ FIGURE IT OUT

Why do the 1s orbitals of the Li atoms not contribute to the bonding in Li₂?

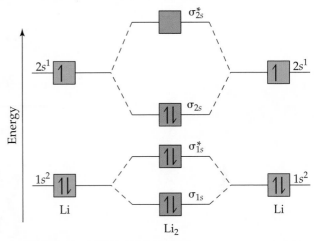

▲ **FIGURE 9.25** Energy-level diagram for the Li_2 molecule.

▲ CONCEPT CHECK 10

Would you expect Be_2^+ to be a stable ion?

Molecular Orbitals from 2p Atomic Orbitals

Before we can consider the remaining second-period molecules, we must look at the MOs that result from combining 2p atomic orbitals. The interactions between p orbitals are shown in ▼ **FIGURE 9.26**, where we have arbitrarily

FIGURE IT OUT

In what types of MOs do we find nodal planes?

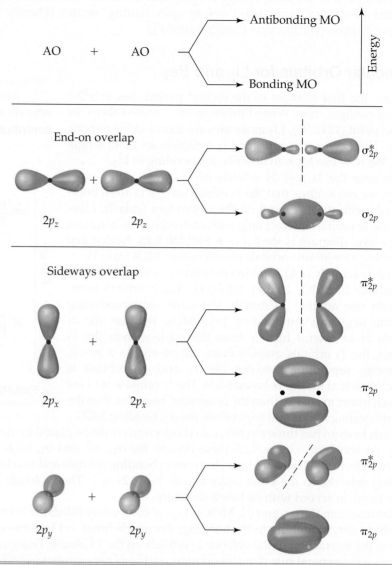

▶ **FIGURE 9.26** **Contour representations of the MOs formed by 2p orbitals.** Each time we combine two atomic orbitals (AOs), we obtain two MOs: one bonding and one antibonding.

chosen the internuclear axis to be the *z*-axis. The $2p_z$ orbitals face each other in a 'head-to-head' fashion. Just as we did for the *s* orbitals, we can combine the $2p_z$ orbitals in two ways. One combination concentrates electron density between the nuclei and is therefore a bonding MO. The other combination excludes electron density from the bonding region; it is an antibonding MO. In each of these MOs the electron density lies along the line through the nuclei, so they are σ MOs: σ_{2p} and σ_{2p}^*.

The other 2*p* orbitals overlap sideways and thus concentrate electron density on opposite sides of the line through the nuclei. MOs of this type are called *pi (π) molecular orbitals*. We get one π-bonding MO by combining the $2p_x$ atomic orbitals and another from the $2p_y$ atomic orbitals. These two π_{2p} MOs have the same energy; in other words, they are degenerate. Likewise, we get two degenerate π_{2p}^*-antibonding MOs.

The $2p_z$ orbitals on two atoms point directly at each other. Hence, the overlap of two $2p_z$ orbitals is greater than that for two $2p_x$ or $2p_y$ orbitals. From rule 3 we therefore expect the σ_{2p} MO to be lower in energy (more stable) than the π_{2p} MOs. Similarly, the σ_{2p}^* MO should be higher in energy (less stable) than the π_{2p}^* MOs.

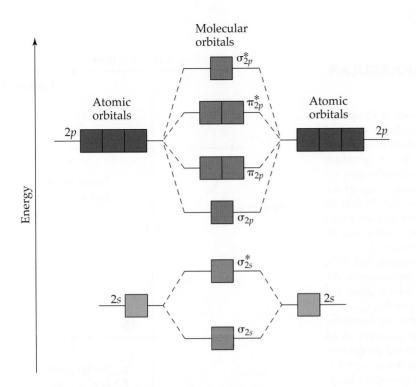

Molecular orbitals

Atomic orbitals

$2p$

σ_{2p}^*

π_{2p}^*

π_{2p}

σ_{2p}

Atomic orbitals

$2p$

Energy

σ_{2s}^*

$2s$

σ_{2s}

$2s$

◀ **FIGURE 9.27 Energy-level diagram for MOs of second-period homonuclear diatomic molecules.** The diagram assumes no interaction between the 2s atomic orbital on one atom and the 2p atomic orbitals on the other atom, and experiment shows that it fits only for O_2, F_2 and Ne_2.

Electron Configurations for B_2 to Ne_2

So far we have considered independently the MOs that result from s orbitals (Figure 9.25) and from p orbitals (Figure 9.26). We can combine these results to construct an energy-level diagram (▲ **FIGURE 9.27**) for homonuclear diatomic molecules of the elements lithium to neon, all of which have valence 2s and 2p atomic orbitals. The following features of the diagram are notable.

1. The 2s atomic orbitals are lower in energy than the 2p atomic orbitals (∞ Section 6.7, 'Many-Electron Atoms'). Consequently, both of the MOs that result from the 2s orbitals, the bonding σ_{2s} and the antibonding σ_{2s}^* are lower in energy than the lowest-energy MO derived from the 2p atomic orbitals.

2. The overlap of the two $2p_z$ orbitals is greater than that of the two $2p_x$ or $2p_y$ orbitals. As a result, the bonding σ_{2p} MO is lower in energy than the π_{2p} MOs, and the antibonding σ_{2p}^* MO is higher in energy than the π_{2p}^* MOs.

3. Both the π_{2p} and π_{2p}^* MOs are *doubly degenerate*; that is, there are two degenerate MOs of each type.

∞ Review this on page 201

Before we can add electrons to the energy-level diagram in Figure 9.27, there is one more effect to consider. We have constructed the diagram assuming there is no interaction between the 2s orbital on one atom and the 2p orbitals on the other. In fact, such interactions can and do take place. ▶ **FIGURE 9.28** shows the overlap of a 2s orbital on one of the atoms with a $2p_z$ orbital on the other.

These interactions affect the energies of the σ_{2s} and σ_{2p} MOs in such a way that these MOs move further apart in energy, the σ_{2s} falling and the σ_{2p} rising in energy ▼ **FIGURE 9.32** on page 324. These 2s–2p interactions are strong enough that the energetic ordering of the MOs can be altered: for B_2, C_2 and N_2, the σ_{2p} MO is above the π_{2p} MOs in energy. For O_2, F_2 and Ne_2, the σ_{2p} MO is below the π_{2p} MOs.

Given the energy ordering of the MOs, it is a simple matter to determine the electron configurations for the second-period diatomic molecules B_2 to Ne_2. For example, a boron atom has three valence electrons. (Remember that we are ignoring the inner-shell 1s electrons.) Thus for B_2 we must place six electrons in

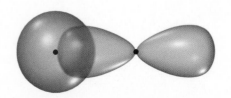

▲ **FIGURE 9.28 Interaction of 2s and 2p atomic orbitals.** The 2s orbital on one atom of a diatomic molecule can overlap with the $2p_z$ orbital on the other atom.

A CLOSER LOOK

PHASES IN ATOMIC AND MOLECULAR ORBITALS

Our discussion of atomic orbitals in Chapter 6 and MOs in this chapter highlights some of the most important applications of quantum mechanics in chemistry. In the quantum mechanical treatment of electrons in atoms and molecules, we are mainly interested in obtaining two characteristics of the electrons—namely, their energies and their distribution in space. Recall that solving Schrödinger's wave equation yields the electron's energy, E, and wave function, ψ, but that ψ itself does not have a direct physical meaning (Section 6.5).

The contour representations of atomic orbitals and MOs that we have presented so far are based on the square of the wave function, ψ^2 (the *probability density*), which gives the probability of finding the electron at a given point in space.

Because probability densities are squares of functions, their values must be non-negative (zero or positive) at all points in space. Remember that points where the probability density is zero are called *nodes*: at a node there is a zero probability of finding the electron. Consider, for example, the p_x orbital presented in Figure 6.19. For this orbital, the probability density at any point on the plane defined by the y- and z-axes is zero—we say that the yz plane is a *nodal plane* of the orbital. Likewise, the σ_{1s}^* MO of H_2 shown in Figure 9.22 has a nodal plane that is perpendicular to the line connecting the H atoms and midway between them. Why do these nodes arise? To answer this question, we must take a closer look at the wave functions for the orbitals.

In order to understand the nodes in p orbitals, we can draw an analogy to a sine function. ▶ FIGURE 9.29 shows one cycle of the function $\sin x$ centred on the origin. Notice that the two halves of the wave have the same shape except that one has positive values and the other negative values—the two halves of the function differ in their sign, or *phase*. The origin, which is where the function changes sign, is a node ($\sin 0 = 0$). What happens if we square this function? Figure 9.27(b) shows that when we square the function, we get two peaks that look the same on each side of the origin. Both peaks are positive because squaring a negative number produces a positive number—we lose the phase information of the function upon squaring it. Note also that because $\sin 0 = 0$, it follows that $\sin^2 0 = 0$. Thus the origin is still a node even though the function is positive on both sides of the origin.

The wave function for a p orbital is much like a sine function inasmuch as it has two equal parts that have opposite

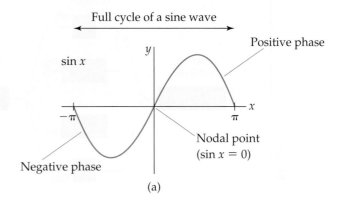

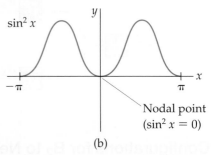

▲ **FIGURE 9.29** **Graphs for a sine function (a) and the same function squared (b).**

phases. ▶ FIGURE 9.30 gives a typical representation used by chemists of the wave function for a p_x orbital.* We typically use two different colours for the lobes to indicate the different phases. Like the sine function, the origin is a node. When we square the wave function for the p_x orbital, we get the probability density for the orbital, which is given as a contour representation Figure 9.31(b). For this function, both lobes have the same phase and therefore the same colour.

The lobes of the wave functions for the d orbitals also have different phases. For example, the wave function for a d_{xy} orbital has four lobes that alternate in phase, as shown in Figure 9.31(c). The wave functions for the d_{xz}, d_{yz} and $d_{x^2-y^2}$ orbitals likewise have lobes of alternating phase. In the wave function for the d_{z^2} orbital, the 'doughnut' has a phase opposite that of the two large lobes.

We encounter a very similar situation with MOs. The wave function of the σ_{1s}^* MO of H_2 is constructed by adding the wave function for a $1s$ orbital on one atom to the wave function for a $1s$ orbital on the other atom, but with the two orbitals having

MOs. Four of these fully occupy the σ_{2s} and σ_{2s}^* MOs, leading to no net bonding. The last two electrons are put in the π_{2p} bonding MOs; one electron is put in one π_{2p} MO and the other electron is put in the other π_{2p} MO, with the two electrons having the same spin. Therefore, B_2 has a bond order of 1. Each time we move one element to the right in the second period, two more electrons must be placed in the diagram. For example, on moving to C_2, we have two more electrons than in B_2, and these electrons are also placed in the π_{2p} MOs, completely filling them. The electron configurations and bond orders for the diatomic molecules B_2 to Ne_2 are given in ▼ FIGURE 9.33.

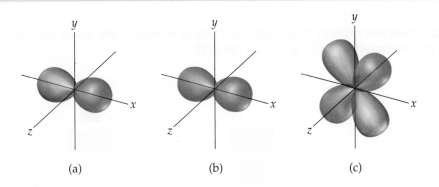

◀ **FIGURE 9.30** **Atomic orbital wave functions.** (a) The wave function for a *p* orbital has two equivalent lobes that have opposite signs. (b) The probability density for a *p* orbital has two lobes that have the same sign. (c) The wave function for the d_{xy} orbital has lobes of alternating sign.

(a)   (b)   (c)

different phases. Chemists generally sketch this wave function by simply drawing the contour representations of the $1s$ orbitals in different colours, as shown in ▶ **FIGURE 9.31(a)**. The fact that the $1s$ orbitals have different phases causes a node halfway between the atoms. Notice how this wave function has similarities to the sine function and to a *p* orbital—in each case we have two parts of the function of opposite phase separated by a node. When we square the wave function of the σ_{1s}^* MO, we get the probability density representation given in Figure 9.22—notice that we once again lose the phase information when we look at the probability density.

The same phenomenon occurs when *p* orbitals form bonding and antibonding MOs. Figure 9.31(b) and 9.31(c) show the atomic orbital wave functions oriented to form the σ_{2p} and σ_{2p}^* MOs of a homonuclear diatomic molecule. In the bonding σ_{2p} orbital, the lobes of the same phase point at one another, which concentrates the electron density between the nuclei. In contrast, to construct the antibonding σ_{2p}^* MO, the $2p$ orbitals are lined up so that the lobes that point at one another have different phases. Because they have a different phase, the mixing of these orbitals leads to the exclusion of electron density in the region between the nuclei and the formation of a node halfway between the nuclei, both of which are important characteristics of antibonding MOs. We present the probability densities for these MOs in Figure 9.26.

This brief discussion has given you only an introduction to the mathematical subtleties of atomic and molecular orbitals. The wave functions of atomic and molecular orbitals are used by chemists to understand many aspects of chemical bonding and spectroscopy, and you will probably see orbitals drawn in colour to show phases if you take a course in organic chemistry.

RELATED EXERCISES: 9.63, 9.74

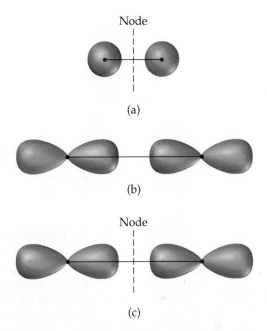

Node

(a)

(b)

Node

(c)

▲ **FIGURE 9.31** **Molecular orbitals from atomic orbital wave functions.** (a) The mixing of two $1s$ orbitals that are opposite in phase produces a σ_{1s}^* MO. (b) The mixing of two $2p$ orbitals pointing at each other with lobes of the same phase produces a σ_{2p} MO. (c) The mixing of two $2p$ orbitals pointing at each other with lobes of opposite phase produces a σ_{2p}^* MO. Notice that a node is generated in (a) and (c).

---

* The mathematical development of this three-dimensional function (and its square) is beyond the scope of this book and, as is typically done by chemists, we have used lobes that are the same shape as in Figure 6.19.

Electron Configurations and Molecular Properties

The way a substance behaves in a magnetic field provides an important insight into the arrangements of its electrons. Molecules with one or more unpaired electrons are attracted into a magnetic field. The more unpaired electrons in a species, the stronger the force of attraction. This type of magnetic behaviour is called **paramagnetism**.

Substances with no unpaired electrons are weakly repelled from a magnetic field. This property is called **diamagnetism**. Diamagnetism is a much weaker effect than paramagnetism. A straightforward method for measuring

FIGURE IT OUT

Which molecular orbitals have switched relative energy in the group on the right as compared with the group on the left?

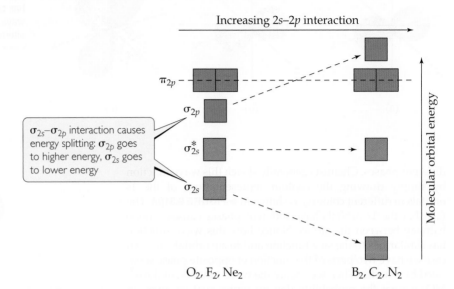

▲ **FIGURE 9.32** **Molecular orbital representation of the effect of interaction of 2s and 2p atomic orbitals.** These 2s–2p interactions can alter the energetic ordering of the MOs of the molecule.

the magnetic properties of a substance, illustrated in ▶ **FIGURE 9.34**, involves weighing the substance in the presence and absence of a magnetic field. If the substance is paramagnetic, it will appear to weigh more in the magnetic field; if it is diamagnetic, it will appear to weigh less. The magnetic behaviours observed for the diatomic molecules of the second-period elements agree with the electron configurations shown in Figure 9.33.

	Large 2s–2p interaction			Small 2s–2p interaction		
	B₂	**C₂**	**N₂**	**O₂**	**F₂**	**Ne₂**
σ^*_{2p}	☐	☐	☐	σ^*_{2p} ☐	☐	⇅
π^*_{2p}	☐☐	☐☐	☐☐	π^*_{2p} ↑ ↑	⇅ ⇅	⇅ ⇅
σ_{2p}	☐	☐	⇅	π_{2p} ⇅ ⇅	⇅ ⇅	⇅ ⇅
π_{2p}	↑ ↑	⇅ ⇅	⇅ ⇅	σ_{2p} ⇅	⇅	⇅
σ^*_{2s}	⇅	⇅	⇅	σ^*_{2s} ⇅	⇅	⇅
σ_{2s}	⇅	⇅	⇅	σ_{2s} ⇅	⇅	⇅
Bond order	1	2	3	2	1	0
Bond enthalpy (kJ mol⁻¹)	290	620	941	495	155	—
Bond length (pm)	159	131	110	121	143	—
Magnetic behaviour	Paramagnetic	Diamagnetic	Diamagnetic	Paramagnetic	Diamagnetic	—

▲ **FIGURE 9.33** **The second-period diatomic molecules.** Molecular orbital electron configurations and some experimental data for several second-period diatomic molecules.

Weigh sample in absence of a magnetic field

A diamagnetic sample appears to weigh less in a magnetic field (weak effect)

A paramagnetic sample appears to weigh more in a magnetic field

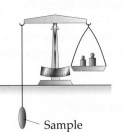
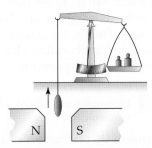

Sample

N S

N S

◀ **FIGURE 9.34 Determining the magnetic properties of a sample.**

▲ **CONCEPT CHECK 11**

Figure 9.33 indicates that the C_2 molecule is diamagnetic. Would that be expected if the σ_{2p} MO were lower in energy than the π_{2p} MOs?

The electron configurations can also be related to the bond distances and bond enthalpies of the molecules (∞ Section 8.8, 'Strengths of Covalent Bonds').

As bond orders increase, bond distances decrease and bond enthalpies increase. N_2, for example, whose bond order is 3, has a short bond distance and a large bond enthalpy. The N_2 molecule does not react readily with other substances to form nitrogen compounds. The high bond order of the molecule helps explain its exceptional stability. We should also note, however, that molecules with the same bond orders do *not* have the same bond distances and bond enthalpies. Bond order is only one factor influencing these properties. Other factors include the nuclear charges and the extent of orbital overlap.

Bonding in the dioxygen molecule, O_2 is especially interesting. Its Lewis structure shows a double bond and complete pairing of electrons

$$\ddot{\text{O}}=\ddot{\text{O}}$$

The short O–O bond distance (121 pm) and the relatively high bond enthalpy (495 kJ mol^{-1}) are in agreement with the presence of a double bond. However, the molecule is found to contain two unpaired electrons. The paramagnetism of O_2 is demonstrated in ▼ **FIGURE 9.35**. Although the Lewis structure fails to

∞ Review this on page 282

▲ **FIGURE IT OUT**

What would you expect to see if liquid nitrogen were poured between the poles of the magnet?

Because O_2 molecules are paramagnetic ...

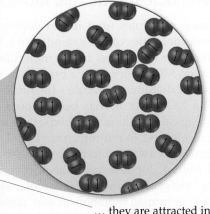

... they are attracted into the magnetic field.

◀ **FIGURE 9.35 Paramagnetism of O_2.**
Liquid O_2 being poured between the poles of a magnet. Because each O_2 molecule contains two unpaired electrons, O_2 is paramagnetic. It is therefore attracted into the magnetic field and 'sticks' between the magnetic poles.

account for the paramagnetism of O_2, MO theory correctly predicts that there are two unpaired electrons in the π_{2p}^* orbital of the molecule (Figure 9.33). The MO description also correctly indicates a bond order of 2.

Going from O_2 to F_2, we add two more electrons, completely filling the π_{2p}^* MOs. Thus F_2 is expected to be diamagnetic and have an F–F single bond, in accord with its Lewis structure. Finally, the addition of two more electrons to make Ne_2 fills all the bonding and antibonding MOs; therefore, the bond order of Ne_2 is zero, and the molecule is not expected to exist.

SAMPLE EXERCISE 9.7 | **Molecular orbitals of a second-period diatomic ion**

Predict the following properties of O_2^+: **(a)** number of unpaired electrons, **(b)** bond order, **(c)** bond enthalpy and bond length.

SOLUTION

Analyse We use the MO description of O_2^+ to determine the number of unpaired electrons and the bond order of O_2^+. Comparison of the bond order to compounds of similar bond order and known bond enthalpies enables us to estimate the bond enthalpy of the O_2^+ ion.

Plan We must first determine the number of electrons in O_2^+ and then draw its MO energy diagram. The unpaired electrons are those without a partner of opposite spin. The bond order is one-half the difference between the number of bonding and antibonding electrons. After calculating the bond order, we can use the data in Figure 9.33 to estimate the bond enthalpy and bond length.

Solve **(a)** The O_2^+ ion has 11 valence electrons, one fewer than O_2. The electron removed from O_2 to form O_2^+ is one of the two unpaired π^* electrons (see Figure 9.33). Therefore, O_2^+ has just one unpaired electron.

(b) The molecule has eight bonding electrons (the same as O_2) and three antibonding electrons (one fewer than O_2). Thus its bond order is

$$\tfrac{1}{2}(8 - 3) = 2\tfrac{1}{2}$$

(c) The bond order of O_2^+ is between that for O_2 (bond order 2) and N_2 (bond order 3). Thus the bond enthalpy and bond length should be about midway between those for O_2 and N_2, approximately 700 kJ mol^{-1} and 115 pm, respectively. The experimental bond enthalpy and bond length of the ion are 625 kJ mol^{-1} and 112.3 pm, respectively.

PRACTICE EXERCISE

Predict the magnetic properties and bond orders of **(a)** the peroxide ion, O_2^{2-}; **(b)** the acetylide ion, C_2^{2-}.

Answers: **(a)** diamagnetic, 1; **(b)** diamagnetic, 3

(See also Exercises 9.49, 9.50.)

Heteronuclear Diatomic Molecules

The same principles used in developing an MO description of homonuclear diatomic molecules can be extended to *heteronuclear* diatomic molecules—those in which the two atoms in the molecule are not the same. We conclude this section on MO theory with a brief discussion of the MOs of a fascinating heteronuclear diatomic molecule—the nitric oxide, NO, molecule.

The NO molecule has been shown to control several important human physiological functions. Our bodies use it, for example, to relax muscles, to kill foreign cells and to reinforce memory. The 1998 Nobel Prize in Physiology or Medicine was awarded to three scientists, Robert Furchgott, Louis Ignarro and Ferid Murad, for research that uncovered the importance of NO as a 'signalling' molecule in the cardiovascular system. That NO plays such an important role in human metabolism was unsuspected before 1987 because NO has an odd number of electrons and is highly reactive. The molecule has 11 valence electrons, and two possible Lewis structures can be drawn. The one with the lower formal charges places the odd electron on the N atom

$$\overset{0}{\underset{\cdot\cdot}{\dot{N}}}=\overset{0}{\underset{\cdot\cdot}{\dot{O}}} \longleftrightarrow \overset{-1}{\underset{\cdot\cdot}{\ddot{N}}}=\overset{+1}{\underset{\cdot\cdot}{\dot{O}}}$$

Both structures indicate the presence of a double bond but, when compared with the molecules in Figure 9.33, the experimental bond length of NO (115 pm) suggests a bond order greater than two. How do we treat NO using the MO model?

If the atoms in a heteronuclear diatomic molecule do not differ too greatly in their electronegativities, the description of their MOs will resemble those for homonuclear diatomics, with one important modification: the atomic energies of the more electronegative atom will be lower in energy than those of the less electronegative element. The MO diagram for NO is shown in ▶ **FIGURE 9.36**, and you can see that the 2s and 2p atomic orbitals of oxygen (more electronegative) are slightly lower than those of nitrogen (less electronegative). We see that the MO energy-level diagram is much like that of a homonuclear diatomic molecule—because the 2s and 2p orbitals on the two atoms interact, the same types of MOs are produced.

There is one other important change in the MOs when we consider heteronuclear molecules. The MOs that result are still a mix of the atomic orbitals from both atoms, but in general *an MO will have a greater contribution from the atomic orbital to which it is closer in energy*. In the case of NO, for example, the σ_{2s} bonding MO is closer in energy to the O 2s atomic orbital than to the N 2s atomic orbital. As a result, the σ_{2s} MO has a slightly greater contribution from O than from N—the orbital is no longer an equal mixture of the two atoms, as was the case for the homonuclear diatomic molecules. Similarly, the σ_{2s}^* antibonding MO is weighted more heavily towards the N atom because that MO is closest in energy to the N 2s atomic orbital.

We complete the MO diagram for NO by filling the MOs in Figure 9.36 with the 11 valence electrons. We see that there are eight bonding and three antibonding electrons, giving a bond order of $\frac{1}{2}(8 - 3) = 2\frac{1}{2}$, which agrees more closely with the experimental value than that derived from Lewis structures. The unpaired electron resides in one of the π_{2p}^* MOs, which are more heavily weighted towards the N atom. This description is consistent with the left Lewis structure above (the one preferred on the basis of formal charge), which places the unpaired electron on the N atom.

▲ **FIGURE IT OUT**

How many valence-shell electrons are there in NO?

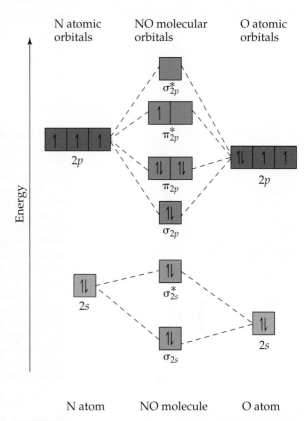

▲ **FIGURE 9.36** The energy-level diagram for atomic and molecular orbitals in NO.

SAMPLE INTEGRATIVE EXERCISE Putting concepts together

Elemental sulfur is a yellow solid that consists of S_8 molecules. The structure of the S_8 molecule is a puckered eight-membered ring. Heating elemental sulfur to high temperatures produces gaseous S_2 molecules:

$$S_8(s) \longrightarrow 4\,S_2(g)$$

(a) With respect to electronic structure, which element in the second period of the periodic table is most similar to sulfur? **(b)** Use the VSEPR model to predict the S–S–S bond angles in S_8 and the hybridisation at S in S_8. **(c)** Use MO theory to predict the sulfur–sulfur bond order in S_2. Is the molecule expected to be diamagnetic or paramagnetic? **(d)** Use average bond enthalpies (Table 8.4) to estimate the enthalpy change for the reaction just described.

SOLUTION

Analyse We use a similar valence electron configuration to suggest a similar element. Drawing a Lewis structure of S_8 allows us to use the VSEPR model to predict the bond angle within the ring and the hybridisation of each sulfur atom. The bond order is determined from the MO diagram and reaction enthalpy is estimated from the difference in enthalpy of the bonds broken and the bonds made in the reaction.

∞ Review this on page 207

Solve **(a)** Sulfur is a group 16 element with an $[Ne]3s^2 3p^4$ electron configuration. It is expected to be most similar electronically to oxygen (electron configuration, $[He]2s^2 2p^4$), which is immediately above it in the periodic table (∞ Section 6.9, 'Electron Configuration and the Periodic Table').

(b) The Lewis structure of S_8 is

There is a single bond between each pair of S atoms and two non-bonding electron pairs on each S atom. Thus we see four electron domains around each S atom, and we would expect a tetrahedral electron-domain geometry corresponding to sp^3 hybridisation (Sections 9.2 and 9.5).

Because of the non-bonding pairs, we would expect the S–S–S angles to be somewhat less than 109°, the tetrahedral angle. Experimentally, the S–S–S angle in S_8 is 108°, in good agreement with this prediction. Further, if S_8 were a planar ring it would have S–S–S angles of 135°. Instead, the S_8 ring puckers to accommodate the smaller angles dictated by sp^3 hybridisation. **(c)** The MOs of S_2 are entirely analogous to those of O_2, although the MOs for S_2 are constructed from the $3s$ and $3p$ atomic orbitals of sulfur. Further, S_2 has the same number of valence electrons as O_2. Thus by analogy with our discussion of O_2, we would expect S_2 to have a bond order of 2 (a double bond) and to be paramagnetic with two unpaired electrons in the π^*_{3p} molecular orbitals of S_2 (Section 9.8). **(d)** We are considering the reaction in which an S_8 molecule falls apart into four S_2 molecules. From parts (b) and (c), we see that S_8 has S–S single bonds and S_2 has S–S double bonds. During the course of the reaction, therefore, we are breaking eight S–S single bonds and forming four S–S double bonds. We can estimate the enthalpy of the reaction by using Equation 8.6 and the average bond enthalpies in Table 8.4:

$$\Delta_{rxn}H = 8\,D(S{-}S) - 4\,D(S{=}S) = 8(266\text{ kJ}) - 4(418\text{ kJ}) = +456\text{ kJ}$$

Comment The very positive value of $\Delta_{rxn}H$ suggests that high temperatures are required to cause the reaction to occur. Reactions in which $\Delta_{rxn}H > 0$ are said to be endothermic (∞ Section 14.4, 'Enthalpies of Reaction').

∞ Find out more on page 513

CHAPTER SUMMARY AND KEY TERMS

SECTION 9.1 **Molecular geometries** are the three-dimensional shapes and sizes of molecules. The molecular geometry of a molecule is determined by the **bond angles** and **bond lengths** of a molecule. Molecules with a central atom A surrounded by n atoms B, denoted AB_n, adopt a number of different geometric shapes, depending on the value of n and on the particular atoms involved. In the overwhelming majority of cases, the geometries are related to only five basic shapes (linear, trigonal pyramidal, tetrahedral, trigonal bypyramidal and octahedral).

SECTION 9.2 Using the **valence-shell electron-pair repulsion (VSEPR) model** makes it possible to predict the geometries of molecules by considering the repulsions between regions of high electron density called **electron domains**. The electron domains are created by **bonding pairs** of electrons or by **non-bonding pairs** (lone pairs) of electrons. According to the VSEPR model, electron domains orient themselves to minimise electrostatic repulsion; that is, they remain as far apart as possible. Electron domains from non-bonding pairs exert slightly greater repulsions than those from bonding pairs, which leads to certain preferred positions for non-bonding pairs and to the departure of bond angles from idealised values. Electron domains from multiple bonds exert slightly greater repulsions than those from single bonds. The arrangement of electron domains around a central atom is called the **electron-domain geometry** whereas the arrangement of atoms is called the molecular geometry.

SECTION 9.3 Knowing the types of bonds in a molecule and its molecular geometry makes it possible to determine whether the molecule is polar or non-polar. The dipole moment of a polyatomic molecule can be determined from the vector sum of the dipole moments associated with the individual bonds called the **bond dipoles**. When the bond dipoles cancel, the molecule is non-polar.

SECTION 9.4 In **valence-bond theory**, covalent bonds are formed when atomic orbitals on adjacent atoms **overlap** one another. The greater the overlap between the two orbitals, the stronger is the bond that is formed.

SECTION 9.5 To account for molecular shape, we consider how the orbitals of an atom (s, p and d) mix with one another in a process called **hybridisation** to form differently shaped orbitals called **hybrid orbitals**, which can then overlap to form a bond.

SECTION 9.6 Atomic orbitals overlap so that the result is a covalent bond in which the electron density lies along the internuclear axis. These are called **sigma (σ) bonds**. Bonds can also be formed from the sideways overlap of p orbitals which results in a bond called a **pi (π) bond**. A single bond usually consists of a single sigma bond between the bonded atoms and a multiple bond consists of a sigma bond and one or more pi bonds. The formation of a π bond requires that molecules adopt a specific orientation; the two CH_2 groups in C_2H_4, for example, must lie in the same plane.

SECTION 9.7 **Molecular orbital theory** is another model used to describe the bonding in molecules. In this model, electrons in a molecule exist in allowed energy levels called **molecular orbitals (MOs)** which can be spread among all the atoms of the molecule. That is, they are delocalised around the molecule as opposed to the valance-bond theory in which the electrons forming a bond are localised in one position. Like an atomic orbital, a molecular orbital has a definite energy and can hold two electrons of opposite spin. We see that when two atomic orbitals combine, the result is the formation of two molecular orbitals. One of these orbitals is of low energy (the **bonding molecular orbital**) and the other is of higher energy (the **antibonding molecular orbital**). The electron density is distributed around the internuclear axis resulting in **σ-bonding and σ-antibonding molecular orbitals**. The combination of atomic orbitals and the relative energies of the resulting MOs can be shown by an **energy-level** (or **molecular orbital**) diagram. Occupation of bonding MOs favours bond formation, whereas occupation of antibonding MOs is unfavourable. When we put in the appropriate number of electrons into the MO diagram we can calculate the **bond order** of a bond, which is half the difference between the number of electron in bonding MOs and the number of electrons in antibonding MOs. A bond order of one corresponds to a single bond and so forth.

SECTION 9.8 Electrons in core orbitals do not contribute to the bonding between atoms so we see that to describe MOs we need to consider only electrons in the outermost electron subshells. We also see that in the case of p orbitals, those that point directly at one another can form σ-bonding and σ*-antibonding MOs whereas the p orbitals which are oriented perpendicular to the internuclear axis combine to form **π-bonding** and **π*-antibonding molecular orbitals**. We find that the bond orders using the MO approach are in accord with the Lewis structures of these molecules. Further, the model predicts correctly that O_2 should exhibit **paramagnetism**, which leads to attraction of a molecule into a magnetic field due to the influence of unpaired electrons. Molecules in which all the electrons are paired exhibit **diamagnetism**, which leads to weak repulsion from a magnetic field.

KEY SKILLS

- Understand the VSEPR model and be able to use it to describe the three-dimensional shape of molecules. (Sections 9.1 and 9.2)
- Determine whether a molecule is polar or non-polar based on its geometry and the individual bond dipole moments. (Section 9.3)
- Be able to identify the hybridisation state of atoms in molecules. (Section 9.5)
- Be able to visualise how orbitals overlap to form sigma and pi bonds. (Section 9.6)
- Be able to explain the concept of bonding and antibonding molecular orbitals. (Section 9.7)
- Be able to draw molecular orbital energy-level diagrams and use them to obtain bond orders and electron configurations of diatomic molecules. (Section 9.8)
- Understand the relationships between bond order, bond length and bond strength. (Section 9.8)

KEY EQUATIONS

- The stability of a convalent bond is related to its bond order:

$$\text{Bond order} = \frac{1}{2}(\text{no. of bonding electrons} - \text{no. of antibonding electrons}) \qquad [9.1]$$

EXERCISES

VISUALISING CONCEPTS

9.1 A certain AB_4 molecule has a 'seesaw' shape:

From which of the fundamental geometries shown in Figure 9.3 could you remove one or more atoms to create a molecule having this seesaw shape? [Section 9.1]

9.2 (a) If the three balloons shown below are all the same size, what angle is formed between the red one and the green one? **(b)** If additional air is added to the blue balloon so that it gets larger, what happens to the angle between the red and green balloons? **(c)** What aspect of the VSEPR model is illustrated by part **(b)**? [Section 9.2]

9.3 An AB_5 molecule adopts the geometry shown below. **(a)** What is the name of this geometry? **(b)** Do you think there are any non-bonding electron pairs on atom A? Why or why not? **(c)** Suppose the atoms B are halogen atoms. Can you determine uniquely to which group in the periodic table atom A belongs? [Section 9.2]

9.4 The molecule shown here is *difluoromethane* (CH_2F_2), which is used as a refrigerant called R-32. **(a)** Based on the structure, how many electron domains surround the C atom in this molecule? **(b)** Would the molecule have a non-zero dipole moment? **(c)** If the molecule is polar, in what direction will the overall dipole moment vector point in the molecule? [Sections 9.2 and 9.3]

9.5 The plot here shows the potential energy of two Cl atoms as a function of the distance between them. **(a)** To what does an energy of zero correspond in this diagram? **(b)** According to the valence-bond model, why does the energy decrease as the Cl atoms move from a large separation to a smaller one? **(c)** What is the significance of the Cl–Cl distance at the minimum point in the plot? **(d)** Why does the energy rise at Cl–Cl distances less than that at the minimum point in the plot? [Section 9.4]

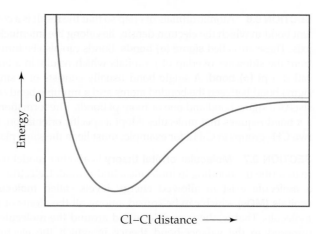

9.6 For each of the following contour representations of MOs, identify **(a)** the atomic orbitals (*s* or *p*) used to construct the MO, **(b)** the type of MO (σ or π), and **(c)** whether the MO is bonding or antibonding. [Sections 9.7 and 9.8]

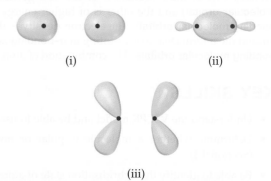

9.7 The diagram below shows the highest occupied MOs of a neutral molecule CX, where element X is in the same period of the periodic table as C. **(a)** Based on the number of electrons, can you determine the identity of X? **(b)** Would the molecule be diamagnetic or paramagnetic? **(c)** Consider the π_{2p} MOs of the molecule. Would you expect them to have a greater atomic orbital contribution from C, a greater atomic orbital contribution from X, or be an equal mixture of atomic orbitals from the two atoms? [Section 9.8]

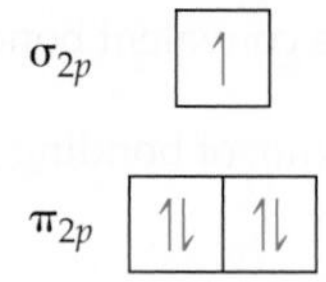

MOLECULAR SHAPES; THE VSEPR MODEL (Sections 9.1 and 9.2)

9.8 **(a)** An AB_2 molecule is described as linear, and the A–B bond length is specified. Does this information completely describe the geometry of the molecule? **(b)** The molecules BF_3 and SO_3 are both described as trigonal planar. Does this information completely define the bond angles of these molecules?

9.9 **(a)** What is meant by the term *electron domain*? **(b)** Explain in what way electron domains behave like the balloons in Figure 9.5. Why do they do so?

9.10 **(a)** How do we determine the number of electron domains in a molecule or ion? **(b)** What is the difference between a *bonding electron domain* and a *non-bonding electron domain*?

9.11 Describe the characteristic electron-domain geometry of each of the following numbers of electron domains about a central atom: **(a)** 3, **(b)** 4, **(c)** 5, **(d)** 6.

9.12 Indicate the number of electron domains about a central atom, given the following angles between them: **(a)** 120°, **(b)** 180°, **(c)** 90°.

9.13 What is the difference between the electron-domain geometry and the molecular geometry of a molecule? Use the water molecule as an example in your discussion.

9.14 An AB_3 molecule is described as having a trigonal-bipyramidal electron-domain geometry. How many non-bonding domains are on atom A? Explain.

9.15 Draw the Lewis structure for each of the following molecules or ions, and predict their electron-domain and molecular geometries: **(a)** PF_3, **(b)** CH_3^+, **(c)** BrF_3, **(d)** ClO_4^-, **(e)** XeF_2, **(f)** BrO_2^-.

9.16 Give the electron-domain and molecular geometries for the following molecules and ions: **(a)** HCN, **(b)** SO_3^{2-}, **(c)** SF_4, **(d)** PF_6^-, **(e)** NH_3Cl^+, **(f)** N_3^-.

9.17 The figure that follows shows ball-and-stick drawings of three possible shapes of an AF_3 molecule. **(a)** For each shape, give the electron-domain geometry on which the molecular geometry is based. **(b)** For each shape, how many non-bonding electron domains are there on atom A? **(c)** Which of the following elements will lead to an AF_3 molecule with the shape in (ii): Li, B, N, Al, P, Cl? **(d)** Name an element A that is expected to lead to the AF_3 structure shown in (iii). Explain your reasoning.

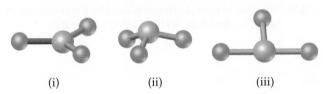

(i)   (ii)   (iii)

9.18 The figure that follows contains ball-and-stick drawings of three possible shapes of an AF_4 molecule. **(a)** For each shape, give the electron-domain geometry on which the molecular geometry is based. **(b)** For each shape, how many non-bonding electron domains are there on atom A? **(c)** Which of the following elements will lead to an AF_4 molecule with the shape in (iii): Be, C, S, Se, Si, Xe? **(d)** Name an element A that is expected to lead to the AF_4 structure shown in (i).

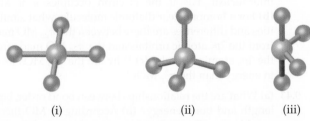

(i)   (ii)   (iii)

9.19 Give the approximate values for the indicated bond angles in the following molecules:

(a) 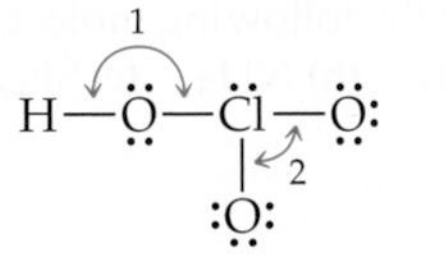 H—Ö—Cl—Ö: with angles 1, 2 and :O: below

(b) 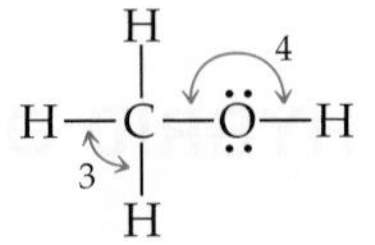 H—C—Ö—H with H's on C and angles 3, 4

(c) H—C≡C—H with angle 5

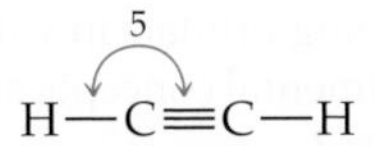

(d) 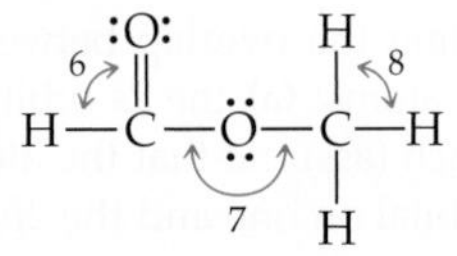 H—C—Ö—C—H with :O: above, angles 6, 7, 8 and H below

9.20 Give approximate values for the indicated bond angles in the following molecules:

(a) 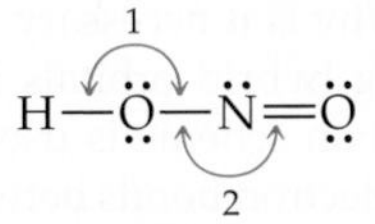 H—Ö—N=Ö with angles 1, 2

(b) 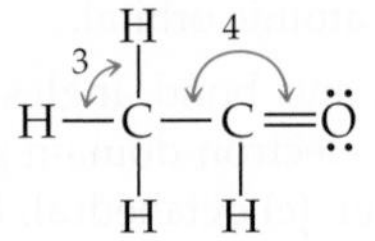 H—C—C=Ö with H's, angles 3, 4

(c) H—N—Ö—H with H's, angles 5, 6

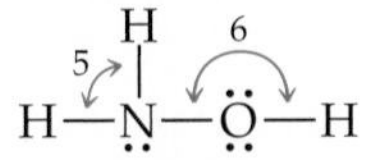

(d) 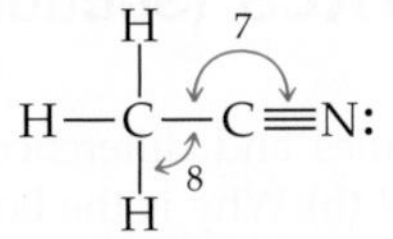 H—C—C≡N: with H's, angles 7, 8

9.21 The three species NH_2^-, NH_3 and NH_4^+ have H–N–H bond angles of 105°, 107° and 109°, respectively. Explain this variation in bond angles.

9.22 Predict the trend in the F(axial)l–A–F(equatorial) bond angle in the following AF_n molecules: PF_5, SF_4 and ClF_3.

9.23 **(a)** Explain why BrF_4^- is square planar, whereas BF_4^- is tetrahedral. **(b)** In which of these molecules, CF_4 or SF_4, do you think the actual bond angle is closest to the ideal angle predicted by the VSEPR model? Explain briefly.

9.24 **(a)** Explain why the following ions have different bond angles: ClO_2^- and NO_2^-. Predict the bond angle in each case. **(b)** Given that the spatial requirement of a non-bonding pair of electrons is greater than that of a bonding pair, explain why the XeF_2 molecule is linear and not bent.

POLARITY OF POLYATOMIC MOLECULES (Section 9.3)

9.25 **(a)** Consider the AF_3 molecules in Exercise 9.17. Which of these will have a non-zero dipole moment? Explain. **(b)** Which of the AF_4 molecules in Exercise 9.18 will have a zero dipole moment?

9.26 **(a)** What conditions must be met if a molecule with polar bonds is non-polar? **(b)** What geometries will give non-polar molecules for AB_2, AB_3 and AB_4 geometries?

9.27 Predict whether each of the following molecules is polar or non-polar: **(a)** CCl_4, **(b)** NH_3, **(c)** SF_4, **(d)** XeF_4, **(e)** CH_3Br, **(f)** GaH_3.

9.28 Predict whether each of the following molecules is polar or non-polar: **(a)** IF, **(b)** CS_2, **(c)** SO_3, **(d)** PCl_3, **(e)** SF_6, **(f)** IF_5.

ORBITAL OVERLAP; HYBRID ORBITALS (Sections 9.4 and 9.5)

9.29 (a) What is meant by the term *orbital overlap*? **(b)** What is the significance of overlapping orbitals in valence-bond theory? **(c)** What two fundamental concepts are incorporated in valence-bond theory?

9.30 Draw sketches illustrating the overlap between the following orbitals on two atoms: **(a)** the 2s orbital on each, **(b)** the $2p_z$ orbital on each (assume that the atoms are on the z-axis), **(c)** the 2s orbital on one and the $2p_z$ orbital on the other.

9.31 Consider the bonding in an MgH_2 molecule. **(a)** Draw a Lewis structure for the molecule, and predict its molecular geometry. **(b)** Why is it necessary to promote an electron before forming hybrid orbitals for the Mg atom? **(c)** What hybridisation scheme is used in MgH_2 **(d)** Sketch one of the two-electron bonds between an Mg hybrid orbital and an H 1s atomic orbital.

9.32 Indicate the hybridisation and bond angles associated with each of the following electron-domain geometries: **(a)** linear, **(b)** trigonal planar, **(c)** octahedral, **(d)** trigonal bipyramidal.

9.33 What is the designation for the hybrid orbitals formed from each of the following combinations of atomic orbitals: **(a)** one s and two p; **(b)** one s, three p and one d; **(c)** one s, three p and two d? What characteristic bond angles are associated with each?

9.34 (a) Starting with the orbital diagram of a boron atom, describe the steps needed to construct hybrid orbitals appropriate to describe the bonding in BF_3. **(b)** What is the name given to the hybrid orbitals constructed in (a)? **(c)** On one origin, sketch the large lobes of the hybrid orbitals constructed in part (a). **(d)** Are there any valence atomic orbitals of B that are left unhybridised? If so, how are they oriented relative to the hybrid orbitals?

9.35 Indicate the hybridisation of the central atom in **(a)** BCl_3, **(b)** $AlCl_4^-$, **(c)** CS_2, **(d)** KrF_2, **(e)** PF_6^-.

9.36 What set of hybrid orbitals is used by the central atom in **(a)** $SiCl_4$, **(b)** HCN, **(c)** SO_3, **(d)** ICl_2^-, **(e)** BrF_4^-?

MOLECULAR ORBITALS (Sections 9.7 and 9.8)

9.37 (a) What are the similarities and differences between atomic orbitals and MOs? **(b)** Why is the bonding molecular orbital of H_2 at lower energy than the electron in a hydrogen atom? **(c)** How many electrons can be placed into each MO of a molecule?

9.38 (a) Why is the antibonding MO of H_2 at higher energy than the electron in a hydrogen atom? **(b)** Does the Pauli exclusion principle (Section 6.7) apply to MOs? Explain. **(c)** If two p orbitals of one atom combine with two p orbitals of another atom, how many MOs result? Explain.

9.39 Consider the H_2^+ ion. **(a)** Sketch the MOs of the ion, and draw its energy-level diagram. **(b)** How many electrons are there in the H_2^+ ion? **(c)** Write the electron configuration of the ion in terms of its MOs. **(d)** What is the bond order in H_2^+? **(e)** Suppose that the ion is excited by light so that an electron moves from a lower-energy to a higher-energy MO. Would you expect the excited-state H_2^+ ion to be stable or to fall apart? Explain.

9.40 (a) Sketch the MOs of the H_2^- ion, and draw its energy-level diagram. **(b)** Write the electron configuration of the ion in terms of its MOs. **(c)** Calculate the bond order in H_2^-. **(d)** Suppose that the ion is excited by light, so that an electron moves from a lower-energy to a higher-energy MO. Would you expect the excited-state H_2^- ion to be stable? Explain.

9.41 (a) Sketch the σ and σ* MOs that can result from the combination of two $2p_z$ atomic orbitals. **(b)** Sketch the

π and π* MOs that result from the combination of two $2p_x$ atomic orbitals. **(c)** Place the MOs from parts (a) and (b) in order of increasing energy, assuming no mixing of 2s and 2p orbitals.

9.42 (a) What is the probability of finding an electron on the internuclear axis if the electron occupies a π MO? **(b)** For a homonuclear diatomic molecule, what similarities and differences are there between the π_{2p} MO made from the $2p_x$ atomic orbitals and the π_{2p} MO made from the $2p_y$ atomic orbitals? **(c)** Why are the π_{2p} MOs lower in energy than the π_{2p}^* MOs?

9.43 (a) What are the relationships between bond order, bond length and bond energy? **(b)** According to MO theory, would either Be_2 or Be_2^+ be expected to exist? Explain.

9.44 Explain the following: **(a)** the peroxide ion, O_2^{2-}, has a longer bond than the superoxide ion, O_2^-; **(b)** the magnetic properties of B_2 are consistent with the π_{2p} MOs being lower in energy than the σ_{2p} MO.

9.45 (a) What is meant by the term *diamagnetism*? **(b)** How does a diamagnetic substance respond to a magnetic field? **(c)** Which of the following ions would you expect to be diamagnetic: N_2^{2-}, O_2^{2-}, Be_2^{2+}, C_2^-?

9.46 (a) What is meant by the term *paramagnetism*? **(b)** How can we determine experimentally whether a substance is paramagnetic? **(c)** Which of the following ions would you expect to be paramagnetic: O_2^+, N_2^{2-}, Li_2^+, O_2^{2-}?

If the ion is paramagnetic, how many unpaired electrons does it possess?

9.47 Using Figures 9.27 and 9.33 as guides, give the MO electron configuration for **(a)** B_2^+, **(b)** Li_2^+, **(c)** N_2^+, **(d)** Ne_2^{2+}. In each case, indicate whether the addition of an electron to the ion would increase or decrease the bond order of the species.

9.48 If we assume that the energy-level diagrams for homonuclear diatomic molecules shown in Figure 9.27 can be applied to heteronuclear diatomic molecules and ions, predict the bond order and magnetic behaviour of **(a)** CO^+, **(b)** NO^-, **(c)** OF^+, **(d)** NeF^+.

9.49 Determine the electron configurations for CN^+, CN and CN^-. Calculate the bond order for each one, and indicate which are paramagnetic.

9.50 **(a)** The nitric oxide molecule, NO, readily loses one electron to form the NO^+ ion. Why is this consistent with the electronic structure of NO? **(b)** Predict the order of the N–O bond strengths in NO, NO^+ and NO^- and describe the magnetic properties of each. **(c)** With what neutral homonuclear diatomic molecules are the NO^+ and NO^- ions isoelectronic (same number of electrons)?

9.51 Consider the MOs of the P_2 molecule. (Assume that the MOs of diatomics from the third period of the periodic table are analogous to those from the second period.)

(a) Which valence atomic orbitals of P are used to construct the MOs of P_2? **(b)** The figure below shows a sketch of one of the MOs for P_2. What is the label for this MO? **(c)** For the P_2 molecule, how many electrons occupy the MO in the figure? **(d)** Is P_2 expected to be diamagnetic or paramagnetic? Explain.

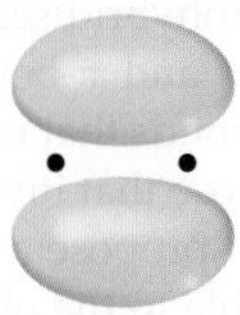

9.52 The iodine bromide molecule, IBr, is an *interhalogen compound*. Assume that the MOs of IBr are analogous to the homonuclear diatomic molecule F_2. **(a)** Which valence atomic orbitals of I and of Br are used to construct the MOs of IBr? **(b)** What is the bond order of the IBr molecule? **(c)** One of the valence MOs of IBr is sketched below. Why are the atomic orbital contributions to this MO different in size? **(d)** What is the label for the MO? **(e)** For the IBr molecule, how many electrons occupy the MO?

ADDITIONAL EXERCISES

9.53 **(a)** What is the physical basis for the VSEPR model? **(b)** When applying the VSEPR model, we count a double or triple bond as a single electron domain. Why is this justified?

9.54 The molecules SiF_4, SF_4 and XeF_4 all have molecular formulae of the type AF_4, but the molecules have different molecular geometries. Predict the shape of each molecule, and explain why the shapes differ.

9.55 Consider the molecule PF_4Cl. **(a)** Draw a Lewis structure for the molecule, and predict its electron-domain geometry. **(b)** Which would you expect to produce a larger electron domain, a P–F bond or a P–Cl bond? Explain. **(c)** Predict the molecular geometry of PF_4Cl. How did your answer for part (b) influence your answer here in part (c)? **(d)** Would you expect the molecule to distort from its ideal electron domain geometry? If so, how would it distort?

9.56 From their Lewis structures, determine the number of σ-bonds and π-bonds in each of the following molecules or ions: **(a)** CO_2; **(b)** thiocyanate ion, NCS^-; **(c)** formaldehyde, HCHO; **(d)** formic acid, HCOOH, which has one H and two O atoms attached to C.

9.57 The lactic acid molecule, $CH_3CH(OH)COOH$, gives sour milk its unpleasant taste. **(a)** Draw the Lewis structure for the molecule, assuming that carbon always forms four bonds in its stable compounds. **(b)** How many σ-bonds and how many π-bonds are in the molecule? **(c)** Which CO bond is shortest in the molecule? **(d)** What is the hybridisation of atomic orbitals around each car-

bon atom associated with that short bond? **(e)** What are the approximate bond angles around each carbon atom in the molecule?

9.58 The PF_3 molecule has a dipole moment of 1.03 D, but BF_3 has a dipole moment of zero. How can you explain the difference?

9.59 There are two compounds of the formula $Pt(NH_3)_2Cl_2$.

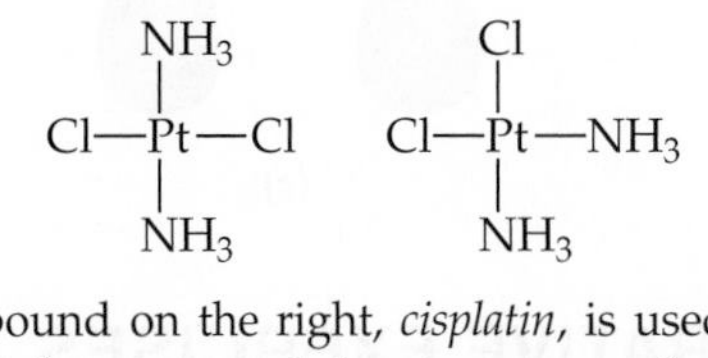

$$
\begin{array}{cc}
& \text{NH}_3 & & \text{Cl} \\
& | & & | \\
\text{Cl}-\text{Pt}-\text{Cl} & & \text{Cl}-\text{Pt}-\text{NH}_3 \\
& | & & | \\
& \text{NH}_3 & & \text{NH}_3
\end{array}
$$

The compound on the right, *cisplatin*, is used in cancer therapy. Both compounds have a square planar geometry. Which compound has a non-zero dipole moment?

9.60 The reaction of three molecules of fluorine gas with an Xe atom produces the substance xenon hexafluoride, XeF_6

$$Xe(g) + 3\,F_2(g) \longrightarrow XeF_6(s)$$

(a) Draw a Lewis structure for XeF_6. **(b)** If you try to use the VSEPR model to predict the molecular geometry of XeF_6, you run into a problem. What is it? **(c)** What could you do to resolve the difficulty in part (b)? **(d)** Suggest a hybridisation scheme for the Xe atom in XeF_6. **(e)** The molecule IF_7 has a pentagonal-bipyramidal structure (five equatorial fluorine atoms at the vertices of a regular pentagon and two axial fluorine atoms). Based on the structure of IF_7, suggest a structure for XeF_6.

9.61 The azide ion, N_3^-, is linear with two N–N bonds of equal length, 116 pm. **(a)** Draw a Lewis structure for the azide ion. **(b)** With reference to Table 8.5, is the observed N–N bond length consistent with your Lewis structure? **(c)** What hybridisation scheme would you expect at each of the nitrogen atoms in N_3^-? **(d)** Show which hybridised and unhybridised orbitals are involved in the formation of σ- and π-bonds in N_3^-. **(e)** It is often observed that σ-bonds that involve an *sp* hybrid orbital are shorter than those that involve only sp^2 or sp^3 hybrid orbitals. Can you propose a reason for this? Is this observation applicable to the observed bond lengths in N_3^-?

9.62 In ozone, O_3, the two oxygen atoms on the ends of the molecule are equivalent to one another. **(a)** What is the best choice of hybridisation scheme for the atoms of ozone? **(b)** For one of the resonance forms of ozone, which of the orbitals are used to make bonds and which are used to hold non-bonding pairs of electrons? **(c)** Which of the orbitals can be used to delocalise the π electrons? **(d)** How many electrons are delocalised in the π system of ozone?

9.63 The sketches below show the atomic orbital wave functions (with phases) used to construct some of the MOs of a homonuclear diatomic molecule. For each sketch, determine the MO that will result from mixing the atomic orbital wave functions as drawn. Use the same labels for the MOs as in Figure 9.27.

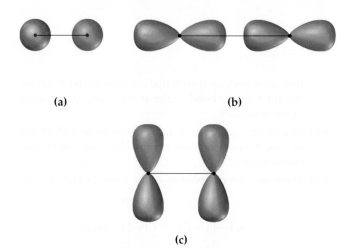

(a)   (b)

(c)

9.64 Write the electron configuration for the first excited state for N_2 (that is, the state with the highest-energy electron moved to the next available energy level). What differences do you expect in the properties of N_2 in its ground state and its first excited state?

9.65 Figure 9.34 shows how the magnetic properties of a compound can be measured experimentally. When such measurements are made, the sample is generally covered by an atmosphere of pure nitrogen gas rather than air. Why do you suppose this is done?

9.66 **(a)** Using only the valence atomic orbitals of a hydrogen atom and a fluorine atom, how many MOs would you expect for the HF molecule? **(b)** How many of the MOs from part (a) would be occupied by electrons? **(c)** Do you think the MO diagram shown in Figure 9.36 could be used to describe the MOs of the HF molecule? Why or why not?

9.67 Carbon monoxide, CO, is isoelectronic to N_2. **(a)** Draw a Lewis structure for CO that satisfies the octet rule. **(b)** Assume that the diagram in Figure 9.36 can be used to describe the MOs of CO. What is the predicted bond order for CO? Is this answer in accord with the Lewis structure you drew in part (a)? **(c)** Experimentally, it is found that the highest-energy electrons in CO reside in a σ-type MO. Is that observation consistent with Figure 9.36? If not, what modification needs to be made to the diagram? How does this modification relate to Figure 9.32? **(d)** Would you expect the π_{2p} MOs of CO to have equal atomic orbital contributions from the C and O atoms? If not, which atom would have the greater contribution?

INTEGRATIVE EXERCISES

9.68 A compound composed of 2.1% H, 29.8% N and 68.1% O has a molar mass of approximately 50 g mol^{-1}. **(a)** What is the molecular formula of the compound? **(b)** What is its Lewis structure if H is bonded to O? **(c)** What is the geometry of the molecule? **(d)** What is the hybridisation of the orbitals around the N atom? **(e)** How many σ-bonds and how many π-bonds are there in the molecule?

9.69 Sulfur tetrafluoride (SF_4) reacts slowly with O_2 to form sulfur tetrafluoride monoxide (OSF_4), according to the following unbalanced reaction

$$SF_4(g) + O_2(g) \longrightarrow OSF_4(g)$$

The O atom and the four F atoms in OSF_4 are bonded to a central S atom. **(a)** Balance the equation. **(b)** Write a Lewis structure of OSF_4 in which the formal charges of all atoms are zero. **(c)** Use average bond enthalpies (Table 8.4) to estimate the enthalpy of the reaction. Is it endothermic or exothermic? **(d)** Determine the electron-domain geometry of OSF_4, and write two possible molecular geometries for the molecule based on this electron-domain geometry. **(e)** Which of the molecular geometries in part (d) is more likely to be observed for the molecule? Explain.

9.70 The phosphorus trihalides (PX_3) show the following variation in the bond angle X–P–X: PF_3, 96.3°; PCl_3,

100.3°; PBr$_3$, 101.0°; PI$_3$, 102°. The trend is generally attributed to the change in the electronegativity of the halogen. **(a)** Assuming that all electron domains exhibit the same repulsion, what value of the X–P–X angle is predicted by the VSEPR model? **(b)** What is the general trend in the X–P–X angle as the electronegativity increases? **(c)** Using the VSEPR model, explain the observed trend in X–P–X angle as the electronegativity of X changes. **(d)** Based on your answer to part (c), predict the structure of PBrCl$_4$.

9.71 **(a)** Compare the bond enthalpies (Table 8.4) of the carbon–carbon single, double and triple bonds to deduce an average π-bond contribution to the enthalpy. What fraction of a single bond does this quantity represent? **(b)** Make a similar comparison of nitrogen–nitrogen bonds. What do you observe? **(c)** Write Lewis structures of N$_2$H$_4$, N$_2$H$_2$ and N$_2$ and determine the hybridisation around nitrogen in each case. **(d)** Propose a reason for the large difference in your observations of parts (a) and (b).

9.72 For both atoms and molecules, ionisation energies (Section 7.4) are related to the energies of orbitals: the lower the energy of the orbital, the greater the ionisation energy. The first ionisation energy of a molecule is therefore a measure of the energy of the highest occupied MO. The first ionisation energies of several diatomic molecules are given in electron-volts in the following table:

Molecule	I_1 (eV)
H$_2$	15.4
N$_2$	15.6
O$_2$	12.1
F$_2$	15.7

(a) Convert these ionisation energies to kJ mol^{-1}. **(b)** On the same piece of graph paper, plot I_1 for the H, N, O and F atoms (estimated from Figure 7.9) and I_1 for the molecules listed. **(c)** Do the ionisation energies of the molecules follow the same periodic trends as the ionisation energies of the atoms? **(d)** Use MO energy-level diagrams to explain the trends in the ionisation energies of the molecules.

9.73 Many compounds of the transition metal elements contain direct bonds between metal atoms. We will assume that the z-axis is defined as the metal–metal bond axis. **(a)** Which of the 3d orbitals (Figure 6.20) can be used to make a σ-bond between metal atoms? **(b)** Sketch the σ$_{3d}$ bonding and σ$_{3d}^*$ antibonding MOs. **(c)** With reference to the A Closer Look box on the phases of orbitals, explain why a node is generated in the σ$_{3d}^*$ MO. **(d)** Sketch the energy-level diagram for the Sc$_2$ molecule, assuming that only the 3d orbital from part (a) is important. **(e)** What is the bond order in Sc$_2$?

9.74 The chemistry of astatine (At) is far less developed than that of the other halogen elements. **(a)** In the periodic table the atomic mass of At is written as (210). Why are the parentheses used? How is this nomenclature related to the difficulty of studying astatine? **(b)** Write the complete electron configuration for a neutral At atom. **(c)** Although At$_2$ is not known, the interhalogen compound AtI has been characterised. Would this compound be expected to have a covalent, polar covalent or ionic bond? Explain. **(d)** The reaction of AtI with I$^-$ forms the AtI$_2^-$ ion. Use the VSEPR method to predict the geometry of this ion. **(e)** Suppose we construct the MOs of the unknown At$_2$ molecule. What bond order is predicted for the molecule? What type of MO is the highest-occupied MO of the molecule?

MasteringChemistry (www.pearson.com.au/masteringchemistry)

Make learning part of the grade. Access:

- tutorials with peronalised coaching
- study area
- Pearson eText

PHOTO/ART CREDITS

292 © Hans & Annie Wapstra; **296** Kristen Brochmann, Fundamental Photographs, NYC; **325** Richard Megna, Fundamental Photographs, NYC.

INTERMOLECULAR FORCES: GASES

A cyclone as seen from space gathers intensity. The swirling gases of the atmosphere (mainly nitrogen and oxygen), with a diameter of over 1000 km, can reach wind speeds of over 300 km h^{-1}.

KEY CONCEPTS

10.1 CHARACTERISTICS OF GASES
We compare the distinguishing characteristics of gases with those of liquids and solids.

10.2 PRESSURE AND ITS MEASUREMENT
We examine gas *pressure*, how it is measured and the units used to express it, as well as consider Earth's atmosphere and the pressure it exerts.

10.3 THE GAS LAWS
The state of a gas can be expressed in terms of its volume, pressure, temperature and quantity. We examine several *gas laws*, which are empirical relationships between these four variables.

10.4 THE IDEAL-GAS EQUATION
The gas laws yield the *ideal-gas equation*, $PV = nRT$. Although this equation is not obeyed exactly by any real gas, most gases come very close to obeying it at ordinary temperatures and pressures.

10.5 FURTHER APPLICATIONS OF THE IDEAL-GAS EQUATION
We see that the ideal-gas equation has many uses, such as the calculation of the density or molar mass of a gas.

10.6 GAS MIXTURES AND PARTIAL PRESSURES
In a mixture of gases, each gas exerts a pressure that is part of the total pressure. This *partial pressure* is the pressure the gas would exert if it were by itself.

10.7 KINETIC-MOLECULAR THEORY
This theory helps us understand gas behaviour at the molecular level. According to the theory, the atoms or molecules that make up a gas move with an average kinetic energy that is proportional to the gas temperature.

10.8 MOLECULAR EFFUSION AND DIFFUSION
The kinetic-molecular theory helps us account for gas properties such as *effusion*, movement through tiny openings, and *diffusion*, movement through another substance.

10.9 REAL GASES: DEVIATIONS FROM IDEAL BEHAVIOUR
We learn that real gases deviate from ideal behaviour because the gas molecules have finite volume and because attractive forces exist between molecules. The *van der Waals equation* gives an accurate account of real gas behaviour at high pressures and low temperatures.

I n the past several chapters we have learned about the electronic structures of atoms and how atoms combine to form molecules and ionic substances. In everyday life, however, we don't have direct experiences with atoms. Instead, we encounter matter as collections of enormous numbers of atoms or molecules that make up gases, liquids and solids. In the atmosphere it's the action of such large collections of atoms and molecules that is responsible for our weather—the gentle breezes and the gales, the humidity and the rain. Cyclones, such as the one shown in the chapter-opening photo, form when moist, warm air at lower elevations converges with cooler, dry air above. The resultant air flows produce winds that can approach speeds of up to 300 km h^{-1}.

We now know that the properties of gases, liquids and solids are readily understood in terms of the behaviour of their component atoms, ions and molecules. In this chapter we examine the physical properties of gases and consider how we can understand these properties in terms of the behaviour of gas molecules. In Chapter 11 we turn our attention to the physical properties of liquids and solids.

In many ways gases are the most easily understood form of matter. Even though different gaseous substances may have very different *chemical* properties, they behave quite similarly as far as their *physical* properties are concerned. For example, we live in an atmosphere composed of a mixture of gases that we refer to as air. We breathe air to absorb oxygen, O_2, which supports human life. Air also contains nitrogen, N_2, which has very different chemical properties from oxygen, yet this mixture behaves physically as one gaseous material. The relative simplicity of the gas state affords a good starting point as we seek to understand the properties of the states of matter.

10.1 | CHARACTERISTICS OF GASES

There are relatively few elements that are gases under ordinary conditions of temperature and pressure. The noble gases, He, Ne, Ar, Kr and Xe, are all monoatomic gases, whereas H_2, N_2, O_2, F_2 and Cl_2 are diatomic gases. Many molecular compounds are also gases. ▼ TABLE 10.1 lists a few of the more common gaseous compounds. Notice that all these gases are composed entirely of non-metallic elements. Furthermore, all have simple molecular formulae and therefore low molar masses. Substances that are solids or liquids under ordinary conditions can also exist in the gaseous state, where they are often referred to as **vapours**. The substance H_2O, for example, can exist as solid ice, liquid water or water vapour.

Gases differ significantly from solids and liquids in several respects. For example, a gas expands spontaneously to fill its container. Consequently, the volume of a gas equals the volume of the container in which it is held. Gases also are highly compressible: when pressure is applied to a gas, its volume readily decreases. Solids and liquids, on the other hand, do not expand to fill their containers, and solids and liquids are not readily compressible.

Gases form homogeneous mixtures with each other regardless of the identities or relative proportions of the component gases. The atmosphere serves as an excellent example. As a further example, when water and petrol are mixed, the two liquids remain as separate layers. In contrast, the water vapour and petrol vapours above the liquids form a homogeneous gas mixture.

The characteristic properties of gases arise because the individual molecules are relatively far apart. In the air we breathe, which is composed roughly of 78% N_2, 21% O_2 and small amounts of other gases, the molecules take up only about 0.1% of the total volume, with the rest being empty space. *Thus each molecule behaves largely as though the others were not present.* As a result, different gases behave similarly, even though they are made up of different molecules. In contrast, the individual molecules in a liquid are close together and occupy perhaps 70% of the total space. The attractive forces between the molecules keep the liquid together.

 CONCEPT CHECK 1

What is the major reason that physical properties do not differ much from one gaseous substance to another?

10.2 | PRESSURE AND ITS MEASUREMENT

One of the easily measured properties of a gas is its **pressure**, P, which is defined as the force, F, exerted by the gas on a given area, A:

$$P = \frac{F}{A}$$

[10.1]

TABLE 10.1 • **Some common compounds that are gases at room temperature**

Formula	Name	Characteristics
HCN	Hydrogen cyanide	Very toxic, slight odour of bitter almonds
H_2S	Hydrogen sulfide	Very toxic, odour of rotten eggs
CO	Carbon monoxide	Toxic, colourless, odourless
CO_2	Carbon dioxide	Colourless, odourless
CH_4	Methane	Colourless, odourless, flammable
C_2H_4	Ethylene	Colourless, ripens fruit
C_3H_8	Propane	Colourless, bottled heating gas
N_2O	Nitrous oxide	Colourless, sweet odour, laughing gas
NO_2	Nitrogen dioxide	Toxic, red-brown, irritating odour
NH_3	Ammonia	Colourless, pungent odour
SO_2	Sulfur dioxide	Colourless, irritating odour

Pressure conveys the idea of a force, a push that tends to move something in a given direction, as can be felt by standing in a strong wind. The molecules that make up the Earth's atmosphere all experience a gravitational force that pulls towards the centre of the Earth. This is counteracted by the thermal motion of the molecules so the particles that make up the Earth's atmosphere do not pile up on the surface of the Earth. Nevertheless, the gravitational force causes the atmosphere as a whole to press down on Earth's surface, creating *atmospheric pressure*, defined as the force exerted by the atmosphere on a given surface area.

We can calculate the magnitude of this atmospheric pressure as follows: The force, F, exerted by any object is the product of its mass, m, and its acceleration, a: $F = ma$. The acceleration given by Earth's gravitational force to any object located near Earth's surface is 9.8 m s^{-2}. Now imagine a column of air 1 m^2 in cross-section extending through the entire atmosphere (▶ FIGURE 10.1). That column has a mass of roughly 10 000 kg. The downward gravitational force exerted on this column is

$$F = (10\,000\ \text{kg})\,(9.8\ \text{m s}^{-2}) = 1 \times 10^5\ \text{kg m s}^{-2} = 1 \times 10^5\ \text{N}$$

where N is the abbreviation for *newton*, the SI unit for force: $1\ \text{N} = 1\ \text{kg m s}^{-2}$. The pressure exerted by the column is this force divided by the cross-sectional area, A, over which the force is applied. Because our air column has a cross-sectional area of 1 m^2, we have for the magnitude of atmospheric pressure at sea level

$$P = \frac{F}{A} = \frac{1 \times 10^5\ \text{N}}{1\ \text{m}^2} = 1 \times 10^5\ \text{N m}^{-2} = 1 \times 10^5\ \text{Pa} = 1 \times 10^2\ \text{kPa}$$

The SI unit of pressure is the **pascal** (Pa), named for Blaise Pascal (1623–1662), a French scientist who studied pressure: $1\ \text{Pa} = 1\ \text{N m}^{-2}$. A related pressure unit is the **bar**: $1\ \text{bar} = 10^5\ \text{Pa} = 10^5\ \text{N m}^{-2}$. Thus the atmospheric pressure at sea level we just calculated, 100 kPa, can be reported as 1 bar. (The actual atmospheric pressure at any location depends on weather conditions and altitude.)

In the early part of the seventeenth century it was widely believed by scientists and philosophers that the atmosphere had no weight. Evangelista Torricelli (1608–1647), a student of Galileo's, proved this to be untrue by inventing the *barometer* (▶ FIGURE 10.2). A glass tube more than 760 mm long that is closed at one end is completely filled with mercury and inverted into a dish that contains additional mercury. (Care must be taken so that no air gets into the tube.) Some of the mercury flows out when the tube is inverted, but a column of mercury remains in the tube. Torricelli argued that this was due to Earth's atmosphere pushing on the surface of the mercury in the dish thus pushing the mercury up the tube until the pressure exerted by the mercury column downward, due to gravity, equals the atmospheric pressure at the base of the tube. Thus the height, h, of the mercury column is a measure of the atmosphere's pressure, so it will change as the atmospheric pressure changes. Since pressure is defined as *force per unit area*, the diameter of the tube does not matter. A larger tube containing more mercury will exert a larger downward force, but the pressure will be the same because the area is larger.

Although Torricelli's explanation met with fierce opposition, it also had its supporters. Blaise Pascal, for example, had one of the barometers carried to the top of Puy de Dome, a volcanic mountain in central France, and compared its readings with a duplicate barometer kept at the foot of the mountain. As the barometer ascended, the height of the mercury column diminished, as expected, because the amount of atmosphere pressing down on the surface decreases as one moves higher.

Standard atmospheric pressure (which corresponds to the typical atmospheric pressure at sea level) was defined as the pressure sufficient to support a column of mercury 760 mm high because, over a long series of experiments, the average height h, of the column of mercury turned out to be 760 mm. In SI units this pressure of 760 mm Hg = 1.01325×10^5 Pa or 1 mm Hg = 133.322 Pa.

The commonly used quantity in chemical calculations is the kilopascal (kPa) where $1\ \text{kPa} = 1 \times 10^3\ \text{Pa}$.

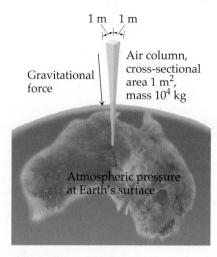

▲ **FIGURE 10.1**   **Calculating atmospheric pressure.**

FIGURE IT OUT

What happens to h, the height of the mercury column, if the atmospheric pressure increases?

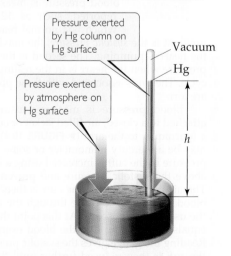

▲ **FIGURE 10.2**   **A mercury barometer.**

Standard atmospheric pressure is now defined as 1 bar, where

$$1 \text{ bar} = 100 \text{ kPa} = 1 \times 10^5 \text{ Pa} = 1 \times 10^5 \text{ N m}^{-2}$$

We can use various devices to measure the pressures of enclosed gases. Tyre gauges, for example, measure the pressure of air in car and bicycle tyres. In laboratories we sometimes use a device called a *manometer*. A manometer operates on a principle similar to that of a barometer, as shown in Sample Exercise 10.1.

SAMPLE EXERCISE 10.1 **Using a manometer to measure gas pressure**

On a certain day the barometer in a laboratory indicates that the atmospheric pressure is 102 kPa. A sample of gas is placed in a flask attached to an open-end mercury manometer, shown in ▶ FIGURE 10.3. A metre stick is used to measure the height of the mercury above the bottom of the manometer. The level of mercury in the open-end arm of the manometer has a height of 136.4 mm, and the mercury in the arm that is in contact with the gas has a height of 103.8 mm. What is the pressure of the gas in kPa and in bar?

SOLUTION

Analyse We are asked to use the height of mercury that a gas supports to determine its pressure.

Plan We use the difference in height between the two arms (*h* in Figure 10.3) to obtain the amount by which the pressure of the gas exceeds atmospheric pressure. Because an open-end mercury manometer is used, the height difference directly measures the pressure difference in mm Hg between the gas and the atmosphere.

Solve The pressure of the gas equals the atmospheric pressure plus *h*:

$$P_{gas} = P_{atm} + h$$
$$= 102 \text{ kPa} + (136.4 \text{ mm Hg} - 103.8 \text{ mm Hg})$$
$$= 102 \text{ kPa} + 32.6 \text{ mm Hg}$$
$$= 102 \text{ kPa} + (32.6 \times 133.322 \times 10^{-3} \text{ kPa})$$
$$= 106 \text{ kPa}$$
$$= 1.06 \text{ bar}$$

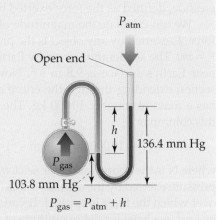

▲ **FIGURE 10.3** **A mercury manometer.** This device is sometimes employed in the laboratory to measure gas pressures near atmospheric pressure.

PRACTICE EXERCISE

Convert a pressure of 740 mm Hg into kPa and bar.

Answer: 98.6 kPa and 0.986 bar.

(See also Exercise 10.11.)

MY WORLD OF CHEMISTRY

BLOOD PRESSURE

The human heart pumps blood to the parts of the body through arteries, and the blood returns to the heart through veins. When your blood pressure is measured, two values are reported, such as 120/80 (120 over 80), which is a normal reading. The first measurement is the *systolic pressure*, the maximum pressure when the heart is pumping. The second is the *diastolic pressure*, the pressure when the heart is in the resting part of its pumping cycle. The units associated with these pressure measurements are torr.

Blood pressure is measured using a pressure gauge attached to a closed, air-filled jacket or cuff that is applied like a tourniquet to the arm (▶ FIGURE 10.4). The pressure gauge may be a mercury manometer or some other device. The air pressure in the cuff is increased using a small pump until it is above the systolic pressure and prevents the flow of blood. The air pressure inside the cuff is then slowly reduced until blood just begins to pulse through the artery, as detected by the use of a stethoscope. At this point the pressure in the cuff equals the pressure that the blood exerts inside the arteries. Reading the gauge gives the systolic pressure. The pressure in the cuff is then reduced further until the blood flows freely. The pressure at this point is the diastolic pressure.

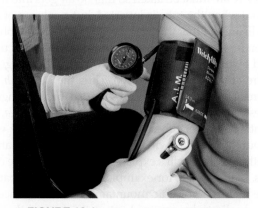

▲ **FIGURE 10.4** **Measuring blood pressure.**

Hypertension is the presence of abnormally high blood pressure. The usual criterion for hypertension is a blood pressure greater than 140/90, although recent studies suggest that health risks increase for systolic readings above 120. Hypertension significantly increases the workload on the heart and also places a stress on the walls of the blood vessels throughout the body. These effects increase the risk of aneurysms, heart attacks and strokes.

10.3 | THE GAS LAWS

Four variables are needed to define the physical condition, or *state*, of a gas: temperature, T, pressure, P, volume, V, and the amount of gas, which is usually expressed as the number of moles, n. The equations that express the relationships between T, P, V and n are known as the *gas laws*.

The Pressure–Volume Relationship: Boyle's Law

If the pressure on a balloon is decreased, the balloon expands. That is why weather balloons expand as they rise through the atmosphere. Conversely, when a volume of gas is compressed, the pressure of the gas increases. British chemist Robert Boyle (1627–1691) first investigated the relationship between the pressure of a gas and its volume.

To perform his gas experiments, Boyle used a J-shaped tube like that shown in ▶ FIGURE 10.5. In the tube on the left, a quantity of gas is trapped behind a column of mercury. Boyle changed the pressure on the gas by adding mercury to the tube. He found that the volume of the gas decreased as the pressure increased. For example, doubling the pressure caused the gas volume to decrease to half its original value.

Boyle's law, which summarises these observations, states that *the volume of a fixed quantity of gas maintained at constant temperature is inversely proportional to the pressure.* When two measurements are inversely proportional, one gets smaller as the other gets larger. Boyle's law can be expressed mathematically as

$$V = \text{constant} \times \frac{1}{P} \quad \text{or} \quad PV = \text{constant} \qquad [10.2]$$

The value of the constant depends on the temperature and the amount of gas in the sample. The graph of V versus P in ▼ FIGURE 10.6(a) shows the type of curve obtained for a given quantity of gas at a fixed temperature. A linear relationship is obtained when V is plotted against $1/P$ (Figure 10.6(b)).

We apply Boyle's law every time we breathe. The volume of the lungs is governed by the rib cage, which can expand and contract, and the diaphragm, a muscle beneath the lungs. Inhalation occurs when the rib cage expands and the

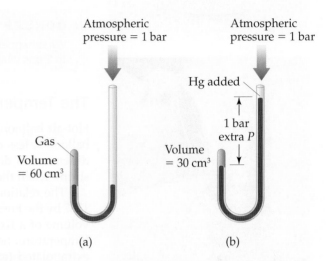

▲ **FIGURE 10.5** **An illustration of Boyle's experiment relating pressure and volume.** In (a) the volume of the gas trapped in the J-tube is 60 cm³ when the gas pressure is 1 bar. When additional mercury is added, as shown in (b), the trapped gas is compressed. The volume is 30 cm³ when its total pressure is 2 bar, corresponding to atmospheric pressure plus the pressure exerted by the column of mercury.

━━━━━━━━━━━━━━━━━━━━

△ **FIGURE IT OUT**

What would a plot of P versus $1/V$ look like for a fixed quantity of gas at a fixed temperature?

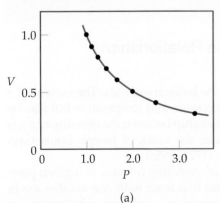

(a)

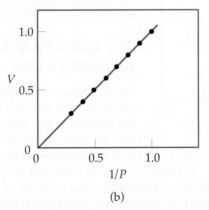

(b)

▲ **FIGURE 10.6** **Boyle's Law.** For a fixed quantity of gas at constant temperature, the volume of the gas is inversely proportional to its pressure.

▲ **FIGURE 10.7   An illustration of the effect of temperature on volume.** As liquid nitrogen (−196 °C) is poured over a balloon, the gas in the balloon is cooled and its volume decreases.

∞ Review this on page 13

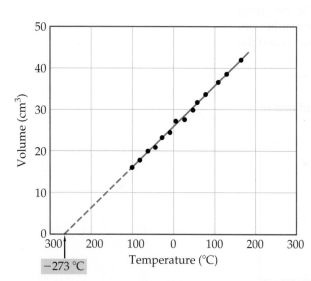

▲ **FIGURE 10.8   Graph based on Charles's law.** At constant pressure, the volume of an enclosed gas increases as the temperature increases. The dashed line is an extrapolation to temperatures at which the substance is no longer a gas.

diaphragm moves downward. Both of these actions increase the volume of the lungs, thus decreasing the gas pressure inside the lungs. The atmospheric pressure then forces air into the lungs until the pressure in the lungs equals atmospheric pressure. Exhalation reverses the process: the rib cage contracts and the diaphragm moves up, both of which decrease the volume of the lungs. Air is forced out of the lungs by the resulting increase in pressure.

> ⚠ **CONCEPT CHECK 2**
>
> What happens to the volume of a gas if you double its pressure, say, from 1 bar to 2 bar, while its temperature is held constant?

The Temperature–Volume Relationship: Charles's Law

Hot-air balloons rise because air expands as it is heated. The warm air in the balloon is less dense than the surrounding cool air at the same pressure. This difference in density causes the balloon to ascend. Conversely, a balloon will shrink when the gas in it is cooled, as seen in ◀ **FIGURE 10.7**.

The relationship between gas volume and temperature was discovered in 1787 by the French scientist Jacques Charles (1746–1823). Charles found that the volume of a fixed quantity of gas at constant pressure increases linearly with temperature. Some typical data are shown in ◀ FIGURE 10.8. Notice that the extrapolated (extended) line (which is dashed) passes through −273 °C. Note also that the gas is predicted to have zero volume at this temperature. This condition is never realised, however, because all gases liquefy or solidify before reaching this temperature.

In 1848 William Thomson (1824–1907), a British physicist whose title was Lord Kelvin, proposed an absolute-temperature scale, now known as the Kelvin scale. On this scale 0 K, which is called *absolute zero*, equals −273.15 °C (∞ Section 1.4, 'Units of Measurement'). In terms of the Kelvin scale, **Charles's law** can be stated as follows: *The volume of a fixed amount of gas maintained at constant pressure is directly proportional to its absolute temperature.* Thus doubling the absolute temperature, for example, from 200 K to 400 K, causes the gas volume to double. Mathematically, Charles's law takes the following form:

$$V = \text{constant} \times T \quad \text{or} \quad \frac{V}{T} = \text{constant} \qquad [10.3]$$

The value of the constant depends on the pressure and amount of gas.

> ⚠ **CONCEPT CHECK 3**
>
> Does the volume of a fixed quantity of gas decrease to half its original value when the temperature is lowered from 100 °C to 50 °C?

The Quantity–Volume Relationship: Avogadro's Law

As we add gas to a balloon the balloon expands. The volume of a gas is affected not only by pressure and temperature but also by the amount of gas. The relationship between the quantity of a gas and its volume follows from the work of Joseph Louis Gay-Lussac (1778–1823) and Amedeo Avogadro (1776–1856).

In 1808 Gay-Lussac observed the *law of combining volumes*: at a given pressure and temperature, the volumes of gases that react with one another are in the ratios of small whole numbers. For example, two volumes of hydrogen gas react with one volume of oxygen gas to form two volumes of water vapour, as shown in ▶ FIGURE 10.9.

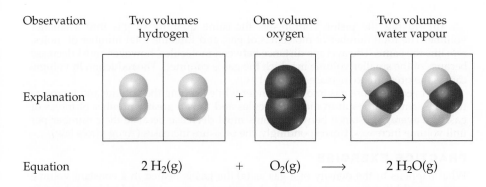

Observation	Two volumes hydrogen		One volume oxygen		Two volumes water vapour	
Explanation		+		→		
Equation	2 H₂(g)	+	O₂(g)	→	2 H₂O(g)	

◀ **FIGURE 10.9   The law of combining volumes.** Gay-Lussac's experimental observation of combining volumes is shown together with Avogadro's explanation of this phenomenon.

Three years later Avogadro interpreted Gay-Lussac's observation by proposing what is now known as **Avogadro's hypothesis**: *Equal volumes of gases at the same temperature and pressure contain equal numbers of molecules.* For example, experiments show that 22.7 dm³ of any gas at 0 °C and 1 bar (1 × 10² kPa) contain 6.02 × 10²³ gas molecules (that is, 1 mol), as depicted in ▼ **FIGURE 10.10**.

Avogadro's law follows from Avogadro's hypothesis: *The volume of a gas maintained at constant temperature and pressure is directly proportional to the number of moles of the gas.* That is,

$$V = \text{constant} \times n \qquad [10.4]$$

Thus doubling the number of moles of gas will cause the volume to double if T and P remain constant.

FIGURE IT OUT

How many moles of gas are in each vessel?

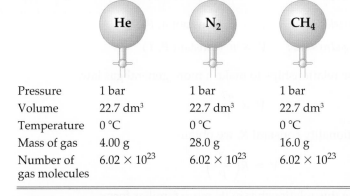

	He	N₂	CH₄
Pressure	1 bar	1 bar	1 bar
Volume	22.7 dm³	22.7 dm³	22.7 dm³
Temperature	0 °C	0 °C	0 °C
Mass of gas	4.00 g	28.0 g	16.0 g
Number of gas molecules	6.02 × 10²³	6.02 × 10²³	6.02 × 10²³

◀ **FIGURE 10.10   A comparison illustrating Avogadro's hypothesis.** Note that helium gas consists of helium atoms. Each gas has the same pressure, volume and temperature and thus contains the same number of molecules. Because a molecule of one substance differs in mass from a molecule of another, the masses of gas in the three containers differ.

SAMPLE EXERCISE 10.2	**Evaluating the effects of changes in *P*, *V*, *n* and *T* on a gas**

Suppose we have a gas confined to a cylinder, as shown in ▶ **FIGURE 10.11**. Consider the following changes: **(a)** Heat the gas from 298 K to 360 K, while maintaining the piston in the position shown in the drawing. **(b)** Move the piston to reduce the volume of gas from 1 dm³ to 0.5 dm³. **(c)** Inject additional gas through the gas inlet valve. Indicate how each of these changes will affect the average distance between molecules, the pressure of the gas and the number of moles of gas present in the cylinder.

SOLUTION

Analyse We need to think how each change affects (1) the distance between molecules, (2) the pressure of the gas and (3) the number of moles of gas in the cylinder.

Plan We will use the gas laws and the general properties of gases to analyse the situation.

Solve **(a)** Heating the gas while maintaining the position of the piston will cause no change in the number of molecules per unit volume. Thus the distance between molecules and the total moles of gas remain the same. The increase in temperature, however, will cause the pressure to increase (Charles's law).

▲ **FIGURE 10.11   Cylinder with piston and gas inlet valve.**

(b) Moving the piston compresses the same quantity of gas into a smaller volume. The total number of molecules of gas, and thus the total number of moles, remains the same. The average distance between molecules, however, must decrease because of the smaller volume in which the gas is confined. The reduction in volume causes the pressure to increase (Boyle's law).

(c) Injecting more gas into the cylinder while keeping the volume and temperature the same will result in more molecules and thus a greater number of moles of gas. The average distance between atoms must decrease because their number per unit volume increases. Correspondingly, the pressure increases (Avogadro's law).

PRACTICE EXERCISE

What happens to the density of a gas as **(a)** the gas is heated in a constant-volume container; **(b)** the gas is compressed at constant temperature; **(c)** additional gas is added to a constant-volume container?

Answers: **(a)** no change, **(b)** increase, **(c)** increase

(See also Exercises 10.13–10.16.)

10.4 THE IDEAL-GAS EQUATION

In Section 10.3 we examined three historically important gas laws that describe the relationships between the four variables P, V, T and n that define the state of a gas. Each law was obtained by holding two variables constant in order to see how the remaining two variables affect each other. We can express each law as a proportionality relationship. Using the symbol $\propto$, which is read 'is proportional to', we have

$$Boyle's\ law:\quad V \propto \frac{1}{P}\ \ (\text{constant } n, T)$$

$$Charles's\ law:\quad V \propto T\ \ (\text{constant } n, P)$$

$$Avogadro's\ law:\quad V \propto n\ \ (\text{constant } P, T)$$

We can combine these relationships to make a more general gas law.

$$V \propto \frac{nT}{P}$$

If we call the proportionality constant R, we obtain

$$V = R\left(\frac{nT}{P}\right)$$

Rearranging, we have this relationship in its more familiar form:

$$PV = nRT \qquad\qquad [10.5]$$

This equation is known as the **ideal-gas equation**. An **ideal gas** is a hypothetical gas whose pressure, volume and temperature behaviour is completely described by the ideal-gas equation.

In deriving the ideal-gas equation, we assume that (1) the molecules of an ideal gas do not interact with one another and (2) the combined volume of the molecules is much smaller than the volume the gas occupies; for this reason, we consider the molecules as taking up no space in the container. In many cases, the small error introduced by these assumptions is acceptable. If more accurate calculations are needed, we can correct for the assumptions if we know something about the attraction molecules have for one another and if we know the diameter of the molecules.

The term R in the ideal-gas equation is called the **gas constant**. The value and units of R depend on the units of P, V, T and n. Temperature must *always* be expressed as an absolute temperature (K) when used in the ideal-gas equation. The quantity of gas, n, is normally expressed in moles. The units of pressure and volume using SI units are pascals and cubic metres (m^3), respectively.

Rearranging the ideal-gas equation, and remembering that 1 m = 10 dm, we have

$$R = \frac{PV}{nT} = \frac{Pa\,m^3}{mol\,K} = \frac{Pa\,(1000\,dm^3)}{mol\,K} = \frac{1000\,Pa\,dm^3}{mol\,K} = \frac{kPa\,dm^3}{mol\,K}$$

$$= kPa\,dm^3\,mol^{-1}\,K^{-1}$$

The units, pascals and m^3, are converted to kPa and dm^3 as these are the units commonly used in chemistry. The numerical value of R in this system of units is 8.314 kPa dm^3 mol^{-1} K^{-1} (8.314 Pa dm^3 mol^{-1} K^{-1}).

It is useful at this point to derive another set of units of R which have the dimensions of energy.

$$R = \frac{Pa\,m^3}{mol\,K} \text{ and since } 1\,Pa = 1\,N\,m^{-2} = 1\,kg\,m^{-1}\,s^{-2}$$

$$R = \frac{1\,kg\,m^{-1}\,s^{-2}\,m^3}{mol\,K} = \frac{1\,kg\,m^2\,s^{-2}}{mol\,K} = J\,mol^{-1}\,K^{-1}$$

The numerical value is the same as above, 8.314 J mol^{-1} K^{-1}.

If we define a set of conditions where the temperature is 0.00 °C (273.15 K) and pressure is 1.00 bar (100 kPa), this is known as the **standard temperature and pressure (STP)**. If we have 1.000 mol of an ideal gas at STP, then, according to the ideal-gas equation,

$$V = \frac{nRT}{P} = \frac{(1.000\,mol)(8.314\,kPa\,dm^3\,mol^{-1}\,K^{-1})(273.15\,K)}{100\,kPa} = 22.71\,dm^3$$

This is the **molar volume** (V_m) of any ideal gas at STP.

CONCEPT CHECK 4

How many molecules are in 22.71 dm^3 of an ideal gas at STP?

The ideal-gas equation accounts adequately for the properties of most gases under a wide variety of circumstances. It is not exactly correct, however, for any real gas. Thus the measured volume, V, for given conditions of P, T and n might differ from the volume calculated from $PV = nRT$. To illustrate, the measured molar volumes of real gases at STP are compared with the calculated volume of an ideal gas in ▼ FIGURE 10.12. Although these real gases don't match the ideal-gas behaviour exactly, the differences are so small that we can ignore them for all but the most accurate work. We have more to say about the differences between ideal and real gases in Section 10.9.

◢ FIGURE IT OUT

Suggest an explanation for the 'ideal' nature of helium compared to the other gases.

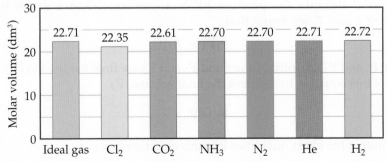

▲ FIGURE 10.12   **Comparison of molar volumes at STP.** One mole of an ideal gas at STP occupies a volume of 22.71 dm^3. One mole of various real gases at STP occupies close to this ideal volume.

SAMPLE EXERCISE 10.3 **Using the ideal-gas equation**

Calcium carbonate, $CaCO_3(s)$, decomposes upon heating to give $CaO(s)$ and $CO_2(g)$. A sample of $CaCO_3$ is decomposed and the carbon dioxide is collected in a 250 cm³ flask. After the decomposition is complete, the gas has a pressure of 132 kPa at a temperature of 31 °C. How many moles of CO_2 gas were generated?

SOLUTION

Analyse We are given the volume (250 cm³), pressure 132 kPa) and temperature (31 °C) of a sample of CO_2 gas and asked to calculate the number of moles of CO_2 in the sample.

Plan Because we are given V, P and T, we can solve the ideal-gas equation for the unknown quantity, n.

Solve In analysing and solving gas-law problems, it is helpful to tabulate the information given in the problems and then to convert the values to units that are consistent with those for R (8.314 kPa dm³ mol⁻¹ K⁻¹). In this case the given values are

$V = 250$ cm³ $= 0.250$ dm³
$P = 132$ kPa
$T = 31$ °C $= (31 + 273)$ K $= 304$ K

Remember: *Absolute temperature must always be used when the ideal-gas equation is solved.*

We now rearrange the ideal-gas equation (Equation 10.5) to solve for n.

$$n = \frac{PV}{RT}$$

$$n = \frac{(132 \text{ kPa})(0.250 \text{ dm}^3)}{(8.314 \text{ kPa dm}^3 \text{ mol}^{-1} \text{ K}^{-1})(304 \text{ K})} = 0.0131 \text{ mol } CO_2$$

Check Appropriate units cancel, thus ensuring that we have properly rearranged the ideal-gas equation and have converted to the correct units.

PRACTICE EXERCISE

Tennis balls are usually filled with air or N_2 gas to a pressure above atmospheric pressure to increase their bounce. If a particular tennis ball has a volume of 144 cm³ and contains 0.330 g of N_2 gas, what is the pressure inside the ball at 24 °C?

Answer: 202 kPa or 2.02 bar

(See also Exercises 10.22–10.24.)

Relating the Ideal-Gas Equation and the Gas Laws

The simple gas laws discussed in Section 10.3, such as Boyle's law, are special cases of the ideal-gas equation. For example, when n and T are held constant, the product nRT contains three constants and so must itself be a constant.

$$PV = nRT = \text{constant} \quad \text{or} \quad PV = \text{constant} \qquad [10.6]$$

Thus we have Boyle's law. We see that if n and T are constant, the individual values of P and V can change, but the product PV must remain constant.

We can use Boyle's law to determine how the volume of a gas changes when its pressure changes. For example, if a metal cylinder holds 50.0 dm³ of O_2 gas at 1875 kPa and 21 °C, what volume will the gas occupy if the temperature is maintained at 21 °C while the pressure is reduced to 101.3 kPa? Because the product PV is a constant when a gas is held at constant n and T, we know that

$$P_1V_1 = P_2V_2 \qquad [10.7]$$

where P_1 and V_1 are initial values and P_2 and V_2 are final values. Dividing both sides of this equation by P_2 gives the final volume, V_2.

$$V_2 = V_1 \times \frac{P_1}{P_2}$$

Substituting the given quantities into this equation gives

$$V_2 = (50.0 \text{ dm}^3)\left(\frac{1875 \text{ kPa}}{101.3 \text{ kPa}}\right) = 925 \text{ dm}^3$$

The answer is reasonable because gases expand as their pressures are decreased.

In a similar way we can start with the ideal-gas equation and derive relationships between any other two variables—V and T (Charles's law), n and V (Avogadro's law) or P and T.

SAMPLE EXERCISE 10.4   Calculating the effect of temperature changes on pressure

The gas pressure in an aerosol can is 152 kPa at 25 °C. Assuming that the gas inside obeys the ideal-gas equation, what will the pressure be if the can is heated to 450 °C?

SOLUTION

Analyse We are given the initial pressure (152 kPa) and temperature (25 °C) of the gas and asked for the pressure at a higher temperature (450 °C).

Plan The volume and number of moles of gas do not change, so we must use a relationship connecting pressure and temperature. Converting temperature to the Kelvin scale and tabulating the given information, we have

	P	T
Initial	152 kPa	298 K
Final	P_2	723 K

Solve In order to determine how P and T are related, we start with the ideal-gas equation and isolate the quantities that don't change (n, V and R) on one side and the variables (P and T) on the other side.

$$\frac{P}{T} = \frac{nR}{V} = \text{constant}$$

Because the quotient P/T is a constant, we can write

$$\frac{P_1}{T_1} = \frac{P_2}{T_2}$$

(where the subscripts 1 and 2 represent the initial and final states, respectively). Rearranging to solve for P_2 and substituting the given data gives

$$P_2 = P_1 \times \frac{T_2}{T_1}$$

$$P_2 = (152\ \text{kPa})\left(\frac{723\ \text{K}}{298\ \text{K}}\right) = 369\ \text{kPa}$$

Check This answer is intuitively reasonable—increasing the temperature of a gas increases its pressure.

Comment It is evident from this example why aerosol cans carry a warning not to incinerate.

PRACTICE EXERCISE

A large natural-gas storage tank is arranged so that the pressure is maintained at 223 kPa. On a cold day in July when the temperature is −15 °C, the volume of gas in the tank is 750 m^3. What is the volume of the same quantity of gas on a hot December day when the temperature is 31 °C?

Answer: 884 m^3

(See also Exercise 10.25.)

We are often faced with the situation in which P, V and T all change for a fixed number of moles of gas. Because n is constant under these circumstances, the ideal-gas equation gives

$$\frac{PV}{T} = nR = \text{constant}$$

If we represent the initial and final conditions of pressure, temperature and volume by subscripts 1 and 2, respectively, we can write

$$\frac{P_1 V_1}{T_1} = \frac{P_2 V_2}{T_2} \qquad [10.8]$$

This is often called the *combined gas law*.

SAMPLE EXERCISE 10.5   Calculating the effect of changing *P* and *T* on the volume of a gas

An inflated balloon has a volume of 6.0 dm^3 at sea level (101.3 kPa) and is allowed to ascend in altitude until the pressure is 45.6 kPa. During ascent the temperature of the gas falls from 22 °C to −21 °C. Calculate the volume of the balloon at its final altitude.

SOLUTION

Analyse We need to determine a new volume for a gas sample when both gas pressure and temperature change.

Plan Let's again proceed by converting temperature to the Kelvin scale and tabulating the given information.

	P	V	T
Initial	101.3 kPa	6.0 dm^3	295 K
Final	45.6 kPa	V_2	252 K

Solve Because n is constant, we can use Equation 10.8. Rearranging Equation 10.8 to solve for V_2 gives

$$V_2 = V_1 \times \frac{P_1}{P_2} \times \frac{T_2}{T_1} = (6.0 \text{ dm}^3)\left(\frac{101.3 \text{ kPa}}{45.6 \text{ kPa}}\right)\left(\frac{252 \text{ K}}{295 \text{ K}}\right) = 11.4 \text{ dm}^3 = 11 \text{ dm}^3$$

PRACTICE EXERCISE

A 0.50 mol sample of oxygen gas is confined at 0 °C in a cylinder with a movable piston, such as that shown in Figure 10.11. The gas has an initial pressure of 1 bar. The gas is then compressed by the piston so that its final volume is half the initial volume. The final pressure of the gas is 2 bar. What is the final temperature of the gas in degrees Celsius?

Answer: 0 °C

(See also Exercise 10.26(b).)

10.5 | FURTHER APPLICATIONS OF THE IDEAL-GAS EQUATION

The ideal-gas equation can be used to determine many relationships involving the physical properties of gases. In this section we use it first to define the relationship between the density of a gas and its molar mass, and then to calculate the volumes of gases formed or consumed in chemical reactions.

Gas Densities and Molar Mass

∞ Review this on page 14

The ideal-gas equation allows us to calculate gas density from the molar mass, pressure and temperature of the gas. Recall that density has the units of mass per unit volume ($\rho = m/V$) (∞ Section 1.4, 'Units of Measurement'). We can arrange the gas equation to obtain similar units, moles per unit volume, n/V:

$$\frac{n}{V} = \frac{P}{RT}$$

If we multiply both sides of this equation by the molar mass, M, which is the number of grams in one mole of a substance, we obtain the following relationship:

$$\frac{nM}{V} = \frac{PM}{RT} \tag{10.9}$$

The product of the quantities n/V and M equals the density in g dm^{-3}, as seen from their units:

$$\frac{\text{mol}}{\text{dm}^3} \times \frac{\text{grams}}{\text{mol}} = \frac{\text{grams}}{\text{dm}^3} = \text{g dm}^{-3}$$

Thus the density, ρ, of the gas is given by the expression on the right in Equation 10.9:

$$\rho = \frac{PM}{RT} \tag{10.10}$$

From Equation 10.10 we see that the density of a gas depends on its pressure, molar mass and temperature. The higher the molar mass and pressure, the more dense the gas; the higher the temperature, the less dense the gas. Although gases form homogeneous mixtures regardless of their identities, a less dense gas will lie above a more dense one in the absence of mixing. For example, CO_2 has a higher molar mass than N_2 or O_2 and is therefore more dense than air. When CO_2 is released from a CO_2 fire extinguisher, it blankets a fire, preventing O_2 from reaching the combustible material. 'Dry ice' which is solid CO_2, converts directly to CO_2 gas at room temperature, and the resulting 'fog' which is actually condensed water droplets cooled by the CO_2 flows downhill in air (▶ FIGURE 10.13).

The fact that a hotter gas is less dense than a cooler one explains why hot air rises. The difference between the densities of hot and cold air is responsible for the lift of hot-air balloons. It is also responsible for many phenomena in weather, such as the formation of large thunder clouds during thunderstorms.

▲ **FIGURE 10.13** **Carbon dioxide gas flows downhill because it is denser than air.** The CO_2 'fog' is not the gas made visible but rather is made up of drops of water that have condensed from water vapour in the air.

 CONCEPT CHECK 5

Is water vapour more or less dense than N_2 under the same conditions of temperature and pressure?

SAMPLE EXERCISE 10.6 Calculating gas density

What is the density of carbon tetrachloride vapour at 0.952 bar and 125 °C?

SOLUTION

Analyse We are asked to calculate the density of a gas given its name, its pressure and its temperature. From the name we can write the chemical formula of the substance and determine its molar mass.

Plan We can use Equation 10.10 to calculate the density. Before we can do that, however, we must convert the given quantities to the appropriate units, degrees Celsius to Kelvin and pressure to atmospheres. We must also calculate the molar mass of CCl_4.

Solve The absolute temperature is $125 + 273 = 398$ K. The molar mass of CCl_4 is $12.01 + (4)(35.45) = 153.8$ g mol^{-1}. Therefore,

$$\rho = \frac{(95.2 \text{ kPa})(154.0 \text{ g mol}^{-1})}{(8.314 \text{ kPa dm}^3 \text{ mol}^{-1} \text{ K}^{-1})(398 \text{ K})} = 4.43 \text{ g dm}^{-3}$$

PRACTICE EXERCISE

The mean molar mass of the atmosphere at the surface of Titan, Saturn's largest moon, is 28.6 g mol^{-1}. The surface temperature is 95 K, and the pressure is 0.162 bar. Assuming ideal behaviour, calculate the density of Titan's atmosphere.

Answer: 0.59 g dm^{-3}

(See also Exercise 10.32(a).)

Equation 10.10 can be rearranged to solve for the molar mass of a gas:

$$M = \frac{\rho RT}{P} \qquad\qquad [10.11]$$

Thus we can use the experimentally measured density of a gas to determine the molar mass of the gas molecules, as shown in Sample Exercise 10.7.

SAMPLE EXERCISE 10.7 | **Calculating the molar mass of a gas**

A series of measurements are made in order to determine the molar mass of an unknown gas. First, a large flask is evacuated and found to weigh 134.567 g. It is then filled with the gas to a pressure of 98.0 kPa at 31 °C and reweighed; its mass is now 137.456 g. Finally, the flask is filled with water at 31 °C and found to weigh 1067.9 g. (The density of the water at this temperature is 0.997 g cm^3.) Assuming that the ideal-gas equation applies, calculate the molar mass of the unknown gas.

SOLUTION

Analyse We are given the temperature (31 °C) and pressure (98.0 kPa) for a gas together with information to determine its volume and mass, and we are asked to calculate its molar mass.

Plan We need to use the mass information given to calculate the volume of the container and the mass of the gas within it. From this we calculate the gas density and then apply Equation 10.11 to calculate the molar mass of the gas.

Solve The mass of the gas is the difference between the mass of the flask filled with gas and that of the empty (evacuated) flask:

$$137.456 \text{ g} - 134.567 \text{ g} = 2.889 \text{ g}$$

The volume of the gas equals the volume of water that the flask can hold. The volume of water is calculated from its mass and density. The mass of the water is the difference between the masses of the full and empty flask:

$$1067.9 \text{ g} - 134.567 \text{ g} = 933.3 \text{ g}$$

By rearranging the equation for density ($\rho = m/V$), we have

$$V = \frac{m}{\rho} = \frac{(933.3 \text{ g})}{(0.997 \text{ g cm}^{-3})} = 936 \text{ cm}^3$$

Knowing the mass of the gas (2.889 g) and its volume (936 cm^3), we can calculate the density of the gas:

$$2.889 \text{ g}/0.936 \text{ dm}^3 = 3.09 \text{ g dm}^{-3}$$

After converting pressure to atmospheres and temperature to Kelvin, we can use Equation 10.11 to calculate the molar mass:

$$M = \frac{\rho RT}{P}$$

$$= \frac{(3.09 \text{ g dm}^{-3})(8.314 \text{ kPa dm}^3 \text{ mol}^{-1} \text{ K}^{-1})(304 \text{ K})}{98.0 \text{ kPa}}$$

$$= 79.7 \text{ g mol}^{-1}$$

Check The units work out appropriately, and the value of molar mass obtained is reasonable for a substance that is gaseous near room temperature.

PRACTICE EXERCISE

Calculate the average molar mass of dry air if it has a density of 1.17 g dm^{-3} at 21 °C and 98.7 kPa.

Answer: 28.6 g mol^{-1}

(See also Exercise 10.32(b).)

Volumes of Gases in Chemical Reactions

Understanding the properties of gases is important because gases are often reactants or products in chemical reactions. For this reason we are often faced with calculating the volumes of gases consumed or produced in reactions. We have seen that the coefficients in balanced chemical equations tell us the relative amounts (in moles) of reactants and products in a reaction. The number of moles of a gas, in turn, is related to P, V and T.

SAMPLE EXERCISE 10.8 | **Relating the volume of a gas to the amount of another substance in a reaction**

The safety air bags in vehicles are inflated by nitrogen gas generated by the rapid decomposition of sodium azide, NaN$_3$:

$$2 \text{ NaN}_3(s) \longrightarrow 2 \text{ Na}(s) + 3 \text{ N}_2(g)$$

If an air bag has a volume of 36 dm^3 and is to be filled with nitrogen gas at a pressure of 116.5 kPa at a temperature of 26.0 °C, how many grams of NaN$_3$ must be decomposed?

SOLUTION

Analyse This is a multistep problem. We are given the volume, pressure and temperature of the N_2 gas and the chemical equation for the reaction by which the N_2 is generated. We must use this information to calculate the number of grams of NaN_3 needed to obtain the necessary N_2.

Plan We need to use the gas data (P, V and T) and the ideal-gas equation to calculate the number of moles of N_2 gas that should be formed for the air bag to operate correctly. We can then use the balanced equation to determine the number of moles of NaN_3. Finally, we can convert the moles of NaN_3 to grams.

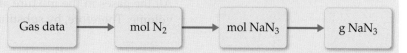

Gas data ⟶ mol N_2 ⟶ mol NaN_3 ⟶ g NaN_3

Solve The number of moles of N_2 is determined using the ideal-gas equation:

$$n = \frac{PV}{RT} = \frac{(116.5 \text{ kPa})(36 \text{ dm}^3)}{(8.314 \text{ kPa dm}^3 \text{ mol}^{-1} \text{ K}^{-1})(299 \text{ K})} = 1.7 \text{ mol } N_2$$

From here we use the coefficients in the balanced equation to calculate the number of moles of NaN_3.

$$(1.7 \text{ mol } N_2)\left(\frac{2 \text{ mol } NaN_3}{3 \text{ mol } N_2}\right) = 1.1 \text{ mol } NaN_3$$

Finally, using the molar mass of NaN_3, we convert moles of NaN_3 to grams:

$$(1.1 \text{ mol } NaN_3)(65.0 \text{ g mol}^{-1} NaN_3) = 73 \text{ g } NaN_3$$

Check The best way to check our approach is to make sure the units cancel properly at each step in the calculation, leaving us with the correct units in the answer, g NaN_3.

PRACTICE EXERCISE

In the first step in the industrial process for making nitric acid, ammonia reacts with oxygen in the presence of a suitable catalyst to form nitric oxide and water vapour:

$$4 \text{ NH}_3(g) + 5 \text{ O}_2(g) \longrightarrow 4 \text{ NO}(g) + 6 \text{ H}_2O(g)$$

How many litres of $NH_3(g)$ at 850 °C and 507 kPa are required to react with 1.00 mol of $O_2(g)$ in this reaction?

Answer: 14.7 dm^3

(See also Exercises 10.33, 10.34.)

10.6 GAS MIXTURES AND PARTIAL PRESSURES

So far we have considered only the behaviour of pure gases—those that consist of only one substance in the gaseous state. How do we deal with gases composed of a mixture of two or more different substances? While studying the properties of air, John Dalton (Section 2.1) observed that *the total pressure of a mixture of gases equals the sum of the pressures that each would exert if it were present alone*. The pressure exerted by a particular component of a mixture of gases is called the **partial pressure** of that gas, and Dalton's observation is known as **Dalton's law of partial pressures**.

CONCEPT CHECK 6

How is the pressure exerted by N_2 gas affected when some O_2 is introduced into a container if the temperature and volume remain constant?

If we let P_t be the total pressure of a mixture of gases and P_1, P_2, P_3 and so forth be the partial pressures of the individual gases, we can write Dalton's law as follows:

$$P_t = P_1 + P_2 + P_3 + \dots \qquad [10.12]$$

This equation implies that each gas in the mixture behaves independently of the others, as we can see by the following analysis. Let n_1, n_2, n_3 and so forth be the number of moles of each of the gases in the mixture and n_t be the total number of moles of gas ($n_t = n_1 + n_2 + n_3 + ...$).

If each of the gases obeys the ideal-gas equation, we can write

$$P_1 = n_1\left(\frac{RT}{V}\right); \qquad P_2 = n_2\left(\frac{RT}{V}\right); \qquad P_3 = n_3\left(\frac{RT}{V}\right); \qquad \text{and so forth.}$$

All the gases in the mixture are at the same temperature and occupy the same volume. Therefore, by substituting into Equation 10.12, we obtain

$$P_t = (n_1 + n_2 + n_3 + ...)\frac{RT}{V} = n_t\left(\frac{RT}{V}\right) \qquad [10.13]$$

That is, the total pressure at constant temperature and constant volume is determined by the total number of moles of gas present, whether that total represents just one substance or a mixture.

SAMPLE EXERCISE 10.9   **Applying Dalton's law of partial pressures**

A gaseous mixture made from 6.00 g O_2 and 9.00 g CH_4 is placed in a 15.0 dm^3 vessel at 0 °C. What is the partial pressure of each gas, in bar, and what is the total pressure in the vessel?

SOLUTION

Analyse We need to calculate the pressure for two gases in the same volume and at the same temperature.

Plan Because each gas behaves independently, we can use the ideal-gas equation to calculate the pressure that each would exert if the other were not present. The total pressure is the sum of these two partial pressures.

Solve We must first convert the mass of each gas to moles:

$$n_{O_2} = \frac{6.00 \text{ g}}{32.0 \text{ g mol}^{-1}} = 0.188 \text{ mol } O_2$$

$$n_{CH_4} = \frac{9.0 \text{ g}}{16.0 \text{ g mol}^{-1}} = 0.563 \text{ mol } CH_4$$

We can now use the ideal-gas equation to calculate the partial pressure of each gas:

$$P_{O_2} = \frac{n_{O_2}RT}{V} = \frac{(0.188 \text{ mol})(8.314 \text{ kPa dm}^3 \text{ mol}^{-1} \text{ K}^{-1})(273 \text{ K})}{15.0 \text{ dm}^3} = 28.4 \text{ kPa} = 0.284 \text{ bar}$$

$$P_{CH_4} = \frac{n_{CH_4}RT}{V} = \frac{(0.563 \text{ mol})(8.314 \text{ kPa dm}^3 \text{ mol}^{-1} \text{ K}^{-1})(273 \text{ K})}{15.0 \text{ dm}^3} = 85.2 \text{ kPa} = 0.852 \text{ bar}$$

According to Dalton's law (Equation 10.12), the total pressure in the vessel is the sum of the partial pressures:

$$P_t = P_{O_2} + P_{CH_2} = 28.4 \text{ kPa} + 85.2 \text{ kPa} = 113.6 \text{ kPa} = 1.14 \text{ bar}$$

PRACTICE EXERCISE

What is the total pressure (bar) exerted by a mixture of 2.00 g of H_2 and 8.00 g of N_2 at 273 K in a 10.0 dm^3 vessel?

Answer: 292 kPa = 2.92 bar

(See also Exercises 10.39, 10.40.)

Partial Pressures and Mole Fractions

Because each gas in a mixture behaves independently, we can relate the amount of a given gas in a mixture to its partial pressure. For an ideal gas, $P = nRT/V$, and so we can write

$$\frac{P_1}{P_t} = \frac{n_1 RT/V}{n_t RT/V} = \frac{n_1}{n_t} = X_1 \qquad [10.14]$$

The ratio n_1/n_t is called the mole fraction of gas 1, which we denote X_1. The **mole fraction**, X, is a dimensionless number that expresses the ratio of the number of moles of one component to the total number of moles in the mixture. We can rearrange Equation 10.14 to give

$$P_1 = \left(\frac{n_1}{n_t}\right)P_t = X_1 P_t \qquad\qquad [10.15]$$

Thus the partial pressure of a gas in a mixture is its mole fraction times the total pressure.

The mole fraction of N_2 in air is 0.78 (that is, 78% of the molecules in air are N_2). If the total barometric pressure is 1 bar, then the partial pressure of N_2 is

$$P_{N_2} = (0.78)(1 \text{ bar}) = 0.78 \text{ bar}$$

This result makes intuitive sense: because N_2 makes up 78% of the molecules in the mixture, it contributes 78% of the total pressure.

SAMPLE EXERCISE 10.10   Relating mole fractions and partial pressures

A study of the effects of certain gases on plant growth requires a synthetic atmosphere composed of 1.5 mol percent CO_2, 18.0 mol percent O_2 and 80.5 mol percent Ar. **(a)** Calculate the partial pressure of O_2 in the mixture if the total pressure of the atmosphere is to be 99.3 kPa. **(b)** If this atmosphere is to be held in a 120 dm^3 space at 295 K, how many moles of O_2 are needed?

SOLUTION

Analyse For (a) we need to calculate the partial pressure of O_2 given its mole percent and the total pressure of the mixture. For (b) we need to calculate the number of moles of O_2 in the mixture given its volume (120 dm^3), temperature (295 K) and partial pressure from part (a).

Plan We calculate the partial pressures using Equation 10.15. We then use P_{O_2}, V and T together with the ideal-gas equation to calculate the number of moles of O_2, n_{O_2}.

Solve
(a) The mole percent is just the mole fraction times 100. Therefore, the mole fraction of O_2 is 0.180. Using Equation 10.15, we have

$$P_{O_2} = (0.180)(99.3 \text{ kPa}) = 17.9 \text{ kPa}$$

(b) Listing the given variables and changing them to appropriate units, we have

$$P_{O_2} = 17.9 \text{ kPa}$$
$$V = 120 \text{ dm}^3$$
$$n_{O_2} = ?$$
$$R = 8.314 \text{ kPa dm}^3 \text{ mol}^{-1} \text{ K}^{-1}$$
$$T = 295 \text{ K}$$

Solving the ideal-gas equation for n_{O_2}, we have

$$n_{O_2} = P_{O_2}\left(\frac{V}{RT}\right) = (17.9 \text{ kPa})\left(\frac{120 \text{ dm}^3}{(8.314 \text{ kPa dm}^3 \text{ mol}^{-1} \text{ K}^{-1})(295 \text{ K})}\right) = 0.876 \text{ mol}$$

Check The units check out satisfactorily and the answer seems to be the right order of magnitude.

PRACTICE EXERCISE

From data gathered by the Cassini-Huygens probe, the composition of the atmosphere of Titan, Saturn's largest moon, is known. The total pressure on the surface of Titan is about 160 kPa. The atmosphere consists of 94.1 mol percent nitrogen and 5.8 mol percent methane. Calculate the partial pressure of each of these gases in Titan's atmosphere.

Answer: 1.5×10^2 kPa N_2 and 9.3 kPa CH_4

(See also Exercises 10.44, 10.55.)

Collecting Gases over Water

An experiment that is often encountered in general chemistry laboratories involves determining the number of moles of gas collected from a chemical reaction. Sometimes this gas is collected over water. For example, solid potassium chlorate, $KClO_3$, can be decomposed by heating it in a test tube in an

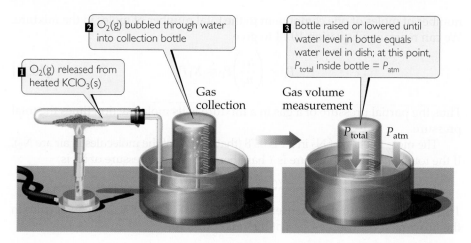

① $O_2(g)$ released from heated $KClO_3(s)$

② $O_2(g)$ bubbled through water into collection bottle

③ Bottle raised or lowered until water level in bottle equals water level in dish; at this point, P_{total} inside bottle = P_{atm}

Gas collection

Gas volume measurement

P_{total} P_{atm}

▶ **FIGURE 10.14   Collecting a water-insoluble gas over water.**

arrangement such as that shown in ▲ FIGURE 10.14. The balanced equation for the reaction is

$$2\,KClO_3(s) \longrightarrow 2\,KCl(s) + 3\,O_2(g) \qquad [10.16]$$

The oxygen gas is collected in a bottle that is initially filled with water and inverted in a water pan.

The volume of gas collected is measured by raising or lowering the bottle as necessary until the water levels inside and outside the bottle are the same. When this condition is met, the pressure inside the bottle is equal to the atmospheric pressure outside. The total pressure inside is the sum of the pressure of gas collected and the pressure of water vapour in equilibrium with liquid water.

$$P_{total} = P_{gas} + P_{H_2O} \qquad [10.17]$$

The pressure exerted by water vapour, P_{H_2O}, at various temperatures is listed in Appendix B.

SAMPLE EXERCISE 10.11   Calculating the amount of gas collected over water

A sample of $KClO_3$ is partially decomposed (Equation 10.16), producing O_2 gas that is collected over water as in Figure 10.14. The volume of gas collected is 0.250 dm³ at 26 °C and 102 kPa total pressure. **(a)** How many moles of O_2 are collected? **(b)** How many grams of $KClO_3$ were decomposed?

SOLUTION

Analyse **(a)** We need to calculate the number of moles of O_2 gas in a container that also contains water vapour. **(b)** We need to calculate the number of moles of reactant $KClO_3$, decomposed.

Plan **(a)** If we tabulate the information presented, we see that values are given for V and T. In order to use the ideal-gas equation to calculate the unknown, n_{O_2}, we must know also the partial pressure of O_2 in the system. We can calculate the partial pressure of O_2 from the total pressure (102 kPa) and the vapour pressure of water.

Solve **(a)** The partial pressure of the O_2 gas is the difference between the total pressure, 102 kPa, and the pressure of the water vapour at 26 °C, 3.33 kPa (Appendix B):

$$P_{O_2} = 102\ \text{kPa} - 3.33\ \text{kPa} = 98.7\ \text{kPa}$$

We can use the ideal-gas equation to calculate the number of moles of O_2:

$$n_{O_2} = \frac{P_{O_2}V}{RT} = \frac{(98.7\ \text{kPa})(0.250\ \text{dm}^3)}{(8.314\ \text{kPa dm}^3\,\text{mol}^{-1}\,\text{K}^{-1})(299\ \text{K})} = 9.92 \times 10^{-3}\ \text{mol}\ O_2$$

(b) We now need to calculate the number of moles of reactant $KClO_3$ decomposed. We can use the number of moles of O_2 formed and the balanced chemical equation to determine the number of moles of $KClO_3$ decomposed, which we can then convert to grams of $KClO_3$.

From Equation 10.16, we have 2 mol $KClO_3$ gives 3 mol O_2. The molar mass $KClO_3$ is 122.6 g mol⁻¹. Thus we can convert the moles of O_2 that we found in part (a) to moles of $KClO_3$ and then to grams of $KClO_3$:

$$\text{moles KClO}_3 = \frac{2}{3} \text{ mol O}_2$$

$$= \frac{2}{3} (9.92 \times 10^{-3} \text{ mol})$$

$$= 6.61 \times 10^{-3} \text{ mol}$$

$$\text{grams KClO}_3 = (6.61 \times 10^{-3} \text{ mol})(122.6 \text{ g mol}^{-1})$$

$$= 0.811 \text{ g}$$

PRACTICE EXERCISE

Ammonium nitrite, NH_4NO_2, decomposes upon heating to form N_2 gas:

$$NH_4NO_2(s) \longrightarrow N_2(g) + 2 H_2O(l)$$

When a sample of NH_4NO_2 is decomposed in a test tube, as in Figure 10.14, 511 cm³ of gas is collected over water at 26 °C and 99.3 kPa total pressure. How many grams of are decomposed?

Answer: 1.26 g
(See also Exercise 10.65.)

10.7 | KINETIC-MOLECULAR THEORY

The ideal-gas equation describes *how* gases behave, but it doesn't explain *why* they behave as they do. Why does a gas expand when heated at constant pressure? Or why does its pressure increase when the gas is compressed at constant temperature? To understand the physical properties of gases, we need a model that helps us picture what happens to gas particles as experimental conditions such as pressure or temperature change. Such a model, known as the **kinetic-molecular theory**, was developed over a period of about 100 years, culminating in 1857 when Rudolf Clausius (1822–1888) published a complete and satisfactory form of the theory.

The kinetic-molecular theory (the theory of moving molecules) is summarised by the following statements.

1. Gases consist of large numbers of molecules that are in continuous, random motion. (The word *molecule* is used here to designate the smallest particle of any gas; some gases, such as the noble gases, consist of individual atoms.)
2. The combined volume of all the molecules of the gas is negligible relative to the total volume in which the gas is contained.
3. Attractive and repulsive forces between gas molecules are negligible.
4. Energy can be transferred between molecules during collisions, but the *average* kinetic energy of the molecules does not change with time, as long as the temperature of the gas remains constant. In other words, the collisions are perfectly elastic.
5. The average kinetic energy of the molecules is proportional to the absolute temperature. At any given temperature the molecules of all gases have the same average kinetic energy.

The kinetic-molecular theory explains both pressure and temperature at the molecular level. The pressure of a gas is caused by collisions of the molecules with the walls of the container, as shown in ▶ **FIGURE 10.15**. The magnitude of the pressure is determined by how often and how forcefully the molecules strike the walls.

The absolute temperature of a gas is a measure of the *average* kinetic energy of its molecules. If two different gases are at the same temperature, their molecules have the same average kinetic energy (statement 5 of the kinetic-molecular theory). If the absolute temperature of a gas is doubled, the average kinetic energy of its molecules doubles. Thus molecular motion increases with increasing temperature.

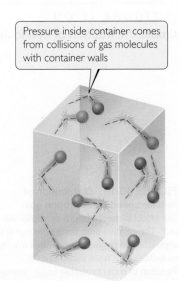

Pressure inside container comes from collisions of gas molecules with container walls

▲ **FIGURE 10.15**   **The molecular origin of gas pressure.**

Distribution of Molecular Speed

Although the molecules in a sample of gas have an *average* kinetic energy and hence an average speed, the individual molecules move at varying speeds. The moving molecules collide frequently with other molecules. Momentum is conserved in each collision, but one of the colliding molecules might be deflected off at high speed while the other is nearly stopped. The result is that the molecules at any instant have a wide range of speeds.

In ▼ FIGURE 10.16(a), which shows the distribution of molecular speeds for nitrogen gas at 0 °C and 100 °C, we see that a larger fraction of the 100 °C molecules moves at the higher speeds. This means that the 100 °C sample has the higher average kinetic energy of the two temperatures.

In any graph of the distribution of molecular speeds in a gas sample, the peak of the curve represents the most probable speed, u_{mp}, which is the speed of the largest number of molecules (Figure 10.16(b)). The most probable speeds in Figure 10.16(a), for instance, are 4×10^2 m s^{-1} for the 0 °C sample and 5×10^2 m s^{-1} for the 100 °C sample. Figure 10.16(b) also shows the **root-mean-square (rms) speed**, u_{rms}, of the molecules. This is the speed of a molecule possessing a kinetic energy identical to the average kinetic energy of the sample. The rms speed is not quite the same as the average (mean) speed, u_{av}. However, the difference between the two is small. In Figure 10.16(b), for example, the root-mean-square speed is almost 5×10^2 m s^{-1} and the average speed is about 4.5×10^2 m s^{-1}.

If you calculate the rms speeds as we will in Section 10.8, you will find that the rms speed is almost 6×10^2 m s^{-1} for the 100 °C sample but slightly less than 5×10^2 m s^{-1} for the 0 °C sample. Notice that the distribution curve broadens as we go to a higher temperature, which tells us that the range of molecular speeds increases with temperature.

∞ Find out more on page 502

The rms speed is important because the average kinetic energy of the gas molecules in a sample is equal to $\frac{1}{2} m (u_{rms})^2$ (∞ Section 14.1, 'The Nature of Energy'). Because mass does not change with temperature, the increase in the average kinetic energy $\frac{1}{2} m (u_{rms})^2$ as the temperature increases implies that the rms speed of the molecules (as well as their average speed) increases as temperature increases.

🔺 **FIGURE IT OUT**

Estimate the fraction of molecules at 100 °C with speeds of less than 300 m s^{-1}.

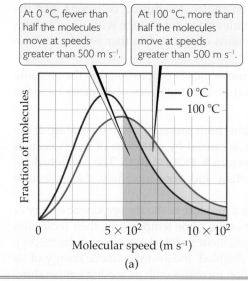

At 0 °C, fewer than half the molecules move at speeds greater than 500 m s^{-1}.

At 100 °C, more than half the molecules move at speeds greater than 500 m s^{-1}.

▶ **FIGURE 10.16 Distribution of molecular speeds for nitrogen gas.**
(a) The effect of temperature on molecular speed. The relative area under the curve for a range of speeds gives the relative fraction of molecules that have those speed.
(b) Position of most probable (u_{mp}), average (u_{av}) and root-mean-square (u_{rms}) speeds of gas molecules. The data shown here are for nitrogen gas at 0 °C.

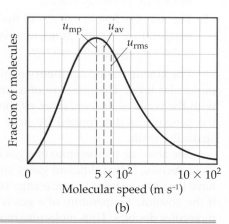

CONCEPT CHECK 7

Consider three gases all at 298 K: HCl, H_2 and O_2. List the gases in order of increasing average speed.

Application of Kinetic-Molecular Theory to the Gas Laws

The empirical observations of gas properties as expressed in the various gas laws are readily understood in terms of the kinetic-molecular theory. The following examples illustrate this point.

1. *Effect of a volume increase at constant temperature.* A constant temperature means that the average kinetic energy of the gas molecules remains unchanged. This, in turn, means that the rms speed of the molecules, u_{rms}, is unchanged. If the volume is increased, however, the molecules must move a longer distance between collisions. Consequently, there are fewer collisions per unit time with the container walls, and pressure decreases. Thus the model accounts in a simple way for Boyle's law.

2. *Effect of a temperature increase at constant volume.* An increase in temperature means an increase in the average kinetic energy of the molecules and thus an increase in u_{rms}. If there is no change in volume, there will be more collisions with the walls per unit time. Furthermore, the change in momentum in each collision increases (the molecules strike the walls more forcefully). Hence the model explains the observed pressure increase.

A CLOSER LOOK

THE IDEAL-GAS EQUATION

Beginning with the five statements given in the text for the kinetic-molecular theory, it is possible to derive the ideal-gas equation. As we have seen, pressure is force per unit area (Section 10.2). The total force of the molecular collisions on the walls and hence the pressure produced by these collisions depend both on how strongly the molecules strike the walls (impulse imparted per collision) and on the rate at which these collisions occur:

$$P \propto \text{impulse imparted per collision} \times \text{rate of collisions}$$

For a molecule travelling at the rms speed, u, the impulse imparted by a collision with a wall depends on the momentum of the molecule; that is, it depends on the product of its mass and speed, mu. The rate of collisions is proportional to both the number of molecules per unit volume, n/V, and their speed, u. If there are more molecules in a container, there will be more frequent collisions with the container walls. As the molecular speed increases or the volume of the container decreases, the time required for molecules to traverse the distance from one wall to another is reduced, and the molecules collide more frequently with the walls. Thus we have

$$P \propto mu_{rms} \times \frac{n}{V} \times u_{rms} \propto \frac{nmu_{rms}^2}{V} \qquad [10.19]$$

Because the average kinetic energy, $\frac{1}{2}mu_{rms}^2$, is proportional to temperature, we have $mu_{rms}^2 \propto T$. Making this substitution into Equation 10.19 gives

$$P \propto \frac{n(mu_{rms}^2)}{V} \propto \frac{nT}{V} \qquad [10.20]$$

Let's now convert the proportionality sign to an equal sign by expressing n as the number of moles of gas; we then insert a proportionality constant—R, the gas constant:

$$P = \frac{nRT}{V} \qquad [10.21]$$

This expression is the ideal-gas equation.

An eminent Swiss mathematician, Daniel Bernoulli (1700–1782), conceived of a model for gases that was, for all practical purposes, the same as the kinetic-molecular model. From this model, Bernoulli derived Boyle's law and the ideal-gas equation. His was one of the first examples in science of developing a mathematical model from a set of assumptions, or hypothetical statements. However, Bernoulli's work on this subject was completely ignored, only to be rediscovered a hundred years later by Clausius and others. It was ignored because the equivalence of temperature and the kinetic energy of particles was not yet understood—the popular idea was that heat was a substance, called 'caloric'. Thus the idea of a gas being composed of particles in perpetual motion seemed like any other perpetual motion machine—a fantasy. Growth of a better understanding of heat went hand in hand with the development and acceptance of the kinetic-molecular model. As this story illustrates, science is not a straight road running from ignorance to the 'truth'.

RELATED EXERCISES: 10.47, 10.48

SAMPLE EXERCISE 10.12 Applying the kinetic-molecular theory

A sample of O_2 gas initially at STP is compressed to a smaller volume at constant temperature. What effect does this change have on (a) the average kinetic energy of O_2 molecules, (b) the average speed of O_2 molecules, (c) the total number of collisions of O_2 molecules with the container walls in a unit time, (d) the number of collisions of O_2 molecules with a unit area of container wall per unit time?

SOLUTION

Analyse We need to apply the concepts of the kinetic-molecular theory of gases to a gas compressed at constant temperature.

Plan We will determine how each of the quantities in (a)–(d) is affected by the change in volume at constant temperature.

Solve
(a) The average kinetic energy of the O_2 molecules is determined only by temperature. Thus the average kinetic energy is unchanged by the compression of O_2 at constant temperature. (b) If the average kinetic energy of O_2 molecules doesn't change, the average speed remains constant. (c) The total number of collisions with the container walls per unit time must increase because the molecules are moving within a smaller volume but with the same average speed as before. Under these conditions they must encounter a wall more frequently. (d) The number of collisions with a unit area of wall per unit time increases because the total number of collisions with the walls per unit time increases and the area of the walls decreases.

PRACTICE EXERCISE

How is the rms speed of N_2 molecules in a gas sample changed by (a) an increase in temperature at constant volume, (b) an increase in volume at constant temperature, (c) mixing with a sample of Ar at the same temperature?

Answers: (a) increases, (b) no effect, (c) no effect

(See also Exercises 10.47–10.49.)

10.8 | MOLECULAR EFFUSION AND DIFFUSION

According to the kinetic-molecular theory, the average kinetic energy of *any* collection of gas molecules, $\frac{1}{2}mu_{rms}^2$, has a specific value at a given temperature. Thus a gas composed of light particles, such as He, will have the same average kinetic energy as one composed of much heavier particles, such as Xe, provided the two gases are at the same temperature. The mass, m, of the particles in the lighter gas is smaller than that in the heavier gas. Consequently, the particles of the lighter gas must have a higher rms speed, u_{rms}, than the particles of the heavier one. The following equation, which expresses this fact quantitatively, can be derived from kinetic-molecular theory:

$$u_{rms} = \sqrt{\frac{3RT}{M}} \qquad [10.22]$$

Because the molar mass, M, appears in the denominator, the less massive the gas molecules, the higher the rms speed, u_{rms}. ▶ FIGURE 10.17 shows the distribution of molecular speeds for several gases at 25 °C. Notice how the distributions are shifted towards higher speeds for gases of lower molar masses.

▲ **CONCEPT CHECK 8**
Consider three samples of gas: HCl at 298 K, H_2 at 298 K and O_2 at 350 K. Compare the average kinetic energies of the molecules in the three samples.

⚠ FIGURE IT OUT

How does root-mean-square speed vary with molar mass?

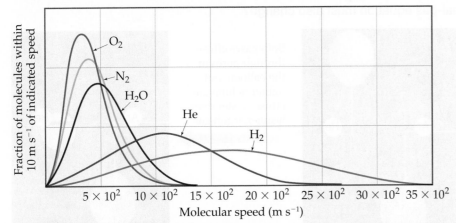

▲ **FIGURE 10.17** **The effect of molecular mass on molecular speeds.** The distributions of molecular speeds for different gases are compared at 25 °C. The molecules with lower molecular masses have higher rms speeds.

SAMPLE EXERCISE 10.13 **Calculating a root-mean-square speed**

Calculate the rms speed, u_{rms}, of an N_2 molecule at 25 °C.

SOLUTION

Analyse We are given the identity of a gas and the temperature, which are the two quantities we need to calculate the rms speed.

Plan We calculate the rms speed using Equation 10.22.

Solve We must convert each quantity to SI units so that all the units are compatible. We also use R in units of J mol^{-1} K^{-1} in order to make the units cancel correctly.

$$T = 25 + 273 = 298 \text{ K}$$

$$M = 28.0 \text{ g mol}^{-1} = 28.0 \times 10^{-3} \text{ kg mol}^{-1}$$

$$R = 8.314 \text{ J mol}^{-1} \text{K}^{-1} = 8.314 \text{ kg m}^2 \text{ s}^{-2} \text{ mol}^{-1} \text{K}^{-1} \quad \text{(These units follow from the fact that 1 J = 1 kg m}^2 \text{ s}^{-2})$$

$$u_{rms} = \sqrt{\frac{3(8.314 \text{ kg m}^2 \text{ s}^{-2} \text{ mol}^{-1} \text{K}^{-1})(298 \text{ K})}{28.0 \times 10^{-3} \text{ kg mol}^{-1}}} = 5.15 \times 10^2 \text{ m s}^{-1}$$

Comment Because the average molecular weight of air molecules is slightly greater than that of N_2, the rms speed of air molecules is a little slower than that for N_2. The speed at which sound propagates through air is about 350 m s^{-1}, a value about two-thirds the average rms speed for air molecules.

PRACTICE EXERCISE

What is the rms speed of an He atom at 25 °C?

Answer: 1.36×10^3 m s^{-1}

The dependence of molecular speeds on mass has several interesting consequences. The first phenomenon is **effusion**, which is the escape of gas molecules through a tiny hole into an evacuated space, as shown in ▶ **FIGURE 10.18**. The second is **diffusion**, which is the spread of one substance throughout a space or throughout a second substance. For example, the molecules of a perfume diffuse throughout a room.

Graham's Law of Effusion

In 1846 Thomas Graham (1805–1869) discovered that the effusion rate of a gas is inversely proportional to the square root of its molar mass. Assume that we

Gas molecules in top half effuse through pinhole only when they happen to hit pinhole

▲ **FIGURE 10.18** **Effusion.**

Because pressure and temperature are constant in this figure but volume changes, which other quantity in the ideal-gas equation must also change?

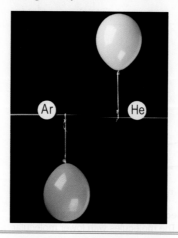

Both gases effuse through pores in the balloon, but lighter helium gas effuses faster than heavier argon gas

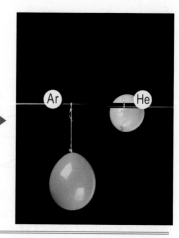

▶ **FIGURE 10.19** An illustration of Graham's law of effusion.

have two gases at the same temperature and pressure in containers with identical pinholes. If the rates of effusion of the two substances are r_1 and r_2 and their respective molar masses are M_1 and M_2, **Graham's law** states

$$\frac{r_1}{r_2} = \sqrt{\frac{M_2}{M_1}}$$ [10.23]

Equation 10.23 compares the *rates* of effusion of two different gases under identical conditions, and it indicates that the lighter gas effuses more rapidly.

Figure 10.18 illustrates the basis of Graham's law. The only way for a molecule to escape from its container is for it to 'hit' the hole in the partitioning wall. The faster the molecules are moving, the greater the likelihood that a molecule will hit the hole and effuse. This implies that the rate of effusion is directly proportional to the rms speed of the molecules. Because R and T are constant, we have, from Equation 10.22

$$\frac{r_1}{r_2} = \frac{u_{\text{rms}(1)}}{u_{\text{rms}(2)}} = \sqrt{\frac{3RT/M_1}{3RT/M_2}} = \sqrt{\frac{M_2}{M_1}}$$ [10.24]

As expected from Graham's law, helium escapes from containers through tiny pinhole leaks more rapidly than other gases of higher molecular weight (▲ **FIGURE 10.19**).

SAMPLE EXERCISE 10.14 | **Applying Graham's law**

An unknown gas composed of homonuclear diatomic molecules effuses at a rate that is only 0.355 times that of O_2 at the same temperature. Calculate the molar mass of the unknown gas, and identify it.

SOLUTION

Analyse We are given the rate of effusion of an unknown gas relative to that of O_2 and asked to find the molar mass and the identity of the unknown. Thus we need to connect relative rates of effusion to relative molar masses.

Plan We can use Graham's law of effusion, Equation 10.23, to determine the molar mass of the unknown gas. If we let r_x and M_x represent the rate of effusion and molar mass of the unknown gas, Equation 10.23 can be written as follows:

$$\frac{r_x}{r_{O_2}} = \sqrt{\frac{M_{O_2}}{M_x}}$$

Solve From the information given,

$$r_x = 0.355 \times r_{O_2}$$

Thus

$$\frac{r_x}{r_{O_2}} = 0.355 = \sqrt{\frac{32.0 \text{ g mol}^{-1}}{M_x}}$$

We now solve for the unknown molar mass, M_x.

$$\frac{32.0 \text{ g mol}^{-1}}{M_x} = (0.355)^2 = 0.126$$

$$M_x = \frac{32.0 \text{ g mol}^{-1}}{0.126} = 254 \text{ g mol}^{-1}$$

Because we are told that the unknown gas is composed of homonuclear diatomic molecules, it must be an element. The molar mass must represent twice the atomic weight of the atoms in the unknown gas. We conclude that the unknown gas is I_2.

PRACTICE EXERCISE

Calculate the ratio of the effusion rates of N_2 and O_2, r_{N_2}/r_{O_2}.

Answer: $r_{N_2}/r_{O_2} = 1.07$

(See also Exercises 10.54, 10.55.)

Diffusion and Mean Free Path

Diffusion, like effusion, is faster for lower-mass molecules than for higher-mass ones. In fact, the ratio of rates of diffusion of two gases under identical experimental conditions is approximated by Graham's law, Equation 10.23. Nevertheless, molecular collisions make diffusion more complicated than effusion.

We can see from the horizontal scale in Figure 10.16 that the speeds of molecules are quite high. For example, the average speed of N_2 at room temperature is 515 m s^{-1}. Despite this high speed, if someone opens a vial of perfume at one end of a room, some time elapses—perhaps a few minutes—before the odour is detected at the other end of the room. The diffusion of gases is much slower than molecular speeds because of molecular collisions.* These collisions occur quite frequently for a gas at atmospheric pressure—about 10^{10} times per second for each molecule. Collisions occur because real gas molecules have finite volumes.

Because of molecular collisions, the direction of motion of a gas molecule is constantly changing. Therefore, the diffusion of a molecule from one point to another consists of many short, straight-line segments as collisions buffet it around in random directions, as depicted in ▶ **FIGURE 10.20**. First the molecule moves in one direction, then in another; one instant at high speed, the next at low speed.

The average distance travelled by a molecule between collisions is called the **mean free path** of the molecule. The mean free path varies with pressure as the following analogy illustrates. Imagine walking blindfolded through a shopping mall. When the mall is very crowded (high pressure), the average distance you can walk before bumping into someone is short (short mean free path). When the mall is empty (low pressure), you can walk a long way (long mean free path) before bumping into someone. The mean free path for air molecules at sea level is about 60 nm (6×10^{-8} m). At about 100 km in altitude, where the air density is much lower, the mean free path is about 10 cm, about 1 million times longer than at the Earth's surface.

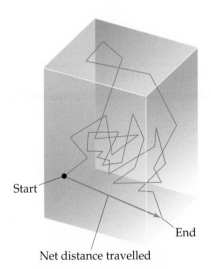

Start

End

Net distance travelled

▲ **FIGURE 10.20 Diffusion of a gas molecule.** For clarity, no other gas molecules in the container are shown. The path of the molecule of interest begins at the dot. Each short segment of line represents travel between collisions. The blue arrow indicates the net distance travelled by the molecule.

---

* The rate at which the perfume moves across the room also depends on how well stirred the air is from, for example, temperature gradients and the movement of people. Nevertheless, even with the aid of these factors, it still takes much longer for the molecules to traverse the room than one would expect from the rms speed alone.

MY WORLD OF CHEMISTRY

GAS SEPARATIONS

The fact that lighter molecules move at higher average speeds than more massive ones has many interesting consequences and applications. For example, the effort to develop the atomic bomb during World War II required scientists to separate the relatively low abundance uranium isotope ^{235}U (0.7%) from the much more abundant ^{238}U (99.3%). This was accomplished by converting the uranium into a volatile compound, UF_6, that was then allowed to pass through porous barriers. Because of the diameters of the pores, this is not a simple effusion. Nevertheless, the depend-

ence on molar mass is essentially the same. The slight difference in molar mass between the two hexafluorides, $^{235}UF_6$ and $^{238}UF_6$, caused the molecules to move at slightly different rates:

$$\frac{r_{235}}{r_{238}} = \sqrt{\frac{352.04}{349.03}} = 1.0043$$

Thus the gas initially appearing on the opposite side of the barrier was very slightly enriched in the lighter molecule. The diffusion process was repeated thousands of times, leading to a nearly complete separation of the two isotopes of uranium.

RELATED EXERCISES: 10.52, 10.53

CONCEPT CHECK 9

Will the following changes increase, decrease or have no effect on the mean free path of the gas molecules in a sample of gas?
a. Increasing pressure.
b. increasing temperature.

10.9 | REAL GASES: DEVIATIONS FROM IDEAL BEHAVIOUR

Although the ideal-gas equation is a very useful description of gases, all real gases fail to obey the relationship to some degree. The extent to which a real gas departs from ideal behaviour can be seen by rearranging the ideal-gas equation to solve for n:

$$\frac{PV}{RT} = n \qquad [10.25]$$

For one mole of ideal gas ($n = 1$) the quantity PV/RT equals 1 at all pressures. In ▼ **FIGURE 10.21** PV/RT is plotted as a function of P for one mole of several

---

▲ FIGURE IT OUT

Does molar mass correlate with non-ideal-gas behaviour below 200 bar?

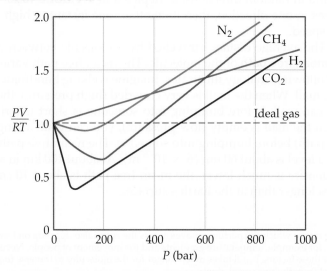

▶ **FIGURE 10.21** **The effect of pressure on the behaviour of several real gases.** Data for 1 mol of gas in all cases. Data for N_2, CH_4 and H_2 are at 300 K; for CO_2 data are at 313 K because under high pressure CO_2 liquefies at 300 K.

True or false: Nitrogen gas behaves more like an ideal gas as the temperature increases.

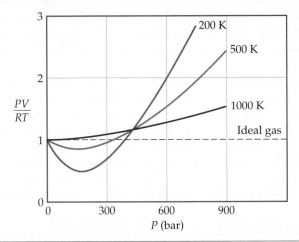

◀ **FIGURE 10.22   The effect of temperature and pressure on the behaviour of nitrogen gas.** The ratios of *PV/RT* against pressure are shown for 1 mol of nitrogen gas at three temperatures. As temperature increases, the gas more closely approaches ideal behaviour, which is represented by the dashed horizontal line.

different gases. At high pressures the deviation from ideal behaviour ($PV/RT = 1$) is large and is different for each gas. *Real gases, therefore, do not behave ideally at high pressure.* At lower pressures (usually below 10 bar), however, the deviation from ideal behaviour is small, and we can use the ideal-gas equation without generating serious error.

The deviation from ideal behaviour also depends on temperature. ▲ FIGURE 10.22 shows graphs of PV/RT against P for 1 mol of N_2 at three temperatures. As temperature increases, the behaviour of the gas more nearly approaches that of the ideal gas. In general, *the deviations from ideal behaviour increase as temperature decreases*, becoming significant near the temperature at which the gas is converted into a liquid.

◢ **CONCEPT CHECK 10**

Would you expect helium gas to deviate from ideal behaviour more at
a. 100 K and 1 bar?
b. 100 K and 5 bar?
c. 300 K and 2 bar?

The basic assumptions of the kinetic-molecular theory give us insight into why real gases deviate from ideal behaviour. The molecules of an ideal gas are assumed to occupy no space and have no attractions for one another. *Real molecules, however, do have finite volumes and they do attract one another.* As shown in ▶ FIGURE 10.23, the free, unoccupied space in which molecules can move is somewhat less than the container volume. At relatively low pressures the volume of the gas molecules is negligible relative to the container volume. Thus the free volume available to the molecules is essentially the entire volume of the container. As the pressure increases, however, the free space in which the molecules can move becomes a smaller fraction of the container volume. Under these conditions, therefore, gas volumes tend to be slightly greater than those predicted by the ideal-gas equation.

In addition, the attractive forces between molecules come into play at short distances, as when molecules are crowded together at high pressures. Because of these attractive forces, the impact of a given molecule with the wall of the container is lessened. If we could stop the action in a gas, the positions of the

Low pressure   High pressure

▲ **FIGURE 10.23   Gases behave more ideally at low pressure than at high pressure.** The combined volume of the molecules can be neglected at low pressure but not at high pressure.

molecules might resemble the illustration in ▼ **FIGURE 10.24**. The molecule about to make contact with the wall experiences the attractive forces of nearby molecules. These attractions lessen the force with which the molecule hits the wall. As a result, the pressure is less than that of an ideal gas. This effect serves to decrease PV/RT below its ideal value, as seen in Figure 10.21. When the pressure is sufficiently high, however, the volume effects dominate and PV/RT increases to above the ideal value.

Temperature determines how effective attractive forces between gas molecules are. As a gas is cooled, the average kinetic energy of the molecules decreases but intermolecular attractions remain constant. In a sense, cooling a gas deprives molecules of the energy they need to overcome their mutual attractive influence. The effects of temperature shown in Figure 10.22 illustrate this point very well. As temperature increases, the negative departure of PV/RT from ideal-gas behaviour disappears. The difference that remains at high temperature stems mainly from the effect of the finite volumes of the molecules.

 CONCEPT CHECK 11
List two reasons why gases deviate from ideal behaviour.

The van der Waals Equation

Engineers and scientists who work with gases at high pressures often cannot use the ideal-gas equation to predict the pressure–volume properties of gases because departures from ideal behaviour are too large. One useful equation developed to predict the behaviour of real gases was proposed by the Dutch scientist Johannes van der Waals (1837–1923).

The ideal-gas equation predicts that the pressure of a gas is

$$P = \frac{nRT}{V} \qquad \text{(ideal gas)}$$

van der Waals recognised that for a real gas this expression would have to be corrected for the finite volume occupied by the gas molecules and for the attractive forces between the gas molecules. He introduced two constants, a and b, to make these corrections.

 FIGURE IT OUT

How would you expect the pressure of a gas to change if suddenly the intermolecular forces were repulsive rather than attractive?

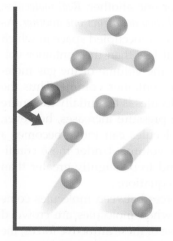
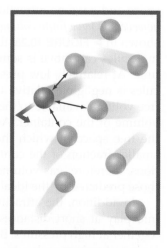

Ideal gas · Real gas

▶ **FIGURE 10.24** In any real gas, attractive intermolecular forces reduce pressure to values lower than in an ideal gas.

$$P = \frac{nRT}{V - nb} - \frac{n^2a}{V^2} \qquad [10.26]$$

Correction for   Correction for
volume of molecules   molecular attraction

The volume of the container, V, is decreased by the factor nb, to account for the small but finite volume occupied by the gas molecules themselves (Figure 10.23). Thus the free volume available to the gas molecules is $V - nb$. The van der Waals constant b is a measure of the actual intrinsic volume occupied by a mole of gas molecules; b has units of $dm^3\ mol^{-1}$. The pressure is in turn decreased by the factor n^2a/V^2, which accounts for the attractive forces between the gas molecules (Figure 10.24). The unusual form of this correction results because the attractive forces between pairs of molecules increase as the square of the number of molecules per unit volume $(n/V)^2$. Hence, the van der Waals constant a has units of $kPa\ dm^6\ mol^{-2}$. The magnitude of a reflects how strongly the gas molecules attract each other.

Equation 10.26 is generally rearranged to give the following form of the **van der Waals equation**:

$$\left(P + \frac{n^2a}{V^2} \right)(V - nb) = nRT \qquad [10.27]$$

The van der Waals constants a and b are different for each gas. Values of these constants for several gases are listed in ▼ TABLE 10.2. Note that the values of both a and b generally increase with an increase in mass of the molecule and with an increase in the complexity of its structure. Larger, more massive molecules not only have larger volumes, they also tend to have greater intermolecular attractive forces.

TABLE 10.2 • **van der Waals constants for gas molecules**

Substance	a ($kPa\ dm^6\ mol^{-2}$)	b ($dm^3\ mol^{-1}$)
He	3.46	0.0237
Ne	21.4	0.0171
Ar	136	0.0322
Kr	235	0.0398
Xe	425	0.0510
H_2	24.7	0.0266
N_2	141	0.0391
O_2	138	0.0318
Cl_2	658	0.0562
H_2O	553	0.0305
CH_4	228	0.0428
CO_2	364	0.0427
CCl_4	2070	0.1383

SAMPLE EXERCISE 10.15 **Using the van der Waals equation**

If 1.000 mol of an ideal gas is confined to 22.71 dm^3 at 0.0 °C, it will exert a pressure of 1 bar (100 kPa). Use the van der Waals equation and the constants in Table 10.2 to estimate the pressure exerted by 1.000 mol of Cl_2(g) in 22.71 dm^3 at 0.0 °C.

SOLUTION

Analyse We need to determine a pressure. Because we will use the van der Waals equation, we must identify the appropriate values for the constants that appear there.

Plan Using Equation 10.26, we have

$$P = \frac{nRT}{V - nb} - \frac{n^2a}{V^2}$$

Solve Substituting $n = 1.000$ mol, $R = 8.314$ kPa dm^3 mol^{-1} K^{-1}, $T = 273.2$ K, $V = 22.71$ dm^3, $a = 6.58 \times 10^2$ kPa dm^6 mol^{-2} and $b = 0.0562$ dm^3 mol^{-1}:

$$P = \frac{(1.000 \text{ mol})(8.314 \text{ kPa dm}^3 \text{ mol}^{-1} \text{ K}^{-1})(273.2 \text{ K})}{22.71 \text{ dm}^3 - (1.000 \text{ mol})(0.0562 \text{ dm}^3 \text{ mol}^{-1})} - \frac{(1.000 \text{ mol})^2(6.58 \times 10^2 \text{ kPa dm}^6 \text{ mol}^{-2})}{(22.71 \text{ dm}^3)^2}$$

$$= 100.3 \text{ kPa} - 1.28 \text{ kPa} = 99.0 \text{ kPa}$$

Check We expect a pressure not far from 100 kPa, which would be the value for an ideal gas, so our answer seems very reasonable.

Comment Notice that the first term, 100.3 kPa, is the pressure corrected for molecular volume. This value is higher than the ideal value, 100 kPa, because the volume in which the molecules are free to move is smaller than the container volume, 22.71 dm^3. Thus the molecules must collide more frequently with the container walls. The second factor, 1.28 kPa, corrects for intermolecular forces. The intermolecular attractions between molecules reduce the pressure to 99.02 kPa. We can conclude, therefore, that the intermolecular attractions are the main cause of the slight deviation of Cl$_2$(g) from ideal behaviour under the stated experimental conditions.

PRACTICE EXERCISE

Consider a sample of 1.00 mol of CO$_2$(g) confined to a volume of 3.00 dm^3 at 0.0 °C. Calculate the pressure of the gas using **(a)** the ideal-gas equation and **(b)** the van der Waals equation.

Answers: **(a)** 757 kPa, **(b)** 728 kPa

(See also Exercises 10.60, 10.61.)

SAMPLE INTEGRATIVE EXERCISE **Putting concepts together**

Cyanogen, a highly toxic gas, is composed of 46.2% C and 53.8% N by mass. At 25 °C and 100 kPa, 1.05 g of cyanogen occupies 0.500 dm^3. **(a)** What is the molecular formula of cyanogen? **(b)** Predict its molecular structure. **(c)** Predict the polarity of the compound.

SOLUTION

(a) We can use the percentage composition of the compound to calculate its empirical formula (∞ Section 3.5, 'Empirical Formulae from Analyses'). Then we can determine the molecular formula by comparing the mass of the empirical formula with the molar mass (∞ Section 3.5, 'Empirical Formulae from Analyses').

> ∞ Review this on pages 83 and 85

To determine the empirical formula, we assume that we have a 100 g sample of the compound and then calculate the number of moles of each element in the sample:

$$\text{Moles C} = \frac{46.2 \text{ g}}{12.01 \text{ g mol}^{-1}} = 3.85 \text{ mol C}$$

$$\text{Moles N} = \frac{53.8 \text{ g}}{14.01 \text{ g mol}^{-1}} = 3.84 \text{ mol N}$$

Because the ratio of the moles of the two elements is essentially 1:1, the empirical formula is CN.

To determine the molar mass of the compound, we use Equation 10.11.

$$M = \frac{\rho RT}{P} = \frac{(1.05 \text{ g}/0.500 \text{ dm}^3)(8.314 \text{ kPa dm}^3 \text{ mol}^{-1} \text{ K})(298 \text{ K})}{100 \text{ kPa}} = 52.0 \text{ g mol}^{-1}$$

The molar mass associated with the empirical formula, CN, is $12.0 + 14.0 = 26.0$ g mol^{-1}. Dividing the molar mass of the compound by that of its empirical formula gives $(52.0 \text{ g mol}^{-1})/(26.0 \text{ g mol}^{-1}) = 2.00$. Thus the molecule has twice as many atoms of each element as the empirical formula, giving the molecular formula C$_2$N$_2$.

> ∞ Review this on pages 267 and 296

(b) To determine the molecular structure of the molecule, we must first determine its Lewis structure (∞ Section 8.5, 'Drawing Lewis Structures'). We can then use the VSEPR model to predict the structure (∞ Section 9.2, 'The VSEPR Model').

The molecule has $2(4) + 2(5) = 18$ valence-shell electrons. We seek a Lewis structure in which each atom has an octet and in which the formal charges are as low as possible. Usually the atom(s) that are capable of forming the higher number of bonds will be at the centre of the optimal Lewis structure, and we can obtain the following structure:

$$:N \equiv C - C \equiv N:$$

(This structure has zero formal charges on each atom.)

The Lewis structure shows that each atom has two electron domains. (Each nitrogen has a non-bonding pair of electrons and a triple bond, whereas each carbon has a triple bond and a single bond.) Thus the electron-domain geometry around each atom is linear, causing the overall molecule to be linear.

(c) To determine the polarity of the molecule, we must examine the polarity of the individual bonds and the overall geometry of the molecule. Because the molecule is linear, we expect the two dipoles created by the polarity in the carbon–nitrogen bond to cancel each other, leaving the molecule with no dipole moment.

MY WORLD OF CHEMISTRY

GAS PIPELINES

Most people are quite unaware of the vast network of underground pipelines that undergirds the developed world and which is used to move massive quantities of liquids and gases over considerable distances. The pipeline system consists of large-diameter (120 cm) trunk lines at high pressure, with branch lines of smaller diameter (about 35 cm) and lower pressure for local transport to and from the trunk lines. Australia has large reserves of natural gas and LP gas, of which natural gas makes up the largest component to be transported by pipeline (▶ FIGURE 10.25). The methane-rich gas from oil and gas wells is processed to remove particulates, water and various gaseous impurities such as hydrogen sulfide and carbon dioxide. The gas is then compressed to pressures ranging from 3.5 MPa (35 bar) to 10 MPa (100 bar) depending on the diameter of the pipe being used. Pressure is maintained by compressor stations along the pipeline spaced at roughly 100-km intervals. Australia's gas supplies are linked to the major industrial centres and other markets by more than 20 000 kilometres of high-pressure pipelines. The supplies from the Carnarvon Basin in northwest Western Australia are connected to Perth and Kalgoorlie, and those off the coast of Darwin link this city and Alice Springs. Gas from the Cooper/Eromanga basin on the border of South Australia and Queensland connects Mount Isa, Brisbane, Sydney and Adelaide. In addition, Victoria and Tasmania are supplied with Bass Strait gas from the Gippsland Basin by a pipeline running 301 km from Longford in Victoria to Five Mile Bluff in Tasmania, continuing on to Hobart and Launceston. In the near future it is hoped to link the gas fields of Papua New Guinea to Australia across the Torres Strait.

▲ FIGURE 10.25   **High-pressure pipelines are used for the transportation of natural gas in Australia.**

CHAPTER SUMMARY AND KEY TERMS

SECTION 10.1   Substances that are gases at room temperature tend to be molecular substances with low molar masses. Air, a mixture composed mainly of N_2 and O_2, is the most common gas we encounter. Some liquids and solids can also exist in the gaseous state, where they are known as **vapours**. Gases are compressible; they mix in all proportions because their component molecules are far apart from each other.

SECTION 10.2   To describe the state or condition of a gas, we must specify four variables: pressure (P), volume (V), temperature (T) and quantity (n). Volume is usually measured in litres, temperature in kelvins and quantity of gas in moles. **Pressure** is the force per unit area. It is expressed in SI units as **pascals**, Pa (1 Pa $= 1$ N m^{-2}). In chemistry, **standard atmospheric pressure** is defined as 1 **bar** $= 100$ kPa. A barometer is often used to measure the atmospheric pressure. A manometer can be used to measure the pressure of enclosed gases.

SECTIONS 10.3 and 10.4   Studies have revealed several simple gas laws. For a constant quantity of gas at constant temperature, the volume of the gas is inversely proportional to the pressure (**Boyle's law**). For a fixed quantity of gas at constant pressure, the volume is directly proportional to its absolute temperature (**Charles's law**). Equal volumes of gases at the same temperature and pressure contain equal numbers of molecules (**Avogadro's hypothesis**). For a gas at constant temperature and pressure, the volume of the gas is directly proportional to the number of moles of gas (**Avogadro's law**). Each of these gas laws is a special case of the **ideal-gas equation**, $PV = nRT$, which is the equation of state for an **ideal gas**. The term R in this equation is the **gas constant**. We can use the ideal-gas equation to calculate variations in one variable when one or more of the others are changed. Most gases at pressures less than 10 bar and temperatures near 273 K and above obey the ideal-gas equation reasonably well. The conditions of 273 K

(0 °C) and 1 bar are known as the **standard temperature and pressure (STP)**.

SECTIONS 10.5 and 10.6 Using the ideal-gas equation, we can relate the density of a gas to its molar mass: $M = \dfrac{PRT}{\rho}$. We can also use the ideal-gas equation to solve problems involving gases as reactants or products in chemical reactions.

In gas mixtures the total pressure is the sum of the **partial pressures** that each gas would exert if it were present alone under the same conditions (**Dalton's law of partial pressures**). The partial pressure of a component of a mixture is equal to its mole fraction times the total pressure: $P_1 = X_1 P_t$. The **mole fraction** is the ratio of the moles of one component of a mixture to the total moles of all components. In calculating the quantity of a gas collected over water, correction must be made for the partial pressure of water vapour in the gas mixture.

SECTION 10.7 The **kinetic-molecular theory** of gases accounts for the properties of an ideal gas in terms of a set of statements about the nature of gases. Briefly, these statements are as follows: Molecules are in continuous chaotic motion. The volume of gas molecules is negligible compared to the volume of their container. The gas molecules neither attract nor repel each other. The average kinetic energy of the gas molecules is proportional to the absolute temperature and does not change if the temperature remains constant.

The individual molecules of a gas do not all have the same kinetic energy at a given instant. Their speeds are distributed over a wide range; the distribution varies with the molar mass of the gas and with temperature. The **root-mean-square (rms) speed**, u_{rms}, varies in proportion to the square root of the absolute temperature and inversely with the square root of the molar mass: $u_{rms} = \sqrt{3RT/M}$.

SECTION 10.8 It follows from kinetic-molecular theory that the rate at which a gas undergoes **effusion** (escapes through a tiny hole) is inversely proportional to the square root of its molar mass (**Graham's law**). The **diffusion** of one gas through the space occupied by a second gas is another phenomenon related to the speeds at which molecules move. Because molecules undergo frequent collisions with one another, the **mean free path**—the mean distance travelled between collisions—is short. Collisions between molecules limit the rate at which a gas molecule can diffuse.

SECTION 10.9 Departures from ideal behaviour increase in magnitude as pressure increases and as temperature decreases. The extent of non-ideality of a real gas can be seen by examining the quantity PV/RT for one mole of the gas as a function of pressure; for an ideal gas, this quantity is exactly 1 at all pressures. Real gases depart from ideal behaviour because the molecules possess finite volume and because the molecules experience attractive forces for one another. The **van der Waals equation** is an equation of state for gases that modifies the ideal-gas equation to account for intrinsic molecular volume and intermolecular forces. It is represented by

$$\left(P + \frac{n^2 a}{V^2}\right)(V - nb) = nRT.$$

KEY SKILLS

- Understand how the gas laws relate to the ideal-gas equation and apply the gas laws in calculations. (Sections 10.3 and 10.4)

- Calculate P, V, n or T using the ideal-gas equation. (Section 10.4)

- Calculate the density or molecular weight of a gas. (Section 10.5)

- Calculate the volume of gas consumed or formed in a chemical reaction. (Section 10.5)

- Calculate the total pressure of a gas mixture given its partial pressures or given information for calculating partial pressures. (Section 10.6)

- Describe the kinetic-molecular theory of gases and how it explains the pressure and temperature of a gas, the gas laws and the rates of effusion and diffusion. (Sections 10.7 and 10.8)

- Explain why intermolecular attractions and molecular volumes cause real gases to deviate from ideal behaviour at high pressure or low temperature. (Section 10.9)

KEY EQUATIONS

- Ideal-gas equation $$PV = nRT \qquad [10.5]$$

- The combined gas law, showing how P, V and T are related for a constant n $$\frac{P_1 V_1}{T_1} = \frac{P_2 V_2}{T_2} \qquad [10.8]$$

- Calculating the density or molar mass of a gas

$$\rho = \frac{PM}{RT} \qquad [10.10]$$

- Relating the total pressure of a gas mixture to the partial pressures of its components (Dalton's law of partial pressures)

$$P_t = P_1 + P_2 + P_3 + \ldots \qquad [10.12]$$

- Relating partial pressure to mole fraction

$$P_1 = \left(\frac{n_1}{n_t}\right) P_t = X_1 P_t \qquad [10.15]$$

- Definition of the root-mean-square (rms) speed of gas molecules

$$u_{rms} = \sqrt{\frac{3RT}{M}} \qquad [10.22]$$

- Relating the relative rates of effusion of two gases to their molar masses

$$\frac{r_1}{r_2} = \sqrt{\frac{M_2}{M_1}} \qquad [10.23]$$

- The van der Waals equation

$$\left(P + \frac{n^2 a}{V^2}\right)(V - nb) = nRT \qquad [10.27]$$

EXERCISES

VISUALISING CONCEPTS

10.1 Assume that you have a sample of gas in a container with a movable piston such as the one in the drawing. **(a)** Redraw the container to show what it might look like if the temperature of the gas is increased from 300 K to 500 K while the pressure is kept constant. **(b)** Redraw the container to show what it might look like if the pressure on the piston is increased from 1 bar to 2 bar while the temperature is kept constant. [Section 10.3]

10.2 Consider the sample of gas depicted below. What would the drawing look like if the volume and temperature remained constant while you removed enough of the gas to decrease the pressure by a factor of 2? [Section 10.3]

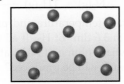

10.3 Consider the following reaction:

$$2\,CO(g) + O_2(g) \longrightarrow 2\,CO_2(g)$$

Imagine that this reaction occurs in a container that has a piston that moves to allow a constant pressure to be maintained when the reaction occurs at constant temperature. **(a)** What happens to the volume of the container as a result of the reaction? Explain. **(b)** If the piston is not allowed to move, what happens to the pressure as a result of the reaction? [Sections 10.3 and 10.5]

10.4 Consider the apparatus below, which shows gases in two containers and one empty container. When the stopcocks are opened and the gases allowed to mix at constant temperature, what is the distribution of atoms in each container? Assume that the containers are of equal volume and ignore the volume of the tubing connecting them. Which gas has the greater partial pressure after the stopcocks are opened? [Section 10.6]

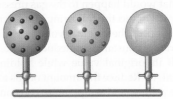

10.5 The drawing below represents a mixture of three different gases. **(a)** Rank the three components in order of increasing partial pressure. **(b)** If the total pressure of the mixture is 91.2 kPa, calculate the partial pressure of each gas. [Section 10.6]

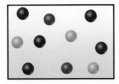

10.6 On a single plot, qualitatively sketch the distribution of molecular speeds for **(a)** Kr(g) at −50 °C, **(b)** Kr(g) at 0 °C, **(c)** Ar(g) at 0 °C. [Section 10.7]

10.7 Consider the drawing below. **(a)** If the curves A and B refer to two different gases, He and O_2, at the same temperature, which is which? Explain. **(b)** If A and B refer to the same gas at two different temperatures, which represents the higher temperature? [Section 10.7]

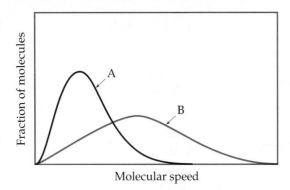

10.8 Consider the following samples of gases:

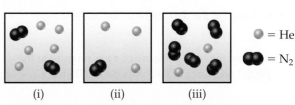

If the three samples are all at the same temperature, rank them with respect to **(a)** total pressure, **(b)** partial pressure of helium, **(c)** density, **(d)** average kinetic energy of particles. [Section 10.7]

GAS CHARACTERISTICS; PRESSURE (Sections 10.1 and 10.2)

10.9 How does a gas differ from a liquid with respect to each of the following properties: **(a)** density, **(b)** compressibility, **(c)** ability to mix with other substances of the same phase to form homogeneous mixtures?

10.10 **(a)** Both a liquid and a gas are moved to larger containers. How does their behaviour differ? Explain the difference in molecular terms. **(b)** Although water and carbon tetrachloride, CCl_4(l), do not mix, their vapours form homogeneous mixtures. Explain. **(c)** The densities of gases are generally reported in units of g dm^{-3}, whereas those for liquids are reported as g cm^{-3}. Explain the molecular basis for this difference.

10.11 How high in metres must a column of water be to exert a pressure equal to that of a 760 mm column of mercury? The density of water is 1.0 g cm^{-3}, whereas that of mercury is 13.6 g cm^{-3}.

10.12 Each of the following statements concerns a mercury barometer such as that shown in Figure 10.2. Identify any incorrect statements, and correct them. **(a)** The tube must be 1 cm^2 in cross-sectional area. **(b)** At equilibrium the force of gravity per unit area acting on the mercury column at the level of the outside mercury equals the force of gravity per unit area acting on the atmosphere. **(c)** The column of mercury is held up by the vacuum at the top of the column.

THE GAS LAWS (Section 10.3)

10.13 Assume that you have a cylinder with a movable piston. What would happen to the gas pressure inside the cylinder if you do the following? **(a)** Decrease the volume to one-fourth the original volume while holding the temperature constant. **(b)** Reduce the Kelvin temperature to half its original value while holding the volume constant. **(c)** Reduce the amount of gas to half while keeping the volume and temperature constant.

10.14 A fixed quantity of gas at 23 °C exhibits a pressure of 98 kPa and occupies a volume of 5.2 dm^3. **(a)** Use Boyle's law to calculate the volume the gas will occupy if the pressure is increased to 190 kPa while the temperature is held constant. **(b)** Use Charles's law to calculate the volume the gas will occupy if the temperature is increased to 165 °C while the pressure is held constant.

10.15 **(a)** How is the law of combining volumes explained by Avogadro's hypothesis? **(b)** Consider a 1.0 dm^3 flask containing neon gas and a 1.5 dm^3 flask containing xenon gas. Both gases are at the same pressure and temperature. According to Avogadro's law, what can be said about the ratio of the number of atoms in the two flasks?

10.16 Nitrogen and hydrogen gases react to form ammonia gas as follows:

$$N_2(g) + 3 H_2(g) \longrightarrow 2 NH_3(g)$$

At a certain temperature and pressure, 1.2 dm^3 of N_2 reacts with 3.6 dm^3 of H_2. If all the N_2 and H_2 are consumed, what volume of NH_3, at the same temperature and pressure, will be produced?

THE IDEAL-GAS EQUATION (Section 10.4)

10.17 **(a)** Write the ideal-gas equation, and give the units used for each term in the equation when $R = 8.314$ kPa dm^3 mol^{-1} K^{-1}. **(b)** What is an ideal gas?

10.18 **(a)** What conditions are represented by the abbreviation STP? **(b)** What is the molar volume of an ideal gas at STP? **(c)** Room temperature is often assumed to be 25 °C. Calculate the molar volume of an ideal gas at room temperature and 1 atm pressure.

10.19 Suppose you are given two 1 dm^3 flasks and told that one contains a gas of molar mass 30, the other a gas of molar mass 60, both at the same temperature. The pressure in flask A is X kPa, and the mass of gas in the flask is 1.2 g. The pressure in flask B is 0.5 X kPa, and the mass of gas in that flask is 1.2 g. Which flask contains gas of molar mass 30, and which contains gas of molar mass 60?

10.20 Suppose you are given two flasks at the same temperature, one of volume 2 dm^3 and the other of volume 3 dm^3. The 2 dm^3 flask contains 4.8 g of gas, and the gas pressure is X kPa. The 3 dm^3 flask contains 0.36 g of gas, and the gas pressure is 0.1 X. Do the two gases have the same molar mass? If not, which contains the gas of higher molar mass?

10.21 The *Hindenburg* was a hydrogen-filled dirigible that exploded in 1937. If the *Hindenburg* held 2.0×10^5 m^3 of hydrogen gas at 23 °C and 1 bar, what mass of hydrogen was present?

10.22 Calculate the number of molecules in a deep breath of air whose volume is 2.50 dm^3 at body temperature, 37 °C, and a pressure of 0.980 bar.

10.23 If the pressure exerted by ozone, O$_3$, in the stratosphere is 0.304 kPa and the temperature is 250 K, how many ozone molecules are in a litre?

10.24 A scuba diver's tank contains 0.29 kg of O$_2$ compressed into a volume of 2.3 dm^3. **(a)** Calculate the gas pressure inside the tank at 9 °C. **(b)** What volume would this oxygen occupy at 26 °C and 96 kPa?

10.25 An aerosol spray can with a volume of 250 cm^3 contains 2.30 g of propane gas as a propellant. **(a)** If the can is at 23 °C, what is the pressure in the can? **(b)** What volume would the propane occupy at STP? **(c)** The can says that exposure to temperatures above 55 °C may cause the can to burst. What is the pressure in the can at this temperature?

10.26 Chlorine is widely used to purify municipal water supplies and to treat swimming pool water. Suppose that the volume of a particular sample of Cl$_2$ gas is 8.70 dm^3 at 120 kPa and 24 °C. **(a)** How many grams of Cl$_2$ are in the sample? **(b)** What volume will the Cl$_2$ occupy at STP? **(c)** At what temperature will the volume be 15.0 dm^3 if the pressure is 118 kPa? **(d)** At what pressure will the volume equal 6.00 dm^3 if the temperature is 58 °C?

10.27 Many gases are shipped in high-pressure containers. Consider a steel tank of volume 65.0 dm^3, containing O$_2$ gas at a pressure of 165 bar at 23 °C. **(a)** What mass of O$_2$ does the tank contain? **(b)** What volume would the gas occupy at STP? **(c)** At what temperature would the pressure in the tank equal 1.52×10^4 kPa? **(d)** What would be the pressure of the gas, in kPa, if it were transferred to a container at 24 °C whose volume is 55.0 dm^3?

FURTHER APPLICATIONS OF THE IDEAL-GAS EQUATION (Section 10.5)

10.28 Which gas is most dense at 1 bar and 298 K: **(a)** CO$_2$, **(b)** N$_2$O, **(c)** Cl$_2$? Explain.

10.29 Which gas is least dense at 1 bar and 298 K: **(a)** SO$_2$, **(b)** HBr, **(c)** CO$_2$? Explain.

10.30 Which of the following statements best explains why a closed balloon filled with helium gas rises in air?

(a) Helium is a monatomic gas, whereas nearly all the molecules that make up air, such as nitrogen and oxygen, are diatomic.

(b) The average speed of helium atoms is higher than the average speed of air molecules, and the higher speed of collisions with the balloon walls propels the balloon upwards.

(c) Because the helium atoms are of lower mass than the average air molecule, the helium gas is less dense than air. Thus the balloon weighs less than the air displaced by its volume.

(d) Because helium has a lower molar mass than the average air molecule, the helium atoms are in faster motion. This means that the temperature of the helium is higher than the air temperature. Hot gases tend to rise.

10.31 Which of the following statements best explains why nitrogen gas at STP is less dense than Xe gas at STP?

(a) Because Xe is a noble gas, there is less tendency for the Xe atoms to repel one another, so they pack more densely in the gas state.

(b) Xe atoms have a higher mass than N$_2$ molecules. Because both gases at STP have the same number of molecules per unit volume, the Xe gas must be denser.

(c) The Xe atoms are larger than N$_2$ molecules and thus take up a larger fraction of the space occupied by the gas.

(d) Because the Xe atoms are much more massive than the molecules, they move more slowly and thus exert less upward force on the gas container and make the gas appear denser.

10.32 **(a)** Calculate the density of NO$_2$ gas at 98.0 kPa and 35 °C. **(b)** Calculate the molar mass of a gas if 2.50 g occupies 0.875 dm^3 at 91.3 kPa and 35 °C.

10.33 Magnesium can be used as a 'getter' in evacuated enclosures, to react with the last traces of oxygen. (The magnesium is usually heated by passing an electric current through a wire or ribbon of the metal.) If an enclosure of 0.382 dm^3 has a partial pressure of O$_2$ of 4.66×10^{-4} kPa at 27 °C, what mass of magnesium will react according to the following equation?

$$2\,\text{Mg(s)} + \text{O}_2(\text{g}) \longrightarrow 2\,\text{MgO(s)}$$

10.34 Calcium hydride, CaH$_2$, reacts with water to form hydrogen gas:

$$\text{CaH}_2(\text{s}) + 2\,\text{H}_2\text{O(l)} \longrightarrow \text{Ca(OH)}_2(\text{aq}) + 2\,\text{H}_2(\text{g})$$

This reaction is sometimes used to inflate life rafts, weather balloons and the like, where a simple, compact means of generating H_2 is desired. How many grams of CaH_2 are needed to generate 53.5 dm^3 of H_2 gas if the pressure of H_2 is 108.5 kPa at 21 °C?

10.35 The metabolic oxidation of glucose, $C_6H_{12}O_6$, in our bodies produces CO_2, which is expelled from our lungs as a gas:

$$C_6H_{12}O_6(aq) + 6\,O_2(g) \longrightarrow 6\,CO_2(g) + 6\,H_2O(l)$$

Calculate the volume of dry CO_2 produced at body temperature (37 °C) and 98.3 kPa when 24.5 g of glucose is consumed in this reaction.

10.36 Acetylene gas, $C_2H_2(g)$, can be prepared by the reaction of calcium carbide with water:

$$CaC_2(s) + 2\,H_2O(l) \longrightarrow Ca(OH)_2(s) + C_2H_2(g)$$

Calculate the volume of C_2H_2 that is collected over water at 26 °C by reaction of 0.887 g of CaC_2 if the total pressure of the gas is 96.8 kPa. (The vapour pressure of water is tabulated in Appendix B.)

PARTIAL PRESSURES (Section 10.6)

10.37 Consider the apparatus shown in the drawing. **(a)** When the stopcock between the two containers is opened and the gases allowed to mix, how does the volume occupied by the N_2 gas change? What is the partial pressure of N_2 after mixing? **(b)** How does the volume of the O_2 gas change when the gases mix? What is the partial pressure of O_2 in the mixture? **(c)** What is the total pressure in the container after the gases mix?

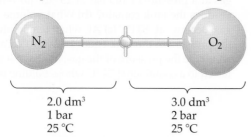

2.0 dm^3	3.0 dm^3
1 bar	2 bar
25 °C	25 °C

10.38 Consider a mixture of two gases, A and B, confined to a closed vessel. A quantity of a third gas, C, is added to the same vessel at the same temperature. How does the addition of gas C affect the following: **(a)** the partial pressure of gas A, **(b)** the total pressure in the vessel, **(c)** the mole fraction of gas B?

10.39 A mixture containing 0.538 mol He(g), 0.315 mol Ne(g) and 0.103 mol Ar(g) is confined in a 7.00 dm^3 vessel at 25 °C. **(a)** Calculate the partial pressure of each of the gases in the mixture. **(b)** Calculate the total pressure of the mixture.

10.40 A piece of solid carbon dioxide with a mass of 5.50 g is placed in a 10.0 dm^3 vessel that already contains air at 94.0 kPa and 24 °C. After the carbon dioxide has totally vapourised, what is the partial pressure of carbon dioxide and the total pressure in the container at 24 °C?

10.41 A sample of 4.00 cm^3 of diethylether ($C_2H_5OC_2H_5$; density = 0.7134 g cm^{-3} is introduced into a 5.00 dm^3 vessel that already contains a mixture of N_2 and O_2, whose partial pressures are $P_{N_2} = 76.1$ kPa and $P_{O_2} = 21.1$ kPa. The temperature is held at 35.0 °C and the diethylether totally evaporates. **(a)** Calculate the partial pressure of the diethylether. **(b)** Calculate the total pressure in the container.

10.42 A mixture of gases contains 0.75 mol N_2, 0.30 mol O_2 and 0.15 mol CO_2. If the total pressure of the mixture is 158 kPa, what is the partial pressure of each component?

10.43 A mixture of gases contains 10.25 g of N_2, 2.05 g of H_2 and 7.63 g of NH_3. If the total pressure of the mixture is 238 kPa, what is the partial pressure of each component?

10.44 At an underwater depth of 90 m, the pressure is 849 kPa. What should the mole percent of oxygen be in the diving gas for the partial pressure of oxygen in the mixture to be 21.3 kPa, the same as in air at 1 bar?

10.45 **(a)** What are the mole fractions of each component in a mixture of 5.08 g of O_2, 7.17 g of N_2 and 1.32 g of H_2? **(b)** What is the partial pressure in kPa of each component of this mixture if it is held in a 12.40 dm^3 vessel at 15 °C?

10.46 A quantity of N_2 gas originally held at 481 kPa pressure in a 1.00 dm^3 container at 26 °C is transferred to a 10.0 dm^3 container at 20 °C. A quantity of O_2 gas, originally at 532 kPa and 26 °C in a 5.00 dm^3 container, is transferred to the same container. What is the total pressure in the new container?

KINETIC-MOLECULAR THEORY OF GASES: EFFUSION AND DIFFUSION (Sections 10.7 and 10.8)

10.47 What change or changes in the state of a gas bring about each of the following effects? **(a)** The number of impacts per unit time on a given container wall increases. **(b)** The average energy of impact of molecules with the wall of the container decreases. **(c)** The average distance between gas molecules increases. **(d)** The average speed of molecules in the gas mixture is increased.

10.48 Indicate which of the following statements regarding the kinetic-molecular theory of gases are correct. For those that are false, formulate a correct version of the statement. **(a)** The average kinetic energy of a collection of gas molecules at a given temperature is proportional to $m^{\frac{1}{2}}$. **(b)** The gas molecules are assumed to exert no forces on each other. **(c)** All the molecules of a gas at a given temperature have the same kinetic energy. **(d)** The volume of the gas molecules is negligible in comparison with the total volume in which the gas is contained.

10.49 Vessel A contains CO(g) at 0 °C and 1 bar. Vessel B contains SO_2(g) at 20 °C and 0.5 bar. The two vessels have the same volume. **(a)** Which vessel contains more molecules? **(b)** Which contains more mass? **(c)** In which vessel is the average kinetic energy of molecules higher? **(d)** In which vessel is the rms speed of molecules higher?

10.50 Suppose you have two 1 dm^3 flasks, one containing N_2 at STP, the other containing CH_4 at STP. How do these systems compare with respect to **(a)** number of molecules, **(b)** density, **(c)** average kinetic energy of the molecules, **(d)** rate of effusion through a pinhole leak?

10.51 **(a)** Place the following gases in order of increasing average molecular speed at 25 °C: Ne, HBr, SO_2, NF_3, CO. **(b)** Calculate the rms speed of NF_3 molecules at 25 °C.

10.52 Hydrogen has two naturally occurring isotopes, ^{1}H and ^{2}H. Chlorine also has two naturally occurring isotopes, ^{35}Cl and ^{37}Cl. Thus hydrogen chloride gas consists of four distinct types of molecules: ^{1}H^{35}Cl, ^{1}H^{37}Cl, ^{2}H^{35}Cl and ^{2}H^{37}Cl. Place these four molecules in order of increasing rate of effusion.

10.53 As discussed in the My World of Chemistry box in Section 10.8, enriched uranium is produced via gaseous diffusion of UF_6. Suppose a process is developed to allow diffusion of gaseous uranium atoms, U(g). Calculate the ratio of diffusion rates for ^{235}U and ^{238}U, and compare it with the ratio for UF_6 given in the essay.

10.54 Arsenic(III) sulfide sublimes readily, even below its melting point of 320 °C. The molecules of the vapour phase are found to effuse through a tiny hole at 0.285 times the rate of effusion of argon atoms under the same conditions of temperature and pressure. What is the molecular formula of arsenic(III) sulfide in the gas phase?

10.55 A gas of unknown molecular mass was allowed to effuse through a small opening under constant-pressure conditions. It required 105 s for 1.0 dm^3 of the gas to effuse. Under identical experimental conditions it required 31 s for 1.0 dm^3 of O_2 gas to effuse. Calculate the molar mass of the unknown gas. (Remember that the faster the rate of effusion, the shorter the time required for effusion of 1.0 dm^3; that is, rate and time are inversely proportional.)

NON-IDEAL-GAS BEHAVIOUR (Section 10.9)

10.56 **(a)** List two experimental conditions under which gases deviate from ideal behaviour. **(b)** List two reasons why the gases deviate from ideal behaviour. **(c)** Explain how the function PV/RT can be used to show how gases behave non-ideally.

10.57 The planet Jupiter has a mass 318 times that of Earth, and its surface temperature is 140 K. Mercury has a mass 0.05 times that of Earth, and its surface temperature is between 600 K and 700 K. On which planet is the atmosphere more likely to obey the ideal-gas law? Explain.

10.58 Based on their respective van der Waals constants (Table 10.2), is Ar or CO_2 expected to behave more nearly like an ideal gas at high pressures? Explain.

10.59 Briefly explain the significance of the constants a and b in the van der Waals equation.

10.60 Calculate the pressure that CCl_4 will exert at 40 °C if 1.00 mol occupies 28.0 dm^3, assuming that **(a)** CCl_4 obeys the ideal-gas equation, **(b)** CCl_4 obeys the van der Waals equation. (Values for the van der Waals constants are given in Table 10.2.)

10.61 It turns out that the van der Waals constant b equals four times the total volume actually occupied by the molecules of a mole of gas. Using this figure, calculate the fraction of the volume in a container actually occupied by Ar atoms **(a)** at STP, **(b)** at 100 bar pressure and 0 °C. (Assume for simplicity that the ideal-gas equation still holds.)

ADDITIONAL EXERCISES

10.62 A gas bubble with a volume of 1.0 mm^3 originates at the bottom of a lake where the pressure is 3 bar. Calculate its volume when the bubble reaches the surface of the lake where the pressure is 0.93 bar, assuming that the temperature doesn't change.

10.63 A 15.0 dm^3 tank is filled with helium gas at a pressure of 1.01×10^4 kPa. How many balloons (each 2.0 dm^3) can be inflated to a pressure of 1.01×10^2 kPa, assuming that the temperature remains constant and that the tank cannot be emptied below 1 bar?

10.64 Nickel carbonyl, $Ni(CO)_4$, is one of the most toxic substances known. The present maximum allowable concentration in laboratory air during an 8 hour working day is 1 part in 10^9 by volume, which means that there is one mole of $Ni(CO)_4$ for every 10^9 moles of gas. Assume 24 °C and 1 bar pressure. What mass of $Ni(CO)_4$ is allowable in a laboratory that is 54 m^2 in area, with a ceiling height of 3.1 m?

10.65 Assume that a single cylinder of a car engine has a volume of 524 cm^3. **(a)** If the cylinder is full of air at 74 °C and 910.3 kPa, how many moles of O_2 are present? (The mole fraction of O_2 in dry air is 0.2095.) **(b)** How many grams of C_8H_{18} could be combusted by this quantity of O_2, assuming complete combustion with formation of CO_2 and H_2O?

10.66 Ammonia, NH_3(g), and hydrogen chloride, HCl(g), react to form solid ammonium chloride, NH_4Cl(s):

$$NH_3(g) + HCl(g) \longrightarrow NH_4Cl(s)$$

Two 2.00 dm^3 flasks at 25 °C are connected by a stopcock, as shown in the drawing. One flask contains 5.00 g NH_3(g) and the other contains 5.00 g HCl(g).

When the stopcock is opened, the gases react until one is completely consumed. **(a)** Which gas will remain in the system after the reaction is complete? **(b)** What will be the final pressure of the system after the reaction is complete? (Neglect the volume of the ammonium chloride formed.)

10.67 A gaseous mixture of O_2 and Kr has a density of 1.104 g dm^{-3} at 58 kPa and 300 K. What is the mole percent of O_2 in the mixture?

10.68 A glass vessel fitted with a stopcock has a mass of 337.428 g when evacuated. When filled with Ar, it has a mass of 3310.854 g. When evacuated and refilled with a mixture of Ne and Ar, under the same conditions of temperature and pressure, it weighs 3310.076 g. What is the mole percent of Ne in the gas mixture?

10.69 Does the effect of intermolecular attraction on the properties of a gas become more significant or less significant if **(a)** the gas is compressed to a smaller volume at constant temperature, **(b)** the temperature of the gas is increased at constant volume?

10.70 For nearly all real gases, the quantity PV/RT decreases below the value of 1, which characterises an ideal gas, as pressure on the gas increases. At much higher pressures, however, PV/RT increases and rises above the value of 1. **(a)** Explain the initial drop in value of PV/RT below 1 and the fact that it rises above 1 for still higher pressures. **(b)** The effects we have just noted are smaller for gases at higher temperature. Why is this so?

INTEGRATIVE EXERCISES

10.71 A 4.00 g sample of a mixture of CaO and BaO is placed in a 1.00 dm^3 vessel containing CO_2 gas at a pressure of 97.3 kPa at a temperature of 25 °C. The CO_2 reacts with the CaO and BaO, forming $CaCO_3$ and $BaCO_3$. When the reaction is complete, the pressure of the remaining CO_2 is 20.0 kPa. **(a)** Calculate the number of moles of CO_2 that have reacted. **(b)** Calculate the mass percentage of CaO in the mixture.

10.72 A gas forms when elemental sulfur is heated carefully with AgF. The initial product boils at 15 °C. Experiments on several samples yielded a gas density of 0.803 ± 0.010 g dm^{-3} for the gas at 20.0 kPa pressure and 32 °C. When the gas reacts with water, all the fluorine is converted to aqueous HF. Other products are elemental sulfur, S_8, and other sulfur-containing compounds. A 480 cm^3 sample of the dry gas at 16.8 kPa pressure and 28 °C, when reacted with 80 cm^3 of water, yielded a 0.081 M solution of HF. The initial gaseous product undergoes a transformation over a period of time to a second compound with the same empirical and molecular formula, which boils at −10 °C. **(a)** Determine the empirical and molecular formulae of the first compound formed. **(b)** Draw at least two reasonable Lewis structures that represent the initial compound and the one into which it is transformed over time. **(c)** Describe the likely geometries of these compounds, and estimate the single bond distances, given that the S–S bond distance in S_8 is 240 pm and the F–F distance in F_2 is 143 pm.

10.73 Gaseous iodine pentafluoride, IF_5, can be prepared by the reaction of solid iodine and gaseous fluorine:

$$I_2(s) + 5\ F_2(g) \longrightarrow 2\ IF_5(g)$$

A 5.00 dm^3 flask containing 10.0 g I_2 is charged with 10.0 g F_2 and the reaction proceeds until one of the reagents is completely consumed. After the reaction is complete, the temperature in the flask is 125 °C. **(a)** What is the partial pressure of IF_5 in the flask? **(b)** What is the mole fraction of IF_5 in the flask?

10.74 A 6.53 g sample of a mixture of magnesium carbonate and calcium carbonate is treated with excess hydrochloric acid. The resulting reaction produces 1.72 dm^3 of carbon dioxide gas at 28 °C and 910.0 kPa pressure. **(a)** Write balanced chemical equations for the reactions that occur between hydrochloric acid and each component of the mixture. **(b)** Calculate the total number of moles of carbon dioxide that forms from these reactions. **(c)** Assuming that the reactions are complete, calculate the percentage by mass of magnesium carbonate in the mixture.

PHOTO/ART CREDITS

336 Jeff Schmaltz MODIS Rapid Response Team, NASA GSFC. Courtesy of nasaimages.org; **339** © NASA. Adapted from NASA_world_topo_bathy_200401, The Visible Earth http://visibleearth.nasa.gov; **349** Getty Images Australia; **360** Richard Megna, Fundamental Photographs, NYC.

INTERMOLECULAR FORCES: LIQUIDS AND SOLIDS

The water in hot springs is at least 5–10 °C warmer than the mean annual air temperature where they are located.

KEY CONCEPTS

11.1 A MOLECULAR COMPARISON OF GASES, LIQUIDS AND SOLIDS

We compare solids, liquids and gases from a molecular perspective. This comparison reveals the important roles that temperature and *intermolecular forces* play in determining the physical state of a substance.

11.2 INTERMOLECULAR FORCES

We examine the four intermolecular forces: *dispersion forces*, *dipole–dipole forces*, *hydrogen bonds* and *ion–dipole forces*.

11.3 SOME PROPERTIES OF LIQUIDS

We learn that the nature and strength of the intermolecular forces between molecules are largely responsible for many properties of liquids, including *viscosity* and *surface tension*.

11.4 PHASE CHANGES

We examine *phase changes*—the transitions of matter between the gaseous, liquid and solid states—and their associated energies.

11.5 VAPOUR PRESSURE

We examine the *dynamic equilibrium* that exists between a liquid and its gaseous state and introduce *vapour pressure*.

11.6 PHASE DIAGRAMS

We introduce *phase diagrams*, which are graphic representations of the equilibria among the gaseous, liquid and solid phases, and learn how to read them.

11.7 LIQUID CRYSTALS

We learn about substances that pass into a liquid crystalline phase, which is an intermediate phase between the solid and liquid states. A substance in the liquid crystalline phase has some of the structural order of a solid and some of the freedom of motion of a liquid.

11.8 STRUCTURES OF SOLIDS

We see that solids can be *crystalline*, where atoms are arranged in an orderly repeating pattern, or *amorphous*, where the atoms have no regular, predictable pattern. We further learn about *lattices* and *unit cells*, which define the repeating patterns that characterise crystalline solids.

11.9 BONDING IN SOLIDS

We learn about the different kinds of solids, such as *molecular solids*, *covalent-network solids*, *ionic solids* and *metallic solids* and the characteristics that define them.

Sitting in a hot spring on a snowy day is not something many of us have experienced. If we were in a hot spring, however, we would be surrounded simultaneously by all three phases of water—gas, liquid and solid. The water vapour—or humidity—in the air, the water in the hot spring and the surrounding snow are all forms of the same substance, H₂O. They all have the same chemical properties. Their physical properties differ greatly, however, because the physical properties of a substance depend on its physical state. In Chapter 10 we discussed the gaseous state in some detail. In this chapter we turn our attention to the physical properties of liquids and solids and to the phase changes that occur between the three states of matter.

Many of the substances that we consider in this chapter are molecular. In fact, virtually all substances that are liquids at room temperature are molecular substances. The intramolecular forces within molecules that give rise to covalent bonding influence molecular shape, bond energies and many aspects of chemical behaviour. The physical properties of molecular liquids and solids, however, are due largely to **intermolecular forces**, the forces that exist between molecules. We learned in Section 10.9 that attractions between gas molecules lead to deviations from ideal-gas behaviour. But how do these intermolecular attractions arise? By understanding the nature and strength of intermolecular forces, we can begin to relate the composition and structure of molecules to their physical properties.

11.1 | A MOLECULAR COMPARISON OF GASES, LIQUIDS AND SOLIDS

We learnt in Chapter 10 that the molecules in a gas are widely separated and in a state of constant, chaotic motion. One of the key tenets of kinetic-molecular theory is the assumption that we can neglect the interactions between molecules (∞ Section 10.7, 'Kinetic-Molecular Theory'). The properties of liquids and solids are quite different from gases largely because the intermolecular forces in liquids and solids are much stronger. A comparison of the properties of gases, liquids and solids is given in ▼ TABLE 11.1.

∞ Review this on page 355

In liquids the intermolecular attractive forces are strong enough to hold molecules close together. Thus liquids are much denser and far less compressible than gases. Unlike gases, liquids have a definite volume, independent of the size and shape of their container.

In solids the intermolecular attractive forces are strong enough not only to hold molecules close together, but to actually lock them in place. Solids, like liquids, are not very compressible because the molecules have little free space between them. Because the particles in a solid or liquid are fairly close together compared with those of a gas, we often refer to solids and liquids as *condensed phases*. Often the molecules of a solid take up positions in a highly regular pattern. Solids that possess regular repeating structures are said to be *crystalline*. (The transition from a liquid to a crystalline solid can be likened to the change that occurs on a military parade ground when the soldiers are called to formation.) Because the particles of a solid are not free to undergo long-range movement, solids are rigid. Keep in mind, however, that the units that form the solid, whether ions or molecules, possess thermal energy and vibrate about fixed mean positions. This vibrational motion increases in amplitude as a solid is heated. In fact, the energy may increase to the point that the solid either melts or sublimes.

▶ FIGURE 11.1 compares the three states of matter. The particles that compose the substance can be individual atoms, as in Ar; molecules, as in H_2O; or ions, as in NaCl. *The state of a substance depends largely on the balance between the kinetic energies of the particles and the interparticle energies of attraction.* The kinetic energies, which depend on temperature, tend to keep the particles apart and moving. The interparticle attractions tend to draw the particles together. Substances that are gases at room temperature have weaker interparticle attractions than those that are liquids; substances that are liquids have weaker interparticle attractions than those that are solids.

We can change a substance from one state to another by heating or cooling, which changes the average kinetic energy of the particles. NaCl, for example, which is a solid at room temperature, melts at 801 °C and boils at 1413 °C under 1 bar pressure. N_2O, however, which is a gas at room tempera-

TABLE 11.1 • Some characteristic properties of the states of matter	
Gas	Assumes both the volume and shape of its container
	Is compressible
	Flows readily
	Diffusion within a gas occurs rapidly
Liquid	Assumes the shape of the portion of the container it occupies
	Does not expand to fill container
	Is virtually incompressible
	Flows readily
	Diffusion within a liquid occurs slowly
Solid	Retains its own shape and volume
	Is virtually incompressible
	Does not normally flow
	Diffusion within a solid occurs extremely slowly

FIGURE IT OUT

For a given substance, do you expect the density of the substance in its liquid state to be closer to the density in the gaseous state or in the solid state?

Strength of intermolecular attractions increasing

Gas

Liquid

Crystalline solid

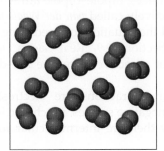

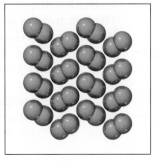

Chlorine, Cl$_2$
Particles far apart; possess complete freedom of motion

Bromine, Br$_2$
Particles are closely packed but randomly oriented; retain freedom of motion; rapidly change neighbours

Iodine, I$_2$
Particles are closely packed in an ordered array; positions are essentially fixed

▲ **FIGURE 11.1   Gases, liquids and solids.** Chlorine, bromine and iodine are all made up of diatomic molecules as a result of covalent bonding. However, due to differences in the strength of the intermolecular forces, they exist in three different states at room temperature and standard pressure: Cl$_2$ gaseous, Br$_2$ liquid, I$_2$ solid.

ture, liquefies at −88.5 °C and solidifies at −90.8 °C at 1 bar pressure. As the temperature of a gas decreases, the average kinetic energy of its particles decreases, allowing the attractions between the particles first to draw the particles close together, forming a liquid, and then virtually locking them in place, forming a solid.

We can also change the state of a substance by changing the pressure. Increasing the pressure on a gas forces the molecules closer together, which in turn increases the strength of the intermolecular forces of attraction. Propane (C$_3$H$_8$) is a gas at room temperature and 1 bar pressure, whereas liquefied propane gas (LPG) is a liquid at room temperature because it is stored under much higher pressure.

CONCEPT CHECK 1

How does the energy of attraction between particles compare with their kinetic energies in
a. a gas?
b. a solid?

11.2 | INTERMOLECULAR FORCES

The strengths of intermolecular forces in different substances vary over a wide range but are generally much weaker than intramolecular forces—ionic, metallic or covalent bonds (▼ FIGURE 11.2). Less energy, therefore, is required to vaporise a liquid or melt a solid than to break covalent bonds. For example, only 16 kJ mol^{-1} is required to overcome the intermolecular attractions in liquid HCl in order to vaporise it. In contrast, the energy required to break the covalent bond in HCl is 431 kJ mol^{-1}. Thus when a molecular substance such as HCl changes from solid to liquid to gas, the molecules remain intact.

Many properties of liquids, including *boiling points*, reflect the strength of the intermolecular forces. A liquid boils when bubbles of its vapour form within the liquid. The molecules of the liquid must overcome their attractive forces in order to separate and form a vapour. The stronger the attractive forces, the higher the temperature at which the liquid boils. Similarly, the *melting points* of solids increase as the strengths of the intermolecular forces increase. As shown in ▼ TABLE 11.2, the melting and boiling points of substances in which the particles are held together by chemical bonds tend to be much higher than those of substances in which the particles are held together by intermolecular forces.

Three types of intermolecular attractions exist between electrically neutral molecules: dispersion forces, dipole–dipole attractions and hydrogen bonding. The first two are collectively called *van der Waals forces* after Johannes van der Waals, who developed the equation for predicting the deviation of gases from ideal behaviour (∞ Section 10.9, 'Real Gases: Deviations from Ideal Behaviour'). Another kind of attractive force, the ion–dipole force, is important in solutions.

All intermolecular interactions are electrostatic, involving attractions between positive and negative species, much like ionic bonds (∞ Section 8.2, 'Ionic Bonding'). Why then are intermolecular forces so much weaker than

∞ Review this on page 364

∞ Review this on page 254

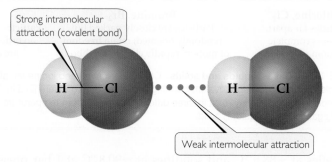

Strong intramolecular attraction (covalent bond)

Weak intermolecular attraction

▲ FIGURE 11.2 **Intermolecular and intramolecular forces.**

TABLE 11.2 • Melting and boiling points of representative substances			
Force holding particles together	**Substance**	**Melting point (K)**	**Boiling point (K)**
Chemical bonds			
Covalent bonds	Diamond (C)	3800	4300
Metallic bonds	Beryllium (Be)	1560	2742
Ionic bonds	Lithium fluoride (LiF)	1118	1949
Intermolecular forces			
Dispersion force	Nitrogen (N_2)	63	77
Dipole–dipole force	Hydrogen chloride (HCl)	158	188
Hydrogen bonding force	Hydrogen fluoride (HF)	190	293

ionic bonds? Recall from Equation 8.4 that electrostatic interactions get stronger as the magnitude of the charges increases and weaker as the distance between charges increases. The charges responsible for intermolecular forces are generally much smaller than the charges in ionic compounds. For example, from its dipole moment it is possible to estimate charges of $+0.178$ and -0.178 for the hydrogen and chlorine ends of the HCl molecule (see Sample Exercise 8.5). Furthermore, the distances between molecules are often larger than the distances between atoms held together by chemical bonds.

Dispersion Forces

You might think there would be no electrostatic interactions between electrically neutral, non-polar atoms and/or molecules. Yet some kind of attractive interactions must exist because non-polar gases like helium, argon and nitrogen can be liquefied. Fritz London (1900–1954), a German-American physicist, first proposed the origin of this attraction in 1930. London recognised that the motion of electrons in an atom or molecule can create an *instantaneous*, or momentary, dipole moment.

In a collection of helium atoms, for example, the *average* distribution of the electrons about each nucleus is spherically symmetrical, as shown in ▼ FIGURE 11.3(a). The atoms are non-polar and so possess no permanent dipole moment. The *instantaneous* distribution of the electrons, however, can be different from the average distribution. If we could freeze the motion of the electrons at any given instant, both electrons could be on one side of the nucleus. At just that instant, the atom has an instantaneous dipole moment as shown in Figure 11.3(b). The motions of electrons in one atom influence the motions of electrons in its neighbours. The instantaneous dipole on one atom can induce an instantaneous dipole on an adjacent atom, causing the atoms to be attracted to each other as shown in Figure 11.3(c). This attractive interaction is called the **dispersion force** (or the *London dispersion force* in some texts). It is significant only when molecules are very close together.

The strength of the dispersion force depends on the ease with which the charge distribution in a molecule can be distorted to induce an instantaneous dipole. The ease with which the charge distribution is distorted is called the molecule's **polarisability**. We can think of the polarisability of a molecule as a

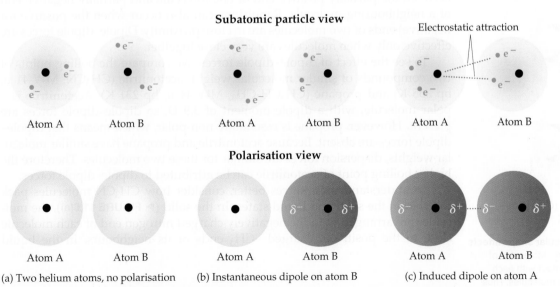

(a) Two helium atoms, no polarisation   (b) Instantaneous dipole on atom B   (c) Induced dipole on atom A

▲ **FIGURE 11.3** **Dispersion forces.** 'Snapshots' of the charge distribution for a pair of helium atoms at three instants.

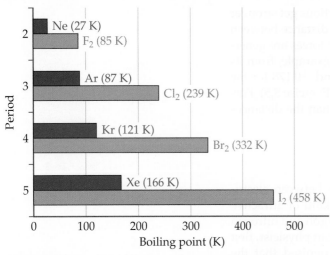

▲ **FIGURE 11.4   Boiling points of the halogens and noble gases.** This plot shows how the boiling points increase as the molecular weight increases due to stronger dispersion forces.

measure of the 'squashiness' of its electron cloud: the greater the polarisability, the more easily the electron cloud can be distorted to give an instantaneous dipole. Therefore, more polarisable molecules have larger dispersion forces.

In general, polarisability increases as the number of electrons in an atom or molecule increases. The strength of dispersion forces therefore tends to increase with increasing atomic or molecular size. Because molecular size and mass generally parallel each other, *dispersion forces tend to increase in strength with increasing molecular weight*. We can see this in the boiling points of the halogens and noble gases (◄ **FIGURE 11.4**), where dispersion forces are the only intermolecular forces at work. In both families the molecular weight increases on moving down the periodic table. The higher molecular weights translate into stronger dispersion forces, which in turn lead to higher boiling points.

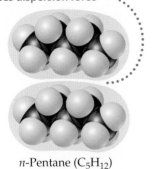

Linear molecule, larger surface area enhances intermolecular contact and increases dispersion force ·····

n-Pentane (C_5H_{12})
bp = 309.4 K

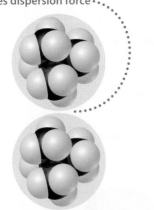

Spherical molecule, smaller surface area diminishes intermolecular contact and decreases dispersion force ·····

Neopentane (C_5H_{12})
bp = 282.7 K

▲ **FIGURE 11.5   Molecular shape affects intermolecular attraction.** Molecules of *n*-pentane make more contact with each other than do neopentane molecules. Thus *n*-pentane has stronger intermolecular attractive forces and a higher boiling point.

CONCEPT CHECK 2

List the substances CCl_4, CBr_4 and CH_4 in order of increasing boiling point.

Molecular shape also influences the magnitudes of dispersion forces. For example, *n*-pentane* and neopentane (◄ **FIGURE 11.5**) have the same molecular formula (C_5H_{12}), yet the boiling point of *n*-pentane is about 27 K higher than that of neopentane. The difference can be traced to the different shapes of the two molecules. Intermolecular attraction is greater for *n*-pentane because the molecules can come in contact over the entire length of the long, somewhat cylindrical molecules. Less contact is possible between the more compact and nearly spherical neopentane molecules.

Dipole–Dipole Forces

The presence of a permanent dipole moment in polar molecules gives rise to **dipole–dipole forces**. These forces originate from electrostatic attractions between the partially positive end of one molecule and partially negative end of a neighbouring molecule. Repulsions can also occur when the positive (or negative) ends of two molecules are in close proximity. Dipole–dipole forces are effective only when molecules are very close together.

To see the effect of dipole–dipole forces, we compare the boiling points of two compounds of similar molecular weight: acetonitrile (CH_3CN, MW 41 u, bp 355 K) and propane ($CH_3CH_2CH_3$, MW 44 u, bp 231 K). Acetonitrile is a polar molecule, with a dipole moment of 3.9 D, so dipole–dipole forces are present. However, propane is essentially non-polar, which means that dipole–dipole forces are absent. Because acetonitrile and propane have similar molecular weights, dispersion forces are similar for these two molecules. Therefore the higher boiling point of acetonitrile can be attributed to dipole–dipole forces.

To understand these forces better, consider how CH_3CN molecules pack together in the solid and liquid states. In the solid (▶ **FIGURE 11.6(a)**), the molecules are arranged with the negatively charged nitrogen end of each molecule close to the positively charged –CH_3 ends of its neighbours. In the liquid

---
* The *n* in *n*-pentane is an abbreviation for the word *normal*. A normal hydrocarbon is one in which the carbon atoms are arranged in a straight chain (Section 2.9).

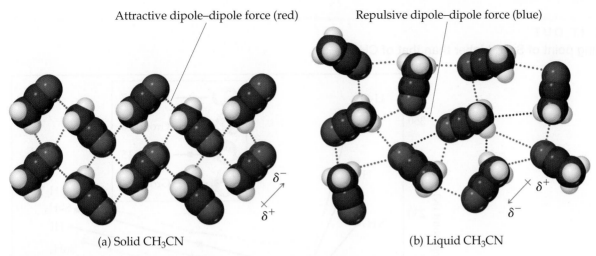

Attractive dipole–dipole force (red)   Repulsive dipole–dipole force (blue)

(a) Solid CH_3CN (b) Liquid CH_3CN

▲ **FIGURE 11.6   Dipole–dipole interactions.** The dipole–dipole interactions in (a) crystalline CH_3CN and (b) liquid CH_3CN.

(Figure 11.6(b)), the molecules are free to move with respect to one another, and their arrangement becomes more disordered. This means that, at any given instant, both attractive and repulsive dipole–dipole interactions are present. However, not only are there more attractive interactions than repulsive ones, but also molecules that are attracting each other spend more time near each other than do molecules that are repelling each other. The overall effect is a net attraction strong enough to keep the molecules in liquid CH_3CN from moving apart to form a gas.

For molecules of approximately equal mass and size, the strength of intermolecular attractions increases with increasing polarity, a trend we see in ▼ **FIGURE 11.7**. Notice how the boiling point increases as the dipole moment increases.

Hydrogen Bonding

▼ **FIGURE 11.8** shows the boiling points of the binary compounds that form between hydrogen and the elements in groups 14 to 17. The boiling points of the compounds containing group 14 elements (CH_4 to SnH_4, all non-polar) increase systematically moving down the group. This is the expected trend because polarisability and, hence, dispersion forces generally increase as molecular weight increases. The three heavier members of groups 15, 16 and 17

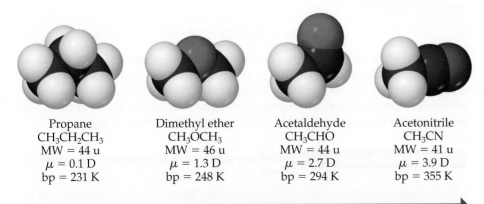

Propane	Dimethyl ether	Acetaldehyde	Acetonitrile
$CH_3CH_2CH_3$	CH_3OCH_3	CH_3CHO	CH_3CN
MW = 44 u	MW = 46 u	MW = 44 u	MW = 41 u
$\mu = 0.1$ D	$\mu = 1.3$ D	$\mu = 2.7$ D	$\mu = 3.9$ D
bp = 231 K	bp = 248 K	bp = 294 K	bp = 355 K

Increasing polarity
Increasing strength of dipole–dipole forces

◀ **FIGURE 11.7   Molecular weights, dipole moments and boiling points of several simple organic substances.**

Why is the boiling point of SnH₄ higher than that of CH₄?

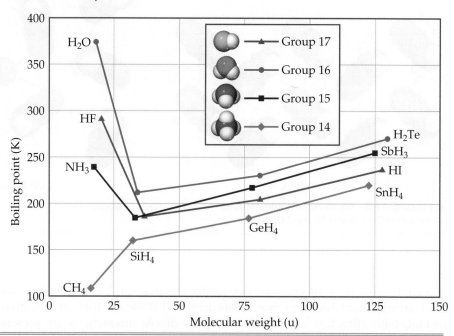

▶ **FIGURE 11.8** Boiling points of the covalent hydrides of the elements in groups 14–17 as a function of molecular weight.

⚠ **FIGURE IT OUT**

To form a hydrogen bond what must the non-hydrogen atom (N, O or F) involved in the bond possess?

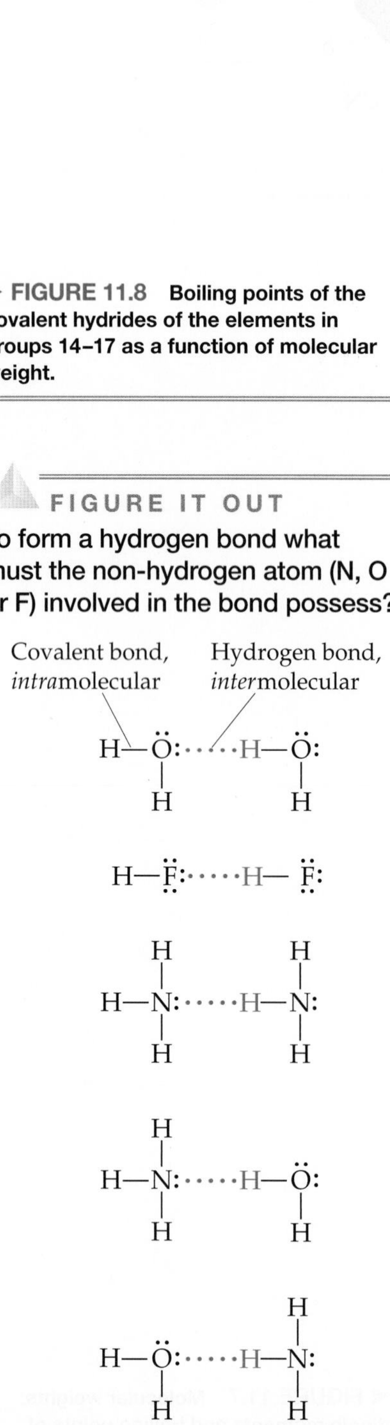

Covalent bond, *intra*molecular	Hydrogen bond, *inter*molecular

H—Ö:⋯⋯H—Ö:
    | |
    H   H

H—F̈:⋯⋯H— F̈:

    H   H
    | |
H—N:⋯⋯H—N:
    | |
    H   H

    H
    |
H—N:⋯⋯H—Ö:
    | |
    H   H

              H
              |
H—Ö:⋯⋯H—N:
    | |
    H   H

▲ **FIGURE 11.9** Hydrogen bonding.

follow the same trend, but NH_3, H_2O and HF have boiling points that are much higher than expected. In fact, these three compounds also have many other characteristics that distinguish them from other substances of similar molecular weight and polarity. For example, water has a high melting point, a high specific heat and a high heat of vaporisation. Each of these properties indicates that the intermolecular forces are abnormally strong.

The strong intermolecular attractions in HF, H_2O and NH_3 result from hydrogen bonding. **Hydrogen bonding** *is a special type of intermolecular attraction between the hydrogen atom in a polar bond (particularly H–F, H–O and H–N) and non-bonding electron pair on a nearby small electronegative ion or atom, usually F, O or N (in another molecule).* For example, a hydrogen bond exists between the H atom in an HF molecule and the F atom of an adjacent HF molecule, as shown in ◀ **FIGURE 11.9** along with several additional examples.

Hydrogen bonds can be considered a type of dipole–dipole attraction. Because N, O and F are so electronegative, a bond between hydrogen and any of these elements is quite polar, with hydrogen at the positive end (remember that the + on the right-hand side of the dipole symbol represents the positive end of the dipole):

$$\overset{\longleftarrow\ +}{N-H} \quad \overset{\longleftarrow\ +}{O-H} \quad \overset{\longleftarrow\ +}{F-H}$$

The hydrogen atom has no inner electrons. Thus the positive side of the dipole has the concentrated charge of the nearly bare hydrogen nucleus. This positive charge is attracted to the negative charge of an electronegative atom in a nearby molecule. Because the electron-poor hydrogen is so small, it can approach an electronegative atom very closely and thus interact strongly with it.

SAMPLE EXERCISE 11.1	**Identifying substances that can form hydrogen bonds**

In which of these substances is hydrogen bonding likely to play an important role in determining physical properties: methane (CH_4), hydrazine (H_2NNH_2), fluoromethane (CH_3F), hydrogen sulfide (H_2S)?

SOLUTION

Analyse We are given the chemical formulae of four compounds and asked to predict whether they can participate in hydrogen bonding. All the compounds contain H, but hydrogen bonding usually occurs only when the hydrogen is covalently bonded to N, O or F.

Plan We analyse each formula to see if it contains N, O or F directly bonded to H. There also needs to be a non-bonding pair of electrons on an electronegative atom (usually N, O or F) in a nearby molecule, which can be revealed by drawing the Lewis structure for the molecule.

Solve The foregoing criteria eliminate CH_4 and H_2S, which do not contain H bonded to N, O or F. They also eliminate CH_3F, whose Lewis structure shows a central C atom surrounded by three H atoms and an F atom. (Carbon always forms four bonds, whereas hydrogen and fluorine form one each.) Because the molecule contains a C–F bond and not an H–F bond, it does not form hydrogen bonds. In H_2NNH_2, however, we find N–H bonds, and the Lewis structure shows a nonbonding pair of electrons on each N atom, telling us hydrogen bonds can exist between the molecules:

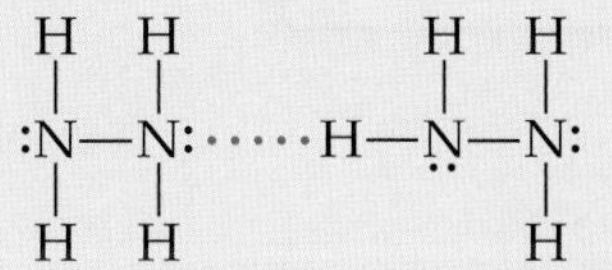

$$
\begin{array}{ccccc}
\text{H} & \text{H} & & \text{H} & \text{H} \\
| & | & & | & | \\
:\text{N}\!-\!\text{N}: & \cdots\cdots & \text{H}\!-\!\text{N}\!-\!\text{N}: \\
| & | & & \overset{..}{} & | \\
\text{H} & \text{H} & & \text{H} &
\end{array}
$$

Check Although we can generally identify substances that participate in hydrogen bonding based on their containing N, O or F covalently bonded to H, drawing the Lewis structure for the interaction provides a way to check the prediction.

PRACTICE EXERCISE

In which of these substances is significant hydrogen bonding possible: dichloromethane (CH_2Cl_2), phosphine (PH_3), hydrogen peroxide (HOOH), acetone (CH_3COCH_3)?

Answer: HOOH

(See also Exercise 11.17.)

The energies of hydrogen bonds vary from about 5 kJ mol^{-1} to 25 kJ mol^{-1}, although there are isolated examples of hydrogen bond energies close to 100 kJ mol^{-1}. Thus hydrogen bonds are typically much weaker than covalent bonds, which have bond enthalpies of 150–1100 kJ mol^{-1} (see Table 8.4). Nevertheless, because hydrogen bonds are generally stronger than dipole–dipole or dispersion forces, they play important roles in many chemical systems, including those of biological significance. For example, hydrogen bonds help stabilise the structures of proteins and are also responsible for the way that DNA is able to carry genetic information.

One remarkable consequence of hydrogen bonding is seen in the densities of ice and liquid water. In most substances the molecules in the solid are more densely packed than in the liquid, making the solid phase denser than the liquid phase. By contrast, the density of ice at 0 °C (0.917 g cm^{-3}) is less than that of liquid water at 0 °C (1.00 g cm^{-3}) so ice floats on liquid water.

The lower density of ice can be understood in terms of hydrogen bonding. In ice, the H_2O molecules assume the ordered, open arrangement shown in ▼ FIGURE 11.10. This arrangement optimises hydrogen bonding between molecules, with each H_2O molecule forming hydrogen bonds to four neighbouring H_2O molecules. These hydrogen bonds, however, create the cavities seen in the middle image of Figure 11.10. When ice melts, the motions of the molecules cause the structure to collapse. The hydrogen bonding in the liquid is more random than in the solid but is strong enough to hold the molecules close together. Consequently, liquid water has a denser structure than ice, meaning that a given mass of water occupies a smaller volume than the same mass of ice.

What is the approximate H—O····H bond angle in ice, where H—O is the covalent bond and O····H is the hydrogen bond?

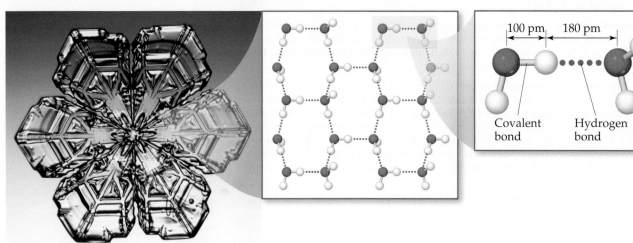

▲ **FIGURE 11.10** **Hydrogen bonding in ice.** The empty channels in the structure of ice make water less dense as a solid than as a liquid.

▲ **FIGURE 11.11** **Expansion of water upon freezing.** Liquid on the left, solid on the right.

The expansion of water upon freezing (◀ **FIGURE 11.11**) is responsible for many phenomena we take for granted. It causes icebergs to float and water pipes to burst in cold weather. The lower density of ice compared with liquid water also profoundly affects life on Earth. Because ice floats, it covers the top of the water when a lake freezes, thereby insulating the water. If ice were denser than water, ice forming at the top of a lake would sink to the bottom, and the lake could freeze solid. Most aquatic life could not survive under these conditions.

Ion–Dipole Forces

An **ion–dipole force** exists between an ion and a polar molecule (▼ **FIGURE 11.12**). Cations are attracted to the negative end of a dipole, and anions are attracted to the positive end. The magnitude of the attraction increases as either the ionic charge or the magnitude of the dipole moment increases. Ion–dipole forces are especially important for solutions of ionic

▶ **FIGURE 11.12** **Ion–dipole forces.**

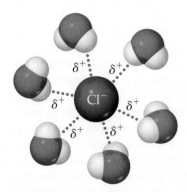

Positive ends of polar molecules are oriented towards negatively charged anion

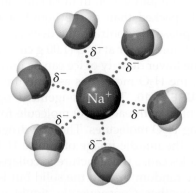

Negative ends of polar molecules are oriented towards positively charged cation

substances in polar liquids, such as a solution of NaCl in water (Section 4.1, 'General Properties of Aqueous Solutions').

Review this on page 105

CONCEPT CHECK 3

In which mixture do you expect to find ion–dipole forces: CH_3OH in water or $Ca(NO_3)_2$ in water?

Comparing Intermolecular Forces

We can identify the intermolecular forces operative in a substance by considering its composition and structure. *Dispersion forces are found in all substances.* The strength of these attractive forces increases with increasing molecular weight and depends on molecular shapes. With polar molecules dipole–dipole forces are also operative, but these forces often make a smaller contribution to the total intermolecular attraction than dispersion forces. For example, in liquid HCl dispersion forces are estimated to account for more than 80% of the total attraction between molecules, whereas dipole–dipole attractions account for the rest. Hydrogen bonds, when present, make an important contribution to the total intermolecular interaction. In general, the energies associated with dispersion and dipole–dipole forces are 2–10 kJ mol^{-1}, whereas the energies of hydrogen bonds are 5–25 kJ mol^{-1}. Ion–dipole attractions have energies of approximately 15 kJ mol^{-1}. All these interactions are considerably weaker than covalent and ionic bonds, which have energies that are hundreds of kilojoules per mole.

When comparing the relative strengths of intermolecular attractions, consider the following generalisations.

1. When the molecules of two substances have comparable molecular weights and shapes, dispersion forces are approximately equal in the two substances. Differences in the magnitudes of the intermolecular forces are due to differences in the strengths of dipole–dipole attractions. The intermolecular forces get stronger as molecule polarity increases, with those molecules capable of hydrogen bonding having the strongest interactions.

2. When the molecules of two substances differ widely in molecular weights, dispersion forces tend to determine which substance has the stronger intermolecular attractions. Intermolecular attractive forces are generally higher in the substance with higher molecular weight.

▶ **FIGURE 11.13** presents a systematic way of identifying the intermolecular forces in a particular system.

It is important to realise that the effects of all these attractions are additive. For example, acetic acid, CH_3COOH, and 1-propanol, $CH_3CH_2CH_2OH$, have the same molecular weight, 60 u, and both are capable of forming hydrogen bonds. However, a pair of acetic acid

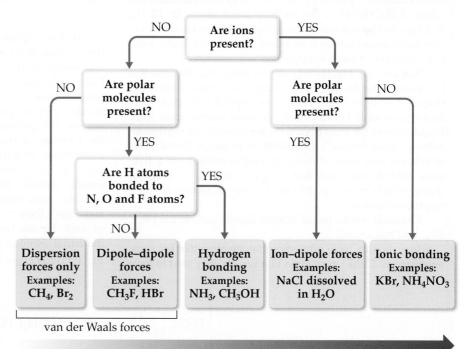

▲ **FIGURE 11.13 Flowchart for determining intermolecular forces.** Multiple types of intermolecular forces can be at work in a given substance or mixture. In particular, dispersion forces occur in all substances.

Each molecule can form two
hydrogen bonds with a neighbour

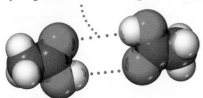

Acetic acid, CH_3COOH
MW = 60 u
bp = 391 K

▲ **FIGURE 11.14** **Hydrogen bonding in
acetic acid and 1-propanol.** The greater
the number of hydrogen bonds possible, the
more tightly the molecules are held together
and therefore the higher the boiling point.

Each molecule can form one
hydrogen bond with a neighbour

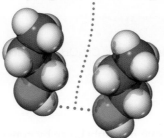

1-Propanol, $CH_3CH_2CH_2OH$
MW = 60 u
bp = 370 K

molecules can form two hydrogen
bonds, whereas a pair of 1-propanol
molecules can form only one
(◄ **FIGURE 11.14**). Hence, the boiling
point of acetic acid is higher than
1-propanol. These effects can be important, especially for very large polar
molecules such as proteins, which have
multiple dipoles over their surfaces.
These molecules can be held together in
solution to a surprisingly high degree
due to the presence of multiple dipole–
dipole attractions.

MY WORLD OF CHEMISTRY

IONIC LIQUIDS

The strong electrostatic attractions between
cations and anions are responsible for the fact
that most ionic compounds are solids at
room temperature, with high melting and
boiling points. However, the melting point
of an ionic compound can be low if the ionic charges are not
too high and the cation–anion distance is sufficiently large.
For example, the melting point of NH_4NO_3, where both cation
and anion are larger polyatomic ions, is 170 °C. If the ammonium cation is replaced by the even larger ethylammonium
cation, $CH_3CH_2NH_3^+$, the melting point drops to 12 °C,
making ethylammonium nitrate a liquid at room temperature!
Ethylammonium nitrate is an example of an *ionic liquid*: a salt
that is a liquid at room temperature.

Not only is $CH_3CH_2NH_3^+$ larger than NH_4^+ but also it is
less symmetric. In general, the larger and more irregularly
shaped the ions in an ionic substance, the better the chances of
forming an ionic liquid. Although many cations form ionic
liquids, one of the most popular is the 1-butyl-3-methylimida-
zolium cation (bmim$^+$), (▼ **FIGURE 11.15** and ► **TABLE 11.3**),
which has two arms of different lengths coming off a five-
atom central ring. This feature gives bmim$^+$ an irregular
shape, which makes it difficult for the molecules to pack
together in a solid.

Common anions found in ionic liquids include the PF_6^+,
BF_4^- and halide ions.

Ionic liquids have properties that are attractive for some
applications. Unlike most molecular liquids, they tend to have

TABLE 11.3 • **Melting point and decomposition
temperature of four 1-butyl-3-methylimidazolium
(bmim$^+$) salts**

Cation	Anion	Melting point (°C)	Decomposition temperature (°C)
bmim$^+$	Cl^-	41	254
bmim$^+$	I^-	−72	265
bmim$^+$	PF_6^-	10	349
bmim$^+$	BF_4^-	−81	403

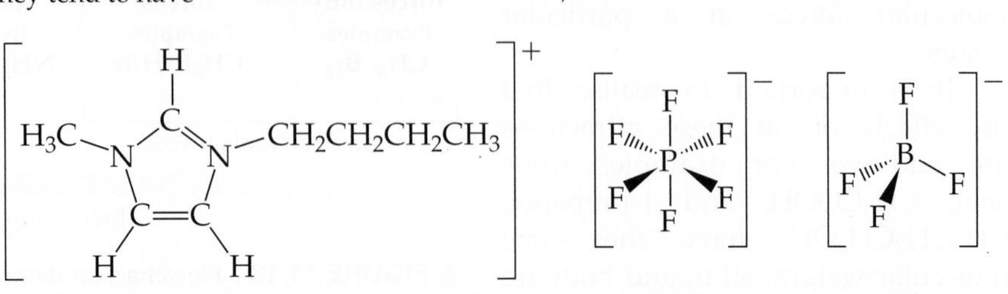

a very low vapour pressure. Because they are non-volatile
(that is, they don't evaporate), they tend to be non-flammable
and remain in the liquid state at temperatures up to 673 K.
Most molecular substances are liquids only at much lower
temperatures, for example 373 K or less in many cases.
Because they are good solvents for a wide range of inorganic,
organic and polymeric substances, ionic liquids can be used
for a variety of reactions and separations. These properties
make them attractive replacements for volatile organic solvents in many industrial processes. Relative to traditional
organic solvents, ionic liquids offer the promise of reduced
volumes, safer handling and easier reuse. For these reasons
and others, there is considerable excitement about the promise
of ionic liquids for reducing the environmental impact of
industrial chemical processes.

RELATED EXERCISES: 11.21, 11.22

► **FIGURE 11.15** **Representative ions
found in ionic liquids.**

$$\left[\begin{array}{c} H \\ | \\ C \\ H_3C-N \diagdown \diagup N-CH_2CH_2CH_2CH_3 \\ C=C \\ | \quad | \\ H \quad H \end{array} \right]^+$$

1-Butyl-3-methylimidazolium (bmim$^+$)
cation

$$\left[\begin{array}{c} F \\ F_{\diagup} P_{\diagdown} F \\ F \diagup | \diagdown F \\ F \end{array} \right]^-$$

PF_6^-
anion

$$\left[\begin{array}{c} F \\ B \\ F \diagup | \diagdown F \\ F \end{array} \right]^-$$

BF_4^-
anion

SAMPLE EXERCISE 11.2   **Predicting types and relative strengths of intermolecular attractions**

List the substances $BaCl_2$, H_2, CO, HF and Ne in order of increasing boiling point.

SOLUTION

Analyse We need to assess the intermolecular forces in these substances and use that information to determine the relative boiling points.

Plan The boiling point depends in part on the attractive forces in each substance. We need to order these according to the relative strengths of the different kinds of intermolecular attractions.

Solve The attractive forces are stronger for ionic substances than for molecular ones, so $BaCl_2$ should have the highest boiling point. The intermolecular forces of the remaining substances depend on molecular weight, polarity and hydrogen bonding. The molecular weights are H_2 (2), CO (28), HF (20) and Ne (20). The boiling point of H_2 should be the lowest because it is non-polar and has the lowest molecular weight. The molecular weights of CO, HF and Ne are similar. Because HF can hydrogen bond, however, it should have the highest boiling point of the three. Next is CO, which is slightly polar and has the highest molecular weight. Finally, Ne, which is non-polar, should have the lowest boiling point of these three. The predicted order of boiling points is, therefore,

$$H_2 < Ne < CO < HF < BaCl_2$$

Check The boiling points reported in the literature are H_2 (20 K), Ne (27 K), CO (83 K), HF (293 K) and $BaCl_2$ (1813 K)—in agreement with our predictions.

PRACTICE EXERCISE

(a) Identify the intermolecular attractions present in the following substances, and **(b)** select the substance with the highest boiling point: CH_3CH_3, CH_3OH and CH_3CH_2OH.

Answers: **(a)** CH_3CH_3 has only dispersion forces, whereas the other two substances have both dispersion forces and hydrogen bonds, **(b)** CH_3CH_2OH

(See also Exercise 11.18.)

11.3 | SOME PROPERTIES OF LIQUIDS

The intermolecular forces we have just discussed can help us understand many familiar properties of liquids and solids. In this section we examine two important properties of liquids: viscosity and surface tension.

Viscosity

Some liquids, such as molasses and motor oil, flow very slowly; others, such as water and petrol, flow easily. The resistance of a liquid to flow is called its **viscosity**. The greater a liquid's viscosity, the more slowly it flows. Viscosity can be measured by timing how long it takes a certain amount of the liquid to flow through a thin tube under gravitational force. More viscous liquids take longer (▶ FIGURE 11.16). Viscosity can also be determined by measuring the rate at which steel spheres fall through the liquid. The spheres fall more slowly as the viscosity increases.

Viscosity is related to the ease with which individual molecules of the liquid can move with respect to one another. It thus depends on the attractive forces between molecules and on whether structural features exist that cause the molecules to become entangled. For a series of related compounds, therefore, viscosity increases with molecular mass, as illustrated in ▼ TABLE 11.4. (The SI units for viscosity are $kg\ m^{-1}\ s^{-1}$.) For any given substance, viscosity decreases with increasing temperature. Octane, for example, has a viscosity of $7.06 \times 10^{-4}\ kg\ m^{-1}\ s^{-1}$ at 0 °C, and of $4.33 \times 10^{-4}\ kg\ m^{-1}\ s^{-1}$ at 40 °C. At higher

▲ FIGURE 11.16 **Comparing viscosities.** The Society of Automotive Engineers (*SAE International*) has established numbers to indicate the viscosity of motor oils. The higher the number, the greater the viscosity is at any given temperature. The SAE 40 motor oil on the left is more viscous and flows more slowly than the less viscous SAE 10 oil on the right.

TABLE 11.4 • Viscosities of a series of hydrocarbons at 20 °C		
Substance	**Formula**	**Viscosity ($kg\,m^{-1}\,s^{-1}$)**
Hexane	$CH_3CH_2CH_2CH_2CH_2CH_3$	3.26×10^{-4}
Heptane	$CH_3CH_2CH_2CH_2CH_2CH_2CH_3$	4.09×10^{-4}
Octane	$CH_3CH_2CH_2CH_2CH_2CH_2CH_2CH_3$	5.42×10^{-4}
Nonane	$CH_3CH_2CH_2CH_2CH_2CH_2CH_2CH_2CH_3$	7.11×10^{-4}
Decane	$CH_3CH_2CH_2CH_2CH_2CH_2CH_2CH_2CH_2CH_3$	1.42×10^{-3}

temperatures the greater average kinetic energy of the molecules more easily overcomes the attractive forces between molecules.

Surface Tension

The surface of water behaves almost as if it had an elastic skin, as evidenced by the ability of certain insects to 'walk' on water. This behaviour is due to an imbalance of intermolecular forces at the surface of the liquid, as shown in ◀ **FIGURE 11.17**. Notice that molecules in the interior are attracted equally in all directions, whereas those at the surface experience a net inward force. The resultant inward force pulls molecules from the surface into the interior, thereby reducing the surface area and making the molecules at the surface pack closely together. Because spheres have the smallest surface area for their volume, water droplets assume an almost spherical shape. Similarly, water tends to 'bead up' on a newly waxed car because there is little or no attraction between the polar water molecules and the non-polar wax molecules.

A measure of the inward forces that must be overcome in order to expand the surface area of a liquid is given by its surface tension. **Surface tension** is the energy required to increase the surface area of a liquid by a unit amount. For example, the surface tension of water at 20 °C is $7.29 \times 10^{-2}\,J\,m^{-2}$, which means that an energy of 7.29×10^{-2} J must be supplied to increase the surface area of a given amount of water by 1 m^2. Water has a high surface tension because of its strong hydrogen bonds. The surface tension of mercury is even higher (4.6×10^{-1} J m^{-2}) because of even stronger metallic bonds between the atoms of mercury.

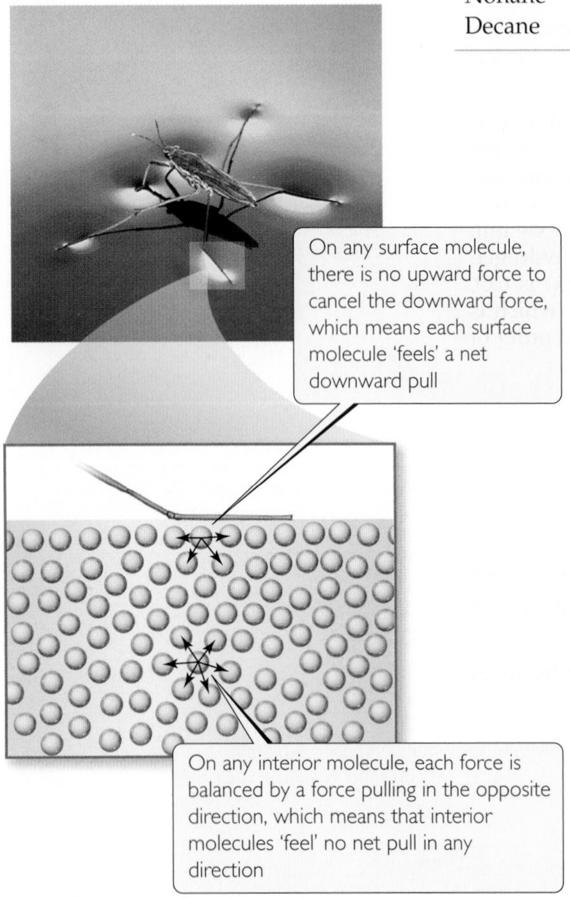

On any surface molecule, there is no upward force to cancel the downward force, which means each surface molecule 'feels' a net downward pull

On any interior molecule, each force is balanced by a force pulling in the opposite direction, which means that interior molecules 'feel' no net pull in any direction

▲ **FIGURE 11.17 Molecular-level view of surface tension.** A water strider does not sink because of the high surface tension of water.

⚠ **CONCEPT CHECK 4**

How do viscosity and surface tension change
a. as temperature increases?
b. as intermolecular forces of attraction become stronger?

Intermolecular forces that bind similar molecules to one another, such as the hydrogen bonding in water, are called *cohesive forces*. Intermolecular forces that bind a substance to a surface are called *adhesive forces*. Water placed in a glass tube adheres to the glass because the adhesive forces between the water and glass are even greater than the cohesive forces between water molecules. If the tube is narrow enough, water will rise in the tube until the force of gravity stops it. This is called **capillary action**. Any liquid in which the adhesive forces are greater than the cohesive forces will exhibit this phenomenon. The curved upper surface, or *meniscus*, of the water is therefore U-shaped (▶ **FIGURE 11.18**). For mercury, however, the meniscus is curved downward where the mercury contacts the glass. In this case the cohesive forces between the mercury atoms are much greater than the adhesive forces between the mercury atoms and the glass.

△ **FIGURE IT OUT**

If the inside surface of each tube were coated with wax, would the general shape of the water meniscus change? Would the general shape of the mercury meniscus change?

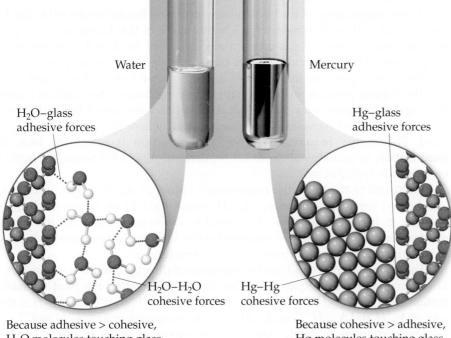

Water   Mercury

H₂O–glass
adhesive forces

Hg–glass
adhesive forces

H₂O–H₂O
cohesive forces

Hg–Hg
cohesive forces

Because adhesive > cohesive, H₂O molecules touching glass adhere to the wall more than to each other, forming concave surface

Because cohesive > adhesive, Hg molecules touching glass adhere to the wall less than to each other, forming convex surface

◀ **FIGURE 11.18** **Meniscus shapes for water and mercury in glass tubes.**

△ **CONCEPT CHECK 5**

Do the viscosity and surface tension of a substance reflect adhesive forces or cohesive forces of attraction?

11.4 | PHASE CHANGES

Water left uncovered in a glass for several days evaporates. An ice cube left in a warm room quickly melts. Solid CO_2 (sold as dry ice) *sublimes* at room temperature; that is, it changes directly from the solid to the vapour state. In general, each state of matter can change into either of the other two states. ▶ **FIGURE 11.19** shows the name associated with each of these transformations. These transformations are called either **phase changes** or changes of state.

Energy Changes Accompanying Phase Changes

Every phase change is accompanied by a change in the energy of the system. In a solid lattice, for example, the molecules or ions are in more or less fixed positions with respect to one another and closely arranged to minimise the energy of the system. As the temperature of the solid increases, the units of the solid vibrate about their equilibrium positions with

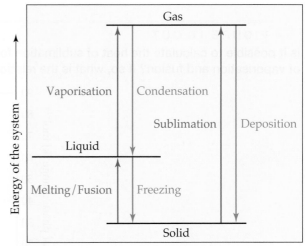

—— Endothermic process (energy added to substance)

—— Exothermic process (energy released from substance)

▲ **FIGURE 11.19** **Phase changes and the names associated with them.**

increasingly energetic motion. When the solid melts, the units that made up the solid are freed to move with respect to one another, which ordinarily means that their average separations increase. This melting process is called *fusion*. The increased freedom of motion of the molecules or ions requires energy, measured by the **heat of fusion**, or enthalpy of fusion, denoted $\Delta_{fus}H$. The heat of fusion of ice, for example, is 6.01 kJ mol^{-1}.

As the temperature of the liquid phase increases, the molecules of the liquid move about with increasing energy. One measure of this increasing energy is that the concentration of gas-phase molecules over the liquid increases with temperature. These molecules exert a pressure called the vapour pressure. We explore vapour pressure in Section 11.5. For now we just need to understand that the vapour pressure increases with increasing temperature until it equals the external pressure over the liquid. At this point the liquid boils; the molecules of the liquid move into the gaseous state, where they are widely separated. The energy required to cause this transition is called the **heat of vaporisation**, or enthalpy of vaporisation, denoted $\Delta_{vap}H$. For water, the heat of vaporisation is 40.7 kJ mol^{-1}.

▼ FIGURE 11.20 shows the comparative values of $\Delta_{fus}H$, $\Delta_{vap}H$ and $\Delta_{sub}H$ (see below) for four substances. $\Delta_{vap}H$ values tend to be larger than $\Delta_{fus}H$ because, in the transition from the liquid to the vapour state, the molecules must essentially sever all their intermolecular attractive interactions; whereas, in melting, many of these attractive interactions remain.

The molecules of a solid can be transformed directly into the gaseous state. The enthalpy change required for this transition is called the **heat of sublimation**, denoted $\Delta_{sub}H$. For the substances shown in Figure 11.20, $\Delta_{sub}H$ is the sum of $\Delta_{fus}H$ and $\Delta_{vap}H$. Thus $\Delta_{sub}H$ for water is approximately 47 kJ mol^{-1}.

Phase changes of matter show up in important ways in our everyday experiences. We use ice cubes to cool our liquid drinks; the heat of fusion of ice cools the liquid in which the ice is immersed. We feel cool when we step out of a swimming pool or a warm shower because the heat of vaporisation is drawn from our bodies as the water evaporates from our skin. Our bodies use the evaporation of water from skin to regulate body temperature, especially when we exercise vigorously in warm weather. A refrigerator also relies on the cooling effects of vaporisation. Its mechanism contains an enclosed gas that can be liquefied under pressure. The liquid absorbs heat as it subsequently evaporates, thereby cooling the interior of the refrigerator. The vapour is then recycled through a compressor.

◢ **FIGURE IT OUT**

Is it possible to calculate the heat of sublimation for a substance given its heats of vaporisation and fusion? If so, what is the relationship?

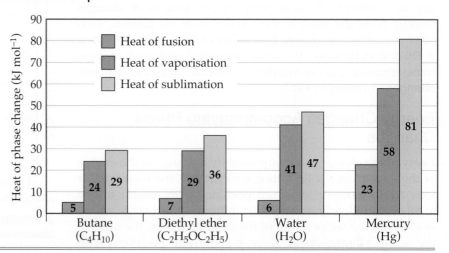

▶ **FIGURE 11.20 Heats of fusion, vaporisation and sublimation.**

What happens to the heat absorbed when the liquid refrigerant vaporises? According to the first law of thermodynamics, the heat absorbed by the liquid in vaporising must be released when the reverse process, condensation of the vapour to form the liquid, occurs. As the refrigerator compresses the vapour and liquid is formed, the heat released is dissipated through cooling coils in the back of the refrigerator. Just as the heat of condensation is equal in magnitude to the heat of vaporisation and has the opposite sign, so also the *heat of deposition* is exothermic to the same degree that the heat of sublimation is endothermic; and the *heat of freezing* is exothermic to the same degree that the heat of fusion is endothermic. These relationships, shown in Figure 11.19, are consequences of the first law of thermodynamics.

CONCEPT CHECK 6

What is the name of the phase change that occurs when ice left at room temperature changes to liquid water? Is this change exothermic or endothermic?

Heating Curves

What happens when we heat a sample of ice that is initially at −25 °C and 1 bar pressure? The addition of heat causes the temperature of the ice to increase. As long as the temperature is below 0 °C, the sample remains frozen. When the temperature reaches 0 °C, the ice begins to melt. Because melting is an endothermic process, the heat we add at 0 °C is used to convert ice to water and the temperature remains constant until all the ice has melted. Once we reach this point, the further addition of heat causes the temperature of the liquid water to increase.

A graph of the temperature of the system versus the amount of heat added is called a *heating curve*. ▼ **FIGURE 11.21** shows a heating curve for transforming ice at −25 °C to steam at 125 °C under a constant pressure of 1 bar. Heating the ice from −25 °C to 0 °C is represented by the line segment *AB* in Figure 11.21, and converting the ice at 0 °C to water at 0 °C is the horizontal segment *BC*. Additional heat increases the temperature of the water until the temperature reaches 100 °C (segment *CD*). The heat is then used to convert water to steam at a constant temperature of 100 °C (segment *DE*). Once all the water has been converted to steam, the steam is heated to its final temperature of 125 °C (segment *EF*).

We can calculate the enthalpy change of the system for each of the segments of the heating curve. In segments *AB*, *CD* and *EF* we are heating a single phase from one temperature to another. As we shall see in Section 14.5, the amount of heat needed to raise the temperature of a substance is given by the product of the specific heat, mass and temperature change (Equation 14.18). The greater the specific heat of a substance, the more heat we must add to accomplish a certain temperature increase. Because the specific heat of water is greater than that of ice, the slope of segment *CD* is less than that of segment *AB*; we must add more heat to water to achieve a 1 °C temperature change than is needed to warm the same quantity of ice by 1 °C.

In segments *BC* and *DE* we are converting one phase to another at a constant temperature. The temperature remains constant during these phase changes because the added energy is used to overcome the attractive forces between molecules rather than to increase their average kinetic energy. For segment *BC*, in which ice is converted to water, the enthalpy change can be calculated by using $\Delta_{fus}H$, and for segment *DE* we can use $\Delta_{vap}H$. In Sample Exercise 11.3 we calculate the total enthalpy change for the heating curve in Figure 11.21.

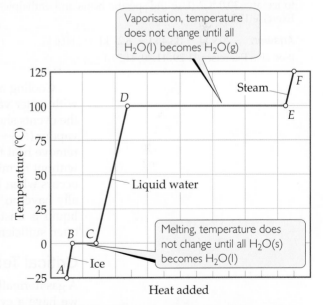

▲ **FIGURE 11.21   Heating curve for water.** Changes that occur when 1.00 mol of H_2O is heated from $H_2O(s)$ at −25 °C to $H_2O(g)$ at 125 °C at a constant pressure of 1 bar. Heat is added over the entire temperature range, but the temperature of the system increases only when the H_2O is either all solid or all liquid or all gas (blue lines). Even though heat is being added continuously, the system temperature does not change during the two phase changes (red lines).

SAMPLE EXERCISE 11.3 Calculating Δ*H* for temperature and phase changes

Calculate the enthalpy change upon converting 1.00 mol of ice at –25 °C to water vapour (steam) at 125 °C under a constant pressure of 1 bar. The specific heats of ice, water and steam are 2.09 J g⁻¹ K⁻¹, 4.18 J g⁻¹ K⁻¹ and 1.84 J g⁻¹ K⁻¹, respectively. For H_2O, $\Delta_{fus}H$ = 6.01 kJ mol⁻¹ and $\Delta_{vap}H$ = 40.67 kJ mol⁻¹.

SOLUTION

Analyse Our goal is to calculate the total heat required to convert 1 mol of ice at −25 °C to steam at 125 °C.

Plan We can calculate the enthalpy change for each segment and then sum them to get the total enthalpy change.

Solve For segment *AB* in Figure 11.21, we are adding enough heat to ice to increase its temperature by 25 °C. A temperature change of 25 °C is the same as a temperature change of 25 K, so we can use the specific heat of ice to calculate the enthalpy change during this process:

AB: ΔH = (1.00 mol)(18.0 g mol⁻¹)(2.09 J g⁻¹ K⁻¹)(25 K) = 940 J = 0.94 kJ

For segment *BC* in Figure 11.21, in which we convert ice to water at 0 °C, we can use the molar enthalpy of fusion directly:

BC: ΔH = (1.00 mol)(6.01 kJ mol⁻¹) = 6.01 kJ

The enthalpy changes for segments *CD*, *DE* and *EF* can be calculated in similar fashion:

CD: ΔH = (1.00 mol)(18.0 g mol⁻¹)(4.18 J g⁻¹ K⁻¹)(100 K) = 7520 J = 7.52 kJ

DE: ΔH = (1.00 mol)(40.67 kJ mol⁻¹) = 40.7 kJ

EF: ΔH = (1.00 mol)(18.0 g mol⁻¹)(1.84 J g⁻¹ K⁻¹)(25 K) = 830 J = 0.83 kJ

The total enthalpy change is the sum of the changes of the individual steps:

ΔH = 0.94 kJ + 6.01 kJ + 7.52 kJ + 40.7 kJ + 0.83 kJ = 56.0 kJ

Check The components of the total energy change are reasonable in comparison with the lengths of the horizontal segments of the lines in Figure 11.21. Notice that the largest component is the heat of vaporisation.

PRACTICE EXERCISE

What is the enthalpy change during the process in which 100.0 g of water at 50.0 °C is cooled to ice at −30.0 °C? (Use the specific heats and enthalpies for phase changes given in Sample Exercise 11.3.)

Answer: −20.9 kJ − 33.4 kJ − 6.27 kJ = −60.6 kJ
(See also Exercises 11.29, 11.30, 11.31.)

Cooling a substance has the opposite effect of heating it. Thus if we start with water vapour and begin to cool it, we would move right to left through the events shown in Figure 11.21. We would first lower the temperature of the vapour (*F* ⟶ *E*), then condense it (*E* ⟶ *D*) and so forth. Sometimes, as we remove heat from a liquid, we can temporarily cool it below its freezing point without forming a solid. This phenomenon is called *supercooling*. Supercooling occurs when heat is removed from a liquid so rapidly that the molecules literally have no time to assume the ordered structure of a solid. A supercooled liquid is unstable; particles of dust entering the solution or gentle stirring is often sufficient to cause the substance to solidify quickly.

Critical Temperature and Pressure

A gas normally liquefies at some point when pressure is applied to it. Suppose we have a cylinder with a piston, containing water vapour at 100 °C. If we increase the pressure on the water vapour, liquid water will form when the pressure is 101.3 kPa. However, if the temperature is 110 °C, the liquid phase does not form until the pressure is 143.2 kPa. At 374 °C the liquid phase forms only at 2.206 × 10⁴ kPa. Above this temperature no amount of pressure will cause a distinct liquid phase to form. Instead, as pressure increases, the gas merely becomes steadily more compressed. The highest temperature at which a distinct liquid phase can form is referred to as the **critical temperature**.

TABLE 11.5 • Critical temperatures and pressures of selected substances

Substance	Critical temperature (K)	Critical pressure (kPa)
Ammonia, NH_3	405.6	1.13×10^4
Phosphine, PH_3	324.4	6.53×10^3
Argon, Ar	150.9	4.86×10^3
Carbon dioxide, CO_2	304.3	7.39×10^3
Nitrogen, N_2	126.1	3.39×10^3
Oxygen, O_2	154.4	5.03×10^3
Propane, $CH_3CH_2CH_3$	370.0	4.25×10^3
Water, H_2O	647.6	2.21×10^4
Hydrogen sulfide, H_2S	373.5	9.01×10^3

The **critical pressure** is the pressure required to bring about liquefaction at this critical temperature.

The critical temperature is the highest temperature at which a liquid can exist. Above the critical temperature, the motional energies of the molecules are greater than the attractive forces that lead to the liquid state, regardless of how much the substance is compressed to bring the molecules closer together. The greater the intermolecular forces, the greater the critical temperature of a substance.

The critical temperatures and pressures are listed for several substances in ▲ TABLE 11.5. Notice that non-polar, low molecular mass substances, which have weak intermolecular attractions, have lower critical temperatures and pressures than those that are polar or of higher molecular mass. Notice also that water and ammonia have exceptionally high critical temperatures and pressures as a consequence of strong intermolecular hydrogen-bonding forces.

The critical temperatures and pressures of substances are often of considerable importance to engineers and other people working with gases, because they provide information about the conditions under which gases liquefy. Sometimes we want to liquefy a gas; other times we want to avoid liquefying it. It is useless to try to liquefy a gas by applying pressure if the gas is above its critical temperature. For example, O_2 has a critical temperature of 154.4 K. It must be cooled below this temperature before it can be liquefied by pressure. In contrast, ammonia has a critical temperature of 405.6 K. Thus it can be liquefied at room temperature (approximately 295 K) by compressing the gas to a sufficient pressure.

When the temperature exceeds the critical temperature and the pressure exceeds the critical pressure, the liquid and gas phases are indistinguishable from each other, and the substance is in a state called a **supercritical fluid**. Like liquids, supercritical fluids can behave as solvents dissolving a wide range of substances. Using *supercritical fluid extraction*, the components of mixtures can be separated from one another. Supercritical fluid extraction has been successfully used to separate complex mixtures in the chemical, food, pharmaceutical and energy industries. Supercritical CO_2 is a popular choice because it is relatively inexpensive and there are no problems associated with disposing of solvent, and there are no toxic residues resulting from the process.

11.5 | VAPOUR PRESSURE

Molecules can escape from the surface of a liquid into the gas phase by evaporation. Suppose we conduct an experiment in which we place a quantity of ethanol (C_2H_5OH) in an evacuated, closed container such as that in ▼ FIGURE 11.22. The ethanol will quickly begin to evaporate. As a result, the pressure exerted by the vapour in the space above the liquid will begin to increase. After a short time the pressure of the vapour will attain a constant value, which we call the **vapour pressure** of the substance.

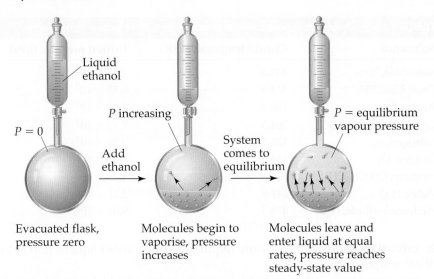

▶ **FIGURE 11.22** **Equilibrium vapour pressure over a liquid.**

Explaining Vapour Pressure at the Molecular Level

The molecules of a liquid move at various speeds. ▼ FIGURE 11.23 shows the distribution of kinetic energies of the particles at the surface of a liquid at two temperatures. The distribution curves are like those shown earlier for gases (Figure 10.16) (∞ Section 10.7, 'Kinetic-Molecular Theory'). At any instant some of the molecules on the surface of the liquid possess sufficient kinetic energy to overcome the attractive forces of their neighbours and escape into the gas phase. The weaker the attractive forces, the larger is the number of molecules that are able to escape and therefore the higher is the vapour pressure.

∞ Review this on page 356

At any particular temperature the movement of molecules from the liquid to the gas phase goes on continuously. As the number of gas-phase molecules increases, however, the probability increases that a molecule in the gas phase will strike the liquid surface and be recaptured by the liquid, as shown in the flask on the right in Figure 11.22. Eventually, the rate at which molecules return to the liquid exactly equals the rate at which they escape. The number of molecules in the gas phase then reaches a steady value, and the pressure of the vapour at this stage becomes constant.

The condition in which two opposing processes are occurring simultaneously at equal rates is called a **dynamic equilibrium**, but is usually referred to merely as an *equilibrium*. A liquid and its vapour are in dynamic equilibrium when evaporation and condensation occur at equal rates. It may appear that nothing is occurring at equilibrium, because there is no net change in the system. In fact, a great deal is happening; molecules continuously pass from the liquid state to the gas state and from the gas state to the liquid state. All equilibria between different states of matter possess this dynamic character. *The vapour pressure of a liquid is the pressure exerted by its vapour when the liquid and vapour states are in dynamic equilibrium.*

▲ FIGURE IT OUT

As the temperature increases, does the rate of molecules escaping into the gas phase increase or decrease?

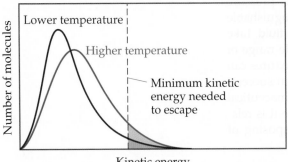

Blue area = number of molecules having enough energy to evaporate at lower temperature

Red + blue areas = number of molecules having enough energy to evaporate at higher temperature

▲ **FIGURE 11.23** **The effect of temperature on the distribution of kinetic energies in a liquid.**

Volatility, Vapour Pressure and Temperature

When vaporisation occurs in an open container, as when water evaporates from a bowl, the vapour spreads away from the liquid. Little, if any, is recaptured at the surface of the liquid. Equilibrium never occurs, and the vapour continues to form until the liquid evaporates to dryness. Substances with high vapour pressure (such as petrol) evaporate more quickly than substances with low vapour pressure (such as motor oil). Liquids that evaporate readily are said to be **volatile**.

Hot water evaporates more quickly than cold water because vapour pressure increases with increasing temperature. We see this effect in Figure 11.23: as the temperature of a liquid is increased, the molecules move more energetically and a greater fraction can therefore escape more readily from their neighbours. ▶ **FIGURE 11.24** depicts the variation in vapour pressure with temperature for four common substances that differ greatly in volatility. Note that the vapour pressure in all cases increases non-linearly with increasing temperature.

CONCEPT CHECK 7

a. Predict the relative vapour pressures of CCl₄ and CBr₄ at 25 °C.
b. Predict the relative volatilities of these two substances at this temperature.

Vapour Pressure and Boiling Point

A liquid boils when its vapour pressure equals the external pressure acting on the surface of the liquid. At this point bubbles of vapour are able to form within the liquid. The temperature at which a given liquid boils increases with increasing external pressure. The boiling point of a liquid at 1 bar pressure (101.3 kPa) is called its **normal boiling point**. From Figure 11.24 we see that the normal boiling point of water is 100 °C.

The boiling point is important to many processes that involve heating liquids, including cooking. The time required to cook food depends on the temperature. As long as water is present, the maximum temperature of the food being cooked is the boiling point of water. Pressure cookers work by allowing steam to escape only when it exceeds a predetermined pressure; the pressure above the water can therefore increase above atmospheric pressure. The higher pressure causes the water to boil at a higher temperature, thereby allowing the food to get hotter and to cook more rapidly. The effect of pressure on boiling point also explains why it takes longer to cook food at higher elevations than at sea level. The atmospheric pressure is lower at higher altitudes, so water boils at a lower temperature.

FIGURE IT OUT

What is the vapour pressure of ethylene glycol at its normal boiling point?

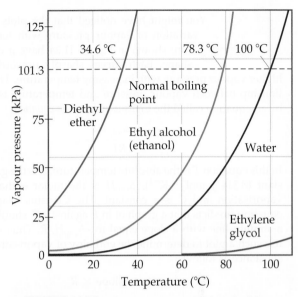

▲ **FIGURE 11.24**   **Vapour pressure for four liquids as a function of temperature.**

SAMPLE EXERCISE 11.4   **Relating boiling point to vapour pressure**

Use Figure 11.24 to estimate the boiling point of diethyl ether under an external pressure of 80 kPa.

SOLUTION

Analyse We are asked to read a graph of vapour pressure versus temperature to determine the boiling point of a substance at a particular pressure. The boiling point is the temperature at which the vapour pressure is equal to the external pressure.

Plan We estimate the location of that pressure on the graph, move horizontally to the vapour pressure curve, and then drop vertically from the curve to estimate the temperature.

Solve From Figure 11.24 we see that the boiling point at this pressure is about 27 °C, which is close to room temperature.

PRACTICE EXERCISE

At what external pressure will ethanol have a boiling point of 60 °C?

Answer: about 45 kPa

(See also Exercise 11.37.)

A CLOSER LOOK

THE CLAUSIUS–CLAPEYRON EQUATION

You might have noticed that the plots of the variation of vapour pressure with temperature shown in Figure 11.24 have a distinct shape. Each curves sharply upwards to a higher vapour pressure with increasing temperature. The relationship between vapour pressure and temperature is given by an equation called the *Clausius–Clapeyron equation*:

$$\ln P = \frac{-\Delta_{vap}H}{RT} + C \qquad [11.1]$$

In this equation T is the absolute temperature, R is the gas constant (8.314 J mol^{-1} K^{-1}), $\Delta_{vap}H$ is the molar enthalpy of vaporisation and C is a constant. The Clausius–Clapeyron equation predicts that a graph of $\ln P$ against $1/T$ should give a straight line with a slope equal to $-\Delta_{vap}H/R$. Thus we can use such a plot to determine the enthalpy of vaporisation of a substance as follows:

$$\Delta_{vap}H = -\text{slope} \times R \qquad [11.2]$$

As an example of the application of the Clausius–Clapeyron equation, the vapour pressure data for ethanol shown in Figure 11.24 are graphed as $\ln P$ against $1/T$ in ▶ **FIGURE 11.25**. The data lie on a straight line with a negative slope. We can use the slope of the line to determine $\Delta_{vap}H$ for ethanol. We can also extrapolate the line to estimate values for the vapour pressure of ethanol at temperatures above and below the temperature range for which we have data.

RELATED EXERCISES: 11.76–11.78

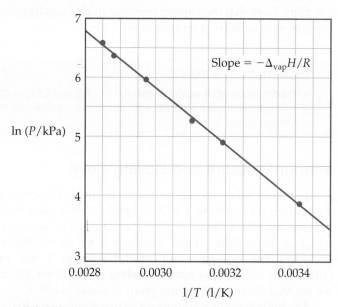

▲ **FIGURE 11.25** **The natural logarithm of vapour pressure versus 1/T for ethanol.**

11.6 │ PHASE DIAGRAMS

The equilibrium between a liquid and its vapour is not the only dynamic equilibrium that can exist between states of matter. Under appropriate conditions of temperature and pressure, a solid can be in equilibrium with its liquid state or even with its vapour state. A **phase diagram** is a graphical way to summarise the conditions under which equilibria exist between the different states of matter. Such a diagram also allows us to predict the phase of a substance that is stable at any given temperature and pressure.

The general form of a phase diagram for a substance is shown in ▶ **FIGURE 11.26**. The diagram is a two-dimensional graph, with pressure and temperature as the axes. It contains three important curves, each of which represents the conditions of pressure and temperature at which the various phases can coexist at equilibrium. The only substance present in the system is the one whose phase diagram is under consideration. The pressure shown in the diagram is either the pressure applied to the system or the pressure generated by the substance itself. The curves in Figure 11.26 may be described as follows.

1. The red curve is the *vapour pressure curve* of the liquid, representing equilibrium between the liquid and gas phases. The point on this curve where the vapour pressure is 1 bar is the normal boiling point of the substance. The vapour pressure curve ends at the *critical point* (cp), which corresponds to the critical temperature and critical pressure of the substance. Beyond the critical point, the liquid and gas phases are indistinguishable from each other, and the substance is a *supercritical fluid*.

2. The green curve, the *sublimation curve*, separates the solid phase from the gas phase and represents the change in the vapour pressure of the solid as it sublimes at different temperatures.

FIGURE IT OUT

If the pressure exerted on a liquid is increased, while the temperature is held constant, what type of phase transition will eventually occur?

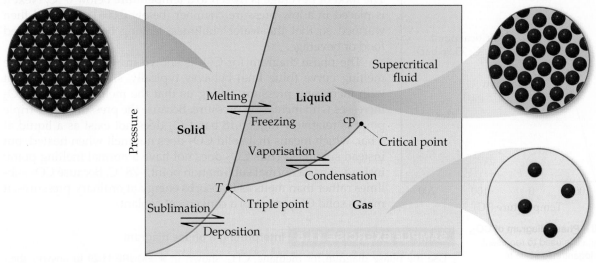

▲ **FIGURE 11.26   Generic phase diagram for a pure substance.**
The green line is the sublimation curve, the blue line is the melting curve and the red line is the vapour pressure curve.

3. The blue curve, the *melting curve*, separates the solid phase from the liquid phase and represents the change in melting point of the solid with increasing pressure. This curve usually slopes slightly to the right as pressure increases because for most substances the solid form is denser than the liquid form. An increase in pressure usually favours the more compact solid phase; thus higher temperatures are required to melt the solid at higher pressures. The melting point at 1 bar is the **normal melting point**.

Point T, where the three curves intersect, is the **triple point**, and here all three phases are in equilibrium. Any other point on any of the three curves represents equilibrium between two phases. Any point on the diagram that does not fall on one of the curves corresponds to conditions under which only one phase is present. The gas phase, for example, is stable at low pressures and high temperatures, whereas the solid phase is stable at low temperatures and high pressures. Liquids are stable in the region between the other two.

The Phase Diagrams of H_2O and CO_2

▶ **FIGURE 11.27** shows the phase diagram of H_2O. Because of the large range of pressures covered in the diagram, a logarithmic scale is used to represent pressure. The melting curve (blue line) of H_2O is atypical, slanting slightly to the left with increasing pressure, indicating that for water the melting point *decreases* with increasing pressure. This unusual behaviour occurs because water is among the very few substances whose liquid form is more compact than its solid form, as we learned in Section 11.2.

If the pressure is held constant at 1 bar, it is possible to move from the solid to liquid to gaseous regions of the phase diagram by changing the temperature, as we expect from our

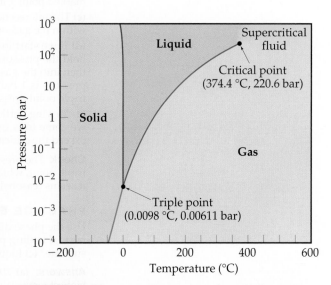

▲ **FIGURE 11.27   Phase diagram of H_2O.** Note that a linear scale is used to represent temperature and a logarithmic scale to represent pressure.

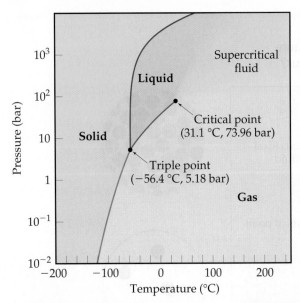

▲ **FIGURE 11.28** **Phase diagram of CO₂.**
Note that a linear scale is used to represent temperature and a logarithmic scale to represent pressure.

everyday encounters with water. The triple point of H_2O falls at a relatively low pressure, 0.00611 bar. Below this pressure, liquid water is not stable and ice sublimes to water vapour on heating. This property of water is used to 'freeze-dry' foods and beverages. The food or beverage is frozen to a temperature below 0 °C. Next it is placed in a low-pressure chamber (below 0.00611 bar) and then warmed so that the water sublimes, leaving behind dehydrated food or beverage.

The phase diagram for CO_2 is shown in ◄ **FIGURE 11.28.** The melting curve (blue line) behaves typically, slanting to the right with increasing pressure, telling us that the melting point of CO_2 increases with increasing pressure. Because the pressure at the triple point is relatively high, 5.18 bar, CO_2 does not exist as a liquid at 1 bar, which means that solid CO_2 does not melt when heated, but instead sublimes. Thus CO_2 does not have a normal melting point; instead, it has a normal sublimation point, –78 °C. Because CO_2 sublimes rather than melts as it absorbs energy at ordinary pressures, it makes solid CO_2 (dry ice) a convenient coolant.

SAMPLE EXERCISE 11.5 | **Interpreting a phase diagram**

Use the phase diagram for methane, CH_4, shown in ▶ FIGURE 11.29 to answer the following questions. **(a)** What are the approximate temperature and pressure of the critical point? **(b)** What are the approximate temperature and pressure of the triple point? **(c)** Is methane a solid, liquid or gas at 1 bar and 0 °C? **(d)** If solid methane at 1 bar is heated while the pressure is held constant, will it melt or sublime? **(e)** If methane at 1 bar and 0 °C is compressed until a phase change occurs, in which state is the methane when the compression is complete?

SOLUTION

Analyse We are asked to identify key features of the phase diagram and to use it to deduce what phase changes occur when specific pressure and temperature changes take place.

Plan We must identify the triple and critical points on the diagram and also identify which phase exists at specific temperatures and pressures.

Solve

(a) The critical point is the point where the liquid, gaseous and supercritical fluid phases coexist. It is marked point 3 in the phase diagram and located at approximately –80 °C and 50 bar.

(b) The triple point is the point where the solid, liquid and gaseous phases coexist. It is marked point 1 in the phase diagram and located at approximately –180 °C and 0.1 bar.

(c) The intersection of 0 °C and 1 bar is marked point 2 in the phase diagram. It is well within the gaseous region of the phase diagram.

(d) If we start in the solid region at P = 1 bar and move horizontally (this means we hold the pressure constant), we cross first into the liquid region, at $T \approx$ –180 °C, and then into the gaseous region, at $T \approx$ –160 °C. Therefore solid methane melts when the pressure is 1 bar. (In order for methane to sublime, the pressure must be below the triple point pressure.)

(e) Moving vertically up from point 2, which is 1 bar and 0 °C, the first phase change we come to is from gas to supercritical fluid. This phase change happens when we exceed the critical pressure (~ 50 bar).

Check The pressure and temperature at the critical point are higher than those at the triple point, which is expected. Methane is the principal component of natural gas. So it seems reasonable that it exists as a gas at 1 bar and 0 °C.

PRACTICE EXERCISE

Use the phase diagram of methane to answer the following questions. **(a)** What is the normal boiling point of methane? **(b)** Over what pressure range does solid methane sublime? **(c)** Liquid methane does not exist above what temperature?

Answers: **(a)** –162 °C. **(b)** It sublimes whenever the pressure is less than 0.1 bar; **(c)** The highest temperature at which a liquid can exist is defined by the critical temperature. So we do not expect to find liquid methane when the temperature is higher than –80 °C.

(See also Exercises 11.42, 11.43.)

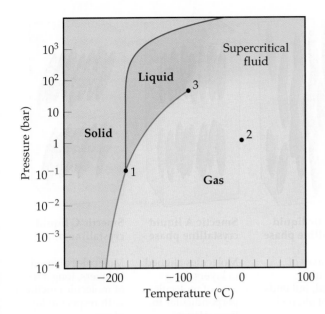

◀ FIGURE 11.29   **Phase diagram of CH₄.** Note that a linear scale is used to represent temperature and a logarithmic scale to represent pressure.

11.7 | LIQUID CRYSTALS

In 1888 Frederick Reinitzer (1857–1927), an Austrian botanist, discovered that the organic compound cholesteryl benzoate has an interesting and unusual property, shown in ▶ FIGURE 11.30. Solid cholesteryl benzoate melts at 145 °C, forming a viscous milky liquid; then at 179 °C the milky liquid becomes clear and remains that way at temperatures above 179 °C. When cooled, the clear liquid turns viscous and milky at 179 °C, and the milky liquid solidifies at 145 °C. Reinitzer's work represents the first systematic report of what we call a **liquid crystal**, the term we use today for the viscous, milky state.

Instead of passing directly from the solid phase to the liquid phase when heated, some substances, such as cholesteryl benzoate, pass through an intermediate liquid crystalline phase that has some of the structure of solids and some of the freedom of motion of liquids. Because of the partial ordering, liquid crystals may be viscous and possess properties intermediate between those of solids and those of liquids. The region in which they exhibit these properties is marked by sharp transition temperatures, as in Reinitzer's sample.

Today liquid crystals are used as pressure and temperature sensors and as the display element in such devices as digital watches and laptop computers. They can be used for these applications because the weak intermolecular forces that hold the molecules together in the liquid crystalline phase are easily affected by changes in temperature, pressure and electric fields.

Types of Liquid Crystals

Substances that form liquid crystals are often composed of rod-shaped molecules that are somewhat rigid in the middle. In the liquid phase these molecules are oriented randomly. In the liquid crystalline phase, by contrast, the molecules are arranged in specific patterns as illustrated in ▼ FIGURE 11.31. Depending on the nature of the ordering, liquid crystals are classified as nematic, smectic A, smectic C or cholesteric.

In a **nematic liquid crystal** the molecules are aligned so that their long axes tend to point in the same direction but the ends are not aligned with one another. In **smectic A** and **smectic C liquid crystals**, the molecules maintain the long-axis alignment seen in nematic crystals, but in addition they pack into layers.

Two molecules that exhibit liquid crystalline phases are shown in ▼ FIGURE 11.32. The lengths of these molecules are much greater than their widths. The double bonds, including those in the benzene rings, add rigidity to the molecules, and the rings, because they are flat, help the molecules stack with

145 °C < T < 179 °C
Liquid crystalline phase

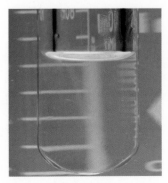

T > 179 °C
Liquid phase

▲ FIGURE 11.30   **Cholesteryl benzoate in its liquid and liquid crystalline states.**

Liquid phase

Molecules arranged randomly

Nematic liquid crystalline phase

Long axes of molecules aligned, but ends are not aligned

Smectic A liquid crystalline phase

Molecules aligned in layers, long axes of molecules perpendicular to layer planes

Smectic C liquid crystalline phase

Molecules aligned in layers, long axes of molecules inclined with respect to layer planes

▶ **FIGURE 11.31 Molecular order in nematic and smectic liquid crystals.** In the liquid phase of any substance, the molecules are arranged randomly, whereas in the liquid crystalline phases the molecules are arranged in a partially ordered way.

Double bonds provide rigidity

Benzene rings allow molecules to stack easily

CH_3O—[ring]—$\overset{\overset{H}{|}}{C}$=N—[ring]—$C_4H_9$ 21–47 °C

Polar groups create dipole moments

$CH_3(CH_2)_7$—O—[ring]—$\overset{\overset{O}{||}}{C}$—OH   108–147 °C

▲ **FIGURE 11.32 Molecular structure and liquid crystal temperature range for two typical liquid crystalline materials.**

one another. The polar CH_3O and COOH groups give rise to dipole–dipole interactions and promote alignment of the molecules. Thus the molecules order themselves quite naturally along their long axes. They can, however, rotate around their axes and slide parallel to one another. In smectic liquid crystals, the intermolecular forces (dispersion forces, dipole–dipole attractions and hydrogen bonding) limit the ability of the molecules to slide past one another.

In a **cholesteric liquid crystal** the molecules are arranged in layers, with their long axes parallel to the other molecules within the same layer.* Upon moving from one layer to the next, the orientation of the molecules rotates, resulting in the spiral pattern shown in ▶ **FIGURE 11.33**. These liquid crystals are so named because many derivatives of cholesterol adopt this structure.

The molecular arrangement in cholesteric liquid crystals produces unusual colouring patterns with visible light. Changes in temperature and pressure change the order and hence the colour. Cholesteric liquid crystals are used to monitor temperature changes in situations where conventional methods are not

---

* Cholesteric liquid crystals are sometimes called chiral nematic phases because the molecules within each plane adopt an arrangement similar to a nematic liquid crystal.

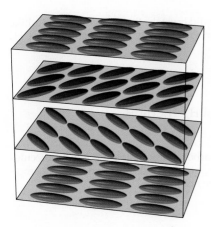

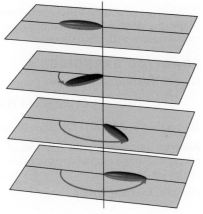

In a cholesteric liquid crystal the molecules pack into layers; the long axis of each molecule is oriented parallel to its neighbours within the same layer

The direction along which the molecules point rotates from one layer to the next, resulting in a spiralling pattern resembling the threads of a screw

▲ **FIGURE 11.33**   **Molecular order in a cholesteric liquid crystal.**

feasible. For example, they can detect hot spots in microelectronic circuits, which may signal the presence of flaws. They can also be fashioned into thermometers for measuring the skin temperature of infants. Because cholesteric liquid crystal displays can be built that draw very little power, they are also being investigated for use in electronic paper (▶ **FIGURE 11.34**).

▲ **FIGURE 11.34**   **Electronic paper (e-paper) based on cholesteric liquid crystal technology.**

SAMPLE EXERCISE 11.6   **Properties of liquid crystals**

Which of these substances is most likely to exhibit liquid crystalline behaviour?

$$CH_3-CH_2-\underset{\underset{CH_3}{|}}{\overset{\overset{CH_3}{|}}{C}}-CH_2-CH_3$$

(i)

$$CH_3CH_2-\langle\rangle-N=N-\langle\rangle-\overset{\overset{O}{\|}}{C}-OCH_3$$

(ii)

$$\langle\rangle-CH_2-\overset{\overset{O}{\|}}{C}-O^-\ Na^+$$

(iii)

SOLUTION

Analyse We have three molecules with different structures, and we are asked to determine which one is most likely to be a liquid crystalline substance.

Plan We need to identify all structural features that might induce liquid crystalline behaviour.

Solve Molecule (i) is not likely to be liquid crystalline because the absence of double and/or triple bonds make this molecule flexible rather than rigid. Molecule (iii) is ionic and the generally high melting points of ionic materials make it unlikely that this substance is liquid crystalline. Molecule (ii) possesses the characteristic long axis and the kinds of structural features often seen in liquid crystals: the molecule has a rodlike shape, the double bonds and benzene rings provide rigidity and the polar COOCH$_3$ group creates a dipole moment.

PRACTICE EXERCISE

Suggest a reason why decane

$$CH_3CH_2CH_2CH_2CH_2CH_2CH_2CH_2CH_2CH_3$$

does not exhibit liquid crystalline behaviour.

Answer: Because rotation can occur about carbon–carbon single bonds, molecules whose backbone consists predominantly of C–C single bonds are too flexible; the molecules tend to coil in random ways and thus are not rodlike.

(See also Exercise 11.42.)

MY WORLD OF CHEMISTRY

LIQUID CRYSTAL DISPLAYS

Liquid crystals displays (LCDs) are widely used in electronic devices such as watches, calculators and computer screens. These applications are possible because an applied electrical field changes the orientation of liquid crystal molecules and thus affects the optical properties of the device.

LCDs come in a variety of designs, but the structure illustrated in ▼ FIGURE 11.35 is typical. A thin layer (5–20 μm) of liquid crystalline material is placed between electrically conducting, transparent glass electrodes. Ordinary light passes through a vertical polariser that permits light in only the vertical plane to pass. Using a special process during fabrication, the liquid crystal molecules are oriented so that the molecules at the front electrode are oriented vertically and those at the back electrode horizontally. The orientation of the molecules in between the two electrodes varies systematically from vertical to horizontal, as shown in Figure 11.35(a). The plane of polarisation of the light is turned by 90° as it passes through the liquid crystal layer and is thus in the correct orientation to pass

through the horizontal polariser. In a watch display, a mirror reflects the light back, and the light retraces its path, allowing the device to look bright. When a voltage is applied to the plates the liquid crystalline molecules align with the voltage, as shown in Figure 11.35(b). The light rays thus are not properly oriented to pass through the horizontal polariser, and the device appears dark. Displays of this kind are called 'twisted nematic'. As the name implies, materials that order as nematic liquid crystals are used for this application.

Liquid crystal displays for computer and televisions employ a light source in place of the reflector, but the principle is the same. The screen is divided into a large number of tiny cells, with the voltages at points on the screen surface controlled by transistors made from thin films of amorphous silicon. Red-green-blue colour filters are employed to provide full colour. The entire display is refreshed at a frequency of about 60 Hz, so the display can change rapidly with respect to the response time of the human eye. Displays of this kind are remarkable technical achievements based on a combination of basic scientific discovery and creative engineering.

RELATED EXERCISES: 11.79, 11.80

▶ FIGURE 11.35 **Schematic illustration of the operation of a twisted nematic liquid crystal display (LCD).** (a) When the voltage is off, the molecules in the liquid crystal are aligned so that they rotate the polarisation of the light by 90°. This alignment allows light to pass through both the vertical and horizontal polarisers before being reflected and retracing its path to give a bright display. (b) When a voltage is applied to the electrodes the liquid crystal molecules align parallel to the light path. In this state the light retains the vertical polarisation and cannot pass through the horizontal polariser. The area covered by the front electrode therefore appears dark.

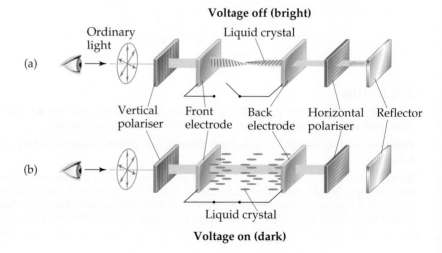

Voltage off (bright)

Ordinary light — Liquid crystal

(a)

Vertical polariser — Front electrode — Back electrode — Horizontal polariser — Reflector

(b)

Liquid crystal

Voltage on (dark)

11.8 | STRUCTURES OF SOLIDS

Throughout the rest of this chapter we focus on how the properties of solids relate to their structures and bonding. Solids can be either crystalline or amorphous (non-crystalline). In a **crystalline solid** (or crystal) the atoms, ions or molecules are ordered in well-defined arrangements. These solids usually have flat surfaces, or *faces*, that make definite angles with one another. The orderly stacks of particles that produce these faces also cause the solids to have highly regular shapes (▼ FIGURE 11.36). Quartz, sodium chloride and diamond are crystalline solids.

An **amorphous solid** is a solid whose particles have no orderly structure. These solids lack well-defined faces and shapes. Many amorphous solids are mixtures of molecules that do not stack together well. Most others are composed of large, complicated molecules. Familiar amorphous solids include rubber and glass.

Quartz (SiO_2) is a crystalline solid with a three-dimensional structure like that shown in ▼ FIGURE 11.37(a). When quartz melts (at about 1600 °C) it becomes a viscous, tacky liquid. Although the silicon–oxygen network remains largely intact, many Si–O bonds are broken, and the rigid order of the quartz is lost. If the liquid is rapidly cooled, the atoms are unable to return to an orderly arrangement. As a result, an amorphous solid known either as quartz glass or as silica glass results (Figure 11.37(b)).

Because the particles of an amorphous solid lack any long-range order, intermolecular forces vary in strength throughout a sample. Thus amorphous solids do not melt at specific temperatures. Instead, they soften over a temperature range as intermolecular forces of various strengths are overcome. A crystalline solid, in contrast, melts at a specific temperature.

CONCEPT CHECK 8

What is the general difference in the melting behaviours of crystalline and amorphous solids?

Unit Cells

The characteristic order of crystalline solids allows us to convey a picture of an entire crystal by looking at only a small part of it. We can think of the solid as being built up by stacking together identical building blocks, much as a brick wall is formed by stacking rows of individual 'identical' bricks. The repeating unit of a solid, the crystalline 'brick', is known as the **unit cell**. A simple two-dimensional example appears in the sheet of wallpaper shown in ▼ FIGURE 11.38. There are several ways of choosing a unit cell, but the choice is usually the smallest unit cell that shows clearly the symmetry characteristic of the entire pattern.

(a)   (b)   (c)

▲ **FIGURE 11.36 Crystalline solids.** Crystalline solids come in a variety of forms and colours: (a) pyrite (fool's gold), (b) fluorite, and (c) amethyst.

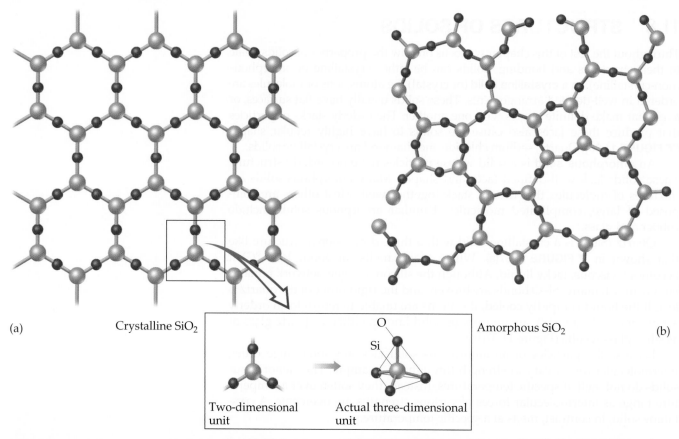

(a) Crystalline SiO$_2$

O

Si

Two-dimensional unit

Actual three-dimensional unit

Amorphous SiO$_2$ (b)

▲ **FIGURE 11.37** **Schematic comparisons of crystalline SiO$_2$ (quartz) and amorphous SiO$_2$ (quartz glass).** The structures are actually three-dimensional and not planar as drawn. The two-dimensional unit shown as the basic building block of the structure (one silicon and three oxygens) actually has four oxygens, the fourth coming out of the plane of the paper and capable of bonding to other silicon atoms. The actual three-dimensional building block is shown.

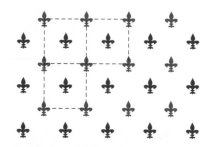

▲ **FIGURE 11.38** **A two-dimensional analogue of a lattice and its unit cell.** The wallpaper design shows a characteristic repeat pattern. Each dashed blue square denotes a unit cell of the pattern. The unit cell could equally well be selected with red figures at the corners.

▶ **FIGURE 11.39** **Part of a simple crystal lattice and its associated unit cell.** A lattice is an array of points that define the positions of particles in a crystalline solid. Each lattice point represents an identical environment in the solid. The points here are shown connected by lines to help convey the three-dimensional character of the lattice and to show the unit cell.

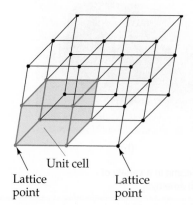

Unit cell

Lattice point

Lattice point

A crystalline solid can be represented by a three-dimensional array of points called a **crystal lattice**. Each point in the lattice is called a *lattice point*, and it represents an identical environment within the solid. The crystal lattice is, in effect, an abstract scaffolding for the crystal structure. We can imagine forming the entire crystal structure by arranging the contents of the unit cell repeatedly on the crystal lattice. In the simplest case the crystal structure would consist of identical atoms, and each atom would be centred on a lattice point. This is the case for most metals.

▼ **FIGURE 11.39** shows a crystal lattice and its associated unit cell. In general, unit cells are parallelepipeds (six-sided figures whose faces are parallelograms). Each unit cell can be described by the lengths of the edges of the cell and by the angles between these edges. The lattices of all crystalline compounds can be described by seven basic types of unit cells. The simplest of these is the cubic unit cell, in which all the sides are equal in length and all the angles are 90°.

There are three kinds of cubic unit cells, as illustrated in ▶ **FIGURE 11.40**. When lattice points are at the corners only, the unit cell is called *primitive cubic* (or *simple cubic*). When a lattice point also occurs at the centre of the unit cell, the cell is *body-centred cubic*. When the cell has lattice points at the centre of each face, as well as at each corner, it is *face-centred cubic*.

Lattice points only at corners

(a)
Primitive cubic lattice

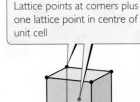
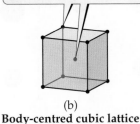

Lattice points at corners plus one lattice point in centre of unit cell

Lattice points at corners plus one lattice point at the centre of each face

(b)
Body-centred cubic lattice

(c)
Face-centred cubic lattice

◀ **FIGURE 11.40** The three types of cubic lattices.

The simplest crystal structures are cubic unit cells with only one atom centred at each lattice point. Most metals have such structures. Nickel, for example, has a face-centred cubic unit cell, whereas sodium has a body-centred cubic one. ▼ **FIGURE 11.41** shows how atoms fill the cubic unit cells. Notice that the atoms on the corners and faces do not lie wholly within the unit cell. Instead, these atoms are shared between unit cells. ▶ **TABLE 11.6** summarises the fraction of an atom that occupies a unit cell when atoms are shared between unit cells.

TABLE 11.6 • **Fraction of an atom that occupies a unit cell for various positions in the unit cell**

Position in unit cell	Fraction in unit cell
Centre	1
Face	$\frac{1}{2}$
Edge	$\frac{1}{4}$
Corner	$\frac{1}{8}$

 CONCEPT CHECK 9
For a structure consisting of identical atoms, how many atoms are contained in the body-centred cubic unit cell?

The Crystal Structure of Sodium Chloride

In the crystal structure of NaCl (▼ **FIGURE 11.42**) we can centre either the Na^+ ions or the Cl^- ions on the lattice points of a face-centred cubic unit cell. Thus we can describe the structure as being face-centred cubic.

In Figure 11.42 the Na^+ and Cl^- ions have been moved apart so the symmetry of the structure can be seen more clearly. In this representation no attention is paid to the relative sizes of the ions. The representation in ▼ **FIGURE 11.43**, on the other hand, shows the relative sizes of the ions and how they fill the unit cell. Notice that the particles at corners, edges and faces are shared by other unit cells.

The total cation to anion ratio of a unit cell must be the same as that for the entire crystal. Therefore, within the unit cell of NaCl there must be an equal number of Na^+ and Cl^- ions. Similarly, the unit cell for $CaCl_2$ would have one Ca^{2+} for every two Cl^- and so forth.

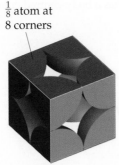

$\frac{1}{8}$ atom at 8 corners

(a) Primitive cubic metal
1 atom per unit cell

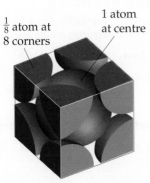

$\frac{1}{8}$ atom at 8 corners

1 atom at centre

(b) Body-centred cubic metal
2 atoms per unit cell

$\frac{1}{8}$ atom at 8 corners

$\frac{1}{2}$ atom at 6 faces

(c) Face-centred cubic metal
4 atoms per unit cell

▲ **FIGURE 11.41** **A space-filling view of unit cells for metals with a cubic structure.** Only the portion of each atom that falls within the unit cell is shown.

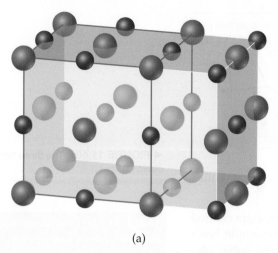

(a)

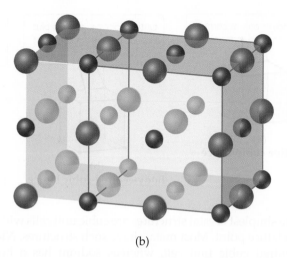

(b)

▲ **FIGURE 11.42** **Two ways of defining the unit cell of NaCl.** A representation of an NaCl crystal lattice can show either (a) Cl^- ions (green spheres) or (b) Na^+ ions (purple spheres) at the lattice points of the unit cell. In both cases, the red lines define the unit cell. Both of these choices for the unit cell are acceptable; both have the same volume and, in both cases, identical points are arranged in a face-centred cubic fashion.

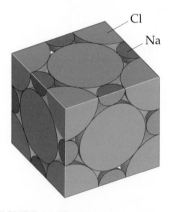

Cl

Na

▲ **FIGURE 11.43** **Relative size of ions in an NaCl unit cell.** As in Figure 11.42, purple represents Na^+ ions and green represents Cl^- ions. Only portions of most of the ions lie within the boundaries of the single unit cell.

SAMPLE EXERCISE 11.7 **Determining the contents of a unit cell**

Determine the net number of Na^+ and Cl^- ions in the NaCl unit cell (Figure 11.43).

SOLUTION

Analyse We are given a unit all of NaCl and asked to determine the number of Na^+ ions and Cl^- ions contained in it.

Plan To find the total number of ions of each type, we must identify the different locations within the unit cell and determine the fraction of the ion that lies within the unit cell boundaries.

Solve There is one-fourth of an Na^+ on each edge, a whole Na^+ in the centre of the cube (refer also to Figure 11.42), one-eighth of a Cl^- on each corner and one-half of a Cl^- on each face. Thus we have the following:

Na^+: $\left(\frac{1}{4} Na^+ \text{ per edge}\right)(12 \text{ edges}) = 3 Na^+$

$(1 Na^+ \text{ per centre})(1 \text{ centre}) = 1 Na^+$

Cl^-: $\left(\frac{1}{8} Cl^- \text{ per corner}\right)(8 \text{ corners}) = 1 Cl^-$

$\left(\frac{1}{2} Cl^- \text{ per face}\right)(6 \text{ faces}) = 3 Cl^-$

Thus the unit cell contains

$4 Na^+$ and $4 Cl^-$

Check This result agrees with the compound's stoichiometry:

$1 Na^+$ for each Cl^-

PRACTICE EXERCISE

The element iron crystallises in a form called α-iron, which has a body-centred cubic unit cell. How many iron atoms are in the unit cell?

Answer: Two

(See also Exercise 11.55.)

SAMPLE EXERCISE 11.8 **Using the contents and dimensions of a unit cell to calculate density**

The geometric arrangement of ions in crystals of LiF is the same as that in NaCl. The unit cell of LiF is 402 pm on an edge. Calculate the density of LiF.

SOLUTION

Analyse We are given a unit cell of LiF and are asked to calculate the density of the cell.

Plan We need to determine the number of formula units of LiF within the unit cell. From that we can calculate the total mass within the unit cell. Because we know the mass and can calculate the volume of the unit cell, we can then calculate density.

Solve The arrangement of ions in LiF is the same as that in NaCl (Sample Exercise 11.7), so a unit cell of LiF contains

4 Li^+ ions and 4 F^- ions.

Density measures mass per unit volume. Thus we can calculate the density of LiF from the mass contained in a unit cell and the volume of the unit cell. The mass contained in one unit cell is

$4(6.94\ u) + 4(19.0\ u) = 103.8\ u$

The volume of a cube of length a on an edge is a^3, so the volume of the unit cell is $(402\ pm)^3$. We can now calculate the density, converting to the common units of g cm^{-3}:

$1\ u = 1.66 \times 10^{-24}\ g$ and $1\ pm = 1 \times 10^{-10}\ cm$

$$\text{Density} = \frac{103.8\ u}{(402\ pm)^3} = \frac{(103.8 \times 1.66 \times 10^{-24})\ g}{(402 \times 10^{-10}\ cm)^3} = 2.65\ g\ cm^{-3}$$

Check This value agrees with that found by simple density measurements, 2.640 g cm^{-3} at 20 °C. The size and contents of the unit cell are therefore consistent with the macroscopic density of the substance.

PRACTICE EXERCISE

The body-centred cubic unit cell of a particular crystalline form of iron is 286.64 pm on each side. Calculate the density of this form of iron.

Answer: 7.88 g cm^{-3}

(See also Exercises 11.55, 11.58, 11.59.)

Close Packing of Spheres

The structures adopted by crystalline solids are those that bring particles into closest contact to maximise the attractive forces between them. In many cases, the particles that make up the solids are spherical or approximately so. Such is the case for atoms in metallic solids. It is therefore instructive to consider how equal-sized spheres can pack most efficiently (that is, with the minimum amount of empty space).

The most efficient arrangement of a layer of equal-sized spheres is shown in ▼ **FIGURE 11.44**. Each sphere is surrounded by six others in the layer. A second layer of spheres can be placed in the depressions on top of the first layer. A third layer can then be added above the second with the spheres sitting in the depressions of the second layer. However, there are two types of depressions for this third layer and they result in different structures, as shown in Figure 11.44.

If the spheres of the third layer are placed in line with those of the first layer the structure is known as *hexagonal close packing*. The third layer repeats the first layer, the fourth layer repeats the second layer and so forth, giving a layer sequence that we denote ABAB.

The spheres of the third layer, however, can be placed so they do not sit above the spheres in the first layer. The resulting structure, shown in Figure 11.44(c), is known as *cubic close packing*. In this case it is the fourth layer that repeats the first layer, and the layer sequence is ABCA. Although it cannot be seen in Figure 11.44, the unit cell of the cubic close-packed structure is face-centred cubic.

In both close-packed structures, each sphere has 12 equidistant nearest neighbours: six in one plane, three above that plane and three below. We say that each sphere has a *coordination number* of 12. The coordination number is the number of particles immediately surrounding a particle in the crystal structure. In both types of close packing, 74% of the total volume of the structure is occupied by spheres; 26% is empty space between the spheres. By comparison, each sphere in the body-centred cubic structure has a coordination

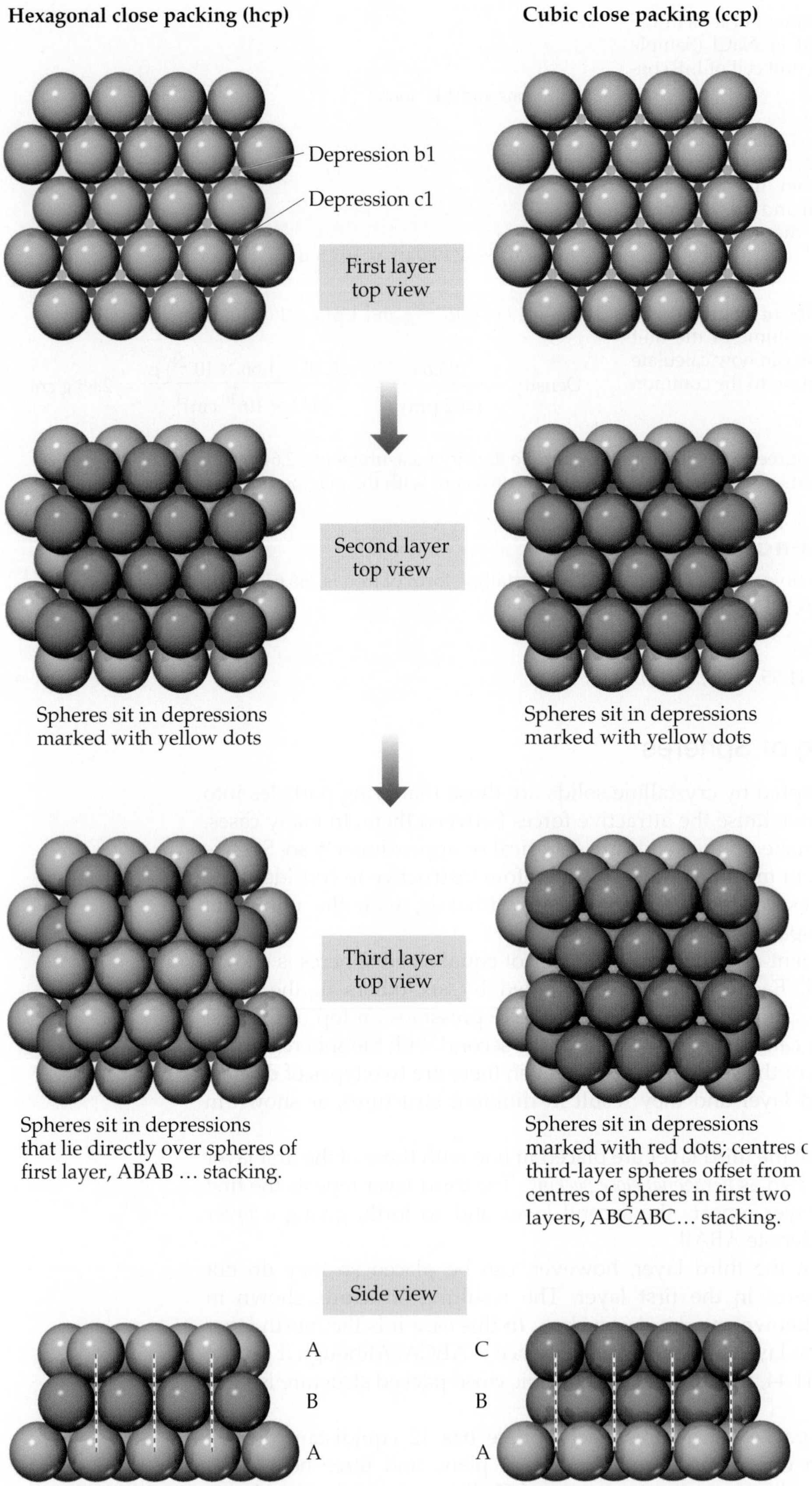

Hexagonal close packing (hcp)   **Cubic close packing (ccp)**

Depression b1

Depression c1

First layer
top view

Second layer
top view

Spheres sit in depressions
marked with yellow dots

Spheres sit in depressions
marked with yellow dots

Third layer
top view

Spheres sit in depressions
that lie directly over spheres of
first layer, ABAB … stacking.

Spheres sit in depressions
marked with red dots; centres of
third-layer spheres offset from
centres of spheres in first two
layers, ABCABC… stacking.

Side view

A   C

B   B

A   A

▲ FIGURE 11.44 **Close packing of equal-sized spheres.** Hexagonal (left) close packing
and cubic (right) close packing are equally efficient ways of packing spheres.

number of 8, and only 68% of the space is occupied. In the primitive cubic structure the coordination number is 6, and only 52% of the space is occupied.

When unequal-sized spheres are packed in a lattice, the larger particles sometimes assume one of the close-packed arrangements, with smaller particles occupying the holes between the large spheres. In Li_2O, for example, the larger oxide ions assume a cubic close-packed structure, and the smaller Li^+ ions occupy small cavities that exist between oxide ions.

 CONCEPT CHECK 10

Based on the information given above for close-packed structures and structures with cubic unit cells, what qualitative relationship exists between coordination numbers and packing efficiencies?

11.9 | BONDING IN SOLIDS

The physical properties of crystalline solids, such as melting point and hardness, depend both on the arrangements of particles and on the attractive forces between them. ▼ TABLE 11.7 classifies solids according to the types of forces between particles in solids.

Molecular Solids

Molecular solids consist of atoms or molecules held together by intermolecular forces (dipole–dipole forces, London dispersion forces and hydrogen bonds). Because these forces are weak, molecular solids are soft. Furthermore, they normally have relatively low melting points (usually below 200 °C). Most substances that are gases or liquids at room temperature form molecular solids at low temperature. Examples include Ar, H_2O and CO_2.

The properties of molecular solids depend not only on the strengths of the forces that exist between molecules, but also on the abilities of the molecules to pack efficiently in three dimensions. Benzene (C_6H_6), for example, is a highly symmetrical planar molecule. It has a higher melting point than toluene, a compound in which one of the hydrogen atoms of benzene has been replaced by a

TABLE 11.7 • Types of crystalline solids				
Type of solid	Form of unit particles	Forces between particles	Properties	Examples
Molecular	Atoms or molecules	Dipole–dipole forces, London dispersion forces, hydrogen bonds	Fairly soft, low to moderately high melting point, poor thermal and electrical conduction	Argon, Ar; methane, CH_4; sucrose, $C_{12}H_{22}O_{11}$; dry ice, CO_2
Covalent-network	Atoms connected in a network of covalent bonds	Covalent bonds	Very hard, very high melting point, often poor thermal and electrical conduction	Diamond, C; quartz, SiO_2
Ionic	Positive and negative ions	Electrostatic attractions	Hard and brittle, high melting point, poor thermal and electrical conduction	Typical salts—for example, NaCl, $Ca(NO_3)_2$
Metallic	Atoms	Metallic bonds	Soft to very hard, low to very high melting point, excellent thermal and electrical conduction, malleable and ductile	All metallic elements—for example, Cu, Fe, Al, Pt

A CLOSER LOOK

X-RAY DIFFRACTION BY CRYSTALS

When light waves pass through a narrow slit, they are scattered in such a way that the wave seems to spread out. This physical phenomenon is called diffraction. When light passes through many evenly spaced narrow slits (a diffraction grating), the scattered waves interact to form a series of light and dark bands, known as a diffraction pattern. The most effective diffraction of light occurs when the wavelength of the light and the width of the slits are similar in magnitude.

The spacing of the layers of atoms in solid crystals is usually about 200–2000 pm. The wavelengths of X-rays are also in this range. Thus a crystal can serve as an effective diffraction grating for X-rays. X-ray diffraction results from the scattering of X-rays by a regular arrangement of atoms, molecules or ions. Much of what we know about crystal structures has been obtained from studies of X-ray diffraction by crystals, a technique known as *X-ray crystallography*. ▼ FIGURE 11.45 depicts the diffraction of a beam of X-rays as it passes through a crystal. The diffracted X-rays were formerly detected by a photographic film. Today, crystallographers use an *array detector*, a device analogous to that used in digital cameras, to capture and measure the intensities of the diffracted rays. The diffraction pattern of spots on the detector in Figure 11.45 depends on the particular arrangement of atoms in the crystal. Thus different types of crystals give rise to different diffraction patterns.

In 1913 the New Zealand born scientists William and Lawrence Bragg (father and son) determined for the first time how the spacing of layers in crystals leads to different X-ray diffraction patterns. By measuring the intensities of the diffracted beams and the angles at which they are diffracted, it is possible to reason backwards to the structure that must have given rise to the pattern. One of the most famous X-ray diffraction patterns is the one for crystals of the genetic material DNA (▼ FIGURE 11.46), first obtained in the early 1950s. Working from photographs such as this one, Francis Crick, Rosalind Franklin, James Watson and Maurice Wilkins determined the double-helix structure of DNA, one of the most important discoveries in molecular biology.

Today X-ray crystallography is used extensively to determine the structures of molecules in crystals. The instruments used to measure X-ray diffraction, known as *X-ray diffractometers*, are now computer-controlled, making the collection of diffraction data highly automated. The diffraction pattern of a crystal can be determined very accurately and quickly even though thousands of diffraction points are measured. Computer programs are then used to analyse the diffraction data and determine the arrangement and structure of the molecules in the crystal.

RELATED EXERCISES: 11.83, 11.84

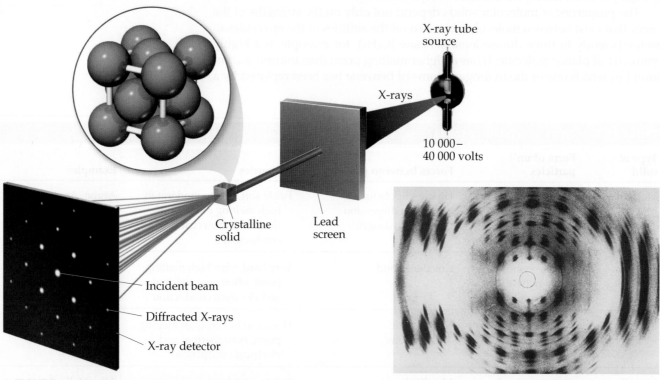

X-ray tube source

X-rays

10 000–
40 000 volts

Crystalline solid

Lead screen

Incident beam

Diffracted X-rays

X-ray detector

▲ **FIGURE 11.45 Diffraction of X-rays by a crystal.** In X-ray crystallography a monochromatic X-ray beam is passed through a crystal. The X-rays are diffracted, and the resulting interference pattern is recorded. The crystal is rotated and another diffraction pattern recorded. Analysis of many diffraction patterns gives the positions of the atoms in the crystal.

▲ **FIGURE 11.46 The X-ray diffraction photograph of one form of crystalline DNA.** This photograph was taken in the early 1950s. From the pattern of dark spots, the double-helical shape of the DNA molecule was deduced.

CH$_3$ group (▶ FIGURE 11.47). The lower symmetry of toluene molecules prevents them from packing as efficiently as benzene molecules. As a result, the intermolecular forces that depend on close contact are not as effective and the melting point is lower. In contrast, the boiling point of toluene is higher than that of benzene, indicating that the intermolecular attractive forces are larger in liquid toluene than in liquid benzene. Both the melting and boiling points of phenol, another substituted benzene shown in Figure 11.47, are higher than those of benzene because the OH group of phenol can form hydrogen bonds.

> **CONCEPT CHECK 11**
> Which of the following substances would you expect to form molecular solids: Co, C$_6$H$_6$ or K$_2$O?

Covalent-Network Solids

Covalent-network solids consist of atoms held together in large networks or chains by covalent bonds. Because covalent bonds are much stronger than intermolecular forces, these solids are much harder and have higher melting points than molecular solids. Diamond and graphite, two allotropes of carbon, are covalent-network solids. Other examples include quartz, SiO$_2$; silicon carbide, SiC; and boron nitride, BN.

In diamond each carbon atom is bonded to four other carbon atoms, as shown in ▼ FIGURE 11.48(a). This interconnected three-dimensional array of strong carbon–carbon single bonds contributes to diamond's unusual hardness. Industrial-grade diamonds are employed in the blades of saws for the most demanding cutting jobs. Diamond also has a high melting point of 3550 °C.

In graphite the carbon atoms are arranged in layers of interconnected hexagonal rings, as shown in Figure 11.48(b). Each carbon atom is bonded to three others in the layer. The distance between adjacent carbon atoms in the plane is 142 pm. Electrons move freely through the delocalised orbitals, making graphite a good conductor of electricity along the layers. (If you have ever taken apart a torch battery, you know that the central electrode in the battery is made of graphite.) The layers, which are separated by 341 pm, are held together by weak dispersion forces. The layers readily slide past one another when rubbed, giving graphite a greasy feel. Graphite is used as a lubricant and in the 'lead' in pencils.

▲ **FIGURE IT OUT**

In which substance, benzene or toluene, are the intermolecular forces stronger? In which substance do the molecules pack more efficiently?

	Benzene	Toluene	Phenol
Melting point (°C)	5	−95	43
Boiling point (°C)	80	111	182

▲ **FIGURE 11.47** Melting and boiling points for benzene, toluene and phenol.

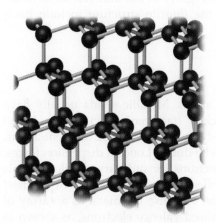

(a) Diamond

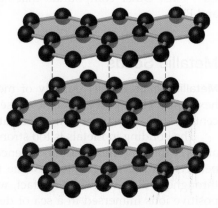

(b) Graphite

◀ **FIGURE 11.48 The structures of (a) diamond and (b) graphite.** The blue colour in (b) is added to emphasise the planarity of the carbon layers.

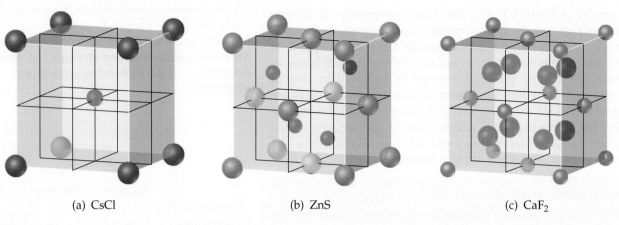

(a) CsCl   (b) ZnS   (c) CaF₂

▲ FIGURE 11.49   Unit cells of some common ionic structures.

Ionic Solids

∞∞ Review this on page 253

Ionic solids consist of ions held together by ionic bonds (∞∞ Section 8.2, 'Ionic Bonding'). The strength of an ionic bond depends greatly on the charges of the ions. Thus NaCl, in which the ions have charges of $+$ and $-$, has a melting point of 801 °C, whereas MgO, in which the charges are $2+$ and $2-$, melts at 2852 °C.

The structures of simple ionic solids can be classified as a few basic types. The NaCl structure is a representative example of one type. Other compounds that possess this same structure include LiF, KCl, AgCl and CaO. Three other common types of crystal structures are shown in ▲ **FIGURE 11.49**.

The structure adopted by an ionic solid depends largely on the charges and relative sizes of the ions. In the NaCl structure, for example, the Na^+ ions have a coordination number of 6 because each Na^+ ion is surrounded by six nearest neighbour Cl^- ions. In the CsCl structure (Figure 11.49(a)), by comparison, the Cl^- ions adopt a primitive cubic arrangement with each Cs^+ ion surrounded by eight Cl^- ions. The increase in the coordination number as the alkali metal ion changes from Na^+ to Cs^+ is a consequence of the larger size of Cs^+ compared with Na^+.

In the zinc blende (ZnS) structure (Figure 11.49(b)), the S^{2-} ions adopt a face-centred cubic arrangement, with the smaller Zn^{2+} ions arranged so they are each surrounded tetrahedrally by four S^{2-} ions (compare with Figure 11.40). CuCl also adopts this structure.

In the fluorite (CaF_2) structure (Figure 11.49(c)), the Ca^{2+} ions are shown in a face-centred cubic arrangement. As required by the chemical formula of the substance, there are twice as many F^- ions (grey) in the unit cell as there are Ca^{2+} ions. Other compounds that have the fluorite structure include $BaCl_2$ and PbF_2.

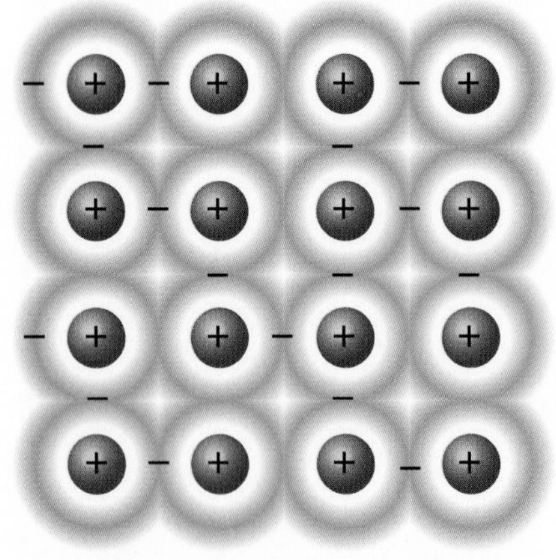

Metallic Solids

Metallic solids consist entirely of metal atoms. Metallic solids usually have hexagonal close-packed, cubic close-packed (face-centred cubic) or body-centred cubic structures. Thus, each atom typically has 8 or 12 adjacent atoms.

The bonding in metals is too strong to be due to London dispersion forces, and yet there are not enough valence electrons for ordinary covalent bonds between atoms. The bonding is due to valence electrons that are delocalised throughout the entire solid. In fact, we can visualise the metal as an array of positive ions immersed in a sea of delocalised valence electrons, as shown in ◄ **FIGURE 11.50**.

▲ FIGURE 11.50   **Representation of a cross-section of a metal.** Each sphere represents the nucleus and inner-core electrons of a metal atom. The surrounding blue 'fog' represents the mobile sea of electrons that binds the atoms together.

Metals vary greatly in the strength of their bonding, as shown by their wide range of physical properties such as hardness and melting point. In general, however, the strength of the bonding increases as the number of electrons available for bonding increases. Thus sodium, which has only one valence electron per atom, melts at 97.5 °C, whereas chromium, with six electrons beyond the noble gas core, melts at 1890 °C. The mobility of the electrons explains why metals are good conductors of heat and electricity.

SAMPLE INTEGRATIVE EXERCISE Putting concepts together

The substance CS_2 has a melting point of −110.8 °C and a boiling point of 46.3 °C. Its density at 20 °C is 1.26 g cm^{-3}. It is highly flammable. **(a)** What is the name of this compound? **(b)** List the intermolecular forces that CS_2 molecules exert on one other. **(c)** Write a balanced equation for the combustion of this compound in air. (You will have to decide on the most likely oxidation products.) **(d)** The critical temperature and pressure for CS_2 are 552 K and 7.8×10^3 kPa, respectively. Compare these values with those for CO_2 (Table 11.5) and discuss the possible origins of the differences. **(e)** Would you expect the density of CS_2 at 40 °C to be greater or less than at 80 °C? What accounts for the difference?

SOLUTION

(a) The compound is named carbon disulfide, in analogy with the naming of other binary molecular compounds such as carbon dioxide (∞ Section 2.8, 'Naming Inorganic Compounds').

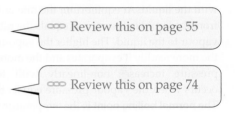
∞ Review this on page 55

∞ Review this on page 74

(b) Only dispersion forces affect CS_2; it does not have a dipole moment, based upon its molecular shape, and obviously cannot undergo hydrogen bonding.

(c) The most likely products of the combustion will be CO_2 and SO_2 (∞ Section 3.2, 'Some Simple Patterns of Chemical Reactivity'). Under some conditions SO_3 might be formed, but this would be the less likely outcome. Thus we have the following equation for combustion:

$$CS_2(l) + 3\,O_2(g) \rightarrow CO_2(g) + 2\,SO_2(g)$$

(d) The critical temperature and pressure of CS_2 (552 K and 7.8×10^3 kPa) are both higher than those given for CO_2 in Table 11.5 (304 K and 7.39×10^3 kPa). The difference in critical temperatures is especially notable. The higher values for CS_2 arise from the greater dispersion attractions between the CS_2 molecules compared with CO_2. These greater attractions are due to the larger size of the sulfur compared to oxygen and therefore its greater polarisability.

(e) The density would be lower at the higher temperature. Density decreases with increasing temperature because the molecules possess higher kinetic energies. Their more energetic movements result in larger average distances between molecules, which translate into lower densities.

CHAPTER SUMMARY AND KEY TERMS

SECTION 11.1 Substances that are gases or liquids at room temperature are usually composed of molecules. In gases the intermolecular attractive forces are negligible compared to the kinetic energies of the molecules; thus the molecules are widely separated and undergo constant, chaotic motion. In liquids the **intermolecular forces** are strong enough to keep the molecules in close proximity; nevertheless, the molecules are free to move with respect to one another. In solids the intermolecular attractive forces are strong enough to restrain molecular motion and to force the particles to occupy specific locations in a three-dimensional arrangement.

SECTION 11.2 Three types of intermolecular forces exist between neutral molecules: **dispersion forces, dipole–dipole forces** and **hydrogen bonding**. Dispersion forces operate between all molecules (and atoms, for atomic substances such as He, Ne, Ar). As molecular weight increases, the **polarisability**

of a molecule increases, which results in stronger dispersion forces. Molecular shape is also an important factor. Dipole–dipole forces increase in strength as the polarity of the molecule increases. Hydrogen bonding occurs in compounds containing O–H, N–H and F–H bonds. Hydrogen bonds are generally stronger than dipole–dipole or dispersion forces. **Ion–dipole forces** are important in solutions in which ionic compounds are dissolved in polar solvents.

SECTION 11.3 The stronger the intermolecular forces, the greater is the **viscosity**, or resistance to flow, of a liquid. The surface tension of a liquid also increases as intermolecular forces increase in strength. **Surface tension** is a measure of the tendency of a liquid to maintain a minimum surface area. The adhesion of a liquid to the walls of a narrow tube and the cohesion of the liquid account for **capillary action** and the formation of a meniscus at the surface of a liquid.

SECTION 11.4 A substance may exist in more than one state of matter, or phase. **Phase changes** are transformations from one phase to another. Changes of a solid to liquid (melting), solid to gas (sublimation) and liquid to gas (vaporisation) are all endothermic processes. Thus the **heat of fusion** (melting), the **heat of sublimation** and the **heat of vaporisation** are all positive quantities. The reverse processes (freezing, deposition and condensation) are exothermic. A gas cannot be liquefied by application of pressure if the temperature is above its **critical temperature**. The pressure required to liquefy a gas at its critical temperature is called the **critical pressure**. When the temperature exceeds the critical temperature and the pressure exceeds the critical pressure, the liquid and gas phases cannot be distinguished and the substance is in a state called a **supercritical fluid**.

SECTION 11.5 The **vapour pressure** of a liquid indicates the tendency of the liquid to evaporate. The vapour pressure is the partial pressure of the vapour when it is in **dynamic equilibrium** with the liquid. At equilibrium the rate of transfer of molecules from the liquid to the vapour equals the rate of transfer from the vapour to the liquid. The higher the vapour pressure of a liquid, the more readily it evaporates and the more **volatile** it is. Vapour pressure increases non-linearly with temperature. Boiling occurs when the vapour pressure equals the external pressure. The **normal boiling point** is the temperature at which the vapour pressure equals 1 bar.

SECTION 11.6 The equilibria between the solid, liquid and gas phases of a substance as a function of temperature and pressure are displayed on a **phase diagram**. A line indicates equilibria between any two phases. The line through the melting point usually slopes slightly to the right as pressure increases, because the solid is usually more dense than the liquid. The melting point at 1 bar is the **normal melting point**. The point on the diagram at which all three phases coexist in equilibrium is called the **triple point**.

SECTION 11.7 A **liquid crystal** is a substance that exhibits one or more ordered phases at a temperature above the melting point of the solid. In a **nematic liquid crystal** the molecules are aligned along a common direction, but the ends of the molecules are not lined up. In a smectic liquid crystal the ends of the molecules are lined up so that the molecules form layers. In **smectic A liquid crystals** the long axes of the molecules line up perpendicular to the layers. In **smectic C liquid crystals** the long axes of molecules are inclined with respect to the layers. A **cholesteric liquid crystal** is composed of molecules that align parallel to each other within a layer, as they do in nematic liquid crystalline phases, but the direction along which the long axes of the molecules align rotates from one layer to the next to form a helical structure. Substances that form liquid crystals are generally composed of molecules with fairly rigid, elongated shapes, as well as polar groups to help align molecules through dipole–dipole interactions.

SECTION 11.8 In solids regular arrangements of molecules or ions in three dimensions characterise **crystalline solids** as opposed to **amorphous solids** in which no such ordered arrangement are found. The three-dimensional structure of a crystalline solid can be represented by its **crystal lattice** but can also be conveyed by its **unit cell**, which is the smallest part of a crystal that can, by simple displacement, reproduce the three-dimensional structure.

SECTION 11.9 Solids are characterised and classified according to the attractive forces between their component atoms, molecules or ions. We consider four such classes: **molecular solids**, **covalent-network solids**, **ionic solids** and **metallic solids**.

KEY SKILLS

- Identify the intermolecular attractive interactions (dispersion, dipole–dipole, hydrogen bonding, ion–dipole) that exist between molecules or ions based on their composition and molecular structure and be able to compare the relative strengths of these intermolecular forces. (Section 11.2)

- Explain the concept of polarisability and how it relates to dispersion forces. (Section 11.2)

- Explain the concepts of viscosity and surface tension in liquids. (Section 11.3)

- Know the names of the various changes of state for a pure substance. (Section 11.4)

- Interpret heating curves and be able to calculate quantities related to temperature and enthalpies of phase changes. (Section 11.4)

- Define critical pressure, critical temperature, vapour pressure, normal boiling point, normal melting point, critical point and triple point. (Sections 11.5 and 11.6)

- Be able to interpret and sketch phase diagrams. Explain how water's phase diagram differs from most other substances, and why. (Section 11.6)

- Understand how the molecular arrangements characteristic of nematic, smectic and cholesteric liquid crystals differ from ordinary liquids and from each other. Be able to recognise the features of molecules that favour formation of liquid crystalline phases. (Section 11.7)

- Know the difference between crystalline and amorphous solids. [Section 11.8]

- Understand the concept of the unit cell and the crystal lattice. [Section 11.8]

- Classify solids based on their bonding/intermolecular forces and understand how difference in bonding relates to physical properties. [Section 11.9]

- Explain the difference between molecular, covalent-network, ionic and metallic solids. [Section 11.9]

KEY EQUATIONS

- The Clausius–Clapeyron equation which relates the pressure of heat to vaporisation.

$$\ln P = \frac{-\Delta_{vap}H}{RT} + C \qquad [11.1]$$

- Method of graphically evaluating the heat of vaporisation.

$$\Delta_{vap}H = -\text{slope} \times R \qquad [11.2]$$

EXERCISES

VISUALISING CONCEPTS

11.1 Does the following diagram best depict a crystalline solid, liquid or gas? Explain. [Section 11.1]

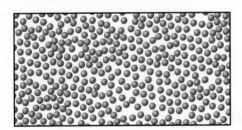

11.2 What kind of intermolecular attractive force is shown in each of the following cases? [Section 11.2]

(i)   (iii)

(ii)   (iv)

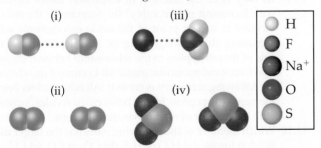

○	H
●	F
●	Na$^+$
●	O
○	S

11.3 Using the following graph, determine **(a)** the approximate vapour pressure of CS$_2$ at 30 °C, **(b)** the temperature at which the vapour pressure equals 0.45 bar, **(c)** the normal boiling point of CS$_2$. [Section 11.5]

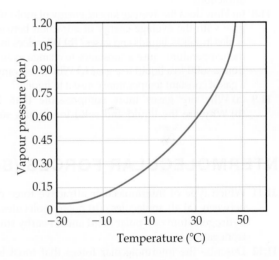

11.4 The following molecules have the same molecular formula (C$_3$H$_8$O), yet they have different normal boiling points, as shown. Rationalise the difference in boiling points. [Sections 11.2 and 11.5]

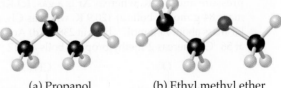

(a) Propanol
97.2 °C

(b) Ethyl methyl ether
10.8 °C

11.5 The phase diagram of a hypothetical substance is shown below.

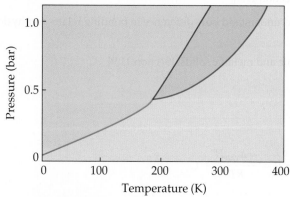

(a) Estimate the normal boiling point and freezing point of the substance.

(b) What is the physical state of the substance under the following conditions?
(i) $T = 150$ K, $P = 0.2$ bar,
(ii) $T = 100$ K, $P = 0.8$ bar,
(iii) $T = 300$ K, $P = 1.0$ bar. [Section 11.6]

11.6 Niobium(II) oxide crystallises in the following cubic unit cell.

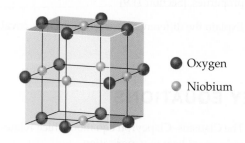

● Oxygen
● Niobium

(a) How many niobium atoms and how many oxygen atoms are within the unit cell?

(b) What is the empirical formula of niobium oxide?

(c) Is this a molecular, covalent-network or ionic solid? [Sections 11.8 and 11.9]

KINETIC-MOLECULAR THEORY (Section 11.1)

11.7 List the three states of matter in order of **(a)** increasing molecular disorder, and **(b)** increasing intermolecular attractions.

11.8 **(a)** How does the average kinetic energy of molecules compare with the average energy of attraction between molecules in solids, liquids and gases? **(b)** Why does increasing the temperature cause a substance to change in succession from a solid to a liquid to a gas? **(c)** Why does compressing a gas at constant temperature cause it to liquefy?

11.9 **(a)** Why are gases more compressible than liquids? **(b)** Why are the liquid and solid forms of a substance referred to as *condensed phases*? **(c)** Why do liquids have a greater ability to flow than solids?

11.10 Benzoic acid, C_6H_5COOH, melts at 122 °C. The density in the liquid state at 130 °C is 1.08 g cm^{-3}. The density of solid benzoic acid at 15 °C is 1.266 g cm^{-3}. **(a)** In which of these two states is the average distance between molecules the greater? **(b)** Explain the difference in densities at the two temperatures in terms of the kinetic-molecular theory.

INTERMOLECULAR FORCES (Section 11.2)

11.11 Which type of intermolecular attractive force operates between **(a)** all molecules, **(b)** polar molecules, **(c)** the hydrogen atom of a polar bond and a nearby small electronegative atom?

11.12 Describe the intermolecular forces that must be overcome to convert each of the following from a liquid to a gas: **(a)** Br_2, **(b)** CH_3OH, **(c)** H_2S.

11.13 What type of intermolecular force accounts for the following differences in each case? **(a)** CH_3OH boils at 65 °C, CH_3SH boils at 6 °C. **(b)** Xe is liquid at atmospheric pressure and 120 K, whereas Ar is a gas. **(c)** Kr, atomic mass 84 g mol^{-1}, boils at 120.9 K, whereas Cl_2, molecular mass about 71 g mol^{-1}, boils at 238 K. **(d)** Acetone boils at 56 °C, whereas 2-methylpropane boils at –12 °C.

$$\underset{\text{Acetone}}{CH_3-\overset{\overset{\displaystyle O}{\|}}{C}-CH_3} \qquad \underset{\text{2-methylpropane}}{CH_3-\overset{\overset{\displaystyle CH_3}{|}}{CH}-CH_3}$$

11.14 **(a)** What is meant by the term *polarisability*? **(b)** Which of the following atoms would you expect to be most polarisable: O, S, Se or Te? Explain. **(c)** Put the following molecules in order of increasing polarisability: $GeCl_4$, CH_4, $SiCl_4$, SiH_4 and $GeBr_4$. **(d)** Predict the order of boiling points of the substances in part (c).

11.15 **(a)** Why does the strength of dispersion forces increase with increasing polarisability? **(b)** Account for the steady increase in boiling point of the noble gas elements with increasing atomic weight (Figure 11.4). **(c)** What general rule of thumb applies to the relationship between dispersion forces and molecular mass? **(d)** Comment on whether the following statement is correct: 'All other factors being the same, dispersion forces between molecules increase with the number of electrons in the molecules.'

11.16 Which member of the following pairs has the larger dispersion forces: **(a)** H_2O or H_2S, **(b)** CO_2 or CO, **(c)** CH_4 or SiH_4?

11.17 **(a)** What molecular features must a molecule have to participate in hydrogen bonding with other molecules of the same kind? **(b)** Which of the following molecules can form hydrogen bonds with other molecules of the same kind: CH_3F, CH_3NH_2, CH_3OH, CH_3Br?

11.18 Rationalise the difference in boiling points between the members of the following pairs of substances: **(a)** HF (20 °C) and HCl (–85 °C), **(b)** $CHCl_3$ (61 °C) and $CHBr_3$ (150 °C), **(c)** Br_2 (59 °C) and ICl (97 °C).

11.19 How are the following observations related to the ability of water to form hydrogen bonds? **(a)** Ice is less dense than liquid water. **(b)** Water has a high specific heat, meaning it requires a large amount of heat to produce a temperature increase of 1 K.

11.20 The following statement about ammonia (NH_3) is from a textbook of inorganic chemistry: 'It is estimated that 26% of the hydrogen bonding in NH_3 breaks down on melting, 7% on warming from the melting to the boiling point and the final 67% on transfer to the gas phase at the boiling point.' From the standpoint of the kinetic-molecular theory, explain **(a)** why there is a decrease of hydrogen-bonding energy on melting and **(b)** why most of the loss in hydrogen bonding occurs in the transition from the liquid to the vapour state.

11.21 A number of salts containing the tetrahedral polyatomic anion, BF_4^-, are ionic liquids, whereas salts containing the somewhat larger tetrahedral ion SO_4^{-2}, do not form ionic liquids. Explain this observation.

11.22 The generic structural formula for 1-alkyl-3-methyl-imidozolium cation is

where R is a —$CH_2(CH_2)_nCH_3$ alkyl group. The melting points of the salts that form between 1-alkyl-3-methylimidazolium cation and the PF_6^- anion are as follows:
R = CH_2CH_3 (mp = 60 °C), R = $CH_2CH_2CH_3$ (mp = 40 °C),
R = $CH_2CH_2CH_2CH_3$ (mp = 10 °C) and
R = $CH_2CH_2CH_2CH_2CH_2CH_3$ (mp = −61 °C).
Why does the melting point decrease as the length of alkyl group increases?

VISCOSITY AND SURFACE TENSION (Section 11.3)

11.23 **(a)** Why do surface tension and viscosity decrease with increasing temperature? **(b)** Why do substances with high surface tensions also tend to have high viscosities?

11.24 **(a)** Distinguish between adhesive forces and cohesive forces. **(b)** What adhesive and what cohesive forces are involved when a paper towel absorbs water? **(c)** Explain the cause for the U-shaped meniscus formed when water is in a glass tube.

11.25 Explain the following observations. **(a)** The surface tension of $CHBr_3$ is greater than that of $CHCl_3$. **(b)** As temperature increases, oil flows faster through a narrow tube. **(c)** Raindrops that collect on a waxed car roof take on a nearly spherical shape.

11.26 Hydrazine (NH_2NH_2), hydrogen peroxide (HOOH) and water (H_2O) all have exceptionally high surface tensions in comparison with other substances of comparable molecular weights. **(a)** Draw the Lewis structures for these three compounds. **(b)** What structural property do these substances have in common, and how might that account for the high surface tensions?

PHASE CHANGES (Section 11.4)

11.27 Name the phase transition in each of the following situations and indicate whether it is exothermic or endothermic. **(a)** When water is cooled, it turns to ice. **(b)** Wet clothes dry on a warm summer day. **(c)** Frost appears on a window on a cold winter day.

11.28 Name the phase transition in each of the following situations, and indicate whether it is exothermic or endothermic. **(a)** Bromine vapour turns to bromine liquid as it is cooled. **(b)** Crystals of iodine disappear from an evaporating dish as they stand in a fume cupboard. **(c)** Rubbing alcohol in an open container slowly disappears. **(d)** Molten lava from a volcano turns into solid rock.

11.29 Explain why the heat of fusion of any substance is generally lower than its heat of vaporisation.

11.30 Chloroethane (C_2H_5Cl) boils at 12 °C. When liquid C_2H_5Cl under pressure is sprayed on a room-temperature surface in air, the surface is cooled considerably. **(a)** What does this observation tell us about the enthalpy content of C_2H_5Cl(g) as compared with C_2H_5Cl(l)? **(b)** In terms of the kinetic-molecular theory, what is the origin of this difference?

11.31 For many years drinking water has been cooled in hot climates by evaporating it from the surfaces of canvas bags or porous clay pots. How many grams of water can be cooled from 35 °C to 22 °C by the evaporation of 50 g of water? (The heat of vaporisation of water in this temperature range is 2.4 kJ g^{-1}. The specific heat of water is 4.18 J g^{-1} K^{-1}.)

11.32 Compounds like CCl_2F_2 are known as chlorofluorocarbons, or CFCs. These compounds were once widely used as refrigerants but are now being replaced by compounds that are believed to be less harmful to the environment. The heat of vaporisation of CCl_2F_2 is 289 J g^{-1}. What mass of this substance must evaporate in order to freeze 100 g of water initially at 18 °C? (The heat of fusion of water is 334 J g^{-1}; the specific heat of water is 4.18 J g^{-1} K^{-1}.)

11.33 **(a)** What is the significance of the critical pressure of a substance? **(b)** What happens to the critical temperature of a series of compounds as the force of attraction between molecules increases? **(c)** Which of the substances listed in Table 11.5 can be liquefied at the temperature of liquid nitrogen (−196 °C)?

11.34 The critical temperatures (K) and pressures (bar) of a series of halogenated methanes are as follows:

Compound	CCl_3F	CCl_2F_2	$CClF_3$	CF_4
Critical temperature	471	385	302	227
Critical pressure	42.9	40.1	37.7	36.5

(a) What in general can you say about the variation in intermolecular forces in this series? (b) What specific kinds of intermolecular forces are most likely to account for most of the variation in critical parameters in this series?

VAPOUR PRESSURE AND BOILING POINT (Section 11.5)

11.35 Explain how each of the following affects the vapour pressure of a liquid: (a) volume of the liquid, (b) surface area, (c) intermolecular attractive forces, (d) temperature.

11.36 A liquid that has an equilibrium vapour pressure of 17.3 kPa at 25 °C is placed into a 1 dm³ vessel like that shown in Figure 11.22. What is the pressure difference shown on the manometer, and what is the composition of the gas in the vessel, under each of the following conditions. (a) 200 cm³ of the liquid is introduced into the vessel and frozen at the bottom. The vessel is evacuated and sealed, and the liquid is allowed to warm to 25 °C. (b) 200 cm³ of the liquid is added to the vessel at 25 °C under atmospheric pressure, and after a few minutes the vessel is closed off. (c) A few cm³ of the liquid is introduced into the vessel at 25 °C while it has a pressure of 1 bar of air in it, without allowing any of the air to escape. After a few minutes a few drops of liquid remain in the vessel.

11.37 (a) Two pans of water are on different burners of a stove. One pan of water is boiling vigorously, whereas the other is boiling gently. What can be said about the temperature of the water in the two pans? (b) A large container of water and a small one are at the same temperature. What can be said about the relative vapour pressures of the water in the two containers?

11.38 Explain the following observations. (a) Water evaporates more quickly on a hot, dry day than on a hot, humid day. (b) It takes longer to cook hard-boiled eggs at high altitudes than at lower altitudes.

11.39 (a) Use the vapour pressure curve in Figure 11.24 to estimate the boiling point of diethyl ether at 50 kPa. (b) Use the vapour pressure curve in Figure 11.24 to estimate the external pressure under which ethanol will boil at 70 °C. (c) Using the vapour pressure table in Appendix B, determine the boiling point of water when the external pressure is 3.33 kPa. (d) Suppose the pressure inside a pressure cooker reaches 1.2 bar. By using the vapour pressure table in Appendix B, estimate the temperature at which water will boil in this cooker.

PHASE DIAGRAMS (Section 11.6)

11.40 (a) What is the significance of the critical point in a phase diagram? (b) Why does the line that separates the gas and liquid phases end at the critical point?

11.41 (a) What is the significance of the triple point in a phase diagram? (b) Could you measure the triple point of water by measuring the temperature in a vessel in which water vapour, liquid water and ice are in equilibrium under one bar pressure of air? Explain.

11.42 Refer to Figure 11.27 and describe all the phase changes that would occur in each of the following cases. (a) Water vapour, originally at 1.0×10^{-3} bar and −0.10 °C, is slowly compressed at constant temperature until the final pressure is 10 bar. (b) Water, originally at 100.0 °C and 0.50 bar, is cooled at constant pressure until the temperature is −10 °C.

11.43 Refer to Figure 11.28 and describe the phase changes (and the temperatures at which they occur) when CO_2 is heated from −80 °C to −20 °C at (a) a constant pressure of 3 bar, (b) a constant pressure of 6 bar.

11.44 The normal melting and boiling points of xenon are −112 °C and −107 °C, respectively. Its triple point is at −121 °C and 37.5 kPa, and its critical point is at 16.6 °C and 5.83×10^3 kPa. (a) Sketch the phase diagram for Xe, showing the four points given and indicating the area in which each phase is stable. (b) Which is denser, Xe(s) or Xe(l)? Explain. (c) If Xe gas is cooled under an external pressure of 13 kPa, will it undergo condensation or deposition? Explain.

11.45 The normal melting and boiling points of O_2 are −218 °C and −183 °C, respectively. Its triple point is at −219 °C and 0.152 kPa, and its critical point is at −119 °C and 5.04×10^3 kPa. (a) Sketch the phase diagram for O_2, showing the four points given and indicating the area in which each phase is stable. (b) Will $O_2(s)$ float on $O_2(l)$? Explain. (c) As it is heated, will solid O_2 sublime or melt under a pressure of 1 bar?

LIQUID CRYSTALS (SECTION 11.7)

11.46 In terms of the arrangement and freedom of motion of the molecules, how are the nematic liquid crystalline phase and an ordinary liquid phase similar? How are they different?

11.47 What observations made by Reinitzer on cholesteryl

benzoate suggested that this substance possesses a liquid crystalline phase?

11.48 The molecules shown in Figure 11.32 possess polar groups (that is, groupings of atoms that give rise to sizeable dipole moments within the molecules). How might the presence of polar groups enhance the tendency towards liquid crystal formation?

11.49 One of the more effective liquid crystalline substances employed in LCDs is the molecule

$$CH_3(CH_2)_2CH{=}CH{-}\overset{\displaystyle CH_2{-}CH_2}{\underset{\displaystyle CH_2{-}CH_2}{CH}}\ \ \overset{\displaystyle CH_2{-}CH_2}{\underset{\displaystyle CH_2{-}CH_2}{CH{-}CH}}\ \ \overset{\displaystyle CH_2{-}CH_2}{\underset{\displaystyle CH_2{-}CH_2}{CH{-}C{\equiv}N}}$$

(a) How many double bonds are there in this molecule?
(b) Describe the features of the molecule that make it prone to show liquid crystalline behaviour.

11.50 For a given substance, the liquid crystalline phase tends to be more viscous than the liquid phase. Why?

11.51 Describe how a cholesteric liquid crystal phase differs from a nematic phase.

11.52 It often happens that a substance possessing a smectic liquid crystalline phase just above the melting point passes into a nematic liquid crystalline phase at a higher temperature. Account for this type of behaviour.

11.53 The smectic liquid crystalline phase can be said to be more highly ordered than the nematic phase. In what sense is this true?

STRUCTURES OF SOLIDS (Section 11.8)

11.54 How does an amorphous solid differ from a crystalline one? Give an example of an amorphous solid.

11.55 Amorphous silica has a density of about 2.2 g cm^{-3}, whereas the density of crystalline quartz is 2.65 g cm^{-3}. Account for this difference in densities.

11.56 Perovskite, a mineral composed of Ca, O and Ti, has the cubic unit cell shown in the drawing. What is the chemical formula of this mineral?

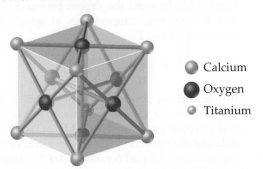

○ Calcium
● Oxygen
○ Titanium

11.57 Rutile is a mineral composed of Ti and O. Its unit cell, shown in the drawing, contains Ti atoms at each corner and a Ti atom at the centre of the cell. Four O atoms are on the opposite faces of the cell, and two are entirely within the cell. (a) What is the chemical formula of this mineral? (b) What is the nature of the bonding that holds the solid together?

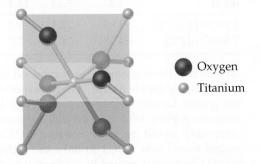

● Oxygen
○ Titanium

11.58 An element crystallises in a body-centred cubic lattice. The edge of the unit cell is 286 pm, and the density of the crystal is 7.92 g cm^{-3}. Calculate the atomic mass of the element.

11.59 KCl has the same structure as NaCl. The length of the unit cell is 628 pm. The density of KCl is 1.984 g cm^{-3}, and its formula mass is 74.55 u. Using this information, calculate Avogadro's number.

11.60 What is the coordination number of each sphere in (a) a three-dimensional, close-packed array of equal-sized spheres; (b) a primitive cubic structure; (c) a body-centred cubic lattice?

11.61 What is the coordination number of (a) Na^+ in the NaCl structure, Figure 11.42; (b) Zn^{2+} in the ZnS unit cell, Figure 11.49(b); (c) Ca^{2+} in the CaF_2 unit cell, Figure 11.49(c)?

11.62 Clausthalite is a mineral composed of lead selenide (PbSe). The mineral adopts an NaCl-type structure. The density of PbSe at 25 °C is 8.27 g cm^{-3}. Calculate the length of an edge of the PbSe unit cell.

11.63 Nickel oxide (NiO) crystallises in the NaCl type of crystal structure (Figure 11.42). The length of the unit cell of NiO is 418 pm. Calculate the density of NiO.

11.64 The mineral uraninite (UO_2) adopts a fluorite structure (Figure 11.49(c)) in which the length of an edge of the unit cell is 546.8 pm. (a) Will the uranium ions be represented by the larger or the smaller spheres in Figure 11.49(c)? Explain. (b) Calculate the density of uraninite.

11.65 A particular form of cinnabar (HgS) adopts the zinc blende structure, Figure 11.49(b). The length of the unit cell side is 585.2 pm. (a) Calculate the density of HgS in this form. (b) The mineral tiemmanite (HgSe) also forms a solid phase with the zinc blende structure. The length of the unit cell side in this mineral is 608.5 pm. What accounts for the larger unit cell length in tiemmanite? (c) Which of the two substances has the higher density? How do you account for the difference in densities?

BONDING IN SOLIDS (Section 11.9)

11.66 What kinds of attractive forces exist between particles in **(a)** molecular crystals, **(b)** covalent-network crystals, **(c)** ionic crystals, **(d)** metallic crystals?

11.67 Indicate the type of crystal (molecular, metallic, covalent-network or ionic) each of the following would form upon solidification: **(a)** $CaCO_3$, **(b)** Pt, **(c)** ZrO_2 (melting point, 2677 °C), **(d)** Kr, **(e)** benzene (C_6H_6), **(f)** I_2.

11.68 Covalent bonding occurs in both molecular and covalent-network solids. Why do these two kinds of solids differ so greatly in their hardness and melting points?

11.69 Which type (or types) of crystalline solid is characterised by each of the following: **(a)** high mobility of electrons throughout the solid; **(b)** softness, relatively low melting point; **(c)** high melting point and poor electrical conductivity; **(d)** network of covalent bonds; **(e)** charged particles throughout the solid.

11.70 A white substance melts with some decomposition at 730 °C. As a solid, it is a non-conductor of electricity, but it dissolves in water to form a conducting solution. Which type of solid (Table 11.7) might the substance be?

11.71 You are given a white substance that sublimes at 3000 °C; the solid is a non-conductor of electricity and is insoluble in water. Which type of solid (Table 11.7) might this substance be?

ADDITIONAL EXERCISES

11.72 Suppose you have two colourless molecular liquids, one boiling at –84 °C, the other at 34 °C, and both at atmospheric pressure. Which of the following statements is correct? For those that are not correct, modify the statement so that it is correct. **(a)** The higher-boiling liquid has greater total intermolecular forces than the other. **(b)** The lower-boiling liquid must consist of non-polar molecules. **(c)** The lower-boiling liquid has a lower molecular mass than the higher-boiling liquid. **(d)** The two liquids have identical vapour pressures at their normal boiling points. **(e)** At 34 °C both liquids have vapour pressures of 101.3 kPa.

11.73 In dichloromethane, CH_2Cl_2 ($\mu = 1.60$ D), the dispersion force contribution to the intermolecular attractive forces is about five times larger than the dipole-dipole contribution. Would you expect the relative importance of the two kinds of intermolecular attractive forces to differ **(a)** in dibromomethane ($\mu = 1.43$ D), **(b)** in difluoromethane ($\mu = 1.93$ D) Explain.

11.74 When an atom or group of atoms is substituted for an H atom in benzene (C_6H_6), the boiling point changes. Explain the order of the following boiling points: C_6H_6 (80 °C), C_6H_5Cl (132 °C), C_6H_5Br (156 °C), C_6H_5OH (182 °C).

11.75 **(a)** When you exercise vigorously, you sweat. How does this help your body cool? **(b)** A flask of water is connected to a vacuum pump. A few moments after the pump is turned on, the water begins to boil. After a few minutes, the water begins to freeze. Explain why these processes occur.

11.76 The following table gives the vapour pressure of hexafluorobenzene (C_6F_6) as a function of temperature:

Temperature (K)	Vapour pressure (kPa)
280.0	4.321
300.0	12.32
320.0	30.00
330.0	44.57
340.0	64.36

(a) By plotting these data in a suitable fashion, determine whether the Clausius–Clapeyron equation is obeyed. If it is obeyed, use your plot to determine $\Delta_{vap}H$ for C_6F_6. **(b)** Use these data to determine the boiling point of the compound.

11.77 Suppose the vapour pressure of a substance is measured at two different temperatures. **(a)** By using the Clausius–Clapeyron equation, Equation 11.1, derive the following relationship between the vapour pressures, P_1 and P_2, and the absolute temperatures at which they were measured, T_1 and T_2:

$$\ln \frac{P_1}{P_2} = -\frac{\Delta_{vap}H}{R}\left(\frac{1}{T_1} - \frac{1}{T_2}\right)$$

(b) The melting point of potassium is 63.2 °C. Molten potassium has a vapour pressure of 1.33 kPa at 443 °C and a vapour pressure of 53.32 kPa at 708 °C. Use these data and the equation in part (a) to calculate the heat of vaporisation of liquid potassium. **(c)** By using the equation in part (a) and the data given in part (b), calculate the boiling point of potassium. **(d)** Calculate the vapour pressure of liquid potassium at 100 °C.

11.78 The following data present the temperatures at which certain vapour pressures are achieved for dichloromethane (CH_2Cl_2) and iodomethane (CH_3I):

Vapour pressure (kPa)	1.33	5.33	13.3	53.3
T for CH_2Cl_2 (°C)	–43.3	–22.3	–6.3	24.1
T for CH_3I (°C)	–45.8	–24.2	–7.0	25.3

(a) Which of the two substances is expected to have the greater dipole–dipole forces? Which is expected to have the greater London dispersion forces? Based on your answers, explain why it is difficult to predict which compound would be more volatile. **(b)** Which compound would you expect to have the higher boiling point? Check your answer in a reference book such as the *CRC Handbook of Chemistry and Physics*. **(c)** The order of volatility of these two substances changes as the temperature is increased. What quantity must be different for the two substances in order for this phenomenon to occur? **(d)** Substantiate your answer for part (c) by drawing an appropriate graph.

11.79 A watch with a liquid crystal display (LCD) does not function properly when it is exposed to low temperatures during a trip to Antarctica. Explain why the LCD might not function well at low temperature.

11.80 The elements xenon and gold both have solid-state structures with face-centred cubic unit cells, yet Xe melts at –112 °C and gold melts at 1064 °C. Account for these greatly different melting points.

11.81 (a) Consider the cubic unit cells (Figure 11.40) with an atom located at each lattice point. Calculate the net number of atoms in (i) a primitive cubic unit cell, (ii) a body-centred cubic unit cell, (iii) a face-centred cubic unit cell. (b) Why can't $CaCl_2$ have the same crystal structure as NaCl?

11.82 In a typical X-ray crystallography experiment, X-rays of wavelength λ = 71 pm are generated by bombarding molybdenum metal with an energetic beam of electrons. Why are these X-rays more effectively diffracted by crystals than is visible light?

11.83 In their study of X-ray diffraction, William and Lawrence Bragg determined that the relationship between the wavelength of the radiation (λ), the angle at which the radiation is diffracted (θ) and the distance between the layers of atoms in the crystal that cause the diffraction (d) is given by $n\lambda = 2d \sin \theta$. X-rays from a copper X-ray tube (λ = 154 pm) are diffracted at an angle of 14.22 degrees by crystalline silicon. Using the Bragg equation, calculate the interplanar spacing in the crystal, assuming n = 1 (first-order diffraction).

INTEGRATIVE EXERCISES

11.84 Spinel is a mineral that contains 37.9% Al, 17.1% Mg and 45.0% O, by mass, and has a density of 3.57 g cm^{-3}. The unit cell is cubic, with an edge length of 809 pm. How many atoms of each type are in the unit cell?

11.85 (a) At the molecular level, what factor is responsible for the steady increase in viscosity with increasing molecular weight in the hydrocarbon series shown in Table 11.4? (b) Although the viscosity varies over a factor of more than two in the series from hexane to nonane, the surface tension at 25 °C increases by only about 20% in the same series. How do you account for this? (c) *l*-octanol, $CH_3CH_2CH_2CH_2CH_2CH_2CH_2CH_2OH$ has a viscosity of 1.01×10^{-2} kg m^{-1} s^{-1}, much higher than nonane, which has about the same molecular weight. What accounts for this difference? How does your answer relate to the difference in normal boiling points for these two substances?

11.86 The table shown here lists the molar heats of vaporisation for several organic compounds. Use specific examples from this list to illustrate how the heat of vaporisation varies with (a) molar mass, (b) molecular shape, (c) molecular polarity, (d) hydrogen-bonding interactions. Explain these comparisons in terms of the nature of the intermolecular forces at work. (You may find it helpful to draw out the structural formula for each compound.)

Compound	Heat of vaporisation (kJ mol^{-1})
$CH_3CH_2CH_3$	19.0
$CH_3CH_2CH_2CH_2CH_3$	27.6
$CH_3CHBrCH_3$	31.8
CH_3COCH_3	32.0
$CH_3CH_2CH_2Br$	33.6
$CH_3CH_2CH_2OH$	47.3

11.87 Liquid butane, C_4H_{10}, is stored in cylinders, to be used as a fuel. The normal boiling point of butane is listed as -0.5 °C. (a) Suppose the tank is standing in the sun and reaches a temperature of 46 °C. Would you expect the pressure in the tank to be greater or less than atmospheric pressure? How does the pressure within the tank depend on how much liquid butane is in it? (b) Suppose the valve to the tank is opened and a few litres of butane are allowed to escape rapidly. What do you expect would happen to the temperature of the remaining liquid butane in the tank? Explain. (c) How much heat must be added to vaporise 155 g of butane if its heat of vaporisation is 21.3 kJ mol^{-1}? What volume does this much butane occupy at 100.6 kPa and 35 °C?

11.88 Using information in Appendices B and C, calculate the minimum number of grams of $C_3H_8(g)$ that must be combusted to provide the energy necessary to convert 2.50 kg of H_2O from its solid form at -14.0 °C to its liquid form at 60.0 °C.

11.89 In a certain type of nuclear reactor, liquid sodium metal is employed as a circulating coolant in a closed system, protected from contact with air or water. Much like the coolant that circulates in a car engine, the liquid sodium carries heat from the hot reactor core to heat exchangers. (a) What properties of the liquid sodium are of special importance in this application? (b) The viscosity of liquid sodium varies with temperature as follows:

Temperature (°C)	Viscosity (kg m^{-1} s^{-1})
100	7.05×10^{-4}
200	4.50×10^{-4}
300	3.45×10^{-4}
600	2.10×10^{-4}

What forces within the liquid sodium are likely to be the major contributors to the viscosity? Why does viscosity decrease with increasing temperature?

11.90 The vapour pressure of a volatile liquid can be determined by slowly bubbling a known volume of gas through it at a known temperature and pressure. In an experiment, 5.00 dm^3 of N_2 gas is passed through 7.2146 g of liquid benzene, C_6H_6, at 26.0 °C. The liquid remaining after the experiment weighs 5.1493 g. Assuming that the gas becomes saturated with benzene vapour and that the total gas volume and temperature remain constant, what is the vapour pressure of the benzene in kPa?

MasteringChemistry (www.pearson.com.au/masteringchemistry)

Make learning part of the grade. Access:

- tutorials with peronalised coaching
- study area
- Pearson eText

PHOTO/ART CREDITS

376, 412 Getty Images Australia; **379 (top left)** Alamy Limited, **(top middle)** Clive Streeter © Dorling Kindersley, Courtesy of The Science Museum, London; **386 (Figure 11.10)** Ted Kinsman/Photo Researchers, Inc, **(Figure 11.11)** Leslie Garland Picture Library/Alamy; **389** Kristen Brochmann, Fundamental Photographs, NYC; **390** Hermann Eisenbeiss/Photo Researchers, Inc; **391, 401** Richard Megna, Fundamental Photographs, NYC; **403** Yuriko Nakao, Reuters Limited/Picture Media.

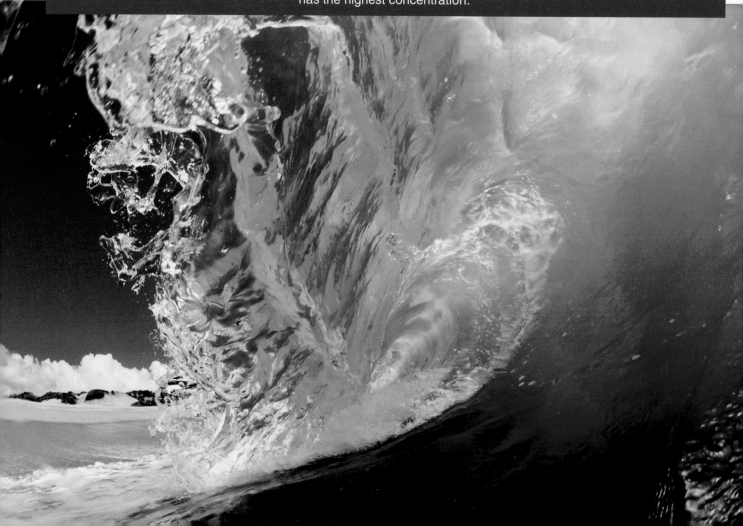

12

PROPERTIES OF SOLUTIONS

The water in the ocean is an aqueous solution of many dissolved substances, of which sodium chloride has the highest concentration.

KEY CONCEPTS

12.1 THE SOLUTION PROCESS
We consider what happens at the molecular level when a substance dissolves, paying particular attention to the role of *intermolecular forces*. Two important aspects of the solution process are the natural tendency of particles to mix and changes in *energy*.

12.2 SATURATED SOLUTIONS AND SOLUBILITY
We learn that when a *saturated solution* is in contact with undissolved solute, the dissolved and undissolved solutes are in *equilibrium*. The amount of solute in a saturated solution defines the *solubility* of the solute, the extent to which a particular solute dissolves in a particular solvent.

12.3 FACTORS AFFECTING SOLUBILITY
We consider the major factors affecting solubility. The nature of the solute and solvent determines the kinds of intermolecular forces between solute and solvent particles and strongly influences solubility. Temperature also affects solubility: most solids are more soluble in water at higher temperatures, whereas gases are less soluble in water at higher temperatures. The solubility of gases increases with increasing pressure.

12.4 WAYS OF EXPRESSING CONCENTRATION
We discuss several common ways of expressing concentration, including *mole fraction*, *molarity* and *molality*.

12.5 COLLIGATIVE PROPERTIES
We observe that some physical properties of solutions depend only on concentration and not on the identity of the solute. These *colligative properties* include the extent to which the solute lowers the vapour pressure, increases the boiling point and decreases the freezing point of the solvent. The *osmotic pressure* of a solution is also a colligative property.

12.6 COLLOIDS
We investigate *colloids*, mixtures that are not true solutions but consist of a solute-like phase (the dispersed phase) and a solvent-like phase (the dispersion medium). The dispersed phase consists of particles larger than typical molecular sizes.

When we think of solutions, we tend to think of liquids, such as that of the breaking ocean wave in the chapter opening photograph. However, solutions can be gaseous, such as the air we breathe (a homogeneous mixture of nitrogen and oxygen) or solids, such as sterling silver, a homogeneous mixture of about 7% copper in silver (▼ TABLE 12.1).

Each of the substances in a mixture is called a *component* of the solution with the *solvent* being the component present in greatest amount. The other component(s) in the solution are called the *solute(s)*. In this chapter we deal primarily with liquid solutions and particularly with aqueous solutions of ionic substances. Our goal is to examine the physical properties of solutions and compare them with the physical properties of their components.

TABLE 12.1 • Examples of solutions

State of solution	State of solvent	State of solute	Example
Gas	Gas	Gas	Air
Liquid	Liquid	Gas	Oxygen in water
Liquid	Liquid	Liquid	Alcohol in water
Liquid	Liquid	Solid	Salt in water
Solid	Solid	Gas	Hydrogen in palladium
Solid	Solid	Liquid	Mercury in silver
Solid	Solid	Solid	Silver in gold

12.1 | THE SOLUTION PROCESS

A solution is formed when one substance disperses uniformly throughout another. We learned in Chapter 11 that the molecules or ions of substances in the liquid and solid states experience intermolecular attractive forces that hold them together. Intermolecular forces also operate between solute particles and solvent molecules.

Any of the various kinds of intermolecular forces that we discussed in Chapter 11 can operate between solute and solvent particles in a solution. Ion–dipole forces, for example, dominate in solutions of ionic substances in water. Dispersion forces dominate when a non-polar substance like hexane dissolves in another non-polar one like carbon tetrachloride. Indeed, a major factor determining whether a solution forms is the relative strengths of intermolecular forces between and among the solute and solvent particles.

Solutions form when the attractive forces between solute and solvent particles are comparable in magnitude with those that exist between the solute particles themselves or between the solvent particles themselves. For example, the ionic substance NaCl dissolves readily in water because the attractive interactions between the ions and the polar H_2O molecules overcome the lattice energy of NaCl(s) (∞ Section 4.1, 'General Properites of Aqueous Solutions'). Let's examine this solution process more closely, paying attention to these attractive forces.

 Review this on page 105

When NaCl is added to water (▼ FIGURE 12.1), the water molecules orient themselves on the surface of the NaCl crystals. The positive end of the water dipole is oriented towards the Cl^- ions, and the negative end of the water dipole is oriented towards the Na^+ ions. The ion–dipole attractions between the ions and water molecules are sufficiently strong to pull the ions from their positions in the crystal.

▲ **FIGURE IT OUT**

How does the orientation of H₂O molecules around Na⁺ differ from that around Cl⁻?

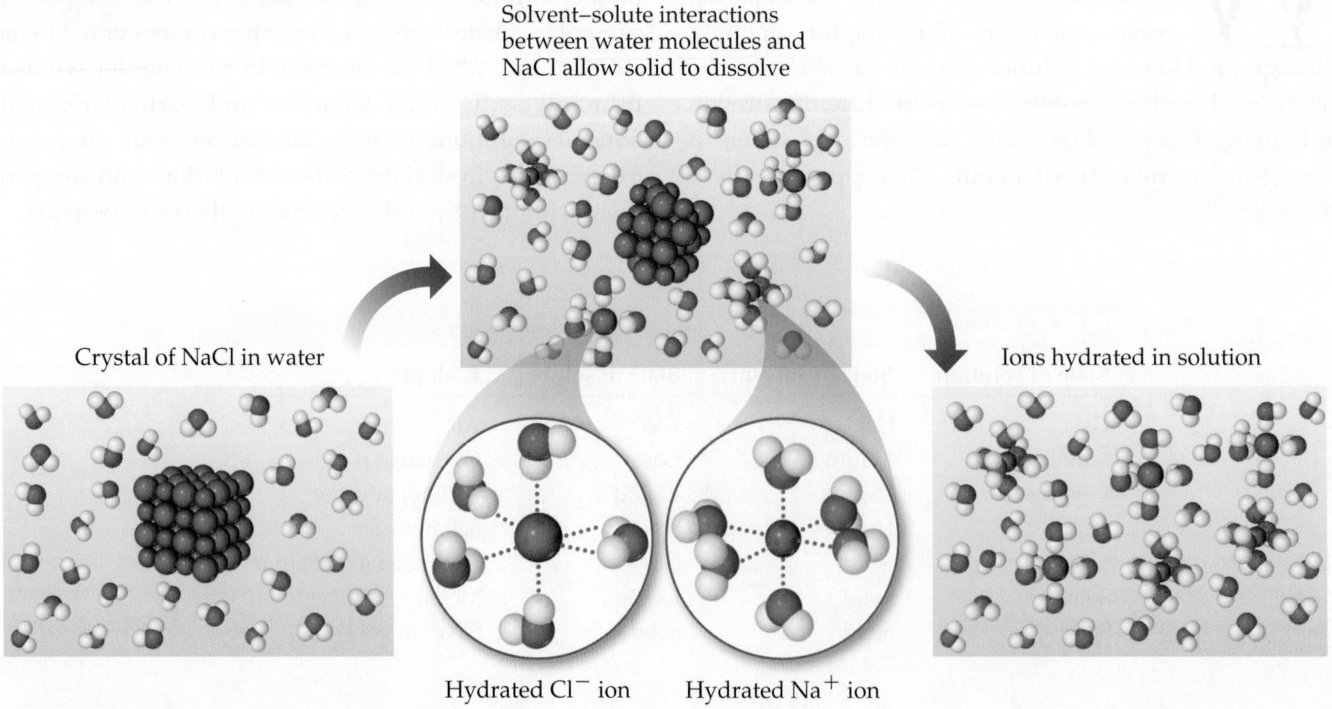

Solvent–solute interactions between water molecules and NaCl allow solid to dissolve

Crystal of NaCl in water

Ions hydrated in solution

Hydrated Cl⁻ ion   Hydrated Na⁺ ion

▲ **FIGURE 12.1** Dissolution of an ionic solid in water.

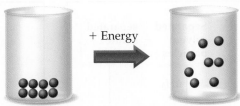

ΔH_1: Separation of solute molecules

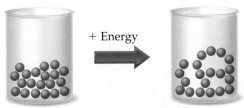

ΔH_2: Separation of solvent molecules

ΔH_3: Formation of solute–solvent interactions

◄ **FIGURE 12.2   Enthalpy contributions to $\Delta_{soln}H$.** The enthalpy changes ΔH_1 and ΔH_2 represent endothermic processes, requiring an input of energy, whereas ΔH_3 represents an exothermic process.

Once separated from the crystal, the Na^+ and Cl^- ions are surrounded by water molecules, as shown in Figure 12.1. We learned in Section 4.1 that interactions like this between solute and solvent molecules are known as **solvation**. When the solvent is water, the interactions are also referred to as **hydration**.

Energy Changes and Solution Formation

Sodium chloride dissolves in water because the water molecules have a sufficient attraction for the Na^+ and Cl^- ions to overcome the attraction of these two ions for one another in the crystal. To form an aqueous solution of NaCl, water molecules must also separate from one another to form spaces in the solvent that will be occupied by the Na^+ and Cl^- ions. Thus we can think of the overall energetics of solution formation as having three components, each with an associated enthalpy change, illustrated in ▲ **FIGURE 12.2.** The overall enthalpy change in forming a solution, $\Delta_{soln}H$, is the sum of three terms:

$$\Delta_{soln}H = \Delta H_1 + \Delta H_2 + \Delta H_3 \qquad [12.1]$$

or $\qquad \Delta_{soln}H = \Delta_{solute}H + \Delta_{solvent}H + \Delta_{mix}H$

▼ **FIGURE 12.3** depicts the enthalpy change associated with each of these components. Separation of the solute particles from one another requires an input of energy to overcome their attractive interactions (for example, separating Na^+ and Cl^- ions). The process is therefore endothermic ($\Delta H_1 > 0$). Separation of solvent molecules to accommodate the solute also requires energy ($\Delta H_2 > 0$). The third component arises from the attractive interactions between solute and solvent and is exothermic ($\Delta H_3 < 0$).

As shown in Figure 12.3, the three enthalpy terms in Equation 12.1 can add together to give either a negative or a positive sum. Thus the formation of a solution can be either exothermic or endothermic. For example, when magnesium sulfate, $MgSO_4$, is added to water, the resultant solution gets quite

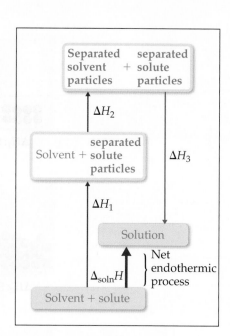

▶ **FIGURE 12.3 Enthalpy changes accompanying the solution process.** The three processes are illustrated in Figure 12.2. In this figure, the diagram on the left illustrates a net exothermic process ($\Delta_{soln}H < 0$); that on the right shows a net endothermic process ($\Delta_{soln}H > 0$).

∞ Review this on page 383

warm: $\Delta_{soln}H = -91.2$ kJ mol^{-1}. In contrast, the dissolution of ammonium nitrate (NH$_4$NO$_3$) is endothermic: $\Delta_{soln}H = 26.4$ kJ mol^{-1}. These particular substances have been used to make the instant heat packs and ice packs that are used to treat athletic injuries. The packs consist of a pouch of water and a dry chemical, MgSO$_4$ for hot packs and NH$_4$NO$_3$ for cold packs. When the pack is squeezed, the seal separating the solid from the water is broken and a solution forms, either increasing or decreasing the temperature.

Processes that are exothermic tend to proceed spontaneously. Although an endothermic solution process, such as the dissolution of NH$_4$NO$_3$ in water, may also proceed spontaneously, the solute–solvent interaction must be strong enough that ΔH_3 is comparable in magnitude with $\Delta H_1 + \Delta H_2$ (see Figure 12.3). This is why ionic solutes like NaCl do not dissolve in non-polar liquids such as petrol. The non-polar hydrocarbon molecules of the petrol would experience only weak attractive interactions with the ions, and these interactions would not compensate for the energies required to separate the ions from one another.

By similar reasoning, a polar liquid such as water does not form solutions with a non-polar liquid such as octane (C$_8$H$_{18}$). The water molecules experience strong hydrogen-bonding interactions with one another (ΔH_1 is large) (∞ Section 11.2, 'Intermolecular Forces'). These attractive forces must be overcome to disperse the water molecules throughout the non-polar liquid. The energy required to separate the H$_2$O molecules is not recovered in the form of attractive interactions between H$_2$O and C$_8$H$_{18}$ molecules (ΔH_3 is small).

Solution Formation, Spontaneity and Disorder

When any amounts of carbon tetrachloride (CCl$_4$) and hexane (C$_6$H$_{14}$) are mixed, they dissolve in one another. We therefore say they are **miscible** in all proportions. Both substances are non-polar and they have similar boiling points (77 °C for CCl$_4$ and 69 °C for C$_6$H$_{14}$). It is therefore reasonable to suppose that the magnitudes of the attractive forces (dispersion forces) between molecules in the two substances and in their solution are comparable. When the two are mixed, dissolving occurs spontaneously; that is, it occurs without any extra input of energy from outside the system. Two distinct factors are involved in processes that occur spontaneously. The most obvious is energy; the other is the distribution of each component into a larger volume.

As we have just seen, *processes in which the energy content of the system decreases tend to occur spontaneously* and such processes are exothermic. Endothermic processes (such as ammonium nitrate dissolving in water) that occur spontaneously are characterised by a more dispersed state of one or more of the components. This results in an overall increase in randomness of the system. In the above case, the densely ordered solid NH_4NO_3 separates into its component ions, NH_4^+ and NO_3^-, which are then dispersed throughout the water solvent. The effect of this increase in randomness of the system is enough to offset the fact that $\Delta H_1 + \Delta H_2 > \Delta H_3$ (see Figure 12.3).

Processes such as this are characterised by a more dispersed state of one or more components, resulting in an overall increase in the randomness of the system. The mixing of CCl_4 and C_6H_{14} provides another simple example. Suppose that we could suddenly remove a barrier that separates 500 cm³ of CCl_4 from 500 cm³ of C_6H_{14}, as in ▶ **FIGURE 12.4(a)**. Before the barrier is removed, each liquid occupies a volume of 500 cm³. All the CCl_4 molecules are in the 500 cm³ to the left of the barrier and all the C_6H_{14} molecules are in the 500 cm³ to the right. When equilibrium is established after the barrier has been removed, the two liquids together occupy a volume of about 1000 cm³. Formation of a homogeneous solution has increased the degree of dispersal, or randomness, because the molecules of each substance are now mixed and distributed in a volume twice as large as that which they individually occupied before mixing. This example illustrates our second basic principle: *Processes occurring at a constant temperature in which the randomness or dispersal in space of the system increases tend to occur spontaneously.*

In most cases, *formation of solutions is favoured by the increase in randomness that accompanies mixing*. Consequently, a solution will form unless solute–solute or solvent–solvent interactions are too strong relative to the solute–solvent interactions.

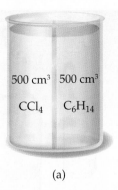

(a)

(b)

▲ **FIGURE 12.4** **Increasing randomness in a solution process.** A homogeneous solution of CCl_4 and C_6H_{14} forms when a barrier separating the two liquids is removed. Each CCl_4 molecule of the solution in (b) is more dispersed in space than it was in the left compartment in (a), and each C_6H_{14} molecule in (b) is more dispersed than it was in the right compartment in (a).

 CONCEPT CHECK 1

Label the following processes as exothermic or endothermic:
a. breaking solvent–solvent interactions to form separated particles,
b. forming solvent–solute interactions from separated particles.

Solution Formation and Chemical Reactions

In all our discussions of solutions, we must be careful to distinguish the physical process of solution formation from chemical reactions that lead to a solution. For example, nickel metal is dissolved on contact with hydrochloric acid solution because the following chemical reaction occurs:

$$Ni(s) + 2\,HCl(aq) \longrightarrow NiCl_2(aq) + H_2(g) \qquad [12.2]$$

In this instance, the chemical form of the substance being dissolved is changed from Ni to $NiCl_2$. If the solution is evaporated to dryness, $NiCl_2 \cdot 6H_2O(s)$, not Ni(s), is recovered (▼ **FIGURE 12.5**). When NaCl(s) is dissolved in water, on the other hand, no chemical reaction occurs. If the solution is evaporated to dryness, NaCl is recovered. Our focus throughout this chapter is on solutions from which the solute can be recovered unchanged from the solution.

 CONCEPT CHECK 2

Silver chloride, AgCl, is essentially insoluble in water. Would you expect a significant change in the randomness of the system when 10 g of AgCl is added to 500 cm³ of water?

Nickel metal and hydrochloric acid

Nickel reacts with hydrochloric acid, forming $NiCl_2(aq)$ and $H_2(g)$. The solution is of $NiCl_2$, not Ni metal

$NiCl_2 \cdot 6H_2O(s)$ remains when solvent evaporated

▲ FIGURE 12.5 **The reaction between nickel metal and hydrochloric acid is *not* a simple dissolution.**

A CLOSER LOOK

HYDRATES

Frequently, hydrated ions remain in crystalline salts that are obtained by evaporation of water from aqueous solutions. Common examples include $FeCl_3 \cdot 6H_2O$ (iron(III) chloride hexahydrate) and $CuSO_4 \cdot 5H_2O$ (copper(II) sulfate pentahydrate). The $FeCl_3 \cdot 6H_2O$ consists of $Fe(H_2O)_6^{3+}$ and Cl^- ions; the $CuSO_4 \cdot 5H_2O$ consists of $Cu(H_2O)_4^{2+}$ and $SO_4(H_2O)^{2-}$ ions. Water molecules can also occur in positions in the crystal lattice that are not specifically associated with either a cation or an anion. $BaCl_2 \cdot 2H_2O$ (barium chloride dihydrate) is an example. Compounds such as $FeCl_3 \cdot 6H_2O$, $CuSO_4 \cdot 5H_2O$ and $BaCl_2 \cdot 2H_2O$, which contain a salt and water combined in definite proportions, are known as *hydrates*; the water associated with them is called *water of hydration*. ▶ FIGURE 12.6 shows an example of a hydrate and the corresponding anhydrous (water-free) substance.

RELATED EXERCISES: 12.2, 12.72

$CuSO_4 \cdot 5H_2O$ $CuSO_4$

▲ FIGURE 12.6 **A hydrate and its anhydrous salt.** The anhydrous salt is the white substance, which turns blue upon addition of water.

12.2 | SATURATED SOLUTIONS AND SOLUBILITY

As a solid solute begins to dissolve in a solvent, the concentration of solute particles in solution increases and so do their chances of colliding with the surface of the solid (▶ FIGURE 12.7). Such a collision may result in the solute particles becoming reattached to the solid. This process, which is the opposite of the solution process, is called **crystallisation**. Thus two opposing processes occur in a solution in contact with undissolved solute. This situation is represented in a chemical equation by use of a double arrow:

$$\text{Solute} + \text{solvent} \underset{\text{crystallise}}{\overset{\text{dissolve}}{\rightleftharpoons}} \text{solution} \qquad [12.3]$$

FIGURE IT OUT

What two processes are represented in this figure, and what are their relative rates at equilibrium?

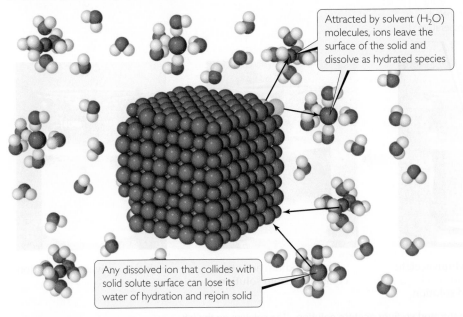

Attracted by solvent (H_2O) molecules, ions leave the surface of the solid and dissolve as hydrated species

Any dissolved ion that collides with solid solute surface can lose its water of hydration and rejoin solid

▲ **FIGURE 12.7** **Dynamic equilibrium in a saturated solution with excess ionic solute.**

When the rates of these opposing processes become equal, no further net increase in the amount of solute in solution occurs. A dynamic equilibrium is established similar to the one between evaporation and condensation discussed in Section 11.5.

A solution that is in equilibrium with undissolved solute is a **saturated solution**. Additional solute will not dissolve if added to a saturated solution. The amount of solute needed to form a saturated solution in a given quantity of solvent is known as the **solubility** of that solute. For example, the solubility of NaCl in water at 0 °C is 35.7 g per 100 cm^3 of water. This is the maximum amount of NaCl that can be dissolved in water to give a stable equilibrium solution at that temperature.

If we dissolve less solute than that needed to form a saturated solution, the solution is **unsaturated**. Thus a solution containing only 10.0 g of NaCl per 100 cm^3 of water at 0 °C is unsaturated because it has the capacity to dissolve more solute.

Under suitable conditions it is sometimes possible to form solutions that contain a greater amount of solute than that needed to form a saturated solution. Such solutions are **supersaturated**. For example, considerably more sodium acetate (CH_3COONa) can dissolve in water at high temperatures than at low temperatures. When a saturated solution of sodium acetate is made at a high temperature and then slowly cooled, all the solute may remain dissolved even though the solubility decreases as the temperature is reduced. Because the solute in a supersaturated solution is present in a concentration higher than the equilibrium concentration, supersaturated solutions are unstable. In order for crystallisation to occur, the molecules or ions of solute must arrange themselves properly to form crystals. Addition of a small crystal of the solute (a seed crystal) provides a template for crystallisation of the excess solute, leading to a saturated solution in contact with excess solid (▼ **FIGURE 12.8**).

FIGURE IT OUT

What evidence is there that the solution is supersaturated?

Amount of sodium acetate dissolved is greater than its solubility at this temperature

Seed crystal of sodium acetate added to supersaturated solution

Excess sodium acetate crystallises from solution

Solution arrives at saturation

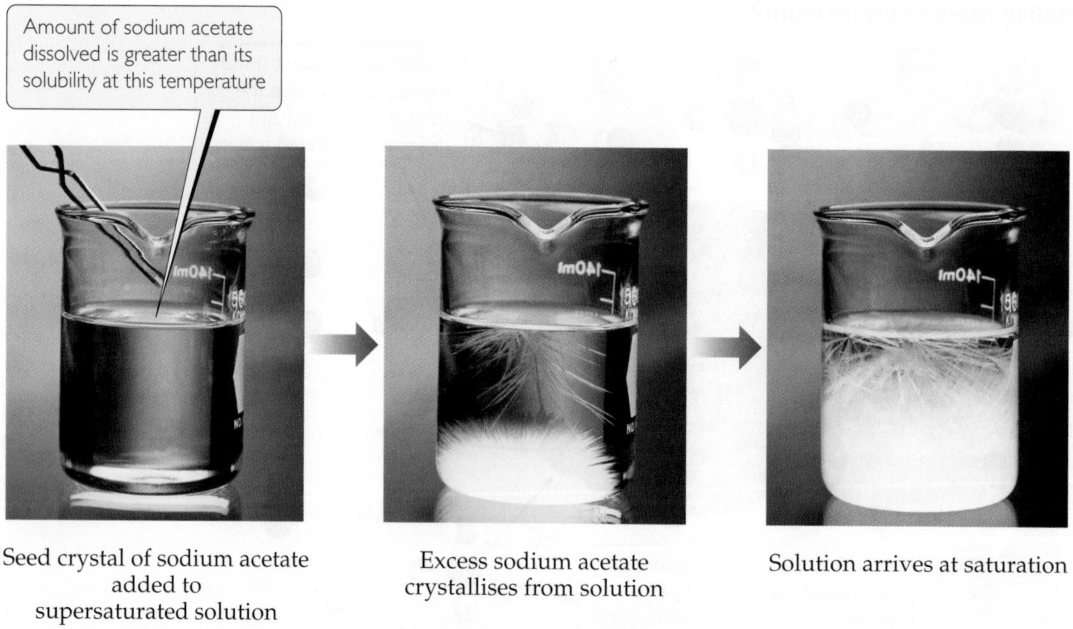

▲ **FIGURE 12.8** **Precipitation from a supersaturated sodium acetate solution.** The solution on the left was formed by dissolving about 170 g of the salt in 100 cm^{-3} of water at 100 °C and then slowly cooling it to 20 °C. Because the solubility of sodium acetate in water at 20 °C is 46 g per 100 cm^{-3} of water, the solution is supersaturated. Addition of a sodium acetate crystal causes the excess solute to crystallise from solution.

CONCEPT CHECK 3
What happens if solute is added to a saturated solution?

12.3 | FACTORS AFFECTING SOLUBILITY

The extent to which one substance dissolves in another depends on the nature of both the solute and the solvent. It also depends on temperature and, at least for gases, on pressure. Let's consider these factors more closely.

Solute–Solvent Interactions

One factor determining solubility is the natural tendency of substances to mix (the tendency of systems to move towards a more dispersed, or random, state). If this were all that was involved, however, we would expect substances to be completely soluble in one another. This is clearly not the case. So what other factors are involved? As we saw in Section 12.1, the relative forces of attraction between the solute and solvent molecules also play very important roles in the solution process.

Although solute–solute and solute–solvent interactions are important in determining the solubilities, considerable insight can often be gained by focusing on the interaction between the solute and solvent. The data in ◀ TABLE 12.2 show, for example, that the solubilities of various simple gases in water increase with increasing molecular mass or polarity. The attractive forces between the gas and solvent molecules are mainly dispersion forces, which increase with increasing size and mass of the gas molecules (∞ Section 11.2, 'Intermolecular Forces'). Thus the data indicate that the solubilities of gases in water increase as the attraction between the solute (gas) and solvent (water) increases. In general, when other

TABLE 12.2 • Solubilities of gases in water at 20 °C, at 1 bar gas pressure

Gas	Solubility (M)
N_2	0.68×10^{-3}
CO	1.03×10^{-3}
O_2	1.36×10^{-3}
Ar	1.48×10^{-3}
Kr	2.75×10^{-3}

 Review this on page 381

factors are comparable, *the stronger the attractions between solute and solvent molecules, the greater the solubility.*

As a result of favourable dipole–dipole attractions between solute molecules and solvent molecules, *polar liquids tend to dissolve readily in polar solvents.* Water is not only polar, but also able to form hydrogen bonds (∞ Section 11.2, 'Intermolecular Forces'). Thus polar molecules, and especially those that can form hydrogen bonds with water molecules, tend to be soluble in water. For example, acetone, a polar molecule whose structural formula is shown in the margin, mixes in all proportions with water. Acetone has a strongly polar C–O double bond and pairs of non-bonding electrons on the O atom that can form hydrogen bonds with water.

Pairs of liquids such as acetone and water that mix in all proportions are miscible, whereas those that do not dissolve in one another are **immiscible**. Petrol, which is a mixture of hydrocarbons, such as hexane (C_6H_{14}), is immiscible in water. Hydrocarbons are non-polar substances because of several factors: the C–C bonds are non-polar, the C–H bonds are nearly non-polar, and the shapes of the molecules are symmetrical enough to cancel much of the weak C–H bond dipoles. The attraction between the polar water molecules and the non-polar hydrocarbon molecules is not sufficiently strong to allow the formation of a solution. *Non-polar liquids tend to be insoluble in polar liquids* (▶ FIGURE 12.9). As a result, hexane (C_6H_{14}) does not dissolve in water.

Many organic compounds have polar groups attached to a non-polar framework of carbon and hydrogen atoms. For example, the series of organic compounds in ▼ TABLE 12.3 all contain the polar OH group. Organic compounds with this molecular feature are called *alcohols*. The O–H bond is able to form hydrogen bonds. For example, ethanol (CH_3CH_2OH) molecules can form hydrogen bonds with water molecules as well as with each other (▼ FIGURE 12.10). As a result, the solute–solute, solvent–solvent and solute–solvent forces are not greatly different in a mixture of CH_3CH_2OH and H_2O. No major change occurs in the environments of the molecules as they are mixed. Therefore the increased entropy when the components mix plays a significant role in solution formation, and ethanol is completely miscible with water.

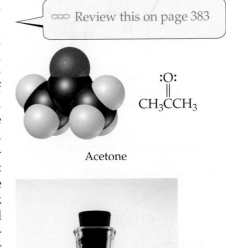

∞ Review this on page 383

$$:\!O\!:$$
$$\parallel$$
$$CH_3CCH_3$$

Acetone

▲ **FIGURE 12.9   Hexane is immiscible with water.** Hexane is the top layer because it is less dense than water.

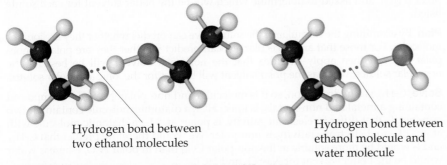

Hydrogen bond between two ethanol molecules

Hydrogen bond between ethanol molecule and water molecule

▲ **FIGURE 12.10   Hydrogen bonding involving OH groups.**

TABLE 12.3 • Solubilities of some alcohols in water and in hexane*

Alcohol	Solubility in H_2O	Solubility in C_6H_{14}
CH_3OH (methanol)	∞	0.12
CH_3CH_2OH (ethanol)	∞	∞
$CH_3CH_2CH_2OH$ (propanol)	∞	∞
$CH_3CH_2CH_2CH_2OH$ (butanol)	0.11	∞
$CH_3CH_2CH_2CH_2CH_2OH$ (pentanol)	0.030	∞
$CH_3CH_2CH_2CH_2CH_2CH_2OH$ (hexanol)	0.0058	∞

* Expressed in mol alcohol/100 g solvent at 20 °C. The infinity symbol (∞) indicates that the alcohol is completely miscible with the solvent.

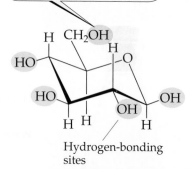

Cyclohexane, C_6H_{12}, which has no polar OH groups, is essentially insoluble in water

OH groups enhance the aqueous solubility because of their ability to hydrogen bond with H_2O

H CH_2OH
HO   H O
HO   OH
    H H   OH   H

Hydrogen-bonding sites

Glucose, $C_6H_{12}O_6$, has five OH groups and is highly soluble in water

▲ **FIGURE 12.11** **Structure and solubility.**

Notice in Table 12.3 that the number of carbon atoms in an alcohol affects its solubility in water. As this number increases, the polar OH group becomes an ever smaller part of the molecule, and the molecule behaves more like a hydrocarbon. The solubility of the alcohol in water decreases correspondingly. In contrast, the solubility of alcohols in a non-polar solvent like hexane (C_6H_{14}) increases as the non-polar hydrocarbon chain lengthens.

One way to enhance the solubility of a substance in water is to increase the number of polar groups the substance contains. For example, increasing the number of OH groups in a solute increases the extent of hydrogen bonding between that solute and water, thereby increasing solubility. Glucose ($C_6H_{12}O_6$, ◀ FIGURE 12.11) has five OH groups on a six-carbon framework, which makes the molecule very soluble in water (83 g dissolves in 100 cm^3 of water at 17.5 °C).

Over the years, examination of different solvent–solute combinations has led to an important generalisation: *Substances with similar intermolecular attractive forces tend to be soluble in one another.* This generalisation is often simply stated as *'like dissolves like'.* Non-polar substances are more likely to be soluble in non-polar solvents; ionic and polar solutes are more likely to be soluble in polar solvents. Network solids such as diamond and quartz are not soluble in either polar or non-polar solvents because of the strong bonding forces within the solid.

CONCEPT CHECK 4

Suppose the hydrogens on the OH groups in glucose (Figure 12.11) were replaced with methyl groups, CH_3. Would you expect the water solubility of the resulting molecule to be higher than, lower than or about the same as the solubility of glucose?

SAMPLE EXERCISE 12.1 Predicting solubility patterns

Predict whether each of the following substances is more likely to dissolve in carbon tetrachloride (CCl_4) or in water: C_7H_{16}, Na_2SO_4, HCl and I_2.

SOLUTION

Analyse We are given two solvents, one that is non-polar (CCl_4) and the other that is polar (H_2O), and asked to determine which will be the better solvent for each solute listed.

Plan By examining the formulae of the solutes, we can predict whether they are ionic or molecular. For those that are molecular, we can predict whether they are polar or non-polar. We can then apply the idea that the non-polar solvent will be best for the non-polar solutes, whereas the polar solvent will be best for the ionic and polar solutes.

Solve C_7H_{16} is a hydrocarbon, so it is molecular and non-polar. Na_2SO_4, a compound containing a metal and non-metals, is ionic; HCl, a diatomic molecule containing two non-metals that differ in electronegativity, is polar; and I_2, a diatomic molecule with atoms of equal electronegativity, is non-polar. We would therefore predict that C_7H_{16} and I_2 would be more soluble in the non-polar CCl_4 than in polar H_2O, whereas water would be the better solvent for Na_2SO_4 and HCl.

PRACTICE EXERCISE

Arrange the following substances in order of increasing solubility in water:

$$H-\overset{H}{\underset{H}{C}}-\overset{H}{\underset{H}{C}}-\overset{H}{\underset{H}{C}}-\overset{H}{\underset{H}{C}}-\overset{H}{\underset{H}{C}}-H \qquad HO-\overset{H}{\underset{H}{C}}-\overset{H}{\underset{H}{C}}-\overset{H}{\underset{H}{C}}-\overset{H}{\underset{H}{C}}-\overset{H}{\underset{H}{C}}-OH$$

$$H-\overset{H}{\underset{H}{C}}-\overset{H}{\underset{H}{C}}-\overset{H}{\underset{H}{C}}-\overset{H}{\underset{H}{C}}-\overset{H}{\underset{H}{C}}-OH \qquad H-\overset{H}{\underset{H}{C}}-\overset{H}{\underset{H}{C}}-\overset{H}{\underset{H}{C}}-\overset{H}{\underset{H}{C}}-\overset{H}{\underset{H}{C}}-Cl$$

Answer: $C_5H_{12} < C_5H_{11}Cl < C_5H_{11}OH < C_5H_{10}(OH)_2$ (in order of increasing polarity and hydrogen-bonding ability)
(See also Exercise 12.10.)

MY WORLD OF CHEMISTRY

FAT-SOLUBLE AND WATER-SOLUBLE VITAMINS

Vitamins have unique chemical structures that affect their solubilities in different parts of the human body. Vitamin C and the B vitamins are soluble in water, for example, whereas vitamins A, D, E and K are soluble in non-polar solvents and in fatty tissue (which is non-polar). Because of their water solubility, vitamins B and C are not stored to any appreciable extent in the body, and so foods containing these vitamins should be included in the daily diet. In contrast, the fat-soluble vitamins are stored in sufficient quantities to keep

vitamin-deficiency diseases from appearing even after a person has subsisted for a long period on a vitamin-deficient diet.

That some vitamins are soluble in water and others are not can be explained in terms of their structures. Notice in ▼ **FIGURE 12.12** that vitamin A (retinol) is an alcohol with a very long carbon chain. Because the OH group is such a small part of the molecule, the molecule resembles the long-chain alcohols listed in Table 12.3. This vitamin is nearly non-polar. In contrast, the vitamin C molecule is smaller and has several OH groups that can form hydrogen bonds with water. In this regard, it is somewhat like glucose.

RELATED EXERCISES: 12.5, 12.20, 12.21

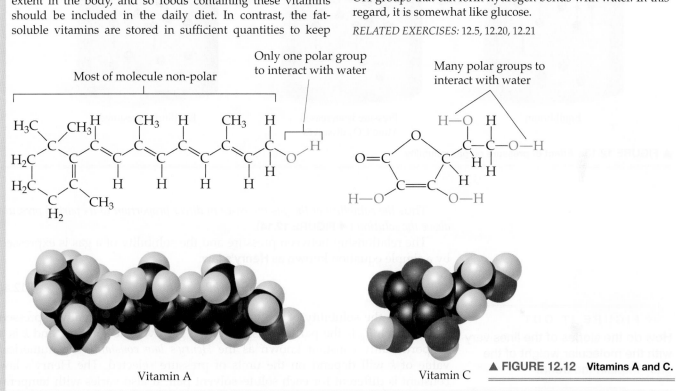

▲ FIGURE 12.12 Vitamins A and C.

Pressure Effects

The solubilities of solids and liquids are not appreciably affected by pressure, whereas the solubility of a gas in any solvent is increased as the pressure over the solvent increases. We can understand the effect of pressure on the solubility of a gas by considering ▼ **FIGURE 12.13**, which shows carbon dioxide gas distributed between the gas and solution phases. When equilibrium is established, the rate at which gas molecules enter the solution equals the rate at which solute molecules escape from the solution to enter the gas phase. The equal number of up and down arrows in the left container of Figure 12.13 represent these opposing processes. Now suppose that we exert added pressure on the piston and compress the gas above the solution, as shown in the middle container of Figure 12.13. If we reduced the volume to half its original value, the pressure of the gas would increase to about twice its original value. As a result the rate at which gas molecules strike the surface to enter the solution phase would therefore increase. Consequently, the solubility of the gas in the solution would increase until an equilibrium is again established; that is, solubility increases until the rate at which gas molecules enter the solution equals the rate at which solute molecules escape from the solution.

FIGURE IT OUT

If the partial pressure of a gas over a solution is doubled, how has the concentration of gas in the solution changed after equilibrium is restored?

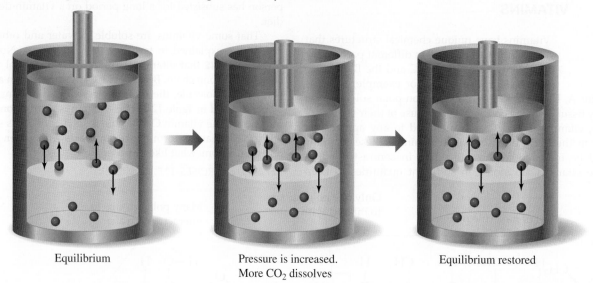

Equilibrium　　　　　Pressure is increased.　　　　Equilibrium restored
　　　　　　　　　　　More CO_2 dissolves

▲ **FIGURE 12.13**　**Effect of pressure on gas solubility.**

Thus *the solubility of the gas increases in direct proportion to its partial pressure above the solution* (◄ **FIGURE 12.14**).

The relationship between pressure and the solubility of a gas is expressed by a simple equation known as **Henry's law**:

$$S_g = kP_g \qquad [12.4]$$

Here, S_g is the solubility of the gas in the solution phase (usually expressed as molarity), P_g is the partial pressure of the gas over the solution, and k is a proportionality constant known as the *Henry's law constant*. The numerical value of k will depend on the units of pressure selected. The Henry's law constant is different for each solute–solvent pair. It also varies with temperature. As an example, the solubility of N_2 gas in water at 25 °C and 79.0 kPa pressure is 5.3×10^{-4} M. The Henry's law constant for N_2 in water at 25 °C is thus given by $(5.3 \times 10^{-4}$ mol dm^{-3})/ (79.0 kPa) $= 6.7 \times 10^{-6}$ mol dm^{-3} kPa^{-1}. If the partial pressure of N_2 is doubled at this temperature, Henry's law predicts that the solubility of N_2 in water will also double to 1.06×10^{-3} M.

Bottlers use the effect of pressure on solubility in producing carbonated beverages such as beer and many soft drinks. These are bottled under a carbon dioxide pressure greater than 1 bar. When the bottles are opened to the air, the partial pressure of CO_2 above the solution decreases. Hence, the solubility of CO_2 decreases and CO_2 bubbles out of the solution (◄ **FIGURE 12.15**).

FIGURE IT OUT

How do the slopes of the lines vary with the molecular weight of the gas? Explain the trend.

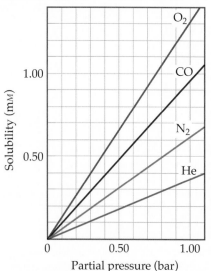

▲ **FIGURE 12.14**　**The solubility of a gas in water is directly proportional to the partial pressure of the gas.** The solubilities are in millimoles per dm^3 of solution.

▲ **FIGURE 12.15**　**Gas solubility decreases as pressure decreases.** CO_2 bubbles out of solution when a carbonated beverage is opened because the CO_2 partial pressure above the solution is reduced.

MY WORLD OF CHEMISTRY

BLOOD GASES AND DEEP-SEA DIVING

Because the solubility of gases increases with increasing pressure, divers who breathe compressed air (▶ FIGURE 12.16) must be concerned about the solubility of gases in their blood. Although the gases are not very soluble at sea level, their solubilities can become appreciable at deep levels where their partial pressures are greater. Thus deep-sea divers must ascend slowly to prevent dissolved gases from being released rapidly from blood and other fluids in the body. These bubbles affect nerve impulses and give rise to the affliction known as decompression sickness, or 'the bends', which is painful and can be fatal. Nitrogen is the main problem because it has the highest partial pressure in air and because it can be removed only through the respiratory system. Oxygen, in contrast, is consumed in metabolism.

Deep-sea divers sometimes substitute helium for nitrogen in the air that they breathe, because helium has a much lower solubility in biological fluids than N_2. For example, divers working at a depth of 30 m experience a pressure of about 400 kPa. At this pressure a mixture of 95% helium and 5% oxygen will give an oxygen partial pressure of about 20 kPa, which is the partial pressure of oxygen in normal air at 1 bar. If

▲ FIGURE 12.16 Solubility increases as pressure increases.

the oxygen partial pressure becomes too great, the urge to breathe is reduced, CO_2 is not removed from the body and CO_2 poisoning occurs. At excessive concentrations in the body, carbon dioxide acts as a neurotoxin, interfering with nerve conduction and transmission.

RELATED EXERCISES: 12.23, 12.24, 12.71

SAMPLE EXERCISE 12.2   **A Henry's law calculation**

Calculate the concentration of CO_2 in a soft drink that is bottled with a partial pressure of CO_2 of 405.3 kPa over the liquid at 25 °C. The Henry's law constant for CO_2 in water at this temperature is 3.06×10^{-4} mol dm^{-3} kPa^{-1}.

SOLUTION

Analyse We are given the partial pressure of CO_2, P_{CO_2} and the Henry's law consant k, and asked to calculate the concentration of CO_2 in the solution.

Plan Use Henry's law, Equation 12.4, to calculate the solubility, S_{CO_2}.

$$S_{CO_2} = kP_{CO_2} = (3.06 \times 10^{-4} \text{ mol dm}^{-3} \text{ kPa}^{-1})(405.3 \text{ kPa})$$

Solve $= 0.124 \text{ mol dm}^{-3} = 0.124 \text{ M}$

PRACTICE EXERCISE

Calculate the concentration of CO_2 in a soft drink after the bottle is opened and equilibrates at 25 °C under a CO_2 partial pressure of 3.04×10^{-2} kPa.

Answer: 9.30×10^{-6} M

(See also Exercises 12.25, 12.63.)

Temperature Effects

The solubility of most solid solutes in water increases as the temperature of the solution increases. ▼ FIGURE 12.17 shows this effect for several ionic substances in water. There are exceptions to this rule, however, as seen for $Ce_2(SO_4)_3$, whose solubility curve slopes downward with increasing temperature. Salts such as these must have a negative entropy of solution, so the order they induce in the water must be greater than the order lost by destroying the crystal lattice.

How does the solubility of KCl at 80 °C compare with that of NaCl at the same temperature?

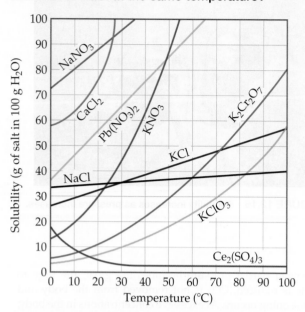

▲ FIGURE 12.17 **Solubilities of several ionic compounds in water as a function of temperature.**

Where would you expect N_2 to fit on this graph?

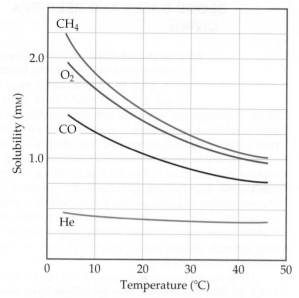

▲ FIGURE 12.18 **Variation of gas solubility with temperature.** Note that solubilities are in units of millimoles per dm^3 (mM), for a constant total pressure of 1 bar in the gas phase.

In contrast to solid solutes, *the solubility of gases in water decreases with increasing temperature* (▲ **FIGURE 12.18**). If a glass of cold tap water is warmed, bubbles of air are seen on the inside of the glass. Similarly, carbonated beverages go flat as they are allowed to warm; as the temperature of the solution increases, the solubility of CO_2 decreases and $CO_2(g)$ escapes from the solution. The decreased solubility of O_2 in water as temperature increases is one of the effects of *thermal pollution* of lakes and streams. The effect is particularly serious in deep lakes because warm water is less dense than cold water. It therefore tends to remain on top of cold water, at the surface. This situation impedes the dissolving of oxygen into the deeper layers, thus stifling the respiration of all aquatic life needing oxygen. Fish may suffocate and die under these conditions.

CONCEPT CHECK 5

Why do bubbles form on the inside wall of a cooking pot when water is heated on the stove, even though the temperature is well below the boiling point of water?

12.4 | WAYS OF EXPRESSING CONCENTRATION

The concentration of a solution can be expressed either qualitatively or quantitatively. The terms *dilute* and *concentrated* are used to describe a solution qualitatively. A solution with a relatively small concentration of solute is said to be dilute; one with a large concentration is said to be concentrated. We use several different ways to express concentration in quantitative terms, and we examine four of these in this section: mass percentage, mole fraction, molarity and molality.

Mass Percentage, ppm and ppb

One of the simplest quantitative expressions of concentration is the **mass percentage** of a component in a solution, given by

$$\text{Mass \% of component} = \frac{\text{mass of component in solution}}{\text{total mass of solution}} \times 100 \qquad [12.5]$$

Thus a solution of hydrochloric acid that is 36% HCl by mass contains 36 g of HCl for each 100 g of solution.

We often express the concentrations of very dilute solution in **parts per million (ppm)**, defined as

$$\text{ppm of component} = \frac{\text{mass of component in solution}}{\text{total mass of solution}} \times 10^6 \qquad [12.6]$$

A solution whose solute concentration is 1 ppm contains 1 g of solute for each million (10^6) grams of solution or, equivalently, 1 mg of solute per kilogram of solution. Because the density of water is 1 g cm^{-3}, 1 kg of a dilute aqueous solution will have a volume very close to 1 dm^3. Thus 1 ppm also corresponds to 1 mg of solute per litre of solution. Consequently,

$$\text{ppm of component} = \frac{\text{mg of solute in solution}}{\text{volume of solution in litres}}$$

The acceptable maximum concentrations of toxic or carcinogenic substances in the environment are often expressed in ppm. For example, the maximum allowable concentration of arsenic in drinking water is 0.010 ppm that is, 0.010 mg of arsenic per litre of water.

For solutions that are even more dilute, **parts per billion (ppb)** is used. A concentration of 1 ppb represents 1 g of solute per billion (10^9) grams of solution, or 1 microgram (μg) of solute per litre of solution. Thus the allowable concentration of arsenic in water can be expressed as 10 ppb.

 CONCEPT CHECK 6

A solution of SO_2 in water contains 0.00023 g of SO_2 per litre of solution. What is the concentration of SO_2 in ppm? In ppb?

SAMPLE EXERCISE 12.3 Calculation of mass-related concentrations

(a) A solution is made by dissolving 13.5 g of glucose ($C_6H_{12}O_6$) in 0.100 kg of water. What is the mass percentage of solute in this solution? **(b)** A 2.5 g sample of groundwater was found to contain 5.4 μg of Zn^{2+}. What is the concentration of Zn^{2+} in parts per million (ppm)?

SOLUTION

Analyse (a) We are given the number of grams of solute (13.5 g) and the number of grams of solvent (0.100 kg = 100 g). From this we must calculate the mass percentage of solute.

Plan We can calculate the mass percentage by using Equation 12.5. The mass of the solution is the sum of the mass of solute (glucose) and the mass of solvent (water).

Solve $\text{Mass \% of glucose} = \dfrac{\text{mass glucose}}{\text{total soln}} \times 100 = \dfrac{13.5\ \text{g}}{(13.5\ \text{g} + 100\ \text{g})} \times 100 = 11.9\%$

Comment The mass percentage of water in this solution is $(100 - 11.9)\% = 88.1\%$.

Analyse (b) In this case we are given the number of micrograms of solute. Because 1 μg is 1×10^{-6} g, 5.4 μg = 5.4×10^{-6} g.

Plan We calculate the parts per million using Equation 12.6.

Solve $\text{ppm} = \dfrac{\text{mass of solute}}{\text{mass of soln}} \times 10^6 = \dfrac{5.4 \times 10^{-6}\ \text{g}}{2.5\ \text{g}} \times 10^6 = 2.2\ \text{ppm}$

PRACTICE EXERCISE

(a) Calculate the mass percentage of NaCl in a solution containing 1.50 g of NaCl in 50.0 g of water. **(b)** A commercial bleaching solution contains 3.62 mass % sodium hypochlorite, NaOCl. What is the mass of NaOCl in a bottle containing 2.50 kg of bleaching solution?

Answers: **(a)** 2.91%, **(b)** 90.5 g of NaOCl

(See also Exercises 12.26(a), 12.27(b).)

Mole Fraction, Molarity and Molality

Concentration expressions are often based on the number of moles of one or more components of the solution. The three most commonly used are mole fraction, molarity and molality.

Recall from Section 10.6 that the *mole fraction* of a component of a solution is given by

$$\text{Mole fraction of component} = \frac{\text{moles of component}}{\text{total moles of all components}} \quad [12.7]$$

The symbol X is commonly used for mole fraction, with a subscript to indicate the component of interest. For example, the mole fraction of HCl in a hydrochloric acid solution is represented as X_{HCl}. Thus a solution containing 1.00 mol of HCl (36.5 g) and 8.00 mol of water (144 g) has a mole fraction of HCl of $X_{HCl} = (1.00 \text{ mol})/(1.00 \text{ mol} + 8.00 \text{ mol}) = 0.111$. Mole fractions have no units because the units in the numerator and the denominator cancel. The sum of the mole fractions of all components of a solution must equal 1. Thus in the aqueous HCl solution, $X_{H_2O} = 1.000 - 0.111 = 0.889$. Mole fractions are very useful when dealing with gases, as we saw in Section 10.6 (Section 10.6, 'Gas Mixtures and Partial Pressures'), but have limited use when dealing with liquid solutions.

Review this on page 351

Recall from Section 4.5 that the *molarity* (M) of a solute in a solution is defined as

$$\text{Molarity} = \frac{\text{moles solute}}{\text{litres of solution}} \quad [12.8]$$

For example, if you dissolve 0.500 mol of Na_2CO_3 in enough water to form 0.250 dm^3 of solution, then the solution has a concentration of $(0.500 \text{ mol})/(0.250 \text{ dm}^3) = 2.00$ M in Na_2CO_3.

The **molality** of a solution, denoted m, is a unit that we haven't encountered in previous chapters. This concentration unit equals the number of moles of solute per kilogram of solvent:

$$\text{Molality} = \frac{\text{moles of solute}}{\text{kilograms of solvent}} \quad [12.9]$$

Thus if you form a solution by mixing 0.200 mol of NaOH and 0.500 kg of water (500 g), the concentration of the solution is $(0.200 \text{ mol})/(0.500 \text{ kg}) = 0.400\ m$ (that is, 0.400 molal) in NaOH.

The definitions of molarity and molality are similar enough that they can be easily confused. Molarity depends on the *volume* of *solution*, whereas molality depends on the *mass* of *solvent*. When water is the solvent, the molality and molarity of dilute solutions are numerically about the same because 1 kg of solvent is nearly the same as 1 kg of solution, and 1 kg of the solution has a volume of about 1 dm^3.

The molality of a given solution does not vary with temperature because masses do not vary with temperature. Molarity, however, changes with temperature because the expansion or contraction of the solution changes its volume. Thus molality is often the concentration unit of choice when a solution is to be used over a range of temperatures.

SAMPLE EXERCISE 12.4 Calculation of molality

A solution is made by dissolving 4.35 g glucose ($C_6H_{12}O_6$) in 25.0 cm^3 of water at 25 °C.
Calculate the molality of glucose in the solution.

SOLUTION

Analyse We are asked to calculate a molality. To do this, we must determine the number of moles of solute (glucose) and the number of kilograms of solvent (water).

Plan We use the molar mass of $C_6H_{12}O_6$ to convert grams to moles. We use the density of water to convert millilitres to kilograms. The molality equals the number of moles of solute divided by the number of kilograms of solvent (Equation 12.9).

Solve Use the molar mass of glucose, 180.2 g mol^{-1}, to convert grams to moles:

$$\text{Mol } C_6H_{12}O_6 = \left(\frac{4.35 \text{ g}}{180.2 \text{ g mol}^{-1}}\right) = 0.0241 \text{ mol}$$

Because water has a density of 1.00 g cm^{-1}, the mass of the solvent is

$$(25.0 \text{ cm}^3)(1.00 \text{ g cm}^{-3}) = 25.0 \text{ g} = 0.0250 \text{ kg}$$

Finally, use Equation 12.9 to obtain the molality:

$$\text{Molality of } C_6H_{12}O_6 = \frac{0.0241 \text{ mol } C_6H_{12}O_6}{0.0250 \text{ kg } H_2O} = 0.964 \text{ } m$$

PRACTICE EXERCISE

What is the molality of a solution made by dissolving 36.5 g of naphthalene ($C_{10}H_8$) in 425 g of toluene (C_7H_8)?

Answer: 0.670 m
(See also Exercises 12.30, 12.31.)

Conversion of Concentration Units

Sometimes the concentration of a given solution needs to be known in several different concentration units. It is possible to interconvert concentration units, as shown in Sample Exercise 12.5.

SAMPLE EXERCISE 12.5 Calculation of mole fraction and molality

An aqueous solution of hydrochloric acid contains 36% HCl by mass.
(a) Calculate the mole fraction of HCl in the solution. **(b)** Calculate the molality of HCl in the solution.

SOLUTION

Analyse We are asked to calculate the concentration of the solute, HCl, in two related concentration units, given only the percentage by mass of the solute in the solution.

Plan In converting concentration units based on the mass or moles of solute and solvent (mass percentage, mole fraction and molality), *it is useful to assume a certain total mass of*

solution. Let's assume that there is exactly 100 g of solution. Because the solution is 36% HCl, it contains 36 g of HCl and $(100 - 36)$ g = 64 g of H_2O. We must convert grams of solute (HCl) to moles in order to calculate either mole fraction or molality. We must convert grams of solvent (H_2O) to moles to calculate mole fractions, and to kilograms to calculate molality.

Solve
(a) To calculate the mole fraction of HCl, we convert the masses of HCl and H_2O to moles and then use Equation 12.7:

$$\text{Moles HCl} = \left(\frac{36 \text{ g}}{36.5 \text{ g mol}^{-1}}\right) = 0.99 \text{ mol}$$

$$\text{Moles } H_2O = \left(\frac{64 \text{ g}}{18 \text{ g mol}^{-1}}\right) = 3.6 \text{ mol}$$

$$X_{HCl} = \frac{\text{moles HCl}}{\text{moles } H_2O + \text{moles HCl}} = \frac{0.99}{3.6 + 0.99} = \frac{0.99}{4.6} = 0.22$$

(b) To calculate the molality of HCl in the solution, we use Equation 12.9. We calculated the number of moles of HCl in part (a), and the mass of solvent is 64 g = 0.064 kg:

$$\text{Molality of HCl} = \frac{0.99 \text{ mol HCl}}{0.064 \text{ kg } H_2O} = 15 \text{ } m$$

PRACTICE EXERCISE

A commercial bleach solution contains 3.62 mass % NaOCl in water. Calculate **(a)** the molality and **(b)** the mole fraction of NaOCl in the solution.

Answers: (a) 0.504 *m*, **(b)** 9.00×10^{-3}
(See also Exercises 12.31, 12.32.)

In order to interconvert molality and molarity, we need to know the density of the solution. ▼ FIGURE 12.19 outlines the calculation of the molality and molarity of a solution from the mass of solute and the mass of solvent. The mass of the solution is the sum of masses of the solvent and solute. The volume of the solution can be calculated from its mass and density.

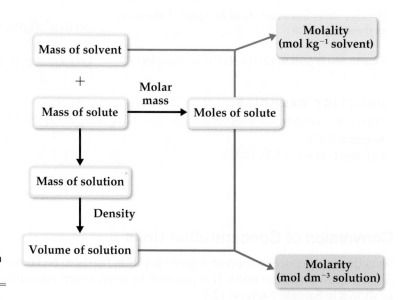

▶ **FIGURE 12.19 Calculating molality and molarity from solute mass, solvent mass and solution density.**

SAMPLE EXERCISE 12.6 | Calculation of molality

A solution contains 5.0 g of toluene (C_7H_8) and 225 g of benzene and has a density of 0.876 g cm^{-3}. Calculate the molarity of the solution.

SOLUTION

Analyse Our goal is to calculate the molarity of a solution, given the masses of solute (5.0 g) and solvent (225 g) and the density of the solution 0.876 g cm^{-3}.

Plan The molarity of a solution is the number of moles of solute divided by the number of litres of solution (Equa-

tion 12.8). The number of moles of solute (C_7H_8) is calculated from the number of grams of solute and its molar mass. The volume of the solution is obtained from the mass of the solution (mass of solute + mass of solvent = 5.0 g + 225 g = 230 g) and its density.

Solve
The number of moles of solute is

$$\text{Moles } C_7H_8 = \left(\frac{5.0 \text{ g}}{92 \text{ g mol}^{-1}}\right) = 0.054 \text{ mol}$$

The density of the solution is used to convert the mass of the solution to its volume:

$$\text{Millilitres soln} = \left(\frac{230 \text{ g}}{0.876 \text{ g cm}^{-3}}\right) = 263 \text{ cm}^3 = 0.263 \text{ dm}^3$$

Molarity is moles of solute per litre of solution:

$$\text{Molarity} = \left(\frac{\text{moles}}{\text{litre soln}}\right) = \left(\frac{0.054 \text{ mol}}{0.263 \text{ dm}^3}\right) = 0.21 \text{ M}$$

Check The magnitude of our answer is reasonable. Rounding moles to 0.05 and litres to 0.25 gives a molarity of

$$\left(\frac{0.05 \text{ mol}}{0.25 \text{ dm}^3}\right) = 0.2 \text{ M}$$

The units for our answer (mol dm^{-3}) are correct and the answer, 0.21 M, has two significant figures, corresponding to the number of significant figures in the mass of solute (2).

Comment Because the mass of the solvent (0.225 kg) and the volume of the solution (0.263 dm^3) are similar in magnitude, the molarity and molality are also similar in magnitude:

$$\text{Molality} = \left(\frac{0.054 \text{ mol}}{0.225 \text{ kg}}\right) = 0.24 \ m$$

PRACTICE EXERCISE

A solution containing equal masses of glycerol (C$_3$H$_8$O$_3$) and water has a density of 1.10 g cm^{-3}. Calculate **(a)** the molality of glycerol, **(b)** the mole fraction of glycerol, **(c)** the molarity of glycerol in the solution.

Answers: **(a)** 10.9 m, **(b)** $X_{C_3H_8O_3} = 0.163$, **(c)** 5.97 M

(See also Exercises 12.32, 12.33.)

12.5 │ COLLIGATIVE PROPERTIES

Some physical properties of solutions differ in important ways from those of the pure solvent. For example, pure water freezes at 0 °C, but aqueous solutions freeze at lower temperatures. Ethylene glycol is added to the water in radiators of cars as an antifreeze to lower the freezing point of the solution. It also raises the boiling point of the solution above that of pure water, making it possible to operate the engine at a higher temperature.

The lowering of the freezing point and the raising of the boiling point are physical properties of solutions that depend on the *quantity* (concentration) but not the *kind* or *identity* of the solute particles. Such properties are called **colligative properties**. (*Colligative* means 'depending on the collection': colligative properties depend on the collective effect of the number of solute particles.) In addition to the decrease in freezing point and the increase in boiling point, vapour pressure reduction and osmotic pressure are colligative properties. As we examine each of these, notice how the concentration of the solute affects the property relative to that of the pure solvent.

Lowering the Vapour Pressure

We learned in Section 11.5 that a liquid in a closed container will establish an equilibrium with its vapour. When that equilibrium is reached, the pressure exerted by the vapour is called the *vapour pressure*. A substance that has no measurable vapour pressure is *non-volatile*, whereas one that exhibits a vapour pressure is *volatile*.

When we compare the vapour pressures of various solvents with those of their solutions, we find that adding a non-volatile solute to a solvent always lowers the vapour pressure. This effect is illustrated in ▼ FIGURE 12.20. The

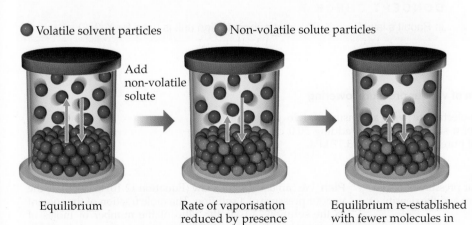

● Volatile solvent particles   ● Non-volatile solute particles

Equilibrium   Rate of vaporisation   Equilibrium re-established
                     reduced by presence   with fewer molecules in
                     of non-volatile solute   gas phase

◀ **FIGURE 12.20   Vapour pressure lowering.** The presence of non-volatile solute particles in a liquid solvent results in a reduction of the vapour pressure above the liquid.

extent to which a non-volatile solute lowers the vapour pressure is proportional to its concentration. This relationship is expressed by **Raoult's law**, which states that *the partial pressure exerted by solvent vapour above a solution, $P_{solution}$, equals the product of the mole fraction of the solvent, times the vapour pressure of the pure solvent, $P°_{solvent}$:*

$$P_{solution} = X_{solvent} P°_{solvent} \qquad [12.10]$$

For example, the vapour pressure of water is 2.33 kPa at 20 °C. Imagine holding the temperature constant while adding glucose ($C_6H_{12}O_6$) to the water so that the resulting solution has $X_{H_2O} = 0.800$ and $X_{C_6H_{12}O_6} = 0.200$. According to Equation 12.10, the vapour pressure of water over the solution will be 80.0% of that of pure water:

$$P_{solution} = (0.800)(2.33 \text{ kPa}) = 1.87 \text{ kPa}$$

In other words, the presence of the non-volatile solute lowers the vapour pressure of the volatile solvent by 2.33 kPa – 1.87 kPa = 0.46 kPa.

The vapour pressure lowering, ΔP, is directly proportional to the mole fraction of the solute, X_{solute}:

$$\Delta P = X_{solute} P°_{solvent} \qquad [12.11]$$

The reduction in vapour pressure depends on the total concentration of solute particles, regardless of whether they are molecules or ions. Remember that vapour pressure lowering is a colligative property, so it depends on the concentration of solute particles and not on their kind. The simplest applications of Raoult's law concern solutes that are non-volatile and non-electrolytes. We consider the effects of *volatile* substances on vapour pressure in the Closer Look box in this section, and we look at the effects of *electrolytes* in our discussion of freezing points and boiling points.

 Review this on page 344

An ideal gas obeys the ideal-gas equation (∞ Section 10.4, 'The Ideal-Gas Equation') and an **ideal solution** obeys Raoult's law. Real solutions best approximate ideal behaviour when the solute concentration is low and when the solute and solvent have similar molecular sizes and similar types of intermolecular attractions.

Many solutions do not obey Raoult's law exactly: they are not ideal solutions. If the intermolecular forces between solute and solvent are weaker than those between solvent and solvent and between solute and solute, then the solvent vapour pressure tends to be greater than predicted by Raoult's law. Conversely, when the interactions between solute and solvent are exceptionally strong, as might be the case when hydrogen bonding exists, the solvent vapour pressure is lower than Raoult's law predicts. Although you should be aware that these departures from an ideal solution occur, we ignore them for the remainder of this chapter.

⚠ CONCEPT CHECK 7

In Raoult's law, $P_A = X_A P°_A$, what concentration unit is used for X_A?

SAMPLE EXERCISE 12.7 | Calculation of vapour pressure lowering

Glycerol ($C_3H_8O_3$) is a non-volatile non-electrolyte with a density of 1.26 g cm^{-3} at 25 °C. Calculate the vapour pressure at 25 °C of a solution made by adding 50.0 cm^3 of glycerol to 500.0 cm^3 of water. The vapour pressure of pure water at 25 °C is 3.17 kPa.

SOLUTION

Analyse Our goal is to calculate the vapour pressure of a solution, given the volumes of solute and solvent and the density of the solute.

Plan We can use Raoult's law (Equation 12.10) to calculate the vapour pressure of a solution. The mole fraction of the solvent in the solution, X_A, is the ratio of the number of moles of solvent (H_2O) to total solution (moles $C_3H_8O_3$ + moles H_2O).

Solve To calculate the mole fraction of water in the solution, we must determine the number of moles of $C_3H_8O_3$ and H_2O:

$$\text{Moles } C_3H_8O_3 = \frac{(50.0 \text{ cm}^3)(1.26 \text{ g cm}^{-3})}{(92.1 \text{ g mol}^{-1})} = 0.684 \text{ mol}$$

$$\text{Moles } H_2O = \frac{(500.0 \text{ cm}^3)(1.00 \text{ g mol}^{-1})}{(18.0 \text{ g mol}^{-1})} = 27.8 \text{ mol}$$

$$X_{H_2O} = \frac{\text{mol } H_2O}{\text{mol } H_2O + \text{mol } C_3H_8O_3} = \frac{27.8}{27.8 + 0.684} = 0.976$$

We now use Raoult's law to calculate the vapour pressure of water for the solution:

$$P_{H_2O} = X_{H_2O}P^\circ_{H_2O} = (0.976)(3.17 \text{ kPa}) = 3.09 \text{ kPa}$$

The vapour pressure of the solution has been lowered by 0.08 kPa relative to that of pure water.

PRACTICE EXERCISE

The vapour pressure of pure water at 110 °C is 142.7 kPa. A solution of ethylene glycol and water has a vapour pressure of 101.3 kPa at 110 °C. Assuming that Raoult's law is obeyed, what is the mole fraction of ethylene glycol in the solution?

Answer: 0.290

(See also Exercise 12.44.)

Boiling-Point Elevation

In Sections 11.5 and 11.6 we examined the vapour pressures of pure substances and how they can be used to construct phase diagrams. How will the phase diagram of a solution, and hence its boiling and freezing points, differ from those of the pure solvent? The addition of a non-volatile solute lowers the vapour pressure of the solution. Thus as shown in ▼ **FIGURE 12.21**, the vapour pressure curve of the solution (blue line) will be shifted downwards relative to the vapour pressure curve of the pure liquid (black line); at any given temperature the vapour pressure of the solution is lower than that of the pure liquid. Recall that the normal boiling point of a liquid is the temperature at which its vapour pressure equals 1 bar (∞ Section 11.5, 'Vapour Pressure'). At the normal boiling point of the pure liquid, the vapour pressure of the solution will be less than 1 bar (Figure 12.21). Therefore, a higher temperature is required to attain a vapour pressure of 1 bar. Thus *the boiling point of the solution is higher than that of the pure liquid.*

∞ Review this on page 397

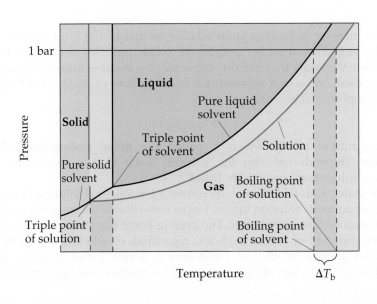

◀ **FIGURE 12.21   Phase diagram illustrating boiling-point elevation.**

A CLOSER LOOK

IDEAL SOLUTIONS WITH TWO OR MORE VOLATILE COMPONENTS

Solutions sometimes have two or more volatile components. Petrol, for example, is a complex solution containing several volatile substances. To gain some understanding of such mixtures, consider an ideal solution containing two components, A and B. The partial pressures of A and B vapours above the solution are given by Raoult's law:

$$P_A = X_A P_A^\circ \quad \text{and} \quad P_B = X_B P_B^\circ$$

The total vapour pressure over the solution is the sum of the partial pressures of each volatile component:

$$P_{total} = P_A + P_B = X_A P_A^\circ + X_B P_B^\circ$$

Consider, for example, a mixture of benzene (C_6H_6) and toluene (C_7H_8) containing 1.0 mol of benzene and 2.0 mol of toluene ($X_{ben} = 0.33$ and $X_{tol} = 0.67$). At 20 °C the vapour pressures of the pure substances are

$$\text{Benzene:} \quad P_{ben}^\circ = 10 \text{ kPa}$$

$$\text{Toluene:} \quad P_{tol}^\circ = 2.9 \text{ kPa}$$

Thus the partial pressures of benzene and toluene above the solution are

$$P_{ben} = (0.33)(10 \text{ kPa}) = 3.3 \text{ kPa}$$

$$P_{tol} = (0.67)(2.9 \text{ kPa}) = 1.9 \text{ kPa}$$

The total vapour pressure is

$$P_{total} = 3.3 \text{ kPa} + 1.9 \text{ kPa} = 5.2 \text{ kPa}$$

The vapour, therefore, is richer in benzene, the more volatile component. The mole fraction of benzene in the vapour is given by the ratio of its vapour pressure to the total pressure (Equation 10.15):

$$X_{ben} \text{ in vapour} = \frac{P_{ben}}{P_{total}} = \frac{3.3 \text{ kPa}}{5.2 \text{ kPa}} = 0.63$$

Although benzene constitutes only 33% of the molecules in the solution, it makes up 63% of the molecules in the vapour.

When ideal solutions are in equilibrium with their vapour, the more volatile component of the mixture will be relatively richer in the vapour. This fact forms the basis of *distillation*, a technique used to separate (or partially separate) mixtures containing volatile components. Distillation is the procedure by which petrochemical plants achieve the separation of crude petroleum into petrol, diesel fuel, lubricating oil and so forth (▼ FIGURE 12.22). It is also used routinely on a small scale in the laboratory. A specially designed *fractional distillation* apparatus can achieve in a single operation a degree of separation that would be equivalent to several successive simple distillations.

RELATED EXERCISES: 12.44, 12.45

▲ FIGURE 12.22 **Separating volatile components.** In an industrial distillation tower, such as the ones shown here, the components of a volatile organic mixture are separated according to boiling-point range.

The increase in boiling point relative to that of the pure solvent, ΔT_b, is directly proportional to the number of solute particles per mole of solvent molecules. We know that molality expresses the number of moles of solute per 1000 g of solvent, which represents a fixed number of moles of solvent. Thus ΔT_b is proportional to molality:

$$\Delta T_b = K_b \, m \qquad\qquad [12.12]$$

The magnitude of K_b, which is called the **molal boiling-point-elevation constant**, depends only on the solvent. Some typical values for several common solvents are given in ▶ TABLE 12.4.

For water, K_b is 0.51 °C m^{-1}; therefore, a 1 m aqueous solution of sucrose or any other aqueous solution that is 1 m in non-volatile solute particles will boil 0.51 °C higher than pure water. The boiling-point elevation is proportional to the concentration of solute particles, regardless of whether the particles are molecules or ions. When NaCl dissolves in water, 2 mol of solute particles (1 mol of Na^+ and 1 mol of Cl^-) are formed for each mole of NaCl that dissolves.

TABLE 12.4 • Molal boiling-point-elevation and freezing-point-depression constants

Solvent	Normal boiling point (°C)	K_b (°C m^{-1})	Normal freezing point (°C)	K_f (°C m^{-1})
Water, H_2O	100.0	0.51	0.0	1.86
Benzene, C_6H_6	80.1	2.64	5.5	5.07
Ethanol, C_2H_5OH	78.4	1.23	−114.6	1.96*
Carbon tetrachloride, CCl_4	76.8	5.26	−22.3	31.3*
Chloroform, $CHCl_3$	61.2	3.80	−63.5	4.60*

* Cryoscopic constants calculated from enthalpy of fusion data in 2008 *CRC Handbook*.

Therefore, a 1 m aqueous solution of NaCl is 1 m in Na^+ and 1 m in Cl^-, making it 2 m in total solute particles. As a result, the boiling-point elevation of a 1 m aqueous solution of NaCl is approximately $(2\ m)(0.51\ °C\ m^{-1}) = 1\ °C$, twice as large as a 1 m solution of a non-electrolyte such as sucrose. Thus to predict the effect of a particular solute on the boiling point properly (or any other colligative property), it is important to know whether the solute is an electrolyte or a non-electrolyte (∞ Section 4.1, 'General Properties of Aqueous Solutions' and Section 4.3, 'Acids, Bases and Neutralisation Reactions').

∞ Review this on pages 104 and 112

CONCEPT CHECK 8

An unknown solute dissolved in water causes the boiling point to increase by 0.51 °C. Does this mean that the concentration of the solute is 1.0 m?

Freezing-Point Depression

When a solution freezes, crystals of pure solvent usually separate out; the solute molecules are not normally soluble in the solid phase of the solvent. When aqueous solutions are partially frozen, for example, the solid that separates out is almost always pure ice. As a result, the part of the phase diagram in ▼ **FIGURE 12.23** that represents the vapour pressure of the solid is

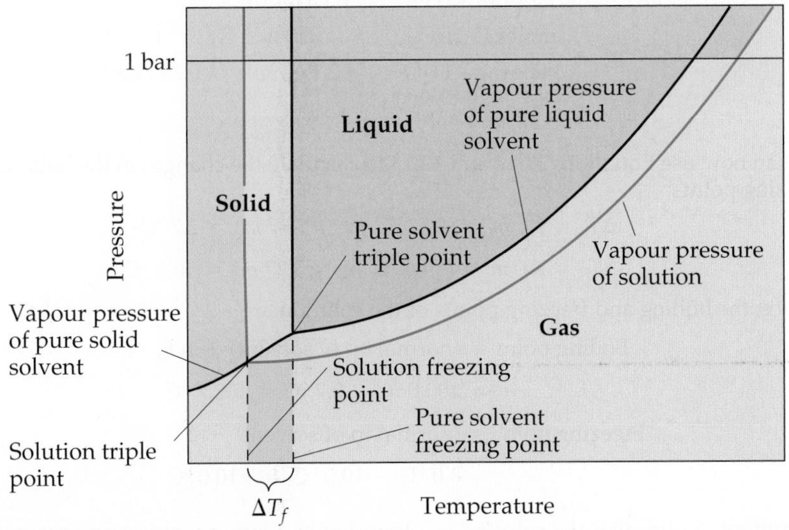

▲ **FIGURE 12.23   Phase diagram illustrating freezing-point depression.**

∞ Review this on page 399

the same as that for the pure liquid. The vapour pressure curves for the liquid and solid phases meet at the triple point (∞ Section 11.6, 'Phase Diagrams'). In Figure 12.23 we see that the triple point of the solution must be at a lower temperature than that in the pure liquid because the solution has a lower vapour pressure than the pure liquid.

The freezing point of a solution is the temperature at which the first crystals of pure solvent begin to form in equilibrium with the solution. Recall from Section 11.6 that the line representing the solid–liquid equilibrium rises nearly vertically from the triple point. Because the triple-point temperature of the solution is lower than that of the pure liquid, *the freezing point of the solution is lower than that of the pure liquid*.

Like the boiling-point elevation, the decrease in freezing point, ΔT_f, is directly proportional to the molality of the solute:

$$\Delta T_f = K_f\, m \qquad [12.13]$$

The values of K_f, the **molal freezing-point-depression constant**, for several common solvents are given in Table 12.4. For water, K_f is 1.86 °C m^{-1}; therefore, a 1 m aqueous solution of sucrose or any other aqueous solution that is 1 m in non-volatile solute particles (such as 0.5 m NaCl) will freeze 1.86 °C lower than pure water.

SAMPLE EXERCISE 12.8 | **Calculation of boiling-point elevation and freezing-point lowering**

Ethylene glycol ($C_2H_6O_2$) is a non-volatile non-electrolyte. Calculate the boiling point and freezing point of a 25.0 mass % solution of ethylene glycol in water.

SOLUTION

Analyse We are given that a solution contains 25.0 mass % of a non-volatile, non-electrolyte solute and asked to calculate the boiling and freezing points of the solution. To do this, we need to calculate the boiling-point elevation and freezing-point depression.

Plan In order to calculate the boiling-point elevation and the freezing-point depression using Equations 12.12 and 12.13, we must express the concentration of the solution as molality. Let's assume for convenience that we have 1000 g of solution. Because the solution is 25.0 mass % ethylene glycol, the masses of ethylene glycol and water in the solution are 250 g and 750 g, respectively. Using these quantities, we can calculate the molality of the solution, which we use with the molal boiling-point-elevation and freezing-point-depression constants (Table 12.4) to calculate ΔT_b and ΔT_f. We add ΔT_b to the boiling point and subtract ΔT_f from the freezing point of the solvent to obtain the boiling point and freezing point of the solution.

Solve The molality of the solution is calculated as follows:

$$\text{Molality} = \frac{\text{moles } C_2H_6O_2}{\text{kilograms } H_2O} = \left(\frac{250\text{ g}}{62.1\text{ g mol}^{-1}}\right)\left(\frac{1}{0.750\text{ kg}}\right)$$

$$= 5.37\ m$$

We can now use Equations 12.12 and 12.13 to calculate the changes in the boiling and freezing points:

$$\Delta T_b = K_b\, m = (0.51\,°C\ m^{-1})(5.37\ m) = 2.7\ °C$$

$$\Delta T_f = K_f\, m = (1.86\,°C\ m^{-1})(5.37\ m) = 10.0\ °C$$

Hence, the boiling and freezing points of the solution are

$$\text{Boiling point} = (\text{normal bp of solvent}) + \Delta T_b$$

$$= 100.0\,°C + 2.7\,°C = 102.7\ °C$$

$$\text{Freezing point} = (\text{normal fp of solvent}) - \Delta T_f$$

$$= 0.0\,°C - 10.0\,°C = -10.0\ °C$$

Comment Notice that the solution is a liquid over a larger temperature range than the pure solvent.

PRACTICE EXERCISE

Calculate the freezing point of a solution containing 0.600 kg of $CHCl_3$ and 42.0 g of eucalyptol ($C_{10}H_{18}O$), a fragrant substance found in the leaves of eucalyptus trees. (See Table 12.4.)

Answer: $-65.6\,°C$

(See also Exercises 12.49, 12.50.)

SAMPLE EXERCISE 12.9   **Freezing-point depression in aqueous solutions**

List the following aqueous solutions in order of their expected freezing point: 0.050 m $CaCl_2$, 0.15 m NaCl, 0.10 m HCl, 0.050 m CH_3COOH, 0.10 m $C_{12}H_{22}O_{11}$.

SOLUTION

Analyse We must order five aqueous solutions according to expected freezing points, based on molalities and the solute formulae.

Plan The lowest freezing point will correspond to the solution with the greatest concentration of solute particles. To determine the total concentration of solute particles in each case, we must determine whether the substance is a non-electrolyte or an electrolyte and consider the number of ions formed when it ionises.

Solve $CaCl_2$, NaCl and HCl are strong electrolytes, CH_3COOH is a weak electrolyte and $C_{12}H_{22}O_{11}$ is a non-electrolyte. The molality of each solution in total particles is as follows:

$$0.050\ m\ CaCl_2 \Rightarrow 0.050\ m\ \text{in}\ Ca^{2+}\ \text{and}\ 0.10\ m\ \text{in}\ Cl^- \Rightarrow 0.15\ m\ \text{in particles}$$

$$0.15\ m\ NaCl \Rightarrow 0.15\ m\ Na^+\ \text{and}\ 0.15\ m\ \text{in}\ Cl^- \Rightarrow 0.30\ m\ \text{in particles}$$

$$0.10\ m\ HCl \Rightarrow 0.10\ m\ \text{in}\ H^+\ \text{and}\ 0.10\ m\ \text{in}\ Cl^- \Rightarrow 0.20\ m\ \text{in particles}$$

$$0.050\ m\ CH_3COOH \Rightarrow \text{weak electrolyte} \Rightarrow \text{between 0.050}\ m\ \text{and 0.10}\ m\ \text{in particles}$$

$$0.10\ m\ C_{12}H_{22}O_{11} \Rightarrow \text{non-electrolyte} \Rightarrow 0.10\ m\ \text{in particles}$$

Because the freezing points depend on the total molality of particles in solution, the expected ordering is 0.15 m NaCl (lowest freezing point), 0.10 m HCl, 0.050 m $CaCl_2$, 0.10 m $C_{12}H_{22}O_{11}$, 0.050 m CH_3COOH (highest freezing point).

PRACTICE EXERCISE

Which of the following solutes will produce the largest increase in boiling point upon addition to 1 kg of water: 1 mol of $Co(NO_3)_2$, 2 mol of KCl, 3 mol of ethylene glycol ($C_2H_6O_2$)?

Answer: 2 mol of KCl because it contains the highest concentration of particles, 2 m K^+ and 2 m Cl^-, giving 4 m in all

(See also Exercises 12.46, 12.48.)

Osmosis

Certain materials, including many membranes in biological systems and synthetic substances such as cellophane, are *semi-permeable*. When in contact with a solution, they allow some molecules to pass through their network of tiny pores, but not others. Most importantly, they generally allow small solvent molecules such as water to pass through, but block larger solute molecules or ions.

Consider a situation in which only solvent molecules are able to pass through a membrane. If such a membrane is placed between two solutions of different concentration, solvent molecules move in both directions through the membrane. The concentration of *solvent* is higher in the solution containing less solute, however, so the rate with which solvent passes from the less concentrated to the more concentrated solution is greater than the rate in the opposite direction. Thus there is a net movement of solvent molecules from the less concentrated solution into the more concentrated one. In this process, called **osmosis**, *the net movement of solvent is always towards the solution with the higher solute concentration.*

⬛ FIGURE IT OUT

If the pure water in the left arm of the U-tube is replaced by a solution more concentrated than the one in the right arm, what will happen?

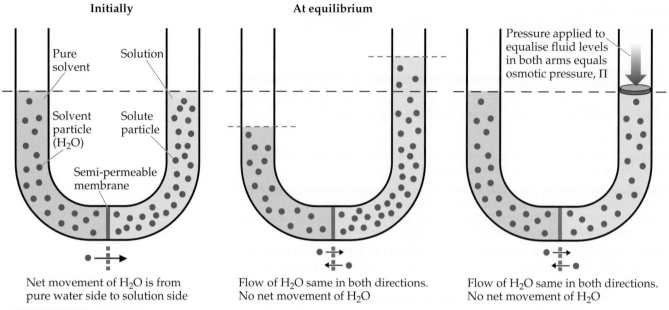

▲ FIGURE 12.24   Osmosis.

▲ **FIGURE 12.24** shows the osmosis that occurs between an aqueous solution and pure water. The U-tube contains water on the left and an aqueous solution on the right. There is a net movement of water through the membrane from left to right. As a result, the liquid levels in the two arms become unequal. Eventually, the pressure difference resulting from the unequal liquid heights becomes so large that the net flow of water ceases. The pressure required to stop osmosis from a pure solvent to a solution is the **osmotic pressure** of the solution. If an external pressure equal to the osmotic pressure is applied to the solution, the liquid levels in the two arms can be equalised, as shown in the right panel of Figure 12.24.

The osmotic pressure obeys a law similar in form to the ideal-gas law, $\Pi V = nRT$, where Π is the osmotic pressure, V is the volume of the solution, n is the number of moles of solute, R is the ideal-gas constant and T is the Kelvin temperature. From this equation, we can write

$$\Pi = \left(\frac{n}{V}\right)RT = \text{M}RT \qquad [12.14]$$

where M is the molarity of the solution. Because the osmotic pressure for any solution depends on the solution concentration, osmotic pressure is a colligative property.

If two solutions of identical osmotic pressure are separated by a semi-permeable membrane, no osmosis will occur. The two solutions are *isotonic* with respect to each other. If one solution is of lower osmotic pressure, it is *hypotonic* with respect to the more concentrated solution. The more concentrated solution is *hypertonic* with respect to the dilute solution.

⬛ CONCEPT CHECK 9

Of two KBr solutions, one 0.5 *m* and the other 0.20 *m*, which is hypotonic with respect to the other?

FIGURE IT OUT

If the fluid surrounding a patient's red blood cells is depleted in electrolytes, is crenation or haemolysis more likely to occur?

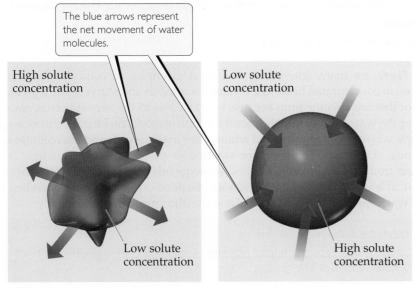

> The blue arrows represent the net movement of water molecules.

High solute concentration

Low solute concentration

Low solute concentration

High solute concentration

Crenation of red blood cell placed in hypertonic environment

Haemolysis of red blood cell placed in hypotonic environment

◀ **FIGURE 12.25**   **Osmosis through red blood cell wall.**

Osmosis plays a very important role in living systems. The membranes of red blood cells, for example, are semi-permeable. Placing a red blood cell in a solution that is *hyper*tonic relative to the intracellular solution (the solution within the cells) causes water to move out of the cell, as shown in ▲ FIGURE 12.25. This causes the cell to shrivel, a process called *crenation*. Placing the cell in a solution that is *hypo*tonic relative to the intracellular fluid causes water to move into the cell. This may cause the cell to rupture, a process called *haemolysis*. People who need body fluids or nutrients replaced but cannot be fed orally are given solutions by intravenous (IV) infusion, which feeds nutrients directly into the veins. To prevent crenation or haemolysis of red blood cells, the IV solutions must be isotonic with the intracellular fluids of the cells.

| SAMPLE EXERCISE 12.10 | Calculations involving osmotic pressure |

The average osmotic pressure of blood is 780.2 kPa at 25 °C. What molarity of glucose ($C_6H_{12}O_6$) will be isotonic with blood?

SOLUTION

Analyse We are asked to calculate the concentration of glucose in water that would be isotonic with blood, given that the osmotic pressure of blood at 25 °C is 780.3 kPa.

Plan Because we are given the osmotic pressure and temperature, we can solve for the concentration, using Equation 12.14.

Solve
$$\Pi = MRT$$

$$M = \frac{\Pi}{RT} = \frac{(780.2 \text{ kPa})}{(8.314 \text{ kPa dm}^3 \text{ mol}^{-1} \text{ K}^{-1})(298 \text{ K})}$$

$$= 0.315 \text{ M}$$

Comment In clinical situations the concentrations of solutions are generally expressed as mass percentages. The mass percentage of a 0.31 M solution of glucose is 5.7%. The concentration of NaCl that is isotonic with blood is 0.16 M because NaCl ionises to form two particles, Na^+ and Cl^- (a 0.155 M solution of NaCl is 0.310 M in particles). A 0.16 M solution of NaCl is 0.9 mass % in NaCl. This kind of solution is known as a physiological saline solution.

PRACTICE EXERCISE

What is the osmotic pressure at 20 °C of a 0.0020 M sucrose ($C_{12}H_{22}O_{11}$) solution?

Answer: 4.9 kPa

(See also Exercises 12.51, 12.52.)

There are many interesting biological examples of osmosis. A cucumber placed in concentrated brine loses water via osmosis and shrivels into a pickle. If a carrot that has become limp because of water loss to the atmosphere is placed in water, the water moves into the carrot through osmosis, making it firm once again. People who eat a lot of salty food retain water in tissue cells and intercellular space because of osmosis. The resultant swelling or puffiness is called *oedema*. Water moves from soil into plant roots and subsequently into the upper portions of the plant, at least in part because of osmosis. Bacteria on salted meat or candied fruit lose water through osmosis, shrivel and die, thus preserving the food.

 CONCEPT CHECK 10

Which would have the higher osmotic pressure: a 0.10 M solution of NaCl or a 0.10 M solution of KBr?

Determination of Molar Mass

The colligative properties of solutions provide a useful means of experimentally determining molar mass. Any of the four colligative properties can be used, as shown in Sample Exercises 12.11 and 12.12.

SAMPLE EXERCISE 12.11   **Molar mass from freezing-point depression**

A solution of an unknown non-volatile electrolyte was prepared by dissolving 0.250 g of the substance in 40.0 g of CCl_4. The boiling point of the resultant solution was 0.357 °C higher than that of the pure solvent. Calculate the molar mass of the solute (K_b CCl_4 = 5.26 °C m^{-1}).

SOLUTION

Analyse Our goal is to calculate the molar mass of a solute based on knowledge of the boiling-point elevation of the solution in CCl_4, ΔT_b = 0.357 °C, and the masses of solute and solvent. We are given that K_b for the solvent (CCl_4) = 5.26 °C m^{-1}.

Plan We can use Equation 12.12, $\Delta T_b = K_b\, m$, to calculate the molality of the solution. Then we can use molality and the quantity of solvent (40.0 g CCl_4) to calculate the number of moles of solute. Finally, the molar mass of the solute equals the number of grams per mole, so we divide the number of grams of solute (0.250 g) by the number of moles we have just calculated.

Solve From Equation 12.12 we have

$$\text{Molality} = \frac{\Delta T_b}{K_b} = \frac{0.357\ °C}{5.26\ °C\ m^{-1}} = 0.0679\ m$$

Thus, the solution contains 0.0679 mol of solute per kilogram of solvent. The solution was prepared using 40.0 g = 0.0400 kg of solvent (CCl_4). The number of moles of solute in the solution is therefore

$$(0.0400\ \text{kg}\ CCl_4)\left(0.0679\ \frac{\text{mol solute}}{\text{kg}\ CCl_4}\right) = 2.71 \times 10^{-3}\ \text{mol solute}$$

The molar mass of the solute is the number of grams per mole of the substance:

$$\text{Molar mass} = \frac{0.250\ \text{g}}{2.71 \times 10^{-3}\ \text{mol}} = 92.1\ \text{g mol}^{-1}$$

PRACTICE EXERCISE

Camphor ($C_{10}H_{16}O$) melts at 179.8 °C, and it has a particularly large freezing-point-depression constant, K_f = 40.0 °C m^{-1}. When 0.186 g of an organic substance of unknown molar mass is dissolved in 22.01 g of liquid camphor, the freezing point of the mixture is found to be 176.7 °C. What is the molar mass of the solute?

Answer: 110 g mol^{-1}

(See also Exercise 12.53.)

SAMPLE EXERCISE 12.12   **Molar mass from osmotic pressure**

The osmotic pressure of an aqueous solution of a certain protein was measured in order to determine the protein's molar mass. The solution contained 3.50 mg of protein dissolved in sufficient water to form 5.00 cm^3 of solution. The osmotic pressure of the solution at 25 °C was found to be 0.205 kPa. Calculate the molar mass of the protein.

SOLUTION

Analyse Our goal is to calculate the molar mass of a high-molecular-mass protein, based on its osmotic pressure and a knowledge of the mass of protein and solution volume.

Plan The temperature ($T = 25$ °C) and osmotic pressure ($\Pi = 0.205$ kPa) are given and we know the value of R, so we can use Equation 12.14 to calculate the molarity of the solution, M. We then use the molarity and the volume of the solution (5.00 cm^3) to determine the number of moles of solute. Finally, we obtain the molar mass by dividing the mass of the solute (3.50 mg) by the number of moles of solute.

Solve Solving Equation 12.14 for molarity gives

$$\text{Molarity} = \frac{\Pi}{RT} = \frac{(0.205 \text{ kPa})}{(8.314 \text{ kPa dm}^3 \text{ mol}^{-1} \text{ K}^{-1})(298 \text{ K})}$$

$$= 8.27 \times 10^{-5} \text{ mol dm}^{-3}$$

Because the volume of the solution is 5.00 cm^3 = 5.00 × 10^{-3} dm^3, the number of moles of protein must be

$$\text{Moles} = (8.27 \times 10^{-5} \text{ mol dm}^{-3})(5.00 \times 10^{-3} \text{ dm}^3) = 4.14 \times 10^{-7} \text{ mol}$$

The molar mass is the number of grams per mole of the substance. The sample has a mass of 3.50 mg = 3.50 × 10^{-3} g. The molar mass is the number of grams divided by the number of moles:

$$\text{Molar mass} = \frac{\text{grams}}{\text{moles}} = \frac{3.50 \times 10^{-3} \text{ g}}{4.14 \times 10^{-7} \text{ mol}} = 8.45 \times 10^{3} \text{ g mol}^{-1}$$

Comment Because small pressures can be measured easily and accurately, osmotic pressure measurements provide a useful way to determine the molar masses of large molecules.

PRACTICE EXERCISE

A sample of 2.05 g of polystyrene of uniform polymer chain length was dissolved in enough toluene to form 0.100 dm^3 of solution. The osmotic pressure of this solution was found to be 1.21 kPa at 25 °C. Calculate the molar mass of the polystyrene.

Answer: 4.20 × 10^4 g mol^{-1}

(See also Exercises 12.54, 12.55.)

12.6 | COLLOIDS

When finely divided clay particles are dispersed throughout water, they eventually settle out of the water because of gravity. The dispersed clay particles are much larger than molecules and consist of many thousands or even millions of atoms. In contrast, the dispersed particles of a solution are of molecular size. Between these extremes lie dispersed particles that are larger than molecules, but not so large that the components of the mixture separate under the influence of gravity. These intermediate types of dispersions or suspensions are called **colloidal dispersions**, or simply **colloids**. Colloids form the dividing line between solutions and heterogeneous mixtures. Like solutions, colloids can be gases, liquids or solids. If a colloid consists of a solid material dispersed in a liquid medium it is referred to as a *sol*. Examples of each are listed in ▼ TABLE 12.5.

The size of the dispersed particles is used to classify a mixture as a colloid. Colloid particles range in diameter from approximately 5 nm to 1000 nm. Solute particles are smaller. The colloid particle may consist of many atoms, ions or molecules, or it may even be a single giant molecule. The haemoglobin

Phase of colloid	Dispersing (solvent-like) substance	Dispersed (solute-like) substance	Colloid type	Example
Gas	Gas	Gas	—	None (all are solutions)
Gas	Gas	Liquid	Aerosol	Fog
Gas	Gas	Solid	Aerosol	Smoke
Liquid	Liquid	Gas	Foam	Whipped cream
Liquid	Liquid	Liquid	Emulsion	Milk
Liquid	Liquid	Solid	Sol	Paint
Solid	Solid	Gas	Solid foam	Marshmallow
Solid	Solid	Liquid	Solid emulsion	Butter
Solid	Solid	Solid	Solid sol	Ruby glass

TABLE 12.5 • Types of colloids

▶ **FIGURE 12.26 Tyndall effect in the laboratory.** The glass on the right contains a colloidal dispersion; that on the left contains a solution.

(a)

(b)

▲ **FIGURE 12.27 Tyndall effect in nature.** (a) Scattering of sunlight by colloidal particles in the misty air of a forest. (b) The scattering of light by smoke or dust particles produces a rich red sunset.

molecule, for example, which carries oxygen in blood, has molecular dimensions of 6500 pm × 5500 pm × 5000 pm and a molecular weight of 64 500 u.

Although colloid particles may be so small that the dispersion appears uniform even under a microscope, they are large enough to scatter light very effectively. Consequently, most colloids appear cloudy or opaque unless they are very dilute. (Homogenised milk is a colloid.) Furthermore, because they scatter light, a light beam can be seen as it passes through a colloidal suspension, as shown in ▲ **FIGURE 12.26**. This scattering of light by colloidal particles, known as the **Tyndall effect**, makes it possible to see the light beam of a car on a dusty dirt road or the sunlight coming through a forest canopy (◀ **FIGURE 12.27(a)**). Not all wavelengths are scattered to the same extent. As a result, brilliant red sunsets are seen when the sun is near the horizon and the air contains dust, smoke or other particles of colloidal size (Figure 12.27(b)).

Hydrophilic and Hydrophobic Colloids

The most important colloids are those in which the dispersing medium is water. These colloids may be **hydrophilic** (water loving) or **hydrophobic** (water fearing). Hydrophilic colloids are most like the solutions that we have previously examined. In the human body the extremely large molecules that make up such important substances as enzymes and antibodies are kept in suspension by interaction with surrounding water molecules. The molecules fold in such a way that the hydrophobic groups are away from the water molecules, on the 'inside' of the folded molecule, while the hydrophilic, polar groups are on the surface, interacting with the water molecules. These hydrophilic groups generally contain

What is the chemical composition of the groups that carry a negative charge?

Hydrophilic polar and charged groups on molecule surface help molecule remain dispersed in water and other polar solvents

▲ **FIGURE 12.28** **Hydrophilic colloidal particle.** Examples of the hydrophilic groups that help to keep a giant molecule (macromolecule) suspended in water.

oxygen or nitrogen and often carry a charge. Some examples are shown in ▲ **FIGURE 12.28**.

Hydrophobic colloids can be prepared in water only if they are stabilised in some way. Otherwise, their natural lack of affinity for water causes them to separate from the water. Hydrophobic colloids can be stabilised by adsorption of ions on their surface, as shown in ▼ **FIGURE 12.29**. (*Adsorption* means to adhere to a surface. It differs from *absorption*, which means to pass into the interior, as when a sponge absorbs water.) These adsorbed ions can interact with water, thereby stabilising the colloid. At the same time, the mutual repulsion between colloid particles with adsorbed ions of the same charge keeps the particles from colliding and getting larger.

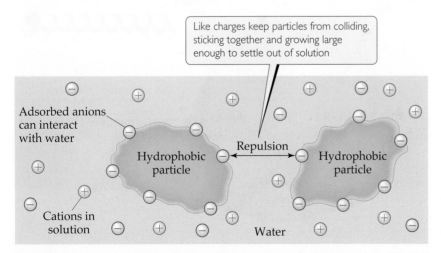

Like charges keep particles from colliding, sticking together and growing large enough to settle out of solution

▲ **FIGURE 12.29** **Hydrophobic colloids stabilised by adsorbed anions.**

Hydrophobic colloids can also be stabilised by the presence of hydrophilic groups on their surfaces. Small droplets of oil are hydrophobic, for example, so they do not remain suspended in water. Instead, they aggregate, forming an oil slick on the surface of the water. Sodium stearate (◀ **FIGURE 12.30**), or any similar substance having one end that is hydrophilic (polar, or charged) and one that is hydrophobic (non-polar), will stabilise a suspension of oil in water. Stabilisation results from the interaction of the hydrophobic ends of the stearate ions with the oil droplet and the hydrophilic ends with the water as shown in ▼ **FIGURE 12.31**.

⚠️ **CONCEPT CHECK 11**
Why don't the oil droplets emulsified by sodium stearate coagulate to form larger oil droplets?

The stabilisation of colloids has an interesting application in our own digestive system. When fats in our diet reach the small intestine, a hormone causes the gall bladder to excrete a fluid called bile. Among the components of bile are compounds that have chemical structures similar to sodium stearate; that is, they have a hydrophilic (polar) end and a hydrophobic (non-polar) end. These compounds emulsify the fats present in the intestine and thus permit digestion and absorption of fat-soluble vitamins through the intestinal wall. The term *emulsify* means 'to form an emulsion', a suspension of one liquid in another, as in milk, for example (Table 12.5). A substance that aids in the formation of an emulsion is called an emulsifying agent. If you read the labels on foods and other materials, you will find that a variety of chemicals are used as emulsifying agents. These chemicals typically have a hydrophilic end and a hydrophobic end.

Sodium stearate

▲ FIGURE 12.30 **Sodium stearate.**

Removal of Colloidal Particles

Colloidal particles must often be removed from a dispersing medium, as in the removal of smoke from stacks or butterfat from milk. Because colloidal particles are so small, they cannot be separated by simple filtration. Instead, the colloidal particles must be enlarged in a process called *coagulation*. The resultant larger particles can then be separated by filtration or merely by allowing them to settle out of the dispersing medium.

Heating the mixture or adding an electrolyte may bring about coagulation. Heating the colloidal dispersion increases the particle motion and so the number of collisions. The particles increase in size as they stick together after colliding. The addition of electrolytes neutralises the surface charges of the

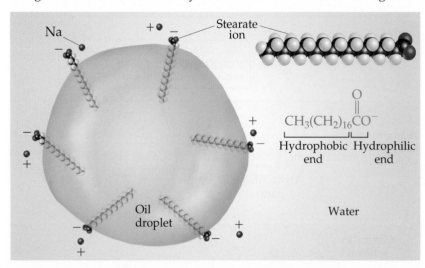

▲ FIGURE 12.31 **Stabilisation of an emulsion of oil in water by stearate ions.**

particles, thereby removing the electrostatic repulsions that prevent them from coming together. Wherever rivers empty into oceans or other salty bodies of water, for example, the suspended clay in the river is deposited as a delta when the electrolytes in the salt water cause it to coagulate.

Semi-permeable membranes can also be used to separate ions from colloidal particles because the ions can pass through the membrane but the colloidal particles cannot. This type of separation is known as *dialysis* and is used to purify blood in artificial kidney machines. Our kidneys normally remove the waste products of metabolism from blood. In a kidney machine, blood is circulated through a dialysing tube immersed in a washing solution. The washing solution is isotonic in ions that must be retained by the blood, but is lacking the waste products. Wastes therefore dialyse out of the blood, but the ions do not.

SAMPLE INTEGRATIVE EXERCISE   **Putting concepts together**

A 0.100 dm^3 solution is made by dissolving 0.441 g of $CaCl_2(s)$ in water. **(a)** Calculate the osmotic pressure of this solution at 27 °C, assuming that it is completely dissociated into its component ions. **(b)** The enthalpy of solution for $CaCl_2$ is $\Delta H = -81.3$ kJ mol^{-1}. If the final temperature of the solution was 27.0 °C, what was its initial temperature? (Assume that the density of the solution is 1.00 g cm^{-3}, that its specific heat is 4.18 J g^{-1} K^{-1} and that the solution loses no heat to its surroundings.)

Solution **(a)** The osmotic pressure is given by Equation 12.14, $\Pi = MRT$. We know the temperature, $T = 27$ °C $= 300$ K, and the gas constant, $R = 8.314$ kPa dm^3 mol^{-1} K^{-1}. We can calculate the molarity of the solution from the mass of $CaCl_2$ and the volume of the solution:

$$\text{Molarity} = \left(\frac{0.441 \text{ g}}{111.0 \text{ g mol}^{-1}}\right)\left(\frac{1}{0.100 \text{ dm}^3}\right) = 0.0397 \text{ mol dm}^{-3}$$

Soluble ionic compounds are strong electrolytes. Thus $CaCl_2$ consists of metal cations (Ca^{2+}) and non-metal anions (Cl^-). When completely dissociated, each $CaCl_2$ unit forms three ions (one Ca^{2+} and two Cl^-). Hence, the total concentration of ions in the solution is $(3)(0.0397$ M$) = 0.119$ M, and the osmotic pressure is

$$\Pi = MRT = (0.119 \text{ mol dm}^{-3})(8.314 \text{ kPa dm}^3 \text{ mol}^{-1} \text{ K}^{-1})(300 \text{ K}) = 2.97 \times 10^2 \text{ kPa}$$

(b) If the solution is 0.0397 M in $CaCl_2$ and has a total volume of 0.100 dm^3, the number of moles of solute is $(0.100 \text{ dm}^3)(0.0397 \text{ mol dm}^{-3}) = 0.00397$ mol. Hence the quantity of heat generated in forming the solution is $(0.00397 \text{ mol})(-81.3 \text{ kJ mol}^{-1}) = -0.323$ kJ. The solution absorbs this heat, causing its temperature to increase. The relationship between temperature change and heat is given by Equation 14.19:

$$q = (\text{specific heat})(\text{grams})(\Delta T)$$

The heat absorbed by the solution is $q = +0.323$ kJ $= 323$ J. The mass of the 0.100 dm^3 of solution is $(100 \text{ cm}^3)(1.00 \text{ g cm}^{-3}) = 100$ g (to three significant figures). Thus the temperature change is

$$\Delta T = \frac{q}{(\text{specific heat of solution})(\text{grams of solution})}$$

$$= \frac{323 \text{ J}}{(4.18 \text{ J g}^{-1} \text{ K}^{-1})(100 \text{ g})} = 0.773 \text{ K}$$

A kelvin has the same size as a degree Celsius. Because the solution temperature increases by 0.773 °C, the initial temperature was 27.0 °C $-$ 0.773 °C $= 26.2$ °C.

CHAPTER SUMMARY AND KEY TERMS

SECTION 12.1 Solutions form when one substance disperses uniformly throughout another. The attractive interaction of solvent molecules with solute is called **solvation**. When the solvent is water, the interaction is called **hydration**. The dissolution of ionic substances in water is promoted by hydration of the separated ions by the polar water molecules. The overall enthalpy change upon solution formation may be either positive or negative. Solution formation is favoured both by a positive entropy change, corresponding to an increased dispersal of the components of the solution, and by a negative enthalpy change, indicating an exothermic process.

SECTION 12.2 The equilibrium between a saturated solution and undissolved solute is dynamic; the process of solution and the reverse process, **crystallisation**, occur simultaneously. In a solution in equilibrium with undissolved solute, the two processes occur at equal rates, giving a **saturated** solution. If there is less solute present than is needed to saturate the solution, the solution is **unsaturated**. When solute concentration is greater than the equilibrium concentration value, the solution is **supersaturated**. This is an unstable condition, and separation of some solute from the solution will occur if the process is initiated with a solute seed crystal. The amount of solute needed to form a saturated solution at any particular temperature is the **solubility** of that solute at that temperature.

SECTION 12.3 The solubility of one substance in another depends on the tendency of systems to become more random, by becoming more dispersed in space, and on the relative intermolecular solute–solute and solvent–solvent energies compared with solute–solvent interactions. Polar and ionic solutes tend to dissolve in polar solvents, and non-polar solutes tend to dissolve in non-polar solvents ('like dissolves like'). Liquids that mix in all proportions are **miscible**; those that do not dissolve significantly in one another are **immiscible**. Hydrogen-bonding interactions between solute and solvent often play an important role in determining solubility; for example, ethanol and water, whose molecules form hydrogen bonds with each other, are miscible. The solubilities of gases in a liquid are generally proportional to the pressure of the gas over the solution, as expressed by **Henry's law**: $S_g = kP_g$. The solubilities of most solid solutes in water increase as the temperature of the solution increases. In contrast, the solubilities of gases in water generally decrease with increasing temperature.

SECTION 12.4 Concentrations of solutions can be expressed quantitatively by several different measures, including **mass percentage** ((mass solute/mass solution) $\times$ 10^2), **parts per million (ppm)**, **parts per billion (ppb)** and mole fraction. Molarity, M, is defined as moles of solute per litre of solution; **molality**, m, is defined as moles of solute per kg of solvent. Molarity can

be converted to these other concentration units if the density of the solution is known.

SECTION 12.5 A physical property of a solution that depends on the concentration of solute particles present, regardless of the nature of the solute, is a **colligative property**. Colligative properties include vapour pressure lowering, freezing-point lowering, boiling-point elevation and osmotic pressure. **Raoult's law** expresses the lowering of vapour pressure. An **ideal solution** obeys Raoult's law. Differences in solvent–solute compared with solvent–solvent and solute–solute intermolecular forces cause many solutions to depart from ideal behaviour.

A solution containing a non-volatile solute possesses a higher boiling point than the pure solvent. The **molal boiling-point-elevation constant**, K_b, represents the increase in boiling point for a 1 m solution of solute particles as compared with the pure solvent. Similarly, the **molal freezing-point-depression constant**, K_f, measures the lowering of the freezing point of a solution for a 1 m solution of solute particles. The temperature changes are given by the equations $\Delta T_b = K_b m$ and $\Delta T_f = K_f m$. When NaCl dissolves in water, two moles of solute particles are formed for each mole of dissolved salt. The boiling point or freezing point is thus elevated or depressed, respectively, approximately twice as much as that of a non-electrolyte solution of the same concentration. Similar considerations apply to other strong electrolytes.

Osmosis is the movement of solvent molecules through a semi-permeable membrane from a less concentrated to a more concentrated solution. This net movement of solvent generates an **osmotic pressure**, Π, which can be measured in units of gas pressure, such as kilopascals. The osmotic pressure of a solution is proportional to the solution molarity: $\Pi = MRT$. Osmosis is a very important process in living systems, in which cell walls act as semi-permeable membranes, permitting the passage of water but restricting the passage of ionic and macromolecular components.

SECTION 12.6 Particles that are large on the molecular scale but still small enough to remain suspended indefinitely in a solvent system form **colloids**, or **colloidal dispersions**. Colloids, which are intermediate between solutions and heterogeneous mixtures, have many practical applications. One useful physical property of colloids, the scattering of visible light, is referred to as the **Tyndall effect**. Aqueous colloids are classified as **hydrophilic** or **hydrophobic**. Hydrophilic colloids are common in living organisms, in which large molecular aggregates (enzymes, antibodies) remain suspended because they have many polar, or charged, atomic groups on their surfaces that interact with water. Hydrophobic colloids, such as small droplets of oil, may remain in suspension through adsorption of charged particles on their surfaces.

KEY SKILLS

- Understand how enthalpy and entropy changes affect solution formation. (Section 12.1)
- Be aware of how intermolecular forces affect solubility. (Sections 12.1 and 12.3)
- Understand the 'like dissolves like' rule. (Section 12.3)
- Describe the effect of temperature on the solubility of solids and gases. (Section 12.3)
- Describe the relationship between partial pressure of a gas and its solubility. (Section 12.3)
- Be able to express concentration in terms of molarity, molality, mole fraction, percent composition, parts per million and parts per billion and be able to interconvert between them. (Section 12.4)
- Understand colligative properties and the difference between the effects of electrolytes and non-electrolytes on these properties. (Section 12.5)
- Be able to calculate the vapour pressure of a solvent over a solution. (Section 12.5)
- Be able to calculate the boiling-point elevation and the freezing-point depression of a solution. (Section 12.5)
- Be able to calculate the osmotic pressure of a solution. (Section 12.5)
- Explain the difference between a solution and a colloid. (Section 12.6)

KEY EQUATIONS

- Henry's law relating gas solubility to partial pressure

$$S_g = kP_g \tag{12.4}$$

- Concentration expressed as mass percent

$$\text{Mass \% of component} = \frac{\text{mass of component in soln}}{\text{total mass of soln}} \times 100 \tag{12.5}$$

- Concentration expressed as parts per million

$$\text{ppm of component} = \frac{\text{mass of component in soln}}{\text{total mass of soln}} \times 10^6 \tag{12.6}$$

- Concentration expressed as mole fraction

$$\text{Mole fraction of component} = \frac{\text{moles of component}}{\text{total moles of all components}} \tag{12.7}$$

- Concentration expressed as molarity

$$\text{Molarity} = \frac{\text{moles solute}}{\text{litres of solution}} \tag{12.8}$$

- Concentration expressed as molality

$$\text{Molality} = \frac{\text{moles of solute}}{\text{kilograms of solvent}} \tag{12.9}$$

- Raoult's law for calculating vapour pressure of solvent over a solution

$$P_{\text{solution}} = X_{\text{solvent}} P^\circ_{\text{solvent}} \tag{12.10}$$

- Calculating the boiling-point elevation of a solution

$$\Delta T_b = K_b\, m \tag{12.12}$$

- Calculating the freezing-point depression of a solution

$$\Delta T_f = K_f\, m \tag{12.13}$$

- Calculating the osmotic pressure of a solution

$$\Pi V = nRT \quad \text{or} \quad \Pi = \left(\frac{n}{V}\right)RT = MRT \tag{12.14}$$

EXERCISES

VISUALISING CONCEPTS

12.1 This figure shows the interaction of a cation with surrounding water molecules.

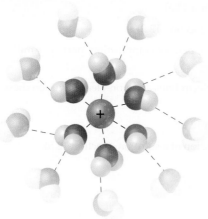

Would you expect the energy of ion–solvent interaction to be greater for Na^+ or Li^+? Explain. [Section 12.1]

12.2 A quantity of the blue solid on the left in Figure 12.6 is placed in a warming oven and heated for a time. It slowly turns from blue to the white colour of the solid on the right. What has occurred? [Section 12.1]

12.3 Which of the following is the best representation of a saturated solution? Explain your reasoning. [Section 12.2]

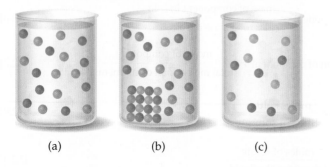

    (a)   (b)   (c)

12.4 The solubility of Xe in water at 20 °C is approximately 5×10^{-3} M. Compare this with the solubilities of Ar and Kr in water, Table 12.2, and explain what properties of the rare gas atoms account for the variation in solubility. [Section 12.3]

12.5 The structures of vitamins B_6 and E are shown below. Predict which is largely water soluble and which is largely fat soluble. Explain. [Section 12.3]

Vitamin B_6

Vitamin E

12.6 The figure shows two volumetric flasks containing the same solution at two temperatures.

    25 °C   55 °C

(a) Does the molarity of the solution change with the change in temperature? Explain.

(b) Does the molality of the solution change with the change in temperature? Explain. [Section 12.4]

12.7 Suppose you had a balloon made of some highly flexible semi-permeable membrane. The balloon is filled completely with a 0.2 M solution of some solute and is submerged in a 0.1 M solution of the same solute:

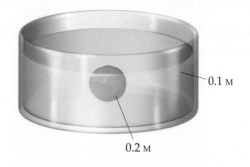

0.1 M

0.2 M

Initially, the volume of solution in the balloon is 0.25 dm^3. Assuming the volume outside the semi-permeable membrane is large, as the illustration shows, what would you expect for the solution volume inside the balloon once the system has come to equilibrium through osmosis? [Section 12.5]

12.8 If you wanted to prepare a solution of CO in water at 25 °C in which the CO concentration was 2.5 mM, what pressure of CO would you need to use? (See Figure 12.18.) [Section 12.3]

THE SOLUTION PROCESS (Section 12.1)

12.9 In general, the attractive intermolecular forces between solvent and solute particles must be comparable or greater than solute–solute interactions if significant solubility is to occur. Explain this statement in terms of the overall energetics of solution formation.

12.10 **(a)** Considering the energetics of solute–solute, solvent–solvent and solute–solvent interactions, explain why NaCl dissolves in water but not in benzene (C$_6$H$_6$). **(b)** Why do ionic substances with higher lattice energies tend to be less soluble in water than those with lower lattice energies? **(c)** What factors cause a cation to be strongly hydrated?

12.11 Indicate the principal type of solute–solvent interaction in each of the following solutions, and rank the solutions from weakest to strongest solute–solvent interaction: **(a)** KCl in water, **(b)** dichloroethane (CH$_2$Cl$_2$) in benzene (C$_6$H$_6$), **(c)** methanol (CH$_3$OH) in water.

12.12 **(a)** In Equation 12.1 which of the energy terms for dissolving an ionic solid would correspond to the lattice energy? **(b)** Which energy terms in this equation are always exothermic?

12.13 When two non-polar organic liquids such as hexane (C$_6$H$_{14}$) and heptane (C$_7$H$_{16}$) are mixed, the enthalpy change that occurs is generally quite small. **(a)** Use the energy diagram in Figure 12.3 to explain why. **(b)** Given that $\Delta_{soln}H \approx 0$, explain why hexane and heptane spontaneously form a solution.

12.14 The enthalpy of solution of KBr in water is about +198 kJ mol^{-1}. Nevertheless, the solubility of KBr in water is relatively high. Why does the solution process occur even though it is endothermic?

SATURATED SOLUTIONS; FACTORS AFFECTING SOLUBILITY (Sections 12.2 and 12.3)

12.15 The solubility of Cr(NO$_3$)$_3 \cdot$ 9H$_2$O in water is 208 g per 100 g of water at 15 °C. A solution of Cr(NO$_3$)$_3 \cdot$9H$_2$O in water at 35 °C is formed by dissolving 324 g in 100 g water. When this solution is slowly cooled to 15 °C, no precipitate forms. **(a)** What term describes this solution? **(b)** What action might you take to initiate crystallisation? Use molecular-level processes to explain how your suggested procedure works.

12.16 The solubility of MnSO$_4 \cdot$ H$_2$O in water at 20 °C is 70 g per 100 cm^3 of water. **(a)** Is a 1.22 M solution of MnSO$_4 \cdot$H$_2$O in water at 20 °C saturated, supersaturated or unsaturated? **(b)** Given a solution of MnSO$_4 \cdot$ H$_2$O of unknown concentration, what experiment could you perform to determine whether the new solution is saturated, supersaturated or unsaturated?

12.17 By referring to Figure 12.17, determine whether the addition of 40.0 g of each of the following ionic solids to 100 g of water at 40 °C will lead to a saturated solution: **(a)** NaNO$_3$, **(b)** KCl, **(c)** K$_2$Cr$_2$O$_7$, **(d)** Pb(NO$_3$)$_2$.

12.18 Water and glycerol, CH$_2$(OH)CH(OH)CH$_2$OH, are miscible in all proportions. What does this mean? How do the OH groups of the alcohol molecule contribute to this miscibility?

12.19 Oil and water are immiscible. What does this mean? Explain in terms of the structural features of their respective molecules and the forces between them.

12.20 Consider a series of carboxylic acids whose general formula is CH$_3$(CH$_2$)$_n$COOH. How would you expect the solubility of these compounds in water and in hexane to change as n increases? Explain.

12.21 Which of the following in each pair is likely to be more soluble in water: **(a)** cyclohexane (C$_6$H$_{12}$) or glucose (C$_6$H$_{12}$O$_6$); **(b)** propionic acid (CH$_3$CH$_2$COOH) or sodium propionate (CH$_3$CH$_2$COONa); **(c)** HCl or chloroethane (CH$_3$CH$_2$Cl)? Explain in each case.

12.22 **(a)** Explain why carbonated beverages must be stored in sealed containers. **(b)** Once the beverage has been opened, why does it maintain more carbonation when refrigerated than at room temperature?

12.23 Explain why pressure affects the solubility of O$_2$ in water, but not the solubility of NaCl in water.

12.24 The Henry's law constant for helium gas in water at 30 °C is 3.7×10^{-6} M kPa^{-1}; the constant for N$_2$ at 30 °C is 5.9×10^{-6} M kPa^{-1}. If the two gases are each present at 152 kPa pressure, calculate the solubility of each gas.

12.25 The partial pressure of O$_2$ in air at sea level is 21.3 kPa. Using the data in Table 12.2, together with Henry's law, calculate the molar concentration of O$_2$ in the surface water of a mountain lake saturated with air at 20 °C and an atmospheric pressure of 88.7 kPa.

CONCENTRATIONS OF SOLUTIONS (Section 12.4)

12.26 (a) Calculate the mass percentage of Na_2SO_4 in a solution containing 10.6 g Na_2SO_4 in 483 g water. **(b)** An ore contains 2.86 g of silver per tonne of ore. What is the concentration of silver in ppm?

12.27 A solution is made containing 14.6 g of CH_3OH in 184 g H_2O. Calculate **(a)** the mole fraction of CH_3OH, **(b)** the mass percent of CH_3OH, **(c)** the molality of CH_3OH.

12.28 A solution is made containing 25.5 g phenol (C_6H_5OH) in 495 g ethanol (CH_3CH_2OH). Calculate **(a)** the mole fraction of phenol, **(b)** the mass percent of phenol, **(c)** the molality of phenol.

12.29 Calculate the molarity of the following aqueous solutions: **(a)** 0.540 g $Mg(NO_3)_2$ in 250.0 cm^3 of solution, **(b)** 22.4 g $LiClO_4\cdot3\,H_2O$ in 125 cm^3 of solution, **(c)** 25.0 cm^3 of 3.50 M HNO_3 diluted to 0.250 dm^3.

12.30 Calculate the molality of each of the following solutions: **(a)** 8.66 g benzene (C_6H_6) dissolved in 23.6 g carbon tetrachloride (CCl_4), **(b)** 4.80 g NaCl dissolved in 0.350 dm^3 of water.

12.31 (a) What is the molality of a solution formed by dissolving 1.50 mol of KCl in 16.0 mol of water at 25 °C? **(b)** How many grams of sulfur (S_8) must be dissolved in 100.0 g naphthalene ($C_{10}H_8$) to make a 0.12 m solution?

12.32 A sulfuric acid solution containing 571.6 g of H_2SO_4 per litre of solution has a density of 1.329 g cm^{-3}. Calculate **(a)** the mass percentage, **(b)** the mole fraction, **(c)** the molality, **(d)** the molarity of H_2SO_4 in this solution.

12.33 Ascorbic acid (vitamin C, $C_6H_8O_6$) is a water-soluble vitamin. A solution containing 80.5 g of ascorbic acid dissolved in 210 g of water has a density of 1.22 g cm^{-3} at 55 °C. Calculate **(a)** the mass percentage, **(b)** the mole fraction, **(c)** the molality, **(d)** the molarity of ascorbic acid in this solution.

12.34 Calculate the number of moles of solute present in each of the following aqueous solutions: **(a)** 600 cm^3 of 0.250 M $SrBr_2$, **(b)** 86.4 g of 0.180 m KCl, **(c)** 124.0 g of a solution that is 6.45% glucose ($C_6H_{12}O_6$) by mass.

12.35 Describe how you would prepare each of the following aqueous solutions, starting with solid KBr: **(a)** 0.75 dm^3 of 1.5×10^{-2} M KBr, **(b)** 125 g of 0.180 m KBr, **(c)** 1.85 dm^3 of a solution that is 12.0% KBr by mass (the density of the solution is 1.10 g cm^{-3}), **(d)** a 0.150 M solution of KBr that contains just enough KBr to precipitate 16.0 g of AgBr from a solution containing 0.480 mol of $AgNO_3$.

12.36 Commercial concentrated aqueous ammonia is 28% NH_3 by mass and has a density of 0.90 g cm^{-3}. What is the molarity of this solution?

12.37 Brass is a substitutional alloy consisting of a solution of copper and zinc. A particular sample of red brass consisting of 80.0% Cu and 20.0% Zn by mass has a density of 8750 kg m^{-3}. **(a)** What is the molality of Zn in the solid solution? **(b)** What is the molarity of Zn in the solution?

12.38 During a typical breathing cycle the CO_2 concentration in the expired air rises to a peak of 4.6% by volume. Calculate the partial pressure of the CO_2 at this point, assuming 1 bar pressure. What is the molarity of the CO_2 in air at this point, assuming a body temperature of 37 °C?

12.39 Breathing air that contains 4.0% by volume CO_2 over a period of time causes rapid breathing, throbbing headache and nausea, among other symptoms. What is the concentration of CO_2 in such air in terms of **(a)** mol percentage, **(b)** molarity, assuming 1 bar pressure and a body temperature of 37 °C?

COLLIGATIVE PROPERTIES (Section 12.5)

12.40 List four properties of a solution that depend on the total concentration but not the type of particle or particles present as solute. Write the mathematical expression that describes how each of these properties depends on concentration.

12.41 How does increasing the concentration of a non-volatile solute in water affect the following properties: **(a)** vapour pressure, **(b)** freezing point, **(c)** boiling point, **(d)** osmotic pressure?

12.42 Consider two solutions, one formed by adding 10 g of glucose ($C_6H_{12}O_6$) to 1 dm^3 of water and the other formed by adding 10 g of sucrose ($C_{12}H_{22}O_{11}$) to 1 dm^3 of water. Are the vapour pressures over the two solutions the same? Why or why not?

12.43 (a) What is an *ideal solution*? **(b)** The vapour pressure of pure water at 60 °C is 19.9 kPa. The vapour pressure of water over a solution at 60 °C containing equal numbers of moles of water and ethylene glycol (a non-volatile solute) is 8.93 kPa. Is the solution ideal according to Raoult's law? Explain.

12.44 (a) Calculate the vapour pressure of water above a solution prepared by dissolving 35.0 g of glycerol ($C_3H_8O_3$) in 125 g of water at 343 K. (The vapour pressure of water is given in Appendix B.) **(b)** Calculate the mass of ethylene glycol ($C_2H_6O_2$) that must be added to 1.00 kg of ethanol (C_2H_5OH) to reduce its vapour pressure by 1.33 kPa at 35 °C. The vapour pressure of pure ethanol at 35 °C is 13.3 kPa.

12.45 At 63.5 °C the vapour pressure of H_2O is 23.3 kPa and that of ethanol (C_2H_5OH) is 53.3 kPa. A solution is made by mixing equal masses of H_2O and C_2H_5OH. **(a)** What is the mole fraction of ethanol in the solution? **(b)** Assuming ideal-solution behaviour, what is the vapour pressure of the solution at 63.5 °C? **(c)** What is the mole fraction of ethanol in the vapour above the solution?

12.46 (a) Why does a 0.10 m aqueous solution of NaCl have a higher boiling point than a 0.10 m aqueous solution of $C_6H_{12}O_6$? **(b)** Calculate the boiling point of each solution. **(c)** The experimental boiling point of the NaCl solution is lower than that calculated, assuming that NaCl is completely dissociated in solution. Why is this the case?

12.47 Arrange the following aqueous solutions, each 10% by mass in solute, in order of increasing boiling point: glucose ($C_6H_{12}O_6$), sucrose ($C_{12}H_{22}O_{11}$), sodium nitrate ($NaNO_3$).

12.48 List the following aqueous solutions in order of decreasing freezing point: 0.040 m glycerol 0.020 m KBr, 0.030 m phenol (C_6H_5OH).

12.49 Using data from Table 12.4, calculate the freezing and boiling points of each of the following solutions: **(a)** 0.22 m glycerol ($C_3H_8O_3$) in ethanol, **(b)** 0.240 mol of naphthalene ($C_{10}H_8$) in 2.45 mol of chloroform, **(c)** 2.04 g KBr and 4.82 g glucose ($C_6H_{12}O_6$) in 188 g of water.

12.50 Using data from Table 12.4, calculate the freezing and boiling points of each of the following solutions: **(a)** 0.40 m glucose in ethanol, **(b)** 20.0 g of decane, $C_{10}H_{22}$, in 45.5 g CHCl$_3$, **(c)** 0.45 mol ethylene glycol and 0.15 mol KBr in 150 g H_2O.

12.51 What is the osmotic pressure of a solution formed by dissolving 44.2 mg of aspirin ($C_9H_8O_4$) in 0.358 dm^3 of water at 25 °C?

12.52 Seawater contains 3.4 g of salts for every litre of solution. Assuming that the solute consists entirely of NaCl (over 90%), calculate the osmotic pressure of seawater at 20 °C.

12.53 Lauryl alcohol is obtained from coconut oil and is used to make detergents. A solution of 5.00 g of lauryl alcohol in 0.100 kg of benzene freezes at 4.1 °C. What is the approximate molar mass of lauryl alcohol?

12.54 Lysozyme is an enzyme that breaks bacterial cell walls. A solution containing 0.150 g of this enzyme in 210 mL of solution has an osmotic pressure of 0.127 kPa at 25 °C. What is the molar mass of lysozyme?

12.55 A dilute aqueous solution of an organic compound soluble in water is formed by dissolving 2.35 g of the compound in water to form 0.250 dm^3 solution. The resulting solution has an osmotic pressure of 61.3 kPa at 25 °C. Assuming that the organic compound is a non-electrolyte, what is its molar mass?

COLLOIDS (Section 12.6)

12.56 **(a)** Why is there no colloid in which both the dispersed substance and the dispersing substance are gases? **(b)** Michael Faraday first prepared ruby-red colloids of gold particles in water that were stable for indefinite times. To the unaided eye these brightly coloured colloids are not distinguishable from solutions. How could you determine whether a given coloured preparation is a solution or a colloid?

12.57 **(a)** Many proteins that remain homogeneously distributed in aqueous medium have molecular masses in the range of 30 000 u and larger. In what sense is it appropriate to consider such suspensions to be colloids rather than solutions? Explain. **(b)** What general name is given to a colloidal dispersion of one liquid in another? What is an emulsifying agent?

12.58 Indicate whether each of the following is a hydrophilic or a hydrophobic colloid: **(a)** butterfat in homogenised

milk, **(b)** haemoglobin in blood, **(c)** vegetable oil in a salad dressing, **(d)** colloidal gold particles in water.

12.59 Explain how each of the following factors helps to determine the stability or instability of a colloidal dispersion: **(a)** particulate mass, **(b)** hydrophobic character, **(c)** charges on colloidal particles.

12.60 Colloidal suspensions of proteins, such as a gelatin, can often be caused to separate into two layers by addition of a solution of an electrolyte. Given that protein molecules may carry electrical charges on their outer surface, as illustrated in Figure 12.28, what do you believe happens when the electrolyte solution is added?

12.61 Explain how **(a)** a soap such as sodium stearate stabilises a colloidal dispersion of oil droplets in water; **(b)** milk curdles upon addition of an acid.

ADDITIONAL EXERCISES

12.62 A saturated solution of sucrose ($C_{12}H_{22}O_{11}$) is made by dissolving excess table sugar in a flask of water. There are 50 g of undissolved sucrose crystals at the bottom of the flask, in contact with the saturated solution. The flask is stoppered and set aside. A year later a single large crystal of mass 50 g is at the bottom of the flask. Explain how this experiment provides evidence for a dynamic equilibrium between the saturated solution and the undissolved solute.

12.63 Fish need at least 4 ppm dissolved O_2 for survival. **(a)** What is this concentration in mol dm^{-3}? **(b)** What partial pressure of O_2 above the water is needed to obtain this concentration at 10 °C? (The Henry's law constant for O_2 at this temperature is 1.70×10^{-5} mol dm^{-3} kPa^{-1}.)

12.64 Glucose makes up about 0.10% by mass of human blood. Calculate the concentration in **(a)** ppm, **(b)** molality. What further information would you need to determine the molarity of the solution?

12.65 A solution of KCl in water has a concentration of 260 ppm KCl. What is the concentration in ppm of K^+?

12.66 One aqueous solution is 15.0% by weight in propanol ($CH_3CH_2CH_2OH$) and another is 15.0% by weight in

ethanol (CH_3CH_2OH). Assuming the two solutions have the same density, which has the higher **(a)** molarity, **(b)** molality, **(c)** mole fraction?

12.67 Sodium metal dissolves in liquid mercury to form a solution called a sodium amalgam. The densities of Na(s) and Hg(l) are 0.97 g cm^{-3} and 13.6 g cm^{-3}, respectively. A sodium amalgam is made by dissolving 1.0 cm^3 Na(s) in 20.0 cm^3 Hg(l). Assume that the final volume of the solution is 21.0 cm^3. **(a)** Calculate the molality of Na in the solution. **(b)** Calculate the molarity of Na in the solution. **(c)** For dilute aqueous solutions, the molality and molarity are generally nearly equal in value. Is that the case for the sodium amalgam described here? Explain.

12.68 When 10.0 g of mercuric nitrate, $Hg(NO_3)_2$, is dissolved in 1.00 kg of water, the freezing point of the solution is −0.162 °C. When 10.0 g of mercuric chloride, $HgCl_2$, is dissolved in 1.00 kg of water, the solution freezes at −0.0685 °C. Use these data to determine which is the stronger electrolyte, $Hg(NO_3)_2$ or $HgCl_2$.

12.69 Carbon disulfide (CS_2) boils at 46.30 °C and has a density of 1.261 g cm^{-3}. **(a)** When 0.250 mol of a non-dissociating solute is dissolved in 400.0 cm^3 of CS_2, the solution boils at 47.46 °C. What is the molal boiling-point-elevation constant for CS_2? **(b)** When 5.39 g of a non-dissociating unknown is dissolved in 50.0 cm^3 of CS_2, the solution boils at 47.08 °C. What is the molecular weight of the unknown?

INTEGRATIVE EXERCISES

12.70 Fluorocarbons (compounds that contain both carbon and fluorine) were, until recently, used as refrigerants. The compounds listed in the following table are all gases at 25 °C, and their solubilities in water at 25 °C and 1 bar fluorocarbon pressure are given as mass percentages. **(a)** For each fluorocarbon, calculate the molality of a saturated solution. **(b)** Explain why the molarity of each of the solutions should be very close numerically to the molality. **(c)** Based on their molecular structures, account for the differences in solubility of the four fluorocarbons. **(d)** Calculate the Henry's law constant at 25 °C for $CHClF_2$, and compare its magnitude with that for N_2 (6.7×10^{-6} mol dm^{-3} kPa^{-1}). Can you account for the difference in magnitude?

Fluorocarbon	Solubility (mass %)
CF_4	0.0015
$CClF_3$	0.009
CCl_2F_2	0.028
$CHClF_2$	0.30

12.71 At ordinary body temperature (37 °C) the solubility of N_2 in water in contact with air at approximately atmospheric pressure (1.01×10^2 kPa) is 0.015 g dm^{-3}. Air is approximately 78 mol % N_2. Calculate the number of moles of N_2 dissolved per litre of blood, which is essentially an aqueous solution. At a depth of 30 m in water, the pressure is 405 kPa. What is the solubility of N_2 from air in blood at this pressure? If a scuba diver suddenly surfaces from this depth, how many millilitres of N_2 gas, in the form of tiny bubbles, are released into the bloodstream from each litre of blood?

12.72 The enthalpies of solution of hydrated salts are generally more positive than those of anhydrous materials. For example, ΔH of solution for KOH is -57.3 kJ mol^{-1}, whereas that for KOH $\cdot$ H_2O is -14.6 kJ mol^{-1}. Similarly, $\Delta_{soln}H$ for $NaClO_4$ is $+13.8$ kJ mol^{-1}, whereas that for $NaClO_4 \cdot H_2O$ is $+22.5$ kJ mol^{-1}. Use the enthalpy contributions to the solution process depicted in Figure 12.3 to explain this effect.

12.73 **(a)** A sample of hydrogen gas is generated in a closed container by reacting 2.050 g of zinc metal with 15.0 cm^3 of 1.00 M sulfuric acid. Write the balanced equation for the reaction, and calculate the number of moles of hydrogen formed, assuming that the reaction is complete. **(b)** The volume over the solution is 122 cm^3. Calculate the partial pressure of the hydrogen gas in this volume at 25 °C, ignoring any solubility of the gas in the solution. **(c)** The Henry's law constant for hydrogen in water at 25 °C is 7.7×10^{-6} mol dm^{-3} kPa^{-1}. Estimate the number of moles of hydrogen gas that remain dissolved in the solution. What fraction of the gas molecules in the system is dissolved in the solution? Was it reasonable to ignore any dissolved hydrogen in part (b)?

PHOTO/ART CREDITS

426 © Mark A. Johnson/Corbis; 432, 434 (Figure 12.8a & c) Fundamental Photographs, NYC; 434 (Figure 12.8b), 435, 456 (Figure 12.26) Richard Megna, Fundamental Photographs, NYC; 438 Charles Winters/Photo Researchers, Inc.

13

ENVIRONMENTAL CHEMISTRY

The pristine Daintree rainforest may be adversely affected by climate change and human activity.

KEY CONCEPTS

13.1 EARTH'S ATMOSPHERE

We consider the temperature profile, pressure profile and chemical composition of Earth's atmosphere and examine *photoionisation* and *photodissociation*—reactions that result from atmospheric absorption of solar radiation.

13.2 HUMAN ACTIVITIES AND EARTH'S ATMOSPHERE

We next examine the effect human activities have on the atmosphere. We discuss how atmospheric ozone is depleted by reactions involving human-made gases and how acid rain and smog are the result of atmospheric reactions involving compounds produced by human activity.

13.3 EARTH'S WATER

We examine the global water cycle, which describes how water moves from the ground to surface water to the atmosphere and back into the ground. We compare the chemical compositions of *seawater*, *freshwater* and *groundwater*.

13.4 HUMAN ACTIVITIES AND EARTH'S WATER

We consider how Earth's water is connected to the global climate and examine one measure of water quality: dissolved oxygen concentration. Water for drinking and for irrigation must be free of salts and pollutants.

13.5 GREEN CHEMISTRY

We conclude by examining *green chemistry* and highlighting the drive towards making industrial products, processes and chemical reactions sustainable and of low impact on the environment.

Life on Earth is rich, diversified and, as far as we know, unique. Earth's atmosphere, the energy received from the sun and the abundance of water on our planet are all features currently believed to be necessary for life.

As technology has advanced and the world human population has increased, humans have put new and greater stresses on the environment. Paradoxically, the very technology that can cause pollution also provides the tools to help understand and manage the environment in a beneficial way. Chemistry is often at the heart of environmental issues. The economic growth of both developed and developing nations depends critically on chemical processes that range from treatment of water supplies to industrial processes. Some of these processes produce products or by-products that are harmful to the environment.

We will now examine how our environment operates and how human activities affect it. To understand and protect the environment in which we live, we must understand how human-made and natural chemical compounds interact on land and in the sea and sky. Our daily decisions as consumers mirror those of leading experts and governmental leaders: in making each decision, we must weigh the costs against the benefits of our actions. Unfortunately, the environmental impacts of our decisions are often subtle and not immediately evident.

13.1 | EARTH'S ATMOSPHERE

Because most of us have never been very far from the Earth's surface, we tend to take for granted the many ways in which the atmosphere determines the environment in which we live. In this section we examine some of the important characteristics of our planet's atmosphere.

The temperature of the atmosphere varies in a complex manner as a function of altitude, as shown in ▼ **FIGURE 13.1**. The atmosphere is divided into four regions based on this temperature profile. Just above the surface, in the **troposphere**, the temperature normally decreases with increasing altitude, reaching a minimum of about 215 K at about 12 km. Nearly all of us live out our entire lives in the troposphere. Howling winds and soft breezes, rain, sunny skies—all that we normally think of as 'weather'—occur in this region. Commercial jet aircraft typically fly about 10 km above the Earth, an altitude that approaches the upper limit of the troposphere, which we call the *tropopause*.

Above the tropopause the temperature increases with altitude, reaching a maximum of about 275 K at about 50 km. The region from 10–50 km is called the **stratosphere**. Beyond the stratosphere are the **mesosphere** and the **thermosphere**. The boundaries of these regions are defined by temperature extremes where the direction of temperature change with increasing height reverses, denoted by the suffix *-pause* (Figure 13.1). The boundaries are important because gases mix across them relatively slowly; for example, pollutant gases generated in the troposphere find their way into the stratosphere only very slowly.

Unlike temperature, the pressure of the atmosphere decreases in a regular way with increasing elevation, as shown in Figure 13.1. Atmospheric pressure drops off much more rapidly at lower elevations than at higher ones because of the atmosphere's compressibility. Thus the pressure decreases from an average value of 1 bar at sea level to 3.1×10^{-6} bar at 100 km, to only 1.3×10^{-9} bar at 200 km. The troposphere and stratosphere together account for 99.9% of the mass of the atmosphere, of which 75% of the mass is in the troposphere.

▲ **FIGURE IT OUT**

At what altitude is the atmospheric temperature lowest?

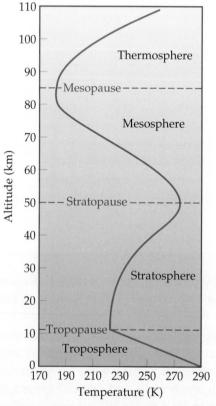
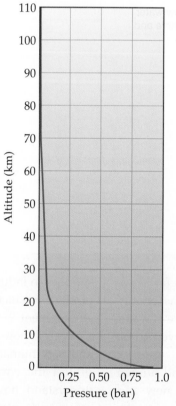

▲ **FIGURE 13.1** **Temperature and pressure in the atmosphere vary as a function of altitude above sea level.**

Composition of the Atmosphere

The atmosphere is an extremely complex system. Its temperature and pressure change over a wide range with altitude. The atmosphere is bombarded by radiation and energetic particles from the sun. This barrage of energy has profound chemical and physical effects, especially on the outer reaches of the atmosphere above 80 km. In addition, because of Earth's gravitational field, heavier atoms and molecules tend to sink in the atmosphere leaving lighter atoms and molecules at the top of the atmosphere. As a result of all these factors, the composition of the atmosphere is not uniform.

▶ **TABLE 13.1** shows the composition of dry air near sea level. Although traces of many substances are present, N_2 and O_2 make up about 99% of the atmosphere. The noble gases and CO_2 make up most of the remainder.

When speaking of trace constituents, we commonly use *parts per million* (ppm) as the unit of concentration. When applied to substances in aqueous solution, parts per million refers to grams of substance per million grams of solution (∞ Section 12.4, 'Ways of Expressing Concentration'). When dealing with gases, however, one part per million refers to one part by *volume* in 1 million volume units of the whole. Because volume (V) is proportional to the number of moles, n, of gas via the ideal-gas equation ($PV = nRT$), volume fraction and mole fraction are the same. Thus 1 ppm of a trace constituent of the atmosphere amounts to one mole of that constituent in 1 million moles of total gas; that is, the concentration in ppm is equal to the mole fraction times 10^6. Table 13.1 lists the mole fraction of CO_2 in the atmosphere as 0.000375. Its concentration in ppm is therefore $0.000375 \times 10^6 = 375$ ppm.

∞ Review this on page 441

Before we consider the chemical processes that occur in the atmosphere, let's review some of the important chemical properties of the two major components, N_2 and O_2. Recall that the N_2 molecule possesses a triple bond between the nitrogen atoms (∞ Section 8.3, 'Covalent Bonding'). This very strong bond (bond energy 941 kJ mol^{-1}) is largely responsible for the very low reactivity of N_2, which undergoes reaction only under extreme conditions. The bond energy in O_2, 495 kJ mol^{-1}, is much lower than that in N_2, making O_2 much more reactive than N_2. For example, oxygen reacts with many substances to form oxides.

∞ Review this on page 260

TABLE 13.1 • Composition of dry air near sea level

Component*	Content (mole fraction)	Molar mass
Nitrogen	0.78084	28.013
Oxygen	0.20948	31.998
Argon	0.00934	39.948
Carbon dioxide	0.000375	44.0099
Neon	0.00001818	20.183
Helium	0.00000524	4.003
Methane	0.000002	20.043
Krypton	0.00000114	83.80
Hydrogen	0.0000005	2.0159
Nitrous oxide	0.0000005	44.0128
Xenon	0.000000087	131.30

* Ozone, sulfur dioxide, nitrogen dioxide, ammonia and carbon monoxide are present as trace gases in variable amounts.

Photochemical Reactions in the Atmosphere

Although the outer portion of the atmosphere, beyond the stratosphere, contains only a small fraction of the atmospheric mass, it forms the outer defence against the hail of radiation and high-energy particles that continuously bombard the Earth. As the bombarding molecules and atoms pass through the upper atmosphere, they undergo chemical changes.

SAMPLE EXERCISE 13.1   Calculating the concentration of water in air

What is the concentration, in parts per million, of water vapour in a sample of air if the partial pressure of the water is 0.107 kPa and the total pressure of the air is 98 kPa?

SOLUTION

Analyse We are given the partial pressure of water vapour and the total pressure of an air sample and asked to determine the water vapour concentration.

Plan The partial pressure of a component in a mixture of gases is given by the product of its mole fraction and the total pressure of the mixture (∞ Section 10.6, 'Gas Mixtures and Partial Pressures'):

∞ Review this on page 351

$$P_{H_2O} = X_{H_2O} P_T$$

Solve Solving for the mole fraction of water vapour in the mixture, X_{H_2O}, gives:

$$X_{H_2O} = \frac{P_{H_2O}}{P_T} = \frac{0.107 \text{ kPa}}{98 \text{ kPa}} = 0.0011$$

The concentration in ppm is the mole fraction times 10^6:

$$0.0011 \times 10^6 = 1100 \text{ ppm}$$

PRACTICE EXERCISE

The concentration of CO in a sample of air is found to be 4.3 ppm. What is the partial pressure of the CO if the total air pressure is 92.7 kPa?

Answer: 4.0×10^{-4} kPa

FIGURE IT OUT

Why doesn't the solar spectrum at sea level perfectly match the solar spectrum outside the atmosphere?

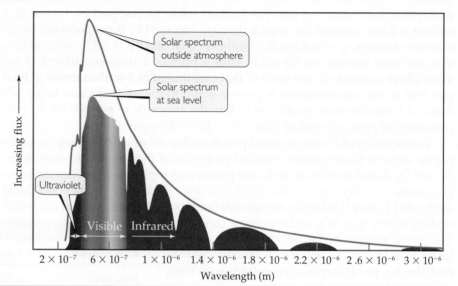

▶ **FIGURE 13.2** **The solar spectrum above Earth's atmosphere compared to that at sea level.** The more structured curve at sea level is due to gases in the atmosphere absorbing specific wavelengths of light. 'Flux,' the unit on the vertical axis, is light energy per area per unit of time.

∞ Review this on page 184

The sun emits radiant energy over a wide range of wavelengths (▲ **FIGURE 13.2**). The shorter-wavelength, higher-energy radiations in the ultraviolet range of the spectrum are sufficiently energetic to cause chemical changes even to relatively unreactive molecules. Recall that electromagnetic radiation can be pictured as a stream of photons (∞ Section 6.2, 'Quantised Energy and Photons'). The energy of each photon is given by the relationship $E = h\nu$, where h is Planck's constant and ν is the frequency of the radiation. For a chemical change to occur when radiation falls on the Earth's atmosphere, two conditions must be met. First, there must be photons with energy sufficient to break a chemical bond or remove an electron from the atom or molecule. Second, the atoms or molecules being bombarded must absorb these photons. When these requirements are met, the energy of the photons is used to do the work associated with some chemical change.

The rupture of a chemical bond resulting from absorption of a photon by a molecule is called **photodissociation**. No ions are formed when the bond between two atoms is cleaved by photodissociation. Instead, half the bonding electrons stay with one of the atoms and half stay with the other atom. The result is two neutral particles. Such neutral particles with unfilled valence shells are called **free radicals** and are important in many fields of chemistry.

One of the most important processes occurring in the upper atmosphere above about 120 km elevation is the photodissociation of the oxygen molecule:

$$:\!\overset{..}{O}\!=\!\overset{..}{O}: + h\nu \longrightarrow :\overset{..}{\underset{..}{O}} + \overset{..}{\underset{..}{O}}: \qquad [13.1]$$

The minimum energy required to cause this change is determined by the bond energy (or *dissociation energy*) of O_2, 495 kJ mol^{-1}. In Sample Exercise 13.2 we calculate the longest wavelength photon having sufficient energy to photodissociate the O_2 molecule.

Fortunately for us, O_2 absorbs much of the high-energy, short-wavelength radiation from the solar spectrum before that radiation reaches the lower atmosphere. As it does, atomic oxygen, $:\overset{..}{\underset{.}{O}}$, is formed. At higher elevations the dissociation of O_2 is very extensive. At 400 km, for example, only 1% of the oxygen is in the form of O_2; the other 99% is atomic oxygen. At 130 km, O_2 and $:\overset{..}{\underset{.}{O}}$ are just about equally abundant. Below 130 km, O_2 is more abundant than atomic oxygen because most of the solar energy has been absorbed in the upper atmosphere.

SAMPLE EXERCISE 13.2 Calculating the wavelength required to break a bond

What is the maximum wavelength of light, in nanometres, that has enough energy per photon to dissociate the O_2 molecule?

SOLUTION

Analyse We are asked to determine the wavelength of a photon that has just enough energy to break the O=O double bond in O_2.

Plan We first need to calculate the energy required to break the O=O double bond in one molecule, then find the wavelength of a photon of this energy.

Solve The dissociation energy of O_2 is 495 kJ mol^{-1}. Using this value and Avogadro's number, we can calculate the amount of energy needed to break the bond in a single O_2 molecule:

$$(495 \times 10^3 \text{ J mol}^{-1})\left(\frac{1}{6.022 \times 10^{23} \text{ molecules mol}^{-1}}\right) = 8.22 \times 10^{-19} \text{ J molecule}^{-1}$$

We next use the Planck relationship, $E = h\nu$ (Equation 6.2) to calculate the frequency, ν, of a photon that has this amount of energy:

$$\nu = \frac{E}{h} = \frac{8.22 \times 10^{-19} \text{ J}}{6.626 \times 10^{-34} \text{ J s}} = 1.24 \times 10^{15} \text{ s}^{-1}$$

Finally, we use the relationship between the frequency and wavelength of light (Equation 6.1) to calculate the wavelength of the light:

$$\lambda = \frac{c}{\nu} = \left(\frac{3.00 \times 10^8 \text{ m s}^{-1}}{1.24 \times 10^{15} \text{ s}^{-1}}\right) \times 10^9 \text{ nm m}^{-1} = 242 \text{ nm}$$

Thus light of wavelength 242 nm, which is in the ultraviolet region of the electromagnetic spectrum, has sufficient energy per photon to photodissociate an O_2 molecule. Because photon energy increases as wavelength *decreases*, any photon of wavelength *shorter* than 242 nm will have sufficient energy to dissociate O_2.

PRACTICE EXERCISE

The bond energy in N_2 is 941 kJ mol^{-1}. What is the longest wavelength a photon can have and still have sufficient energy to dissociate N_2?

Answer: 127 nm

(See also Exercises 13.14, 13.15.)

The dissociation energy of N_2 is very high, 941 kJ mol^{-1}. Analogous to Practice Exercise 13.2, only photons having a wavelength shorter than 127 nm possess sufficient energy to dissociate N_2. Furthermore, N_2 does not readily absorb photons, even when they possess sufficient energy. As a result, very little atomic nitrogen is formed in the upper atmosphere by photodissociation of N_2.

Photoionisation

In about 1924 the existence of electrons in the upper atmosphere was established by experimental studies. For each electron present in the upper atmosphere, there must be a corresponding positively charged particle. The electrons in the upper atmosphere result mainly from the **photoionisation** of molecules, caused by solar radiation. Photoionisation occurs when a molecule absorbs radiation and the absorbed energy causes an electron to be ejected from the molecule. The molecule then becomes a positively charged ion. For photoionisation to occur, therefore, a molecule must absorb a photon and the photon must have enough energy to remove an electron (∞ Section 7.4, 'Ionisation Energy').

∞ Review this on page 230

Some of the more important ionisation processes occurring in the atmosphere above about 90 km are shown in ▶ TABLE 13.2, together with the ionisation energies and λ_{max}, the maximum wavelength of a photon capable of causing ionisation. Photons with

TABLE 13.2 • Ionisation processes, ionisation energies and maximum wavelengths capable of causing ionisation

Process	Ionisation energy (kJ mol^{-1})	λ_{max}(nm)
$N_2 + h\nu \longrightarrow N_2^+ + e^-$	1495	80.1
$O_2 + h\nu \longrightarrow O_2^+ + e^-$	1205	99.3
$O + h\nu \longrightarrow O^+ + e^-$	1313	91.2
$NO + h\nu \longrightarrow NO^+ + e^-$	890	134.5

energies sufficient to cause ionisation have wavelengths in the high-energy end of the ultraviolet region of the electromagnetic spectrum. These wavelengths are completely filtered out of the radiation reaching the Earth, because they are absorbed by the upper atmosphere.

⚠ **CONCEPT CHECK 1**

Explain the difference between photoionisation and photodissociation.

Ozone in the Stratosphere

Although N_2, O_2 and atomic oxygen absorb photons having wavelengths shorter than 240 nm, ozone, O_3, is the key absorber of photons having wavelengths ranging from 240 nm to 310 nm, in the ultraviolet region of the electromagnetic spectrum. Ozone in the upper atmosphere protects us from these harmful high-energy photons, which would otherwise penetrate to Earth's surface. Let's consider how ozone forms in the upper atmosphere and how it absorbs photons.

By the time radiation from the sun reaches an altitude of 90 km above the Earth's surface, most of the short-wavelength radiation capable of photoionisation has been absorbed. Radiation capable of dissociating the O_2 molecule is sufficiently intense, however, for photodissociation of O_2 (Equation 13.1) to remain important down to an altitude of 30 km. In the region between 30 km and 90 km, the concentration of O_2 is much greater than that of atomic oxygen. Therefore the $:\ddot{O}$ atoms that form in this region undergo frequent collisions with O_2 molecules, resulting in the formation of ozone, O_3:

$$:\ddot{O} + O_2 \longrightarrow O_3{}^* \qquad [13.2]$$

The asterisk beside the O_3 denotes that the ozone molecule contains an excess of energy. The reaction in Equation 13.2 releases 105 kJ mol^{-1}. This energy must be transferred away from the $O_3{}^*$ molecule in a very short time or else the molecule will fly apart again into O_2 and $:\ddot{O}$—a decomposition that is the reverse of the process by which $O_3{}^*$ is formed.

An energy-rich $O_3{}^*$ molecule can release its excess energy by colliding with another atom or molecule and transferring some of the excess energy to it. Let's represent the atom or molecule with which $O_3{}^*$ collides as M. (Usually M is N_2 or O_2 because these are the most abundant molecules in the atmosphere.) The formation of $O_3{}^*$ and the transfer of excess energy to M are summarised by these equations (where O atoms are shown without valence electrons):

$$O(g) + O_2(g) \rightleftharpoons O_3{}^*(g) \qquad [13.3]$$

$$O_3{}^*(g) + M(g) \longrightarrow O_3(g) + M^*(g) \qquad [13.4]$$

$$\overline{O(g) + O_2(g) + M(g) \longrightarrow O_3(g) + M^*(g)} \qquad [13.5]$$

The rate at which O_3 forms, according to Equations 13.3 and 13.4, depends on two factors that vary in opposite directions with increasing altitude. First, the formation of $O_3{}^*$, according to Equation 13.3, depends on the presence of O atoms. At low altitudes most of the radiation energetic enough to dissociate O_2 has been absorbed; thus the formation of O is favoured at higher altitudes. Second, both Equations 13.3 and 13.4 depend on molecular collisions (∞ Section 15.5, 'Temperature and Rate'). The concentration of molecules is greater at low altitudes, however, and so the frequency of collisions between O and O_2 (Equation 13.3) and between $O_3{}^*$ and M (Equation 13.4) are both greater at lower altitudes. Because these processes vary with altitude in opposite directions, the highest rate of O_3 formation occurs in a band at an altitude of about 50 km, near the stratopause (Figure 13.1). Overall, roughly 90% of the Earth's ozone is found in the stratosphere, between the altitudes of 10 km and 50 km.

∞ Find out more on page 584

Why don't O_2 and N_2 molecules filter out ultraviolet light with wavelengths between 240 nm and 310 nm?

The photodissociation of ozone reverses the reaction that forms it. We thus have a cyclic process of ozone formation and decomposition, summarised as follows:

$$O_2(g) + h\nu \longrightarrow O(g) + O(g)$$

$$O(g) + O_2(g) + M(g) \longrightarrow O_3(g) + M^*(g) \text{ (heat released)}$$

$$O_3(g) + h\nu \longrightarrow O_2(g) + O(g)$$

$$O(g) + O(g) + M(g) \longrightarrow O_2(g) + M^*(g) \text{ (heat released)}$$

The first and third processes are photochemical; they use a solar photon to initiate a chemical reaction. The second and fourth processes are exothermic chemical reactions. The net result of all four processes is a cycle in which solar radiant energy is converted into thermal energy. The ozone cycle in the stratosphere is largely responsible for the rise in temperature that reaches its maximum at the stratopause, as illustrated in Figure 13.1.

The reactions of the ozone cycle account for some, but not all, of the facts about the ozone layer. Many chemical reactions occur that involve substances other than oxygen. We must also consider the effects of turbulence and winds that mix up the stratosphere. A complicated picture results. The overall result of ozone formation and removal reactions, coupled with atmospheric turbulence and other factors, is to produce the upper-atmosphere ozone profile shown in ▶ FIGURE 13.3, with a maximum ozone concentration occurring at an altitude of about 25 km. This band of relatively high ozone concentration is referred to as the 'ozone layer' or the 'ozone shield'.

Photons with wavelengths shorter than about 300 nm are energetic enough to break many kinds of single chemical bonds. Thus the 'ozone shield' is essential for our continued well-being. The ozone molecules that form this essential shield against high-energy radiation, however, represent only a tiny fraction of the oxygen atoms present in the stratosphere.

Estimate the ozone concentration in ppm in the atmosphere as a function of altitude.

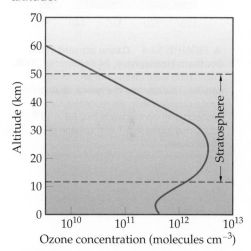

▲ **FIGURE 13.3**   **Variation in ozone concentration in the atmosphere as a function of altitude.**

13.2 | HUMAN ACTIVITIES AND EARTH'S ATMOSPHERE

Both natural and anthropogenic (human-caused) events can modify Earth's atmosphere. One impressive natural event was the eruption of Mount Pinatubo in June 1991. The volcano ejected approximately 10 km^3 of material into the stratosphere, causing a 10% drop in the amount of sunlight reaching Earth's surface during the next 2 years. That drop in sunlight led to a temporary 0.5 °C drop in the Earth's surface temperature. The volcanic particles that made it to the stratosphere remained there for approximately three years, raising the temperature of the stratosphere by several degrees due to light absorption. Measurements of the stratospheric ozone concentration showed significantly increased ozone decomposition in this three-year period.

Eruption of the Icelandic volcano Eyjafjallajökull in 2010, though not as large as the Pinatubo eruption, has similarly affected the atmosphere over large regions of the Northern Hemisphere.

The Ozone Layer and its Depletion

The ozone layer protects Earth's surface from damaging ultraviolet (UV) radiation. Therefore, if the concentration of ozone in the stratosphere decreases substantially, more UV radiation will reach the Earth's surface, causing unwanted photochemical reactions, including reactions correlated with skin

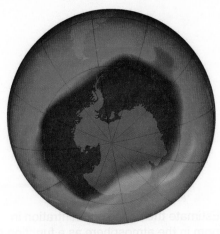

Total ozone (Dobson units)

110   220   330   440   550

▲ FIGURE 13.4 **Ozone present in the Southern Hemisphere, 24 September 2006.** These data were taken from an orbiting satellite. This day had the lowest stratospheric ozone concentration yet recorded. One 'Dobson unit' corresponds to 2.69×10^{16} ozone molecules in a 1 cm² column of atmosphere.

∞ Find out more on page 571

cancer. Satellite monitoring of ozone, which began in 1978, has revealed a depletion of ozone in the stratosphere that is particularly severe over Antarctica, a phenomenon known as the ozone hole (◀ FIGURE 13.4).

In 1995 the Nobel Prize in Chemistry was awarded to F. Sherwood Rowland, Mario Molina and Paul Crutzen for their studies of ozone depletion. In 1970 Crutzen showed that naturally occurring nitrogen oxides catalytically destroy ozone. Rowland and Molina recognised in 1974 that chlorine from **chlorofluorocarbons** (CFCs) may deplete the ozone layer. These substances, principally $CFCl_3$ and CF_2Cl_2, do not occur in nature and have been widely used as propellants in spray cans, as refrigerant and air-conditioner gases and as foaming agents for plastics. They are virtually unreactive in the lower atmosphere. Furthermore, they are relatively insoluble in water and are therefore not removed from the atmosphere by rainfall or by dissolution in the oceans. Unfortunately, the lack of reactivity that makes them commercially useful also allows them to survive in the atmosphere and to diffuse into the stratosphere. It is estimated that several million tons of chlorofluorocarbons are now present in the atmosphere.

As CFCs diffuse into the stratosphere, they are exposed to high-energy radiation, which can cause photodissociation. The C–Cl bonds are considerably weaker than the C–F bonds. As a result, free chlorine atoms are formed readily in the presence of light with wavelengths in the range of 190 nm to 225 nm, as shown in this typical reaction:

$$CF_2Cl_2(g) + h\nu \longrightarrow CF_2Cl(g) + Cl(g) \qquad [13.6]$$

Calculations suggest that chlorine atom formation occurs at the greatest rate at an altitude of about 30 km, the altitude at which ozone is at its highest concentration.

Atomic chlorine reacts rapidly with ozone to form chlorine monoxide (ClO) and molecular oxygen (O_2):

$$Cl(g) + O_3(g) \longrightarrow ClO(g) + O_2(g) \qquad [13.7]$$

Equation 13.7 follows a second-order rate law (∞ Section 15.3, 'Concentration and Rate Laws') with a very large rate constant:

$$\text{Rate} = k[Cl][O_3] \quad k = 7.2 \times 10^9 \text{ dm}^3 \text{ mol}^{-1} \text{ s}^{-1} \text{ at 298 K} \qquad [13.8]$$

Under certain conditions the ClO generated in Equation 13.7 can react to regenerate free Cl atoms. One way that this can happen is by photodissociation of the ClO:

$$ClO(g) + h\nu \longrightarrow Cl(g) + O(g) \qquad [13.9]$$

The Cl atoms generated in Equation 13.9 can react with more O_3, according to Equation 13.7. These two equations form a cycle for the Cl atom-catalysed decomposition of O_3 to O_2 as we see when we add the equations:

$$2\,Cl(g) + 2\,O_3(g) \longrightarrow 2\,ClO(g) + 2\,O_2(g)$$

$$2\,ClO(g) + h\nu \longrightarrow 2\,Cl(g) + 2\,O(g)$$

$$O(g) + O(g) \longrightarrow O_2(g)$$

$$2\,Cl(g) + 2\,O_3(g) + 2\,ClO(g) + 2\,O(g) \longrightarrow 2\,Cl(g) + 2\,ClO(g) + 3\,O_2(g) + 2\,O(g)$$

The equation can be simplified by eliminating like species from each side to give

$$2\,O_3(g) \xrightarrow{\;Cl\;} 3\,O_2(g) \qquad [13.10]$$

Because the rate of Equation 13.7 increases linearly with [Cl], the rate at which ozone is destroyed increases as the quantity of Cl atoms increases. Thus the greater the amount of CFCs that diffuse into the stratosphere, the faster the destruction of the ozone layer. Rates of diffusion of molecules from the troposphere into the stratosphere are slow. Nevertheless a significant thinning of the ozone layer over the poles (where $ClO(g)$ can accumulate during the long polar nights) has already been observed.

Because of the environmental problems associated with CFCs, steps have been taken to limit their manufacture and use. A major step was the signing in 1987 of the Montreal Protocol on Substances That Deplete the Ozone Layer, in which participating nations agreed to reduce CFC production. More stringent limits were set in 1992, when representatives of approximately 100 nations agreed to ban the production and use of CFCs by 1996. Because CFCs are so unreactive and diffuse so slowly into the stratosphere, ozone depletion will continue for many years to come, but there are already encouraging signs of a reversal in the trend towards the thinning of the ozone layer at the poles.

What substances have replaced CFCs? At this time the main alternatives are hydrofluorocarbons, compounds in which C–H bonds replace the C–Cl bonds of CFCs. One such compound in current use is CF_3CH_2F (1,1,1,2-tetrafluoroethane) known as HFC-134a.

There are no naturally occurring CFCs, but there are some natural sources that contribute chlorine and bromine to the atmosphere and, just like halogens from CFC, these naturally occurring Cl and Br atoms can participate in ozone-depleting reactions. The principal natural sources are bromomethane and chloromethane, CH_3Br and CH_3Cl, which are emitted from the oceans. It is estimated that these molecules contribute less than a third to the total Cl and Br in the upper atmosphere; the remaining two-thirds is a result of human activities. Volcanos are a source of Cl atoms, but generally they release HCl which reacts with water in the troposphere and does not make it to the upper atmosphere.

The Troposphere

The troposphere consists primarily of N_2 and O_2, which together make up 99% of the Earth's atmosphere at sea level (Table 13.1). Other gases, although present only at very low concentrations, can have major effects on our environment. ▼ TABLE 13.3 lists the major sources and typical concentrations of some of the important minor constituents of the troposphere. Many of these substances occur to only a slight extent in the natural environment but exhibit much higher concentrations in certain areas as a result of human activities. In this section we discuss the most important characteristics of a few of these substances and their chemical roles as air pollutants. Most of them are formed either directly or indirectly from our widespread use of combustion reactions.

Sulfur Compounds and Acid Rain

Sulfur-containing compounds are present to some extent in the natural, unpolluted atmosphere. They originate in the bacterial decay of organic matter,

TABLE 13.3 • Sources and typical concentrations of some minor tropospheric constituents

Minor constituent	Sources	Typical concentrations
Carbon dioxide, CO_2	Decomposition of organic matter; release from the oceans; fossil-fuel combustion	375 ppm throughout the troposphere
Carbon monoxide, CO	Decomposition of organic matter; industrial processes; fossil-fuel combustion	0.05 ppm in unpolluted air; 1–50 ppm in urban traffic areas
Methane, CH_4	Decomposition of organic matter; natural-gas seepage	1.77 ppm throughout the troposphere
Nitric oxide, NO	Electrical discharges; internal combustion engines; combustion of organic matter	0.01 ppm in unpolluted air; 0.2 ppm in smog
Ozone, O_3	Electrical discharges; diffusion from the stratosphere; photochemical smog	0–0.01 ppm in unpolluted air; 0.5 ppm in photochemical smog
Sulfur dioxide, SO_2	Volcanic gases; forest fires; bacterial action; fossil-fuel combustion; industrial processes	0–0.01 ppm in unpolluted air; 0.1–2 ppm in polluted urban environment

TABLE 13.4 • Median concentrations of atmospheric pollutants in a typical urban atmosphere	
Pollutant	Concentration (ppm)
Carbon monoxide	10
Hydrocarbons	3
Sulfur dioxide	0.08
Nitrogen oxides	0.05
Total oxidants (ozone and others)	0.02

(a)

(b)

▲ **FIGURE 13.5 Damage from acid rain.** Photograph (b), recently taken, shows how the statue has lost detail in its carvings.

in volcanic gases and from other sources listed in Table 13.3. The amount of sulfur-containing compounds released into the atmosphere from natural sources is about 24×10^6 tonnes per year, which is less than the amount from human activities (about 79×10^6 tonnes per year). Sulfur compounds, chiefly sulfur dioxide, SO_2, are among the most unpleasant and harmful of the common pollutant gases. ◀ TABLE 13.4 shows the concentrations of several pollutant gases in a *typical* urban environment (not one that is particularly affected by smog). According to these data, the level of sulfur dioxide is 0.08 ppm or higher about half the time. This concentration is considerably lower than that of other pollutants, notably carbon monoxide. Nevertheless, SO_2 is regarded as the most serious health hazard among the pollutants shown, especially for people with respiratory difficulties.

Combustion of coal, oil and other petroleum products containing sulfur accounts for a large percentage of the total SO_2 released in the world.

Sulfur dioxide is harmful to both human health and property; furthermore, atmospheric SO_2 can be oxidised to SO_3 by several pathways (such as reaction with O_2 or O_3). When SO_3 dissolves in water, it produces sulfuric acid, H_2SO_4:

$$SO_3(g) + H_2O(l) \longrightarrow H_2SO_4(aq)$$

Many of the environmental effects ascribed to SO_2 are actually due to H_2SO_4.

The presence of SO_2 in the atmosphere and the sulfuric acid that it produces result in the phenomenon of **acid rain**. (Nitrogen oxides, which form nitric acid, are also major contributors to acid rain.) Uncontaminated rainwater is naturally acidic and generally has a pH value of about 5.6. The primary source of this natural acidity is CO_2, which reacts with water to form carbonic acid, H_2CO_3. Acid rain typically has a pH value of about 4. This acidity has affected many lakes in northern Europe, northern United States and Canada, reducing fish populations and affecting other organisms within the lakes and surrounding forests.

The pH of most natural waters containing living organisms is between 6.5 and 8.5, but freshwater pH values are far below 6.5 in many parts of the industrialised world. At pH levels below 4.0, all vertebrates, most invertebrates and many microorganisms are destroyed. The lakes that are most susceptible to damage are those with low concentrations of basic ions, such as HCO_3^-, that buffer them against changes in pH.

Because acids react with metals and with carbonates, acid rain is corrosive both to metals and to stone building materials. Marble and limestone, for example, whose major constituent is $CaCO_3$, are readily attacked by acid rain (◀ FIGURE 13.5). Billions of dollars each year are lost as a result of corrosion that is due to SO_2 pollution.

One way to reduce the quantity of SO_2 released into the environment is to remove sulfur from coal and oil before it is burned. Although difficult and expensive, several methods have been developed for removing SO_2 from the gases formed when coal and oil are combusted. Powdered limestone ($CaCO_3$), for example, can be injected into the furnace of a power plant where it decomposes into lime (CaO) and carbon dioxide:

$$CaCO_3(s) \longrightarrow CaO(s) + CO_2(g)$$

The CaO then reacts with SO_2 to form calcium sulfite:

$$CaO(s) + SO_2(g) \longrightarrow CaSO_3(s)$$

The solid particles of $CaSO_3$ as well as much of the unreacted SO_2 can be removed from the furnace gas by passing it through an aqueous suspension of lime (▶ FIGURE 13.6). Not all the SO_2 is removed, however, and, given the enormous quantities of coal and oil burned worldwide, controllable pollution by SO_2 will probably remain a problem for some time.

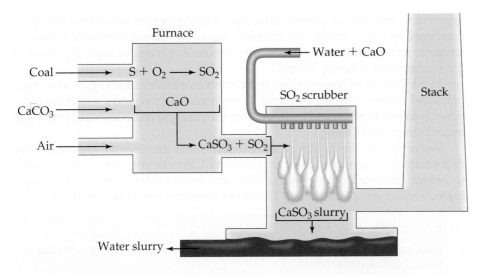

◀ **FIGURE 13.6**   **Common method for** removing SO$_2$ from combusted fuel.

⚠ CONCEPT CHECK 3
What chemical behaviour associated with sulfur oxides gives rise to acid rain?

Nitrogen Oxides and Photochemical Smog

Nitrogen oxides are primary components of smog, a phenomenon with which city dwellers are all too familiar. The term *smog* refers to a particularly unpleasant condition of pollution in certain urban environments that occurs when weather conditions produce a relatively stagnant air mass. Smog is more accurately described as **photochemical smog** because photochemical processes play a major role in its formation (▶ **FIGURE 13.7**).

Nitric oxide, NO, forms in small quantities in the cylinders of internal combustion engines by the direct combination of nitrogen and oxygen:

$$N_2(g) + O_2(g) \rightleftharpoons 2\,NO(g) \qquad \Delta H = +180.8 \text{ kJ} \qquad [13.11]$$

This reaction is more favourable at higher temperatures. The equilibrium constant K (∞ Section 16.2, 'The Equilibrium Constant') for this reaction increases from about 10^{-15} at 300 K (near room temperature) to about 0.05 at 2400 K (approximately the temperature in the cylinder of an engine during combustion).

In air, nitric oxide (NO) is rapidly oxidised to nitrogen dioxide (NO$_2$):

$$2\,NO(g) + O_2(g) \rightleftharpoons 2\,NO_2(g) \qquad \Delta H = -113.1 \text{ kJ} \qquad [13.12]$$

The equilibrium constant for this reaction decreases from about 10^{12} at 300 K to about 10^{-5} at 2400 K. The photodissociation of NO$_2$ initiates the reactions associated with photochemical smog. The dissociation of NO$_2$ into NO and O requires 304 kJ mol^{-1}, which corresponds to a photon wavelength of 393 nm. In sunlight, therefore, NO$_2$ undergoes dissociation to NO and O:

$$NO_2(g) + h\nu \longrightarrow NO(g) + O(g) \qquad [13.13]$$

The atomic oxygen formed undergoes several possible reactions, one of which gives ozone, as described earlier:

$$O(g) + O_2(g) + M(g) \longrightarrow O_3(g) + M^*(g) \qquad [13.14]$$

Ozone is a key component of photochemical smog. Although it is an essential UV screen in the upper atmosphere, it is an undesirable pollutant in the troposphere. It is extremely reactive and toxic, and breathing air that contains

▲ **FIGURE 13.7**   **Photochemical smog.** Photochemical smog is produced largely by the action of sunlight on vehicle exhaust gases.

∞ Find out more on page 618

appreciable amounts of ozone can be especially dangerous for asthma sufferers, exercisers and the elderly. We therefore have two ozone problems: excessive amounts in many urban environments, where it is harmful, and depletion in the stratosphere, where it is vital.

In addition to nitrogen oxides and carbon monoxide, a car engine also emits unburned *hydrocarbons* as pollutants. These organic compounds, which are composed entirely of carbon and hydrogen, are the principal components of petrol and major ingredients of smog.

Reduction or elimination of smog requires that the essential ingredients for its formation are removed from car exhausts. Catalytic converters are designed to reduce drastically the levels of NO_x and hydrocarbons, two of the major ingredients of smog (see the My World of Chemistry box in Section 15.7).

 CONCEPT CHECK 4

What photochemical reaction involving nitrogen oxides initiates the formation of photochemical smog?

Greenhouse Gases: Water Vapour, Carbon Dioxide and Climate

We have seen how the atmosphere makes life as we know it possible on the Earth by screening out harmful short-wavelength radiation. In addition, the atmosphere is essential in maintaining a reasonably uniform and moderate temperature on the surface of the planet. The two atmospheric components of greatest importance in maintaining the Earth's surface temperature are carbon dioxide and water.

The Earth is in overall thermal balance with its surroundings. This means that the Earth radiates energy into space at a rate equal to the rate at which it absorbs energy from the sun. ▼ **FIGURE 13.8** shows the distribution of radia-

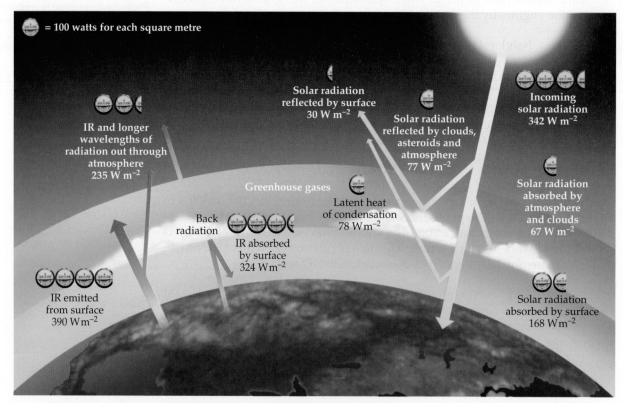

▲ **FIGURE 13.8   Earth's thermal balance.** The amount of radiation reaching the surface of the planet is approximately equal to the amount radiated back into space.

tion to and from the Earth's surface. ▶ FIGURE 13.9 shows which portion of the infrared radiation leaving the surface is absorbed by atmospheric water vapour and carbon dioxide. In doing so, they help to maintain a livable uniform temperature at the surface by holding in, as it were, the infrared radiation from the surface, which we feel as heat. The influence of H_2O, CO_2 and certain other atmospheric gases on the Earth's temperature is often called the *greenhouse effect* because in trapping infrared radiation these gases act much like the glass of a greenhouse. The gases themselves are called **greenhouse gases**.

Water vapour makes the largest contribution to the greenhouse effect. The partial pressure of water vapour in the atmosphere varies greatly from place to place and time to time, but it is generally highest near the Earth's surface and drops off very sharply with increased elevation. Because water vapour absorbs infrared radiation so strongly, it plays the major role in maintaining the atmospheric temperature at night, when the surface is emitting radiation into space and not receiving energy from the sun. In very dry desert climates, where the water vapour concentration is unusually low, it may be extremely hot during the day, but very cold at night. In the absence of an extensive layer of water vapour to absorb and then radiate part of the infrared radiation back to the Earth, the surface loses this radiation into space and cools off very rapidly.

Carbon dioxide plays a secondary, but very important, role in maintaining the surface temperature. The worldwide combustion of fossil fuels, principally coal and oil, on a prodigious scale in the modern era has sharply increased the carbon dioxide level of the atmosphere. Measurements conducted over several decades show that the CO_2 concentration in the atmosphere is steadily increasing, currently at a high of about 390 ppm (▶ **FIGURE 13.10**). In comparison, it is estimated that CO_2 concentrations in the atmosphere fluctuated between 200 ppm and 300 ppm over the past 150 000 years, based on data from air bubbles trapped in ice cores. A consensus is emerging among scientists that this increase is already perturbing the Earth's climate and that it may be responsible for the observed increase in the average global air temperature of 0.3 °C to 0.6 °C over the past century. Scientists often use the term *climate change* instead of *global warming* to refer to this effect because as the Earth's temperature increases, it affects winds and ocean currents in ways that can cool some areas and warm others.

On the basis of present and expected future rates of fossil-fuel use, the atmospheric CO_2 level is expected to double from its present level some time between 2050 and 2100. Computer models predict that this increase will result in an average global temperature increase of 1 °C to 3 °C. We cannot predict with certainty what changes will occur. Clearly, however, humanity has acquired the potential, by changing

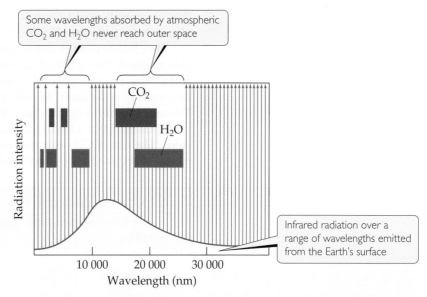

▲ **FIGURE 13.9**   **Portions of the infrared radiation emitted by the Earth's surface that are absorbed by atmospheric CO_2 and H_2O.**

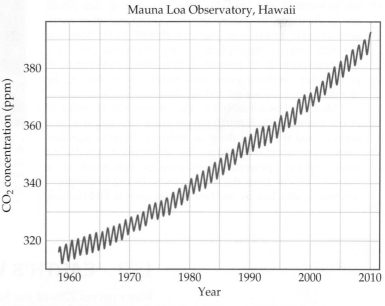

▲ **FIGURE 13.10**   **Rising CO_2 levels.** The sawtooth shape of the graph is due to regular seasonal variations in CO_2 concentration for each year.

the concentration of CO_2 and other heat-trapping gases in the atmosphere, to alter substantially the climate of the planet.

The climate change threat posed by atmospheric CO_2 has sparked considerable research into ways of capturing the gas at its largest combustion sources and storing it under ground or under the seafloor. There is also much interest in developing new ways to use CO_2 as a chemical feedstock. However, the 115 million tonnes (approximately) of CO_2 used annually by the global chemical industry is but a small fraction of the 24 billion tonnes of annual CO_2 emissions. The use of CO_2 as a raw material will probably never be great enough to reduce its atmospheric concentration.

CONCEPT CHECK 5

Explain why night-time temperatures remain higher in locations where there is higher humidity.

MY WORLD OF CHEMISTRY

METHANE AS A GREENHOUSE GAS

Although CO_2 receives most of the attention, other gases in total make an equal contribution to the greenhouse effect. Chief among these is methane, CH_4. Each methane molecule has about 25 times the greenhouse effect of a CO_2 molecule. Studies of atmospheric gas trapped long ago in the Greenland and Antarctic ice sheets show that the concentration of methane in the atmosphere has increased during the pre-industrial age, from pre-industrial values in the range of 0.3 ppm to 0.7 ppm to the present value of about 1.8 ppm.

Methane is formed in biological processes that occur in low-oxygen environments. Anaerobic bacteria, which flourish in swamps and landfills, near the roots of rice plants and in the digestive systems of cows, sheep and other ruminant animals, produce methane (▶ FIGURE 13.11). Methane also leaks into the atmosphere during natural-gas extraction and transport. It is estimated that about two-thirds of present-day methane emissions, which are increasing by about 1% per year, are related to human activities.

Methane has a half-life in the atmosphere of about 10 years, whereas CO_2 is much longer lived. This might at first seem a good thing, but there are indirect effects to consider. Some methane is oxidised in the stratosphere, producing water vapour, a powerful greenhouse gas that is otherwise virtually absent from the stratosphere. In the troposphere methane is attacked by reactive species such as OH radicals

and nitrogen oxides, eventually producing other greenhouse gases such as O_3. It has been estimated that the climate-changing effects of CH_4 are at least one-third those of CO_2, and perhaps even half as large. Given this large contribution, important reductions in the greenhouse effect could be achieved by reducing methane emissions or capturing the emissions for use as a fuel.

▲ FIGURE 13.11 **Methane production.** Ruminant animals, such as cows and sheep, produce methane in their digestive systems. In Australia, cattle and sheep produce about 14% of the country's total greenhouse emissions.

∞ Review this on page 383

13.3 | EARTH'S WATER

Water covers 72% of the Earth's surface and is essential to life. Our bodies are about 65% water by mass. Because of extensive hydrogen bonding, water has unusually high melting and boiling points, and a high heat capacity (∞ Section 11.2, 'Intermolecular Forces'). Its highly polar character is responsible for its exceptional ability to dissolve a wide range of ionic and polar-covalent substances. Many reactions occur in water, including reactions in which H_2O itself is a reactant. Recall, for example, that H_2O can participate in acid–base reactions as either a proton donor or a proton acceptor.

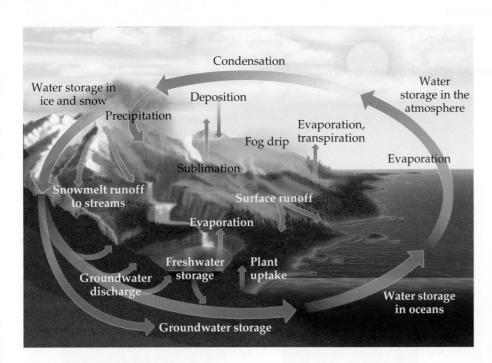

◀ FIGURE 13.12   **The global water cycle.**

The Global Water Cycle

All the water on Earth is connected in a global water cycle (▲ FIGURE 13.12). Most of the processes depicted here rely on the phase changes of water. For instance, warmed by the sun, liquid water in the oceans evaporates into the atmosphere as water vapour and condenses into liquid water droplets that we see as clouds. Water droplets in the clouds can crystallise to ice, which can precipitate as hail or snow. Once on the ground, the hail or snow melts to liquid water, which soaks into the ground. If conditions are right, it is also possible for ice on the ground to sublime to water vapour in the atmosphere.

Salt Water: Earth's Oceans and Seas

The vast layer of salty water that covers so much of the planet is connected and is generally constant in composition. For this reason, oceanographers speak of a *world ocean* rather than of the separate oceans we learn about in geography books. The world ocean is huge. Its volume is $1.35 \times 10^9 \ km^3$. Almost all the water on the Earth, 97.2%, is in the world ocean. Of the remaining 2.8%, 2.1% is in the form of ice caps and glaciers. All the freshwater—in lakes, rivers and groundwater—amounts to only 0.6%.

Seawater is often referred to as saline water. The **salinity** of seawater is the mass in grams of dry salts present in 1 kg of seawater. In the world ocean the salinity averages about 35 g. To put it another way, seawater contains about 3.5% dissolved salts by mass. The list of elements present in seawater is very long. Most, however, are present only in very low concentrations. ▶ TABLE 13.5 lists the 11 ionic species that are most abundant in seawater.

The properties of seawater—its temperature, salinity and density—vary as a function of depth (▼ FIGURE 13.13). Sunlight penetrates well only 200 m into the sea; the region between 200 m and 1000 m deep is the 'twilight zone', where visible light is faint. Below 1000 m in depth, the ocean is pitch-black and cold, about 4 °C. The transport of heat, salt and other chemicals throughout the ocean is influenced by these changes in the physical properties of seawater and, in turn, the changes in the way heat and substances are transported affects ocean currents and the global climate.

TABLE 13.5 • **Ionic constituents of seawater present in concentrations greater than 0.001 g kg^{-1} (1 ppm)**

Ionic constituent	g kg^{-1} Seawater	Concentration (M)
Chloride, Cl^-	19.35	0.55
Sodium, Na^+	10.76	0.47
Sulfate, SO_4^{2-}	2.71	0.028
Magnesium, Mg^{2+}	1.29	0.054
Calcium, Ca^{2+}	0.412	0.010
Potassium, K^+	0.40	0.010
Carbon dioxide*	0.106	2.3×10^{-3}
Bromide, Br^-	0.067	8.3×10^{-4}
Boric acid, H_3BO_3	0.027	4.3×10^{-4}
Strontium, Sr^{2+}	0.0079	9.1×10^{-5}
Fluoride, F^-	0.0013	7.0×10^{-5}

*CO_2 is present in seawater as HCO_3^- and CO_3^{2-}.

∞∞ Find out more on page 578

FIGURE IT OUT

Look at the trend in density as a function of depth; does it mirror the trend better in salinity or in temperature?

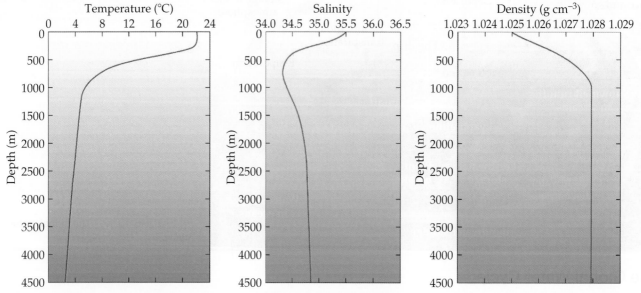

▲ **FIGURE 13.13** **Average temperature, salinity and density of seawater as a function of depth.** (From Windows to the Universe, University Corporation for Atmospheric Research. Copyright © 2004 University Corporation for Atmospheric Research. All rights reserved.)

The sea is so vast that, if a substance is present in seawater to the extent of only 1 part per billion (ppb, that is, 1×10^{-6} g per kg of water), there is still 5×10^9 kg of it in the world ocean. Despite this, the ocean is rarely used as a source of raw materials because the cost of extracting the desired substances is too high. Only three substances are obtained from seawater in commercially important amounts: sodium chloride, bromine (from bromide salts) and magnesium (from magnesium salts).

Absorption of CO_2 by the ocean plays a large role in global climate. Carbon dioxide reacts with water to form carbonic acid, H_2CO_3, and so, as CO_2 from the atmosphere is absorbed by the world ocean, the concentration of H_2CO_3 in the ocean increases. Most of the carbon in the ocean, however, is in the form of HCO_3^- and CO_3^{2-} ions. These ions form a buffer system that maintains the ocean's average pH between 8.0 and 8.3. The buffering capacity of the world ocean is predicted to decrease as the concentration of CO_2 in the atmosphere increases, because of the increase in H_2CO_3 concentration resulting in a drop in pH of the ocean. In more acidic waters the equilibrium between HCO_3^- and CO_3^{2-} lies further towards HCO_3^-, and the precipitation of $CaCO_3$, the main constituent of seashells and coral skeletons, becomes less thermodynamically favourable. Thus both acid–base equilibrium reactions and solubility equilibrium reactions (∞∞ Section 15.4, 'The Change of Concentration with Time') form a complicated web of interactions that ties the ocean to the atmosphere and to the global climate.

Freshwater and groundwater

Freshwater is the term used to denote natural waters that have low concentrations (less than 500 ppm) of dissolved salts and solids.

The total amount of freshwater on the Earth is not a very large fraction of the total water present. Indeed, freshwater is one of our most precious resources. It forms by evaporation from the oceans and the land. The water vapour that accumulates in the atmosphere is transported by global atmospheric circulation, eventually returning to the Earth as rain and snow (Figure 13.12).

As rain falls and as water runs off the land on its way to the oceans, it dissolves a variety of cations (mainly Na^+, K^+, Mg^{2+}, Ca^{2+} and Fe^{2+}), anions (mainly Cl^-, SO_4^{2-} and HCO_3^-) and gases (principally O_2, N_2 and CO_2). As we use water, it becomes laden with additional dissolved material, including the wastes of human society. As our population and output of environmental pollutants increase, we find that we must spend ever-increasing amounts of money and resources to guarantee a supply of freshwater.

Approximately 20% of the world's freshwater is under the soil, in the form of *groundwater*. Groundwater resides in *aquifers*, which are layers of porous rock that hold water. The water in aquifers can be very pure and accessible for human consumption if near the surface (▼ FIGURE 13.14). Dense rock that does not allow water to penetrate readily can hold groundwater for years or even millennia.

The nature of the rock that contains the groundwater has a large influence on the water's chemical composition. If minerals in the rock are water soluble to some extent, ions can leach out of the rock and remain dissolved in the groundwater. Arsenic in the form of $HAsO_4^{2-}$, $H_2AsO_4^-$ and H_3AsO_4 are found in groundwater across the world, most infamously in Bangladesh, at concentrations poisonous to humans. In some reducing waters, the arsenic is found as H_3AsO_3.

13.4 | HUMAN ACTIVITIES AND EARTH'S WATER

All life on Earth depends on the availability of suitable water. Some organisms can thrive under temperature, pH and ionic conditions where other organisms would die. Many human activities rely on waste disposal via Earth's waters, even today, and this practice can be detrimental to aquatic organisms.

Dissolved Oxygen and Water Quality

The amount of dissolved O_2 in water is an important indicator of water quality. Water fully saturated with air at 1 bar and 20 °C contains about 8.5 ppm of O_2. Oxygen is necessary for fish and much other aquatic life. Cold-water fish require that the water contain at least 5 ppm of dissolved oxygen

FIGURE IT OUT

What factors influence how long it takes for water to migrate from a deep aquifer to the surface?

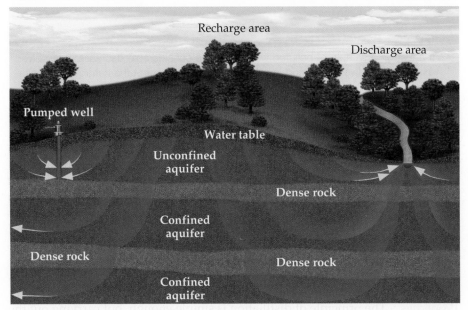

◀ **FIGURE 13.14 Groundwater is water located in aquifers below the soil.** An unconfined aquifer, that has no dense rock between it and the water table, can hold water for days or years. Confined aquifers can hold water for centuries or millennia, depending on their depth. Aquifers are discharged through wells or rivers, and are recharged from water flowing through the soil (e.g. from rain).

▶ **FIGURE 13.15 Eutrophication.** The growth of algae and duckweed in this pond is due to agricultural wastes. The wastes feed the growth of the algae and weeds which deplete the oxygen in the water, a process called eutrophication. A eutrophic lake cannot support fish.

for survival. Aerobic bacteria consume dissolved oxygen in order to oxidise organic materials and so meet their energy requirements. The organic material that the bacteria are able to oxidise is said to be **biodegradable**. This oxidation occurs by a complex set of chemical reactions, and the organic material disappears gradually.

Excessive quantities of biodegradable organic materials in water are detrimental because they deplete the water of the oxygen necessary to sustain normal animal life. Typical sources of these biodegradable materials, which are called *oxygen-demanding wastes*, include sewage, industrial wastes from food-processing plants and paper mills and effluent (liquid waste) from meatpacking plants.

In the presence of oxygen, the carbon, hydrogen, nitrogen, sulfur and phosphorus in biodegradable material end up mainly as CO_2, HCO_3^-, H_2O, NO_3^-, SO_4^{2-} and phosphates. The formation of these oxidation products sometimes reduces the amount of dissolved oxygen to the point where aerobic bacteria can no longer survive. Anaerobic bacteria then take over the decomposition process, forming CH_4, NH_3, H_2S, PH_3 and other products, several of which contribute to the offensive odours of some polluted waters.

Plant nutrients, particularly nitrogen and phosphorus, contribute to water pollution by stimulating excessive growth of aquatic plants. The most visible results of excessive plant growth are floating algae and murky water. More significantly, however, as plant growth becomes excessive, the amount of dead and decaying plant matter increases rapidly, a process called *eutrophication* (▲ **FIGURE 13.15**). The decay of plants consumes O_2 as the plants are biodegraded, leading to the depletion of oxygen in the water. Without sufficient supplies of oxygen the water, in turn, cannot sustain any form of animal life. The most significant sources of nitrogen and phosphorus compounds in water are domestic sewage (phosphate-containing detergents and nitrogen-containing body wastes), runoff from agricultural land (fertilisers containing both nitrogen and phosphorus) and runoff from livestock areas (animal wastes containing nitrogen).

CONCEPT CHECK 6

If a test on a sample of polluted water shows a considerable decrease in dissolved oxygen over a five-day period, what can we conclude about the nature of the pollutants present?

Water Purification: Desalination

Because of its high salt content, seawater is unfit for human consumption and for most of the uses to which we put water. In Australia the salt content of municipal water supplies is restricted by health codes to no more than 0.05% by mass. This amount is much lower than the 3.5% dissolved salts present in seawater and the 0.5% or so present in brackish water found underground in some regions. The removal of salts from seawater or brackish water to make the water usable is called **desalination**.

Water can be separated from dissolved salts by *distillation* because water is a volatile substance and the salts are non-volatile (∞ Section 12.5, 'Colligative Properties'). The principle of distillation is simple enough, but carrying out the

∞ Review this on page 448

process on a large scale presents many problems. As water is distilled from seawater, for example, the salts become more and more concentrated and eventually precipitate out. Distillation is also an energy-intensive process.

Seawater can also be desalinated using **reverse osmosis**. Recall that osmosis is the net movement of solvent molecules, but not solute molecules, through a semi-permeable membrane (∞ Section 12.5, 'Colligative Properties'). In osmosis, the solvent passes from the more dilute solution into the more concentrated one. However, if sufficient external pressure is applied, osmosis can be stopped and, at still higher pressures, reversed. When reverse osmosis occurs, solvent passes from the more concentrated into the more dilute solution. In a modern reverse-osmosis facility, hollow fibres are used as the semi-permeable membrane (▶ FIGURE 13.16). Water is introduced under pressure into the fibres, and desalinated water is recovered.

The world's largest desalination plant, in Jubail, Saudi Arabia, provides 50% of that country's drinking water by using reverse osmosis to desalinate seawater from the Persian Gulf. Such plants are becoming common in Australia. The Sydney desalination plant provides 15% of Sydney's water supply. Small-scale, manually operated reverse-osmosis desalinators are used in camping, travelling and at sea.

Water Purification: Municipal Treatment

The water needed for domestic uses, agriculture and industrial processes is taken either from naturally occurring lakes, rivers and underground sources or from reservoirs. This water must be treated before it is distributed. Municipal water treatment usually involves five steps: coarse filtration, sedimentation, sand filtration, aeration and sterilisation. ▼ FIGURE 13.17 shows a typical treatment process.

After coarse filtration through a screen, the water is allowed to stand in large settling tanks in which finely divided sand and other minute particles can settle out. To aid in removing very small particles, the water may first be made slightly basic by adding CaO. Then $Al_2(SO_4)_3$ is added. The aluminium sulfate reacts with OH^- ions to form a spongy, gelatinous precipitate of $Al(OH)_3$. This precipitate settles slowly, carrying suspended particles down with it, thereby removing nearly all finely divided matter and most bacteria. The water is then filtered through a sand bed. Following filtration, the water may be sprayed into the air to hasten the oxidation of dissolved organic substances.

 ∞ Review this on page 451

▲ **FIGURE IT OUT**

What feature of this process is responsible for it being called *reverse* osmosis?

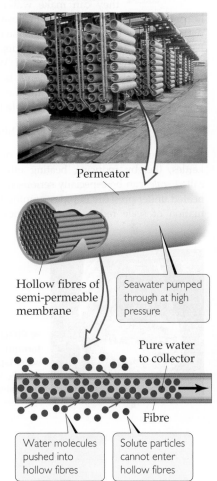

Permeator

Hollow fibres of semi-permeable membrane

Seawater pumped through at high pressure

Pure water to collector

Fibre

Water molecules pushed into hollow fibres

Solute particles cannot enter hollow fibres

▲ FIGURE 13.16 **Reverse osmosis.**

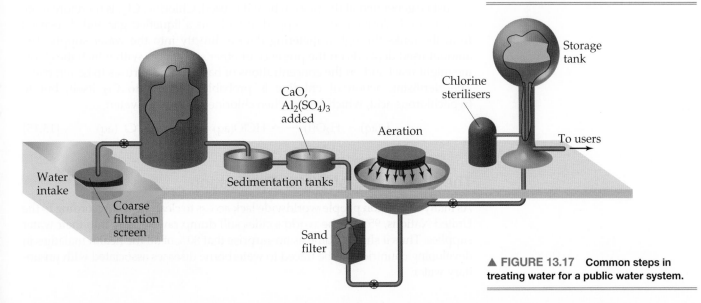

▲ FIGURE 13.17 **Common steps in treating water for a public water system.**

Storage tank

Chlorine sterilisers

CaO, $Al_2(SO_4)_3$ added

Aeration

To users

Water intake

Coarse filtration screen

Sedimentation tanks

Sand filter

A CLOSER LOOK

WATER SOFTENING

Water containing a relatively high concentration of Ca^{2+}, Mg^{2+} and other divalent cations is called **hard water**. Although the presence of these ions is generally not a health threat, they can make water unsuitable for some household and industrial uses. For example, these ions react with soaps to form an insoluble soap scum, the stuff of bathtub rings. In addition, mineral deposits may form when water containing these ions is heated. When water containing calcium ions and bicarbonate ions is heated, some carbon dioxide is driven off. As a result, the solution becomes less acidic and insoluble calcium carbonate forms:

$$Ca^{2+}(aq) + 2\,HCO_3^-(aq) \longrightarrow CaCO_3(s) + CO_2(g) + H_2O(l)$$

The solid $CaCO_3$ coats the surface of hot-water systems and kettles, thereby reducing heating efficiency. These deposits, called *scale*, can be especially serious in boilers where water is heated under pressure in pipes running through a furnace. Formation of scale reduces the efficiency of heat transfer and

◄ **FIGURE 13.18 Scale formation.** The interior of this water pipe has been coated with $CaCO_3$ and other insoluble salts deposited from hard water.

reduces the flow of water through pipes (▼ FIGURE 13.18). The removal of the ions that cause hard water is called water *softening*.

Ion exchange is a typical household method for water softening. In this procedure the hard water is passed through a bed of an ion-exchange resin: plastic beads with covalently bound anion groups such as $-COO^-$ or $-SO_3^-$. These negatively charged groups have Na^+ ions attached to balance their charges. The Ca^{2+} ions and other cations in the hard water are attracted to the anionic groups and displace the lower-charged Na^+ ions into the water. Thus one type of ion is exchanged for another. To maintain charge balance, $2\,Na^+$ ions enter the water for each Ca^{2+} removed. If we represent the resin with its anionic site as $R-COO^-$, we can write the equation for the process as follows:

$$2\,Na(R-COO)(s) + Ca^{2+}(aq) \rightleftharpoons$$
$$Ca(R-COO)_2(s) + 2\,Na^+(aq)$$

Water softened in this way contains an increased concentration of Na^+ ions. Although Na^+ ions do not form precipitates or cause other problems associated with hard-water cations, individuals concerned about their sodium intake, such as those who have high blood pressure (hypertension), should avoid drinking water softened in this way.

When all the available Na^+ ions have been displaced from the ion-exchange resin, the resin is regenerated by flushing it with a concentrated solution of NaCl. Home owners can do this by charging their units with large amounts of NaCl(s), which can be purchased at most grocery stores. The high concentration of Na^+ forces the equilibrium shown in the earlier equation to shift to the left, causing the Na^+ ions to displace the hard-water cations, which are flushed down the drain.

The final stage of the operation normally involves treating the water with a chemical agent to ensure the destruction of bacteria. Ozone is most effective, but it must be generated at the place where it is used. Chlorine, Cl_2, is therefore more convenient. Chlorine can be shipped in tanks as a liquefied gas and dispensed from the tanks through a metering device directly into the water supply. The amount used depends on the presence of other substances with which the chlorine might react and on the concentrations of bacteria and viruses to be removed. The sterilising action of chlorine is probably due not to Cl_2 itself, but to hypochlorous acid, which forms when chlorine reacts with water:

$$Cl_2(aq) + H_2O(l) \longrightarrow HClO(aq) + H^+(aq) + Cl^-(aq) \qquad [13.15]$$

The Challenges of Water Purification

As many as a billion people worldwide lack access to clean water. According to the United Nations, 95% of the world's cities still dump raw sewage into their water supplies. Thus it should come as no surprise that 80% of all the health maladies in developing countries can be traced to waterborne diseases associated with unsanitary water.

One promising development is a device called the LifeStraw (▶ FIGURE 13.19). When a person sucks water through the straw, the water first encounters a textile filter with a mesh opening of 100 μm followed by a second textile filter with a mesh opening of 15 μm. These filters remove debris and even clusters of bacteria. The water next encounters a chamber of iodine-impregnated beads, where bacteria, viruses and parasites are killed. Finally, the water passes through granulated active carbon, which removes the smell of iodine as well as the parasites that have not been taken by the filters or killed by the iodine.

Access to clean water is essential to the workings of a stable, thriving society. We saw earlier that disinfection of water is an important step in water treatment for human consumption. Water disinfection is one of the greatest public health innovations in human history. It has dramatically decreased the incidences of waterborne bacterial diseases such as cholera and typhus. But this great benefit comes at a price.

In 1974 scientists in both Europe and the United States discovered that chlorination of water produces a group of by-products that had previously gone undetected. These by-products are called *trihalomethanes* (THMs) because all have a single carbon atom and three halogen atoms: $CHCl_3$, $CHCl_2Br$, $CHClBr_2$ and $CHBr_3$. These and many other chlorine- and bromine-containing organic substances are produced by the reaction of aqueous chlorine with organic materials present in nearly all natural waters, as well as with substances that are by-products of human activity. Recall that chlorine dissolves in water to form HClO, which is the active oxidising agent:

$$Cl_2(g) + H_2O(l) \longrightarrow HClO(aq) + HCl(aq) \qquad [13.16]$$

HClO in turn reacts with organic substances to form the THMs. Bromine enters through the reaction of HClO with dissolved bromide ion:

$$HOCl(aq) + Br^-(aq) \longrightarrow HBrO(aq) + Cl^-(aq) \qquad [13.17]$$

HBrO(aq) halogenates organic substances analogously to HClO(aq).

Some THMs and other halogenated organic substances are suspected carcinogens. As a result, the World Health Organization has placed concentration limits of 0.080 mg dm^{-3} (80 ppm) on the total quantity of such substances in drinking water. The goal is to reduce the levels of THMs and related substances in the drinking water supply while preserving the antibacterial effectiveness of the water treatment. In some cases, simply lowering the concentration of chlorine may provide adequate disinfection while reducing the concentrations of THMs formed. Alternative oxidising agents, such as ozone (O_3) or chlorine dioxide (ClO_2), produce less of the halogenated substances, but they have their own disadvantages. Each is capable of oxidising aqueous bromide, as shown, for example, for ozone:

$$O_3(aq) + Br^-(aq) + H_2O(l) \longrightarrow HBrO(aq) + O_2(aq) + OH^-(aq) \qquad [13.18]$$

$$HBrO(aq) + 2 O_3(aq) \longrightarrow BrO_3^-(aq) + 2 O_2(aq) + H^+(aq) \qquad [13.19]$$

As we have seen, HBrO(aq) is capable of reacting with dissolved organic substances to form halogenated organic compounds. Furthermore, bromate ions (BrO_3^-) have been shown to cause cancer in animal tests.

There seem to be no completely satisfactory alternatives to chlorination at present. The risks of possible cancer from THMs and related substances in municipal water are very low compared with the risks of cholera, typhus and gastrointestinal disorders from untreated water. When the water supply is cleaner to begin with, less disinfectant is needed; thus the danger of contamination through disinfection is reduced. Once the THMs are formed, their concentrations in the water supply can be reduced by aeration because the

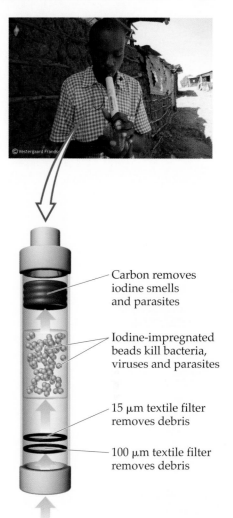

Carbon removes iodine smells and parasites

Iodine-impregnated beads kill bacteria, viruses and parasites

15 μm textile filter removes debris

100 μm textile filter removes debris

▲ FIGURE 13.19   **A LifeStraw purifies water as it is drunk.**

THMs are more volatile than water. Alternatively, they can be removed by adsorption onto activated charcoal or other adsorbents.

13.5 | GREEN CHEMISTRY

More people are beginning to understand that the planet on which we live is, to a large extent, a closed system. There is a growing realisation that we must create a sustainable society in order to survive—that is, a society in which all the processes we carry out are in balance with the Earth's natural processes; a society in which minimal toxic material is released into the environment and in which our needs are met with renewable resources. Finally, all of this must be accomplished using the smallest possible amount of energy.

Although the chemical industry is only a small part of the whole, chemical processes are involved in nearly all aspects of modern life. Chemistry is therefore at the heart of efforts to accomplish the goals of sustainability. The **green chemistry initiative** promotes the design and application of chemical products and processes that are compatible with human health and preserve the environment. Some of the major principles that govern green chemistry are:

- **Prevention.** It is better to prevent waste than to treat it or clean it up after it has been created.
- **Atom economy.** The synthesis of new substances should generate as little waste product as possible by incorporation of all starting atoms into the final product.
- **Safer chemicals.** Those substances that are generated should possess little or no toxicity to human health or the environment.
- **Energy efficiency.** Chemical processes should be as energy efficient as possible, avoiding high temperatures and pressures.
- **Catalysis.** Catalysts that permit the use of common and safe reagents while minimising energy costs should be employed whenever possible.
- **Renewable feedstock.** The raw materials for chemical processes should be renewable feedstocks when technically and economically feasible.
- **Solvent choice.** Auxiliary substances, such as solvents, should be eliminated or made as innocuous as possible, for example, use water.

> **⚠ CONCEPT CHECK 7**
>
> Explain how a consideration of atom economy can result in a 'greener' chemical process.

Let's consider some of the areas in which green chemistry can operate to improve environmental quality.

A major area of concern in chemical processes is the use of volatile organic compounds as solvents for reaction. Generally, the solvent in which a reaction is undertaken is not consumed in the reaction and has to be removed to isolate the product. Even in the most carefully controlled processes there are unavoidable releases of solvent into the atmosphere. The use of supercritical fluids represents a way to replace conventional solvents with other reagents. A supercritical fluid is an unusual state of matter that has properties of both a gas and a liquid (∞ Section 11.6, 'Phase Diagrams').

∞ Review this on page 398

Water and carbon dioxide are the two most popular choices as supercritical fluid solvents. One recently developed industrial process, for example, replaces chlorofluorocarbon solvents with liquid or supercritical CO_2 in the production of Teflon®. The chlorofluorocarbon solvents, aside from their costs, have harmful effects on the Earth's ozone layer (Section 13.2). However, CO_2 does participate in global-warming processes. Therefore, in making choices about 'being green', there are trade-offs.

As a further example of green chemistry, *p*-xylene is oxidised industrially to form terephthalic acid, which in turn is used to make polyethylene terephthalate (PET) plastic for applications such as magnetic tape, soft-drink bottles and polyester fibre:

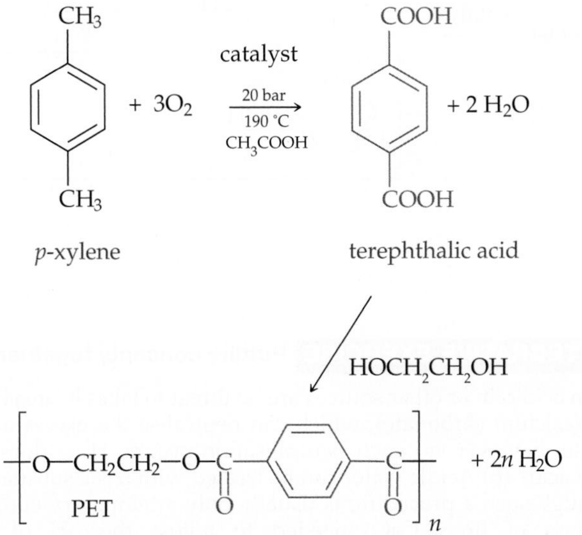

This commercial oxidation process usually requires pressurisation (20 bar) and a relatively high temperature. The catalyst is a manganese–cobalt mixture, oxygen is the oxidising agent and the solvent is acetic acid (CH_3COOH). A research group at the University of Nottingham in England has developed an alternative route that employs supercritical water as the solvent and hydrogen peroxide as the oxidant. This alternative process has several advantages, most particularly the elimination of acetic acid as solvent and the use of an oxidising agent whose reduction product is water. Whether it can successfully replace the existing commercial process depends on many factors, including infrastructure and the cost of installing a new process.

An example of atom economy is the new method for the manufacture of hydroquinone, HOC_6H_4OH. The standard industrial route yields many by-products that are treated as waste (shown in red):

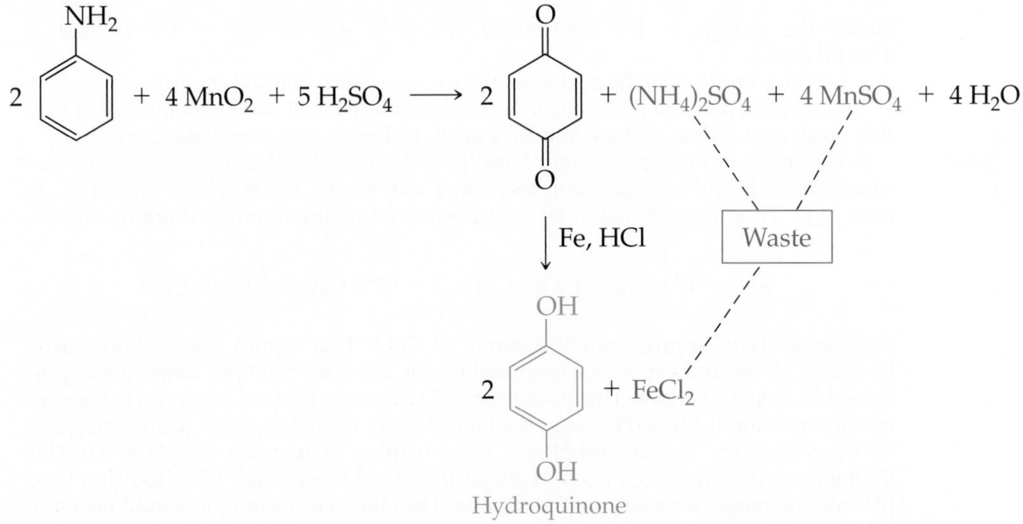

The new process uses a new starting material and also yields by-products (shown in green). However, these by-products may be isolated and then reacted together to form more starting material.

SAMPLE INTEGRATIVE EXERCISE | Putting concepts together

(a) Acids from acid rain or other sources are no threat to lakes in areas where the rock is limestone (calcium carbonate), which can neutralise the excess acid. Where the rock is granite, however, no such neutralisation occurs. How does the limestone neutralise the acid? (b) Acidic water can be treated with basic substances to increase the pH, although such a procedure is usually only a temporary cure. Calculate the minimum mass of lime, CaO, needed to adjust the pH of a small lake ($V = 4 \times 10^9$ dm^3) from 5.0 to 6.5. Why might more lime be needed?

SOLUTION

(a) The carbonate ion, which is the anion of a weak acid, is basic (∞ Section 17.2, 'Brønsted–Lowry Acids And Bases' and Section 17.7, 'Weak Bases'). Thus the carbonate ion, CO_3^{2-}, reacts with $H^+(aq)$. If the concentration of $H^+(aq)$ is small, the major product is the bicarbonate ion, HCO_3^-. If the concentration of $H^+(aq)$ is higher, however, H_2CO_3 forms and decomposes to CO_2 and H_2O (∞ Section 4.3, 'Acids, Bases and Neutralisation Reactions').

(b) The initial and final concentrations of $H^+(aq)$ in the lake are obtained from their pH values (∞ Section 17.4, 'The pH Scale'):

$$[H^+]_{initial} = 10^{-5.0} = 1 \times 10^{-5} \text{ M and } [H^+]_{final} = 10^{-6.5} = 3 \times 10^{-7} \text{ M}$$

Using the volume of the lake, we can calculate the number of moles of $H^+(aq)$ at both pH values:

$$(1 \times 10^{-5} \text{M}) (4.0 \times 10^9 \text{ dm}^3) = 4 \times 10^4 \text{ mol}$$
$$(3 \times 10^{-7} \text{M}) (4.0 \times 10^9 \text{ dm}^3) = 1 \times 10^3 \text{ mol}$$

Hence the change in the amount of $H^+(aq)$ is 4×10^4 mol $- 1 \times 10^3$ mol $\approx 4 \times 10^4$ mol.

Let's assume that all the acid in the lake is completely ionised, so that only the free $H^+(aq)$ measured by the pH needs to be neutralised. We will need to neutralise at least that much acid, although there may be a great deal more acid in the lake than that.

The oxide ion of CaO is very basic (∞ Section 7.6, 'Metals, Non-metals and Metalloids'). In the neutralisation reaction 1 mol of O^{2-} reacts with 2 mol of H^+ to form H_2O. Thus 4×10^4 mol of H^+ requires the following number of grams of CaO:

$$\frac{1}{2}(4 \times 10^4 \text{ mol})(56.1 \text{ g mol}^{-1}) = 1 \times 10^6 \text{ g CaO} = 1 \text{ tonne CaO}$$

This amounts to approximately a tonne of CaO. That would not be very costly because CaO is an inexpensive base, selling for less than $100 per tonne when purchased in large quantities. The amount of CaO calculated above, however, is the very minimum amount needed because the lake is likely to contain weak acids acting as a buffer system (e.g. H_2CO_3 and HCO_3^-) and helping to maintain the pH at 5.0. This liming procedure has been used to adjust the pH of some small lakes to bring their pH into the range necessary for fish to live. The lake in our example would be about 0.3 km long and 0.3 km wide, and have an average depth of 6 m.

∞ Find out more on pages 656 and 680

∞ Review this on pages 116–117

∞ Find out more on page 664

∞ Review this on page 240

CHAPTER SUMMARY AND KEY TERMS

SECTION 13.1   This section examined the physical and chemical properties of the Earth's atmosphere and noted that there are four regions in the atmosphere, each with characteristic properties. The lowest of these regions, the **troposphere**, extends from the Earth's surface up to an altitude of about 12 km. Above the troposphere in order of increasing altitude, are the **stratosphere**, **mesosphere** and **thermosphere**. In the upper reaches of the atmosphere, only the simplest chemical species can survive the bombardment of highly energetic particles and radiation from the sun. The average molecular weight of the atmosphere at high elevations is lower than that at the Earth's surface because the lightest atoms and molecules diffuse upwards and also because of **photodissociation**, which is the breaking of bonds in molecules because of the absorption of light to form neutral particles with unfilled valence shells called **free radicals**. Absorption of radiation may also lead to the formation of ions via **photoionisation**.

SECTION 13.2   Ozone is produced in the upper atmosphere from the reaction of atomic oxygen with O_2. Ozone is itself decomposed by absorption of a photon or by reaction with an active species such as Cl. **Chlorofluorocarbons** can undergo photodissociation in the stratosphere, introducing atomic chlorine, which is capable of catalytically destroying ozone. A marked reduction in the ozone level in the upper atmosphere would have serious adverse consequences because the ozone layer filters out certain wavelengths of ultraviolet light that are not removed by any other atmospheric component. In the troposphere the chemistry of trace atmospheric components is of major importance. Many of these minor components are pollutants. Sulfur dioxide is one of the more noxious and prevalent examples. It is oxidised in air to form sulfur trioxide, which, upon dissolving in water, forms sulfuric acid. The oxides of sulfur are major contributors to **acid rain**. One method of preventing the escape of SO_2 from industrial operations is to react it with CaO to form calcium sulfite ($CaSO_3$).

Photochemical smog is a complex mixture in which both nitrogen oxides and ozone play important roles. Smog components are generated mainly in car engines, and smog control consists largely of controlling car emissions.

Carbon dioxide and water vapour are the major components of the atmosphere that strongly absorb infrared radiation. CO_2 and H_2O are therefore critical in maintaining the Earth's surface temperature. The concentrations of CO_2 and other so-called **greenhouse gases** in the atmosphere are thus important in determining worldwide climate. Because of the extensive combustion of fossil fuels (coal, oil and natural gas), the concentration of carbon dioxide in the atmosphere is steadily increasing.

SECTION 13.3   Earth's water is largely in the oceans and seas; only a small fraction is freshwater. Seawater contains about 3.5% by mass of dissolved salts and is described as having a **salinity** (grams of dry salts per 1 kg seawater) of 35. Seawater's density and salinity vary with depth. Because most of the world's water is in the oceans, humans may eventually need to recover freshwater from seawater. The global water cycle involves continuous phase changes of water.

SECTION 13.4   Freshwater contains many dissolved substances including dissolved oxygen, which is necessary for fish and other aquatic life. Substances that are decomposed by bacteria are said to be **biodegradable**. Because the oxidation of biodegradable substances by aerobic bacteria consumes dissolved oxygen, these substances are called oxygen-demanding wastes. The presence of an excess amount of oxygen-demanding wastes in water can sufficiently deplete the dissolved oxygen to kill fish and produce offensive odours. Plant nutrients can contribute to the problem by stimulating the growth of plants that become oxygen-demanding wastes when they die.

Desalination is the removal of dissolved salts from seawater or brackish water to make it fit for human consumption. Desalination may be accomplished by distillation or by **reverse osmosis**.

The water available from freshwater sources may require treatment before it can be used domestically. The several steps generally used in municipal water treatment include coarse filtration, sedimentation, sand filtration, aeration, sterilisation and sometimes water softening. Water softening is required when the water contains significant concentrations of ions such as Mg^{2+} and Ca^{2+}, which react with soap to form soap scum. Water containing such ions is called **hard water**. Individual homes usually rely on **ion exchange**, a process for water softening by which hard-water ions are exchanged for Na^+ ions.

SECTION 13.5   The **green chemistry initiative** promotes the design and application of chemical products and processes that are compatible with human health and that preserve the environment. The areas in which the principles of green chemistry can operate to improve environmental quality include choices of solvents and reagents for chemical reactions, development of atom and energy-efficient processes and improvements in existing systems and practices.

KEY SKILLS

- Describe the regions of the Earth's atmosphere in terms of how temperature varies with altitude. (Section 13.1)

- Describe the composition of the atmosphere in terms of the major components in dry air at sea level. (Section 13.1)

- Calculate the concentrations of gases in parts per million (ppm). (Section 13.1)

- Describe the processes of photodissociation and photoionisation. (Section 13.1)

- Calculate the maximum wavelength needed to cause photodissociation or photoionisation using bond energies and ionisation energies. (Section 13.1)

- Explain the role of ozone in the upper atmosphere. (Section 13.1)

- Explain how chlorofluorocarbons (CFCs) are involved in depleting the ozone layer. (Section 13.2)

- Describe the origins and behaviour of sulfur oxides, carbon monoxide and nitrogen oxides as air pollutants, including the generation of acid rain and photochemical smog. (Section 13.2)

- Describe how water and carbon dioxide in the atmosphere affect atmospheric temperature via the greenhouse effect. (Section 13.2)

- Describe the global water cycle. (Section 13.3)

- Explain what is meant by the salinity of water and describe the process of reverse osmosis as a method of desalination. (Section 13.4)

- List the major cations, anions and gases present in natural waters and describe the relationship between dissolved oxygen and water quality. (Section 13.4)

- List the main steps in treating water for domestic use. (Section 13.4)

- Describe the main goals of green chemistry. (Section 13.5)

- Compare reactions and decide which reaction is environmentally more appropriate. (Section 13.5)

EXERCISES

VISUALISING CONCEPTS

13.1 At room temperature (298 K) and 1 bar pressure (which corresponds to the atmospheric pressure at sea level), one mole of an ideal gas occupies a volume of 24.8 dm^3. **(a)** Looking back at Figure 13.1, do you predict that 1 mole of an ideal gas in the middle of the stratosphere would occupy a greater or smaller volume than 24.8 dm^3? **(b)** Looking at Figure 13.1, we see that the temperature is lower at 85 km altitude than at 50 km. Does this mean that one mole of an ideal gas would occupy less volume at 85 km than at 50 km? Explain. [Section 13.1]

13.2 Molecules in the upper atmosphere tend to contain double and triple bonds rather than single bonds. Suggest an explanation. [Section 13.1]

13.3 Why does ozone concentration in the atmosphere vary as a function of altitude? [Section 13.2]

13.4 How does carbon dioxide interact with the world ocean? [Section 13.4]

13.5 The following picture represents an ion-exchange column, in which water containing 'hard' ions, like Ca^{2+}, is added to the top of the column and out of the bottom comes 'softened' water that has Na$^+$ instead of Ca^{2+} in it. Explain what is happening in the column. [Section 13.4]

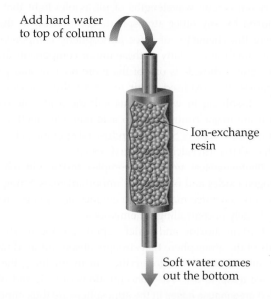

Add hard water to top of column

Ion-exchange resin

Soft water comes out the bottom

13.6 In areas with large deposits of granite, bore water can contain significant amounts of arsenic in the form of the water-soluble arsenate ion, $HAsO_4^-$. This ion can be easily reduced, in slightly acid solution, to the much less soluble arseneous ion, $H_2AsO_3^-$.

Draw a diagram of a device for removing the arsenic from bore water if the reaction is represented by $HAsO_4^{2-} + 4\,H^+ + Fe \longrightarrow H_3O^+ + H_2AsO_3^- + Fe^{2+}$. [Section 13.4]

13.7 One mystery in environmental science is the imbalance in the 'carbon dioxide budget'. Considering only human activities, scientists have estimated that 1.6 billion tonnes of CO_2 is added to the atmosphere every year because of deforestation (plants use CO_2 to make glucose, and fewer plants will leave more CO_2 in the atmosphere) and another 5.5 billion tonnes per year is put into the atmosphere as a result of fossil-fuel burning. It is further estimated (again, considering only human activities) that the atmosphere has actually taken up about 3.3 billion tonnes of CO_2 per year and the oceans have taken up 2 billion tonnes per year, leaving about 1.8 billion tonnes of CO_2 per year unaccounted for. This 'missing' CO_2 is assumed to be taken up by the 'land'. What do you think might be happening? [Sections 13.1–13.3]

13.8 Describe the basic goals of green chemistry. [Section 13.5]

EARTH'S ATMOSPHERE (Section 13.1)

13.9 **(a)** What is the primary basis for the division of the atmosphere into different regions? **(b)** Name the regions of the atmosphere, indicating the altitude interval for each one.

13.10 **(a)** How are the boundaries between the regions of the atmosphere determined? **(b)** Explain why the stratosphere, which is more than 32 km thick, has a smaller total mass than the troposphere, which is less than 16 km thick.

13.11 From the data in Table 13.1, calculate the partial pressures of carbon dioxide and argon when the total atmospheric pressure is 98 kPa.

13.12 If the average concentration of carbon monoxide in air is 3.4 ppm, calculate the number of CO molecules in 1.0 dm^3 of this air at a pressure of 1 bar and a temperature of 22 °C.

13.13 **(a)** From the data in Table 13.1, what is the concentration of neon in the atmosphere in ppm? **(b)** What is the concentration of neon in the atmosphere in molecules per dm^3, assuming an atmospheric pressure of 99 kPa and a temperature of 300 K?

13.14 The dissociation energy of a carbon–bromine bond is typically about 210 kJ mol^{-1}. What is the maximum wavelength of photons that can cause C–Br bond dissociation?

13.15 In CF_3Cl the C–Cl bond-dissociation energy is 339 kJ mol^{-1}. In CCl_4 the C–Cl bond-dissociation energy is 293 kJ mol^{-1}. What is the range of wavelengths of photons that can cause C–Cl bond rupture in one molecule but not in the other?

13.16 **(a)** Distinguish between *photodissociation* and *photoionisation*. **(b)** Use the energy requirements of these two processes to explain why photodissociation of oxygen is more important than photoionisation of oxygen at altitudes below about 90 km.

13.17 Why is the photodissociation of N_2 relatively unimportant compared with the photodissociation of O_2?

HUMAN ACTIVITIES AND EARTH'S ATMOSPHERE (Section 13.2)

13.18 What is a hydrofluorocarbon? Why are these compounds potentially less harmful to the ozone layer than CFCs?

13.19 Draw the Lewis structure for the chlorofluorocarbon CFC-11, $CFCl_3$. What chemical characteristics of this substance allow it to effectively deplete stratospheric ozone?

13.20 **(a)** Why is the fluorine present in chlorofluorocarbons not a major contributor to depletion of the ozone layer? **(b)** What are the chemical forms in which chlorine exists in the stratosphere following cleavage of the carbon–chlorine bond?

13.21 Would you expect the substance $CFBr_3$ to be effective in depleting the ozone layer, assuming that it is present in the stratosphere? Explain.

13.22 Why is rainwater naturally acidic, even in the absence of polluting gases such as SO_2?

13.23 **(a)** Write a chemical equation that describes the attack of acid rain on limestone, $CaCO_3$. **(b)** If a limestone sculpture is treated to form a surface layer of calcium sulfate, will this help to slow down the effects of acid rain? Explain.

13.24 The first stage in corrosion of iron upon exposure to air is oxidation to Fe^{2+}. **(a)** Write a balanced chemical equation to show the reaction of iron with oxygen and protons from acid rain. **(b)** Would you expect the same sort of reaction to occur with a silver surface? Explain.

13.25 Alcohol-based fuels for cars lead to the production of formaldehyde (CH_2O) in exhaust gases. Formaldehyde undergoes photodissociation, which contributes to photochemical smog:

$$CH_2O + h\nu \longrightarrow CHO + H$$

The maximum wavelength of light that can cause this reaction is 335 nm. **(a)** In what part of the electromagnetic spectrum is light with this wavelength found? **(b)** What is the maximum strength of a bond, in kJ mol^{-1}, that can be broken by absorption of a photon of 335 nm light? **(c)** Compare your answer from part (b) with the appropriate value from Table 8.4. What do you conclude about the C–H bond energy in formaldehyde? **(d)** Write out the formaldehyde photodissociation reaction, showing Lewis-dot structures.

13.26 An important reaction in the formation of photochemical smog is the photodissociation of NO_2:

$$NO_2 + h\nu \longrightarrow NO(g) + O(g)$$

The maximum wavelength of light that can cause this reaction is 420 nm. **(a)** In what part of the electromagnetic spectrum is light with this wavelength found? **(b)** What is the maximum strength of a bond, in kJ mol^{-1}, that can be broken by absorption of a photon of 420 nm light? **(c)** Write out the photodissociation reaction showing Lewis-dot structures.

EARTH'S WATER (Section 13.3)

13.27 What is the molarity of Na^+ in a solution of NaCl whose salinity is 5.6 if the solution has a density of 1.03 g cm^{-3}?

13.28 Phosphorus is present in seawater to the extent of 0.07 ppm by mass. If the phosphorus is present as phosphate, PO_4^{3-}, calculate the corresponding molar concentration of phosphate in seawater.

13.29 A first-stage recovery of magnesium from seawater is precipitation of $Mg(OH)_2$ with CaO:

$$Mg^{2+}(aq) + CaO(s) + H_2O(l) \longrightarrow$$
$$Mg(OH)_2(s) + Ca^{2+}(aq)$$

What mass of CaO, in grams, is needed to precipitate 454 kg of $Mg(OH)_2$?

13.30 Gold is found in seawater at very low levels, about 0.05 ppb by mass. Assuming that gold is worth about $1500 per troy ounce (1 troy ounce = 31.1 g), how many litres of seawater would you have to process to obtain $1 000 000 worth of gold? Assume the density of seawater is 1.03 g cm^{-3} and that your gold recovery process is 50% efficient.

13.31 Suppose that we want to use reverse osmosis to reduce the salt content of brackish water containing 0.22 M total salt concentration to a value of 0.01 M, thus making it suitable for human consumption. What is the minimum pressure that needs to be applied in the permeators (Figure 13.16) to achieve this goal, assuming that the operation occurs at 298 K? [*Hint:* Refer to Section 12.5.]

HUMAN ACTIVITIES AND EARTH'S WATER (Section 13.4)

13.32 List the common products formed when an organic material containing the elements carbon, hydrogen, oxygen, sulfur and nitrogen decomposes **(a)** under aerobic conditions, **(b)** under anaerobic conditions.

13.33 **(a)** Explain why the concentration of dissolved oxygen in freshwater is an important indicator of the quality of the water. **(b)** How is the solubility of oxygen in water affected by increasing temperature?

13.34 The organic anion

is found in most detergents. Assume that the anion undergoes aerobic decomposition in the following manner:

$$2\, C_{18}H_{29}SO_3^-(aq) + 51\, O_2(aq) \longrightarrow$$
$$36\, CO_2(aq) + 28\, H_2O(l) + 2\, H^+(aq) + 2\, SO_4^{2-}(aq)$$

What is the total mass of O_2 required to biodegrade 1.0 g of this substance?

13.35 The average daily mass of O_2 taken up by sewage discharged in a city is 59 g per person. How many litres of water at 9 ppm O_2 are totally depleted of oxygen in 1 day by a population of 120 000 people?

13.36 Write a balanced chemical equation to describe how magnesium ions are removed in water treatment by the addition of slaked lime, $Ca(OH)_2$.

13.37 **(a)** Which of the following ionic species is, or could be, responsible for hardness in a water supply: Ca^{2+}, K^+, Mg^{2+}, Fe^{2+}, Na^+? **(b)** What properties of an ion determine whether it will contribute to water hardness?

13.38 Ferrous sulfate ($FeSO_4$) is often used as a coagulant in water purification. The iron(II) salt is dissolved in the water to be purified, then oxidised to the iron(III) state by dissolved oxygen, at which time gelatinous $Fe(OH)_3$ forms, assuming the pH is above approximately 6. Write balanced chemical equations for the oxidation of Fe^{2+} to Fe^{3+} by dissolved oxygen, and for the formation of $Fe(OH)_3(s)$ by reaction of $Fe^{3+}(aq)$ with $HCO_3^-(aq)$.

13.39 What properties make a substance a good coagulant for water purification?

13.40 Lakes and dam catchments in limestone areas are fairly resistant to the harmful effects of acid rain. **(a)** Write the chemical equation to show how this occurs. **(b)** Suggest what could be added to small, acid-rain-affected areas of static water in non-limestone areas to help revive them.

13.41 An alternative to using chlorine to purify drinking water is the use of ozone, an established method in large parts of Europe. Write the chemical equations that describe the possible disadvantages of this method.

13.42 Although marine life is tolerant of even relatively large changes in the pH of their environment, in granite (aluminium silicate) areas fish kills can occur very readily with decrease in pH of the water. Explain this phenomenon. (*Hint:* Aluminium ion is leached from rock by H^+ and is soluble as the hydrated species until it comes in contact with a higher-pH environment of 6 or 7.)

GREEN CHEMISTRY (Section 13.5)

13.43 One of the principles of green chemistry is that it is better to use as few steps as possible in making new chemicals. How, if at all, does this principle relate to energy efficiency?

13.44 The Baeyer–Villiger reaction is a classic organic oxidation reaction for converting ketones to lactones, as in this reaction:

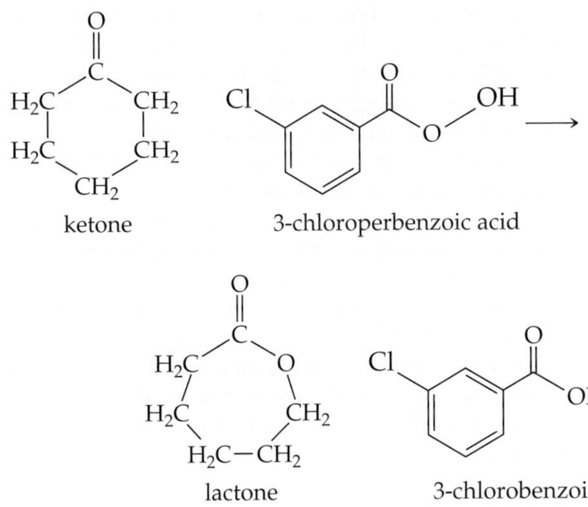

ketone   3-chloroperbenzoic acid $\longrightarrow$

lactone   3-chlorobenzoic acid

The reaction is used in the manufacture of plastics and pharmaceuticals. The reactant, 3-chloroperbenzoic acid, is somewhat shock-sensitive, however, and prone to explode. Also, 3-chlorobenzoic acid is a waste product. An alternative process being developed uses hydrogen peroxide and a catalyst consisting of tin deposited within a solid support. The catalyst is readily recovered from the reaction mixture. **(a)** What would you expect to be the other product of oxidation of the ketone to lactone by hydrogen peroxide? **(b)** What principles of green chemistry are addressed by use of the proposed process?

13.45 The reaction shown here was performed with an iridium catalyst, both in supercritical CO_2 ($scCO_2$) and in the chlorinated solvent CH_2Cl_2. The kinetic data for the re-

action in both solvents are plotted in the graph. Why is this a good example of a green chemical reaction?

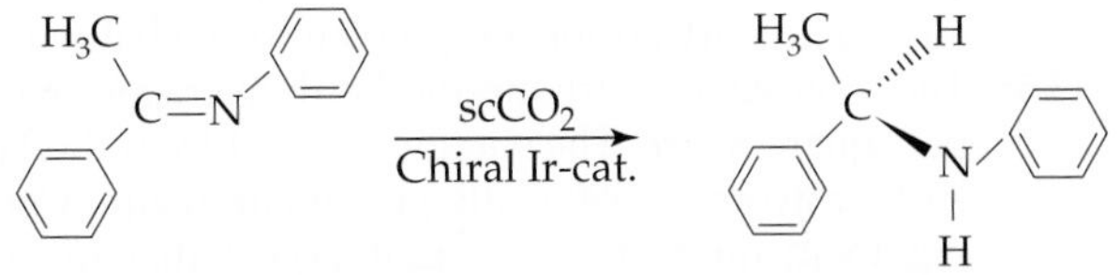

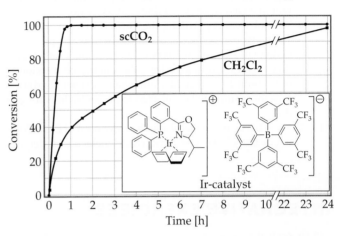

13.46 Which choice is greener in a chemical process? Explain. **(a)** A reaction that can be run at 350 K for 12 hours without a catalyst or one that can be run at 300 K for 1 hour with a catalyst. **(b)** A reagent for the reaction that can be obtained from corn husks or one that can be obtained from petroleum. **(c)** A process that produces no by-products or one in which the by-products are recycled for another process.

ADDITIONAL EXERCISES

13.47 Suppose that, on another planet, the atmosphere consists of 17% Kr, 38% CH_4 and 45% O_2. What is the average molar mass at the surface? What is the average molar mass at an altitude at which all the O_2 is photodissociated? Assume the atmosphere is well mixed so the effects of gravity on composition can be ignored.

13.48 If an average O_3 molecule 'lives' only 100–200 seconds in the stratosphere before undergoing dissociation, how can O_3 offer any protection from ultraviolet radiation?

13.49 Show how Equations 13.7 and 13.9 can be added to give Equation 13.10. (You may need to multiply one of the reactions by a factor to have them add properly.)

13.50 *Halons* are fluorocarbons that contain bromine, such as $CBrF_3$. They are used extensively as foaming agents for fighting fires. Like CFCs, halons are very unreactive and may ultimately diffuse into the stratosphere. **(a)** Based on the data in Table 8.4, would you expect photodissociation of Br atoms to occur in the stratosphere? **(b)** Propose a mechanism by which the presence of halons in the stratosphere could lead to the depletion of stratospheric ozone.

13.51 The *hydroxyl radical*, OH, is formed at low altitudes via the reaction of excited oxygen atoms with water:

$$O^*(g) + H_2O(g) \longrightarrow 2\,OH(g)$$

(a) Write the Lewis structure for the hydroxyl radical. [*Hint:* It has one unpaired electron.]

Once produced, the hydroxyl radical is very reactive. Explain why each of the following series of reactions affects the pollution in the troposphere:
(b) $OH + NO_2 \longrightarrow HNO_3$
(c) $OH + CO + O_2 \longrightarrow CO_2 + OOH$
    $OOH + NO \longrightarrow OH + NO_2$
(d) $OH + CH_4 \longrightarrow H_2O + CH_3$
    $CH_3 + O_2 \longrightarrow OOCH_3$
    $OOCH_3 + NO \longrightarrow OCH_3 + NO_2$

13.52 The affinity of carbon monoxide for haemoglobin is about 210 times that of O_2. Assume a person is inhaling air that contains 125 ppm of CO. If all the haemoglobin leaving the lungs carries either oxygen or CO, calculate the fraction in the form of carboxyhaemoglobin.

13.53 Natural gas consists primarily of methane, $CH_4(g)$. **(a)** Write a balanced chemical equation for the complete combustion of methane to produce $CO_2(g)$ as the only carbon-containing product. **(b)** Write a balanced chemical equation for the incomplete combustion of methane to produce $CO(g)$ as the only carbon-containing product. **(c)** At 25 °C and 1.0 bar pressure, what is the minimum quantity of dry air needed to combust 1.0 dm^3 of $CH_4(g)$ completely to $CO_2(g)$?

13.54 One of the possible consequences of global warming is an increase in the temperature of ocean water. The oceans serve as a 'sink' for CO_2 by dissolving large amounts of it. **(a)** How would the solubility of CO_2 in the oceans be affected by an increase in the temperature of the water? **(b)** Discuss the implications of your answer to part (a) for the problem of global warming.

13.55 The solar energy striking the Earth averages 169 watts per square metre. The energy radiated from the Earth's surface averages 390 watts per square metre. Comparing these numbers, we might expect that the planet would cool quickly, yet it does not. Why not?

13.56 The solar energy striking the Earth every day averages 169 watts per square metre, and each day the people of Sydney use about 3 watts per square metre. Considering that present technology for solar energy conversion is only about 10% efficient, from what fraction of Sydney's area must sunlight be collected to power Sydney for one day?

13.57 Write balanced chemical equations for each of the following reactions. **(a)** The nitric oxide molecule undergoes photodissociation in the upper atmosphere. **(b)** The nitric oxide molecule undergoes photoionisation in the upper atmosphere. **(c)** Nitric oxide undergoes oxidation by ozone in the stratosphere. **(d)** Nitrogen dioxide dissolves in water to form nitric acid and nitric oxide.

13.58 Explain why $Mg(OH)_2$ precipitates when CO_3^{2-} ion is added to a solution containing Mg^{2+}.

13.59 It has recently been pointed out that there may be increased amounts of NO in the troposphere compared with the past because of massive use of nitrogen-containing compounds in fertilisers. Assuming that NO can eventually diffuse into the stratosphere, how might it affect the conditions of life on Earth? Using the index to this text, look up the chemistry of nitrogen oxides. What chemical pathways might NO in the troposphere follow?

INTEGRATIVE EXERCISES

13.60 **(a)** If the estimated average concentration of NO_2 in air is 0.019 ppm, calculate the partial pressure of the NO_2 in a sample of this air when the atmospheric pressure is 99.1 kPa. **(b)** How many molecules of NO_2 are present under these conditions at 20 °C in a room that measures 47.6 m^3?

13.61 The water supply for a city contains the following impurities: coarse sand, finely divided particulates, nitrate ion, trihalomethanes, dissolved phosphorus in the form of phosphates, potentially harmful bacterial strains, dissolved organic substances. Which of the following processes or agents, if any, is effective in removing each of these impurities: coarse sand filtration, activated carbon filtration, aeration, ozonisation, precipitation with aluminium hydroxide?

13.62 The concentration of H_2O in the stratosphere is about 5 ppm. It undergoes photodissociation as follows:

$$H_2O(g) \longrightarrow H(g) + OH(g)$$

(a) Write out the Lewis-dot structures for both products and reactant.

(b) Using Table 8.4, calculate the wavelength required to cause this dissociation.

(c) The hydroxyl radicals, OH, can react with ozone, giving the following reactions:

$$OH(g) + O_3(g) \longrightarrow HO_2(g) + O_2(g)$$
$$HO_2(g) + O(g) \longrightarrow OH(g) + O_2(g)$$

What overall reaction results from these two elementary reactions? What is the catalyst in the overall reaction? Explain.

13.63 Nitrogen dioxide (NO_2) is the only important gaseous species in the lower atmosphere that absorbs visible light. **(a)** Write the Lewis structure(s) for NO_2. **(b)** How does this structure account for the fact that NO_2 dimerises to form N_2O_4? Based on what you can find about this dimerisation reaction in the text, would you expect to find the NO_2 that forms in an urban environment to be in the form of dimer? Explain. **(c)** What would you expect as products, if any, for the reaction of NO_2 with CO? **(d)** Would you expect NO_2 generated in an urban environment to migrate to the stratosphere? Explain.

13.64 The Bayswater Power Station in the Hunter Valley burns 7.5 million tonnes of coal in a year. **(a)** Assuming that the coal was 83% carbon and 2.5% sulfur and that combustion was complete, calculate the number of tonnes of carbon dioxide and sulfur dioxide produced by the plant during the year. **(b)** If 55% of the SO_2 could be removed by reaction with powdered CaO to form $CaSO_3$, how many tonnes of $CaSO_3$ would be produced?

13.65 The valuable polymer polyurethane is made by a condensation reaction of alcohols (ROH) with compounds that contain an isocyanate group (RNCO). Two reactions that can generate a urethane monomer are shown here:

(i) $RNH_2 + CO_2 \longrightarrow R-N{=}C{=}O + 2H_2O$

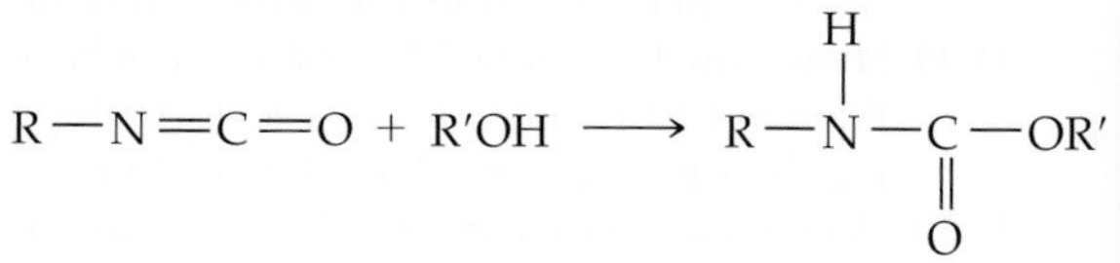

$$R-N{=}C{=}O + R'OH \longrightarrow R-\overset{\overset{\displaystyle H}{|}}{\underset{\underset{\displaystyle O}{\|}}{N-C}}-OR'$$

(ii) $RNH_2 + \underset{Cl\quad Cl}{\overset{\overset{\displaystyle O}{\|}}{C}} \longrightarrow R-N{=}C{=}O + 2HCl$

$$R-N{=}C{=}O + R'OH \longrightarrow R-\overset{\overset{\displaystyle H}{|}}{\underset{\underset{\displaystyle O}{\|}}{N-C}}-OR'$$

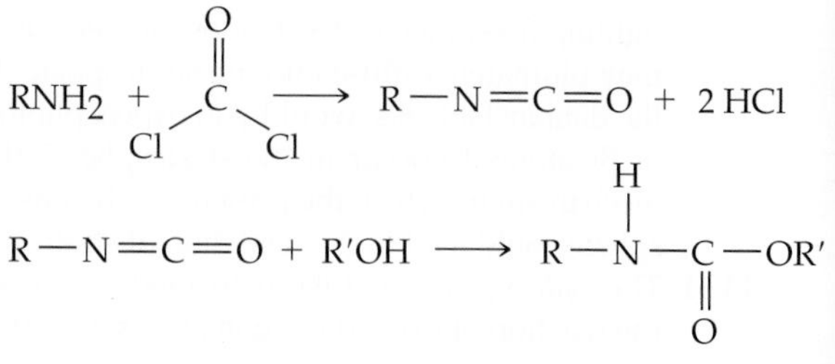

(a) Which process, (i) or (ii), is greener? Explain.
(b) What are the hybridisation and geometry of the carbon atoms in each C-containing compound in each reaction?

MasteringChemistry (www.pearson.com.au/masteringchemistry)

Make learning part of the grade. Access:

- tutorials with peronalised coaching
- study area
- Pearson eText

PHOTO/ART CREDITS

468 © Ben Goode/Dreamstime.com; **478 (a), 488** Getty Images Australia, **478 (b)** Jose Ignacio Soto/Shutterstock; **486** River Alliance of Wisconsin; **487** DuPont; **489** © Vestergaard Frandsen.

14

THERMODYNAMICS

There is a tremendous amount of energy associated with rushing water that can be harnessed to produce electricity.

KEY CONCEPTS

14.1 THE NATURE OF ENERGY
We consider the nature of *energy* and the forms it takes, notably *kinetic energy* and *potential energy* and discuss the fact that energy can be used to do *work* or to transfer *heat*. To study energy changes, we focus on a particular part of the universe, which we call the *system*. Everything else is called the *surroundings*.

14.2 THE FIRST LAW OF THERMODYNAMICS
The *first law of thermodynamics* states: energy cannot be created or destroyed but can be transformed from one form to another or transferred between systems and surroundings. The energy possessed by a system is called its *internal energy*. Internal energy is a *state function,* a quantity whose value depends only on the current state of a system, not on how the system came to be in that state.

14.3 ENTHALPY
We encounter a state function called *enthalpy* that is useful because the change in enthalpy measures the quantity of heat energy gained or lost by a system in a process occurring under constant pressure.

14.4 ENTHALPIES OF REACTION
The enthalpy change associated with a chemical reaction is the enthalpies of the products minus the enthalpies of the reactants. This quantity is directly proportional to the amount of reactant consumed in the reaction.

14.5 CALORIMETRY
We examine *calorimetry*, an experimental technique used to measure heat changes in chemical processes.

14.6 HESS'S LAW
The enthalpy change for a given reaction can be calculated using appropriate enthalpy changes for related reactions. To do so, we apply *Hess's law.*

14.7 ENTHALPIES OF FORMATION
We discuss how to establish standard values for enthalpy changes in chemical reactions and how to use them to calculate enthalpy changes for reactions.

14.8 SPONTANEOUS PROCESSES
Changes that occur in nature have a directional character. They move *spontaneously* in one direction but not in the reverse direction.

14.9 ENTROPY AND THE SECOND LAW OF THERMODYNAMICS
We examine *entropy*, a thermodynamic state function that is important in determining whether a process is spontaneous. The *second law of thermodynamics* tells us that in any spontaneous process the entropy of the universe (system plus surroundings) increases.

14.10 MOLECULAR INTERPRETATION OF ENTROPY
On the molecular level, the entropy of a system is related to the number of accessible *microstates*. The entropy of the system increases as the randomness of the system increases. The *third law of thermodynamics* states that, at 0 K, the entropy of a perfect crystalline solid is zero.

14.11 ENTROPY CHANGES IN CHEMICAL REACTIONS
Using tabulated *standard molar entropies*, we can calculate the standard entropy changes for systems undergoing reaction.

14.12 GIBBS FREE ENERGY
We encounter another thermodynamic state function, *free energy* (or *Gibbs free energy*), a measure of how far removed a system is from equilibrium. The change in free energy measures the maximum amount of useful work obtainable from a process and tells us the direction in which a chemical reaction is spontaneous.

14.13 GIBBS ENERGY AND TEMPERATURE
Finally, we consider how the standard free energy change for a chemical reaction can be used to calculate the equilibrium constant for the reaction.

W e live in a world where we discuss 'energy' daily. Modern society depends on energy for its existence. Energy is used to drive our machinery, to operate our computing and communications systems, to power our transportation vehicles and to keep us warm in winter and cool in summer.

However, it is sometimes difficult to understand just exactly what this 'energy' is that we so frequently talk about. There are many manifestations of energy such as solar energy, geothermal energy, wave energy, chemical energy and so on. In fact, we often use a chemical reaction to obtain energy, for example when we burn fossil fuels (coal, oil, petrol, natural gas). Indeed, nearly all the energy we depend on ultimately comes from the sun, involving many transactions and transformations in its passage from sun to plants to animals. In this chapter we will examine the nature of energy and its transformations as well as how fast and how completely these transformations occur. In so doing, we explore the connection between energy and the extent of a reaction.

The study of energy is known as **thermodynamics** (Greek: *thérme*- 'heat'; *dynamis*, 'power') and we will start with **thermochemistry**, which is the study of the

(a)

(b)

▲ **FIGURE 14.1 Work and heat.** Energy can be used to achieve two basic types of tasks: (a) work is energy used to cause an object with mass to move; (b) heat is energy used to cause the temperature of an object to increase.

relationships between chemical reactions and energy changes involving heat. We will continue by examining the relationship between energy and the extent of chemical reactions, that is, between energy and chemical equilibria. By introducing a concept called *entropy* we will be able to provide an insight into why physical and chemical changes tend to favour one direction and not another. For example, we would not expect a half burnt candle to spontaneously regenerate itself even if we had kept all the gases produced when the candle burned Thermodynamics helps us to understand not only the energy requirements of chemical reactions but also the extent to which they proceed and the reasons for their directional character.

14.1 | THE NATURE OF ENERGY

Although the idea of energy is a familiar one, it is a bit challenging to deal with the concept in a precise way. **Energy** is commonly defined as *the capacity to do work or to transfer heat*. This definition requires us to understand the concepts of work and heat. We can think of **work** as being *energy used to cause an object with mass to move*, and **heat** as being *the energy used to cause the temperature of an object to increase* (◄ **FIGURE 14.1**). We will consider each of these concepts more closely to give them fuller meaning. We begin by examining the ways in which matter can possess energy and how that energy can be transferred from one piece of matter to another.

Kinetic Energy and Potential Energy

Moving objects possess **kinetic energy**, the energy of *motion*, whose magnitude depends on the object's mass, m, and velocity, u:

$$E_k = \tfrac{1}{2}mu^2 \qquad [14.1]$$

Not only do large moving objects such as cars, balls and so on possess kinetic energy but atoms and molecules which also have mass and are in motion possess this energy of motion.

An object can also possess **potential energy**, by virtue of its *position* relative to other objects. Potential energy arises when there is a force operating on an object. A **force** is any kind of push or pull exerted on an object. The most familiar force is the pull of gravity. Think of a cyclist poised at the top of a hill, as illustrated in ▼ **FIGURE 14.2**. Gravity acts upon her and her bicycle, exerting a

▶ **FIGURE 14.2 Potential energy and kinetic energy.** The potential energy initially stored in the motionless bicycle and rider at the top of the hill is converted to kinetic energy as the bicycle moves down the hill and loses potential energy.

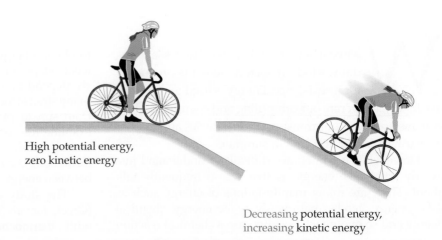

High potential energy, zero kinetic energy

Decreasing potential energy, increasing kinetic energy

force directed towards the centre of the Earth. At the top of the hill the cyclist and her bicycle possess a certain potential energy by virtue of their elevation. The potential energy, E_p, is given by Equation 14.2.

$$E_p = mgh \qquad [14.2]$$

Here m is the mass of the object in question (in this case, the cyclist and bicycle), h is the height of the object relative to some reference height and g is the gravitational constant, 9.8 m s^{-2}. Once in motion, without any further effort on her part, the cyclist gains speed as the bicycle rolls down the hill. Her potential energy decreases as she moves downwards, but the energy does not simply disappear. It is converted to other forms of energy, principally kinetic energy, the energy of motion. This example illustrates that different forms of energy are interconvertible.

Chemistry deals mostly with extremely small objects—atoms and molecules—and gravitational forces play a negligible role in the ways these submicroscopic objects interact with one another. More important are forces that arise from electrical charges. One of the most important forms of potential energy in chemistry is *electrostatic potential energy*, which arises from the interactions between charged particles. The electrostatic potential energy, E_{el} is proportional to the electrical charges on the two interacting objects, q_1 and q_2, and inversely proportional to the square of the distance, r, separating them:

$$E_{el} = k\frac{q_1 q_2}{r^2} \qquad [14.3]$$

Here k is simply a constant of proportionality, 8.99×10^9 J m C^{-2} (C is the coulomb, a unit of electrical charge) (∞ Section 2.2, 'The Discovery of Atomic Structure').

∞ Review this on page 31

When dealing with molecular-level objects, the electrical charges q_1 and q_2 are typically of the order of magnitude of the charge of the electron. When q_1 and q_2 have the same sign (for example, both are positive) the two charges repel one another, pushing them apart, and E_{el} is positive. When they have opposite signs they attract one another, pulling them towards each other, and E_{el} is negative. The lower the energy of a system (that is, the more negative is E_{el}), the more stable it is. Thus the more strongly opposite charges interact, the more stable the system.

One of our goals in chemistry is to relate the energy changes that we see in our macroscopic world to the kinetic or potential energy of substances at the atomic or molecular level. Many substances—fuels, for example—release energy when they react. The *chemical energy* of these substances is due to the potential energy stored in the arrangements of the atoms of the substance. Likewise, we will see that the energy a substance possesses because of its temperature (its *thermal energy*) is associated with the kinetic energy of the molecules in the substance.

 CONCEPT CHECK 1

When the cyclist and bicycle in Figure 14.2 come to a stop at the bottom of the hill:
a. Is the potential energy the same as it was at the top of the hill?
b. Is the kinetic energy the same as it was at the top of the hill?

Units of Energy

The SI unit for energy is the **joule, J**, in honour of James Joule (1818–1889), a British scientist who investigated work and heat: 1 J = 1 kg m^2 s^{-2}. A mass of 2 kg moving at a speed of 1 m s^{-1} possesses a kinetic energy of 1 J:

$$E_k = \tfrac{1}{2}mu^2 = \tfrac{1}{2}(2 \text{ kg})(1 \text{ m s}^{-1})^2 = 1 \text{ kg m}^2\text{s}^{-2} = 1 \text{ J}$$

A joule is not a large amount of energy, so we will often use *kilojoules* (kJ) in discussing the energies associated with chemical reactions.

System and Surroundings

When we analyse energy changes, we need to focus our attention on a limited and well-defined part of the universe to keep track of the energy changes that occur. The part of the universe we single out for study is called the **system**; everything else is called the **surroundings**. When we study the energy change that accompanies a chemical reaction in the laboratory, the reactants and products usually constitute the system. The container and everything beyond it are then considered the surroundings. Systems may be open, closed or isolated. The systems we can most readily study in thermodynamics are called *closed systems*. A closed system can exchange energy but not matter with its surroundings. For example, consider a mixture of hydrogen gas and oxygen gas in a cylinder, as illustrated in ◀ **FIGURE 14.3**. The system in this case is just the hydrogen and oxygen; the cylinder, piston and everything beyond them (including us) are the surroundings. If the hydrogen and oxygen react to form water, energy is liberated:

$$2 \, H_2(g) + O_2(g) \longrightarrow 2 \, H_2O(g) + energy$$

Although the chemical form of the hydrogen and oxygen atoms in the system is changed by this reaction, the system has not lost or gained mass; it undergoes no exchange of matter with its surroundings. However, it does exchange energy with its surroundings in the form of *work* and *heat*. These are quantities that we can measure, as we will now discuss.

An *isolated system* is one in which neither energy nor matter can be exchanged with the surroundings. An insulated thermos flask containing hot tea approximates an isolated system. We know, however, that the tea eventually cools, so it is not perfectly isolated.

Transferring Energy: Work and Heat

Since a force is any push or pull exerted on an object we can define *work* as the energy transferred when an object is moved by a force. The magnitude of this work equals the product of the force, F, and the distance, d, that the object is moved:

$$w = F \times d \qquad [14.4]$$

We perform work, for example, when we lift an object against the force of gravity or when we bring two like charges closer together. If we define the object as the system, then we—as part of the surroundings—are performing work on that system, transferring energy to it.

The other way in which energy is transferred is as heat. *Heat* is the energy transferred from a hotter object to a colder one. Or stating this idea in a slightly more abstract but nevertheless useful way, we can define heat as the energy transferred between a system and its surroundings as a result of their difference in temperature. A combustion reaction, such as the burning of natural gas illustrated in Figure 14.1(b), releases the chemical energy stored in the molecules of the fuel (∞ Section 3.2, 'Some Simple Patterns of Chemical Reactivity').

If we define the substances involved in the reaction as the system and everything else as the surroundings, we find that the released energy causes the temperature of the system to increase. Energy in the form of heat is then transferred from the hotter system to the cooler surroundings.

∞ Review this on page 74

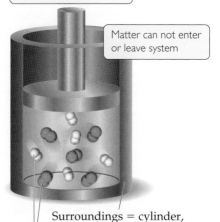

FIGURE IT OUT

If the piston is pulled upwards so that it sits halfway between the position shown and the top of the cylinder, is the system still closed?

Energy can enter or leave system as heat or as work done on piston

Matter can not enter or leave system

Surroundings = cylinder, piston and everything beyond

System = $H_2(g)$ and $O_2(g)$

▲ **FIGURE 14.3** **A closed system.**

SAMPLE EXERCISE 14.1	Describing and calculating energy changes

A bowler lifts a 5.4 kg bowling ball from ground level to a height of 1.6 m and then drops the ball back to the ground. **(a)** What happens to the potential energy of the bowling ball as it is raised from the ground? **(b)** What quantity of work, in J, is used to raise the ball? **(c)** After the ball is dropped, it gains kinetic energy. If we assume that all of the work done in part (b) has been converted to kinetic energy by the time the ball strikes the ground, what is the speed of the ball at the instant just before it hits the ground? (Note: The force due to gravity is $F = m \times g$, where m is the mass of the object and g is the gravitational constant, $g = 9.8$ m s^{-2}.)

SOLUTION

Analyse We need to relate the potential energy of the bowling ball to its position relative to the ground. We then need to establish the relationship between work and the change in the ball's potential energy. Finally, we need to connect the change in potential energy when the ball is dropped with the kinetic energy attained by the ball.

Plan We can calculate the work done in lifting the ball by using Equation 14.4: $w = F \times d$. The kinetic energy of the ball just before it hits the ground equals its initial potential energy. We can use the kinetic energy and Equation 14.1 to calculate the speed, u, just before impact.

Solve
(a) Because the bowling ball is raised to a greater height above the ground, its potential energy increases.

(b) The ball has a mass of 5.4 kg and it is lifted a distance of 1.6 m. To calculate the work performed to raise the ball, we use both Equation 14.4 and $F = m \times g$ for the force that is due to gravity:

$$w = F \times d = m \times g \times d = (5.4 \text{ kg})(9.8 \text{ m s}^{-2})(1.6 \text{ m}) = 85 \text{ kg m}^2 \text{s}^{-2} = 85 \text{ J}$$

Thus the bowler has done 85 J of work to lift the ball to a height of 1.6 m.

(c) When the ball is dropped its potential energy is converted to kinetic energy. At the instant just before the ball hits the ground, we assume that the kinetic energy is equal to the work done in part (b), 85 J:

$$E_k = \tfrac{1}{2}mu^2 = 85 \text{ J} = 85 \text{ kg m}^2 \text{s}^{-2}$$

We can now solve this equation for u:

$$u^2 = \left(\frac{2E_k}{m}\right) = \left(\frac{2(85 \text{ kg m}^2 \text{ s}^{-2})}{5.4 \text{ kg}}\right) = 31.5 \text{ m}^2 \text{s}^{-2}$$

$$u = \sqrt{31.5 \text{ m}^2 \text{s}^{-2}} = 5.6 \text{ m s}^{-1}$$

Check Work must be done in part (b) to increase the potential energy of the ball, which is in accord with our experiences. The units are appropriate in both parts (b) and (c). The work is in units of J and the speed in units of m s^{-1}. In part (c) we have carried an additional digit in the intermediate calculation involving the square root, but we report the final value to only two significant figures, as appropriate.

PRACTICE EXERCISE

What is the kinetic energy, in J, of **(a)** an Ar atom moving with a speed of 650 m s^{-1}, **(b)** a mole of Ar atoms moving with a speed of 650 m s^{-1}? (Hint: 1 u = 1.66×10^{-27} kg)

Answers: (a) 1.4×10^{-20} J, **(b)** 8.4×10^3 J

(See also Exercises 14.9–14.11.)

14.2 | THE FIRST LAW OF THERMODYNAMICS

In general, energy can be converted from one form to another, and it can be transferred back and forth between a system and its surroundings in the form of work and heat. All these transactions proceed in accord with one of the most important observations in science—that energy can be neither created nor destroyed. This universal truth, known as the **first law of thermodynamics**, can be summarised by a simple statement: *Energy is conserved.* Any energy that is lost by the system must be gained by the surroundings, and vice versa. In order to apply the first law quantitatively, we must first define the energy of a system more precisely.

Internal Energy

We will use the first law of thermodynamics to analyse energy changes in chemical systems. In order to do so, we must consider all the sources of kinetic and potential energy in the system we are studying.

We define the **internal energy**, U, of a system as the sum of *all* the kinetic and potential energies of all its components. For the system in Figure 14.3, for example, the internal energy includes the kinetic energy of motions of the H_2 and O_2 molecules through space, their rotations and internal vibrations. It also includes the potential energy of the electrostatic interactions between the nuclei and electrons.

We can't measure the actual numerical value of U. However, in thermodynamics we are mainly concerned with the *change* in U, denoted ΔU (read 'delta U') that accompanies a change in the system as the difference between U_{final} and U_{initial}:

$$\Delta U = U_{\text{final}} - U_{\text{initial}} \quad\quad [14.5]$$

To apply the first law of thermodynamics, we need only the value of ΔU which we can determine even though we don't know the specific values of U_{final} and U_{initial}.

Thermodynamic quantities such as ΔU have three parts: (1) a number and (2) a unit that together give the magnitude of the change, and (3) a sign that gives the direction. A *positive* value of ΔU results when $U_{\text{final}} > U_{\text{initial}}$, indicating the system has gained energy from its surroundings. A *negative* value of ΔU is obtained when $U_{\text{final}} < U_{\text{initial}}$, indicating the system has lost energy to its surroundings. Notice that we are taking the system's point of view rather than that of the surroundings in discussing the energy changes. We need to remember, however, that any change in the energy of the system is accompanied by an opposite change in the energy of the surroundings. These features of energy changes are summarised in ▼ **FIGURE 14.4**.

In a chemical reaction the initial state of the system refers to the reactants, and the final state refers to the products. When hydrogen and oxygen form water at a given temperature, the system loses energy to the surroundings as heat.

◢ **FIGURE IT OUT**

What is the value of ΔU if U_{final} equals U_{initial}?

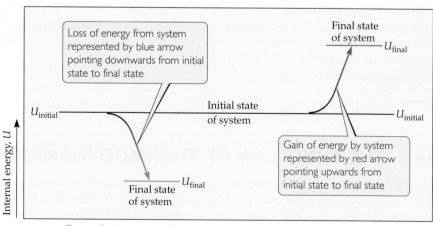

▶ **FIGURE 14.4   Changes in internal energy.**

Because heat is lost from the system, the internal energy of the products (final state) is less than that of the reactants (initial state), and ΔU for the process is negative. Thus the *energy diagram* in ▶ **FIGURE 14.5** shows that the internal energy of the mixture of H_2 and O_2 is greater than that of H_2O.

Relating ΔU to Heat and Work

As we noted in Section 14.1, a system may exchange energy with its surroundings as heat or as work. The internal energy of a system changes in magnitude as heat is added to or removed from the system or as work is done on or by the system.

We can use these ideas to write a very useful algebraic expression of the first law of thermodynamics. When a system undergoes any chemical or physical change, the magnitude and sign of the accompanying change in internal energy, ΔU, is given by the heat added to or liberated from the system, q, plus the work done on or by the system, w:

$$\Delta U = q + w \qquad [14.6]$$

When heat is added to a system or work is done on a system, its internal energy increases. Therefore, when heat is transferred to the system from the surroundings, q has a positive value: the total amount of energy goes up. Likewise, when work is done on the system by the surroundings, w has a positive value (▼ **FIGURE 14.6**). That too increases the internal energy of the system. Conversely, both the heat lost by the system to the surroundings and the work done

▲ **FIGURE IT OUT**

The internal energy for Mg(s) and $Cl_2(g)$ is greater than that of $MgCl_2(s)$. Sketch an energy diagram that represents the reaction $MgCl_2(s) \longrightarrow Mg(s) + Cl_2(g)$.

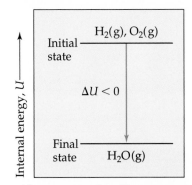

$U_{initial}$ greater than U_{final}, energy released from system to surroundings during reaction, $\Delta U < 0$

▲ **FIGURE 14.5**   **Energy diagram for the reaction** $2\,H_2(g) + O_2(g) \longrightarrow 2\,H_2O(g)$.

▲ **FIGURE IT OUT**

Suppose a system receives a 'deposit' of work from the surroundings and loses a 'withdrawal' of heat to the surroundings. Can we determine the sign of ΔU for this process?

System is interior of vault

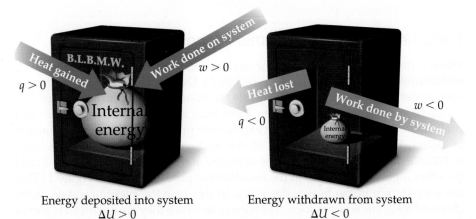

Energy deposited into system
$\Delta U > 0$

Energy withdrawn from system
$\Delta U < 0$

◀ **FIGURE 14.6**   **Sign conventions for heat and work.** Heat, q, gained by a system and work, w, done on a system are both positive quantities, corresponding to 'deposits' of internal energy into the system. Conversely, heat transferred from the system to the surroundings and work done by the system on the surroundings are both 'withdrawals' of internal energy from the system.

TABLE 14.1 • Sign conventions for *q*, *w* and Δ*U*

For *q*	+ means system *gains* heat	− means system *loses* heat
For *w*	+ means work done *on* system	− means work done *by* system
For Δ*U*	+ means *net gain* of energy by system	− means *net loss* of energy by system

by the system on the surroundings have negative values; that is, they lower the internal energy of the system. The sign conventions for *q*, *w* and Δ*U* are summarised in ▲ TABLE 14.1. Notice that any energy entering the system as either heat or work carries a positive sign.

SAMPLE EXERCISE 14.2 **Relating heat and work to changes of internal energy**

Two gases, A(g) and B(g), are confined in a cylinder-and-piston arrangement like that in Figure 14.3. Substances A and B react to form a solid product: A(g) + B(g) ⟶ C(s). As the reaction occurs, the system loses 1150 J of heat to the surroundings. The piston moves downwards as the gases react to form a solid. As the volume of the gas decreases under the constant pressure of the atmosphere, the surroundings do 480 J of work on the system. What is the change in the internal energy of the system?

SOLUTION

Analyse The question asks us to determine Δ*U*, given information about *q* and *w*.

Plan We first determine the signs of *q* and *w* (Table 14.1) and then use Equation 14.6, Δ*U* = *q* + *w*, to calculate Δ*U*.

Solve Heat is transferred from the system to the surroundings, and work is done on the system by the surroundings, so *q* is negative and *w* is positive: *q* = −1150 J and *w* = 480 kJ. Thus Δ*U* is

$$\Delta U = q + w = (-1150 \text{ J}) + (480 \text{ J}) = -670 \text{ J}$$

The negative value of Δ*U* tells us that a net quantity of 670 J of energy has been transferred from the system to the surroundings.

Comment You can think of this change as a decrease of 670 J in the net value of the system's energy bank account (hence the negative sign); 1150 J is withdrawn in the form of heat, while 480 J is deposited in the form of work. Notice that as the volume of the gases decreases, work is being done on the system by the surroundings, resulting in a deposit of energy.

PRACTICE EXERCISE

Calculate the change in the internal energy of the system for a process in which the system absorbs 140 J of heat from the surroundings and does 85 J of work on the surroundings.

Answer: + 55 J

(See also Exercises 14.13, 14.14.)

Endothermic and Exothermic Processes

When a process occurs in which the system absorbs heat, the process is called **endothermic** (*endo-* is a prefix meaning 'into'). During an endothermic process, such as the melting of ice, heat flows *into* the system from its surroundings.

A process in which the system loses heat is called **exothermic** (*exo-* is a prefix meaning 'out of'). During an exothermic process, such as the combustion of petrol, heat *exits* or flows *out* of the system and into the surroundings. ▶ FIGURE 14.7 shows two examples of chemical reactions, one endothermic and the other highly exothermic. In the endothermic process shown in Figure 14.7(a), the temperature in the beaker decreases. In this case the system consists of the chemical reactants and products. The solvent in which they are dissolved is part of the surroundings. Heat flows from the solvent, as part of the surroundings, into the system as reactants are converted to products. Thus the temperature of the solution drops.

State Functions

Although we usually have no way of knowing the precise value of the internal energy of a system, U, it does have a fixed value for a given set of conditions. The conditions that influence internal energy include the temperature and pressure. Furthermore, the total internal energy of a system is proportional to the total quantity of matter in the system because energy is an extensive property (Section 1.3 'Properties of Matter').

Review this on page 9

Suppose we define our system as 50 g of water at 25 °C, as in ▼ **FIGURE 14.8**. The system could have arrived at this state by cooling 50 g of water from 100 °C or by melting 50 g of ice and subsequently warming the water to 25 °C. The internal energy of the water at 25 °C is the same in either case. Internal energy is an example of a **state function**, a property of a system that is determined by specifying the system's condition, or state (in terms of temperature, pressure, location and so forth). *The value of a state function depends only on the present state of the system, not on the path the system took to reach that state.* Because U is a state function, ΔU depends only on the initial and final states of the system, not on how the change occurs. By analogy, if you drive from Sydney, which is at sea level, to Katoomba in the Blue Mountains, which is 1017 m above sea level, no matter which route you take the final altitude change is still 1017 m. The distance you travel will of course depend on the route you take. So the altitude is analogous to a state function whereas the distance travelled is not.

▲ **FIGURE 14.7 Examples of endothermic and exothermic reactions.**
(a) When ammonium thiocyanate and barium hydroxide octahydrate are mixed at room temperature, an endothermic reaction occurs:
$2\,NH_4SCN(s) + Ba(OH)_2.8\,H_2O(s) \longrightarrow Ba(SCN)_2(aq) + 2\,NH_3(aq) + 10\,H_2O(l)$.
As a result, the temperature of the system drops from about 20 °C to −9 °C.
(b) The reaction of powdered aluminium with Fe_2O_3 (the thermite reaction) is highly exothermic. The reaction proceeds vigorously to form Al_2O_3 and molten iron:
$2\,Al(s) + Fe_2O_3(s) \longrightarrow Al_2O_3(s) + 2\,Fe(l)$.

Some thermodynamic quantities, such as U, are state functions. Others, such as q and w, are not. Although $\Delta U = q + w$ does not depend on how the change occurs, the specific amounts of heat and work produced depend on the way in which the change is carried out (analogous to the choice of route from Sydney to Katoomba). Nevertheless, if changing the path by which a system goes from an initial state to a final state increases the value of q, that path change will also decrease the value of w by exactly the same amount. The result is that the value for ΔU for the two paths will be the same.

We can illustrate this principle with the example of a torch battery as our system. In ▼ **FIGURE 14.9**, we consider two possible ways of discharging the battery at constant temperature. If the battery is shorted out by a coil of wire, no work is accomplished because nothing is moved against a force. All the energy

50 g
$H_2O(l)$
100 °C

Initially hot water cools to water at 25 °C; once this temperature is reached, system has internal energy U

50 g
$H_2O(l)$
25 °C

Ice warms up to water at 25 °C; once this temperature is reached, system has internal energy U

50 g
$H_2O(s)$
0 °C

◀ **FIGURE 14.8 Internal energy, U, a state function.** Any state function depends only on the present state of the system and not on the path by which the system arrived at that state.

▲ FIGURE IT OUT

If the battery is defined as the system, what is the sign on *w* in part (b)?

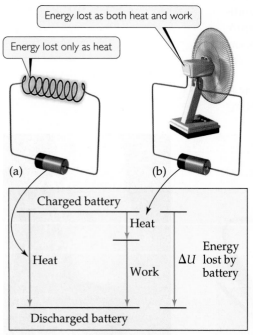

▲ **FIGURE 14.9 Internal energy is a state function but heat and work are not.** (a) A battery shorted out by a wire loses energy to the surroundings only as heat; no work is performed. (b) A battery discharged through a motor loses energy as work (to make the fan turn) and also loses some energy as heat. The value of ΔU is the same for both processes even though the values of q and w in (a) are different from those in (b).

is lost from the battery in the form of heat. (The wire coil will get warmer and release heat to the surrounding air.) In contrast though, if the battery is used to make a small motor turn, the discharge of the battery produces work. Some heat will be released as well, although not as much as when the battery is shorted out. The magnitudes of q and w are different for these two cases. If the initial and final states of the battery are identical in both cases, however, then $\Delta U = q + w$ must be the same in both cases because U is a state function. Thus ΔU depends only on the initial and final states of the system, regardless of how the transfers of energy occur in terms of heat and work.

> ▲ **CONCEPT CHECK 2**
> In what ways is the balance in your bank account a state function?

14.3 | ENTHALPY

The chemical and physical changes that occur around us, such as photosynthesis in the leaves of a plant, the evaporation of water from a lake or a reaction in an open beaker in a laboratory, occur under the essentially constant pressure of the Earth's atmosphere.

The changes can result in the release or absorption of heat or can be accompanied by work that is done by or on the system. In exploring these changes, we have a number of experimental means to measure the flow of heat into and out of the system and we therefore focus much of our discussion on what we can learn from heat flow. Of course in order to apply the first law of thermodynamics to these processes, we still need to account for any work that accompanies the process.

A system that consists of a gas confined to a container can be characterised by several different properties. Among the most important are the *pressure* of the gas, P, and the *volume* of the container, V. Like internal energy U, both P and V are state functions—they depend only on the current state of the system and not on the path taken to that state.

We can combine these three state functions—U, P and V—to define a new state function called **enthalpy** (from the Greek *enthalpein*, 'to warm'). This new function is particularly useful for discussing heat flow in processes that occur under constant (or nearly constant) pressure. Enthalpy, which we denote by the symbol H, is defined as the internal energy plus the product of the pressure and volume of the system:

$$H = U + PV \qquad [14.7]$$

> ▲ **CONCEPT CHECK 3**
> Given the definition of enthalpy in Equation 14.7, why must H be a state function?

You might be asking yourself why it is convenient to define a new function H. To answer that question, recall from Equation 14.6 that ΔU involves not only the heat q added to or removed from the system but also the work w done by or on the system.

The main kind of work produced by chemical or physical changes open to the atmosphere is the mechanical work associated with a change in the volume of the system. Consider, for example, the reaction of zinc metal with hydrochloric acid solution:

$$\text{Zn(s)} + 2\,\text{H}^+\text{(aq)} \longrightarrow \text{Zn}^{2+}\text{(aq)} + \text{H}_2\text{(g)} \qquad [14.8]$$

If we conduct this reaction at constant pressure in the apparatus illustrated in ▶ **FIGURE 14.10** the piston moves up or down to maintain a constant pres-

If the amount of zinc used in the reaction is increased, will more work be done by the system? Is there additional information you need in order to answer this question?

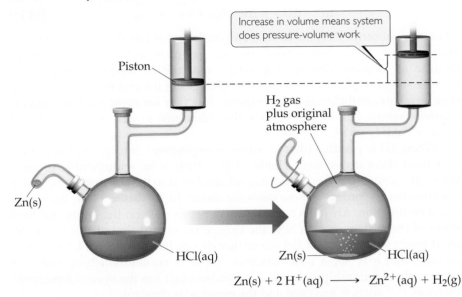

▲ FIGURE 14.10 **A system that does work on its surroundings.**

sure in the reaction vessel. If we assume for simplicity that the piston has no mass, the pressure in the apparatus is the same as the atmospheric pressure outside the apparatus. As the reaction proceeds, gas forms, and the piston rises. The gas within the flask is thus doing work on the surroundings by lifting the piston against the force of atmospheric pressure that presses down on it.

The work (w) involved in the expansion or compression of gases is called **pressure-volume work** (or *P-V* work). When the pressure is constant, as in our example, the sign and magnitude of the pressure-volume work is given by

$$w = -P\,\Delta V \qquad\qquad [14.9]$$

where P is pressure and ΔV is the change in volume of the system ($\Delta V = V_{final} - V_{initial}$). The negative sign in Equation 14.9 is necessary to conform to the sign conventions given in Table 14.1. The pressure P is always a positive number or zero. If the volume of the system expands, then ΔV is positive as well. Because the expanding system does work on the surroundings, energy leaves the system as work and w is negative. If the gas is compressed, ΔV is a negative quantity (the volume decreases), and Equation 14.9 indicates that w is positive, That is, energy enters the system as work, indicating that work is done on the system by the surroundings.

⚠ **CONCEPT CHECK 4**
If a system does not change its volume during the course of a process, does it do pressure-volume work?

We now return to our discussion of enthalpy.

When a change occurs at constant pressure, the change in enthalpy is given by the following relationship:

$$\Delta H = \Delta(U + PV)$$

$$= \Delta U + P\,\Delta V \quad \text{(constant pressure)} \qquad\qquad [14.10]$$

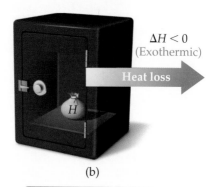

Constant pressure maintained in system

$\Delta H > 0$
(Endothermic)
Heat gain

(a)

$\Delta H < 0$
(Exothermic)
Heat loss

(b)

ΔH is amount of heat that flows into or out of system under constant pressure

▲ **FIGURE 14.11** **Endothermic and exothermic processes.** (a) An endothermic process ($\Delta H > 0$) deposits heat into the system. (b) An exothermic process ($\Delta H < 0$) withdraws heat from the system.

That is, the change in enthalpy equals the change in internal energy plus the product of the constant pressure times the change in volume. Remember that $\Delta U = q + w$ (Equation 14.6) and that the work involved in the expansion or compression of gases is $w = -P \,\Delta V$. If we substitute $-w$ for $P \,\Delta V$ and $q + w$ for ΔU into Equation 14.9, we have

$$\Delta H = \Delta U + P \,\Delta V = (q_P + w) - w = q_P \qquad [14.11]$$

where the subscript P on the heat, q, emphasises changes at constant pressure. Thus *the change in enthalpy equals the heat gained or lost at constant pressure.* Because q_P is something we can either measure or readily calculate and because so many physical and chemical changes of interest to us occur at constant pressure, enthalpy is a more useful function for most reactions than is internal energy. For most reactions the difference in ΔH and ΔU is small because $P \,\Delta V$ is small.

When ΔH is positive (that is, when q_P is positive), the system has gained heat from the surroundings (Table 14.1), which is an endothermic process. When ΔH is negative, the system has released heat to the surroundings, which is an exothermic process. These cases are shown in ◀ **FIGURE 14.11**. Because H is a state function, ΔH (which equals q_P) depends only on the initial and final states of the system, not on how the change occurs. At first glance this statement might seem to contradict our earlier discussion in Section 14.2, in which we said that q is *not* a state function. There is no contradiction, however, because the relationship between ΔH and heat (q_P) has the special limitations that only *P-V* work is involved and the pressure is constant.

▲ **CONCEPT CHECK 5**

What common laboratory measuring device are we likely always to use in experiments that measure enthalpy changes?

SAMPLE EXERCISE 14.3 **Determining the sign of ΔH**

Indicate the sign of the enthalpy change, in each of the following processes conducted under atmospheric pressure, and indicate whether the process is endothermic or exothermic: **(a)** an ice cube melts; **(b)** 1 g of butane is combusted in sufficient oxygen to give complete combustion to CO_2 and H_2O.

SOLUTION

Analyse Our goal is to determine whether ΔH is positive or negative for each process. Because each process occurs at constant pressure, the enthalpy change equals the quantity of heat absorbed or released, $\Delta H = q_P$.

Plan We must predict whether heat is absorbed or released by the system in each process. Processes in which heat is absorbed are endothermic and have a positive sign for ΔH; those in which heat is released are exothermic and have a negative sign for ΔH.

Solve In (a) the water that makes up the ice cube is the system. The ice cube absorbs heat from the surroundings as it melts, so is positive and the process is endothermic. In (b) the system is the 1 g of butane and the oxygen required to combust it. The combustion of butane in oxygen gives off heat, so is negative and the process is exothermic.

PRACTICE EXERCISE

Molten gold poured into a mould solidifies at atmospheric pressure. With the gold defined as the system, is the solidification an exothermic or endothermic process?

Answer: In order to solidify, the gold must cool to below its melting temperature. It cools by transferring heat to its surroundings. The air around the sample would feel hot because heat is transferred to it from the molten gold, meaning the process is exothermic.

You may notice that solidification of a liquid is the reverse of the melting we analysed in the exercise. As we will see, reversing the direction of a process changes the sign of the heat transferred.

(See also Exercises 14.22, 14.23.)

14.4 ENTHALPIES OF REACTION

Because $\Delta H = H_{final} - H_{initial}$, the enthalpy change for a chemical reaction is given by the enthalpy of the products minus the enthalpy of the reactants:

$$\Delta H = H_{products} - H_{reactants} \quad [14.12]$$

The enthalpy change that accompanies a reaction is called the **enthalpy of reaction**, or merely the *heat of reaction*, and is sometimes written $\Delta_{rxn}H$, where 'rxn' is a commonly used abbreviation for 'reaction'.

The combustion of hydrogen is shown in ▼ **FIGURE 14.12**. When the reaction is controlled so that 2 mol $H_2(g)$ burn to form 2 mol $H_2O(g)$ at a constant pressure, the system releases 483.6 kJ of heat. We can summarise this information as

$$2\,H_2(g) + O_2(g) \longrightarrow 2\,H_2O(g) \qquad \Delta H = -483.6\text{ kJ} \quad [14.13]$$

ΔH is negative, so this reaction is exothermic. Notice that ΔH is reported at the end of the balanced equation, without explicitly mentioning the amounts of chemicals involved. In such cases, the coefficients in the balanced equation represent the number of moles of reactants and products producing the associated enthalpy change. Balanced chemical equations that show the associated enthalpy change in this way are called *thermochemical equations*.

CONCEPT CHECK 6

If the reaction to form water were written $H_2(g) + \frac{1}{2} O_2(g) \longrightarrow H_2O(g)$, would you expect the same value of ΔH as in Equation 14.13? Why or why not?

The enthalpy change accompanying a reaction may also be represented in an *enthalpy diagram* such as that shown in Figure 14.12. Because the combustion of $H_2(g)$ is exothermic, the enthalpy of the products in the reaction is lower than the enthalpy of the reactants. The enthalpy of the system is lower after the reaction because energy has been lost in the form of heat released to the surroundings.

The reaction of hydrogen with oxygen is highly exothermic (ΔH is negative and has a large magnitude) and it occurs rapidly once it starts. It can occur with explosive violence, too, as demonstrated by the disastrous explosions of the German airship *Hindenburg* in 1937 (▼ **FIGURE 14.13**) and of the US space shuttle *Challenger* in 1986.

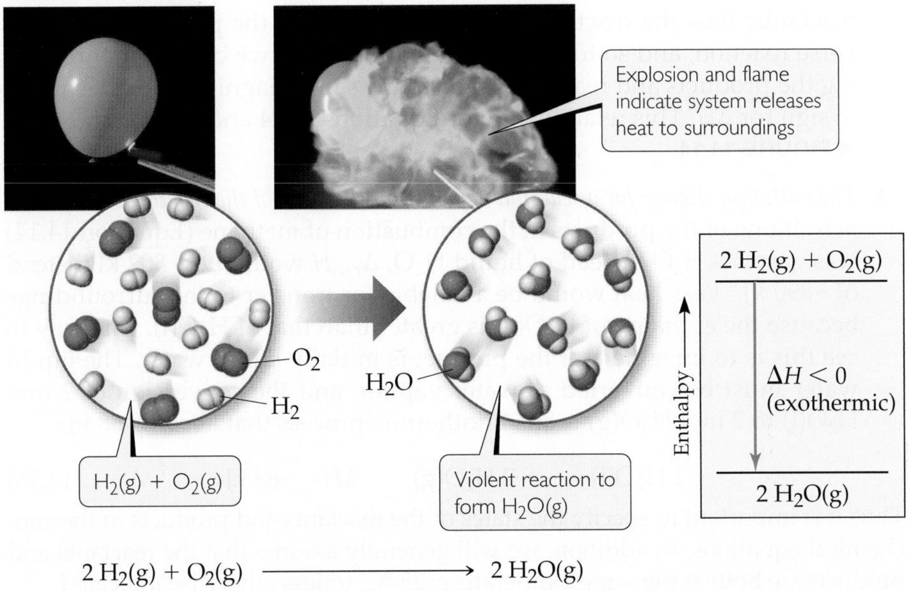

◀ **FIGURE 14.12 Exothermic reaction of hydrogen with oxygen.** When a mixture of $H_2(g)$ and $O_2(g)$ is ignited to form $H_2O(g)$, the resultant explosion produces a ball of flame. Because the system releases heat to the surroundings, the reaction is exothermic as indicated in the enthalpy diagram.

▶ **FIGURE 14.13 The burning of the hydrogen-filled airship *Hindenburg*.** This photograph was taken only 22 seconds after the first explosion occurred. This tragedy, which occurred in Lakehurst, New Jersey, on 6 May 1937, led to the discontinuation of hydrogen as a buoyant gas in such craft. Modern-day blimps are filled with helium, which is not as buoyant as hydrogen but is not flammable.

The following guidelines are helpful when using thermochemical equations and enthalpy diagrams.

1. *Enthalpy is an extensive property.* The magnitude of ΔH therefore is directly proportional to the amount of reactant consumed in the process. For the combustion of 1 mol of methane to form carbon dioxide and liquid water, 890 kJ of heat is produced when it is burned in a constant-pressure system:

$$CH_4(g) + 2\,O_2(g) \longrightarrow CO_2(g) + 2\,H_2O(l) \qquad \Delta H = -890 \text{ kJ} \qquad [14.14]$$

Because the combustion of 1 mol of CH_4 with 2 mol of O_2 releases 890 kJ of heat, the combustion of 2 mol of CH_4 with 4 mol O_2 of releases twice as much heat, 1780 kJ.

2. *The enthalpy change for a reaction is equal in magnitude, but opposite in sign, to ΔH for the reverse reaction.* For example, if we could reverse Equation 14.14 so that $CH_4(g)$ and $O_2(g)$ formed from $CO_2(g)$ and $H_2O(l)$ ΔH for the process would be

$$CO_2(g) + 2\,H_2O(l) \longrightarrow CH_4(g) + 2\,O_2(g) \qquad \Delta H = +890 \text{ kJ} \qquad [14.15]$$

When we reverse a reaction, we reverse the roles of the products and the reactants; thus the reactants in a reaction become the products of the reverse reaction, and so forth. From Equation 14.12, we can see that reversing the products and reactants leads to the same magnitude, but a change in sign for ΔH. This relationship for Equations 14.14 and 14.15 is shown in ◀ **FIGURE 14.14.**

3. *The enthalpy change for a reaction depends on the state of the reactants and products.* If one of the products in the combustion of methane (Equation 14.14) was gaseous H_2O instead of liquid H_2O, $\Delta_{rxn}H$ would be -802 kJ instead of -890 kJ. Less heat would be available for transfer to the surroundings because the enthalpy of $H_2O(g)$ is greater than that of $H_2O(l)$. One way to see this is to imagine that the product is initially liquid water. The liquid water must be converted to water vapour, and the conversion of 2 mol $H_2O(l)$ to 2 mol $H_2O(g)$ is an endothermic process that absorbs 88 kJ:

$$2\,H_2O(l) \longrightarrow 2\,H_2O(g) \qquad \Delta H = +88 \text{ kJ} \qquad [14.16]$$

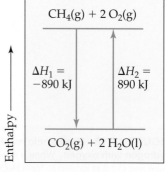

▲ **FIGURE 14.14 ΔH for a reverse reaction.** Reversing a reaction changes the sign but not the magnitude of the enthalpy change: $\Delta H_2 = -\Delta H_1$.

Thus it is important to specify the states of the reactants and products in thermochemical equations. In addition, we will generally assume that the reactants and products are both at the same temperature, 25 °C, unless otherwise indicated.

SAMPLE EXERCISE 14.4	Relating ΔH to quantities of reactants and products

How much heat is released when 4.50 g of methane gas is burned in a constant-pressure system? (Use the information given in Equation 14.14.)

SOLUTION

Analyse Our goal is to use a thermochemical equation to calculate the heat produced when a specific amount of methane gas is combusted. According to Equation 14.14, 890 kJ is released by the system when 1 mol CH_4 is burned at constant pressure.

Plan Equation 14.14 provides us with a stoichiometric conversion factor: (1 mol CH_4 yields -890 kJ). Thus we can convert moles of CH_4 to kJ of energy. First, however, we must convert grams of CH_4 to moles of CH_4. Thus the conversion sequence is grams CH_4 (given) $\longrightarrow$ moles CH_4 $\longrightarrow$ kJ (unknown to be found).
The molar mass of CH_4 is 16 g mol^{-1}. Hence 4.50 g of CH_4 is 0.281 mol.

$$\text{Heat} = (0.281 \times -890) \text{ kJ} = -250 \text{ kJ}$$

The negative sign indicates that 250 kJ is released by the system into the surroundings.

PRACTICE EXERCISE

Hydrogen peroxide can decompose to water and oxygen by the following reaction:

$$2\,H_2O_2(l) \longrightarrow 2\,H_2O(l) + O_2(g) \qquad \Delta H = -196 \text{ kJ}$$

Calculate the amount of heat produced when 5.00 g of $H_2O_2(l)$ decomposes at constant pressure.
Answer: -14.4 kJ
(See also Exercises 14.24–14.26.)

A CLOSER LOOK

USING ENTHALPY AS A GUIDE

If you hold a brick in the air and let it go, you know what happens: it falls as the force of gravity pulls it towards the Earth. A process that is thermodynamically favoured to happen, such as a brick falling to the ground, is called a *spontaneous* process. A spontaneous process can be either fast or slow; the rate at which processes occur is not governed by thermodynamics.

Chemical processes can be thermodynamically favoured, or spontaneous, too. By spontaneous, however, we do not mean that the reaction will form products without any intervention. That can be the case, but often some energy must be imparted to get the process started. The enthalpy change in a reaction gives one indication as to whether the reaction is likely to be spontaneous. The combustion of $H_2(g)$ and $O_2(g)$, for example, is highly exothermic:

$$H_2(g) + \tfrac{1}{2}O_2(g) \longrightarrow H_2O(g) \qquad \Delta H = -242 \text{ kJ}$$

Hydrogen gas and oxygen gas can exist together in a volume indefinitely without noticeable reaction occurring. Once the reaction is initiated, however, energy is rapidly transferred from the system (the reactants) to the surroundings as heat. The system thus loses enthalpy by transferring the heat to the surroundings. (Recall that the first law of thermodynamics

tells us that the total energy of the system plus the surroundings does not change; energy is conserved.)

Enthalpy change is not the only consideration in the spontaneity of reactions, however, nor is it a foolproof guide. For example, even though ice melting is an endothermic process,

$$H_2O(s) \longrightarrow H_2O(l) \qquad \Delta H = +6.01 \text{ kJ}$$

this process is spontaneous at temperatures above the freezing point of water (0 °C). The reverse process, water freezing, is spontaneous at temperatures below 0 °C. Thus we know that ice at room temperature melts and water put into a freezer at –20 °C turns into ice. Both processes are spontaneous under different conditions even though they are the reverse of one another. We will address the spontaneity of processes more fully later in this chapter. We will see why a process can be spontaneous at one temperature but not at another, as is the case for the conversion of water to ice.

Despite these complicating factors, you should pay attention to the enthalpy changes in reactions. As a general observation, when the enthalpy change is large it is the dominant factor in determining spontaneity. Thus reactions for which ΔH is *large* and *negative* tend to be spontaneous. Reactions for which ΔH is *large* and *positive* tend to be spontaneous only in the reverse direction.

RELATED EXERCISE: 14.32

In many situations we will find it valuable to know the sign and magnitude of the enthalpy change associated with a given chemical process. As we see in the following sections, ΔH can be either determined directly by experiment or calculated from known enthalpy changes of other reactions.

14.5 | CALORIMETRY

The value of ΔH can be determined experimentally by measuring the heat flow accompanying a reaction at constant pressure. Typically, we can determine the magnitude of the heat flow by measuring the magnitude of the temperature change the heat flow produces. The measurement of heat flow is **calorimetry**; a device used to measure heat flow is a **calorimeter**.

Heat Capacity and Specific Heat Capacity

All substances change temperature when they gain heat, but the magnitude of the temperature change produced by a given quantity of heat varies from substance to substance. The temperature change experienced by an object when it absorbs a certain amount of heat is determined by its **heat capacity**, C. The heat capacity of an object is the amount of heat required to raise its temperature by 1 K (or 1 °C). The greater the heat capacity of a substance, the greater the heat required to produce a given increase in temperature.

For pure substances the heat capacity is usually given for a specified amount of the substance. The heat capacity of one mole of a substance is called its **molar heat capacity**, C_{molar} or simply C_m. The heat capacity of one gram of a substance is called its **specific heat capacity**, C_s. The specific heat capacity of a substance can be determined experimentally by measuring the temperature change that a known mass, m, of the substance undergoes when it gains or loses a specific quantity of heat, q:

$$\text{Specific heat capacity} = \frac{\text{(quantity of heat transferred)}}{\text{(grams of substance)} \times \text{(temperature change)}}$$

$$C_s = \frac{q}{m \times \Delta T} \qquad [14.17]$$

For example, 209 J is required to increase the temperature of 50.0 g of water by 1.00 K. Thus the specific heat capacity of water is

$$C_s = \frac{209 \text{ J}}{(50.1 \text{ g})(1.00 \text{ K})} = 4.18 \text{ J g}^{-1} \text{ K}^{-1}$$

When the sample gains heat (positive q), the temperature of the sample increases (positive ΔT). Rearranging Equation 14.17 we get

$$q = C_s \times m \times \Delta T \qquad [14.18]$$

Thus we can calculate the quantity of heat that a substance has gained or lost by using its specific heat capacity together with its measured mass and temperature change.

The specific heat capacities of several substances are listed in ▶ TABLE 14.2. Notice that the specific heat capacity of liquid water is higher than those of the other substances listed. For example, it is about five times as great as that of aluminium metal. The high specific heat capacity of water affects the Earth's climate because it keeps the temperatures of the oceans relatively resistant to change.

TABLE 14.2 • Specific heat capacities of some substances at 298 K

Elements		Compounds	
Substance	Specific heat capacity $(J\,g^{-1}\,K^{-1})$	Substance	Specific heat capacity $(J\,g^{-1}\,K^{-1})$
$N_2(g)$	1.04	$H_2O(l)$	4.18
$Al(s)$	0.90	$CH_4(g)$	2.20
$Fe(s)$	0.45	$CO_2(g)$	0.84
$Hg(l)$	0.14	$CaCO_3(s)$	0.82

SAMPLE EXERCISE 14.5 **Relating heat, temperature change and heat capacity**

(a) How much heat is needed to warm 250 g of water from 22 °C (about room temperature) to near its boiling point, 98 °C? The specific heat capacity of water is 4.18 J g^{-1} K^{-1}. **(b)** What is the molar heat capacity of water?

SOLUTION

Analyse In part (a) we must find the quantity of heat (q) needed to warm the water, given the mass of water (m), its temperature change (ΔT) and its specific heat (C_s). In part (b) we must calculate the molar heat capacity (heat capacity per mole, C_m) of water from its specific heat (heat capacity per gram).

Plan (a) Given C_s, m and ΔT, we can calculate the quantity of heat, q, using Equation 14.18. (b) We can use the molar mass of water and dimensional analysis to convert from heat capacity per gram to heat capacity per mole.

Solve

(a) The water undergoes a temperature change of

$$\Delta T = 98\ ^\circ C - 22\ ^\circ C = 76\ ^\circ C = 76\ K$$

Using Equation 14.18, we have

$$q = C_s \times m \times \Delta T$$
$$= (4.18\ J\,g^{-1}\,K^{-1})(250\ g)(76\ K) = 7.9 \times 10^4\ J$$

(b) The molar heat capacity is the heat capacity of one mole of substance. Using the atomic weights of hydrogen and oxygen, we have

$$1\ mol\ H_2O = 18.0\ g\ H_2O$$

From the specific heat capacity given in part (a), we have

$$C_m = (4.18\ J\,g^{-1}\,K^{-1})\,(18\ g\,mol^{-1}) = 75.2\ J\,mol^{-1}\,K^{-1}$$

PRACTICE EXERCISE

(a) Large beds of rocks are used in some solar-heated homes to store heat. Assume that the specific heat capacity of the rocks is 0.82 J g^{-1} K^{-1}. Calculate the quantity of heat absorbed by 50.0 kg of rocks if their temperature increases by 12.0 °C. **(b)** What temperature change would these rocks undergo if they emitted 450 kJ of heat?

Answers: **(a)** 490 kJ, **(b)** 11 K = 11 °C decrease

(See also Exercises 14.33, 14.34, 14.97.)

Constant-Pressure Calorimetry

The techniques and equipment employed in calorimetry depend on the nature of the process being studied. For many reactions, such as those occurring in solution, it is easy to control pressure so that ΔH is measured directly. (Recall that $\Delta H = q_P$.) Although the calorimeters used for highly accurate work are precision instruments, a very simple 'coffee-cup' calorimeter, as shown in ▼ **FIGURE 14.15**, is often used in general chemistry labs to illustrate the principles of calorimetry. Because the calorimeter is not sealed, the reaction occurs under the essentially constant pressure of the atmosphere.

Imagine adding two aqueous solutions, each containing a reactant, to a coffee-cup calorimeter. Once mixed, the reactants can react to form products. In this case there is no physical boundary between the system and the surroundings. The reactants and products of the reaction are the system, and the

Propose a reason for why two Styrofoam® cups are often used instead of just one.

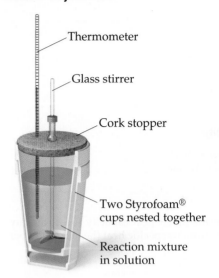

Thermometer

Glass stirrer

Cork stopper

Two Styrofoam® cups nested together

Reaction mixture in solution

▲ **FIGURE 14.15 Coffee-cup calorimeter.** This simple apparatus is used to measure temperature changes of reactions at constant pressure.

water in which they dissolve and the calorimeter are part of the surroundings. If we assume that the calorimeter perfectly prevents the gain or loss of heat from the solution, the heat gained by the solution must be produced from the chemical reaction under study. In other words, the heat produced by the reaction is entirely absorbed by the solution; it does not escape the calorimeter. (We also assume that the calorimeter itself does not absorb heat. In the case of the coffee-cup calorimeter this is a reasonable approximation because the calorimeter has a very low thermal conductivity and heat capacity.) For an exothermic reaction, heat is 'lost' by the reaction and 'gained' by the solution, so the temperature of the solution rises. The opposite occurs for an endothermic reaction. The heat gained or lost by the solution, q_{soln}, is therefore equal in magnitude to but opposite in sign to the heat absorbed or released by the reaction, q_{rxn}. Therefore, $q_{soln} = -q_{rxn}$. The value of q_{soln} is readily calculated from the mass of the solution, its specific heat and the temperature change:

$$q_{soln} = \text{(specific heat capacity of solution)}$$
$$\times \text{(grams of solution)} \times \Delta T = -q_{rxn} \quad [14.19]$$

For dilute aqueous solutions the specific heat capacity of the solution will be approximately the same as that of water, $4.18\ \text{J g}^{-1}\ \text{K}^{-1}$.

Equation 14.19 makes it possible to calculate q_{rxn} from the temperature change of the solution in which the reaction occurs. A temperature increase means the reaction is exothermic ($q_{rxn} < 0$).

SAMPLE EXERCISE 14.6 Measuring ΔH using a coffee-cup calorimeter

When a student mixes 50 cm³ of 1.0 M HCl and 50 cm³ of 1.0 M NaOH in a coffee-cup calorimeter, the temperature of the resultant solution increases from 21.0 °C to 27.5 °C. Calculate the enthalpy change for the reaction in kJ per mol of HCl, assuming that the calorimeter loses only a negligible quantity of heat, that the total volume of the solution is 100 cm³, that its density is 1.0 g cm⁻³ and that its specific heat capacity is 4.18 J g⁻¹ K⁻¹.

SOLUTION

Analyse Mixing solutions of HCl and NaOH results in an acid–base reaction:

$$HCl(aq) + NaOH(aq) \longrightarrow H_2O(l) + NaCl(aq)$$

We need to calculate the heat produced per mole of HCl, given the temperature increase of the solution, the number of moles of HCl and NaOH involved, and the density and specific heat of the solution.

Plan The total heat produced can be calculated using Equation 14.19. The number of moles of HCl consumed in the reaction must be calculated from the volume and molarity of this substance, and this amount is then used to determine the heat produced per mol HCl.

Solve

Because the total volume of the solution is 100 cm³, its mass is $(100\ \text{cm}^3)(1.0\ \text{g cm}^{-3}) = 100\ \text{g}$

The temperature change is $\Delta T = 27.5\ °C - 21.0\ °C = 6.5\ °C = 6.5\ K$

Using Equation 14.19, we have
$$q_{rxn} = -C_s \times m \times \Delta T$$
$$-(4.18\ \text{J g}^{-1}\ \text{K}^{-1})(100\ \text{g})(6.5\ \text{K}) = -2.7 \times 10^3\ \text{J} = -2.7\ \text{kJ}$$

Because the process occurs at constant pressure, $\Delta H = q_P = -2.7\ \text{kJ}$

To express the enthalpy change on a molar basis, we use the fact that the number of moles of HCl and NaOH is given by the product of the respective solution volumes (50 cm³ = 0.050 dm³) and concentrations: $(0.050\ \text{dm}^3)(1.0\ \text{mol dm}^{-3}) = 0.050\ \text{mol}$

Thus the enthalpy change per mole of HCl is $\Delta H = -2.7\ \text{kJ}/0.050\ \text{mol} = -54\ \text{kJ mol}^{-1}$

Check: ΔH is negative (exothermic), which is expected for the reaction of an acid with a base. The molar magnitude of the heat evolved seems reasonable.

PRACTICE EXERCISE

When 50.0 cm^3 of 0.100 M AgNO$_3$ and 50.0 cm^3 of 0.100 M HCl are mixed in a constant-pressure calorimeter, the temperature of the mixture increases from 22.20 °C to 23.11 °C. The temperature increase is caused by the following reaction:

$$AgNO_3(aq) + HCl(aq) \longrightarrow AgCl(s) + HNO_3(aq)$$

Calculate ΔH for this reaction in kJ mol^{-1} AgNO$_3$, assuming that the combined solution has a mass of 100.0 g and a specific heat capacity of 4.18 J g^{-1} K^{-1}.

Answer: -68000 J mol$^{-1} => -76$ kJ mol^{-1}

(See also Exercise 14.36.)

Bomb Calorimetry (Constant-Volume Calorimetry)

One of the most important types of reactions studied using calorimetry is combustion, in which a compound reacts completely with excess oxygen (∞ Section 3.2, 'Some Simple Patterns of Chemical Reactivity').

∞ Review this on page 74

Combustion reactions are most conveniently studied using a *bomb calorimeter* (▶ FIGURE 14.16). The substance to be studied is placed in a small cup within an insulated vessel called a *bomb*. The bomb, which is designed to withstand high pressures, has an inlet valve for adding oxygen and electrical contacts to initiate the combustion. After the sample has been placed in the bomb, the bomb is sealed and pressurised with oxygen. It is then placed in the calorimeter, which is essentially an insulated container, and covered with an accurately measured quantity of water. When all the components within the calorimeter have come to the same temperature, the combustion reaction is initiated by passing an electrical current through a fine wire that is in contact with the sample. When the wire gets sufficiently hot, the sample ignites.

Heat is released when combustion occurs. This heat is absorbed by the water and the various components of the calorimeter, causing a rise in the temperature of the water. The change in water temperature caused by the reaction is measured very precisely.

To calculate the heat of combustion from the measured temperature increase, we must know the total heat capacity of the calorimeter, C_{cal}. This quantity is determined by combusting a sample that releases a known quantity of heat and measuring the resulting temperature change. For example, the combustion of exactly 1 g of benzoic acid, C_6H_5COOH, in a bomb calorimeter produces 26.38 kJ of heat. Suppose 1.000 g of benzoic acid is combusted in a calorimeter, increasing the temperature by 4.857°C. The heat capacity of the calorimeter is then given by $C_{cal} = 26.38$ kJ/4.857°C $= 5.431$ kJ °C^{-1}. Once we know the value of C_{cal} we can measure temperature changes produced by other reactions, and from these we can calculate the heat evolved in the reaction:

$$q_{rxn} = -C_{cal}\Delta T \qquad [14.20]$$

FIGURE IT OUT

Why is a stirrer used in calorimeters?

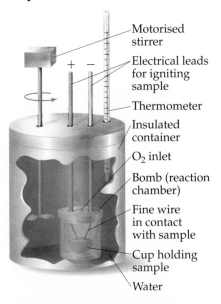

Motorised stirrer

Electrical leads for igniting sample

Thermometer

Insulated container

O$_2$ inlet

Bomb (reaction chamber)

Fine wire in contact with sample

Cup holding sample

Water

▲ **FIGURE 14.16** Bomb calorimeter.

SAMPLE EXERCISE 14.7 | Measuring q_{rxn} using a bomb calorimeter

Methylhydrazine is commonly used as a liquid rocket fuel. The combustion of methylhydrazine with oxygen produces N$_2$(g), CO$_2$(g) and H$_2$O(l):

$$2\,CH_3N_2H_3(l) + 5\,O_2(g) \longrightarrow 2\,N_2(g) + 2\,CO_2(g) + 6\,H_2O(l)$$

When 4.00 g of methylhydrazine is combusted in a bomb calorimeter, the temperature of the calorimeter increases from 25.00 °C to 39.50 °C. In a separate experiment the heat capacity of the calorimeter is measured to be 7.794 kJ °C^{-1}. What is the heat of reaction for the combustion of a mole of methylhydrazine in this calorimeter?

SOLUTION

Analyse We are given a temperature change and the total heat capacity of the calorimeter. We are also given the amount of reactant combusted. Our goal is to calculate the enthalpy change per mole for combustion of the reactant.

Plan We will first calculate the heat evolved for the combustion of the 4.00 g sample. We will then convert this heat to a molar quantity.

Solve

For combustion of the 4.00 g sample of methylhydrazine, the temperature change of the calorimeter is

$$\Delta T = (39.50\ ^{\circ}C - 25.00\ ^{\circ}C) = 14.50\ ^{\circ}C$$

We can use this value and the value for C_{cal} to calculate the heat of reaction (Equation 14.20):

$$q_{rxn} = -C_{cal} \times \Delta T = -(7.794\ \text{kJ}\ ^{\circ}C^{-1})(14.50\ ^{\circ}C) = -113.0\ \text{kJ}$$

We can readily convert this value to the heat of reaction for a mole of $CH_3N_2H_3$

$$\text{mol } CH_3N_2H_3 = \frac{4.00\ \text{g}}{46.1\ \text{g mol}^{-1}} = 0.0868\ \text{mol}$$

$$q_{rxn}\ \text{per mol } CH_3N_2H_3 = \frac{-113.0\ \text{kJ}}{0.0868\ \text{mol}} = 1.30 \times 10^3\ \text{kJ mol}^{-1}\ CH_3N_2H_3$$

Check The units cancel properly, and the sign of the answer is negative as it should be for an exothermic reaction.

PRACTICE EXERCISE

A 0.5865 g sample of lactic acid is burned in a calorimeter whose heat capacity is 4.812 kJ $^{\circ}C^{-1}$. The temperature increases from 23.10°C to 24.95°C. Calculate the heat of combustion of lactic acid **(a)** per gram and **(b)** per mole.

Answers: (a) -15.2 kJ g^{-1}, **(b)** -1370 kJ mol^{-1}

(See also Exercises 14.37, 14.38.)

Because the reactions in a bomb calorimeter are conducted under constant-volume conditions, the heat transferred corresponds to the change in internal energy, $\Delta U = q_V$, where q_V is the heat change at constant volume, rather than the change in enthalpy, $\Delta H = q_P$. For most reactions, however, the difference between ΔU and ΔH is very small. For the reaction discussed in Sample Exercise 14.7, for example, the difference between ΔU and ΔH is only about 1 kJ mol^{-1}, a difference of less than 0.1%. It is possible to correct the measured heat changes to obtain ΔH values, and these form the basis of the tables of enthalpy change that we see in the following sections. We need not concern ourselves with how these small corrections are made.

14.6 | HESS'S LAW

It is often possible to calculate the ΔH for a reaction from the tabulated ΔH values of other reactions. Thus it is not necessary to make calorimetric measurements for all reactions.

Because enthalpy is a state function, the enthalpy change, ΔH, associated with any chemical process depends only on the amount of matter that undergoes change and on the nature of the initial state of the reactants and the final state of the products. This means that if a particular reaction can be conducted in one step or in a series of steps, the sum of the enthalpy changes associated with the individual steps must be the same as the enthalpy change associated with the one-step process. As an example, the combustion of methane gas, $CH_4(g)$, to form $CO_2(g)$ and liquid water can be thought of as occurring in two steps: (1) the combustion of $CH_4(g)$ to form $CO_2(g)$ and gaseous water, $H_2O(g)$, and (2) the condensation of gaseous water to form liquid water, $H_2O(l)$. The enthalpy change for the overall process is simply the sum of the enthalpy changes for these two steps:

$$CH_4(g) + 2\,O_2(g) \longrightarrow CO_2(g) + 2\,H_2O(g) \qquad \Delta H = -802 \text{ kJ}$$

(Add) $\qquad\qquad 2\,H_2O(g) \longrightarrow 2\,H_2O(l) \qquad\qquad\qquad \Delta H = -88 \text{ kJ}$

$$\overline{CH_4(g) + 2\,O_2(g) + 2\,H_2O(g) \longrightarrow CO_2(g) + 2\,H_2O(l) + 2\,H_2O(g)} \quad \Delta H = -890 \text{ kJ}$$

The net equation is

$$CH_4(g) + 2\,O_2(g) \longrightarrow CO_2(g) + 2\,H_2O(l) \qquad \Delta H = -890 \text{ kJ}$$

Hess's law states that *if a reaction is conducted in a series of steps, ΔH for the overall reaction will equal the sum of the enthalpy changes for the individual steps.* This means that a relatively small number of experimental measurements can be used to calculate ΔH for a vast number of different reactions.

Hess's law provides a useful means of calculating energy changes that are difficult to measure directly. For instance, it is impossible to measure directly the enthalpy for the combustion of carbon to form carbon monoxide. Combustion of 1 mol of carbon with 0.5 mol of O_2 produces not only CO but also leaves some carbon unreacted. However, solid carbon and carbon monoxide can both be completely burned in O_2 to produce CO_2. We can use the enthalpy changes of these reactions to calculate the heat of combustion of C to CO, as shown in Sample Exercise 14.8.

CONCEPT CHECK 8

What effect do the following changes have on ΔH for a reaction:
a. Reversing a reaction?
b. Multiplying coefficients by 2?

SAMPLE EXERCISE 14.8 | **Using Hess's law to calculate ΔH**

The enthalpy of reaction for the combustion of C to CO_2 is -393.5 kJ mol^{-1} and the enthalpy for the combustion of CO to CO_2 to is -283.0 kJ mol^{-1}:

(1) $\qquad C(s) + O_2(g) \longrightarrow CO_2(g) \qquad \Delta H_1 = -393.5 \text{ kJ}$

(2) $\quad CO(g) + \frac{1}{2} O_2(g) \longrightarrow CO_2(g) \qquad \Delta H_2 = -283.0 \text{ kJ}$

Using these data, calculate the enthalpy for the combustion of C to CO:

(3) $\qquad C(s) + \frac{1}{2} O_2(g) \longrightarrow CO(g) \qquad \Delta H_3 =?$

SOLUTION

Analyse We are given two thermochemical equations, and our goal is to combine them in such a way as to obtain the third equation and its enthalpy change.

Plan We will use Hess's law. In doing so, we first note the numbers of moles of substances among the reactants and products in the target equation, (3). We then manipulate equations (1) and (2) to give the same number of moles of these substances, so that when the resulting equations are added, we obtain the target equation. At the same time, we keep track of the enthalpy changes, which we add.

Solve In order to use equations (1) and (2), we arrange them so that C(s) is on the reactant side and CO(g) is on the product side of the arrow, as in the target reaction, equation (3). Because equation (1) has C(s) as a reactant, we can use that equation just as it is. We need to turn equation (2) around, however, so that CO(g) is a product. Remember that when reactions are turned around, the sign of ΔH is reversed. We arrange the two equations so that they can be added to give the desired equation:

$$C(s) + O_2(g) \longrightarrow CO_2(g) \qquad\qquad \Delta H_1 = -393.5 \text{ kJ}$$

$$\underline{CO_2(g) \longrightarrow CO(g) + \tfrac{1}{2} O_2(g) \qquad\qquad -\Delta H_2 = 283.0 \text{ kJ}}$$

$$C(s) + \tfrac{1}{2} O(g) \longrightarrow CO(g) \qquad\qquad \Delta H_3 = -110.5 \text{ kJ}$$

When we add the two equations, $CO_2(g)$ appears on both sides of the arrow and therefore cancels out. Likewise, $\frac{1}{2} O_2(g)$ is eliminated from each side.

Comment It is sometimes useful to add subscripts to the enthalpy changes, as we have done here, to keep track of the associations between the chemical reactions and their ΔH values.

PRACTICE EXERCISE

Carbon occurs in two forms, graphite and diamond. The enthalpy of the combustion of graphite is -393.5 kJ mol^{-1} and that of diamond is -395.4 kJ mol^{-1}.

$$C(\text{graphite}) + O_2(g) \longrightarrow CO_2(g) \qquad \Delta H_1 = -393.5 \text{ kJ}$$

$$C(\text{diamond}) + O_2(g) \longrightarrow CO_2(g) \qquad \Delta H_2 = -395.4 \text{ kJ}$$

Calculate for the conversion of graphite to diamond:

$$C(\text{graphite}) \longrightarrow C(\text{diamond}) \qquad \Delta H_3 = ?$$

Answer: $\Delta H_3 = +1.9$ kJ

(See also Exercises 14.40, 14.41.)

SAMPLE EXERCISE 14.9 | **Using three equations with Hess's law to calculate ΔH**

Calculate ΔH for the reaction

$$2\,C(s) + H_2(g) \longrightarrow C_2H_2(g)$$

given the following chemical equations and their respective enthalpy changes:

$$C_2H_2(g) + \tfrac{5}{2} O_2(g) \longrightarrow 2\,CO_2(g) + H_2O(l) \qquad \Delta H = -1299.6 \text{ kJ}$$

$$C(s) + O_2(g) \longrightarrow CO_2(g) \qquad \Delta H = -393.5 \text{ kJ}$$

$$H_2(g) + \tfrac{1}{2} O_2(g) \longrightarrow H_2O(l) \qquad \Delta H = -285.8 \text{ kJ}$$

SOLUTION

Analyse We are given a chemical equation and asked to calculate its ΔH using three chemical equations and their associated enthalpy changes.

Plan We will use Hess's law, summing the three equations or their reverses and multiplying each by an appropriate coefficient so that they add to give the net equation for the reaction of interest. At the same time, we keep track of the ΔH values, reversing their signs if the reactions are reversed and multiplying them by whatever coefficient is employed in the equation.

Solve Because the target equation has C_2H_2 as a product, we turn the first equation around; the sign of ΔH is therefore changed. The desired equation has $2\,C(s)$ as a reactant, so we multiply the second equation and its ΔH by 2. Because the target equation has H_2 as a reactant, we keep the third equation as it is. We then add the three equations and their enthalpy changes in accordance with Hess's law:

$$2\,CO_2(g) + H_2O(l) \longrightarrow C_2H_2(g) + \tfrac{5}{2} O_2(g) \qquad \Delta H = +1299.6 \text{ kJ}$$

$$2\,C(s) + 2\,O_2(g) \longrightarrow 2\,CO_2(g) \qquad \Delta H = -787.0 \text{ kJ}$$

$$\underline{H_2(g) + \tfrac{1}{2} O_2(g) \longrightarrow H_2O(l) \qquad \Delta H = -285.8 \text{ kJ}}$$

$$2\,C(s) + H_2(g) \longrightarrow C_2H_2(g) \qquad \Delta H = 226.8 \text{ kJ}$$

When the equations are added, there are $2\,CO_2$, $\tfrac{5}{2}\,O_2$ and H_2O on both sides of the arrow. These are cancelled in writing the net equation.

Check The procedure must be correct because we obtained the correct net equation. In cases like this you should go back over the numerical manipulations of the values to ensure that you did not make an inadvertent error with signs.

PRACTICE EXERCISE

Calculate ΔH for the reaction

$$NO(g) + O(g) \longrightarrow NO_2(g)$$

given the following information:

$$NO(g) + O_3(g) \longrightarrow NO_2(g) + O_2(g) \qquad \Delta H = -198.9 \text{ kJ}$$
$$O_3(g) \longrightarrow \tfrac{3}{2} O_2(g) \qquad \Delta H = -142.3 \text{ kJ}$$
$$O_2(g) \longrightarrow 2 O(g) \qquad \Delta H = +495.0 \text{ kJ}$$

Answer: −304.1 kJ
(See also Exercises 14.32, 14.33.)

The key point of these examples is that H is a state function, so *for a particular set of reactants and products, ΔH is the same whether the reaction takes place in one step or in a series of steps.* For example, consider the reaction of methane, CH_4, and oxygen, O_2, to form CO_2 and H_2O. We can envisage the reaction forming CO_2 directly or with the initial formation of CO, which is then combusted to CO_2. These two paths are compared in ▶ **FIGURE 14.17**. Because ΔH is a state function, both paths *must* produce the same value of ΔH. In the enthalpy diagram, that means $\Delta H_1 = \Delta H_2 + \Delta H_3$.

14.7 | ENTHALPIES OF FORMATION

By using the methods we have just discussed, we can calculate the enthalpy changes for a great many reactions from tabulated ΔH values. Many experimental data are tabulated according to the type of process. For example, extensive tables exist of *enthalpies of vaporisation* (ΔH for converting liquids to gases), *enthalpies of fusion* (ΔH for melting solids), *enthalpies of combustion* (ΔH for combusting a substance in oxygen) and so forth. A particularly important process used for tabulating thermochemical data is the formation of a compound from its constituent elements. The enthalpy change associated with this process is called the **enthalpy of formation** (or *heat of formation*) and is labelled $\Delta_f H$.

The magnitude of any enthalpy change depends on the conditions of temperature, pressure and state (gas, liquid or solid crystalline form) of the reactants and products. To compare the enthalpies of different reactions, we must define a set of conditions, called a *standard state*, at which most enthalpies are tabulated. The standard state of a substance is its pure form at 1 bar pressure and the temperature of interest, which we usually choose to be 298 K (25 °C). The older definition of standard state used 1 atmosphere of pressure for gases. However since 1 bar = 0.9872 atm, for most purposes this makes very little difference in the standard enthalpy changes. The **standard enthalpy change** of a reaction is defined as the enthalpy change when all reactants and products are in their standard states. We denote a standard enthalpy change as $\Delta H°$, where the superscript ° indicates standard-state conditions.

The **standard enthalpy of formation** of a compound, $\Delta_f H°$, is the change in enthalpy for the reaction that forms *one mole of the compound from its elements*, with all substances in their standard states. If an element exists in more than one form under standard conditions, the most stable form of the element is usually used for the formation reaction. For example, the standard enthalpy of formation for ethanol is the enthalpy change for the following reaction:

$$2 \text{ C(graphite)} + 3 \text{ H}_2(g) + \tfrac{1}{2} \text{ O}_2(g) \longrightarrow \text{C}_2\text{H}_5\text{OH(l)} \qquad \Delta_f H° = -277.7 \text{ kJ} \qquad [14.21]$$

The elemental source of oxygen is O_2 not O or O_3 because O_2 is the stable form of oxygen at 298 K and standard pressure. Similarly, the elemental source of carbon is graphite and not diamond, because graphite is more stable (lower energy) at 298 K and standard pressure (see Practice Exercise 14.8). Likewise, the most stable form of hydrogen under standard conditions is $H_2(g)$, so this is used as the source of hydrogen in Equation 14.21.

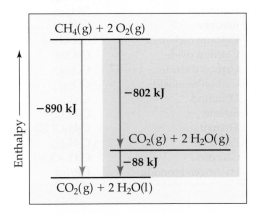

▲ **FIGURE IT OUT**

What process corresponds to the −88 kJ enthalpy change?

▲ **FIGURE 14.17 Enthalpy diagram for combustion of 1 mol of methane.** The enthalpy change of the one-step reaction equals the sum of the enthalpy changes of the reaction run in two steps: −890 kJ = −802 kJ + (−88 kJ).

TABLE 14.3 • Standard enthalpies of formation, $\Delta_f H°$ at 298 K

Substance	Formula	$\Delta_f H°$ (kJ mol^{-1})	Substance	Formula	$\Delta_f H°$ (kJ mol^{-1})
Acetylene	$C_2H_2(g)$	226.8	Hydrogen chloride	$HCl(g)$	−92.30
Ammonia	$NH_3(g)$	−46.19	Hydrogen fluoride	$HF(g)$	−268.61
Benzene	$C_6H_6(l)$	49.0	Hydrogen iodide	$HI(g)$	25.9
Calcium carbonate	$CaCO_3(s)$	−1207.1	Methane	$CH_4(g)$	−74.80
Calcium oxide	$CaO(s)$	−635.5	Methanol	$CH_3OH(l)$	−238.6
Carbon dioxide	$CO_2(g)$	−393.5	Propane	$C_3H_8(g)$	−103.85
Carbon monoxide	$CO(g)$	−110.5	Silver chloride	$AgCl(s)$	−127.0
Diamond	$C(s)$	1.88	Sodium bicarbonate	$NaHCO_3(s)$	−947.7
Ethane	$C_2H_6(g)$	−84.68	Sodium carbonate	$Na_2CO_3(s)$	−1130.9
Ethanol	$C_2H_5OH(l)$	−277.7	Sodium chloride	$NaCl(s)$	−410.9
Ethylene	$C_2H_4(g)$	52.30	Sucrose	$C_{12}H_{22}O_{11}(s)$	−2221
Glucose	$C_6H_{12}O_6(s)$	−1273	Water	$H_2O(l)$	−285.8
Hydrogen bromide	$HBr(g)$	−36.23	Water vapour	$H_2O(g)$	−241.8

The stoichiometry of formation reactions always indicates that one mole of the desired substance is produced, as in Equation 14.21. As a result, enthalpies of formation are reported in kJ mol^{-1} of the substance. Several standard enthalpies of formation are given in ▲ TABLE 14.3. A more complete table is provided in Appendix C. By definition, *the standard enthalpy of formation of the most stable form of any element is zero* because no formation reaction is needed when the element is already in its standard state. Thus the values of $\Delta_f H°$ for C(graphite), $H_2(g)$, $O_2(g)$ and the standard states of other elements are zero by definition.

 CONCEPT CHECK 9

In Table 14.3 the standard enthalpy of formation of $C_2H_2(g)$ is listed as 226.8 kJ mol^{-1}. Write the thermochemical equation associated with $\Delta_f H°$ for this substance.

Using Enthalpies of Formation to Calculate Enthalpies of Reaction

Tabulations of $\Delta_f H°$, such as those in Table 14.3 and Appendix C, have many important uses. As we see in this section, we can use Hess's law to calculate the standard enthalpy change for any reaction for which we know the $\Delta_f H°$ values for all reactants and products. For example, consider the combustion of propane gas, $C_3H_8(g)$, with oxygen to form $CO_2(g)$ and $H_2O(l)$ under standard conditions:

$$C_3H_8(g) + 5\,O_2(g) \longrightarrow 3\,CO_2(g) + 4\,H_2O(l)$$

We can write this equation as the sum of three formation reactions:

$$C_3H_8(g) \longrightarrow 3\,C(s) + 4\,H_2(g) \qquad \Delta H_1 = -\Delta_f H°[C_3H_8(g)] \quad [14.22]$$

$$3\,C(s) + 3\,O_2(g) \longrightarrow 3\,CO_2(g) \qquad \Delta H_2 = 3\Delta_f H°[CO_2(g)] \quad [14.23]$$

$$4\,H_2(g) + 2\,O_2(g) \longrightarrow 4\,H_2O(l) \qquad \Delta H_3 = 4\Delta_f H°[H_2O(l)] \quad [14.24]$$

$$C_3H_8(g) + 5\,O_2(g) \longrightarrow 3\,CO_2(g) + 4\,H_2O(l) \quad \Delta_{rxn} H° = \Delta H_1 + \Delta H_2 + \Delta H_3 \;[14.25]$$

Note that ΔH_1 is the reverse of $\Delta_f H°$ for the formation of $C_3H_8(g)$, ΔH_2 is three times $\Delta_f H°$ for the formation of $CO_2(g)$ and ΔH_3 is four times $\Delta_f H°$ for the formation of $H_2O(l)$, since the heat of formation is defined *per mole* of product.

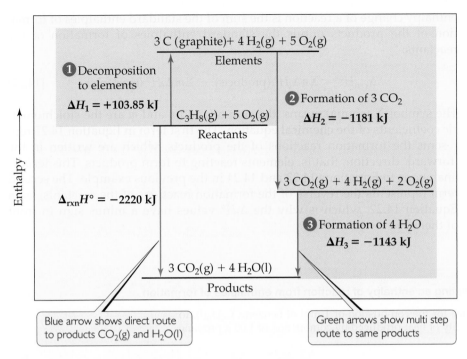

▲ **FIGURE 14.18** **Enthalpy diagram for propane combustion.**

From Hess's law we can write the standard enthalpy change for the overall reaction, Equation 14.25, as the sum of the enthalpy changes for the processes in Equations 14.22 to 14.24. We can then use values from Table 14.3 to calculate a numerical value for $\Delta H°$ for the overall reaction:

$$\Delta_{rxn}H° = \Delta H_1 + \Delta H_2 + \Delta H_3$$
$$= -\Delta_f H°[C_3H_8(g)] + 3\Delta_f H°[CO_2(g)] + 4\Delta_f H°[H_2O(l)]$$
$$= -(-103.85 \text{ kJ}) + 3(-393.5 \text{ kJ}) + 4(-285.8 \text{ kJ}) = -2220 \text{ kJ} \qquad [14.26]$$

An enthalpy diagram (▲ **FIGURE 14.18**) for Equation 14.25 shows the propane reaction broken down to the three reactions. Several aspects of this calculation depend on the guidelines we discussed in Section 14.4.

1. **Decomposition.** Equation 14.22 is the reverse of the formation reaction for $C_3H_8(g)$ so the enthalpy change for this reaction is $-\Delta_f H°[C_3H_8(g)]$.

2. **Formation of CO_2.** Equation 14.23 is the formation reaction for 3 mol of $CO_2(g)$. Because enthalpy is an extensive property, the enthalpy change for this step is $3\Delta_f H°[CO_2(g)]$.

3. **Formation of H_2O.** The enthalpy change for Equation 14.24, formation of 4 mol of $H_2O(l)$, is $4\Delta_f H°[H_2O(l)]$. The reaction specifies that $H_2O(l)$ is produced, so be careful to use the value of $\Delta_f H°$ for $H_2O(l)$ not $H_2O(g)$.

Note that in this analysis we assume that the stoichiometric coefficients in the balanced equation represent moles. For Equation 14.25, therefore, $\Delta_{rxn}H° = -2220$ kJ represents the enthalpy change for the reaction of 1 mol C_3H_8 and 5 mol O_2 to form 3 mol CO_2 and 4 mol H_2O. The product of the number of moles and the enthalpy change in kJ mol^{-1} has the units kJ: (number of moles) $\times \Delta_f H°$ (in kJ mol^{-1}) = kJ. We therefore report $\Delta_{rxn}H°$ in kJ.

We can break down any reaction into formation reactions as we have done here. When we do, we obtain the general result that the standard enthalpy change of a reaction is the sum of the standard enthalpies of formation of the products minus the standard enthalpies of formation of the reactants:

$$\Delta_{rxn}H° = \Sigma n\Delta_f H° \text{ (products)} - \Sigma m\Delta_f H° \text{ (reactants)} \qquad [14.27]$$

The symbol Σ (sigma) means 'the sum of', and n and m are the stoichiometric coefficients of the chemical equation. The first term in Equation 14.27 represents the formation reactions of the products, which are written in the 'forward' direction, that is, elements reacting to form products. This term is analogous to Equations 14.23 and 14.24 in the previous example. The second term represents the reverse of the formation reactions of the reactants, as in Equation 14.22, which is why the $\Delta_f H°$ values have a minus sign in front of them.

SAMPLE EXERCISE 14.10 **Calculating an enthalpy of reaction from enthalpies of formation**

(a) Calculate the standard enthalpy change for the combustion of 1 mol of benzene, $C_6H_6(l)$, to $CO_2(g)$ and $H_2O(l)$. **(b)** Compare the quantity of heat produced by combustion of 1.00 g propane to that produced by 1.00 g benzene.

SOLUTION

Analyse (a) We are given a combustion reaction of $C_6H_6(l)$ to form $CO_2(g)$ and $H_2O(l)$ and asked to calculate its standard enthalpy change, $\Delta H°$ (b) We then need to compare the quantity of heat produced by combustion of 1.00 g C_6H_6 with that produced by 1.00 g C_3H_8, whose combustion was treated previously in the text. (See Equations 14.25 and 14.26.)

Plan (a) We need to write the balanced equation for the combustion of C_6H_6. We then look up $\Delta_f H°$ values in Appendix C or in Table 14.3 and apply Equation 14.27 to calculate the enthalpy change for the reaction. (b) We use the molar mass of C_6H_6 to change the enthalpy change per mole to that per gram. We similarly use the molar mass of C_3H_8 and the enthalpy change per mole calculated in the text previously to calculate the enthalpy change per gram of that substance.

Solve

(a) We know that a combustion reaction involves $O_2(g)$ as a reactant. Thus the balanced equation for the combustion reaction of 1 mol $C_6H_6(l)$ is

We can calculate $\Delta_{rxn}H°$ for this reaction by using Equation 14.27 and data in Table 14.3. Remember to multiply the $\Delta_f H°$ value for each substance in the reaction by that substance's stoichiometric coefficient. Recall also that $\Delta_f H° = 0$ for any element in its most stable form under standard conditions, so $\Delta_f H°[O_2(g)] = 0$:

$$C_6H_6(l) + \tfrac{15}{2} O_2(g) \longrightarrow 6\,CO_2(g) + 3\,H_2O(l)$$

$$\Delta_{rxn}H° = [6\Delta_f H°(CO_2) + 3\Delta_f H°(H_2O)] - [\Delta_f H°(C_6H_6) + \tfrac{15}{2}\Delta_f H°(O_2)]$$

$$= [6(-393.5\text{ kJ}) + 3(-285.8\text{ kJ}) - (49.0\text{ kJ}) + \tfrac{15}{2}(0\text{ kJ})]$$

$$= (-2361 - 857.4 - 49.0)\text{ kJ}$$

$$= -3267\text{ kJ}$$

(b) From the example worked in the text, $\Delta H° = -2220$ kJ for the combustion of 1 mol of propane. In part (a) of this exercise we determined that $\Delta H° = -3267$ kJ for the combustion of 1 mol benzene. To determine the heat of combustion per gram of each substance, we use the molar masses to convert moles to grams:

$C_3H_8(g)$: $(-2220\text{ kJ mol}^{-1})(1\text{ mol}/44.1\text{ g}) = -50.3\text{ kJ g}^{-1}$

$C_6H_6(l)$: $(-3267\text{ kJ mol}^{-1})(1\text{ mol}/78.1\text{ g}) = -41.8\text{ kJ g}^{-1}$

Comment Both propane and benzene are hydrocarbons. As a rule, the energy obtained from the combustion of a gram of hydrocarbon is between 40 kJ and 50 kJ.

PRACTICE EXERCISE

Using the standard enthalpies of formation listed in Table 14.3, calculate the enthalpy change for the combustion of 1 mol of ethanol:

$$C_2H_5OH(l) + 3\,O_2(g) \longrightarrow 2\,CO_2(g) + 3\,H_2O(l)$$

Answer: −1367 kJ

(See also Exercises 14.47, 14.48.)

| **SAMPLE EXERCISE 14.11** | **Calculating an enthalpy of formation using an enthalpy of reaction** |

The standard enthalpy change for the reaction $CaCO_3(s) \longrightarrow CaO(s) + CO_2(g)$ is 178.1 kJ. From the values for the standard enthalpies of formation of CaO(s) and $CO_2(g)$ given in Table 14.3, calculate the standard enthalpy of formation of $CaCO_3(s)$.

SOLUTION

Analyse Our goal is to obtain $\Delta_f H°$ [$CaCO_3(s)$].

Plan We begin by writing the expression for the standard enthalpy change for the reaction:

$$\Delta_{rxn}H° = [\Delta_f H°(CaO) + \Delta_f H°(CO_2)] - \Delta_f H°(CaCO_3)$$

Solve Inserting the known values from Table 14.3 or Appendix C, we have

$$178.1 \text{ kJ} = -635.5 \text{ kJ} - 393.5 \text{ kJ} - \Delta_f H°(CaCO_3)$$

Solving for $\Delta_f H°(CaCO_3)$ gives

$$\Delta_f H°(CaCO_3) = -1207.1 \text{ kJ mol}^{-1}$$

Check We expect the enthalpy of formation of a stable solid such as calcium carbonate to be negative, as obtained.

PRACTICE EXERCISE

Given the following standard enthalpy change, use the standard enthalpies of formation in Table 14.3 to calculate the standard enthalpy of formation of CuO(s):

$$CuO(s) + H_2(g) \longrightarrow Cu(s) + H_2O(l) \qquad \Delta_{rxn}H° = -129.7 \text{ kJ}$$

Answer: $-156.1 \text{ kJ mol}^{-1}$
(See also Exercise 14.51.)

14.8 | SPONTANEOUS PROCESSES

We have seen from the first law of thermodynamics that energy cannot be created or destroyed (Section 14.2) in any process, but can be converted from one form to another or transferred between a system and its surroundings. The first law is expressed mathematically as $\Delta U = q + w$ (Equation 14.6) and helps us to balance the books, so to speak, on the heat transferred between a system and its surroundings and the work done by a particular process or reaction. However, the first law does not tell us the *extent* to which a reaction occurs. For example, sodium metal and chlorine gas combine readily to form sodium chloride. But we will never find sodium chloride decomposing of its own accord into sodium and chlorine. A process that occurs of its own accord, without any ongoing outside intervention, is a **spontaneous process**. A spontaneous process occurs in a definite direction. A dropped brick falls to the ground and bricks do not rise from the ground into a waiting hand. A brick falling is a spontaneous process whereas the reverse is *non-spontaneous*.

A gas will expand into a vacuum, as shown in ▶ **FIGURE 14.19**, but the process will never reverse itself. The expansion of the gas is spontaneous. Likewise, a nail left out in the weather will rust (▼ **FIGURE 14.20**). In this process the iron in the nail reacts with oxygen from the air to form an iron oxide. We would never expect the rusty nail to reverse this process and become shiny. The rusting process is spontaneous, whereas the reverse process is non-spontaneous. There are countless other examples we could cite that illustrate the same idea: *processes that are spontaneous in one direction are non-spontaneous in the opposite direction.*

Experimental conditions, such as temperature and pressure, are often important in determining whether a process is spontaneous. Consider, for example, the melting of ice. When the temperature of the surroundings is above

FIGURE IT OUT

If flask B were smaller than flask A, would the final pressure after the stopcock is opened be greater than, equal to, or less than 0.5 bar?

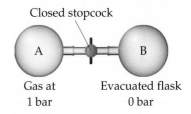

Closed stopcock

A   B

Gas at   Evacuated flask
1 bar   0 bar

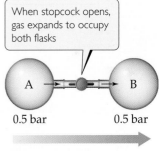

When stopcock opens, gas expands to occupy both flasks

A   B

0.5 bar   0.5 bar

This process is spontaneous

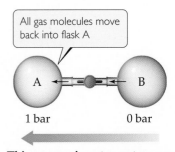

All gas molecules move back into flask A

A   B

1 bar   0 bar

This process is not spontaneous

▲ **FIGURE 14.19 Expansion of a gas into an evacuated space is a spontaneous process.** The reverse process—gas molecules initially distributed evenly in two flasks all moving into one flask—is not spontaneous.

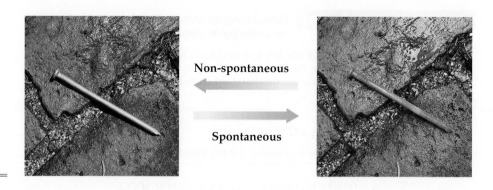

▶ FIGURE 14.20 **A spontaneous process.** Elemental iron in the shiny nail in the left-hand photograph spontaneously combines with H_2O and O_2 in the surrounding air to form a layer of rust (Fe_2O_3) on the nail surface.

Non-spontaneous

Spontaneous

0 °C at ordinary atmospheric pressures, ice melts spontaneously and the reverse process—liquid water turning into ice—is not spontaneous. However, when the surroundings are below 0 °C, the opposite is true. Liquid water converts into ice spontaneously, and the conversion of ice into water is *not* spontaneous (▼ FIGURE 14.21).

What happens at $T = 0$ °C, the normal melting point of water, when the flask in Figure 14.21 contains both water and ice? At the normal melting point of a substance, the solid and liquid phases are in equilibrium.

At this temperature the two phases are interconverting at the same rate and there is no preferred direction for the process.

It is important to realise that the fact that a process is spontaneous does not necessarily mean that it will occur at an observable rate. A chemical reaction is spontaneous if it occurs on its own accord, regardless of its speed. A spontaneous reaction can be very fast, as in the case of acid–base neutralisation, or very slow, as in the rusting of iron. Thermodynamics can tell us the *direction* and *extent* of a reaction but tells us nothing about the *speed* of the reaction. The study of the speed of reaction is chemical kinetics (∞ Chapter 15, 'Chemical Kinetics').

∞ Find out more on page 564

△ **FIGURE IT OUT**

In which direction is this process exothermic?

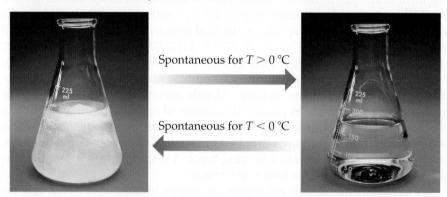

Spontaneous for $T > 0$ °C

Spontaneous for $T < 0$ °C

▲ FIGURE 14.21 **Spontaneity can depend on temperature.** At $T > 0$ °C, ice melts spontaneously to liquid water. At $T < 0$ °C, the reverse process, water freezing to ice, is spontaneous. At $T = 0$ °C the two states are in equilibrium.

SAMPLE EXERCISE 14.12 **Identifying spontaneous processes**

Predict whether the following processes are spontaneous as described, spontaneous in the reverse direction or in equilibrium. **(a)** When a piece of metal heated to 150 °C is added to water at 40 °C, the water gets hotter. **(b)** Water at room temperature decomposes into $H_2(g)$ and $O_2(g)$. **(c)** Benzene vapour $C_6H_6(g)$, at a pressure of 1 bar, condenses to liquid benzene at the normal boiling point of benzene, 80.1 °C.

SOLUTION

Analyse We are asked to judge whether each process is spontaneous in the direction indicated, in the reverse direction or in neither direction.

Plan We need to think about whether each process is consistent with our experience about the natural direction of events or whether we expect the reverse process to occur.

Solve

(a) This process is spontaneous. Whenever two objects at different temperatures are brought into contact, heat is transferred from the hotter object to the colder one. In this instance, heat is transferred from the hot metal to the cooler water. The final temperature, after the metal and water achieve the same temperature (thermal equilibrium), will be somewhere between the initial temperatures of the metal and the water.

(b) Experience tells us that this process is not spontaneous—we certainly have never seen hydrogen and oxygen gas bubbling up out of water! Rather, the *reverse* process (the reaction of H_2 and O_2 to form H_2O) is spontaneous once initiated by a spark or flame.

(c) By definition, the normal boiling point is the temperature at which a vapour at 1 bar is in equilibrium with its liquid. Thus this is an equilibrium situation. Neither the condensation of benzene vapour nor the reverse process is spontaneous. If the temperature were below 80.1 °C, condensation would be spontaneous.

PRACTICE EXERCISE

Under 1 bar pressure $CO_2(s)$ sublimes at −78 °C. Is the transformation of $CO_2(s)$ to $CO_2(g)$ a spontaneous process at −100 °C and 1 bar pressure?

Answer: No, the reverse process is spontaneous at this temperature.

(See also Exercises 14.54, 14.55.)

 CONCEPT CHECK 10

If a process is non-spontaneous, does that mean it cannot occur under any circumstances?

Seeking a Criterion for Spontaneity

A marble rolling down an incline or a brick falling from your hand loses potential energy. The loss of some form of energy is a common feature of spontaneous change in mechanical systems. During the 1870s Marcellin Bertholet (1827–1907), a famous chemist of that era, suggested that the direction of spontaneous changes in chemical systems is determined by the loss of energy. He proposed that all spontaneous chemical and physical changes are exothermic. It takes only a few moments, however, to find exceptions to this generalisation. For example, the melting of ice at room temperature is spontaneous and endothermic. Similarly, many spontaneous dissolution processes, such as the dissolving of NH_4NO_3, are endothermic. We conclude that although the majority of spontaneous reactions are exothermic, there are spontaneous endothermic ones as well. Clearly, some other factor must be at work in determining the natural direction of processes.

To understand why certain processes are spontaneous, we need to consider more closely the ways in which the state of a system can change. Recall from Section 14.2 that quantities such as temperature, internal energy and enthalpy are *state functions*, properties that define a state and do not depend on how we reach that state. The heat transferred between a system and its surroundings, q, and the work done by or on the system, w, are *not* state functions—their values depend on the specific path taken between states. One key to understanding spontaneity is understanding differences in the paths between states.

Reversible and Irreversible Processes

In 1824 a 28-year-old French engineer by the name of Sadi Carnot (1796–1832) published an analysis of the factors that determine how efficiently a steam engine can convert heat to work. Carnot considered what an *ideal engine*, one with the highest possible efficiency, would be like. He observed that it is impossible to convert the energy content of a fuel completely to work because a significant amount of heat is always lost to the surroundings.

An ideal engine, one with the maximum efficiency, operates under an ideal set of conditions in which all the processes are reversible. In a **reversible process**, a system is changed in such a way that the system and surroundings can be restored to their original state by *exactly* reversing the change. In other words, we can completely restore the system to its original condition with no net change to either the system or its surroundings. An **irreversible process** is one that cannot simply be reversed to restore the system and its surroundings to their original states. What Carnot discovered is that the amount of work we can extract from any process depends on the manner in which the process is carried out. *A reversible change produces the maximum amount of work that can be achieved by the system on the surroundings* $(w_{rev} = w_{max})$.

>
> **CONCEPT CHECK 11**
> If you evaporate water and then condense it back into its original container, have you necessarily performed a reversible process?

Let's examine some examples of reversible and irreversible processes. If two objects are at different temperatures, heat will flow spontaneously from the hotter object to the colder one. Because it is impossible to make heat flow in the opposite direction, the flow of heat is irreversible. Given these facts, can we imagine any conditions under which heat transfer can be made reversible? To answer this question, we must consider temperature differences that are infinitesimally small, as opposed to the discrete temperature differences with which we are most familiar. For example, consider a system and its surroundings at essentially the same temperature, with just an infinitesimal temperature difference δT between them (▼ **FIGURE 14.22**). If the surroundings are at temperature T and the system is at the infinitesimally higher temperature $T + \delta T$, then an infinitesimal amount of heat flows from system to surroundings. We can reverse the direction of heat flow by making an infinitesimal change of temperature in the opposite direction, lowering the system temperature to $T - \delta T$. Now the direction of heat flow is from surroundings to system.

Reversible processes are those that reverse direction whenever an infinitesimal change is made in some property of the system. For a process to be truly reversible, the amounts of heat must be infinitesimally small and the transfer of heat must occur infinitely slowly; thus no process that we can observe is truly reversible. The notion of infinitesimal amounts are related to the infinitesimals that you may have studied in a calculus course.

FIGURE IT OUT

If the flow of heat in to or out of the system is to be reversible, what must be true of δT?

▶ **FIGURE 14.22 Reversible flow of heat.** Heat can flow reversibly between a system and its surroundings only if the two have an infinitesimally small difference in temperature, δT. (a) Increasing the temperature of the system by δT causes heat to flow from the hotter system to the colder surroundings. (b) Decreasing the temperature of the system by δT causes heat to flow from the hotter surroundings to the colder system.

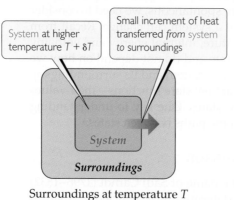

System at higher temperature $T + \delta T$

Small increment of heat transferred *from* system *to* surroundings

System

Surroundings

Surroundings at temperature T

(a)

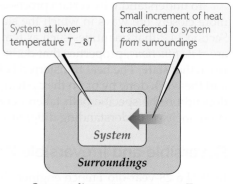

System at lower temperature $T - \delta T$

Small increment of heat transferred *to* system *from* surroundings

System

Surroundings

Surroundings at temperature T

(b)

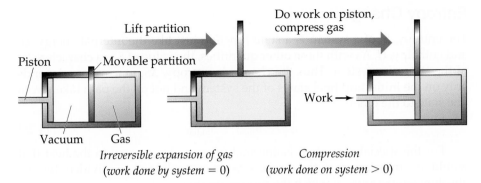

◀ **FIGURE 14.23 An irreversible process.** Initially an ideal gas is confined to the right half of a cylinder. When the partition is removed, the gas spontaneously expands to fill the whole cylinder. No work is done by the system during this expansion. Using the piston to compress the gas back to its original state requires the surroundings to do work on the system.

Now consider another example, the expansion of an ideal gas at constant temperature, referred to as an **isothermal process**. To keep the example simple, consider the gas in the cylinder and piston arrangement shown in ▲ FIGURE 14.23; when the partition is removed, the gas expands spontaneously to fill the evacuated space. Because the gas is expanding into a vacuum with no external pressure, it does no P-V work on the surroundings. So, for the expansion, $w = 0$ (∞ Section 14.3, 'Enthalpy').

We can use the piston to compress the gas back to its original state, but doing so requires that the surroundings do work on the system, meaning that $w > 0$ for the compression. That is, reversing the process has produced a change in the surroundings as energy is used to do work on the system. The fact that the same path can't be followed to restore the system to its original state indicates that the system is irreversible.

What might a reversible, isothermal expansion of an ideal gas be like? It will occur only if the external pressure acting on the piston exactly balances the pressure exerted by the gas. Under these conditions, the piston will not move unless the external pressure is reduced infinitely slowly, allowing the pressure of the confined gas to readjust to maintain a balance in the two pressures. This gradual, infinitely slow process in which the external pressure and internal pressure are always equal, is reversible. If we reverse the process and compress the gas in the same infinitely slow manner, we can return the gas to its original volume. The complete cycle of expansion and compression in this hypothetical process, moreover, is accomplished without any net change to the surroundings.

Because real processes can, at best, only approximate the slow, ever-in-equilibrium change associated with reversible processes, *all real processes are irreversible*. Further, the reverse of any spontaneous process is a non-spontaneous process. A non-spontaneous process can occur only if the surroundings do work on the system. Thus *any spontaneous process is irreversible*; even if we return the system to the original condition, the surroundings will have changed.

∞ Review this on pages 510–12

14.9 | ENTROPY AND THE SECOND LAW OF THERMODYNAMICS

To understand spontaneity, we must examine the thermodynamic quantity called **entropy**. Entropy has been variously associated with the extent of *randomness* in a system or with the extent to which energy is distributed between the various motions of the molecules of the system. In fact, entropy is a multifaceted concept whose interpretations are not so quickly summarised by a simple definition. In this section we consider how we can relate entropy changes to heat transfer and temperature. Our analysis will bring us to a profound statement about spontaneity that we call the second law of thermodynamics.

Entropy Change

The entropy, S, of a system is a state function just like the internal energy, U, and enthalpy, H. As with these other quantities, the value of S is a characteristic of the state of a system. Thus the change in entropy, ΔS, in a system depends only on the initial and final states of the system and not on the path taken from one state to the other:

$$\Delta S = S_{\text{final}} - S_{\text{initial}} \qquad [14.28]$$

For the special case of an isothermal process, ΔS is equal to the heat that would be transferred if the process were reversible, q_{rev}, divided by the absolute temperature at which the process occurs:

$$\Delta S = \frac{q_{\text{rev}}}{T} \text{ (constant } T) \qquad [14.29]$$

Although there are many possible paths that can take the system from one state to another, only one path is associated with a reversible process. Thus the value of q_{rev} is uniquely defined for any two states of the system. Because S is a state function, we can use Equation 14.29 to calculate ΔS for *any* isothermal process, not just those that are reversible. If a change between two states is irreversible, we calculate ΔS by using a reversible path between the states.

 CONCEPT CHECK 12

How do we reconcile the fact that S is a state function but that ΔS depends on q, which is not a state function?

ΔS for Phase Changes

The melting of a substance at its melting point and the vaporisation of a substance at its boiling point are isothermal processes. Consider the melting of ice. At 1 bar pressure, ice and liquid water are in equilibrium with each other at 0 °C. Imagine that we melt one mole of ice at 0 °C and 1 bar to form one mole of liquid water at 0 °C, 1 bar. We can achieve this change by adding a certain amount of heat to the system from the surroundings: $q = \Delta_{\text{fusion}}H$. Now imagine that we carry out the change by adding the heat infinitely slowly, raising the temperature of the surroundings only infinitesimally above 0 °C. When we do it this way, the process is reversible because we can reverse the process simply by infinitely slowly removing the same amount of heat, $\Delta_{\text{fusion}}H$, from the system, using immediate surroundings that are infinitesimally below 0 °C. Thus for the melting of ice at $T = 0$ °C $= 273$ K, $q_{\text{rev}} = \Delta_{\text{fusion}}H$.

The enthalpy of fusion for H_2O is $q = \Delta_{\text{fusion}}H = 6.01$ kJ mol^{-1}. (The melting is an endothermic process and so the sign of ΔH is positive.) Thus we can use Equation 14.29 to calculate $\Delta_{\text{fusion}}S$ for melting one mole of ice at 273 K.

Notice that the units for ΔS, J K^{-1}, are energy divided by absolute temperature, as we expect from Equation 14.29.

$$\Delta_{\text{fusion}}S = \frac{q_{\text{rev}}}{T} = \frac{\Delta_{\text{fusion}}H}{T} = \frac{(1\,\text{mol})(6.01 \times 10^3\,\text{J mol}^{-1})}{273\,\text{K}} = 22.0\,\text{J K}^{-1}$$

SAMPLE EXERCISE 14.13   Calculating ΔS for a phase change

The element mercury, Hg, is a silvery liquid at room temperature. The normal freezing point of mercury is −38.98 °C, and its molar enthalpy of fusion is $\Delta_{\text{fusion}}H = 2.29$ kJ mol^{-1}. What is the entropy change of the system when 50.0 g of Hg(l) freezes at the normal freezing point?

SOLUTION

Analyse We first recognise that freezing is an *exothermic* process, which means heat is transferred from system to surroundings and $q < 0$. The enthalpy of fusion refers to the process of melting. Because freezing is the reverse of melting, the enthalpy change that accompanies the freezing of 1 mol of Hg is $-\Delta_{fusion}H = -2.29$ kJ mol^{-1}.

Plan We can use $-\Delta_{fusion}H$ and the atomic weight of Hg to calculate q for freezing 50.0 g of Hg. Then we use this value of q as q_{rev} in Equation 14.29 to determine ΔS for the system.

Solve

$$n_{Hg} = \frac{50.0 \text{ g}}{200.59 \text{ g mol}^{-1}} = 0.24926 \text{ mol} = 0.249 \text{ mol}$$

$$q = 0.2496 \text{ mol} \times -2.29 \text{ kJ mol}^{-1} = -0.571 \text{ kJ} = -571 \text{ J}$$

We can use this value of q as q_{rev} in Equation 14.29. We must first, however, convert the temperature to K:

$$-38.9\ ^\circ C = (-38.9 + 273.15)\text{ K} = 234.3 \text{ K}$$

We can now calculate the value of $\Delta_{sys}S$:

$$\Delta_{sys}S = \frac{q_{rev}}{T} = \frac{-571 \text{ J}}{234.3 \text{ K}} = -2.44 \text{ J K}^{-1}$$

Check The entropy change is negative because heat flows from the system, making q_{rev} negative.

Comment The procedure used here can also be used to calculate ΔS for other isothermal phase changes, such as the vaporisation of a liquid at its boiling point.

PRACTICE EXERCISE

The normal boiling point of ethanol, C_2H_5OH, is 78.3 °C and its molar enthalpy of vaporisation is 38.56 kJ mol^{-1}. What is the change in entropy in the system when 68.3 g of $C_2H_5OH(g)$ at 1 bar condenses to liquid at the normal boiling point?

Answer: -163 J K^{-1}

A CLOSER LOOK

THE ENTROPY CHANGE WHEN A GAS EXPANDS ISOTHERMALLY

In general, the entropy of any system increases as the system becomes more random or more spread out. Thus we expect the spontaneous expansion of a gas to result in an increase in entropy. To see how this entropy increase can be calculated, consider the expansion of an ideal gas that is initially constrained by a piston, as in the rightmost part of Figure 14.23. Imagine that we allow the gas to undergo a reversible isothermal expansion by infinitesimally decreasing the external pressure on the piston. The work done on the surroundings by the reversible expansion of the system against the piston can be calculated with the aid of calculus (we do not show the derivation):

$$w_{rev} = -nRT \ln \frac{V_2}{V_1}$$

In this equation, n is the number of moles of gas, R is the gas constant (Section 10.4), T is the absolute temperature, V_1 is the initial volume and V_2 is the final volume. Notice that if $V_2 > V_1$, as it must be in our expansion, then $w_{rev} < 0$, meaning that the expanding gas does work on the surroundings.

One characteristic of an ideal gas is that its internal energy depends only on temperature, not on pressure. Thus when an ideal gas expands isothermally, $\Delta U = 0$. Because $\Delta U = q_{rev} + w_{rev} = 0$, we see that $q_{rev} = -w_{rev} = nRT \ln(V_2/V_1)$. Then, using Equation 14.29, we can calculate the entropy change in the system:

$$\Delta_{sys}S = \frac{q_{rev}}{T} = \frac{nRT \ln \dfrac{V_2}{V_1}}{T} = nR \ln \frac{V_2}{V_1}$$

From the ideal-gas equation, we can calculate the number of moles in 1.00 dm^3 of an ideal gas at 1.00 bar (100 kPa) and 0 °C by using the value 8.314 kPa dm^3 mol^{-1} K^{-1} for R:

$$n = \frac{PV}{RT} = \frac{(100 \text{ kPa})(1.00 \text{ dm}^3)}{(8.314 \text{ kPa dm}^3 \text{ mol}^{-1} \text{ K}^{-1})(273 \text{ K})} = 4.41 \times 10^{-2} \text{ mol}$$

Thus for the expansion of the gas from 1.00 dm^3 to 2.00 dm^3, we have

$$\Delta_{sys}S = (4.41 \times 10^{-2} \text{ mol})(8.314 \text{ J mol}^{-1} \text{ K}^{-1})\left(\ln \frac{2.00}{1.00} \right)$$
$$= 0.254 \text{ J K}^{-1}$$

This increase in entropy is a measure of the increased randomness of the molecules because of the expansion, as will be seen in Section 14.11.

RELATED EXERCISE: 14.67

⚬⚬ Review this on page 504

The Second Law of Thermodynamics

The key idea of the first law of thermodynamics is that energy is conserved in any process. Thus the quantity of energy lost by a system equals the quantity gained by its surroundings, (⚬⚬ Section 14.1, 'The Nature of Energy').

Entropy, however, is not conserved, because it actually increases in any spontaneous process. Thus the sum of the entropy change of the system and surroundings for any spontaneous process is always greater than zero.

Let's illustrate this generalisation by calculating the entropy change of a system and the entropy change of its surroundings when our system is 1 mol of ice (a piece roughly the size of an ice cube) melting in the palm of your hand, which is part of the surroundings. The process is not reversible because the system and surroundings are at different temperatures. Nevertheless, because ΔS is a state function, its value is the same regardless of whether the process is reversible or irreversible. We have calculated the entropy change of the system (above):

$$(\Delta_{fusion}S) = \Delta_{sys}S = \frac{q_{rev}}{T} = \frac{(1 \text{ mol})(6.01 \times 10^3 \text{ J mol}^{-1})}{273 \text{ K}} = 22.0 \text{ J K}^{-1}$$

The surroundings immediately in contact with the ice are your hand, which we assume is at body temperature, $37\,°C = 310$ K. The quantity of heat lost by your hand is -6.01×10^3 J mol^{-1}, which is equal in magnitude to the quantity of heat gained by the ice but has the opposite sign. Hence, the entropy change of the surroundings is

$$\Delta_{surr}S = \frac{q_{rev}}{T} = \frac{(1 \text{ mol})(-6.01 \times 10^3 \text{ J mol}^{-1})}{310 \text{ K}} = -19.4 \text{ J K}^{-1}$$

Thus the total entropy change is positive:

$$\Delta_{total}S = \Delta_{sys}S + \Delta_{surr}S = (22.0 \text{ J K}^{-1}) + (-19.4 \text{ J K}^{-1}) = 2.6 \text{ J K}^{-1}$$

If the temperature of the surroundings were not 310 K but rather some temperature infinitesimally above 273 K, the melting would be reversible instead of irreversible. In that case, the entropy change of the surroundings would equal -22.0 J K^{-1} and $\Delta_{total}S$ would be zero.

In general, any irreversible process results in an overall increase in entropy, whereas a reversible process results in no overall change in entropy. This general statement is known as the **second law of thermodynamics**. The sum of the entropy of a system plus the entropy of the surroundings is everything there is, and so we refer to the total entropy change as the entropy change of the universe, $\Delta_{univ}S$. We can therefore state the second law of thermodynamics in terms of the following equations:

Reversible process: $\Delta_{univ}S = \Delta_{sys}S + \Delta_{surr}S = 0$

Irreversible process: $\Delta_{univ}S = \Delta_{sys}S + \Delta_{surr}S > 0$ [14.30]

All real processes that occur of their own accord are irreversible (with reversible processes being a useful idealisation). These processes are also spontaneous. Thus *the total entropy of the universe increases in any spontaneous process.* This profound generalisation is yet another way of expressing the second law of thermodynamics.

> ⚠ **CONCEPT CHECK 13**
>
> The rusting of iron is spontaneous and is accompanied by a decrease in the entropy of the system (the iron and oxygen). What can we conclude about the entropy change of the surroundings?

The second law of thermodynamics tells us the essential character of any spontaneous change—it is always accompanied by an overall increase in entropy.

We can, in fact, use this criterion to predict whether processes will be spontaneous. Before beginning to use the second law to predict spontaneity, however, it will be useful to explore further the meaning of entropy from a molecular perspective.

Throughout most of the remainder of this chapter, we focus mainly on the systems we encounter rather than on their surroundings. To simplify the notation, we usually refer to the entropy change of the system merely as ΔS rather than explicitly indicating $\Delta_{sys}S$.

14.10 | MOLECULAR INTERPRETATION OF ENTROPY

As chemists, we are interested in molecules. What does entropy have to do with them and their transformations? What molecular property does entropy reflect? It was Ludwig Boltzmann (1844–1906) who gave conceptual meaning to entropy. To understand Boltzmann's contribution, we need to examine the ways in which molecules can store energy.

Expansion of a Gas at the Molecular Level

In discussing Figure 14.19, we talked about the expansion of a gas into a vacuum as a spontaneous process. We now understand that it is an irreversible process and that the entropy of the universe increases during the expansion. How can we explain the spontaneity of this process at the molecular level? We can get a sense of what makes this expansion spontaneous by envisioning the gas as a collection of particles in constant motion, as we did in discussing the kinetic-molecular theory of gases (∞ Section 10.7, 'Kinetic-Molecular Theory'). When the stopcock in Figure 14.19 is opened, we can view the expansion of the gas as the ultimate result of the gas molecules moving randomly throughout the larger volume.

∞ Review this on page 355

Let's look at this idea more closely by tracking two of the gas molecules as they move around. Before the stopcock is opened, both molecules are confined to the left flask, as shown in ▼ FIGURE 14.24(a). After the stopcock is opened, the molecules travel randomly throughout the entire apparatus. As Figure 14.24(b) shows, there are four possible arrangements for the two molecules once both flasks are available to them. Because the molecular motion is random, all four arrangements are equally likely. Note that now only one arrangement corresponds to the situation before the stopcock was opened: both molecules in the left flask.

Figure 14.24(b) shows that with both flasks available to the molecules, the probability of the red molecule being in the left flask is two in four (top right and bottom left arrangements), and the probability of the blue molecule being

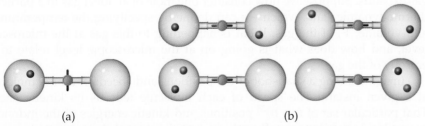

(a)	(b)
The two molecules are coloured red and blue to keep track of them.	Four possible arrangements (microstates) once the stopcock is opened.

▲ **FIGURE 14.24   Possible arrangements of two gas molecules in two flasks.** (a) Before the stopcock is opened, both molecules are in the left flask. (b) After the stopcock is opened, there are four possible arrangements of the two molecules.

in the left flask is the same (top left and bottom left arrangements). Because the probability is $\frac{2}{4} = \frac{1}{2}$ that each molecule is in the left flask, the probability that *both* are there is $\left(\frac{1}{2}\right)^2 = \frac{1}{4}$. If we apply the same analysis to *three* gas molecules, we find that the probability that all three are in the left flask at the same time is $\left(\frac{1}{2}\right)^3 = \frac{1}{8}$.

Now let's consider a *mole* of gas. The probability that all the molecules are in the left flask at the same time is $\left(\frac{1}{2}\right)^N$, where $N = 6.02 \times 10^{23}$. This is a vanishingly small number! Thus there is essentially zero likelihood that all the gas molecules will be in the left flask at the same time. This analysis of the microscopic behaviour of the gas molecules leads to the expected macroscopic behaviour; the gas spontaneously expands to fill both the left and right flasks, and it does not spontaneously all go back in the left flask.

This molecular view of gas expansion shows the tendency of the molecules to 'spread out' among the different arrangements they can take. Before the stopcock is opened, there is only one possible arrangement: all molecules in the left flask. When the stopcock is opened, the arrangement in which all the molecules are in the left flask is but one of an extremely large number of possible arrangements. The most probable arrangements by far are those in which there are essentially equal numbers of molecules in the two flasks. When the gas spreads throughout the apparatus, any given molecule can be in either flask rather than confined to the left flask. We say that with the stopcock opened, the arrangement of gas molecules is more random or disordered than when the molecules are all confined in the left flask.

We will see that this notion of increasing randomness helps us understand entropy at the molecular level.

Boltzmann's Equation and Microstates

The science of thermodynamics developed as a means of describing the properties of matter in our macroscopic world without regard to microscopic structure. In fact, thermodynamics was a well-developed field before the modern view of atomic and molecular structure was even known. The thermodynamic properties of water, for example, addressed the behaviour of bulk water (or ice or water vapour) as a substance without considering any specific properties of individual H_2O molecules.

To connect the microscopic and macroscopic descriptions of matter, scientists have developed the field of *statistical thermodynamics*, which uses the tools of statistics and probability to link the microscopic and macroscopic worlds. Here we show how entropy, which is a property of bulk matter, can be connected to the behaviour of atoms and molecules. Because the mathematics of statistical thermodynamics is complex, our discussion will be largely conceptual.

In our discussion of two gas molecules in the two-flask system in Figure 14.24 we saw that the number of possible arrangements helped explain why the gas expands. Suppose we now consider one mole of an ideal gas in a particular thermodynamic state, which we can define by specifying the temperature, T, and volume, V, of the gas. What is happening to this gas at the microscopic level, and how does what is going on at the microscopic level relate to the entropy of the gas?

Imagine taking a snapshot of the positions and speeds of all the molecules at a given instant. The speed of each molecule tells us its kinetic energy. That particular set of 6×10^{23} positions and kinetic energies of the individual gas molecules is what we call a *microstate* of the system. A **microstate** is a single possible arrangement of the positions and kinetic energies of the gas molecules when the gas is in a specific thermodynamic state. We could envisage continuing to take snapshots of our system to see other possible microstates.

As you no doubt see, there would be such a staggeringly large number of microstates that taking individual snapshots of all of them is not feasible.

Because we are examining such a large number of particles, however, we can use the tools of statistics and probability to determine the total number of microstates for the thermodynamic state. (That is where the *statistical* part of the name *statistical thermodynamics* comes in.) Each thermodynamic state has a characteristic number of microstates associated with it, and we will use the symbol W for that number.

Students sometimes have difficulty distinguishing between the state of a system and the microstates associated with the state. The difference is that *state* is used to describe the macroscopic view of our system as characterised, for example, by the pressure or temperature of a sample of gas. A *microstate* is a particular microscopic arrangement of the atoms or molecules of the system that corresponds to the given state of the system. Each of the snapshots we described is a microstate—the positions and kinetic energies of individual gas molecules will change from snapshot to snapshot, but each one is a possible arrangement of the collection of molecules corresponding to a single state. For macroscopically sized systems, such as a mole of gas, there is a very large number of microstates for each state—that is, W is generally an extremely large number.

The connection between the number of microstates of a system, W, and the entropy of the system, S, is expressed in a beautifully simple equation developed by Boltzmann and engraved on his tombstone (▶ **FIGURE 14.25**):

$$S = k \ln W \qquad [14.31]$$

In this equation, k is Boltzmann's constant, 1.38×10^{-23} J K^{-1}. Thus *entropy is a measure of how many microstates are associated with a particular macroscopic state.*

▲ **FIGURE 14.25   Ludwig Boltzmann's gravestone.** Boltzmann's gravestone in Vienna is inscribed with his famous relationship between the entropy of a state and the number of available microstates. (In Boltzmann's time, 'log' was used to represent the natural logarithm.)

 CONCEPT CHECK 14

What is the entropy of a system that has only a single microstate?

From Equation 14.31, we see that the entropy change accompanying any process is

$$\Delta S = k \ln W_{\text{final}} - k \ln W_{\text{initial}} = k \ln \frac{W_{\text{final}}}{W_{\text{initial}}} \qquad [14.32]$$

Any change in the system that leads to an increase in the number of microstates ($W_{\text{final}} > W_{\text{initial}}$) leads to a positive value of ΔS: *entropy increases with the number of microstates of the system.*

Let's consider two modifications to our ideal-gas sample and see how the entropy changes in each case. First, suppose we increase the volume of the system, which is analogous to allowing the gas to expand isothermally. A greater volume means a greater number of positions available to the gas atoms and therefore a greater number of microstates. The entropy therefore increases as the volume increases.

Second, suppose we keep the volume fixed but increase the temperature. How does this change affect the entropy of the system? Recall the distribution of molecular speeds presented in Figure 10.16. An increase in temperature increases the most probable speed of the molecules and also broadens the distribution of speeds. Hence the molecules have a greater number of possible kinetic energies, and the number of microstates increases. Thus the entropy of the system increases with increasing temperature.

Molecular Motions and Energy

When a substance is heated, the motion of its molecules increases. In Section 10.7, we found that the average kinetic energy of the molecules of an ideal gas is directly proportional to the absolute temperature of the gas. That means the higher the temperature, the faster the molecules move and the more kinetic energy they possess. Moreover, hotter systems have a *broader distribution* of molecular speeds, as Figure 10.16 shows.

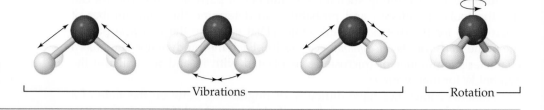

▶ **FIGURE 14.26** **Vibrational and rotational motions in a water molecule.**

The particles of an ideal gas are idealised points with no volume and no bonds, however; points that we visualise as flitting around through space. Any real molecule can undergo three kinds of more complex motion. The entire molecule can move in one direction, which is the simple motion we visualise for an ideal particle and see in a macroscopic object, such as a moving football. We call such movement **translational motion**. The molecules in a gas have more freedom of translational motion than those in a liquid, which have more freedom of translational motion than the molecules of a solid.

A real molecule can also undergo **vibrational motion**, in which the atoms in the molecule move periodically towards and away from one another, and **rotational motion**, in which the molecule spins about an axis. ▲ **FIGURE 14.26** shows the vibrational motions and one of the rotational motions possible for the water molecule. These different forms of motion are ways in which a molecule can store energy, and we refer to the various forms collectively as the *motional energy* of the molecule.

> ◣ **CONCEPT CHECK 15**
> What kinds of motion can a molecule undergo that a single atom cannot?

The vibrational and rotational motions possible in real molecules lead to arrangements that a single atom can't have. A collection of real molecules therefore has a greater number of possible microstates than does the same number of ideal-gas particles. In general, *the number of microstates possible for a system increases with an increase in volume, an increase in temperature or an increase in the number of molecules because any of these changes increases the possible positions and kinetic energies of the molecules making up the system.* We will also see that the number of microstates increases as the complexity of the molecule increases because there are more vibrational motions available.

Chemists have several ways of describing an increase in the number of microstates possible for a system and therefore an increase in the entropy for the system. Each way seeks to capture a sense of the increased freedom of motion that causes molecules to spread out when not restrained by physical barriers or chemical bonds.

The most common way for describing an increase in entropy is as an increase in the *randomness* or *disorder*, of the system. Another way likens an entropy increase to an increased *dispersion (spreading out) of energy* because there is an increase in the number of ways the positions and energies of the molecules can be distributed throughout the system. Each description (randomness or energy dispersal) is conceptually helpful if applied correctly.

Making Qualitative Predictions about ΔS

It is usually not difficult to construct a mental picture to estimate qualitatively how the entropy of a system changes during a simple process. In most instances, an increase in the number of microstates and, hence, an increase in entropy parallels an increase in temperature, volume and the number of independently moving particles.

◢ **FIGURE IT OUT**

In which phase are water molecules least able to have rotational motion?

Increasing entropy

| Ice | Liquid water | Water vapour |

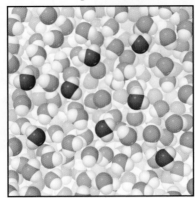
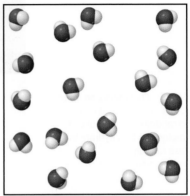

Rigid, crystalline structure

Motion restricted to **vibration** only

Smallest number of microstates

Increased freedom with respect to **translation**

Free to **vibrate** and **rotate**

Larger number of microstates

Molecules spread out, essentially independent of one another

Complete freedom for **translation**, **vibration** and **rotation**

Largest number of microstates

▲ **FIGURE 14.27   Entropy and the phases of water.** The larger the number of possible microstates, the higher the entropy of the system.

We can usually make qualitative predictions about entropy changes by focusing on these factors. For example, when water vaporises, the molecules spread out into a larger volume. Because they occupy a larger space, there is an increase in their freedom of motion, giving rise to more accessible microstates and hence to an increase in entropy.

Now consider the phases of water. In ice, hydrogen bonding leads to the rigid structure shown in ▲ **FIGURE 14.27**. Each molecule in the ice is free to vibrate, but its translational and rotational motions are much more restricted than in liquid water. Although there are hydrogen bonds in liquid water, the molecules can more readily move about relative to one another (translation) and tumble around (rotation). During melting, therefore, the number of possible microstates increases and so does the entropy. In water vapour, the molecules are essentially independent of one another and have their full range of translational, vibrational and rotational motions. Thus water vapour has an even greater number of possible microstates and therefore a higher entropy than liquid water or ice.

When an ionic solid, such as KCl, dissolves in water, the pure solid and pure water are replaced by a mixture of water and ions (▼ **FIGURE 14.28**). The ions in the liquid now move in a larger volume and possess more motional energy than in the rigid solid. This increased motion might lead us to conclude that the entropy of the system had increased. We have to be careful, however, because some of the water molecules have lost some freedom of motion because they are now held around the ions as water of hydration (Section 12.1, 'The Solution Process').

Review this on page 428

These water molecules have less motional energy than before because they are now confined to the immediate environment of the ions and so are in a *more* ordered state than before. Therefore, the dissolving of a salt involves both a disordering process (the ions become less confined) and an ordering process (some water molecules become more confined). The disordering processes are usually

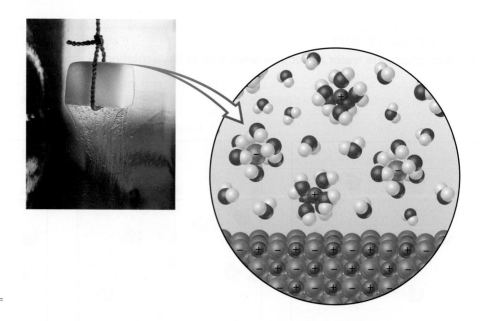

▶ **FIGURE 14.28 Dissolving an ionic solid in water.** The ions become more spread out and random in their motions, but the water molecules that hydrate the ions become less random.

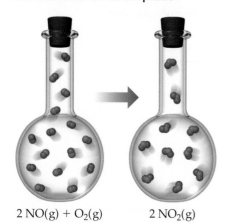

FIGURE IT OUT

What major factor leads to a decrease in entropy as the reaction shown takes place?

2 NO(g) + O₂(g)   2 NO₂(g)

▲ **FIGURE 14.29 Entropy decreases when NO(g) is oxidised by O₂(g) to NO₂(g).** A decrease in the number of gaseous molecules leads to a decrease in the entropy of the system.

dominant, so the overall effect is an increase in the randomness of the system when most salts dissolve in water.

The same ideas apply to systems involving chemical reactions. Consider the reaction between nitric oxide gas and oxygen gas to form nitrogen dioxide gas:

$$2\,NO(g) + O_2(g) \longrightarrow 2\,NO_2(g) \qquad [14.33]$$

In this case the reaction results in a decrease in the number of molecules—three molecules of gaseous reactants form two molecules of gaseous products (◀ **FIGURE 14.29**). The formation of new N–O bonds reduces the motions of the atoms in the system. The formation of new bonds decreases the *number of degrees of freedom*, or forms of motion, available to the atoms. That is, the atoms are less free to move in random fashion because of the formation of new bonds. The decrease in the number of molecules and the resultant decrease in motion result in fewer possible microstates and therefore a decrease in the entropy of the system.

In summary, we generally expect the entropy of the system to increase for processes in which:

1. gases are formed from either solids or liquids,
2. liquids or solutions are formed from solids, and
3. the number of gas molecules increases during a chemical reaction.

SAMPLE EXERCISE 14.14   **Predicting the sign of ΔS**

Predict whether ΔS is positive or negative for each of the following processes, assuming each occurs at constant temperature:

(a) $H_2O(l) \longrightarrow H_2O(g)$

(b) $Ag^+(aq) + Cl^-(aq) \longrightarrow AgCl(s)$

(c) $4\,Fe(s) + 3\,O_2(g) \longrightarrow 2\,Fe_2O_3(s)$

(d) $N_2(g) + O_2(g) \longrightarrow 2\,NO(g)$

SOLUTION

Analyse We are given four reactions and asked to predict the sign of ΔS for each.

Plan We expect ΔS to be positive if there is an increase in temperature, increase in volume or increase in number of gas particles. The question states that the temperature is constant, and so we need to concern ourselves only with volume and number of particles.

Solve

(a) The evaporation of a liquid is accompanied by a large increase in volume. One mole of water (18 g) occupies about 18 cm^3 as a liquid and 22.7 dm^3 as a gas at 0 °C and 1 bar. Because the molecules are distributed throughout a much larger volume in the gaseous state than in the liquid state, an increase in motional freedom accompanies vaporisation. Therefore ΔS is positive.

(b) In this process the ions, which are free to move throughout the volume of the solution, form a solid in which they are confined to a smaller volume and restricted to more highly constrained positions. Thus ΔS is negative.

(c) The particles of a solid are confined to specific locations and have fewer ways to move (fewer microstates) than do the molecules of a gas. Because O_2 gas is converted into part of the solid product Fe_2O_3, ΔS is negative.

(d) The number of moles of gases is the same on both sides of the equation, and so the entropy change will be small. The sign of ΔS is impossible to predict based on our discussions so far, but we can predict that ΔS will be close to zero.

PRACTICE EXERCISE

Indicate whether each of the following processes produces an increase or decrease in the entropy of the system:

(a) $CO_2(s) \longrightarrow CO_2(g)$

(b) $CaO(s) + CO_2(g) \longrightarrow CaCO_2(s)$

(c) $HCl(g) + NH_3(g) \longrightarrow NH_4Cl(s)$

(d) $2\,SO_2(g) + O_2(g) \longrightarrow 2\,SO_3(g)$

Answers: **(a)** increase, **(b)** decrease, **(c)** decrease, **(d)** decrease

(See also Exercise 14.76.)

The Third Law of Thermodynamics

If we decrease the thermal energy of a system by lowering the temperature, the energy stored in translational, vibrational and rotational forms of motion decreases. As less energy is stored, the entropy of the system decreases. If we keep lowering the temperature, do we reach a state in which these motions are essentially shut down, a point described by a single microstate? This question is addressed by the **third law of thermodynamics**, which states that *the entropy of a pure crystalline substance at absolute zero is zero: S* (0 K) *= 0.*

Consider a pure crystalline solid. At absolute zero the individual atoms or molecules in the lattice have no thermal motion. There is, therefore, only one microstate. As a result $S = k \ln W = k \ln 1 = 0$. As the temperature is increased from absolute zero, the atoms or molecules in the crystal gain energy in the form of vibrational motion about their lattice positions. Thus the degrees of freedom of the crystal increase. The entropy of the lattice therefore increases with temperature because vibrational motion causes the atoms or molecules to have a greater number of accessible microstates.

What happens to the entropy of the substance as we continue to heat it? ▼ **FIGURE 14.30** is a plot of how the entropy of a typical substance varies with temperature. We see that the entropy of the solid continues to increase steadily with increasing temperature up to the melting point of the solid. When the solid melts, the bonds holding the atoms or molecules are broken and the particles are free to move about the entire volume of the substance. The added degrees of freedom for the individual molecules allow greater dispersal of the substance's energy, thereby increasing its entropy. We therefore see a sharp increase in the entropy at the melting point. After all the solid has melted to liquid, the temperature again increases and, with it, the entropy.

At the boiling point of the liquid, another abrupt increase in entropy occurs. We can understand this increase as resulting from the increased volume in which the molecules may be found. When the gas is heated further, the entropy increases steadily as more energy is stored in the translational motion of the gas molecules. At higher temperatures, the distribution of molecular

Why does the plot show vertical jumps at the melting and boiling points?

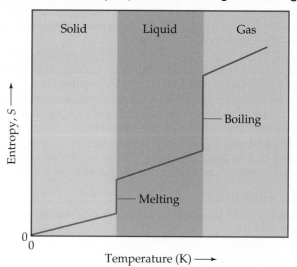

▲ **FIGURE 14.30   Entropy increases with increasing temperature.**

speeds is spread out towards higher values (Figure 10.16). More of the molecules have speeds that differ greatly from the most probable value. The expansion of the range of speeds of the gas molecules leads to an increased entropy.

The general conclusions we reach in examining Figure 14.30 are consistent with what we noted earlier: entropy generally increases with increasing temperature because the increased motional energy can be dispersed in more ways. Further, the entropies of the phases of a given substance follow the order $S_{solid} < S_{liquid} < S_{gas}$. This ordering fits in nicely with our picture of the number of microstates available to solids, liquids and gases.

CONCEPT CHECK 16

If you are told that the entropy of a certain system is zero, what do you know about the system?

14.11 | ENTROPY CHANGES IN CHEMICAL REACTIONS

In Section 14.5 we discussed how calorimetry can be used to measure ΔH for chemical reactions. No comparable, easy method exists for measuring ΔS for a reaction. Entropy plots such as Figure 14.30 can be obtained by carefully measuring how the heat capacity of a substance (Section 14.5) varies with temperature, and we can use the data to obtain the absolute entropies at different temperatures, which are based on the reference point of zero for perfect crystalline solids at 0 K (the third law). (The theory and methods used for these measurements and calculations are beyond the scope of this text.) Entropies are usually tabulated as molar quantities, in units of joules per mole per kelvin ($J\ mol^{-1}\ K^{-1}$).

The molar entropy values of substances in their standard states are known as **standard molar entropies** and are denoted $S°$. The standard state for any substance is defined as the pure substance at 1 bar pressure. ◀ **TABLE 14.4** lists the values of $S°$ for several substances at 298 K; Appendix C gives a more extensive list.

TABLE 14.4 • Standard molar entropies of selected substances at 298 K

Substance	$S°\ J\ mol^{-1}\ K^{-1}$
Gases	
$H_2(g)$	130.6
$N_2(g)$	191.5
$O_2(g)$	205.0
$H_2O(g)$	188.8
$NH_3(g)$	192.5
$CH_3OH(g)$	237.6
$C_6H_6(g)$	269.2
Liquids	
$H_2O(l)$	69.9
$CH_3OH(l)$	126.8
$C_6H_6(l)$	172.8
Solids	
$Li(s)$	29.1
$Na(s)$	51.4
$K(s)$	64.7
$Fe(s)$	27.23
$FeCl_3(s)$	142.3
$NaCl(s)$	72.3

FIGURE IT OUT

What might you expect for the value of $S°$ for butane, C_4H_{10}?

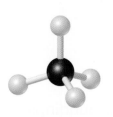

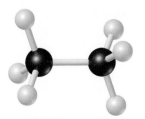

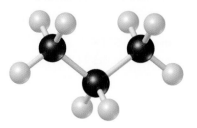

Methane, CH_4
$S° = 186.3 \text{ J mol}^{-1}\text{K}^{-1}$

Ethane, C_2H_6
$S° = 229.6 \text{ J mol}^{-1}\text{K}^{-1}$

Propane, C_3H_8
$S° = 270.3 \text{ J mol}^{-1}\text{K}^{-1}$

▲ **FIGURE 14.31   Entropy increases with increasing molecular complexity.**

We can make several observations about the $S°$ values in Table 14.4.

1. Unlike enthalpies of formation, the standard molar entropies of elements at the reference temperature of 298 K are *not* zero.

2. The standard molar entropies of gases are greater than those of liquids and solids, consistent with our interpretation of experimental observations, as represented in Figure 14.30.

3. Standard molar entropies generally increase with increasing molar mass. (Compare Li(s), Na(s) and K(s).)

4. Standard molar entropies generally increase with an increasing number of atoms in the formula of a substance.

Point 4 is related to molecular motion (Section 14.10). In general, the number of degrees of freedom for a molecule increases with increasing number of atoms, and thus the number of accessible microstates increases also. ▲ **FIGURE 14.31** compares the standard molar entropies of three hydrocarbons. Notice how the entropy increases as the number of atoms in the molecule increases.

The entropy change in a chemical reaction equals the sum of the entropies of the products less the sum of the entropies of the reactants:

$$\Delta S° = \Sigma nS°(\text{products}) - \Sigma mS°(\text{reactants}) \qquad [14.34]$$

As in Equation 14.27, the coefficients n and m are the coefficients in the chemical equation, as illustrated in Sample Exercise 14.15.

SAMPLE EXERCISE 14.15   Calculating $\Delta S°$ from tabulated entropies

Calculate $\Delta S°$ for the synthesis of ammonia from $N_2(g)$ and $H_2(g)$ at 298 K:

$$N_2(g) + 3\,H_2(g) \longrightarrow 2\,NH_3(g)$$

SOLUTION

Analyse We are asked to calculate the standard entropy change for the synthesis of $NH_3(g)$ from its constituent elements.

Plan We can make this calculation using Equation 14.34 and the standard molar entropy values in Table 14.4 and Appendix C.

$$\Delta S° = 2S°(NH_3) - [S°(N_2) + 3S°(H_2)]$$

Solve Substituting the appropriate $S°$ values from Table 14.4 yields

$$\Delta S° = (2\text{ mol})(192.5\text{ J mol}^{-1}\text{K}^{-1}) - [(1\text{ mol})(191.5\text{ J mol}^{-1}\text{K}^{-1}) + (3\text{ mol})(130.6\text{ J mol}^{-1}\text{K}^{-1})]$$

$$= -198.3\text{ J K}^{-1}$$

Check The value for $\Delta S°$ is negative, in agreement with our qualitative prediction based on the decrease in the number of molecules of gas during the reaction.

PRACTICE EXERCISE

Using the standard entropies in Appendix C, calculate the standard entropy change, $\Delta S°$, for the following reaction at 298 K:

$$Al_2O_3(s) + 3\ H_2(g) \rightarrow 2\ Al(s) + 3\ H_2O(g)$$

Answer: 180.39 J K^{-1}

(See also Exercises 14.78, 14.79.)

Entropy Changes in the Surroundings

Tabulated absolute entropy values can be used to calculate the standard entropy change in a system, such as a chemical reaction, as just described, but what about the entropy change in the surroundings? We encountered this situation in Section 14.9, but it is good to revisit it now that we are examining chemical reactions.

We should recognise that the surroundings serve essentially as a large, constant temperature heat source (or heat sink if the heat flows from the system to the surroundings). The change in entropy of the surroundings will depend on how much heat is absorbed or given off by the system. For an isothermal process, the entropy change of the surroundings is given by

$$\Delta_{surr}S = \frac{-q_{sys}}{T} = \frac{-\Delta_{sys}H}{T} \qquad [14.35]$$

For a reaction occurring at constant pressure, q_{sys} is simply the enthalpy change for the reaction, ΔH. For the reaction in Sample Exercise 14.15, the formation of ammonia from $H_2(g)$ and $N_2(g)$ at 298 K, q_{sys} is the enthalpy change for reaction under standard conditions, $\Delta H°$. Using the procedures described in Section 14.7, we have

$$\Delta_{rxn}H° = 2\Delta_f H°[NH_3(g)] - 3\Delta_f H°[H_2(g)] - \Delta_f H°[N_2(g)]$$
$$= 2(-46.19\ kJ) - 3(0)\ kJ - (0)\ kJ = -92.38\ kJ$$

Thus at 298 K the formation of ammonia from $H_2(g)$ and $N_2(g)$ is exothermic. Absorption of the heat given off by the system results in an increase in the entropy of the surroundings:

$$\Delta_{surr}S° = \frac{92.38\ kJ}{298\ K} = 0.310\ kJ\ K^{-1} = 310\ J\ K^{-1}$$

Notice that the magnitude of the entropy gained by the surroundings (310 J K^{-1}) is greater than that lost by the system (198.3 J K^{-1}, as calculated in Sample Exercise 14.15):

$$\Delta_{univ}S° = \Delta_{sys}S° + \Delta_{surr}S° = -198.3\ J\ K^{-1} + 310\ J\ K^{-1} = 112\ J\ K^{-1}$$

Because $\Delta_{univ}S°$ is positive for any spontaneous reaction, this calculation indicates that, when $NH_3(g)$, $H_2(g)$ and $N_2(g)$ are together at 298 K in their standard states (each at 1 bar pressure), the reaction system will move spontaneously towards formation of $NH_3(g)$. Keep in mind that, although the thermodynamic calculations indicate that formation of ammonia is spontaneous, they do not tell us anything about the rate at which ammonia is formed.

CONCEPT CHECK 17

If a process is exothermic, does the entropy of the surroundings (1) always increase, (2) always decrease or (3) sometimes increase and sometimes decrease, depending on the process?

14.12 | GIBBS FREE ENERGY (GIBBS ENERGY)

Although the IUPAC convention now assigns Gibbs energy as the preferred name for Gibbs free energy, the term free energy will be employed due to common usage. There are examples of endothermic processes that are spontaneous, such as the dissolution of ammonium nitrate in water (∞ Section 12.1, 'The Solution Process').

The solution process is a spontaneous, endothermic process and must be accompanied by an increase in the entropy of the system. However, we may also encounter processes that are spontaneous and yet proceed with a *decrease* in the entropy of the system, such as the highly exothermic formation of sodium chloride from its constituent elements (∞ Section 8.2, 'Ionic Bonding').

Spontaneous processes that result in a decrease in the system's entropy are always exothermic. Thus the spontaneity of a reaction seems to involve two thermodynamic concepts, enthalpy and entropy.

How can we use use ΔH and ΔS to predict whether a given reaction occurring at constant temperature and pressure will be spontaneous? The means for doing so was first developed by the American mathematician J. Willard Gibbs (1839–1903). Gibbs (▶ FIGURE 14.32) proposed a new state function, now called the **Gibbs free energy** (or **free energy**). The Gibbs free energy, G, of a state is defined as

$$G = H - TS \qquad [14.36]$$

where T is the absolute temperature. For a process occurring at constant temperature, the change in Gibbs free energy of the system, ΔG, is given by the expression

$$\Delta G = \Delta H - T\Delta S \qquad [14.37]$$

Under standard conditions, this equation becomes

$$\Delta G° = \Delta H° - T\Delta S° \qquad [14.38]$$

To see how the state function G relates to reaction spontaneity, recall that for a reaction occurring at constant temperature and pressure

$$\Delta_{univ}S = \Delta_{sys}S + \Delta_{surr}S = \Delta_{sys}S + \left(\frac{-\Delta_{sys}H}{T}\right)$$

where we have used Equation 14.35 to substitute for $\Delta_{surr}S$. Multiplying both sides by $(-T)$ gives us

$$-T\Delta_{univ}S = \Delta_{sys}H - T\Delta_{sys}S \qquad [14.39]$$

Comparing Equation 14.39 with Equation 14.37 we see that the Gibbs free energy change in a process occurring at constant temperature and pressure, ΔG, is equal to $-T\Delta_{univ}S$. We know that, for spontaneous processes, $\Delta_{univ}S$ is positive. Thus the sign of ΔG provides us with extremely valuable information about the spontaneity of processes that occur at constant temperature and pressure. If both T and P are constant, the relationship between the sign of ΔG and the spontaneity of a reaction is as follows.

1. If ΔG is negative, the reaction is spontaneous in the forward direction.

2. If ΔG is zero, the reaction is at equilibrium (∞ Section 10.1, 'The Concept of Equilibrium').

∞ Review this on page 430

∞ Review this on page 253

▲ **FIGURE 14.32   Josiah Willard Gibbs.** Gibbs was the first person to be awarded a PhD in science from an American university (Yale, 1863). From 1871 until his death, he held the chair of mathematical physics at Yale. He developed much of the theoretical foundation that led to the development of chemical thermodynamics.

∞ Find out more on page 616

Are the processes that move a system towards equilibrium spontaneous or non-spontaneous?

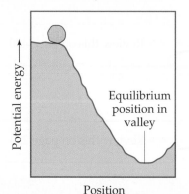

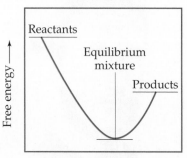

▲ **FIGURE 14.33 Potential energy and free energy.** An analogy is shown between the gravitational potential energy change of a boulder rolling down a hill and the free energy change in a spontaneous reaction.

TABLE 14.5 • Conventions used in establishing standard free energies

State of matter	Standard state
Solid	Pure solid
Liquid	Pure liquid
Gas	1 bar pressure
Solution	1 M concentration
Elements	Standard free energy of formation of an element in its standard state is defined as zero

3. If ΔG is positive, the reaction in the forward direction is non-spontaneous; work must be supplied from the surroundings to make it occur. However, the reverse reaction will be spontaneous.

It is more convenient to use ΔG as a criterion for spontaneity than to use $\Delta_{univ}S$, because ΔG relates to the system alone and avoids the complication of having to examine the surroundings.

An analogy is often drawn between the free energy change during a spontaneous reaction and the potential energy change when a boulder rolls down a hill. Potential energy in a gravitational field 'drives' the boulder until it reaches a state of minimum potential energy in the valley (◀ FIGURE 14.33(a)). Similarly, the free energy of a chemical system decreases until it reaches a minimum value (Figure 14.33(b)). When this minimum is reached, a state of equilibrium exists. *In any spontaneous process at constant temperature and pressure, the free energy always decreases.*

 CONCEPT CHECK 18

Give the criterion for spontaneity
a. in terms of entropy,
b. in terms of free energy.

In fact, thermodynamics tells us that *the change in free energy for a process,* ΔG, *equals the maximum useful work that can be done by a system on its surroundings in a spontaneous process occurring at constant temperature and pressure:*

$$\Delta G = -w_{max} \qquad [14.40]$$

This relationship explains why ΔG is called *free* energy. It is the portion of the energy of a spontaneous reaction that is free to do useful work. The remainder of the energy enters the environment as heat.

Standard Free Energy Changes

Like enthalpy, free energy is a state function. We can tabulate **standard free energies of formation**, $\Delta_f G°$, for substances, just as we can tabulate standard enthalpies of formation (Section 14.7). It is important to remember that standard values for these functions imply a particular set of conditions, or standard states. The standard state for gaseous substances is 1 bar pressure. For solid substances, the standard state is the pure solid; for liquids, the pure liquid. For substances in solution, the standard state is normally a concentration of 1 M. The temperature usually chosen for purposes of tabulating data is 25 °C, but we will calculate $\Delta G°$ at other temperatures as well. Just as for the standard heats of formation, the free energies of elements in their standard states are set to zero. This arbitrary choice of a reference point has no effect on the quantity in which we are really interested, namely, the *difference* in free energy between reactants and products. The rules about standard states are summarised in ◀ TABLE 14.5. A listing of standard free energies of formation is given in Appendix C.

CONCEPT CHECK 19

What does the superscript ° indicate when associated with a thermodynamic quantity, as in $\Delta H°$, $\Delta S°$ or $\Delta G°$?

The standard free energies of formation are useful in calculating the *standard Gibbs energy change* for chemical processes. The procedure is analogous to the calculation of $\Delta H°$ (Equation 14.27) and $\Delta S°$ (Equation 14.34):

$$\Delta G° = \sum n\Delta_f G°(\text{products}) - \sum m\Delta_f G°(\text{reactants}) \qquad [14.41]$$

SAMPLE EXERCISE 14.16   Calculating standard free energy change from free energies of formation

(a) By using data from Appendix C, calculate the standard free energy change for the following reaction at 298 K:

$$P_4(g) + 6\,Cl_2(g) \longrightarrow 4\,PCl_3(g)$$

(b) What is $\Delta G°$ for the reverse of the above reaction?

SOLUTION

Analyse We are asked to calculate the free energy change for a reaction and then to determine the free energy change for the reverse reaction.

Plan We look up the free energy values for the products and reactants and use Equation 14.41. We multiply the molar quantities by the coefficients in the balanced equation and subtract the total for the reactants from that for the products.

Solve

(a) $Cl_2(g)$ is in its standard state, so $\Delta_f G°$ is zero for this reactant. $P_4(g)$, however, is not in its standard state, so $\Delta_f G°$ is not zero for this reactant. From the balanced equation and using Appendix C, we have:

$$\Delta_{rxn}G° = 4\Delta_f G°[PCl_3(g)] - \Delta_f G°[P_4(g)] - 6\Delta_f G°\,[Cl_2(g)]$$

$$= (4\text{ mol})(-269.6\text{ kJ mol}^{-1}) - (1\text{ mol})(24.4\text{ kJ mol}^{-1}) - 0$$

$$= -1102.8\text{ kJ}$$

The fact that $\Delta G°$ is negative tells us that a mixture of $P_4(g)$, $Cl_2(g)$ and $PCl_3(g)$ at 25 °C, each present at a partial pressure of 1 bar, would react spontaneously in the forward direction to form more PCl_3. Remember, however, that the value of $\Delta G°$ tells us nothing about the rate at which the reaction occurs.

(b) Remember that $\Delta G = G$ (products) $- G$ (reactants). If we reverse the reaction, we reverse the roles of the reactants and products. Thus reversing the reaction changes the sign of ΔG, just as reversing the reaction changes the sign of ΔH (Section 14.4). Hence, using the result from part (a):

$$4\,PCl_3(g) \longrightarrow P_4(g) + 6\,Cl_2(g) \qquad \Delta G° = +1102.8\text{ kJ}$$

PRACTICE EXERCISE

By using data from Appendix C, calculate $\Delta G°$ at 298 K for the combustion of methane: $CH_4(g) + 2\,O_2(g) \longrightarrow CO_2(g) + 2\,H_2O(g)$.

Answer: -800.7 kJ

(See also Exercises 14.83, 14.84.)

SAMPLE EXERCISE 14.17   Estimating and calculating $\Delta G°$

In Section 14.7 we used Hess's law to calculate $\Delta H°$ for the combustion of propane gas at 298 K:

$$C_3H_8(g) + 5\,O_2(g) \longrightarrow 3\,CO_2(g) + 4\,H_2O(l) \qquad \Delta H° = -2220\text{ kJ}$$

(a) *Without using data from Appendix C*, predict whether $\Delta G°$ for this reaction is more negative or less negative than $\Delta H°$. **(b)** Use data from Appendix C to calculate $\Delta G°$ for the reaction at 298 K. Is your prediction from part (a) correct?

SOLUTION

Analyse In part (a) we must predict the value for $\Delta G°$ relative to that for $\Delta H°$ on the basis of the balanced equation for the reaction. In part (b) we must calculate the value for $\Delta G°$ and compare this value with our qualitative prediction.

Plan The free energy change incorporates both the change in enthalpy and the change in entropy for the reaction (Equation 14.38), so under standard conditions

$$\Delta G° = \Delta H° - T\Delta S°$$

To determine whether $\Delta G°$ is more negative or less negative than $\Delta H°$, we need to determine the sign of the term $T\Delta S°$. Because T is the absolute temperature, 298 K, it is always a positive number. We can predict the sign of $\Delta S°$ by looking at the reaction.

Solve

(a) The reactants are six molecules of gas, and the products are three molecules of gas and four molecules of liquid. Thus the number of molecules of gas has decreased significantly during the reaction. By using the general rules discussed in Section 14.10, we would expect a decrease in the number of gas molecules to lead to a decrease in the entropy of the system—the products have fewer accessible microstates than the reactants. We therefore expect $\Delta S°$ and $T\Delta S°$ to be negative numbers. Because we are subtracting $T\Delta S°$, which is a negative number, we would predict that $\Delta G°$ is *less negative* than $\Delta H°$.

(b) Using Equation 14.41 and values from Appendix C, we can calculate the value of $\Delta G°$:

$$\Delta G° = 3\Delta_f G°[CO_2(g)] + 4\Delta_f G°[H_2O(l)] - \Delta_f G°[C_3H_8(g)] - 5\Delta_f G°[O_2(g)]$$

$$= 3\ mol(-394.4\ kJ\ mol^{-1}) + 4\ mol(-237.13\ kJ\ mol^{-1})$$

$$- 1\ mol(-23.47\ kJ\ mol^{-1}) - 5\ mol(0\ kJ\ mol^{-1}) = -2108\ kJ$$

Notice that we have been careful to use the value of $\Delta_f G°$ for $H_2O(l)$; as in the calculation of ΔH values, the phases of the reactants and products are important. As we predicted, $\Delta G°$ is less negative than $\Delta H°$ because of the decrease in entropy during the reaction.

PRACTICE EXERCISE

Consider the combustion of propane to form $CO_2(g)$ and $H_2O(g)$ at 298 K: $C_3H_8(g) + 5\ O_2(g) \longrightarrow 3\ CO_2(g) + 4\ H_2O(g)$. Would you expect $\Delta G°$ to be more negative or less negative than $\Delta H°$?

Answer: More negative

(See also Exercise 14.85.)

14.13 | GIBBS ENERGY AND TEMPERATURE

We have seen that tabulations of $\Delta_f G°$, such as those in Appendix C, make it possible to calculate $\Delta G°$ for reactions at the standard temperature of 25 °C. However, we are often interested in examining reactions at other temperatures. How is the change in free energy affected by the change in temperature? Let's look again at Equation 14.37:

$$\Delta G = \Delta H - T\Delta S = \Delta H + (-T\Delta S)$$

$$\underset{\substack{\text{Enthalpy} \\ \text{term}}}{} \quad \underset{\substack{\text{Entropy} \\ \text{term}}}{}$$

Note that we have written the expression for ΔG as a sum of two contributions, an enthalpy term, ΔH, and an entropy term, $-T\Delta S$. Because the value of $-T\Delta S$ depends directly on the absolute temperature T, ΔG will vary with temperature. T is a positive number at all temperatures other than absolute zero. We know that the enthalpy term, ΔH, can be positive or negative. The entropy term, $-T\Delta S$, can also be positive or negative. When ΔS is positive, which means that the final state has greater randomness (a greater number of microstates) than the initial state, the term $-T\Delta S$ is negative. When ΔS is negative, the term $-T\Delta S$ is positive.

The sign of ΔG, which tells us whether a process is spontaneous, will depend on the signs and magnitudes of ΔH and $-T\Delta S$. When both ΔH and $-T\Delta S$ are negative, ΔG will always be negative and the process will be spontaneous at all temperatures. Likewise, when both ΔH and $-T\Delta S$ are positive, ΔG will always be positive and the process will be non-spontaneous at all temperatures (the reverse process will be spontaneous at all temperatures). When ΔH and $-T\Delta S$ have opposite signs, however, the sign of ΔG will depend on the magnitudes of these two terms. In these instances, temperature is an important consideration. Generally, ΔH and ΔS change very little with temperature.

TABLE 14.6 • **Effect of temperature on the spontaneity of reactions**

ΔH	ΔS	$-T\Delta S$	$\Delta G = \Delta H - T\Delta S$	Reaction characteristics	Example
−	+	−	−	Spontaneous at all temperatures	$2\,O_3(g) \longrightarrow 3\,O_2(g)$
+	−	+	+	Non-spontaneous at all temperatures	$3\,O_2(g) \longrightarrow 2\,O_3(g)$
−	−	+	+ or −	Spontaneous at low T; non-spontaneous at high T	$H_2O(l) \longrightarrow H_2O(s)$
+	+	−	+ or −	Spontaneous at high T; non-spontaneous at low T	$H_2O(s) \longrightarrow H_2O(l)$

However, the value of T directly affects the magnitude of $-T\Delta S$. As the temperature increases, the magnitude of the term $-T\Delta S$ increases and it will become relatively more important in determining the sign and magnitude of ΔG.

For example, let's consider once more the melting of ice to liquid water at 1 bar pressure:

$$H_2O(s) \longrightarrow H_2O(l) \qquad \Delta H > 0, \Delta S > 0$$

This process is endothermic, which means that ΔH is positive. We also know that the entropy increases during this process, so ΔS is positive and $-T\Delta S$ is negative. At temperatures below 0 °C (273 K) the magnitude of ΔH is greater than that of $-T\Delta S$. Hence the positive enthalpy term dominates, leading to a positive value for ΔG. The positive value of ΔG means that the melting of ice is not spontaneous at $T < 0$ °C; rather, the reverse process, the freezing of liquid water into ice, is spontaneous at these temperatures.

What happens at temperatures greater than 0 °C? As the temperature increases, so does the magnitude of the entropy term $-T\Delta S$. When $T > 0$ °C, the magnitude of $-T\Delta S$ is greater than the magnitude of ΔH. At these temperatures the negative entropy term dominates, which leads to a negative value for ΔG. The negative value of ΔG tells us that the melting of ice is spontaneous at $T > 0$ °C. At the normal melting point of water, $T = 0$ °C, the two phases are in equilibrium. At equilibrium, $\Delta G = 0$; at $T = 273$ K, ΔH and $-T\Delta S$ are equal in magnitude and opposite in sign, so they cancel one another and give $\Delta G = 0$.

 CONCEPT CHECK 20

The normal boiling point of benzene is 80 °C. At 100 °C and 1 bar, which term is greater for the vaporisation of benzene, ΔH or $T\Delta S$?

The possible situations for the relative signs of ΔH and ΔS are given in ▲ TABLE 14.6, along with examples of each. By applying the concepts we have developed for predicting entropy changes, we can often predict how ΔG will change with temperature.

Our discussion of the temperature dependence of ΔG is also relevant to standard free energy changes.

We can readily calculate the values of $\Delta H°$ and $\Delta S°$ at 298 K from the data tabulated in Appendix C. If we assume that the values of $\Delta H°$ and $\Delta S°$ do not change with temperature, we can use Equation 14.38 to estimate the value of $\Delta G°$ at temperatures other than 298 K.

SAMPLE EXERCISE 14.18 | **Determining the effect of temperature on spontaneity**

The Haber process for the production of ammonia involves the equilibrium

$$N_2(g) + 3\,H_2(g) \rightleftharpoons 2\,NH_3(g)$$

Assume that $\Delta H°$ and $\Delta S°$ for this reaction do not change with temperature. **(a)** Predict the direction in which $\Delta G°$ for this reaction changes with increasing temperature. **(b)** Calculate the values of $\Delta G°$ for the reaction at 25 °C and 500 °C.

SOLUTION

Analyse In part (a) we are asked to predict the direction in which $\Delta G°$ changes as temperature increases. In part (b) we need to determine $\Delta G°$ for the reaction at two temperatures.

Plan We can answer part (a) by determining the sign of ΔS for the reaction and then using that information to analyse Equation 14.38. In part (b) we first calculate $\Delta H°$ and $\Delta S°$ for the reaction using data in Appendix C and then use Equation 14.38 to calculate $\Delta G°$.

Solve

(a) Equation 14.38 tells us that $\Delta G°$ is the sum of the enthalpy term $\Delta H°$ and the entropy term $-T\Delta S$. The temperature dependence of $\Delta G°$ comes from the entropy term. We expect $\Delta S°$ for this reaction to be negative because the number of molecules of gas is smaller in the products. Because $\Delta S°$ is negative, the term $-T\Delta S$ is positive and grows larger with increasing temperature. As a result, $\Delta G°$ becomes less negative (or more positive) with increasing temperature. Thus the driving force for the production of NH_3 becomes smaller with increasing temperature.

(b) $\Delta H° = -92.38$ kJ and $\Delta S° = -198.4$ J K^{-1}. If we assume that these values don't change with temperature, we can calculate $\Delta G°$ at any temperature by using Equation 14.38. At $T = 298$ K we have:

$$\Delta G° = -92.38 \text{ kJ} - (298 \text{ K})\left(\frac{-198.4 \text{ kJ K}^{-1}}{1000}\right)$$

$$= -92.38 \text{ kJ} + 59.1 \text{ kJ} = -33.3 \text{ kJ}$$

At $T = 500 + 273 = 773$ K we have

$$\Delta G° = -92.38 \text{ kJ} - (773 \text{ K})\left(\frac{-198.4 \text{ kJ K}^{-1}}{1000}\right)$$

$$= -92.38 \text{ kJ} + 153 \text{ kJ} = 61 \text{ kJ}$$

Notice that we have been careful to convert $-T\Delta S$ into units of kJ so that it can be added to $\Delta H°$, which has units of kJ.

Comment Increasing the temperature from 298 K to 773 K changes $\Delta G°$ from -33.3 kJ to $+61$ kJ. Of course, the result at 773 K depends on the assumption that $\Delta H°$ and $\Delta S°$ do not change with temperature. In fact, these values do change slightly with temperature. Nevertheless, the result at 773 K should be a reasonable approximation. The positive increase in $\Delta G°$ with increasing T agrees with our prediction in part (a) of this exercise. Our result indicates that a mixture of $N_2(g)$, $H_2(g)$ and $NH_3(g)$, each present at a partial pressure of 1 bar, will react spontaneously at 298 K to form more $NH_3(g)$. In contrast, at 773 K the positive value of $\Delta G°$ tells us that the reverse reaction is spontaneous. Thus when the mixture of three gases, each at a partial pressure of 1 bar, is heated to 773 K, some of the $NH_3(g)$ spontaneously decomposes into $N_2(g)$ and $H_2(g)$.

PRACTICE EXERCISE

(a) Using standard enthalpies of formation and standard entropies in Appendix C, calculate $\Delta H°$ and $\Delta S°$ at 298 K for the following reaction: $2 SO_2(g) + O_2(g) \longrightarrow 2 SO_3(g)$. **(b)** Using the values obtained in part (a), estimate $\Delta G°$ at 400 K.

Answers: **(a)** $\Delta H° = -196.6$ kJ, $\Delta S° = -189.6$ J K^{-1}; **(b)** $\Delta G° = -120.8$ kJ

(See also Exercise 14.91.)

MY WORLD OF CHEMISTRY

DRIVING NON-SPONTANEOUS REACTIONS

Many desirable chemical reactions, including a large number that are central to living systems, are non-spontaneous as written. For example, consider the extraction of copper metal from the mineral *chalcocite*, which contains Cu_2S. The decomposition of Cu_2S to its elements is non-spontaneous:

$$Cu_2S(s) \longrightarrow 2\,Cu(s) + S(s) \qquad \Delta G° = +186.2 \text{ kJ}$$

Because $\Delta G°$ is very positive, we cannot obtain Cu(s) directly via this reaction. Instead, we must find some way to 'do work' on the reaction to force it to occur as we wish. We can do this by coupling the reaction to another one so that the overall reaction *is* spontaneous. For example, we can imagine the S(s) reacting with $O_2(g)$ to form $SO_2(g)$:

$$S(s) + O_2(g) \longrightarrow SO_2(g) \qquad \Delta G° = -300.4 \text{ kJ}$$

By coupling these reactions together, we can extract much of the copper metal via a spontaneous reaction:

$$Cu_2S(s) + O_2(g) \longrightarrow 2\,Cu(s) + SO_2(g)$$

$$\Delta G° = (+86.2 \text{ kJ}) + (-300.4 \text{ kJ}) = -214.2 \text{ kJ}$$

In essence, we have used the spontaneous reaction of S(s) with $O_2(g)$ to provide the free energy needed to extract the copper metal from the mineral.

Biological systems employ the same principle of using spontaneous reactions to drive non-spontaneous ones. Many of the biochemical reactions that are essential for the forma-

tion and maintenance of highly ordered biological structures are not spontaneous. These necessary reactions are made to occur by coupling them with spontaneous reactions that release energy. The metabolism of food is the usual source of the free energy needed to do the work of maintaining biological systems. For example, complete oxidation of the sugar *glucose*, $C_6H_{12}O_6$, to CO_2 and H_2O yields substantial free energy:

$$C_6H_{12}O_6(s) + 6\,O_2(g) \longrightarrow 6\,CO_2(g) + 6\,H_2O(l)$$

$$\Delta G° = -2880 \text{ kJ}$$

This energy can be used to drive non-spontaneous reactions in the body. However, a means is necessary to transport the energy released by glucose metabolism to the reactions that require energy. One way, shown in ▼ FIGURE 14.34, involves the interconversion of adenosine triphosphate (ATP) and adenosine diphosphate (ADP), molecules that are related to the building blocks of nucleic acids. The conversion of ATP to ADP releases free energy ($\Delta G° = -30.5$ kJ) that can be used to drive other reactions.

In the human body the metabolism of glucose occurs via a complex series of reactions, most of which release free energy. The free energy released during these steps is used in part to reconvert lower-energy ADP back to higher-energy ATP. Thus the ATP–ADP interconversions are used to store energy during metabolism and to release it as needed to drive non-spontaneous reactions in the body. If you take a course in biochemistry, you will have the opportunity to learn more about the remarkable sequence of reactions used to transport free energy throughout the human body.

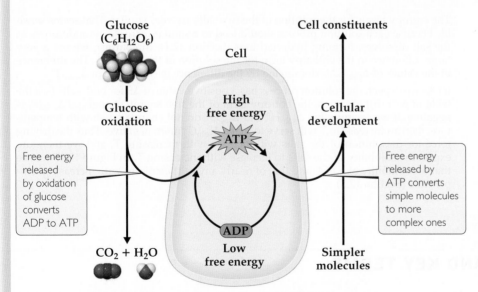

◀ **FIGURE 14.34 Schematic representation of Gibbs energy changes during cell metabolism.** The oxidation of glucose to CO_2 and H_2O produces free energy that is then used to convert ADP into the more energetic ATP. The ATP is then used, as needed, as an energy source to drive non-spontaneous reactions, such as the conversion of simple molecules into more complex cell constituents.

Consider the simple salts NaCl(s) and AgCl(s). We examine the equilibria in which these salts dissolve in water to form aqueous solutions of ions:

$$NaCl(s) \rightleftharpoons Na^+(aq) + Cl^-(aq)$$

$$AgCl(s) \rightleftharpoons Ag^+(aq) + Cl^-(aq)$$

(a) Calculate the value of $\Delta G°$ at 298 K for each of the preceding reactions. **(b)** The two values from part (a) are very different. Is this difference primarily due to the enthalpy term or the entropy term of the standard free energy change? **(c)** How will $\Delta G°$ for the solution process of these salts change with increasing T? What effect should this change have on the solubility of the salts?

SOLUTION

(a) We use Equation 14.41 together with $\Delta_f G°$ values from Appendix C to calculate the $\Delta_{soln}G°$ values for each equilibrium. (We use the subscript 'soln' to indicate that these are thermodynamic quantities for the formation of a solution.) We find

$$\Delta_{soln}G°(NaCl) = (-261.9\,kJ\,mol^{-1}) + (-131.2\,kJ\,mol^{-1}) - (-384.0\,kJ\,mol^{-1})$$
$$= -9.1\,kJ\,mol^{-1}$$
$$\Delta_{soln}G°(AgCl) = (+77.11\,kJ\,mol^{-1}) + (-131.2\,kJ\,mol^{-1}) - (-109.70\,kJ\,mol^{-1})$$
$$= +55.6\,kJ\,mol^{-1}$$

(b) We can write $\Delta_{soln}G°$ as the sum of an enthalpy term, $\Delta_{soln}H°$, and an entropy term, $-T\Delta_{soln}S°$: $\Delta_{soln}G° = \Delta_{soln}H° + (-T\Delta_{soln}S°)$. We can calculate the values of $\Delta_{soln}H°$ and $\Delta_{soln}S°$ by using Equations 14.27 and 14.34. We can then calculate $-T\Delta_{soln}S°$ at $T = 298$ K. All these calculations are now familiar to us. The results are summarised in the following table:

Salt	$\Delta_{soln}H°$ (kJ mol^{-1})	$\Delta_{soln}S°$ (kJ mol^{-1})	$-T\Delta_{soln}S°$ (kJ mol^{-1})
NaCl	+ 3.6	+ 43.2	-12.9
AgCl	+ 65.7	+ 34.3	-10.2

The entropy terms for the solution of the two salts are very similar. That seems sensible because each solution process should lead to a similar increase in randomness as the salt dissolves, forming hydrated ions (Section 12.1). In contrast, we see a very large difference in the enthalpy term for the solution of the two salts. The difference in the values of $\Delta_{soln}G°$ is dominated by the difference in the values of $\Delta_{soln}H°$.

(c) As we expect, the solution process has a positive value of ΔS for both salts (see the table in part (b)). As such, the entropy term of the free energy change, $-T\Delta_{soln}S°$, is negative. If we assume that $\Delta_{soln}H°$ and $\Delta_{soln}S°$ do not change much with temperature, then an increase in T will serve to make $\Delta_{soln}G°$ more negative. Thus the driving force for dissolution of the salts will increase with increasing T, and we therefore expect the solubility of the salts to increase with increasing T. In Figure 12.17 we see that the solubility of NaCl (and that of nearly any salt) increases with increasing temperature (Section 12.3).

CHAPTER SUMMARY AND KEY TERMS

SECTION 14.1 **Thermodynamics** is the study of energy and its transformations and when applied to chemical reactions is known as **thermochemistry**. We consider the nature of energy and some of the forms it takes such as **kinetic energy** (E_k) and **potential energy** (E_p), which arises when there is a **force** acting on a body. We see that **energy** is the capacity to do **work** (w) or to transfer **heat** (q). The focus of our studies is the **system**, which for chemical reactions is the reactants and products. Everything else is the **surroundings**. The SI unit of energy is the **joule (J)** where $1\,J = 1\,kg\,m^2\,s^{-2}$.

SECTION 14.2 The **internal energy** (U) of the system is the sum of all the kinetic and potential energies of its component parts. The **first law of thermodynamics** states that the change in the internal energy of the system is given by $\Delta U = q + w$. If the system absorbs heat from the surroundings, the process is said to be **endothermic** and if it delivers heat to the surroundings it is **exothermic**. The internal energy of the system is a **state function**. The value of any state function depends only on the state or condition of the system and not on the details of how it came to be in that state.

SECTIONS 14.3 and 14.4 We introduce a new state function, **enthalpy** (H), where $H = U + PV$, to account for **pressure-volume (P-V) work**, done by a system in which a gas is produced and expands against the prevailing pressure. At constant pressure the change in enthalpy of the system is given by $\Delta H = \Delta U + P\Delta V$ and $\Delta H = q_P$, that is, the heat gained or lost by the system at constant pressure. For an endothermic process $\Delta H > 0$ and for an endothermic process $\Delta H < 0$. The change in enthalpy of the system in a chemical reaction is calculated from $\Delta_{rxn}H = \sum \Delta H(\text{products}) - \sum \Delta H(\text{reactants})$, where $\Delta_{rxn}H$ is known as the **enthalpy of reaction**.

SECTION 14.5 The amount of heat transferred between the system and the surroundings is measured experimentally by **calorimetry**. A **calorimeter** measures the temperature change during a chemical reaction. The temperature change of a calorimeter depends on its **heat capacity** (C), which is the amount of heat required to raise its temperature by 1 K. The amount of heat (q) absorbed by an object is related to the heat capacity (C) of the object, the mass of the object (m) and the temperature change (ΔT) by the equation $q = C \times m \times \Delta T$. The heat capacity of one gram of a substance is its **specific heat capacity** whereas that of one mole is the **molar heat capacity**.

SECTION 14.6 Since enthalpy is a state function, ΔH depends only on the initial and final states of the system and so the enthalpy change of a process is the same whether the process is carried out in one step or in a series of steps. **Hess's law** states that if a reaction is carried out in a series of steps, ΔH for the reaction will be equal to the sum of the enthalpy changes for the steps.

SECTION 14.7 The **enthalpy of formation**, $\Delta_f H$, of a substance is the enthalpy change for the reaction in which the substance is formed from its constituent elements. The **standard enthalpy change**, $\Delta H°$, of a reaction is the enthalpy change when all reactants and products are at 1 bar pressure and 298 K (25 °C). The **standard enthalpy of formation**, $\Delta_f H°$, of a substance is the change in enthalpy for the reaction that forms one mole of the substance from its elements in their most stable form with all reactants and products at 1 bar pressure and 298 K.

SECTION 14.8 Since most chemical reactions are **spontaneous** in one direction and non-spontaneous in the reverse direction, the spontaneity of chemical reactions is examined. **Reversible** processes are studied in detail with a view to explaining the maximum amount of work that can be obtained from a system. It is noted that any spontaneous process is **irreversible**. A process that occurs at constant temperature is said to be **isothermal**.

SECTION 14.9 The relation between spontaneity of a process and a thermodynamic state function, **entropy** (S), is discussed. It is determined that for an isothermal process entropy can be defined as the heat absorbed by the system along a reversible path, divided by the temperature; $S = q_{rev}/T$. The **second law of thermodynamics** is used to explain how entropy controls the spontaneity of processes. This yields the equation $\Delta_{univ}S° = \Delta_{sys}S° + \Delta_{surr}S°$. In a reversible process $\Delta_{univ}S = 0$; in an irreversible (spontaneous) process $\Delta_{univ}S > 0$. Entropy values are usually expressed in J K^{-1}.

SECTION 14.10 The kinetic energy of molecules is discussed in terms of the their possible types of motion, that is **translational**, **vibrational** and **rotational**. The concept of a **microstate** (W) is introduced, relating entropy to the number of microstates associated with a particular macroscopic state. This is summarised by the equation $S = k \ln W$ where k is a constant. The **third law of thermodynamics** concludes that at 0 K the entropy of a pure crystalline solid would be zero as at that temperature all motion ceases and consequently the system would possess only one microstate.

SECTION 14.11 The concept of standard entropy ($S°$) is introduced and it is shown that from tabulated values of $S°$ we can calculate the entropy change for any process under standard conditions. For an isothermal process, $\Delta_{surr}S = -\Delta_{sys}H/T$. The entropy of a mole of a substance is known as the **standard molar entropy**.

SECTION 14.12 The **Gibbs free energy** (G) is discussed. This relates enthalpy and entropy changes. For an isothermal process the change in Gibbs free energy is given by the equation $\Delta G = \Delta H - T\Delta S$. If $\Delta G < 0$, then the process is spontaneous. If standard states of reactants and products are employed, the free energy of formation is known as the **standard free energy of formation**, $\Delta_f G°$.

SECTION 14.13 It is noted that because changes in enthalpy and entropy are small as the temperature changes, it is in fact the temperature of a process which governs the value of ΔG. Therefore, the dependence of ΔG on temperature is governed mainly by the value of T in the expression $\Delta G = -\Delta H - T\Delta S$. The entropy term $-T\Delta S$ has a greater effect on the temperature dependence of ΔG and hence on the spontaneity of the process. The standard Gibbs free energy ($\Delta G°$) is related to the equilibrium constant of a reaction as given by the equation $\Delta G° = -RT \ln K$.

KEY SKILLS

- Distinguish between the system and the surroundings in thermodynamics. (Section 14.1)
- State the first law of thermodynamics. (Section 14.2)
- Understand the concept of a state function and be able to give examples. (Section 14.2)
- Express the relationship between the quantities q, w, ΔU and ΔH. Understand their sign conventions, including how the signs of q and ΔH relate to whether a process is exothermic or endothermic. (Sections 14.2 and 14.3)
- Use thermochemical equations to relate the amount of heat energy transferred in reactions at constant pressure (ΔH) to the amount of substance involved in the reaction. (Section 14.4)
- Calculate the heat transferred in a process from temperature measurements together with heat capacities or specific heats (calorimetry). (Section 14.5)
- Use Hess's law to determine enthalpy changes for reactions. (Section 14.6)
- Use standard enthalpies of formation to calculate $\Delta H°$ for reactions. (Section 14.7)

- Understand the meaning of spontaneous process, reversible process, irreversible process and isothermal process. (Section 14.8)
- State the second law of thermodynamics. (Section 14.9)
- Describe the kinds of molecular motion that a molecule can possess. (Section 14.10)
- Understand the concept of microstates and how they are related to entropy. (Section 14.10)
- State the third law of thermodynamics. (Section 14.10)
- Calculate the standard entropy changes for a system from standard molar entropies. (Section 14.11)
- Calculate entropy changes in the surroundings for an isothermal process. (Section 14.12)
- Calculate the Gibbs free energy from the enthalpy change and the entropy change at constant temperature. (Section 14.12)
- Use free energy changes to predict whether a reaction is spontaneous. (Section 14.12)
- Predict the effect of temperature on spontaneity given ΔH and ΔS. (Section 14.13)
- Calculate the Gibbs free energy under non-standard conditions. (Section 14.13)

KEY EQUATIONS

- Kinetic energy

$$E_k = \frac{1}{2}mu^2 \qquad [14.1]$$

- Potential energy

$$E_p = mgh \qquad [14.2]$$

- The change in internal energy

$$\Delta U = \Delta U_{final} - \Delta U_{initial} \qquad [14.5]$$

- Relates the change in internal energy to heat and work (first law of thermodynamics)

$$\Delta U = q + w \qquad [14.6]$$

- The work done by an expanding gas at constant pressure

$$w = -P\,\Delta V \qquad [14.9]$$

- Enthalpy change at constant pressure

$$\Delta H = \Delta U + P\,\Delta V = q_P \qquad [14.11]$$

- Heat gained or lost by the system based on specific heat, mass and change in temperature

$$q = C_s \times m \times \Delta T \qquad [14.18]$$

- Standard enthalpy change of a reaction from standard enthalpies of formation

$$\Delta_{rxn}H° = \sum n\Delta_f H°(\text{products}) - \sum m\Delta_f H°(\text{reactants}) \qquad [14.27]$$

- Relates entropy change to the heat absorbed or released in a reversible process

$$\Delta S = \frac{q_{rev}}{T}(\text{constant } T) \qquad [14.29]$$

- The second law of thermodynamics

  Reversible process

$$\Delta_{univ}S = \Delta_{sys}S + \Delta_{surr}S = 0 \qquad [14.30]$$

  Irreversible process

$$\Delta_{univ}S = \Delta_{sys}S + \Delta_{surr}S > 0$$

- Relates entropy to the number of microstates

$$S = k\ln W \qquad [14.31]$$

- To calculate the standard entropy change from standard molar entropies

$$\Delta S° = \sum nS°(\text{products}) - \sum mS°(\text{reactants}) \qquad [14.34]$$

- The entropy change of the surroundings at constant pressure and temperature

$$\Delta_{surr}S = \frac{-\Delta_{sys}H}{T} \qquad [14.35]$$

- Relates the Gibbs free energy change to enthalpy and entropy changes at constant temperature

$$\Delta G = \Delta H - T\Delta S \qquad [14.37]$$

- Relates the free energy change to the maximum work a process can perform

$$\Delta G = -w_{max} \qquad [14.40]$$

- To calculate standard free energy changes from standard free energies of formation

$$\Delta G° = \sum n\Delta_f G°(\text{products}) - \sum m\Delta_f G°(\text{reactants}) \qquad [14.41]$$

EXERCISES

VISUALISING CONCEPTS

14.1 Imagine a book that is falling from a shelf. At a particular moment during its fall, it has a kinetic energy of 13 J and a potential energy with respect to the floor of 72 J. How do its kinetic energy and its potential energy change as it continues to fall? What is its total kinetic energy at the instant just before it strikes the floor? [Section 14.1]

14.2 The contents of the closed box in each of the following illustrations represent a system, and the arrows show the changes to the system during some process. The lengths of the arrows represent the relative magnitudes of q and w. **(a)** Which of these processes is endothermic? **(b)** For which of these processes, if any, is $\Delta U < 0$? **(c)** For which process, if any, is there a net gain in internal energy? [Section 14.2]

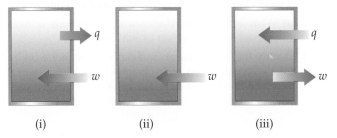

(i)   (ii)   (iii)

14.3 In the cylinder shown below, a chemical process occurs at constant temperature and pressure. Is the sign of w indicated by this change positive or negative? If the process is endothermic, does the internal energy of the system within the cylinder increase or decrease during the change? Is ΔU positive or negative? [Sections 14.2 and 14.3]

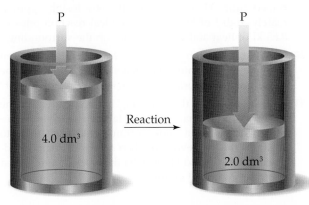

14.4 Imagine a container placed in a tub of water, as depicted below. **(a)** If the contents of the container are the system and heat is able to flow through the container walls, what qualitative changes will occur in the temperatures of the system and in its surroundings? What is the sign of q associated with each change? From the system's perspective, is the process exothermic or endothermic? **(b)** If neither the volume nor the pressure of the system changes during the process, how is the change in internal energy related to the change in enthalpy? [Sections 14.2 and 14.3]

14.5 Two different gases occupy two separate bulbs. Consider the process that occurs when the stopcock separating the gases is opened, assuming the gases behave ideally. **(a)** Draw the final (equilibrium) state. **(b)** Predict the signs of ΔH and ΔS for the process. **(c)** Is the process that occurs when the stopcock is opened a reversible one? **(d)** How does the process affect the entropy of the surroundings? [Sections 14.8 and 14.9]

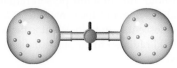

14.6 **(a)** What are the signs of ΔS, ΔH and ΔG for the process depicted below? **(b)** If energy can flow in and out of the system, what can you say about the entropy change of the surroundings as a result of this process? [Sections 14.9 and 14.12]

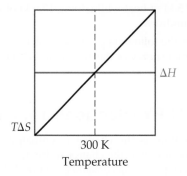

14.7 The diagram below shows how ΔH (red line) and $T\Delta S$ (blue line) change with temperature for a hypothetical reaction. **(a)** What is the significance of the point at 300 K, where ΔH and $T\Delta S$ are equal? **(b)** In what temperature range is this reaction spontaneous? [Section 14.13]

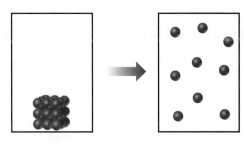

14.8 Consider a reaction $A_2(g) + B_2(g) \rightleftharpoons 2\ AB(g)$, with atoms of A shown in red here and atoms of B shown in blue. **(a)** If box 1 represents an equilibrium mixture, what is the sign of ΔG for the process in which the contents of a reaction vessel change from what is shown in box 2 to what is shown in box 1? **(b)** Rank the boxes in order of increasing magnitude of ΔG for the reaction. [Section 14.12]

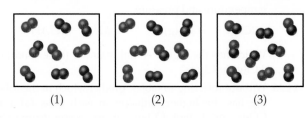

(1)   (2)   (3)

THE NATURE OF ENERGY (Section 14.1)

14.9 (a) Calculate the kinetic energy in joules of a 45 g golf ball moving at 61 m s^{-1}. (b) What happens to this energy when the ball lands in a sand trap?

14.10 A watt is a measure of power (the rate of energy change) equal to 1 J s^{-1}. (a) Calculate the number of joules in a kilowatt-hour. (b) An adult person radiates heat to the surroundings at about the same rate as a 100 watt electric incandescent light bulb. What is the total amount of energy in kJ radiated to the surroundings by an adult in 24 hours?

14.11 Describe the source of the energy and the nature of the energy conversions involved when a 100 watt electric light bulb radiates energy to its surroundings. Compare this with the energy source and energy conversions in-

volved when an adult person radiates energy to the surroundings.

14.12 (a) What is meant by the term *system* in thermodynamics? (b) What is a *closed system*?

14.13 (a) What is work? (b) How do we determine the amount of work done, given the force associated with the work?

14.14 (a) What is heat? (b) Under what conditions is heat transferred from one object to another?

14.15 Identify the force present, and explain whether work is done when (a) a positively charged particle moves in a circle at a fixed distance from a negatively charged particle, and (b) an iron nail is pulled off a magnet.

THE FIRST LAW OF THERMODYNAMICS (Section 14.2)

14.16 (a) State the first law of thermodynamics. (b) What is meant by the *internal energy* of a system? (c) By what means can the internal energy of a system increase?

14.17 (a) Write an equation that expresses the first law of thermodynamics. (b) Under what conditions will the quantities q and w be negative numbers?

14.18 Calculate ΔU, and determine whether the process is endothermic or exothermic for the following cases: (a) A system absorbs 85 kJ of heat from its surroundings while doing 29 kJ of work on the surroundings; (b) $q = 1.50$ kJ and $w = -657$ J; (c) the system releases 57.5 kJ of heat while doing 13.5 kJ of work on the surroundings.

14.19 For the following processes, calculate the change in internal energy of the system and determine whether the

process is endothermic or exothermic: (a) A balloon is heated by adding 900 J of heat. It expands, doing 422 J of work on the atmosphere. (b) A 50 g sample of water is cooled from 30 °C to 15 °C, thereby losing approximately 3140 J of heat. (c) A chemical reaction releases 8.65 kJ of heat and does no work on the surroundings.

14.20 (a) What is meant by the term *state function*? (b) Give an example of a quantity that is a state function and one that is not. (c) Is work a state function? Why or why not?

14.21 Indicate which of the following is independent of the path by which a change occurs: (a) the change in potential energy when a book is transferred from table to shelf, (b) the heat evolved when a cube of sugar is oxidised to $CO_2(g)$ and $H_2O(g)$, (c) the work accomplished in burning a litre of petrol.

ENTHALPY (Sections 14.3 and 14.4)

14.22 (a) Why is the change in enthalpy usually easier to measure than the change in internal energy? (b) For a given process at constant pressure, ΔH is negative. Is the process endothermic or exothermic?

14.23 (a) Under what condition will the enthalpy change of a process equal the amount of heat transferred into or out of the system? (b) During a constant-pressure process the system absorbs heat from the surroundings. Does the enthalpy of the system increase or decrease during the process?

14.24 Consider the following reaction which occurs at room temperature and pressure:

$$2Cl(g) \longrightarrow Cl_2(g) \qquad \Delta H = -243.4\,\text{kJ}$$

Which has the higher enthalpy under these conditions: 2 Cl(g) or $Cl_2(g)$?

14.25 Without referring to tables, predict which of the following has the higher enthalpy in each case: (a) 1 mol $CO_2(s)$ or 1 mol $CO_2(g)$ at the same temperature, (b) 2 mol of hydrogen atoms or 1 mol of H_2, (c) 1 mol

$H_2(g)$ and 0.5 mol $O_2(g)$ at 25 °C or 1 mol $H_2O(g)$ at 25 °C, (d) 1 mol $N_2(g)$ at 100 °C or 1 mol $N_2(g)$ at 300 °C.

14.26 Consider the following reaction:

$$CH_3OH(g) \longrightarrow CO(g) + 2H_2(g)$$
$$\Delta H = +90.7 \text{ kJ}$$

(a) Is heat absorbed or evolved in the course of this reaction? (b) Calculate the amount of heat transferred when 45.0 g of $CH_3OH(g)$ is decomposed by this reaction at constant pressure. (c) For a given sample of CH_3OH, the enthalpy change on reaction is 18.5 kJ. How many grams of hydrogen gas are produced? (d) What is the value of ΔH for the reverse of the previous reaction? How many kilojoules of heat are released when 27.0 g of CO(g) reacts completely with $H_2(g)$ to form $CH_3OH(g)$ at constant pressure?

14.27 When solutions containing silver ions and chloride ions are mixed, silver chloride precipitates:

$$Ag^+(aq) + Cl^-(aq) \longrightarrow AgCl(s) \qquad \Delta H = -65.5 \text{ kJ}$$

(a) Calculate ΔH for formation of 0.200 mol of AgCl by this reaction. **(b)** Calculate ΔH for the formation of 2.50 g of AgCl. **(c)** Calculate ΔH when 0.150 mol of AgCl dissolves in water.

14.28 At one time, a common means of forming small quantities of oxygen gas in the laboratory was to heat $KClO_3$:

$$2\ KClO_3(s) \longrightarrow 2\ KCl(s) + 3\ O_2(g) \quad \Delta H = -89.4\ kJ$$

For this reaction, calculate ΔH for the formation of **(a)** 0.855 mol of O_2 and **(b)** 10.75 g of KCl. **(c)** The decomposition of $KClO_3$ proceeds spontaneously when it is heated. Do you think that the reverse reaction, the formation of $KClO_3$ from KCl and O_2, is likely to be feasible under ordinary conditions? Explain your answer.

14.29 You are given ΔH for a process that occurs at constant pressure. What additional information is needed to determine ΔU for the process?

14.30 Suppose the gas-phase reaction $2\ NO(g) + O_2(g) \longrightarrow 2\ NO_2(g)$ is conducted in a constant-volume container at constant temperature. Would the measured heat change represent ΔH or ΔU? If there is a difference, which quantity is larger for this reaction? Explain.

14.31 A gas is confined to a cylinder under constant atmospheric pressure, as illustrated in Figure 14.3. When the gas undergoes a particular chemical reaction, it releases 79 kJ of heat to its surroundings and does 18 kJ of P-V work on its surroundings. What are the values of ΔH and ΔU for this process?

14.32 Consider the combustion of liquid methanol, $CH_3OH(l)$:

$$CH_3OH(l) + \tfrac{3}{2} O_2(g) \longrightarrow CO_2(g) + 2\ H_2O(l)$$

$$\Delta H = -726.5\ kJ$$

(a) What is the enthalpy change for the reverse reaction? **(b)** Balance the forward reaction with whole-number coefficients. What is ΔH for the reaction represented by this equation? **(c)** Which is more likely to be thermodynamically favoured, the forward reaction or the reverse reaction? **(d)** If the reaction were written to produce $H_2O(g)$ instead of $H_2O(l)$, would you expect the magnitude of ΔH to increase, decrease or stay the same? Explain.

CALORIMETRY (Section 14.5)

14.33 **(a)** What are the units of heat capacity? **(b)** What are the units of specific heat capacity? **(c)** If you know the specific heat capacity of copper, what additional information do you need to calculate the heat capacity of a particular piece of copper pipe?

14.34 **(a)** What is the specific heat capacity of liquid water? **(b)** What is the molar heat capacity of liquid water? **(c)** What is the heat capacity of 185 g of liquid water? **(d)** How many kJ of heat are needed to raise the temperature of 10.00 kg of liquid water from 24.6 °C to 46.2 °C?

14.35 When a 9.55 g sample of solid sodium hydroxide dissolves in 100.0 g of water in a coffee-cup calorimeter (Figure 14.15), the temperature rises from 23.6 °C to 47.4 °C. Calculate ΔH (in $kJ\ mol^{-1}$ NaOH) for the solution process

$$NaOH(s) \longrightarrow Na^+(aq) + OH^-(aq)$$

Assume that the specific heat capacity of the solution is the same as that of pure water.

14.36 **(a)** When a 3.88 g sample of solid ammonium nitrate dissolves in 60.0 g of water in a coffee-cup calorimeter (Figure 14.15), the temperature drops from 23.0 °C to 18.4 °C. Calculate ΔH (in $kJ\ mol^{-1}$ NH_4NO_3) for the solution process

$$NH_4NO_3(s) \longrightarrow NH_4^+(aq) + NO_3^-(aq)$$

Assume that the specific heat capacity of the solution is the same as that of pure water. **(b)** Is this process endothermic or exothermic?

14.37 A 2.200 g sample of quinone ($C_6H_4O_2$) is burned in a bomb calorimeter whose total heat capacity is 7.854 $kJ\ °C^{-1}$. The temperature of the calorimeter increases from 23.44 °C to 30.57 °C. What is the heat of combustion per gram of quinone? Per mole of quinone?

14.38 Under constant-volume conditions the heat of combustion of benzoic acid ($C_7H_6O_2$) is 26.38 $kJ\ g^{-1}$. A 1.640 g sample of benzoic acid is burned in a bomb calorimeter. The temperature of the calorimeter increases from 22.25 °C to 27.20 °C. **(a)** What is the total heat capacity of the calorimeter? **(b)** A 1.320 g sample of a new organic substance is combusted in the same calorimeter. The temperature of the calorimeter increases from 22.14 °C to 26.82 °C. What is the heat of combustion per gram of the new substance? **(c)** Suppose that, in changing samples, a portion of the water in the calorimeter is lost. In what way, if any, would this change the heat capacity of the calorimeter?

HESS'S LAW (Section 14.6)

14.39 What is the connection between Hess's law and the fact that H is a state function?

14.40 Consider the following hypothetical reactions:

$$A \longrightarrow B \quad \Delta H = +30\ kJ$$
$$B \longrightarrow C \quad \Delta H = +60\ kJ$$

(a) Use Hess's law to calculate the enthalpy change for the reaction $A \longrightarrow C$. **(b)** Construct an enthalpy diagram for substances A, B and C, and show how Hess's law applies.

14.41 Calculate the enthalpy change for the reaction:

$$P_4O_6(s) + 2\ O_2(g) \longrightarrow P_4O_{10}(s)$$

given the following enthalpies of reaction:

$$P_4(s) + 3\ O_2(g) \longrightarrow P_4O_6(s) \quad \Delta H = -1640.1\ kJ$$
$$P_4(s) + 5\ O_2(g) \longrightarrow P_4O_{10}(s) \quad \Delta H = -2940.1\ kJ$$

14.42 From the enthalpies of reaction:

$$H_2(g) + F_2(g) \longrightarrow 2\,HF(g) \qquad \Delta H = -537\text{ kJ}$$
$$C(s) + 2\,F_2(g) \longrightarrow CF_4(g) \qquad \Delta H = -680\text{ kJ}$$
$$2\,C(s) + 2\,H_2(g) \longrightarrow C_2H_4(g) \qquad \Delta H = +52.3\text{ kJ}$$

calculate ΔH for the reaction of ethylene with F_2:

$$C_2H_4(g) + 6\,F_2(g) \longrightarrow 2\,CF_4(g) + 4\,HF(g).$$

14.43 Given the data:

$$N_2(g) + O_2(g) \longrightarrow 2\,NO(g) \qquad \Delta H = +180.7\text{ kJ}$$
$$2\,NO(g) + O_2(g) \longrightarrow 2\,NO_2(g) \quad \Delta H = -113.1\text{ kJ}$$
$$2\,N_2O(g) \longrightarrow 2\,N_2(g) + O_2(g) \quad \Delta H = -163.2\text{ kJ}$$

use Hess's law to calculate ΔH for the reaction:

$$N_2O(g) + NO_2(g) \longrightarrow 3\,NO(g).$$

ENTHALPIES OF FORMATION (Section 14.7)

14.44 (a) What is meant by the term *standard conditions*, with reference to enthalpy changes? (b) What is meant by the term *enthalpy of formation*? (c) What is meant by the term *standard enthalpy of formation*?

14.45 (a) Why are tables of standard enthalpies of formation so useful? (b) What is the value of the standard enthalpy of formation of an element in its most stable form? (c) Write the chemical equation for the reaction whose enthalpy change is the standard enthalpy of formation of glucose, $C_6H_{12}O_6(s)$, $\Delta_f H°[C_6H_{12}O_6]$.

14.46 For each of the following compounds, write a balanced thermochemical equation depicting the formation of one mole of the compound from its elements in their standard states, and use Appendix C to obtain the value of $\Delta_f H°$: (a) $NH_3(g)$, (b) $SO_2(g)$, (c) $RbClO_3(s)$, (d) $NH_4NO_3(s)$.

14.47 The following is known as the thermite reaction (Figure 14.7(b)):

$$2\,Al(s) + Fe_2O_3(s) \longrightarrow Al_2O_3(s) + 2\,Fe(l)$$

This highly exothermic reaction is used for welding massive units, such as propellers for large ships. Using enthalpies of formation in Appendix C, calculate $\Delta H°$ for this reaction.

14.48 Many cigarette lighters contain liquid butane, $C_4H_{10}(l)$. Using enthalpies of formation, calculate the quantity of heat produced when 1.0 g of butane is completely combusted in air.

14.49 Using values from Appendix C, calculate the standard enthalpy change for each of the following reactions:

(a) $SO_2(g) + O_2(g) \longrightarrow 2\,SO_3(g)$
(b) $Mg(OH)_2(s) \longrightarrow MgO(s) + H_2O(l)$
(c) $CH_4(g) + 4\,Cl_2(g) \longrightarrow CCl_4(l) + 4\,HCl(g)$
(d) $SiCl_4(l) + 2\,H_2O(l) \longrightarrow SiO_2(s) + 4\,HCl(g)$

14.50 Complete combustion of 1 mol of acetone (C_3H_6O) liberates 1790 kJ:

$$C_3H_6O(l) + 4\,O_2(g) \longrightarrow 3\,CO_2(g) + 3\,H_2O(l)$$
$$\Delta H° = -1790\text{ kJ}$$

Using this information together with data from Appendix C, calculate the enthalpy of formation of acetone.

14.51 Calcium carbide (CaC_2) reacts with water to form acetylene (C_2H_2) and $Ca(OH)_2$. From the following enthalpy of reaction data and data in Appendix C, calculate $\Delta_f H°$ for $CaC_2(s)$:

$$CaC_2(s) + 2\,H_2O(l) \longrightarrow Ca(OH)_2(s) + C_2H_2(g)$$
$$\Delta H° = -127.2\text{ kJ}$$

14.52 Calculate the standard enthalpy of formation of solid $Mg(OH)_2$, given the following data:

$$2\,Mg(s) + O_2(g) \longrightarrow 2\,MgO(s) \qquad \Delta H° = -1203.6\text{ kJ}$$
$$Mg(OH)_2(s) \longrightarrow MgO(s) + H_2O(l) \quad \Delta H° = +37.1\text{ kJ}$$
$$2\,H_2(g) + O_2(g) \longrightarrow 2\,H_2O(l) \qquad \Delta H° = -571.7\text{ kJ}$$

14.53 Naphthalene ($C_{10}H_8$) is a solid aromatic compound often sold as mothballs. The complete combustion of this substance to yield $CO_2(g)$ and $H_2O(l)$ at 25 °C yields 5154 kJ mol^{-1}. (a) Write balanced equations for the formation of naphthalene from the elements and for its combustion. (b) Calculate the standard enthalpy of formation of naphthalene.

SPONTANEOUS PROCESSES (Section 14.8)

14.54 Which of the following processes are spontaneous and which are non-spontaneous? (a) The melting of ice cubes at −5 °C and 1 bar pressure; (b) dissolution of sugar in a cup of hot coffee; (c) the reaction of nitrogen atoms to form N_2 molecules at 25 °C and 1 bar; (d) alignment of iron filings in a magnetic field, (e) formation of CH_4 and O_2 molecules from CO_2 and H_2O at room temperature and 1 bar of pressure.

14.55 Which of the following processes are spontaneous? (a) Spreading of the fragrance of perfume through a room; (b) separating a mixture of N_2 and O_2 into two separate samples, one that is pure N_2 and one that is pure O_2; (c) the bursting of an inflated balloon; (d) the reaction of sodium metal with chlorine gas to form sodium chloride; (e) the dissolution of $HCl(g)$ in water to form concentrated hydrochloric acid.

14.56 (a) Give two examples of endothermic processes that are spontaneous. (b) Give an example of a process that is spontaneous at one temperature but non-spontaneous at a different temperature.

14.57 Consider the vaporisation of liquid water to steam at a pressure of 1 bar. (a) Is this process endothermic or exothermic? (b) In what temperature range is it a spontaneous process? (c) In what temperature range is it a non-spontaneous process? (d) At what temperature are the two phases in equilibrium?

14.58 (a) What is special about a *reversible* process? (b) Suppose a reversible process is reversed, restoring the system to its original state. What can be said about the surroundings after the process is reversed? (c) Under what circumstances will the vaporisation of water to steam be a reversible process?

14.59 **(a)** What is meant by calling a process *irreversible*? **(b)** After an irreversible process the system is restored to its original state. What can be said about the condition of the surroundings after the system is restored to its original state? **(c)** Under what conditions will the condensation of a liquid be an irreversible process?

14.60 Consider a process in which an ideal gas changes from state 1 to state 2 in such a way that its temperature changes from 300 K to 200 K. Does the change in ΔU depend on the particular pathway taken to carry out this change of state? Explain.

14.61 A system goes from state 1 to state 2 and back to state 1. **(a)** What is the relationship between the value of ΔU for going from state 1 to state 2 and that for going from state 2 back to state 1? **(b)** Without further information, can you conclude anything about the amount of heat transferred to the system as it goes from state 1 to state 2 compared with that as it goes from state 2 back to state 1? **(c)** Suppose the changes in state are reversible processes. Can you conclude anything about the work done by the system as it goes from state 1 to state 2 compared with that as it goes from state 2 back to state 1?

14.62 Consider a system consisting of an ice cube. **(a)** Under what conditions can the ice cube melt reversibly? **(b)** If the ice cube melts reversibly, is ΔU zero for the process? Explain.

ENTROPY AND THE SECOND LAW OF THERMODYNAMICS (Section 14.9)

14.63 **(a)** How can we calculate ΔS for an isothermal process? **(b)** Does ΔS for a process depend on the path taken from the initial to the final state of the system? Explain.

14.64 **(a)** Give an example of a process in which the entropy of the system decreases. **(b)** What is the sign of ΔS for the process? **(c)** What is the significance of the statement that entropy is a state function?

14.65 The element caesium (Cs) freezes at 28.4 °C, and its molar enthalpy of fusion is $\Delta_{fus}H = 2.09$ kJ mol^{-1} **(a)** When molten caesium solidifies to Cs(s) at its normal melting point, is ΔS positive or negative? **(b)** Calculate the value of ΔS when 15.0 g of Cs(l) solidifies at 28.4 °C.

14.66 **(a)** Express the second law of thermodynamics in words. **(b)** If the entropy of the system increases during a reversible process, what can you say about the entropy change of the surroundings? **(c)** In a certain spontaneous process the system undergoes an entropy change, $\Delta S = 42$ J K^{-1}. What can you conclude about $\Delta_{surr}S$?

14.67 The volume of 0.100 mol of helium gas at 27 °C is increased isothermally from 2.00 dm^3 to 5.00 dm^3. Assuming the gas to be ideal, calculate the entropy change for the process.

THE MOLECULAR INTERPRETATION OF ENERGY (Section 14.10)

14.68 How does each of the following affect the number of microstates available to a system: **(a)** temperature, **(b)** volume, **(c)** state of matter?

14.69 **(a)** Using the heat of vaporisation in Appendix B, calculate the entropy change for the vaporisation of water at 25 °C and at 100 °C. **(b)** From your knowledge of microstates and the structure of liquid water, explain the difference in these two values.

14.70 **(a)** What do you expect for the sign of ΔS in a chemical reaction in which two moles of gaseous reactants are converted to three moles of gaseous products? **(b)** For which of the processes in Exercise 14.43 does the entropy of the system increase?

14.71 How does the entropy of the system change when **(a)** a solid melts, **(b)** a gas liquefies, **(c)** a solid sublimes?

14.72 How does the entropy of the system change when **(a)** the temperature of the system increases, **(b)** the volume of a gas increases, **(c)** a solid dissolves in water?

14.73 **(a)** State the third law of thermodynamics. **(b)** Distinguish between translational motion, vibrational motion and rotational motion of a molecule. **(c)** Illustrate these three kinds of motion with sketches for the HCl molecule.

14.74 **(a)** The energy of a gas is increased by heating it. Using CO_2 as an example, illustrate the different ways in which additional energy can be distributed between the molecules of the gas. **(b)** You are told that the number of microstates for a system increases. What do you know about the entropy of the system?

14.75 Propanol (C_3H_7OH) melts at –126.5 °C and boils at 97.4 °C at 1 bar. Draw a qualitative sketch of how the entropy changes as propanol vapour at 150 °C and 1 bar is cooled to solid propanol at –150 °C and 1 bar.

14.76 Predict the sign of the entropy change of the system for each of the following reactions:

(a) $2\,SO_2(g) + O_2(g) \longrightarrow 2\,SO_3(g)$

(b) $Ba(OH)_2(s) \longrightarrow BaO(s) + H_2O(g)$

(c) $CO(g) + 2\,H_2(g) \longrightarrow CH_3OH(l)$

(d) $FeCl_2(s) + H_2(g) \longrightarrow Fe(s) + 2\,HCl(g)$

ENTROPY CHANGES IN CHEMICAL REACTIONS (Section 14.11)

14.77 In each of the following pairs, which compound would you expect to have the higher standard molar entropy: **(a)** $C_2H_2(g)$ or $C_2H_6(g)$; **(b)** $CO_2(g)$ or $CO(g)$?

14.78 The standard entropies at 298 K for certain of the group 14 elements are as follows: C(s, diamond) = 2.43 J mol^{-1} K^{-1}; Si(s) = 18.81 J mol^{-1} K^{-1}; Ge(s) = 31.09 J mol^{-1} K^{-1};

and $Sn(s) = 51.18$ J mol^{-1} K^{-1}. All but Sn have the diamond structure. How do you account for the trend in the $S°$ values?

14.79 Using $S°$ values from Appendix C, calculate $\Delta S°$ values for the following reactions. In each case, account for the sign of $\Delta S°$.

(a) $C_2H_4(g) + H_2(g) \longrightarrow C_2H_6(g)$

(b) $N_2O_4(g) \longrightarrow 2 NO_2(g)$

(c) $Be(OH)_2(s) \longrightarrow BeO(s) + H_2O(g)$

(d) $2 CH_3OH(g) + 3 O_2(g) \longrightarrow 2 CO_2(g) + 4 H_2O(g)$

GIBBS FREE ENERGY (Sections 14.12 and 14.13)

14.80 **(a)** For a process that occurs at constant temperature, express the change in Gibbs free energy in terms of changes in the enthalpy and entropy of the system. **(b)** For a certain process that occurs at constant T and P, the value of ΔG is positive. What can you conclude? **(c)** What is the relationship between ΔG for a process and the rate at which it occurs?

14.81 **(a)** What is the meaning of the standard free energy change, $\Delta G°$, as compared with ΔG? **(b)** For any process that occurs at constant temperature and pressure, what is the significance of $\Delta G = 0$? **(c)** For a certain process, ΔG is large and negative. Does this mean that the process necessarily occurs rapidly?

14.82 For a certain chemical reaction, $\Delta H° = -35.4$ kJ and $\Delta S° = -85.5$ J K^{-1}. **(a)** Is the reaction exothermic or endothermic? **(b)** Does the reaction lead to an increase or decrease in the disorder of the system? **(c)** Calculate $\Delta G°$ for the reaction at 298 K. **(d)** Is the reaction spontaneous at 298 K under standard conditions?

14.83 Use data in Appendix C to calculate $\Delta H°$, $\Delta S°$ and $\Delta G°$ at 25 °C for each of the following reactions. In each case, show that $\Delta G° = \Delta H° - T\Delta S°$.

(a) $Ni(s) + Cl_2(g) \longrightarrow NiCl_2(s)$

(b) $CaCO_3(s, calcite) \longrightarrow CaO(s) + CO_2(g)$

(c) $P_4O_{10}(s) + 6 H_2O(l) \longrightarrow 4 H_3PO_4(aq)$

(d) $2 CH_3OH(l) + 3 O_2(g) \longrightarrow 2 CO_2(g) + 4 H_2O(l)$

14.84 Using data from Appendix C, calculate $\Delta G°$ for the following reactions. Indicate whether each reaction is spontaneous under standard conditions.

(a) $2 SO_2(g) + O_2(g) \longrightarrow 2 SO_3(g)$

(b) $NO_2(g) + N_2O(g) \longrightarrow 3 NO(g)$

(c) $6 Cl_2(g) + 2 Fe_2O_3(s) \longrightarrow 4 FeCl_3(s) + 3 O_2(g)$

(d) $SO_2(g) + 2 H_2(g) \longrightarrow S(s) + 2 H_2O(g)$

14.85 Cyclohexane (C_6H_{12}) is a liquid hydrocarbon at room temperature. **(a)** Write a balanced equation for the combustion of $C_6H_{12}(l)$ to form $CO_2(g)$ and $H_2O(l)$. **(b)** Without using thermochemical data, predict whether $\Delta G°$ for this reaction is more negative or less negative than $\Delta H°$.

14.86 Classify each of the following reactions as one of the four possible types summarised in Table 14.6.

(a) $N_2(g) + 3 F_2(g) \longrightarrow 2 NF_3(g)$
$$\Delta H° = -249 \text{ kJ}; \Delta S° = -278 \text{ J K}^{-1}$$

(b) $N_2(g) + 3 Cl_2(g) \longrightarrow 2 NCl_3(g)$
$$\Delta H° = 460 \text{ kJ}; \Delta S° = -275 \text{ J K}^{-1}$$

(c) $N_2F_4(g) \longrightarrow 2 NF_2(g)$
$$\Delta H° = 85 \text{ kJ}; \Delta S° = 198 \text{ J K}^{-1}$$

14.87 From the values given for $\Delta H°$ and $\Delta S°$, calculate $\Delta G°$ for each of the following reactions at 298 K. If the reaction is not spontaneous under standard conditions at 298 K, at what temperature (if any) would the reaction become spontaneous?

(a) $2 PbS(s) + 3 O_2(g) \longrightarrow 2 PbO(s) + 2 SO_2(g)$
$$\Delta H° = -844 \text{ kJ}; \Delta S° = -165 \text{ J K}^{-1}$$

(b) $2 POCl_3(g) \longrightarrow 2 PCl_3(g) + O_2(g)$
$$\Delta H° = 572 \text{ kJ}; \Delta S° = 179 \text{ J K}^{-1}$$

14.88 A particular reaction is spontaneous at 450 K. The enthalpy change for the reaction is +34.5 kJ. What can you conclude about the sign and magnitude of ΔS for the reaction?

14.89 A certain reaction is non-spontaneous at -25 °C. The entropy change for the reaction is 95 J K^{-1}. What can you conclude about the sign and magnitude of ΔH?

14.90 For a particular reaction, $\Delta H = -32$ kJ and $\Delta S = -98$ J K^{-1}. Assume that ΔH and ΔS do not vary with temperature. **(a)** At what temperature will the reaction have $\Delta G = 0$? **(b)** If T is increased from that in part (a), will the reaction be spontaneous or non-spontaneous?

14.91 Consider the following reaction between oxides of nitrogen:

$NO_2(g) + N_2O(g) \longrightarrow 3 NO(g)$

(a) Use data in Appendix C to predict how $\Delta G°$ for the reaction varies with increasing temperature. **(b)** Calculate $\Delta G°$ at 800 K, assuming that $\Delta H°$ and $\Delta S°$ do not change with temperature. Under standard conditions is the reaction spontaneous at 800 K? **(c)** Calculate $\Delta G°$ at 1000 K. Is the reaction spontaneous under standard conditions at this temperature?

14.92 Acetylene gas, $C_2H_2(g)$, is used in welding. **(a)** Write a balanced equation for the combustion of acetylene gas to $CO_2(g)$ and $H_2O(l)$. **(b)** Use data in Appendix C to determine how much heat is produced in burning 1 mol of C_2H_2 under standard conditions if both reactants and products are brought to 298 K? **(c)** What is the maximum amount of useful work that can be accomplished under standard conditions by this reaction?

ADDITIONAL EXERCISES

14.93 Suppose an Olympic diver who weighs 52.0 kg executes a straight dive from a 10 m platform. At the apex of the dive, the diver is 10.8 m above the surface of the water. **(a)** What is the potential energy of the diver at the apex of the dive, relative to the surface of the water? **(b)** Assuming that all the potential energy of the diver is converted into kinetic energy at the surface of the water, at what speed in m s^{-1} will the diver enter the water? **(c)** Does the diver do work on entering the water? Explain.

14.94 The air bags that provide protection in cars in the event of an accident expand as a result of a rapid chemical reaction. From the viewpoint of the chemical reactants as the system, what do you expect for the signs of q and w in this process?

14.95 Limestone stalactites and stalagmites are formed in caves by the following reaction:

$$Ca^{2+}(aq) + 2\,HCO_3^-(aq) \longrightarrow$$
$$CaCO_3(s) + CO_2(g) + H_2O(l)$$

If 1 mol of $CaCO_3$ forms at 298 K under 1 bar pressure, the reaction performs 2.47 kJ of P-V work, pushing back the atmosphere as the gaseous CO_2 forms. At the same time, 38.95 kJ of heat is absorbed from the environment. What are the values of ΔH and ΔU for this reaction?

14.96 Consider the systems shown in Figure 14.9. In one case the battery becomes completely discharged by running the current through a heater, and in the other by running a fan. Both processes occur at constant pressure. In both cases the change in state of the system is the same: the battery goes from being fully charged to being fully discharged. Yet in one case the heat evolved is large, and in the other it is small. Is the enthalpy change the same in the two cases? If not, how can enthalpy be considered a state function? If it is, what can you say about the relationship between enthalpy change and q in this case compared with others that we have considered?

14.97 Use data in Appendix C to determine how many grams of methane ($CH_4(g)$) must be combusted to heat 1.00 kg of water from 25.0 °C to 90.0 °C, assuming $H_2O(l)$ as a product and 100% efficiency in heat transfer?

14.98 Burning methane in oxygen can produce three different carbon-containing products: soot (very fine particles of graphite), $CO(g)$ and $CO_2(g)$. **(a)** Write three balanced equations for the reaction of methane gas with oxygen to produce these three products. In each case assume that $H_2O(l)$ is the only other product. **(b)** Use data in Appendix C to determine the standard enthalpies for the reactions in part (a). **(c)** Why, when the oxygen supply is adequate, is $CO_2(g)$ the predominant carbon-containing product of the combustion of methane?

14.99 From the following data for three prospective fuels, calculate which would provide the most energy per unit volume:

Fuel	Density at 20 °C (g cm^{-3})	Molar enthalpy of combustion (kJ mol^{-1})
Nitroethane, $C_2H_5NO_2$(l)	1.052	−1368
Ethanol, C_2H_5OH(l)	0.789	−1367
Methylhydrazine, CH_6N_2(l)	0.874	−1305

14.100 The hydrocarbons acetylene (C_2H_2) and benzene (C_6H_6) have the same empirical formula. Benzene is an 'aromatic' hydrocarbon, one that is unusually stable because of its structure. **(a)** By using the data in Appendix C, determine the standard enthalpy change for the reaction $3\,C_2H_2(g) \longrightarrow C_6H_6(l)$. **(b)** Which has greater enthalpy, 3 mol of acetylene gas or 1 mol of liquid benzene? **(c)** Determine the fuel value in kJ g^{-1} for acetylene and benzene.

14.101 The two common sugars, glucose ($C_6H_{12}O_6$) and sucrose ($C_{12}H_{22}O_{11}$), are both carbohydrates. Their standard enthalpies of formation are given in Table 14.3. Using these data together with data in Appendix C, **(a)** calculate the molar enthalpy of combustion to $CO_2(g)$ and $H_2O(l)$ for the two sugars and **(b)** calculate the enthalpy of combustion per gram of each sugar.

14.102 It is estimated that the net amount of carbon dioxide fixed by photosynthesis on the landmass of the Earth is 55×10^{16} g year^{-1} of CO_2. All this carbon is converted into glucose. **(a)** Calculate the energy stored by photosynthesis on land per year in kJ. **(b)** Calculate the average rate of conversion of solar energy into plant energy in MW (1 W = 1 J s^{-1}). A large nuclear power plant produces about 10^3 MW. The energy of how many such nuclear power plants is equivalent to the solar energy conversion?

14.103 For each of the following processes, indicate whether the signs of ΔS and ΔH are expected to be positive, negative or about zero. **(a)** A solid sublimes. **(b)** The temperature of a sample of $Co(s)$ is lowered from 60 °C to 25 °C. **(c)** Ethanol evaporates from a beaker. **(d)** A diatomic molecule dissociates into atoms. **(e)** A piece of charcoal is combusted to form $CO_2(g)$ and $H_2O(g)$.

14.104 The reaction $2\,Mg(s) + O_2(g)\ 2\,MgO(s)$ is highly spontaneous and has a negative value for $\Delta S°$. The second law of thermodynamics states that in any spontaneous process there is always an increase in the entropy of the universe. Is there an inconsistency between the above reaction and the second law?

14.105 Ammonium nitrate dissolves spontaneously and endothermally in water at room temperature. What can you deduce about the sign of ΔS for this solution process?

14.106 For the majority of the compounds listed in Appendix C, the value of $\Delta_f G°$ is more positive (or less negative) than the value of $\Delta_f H°$. **(a)** Explain this observation, using $NH_3(g)$, $CCl_4(l)$ and $KNO_3(s)$ as examples. **(b)** An exception to this observation is $CO(g)$. Explain the trend in the $\Delta_f H°$ and $\Delta_f G°$ values for this molecule.

14.107 The potassium ion concentration in blood plasma is about 5.0×10^{-3} M, whereas the concentration in muscle-cell fluid is much greater (0.15 M). The plasma and intracellular fluid are separated by the cell membrane, which we assume is permeable only to K^+. **(a)** What is ΔG for the transfer of 1 mol of K^+ from blood plasma to the cellular fluid at body temperature (37 °C)? **(b)** What is the minimum amount of work that must be used to transfer this K^+?

INTEGRATIVE EXERCISES

14.108 Using data in Appendix C consider the combustion of a single molecule of $CH_4(g)$ forming $H_2O(l)$ as a product. **(a)** How much energy, in J, is produced during this reaction? **(b)** A typical X-ray photon has an energy of 8×10^3 eV. How does the energy of combustion compare with the energy of the X-ray photon ($1 \text{ eV} = 1.602 \times 10^{-19}$ J)?

14.109 Consider the dissolving of NaCl in water, illustrated in Figure 4.3. Let's say that the system consists of 0.1 mol NaCl and 1 dm^3 of water. Considering that the NaCl readily dissolves in the water and that the ions are strongly stabilised by the water molecules, as shown in the figure, is it safe to conclude that the dissolution of NaCl in water results in a lower enthalpy for the system? Explain your response. What experimental evidence would you examine to test this question?

14.110 Consider the following unbalanced oxidation–reduction reactions in aqueous solution:

$$Ag^+(aq) + Li(s) \longrightarrow Ag(s) + Li^+(aq)$$
$$Fe(s) + Na^+(aq) \longrightarrow Fe^{2+}(aq) + Na(s)$$
$$K(s) + H_2O(l) \longrightarrow KOH(aq) + H_2(g)$$

(a) Balance each of the reactions. **(b)** By using data in Appendix C, calculate $\Delta H°$ for each of the reactions. **(c)** Based on the values you obtain for $\Delta H°$, which of the reactions would you expect to be favourable? Which would you expect to be unfavourable? **(d)** Use the activity series to predict which of these reactions should occur (Table 4.4). Are these results in accord with your conclusion in part (c) of this problem?

14.111 The metathesis reaction between $AgNO_3(aq)$ and $NaCl(aq)$ proceeds as follows:

$$AgNO_3(aq) + NaCl(aq) \longrightarrow NaNO_3(aq) + AgCl(s)$$

(a) By using Appendix C, calculate $\Delta H°$ for the net ionic equation of this reaction. **(b)** What would you expect for the value of $\Delta H°$ of the overall molecular equation compared with that for the net ionic equation? Explain. **(c)** Use the results from (a) and (b) along with data in Appendix C to determine the value of $\Delta_f H°$ for $AgNO_3(aq)$.

14.112 A sample of a hydrocarbon is combusted completely in $O_2(g)$ to produce 21.83 g $CO_2(g)$, 4.47 g $H_2O(g)$ and 311 kJ of heat. **(a)** What is the mass of the hydrocarbon sample that was combusted? **(b)** What is the empirical formula of the hydrocarbon? **(c)** Calculate the value of $\Delta_f H°$ per empirical formula unit of the hydrocarbon. **(d)** Do you think that the hydrocarbon is one of those listed in Appendix C? Explain your answer.

14.113 Most liquids follow Trouton's rule, which states that the molar entropy of vaporisation lies in the range of 88 ± 5 J mol^{-1} K^{-1}. The normal boiling points and enthalpies of vaporisation of several organic liquids are as follows:

Substance	Normal boiling point (°C)	$\Delta_{vap}H$ (kJ mol^{-1})
Acetone, $(CH_3)_2CO$	56.1	29.1
Dimethyl ether, $(CH_3)_2O$	−24.8	21.5
Ethanol, C_2H_5OH	78.4	38.6
Octane, C_8H_{18}	125.6	34.4
Pyridine, C_5H_5N	115.3	35.1

(a) Calculate $\Delta_{vap}S$ for each of the liquids. Do all the liquids obey Trouton's rule? **(b)** With reference to inter-molecular forces (Section 11.2), can you explain any exceptions to the rule? **(c)** Would you expect water to obey Trouton's rule? By using data in Appendix B, check the accuracy of your conclusion. **(d)** Chlorobenzene (C_6H_5Cl) boils at 131.8 °C. Use Trouton's rule to estimate $\Delta_{vap}H$ for this substance.

14.114 The following data compare the standard enthalpies and free energies of formation of some crystalline ionic substances and aqueous solutions of the substances:

Substance	$\Delta_f H°$ (kJ mol^{-1})	$\Delta_f G°$ (kJ mol^{-1})
$AgNO_3(s)$	−124.4	−33.4
$AgNO_3(aq)$	−101.7	−34.2
$MgSO_4(s)$	−1283.7	−1169.6
$MgSO_4(aq)$	−1374.8	−1198.4

(a) Write the formation reaction for $AgNO_3(s)$. Based on this reaction, do you expect the entropy of the system to increase or decrease upon the formation of $AgNO_3(s)$? **(b)** Use $\Delta_f H°$ and $\Delta_f G°$ of $AgNO_3(s)$ to determine the entropy change upon formation of the substance. Is your answer consistent with your reasoning in part (a)? **(c)** Is dissolving $AgNO_3$ in water an exothermic or endothermic process? What about dissolving $MgSO_4$ in water? **(d)** For both $AgNO_3$ and $MgSO_4$, use the data to calculate the entropy change when the solid is dissolved in water.

14.115 When most elastomeric polymers (e.g. a rubber band) are stretched, the molecules become more ordered, as illustrated here:

Suppose you stretch a rubber band. **(a)** Do you expect the entropy of the system to increase or decrease? **(b)** If the rubber band were stretched isothermally, would heat need to be absorbed or emitted to maintain constant temperature?

MasteringChemistry (www.pearson.com.au/masteringchemistry)

Make learning part of the grade. Access:

- tutorials with peronalised coaching
- study area
- Pearson eText

PHOTO/ART CREDITS

500 © Du Huaju/Xinhua Press/Corbis; **502** Tom Pantages; **509 (b), 528 (Figure 14.20), 540** Richard Megna, Fundamental Photographs, NYC; **513** Charles Winters/Photo Researchers, Inc; **514** Corbis; **528 (Figure 14.21) (a)** Fundamental Photographs, NYC, **(Figure 42.1) (b), 537** © 1996 Michael Dalton, Fundamental Photographs, NYC.

15

CHEMICAL KINETICS

Fireworks rely on rapid chemical reactions both to propel them skyward and to produce their burst of light.

KEY CONCEPTS

15.1 FACTORS THAT AFFECT REACTION RATES
We begin by examining four variables that affect reaction rates: concentration, physical states of reactants, temperature and presence of catalysts. These factors can be understood in terms of the collisions between reactant molecules that lead to reaction.

15.2 REACTION RATES
We discuss the way to express *reaction rates* and how reactant disappearance rates and product appearance rates are related to the reaction stoichiometry.

15.3 CONCENTRATION AND RATE LAWS
We show the effect of concentration on rate is expressed quantitatively by *rate laws* and how rate laws and *rate constants* can be determined experimentally.

15.4 THE CHANGE OF CONCENTRATION WITH TIME
We learn that rate equations can be written to express how concentrations change with time and look at several examples of rate equations, namely for *zero-order*, *first-order* and *second-order* reactions.

15.5 TEMPERATURE AND RATE
We consider the effect of temperature on rate. We see that in order to occur, most reactions require a minimum input of energy called the *activation energy*.

15.6 REACTION MECHANISMS
We examine *reaction mechanisms*, the step-by-step molecular pathways leading from reactants to products.

15.7 CATALYSIS
We end by discussing how *catalysts* increase reaction rates, including catalysts in biological systems called *enzymes*.

Chemistry is, by its very nature, concerned with change. Chemical reactions convert substances with well-defined properties into other materials with different well-defined properties. Much of our study of chemical reactivity is concerned with the formation of new substances from a given set of reactants. However, it is equally important to understand how rapidly chemical reactions occur.

The rates of reactions span an enormous range, from those that are complete within fractions of seconds, such as explosions, to those that take thousands or even millions of years, such as the formation of diamonds or other minerals in the Earth's crust (▼ FIGURE 15.1). The fireworks shown in the chapter-opening photograph require very rapid reactions both to propel them skyward and to produce their colourful bursts of light. The chemicals used in the fireworks are chosen to give the desired colours and to do so very rapidly. The characteristic red, blue and green colours are produced by salts of strontium, copper and barium, respectively.

The area of chemistry that is concerned with the speeds, or rates, of reactions is called **chemical kinetics**. Chemical kinetics is a subject of broad importance. It relates, for example, to how quickly a medicine is able to work, to whether the formation and depletion of ozone in the upper atmosphere are in balance, and to industrial challenges such as the development of catalysts to synthesise new materials.

Our goal in this chapter is not only to understand how to determine the rates at which reactions occur but also to consider the factors that control these rates. For example, what factors determine how rapidly food spoils? How does one design a fast-setting material for dental fillings? What determines the rate at which steel rusts? What controls the rate at which fuel burns in a combustion engine? Although we won't address these specific questions directly, we will see that the rates of all chemical reactions are subject to the same basic principles.

▲ FIGURE 15.1 **Reaction rates.** The rates of chemical reactions span a range of time scales. For example, explosions are rapid, occurring in seconds or fractions of seconds; corrosion can take years; and the weathering of rocks takes place over thousands or even millions of years.

Steel wool heated in air (about 20% O_2) glows red-hot but oxidises to Fe_2O_3 slowly

Red-hot steel wool in 100% O_2 burns vigorously, forming Fe_2O_3 quickly

▲ **FIGURE 15.2** **Effect of concentration on reaction rate.** The difference in behaviour is due to the different concentrations of O_2 in the two environments.

∞ Review this on page 356

15.1 FACTORS THAT AFFECT REACTION RATES

Before we examine the quantitative aspects of chemical kinetics, such as how rates are measured, let's examine the key factors that influence the rates of reactions. Because reactions involve the breaking and forming of bonds, their speeds depend on the nature of the reactants themselves. There are, however, four factors that allow us to change the rates at which particular reactions occur.

1. **The physical state of the reactants.** For a reaction involving two or more reactants, the reactants must come together in order to react. The more readily molecules collide with each other, the more rapidly they react. Most of the reactions we consider are homogeneous, involving either gases or liquid solutions. When reactants are in different phases, such as when one is a gas and another a solid, the reaction is limited to their area of contact. Thus reactions that involve solids tend to proceed faster if the surface area of the solid is increased. For example, a medicine in the form of a tablet will dissolve in the stomach and enter the bloodstream more slowly than the same medicine in the form of a fine powder.

2. **The concentrations of the reactants.** Most chemical reactions proceed faster if the concentration of one or more of the reactants is increased. For example, steel wool burns with difficulty in air, which contains 20% O_2, (◄ FIGURE 15.2(a)) but bursts into a brilliant white flame in pure oxygen (Figure 52.2(b)). As concentration increases, the frequency with which the reactant molecules collide increases, leading to increased rates.

3. **The temperature at which the reaction occurs.** The rates of chemical reactions increase as temperature is increased. It is for this reason that we refrigerate perishable foods such as milk. The bacterial reactions that lead to the spoiling of milk proceed much more rapidly at room temperature than they do at the lower temperature of a refrigerator. Increasing temperature increases the kinetic energies of molecules. (∞ Section 10.7, 'Kinetic-Molecular Theory'). As molecules move more rapidly, they collide more frequently and also with higher energy, leading to increased reaction rates.

4. **The presence of a catalyst.** Catalysts are agents that increase reaction rates without being consumed during the reaction. They affect the mechanism of reaction, the individual bond-forming and bond-breaking steps that need to happen to get from the reactants to the products. Catalysts play a crucial role in our lives. The physiology of most living species depends on *enzymes*, protein molecules that act as catalysts, increasing the rates of selected biochemical reactions.

On a molecular level, reaction rates depend on the frequency of collisions between molecules. The greater the frequency of collisions, the greater the rate of reaction. For a collision to lead to reaction, however, it must occur with sufficient energy to allow interpenetration of the electron clouds, leading to a breaking of old bonds and the formation of new bonds. The orientation of the colliding sites must also be favourable so as to form the new bonds in the correct locations. We consider these factors as we proceed through the chapter.

▲ **CONCEPT CHECK 1**

How does increasing the partial pressures of the reactive components of a gaseous mixture affect the rate at which the components react with one another?

15.2 | REACTION RATES

The *speed* of an event is defined as the *change* that occurs in a given interval of *time*. Whenever we talk about speed, we necessarily bring in the notion of time. For example, the speed of a car is expressed as the change in the car's position over a certain period of time. The units of this speed are usually kilometres per hour (km h^{-1})—that is, the quantity that is changing (position, measured in kilometres) divided by a time interval (hours).

Similarly, the speed of a chemical reaction—its **reaction rate**—is the change in the concentration of reactants or products per unit time. Thus the units for reaction rate are usually molarity per second ($M \, s^{-1}$)—that is, the change in concentration (measured in molarity) divided by a time interval (seconds).

Let's consider a simple hypothetical reaction, A $\longrightarrow$ B, depicted in ▼ **FIGURE 15.3**. Each red sphere represents 0.01 mol of A and each blue sphere represents 0.01 mol of B. The container has a volume of 1.00 dm^3. At the beginning of the reaction there is 1.00 mol A. After 20 s the concentration of A has fallen to 0.54 M, whereas that of B has risen to 0.46 M. The sum of the concentrations is still 1.00 M because 1 mol of B is produced for each mol of A that reacts. After 40 s the concentration of A is 0.30 M and that of B is 0.70 M.

The rate of this reaction can be expressed either as the rate of disappearance of reactant A or as the rate of appearance of product B. The *average* rate of appearance of B over a particular time interval is given by the change in concentration of B divided by the change in time:

$$\text{Average rate of appearance of B} = \frac{\text{change in concentration of B}}{\text{change in time}}$$

$$= \frac{[B]t_2 - [B]t_1}{t_2 - t_1} = \frac{\Delta[B]}{\Delta t} \qquad [15.1]$$

We use square brackets around a chemical formula, as in [B], to indicate the concentration of the substance in molarity. The average rate of appearance of B over the 20 s interval from the beginning of the reaction ($t_1 = 0$ s to $t_2 = 20$ s) is given by

$$\text{Average rate} = \frac{0.46 \, M - 0.00 \, M}{20 \, s - 0 \, s} = 2.3 \times 10^{-2} \, M \, s^{-1}$$

FIGURE IT OUT

If A converts completely to B, what type of molecules will the container hold?

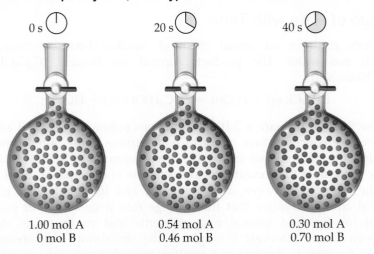

0 s	20 s	40 s
1.00 mol A	0.54 mol A	0.30 mol A
0 mol B	0.46 mol B	0.70 mol B

▲ **FIGURE 15.3   Progress of a hypothetical reaction A $\longrightarrow$ B.**

We could equally well express the rate of the reaction with respect to the change of concentration of the reactant, A. In this case we would be describing the rate of disappearance of A, which we express as

$$\text{Average rate of disappearance of A} = -\frac{\Delta[A]}{\Delta t} \qquad [15.2]$$

Notice the minus sign in this equation. By convention, *rates are always expressed as positive quantities*. Because [A] is decreasing with time, $\Delta[A]$ is a negative number. We use the negative sign to convert the negative $\Delta[A]$ to a positive rate. Because one molecule of A is consumed for every molecule of B that forms, the average rate of disappearance of A equals the average rate of appearance of B, as the following calculation shows:

$$\text{Average rate} = -\frac{\Delta[A]}{\Delta t} = -\frac{0.54\ \text{M} - 1.00\ \text{M}}{20\ \text{s} - 0\ \text{s}} = 2.3 \times 10^{-2}\ \text{M}\ \text{s}^{-1}$$

SAMPLE EXERCISE 15.1 **Calculating an average rate of reaction**

From the data given in the caption to Figure 15.3, calculate the average rate at which A disappears over the time interval from 20 s to 40 s.

SOLUTION

Analyse We are given the concentration of A at 20 s (0.54 M) and at 40 s (0.30 M) and asked to calculate the average rate of reaction over this time interval.

Plan The average rate is given by the change in concentration, $\Delta[A]$, divided by the corresponding change in time, Δt. Because A is a reactant, a minus sign is used in the calculation to make the rate a positive quantity—that is, we are calculating the rate of disappearance of A.

Solve $\text{Average rate} = -\dfrac{\Delta[A]}{\Delta t} = -\dfrac{0.30\ \text{M} - 0.54\ \text{M}}{40\ \text{s} - 20\ \text{s}} = 1.2 \times 10^{-2}\ \text{M}\ \text{s}^{-1}$

PRACTICE EXERCISE

For the reaction pictured in Figure 15.3, calculate the average rate of appearance of B over the time interval from 0 to 40 s. (The necessary data are given in the figure caption.)

Answer: $1.8 \times 10^{-2}\ \text{M s}^{-1}$

(See also Exercise 15.15.)

Change of Rate with Time

Now let's consider an actual chemical reaction between chlorobutane (C_4H_9Cl) and water. The products formed are butanol (C_4H_9OH) and hydrochloric acid:

$$C_4H_9Cl(aq) + H_2O(l) \longrightarrow C_4H_9OH(aq) + HCl(aq) \qquad [15.3]$$

Suppose that we prepare a 0.1000 M aqueous solution of C_4H_9Cl and then measure the concentration of C_4H_9Cl at various times after time zero, collecting the data shown in the first two columns of ▶ TABLE 15.1. We can use these data to calculate the average rate of disappearance of C_4H_9Cl over the intervals between measurements, and these rates are given in the third column. Notice that the average rate decreases over each 50 s interval for the first several measurements and continues to decrease over even larger intervals throughout the remaining measurements. *It is typical for rates to decrease as a reaction proceeds because the concentration of reactants decreases.* As the concentrations decrease the number of effective

TABLE 15.1 • Rate data for reaction of C_4H_9Cl with water

Time, t (s)	$[C_4H_9Cl]$(M)	Average rate (M s^{-1})
0.0	0.1000	1.9×10^{-4}
50.0	0.0905	1.7×10^{-4}
100.0	0.0820	1.6×10^{-4}
150.0	0.0741	1.4×10^{-4}
200.0	0.0671	1.22×10^{-4}
300.0	0.0549	1.01×10^{-4}
400.0	0.0448	0.80×10^{-4}
500.0	0.0368	0.560×10^{-4}
800.0	0.0200	
10 000	0	

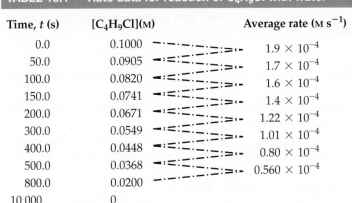

FIGURE IT OUT

How does the instantaneous rate of reaction change as the reaction proceeds?

$$C_4H_9Cl(aq) + H_2O(l) \longrightarrow C_4H_9OH(aq) + HCl(aq)$$

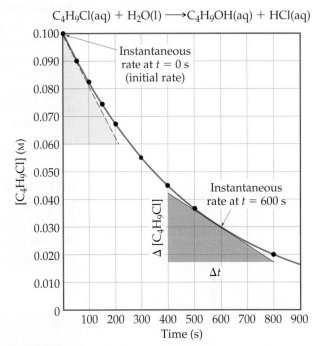

▲ **FIGURE 15.4   Concentration of butyl chloride (C_4H_9Cl) as a function of time.**

collisions decreases. The change in rate as the reaction proceeds is also seen in a graph of the concentration of C_4H_9Cl against time (▶ **FIGURE 15.4**). Notice how the steepness of the curve decreases with time, indicating a decreasing rate of reaction.

The graph shown in Figure 15.4 is particularly useful because it allows us to evaluate the **instantaneous rate**, the rate at a particular moment in the reaction. The instantaneous rate is determined from the slope (or tangent) of this curve at the point of interest. We have drawn two tangents in Figure 15.4, one at $t = 0$ and the other at $t = 600$ s. The slopes of these tangents give the instantaneous rates at these times.* For example, to determine the instantaneous rate at 600 s, we draw the tangent to the curve at this time, then construct horizontal and vertical lines to form the right-angled triangle shown. The slope is the ratio of the height of the vertical side to the length of the horizontal side:

$$\text{Instantaneous rate} = -\frac{\Delta[C_4H_9Cl]}{\Delta t} = -\frac{(0.017 - 0.042)\,\text{M}}{(800 - 400)\,\text{s}}$$

$$= 6.3 \times 10^{-5}\,\text{M s}^{-1}$$

In what follows, the term 'rate' means 'instantaneous rate' unless indicate otherwise. The instantaneous rate at $t = 0$ is called the *initial rate* of the reaction.

To understand better the difference between average rate and instantaneous rate, imagine that you have just driven 98 km in 2.0 hours. Your average speed is 49 km h^{-1}, whereas your instantaneous speed at any moment is the speedometer reading at that time.

---

* You may wish to review briefly the idea of graphical determination of slopes by referring to Appendix A. If you are familiar with calculus, you may recognise that the average rate approaches the instantaneous rate as the time interval approaches zero. This limit, in the notation of calculus, is represented as $-d[C_4H_9Cl]/dt$.

SAMPLE EXERCISE 15.2   Calculating an instantaneous rate of reaction

Using Figure 15.4, calculate the instantaneous rate of disappearance of C_4H_9Cl at $t = 0$ (the initial rate).

SOLUTION

Analyse We are asked to determine an instantaneous rate from a graph of reactant concentration versus time.

Plan To obtain the instantaneous rate at $t = 0$, we must determine the slope of the curve at $t = 0$. The tangent is drawn on the graph.

Solve The tangent line falls from $[C_4H_9Cl] = 0.100$ M to 0.060 M in the time change from 0 s to 200 s, as indicated by the upper triangle shown in Figure 15.4. Thus the initial rate is

$$\text{Rate} = -\frac{\Delta[C_4H_9Cl]}{\Delta t} = -\frac{(0.060 - 0.100)\text{ M}}{(200 - 0)\text{ s}} = 2.0 \times 10^{-4}\text{ M s}^{-1}$$

PRACTICE EXERCISE

Using Figure 15.4, determine the instantaneous rate of disappearance of C_4H_9Cl at $t = 300$ s.

Answer: 1.1×10^{-4} M s^{-1}

(See also Exercise 15.16.)

CONCEPT CHECK 2

What is the difference between average rate and instantaneous rate? In a given reaction, can these two rates ever have the same numerical values?

Reaction Rates and Stoichiometry

During our earlier discussion of the hypothetical reaction A ⟶ B, we saw that the stoichiometry requires that the rate of disappearance of A equals the rate of appearance of B. Likewise, the stoichiometry of Equation 15.3 indicates that 1 mol of C_4H_9OH is produced for each mol of C_4H_9Cl consumed. Therefore the rate of appearance of C_4H_9OH equals the rate of disappearance of C_4H_9Cl:

$$\text{Rate} = -\frac{\Delta[C_4H_9Cl]}{\Delta t} = \frac{\Delta[C_4H_9OH]}{\Delta t}$$

What happens when the stoichiometric relationships are not one to one? For example, consider the following reaction:

$$2\,HI(g) \longrightarrow H_2(g) + I_2(g)$$

We can measure the rate of disappearance of HI or the rate of appearance of either H_2 or I_2. Because 2 mol of HI disappear for each mol of H_2 or I_2 that forms, the rate of disappearance of HI is twice the rate of appearance of either H_2 or I_2. To equate the rates, we must therefore divide the rate of disappearance of HI by 2 (its coefficient in the balanced chemical equation):

$$\text{Rate} = -\frac{1}{2}\frac{\Delta[HI]}{\Delta t} = \frac{\Delta[H_2]}{\Delta t} = \frac{\Delta[I_2]}{\Delta t}$$

In general, for the reaction

$$a\,A + b\,B \longrightarrow c\,C + d\,D$$

the relationship between the rate of appearance or disappearance of a particular component is given by

$$\text{Rate} = -\frac{1}{a}\frac{\Delta[A]}{\Delta t} = -\frac{1}{b}\frac{\Delta[B]}{\Delta t} = \frac{1}{c}\frac{\Delta[C]}{\Delta t} = \frac{1}{d}\frac{\Delta[D]}{\Delta t} \qquad [15.4]$$

When we speak of the rate of a reaction without specifying a particular reactant or product, we mean it in this sense.*

SAMPLE EXERCISE 15.3 Relating rates at which products appear and reactants disappear

(a) How is the rate at which ozone disappears related to the rate at which oxygen appears in the reaction 2 O_3(g) $\longrightarrow$ 3 O_2(g)? **(b)** If the rate at which O_2 appears, $\Delta[O_2]/\Delta t$, is 6.0 $\times$ 10^{-5} M s^{-1} at a particular instant, at what rate is O_3 disappearing at this same time, $-\Delta[O_3]/\Delta t$?

SOLUTION

Analyse We are given a balanced chemical equation and asked to relate the rate of appearance of the product to the rate of disappearance of the reactant.

Plan We can use the coefficients of the chemical equation as shown in Equation 15.4 to express the relative rates of reactions.

Solve
(a) Using the coefficients in the balanced equation and the relationship given by Equation 15.4, we have:

$$\text{Rate} = -\frac{1}{2}\frac{\Delta[O_3]}{\Delta t} = \frac{1}{3}\frac{\Delta[O_2]}{\Delta t}$$

(b) Solving the equation from part (a) for the rate at which O_3 disappears, $-\Delta[O_3]/\Delta t$, we have:

$$-\frac{\Delta[O_3]}{\Delta t} = \frac{2}{3}\frac{\Delta[O_2]}{\Delta t} = \frac{2}{3}(6.0 \times 10^{-5} \text{ M s}^{-1}) = 4.0 \times 10^{-5} \text{ M s}^{-1}$$

PRACTICE EXERCISE
The decomposition of N_2O_5 proceeds according to the following equation:

$$2\,N_2O_5(g) \longrightarrow 4\,NO_2(g) + O_2(g)$$

If the rate of decomposition of N_2O_5 at a particular instant in a reaction vessel is 4.2 $\times$ 10^{-7} M s^{-1}, what is the rate of appearance of **(a)** NO_2, **(b)** O_2?

Answers: **(a)** 8.4 $\times$ 10^{-7} M s^{-1}, **(b)** 2.1 $\times$ 10^{-7} M s^{-1}

(See also Exercise 15.19.)

15.3 | CONCENTRATION AND RATE LAWS

One way of studying the effect of concentration on reaction rate is to determine the way in which the rate at the beginning of a reaction (the initial rate) depends on the starting concentrations. To illustrate this approach, consider the following reaction:

$$NH_4^+(aq) + NO_2^-(aq) \longrightarrow N_2(g) + 2\,H_2O(l)$$

We might study the rate of this reaction by measuring the concentration of NH_4^+ or NO_2^- as a function of time or by measuring the volume of N_2 collected. Because the stoichiometric coefficients on NH_4^+, NO_2^- and N_2 are all the same, all these rates will be equal.

▼ TABLE 15.2 shows the initial reaction rate for various starting concentrations of NH_4^+ and NO_2^-. These data indicate that changing either $[NH_4^+]$ or $[NO_2^-]$ changes the reaction rate. If we double $[NH_4^+]$ while holding $[NO_2^-]$ constant, the rate doubles (compare experiments 1 and 2). If $[NH_4^+]$ is increased by a factor of 4 with $[NO_2^-]$ left unchanged (compare experiments 1 and 3), the rate changes by a factor of 4 and so forth. These results indicate that the rate is directly proportional to $[NH_4^+]$. When $[NO_2^-]$ is similarly varied while $[NH_4^+]$ is held constant, the rate is affected in the same manner. Thus the rate is also directly proportional to the concentration of NO_2^-.

---

* Equation 15.4 does not hold true if substances other than C and D are formed in significant amounts during the course of the reaction. For example, sometimes intermediate substances build in concentration before forming the final products. In that case, the relationship between the rate of disappearance of reactants and the rate of appearance of products will not be given by Equation 15.4. All reactions whose rates we consider in this chapter obey Equation 15.4.

TABLE 15.2 • **Rate data for the reaction of ammonium and nitrite ions in water at 25 °C**

Experiment number	Initial NH_4^+ concentration (M)	Initial NO_2^- concentration (M)	Observed initial rate (M s^{-1})
1	0.0100	0.200	5.4×10^{-7}
2	0.0200	0.200	10.8×10^{-7}
3	0.0400	0.200	21.5×10^{-7}
4	0.200	0.0202	10.8×10^{-7}
5	0.200	0.0404	21.6×10^{-7}
6	0.200	0.0808	43.3×10^{-7}

We can express the way in which the rate depends on the concentrations of the reactants, NH_4^+ and NO_2^-, in terms of the following equation:

$$\text{Rate} = k[NH_4^+][NO_2^-] \qquad [15.5]$$

An equation such as Equation 15.5, which shows how the rate depends on the concentrations of reactants, is called a **rate law**. For a general reaction,

$$a\,A + b\,B \longrightarrow c\,C + d\,D$$

the rate law generally has the form

$$\text{Rate} = k[A]^m[B]^t \qquad [15.6]$$

The constant k in the rate law is called the **rate constant** or rate coefficient. The magnitude of k changes with temperature and therefore reflects how temperature affects the rate, as we see in Section 15.5. The exponents m and n are typically small whole numbers (usually 0, 1 or 2). We consider these exponents in more detail later.

> ⚠️ **CONCEPT CHECK 3**
>
> How do reaction rate, rate constant and rate law differ?

If we know the rate law for a reaction and its rate for a set of reactant concentrations, we can calculate the value of the rate constant, k. For example, using the data in Table 15.2 and the results from experiment 1, we can substitute into Equation 15.6

$$5.4 \times 10^{-7}\,M\ s^{-1} = k(0.0100\,M)(0.200\,M)$$

Solving for k gives

$$k = \frac{5.4 \times 10^{-7}\,M^{-1}\,s^{-1}}{(0.0100\,M)(0.200\,M)} = 2.7 \times 10^{-4}\,M^{-1}\,s^{-1}$$

You may wish to verify that this same value of k is obtained using any of the other experimental results given in Table 15.2.

Once we have both the rate law and the value of the rate constant for a reaction, we can calculate the rate of reaction for any set of concentrations. For example, using Equation 15.6 and $k = 2.7 \times 10^{-4}\,M^{-1}\,s^{-1}$, we can calculate the rate for $[NH_4^+] = 0.100\,M$ and $[NO_2^-] = 0.100\,M$:

$$\text{Rate} = (2.7 \times 10^{-4}\,M^{-1}\,s^{-1})(0.100\,M)(0.100\,M) = 2.7 \times 10^{-6}\,M\ s^{-1}$$

> ⚠️ **CONCEPT CHECK 4**
>
> Does the rate constant have the same units as the rate? Explain your answer.

A CLOSER LOOK

USING SPECTROSCOPIC METHODS TO MEASURE REACTION RATES

A variety of techniques can be used to monitor reactant and product concentration during a reaction, including spectroscopic methods, which rely on the ability of substances to absorb (or emit) light. Spectroscopic kinetic studies are often performed with the reaction mixture in the sample compartment of a *spectrometer*, an instrument that measures the amount of light transmitted or absorbed by a sample at different wavelengths. For kinetic studies, the spectrometer is set to measure the light absorbed at a wavelength characteristic of one of the reactants or products. In the decomposition of $HI(g)$ into $H_2(g)$ and $I_2(g)$, for example, both HI and H_2 are colourless, whereas I_2 is violet. During the reaction, the violet colour of the reaction mixture gets deeper as I_2 forms. Thus visible light of appropriate wavelength can be used to monitor the reaction (▶ FIGURE 15.5).

▼ FIGURE 15.6 shows the components of a spectrometer. The spectrometer measures the amount of light absorbed by the sample by comparing the intensity of the light emitted from the light source with the intensity of the light transmitted through the sample, for various wavelengths. As the concentration of I_2 increases and its colour becomes more intense, the amount of light absorbed by the reaction mixture increases, as Figure 15.5 shows, causing less light to reach the detector.

Beer's law relates the amount of light absorbed to the concentration of the absorbing substance:

$$A = \varepsilon\, bc \qquad [15.7]$$

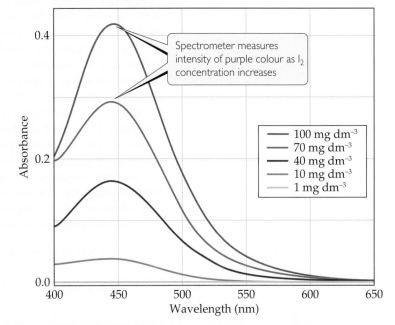

▲ **FIGURE 15.5**   **Visible spectra of I_2 at different concentrations.**

In this equation, A is the measured absorbance, ε is the extinction coefficient (a characteristic of the substance being monitored at a given wavelength of light), b is the path length through which the light passes, and c is the molar concentration of the absorbing substance. Thus the concentration is directly proportional to absorbance. Many chemical and pharmaceutical companies routinely use Beer's law to calculate the concentration of purified solutions of the compounds that they make.

RELATED EXERCISE: 15.6

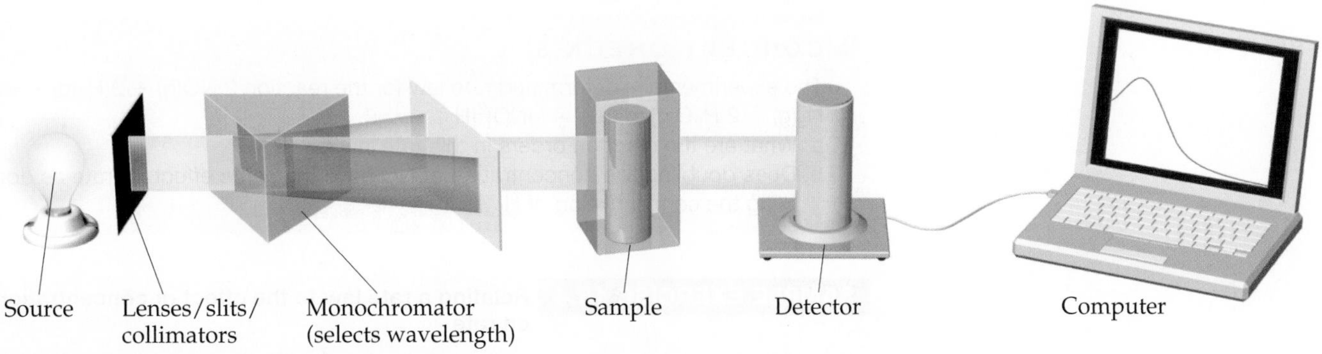

| Source | Lenses/slits/ collimators | Monochromator (selects wavelength) | Sample | Detector | Computer |

▲ **FIGURE 15.6**   **Components of a spectrometer.**

Reaction Orders: Exponents in the Rate Law

The rate laws for most reactions have the general form

$$\text{Rate} = k[\text{reactant 1}]^{m}[\text{reactant 2}]^{n}\ldots \qquad [15.8]$$

The exponents m and n in a rate law are called **reaction orders**. For example, consider again the rate law for the reaction of NH_4^+ with NO_2^-:

$$\text{Rate} = k[NH_4^+][NO_2^-]$$

Because the exponent of $[NH_4^+]$ is 1, the rate is *first order* in NH_4^+. The rate is also first order in NO_2^-. (The exponent '1' is not shown explicitly in rate laws.) The **overall reaction order** is the sum of the orders with respect to each reactant in the rate law. Thus the rate law has an overall reaction order of $1 + 1 = 2$, and the reaction is *second order overall*.

The exponents in a rate law indicate how the rate is affected by the concentration of each reactant. Because the rate at which NH_4^+ reacts with NO_2^- depends on $[NH_4^+]$ raised to the first power, the rate doubles when $[NH_4^+]$ doubles, triples when $[NH_4^+]$ triples and so forth. Doubling or tripling $[NO_2^-]$ likewise doubles or triples the rate. If a rate law is second order with respect to a reactant, $[A]^2$, then doubling the concentration of that substance causes the reaction rate to quadruple ($[2]^2 = 4$), whereas tripling the concentration causes the rate to increase 9-fold ($[3]^2 = 9$).

In some rare cases a change in concentration of a particular reactant does not affect the rate. The rate law is zero order with respect to that component and the concentration of that component does not appear in the rate law (Equation 15.9).

The following are some additional examples of rate laws:

$$2\,NH_3(g) \longrightarrow N_2(g) + 3H_2(g) \qquad \text{Rate} = k \qquad\qquad [15.9]$$

$$2\,N_2O_5(g) \longrightarrow 4\,NO_2(g) + O_2(g) \qquad \text{Rate} = k[N_2O_5] \qquad [15.10]$$

$$CHCl_3(g) + Cl_2(g) \longrightarrow CCl_4(g) + HCl(g) \qquad \text{Rate} = k[CHCl_3][Cl_2]^{\frac{1}{2}} \quad [15.11]$$

$$H_2(g) + I_2(g) \longrightarrow 2\,HI(g) \qquad \text{Rate} = k[H_2][I_2] \qquad [15.12]$$

Although the exponents in a rate law are sometimes the same as the coefficients in the balanced equation, this is not necessarily the case, as seen in Equations 15.10 and 15.11. *The values of these exponents must be determined experimentally.* In most rate laws, reaction orders are 0, 1 or 2. However, we also occasionally encounter rate laws in which the reaction order is fractional (such as Equation 15.11) or even negative. To review the meaning of negative and fractional exponents, see Appendix A. The values of the experimental exponents can also give insights into the nature of the mechanism of the reaction, as we see later.

CONCEPT CHECK 5

The experimentally determined rate law for the reaction $2\,NO(g) + 2\,H_2(g) \longrightarrow N_2(g) + 2\,H_2O(g)$ is rate $= k[NO]^2[H_2]$.
a. What are the reaction orders in this rate law?
b. Does doubling the concentration of NO have the same effect on rate as doubling the concentration of H_2?

SAMPLE EXERCISE 15.4   **Relating a rate law to the effect of concentration on rate**

Consider a reaction $A + B \longrightarrow C$ for which rate $= k[A][B]^2$. Each of the boxes below represents a reaction mixture in which A is shown as red spheres and B as purple spheres. Rank these mixtures in order of increasing rate of reaction.

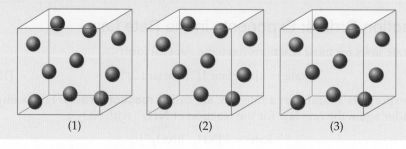

(1)   (2)   (3)

SOLUTION

Analyse We are given three boxes containing different numbers of spheres representing mixtures containing different reactant concentrations. We are asked to use the given rate law and the compositions of the boxes to rank the mixtures in order of increasing reaction rates.

Plan Because all three boxes have the same volume, we can put the number of spheres of each kind into the rate law and calculate the rate for each box.

Solve Box 1 contains five red spheres and five purple spheres, giving the following rate:

$$\text{Box 1: Rate} = k(5)(5)^2 = 125k$$

Box 2 contains seven red spheres and three purple spheres:

$$\text{Box 2: Rate} = k(7)(3)^2 = 63k$$

Box 3 contains three red spheres and seven purple spheres:

$$\text{Box 3: Rate} = k(3)(7)^2 = 147k$$

The slowest rate is $63k$ (box 2) and the highest is $147k$ (box 3). Thus the rates vary in the order $2 < 1 < 3$.

Check Each box contains 10 spheres. The rate law indicates that in this case [B] has a greater influence on rate than [A] because B has a higher reaction order. Hence the mixture with the highest concentration of B (most purple spheres) should react fastest. This analysis confirms the order $2 < 1 < 3$.

PRACTICE EXERCISE

Assuming that rate = k[A][B], rank the mixtures represented in Sample Exercise 15.4 in order of increasing rate.

Answer: $2 = 3 < 1$

Units of Rate Constants

The units of the rate constant depend on the overall reaction order of the rate law. In a reaction that is second order overall, for example, the units of the rate constant must satisfy the equation:

$$\text{Units of rate} = (\text{units of rate constant})(\text{units of concentration})^2$$

Hence, in our usual units of concentration and time

$$\text{Units of rate constant} = \frac{\text{units of rate}}{(\text{units of concentration})^2} = \frac{\text{M s}^{-1}}{\text{M}^2} = \text{M}^{-1}\,\text{s}^{-1}$$

SAMPLE EXERCISE 15.5 | **Determining reaction orders and units for rate constants**

(a) What are the overall reaction orders for the reactions described in Equations 15.10 and 15.11? **(b)** What are the units of the rate constant for the rate law for Equation 15.10?

SOLUTION

Analyse We are given two rate laws and asked to express **(a)** the overall reaction order for each and **(b)** the units for the rate constant for the first reaction.

Plan The overall reaction order is the sum of the exponents in the rate law. The units for the rate constant, k, are found by using the normal units for rate (M s^{-1}) and concentration (M) in the rate law and applying algebra to solve for k.

Solve
(a) The rate of the reaction in Equation 15.10 is first order in N_2O_5 and first order overall. The reaction in Equation 15.11 is first order in $CHCl_3$ and one-half order in Cl_2. The overall reaction order is three halves.

(b) For the rate law for Equation 15.10, we have

$$\text{Units of rate} = (\text{units of rate constant})(\text{units of concentration})$$

So,

$$\text{Units of rate constant} = \frac{\text{units of rate}}{\text{units of concentration}} = \frac{\text{M s}^{-1}}{\text{M}} = \text{s}^{-1}$$

Notice that the units of the rate constant change as the overall order of the reaction changes.

PRACTICE EXERCISE

(a) What is the reaction order of the reactant H_2 in Equation 15.12? **(b)** What are the units of the rate constant for Equation 15.12?

Answers: **(a)** 1, **(b)** $M^{-1} s^{-1}$

(See also Exercises 15.22, 15.23.)

Summary of units for the rate constant, k

Overall reaction order	Units of k (t in seconds)
0	$M s^{-1}$
1	s^{-1}
2	$M^{-1} s^{-1}$
3	$M^{-2} s^{-1}$

Using Initial Rates to Determine Rate Laws

We have seen that the rate law for most reactions has the general form (Equation 15.8)

$$\text{Rate} = k[\text{reactant 1}]^m[\text{reactant 2}]^n \ldots$$

So, to determine the rate law, we must determine the reaction orders m and n.

In most reactions the exponents in the rate law are 0, 1 or 2. If a reaction is zero order in a particular reactant, changing its concentration will have no effect on rate (as long as some of the reactant is present) because any concentration raised to the zero power equals 1. In contrast, we have seen that when a reaction is first order in a reactant, changes in the concentration of that reactant will produce proportional changes in the rate. Thus doubling the concentration will double the rate and so forth. Finally, when the rate law is second order in a particular reactant, doubling its concentration increases the rate by a factor of $2^2 = 4$, tripling its concentration causes the rate to increase by a factor of $3^2 = 9$ and so forth.

In working with rate laws, it is important to realise that the *rate* of a reaction depends on concentration, but the *rate constant* does not. As we see later in this chapter, the rate constant (and hence the reaction rate) is affected by temperature and by the presence of a catalyst.

SAMPLE EXERCISE 15.6 Determining a rate law from initial rate data

The initial rate of a reaction A + B ⟶ C was measured for several different starting concentrations of A and B, and the results are as follows:

Experiment number	[A] (M)	[B] (M)	Initial rate (M s^{-1})
1	0.100	0.100	4.0×10^{-5}
2	0.100	0.200	4.0×10^{-5}
3	0.200	0.100	16.0×10^{-5}

Using these data, determine **(a)** the rate law for the reaction, **(b)** the magnitude of the rate constant, **(c)** the rate of the reaction when [A] = 0.050 M and [B] = 0.100 M.

SOLUTION

Analyse We are given a table of data that relates concentrations of reactants with initial rates of reaction and asked to determine **(a)** the rate law, **(b)** the rate constant and **(c)** the rate of reaction for a set of concentrations not listed in the table.

Plan (a) We assume that the rate law has the following form: Rate $= k[A]^m[B]^n$. So we must use the given data to deduce the reaction orders m and n by determining how changes in the concentration change the rate. (b) Once we know m and n, we can use the rate law and one of the sets of data to determine the rate constant k. (c) Now that we know both the rate constant and the reaction orders, we can use the rate law with the given concentrations to calculate rate.

Solve
(a) As we move from experiment 1 to experiment 2, [A] is held constant and [B] is doubled. Because the rate remains the same when [B] is doubled, the concentration of B has no effect on the reaction rate. The rate law is therefore zero order in B (that is, $n = 0$).

In experiments 1 and 3, [B] is held constant while doubling [A] increases the rate fourfold. This result indicates that rate is proportional to $[A]^2$ (that is, the reaction is second order in A). Hence, the rate law is

$$\text{Rate} = k[A]^2[B]^0 = k[A]^2$$

This rate law could be reached in a more formal way by taking the ratio of the rates from two experiments:

$$\frac{\text{Rate 2}}{\text{Rate 1}} = \frac{4.0 \times 10^{-5}\,\text{M s}^{-1}}{4.0 \times 10^{-5}\,\text{M s}^{-1}} = 1$$

Using the rate law, we have, after cancelling like terms

$$1 = \frac{\text{Rate 2}}{\text{Rate 1}} = \frac{[0.200\,\text{M}]^n}{[0.100\,\text{M}]^n} = \frac{[0.200]^n}{[0.100]^n} = 2^n$$

2^n equals 1 under only one condition:

$$n = 0$$

We can deduce the value of m in a similar fashion:

$$\frac{\text{Rate 3}}{\text{Rate 1}} = \frac{16.0 \times 10^{-5}\,\text{M s}^{-1}}{4.0 \times 10^{-5}\,\text{M s}^{-1}} = 4$$

Using the rate law gives

$$4 = \frac{\text{Rate 3}}{\text{Rate 1}} = \frac{[0.200\,\text{M}]^m}{[0.100\,\text{M}]^m} = \frac{[0.200]^m}{[0.100]^m} = 2^m$$

Because $2^m = 4$, we conclude that

$$m = 2$$

(b) Using the rate law and the data from experiment 1, we have

$$k = \frac{\text{Rate}}{[A]^2} = \frac{4.0 \times 10^{-5}\,\text{M s}^{-1}}{(0.100\,\text{M})^2} = 4.0 \times 10^{-3}\,\text{M}^{-1}\,\text{s}^{-1}$$

(c) Using the rate law from part (a) and the rate constant from part (b), we have

$$\text{Rate} = k[A]^2 = (4.0 \times 10^{-3}\,\text{M}^{-1}\,\text{s}^{-1})(0.050\,\text{M})^2 = 1.0 \times 10^{-5}\,\text{M s}^{-1}$$

Because [B] is not part of the rate law, it is irrelevant to the rate, provided there is at least some B present to react with A.

Check A good way to check our rate law is to use the concentrations in experiment 2 or 3 and see if we can correctly calculate the rate. Using data from experiment 3, we have

$$\text{Rate} = k[A]^2 = (4.0 \times 10^{-3}\,\text{M}^{-1}\,\text{s}^{-1})(0.200\,\text{M})^2 = 16.0 \times 10^{-5}\,\text{M s}^{-1}$$

Thus the rate law correctly reproduces the data, giving the correct number and the correct units for the rate.

PRACTICE EXERCISE
The following data were measured for the reaction of nitric oxide with hydrogen:

$$2\,NO(g) + 2\,H_2(g) \longrightarrow N_2(g) + 2\,H_2O(g)$$

Experiment number	[NO] (M)	[H$_2$] (M)	Initial rate (M s^{-1})
1	0.100	0.100	1.23×10^{-3}
2	0.100	0.200	2.46×10^{-3}
3	0.200	0.100	4.92×10^{-3}

(a) Determine the rate law for this reaction. (b) Calculate the rate constant. (c) Calculate the rate when [NO] = 0.050 M and [H$_2$] = 0.150 M.

Answers: (a) rate $= k[NO]^2[H_2]$; (b) $k = 1.23\,\text{M}^{-2}\,\text{s}^{-1}$; (c) rate $= 4.6 \times 10^{-4}\,\text{M s}^{-1}$

(See also Exercise 15.24.)

15.4 | THE CHANGE OF CONCENTRATION WITH TIME (INTEGRATED RATE EQUATIONS)

From a rate law we are able to calculate the rate of reaction from the rate constant and reactant concentrations. However, it would be useful to have an equation that also allows the concentration of reactants or products to be determined at any particular time during the reaction. To derive such an equation we integrate the differential form of the rate equation.

Zero-Order Reactions

A **zero-order reaction** is one whose rate is independent of the concentration of the reacting species. As shown in Equation 15.9, ammonia in contact with a hot platinum wire decomposes at a constant rate until it has all disappeared. For a reaction of the type A $\longrightarrow$ products, the rate law is given by,

$$\text{Rate} = -\frac{\Delta[A]}{\Delta t} = k[A]^0 = k \qquad [15.13]$$

By integrating this *differential rate law* we obtain an equation which relates the concentration of A at the start of the reaction, $[A]_0$, to the concentration at any other time t, $[A]_t$. This form of the rate law is called the *integrated rate law*.

$$[A]_t = -kt + [A]_0 \qquad [15.14]$$

First-Order Reactions

A **first-order reaction** is one whose rate depends on the concentration of a single reactant raised to the first power. For a reaction of the type A $\longrightarrow$ products, the rate law may be first order:

$$\text{Rate} = -\frac{\Delta[A]}{\Delta t} = k[A]$$

Integration of the differential rate law yields an equation that relates the concentration of A at the start of the reaction, $[A]_0$, to its concentration at any other time t, $[A]_t$:

$$\ln[A]_t - \ln[A]_0 = -kt \quad \text{or} \quad \ln\frac{[A]_t}{[A]_0} = -kt \qquad [15.15]$$

The function 'ln' in Equation 15.15 is the natural logarithm (Appendix A). Equation 15.15 can also be rearranged and written as follows:

$$\ln[A]_t = -kt + \ln[A]_0 \qquad [15.16]$$

Equations 15.15 and 15.16 can be used with any concentration units, as long as the units are the same for both $[A]_0$ and $[A]_t$.

For a first-order reaction, Equation 15.15 or 15.16 can be used in several ways. Given any three of the quantities k, t, $[A]_0$ and $[A]_t$, we can solve for the fourth. Thus these equations can be used, for example, to determine (1) the concentration of a reactant remaining at any time after the reaction has started, (2) the time required for a given fraction of a sample to react, or (3) the time required for a reactant concentration to fall to a certain level (◀ FIGURE 15.7).

Equation 15.16 can be used to verify whether a reaction is first order and to determine its rate constant. This equation has the form of the general equation for a straight line, $y = mx + b$, in which m is the slope and b is the y-intercept of the line (Appendix A):

$$\underset{y}{\ln[A]_t} = \underset{m}{-k} \; \underset{x}{t} + \underset{b}{\ln[A]_0}$$

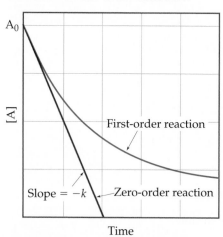

FIGURE IT OUT

At which times during the reaction would you have trouble distinguishing a zero-order reaction from a first-order reaction?

▲ **FIGURE 15.7** Comparison of first-order and zero-order reactions for the disappearance of reactant A with time.

SAMPLE EXERCISE 15.7 | Using the integrated first-order rate law

The decomposition of a certain insecticide in water follows first-order kinetics with a rate constant of 1.45 yr^{-1} at 12 °C. A quantity of this insecticide is washed into a lake on 1 June, leading to a concentration of 5.0×10^{-7} g cm^{-3}.

Assume that the average temperature of the lake is 12 °C. **(a)** What is the concentration of the insecticide on 1 June of the following year? **(b)** How long will it take for the concentration of the insecticide to drop to 3.0×10^{-7} g cm^{-3}?

SOLUTION

Analyse We are given the rate constant for a reaction that obeys first-order kinetics, as well as information about concentrations and times, and asked to calculate how much reactant (insecticide) remains after one year. We must also determine the time interval needed to reach a particular insecticide concentration. Because the exercise gives time in **(a)** and asks for time in **(b)**, we know that the integrated rate law, Equation 15.16, is required.

Plan (a) We are given $k = 1.45$ yr^{-1}, $t = 1.00$ yr and [insecticide]$_0$ = 5.0×10^{-7} g cm^{-3} and so Equation 15.16 can be solved for [insecticide]$_t$. **(b)** We have $k = 1.45$ yr^{-1}, [insecticide]$_0$ = 5.0×10^{-7} g cm^{-3} and [insecticide]$_t$ = 3.0×10^{-7} g cm^{-3} and so we can solve Equation 15.16 for time, t.

Solve

(a) Substituting the known quantities into Equation 15.16, we have

$$\ln[\text{insecticide}]_{t=1\text{ yr}} = -(1.45 \text{ yr}^{-1})(1.00 \text{ yr}) + \ln(5.0 \times 10^{-7})$$

We use the ln function on a calculator to evaluate the second term on the right, giving

$$\ln[\text{insecticide}]_{t=1\text{ yr}} = -1.45 + (-14.51) = -15.96$$

To obtain [insecticide]$_{t=1\text{ yr}}$, we use the inverse natural logarithm, or e^x, function on the calculator:

$$[\text{insecticide}]_{t=1\text{ yr}} = e^{-15.96} = 1.2 \times 10^{-7} \text{ g cm}^{-3}$$

Note that the concentration units for [A]$_t$ and [A]$_0$ must be the same.

(b) Again substituting into Equation 15.16, with [insecticide]$_t$ = 3.0×10^{-7} g cm^{-3} gives

Solving for t gives

$$\ln(3.0 \times 10^{-7}) = -(1.45 \text{ year}^{-1})(t) + \ln(5.0 \times 10^{-7})$$
$$t = -[\ln(3.0 \times 10^{-7}) - \ln(5.0 \times 10^{-7})]/1.45 \text{ yr}^{-1}$$
$$= -(-15.02 + 14.51)/1.45 \text{ yr}^{-1} = 0.35 \text{ yr}$$

Check In part **(a)** the concentration remaining after 1.00 yr (that is, 1.2×10^{-7} g cm^{-3}) is less than the original concentration (5.0×10^{-7} g cm^{-3}), as it should be. In **(b)** the given concentration (3.0×10^{-7} g cm^{-3}) is greater than that remaining after 1.00 year, indicating that the time must be less than a year. Thus $t = 0.35$ year is a reasonable answer.

PRACTICE EXERCISE

The decomposition of dimethyl ether, (CH$_3$)$_2$O, at 510 °C is a first-order process with a rate constant of 6.8×10^{-4} s^{-1}:

$$(CH_3)_2O(g) \longrightarrow CH_4(g) + H_2(g) + CO(g)$$

If the initial pressure of (CH$_3$)$_2$O is 18 kPa, what is its partial pressure after 1420 s? [*Hint:* From the ideal-gas equation, $P \propto n$.]

Answer: 6.8 kPa

(See also Exercise 15.32.)

For a first-order reaction, therefore, a graph of ln[A]$_t$ versus time gives a straight line with a slope of $-k$ and a y-intercept of ln[A]$_0$. A reaction that is not first order will not yield a straight line.

As an example, consider the conversion of methyl isonitrile (CH$_3$NC) to acetonitrile (CH$_3$CN) (▼ **FIGURE 15.8**). ▼ **FIGURE 15.9 (a)** shows how the partial

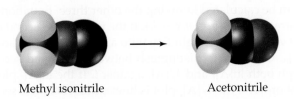

Methyl isonitrile   Acetonitrile

◀ **FIGURE 15.8   A first-order reaction.** The transformation of methyl isonitrile (CH$_3$NC) to acetonitrile (CH$_3$CN) is a first-order process.

What can you conclude from the fact that the plot of ln P versus t is linear?

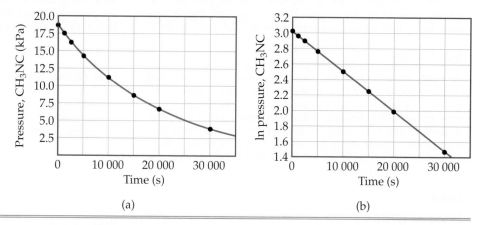

▶ **FIGURE 15.9 Kinetic data for conversion of methyl isonitrile into acetonitrile.**

pressure of methyl isonitrile varies with time as it rearranges in the gas phase at 198.9 °C. We can use pressure as a unit of concentration for a gas because from the ideal-gas law the pressure is directly proportional to the number of moles per unit volume. Figure 15.9(b) shows a plot of the natural logarithm of the pressure versus time, a plot that yields a straight line. The slope of this line is $-5.1 \times 10^{-5}\,\text{s}^{-1}$. (You should verify this for yourself, remembering that your result may vary slightly from ours because of inaccuracies associated with reading the graph.) This straight line shows that the reaction is first order, so that we can write the rate equation:

$$\ln[CH_3NC]_t = -kt + \ln[CH_3NC]_0.$$

Because the slope of the line equals $-k$, the rate constant for this reaction equals $5.1 \times 10^{-5}\,\text{s}^{-1}$.

 CONCEPT CHECK 6

What do the y-intercepts in Figure 15.9(a) and (b) represent?

Second-Order Reactions

A **second-order reaction** is one whose rate depends on the reactant concentration raised to the second power or on the concentrations of two different reactants, each raised to the first power. For simplicity, let's consider reactions of the type A ⟶ products or A + B ⟶ products that are second order in just one reactant, A:

$$\text{Rate} = -\frac{\Delta[A]}{\Delta t} = k[A]^2$$

With the use of calculus, this differential rate law can be used to derive the following integrated rate law:

$$\frac{1}{[A]_t} = kt + \frac{1}{[A]_0} \qquad \text{[15.17]}$$

This equation, like Equation 15.16, has four variables, k, t, $[A]_0$ and $[A]_t$, and any one of these can be calculated knowing the other three. Equation 15.17 also has the form of a straight line ($y = mx + b$). If the reaction is second order, a plot of $1/[A]_t$ versus t will yield a straight line with a slope equal to k and a y-intercept equal to $1/[A]_0$. One way to distinguish between first- and second-order rate laws is to graph both $\ln[A]_t$ and $1/[A]_t$ against t. If the $\ln[A]_t$ plot is linear, the reaction is first order; if the $1/[A]_t$ plot is linear, the reaction is second order.

SAMPLE EXERCISE 15.8 **Determining reaction order from the integrated rate law**

The following data were obtained for the gas-phase decomposition of nitrogen dioxide at 300 °C, $NO_2(g) \longrightarrow NO(g) + \frac{1}{2} O_2(g)$:

Time (s)	$[NO_2]$ (M)
0.0	0.01000
50.0	0.00787
100.0	0.00649
200.0	0.00481
300.0	0.00380

Is the reaction first or second order in NO_2?

SOLUTION

Analyse We are given the concentrations of a reactant at various times during a reaction and asked to determine whether the reaction is first order or second order.

Plan We can plot $\ln[NO_2]$ and $1/[NO_2]$ against time. One or the other will be linear, indicating whether the reaction is first order or second order.

Solve In order to graph $\ln[NO_2]$ and $1/[NO_2]$ against time, we first prepare the following table from the data given:

Time (s)	$[NO_2]$ (M)	$\ln[NO_2]$	$1/[NO_2]$
0.0	0.01000	−4.605	100
50.0	0.00787	−4.845	127
100.0	0.00649	−5.038	154
200.0	0.00481	−5.337	208
300.0	0.00380	−5.573	263

As ▼ **FIGURE 15.10** shows, only the plot of $1/[NO_2]$ versus time is linear. Thus the reaction obeys a second-order rate law: Rate = $k[NO_2]^2$. From the slope of this straight-line graph, we determine that $k = 0.543$ M^{-1} s^{-1} for the disappearance of NO_2.

PRACTICE EXERCISE

Consider again the decomposition of $[NO_2]$ discussed in Sample Exercise 15.8. The reaction is second order in NO_2 with $k = 0.543$ M^{-1} s^{-1}. If the initial concentration of NO_2 in a closed vessel is 0.0500 M, what is the remaining concentration after 0.500 h?

Answer: Using Equation 15.17, we find $[NO_2] = 1.00 \times 10^{-3}$ M

(See also Exercise 15.35.)

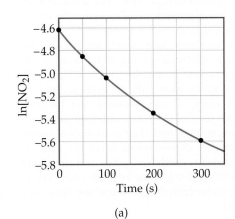

(a)

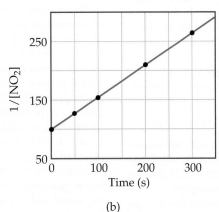

(b)

◀ **FIGURE 15.10   Kinetic data for decomposition of NO_2.**

Half-Life

The **half-life** of a reaction, $t_{\frac{1}{2}}$, is the time required for the concentration of a reactant to reach one-half of its initial value, $[A]_{t_{\frac{1}{2}}} = \frac{1}{2}[A]_0$. In essence, the half-life is a convenient way to describe how fast a reaction occurs. A fast reaction will have a short half-life and *vice versa*. Many applications of chemical kinetics require knowledge of the half-life of a substance. Such applications include rates of drug uptake and the use of radioactive isotopes.

We can determine the half lives of reactions by substituting $[A]_{t_{\frac{1}{2}}}$, that is $\frac{1}{2}[A]_0$, into the integrated forms of the rate equations.

For a zero-order reaction, by substituting $[A]_{t_{\frac{1}{2}}}$ into Equation 15.14 we obtain,

$$\frac{1}{2}[A]_0 = -kt_{\frac{1}{2}} + [A]_0$$

$$kt_{\frac{1}{2}} = \frac{1}{2}[A]_0$$

$$t_{\frac{1}{2}} = \frac{[A]_0}{2k} \qquad [15.18]$$

MY WORLD OF CHEMISTRY

BROMOMETHANE IN THE ATMOSPHERE

Several small molecules containing carbon–chlorine or carbon–bromine bonds, when present in the stratosphere, are capable of reacting with ozone (O_3) and thus contributing to the destruction of the Earth's ozone layer. Whether a halogen-containing molecule contributes significantly to destruction of the ozone layer depends in part on the molecule's average lifetime in the atmosphere. It takes quite a long time for molecules formed at the surface of the Earth to diffuse through the lower atmosphere (called the troposphere) and move into the stratosphere, where the ozone layer is located (▶ FIGURE 15.11). Decomposition in the lower atmosphere competes with diffusion into the stratosphere.

The much-discussed chlorofluorocarbons, or CFCs, contribute to the destruction of the ozone layer because they have long lifetimes in the troposphere. Thus they persist long enough for a substantial fraction of the molecules to find their way to the stratosphere.

Another simple molecule that has the potential to destroy the stratospheric ozone layer is bromomethane (CH_3Br). This substance has a wide range of uses, including antifungal treatment of plant seeds, and has therefore been produced in large quantities (about 70 million kilograms per year). In the stratosphere, the C–Br bond is broken through absorption of short-wavelength radiation. The resultant Br atoms then catalyse decomposition of O_3.

Bromomethane is removed from the lower atmosphere by a variety of mechanisms, including a slow reaction with ocean water:

$$CH_3Br(g) + H_2O(l) \longrightarrow CH_3OH(aq) + HBr(aq) \qquad [15.20]$$

To determine the potential importance of CH_3Br in destruction of the ozone layer, it is important to know how rapidly Equation 15.20 and all other mechanisms together remove CH_3Br from the atmosphere before it can diffuse into the stratosphere.

Scientists have conducted research to estimate the average lifetime of CH_3Br in the Earth's atmosphere. Such an estimate

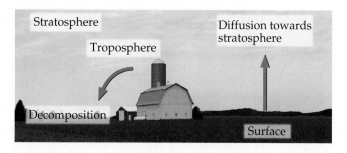

▲ **FIGURE 15.11 Distribution and fate of bromomethane in the atmosphere.** Some CH_3Br is removed from the atmosphere by decomposition, and some diffuses upwards into the stratosphere, where it contributes to destruction of the ozone layer. The relative rates of decomposition and diffusion determine how extensively bromomethane is involved in destruction of the ozone layer.

is difficult to make. It cannot be done in laboratory-based experiments because the conditions that exist in the atmosphere above the planet are too complex to be simulated in the laboratory. Instead, scientists gathered nearly 4000 samples of the atmosphere during aircraft flights all over the Pacific Ocean and analysed them for the presence of several trace organic substances, including bromomethane. From a detailed analysis of the concentrations, it was possible to estimate that the *atmospheric residence time* for CH_3Br is 0.8 ± 0.1 year.

The atmospheric residence time equals the half-life for CH_3Br in the lower atmosphere, assuming that it decomposes by a first-order process. That is, a collection of CH_3Br molecules present at any given time will, on average, be 50% decomposed after 0.8 years, 75% decomposed after 1.6 years and so on. A residence time of 0.8 years, although comparatively short, is still long enough for CH_3Br to contribute significantly to the destruction of the ozone layer. In 1997 an international agreement was reached to phase out the use of bromomethane in developed countries by 2005. However, in recent years exemptions for critical agricultural use have been requested and granted.

RELATED EXERCISE: 15.61

So, for a zero-order reaction, the half-life is directly proportional to the initial concentration.

For a first-order reaction, we substitute $[A]_{t_{\frac{1}{2}}}$ into Equation 15.15 and obtain,

$$\ln \frac{\frac{1}{2}[A]_0}{[A]_0} = -kt_{\frac{1}{2}}$$

$$\ln \frac{1}{2} = -kt_{\frac{1}{2}}$$

$$t_{\frac{1}{2}} = -\frac{\ln \frac{1}{2}}{k} = \frac{0.693}{k} \qquad [15.19]$$

From Equation 15.19, we see that $t_{\frac{1}{2}}$ for a first-order rate law does not depend on the starting concentration. Consequently, the half-life remains constant throughout the reaction. If, for example, the concentration of the reactant is 0.120 M at some moment in the reaction, it will be 0.06 after one half-life. After one more half-life passes, the concentration will drop to 0.030 M and so on. Equation 15.19 also indicates that we can calculate $t_{\frac{1}{2}}$ for a first-order reaction if k is known or k if $t_{\frac{1}{2}}$ is known.

The change in concentration over time for the first-order rearrangement of methyl isonitrile at 198.9 °C is graphed in ▶ FIGURE 15.12. The first half-life is shown at 13 600 s (that is, 3.78 h). At a time 13 600 s later, the isonitrile concentration has decreased to one-half of one-half, or one-fourth the original concentration. *In a first-order reaction, the concentration of the reactant decreases by a factor of a $\frac{1}{2}$ in each of a series of regularly spaced time intervals, namely,* $t_{\frac{1}{2}}$.

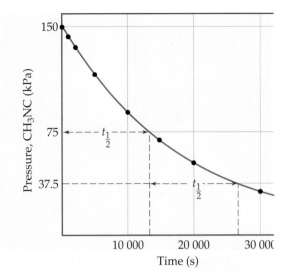

▲ FIGURE 15.12   **Half-life of a first-order reaction.** Pressure of methyl isonitrile as a function of time showing two successive half-lives of the isomerisation reaction depicted in Figure 15.8.

SAMPLE EXERCISE 15.9   **Determining the half-life of a first-order reaction**

The reaction of C_4H_9Cl with water is a first-order reaction. Figure 15.4 shows how the concentration of C_4H_9Cl changes with time at a particular temperature. **(a)** From that graph, estimate the half-life for this reaction. **(b)** Use the half-life from (a) to calculate the rate constant.

SOLUTION

Analyse We are asked to estimate the half-life of a reaction from a graph of concentration versus time and then to use the half-life to calculate the rate constant for the reaction.

Plan (a) To estimate a half-life, we can select a concentration and then determine the time required for the concentration to decrease to half of that value. **(b)** Equation 15.19 is used to calculate the rate constant from the half-life.

Solve
(a) From the graph, we see that the initial value of $[C_4H_9Cl]$ is 0.100 M. The half-life for this first-order reaction is the time required for $[C_4H_9Cl]$ to decrease to 0.050 M, which we can read off the graph. This point occurs at approximately 340 s.
(b) Solving Equation 15.19 for k, we have

$$k = \frac{0.693}{t_{\frac{1}{2}}} = \frac{0.693}{340 \text{ s}} = 2.0 \times 10^{-3} \text{ s}^{-1}$$

Check At the end of the second half-life, which should occur at 680 s, the concentration should have decreased by yet another factor of 2, to 0.025 M. Inspection of the graph shows that this is indeed the case.

PRACTICE EXERCISE
(a) Using Equation 15.19, calculate $t_{\frac{1}{2}}$ for the decomposition of the insecticide described in Sample Exercise 15.7. **(b)** How long does it take for the concentration of the insecticide to reach one-quarter of the initial value?

Answers: **(a)** 0.478 yr = 1.51×10^7 s; **(b)** it takes two half-lives, 2(0.478 yr) = 0.956 yr

(See also Exercise 15.33.)

Hot water   Cold water

▲ FIGURE 15.13 **Temperature affects the rate of the chemiluminescence reaction in light sticks.**

For a second-order reaction by substituting $[A]_{t_{\frac{1}{2}}}$ into Equation 15.17 we obtain,

$$t_{\frac{1}{2}} = \frac{1}{k[A]_0}$$ [15.21]

In this case, the half-life depends on the initial concentration of reactant—the lower the initial concentration, the greater the half-life. If fewer reactant particles are available the collision frequency will be less, hence the rate of reaction will be less.

> ▲ **CONCEPT CHECK 7**
>
> How does the half-life of a second-order reaction change as the reaction proceeds?

15.5 | TEMPERATURE AND RATE

The rates of most chemical reactions increase as the temperature rises. For example, dough rises faster at room temperature than when refrigerated, and plants grow more rapidly in warm weather than in cold. We can literally see the effect of temperature on reaction rate by observing a chemiluminescent reaction (one that produces light). The characteristic glow of fireflies is a familiar example of chemiluminescence. Another is the light produced by Cyalume® light sticks, which contain chemicals that produce chemiluminescence when mixed. As seen in ◄ FIGURE 15.13, these light sticks produce a brighter light at higher temperature. The amount of light produced is greater because the rate of the reaction is faster at the higher temperature. Although the light stick glows more brightly initially, its luminescence also dies out more rapidly.

How is this experimentally observed temperature effect reflected in the rate expression? The faster rate at higher temperature is due to an increase in the rate constant with increasing temperature. For example, let's reconsider the first-order reaction $CH_3NC \longrightarrow CH_3CN$ (Figure 15.8). ▼ FIGURE 15.14 shows the rate constant for this reaction as a function of temperature. The rate constant, and hence the rate of the reaction, increases rapidly with temperature, approximately doubling for each 10 °C rise.

▶ FIGURE 15.14 **Dependence of rate constant on temperature.** The data show the variation in the first-order rate constant for the rearrangement of methyl isonitrile as a function of temperature. The four points indicated are used in connection with Sample Exercise 15.11.

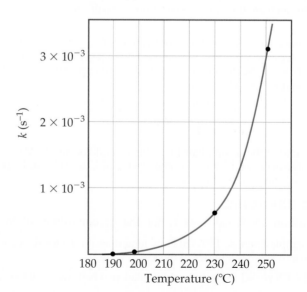

The Collision Model

We have seen that reaction rates are affected both by the concentrations of reactants and by temperature. The **collision model**, which is based on the kinetic-molecular theory (∞ Section 10.7, 'Kinetic-Molecular Theory'), accounts for both of these effects at the molecular level. The central idea of the collision model is that molecules must collide to react. The greater the number of collisions occurring per second, the greater the reaction rate. As the concentration of reactant molecules increases, therefore, the number of collisions increases, leading to an increase in reaction rate. As molecules move faster, they collide more forcefully (with more energy) and more frequently, increasing reaction rates.

For most reactions, only a tiny fraction of the collisions leads to reaction. For example, in a mixture of H_2 and I_2 at ordinary temperatures and pressures, each molecule undergoes about 10^{10} collisions per second. If every collision between H_2 and I_2 resulted in the formation of HI, the reaction would be over in much less than a second. Instead, at room temperature the reaction proceeds very slowly. Only about one in every 10^{13} collisions produces a reaction. What keeps the reaction from occurring more rapidly?

∞ Review this on page 355

 CONCEPT CHECK 8

What is the central idea of the collision model?

The Orientation Factor

In most reactions, molecules must be oriented in a certain way during collisions if a reaction is to occur. The relative orientations of the molecules during their collisions determine whether the atoms are suitably positioned to form new bonds. For example, consider the reaction of Cl atoms with NOCl:

$$Cl + NOCl \longrightarrow NO + Cl_2$$

The reaction will take place if the collision brings Cl atoms together to form Cl_2, as shown in the top part of ▼ **FIGURE 15.15**. In contrast, the collision shown in the bottom part of Figure 15.15 will be ineffective and will not yield products. Indeed, a great many collisions do not lead to reaction, merely because the molecules are not suitably oriented. There is, however, another factor that is usually even more important in determining whether particular collisions result in reaction.

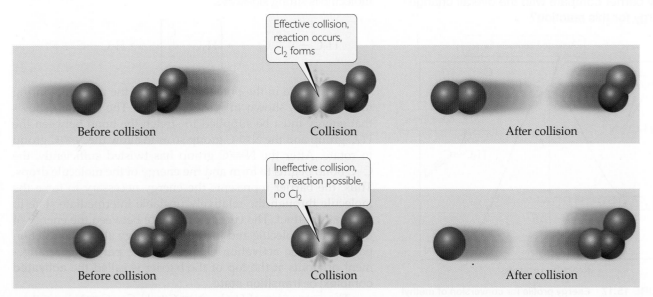

Effective collision, reaction occurs, Cl_2 forms

Before collision   Collision   After collision

Ineffective collision, no reaction possible, no Cl_2

Before collision   Collision   After collision

▲ **FIGURE 15.15   Molecular collisions may or may not lead to a chemical reaction between Cl and NOCl.**

▶ **FIGURE 15.16** **Energy is needed to overcome a barrier between initial and final states.**

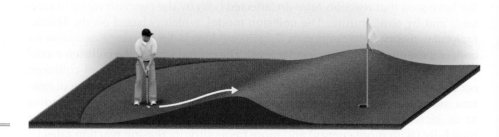

Activation Energy

In 1888 the Swedish chemist Svante Arrhenius (1859–1927) suggested that molecules must possess a certain minimum amount of energy in order to react. According to the collision model, this energy comes from the kinetic energies of the colliding molecules. Upon collision, the kinetic energy of the molecules can be used to stretch, bend and ultimately break bonds, leading to chemical reactions. That is, the kinetic energy is used to change the potential energy of the molecule. If molecules are moving too slowly, with too little kinetic energy, they merely bounce off one another without changing. For new bonds to form, the electron clouds of the reactants must interact in such a way as to allow electronic rearrangements to occur. If the particles have insufficient energy, the electron clouds will simply repel each other. In order to react, colliding molecules must have a total kinetic energy equal to or greater than some minimum value. The minimum energy required to initiate a chemical reaction is called the **activation energy, E_a**. The value of E_a varies from reaction to reaction.

The situation during reactions is rather like that shown in ▲ **FIGURE 15.16**. The player on the putting green needs to move his ball over the hill to the vicinity of the putting hole. To do this, he must impart enough kinetic energy with the putter to move the ball to the top of the hill. If he doesn't impart enough energy, the ball will roll partway up the hill and then back down. In the same way, molecules may require a certain minimum energy to break existing bonds during a chemical reaction. In the rearrangement of methyl isonitrile to acetonitrile, for example, we might imagine the reaction passing through an intermediate state in which the N≡C portion of the molecule is sitting sideways:

$$H_3C-N\equiv C: \longrightarrow \left[H_3C\cdots \overset{\ddot{C}}{\underset{\ddot{N}}{|||}} \right] \longrightarrow H_3C-C\equiv N:$$

The change in the potential energy of the molecule during the reaction is shown in ◀ **FIGURE 15.17**. The diagram shows that energy must be supplied to stretch the bond between the H_3C group and the N≡C group so as to allow the N≡C group to rotate. After the N≡C group has twisted sufficiently, the C–C bond begins to form and the energy of the molecule drops. Thus, the barrier represents the energy necessary to force the molecule through the relatively unstable intermediate state to the final product. The energy difference between the energy of the starting molecule and the highest energy along the reaction pathway is the activation energy, E_a. The particular arrangement of atoms at the top of the barrier is called the **activated complex**, or **transition state**.

The conversion of $H_3C-N\equiv C$ to $H_3C-C\equiv N$ is exothermic. Figure 15.17 therefore shows the product as having a lower

▲ **FIGURE IT OUT**

How does the energy needed to overcome the energy barrier compare with the overall change in energy for this reaction?

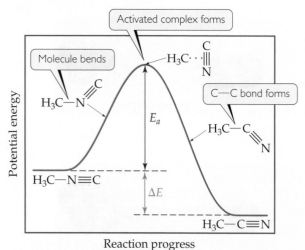

▲ **FIGURE 15.17** **Energy profile for conversion of methyl isonitrile (H_3CNC) to its isomer acetonitrile (H_3CCN).**

internal energy than the reactant. The energy change for the reaction, ΔE, has no effect on the rate of the reaction. The rate depends on the magnitude of E_a. Generally, the lower E_a is the faster the reaction. Notice that the reverse reaction is endothermic. The activation barrier for the reverse reaction is equal to the sum of ΔE and E_a for the forward reaction.

How does any particular methyl isonitrile molecule acquire sufficient energy to overcome the activation barrier? It does so through collisions with other molecules. Recall from the kinetic-molecular theory of gases that, at any given instant, gas molecules are distributed in energy over a wide range. (∞ Section 10.7, 'Kinetic-Molecular Theory'). ▶ FIGURE 15.18 shows the distribution of kinetic energies for two different temperatures, comparing them with the minimum energy needed for reaction, E_a. At the higher temperature a much greater fraction of the molecules has kinetic energy greater than E_a, which leads to a much greater rate of reaction.

The fraction of molecules that has an energy equal to or greater than E_a is given by the expression

$$f = e^{-E_a/RT} \qquad [15.22]$$

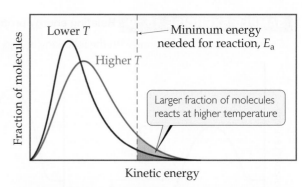

▶ **FIGURE 15.18**   **The effect of temperature on the distribution of kinetic energies of molecules in a sample.**

∞ Review this on page 356

In this equation, R is the gas constant (8.314 J mol^{-1} K^{-1}) and T is the absolute temperature. To get an idea of the magnitude of f, let's suppose that E_a is 100 kJ mol^{-1}, a value typical of many reactions, and that T is 300 K, around room temperature. The calculated value of f is 3.8×10^{-18}, an extremely small number! At 310 K the fraction is $f = 1.4 \times 10^{-17}$. Thus a 10 degree increase in temperature produces a 3.7-fold increase in the fraction of molecules possessing at least 100 kJ mol^{-1} of energy.

 CONCEPT CHECK 9

In a chemical reaction, why does not every collision between reactant molecules result in formation of a product molecule?

The Arrhenius Equation

Arrhenius noted that for most reactions the increase in rate with increasing temperature is non-linear, as shown in Figure 15.14. He found that most reaction-rate data obeyed an equation based on three factors: (1) the fraction of molecules possessing an energy of E_a or greater; (2) the number of collisions occurring per second; and (3) the fraction of collisions that have the appropriate orientation. These three factors are incorporated into the **Arrhenius equation**:

$$k = Ae^{-E_a/RT} \qquad [15.23]$$

In this equation, k is the rate constant, E_a is the activation energy, R is the gas constant (8.314 J mol^{-1} K^{-1}) and T is the absolute temperature. The **frequency factor**, A, is constant, or nearly so, as temperature is varied. It is related to the frequency of collisions favourably oriented for reaction.* As the magnitude of E_a increases, k decreases because the fraction of molecules that possess the required energy is smaller. Thus *reaction rates decrease as* E_a *increases*.

---

*Because the frequency of collisions increases with temperature, A also has some temperature dependence, but it is small compared with the exponential term. Therefore, A is considered approximately constant.

SAMPLE EXERCISE 15.10   **Relating energy profiles to activation energies and speeds of reaction**

Consider a series of reactions with the following energy profiles:

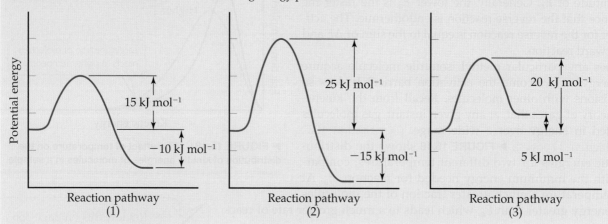

Reaction pathway
(1)

Reaction pathway
(2)

Reaction pathway
(3)

Assuming that all three reactions have nearly the same frequency factors, rank the reactions from slowest to fastest.

Solution   The lower the activation energy, the faster the reaction. The value of ΔE does not affect the rate. Hence, the order is (2) < (3) < (1).

PRACTICE EXERCISE

Imagine that these reactions are reversed. Rank these reverse reactions from slowest to fastest.

Answer: (2) < (1) < (3) because E_a values are 40, 25 and 15 kJ mol^{-1}, respectively

(See also Exercise 15.41.)

Determining the Activation Energy

Taking the natural log of both sides of Equation 15.23, we have

$$\ln k = -\frac{E_a}{RT} + \ln A \qquad [15.24]$$

Equation 15.24 has the form of a straight line; it predicts that a graph of $\ln k$ versus $1/T$ will be a line with a slope equal to $-E_a/R$ and a y-intercept equal to $\ln A$. Thus the activation energy can be determined by measuring k at a series of temperatures, graphing $\ln k$ versus $1/T$ and then calculating E_a from the slope of the resultant line.

We can also use Equation 15.24 to evaluate E_a in a non-graphical way if we know the rate constant of a reaction at two or more temperatures. For example, suppose that at two different temperatures, T_1 and T_2, a reaction has rate constants k_1 and k_2. For each condition, we have

$$\ln k_1 = -\frac{E_a}{RT_1} + \ln A \qquad \text{and} \qquad \ln k_2 = -\frac{E_a}{RT_2} + \ln A$$

Assuming that A is constant over the temperature range under study allows rearrangement as follows:

$$\ln A = \ln k_1 + \frac{E_a}{RT_1} \qquad \text{and} \qquad \ln A = \ln k_2 + \frac{E_a}{RT_2}$$

Hence, $$\ln k_1 + \frac{E_a}{RT_1} = \ln k_2 + \frac{E_a}{RT_2}$$

and $$\ln k_1 - \ln k_2 = \frac{E_a}{RT_2} - \frac{E_a}{RT_1}$$

Simplifying this equation and rearranging it gives

$$\ln \frac{k_1}{k_2} = \frac{E_a}{R}\left(\frac{1}{T_2} - \frac{1}{T_1}\right)$$ [15.25]

Equation 15.25 provides a convenient way to calculate the rate constant, k_1, at some temperature, T_1, when we know the activation energy and the rate constant, k_2, at some other temperature, T_2.

SAMPLE EXERCISE 15.11 | Determining the energy of activation

The following table shows the rate constants for the rearrangement of methyl isonitrile at various temperatures (these are the data in Figure 15.14):

Temperature (°C)	k (s^{-1})
189.7	2.52×10^{-5}
198.9	5.25×10^{-5}
230.3	6.30×10^{-4}
251.2	3.16×10^{-3}

(a) From these data, calculate the activation energy for the reaction. **(b)** What is the value of the rate constant at 430.0 K?

SOLUTION

Analyse We are given rate constants, k, measured at several temperatures and asked to determine the activation energy, E_a, and the rate constant, k, at a particular temperature.

Plan We can obtain E_a from the slope of a graph of $\ln k$ versus $1/T$. Once we know E_a, we can use Equation 15.25 together with the given rate data to calculate the rate constant at 430.0 K.

Solve

(a) We must first convert the temperatures from degrees celsius to kelvin. We then take the inverse of each temperature, $1/T$, and the natural log of each rate constant, $\ln k$. This gives us the table shown below:

T (K)	$1/T$ (K^{-1})	$\ln k$
462.9	2.160×10^{-3}	-10.589
472.1	2.118×10^{-3}	-9.855
503.5	1.986×10^{-3}	-7.370
524.4	1.907×10^{-3}	-5.757

A graph of $\ln k$ versus $1/T$ results in a straight line, as shown in ▼ **FIGURE 15.19**.

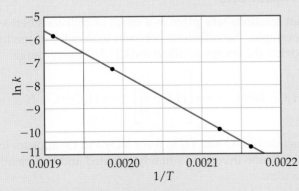

◀ **FIGURE 15.19 Graphical determination of activation energy.** The natural logarithm of the rate constant for the rearrangement of methyl isonitrile is plotted as a function of $1/T$. The linear relationship is predicted by the Arrhenius equation, giving a slope equal to $-E_a/R$.

The slope of the line is obtained by choosing two well-separated points, as shown, and using the coordinates of each:

$$\text{Slope} = \frac{\Delta y}{\Delta x} = \frac{-6.6 - (-10.4)}{0.00195 - 0.00215} = -1.9 \times 10^4$$

Because logarithms have no units, the numerator in this equation is dimensionless. The denominator has the units of $1/T$, namely, K^{-1}. Thus the overall units for the slope are K. The slope equals $-E_a/R$. R in $J\ mol^{-1}\ K^{-1}$ is used:

$$Slope = -\frac{E_a}{R}$$

$$E_a = -(slope)(R) = -(-1.9 \times 10^4\ K)(8.314\ J\ mol^{-1}\ K^{-1})$$

$$= 1.6 \times 10^5\ J\ mol^{-1}$$

$$= 160\ kJ\ mol^{-1}$$

We report the activation energy to only two significant figures because we are limited by the precision with which we can read the graph in Figure 15.19.

(b) To determine the rate constant, k_1, at $T_1 = 430.0$ K, we can use Equation 15.25 with $E_a = 160$ kJ mol^{-1}, and one of the rate constants and temperatures from the given data, such as $k_2 = 2.52 \times 10^{-5}\ s^{-1}$ and $T_2 = 462.9$ K:

$$\ln\left(\frac{k_1}{2.52 \times 10^{-5}\ s^{-1}}\right) = \left(\frac{160 \times 1000\ J\ mol^{-1}}{8.314\ J\ mol^{-1}\ K^{-1}}\right)\left(\frac{1}{462.9\ K} - \frac{1}{430.0\ K}\right) = -3.18$$

Thus,

$$\frac{k_1}{2.52 \times 10^{-5}\ s^{-1}} = e^{-3.18} = 4.15 \times 10^{-2}$$

$$k_1 = (4.15 \times 10^{-2})(2.52 \times 10^{-5}\ s^{-1}) = 1.0 \times 10^{-6}\ s^{-1}$$

Note that the units of k_1 are the same as those of k_2.

PRACTICE EXERCISE

Using the data in Sample Exercise 15.11, calculate the rate constant for the rearrangement of methyl isonitrile at 280 °C.

Answer: $2.0 \times 10^{-2}\ s^{-1}$

(See also Exercises 15.43, 15.44.)

15.6 | REACTION MECHANISMS

A balanced equation for a chemical reaction indicates the substances present at the start of the reaction and those produced as the reaction proceeds. It provides no information, however, about how the reaction occurs. The process by which a reaction occurs is called the **reaction mechanism**. At the most sophisticated level, a reaction mechanism will describe in great detail the order in which bonds are broken and formed and the changes in relative positions of the atoms in the course of the reaction. We begin with more rudimentary descriptions of how reactions occur, considering further the nature of the collisions leading to reaction.

Elementary Reactions

We have seen that reactions take place as a result of collisions between reacting molecules. For example, the collisions between molecules of methyl isonitrile (CH_3NC) can provide the energy to allow the CH_3NC to rearrange:

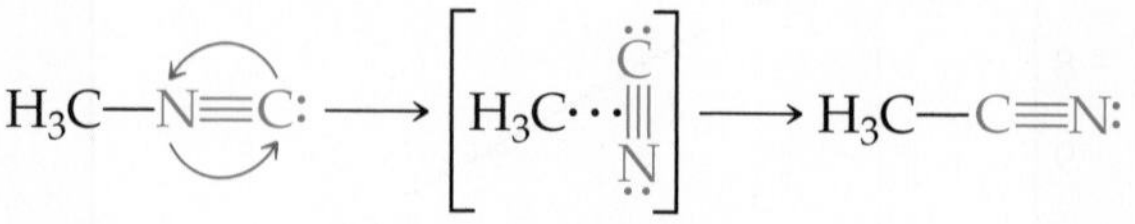

$$H_3C-N\equiv C: \longrightarrow \left[H_3C\cdots\overset{\ddot{C}}{\underset{\ddot{N}}{\vert\vert\vert}} \right] \longrightarrow H_3C-C\equiv N:$$

Similarly, the reaction of NO and O_3 to form NO_2 and O_2 appears to occur as a result of a single collision involving suitably oriented and sufficiently energetic NO and O_3 molecules:

$$NO(g) + O_3(g) \longrightarrow NO_2(g) + O_2(g) \qquad [15.26]$$

Both of these processes occur in a single event or step and are called **elementary reactions** (or elementary processes).

The number of molecules that participate as reactants in an elementary reaction defines the **molecularity** of the reaction. If a single molecule is involved, the reaction is **unimolecular**. The rearrangement of methyl isonitrile is a unimolecular process. Elementary reactions involving the collision of two reactant molecules are **bimolecular**. The reaction between NO and O_3 (Equation 15.26) is bimolecular. Elementary reactions involving the simultaneous collision of three molecules are **termolecular**. Termolecular reactions are far less probable than unimolecular or bimolecular processes and are rarely encountered. The chance that four or more molecules will collide simultaneously with any regularity is even more remote; consequently such collisions are never proposed as part of a reaction mechanism.

CONCEPT CHECK 10

What is the molecularity of this elementary reaction?
$$NO(g) + Cl_2(g) \longrightarrow NOCl(g) + Cl(g)$$

Multistep Mechanisms

The net change represented by a balanced chemical equation often occurs by a *multistep mechanism*, which consists of a sequence of elementary reactions. For example, consider the reaction of NO_2 and CO:

$$NO_2(g) + CO(g) \longrightarrow NO(g) + CO_2(g) \qquad [15.27]$$

Below 225 °C, this reaction appears to proceed in two elementary reactions (or two *elementary steps*), each of which is bimolecular. First, two NO_2 molecules collide, and an oxygen atom is transferred from one to the other. The resultant NO_3 then collides with a CO molecule and transfers an oxygen atom to it:

$$NO_2(g) + NO_2(g) \longrightarrow NO_3(g) + NO(g)$$

$$NO_3(g) + CO(g) \longrightarrow NO_2(g) + CO_2(g)$$

Thus we say that the reaction occurs by a two-step mechanism.

The chemical equations for the elementary reactions in a multistep mechanism must always add to give the chemical equation of the overall process.

In the present example the sum of the two elementary reactions is

$$2\,NO_2(g) + NO_3(g) + CO(g) \longrightarrow NO_2(g) + NO_3(g) + NO(g) + CO_2(g)$$

Simplifying this equation by eliminating substances that appear on both sides gives Equation 15.27, the net equation for the process.

Because NO_3 is neither a reactant nor a product in the overall reaction—it is formed in one elementary reaction and consumed in the next—it is called an **intermediate**. Multistep mechanisms involve one or more intermediates.

Intermediates are not the same as transition states, as shown in ▶ FIGURE 15.20. Intermediates can be stable and can therefore sometimes be identified and even isolated. Transition states, on the other hand, are always inherently unstable and as such can never be isolated. Nevertheless, the use of advanced 'ultrafast' techniques sometimes allows us to characterise them.

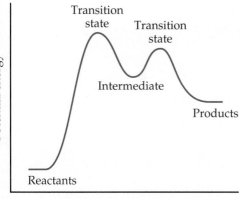

▲ **FIGURE 15.20**   **The energy profile of a reaction, showing transition states and an intermediate.**

| **SAMPLE EXERCISE 15.12** | **Determining molecularity and identifying intermediates** |

It has been proposed that the conversion of ozone into O_2 proceeds by a two-step mechanism:

$$O_3(g) \longrightarrow O_2(g) + O(g)$$

$$O_3(g) + O(g) \longrightarrow 2\,O_2(g)$$

(a) Describe the molecularity of each elementary reaction in this mechanism. **(b)** Write the equation for the overall reaction. **(c)** Identify the intermediate(s).

SOLUTION

Analyse We are given a two-step mechanism and asked for **(a)** the molecularities of each of the two elementary reactions, **(b)** the equation for the overall process, and **(c)** the intermediate.

Plan The molecularity of each elementary reaction depends on the number of reactant molecules in the equation for that reaction. The overall equation is the sum of the equations for the elementary reactions. The intermediate is a substance formed in one step of the mechanism and used in another and therefore not part of the equation for the overall reaction.

Solve
(a) The first elementary reaction involves a single reactant and is consequently unimolecular. The second reaction, which involves two reactant molecules, is bimolecular.

(b) Adding the two elementary reactions gives

$$2\,O_3(g) + O(g) \longrightarrow 3\,O_2(g) + O(g)$$

Because O(g) appears in equal amounts on both sides of the equation, it can be eliminated to give the net equation for the chemical process:

$$2\,O_3(g) \longrightarrow 3\,O_2(g)$$

(c) The intermediate is O(g). It is neither an original reactant nor a final product, but is formed in the first step of the mechanism and consumed in the second.

PRACTICE EXERCISE

The following mechanism has been proposed for the reaction of NO with H_2 to form N_2O and H_2O.

$$NO(g) + NO(g) \longrightarrow N_2O_2(g)$$

$$N_2O_2(g) + H_2(g) \longrightarrow N_2O(g) + H_2O(g)$$

(a) Show that the elementary reactions of the proposed mechanism add to provide a balanced equation for the reaction. **(b)** Write a rate law for each elementary reaction in the mechanism. **(c)** Identify any intermediates in the mechanism. **(d)** The observed rate law is rate = $k[NO]^2[H_2]$. If the proposed mechanism is correct, what can we conclude about the relative speeds of the first and second reactions?

Answers: **(a)** $2\,NO(g) + H_2(g) \longrightarrow N_2O(g) + H_2O(g)$. **(b)** rate = $k[NO]^2$; rate = $k[N_2O_2][H_2]$. **(c)** $N_2O_2(g)$. **(d)** The second step is the rate-determining step.

(See also Exercise 15.68.)

Rate Laws for Elementary Reactions

In Section 15.3 we stressed that rate laws must be determined experimentally; they cannot be predicted from the coefficients of balanced chemical equations. We are now in a position to understand why this is so. Every reaction is made up of a series of one or more elementary steps, and the rate laws and relative speeds of these steps will dictate the overall rate law. Indeed, the rate law for a reaction can be determined from its mechanism, as we will see shortly. Thus our next challenge in kinetics is to arrive at reaction mechanisms that lead to rate laws that are consistent with those observed experimentally. We start by examining the rate laws of elementary reactions.

Elementary reactions are significant in a very important way. *If we know that a reaction is an elementary reaction, then we know its rate law.* The rate law of any elementary reaction is based directly on its molecularity. For example, consider the general unimolecular process:

$$A \longrightarrow products$$

As the number of A molecules increases, the number that decompose in a given interval of time will increase proportionally. Thus the rate of a unimolecular process will be first order:

$$Rate = k[A]$$

TABLE 15.3 • **Elementary reactions and their rate laws**

Molecularity	Elementary reaction	Rate law
*Uni*molecular	A $\longrightarrow$ products	Rate = $k[A]$
*Bi*molecular	A + A $\longrightarrow$ products	Rate = $k[A]^2$
*Bi*molecular	A + B $\longrightarrow$ products	Rate = $k[A][B]$
*Ter*molecular	A + A + A $\longrightarrow$ products	Rate = $k[A]^3$
*Ter*molecular	A + A + B $\longrightarrow$ products	Rate = $k[A]^2[B]$
*Ter*molecular	A + B + C $\longrightarrow$ products	Rate = $k[A][B][C]$

In the case of bimolecular elementary steps, the rate law is second order, as in the following example:

$$A + B \longrightarrow \text{products} \qquad \text{Rate} = k[A][B]$$

The second-order rate law follows directly from the collision theory. If we double the concentration of A, the number of collisions between molecules of A and B will double; likewise, if we double [B], the number of collisions will double. Therefore the rate law will be first order in both [A] and [B], and second order overall.

The rate laws for all feasible elementary reactions are given in ▲ **TABLE 15.3**. Notice how the rate law for each kind of elementary reaction follows directly from the molecularity of that reaction. It is important to remember, however, that we cannot tell by merely looking at a balanced chemical equation whether the reaction involves one or several elementary steps. It is also important to note that a mechanism based on only unimolecular and bimolecular reactions is far more probable than a single termolecular step.

SAMPLE EXERCISE 15.13 | **Predicting the rate law for an elementary reaction**

If the following reaction occurs in a single elementary reaction, predict the rate law:

$$H_2(g) + Br_2(g) \longrightarrow 2\,HBr(g)$$

SOLUTION

Analyse We are given the equation and asked for its rate law, assuming that it is an elementary process.

Plan Because we are assuming that the reaction occurs as a single elementary reaction, we are able to write the rate law using the coefficients for the reactants in the equation as the reaction orders.

Solve The reaction is bimolecular, involving one molecule of H_2 with one molecule of Br_2. Thus the rate law is first order in each reactant and second order overall:

$$\text{Rate} = k[H_2][Br_2]$$

Comment Experimental studies of this reaction show that the reaction actually has a very different rate law:

$$\text{Rate} = k[H_2][Br_2]^{\frac{1}{2}}$$

Because the experimental rate law differs from the one obtained by assuming a single elementary reaction, we can conclude that the mechanism must involve two or more elementary steps.

PRACTICE EXERCISE

Consider the following reaction: $2\,NO(g) + Br_2(g) \longrightarrow 2\,NOBr(g)$. **(a)** Write the rate law for the reaction, assuming it involves a single elementary reaction. **(b)** Is a single-step mechanism likely for this reaction?

Answers: **(a)** Rate = $k[NO]^2[Br_2]$ **(b)** No, because termolecular reactions require three molecules to strike simultaneously with enough energy and the correct geometry, which is extremely improbable.

(See also Exercise 15.51.)

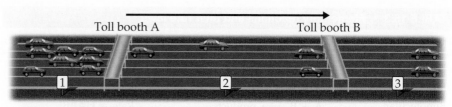

(a) Cars slowed at toll booth A, rate-determining step is passage through A

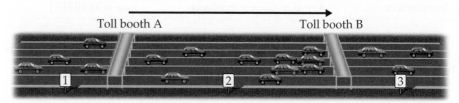

▶ FIGURE 15.21 **Rate-determining steps in traffic flow on a toll road.**

(b) Cars slowed at toll booth B, rate-determining step is passage through B

The Rate-Determining Step for a Multistep Mechanism

As with the reaction in Sample Exercise 15.12, most chemical reactions occur by mechanisms that involve two or more elementary reactions. Each of these steps of the mechanism has its own rate constant and activation energy. Often one of the steps is much slower than the others. The overall rate of a reaction cannot exceed the rate of the slowest elementary step of its mechanism. Because the slow step limits the overall reaction rate, it is called the **rate-determining step** (or *rate-limiting step*).

To understand the concept of the rate-determining step for a reaction, consider a toll road with two toll booths (▲ **FIGURE 15.21**). Cars enter the toll road at point 1 and pass through toll booth A. They then pass an intermediate point 2 before passing through toll booth B and arriving at point 3. We can envisage this trip along the toll road as occurring in two elementary steps:

$$\text{Step 1:} \quad \text{Point 1} \longrightarrow \text{Point 2} \quad \text{(through toll booth A)}$$

$$\underline{\text{Step 2:} \quad \text{Point 2} \longrightarrow \text{Point 3} \quad \text{(through toll booth B)}}$$

$$\text{Overall:} \quad \text{Point 1} \longrightarrow \text{Point 3} \quad \text{(through both toll booths)}$$

Now suppose that one or more gates at toll booth A are malfunctioning, so that traffic backs up behind the gates, as depicted in Figure 15.21(a). The rate at which cars can get to point 3 is limited by the rate at which they can get through the traffic jam at booth A. Thus step 1 is the rate-determining step of the journey along the toll road. If, however, all gates at A are functioning but one or more at B are not, traffic flows quickly through A but gets backed up at B, as depicted in Figure 15.21(b). In this case step 2 is the rate-determining step.

In the same way, *the slowest step in a multistep reaction limits the overall rate.* By analogy to Figure 15.21(a), the rate of a fast step following the rate-determining step does not speed up the overall rate. If the slow step is not the first one, as is the case in Figure 15.21(b), the faster preceding steps produce intermediate products that accumulate before being consumed in the slow step. In either case, *the rate-determining step governs the rate law for the overall reaction.*

 CONCEPT CHECK 11

Why can't the rate law for a reaction generally be deduced from the balanced equation for the reaction?

Mechanisms with a Slow Initial Step

The relationship between the slow step in a mechanism and the rate law for the overall reaction is most easily seen by considering an example in which the first step in a multistep mechanism is the slow rate-determining step. As an example, consider the reaction of NO_2 and CO to produce NO and CO_2 (Equation 15.27). Below 225 °C, it is found experimentally that the rate law for this reaction is second order in NO_2 and zero order in CO: Rate $= k[NO_2]^2$. Can we propose a reaction mechanism that is consistent with this rate law? Consider the following two-step mechanism:*

$$\text{Step 1:   } NO_2(g) + NO_2(g) \xrightarrow{k_1} NO_3(g) + NO(g) \quad \text{(slow)}$$

$$\text{Step 2:   } \underline{NO_3(g) + CO(g) \xrightarrow{k_2} NO_2(g) + CO_2(g)} \quad \text{(fast)}$$

$$\text{Overall: } NO_2(g) + CO(g) \longrightarrow NO(g) + CO_2(g)$$

Step 2 is much faster than step 1; that is, $k_2 \gg k_1$. The intermediate $NO_3(g)$ is slowly produced in step 1 and immediately consumed in step 2.

Because step 1 is slow and step 2 is fast, step 1 is rate determining. Thus the rate of the overall reaction equals the rate of step 1, and the rate law of the overall reaction equals the rate law of step 1. Step 1 is a bimolecular process that has the rate law

$$\text{Rate} = k_1[NO_2]^2$$

Thus the rate law predicted by this mechanism agrees with the one observed experimentally. The reactant CO is absent from the rate law, because it reacts in a step that follows the rate-determining step.

---

* The subscript on the rate constant identifies the elementary step involved. Thus k_1 is the rate constant for step 1, k_2 is the rate constant for step 2 and so forth. A negative subscript refers to the rate constant for the reverse of an elementary step. For example, k_{-1} is the rate constant for the reverse of the first step.

SAMPLE EXERCISE 15.14 **Determining the rate law for a multistep mechanism**

The decomposition of nitrous oxide, N_2O, is believed to occur by a two-step mechanism:

$$N_2O(g) \longrightarrow N_2(g) + O(g) \quad \text{(slow)}$$

$$N_2O(g) + O(g) \longrightarrow N_2(g) + O_2(g) \quad \text{(fast)}$$

(a) Write the equation for the overall reaction. **(b)** Write the rate law for the overall reaction.

SOLUTION

Analyse Given a multistep mechanism with the relative speeds of the steps, we are asked to write the overall reaction and the rate law for that overall reaction.

Plan **(a)** Find the overall reaction by adding the elementary steps and eliminating the intermediates. **(b)** The rate law for the overall reaction will be that of the slow, rate-determining step.

Solve

(a) Adding the two elementary reactions gives

$$2\,N_2O(g) + O(g) \longrightarrow 2\,N_2(g) + O_2(g) + O(g)$$

Omitting the intermediate, $O(g)$, which occurs on both sides of the equation, gives the overall reaction:

$$2\,N_2O(g) \longrightarrow 2\,N_2(g) + O_2(g)$$

(b) The rate law for the overall reaction is just the rate law for the slow rate-determining elementary reaction. Because that slow step is a unimolecular elementary reaction, the rate law is first order:

$$\text{Rate} = k[N_2O]$$

PRACTICE EXERCISE
Ozone reacts with nitrogen dioxide to produce dinitrogen pentoxide and oxygen:

$$O_3(g) + 2\,NO_2(g) \longrightarrow N_2O_5(g) + O_2(g)$$

The reaction is believed to occur in two steps:

$$O_3(g) + NO_2(g) \longrightarrow NO_3(g) + O_2(g)$$
$$NO_3(g) + NO_2(g) \longrightarrow N_2O_5(g)$$

The experimental rate law is rate $= k[O_3][NO_2]$. What can you say about the relative rates of the two steps of the mechanism?

Answer: Because the rate law conforms to the molecularity of the first step, that must be the rate-determining step. The second step must be much faster than the first one.

(See also Exercise 15.52.)

Mechanisms with a Fast Initial Step

It is difficult to derive the rate law for a mechanism in which an intermediate is a reactant in the rate-determining step. This situation arises in multistep mechanisms when the first step is *not* rate determining. Let's consider one example: the gas-phase reaction of nitric oxide (NO) with bromine (Br_2).

$$2\,NO(g) + Br_2(g) \longrightarrow 2\,NOBr(g) \qquad [15.28]$$

The experimentally determined rate law for this reaction is second order in NO and first order in Br_2:

$$\text{Rate} = k[NO]^2[Br_2] \qquad [15.29]$$

We seek a reaction mechanism that is consistent with this rate law. One possibility is that the reaction occurs in a single termolecular step:

$$NO(g) + NO(g) + Br_2(g) \longrightarrow 2\,NOBr(g) \qquad \text{Rate} = k[NO]^2[Br_2] \qquad [15.30]$$

As noted in Practice Exercise 15.13, this does not seem likely because termolecular processes are so rare.

> ### CONCEPT CHECK 12
> Why are termolecular elementary steps rare in gas-phase reactions?

Let's consider an alternative mechanism that does not invoke termolecular steps:

Step 1: $\quad NO(g) + Br_2(g) \underset{k_{-1}}{\overset{k_1}{\rightleftharpoons}} NOBr_2(g) \quad$ (fast)

Step 2: $\quad NOBr_2(g) + NO(g) \overset{k_2}{\longrightarrow} 2\,NOBr(g) \quad$ (slow)

In this mechanism, step 1 actually involves two processes: a forward reaction and its reverse.

Because step 2 is the slow rate-determining step, the rate of the overall reaction is governed by the rate law for that step:

$$\text{Rate} = k[NOBr_2][NO] \qquad [15.31]$$

However, $NOBr_2$ is an intermediate generated in step 1. Intermediates are usually unstable molecules that have a low, unknown concentration. Thus our rate law depends on the unknown concentration of an intermediate.

Fortunately, with the aid of some assumptions, we can express the concentration of the intermediate ($NOBr_2$) in terms of the concentrations of the starting reactants (NO and Br_2). We first assume that $NOBr_2$ is intrinsically unstable and that it does not accumulate to a significant extent in the reaction mixture. There are two ways for $NOBr_2$ to be consumed once it is formed: it can either react with NO to form NOBr or fall back apart into NO and Br_2. The first of these possibilities is step 2, a slow process. The second is the reverse of step 1, a unimolecular process:

$$NOBr_2(g) \xrightarrow{k_{-1}} NO(g) + Br_2(g) \qquad [15.32]$$

Because step 2 is slow, we assume that most of the $NOBr_2$ falls apart according to Equation 15.32. Thus we have both the forward and reverse reactions of step 1 occurring much faster than step 2. Because they occur rapidly with respect to the reaction in step 2, the forward and reverse processes of step 1 establish an equilibrium. We have seen examples of dynamic equilibrium before, in the equilibrium between a liquid and its vapour (∞ Section 11.5, 'Vapour Pressure') and between a solid solute and its solution (∞ Section 12.2, 'Saturated Solutions and Solubility'). As in any dynamic equilibrium, the rates of the forward and reverse reactions are equal. Thus we can equate the rate expression for the forward reaction in step 1 with the rate expression for the reverse reaction:

∞ Review this on pages 395 and 432

$$\underset{\text{Rate of forward reaction}}{k_1[NO][Br_2]} = \underset{\text{Rate of reverse reaction}}{k_{-1}[NOBr_2]}$$

Solving for $[NOBr_2]$, we have

$$[NOBr_2] = \frac{k_1}{k_{-1}}[NO][Br_2]$$

Substituting this relationship into the rate law for the rate-determining step (Equation 15.31), we have

$$\text{Rate} = k_2\frac{k_1}{k_{-1}}[NO][Br_2][NO] = k[NO]^2[Br_2]$$

This is consistent with the experimental rate law (Equation 15.29). The experimental rate constant, k, equals k_2k_1/k_{-1}.

In general, *whenever a fast step precedes a slow one, we can solve for the concentration of an intermediate by assuming that an equilibrium is established in the fast step.*

SAMPLE EXERCISE 15.15 **Deriving the rate law for a mechanism with a fast initial step**

Show that the following mechanism for Equation 15.28 also produces a rate law consistent with the experimentally observed one:

Step 1: $NO(g) + NO(g) \underset{k_{-1}}{\overset{k_1}{\rightleftharpoons}} N_2O_2(g)$ (fast, equilibrium)

Step 2: $N_2O_2(g) + Br_2(g) \xrightarrow{k_2} 2\,NOBr(g)$ (slow)

SOLUTION

Analyse We are given a mechanism with a fast initial step and asked to write the rate law for the overall reaction.

Plan The rate law of the slow elementary step in a mechanism determines the rate law for the overall reaction. Thus we first write the rate law based on the molecularity of the slow step. In this case the slow step involves the intermediate N_2O_2 as a reactant. Experimental rate laws, however, do not contain the concentrations of intermediates; instead they are expressed in terms of the concentrations of starting substances. Thus we must relate the concentration of N_2O_2 to the concentration of NO by assuming that an equilibrium is established in the first step.

Solve The second step is rate determining, so the overall rate is

$$\text{Rate} = k_2[\text{N}_2\text{O}_2][\text{Br}_2]$$

We solve for the concentration of the intermediate, N_2O_2, by assuming that an equilibrium is established in step 1; thus the rates of the forward and reverse reactions in step 1 are equal:

$$k_1[\text{NO}]^2 = k_{-1}[\text{N}_2\text{O}_2]$$

$$[\text{N}_2\text{O}_2] = \frac{k_1}{k_{-1}}[\text{NO}]^2$$

Substituting this expression into the rate expression gives

$$\text{Rate} = k_2\frac{k_1}{k_{-1}}[\text{NO}]^2[\text{Br}_2] = k[\text{NO}]^2[\text{Br}_2]$$

Thus this mechanism also yields a rate law consistent with the experimental one.

PRACTICE EXERCISE

The first step of a mechanism involving the reaction of bromine is

$$\text{Br}_2(g) \underset{k_{-1}}{\overset{k_1}{\rightleftharpoons}} 2\,\text{Br}(g) \quad \text{(fast, equilibrium)}$$

What is the expression relating the concentration of $\text{Br}(g)$ to that of $\text{Br}_2(g)$?

Answer: $[\text{Br}] = \left(\dfrac{k_1}{k_{-1}}[\text{Br}_2]\right)^{\frac{1}{2}}$

(See also Exercise 15.68.)

15.7 | CATALYSIS

A **catalyst** is a substance that changes the speed of a chemical reaction without undergoing a permanent chemical change itself in the process. Catalysts are very common; most reactions in the body, the atmosphere and the oceans occur with the help of catalysts. Much industrial chemical research is devoted to the search for new and more effective catalysts for reactions of commercial importance. Extensive research efforts are also devoted to finding means of inhibiting or removing certain catalysts that promote undesirable reactions, such as those that corrode metals, age our bodies and cause tooth decay.

Homogeneous Catalysis

A catalyst that is present in the same phase as the reacting molecules is called a **homogeneous catalyst**. Examples abound both in solution and in the gas phase. Consider, for example, the decomposition of aqueous hydrogen peroxide, $\text{H}_2\text{O}_2(aq)$, into water and oxygen:

$$2\,\text{H}_2\text{O}_2(aq) \longrightarrow 2\,\text{H}_2\text{O}(l) + \text{O}_2(g) \qquad [15.33]$$

In the absence of a catalyst or light, this reaction occurs extremely slowly.

Many different substances are capable of catalysing the reaction represented by Equation 15.33, including bromide ion, $\text{Br}^-(aq)$, as shown in ▶ FIGURE 15.22. The bromide ion reacts with hydrogen peroxide in acidic solution, forming aqueous bromine and water:

$$2\,\text{Br}^-(aq) + \text{H}_2\text{O}_2(aq) + 2\,\text{H}^+ \longrightarrow \text{Br}_2(aq) + 2\,\text{H}_2\text{O}(l) \qquad [15.34]$$

The brown colour observed in the middle photograph of Figure 15.22 indicates the formation of $\text{Br}_2(aq)$. If this were the complete reaction, bromide ion would not be

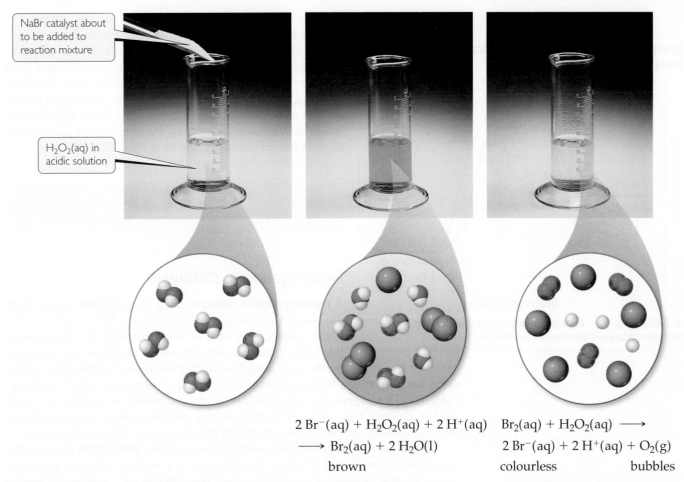

NaBr catalyst about to be added to reaction mixture

$H_2O_2(aq)$ in acidic solution

$$2\,Br^-(aq) + H_2O_2(aq) + 2\,H^+(aq)$$
$$\longrightarrow Br_2(aq) + 2\,H_2O(l)$$
brown

$$Br_2(aq) + H_2O_2(aq) \longrightarrow$$
$$2\,Br^-(aq) + 2\,H^+(aq) + O_2(g)$$
colourless   bubbles

▲ **FIGURE 15.22   Homogeneous catalysis.** Effect of catalyst on the speed of hydrogen peroxide decomposition to water and oxygen gas.

a catalyst because it undergoes chemical change during the reaction. However, hydrogen peroxide also reacts with the $Br_2(aq)$ generated in Equation 15.34:

$$Br_2(aq) + H_2O_2(aq) \longrightarrow 2\,Br^-(aq) + 2\,H^+(aq) + O_2(g) \qquad [15.35]$$

The sum of Equations 15.34 and 15.35 is Equation 15.33:

$$2\,H_2O_2(aq) \longrightarrow 2\,H_2O(l) + O_2(g)$$

When H_2O_2 has been totally decomposed, we are left with a colourless solution of $Br^-(aq)$, as seen in the photograph on the right in Figure 15.22. Bromide ion, therefore, is indeed a catalyst of the reaction because it speeds the overall reaction without itself undergoing any net change. In contrast, Br_2 is an intermediate because it is first formed (Equation 15.34) and then consumed (Equation 15.35). Neither the catalyst nor the intermediate appears in the chemical equation for the overall reaction. Notice, however, that *the catalyst is there at the start of the reaction, whereas the intermediate is formed during the course of the reaction.*

How does a catalyst work? If we think about the general form of rate laws (rate = $k[A]^m[B]^n$), we must conclude that the catalyst must affect the numerical value of k, the rate constant.

On the basis of the Arrhenius equation (Equation 15.23), the rate constant (k) is determined by the activation energy (E_a) and the frequency factor (A). A catalyst may affect the rate of reaction by altering the value of either E_a or A. The most dramatic catalytic effects come from lowering E_a. As a general rule, *a catalyst provides an alternative mechanism of lower activation energy for a chemical reaction.*

A catalyst usually affects the activation energy and/or increases the frequency factor for a reaction by providing a completely different mechanism for

Where are the intermediates and transition states in this diagram?

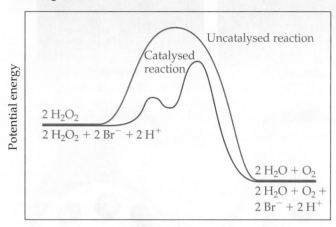

▲ **FIGURE 15.23   Energy profiles for the uncatalysed and bromide-catalysed decomposition of H$_2$O$_2$.**

∞ Review this on page 457

∞ Review this on page 272

the reaction. The examples given above involve a reversible, cyclic reaction of the catalyst with the reactants. In the decomposition of hydrogen peroxide, for example, two successive reactions of H$_2$O$_2$, with bromide and then with bromine, take place. Because these two reactions together serve as a catalytic pathway for hydrogen peroxide decomposition, *both* of them must have significantly lower activation energies than the uncatalysed decomposition, as shown schematically in ◀ **FIGURE 15.23**.

> **CONCEPT CHECK 13**
>
> How does a catalyst increase the rate of a reaction?

Heterogeneous Catalysis

A **heterogeneous catalyst** exists in a different phase from the reactant molecules, usually as a solid in contact with either gaseous reactants or with reactants in a liquid solution. Many industrially important reactions are catalysed by the surfaces of solids. For example, hydrocarbon molecules are rearranged to form petrol with the aid of what are called 'cracking' catalysts. Heterogeneous catalysts are often composed of metals or metal oxides. Because the catalysed reaction occurs on the surface, special methods are often used to prepare catalysts so that they have very large surface areas.

The initial step in heterogeneous catalysis is usually **adsorption** of reactants. *Adsorption* refers to the binding of molecules to a surface, whereas *absorption* refers to the uptake of molecules into the interior of another substance (∞ Section 12.6, 'Colloids'). Adsorption occurs because the atoms or ions at the surface of a solid are extremely reactive. Unlike their counterparts in the interior of the substance, surface atoms and ions have unused bonding capacity, that can be used to bond molecules from the gas or solution phase to the surface of the solid.

The reaction of hydrogen gas with ethene gas to form ethane gas provides an example of heterogeneous catalysis:

$$C_2H_4(g) + H_2(g) \longrightarrow C_2H_6(g) \qquad \Delta H° = -137 \text{ kJ mol}^{-1} \qquad [15.36]$$

Ethene   Ethane

Even though this reaction is exothermic, it occurs very slowly in the absence of a catalyst. In the presence of a finely powdered metal, however, such as nickel, palladium or platinum, the reaction occurs rather easily at room temperature, via the mechanism shown in ▶ **FIGURE 15.24**. Both ethylene and hydrogen are adsorbed on the metal surface. Upon adsorption, the H–H bond of H$_2$ breaks, leaving two H atoms initially bonded to the metal surface but relatively free to move. When a hydrogen encounters an adsorbed ethene molecule, it can form a σ-bond to one of the carbon atoms, effectively destroying the C–C π-bond and leaving an *ethyl group* (C$_2$H$_5$) bonded to the surface via a metal-to-carbon σ-bond. This σ-bond is relatively weak, so when the other carbon atom also encounters a hydrogen atom, a sixth C–H σ-bond is readily formed, and an ethane molecule (C$_2$H$_6$) is released from the metal surface.

We can understand the role of the catalyst in this process by considering the bond enthalpies involved (∞ Section 8.8, 'Strength of Covalent Bonds'). In the course of the reaction, the H–H σ-bond and the C–C π-bond must be broken, and to do so requires the input of energy, which is essentially the activation energy of the reaction. The formation of the new C–H σ-bonds *releases* an even greater amount of energy, making the reaction exothermic. When H$_2$ and

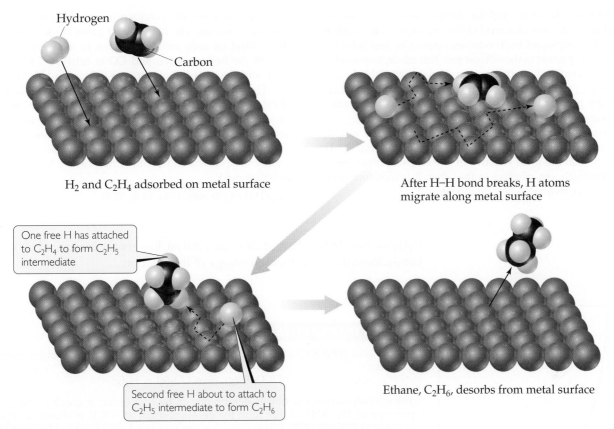

Hydrogen

Carbon

H_2 and C_2H_4 adsorbed on metal surface

After H–H bond breaks, H atoms migrate along metal surface

One free H has attached to C_2H_4 to form C_2H_5 intermediate

Second free H about to attach to C_2H_5 intermediate to form C_2H_6

Ethane, C_2H_6, desorbs from metal surface

▲ FIGURE 15.24 **Heterogeneous catalysis.** Mechanism for reaction of ethylene with hydrogen on a catalytic surface.

MY WORLD OF CHEMISTRY

CATALYTIC CONVERTERS

Heterogeneous catalysis plays a major role in the fight against urban air pollution. Two components of car exhausts that help form photochemical smog are nitrogen oxides and unburned hydrocarbons. In addition, car exhaust may contain considerable quantities of carbon monoxide. Even with the most careful attention to engine design, it is impossible under normal driving conditions to reduce the quantity of these pollutants to an acceptable level in the exhaust gases. It is therefore necessary to remove them from the exhaust before they are vented to the air. This removal is accomplished in the *catalytic converter*.

The catalytic converter, which is part of a car's exhaust system, must perform two functions: (1) oxidation of CO and unburned hydrocarbons (C_xH_y) to carbon dioxide and water, and (2) reduction of nitrogen oxides to nitrogen gas:

$$CO, C_xH_y \xrightarrow{O_2} CO_2 + H_2O$$

$$NO, NO_2 \longrightarrow N_2$$

These two functions require different catalysts, so the development of a successful catalyst system is a difficult challenge. The catalysts must be effective over a wide range of operating temperatures. They must continue to be active despite the fact that various components of the exhaust can block the active sites of the catalyst. The catalysts must also be sufficiently rugged to withstand exhaust gas turbulence and the mechanical shocks of driving under various conditions for thousands of miles.

Catalysts that promote the combustion of CO and hydrocarbons are, in general, the transition metal oxides and the noble metals. These materials are supported on a structure (▼ FIGURE 15.25) that allows the best possible contact between

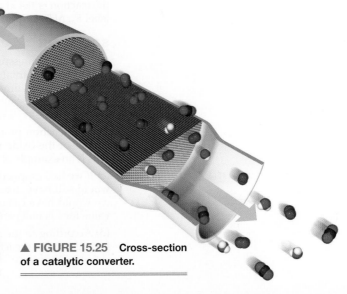

▲ FIGURE 15.25 **Cross-section of a catalytic converter.**

the flowing exhaust gas and the catalyst surface. A honeycomb structure made from alumina (Al_2O_3) and impregnated with the catalyst is employed. Such catalysts operate by first adsorbing oxygen gas present in the exhaust gas. This adsorption weakens the O–O bond in O_2, so that oxygen atoms are available for reaction with adsorbed CO to form CO_2. Hydrocarbon oxidation probably proceeds somewhat similarly, with the hydrocarbons first being adsorbed followed by rupture of a C–H bond.

Transition metal oxides and noble metals are also the most effective catalysts for reduction of NO to N_2 and O_2. The catalysts that are most effective in one reaction, however, are usually much less effective in the other. It is therefore necessary to have two catalytic components.

Catalytic converters contain remarkably efficient heterogeneous catalysts. The car exhaust gases are in contact with the catalyst for only 100–400 ms, but in this very short time 96% of the hydrocarbons and CO is converted to CO_2 and H_2O, and the emission of nitrogen oxides is reduced by 76%.

There are costs as well as benefits associated with the use of catalytic converters, one being that some of the metals are very expensive. Catalytic converters currently account for about 35% of the platinum, 65% of the palladium and 95% of the rhodium used annually. All of these metals, which come mainly from Russia and South Africa, can be far more expensive than gold.

RELATED EXERCISES: 15.44, 15.58, 15.59

C_2H_4 are bonded to the surface of the catalyst, less energy is required to break the bonds, lowering the activation energy of the reaction.

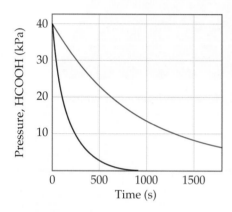

▲ **FIGURE 15.26**   **Variation in pressure of HCOOH(g) as a function of time at 838 K.** The red line corresponds to decomposition when only gaseous HCOOH is present. The blue line corresponds to decomposition in the presence of added ZnO(s).

SAMPLE INTEGRATIVE EXERCISE | Putting concepts together

Formic acid (HCOOH) decomposes in the gas phase at elevated temperatures as follows:

$$HCOOH(g) \longrightarrow CO_2(g) + H_2(g)$$

The decomposition reaction is determined to be first order. A graph of the partial pressure of HCOOH versus time for decomposition at 838 K is shown as the red curve in ◄ FIGURE 15.26. When a small amount of solid ZnO is added to the reaction chamber the partial pressure of acid versus time varies as shown by the blue curve in Figure 15.26.

(a) Estimate the half-life and first-order rate constant for formic acid decomposition in the absence of ZnO.

(b) What can you conclude from the effect of added ZnO on the decomposition of formic acid?

(c) The progress of the reaction was followed by measuring the partial pressure of formic acid vapour at selected times. Suppose that, instead, we had plotted the concentration of formic acid in units of mol dm^{-3}. What effect would this have had on the calculated value of k?

(d) The pressure of formic acid vapour at the start of the reaction is 40.0 kPa. Assuming constant temperature and ideal-gas behaviour, what is the pressure in the system at the end of the reaction? If the volume of the reaction chamber is 436 cm^3, how many moles of gas occupy the reaction chamber at the end of the reaction?

(e) The standard heat of formation of formic acid vapour is $\Delta_f H° = -378.6$ kJ mol^{-1}. Calculate $\Delta H°$ for the overall reaction. Assuming that the activation energy (E_a) for the reaction is 184 kJ mol^{-1} sketch an approximate energy profile for the reaction, and label E_a, $\Delta H°$ and the transition state.

SOLUTION

(a) The initial pressure of HCOOH is 40.0 kPa. On the graph we move to the level at which the partial pressure of HCOOH is 20.0 kPa, half the initial value. This corresponds to a time of about 660×10^2 s, which is therefore the half-life. The first-order rate constant is given by Equation 15.19: $k = 0.693/t_{\frac{1}{2}} = 0.693/660$ s $= 1.05 \times 10^{-3}$ s^{-1}.

(b) The reaction proceeds much more rapidly in the presence of solid ZnO, so the surface of the oxide must be acting as a catalyst for the decomposition of the acid. This is an example of heterogeneous catalysis.

(c) If we had graphed the concentration of formic acid in units of moles per litre, we would still have determined that the half-life for decomposition is 660 seconds, and we would have calculated the same value for k. Because the units for k are s^{-1}, the value for k is independent of the units used for concentration.

(d) According to the stoichiometry of the reaction, two moles of product are formed for each mole of reactant. When reaction is completed, therefore, the pressure will be

80 kPa, just twice the initial pressure, assuming ideal-gas behaviour. (Because we are working at quite high temperature and fairly low gas pressure, assuming ideal-gas behaviour is reasonable.) The number of moles of gas present can be calculated using the ideal-gas equation (Section 10.4):

$$n = \frac{PV}{RT} = \frac{(80 \text{ kPa})(0.436 \text{ dm}^3)}{(8.314 \text{ kPa dm}^3 \text{ mol}^{-1} \text{ K}^{-1})(838 \text{ K})} = 5.00 \times 10^{-3} \text{ moles}$$

(e) We first calculate the overall change in energy, $\Delta H°$ (Section 14.7 and Appendix C), as in

$$\Delta H° = \Delta_f H°(CO_2(g)) + \Delta_f H°(H_2(g)) - \Delta_f H°(HCOOH(g))$$
$$= -393.5 \text{ kJ mol}^{-1} + 0 - (-378.6 \text{ kJ mol}^{-1})$$
$$= -14.9 \text{ kJ mol}^{-1}$$

From this and the given value for E_a, we can draw an approximate energy profile for the reaction, in analogy to Figure 15.17.

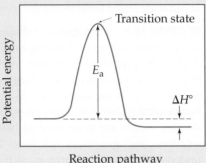

CHAPTER SUMMARY AND KEY TERMS

SECTION 15.1   Chemical kinetics is the study of the rates of chemical reactions and the factors that affect them, such as concentration, physical states of reactants, temperature and catalysts.

SECTION 15.2   Reaction rates are usually expressed as the decrease in concentration per unit time of the reactants or an increase in concentration per unit time of the products. Typically the rates are given in units of molarity per second ($M \text{ s}^{-1}$). We find that the stoichiometry of the reaction dictates the relationship between rates of appearance of products and rates of disappearance of reactants. The **instantaneous rate** is the slope of a line drawn tangent to the concentration versus time curve at a specific time.

SECTION 15.3   The quantitative relationship between rate and concentration is expressed by a **rate law**, which usually has the following form:

$$\text{Rate} = k[\text{reactant 1}]^m[\text{reactant 2}]^n$$

The constant k in the rate law is called the **rate constant**; the exponents m, n and so forth are called **reaction orders** for the reactants. The sum of the reaction orders gives the **overall reaction order**. Reaction orders must be determined experimentally. The units of the rate constant depend on the overall reaction order. For a reaction in which the overall reaction order is 0, k has units of $M \text{ s}^{-1}$; for one in which the overall reaction order is 1, k has units of s^{-1}; for one in which the overall reaction order is 2, k has units of $M^{-1} \text{ s}^{-1}$. Spectroscopy is one technique that can be used to monitor the course of a reaction. According to Beer's law, the absorption of electromagnetic radiation by a substance at a particular wavelength is directly proportional to its concentration.

SECTION 15.4   Rate laws can be used to determine the concentrations of reactants or products at any time during a reaction. A **zero-order reaction** is one for which the overall reaction order is 0. Rate = k if the reaction is zero order. In a **first-order reaction** the rate is proportional to the concentration of a single reactant raised to the first power: Rate = $k[A]$. In such cases the integrated form of the rate law is $\ln[A]_t = -kt + \ln[A]_0$, where $[A]_t$ is the concentration of reactant A at time t, k is the rate constant, and $[A]_0$ is the initial concentration of A. Thus for a first-order reaction, a graph of $\ln[A]$ versus time yields a straight line of slope $-k$.

A **second-order reaction** is one for which the overall reaction order is 2. If a second-order rate law depends on the concentration of only one reactant, then rate = $k[A]^2$, and the time dependence of $[A]$ is given by the integrated form of the rate law: $1/[A]_t = 1/[A]_0 + kt$. In this case a graph of $1/[A]_t$ versus time yields a straight line.

The **half-life** of a reaction, $t_{\frac{1}{2}}$, is the time required for the concentration of a reactant to drop to one-half of its original value. For a first-order reaction, the half-life depends only on the rate constant and not on the initial concentration: $t_{\frac{1}{2}} = 0.693/k$. The half-life of a second-order reaction depends on both the rate constant and the initial concentration of A: $t_{\frac{1}{2}} = 1/k[A]_0$.

SECTION 15.5 The **collision model**, which assumes that reactions occur as a result of collisions between molecules, helps explain why the magnitudes of rate constants increase with increasing temperature. The greater the kinetic energy of the colliding molecules, the greater is the energy of collision. The minimum energy required for a reaction to occur is called the **activation energy, E_a**. A collision with energy E_a or greater can cause the atoms of the colliding molecules to reach the **activated complex** (or **transition state**), which is the highest energy arrangement in the pathway from reactants to products. Even if a collision is energetic enough, it may not lead to reaction; the reactants must also be correctly oriented relative to one another in order for a collision to be effective.

Because the kinetic energy of molecules depends on temperature, the rate constant of a reaction is very dependent on temperature. The relationship between k and temperature is given by the **Arrhenius equation**: $k = Ae^{-E_a/RT}$. The term A is called the **frequency factor**; it relates to the number of collisions that are favourably oriented for reaction. The Arrhenius equation is often used in logarithmic form: $\ln k = \ln A - E_a/RT$. Thus a graph of $\ln k$ versus $1/T$ yields a straight line with slope $-E_a/R$.

SECTION 15.6 A **reaction mechanism** details the individual steps that occur in the course of a reaction. Each of these steps, called **elementary reactions**, has a well-defined rate law that depends on the number of molecules (the **molecularity**) of the step. Elementary reactions are defined as either **unimolecular**, **bimolecular** or **termolecular**, depending on whether one, two or three reactant molecules are involved, respectively. Termolecular elementary reactions are very rare. Unimolecular, bimolecular and termolecular reactions follow rate laws that are first order overall, second order overall and third order overall, respectively. Many reactions occur by a multistep mechanism, involving two or more elementary reactions, or steps. An **intermediate** that is produced in one elementary step, is consumed in a later elementary step, and therefore does not appear in the overall equation for the reaction. When a mechanism has several elementary steps, the overall rate is limited by the slowest elementary step, called the **rate-determining step**. A fast elementary step that follows the rate-determining step will have no effect on the rate law of the reaction. A fast step that precedes the rate-determining step often creates an equilibrium that involves an intermediate. For a mechanism to be valid, the rate law predicted by the mechanism must be the same as that observed experimentally.

SECTION 15.7 A **catalyst** is a substance that increases the rate of a reaction without undergoing a net chemical change itself. It does so by providing a different mechanism for the reaction, one that has a lower activation energy. A **homogeneous catalyst** is one that is in the same phase as the reactants. A **heterogeneous catalyst** has a different phase from the reactants. Finely divided metals are often used as heterogeneous catalysts for solution- and gas-phase reactions. Reacting molecules can undergo binding, or adsorption, at the surface of the catalyst. The **adsorption** of a reactant at specific sites on the surface makes bond breaking easier, lowering the activation energy.

KEY SKILLS

- Understand the factors that affect the rate of chemical reactions. (Section 15.1)

- Determine the rate of a chemical reaction given time and concentration. (Section 15.2)

- Relate the rate of disappearance of reactants and the rate of appearance of products from the stoichiometry of the reaction. (Section 15.2)

- Understand the meaning of a rate law, reaction order and rate constant. (Section 15.3)

- Determine the rate law and rate constant, given the experimentally determined rates for a series of concentrations of reactants. (Section 15.3)

- Use the integrated form of a rate law to determine the concentration of a reactant at a given time. (Section 15.4)

- Understand the Arrhenius equation and how it relates temperature to the rate constant and be able to use it. (Section 15.5)

- Predict a rate law for a reaction having a multistep mechanism given the individual steps of the mechanism. (Section 15.6)

- Explain the action of a catalyst. (Section 15.7)

KEY EQUATIONS

- Relating rates to the change in concentration of components with time for a balanced chemical equation

$$\text{Rate} = -\frac{1}{a}\frac{\Delta[A]}{\Delta t} = -\frac{1}{b}\frac{\Delta[B]}{\Delta t} = \frac{1}{c}\frac{\Delta[C]}{\Delta t} = \frac{1}{d}\frac{\Delta[D]}{\Delta t} \qquad [15.4]$$

- General form of the rate law

$$\text{Rate} = k[\text{reactant 1}]^m[\text{reactant 2}]^n \ldots \qquad [15.8]$$

- The integrated form of a zero-order rate law

$$[A]_t = -kt + [A]_0 \qquad [15.14]$$

- The integrated form of a first-order rate law

$$\ln[A]_t = -kt + \ln[A]_0 \qquad [15.16]$$

- The integrated form of a second-order rate law

$$\frac{1}{[A]_t} = kt + \frac{1}{[A]_0} \qquad [15.17]$$

- Relating the half-life and rate constant for a zero-order reaction

$$t_{\frac{1}{2}} = \frac{[A]_0}{2k} \qquad [15.18]$$

- Relating the half-life and rate constant for a first-order reaction

$$t_{\frac{1}{2}} = \frac{0.693}{k} \qquad [15.19]$$

- Relating the half-life and rate constant for a second-order reaction

$$t_{\frac{1}{2}} = \frac{1}{k[A]_0} \qquad [15.21]$$

- Relating the rate constant and temperature

$$k = Ae^{-E_a/RT} \qquad [15.23]$$

- Linear form of the Arrhenius equation

$$\ln k = -\frac{E_a}{RT} + \ln A \qquad [15.24]$$

EXERCISES

VISUALISING CONCEPTS

15.1 For which one of the following vessels for the reaction A + B $\longrightarrow$ C is the reaction the fastest? Assume all vessels are at the same temperature. [Section 15.1]

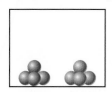
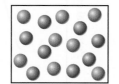

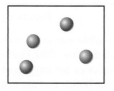

15.2 Consider the following graph of the concentration of a substance over time. **(a)** Is X a reactant or product of the reaction? **(b)** Why is the average rate of the reaction greater between points 1 and 2 than between points 2 and 3? [Section 15.2]

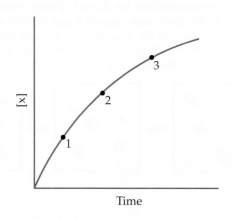

15.3 You study the rate of a reaction, measuring both the concentration of the reactant and the concentration of the product as a function of time, and obtain the following results:

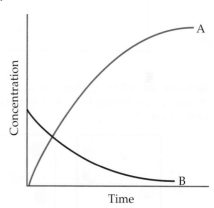

What chemical equation is consistent with these data: **(a)** A $\longrightarrow$ B, **(b)** B $\longrightarrow$ A, **(c)** A $\longrightarrow$ 2 B, **(d)** B $\longrightarrow$ 2 A. Explain your choice. [Section 15.2]

15.4 You perform a series of experiments for the reaction A $\longrightarrow$ B + C and find that the rate law has the form rate = $k[A]^x$. Determine the value of x in each of the following cases. **(a)** There is no rate change when [A] is tripled. **(b)** The rate increases by a factor of 9 when [A] is tripled. **(c)** When [A] is doubled, the rate increases by a factor of 8. [Section 15.3]

15.5 The following diagrams represent mixtures of NO(g) and O_2(g). These two substances react as follows:

$$2\,NO(g) + O_2(g) \longrightarrow 2\,NO_2(g)$$

It has been determined experimentally that the rate is second order in NO and first order in O_2. Based on this fact, which of the following mixtures will have the fastest initial rate? [Section 15.3]

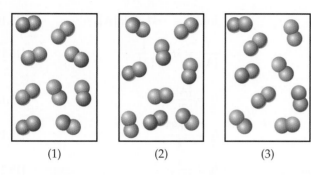

(1)   (2)   (3)

15.6 A friend studies a first-order reaction and obtains the following three graphs for experiments done at two different temperatures. **(a)** Which two lines represent experiments conducted at the same temperature? What accounts for the difference in these two lines? In what way are they the same? **(b)** Which two lines represent experiments done with the same starting concentration but at different temperatures? Which line probably represents the lower temperature? How do you know? [Section 15.4]

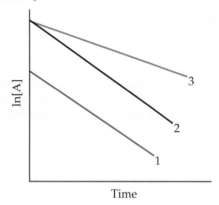

15.7 (a) Given the following diagrams at $t = 0$ min and $t = 30$ min, what is the half-life of the reaction if it follows first-order kinetics?

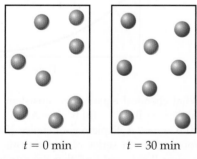

$t = 0$ min $t = 30$ min

(b) After four half-life periods for a first-order reaction, what fraction of reactant remains? [Section 15.4]

15.8 You study the effect of temperature on the rate of two reactions and graph the natural logarithm of the rate constant for each reaction as a function of $1/T$. How do the two graphs compare **(a)** if the activation energy of the second reaction is higher than the activation energy of the first reaction but the two reactions have the same frequency factor, and **(b)** if the frequency factor of the second reaction is higher than the frequency factor of the first reaction but the two reactions have the same activation energy? [Section 15.5]

15.9 Consider the diagram that follows, which represents two steps in an overall reaction. The red spheres are oxygen, the blue ones nitrogen and the green ones fluorine. **(a)** Write the chemical equation for each step in the reaction. **(b)** Write the equation for the overall reaction. **(c)** Identify the intermediate in the mechanism. **(d)** Write the rate law for the overall reaction if the first step is the slow, rate-determining step. [Section 15.6]

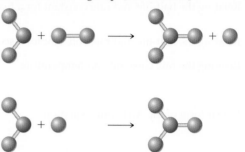

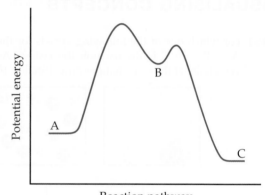

15.10 Based on the following reaction profile, how many intermediates are formed in the reaction A ⟶ C? How many transition states are there? Which step is the fastest? Is the reaction A ⟶ C exothermic or endothermic? [Section 15.6]

Reaction pathway

15.11 Draw a possible transition state for the bimolecular reaction depicted below. (The blue spheres are nitrogen atoms and the red ones are oxygen atoms.) Use dashed lines to represent the bonds that are in the process of being broken or made in the transition state. [Section 15.6]

15.12 The following diagram represents an imaginary two-step mechanism. Let the red spheres represent element A, the green ones element B and the blue ones element C. **(a)** Write the equation for the net reaction that is occurring. **(b)** Identify the intermediate. **(c)** Identify the catalyst. [Sections 15.6 and 15.7]

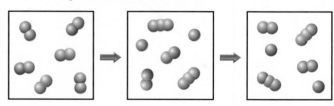

REACTION RATES (Sections 15.1 and 15.2)

15.13 **(a)** What is meant by the term *reaction rate*? **(b)** Name three factors that can affect the rate of a chemical reaction. **(c)** What information is necessary to relate the rate of disappearance of reactants to the rate of appearance of products?

15.14 **(a)** What are the units usually used to express the rates of reactions occurring in solution? **(b)** From your everyday experience, give two examples of the effects of temperature on the rates of reactions. **(c)** What is the difference between average rate and instantaneous rate?

15.15 Consider the following hypothetical aqueous reaction: A(aq) $\longrightarrow$ B(aq). A flask is charged with 0.065 mol of A in a total volume of 100.0 cm^3. The following data are collected:

Time (min)	0	10	20	30	40
Moles of A	0.065	0.051	0.042	0.036	0.031

(a) Calculate the number of moles of B at each time in the table, assuming that there are no molecules of B at time zero. **(b)** Calculate the average rate of disappearance of A for each 10 min interval, in units of M s^{-1}. **(c)** Between t = 10 min and t = 30 min, what is the average rate of appearance of B in units of M s^{-1}? Assume that the volume of the solution is constant.

15.16 The rate of disappearance of HCl was measured for the following reaction:

$$CH_3OH(aq) + HCl(aq) \longrightarrow CH_3Cl(aq) + H_2O(l)$$

The following data were collected:

Time (min)	[HCl] (M)
0.0	1.85
54.0	1.58
107.0	1.36
215.0	1.02
430.0	0.580

(a) Calculate the average rate of reaction, in M s^{-1}, for the time interval between each measurement. **(b)** Graph [HCl] versus time, and determine the instantaneous rates in M min^{-1} and M s^{-1} at t = 75.0 min and t = 250 min.

15.17 For each of the following gas-phase reactions, write the rate expression in terms of the appearance of each product or disappearance of each reactant:
(a) $2\,HBr(g) \longrightarrow H_2(g) + Br_2(g)$
(b) $2\,SO_2(g) + O_2(g) \longrightarrow 2\,SO_3(g)$
(c) $2\,NO(g) + 2\,H_2(g) \longrightarrow N_2(g) + 2\,H_2O(g)$

15.18 **(a)** Consider the combustion of $H_2(g)$: $2\,H_2(g) + O_2(g) \longrightarrow 2\,H_2O(g)$. If hydrogen is burning at the rate of 0.85 mol s^{-1} what is the rate of consumption of oxygen? What is the rate of formation of water vapour? **(b)** The reaction $2\,NO(g) + Cl_2(g) \longrightarrow 2\,NOCl(g)$ is conducted in a closed vessel. If the partial pressure of NO is decreasing at the rate of 23 Pa min^{-1}, what is the rate of change of the total pressure of the vessel?

15.19 **(a)** Consider the combustion of ethylene, $C_2H_4(g) + 3\,O_2(g) \longrightarrow 2\,CO_2(g) + 2\,H_2O(g)$. If the concentration of C_2H_4 is decreasing at the rate of 0.37 M s^{-1}, what are the rates of change in the concentrations of CO_2 and H_2O? **(b)** The rate of decrease in N_2H_4 partial pressure in a closed reaction vessel from the reaction $N_2H_4(g) + H_2(g) \longrightarrow 2\,NH_3(g)$ is 63 Pa h^{-1}. What are the rates of change of NH_3 partial pressure and total pressure in the vessel?

RATE LAWS (Section 15.3)

15.20 A reaction A + B $\longrightarrow$ C obeys the following rate law: Rate = $k[A]^2[B]$. **(a)** If [A] is doubled, how will the rate change? Will the rate constant change? Explain. **(b)** What are the reaction orders for A and B? What is the overall reaction order? **(c)** What are the units of the rate constant?

15.21 The decomposition of N_2O_5 in carbon tetrachloride proceeds as follows: $2\,N_2O_5 \longrightarrow 4\,NO_2 + O_2$. The rate law is first order in N_2O_5. At 64 °C the rate constant is 4.82×10^{-3} s^{-1}. **(a)** Write the rate law for the reaction. **(b)** What is the rate of reaction when $[N_2O_5]$ = 0.0240 M? **(c)** What happens to the rate when the concentration of N_2O_5 is doubled to 0.0480 M?

15.22 Consider the following reaction:

$$CH_3Br(aq) + OH^-(aq) \longrightarrow CH_3OH(aq) + Br^-(aq)$$

The rate law for this reaction is first order in CH_3Br and first order in OH$^-$. When $[CH_3Br]$ is 5.0×10^{-3} M and $[OH^-]$ is 0.050 M, the reaction rate at 298 K is 0.0432 M s^{-1}. **(a)** What is the value of the rate constant? **(b)** What are the units of the rate constant? **(c)** What would happen to the rate if the concentration of OH$^-$ were tripled?

15.23 The reaction between ethyl bromide (C_2H_5Br) and hydroxide ion in ethyl alcohol at 330 K, $C_2H_5Br + OH^- \longrightarrow C_2H_5OH + Br^-$, is first order each in bromoethane and hydroxide ion. When $[C_2H_5Br]$ is 0.0477 M and $[OH^-]$ is 0.100 M, the rate of disappearance of bromoethane is 1.7×10^{-7} M s^{-1}. **(a)** What is the value of the rate constant? **(b)** What are the units of the rate constant? **(c)** How would the rate of disappearance of bromoethane change if the solution were diluted by adding an equal volume of pure ethyl alcohol to the solution?

15.24 The iodide ion reacts with hypochlorite ion (the active ingredient in chlorine bleaches) in the following way: $OCl^- + I^- \longrightarrow OI^- + Cl^-$. This rapid reaction gives the following rate data:

[OCl⁻] (M)	[I⁻] (M)	Initial rate (M s⁻¹)
1.5×10^{-3}	1.5×10^{-3}	1.36×10^{-4}
3.0×10^{-3}	1.5×10^{-3}	2.72×10^{-4}
1.5×10^{-3}	3.0×10^{-3}	2.72×10^{-4}

(a) Write the rate law for this reaction. **(b)** Calculate the rate constant. **(c)** Calculate the rate when $[OCl^-] = 2.0 \times 10^{-3}$ M and $[I^-] = 5.0 \times 10^{-4}$ M.

15.25 The following data were measured for the reaction $BF_3(g) + NH_3(g) \longrightarrow F_3BNH_3(g)$:

Experiment	[BF₃] (M)	[NH₃] (M)	Initial rate (M s⁻¹)
1	0.250	0.250	0.2130
2	0.250	0.125	0.1065
3	0.200	0.100	0.0682
4	0.350	0.100	0.1193
5	0.175	0.100	0.0596

(a) What is the rate law for the reaction? **(b)** What is the overall order of the reaction? **(c)** What is the value of the rate constant for the reaction? **(d)** What is the rate when $[BF_3] = 0.100$ M and $[NH_3] = 0.500$ M?

15.26 Consider the gas-phase reaction between nitric oxide and bromine at 273 °C: $2 NO(g) + Br_2(g) \longrightarrow 2 NOBr(g)$. The following data for the initial rate of appearance of NOBr were obtained:

Experiment	[NO] (M)	[Br₂] (M)	Initial rate (M s⁻¹)
1	0.10	0.20	24
2	0.25	0.20	150
3	0.10	0.50	60
4	0.35	0.50	735

(a) Determine the rate law. **(b)** Calculate the average value of the rate constant for the appearance of NOBr from the four data sets. **(c)** How is the rate of appearance of NOBr related to the rate of disappearance of Br_2? **(d)** What is the rate of disappearance of Br_2 when $[NO] = 0.075$ M and $[Br_2] = 0.25$ M?

15.27 Consider the reaction of peroxydisulfate ion ($S_2O_8^{2-}$) with iodide ion (I^-) in aqueous solution:

$$S_2O_8^{2-}(aq) + 3 I^-(aq) \longrightarrow 2 SO_4^{2-}(aq) + I_3^-(aq)$$

At a particular temperature the rate of disappearance of $S_2O_8^{2-}$ varies with reactant concentrations in the following manner:

Experiment	[S₂O₈²⁻] (M)	[I⁻] (M)	Initial rate (M s⁻¹)
1	0.018	0.036	2.6×10^{-6}
2	0.027	0.036	3.9×10^{-6}
3	0.036	0.054	7.8×10^{-6}
4	0.050	0.072	1.4×10^{-5}

(a) Determine the rate law for the reaction. **(b)** What is the average value of the rate constant for the disappearance of $S_2O_8^{2-}$ based on the four sets of data? **(c)** How is the rate of disappearance of $S_2O_8^{2-}$ related to the rate of disappearance of I^-? **(d)** What is the rate of disappearance of I^- when $[S_2O_8^{2-}] = 0.015$ M and $[I^-] = 0.040$ M?

CHANGE OF CONCENTRATION WITH TIME (Section 15.4)

15.28 **(a)** Define the following symbols that are encountered in rate equations: $[A]_0$, $t_{\frac{1}{2}}$, $[A]_t$, k. **(b)** What quantity, when graphed versus time, will yield a straight line for a first-order reaction?

15.29 **(a)** For a second-order reaction, what quantity, when graphed versus time, will yield a straight line? **(b)** How do the half-lives of first-order and second-order reactions differ?

15.30 **(a)** The gas-phase decomposition of SO_2Cl_2, $SO_2Cl_2(g) \longrightarrow SO_2(g) + Cl_2(g)$, is first order in SO_2Cl_2. At 600 K the half-life for this process is 2.3×10^5 s. What is the rate constant at this temperature? **(b)** At 320 °C the rate constant is 2.2×10^{-5} s⁻¹. What is the half-life at this temperature?

15.31 Americium-241 is used in smoke detectors. It has a rate constant for radioactive decay of $k = 1.6 \times 10^{-3}$ year⁻¹. By contrast, iodine-125, which is used to test for thyroid functioning, has a rate constant for radioactive decay of $k = 0.011$ day⁻¹. **(a)** What are the half-lives of these two isotopes? **(b)** Which one decays at a faster rate? **(c)** How much of a 1.00 mg sample of either isotope remains after three half-lives?

15.32 As described in Exercise 15.30, the decomposition of sulfuryl chloride (SO_2Cl_2) is a first-order process. The rate constant for the decomposition at 660 K is 4.5×10^{-2} s⁻¹. **(a)** If we begin with an initial SO_2Cl_2 pressure of 375 Pa, what is the pressure of this substance after 65 s? **(b)** At what time will the pressure of SO_2Cl_2 decline to one-tenth its initial value?

15.33 From the following data for the first-order gas-phase isomerisation of CH_3NC at 215 °C, calculate the first-order rate constant and half-life for the reaction.

Time (s)	Pressure CH₃NC (kPa)
0	0.661
2 000	0.441
5 000	0.237
8 000	0.126
12 000	0.0549
15 000	0.0295

15.34 Consider the data presented in Exercise 15.15. **(a)** By using appropriate graphs, determine whether the reaction is first order or second order. **(b)** What is the value of the rate constant for the reaction? **(c)** What is the half-life for the reaction?

15.35 Sucrose ($C_{12}H_{22}O_{11}$), which is commonly known as table sugar, reacts in dilute acid solutions to form two simpler sugars, glucose and fructose, both of which have the formula $C_6H_{12}O_6$. At 23 °C and in 0.5 M HCl, the following data were obtained for the disappearance of sucrose:

Time (min)	$[C_{12}H_{22}O_{11}]$ (M)
0	0.316
39	0.274
80	0.238
140	0.190
210	0.146

(a) Is the reaction first order or second order with respect to $[C_{12}H_{22}O_{11}]$? **(b)** What is the value of the rate constant?

TEMPERATURE AND RATE (Section 15.5)

15.36 **(a)** What factors determine whether a collision between two molecules will lead to a chemical reaction? **(b)** According to the collision model, why does temperature affect the value of the rate constant?

15.37 **(a)** Explain the rate of a unimolecular (that is, one-molecule) reaction, such as the isomerisation of methyl isonitrile (Figure 15.8), in terms of the collision model. **(b)** In a reaction of the form $A(g) + B(g) \longrightarrow$ products, are all collisions of A with B that are sufficiently energetic likely to lead to reaction? Explain. **(c)** How does the kinetic-molecular theory help us understand the temperature dependence of chemical reactions?

15.38 Calculate the fraction of atoms in a sample of argon gas at 400 K that have an energy of 10.0 kJ or greater.

15.39 **(a)** The activation energy for the isomerisation of methyl isonitrile (Figure 15.8) is 160 kJ mol^{-1}. Calculate the fraction of methyl isonitrile molecules that have an energy of 160.0 kJ or greater at 500 K. **(b)** Calculate this fraction for a temperature of 510 K. What is the ratio of the fraction at 510 K to that at 500 K?

15.40 The gas-phase reaction $Cl(g) + HBr(g) \longrightarrow HCl(g) + Br(g)$ has an overall enthalpy change of -66 kJ. The activation energy for the reaction is 7 kJ. **(a)** Sketch the energy profile for the reaction, and label E_a and ΔE. **(b)** What is the activation energy for the reverse reaction?

15.41 Based on their activation energies and energy changes and assuming that all collision factors are the same, which of the following reactions would be fastest and which would be slowest? Explain your answer.

(a) $E_a = 45$ kJ mol^{-1}; $\Delta E = -25$ kJ mol^{-1}
(b) $E_a = 35$ kJ mol^{-1}; $\Delta E = -10$ kJ mol^{-1}
(c) $E_a = 55$ kJ mol^{-1}; $\Delta E = 10$ kJ mol^{-1}

15.42 Which of the reactions in Exercise 15.41 will be fastest in the reverse direction? Which will be slowest? Explain.

15.43 A certain first-order reaction has a rate constant of 2.75×10^{-2} s^{-1} at 20 °C. What is the value of k at 60 °C if **(a)** $E_a = 75.5$ kJ mol^{-1}, **(b)** $E_a = 105$ kJ mol^{-1}?

15.44 Understanding the high-temperature behaviour of nitrogen oxides is essential for controlling pollution generated in vehicle engines. The decomposition of nitric oxide (NO) to N_2 and O_2 is second order with a rate constant of 0.0796 M^{-1} s^{-1} at 737 °C and 0.0815 M^{-1} s^{-1} at 947 °C. Calculate the activation energy for the reaction.

15.45 The rate of the reaction

$$CH_3COOC_2H_5(aq) + OH^-(aq) \longrightarrow$$
$$CH_3COO^-(aq) + C_2H_5OH(aq)$$

was measured at several temperatures, and the following data were collected:

Temperature (°C)	k (M^{-1} s^{-1})
15	0.0521
25	0.101
35	0.184
45	0.332

Using these data, graph ln k versus $1/T$. Using your graph, determine the value of E_a and A.

REACTION MECHANISMS (Section 15.6)

15.46 **(a)** What is meant by the term *elementary reaction*? **(b)** What is the difference between a *unimolecular* and a *bimolecular* elementary reaction? **(c)** What is a *reaction mechanism*?

15.47 **(a)** What is meant by the term *molecularity*? **(b)** Why are termolecular elementary reactions so rare? **(c)** What is an *intermediate* in a mechanism?

15.48 What is the molecularity of each of the following elementary reactions? Write the rate law for each.
(a) $Cl_2(g) \longrightarrow 2 Cl(g)$
(b) $OCl^-(g) + H_2O(g) \longrightarrow HOCl(g) + OH^-(g)$
(c) $NO(g) + Cl_2(g) \longrightarrow NOCl_2(g)$

15.49 **(a)** Based on the following reaction profile, how many intermediates are formed in the reaction $A \longrightarrow D$? **(b)** How many transition states are there? **(c)** Which step

is the fastest? **(d)** Is the reaction $A \longrightarrow D$ exothermic or endothermic? **(e)** Which is the rate-limiting step?

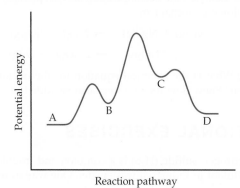

Reaction pathway

15.50 The following mechanism has been proposed for the gas-phase reaction of H_2 with ICl:

$$H_2(g) + ICl(g) \longrightarrow HI(g) + HCl(g)$$
$$HI(g) + ICl(g) \longrightarrow I_2(g) + HCl(g)$$

(a) Write the balanced equation for the overall reaction. (b) Identify any intermediates in the mechanism. (c) Write rate laws for each elementary reaction in the mechanism. (d) If the first step is slow and the second one is fast, what rate law do you expect to be observed for the overall reaction?

15.51 The decomposition of hydrogen peroxide is catalysed by iodide ion. The catalysed reaction is thought to proceed by a two-step mechanism:

$$H_2O_2(aq) + I^-(aq) \longrightarrow H_2O(l) + IO^-(aq) \quad \text{(slow)}$$
$$IO^-(aq) + H_2O_2(aq) \longrightarrow H_2O(l) + O_2(g) + I^-(aq) \quad \text{(fast)}$$

(a) Write the rate law for each of the elementary reactions of the mechanism. (b) Write the chemical equation for the overall process. (c) Identify the intermediate, if any, in the mechanism. (d) Assuming that the first step of the mechanism is rate determining, predict the rate law for the overall process.

15.52 The reaction $2\,NO(g) + Cl_2(g) \longrightarrow 2\,NOCl(g)$ obeys the rate law, rate $= k[NO]^2[Cl_2]$. The following mechanism has been proposed for this reaction:

$$NO(g) + Cl_2(g) \longrightarrow NOCl_2(g)$$
$$NOCl_2(g) + NO(g) \longrightarrow 2\,NOCl(g)$$

(a) What would the rate law be if the first step were rate determining? (b) Based on the observed rate law, what can we conclude about the relative rates of the two steps?

15.53 You have studied the gas-phase oxidation of HBr by O_2:

$$4\,HBr(g) + O_2(g) \longrightarrow 2\,H_2O(g) + 2\,Br_2(g)$$

You find the reaction to be first order with respect to HBr and first order with respect to O_2. You propose the following mechanism:

$$HBr(g) + O_2(g) \longrightarrow HOOBr(g)$$
$$HOOBr(g) + HBr(g) \longrightarrow 2\,HOBr(g)$$
$$HOBr(g) + HBr(g) \longrightarrow H_2O(g) + Br_2(g)$$

(a) Indicate how the elementary reactions add to give the overall reaction. (b) Based on the rate law, which step is rate determining? (c) What are the intermediates in this mechanism? (d) If you are unable to detect HOBr or HOOBr among the products, does this disprove your mechanism?

CATALYSIS (Section 15.7)

15.54 (a) What part of the energy profile of a reaction is affected by a catalyst? (b) What is the difference between a homogeneous and a heterogeneous catalyst?

15.55 (a) Most heterogeneous catalysts of importance are extremely finely divided solid materials. Why is particle size important? (b) What role does adsorption play in the action of a heterogeneous catalyst?

15.56 The oxidation of SO_2 to SO_3 is catalysed by NO_2. The reaction proceeds as follows:

$$NO_2(g) + SO_2(g) \longrightarrow NO(g) + SO_3(g)$$
$$2\,NO(g) + O_2(g) \longrightarrow 2\,NO_2(g)$$

(a) Show that the two reactions can be summed to give the overall oxidation of SO_2 by O_2 to give SO_3. (b) Why do we consider NO_2 a catalyst and not an intermediate in this reaction? (c) Is this an example of homogeneous catalysis or heterogeneous catalysis?

15.57 NO catalyses the decomposition of N_2O, possibly by the following mechanism:

$$NO(g) + N_2O(g) \longrightarrow N_2(g) + NO_2(g)$$
$$2\,NO_2(g) \longrightarrow 2\,NO(g) + O_2(g)$$

(a) What is the chemical equation for the overall reaction? Show how the two steps can be added to give the overall equation. (b) Why is NO considered a catalyst and not an intermediate? (c) If experiments show that during the decomposition of N_2O, NO_2 does not accumulate in measurable quantities, does this rule out the proposed mechanism? If you think not, suggest what might be going on.

15.58 Many metallic catalysts, particularly the precious-metal ones, are often deposited as very thin films on a substance of high surface area per unit mass, such as alumina (Al_2O_3) or silica (SiO_2). Why is this an effective way of utilising the catalyst material?

15.59 (a) If you were going to build a system to check the effectiveness of catalytic converters on cars, what substances would you want to look for in the car exhaust? (b) Vehicle catalytic converters have to work at high temperatures as hot exhaust gases stream through them. In what ways could this be an advantage? In what ways a disadvantage? (c) Why is the rate of flow of exhaust gases over a catalytic converter important?

15.60 The activation energy of an uncatalysed reaction is $95\ kJ\ mol^{-1}$. The catalyst reaction has an activation energy of $55\ kJ\ mol^{-1}$. Assuming that the collision factor remains the same, by what factor will the catalyst increase the rate of the reaction at (a) 25 °C, (b) 125 °C?

ADDITIONAL EXERCISES

15.61 Hydrogen sulfide (H_2S) is a common and troublesome pollutant in industrial wastewater. One way to remove H_2S is to treat the water with chlorine, in which case the following reaction occurs:

$$H_2S(aq) + Cl_2(aq) \longrightarrow S(s) + 2\,H^+(aq) + 2\,Cl^-(aq)$$

The rate of this reaction is first order in each reactant. The rate constant for the disappearance of H_2S at 28 °C is 3.5×10^{-2} M^{-1} s^{-1}. If at a given time the concentration of H_2S is 2.0×10^{-4} M and that of Cl_2 is 0.050 M, what is the rate of formation of Cl^-?

15.62 For the reaction of iodide ion with hypochlorite ion, $I^-(aq) + OCl^-(aq) \longrightarrow OI^-(aq) + Cl^-(aq)$, the reaction is found to be first order each in iodide and hypochlorite ions, and inversely proportional to the concentration of hydroxide ion present in the solution. **(a)** Write the rate law for the reaction. **(b)** By what factor will the rate change if the concentration of iodide ion is tripled? **(c)** By what factor will the rate change if the hydroxide ion concentration is doubled?

15.63 Consider the following reaction between mercury(II) chloride and oxalate ion:

$$2\,HgCl_2(aq) + C_2O_4{}^{2-}(aq) \longrightarrow$$
$$2\,Cl^-(aq) + 2\,CO_2(g) + Hg_2Cl_2(s)$$

The initial rate of this reaction was determined for several concentrations of $HgCl_2$ and $C_2O_4{}^{2-}$, and the following rate data were obtained for the rate of disappearance of $C_2O_4{}^{2-}$:

Experiment	[HgCl$_2$] (M)	[C$_2$O$_4{}^{2-}$] (M)	Rate (M s^{-1})
1	0.164	0.15	3.2×10^{-5}
2	0.164	0.45	2.9×10^{-4}
3	0.082	0.45	1.4×10^{-4}
4	0.246	0.15	4.8×10^{-5}

(a) What is the rate law for this reaction? **(b)** What is the value of the rate constant? **(c)** What is the reaction rate when the concentration of $HgCl_2$ is 0.050 M and that of $C_2O_4{}^{2-}$ is 0.10 M, if the temperature is the same as that used to obtain the data shown?

15.64 Urea (NH_2CONH_2) is the end product in protein metabolism in animals. The decomposition of urea in 0.1 M HCl occurs according to the reaction

$$NH_2CONH_2(aq) + H^+(aq) + 2\,H_2O(l) \longrightarrow$$
$$2\,NH_4{}^+(aq) + HCO_3{}^-(aq)$$

The reaction is first order in urea and first order overall. When $[NH_2CONH_2] = 0.200$ M, the rate at 61.05 °C is 8.56×10^{-5} M s^{-1}. **(a)** What is the value for the rate constant, k? **(b)** What is the concentration of urea in this solution after 4.00×10^3 s if the starting concentration is

0.500 M? **(c)** What is the half-life for this reaction at 61.05 °C?

15.65 The rate of a first-order reaction is followed by spectroscopy, monitoring the absorbance of a coloured reactant at 520 nm. The reaction occurs in a 1.00 cm sample cell, and the only coloured species in the reaction has an extinction coefficient of at 5.60×10^3 M^{-1} cm^{-1}. **(a)** Calculate the initial concentration of the coloured reactant if the absorbance is 0.605 at the beginning of the reaction. **(b)** The absorbance falls to 0.250 at 30.0 min. Calculate the rate constant in units of s^{-1}. **(c)** Calculate the half-life of the reaction. **(d)** How long does it take for the absorbance to fall to 0.100?

15.66 Cyclopentadiene (C_5H_6) reacts with itself to form dicyclopentadiene ($C_{10}H_{12}$). A 0.0400 M solution of C_5H_6 was monitored as a function of time as the reaction $2\,C_5H_6 \longrightarrow C_{10}H_{12}$ proceeded. The following data were collected:

Time (s)	[C$_5$H$_6$] (M)
0.0	0.0400
50.0	0.0300
100.0	0.0240
150.0	0.0200
200.0	0.0174

Plot $[C_5H_6]$ against time, $\ln[C_5H_6]$ against time and $1/[C_5H_6]$ against time. What is the order of the reaction? What is the value of the rate constant?

15.67 **(a)** Two reactions have identical values for E_a. Does this ensure that they will have the same rate constant if run at the same temperature? Explain. **(b)** Two similar reactions have the same rate constant at 25 °C, but at 35 °C one of the reactions has a higher rate constant than the other. Account for these observations.

15.68 The following mechanism has been proposed for the gas-phase reaction of chloroform ($CHCl_3$) and chlorine:

Step 1:
$$Cl_2(g) \underset{k_{-1}}{\overset{k_1}{\rightleftharpoons}} 2\,Cl(g) \text{ (fast)}$$

Step 2: $Cl(g) + CHCl_3(g) \xrightarrow{k_2} HCl(g) + CCl_3(g)$ (slow)

Step 3:
$$Cl(g) + CCl_3(g) \xrightarrow{k_3} CCl_4(g) \text{ (fast)}$$

(a) What is the overall reaction? **(b)** What are the intermediates in the mechanism? **(c)** What is the molecularity of each of the elementary reactions? **(d)** What is the rate-determining step? **(e)** What is the rate law predicted by this mechanism?

INTEGRATIVE EXERCISES

15.69 Dinitrogen pentoxide (N_2O_5) decomposes in chloroform as a solvent to yield NO_2 and O_2. The decomposition is first order with a rate constant at 45 °C of 1.00×10^{-5} s^{-1}. Calculate the partial pressure of O_2 produced from 1.00 dm^3 of 0.600 M N_2O_5 solution at 45 °C over a period of 20.0 h if the gas is collected in a 10.00 dm^3 container. (Assume that the products do not dissolve in chloroform.)

15.70 The reaction between iodoethane and hydroxide ion in ethanol (C_2H_5OH) solution, $C_2H_5I + OH^- \longrightarrow C_2H_5OH + I^-$, has an activation energy of 86.8 kJ mol^{-1} and a frequency factor of 2.10×10^{11} M^{-1} s^{-1}. **(a)** Predict the rate constant for the reaction at 35 °C. **(b)** A solution of KOH in ethanol is made up by dissolving 0.335 g KOH in ethanol to form 250.0 cm^3 of solution. Similarly, 1.453 g of C_2H_5I is dissolved in ethanol to form 250.0 cm^3 of solution. Equal volumes of the two solutions are mixed. Assuming the reaction is first order in each reactant, what is the initial rate at 35 °C? **(c)** Which reagent in the reaction is limiting, assuming the reaction proceeds to completion?

15.71 Zinc metal dissolves in hydrochloric acid according to the reaction

$$Zn(s) + 2\,HCl(aq) \longrightarrow ZnCl_2(aq) + H_2(g)$$

Suppose you are asked to study the kinetics of this reaction by monitoring the rate of production of $H_2(g)$. **(a)** By using a reaction flask, a manometer and any other common laboratory equipment, design an experimental apparatus that would allow you to monitor the partial pressure of $H_2(g)$ produced as a function of time. **(b)** Explain how you would use the apparatus to determine the rate law of the reaction. **(c)** Explain how you would use the apparatus to determine the reaction order for [H$^+$] for the reaction. **(d)** How could you use the apparatus to determine the activation energy of the reaction? **(e)** Explain how you would use the apparatus to determine the effects of changing the form of Zn(s) from metal strips to granules.

15.72 The gas-phase reaction of NO with F_2 to form NOF and F has an activation energy of $E_a = 6.3$ kJ mol^{-1} and a frequency factor of $A = 6.0 \times 10^8$ M^{-1} s^{-1}. The reaction is believed to be bimolecular:

$$NO(g) + F_2(g) \longrightarrow NOF(g) + F(g)$$

(a) Calculate the rate constant at 100 °C. **(b)** Draw the Lewis structures for the NO and the NOF molecules, given that the chemical formula for NOF is misleading because the nitrogen atom is actually the central atom in the molecule. **(c)** Predict the structure for the NOF molecule. **(d)** Draw a possible transition state for the formation of NOF, using dashed lines to indicate the weak bonds that are beginning to form. **(e)** Suggest a reason for the low activation energy for the reaction.

15.73 The rates of many atmospheric reactions are accelerated by the absorption of light by one of the reactants. For example, consider the reaction between methane and chlorine to produce chloromethane and hydrogen chloride:

Reaction 1: $CH_4(g) + Cl(g) \longrightarrow CH_3Cl(g) + HCl(g)$

This reaction is very slow in the absence of light. However, $Cl_2(g)$ can absorb light to form Cl atoms:

Reaction 2: $Cl_2(g) + h\nu \longrightarrow 2\,Cl(g)$

Once the Cl atoms are generated, they can catalyse the reaction of CH_4 and Cl_2 according to the following proposed mechanism:

Reaction 3: $\quad CH_4(g) + Cl(g) \longrightarrow CH_3(g) + HCl(g)$

Reaction 4: $\quad CH_3(g) + Cl_2(g) \longrightarrow CH_3Cl(g) + Cl(g)$

The enthalpy changes and activation energies for these two reactions are tabulated as follows:

Reaction	$\Delta_{rxn}H°$ (kJ mol^{-1})	E_a (kJ mol^{-1})
3	+4	17
4	−109	4

(a) By using the bond enthalpy for Cl_2 (Table 8.4), determine the longest wavelength of light that is energetic enough to cause reaction 2 to occur. In which portion of the electromagnetic spectrum is this light found? **(b)** By using the data tabulated here, sketch a quantitative energy profile for the catalysed reaction represented by reactions 3 and 4. **(c)** By using bond enthalpies, estimate where the reactants, $CH_4(g) + Cl_2(g)$, should be placed on your diagram in part (b). Use this result to estimate the value of E_a for the reaction $CH_4(g) + Cl_2(g) \longrightarrow CH_3(g) + HCl(g) + Cl(g)$. **(d)** The species Cl(g) and $CH_3(g)$ in reactions 3 and 4 are radicals, atoms or molecules with unpaired electrons. Draw a Lewis structure of CH_3 and verify that it is a radical. **(e)** The sequence of reactions 3 and 4 comprises a radical chain mechanism. Why do you think this is called a 'chain reaction'? Propose a reaction that will terminate the chain reaction.

MasteringChemistry (www.pearson.com.au/masteringchemistry)

Make learning part of the grade. Access:

- tutorials with peronalised coaching
- study area
- Pearson eText

PHOTO/ART CREDITS

564 © R. Gino Santa Maria/Dreamstime.com; **565 (left)** S. Yamashita/Corbis Australia Pty Ltd; **566 (a)** Michael Dalton, Fundamental Photographs, NYC; **566 (b), 584, 599** Richard Megna, Fundamental Photographs, NYC.

16

CHEMICAL EQUILIBRIUM

The number of players in an Australian Rules football team is always 18, although individual players may be changed.

KEY CONCEPTS

16.1 THE CONCEPT OF EQUILIBRIUM
We examine reversible reactions and the concept of equilibrium.

16.2 THE EQUILIBRIUM CONSTANT
We define the *equilibrium constant* based on rates of forward and reverse reactions, and learn how to write *equilibrium constant expressions* for homogeneous reactions.

16.3 INTERPRETING AND WORKING WITH EQUILIBRIUM CONSTANTS
We learn how to interpret the magnitude of an equilibrium constant and how its value depends on the way the corresponding chemical equation is expressed.

16.4 HETEROGENEOUS EQUILIBRIA
We learn how to write equilibrium constant expressions for heterogeneous reactions.

16.5 CALCULATING EQUILIBRIUM CONSTANTS
We see that the value of an equilibrium constant can be calculated from equilibrium concentrations of reactants and products.

16.6 APPLICATIONS OF EQUILIBRIUM CONSTANTS
We see that equilibrium constants can be used to predict equilibrium concentrations of reactants and products and to determine the direction in which a reaction mixture must proceed to achieve equilibrium.

16.7 THE EQUILIBRIUM CONSTANT AND FREE ENERGY
We consider how the standard free energy change for a chemical reaction can be used to calculate the equilibrium constant for the reaction.

16.8 LE CHÂTELIER'S PRINCIPLE
We discuss *Le Châtelier's principle*, which predicts how a system at equilibrium responds to changes in concentration, volume, pressure and temperature.

To be in equilibrium is to be in a state of balance. A tug of war in which the two sides are pulling with equal force so that the rope doesn't move is an example of a *static* equilibrium, one in which an object is at rest. Team sports that allow player substitutions (rugby, water polo, soccer and Australian Rules football) are examples of *dynamic* equilibria. The number of players 'in play' is always constant but they are not always the same players because of the substitutions that are allowed.

We have already encountered several instances of dynamic equilibrium. For example, the vapour above a liquid is in equilibrium with the liquid phase (∞ Section 11.5). The rate at which molecules escape from the liquid into the gas phase equals the rate at which molecules in the gas phase strike the surface and become part of the liquid. Similarly, in a saturated solution of sodium chloride the solid sodium chloride is in equilibrium with the ions dispersed in water (∞ Section 12.2). The rate at which ions leave the solid surface equals the rate at which other ions are removed from the liquid to become part of the solid. Both of these examples involve a pair of opposing processes. At equilibrium these opposing processes are occurring at the same rate.

In this chapter we will consider yet another type of dynamic equilibrium, one involving chemical reactions. **Chemical equilibrium** *occurs when opposing reactions are proceeding at equal rates*: the rate at which the products are formed from the reactants equals the rate at which the reactants are formed from the products. As a result, concentrations cease to change, making it appear as if the reaction has stopped. When chemical reactions occur in closed systems, the reactions will sooner or later reach an equilibrium state—that is, a mixture in which the concentrations of reactants and products no longer change with time. Consequently, closed systems will always be either at equilibrium or approaching equilibrium. How fast a reaction reaches equilibrium is a matter of kinetics.

16.1 | THE CONCEPT OF EQUILIBRIUM

Let's examine a simple chemical reaction to see how it reaches an *equilibrium state*—that is, a mixture of reactants and products whose concentrations no longer change with time. We begin with N_2O_4, a colourless substance that dissociates to form brown NO_2.

▼ FIGURE 16.1 shows a sample of frozen N_2O_4, a colourless substance, inside a sealed tube resting in a beaker. When this solid is warmed until the substance is above its boiling point (21.2 °C), the gas in the sealed tube turns progressively darker as the colourless N_2O_4 gas dissociates into brown NO_2 gas. Eventually, even though there is still N_2O_4 in the tube, the colour stops getting darker because the system reaches equilibrium. We are left with an *equilibrium mixture* of N_2O_4 and NO_2 in which the concentrations of the gases no longer change as time passes. The equilibrium mixture results because the reaction is *reversible*. Not only can N_2O_4 react to form NO_2, but NO_2 can also react to form N_2O_4. This situation is represented by writing the equation for the reaction with a double arrow (∞ Section 4.1, 'General Properties of Aqueous Solutions').

∞ Review this on page 106

$$N_2O_4(g) \rightleftharpoons 2\,NO_2(g) \qquad\qquad [16.1]$$
$$\text{colourless} \qquad\qquad \text{brown}$$

◢ **FIGURE IT OUT**

How can you tell if you are at equilibrium?

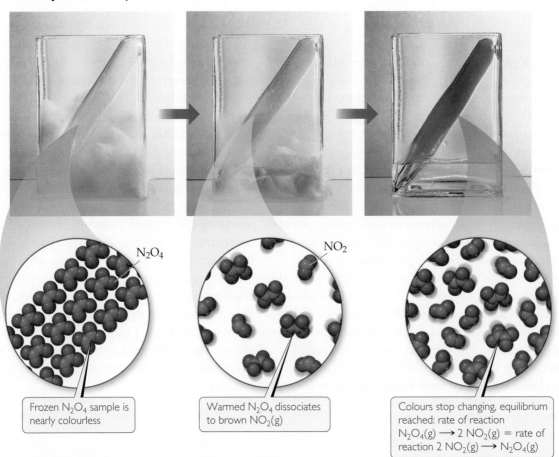

Frozen N_2O_4 sample is nearly colourless

Warmed N_2O_4 dissociates to brown $NO_2(g)$

N_2O_4

NO_2

Colours stop changing, equilibrium reached: rate of reaction $N_2O_4(g) \longrightarrow 2\,NO_2(g)$ = rate of reaction $2\,NO_2(g) \longrightarrow N_2O_4(g)$

▲ FIGURE 16.1 **The equilibrium between NO_2 and N_2O_4 in a sealed tube.**

We can analyse this equilibrium using our knowledge of kinetics. Let's call the decomposition of N_2O_4 to form NO_2 the forward reaction and the reaction of NO_2 to re-form N_2O_4 the reverse reaction. In this case both the forward reaction and the reverse reaction are elementary reactions. As we learned in Section 15.6, the rate laws for elementary reactions can be written from their chemical equations:

Forward reaction: $N_2O_4(g) \longrightarrow 2\,NO_2(g)$ $\text{Rate}_f = k_f[N_2O_4]$ [16.2]

Reverse reaction: $2\,NO_2(g) \longrightarrow N_2O_4(g)$ $\text{Rate}_r = k_r[NO_2]^2$ [16.3]

where k_f and k_r are the rate constants for the forward and reverse reactions, respectively. At equilibrium the rate at which products are produced from reactants equals the rate at which reactants are produced from products:

$$k_f[N_2O_4] = k_r[NO_2]^2 \qquad\qquad [16.4]$$

Rearranging this equation gives

$$\frac{[NO_2]^2}{[N_2O_4]} = \frac{k_f}{k_r} = \text{a constant} \qquad\qquad [16.5]$$

As shown in Equation 16.5, the quotient of two constants, such as k_f and k_r, is itself a constant called the *equilibrium constant*. Thus at equilibrium the ratio of the concentration terms involving N_2O_4 and NO_2 equals a constant. If the temperature of the reaction is changed, the values k_f and k_r will change and the value of the equilibrium constant will change. As the value of the equilibrium constant will depend on the temperature of the reaction, so temperature must always be specified. This point is discussed further in Section 16.7.

It makes no difference whether we start with N_2O_4 or with NO_2, or even with some mixture of the two. At equilibrium the ratio equals a specific value. Thus there is an important constraint on the proportions of N_2O_4 and NO_2 at equilibrium.

Once equilibrium is established, the concentrations of N_2O_4 and NO_2 no longer change, as shown in ▼ FIGURE 16.2. Just because the composition of the equilibrium mixture remains constant with time does not mean, however, that N_2O_4 and NO_2 stop reacting. On the contrary, the equilibrium is dynamic— some N_2O_4 is still converting to NO_2, and some NO_2 is still converting to N_2O_4. At equilibrium, however, the two processes occur at the same rate and so there is no *net* change in their amounts.

We learn several important lessons about equilibrium from this example.

- A mixture of reactants and products is formed in which concentrations no longer change with time, which indicates that the reaction has reached a state of equilibrium.

⚠ **FIGURE IT OUT**

At equilibrium, are the concentrations of NO_2 and N_2O_4 equal?

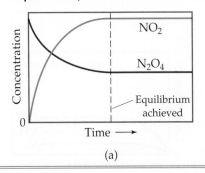

(a)

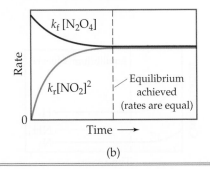

(b)

◀ **FIGURE 16.2 Achieving chemical equilibrium in the $N_2O_4(g) \rightleftharpoons 2\,NO_2(g)$ reaction.** Equilibrium occurs when the rate of the forward reaction equals the rate of the reverse reaction.

- For equilibrium to occur, neither reactants nor products can escape from the system.
- At equilibrium the particular ratio of concentration terms equals a constant.

> **CONCEPT CHECK 1**
> a. Which quantities are equal in a dynamic equilibrium?
> b. If the rate constant for the forward reaction in Equation 16.1 is larger than the rate constant for the reverse reaction, will the constant in Equation 16.5 be greater than 1 or smaller than 1?

16.2 | THE EQUILIBRIUM CONSTANT

A reaction in which reactants convert to products and products convert to reactants in the same reaction vessel naturally leads to an equilibrium, regardless of how complicated the reaction might be and regardless of the nature of the kinetic processes for the forward and reverse reactions. Consider the synthesis of ammonia from nitrogen and hydrogen:

$$N_2(g) + 3H_2(g) \rightleftharpoons 2NH_3(g) \qquad [16.6]$$

This reaction is the basis for the **Haber process**, which, in the presence of a catalyst, combines N_2 and H_2 at a pressure of several hundred bar and a temperature of several hundred degrees Celsius. The two gases react to form ammonia under these conditions, but the reaction does not lead to complete consumption of the N_2 and H_2. Rather, at some point the reaction appears to stop, with all three components of the reaction mixture present at the same time.

The manner in which the concentrations of N_2, H_2 and NH_3 vary with time is shown in ▼ **FIGURE 16.3**. Notice that an equilibrium mixture is obtained regardless of whether we begin with N_2 and H_2 or only with NH_3. At equilibrium the relative concentrations of N_2, H_2 and NH_3 are the same, regardless of whether the starting mixture was a 1 : 3 molar ratio of N_2 and H_2 or pure NH_3. *The equilibrium condition can be reached from either direction.*

> **CONCEPT CHECK 2**
> How do we know when equilibrium has been reached in a chemical reaction?

Earlier, we saw that when the reaction $N_2O_4(g) \rightleftharpoons 2NO_2(g)$ reaches equilibrium, a ratio based on the equilibrium concentrations of N_2O_4 and NO_2 has a constant value (Equation 16.5). A similar relationship governs the concentrations of N_2, H_2 and NH_3 at equilibrium. If we were systematically to change the relative amounts of the three gases in the starting mixture and then analyse each equilibrium mixture, we could determine the relationship between the equilibrium concentrations.

Chemists conducted studies of this kind on other chemical systems in the nineteenth century, before Haber's work. In 1864 Cato Maximilian Guldberg

▶ **FIGURE 16.3** The same equilibrium is reached whether we start with only reactants (N_2 and H_2) or with only product (NH_3).

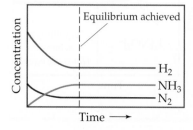

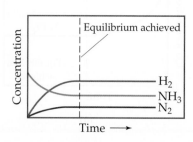

MY WORLD OF CHEMISTRY

THE HABER PROCESS

The My World of Chemistry box in Section 20.7 discusses *nitrogen fixation*, the processes that convert N_2 gas into ammonia, which can then be incorporated into living organisms. It shows that the enzyme nitrogenase is responsible for generating most of the fixed nitrogen essential for plant growth. However, the quantity of food required to feed the ever-increasing human population far exceeds that provided by nitrogen-fixing plants, so human agriculture requires substantial amounts of ammonia-based fertilisers that can be applied directly to croplands. Thus of all the chemical reactions that humans have learned to conduct and control for their own purposes, the synthesis of ammonia from hydrogen and atmospheric nitrogen is one of the most important.

In 1912 the German chemist Fritz Haber (1868–1934) developed a process for synthesising ammonia directly from nitrogen and hydrogen (▶ **FIGURE 16.4**). The process is sometimes called the *Haber–Bosch process*, also honouring Karl Bosch, the engineer who developed the equipment for the industrial production of ammonia. The engineering needed to implement the Haber process requires the use of temperatures and pressures (approximately 500 °C and 200 bar) that were difficult to achieve at that time.

The Haber process provides a historically interesting example of the complex impact of chemistry on our lives. At the start of World War I, in 1914, Germany depended on nitrate deposits in Chile for the nitrogen-containing compounds needed to manufacture explosives. During the war the Allied naval blockade of South America cut off this supply. However, by fixing nitrogen from air, Germany was able to continue to produce explosives. Experts have estimated that

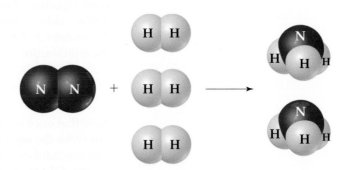

▲ **FIGURE 16.4   The Haber process.** Used to convert $N_2(g)$ and $H_2(g)$ to $NH_3(g)$, this process, although exothermic, requires breaking the very strong triple bond in N_2.

World War I would have ended before 1918 had it not been for the Haber process.

From these unhappy beginnings as a major factor in international warfare, the Haber process has become the world's principal source of fixed nitrogen. The same process that prolonged World War I has enabled scientists to manufacture fertilisers that have increased crop yields, thereby saving millions of people from starvation. About 100 million tonnes of ammonia are produced each year.

Haber was a patriotic German who gave enthusiastic support to his nation's war effort. He served as chief of Germany's Chemical Warfare Service during World War I and developed the use of chlorine as a poison gas weapon. Consequently, the decision to award him the Nobel Prize in Chemistry in 1918 was the subject of considerable controversy and criticism. The ultimate irony, however, came in 1933 when Haber was expelled from Germany because he was Jewish.

RELATED EXERCISE: 16.44

(1836–1902) and Peter Waage (1833–1900) postulated their **law of mass action**, which expresses, for any reaction, the relationship between the concentrations of the reactants and products present at equilibrium. Suppose we have the following general equilibrium equation:

$$a\,A + b\,B \rightleftharpoons d\,D + e\,E \qquad [16.7]$$

where A, B, D and E are the chemical species involved and a, b, d and e are their coefficients in the balanced chemical equation. According to the law of mass action, the equilibrium condition is given by the expression

$$K_c = \frac{[D]^d[E]^e}{[A]^a[B]^b} \qquad [16.8]$$

We call this relationship the **equilibrium constant expression** (or merely the **equilibrium expression**) for the reaction. The constant K_c, which we call the **equilibrium constant**, is the numerical value obtained when we substitute molar equilibrium concentrations into the equilibrium constant expression. The subscript c on the K indicates that concentrations expressed in molarity are used to evaluate the constant.

The numerator of the equilibrium constant expression is the product of the concentrations of all substances on the product side of the equilibrium equation, each raised to a power equal to its coefficient in the balanced equation. The denominator is similarly derived from the reactant side of the equilibrium equation. (Remember, the convention is to write the substances on the *product* side in the *numerator* and the substances on the *reactant* side in the *denominator*.) Thus for the Haber process, $N_2(g) + 3 H_2(g) \rightleftharpoons 2 NH_3(g)$, the equilibrium constant expression is

$$K_c = \frac{[NH_3]^2}{[N_2][H_2]^3} \qquad [16.9]$$

Note that once we know the balanced chemical equation for an equilibrium, we can write the equilibrium constant expression even if we don't know the reaction mechanism. *The equilibrium constant expression depends only on the stoichiometry of the reaction, not on its mechanism.*

The value of the equilibrium constant at any given temperature does not depend on the initial amounts of reactants and products. Nor does it matter whether other substances are present, as long as they do not react with a reactant or a product. The value of the equilibrium constant depends only on the particular reaction and on the temperature.

SAMPLE EXERCISE 16.1 | **Writing equilibrium constant expressions**

Write the equilibrium expression for K_c for the following reactions:
    (a) $2 O_3(g) \rightleftharpoons 3 O_2(g)$
    (b) $2 NO(g) + Cl_2(g) \rightleftharpoons 2 NOCl(g)$
    (c) $Ag^+(aq) + 2 NH_3(aq) \rightleftharpoons Ag(NH_3)_2^+(aq)$

SOLUTION

Analyse We are given three equations and are asked to write an equilibrium constant expression for each.

Plan Using the law of mass action, we write each expression as a quotient having the product concentration terms in the numerator and the reactant concentration terms in the denominator. Each concentration term is raised to the power of its coefficient in the balanced chemical equation.

Solve

(a) $K_c = \dfrac{[O_2]^3}{[O_3]^2}$ **(b)** $K_c = \dfrac{[NOCl]^2}{[NO]^2[Cl_2]}$ **(c)** $K_c = \dfrac{[Ag(NH_3)_2^+]}{[Ag^+][NH_3]^2}$

PRACTICE EXERCISE

Write the equilibrium constant expression, K_c, for   **(a)** $H_2(g) + I_2(g) \rightleftharpoons 2 HI(g)$, **(b)** $Cd^{2+}(aq) + 4 Br^-(aq) \rightleftharpoons CdBr_4^{2-}(aq)$.

Answers: **(a)** $K_c = \dfrac{[HI]^2}{[H_2][I_2]}$ **(b)** $K_c = \dfrac{[CdBr_4^{2-}]}{[Cd^{2+}][Br^-]^4}$

(See also Exercise 16.12.)

Evaluating K_c

We can illustrate how the law of mass action was discovered empirically and demonstrate that the equilibrium constant is independent of starting concentrations by examining some equilibrium concentrations for the gas-phase reaction between dinitrogen tetroxide and nitrogen dioxide:

$$N_2O_4(g) \rightleftharpoons 2 NO_2(g) \qquad K_c = \frac{[NO_2]^2}{[N_2O_4]} \qquad [16.10]$$

Figure 16.1 shows the reaction proceeding to equilibrium starting with pure N_2O_4. Because NO_2 is a dark brown gas and N_2O_4 is colourless, the amount of

TABLE 16.1 • Initial and equilibrium concentrations of N_2O_4 and NO_2 in the gas phase at 100 °C

Experiment	Initial [N_2O_4] (M)	Initial [NO_2] (M)	Equilibrium [N_2O_4] (M)	Equilibrium [NO_2] (M)	K_c
1	0.0	0.0200	0.00140	0.0172	0.211
2	0.0	0.0300	0.00280	0.0243	0.211
3	0.0	0.0400	0.00452	0.0310	0.213
4	0.0200	0.0	0.00452	0.0310	0.213

NO_2 in the mixture can be determined by measuring the intensity of the brown colour of the gas mixture.

We can determine the numerical value for K_c and verify that it is constant, regardless of the starting amounts of NO_2 and N_2O_4, by performing experiments in which we start with several sealed tubes containing different concentrations of NO_2 and N_2O_4, as summarised in ▲ **TABLE 16.1**. The tubes are kept at 100 °C until no further change in the colour of the gas is noted. We then analyse the mixtures and determine the equilibrium concentrations of NO_2 and N_2O_4, as shown in Table 16.1.

To evaluate the equilibrium constant, K_c, we insert the equilibrium concentrations into the equilibrium constant expression. For example, using the experiment 1 data, [NO_2] = 0.0172 M and [N_2O_4] = 0.00140 M, we find

$$K_c = \frac{[NO_2]^2}{[N_2O_4]} = \frac{(0.0172)^2}{0.00140} = 0.211$$

Proceeding in the same way, the values of K_c for the other samples are calculated, as listed in Table 16.1. Note that the value for K_c is constant ($K_c = 0.212$, within the limits of experimental error) even though the initial concentrations vary. Furthermore, the results of experiment 4 show that equilibrium can be achieved beginning with N_2O_4 rather than with NO_2. That is, equilibrium can be approached from either direction. ▼ **FIGURE 16.5** shows how both experiments 3 and 4 result in the same equilibrium mixture even though one begins with 0.400 M NO_2 and the other with 0.0200 M N_2O_4.

Notice that no units are given for the values of K_c either in Table 16.1 or in our calculation using the experiment 1 data. Equilibrium constants are written without units for reasons that we address later in this section.

 CONCEPT CHECK 3

How does the value of K_c in Equation 16.10 depend on the starting concentrations of NO_2 and N_2O_4?

Equilibrium Constants in Terms of Pressure, K_P

When the reactants and products in a chemical reaction are gases, we can formulate the equilibrium constant expression in terms of partial pressures instead of molar concentrations. When partial pressures in bar* are used in the equilibrium constant expression, we can denote the equilibrium constant as K_P (where the subscript P stands for pressure). For the general reaction in Equation 16.7, the expression for K_P is

$$K_P = \frac{(P_D)^d (P_E)^e}{(P_A)^a (P_B)^b} \qquad [16.11]$$

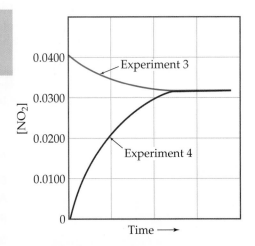

▲ FIGURE 16.5 **Concentration changes approaching equilibrium.** As seen in Table 16.1, the same equilibrium mixture is produced starting with either NO_2 (experiment 3) or N_2O_4 (experiment 4).

---

* The standard SI pressure unit is the pascal, Pa. Thermodynamic data are now calculated for a standard pressure of 1 bar = 10^5 Pa, which is approximately atmospheric pressure.

where P_A is the partial pressure of A in bar and so forth. For example, for $N_2O_4(g) \rightleftharpoons 2\,NO_2(g)$ we have

$$K_P = \frac{(P_{NO_2})^2}{P_{N_2O_4}}$$

⚠ **CONCEPT CHECK 4**

What is the difference between the equilibrium constant, K_c, and the equilibrium constant, K_P?

For a given reaction, the numerical value of K_c is generally different from the numerical value of K_P. We must therefore take care to indicate, via the subscripts, which of these equilibrium constants we are using. It is possible, however, to calculate one from the other using the ideal-gas equation (∞ Section 10.4, 'The Ideal-Gas Equation') to convert between concentration (in mol dm^{-3}) and pressure (in bar):

∞ Review this on page 544

$$PV = nRT, \text{ so } P = \frac{n}{V}RT \qquad [16.12]$$

We see that n/V has the units of mol dm^{-3} and therefore equals molarity, M. For substance A, therefore

$$P_A = \frac{n_A}{V}RT = [A]RT \qquad [16.13]$$

Substituting Equation 16.13 and like expressions for the other gaseous components of the reaction into the expression for K_P (Equation 16.11), we obtain a general expression relating K_P and K_c:

$$K_P = K_c(RT)^{\Delta n} \qquad [16.14]$$

The quantity Δn is the change in the number of moles of gas in the balanced chemical equation for the reaction.

$$\Delta n = \text{(moles of gaseous product)} - \text{(moles of gaseous reactant)} \qquad [16.15]$$

For example, in the reaction $N_2O_4(g) \rightleftharpoons 2\,NO_2(g)$ there are two moles of the product NO_2 (the coefficient in the balanced equation) and one mole of the reactant, N_2O_4. Therefore, $\Delta n = 2 - 1 = 1$ and $K_P = K_c(RT)$ for this reaction. From Equation 16.14, we see that $K_P = K_c$ only when the same number of moles of gas appears on both sides of the balanced chemical equation, which means that $\Delta n = 0$.

SAMPLE EXERCISE 16.2 **Converting between K_c and K_P**

In the synthesis of ammonia from nitrogen and hydrogen,

$$N_2(g) + 3\,H_2(g) \rightleftharpoons 2\,NH_3(g)$$

$K_c = 9.60$ at 300 °C. Calculate K_P for this reaction at this temperature.

SOLUTION

Analyse We are given K_c for a reaction and asked to calculate K_P.

Plan The relationship between K_c and K_P is given by Equation 16.14.

Solve There are two moles of gaseous products ($2\,NH_3$) and four moles of gaseous reactants ($1\,N_2 + 3\,H_2$). Therefore, $\Delta n = 2 - 4 = -2$. (Remember that Δ functions are always based on *products minus reactants*.) The temperature, T, is $273 + 300 = 573$ K. The value for the ideal-gas constant, R, is 8.314 kPa dm^3 mol^{-1} K^{-1}. Remembering that K_P is calculated in terms of bar (100 kPa), and concentration is in mol dm^{-3},

$$K_P = K_c(RT)^{\Delta n} = 9.60 \left(\frac{8.314 \text{ kPa dm}^3 \text{ mol}^{-1} \text{ K}^{-1} \times 573 \text{ K} \times 1 \text{ mol dm}^{-3}}{100 \text{ kPa}} \right)^{-2}$$

$$= 9.60 \left(\frac{100}{8.314 \times 573} \right)^2 = 4.23 \times 10^{-3}$$

PRACTICE EXERCISE

For the equilibrium $2 \text{SO}_3(g) \rightleftharpoons 2 \text{SO}_2(g) + \text{O}_2(g)$, K_c is 4.08×10^{-3} at 1000 K. Calculate the value for K_P.

Answer: 0.339

(See also Exercises 16.14, 16.15.)

Equilibrium Constants and Units

You may wonder why equilibrium constants are reported without units. The equilibrium constant is related not only to the kinetics of a reaction but also to the thermodynamics of the process. Equilibrium constants derived from thermodynamic measurements are defined in terms of *activities* rather than concentrations or partial pressures.

The activity of any substance in an *ideal* mixture is the ratio of the concentration or pressure of the substance to a reference concentration (1 M) or to a reference pressure (1 bar). For example, if the concentration of a substance in an equilibrium mixture is 0.10 M, its activity is 0.10 M/1 M = 0.10. The units of such ratios always cancel and, consequently, activities have no units. Furthermore, the numerical value of the activity equals the concentration because we have divided by 1.

In real systems, activities are not exactly numerically equal to concentrations. In some cases the differences are significant; however, we will ignore these differences. For pure solids and pure liquids, the situation is even simpler because the activities then merely equal 1 (again with no units). Because activities have no units, the *thermodynamic equilibrium constant* derived from them also has no units. It is therefore common practice to write all types of equilibrium constants without units as well, a practice that we adhere to in this text.

16.3 | INTERPRETING AND WORKING WITH EQUILIBRIUM CONSTANTS

Before doing calculations with equilibrium constants, it is valuable to understand what the magnitude of an equilibrium constant can tell us about the relative concentrations of reactants and products in an equilibrium mixture. It is also useful to consider how the magnitude of any equilibrium constant depends on how the chemical equation is expressed.

The Magnitude of Equilibrium Constants

Equilibrium constants can vary from very large to very small. The magnitude of the constant provides us with important information about the composition of an equilibrium mixture. For example, consider the reaction of carbon monoxide gas and chlorine gas at 100 °C to form phosgene ($COCl_2$), a toxic gas used in the manufacture of certain polymers and insecticides:

$$CO(g) + Cl_2(g) \rightleftharpoons COCl_2(g) \qquad K_c = \frac{[COCl_2]}{[CO][Cl_2]} = 4.56 \times 10^9$$

For the equilibrium constant to be so large, the numerator of the equilibrium constant expression must be much larger than the denominator. Thus the equilibrium concentration of $COCl_2$ must be much greater than that of CO or Cl_2, and in

What would this figure look like for a reaction in which $K \approx 1$?

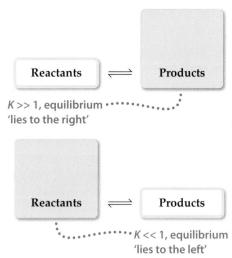

$K \gg 1$, equilibrium 'lies to the right'

$K \ll 1$, equilibrium 'lies to the left'

▲ **FIGURE 16.6** Relationship between magnitude of K and composition of an equilibrium mixture.

fact this is just what we find experimentally. We say that this equilibrium *lies to the right* (that is, towards the product side). Likewise, a very small equilibrium constant indicates that the equilibrium mixture contains mostly reactants. We then say that the equilibrium *lies to the left*. In general:

If $K \gg 1$ (*large K*):   Equilibrium lies to the right; products predominate.

If $K \ll 1$ (*small K*):   Equilibrium lies to the left; reactants predominate.

These situations are summarised in ◀ **FIGURE 16.6**. Note that the equilibrium constant does not tell us anything about how fast the equilibrium is achieved.

SAMPLE EXERCISE 16.3 | **Interpreting the magnitude of an equilibrium constant**

The following diagrams represent three systems at equilibrium, all in the same size of container. **(a)** Without doing any calculations, rank the systems in order of increasing K_c. **(b)** If the volume of the containers is 1.0 dm^3 and each sphere represents 0.10 mol, calculate K_c for each system.

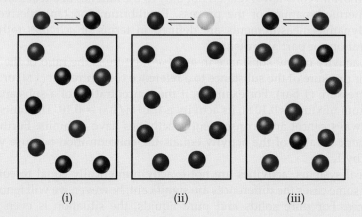

(i)   (ii)   (iii)

SOLUTION

Analyse We are asked to judge the relative magnitudes of three equilibrium constants and then to calculate them.

Plan (a) The more product present at equilibrium, relative to reactant, the larger the equilibrium constant. **(b)** The equilibrium constant is given by Equation 16.8.

Solve
(a) Each box contains 10 spheres. The amount of product in each varies as follows: (i) 6, (ii) 1, (iii) 8. Therefore the equilibrium constant varies in the order (ii) < (i) < (iii), from smallest (most reactant) to largest (most products).

(b) In (i) we have 0.60 mol dm^{-3} product and 0.40 mol dm^{-3} reactant, giving $K_c = 0.60/0.40 = 1.5$. (You will get the same result by merely dividing the number of spheres of each kind: 6 spheres/4 spheres = 1.5.) In (ii) we have 0.10 mol dm^{-3} product and 0.90 mol dm^{-3} reactant, giving $K_c = 0.10/0.90 = 0.11$ (or 1 sphere/ 9 spheres = 0.11). In (iii) we have 0.80 mol dm^{-3} product and 0.20 mol dm^{-3} reactant, giving $K_c = 0.80/0.20 = 4.0$ (or 8 spheres/2 spheres = 4.0). These calculations verify the order in (a).

Comment Imagine a drawing that represents a reaction with a very small or very large value of K_c. For example, what would the drawing look like if $K_c = 1 \times 10^{-5}$? In that case there would need to be 100 000 reactant molecules for only 1 product molecule. But then, that would be impractical to draw.

PRACTICE EXERCISE

For the reaction $H_2(g) + I_2(g) \rightleftharpoons 2 HI(g)$, $K_P = 794$ at 298 K and $K_P = 55$ at 700 K. Is the formation of HI favoured more at the higher or lower temperature?

Answer: The formation of product, HI, is favoured at the lower temperature because K_P is larger at the lower temperature.

(See also Exercise 16.13.)

The Direction of the Chemical Equation and *K*

Because an equilibrium can be approached from either direction, the direction in which we write the chemical equation for an equilibrium is arbitrary. For example, we have seen that we can represent the N_2O_4:NO_2 equilibrium as

$$N_2O_4(g) \rightleftharpoons 2\,NO_2(g) \quad K_c = \frac{[NO_2]^2}{[N_2O_4]} = 0.212 \quad \text{(at 100 °C)} \quad [16.16]$$

We could equally well consider this same equilibrium in terms of the reverse reaction:

$$2\,NO_2(g) \rightleftharpoons N_2O_4(g)$$

The equilibrium expression is then

$$K_c = \frac{[N_2O_4]}{[NO_2]^2} = \frac{1}{0.212} = 4.72 \quad \text{(at 100 °C)} \quad [16.17]$$

Equation 16.17 is the reciprocal of the equilibrium constant expression in Equation 16.16. *The equilibrium constant expression for a reaction written in one direction is the reciprocal of the one for the reaction written in the reverse direction.* Consequently, the numerical value of the equilibrium constant for the reaction written in one direction is the reciprocal of that for the reverse reaction. Both expressions are equally valid, but it is meaningless to say that the equilibrium constant for the equilibrium between NO_2 and N_2O_4 is 0.212 or 4.72. We need to specify that 0.212 is the equilibrium constant for the decomposition of N_2O_4 at 100 °C, whereas 4.72 is the value for the formation of N_2O_4 at 100 °C.

Relating Chemical Equation Stoichiometry and Equilibrium Constants

Just as the equilibrium constants of forward and reverse reactions are reciprocals of each other, the equilibrium constants of reactions associated in other ways are also related. For example, if we multiply the original N_2O_4:NO_2 equilibrium by 2, we have

$$2\,N_2O_4(g) \rightleftharpoons 4\,NO_2(g)$$

The equilibrium constant expression, K_c, for this equation is

$$K_c = \frac{[NO_2]^4}{[N_2O_4]^2}$$

which is simply the square of the equilibrium constant expression for the original equation, given in Equation 16.10. Because the new equilibrium constant expression equals the original expression squared, the new equilibrium constant equals the original constant squared: in this case, $0.212^2 = 0.0449$ (at 100 °C).

> ⚠️ **CONCEPT CHECK 5**
>
> How does the magnitude of the equilibrium constant, K_P, for the reaction $2\,HI(g) \rightleftharpoons H_2(g) + I_2(g)$ change if the equilibrium is written $6\,HI(g) \rightleftharpoons 3\,H_2(g) + 3\,I_2(g)$?

Sometimes, as in problems that utilise Hess's law (∞ Section 14.6, 'Hess's Law'), we must use equations made up of two or more steps in the overall process. We obtain the net equation by adding the individual equations and cancelling identical species. Consider the following two reactions, their equilibrium-constant expressions and their equilibrium constants at 100 °C:

∞ Review this on page 520

$$2 \, NOBr(g) \rightleftharpoons 2 \, NO(g) + Br_2(g) \qquad K_c = \frac{[NO]^2[Br_2]}{[NOBr]^2} = 0.014$$

$$Br_2(g) + Cl_2(g) \rightleftharpoons 2 \, BrCl(g) \qquad K_c = \frac{[BrCl]^2}{[Br_2][Cl_2]} = 7.2$$

The net sum of these two equations is

$$2 \, NOBr(g) + Cl_2(g) \rightleftharpoons 2 \, NO(g) + 2 \, BrCl(g)$$

and the equilibrium constant expression for the net equation is the product of the expressions for the individual steps:

$$K_c = \frac{[NO]^2[BrCl]^2}{[NOBr]^2[Cl_2]} = \frac{[NO]^2[Br_2]}{[NOBr]^2} \times \frac{[BrCl]^2}{[Br_2][Cl_2]}$$

Because the equilibrium constant expression for the net equation is the product of two equilibrium constant expressions, the equilibrium constant for the net equation is the product of the two individual equilibrium constants:

$$K_c = 0.014 \times 7.2 = 0.10$$

To summarise:

1. The equilibrium constant of a reaction in the *reverse* direction is the *inverse* of the equilibrium constant of the reaction in the forward direction.

$$A + B \rightleftharpoons C + D \qquad K_1$$
$$C + D \rightleftharpoons A + B \qquad K = 1/K_1$$

2. The equilibrium constant of a reaction that has been *multiplied* by a number is the equilibrium constant raised to a *power* equal to that number.

$$A + B \rightleftharpoons C + D \qquad K_1$$
$$nA + nB \rightleftharpoons nC + nD \qquad K = K_1{}^n$$

3. The equilibrium constant for a net reaction made up of *two or more steps* is the *product* of the equilibrium constants for the individual steps.

$$1. \quad A + B \rightleftharpoons C + D \quad K_1$$
$$2. \quad C + F \rightleftharpoons G + A \quad K_2$$
$$\overline{\quad\quad\quad\quad\quad\quad\quad\quad\quad\quad\quad\quad\quad\quad\quad}$$
$$3. \quad B + F \rightleftharpoons D + G \quad K_3 = (K_1)(K_2)$$

SAMPLE EXERCISE 16.4 **Combining equilibrium expressions**

Given the following information,

$$HF(aq) \rightleftharpoons H^+(aq) + F^-(aq) \qquad K_c = 6.8 \times 10^{-4}$$

$$H_2C_2O_4(aq) \rightleftharpoons 2 \, H^+(aq) + C_2O_4{}^{2-}(aq) \quad K_c = 3.8 \times 10^{-6}$$

determine the value of K_c for the reaction

$$2 \, HF(aq) + C_2O_4{}^{2-}(aq) \rightleftharpoons 2 \, F^-(aq) + H_2C_2O_4(aq)$$

SOLUTION

Analyse We are given two equilibrium equations and the corresponding equilibrium constants and are asked to determine the equilibrium constant for a third equation, which is related to the first two.

Plan We cannot simply add the first two equations to get the third. Instead, we need to determine how to manipulate the equations to come up with the steps that will add to give us the desired equation.

Solve If we multiply the first equation by 2 and make the corresponding change to its equilibrium constant (raising to the power 2), we get

$$2 \, HF(aq) \rightleftharpoons 2 \, H^+(aq) + 2 \, F^-(aq) \quad K_c = (6.8 \times 10^{-4})^2 = 4.6 \times 10^{-7}$$

Reversing the second equation and again making the corresponding change to its equilibrium constant (taking the reciprocal) gives

$$2\,H^+(aq) + C_2O_4^{2-}(aq) \rightleftharpoons H_2C_2O_4(aq)$$

$$K_c = \frac{1}{3.8 \times 10^{-6}} = 2.6 \times 10^5$$

Now we have two equations that sum to give the net equation, and we can multiply the individual K_c values to get the desired equilibrium constant.

$$2\,HF(aq) \rightleftharpoons 2\,H^+(aq) + 2\,F^-(aq) \qquad K_c = 4.6 \times 10^{-7}$$

$$2\,H^+(aq) + C_2O_4^{2-}(aq) \rightleftharpoons H_2C_2O_4(aq) \qquad K_c = 2.5 \times 10^5$$

$$\overline{2\,HF(aq) + C_2O_4^{2-}(aq) \rightleftharpoons 2\,F^-(aq) + H_2C_2O_4(aq)} \quad K_c = (4.6 \times 10^{-7})(2.6 \times 10^5) = 0.12$$

PRACTICE EXERCISE

Given that, at 700 K, $K_P = 54.0$ for the reaction $H_2(g) + I_2(g) \rightleftharpoons 2\,HI(g)$ and $K_P = 1.04 \times 10^{-4}$ for the reaction $N_2(g) + 3\,H_2(g) \rightleftharpoons 2\,NH_3(g)$, determine the value of K_P for the reaction $2\,NH_3(g) + 3\,I_2(g) \rightleftharpoons 6\,HI(g) + N_2(g)$ at 700 K.

Answer: $\dfrac{(54.0)^3}{1.04 \times 10^{-4}} = 1.51 \times 10^9$

(See also Exercise 16.19.)

16.4 | HETEROGENEOUS EQUILIBRIA

Many equilibria, such as the hydrogen–nitrogen–ammonia system, involve substances all in the same phase. Such equilibria are called **homogeneous equilibria**. In other cases the substances in equilibrium are in different phases, giving rise to **heterogeneous equilibria**. As an example, consider the equilibrium that occurs when solid lead(II) chloride ($PbCl_2$) dissolves in water to form a saturated solution:

$$PbCl_2(s) \rightleftharpoons Pb^{2+}(aq) + 2\,Cl^-(aq) \qquad [16.18]$$

This system consists of a solid in equilibrium with two aqueous species. If we write the equilibrium constant expression for this process, we encounter a new problem: How do we express the concentration of a solid substance? Although it is possible to express the concentration of a solid in terms of moles per unit volume, it is unnecessary to do so in writing equilibrium constant expressions since the concentration remains constant as long as some solid remains. The same will be true for a pure liquid that does not mix with the other reactants. *Whenever a pure solid or a pure liquid is involved in a heterogeneous equilibrium, its concentration is not included in the equilibrium constant expression for the reaction.* Thus the equilibrium constant expression for Equation 16.18 is

$$K_c = [Pb^{2+}][Cl^-]^2 \qquad [16.19]$$

Even though $PbCl_2(s)$ does not appear in the equilibrium constant expression, it must be present for equilibrium to occur.

Because equilibrium constant expressions include terms only for reactants and products whose concentrations can change during a chemical reaction, the concentrations of pure solids and pure liquids are omitted.

CONCEPT CHECK 6

Write the equilibrium constant expression for the evaporation of water, $H_2O(l) \rightleftharpoons H_2O(g)$, in terms of partial pressures, K_P.

As a further example of a heterogeneous reaction, consider the decomposition of calcium carbonate:

$$CaCO_3(s) \rightleftharpoons CaO(s) + CO_2(g)$$

Omitting the concentrations of solids from the equilibrium constant expression gives

$$K_c = [CO_2] \quad \text{and} \quad K_P = P_{CO_2}$$

◣ FIGURE IT OUT

Imagine starting with only CaO in a bell jar and adding $CO_2(g)$ to make its pressure the same as it is in these two bell jars. How does the equilibrium concentration of $CO_2(g)$ in your jar compare with the $CO_2(g)$ equilibrium concentration in these two jars?

$$CaCO_3(s) \rightleftharpoons CaO(s) + CO_2(g)$$

CaCO₃   CaO

Large amount of CaCO₃, small amount of CaO, gas pressure P

CaCO₃   CaO

Small amount of CaCO₃, large amount of CaO, gas pressure still P

▲ **FIGURE 16.7** **At a given temperature, the equilibrium pressure of CO_2 in the bell jars is the same no matter how much of each solid is present.**

These equations tell us that, at a given temperature, an equilibrium among $CaCO_3$, CaO and CO_2 will always lead to the same partial pressure of CO_2 as long as all three components are present. As shown in ◀ **FIGURE 16.7**, we would have the same pressure of CO_2 regardless of the relative amounts of $CaCO_3$ and CaO.

When a solvent is involved as a reactant or product in an equilibrium, its concentration is also excluded from the equilibrium constant expression, provided the concentrations of reactants and products are low, so that the solvent is essentially a pure substance. Applying this guideline to an equilibrium involving water as a solvent,

$$H_2O(l) + CO_3^{2-}(aq) \rightleftharpoons OH^-(aq) + HCO_3^-(aq) \qquad [16.20]$$

gives an equilibrium constant expression in which $[H_2O]$ is excluded:

$$K_c = \frac{[OH^-][HCO_3^-]}{[CO_3^{2-}]} \qquad [16.21]$$

SAMPLE EXERCISE 16.5 | Writing equilibrium constant expressions for heterogeneous reactions

Write the equilibrium constant expression for K_c for each of the following reactions:
(a) $CO_2(g) + H_2(g) \rightleftharpoons CO(g) + H_2O(l)$
(b) $SnO_2(s) + 2\,CO(g) \rightleftharpoons Sn(s) + 2\,CO_2(g)$.

SOLUTION

Analyse We are given two chemical equations, both for heterogeneous equilibria, and asked to write the corresponding equilibrium constant expressions.

Plan We use the law of mass action, remembering to omit any pure solids and pure liquids from the expressions.

Solve
(a) The equilibrium constant expression is $\quad K_c = \dfrac{[CO]}{[CO_2][H_2]}$

Because H_2O appears in the reaction as a pure liquid, its concentration does not appear in the equilibrium constant expression.

(b) The equilibrium constant expression is $\quad K_c = \dfrac{[CO_2]^2}{[CO]^2}$

Because SnO_2 and Sn are both pure solids, their concentrations do not appear in the equilibrium constant expression.

PRACTICE EXERCISE

Write the following equilibrium constant expressions:
(a) K_c for $Cr(s) + 3\,Ag^+(aq) \rightleftharpoons Cr^{3+}(aq) + 3\,Ag(s)$
(b) K_P for $3\,Fe(s) + 4\,H_2O(g) \rightleftharpoons Fe_3O_4(s) + 4\,H_2(g)$

Answers: (a) $K_c = \dfrac{[Cr^{3+}]}{[Ag^+]^3}$ (b) $K_P = \dfrac{(P_{H_2})^4}{(P_{H_2O})^4}$

(See also Exercise 16.20.)

SAMPLE EXERCISE 16.6 | Analysing a heterogeneous equilibrium

Each of these mixtures was placed in a closed container and allowed to stand:
(a) $CaCO_3(s)$
(b) CaO(s) and $CO_2(g)$ at a pressure greater than the value of K_P
(c) $CaCO_3(s)$ and $CO_2(g)$ at a pressure greater than the value of K_P
(d) $CaCO_3(s)$ and CaO(s)

Determine whether or not each mixture can attain the equilibrium
$$CaCO_3(s) \rightleftharpoons CaO(s) + CO_2(g)$$

SOLUTION

Analyse We are asked which of several combinations of species can establish an equilibrium between calcium carbonate and its decomposition products, calcium oxide and carbon dioxide.

Plan For equilibrium to be achieved, it must be possible for both the forward process and the reverse process to occur. For the forward process to occur, there must be some calcium carbonate present. For the reverse process to occur, there must be both calcium oxide and carbon dioxide. In both cases, either the necessary compounds may be present initially or they may be formed by reaction of the other species.

Solve Equilibrium can be reached in all cases except (c) as long as sufficient quantities of solids are present. **(a)** $CaCO_3$ simply decomposes, forming $CaO(s)$ and $CO_2(g)$, until the equilibrium pressure of CO_2 is attained. There must be enough $CaCO_3$, however, to allow the CO_2 pressure to reach equilibrium. **(b)** CO_2 continues to combine with CaO until the partial pressure of the CO_2 decreases to the equilibrium value. **(c)** There is no CaO present, so equilibrium can't be attained because there is no way the CO_2 pressure can decrease to its equilibrium value (which would require some of the CO_2 to react with CaO). **(d)** The situation is essentially the same as in (a): $CaCO_3$ decomposes until equilibrium is attained. The presence of CaO initially makes no difference.

PRACTICE EXERCISE

When added to $Fe_3O_4(s)$ in a closed container, which one of the following substances: $H_2(g)$, $H_2O(g)$, $O_2(g)$, will allow equilibrium to be established in the reaction $3\,Fe(s) + 4\,H_2O(g) \rightleftharpoons Fe_3O_4(s) + 4\,H_2(g)$?

Answer: Only $H_2(g)$

16.5 | CALCULATING EQUILIBRIUM CONSTANTS

If we can measure the equilibrium concentrations of all the reactants and products in a chemical reaction, as we did with the data in Table 16.1, calculating the value of the equilibrium constant is straightforward. We simply insert all the equilibrium concentrations into the equilibrium constant expression for the reaction.

SAMPLE EXERCISE 16.7 | **Calculating K when all equilibrium concentrations are known**

A mixture of hydrogen and nitrogen in a reaction vessel is allowed to attain equilibrium at 472 °C. The equilibrium mixture of gases is analysed and found to contain 2.46 bar N_2, 7.38 bar H_2 and 0.166 bar NH_3. From these data, calculate the equilibrium constant K_P for the reaction

$$N_2(g) + 3\,H_2(g) \rightleftharpoons 2\,NH_3(g)$$

SOLUTION

Analyse We are given a balanced equation and equilibrium partial pressures and are asked to calculate the value of the equilibrium constant.

Plan First, we write the equilibrium constant expression. We then substitute the equilibrium partial pressures into the expression and solve for K_P.

Solve

$$K_P = \frac{(P_{NH_3})^2}{P_{N_2}(P_{H_2})^3} = \frac{(0.166)^2}{(2.46)(7.38)^3} = 2.79 \times 10^{-5}$$

PRACTICE EXERCISE

An aqueous solution of acetic acid is found to have the following equilibrium concentrations at 25 °C: $[CH_3COOH] = 1.65 \times 10^{-2}$ M; $[H^+] = 5.44 \times 10^{-4}$ M; and

$[CH_3COO^-] = 5.44 \times 10^{-4}$ M. Calculate the equilibrium constant K_c for the ionisation of acetic acid at 25 °C. The reaction is

$$CH_3COOH(aq) \rightleftharpoons H^+(aq) + CH_3COO^-(aq)$$

Answer: 1.79×10^{-5}

(See also Exercises 16.22, 16.24.)

We often don't know the equilibrium concentrations of all chemical species in an equilibrium mixture. If we know the equilibrium concentration of at least one species, however, we can generally use the stoichiometry of the reaction to deduce the equilibrium concentrations of the others. The following steps outline the procedure.

1. Tabulate all the known initial and equilibrium concentrations of the species that appear in the equilibrium constant expression.
2. For those species for which both the initial and equilibrium concentrations are known, calculate the change in concentration that occurs as the system reaches equilibrium.
3. Use the stoichiometry of the reaction (that is, use the coefficients in the balanced chemical equation) to calculate the changes in concentration for all the other species in the equilibrium.
4. Use the initial concentrations from step 1 and the changes in concentration from step 3 to calculate any equilibrium concentrations not tabulated in step 1.
5. Determine the value of the equilibrium constant.

SAMPLE EXERCISE 16.8 Calculating K from initial and equilibrium concentrations

A closed system initially containing 1.000×10^{-3} M H_2 and 2.000×10^{-3} M I_2 at 448 °C is allowed to reach equilibrium. Analysis of the equilibrium mixture shows that the concentration of HI is 1.8×10^{-3} M. Calculate K_c at 448 °C for the reaction taking place, which is

$$H_2(g) + I_2(g) \rightleftharpoons 2\,HI(g)$$

SOLUTION

Analyse We are given the initial concentrations of H_2 and I_2 and the equilibrium concentration of HI. We are asked to calculate the equilibrium constant K_c for $H_2(g) + I_2(g) \rightleftharpoons 2\,HI(g)$.

Plan We construct a table to find equilibrium concentrations of all species and then use the equilibrium concentrations to calculate the equilibrium constant.

Solve First, we tabulate the initial and equilibrium concentrations of as many species as we can. We also provide space in our table for listing the changes in concentrations. As shown, it is convenient to use the chemical equation as the heading for the table.

	$H_2(g)$	+	$I_2(g)$	$\rightleftharpoons$	$2\,HI(g)$
Initial	1.000×10^{-3} M		2.000×10^{-3} M		0 M
Change					
Equilibrium					1.87×10^{-3} M

Second, we calculate the change in concentration of HI, which is the difference between the equilibrium values and the initial values:

Change in [HI] = 1.87×10^{-3} M $- 0 = 1.87 \times 10^{-3}$ M

Third, we use the coefficients in the balanced equation to relate the change in [HI] to the changes in [H_2] and [I_2]:

$\frac{1}{2}(1.87 \times 10^{-3}$ mol HI dm$^{-3}) = 0.935 \times 10^{-3}$ mol H_2 dm^{-3}

$\frac{1}{2}(1.87 \times 10^{-3}$ mol HI dm$^{-3}) = 0.935 \times 10^{-3}$ mol I_2 dm^{-3}

Fourth, we calculate the equilibrium concentrations of H_2 and I_2, using the initial concentrations and the changes. The equilibrium concentration equals the initial concentration minus that consumed:

$$[H_2] = 1.000 \times 10^{-3}\,M - 0.935 \times 10^{-3}\,M = 0.065 \times 10^{-3}\,M$$
$$[I_2] = 2.000 \times 10^{-3}\,M - 0.935 \times 10^{-3}\,M = 1.065 \times 10^{-3}\,M$$

The completed table now looks like this (with equilibrium concentrations in blue for emphasis):

	$H_2(g)$	$+$	$I_2(g)$	$\rightleftharpoons$	$2\,HI(g)$
Initial	$1.000 \times 10^{-3}\,M$		$2.000 \times 10^{-3}\,M$		$0\,M$
Change	$-0.935 \times 10^{-3}\,M$		$-0.935 \times 10^{-3}\,M$		$+1.87 \times 10^{-3}\,M$
Equilibrium	$0.065 \times 10^{-3}\,M$		$1.065 \times 10^{-3}\,M$		$1.87 \times 10^{-3}\,M$

Notice that the entries for the changes are negative when a reactant is consumed and positive when a product is formed.

Finally, now that we know the equilibrium concentration of each reactant and product, we can use the equilibrium constant expression to calculate the equilibrium constant.

$$K_c = \frac{[HI]^2}{[H_2][I_2]} = \frac{(1.87 \times 10^{-3})^2}{(0.065 \times 10^{-3})(1.065 \times 10^{-3})} = 51$$

Comment The same method can be applied to gaseous equilibrium problems, in which case partial pressures are used as table entries in place of molar concentrations.

PRACTICE EXERCISE

Sulfur trioxide decomposes at high temperature in a sealed container: $2\,SO_3(g) \rightleftharpoons 2\,SO_2(g) + O_2(g)$. Initially, the vessel is charged at 1000 K with $SO_3(g)$ at a partial pressure of 0.500 bar. At equilibrium the SO_3 partial pressure is 0.200 bar. Calculate the value of K_P at 1000 K.

Answer: 0.338

(See also Exercises 16.25, 16.26.)

16.6 | APPLICATIONS OF EQUILIBRIUM CONSTANTS

We have seen that the magnitude of K indicates the extent to which a reaction will proceed. If K is very large, the reaction will tend to proceed far to the right; if K is very small (that is, much less than 1), the equilibrium mixture will contain mainly reactants. The equilibrium constant also allows us to (1) predict the direction in which a reaction mixture will proceed to achieve equilibrium, and (2) calculate the concentrations of reactants and products when equilibrium has been reached.

Predicting the Direction of Reaction

For the formation of NH_3 from N_2 and H_2 (Equation 16.6), $K_c = 0.105$ at 472 °C. Suppose we place a mixture of 1.00 mol of N_2, 2.00 mol of H_2 and 2.00 mol of NH_3 in a 1.00 dm^3 container at 472 °C. How will the mixture react to reach equilibrium? Will N_2 and H_2 react to form more NH_3, or will NH_3 decompose to form N_2 and H_2?

To answer this question, we can substitute the starting concentrations of N_2, H_2 and NH_3 into the equilibrium constant expression and compare its value with the equilibrium constant:

$$\frac{[NH_3]^2}{[N_2][H_2]^3} = \frac{(2.00)^2}{(1.00)(2.00)^3} = 0.500 \quad \text{whereas} \quad K_c = 0.105 \qquad [16.22]$$

To reach equilibrium, the quotient $[NH_3]^2/[N_2][H_2]^3$ will need to decrease from the starting value of 0.500 to the equilibrium value of 0.105. This change can happen only if the concentration of NH_3 decreases and the concentrations of N_2 and H_2 increase. Thus the reaction proceeds towards equilibrium by forming N_2 and H_2 from NH_3; that is, the reaction as written in Equation 16.6 proceeds from right to left.

The approach we have illustrated can be formalised by defining a quantity called the reaction quotient. The **reaction quotient**, Q, is *a number obtained by substituting reactant and product concentrations or partial pressures at any point during a reaction into an equilibrium constant expression.*

Therefore, for the general reaction

$$a\,A + b\,B \rightleftharpoons d\,D + e\,E$$

the reaction quotient in terms of molar concentrations is

$$Q_c = \frac{[D]^d[E]^e}{[A]^a[B]^b} \qquad [16.23]$$

(A related quantity, Q_P, can be written for any reaction that involves gases by using partial pressures instead of concentrations.)

Although we use what looks like the equilibrium constant expression to calculate the reaction quotient, the concentrations we use may or may not be the equilibrium concentrations. For example, when we substituted the starting concentrations into the equilibrium constant expression of Equation 16.22, we obtained $Q_c = 0.500$ whereas $K_c = 0.105$. The equilibrium constant has only one value at each temperature. The reaction quotient, however, varies as the reaction proceeds.

Of what use is Q? One practical thing we can do with Q is tell whether our reaction really is at equilibrium, which is an especially valuable option when a reaction is very slow. We can take samples of our reaction mixture as the reaction proceeds, separate the components and measure their concentrations. Then we insert these numbers into Equation 16.23 for our reaction. To determine whether or not we are at equilibrium, or in which direction the reaction proceeds to achieve equilibrium, we compare the values of Q_c and K_c or Q_P and K_P. Three possible situations arise:

1. $Q = K$: the reaction quotient will equal the equilibrium constant only if the system is already at equilibrium.
2. $Q > K$: the concentration of products is too large and that of reactants too small. Thus substances on the right side of the chemical equation will react to form substances on the left; the reaction moves from right to left in approaching equilibrium.
3. $Q < K$: the concentration of products is too small and that of reactants too large. Thus the reaction will achieve equilibrium by forming more products; it moves from left to right.

These relationships are summarised in ◀ FIGURE 16.8.

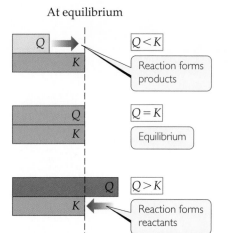

At equilibrium

$Q < K$

Reaction forms products

$Q = K$

Equilibrium

$Q > K$

Reaction forms reactants

▲ **FIGURE 16.8** **Predicting the direction of a reaction by comparing Q and K at a given temperature.**

SAMPLE EXERCISE 16.9 **Predicting the direction of approach to equilibrium**

At 448 °C the equilibrium constant K_c for the reaction

$$H_2(g) + I_2(g) \rightleftharpoons 2\,HI(g)$$

is 50.5. Predict in which direction the reaction will proceed to reach equilibrium at 448 °C if we start with 2.0×10^{-2} mol of HI, 1.0×10^{-2} mol of H_2 and 3.0×10^{-2} mol of I_2 in a 2.00 dm^3 container.

SOLUTION

Analyse We are given a volume and initial molar amounts of the species in a reaction and asked to determine in which direction the reaction must proceed to achieve equilibrium.

Plan We can determine the starting concentration of each species in the reaction mixture. We can then substitute the starting concentrations into the equilibrium constant expression to calculate the reaction quotient, Q_c. Comparing the magnitudes of the equilibrium constant, which is given, and the reaction quotient will tell us in which direction the reaction will proceed.

Solve The initial concentrations are

$$[HI] = 2.0 \times 10^{-2} \, \text{mol per } 2.00 \, \text{dm}^3 = 1.0 \times 10^{-2} \, \text{M}$$

$$[H_2] = 1.0 \times 10^{-2} \, \text{mol per } 2.00 \, \text{dm}^3 = 5.0 \times 10^{-3} \, \text{M}$$

$$[I_2] = 3.0 \times 10^{-2} \, \text{mol per } 2.00 \, \text{dm}^3 = 1.5 \times 10^{-2} \, \text{M}$$

The reaction quotient is therefore

$$Q_c = \frac{[HI]^2}{[H_2][I_2]} = \frac{(1.0 \times 10^{-2})^2}{(5.0 \times 10^{-3})(1.5 \times 10^{-2})} = 1.3$$

Because $Q_c < K_c$, the concentration of HI must increase and the concentrations of H_2 and I_2 must decrease to reach equilibrium; the reaction will proceed from left to right as it moves towards equilibrium.

PRACTICE EXERCISE

At 1000 K the value of K_P for the reaction $2\,SO_3(g) \rightleftharpoons 2\,SO_2(g) + O_2(g)$ is 0.338. Calculate the value for Q_P and predict the direction in which the reaction will proceed towards equilibrium if the initial partial pressures are $P_{SO_3} = 0.16$ bar; $P_{SO_2} = 0.41$ bar; $P_{O_2} = 2.5$ bar.

Answer: $Q_P = 16$; $Q_P > K_P$ and so the reaction will proceed from right to left, forming more SO_3.

(See also Exercise 16.29.)

Calculating Equilibrium Concentrations

Chemists frequently need to calculate the amounts of reactants and products present at equilibrium. Our approach in solving problems of this type is similar to that used for evaluating equilibrium constants. We tabulate the initial concentrations or partial pressures, the changes therein and the final equilibrium concentrations or partial pressures. Usually, we end up using the equilibrium constant expression to derive an equation that must be solved for an unknown quantity, as demonstrated in Sample Exercise 16.10.

In many situations we will know the value of the equilibrium constant and the initial amounts of all species. We must then solve for the equilibrium amounts. Solving this type of problem usually entails treating the change in concentration as a variable. The stoichiometry of the reaction gives us the relationship between the changes in the amounts of all the reactants and products, as illustrated in Sample Exercise 16.11.

SAMPLE EXERCISE 16.10 | **Calculating equilibrium concentrations**

For the Haber process, $N_2(g) + 3\,H_2(g) \rightleftharpoons 2\,NH_3(g)$, $K_P = 1.45 \times 10^{-5}$ at 500 °C. In an equilibrium mixture of the three gases at 500 °C, the partial pressure of N_2 is 0.432 bar and that of H_2 is 0.928 bar. What is the partial pressure of NH_3 in this equilibrium mixture?

SOLUTION

Analyse We are given an equilibrium constant, K_P, and the equilibrium partial pressures of two of the three substances in the equation (N_2 and H_2), and we are asked to calculate the equilibrium partial pressure for the third substance (NH_3).

Plan We can set K_P equal to the equilibrium constant expression and substitute in the partial pressures that we know. Then we can solve for the only unknown in the equation.

Solve We tabulate the equilibrium pressures as follows:

	$N_2(g)$	+	$3\,H_2(g)$	$\rightleftharpoons$	$2\,NH_3(g)$
Equilibrium pressure (bar)	0.432		0.928		x

Because we do not know the equilibrium pressure of NH_3, we represent it with a variable, x. At equilibrium the pressures must satisfy the equilibrium constant expression:

$$K_P = \frac{(P_{NH_3})^2}{P_{N_2}(P_{H_2})^3} = \frac{x^2}{(0.432)(0.928)^3} = 1.45 \times 10^{-5}$$

We now rearrange the equation to solve for x:

$$x^2 = (1.45 \times 10^{-5})(0.432)(0.928)^3 = 5.01 \times 10^{-6}$$
$$x = \sqrt{5.01 \times 10^{-6}} = 2.24 \times 10^{-3} \text{ bar } = P_{NH_3}$$

Comment We can always check our answer by using it to recalculate the value of the equilibrium constant:

$$K_P = \frac{(2.24 \times 10^{-3})^2}{(0.432)(0.928)^3} = 1.45 \times 10^{-5}$$

PRACTICE EXERCISE

At 500 K the reaction $PCl_5(g) \rightleftharpoons PCl_3(g) + Cl_2(g)$ has $K_P = 0.497$. In an equilibrium mixture at 500 K, the partial pressure of PCl_5 is 0.860 bar and that of PCl_3 is 0.350 bar. What is the partial pressure of Cl_2 in the equilibrium mixture?

Answer: 1.22 bar

(See also Exercises 16.35, 16.36.)

SAMPLE EXERCISE 16.11 Calculating equilibrium concentrations from initial concentrations

A 1.000 dm^3 flask is filled with 1.000 mol of H_2 and 2.000 mol of I_2 at 448 °C. The value of the equilibrium constant K_c for the reaction

$$H_2(g) + I_2(g) \rightleftharpoons 2 HI(g)$$

at 448 °C is 50.5. What are the equilibrium concentrations of H_2, I_2 and HI in moles per litre?

SOLUTION

Analyse We are given the volume of a container, an equilibrium constant and starting amounts of reactants in the container and are asked to calculate the equilibrium concentrations of all species.

Plan In this case we are not given any of the equilibrium concentrations. We must develop some relationships that relate the initial concentrations to those at equilibrium. The procedure is similar in many regards to that outlined in Sample Exercise 16.8, where we calculated an equilibrium constant using initial concentrations.

Solve First, we note the initial concentrations of H_2 and I_2 in the 1.000 dm^3 flask:

$[H_2] = 1.000$ M and $[I_2] = 2.000$ M

Second, we construct a table in which we tabulate the initial concentrations:

	$H_2(g)$	+	$I_2(g)$	$\rightleftharpoons$	2 HI(g)
Initial	1.000 M		2.000 M		0 M
Change					
Equilibrium					

Third, we use the stoichiometry of the reaction to determine the changes in concentration that occur as the reaction proceeds to equilibrium. The concentrations of H_2 and I_2 will decrease as equilibrium is established, and that of HI will increase. We represent the change in concentration of H_2 by the variable x. The balanced chemical equation tells us the relationship between the changes in the concentrations of the three gases:

For each x mol of H_2 that reacts, x mol of I_2 are consumed and $2x$ mol of HI are produced:

	$H_2(g)$	+	$I_2(g)$	$\rightleftharpoons$	2 HI(g)
Initial	1.000 M		2.000 M		0 M
Change	$-x$		$-x$		$+2x$
Equilibrium					

Fourth, we use the initial concentrations and the changes in concentrations, as dictated by stoichiometry, to express the equilibrium concentrations. With all our entries, the table now looks like this:

	$H_2(g)$	+	$I_2(g)$	$\rightleftharpoons$	2 HI(g)
Initial	1.000 M		2.000 M		0 M
Change	$-x$		$-x$		$+2x$
Equilibrium	$(1.000 - x)$ M		$(2.000 - x)$ M		$2x$ M

Fifth, we substitute the equilibrium concentrations into the equilibrium constant expression and solve for the single unknown, x:

$$K_c = \frac{[HI]^2}{[H_2][I_2]} = \frac{(2x)^2}{(1.000 - x)(2.000 - x)} = 50.5$$

If you have an equation-solving calculator, you can solve this equation directly for x. If not, expand this expression to obtain a quadratic equation in x:

$$4x^2 = 50.5(x^2 - 3.000x + 2.000)$$

$$46.5x^2 - 151.5x + 101.0 = 0$$

Solving the quadratic equation (Appendix A) leads to two solutions for x:

$$x = \frac{-(-151.5) \pm \sqrt{(-151.5)^2 - 4(46.5)(101.0)}}{2(46.5)} = 2.323 \text{ or } 0.935$$

When we substitute $x = 2.323$ into the expressions for the equilibrium concentrations, we find *negative* concentrations of H_2 and I_2. Because a negative concentration is not chemically meaningful, we reject this solution. We then use $x = 0.935$ to find the equilibrium concentrations:

$$[H_2] = 1.000 - x = 0.065 \text{ M}$$
$$[I_2] = 2.000 - x = 1.065 \text{ M}$$
$$[HI] = 2x = 1.870 \text{ M}$$

Check We can check our solution by putting these numbers into the equilibrium constant expression:

$$K_c = \frac{[HI]^2}{[H_2][I_2]} = \frac{(1.870)^2}{(0.065)(1.065)} = 51$$

Comment Whenever you use a quadratic equation to solve an equilibrium problem, one of the solutions will not be chemically meaningful and should be rejected.

PRACTICE EXERCISE

For the equilibrium $PCl_5(g) \rightleftharpoons PCl_3(g) + Cl_2(g)$, the equilibrium constant K_P has the value 0.497 at 500 K. A gas cylinder at 500 K is charged with $PCl_5(g)$ at an initial pressure of 1.66 bar. What are the equilibrium pressures of PCl_5, PCl_3 and Cl_2 at this temperature?

Answers: $P_{PCl_5} = 0.967$ bar; $P_{PCl_3} = P_{Cl_2} = 0.693$ bar

(See also Exercise 16.34.)

In Sample Exercise 16.11 the quadratic expression has been solved exactly, because the initial concentrations and equilibrium concentrations are quite different. If, however, the equilibrium constant is either very small or very large then the change in concentrations on reaching equilibrium will be very small. If this is the case, a simplifying approximation can be made, as shown in Section 17.6.

16.7 | THE EQUILIBRIUM CONSTANT AND FREE ENERGY

We have seen in Section 16.3 that the magnitude of the equilibrium constant allows us to predict the direction in which a reaction mixture achieves equilibrium. In Section 14.13 we saw a special relationship between ΔG, the free energy of reaction of a process, and the spontaneity of a reaction. We have seen that if $\Delta G < 0$ the reaction is spontaneous in the forward direction and if $\Delta G > 0$ the forward reaction is non-spontaneous and work must be done on it to make it occur. For a system at equilibrium, $\Delta G = 0$. By using data such as that in Appendix C, we can calculate values of the *standard* free energy change for a process, $\Delta G°$ (Section 14.13, 'Gibbs Free Energy'), and then use these values to calculate ΔG, the free energy change under *non-standard* conditions because most chemical reactions occur under non-standard conditions. The set of standard conditions for which $\Delta G°$ values pertain is given in Table 14.5

Review this on page 546

(Section 14.12). We can also directly relate the value of $\Delta G°$ for a reaction to the value of the equilibrium constant for the reaction.

For any chemical process, the free energy of reaction, ΔG, is proportional to $\ln \dfrac{Q}{K}$, where Q represents the concentrations or pressures of the components of a system at any time during the reaction and K represents the equilibrium concentrations or pressures.

Introducing the proportionality constant RT, we have,

$$\Delta G = RT \ln \frac{Q}{K}$$

or
$$\Delta G = RT \ln Q - RT \ln K \qquad \text{[16.24]}$$

In this equation R is the ideal-gas constant, $8.314 \text{ J mol}^{-1}\text{ K}^{-1}$; T is the absolute temperature and Q is the reaction quotient that corresponds to the particular reaction mixture of interest. If Q is very different from K, the reaction absorbs or releases a large amount of energy, whereas if Q and K are similar, very little energy is absorbed or released. And of course, if the system is at equilibrium, $\Delta G = 0$.

If we consider standard conditions for Q, that is, 1 bar for gases, 1 M for solutions, then $Q = 1$ and ΔG becomes $\Delta G°$ by definition. Equation 16.24 then becomes

$$\Delta G° = RT \ln 1 - RT \ln K$$

and since $\ln 1 = 0$,
$$\Delta G° = - RT \ln K \qquad \text{[16.25]}$$

or
$$- RT \ln K = \Delta G° \qquad \text{[16.26]}$$

This provides the relationship between the magnitude of $\Delta G°$ and the equilibrium constant for a reaciton. By substituting Equation 16.26 in Equation 16.24 we obtain

$$\Delta G = \Delta G° + RT \ln Q \qquad \text{[16.27]}$$

which relates the free energy change under standard conditions and the free energy change under any other conditions. We see from Equation 16.25 that if $\Delta G°$ is negative then $\ln K$ must be positive which means that $K > 1$. So the more negative ΔG is, the larger is the value of the equilibrium constant K. Conversely if $\Delta G°$ is positive, then $\ln K$ is negative and $K < 1$. ◀ TABLE 16.2 summarises the conclusions by comparing $\Delta G°$ and K for both positive and negative values of $\Delta G°$.

TABLE 16.2 • Relationship between $\Delta G°$ and K at 298 K

$\Delta G°$, kJ mol^{-1}	K
+200	9.1×10^{-36}
+100	3.0×10^{-18}
+50	1.7×10^{-9}
+10	1.8×10^{-2}
+1.0	6.7×10^{-1}
0	1.0
−1.0	1.5
−10	5.6×10^{1}
−50	5.8×10^{8}
−100	3.3×10^{17}
−200	1.1×10^{35}

SAMPLE EXERCISE 16.12 Calculating the free-energy change under non-standard conditions

We continue to explore the Haber process for the synthesis of ammonia for which

$$\Delta G° = -33.3 \text{ kJ:}$$

$$N_2(g) + 3 H_2(g) \rightleftharpoons 2 NH_3(g)$$

Calculate ΔG at 298 K for a reaction mixture that consists of 1.0 bar N_2, 3.0 bar H_2 and 0.50 bar NH_3.

SOLUTION

Analyse We are asked to calculate ΔG under non-standard conditions.

Plan We can use Equation 16.27 to calculate ΔG. To do this we have to calculate the value of the reaction quotient, Q, using the values of the partial pressures given. The value of $\Delta G°$ is given although it could be calculated from heats of formation data in Appendix C.

Solve Solving for the reaction quotient gives:

$$Q = \frac{(P_{NH_3})^2}{P_{N_2}(P_{H_2})^3} = \frac{(0.50)^2}{(1.0)(3.0)^3} = 9.3 \times 10^{-3}$$

We can now use Equation 16.27 to calculate ΔG for these non-standard conditions:

$$\Delta G = \Delta G^\circ + RT \ln Q$$

$$= (-33.3 \text{ kJ mol}^{-1}) + \left(\frac{8.314 \text{ J mol}^{-1}\text{K}^{-1}}{1000}\right)(298 \text{ K}) \ln (9.3 \times 10^{-3})$$

$$= (-33.3 \text{ kJ mol}^{-1}) + (-11.6 \text{ kJ mol}^{-1}) = -44.9 \text{ kJ mol}^{-1}$$

Note that the units of R are J mol^{-1} and have to be converted to kJ mol^{-1}.

Comment We see that ΔG becomes more negative, changing from -33.3 kJ mol^{-1} to -44.9 kJ mol^{-1}, as the pressures of N$_2$, H$_2$ and NH$_3$ are changed from 1.0 bar each (standard conditions, ΔG°) to 1.0 bar, 3.0 bar and 0.50 bar, respectively. The larger negative value for ΔG indicates a larger 'driving force' to produce NH$_3$.

PRACTICE EXERCISE

Calculate ΔG at 298 K for the reaction of nitrogen and hydrogen to form ammonia if the reaction mixture consists of 0.50 bar N$_2$, 0.75 bar H$_2$ and 2.0 bar NH$_3$.

Answer: -26.0 kJ mol^{-1}

(See also Exercises 16.38, 16.39.)

SAMPLE EXERCISE 16.13 **Calculating an equilibrium constant from ΔG°**

Use standard free energies of formation to calculate the equilibrium constant, K, at 25 °C for the reaction involved in the Haber process:

$$N_2(g) + 3 H_2(g) \rightleftharpoons 2 NH_3(g)$$

The standard free energy change for this reaction is
$\Delta G^\circ = -33.3$ kJ mol^{-1} = $-33\,300$ J mol^{-1}.

SOLUTION

Analyse We are asked to calculate K for a reaction given ΔG°.

Plan We have to solve Equation 16.26 for K.

Solve Solving Equation 16.26 for the exponent $-\Delta G^\circ/RT$ we have

$$\frac{\Delta G^\circ}{RT} = \frac{-(-33\,300 \text{ J mol}^{-1})}{(8.314 \text{ J mol}^{-1} \text{ K}^{-1})(298 \text{ K})} = 13.4$$

We insert this value into Equation 16.26 to obtain K:

$$K = e^{-\Delta G^\circ/RT} = e^{13.4} = 6.6 \times 10^5$$

Comment This is a large equilibrium constant, which indicates that the product, NH$_3$, is greatly favoured in the equilibrium mixture at 25 °C. The equilibrium constants for temperatures in the range of 300 °C to 600 °C, given in Table 16.3, are much smaller than the value at 25 °C. Clearly, a low-temperature equilibrium favours the production of ammonia more than a high-temperature one. Nevertheless, the Haber process is conducted at high temperatures because the reaction is extremely slow at room temperature.

Remember Thermodynamics can tell us the direction and extent of a reaction, but tells us nothing about the rate at which it will occur. If a catalyst were found that would permit the reaction to proceed at a rapid rate at room temperature, high pressures would not be needed to force the equilibrium towards NH$_3$.

PRACTICE EXERCISE

Use data from Appendix C to calculate the standard free energy change, ΔG°, and the equilibrium constant, K, at 298 K for the reaction H$_2$(g) + Br$_2$(l) $\rightleftharpoons$ 2 HBr(g).

Answer: $\Delta G^\circ = -106.4$ kJ mol^{-1}, $K = 4 \times 10^{18}$

16.8 | LE CHÂTELIER'S PRINCIPLE

Many of the products we use in everyday life are obtained from the chemical industry. Chemists and chemical engineers in industry spend a great deal of time and effort to maximise the yield of valuable products and minimise waste. For example, when Haber developed his process for making ammonia from N_2 and H_2, he examined how reaction conditions might be varied to increase yield. Using the values of the equilibrium constant at various temperatures, he calculated the equilibrium amounts of NH_3 formed under a variety of conditions. Some of Haber's results are shown in ▼ FIGURE 16.9.

Notice that the percentage of NH_3 present at equilibrium decreases with increasing temperature and increases with increasing pressure.

We can understand these effects in terms of a principle first put forward by Henri-Louis Le Châtelier* (1850–1936), a French industrial chemist, Le Châtelier's principle: *If a system at equilibrium is disturbed by a change in temperature, pressure or a component concentration, the system will shift its equilibrium position so as to counteract the effect of the disturbance.*

In this section we use Le Châtelier's principle to make qualitative predictions about how a system at equilibrium responds to various changes in external conditions. We consider three ways in which a chemical equilibrium can be disturbed: (1) adding or removing a reactant or product, (2) changing the pressure by changing the volume and (3) changing the temperature.

Change in Reactant or Product Concentration

A system at dynamic equilibrium is in a state of balance. When the concentrations of species in the reaction are altered, the equilibrium shifts until a new state of balance is attained. What does *shift* mean? It means that reactant and product concentrations change over time to accommodate the new situation. *Shift* does *not* mean that the equilibrium constant itself is altered; the equilibrium constant remains the same. Le Châtelier's principle states that the shift is in the direction that minimises or reduces the effect of the change. Therefore, *if a chemical system is already at equilibrium and the concentration of any substance in the mixture is increased (either reactant or product), the system reacts to consume some of that substance. Conversely, if the concentration of a substance is decreased, the system reacts to produce some of that substance.*

◢ **FIGURE IT OUT**

At what combination of pressure and temperature should you run the reaction to maximise NH$_3$ yield?

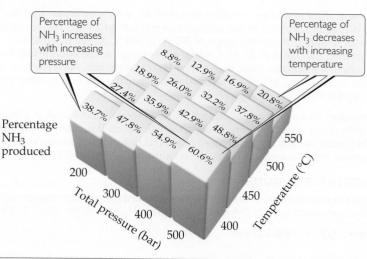

▶ **FIGURE 16.9 Effect of temperature and pressure on NH$_3$ yield in the Haber process.** Each mixture was produced by starting with a 3 : 1 molar mixture of H_2 and N_2.

---

* Pronounced 'le-SHOT-l-yay.'

Le Châtelier's principle

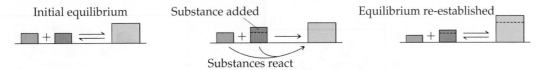

If a system at equilibrium is disturbed by a change in **concentration**, **pressure** or **temperature**, the system will shift its equilibrium position so as to counter the effect of the disturbance.

Concentration: adding or removing a reactant or product

If a substance is added to a system at equilibrium, the system reacts to consume some of the substance. If a substance is removed from a system, the system reacts to produce more of the substance.

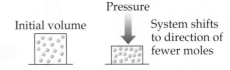

Pressure: changing the pressure by changing the volume

At constant temperature, reducing the volume of a gaseous equilibrium mixture causes the system to shift in the direction that reduces the number of moles of gas.

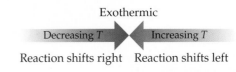

Temperature:

If the temperature of a system at equilibrium is increased, the system reacts as if we added a reactant to an endothermic reaction or a product to an exothermic reaction. The equilibrium shifts in the direction that consumes the 'excess reactant', namely heat.

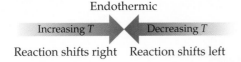

There is no change in the equilibrium constant when we change the concentrations of reactants or products. As an example, consider our familiar equilibrium mixture of N_2, H_2 and NH_3:

$$N_2(g) + 3 H_2(g) \rightleftharpoons 2 NH_3(g)$$

Adding H_2 causes the system to shift so as to reduce the increased concentration of H_2 (▼ **FIGURE 16.10**). This change can occur only if the reaction consumes H_2 and simultaneously consumes N_2 to form more NH_3. Adding N_2 to the equilibrium mixture likewise causes the reaction to shift towards forming more NH_3. Removing NH_3 also causes a shift towards producing more NH_3, whereas *adding* NH_3 to the system at equilibrium causes the reaction to shift in the direction that reduces the increased NH_3 concentration: some of the added ammonia decomposes to form N_2 and H_2.

In the Haber reaction, therefore, removing NH_3 from an equilibrium mixture of N_2, H_2 and NH_3 causes the reaction to shift right to form more NH_3. If the NH_3 can be removed continuously as it is produced, the yield can be increased dramatically. In the industrial production of ammonia, the NH_3 is continuously removed by selectively liquefying it (▼ **FIGURE 16.11**). (The boiling point of NH_3, $-33\,°C$ is much higher than those of N_2, $-196\,°C$, and H_2, $-253\,°C$.) The liquid NH_3 is removed, and the N_2 and H_2 are recycled to form more NH_3. As a result of the product being continuously removed, the reaction is driven essentially to completion.

 CONCEPT CHECK 7

What happens to the equilibrium $2\,NO(g) + O_2(g) \rightleftharpoons 2\,NO_2(g)$ if:
a. O_2 is added to the system,
b. NO is removed?

◢ FIGURE IT OUT

Why does the nitrogen concentration decrease after hydrogen is added?

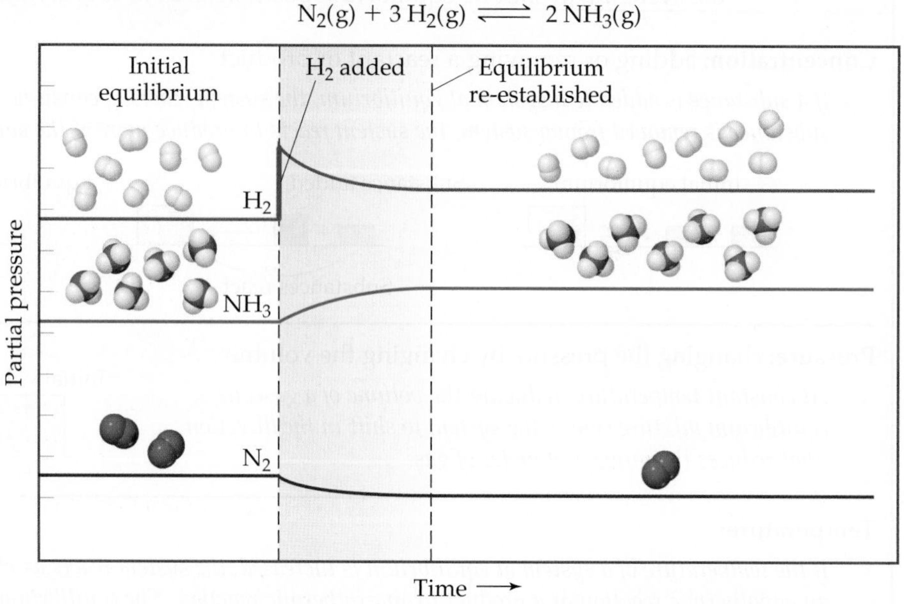

$$N_2(g) + 3 H_2(g) \rightleftharpoons 2 NH_3(g)$$

▶ **FIGURE 16.10** **Effect of adding H₂ to an equilibrium mixture of N₂, H₂ and NH₃.** Adding H₂ causes the reaction as written to shift to the right, consuming some N₂ to produce more NH₃.

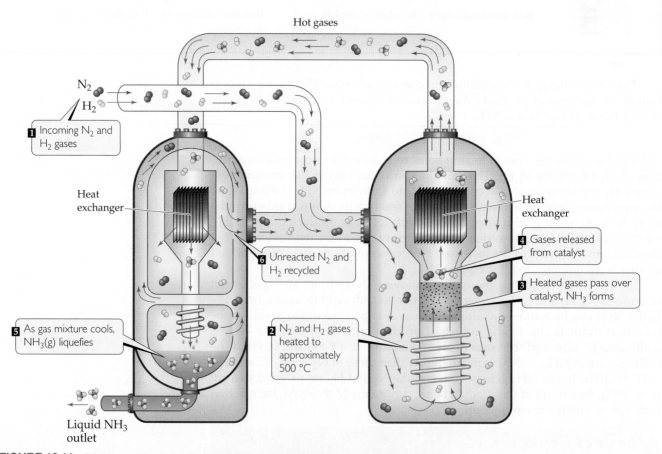

▲ **FIGURE 16.11** **Diagram of the industrial production of ammonia.** Incoming N₂(g) and H₂(g) are heated to approximately 500 °C and passed over a catalyst. When the resultant N₂, H₂ and NH₃ mixture is cooled, the NH₃ liquefies and is removed from the mixture, shifting the reaction to produce more NH₃.

Effects of Volume and Pressure Changes

If a system is at equilibrium and its volume is decreased, thereby increasing its total pressure, Le Châtelier's principle indicates that the system will respond by shifting its equilibrium position to reduce the pressure. A system can reduce its pressure by reducing the total number of gas molecules (fewer molecules of gas exert a lower pressure). Thus at constant temperature, *reducing the volume of a gaseous equilibrium mixture causes the system to shift in the direction that reduces the number of moles of gas.* Conversely, increasing the volume causes a shift in the direction that produces more gas molecules (▼ **FIGURE 16.12**).

> ### ⚠ CONCEPT CHECK 8
> What happens to the equilibrium $2\ SO_2(g) + O_2(g) \rightleftharpoons 2\ SO_3(g)$ if the volume of the system is increased?

For the reaction $N_2(g) + 3\ H_2(g) \rightleftharpoons 2\ NH_3(g)$, there are four molecules of reactant consumed for every two molecules of product produced. Consequently, an increase in pressure (decrease in volume) causes a shift towards the side with fewer gas molecules, which leads to the formation of more NH_3, as indicated in Figure 16.9. In the case of the reaction $H_2(g) + I_2(g) \rightleftharpoons 2\ HI(g)$, the number of molecules of gaseous products (two) equals the number of molecules of gaseous reactants; therefore changing the pressure will not influence the position of the equilibrium.

Keep in mind that pressure-volume changes do *not* change the value of K as long as the temperature remains constant. Rather, they change the partial pressures of the gaseous substances. In Sample Exercise 16.7 we calculated K_P for an equilibrium mixture at 472 °C that contained 2.46 bar N_2, 7.38 bar H_2 and 0.166 bar NH_3. The value of K_P is 2.79×10^{-5}. Consider what happens when we suddenly reduce the volume of the system by one-half. If there were no shift in equilibrium, this volume change would cause the partial pressures of all substances to double, giving $P_{H_2} = 14.76$ bar, $P_{N_2} = 4.92$ bar and $P_{NH_3} = 0.332$ bar. The reaction quotient would then no longer equal the equilibrium constant.

$$Q_P = \frac{(P_{NH_3})^2}{P_{N_2}(P_{H_2})^3} = \frac{(0.332)^2}{(4.92)(14.76)^3} = 6.97 \times 10^{-6} \neq K_P$$

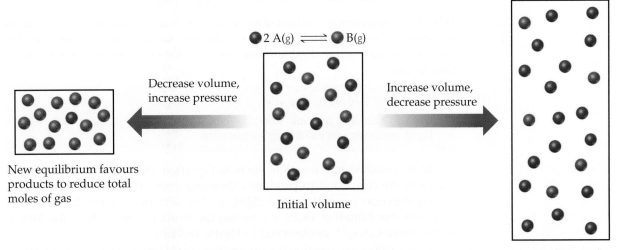

New equilibrium favours products to reduce total moles of gas

Decrease volume, increase pressure

$2\ A(g) \rightleftharpoons B(g)$

Initial volume

Increase volume, decrease pressure

New equilibrium favours reactants to increase total moles of gas

▲ **FIGURE 16.12**   **Pressure and Le Châtelier's principle.**

Because $Q_P < K_P$, the system is no longer at equilibrium. Equilibrium will be re-established by increasing P_{NH_3} and decreasing P_{N_2} and P_{H_2} until $Q_P = K_P = 2.79 \times 10^{-5}$. Therefore the equilibrium shifts to the right as Le Châtelier's principle predicts.

It is possible to change the total pressure of the system without changing its volume. For example, pressure increases if additional amounts of any of the reacting components are added to the system. We have already seen how to deal with a change in concentration of a reactant or product. The total pressure within the reaction vessel might also be increased by adding a gas that is not involved in the equilibrium. For example, argon might be added to the ammonia equilibrium system. The argon would not alter the partial pressures of any of the reacting components and therefore would not cause a shift in equilibrium.

Effect of Temperature Changes

Changes in concentrations or partial pressures cause shifts in equilibrium without changing the value of the equilibrium constant. In contrast, almost every equilibrium constant changes in value as the temperature changes. For example, consider the equilibrium established when cobalt(II) chloride ($CoCl_2$) is dissolved in hydrochloric acid, HCl(aq):

$$Co(H_2O)_6{}^{2+}(aq) + 4\,Cl^-(aq) \rightleftharpoons CoCl_4{}^{2-}(aq) + 6\,H_2O(l) \qquad \Delta H > 0$$
$$\text{pale pink} \qquad\qquad\qquad \text{deep blue} \qquad\qquad\qquad\qquad\qquad [16.28]$$

The formation of $CoCl_4{}^{2-}$ from $Co(H_2O)_6{}^{2+}$ is an endothermic process. We discuss the significance of this enthalpy change shortly. Because $Co(H_2O)_6{}^{2+}$ is pink and $CoCl_4{}^{2-}$ is blue, the position of this equilibrium is readily apparent from the colour of the solution ▶ **FIGURE 16.13(a)**. When the solution is heated (Figure 16.13(b)) it becomes more blue in colour, indicating that the equilibrium has shifted to form more $CoCl_4{}^{2-}$. Cooling the solution (Figure 16.13(c)) leads to a pink solution, indicating that the equilibrium has shifted to produce more $Co(H_2O)_6{}^{2+}$. How can we explain the dependence of this equilibrium on temperature?

We can deduce the rules for the temperature dependence of the equilibrium constant by applying Le Châtelier's principle. A simple way to do this is to treat heat as if it were a chemical reagent. In an *endothermic* reaction we can consider heat as a *reactant*, whereas in an *exothermic* reaction we can consider heat as a *product*.

$$\textit{Endothermic:} \qquad \text{Reactants} + \textit{heat} \rightleftharpoons \text{products}$$
$$\textit{Exothermic:} \qquad \text{Reactants} \rightleftharpoons \text{products} + \textit{heat}$$

When the temperature of a system at equilibrium is increased, it is as if we have added a reactant to an endothermic reaction or a product to an exothermic reaction. The equilibrium shifts in the direction that consumes the excess reactant (or product), namely heat.

 CONCEPT CHECK 9

Use Le Châtelier's principle to explain why the equilibrium vapour pressure of a liquid increases with increasing temperature.

In an endothermic reaction, such as Equation 16.28, heat is absorbed as reactants are converted to products; thus increasing the temperature causes the equilibrium to shift to the right, in the direction of products, and K increases. For Equation 16.28, increasing the temperature leads to the formation of more $CoCl_4{}^{2-}$, as observed in Figure 16.13(b).

In an exothermic reaction the opposite occurs. Heat is absorbed as products are converted to reactants; therefore the equilibrium shifts to the left and K decreases. We can summarise these results as follows:

Endothermic: Increasing T results in an increase in K.

Exothermic: Increasing T results in a decrease in K.

$$\text{Heat} + Co(H_2O)_6{}^{2+}(aq) + 4\,Cl^-(aq) \rightleftharpoons CoCl_4{}^{2-}(aq) + 6\,H_2O(l)$$

pink   blue

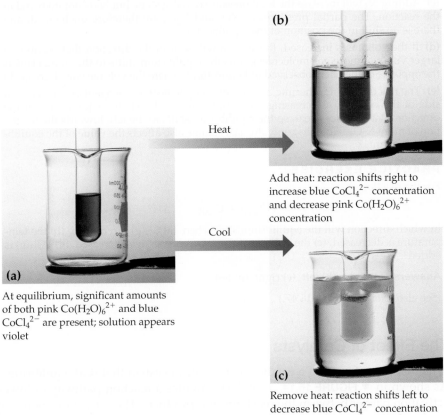

(b)

Heat →

Add heat: reaction shifts right to increase blue $CoCl_4{}^{2-}$ concentration and decrease pink $Co(H_2O)_6{}^{2+}$ concentration

Cool →

(a)

At equilibrium, significant amounts of both pink $Co(H_2O)_6{}^{2+}$ and blue $CoCl_4{}^{2-}$ are present; solution appears violet

(c)

Remove heat: reaction shifts left to decrease blue $CoCl_4{}^{2-}$ concentration and increase pink $Co(H_2O)_6{}^{2+}$ concentration

◀ **FIGURE 16.13   Temperature and Le Châtelier's principle for an endothermic reaction where $\Delta H > 0$.**

Cooling a reaction has the opposite effect. As we lower the temperature the equilibrium shifts to the side that produces heat. Thus cooling an endothermic reaction shifts the equilibrium to the left, decreasing K. We observe this effect in Figure 16.13(c). Cooling an exothermic reaction shifts the equilibrium to the right, increasing K.

SAMPLE EXERCISE 16.14   **Using Le Châtelier's principle to predict shifts in equilibrium**

Consider the equilibrium

$$N_2O_4(g) \rightleftharpoons 2\,NO_2(g) \qquad \Delta H^\circ = 58.0\,kJ$$

In which direction will the equilibrium shift when **(a)** N_2O_4 is added, **(b)** NO_2 is removed, **(c)** the total pressure is increased by addition of $N_2(g)$, **(d)** the volume is increased, **(e)** the temperature is decreased?

SOLUTION

Analyse We are given a series of changes to be made to a system at equilibrium and are asked to predict what effect each change will have on the position of the equilibrium.

Plan Le Châtelier's principle can be used to determine the effects of each of these changes.

Solve

(a) The system will adjust to decrease the concentration of the added N_2O_4, so the equilibrium shifts to the right, in the direction of products.

(b) The system will adjust to the removal of NO_2 by shifting to the side that produces more NO_2; thus the equilibrium shifts to the right.

(c) Adding N_2 will increase the total pressure of the system, but N_2 is not involved in the reaction. The partial pressures of NO_2 and N_2O_4 are therefore unchanged, and there is no shift in the position of the equilibrium.

(d) If the volume is increased, the system will shift in the direction that occupies a larger volume (more gas molecules); thus the equilibrium shifts to the right. (This is the opposite of the effect observed in Figure 16.12, where the volume was decreased.)

(e) The reaction is endothermic, so we can imagine heat as a reagent on the reactant side of the equation. Decreasing the temperature will shift the equilibrium in the direction that produces heat, so the equilibrium shifts to the left, towards the formation of more N_2O_4. Note that only this last change also affects the value of the equilibrium constant, K.

PRACTICE EXERCISE

For the reaction

$$PCl_5(g) \rightleftharpoons PCl_3(g) + Cl_2(g) \qquad \Delta H° = 87.9 \text{ kJ}$$

in which direction will the equilibrium shift when **(a)** $Cl_2(g)$ is removed, **(b)** the temperature is decreased, **(c)** the volume of the reaction system is increased, **(d)** $PCl_3(g)$ is added?

Answers: **(a)** right, **(b)** left, **(c)** right, **(d)** left

(See also Exercises 16.44–16.47.)

The Effect of Catalysts

What happens if we add a catalyst to a chemical system that is at equilibrium? As shown in ▼ **FIGURE 16.14**, a catalyst provides a reaction pathway of lower activation barrier between the reactants and products. The activation energy of the forward reaction is lowered to the same extent as that for the reverse reaction. The catalyst thereby increases the rates of both the forward and reverse reactions. As a result, *a catalyst increases the rate at which equilibrium is achieved, but it does not change the composition of the equilibrium mixture*. The value of the equilibrium constant for a reaction is not affected by the presence of a catalyst.

The rate at which a reaction approaches equilibrium is an important practical consideration. As an example, let's again consider the synthesis of ammonia

 FIGURE IT OUT

How much faster is the catalysed reaction compared to the uncatalysed reaction?

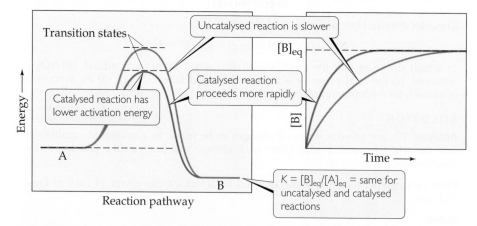

▲ **FIGURE 16.14** A catalyst increases the rate at which equilibrium is reached but does not change the overall composition of the mixture at equilibrium.

from N_2 and H_2. In designing a process for ammonia synthesis, Haber had to deal with a rapid decrease in the equilibrium constant with increasing temperature, as shown in ▶ TABLE 16.3. At temperatures sufficiently high to give a satisfactory reaction rate, the amount of ammonia formed was too small. The solution to this dilemma was to develop a catalyst that would produce a reasonably rapid approach to equilibrium at a sufficiently low temperature, so that the equilibrium constant was still reasonably large. The development of a suitable catalyst thus became the focus of Haber's research efforts.

After trying different substances to see which would be most effective, Haber finally settled on iron mixed with metal oxides. Variants of the original catalyst formulations are still used. These catalysts make it possible to obtain a reasonably rapid approach to equilibrium at temperatures of 400–500 °C and with gas pressures of 200–600 bar. The high pressures are needed to obtain a satisfactory degree of conversion at equilibrium. You can see from Figure 16.9 that if an improved catalyst could be found, one that would lead to sufficiently rapid reaction at temperatures lower than 400–500 °C, it would be possible to obtain the same degree of equilibrium conversion at much lower pressures. This would result in great savings in the cost of equipment for ammonia synthesis. In view of the growing need for nitrogen as fertiliser, the fixation of nitrogen is a process of ever-increasing importance.

TABLE 16.3 • Variation in K_P with temperature for $N_2 + 3 H_2 \rightleftharpoons 2 NH_3$	
Temperature (°C)	K_P
300	4.34×10^{-3}
400	1.64×10^{-4}
450	4.51×10^{-5}
500	1.45×10^{-5}
550	5.38×10^{-6}
600	2.25×10^{-6}

CONCEPT CHECK 10

Does the addition of a catalyst have any effect on the position of an equilibrium?

SAMPLE INTEGRATIVE EXERCISE   Putting concepts together

At temperatures near 800 °C, steam passed over hot coke (a form of carbon obtained from coal) reacts to form CO and H_2:

$$C(s) + H_2O(g) \rightleftharpoons CO(g) + H_2(g)$$

The mixture of gases that results is an important industrial feedstock called *water gas*. **(a)** At 800 °C the equilibrium constant for this reaction is $K_P = 14.1$. What are the equilibrium partial pressures of H_2O, CO and H_2 in the equilibrium mixture at this temperature if we start with solid carbon and 0.100 mol of H_2O in a 1.00 dm³ vessel? **(b)** What is the minimum amount of carbon required to achieve equilibrium under these conditions? **(c)** What is the total pressure in the vessel at equilibrium? **(d)** At 25 °C the value of K_P for this reaction is 1.7×10^{-21}. Is the reaction exothermic or endothermic? **(e)** To produce the maximum amount of CO and H_2 at equilibrium, should the pressure of the system be increased or decreased?

SOLUTION

(a) To determine the equilibrium partial pressures, we use the ideal-gas equation, first determining the starting partial pressure of water.

$$P_{H_2O} = \frac{n_{H_2O}RT}{V} = \frac{(0.100 \text{ mol})(8.314 \text{ kPa dm}^3 \text{ mol}^{-1} \text{ K}^{-1})(1073 \text{ K})}{1.00 \text{ dm}^3} = 892 \text{ kPa} = 8.92 \text{ bar}$$

We then construct a table of starting partial pressures and their changes as equilibrium is achieved:

	C(s) +	$H_2O(g) \rightleftharpoons$	CO(g) +	$H_2(g)$
Initial		8.92 bar	0 bar	0 bar
Change		$-x$	$+x$	$+x$
Equilibrium		$8.92 - x$ bar	x bar	x bar

There are no entries in the table under C(s) because the reactant, being a solid, does not appear in the equilibrium constant expression. Substituting the equilibrium partial pressures of the other species into the equilibrium constant expression for the reaction gives

$$K_P = \frac{P_{CO}P_{H_2}}{P_{H_2O}} = \frac{(x)(x)}{(8.92 - x)} = 14.1$$

Multiplying through by the denominator gives a quadratic equation in x:

$$x^2 = (14.1)(8.92 - x)$$
$$x^2 + 14.1x - 125.77 = 0$$

Solving this equation for x using the quadratic formula yields $x = 6.20$ bar. Hence the equilibrium partial pressures are $P_{CO} = x = 6.20$ bar, $P_{H_2} = x = 6.20$ bar and $P_{H_2O} = (8.92 - x) = 2.72$ bar.

(b) Part (a) shows that $x = 6.20$ bar of H_2O must react in order for the system to achieve equilibrium. We can use the ideal-gas equation to convert this partial pressure into a mole amount.

$$n = \frac{PV}{RT} = \frac{(6.20 \text{ bar})(1.00 \text{ dm}^3)}{(0.08314 \text{ bar dm}^3 \text{ mol}^{-1} \text{ K}^{-1})(1073 \text{ K})} = 0.0695 \text{ mol}$$

Thus 0.0695 mol of H_2O and the same amount of C must react to achieve equilibrium. As a result, there must be at least 0.0695 mol of C (0.835 g of carbon) present among the reactants at the start of the reaction.

(c) The total pressure in the vessel at equilibrium is simply the sum of the equilibrium partial pressures:

$$P_{total} = P_{H_2O} + P_{CO} + P_{H_2} = 2.72 \text{ bar} + 6.2 \text{ bar} + 6.2 \text{ bar} = 15.12 \text{ bar}$$

(d) In discussing Le Châtelier's principle, we saw that endothermic reactions exhibit an increase in K_P with increasing temperature. Because the equilibrium constant for this reaction increases as temperature increases, the reaction must be endothermic. From the enthalpies of formation given in Appendix C, we can verify our prediction by calculating the enthalpy change for the reaction, $\Delta H° = \Delta_f H°(CO) + \Delta_f H°(H_2) - \Delta H°(C) - \Delta_f H°(H_2O) = +131.3$ kJ. The positive sign for $\Delta H°$ indicates that the reaction is endothermic.

(e) According to Le Châtelier's principle, a decrease in the pressure causes a gaseous equilibrium to shift towards the side of the equation with the greater number of moles of gas. In this case, there are two moles of gas on the product side and only one on the reactant side. Therefore the pressure should be reduced to maximise the yield of the CO and H_2.

A CLOSER LOOK

CONTROLLING NITRIC OXIDE EMISSIONS

The formation of NO from N_2 and O_2,

$$\tfrac{1}{2}N_2(g) + \tfrac{1}{2}O_2(g) \rightleftharpoons NO(g) \quad \Delta H° = 90.4 \text{ kJ}$$
$$[16.29]$$

provides an interesting example of the practical importance of the fact that equilibrium constants and reaction rates change with temperature. By applying Le Châtelier's principle to this endothermic reaction and treating heat as a reactant, we deduce that an increase in temperature shifts the equilibrium in the direction of more NO. The equilibrium constant, K_P, for formation of 1 mol of NO from its elements at 300 K is only about 1×10^{-15} (▶ **FIGURE 16.15**). At 2400 K, however, the equilibrium constant is about 0.05, which is 10^{13} times larger than the 300 K value.

Figure 16.15 helps explain why NO is a pollution problem. In the cylinder of a modern high-compression car engine, the temperature during the fuel-burning part of the cycle is approximately 2400 K. Also, there is a fairly large excess of air in the cylinder. These conditions favour the formation of NO. After combustion, however, the gases cool quickly. As the temperature drops, the equilibrium in Equation 16.29 shifts to the left (because the reactant heat is being

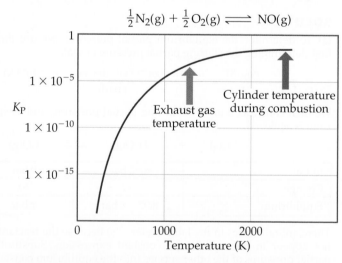

△ **FIGURE IT OUT**

Estimate the value of K_P at 1200 K, the exhaust gas temperature.

$$\tfrac{1}{2}N_2(g) + \tfrac{1}{2}O_2(g) \rightleftharpoons NO(g)$$

▲ **FIGURE 16.15 Equilibrium and temperature.** The equilibrium constant increases with increasing temperature because the reaction is endothermic. It is necessary to use a log scale for K_P because the values vary over such a large range.

removed). The lower temperature also means that the reaction rate decreases, however, so the NO formed at 2400 K is essentially 'frozen' in that form as the gas cools.

The gases exhausting from the cylinder are still quite hot, perhaps 1200 K. At this temperature, as shown in Figure 16.15, the equilibrium constant for formation of NO is about 5×10^{-4}, much smaller than the value at 2400 K. However, the rate of conversion of NO to N_2 and O_2 is too slow to permit much loss of NO before the gases are cooled further.

As discussed in the My World of Chemistry box in Section 15.7, one of the goals of car catalytic converters is to achieve rapid conversion of NO to N_2 and O_2 at the temperature of the exhaust gas. Some catalysts developed for this reaction are reasonably effective under the gruelling conditions in car exhaust systems. Nevertheless, scientists and engineers are continuously searching for new materials that provide even more effective catalysis of the decomposition of nitrogen oxides.

CHAPTER SUMMARY AND KEY TERMS

SECTION 16.1 When a chemical reaction is in a state in which the forward and reverse processes occur at the same rate, this condition is called **chemical equilibrium** and results in the formation of an equilibrium mixture of reactants and products of the reaction. The composition of an equilibrium mixture does not change with time if the temperature is held constant.

SECTION 16.2 An equilibrium that is used throughout this chapter is the reaction $N_2(g) + 3 H_2(g) \rightleftharpoons 2 NH_3(g)$. This reaction is the basis of the **Haber process** for the production of ammonia. The relationship between the concentrations of the reactants and products of a system at equilibrium is given by the **law of mass action**. For an equilibrium equation of the form $a A + b B \rightleftharpoons d D + e E$, the **equilibrium constant expression** is written as

$$K_c = \frac{[D]^d[E]^e}{[A]^a[B]^b}$$

where K_c is a constant called the **equilibrium constant**. When the equilibrium system of interest consists of gases, it is often convenient to express the concentrations of reactants and products in terms of gas pressures:

$$K_P = \frac{(P_D)^d(P_E)^e}{(P_A)^a(P_B)^b}$$

K_c and K_P are related by the expression $K_P = K_c (RT)^{\Delta n}$.

SECTION 16.3 The value of the equilibrium constant changes with temperature. A large value of K_c indicates that the equilibrium mixture contains more products than reactants and therefore lies towards the product side of the equation. A small value for the equilibrium constant means that the equilibrium mixture contains less products than reactants and therefore lies towards the reactant side. The equilibrium constant expression and the equilibrium constant of the reverse of a reaction are the reciprocals of those of the forward reaction. If a reaction is the sum of two or more reactions, its equilibrium constant will be the product of the equilibrium constants for the individual reactions.

SECTION 16.4 Equilibria for which all substances are in the same phase are called **homogeneous equilibria**; in **heterogeneous equilibria** two or more phases are present. The concentrations of pure solids and liquids are left out of the equilibrium constant expression for a heterogeneous equilibrium.

SECTION 16.5 If the concentrations of all species in an equilibrium are known, the equilibrium constant expression can be used to calculate the equilibrium constant. The changes in the concentrations of reactants and products on the way to achieving equilibrium are governed by the stoichiometry of the reaction.

SECTION 16.6 The **reaction quotient (Q)** is found by substituting reactant and product concentrations or partial pressures at any point during a reaction into the equilibrium constant expression. If the system is at equilibrium, $Q = K$. If $Q \neq K$, however, the system is not at equilibrium. When $Q < K$, the reaction will move towards equilibrium by forming more products (the reaction proceeds from left to right); when $Q > K$, the reaction will proceed from right to left. Knowing the value of K makes it possible to calculate the equilibrium amounts of reactants and products, often by the solution of an equation in which the unknown is the change in a partial pressure or concentration.

SECTION 16.7 For a reaction at equilibrium, $\Delta G = 0$. Under non-standard conditions ΔG is related to $\Delta G°$ and the value of the reaction quotient, Q: $\Delta G = \Delta G° + RT \ln Q$. At equilibrium ($\Delta G = 0$, $Q = K$), $\Delta G° = -RT \ln K$. Thus the standard free energy change is directly related to the equilibrium constant for the reaction. This relationship expresses the temperature dependence of equilibrium constants.

SECTION 16.8 **Le Châtelier's principle** states that if a system at equilibrium is disturbed, the equilibrium will shift to minimise the disturbing influence. By this principle, if a reactant or product is added to a system at equilibrium, the equilibrium will shift to consume the added substance. The effects of removing reactants or products and of changing the pressure or volume of a reaction can be similarly deduced. For example, if the volume of the system is reduced, the equilibrium will shift in the direction that decreases the number of gas molecules. The enthalpy change for a reaction indicates how an increase in temperature affects the equilibrium: for an endothermic reaction, an increase in temperature shifts the equilibrium to the right; for an exothermic reaction, a temperature increase shifts the equilibrium to the left. Catalysts affect the speed at which equilibrium is reached but do not affect the magnitude of K.

KEY SKILLS

- Understand what is meant by chemical equilibrium and how it relates to reaction rates. (Section 16.1)

- Write the equilibrium constant expression for any reaction. (Section 16.2)

- Relate K_c and K_P. (Section 16.2)

- Understand the significance of the value of the equilibrium constant as it relates to the amounts of reactants and products in the equilibrium mixture. (Section 16.3)

- Write the equilibrium constant expression for a heterogeneous reaction. (Section 16.4)

- Calculate the equilibrium constant from concentration measurements. (Section 16.5)

- Predict the direction of a reaction given the equilibrium constant and the concentrations of reactants and products. (Section 16.6)

- Calculate an unknown concentration given the equilibrium constant and all other concentrations. (Section 16.6)

- Calculate the equilibrium concentrations given the equilibrium constant and the starting concentrations. (Section 16.6)

- Calculate ΔG under non-standard conditions. (Section 16.7)

- Relate $\Delta G°$ and equilibrium constant. (Section 16.7)

- Understand how changing the concentrations, volume, pressure or temperature of a system at equilibrium affects the equilibrium position. (Section 16.8)

KEY EQUATIONS

- The equilibrium constant expression for a general reaction of the type $a\,A + b\,B \rightleftharpoons d\,D + e\,E$; the concentrations are equilibrium concentrations only

$$K_c = \frac{[D]^d[E]^e}{[A]^a[B]^b}$$ [16.8]

- The equilibrium constant expression in terms of equilibrium partial pressures

$$K_P = \frac{(P_D)^d(P_E)^e}{(P_A)^a(P_B)^b}$$ [16.11]

- Relating the equilibrium constant based on pressures to the equilibrium constant based on concentration

$$K_P = K_c\,(RT)^{\Delta n}$$ [16.14]

- The reaction quotient. The concentrations are for any time during a reaction. If the concentrations are equilibrium concentrations, then $Q_c = K_c$

$$Q_c = \frac{[D]^d[E]^e}{[A]^a[B]^b}$$ [16.23]

- Relating the standard free energy change for a reaction and the equilibrium constant

$$\Delta G° = -RT \ln K$$ [16.25]

EXERCISES

VISUALISING CONCEPTS

16.1 (a) Based on the following energy profile, predict whether $k_f > k_r$ or $k_f < k_r$, where k_f is the reaction coefficient for the forward reaction and k_r is the reaction coefficient for the reverse reaction. **(b)** Using Equation 16.5, predict whether the equilibrium constant for the process is greater than 1 or less than 1. [Section 16.1]

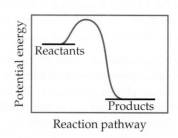

16.2 The following diagrams represent a hypothetical reaction A $\longrightarrow$ B, with A represented by red spheres and B represented by blue spheres. The sequence from left to right represents the system as time passes. Do the diagrams indicate that the system reaches an equilibrium state? Explain. [Sections 16.1 and 16.2]

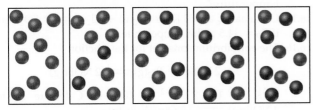

16.3 The following diagram represents a reaction shown going to completion. **(a)** Letting A = red spheres and B = blue spheres, write a balanced equation for the reaction. **(b)** Write the equilibrium constant expression for the reaction. **(c)** Assuming that all the molecules are in the gas phase, calculate Δn, the change in the number of gas molecules that accompanies the reaction. **(d)** How can you calculate K_P if you know K_c at a particular temperature? [Section 16.2]

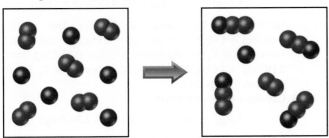

16.4 The reaction $A_2 + B_2 \rightleftharpoons 2\,AB$ has an equilibrium constant $K_c = 1.5$. The following diagrams represent reaction mixtures containing A_2 molecules (red), B_2 molecules (blue) and AB molecules. **(a)** Which reaction mixture is at equilibrium? **(b)** For those mixtures that are not at equilibrium, how will the reaction proceed to reach equilibrium? [Sections 16.5 and 16.6]

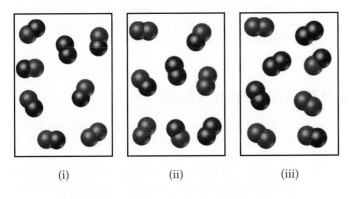

(i)　　　　　　(ii)　　　　　　(iii)

16.5 The reaction $A_2(g) + B(g) \rightleftharpoons A(g) + AB(g)$ has an equilibrium constant of $K_P = 2$. The diagram below shows a mixture containing A atoms (red), A_2 molecules and AB molecules (red and blue). How many B atoms should be added to the diagram if the system is at equilibrium? [Section 16.6]

16.6 The following diagram represents the equilibrium state for the reaction $A_2(g) + 2\,B(g) \rightleftharpoons 2\,AB(g)$. **(a)** Assuming the volume is 1 dm^3, calculate the equilibrium constant, K_c, for the reaction. **(b)** If the volume of the equilibrium mixture is decreased, will the number of AB molecules increase or decrease? [Sections 16.5 and 16.7]

16.7 The diagrams below represent equilibrium mixtures for the reaction $A_2 + B \rightleftharpoons A + AB$ at (i) 300 K and (ii) 500 K. The A atoms are red and the B atoms are blue. Is the reaction exothermic or endothermic? [Section 16.7]

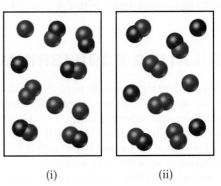

(i)　　　　　　　　　(ii)

EQUILIBRIUM; THE EQUILIBRIUM CONSTANT (Sections 16.1, 16.2, 16.3 and 16.4)

16.8 Suppose that the gas-phase reactions A $\longrightarrow$ B and B $\longrightarrow$ A are both elementary processes with rate constants of $3.8 \times 10^{-2}\,\text{s}^{-1}$ and $3.1 \times 10^{-1}\,\text{s}^{-1}$, respectively. **(a)** What is the value of the equilibrium constant for the equilibrium A(g) $\rightleftharpoons$ B(g)? **(b)** Which is greater at equilibrium, the partial pressure of A or the partial pressure of B? Explain.

16.9 Consider the reaction A + B $\rightleftharpoons$ C + D. Assume that both the forward reaction and the reverse reaction are elementary processes and that the value of the equilibrium constant is very large. **(a)** Which species predominate at equilibrium, reactants or products? **(b)** Which reaction has the larger rate constant, the forward or the reverse? Explain.

16.10 **(a)** What is the *law of mass action*? Illustrate the law by using the reaction NO(g) + Br$_2$(g) $\rightleftharpoons$ NOBr$_2$(g). **(b)** What is the difference between the *equilibrium constant expression* and the *equilibrium constant* for a given equilibrium? **(c)** Describe an experiment that could be used to determine the value of the equilibrium constant for the reaction in part (a).

16.11 **(a)** The mechanism for a certain reaction A + B $\rightleftharpoons$ C + D is unknown. Is it still possible to apply the law of mass action to the reaction? Explain. **(b)** Write the chemical reaction involved in the Haber process. Why is this reaction important to humanity? **(c)** Write the equilibrium constant expression for the reaction in part (b).

16.12 Write the expression for K_c for the following reactions. In each case, indicate whether the reaction is homogeneous or heterogeneous.
(a) 3 NO(g) $\rightleftharpoons$ N$_2$O(g) + NO$_2$(g)
(b) CH$_4$(g) + 2 H$_2$S(g) $\rightleftharpoons$ CS$_2$(g) + 4 H$_2$(g)
(c) Ni(CO)$_4$(g) $\rightleftharpoons$ Ni(s) + 4 CO(g)
(d) HF(aq) $\rightleftharpoons$ H$^+$(aq) + F$^-$(aq)
(e) 2 Ag(s) + Zn^{2+}(aq) $\rightleftharpoons$ 2 Ag$^+$(aq) + Zn(s)

16.13 Which of the following reactions lies to the right, favouring the formation of products, and which lies to the left, favouring formation of reactants?
(a) 2 NO(g) + O$_2$(g) $\rightleftharpoons$ 2 NO$_2$(g); $K_P = 5.0 \times 10^{12}$
(b) 2 HBr(g) $\rightleftharpoons$ H$_2$(g) + Br$_2$(g); $K_c = 5.8 \times 10^{-18}$

16.14 If $K_c = 0.042$ for PCl$_3$(g) + Cl$_2$(g) $\rightleftharpoons$ PCl$_5$(g) at 500 K, what is the value of K_P for this reaction at this temperature?

16.15 Calculate K_c at 303 K for SO$_2$(g) + Cl$_2$(g) $\rightleftharpoons$ SO$_2$Cl$_2$(g) if $K_P = 34.5$ at this temperature.

16.16 The equilibrium constant for the reaction

$$2\,\text{NO(g)} + \text{Br}_2\text{(g)} \rightleftharpoons 2\,\text{NOBr(g)}$$

is $K_c = 1.3 \times 10^{-2}$ at 1000 K. Calculate K_c for 2 NOBr(g) $\rightleftharpoons$ 2 NO(g) + Br$_2$(g).

16.17 The equilibrium constant for the reaction

$$2\,\text{NO(g)} + \text{O}_2\text{(g)} \rightleftharpoons 2\,\text{NO}_2\text{(g)}$$

is $K_P = 1.48 \times 10^4$ at 184 °C. **(a)** Calculate K_P for 2 NO$_2$(g) $\rightleftharpoons$ 2 NO(g) + O$_2$(g). **(b)** Does the equilibrium favour NO and O$_2$, or does it favour NO$_2$ at this temperature?

16.18 Consider the following equilibrium, for which $K_P = 0.0752$ at 480 °C:

$$2\,\text{Cl}_2\text{(g)} + 2\,\text{H}_2\text{O(g)} \rightleftharpoons 4\,\text{HCl(g)} + \text{O}_2\text{(g)}$$

(a) What is the value of K_P for the reaction 4 HCl(g) + O$_2$(g) $\rightleftharpoons$ 2 Cl$_2$(g) + 2 H$_2$O(g)? **(b)** What is the value of K_P for the reaction Cl$_2$(g) + H$_2$O(g) $\rightleftharpoons$ 2 HCl(g) + $\frac{1}{2}$ O$_2$(g)? **(c)** What is the value of K_c for the reaction in part (b)?

16.19 Consider the equilibrium

$$\text{N}_2\text{(g)} + \text{O}_2\text{(g)} + \text{Br}_2\text{(g)} \rightleftharpoons 2\,\text{NOBr(g)}$$

Calculate the equilibrium constant K_P for this reaction, given the following information (at 298 K):

$$2\,\text{NO(g)} + \text{Br}_2\text{(g)} \rightleftharpoons 2\,\text{NOBr(g)} \qquad K_c = 2.0$$
$$2\,\text{NO(g)} \rightleftharpoons \text{N}_2\text{(g)} + \text{O}_2\text{(g)} \qquad K_c = 2.1 \times 10^{30}$$

16.20 Mercury(I) oxide decomposes into elemental mercury and elemental oxygen: 2 Hg$_2$O(s) $\rightleftharpoons$ 4 Hg(l) + O$_2$(g). **(a)** Write the equilibrium constant expression for this reaction in terms of partial pressures. **(b)** Explain why we normally exclude pure solids and liquids from equilibrium constant expressions.

CALCULATING EQUILIBRIUM CONSTANTS (Section 16.5)

16.21 Gaseous hydrogen iodide is placed in a closed container at 425 °C, where it partially decomposes to hydrogen and iodine: 2 HI(g) $\rightleftharpoons$ H$_2$(g) + I$_2$(g). At equilibrium it is found that [HI] = 3.53×10^{-3} M, [H$_2$] = 4.79×10^{-4} M and [I$_2$] = 4.79×10^{-4} M. What is the value of K_c at this temperature?

16.22 Methanol (CH$_3$OH) is produced commercially by the catalysed reaction of carbon monoxide and hydrogen: CO(g) + 2 H$_2$(g) $\rightleftharpoons$ CH$_3$OH(g). An equilibrium mixture in a 2.00 dm^3 vessel is found to contain 0.0406 mol CH$_3$OH, 0.170 mol CO and 0.302 mol H$_2$ at 500 K. Calculate K_c at this temperature.

16.23 Phosphorus trichloride gas and chlorine gas react to form phosphorus pentachloride gas: PCl$_3$(g) + Cl$_2$(g) $\rightleftharpoons$

PCl$_5$(g). A gas vessel is charged with a mixture of PCl$_3$(g) and Cl$_2$(g), which is allowed to equilibrate at 450 K. At equilibrium the partial pressures of the three gases are $P_{\text{PCl}_3} = 0.124$ bar, $P_{\text{Cl}_3} = 0.157$ bar and $P_{\text{PCl}_5} = 1.30$ bar. **(a)** What is the value of K_P at this temperature? **(b)** Does the equilibrium favour reactants or products?

16.24 A mixture of 0.10 mol of NO, 0.050 mol of H$_2$ and 0.10 mol of H$_2$O is placed in a 1.0 dm^3 vessel at 300 K. The following equilibrium is established:

$$2\,\text{NO(g)} + 2\,\text{H}_2\text{(g)} \rightleftharpoons \text{N}_2\text{(g)} + 2\,\text{H}_2\text{O(g)}$$

At equilibrium [NO] = 0.062 M. **(a)** Calculate the equilibrium concentrations of H$_2$, N$_2$ and H$_2$O. **(b)** Calculate K_c.

16.25 A mixture of 1.374 g of H_2 and 70.31 g of Br_2 is heated in a 2.00 dm^3 vessel at 700 K. These substances react as follows:

$$H_2(g) + Br_2(g) \rightleftharpoons 2\,HBr(g)$$

At equilibrium the vessel is found to contain 0.566 g of H_2. **(a)** Calculate the equilibrium concentrations of H_2, Br_2 and HBr. **(b)** Calculate K_c.

16.26 A mixture of 0.2000 mol of CO_2, 0.1000 mol of H_2 and 0.1600 mol of H_2O is placed in a 2.000 dm^3 vessel. The following equilibrium is established at 500 K:

$$CO_2(g) + H_2(g) \rightleftharpoons CO(g) + H_2O(g)$$

(a) Calculate the initial partial pressures of CO_2, H_2 and H_2O. **(b)** At equilibrium $P_{H_2O} = 3.51$ bar. Calculate the equilibrium partial pressures of CO_2, H_2 and CO. **(c)** Calculate K_P for the reaction.

APPLICATIONS OF EQUILIBRIUM CONSTANTS (Section 16.6)

16.27 **(a)** How does a reaction quotient differ from an equilibrium constant? **(b)** If $Q_c < K_c$, in which direction will a reaction proceed in order to reach equilibrium? **(c)** What condition must be satisfied so that $Q_c = K_c$?

16.28 **(a)** How is a reaction quotient used to determine whether a system is at equilibrium? **(b)** If $Q_c > K_c$, how must the reaction proceed to reach equilibrium? **(c)** At the start of a certain reaction, only reactants are present; no products have been formed. What is the value of Q_c at this point in the reaction?

16.29 At 100 °C the equilibrium constant for the reaction $COCl_2(g) \rightleftharpoons CO(g) + Cl_2(g)$ has the value $K_c = 2.19 \times 10^{-10}$. Are the following mixtures of $COCl_2$, CO and Cl_2 at 100 °C at equilibrium? If not, indicate the direction that the reaction must proceed to achieve equilibrium. **(a)** $[COCl_2] = 2.00 \times 10^{-3}$ M, $[CO] = 3.3 \times 10^{-6}$ M, $[Cl_2] = 6.62 \times 10^{-6}$ M; **(b)** $[COCl_2] = 4.50 \times 10^{-2}$ M, $[CO] = 1.1 \times 10^{-7}$ M, $[Cl_2] = 2.25 \times 10^{-6}$ M; **(c)** $[COCl_2] = 0.0100$ M, $[CO] = [Cl_2] = 1.48 \times 10^{-6}$ M.

16.30 At 900 K the following reaction has $K_P = 0.345$:

$$2\,SO_2(g) + O_2(g) \rightleftharpoons 2\,SO_3(g)$$

In an equilibrium mixture the partial pressures of SO_2 and O_2 are 0.165 bar and 0.755 bar, respectively. What is the equilibrium partial pressure of SO_3 in the mixture?

16.31 **(a)** At 1285 °C the equilibrium constant for the reaction $Br_2(g) \rightleftharpoons 2\,Br(g)$ is $K_c = 1.04 \times 10^{-3}$. A 0.200 dm^3 vessel containing an equilibrium mixture of the gases has 0.245 g $Br_2(g)$ in it. What is the mass of Br(g) in the vessel? **(b)** For the reaction $H_2(g) + I_2(g) \rightleftharpoons 2\,HI(g)$,

$K_c = 55.3$ at 700 K. In a 2.00 dm^3 flask containing an equilibrium mixture of the three gases, there are 0.056 g H_2 and 4.36 g I_2. What is the mass of HI in the flask?

16.32 At 2000 °C the equilibrium constant for the reaction

$$2\,NO(g) \rightleftharpoons N_2(g) + O_2(g)$$

is $K_c = 2.4 \times 10^3$. If the initial concentration of NO is 0.200 M, what are the equilibrium concentrations of NO, N_2 and O_2?

16.33 For the equilibrium

$$Br_2(g) + Cl_2(g) \rightleftharpoons 2\,BrCl(g)$$

at 400 K, $K_c = 7.0$. If 0.30 mol of Br_2 and 0.30 mol Cl_2 are introduced into a 1.0 dm^3 container at 400 K, what will be the equilibrium concentrations of Br_2, Cl_2 and BrCl?

16.34 At 373 K, $K_P = 0.416$ for the equilibrium

$$2\,NOBr(g) \rightleftharpoons 2\,NO(g) + Br_2(g)$$

If the pressures of NOBr(g) and NO(g) are equal, what is the equilibrium pressure of $Br_2(g)$?

16.35 At 218 °C, $K_c = 1.2 \times 10^{-4}$ for the equilibrium

$$NH_4HS(s) \rightleftharpoons NH_3(g) + H_2S(g)$$

Calculate the equilibrium concentrations of NH_3 and H_2S if a sample of solid NH_4HS is placed in a closed vessel and decomposes until equilibrium is reached.

16.36 For the reaction $I_2(g) + Br_2(g) \rightleftharpoons 2\,IBr(g)$, $K_c = 280$ at 150 °C. Suppose that 0.500 mol IBr in a 1.00 dm^3 flask is allowed to reach equilibrium at 150 °C. What are the equilibrium concentrations of I_2, Br_2 and IBr?

EQUILIBRIUM CONSTANT AND FREE ENERGY (Section 16.7)

16.37 Consider the reaction $2\,NO_2(g) + O_2(g) \rightleftharpoons 2\,NO_2(g)$. **(a)** Using data from Appendix C, calculate $\Delta G°$ at 298 K. **(b)** Calculate ΔG at 298 K if the partial pressures of NO_2 and N_2O_4 are 0.40 bar and 1.60 bar, respectively.

16.38 Consider the reaction $3\,CH_4(g) \longrightarrow C_3H_8(g) + 2\,H_2(g)$. **(a)** Using data from Appendix C, calculate $\Delta G°$ at 298 K. **(b)** Calculate ΔG at 298 K if the reaction mixture consists of 40.0 bar of CH_4, 0.0100 bar of $C_3H_8(g)$ and 0.0180 bar of H_2.

16.39 Use data from Appendix C to calculate the equilibrium constant, K, at 298 K for each of the following reactions: **(a)** $H_2(g) + I_2(g) \rightleftharpoons 2\,HI(g)$ **(b)** $C_2H_5OH(g) \rightleftharpoons C_2H_4(g) + H_2O(g)$ **(c)** $3\,C_2H_2(g) \rightleftharpoons C_6H_6(g)$

16.40 Using data from Appendix C, write the equilibrium constant expression and calculate the value of the equilib-

rium constant for these reactions at 298 K: **(a)** $NaHCO_3(s) \rightleftharpoons NaOH(s) + CO_2(g)$ **(b)** $2\,HBr(g) + Cl_2(g) \rightleftharpoons 2\,HCl(g) + Br_2(g)$ **(c)** $2\,SO_2(g) + O_2(g) \rightleftharpoons 2\,SO_3(g)$

16.41 Consider the decomposition of barium carbonate:

$$BaCO_3(s) \rightleftharpoons BaO(s) + CO_2(g)$$

Using data from Appendix C, calculate the equilibrium pressure of CO_2 at **(a)** 298 K and **(b)** 1100 K.

16.42 Consider the reaction

$$PbCO_3(s) \rightleftharpoons PbO(s) + CO_2(g)$$

Using data in Appendix C, calculate the equilibrium pressure of CO_2 in the system at **(a)** 400 °C and **(b)** 180 °C.

16.43 The value of K_a for nitrous acid (HNO_2) at 25 °C is given in Appendix D. **(a)** Write the chemical equation for the equilibrium that corresponds to K_a. **(b)** By using the value of K_a, calculate $\Delta G°$ for the dissociation of nitrous acid in aqueous solution. **(c)** What is the value of ΔG at equilibrium? **(d)** What is the value of ΔG when $[H^+] = 5.0 \times 10^{-2}$ M, $[NO_2^-] = 6.0 \times 10^{-4}$ M and $[HNO_2] = 0.20$ M?

LE CHÂTELIER'S PRINCIPLE (Section 16.8)

16.44 Consider the following equilibrium, for which $\Delta H < 0$:

$$2 SO_2(g) + O_2(g) \rightleftharpoons 2 SO_3(g)$$

How will each of the following changes affect an equilibrium mixture of the three gases? **(a)** $O_2(g)$ is added to the system; **(b)** the reaction mixture is heated; **(c)** the volume of the reaction vessel is doubled; **(d)** a catalyst is added to the mixture; **(e)** the total pressure of the system is increased by adding a noble gas; **(f)** $SO_3(g)$ is removed from the system.

16.45 For the following reaction, $\Delta H° = 2816$ kJ:

$$6 CO_2(g) + 6 H_2O(l) \rightleftharpoons C_6H_{12}O_6(s) + 6 O_2(g)$$

How is the equilibrium yield of $C_6H_{12}O_6$ affected by **(a)** increasing P_{CO_2}, **(b)** increasing temperature, **(c)** removing CO_2, **(d)** decreasing the total pressure, **(e)** removing part of the $C_6H_{12}O_6$, **(f)** adding a catalyst?

16.46 How do the following changes affect the value of the equilibrium constant for a gas-phase exothermic reaction: **(a)** removal of a reactant or product, **(b)** decrease in the volume, **(c)** decrease in the temperature, **(d)** addition of a catalyst?

16.47 For a certain gas-phase reaction, the fraction of products in an equilibrium mixture is increased by increasing the temperature and increasing the volume of the reaction vessel. **(a)** What can you conclude about the reaction from the influence of temperature on the equilibrium? **(b)** What can you conclude from the influence of increasing the volume?

ADDITIONAL EXERCISES

16.48 Both the forward reaction and the reverse reaction in the following equilibrium are believed to be elementary steps:

$$CO(g) + Cl_2(g) \rightleftharpoons COCl(g) + Cl(g)$$

At 25 °C the rate constants for the forward and reverse reactions are 1.4×10^{-28} $M^{-1}s^{-1}$ and 9.3×10^{10} $M^{-1}s^{-1}$, respectively. **(a)** What is the value for the equilibrium constant at 25 °C? **(b)** Are reactants or products more plentiful at equilibrium?

16.49 A mixture of CH_4 and H_2O is passed over a nickel catalyst at 1000 K. The emerging gas is collected in a 5.00 dm^3 flask and is found to contain 8.62 g of CO, 2.60 g of H_2, 43.0 g of CH_4 and 48.4 g of H_2O. Assuming that equilibrium has been reached, calculate K_c and K_P for the reaction.

16.50 When 2.00 mol of SO_2Cl_2 is placed in a 2.00 dm^3 flask at 303 K, 56% of the SO_2Cl_2 decomposes to SO_2 and Cl_2:

$$SO_2Cl_2(g) \rightleftharpoons SO_2(g) + Cl_2(g)$$

Calculate K_c for this reaction at this temperature.

16.51 As shown in Table 16.3, the equilibrium constant for the reaction $N_2(g) + 3 H_2(g) \rightleftharpoons 2 NH_3(g)$ is $K_P = 4.34 \times 10^{-3}$ at 300 °C. Pure NH_3 is placed in a 1.00 dm^3 flask and allowed to reach equilibrium at this temperature. There are 1.05 g NH_3 in the equilibrium mixture. **(a)** What are the masses of N_2 and H_2 in the equilibrium mixture? **(b)** What was the initial mass of ammonia placed in the vessel? **(c)** What is the total pressure in the vessel?

16.52 For the equilibrium

$$2 IBr(g) \rightleftharpoons I_2(g) + Br_2(g)$$

$K_P = 8.5 \times 10^{-3}$ at 150 °C. If 0.025 bar of IBr is placed in a 2.0 dm^3 container, what is the partial pressure of this substance after equilibrium is reached?

16.53 Solid NH_4HS is introduced into an evacuated flask at 24 °C. The following reaction takes place:

$$NH_4HS(s) \rightleftharpoons NH_3(g) + H_2S(g)$$

At equilibrium the total pressure (for NH_3 and H_2S taken together) is 0.614 bar. What is K_P for this equilibrium at 24 °C?

16.54 A 0.831 g sample of SO_3 is placed in a 1.00 dm^3 container and heated to 1100 K. The SO_3 decomposes to SO_2 and O_2:

$$2 SO_3(g) \rightleftharpoons 2 SO_2(g) + O_2(g)$$

At equilibrium the total pressure in the container is 1.300 bar. Find the values of K_P and K_c for this reaction at 1100 K.

16.55 Nitric oxide (NO) reacts readily with chlorine gas as follows:

$$2 NO(g) + Cl_2(g) \rightleftharpoons 2 NOCl(g)$$

At 700 K the equilibrium constant K_P for this reaction is 0.26. Predict the behaviour of each of the following mixtures at this temperature: **(a)** $P_{NO} = 0.15$ bar, $P_{Cl_2} = 0.31$ bar, $P_{NOCl} = 0.11$ bar; **(b)** $P_{NO} = 0.12$ bar, $P_{Cl_2} = 0.10$ bar, $P_{NOCl} = 0.050$ bar; **(c)** $P_{NO} = 0.15$ bar, $P_{Cl_2} = 0.20$ bar, $P_{NOCl} = 5.10 \times 10^{-3}$ bar.

16.56 NiO is to be reduced to nickel metal in an industrial process by use of the reaction

$$NiO(s) + CO(g) \rightleftharpoons Ni(s) + CO_2(g)$$

At 1600 K the equilibrium constant for the reaction is $K_P = 6.0 \times 10^2$. If a CO pressure of 0.2 bar is to be employed in the furnace and total pressure never exceeds 1 bar, will reduction occur?

16.57 At 700 K the equilibrium constant for the reaction

$$CCl_4(g) \rightleftharpoons C(s) + 2 Cl_2(g)$$

is $K_P = 0.76$. A flask is charged with 2.0 bar of CCl_4, which then reaches equilibrium at 700 K. **(a)** What fraction of the CCl_4 is converted into C and Cl_2? **(b)** What are the partial pressures of CCl_4 and Cl_2 at equilibrium?

16.58 The reaction $PCl_3(g) + Cl_2(g) \rightleftharpoons PCl_5(g)$ has $K_P = 0.0870$ at 300 °C. A flask is charged with 0.50 bar PCl_3, 0.50 bar Cl_2 and 0.20 bar PCl_5 at this temperature.

(a) Use the reaction quotient to determine the direction the reaction must proceed in order to reach equilibrium. **(b)** Calculate the equilibrium partial pressures of the gases. **(c)** What effect will increasing the volume of the system have on the mole fraction of Cl_2 in the equilibrium mixture? **(d)** The reaction is exothermic. What effect will increasing the temperature of the system have on the mole fraction of Cl_2 in the equilibrium mixture?

16.59 An equilibrium mixture of H_2, I_2 and HI at 458 °C contains 0.112 mol H_2, 0.112 mol I_2 and 0.775 mol HI in a 5.00 dm^3 vessel. What are the equilibrium partial pressures when equilibrium is re-established following the addition of 0.100 mol of HI?

16.60 At 1200 K, the approximate temperature of vehicle exhaust gases K_P for the reaction

$$2 CO_2(g) \rightleftharpoons 2 CO(g) + O_2(g)$$

is about 1×10^{-13}. Assuming that the exhaust gas (total pressure 1 bar) contains 0.2% CO, 12% CO_2 and 3% O_2 by volume, is the system at equilibrium with respect to the above reaction? Based on your conclusion, would the CO concentration in the exhaust be decreased or increased by a catalyst that speeds up the reaction above?

16.61 Suppose that you worked at the Patent Office and a patent application came across your desk claiming that a newly developed catalyst was much superior to the Haber catalyst for ammonia synthesis because the catalyst led to much greater equilibrium conversion of N_2 and H_2 into NH_3 than the Haber catalyst under the same conditions. What would be your response?

INTEGRATIVE EXERCISES

16.62 The hypothetical reaction $A + B \rightleftharpoons C$ occurs in the forward direction in a single step. The energy profile of the reaction is shown in the drawing. **(a)** Is the forward or reverse reaction faster at equilibrium? **(b)** Would you expect the equilibrium to favour reactants or products? **(c)** In general, how would a catalyst affect the energy profile shown? **(d)** How would a catalyst affect the ratio of the rate constants for the forward and reverse reactions? **(e)** How would you expect the equilibrium constant of the reaction to change with increasing temperature?

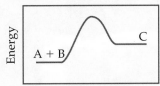

Reaction pathway

16.63 At 25 °C the reaction

$$NH_4HS(s) \rightleftharpoons NH_3(g) + H_2S(g)$$

has $K_P = 0.120$. A 5.00 dm^3 flask is charged with 0.300 g of pure $H_2S(g)$ at 25 °C. Solid NH_4HS is then added until there is excess unreacted solid remaining. **(a)** What is the

initial pressure of $H_2S(g)$ in the flask? **(b)** Why does no reaction occur until NH_4HS is added? **(c)** What are the partial pressures of NH_3 and H_2S at equilibrium? **(d)** What is the mole fraction of H_2S in the gas mixture at equilibrium? **(e)** What is the minimum mass, in grams, of NH_4HS that must be added to the flask to achieve equilibrium?

16.64 Write the equilibrium constant expression for the equilibrium

$$C(s) + CO_2(g) \rightleftharpoons 2 CO(g)$$

The table included below shows the relative mole percentages of $CO_2(g)$ and $CO(g)$ at a total pressure of 1 bar for several temperatures. Calculate the value of K_P at each temperature. Is the reaction exothermic or endothermic? Explain.

Temperature (°C)	CO_2 (mol %)	CO (mol %)
850	6.23	93.77
950	1.32	98.68
1050	0.37	99.63
1200	0.06	99.94

PHOTO/ART CREDITS

614 Dave Hunt/AAP Image; **616, 643** Richard Megna, Fundamental Photographs, NYC.

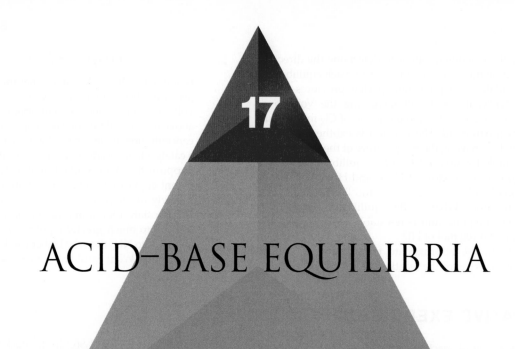

ACID–BASE EQUILIBRIA

17

Rhubarb contains varying amounts of weak acids such as malic and citric acid. In particular, the leaves contain oxalic acid, which is poisonous to animals and humans.

KEY CONCEPTS

17.1 ACIDS AND BASES: A BRIEF REVIEW
The field of acids and bases begins with a review of the *Arrhenius* definition of acids and bases.

17.2 BRØNSTED–LOWRY ACIDS AND BASES
We learn that a Brønsted–Lowry acid is a *proton donor* and a Brønsted–Lowry base is a *proton acceptor*. Two species that differ by the presence or absence of a proton are known as a *conjugate acid–base pair*.

17.3 THE AUTOIONISATION OF WATER
Water can undergo *autoionisation* producing small quantities of H_3O^+ and OH^- ions. The *equilibrium constant* for autoionisation, $K_w = [H_3O^+][OH^-]$, defines the relationship between H_3O^+ and OH^- concentrations in aqueous solutions.

17.4 THE pH SCALE
The pH scale is used to describe the acidity or basicity of an aqueous solution. Neutral solutions have a pH = 7, acidic solutions have a pH below 7 and basic solutions have a pH above 7.

17.5 STRONG ACIDS AND BASES
We learn that acids and bases are categorised as being either strong or weak electrolytes. *Strong* acids and bases are strong electrolytes, ionising or dissociating completely in aqueous solution. *Weak* acids and bases are weak electrolytes and ionise only partially.

17.6 WEAK ACIDS
We see that the ionisation of a weak acid in water is an equilibrium process with an equilibrium constant, K_a, that can be used to calculate the pH of a weak acid solution.

17.7 WEAK BASES
We see that the ionisation of a weak base in water is also an equilibrium process with an equilibrium constant, K_b, that can be used to calculate the pH of a weak base solution.

17.8 RELATIONSHIP BETWEEN K_a AND K_b
We learn that the relationship $K_a \times K_b = K_w$ indicates that the stronger an acid is, the weaker is its conjugate base.

17.9 ACID–BASE PROPERTIES OF SALT SOLUTIONS
We explore the fact that the ions of a soluble ionic compound can serve as Brønsted–Lowry acids or bases.

17.10 ACID–BASE BEHAVIOUR AND CHEMICAL STRUCTURE
We explore the relationship between chemical structure and acid–base behaviour.

17.11 LEWIS ACIDS AND BASES
We see that a Lewis acid is an *electron-pair acceptor* and a Lewis base is an *electron-pair donor*.

Taste is one of the five senses we use to experience the world around us. Receptors on the tongue are sensitive to chemical stimuli that lead to five basic taste sensations: sweet, sour, salty, bitter and umami (from the Japanese word for 'delicious' and triggered by the amino acid glutamic acid). The sensation of sour is a response to the presence of acids, and we associate a sour taste with certain fruits and vegetables because they contain acids. For example, lemons, limes and grapefruits contain citric acid ($C_6H_8O_7$), and green apples and grapes contain malic acid ($C_4H_6O_5$). The vegetable rhubarb is among the sourest of foods, so sour that eating a fresh stalk of rhubarb is sure to elicit a pucker on the first bite. The sour taste comes from the high acid content of the stalks. Acids and bases are important in numerous chemical processes that occur around us, from industrial processes to biological ones, from reactions in the laboratory to those in our environment. The time required for a metal object immersed in water to corrode, the ability of an aquatic environment to support fish and plant life, the fate of pollutants washed out of the air by rain, and even the rates of reactions that maintain our lives all depend critically upon the acidity or basicity of solutions. Indeed, an enormous amount of chemistry can be understood in terms of acid–base reactions.

We have encountered acids and bases many times in earlier discussions. For example, a portion of Chapter 4 focused on their reactions. But what makes a substance behave as an acid or as a base? In this chapter we re-examine acids and bases, taking a closer look at how they are identified and characterised. In doing so, we consider their behaviour not only in terms of their structure and bonding but also in terms of the chemical equilibria in which they participate.

17.1 | ACIDS AND BASES: A BRIEF REVIEW

From the earliest days of experimental chemistry, scientists have recognised acids and bases by their characteristic properties. Acids have a sour taste and cause certain dyes to change colour (for example, litmus turns red on contact with acids). The word *acid* comes from the Latin word *acidus*, meaning sour or tart. Bases, in contrast, have a bitter taste and feel slippery (soap is a good example). The word *base* comes from an old English meaning of the word, which is 'to bring low'. When bases are added to acids, they lower the amount of acid. Indeed, when acids and bases are mixed in certain proportions, their characteristic properties disappear altogether (co Section 4.3, 'Acids, Bases and Neutralisation Reactions').

co Review this on page 114

By 1830 it appeared that all acids contained hydrogen but not all hydrogen-containing substances are acids. In the 1880s the Swedish chemist Svante Arrhenius defined acids as substances that produce H^+ ions in water, and bases as substances that produce OH^- ions in water. Indeed, the properties of aqueous solutions of acids, such as sour taste, are due to $H^+(aq)$, whereas the properties of aqueous solutions of bases are due to $OH^-(aq)$. Over time, the Arrhenius concept of acids and bases came to be stated in the following way:

- An **acid** is a substance that, when dissolved in water, increases the concentration of H^+ ions.

- A **base** is a substance that, when dissolved in water, increases the concentration of OH^- ions.

Hydrogen chloride is an Arrhenius acid. Hydrogen chloride gas is highly soluble in water because of its chemical reaction with water, which produces hydrated H^+ and Cl^- ions:

$$HCl(g) \xrightarrow{\text{H}_2\text{O}} H^+(aq) + Cl^-(aq) \qquad [17.1]$$

The aqueous solution of HCl is known as hydrochloric acid. Concentrated hydrochloric acid is about 37% HCl by mass and is 12 M in HCl. Sodium hydroxide is an Arrhenius base. Because NaOH is an ionic compound, it dissociates into Na^+ and OH^- ions when it dissolves in water, thereby releasing OH^- ions into the solution.

> ⚠ **CONCEPT CHECK 1**
>
> What two ions are central to the Arrhenius definitions of acids and bases?

17.2 | BRØNSTED–LOWRY ACIDS AND BASES

The Arrhenius concept of acids and bases, although useful, has limitations. For one thing, it is restricted to aqueous solutions. In 1923 the Danish chemist Johannes Brønsted (1879–1947) and the English chemist Thomas Lowry (1874–1936) proposed independently a more general definition of acids and bases. Their concept is based on the fact that *acid–base reactions involve the transfer of H^+ ions from one substance to another.*

The H^+ Ion in Water

In Equation 17.1 hydrogen chloride is shown ionising in water to form $H^+(aq)$. *An H^+ ion is simply a proton with no surrounding valence electron.* This small, positively charged particle interacts strongly with the non-bonding electron pairs of water molecules to form hydrated hydrogen ions. For example, the interaction of a proton with one water molecule forms the **hydronium ion**, $H_3O^+(aq)$:

$$H^+ \;+\; :\!\overset{\displaystyle ..}{\underset{\displaystyle H}{O}}\!-\!H \;\longrightarrow\; \left[H\!-\!\overset{\displaystyle ..}{\underset{\displaystyle H}{O}}\!-\!H \right]^+$$

[17.2]

The formation of hydronium ions is one of the complex features of the interaction of the H^+ ion with liquid water. In fact, the H_3O^+ ion can form hydrogen bonds to additional H_2O molecules to generate larger clusters of hydrated hydrogen ions, such as $H_5O_2^+$ and $H_9O_4^+$ (▶ **FIGURE 17.1**).

Chemists use $H^+(aq)$ and $H_3O^+(aq)$ interchangeably to represent the same thing—namely the hydrated proton that is responsible for the characteristic properties of aqueous solutions of acids. We often use the $H^+(aq)$ ion for simplicity and convenience, as we did in Equation 17.1. The $H_3O^+(aq)$ ion, however, more closely represents reality.

Proton-Transfer Reactions

In the reaction that occurs when HCl dissolves in water, the HCl molecule transfers an H^+ ion (a proton) to a water molecule. Thus we can represent the reaction as occurring between an HCl molecule and a water molecule to form hydronium and chloride ions:

$$HCl(g) \;+\; H_2O(l) \;\longrightarrow\; Cl^-(aq) \;+\; H_3O^+(aq)$$

[17.3]

$$:\!\overset{\displaystyle ..}{\underset{\displaystyle ..}{Cl}}\!-\!H \;+\; :\!\overset{\displaystyle ..}{\underset{\displaystyle H}{O}}\!-\!H \;\longrightarrow\; :\!\overset{\displaystyle ..}{\underset{\displaystyle ..}{Cl}}\!:^- \;+\; \left[H\!-\!\overset{\displaystyle ..}{\underset{\displaystyle H}{O}}\!-\!H \right]^+$$

Acid   Base

The polar H_2O molecule promotes the ionisation of acids in water solution by accepting a proton to form H_3O^+.

Brønsted and Lowry proposed definitions of acids and bases in terms of their ability to transfer protons:

- An *acid* is a substance (molecule or ion) that donates a proton to another substance.
- A *base* is a substance that accepts a proton.

Thus when HCl dissolves in water (Equation 17.3), HCl acts as a **Brønsted–Lowry acid** (it donates a proton to H_2O) and H_2O acts as a **Brønsted–Lowry base** (it accepts a proton from HCl).

Because the emphasis in the Brønsted–Lowry concept is on proton transfer, the concept also applies to reactions that do not occur in aqueous solution. In the reaction between HCl and NH_3, for example, a proton is transferred from the acid HCl to the base NH_3:

$$:\!\overset{\displaystyle ..}{\underset{\displaystyle ..}{Cl}}\!-\!H \;+\; :\!\overset{\displaystyle }{\underset{\displaystyle H}{N}}\!-\!H \;\longrightarrow\; :\!\overset{\displaystyle ..}{\underset{\displaystyle ..}{Cl}}\!:^- \;+\; \left[H\!-\!\overset{\displaystyle H}{\underset{\displaystyle H}{N}}\!-\!H \right]^+$$

[17.4]

Acid   Base

FIGURE IT OUT

Which type of intermolecular force do the dotted lines in this figure represent?

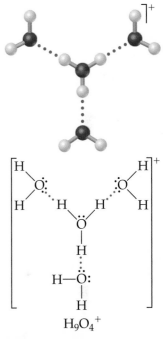

▲ FIGURE 17.1 **Ball-and-stick models and Lewis structures for two hydrated hydronium ions.**

Let us consider another example that compares the relationship between the Arrhenius definitions and the Brønsted–Lowry definitions of acids and bases—an aqueous solution of ammonia, in which the following equilibrium occurs:

$$NH_3(aq) + H_2O(l) \rightleftharpoons NH_4^+(aq) + OH^-(aq) \qquad [17.5]$$

$$\underset{\text{Base}}{}\ \underset{\text{Acid}}{}$$

Ammonia is an Arrhenius base because adding it to water leads to an increase in the concentration of $OH^-(aq)$. It is a Brønsted–Lowry base because it accepts a proton from H_2O. The H_2O molecule in Equation 17.5 acts as a Brønsted–Lowry acid because it donates a proton to the NH_3 molecule.

An acid and a base always work together to transfer a proton. In other words, a substance can function as an acid only if another substance simultaneously behaves as a base. To be a Brønsted–Lowry acid, a molecule or ion must have a hydrogen atom that it can lose as an H^+ ion. To be a Brønsted–Lowry base, a molecule or ion must have a non-bonding pair of electrons that it can use to bind the H^+ ion.

Some substances can act as an acid in one reaction and as a base in another. For example, H_2O is a Brønsted–Lowry base in its reaction with HCl (Equation 17.3) and a Brønsted–Lowry acid in its reaction with NH_3 (Equation 17.5). A substance that is capable of acting as either an acid or a base is called **amphiprotic**. An amphiprotic substance acts as a base when combined with something more strongly acidic than itself, and as an acid when combined with something more strongly basic than itself.

> ### CONCEPT CHECK 2
> In the forward reaction, which substance acts as the Brønsted–Lowry base: $HSO_4^-(aq) + NH_3(aq) \rightleftharpoons SO_4^{2-}(aq) + NH_4^+(aq)$?

Conjugate Acid–Base Pairs

In any acid–base equilibrium both the forward reaction (to the right) and the reverse reaction (to the left) involve proton transfers. For example, consider the reaction of an acid (HX) with water.

$$HX(aq) + H_2O(l) \rightleftharpoons X^-(aq) + H_3O^+(aq) \qquad [17.6]$$

In the forward reaction HX donates a proton to H_2O. Therefore HX is the Brønsted–Lowry acid and H_2O is the Brønsted–Lowry base. In the reverse reaction the H_3O^+ ion donates a proton to the X^- ion, so H_3O^+ is the acid and X^- is the base. When the acid HX donates a proton, it leaves behind a substance, X^-, which can act as a base. Likewise, when H_2O acts as a base, it generates H_3O^+, which can act as an acid.

An acid and a base such as HX and X^- that differ only in the presence or absence of a proton are called a **conjugate acid–base pair**. Every acid has a conjugate base, formed by removing a proton from the acid. For example, OH^- is the conjugate base of H_2O and X^- is the conjugate base of HX. Similarly, every base has associated with it a **conjugate acid**, formed by adding a proton to the base. Thus, H_3O^+ is the conjugate acid of H_2O and HX is the conjugate acid of X^-.

In any acid–base (proton-transfer) reaction we can identify two sets of conjugate acid–base pairs. For example, consider the reaction between nitrous acid (HNO_2) and water:

$$HNO_2(aq) + H_2O(l) \rightleftharpoons NO_2^-(aq) + H_3O^+(aq) \qquad [17.7]$$

$$\underset{\text{Acid}}{}\quad \underset{\text{Base}}{}\quad \underset{\substack{\text{Conjugate}\\\text{base}}}{}\quad \underset{\substack{\text{Conjugate}\\\text{acid}}}{}$$

Likewise for the reaction between ammonia (NH_3) and water (Equation 17.5), we have

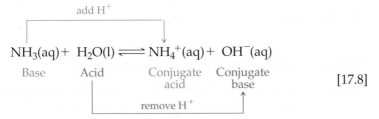

$$NH_3(aq) + H_2O(l) \rightleftharpoons NH_4^+(aq) + OH^-(aq)$$

Base   Acid   Conjugate   Conjugate   [17.8]
acid   base

SAMPLE EXERCISE 17.1   Identifying conjugate acids and bases

(a) What is the conjugate base of $HClO_4$, H_2S, PH_4^+, HCO_3^-? **(b)** What is the conjugate acid of CN^-, SO_4^{2-}, H_2O, HCO_3^-?

SOLUTION

Analyse We are asked to give the conjugate base for several acids and the conjugate acid for several bases.

Plan The conjugate base of a substance is simply the parent substance minus one proton, and the conjugate acid of a substance is the parent substance plus one proton.

Solve
(a) $HClO_4$ less one proton H^+ is ClO_4^-. The other conjugate bases are HS^-, PH_3 and CO_3^{2-}.

(b) CN^- plus one proton H^+ is HCN. The other conjugate acids are HSO_4^-, H_3O^+ and H_2CO_3. Notice that the hydrogen carbonate ion (HCO_3^-) is amphiprotic. It can act as either an acid or a base.

PRACTICE EXERCISE

Write the formula for the conjugate acid of each of the following: HSO_3^-, F^-, PO_4^{3-}, CO.

Answers: H_2SO_3, HF, HPO_4^{2-}, HCO^+

(See also Exercises 17.14, 17.15.)

SAMPLE EXERCISE 17.2   Writing equations for proton-transfer reactions

The hydrogen sulfite ion (HSO_3^-) is amphiprotic. **(a)** Write an equation for the reaction of HSO_3^- with water in which the ion acts as an acid. **(b)** Write an equation for the reaction of HSO_3^- with water in which the ion acts as a base. In both cases, identify the conjugate acid–base pairs.

SOLUTION

Analyse and Plan We are asked to write two equations representing reactions between HSO_3^- and water, one in which HSO_3^- should donate a proton to water, thereby acting as a Brønsted–Lowry acid, and one in which HSO_3^- should accept a proton from water, thereby acting as a base. We are also asked to identify the conjugate pairs in each equation.

Solve
(a) $HSO_3^-(aq) + H_2O(l) \rightleftharpoons SO_3^{2-}(aq) + H_3O^+(aq)$

The conjugate pairs in this equation are HSO_3^- (acid) and SO_3^{2-} (conjugate base); and H_2O (base) and H_3O^+ (conjugate acid).

(b) $HSO_3^-(aq) + H_2O(l) \rightleftharpoons H_2SO_3(aq) + OH^-(aq)$

The conjugate pairs in this equation are H_2O (acid) and OH^- (conjugate base); and HSO_3^- (base) and H_2SO_3 (conjugate acid).

PRACTICE EXERCISE

When lithium oxide (Li_2O) is dissolved in water the solution turns basic from the reaction of the oxide ion (O^{2-}) with water. Write the reaction that occurs, and identify the conjugate acid–base pairs.

Answer: $O^{2-}(aq) + H_2O(l) \rightleftharpoons OH^-(aq) + OH^-(aq)$. OH^- is the conjugate acid of the base O^{2-}. OH^- is also the conjugate base of the acid H_2O.

(See also Exercise 17.17.)

Relative Strengths of Acids and Bases

Some acids are better proton donors than others; likewise, some bases are better proton acceptors than others. If we arrange acids in order of their ability to donate a proton we find that the more easily a substance gives up a proton, the less easily its conjugate base accepts a proton. Similarly, the more easily a base accepts a proton, the less easily its conjugate acid gives up a proton. In other words, *the stronger an acid, the weaker is its conjugate base; the stronger a base, the weaker is its conjugate acid*. Thus if we know something about the strength of an acid (its ability to donate protons), we also know something about the strength of its conjugate base (its ability to accept protons).

The inverse relationship between the strengths of acids and the strengths of their conjugate bases is illustrated in ▼ **FIGURE 17.2**. Here we have grouped acids and bases into three broad categories based on their behaviour in water.

∞ Review this on page 113

1. The *strong acids* completely transfer their protons to water, leaving no undissociated molecules in solution (∞ Section 4.3, 'Acids, Bases and Neutralisation Reactions'). Their conjugate bases have a negligible tendency to be protonated (to abstract protons) in aqueous solution.

2. The *weak acids* only partially dissociate in aqueous solution and therefore exist in the solution as a mixture of acid molecules and their constituent ions. The conjugate bases of weak acids show a slight ability to remove protons from water. (*The conjugate bases of weak acids are weak bases.*)

3. The substances with *negligible acidity* are those such as CH_4 that contain hydrogen but do not demonstrate any acidic behaviour in water. Their conjugate bases are strong bases, reacting completely with water, abstracting protons to form OH^- ions.

FIGURE IT OUT

If O_2^- ions are added to water, what reaction, if any, occurs?

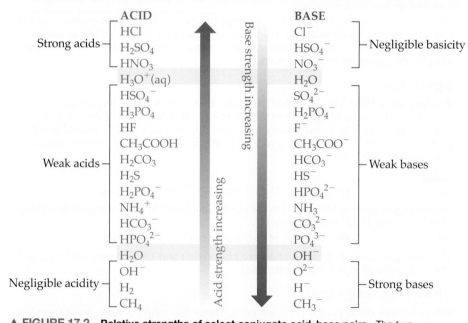

▲ **FIGURE 17.2** **Relative strengths of select conjugate acid–base pairs.** The two members of each pair are listed opposite each other in the two columns.

The ions $H^+(aq)$ and $OH^-(aq)$ are, respectively, the strongest possible acid and strongest possible base that can exist at equilibrium in aqueous solution. Stronger acids react with water to produce $H^+(aq)$ ions, and stronger bases react with water to produce $OH^-(aq)$ ions. This effect is known as the *levelling effect*.

> **⚠ CONCEPT CHECK 3**
>
> Using the three categories above, specify the strength of HNO_3 and the strength of its conjugate base, NO_3^-.

We can think of proton-transfer reactions as being governed by the relative abilities of two bases to abstract protons. For example, consider the proton transfer that occurs when an acid HX dissolves in water:

$$HX(aq) + H_2O(l) \rightleftharpoons H_3O^+(aq) + X^-(aq) \qquad [17.9]$$

If H_2O (the base in the forward reaction) is a stronger base than X^- (the conjugate base of HX), then H_2O will abstract the proton from HX to produce H_3O^+ and X^-. As a result, the equilibrium will lie to the right. This describes the behaviour of a strong acid in water. For example, when HCl dissolves in water the solution consists almost entirely of H_3O^+ and Cl^- ions with a negligible concentration of HCl molecules, as seen in Equation 17.3.

When X^- is a stronger base than H_2O, the equilibrium will lie to the left. This situation occurs when HX is a weak acid. For example, an aqueous solution of acetic acid (CH_3COOH) consists mainly of CH_3COOH molecules with only a relatively few H_3O^+ ions.

$$CH_3COOH(aq) + H_2O(l) \rightleftharpoons H_3O^+(aq) + CH_3COO^-(aq) \qquad [17.10]$$

CH_3COO^- is a stronger base than H_2O (Figure 17.2) and therefore abstracts the proton from H_3O^+. From these examples, we conclude that *in every acid–base reaction the position of the equilibrium favours transfer of the proton to the stronger base*. That is, the position of the equilibrium favours the reaction of the stronger acid and the stronger base to form the weaker acid and the weaker base. As a result, the equilibrium mixture contains more of the weaker acid and weaker base and less of the stronger acid and stronger base.

SAMPLE EXERCISE 17.3 | **Predicting the position of a proton-transfer equilibrium**

For the following proton-transfer reaction, use Figure 17.2 to predict whether the equilibrium lies predominantly to the left (that is, $K_c < 1$) or to the right ($K_c > 1$):

$$HSO_4^-(aq) + CO_3^{2-}(aq) \rightleftharpoons SO_4^{2-}(aq) + HCO_3^-(aq)$$

SOLUTION

Analyse We are asked to predict whether an equilibrium lies to the right, favouring products, or to the left, favouring reactants.

Plan This is a proton-transfer reaction, and the position of the equilibrium will favour the proton going to the stronger of two bases. The two bases in the equation are CO_3^{2-}, the base in the forward reaction, and SO_4^{2-}, the conjugate base of HSO_4^-. We can find the relative positions of these two bases in Figure 17.2 to determine which is the stronger base.

Solve CO_3^{2-} appears lower in the right-hand column in Figure 17.2 and is therefore a stronger base than SO_4^{2-}. CO_3^{2-}, therefore, will get the proton preferentially to become HCO_3^-, whereas SO_4^{2-} will remain mostly unprotonated. The resulting equilibrium will lie to the right, favouring products (that is, $K_c > 1$).

$$\underset{\text{Acid}}{HSO_4^-(aq)} + \underset{\text{Base}}{CO_3^{2-}(aq)} \rightleftharpoons \underset{\substack{\text{Conjugate} \\ \text{base}}}{SO_4^{2-}(aq)} + \underset{\substack{\text{Conjugate} \\ \text{acid}}}{HCO_3^-(aq)} \qquad K_c > 1$$

Comment Of the two acids in the equation, HSO_4^- and HCO_3^-, the stronger one gives up a proton whereas the weaker one retains its proton. Thus the equilibrium favours the direction in which the proton moves from the stronger acid and becomes bonded to the stronger base.

PRACTICE EXERCISE

For each of the following reactions, use Figure 17.2 to predict whether the equilibrium lies predominantly to the left or to the right:

(a) $HPO_4^{2-}(aq) + H_2O(l) \rightleftharpoons H_2PO_4^-(aq) + OH^-(aq)$

(b) $NH_4^+(aq) + OH^-(aq) \rightleftharpoons NH_3(aq) + H_2O(l)$

Answers: (a) left, (b) right

(See also Exercise 17.21.)

17.3 | THE AUTOIONISATION OF WATER

One of the most important chemical properties of water is its ability to act as either a Brønsted–Lowry acid or a Brønsted–Lowry base, depending on the circumstances. In the presence of an acid, water acts as a proton acceptor; in the presence of a base, water acts as a proton donor. In fact, one water molecule can donate a proton to another water molecule:

$$H_2O(l) \quad + \quad H_2O(l) \quad \rightleftharpoons \quad OH^-(aq) \quad + \quad H_3O^+(aq)$$

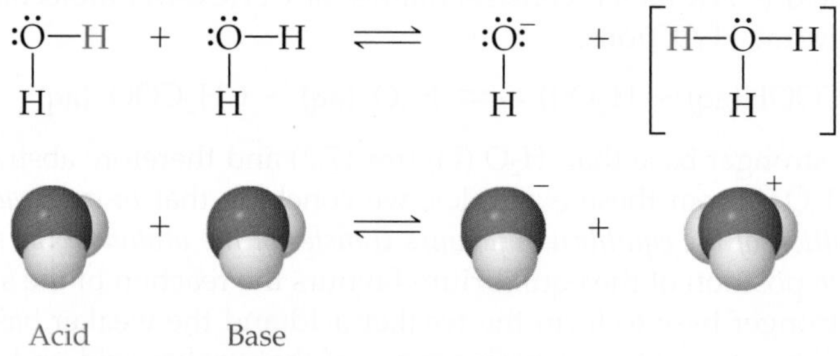

[17.11]

We call this process the **autoionisation** of water.

Because the forward and reverse reactions in Equation 17.11 are extremely rapid no water molecule remains ionised for long. At room temperature only about two out of every 10^9 water molecules are ionised at any given instant. Thus pure water consists almost entirely of H_2O molecules and is an extremely poor conductor of electricity. Nevertheless, the autoionisation of water is very important, as we will soon see.

The Ion Product of Water

Because the autoionisation of water (Equation 17.11) is an equilibrium process, we can write the following equilibrium constant expression for it:

$$K_c = [H_3O^+][OH^-] \qquad [17.12]$$

∞ Review this on page 627

The $[H_2O]$ is excluded from the equilibrium constant expression as it is a pure liquid (∞ Section 16.4, 'Heterogeneous Equilibria'). Because this equilibrium constant expression refers specifically to the autoionisation of water, we use the symbol K_w to denote the equilibrium constant, which we call the **ion product constant** for water. At 25 °C, K_w equals 1.0×10^{-14}. Thus we have

$$K_w = [H_3O^+][OH^-] = 1.0 \times 10^{-14} \text{ (at 25 °C)} \qquad [17.13]$$

Because we use $H^+(aq)$ and $H_3O(aq)^+$ interchangeably to represent the hydrated proton, the autoionisation reaction for water can also be written as

$$H_2O(l) \rightleftharpoons H^+(aq) + OH^-(aq) \qquad [17.14]$$

Similairly the expression for K_w can be written in terms of either H_3O^+ or H^+ and K_w has the same value in either case:

$$K_w = [H_3O^+][OH^-] = [H^+][OH^-] = 1.0 \times 10^{-14} \text{ (at 25 °C)} \quad [17.15]$$

This equilibrium constant expression and the value of K_w at 25 °C are extremely important, and you should commit them to memory.

A solution in which $[H^+] = [OH^-]$ is said to be *neutral*. In most solutions, H^+ and OH^- concentrations are not equal. As the concentration of one of these ions increases, the concentration of the other must decrease, so that the product of their concentrations equals 1.0×10^{-14}. In acidic solutions $[H^+]$ exceeds $[OH^-]$. In basic solutions $[OH^-]$ exceedes $[OH^+]$ (▼ FIGURE 17.3).

SAMPLE EXERCISE 17.4   **Calculating [H⁺] for pure water**

Calculate the values of $[H^+]$ and $[OH^-]$ in a neutral solution at 25 °C.

SOLUTION

Analyse We are asked to determine the concentrations of H^+ and OH^- ions in a neutral solution at 25 °C.

Plan We will use Equation 17.15 and the fact that, by definition, $[H^+] = [OH^-]$ in a neutral solution.

Solve We will represent the concentration of H^+ and OH^- in neutral solution with x. This gives

$$[H^+][OH^-] = (x)(x) = 1.0 \times 10^{-14}$$

$$x^2 = 1.0 \times 10^{-14}$$

$$x = 1.0 \times 10^{-7} \text{ M} = [H^+] = [OH^-]$$

In an acid solution $[H^+]$ is greater than 1.0×10^{-7} M; in a basic solution $[H^+]$ is less than 1.0×10^{-7} M.

PRACTICE EXERCISE

Indicate whether solutions with each of the following ion concentrations are neutral, acidic or basic: **(a)** $[H^+] = 4 \times 10^{-9}$ M; **(b)** $[OH^-] = 1 \times 10^{-7}$ M; **(c)** $[OH^-] = 7 \times 10^{-13}$ M.

Answers: **(a)** basic, **(b)** neutral, **(c)** acidic

(See also Exercise 17.24.)

Acidic solution	**Neutral solution**	**Basic solution**
Hydrochloric acid HCl(aq)	Water H₂O	Sodium hydroxide NaOH(aq)
$[H^+] > [OH^-]$	$[H^+] = [OH^-]$	$[H^+] < [OH^-]$
$[H^+][OH^-] = 1.0 \times 10^{-14}$	$[H^+][OH^-] = 1.0 \times 10^{-14}$	$[H^+][OH^-] = 1.0 \times 10^{-14}$

◀ FIGURE 17.3   Relative concentrations of H^+ and HO^- in aqueous solutions at 25 °C.

SAMPLE EXERCISE 17.5   **Calculating [H⁺] from [OH⁻]**

Calculate the concentration of H^+(aq) in **(a)** a solution in which $[OH^-]$ is 0.010 M, **(b)** a solution in which $[OH^-]$ is 1.8×10^{-9} M. *Note:* In this problem and all that follow, we assume, unless stated otherwise, that the temperature is 25 °C.

SOLUTION

Analyse We are asked to calculate the $[H^+]$ concentration in an aqueous solution where the hydroxide concentration is known.

Plan We can use the equilibrium constant expression for the autoionisation of water and the value of K_w to solve for each unknown concentration.

Solve

$$[H^+][OH^-] = 1.0 \times 10^{-14}$$

(a) Using Equation 17.15, we have:

$$[H^+] = \frac{1.0 \times 10^{-14}}{[OH^-]} = \frac{1.0 \times 10^{-14}}{0.010} = 1.0 \times 10^{-12} \text{ M}$$

This solution is basic because

$$[OH^-] > [H^+]$$

(b) In this instance,

$$[H^+] = \frac{1.0 \times 10^{-14}}{[OH^-]} = \frac{1.0 \times 10^{-14}}{1.8 \times 10^{-9}} = 5.6 \times 10^{-6} \text{ M}$$

This solution is acidic because

$$[H^+] > [OH^-]$$

PRACTICE EXERCISE

Calculate the concentration of OH^-(aq) in a solution in which **(a)** $[H^+] = 2 \times 10^{-6}$ M; **(b)** $[H^+] = [OH^-]$; **(c)** $[H^+] = 100 \times [OH^-]$.

Answers: **(a)** 5×10^{-9} M, **(b)** 1.0×10^{-7} M, **(c)** 1.0×10^{-8} M

(See also Exercise 17.25.)

17.4 | THE pH SCALE

The molar concentration of H^+(aq) in an aqueous solution is usually very small. So for convenience we usually express $[H^+]$ in terms of **pH**, which is the negative logarithm in base 10 of $[H^+]$.*

$$pH = -\log[H^+] \qquad [17.16]$$

If you need to review the use of logs see Appendix A.

We can use Equation 17.16 to calculate the pH of a neutral solution at 25 °C (that is, one in which $[H^+] = 1.0 \times 10^{-7}$ M):

$$pH = -\log(1.0 \times 10^{-7}) = -(-7.00) = 7.00$$

The pH of a neutral solution is 7.00 at 25 °C.

What happens to the pH of a solution as we make the solution acidic? An acidic solution is one in which $[H^+] > 1.0 \times 10^{-7}$ M. Because of the negative sign in Equation 17.16, *the pH decreases as $[H^+]$ increases*. For example, the pH of an acidic solution in which $[H^+] = 1.0 \times 10^{-3}$ M is

$$pH = -\log(1.0 \times 10^{-3}) = -(-3.00) = 3.00$$

At 25 °C the pH of an acidic solution is less than 7.00.

We can also calculate the pH of a basic solution, one in which $[OH^-] > 1.0 \times 10^{-7}$ M. Suppose $[OH^-] = 2.0 \times 10^{-3}$ M. We can use Equation 17.15 to calculate $[H^+]$ for this solution, and Equation 17.16 to calculate the pH:

$$[H^+] = \frac{K_w}{[OH^-]} = \frac{1.0 \times 10^{-14}}{2.0 \times 10^{-3}} = 5.0 \times 10^{-12} \text{ M}$$

$$pH = -\log(5.0 \times 10^{-12}) = 11.30$$

At 25 °C the pH of a basic solution is greater than 7.00.

The relationships between $[H^+]$, $[OH^-]$ and pH are summarised in ▶ TABLE 17.1.

 CONCEPT CHECK 4

Is it possible for a solution to have a negative pH? If so, would that pH signify a basic or acidic solution?

---

* Because $[H^+]$ and $[H_3O^+]$ are used interchangeably, you might also see pH defined as $-\log[H_3O^+]$.

TABLE 17.1 • Relationships between (H⁺), (OH⁻) and pH at 25 °C			
Solution type	**[H⁺] (M)**	**[OH⁻] (M)**	**pH**
Acidic	$>1.0 \times 10^{-7}$	$<1.0 \times 10^{-7}$	<7.00
Neutral	1.0×10^{-7}	1.0×10^{-7}	7.00
Basic	$<1.0 \times 10^{-7}$	$>1.0 \times 10^{-7}$	>7.00

You might think that, when $[H^+]$ is very small, as it is for some of the examples shown in Figure 17.4, it would be unimportant. Nothing is further from the truth. If $[H^+]$ is part of a kinetic rate law, then changing its concentration will change the rate (Section 15.3, 'Concentration and Rate Laws'). Thus if the rate law of a chemical reaction is first order in $[H^+]$, doubling its concentration will double the rate even if the change is merely from 1×10^{-7} M to 2×10^{-7} M. In biological systems many reactions involve proton transfers and have rates that depend on $[H^+]$. Because the speeds of these reactions are crucial, the pH of biological fluids must be maintained within narrow limits. For example, human blood has a normal pH range of 7.35 to 7.45. Illness and even death can result if the pH varies much from this narrow range.

Review this on page 572

SAMPLE EXERCISE 17.6   Calculating pH from [H⁺]

Calculate the pH values for the two solutions described in Sample Exercise 17.5.

SOLUTION

Analyse We are asked to determine the pH of aqueous solutions for which we have already calculated $[H^+]$.

Plan We can calculate pH using the defining equation, Equation 17.16.

Solve

(a) In the first instance we found $[H^+]$ to be 1.0×10^{-12} M.

$$pH = -\log(1.0 \times 10^{-12}) = -(-12.00) = 12.00$$

The rule for using significant figures with logs is that *the number of decimal places in the log equals the number of significant figures in the original number* (see Appendix A). Because 1.0×10^{-12} has two significant figures, the pH has two decimal places, 12.00.

(b) For the second solution, $[H^+] = 5.6 \times 10^{-6}$ M.

$$pH = -\log(5.6 \times 10^{-6}) = 5.25$$

Check Always check that your calculated pH is consistent with your expectations. If you expect an acidic solution, then pH < 7. You should be able to estimate fairly closely whether the pH is about, say, 5 or 8 from the magnitude of $[H^+]$.

PRACTICE EXERCISE

(a) In a sample of lemon juice $[H^+]$ is 3.8×10^{-4} M. What is the pH? **(b)** A commonly available window-cleaning solution has a $[H^+]$ of 5.3×10^{-9} M. What is the pH?

Answers: **(a)** 3.42, **(b)** 8.28

(See also Exercise 17.35.)

pOH and Other 'p' Scales

The negative log is also a convenient way of expressing the magnitudes of other small quantities. We use the convention that the negative log of a quantity is labelled 'p' (quantity). For example, we can express the concentration of OH^- as pOH:

$$pOH = -\log[OH^-] \qquad [17.17]$$

By taking the negative log of both sides of Equation 17.15,

$$-\log K_w = -\log[H^+] + (-\log[OH^-]) \qquad [17.18]$$

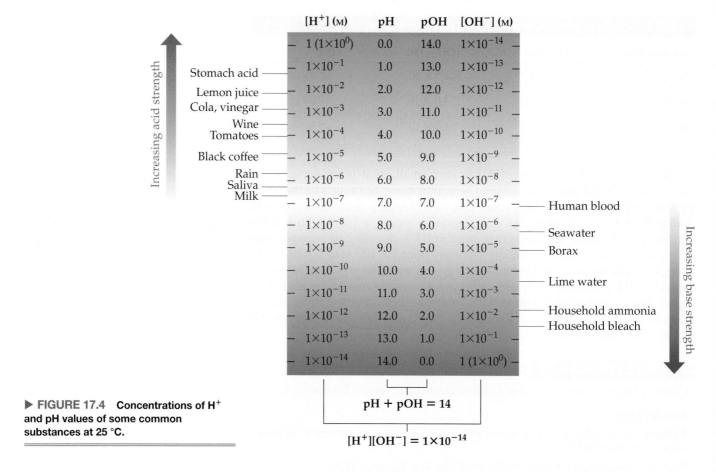

▶ FIGURE 17.4 Concentrations of H⁺ and pH values of some common substances at 25 °C.

we obtain the following useful expression:

$$\text{pH} + \text{pOH} = 14.00 \text{ (at 25 °C)} \qquad [17.19]$$

The pH and pOH values characteristic of a number of familiar solutions are shown in ▲ FIGURE 17.4. Notice that a change in [H⁺] by a factor of 10 causes the pH to change by 1. Thus the concentration of H⁺(aq) in a solution of pH 6 is 10 times the H⁺(aq) concentration in a solution of pH 7.

SAMPLE EXERCISE 17.7 Calculating [H⁺] from pOH

A sample of freshly pressed apple juice has a pOH of 10.24. Calculate [H⁺].

SOLUTION

Analyse We need to calculate [H⁺] from pOH.

Plan We will first use Equation 17.19, pH + pOH = 14.00, to calculate pH from pOH. Then we will use Equation 17.16 to determine the concentration of H⁺.

Solve From Equation 17.19, we have

$$\text{pH} = 14.00 - \text{pOH}$$
$$\text{pH} = 14.00 - 10.24 = 3.76$$

Next we use Equation 17.16:

$$\text{pH} = -\log[\text{H}^+] = 3.76$$

Thus,

$$\log[\text{H}^+] = -3.76$$

To find [H⁺], we need to determine the *antilogarithm* of −3.76. Scientific calculators have an antilogarithm function (sometimes labelled INV log or 10ˣ) that allows us to perform the calculation:

$$[\text{H}^+] = \text{antilog}(-3.76) = 10^{-3.76} = 1.7 \times 10^{-4} \text{ M}$$

Comment The number of significant figures in [H⁺] is two because the number of decimal places in the pH is two.

Check Because the pH is between 3.0 and 4.0, we know that $[H^+]$ will be between 1.0×10^{-3} M and 1.0×10^{-4} M. Our calculated $[H^+]$ falls within this estimated range.

PRACTICE EXERCISE

A solution formed by dissolving an antacid tablet has a pH of 9.18. Calculate $[OH^-]$.

Answer: $[OH^-] = 1.5 \times 10^{-5}$

(See also Exercises 17.30, 17.31.)

 CONCEPT CHECK 5

If the pOH for a solution is 3.00, what is the pH of the solution? Is the solution acidic or basic?

▲ **FIGURE 17.5 A digital pH meter.** The device is a millivoltmeter, and the electrodes immersed in a solution produce a voltage that depends on the pH of the solution.

Measuring pH

The pH of a solution can be measured with a *pH meter* (▶ **FIGURE 17.5**). A complete understanding of how this important device works requires a knowledge of electrochemistry, a subject we take up in Chapter 19. In brief, a pH meter consists of a pair of electrodes connected to a meter capable of measuring small voltages, on the order of millivolts. A voltage, which varies with pH, is generated when the electrodes are placed in a solution. This voltage is read by the meter, which is calibrated to give pH.

Although less precise, acid–base indicators can be used to measure pH. An acid–base indicator is a coloured substance that can exist in either an acid or a base form. The two forms have different colours. Thus the indicator has one colour at lower pH and another at higher pH. If you know the pH at which the indicator turns from one form to the other, you can determine whether a solution has a higher or lower pH than this value. Litmus, for example, changes colour in the vicinity of pH 7. The colour change, however, is not very sharp. Red litmus indicates a pH of about 5 or lower, and blue litmus indicates a pH of about 8 or higher.

Some common indicators are listed in ▼ **FIGURE 17.6**. The chart tells us, for instance, that methyl red changes colour over the pH interval from about 4.5

 FIGURE IT OUT

If a colourless solution turns pink when we add phenolphthalein, what can we conclude about the pH of the solution?

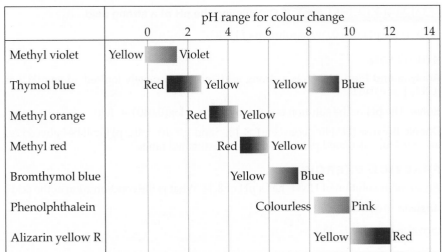

	pH range for colour change							
	0	2	4	6	8	10	12	14
Methyl violet	Yellow — Violet							
Thymol blue	Red — Yellow			Yellow — Blue				
Methyl orange		Red — Yellow						
Methyl red			Red — Yellow					
Bromthymol blue				Yellow — Blue				
Phenolphthalein					Colourless — Pink			
Alizarin yellow R						Yellow — Red		

▲ **FIGURE 17.6 pH ranges for common acid–base indicators.** Most indicators have a useful range of about 2 pH units.

to 6.0. Below pH 4.5 it is in the acid form, which is red. In the interval between 4.5 and 6.0, it is gradually converted to its basic form, which is yellow. Once the pH rises above 6 the conversion is complete, and the solution is yellow. This colour change, along with that of the indicators bromthymol blue and phenolphthalein, is shown in ▶ FIGURE 17.7. Paper tape impregnated with several indicators is widely used for determining approximate pH values.

17.5 | STRONG ACIDS AND BASES

The chemistry of an aqueous solution often depends critically on the pH of the solution. It is therefore important to examine how the pH of solutions relates to the concentrations of acids and bases. The simplest cases are those involving strong acids and strong bases. Strong acids and bases are *strong electrolytes*, existing in aqueous solution entirely as ions. There are relatively few common strong acids and bases (see Table 4.2).

Strong Acids

The seven most common strong acids include six monoprotic acids (HCl, HBr, HI, HNO_3, $HClO_3$ and $HClO_4$) and one diprotic acid (H_2SO_4). Their Lewis structures are shown in the margin. Nitric acid (HNO_3) exemplifies the behaviour of the monoprotic strong acids. For all practical purposes, an aqueous solution of HNO_3 consists entirely of H_3O^+ and NO_3^- ions.

$$HNO_3(aq) + H_2O(l) \longrightarrow H_3O^+(aq) + NO_3^-(aq) \text{ (complete ionisation) [17.20]}$$

We have not used equilibrium arrows for Equation 17.20 because the reaction lies entirely to the right, the side with the ions (∞ Section 4.1, 'General Properties of Aqueous Solutions'). As noted in Section 17.3, we use $H_3O^+(aq)$ and $H^+(aq)$ interchangeably to represent the hydrated proton in water. Thus we often simplify the equations for the ionisation reactions of acids as follows:

$$HNO_3(aq) \longrightarrow H^+(aq) + NO_3^-(aq)$$

In an aqueous solution of a strong acid, the acid is normally the only significant source of H^+ ions.* As a result, calculating the pH of a solution of a strong monoprotic acid is straightforward because $[H^+]$ equals the original concentration of acid. In a 0.20 M solution of $HNO_3(aq)$, for example, $[H^+]$ = $[NO_3^-]$ = 0.20 M and pH = 0.70. The situation with the diprotic acid, H_2SO_4, is more complex, as we see in Section 17.6.

H—Cl

H—Br

H—I

H—Ö—N—Ö: (with :O: double bonded to N)

H—Ö—Cl—Ö: (with :O: above Cl)

H—Ö—Cl—Ö: (with :O: above and :O: below Cl)

H—Ö—S—Ö—H (with :O: above and :O: below S)

∞ Review this on page 106

SAMPLE EXERCISE 17.8 Calculating the pH of a strong acid

What is the pH of a 0.040 M solution of $HClO_4$?

SOLUTION

Analyse and Plan $HClO_4$ is a strong acid and is completely ionised, giving $[H^+]$ = $[ClO_4^-]$ = 0.040 M.

Solve The pH of the solution is given by pH = $-\log(0.040)$ = 1.40.

Check Because $[H^+]$ lies between 1×10^{-2} and 1×10^{-1}, the pH will be between 2.0 and 1.0. Our calculated pH falls within the estimated range.

PRACTICE EXERCISE

An aqueous solution of HNO_3 has a pH of 2.34. What is the concentration of the acid?

Answer: 0.0046 M

(See also Exercise 17.35.)

---

* If the concentration of the acid is 10^{-6} M or less, we also need to consider the H^+ ions that result from the autoionisation of H_2O. Normally, the concentration of H^+ from H_2O is so small that it can be neglected.

Strong Bases

There are relatively few common strong bases. The most common soluble strong bases are the ionic hydroxides of the alkali metals (group 1) and the heavier alkaline earth metals (group 2), such as NaOH, KOH and $Ca(OH)_2$. These compounds completely dissociate into ions in aqueous solution. Thus a solution labelled 0.30 M NaOH consists of 0.30 M Na^+(aq) and 0.30 M OH^-(aq); there is essentially no undissociated NaOH. Because these strong bases dissociate entirely into ions in aqueous solution, calculating the pH of their solutions is also straightforward, as shown in Sample Exercise 17.9.

CONCEPT CHECK 6

Which solution has the higher pH, a 0.001 M solution of NaOH or a 0.001 M solution of $Ba(OH)_2$?

SAMPLE EXERCISE 17.9 **Calculating the pH of a strong base**

What is the pH of **(a)** a 0.028 M solution of NaOH, **(b)** a 0.0011 M solution of $Ca(OH)_2$?

SOLUTION

Analyse We are asked to calculate the pH of two solutions of strong bases.

Plan We can calculate each pH by either of two equivalent methods. First, we could use Equation 17.15 to calculate $[H^+]$ and then use Equation 17.16 to calculate the pH. Alternatively, we could use $[OH^-]$ to calculate pOH and then use Equation 17.19 to calculate the pH.

Solve
(a) NaOH dissociates in water to give one OH^- ion per formula unit. Therefore the OH^- concentration for the solution in (a) equals the stated concentration of NaOH, namely 0.028 M.
Method 1:

$$[H^+] = \frac{1.0 \times 10^{-14}}{0.028} = 3.57 \times 10^{-13} \text{ M} \quad pH = -\log(3.57 \times 10^{-13}) = 12.45$$

Method 2:

$$pOH = -\log(0.028) = 1.55 \qquad pH = 14.00 - pOH = 12.45$$

(b) $Ca(OH)_2$ is a strong base that dissociates in water to give two OH^- ions per formula unit. Thus the concentration of OH^-(aq) for the solution in part (b) is $2 \times (0.0011 \text{ M}) = 0.0022$ M.

Method 1:

$$[H^+] = \frac{1.0 \times 10^{-14}}{0.0022} = 4.55 \times 10^{-12} \text{ M} \quad pH = -\log(4.55 \times 10^{-12}) = 11.34$$

Method 2:

$$pOH = -\log(0.0022) = 2.66 \qquad pH = 14.00 - pOH = 11.34$$

PRACTICE EXERCISE

What is the concentration of a solution of **(a)** KOH, for which the pH is 11.89 and **(b)** $Ca(OH)_2$, for which the pH is 11.68?

Answers: **(a)** 7.8×10^{-3} M, **(b)** 2.4×10^{-3} M

(See also Exercise 17.36.)

Although all the hydroxides of the alkali metals (group 1) are strong electrolytes, LiOH, RbOH and CsOH are not commonly encountered in the laboratory. The hydroxides of the heavier alkaline earth metals, $Ca(OH)_2$, $Sr(OH)_2$ and $Ba(OH)_2$, are also strong electrolytes. They have limited solubilities, however, so they are used only when high solubility is not critical.

FIGURE IT OUT

Which of these indicators is best suited to distinguish between a solution that is slightly acidic and one that is slightly basic?

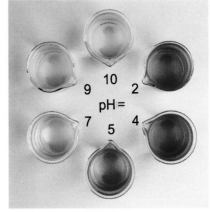

Methyl red

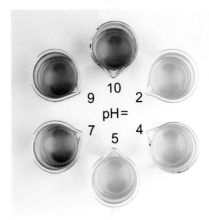

Bromthymol blue

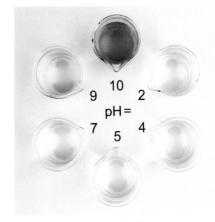

Phenolphthalein

▲ FIGURE 17.7 **Solutions containing three common acid–base indicators at various pH values.**

Strong basic solutions are also created by certain substances that react with water to form $OH^-(aq)$. The most common of these contain the oxide ion. Ionic metal oxides, especially Na_2O and CaO, are often used in industry when a strong base is needed. Each mole of O^{2-} reacts with water to form 2 mol of OH^-, leaving virtually no O^{2-} remaining in the solution:

$$O^{2-}(aq) + H_2O(l) \longrightarrow 2\,OH^-(aq) \qquad [17.21]$$

Thus a solution formed by dissolving 0.010 mol of $Na_2O(s)$ in enough water to form 1.0 dm^3 of solution will have $[OH^-] = 0.020$ M and a pH of 12.30.

Ionic hydrides and nitrides also react with H_2O to form OH^-:

$$H^-(aq) + H_2O(l) \longrightarrow H_2(g) + OH^-(aq) \qquad [17.22]$$

$$N^{3-}(aq) + 3\,H_2O(l) \longrightarrow NH_3(aq) + 3\,OH^-(aq) \qquad [17.23]$$

Because the anions O^{2-}, H^- and N^{3-} are stronger bases than OH^- (the conjugate base of H_2O), they are able to remove a proton from H_2O.

CONCEPT CHECK 7
The CH_3^- ion is the conjugate base of CH_4, and CH_4 shows no evidence of being an acid in water. What happens when CH_3^- is added to water?

17.6 | WEAK ACIDS

Most acidic substances are weak acids and are therefore only partially ionised in aqueous solution (▼ **FIGURE 17.8**). We can use the equilibrium constant for the ionisation reaction to express the extent to which a weak acid ionises. If we represent a general weak acid as HA, we can write the equation for its ionisation reaction in either of the following ways, depending on whether the hydrated proton is represented as $H_3O^+(aq)$ or $H^+(aq)$:

$$HA(aq) + H_2O(l) \rightleftharpoons H_3O^+(aq) + A^-(aq) \qquad [17.24]$$

or

$$HA(aq) \rightleftharpoons H^+(aq) + A^-(aq) \qquad [17.25]$$

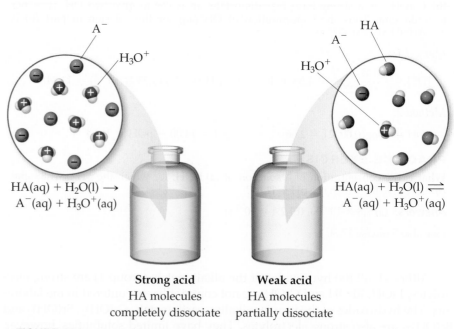

A^- H_3O^+

$HA(aq) + H_2O(l) \longrightarrow$
$A^-(aq) + H_3O^+(aq)$

HA A^- H_3O^+

$HA(aq) + H_2O(l) \rightleftharpoons$
$A^-(aq) + H_3O^+(aq)$

Strong acid
HA molecules
completely dissociate

Weak acid
HA molecules
partially dissociate

▲ **FIGURE 17.8** Species present in a solution of a strong acid and a weak acid.

Because $[H_2O]$ is the solvent, it is omitted from the equilibrium constant expression, which can be written as either

$$K_c = \frac{[H_3O^+][A^-]}{[HA]} \quad \text{or} \quad K_c = \frac{[H^+][A^-]}{[HA]}$$

As we did for the ion product constant for the autoionisation of water, we change the subscript on this equilibrium constant to indicate the type of equation to which it corresponds.

$$K_a = \frac{[H_3O^+][A^-]}{[HA]} \quad \text{or} \quad K_a = \frac{[H^+][A^-]}{[HA]} \qquad [17.26]$$

The subscript a on K_a denotes that it is an equilibrium constant for the ionisation of an acid, so K_a is called the **acid-dissociation constant**.

▼ TABLE 17.2 shows the names, structures and K_a values for several weak acids. A more complete listing is given in Appendix D. Many weak acids are organic compounds composed entirely of carbon, hydrogen and oxygen. These compounds usually contain some hydrogen atoms bonded to carbon atoms and some bonded to oxygen atoms. In almost all cases the hydrogen atoms bonded to carbon do not ionise in water; instead, the acidic behaviour of these compounds is due to the hydrogen atoms attached to oxygen atoms.

The magnitude of K_a indicates the tendency of the acid to ionise in water: *the larger the value of* K_a, *the stronger the acid*. Hydrofluoric acid (HF), for example, is the strongest acid listed in Table 17.2, and phenol (C_6H_5OH) is the weakest. Notice that K_a is typically less than 10^{-3}.

Calculating K_a from pH

To calculate either the K_a value for a weak acid or the pH of its solutions, we use many of the skills for solving equilibrium problems that we developed in Section 16.5 (∞ Section 16.5, 'Calculating Equilibrium Constants'). In many cases the small magnitude of K_a allows us to use approximations to simplify the problem. In doing these calculations, it is important to realise that proton-transfer reactions are generally very rapid. As a result, the measured or calculated pH for a weak acid almost always represents an equilibrium condition.

∞ Review this on page 629

TABLE 17.2 • Some weak acids in water at 25 °C*

Acid	Structural formula	Conjugate base	Equilibrium reaction	K_a
Hydrofluoric (HF)	H—F	F^-	$HF(aq) + H_2O(l) \rightleftharpoons H_3O^+(aq) + F^-(aq)$	6.8×10^{-4}
Nitrous (HNO_2)	H—O—N=O	NO_2^-	$HNO_2(aq) + H_2O(l) \rightleftharpoons H_3O^+(aq) + NO_2^-(aq)$	4.6×10^{-4}
Benzoic (C_6H_5COOH)	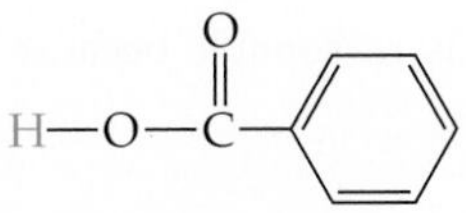 H—O—C(=O)—⬡	$C_6H_5OO^-$	$C_6H_5COOH(aq) + H_2O(l) \rightleftharpoons H_3O^+(aq) + C_6H_5COO^-(aq)$	6.5×10^{-5}
Acetic (CH_3COOH)	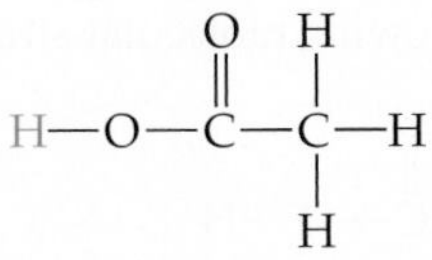 H—O—C(=O)—C(H)(H)—H	CH_3COO^-	$CH_3COOH(aq) + H_2O(l) \rightleftharpoons H_3O^+(aq) + CH_3COO^-(aq)$	1.8×10^{-5}
Hypochlorous (HClO)	H—O—Cl	ClO^-	$HClO(aq) + H_2O(l) \rightleftharpoons H_3O^+(aq) + ClO^-(aq)$	3.0×10^{-8}
Hydrocyanic (HCN)	H—C≡N	CN^-	$HCN(aq) + H_2O(l) \rightleftharpoons H_3O^+(aq) + CN^-(aq)$	4.9×10^{-10}
Phenol (C_6H_5OH)	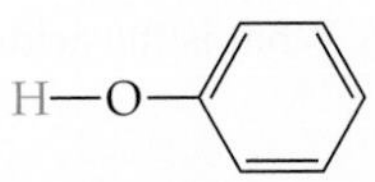 H—O—⬡	$C_6H_5O^-$	$C_6H_5OH(aq) + H_2O(l) \rightleftharpoons H_3O^+(aq) + C_6H_5O^-(aq)$	1.3×10^{-10}

* The proton that ionises is shown in blue.

SAMPLE EXERCISE 17.10 Calculating K_a from measured pH

A student prepared a 0.10 M solution of formic acid (HCOOH) and determined its pH. The pH at 25 °C was found to be 2.38. Calculate K_a for formic acid at this temperature.

SOLUTION

Analyse We are given the molar concentration of an aqueous solution of weak acid and the pH of the solution, and we are asked to determine the value of K_a for the acid.

Plan Although we are dealing specifically with the ionisation of a weak acid, this problem is very similar to the equilibrium problems we encountered in Chapter 16. We can solve this problem using the method first outlined in Sample Exercise 16.8, starting with the chemical reaction and a tabulation of initial and equilibrium concentrations.

Solve The first step in solving any equilibrium problem is to write the equation for the equilibrium reaction. The ionisation equilibrium for formic acid can be written as follows:

$$HCOOH(aq) \rightleftharpoons H^+(aq) + HCOO^-(aq)$$

The equilibrium constant expression is

$$K_a = \frac{[H^+][HCOO^-]}{[HCOOH]}$$

From the measured pH, we can calculate $[H^+]$:

$$pH = -\log[H^+] = 2.38$$

$$\log[H^+] = -2.38$$

$$[H^+] = 10^{-2.38} = 4.2 \times 10^{-3}\ M$$

We can do a little accounting to determine the concentrations of the species involved in the equilibrium. We imagine that the solution is initially 0.10 M in HCOOH molecules. We then consider the ionisation of the acid into H^+ and $HCOO^-$. For each HCOOH molecule that ionises, one H^+ ion and one $HCOO^-$ ion are produced in solution. Because the pH measurement indicates that $[H^+] = 4.2 \times 10^{-3}$ M at equilibrium, we can construct this table:

	HCOOH(aq) $\rightleftharpoons$	H$^+$(aq) +	HCOO$^-$(aq)
Initial	0.10 M	0	0
Change	-4.2×10^{-3} M	$+4.2 \times 10^{-3}$ M	$+4.2 \times 10^{-3}$ M
Equilibrium	$(0.10 - 4.2 \times 10^{-3})$ M	4.2×10^{-3} M	4.2×10^{-3} M

Notice that we have neglected the very small concentration of $H^+(aq)$ that is due to the autoionisation of H_2O. Notice also that the amount of HCOOH that ionises is very small compared with the initial concentration of the acid. To the number of significant figures we are using, the subtraction yields 0.10 M:

$$(0.10 - 4.2 \times 10^{-3})\ M = 0.10\ M$$

We can now insert the equilibrium concentrations into the expression for K_a:

$$K_a = \frac{(4.2 \times 10^{-3})(4.2 \times 10^{-3})}{0.10} = 1.8 \times 10^{-4}$$

Check The magnitude of our answer is reasonable because K_a for a weak acid is usually between 10^{-3} and 10^{-10}.

PRACTICE EXERCISE

Niacin, one of the B vitamins, has the following molecular structure:

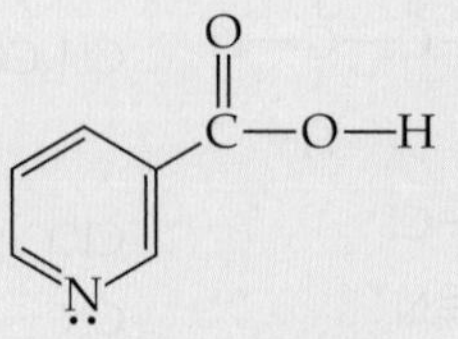

A 0.020 M solution of niacin has a pH of 3.26. What is the acid-dissociation constant, K_a, for niacin?

Answer: 1.5×10^{-5}

(See also Exercise 17.43.)

Percent Ionisation

We have seen that the magnitude of K_a indicates the strength of a weak acid. Another measure of acid strength is **percent ionisation**, defined as

$$\text{Percent ionisation} = \frac{\text{concentration ionised}}{\text{original concentration}} \times 100\% \qquad [17.27]$$

The stronger the acid, the greater the percent ionisation.

For any acid, the concentration of acid that ionises equals the concentration of $H^+(aq)$ that forms, assuming that H_2O autoionisation is negligible. Thus the percent ionisation for an acid HA is also given by

$$\text{Percent ionisation} = \frac{[H^+]_{\text{equilibrium}}}{[HA]_{\text{initial}}} \times 100\% \qquad [17.28]$$

For example, a 0.035 M solution of HNO_2 contains 3.7×10^{-3} M $H^+(aq)$ and its percent ionisation is

$$\text{Percent ionisation} = \frac{[H^+]_{\text{equilibrium}}}{[HNO_2]_{\text{initial}}} \times 100\% = \frac{3.7 \times 10^{-3} \text{ M}}{0.035 \text{ M}} \times 100\% = 11\%$$

SAMPLE EXERCISE 17.11 **Calculating percent ionisation**

As calculated in Sample Exercise 17.10, a 0.10 M solution of formic acid (HCOOH) contains 4.2×10^{-3} M $H^+(aq)$. Calculate the percentage of the acid that is ionised.

SOLUTION

Analyse We are given the molar concentration of an aqueous solution of weak acid and the equilibrium concentration of $H^+(aq)$ and asked to determine the percent ionisation of the acid.

Plan The percent ionisation is given by Equation 17.28.

Solve

$$\text{Percent ionisation} = \frac{[H^+]_{\text{equilibrium}}}{[HCOOH]_{\text{initial}}} \times 100\% = \frac{4.2 \times 10^{-3} \text{ M}}{0.10 \text{ M}} \times 100\% = 4.2\%$$

PRACTICE EXERCISE

A 0.020 M solution of niacin has a pH of 3.26. Calculate the percent ionisation of the niacin.

Answer: 2.7%

(See also Exercise 17.44.)

Using K_a to Calculate pH

Knowing the value of K_a and the initial concentration of the weak acid, we can calculate the concentration of $H^+(aq)$ in a solution of a weak acid. Let's calculate the pH of a 0.30 M solution of acetic acid (CH_3COOH), the weak acid responsible for the characteristic odour and acidity of vinegar, at 25 °C.

1. The first step is to write the ionisation equilibrium for acetic acid:

$$CH_3COOH(aq) \rightleftharpoons H^+(aq) + CH_3COO^-(aq) \qquad [17.29]$$

According to the structural formula of acetic acid, shown in Table 17.2, the hydrogen that ionises is the one attached to an oxygen atom. We write this hydrogen separate from the others in the formula to emphasise that only this one hydrogen is readily ionised.

2. The second step is to write the equilibrium constant expression and the value for the equilibrium constant. From Table 17.2, we have $K_a = 1.8 \times 10^{-5}$. Thus we can write the following:

$$K_a = \frac{[H^+][CH_3COO^-]}{[CH_3COOH]} = 1.8 \times 10^{-5} \qquad [17.30]$$

3. As the third step, we need to express the concentrations that are involved in the equilibrium reaction. This can be done with a little accounting, as described in Sample Exercise 17.10. Because we want to find the equilibrium value for $[H^+]$, let's call this quantity x. The concentration of acetic acid before any of it ionises is 0.30 M. The chemical equation tells us that, for each molecule of CH_3COOH that ionises, one $H^+(aq)$ and one $CH_3COO^-(aq)$ are formed. Consequently, if x moles per litre of $H^+(aq)$ form at equilibrium, x moles per litre of $CH_3COO^-(aq)$ must also form, and x moles per litre of CH_3COOH must be ionised. This gives rise to the following table with the equilibrium concentrations shown on the last line:

	$CH_3COOH(aq)$ $\rightleftharpoons$	$H^+(aq)$ +	$CH_3COO^-(aq)$
Initial	0.30 M	0	0
Change	$-x$ M	$+x$ M	$+x$ M
Equilibrium	$(0.30 - x)$ M	x M	x M

4. As the fourth step of the problem, we need to substitute the equilibrium concentrations into the equilibrium constant expression. The substitutions give the following equation:

$$K_a = \frac{[H^+][CH_3COO^-]}{[CH_3COOH]} = \frac{(x)(x)}{0.30 - x} = 1.8 \times 10^{-5} \qquad [17.31]$$

This expression leads to a quadratic equation in x, which we can solve by using the quadratic formula (Appendix A). We can simplify the problem, however, by noting that the value of K_a is quite small. As a result, we anticipate that the equilibrium will lie far to the left and that x will be very small compared with the initial concentration of acetic acid. Thus we *assume* that x is negligible compared with 0.30, so that $0.30 - x$ is essentially equal to 0.30.

$$0.30 - x \approx 0.30$$

We can (and should) check the validity of this assumption when we finish the problem. By using this assumption, Equation 17.31 now becomes

$$K_a = \frac{x^2}{0.30} = 1.8 \times 10^{-5}$$

Solving for x, we have

$$x^2 = (0.30)(1.8 \times 10^{-5}) = 5.4 \times 10^{-6}$$
$$x = \sqrt{5.4 \times 10^{-6}} = 2.3 \times 10^{-3}$$
$$[H^+] = x = 2.3 \times 10^{-3} \text{ M}$$
$$pH = -\log(2.3 \times 10^{-3}) = 2.64$$

We should now go back and check the validity of our simplifying assumption that $0.30 - x \approx 0.30$. The value of x we determined is so small that, for this number of significant figures, the assumption is entirely valid. We are thus satisfied that the assumption was a reasonable one to make. Because x represents the moles per litre of acetic acid that ionise, we see that, in this particular case, less than 1% of the acetic acid molecules ionise:

$$\text{Percent ionisation of } CH_3COOH = \frac{0.0023 \text{ M}}{0.30 \text{ M}} \times 100\% = 0.77\%$$

As a general rule, if the quantity x is more than about 5% of the initial value, it is better to use the quadratic formula. You should always check the validity of any simplifying assumptions when you have finished solving a problem.

 CONCEPT CHECK 8

> Why can we generally assume that the equilibrium concentration of a weak acid equals its initial concentration?

Finally, we can compare the pH value of this weak acid with a solution of a strong acid of the same concentration. The pH of the 0.30 M solution of acetic acid is 2.64. By comparison, the pH of a 0.30 M solution of a strong acid such as HCl is $-\log(0.30) = 0.52$. As expected, the pH of a solution of a weak acid is higher than that of a solution of a strong acid of the same molarity. (Remember, the higher the pH value, the *less* acidic the solution.)

SAMPLE EXERCISE 17.12 | Using K_a to calculate pH

Calculate the pH of a 0.20 M solution of HCN. K_a for HCN is 4.9×10^{-10}.

SOLUTION

Analyse We are given the molarity of a weak acid and are asked for the pH. From Table 17.2, K_a for HCN is 4.9×10^{-10}.

Plan We proceed as in the example just worked in the text, writing the chemical equation and constructing a table of initial and equilibrium concentrations in which the equilibrium concentration of H^+ is our unknown.

Solve Writing both the chemical equation for the ionisation reaction that forms $H^+(aq)$ and the equilibrium constant (K_a) expression for the reaction:

$$HCN(aq) \rightleftharpoons H^+(aq) + CN^-(aq)$$

$$K_a = \frac{[H^+][CN^-]}{[HCN]} = 4.9 \times 10^{-10}$$

Next, we tabulate the concentration of the species involved in the equilibrium reaction, letting $x = [H^+]$ at equilibrium:

	$HCN(aq)$	$\rightleftharpoons$ $H^+(aq)$	+ $CN^-(aq)$
Initial	0.20 M	0	0
Change	$-x$ M	$+x$ M	$+x$ M
Equilibrium	$(0.20 - x)$ M	x M	x M

Substituting the equilibrium concentrations from the table into the equilibrium constant expression yields

$$K_a = \frac{(x)(x)}{0.20 - x} = 4.9 \times 10^{-10}$$

We next make the simplifying approximation that x, the amount of acid that dissociates, is small compared with the initial concentration of acid; that is,

$$0.20 - x \approx 0.20$$

Thus

$$\frac{x^2}{0.20} = 4.9 \times 10^{-10}$$

Solving for x, we have

$$x^2 = (0.20)(4.9 \times 10^{-10}) = 0.98 \times 10^{-10}$$

$$x = \sqrt{0.98 \times 10^{-10}} = 9.9 \times 10^{-6}\,M = [H^+]$$

A concentration of 9.9×10^{-6} M is much smaller than 5% of 0.20, the initial HCN concentration. Our simplifying approximation is therefore appropriate. We now calculate the pH of the solution:

$$pH = -\log[H^+] = -\log(9.9 \times 10^{-6}) = 5.00$$

PRACTICE EXERCISE

The K_a for niacin (Practice Exercise 17.10) is 1.5×10^{-5}. What is the pH of a 0.010 M solution of niacin?

Answer: 3.42

(See also Exercise 17.53.)

The result obtained in Sample Exercise 17.12 is typical of the behaviour of weak acids; the concentration of $H^+(aq)$ is only a small fraction of the concentration of the acid in solution. Those properties of the acid solution that relate directly to the concentration of $H^+(aq)$, such as electrical conductivity and rate of reaction with an active metal, are much less evident for a solution of a weak acid than for a solution of a strong acid. ▼ **FIGURE 17.9** presents an experiment that demonstrates this difference with 1 M CH_3COOH and 1 M HCl. The 1 M CH_3COOH contains only 0.004 M $H^+(aq)$, whereas the 1 M HCl solution contains 1 M $H^+(aq)$. As a result, the reaction rate with the metal is much faster in the HCl solution.

As the concentration of a weak acid increases, the equilibrium concentration of $H^+(aq)$ increases, as expected. However, as shown in ▼ **FIGURE 17.10**, *the percent ionisation decreases as the concentration increases*. Thus the concentration of

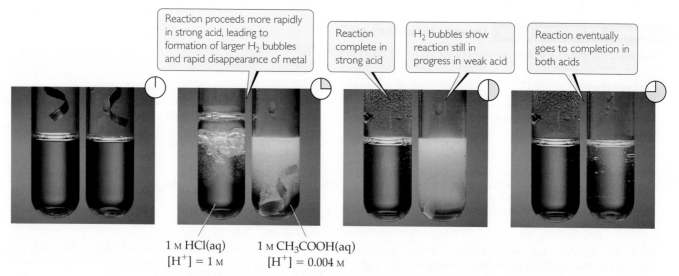

Reaction proceeds more rapidly in strong acid, leading to formation of larger H_2 bubbles and rapid disappearance of metal

Reaction complete in strong acid

H_2 bubbles show reaction still in progress in weak acid

Reaction eventually goes to completion in both acids

1 M HCl(aq)
$[H^+]$ = 1 M

1 M CH_3COOH(aq)
$[H^+]$ = 0.004 M

▲ **FIGURE 17.9** **Rates of the same reaction run in a weak acid and a strong acid.** The bubbles are H_2 gas, which, along with metal cations, is produced when a metal is oxidised by an acid (Section 4.4).

◢ FIGURE IT OUT

Is the trend observed in this graph consistent with Le Châtelier's principle? Explain.

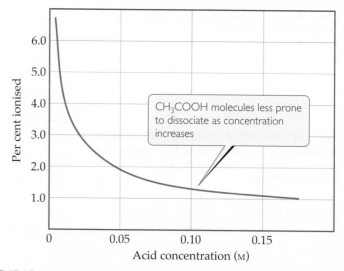

CH₃COOH molecules less prone to dissociate as concentration increases

▲ **FIGURE 17.10** **Effect of concentration on percent ionisation in an acetic acid solution.**

$H^+(aq)$ is not directly proportional to the concentration of the weak acid. For example, doubling the concentration of a weak acid does not double the concentration of $H^+(aq)$.

SAMPLE EXERCISE 17.13 | Using K_a to calculate percent ionisation

Calculate the percentage of HF molecules ionised in **(a)** a 0.10 M HF solution, **(b)** a 0.010 M HF solution.

SOLUTION

Analyse We are asked to calculate the percent ionisation of two HF solutions of different concentration. From Appendix D, we find $K_a = 6.8 \times 10^{-4}$.

Plan We approach this problem as we have previous equilibrium problems. We begin by writing the chemical equation for the equilibrium and tabulating the known and unknown concentrations of all species. We then substitute the equilibrium concentrations into the equilibrium constant expression and solve for the unknown concentration, that of H^+.

Solve
(a) The equilibrium reaction and equilibrium concentrations are as follows:

	$HF(aq)$	$\rightleftharpoons$	$H^+(aq)$	+	$F^-(aq)$
Initial	0.10 M		0		0
Change	$-x$ M		$+x$ M		$+x$ M
Equilibrium	$(0.10 - x)$ M		x M		x M

The equilibrium constant expression is

$$K_a = \frac{[H^+][F^-]}{[HF]} = \frac{(x)(x)}{0.10 - x} = 6.8 \times 10^{-4}$$

When we try solving this equation using the approximation $0.10 - x = 0.10$ (that is, by neglecting the concentration of acid that ionises in comparison with the initial concentration), we obtain

$$x = 8.2 \times 10^{-3}\,M$$

Because this value is greater than 5% of 0.10 M, we should work the problem without the approximation, using the quadratic formula. Rearranging our equation and writing it in standard quadratic form, we have

$$x^2 = (0.10 - x)(6.8 \times 10^{-4})$$
$$= 6.8 \times 10^{-5} - (6.8 \times 10^{-4})x$$
$$x^2 + (6.8 \times 10^{-4})x - 6.8 \times 10^{-5} = 0$$

This equation can be solved using the standard quadratic formula.

$$x = \frac{-b \pm \sqrt{b^2 - 4ac}}{2a}$$

Substituting the appropriate numbers gives

$$x = \frac{-6.8 \times 10^{-4} \pm \sqrt{(6.8 \times 10^{-4})^2 + 4(6.8 \times 10^{-5})}}{2}$$

$$= \frac{-6.8 \times 10^{-4} \pm 1.6 \times 10^{-2}}{2}$$

Of the two solutions, only the one that gives a positive value for x is chemically reasonable. Thus

$$x = [H^+] = [F^-] = 7.9 \times 10^{-3}\,M$$

From our result, we can calculate the percent of molecules ionised:

$$\text{Percent ionisation of HF} = \frac{\text{concentration ionised}}{\text{original concentration}} \times 100\%$$

$$= \frac{7.9 \times 10^{-3}\,M}{0.10\,M} \times 100\% = 7.9\%$$

(b) Proceeding similarly for the 0.010 M solution, we have

$$\frac{x^2}{0.010 - x} = 6.8 \times 10^{-4}$$

Solving the resultant quadratic expression, we obtain

$$x = [H^+] = [F^-] = 2.3 \times 10^{-3}\,M$$

The percentage of molecules ionised is

$$\frac{0.0023}{0.010} \times 100\% = 23\%$$

Comment Notice that if we do not use the quadratic formula to solve the problem properly, we calculate 8.2% ionisation for part (a) and 26% ionisation for part (b). Notice also that in diluting the solution by a factor of 10, the percentage of molecules ionised increases by a factor of 3. This result is what we would expect from Le Châtelier's principle (Section 16.8, 'Le Châtelier's Principle'). There are more 'particles' or reaction components on the right side of the equation than on the left. Dilution causes the reaction to shift in the direction of the larger number of particles because this counters the effect of the decreasing concentration of particles.

PRACTICE EXERCISE

In Practice Exercise 17.10, we found that the percent ionisation of niacin ($K_a = 1.5 \times 10^{-5}$) in a 0.020 M solution is 2.7%. Calculate the percentage of niacin molecules ionised in a solution that is **(a)** 0.010 M, **(b)** 1.0×10^{-3} M.

Answers: **(a)** 3.8%, **(b)** 12%

(See also Exercise 17.50.)

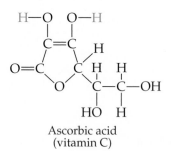

Ascorbic acid
(vitamin C)

Citric acid

Polyprotic Acids

Many acids have more than one ionisable H atom. These acids are known as **polyprotic acids**. For example, each of the H atoms in sulfurous acid (H_2SO_3) can ionise in successive steps:

$$H_2SO_3(aq) \rightleftharpoons H^+(aq) + HSO_3^-(aq) \qquad K_{a1} = 1.7 \times 10^{-2} \qquad [17.32]$$

$$HSO_3^-(aq) \rightleftharpoons H^+(aq) + SO_3^{2-}(aq) \qquad K_{a2}' = 6.4 \times 10^{-8} \qquad [17.33]$$

The acid-dissociation constants for these equilibria are labelled K_{a1} and K_{a2}. The numbers on the constants refer to the particular proton of the acid that is ionising. Thus K_{a2} always refers to the equilibrium involving removal of the second proton of a polyprotic acid.

In the preceding example K_{a2} is much smaller than K_{a1}. On the basis of electrostatic attractions, we would expect a positively charged proton to be lost more readily from the neutral H_2SO_3 molecule than from the negatively charged HSO_3^- ion. This observation is general: *It is always easier to remove the first proton from a polyprotic acid than to remove the second.* Similarly, for an acid with three ionisable protons, it is easier to remove the second proton than the third. Thus the K_a values become successively smaller as successive protons are removed.

CONCEPT CHECK 9

What is meant by the symbol K_{a3} for H_3PO_4?

The acid-dissociation constants for a few common polyprotic acids are listed in ▼ TABLE 17.3. A more complete list is given in Appendix D. The structures for ascorbic and citric acids are shown in the margin. Notice that

TABLE 17.3 • **Acid-dissociation constants of some common polyprotic acids**

Name	Formula	K_{a1}	K_{a2}	K_{a3}
Ascorbic	$C_6H_8O_6$	8.0×10^{-5}	1.6×10^{-12}	
Carbonic	H_2CO_3	4.3×10^{-7}	5.6×10^{-11}	
Citric	$C_6H_8O_7$	7.4×10^{-4}	1.7×10^{-5}	4.0×10^{-7}
Oxalic	$H_2C_2O_4$	5.9×10^{-2}	6.4×10^{-5}	
Phosphoric	H_3PO_4	7.5×10^{-3}	6.2×10^{-8}	4.2×10^{-13}
Sulfurous	H_2SO_3	1.7×10^{-2}	6.4×10^{-8}	
Sulfuric	H_2SO_4	Large	1.2×10^{-2}	
Tartaric	$C_4H_6O_6$	1.0×10^{-3}	4.6×10^{-5}	

the K_a values for successive losses of protons from these acids usually differ by a factor of at least 10^3. Notice also that the value of K_{a1} for sulfuric acid is listed simply as 'large'. Sulfuric acid is a strong acid with respect to the removal of the first proton. Thus the reaction for the first ionisation step lies completely to the right:

$$H_2SO_4(aq) \longrightarrow H^+(aq) + HSO_4^-(aq) \qquad \text{(complete ionisation)}$$

HSO_4^-, however, is a weak acid for which $K_{a2} = 1.2 \times 10^{-2}$.

Because K_{a1} is so much larger than subsequent dissociation constants for these polyprotic acids, almost all of the $H^+(aq)$ in the solution comes from the first ionisation reaction. As long as successive K_a values differ by a factor of 10^3 or more, it is possible to obtain a satisfactory estimate of the pH of polyprotic acid solutions by considering only K_{a1}.

SAMPLE EXERCISE 17.14   Calculating the pH of a polyprotic acid solution

The solubility of CO_2 in pure water at 25 °C and 0.1 bar pressure is 0.0037 M. The common practice is to assume that all the dissolved CO_2 is in the form of carbonic acid (H_2CO_3), which is produced by reaction between the CO_2 and H_2O:

$$CO_2(aq) + H_2O(l) \rightleftharpoons H_2CO_3(aq)$$

What is the pH of a 0.0037 M solution of H_2CO_3?

SOLUTION

Analyse We are asked to determine the pH of a 0.0037 M solution of a polyprotic acid.

Plan H_2CO_3 is a diprotic acid; the two acid-dissociation constants, K_{a1} and K_{a2} (Table 17.3), differ by more than a factor of 10^3. Consequently, the pH can be determined by considering only K_{a1}, thereby treating the acid as if it were a monoprotic acid.

Solve Proceeding as in Sample Exercises 17.12 and 17.13, we can write the equilibrium reaction and equilibrium concentrations as follows:

	$H_2CO_3(aq)$	$\rightleftharpoons$ $H^+(aq)$	$+$ $HCO_3^-(aq)$
Initial	0.0037 M	0	0
Change	$-x$ M	$+x$ M	$+x$ M
Equilibrium	$(0.0037 - x)$ M	x M	x M

The equilibrium constant expression is as follows:

$$K_{a1} = \frac{[H^+][HCO_3^-]}{[H_2CO_3]} = \frac{(x)(x)}{0.0037 - x} = 4.3 \times 10^{-7}$$

Solving this equation we get

$$x = 4.0 \times 10^{-5} \, M$$

Alternatively, because K_{a1} is small, we can make the simplifying approximation that x is small, so that

$$0.0037 - x \approx 0.0037$$

Thus

$$\frac{(x)(x)}{0.0037} = 4.3 \times 10^{-7}$$

Solving for x, we have

$$x^2 = (0.0037)(4.3 \times 10^{-7}) = 1.6 \times 10^{-9}$$
$$x = [H^+] = [HCO_3^-] = \sqrt{1.6 \times 10^{-9}} = 4.0 \times 10^{-5} \, M$$

The small value of x indicates that our simplifying assumption was justified. The pH is therefore

$$pH = -\log[H^+] = -\log(4.0 \times 10^{-5}) = 4.40$$

Comment If we were asked to solve for $[CO_3^{2-}]$, we would need to use K_{a2}. Let's illustrate that calculation. Using the values of $[HCO_3^-]$ and $[H^+]$ calculated above, and setting $[CO_3^{2-}] = y$, we have the following initial and equilibrium concentration values:

	$HCO_3^-(aq)$	$\rightleftharpoons$ $H^+(aq)$	$+$ $CO_3^{2-}(aq)$
Initial	4.0×10^{-5} M	4.0×10^{-5} M	0
Change	$-y$ M	$+y$ M	$+y$ M
Equilibrium	$(4.0 \times 10^{-5} - y)$ M	$(4.0 \times 10^{-5} + y)$ M	y M

Assuming that y is small compared with 4.0×10^{-5}, we have

$$K_{a2} = \frac{[H^+][CO_3^{2-}]}{[HCO_3^-]} = \frac{(4.0 \times 10^{-5})(y)}{4.0 \times 10^{-5}} = 5.6 \times 10^{-11}$$

$$y = 5.6 \times 10^{-11} \, \text{M} = [CO_3^{2-}]$$

The value calculated for y is indeed very small compared with 4.0×10^{-5}, showing that our assumption was justified. It also shows that the ionisation of HCO_3^- is negligible compared with that of H_2CO_3, as far as production of H^+ is concerned. However, it is the *only* source of CO_3^{2-}, which has a very low concentration in the solution. Our calculations thus tell us that in a solution of carbon dioxide in water, most of the CO_2 is in the form of CO_2 or H_2CO_3, a small fraction ionises to form H^+ and HCO_3^- and an even smaller fraction ionises to give CO_3^{2-}. Notice also that $[CO_3^{2-}]$ is numerically equal to K_{a2}.

PRACTICE EXERCISE

(a) Calculate the pH of a 0.020 M solution of oxalic acid $H_2C_2O_4$. (See Table 17.3 for K_{a1} and K_{a2}.)
(b) Calculate the concentration of oxalate ion, $[C_2O_4^{2-}]$, in this solution.

Answers: (a) pH = 1.80, (b) $[C_2O_4^{2-}] = 6.4 \times 10^{-5}$ M

(See also Exercise 17.53.)

17.7 | WEAK BASES

Many substances behave as weak bases in water. Weak bases react with water, abstracting protons from H_2O, thereby forming the conjugate acid of the base and OH^- ions.

$$B(aq) + H_2O(l) \rightleftharpoons HB^+(aq) + OH^-(aq) \qquad [17.34]$$

The equilibrium constant for the reaction can be written as

$$K_b = \frac{[BH^+][OH^-]}{[B]} \qquad [17.35]$$

The most commonly encountered weak base is ammonia.

$$NH_3(aq) + H_2O(l) \rightleftharpoons NH_4^+(aq) + OH^-(aq) \qquad [17.36]$$

The equilibrium constant expression for this reaction can be written as

$$K_b = \frac{[NH_4^+][OH^-]}{[NH_3]} \qquad [17.37]$$

Water is the solvent, so it is omitted from the equilibrium constant expression.

As with K_w and K_a, the subscript b denotes that this equilibrium constant refers to a particular type of reaction, namely the ionisation of a weak base in water. The constant K_b is called the **base-dissociation constant**. *The constant K_b always refers to the equilibrium in which a base reacts with H_2O to form the corresponding conjugate acid and OH^-.* ▶ **TABLE 17.4** lists the names, formulae, Lewis structures, equilibrium reactions and values of K_b for several weak bases in water. Appendix D includes a more extensive list. These bases contain one or more lone pairs of electrons because a lone pair is necessary to form the bond with H^+. Notice that in the neutral molecules in Table 17.4, the lone pairs are on nitrogen atoms. The other bases listed are anions derived from weak acids.

Types of Weak Bases

How can we recognise from a chemical formula whether a molecule or ion is able to behave as a weak base? Weak bases fall into two general categories. The first category contains neutral substances that have an atom with a non-bonding pair of electrons that can serve as a proton acceptor. Most of these bases, including all the uncharged bases listed in Table 17.4, contain a nitrogen atom. These substances include ammonia and a related class of compounds called **amines** (▼ **FIGURE 17.11**). In organic amines, one or more of the N–H bonds in NH_3 is replaced with a bond between N and C. Thus the replacement of one N–H bond in NH_3 with a N–CH$_3$ bond gives methylamine, NH_2CH_3

TABLE 17.4 • **Some weak bases and their aqueous solution equilibria**

Base	Lewis structure	Conjugate acid	Equilibrium reaction	K_b
Ammonia (NH_3)	$H\!-\!\overset{\displaystyle ..}{N}\!-\!H$ with H below	NH_4^+	$NH_3 + H_2O \rightleftharpoons NH_4^+ + OH^-$	1.8×10^{-5}
Pyridine (C_5H_5N)	(pyridine ring) $N:$	$C_5H_5NH^+$	$C_5H_5N + H_2O \rightleftharpoons C_5H_5NH^+ + OH^-$	1.7×10^{-9}
Hydroxylamine ($HONH_2$)	$H\!-\!\overset{\displaystyle ..}{N}\!-\!\overset{\displaystyle ..}{O}H$ with H below	$HONH_3^+$	$H_2NOH + H_2O \rightleftharpoons H_3NOH^+ + OH^-$	1.1×10^{-8}
Methylamine (CH_3NH_2)	$H\!-\!\overset{\displaystyle ..}{N}\!-\!CH_3$ with H below	$CH_3NH_3^+$	$CH_3NH_2 + H_2O \rightleftharpoons CH_3NH_3^+ + OH^-$	4.4×10^{-4}
Hydrosulfide ion (HS^-)	$\left[H\!-\!\overset{\displaystyle ..}{\underset{\displaystyle ..}{S}}: \right]^-$	H_2S	$HS^- + H_2O \rightleftharpoons H_2S + OH^-$	1.8×10^{-7}
Carbonate ion (CO_3^{2-})	$\left[\overset{\displaystyle :\ddot{O}:}{\underset{\displaystyle :\ddot{O}\!-\!C\!=\!\ddot{O}:}{} } \right]^{2-}$	HCO_3^-	$CO_3^{2-} + H_2O \rightleftharpoons HCO_3^- + OH^-$	1.8×10^{-4}
Hypochlorite ion (ClO^-)	$\left[:\ddot{Cl}\!-\!\ddot{O}: \right]^-$	$HClO$	$ClO^- + H_2O \rightleftharpoons HClO + OH^-$	3.3×10^{-7}

SAMPLE EXERCISE 17.15 | **Using K_b to calculate [OH$^-$]**

Calculate the concentration of OH$^-$ in a 0.15 M solution of NH_3.

SOLUTION

Analyse We are given the concentration of a weak base and asked to determine the concentration of OH$^-$.

Plan We will use essentially the same procedure here as used in solving problems involving the ionisation of weak acids, that is, write the chemical equation and tabulate initial and equilibrium concentrations.

Solve We first write the ionisation reaction and the corresponding equilibrium constant (K_b) expression:

$$NH_3(aq) + H_2O(l) \rightleftharpoons NH_4^+(aq) + OH^-(aq)$$

$$K_b = \frac{[NH_4^+][OH^-]}{[NH_3]} = 1.8 \times 10^{-5}$$

We then tabulate the equilibrium concentrations involved in the equilibrium:

	$NH_3(aq)$	+	$H_2O(l)$	$\rightleftharpoons$	$NH_4^+(aq)$	+	$OH^-(aq)$
Initial	0.15 M		—		0		0
Change	$-x$ M		—		$+x$ M		$+x$ M
Equilibrium	$(0.15 - x)$ M		—		x M		x M

Inserting these quantities into the equilibrium constant expression gives the following:

$$K_b = \frac{[NH_4^+][OH^-]}{[NH_3]} = \frac{(x)(x)}{0.15 - x} = 1.8 \times 10^{-5}$$

Because K_b is small, we can neglect the small amount of NH_3 that reacts with water, as compared with the total NH_3 concentration; that is, we can neglect x relative to 0.15 M. Then we have

$$\frac{x^2}{0.15} = 1.8 \times 10^{-5}$$

$$x^2 = (0.15)(1.8 \times 10^{-5}) = 2.7 \times 10^{-6}$$

$$x = [NH_4^+] = [OH^-] = \sqrt{2.7 \times 10^{-6}} = 1.6 \times 10^{-3} \text{ M}$$

Check The value obtained for x is only about 1% of the NH_3 concentration, 0.15 M. Therefore, neglecting x relative to 0.15 is justified.

PRACTICE EXERCISE

Which of the following compounds would produce the highest pH as a 0.05 M solution: pyridine, methylamine or nitrous acid?

Answer: Methylamine (because it has the largest K_b value).

(See also Exercise 17.58.)

▲ **FIGURE IT OUT**

When hydroxylamine acts as a base, which atom accepts the proton?

Ammonia
NH_3

Methylamine
CH_3NH_2

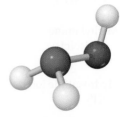
Hydroxylamine
NH_2OH

▲ **FIGURE 17.11** Structures of ammonia and two simple amines.

(usually written CH_3NH_2). Like NH_3, amines can abstract a proton from a water molecule by forming an additional N–H bond, as shown here for methylamine:

$$CH_3{-}\overset{\displaystyle ..}{\underset{\displaystyle H}{N}}{-}H(aq) \; + H_2O(l) \rightleftharpoons \left[CH_3{-}\overset{\displaystyle H}{\underset{\displaystyle H}{N}}{-}H \right]^{+} (aq) + OH^-(aq) \qquad [17.38]$$

The chemical formula for the conjugate acid of methylamine is usually written $CH_3NH_3{}^+$.

The second general category of weak bases consists of the anions of weak acids. In an aqueous solution of sodium hypochlorite (NaClO), for example, NaClO dissociates to give Na^+ and ClO^- ions. The Na^+ ion is always a spectator ion in acid–base reactions (Section 4.3, 'Acids, Bases and Neutralisation Reactions'). The ClO^- ion, however, is the conjugate base of a weak acid, hypochlorous acid. Consequently, the ClO^- ion acts as a weak base in water:

$$ClO^-(aq) + H_2O(l) \rightleftharpoons HClO(aq) + OH^-(aq) \qquad K_b = 3.3 \times 10^{-7} \qquad [17.39]$$

∞ Review this on page 112

SAMPLE EXERCISE 17.16 **Using pH to determine the concentration of a salt solution**

A solution made by adding solid sodium hypochlorite (NaClO) to enough water to make $2.00 \, dm^3$ of solution has a pH of 10.50. Using the information in Equation 17.39, calculate the number of moles of NaClO that were added to the water.

SOLUTION

Analyse We are given the pH of a weak base and asked to determine the concentration of the base.

Plan We will use essentially the same procedure here as used in solving problems involving the ionisation of weak acids, that is, write the chemical equation and tabulate initial and equilibrium concentrations.

Solve We can calculate $[OH^-]$ by using either Equation 17.15 or Equation 17.19; we use the latter method here:

$$pOH = 14.00 - pH = 14.00 - 10.50 = 3.50$$
$$[OH^-] = 10^{-3.50} = 3.16 \times 10^{-4} \, M$$

This concentration is high enough for us to assume that Equation 17.39 is the only source of OH^-; that is, we can neglect any OH^- produced by the autoionisation of H_2O. We now assume a value of x for the initial concentration of ClO^- and solve the equilibrium problem in the usual way.

$$ClO^-(aq) \quad + \quad H_2O(l) \rightleftharpoons HClO(aq) \quad + \quad OH^-(aq)$$

Initial	x M	—	0	0
Change	-3.16×10^{-4} M	—	$+3.16 \times 10^{-4}$ M	$+3.16 \times 10^{-4}$ M
Final	$(x - 3.16 \times 10^{-4})$ M	—	3.16×10^{-4} M	3.16×10^{-4} M

We now use the expression for the base-dissociation constant to solve for x:

$$K_b = \frac{[HClO][OH^-]}{[ClO^-]} = \frac{(3.16 \times 10^{-4})^2}{x - 3.16 \times 10^{-4}} = 3.3 \times 10^{-7}$$

Thus

$$x = \frac{(3.16 \times 10^{-4})^2}{3.3 \times 10^{-7}} + (3.16 \times 10^{-4}) = 0.30 \text{ M}$$

We say that the solution is 0.30 M in NaClO, even though some of the ClO^- ions have reacted with water. Because the solution is 0.30 M in NaClO and the total volume of solution is 2.00 dm^3, 0.60 mol of NaClO is the amount of the salt added to the water.

PRACTICE EXERCISE
A solution of NH_3 in water has a pH of 11.17. What is the molarity of the solution?

Answer: 0.12 M

(See also Exercise 17.59.)

17.8 | RELATIONSHIP BETWEEN K_a AND K_b

We've seen in a qualitative way that the stronger acids have the weaker conjugate bases. To see whether we can find a corresponding *quantitative* relationship, we consider the NH_4^+ and NH_3 conjugate acid–base pair. Each species reacts with water:

$$NH_4^+(aq) \rightleftharpoons NH_3(aq) + H^+(aq) \qquad [17.40]$$

$$NH_3(aq) + H_2O(l) \rightleftharpoons NH_4^+(aq) + OH^-(aq) \qquad [17.41]$$

Each of these equilibria is expressed by a characteristic dissociation constant:

$$K_a = \frac{[NH_3][H^+]}{[NH_4^+]}$$

$$K_b = \frac{[NH_4^+][OH^-]}{[NH_3]}$$

When Equations 17.40 and 17.41 are added together, the NH_4^+ and NH_3 species cancel and we are left with just the autoionisation of water.

$$NH_4^+(aq) \rightleftharpoons NH_3(aq) + H^+(aq)$$
$$\underline{NH_3(aq) + H_2O(l) \rightleftharpoons NH_4^+(aq) + OH^-(aq)}$$
$$H_2O(l) \rightleftharpoons H^+(aq) + OH^-(aq)$$

Recall that, when two equations are added to give a third, the equilibrium constant associated with the third equation equals the product of the equilibrium constants for the two equations added together (∞ Section 16.3, 'Interpreting and Working with Equilibrium Constants').

∞ Review this on page 626

Applying this rule to our present example, when we multiply K_a and K_b, we obtain the following:

$$K_a \times K_b = \left(\frac{[NH_3][H^+]}{[NH_4^+]} \right) \left(\frac{[NH_4^+][OH^-]}{[NH_3]} \right)$$

$$= [H^+][OH^-] = K_w$$

Thus the result of multiplying K_a times K_b is just the ion product constant for water, K_w (Equation 17.15). This is what we would expect, because adding Equations 17.40 and 17.41 gave us the autoionisation equilibrium for water, for which the equilibrium constant is K_w.

This relationship is so important that it should receive special attention. *The product of the acid-dissociation constant for an acid and the base-dissociation constant for its conjugate base is the ion product constant for water.*

$$K_a \times K_b = K_w \qquad [17.42]$$

As the strength of an acid increases (larger K_a), the strength of its conjugate base must decrease (smaller K_b) so that the product $K_a \times K_b$ equals 1.0×10^{-14} at 25 °C. The K_a and K_b data in ▼ TABLE 17.5 demonstrate this relationship.

By using Equation 17.42, we can calculate K_b for any weak base if we know K_a for its conjugate acid. Similarly, we can calculate K_a for a weak acid if we know K_b for its conjugate base. As a practical consequence, ionisation constants are often listed for only one member of a conjugate acid–base pair. For example, Appendix D does not contain K_b values for the anions of weak acids because these can be readily calculated from the tabulated K_a values for their conjugate acids.

If you look up the values for acid- or base-dissociation constants in a chemistry handbook, you may find them expressed as pK_a or pK_b (that is, as $-\log K_a$ or $-\log K_b$) (Section 17.4). Equation 17.42 can be written in terms of pK_a and pK_b by taking the negative log of both sides:

$$pK_a + pK_b = pK_w = 14.00 \quad \text{at 25 °C} \qquad [17.43]$$

TABLE 17.5 • Some conjugate acid–base pairs

Acid	K_a	Base	K_b
HNO_3	(Strong acid)	NO_3^-	(Negligible basicity)
HF	6.8×10^{-4}	F^-	1.5×10^{-11}
CH_3COOH	1.8×10^{-5}	CH_3COO^-	5.6×10^{-10}
H_2CO_3	4.3×10^{-7}	HCO_3^-	2.3×10^{-8}
NH_4^+	5.6×10^{-10}	NH_3	1.8×10^{-5}
HCO_3^-	5.6×10^{-11}	CO_3^{2-}	1.8×10^{-4}
OH^-	(Negligible acidity)	O^{2-}	(Strong base)

SAMPLE EXERCISE 17.17 | Calculating K_a or K_b for a conjugate acid–base pair

Calculate **(a)** the base-dissociation constant, K_b, for the fluoride ion (F^-); **(b)** the acid-dissociation constant, K_a, for the ammonium ion (NH_4^+).

SOLUTION

Analyse We are asked to determine the dissociation constants for F^-, the conjugate base of HF, and NH_4^+, the conjugate acid of NH_3.

Plan We can use the tabulated K values for HF and NH_3 and the relationship between K_a and K_b to calculate the ionisation constants for their conjugates, F^- and NH_4^+.

Solve

(a) K_a for the weak acid, HF, is given in Table 17.5 as $K_a = 6.8 \times 10^{-4}$. We can use Equation 17.42 to calculate K_b for the conjugate base, F^-:

$$K_b = \frac{K_w}{K_a} = \frac{1.0 \times 10^{-14}}{6.8 \times 10^{-4}} = 1.5 \times 10^{-11}$$

(b) K_b for NH_3 is listed in Table 17.5 as $K_b = 1.8 \times 10^{-5}$. Using Equation 17.42, we can calculate K_a for the conjugate acid, NH_4^+:

$$K_a = \frac{K_w}{K_b} = \frac{1.0 \times 10^{-14}}{1.8 \times 10^{-5}} = 5.6 \times 10^{-10}$$

PRACTICE EXERCISE

(a) Which of the following anions has the largest base-dissociation constant: NO_2^-, PO_4^{3-} or N_3^-? **(b)** The base quinoline has the following structure:

Its conjugate acid is listed in handbooks as having a pK_a of 4.90. What is the base-dissociation constant for quinoline?

Answers: **(a)** PO_4^{3-} ($K_b = 2.4 \times 10^{-2}$), **(b)** 7.9×10^{-10}

(See also Exercise 17.62.)

MY WORLD OF CHEMISTRY

AMINES AND AMINE HYDROCHLORIDES

Many low-molecular-weight amines have a fishy odour. Amines and NH_3 are produced by the anaerobic (absence of O_2) decomposition of dead animal or plant matter. Two such amines with very disagreeable odours are $H_2N(CH_2)_4NH_2$, *putrescine*, and $H_2N(CH_2)_5NH_2$, *cadaverine*.

Many drugs, including quinine, codeine, caffeine and amphetamine, are amines. Like other amines, these substances are weak bases; the amine nitrogen is readily protonated upon treatment with an acid. The resulting products are called *acid salts*. If we use A as the abbreviation for an amine, the acid salt formed by reaction with hydrochloric acid can be written AH^+Cl^-. It can also be written as $A \cdot HCl$ and referred to as a hydrochloride. Amphetamine hydrochloride, for example, is the acid salt formed by treating amphetamine with HCl:

Amphetamine

Amphetamine hydrochloride

Acid salts are much less volatile, more stable and generally more water soluble than the corresponding amines. For this reason, many drugs that are amines are sold and administered as acid salts. Some examples of over-the-counter medications that contain amine hydrochlorides as active ingredients are shown in ▼ FIGURE 17.12.

▲ FIGURE 17.12 **Some over-the-counter medications in which an amine hydrochloride is a major active ingredient.**

17.9 | ACID–BASE PROPERTIES OF SALT SOLUTIONS

Even before you began this chapter, you were undoubtedly aware of many substances that are acidic, such as HNO_3, HCl and H_2SO_4, and others that are basic, such as NaOH and NH_3. However, our recent discussions have indicated that ions can also exhibit acidic or basic properties. For example, we calculated K_a for NH_4^+ and K_b for F^- in Sample Exercise 17.17. Such behaviour implies that salt solutions can be acidic or basic. Before discussing acids and bases further, let's examine the way dissolved salts can affect pH.

We can assume that, when salts dissolve in water, they are completely dissociated; nearly all salts are strong electrolytes. Consequently, the acid–base properties of salt solutions are due to the behaviour of their constituent cations and anions. Many ions are able to react with water to generate $H^+(aq)$ or $OH^-(aq)$. This type of reaction is often called **hydrolysis**. The pH of an aqueous salt solution can be predicted qualitatively by considering the salt's cations and anions.

An Anion's Ability to React with Water

In general, an anion X^- in solution can be considered the conjugate base of an acid. For example, Cl^- is the conjugate base of HCl, and CH_3COO^- is the conjugate base of CH_3COOH. Whether or not an anion reacts with water to produce hydroxide ions depends on the strength of the anion's conjugate acid. To identify the acid and assess its strength, we add a proton to the anion's formula. If the acid HX determined in this way is one of the seven strong acids listed in Section 17.5, the anion has a negligible tendency to abstract protons from water and does not affect the pH of the solution. The presence of Cl^- in an aqueous solution, for example, does not result in the production of any OH^- and does not affect the pH. Thus Cl^- is always a spectator ion in acid–base chemistry.

If HX is *not* one of the seven strong acids, it is a weak acid. In this case, the conjugate base X^- is a weak base and it reacts to a small extent with water to produce the weak acid and hydroxide ions:

$$X^-(aq) + H_2O(l) \rightleftharpoons HX(aq) + OH^-(aq) \qquad [17.44]$$

The OH^- ion generated in this way increases the pH of the solution, making it basic. Acetate ion, for example, being the conjugate base of a weak acid, reacts with water to produce acetic acid and hydroxide ions, thereby increasing the pH of the solution:

$$CH_3COO^-(aq) + H_2O(l) \rightleftharpoons CH_3COOH(aq) + OH^-(aq) \qquad [17.45]$$

> ### CONCEPT CHECK 10
> What effect will NO_3^- ions have on the pH of a solution? What effect will CO_3^{2-} ions have?

The situation is more complicated for salts containing anions that have ionisable protons, such as HSO_3^-. These salts are amphiprotic (Section 17.2), and how they behave in water is determined by the relative magnitudes of K_a and K_b for the ion, as will be shown in Sample Exercise 17.19. If $K_a > K_b$, the ion causes the solution to be acidic. If $K_b > K_a$, the solution is made basic by the ion.

A Cation's Ability to React with Water

Polyatomic cations containing one or more protons can be considered the conjugate acids of weak bases. The NH_4^+ ion, for example, is the conjugate acid of the weak base NH_3. Thus, NH_4^+ is a weak acid and will donate a proton to water, producing hydronium ions and thereby lowering the pH:

$$NH_4^+(aq) + H_2O(l) \rightleftharpoons NH_3(aq) + H_3O^+(aq) \qquad [17.46]$$

Many metal ions react with water to decrease the pH of an aqueous solution. This effect is most pronounced for small, highly charged cations like Fe^{3+} and Al^{3+}, as illustrated by the K_a values for metal cations in ◀ TABLE 17.6. A comparison of Fe^{2+} and Fe^{3+} values in the table illustrates how acidity increases as ionic charge increases.

TABLE 17.6 • Acid-dissociation constants for metal cations in aqueous solution at 25 °C

Cation	K_a
Fe^{2+}	3.2×10^{-10}
Zn^{2+}	2.5×10^{-10}
Ni^{2+}	2.5×10^{-11}
Fe^{3+}	6.3×10^{-3}
Cr^{3+}	1.6×10^{-4}
Al^{3+}	1.4×10^{-5}

Notice that K_a values for the 3+ ions in Table 17.6 are comparable to the values for familiar weak acids, such as acetic acid ($K_a = 1.8 \times 10^{-5}$). The ions of alkali and alkaline earth metals, being relatively large and not highly charged, do not react with water and therefore do not affect pH. Note that these are the same cations found in the strong bases (Section 17.5). The different tendencies of four cations to lower the pH of a solution are illustrated in ▼ FIGURE 17.13.

The mechanism by which metal ions produce acidic solutions is shown in ▼ FIGURE 17.14. Because metal ions are positively charged, they attract the unshared electron pairs of water molecules and become hydrated (∞ Section 12.1, 'The Solution Process'). The larger the charge on the metal ion, the stronger the interaction between the ion and the oxygen of its hydrating water molecules. As the strength of this interaction increases, the O–H bonds in the hydrating water molecules become weaker. This facilitates transfer of protons from the hydration water molecules to solvent water molecules.

∞ Review this on page 428

NaNO₃	Ca(NO₃)₂	Zn(NO₃)₂	Al(NO₃)₃
Bromothymol blue	Bromothymol blue	Methyl red	Methyl orange
pH = 7.0	pH = 6.9	pH = 5.5	pH = 3.5

◄ FIGURE 17.13   **Effect of cations on solution pH.** The pH values of 1.0 M solutions of four nitrate salts are estimated using acid–base indicators.

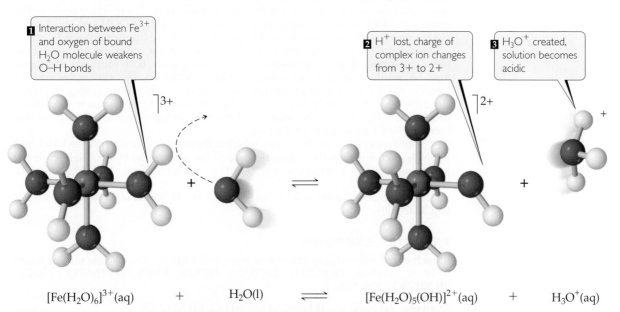

1 Interaction between Fe^{3+} and oxygen of bound H_2O molecule weakens O–H bonds

2 H^+ lost, charge of complex ion changes from 3+ to 2+

3 H_3O^+ created, solution becomes acidic

$[Fe(H_2O)_6]^{3+}$(aq)   + H_2O(l) ⇌ $[Fe(H_2O)_5(OH)]^{2+}$(aq)   + H_3O^+(aq)

▲ FIGURE 17.14   **A hydrated Fe^{3+} ion acts as an acid by donating an H^+ to a free H_2O molecule, forming H_3O^+.**

Combined Effect of Cation and Anion in Solution

To determine whether a salt forms an acidic, a basic or a neutral solution when dissolved in water, we must consider the action of both cation and anion. There are four possible combinations.

1. If the salt contains an anion that does not react with water and a cation that does not react with water, we expect the pH to be neutral. Such is the case when the anion is a conjugate base of a strong acid and the cation is either from group 1 or one of the heavier members of group 2 (Ca^{2+}, Sr^{2+} and Ba^{2+}). *Examples*: $NaCl$, $Ba(NO_3)_2$, $RbClO_4$.

2. If the salt contains an anion that reacts with water to produce hydroxide ions and a cation that does not react with water, we expect the pH to be basic. Such is the case when the anion is the conjugate base of a weak acid and the cation is either from group 1 or one of the heavier members of group 2 (Ca^{2+}, Sr^{2+} and Ba^{2+}). *Examples*: $NaClO$, RbF, $BaSO_3$.

3. If the salt contains a cation that reacts with water to produce hydronium ions and an anion that does not react with water, we expect the pH to be acidic. Such is the case when the cation is a conjugate acid of a weak base or a small cation with a charge of 2+ or greater. *Examples*: NH_4NO_3, $AlCl_3$, $Fe(NO_3)_3$.

4. If the salt contains an anion and a cation *both* capable of reacting with water, both hydroxide ions and hydronium ions are produced. Whether the solution is basic, neutral or acidic depends on the relative abilities of the ions to react with water. *Examples*: NH_4ClO, $Al(CH_3COO)_3$, CrF_3.

SAMPLE EXERCISE 17.18 | **Predicting whether salt solutions are acidic, basic or neutral**

List the following solutions in order of increasing pH: (i) 0.1 M $Ba(CH_3COO)_2$, (ii) 0.1 M NH_4Cl, (iii) 0.1 M CH_3NH_3Br, (iv) 0.1 M KNO_3.

SOLUTION

Analyse We are given the chemical formulae of four ionic compounds (salts) and asked whether their aqueous solutions will be acidic, basic or neutral.

Plan We can determine whether a solution of a salt is acidic, basic or neutral by identifying the ions in solution and by assessing how each ion will affect the pH.

Solve Solution (i) contains barium ions and acetate ions. Ba^{2+} is an ion of one of the heavy alkaline earth metals and will therefore not affect the pH (summary point 4). The anion, CH_3COO^-, is the conjugate base of the weak acid CH_3COOH and will hydrolyse to produce OH^- ions, thereby making the solution basic (summary point 2).

Solutions (ii) and (iii) both contain cations that are conjugate acids of weak bases and anions that are conjugate bases of strong acids. Both solutions will therefore be acidic. Solution (ii) contains NH_4^+, which is the conjugate acid of NH_3 ($K_b = 1.8 \times 10^{-5}$). Solution (iii) contains $CH_3NH_3^+$, which is the conjugate acid of CH_3NH_2 ($K_b = 4.4 \times 10^{-4}$). Because NH_3 has the smaller K_b and is the weaker of the two bases, NH_4^+ will be the stronger of the two conjugate acids. Solution (ii) will therefore be the more acidic of the two.

Solution (iv) contains the K^+ ion, which is the cation of the strong base KOH, and the NO_3^- ion, which is the conjugate base of the strong acid HNO_3. Neither of the ions in solution (iv) will react with water to any appreciable extent, making the solution neutral.

Thus the order of pH is 0.1 M NH_4Cl < 0.1 M CH_3NH_3Br < 0.1 M KNO_3 < 0.1 M $Ba(CH_3COO)_2$.

PRACTICE EXERCISE

In each of the following, indicate which salt will form the more acidic (or less basic) 0.010 M solution: **(a)** $NaNO_3$, $Fe(NO_3)_3$; **(b)** KBr, $KBrO$; **(c)** CH_3NH_3Cl, $BaCl_2$; **(d)** NH_4NO_2, NH_4NO_3.

Answers: **(a)** $Fe(NO_3)_3$, **(b)** KBr, **(c)** CH_3NH_3Cl, **(d)** NH_4NO_3

(See also Exercises 17.66, 17.67.)

| SAMPLE EXERCISE 17.19 | Predicting whether the solution of an amphiprotic anion is acidic or basic |

Predict whether the salt Na_2HPO_4 will form an acidic solution or a basic solution on dissolving in water.

SOLUTION

Analyse We are asked to predict whether a solution of Na_2HPO_4 is acidic or basic. This substance is an ionic compound composed of Na^+ and HPO_4^{2-} ions.

Plan We need to evaluate each ion, predicting whether it is acidic or basic. Because Na^+ is a cation of group 1, it has no influence on pH. Thus our analysis of whether the solution is acidic or basic must focus on the behaviour of the HPO_4^{2-} ion. We need to consider the fact that HPO_4^{2-} can act as either an acid or a base:

As acid $HPO_4^{2-}(aq) \rightleftharpoons H^+(aq) + PO_4^{3-}(aq)$ [17.47]

As base $HPO_4^{2-}(aq) + H_2O \rightleftharpoons H_2PO_4^-(aq) + OH^-(aq)$ [17.48]

Of these two reactions, the one with the larger equilibrium constant determines whether the solution is acidic or basic.

Solve We need to consider the fact that HPO_4^{2-} can act as either acid or base.

$$HPO_4^{2-}(aq) \rightleftharpoons H^+(aq) + PO_4^{3-}(aq)$$

$$HPO_4^{2-}(aq) + H_2O \rightleftharpoons H_2PO_4^-(aq) + OH^-(aq)$$

The value of K_a for the dissociation of HPO_4^{2-} is 4.2×10^{-13}. We must calculate the value of K_b for Equation 17.42 from the value of K_a for its conjugate acid, $H_2PO_4^-$. We make use of the relationship shown in Equation 17.42.

$$K_a \times K_b = K_w$$

We want to know K_b for the base HPO_4^{2-}, knowing the value of K_a for the conjugate acid $H_2PO_4^-$:

$$K_b(HPO_4^{2-}) \times K_a(H_2PO_4^-) = K_w = 1.0 \times 10^{-14}$$

Because K_a for $H_2PO_4^-$ is 6.2×10^{-8} (Table 17.3), we calculate K_b for HPO_4^{2-} to be 1.6×10^{-7}. This is more than 10^5 times larger than K_a for HPO_4^{2-}; thus the reaction shown in the hydrolysis equation predominates over the first reaction and the solution will be basic.

PRACTICE EXERCISE

Predict whether the dipotassium salt of citric acid ($K_2C_6H_6O_7$) will form an acidic or basic solution in water (see Table 17.3 for data).

Answer: Acidic

17.10 | ACID–BASE BEHAVIOUR AND CHEMICAL STRUCTURE

When a substance is dissolved in water, it may behave as an acid or a base, or exhibit no acid–base properties. How does the chemical structure of a substance determine which of these behaviours is exhibited by the substance? For example, why do some substances that contain OH groups behave as bases, releasing OH^- ions into solution, whereas others behave as acids, ionising to release H^+ ions? Why are some acids stronger than others? In this section we look briefly at the effects of chemical structure on acid–base behaviour.

Factors that Affect Acid Strength

A molecule containing H will transfer a proton only if the H–X bond is polarised in the following way:

$$\overset{\longrightarrow}{H-X}$$

In ionic hydrides, such as NaH, the reverse is true; the H atom possesses a negative charge and behaves as a proton acceptor. Essentially, non-polar H–X bonds, such as the H–C bond in CH_4, produce neither acidic nor basic aqueous solutions.

A second factor that helps determine whether a molecule containing an H–X bond will donate a proton is the strength of the bond. Very strong bonds are less easily dissociated than weaker ones. This factor is important, for example, in the case of the hydrogen halides. The H–F bond is the most polar H–X bond. You might expect, therefore, that HF would be a very strong acid if the first factor was all that mattered. However, the energy required to dissociate HF into H and F atoms is much higher than it is for the other hydrogen halides, as shown in Table 8.4. As a result, HF is a weak acid, whereas all the other hydrogen halides are strong acids in water.

A third factor that affects the ease with which a hydrogen atom ionises from HX is the stability of the conjugate base, X^-. In general, the greater the stability of the conjugate base, the stronger is the acid. The strength of an acid is often a combination of all three factors: the polarity of the H–X bond, the strength of the H–X bond and the stability of the conjugate base, X^-.

Binary Acids

In general, the H–X bond strength is the most important factor determining acid strength among the binary acids (those containing hydrogen and just one other element) in which X is in the same *group* in the periodic table. The strength of an H–X bond tends to decrease as the element X increases in size. As a result, the bond strength decreases and the acidity increases down a group. Thus HCl is a stronger acid than HF, and H_2S is a stronger acid than H_2O.

Bond strengths change less across a row in the periodic table than they do down a group. As a result, bond polarity is the major factor determining acidity for binary acids in the same *row*. Thus acidity increases as the electronegativity of the element X increases, as it generally does moving from left to right in a row. For example, the acidity of the second-row elements varies in the following order: $CH_4 < NH_3 \ll H_2O < HF$. Because the C–H bond is essentially non-polar, CH_4 shows no tendency to form H^+ and CH_3^- ions. Although the N–H bond is polar, NH_3 has a non-bonding pair of electrons on the nitrogen atom that dominates its chemistry, so NH_3 acts as a base rather than as an acid. The periodic trends in the acid strengths of binary compounds of hydrogen and The non-metals of periods 2 and 3 are summarised in ▼ FIGURE 17.15.

> **⚠ CONCEPT CHECK 11**
>
> What is the major factor determining the increase in acidity of binary acids going down a group of the periodic table? What is the major factor going across a period?

	GROUP			
	14	15	16	17
Period 2	CH_4 No acid or base properties	NH_3 Weak base	H_2O Neutral	HF Weak acid
Period 3	SiH_4 No acid or base properties	PH_3 Weak base	H_2S Weak acid	HCl Strong acid

Increasing acid strength →

← Increasing base strength

▶ **FIGURE 17.15** **Trends in acid–base properties of binary hydrides.** The acidity of the binary compounds of hydrogen and non-metals increases moving from left to right across a period and moving from top to bottom down a group.

Oxyacids

Many common acids, such as sulfuric acid, contain one or more O–H bonds:

$$\ddot{\text{O}}:$$

$$\text{H}-\ddot{\text{O}}-\overset{|}{\underset{|}{\text{S}}}-\ddot{\text{O}}-\text{H}$$

$$:\ddot{\text{O}}:$$

Acids in which OH groups and possibly additional oxygen atoms are bound to a central atom are called **oxyacids**. The OH group is also present in bases. What factors determine whether an OH group will behave as a base or as an acid?

Let's consider an OH group bound to some atom, Y, which might in turn have other groups attached to it:

$$\overset{\diagdown}{\underset{\diagup}{\text{Y}}}-\text{O}-\text{H}$$

At one extreme, Y might be a metal, such as Na, K or Mg. Because of their low electronegativities, the pair of electrons shared between Y and O is completely transferred to oxygen, and an ionic compound containing OH^- is formed. Such compounds are therefore sources of OH^- ions and behave as bases, as in NaOH and $Mg(OH)_2$.

When Y is a non-metal, the bond to O is covalent and the substance does not readily lose OH^-. Instead, these compounds are either acidic or neutral. As a general rule, *as the electronegativity of Y increases, so will the acidity of the substance*. This happens for two reasons. First, as electron density is drawn towards Y, the O–H bond becomes weaker and more polar, thereby favouring loss of H^+. Second, because the conjugate base is usually an anion, its stability generally increases as the electronegativity of Y increases.

This trend is illustrated by the K_a values of the hypohalous acids (YOH acids where Y is a halide ion), which decrease as the electronegativity of the halogen atom decreases (▼ FIGURE 17.16).

Many oxyacids contain additional oxygen atoms bonded to the central atom Y. The additional electronegative oxygen atoms pull electron density from the O–H bond, further increasing its polarity. Increasing the number of oxygen atoms also helps stabilise the conjugate base by increasing its ability to 'spread out' its negative charge. Thus the strength of an acid will increase as additional electronegative atoms bond to the central atom Y.

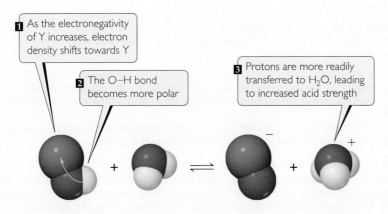

Substance	Y–OH	Electronegativity of Y	Dissociation constant
Hypochlorous acid	Cl–OH	3.0	$K_a = 3.0 \times 10^{-8}$
Hypobromous acid	Br–OH	2.8	$K_a = 2.5 \times 10^{-9}$
Hypoiodous acid	I–OH	2.5	$K_a = 2.3 \times 10^{-11}$
Water	H–OH	2.1	$K_w = 1.0 \times 10^{-14}$

◀ **FIGURE 17.16   Acidity of the hypohalous oxyacids (YOH) as a function of electronegativity of Y.**

We can summarise these ideas as two simple rules that relate the acid strength of oxyacids to the electronegativity of Y and to the number of groups attached to Y.

1. For oxyacids that have the same number of OH groups and the same number of O atoms, acid strength increases with increasing electronegativity of the central atom Y. For example, the strength of the hypohalous acids, which have the structure H–O–Y, increases as the electronegativity of Y increases (Figure 17.16).

2. For oxyacids that have the same central atom Y, acid strength increases as the number of oxygen atoms attached to Y increases. For example, the strength of the oxyacids of chlorine steadily increases from hypochlorous acid (HClO) to perchloric acid (HClO$_4$):

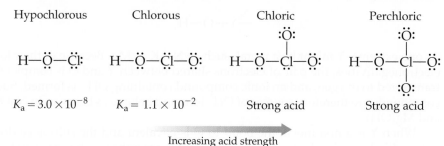

Hypochlorous   Chloric   Chloric   Perchloric

$K_a = 3.0 \times 10^{-8}$ $K_a = 1.1 \times 10^{-2}$ Strong acid   Strong acid

Increasing acid strength

Because the oxidation number of the central atom increases as the number of attached O atoms increases, this correlation can be stated in an equivalent way: *in a series of oxyacids, the acidity increases as the oxidation number of the central atom increases*.

 CONCEPT CHECK 12

Which acid has the larger acid-dissociation constant, HIO$_2$, HBrO$_2$ or HBrO$_3$?

SAMPLE EXERCISE 17.20 | **Predicting relative acidities from composition and structure**

Arrange the compounds in each of the following series in order of increasing acid strength: **(a)** AsH$_3$, HI, NaH, H$_2$O; **(b)** H$_2$SeO$_3$, H$_2$SeO$_4$, H$_2$O.

SOLUTION

Analyse We are asked to arrange two sets of compounds in order from weakest acid to strongest acid. In **(a)**, the substances are binary compounds containing H, and in **(b)** the substances are oxyacids.

Plan For the binary compounds, we will consider the electronegativities of As, Br, K and Se relative to the electronegativity of H. The higher the electronegativity of these atoms, the higher the partial positive charge on H and so the more acidic the compound.

For the oxyacids, we will consider both the electronegativities of the central atom and the number of oxygen atoms bonded to the central atom.

Solve

(a) The elements from the left side of the periodic table form the most basic binary hydrogen compounds because the hydrogen in these compounds carries a negative charge. Thus NaH should be the most basic compound on the list. Because arsenic is less electronegative than oxygen, we might expect that AsH$_3$ would be a weak base towards water. That is also what we would predict by an extension of the trends shown in Figure 17.16. Further, we expect that the binary hydrogen compounds of the halogens, as the most electronegative element in each period, will be acidic relative to water. In fact, HI is one of the strong acids in water. Thus the order of increasing acidity is NaH < AsH$_3$ < H$_2$O < HI.

(b) The acidity of oxyacids increases as the number of oxygen atoms bonded to the central atom increases. Thus H_2SeO_4 will be a stronger acid than H_2SeO_3; in fact, the Se atom in H_2SeO_4 is in its maximum positive oxidation state and so we expect it to be a comparatively strong acid, much like H_2SO_4. H_2SeO_3 is an oxyacid of a non-metal that is similar to H_2SO_3. As such, we expect that H_2SeO_3 is able to donate a proton to H_2O, indicating that H_2SeO_3 is a stronger acid than H_2O. Thus, the order of increasing acidity is $H_2O < H_2SeO_3 < H_2SeO_4$.

PRACTICE EXERCISE

In each of the following pairs choose the compound that leads to the more acidic (or less basic) solution: **(a)** HBr, HF; **(b)** PH_3, H_2S; **(c)** HNO_2, HNO_3; **(d)** H_2SO_3, H_2SeO_3.

Answers: **(a)** HBr, **(b)** H_2S, **(c)** HNO_3, **(d)** H_2SO_3

(See also Exercises 17.70, 17.71.)

Carboxylic Acids

Another large group of acids is illustrated by acetic acid:

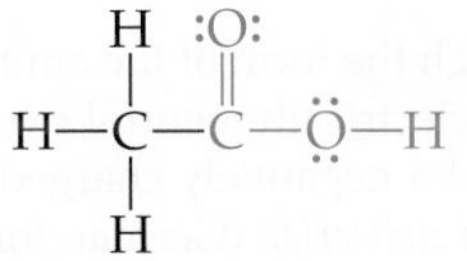

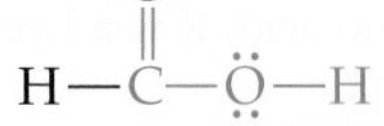

$$\begin{array}{c} \text{H} \quad :\!\text{O}\!: \\ | \quad\quad || \\ \text{H}-\text{C}-\text{C}-\ddot{\text{O}}-\text{H} \\ | \\ \text{H} \end{array}$$

The portion of the structure shown in blue is called the *carboxyl group*, which is often written as COOH. Thus the chemical formula of acetic acid is often written as CH_3COOH, where only the hydrogen atom in the carboxyl group can be ionised. Acids that contain a carboxyl group are called **carboxylic acids**. Formic acid and benzoic acid, whose structures are drawn in the margin, are further examples of this large and important category of acids.

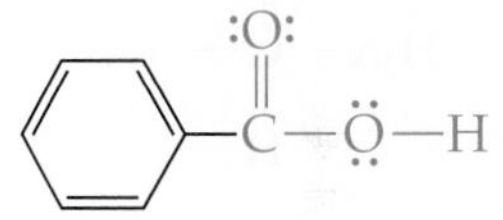

Acetic acid (CH_3COOH) is a weak acid ($K_a = 1.8 \times 10^{-5}$). Methanol (CH_3OH), however, is not an acid in water. Two factors contribute to the acidic behaviour of carboxylic acids. First, the additional oxygen atom attached to the carboxyl group carbon draws electron density from the O–H bond, increasing its polarity and helping to stabilise the conjugate base. Second, the conjugate base of a carboxylic acid (a *carboxylate anion*) can exhibit resonance (∞ Section 8.6, 'Resonance Structures'), which contributes further to the stability of the anion by spreading the negative charge over several atoms:

$$\begin{array}{c} \text{H} \quad :\!\text{O}\!: \\ | \quad\quad || \\ \text{H}-\text{C}-\text{C}-\ddot{\text{O}}\!:^{-} \\ | \\ \text{H} \end{array} \longleftrightarrow \begin{array}{c} \text{H} \quad :\!\ddot{\text{O}}\!:^{-} \\ | \quad\quad | \\ \text{H}-\text{C}-\text{C}=\ddot{\text{O}} \\ | \\ \text{H} \end{array}$$

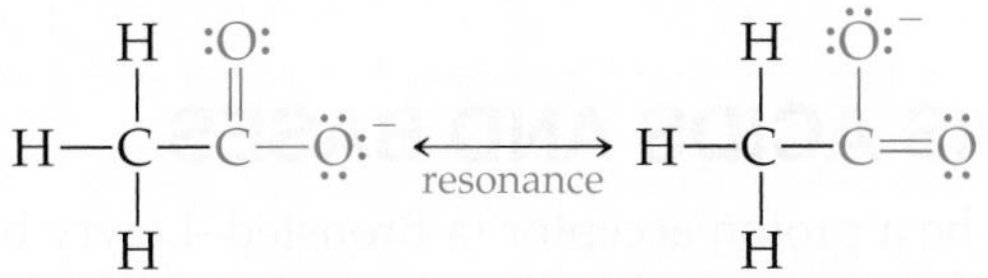

The acid strength of carboxylic acids also increases as the number of electronegative atoms in the acid increases. For example, trifluoroacetic acid (CF_3COOH) has $K_a = 5.0 \times 10^{-1}$; the replacement of three hydrogen atoms of acetic acid with more electronegative fluorine atoms leads to a large increase in acid strength.

$$\begin{array}{c} :\!\text{O}\!: \\ || \\ \text{H}-\text{C}-\ddot{\text{O}}-\text{H} \end{array}$$

Formic acid

$$\begin{array}{c} :\!\text{O}\!: \\ || \\ \text{C}6\text{H}5-\text{C}-\ddot{\text{O}}-\text{H} \end{array}$$

Benzoic acid

∞ Review this on page 272

CONCEPT CHECK 13

What group of atoms is present in all carboxylic acids?

A CLOSER LOOK

THE AMPHIPROTIC BEHAVIOUR OF AMINO ACIDS

The general structure of *amino acids*, the building blocks of proteins, is

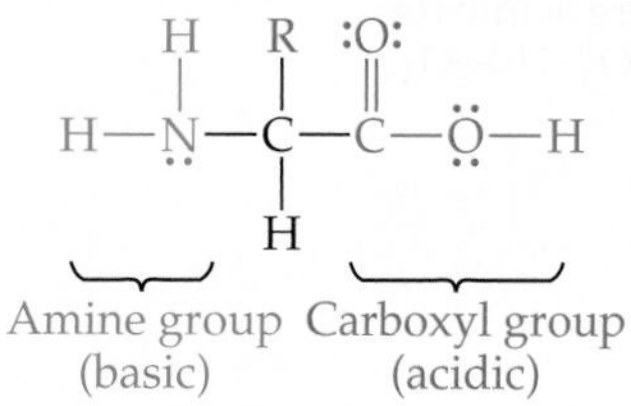

Amine group (basic)   Carboxyl group (acidic)

where different amino acids have different R groups attached to the central carbon atom. For example, in *glycine*, the simplest amino acid, R is a hydrogen atom, and in *alanine* R is a CH_3 group:

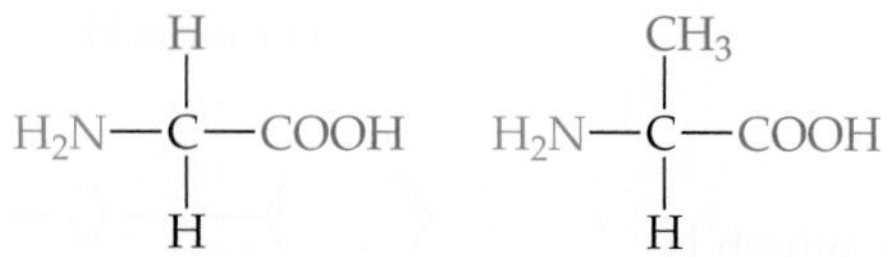

Glycine   Alanine

Amino acids contain a carboxyl group and can therefore serve as acids. They also contain an NH_2 group, characteristic of amines (Section 17.7), and thus they can also act as bases. Amino acids, therefore, are amphiprotic. For glycine, we might expect the acid and base reactions with water to be

Acid: $H_2N—CH_2—COOH(aq) + H_2O(l) \rightleftharpoons$
$\quad H_2N—CH_2—COO^-(aq) + H_3O^+(aq)$ [17.47]

Base: $H_2N—CH_2—COOH(aq) + H_2O(l) \rightleftharpoons$
$\quad {}^+H_3N—CH_2—COOH(aq) + OH^-(aq)$ [17.48]

The pH of a solution of glycine in water is about 6.0, indicating that it is a slightly stronger acid than base. However, the acid–base chemistry of amino acids is more complicated than shown in Equations 17.47 and 17.48. Because the COOH group can act as an acid and the NH_2 group can act as a base, amino acids undergo a 'self-contained' Brønsted–Lowry acid–base reaction in which the proton of the carboxyl group is transferred to the basic nitrogen atom:

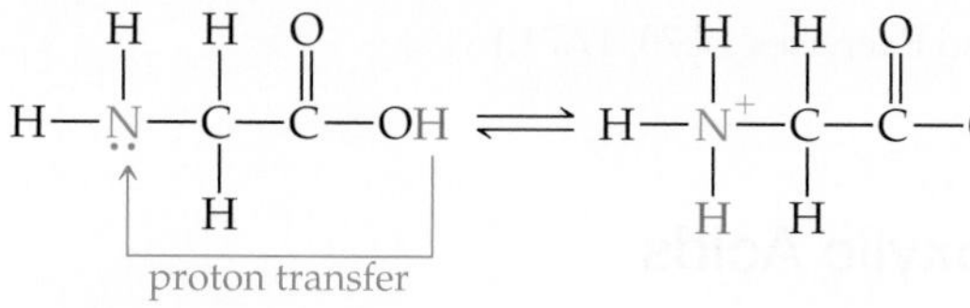

proton transfer

Neutral molecule   Zwitterion

Although the form of the amino acid on the right in this equation is electrically neutral overall, it has a positively charged end and a negatively charged end. A molecule of this type is called a *zwitterion* (German for 'hybrid ion').

Do amino acids exhibit any properties indicating that they behave as zwitterions? If so, their behaviour should be similar to that of ionic substances (∞ Section 11.9, 'Bonding in Solids'). Crystalline amino acids have relatively high melting points, usually above 200 °C, which is characteristic of ionic solids. Amino acids are far more soluble in water than in nonpolar solvents. In addition, the dipole moments of amino acids are large, consistent with a large separation of charge in the molecule. Thus the ability of amino acids to act simultaneously as acids and bases has important effects on their properties.

RELATED EXERCISE: 17.88

17.11 | LEWIS ACIDS AND BASES

For a substance to be a proton acceptor (a Brønsted–Lowry base), it must have an unshared pair of electrons for binding the proton. NH_3, for example, acts as a proton acceptor. Using Lewis structures, we can write the reaction between H^+ and NH_3 as follows:

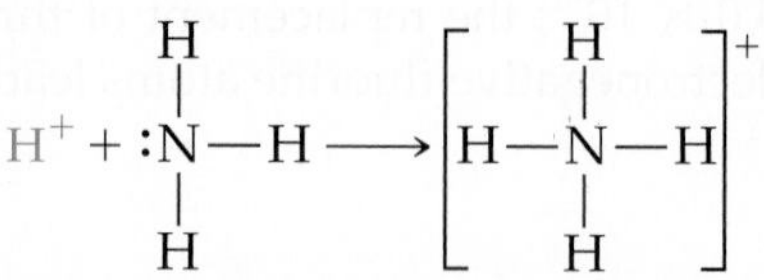

G. N. Lewis was the first to notice this aspect of acid–base reactions. He proposed a definition of acid and base that emphasises the shared electron pair: a **Lewis acid** is an electron-pair acceptor and a **Lewis base** is an electron-pair donor.

Every base that we have discussed so far—whether it be OH^-, H_2O, an amine or an anion—is an electron-pair donor. Everything that is a base in the Brønsted–Lowry sense (a proton acceptor) is also a base in the Lewis sense (an electron-pair donor). In the Lewis theory, however, a base can donate its electron pair to something other than H$^+$. The Lewis definition therefore greatly increases the number of species that can be considered acids; H$^+$ is a Lewis acid, but not the only one. For example, consider the reaction between NH_3 and BF_3 (Equation 17.49). This reaction occurs because BF_3 has a vacant orbital in its valence shell (∞ Section 8.7, 'Exceptions to the Octet Rule'). It therefore acts as an electron-pair acceptor (a Lewis acid) towards NH_3, which donates the electron pair. The curved arrow shows the donation of a pair of electrons from N to B to form a covalent bond:

∞ Review this on page 274

$$
\begin{array}{cc}
\text{H} \quad \text{F} & \text{H} \quad \text{F} \\
| \quad\quad | & | \quad\quad | \\
\text{H—N: + B—F} \longrightarrow \text{H—N—B—F} \\
| \quad\quad | & | \quad\quad | \\
\text{H} \quad \text{F} & \text{H} \quad \text{F}
\end{array}
$$

Lewis base   Lewis acid

[17.49]

CONCEPT CHECK 14

What feature must any molecule or ion have to act as a Lewis base?

Our emphasis throughout this chapter has been on water as the solvent and on the proton as the source of acidic properties. In such cases we find the Brønsted–Lowry definition of acids and bases to be the most useful. In fact, when we speak of a substance as being acidic or basic, we are usually thinking of aqueous solutions and using these terms in the Arrhenius or Brønsted–Lowry sense. The advantage of the Lewis theory is that it allows us to treat a wider variety of reactions, including those that do not involve proton transfer, as acid–base reactions. To avoid confusion, a substance like BF_3 is rarely called an acid unless it is clear from the context that we are using the term in the sense of the Lewis definition. Instead, substances that function as electron-pair acceptors are referred to explicitly as 'Lewis acids'.

Lewis acids include molecules that, like BF_3, have an incomplete octet of electrons. In addition, many simple cations can function as Lewis acids. For example, Fe^{3+} interacts strongly with cyanide ions to form the ferricyanide ion, $Fe(CN)_6{}^{3-}$.

$$Fe^{3+} + 6[:C\equiv N:]^- \longrightarrow [Fe(C\equiv N:)_6]^{3-}$$

The Fe^{3+} ion has vacant orbitals that accept the electron pairs donated by the ions; we learn more in Chapter 21 about just which orbitals are used by the Fe^{3+} ion. The metal ion is highly charged, too, which contributes to the interaction with CN$^-$ ions.

Some compounds with multiple bonds can behave as Lewis acids. For example, the reaction of carbon dioxide with water to form carbonic acid (H_2CO_3) can be pictured as an attack by a water molecule on CO_2, in which the water acts as an electron-pair donor and the CO_2 as an electron-pair acceptor, as shown in the margin. The electron pair of one of the carbon–oxygen double bonds is moved onto the oxygen, leaving a vacant orbital on the carbon that can act as an electron-pair acceptor. The shift of these electrons is indicated with arrows. After forming the initial acid–base product, a proton moves from one oxygen to another, thereby forming carbonic acid.

The hydrated cations we encountered in Section 17.9, such as $[Fe(H_2O)_6]^{3+}$ in Figure 17.14, form through the reaction between the cation acting as a Lewis acid and the water molecules acting as Lewis bases. When a water molecule interacts with the positively charged metal ion, electron density is drawn from the oxygen (▼ FIGURE 17.17). This flow of electron density causes the O–H bond to become more polarised; as a result, water molecules bound to the metal ion are more acidic than those in the bulk solvent. This effect becomes more pronounced as the charge of the cation increases, which explains why 3+ cations are much more acidic than cations with smaller charges.

Lewis base

Lewis acid

Hydration of carbon dioxide to carbonic acid

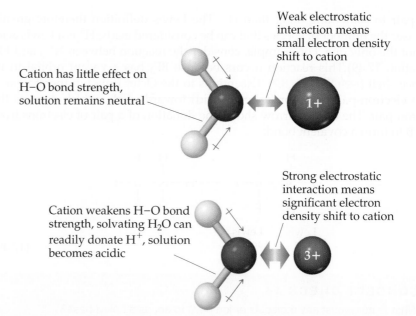

Weak electrostatic interaction means small electron density shift to cation

Cation has little effect on H–O bond strength, solution remains neutral

Strong electrostatic interaction means significant electron density shift to cation

Cation weakens H–O bond strength, solvating H_2O can readily donate H^+, solution becomes acidic

▶ **FIGURE 17.17** **The acidity of a hydrated cation depends on cation charge.**

SAMPLE INTEGRATIVE EXERCISE **Putting concepts together**

Phosphorous acid (H_3PO_3) has the following Lewis structure.

$$\overset{\qquad\;\; H}{\underset{\qquad\;\; \ddot{O}-H}{\ddot{O}=P-\ddot{O}-H}}$$

(a) Explain why H_3PO_3 is diprotic and not triprotic. **(b)** A 25.0 cm³ sample of a solution of H_3PO_3 is titrated with 0.102 M NaOH. It requires 23.3 cm³ of NaOH to neutralise both acidic protons. What is the molarity of the H_3PO_3 solution? **(c)** This solution has a pH of 1.59. Calculate the percent ionisation and K_{a1} for H_3PO_3, assuming that $K_{a1} \gg K_{a2}$. **(d)** How does the osmotic pressure of a 0.050 M solution of HCl compare with that of a 0.050 M solution of H_3PO_3? Explain.

SOLUTION

(a) Acids have polar H–X bonds. The electronegativity of H is 2.1 and that of P is also 2.1. Because the two elements have the same electronegativity, the H–P bond is non-polar (∞ Section 8.4, 'Bond Polarity and Electronegativity'). Thus this H cannot be acidic. The other two H atoms, however, are bonded to O, which has an electronegativity of 3.5. The H–O bonds are therefore polar, with H having a partial positive charge. These two H atoms are consequently acidic.

(b) The chemical equation for the neutralisation reaction is

$$H_3PO_3(aq) + 2\,NaOH(aq) \longrightarrow Na_2HPO_3(aq) + 2\,H_2O(l)$$

From the definition of molarity, M = mol dm⁻³, we see that moles = M × dm³ (∞ Section 4.5, 'Concentrations of Solutions'). Thus the number of moles of NaOH added to the solution is $(0.0233\ dm^3)(0.102\ mol\ dm^{-3}) = 2.38 \times 10^{-3}$ mol NaOH. The balanced equation indicates that 2 mol of NaOH is consumed for each mole of H_3PO_3. Thus, the number of moles of H_3PO_3 in the sample is

$$\frac{1}{2}\,(2.38 \times 10^{-3}\ \text{mol NaOH}) = 1.19 \times 10^{-3}\ \text{mol H}_3\text{PO}_3$$

The concentration of the H_3PO_3 solution, therefore, equals $(1.19 \times 10^{-3}\ \text{mol})/(0.0250\ dm^3)$ = 0.0476 M.

(c) From the pH of the solution, 1.59, we can calculate $[H^+]$ at equilibrium.

$$[H^+] = \text{antilog}(-1.59) = 10^{-1.59} = 0.026\ \text{M (two significant figures)}$$

∞ Review this on page 260

∞ Review this on page 124

Because $K_{a1} \gg K_{a2}$, the vast majority of the ions in solution are from the first ionisation step of the acid.

$$H_3PO_3(aq) \rightleftharpoons H^+(aq) + H_2PO_3^-(aq)$$

Because one $H_2PO_3^-$ ion forms for each H^+ ion formed, the equilibrium concentrations of H^+ and $H_2PO_3^-$ are equal: $[H^+] = [H_2PO_3^-] = 0.026$ M. The equilibrium concentration of H_3PO_3 equals the initial concentration minus the amount that ionises to form H^+ and $H_2PO_3^-$: $[H_3PO_3] = 0.0476$ M $- 0.026$ M $= 0.022$ M (two significant figures). These results can be tabulated as follows:

	$H_3PO_3(aq)$ $\rightleftharpoons$	$H^+(aq)$ +	$H_2PO_3^-(aq)$
Initial	0.0475 M	0	0
Change	−0.026 M	+0.026 M	+0.026 M
Equilibrium	0.022 M	0.026 M	0.026 M

The percent ionisation is

$$\text{Percent ionisation} = \frac{[H^+]_{\text{equilibrium}}}{[H_3PO_3]_{\text{initial}}} \times 100\% = \frac{0.026 \text{ M}}{0.0476 \text{ M}} \times 100\% = 55\%$$

The first acid-dissociation constant is

$$K_{a1} = \frac{[H^+][H_2PO_3^-]}{[H_3PO_3]} = \frac{(0.026)(0.026)}{0.022} = 0.030$$

(d) Osmotic pressure is a colligative property and depends on the total concentration of particles in solution (∞ Section 12.5, 'Colligative Properties'). Because HCl is a strong acid, a 0.050 M solution will contain 0.050 M $H^+(aq)$ and 0.050 M $Cl^-(aq)$, or a total of 0.100 mol dm^{-3} of particles. Because H_3PO_3 is a weak acid, it ionises to a lesser extent than HCl and, hence, there are fewer particles in the H_3PO_3 solution. As a result, the H_3PO_3 solution will have the lower osmotic pressure.

∞ Review this on page 451

CHAPTER SUMMARY AND KEY TERMS

SECTION 17.1 **Acids** and **bases** were first recognised by the properties of their aqueous solutions. For example, acids turn litmus red, whereas bases turn litmus blue. Arrhenius recognised that the properties of acidic solutions are due to $H^+(aq)$ ions and those of basic solutions are due to $OH^-(aq)$ ions.

SECTION 17.2 The Brønsted–Lowry concept of acids and bases is more general than the Arrhenius concept and emphasises the transfer of a proton (H^+) from an acid to a base. The H^+ ion, which is merely a proton with no surrounding valence electrons, is strongly bound to water. For this reason, the **hydronium ion**, $H_3O^+(aq)$, is often used to represent the predominant form of H^+ in water instead of the simpler $H^+(aq)$.

A **Brønsted–Lowry acid** is a substance that donates a proton to another substance; a **Brønsted–Lowry base** is a substance that accepts a proton from another substance. Water is an **amphiprotic** substance, one that can function as either a Brønsted–Lowry acid or base, depending on the substance with which it reacts.

The **conjugate base** of a Brønsted–Lowry acid is the species that remains when a proton is removed from the acid. The **conjugate acid** of a Brønsted–Lowry base is the species formed by adding a proton to the base. Together, an acid and its conjugate base (or a base and its conjugate acid) are called a **conjugate acid–base pair**.

The acid–base strengths of conjugate acid–base pairs are related: the stronger an acid, the weaker is its conjugate base; the weaker an acid, the stronger is its conjugate base. In every acid–base reaction, the position of the equilibrium favours the transfer of the proton from the stronger acid to the stronger base.

SECTION 17.3 Water ionises to a slight degree, forming $H^+(aq)$ and $OH^-(aq)$. The extent of this **autoionisation** is expressed by the **ion product constant** for water: $K_w = [H^+][OH^-] = 1.0 \times 10^{-14}$ (25 °C). This relationship describes both pure water and aqueous solutions. The K_w expression indicates that the product of $[H^+]$ and $[OH^-]$ is a constant. Thus as $[H^+]$ increases, $[OH^-]$ decreases. Acidic solutions are those that contain more $H^+(aq)$ than $OH^-(aq)$, whereas basic solutions contain more $OH^-(aq)$ than $H^+(aq)$.

SECTION 17.4 The concentration of $H^+(aq)$ can be expressed in terms of **pH**: $pH = -\log[H^+]$. At 25 °C the pH of a neutral solution is 7.00, whereas the pH of an acidic solution is below

7.00, and the pH of a basic solution is above 7.00. The pX notation is also used to represent the negative logarithm of other small quantities, as in pOH and pK_w. The pH of a solution can be measured using a pH meter, or it can be estimated using acid–base indicators.

SECTION 17.5 Strong acids are strong electrolytes, ionising completely in aqueous solution. The common strong acids are HCl, HBr, HI, HNO_3, $HClO_3$, $HClO_4$ and H_2SO_4. The conjugate bases of strong acids have negligible basicity.

Common strong bases are the ionic hydroxides of the alkali metals and the heavy alkaline earth metals.

SECTION 17.6 Weak acids are weak electrolytes; only some of the molecules exist in solution in ionised form. The extent of ionisation is expressed by the **acid-dissociation constant**, K_a, which is the equilibrium constant for the reaction $HA(aq) \rightleftharpoons H^+(aq) + A^-(aq)$, which can also be written $HA(aq) + H_2O(l) \rightleftharpoons H_3O^+(aq) + A^-(aq)$. The larger the value of K_a, the stronger is the acid. For solutions of the same concentration, a stronger acid also has a larger **percent ionisation**. The concentration of a weak acid and its K_a value can be used to calculate the pH of a solution.

Polyprotic acids, such as H_2SO_3, have more than one ionisable proton. These acids have acid-dissociation constants that decrease in magnitude in the order $K_{a1} > K_{a2} > K_{a3}$. Because nearly all the $H^+(aq)$ in a polyprotic acid solution comes from the first dissociation step, the pH can usually be estimated satisfactorily by considering only K_{a1}.

SECTION 17.7 Weak bases include NH_3, **amines** and the anions of weak acids. The extent to which a weak base reacts with water to generate the corresponding conjugate acid and OH^- is measured by the **base-dissociation constant**, K_b. This is the equilibrium constant for the reaction $B(aq) + H_2O(l) \rightleftharpoons HB^+(aq) + OH^-(aq)$, where B is the base.

SECTION 17.8 The relationship between the strength of an acid and the strength of its conjugate base is expressed quantitatively by the equation $K_a \times K_b = K_w$, where K_a and K_b are dissociation constants for conjugate acid–base pairs.

SECTION 17.9 The acid–base properties of salts can be ascribed to the behaviour of their respective cations and anions. The reaction of ions with water, with a resultant change in pH, is called **hydrolysis**. The cations of the alkali metals and the alkaline earth metals as well as the anions of strong acids, such as Cl^-, Br^-, I^- and NO_3^-, do not undergo hydrolysis. They are always spectator ions in acid–base chemistry.

SECTION 17.10 The tendency of a substance to show acidic or basic characteristics in water can be correlated with its chemical structure. Acid character requires the presence of a highly polar H–X bond. Acidity is also favoured when the H–X bond is weak and when the X^- ion is stable.

For **oxyacids** with the same number of OH groups and the same number of O atoms, acid strength increases with increasing electronegativity of the central atom. For oxyacids with the same central atom, acid strength increases as the number of oxygen atoms attached to the central atom increases. **Carboxylic acids**, which are organic acids containing the COOH group, are the most important class of organic acids. The presence of resonance structures in the conjugate base is partially responsible for the acidity of these compounds.

SECTION 17.11 The Lewis concept of acids and bases emphasises the shared electron pair rather than the proton. A **Lewis acid** is an electron-pair acceptor, and a **Lewis base** is an electron-pair donor. The Lewis concept is more general than the Brønsted–Lowry concept because it can apply to cases in which the acid is some substance other than H^+.

KEY SKILLS

- Define and identify Arrhenius acids and bases. (Section 17.1)

- Understand the nature of the hydrated proton, represented as either $H^+(aq)$ or $H_3O^+(aq)$. (Section 17.2)

- Define and identify Brønsted–Lowry acids and bases and identify conjugate acid–base pairs. (Section 17.2)

- Relate the strength of an acid to the strength of its conjugate base. (Section 17.2)

- Understand how the equilibrium position of a proton-transfer reaction relates to the strengths of the acids and bases involved. (Section 17.3)

- Describe the autoionisation of water and understand how $[H_3O^+]$ and $[OH^-]$ are related. (Section 17.3)

- Calculate the pH of a solution given $[H_3O^+]$ or $[OH^-]$. (Section 17.4)

- Calculate the pH of a strong acid or strong base given its concentration. (Section 17.5)

- Calculate K_a or K_b for a weak acid or weak base given its concentration and the pH of the solution. (Sections 17.6 and 17.7)

- Calculate the pH of a weak acid or weak base or its percent ionisation given its concentration and K_a or K_b. (Sections 17.6 and 17.7)

- Calculate K_b for a weak base given K_a of its conjugate acid, and similarly calculate K_a from K_b. (Section 17.8)

- Predict whether an aqueous solution of a salt will be acidic, basic or neutral. (Section 17.9)

- Predict the relative strength of a series of acids from their molecular structures. (Section 17.10)
- Define and identify Lewis acids and bases. (Section 17.11)

KEY EQUATIONS

- Ion product of water at 25 °C

$$K_w = [H_3O^+][OH^-] = [H^+][OH^-] = 1.0 \times 10^{-14}$$ [17.15]

- Definition of pH

$$pH = -\log[H^+]$$ [17.16]

- Definition of pOH

$$pOH = -\log[OH^-]$$ [17.17]

- Relationship between pH and pOH

$$pH + pOH = 14.00$$ [17.19]

- Acid-dissociation constant for a weak acid, HA

$$K_a = \frac{[H_3O^+][A^-]}{[HA]} \text{ or } K_a = \frac{[H^+][A^-]}{[HA]}$$ [17.26]

- Percent ionisation of a weak acid

$$\text{Percent ionisation} = \frac{[H^+]_{equilibrium}}{[HA]_{initial}} \times 100\%$$ [17.28]

- Base-dissociation constant for a weak base, B

$$K_b = \frac{[BH^+][OH^-]}{[B]}$$ [17.35]

- Relationship between acid- and base-dissociation constants of a conjugate acid–base pair

$$K_a \times K_b = K_w$$ [17.42]

EXERCISES

VISUALISING CONCEPTS

17.1 (a) Identify the Brønsted–Lowry acid and the Brønsted–Lowry base in the following reaction:

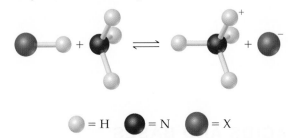

$\bigcirc$ = H   ● = N   ● = X

(b) Identify the Lewis acid and the Lewis base in the reaction. [Sections 17.2 and 17.11]

17.2 The following diagrams represent aqueous solutions of two monoprotic acids, HA (A = X or Y). The water molecules have been omitted for clarity. **(a)** Which is the stronger acid, HX or HY? **(b)** Which is the stronger base, X⁻ or Y⁻? **(c)** If you mix equal concentrations of HX and NaY, will the equilibrium

$$HX(aq) + Y^-(aq) \rightleftharpoons HY(aq) + X^-(aq)$$

lie mostly to the right ($K_c > 1$) or to the left ($K_c < 1$)? [Section 17.2]

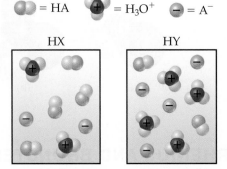

●● = HA   ● = H₃O⁺   ● = A⁻

17.3 The following diagrams represent aqueous solutions of three acids, HX, HY and HZ. The water molecules have been omitted for clarity, and the hydrated proton is represented as a simple sphere rather than as a hydronium ion. **(a)** Which of the acids is a strong acid? Explain. **(b)** Which acid would have the smallest acid-dissociation constant, K_a? **(c)** Which solution would have the highest pH? [Sections 17.5 and 17.6]

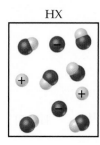
HX

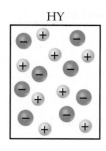
HY

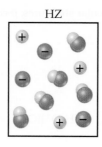
HZ

17.4 Refer to the diagrams accompanying Exercise 17.3. **(a)** Rank the anions X^-, Y^- and Z^- in order of increasing basicity. **(b)** Which of the ions would have the largest base-dissociation constant, K_b? [Sections 17.2 and 17.8]

17.5 **(a)** Draw the Lewis structure for $(CH_3)_2NH$ and explain why it is able to act as a base. **(b)** To what class of organic compounds does this substance belong? [Section 17.7]

17.6 The following diagram represents an aqueous solution formed by dissolving a sodium salt of a weak acid in water. The diagram shows only the Na^+ ions, the X^- ions and the HX molecules. What ion is missing from the diagram? If the drawing is completed by drawing all the ions, how many of the missing ions should be shown? [Section 17.9]

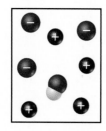

17.7 **(a)** What kinds of acids are represented by the following molecular models? **(b)** Indicate how the acidity of each molecule is affected by increasing the electronegativity of the atom X, and explain the origin of the effect. [Section 17.10]

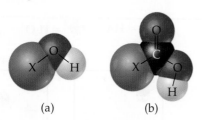

(a)   (b)

17.8 In this model of acetylsalicylic acid (aspirin), identify the carboxyl group in the molecule. [Section 17.10]

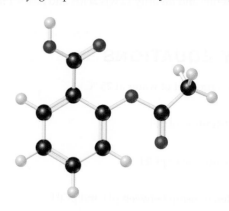

17.9 **(a)** The following diagram represents the reaction of PCl_4^+ with Cl^-. Draw the Lewis structures for the reactants and products, and identify the Lewis acid and the Lewis base in the reaction.

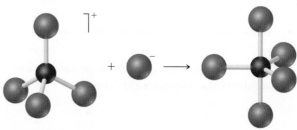

(b) The following reaction represents the acidity of a hydrated cation. How does the equilibrium constant for the reaction change as the charge of the cation increases? [Section 17.11]

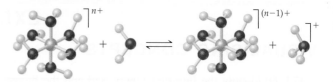

ARRHENIUS AND BRØNSTED–LOWRY ACIDS AND BASES (Sections 17.1 and 17.2)

17.10 Although HCl and H_2SO_4 have very different properties as pure substances, their aqueous solutions possess many common properties. List some general properties of these solutions, and explain their common behaviour in terms of the species present.

17.11 Although pure NaOH and CaO have very different properties, their aqueous solutions possess many common properties. List some general properties of these solutions, and explain their common behaviour in terms of the species present.

17.12 **(a)** What is the difference between the Arrhenius and the Brønsted–Lowry definitions of an acid? **(b)** $NH_3(g)$ and HCl(g) react to form the ionic solid $NH_4Cl(s)$. Which substance is the Brønsted–Lowry acid in this reaction? Which is the Brønsted–Lowry base?

17.13 **(a)** What is the difference between the Arrhenius and the Brønsted–Lowry definitions of a base? **(b)** When ammonia is dissolved in water, it behaves both as an Arrhenius base and as a Brønsted–Lowry base. Explain.

17.14 Give the conjugate base of the following Brønsted–Lowry acids: **(a)** HIO_3, **(b)** NH_4^+, **(c)** $H_2PO_4^-$, **(d)** C_6H_5COOH.

17.15 Give the conjugate acid of the following Brønsted–Lowry bases: **(a)** CN^-, **(b)** O^{2-}, **(c)** HPO_4^{2-}, **(d)** $C_2H_5NH_2$.

17.16 Designate the Brønsted–Lowry acid and the Brønsted–Lowry base on the left side of each of the following equations, and also designate the conjugate acid and conjugate base on the right side:
(a) $NH_4^+(aq) + CN^-(aq) \rightleftharpoons HCN(aq) + NH_3(aq)$
(b) $(CH_3)_3N(aq) + H_2O(l) \rightleftharpoons$
$(CH_3)_3NH^+(aq) + OH^-(aq)$
(c) $HCOOH(aq) + PO_4^{3-}(aq) \rightleftharpoons$
$HCOO^-(aq) + HPO_4^{2-}(aq)$

17.17 **(a)** The hydrogen oxalate ion $HOOC(CH)_2COO^-$ is amphiprotic. Write a balanced chemical equation showing how it acts as an acid towards water and another equation showing how it acts as a base towards water. **(b)** What is the conjugate acid of $HOOC(CH)_2COO^-$ What is its conjugate base?

17.18 Label each of the following as being a strong base, a weak base or a species with negligible basicity. In each case write the formula of its conjugate acid, and indicate whether the conjugate acid is a strong acid, a weak acid or a species with negligible acidity: **(a)** CH_3COO^-, **(b)** HCO_3^-, **(c)** O^{2-}, **(d)** Cl^-, **(e)** NH_3.

17.19 **(a)** Which of the following is the stronger Brønsted–Lowry acid, HBrO or HBr? **(b)** Which is the stronger Brønsted–Lowry base, F^- or Cl^-? Briefly explain your choices.

17.20 **(a)** Which of the following is the stronger Brønsted–Lowry acid, HNO_3 or HNO_2? **(b)** Which is the stronger Brønsted–Lowry base, NH_3 or H_2O? Briefly explain your choices.

17.21 Predict the products of the following acid–base reactions, and also predict whether the equilibrium lies to the left or to the right of the equation:
(a) $O^{2-}(aq) + H_2O(l) \rightleftharpoons ?$
(b) $CH_3COOH(aq) + HS^-(aq) \rightleftharpoons ?$
(c) $NO_3^-(aq) + H_2O(l) \rightleftharpoons ?$

AUTOIONISATION OF WATER (Section 17.3)

17.22 **(a)** What is meant by the term *autoionisation*? **(b)** Explain why pure water is a poor conductor of electricity. **(c)** You are told that an aqueous solution is acidic. What does this statement mean?

17.23 **(a)** Write a chemical equation that illustrates the autoionisation of water. **(b)** Write the expression for the ion product constant for water, K_w. Why is $[H_2O]$ absent from this expression? **(c)** A solution is described as basic. What is meant by this statement?

17.24 Calculate $[H^+]$ for each of the following solutions at 25 °C, and indicate whether the solution is acidic, basic or neutral: **(a)** $[OH^-] = 0.00045$ M; **(b)** $[OH^-] = 8.8 \times 10^{-9}$ M; **(c)** a solution in which $[OH^-]$ is 100 times greater than $[H^+]$.

17.25 Calculate $[OH^-]$ for each of the following solutions, and indicate whether the solution is acidic, basic or neutral: **(a)** $[H^+] = 0.0075$ M; **(b)** $[H^+] = 6.5 \times 10^{-10}$ M; **(c)** a solution in which $[H^+]$ is 10 times greater than $[OH^-]$.

17.26 At the freezing point of water (0 °C), $K_w = 1.2 \times 10^{-15}$. Calculate $[H^+]$ and $[OH^-]$ for a neutral solution at this temperature.

THE pH SCALE (Section 17.4)

17.27 By what factor does $[H^+]$ change for a pH change of **(a)** 2.00 units **(b)** 0.50 units?

17.28 Consider two solutions, solution A and solution B. $[H^+]$ in solution A is 500 times greater than that in solution B. What is the difference in the pH values of the two solutions?

17.29 **(a)** If NaOH is added to water, how does $[H^+]$ change? How does pH change? **(b)** If the pH of a solution is 5.2 at 25 °C, first estimate and then calculate the molar concentrations of $[H^+]$ and $[OH^-]$ in the solution.

17.30 Complete the following table for solutions at 25 °C by calculating the missing entries and indicating whether the solution is acidic or basic.

$[H^+]$	$[OH^-]$	pH	pOH	Acidic or basic?
7.5×10^{-3} M				
	3.6×10^{-10} M			
		8.25		
			5.70	

17.31 Complete the following table by calculating the missing entries. In each case indicate whether the solution is acidic or basic.

pH	pOH	$[H^+]$	$[OH^-]$	Acidic or basic?
4.75				
	11.89			
		6.5×10^{-3} M		
			8.6×10^{-7} M	

17.32 The average pH of normal arterial blood is 7.40. At normal body temperature (37 °C), $K_w = 2.4 \times 10^{-14}$. Calculate $[H^+]$, $[OH^-]$ and pOH for blood at this temperature.

STRONG ACIDS AND BASES (Section 17.5)

17.33 (a) What is a strong acid? (b) A solution is labelled 0.500 M HCl. What is $[H^+]$ for the solution? (c) Which of the following are strong acids: HF, HCl, HBr, HI?

17.34 (a) What is a strong base? (b) A solution is labelled 0.045 M $Sr(OH)_2$. What is $[OH^-]$ for the solution? (c) Is the following statement true or false? Because $Mg(OH)_2$ is not very soluble, it cannot be a strong base. Explain.

17.35 Calculate the pH of each of the following strong acid solutions: (a) 8.5×10^{-3} M HBr, (b) 1.52 g of HNO_3 in 575 cm³ of solution, (c) 5.00 cm³ of 0.250 M $HClO_4$ diluted to 50.0 cm³, (d) a solution formed by mixing 10.0 cm³ of 0.100 M HBr with 20.0 cm³ of 0.200 M HCl.

17.36 Calculate $[OH^-]$ and pH for (a) 1.5×10^{-3} M $Sr(OH)_2$, (b) 2.250 g of LiOH in 250.0 cm³ of solution, (c) 1.00 cm³ of 0.175 M NaOH diluted to 2.00 dm³, (d) a solution formed by adding 5.00 cm³ of 0.105 M KOH to 15.0 cm³ of 9.5×10^{-2} M $Ca(OH)_2$.

17.37 Calculate the concentration of an aqueous solution of NaOH that has a pH of 11.50.

17.38 Calculate the concentration of an aqueous solution of $Ca(OH)_2$ that has a pH of 12.05.

17.39 Calculate the pH of a solution made by adding 15.00 g of sodium hydride (NaH) to enough water to make 2.500 dm³ of solution.

17.40 Calculate the pH of a solution made by adding 2.50 g of lithium oxide (Li_2O) to enough water to make 1.500 dm³ of solution.

WEAK ACIDS (Section 17.6)

17.41 Write the chemical equation and the K_a expression for the ionisation of each of the following acids in aqueous solution. First show the reaction with $H^+(aq)$ as a product and then with the hydronium ion: (a) $HBrO_2$, (b) CH_3CH_2COOH.

17.42 Lactic acid ($CH_3CH(OH)COOH$) has one acidic hydrogen. A 0.10 M solution of lactic acid has a pH of 2.44. Calculate K_a.

17.43 Phenylacetic acid ($C_6H_5CH_2COOH$) is one of the substances that accumulates in the blood of people with phenylketonuria, an inherited disorder that can cause mental retardation or even death. A 0.085 M solution of phenylacetic acid is found to have a pH of 2.68. Calculate the K_a value for this acid.

17.44 A 0.100 M solution of chloroacetic acid ($ClCH_2COOH$) is 11.0% ionised. Using this information, calculate $[ClCH_2COO^-]$, $[H^+]$, $[ClCH_2COOH]$ and K_a for chloroacetic acid.

17.45 How many moles of HF ($K_a = 6.8 \times 10^{-4}$) must be present in 0.200 dm³ to form a solution with a pH of 3.25?

17.46 The acid-dissociation constant for benzoic acid (C_6H_5COOH) is 6.3×10^{-5}. Calculate the equilibrium concentrations of H_3O^+, $C_6H_5COO^-$ and C_6H_5COOH in the solution if the initial concentration of C_6H_5COOH is 0.050 M.

17.47 Calculate the pH of each of the following solutions (K_a and K_b values are given in Appendix D): (a) 0.095 M propionic acid (CH_3CH_2COOH), (b) 0.100 M hydrogen chromate ion ($HCrO_4^-$), (c) 0.120 M pyridine (C_5H_5N).

17.48 Saccharin, a sugar substitute, is a weak acid with $pK_a = 2.32$ at 25 °C. It ionises in aqueous solution as follows:

$$C_7H_5O_3NS(aq) \rightleftharpoons H^+(aq) + C_7H_4O_3NS^-(aq)$$

What is the pH of a 0.10 M solution of this substance?

17.49 The active ingredient in aspirin is acetylsalicylic acid ($C_6H_4(OCOCH_3)COOH$), a monoprotic acid with $K_a = 3.3 \times 10^{-4}$ at 25 °C. What is the pH of a solution obtained by dissolving two extra-strength aspirin tablets, containing 500 mg of acetylsalicylic acid each, in 250 cm³ of water?

17.50 Calculate the percent ionisation of hydrazoic acid (HN_3) in solutions of each of the following concentrations (K_a is given in Appendix D): (a) 0.400 M, (b) 0.100 M, (c) 0.0400 M.

17.51 Show that, for a weak acid, the percent ionisation should vary as the inverse square root of the acid concentration.

17.52 Citric acid, which is present in citrus fruits, is a triprotic acid (Table 17.3). Calculate the pH and the citrate ion $C_3H_5O(COO^-)_3$ concentration for a 0.050 M solution of citric acid. Explain any approximations or assumptions that you make in your calculations.

17.53 Tartaric acid is found in many fruits, including grapes. It is partially responsible for the dry texture of certain wines. Calculate the pH and the tartarate ion $C_2H_4O_2(COO^-)_2$ concentration for a 0.250 M solution of tartaric acid, for which the acid-dissociation constants are listed in Table 17.3. Explain any approximations or assumptions that you make in your calculation.

WEAK BASES (Section 17.7)

17.54 What is the essential structural feature of all Brønsted–Lowry bases?

17.55 What are two kinds of molecules or ions that commonly function as weak bases?

17.56 Write the chemical equation and the K_b expression for the ionisation of each of the following bases in aqueous solution: (a) dimethylamine, $(CH_3)_2NH$; (b) carbonate ion, CO_3^{2-}; (c) formate ion, HCO_2^-.

17.57 Write the chemical equation and the K_b expression for the reaction of each of the following bases with water: (a) propylamine, $C_3H_7NH_2$; (b) monohydrogen phosphate ion, HPO_4^{2-}; (c) benzoate ion, $C_6H_5COO^-$.

17.58 Calculate the molar concentration of OH^- ions in a 0.075 M solution of ethylamine $(C_2H_5NH_2)$ $(K_b = 6.4 \times 10^{-4})$. Calculate the pH of this solution.

17.59 Ephedrine, a central nervous system stimulant, is used in nasal sprays as a decongestant. This compound is a weak organic base:

$$C_{10}H_{15}ON(aq) + H_2O(l) \rightleftharpoons C_{10}H_{15}ONH^+(aq) + OH^-(aq)$$

A 0.035 M solution of ephedrine has a pH of 11.33. **(a)** What are the equilibrium concentrations of $C_{10}H_{15}ON$, $C_{10}H_{15}ONH^+$ and OH^-? **(b)** Calculate K_b for ephedrine.

THE K_a–K_b RELATIONSHIP; ACID–BASE PROPERTIES OF SALTS (Sections 17.8 and 17.9)

17.60 Although the acid-dissociation constant for phenol (C_6H_5OH) is listed in Appendix D, the base-dissociation constant for the phenolate ion $(C_6H_5O^-)$ is not. **(a)** Explain why it is not necessary to list both K_a for phenol and K_b for the phenolate ion. **(b)** Calculate K_b for the phenolate ion. **(c)** Is the phenolate ion a weaker or stronger base than ammonia?

17.61 We can calculate K_b for the carbonate ion if we know the K_a values of carbonic acid (H_2CO_3). **(a)** Is K_{a1} or K_{a2} of carbonic acid used to calculate K_b for the carbonate ion? Explain. **(b)** Calculate K_b for the carbonate ion. **(c)** Is the carbonate ion a weaker or stronger base than ammonia?

17.62 **(a)** Given that K_b for ammonia is 1.8×10^{-5} and for hydroxylamine is 1.1×10^{-8}, which is the stronger base? **(b)** Which is the stronger acid, the ammonium ion or the hydroxylammonium ion? **(c)** Calculate K_a values for NH_4^+ and H_3NOH^+.

17.63 Using data from Appendix D, calculate $[OH^-]$ and pH for each of the following solutions: **(a)** 0.10 M NaCN, **(b)** 0.080 M Na_2CO_3, **(c)** a mixture that is 0.10 M in $NaNO_2$ and 0.20 M in $Ca(NO_2)_2$.

17.64 An unknown salt is NaF, NaCl or NaOCl. When 0.050 mol of the salt is dissolved in water to form 0.500 dm^3 of solution, the pH of the solution is 8.08. What is the identity of the salt?

17.65 Sorbic acid (C_5H_7COOH) is a weak monoprotic acid with $K_a = 1.7 \times 10^{-5}$. Its salt (potassium sorbate) is added to cheese to inhibit the formation of mould. What is the pH of a solution containing 11.25 g of potassium sorbate in 1.75 dm^3 of solution?

17.66 Predict whether aqueous solutions of the following compounds are acidic, basic or neutral: **(a)** NH_4Br, **(b)** $FeCl_3$, **(c)** Na_2CO_3, **(d)** $KClO_4$, **(e)** $NaHC_2O_4$.

17.67 Predict whether aqueous solutions of the following substances are acidic, basic or neutral: **(a)** $AlCl_3$, **(b)** NaBr, **(c)** NaClO, **(d)** $[CH_3NH_3]NO_3$, **(e)** Na_2SO_3.

ACID–BASE CHARACTER AND CHEMICAL STRUCTURE (Section 17.10)

17.68 How does the acid strength of an oxyacid depend on **(a)** the electronegativity of the central atom; **(b)** the number of non-protonated oxygen atoms in the molecule?

17.69 **(a)** How does the strength of an acid vary with the polarity and strength of the H–X bond? **(b)** How does the acidity of the binary acid of an element vary as a function of the electronegativity of the element? How does this relate to the position of the element in the periodic table?

17.70 Explain the following observations: **(a)** HNO_3 is a stronger acid than HNO_2; **(b)** H_2S is a stronger acid than H_2O; **(c)** H_2SO_4 is a stronger acid than HSO_4^-; **(d)** H_2SO_4 is a stronger acid than H_2SeO_4; **(e)** CCl_3COOH is a stronger acid than CH_3COOH.

17.71 Based on their compositions and structures and on conjugate acid–base relationships, select the stronger base in each of the following pairs: **(a)** BrO^- or ClO^-, **(b)** BrO^- or BrO_2^-, **(c)** HPO_4^{2-} or $H_2PO_4^-$.

17.72 Indicate whether each of the following statements is true or false. For each statement that is false, correct the statement so that it is true. **(a)** In general, the acidity of binary acids increases from left to right in a given row of the periodic table. **(b)** In a series of acids that have the same central atom, acid strength increases with the number of hydrogen atoms bonded to the central atom. **(c)** Hydrotelluric acid (H_2Te) is a stronger acid than H_2S because Te is more electronegative than S.

17.73 Indicate whether each of the following statements is true or false. For each statement that is false, correct the statement so that it is true. **(a)** Acid strength in a series of H–X molecules increases with increasing size of X. **(b)** For acids of the same general structure but differing electronegativities of the central atoms, acid strength decreases with increasing electronegativity of the central atom. **(c)** The strongest acid known is HF because fluorine is the most electronegative element.

LEWIS ACIDS AND BASES (Section 17.11)

17.74 If a substance is an Arrhenius base, is it necessarily a Brønsted–Lowry base? Is it necessarily a Lewis base? Explain.

17.75 If a substance is a Lewis acid, is it necessarily a Brønsted–Lowry acid? Is it necessarily an Arrhenius acid? Explain.

17.76 Identify the Lewis acid and Lewis base among the reactants in each of the following reactions:

(a) $Fe(ClO_4)_3(s) + 6 H_2O(l) \rightleftharpoons$
$$Fe(H_2O)_6^{3+}(aq) + 3 ClO_4^-(aq)$$

(b) $CN^-(aq) + H_2O(l) \rightleftharpoons HCN(aq) + OH^-(aq)$

(c) $(CH_3)_3N(g) + BF_3(g) \rightleftharpoons (CH_3)_3NBF_3(s)$

(d) $HIO(lq) + NH_2^-(lq) \rightleftharpoons NH_3(lq) + IO^-(lq)$
(lq denotes liquid ammonia as solvent)

17.77 Predict which member of each pair produces the more acidic aqueous solution: **(a)** K^+ or Cu^{2+}, **(b)** Fe^{2+} or Fe^{3+}, **(c)** Al^{3+} or Ga^{3+}. Explain.

17.78 Which member of each pair produces the more acidic aqueous solution: **(a)** $ZnBr_2$ or $CdCl_2$, **(b)** $CuCl$ or $Cu(NO_3)_2$, **(c)** $Ca(NO_3)_2$ or $NiBr_2$? Explain.

ADDITIONAL EXERCISES

17.79 In your own words, define or explain **(a)** K_w, **(b)** K_a, **(c)** pOH.

17.80 Indicate whether each of the following statements is correct or incorrect. For those that are incorrect, explain why they are wrong.

(a) Every Brønsted–Lowry acid is also a Lewis acid.

(b) Every Lewis acid is also a Brønsted–Lowry acid.

(c) Conjugate acids of weak bases produce more acidic solutions than conjugate acids of strong bases.

(d) K^+ ion is acidic in water because it causes hydrating water molecules to become more acidic.

(e) The percent ionisation of a weak acid in water increases as the concentration of acid decreases.

17.81 Haemoglobin plays a part in a series of equilibria involving protonation–deprotonation and oxygenation–deoxygenation. The overall reaction is approximately as follows:

$$HbH^+(aq) + O_2(aq) \rightleftharpoons HbO_2(aq) + H^+(aq)$$

where Hb stands for haemoglobin and HbO_2 for oxy-haemoglobin. **(a)** What effect does high O_2 concentration have on the position of this equilibrium? **(b)** The normal pH of blood is 7.4. Is the blood acidic, basic or neutral? **(c)** If the blood pH is lowered by the presence of large amounts of acidic metabolism products, a condition known as acidosis results. What effect does lowering blood pH have on the ability of haemoglobin to transport O_2?

17.82 What is the pH of a solution that is 2.5×10^{-9} M in NaOH? Does your answer make sense?

17.83 Which of the following solutions has the higher pH? **(a)** a 0.1 M solution of a strong acid or a 0.1 M solution of a weak acid, **(b)** a 0.1 M solution of an acid with $K_a = 2 \times 10^{-3}$ or one with $K_a = 8 \times 10^{-6}$, **(c)** a 0.1 M solution of a base with $pK_b = 4.5$ or one with $pK_b = 6.5$.

17.84 Caproic acid ($C_5H_{11}COOH$) is found in small amounts in coconut and palm oils and is used in making artificial flavours. A saturated solution of the acid contains 11 g dm^{-3} and has a pH of 2.94. Calculate K_a for the acid.

17.85 Arrange the following 0.10 M solutions in order of increasing acidity (decreasing pH): (i) NH_4NO_3, (ii) $NaNO_3$, (iii) CH_3COONH_4 (iv) NaF, (v) CH_3COONa.

17.86 What are the concentrations of H^+, $H_2PO_4^-$, HPO_4^{2-} and PO_4^{3-} in a 0.0250 M solution of H_3PO_4?

17.87 Many moderately large organic molecules containing basic nitrogen atoms are not very soluble in water as neutral molecules, but they are frequently much more soluble as their acid salts. Assuming that pH in the stomach is 2.5, indicate whether each of the following compounds would be present in the stomach as the neutral base or in the protonated form: nicotine, $K_b = 7 \times 10^{-7}$; caffeine, $K_b = 4 \times 10^{-14}$; strychnine, $K_b = 1 \times 10^{-6}$; quinine, $K_b = 1.1 \times 10^{-6}$.

17.88 The amino acid glycine (H_2N-CH_2-COOH) can participate in the following equilibria in water:

$H_2N-CH_2-COOH + H_2O \rightleftharpoons$
$\quad H_2N-CH_2-COO^- + H_3O^+ \quad K_a = 4.3 \times 10^{-3}$

$H_2N-CH_2-COOH + H_2O \rightleftharpoons$
$\quad {}^+H_3N-CH_2-COOH + OH^- \quad K_b = 6.0 \times 10^{-5}$

(a) Use the values of K_a and K_b to estimate the equilibrium constant for the intramolecular proton transfer to form a zwitterion:

$$H_2N-CH_2-COOH \rightleftharpoons {}^+H_3N-CH_2-COO^-$$

What assumptions did you need to make? **(b)** What is the pH of a 0.050 M aqueous solution of glycine? **(c)** What would be the predominant form of glycine in a solution with pH 13? With pH 1?

INTEGRATIVE EXERCISES

17.89 Calculate the number of $H^+(aq)$ ions in $1.0\ cm^3$ of pure water at 25 °C.

17.90 How many millilitres of concentrated hydrochloric acid solution (36.0% HCl by mass, density = $1.18\ g\ cm^{-3}$) are required to produce $10.0\ dm^3$ of a solution that has a pH of 2.05?

17.91 Atmospheric CO_2 levels have risen by nearly 20% over the past 40 years from 315 ppm to 375 ppm. **(a)** Given that the average pH of clean, unpolluted rain today is 5.4, determine the pH of unpolluted rain 40 years ago. Assume that carbonic acid (H_2CO_3) formed by the reaction of CO_2 and water is the only factor influencing pH.

$$CO_2(g) + H_2O(l) \rightleftharpoons H_2CO_3(aq)$$

(b) What volume of CO_2 at 25 °C and 1.0 bar is dissolved in a $20.0\ dm^3$ bucket of today's rainwater?

17.92 In many reactions the addition of $AlCl_3$ produces the same effect as the addition of H^+. **(a)** Draw a Lewis structure for $AlCl_3$ in which no atoms carry formal charges, and determine its structure using the VSEPR method. **(b)** What characteristic is notable about the structure in part (a) that helps us understand the acidic character of $AlCl_3$? **(c)** Predict the result of the reaction between $AlCl_3$ and NH_3 in a solvent that does not participate as a reactant. **(d)** Which acid–base theory is most suitable for discussing the similarities between $AlCl_3$ and H^+?

17.93 What is the boiling point of a 0.10 M solution of $NaHSO_4$ if the solution has a density of $1.002\ g\ cm^{-3}$?

17.94 The iodate ion is reduced by sulfite according to the following reaction:

$$IO_3^-(aq) + 3\,SO_3^{2-}(aq) \longrightarrow I^-(aq) + 3\,SO_4^{2-}(aq)$$

The rate of this reaction is found to be first order in IO_3^-, first order in SO_3^{2-} and first order in H^+. **(a)** Write the rate law for the reaction. **(b)** By what factor will the rate of the reaction change if the pH is lowered from 5.00 to 3.50? Does the reaction proceed faster or slower at the lower pH? **(c)** By using the concepts discussed in Section 17.6, explain how the reaction can be pH dependent even though H^+ does not appear in the overall reaction.

PHOTO/ART CREDITS

654 © Gynane/Dreamstime.com; **667, 669, 687** Richard Megna, Fundamental Photographs, NYC; **676** Pearson Science; **685** Newspix/Brett Hartwig © Newspix/News Ltd/3rd Party Managed Reproduction & Supply Rights.

18

ADDITIONAL ASPECTS OF AQUEOUS EQUILIBRIA

The Great Barrier Reef stretches for more than 2600 km off the eastern coast of Australia. Coral reefs such as this are among the most diverse ecosystems on the planet.

KEY CONCEPTS

18.1 THE COMMON-ION EFFECT
We consider a specific example of Le Châtelier's principle known as the *common-ion effect*.

18.2 BUFFER SOLUTIONS
We examine the composition of buffer solutions and learn how they resist pH change when small amounts of a strong acid or strong base are added to them.

18.3 ACID–BASE TITRATIONS
We study acid–base titrations and learn how to determine pH at any point in an acid–base titration.

18.4 SOLUBILITY EQUILIBRIA
We learn how to use *solubility-product constants* to determine to what extent a sparingly soluble salt dissolves in water.

18.5 FACTORS THAT AFFECT SOLUBILITY
We investigate some of the factors that affect solubility, including the common-ion effect and the effect of acids.

18.6 PRECIPITATION AND SEPARATION OF IONS
We learn how differences in solubility can be used to separate ions through selective precipitation.

18.7 QUALITATIVE ANALYSIS FOR METALLIC ELEMENTS
We explain how the principles of solubility and complex-ion equilibria can be used to identify ions in solution.

W ater, the most common and most important solvent on Earth, occupies its position of importance because of its abundance and its exceptional ability to dissolve a wide variety of substances. Coral reefs are a striking example of aqueous chemistry at work in nature. Coral reefs are built by tiny animals called stony corals, which secrete a hard calcium carbonate exoskeleton. Over time, the stony corals build up large networks of calcium carbonate upon which a reef is built. The size of such structures can be immense, as illustrated by the Great Barrier Reef.

Stony corals make their exoskeletons from dissolved Ca^{2+} and CO_3^{2-} ions. This process is aided by the fact that the CO_3^{2-} concentration is supersaturated in most parts of the ocean. However, well-documented increases in the amount of CO_2 in the atmosphere threaten to upset the aqueous chemistry that stony corals depend on. As atmospheric CO_2 levels increase, the amount of CO_2 dissolved in the ocean also increases. This lowers the pH of the ocean and leads to a decrease in the CO_3^{2-} concentration. As a result it becomes more difficult for stony corals and other important ocean creatures to maintain their exoskeletons. We will take a closer look at the consequences of ocean acidification later in the chapter.

To understand the chemistry that underlies coral reef formation, we must understand the concepts of aqueous equilibria. In this chapter we take a step towards understanding such complex solutions by looking first at further applications of acid–base equilibria. The idea is to consider not only solutions in which there is a single solute but also those containing a mixture of solutes. We then broaden our discussion to include two additional types of aqueous equilibria: those involving slightly soluble salts and those involving the formation of metal complexes in solution.

18.1 | THE COMMON-ION EFFECT

In Chapter 17 we examined the equilibrium concentrations of ions in solutions containing a weak acid or a weak base. We now consider solutions that contain not only a weak acid, such as acetic acid (CH_3COOH), but also a soluble salt of that acid, such as sodium acetate (CH_3COONa). Notice that these solutions contain two substances that share a *common ion*, CH_3COO^-. It is instructive to view these solutions from the perspective of Le Châtelier's principle (∞ Section 16.8, 'Le Châtelier's Principle').

∞ Review this on page 638

∞ Review this on page 104

Sodium acetate, CH_3COONa, is a soluble ionic compound and therefore a strong electrolyte (∞ Section 4.1, 'General Properties of Aqueous Solutions'). It dissociates completely in aqueous solution to form Na^+ and CH_3COO^- ions:

$$CH_3COONa(aq) \longrightarrow Na^+(aq) + CH_3COO^-(aq)$$

In contrast, CH_3COOH is a weak electrolyte that ionises as follows:

$$CH_3COOH(aq) \rightleftharpoons H^+(aq) + CH_3COO^-(aq) \qquad [18.1]$$

When we have sodium acetate and acetic acid in the same solution, the CH_3COO^- from CH_3COONa causes the equilibrium of Equation 18.1 to shift to the left, thereby decreasing the equilibrium concentration of $H^+(aq)$:

$$CH_3COOH(aq) \rightleftharpoons H^+(aq) + CH_3COO^-(aq)$$

Addition of CH_3COO^- shifts equilibrium, reducing $[H^+]$

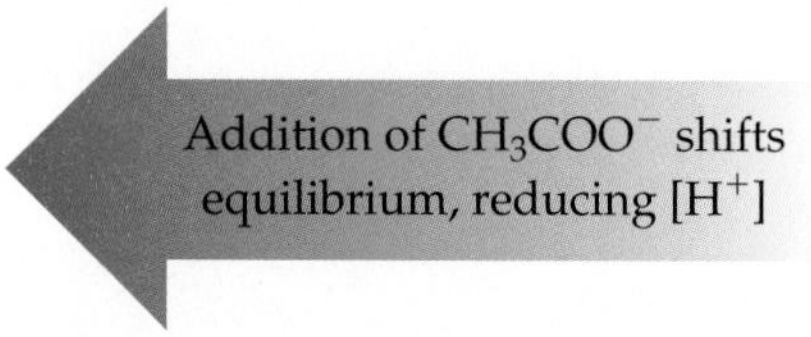

In other words, the presence of the added acetate ion causes the acetic acid to ionise less than it normally would.

Whenever a weak electrolyte and a strong electrolyte containing a common ion are together in solution, the weak electrolyte ionises less than it would if it were alone in solution. We call this observation the **common-ion effect**.

Sample Exercises 18.1 and 18.2 illustrate how equilibrium concentrations may be calculated when a solution contains a mixture of a weak electrolyte and a strong electrolyte that have a common ion. The procedures are similar to those encountered for weak acids and weak bases in Chapter 17.

SAMPLE EXERCISE 18.1 Calculating the pH when a common ion is involved

What is the pH of a solution made by adding 0.30 mol of acetic acid (CH_3COOH) and 0.30 mol of sodium acetate (CH_3COONa) to enough water to make 1.0 dm³ of solution?

SOLUTION

Analyse We are asked to determine the pH of a solution of a weak electrolyte (CH_3COOH) and a strong electrolyte (CH_3COONa) that share a common ion, CH_3COO^-.

Plan In any problem in which we must determine the pH of a solution containing a mixture of solutes, it is helpful to proceed by a series of logical steps:
1. Consider which solutes are strong electrolytes and which are weak electrolytes, and identify the major species in solution.

2. Identify the important equilibrium that is the source of H^+ and therefore determines pH.
3. Tabulate the concentrations of ions involved in the equilibrium.
4. Use the equilibrium constant expression to calculate $[H^+]$ and then pH.

Solve First, the major species in the solution are CH_3COOH (a weak acid), Na^+ (which is neither acidic nor basic and is therefore a spectator in the acid–base chemistry) and CH_3COO^- (which is the conjugate base of CH_3COOH).

Second, [H$^+$] and therefore the pH are controlled by the dissociation equilibrium of CH$_3$COOH:

$$CH_3COOH(aq) \rightleftharpoons H^+(aq) + CH_3COO^-(aq) \quad \text{(weak electrolyte dissociation)}$$
$$CH_3COONa(aq) \longrightarrow Na^+(aq) + CH_3COO^-(aq) \quad \text{(strong electrolyte dissociation)}$$

(We write the equilibrium using H$^+$(aq) rather than H$_3$O$^+$(aq), but both representations of the hydrated hydrogen ion are equally valid.)

Third, we tabulate the initial and equilibrium concentrations much as we did in solving other equilibrium problems in Chapters 16 and 17:

	CH$_3$COOH (aq)	$\rightleftharpoons$	H$^+$(aq)	+	CH$_3$COO$^-$ (aq)
Initial	0.30 M		0		0.30 M
Change	$-x$ M		$+x$ M		$+x$ M
Equilibrium	$(0.30 - x)$ M		x M		$(0.30 + x)$ M

The equilibrium concentration of CH$_3$COO$^-$ (the common ion) is the initial concentration that is due to CH$_3$COONa (0.30 M) plus the change in concentration (x) that is due to the ionisation of CH$_3$COOH.

Now we can use the equilibrium constant expression:

(The dissociation constant for CH$_3$COOH at 25 °C is from Appendix D; addition of CH$_3$COONa does not change the value of this constant.)

$$K_a = 1.8 \times 10^{-5} = \frac{[H^+][CH_3COO^-]}{[CH_3COOH]}$$

Substituting the equilibrium constant concentrations from the table into the equilibrium expression gives

$$K_a = 1.8 \times 10^{-5} = \frac{x(0.30 + x)}{0.30 - x}$$

Because K_a is small, we assume that x is small compared with the original concentrations of CH$_3$COOH and CH$_3$COO$^-$ (0.30 M each). Thus we can ignore the very small x relative to 0.30 M, giving

$$K_a = 1.8 \times 10^{-5} = \frac{x(0.30)}{0.30}$$

The resulting value of x is indeed small relative to 0.30, justifying the approximation made in simplifying the problem.

$$x = 1.8 \times 10^{-5} \text{ M}$$

Finally, we calculate the pH from the equilibrium concentration of H$^+$(aq):

$$pH = -\log(1.8 \times 10^{-5}) = 4.74$$

Comment In Section 17.6 we calculated that a 0.30 M solution of CH$_3$COOH has a pH of 2.64, corresponding to [H$^+$] = 2.3 $\times$ 10^{-3} M. Thus the addition of CH$_3$COONa has substantially decreased [H$^+$], as we would expect from Le Châtelier's principle.

PRACTICE EXERCISE

Calculate the pH of a solution containing 0.085 M nitrous acid (HNO$_2$; K_a = 4.5 $\times$ 10^{-4}) and 0.10 M potassium nitrite (KNO$_2$).

Answer: 3.42

(See also Exercises 18.10, 18.11.)

SAMPLE EXERCISE 18.2   **Calculating ion concentrations when a common ion is involved**

Calculate the fluoride ion concentration and pH of a solution that is 0.20 M in HF and 0.10 M in HCl.

SOLUTION

Analyse We are asked to determine the concentration of F$^-$ and the pH in a solution containing the weak acid HF and the strong acid HCl. In this case the common ion is H$^+$.

Plan We can again use the four steps outlined in Sample Exercise 18.1.

Solve Because HF is a weak acid and HCl is a strong acid, the major species in solution are HF, H^+ and Cl^-. The Cl^-, which is the conjugate base of a strong acid, is merely a spectator ion in any acid–base chemistry. The problem asks for $[F^-]$, which is formed by ionisation of HF.

$$HF(aq) \rightleftharpoons H^+(aq) + F^-(aq) \qquad \text{(weak electrolyte dissociation)}$$

$$HCl(aq) \longrightarrow H^+(aq) + Cl^-(aq) \qquad \text{(strong electrolyte dissociation)}$$

Thus the important equilibrium is the ionisation of HF.

The common ion in this problem is the hydrogen or hydronium ion. Now we can tabulate the initial and equilibrium concentrations of each species involved in this equilibrium:

	$HF(aq)$	$\rightleftharpoons$	$H^+(aq)$	$+$	$F^-(aq)$
Initial	0.20 M		0.10 M		0
Change	$-x$ M		$+x$ M		$+x$ M
Equilibrium	$(0.20 - x)$ M		$(0.10 + x)$ M		x M

The equilibrium constant for the ionisation of HF, from Appendix D, is 6.8×10^{-4}. Substituting the equilibrium constant concentrations into the equilibrium expression gives

$$K_a = 6.8 \times 10^{-4} = \frac{[H^+][F^-]}{[HF]} = \frac{(0.10 + x)(x)}{0.20 - x}$$

If we assume that x is small relative to 0.10 or 0.20 M, this expression simplifies to

$$\frac{(0.10)(x)}{0.20} = 6.8 \times 10^{-4}$$

$$x = \frac{0.20}{0.10}(6.8 \times 10^{-4}) = 1.4 \times 10^{-3} \text{ M} = [F^-]$$

This F^- concentration is substantially smaller than it would be in a 0.20 M solution of HF with no added HCl. The common ion, H^+, suppresses the ionisation of HF. The concentration of $H^+(aq)$ is

$$[H^+] = (0.10 + x) \text{ M} \simeq 0.10 \text{ M}$$

Thus

$$pH = 1.00$$

Comment Notice that for all practical purposes $[H^+]$ is due entirely to the HCl; the HF makes a negligible contribution by comparison.

PRACTICE EXERCISE
Calculate the formate ion concentration and pH of a solution that is 0.050 M in formic acid (HCOOH; $K_a = 1.8 \times 10^{-4}$) and 0.10 M in HNO_3.
Answer: $[HCOO^-] = 9.0 \times 10^{-5}$; pH = 1.00
(See also Exercise 18.11.)

Sample Exercises 18.1 and 18.2 both involve weak acids. The ionisation of a weak base is also decreased by the addition of a common ion. For example, the addition of NH_4^+ (as from the strong electrolyte NH_4Cl) causes the base-dissociation equilibrium of NH_3 to shift to the left, decreasing the equilibrium concentration of OH^- and lowering the pH:

$$NH_3(aq) + H_2O(l) \rightleftharpoons NH_4^+(aq) + OH^-(aq) \qquad [18.2]$$

 CONCEPT CHECK 1

A mixture of 0.10 mol of NH_4Cl and 0.12 mol of NH_3 is added to enough water to make 1.0 dm^3 of solution.
a. What are the initial concentrations of the major species in the solution?
b. Which of the ions in this solution is a spectator ion in any acid–base chemistry occurring in the solution?
c. What equilibrium reaction determines $[OH^-]$ and therefore the pH of the solution?

18.2 | BUFFER SOLUTIONS

Solutions such as those discussed in Section 18.1, which contain a weak conjugate acid–base pair, can resist drastic changes in pH upon the addition of small amounts of strong acid or strong base. These solutions are called **buffer solutions** (or merely **buffers**). Human blood, for example, is a complex aqueous mixture with a pH buffered at about 7.4 (see the My World of Chemistry box near the end of this section). Much of the chemical behaviour of seawater is determined by its pH, buffered at about 8.1 to 8.3 near the surface. Buffer solutions find many important applications in the laboratory and in medicine (▶ FIGURE 18.1).

▲ **FIGURE 18.1   Buffer solutions.** Prepackaged buffer solutions and ingredients for making up buffer solutions of predetermined pH can be purchased.

Composition and Action of Buffer Solutions

A buffer resists changes in pH because it contains both an acidic species to neutralise OH^- ions and a basic species to neutralise H^+ ions. The acidic and basic species that make up the buffer, however, must not consume each other through a neutralisation reaction. These requirements are fulfilled by a weak acid–base conjugate pair such as CH_3COOH/CH_3COO^- or NH_4^+/NH_3. Thus buffers are often prepared by mixing a weak acid or a weak base with a salt of that acid or base. The CH_3COOH/CH_3COO^- buffer can be prepared, for example, by adding CH_3COONa to a solution of CH_3COOH; the NH_4^+/NH_3 buffer can be prepared by adding NH_4Cl to a solution of NH_3. By choosing appropriate components and adjusting their relative concentrations, we can buffer a solution at virtually any pH.

 CONCEPT CHECK 2

Which of the following conjugate acid–base pairs will not function as a buffer: $HCOOH$ and $HCOO^-$; HCO_3^- and CO_3^{2-}; HNO_3 and NO_3^-? Explain.

To understand better how a buffer works, let's consider a buffer composed of a weak acid (HX) and one of its salts (MX, where M^+ could be Na^+, K^+ or another cation). The acid-dissociation equilibrium in this buffer solution involves both the acid and its conjugate base:

$$HX(aq) \rightleftharpoons H^+(aq) + X^-(aq) \qquad [18.3]$$

The corresponding acid-dissociation constant expression is

$$K_a = \frac{[H^+][X^-]}{[HX]} \qquad [18.4]$$

Solving this expression for $[H^+]$, we have

$$[H^+] = K_a \frac{[HX]}{[X^-]} \qquad [18.5]$$

We see from this expression that $[H^+]$, and thus the pH, is determined by two factors: the value of K_a for the weak-acid component of the buffer and the ratio of the concentrations of the conjugate acid–base pair, $[HX]/[X^-]$.

If OH^- ions are added to the buffer solution, they react with the acid component of the buffer to produce water and the base component (X^-):

$$\underset{\text{added base}}{OH^-(aq)} + HX(aq) \longrightarrow H_2O(l) + X^-(aq) \qquad [18.6]$$

This reaction causes [HX] to decrease and $[X^-]$ to increase. As long as the amounts of HX and X^- in the buffer are large compared with the amount of OH^- added, however, the ratio $[HX]/[X^-]$ doesn't change much, and thus the change in pH is small.

If H^+ ions are added, they react with the base component of the buffer:

$$H^+(aq) + X^-(aq) \longrightarrow HX(aq) \qquad [18.7]$$

added acid

This reaction can also be represented using H_3O^+:

$$H_3O^+(aq) + X^-(aq) \longrightarrow HX(aq) + H_2O(l)$$

Using either equation, the reaction causes $[X^-]$ to decrease and $[HX]$ to increase. As long as the change in the ratio $[HX]/[X^-]$ is small, the change in pH will be small.

▼ FIGURE 18.2. shows an HX/MX buffer consisting of equal concentrations of hydrofluoric acid and fluoride ion (centre). The addition of OH^- reduces $[HF]$ and increases $[F^-]$, whereas the addition of $[H^+]$ reduces $[F^-]$ and increases $[HF]$.

⚠ CONCEPT CHECK 3

a. What happens when NaOH is added to a buffer composed of CH_3COOH and CH_3COO^-?

b. What happens when HCl is added to this buffer?

Calculating the pH of a Buffer

Because conjugate acid–base pairs share a common ion, we can use the same procedures to calculate the pH of a buffer that we used to treat the common-ion effect (see Sample Exercise 18.1). However, an alternative approach is sometimes taken that is based on an equation derived from Equation 18.5. Taking the negative log of both sides of Equation 18.5, we have

$$-\log[H^+] = -\log\left(K_a\frac{[HX]}{[X^-]}\right) = -\log K_a - \log\frac{[HX]}{[X^-]}$$

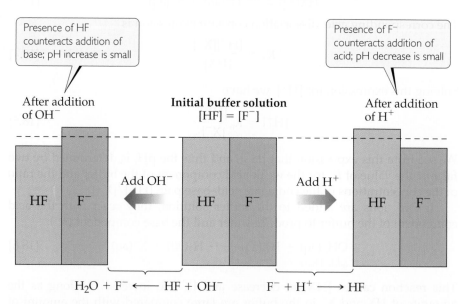

Presence of HF counteracts addition of base; pH increase is small

Presence of F^- counteracts addition of acid; pH decrease is small

After addition of OH^-

Initial buffer solution $[HF] = [F^-]$

After addition of H^+

Add OH^-

Add H^+

HF F^- HF F^- HF F^-

$$H_2O + F^- \longleftarrow HF + OH^- \qquad F^- + H^+ \longrightarrow HF$$

▲ **FIGURE 18.2   Buffer action.** The pH of an HF/F$^-$ buffer solution changes by only a small amount in response to addition of an acid or base.

Because $-\log[H^+] = pH$ and $-\log K_a = pK_a$

we have

$$pH = pK_a - \log\frac{[HX]}{[X^-]} = pK_a + \log\frac{[X^-]}{[HX]} \qquad [18.8]$$

In general,

$$pH = pK_a + \log\frac{[base]}{[acid]} \qquad [18.9]$$

where [base] and [acid] refer to the equilibrium concentrations of the conjugate acid–base pair. Note that when [base] = [acid], pH = pK_a.

Equation 18.9 is known as the **Henderson–Hasselbalch equation**. Biologists, biochemists and others who work frequently with buffers often use this equation to calculate the pH of buffers. In doing equilibrium calculations, we have seen that we can normally neglect the amounts of the acid and base of the buffer that ionise. Therefore we can usually use the starting concentrations of the acid and base components of the buffer directly in Equation 18.9.

SAMPLE EXERCISE 18.3   Calculating the pH of a buffer

What is the pH of a buffer that is 0.12 M in lactic acid ($CH_3CH(OH)COOH$) and 0.10 M in sodium lactate? For lactic acid, $K_a = 1.4 \times 10^{-4}$.

SOLUTION

Analyse We are asked to calculate the pH of a buffer containing lactic acid ($C_3H_6O_3$) and its conjugate base, the lactate ion ($C_3H_5O_3^-$).

Plan We first determine the pH using the method described in Section 18.1. The major species in solution are $CH_3CH(OH)COOH$, Na^+ and $CH_3CH(OH)COO^-$. The Na^+ ion is a spectator ion. The ($CH_3CH(OH)COOH$)/ $CH_3CH(OH)COO^-$ conjugate acid–base pair determines $[H^+]$ and thus pH; $[H^+]$ can be determined using the acid-dissociation equilibrium of lactic acid.

Solve The initial and equilibrium concentrations of the species involved in this equilibrium are

$$CH_3CH(OH)COOH(aq) \rightleftharpoons H^+(aq) + CH_3CH(OH)COO^-(aq)$$

Initial	0.12 M	0	0.10 M
Change	$-x$ M	$+x$ M	$+x$ M
Equilibrium	$(0.12 - x)$ M	x M	$(0.10 + x)$ M

The equilibrium concentrations are governed by the equilibrium expression:

$$K_a = 1.4 \times 10^{-4} = \frac{[H^+][CH_3CH(OH)COO^-]}{[CH_3CH(OH)COOH]} = \frac{x(0.10 + x)}{(0.12 - x)}$$

Because K_a is small and a common ion is present, we expect x to be small relative to either 0.12 or 0.10 M. Thus our equation can be simplified to give

$$K_a = 1.4 \times 10^{-4} = \frac{x(0.10)}{0.12}$$

Solving for x gives a value that justifies our approximation:

$$[H^+] = x = \left(\frac{0.12}{0.10}\right)(1.4 \times 10^{-4}) = 1.7 \times 10^{-4}\,M$$

$$pH = -\log(1.7 \times 10^{-4}) = 3.77$$

Alternatively, we could have used the Henderson–Hasselbalch equation to calculate pH directly:

$$pH = pK_a + \log\left(\frac{[base]}{[acid]}\right) = 3.85 + \log\left(\frac{0.10}{0.12}\right)$$

$$= 3.85 + (-0.08) = 3.77$$

PRACTICE EXERCISE

Calculate the pH of a buffer composed of 0.12 M benzoic acid and 0.20 M sodium benzoate. (Refer to Appendix D.)

Answer: 4.42

(See also Exercises 18.14, 18.15.)

SAMPLE EXERCISE 18.4 **Preparing a buffer**

How many moles of NH_4Cl must be added to $2.0 \ dm^3$ of $0.10 \ M \ NH_3$ to form a buffer whose pH is 9.00? (Assume that the addition of NH_4Cl does not change the volume of the solution: $K_b = 1.8 \times 10^{-5}$.)

SOLUTION

Analyse We are expected to determine the amount of NH_4^+ ion required to prepare a buffer of a specific pH.

Plan The major species in the solution will be NH_4^+, Cl^- and NH_3. Of these, the Cl^- ion is a spectator (it is the conjugate base of a strong acid). Thus the NH_4^+/NH_3 conjugate acid–base pair will determine the pH of the buffer solution. The equilibrium relationship between NH_4^+ and NH_3 is given by the base-dissociation constant for NH_3:

$$NH_3(aq) + H_2O(l) \rightleftharpoons NH_4^+(aq) + OH^-(aq)$$

$$K_b = \frac{[NH_4^+][OH^-]}{[NH_3]} = 1.8 \times 10^{-5}$$

The key to this exercise is to use this K_b expression to calculate $[NH_4^+]$.

Solve We obtain $[OH^-]$ from the given pH:

$$pOH = 14.00 - pH = 14.00 - 9.00 = 5.00$$

and so

$$[OH^-] = 1.0 \times 10^{-5} \ M$$

Because K_b is small and the common ion NH_4^+ is present, the equilibrium concentration of NH_3 will essentially equal its initial concentration:

$$[NH_3] = 0.10 \ M$$

We now use the expression for K_b to calculate $[NH_4^+]$:

$$[NH_4^+] = K_b \frac{[NH_3]}{[OH^-]} = (1.8 \times 10^{-5}) \frac{(0.10 \ M)}{(1.0 \times 10^{-5} \ M)} = 0.18 \ M$$

Thus for the solution to have pH = 9.00, $[NH_4^+]$ must equal 0.18 M. The number of moles of NH_4Cl needed to produce this concentration is given by the product of the volume of the solution and its molarity:

$$(2.0 \ dm^3)(0.18 \ mol \ NH_4Cl \ dm^{-3}) = 0.36 \ mol \ NH_4Cl$$

Comment Because NH_4^+ and NH_3 are a conjugate acid–base pair, we could use the Henderson–Hasselbalch equation (Equation 18.9) to solve this problem. To do so requires first using Equation 17.43 to calculate pK_a for NH_4^+ from the value of pK_b for NH_3. We suggest you try this approach to convince yourself that you can use the Henderson–Hasselbalch equation for buffers for which you are given K_b for the conjugate base rather than K_a for the conjugate acid.

PRACTICE EXERCISE

Calculate the concentration of sodium benzoate that must be present in a 0.20 M solution of benzoic acid (C_6H_5COOH) to produce a pH of 4.00.

Answer: 0.13 M

(See also Exercise 18.19.)

Buffer Capacity and pH Range

Two important characteristics of a buffer are its capacity and its effective pH range. **Buffer capacity** is the amount of acid or base the buffer can neutralise before the pH begins to change to an appreciable degree. The buffer capacity depends on the amount of acid and base from which the buffer is made. The pH of the buffer depends on the K_a for the acid and on the relative concentrations of the acid and base that comprise the buffer. According to Equation 18.5, for example, $[H^+]$ for a $1 \ dm^3$ solution that is 1 M in CH_3COOH and 1 M in

CH_3COONa will be the same as for a 1 dm^3 solution that is 0.1 M in CH_3COOH and 0.1 M in CH_3COONa. The first solution has a greater buffering capacity, however, because it contains more CH_3COOH and CH_3COO^-. The greater the amounts of the conjugate acid–base pair, the more resistant the ratio of their concentrations, and hence the pH, is to change.

The **pH range** of any buffer is the pH range over which the buffer acts effectively. Buffers most effectively resist a change in pH in *either* direction when the concentrations of weak acid and conjugate base are about the same. From Equation 18.9 we see that, when the concentrations of weak acid and conjugate base are equal, pH = pK_a. This relationship gives the optimal pH of any buffer. Thus we usually try to select a buffer whose acid form has a pK_a close to the desired pH. In practice, we find that if the concentration of one component of the buffer is more than 10 times the concentration of the other component, the buffering action is poor. Because log 10 = 1, buffers usually have a usable range within ± 1 pH unit of pK_a.

CONCEPT CHECK 4

What is the optimal pH buffered by a solution containing CH_3COOH and CH_3COONa? (K_a for CH_3COOH is 1.8×10^{-5}.)

Addition of Strong Acids or Bases to Buffers

When a strong acid or a strong base is added to a buffer solution, provided we do not exceed the buffering capacity of the buffer, we can assume that all the acid or base is completely consumed by reaction with the buffer. Consider a typical buffer containing a weak acid, HX, and its conjugate base, X^-:

$$HX(aq) \rightleftharpoons H^+(aq) + X^-(aq)$$

$$MX(aq) \longrightarrow M^+(aq) + X^-(aq)$$

As before, M^+ is a spectator ion and is disregarded. If H^+ is added to the buffer, it reacts with X^- to form HX, and if OH^- is added it reacts with HX to form H_2O and X^-. In other words, the buffer consumes the added acid or base. We can calculate how the pH of a buffer responds to the addition of acid or base as follows.

Consider the buffer formed by adding 0.300 mol of CH_3COOH and 0.300 mol of CH_3COONa to enough water to make 1.00 dm^3 of solution. The pK_a of this buffer is 4.74. How does the pH change if 0.020 mol of NaOH is added to this solution, assuming no change in volume? The pH of the buffer before the addition of NaOH is given by the Henderson–Hasselbalch equation:

$$pH = pK_a + \log \frac{[base]}{[acid]} = 4.74 + \log \frac{[0.300]}{[0.300]} = 4.74 + \log 1 = 4.74$$

If 0.020 mol of NaOH is added, it will react with the acid CH_3COOH according to the equation

$$CH_3COOH(aq) + OH^-(aq) \longrightarrow CH_3COO^-(aq) + H_2O(l)$$

Initially, the concentrations of the above species are 0.300 M, 0.020 M and 0.300 M, respectively. After the reaction of OH^- with CH_3COOH, the concentrations are

$$[CH_3COOH] = [0.300 - 0.020]\ M = 0.280\ M$$

$$[CH_3COO^-] = [0.300 + 0.020]\ M = 0.320\ M$$

BLOOD AS A BUFFER SOLUTION

Many of the chemical reactions that occur in living systems are extremely sensitive to pH. Many of the enzymes that catalyse important biochemical reactions, for example, are effective only within a narrow pH range. For this reason the human body maintains a remarkably intricate system of buffers, both within tissue cells and in the fluids that transport cells. Blood, the fluid that transports oxygen to all parts of the body (▼ FIGURE 18.3), is one of the most prominent examples of the importance of buffers in living beings.

Human blood is slightly basic with a normal pH of 7.35 to 7.45. Any deviation from this normal pH range can have extremely disruptive effects on the stability of cell membranes, the structures of proteins and the activities of enzymes. Death may result if the blood pH falls below 6.8 or rises above 7.8. When the pH falls below 7.35, the condition is called *acidosis*; when it rises above 7.45, the condition is called *alkalosis*. Acidosis is the more common tendency because ordinary metabolism generates several acids within the body.

The major buffer system that is used to control the pH of blood is the *carbonic acid/bicarbonate buffer system*. Carbonic acid (H_2CO_3) and bicarbonate ion (HCO_3^-) are a conjugate acid–base pair. In addition, carbonic acid can decompose into carbon dioxide gas and water. The important equilibria in this buffer system are

$$H^+(aq) + HCO_3^-(aq) \rightleftharpoons H_2CO_3(aq) \rightleftharpoons H_2O(l) + CO_2(g)$$
[18.10]

Several aspects of these equilibria are notable. First, although carbonic acid is a diprotic acid, the carbonate ion (CO_3^{2-}) is unimportant in this system. Second, one of the components of this equilibrium, CO_2, is a gas, which provides a mechanism for the body to adjust the equilibria. Removal of CO_2 via exhalation shifts the equilibria to the right, consuming H^+ ions. Third, the buffer system in blood operates at a pH of 7.4, which is fairly far removed from the pK_{a1} value of H_2CO_3 (6.1 at physiological temperatures). For the buffer to have a pH of 7.4, the ratio [base]/[acid] must have a value of about 20. In normal blood plasma the concentrations of HCO_3^- and H_2CO_3 are about 0.024 M and 0.0012 M, respectively. As a consequence, the buffer has a high capacity to neutralise additional acid, but only a low capacity to neutralise additional base.

The principal organs that regulate the pH of the carbonic acid/bicarbonate buffer system are the lungs and kidneys. Some of the receptors in the brain are sensitive to the concentrations of H^+ and CO_2 in bodily fluids. When the concentration of CO_2 rises, the equilibria in Equation 18.10 shift to the left, which leads to the formation of more H^+. The receptors trigger a reflex to breathe faster and deeper, increasing the rate of elimination of CO_2 from the lungs and shifting the equilibria back to the right. The kidneys absorb or release H^+ and HCO_3^-; much of the excess acid leaves the body in urine, which normally has a pH of 5.0 to 7.0.

The regulation of the pH of blood plasma relates directly to the effective transport of O_2 to bodily tissues. Oxygen is carried by the protein haemoglobin, which is found in red blood cells. Haemoglobin (Hb) reversibly binds both H^+ and O_2. These two substances compete for the Hb, which can be represented approximately by the following equilibrium:

$$HbH^+ + O_2 \rightleftharpoons HbO_2 + H^+$$
[18.11]

Oxygen enters the blood through the lungs, where it passes into the red blood cells and binds to Hb. When the blood reaches tissue in which the concentration of O_2 is low, the equilibrium in Equation 18.11 shifts to the left and O_2 is released. An increase in H^+ ion concentration (decrease in blood pH) also shifts this equilibrium to the left, as does increasing temperature.

During periods of strenuous exertion, three factors work together to ensure the delivery of O_2 to active tissues. (1) As O_2 is consumed, the equilibrium in Equation 18.11 shifts to the left according to Le Châtelier's principle. (2) Exertion raises the temperature of the body, also shifting the equilibrium to the left. (3) Large amounts of CO_2 are produced by metabolism, which shifts the equilibria in Equation 18.10 to the left, thus decreasing the pH. Other acids, such as lactic acid, are also produced during strenuous exertion as tissues become starved for oxygen. The decrease in pH shifts the haemoglobin equilibrium to the left, delivering more O_2. In addition, the decrease in pH stimulates an increase in the rate of breathing, which furnishes more O_2 and eliminates CO_2. Without this elaborate arrangement, the O_2 in tissues would be rapidly depleted, making further activity impossible.

RELATED EXERCISE: 18.69

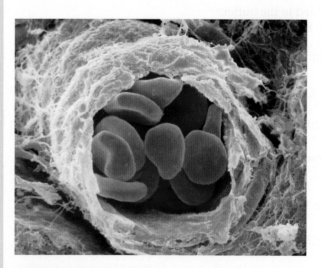

▲ FIGURE 18.3   **Red blood cells.** A scanning electromicrograph of a group of red blood cells travelling through a small branch of an artery. Blood is a buffer solution whose pH is maintained between 7.35 and 7.45.

Consequently, the new pH is given by

$$pH = 4.74 + \log\frac{[0.320]}{[0.280]} = 4.80$$

This is a very small change from the original pH of 4.74. If, however, 0.020 mol NaOH were added to 1.00 dm^3 of pure water at a pH of 7.0, then the new pH would be

$$pH = 14 - pOH = 14 - (-\log 0.20 \text{ mol dm}^{-3}) = 12.30$$

This is a very large change of 5.30 pH units.

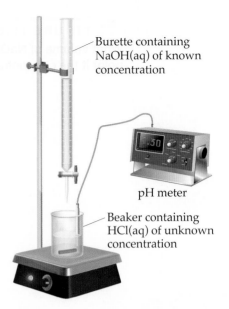

▲ **FIGURE 18.4   Measuring pH during a titration.**

18.3 ACID–BASE TITRATIONS

In Section 4.6 we briefly described *titrations*. In an acid–base titration, a solution containing a known concentration of base is slowly added to an acid (or the acid is added to the base). Acid–base indicators can be used to signal the *equivalence point* of a titration (the point at which stoichiometrically equivalent quantities of acid and base have been brought together). Alternatively, a pH meter can be used to monitor the progress of the reaction and produce a **pH titration curve**, a graph of the pH as a function of the volume of the added titrant. The shape of the titration curve makes it possible to determine the equivalence point in the titration. The titration curve can also be used to select suitable indicators and to determine the K_a of the weak acid or the K_b of the weak base being titrated.

A typical apparatus for measuring pH during a titration is illustrated in ▶ FIGURE 18.4. The titrant is added to the solution from a burette, and the pH is continually monitored using a pH meter. To understand why titration curves have certain characteristic shapes, we examine the curves for three kinds of titrations: (1) strong acid–strong base, (2) weak acid–strong base and (3) polyprotic acid–strong base. We also briefly consider how these curves relate to those involving weak bases.

CONCEPT CHECK 5

For the setup shown in Figure 18.4, will pH increase or decrease as titrant is added?

Strong Acid–Strong Base Titrations

The titration curve produced when a strong base is added to a strong acid has the general shape shown in▼ FIGURE 18.5. This curve depicts the pH change that occurs as 0.100 M NaOH is added to 50.0 cm^3 of 0.100 M HCl. The pH can be calculated at various stages of the titration. To help understand these calculations, we can divide the curve into four regions.

1. **The initial pH.** The pH of the solution before the addition of any base is determined by the initial concentration of the strong acid. For a solution of 0.100 M HCl, [H$^+$] = 0.100 M and hence pH = $-\log(0.100)$ = 1.000. Thus the initial pH is low.

2. **Between the initial pH and the equivalence point.** As NaOH is added, the pH increases, slowly at first and then rapidly in the vicinity of the equivalence point. The pH of the solution before the equivalence point is determined by the concentration of acid that has not yet been neutralised. This calculation is illustrated in Sample Exercise 18.5(a).

⚠ FIGURE IT OUT

What volume of NaOH(aq) would be needed to reach the equivalence point if the concentration of the added base were 0.200 M?

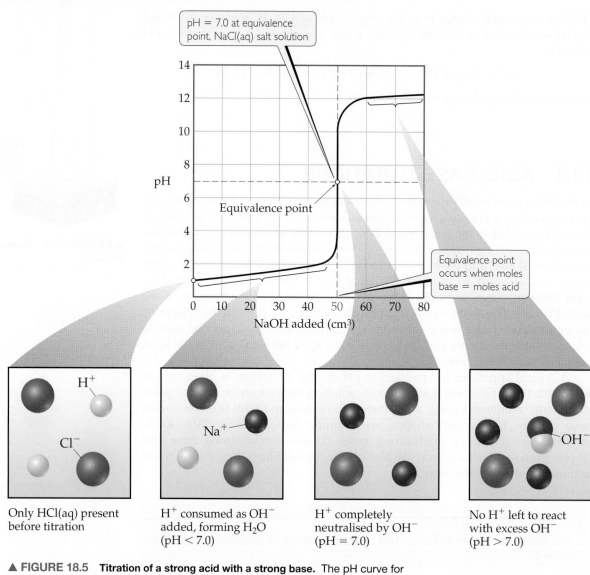

▲ **FIGURE 18.5** **Titration of a strong acid with a strong base.** The pH curve for titration of 50.0 mL of a 0.100 M solution of hydrochloric acid with a 0.100 M solution of NaOH(aq). For clarity, water molecules have been omitted from the molecular art.

3. **The equivalence point.** At the equivalence point an equal number of moles of the NaOH and HCl have reacted, leaving only a solution of their salt, NaCl. The pH of the solution is 7.00 because the cation of a strong base (in this case, Na^+) and the anion of a strong acid (in this case, Cl^-) do not hydrolyse and therefore have no appreciable effect on pH (∞ Section 17.9, 'Acid–Base Properties of Salt Solutions').

∞ Review this on page 685

4. **After the equivalence point.** The pH of the solution after the equivalence point is determined by the concentration of the excess NaOH in the solution. This calculation is shown in Sample Exercise 18.5(b).

Titration of a solution of a strong base with a solution of a strong acid would yield an analogous curve of pH versus added acid. In this case, however, the pH would be high at the outset of the titration and low at its completion, as shown in ▶ **FIGURE 18.6**.

SAMPLE EXERCISE 18.5   **Calculating pH for a strong acid–strong base titration**

Calculate the pH when the following quantities of 0.100 M NaOH solution have been added to 50.0 cm^3 of 0.100 M HCl solution: **(a)** 49.0 cm^3, **(b)** 51.0 cm^3.

SOLUTION

Analyse We are asked to calculate the pH at two points in the titration of a strong acid with a strong base. The first point is just before the equivalence point, so we expect the pH to be determined by the small amount of strong acid that has not yet been neutralised. The second point is just after the equivalence point, so we expect this pH to be determined by the small amount of excess strong base.

Plan As the NaOH solution is added to the HCl solution, $H^+(aq)$ reacts with $OH^-(aq)$ to form H_2O. Both Na^+ and Cl^- are spectator ions, having negligible effect on the pH. To determine the pH of the solution, we must first determine how many moles of H^+ were originally present and how many moles of OH^- were added. We can then calculate how many moles of each ion remain after the neutralisation reaction. To calculate $[H^+]$ and hence pH we must also remember that the volume of the solution increases as we add titrant, thus diluting the concentration of all solutes present.

Solve

(a) Moles H^+ in 50 cm^3 of original solution

$= M \times V = (0.100 \text{ mol dm}^{-3})(0.0500 \text{ dm}^3) = 5.00 \times 10^{-3} \text{ mol}$

Moles OH^- in 49.0 cm^3 of 0.100 M NaOH

$= M \times V = (0.100 \text{ mol dm}^{-3})(0.0490 \text{ dm}^3) = 4.90 \times 10^{-3} \text{ mol}$

Moles of H^+ left after reaction

$= 5.00 \times 10^{-3} \text{ mol } H^+ - 4.90 \times 10^{-3} \text{ mol } OH^- = 0.100 \times 10^{-3} \text{ mol}$

Molarity of H^+

$= \dfrac{(0.100 \times 10^{-3} \text{ mol})}{(0.0500 \text{ dm}^3 + 0.0490 \text{ dm}^3)} = 1.01 \times 10^{-3} \text{ M}$

$pH = -\log(1.01 \times 10^{-3}) = 2.99$

(b) Moles OH^- added
Moles OH^- remaining

$= M \times V = (0.100 \text{ mol dm}^{-3})(0.0510 \text{ dm}^3) = 5.10 \times 10^{-3} \text{ mol}$
$= \text{moles } OH^- \text{ added} - \text{moles } H^+ \text{ initially present}$
$= 5.10 \times 10^{-3} - 5.00 \times 10^{-3} = 0.100 \times 10^{-3} \text{ mol}$

Molarity of OH^- after reaction

$= \dfrac{(0.100 \times 10^{-3} \text{ mol})}{(0.0500 \text{ dm}^3 + 0.0510 \text{ dm}^3)} = 9.9 \times 10^{-4} \text{ M}$

$pOH = -\log(9.9 \times 10^{-4}) = 3.00$

$pH = 14 - pOH = 14 - 3.00 = 11.00$

PRACTICE EXERCISE

Calculate the pH when the following quantities of 0.100 M HNO_3 have been added to 25.0 cm^3 of 0.100 M KOH solution: **(a)** 24.9 cm^3, **(b)** 25.1 cm^3.

Answers: **(a)** 10.30, **(b)** 3.70

(See also Exercise 18.32.)

Weak Acid–Strong Base Titrations

The curve for the titration of a weak acid by a strong base is very similar in shape to that for the titration of a strong acid by a strong base (Figure 18.5). Consider, for example, the titration curve for the titration of 50.0 cm^3 of 0.100 M acetic acid (CH_3COOH) with 0.100 M NaOH

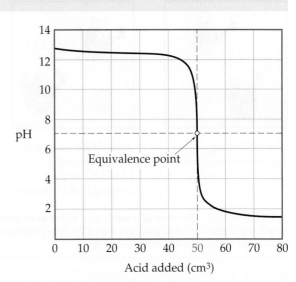

▶ **FIGURE 18.6**   **Titration of a strong base with a strong acid.** The pH curve for titration of 50.0 cm^3 of a 0.100 M solution of a strong base with a 0.100 M solution of a strong acid.

as shown in ▼ FIGURE 18.7. We can calculate the pH at points along this curve, using principles we have discussed earlier. As in the case of the titration of a strong acid by a strong base, we can divide the curve into four regions.

1. **The initial pH.** This pH is just the pH of the 0.100 M CH₃COOH. We performed calculations of this kind in Section 17.6. The calculated pH of 0.100 M CH₃COOH is 2.89.

2. **Between the initial pH and the equivalence point.** To determine pH in this range, we must consider the neutralisation of the acid.

$$CH_3COOH(aq) + OH^-(aq) \longrightarrow CH_3COO^-(aq) + H_2O(l) \qquad [18.12]$$

▲ FIGURE IT OUT

If the acetic acid being titrated here were replaced by hydrochloric acid, would the amount of base needed to reach the equivalence point change? Would the pH at the equivalence point change?

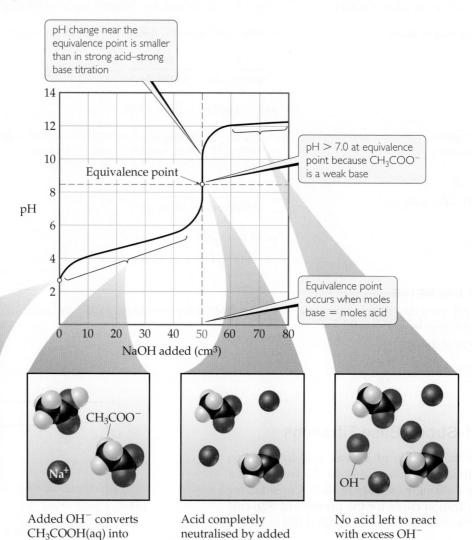

▲ **FIGURE 18.7** **Titration of a weak acid with a strong base.** The pH curve for titration of 50.0 mL of a 0.100 M solution of acetic acid with a 0.100 M solution of NaOH(aq). For clarity, water molecules have been omitted from the molecular art.

▲ **FIGURE 18.8 Procedure for calculating pH when a weak acid is partially neutralised by a strong base.**

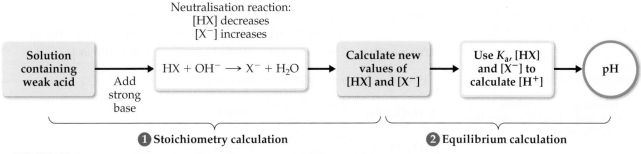

Prior to reaching the equivalence point, part of the CH_3COOH is neutralised to form CH_3COO^-. Thus the solution contains a mixture of CH_3COOH and CH_3COO^-.

The approach we take in calculating the pH in this region of the titration curve involves two main steps. First, we consider the neutralisation reaction between CH_3COOH and OH^- to determine the concentrations of CH_3COOH and CH_3COO^- in the solution. Next, we calculate the pH of this buffer pair using procedures developed in Sections 18.1 and 18.2 and illustrated in Sample Exercise 18.6. The procedure is shown in ▲ FIGURE 18.8.

3. **The equivalence point.** The equivalence point is reached after adding 50.0 cm^3 of 0.100 M NaOH to the 50.0 cm^3 of 0.100 M CH_3COOH. At this point the 5.00×10^{-3} mol of NaOH completely reacts with the 5.00×10^{-3} mol of CH_3COOH to form 5.00×10^{-3} mol of their salt, CH_3COONa. The Na^+ ion of this salt has no significant effect on the pH. The CH_3COO^- ion, however, is a weak base and the pH at the equivalence point is therefore greater than 7. Indeed, the pH at the equivalence point is always above 7 in a weak acid–strong base titration because the anion of the salt formed is a weak base.
4. **After the equivalence point.** In this region of the titration curve, $[OH^-]$ from the reaction of CH_3COO^- with water is negligible compared with $[OH^-]$ from the excess NaOH. Thus the pH is determined by the concentration of OH^- from the excess NaOH. The method for calculating pH in this region is therefore like that for the strong acid–strong base titration illustrated in Sample Exercise 18.5(b). Thus the addition of 51.0 cm^3 of 0.100 M NaOH to 50.0 cm^3 of either 0.100 M HCl or 0.100 M CH_3COOH yields the same pH, 11.00. Notice in Figures 18.5 and 18.7 that the titration curves for the titrations of both the strong acid and the weak acid are the same after the equivalence point.

SAMPLE EXERCISE 18.6 Calculating pH for a weak acid–strong base titration

Calculate the pH of the solution formed when 45.0 cm^3 of 0.100 M NaOH is added to 50.0 cm^3 of 0.100 M CH_3COOH ($K_a = 1.8 \times 10^{-5}$).

SOLUTION

Analyse We are asked to calculate the pH before the equivalence point of the titration of a weak acid with a strong base.

Plan We first must determine the number of moles of CH_3COOH and CH_3COO^- present after the neutralisation reaction. We then calculate pH using K_a, $[CH_3COOH]$ and $[CH_3COO^-]$.

Solve *Stoichiometry calculation:* The product of the volume and concentration of each solution gives the number of moles of each reactant present before the neutralisation:

for CH_3COOH, $(0.0500\ dm^3)(0.100\ mol\ dm^{-3}) = 5.00 \times 10^{-3}$ mol CH_3COOH

for NaOH, $(0.045\ dm^3)(0.100\ mol\ dm^{-3}) = 4.50 \times 10^{-3}$ mol NaOH

The 4.50×10^{-3} mol of NaOH consumes 4.50×10^{-3} mol of CH_3COOH:

Before reaction: 5.00×10^{-3} mol 4.50×10^{-3} mol 0.0 mol
$$CH_3COOH(aq) + OH^-(aq) \longrightarrow CH_3COO^-(aq) + H_2O(l)$$
After reaction: 0.50×10^{-3} mol 0.0 mol 4.50×10^{-3} mol

The total volume of the solution is

$$45.0\ cm^3 + 50.0\ cm^3 = 95.0\ cm^3 = 0.0950\ dm^3$$

The resulting molarities of CH_3COOH and CH_3COO^- after the reaction are therefore

$$[CH_3COOH] = \frac{0.50 \times 10^{-3}\ mol}{0.0950\ dm^3} = 0.0053\ M$$

$$[CH_3COO^-] = \frac{4.50 \times 10^{-3}\ mol}{0.0950\ dm^3} = 0.0474\ M$$

Equilibrium calculation: The equilibrium between CH_3COOH and CH_3COO^- must obey the equilibrium constant expression for CH_3COOH:

$$K_a = \frac{[H^+][CH_3COO^-]}{[CH_3COOH]} = 1.8 \times 10^{-5}$$

Solving for $[H^+]$ gives

$$[H^+] = K_a \times \frac{[CH_3COOH]}{[CH_3COO^-]} = (1.8 \times 10^{-5}) \times \left(\frac{0.0053}{0.0474}\right) = 2.0 \times 10^{-6}\ M$$

$$pH = -\log(2.0 \times 10^{-6}) = 5.70$$

Comment We could have solved for pH equally well using the Henderson–Hasselbalch equation.

PRACTICE EXERCISE

(a) Calculate the pH in the solution formed by adding $10.0\ cm^3$ of $0.050\ M$ NaOH to $40.0\ cm^3$ of $0.0250\ M$ benzoic acid (C_6H_5COOH, $K_a = 6.3 \times 10^{-5}$). **(b)** Calculate the pH in the solution formed by adding $10.0\ cm^3$ of $0.100\ M$ HCl to $20.0\ cm^3$ of $0.100\ M$ NH_3 ($K_b = 1.8 \times 10^{-5}$).

Answers: **(a)** 4.20, **(b)** 9.26

(See also Exercise 18.30(b).)

SAMPLE EXERCISE 18.7   Calculating the pH at the equivalence point

Calculate the pH at the equivalence point in the titration of $50.0\ cm^3$ of $0.100\ M$ CH_3COOH with $0.100\ M$ NaOH.

SOLUTION

Analyse We are asked to determine the pH at the equivalence point of the titration of a weak acid with a strong base. Because the neutralisation of a weak acid produces its anion, which is a weak base, we expect the pH at the equivalence point to be greater than 7.

Plan The initial number of moles of acetic acid equals the number of moles of acetate ion at the equivalence point. We use the volume of the solution at the equivalence point to calculate the concentration of acetate ion. Because the acetate ion is a weak base, we can calculate the pH using K_b and $[CH_3COO^-]$.

Solve The number of moles of acetic acid in the initial solution is obtained from the volume and molarity of the solution:

$$Moles = M \times V = (0.100\ mol\ dm^{-3})(0.0500\ dm^3) = 5.00 \times 10^{-3}\ mol\ CH_3COOH$$

Hence 5.00×10^{-3} mol of CH_3COO^- is formed. It will take $50.0\ cm^3$ of NaOH to reach the equivalence point (Figure 18.7). The volume of this salt solution at the equivalence point is the sum of the volumes of the acid and base, $50.0\ cm^3 + 50.0\ cm^3 = 100.0\ cm^3 = 0.1000\ dm^3$. Thus the concentration of CH_3COO^- is

$$[CH_3COO^-] = \frac{5.00 \times 10^{-3}\ mol}{0.1000\ dm^3} = 0.0500\ M$$

The CH_3COO^- ion is a weak base.

$$CH_3COO^-(aq) + H_2O(l) \rightleftharpoons CH_3COOH(aq) + OH^-(aq)$$

The K_b for CH_3COO^- can be calculated from the K_a value of its conjugate acid, $K_b = K_w/K_a = (1.0 \times 10^{-14})/(1.8 \times 10^{-5}) = 5.6 \times 10^{-10}$. Using the K_b expression, we have

$$K_b = \frac{[CH_3COOH][OH^-]}{[CH_3COO^-]} = \frac{(x)(x)}{0.0500 - x} = 5.6 \times 10^{-10}$$

Making the approximation that $0.0500 - x \approx 0.0500$, and then solving for x, we have $x = [\text{OH}^-] = 5.3 \times 10^{-6}$ M, which gives pOH = 5.28 and pH = 8.72.

Check The pH is above 7, as expected for the salt of a weak acid and strong base.

PRACTICE EXERCISE

Calculate the pH at the equivalence point when **(a)** 40.0 cm^3 of 0.025 M benzoic acid ($\text{C}_6\text{H}_5\text{COOH}$, $K_a = 6.3 \times 10^{-5}$) is titrated with 0.050 M NaOH; **(b)** 40.0 cm^3 of 0.100 M NH$_3$ is titrated with 0.100 M HCl.

Answers: **(a)** 8.21, **(b)** 5.28

(See also Exercise 18.34.)

The pH titration curves for weak acid–strong base titrations (Figure 18.7) differ from those for strong acid–strong base titrations (Figure 18.5) in three noteworthy ways.

1. The solution of the weak acid has a higher initial pH than a solution of a strong acid of the same concentration.

2. The pH change at the rapid-rise portion of the curve near the equivalence point is smaller for the weak acid than it is for the strong acid.

3. The pH at the equivalence point is above 7.00 for the weak acid–strong base titration.

To illustrate these differences further, consider the family of titration curves shown in ▶ **FIGURE 18.9**. As expected, the initial pH of the weak acid solutions is always higher than that of the strong acid solution of the same concentration. Notice also that the pH change near the equivalence point becomes less marked as the acid becomes weaker (that is, as K_a becomes smaller). Finally, the pH at the equivalence point steadily increases as K_a decreases. It is virtually impossible to determine the equivalence point when pK_a is 10 or higher because the pH change is too small and gradual.

Titrations of Polyprotic Acids

When weak acids contain more than one ionisable H atom, the reaction with OH$^-$ occurs in a series of steps. Neutralisation of H$_3$PO$_3$, for example, proceeds in two steps (Chapter 17 Sample Integrative Exercise):

$$\text{H}_3\text{PO}_3(aq) + \text{OH}^-(aq) \longrightarrow \text{H}_2\text{PO}_3^-(aq) + \text{H}_2\text{O}(l) \quad [18.13]$$

$$\text{H}_2\text{PO}_3^-(aq) + \text{OH}^-(aq) \longrightarrow \text{HPO}_3^{2-}(aq) + \text{H}_2\text{O}(l) \quad [18.14]$$

When the neutralisation steps of a polyprotic acid or polybasic base are sufficiently separated, the titration has multiple equivalence points. ▼ **FIGURE 18.10** shows the two equivalence points corresponding to Equations 18.13 and 18.14.

CONCEPT CHECK 6

Sketch an approximate titration curve for the titration of Na$_2$CO$_3$ with HCl.

FIGURE IT OUT

How does the pH at the equivalence point change as the acid being titrated becomes weaker? How does the volume of NaOH(aq) needed to reach the equivalence point change?

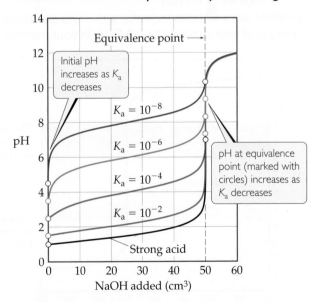

▲ **FIGURE 18.9** A set of curves showing the effect of acid strength on the characteristics of the titration curve when a weak acid is titrated by a strong base. Each curve represents titration of 50.0 cm^3 of 0.10 M acid with 0.10 M NaOH.

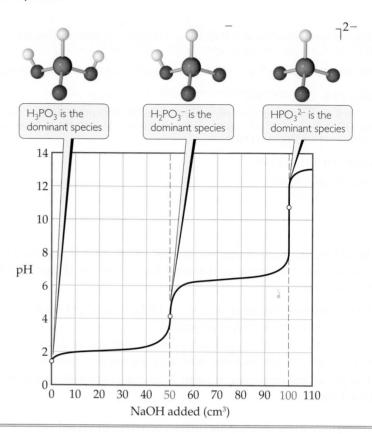

▶ **FIGURE 18.10** **Titration curve for a diprotic acid.** The curve shows the pH change when 50.0 cm³ of 0.10 M H₃PO₃ is titrated with 0.10 M NaOH.

Titrating with an Acid–Base Indicator

Sometimes in an acid–base titration an indicator is used rather than a pH meter. Optimally, an indicator should change colour at the equivalence point in a titration. In practice, however, that is unnecessary. The pH changes very rapidly near the equivalence point, and in this region one drop of titrant can change the pH by several units. Thus an indicator beginning and ending its colour change anywhere on the rapid-rise portion of the titration curve gives a sufficiently accurate measure of the titrant volume needed to reach the equivalence point. The point in a titration where the indicator changes colour is called the *end point* to distinguish it from the equivalence point that it closely approximates.

◀ **FIGURE 18.11** shows the curve for titration of a strong base (NaOH) with a strong acid (HCl). We see from the vertical part of the curve that the pH changes rapidly from roughly 11 to 3 near the equivalence point. Consequently, an indicator for this titration can change colour anywhere in this range. Most strong acid–strong base titrations are conducted using phenolphthalein as an indicator because it changes colour in this range (see Figure 17.7, page 669. Several other indicators would also be satisfactory, including methyl red, which, as the lower colour band in Figure 18.11 shows, changes colour in the pH range from about 4.2 to 6.0. (see Figure 17.7).

As noted in our discussion of Figure 18.9, because the pH change near the equivalence point

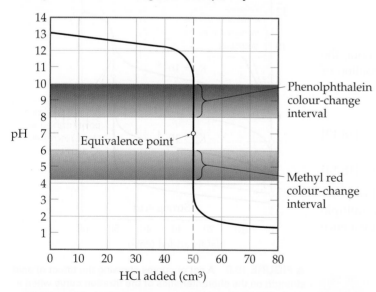

FIGURE IT OUT

Is methyl red a suitable indicator when you are titrating a strong acid with a strong base? Explain your answer.

▲ **FIGURE 18.11** **Using colour indicators for titration of a strong base with a strong acid.** Both phenolphthalein and methyl red change colour in the rapid-rise portion of the titration curve.

becomes smaller as K_a decreases, the choice of indicator for a weak acid–strong base titration is more critical than it is for titrations where both acid and base are strong. When 0.100 M CH_3COOH ($K_a = 1.8 \times 10^{-5}$) is titrated with 0.100 M NaOH, for example, the pH increases rapidly only over the pH range from about 7 to 11 (▼ FIGURE 18.12). Phenolphthalein is therefore an ideal indicator because it changes colour from pH 8.3 to 10.0, close to the pH at the equivalence point. Methyl red is a poor choice, however, because its colour change, from 4.2 to 6.0, begins well before the equivalence point is reached.

Titration of a weak base (such as 0.100 M NH_3) with a strong acid solution (such as 0.100 M HCl) leads to the titration curve shown in ▼ FIGURE 18.13. In this example, the equivalence point occurs at pH 5.28. Thus methyl red is an ideal indicator but phenolphthalein would be a poor choice.

 CONCEPT CHECK 7

Why is the choice of indicator more crucial for a weak acid–strong base titration than for a strong acid–strong base titration?

18.4 | SOLUBILITY EQUILIBRIA

The equilibria considered so far in this chapter have involved acids and bases. Furthermore, they have been homogeneous; that is, all the species have been in the same phase. Through the rest of this chapter we consider the equilibria involved in the dissolution or precipitation of ionic compounds. These reactions are heterogeneous.

The dissolving and precipitating of compounds are phenomena that occur both within us and around us. Tooth enamel dissolves in acidic solutions, for example, causing tooth decay. The precipitation of certain salts in our kidneys produces kidney stones. The waters of the Earth contain salts dissolved as water passes over and through the ground. Precipitation of $CaCO_3$ from groundwater is responsible for the formation of stalactites and stalagmites within limestone caves (Figure 4.1).

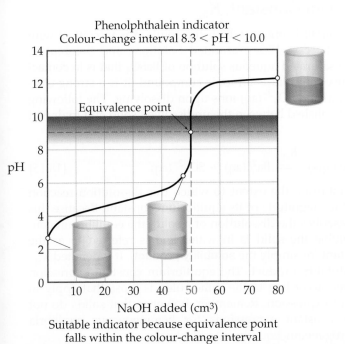

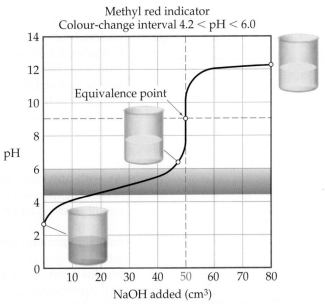

▲ **FIGURE 18.12   Good and poor indicators for titration of a weak acid with a strong base.**

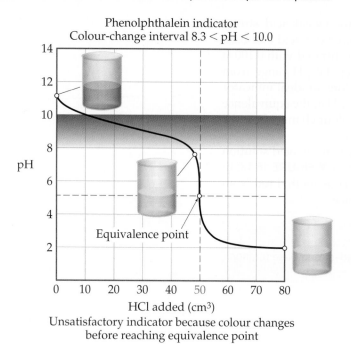

Phenolphthalein indicator
Colour-change interval 8.3 < pH < 10.0

Unsatisfactory indicator because colour changes
before reaching equivalence point

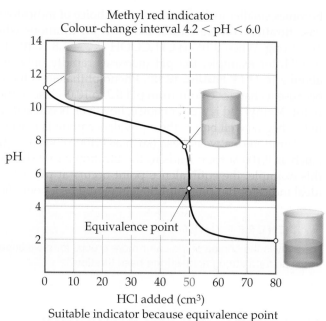

Methyl red indicator
Colour-change interval 4.2 < pH < 6.0

Suitable indicator because equivalence point
falls within the colour-change interval

▲ **FIGURE 18.13**   **Good and poor indicators for titration of a weak base with a strong acid.**

∞ Review this on page 107

In our earlier discussion of precipitation reactions, we considered some general rules for predicting the solubility of common salts in water (∞ Section 4.2, 'Precipitation Reactions'). These rules give us a qualitative sense of whether a compound will have a low or high solubility in water. By considering solubility equilibria, in contrast, we can make quantitative predictions about the amount of a given compound that will dissolve. We can also use these equilibria to analyse the factors that affect solubility.

The Solubility-Product Constant, K_{sp}

∞ Review this on page 432

Recall that a *saturated solution* is one in which the solution is in contact with undissolved solute (∞ Section 12.2, 'Saturated Solutions and Solubility'). Consider, for example, a saturated aqueous solution of $BaSO_4$ that is in contact with solid $BaSO_4$. Because the solid is an ionic compound, it is a strong electrolyte and yields $Ba^{2+}(aq)$ and $SO_4^{2-}(aq)$ ions upon dissolving. The following equilibrium is readily established between the undissolved solid and hydrated ions in solution:

$$BaSO_4(s) \xrightleftharpoons{K_{sp}} Ba^{2+}(aq) + SO_4^{2-}(aq) \qquad [18.15]$$

As with any other equilibrium, the extent to which this dissolution reaction occurs is expressed by the magnitude of its equilibrium constant. Because this equilibrium equation describes the dissolution of a solid, the equilibrium constant indicates how soluble the solid is in water and is referred to as the **solubility-product constant** (or simply the **solubility product**). It is denoted K_{sp}, where sp stands for solubility product. The equilibrium constant expression for this process is written according to the same rules as those that apply to any equilibrium constant expression. Remember, however, that solids do not appear in equilibrium constant expressions for heterogeneous equilibria

∞ Review this on page 627

(∞ Section 16.4, 'Heterogeneous Equilibria').

The solubility product equals the product of the concentration of the ions involved in the equilibrium, each raised to the power of its coefficient in the equilibrium equation.

Thus the solubility-product expression for the equilibrium expressed in Equation 18.15 is

$$K_{sp} = [Ba^{2+}][SO_4^{2-}] \qquad [18.16]$$

In general, the solubility-product constant (K_{sp}) is the equilibrium constant for the equilibrium that exists between a solid ionic solute and its ions in a saturated aqueous solution. The values of K_{sp} at 25 °C for many ionic solids are tabulated in Appendix D. The value of K_{sp} for BaSO$_4$ is 1.1×10^{-10}, a very small number, indicating that only a very small amount of the solid will dissolve in water.

SAMPLE EXERCISE 18.8   Writing solubility-product (K_{sp}) expressions

Write the expression for the solubility-product constant for CaF$_2$, and look up the corresponding K_{sp} value in Appendix D.

SOLUTION

Analyse We are asked to write an equilibrium constant expression for the process by which CaF$_2$ dissolves in water.

Plan We apply the same rules for writing any equilibrium constant expression, making sure to exclude the solid reactant from the expression. We assume that the compound dissociates completely into its component ions.

$$CaF_2(s) \rightleftharpoons Ca^{2+}(aq) + 2\,F^-(aq)$$

Solve Following the italicised rule stated above, the expression for K_{sp} is

$$K_{sp} = [Ca^{2+}][F^-]^2$$

In Appendix D we see that this K_{sp} has a value of 3.9×10^{-11}.

PRACTICE EXERCISE

Give the solubility-product constant expressions and the values of the solubility-product constants (from Appendix D) for the following compounds: **(a)** barium carbonate, **(b)** silver sulfate.

Answers: **(a)** $K_{sp} = [Ba^{2+}][CO_3^{2-}] = 5.0 \times 10^{-9}$; **(b)** $K_{sp} = [Ag^+]^2[SO_4^{2-}] = 1.5 \times 10^{-5}$

(See also Exercise 15.36.)

Solubility and K_{sp}

It is important to distinguish carefully between solubility and the solubility-product constant. The solubility of a substance is the maximum quantity that can dissolve to form a saturated solution (∞ Section 12.2, 'Saturated Solutions and Solubility'). Solubility is often expressed as grams of solute per litre of solution (g dm^{-3}). The molar solubility is the number of moles of the solute that dissolve in forming a litre of saturated solution of the solute (mol dm^{-3}). The solubility-product constant (K_{sp}) is the equilibrium constant for the equilibrium between an ionic solid and its saturated solution, and like all equilibrium constants is a unitless number.

∞ Review this on page 432

> **CONCEPT CHECK 8**
>
> Without doing a calculation, predict which of the following compounds will have the greatest molar solubility in water: AgCl ($K_{sp} = 1.8 \times 10^{-10}$), AgBr ($K_{sp} = 5.0 \times 10^{-13}$) or AgI ($K_{sp} = 8.3 \times 10^{-17}$).

The solubility of a substance can change considerably as the concentrations of other solutes change. The solubility of Mg(OH)$_2$, for example, depends highly on the pH of the solution. The solubility is also affected by the concentrations of other ions in solution, especially common ions. That is, the numerical value of the solubility of a given solute can and does change as

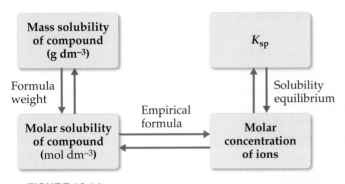

▲ **FIGURE 18.14** **Procedure for converting between solubility and K_{sp}.** Starting from the mass solubility, follow the green arrows to determine K_{sp}. Starting from K_{sp}, follow the red arrows to determine either molar solubility or mass solubility.

the other species in solution change. In contrast, the solubility-product constant, K_{sp}, has only one value for a given solute at any specific temperature.*

In principle, it is possible to use the K_{sp} value of a salt to calculate solubility under a variety of conditions. In practice, great care must be taken in doing so for the reasons indicated in A Closer Look on the limitations of solubility products. Agreement between measured solubility and that calculated from K_{sp} is usually best for salts whose ions have low charges (1+ and 1−) and do not hydrolyse to their conjugate acid or base. ◀ **FIGURE 18.14** summarises the relationships between various expressions of solubility and K_{sp}.

---

* This is strictly true only for very dilute solutions. The values of equilibrium constants change when the total concentration of ionic substances in water is increased. However, we ignore these effects except in work that requires exceptional accuracy.

SAMPLE EXERCISE 18.9 | Calculating K_{sp} from solubility

Solid silver chromate is added to pure water at 25 °C. Some of the solid remains undissolved at the bottom of the flask. The mixture is stirred for several days to ensure that equilibrium is achieved between the undissolved $Ag_2CrO_4(s)$ and the solution. Analysis of the equilibrated solution shows that its silver ion concentration is 1.3×10^{-4} M. Assuming that Ag_2CrO_4 dissociates completely in water and that there are no other important equilibria involving the Ag^+ or CrO_4^{2-} ions in the solution, calculate K_{sp} for this compound.

SOLUTION

Analyse We are given the equilibrium concentration of Ag^+ in a saturated solution of Ag_2CrO_4 and asked to determine the value of K_{sp} for Ag_2CrO_4.

Plan The equilibrium equation and the expression for K_{sp} are

$$Ag_2CrO_4(s) \rightleftharpoons 2\,Ag^+(aq) + CrO_4^{2-}(aq) \qquad K_{sp} = [Ag^+]^2[CrO_4^{2-}]$$

To calculate K_{sp}, we need the equilibrium concentrations of Ag^+ and CrO_4^{2-}. We know that at equilibrium $[Ag^+] = 1.3 \times 10^{-4}$ M. All the Ag^+ and CrO_4^{2-} ions in the solution come from the Ag_2CrO_4 that dissolves. Thus we can use $[Ag^+]$ to calculate $[CrO_4^{2-}]$.

Solve From the stoichiometry of the dissociation reaction

$$\frac{[CrO_4^{2-}]}{[Ag^+]} = \frac{1}{2}$$

so,

$$[CrO_4^{2-}] = \tfrac{1}{2}[Ag^+] = \tfrac{1}{2}(1.3 \times 10^{-4}\ mol\ dm^{-3}) = 6.5 \times 10^{-5}\ M$$

We can now calculate the value of K_{sp}.

$$K_{sp} = [Ag^+]^2[CrO_4^{2-}] = (1.3 \times 10^{-4})^2(6.5 \times 10^{-5}) = 1.1 \times 10^{-12}$$

Check We obtain a small value, as expected for a slightly soluble salt. Furthermore, the calculated value agrees well with the one given in Appendix D, 1.2×10^{-12}.

PRACTICE EXERCISE

A saturated solution of $Mg(OH)_2$ in contact with undissolved solid is prepared at 25 °C. The pH of the solution is found to be 10.17. Assuming that $Mg(OH)_2$ dissociates completely in water and that there are no other simultaneous equilibria involving the Mg^{2+} or OH^- ions in the solution, calculate K_{sp} for this compound.

Answer: 1.6×10^{-12}

(See also Exercise 18.38.)

SAMPLE EXERCISE 18.10 | Calculating solubility from K_{sp}

The K_{sp} for CaF_2 is 3.9×10^{-11} at 25 °C. Assuming that CaF_2 dissociates completely upon dissolving and that there are no other important equilibria affecting its solubility, calculate the solubility of CaF_2 in g dm^{-3}.

SOLUTION

Analyse We are given K_{sp} for CaF_2 and asked to determine its solubility. Recall that the solubility of a substance is the quantity that can dissolve in solvent, whereas the solubility-product constant, K_{sp}, is an equilibrium constant.

Plan To go from K_{sp} to solubility, we follow the steps indicated by the red arrows in Figure 18.14. We first write the chemical equation for the dissolution and set up a table of initial and equilibrium concentrations. We then use the equilibrium constant expression. In this case we know K_{sp}, and so we solve for the concentrations of the ions in solution. Once we know these concentrations, we use the formula weight to determine solubility in $g\ dm^{-3}$.

Solve Assume initially that none of the salt has dissolved, and then allow x mol dm^{-3} of CaF_2 to dissociate completely when equilibrium is achieved.

	$CaF_2(s)$	$\rightleftharpoons$	$Ca^{2+}(aq)$	+	$2\ F^-(aq)$
Initial	—		0		0
Change	—		$+x$ M		$+2x$ M
Equilibrium	—		x M		$2x$ M

The stoichiometry of the equilibrium dictates that $2x$ mol dm^{-3} of F^- are produced for each x mol dm^{-3} of CaF_2 that dissolve. We now use the expression for K_{sp} and substitute the equilibrium concentrations to solve for the value of x

$$K_{sp} = [Ca^{2+}][F^-]^2 = (x)(2x)^2 = 4x^3 = 3.9 \times 10^{-11}$$

$$x = \sqrt[3]{\frac{3.9 \times 10^{-11}}{4}} = 2.1 \times 10^{-4}\ M$$

If you cannot remember how to calculate the cube root of a number, see Appendix A. Thus the molar solubility of CaF_2 is 2.1×10^{-4} mol dm^{-3}. The mass of CaF_2 that dissolves in water to form a litre of solution is

$$(2.1 \times 10^{-4}\ \text{mol dm}^{-3})(78.1\ \text{g mol}^{-1}) = 1.6 \times 10^{-2}\ \text{g } CaF_2\ \text{dm}^{-3}\ \text{soln}$$

Check We expect a small number for the solubility of a slightly soluble salt. If we reverse the calculation, we should be able to recalculate K_{sp}: $K_{sp} = (2.1 \times 10^{-4})(4.2 \times 10^{-4})^2 = 3.7 \times 10^{-11}$, close to the starting value for K_{sp}, 3.9×10^{-11}.

Comment Because F^- is the anion of a weak acid, you might expect that the hydrolysis of the ion would affect the solubility of CaF_2. The basicity of F^- is so small ($K_b = 1.5 \times 10^{-11}$), however, that the hydrolysis occurs to only a slight extent and does not significantly influence the solubility. The reported solubility is 0.017 g dm^{-3} at 25 °C, in good agreement with our calculation.

PRACTICE EXERCISE

The K_{sp} for LaF_3 is 2.0×10^{-19}. What is the solubility of LaF_3 in water in mol dm^{-3}?

Answer: 9.3×10^{-6} mol dm^{-3}

(See also Exercise 18.48.)

A CLOSER LOOK

LIMITATIONS OF SOLUBILITY PRODUCTS

The concentrations of ions calculated from K_{sp} sometimes deviate appreciably from those found experimentally. In part, these deviations are due to electrostatic interactions between ions in solution, which can lead to ion pairs. These interactions increase in magnitude as both the concentrations of the ions and their charges increase. The solubility calculated from K_{sp} tends to be low unless it is corrected to account for these interactions between ions. Chemists have developed procedures for correcting for these 'ionic-strength' or 'ionic-activity' effects, and these procedures are examined in more advanced chemistry courses. As an example of the effect of these interionic interactions, consider $CaCO_3$ (calcite), whose solubility product, $K_{sp} = 4.5 \times 10^{-9}$, gives a calculated solubility of 6.7×10^{-5} mol dm^{-3}. Making corrections for the interionic interactions in the solution yields a higher solubility, 7.3×10^{-5} mol dm^{-3}. The reported solubility, however, is twice as high (1.4×10^{-4} mol dm^{-3}), so there must be one or more additional factors involved.

Another common source of error in calculating ion concentrations from K_{sp} is ignoring other equilibria that occur simultaneously in the solution. For example, acid–base equilibria may take place simultaneously with solubility equilibria. In particular, both basic anions and cations with high charge-to-size ratios undergo hydrolysis reactions that can measurably increase the solubilities of their salts. For example, $CaCO_3$ contains the basic carbonate ion ($K_b = 1.8 \times 10^{-4}$), which hydrolyses in water: $CO_3^{2-}(aq) + H_2O(l) \rightleftharpoons HCO_3^-(aq) + OH^-(aq)$. If we consider both the effect of the interionic interactions in the solution and the effect of the simultaneous solubility and hydrolysis equilibria, we calculate a solubility of 1.4×10^{-4} mol dm^{-3}, in agreement with the measured value.

Finally, we generally assume that ionic compounds dissociate completely into their component ions when they dissolve. This assumption is not always valid. When MgF_2 dissolves, for example, it yields not only Mg^{2+} and F^- ions but also MgF^+ ions in solution. Thus we see that calculating solubility using K_{sp} can be more complicated than it first appears and it requires considerable knowledge of the equilibria occurring in solution.

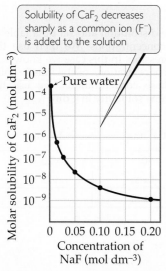

Solubility of CaF_2 decreases sharply as a common ion (F^-) is added to the solution

Pure water

▲ FIGURE 18.15 **Common-ion effect.** Notice that the CaF_2 solubility is on a logarithmic scale.

∞ Review this on page 708

18.5 | FACTORS THAT AFFECT SOLUBILITY

The solubility of a substance is affected not only by temperature but also by the presence of other solutes. The presence of an acid, for example, can have a major influence on the solubility of a substance. In Section 18.4 we considered the dissolving of ionic compounds in pure water. In this section we examine three factors that affect the solubility of ionic compounds: the presence of common ions, the pH of the solution and the presence of complexing agents. We also examine the phenomenon of *amphoterism*, which is related to the effects of both pH and complexing agents.

Common-Ion Effect

The presence of either $Ca^{2+}(aq)$ or $F^-(aq)$ in a solution reduces the solubility of CaF_2, shifting the solubility equilibrium of CaF_2 to the left.

$$CaF_2(s) \rightleftharpoons Ca^{2+}(aq) + 2\,F^-(aq)$$

Addition of Ca^{2+} or F^- shifts equilibrium, reducing solubility

This reduction in solubility is another application of the common-ion effect (∞ Section 18.1, 'The Common-Ion Effect'). In general, *the solubility of a slightly soluble salt is decreased by the presence of a second solute that furnishes a common ion.* ▲ FIGURE 18.15 shows how the solubility of CaF_2 decreases as NaF is added to the solution. Sample Exercise 18.11 shows how the K_{sp} can be used to calculate the solubility of a slightly soluble salt in the presence of a common ion.

SAMPLE EXERCISE 18.11 | **Calculating the effect of a common Ion on solubility**

Calculate the molar solubility of CaF_2 at 25 °C in a solution that is **(a)** 0.010 M in $Ca(NO_3)_2$, **(b)** 0.010 M in NaF.

SOLUTION

Analyse We are asked to determine the solubility of CaF_2 in the presence of two strong electrolytes, each containing an ion common to CaF_2. In (a) the common ion is Ca^{2+}, and NO_3^- is a spectator ion. In (b) the common ion is F^-, and Na^+ is a spectator ion.

Plan Because the slightly soluble compound is CaF_2, we need to use the K_{sp} for this compound, which is available in Appendix D:

$$K_{sp} = [Ca^{2+}][F^-]^2 = 3.9 \times 10^{-11}$$

The value of K_{sp} is unchanged by the presence of additional solutes. Because of the common-ion effect, however, the solubility of the salt will decrease in the presence of common ions. We can again use our standard equilibrium techniques of starting with the equation for CaF_2 dissolution, setting up a table of initial and equilibrium concentrations, and using the K_{sp} expression to determine the concentration of the ion that comes only from CaF_2.

Solve
(a) In this instance the initial concentration of Ca^{2+} is 0.010 M because of the dissolved $Ca(NO_3)_2$:

	$CaF_2(s)$	$\rightleftharpoons$	$Ca^{2+}(aq)$	+	$2\,F^-(aq)$
Initial	—		0.010 M		0
Change	—		$+x$ M		$+2x$ M
Equilibrium	—		$(0.010 + x)$ M		$2x$ M

Substituting into the solubility-product expression gives

$$K_{sp} = 3.9 \times 10^{-11} = [Ca^{2+}][F^-]^2 = (0.010 + x)(2x)^2$$

This would be a messy problem to solve exactly, but fortunately it is possible to simplify matters greatly. Even without the common-ion effect, the solubility of CaF_2 is very small. Assume that the 0.010 M concentration of Ca^{2+} from $Ca(NO_3)_2$ is very much greater than the small additional concentration resulting from the solubility of CaF_2; that is, x is small compared with 0.010 M, and $0.010 + x \approx 0.010$. We then have

$$3.9 \times 10^{-11} = (0.010)(2x)^2$$

$$x^2 = \frac{3.9 \times 10^{-11}}{4(0.010)} = 9.8 \times 10^{-10}$$

$$x = \sqrt{9.8 \times 10^{-10}} = 3.1 \times 10^{-5} \text{ M}$$

The very small value for x validates the simplifying assumption we have made. Our calculation indicates that 3.1×10^{-5} mol of solid CaF_2 dissolves per dm^3 of the 0.010 M $Ca(NO_3)_2$ solution.

(b) In this case the common ion is F^-, and at equilibrium we have

$$[Ca^{2+}] = x \quad \text{and} \quad [F^-] = 0.010 + 2x$$

Assuming that $2x$ is small compared with 0.010 M (that is, $0.010 + 2x \approx 0.010$), we have

$$3.9 \times 10^{-11} = x(0.010)^2$$

$$x = \frac{3.9 \times 10^{-11}}{(0.010)^2} = 3.9 \times 10^{-7} \text{ M}$$

Thus 3.9×10^{-7} mol of solid CaF_2 should dissolve per dm^3 of 0.010 M NaF solution.

Comment The molar solubility of CaF_2 in pure water is 2.1×10^{-4} M (Sample Exercise 18.10). By comparison, our calculations above show that the solubility of CaF_2 in the presence of 0.010 M Ca^{2+} is 3.1×10^{-5} M, and in the presence of 0.010 M F^- ion it is 3.9×10^{-7} M. Thus the addition of either Ca^{2+} or F^- to a solution of CaF_2 decreases the solubility. However, the effect of F^- on the solubility is more pronounced than that of Ca^{2+} because $[F^-]$ appears to the second power in the K_{sp} expression for CaF_2, whereas Ca^{2+} appears to the first power.

PRACTICE EXERCISE

The value for K_{sp} for manganese(II) hydroxide, $Mn(OH)_2$, is 1.6×10^{-13}. Calculate the molar solubility of $Mn(OH)_2$ in a solution that contains 0.020 M NaOH.

Answer: 4.0×10^{-10} M

(See also Exercise 18.71.)

Solubility and pH

The solubility of any substance whose anion is basic will be affected to some extent by the pH of the solution. Consider $Mg(OH)_2$, for example, for which the solubility equilibrium is

$$Mg(OH)_2(s) \rightleftharpoons Mg^{2+}(aq) + 2\,OH^-(aq) \quad K_{sp} = 1.8 \times 10^{-11} \quad [18.17]$$

A saturated solution of $Mg(OH)_2$ has a calculated pH of 10.52 and contains $[Mg^{2+}] = 1.7 \times 10^{-4}$ M. Now suppose that solid $Mg(OH)_2$ is equilibrated with a solution buffered at a more acidic pH of 9.0. The pOH, therefore, is 5.0, so $[OH^-] = 1.0 \times 10^{-5}$. Inserting this value for $[OH^-]$ into the solubility-product expression, we have

$$K_{sp} = [Mg^{2+}][OH^-]^2 = 1.8 \times 10^{-11}$$

$$[Mg^{2+}](1.0 \times 10^{-5})^2 = 1.8 \times 10^{-11}$$

$$[Mg^{2+}] = \frac{1.8 \times 10^{-11}}{(1.0 \times 10^{-5})^2} = 0.18 \text{ M}$$

Thus $Mg(OH)_2$ dissolves in the solution until $[Mg^{2+}] = 0.18$ M. It is apparent that $Mg(OH)_2$ is quite soluble in this solution. If the concentration of OH^- was reduced even further by making the solution more acidic, the Mg^{2+} concentration would have to increase to maintain the equilibrium condition. Thus a sample of $Mg(OH)_2$ will dissolve completely if sufficient acid is added.

MY WORLD OF CHEMISTRY

OCEAN ACIDIFICATION

Seawater is a weakly basic solution, with pH values typically between 8.0 and 8.3. This pH range is maintained through a carbonic acid buffer system similar to the one in blood (see Equation 18.10). Because the pH of seawater is higher than in blood (7.35–7.45), however, the second dissociation of carbonic acid cannot be neglected and CO_3^{2-} becomes an important aqueous species.

The availability of carbonate ions plays an important role in shell formation for a number of marine organisms, including stony corals (▶ FIGURE 18.16). These organisms, which are referred to as marine *calcifiers* and play an important role in the food chains of nearly all oceanic ecosystems, depend on dissolved Ca^{2+} and CO_3^{2-} ions to form their shells and exoskeletons. The relatively low solubility-product constant of $CaCO_3$,

$$CaCO_3(s) \rightleftharpoons Ca^{2+}(aq) + CO_3^{2-}(aq) \quad K_{sp} = 4.5 \times 10^{-9}$$

and the fact that the ocean contains saturated concentrations of Ca^{2+} and CO_3^{2-} mean that $CaCO_3$ is usually quite stable once formed. In fact, calcium carbonate skeletons of creatures that died millions of years ago are not uncommon in the fossil record.

Just as in our bodies, the carbonic acid buffer system can be perturbed by removing or adding $CO_2(g)$. The concentration of dissolved CO_2 in the ocean is sensitive to changes in atmospheric CO_2 levels. As discussed in Chapter 13, the atmospheric CO_2 concentration has risen by approximately 30% over the past three centuries to the present level of 386 ppm. Human activity has played a prominent role in this increase. Scientists estimate that one-third to one-half of the CO_2 emissions resulting from human activity have been absorbed by the Earth's oceans. Although this absorption helps

▲ FIGURE 18.16 **Marine calcifiers.** Many sea-dwelling organisms use $CaCO_3$ for their shells and exoskeletons. Examples include stony coral, crustaceans, some phytoplankton and echinoderms, such as sea urchins and starfish.

mitigate the greenhouse gas effects of CO_2, the extra CO_2 in the ocean produces carbonic acid, which lowers the pH. Because CO_3^{2-} is the conjugate base of the weak acid HCO_3^{-}, the carbonate ion readily combines with the hydrogen ion:

$$CO_3^{2-}(aq) + H^+(aq) \longrightarrow HCO_3^{-}(aq)$$

This consumption of carbonate ion shifts the dissolution equilibrium to the right, increasing the solubility of $CaCO_3$. This can lead to partial dissolution of calcium carbonate shells and exoskeletons. If the amount of atmospheric CO_2 continues to increase at the present rate, scientists estimate that seawater pH will fall to 7.9 sometime over the next 50 years. This change might sound small but it has dramatic ramifications for oceanic ecosystems.

RELATED EXERCISE: 18.77

The solubility of almost any ionic compound is affected if the solution is made sufficiently acidic or basic. The effects are very noticeable, however, only when one or both ions involved are at least moderately acidic or basic. The metal hydroxides, such as $Mg(OH)_2$, are examples of compounds containing a strongly basic ion, the hydroxide ion.

In general, *the solubility of a compound containing a basic anion (that is, the anion of a weak acid) increases as the solution becomes more acidic.* As we have seen, the solubility of $Mg(OH)_2$ greatly increases as the acidity of the solution increases. The solubility of PbF_2 increases as the solution becomes more acidic, too, because F^- is a weak base (it is the conjugate base of the weak acid HF). As a result, the solubility equilibrium of PbF_2 is shifted to the right as the concentration of F^- is reduced by protonation to form HF. Thus the solution process can be understood in terms of two consecutive reactions:

$$PbF_2(s) \rightleftharpoons Pb^{2+}(aq) + 2\,F^-(aq) \qquad [18.18]$$

$$F^-(aq) + H^+(aq) \rightleftharpoons HF(aq) \qquad [18.19]$$

The equation for the overall process is

$$PbF_2(s) + 2\,H^+(aq) \rightleftharpoons Pb^{2+}(aq) + 2\,HF(aq) \qquad [18.20]$$

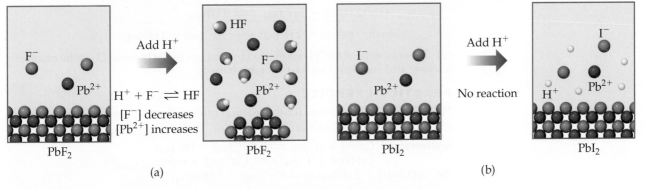

▲ **FIGURE 18.17**   **Response of two ionic compounds to addition of a strong acid.**
(a) The solubility of PbF$_2$ increases on addition of acid. (b) The solubility of PbI$_2$ is not affected by addition of acid. The water molecules and the anion of the strong acid have been omitted for clarity.

The processes responsible for the increase in solubility of PbF$_2$ in acidic solution are illustrated in ▲ **FIGURE 18.17(a)**.

Other salts that contain basic anions, such as CO$_3^{2-}$, PO$_4^{3-}$, CN$^-$, or S^{2-}, behave similarly. These examples illustrate a general rule: *The solubility of slightly soluble salts containing basic anions increases as [H$^+$] increases (as pH is lowered).* The more basic the anion, the more the solubility is influenced by pH. The solubility of salts with anions of negligible basicity (the anions of strong acids), such as Cl$^-$, Br$^-$, I$^-$ and NO$_3^-$, is unaffected by pH changes, as shown in Figure 18.17(b).

SAMPLE EXERCISE 18.12   **Predicting the effect of acid on solubility**

Which of the following substances will be more soluble in acidic solution than in basic solution: **(a)** Ni(OH)$_2$(s), **(b)** CaCO$_3$(s), **(c)** BaF$_2$(s), **(d)** AgCl(s)?

SOLUTION

Analyse The problem lists four sparingly soluble salts, and we are asked to determine which are more soluble at low pH than at high pH.

Plan Ionic compounds that dissociate to produce a basic anion will be more soluble in acid solution.

Solve

(a) Ni(OH)$_2$(s) will be more soluble in acidic solution because of the basicity of OH$^-$; the H$^+$ ion reacts with the OH$^-$ ion, forming water.

$$Ni(OH)_2(s) \rightleftharpoons Ni^{2+}(aq) + 2\,OH^-(aq)$$

$$\underline{2\,OH^-(aq) + 2\,H^+(aq) \rightleftharpoons 2\,H_2O(l)}$$

Overall: $Ni(OH)_2(s) + 2\,H^+(aq) \rightleftharpoons Ni^{2+}(aq) + 2\,H_2O(l)$

(b) Similarly, CaCO$_3$(s) dissolves in acid solutions because CO$_3^{2-}$ is a basic anion.

$$CaCO_3(s) \rightleftharpoons Ca^{2+}(aq) + CO_3^{2-}(aq)$$

$$CO_3^{2-}(aq) + 2\,H^+(aq) \rightleftharpoons H_2CO_3(aq)$$

$$\underline{H_2CO_3(aq) \longrightarrow CO_2(g) + H_2O(l)}$$

Overall: $CaCO_3(s) + 2\,H^+(aq) \longrightarrow Ca^{2+}(aq) + CO_2(g) + H_2O(l)$

The reaction between CO$_3^{2-}$ and H$^+$ occurs in a stepwise fashion, first forming HCO$_3^-$. H$_2$CO$_3$ forms in appreciable amounts only when the concentration of H$^+$ is sufficiently high.

(c) The solubility of BaF_2 is also enhanced by lowering the pH, because F^- is a basic anion.

$$BaF_2(s) \rightleftharpoons Ba^{2+}(aq) + 2\,F^-(aq)$$

$$\underline{2\,F^-(aq) + 2\,H^+(aq) \rightleftharpoons 2\,HF(aq)}$$

Overall: $BaF_2(s) + 2\,H^+(aq) \longrightarrow Ba^{2+}(aq) + 2\,HF(aq)$

(d) The solubility of AgCl is unaffected by changes in pH because Cl^- is the anion of a strong acid and therefore has negligible basicity.

PRACTICE EXERCISE

Write the net ionic equation for the reaction of the following copper(II) compounds with acid: (a) CuS, (b) $Cu(N_3)_2$.

Answers: (a) $CuS(s) + H^+(aq) \rightleftharpoons Cu^{2+}(aq) + HS^-(aq)$
$CuS(s) + 2\,H^+(aq) \rightleftharpoons Cu^{2+}(aq) + H_2S(g)$
(b) $Cu(N_3)_2(s) + 2\,H^+(aq) \rightleftharpoons Cu^{2+}(aq) + 2\,HN_3(aq)$

(See also Exercise 18.45.)

Formation of Complex Ions

∞ Review this on page 694

A characteristic property of metal ions is their ability to act as Lewis acids, or electron-pair acceptors, towards water molecules, which act as Lewis bases, or electron-pair donors (∞ Section 17.11, 'Lewis Acids and Bases'). Lewis bases other than water can also interact with metal ions, particularly with transition metal ions. Such interactions can dramatically affect the solubility of a metal salt. AgCl, for example, which has $K_{sp} = 1.8 \times 10^{-10}$, will dissolve in the presence of aqueous ammonia because Ag^+ interacts with the Lewis base NH_3, as shown in ▼ FIGURE 18.18. This process can be viewed as the sum of two reac-

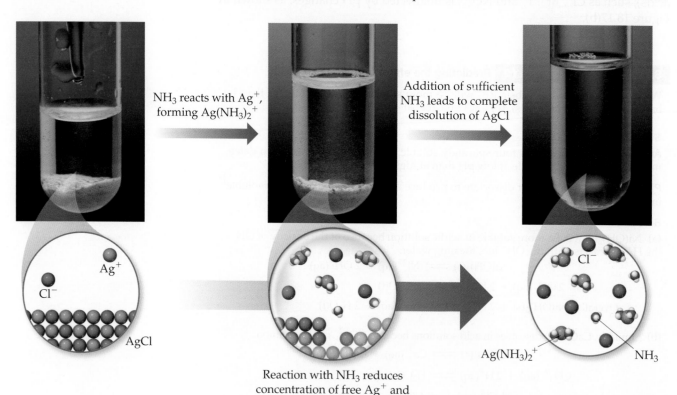

NH$_3$ reacts with Ag$^+$, forming Ag(NH$_3$)$_2$$^+$

Addition of sufficient NH$_3$ leads to complete dissolution of AgCl

Reaction with NH$_3$ reduces concentration of free Ag$^+$ and increases solubility of AgCl

$$AgCl(s) + 2\,NH_3(aq) \rightleftharpoons Ag(NH_3)_2{}^+(aq) + Cl^-(aq)$$

▲ FIGURE 18.18   Using concentrated NH$_3$(aq) to dissolve AgCl(s), which has very low solubility in water.

tions, the dissolution of AgCl and the Lewis acid–base interaction between Ag^+ and NH_3.

$$AgCl(s) \rightleftharpoons Ag^+(aq) + Cl^-(aq) \qquad [18.21]$$

$$Ag^+(aq) + 2\,NH_3(aq) \rightleftharpoons Ag(NH_3)_2^+(aq) \qquad [18.22]$$

$$\text{Overall:} \quad AgCl(s) + 2\,NH_3(aq) \overset{K_f}{\rightleftharpoons} Ag(NH_3)_2^+(aq) + Cl^-(aq) \qquad [18.23]$$

The presence of NH_3 drives the top reaction, the dissolution of AgCl, to the right as $Ag^+(aq)$ is consumed to form $Ag(NH_3)_2^+$.

For a Lewis base such as NH_3 to increase the solubility of a metal salt, it must be able to interact more strongly with the metal ion than water does. The NH_3 must displace solvating H_2O molecules (Sections 12.1 and 17.11) in order to form $Ag(NH_3)_2^+$:

$$Ag^+(aq) + 2\,NH_3(aq) \rightleftharpoons Ag(NH_3)_2^+(aq) \qquad [18.24]$$

An assembly of a metal ion and the Lewis bases bonded to it, such as $Ag(NH_3)_2^+$, is called a **complex ion**. The stability of a complex ion in aqueous solution can be judged by the size of the equilibrium constant for its formation from the hydrated metal ion. For example, the equilibrium constant for formation of $Ag(NH_3)_2^+$ (Equation 18.24) is 1.7×10^7:

$$K_f = \frac{[Ag(NH_3)_2^+]}{[Ag^+][NH_3]^2} = 1.7 \times 10^7$$

$$[18.25]$$

The equilibrium constant for this kind of reaction is called a **formation constant**, K_f. The formation constants for several complex ions are listed in ▼ TABLE 18.1.

SAMPLE EXERCISE 18.13 | Evaluating an equilibrium involving a complex ion

Calculate the concentration of Ag^+ present in solution at equilibrium when concentrated ammonia is added to a 0.010 M solution of $AgNO_3$ to give an equilibrium concentration of $[NH_3] = 0.20$ M. Neglect the small volume change that occurs when NH_3 is added.

SOLUTION

Analyse Addition of $NH_3(aq)$ to $Ag^+(aq)$ forms $Ag(NH_3)_2^+$, as shown in Equation 18.22. We are asked to determine what concentration of $Ag^+(aq)$ remains uncombined when the NH_3 concentration is brought to 0.20 M in a solution originally 0.010 M in $AgNO_3$.

Plan We assume that the $AgNO_3$ is completely dissociated, giving 0.010 M Ag^+. Because K_f for the formation of $Ag(NH_3)_2^+$ is quite large, we assume that essentially all the Ag^+ is converted to $Ag(NH_3)_2^+$ and approach the problem as though we are concerned with the dissociation of $Ag(NH_3)_2^+$ rather than its formation. To facilitate this approach, we need to reverse Equation 18.22 and make the corresponding change to the equilibrium constant:

$$Ag(NH_3)_2^+(aq) \rightleftharpoons Ag^+(aq) + 2\,NH_3(aq)$$

$$\frac{1}{K_f} = \frac{1}{1.7 \times 10^7} = 5.9 \times 10^{-8}$$

Solve If $[Ag^+]$ is 0.010 M initially, $[Ag(NH_3)_2^+]$ will be 0.010 M following addition of the NH_3. We construct a table to solve this equilibrium problem. Note that the NH_3 concentration given in the problem is an equilibrium concentration rather than an initial concentration.

	$Ag(NH_3)_2^+(aq)$ $\rightleftharpoons$	$Ag^+(aq)$ +	$2\,NH_3(aq)$
Initial	0.010 M	0 M	
Change	$-x$ M	$+x$ M	
Equilibrium	$0.010 - x$ M	x M	0.20 M

Because [Ag⁺] is very small, we can ignore x, so that $0.010 - x = 0.010$ M. Substituting these values into the equilibrium constant expression for the dissociation of $Ag(NH_3)_2^+$, we obtain

$$\frac{[Ag^+][NH_3]^2}{[Ag(NH_3)_2^+]} = \frac{(x)(0.20)^2}{0.010} = 5.9 \times 10^{-8}$$

$$x = 1.5 \times 10^{-8} \text{ M} = [Ag^+]$$

Formation of the $Ag(NH_3)_2^+$ complex drastically reduces the concentration of free Ag^+ ion in solution

PRACTICE EXERCISE

Calculate $[Cr^{3+}]$ in equilibrium with $Cr(OH)_4^-$ when 0.010 mol of $Cr(NO_3)_3$ is dissolved in 1 dm³ of solution buffered at pH 10.0.

Answer: $[Cr^{3+}] = 1 \times 10^{-16}$ M

(See also Exercise 18.46.)

TABLE 18.1 • Formation constants for some metal complex ions in water at 25 °C

Complex ion	K_f	Equilibrium equation
$Ag(NH_3)_2^+$	1.7×10^7	$Ag^+(aq) + 2\,NH_3(aq) \rightleftharpoons Ag(NH_3)_2^+(aq)$
$Ag(CN)_2^-$	1×10^{21}	$Ag^+(aq) + 2\,CN^-(aq) \rightleftharpoons Ag(CN)_2^-(aq)$
$Ag(S_2O_3)_2^{3-}$	2.9×10^{13}	$Ag^+(aq) + 2\,S_2O_3^{2-}(aq) \rightleftharpoons Ag(S_2O_3)_2^{3-}(aq)$
$CdBr_4^{2-}$	5×10^3	$Cd^{2+}(aq) + 4\,Br^-(aq) \rightleftharpoons CdBr_4^{2-}(aq)$
$Cr(OH)_4^-$	8×10^{29}	$Cr^{3+}(aq) + 4\,OH^-(aq) \rightleftharpoons Cr(OH)_4^-(aq)$
$Co(SCN)_4^{2-}$	1×10^3	$Co^{2+}(aq) + 4\,SCN^-(aq) \rightleftharpoons Co(SCN)_4^{2-}(aq)$
$Cu(NH_3)_4^{2+}$	5×10^{12}	$Cu^{2+}(aq) + 4\,NH_3(aq) \rightleftharpoons Cu(NH_3)_4^{2+}(aq)$
$Cu(CN)_4^{2-}$	1×10^{25}	$Cu^{2+}(aq) + 4\,CN^-(aq) \rightleftharpoons Cu(CN)_4^{2-}(aq)$
$Ni(NH_3)_6^{2+}$	1.2×10^9	$Ni^{2+}(aq) + 6\,NH_3(aq) \rightleftharpoons Ni(NH_3)_6^{2+}(aq)$
$Fe(CN)_6^{4-}$	1×10^{35}	$Fe^{2+}(aq) + 6\,CN^-(aq) \rightleftharpoons Fe(CN)_6^{4-}(aq)$
$Fe(CN)_6^{3-}$	1×10^{42}	$Fe^{3+}(aq) + 6\,CN^-(aq) \rightleftharpoons Fe(CN)_6^{3-}(aq)$

MY WORLD OF CHEMISTRY

TOOTH DECAY AND FLUORIDATION

Tooth enamel consists mainly of a mineral called hydroxyapatite, $Ca_{10}(PO_4)_6(OH)_2$. It is the hardest substance in the body. Tooth cavities are caused when acids dissolve tooth enamel.

$$Ca_{10}(PO_4)_6(OH)_2(s) + 8\,H^+(aq) \longrightarrow$$
$$10\,Ca^{2+}(aq) + 6\,HPO_4^{2-}(aq) + 2\,H_2O(l)$$

The resultant Ca^{2+} and HPO_4^{2-} ions diffuse out of the tooth enamel and are washed away by saliva. The acids that attack the hydroxyapatite are formed by the action of specific bacteria on sugars and other carbohydrates present in the plaque adhering to the teeth.

Fluoride ion, present in drinking water, toothpaste and other sources, can react with hydroxyapatite to form fluoro-

apatite, $Ca_{10}(PO_4)_6F_2$. This mineral, in which F^- has replaced OH^-, is much more resistant to attack by acids because the fluoride ion is a much weaker Brønsted–Lowry base than the hydroxide ion.

Because the fluoride ion is so effective in preventing cavities, it is added to the public water supply in many places to give a concentration of 1 mg dm⁻³ (1 ppm). The compound added may be NaF or Na_2SiF_6. Na_2SiF_6 reacts with water to release fluoride ions by the following reaction:

$$SiF_6^{2-}(aq) + 2\,H_2O(l) \longrightarrow 6\,F^-(aq) + 4\,H^+(aq) + SiO_2(s)$$

About 80% of all toothpastes now sold contain fluoride compounds, usually at the level of 0.1% fluoride by mass. The most common compounds in toothpastes are sodium fluoride (NaF), sodium monofluorophosphate (Na_2PO_3F) and stannous fluoride (SnF_2).

The general rule is that the solubility of metal salts increases in the presence of suitable Lewis bases, such as NH_3, CN^- or OH^-, if the metal forms a complex with the base. The ability of metal ions to form complexes is an extremely important aspect of their chemistry.

Amphoterism

Some metal oxides and hydroxides that are relatively insoluble in neutral water dissolve in strongly acidic and strongly basic solutions. These substances are soluble in strong acids and bases because they themselves are capable of behaving as either an acid or base; they are **amphoteric oxides and hydroxides**. Amphoteric oxides and hydroxides include those of Al^{3+}, Cr^{3+}, Zn^{2+} and Sn^{2+}. Notice that the term *amphoteric* is applied to the behaviour of insoluble oxides and hydroxides that can be made to dissolve in either acidic or basic solutions. The similar term *amphiprotic*, encountered in Section 17.2, relates more generally to any molecule or ion that can either gain or lose a proton.

These species dissolve in acidic solutions because they contain basic anions.

$$Al(OH)_3(s) + 3\,H^+(aq) \longrightarrow Al^{3+}(aq) + 3\,H_2O(l)$$

What makes amphoteric oxides and hydroxides special, though, is that they also dissolve in strongly basic solutions (▼ **FIGURE 18.19**). This behaviour results from the formation of complex anions containing several (typically four) hydroxides bound to the metal ion.

$$Al(OH)_3(s) + OH^-(aq) \rightleftharpoons Al(OH)_4^-(aq) \qquad [18.26]$$

The extent to which an insoluble metal hydroxide reacts with either acid or base varies with the particular metal ion involved. Many metal hydroxides—such as $Ca(OH)_2$, $Fe(OH)_2$ and $Fe(OH)_3$—are capable of dissolving in acidic solution but do not react with excess base. These hydroxides are not amphoteric.

The purification of aluminium ore in the manufacture of aluminium metal provides an interesting application of the property of amphoterism. As we have seen, $Al(OH)_3$ is amphoteric, whereas $Fe(OH)_3$ is not. Aluminium occurs in large quantities as the ore *bauxite*, which is essentially Al_2O_3 with additional water molecules. The ore is contaminated with Fe_2O_3 as an impurity. When bauxite is added to a strongly basic solution, the Al_2O_3 dissolves because the aluminium forms complex ions, such as $Al(OH)_4^-$. The Fe_2O_3 impurity, however, is not amphoteric and remains as a solid. The solution is filtered, getting rid of the iron impurity. Aluminium hydroxide is then precipitated by addition of acid. The purified hydroxide receives further treatment and eventually yields aluminium metal.

▲ CONCEPT CHECK 9

What is the difference between an amphoteric substance and an amphiprotic substance?

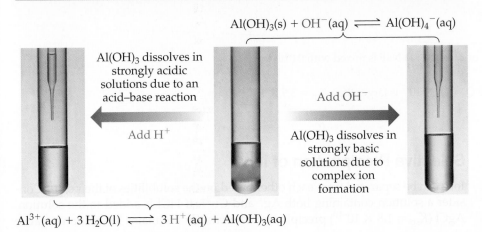

$$Al(OH)_3(s) + OH^-(aq) \rightleftharpoons Al(OH)_4^-(aq)$$

$Al(OH)_3$ dissolves in strongly acidic solutions due to an acid–base reaction

Add H^+

Add OH^-

$Al(OH)_3$ dissolves in strongly basic solutions due to complex ion formation

$$Al^{3+}(aq) + 3\,H_2O(l) \rightleftharpoons 3\,H^+(aq) + Al(OH)_3(aq)$$

◄ **FIGURE 18.19 Amphoterism.** Some metal oxides and hydroxides, such as $Al(OH)_3$, are amphoteric, which means they dissolve in both strongly acidic and strongly basic solutions.

18.6 | PRECIPITATION AND SEPARATION OF IONS

Equilibrium can be achieved starting with the substances on either side of a chemical equation. The equilibrium between $BaSO_4(s)$, $Ba^{2+}(aq)$ and $SO_4^{2-}(aq)$ (Equation 18.15) can be achieved starting with solid $BaSO_4$. It can also be reached starting with solutions of salts containing Ba^{2+} and SO_4^{2-}, say $BaCl_2$ and Na_2SO_4. When these two solutions are mixed, $BaSO_4$ will precipitate if the product of the initial ion concentrations, $Q = [Ba^{2+}][SO_4^{2-}]$, is greater than K_{sp}.

∞ Review this on page 631

The use of the reaction quotient, Q, to determine the direction in which a reaction must proceed to reach equilibrium was discussed earlier (∞ Section 16.6, 'Applications of Equilibrium Constants'). The possible relationships between Q and K_{sp} are summarised as follows:

If $Q > K_{sp}$, precipitation occurs until $Q = K_{sp}$.

If $Q = K_{sp}$, equilibrium exists (saturated solution).

If $Q < K_{sp}$, solid dissolves until $Q = K_{sp}$.

SAMPLE EXERCISE 18.14 | **Predicting whether a precipitate will form**

Will a precipitate form when $0.10 \, dm^3$ of 8.0×10^{-3} M $Pb(NO_3)_2$ is added to $0.40 \, dm^3$ of 5.0×10^{-3} M Na_2SO_4?

SOLUTION

Analyse The problem asks us to determine whether a precipitate forms when two salt solutions are combined.

Plan We should determine the concentrations of all ions immediately upon mixing of the solutions and compare the value of the reaction quotient, Q, to the solubility-product constant, K_{sp}, for any potentially insoluble product. The possible metathesis products are $PbSO_4$ and $NaNO_3$. Sodium salts are quite soluble; $PbSO_4$ has a K_{sp} of 6.3×10^{-7} (Appendix D), however, and will precipitate if the Pb^{2+} and SO_4^{2-} ion concentrations are high enough for Q to exceed K_{sp} for the salt.

Solve When the two solutions are mixed, the total volume becomes $0.10 \, dm^3$ + $0.40 \, dm^3$ = $0.50 \, dm^3$. The number of moles of Pb^{2+} in $0.10 \, dm^3$ of 8.0×10^{-3} M $Pb(NO_3)_2$ is

$$(0.10 \, dm^3)(8.0 \times 10^{-3} \, mol \, dm^{-3}) = 8.0 \times 10^{-4} \, mol$$

The concentration of Pb^{2+} in the $0.50 \, dm^3$ mixture is therefore

$$[Pb^{2+}] = \frac{8.0 \times 10^{-4} \, mol}{0.50 \, dm^3} = 1.6 \times 10^{-3} \, M$$

The number of moles of SO_4^{2-} in $0.40 \, dm^3$ of 5.0×10^{-3} M Na_2SO_4 is

$$(0.40 \, dm^3)(5.0 \times 10^{-3} \, mol \, dm^{-3}) = 2.0 \times 10^{-3} \, mol$$

Therefore, $[SO_4^{2-}]$ in the $0.50 \, dm^3$ mixture is

$$[SO_4^{2-}] = \frac{2.0 \times 10^{-3} \, mol}{0.50 \, dm^3} = 4.0 \times 10^{-3} \, M$$

We then have

$$Q = [Pb^{2+}][SO_4^{2-}] = (1.6 \times 10^{-3})(4.0 \times 10^{-3}) = 6.4 \times 10^{-6}$$

Because $Q > K_{sp}$, $PbSO_4$ will precipitate.

PRACTICE EXERCISE

Will a precipitate form when $0.050 \, dm^3$ of 2.0×10^{-2} M NaF is mixed with $0.010 \, dm^3$ of 1.0×10^{-2} M $Ca(NO_3)_2$?

Answer: Yes, CaF_2 precipitates because $Q = 4.6 \times 10^{-7}$ is larger than $K_{sp} = 3.9 \times 10^{-11}$

(See also Exercise 18.52.)

Selective Precipitation of Ions

Ions can be separated from each other based on the solubilities of their salts. Consider a solution containing both Ag^+ and Cu^{2+}. If HCl is added to the solution, AgCl ($K_{sp} = 1.8 \times 10^{-10}$) precipitates, whereas Cu^{2+} remains in solution because

SAMPLE EXERCISE 18.15   Calculating ion concentrations for precipitation

A solution contains 1.0×10^{-2} M Ag^+ and 2.0×10^{-2} M Pb^{2+}. When Cl^- is added to the solution, both AgCl ($K_{sp} = 1.8 \times 10^{-10}$) and $PbCl_2$ ($K_{sp} = 1.7 \times 10^{-5}$) precipitate from the solution. What concentration of Cl^- is necessary to begin the precipitation of each salt? Which salt precipitates first?

SOLUTION

Analyse We are asked to determine the concentration of Cl^- necessary to begin the precipitation from a solution containing Ag^+ and Pb^{2+} ions, and to predict which metal chloride will begin to precipitate first.

Plan We are given K_{sp} values for the two possible precipitates. Using these and the metal ion concentrations, we can calculate what concentration of Cl^- ion would be necessary to begin precipitation of each. The salt requiring the lower Cl^- ion concentration will precipitate first.

Solve For AgCl we have

$$K_{sp} = [Ag^+][Cl^-] = 1.8 \times 10^{-10}$$

Because $[Ag^+] = 1.0 \times 10^{-2}$ M, the greatest concentration of Cl^- that can be present without causing precipitation of AgCl can be calculated from the K_{sp} expression:

$$K_{sp} = [1.0 \times 10^{-2}][Cl^-] = 1.8 \times 10^{-10}$$

$$[Cl^-] = \frac{1.8 \times 10^{-10}}{1.0 \times 10^{-2}} = 1.8 \times 10^{-8} \text{ M}$$

Any Cl^- in excess of this very small concentration will cause AgCl to precipitate from solution. Proceeding similarly for $PbCl_2$, we have

$$K_{sp} = [Pb^{2+}][Cl^-]^2 = 1.7 \times 10^{-5}$$

$$[2.0 \times 10^{-2}][Cl^-]^2 = 1.7 \times 10^{-5}$$

$$[Cl^-]^2 = \frac{1.7 \times 10^{-5}}{2.0 \times 10^{-2}} = 8.5 \times 10^{-4}$$

$$[Cl^-] = \sqrt{8.5 \times 10^{-4}} = 2.9 \times 10^{-2} \text{ M}$$

Thus a concentration of Cl^- in excess of 2.9×10^{-2} M will cause $PbCl_2$ to precipitate.

Comparing the concentrations of Cl^- required to precipitate each salt, we see that, as Cl^- is added to the solution, AgCl will precipitate first because it requires a much smaller concentration of Cl^-. Thus Ag^+ can be separated from Pb^{2+} by slowly adding Cl^- so $[Cl^-]$ is between 1.8×10^{-8} M and 2.9×10^{-2} M.

PRACTICE EXERCISE

A solution consists of 0.050 M Mg^{2+} and 0.020 M Cu^{2+}. Which ion will precipitate first as OH^- is added to the solution? What concentration of OH^- is necessary to begin the precipitation of each cation? ($K_{sp} = 1.8 \times 10^{-11}$ for $Mg(OH)_2$, and $K_{sp} = 2.2 \times 10^{-20}$ for $Cu(OH)_2$.)

Answer: $Cu(OH)_2$ precipitates first. $Cu(OH)_2$ begins to precipitate when $[OH^-]$ exceeds 1.0×10^{-9} M; $Mg(OH)_2$ begins to precipitate when $[OH^-]$ exceeds 1.9×10^{-5} M.

(See also Exercise 18.53.)

$CuCl_2$ is soluble. Separation of ions in an aqueous solution by using a reagent that forms a precipitate with one or a few of the ions is called *selective precipitation*.

Sulfide ions are often used to separate metal ions because the solubilities of sulfide salts span a wide range and depend greatly on the pH of the solution. Cu^{2+} and Zn^{2+}, for example, can be separated by bubbling H_2S gas through an acidified solution. Because CuS ($K_{sp} = 6 \times 10^{-37}$) is less soluble than ZnS ($K_{sp} = 2 \times 10^{-25}$), CuS precipitates from an acidified solution (pH = 1) while ZnS does not (▼ **FIGURE 18.20**):

$$Cu^{2+}(aq) + H_2S(aq) \rightleftharpoons CuS(s) + 2\,H^+(aq) \qquad \text{[18.27]}$$

◢ FIGURE IT OUT

What would happen if the pH were raised to 8 first and then H_2S were added?

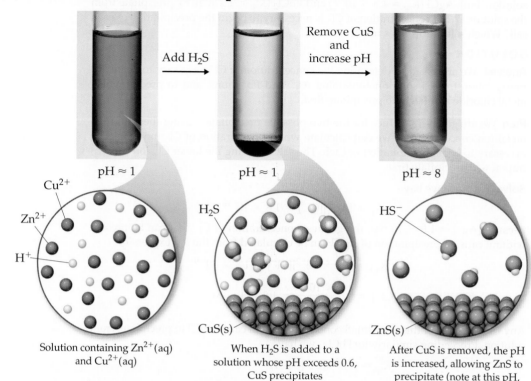

▶ **FIGURE 18.20**
Selective precipitation. In this example Cu^{2+} ions are separated from Zn^{2+} ions.

Solution containing $Zn^{2+}(aq)$ and $Cu^{2+}(aq)$

When H_2S is added to a solution whose pH exceeds 0.6, CuS precipitates

After CuS is removed, the pH is increased, allowing ZnS to precipitate (note at this pH, HS^- rather than H_2S predominates in solution)

The CuS can be separated from the Zn^{2+} solution by filtration. The CuS can then be dissolved by using a high concentration of H^+, shifting the equilibrium shown in Equation 18.27 to the left.

◢ **CONCEPT CHECK 10**

What experimental conditions will leave the smallest concentration of Cu^{2+} ions in solution according to Equation 18.27?

18.7 | QUALITATIVE ANALYSIS FOR METALLIC ELEMENTS

In this chapter we have seen several examples of equilibria involving metal ions in aqueous solution. In this final section we look briefly at how solubility equilibria and complex-ion formation can be used to detect the presence of particular metal ions in solution. Before the development of modern analytical instrumentation, it was necessary to analyse mixtures of metals in a sample by so-called *wet chemical methods*. For example, an ore sample that might contain several metallic elements was dissolved in a concentrated acid solution. This solution was then tested in a systematic way for the presence of various metal ions.

Qualitative analysis determines only the presence or absence of a particular metal ion, whereas **quantitative analysis** determines how much of a given substance is present. Wet methods of qualitative analysis have become less important as a means of analysis. They are frequently used in general chemistry laboratory programs, however, to illustrate equilibria, to teach the properties of

◢ FIGURE IT OUT

If a solution contained a mixture of Cu^{2+} and Zn^{2+} ions, would this separation scheme work? After which step would the first precipitate be observed?

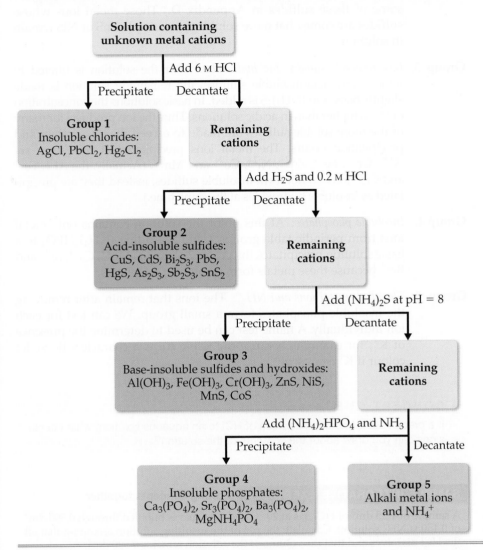

◀ **FIGURE 18.21   Qualitative analysis.**
A flowchart showing a common scheme for identifying cations.

common metal ions in solution and to develop laboratory skills. Typically, such analyses proceed in three stages: (1) the ions are separated into broad groups on the basis of solubility properties; (2) the individual ions within each group are then separated by selectively dissolving members in the group; (3) the ions are then identified by means of specific tests.

A scheme in general use divides the common cations into five groups, as shown in ▲ FIGURE 18.21. The order of addition of reagents is important. The most selective separations, those that involve the smallest number of ions, are carried out first. The reactions that are used must proceed so far towards completion that any concentration of cations remaining in the solution is too small to interfere with subsequent tests. Let's take a closer look at each of these five groups of cations, briefly examining the logic used in this qualitative analysis scheme.

Group 1.   *Insoluble chlorides:* Of the common metal ions, only Ag^+, Hg_2^{2+} and Pb^{2+} form insoluble chlorides. When dilute HCl is added to a mixture of cations, therefore, only AgCl, Hg_2Cl_2 and $PbCl_2$ will precipitate, leaving the other cations in solution. The absence of a precipitate indicates that the starting solution contains no Ag^+, Hg_2^{2+} or Pb^{2+}.

Group 2. *Acid-insoluble sulfides:* After any insoluble chlorides have been removed, the remaining solution, now acidic, is treated with H_2S. Only the most insoluble metal sulfides, CuS, Bi_2S_3, CdS, PbS, HgS, As_2S_3, Sb_2S_3 and SnS_2, can precipitate. (Note the very small values of K_{sp} for some of these sulfides in Appendix D.) Those metal ions whose sulfides are somewhat more soluble, for example ZnS or NiS remain in solution.

Group 3. *Base-insoluble sulfides and hydroxides:* After the solution is filtered to remove any acid-insoluble sulfides, the remaining solution is made slightly basic and $(NH_4)_2S$ is added. In basic solutions the concentration of S^{2-} is higher than in acidic solutions. Thus the ion products for many of the more soluble sulfides are made to exceed their K_{sp} values and precipitation occurs. The metal ions precipitated at this stage are Al^{3+}, Cr^{3+}, Fe^{3+}, Zn^{2+}, Ni^{2+}, Co^{2+} and Mn^{2+}. (Actually, the Al^{3+}, Fe^{3+} and Cr^{3+} ions do not form insoluble sulfides; instead they are precipitated as insoluble hydroxides at the same time.)

Group 4. *Insoluble phosphates:* At this point the solution contains only metal ions from periodic table groups 1 and 2. Adding $(NH_4)_2HPO_4$ to a basic solution precipitates the group 2 elements Mg^{2+}, Ca^{2+}, Sr^{2+} and Ba^{2+} because these metals form insoluble phosphates.

Group 5. *The alkali metal ions and NH_4^+:* The ions that remain after removing the insoluble phosphates form a small group. We can test for each ion individually. A flame test can be used to determine the presence of K^+, for example, because the flame turns a characteristic violet colour if K^+ is present.

> ⚠ **CONCEPT CHECK 11**
>
> If a precipitate forms upon addition of HCl to an aqueous solution, what conclusions can you draw about the contents of the solution?

SAMPLE INTEGRATIVE EXERCISES Putting concepts together

A sample of 1.25 dm^3 of HCl gas at 20 °C and 0.960 bar is bubbled through 0.500 dm^3 of 0.150 M NH_3 solution. Calculate the pH of the resulting solution assuming that all the HCl dissolves and that the volume of the solution remains 0.500 dm^3.

SOLUTION

The number of moles of HCl gas is calculated from the ideal-gas law,

$$n = \frac{PV}{RT} = \frac{(96.0 \text{ kPa})(1.25 \text{ dm}^3)}{(8.314 \text{ kPa dm}^3 \text{ mol}^{-1} \text{ K}^{-1})(293 \text{ K})} = 0.0493 \text{ mol HCl}$$

The number of moles of NH_3 in the solution is given by the product of the volume of the solution and its concentration,

$$\text{Moles NH}_3 = (0.500 \text{ dm}^3)(0.150 \text{ mol NH}_3 \text{ dm}^{-3}) = 0.0750 \text{ mol NH}_3$$

The acid HCl and base NH_3 react, transferring a proton from HCl to NH_3, producing NH_4^+ and Cl^- ions,

$$HCl(g) + NH_3(aq) \longrightarrow NH_4^+(aq) + Cl^-(aq)$$

To determine the pH of the solution, we first calculate the amount of each reactant and each product present at the completion of the reaction,

	HCl(g)	+	NH₃(aq)	⟶	NH₄⁺(aq)	+	Cl
Before addition	0		0.0750 mol		0		0
Addition	0.0493 mol						
After addition	0		0.0258 mol		0.0493 mol		0.0493 mol

Thus the reaction produces a solution containing a mixture of NH_3, NH_4^+ and Cl^-. The NH_3 is a weak base ($K_b = 1.8 \times 10^{-5}$), NH_4^+ is its conjugate acid and Cl^- is neither acidic nor basic. Consequently, the pH depends on $[NH_3]$ and $[NH_4^+]$,

$$[NH_3] = \frac{0.0258 \text{ mol } NH_3}{0.500 \text{ dm}^{-3} \text{ soln}} = 0.0516 \text{ M}$$

$$[NH_4^+] = \frac{0.0493 \text{ mol } NH_4^+}{0.500 \text{ dm}^{-3} \text{ soln}} = 0.0986 \text{ M}$$

We can calculate the pH using either K_b for NH_3 or K_a for NH_4^+. Using the K_b expression, we have

	$NH_3(aq)$	+	$H_2O(l)$	$\longrightarrow$	$NH_4^+(aq)$	+	$OH^-(aq)$
Initial	0.516 M		—		0.0986 M		0
Change	$-x$ M		—		$+x$ M		$+x$ M
Equilibrium	$(0.0516 - x)$ M		—		$(0.0986 + x)$ M		x M

$$K_b = \frac{[NH_4^+][OH^-]}{[NH_3]} = \frac{(0.0986 + x)(x)}{(0.0516 - x)} \simeq \frac{(0.0986)x}{0.0516} = 1.8 \times 10^{-5}$$

$$x = [OH^-] = \frac{(0.0516)(1.8 \times 10^{-5})}{0.0986} = 9.4 \times 10^{-6} \text{ M}$$

Hence, $pOH = -\log(9.4 \times 10^{-6}) = 5.03$, and $pH = 14.00 - pOH = 14.00 - 5.03 = 8.97$.

CHAPTER SUMMARY AND KEY TERMS

SECTION 18.1 In this chapter we have considered several types of important equilibria that occur in aqueous solution. Our primary emphasis has been on acid–base equilibria in solutions containing two or more solutes and on solubility equilibria. The dissociation of a weak acid or weak base is repressed by the presence of a strong electrolyte that provides an ion common to the equilibrium. This phenomenon is called the **common-ion effect**.

SECTION 18.2 A particularly important type of acid–base mixture is that of a weak conjugate acid–base pair. Such mixtures function as **buffer solutions** (**buffers**). Addition of small amounts of a strong acid or a strong base to a buffer solution causes only small changes in pH because the buffer reacts with the added acid or base. (Strong acid–strong base, strong acid–weak base and weak acid–strong base reactions proceed essentially to completion.) Buffer solutions are usually prepared from a weak acid and a salt of that acid or from a weak base and a salt of that base. Two important characteristics of a buffer solution are its **buffer capacity** and its **pH range**. The optimal pH of a buffer is equal to pK_a (or pK_b) of the acid (or base) used to prepare the buffer. The relationship between pH, pK_a and the concentrations of an acid and its conjugate base can be expressed by the **Henderson–Hasselbalch equation**.

SECTION 18.3 The plot of the pH of an acid (or base) as a function of the volume of added base (or acid) is called a **pH titration curve**. The titration curve of a strong acid–strong base titration exhibits a large change in pH in the immediate vicinity of the equivalence point; at the equivalence point for such a titration pH = 7. For strong acid–weak base or weak acid–strong base titrations, the pH change in the vicinity of the equivalence point is not as large. Furthermore, the pH at the equivalence point is not 7 in either of these cases. Rather, it is the pH of the salt solution that results from the neutralisation reaction. For this reason it is important to choose an indicator whose colour change is near the pH at the equivalence point for titrations involving either weak acids or weak bases. It is possible to calculate the pH at any point of the titration curve by first considering the effects of the acid–base reaction on solution concentrations and then examining equilibria involving the remaining solute species.

SECTION 18.4 The equilibrium between a solid compound and its ions in solution provides an example of heterogeneous equilibrium. The **solubility-product constant** (or simply the **solubility product**), K_{sp}, is an equilibrium constant that expresses quantitatively the extent to which the compound dissolves. The K_{sp} can be used to calculate the solubility of an ionic compound, and the solubility can be used to calculate K_{sp}.

SECTION 18.5 Several experimental factors, including temperature, affect the solubilities of ionic compounds in water. The solubility of a slightly soluble ionic compound is decreased by the presence of a second solute that furnishes a common ion (the common-ion effect). The solubility of compounds containing basic anions increases as the solution is made more acidic (as pH decreases). Salts with anions of negligible basicity (the anions of strong acids) are unaffected by pH changes.

The solubility of metal salts is also affected by the presence of certain Lewis bases that react with metal ions to form stable **complex ions**. Complex-ion formation in aqueous solution involves the displacement by Lewis bases (such as NH_3 and CN^-) of water molecules attached to the metal ion. The extent to which such complex formation occurs is expressed quantitatively by the **formation constant** for the complex ion. **Amphoteric oxides and hydroxides** are those that are only slightly soluble in water but dissolve on addition of either acid or base.

SECTION 18.6   Comparison of the ion product, Q, with the value of K_{sp} can be used to judge whether a precipitate will form when solutions are mixed or whether a slightly soluble salt will dissolve under various conditions. Precipitates form when $Q > K_{sp}$. If two salts have sufficiently different solubilities, selec-

tive precipitation can be used to precipitate one ion while leaving the other in solution, effectively separating the two ions.

SECTION 18.7   Metallic elements vary a great deal in the solubilities of their salts, in their acid–base behaviour and in their tendencies to form complex ions. These differences can be used to separate and detect the presence of metal ions in mixtures. **Qualitative analysis** determines the presence or absence of species in a sample, whereas **quantitative analysis** determines how much of each species is present. The qualitative analysis of metal ions in solution can be carried out by separating the ions into groups on the basis of precipitation reactions and then analysing each group for individual metal ions.

KEY SKILLS

- Describe the common-ion effect. (Section 18.1)

- Explain how a buffer solution functions. (Section 18.2)

- Calculate the pH of a buffer solution. (Section 18.2)

- Calculate the pH of a buffer solution after the addition of small amounts of a strong acid or a strong base. (Section 18.2)

- Calculate the pH at any point in an acid–base titration of a strong acid and strong base. (Section 18.3)

- Calculate the pH at any point in a titration of a weak acid with a strong base or a weak base with a strong acid. (Section 18.3)

- Understand the differences between the titration curves for a strong acid–strong base titration and those when either the acid or the base is weak. (Section 18.3)

- Calculate K_{sp} from molar solubility and molar solubility from K_{sp}. (Section 18.4)

- Calculate molar solubility in the presence of a common ion. (Section 18.5)

- Predict the effect of pH on solubility. (Section 18.5)

- Predict whether a precipitate will form when solutions are mixed, by comparing Q and K_{sp}. (Section 18.6)

- Calculate the ion concentrations required to begin precipitation. (Section 18.6)

- Explain the effect of complex-ion formation on solubility. (Section 18.6)

KEY EQUATION

- The Henderson–Hasselbalch equation, used to calculate the pH of a buffer using the concentrations of a conjugate acid–base pair

$$pH = pK_a + \log \frac{[\text{base}]}{[\text{acid}]} \qquad [18.9]$$

EXERCISES

VISUALISING CONCEPTS

18.1 The following boxes represent aqueous solutions containing a weak acid, HX, and its conjugate base, X^-. Water molecules and cations are not shown. Which solution has the highest pH? Explain. [Section 18.1]

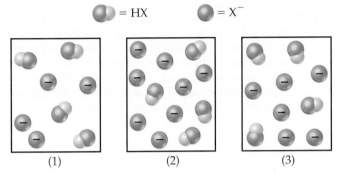

= HX   = X^-

(1)   (2)   (3)

18.2 Drawing (1) below represents a buffer composed of equal concentrations of a weak acid, HX, and its conjugate base, X^-. The heights of the columns are proportional to the concentrations of the components of the buffer. **(a)** Which of the three drawings (1), (2) or (3) represents the buffer after the addition of a strong acid? **(b)** Which of the three represents the buffer after the addition of a strong base? **(c)** Which of the three represents a situation that cannot arise from the addition of either an acid or a base? [Section 18.2]

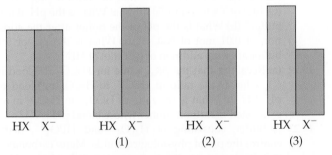

HX X^- HX X^- HX X^- HX X^-
           (1)   (2)   (3)

18.3 The following drawings represent solutions at various stages of the titration of a weak acid, HA, with NaOH. (The Na^+ ions and water molecules have been omitted for clarity.) To which of the following regions of the titration curve does each drawing correspond: **(a)** before addition of NaOH, **(b)** after addition of NaOH but before equivalence point, **(c)** at equivalence point, **(d)** after equivalence point? [Section 18.3]

= HA   = A^- = OH^-

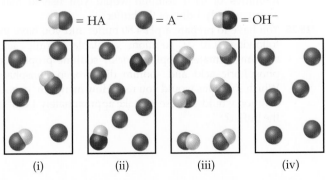

(i)   (ii)   (iii)   (iv)

18.4 Match the following descriptions of titration curves with the diagrams: **(a)** strong acid added to strong base, **(b)** strong base added to weak acid, **(c)** strong base added to strong acid, **(d)** strong base added to polyprotic acid. [Section 18.3]

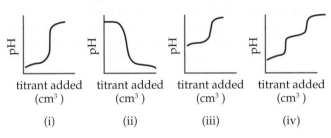

titrant added (cm³)   titrant added (cm³)   titrant added (cm³)   titrant added (cm³)
      (i)   (ii)   (iii)   (iv)

18.5 The following graphs represent the behaviour of $BaCO_3$ under different circumstances. In each case the vertical axis indicates the solubility of the $BaCO_3$ and the horizontal axis represents the concentration of some other reagent. **(a)** Which graph represents what happens to the solubility of $BaCO_3$ as HNO_3 is added? **(b)** Which graph represents what happens to the $BaCO_3$ solubility as Na_2CO_3 is added? **(c)** Which represents what happens to the $BaCO_3$ solubility as $NaNO_3$ is added? [Section 18.5]

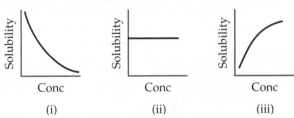

Conc   Conc   Conc
(i)   (ii)   (iii)

18.6 What is the name given to the kind of behaviour demonstrated by a metal hydroxide in this graph? [Section 18.5]

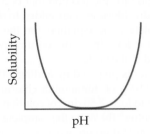

pH

COMMON-ION EFFECT (Section 18.1)

18.7 (a) What is the common-ion effect? **(b)** Give an example of a salt that can decrease the ionisation of HNO_2 in solution.

18.8 (a) Consider the equilibrium $B(aq) + H_2O(l) \rightleftharpoons HB^+(aq) + OH^-(aq)$. Using Le Châtelier's principle, explain the effect of the presence of a salt of HB^+ on the ionisation of B. **(b)** Give an example of a salt that can decrease the ionisation of NH_3 in solution.

18.9 Does the pH increase, decrease or remain the same when each of the following is added: **(a)** $NaNO_2$ to a solution of HNO_2; **(b)** $(CH_3NH_3)Cl$ to a solution of CH_3NH_2; **(c)** sodium formate to a solution of formic acid; **(d)** potassium bromide to a solution of hydrobromic acid; **(e)** HCl to a solution of CH_3COONa?

18.10 Use information from Appendix D to calculate the pH of **(a)** a solution that is 0.060 M in potassium propionate (CH_3CH_2COOK) and 0.085 M in propionic acid (CH_3CH_2COOH); **(b)** a solution that is 0.075 M in trimethylamine, $(CH_3)_3N$, and 0.10 M in trimethylammonium chloride, $(CH_3)_3NHCl$; **(c)** a solution that is made by mixing 50.0 cm³ of 0.15 M acetic acid and 50.0 cm³ of 0.20 M sodium acetate.

18.11 Use information from Appendix D to calculate the pH of **(a)** a solution that is 0.160 M in sodium formate ($HCOONa$) and 0.260 M in formic acid ($HCOOH$); **(b)** a solution that is 0.210 M in pyridine (C_5H_5N) and 0.350 M in pyridinium chloride (C_5H_5NHCl); **(c)** a solution that is made by combining 125 cm³ of 0.050 M hydrofluoric acid with 50.0 cm³ of 0.10 M sodium fluoride.

BUFFER SOLUTIONS (Section 18.2)

18.12 Explain why a mixture of CH_3COOH and CH_3COONa can act as a buffer while a mixture of HCl and NaCl cannot.

18.13 Explain why a mixture formed by mixing 100 cm³ of 0.100 M CH_3COOH and 50 cm³ of 0.100 M NaOH will act as a buffer.

18.14 (a) Calculate the pH of a buffer that is 0.12 M in lactic acid and 0.11 M in sodium lactate. **(b)** Calculate the pH of a buffer formed by mixing 85 cm³ of 0.13 M lactic acid with 95 cm³ of 0.15 M sodium lactate.

18.15 (a) Calculate the pH of a buffer that is 0.120 M in $NaHCO_3$ and 0.105 M in Na_2CO_3. **(b)** Calculate the pH of a solution formed by mixing 65 cm³ of 0.20 M $NaHCO_3$ with 75 cm³ of 0.15 M Na_2CO_3.

18.16 A buffer is prepared by adding 20.0 g of acetic acid (CH_3COOH) and 20.0 g of sodium acetate (CH_3COONa) to enough water to form 2.00 dm³ of solution. **(a)** Determine the pH of the buffer. **(b)** Write the complete ionic equation for the reaction that occurs when a few drops of hydrochloric acid are added to the buffer. **(c)** Write the complete ionic equation for the reaction that occurs when a few drops of sodium hydroxide solution are added to the buffer.

18.17 A buffer is prepared by adding 5.0 g of ammonia (NH_3) and 20.0 g of ammonium chloride (NH_4Cl) to enough water to form 2.50 dm³ of solution. **(a)** What is the pH of this buffer? **(b)** Write the complete ionic equation for the reaction that occurs when a few drops of nitric acid are added to the buffer. **(c)** Write the complete ionic equation for the reaction that occurs when a few drops of potassium hydroxide solution are added to the buffer.

18.18 How many moles of sodium hypobromite (NaBrO) should be added to 1.00 dm³ of 0.050 M hypobromous acid (HBrO) to form a buffer solution of pH 9.15? Assume that no volume change occurs when the NaBrO is added.

18.19 How many grams of sodium lactate ($CH_3CH_3(OH)$ $COONa$) should be added to 1.00 dm³ of 0.150 M lactic

acid ($CH_3CH(OH)COOH$) to form a buffer solution with pH 4.00? Assume that no volume change occurs when the sodium lactate is added.

18.20 A buffer solution contains 0.10 mol of acetic acid and 0.13 mol of sodium acetate in 1.00 dm³. **(a)** What is the pH of this buffer? **(b)** What is the pH of the buffer after the addition of 0.02 mol of KOH? **(c)** What is the pH of the buffer after the addition of 0.02 mol of HNO_3?

18.21 A buffer solution contains 0.12 mol of propionic acid (CH_3CH_2COOH) and 0.10 mol of sodium propionate (CH_3CH_2COONa) in 1.50 dm³. **(a)** What is the pH of this buffer? **(b)** What is the pH of the buffer after the addition of 0.01 mol of NaOH? **(c)** What is the pH of the buffer after the addition of 0.01 mol of HI?

18.22 (a) Recalling that the pK_{a1} value for H_2CO_3 in blood is 6.1, what is the ratio of HCO_3^- to H_2CO_3 in blood of pH 7.4? **(b)** What is the ratio of HCO_3^- to H_2CO_3 in an exhausted marathon runner whose blood pH is 7.1?

18.23 A buffer, consisting of $H_2PO_4^-$ and HPO_4^{2-}, helps control the pH of physiological fluids. Many carbonated soft drinks also use this buffer system. What is the pH of a soft drink in which the major buffer ingredients are 6.5 g of NaH_2PO_4 and 8.0 g of Na_2HPO_4 per 355 cm³ of solution?

18.24 You have to prepare a pH 3.50 buffer, and you have the following 0.10 M solutions available: HCOOH, CH_3COOH, H_3PO_4, HCOONa, CH_3COONa and NaH_2PO_4. Which solutions would you use? How many millilitres of each solution would you use to make approximately 1 dm³ of the buffer?

18.25 You have to prepare a pH 4.80 buffer and you have the following 0.10 M solutions available: formic acid, sodium formate, propionic acid, sodium propionate, phosphoric acid and sodium dihydrogen phosphate. Which solutions would you use? How many cm³ of each solution would you use to make approximately 1 dm³ of the buffer?

ACID–BASE TITRATIONS (Section 18.3)

18.26 The graph below shows the titration curves for two monoprotic acids. **(a)** Which curve is that of a strong acid? **(b)** What is the approximate pH at the equivalence point of each titration?

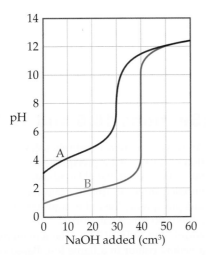

18.27 How does titration of a strong, monoprotic acid with a strong base differ from titration of a weak, monoprotic acid with a strong base with respect to the following: **(a)** quantity of base required to reach the equivalence point, **(b)** pH at the beginning of the titration, **(c)** pH at the equivalence point, **(d)** pH after addition of a slight excess of base, **(e)** choice of indicator for determining the equivalence point?

18.28 Predict whether the equivalence point of each of the following titrations is below, above or at pH = 7: **(a)** $NaHCO_3$ titrated with NaOH, **(b)** NH_3 titrated with HCl, **(c)** KOH titrated with HBr.

18.29 Predict whether the equivalence point of each of the following titrations is below, above or at pH = 7: **(a)** formic acid titrated with NaOH, **(b)** calcium hydroxide titrated with perchloric acid, **(c)** pyridine titrated with nitric acid.

18.30 How many cm^3 of 0.0850 M NaOH are required to titrate each of the following solutions to the equivalence point: **(a)** 40.0 cm^3 of 0.0900 M HNO_3, **(b)** 35.0 cm^3 of 0.0850 M $HC_2H_3O_2$, **(c)** 50.0 cm^3 of a solution that contains 1.85 g of HCl per cm^3?

18.31 How many dm^3 of 0.105 M HCl are needed to titrate each of the following solutions to the equivalence point: **(a)** 55.0 cm^3 of 0.0950 M NaOH, **(b)** 22.5 cm^3 of 0.118 M NH_3, **(c)** 125.0 cm^3 of a solution that contains 1.35 g of NaOH per dm^3?

18.32 A 20.0 cm^3 sample of 0.200 M HBr solution is titrated with 0.200 M NaOH solution. Calculate the pH of the solution after the following volumes of base have been added: **(a)** 18.0 cm^3, **(b)** 19.9 cm^3, **(c)** 20.0 cm^3, **(d)** 20.1 cm^3, **(e)** 35.0 cm^3.

18.33 Consider the titration of 30.0 cm^3 of 0.030 M NH_3 with 0.025 M HCl. Calculate the pH after the following volumes of titrant have been added: **(a)** 0 cm^3, **(b)** 10.0 cm^3, **(c)** 20.0 cm^3, **(d)** 35.0 cm^3, **(e)** 36.0 cm^3, **(f)** 37.0 cm^3.

18.34 Calculate the pH at the equivalence point for titrating 0.200 M solutions of each of the following bases with 0.200 M HBr: **(a)** sodium hydroxide (NaOH), **(b)** hydroxylamine (NH_2OH), **(c)** aniline ($C_6H_5HN_2$).

SOLUBILITY EQUILIBRIA AND FACTORS AFFECTING SOLUBILITY (Sections 18.4 and 18.5)

18.35 **(a)** Why is the concentration of undissolved solid not explicitly included in the expression for the solubility-product constant? **(b)** Write the expression for the solubility-product constant for each of the following strong electrolytes: AgI, $SrSO_4$, $Fe(OH)_2$ and Hg_2Br_2.

18.36 **(a)** Explain the difference between solubility and solubility-product constant. **(b)** Write the expression for the solubility-product constant for each of the following ionic compounds: $MnCO_3$, $Hg(OH)_2$ and $Cu_3(PO_4)_2$.

18.37 **(a)** If the molar solubility of CaF_2 at 35 °C is 1.24×10^{-3} mol dm^{-3}, what is K_{sp} at this temperature? **(b)** It is found that 1.1×10^{-2} g of SrF_2 dissolves per 100 cm^3 of aqueous solution at 25 °C. Calculate the solubility product for SrF_2. **(c)** The K_{sp} of $BA(IO_3)_2$ at 25 °C is 6.0×10^{-10}. What is the molar solubility of $Ba(IO_3)_2$?

18.38 **(a)** The molar solubility of $PbBr_2$ at 25 °C is 1.0×10^{-2} mol dm^{-3}. Calculate K_{sp}. **(b)** If 0.0490 g of $AgIO_3$ dissolves per litre of solution, calculate the solubility-product constant. **(c)** Using the appropriate K_{sp} value from Appendix D, calculate the solubility of $Cu(OH)_2$ in grams per litre of solution.

18.39 A 1.00 dm^3 solution saturated at 25 °C with calcium oxalate (CaC_2O_4) contains 0.0061 g of CaC_2O_4. Calculate the solubility-product constant for this salt at 25 °C.

18.40 A 1.00 dm^3 solution saturated at 25 °C with lead(II) iodide contains 0.54 g of PbI_2. Calculate the solubility-product constant for this salt at 25 °C.

18.41 Using Appendix D, calculate the molar solubility of AgBr in **(a)** pure water, **(b)** 3.0×10^{-2} M $AgNO_3$ solution, **(c)** 0.10 M NaBr solution.

18.42 Calculate the solubility of $Mn(OH)_2$ in g dm^{-3} when buffered at pH **(a)** 9.5, **(b)** 11.8.

18.43 Calculate the molar solubility of $Fe(OH)_2$ when buffered at pH **(a)** 7.0, **(b)** 10.0, **(c)** 12.0.

18.44 Which of the following salts will be substantially more soluble in acidic solution than in pure water: **(a)** $ZnCO_3$, **(b)** ZnS, **(c)** BiI_3, **(d)** AgCN, **(e)** $Ba_3(PO_4)_2$?

18.45 For each of the following slightly soluble salts, write the net ionic equation, if any, for reaction with acid: **(a)** MnS, **(b)** PbF_2, **(c)** $AuCl_3$, **(d)** $Hg_2C_2O_4$, **(e)** CuBr.

18.46 From the value of K_f listed in Table 18.1, calculate the concentration of Cu^{2+} in 1.0 dm^3 of a solution that ontains a total of 1×10^{-3} mol of copper(II) ion and that is 0.10 M in NH_3.

18.47 To what final concentration of NH_3 must a solution be adjusted to just dissolve 0.020 mol of NiC_2O_4 ($K_{sp} = 4 \times 10^{-10}$) in 1.0 dm^3 of solution? [*Hint:* You can neglect the hydrolysis of $C_2O_4^{2-}$ because the solution will be quite basic.]

18.48 By using the values of K_{sp} for AgI (Appendix D) and K_f for $Ag(CN)_2^-$ (Table 18.1), calculate the equilibrium constant for the reaction

$$AgI(s) + 2 CN^-(aq) \rightleftharpoons Ag(CN)_2^-(aq) + I^-(aq)$$

18.49 Using the value of K_{sp} for Ag_2S, K_{a1} and K_{a2} for H_2S and $K_f = 1.1 \times 10^5$ for $AgCl_2^-$, calculate the equilibrium constant for the following reaction:

$$Ag_2S(s) + 4 Cl^-(aq) + 2 H^+(aq) \rightleftharpoons$$
$$2 AgCl_2^-(aq) + H_2S(aq)$$

PRECIPITATION; QUALITATIVE ANALYSIS (Sections 18.6 and 18.7)

18.50 (a) Will $Ca(OH)_2$ precipitate from solution if the pH of a 0.050 M solution of $CaCl_2$ is adjusted to 8.0? (b) Will Ag_2SO_4 precipitate when 100 cm^3 of 0.050 M $AgNO_3$ is mixed with 10 cm^3 of 5.0×10^{-2} M Na_2SO_4 solution?

18.51 Calculate the minimum pH needed to precipitate $Mn(OH)_2$ so completely that the concentration of Mn^{2+} is less than 1 μg per litre (1 part per billion (ppb)).

18.52 A solution contains 2.0×10^{-4} M Ag^+ and 1.5×10^{-3} M Pb^{2+}. If NaI is added, will AgI ($K_{sp} = 8.3 \times 10^{-17}$) or PbI_2 ($K_{sp} = 7.9 \times 10^{-9}$) precipitate first? Specify the concentration of I^- needed to begin precipitation.

18.53 A solution of Na_2SO_4 is added dropwise to a solution that is 0.010 M in Ba^{2+} and 0.010 M in Sr^{2+}. (a) What concentration of SO_4^{2-} is necessary to begin precipitation? (Neglect volume changes. $BaSO_4$: $K_{sp} = 1.1 \times 10^{-10}$; $SrSO_4$: $K_{sp} = 3.2 \times 10^{-7}$.) (b) Which cation precipitates first? (c) What is the concentration of SO_4^{2-} when the second cation begins to precipitate?

18.54 A solution containing an unknown number of metal ions is treated with dilute HCl; no precipitate forms. The pH is adjusted to about 1, and H_2S is bubbled through. Again, no precipitate forms. The pH of the solution is then adjusted to about 8. Again, H_2S is bubbled through. This time a precipitate forms. The filtrate from this solution is treated with $(NH_4)_2HPO_4$. No precipitate forms. Which metal ions discussed in Section 18.7 are possibly present? Which are definitely absent within the limits of these tests?

18.55 An unknown solid is entirely soluble in water. On addition of dilute HCl a precipitate forms. After the precipitate is filtered off, the pH is adjusted to about 1 and H_2S is bubbled in; a precipitate again forms. After filtering off this precipitate, the pH is adjusted to 8 and H_2S is again added; no precipitate forms. No precipitate forms upon addition of $(NH_4)_2HPO_4$. The remaining solution shows a yellow colour in a flame test. Based on these observations, which of the following compounds might be present, which are definitely present and which are definitely absent: CdS, $Pb(NO_3)_2$, HgO, $ZnSO_4$, $Cd(NO_3)_2$ and Na_2SO_4?

18.56 In the course of various qualitative analysis procedures, the following mixtures are encountered: (a) Zn^{2+} and Cd^{2+}, (b) $Cr(OH)_3$ and $Fe(OH)_3$, (c) Mg^{2+} and K^+, (d) Ag^+ and Mn^{2+}. Suggest how each mixture might be separated.

18.57 Suggest how the cations in each of the following solution mixtures can be separated: (a) Na^+ and Cd^{2+}, (b) Cu^{2+} and Mg^{2+}, (c) Pb^{2+} and Al^{3+}, (d) Ag^+ and Hg^{2+}.

18.58 (a) Precipitation of the group 4 cations (Figure 18.21) requires a basic medium. Why is this so? (b) What is the most significant difference between the sulfides precipitated in group 2 and those precipitated in group 3? (c) Suggest a procedure that would serve to redissolve the group 3 cations following their precipitation.

ADDITIONAL EXERCISES

18.59 Furoic acid (C_4H_3OCOOH) has a K_a value of 6.76×10^{-4} at 25 °C. Calculate the pH at 25 °C of (a) a solution formed by adding 35.0 g of furoic acid and 30.0 g of sodium furoate ($C_4H_3OCOONa$) to enough water to form 0.250 dm^3 of solution; (b) a solution formed by mixing 30.0 cm^3 of 0.250 M C_4H_3OCOOH and 20.0 cm^3 of 0.22 M $C_4H_3OCOONa$ and diluting the total volume to 125 cm^3; (c) a solution prepared by adding 50.0 cm^3 of 1.65 M NaOH solution to 0.500 dm^3 of 0.0850 M C_4H_3OCOOH.

18.60 Equal quantities of 0.010 M solutions of an acid HA and a base B are mixed. The pH of the resulting solution is 9.2. (a) Write the equilibrium equation and equilibrium constant expression for the reaction between HA and B.

(b) If K_a for HA is 8.0×10^{-5}, what is the value of the equilibrium constant for the reaction between HA and B? (c) What is the value of K_b for B?

18.61 Two buffers are prepared by adding an equal number of moles of formic acid (HCOOH) and sodium formate (HCOONa) to enough water to make 1.00 dm^3 of solution. Buffer A is prepared using 1.00 mol each of formic acid and sodium formate. Buffer B is prepared by using 0.010 mol of each. (a) Calculate the pH of each buffer, and explain why they are equal. (b) Which buffer will have the greater buffer capacity? Explain. (c) Calculate the change in pH for each buffer upon the addition of 1.0 cm^3 of 1.00 M HCl. (d) Calculate the change in pH for each buffer upon the addition of 10 cm^3 of 1.00 M HCl. **(e)** Discuss your answers for parts (c) and (d) in light of your response to part (b).

18.62 A biochemist needs 750 cm^3 of an acetic acid–sodium acetate buffer with pH 4.50. Solid sodium acetate (CH$_3$COONa) and glacial acetic acid (CH$_3$COOH) are available. Glacial acetic acid is 99% CH$_3$COOH by mass and has a density of 1.05 g cm^{-3}. If the buffer is to be 0.20 M in CH$_3$COOH, how many grams of CH$_3$COONa and how many millilitres of glacial acetic acid must be used?

18.63 A sample of 0.2140 g of an unknown monoprotic acid was dissolved in 25.0 cm^3 of water and titrated with 0.0950 M NaOH. The acid required 27.4 cm^3 of base to reach the equivalence point. **(a)** What is the molar mass of the acid? **(b)** After 15.0 cm^3 of base had been added in the titration, the pH was found to be 6.50. What is the K_a for the unknown acid?

18.64 Show that the pH at the halfway point of a titration of a weak acid with a strong base (where the volume of added base is half of that needed to reach the equivalence point) is equal to pK_a for the acid.

18.65 If 40.00 cm^3 of 0.100 M Na$_2$CO$_3$ is titrated with 0.100 M HCl, calculate **(a)** the pH at the start of the titration; **(b)** the volume of HCl required to reach the first equivalence point and the predominant species present at this point; **(c)** the volume of HCl required to reach the second equivalence point and the predominant species present at this point; **(d)** the pH at the second equivalence point.

18.66 A hypothetical weak acid, HA, was combined with NaOH in the following proportions: 0.20 mol of HA, 0.080 mol of NaOH. The mixture was diluted to a total volume of 1.0 dm^3, and the pH measured. **(a)** If pH = 4.80, what is the pK_a of the acid? **(b)** How many additional moles of NaOH should be added to the solution to increase the pH to 5.00?

18.67 What is the pH of a solution made by mixing 0.30 mol NaOH, 0.25 mol Na$_2$HPO$_4$ and 0.20 mol H$_3$PO$_4$ with water and diluting to 1.00 dm^3?

18.68 How many microlitres of 1.000 M NaOH solution must be added to 25.00 cm^3 of a 0.1000 M solution of lactic acid (CH$_3$CH(OH)COOH) to produce a buffer with pH = 3.75?

18.69 A person suffering from anxiety begins breathing rapidly and as a result suffers alkalosis, an increase in blood pH. **(a)** Using Equation 18.10, explain how rapid breathing can cause the pH of blood to increase. **(b)** One cure for this problem is breathing into a paper bag. Why does this procedure lower blood pH?

18.70 Excess Ca(OH)$_2$ is shaken with water to produce a saturated solution. The solution is filtered, and a 50.00 cm^3 sample titrated with HCl requires 11.23 cm^3 of 0.0983 M HCl to reach the end point. Calculate K_{sp} for Ca(OH)$_2$. Compare your result with that in Appendix D. Do you think the solution was kept at 25 °C?

18.71 The solubility-product constant for barium permanganate, Ba(MnO$_4$)$_2$, is 2.5×10^{-10}. Suppose that solid Ba(MnO$_4$)$_2$ is in equilibrium with a solution of KMnO$_4$. What concentration of KMnO$_4$ is required to establish a concentration of 2.0×10^{-8} M for the Ba^{2+} ion in solution?

18.72 Calculate the ratio of [Ca^{2+}] to [Fe^{2+}] in a lake in which the water is in equilibrium with deposits of both CaCO$_3$ and FeCO$_3$, assuming that the water is quite basic and that the hydrolysis of the carbonate ion can therefore be ignored.

18.73 The solubility products of PbSO$_4$ and SrSO$_4$ are 6.3×10^{-7} and 3.2×10^{-7}, respectively. What are the values of [SO$_4^{2-}$], [Pb^{2+}] and [Sr^{2+}] in a solution at equilibrium with both substances?

18.74 What pH buffer solution is needed to give an Mg^{2+} concentration of 3.0×10^{-2} M in equilibrium with solid magnesium oxalate?

18.75 The value of K_{sp} for Mg$_3$(AsO$_4$)$_2$ is 2.1×10^{-20}. The AsO$_4^{3-}$ ion is derived from the weak acid H$_3$AsO$_4$ (pK_{a1} = 2.22; pK_{a2} = 6.98; pK_{a3} = 11.50). When asked to calculate the molar solubility of Mg$_3$(AsO$_4$)$_2$ in water, a student used the K_{sp} expression and assumed that [Mg^{2+}] = 1.5[AsO$_4^{3-}$]. Why was this a mistake?

18.76 The solubility product for Zn(OH)$_2$ is 3.0×10^{-16}. The formation constant for the hydroxo complex, Zn(OH)$_4^{2-}$, is 4.6×10^{17}. What concentration of OH$^-$ is required to dissolve 0.015 mol of Zn(OH)$_2$ in a litre of solution?

18.77 The solubility of CaCO$_3$ is pH dependent. **(a)** Calculate the molar solubility of CaCO$_3$ ($K_{sp} = 4.5 \times 10^{-9}$) neglecting the acid–base character of the carbonate ion. **(b)** Use the K_b expression for the CO$_3^{2-}$ ion to determine the equilibrium constant for the reaction

$$CaCO_3(s) + H_2O(l) \rightleftharpoons Ca^{2+}(aq) + HCO_3^-(aq) + OH^-(aq)$$

(c) If we assume that the only sources of Ca^{2+}, HCO$_3^-$, OH$^-$ and ions are from the dissolution of CaCO$_3$, what is the molar solubility of CaCO$_3$ using the preceding expression? What is the pH? **(d)** If the pH is buffered at 8.2 (as is historically typical for the ocean), what is the molar solubility of CaCO$_3$? **(e)** If the pH is buffered at 7.5, what is the molar solubility of CaCO$_3$? How much does this drop in pH increase solubility?

INTEGRATIVE EXERCISES

18.78 **(a)** Write the net ionic equation for the reaction that occurs when a solution of hydrochloric acid (HCl) is mixed with a solution of sodium formate (NaCHO$_2$). **(b)** Calculate the equilibrium constant for this reaction. **(c)** Calculate the equilibrium concentrations of Na$^+$, Cl$^-$, H$^+$, HCOO$^-$ and HCOOH when 50.0 cm^3 of 0.15 M HCl is mixed with 50.0 cm^3 of 0.15 M HCOONa.

18.79 **(a)** A 0.1044 g sample of an unknown monoprotic acid requires 22.10 cm^3 of 0.0500 M NaOH to reach the end point. What is the molecular weight of the unknown? **(b)** As the acid is titrated, the pH of the solution after the addition of 11.05 cm^3 of the base is 4.89. What is the K_a for the acid? **(c)** Using Appendix D, suggest the identity of the acid. Do both the molecular weight and K_a value agree with your choice?

18.80 Aspirin has the structural formula

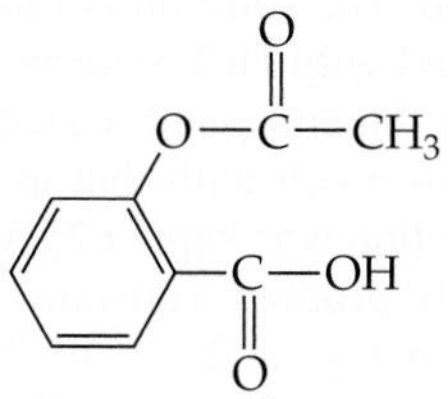

At body temperature (37 °C), K_a for aspirin equals 3×10^{-5}. If two aspirin tablets, each having a mass of 325 mg, are dissolved in a full stomach whose volume is 1 dm^3 and whose pH is 2, what percentage of the aspirin is in the form of neutral molecules?

18.81 The osmotic pressure of a saturated solution of strontium sulfate at 25 °C is 2.80 kPa. What is the solubility product of this salt at 25 °C?

18.82 A concentration of 10–100 parts per billion (by mass) of Ag^+ is an effective disinfectant in swimming pools. However, if the concentration exceeds this range, the Ag^+ can cause adverse health effects. One way to maintain an appropriate concentration of Ag^+ is to add a slightly soluble salt to the pool. Using K_{sp} values from Appendix D, calculate the equilibrium concentration of Ag^+ in parts per billion that would exist in equilibrium with **(a)** AgCl, **(b)** AgBr, **(c)** AgI.

18.83 Fluoridation of drinking water is employed in many places to aid in the prevention of dental caries. Typically the F^- ion concentration is adjusted to about 1 ppb. Some water supplies are also 'hard'; that is, they contain certain cations such as Ca^{2+} that interfere with the action of soap. Consider a case where the concentration of Ca^{2+} is 8 ppb. Could a precipitate of CaF_2 form under these conditions? (Make any necessary approximations.)

MasteringChemistry (www.pearson.com.au/masteringchemistry)

Make learning part of the grade. Access:

- tutorials with peronalised coaching
- study area
- Pearson eText

PHOTO/ART CREDITS

706 © Mark Karrass/Corbis; **711** Donald Clegg and Roxy Wilson Pearson Education/PH College; **716** © Professors P. Motta and S. Correr/Science Photo Library; **732** © Dave Fleetham/Getty Images Australia; **734** Pearson Science; **737, 740** Richard Megna, Fundamental Photographs, NYC.

ELECTROCHEMISTRY

A variety of batteries of different sizes, compositions and voltages.

KEY CONCEPTS

19.1 OXIDATION STATES AND OXIDATION–REDUCTION REACTIONS
We review oxidation states and *oxidation–reduction (redox) reactions*.

19.2 BALANCING REDOX EQUATIONS
We explain how to balance redox equations using the method of *half-reactions*.

19.3 VOLTAIC CELLS
We study *voltaic cells*, which produce electricity from spontaneous redox reactions. In voltaic cells solid electrodes serve as the surfaces at which oxidation and reduction take place. The electrode where oxidation occurs is the *anode*, and the electrode where reduction occurs is the *cathode*.

19.4 CELL POTENTIALS UNDER STANDARD CONDITIONS
We see that an important characteristic of a voltaic cell is its *cell potential*, which is the difference in the electrical potentials at the two electrodes and is measured in units of volts. Half-cell potentials are tabulated for reduction half-reactions under standard conditions (*standard reduction potentials*).

19.5 FREE ENERGY AND REDOX REACTIONS
We explain the relationship between the Gibbs free energy, $\Delta G°$, and cell potential.

19.6 CELL POTENTIALS UNDER NON-STANDARD CONDITIONS
We calculate cell potentials under non-standard conditions by using standard cell potentials and the Nernst equation.

19.7 BATTERIES AND FUEL CELLS
We describe batteries and fuel cells, which are commercially important energy sources that use electrochemical reactions.

19.8 CORROSION
We discuss *corrosion*, a spontaneous electrochemical process involving metals.

19.9 ELECTROLYSIS
Finally, we focus on non-spontaneous redox reactions, examining *electrolytic cells*, which use electricity to perform chemical reactions.

Although we are surrounded by an amazing array of portable electronic devices, such as mobile phones, portable music players, laptop computers, gaming devices and so on, none of these would be of any use in the absence of batteries. So a variety of batteries of different sizes, compositions and voltages have been developed, to provide a source of portable electricity to power these gadgets. Considerable research is in progress to develop new batteries with more power, faster recharging ability, lighter weight or cheaper price. At the heart of such development are the oxidation–reduction reactions that power batteries.

As we discussed in Chapter 4, *oxidation* is the loss of electrons in a chemical reaction, and *reduction* is the gain of electrons (Section 4.4, 'Oxidation–Reduction Reactions'). Thus oxidation–reduction (redox) reactions occur when electrons are transferred from an atom that is oxidised to an atom that is reduced. Redox reactions are involved not only in the operation of batteries but also in a wide variety of important natural processes, including the rusting of iron, the browning of foods and the respiration of animals. **Electrochemistry** is the study of the relationships between electricity and chemical reactions. It includes the study of both spontaneous and non-spontaneous processes.

19.1 | OXIDATION STATES AND OXIDATION–REDUCTION REACTIONS

∞ Review this on page 118

We determine whether a given chemical reaction is an oxidation–reduction reaction by keeping track of the *oxidation numbers* (*oxidation states*) of the elements involved in the reaction (∞ Section 4.4, 'Oxidation–Reduction Reactions'). This procedure identifies whether the oxidation state changes for any elements involved in the reaction. For example, consider the reaction that occurs spontaneously when zinc metal is added to a strong acid (▼ **FIGURE 19.1**):

$$Zn(s) + 2\,H^+(aq) \longrightarrow Zn^{2+}(aq) + H_2(g) \qquad [19.1]$$

The chemical equation for this reaction can be written

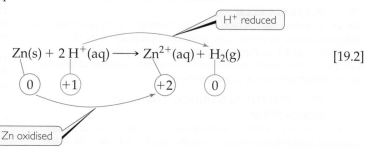

$$Zn(s) + 2\,H^+(aq) \longrightarrow Zn^{2+}(aq) + H_2(g) \qquad [19.2]$$

$$\underset{0}{} \quad \underset{+1}{} \qquad \qquad \underset{+2}{} \quad \underset{0}{}$$

H⁺ reduced

Zn oxidised

From the oxidation numbers below the equation, we see that the oxidation number of Zn changes from 0 to +2 and that of H changes from +1 to 0. Thus this is an oxidation–reduction reaction. Electrons are transferred from zinc atoms to hydrogen ions and therefore Zn is oxidised and H^+ is reduced.

In a reaction such as Equation 19.2, a clear transfer of electrons occurs. In some reactions, however, the oxidation numbers change, but we cannot say that

FIGURE IT OUT

Explain (a) the vigorous bubbling in the beaker on the right and (b) the formation of steam above that beaker.

▶ **FIGURE 19.1** **Oxidation of zinc by hydrochloric acid.**

Zn(s) 2 HCl(aq) ZnCl₂(aq) H₂(g)

any substance literally gains or loses electrons. For example, in the combustion of hydrogen gas,

$$2\,H_2(g) + O_2(g) \longrightarrow 2\,H_2O(g) \qquad [19.3]$$

with oxidation states: H_2 is 0, O_2 is 0, and in H_2O the H is +1 and O is −2.

hydrogen is oxidised from the 0 to the +1 oxidation state and oxygen is reduced from the 0 to the −2 oxidation state. Therefore Equation 19.3 is an oxidation–reduction reaction. Water is not an ionic substance, however, and so there is not a complete transfer of electrons from hydrogen to oxygen as water is formed. Thus keeping track of oxidation states is a convenient form of 'bookkeeping', but you should not generally equate the oxidation state of an atom with its actual charge in a chemical compound (see the A Closer Look box in Section 8.5).

CONCEPT CHECK 1

What are the oxidation numbers of the elements in the nitrite ion, NO_2^-?

In any redox reaction, both oxidation and reduction must occur. If one substance is oxidised, another must be reduced. The substance that makes it possible for another substance to be oxidised is called either the **oxidising agent** or the **oxidant**. The oxidising agent acquires electrons from the other substance and so is itself reduced. A **reducing agent**, or **reductant**, is a substance that gives up electrons, thereby causing another substance to be reduced. The reducing agent is therefore oxidised in the process. In Equation 19.2, $H^+(aq)$, the species that is reduced, is the oxidising agent and $Zn(s)$, the species that is oxidised, is the reducing agent.

SAMPLE EXERCISE 19.1 | **Identifying oxidising and reducing agents**

The nickel-cadmium (nicad) battery uses the following redox reaction to generate electricity:

$$Cd(s) + NiO_2(s) + 2\,H_2O(l) \longrightarrow Cd(OH)_2(s) + Ni(OH)_2(s)$$

Identify the substances that are oxidised and reduced, and indicate which is the oxidising agent and which is the reducing agent.

SOLUTION

Analyse We are given a redox equation and asked to identify the substance oxidised and the substance reduced and to label the oxidising agent and the reducing agent.

Plan First, we assign oxidation states, or numbers, to all the atoms and determine which elements change oxidation state. Second, we apply the definitions of oxidation and reduction.

Solve $Cd(s) + NiO_2(s) + 2\,H_2O(l) \longrightarrow Cd(OH)_2(s) + Ni(OH)_2(s)$

with oxidation states: Cd is 0; Ni is +4 and O is −2; H is +1 and O is −2; in $Cd(OH)_2$ Cd is +2, O is −2, H is +1; in $Ni(OH)_2$ Ni is +2, O is −2, H is +1.

The oxidation state of Cd increases from 0 to +2, and that of Ni decreases from +4 to +2. Thus the Cd atom is oxidised (loses electrons) and is the reducing agent. The oxidation state of Ni decreases as NiO_2 is converted into $Ni(OH)_2$. Thus NiO_2 is reduced (gains electrons) and is the oxidising agent.

Comment A common mnemonic for remembering oxidation and reduction is 'LEO the lion says GER': *losing electrons is oxidation; gaining electrons is reduction*.

PRACTICE EXERCISE

Identify the oxidising and reducing agents in the reaction

$$2\,H_2O(l) + Al(s) + MnO_4^-(aq) \longrightarrow Al(OH)_4^-(aq) + MnO_2(s)$$

Answer: $Al(s)$ is the reducing agent; $MnO_4^-(aq)$ is the oxidising agent.

(See also Exercise 19.16.)

19.2 | BALANCING REDOX EQUATIONS

Whenever we balance a chemical equation, we must obey the law of conservation of mass: The amount of each element must be the same on both sides of the equation. (Atoms are neither created nor destroyed in any chemical reaction.) As we balance oxidation–reduction reactions, there is an additional requirement: the gains and losses of electrons must be balanced. If a substance loses a certain number of electrons during a reaction, another substance must gain that same number of electrons. (Electrons are neither created nor destroyed in any chemical reaction.)

In many simple chemical equations, such as Equation 19.2, balancing the electrons is handled 'automatically' in the sense that we can balance the equation without explicitly considering the transfer of electrons. Many redox equations are more complex than Equation 19.2, however, and cannot be balanced easily without taking into account the number of electrons lost and gained. In this section we examine the *method of half-reactions*, a systematic procedure for balancing redox equations.

Half-Reactions

Although oxidation and reduction must take place simultaneously, it is often convenient to consider them as separate processes. For example, the oxidation of Sn^{2+} by Fe^{3+}:

$$Sn^{2+}(aq) + 2\,Fe^{3+}(aq) \longrightarrow Sn^{4+}(aq) + 2\,Fe^{2+}(aq)$$

can be considered to consist of two processes: (1) the oxidation of Sn^{2+} (Equation 19.4) and (2) the reduction of Fe^{3+} (Equation 19.5):

Oxidation:	$Sn^{2+}(aq) \longrightarrow Sn^{4+}(aq) + 2\,e^-$	[19.4]
Reduction:	$2\,Fe^{3+}(aq) + 2\,e^- \longrightarrow 2\,Fe^{2+}(aq)$	[19.5]

Notice that, in the oxidation, electrons are shown as products whereas, in the reduction, electrons are shown as reactants.

Equations that show either oxidation or reduction alone, such as Equations 19.4 and 19.5, are called **half-reactions**. In the overall redox reaction, the number of electrons lost in the oxidation half-reaction must equal the number of electrons gained in the reduction half-reaction. When this condition is met and each half-reaction is balanced, the electrons on the two sides cancel when the two half-reactions are added to give the overall balanced oxidation–reduction equation.

Balancing Equations by the Method of Half-Reactions

In using the half-reaction method, we usually begin with a 'skeleton' ionic equation showing only the substances undergoing oxidation and reduction. In such cases, we usually do not need to assign oxidation numbers unless we are unsure whether the reaction involves oxidation–reduction. We will find that H^+ (for acidic solutions), OH^- (for basic solutions) and H_2O are often involved as reactants or products in redox reactions. Unless H^+, OH^- or H_2O is being oxidised or reduced, these species do not appear in the skeleton equation. Their presence, however, can be deduced as we balance the equation.

For balancing a redox reaction that occurs *in acidic aqueous* solution, the procedure is as follows:

1. Divide the equation into one oxidation half-reaction and one reduction half-reaction.

2. Balance each half-reaction.
   (a) First, balance elements other than H and O.
   (b) Next, balance O atoms by adding H_2O as needed.
   (c) Then balance H atoms by adding H^+ as needed.
   (d) Finally, balance charge by adding e^- as needed.

This specific sequence (a)–(d) is important, and it is summarised in the diagram in the margin. At this point, you can check whether the number of electrons in each half-reaction corresponds to the changes in oxidation state.

3. Multiply half-reactions by integers as needed to make the number of electrons lost in the oxidation half-reaction equal the number of electrons gained in the reduction half-reaction.

4. Add half-reactions and, if possible, simplify by cancelling species appearing on both sides of the combined equation.

5. Check to make sure that atoms and charges are balanced.

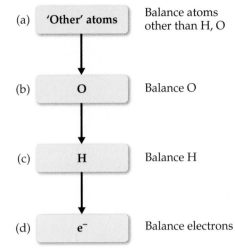

(a) **'Other' atoms** — Balance atoms other than H, O

(b) **O** — Balance O

(c) **H** — Balance H

(d) **e⁻** — Balance electrons

As an example, consider the reaction between permanganate ion (MnO_4^-) and oxalate ion ($C_2O_4^{2-}$) in acidic aqueous solution. When MnO_4^- is added to an acidified solution of $C_2O_4^{2-}$, the deep purple colour of the MnO_4^- ion fades, as illustrated in ▼ FIGURE 19.2. Bubbles of CO_2 form and the solution takes on the pale pink colour of Mn^{2+}. We can therefore write the unbalanced equation as

$$MnO_4^-(aq) + C_2O_4^{2-}(aq) \longrightarrow Mn^{2+}(aq) + CO_2(aq) \qquad [19.6]$$

Experiments show that H^+ is consumed and H_2O is produced in the reaction. These facts can be deduced in the course of balancing the equation.

To complete and balance Equation 19.6, we first write the two half-reactions (step 1). One half-reaction must have Mn on both sides of the arrow, and the other must have C on both sides of the arrow:

$$MnO_4^-(aq) \longrightarrow Mn^{2+}(aq)$$
$$C_2O_4^{2-}(aq) \longrightarrow CO_2(g)$$

We next complete and balance each half-reaction. First, we balance all the atoms except H and O (step 2a). In the permanganate half-reaction, we have one manganese atom on each side of the equation and so need to do nothing. In the oxalate half-reaction, we add a coefficient 2 on the right to balance the two carbons on the left:

$$MnO_4^-(aq) \longrightarrow Mn^{2+}(aq)$$
$$C_2O_4^{2-}(aq) \longrightarrow 2\,CO_2(g)$$

Next we balance O (step 2b). The permanganate half-reaction has four oxygens on the left and none on the right; therefore we need four H_2O molecules on the right to balance the oxygen atoms:

$$MnO_4^-(aq) \longrightarrow Mn^{2+}(aq) + 4\,H_2O(l)$$

▲ **FIGURE IT OUT**

Which species is reduced in this reaction? Which species is the reducing agent?

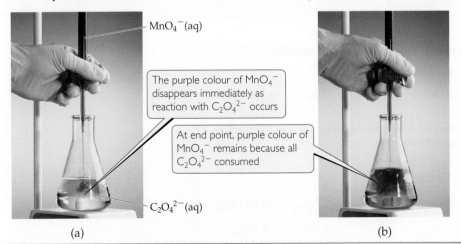

MnO_4^-(aq)

The purple colour of MnO_4^- disappears immediately as reaction with $C_2O_4^{2-}$ occurs

At end point, purple colour of MnO_4^- remains because all $C_2O_4^{2-}$ consumed

$C_2O_4^{2-}$(aq)

(a)

(b)

◀ **FIGURE 19.2 Titration of an acidic solution of $Na_2C_2O_4$ with $KMnO_4$(aq).**

The eight hydrogen atoms now in the products must be balanced by adding $8\,H^+$ to the reactants (step 2c):

$$8\,H^+(aq) + MnO_4^-(aq) \longrightarrow Mn^{2+}(aq) + 4\,H_2O(l)$$

There are now equal numbers of each type of atom on the two sides of the equation, but the charge still needs to be balanced. The total charge of the reactants is $8(1+) + 1(1-) = 7+$, and that of the products is $1(2+) + 4(0) = 2+$. To balance the charge, we add five electrons to the reactant side (step 2d):

$$5\,e^- + 8\,H^+(aq) + MnO_4^-(aq) \longrightarrow Mn^{2+}(aq) + 4\,H_2O(l)$$

We can use oxidation states to check our result. In this half-reaction Mn goes from the +7 oxidation state in MnO_4^- to the +2 oxidation state of Mn^{2+}. Therefore each Mn atom gains five electrons, in agreement with our balanced half-reaction.

In the oxalate half-reaction, we have C and O balanced (step 2a). We balance the charge (step 2d) by adding two electrons to the products:

$$C_2O_4^{2-}(aq) \longrightarrow 2\,CO_2(g) + 2\,e^-$$

We can check this result using oxidation states. Carbon goes from the +3 oxidation state in $C_2O_4^{2-}$ to the +4 oxidation state in CO_2. Thus each C atom loses one electron; therefore the two C atoms in $C_2O_4^{2-}$ lose two electrons, in agreement with our balanced half-reaction.

Now we multiply each half-reaction by an appropriate integer so that the number of electrons gained in one half-reaction equals the number of electrons lost in the other (step 3). We multiply the MnO_4^- half-reaction by 2 and the $C_2O_4^{2-}$ half-reaction by 5:

$$10\,e^- + 16\,H^+(aq) + 2\,MnO_4^-(aq) \longrightarrow 2\,Mn^{2+}(aq) + 8\,H_2O(l)$$
$$\underline{5\,C_2O_4^{2-}(aq) \longrightarrow 10\,CO_2(g) + 10\,e^-}$$

$$16\,H^+(aq) + 2\,MnO_4^-(aq) + 5\,C_2O_4^{2-}(aq) \longrightarrow$$

$$2\,Mn^{2+}(aq) + 8\,H_2O(l) + 10\,CO_2(g)$$

The balanced equation is the sum of the balanced half-reactions (step 4). Note that the electrons on the reactant and product sides of the equation cancel each other.

We can check the balanced equation by counting atoms and charges (step 5): There are 16 H, 2 Mn, 28 O, 10 C and a net charge of 4+ on each side of the equation, confirming that the equation is correctly balanced.

CONCEPT CHECK 2

Do free electrons appear anywhere in the balanced equation for a redox reaction?

SAMPLE EXERCISE 19.2   Balancing redox equations in acidic solution

Complete and balance this equation by the method of half-reactions:

$$Cr_2O_7^{2-}(aq) + Cl^-(aq) \longrightarrow Cr^{3+}(aq) + Cl_2(g) \qquad \text{(acidic aqueous solution)}$$

SOLUTION

Analyse We are given an incomplete, unbalanced (skeleton) equation for a redox reaction occurring in acidic solution and asked to complete and balance it.

Plan We use the half-reaction procedure we just learned.

Solve

Step 1:   We divide the equation into two half-reactions: $Cr_2O_7^{2-}(aq) \longrightarrow Cr^{3+}(aq)$

$$Cl^-(aq) \longrightarrow Cl_2(g)$$

Step 2: We balance each half-reaction. In the first half-reaction the presence of one $Cr_2O_7^{2-}$ among the reactants requires two Cr^{3+} among the products. The seven oxygen atoms in $Cr_2O_7^{2-}$ are balanced by adding seven H_2O to the products. The 14 hydrogen atoms in 7 H_2O are then balanced by adding 14 H^+ to the reactants:

$$14\,H^+(aq) + Cr_2O_7^{2-}(aq) \longrightarrow 2\,Cr^{3+}(aq) + 7\,H_2O(l)$$

Charge is then balanced by adding electrons to the left side of the equation so that the total charge is the same on the two sides:

$$6\,e^- + 14\,H^+(aq) + Cr_2O_7^{2-}(aq) \longrightarrow 2\,Cr^{3+}(aq) + 7\,H_2O(l)$$

We can check this result by looking at the oxidation state changes. Each chromium atom goes from +6 to +3, gaining three electrons; therefore the two Cr atoms in $Cr_2O_7^{2-}$ gain six electrons, in agreement with our half-reaction.

In the second half-reaction, two Cl^- are required to balance one Cl_2:

$$2\,Cl^-(aq) \longrightarrow Cl_2(g)$$

We add two electrons to the right side to attain charge balance:

$$2\,Cl^-(aq) \longrightarrow Cl_2(g) + 2\,e^-$$

This result agrees with the oxidation state changes. Each chlorine atom goes from −1 to 0, losing one electron; therefore the two chlorine atoms lose two electrons.

Step 3: We equalise the number of electrons transferred in the two half-reactions. To do so, we multiply the Cl half-reaction by 3 so that the number of electrons gained in the Cr half-reaction (six) equals the number lost in the Cl half-reaction, allowing the electrons to cancel when the half-reactions are added:

$$6\,Cl^-(aq) \longrightarrow 3\,Cl_2(g) + 6\,e^-$$

Step 4: The equations are added to give the balanced equation:

$$14\,H^+(aq) + Cr_2O_7^{2-}(aq) + 6\,Cl^-(aq) \longrightarrow 2\,Cr^{3+}(aq) + 7\,H_2O(l) + 3\,Cl_2(g)$$

Step 5: There are equal numbers of atoms of each kind on the two sides of the equation (14 H, 2 Cr, 7 O, 6 Cl). In addition, the charge is the same on the two sides (6+). Thus the equation is balanced.

PRACTICE EXERCISE

Complete and balance the following equations using the method of half-reactions. Both reactions occur in acidic solution.

(a) $Cu(s) + NO_3^-(aq) \longrightarrow Cu^{2+}(aq) + NO_2(g)$

(b) $Mn^{2+}(aq) + NaBiO_3(s) \longrightarrow Bi^{3+}(aq) + MnO_4^-(aq)$

Answers: **(a)** $Cu(s) + 4\,H^+(aq) + 2\,NO_3^-(aq) \longrightarrow Cu^{2+}(aq) + 2\,NO_2(g) + 2\,H_2O(l)$

   (b) $2\,Mn^{2+}(aq) + 5\,NaBiO_3(s) + 14\,H^+(aq) \longrightarrow 2\,MnO_4^-(aq) + 5\,Bi^{3+}(aq) + 5\,Na^+(aq) + 7\,H_2O(l)$

(See also Exercises 19.19(b), (c), (f), 19.20(a), (b), (c), (d).)

Balancing Equations for Reactions Occurring in Basic Solution

In basic solution, instead of H^+ and H_2O, we have OH^- and H_2O available. The easiest way to balance a redox equation occurring in basic solution is first to balance it exactly as before, *as if it occurs in acidic solution*. We then count the number of H^+ in the equation and add that number of OH^- *to each side* of the equation. This way the reaction is still mass-balanced because you are adding the same thing to both sides. In essence, what you are doing is 'neutralising' the protons to form water ($H^+ + OH^- \longrightarrow H_2O$) on the side containing H^+, and the other side ends up with the OH^-. The resulting water molecules can be cancelled as needed. This procedure is shown in Sample Exercise 19.3.

SAMPLE EXERCISE 19.3 Balancing redox equations in basic solution

Complete and balance this equation for a redox reaction that takes place in basic solution:

$$CN^-(aq) + MnO_4^-(aq) \longrightarrow CNO^-(aq) + MnO_2(s) \text{ (basic solution)}$$

SOLUTION

Analyse We are given an incomplete equation for a basic redox reaction and asked to balance it.

Plan We go through the first steps of our procedure as if the reaction were occurring in acidic solution. We then add the appropriate number of OH^- ions to each side of the equation, combining H^+ and OH^- to form H_2O. We complete the process by simplifying the equation.

Solve

Step 1: We write the incomplete, unbalanced half-reactions:

$$CN^-(aq) \longrightarrow CNO^-(aq)$$
$$MnO_4^-(aq) \longrightarrow MnO_2(s)$$

Step 2: We balance each half-reaction as if it took place in acidic solution:

$$CN^-(aq) + H_2O(l) \longrightarrow CNO^-(aq) + 2\,H^+(aq) + 2\,e^-$$
$$3\,e^- + 4\,H^+(aq) + MnO_4^-(aq) \longrightarrow MnO_2(s) + 2\,H_2O(l)$$

Now we must take into account that the reaction occurs in basic solution, adding OH^- to both sides of both half-reactions to neutralise H^+:

$$CN^-(aq) + H_2O(l) + 2\,OH^-(aq) \longrightarrow CNO^-(aq) + 2\,H^+(aq) + 2\,e^- + 2\,OH^-(aq)$$
$$3\,e^- + 4\,H^+(aq) + MnO_4^-(aq) + 4\,OH^-(aq) \longrightarrow MnO_2(s) + 2\,H_2O(l) + 4\,OH^-(aq)$$

We 'neutralise' H^+ and OH^- by forming H_2O when they are on the same side of either half-reaction:

$$CN^-(aq) + H_2O(l) + 2\,OH^-(aq) \longrightarrow CNO^-(aq) + 2\,H_2O(l) + 2\,e^-$$
$$3\,e^- + 4\,H_2O(l) + MnO_4^-(aq) \longrightarrow MnO_2(s) + 2\,H_2O(l) + 4\,OH^-(aq)$$

Next, we cancel water molecules that appear as both reactants and products:

$$CN^-(aq) + 2\,OH^-(aq) \longrightarrow CNO^-(aq) + H_2O(l) + 2\,e^-$$
$$3\,e^- + 2\,H_2O(l) + MnO_4^-(aq) \longrightarrow MnO_2(s) + 4\,OH^-(aq)$$

Both half-reactions are now balanced. You can check the atoms and the overall charge.

Step 3: We multiply the cyanide half-reaction by 3, which gives 6 electrons on the product side, and multiply the permanganate half-reaction by 2, which gives 6 electrons on the reactant side::

$$3\,CN^-(aq) + 6\,OH^-(aq) \longrightarrow 3\,CNO^-(aq) + 3\,H_2O(l) + 6\,e^-$$
$$6\,e^- + 4\,H_2O(l) + 2\,MnO_4^-(aq) \longrightarrow 2\,MnO_2(s) + 8\,OH^-(aq)$$

Step 4: We add the two half-reactions together and simplify by cancelling species that appear as both reactants and products:

$$3\,CN^-(aq) + H_2O(l) + 2\,MnO_4^-(aq) \longrightarrow 3\,CNO^-(aq) + 2\,MnO_2(s) + 2\,OH^-(aq)$$

Step 5: Check that the atoms and charges are balanced.

There are 3 C, 3 N, 2 H, 9 O, 2 Mn and a charge of 5− on both sides of the equation.

Comment It is important to remember that this procedure doesn't imply that H^+ ions are involved in the chemical reaction. Recall that in aqueous solutions at 20 °C, $K_w = [H^+][OH^-] = 1.0 \times 10^{-14}$. Thus $[H^+]$ is very small in this basic solution (Section 17.3).

PRACTICE EXERCISE

Complete and balance the following equations for oxidation–reduction reactions that occur in basic solution:

(a) $NO_2^-(aq) + Al(s) \longrightarrow NH_3(aq) + Al(OH)_4^-(aq)$

(b) $Cr(OH)_3(s) + ClO^-(aq) \longrightarrow CrO_4^{2-}(aq) + Cl_2(g)$

Answers: **(a)** $NO_2^-(aq) + 2\,Al(s) + 5\,H_2O(l) + OH^-(aq) \longrightarrow NH_3(aq) + 2\,Al(OH)_4^-(aq)$

(b) $2\,Cr(OH)_3(s) + 6\,ClO^-(aq) \longrightarrow 2\,CrO_4^{2-}(aq) + 3\,Cl_2(g) + 2\,OH^-(aq) + 2\,H_2O(l)$

(See also Exercises 19.19(a), (d), (e), (g), 19.20(e), (f).)

▲ FIGURE IT OUT

Why does the intensity of the blue solution colour lessen as the reaction proceeds?

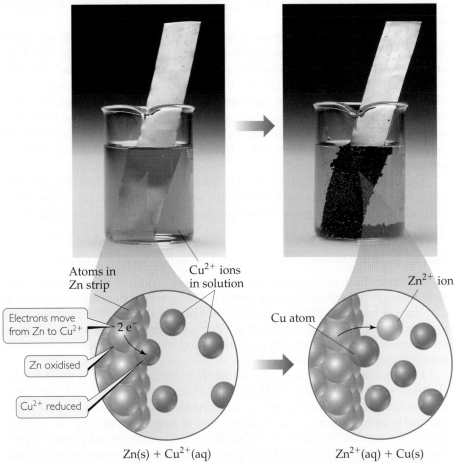

Electrons move from Zn to Cu^{2+} $2\,e^-$

Atoms in Zn strip

Cu^{2+} ions in solution

Zn^{2+} ion

Cu atom

Zn oxidised

Cu^{2+} reduced

$Zn(s) + Cu^{2+}(aq)$ $Zn^{2+}(aq) + Cu(s)$

▲ FIGURE 19.3 **A spontaneous oxidation–reduction reaction involving zinc and copper.**

19.3 | VOLTAIC CELLS

The energy released in a spontaneous redox reaction can be used to perform electrical work. This task is accomplished through a **voltaic** (or **galvanic**) **cell**, a device in which the transfer of electrons takes place through an external pathway rather than directly between reactants present in the same reaction vessel.

One such spontaneous reaction occurs when a strip of zinc is placed in contact with a solution containing Cu^{2+}. As the reaction proceeds, the blue colour of $Cu^{2+}(aq)$ ions fades and copper metal deposits on the zinc. At the same time, the zinc begins to dissolve. These transformations are shown in ▲ FIGURE 19.3 and are summarised by the equation

$$Zn(s) + Cu^{2+}(aq) \longrightarrow Zn^{2+}(aq) + Cu(s) \qquad [19.7]$$

▶ FIGURE 19.4 shows a voltaic cell that uses the redox reaction between Zn and Cu^{2+} given in Equation 19.7. Although the setup shown in Figure 19.4 is more complex than that in Figure 19.3, the reaction is the same in both cases. The significant difference between the two setups is that the Zn metal and $Cu^{2+}(aq)$ are not in direct contact in the voltaic cell. Instead, the Zn metal is placed in contact with $Zn^{2+}(aq)$ in one compartment of the cell and

▲ FIGURE IT OUT

Which metal, Cu or Zn, is oxidised in this voltaic cell?

Zn electrode in 1 M $ZnSO_4$ solution

Cu electrode in 1 M $CuSO_4$ solution

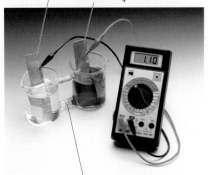

Solutions in contact with each other through porous glass disc

▲ FIGURE 19.4 **A Cu–Zn voltaic cell based on the reaction in Equation 19.7.**

Cu metal is placed in contact with Cu^{2+}(aq) in another compartment. Consequently, the reduction of the Cu^{2+} can occur only by a flow of electrons through an external circuit, namely, the wire that connects the Zn and Cu strips. Electrons flowing through a wire and ions moving in solution both constitute an *electric current*. This flow of electrical charge can be used to accomplish electrical work.

The two solid metals that are connected by the external circuit are called *electrodes*. By definition, the electrode at which oxidation occurs is called the **anode**; the electrode at which reduction occurs is called the **cathode**.* The electrodes can be made of materials that participate in the reaction, as in Figure 19.4. Over the course of the reaction, the Zn electrode will gradually disappear and the copper electrode will gain mass. More typically, the electrodes are made of a conducting material, such as platinum or graphite, that does not gain or lose mass during the reaction but serves as a surface at which electrons are transferred.

Each of the two compartments of a voltaic cell is called a *half-cell*. One half-cell is the site of the oxidation half-reaction, and the other is the site of the reduction half-reaction. In our present example Zn is oxidised and Cu^{2+} is reduced:

Anode (oxidation half-reaction): $\qquad\qquad\qquad\qquad Zn(s) \longrightarrow Zn^{2+}(aq) + 2\,e^-$

Cathode (reduction half-reaction): $\qquad\qquad Cu^{2+}(aq) + 2\,e^- \longrightarrow Cu(s)$

Electrons become available as zinc metal is oxidised at the anode. They flow through the external circuit to the cathode, where they are consumed as Cu^{2+}(aq) is reduced. Because Zn(s) is oxidised in the cell, the zinc electrode loses mass and the concentration of the Zn^{2+} solution increases as the cell operates. Similarly, the Cu electrode gains mass and the Cu^{2+} solution becomes less concentrated as Cu^{2+} is reduced to Cu(s).

For a voltaic cell to work, the solutions in the two half-cells must remain electrically neutral. As Zn is oxidised in the anode compartment, Zn^{2+} ions enter the solution. Thus there must be some means for positive ions to migrate out of the anode compartment or for negative ions to migrate in to keep the solution electrically neutral. Similarly, the reduction of Cu^{2+} at the cathode removes positive charge from the solution, leaving an excess of negative charge in that half-cell. Thus positive ions must migrate into the compartment or negative ions must migrate out. In fact, no measurable electron flow will occur between electrodes unless a means is provided for ions to migrate through the solution from one electrode compartment to the other, thereby completing the circuit.

In Figure 19.4 a porous glass disc separating the two compartments allows a migration of ions that maintains the electrical neutrality of the solutions. In ▶ **FIGURE 19.5** a *salt bridge* serves this purpose. A salt bridge consists of a U-shaped tube that contains an electrolyte solution, such as $NaNO_3$(aq), whose ions will not react with other ions in the cell or with the electrode materials. The electrolyte is often incorporated into a paste or gel so that the electrolyte solution does not pour out when the U-tube is inverted. As oxidation and reduction proceed at the electrodes, ions from the salt bridge migrate to neutralise charge in the cell compartments. Whatever means is used to allow ions to migrate between half-cells, *anions always migrate towards the anode and cations towards the cathode*.

⚠ CONCEPT CHECK 3

Why do anions in a salt bridge migrate towards the anode?

▶ **FIGURE 19.6** summarises the relationships between the anode, the cathode, the chemical process occurring in a voltaic cell, the direction of migration of ions in solution and the motion of electrons between electrodes

---

* To help remember these definitions, note that *anode* and *oxidation* both begin with a vowel, and *cathode* and *reduction* both begin with a consonant. Or, think of two animals: an ox and a red cat.

FIGURE IT OUT

How is electrical balance maintained in the left beaker as Zn^{2+} ions are formed at the anode?

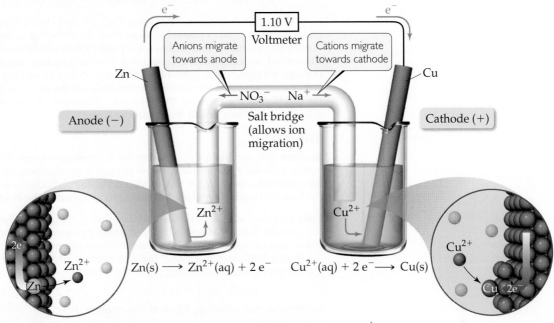

▲ FIGURE 19.5   A voltaic cell that uses a salt bridge to complete the electrical circuit.

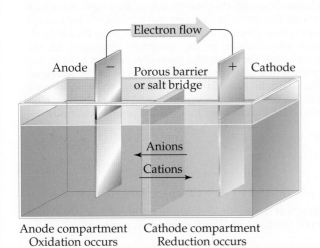

◀ FIGURE 19.6   **A summary of the terminology used to describe voltaic cells.** Oxidation occurs at the anode; reduction occurs at the cathode. The electrons flow spontaneously from the negative anode to the positive cathode. The electrical circuit is completed by the movement of ions in solution. Anions move towards the anode, whereas cations move towards the cathode. The cell compartments can be separated by either a porous glass barrier (as in Figure 19.4) or by a salt bridge (as in Figure 19.5).

in the external circuit. Notice in particular that *in any voltaic cell the electrons flow from the anode through the external circuit to the cathode.* Because the negatively charged electrons flow from the anode to the cathode, the anode in a voltaic cell is labelled with a negative sign and the cathode with a positive sign; we can imagine the electrons as being attracted to the positive cathode from the negative anode through the external circuit.*

---

* Although the anode and cathode in a voltaic cell are labelled with − and + signs, respectively, you should not interpret the labels as charges on the electrodes. The labels simply tell us the electrode at which the electrons are released to the external circuit (the anode) and received from the external circuit (the cathode). The actual charges on the electrodes are essentially zero.

SAMPLE EXERCISE 19.4 | **Reactions in a voltaic cell**

The oxidation–reduction reaction

$$Cr_2O_7^{2-}(aq) + 14\,H^+(aq) + 6\,I^-(aq) \longrightarrow 2\,Cr^{3+}(aq) + 3\,I_2(s) + 7\,H_2O(l)$$

is spontaneous. A solution containing $K_2Cr_2O_7$ and H_2SO_4 is poured into one beaker, and a solution of KI is poured into another. A salt bridge is used to join the beakers. A metallic conductor that will not react with either solution (such as platinum foil) is suspended in each solution, and the two conductors are connected with wires through a voltmeter or some other device to detect an electric current. The resultant voltaic cell generates an electric current. Indicate the reaction occurring at the anode, the reaction at the cathode, the direction of electron migration, the direction of ion migration and the signs of the electrodes.

SOLUTION

Analyse We are given the equation for a spontaneous reaction that takes place in a voltaic cell and a description of how the cell is constructed. We are asked to write the half-reactions occurring at the anode and at the cathode, as well as the directions of electron and ion movements and the signs assigned to the electrodes.

Plan Our first step is to divide the chemical equation into half-reactions so that we can identify the oxidation and the reduction processes. We then use the definitions of anode and cathode and the other terminology summarised in Figure 19.6.

In one half-reaction, $Cr_2O_7^{2-}(aq)$ is converted into $Cr^{3+}(aq)$. Starting with these ions and then completing and balancing the half-reaction, we have

$$Cr_2O_7^{2-}(aq) + 14\,H^+(aq) + 6\,e^- \longrightarrow 2\,Cr^{3+}(aq) + 7\,H_2O(l)$$

In the other half-reaction, $I^-(aq)$ is converted to $I_2(s)$:

$$6\,I^-(aq) \longrightarrow 3\,I_2(s) + 6\,e^-$$

Now we can use the summary in Figure 19.6 to help us describe the voltaic cell. The first half-reaction is the reduction process (electrons shown on the reactant side of the equation) and, by definition, this process occurs at the cathode. The second half-reaction is the oxidation (electrons shown on the product side of the equation), which occurs at the anode. The I^- ions are the source of electrons and the $Cr_2O_7^{2-}$ ions accept the electrons. Hence, the electrons flow through the external circuit from the electrode immersed in the KI solution (the anode) to the electrode immersed in the $K_2Cr_2O_7/H_2SO_4$ solution (the cathode). The electrodes themselves do not react in any way; they merely provide a means of transferring electrons from or to the solutions. The cations move through the solutions towards the cathode and the anions move towards the anode. The anode (from which the electrons move) is the negative electrode and the cathode (towards which the electrons move) is the positive electrode.

PRACTICE EXERCISE

The two half-reactions in a voltaic cell are

$$Zn(s) \longrightarrow Zn^{2+}(aq) + 2\,e^-$$

$$ClO_3^-(aq) + 6\,H^+(aq) + 6\,e^- \longrightarrow Cl^-(aq) + 3\,H_2O(l)$$

(a) Indicate which reaction occurs at the anode and which at the cathode. **(b)** Which electrode is consumed in the cell reaction? **(c)** Which electrode is positive?

Answers: **(a)** The first reaction occurs at the anode, the second reaction at the cathode. **(b)** The anode (Zn) is consumed in the cell reaction. **(c)** The cathode is positive.

(See also Exercises 19.22, 19.23.)

19.4 | CELL POTENTIALS UNDER STANDARD CONDITIONS

Why do electrons transfer spontaneously from a Zn atom to a Cu^{2+} ion, either directly as in the reaction of Figure 19.3 or through an external circuit as in the voltaic cell of Figure 19.4? In this section we examine the 'driving force' that pushes the electrons through an external circuit in a voltaic cell.

The chemical processes that constitute any voltaic cell are spontaneous because they increase the total entropy of the universe as we described spontaneous processes in Chapter 14. In a simple sense, we can compare the electron flow caused by a voltaic cell to the flow of water in a waterfall (▼ **FIGURE 19.7**). Water flows spontaneously over a waterfall because of a difference in potential energy between the top of the falls and the stream below (⚬⚬⚬ Section 14.1, 'The Nature of Energy'). In a similar fashion, electrons flow from the anode of a voltaic cell to the cathode because of a difference in potential energy. The potential energy of electrons is higher in the anode than in the cathode, and they spontaneously flow through an external circuit from the anode to the cathode.

⚬⚬⚬ Review this on page 502

The difference in potential energy per electrical charge (the *potential difference*) between two electrodes is measured in units of volts. One volt (V) is the potential difference required to impart 1 J of energy to a charge of 1 coulomb (C).

$$1\,\mathrm{V} = 1\,\mathrm{J\,C^{-1}}$$

Recall that one electron has a charge of 1.60×10^{-19} C (⚬⚬⚬ Section 2.2, 'The Discovery of Atomic Structure').

⚬⚬⚬ Review this on page 31

The potential difference between the two electrodes of a voltaic cell provides the driving force that pushes electrons through the external circuit. Therefore we call this potential difference the **electromotive** ('causing electron motion') **force**, or **emf**. The emf of a cell, denoted E_{cell}, is also called the **cell potential**. Because E_{cell} is measured in volts, we often refer to it as the *cell voltage*. For any cell reaction that proceeds spontaneously, such as that in a voltaic cell, the cell potential will be *positive*.

The emf of a particular voltaic cell depends on the specific reactions that occur at the cathode and anode, the concentrations of reactants and products, and the temperature, which we will assume to be 25 °C unless otherwise noted. In this section we focus on cells that are operated at 25 °C under *standard conditions*. Recall that standard conditions include 1 M concentrations for reactants and products in solution and 1 bar pressure for those that are gases (Table 14.7). Under standard conditions the emf is called the **standard emf**, or the **standard cell potential**, and is denoted E°_{cell}. For the Zn–Cu voltaic cell in Figure 19.5, for example, the standard cell potential at 25 °C is +1.10 V.

$$\mathrm{Zn(s)} + \mathrm{Cu^{2+}(aq,\ 1\ M)} \longrightarrow \mathrm{Zn^{2+}(aq,\ 1\ M)} + \mathrm{Cu(s)} \qquad E^{\circ}_{cell} = +1.10\ \mathrm{V}$$

Recall that the superscript ° indicates standard-state conditions (⚬⚬⚬ Section 14.7, 'Enthalpies of Formation').

⚬⚬⚬ Review this on page 523

CONCEPT CHECK 4

A certain voltaic cell has a standard cell potential of +0.85 V at 25 °C. Does this mean the redox reaction of the cell is spontaneous?

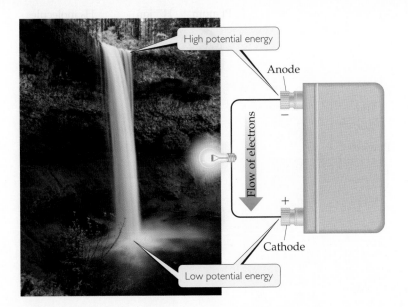

High potential energy

Anode

Flow of electrons

Cathode

Low potential energy

◀ **FIGURE 19.7**   **Water analogy for electron flow.**

Standard Reduction (Half-Cell) Potentials

The standard cell potential of a voltaic cell, E°_{cell}, depends on the particular cathode and anode half-cells involved.

We could, in principle, tabulate the standard cell potentials for all possible cathode/anode combinations. However, it is not necessary to undertake this arduous task. Rather, we can assign a standard potential to each half-cell and then use these half-cell potentials to determine E°_{cell}.

The cell potential is the difference between two electrode potentials, one associated with the cathode and the other associated with the anode. *By convention, the potential associated with each electrode is chosen to be the potential for reduction at that electrode.* Thus standard electrode potentials are tabulated for reduction reactions; they are **standard reduction potentials**, denoted E°_{red}. The cell potential, E°_{cell}, is given by the difference between the standard reduction potential of the cathode reaction, E°_{red} (cathode), and the standard reduction potential of the anode reaction, E°_{red} (anode):

$$E^\circ_{cell} = E^\circ_{red} \text{ (cathode)} - E^\circ_{red} \text{ (anode)} \qquad [19.8]$$

We discuss Equation 19.8 in greater detail shortly.

Because every voltaic cell involves two half-cells, it is not possible to measure the standard reduction potential of a half-reaction directly. If we assign a standard reduction potential to a certain reference half-reaction, however, we can then determine the standard reduction potentials of other half-reactions relative to that reference. The reference half-reaction is the reduction of $H^+(aq)$ to $H_2(g)$ under standard conditions, which is assigned a standard reduction potential of exactly 0 V.

$$2\,H^+(aq, 1\,\text{M}) + 2\,e^- \longrightarrow H_2(g, 1\,\text{bar}) \qquad E^\circ_{red} = 0\,\text{V} \qquad [19.9]$$

An electrode designed to produce this half-reaction is called a **standard hydrogen electrode** (SHE). A SHE consists of a platinum wire connected to a piece of platinum foil covered with finely divided platinum that serves as an inert surface for the reaction. The electrode is encased in a glass tube so that hydrogen gas under standard conditions (1 bar) can bubble over the platinum, and the solution contains $H^+(aq)$ under standard (1 M) conditions (▼ FIGURE 19.8). The SHE can operate as either the anode or cathode of a cell, depending on the nature of the other electrode.

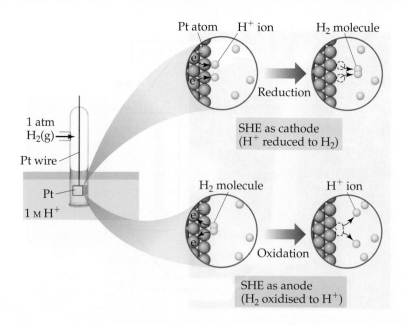

► FIGURE 19.8 **The standard hydrogen electrode (SHE) is used as a reference electrode.**

▼ **FIGURE 19.9** shows a voltaic cell using a SHE and a standard Zn^{2+}/Zn electrode. The spontaneous reaction is the oxidation of Zn and the reduction of H^+.

$$Zn(s) + 2\,H^+(aq) \longrightarrow Zn^{2+}(aq) + H_2(g)$$

Notice that the Zn^{2+}/Zn electrode is the anode and the SHE is the cathode and that the cell potential is +0.76 V when operated under standard conditions. By using the defined standard reduction potential of $H^+ (E^\circ_{red} = 0)$ and Equation 19.8, we can determine the standard reduction potential for the Zn^{2+}/Zn half-reaction:

$$E^\circ_{cell} = E^\circ_{red}\,(\text{cathode}) - E^\circ_{red}\,(\text{anode})$$

$$+0.76\ \text{V} = 0\ \text{V} - E^\circ_{red}\,(\text{anode})$$

$$E^\circ_{red}\,(\text{anode}) = -0.76\ \text{V}$$

Thus a standard reduction potential of −0.76 V can be assigned to the reduction of Zn^{2+} to Zn.

$$Zn^{2+}(aq, 1\ \text{M}) + 2\,e^- \longrightarrow Zn(s) \qquad E^\circ_{red} = -0.76\ \text{V}$$

We write the reaction as a reduction even though it is 'running in reverse' as an oxidation in the cell in Figure 19.9. *Whenever we assign a potential to a half-reaction, we write the reaction as a reduction.*

The standard reduction potentials for other half-reactions can be established from other cell potentials in a fashion analogous to that used for the Zn^{2+}/Zn half-reaction. ▼ **TABLE 19.1** lists some standard reduction potentials; a more complete list is found in Appendix E. These standard reduction potentials, often called *half-cell potentials*, can be combined to calculate the emfs of a large variety of voltaic cells.

CONCEPT CHECK 5

For the half-reaction $Cl_2(g) + 2\,e^- \longrightarrow 2\,Cl^-(aq)$, what are the standard conditions for the reactant and product?

Because electrical potential measures potential energy per electrical charge, standard reduction potentials are *intensive properties* (Section 1.3, 'Properties of Matter'). In other words, if we increased the amount of substances in a redox reaction, we would increase both the energy and charges involved, but

Review this on page 9

FIGURE IT OUT

Why does Na^+ migrate into the cathode half-cell as the cell reaction proceeds?

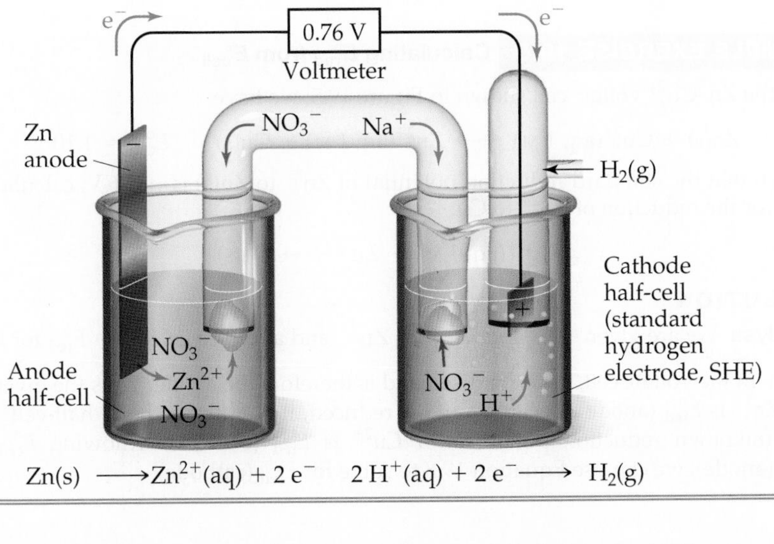

◀ **FIGURE 19.9 A voltaic cell using a standard hydrogen electrode (SHE).** The anode half-cell is Zn metal in a $Zn(NO_3)_2(aq)$ solution, and the cathode half-cell is the SHE in a $HNO_3(aq)$ solution.

TABLE 19.1 • Standard reduction potentials in water at 25 °C

Potential (V)	Reduction half-reaction
+2.87	$F_2(g) + 2\,e^- \longrightarrow 2\,F^-(aq)$
+1.51	$MnO_4^-(aq) + 8\,H^+(aq) + 5\,e^- \longrightarrow Mn^{2+}(aq) + 4\,H_2O(l)$
+1.36	$Cl_2(g) + 2\,e^- \longrightarrow 2\,Cl^-(aq)$
+1.33	$Cr_2O_7^{2-}(aq) + 14\,H^+(aq) + 6\,e^- \longrightarrow 2\,Cr^{3+}(aq) + 7\,H_2O(l)$
+1.23	$O_2(g) + 4\,H^+(aq) + 4\,e^- \longrightarrow 2\,H_2O(l)$
+1.07	$Br_2(l) + 2\,e^- \longrightarrow 2\,Br^-(aq)$
+0.96	$NO_3^-(aq) + 4\,H^+(aq) + 3\,e^- \longrightarrow NO(g) + 2\,H_2O(l)$
+0.80	$Ag^+(aq) + e^- \longrightarrow Ag(s)$
+0.77	$Fe^{3+}(aq) + e^- \longrightarrow Fe^{2+}(aq)$
+0.68	$O_2(g) + 2\,H^+(aq) + 2\,e^- \longrightarrow H_2O_2(aq)$
+0.59	$MnO_4^-(aq) + 2\,H_2O(l) + 3\,e^- \longrightarrow MnO_2(s) + 4\,OH^-(aq)$
+0.54	$I_2(s) + 2\,e^- \longrightarrow 2\,I^-(aq)$
+0.40	$O_2(g) + 2\,H_2O(l) + 4\,e^- \longrightarrow 4\,OH^-(aq)$
+0.34	$Cu^{2+}(aq) + 2\,e^- \longrightarrow Cu(s)$
0 [defined]	$2\,H^+(aq) + 2\,e^- \longrightarrow H_2(g)$
−0.28	$Ni^{2+}(aq) + 2\,e^- \longrightarrow Ni(s)$
−0.44	$Fe^{2+}(aq) + 2\,e^- \longrightarrow Fe(s)$
−0.76	$Zn^{2+}(aq) + 2\,e^- \longrightarrow Zn(s)$
−0.83	$2\,H_2O(l) + 2\,e^- \longrightarrow H_2(g) + 2\,OH^-(aq)$
−1.66	$Al^{3+}(aq) + 3\,e^- \longrightarrow Al(s)$
−2.71	$Na^+(aq) + e^- \longrightarrow Na(s)$
−3.05	$Li^+(aq) + e^- \longrightarrow Li(s)$

the ratio of energy (joules) to electrical charge (coulombs) would remain constant ($V = J\,C^{-1}$). Thus *changing the stoichiometric coefficient in a half-reaction does not affect the value of the standard reduction potential.* For example, E°_{red} for the reduction of 50 Zn^{2+} is the same as that for the reduction of 1 Zn^{2+}:

$$50\,Zn^{2+}(aq, 1\,M) + 100\,e^- \longrightarrow 50\,Zn(s) \qquad E^\circ_{red} = -0.76\,V$$

Earlier, we made an analogy between emf and the flow of water over a waterfall (Figure 19.7). The emf corresponds to the difference in height between the top and bottom of the waterfall. This height remains the same, whether the flow of water is large or small.

SAMPLE EXERCISE 19.5 **Calculating E°_{red} from E°_{cell}**

For the Zn–Cu^{2+} voltaic cell shown in Figure 19.5, we have

$$Zn(s) + Cu^{2+}(aq, 1\,M) \longrightarrow Zn^{2+}(aq, 1\,M) + Cu(s) \qquad E^\circ_{cell} = 1.10\,V$$

Given that the standard reduction potential of Zn^{2+} to Zn(s) is −0.76 V, calculate the E°_{red} for the reduction of Cu^{2+} to Cu:

$$Cu^{2+}(aq, 1\,M) + 2\,e^- \longrightarrow Cu(s)$$

SOLUTION

Analyse We are given E°_{cell} and E°_{red} for Zn^{2+} and asked to calculate E°_{red} for Cu^{2+}.

Plan In the voltaic cell, Zn is oxidised and is therefore the anode. Thus the given E°_{red} for Zn^{2+} is E°_{red} (anode). Because Cu^{2+} is reduced, it is in the cathode half-cell. Thus the unknown reduction potential for Cu^{2+} is E°_{red} (cathode). Knowing E°_{cell} and E°_{red} (anode), we can use Equation 19.8 to solve for E°_{red} (cathode).

$$E^\circ_{cell} = E^\circ_{red} \text{ (cathode)} - E^\circ_{red} \text{ (anode)}$$

$$1.10 \text{ V} = E^\circ_{red} \text{ (cathode)} - (-0.76 \text{ V})$$

$$E^\circ_{red} \text{ (cathode)} = 1.10 \text{ V} - 0.76 \text{ V} = 0.34 \text{ V}$$

Check This standard reduction potential agrees with the one listed in Table 19.1.

Comment The standard reduction potential for Cu^{2+} can be represented as $E^\circ_{Cu^{2+}} = 0.34$ V, and that for Zn^{2+} as $E^\circ_{Zn^{2+}} = -0.76$ V. The subscript identifies the ion that is reduced in the reduction half-reaction.

PRACTICE EXERCISE

A voltaic cell is based on the half-reactions

$$In^+(aq) \longrightarrow In^{3+}(aq) + 2 e^-$$

$$Br_2(l) + 2 e^- \longrightarrow 2 Br^-(aq)$$

The standard emf for this cell is 1.46 V. Using the data in Table 19.1, calculate E°_{red} for the reduction of In^{3+} to In^+.

Answer: −0.40 V

(See also Exercise 19.30.)

SAMPLE EXERCISE 19.6 Calculating E°_{cell} from E°_{red}

Using the standard reduction potentials listed in Table 19.1, calculate the standard emf for the voltaic cell described in Sample Exercise 19.4, which is based on the reaction

$$Cr_2O_7{}^{2-}(aq) + 14 H^+(aq) + 6 I^-(aq) \longrightarrow 2 Cr^{3+}(aq) + 3 I_2(s) + 7 H_2O(l)$$

SOLUTION

Analyse We are given the equation for a redox reaction and asked to use data in Table 19.1 to calculate the standard cell potential for the associated voltaic cell.

Plan Our first step is to identify the half-reactions that occur at the cathode and the anode, which we did in Sample Exercise 19.3. Then we can use data from Table 19.1 and Equation 19.8 to calculate the standard emf.

Solve The half-reactions are

Cathode: $Cr_2O_7{}^{2-}(aq) + 14 H^+(aq) + 6 e^- \longrightarrow 2 Cr^{3+}(aq) + 7 H_2O(l)$

Anode: $6 I^-(aq) \longrightarrow 3 I_2(s) + 6 e^-$

According to Table 19.1, the standard reduction potential for the reduction of $Cr_2O_7{}^{2-}$ to Cr^{3+} is +1.33 V, and the standard reduction potential for the reduction of I_2 to I^- (the reverse of the oxidation half-reaction) is +0.54 V. We then use these values in Equation 19.8.

$$E^\circ_{cell} = E^\circ_{red} \text{ (cathode)} - E^\circ_{red} \text{ (anode)} = 1.33 \text{ V} - 0.54 \text{ V} = 0.79 \text{ V}$$

Although the iodide half-reaction at the anode must be multiplied by 3 in order to obtain a balanced equation for the reaction, the value of E°_{red} is *not* multiplied by 3. As we have noted, the standard reduction potential is an intensive property, so it is independent of the specific stoichiometric coefficients.

Check The cell potential, 0.79 V, is a positive number. As noted earlier, a voltaic cell must have a positive emf in order to operate.

PRACTICE EXERCISE

Using the data in Table 19.1, calculate the standard emf for a cell that employs the following overall cell reaction:

$$2 Al(s) + 3 I_2(s) \longrightarrow 2 Al^{3+}(aq) + 6 I^-(aq)$$

Answer: +2.20 V

(See also Exercise 19.31.)

FIGURE IT OUT

Given E_{red}° values for the two electrodes in a standard voltaic cell, how do you determine which electrode is the cathode?

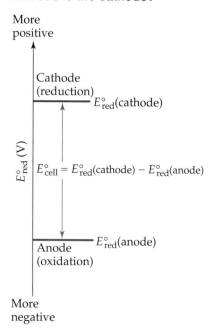

▲ **FIGURE 19.10** Graphical representation of standard cell potential of a voltaic cell.

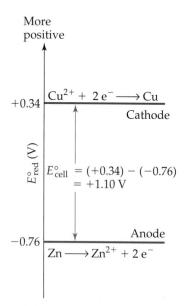

▲ **FIGURE 19.11** Half-cell potentials and standard cell potential for the Zn–Cu voltaic cell.

For each half-cell in a voltaic cell, the standard reduction potential provides a measure of the tendency for reduction to occur: *The more positive the value of* E_{red}°*, the greater the tendency for reduction under standard conditions.* In any voltaic cell operating under standard conditions, the E_{red}° value for the reaction at the cathode is more positive than the E_{red}° value for the reaction at the anode. Thus electrons flow spontaneously through the external circuit from the electrode with the more negative value of E_{red}° to the electrode with the more positive value of E_{red}°.

The fact that the standard cell potential is the difference between the standard reduction potentials of cathode and anode is illustrated graphically in ◄ FIGURE 19.10. The more positive E_{red}° value identifies the cathode, and the difference between the two standard reduction potentials is the standard cell potential. ▼ FIGURE 19.11 shows E_{red}° values for the two half-reactions in the Zn–Cu voltaic cell of Figure 19.5.

▲ **CONCEPT CHECK 6**

Is the following statement true or false? The smaller the difference between cathode and anode standard reduction potentials, the smaller the driving force for the overall redox reaction.

SAMPLE EXERCISE 19.7 | **Determining half-reactions at electrodes and calculating cell potentials**

A voltaic cell is based on the following two standard half-reactions:

$$Cd^{2+}(aq) + 2\,e^- \longrightarrow Cd(s)$$
$$Sn^{2+}(aq) + 2\,e^- \longrightarrow Sn(s)$$

By using the data in Appendix E, determine **(a)** the half-reactions that occur at the cathode and the anode, and **(b)** the standard cell potential.

SOLUTION

Analyse We need to look up E_{red}° for two half-reactions. We then use these values first to determine the cathode and the anode and then to calculate the standard cell potential E_{cell}°.

Plan The cathode will have the reduction with the most positive E_{red}° value. The anode will have the less positive E_{red}°. To write the half-reaction at the anode, we reverse the half-reaction written for the reduction.

Solve

(a) According to Appendix E, $E_{red}^\circ(Cd^{2+}/Cd) = -0.403$ V and $E_{red}^\circ(Sn^{2+}/Sn) = -0.136$ V. The standard reduction potential for Sn^{2+} is more positive (less negative) than that for Cd^{2+}; hence the reduction of Sn^{2+} is the reaction that occurs at the cathode.

Cathode: $\qquad\qquad Sn^{2+}(aq) + 2\,e^- \longrightarrow Sn(s)$

The anode reaction therefore is the loss of electrons by Cd.

Anode: $\qquad\qquad Cd(s) \longrightarrow Cd^{2+}(aq) + 2\,e^-$

(b) The cell potential is given by Equation 19.8:

$$E_{cell}^\circ = E_{red}^\circ(\text{cathode}) - E_{red}^\circ(\text{anode}) = (-0.136\text{ V}) - (-0.403\text{ V}) = 0.267\text{ V}$$

Notice that it is unimportant that the E_{red}° values of both half-reactions are negative; the negative values merely indicate how these reductions compare with the reference reaction, the reduction of $H^+(aq)$.

Check The cell potential is positive, as it must be for a voltaic cell.

PRACTICE EXERCISE

A voltaic cell is based on a Co^{2+}/Co half-cell and an $AgCl/Ag$ half-cell.
(a) What reaction occurs at the anode? **(b)** What is the standard cell potential?

Answers: **(a)** $Co \longrightarrow Co^{2+} + 2\,e^-$; **(b)** $+0.499$ V

(See also Exercise 19.39.)

Strengths of Oxidising and Reducing Agents

Table 19.1 lists half-reactions in order of decreasing tendency to undergo reduction. For example, F_2 is located at the top of the table, having the most positive value for E°_{red}. Thus F_2 is the most easily reduced species in Table 19.1 and therefore the strongest oxidising agent listed.

Among the most frequently used oxidising agents are the halogens, O_2 and oxyanions such as MnO_4^-, $Cr_2O_7^{2-}$ and NO_3^-, whose central atoms have high positive oxidation states. As seen in Table 19.1, all these species have large positive values of E°_{red} and therefore easily undergo reduction.

The lower the tendency for a half-reaction to occur in one direction, the greater the tendency for it to occur in the opposite direction. Thus *the half-reaction with the most negative reduction potential in Table 19.1 is the one most easily reversed and run as an oxidation.* Being at the bottom of Table 19.1, Li^+(aq) is the most difficult species in the list to reduce and is therefore the poorest oxidising agent listed. Although Li^+(aq) has little tendency to gain electrons, the reverse reaction, oxidation of Li(s) to Li^+(aq), is highly favourable. Thus Li is the strongest reducing agent among the substances listed in Table 19.1. (Note that, because Table 19.1 lists half-reactions as reductions, only the substances on the reactant side of these equations can serve as oxidising agents; only those on the product side can serve as reducing agents.)

Commonly used reducing agents include H_2 and the active metals, such as the alkali metals and the alkaline earth metals. Other metals whose cations have negative E°_{red} values—Zn and Fe, for example—are also used as reducing agents. Solutions of reducing agents are difficult to store for extended periods because of the ubiquitous presence of O_2, a good oxidising agent.

The information contained in Table 19.1 is summarised graphically in ▼ **FIGURE 19.12**. The reactants in half-reactions at the top of Table 19.1 are the most readily reduced species in the table and are therefore the strongest oxidising agents. Thus Figure 19.12 shows F_2(g) as the strongest oxidising agent (the position at the top of the red arrow). The products in half-reactions at the top of Table 19.1 are the most difficult to oxidise and are therefore the weakest reducing agents in the table. The reactants in half-reactions at the bottom of Table 19.1 are the most difficult to reduce and so are the weakest oxidising agents. The products in half-reactions at the bottom of Table 19.1 are the most readily oxidised species in the table and so are the strongest reducing agents.

SAMPLE EXERCISE 19.8 | **Determining the relative strengths of oxidising agents**

Using Table 19.1, rank the following ions in order of increasing strength as oxidising agents: NO_3^-(aq), Ag^+(aq), $Cr_2O_7^{2-}$(aq).

SOLUTION

Analyse We are asked to rank the abilities of several ions to act as oxidising agents.

Plan The more readily an ion is reduced (the more positive its E°_{red} value), the stronger it is as an oxidising agent.

Solve From Table 19.1, we have

$$NO_3^-(aq) + 4\,H^+(aq) + 3\,e^- \longrightarrow NO(g) + 2\,H_2O(l) \qquad E^\circ_{red} = +0.96\ V$$

$$Ag^+(aq) + e^- \longrightarrow Ag(s) \qquad E^\circ_{red} = +0.80\ V$$

$$Cr_2O_7^{2-}(aq) + 14\,H^+(aq) + 6\,e^- \longrightarrow 2\,Cr^{3+}(aq) + 7\,H_2O(l) \qquad E^\circ_{red} = +1.33\ V$$

Because the standard reduction potential of $Cr_2O_7^{2-}$ is the most positive, $Cr_2O_7^{2-}$ is the strongest oxidising agent of the three. The rank order is $Ag^+ < NO_3^- < Cr_2O_7^{2-}$.

PRACTICE EXERCISE

Using Table 19.1, rank the following species from the strongest to the weakest reducing agent: I^-(aq), Fe(s), Al(s).

Answer: Al(s) > Fe(s) > I^-(aq)

(See also Exercise 19.38.)

FIGURE IT OUT

Why is a strong oxidising agent easy to reduce?

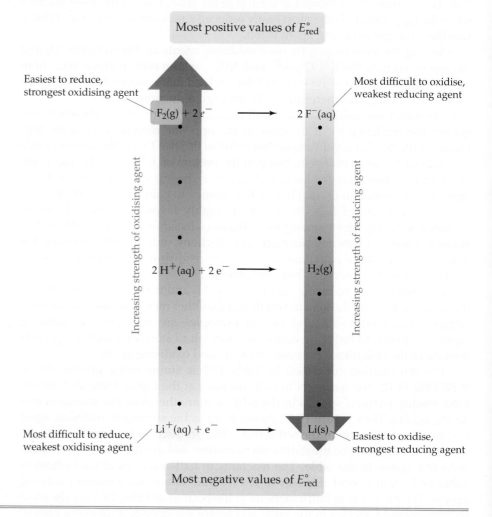

▶ **FIGURE 19.12 Relative strengths of oxidising and reducing agents.** The standard reduction potentials in Table 19.1 are related to the ability of substances to serve as oxidising or reducing agents. Species on the left side of the half-reactions can act as oxidising agents, and those on the right side can act as reducing agents.

∞ Review this on pages 656 and 663

This inverse relationship between oxidising and reducing strength is similar to the inverse relationship between the strengths of conjugate acids and bases (∞ Section 17.2, 'Brønsted–Lowry Acids and Bases' and Figure 17.3).

19.5 | FREE ENERGY AND REDOX REACTIONS

We have observed that voltaic cells use redox reactions that proceed spontaneously. Any reaction that can occur in a voltaic cell to produce a positive emf must be spontaneous. Consequently, it is possible to decide whether a redox reaction will be spontaneous by using half-cell potentials to calculate the emf associated with it.

The following discussion pertains to general redox reactions, not just reactions in voltaic cells. We can make Equation 19.8 more general by writing it as

$$E° = E°_{red} \text{ (reduction process)} - E°_{red} \text{ (oxidation process)} \qquad [19.10]$$

In modifying Equation 19.8, we have dropped the subscript 'cell' to indicate that the calculated emf does not necessarily refer to a voltaic cell. Similarly, we have generalised the standard reduction potentials on the right side of the equation by referring to the *reduction* and *oxidation* processes, rather than the *cathode* and the *anode*. We can now make a general statement about the spontaneity of a reaction and its associated emf, E. *A positive value of* E *indicates a spontaneous process, and a negative value of* E *indicates a non-spontaneous one.* We will use E to represent the emf under non-standard conditions and $E°$ to indicate the standard emf.

SAMPLE EXERCISE 19.9   Determining spontaneity

Using standard reduction potentials (Table 19.1), determine whether the following reactions are spontaneous under standard conditions.
(a) $Cu(s) + 2H^+(aq) \longrightarrow Cu^{2+}(aq) + H_2(g)$
(b) $Cl_2(g) + 2I^-(aq) \longrightarrow 2Cl^-(aq) + I_2(s)$

SOLUTION

Analyse We are given two reactions and must determine whether each is spontaneous.

Plan To determine whether a redox reaction is spontaneous under standard conditions, we first need to write its reduction and oxidation half-reactions. We can then use the standard reduction potentials and Equation 19.10 to calculate the standard emf, $E°$, for the reaction. If a reaction is spontaneous, its standard emf must be a positive number.

Solve

(a) In this reaction Cu is oxidised to Cu^{2+} and H^+ is reduced to H_2. The corresponding half-reactions and associated standard reduction potentials are

Reduction:	$2H^+(aq) + 2e^- \longrightarrow H_2(g)$	$E°_{red} = 0\ V$
Oxidation:	$Cu(s) \longrightarrow Cu^{2+}(aq) + 2e^-$	$E°_{red} = +0.34\ V$

Notice that, for the oxidation, we use the standard reduction potential from Table 19.1 for the reduction of Cu^{2+} to Cu. We now calculate $E°$ by using Equation 19.10:

$$E° = E°_{red}\ (\text{reduction process}) - E°_{red}\ (\text{oxidation process})$$
$$= (0\ V) - (0.34\ V) = -0.34\ V$$

Because $E°$ is negative, the reaction is not spontaneous in the direction written. Copper metal does not react with acids in this fashion. The reverse reaction, however, *is* spontaneous and would have an $E°$ of +0.34 V:

Cu^{2+} can be reduced by H_2.

$$Cu^{2+}(aq) + H_2(g) \longrightarrow Cu(s) + 2H^+(aq) \qquad E° = +0.34\ V$$

(b) We follow a procedure analogous to that in (a):

Reduction:	$Cl_2(g) + 2e^- \longrightarrow 2Cl^-(aq)$	$E°_{red} = +1.36\ V$
Oxidation:	$2I^-(aq) \longrightarrow I_2(s) + 2e^-$	$E°_{red} = +0.54\ V$

In this case

$$E° = (1.36\ V) - (0.54\ V) = +0.82\ V$$

Because the value of $E°$ is positive, this reaction is spontaneous and could be used to build a voltaic cell.

PRACTICE EXERCISE

Using the standard reduction potentials listed in Appendix E, determine which of the following reactions are spontaneous under standard conditions.
(a) $I_2(s) + 5Cu^{2+}(aq) + 6H_2O(l) \longrightarrow 2IO_3^-(aq) + 5Cu(s) + 12H^+(aq)$
(b) $Hg^{2+}(aq) + 2I^-(aq) \longrightarrow Hg(l) + I_2(s)$
(c) $H_2SO_3(aq) + 2Mn(s) + 4H^+(aq) \longrightarrow S(s) + 2Mn^{2+}(aq) + 3H_2O(l)$

Answers: Reactions **(b)** and **(c)** are spontaneous.

(See also Exercise 19.43.)

We can use standard reduction potentials to understand the activity series of metals (∞ Section 4.4, 'Oxidation–Reduction Reactions'). Recall that any metal in the activity series will be oxidised by the ions of any metal below it. We can now interpret this on the basis of standard reduction potentials. The activity series, tabulated in Table 4.4, consists of the oxidation reactions of the

∞ Review this on page 118

metals, ordered from the strongest reducing agent at the top to the weakest reducing agent at the bottom. (Thus the ordering is inverted relative to that in Table 19.1.) For example, nickel lies above silver in the activity series. In a mixture of nickel metal and silver cations, therefore we expect nickel to displace silver, according to the net reaction

$$Ni(s) + 2\,Ag^+(aq) \longrightarrow Ni^{2+}(aq) + 2\,Ag(s)$$

In this reaction Ni is oxidised and Ag^+ is reduced. Using data from Table 19.1, the standard emf for the reaction is

$$E° = E°_{red}(Ag^+/Ag) - E°_{red}(Ni^{2+}/Ni)$$
$$= (+0.80\,V) - (-0.28\,V) = +1.08\,V$$

The positive value of $E°$ indicates that the displacement of silver by nickel is a spontaneous process. Remember that, although the silver half-reaction is multiplied by 2, the reduction potential is not.

CONCEPT CHECK 7

Based on Table 4.4, which is the stronger reducing agent: Hg(l) or Pb(s)?

Emf, Free Energy and the Equilibrium Constant

∞ Review this on page 545

The change in Gibbs free energy, ΔG, is a measure of the spontaneity of a process that occurs at constant temperature and pressure (∞ Section 14.12, 'Gibbs Free Energy (Gibbs Energy)'). Because the emf, E, of a redox reaction indicates whether the reaction is spontaneous, the relationship between emf and the free energy change is

$$\Delta G = -nFE \qquad [19.11]$$

In this equation n is a positive number without units and represents the number of moles of electrons transferred according to the balanced equation for the reaction. The constant, F, is called **Faraday's constant**, named after Michael Faraday, an English scientist (1791–1867). Faraday's constant is the quantity of electrical charge on one mole of electrons. This quantity of charge is called a **faraday** (F).

$$1\,F = 96\,485\,C\,mol^{-1} = 96{,}485\,J\,V^{-1}\,mol^{-1}$$

The units of ΔG calculated by using Equation 19.11 and later using Equation 19.13 are $J\,mol^{-1}$.

Both n and F are positive numbers. Thus a positive value of E in Equation 19.11 leads to a negative value of ΔG. Remember: *A positive value of* E *and a negative value of* ΔG *both indicate that a reaction is spontaneous.* When the reactants and products are all in their standard states, Equation 19.11 can be modified to relate $\Delta G°$ and $E°$.

$$\Delta G° = -nFE° \qquad [19.12]$$

Because $\Delta G°$ is related to the equilibrium constant, K, for a reaction by the expression $\Delta G° = -RT \ln K$ (Equation 16.25), we can relate $E°$ to K by solving Equation 19.12 for $E°$ and then substituting the Equation 16.25 expression for $\Delta G°$:

$$E° = \frac{\Delta G°}{-nF} = \frac{-RT \ln K}{-nF} = \frac{RT}{nF} \ln K \qquad [19.13]$$

◀ **FIGURE 19.13** summarises the relationships between $E°$, $\Delta G°$ and K.

FIGURE IT OUT

What does the variable n represent in the $\Delta G°$ and $E°$ equations?

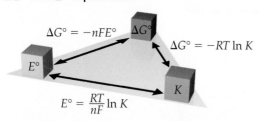

▲ **FIGURE 19.13   Relationships of $E°$, $\Delta G°$ and K.** Any one of these important parameters can be used to calculate the other two. The signs of $E°$ and $\Delta G°$ determine the direction that the reaction proceeds under standard conditions. The magnitude of K determines the relative amounts of reactants and products in an equilibrium mixture.

SAMPLE EXERCISE 19.10   Using standard reduction potentials to calculate $\Delta G°$ and K

(a) Use the standard reduction potentials listed in Table 19.1 to calculate the standard free energy change, $\Delta G°$, and the equilibrium constant, K, at room temperature ($T = 298$ K) for the reaction

$$4\,Ag(s) + O_2(g) + 4\,H^+(aq) \longrightarrow 4\,Ag^+(aq) + 2\,H_2O(l)$$

(b) Suppose the reaction in part (a) was written

$$2\,Ag(s) + \tfrac{1}{2}O_2(g) + 2\,H^+(aq) \longrightarrow 2\,Ag^+(aq) + H_2O(l)$$

What are the values of $E°$, $\Delta G°$ and K when the reaction is written in this way?

SOLUTION

Analyse We are asked to determine $\Delta G°$ and K for a redox reaction, using standard reduction potentials.

Plan We use the data in Table 19.1 and Equation 19.10 to determine $E°$ for the reaction and then use $E°$ in Equation 19.12 to calculate $\Delta G°$. We then use Equation 16.25, $\Delta G° = -RT \ln K$, to calculate K.

Solve

(a) We first calculate $E°$ by breaking the equation into two half-reactions, as in Sample Exercise 19.8, and then obtain $E_{red}°$ values from Table 19.1 (or Appendix E):

Reduction:	$O_2(g) + 4\,H^+(aq) + 4\,e^- \longrightarrow 2\,H_2O(l)$	$E_{red}° = +1.23$ V
Oxidation:	$4\,Ag(s) \longrightarrow 4\,Ag^+(aq) + 4\,e^-$	$E_{red}° = +0.80$ V

Even though the second half-reaction has 4 Ag, we use the $E_{red}°$ value directly from Table 19.1 because emf is an intensive property.

Using Equation 19.10, we have

$$E° = (1.23\text{ V}) - (0.80\text{ V}) = 0.43\text{ V}$$

The half-reactions show the transfer of four electrons. Thus for this reaction $n = 4$. We now use Equation 19.12 to calculate $\Delta G°$:

$$\Delta G° = -nFE°$$
$$= -(4)(96\,485\text{ J V}^{-1}\text{ mol}^{-1})(+0.43)\text{ V}$$
$$= -1.7 \times 10^5\text{ J mol}^{-1} = -166\text{ kJ mol}^{-1}$$

The positive value of $E°$ leads to a negative value of $\Delta G°$.

Now we need to calculate the equilibrium constant, K, using $\Delta G° = -RT \ln K$. Because $\Delta G°$ is a large negative number, which means the reaction is thermodynamically very favourable, we expect K to be large.

$$\Delta G° = -RT \ln K$$
$$-1.7 \times 10^5\text{ J mol}^{-1} = -(8.314\text{ J K}^{-1}\text{ mol}^{-1})\,(298\text{ K}) \ln K$$
$$\ln K = \frac{-1.7 \times 10^5\text{ J mol}^{-1}}{-(8.314\text{ J K}^{-1}\text{ mol}^{-1})(298\text{ K})}$$
$$\ln K = 69$$
$$K = 9.3 \times 10^{29}$$

K is indeed very large. This means that we expect silver metal to oxidise in acidic environments, in air, to Ag^+. Notice that the voltage calculated for the reaction was 0.43 V, which is easy to measure. Directly measuring such a large equilibrium constant by measuring reactant and product concentrations at equilibrium, on the other hand, would be very difficult.

(b) The overall equation is the same as that in part (a), multiplied by $\tfrac{1}{2}$. The half-reactions are:

Reduction:	$\tfrac{1}{2}O_2(g) + 2\,H^+(aq) + 2\,e^- \longrightarrow H_2O(l)$	$E_{red}° = 1.23$ V
Oxidation:	$2\,Ag(s) \longrightarrow 2\,Ag^+(aq) + 2\,e^-$	$E_{red}° = +0.80$ V

The values of $E_{red}°$ are the same as they were in part (a); they are not changed by multiplying the half-reactions by $\tfrac{1}{2}$. Thus $E°$ has the same value as in part (a):

$$E° = +0.43\text{ V}$$

Notice, though, that the value of n has changed to $n = 2$, which is half the value in part (a). Thus $\Delta G°$ is half as large as in part (a).

$$\Delta G° = -(2)(96\,485\ \text{J V}^{-1}\ \text{mol}^{-1})(+0.43\ \text{V}) = -83\ \text{kJ mol}^{-1}$$

Now we can calculate K as before:

$$-8.3 \times 10^4\ \text{J mol}^{-1} = -(8.314\ \text{J K}^{-1}\ \text{mol}^{-1})(298\ \text{K}) \ln K$$

$$K = 3.5 \times 10^{14}$$

Comment $E°$ is an *intensive* quantity, so multiplying a chemical equation by a certain factor will not affect the value of $E°$. Multiplying an equation will change the value of n, however, and hence the value of $\Delta G°$. The change in free energy, in units of J mol^{-1} of reaction as written, is an *extensive* quantity. The equilibrium constant is also an extensive quantity.

PRACTICE EXERCISE

For the reaction

$$3\,Ni^{2+}(aq) + 2\,Cr(OH)_3(s) + 10\,OH^-(aq) \longrightarrow 3\,Ni(s) + 2\,CrO_4{}^{2-}(aq) + 8\,H_2O(l)$$

(a) What is the value of n? **(b)** Use the data in Appendix E to calculate $\Delta G°$. **(c)** Calculate K at $T = 298$ K.

Answers: **(a)** 6, **(b)** +87 kJ mol^{-1}, **(c)** $K = 6 \times 10^{-16}$

(See also Exercise 19.43.)

A CLOSER LOOK

ELECTRICAL WORK

For any spontaneous process, ΔG is a measure of the maximum useful work, w_{max}, that can be extracted from the process: $\Delta G = w_{max}$ (∞ Section 14.13, 'Gibbs Energy and Temperature'). Because $\Delta G = -nFE$, the maximum useful electrical work obtainable from a voltaic cell is

$$w_{max} = -nFE_{cell} \qquad [19.14]$$

Because cell emf, E_{cell}, is always positive for a voltaic cell, w_{max} is negative, indicating that work is done *by* a system *on* its surroundings, as we expect for a voltaic cell (∞ Section 14.2, 'The First Law of Thermodynamics').

As Equation 19.14 shows, the more charge a voltaic cell moves through a circuit (that is, the larger nF is) and the larger the emf pushing the electrons through the circuit (that is, the larger E_{cell} is), the more work the cell can accomplish. In Sample Exercise 19.10, we calculated $\Delta G° = -170\ \text{kJ mol}^{-1}$ for the reaction $4\,Ag(s) + O_2(s) + 4\,H^+(aq) \longrightarrow 4\,Ag^+(aq) + 2\,H_2O(l)$. Thus a voltaic cell utilising this reaction could perform a maximum of 170 kJ of work in consuming 4 mol Ag, 1 mol O_2 and 4 mol H^+.

If a reaction is not spontaneous, ΔG is positive and E is negative. To force a non-spontaneous reaction to occur in an electrochemical cell, we need to apply an external potential, E_{ext}, that exceeds $|E_{cell}|$. For example, if a non-spontaneous process has $E = -0.9$ V, then the external potential, E_{ext}, must be greater than +0.9 V in order for the process to occur. We will examine such non-spontaneous processes in Section 19.9.

Electrical work can be expressed in energy units of watts times time. The *watt* (W) is a unit of electrical power (that is, rate of energy expenditure):

$$1\ \text{W} = 1\ \text{J s}^{-1}$$

Thus a watt-second is a joule. The unit employed by electric utilities is the kilowatt-hour (kWh), which equals 3.6×10^6 J:

$$1\ \text{kWh} = (1000\ \text{W})(1\ \text{h})\left(\frac{3600\ \text{s}}{1\ \text{h}}\right)\left(\frac{1\ \text{J s}^{-1}}{1\ \text{W}}\right) = 3.6 \times 10^6\ \text{J}$$

RELATED EXERCISES: 19.84

19.6 CELL POTENTIALS UNDER NON-STANDARD CONDITIONS

We have seen how to calculate the emf of a cell when the reactants and products are under standard conditions. As a voltaic cell is discharged, however, the reactants of the reaction are consumed and the products are generated, so the concentrations of these substances change. The emf progressively drops until $E = 0$, at which point we say the cell is 'dead'. At that point the concentrations of the reactants and products cease to change: they are at equilibrium. In this section we examine how the cell emf depends on the concentrations of the reactants and products of the cell reaction. The emf generated under non-standard conditions can be calculated by using an equation first derived by Walther Nernst (1864–1941), a German chemist who established many of the theoretical foundations of electrochemistry.

The Nernst Equation

The dependence of the cell emf on concentration can be obtained from the dependence of the free energy change on concentration (∞ Section 16.7, 'The Equilibrium Constant and Free Energy'). Recall that the free energy change, ΔG, is related to the standard free energy change, $\Delta G°$:

Review this on page 635

$$\Delta G = \Delta G° + RT \ln Q \qquad [19.15]$$

The quantity Q is the reaction quotient, which has the form of the equilibrium constant expression except that the concentrations are those that exist in the reaction mixture at a given moment.

Substituting $\Delta G = -nFE$ (Equation 19.11) into Equation 19.15 gives

$$-nFE = -nFE° + RT \ln Q$$

Solving this equation for E gives the **Nernst equation**:

$$E = E° - \frac{RT}{nF} \ln Q \qquad [19.16]$$

At 25 °C (298 K) and converting to base 10 logarithms, Equation 19.16 can be simplified to a form in common use:

$$E = E° - \frac{0.0592}{n} \log Q \qquad [19.17]$$

This is because most electrochemical experiments are made at a standard temperature, which is 298 K. For non-standard conditions, Equation 19.16 is normally used.

To show how Equation 19.16 might be used, consider the following reaction, which we discussed earlier:

$$Zn(s) + Cu^{2+}(aq) \longrightarrow Zn^{2+}(aq) + Cu(s)$$

In this case, $n = 2$ (two electrons are transferred from Zn to Cu^{2+}) and the standard emf is +1.10 V. Thus at 298 K with $R = 8.314$ J mol^{-1} K^{-1} and $F = 96\,485$ C mol^{-1}, the Nernst equation gives

$$E = 1.10 \text{ V} - \frac{(8.314 \text{ J mol}^{-1} \text{ K}^{-1})(298 \text{ K})}{2(96\,485 \text{ J V}^{-1} \text{ mol}^{-1})} \ln \frac{[Zn^{2+}]}{[Cu^{2+}]} \qquad [19.18]$$

Recall that pure solids are excluded from the expression for Q (∞ Section 16.6, 'Applications of Equilibrium Constants'). According to Equation 19.18, the emf increases as $[Cu^{2+}]$ increases and as $[Zn^{2+}]$ decreases. For example, when $[Cu^{2+}]$ is 5.0 M and $[Zn^{2+}]$ is 0.050 M, we have

Review this on page 631

$$E = 1.10 \text{ V} - \frac{0.0252 \text{ V}}{2} \ln \frac{[0.050]}{[5.0]}$$

$$= 1.10 \text{ V} - 0.0126 \text{ V}(-4.61) = 1.16 \text{ V}$$

Thus increasing the concentration of the reactant (Cu^{2+}) and decreasing the concentration of the product (Zn^{2+}) relative to standard conditions increases the emf of the cell ($E = +1.16$ V) relative to standard conditions ($E° = +1.10$ V). We could have anticipated this result by applying Le Châtelier's principle (∞ Section 16.8, 'Le Châtelier's Principle').

Review this on page 638

In general, if the concentrations of reactants increase relative to those of products, the emf increases. Conversely, if the concentrations of products increase relative to reactants, the emf decreases. As a voltaic cell operates, reactants are converted into products, which increases the value of Q and decreases the value of E, eventually reaching $E = 0$. The Nernst equation helps us understand why the emf of a voltaic cell drops as the cell discharges: as reactants are converted to products, the value of Q increases, so the value of E decreases, eventually reaching $E = 0$. Because $\Delta G = -nFE$ (Equation 19.11), it follows that $\Delta G = 0$ when $E = 0$. Recall that a system is at equilibrium when $\Delta G = 0$

∞ Review this on page 635

(∞ Section 16.7, 'The Equilibrium Constant and Free Energy'). Thus when $E = 0$, the cell reaction has reached equilibrium and no net reaction is occurring.

SAMPLE EXERCISE 19.11 Voltaic cell emf under non-standard conditions

Calculate the emf at 298 K generated by the cell described in Sample Exercise 19.4 below, when $[Cr_2O_7^{2-}] = 2.0$ M, $[H^+] = 1.0$ M, $[I^-] = 1.0$ M and $[Cr^{3+}] = 1.0 \times 10^{-5}$ M.

$$Cr_2O_7^{2-}(aq) + 14\,H^+(aq) + 6\,I^-(aq) \longrightarrow 2\,Cr^{3+}(aq) + 3\,I_2(s) + 7\,H_2O(l)$$

SOLUTION

Analyse We are given a chemical equation for a voltaic cell and the concentration of reactants and products under which it operates. We are asked to calculate the emf of the cell under these non-standard conditions.

Plan To calculate the emf of a cell under non-standard conditions, we use the Nernst equation, Equation 19.16.

Solve We first calculate $E°$ for the cell from standard reduction potentials (Table 19.1 or Appendix E). The standard emf for this reaction was calculated in Sample Exercise 19.6: $E° = 0.79$ V. If you refer back to that exercise, you will see that the balanced equation shows six electrons transferred from reducing agent to oxidising agent, so $n = 6$. The reaction quotient is

$$Q = \frac{[Cr^{3+}]^2}{[Cr_2O_7^{2-}][H^+]^{14}[I^-]^6} = \frac{(1.0 \times 10^{-5})^2}{(2.0)(1.0)^{14}(1.0)^6} = 5.0 \times 10^{-11}$$

Using Equation 19.16, we have

$$E = 0.79\,V - \frac{(8.314\,J\,mol^{-1}\,K^{-1})(298\,K)}{6(96\,485\,J\,V^{-1}\,mol^{-1})} \ln(5.0 \times 10^{-11})$$

$$= 0.79\,V - (0.00428\,V)\,(-23.7)$$

$$= 0.79\,V + 0.10\,V = 0.89\,V$$

Check This result is qualitatively what we expect. Because the concentration of $Cr_2O_7^{2-}$ (a reactant) is greater than 1 M and the concentration of Cr^{3+} (a product) is less than 1 M, the emf is greater than $E°$. Q is about 10^{-10}, so $\log Q$ is about -10. Thus the correction to $E°$ is about $0.06 \times (10)/6$, which is 0.1, in agreement with the more detailed calculation.

PRACTICE EXERCISE

Calculate the emf generated by the cell described in the practice exercise accompanying Sample Exercise 19.6 when $[Al^{3+}] = 4.0 \times 10^{-3}$ M and $[I^-] = 0.010$ M.

Answer: $E = +2.37$ V

(See also Exercises 19.50, 19.51.)

SAMPLE EXERCISE 19.12 Calculating concentrations in a voltaic cell

If the voltage of a Zn–H$^+$ cell (like that in Figure 19.4) is 0.45 V at 25 °C when $[Zn^{2+}] = 1.0$ M and $P_{H_2} = 1$ bar, what is the concentration of H$^+$?

SOLUTION

Analyse We are given a description of a voltaic cell, its emf and the concentration of Zn^{2+} and the partial pressure of H_2 (both products in the cell reaction). We are asked to calculate the concentraiton of H^+, a reactant.

Plan First, we write the equation for the cell reaction and use standard reduction potentials from Table 19.1 to calculate $E°$ for the reaction. After determining the value of n from our reaction equation, we solve the Nernst equation for Q. Finally, we use the equation for the cell reaction to write an expression for Q that contains $[H^+]$ to determine $[H^+]$.

Solve

The cell reaction is

$$Zn(s) + 2\,H^+(aq) \longrightarrow Zn^{2+}(aq) + H_2(g)$$

The standard emf is

$$E° = E°_{red}\,(reduction) - E°_{red}\,(oxidation)$$

$$= 0\,V - (-0.76\,V) = +0.76\,V$$

Because each Zn atom loses two electrons,

$$n = 2$$

Using Equation 19.17, we can solve for Q:

$$0.45 \text{ V} = 0.76 \text{ V} - 0.0128 \text{ V} \ln Q$$

$$\ln Q = \frac{(0.76 \text{ V} - 0.45 \text{ V})}{(0.0128 \text{ V})} = 24.2$$

$$Q = e^{24.2} = 3.3 \times 10^{10}$$

Solving for $[H^+]$, we have

$$Q = \frac{[Zn^{2+}]P_{H_2}}{[H^+]^2} = \frac{(1.0)(1.0)}{[H^+]^2} = 3.0 \times 10^{10}$$

Q has the form of the equilibrium constant for the reaction

$$[H^+]^2 = \frac{1.0}{3.0 \times 10^{10}} = 3.3 \times 10^{-11}$$

$$[H^+] = \sqrt{3.3 \times 10^{-11}} = 5.8 \times 10^{-6} \text{ M}$$

Comment A voltaic cell whose cell reaction involves H^+ can be used to measure $[H^+]$ or pH. A pH meter is a specially designed voltaic cell with a voltmeter calibrated to read pH directly (Section 17.4, 'The pH Scale').

PRACTICE EXERCISE

What is the pH of the solution in the cathode compartment of the cell pictured in Figure 19.8 when $P_{H_2} = 1$ bar, $[Zn^{2+}]$ in the anode compartment is 0.10 M and cell emf is 0.542 V?

Answer: pH = 4.19

(See also Exercise 19.54(a).)

Concentration Cells

In each of the voltaic cells that we have looked at so far, the reactive species at the anode has been different from the one at the cathode. Cell emf depends on concentration, however, so a voltaic cell can be constructed using the *same* species in both the anode and cathode compartments as long as the concentrations are different. A cell based solely on the emf generated because of a difference in a concentration is called a **concentration cell**.

A diagram of a concentration cell is shown in ▼ **FIGURE 19.14(a)**. One half-cell consists of a strip of nickel metal immersed in a 1.00 M solution of Ni^{2+}(aq). The other half-cell also has an Ni(s) electrode, but it is immersed in a

 FIGURE IT OUT

Assuming that the solutions are made from $Ni(NO_3)_2$, how do the ions migrate as the cell operates?

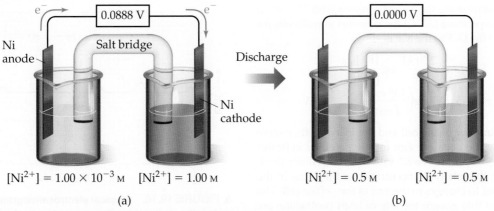

(a)   (b)

▲ **FIGURE 19.14   Concentration cell based on the Ni^{2+}–Ni cell reaction.**   (a) Concentrations of Ni^{2+}(aq) in the two half-cells are unequal, and the cell generates an electrical current. (b) The cell operates until $[Ni^{2+}$(aq)] is the same in the two half-cells, at which point the cell has reached equilibrium and is 'dead'.

1.00 × 10⁻³ M solution of $Ni^{2+}(aq)$. The two half-cells are connected by a salt bridge and by an external wire running through a voltmeter. The half-cell reactions are the reverse of each other:

Anode: $Ni(s) \longrightarrow Ni^{2+}(aq) + 2\,e^-$ $E^{\circ}_{red} = -0.28\ V$

Cathode: $Ni^{2+}(aq) + 2\,e^- \longrightarrow Ni(s)$ $E^{\circ}_{red} = -0.28\ V$

MY WORLD OF CHEMISTRY

HEARTBEATS AND ELECTROCARDIOGRAPHY

The human heart is a marvel of efficiency and dependability. In a typical day an adult's heart pumps more than 7000 dm^3 of blood through the circulatory system, usually with no maintenance required beyond a sensible diet and lifestyle. We generally think of the heart as a mechanical device, a muscle that circulates blood via regularly spaced muscular contractions. However, more than two centuries ago two pioneers in electricity, Luigi Galvani (1729–1787) and Alessandro Volta (1745–1827), discovered that the contractions of the heart are controlled by electrical phenomena, as are nerve impulses throughout the body. The pulses of electricity that cause the heart to beat result from a remarkable combination of electrochemistry and the properties of semi-permeable membranes (∞ Section 12.5, 'Colligative Properties').

Cell walls are membranes with variable permeability with respect to a number of physiologically important ions (especially Na^+, K^+ and Ca^{2+}). The concentrations of these ions are different for the fluids inside the cells (the *intracellular fluid*, or ICF) and outside the cells (the *extracellular fluid*, or ECF). In cardiac muscle cells, for example, the concentrations of K^+ in the ICF and ECF are typically about 135 millimolar (mM) and 4 mM, respectively. For Na^+, however, the concentration difference between the ICF and ECF is opposite that for K^+; typically, $[Na^+]_{ICF} = 10$ mM and $[Na^+]_{ECF} = 145$ mM.

The cell membrane is initially permeable to K^+ ions but is much less so to Na^+ and Ca^{2+}. The difference in concentration of K^+ ions between the ICF and ECF generates a concentration cell: Even though the same ions are present on both sides of the membrane, there is a potential difference between the two fluids that we can calculate using the Nernst equation with $E^\circ = 0$. At physiological temperature (37 °C) the potential in millivolts for moving K^+ from the ECF to the ICF is

$$E = E^\circ - \frac{2.30\,RT}{nF} \log \frac{[K^+]_{ICF}}{[K^+]_{ECF}}$$

$$= 0 - (61.5\ mV) \log\left(\frac{135\ mM}{4\ mM}\right) = -94\ mV$$

In essence, the interior of the cell and the ECF together serve as a voltaic cell. The negative sign for the potential indicates that work is required to move K^+ into the intracellular fluid.

Changes in the relative concentrations of the ions in the ECF and ICF lead to changes in the emf of the voltaic cell. The cells of the heart that govern the rate of heart contraction are called the *pacemaker cells*. The membranes of the cells regulate the concentrations of ions in the ICF, allowing them to change

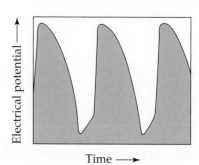

▲ **FIGURE 19.15** **Changes in electrical potential in the human heart.** Variation of the electrical potential caused by changes of ion concentrations in the pacemaker cells of the heart.

in a systematic way. The concentration changes cause the emf to change in a cyclic fashion, as shown in ▲ **FIGURE 19.15.** The emf cycle determines the rate at which the heart beats. If the pacemaker cells malfunction because of disease or injury, an artificial pacemaker can be surgically implanted. The artificial pacemaker contains a small battery that generates the electrical pulses needed to trigger the contractions of the heart.

During the late 1800s scientists discovered that the electrical impulses that cause the contraction of the heart muscle are strong enough to be detected at the surface of the body. This observation formed the basis for *electrocardiography*, non-invasive monitoring of the heart by using a complex array of electrodes on the skin to measure voltage changes during heartbeats. A typical electrocardiogram is shown in ▼ **FIGURE 19.16.** It is quite striking that, although the heart's major function is the *mechanical* pumping of blood, it is most easily monitored by using the *electrical* impulses generated by tiny voltaic cells.

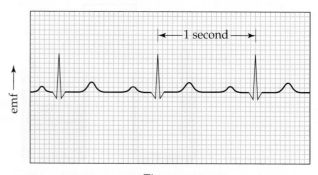

▲ **FIGURE 19.16** **A typical electrocardiogram.** The printout records the electrical events monitored by electrodes attached to the body surface.

Although the *standard* emf for this cell is zero,

$$E^\circ_{cell} = E^\circ_{red}\,(\text{cathode}) - E^\circ_{red}\,(\text{anode}) = (-0.28\ \text{V}) - (-0.28\ \text{V}) = 0\ \text{V}$$

the cell operates under *non-standard* conditions because the concentration of $Ni^{2+}(aq)$ is different in the two compartments. In fact, the cell will operate until the concentrations of Ni^{2+} in both compartments are equal, that is $[Ni^{2+}]_{anode} = [Ni^{2+}]_{cathode}$. Oxidation of $Ni(s)$ occurs in the half-cell containing the *more dilute* solution, thereby increasing the concentration of $Ni^{2+}(aq)$. It is therefore the *anode* compartment of the cell. Reduction of $Ni^{2+}(aq)$ occurs in the half-cell containing the more concentrated solution, thereby decreasing the concentration of $Ni^{2+}(aq)$, making it the cathode compartment. The *overall* cell reaction is therefore

Anode:	$Ni(s) \longrightarrow Ni^{2+}\,(aq,\ dilute) + 2\ e^-$
Cathode:	$Ni^{2+}(aq,\ concentrated) + 2\ e^- \longrightarrow Ni(s)$
Overall:	$Ni^{2+}(aq,\ concentrated) \longrightarrow Ni^{2+}(aq,\ dilute)$

We can calculate the emf of a concentration cell by using the Nernst equation. For this particular cell, we see that $n = 2$. The expression for the reaction quotient for the overall reaction is

$$Q = \frac{[Ni^{2+}]_{dilute}}{[Ni^{2+}]_{concentrated}}$$

Thus the emf at 298 K is

$$E = E^\circ - \frac{RT}{nF}\ln Q$$

$$= 0 - \frac{0.0256\ \text{V}}{2}\ln\frac{[Ni^{2+}]_{dilute}}{[Ni^{2+}]_{concentrated}} = -0.0128\ln\frac{1.00 \times 10^{-3}\ \text{M}}{1.00\ \text{M}}$$

$$= +0.0888\ \text{V}$$

This concentration cell generates an emf of nearly 0.09 V even though $E^\circ = 0$. The difference in concentration provides the driving force for the cell. When the concentrations in the two compartments become the same, the value of $Q = 1$ and $E = 0$.

The idea of generating a potential by a difference in concentration is the basis for the operation of pH meters. It is also a critical aspect in biology; nerve cells in the brain generate a voltage across the cell membrane by having different concentrations of ions on either side of the membrane. The regulation of the heartbeat in mammals is another example of the importance of electrochemistry to living organisms.

SAMPLE EXERCISE 19.13   **Determining pH using a concentration cell**

A voltaic cell is constructed with two hydrogen electrodes. Electrode 1 has $P_{H_2} = 1$ bar and an unknown concentration of $H^+(aq)$. Electrode 2 is a standard hydrogen electrode ($[H^+] = 1.00$ M, $P_{H_2} = 1$ bar. At 298 K the measured cell voltage is 0.211 V, and the electrical current is observed to flow from electrode 1 through the external circuit to electrode 2. Calculate $[H^+]$ for the solution at electrode 1. What is its pH?

SOLUTION

Analyse We are given the potential of a concentration cell and the direction in which the current flows. We also have the concentration or partial pressures of all reactants and products except for $[H^+]$ in half-cell 1, which is our unknown.

Plan We can use the Nernst equation to determine Q and then use Q to calculate the unknown concentration. Because this is a concentration cell, $E^\circ_{cell} = 0$ V.

Solve Using the Nernst equation, we have

$$0.211\ \text{V} = 0 - \frac{0.0256\ \text{V}}{2}\ln Q$$

$$\ln Q = \frac{-0.211\ \text{V}}{0.0128\ \text{V}} = -16.48$$

$$Q = e^{-16.48} = 6.9 \times 10^{-8}$$

Because electrons flow from electrode 1 to electrode 2, electrode 1 is the anode of the cell and electrode 2 is the cathode. The electrode reactions are therefore as follows, with the concentration of $H^+(aq)$ in electrode 1 represented by the unknown x:

Electrode 1: $\qquad H_2(g, 1\text{ bar}) \longrightarrow 2\,H^+(aq, x\text{ M}) + 2\,e^- \qquad E^\circ_{red} = 0\text{ V}$

Electrode 2: $\quad 2\,H^+(aq, 1.00\text{ M}) + 2\,e^- \longrightarrow H_2(g, 1\text{ bar}) \qquad E^\circ_{red} = 0\text{ V}$

Thus

$$Q = \frac{[H^+ \text{ (electrode 1)}]^2\, P_{H_2} \text{ (electrode 2)}}{[H^+ \text{ (electrode 2)}]^2\, P_{H_2} \text{ (electrode 1)}}$$

$$= \frac{x^2(1.00)}{(1.00)^2(1.00)} = x^2 = 6.9 \times 10^{-8}$$

$$x = \sqrt{6.9 \times 10^{-8}} = 2.6 \times 10^{-4}$$

At electrode 1, therefore

$$[H^+] = 2.6 \times 10^{-4}\text{ M}$$

and the pH of the solution is

$$pH = -\log[H^+] = -\log(2.6 \times 10^{-4}) = 3.6$$

Comment The concentration of H^+ at electrode 1 is lower than that at electrode 2, which is why electrode 1 is the anode of the cell: the oxidation of H_2 to $H^+(aq)$ increases $[H^+]$ at electrode 1.

PRACTICE EXERCISE

A concentration cell is constructed with two $Zn(s)$–$Zn^{2+}(aq)$ half-cells. The first half-cell has $[Zn^{2+}] = 1.35$ M and the second half-cell has $[Zn^{2+}] = 3.75 \times 10^{-4}$ M. **(a)** Which half-cell is the anode of the cell? **(b)** What is the emf of the cell?

Answers: **(a)** the second half-cell, **(b)** 0.105 V

(See also Exercise 19.52.)

19.7 | BATTERIES AND FUEL CELLS

A **battery** is a portable, self-contained electrochemical power source that consists of one or more voltaic cells. For example, the common 1.5 V batteries used to power torches and many consumer electronic devices are single voltaic cells. Greater voltages can be achieved by using multiple voltaic cells in a single battery, as is the case in 12 V car batteries. When cells are connected in series (with the cathode of one attached to the anode of another), the battery produces a voltage that is the sum of the emfs of the individual cells. Higher emfs can also be achieved by using multiple batteries in series (◀ **FIGURE 19.17**). The electrodes of batteries are marked following the convention of Figure 19.6—the cathode is labelled with a plus sign and the anode with a minus sign.

Although any spontaneous redox reaction can serve as the basis for a voltaic cell, making a commercial battery that has specific performance characteristics can require considerable ingenuity. The substances that are oxidised at the anode and reduced by the cathode determine the emf of a battery, and the usable life of the battery depends on the quantities of these substances packaged in the battery. Usually the anode and the cathode compartments are separated by a barrier analogous to the porous barrier of Figure 19.6.

Different applications require batteries with different properties. The battery required to start a car, for example, must be capable of delivering a large electrical current for a short time period. The battery that powers a heart pacemaker, on the other hand, must be very small and capable of delivering a small but steady current over an extended time period. Some batteries are *primary* cells, meaning that they can't be recharged. A primary cell must be discarded or recycled after its emf drops to zero. A *secondary* cell can be recharged from an external power source after its emf has dropped.

▲ **FIGURE 19.17 Combining batteries.** When batteries are connected in series, as in most torches, the total emf is the sum of the individual emfs.

Lead-Acid Battery

A 12 V lead-acid car battery consists of six voltaic cells in series, each producing 2 V. The cathode of each cell consists of lead dioxide (PbO_2) packed on a metal grid. The anode of each cell is composed of lead. Both electrodes are immersed in sulfuric acid. The electrode reactions that occur during discharge are

Cathode:

$$PbO_2(s) + HSO_4^-(aq) + 3\,H^+(aq) + 2\,e^- \longrightarrow PbSO_4(s) + 2\,H_2O(l)$$

Anode:

$$Pb(s) + HSO_4^-(aq) \longrightarrow PbSO_4(s) + H^+(aq) + 2\,e^-$$

Overall:

$$PbO_2(s) + Pb(s) + 2\,HSO_4^-(aq) + 2\,H^+(aq) \longrightarrow 2\,PbSO_4(s) + 2\,H_2O(l)$$

The standard cell potential can be obtained from the standard reduction potentials in Appendix E:

$$E^\circ_{cell} = E^\circ_{red}\,(\text{cathode}) - E^\circ_{red}\,(\text{anode}) = (+1.685\,\text{V}) - (-0.356\,\text{V}) = +2.041\,\text{V}$$

The reactants Pb and PbO_2 serve as the electrodes. Because the reactants are solids, there is no need to separate the cell into half-cells; the Pb and PbO_2 cannot come into direct physical contact unless one electrode plate touches another. To keep the electrodes from touching, glass-fibre spacers are placed between them (▶ **FIGURE 19.18**).

Using a reaction whose reactants and products are solids has another benefit. Because solids are excluded from the reaction quotient Q, the relative amounts of $Pb(s)$, $PbO_2(s)$ and $PbSO_4(s)$ have no effect on the emf of the lead storage battery, helping the battery maintain a relatively constant emf during discharge. The emf does vary somewhat with use because the concentration of H_2SO_4 varies with the extent of cell discharge. As the equation for the overall cell reaction indicates, H_2SO_4 is consumed during the discharge.

One advantage of a lead-acid battery is that it can be recharged. During recharging, an external source of energy is used to reverse the direction of the overall cell reaction, regenerating $Pb(s)$ and $PbO_2(s)$.

$$2\,PbSO_4(s) + 2\,H_2O(l) \longrightarrow$$
$$PbO_2(s) + Pb(s) + 2\,HSO_4^-(aq) + 2\,H^+(aq)$$

In a car the energy necessary for recharging the battery is provided by the alternator, driven by the engine. Recharging is possible because $PbSO_4$ formed during discharge adheres to the electrodes. As the external source forces electrons from one electrode to another, the $PbSO_4$ is converted to Pb at one electrode and to PbO_2 at the other.

Alkaline Battery

The most common primary (non-rechargeable) battery is the alkaline battery. More than 10^{10} alkaline batteries are produced annually. The anode of this battery consists of powdered zinc metal immobilised in a gel in contact with a concentrated solution of KOH (hence the name *alkaline* battery). The cathode is a mixture of $MnO_2(s)$ and graphite, separated from the anode by a porous fabric. The battery is sealed in a steel can to reduce the risk of leakage of the concentrated KOH. A schematic view of an alkaline battery is shown in ▶ **FIGURE 19.19**. The cell reactions are complex, but can be approximately represented as follows:

Cathode: $2\,MnO_2(s) + 2\,H_2O(l) + 2\,e^- \longrightarrow 2\,MnO(OH)(s) + 2\,OH^-(aq)$

Anode: $Zn(s) + 2\,OH^-(aq) \longrightarrow Zn(OH)_2(s) + 2\,e^-$

The emf of an alkaline battery is 1.55 V at room temperature. The alkaline battery provides far superior performance over the older 'dry cells' that were also based on MnO_2 and Zn as the electrochemically active substances which used an acidic zinc chloride or ammonium chloride paste as the electrolyte.

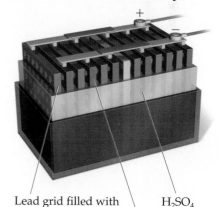

△ **FIGURE IT OUT**

What is the oxidation state of lead in the cathode of this battery?

Lead grid filled with spongy lead (anode)

H_2SO_4 electrolyte

Lead grid filled with PbO_2 (cathode)

▲ **FIGURE 19.18**   **A 12 V car lead-acid battery.** Each anode/cathode pair in this schematic cutaway produces a voltage of about 2 V. Six pairs of electrodes are connected in series, producing 12 V.

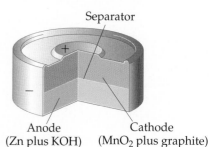

Separator

Anode
(Zn plus KOH)

Cathode
(MnO_2 plus graphite)

▲ **FIGURE 19.19**   **Cutaway view of a miniature alkaline battery.**

Nickel-Cadmium, Nickel-Metal Hydride and Lithium-Ion Batteries

The tremendous growth in high-power-demand portable electronic devices, such as mobile phones, notebook computers and video recorders, has increased the demand for lightweight, readily recharged batteries. One of the most common rechargeable batteries is the nickel-cadmium (nicad) battery. During discharge, cadmium metal is oxidised at the anode of the battery while nickel oxyhydroxide ($NiO(OH)(s)$) is reduced at the cathode.

Cathode: $2\,NiO(OH)(s) + 2\,H_2O(l) + 2\,e^- \longrightarrow 2\,Ni(OH)_2(s) + 2\,OH^-(aq)$

Anode: $Cd(s) + 2\,OH^-(aq) \longrightarrow Cd(OH)_2(s) + 2\,e^-$

As in the lead-acid battery, the solid reaction products adhere to the electrodes, which permits the electrode reactions to be reversed during charging. A single nicad voltaic cell has an emf of 1.30 V. Nicad battery packs typically contain three or more cells in series to produce the higher emfs needed by most electronic devices.

There are drawbacks to nickel-cadmium batteries. Cadmium is a toxic heavy metal. Its use increases the weight of batteries and provides an environmental hazard; roughly 1.5 billion nickel-cadmium batteries are produced annually, and these must eventually be recycled as they lose their ability to be recharged. Some of these problems have been alleviated by the development of nickel-metal hydride (NiMH) batteries. The cathode reaction of NiMH batteries is the same as that for nickel-cadmium batteries, but the anode reaction is very different. The anode consists of a metal *alloy*, such as $ZrNi_2$, that has the ability to absorb hydrogen atoms. During the oxidation at the anode, the hydrogen atoms lose electrons, and the resultant H^+ ions react with OH^- ions to form H_2O, a process that is reversed during charging. Hybrid gas–electric cars, which are powered by both a petrol engine and an electric motor, use NiMH batteries to store the electrical power. The batteries are recharged by the electric motor while braking and, as a result, the batteries can last up to eight years.

The newest rechargeable battery to receive large use in consumer electronic devices is the lithium-ion (Li-ion) battery. This is the battery you will find in mobile phones and laptop computers. Because lithium is a very light element, Li-ion batteries achieve a greater *energy density* (the amount of energy stored per unit mass) than nickel-based batteries. The technology of Li-ion batteries is very different from that of the other batteries described here. It is based on the ability of Li^+ ions to be inserted into and removed from certain layered solids. For example, Li^+ ions can be inserted reversibly into layers of graphite (Figure 11.48). In most commercial cells, one electrode is graphite or some other carbon-based material and the other is usually made of lithium cobalt oxide ($LiCoO_2$). When charged, cobalt ions are oxidised and Li^+ ions migrate into the graphite. During discharge, when the battery is producing electricity for use, the Li^+ ions spontaneously migrate from the graphite anode to the cathode, enabling electrons to flow through the external circuit. An Li-ion battery produces a maximum voltage of 3.7 V, considerably higher than typical 1.5 V alkaline batteries.

Hydrogen Fuel Cells

The thermal energy released by the combustion of fuels can be converted to electrical energy. The heat may convert water to steam, which drives a turbine that in turn drives a generator. Typically, a maximum of only 40% of the energy from combustion is converted to electricity; the remainder is lost as heat. The direct production of electricity from fuels by a voltaic cell could, in principle, yield a higher rate of conversion of the chemical energy of the reaction. Voltaic cells that perform this conversion using conventional fuels, such as H_2 and CH_4, are called **fuel cells**. Strictly speaking, fuel cells are *not* batteries because they are not self-contained systems, as the fuel must be continuously supplied to generate electricity.

The most common fuel-cell system involves the reaction of $H_2(g)$ and $O_2(g)$ to form $H_2O(l)$ as the only product. These cells can generate electricity twice as efficiently as the best internal combustion engine. Under acidic conditions, the electrode reactions are

Cathode:	$O_2(g) + 4\,H^+ + 4\,e^- \longrightarrow 2\,H_2O(l)$
Anode:	$2\,H_2(g) \longrightarrow 4\,H^+ + 4\,e^-$
Overall:	$2\,H_2(g) + O_2(g) \longrightarrow 2\,H_2O(l)$

The standard emf of an H_2/O_2 fuel cell is $+1.23$ V, reflecting the large driving force for the reaction of H_2 and O_2 to form H_2O.

In this fuel cell (known as the PEM fuel cell, for 'proton-exchange membrane') the anode and cathode are separated by a thin polymer membrane that is permeable to protons but not electrons. The polymer membrane therefore acts as a salt bridge. The electrodes are typically made from graphite. A PEM cell operates at a temperature of around 80 °C. At this low temperature the electrochemical reactions would normally occur very slowly, and so they are catalysed by a thin layer of platinum on each electrode.

Under basic conditions the electrode reactions in the hydrogen fuel cell are

Cathode:	$4\,e^- + O_2(g) + 2\,H_2O(l) \longrightarrow 4\,OH^-(aq)$
Anode:	$2\,H_2(g) + 4\,OH^-(aq) \longrightarrow 4\,H_2O(l) + 4\,e^-$
Overall:	$2\,H_2(g) + O_2(g) \longrightarrow 2\,H_2O(l)$

NASA has used the PEM fuel cell as the energy source for its spacecraft. Liquid hydrogen and oxygen are stored as fuel and the product of the reaction, water, is drunk by the spacecraft crew.

A schematic drawing of a low-temperature H_2/O_2 fuel cell is shown in ▶ FIGURE 19.20. Currently, a great deal of research is going into improving fuel cells. Much effort is being directed towards developing fuel cells that use conventional fuels such as hydrocarbons and alcohols, which are not as difficult to handle and distribute as hydrogen gas.

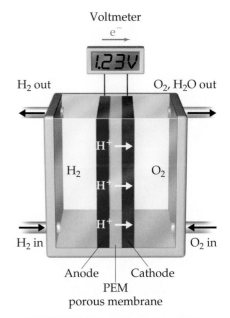

▲ **FIGURE 19.20   A hydrogen-PEM fuel cell.** The proton-exchange membrane (PEM) allows H^+ ions generated by H_2 oxidation at the anode to migrate to the cathode, where H_2O is formed.

Direct Methanol Fuel Cells

The direct methanol fuel cell is similar to the PEM fuel cell, but instead of using hydrogen gas as a reactant it uses methanol, CH_3OH. The reactions are

Cathode:	$\frac{3}{2}O_2(g) + 6\,H^+ + 6\,e^- \longrightarrow 3\,H_2O(g)$
Anode:	$CH_3OH(l) + H_2O(g) \longrightarrow CO_2(g) + 6\,H^+ + 6\,e^-$
Overall:	$CH_3OH(g) + \frac{3}{2}O_2(g) \longrightarrow CO_2(g) + 2\,H_2O(g)$

These cells operate at around 120 °C, which is slightly higher than the operating temperature of the standard PEM fuel cell. One shortcoming of the methanol cell is that the quantity of platinum catalyst needed is greater than in conventional PEM cells. Also, the carbon dioxide produced in the methanol reaction is not as environmentally friendly as water. However, liquid methanol is a far more attractive fuel to store and transport than hydrogen gas.

19.8 | CORROSION

In this section we examine the undesirable redox reactions that lead to the **corrosion** of metals. Corrosion reactions are spontaneous redox reactions in which a metal is attacked by some substance in its environment and converted to an unwanted compound.

For nearly all metals, oxidation is a thermodynamically favourable process in air at room temperature. When the oxidation process is not inhibited in some way, it can be very destructive. For some metals, however, oxidation can also form an insulating protective oxide layer, however, that prevents further reaction of the underlying metal. On the basis of the standard reduction potential for Al^{3+}, for example, we would expect aluminium metal to be very readily oxidised. The many aluminium cans that litter the environment are ample evidence, however, that aluminium undergoes only very slow chemical corrosion. The exceptional stability of this active metal in air is due to the formation of a thin protective coat of oxide (a hydrated form of Al_2O_3) on the surface of the metal. The oxide coat is impermeable to O_2 and H_2O and so protects the underlying metal from further corrosion. Magnesium metal is similarly protected. Some metal alloys, such as stainless steel, likewise form protective impervious oxide coats.

Corrosion of Iron (Rusting)

The rusting of iron requires both oxygen and water. Other factors, such as the pH of the solution, the presence of salts, contact with metals more difficult to oxidise than iron and stress on the iron, can accelerate rusting.

The corrosion of iron is electrochemical in nature. Not only does the corrosion process involve oxidation and reduction, the metal itself conducts electricity. Thus electrons can move through the metal from a region where oxidation occurs to another region where reduction occurs, as in voltaic cells.

Because the standard reduction potential for the reduction of $Fe^{2+}(aq)$ is less positive than that for the reduction of O_2, $Fe(s)$ can be oxidised by $O_2(g)$.

$$\textit{Cathode:}\quad O_2(g) + 4\,H^+(aq) + 4\,e^- \longrightarrow 2\,H_2O(l) \qquad\qquad E^\circ_{red} = 1.23\text{ V}$$

$$\textit{Anode:}\qquad\qquad\qquad\qquad Fe(s) \longrightarrow Fe^{2+}(aq) + 2\,e^- \qquad E^\circ_{red} = -0.44\text{V}$$

A portion of the iron, often associated with a dent or region of strain, can serve as an anode at which the oxidation of Fe to Fe^{2+} occurs (▼ **FIGURE 19.21**). The electrons produced migrate through the metal to another portion of the surface that serves as the cathode, at which O_2 is reduced. The reduction of O_2 requires H^+, so lowering the concentration of H^+ (increasing the pH) makes the reduction of O_2 less favourable. Iron in contact with a solution whose pH is greater than 9 does not corrode.

FIGURE IT OUT

What is the oxidising agent in this corrosion reaction?

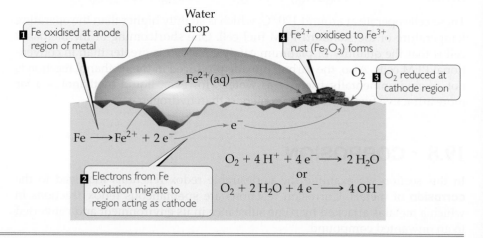

▶ **FIGURE 19.21   Corrosion of iron in contact with water.** One region of the iron acts as the cathode and another region acts as the anode.

The Fe^{2+} formed at the anode is eventually oxidised further to Fe^{3+}, which forms the hydrated iron(III) oxide known as rust:*

$$4\,Fe^{2+}(aq) + O_2(g) + 4\,H_2O(l) + 2\,xH_2O(l) \longrightarrow 2\,Fe_2O_3{\cdot}xH_2O(s) + 8\,H^+(aq)$$

Because the cathode is generally the area with the largest supply of O_2, rust often deposits there. If you look closely at a shovel after it has stood outside in the moist air with wet dirt adhering to its blade, you may notice that pitting has occurred under the dirt but that rust has formed elsewhere, where O_2 is more readily available.

The enhanced corrosion caused by the presence of salts is usually evident on cars in areas on the seaboard. Like a salt bridge in a voltaic cell, the ions of the salt provide the electrolyte necessary to complete the electrical circuit.

Preventing the Corrosion of Iron

Iron is often covered with a coat of paint or another metal such as tin or zinc to protect its surface against corrosion. Covering the surface with paint or tin is simply a means of preventing oxygen and water from reaching the iron surface. If the coating is broken and the iron is exposed to oxygen and water, corrosion will begin.

Galvanised iron, which is iron coated with a thin layer of zinc, uses the principles of electrochemistry to protect the iron from corrosion even after the surface coat is broken. The standard reduction potentials for iron and zinc are

$$Fe^{2+}(aq) + 2\,e^- \longrightarrow Fe(s) \qquad E^{\circ}_{red} = -0.44\ V$$

$$Zn^{2+}(aq) + 2\,e^- \longrightarrow Zn(s) \qquad E^{\circ}_{red} = -0.76\ V$$

Because the E°_{red} value for the reduction of Fe^{2+} is higher than that for the reduction of Zn^{2+}, Fe^{2+} is easier to reduce than Zn^{2+}. Conversely, $Zn(s)$ is easier to oxidise than $Fe(s)$. Thus even if the zinc coating is broken and the galvanised iron is exposed to oxygen and water, the zinc, which is most easily oxidised, serves as the anode and is corroded instead of the iron. The iron serves as the cathode at which O_2 is reduced, as shown in ▼ FIGURE 19.22.

Protecting a metal from corrosion by making it the cathode in an electrochemical cell is known as **cathodic protection**. The metal that is oxidised while protecting the cathode is called the *sacrificial anode*. Underground pipelines are often protected against corrosion by making the pipeline the cathode of a voltaic

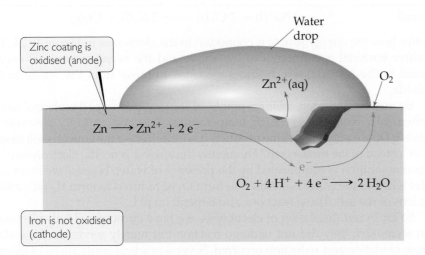

Zinc coating is oxidised (anode)

Water drop

$Zn^{2+}(aq)$

O_2

$Zn \longrightarrow Zn^{2+} + 2\,e^-$

e^-

$O_2 + 4\,H^+ + 4\,e^- \longrightarrow 2\,H_2O$

Iron is not oxidised (cathode)

◀ **FIGURE 19.22   Cathodic protection of iron in contact with zinc.** The standard reduction potentials are $E^{\circ}_{red,\,Fe^{2+}} = -0.440\ V$, $E^{\circ}_{red,\,Zn^{2+}} = -0.763\ V$, making the zinc more readily oxidised.

---

* Frequently, metal compounds obtained from aqueous solution have water associated with them. For example, copper(II) sulfate crystallises from water with five moles of water per mole of $CuSO_4$. We represent this formula as $CuSO_4{\cdot}5\,H_2O$. Such compounds are called hydrates (∞ Section 11.2, 'The Solution Process'). Rust is a hydrate of iron(III) oxide with a variable amount of water of hydration. We represent this variable water content by writing the formula as $Fe_2O_3{\cdot}xH_2O$.

Ground level | Soil electrolyte | Iron pipe, cathode

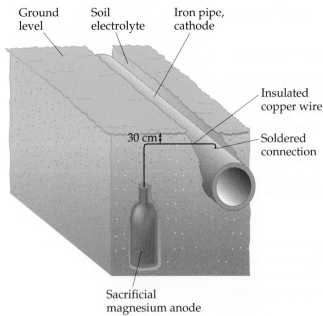

Insulated copper wire

30 cm

Soldered connection

Sacrificial magnesium anode

▲ **FIGURE 19.23** **Cathodic protection of an iron pipe.** A mixture of gypsum, sodium sulfate and clay surrounds the sacrificial magnesium anode to promote conductivity of ions.

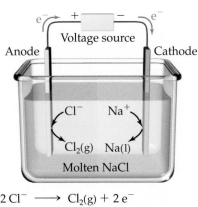

$2\,Cl^- \longrightarrow Cl_2(g) + 2\,e^-$

$2\,Na^+ + 2\,e^- \longrightarrow 2\,Na(l)$

▲ **FIGURE 19.24** **Electrolysis of molten sodium chloride.** Pure NaCl melts at 801 °C.

∞∞ Review this on page 405

cell. Pieces of an active metal such as magnesium are buried along the pipeline and connected to it by wire, as shown in ◀ **FIGURE 19.23**. In moist soil, where corrosion can occur, the active metal serves as the anode and provides cathodic protection for the pipes.

CONCEPT CHECK 8

Based on the standard reduction potentials in Table 19.1, which of the following metals could provide cathodic protection to iron: Al, Cu, Ni, Zn?

19.9 | ELECTROLYSIS

Voltaic cells are based on spontaneous redox reactions. It is also possible for *non-spontaneous* redox reactions to occur, however, by using electrical energy to drive them. For example, electricity can be used to decompose molten sodium chloride into its component elements Na and Cl_2. Such processes driven by an outside source of electrical energy are called **electrolysis reactions** and take place in **electrolytic cells**.

An electrolytic cell consists of two electrodes immersed either in a molten salt or in a solution. A battery or some other source of electrical energy acts as an electron pump, pushing electrons into one electrode and pulling them from the other. Just as in voltaic cells, the electrode at which reduction occurs is called the cathode, and the electrode at which oxidation occurs is called the anode.

In the electrolysis of molten NaCl, Na^+ ions pick up electrons and are reduced to Na at the cathode (◀ **FIGURE 19.24**). As Na^+ ions near the cathode are depleted, additional Na^+ ions migrate in. Similarly, there is net movement of Cl^- ions to the anode where they are oxidised. The electrode reactions for the electrolysis are

Cathode:	$2\,Na^+(l) + 2\,e^- \longrightarrow 2\,Na(l)$
Anode:	$2\,Cl^-(l) \longrightarrow Cl_2(g) + 2\,e^-$
Overall:	$2\,Na^+(l) + 2\,Cl^-(l) \longrightarrow 2\,Na(l) + Cl_2(g)$

Notice how the energy source is connected to the electrodes in Figure 19.24. The positive terminal is connected to the anode and the negative terminal is connected to the cathode, which forces electrons to move from the anode to the cathode.

Because of the high melting points of ionic substances, the electrolysis of molten salts requires very high temperatures (∞∞ Section 11.8, 'Structure of Solids'). Do we obtain the same products if we electrolyse the aqueous solution of a salt instead of the molten salt? Frequently the answer is no: the electrolysis of an aqueous solution is complicated by the presence of water, because we must consider whether the water is oxidised to form O_2 or reduced to form H_2 rather than the ions of the salt. These reactions also depend on pH.

So far in our discussion of electrolysis, we have encountered only electrodes that were *inert*; they did not undergo reaction but merely served as the surface where oxidation and reduction occurred. Several practical applications of electrochemistry, however, are based on *active* electrodes—electrodes that participate in the electrolysis process. *Electroplating*, for example, uses electrolysis to deposit a thin layer of one metal on another metal in order to improve beauty or resistance to corrosion. We can illustrate the principles of electrolysis with active electrodes by describing how to electroplate nickel on a piece of steel.

▶ **FIGURE 19.25** illustrates the electrolytic cell for our electroplating experiment. The anode of the cell is a strip of nickel metal, and the cathode is the piece of steel that will be electroplated. The electrodes are immersed in a solution of $NiSO_4(aq)$. What happens at the electrodes when the external voltage source is turned on? Reduction will occur at the cathode. The standard reduction potential of Ni^{2+} ($E^\circ_{red} = -0.28$ V) is less negative than that of H_2O ($E^\circ_{red} = -0.83$ V), so Ni^{2+} will be preferentially reduced at the cathode.

At the anode we need to consider which substances can be oxidised. For the $NiSO_4(aq)$ solution, only the H_2O solvent is readily oxidised because neither Na^+ nor SO_4^{2-} can be oxidised (both already have their elements in their highest possible oxidation state). The Ni atoms in the anode, however, can undergo oxidation. Thus the two possible oxidation processes are

$$2\,H_2O(l) \longrightarrow O_2(g) + 4\,H^+(aq) + 4\,e^- \qquad E^\circ_{red} = +1.23\ V$$

$$Ni(s) \longrightarrow Ni^{2+}(aq) + 2\,e^- \qquad E^\circ_{red} = -0.28\ V$$

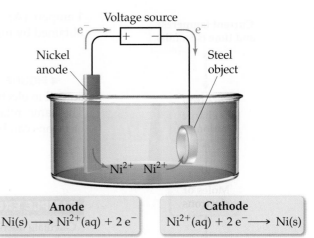

Anode	Cathode
$Ni(s) \longrightarrow Ni^{2+}(aq) + 2\,e^-$	$Ni^{2+}(aq) + 2\,e^- \longrightarrow Ni(s)$

▲ **FIGURE 19.25   Electrolytic cell with an active metal electrode.** Nickel dissolves from the anode to form Ni^{2+}(aq). At the cathode Ni^{2+}(aq) is reduced and forms a nickel 'plate' on the steel cathode.

where the potentials are the standard *reduction* potentials of these reactions. Because this is an oxidation, the half-reaction with the more negative value of E°_{red} is favoured. We therefore expect Ni(s) to be oxidised at the anode. We can summarise the electrode reactions as

Cathode (steel strip): $Ni^{2+}(aq) + 2\,e^- \longrightarrow Ni(s)$ $E^\circ_{red} = -0.28$ V

Anode (nickel strip): $Ni(s) \longrightarrow Ni^{2+}(aq) + 2\,e$ $E^\circ_{red} = -0.28$ V

If we look at the overall reaction, it appears as if nothing has been accomplished. During the electrolysis, however, we are transferring Ni atoms from the Ni anode to the steel cathode, plating the steel electrode with a thin layer of nickel atoms. The standard emf for the overall reaction is $E^\circ_{cell} = E^\circ_{red}$ (cathode) $- E^\circ_{red}$ (anode) $= 0$ V. Only a small emf is needed to provide the 'push' to transfer the nickel atoms from one electrode to the other.

Quantitative Aspects of Electrolysis

The stoichiometry of a half-reaction shows how many electrons are needed to achieve an electrolytic process. For example, the reduction of Na^+ to Na is a one-electron process:

$$Na^+ + e^- \longrightarrow Na$$

Thus one mole of electrons will plate out 1 mol of Na metal, two moles of electrons will plate out 2 mol of Na metal and so forth. Similarly, two moles of electrons are required to produce 1 mol of copper from Cu^{2+}, and three moles of electrons are required to produce 1 mol of aluminium from Al^{3+}.

$$Cu^{2+} + 2\,e^- \longrightarrow Cu$$

$$Al^{3+} + 3\,e^- \longrightarrow Al$$

For any half-reaction, the amount of a substance that is reduced or oxidised in an electrolytic cell is directly proportional to the number of electrons passed into the cell.

The quantity of charge passing through an electrical circuit, such as that in an electrolytic cell, is generally measured in *coulombs*. As noted in Section 19.4, the charge on one mole of electrons is 96 485 C (1 faraday). A coulomb is the quantity of charge passing a point in a circuit in 1 s when the current is

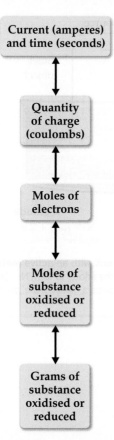

Current (amperes) and time (seconds)

↕

Quantity of charge (coulombs)

↕

Moles of electrons

↕

Moles of substance oxidised or reduced

↕

Grams of substance oxidised or reduced

▲ **FIGURE 19.26** **Relationship between charge and amount of reactant and product in electrolysis reactions.**

1 ampere (A). Therefore the number of coulombs passing through a cell can be obtained by multiplying the amperage and the elapsed time in seconds.

$$\text{Coulombs} = \text{amperes} \times \text{seconds} \qquad [19.19]$$

◀ **FIGURE 19.26** shows how the quantities of substances produced or consumed in electrolysis are related to the quantity of electrical charge that is used. The same relationships can also be applied to voltaic cells. In other words, electrons can be thought of as reagents in electrolysis reactions.

SAMPLE EXERCISE 19.14 **Aluminium electrolysis**

Calculate the number of grams of aluminium produced in 1.00 h by the electrolysis of molten $AlCl_3$ if the electrical current is 10.0 A.

SOLUTION

Analyse We are told that $AlCl_3$ is electrolysed to form Al and asked to calculate the number of grams of Al produced in 1.00 h with 10.0 A.

Plan Figure 19.26 provides a road map of the problem. First, the product of the amperage and the time in seconds gives the number of coulombs of electrical charge being used (Equation 19.19). Second, the coulombs can be converted with the Faraday constant ($F = 96\,485$ C mol⁻¹) to tell us the number of moles of electrons being supplied. Third, reduction of 1 mol of Al^{3+} to Al requires three moles of electrons. Hence we can use the number of moles of electrons to calculate the number of moles of Al metal it produces. Finally, we convert moles of Al into grams.

Solve First, we calculate the coulombs of electrical charge that are passed into the electrolytic cell:

$$\text{Coulombs} = \text{amperes} \times \text{seconds}$$
$$= (10.0 \text{ A})(1.0 \times 3600 \text{ s}) = 3.60 \times 10^4 \text{ C}$$

Second, we calculate the number of moles of electrons that pass into the cell:

$$\text{Moles e}^- = \left(\frac{3.60 \times 10^4 \text{ C}}{96\,485 \text{ C mol}^{-1}} \right) = 0.373 \text{ mol}$$

Third we relate the number of moles of electrons to the number of moles of aluminium being formed, using the half-reaction for the reduction of Al^{3+}:

$$Al^{3+} + 3\,e^- \longrightarrow Al$$

Thus three moles of electrons (3 F of electrical charge) are required to form 1 mol of Al:

$$\frac{\text{mol Al}}{\text{mol e}^-} = \frac{1}{3}$$

$$\text{moles Al} = \tfrac{1}{3}\text{mol e}-$$

$$= \tfrac{1}{3}(0.373 \text{ mol}) = 0.124 \text{ mol}$$

Finally, we convert moles to grams:

$$\text{Grams Al} = (0.124 \text{ mol})(27.0 \text{ g mol}^{-1}) = 3.36 \text{ g Al}$$

PRACTICE EXERCISE

(a) The half-reaction for formation of magnesium metal upon electrolysis of molten is $MgCl_2$ is $Mg^{2+} + 2\,e^- \longrightarrow Mg$. Calculate the mass of magnesium formed upon passage of a current of 60.0 A for a period of 4.00×10^3 s. **(b)** How many seconds would be required to produce 50.0 g of Mg from $MgCl_2$ if the current is 100.0 A?

Answers: **(a)** 30.2 g of Mg, **(b)** 3.97×10^3 s

(See also Exercise 19.72.)

SAMPLE INTEGRATIVE EXERCISE   **Putting concepts together**

The K_{sp} at 298 K for iron(II) fluoride is 2.4×10^{-6}. **(a)** Write a half-reaction that gives the likely products of the two-electron reduction of $FeF_2(s)$ in water. **(b)** Use the K_{sp} value and the standard reduction potential of $Fe^{2+}(aq)$ to calculate the standard reduction potential for the half-reaction in part (a). **(c)** Rationalise the difference in the reduction potential for the half-reaction in part (a) with that for $Fe^{2+}(aq)$.

SOLUTION

Analyse We are going to have to combine what we know about equilibrium constants and electrochemistry to obtain reduction potentials.

Plan For part (a) we need to determine which ion, Fe^{2+} or F^-, is more likely to be reduced by two electrons and write the overall reaction for $FeF_2 + 2\,e^- \longrightarrow$? For part (b) we need to write the K_{sp} reaction and manipulate it to get $E°$ for the reaction in part (a). For part (c) we need to see what we get for parts (a) and (b).

Solve

(a) Iron(II) fluoride is an ionic substance that consists of Fe^{2+} and F^- ions. We are asked to predict where two electrons could be added to FeF_2. We can't imagine adding the electrons to the F^- ions to form F^{3-}, so it seems likely that we can reduce the Fe^{2+} ions to $Fe(s)$. We therefore predict the half-reaction

$$FeF_2(s) + 2\,e^- \longrightarrow Fe(s) + 2\,F^-(aq)$$

(b) The K_{sp} value refers to the following equilibrium (∞ Section 15.4, 'The Change of Concentration with Time (Integrated Rate Equations)'):

∞ Review this on page 578

$$FeF_2(s) \rightleftharpoons Fe^{2+}(aq) + 2\,F^-(aq) \qquad K_{sp} = [Fe^{2+}][F^-]^2 = 2.4 \times 10^{-6}$$

We are asked to use the standard reduction potential of Fe^{2+}, whose half-reaction and standard voltage are listed in Appendix E:

$$Fe^{2+}(aq) + 2\,e^- \longrightarrow Fe(s) \qquad E° = -0.440 \text{ V}$$

Recall that according to Hess's law, we can add reactions to get the one we want and that we can add thermodynamic quantities like ΔH and ΔG to solve for the enthalpy or free energy of the reaction we want. In this case, notice that, if we add the K_{sp} reaction to the standard reduction half-reaction for Fe^{2+}, we get the half-reaction we want:

1.	$FeF_2(s) \longrightarrow Fe^{2+}(aq) + 2\,F^-(aq)$	
2.	$Fe^{2+}(aq) + 2\,e^- \longrightarrow Fe(s)$	
Overall: 3.	$FeF_2(s) + 2\,e^- \longrightarrow Fe(s) + 2\,F^-(aq)$	

Reaction 3 is still a half-reaction, so we do see the free electrons.

If we knew $\Delta G°$ for reactions 1 and 2, we could add them to get $\Delta G°$ for reaction 3. Recall that we can relate $\Delta G°$ to $E°$ by $\Delta G° = -nFE°$ and to K by $\Delta G° = -RT \ln K$. We know K for reaction 1: it is K_{sp}. We know $E°$ for reaction 2. Therefore, we can calculate $\Delta G°$ for reactions 1 and 2:

Reaction 1:

$$\Delta G° = -RT \ln K = -(8.314 \text{ J K}^{-1}\text{ mol}^{-1})(298 \text{ K})\ln(2.4 \times 10^{-6}) = 3.2 \times 10^4 \text{ J mol}^{-1}$$

Reaction 2:

$$\Delta G° = -nFE° = -(2 \text{ mol})(96\,485 \text{ C mol}^{-1})(-0.440 \text{ J C}^{-1}) = 8.49 \times 10^4 \text{ J}$$

(Recall that 1 volt is 1 joule per coulomb.)

Then, $\Delta G°$ for reaction 3, the one we want, is 3.2×10^4 J (for one mole of FeF_2) + 8.49×10^4 J $= 1.2 \times 10^5$ J. We can convert this to $E°$ easily from the relationship $\Delta G° = -nFE°$:

$$1.2 \times 10^5 \text{ J} = -(2 \text{ mol})(96\,485 \text{ C mol}^{-1})\,E°$$

$$E° = -0.61 \text{ J C}^{-1} = -0.61 \text{ V.}$$

(c) The standard reduction potential for FeF_2 (-0.61 V) is more negative than that for Fe^{2+} (-0.440 V), telling us that the reduction of FeF_2 is the less favourable process. When FeF_2 is reduced, we not only reduce the Fe^{2+} ions but also break up the ionic solid. Because this additional energy must be overcome, the reduction of FeF_2 is less favourable than the reduction of Fe^{2+}.

CHAPTER SUMMARY AND KEY TERMS

INTRODUCTION AND SECTION 19.1 In this chapter we have focused on **electrochemistry**, the branch of chemistry that relates electricity and chemical reactions. Electrochemistry involves oxidation–reduction reactions, also called redox reactions. These reactions involve a change in the oxidation state of one or more elements. In every oxidation–reduction reaction one substance is oxidised (its oxidation state, or number, increases) and one substance is reduced (its oxidation state, or number, decreases). The substance that is oxidised is referred to as a **reducing agent**, or **reductant**, because it causes the reduction of some other substance. Similarly, the substance that is reduced is referred to as an **oxidising agent**, or **oxidant**, because it causes the oxidation of some other substance.

SECTION 19.2 An oxidisation–reduction reaction can be balanced by dividing the reaction into two **half-reactions**, one for oxidation and one for reduction. A half-reaction is a balanced chemical equation that includes electrons. In oxidation half-reactions the electrons are on the product (right) side of the equation; we can envision that these electrons are transferred from a substance when it is oxidised. In reduction half-reactions the electrons are on the reactant (left) side of the equation. Each half-reaction is balanced separately, and the two are brought together with proper coefficients to balance the electrons on each side of the equation, so the electrons cancel when the half-reactions are added.

SECTION 19.3 A **voltaic** (or **galvanic**) **cell** uses a spontaneous oxidation–reduction reaction to generate electricity. In a voltaic cell the oxidation and reduction half-reactions often occur in separate half-cells. Each half-cell has a solid surface called an electrode, where the half-reaction occurs. The electrode where oxidation occurs is called the **anode**; reduction occurs at the **cathode**. The electrons released at the anode flow through an external circuit (where they do electrical work) to the cathode. Electrical neutrality in the solution is maintained by the migration of ions between the two half-cells through a device such as a salt bridge.

SECTION 19.4 A voltaic cell generates an **electromotive force** (emf) that moves the electrons from the anode to the cathode through the external circuit. The origin of emf is a difference in the electrical potential energy of the two electrodes in the cell. The emf of a cell is called its **cell potential**, E_{cell}, and is measured in volts ($1 \, V = 1 \, J \, C^{-1}$). The cell potential under standard conditions is called the **standard emf**, or the **standard cell potential**, and is denoted E°_{cell}.

A **standard reduction potential**, E°_{red}, can be assigned for an individual half-reaction. This is achieved by comparing the potential of the half-reaction to that of the **standard hydrogen electrode** (SHE), which is defined to have $E^{\circ}_{red} = 0 \, V$ and is based on the following half-reaction:

$$2 \, H^{+}(aq, 1 \, M) + 2 \, e^{-} \longrightarrow H_2(g, 1 \, bar) \quad E^{\circ}_{red} = 0 \, V$$

The standard cell potential of a voltaic cell is the difference between the standard reduction potentials of the half-reactions that occur at the cathode and the anode: $E^{\circ}_{cell} = E^{\circ}_{red} \text{(cathode)} - E^{\circ}_{red} \text{(anode)}$. The value of E°_{cell} is positive for a voltaic cell.

For a reduction half-reaction, E°_{red} is a measure of the tendency of the reduction to occur; the more positive the value for E°_{red}, the greater the tendency of the substance to be reduced.

Thus E°_{red} provides a measure of the oxidising strength of a substance. Substances that are strong oxidising agents produce products that are weak reducing agents, and *vice versa*.

SECTION 19.5 The emf, E, is related to the change in the Gibbs free energy, $\Delta G : \Delta G = -nFE$, where n is the number of electrons transferred during the redox process and F is **Faraday's constant**, defined as the quantity of electrical charge on one mole of electrons: $F = 96485 \, C \, mol^{-1}$. Because E is related to ΔG, the sign of E indicates whether a redox process is spontaneous: $E > 0$ indicates a spontaneous process and $E < 0$ indicates a non-spontaneous one. Because ΔG is also related to the equilibrium constant for a reaction ($\Delta G = -RT \ln K$), we can relate E to K as well.

The maximum amount of electrical work produced by a voltaic cell is given by the product of the total charge delivered, nF, and the emf, E: $w_{max} = -nFE$. The watt is a unit of power: $1 \, W = 1 \, J \, s^{-1}$. Electrical work is often measured in kilowatt-hours (kWh).

SECTION 19.6 The emf of a redox reaction varies with temperature and with the concentrations of reactants and products. The **Nernst equation** relates the emf under non-standard conditions to the standard emf and the reaction quotient Q:

$$E = E^{\circ} - (RT/nF) \ln Q = E^{\circ} - (0.0592/n) \log Q$$

The factor 0.0592 is valid when $T = 298 \, K$. A **concentration cell** is a voltaic cell in which the same half-reaction occurs at both the anode and cathode but with different concentrations of reactants in each half-cell. At equilibrium, $Q = K$ and $E = 0$.

SECTION 19.7 A **battery** is a self-contained electrochemical power source that contains one or more voltaic cells. Batteries are based on a variety of different redox reactions. Several common batteries were discussed. The lead-acid battery, the nickel-cadmium battery, the nickel-metal hydride battery and the lithium-ion battery are examples of rechargeable batteries. The common alkaline dry cell is not rechargeable. **Fuel cells** are voltaic cells that utilise redox reactions in which reactants such as H_2 have to be continuously supplied to the cell to generate voltage.

SECTION 19.8 Electrochemical principles help us understand **corrosion**, undesirable redox reactions in which a metal is attacked by some substance in its environment. The corrosion of iron into rust is caused by the presence of water and oxygen, and it is accelerated by the presence of electrolytes, such as road salt. The protection of a metal by putting it in contact with another metal that more readily undergoes oxidation is called **cathodic protection**. Galvanised iron, for example, is coated with a thin layer of zinc; because zinc is oxidised more readily than iron, the zinc serves as a sacrificial anode in the redox reaction.

SECTION 19.9 An **electrolysis reaction**, which is carried out in an **electrolytic cell**, employs an external source of electricity to drive a non-spontaneous electrochemical reaction. The current-carrying medium within an electrolytic cell may be either a molten salt or an electrolyte solution. The products of electrolysis can generally be predicted by comparing the reduction potentials associated with possible oxidation and reduction processes. The electrodes in an electrolytic cell can be active, meaning that the electrode can be involved in the electrolysis reaction. Active electrodes are important in electroplating and in metallurgical processes.

The quantity of substances formed during electrolysis can be calculated by considering the number of electrons involved in the redox reaction and the amount of electrical charge that passes into the cell. The amount of electrical charge is measured in coulombs and is related to the magnitude of the current and the time it flows ($1 C = 1 A s$).

KEY SKILLS

- Identify oxidation, reduction, oxidising agent and reducing agent in a chemical equation. (Section 19.1)
- Complete and balance redox equations using the method of half-reactions. (Section 19.2)
- Sketch a voltaic cell and identify its cathode, anode and the directions that electrons and ions move. (Section 19.3)
- Calculate standard emfs (cell potentials), E_{cell}°, from standard reduction potentials. (Section 19.4)
- Use reduction potentials to predict whether a redox reaction is spontaneous. (Section 19.4)
- Relate E_{cell}° to ΔG° and equilibrium constants. (Section 19.5)
- Calculate emf under non-standard conditions. (Section 19.6)
- Describe the components of common batteries and fuel cells. (Section 19.7)
- Explain how corrosion occurs and how it is prevented by cathodic protection. (Section 19.8)
- Describe the reactions in electrolytic cells. (Section 19.9)
- Relate amounts of products and reactants in redox reactions to electrical charge. (Section 19.9)

KEY EQUATIONS

- The electrochemical potential of a cell under standard conditions

$$E_{cell}^{\circ} = E_{red}^{\circ} \text{ (cathode)} - E_{red}^{\circ} \text{ (anode)} \qquad [19.8]$$

- The standard free energy change of a cell at standard conditions

$$\Delta G^{\circ} = -nFE^{\circ} \qquad [19.12]$$

- The simplified form of the Nernst equation

$$E = E^{\circ} - \frac{0.0592}{n} \log Q \quad \text{at 298 K} \qquad [19.17]$$

EXERCISES

VISUALISING CONCEPTS

19.1 In the Brønsted–Lowry concept of acids and bases, acid–base reactions are viewed as proton-transfer reactions. The stronger the acid, the weaker is its conjugate base. In what ways are redox reactions analogous? [Sections 19.1 and 19.2]

19.2 You may have heard that 'antioxidants' are good for your health. Based on what you have learned in this chapter, what do you deduce an 'antioxidant' is? [Sections 19.1 and 19.2]

19.3 The diagram that follows represents a molecular view of a process occurring at an electrode in a voltaic cell.

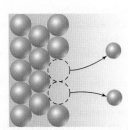

(a) Does the process represent oxidation or reduction? **(b)** Is the electrode the anode or cathode? **(c)** Why are the atoms in the electrode represented by larger spheres than the ions in the solution? [Section 19.3]

19.4 Assume that you want to construct a voltaic cell that uses the following half-reactions:

$$A^{2+}(aq) + 2\,e^- \longrightarrow A(s) \quad E^\circ_{red} = -0.10\ V$$
$$B^{2+}(aq) + 2\,e^- \longrightarrow B(s) \quad E^\circ_{red} = -1.10\ V$$

You begin with the incomplete cell pictured here in which the electrodes are immersed in water.

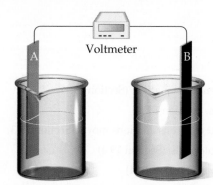

(a) What additions must you make to the cell for it to generate a standard emf? **(b)** Which electrode functions as the cathode? **(c)** Which direction do electrons move through the external circuit? **(d)** What voltage will the cell generate under standard conditions? [Sections 19.3 and 19.4]

19.5 For a spontaneous reaction $A(aq) + B(aq) \longrightarrow A^-(aq) + B^+(aq)$, answer the following questions:

(a) If you made a voltaic cell out of this reaction, what half-reaction would be occurring at the cathode, and what half-reaction would be occurring at the anode?
(b) Which half-reaction from (a) is higher in potential energy?
(c) What is the sign of E°_{cell}? [Section 19.3]

19.6 Consider the following table of standard electrode potentials for a series of hypothetical reactions in aqueous solution:

Reduction half-reaction	E°(V)
$A^+(aq) + e^- \longrightarrow A(s)$	1.33
$B^{2+}(aq) + 2\,e^- \longrightarrow B(s)$	0.87
$C^{3+}(aq) + e^- \longrightarrow C^{2+}(aq)$	-0.12
$D^{3+}(aq) + 3\,e^- \longrightarrow D(s)$	-1.59

(a) Which substance is the strongest oxidising agent? Which is weakest?
(b) Which substance is the strongest reducing agent? Which is weakest?
(c) Which substance(s) can oxidise C^{2+}? [Sections 19.4 and 19.5]

19.7 Consider a redox reaction for which E° is a negative number.
(a) What is the sign of ΔG° for the reaction?
(b) Will the equilibrium constant for the reaction be larger or smaller than 1?
(c) Can an electrochemical cell based on this reaction accomplish work on its surroundings? [Section 19.5]

19.8 Consider the following voltaic cell:

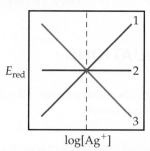

(a) Which electrode is the cathode?
(b) How would you determine the standard emf generated by this cell?
(c) What is the change in the cell voltage when the ion concentrations in the cathode half-cell are increased by a factor of 10?
(d) What is the change in the cell voltage when the ion concentrations in the anode half-cell are increased by a factor of 10? [Sections 19.4 and 19.6]

19.9 Consider the half-reaction $Ag^+(aq) + e^- \longrightarrow Ag(s)$.
(a) Which of the lines in the following diagram indicates how the reduction potential varies with the concentration of Ag^+? **(b)** What is the value of E_{red} when $\log[Ag^+] = 0$? [Section 19.6]

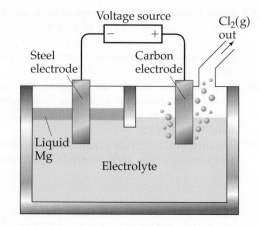

19.10 Draw a generic picture of a fuel cell. What is the main difference between it and a battery, regardless of the redox reactions that occur inside? [Section 19.7]

19.11 How does a zinc coating on iron protect the iron from unwanted oxidation? [Section 19.8]

19.12 Magnesium is produced commercially by electrolysis from a molten salt using a cell similar to the one shown here. **(a)** What salt is used as the electrolyte? **(b)** Which electrode is the anode, and which one is the cathode? **(c)** Write the overall cell reaction and individual half-reactions. **(d)** What precautions would need to be taken with respect to the magnesium formed? [Section 19.9]

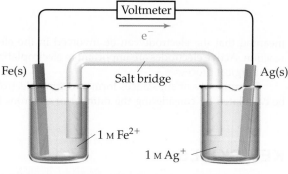

OXIDATION–REDUCTION REACTIONS (Section 19.1)

19.13 **(a)** What is meant by the term *oxidation*? **(b)** On which side of an oxidation half-reaction do the electrons appear? **(c)** What is meant by the term *oxidant*? **(d)** What is meant by the term *oxidising agent*?

19.14 **(a)** What is meant by the term *reduction*? **(b)** On which side of a reduction half-reaction do the electrons appear? **(c)** What is meant by the term *reductant*? **(d)** What is meant by the term *reducing agent*?

19.15 Indicate whether each of the following statements is true or false.
(a) If something is oxidised, it is formally losing electrons.
(b) For the reaction $Fe^{3+}(aq) + Co^{2+}(aq) \longrightarrow Fe^{2+}(aq) + Co^{3+}(aq)$, $Fe^{3+}(aq)$ is the reducing agent and $Co^{2+}(aq)$ is the oxidising agent.

(c) If there are no changes in the oxidation state of the reactants or products of a particular reaction, that reaction is not a redox reaction.

19.16 In each of the following balanced oxidation–reduction equations, identify those elements that undergo changes in oxidation number and indicate the magnitude of the change in each case.
(a) $I_2O_5(s) + 5\,CO(g) \longrightarrow I_2(s) + 5\,CO_2(g)$
(b) $2\,Hg^{2+}(aq) + N_2H_4(aq) \longrightarrow$
$$2\,Hg(l) + N_2(g) + 4\,H^+(aq)$$
(c) $3\,H_2S(aq) + 2\,H^+(aq) + 2\,NO_3^-(aq) \longrightarrow$
$$3\,S(s) + 2\,NO(g) + 4\,H_2O(l)$$
(d) $Ba^{2+}(aq) + 2\,OH^-(aq) + H_2O_2(aq) + 2\,ClO_2(aq)$
$$\longrightarrow Ba(ClO_2)_2(s) + 2\,H_2O(l) + O_2(g)$$

BALANCING OXIDATION–REDUCTION REACTIONS (Section 19.2)

19.17 At 900 °C titanium tetrachloride vapour reacts with molten magnesium metal to form solid titanium metal and molten magnesium chloride. **(a)** Write a balanced equation for this reaction. **(b)** What is being oxidised, and what is being reduced? **(c)** Which substance is the reductant and which is the oxidant?

19.18 Hydrazine (N_2H_4) and dinitrogen tetroxide (N_2O_4) form a self-igniting mixture that has been used as a rocket propellant. The reaction products are N_2 and H_2O. **(a)** Write a balanced chemical equation for this reaction. **(b)** What is being oxidised, and what is being reduced? **(c)** Which substance serves as the reducing agent and which as the oxidising agent?

19.19 Complete and balance the following half-reactions. In each case indicate whether the half-reaction is an oxidation or a reduction.
(a) $Sn^{2+}(aq) \longrightarrow Sn^{4+}(aq)$ (acidic or basic solution)
(b) $TiO_2(s) \longrightarrow Ti^{2+}(aq)$ (acidic solution)
(c) $ClO_3^-(aq) \longrightarrow Cl^-(aq)$ (acidic solution)

(d) $OH^-(aq) \longrightarrow O_2(g)$ (basic solution)
(e) $SO_3^{2-}(aq) \longrightarrow SO_4^{2-}(aq)$ (basic solution)
(f) $N_2(g) \longrightarrow NH_4^+(aq)$ (acidic solution)
(g) $N_2(g) \longrightarrow NH_3(g)$ (basic solution)

19.20 Complete and balance the following equations, and identify the oxidising and reducing agents.
(a) $Cr_2O_7^{2-}(aq) + I^-(aq) \longrightarrow Cr^{3+}(aq) + IO_3^-(aq)$
(acidic solution)
(b) $MnO_4^-(aq) + CH_3OH(aq) \longrightarrow$
$$Mn^{2+}(aq) + HCO_2H(aq)$$ (acidic solution)
(c) $I_2(s) + OCl^-(aq) \longrightarrow IO_3^-(aq) + Cl^-(aq)$
(acidic solution)
(d) $As_2O_3(s) + NO_3^-(aq) \longrightarrow H_3AsO_4(aq) + N_2O_3(aq)$
(acidic solution)
(e) $MnO_4^-(aq) + Br^-(aq) \longrightarrow MnO_2(s) + BrO_3^-(aq)$
(basic solution)
(f) $Pb(OH)_4^{2-}(aq) + ClO^-(aq) \longrightarrow PbO_2(s) + Cl^-(aq)$
(basic solution)

VOLTAIC CELLS (Section 19.3)

19.21 **(a)** What are the similarities and differences between Figure 19.3 and Figure 19.4? **(b)** Why are Na^+ ions drawn into the cathode compartment as the voltaic cell shown in Figure 19.5 operates?

19.22 A voltaic cell similar to that shown in Figure 19.5 is constructed. One electrode compartment consists of a silver strip placed in a solution of $AgNO_3$, and the other has an iron strip placed in a solution of $FeCl_2$. The overall cell reaction is

$$Fe(s) + 2\,Ag^+(aq) \longrightarrow Fe^{2+}(aq) + 2\,Ag(s)$$

(a) What is being oxidised and what is being reduced? **(b)** Write the half-reactions that occur in the two electrode compartments. **(c)** Which electrode is the anode, and which is the cathode? **(d)** Indicate the signs of the electrodes. **(e)** Do electrons flow from the silver electrode to the iron electrode, or from the iron to the silver? **(f)** In which directions do the cations and anions migrate through the solution?

19.23 A voltaic cell similar to that shown in Figure 19.5 is constructed. One electrode compartment consists of an aluminium strip placed in a solution of $Al(NO_3)_3$, and the other has a nickel strip placed in a solution of $NiSO_4$. The overall cell reaction is

$$2\,Al(s) + 3\,Ni^{2+}(aq) \longrightarrow 2\,Al^{3+}(aq) + 3\,Ni(s)$$

(a) What is being oxidised and what is being reduced? **(b)** Write the half-reactions that occur in the two electrode compartments. **(c)** Which electrode is the anode, and which is the cathode? **(d)** Indicate the signs of the electrodes. **(e)** Do electrons flow from the aluminium electrode to the nickel electrode, or from the nickel to the aluminium? **(f)** In which directions do the cations and anions migrate through the solution? Assume the Al is not coated with its oxide.

CELL POTENTIALS UNDER STANDARD CONDITIONS (Section 19.4)

19.24 **(a)** What is meant by the term *electromotive force*? **(b)** What is the definition of the *volt*? **(c)** What is meant by the term *cell potential*?

19.25 **(a)** Which electrode of a voltaic cell, the cathode or the anode, corresponds to the higher potential energy for the electrons? **(b)** What are the units for electrical potential? How does this unit relate to energy expressed in joules? **(c)** What is special about a *standard* cell potential?

19.26 **(a)** Write the half-reaction that occurs at a hydrogen electrode in acidic aqueous solution when it serves as the cathode of a voltaic cell. **(b)** What is *standard* about the standard hydrogen electrode? **(c)** What is the role of the platinum foil in a standard hydrogen electrode?

19.27 **(a)** Write the half-reaction that occurs at a hydrogen electrode in acidic aqueous solution when it serves as the anode of a voltaic cell. **(b)** The platinum electrode in a standard hydrogen electrode is specially prepared to have a large surface area. Why is this important? **(c)** Sketch a standard hydrogen electrode.

19.28 **(a)** What is a *standard reduction potential*? **(b)** What is the standard reduction potential of a standard hydrogen electrode? **(c)** Based on the standard reduction potentials listed in Appendix E, which is the more favourable process, the reduction of $Ag^+(aq)$ to $Ag(s)$ or the reduction of $Sn^{2+}(aq)$ to $Sn(s)$?

19.29 **(a)** Why is it impossible to measure the standard reduction potential of a single half-reaction? **(b)** Describe how the standard reduction potential of a half-reaction can be determined. **(c)** By using data in Appendix E, determine which is the more unfavourable reduction, $Cd^{2+}(aq)$ to $Cd(s)$ or $Ca^{2+}(aq)$ to $Ca(s)$.

19.30 A voltaic cell that uses the reaction

$$Tl^{3+}(aq) + 2\,Cr^{2+}(aq) \longrightarrow Tl^+(aq) + 2\,Cr^{3+}(aq)$$

has a measured standard cell potential of +1.19 V. **(a)** Write the two half-cell reactions. **(b)** By using data from Appendix E, determine $E°_{red}$ for the reduction of $Tl^{3+}(aq)$ to $Tl^+(aq)$. **(c)** Sketch the voltaic cell, label the anode and cathode and indicate the direction of electron flow.

19.31 Using standard reduction potentials (Appendix E), calculate the standard emf for each of the following reactions:

(a) $Cl_2(g) + 2\,I^-(aq) \longrightarrow 2\,Cl^-(aq) + I_2(s)$
(b) $Ni(s) + 2\,Ce^{4+}(aq) \longrightarrow Ni^{2+}(aq) + 2\,Ce^{3+}(aq)$
(c) $Fe(s) + 2\,Fe^{3+}(aq) \longrightarrow 3\,Fe^{2+}(aq)$
(d) $2\,Al^{3+}(aq) + 3\,Ca(s) \longrightarrow 2\,Al(s) + 3\,Ca^{2+}(aq)$

19.32 The standard reduction potentials of the following half-reactions are given in Appendix E:

$$Ag^+(aq) + e^- \longrightarrow Ag(s)$$
$$Cu^{2+}(aq) + 2\,e^- \longrightarrow Cu(s)$$
$$Ni^{2+}(aq) + 2\,e^- \longrightarrow Ni(s)$$
$$Cr^{3+}(aq) + 3\,e^- \longrightarrow Cr(s)$$

(a) Determine which combination of these half-cell reactions leads to the cell reaction with the largest positive cell emf, and calculate the value. **(b)** Determine which combination of these half-cell reactions leads to the cell reaction with the smallest positive cell emf, and calculate the value.

19.33 Given the following half-reactions and associated standard reduction potentials:

$$AuBr_4^-(aq) + 3\,e^- \longrightarrow Au(s) + 4\,Br^-(aq)$$
$$E°_{red} = -0.858\ V$$
$$Eu^{3+}(aq) + e^- \longrightarrow Eu^{2+}(aq)$$
$$E°_{red} = -0.43\ V$$
$$IO^-(aq) + H_2O(l) + 2\,e^- \longrightarrow I^-(aq) + 2\,OH^-(aq)$$
$$E°_{red} = +0.49\ V$$
$$Sn^{2+}(aq) + 2\,e^- \longrightarrow Sn(s)$$
$$E°_{red} = -0.1364\ V$$

(a) Write the cell reaction for the combination of these half-cell reactions that leads to the largest positive cell emf, and calculate the value. **(b)** Write the cell reaction for the combination of half-cell reactions that leads to the smallest positive cell emf, and calculate that value.

STRENGTHS OF OXIDISING AND REDUCING AGENTS (Section 19.4)

19.34 From each of the following pairs of substances, use data in Appendix E to choose the one that is the stronger oxidising agent:
(a) $Cl_2(g)$ or $Br_2(l)$
(b) $Ni^{2+}(aq)$ or $Cd^{2+}(aq)$
(c) $BrO_3^-(aq)$ or $IO_3^-(aq)$
(d) $H_2O_2(aq)$ or $O_3(g)$

19.35 From each of the following pairs of substances, use data in Appendix E to choose the one that is the stronger reducing agent:
(a) $Fe(s)$ or $Mg(s)$
(b) $Ca(s)$ or $Al(s)$
(c) $H_2(g, \text{acidic solution})$ or $H_2S(g)$
(d) $H_2SO_3(aq)$ or $H_2C_2O_4(aq)$

19.36 By using the data in Appendix E, determine whether each of the following substances is likely to serve as an oxidant or a reductant: **(a)** $Cl_2(g)$, **(b)** MnO_4^- (aq, acidic solution), **(c)** $Ba(s)$, **(d)** $Zn(s)$.

19.37 Is each of the following substances likely to serve as an oxidant or a reductant: **(a)** $Na(s)$, **(b)** $O_3(g)$, **(c)** $Ce^{3+}(aq)$, **(d)** $Sn^{2+}(aq)$?

19.38 **(a)** Assuming standard conditions, arrange the following in order of increasing strength as oxidising agents in acidic solution: $Cr_2O_7^{2-}$, H_2O_2, Cu^{2+}, Cl_2, O_2. **(b)** Arrange the following in order of increasing strength as reducing agents in acidic solution: Zn, I^-, Sn^{2+}, H_2O_2, Al.

19.39 The standard reduction potential for the reduction of $Eu^{3+}(aq)$ to $Eu^{2+}(aq)$ is -0.43 V. Using Appendix E, which of the following substances is capable of reducing $Eu^{3+}(aq)$ to $Eu^{2+}(aq)$ under standard conditions: Al, Co, H_2O_2, $N_2H_5^+$, $H_2C_2O_4$?

FREE ENERGY AND REDOX REACTIONS (Section 19.5)

19.40 Given the following reduction half-reactions:

$$Fe^{3+}(aq) + e^- \longrightarrow Fe^{2+}(aq)$$
$$E_{red}^° = +0.77 \text{ V}$$

$$S_2O_6{}^{2-}(aq) + 4\,H^+(aq) + 2\,e^- \longrightarrow 2\,H_2SO_3(aq)$$
$$E_{red}^° = +0.60 \text{ V}$$

$$N_2O(aq) + 2\,H^+(aq) + 2\,e^- \longrightarrow N_2(g) + H_2O(l)$$
$$E_{red}^° = -1.77 \text{ V}$$

$$VO_2{}^+(aq) + 2\,H^+(aq) + e^- \longrightarrow VO^{2+}(aq) + H_2O(l)$$
$$E_{red}^° = +1.00 \text{ V}$$

(a) Write balanced chemical equations for the oxidation of $Fe^{2+}(aq)$ by $S_2O_6{}^{2-}(aq)$, by $N_2O(aq)$ and by $VO_2{}^+(aq)$. **(b)** Calculate $\Delta G^°$ for each reaction at 298 K. **(c)** Calculate the equilibrium constant, K, for each reaction at 298 K.

19.41 If the equilibrium constant for a two-electron redox reaction at 298 K is 1.5×10^{-4}, calculate the corresponding $\Delta G^°$ and $E_{cell}^°$ under standard conditions.

19.42 If the equilibrium constant for a one-electron redox reaction at 298 K is 3.7×10^6, calculate the corresponding $\Delta G^°$ and $E_{cell}^°$ under standard conditions.

19.43 Using the standard reduction potentials listed in Appendix E, calculate the equilibrium constant for each of the following reactions at 298 K:
(a) $Fe(s) + Ni^{2+}(aq) \longrightarrow Fe^{2+}(aq) + Ni(s)$
(b) $Co(s) + 2\,H^+(aq) \longrightarrow Co^{2+}(aq) + H_2(g)$
(c) $10\,Br^-(aq) + 2\,MnO_4{}^-(aq) + 16\,H^+(aq) \longrightarrow$
$$2\,Mn^{2+}(aq) + 8\,H_2O(l) + 5\,Br_2(l)$$

19.44 A cell has a standard emf of $+0.177$ V at 298 K. What is the value of the equilibrium constant for the cell reaction **(a)** if $n = 1$, **(b)** if $n = 2$, **(c)** if $n = 3$?

19.45 At 298 K a cell reaction has a standard emf of $+0.17$ V. The equilibrium constant for the cell reaction is 5.5×10^5. What is the value of n for the cell reaction?

CELL POTENTIALS UNDER NON-STANDARD CONDITIONS (Section 19.6)

19.46 **(a)** Under what circumstances is the Nernst equation applicable? **(b)** What is the numerical value of the reaction quotient, Q, under standard conditions? **(c)** What happens to the emf of a cell if the concentrations of the reactants are increased?

19.47 **(a)** A voltaic cell is constructed with all reactants and products in their standard states. Will this condition hold as the cell operates? Explain. **(b)** Can the Nernst equation be used at temperatures other than room temperature? Explain. **(c)** What happens to the emf of a cell if the concentrations of the products are increased?

19.48 What is the effect on the emf of the cell shown in Figure 19.9, which has the overall reaction $Zn(s) + 2\,H^+(aq) \longrightarrow Zn^{2+}(aq) + H_2(g)$, for each of the following changes? **(a)** The pressure of the H_2 gas is increased in the cathode compartment. **(b)** Zinc nitrate is added to the anode compartment. **(c)** Sodium hydroxide is added to the cathode compartment, decreasing $[H^+]$. **(d)** The surface area of the anode is doubled.

19.49 A voltaic cell utilises the following reaction:

$$Al(s) + 3\,Ag^+(aq) \longrightarrow Al^{3+}(aq) + 3\,Ag(s)$$

What is the effect on the cell emf of each of the following changes? **(a)** Water is added to the anode compartment, diluting the solution. **(b)** The size of the aluminium electrode is increased. **(c)** A solution of $AgNO_3$ is added to the cathode compartment, increasing the quantity of Ag^+ but not changing its concentration. **(d)** HCl is added to the $AgNO_3$ solution, precipitating some of the Ag^+ as AgCl.

19.50 A voltaic cell utilises the following reaction:

$$4\,Fe^{2+}(aq) + O_2(g) + 4\,H^+(g) \longrightarrow 4\,Fe^{3+}(aq) + 2\,H_2O(l)$$

(a) What is the emf of this cell under standard conditions? **(b)** What is the emf of this cell when $[Fe^{2+}] = 1.3$ M, $[Fe^{3+}] = 0.010$ M, $P_{O_2} = 0.51$ bar and the pH of the solution in the cathode compartment is 3.50?

19.51 A voltaic cell utilises the following reaction:

$$2\,Fe^{3+}(aq) + H_2(g) \longrightarrow 2\,Fe^{2+}(aq) + 2\,H^+(aq)$$

(a) What is the emf of this cell under standard conditions? **(b)** What is the emf for this cell when $[Fe^{3+}] = 1.50$ M, $P_{H_2} = 0.050$ bar, $[Fe^{2+}] = 0.0010$ M and the pH in both compartments is 5.00?

19.52 A voltaic cell is constructed with two Zn^{2+}/Zn electrodes. The two cell compartments have $[Zn^{2+}] = 1.8$ M and $[Zn^{2+}] = 1.00 \times 10^{-2}$ M, respectively. **(a)** Which electrode is the anode of the cell? **(b)** What is the standard emf of the cell? **(c)** What is the cell emf for the concentrations given? **(d)** For each electrode, predict whether $[Zn^{2+}]$ will increase, decrease or stay the same as the cell operates.

19.53 A voltaic cell is constructed with two silver–silver chloride electrodes, each of which is based on the following half-reaction:

$$AgCl(s) + e^- \longrightarrow Ag(s) + Cl^-(aq)$$

The two cell compartments have $[Cl^-] = 0.0150$ M and $[Cl^-] = 2.55$ M, respectively. **(a)** Which electrode is the cathode of the cell? **(b)** What is the standard emf of the cell? **(c)** What is the cell emf for the concentrations given? **(d)** For each electrode, predict whether $[Cl^-]$ will increase, decrease or stay the same as the cell operates.

19.54 A voltaic cell is constructed that is based on the following reaction:

$$Sn^{2+}(aq) + Pb(s) \longrightarrow Sn(s) + Pb^{2+}(aq)$$

(a) If the concentration of Sn^{2+} in the cathode compartment is 1.00 M and the cell generates an emf of +0.22 V, what is the concentration of Pb^{2+} in the anode com-

partment? **(b)** If the anode compartment contains $[SO_4^{2-}]$ = 1.00 M in equilibrium with $PbSO_4(s)$, what is the K_{sp} of $PbSO_4$?

BATTERIES AND FUEL CELLS (Section 19.7)

19.55 **(a)** What happens to the emf of a battery as it is used? Why does this happen? **(b)** The AA size and D size alkaline batteries are both 1.5 V batteries that are based on the same electrode reactions. What is the major difference between the two batteries? What performance feature is most affected by this difference?

19.56 Suggest an explanation for why liquid water is needed in an alkaline battery.

19.57 During the discharge of an alkaline battery, 12.6 g of Zn is consumed at the anode of the battery. What mass of MnO_2 is reduced at the cathode during this discharge?

19.58 Mercuric oxide dry-cell batteries are often used where a high-energy density is required, such as in watches and cameras. The two half-cell reactions that occur in the battery are

$$HgO(s) + H_2O(l) + 2\,e^- \longrightarrow Hg(l) + 2\,OH^-(aq)$$
$$Zn(s) + 2\,OH^-(aq) \longrightarrow ZnO(s) + H_2O(l) + 2\,e^-$$

(a) Write the overall cell reaction. **(b)** The value of E°_{red} for the cathode reaction is +0.098 V. The overall cell potential is +1.35 V. Assuming that both half-cells operate under standard conditions, what is the standard reduction potential for the anode reaction? **(c)** Why is the potential of the anode reaction different than would be expected if the reaction occurred in an acidic medium?

19.59 **(a)** Suppose that an alkaline battery was manufactured using cadmium metal rather than zinc. What effect would this have on the cell emf? **(b)** What environmental advantage is provided by the use of nickel-metal hydride batteries over nickel-cadmium batteries?

19.60 The hydrogen–oxygen fuel cell has a standard emf of 1.23 V. What advantages and disadvantages are there to using this device as a source of power, compared with a 1.55 V alkaline battery?

19.61 Can the 'fuel' of a fuel cell be a solid? Explain.

CORROSION (Section 19.8)

19.62 **(a)** Write the anode and cathode reactions that cause the corrosion of iron metal to aqueous iron(II). **(b)** Write the balanced half-reactions involved in the air oxidation of $Fe^{2+}(aq)$ to $Fe_2O_3 \cdot 3\,H_2O$.

19.63 **(a)** Magnesium metal is used as a sacrificial anode to protect underground pipes from corrosion. Why is the magnesium referred to as a 'sacrificial anode'? **(b)** Looking in Appendix E, suggest what metal the underground pipes could be made from in order for magnesium to be successful as a sacrificial anode.

19.64 An iron object is plated with a coating of cobalt to protect against corrosion. Does the cobalt protect iron by cathodic protection? Explain.

19.65 A plumber's handbook states that you should not connect a copper pipe directly to a steel pipe because electrochemical reactions between the two metals will cause corrosion. The handbook recommends you use, instead, an insulating fitting to connect them. What spontaneous redox reaction(s) might cause the corrosion? Justify your answer with standard emf calculations.

ELECTROLYSIS (Section 19.9)

19.66 **(a)** What is *electrolysis*? **(b)** Are electrolysis reactions thermodynamically spontaneous? Explain. **(c)** What process occurs at the anode in the electrolysis of molten NaCl?

19.67 **(a)** What is an *electrolytic cell*? **(b)** The negative terminal of a voltage source is connected to an electrode of an electrolytic cell. Is the electrode the anode or the cathode of the cell? Explain. **(c)** The electrolysis of water is often done with a small amount of sulfuric acid added to the water. What is the role of the sulfuric acid?

19.68 **(a)** A $Cr^{3+}(aq)$ solution is electrolysed, using a current of 7.60 A. What mass of $Cr(s)$ is plated out after 2.00 days? **(b)** What amperage is required to plate out 0.250 mol Cr from a Cr^{3+} solution in a period of 8.00 h?

19.69 Metallic magnesium can be made by the electrolysis of molten $MgCl_2$. **(a)** What mass of Mg is formed by passing a current of 5.25 A through molten $MgCl_2$ for 2.50 days? **(b)** How many minutes are needed to plate out 10.00 g Mg from molten $MgCl_2$, using 3.50 A of current?

19.70 A voltaic cell is based on the reaction

$$Sn(s) + I_2(s) \longrightarrow Sn^{2+}(aq) + 2\,I^-(aq)$$

Under standard conditions, what is the maximum electrical work, in joules, that the cell can accomplish if 75.0 g of Sn is consumed?

19.71 Consider the voltaic cell illustrated in Figure 19.5, which is based on the cell reaction

$$Zn(s) + Cu^{2+}(aq) \longrightarrow Zn^{2+}(aq) + Cu(s)$$

Under standard conditions, what is the maximum electrical work, in joules, that the cell can accomplish if 50.0 g of copper is plated out?

19.72 **(a)** Calculate the mass of Li formed by electrolysis of molten LiCl by a current of 7.5×10^4 A flowing for a period of 24 h. Assume the electrolytic cell is 85% efficient. **(b)** What is the energy requirement for this electrolysis per mole of Li formed if the applied emf is +7.5 V?

19.73 Elemental calcium is produced by the electrolysis of molten $CaCl_2$. **(a)** What mass of calcium can be produced by this process if a current of 6.5×10^3 A is applied for 48 h? Assume that the electrolytic cell is 68% efficient.

ADDITIONAL EXERCISES

19.74 A *disproportionation* reaction is an oxidation–reduction reaction in which the same substance is oxidised and reduced. Complete and balance the following disproportionation reactions:
(a) $Ni^+(aq) \longrightarrow Ni^{2+}(aq) + Ni(s)$ (acidic solution)
(b) $MnO_4^{2-}(aq) \longrightarrow MnO_4^-(aq) + MnO_2(s)$ (acidic solution)
(c) $H_2SO_3(aq) \longrightarrow S(s) + HSO_4^-(aq)$ (acidic solution)
(d) $Cl_2(aq) \longrightarrow Cl^-(aq) + ClO^-(aq)$ (basic solution)

19.75 A common shorthand way to represent a voltaic cell is to list its components as follows:

anode|anode solution||cathode solution|cathode

A double vertical line represents a salt bridge or a porous barrier. A single vertical line represents a change in phase, such as from solid to solution. **(a)** Write the half-reactions and overall cell reaction represented by $Fe|Fe^{2+}||Ag^+|Ag$; sketch the cell. **(b)** Write the half-reactions and overall cell reaction represented by $Zn|Zn^{2+}||H^+|H_2$; sketch the cell. **(c)** Using the notation just described, represent a cell based on the following reaction

$$ClO_3^-(aq) + 3\,Cu(s) + 6\,H^+(aq) \longrightarrow$$
$$Cl^-(aq) + 3\,Cu^{2+}(aq) + 3\,H_2O(l)$$

Pt is used as an inert electrode in contact with the ClO_3^- and Cl^-. Sketch the cell.

19.76 Predict whether the following reactions will be spontaneous in acidic solution under standard conditions: **(a)** oxidation of Sn to Sn^{2+} by I_2 (to form I^-), **(b)** reduction of Ni^{2+} to Ni by I^- (to form I_2), **(c)** reduction of Ce^{4+} to Ce^{3+} by H_2O_2, **(d)** reduction of Cu^{2+} to Cu by Sn^{2+} (to form Sn^{4+}).

19.77 Gold exists in two common positive oxidation states, +1 and +3. The standard reduction potentials for these oxidation states are

$$Au^+(aq) + e^- \longrightarrow Au(s) \qquad E^\circ_{red} = +1.69\text{ V}$$
$$Au^{3+}(aq) + 3\,e^- \longrightarrow Au(s) \qquad E^\circ_{red} = +1.50\text{ V}$$

(a) Can you use these data to explain why gold does not tarnish in the air? **(b)** Suggest several substances that should be strong enough oxidising agents to oxidise gold metal. **(c)** Miners obtain gold by soaking gold-containing ores in an aqueous solution of sodium cyanide. A very soluble complex ion of gold forms in the aqueous solution as a result of the redox reaction

$$4\,Au(s) + 8\,NaCN(aq) + 2\,H_2O(l) + O_2(g) \longrightarrow$$
$$4\,Na[Au(CN)_2](aq) + 4\,NaOH(aq)$$

What is being oxidised, and what is being reduced, in this reaction? **(d)** Gold miners then react the basic aqueous product solution from part (c) with Zn dust to

get gold metal. Write a balanced redox reaction for this process. What is being oxidised, and what is being reduced?

19.78 A voltaic cell is constructed that uses the following half-cell reactions:

$$Cu^+(aq) + e^- \longrightarrow Cu(s)$$
$$I_2(s) + 2\,e^- \longrightarrow 2\,I^-(aq)$$

The cell is operated at 298 K with $[Cu^+] = 2.5$ M and $[I^-] = 3.5$ M. **(a)** Determine E for the cell at these concentrations. **(b)** Which electrode is the anode of the cell? **(c)** Is the answer to part (b) the same as it would be if the cell was operated under standard conditions? **(d)** If $[Cu^+]$ was equal to 1.4 M, at what concentration of I^- would the cell have zero potential?

19.79 Derive an equation that directly relates the standard emf of a redox reaction to its equilibrium constant.

19.80 Using data from Appendix E, calculate the equilibrium constant for the disproportionation of the copper(I) ion at room temperature: $2\,Cu^+(aq) \longrightarrow Cu^{2+}(aq) + Cu(s)$.

19.81 **(a)** Write the reactions for the discharge and charge of a nickel-cadmium rechargeable battery. **(b)** Given the following reduction potentials, calculate the standard emf of the cell:

$$Cd(OH)_2(s) + 2\,e^- \longrightarrow Cd(s) + 2\,OH^-(aq)$$
$$E^\circ_{red} = -0.76\text{ V}$$
$$NiO(OH)(s) + H_2O(l) + e^- \longrightarrow Ni(OH)_2(s) + OH(aq)$$
$$E^\circ_{red} = +0.49\text{ V}$$

(c) A typical nicad voltaic cell generates an emf of +1.30 V. Why is there a difference between this value and the one you calculated in part (b)? **(d)** Calculate the equilibrium constant for the overall nicad reaction based on this typical emf value.

19.82 If you were going to apply a small potential to a steel ship resting in the water as a means of inhibiting corrosion, would you apply a negative or a positive charge? Explain.

19.83 **(a)** How many coulombs are required to plate a layer of chromium metal 0.25 mm thick on a car bumper with a total area of 0.32 m^2 from a solution containing CrO_4^{2-}? The density of chromium metal is 7.20 g cm^{-3}. **(b)** What current flow is required for this electroplating if the bumper is to be plated in 10.0 s? **(c)** If the external source has an emf of +6.0 V and the electrolytic cell is 65% efficient, how much electrical power is expended to electroplate the bumper?

19.84 **(a)** What is the maximum amount of work that a 6 V lead-acid battery of a golf cart can accomplish if it is rated at 300 A h? **(b)** List some of the reasons why this amount of work is never realised.

(b) What is the total energy requirement for this electrolysis if the applied emf is +5.00 V?

INTEGRATIVE EXERCISES

19.85 Copper dissolves in concentrated nitric acid with the evolution of $NO(g)$, which is subsequently oxidised to $NO_2(g)$ in air. In contrast, copper does not dissolve in concentrated hydrochloric acid. Explain these observations by using standard reduction potentials from Table 19.1.

19.86 The Haber process is the principal industrial route for converting nitrogen into ammonia:

$$N_2(g) + 3\,H_2(g) \longrightarrow 2\,NH_3(g)$$

(a) What is being oxidised, and what is being reduced? **(b)** Using the thermodynamic data in Appendix C, calculate the equilibrium constant for the process at room temperature. **(c)** Calculate the standard emf of the Haber process at room temperature.

19.87 In a galvanic cell the cathode is an Ag^+ [1.00 M/Ag(s)] half-cell. The anode is a standard hydrogen electrode immersed in a buffer solution containing 0.10 M benzoic acid (C_6H_5COOH) and 0.050 M sodium benzoate ($C_6H_5COO^-Na^+$). The measured cell voltage is 1.030 V. What is the pK_a of benzoic acid?

19.88 Consider the general oxidation of a species A in solution: $A \longrightarrow A^+ + e^-$. The term 'oxidation potential' is sometimes used to describe the ease with which species A is oxidised—the easier a species is to oxidise, the greater its oxidation potential. **(a)** What is the relationship between the standard oxidation potential of A and the standard reduction potential of A^+? **(b)** Which of the metals listed in Table 4.4 has the highest standard oxidation potential? Which has the lowest? **(c)** For a series of substances, the trend in oxidation potential is often related to the trend in the first ionisation energy. Explain why this relationship makes sense.

19.89 Gold metal dissolves in aqua regia, a mixture of concentrated hydrochloric acid and concentrated nitric acid.

The standard reduction potentials

$$Au^{3+}(aq) + 3\,e^- \longrightarrow Au(s) \qquad E^\circ_{red} = +1.498\text{ V}$$

$$AuCl_4^-(aq) + 3\,e^- \longrightarrow Au(s) + 4\,Cl^-(aq)$$

$$E^\circ_{red} = +1.002\text{ V}$$

are important in gold chemistry. **(a)** Use half-reactions to write a balanced equation for the reaction of Au and nitric acid to produce Au^{3+} and $NO(g)$, and calculate the standard emf of this reaction. Is this reaction spontaneous? **(b)** Use half-reactions to write a balanced equation for the reaction of Au and hydrochloric acid to produce $AuCl_4^-(aq)$ and $H_2(g)$, and calculate the standard emf of this reaction. Is this reaction spontaneous? **(c)** Use half-reactions to write a balanced equation for the reaction of Au and aqua regia to produce $AuCl_4^-(aq)$ and $NO(g)$, and calculate the standard emf of this reaction. Is this reaction spontaneous under standard conditions? **(d)** Use the Nernst equation to explain why aqua regia made from *concentrated* hydrochloric and nitric acids is able to dissolve gold.

19.90 The standard potential for the reduction of AgSCN(s) is +0.0895 V.

$$AgSCN(s) + e^- \longrightarrow Ag(s) + SCN^-(aq)$$

Using this value and the electrode potential for $Ag^+(aq)$, calculate the K_{sp} for AgSCN.

19.91 The K_{sp} value for PbS(s) is 8.0×10^{-28}. Consider the equation $PbS(s) + 2\,e^- \longrightarrow Pb(s) + S^{2-}(aq)$. By using the K_{sp} value together with an electrode potential from Appendix E, determine the value of the standard reduction potential for the reaction.

MasteringChemistry (www.pearson.com.au/masteringchemistry)

Make learning part of the grade. Access:

- tutorials with peronalised coaching
- study area
- Pearson eText

PHOTO/ART CREDITS

752 Angela Waye/Shutterstock; **754, 757, 761** Richard Megna, Fundamental Photographs, NYC.

20

CHEMISTRY OF THE NON-METALS

A modern living space in which the artifacts are composed of a very large variety and type of atoms and molecules.

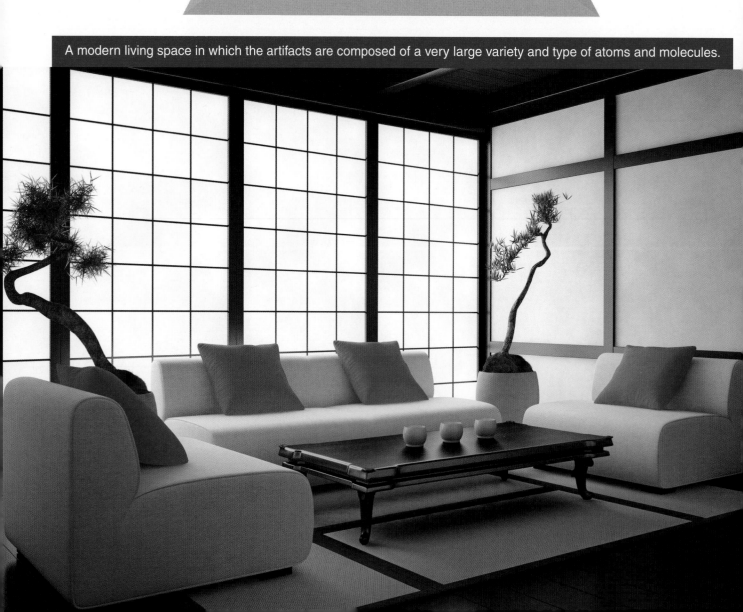

KEY CONCEPTS

20.1 PERIODIC TRENDS AND CHEMICAL REACTIONS
We review periodic trends and types of chemical reactions, which will help us focus on general patterns of behaviour as we examine each group in the periodic table.

20.2 HYDROGEN
The first element in the periodic table, hydrogen, forms compounds with most other non-metals and with many metals.

20.3 GROUP 18: THE NOBLE GASES
We consider the noble gases, the elements of group 18, which exhibit very limited chemical reactivity.

20.4 GROUP 17: THE HALOGENS
We examine the most electronegative elements, the halogens of group 17.

20.5 OXYGEN
Oxygen, the most abundant element by mass in both the Earth's crust and the human body, and the oxide and peroxide compounds it forms are examined in detail.

20.6 THE OTHER GROUP 16 ELEMENTS: S, Se, Te AND Po
We study the other members of group 16 (S, Se, Te and Po), of which sulfur is the most important.

20.7 NITROGEN
We describe nitrogen, a key component of our atmosphere. It forms compounds in which its oxidation number ranges from −3 to +5, including such important compounds as NH_3 and HNO_3.

20.8 THE OTHER GROUP 15 ELEMENTS: P, As, Sb AND Bi
We take a closer look at phosphorus—the most commercially important of the other members of group 15 (P, As, Sb and Bi), and the only one that plays an important and beneficial role in biological systems.

20.9 CARBON
We focus on the inorganic compounds of carbon.

20.10 THE OTHER GROUP 14 ELEMENTS: Si, Ge, Sn, AND Pb
We consider silicon, the most abundant and significant element of the heavier members of group 14.

20.11 BORON
Finally, we examine boron, the sole non-metallic element of group 13.

The chapter-opening photograph shows a living space. A great number of chemical elements have gone into making up the space and all its decorations. One of the first questions we might ask is: What are the relative importances of metallic and non-metallic elements? The carpets or mats are probably made of an organic fibre derived from bamboo. The walls are also made of a cellulososic material that is essentially non-metallic. The chair coverings, the chairs themselves and the ceramic plant pot are all constructed from mainly non-metallic elements as well. Metals are certainly important too, as they are necessary to manufacture the electrical wiring in the walls, perhaps steel girders in the building's frame, perhaps plumbing fixtures. Overall, however, the stuff of which this scene is constructed is largely non-metallic in origin.

Chemical innovation has resulted in a host of materials that form the stuff of modern life. We've learned in earlier chapters about a variety of chemical reaction types that allows chemists to purify materials and polymerise small molecules into much larger ones. We've also learned how the properties of pure substances and mixtures depend on underlying molecular properties (∞ Chapter 11, 'Intermolecular Forces: Liquids and Solids'). But what are the chemical and physical properties of the elements that make up all these materials? In this and the next chapter, we look at the properties of many elements and at how these properties determine the possibilities for useful applications.

In this chapter we start by taking a panoramic view of the descriptive chemistry of the non-metallic elements, starting with hydrogen and then progressing, group by group, from right to left across the periodic table. We emphasise hydrogen, oxygen, nitrogen and carbon. These four non-metals form many commercially important compounds and account for 99% of the atoms required by living cells. We discuss further aspects of the chemistry of these elements when we consider organic and biological chemistry in Chapters 22 to 30.

20.1 | PERIODIC TRENDS AND CHEMICAL REACTIONS

∞ Review this on page 237

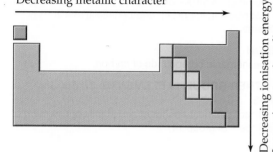

■ Metals ■ Metalloids ■ Non-metals

Increasing ionisation energy
Decreasing atomic radius
Increasing electronegativity
Decreasing metallic character

▲ **FIGURE 20.1** **Trends in elemental properties.**

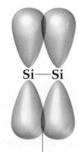

Smaller nucleus-to-nucleus distance, more orbital overlap, stronger π-bond

Larger nucleus-to-nucleus distance, less orbital overlap, weaker π-bond

▲ **FIGURE 20.2** π-bonds in period 2 and period 3 elements.

Recall that we can classify elements as metals, metalloids and non-metals (∞ Sections 7.6, 'Metals, Non-metals and Metalloids'). Except for hydrogen, which is a special case, the non-metals occupy the upper right portion of the periodic table. This division of elements relates nicely to trends in the properties of the elements as summarised in ◄ FIGURE 20.1. Electronegativity, for example, increases as we move left to right across a period and decreases as we move down a group. The non-metals thus have higher electronegativities than the metals. This difference leads to the formation of ionic solids in reactions between metals and non-metals (Sections 7.6, 8.2, 8.4). In contrast, compounds formed between two or more non-metals are usually molecular substances (Sections 7.8 and 8.4).

The chemistry exhibited by the first member of a non-metal group can differ from that of subsequent members. For example, non-metals in period 3 and below can accommodate a larger number of bonded neighbours (Section 8.7). Another important difference is that the first element in any group can more readily form π-bonds. This trend is due, in part, to size because small atoms are able to approach each other more closely. As a result, the overlap of *p* orbitals, which results in the formation of π-bonds, is more effective for the first element in each group (◄ FIGURE 20.2). More effective overlap means stronger π-bonds, reflected in bond enthalpies (Section 8.8). For example, the difference between the enthalpies of the C–C single bond and the C–C double bond is about 270 kJ mol^{-1} (Table 8.4); this value reflects the 'strength' of a carbon–carbon π-bond, and the difference between Si–Si single and Si–Si double bonds is only about 100 kJ mol^{-1}, significantly lower than that for carbon.

As we shall see, π-bonds are particularly important in the chemistry of carbon, nitrogen and oxygen, each the first member in its group. The heavier elements in these groups have a tendency to form only single bonds.

SAMPLE EXERCISE 20.1 Identifying elemental properties

Of the elements Li, K, N, P and Ne, which **(a)** is the most electronegative, **(b)** has the greatest metallic character, **(c)** can bond to more than four atoms in a molecule, **(d)** forms π-bonds most readily?

SOLUTION

Analyse We are given a list of elements and asked to predict several properties that can be related to periodic trends.

Plan We can use Figures 20.1 and 20.2 to guide us to the answers.

Solve

(a) Electronegativity increases as we proceed towards the upper right portion of the periodic table, excluding the noble gases. Thus N is the most electronegative element of our choices.

(b) Metallic character correlates inversely with electronegativity—the less electronegative an element, the greater its metallic character. The element with the greatest metallic character is therefore K, which is closest to the lower left corner of the periodic table.

(c) Non-metals tend to form molecular compounds, so we can narrow our choice to the three non-metals on the list: N, P and Ne. To form more than four bonds, an element must be able to expand its valence shell to allow more than an octet of electrons around it. Valence-shell expansion occurs for period 3 elements and below; N and Ne are both in period 2 and do not undergo valence-shell expansion. Thus the answer is P.

(d) Period 2 non-metals form π-bonds more readily than elements in period 3 and below. There are no compounds known that contain covalent bonds to Ne. Thus N is the element from the list that forms π-bonds most readily.

PRACTICE EXERCISE

Of the elements Be, C, Cl, Sb and Cs, which **(a)** has the lowest electronegativity, **(b)** has the greatest non-metallic character, **(c)** is most likely to participate in extensive π-bonding, **(d)** is most likely to be a metalloid?

Answers: **(a)** Cs, **(b)** Cl, **(c)** C, **(d)** Sb

(See also Exercise 20.12.)

The ready ability of period 2 elements to form π-bonds is an important factor in determining the structures of these elements and their compounds. Compare, for example, the elemental forms of carbon and silicon. Carbon has five major crystalline allotropes: diamond, graphite, buckminsterfullerene, graphene and carbon nanotubes (Section 11.9, 'Bonding in Solids').

Diamond is a covalent-network solid that has C–C σ-bonds but no π-bonds. Graphite, buckminsterfullerene, graphene and carbon nanotubes have π-bonds that result from the sideways overlap of *p* orbitals. Elemental silicon, however, exists only as a diamond-like covalent-network solid with σ-bonds; it has no forms analogous to graphite, buckminsterfullerene, graphene or carbon nanotubes, apparently because Si–Si π-bonds are weak.

Review this on page 413

CONCEPT CHECK 1

Can silicon–silicon double bonds form in elemental silicon?

We likewise see significant differences in the dioxides of carbon and silicon (▶ FIGURE 20.3). CO_2 is a molecular substance with C–O double bonds, whereas SiO_2 contains no double bonds. SiO_2 is a covalent-network solid in which four oxygen atoms are bonded to each silicon atom by single bonds, forming an extended structure that has the empirical formula SiO_2.

CONCEPT CHECK 2

The element nitrogen is found in nature as $N_2(g)$. Would you expect phosphorus to be found in nature as $P_2(g)$? Explain.

Chemical Reactions

Because O_2 and H_2O are abundant in our environment, it is particularly important to consider the possible reactions of these substances with other compounds. About one-third of the reactions discussed in this chapter involve either O_2 (oxidation or combustion reactions) or H_2O (especially proton-transfer reactions).

In combustion reactions with O_2, hydrogen-containing compounds produce H_2O. Carbon-containing compounds produce CO_2 (unless the amount of O_2 is insufficient, in which case CO or even C can form). Nitrogen-containing compounds tend to form N_2, although NO can form in special cases. A reaction illustrating these points is:

$$4\,CH_3NH_2(g) + 9\,O_2(g) \longrightarrow 4\,CO_2(g) + 10\,H_2O(g) + 2\,N_2(g) \qquad [20.1]$$

The formation of H_2O, CO_2 and N_2 reflects the high thermodynamic stabilities of these substances, which are indicated by the large bond energies for the O–H, C–O double and N–N triple bonds that they contain (463, 799 and 941 kJ mol^{-1}, respectively) (Section 8.8, 'Strengths of Covalent Bonds').

When dealing with proton-transfer reactions, remember that the weaker a Brønsted–Lowry acid is, the stronger its conjugate base (Section 17.2, 'Brønsted–Lowry Acids and Bases'). For example, H_2, OH$^-$, NH_3 and CH_4 are

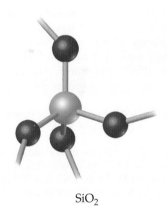

SiO$_2$

CO$_2$

▲ **FIGURE 20.3   Comparison of SiO₂ and CO₂.** SiO₂ has only single bonds, whereas CO₂ has double bonds.

Review this on page 279

Review this on page 660

exceedingly weak proton donors that have *no* tendency to act as acids in water. Thus species formed from them by removing one or more protons (such as H^-, O^{2-} and NH_2^-) are extremely strong bases. All react readily with water, removing protons from H_2O to form OH^-. The following reactions are illustrative:

$$CH_3^-(aq) + H_2O(l) \longrightarrow CH_4(g) + OH^-(aq) \qquad [20.2]$$

$$N^{3-}(aq) + 3\,H_2O(l) \longrightarrow NH_3(aq) + 3\,OH^-(aq) \qquad [20.3]$$

Substances that are stronger proton donors than H_2O, such as HCl, H_2SO_4, CH_3COOH and other acids, also react readily with basic anions.

SAMPLE EXERCISE 20.2   Predicting the products of chemical reactions

Predict the products formed in each of the following reactions, and write a balanced equation:

(a) $CH_3NHNH_2(g) + O_2(g) \longrightarrow$
(b) $Mg_3P_2(s) + H_2O(l) \longrightarrow$
(c) $NaCN(s) + HCl(aq) \longrightarrow$

SOLUTION

Analyse We are given the reactants for three chemical equations and asked to predict the products and balance the equations.

Plan We need to examine the reactants to see if there is a reaction type that we might recognise. In **(a)** the carbon compound is reacting with O_2, which suggests a combustion reaction. In **(b)** water reacts with an ionic compound. The anion, P^{3-}, is a strong base and H_2O is able to act as an acid, so the reactants suggest an acid–base (proton-transfer) reaction. In **(c)** we have an ionic compound and a strong acid. Again, a proton-transfer reaction is suggested.

Solve

(a) Based on the elemental composition of the carbon compound, this combustion reaction should produce CO_2, H_2O and N_2:

$$2\,CH_3NHNH_2(g) + 5\,O_2(g) \longrightarrow 2\,CO_2(g) + 6\,H_2O(g) + 2\,N_2(g)$$

(b) Mg_3P_2 is ionic, consisting of Mg^{2+} and P^{3-} ions. The P^{3-} ion, like N^{3-}, has a strong affinity for protons and reacts with H_2O to form OH^- and PH_3 (PH^{2-}, PH_2^- and PH_3 are all exceedingly weak proton donors).

$$Mg_3P_2(s) + 6\,H_2O(l) \longrightarrow 2\,PH_3(g) + 3\,Mg(OH)_2(s)$$

$Mg(OH)_2$ has low solubility in water and will precipitate.

(c) NaCN consists of Na^+ and CN^- ions. The CN^- ion is basic (HCN is a weak acid). Thus CN^- reacts with protons to form its conjugate acid.

$$NaCN(s) + HCl(aq) \longrightarrow HCN(aq) + NaCl(aq)$$

HCN has limited solubility in water and escapes as a gas. HCN is *extremely* toxic; in fact, this reaction has been used to produce the lethal gas in gas chambers.

PRACTICE EXERCISE

Write a balanced equation for the reaction of solid sodium hydride with water.

Answer: $NaH(s) + H_2O(l) \longrightarrow NaOH(aq) + H_2(g)$

(See also Exercises 20.15, 20.16.)

20.2 | HYDROGEN

The English chemist Henry Cavendish (1731–1810) first isolated pure hydrogen. Because the element produces water when burned in air, the French chemist Antoine Lavoisier gave it the name *hydrogen*, which means 'water producer' (Greek: *hydro*, water; *gennao*, to produce).

Hydrogen is the most abundant element in the universe. Although about 70% of the visible mass of the universe is composed of hydrogen, it constitutes only 0.87% of the Earth's mass. Most of the hydrogen on our planet is found associated with oxygen. Water, which is 11% hydrogen by mass, is the most abundant hydrogen compound. Hydrogen is also an important part of petroleum, cellulose, starch, fats, alcohols, acids and a wide variety of other materials.

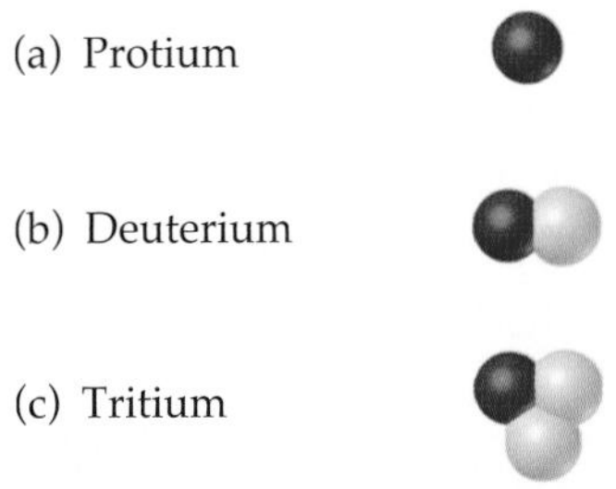

Isotopes of Hydrogen

The most common isotope of hydrogen, $_1^1H$, has a nucleus consisting of a single proton. This isotope, sometimes referred to as **protium**,* comprises 99.9844% of naturally occurring hydrogen.

Two other isotopes are known: $_1^2H$, whose nucleus contains a proton and a neutron, and $_1^3H$, whose nucleus contains a proton and two neutrons (▶ **FIGURE 20.4**). The $_1^2H$ isotope, called **deuterium**, comprises 0.0156% of naturally occurring hydrogen. Deuterium is often given the symbol D in chemical formulae, as in D_2O (deuterium oxide), which is known as *heavy water*.

Because an atom of deuterium is about twice as massive as an atom of protium, the properties of deuterium-containing substances vary somewhat from those of the 'normal' protium-containing analogues. For example, the normal melting and boiling points of D_2O are 3.81 °C and 101.42 °C, respectively, whereas they are 0.00 °C and 100.00 °C for H_2O. Not surprisingly, the density of D_2O at 25 °C (1.104 g cm^{-3}) is greater than that of H_2O (0.997 g cm^{-3}). Replacing protium with deuterium (a process called *deuteration*) can also have a profound effect on the rates of reactions, a phenomenon called a *kinetic isotope effect*. In fact, heavy water can be obtained by the electrolysis of ordinary water because D_2O undergoes electrolysis at a slower rate and therefore becomes concentrated during the electrolysis.

The third isotope, $_1^3H$, is known as **tritium**. It is radioactive, with a half-life of 12.3 years.

$$_1^3H \longrightarrow {}_2^3He + {}_{-1}^0e \qquad t_{\frac{1}{2}} = 12.3 \text{ years} \qquad [20.4]$$

Tritium is formed continuously in the upper atmosphere in nuclear reactions induced by cosmic rays; because of its short half-life, however, only trace quantities exist naturally.

Properties of Hydrogen

Hydrogen is the only element that is not a member of any family in the periodic table. Because of its $1s^1$ electron configuration, it is generally placed above lithium in the periodic table. However, it is definitely *not* an alkali metal. It forms a positive ion much less readily than any alkali metal; the ionisation energy of the hydrogen atom is 1312 kJ mol^{-1}, whereas that of lithium is 520 kJ mol^{-1}.

Hydrogen is sometimes placed above the halogens in the periodic table because the hydrogen atom can pick up one electron to form the *hydride ion*, H$^-$, which has the same electron configuration as helium. The electron affinity of hydrogen ($EA = -73$ kJ mol^{-1}), however, is not as large as that of any halogen; the electron affinity of fluorine is -328 kJ mol^{-1} and that of iodine is -295 kJ mol^{-1}. In general, hydrogen shows no closer resemblance to the halogens than it does to the alkali metals.

Elemental hydrogen exists at room temperature as a colourless, odourless, tasteless gas composed of diatomic molecules. We can call H_2 dihydrogen, but it is more commonly referred to as molecular hydrogen or simply hydrogen.

(a) Protium

(b) Deuterium

(c) Tritium

▲ **FIGURE 20.4   Nuclei of the three isotopes of hydrogen.** (a) Protium, $_1^1H$, has a single proton (depicted as a red sphere) in its nucleus. (b) Deuterium, $_1^2H$, has a proton and a neutron (depicted as a grey sphere). (c) Tritium, $_1^3H$, has a proton and two neutrons. Remember that the nuclei are very small! (∞ Section 2.3, 'The Modern View of Atomic Structure')

---

* Giving unique names to isotopes is limited to hydrogen. Because of the proportionally large differences in their masses, the isotopes of H show appreciably more differences in their chemical and physical properties than isotopes of heavier elements.

Because H_2 is non-polar and has only two electrons, attractive forces between molecules are extremely weak. As a result, the melting point (-259 °C) and boiling point (-253 °C) or H_2 are very low.

The H–H bond enthalpy (436 kJ mol^{-1}) is high for a single bond (Table 8.4). By comparison, the Cl–Cl bond enthalpy is only 242 kJ mol^{-1}. Because H_2 has a strong bond, most reactions of H_2 are slow at room temperature. However, the molecule is readily activated by heat, irradiation or catalysis. The activation process generally produces hydrogen atoms, which are very reactive. Once H_2 is activated, it reacts rapidly and exothermically with a wide variety of substances.

CONCEPT CHECK 3

If H_2 is activated to produce H^+, what must the other product be?

Hydrogen forms strong covalent bonds with many elements, including oxygen; the O–H bond enthalpy is 463 kJ mol^{-1}. The formation of the strong O–H bond makes hydrogen an effective reducing agent for many metal oxides. When H_2 is passed over heated CuO, for example, copper is produced.

$$CuO(s) + H_2(g) \longrightarrow Cu(s) + H_2O(g) \qquad [20.5]$$

When H_2 is ignited in air, a vigorous reaction occurs, forming H_2O.

$$2\,H_2(g) + O_2(g) \longrightarrow 2\,H_2O(g) \qquad \Delta H° = -483.6 \text{ kJ} \qquad [20.6]$$

Air containing as little as 4% H_2 (by volume) is potentially explosive. Combustion of hydrogen–oxygen mixtures is commonly used in liquid-fuel rocket engines such as those of the space shuttles.

Preparation of Hydrogen

When a small quantity of H_2 is needed in the laboratory, it is usually obtained by the reaction between an active metal such as zinc and a dilute strong acid such as HCl or H_2SO_4.

$$Zn(s) + 2\,H^+(aq) \longrightarrow Zn^{2+}(aq) + H_2(g) \qquad [20.7]$$

Large quantities of H_2 are produced by reacting methane (CH_4, the principal component of natural gas) with steam at 1100 °C.

$$CH_4(g) + 2\,H_2O(g) \longrightarrow CO_2(g) + 4\,H_2(g) \qquad [20.8]$$

When heated to about 1000 °C, carbon also reacts with steam to produce a mixture of H_2 and CO gases.

$$C(s) + H_2O(g) \longrightarrow H_2(g) + CO(g) \qquad [20.9]$$

This mixture, known as *water gas*, is used as an industrial fuel.

Simple electrolysis of water can be used to produce hydrogen, but this is a high energy process and is not commercially viable; see A Closer Look: The Hydrogen Economy.

$$2\,H_2O(l) \longrightarrow 2\,H_2(g) + O_2(g) \qquad \Delta H° = +286 \text{ kJ} \qquad [20.10]$$

CONCEPT CHECK 4

What are the oxidation states of H atoms in Equations 20.7–20.10?

Uses of Hydrogen

Hydrogen is a commercially important substance. Over two-thirds of the H_2 produced is used to synthesise ammonia by the Haber process (Section 16.2,

A CLOSER LOOK

THE HYDROGEN ECONOMY

The reaction of hydrogen with oxygen is highly exothermic:

$$2 H_2(g) + O_2(g) \longrightarrow 2 H_2O(g)$$
$$\Delta H = -483.6 \text{ kJ} \quad [20.11]$$

Because the only product of the reaction is water vapour, the prospect of using hydrogen as a fuel in fuel cells is attractive (∞ Section 19.7, 'Batteries and Fuel Cells').

Alternatively, hydrogen could be combusted directly with oxygen from the atmosphere in an internal combustion engine. In either case, it would be necessary to generate elemental hydrogen on a large scale and arrange for its transport and storage.

▼ FIGURE 20.5 illustrates various sources and uses of H_2 fuel. The generation of H_2 through electrolysis of water is in principle the cleanest route, because this process—the reverse of Equation 20.11—produces only hydrogen and oxygen (∞ Section 19.9. 'Electrolysis').

However, the energy required to electrolyse water must come from somewhere. If we burn fossil fuels to generate this energy, we have not advanced very far towards a true hydrogen economy. If the energy for electrolysis came instead from a hydroelectric or nuclear power plant, solar cells or wind generators, consumption of non-renewable energy sources and undesired production of CO_2 could be avoided.

RELATED EXERCISES: 20.20, 20.78

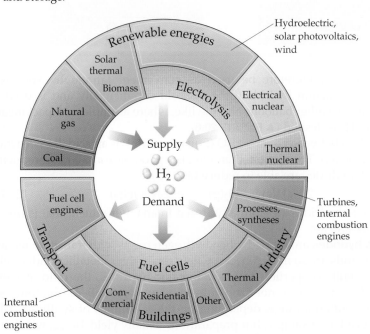

◀ FIGURE 20.5   The 'hydrogen economy' would require hydrogen to be produced from various sources and would use hydrogen in energy-related applications.

'The Equilibrium Constant'). Much of the remaining hydrogen is used to convert oil into hydrocarbons suitable for fuel (petrol, diesel and others) in a process known as *cracking*. Hydrogen is also used to manufacture methanol (CH_3OH) via the catalytic reaction of CO and H_2 at high pressure and temperature.

∞ Review this on page 618

$$CO(g) + 2 H_2(g) \longrightarrow CH_3OH(g) \quad [20.12]$$

Binary Hydrogen Compounds

Hydrogen reacts with other elements to form compounds of three general types: (1) ionic hydrides, (2) metallic hydrides and (3) molecular hydrides.

The **ionic hydrides** are formed by the alkali metals and by the heavier alkaline earths (Ca, Sr and Ba). These active metals are much less electronegative than hydrogen. Consequently, hydrogen acquires electrons from them to form hydride ions (H^-):

$$Ca(s) + H_2(g) \longrightarrow CaH_2(s) \quad [20.13]$$

The hydride ion is very basic and reacts readily with compounds that have even weakly acidic protons to form H_2. For example, H^- reacts readily with H_2O.

$$H^-(aq) + H_2O(l) \longrightarrow H_2(g) + OH^-(aq) \quad [20.14]$$

FIGURE IT OUT

This reaction is exothermic. Is the beaker on the right warmer or colder than the beaker on the left?

▶ **FIGURE 20.6** **The reaction of CaH$_2$ with water.**

Ionic hydrides can therefore be used as convenient (although expensive) sources of H$_2$. Calcium hydride (CaH$_2$) is sold commercially and is used to inflate life rafts, weather balloons and the like, where a simple, compact means of generating H$_2$ is desired (▲ **FIGURE 20.6**).

The reaction between H$^-$ and H$_2$O (Equation 20.14) is an acid–base reaction *and* a redox reaction. The H$^-$ ion, therefore, is a good base *and* a good reducing agent. In fact, hydrides are able to reduce O$_2$ to OH$^-$:

$$2\,NaH(s) \,+\, O_2(g) \longrightarrow 2\,NaOH(s) \qquad [20.15]$$

For this reason, hydrides are normally stored in an environment that is free of both moisture and air.

Metallic hydrides are formed when hydrogen reacts with transition metals. These compounds are so named because they retain their metallic conductivity and other metallic properties. In many metallic hydrides, the ratio of metal atoms to hydrogen atoms is not fixed or in small whole numbers. The composition can vary within a range, depending on the conditions of synthesis. TiH$_2$ can be produced, for example, but preparations usually yield TiH$_{1.8}$, which has about 10% less hydrogen than TiH$_2$. These non-stoichiometric metallic hydrides are sometimes called *interstitial hydrides*. They may be considered to be solutions of hydrogen atoms in the metal, with the hydrogen atoms occupying the holes, or interstices, between metal atoms in the solid lattice. This is an oversimplification, however, because there is evidence for chemical interaction between metal and hydrogen.

The most interesting interstitial metallic hydride is that of palladium. Palladium can take up nearly 900 times its volume of hydrogen, making it very attractive for hydrogen storage in any possible future 'hydrogen economy'. However, to be practical, any hydrogen storage compound will have to contain 75% or more hydrogen by mass and be able to charge and discharge hydrogen quickly and safely near room temperature.

14	15	16	17
CH$_4$(g)	NH$_3$(g)	H$_2$O(l)	HF(g)
−50.8	−16.7	−237	−271
SiH$_4$(g)	PH$_3$(g)	H$_2$S(g)	HCl(g)
+56.9	+18.2	−33.0	−95.3
GeH$_4$(g)	AsH$_3$(g)	H$_2$Se(g)	HBr(g)
+117	+111	+71	−53.2
	SbH$_3$(g)	H$_2$Te(g)	HI(g)
	+187	+138	+1.30

▲ **FIGURE 20.7** **Standard free energies of formation of molecular hydrides.** All values given are in kilojoules per mole of hydride.

CONCEPT CHECK 5

Palladium has a density of 12.023 g cm^{-3}. Can a sample of Pd that has a volume of 1 cm^3 increase its mass to over 900 g by adsorbing hydrogen?

The **molecular hydrides**, formed by non-metals and semi-metals, are either gases or liquids under standard conditions. The simple molecular hydrides are listed in ◀ **FIGURE 20.7**, together with their standard free energies of formation,

$\Delta_f G°$. In each family the thermal stability (measured by $\Delta_f G°$) decreases as we move down the family. (Recall that the more stable a compound is with respect to its elements under standard conditions, the more negative $\Delta_f G°$ is.) We discuss the molecular hydrides further in the course of examining the other non-metallic elements.

∞ Review this on page 380

20.3 | GROUP 18: THE NOBLE GASES

The elements of group 18, the first five of which were isolated and characterised between 1894 and 1898 by Sir William Ramsay (1852–1916), a Scottish chemist, are chemically unreactive. Indeed, most of our references to these elements have been in relation to their physical properties, as when we discussed inter-molecular forces (∞ Section 11.2, 'Intermolecular Forces'). The relative inert-ness of these elements is due to the presence of a completed octet of valence-shell electrons (except He, which has a filled $1s$ shell). The stability of such an arrangement is reflected in the high ionisation energies of the group 18 elements (∞ Section 7.4, 'Ionisation Energy').

∞ Review this on page 232

The group 18 elements are all gases at room temperature. They are compo-nents of the Earth's atmosphere, except for radon, which exists only as a short-lived radioisotope. Only argon is relatively abundant (Table 13.1). Neon, argon, krypton and xenon are recovered from liquid air by distillation. Argon is used as a blanketing atmosphere in electric light bulbs. The gas conducts heat away from the filament but does not react with it. It is also used as a protective atmos-phere to prevent oxidation in welding and certain high-temperature metallurgi-cal processes. Neon and argon are used in electric signs; the gas is caused to radiate by passing an electric discharge through the tube.

Helium is, in many ways, the most important of the noble gases. Liquid helium is used as a coolant to conduct experiments at very low temperatures. Helium boils at 4.2 K at 1 bar pressure, the lowest boiling point of any substance. Fortunately, helium is found in relatively high concentrations in many natural-gas wells.

Noble Gas Compounds

Although the noble gases are exceedingly stable, some will nonetheless undergo reaction, but only under rigorous conditions. Because of their lower ionisation energy, the heavier noble gases such as xenon might be expected to form compounds with highly reactive, highly electronegative elements such as fluorine. Because the group 18 elements already have a full valence shell, any resulting compounds are required to have an expanded valence octet (∞ Section 8.7, 'Exceptions to the Octet Rule', and ▶ FIGURE 20.8).

∞ Review this on page 276

The first noble gas compound was prepared by Neil Bartlett (1932–2008) at the University of British Columbia. He prepared a number of xenon fluoride compounds by direct reaction of xenon and fluorine under varying conditions and varying stoichiometric ratios of the reactants. The oxygen-containing com-pounds are formed when the fluorides are reacted with water and again, by varying the stoichiometry, different compounds can be produced.

$$Xe(g) + 3\,F_2(g) \longrightarrow XeF_6(s) \qquad [20.16]$$

$$XeF_6(s) + H_2O(l) \longrightarrow XeOF_4(l) + 2\,HF(g) \qquad [20.17]$$

$$XeF_6(s) + 3\,H_2O(l) \longrightarrow XeO_3(aq) + 6\,HF(aq) \qquad [20.18]$$

From ▼ TABLE 20.1 it can be seen that the enthalpies of formation for the xenon fluorides are negative, indicating that they should be relatively stable, whereas the oxygen-containing compounds have positive enthalpies of formation and are quite unstable. Krypton is the only other noble gas to form compounds but only one, KrF_2, is known with certainty and it decomposes to its elements at $-10\,°C$.

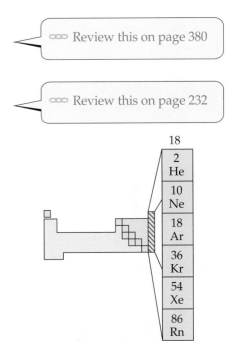

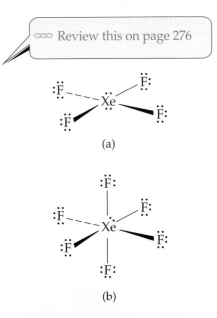

▲ **FIGURE 20.8 Expanded octets in xenon fluorides.** (a) Lewis structure of xenontetrafluoride. (b) Lewis structure of xenonhexafluoride.

TABLE 20.1 • Properties of xenon compounds

Compound	Oxidation state of Xe	Melting point (°C)	$\Delta_f H°$ (kJ mol^{-1})[a]
XeF_2	+2	129	−109(g)
XeF_4	+4	117	−218(g)
XeF_6	+6	49	−298(g)
$XeOF_4$	+6	−41 to −28	+146(l)
XeO_3	+6	—[b]	+402(s)
XeO_2F_2	+6	31	+145(s)
XeO_4	+8	—[c]	—

[a] At 25 °C, for the compound in the state indicated.
[b] A solid; decomposes at 40 °C.
[c] A solid; decomposes at −40 °C.

SAMPLE EXERCISE 20.3 **Predicting a molecular structure**

Use the VSEPR model to predict the structure of XeF_4.

SOLUTION

Analyse We must predict the geometrical structure given only the molecular formula.

Plan We must first write the Lewis structure for the molecule. We then count the number of electron pairs (domains) around the Xe atom and use that number and the number of bonds to predict the geometry.

Solve There are 36 valence-shell electrons (8 from xenon and 7 from each fluorine). If we make four single Xe–F bonds, each fluorine has its octet satisfied. Xe then has 12 electrons in its valence shell, so we expect an octahedral arrangement of six electron pairs. Two of these are non-bonded pairs. Because non-bonded pairs require more volume than bonded pairs (∞ Section 9.2, 'The VSEPR Model'), it is reasonable to expect these non-bonded pairs to be opposite each other. The expected structure is square planar, as shown in ◀ **FIGURE 20.9**.

Comment The experimentally determined structure agrees with this prediction.

PRACTICE EXERCISE

Describe the electron-domain geometry and molecular geometry of XeF_2.

Answer: trigonal bipyramidal, linear

∞ Review this on page 300

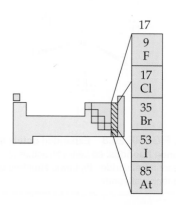

▲ **FIGURE 20.9**   **Xenon tetrafluoride.**

∞ Review this on page 235

17
9 F
17 Cl
35 Br
53 I
85 At

20.4 | GROUP 17: THE HALOGENS

The elements of group 17, the halogens, have outer electron configurations of ns^2np^5, where n ranges from 2 through to 6. The halogens have large negative electron affinities (∞ Section 7.5, 'Electron Affinities'), and they most often achieve a noble gas configuration by gaining an electron, which results in a −1 oxidation state. Fluorine, being the most electronegative element, exists in compounds only in the −1 state. The other halogens also exhibit positive oxidation states up to +7 in combination with more electronegative atoms such as O. In the positive oxidation states the halogens tend to be good oxidising agents, readily accepting electrons.

Chlorine, bromine and iodine are found as the halides in seawater and in salt deposits. The concentration of iodine in these sources is very small, but it is concentrated by certain seaweeds. When they are harvested, dried and burned, iodine can be extracted from the ashes. Fluorine occurs in minerals* such as cryolite (Na_3AlF_6), which is an important commercial source of fluorine.

___

* Minerals are solid substances that occur in nature. They are usually known by their common names rather than by their chemical names. What we know as *rock* is merely an aggregate of different kinds of minerals.

Property	F	Cl	Br	I
TABLE 20.2 • Some properties of the halogens				
Atomic radius (pm)	71	99	114	133
Ionic radius, X^- (pm)	133	181	196	220
First ionisation energy (kJ mol^{-1})	1681	1251	1140	1008
Electron affinity (kJ mol^{-1})	−328	−349	−325	−295
Electronegativity	4.0	3.0	2.8	2.5
X–X single-bond enthalpy (kJ mol^{-1})	155	242	193	151
Reduction potential (V): $\frac{1}{2}X_2(aq) + e^- \longrightarrow X^-(aq)$	2.87	1.36	1.07	0.54

All isotopes of astatine are radioactive. The longest lived isotope is astatine-210, which has a half-life of 8.1 hours and decays mainly by electron capture. Because astatine is so unstable to nuclear decay, very little is known about its chemistry.

Properties and Preparation of the Halogens

Some of the properties of the halogens are summarised in ▲ TABLE 20.2. Most of the properties vary in a regular fashion as we go from fluorine to iodine. The electronegativity steadily decreases, for example, from 4.0 for fluorine to 2.5 for iodine. The halogens have the highest electronegativities in each period of the periodic table. Under ordinary conditions the halogens exist as diatomic molecules. The molecules are held together in the solid and liquid states by London dispersion forces (⚬⚬⚬ Section 11.2, 'Intermolecular Forces'). Because I_2 is the largest and most polarisable of the halogen molecules, the intermolecular forces between I_2 molecules are the strongest. Thus I_2 has the highest melting point and boiling point. At room temperature and 1 bar pressure, I_2 is a solid, Br_2 is a liquid and Cl_2 and F_2 are gases.

The comparatively low bond enthalpy in F_2 (155 kJ mol^{-1}) accounts in part for the extreme reactivity of elemental fluorine. Because of its high reactivity, F_2 is very difficult to work with. Certain metals, such as copper and nickel, can be used to contain F_2 because their surfaces form a protective coating of metal fluoride. Chlorine and the heavier halogens are also reactive, although less so than fluorine. They combine directly with most elements except the noble gases.

Because of their high electronegativities, the halogens tend to gain electrons from other substances and thereby serve as oxidising agents. The oxidising ability of the halogens, which is indicated by their standard reduction potentials, decreases going down the group. As a result, a given halogen is able to oxidise the anions of the halogens below it in the group. For example, Cl_2 will oxidise Br^- and I^-, but not F^-, as seen in ▶ FIGURE 20.10.

⚬⚬⚬ Review this on page 381

SAMPLE EXERCISE 20.4   Predicting chemical reactions between the halogens

Write the balanced equation for the reaction, if any, that occurs between **(a)** $I^-(aq)$ and $Br_2(l)$, **(b)** $Cl^-(aq)$ and $I_2(s)$.

SOLUTION

Analyse We are asked to determine whether a reaction occurs when a particular halide and halogen are combined.

Plan A given halogen is able to oxidise anions of the halogens below it in the periodic table. Thus in each pair, the halogen having the smaller atomic number ends up

FIGURE IT OUT

Do Br_2 and I_2 appear to be more or less soluble in CCl_4 than in H_2O?

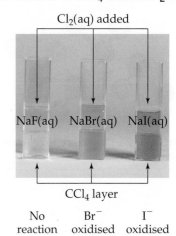

$Cl_2(aq)$ added

NaF(aq)   NaBr(aq)   NaI(aq)

CCl_4 layer

| No reaction | Br^- oxidised to Br_2 | I^- oxidised to I_2 |

▲ FIGURE 20.10   Reaction of Cl_2 with aqueous solutions of NaF, NaBr and NaI. The top liquid layer is water; the bottom liquid layer is carbon tetrachloride.

as the halide ion. If the halogen with the smaller atomic number is already the halide, there is no reaction. Thus the key to determining whether a reaction occurs is locating the elements in the periodic table.

Solve

(a) Br_2 is able to oxidise (remove electrons from) the anions of the halogens below it in the periodic table. Thus it will oxidise I^-.

$$2\,I^-(aq) + Br_2(l) \longrightarrow I_2(s) + 2\,Br^-(aq)$$

(b) Cl^- is the anion of a halogen above iodine in the periodic table. Thus I_2 cannot oxidise Cl^-; there is no reaction.

PRACTICE EXERCISE

Write the balanced chemical equation for the reaction that occurs between $Br^-(aq)$ and $Cl_2(aq)$.

Answer: $2\,Br^-(aq) + Cl_2(aq) \longrightarrow Br_2(l) + 2\,Cl^-(aq)$

(See also Exercise 20.30.)

Notice in Table 20.2 that the reduction potential of F_2 is exceptionally high. Fluorine gas readily oxidises water:

$$F_2(aq) + H_2O(l) \longrightarrow 2\,HF(aq) + \tfrac{1}{2}O_2(g) \qquad E° = 1.80\text{ V} \qquad [20.19]$$

∞ Review this on page 788

Fluorine cannot be prepared by electrolytic oxidation of aqueous solutions of fluoride salts because water itself is oxidised more readily than F^- (∞ Section 19.9, 'Electrolysis'). In practice, the element is formed by electrolytic oxidation of a solution of KF in anhydrous HF. The KF reacts with HF to form a salt, $K^+HF_2^-$, which acts as the current carrier in the liquid. (The HF_2^- ion is stable because of very strong hydrogen bonding.) The overall cell reaction is

$$2\,KHF_2(l) \longrightarrow H_2(g) + F_2(g) + 2\,KF(l) \qquad [20.20]$$

Chlorine is produced mainly by electrolysis of either molten or aqueous sodium chloride, as described in Section 19.9. Both bromine and iodine are obtained commercially from brines containing the halide ions by oxidation with Cl_2.

Uses of the Halogens

Fluorine is an important industrial chemical. It is used, for example, to prepare fluorocarbons—very stable carbon–fluorine compounds used as lubricants and plastics. Teflon® is a polymeric fluorocarbon noted for its high thermal stability and lack of chemical reactivity (◀ **FIGURE 20.11**).

Chlorine is by far the most commercially important halogen. About half the chlorine produced in industrial nations finds its way eventually into the manufacture of chlorine-containing organic compounds such as vinyl chloride used in making polyvinyl chloride (PVC) plastics. Much of the remainder is used as a bleaching agent in the paper and textile industries. When Cl_2 dissolves in a cold dilute base, it disproportionates into Cl^- and hypochlorite, ClO^-.

$$Cl_2(aq) + 2\,OH^-(aq) \rightleftharpoons Cl^-(aq) + ClO^-(aq) + H_2O(l) \qquad [20.21]$$

FIGURE IT OUT

What is the repeating unit in this polymer?

▲ **FIGURE 20.11** Structure of Teflon®, a fluorocarbon polymer.

🔺 **CONCEPT CHECK 6**

What is the oxidation state of Cl on each Cl species in Equation 20.21?

Sodium hypochlorite (NaClO) is the active ingredient in many liquid bleaches. Chlorine is also used in water treatment to oxidise and thereby destroy bacteria (∞ Section 13.4, 'Human Activities and Earth's Water').

A common use of iodine is as KI in table salt. Iodised salt provides the small amount of iodine necessary in our diets; it is essential for the formation of thyroxin, a hormone secreted by the thyroid gland. Lack of iodine in the diet results in an enlarged thyroid gland, a condition called *goitre*.

∞ Review this on page 488

▲ **FIGURE IT OUT**

Are these reactions acid–base reactions or oxidation–reduction reactions?

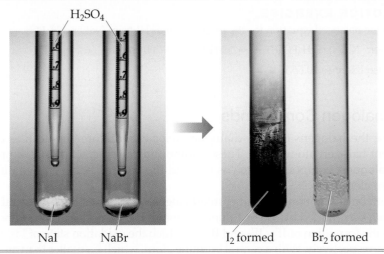

NaI   NaBr I_2 formed Br_2 formed

◀ **FIGURE 20.12**   Reaction of H_2SO_4 with NaI and NaBr.

The Hydrogen Halides

All the halogens form stable diatomic molecules with hydrogen. Aqueous solutions of HCl, HBr and HI are strong acids.

The hydrogen halides can be formed by direct reaction of the elements. The most important means of preparing them, however, is by reacting a salt of the halide with a strong non-volatile acid. Hydrogen fluoride and hydrogen chloride are prepared in this manner by reaction of an inexpensive, readily available salt with concentrated sulfuric acid.

$$CaF_2(s) + H_2SO_4(l) \xrightarrow{\Delta} 2\,HF(g) + CaSO_4(s) \qquad [20.22]$$

$$NaCl(s) + H_2SO_4(l) \xrightarrow{\Delta} HCl(g) + NaHSO_4(s) \qquad [20.23]$$

Neither hydrogen bromide nor hydrogen iodide can be prepared by analogous reactions of salts with H_2SO_4 because H_2SO_4 oxidises Br^- and I^- (▲ **FIGURE 20.12**). This difference in reactivity reflects the greater ease of oxidation of Br^- and I^- relative to F^- and Cl^-. These undesirable oxidations are avoided by using a non-volatile acid, such as H_3PO_4, that is a weaker oxidising agent than H_2SO_4.

The hydrogen halides form hydrohalic acid solutions when dissolved in water. These solutions exhibit the characteristic properties of acids, such as reactions with active metals to produce hydrogen gas (∞ Section 4.4, 'Oxidation–Reduction Reactions'). Hydrofluoric acid also reacts readily with silica (SiO_2) and with various silicates to form hexafluorosilicic acid (H_2SiF_6), as in these examples:

∞ Review this on page 120

$$SiO_2(s) + 6\,HF(aq) \longrightarrow H_2SiF_6(aq) + 2\,H_2O(l) \qquad [20.24]$$

$$CaSiO_3(s) + 8\,HF(aq) \longrightarrow H_2SiF_6(aq) + CaF_2(s) + 3\,H_2O(l) \qquad [20.25]$$

SAMPLE EXERCISE 20.5   **Writing a balanced chemical equation**

Write a balanced equation for the formation of hydrogen bromide gas from the reaction of solid sodium bromide with phosphoric acid.

SOLUTION

Analyse We are asked to write a balanced equation for the reaction between NaBr and H_3PO_4 to form HBr and another product.

Plan As in Equation 20.22, a metathesis reaction takes place (∞ Section 4.2, 'Precipitation Reactions'). Let's assume that only one H in H_3PO_4 reacts. (The actual number depends on the reaction conditions.) The $H_2PO_4^-$ and Na^+ will form NaH_2PO_4 as one product.

∞ Review this on page 109

Solve The balanced equation is

$$NaBr(s) + H_3PO_4(l) \longrightarrow NaH_2PO_4(s) + HBr(g)$$

PRACTICE EXERCISE

Write the balanced equation for the preparation of HI from NaI and H_3PO_4.

Answer: $NaI(s) + H_3PO_4(l) \longrightarrow NaH_2PO_4(s) + HI(g)$

(See also Exercise 20.66.)

Interhalogen Compounds

Because the halogens exist as diatomic molecules, diatomic molecules of two different halogen atoms exist. These compounds are the simplest examples of **interhalogens**, compounds such as ClF and IF_5, formed between two different halogen elements.

With one exception, the higher interhalogen compounds have a central Cl, Br or I atom surrounded by 3, 5 or 7 fluorine atoms. The large size of the iodine atom allows the formation of IF_3, IF_5 and IF_7, in which the oxidation state of I is +3, +5 and +7, respectively. With a central bromine atom, which is smaller than an iodine atom, only BrF_3 and BrF_5 can be formed. Chlorine, which is smaller still, can form ClF_3 and, with difficulty, ClF_5. The only higher interhalogen compound that does not have outer F atoms is ICl_3; the large size of the I atom can accommodate three Cl atoms, whereas Br is not large enough to allow $BrCl_3$ to form.

Because the interhalogen compounds contain a halogen atom in a positive oxidation state, they are exceedingly reactive. They are invariably powerful oxidising agents. When the compound acts as an oxidant, the oxidation state of the central halogen atom is decreased (usually to 0 or −1) as in the reaction:

$$2\,CoCl_2(s) + 2\,ClF_3(g) \longrightarrow 2\,CoF_3(s) + 3\,Cl_2(g) \qquad [20.26]$$

Oxyacids and Oxyanions

∞ Review this on page 54

∞ Review this on page 692

▼ **TABLE 20.3** summarises the formulae of the known oxyacids of the halogens and the way they are named* (∞ Section 2.8, 'Naming Inorganic Compounds'). The acid strengths of the oxyacids increase with increasing oxidation state of the central halogen atom (∞ Section 17.10, 'Acid–Base Behaviour and Chemical Structure'). All the oxyacids are strong oxidising agents. The oxyanions, formed on removal of H^+ from the oxyacids, are generally more stable than the oxyacids. Hypochlorite salts are used as bleaches and disinfectants because of the powerful oxidising capabilities of the ClO^- ion. Sodium chlorite is used as a bleaching agent. Chlorate salts are similarly very reactive. For example, potassium chlorate is used to make matches and fireworks.

⚠ CONCEPT CHECK 7

Which do you expect to be the stronger oxidising agent, $NaBrO_3$ or $NaClO_3$?

TABLE 20.3 • The oxyacids of the halogens

Oxidation state of halogen	Formula of acid			Acid name
	Cl	**Br**	**I**	
+1	HClO	HBrO	HIO	*Hypohalous* acid
+3	$HClO_2$	—	—	*Halous* acid
+5	$HClO_3$	$HBrO_3$	HIO_3	*Halic* acid
+7	$HClO_4$	$HBrO_4$	HIO_4, H_5IO_6	*Perhalic* acid

* Fluorine forms one oxyacid, HOF. Because the electronegativity of fluorine is greater than that of oxygen, we consider fluorine to be in a −1 oxidation state and oxygen to be in the 0 oxidation state in this compound.

Perchloric acid and its salts are the most stable of the oxyacids and oxyanions. Dilute solutions of perchloric acid are quite safe, and most perchlorate salts are stable except when heated with organic materials. When heated, perchlorates can become vigorous, even violent, oxidisers. So considerable caution should be exercised when handling these substances.

There are two oxyacids that have iodine in the +7 oxidation state. These periodic acids are HIO_4 (called metaperiodic acid) and H_5IO_6 (called paraperiodic acid). The two forms exist in equilibrium in aqueous solution.

$$H_5IO_6(aq) \rightleftharpoons H^+(aq) + IO_4^-(aq) + 2\,H_2O(l) \qquad K_{eq} = 0.015 \quad [20.27]$$

HIO_4 is a strong acid and H_5IO_6 is a weak one; the first two acid-dissociation constants for H_5IO_6 are $K_{a1} = 2.8 \times 10^{-2}$ and $K_{a2} = 4.9 \times 10^{-9}$. The structure of H_5IO_6 is given in ▶ **FIGURE 20.13**. The large size of the iodine atom allows it to accommodate six surrounding oxygen atoms. The smaller halogens do not form acids of this type.

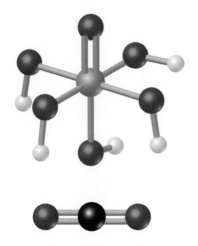

▲ **FIGURE 20.13   Paraperiodic acid (H_5IO_6).**

20.5 | OXYGEN

By the middle of the seventeenth century, scientists recognised that air contained a component associated with burning and breathing. That component was not isolated until 1774, however, when Joseph Priestley (▶ **FIGURE 20.14**) discovered oxygen. Lavoisier subsequently named the element *oxygen*, meaning 'acid former'.

Oxygen is found in combination with other elements in a great variety of compounds. Indeed, oxygen is the most abundant element by mass both in the Earth's crust and in the human body. It is the oxidising agent for the metabolism of our foods and is crucial to human life.

Properties of Oxygen

Oxygen has two allotropes, O_2 and O_3. When we speak of molecular oxygen or simply oxygen, it is usually understood that we are speaking of *dioxygen* (O_2), the normal form of the element; O_3 is called *ozone*.

At room temperature dioxygen is a colourless and odourless gas. It condenses to a liquid at –183 °C and freezes at –218 °C. It is only slightly soluble in water, but its presence in water is essential to marine life.

The electron configuration of the oxygen atom is $[He]2s^2 2p^4$. Thus, oxygen can complete its octet of electrons either by picking up two electrons to form the oxide ion (O^{2-}) or by sharing two electrons. In its covalent compounds it tends to form two bonds: either as two single bonds, as in H_2O, or as a double bond, as in formaldehyde ($H_2C{=}O$). The O_2 molecule itself contains a double bond.

The bond in O_2 is very strong (the bond enthalpy is 495 kJ mol^{-1}). Oxygen also forms strong bonds with many other elements. Consequently, many oxygen-containing compounds are thermodynamically more stable than O_2. In the absence of a catalyst, however, most reactions of O_2 have high activation energies and thus require high temperatures to proceed at a suitable rate. Once a sufficiently exothermic reaction begins, however, it may accelerate rapidly, producing a reaction of explosive violence.

▲ **FIGURE 20.14   Joseph Priestley (1733–1804).** Priestley became interested in chemistry at the age of 39. Priestley lived next door to a brewery from which he could obtain carbon dioxide, so his studies focused on this gas first and were later extended to other gases. Because he was suspected of sympathising with the American and French revolutions, his church, home and laboratory in Birmingham, England, were burned by a mob in 1791. Priestley had to flee in disguise. He eventually emigrated to the United States in 1794, where he lived his remaining years in relative seclusion in Pennsylvania.

Preparation of Oxygen

Nearly all commercial oxygen is obtained from air. The normal boiling point of O_2 is –183 °C, whereas that of N_2, the other principal component of air, is –196 °C. Thus when air is liquefied and then allowed to warm, the N_2 boils off, leaving liquid O_2 contaminated mainly by small amounts of N_2 and Ar.

A common laboratory method for preparing O_2 is the thermal decomposition of potassium chlorate ($KClO_3$), with manganese dioxide (MnO_2) added as a catalyst:

$$2\,KClO_3(s) \xrightarrow{\;MnO_2\;} 2\,KCl(s) + 3\,O_2(g) \qquad [20.28]$$

Much of the O_2 in the atmosphere is replenished through the process of photosynthesis, in which green plants use the energy of sunlight to generate O_2 from atmospheric CO_2. Photosynthesis, therefore, not only regenerates O_2, but also uses up CO_2:

$$6\,CO_2(g) + 6\,H_2O(l) \longrightarrow C_6H_{12}O_6(aq) + 6\,O_2(g)$$

Uses of Oxygen

Oxygen is one of the most widely used industrial chemicals, ranking behind only sulfuric acid (H_2SO_4) and nitrogen (N_2). Oxygen can be shipped and stored either as a liquid or in steel containers as a compressed gas.

Oxygen is by far the most widely used oxidising agent. Over half of the O_2 produced is used in the steel industry, mainly to remove impurities from steel. It is also used to bleach pulp and paper. (Oxidation of coloured compounds often gives colourless products.) In medicine, oxygen eases breathing difficulties. It is also used together with acetylene (C_2H_2) in oxyacetylene welding. The reaction between C_2H_2 and O_2 is highly exothermic, producing temperatures in excess of 3000 °C:

$$2\,C_2H_2(g) + 5\,O_2(g) \longrightarrow 4\,CO_2(g) + 2\,H_2O(g) \qquad \Delta H° = -2510 \text{ kJ} \quad [20.29]$$

Ozone

Ozone is a pale blue poisonous gas with a sharp, irritating odour. Most people can detect about 0.01 ppm in air. Exposure to 0.1 to 1 ppm produces headaches, burning eyes and irritation to the respiratory passages.

The structure of the O_3 molecule is shown in ◀ **FIGURE 20.15**. The molecule possesses a π-bond that is delocalised over the three oxygen atoms (∞ Section 8.6, 'Resonance Structures'). The molecule dissociates readily, forming reactive oxygen atoms:

$$O_3(g) \longrightarrow O_2(g) + O(g) \qquad \Delta H° = 105 \text{ kJ} \qquad [20.30]$$

> 🔺 **CONCEPT CHECK 8**
> What wavelength of light is needed to break an O–O bond in ozone?

Ozone is a stronger oxidising agent than dioxygen. One measure of this oxidising power is the high standard reduction potential of O_3, compared with that of O_2.

$$O_3(g) + 2\,H^+(aq) + 2\,e^- \longrightarrow O_2(g) + H_2O(l) \qquad E° = 2.07 \text{ V} \qquad [20.31]$$

$$O_2(g) + 4\,H^+(aq) + 4\,e^- \longrightarrow 2\,H_2O(l) \qquad E° = 1.23 \text{ V} \qquad [20.32]$$

Ozone forms oxides with many elements under conditions where O_2 will not react; indeed, it oxidises all the common metals except gold and platinum.

Ozone can be prepared by passing electricity through dry O_2 in a flow-through apparatus. The electrical discharge causes rupture of the O_2 bond, resulting in reactions like those described in Section 13.1.

$$3\,O_2(g) \xrightarrow{\;electricity\;} 2\,O_3(g) \qquad \Delta H° = 285 \text{ kJ} \qquad [20.33]$$

Ozone cannot be stored for long, except at low temperature, because it readily decomposes to O_2. The decomposition is catalysed by certain metals, such as Ag, Pt and Pd, and by many transition metal oxides.

∞ Review this on page 272

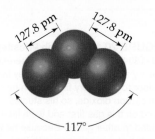

127.8 pm   127.8 pm

117°

🔺 **FIGURE 20.15** **Structure of the ozone molecule.**

SAMPLE EXERCISE 20.6   Calculating an equilibrium constant

Using $\Delta_f G°$ for ozone from Appendix C, calculate the equilibrium constant, K, for Equation 20.33 at 298.0 K, assuming no electrical input.

SOLUTION

Analyse We are asked to calculate the equilibrium constant for the formation of O_3 from O_2, given the temperature and $\Delta_f G°$.

Plan The relationship between the standard free energy change, $\Delta_f G°$, for a reaction and the equilibrium constant for the reaction is given in Section 16.7 as $\Delta_f G° = -RT \ln K$.

Solve From Appendix C we have

$$\Delta_f G°(O_3) = 163.4 \text{ kJ mol}^{-1}$$

Thus for Equation 20.33,

$$\Delta G° = (2 \text{ mol } O_3)(163.4 \text{ kJ mol}^{-1} O_3) = 326.8 \text{ kJ}$$

From Equation 16.24 we have

$$\Delta G° = -RT \ln K$$

Thus

$$\ln K = \frac{-\Delta G°}{RT} = \frac{-326.8 \times 10^3 \text{ J}}{(8.314 \text{ J K}^{-1} \text{ mol}^{-1})(298.0 \text{ K})} = -131.9$$

$$K = e^{-131.19} = 5 \times 10^{-58}$$

Comment Despite the unfavourable equilibrium constant, ozone can be prepared from O_2 as described in the preceding text. The unfavourable free energy of formation is overcome by energy from the electrical discharge, and O_3 is removed before the reverse reaction can occur, so a non-equilibrium mixture results.

PRACTICE EXERCISE

Using the data in Appendix C, calculate $\Delta G°$ and the equilibrium constant (K) for Equation 20.30 at 298.0 K.

Answer: $\Delta G° = 66.7 \text{ kJ}$, $K = 2 \times 10^{-12}$

Ozone is sometimes used to treat domestic water in place of chlorine. Like Cl_2, it kills bacteria and oxidises organic compounds. The largest use of ozone, however, is in the preparation of pharmaceuticals, synthetic lubricants and other commercially useful organic compounds, where O_3 is used to sever carbon–carbon double bonds.

Ozone is an important component of the upper atmosphere, where it screens out ultraviolet radiation. In this way, ozone protects the Earth from the effects of these high-energy rays. For this reason, depletion of stratospheric ozone is a major scientific concern (Section 13.2, 'Human Activities and Earth's Atmosphere'). In the lower atmosphere, however, ozone is considered an air pollutant. It is a major constituent of smog (Section 13.2). Because of its oxidising power, it damages living systems and structural materials, especially rubber.

Review this on page 475

Oxides

The electronegativity of oxygen is second only to that of fluorine. As a result, oxygen exhibits negative oxidation states in all compounds except those with fluorine, OF_2 and O_2F_2. The −2 oxidation state is by far the most common. Compounds in this oxidation state are called *oxides*.

Non-metals form covalent oxides. Most of these oxides are simple molecules with low melting and boiling points. SiO_2 and B_2O_3, however, have polymeric structures (Sections 20.10, 'The Other Group 14 Elements: Si, Ge, Sn and Pb' and 20.11, 'Boron'). Most non-metal oxides combine with water to give oxyacids. Sulfur dioxide (SO_2), for example, dissolves in water to give sulfurous acid (H_2SO_3):

Find out more on pages 838 and 841

$$SO_2(g) + H_2O(l) \longrightarrow H_2SO_3(aq) \qquad [20.34]$$

This reaction and that of SO_3 with H_2O to form H_2SO_4 are largely responsible for acid rain (Section 13.2, 'Human Activities and Earth's Atmosphere').

Review this on page 478

▲

FIGURE IT OUT

Is this reaction a redox reaction?

$$BaO(s) \; + \; H_2O(l) \; \longrightarrow \; Ba(OH)_2(aq)$$

▲ **FIGURE 20.16**   Reaction of a basic oxide with water.

TABLE 20.4 • Acid–base character of chromium oxides

Oxide	Oxidation state of Cr	Nature of oxide
CrO	+2	Basic
Cr_2O_3	+3	Amphoteric
CrO_3	+6	Acidic

∞ Review this on page 737

$$4\,KO_2(s) + 2\,H_2O(l, \text{ from breath}) \longrightarrow$$
$$4\,K^+(aq) + 4\,OH^-(aq) + 3\,O_2(g)$$

$$2\,OH^-(aq) + CO_2(g, \text{ from breath}) \longrightarrow$$
$$H_2O(l) + CO_3^{2-}(aq)$$

▲ **FIGURE 20.17**   A self-contained breathing apparatus.

The analogous reaction of CO_2 with H_2O to form carbonic acid (H_2CO_3) causes the acidity of carbonated water.

Oxides that react with water to form acids are called **acidic anhydrides** (anhydride means 'without water') or **acidic oxides**.

▲

CONCEPT CHECK 9

What acid is produced by the reaction of I_2O_5 with water?

Most metal oxides are ionic compounds. Those ionic oxides that dissolve in water react to form hydroxides and are consequently called **basic anhydrides** or **basic oxides**. Barium oxide (BaO), for example, reacts with water to form barium hydroxide ($Ba(OH)_2$). (▲ **FIGURE 20.16**) These kinds of reactions are due to the high basicity of the O^{2-} ion and its virtually complete hydrolysis in water.

$$O^{2-}(aq) + H_2O(l) \longrightarrow 2\,OH^-(aq) \qquad [20.35]$$

Oxides that can exhibit both acidic and basic characters are said to be *amphoteric* (∞ Section 18.5, 'Factors that Affect Solubility'). If a metal forms more than one oxide, the basic character of the oxide decreases as the oxidation state of the metal increases, as illustrated in ◀ **TABLE 20.4**.

Peroxides and Superoxides

Compounds containing O–O bonds with oxygen in an oxidation state of −1 are called *peroxides*. Oxygen has an oxidation state of $-\frac{1}{2}$ in O_2^-, which is called the *superoxide* ion. The most active metals (K, Rb and Cs) react with O_2 to give superoxides (KO_2, RbO_2 and CsO_2). Their active neighbours in the periodic table (Na, Ca, Sr and Ba) react with O_2, producing peroxides (Na_2O_2, CaO_2, SrO_2 and BaO_2). Less active metals and non-metals produce normal oxides.

When superoxides dissolve in water, O_2 is produced:

$$4\,KO_2(s) + 2\,H_2O(l) \longrightarrow 4\,K^+(aq) + 4\,OH^-(aq) + 3\,O_2(g) \qquad [20.36]$$

Because of this reaction, potassium superoxide (KO_2) is used as an oxygen source in masks worn by rescue workers (◀ **FIGURE 20.17**). Moisture in the breath causes the compound to decompose to form O_2 and KOH. The KOH so formed removes CO_2 from the exhaled breath:

$$2\,OH^-(aq) + CO_2(g) \longrightarrow H_2O(l) + CO_3^{2-}(aq) \qquad [20.37]$$

Hydrogen peroxide (H_2O_2) is the most familiar and commercially important peroxide. The structure of H_2O_2 is shown in ▶ **FIGURE 20.18**. Pure hydrogen peroxide is a clear, syrupy liquid with a density of 1.47 g cm^{-3} at 0 °C. It melts at –0.4 °C, and its normal boiling point is 151 °C. These properties are characteristic of a highly polar, strongly hydrogen-bonded liquid such as water. Concentrated hydrogen peroxide is a dangerously reactive substance because the decomposition to form water and oxygen gas is very exothermic.

$$2\,H_2O_2(l) \longrightarrow 2\,H_2O(l) + O_2(g) \qquad \Delta H^\circ = -196.1\ \text{kJ} \qquad [20.38]$$

Note that Equation 20.38 represents a **disproportionation** reaction. The hydrogen peroxide is simultaneously oxidised to oxygen and reduced to water.

Hydrogen peroxide is marketed as a chemical reagent in aqueous solutions of up to about 30% by mass. A solution containing about 3% H_2O_2 by mass is sold in pharmacies and used as a mild antiseptic. Somewhat more concentrated solutions are used to bleach fabrics.

The peroxide ion is a by-product of metabolism that results from the reduction of O_2. The body disposes of this reactive ion with enzymes such as peroxidase and catalase.

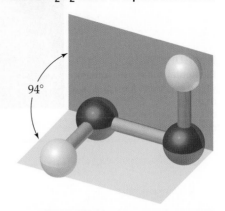

94°

▲ **FIGURE 20.18** Molecular structure of hydrogen peroxide.

20.6 | THE OTHER GROUP 16 ELEMENTS: S, Se, Te AND Po

In addition to oxygen, the group 16 elements are sulfur, selenium, tellurium and polonium. In this section we survey the properties of the group as a whole and then examine the chemistry of sulfur, selenium and tellurium. We do not say much about polonium, which has no stable isotopes and is found only in minute quantities in radium-containing minerals.

General Characteristics of the Group 16 Elements

The group 16 elements possess the general outer-electron configuration ns^2np^4, where n has values ranging from 2 to 6. Thus these elements may attain a noble gas electron configuration by the addition of two electrons, which results in a -2 oxidation state. Because the group 16 elements are non-metals, this is a common oxidation state. Except for oxygen, however, the group 16 elements are also commonly found in positive oxidation states up to $+6$, and they can have expanded valence shells. Thus compounds such as SF_6, SeF_6 and TeF_6 occur in which the central atom is in the $+6$ oxidation state with more than an octet of valence electrons.

▼ **TABLE 20.5** summarises some of the more important properties of the atoms of the group 16 elements. In most of the properties listed in Table 20.5, we

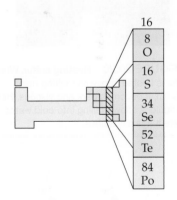

16
8
O
16
S
34
Se
52
Te
84
Po

TABLE 20.5 • Some properties of the group 16 elements				
Property	O	S	Se	Te
Atomic radius (pm)	73	104	117	143
Ionic radius, X^{2-} (pm)	140	184	198	221
First ionisation energy (kJ mol^{-1})	1314	1000	941	869
Electron affinity (kJ mol^{-1})	-141	-200	-195	-190
Electronegativity	3.5	2.5	2.4	2.1
X–X single-bond enthalpy (kJ mol^{-1})	146*	266	172	126
Reduction potential to H_2X in acidic solution (V)	1.23	0.14	-0.40	-0.72

* Based on O–O bond energy in H_2O_2.

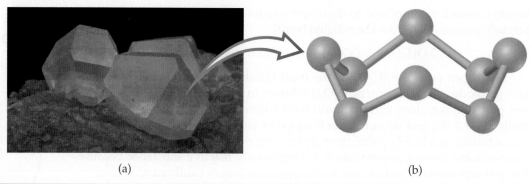

▶ **FIGURE 20.19**
Elemental sulfur. The common yellow crystalline form of rhombic sulfur, S_8, consists of eight-membered puckered rings of S atoms.

(a)　　　　　　　　　　　　　(b)

∞ Review this on page 477

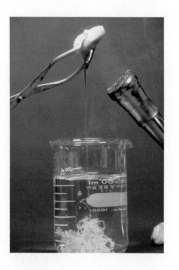

▲ **FIGURE 20.20 Heating sulfur.** When sulfur is heated above its melting point (113 °C), it becomes dark and viscous. Here the liquid is shown falling into cold water, where it again solidifies.

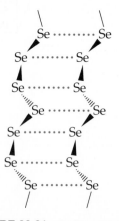

▲ **FIGURE 20.21 Portion of helical chains making up the structure of crystalline selenium.**

see a regular variation as a function of increasing atomic number. For example, atomic and ionic radii increase and ionisation energies decrease, as expected, as we move down the family.

Occurrence and Preparation of S, Se and Te

Large underground deposits are the principal source of elemental sulfur. Sulfur also occurs widely as sulfide and sulfate minerals. Its presence as a minor component of coal and petroleum poses a major problem. Combustion of these 'unclean' fuels leads to serious sulfur oxide pollution (∞ Section 13.2, 'Human Activities and Earth's Atmosphere'). Much effort has been directed at removing this sulfur, and these efforts have increased the availability of sulfur. The sale of this sulfur helps to partially offset the costs of the desulfurising processes and equipment.

Selenium and tellurium occur in rare minerals such as Cu_2Se, $PbSe$, Ag_2Se, Cu_2Te, $PbTe$, Ag_2Te and Au_2Te. They also occur as minor constituents in sulfide ores of copper, iron, nickel and lead.

Properties and Uses of Sulfur, Selenium and Tellurium

As we normally encounter it, sulfur is yellow, tasteless and nearly odourless. It is insoluble in water and exists in several allotropic forms. The thermodynamically stable form at room temperature is rhombic sulfur, which consists of puckered S_8 rings, as shown in ▲ **FIGURE 20.19**. When heated above its melting point (113 °C), sulfur undergoes a variety of changes. The molten sulfur first contains S_8 molecules and is fluid because the rings readily slip over one another. Further heating of this straw-coloured liquid causes the rings to break; the fragments then join to form very long molecules that can become entangled. The sulfur consequently becomes highly viscous. This change is marked by a colour change to dark reddish-brown (◀ **FIGURE 20.20**). Further heating breaks the chains, and the viscosity again decreases.

Most of the sulfur produced worldwide each year is used to manufacture sulfuric acid. Sulfur is also used to vulcanise rubber, a process that toughens rubber by introducing cross-linking between polymer chains (see the My World of Chemistry box in Section 24.6).

The most stable allotropes of both selenium and tellurium are crystalline substances containing helical chains of atoms, as illustrated in ◀ **FIGURE 20.21**. Each atom of the chain is close to atoms in adjacent chains, and it appears that some sharing of electron pairs between these atoms occurs.

The electrical conductivity of selenium is very low in the dark, but increases greatly upon exposure to light. This property of the element is utilised in photoelectric cells and light meters.

Sulfides

Sulfur forms compounds by direct combination with many elements. When the element is less electronegative than sulfur, *sulfides*, which contain S^{2-}, form.

Iron(II) sulfide (FeS) forms, for example, by direct combination of iron and sulfur. Many metallic elements are found in the form of sulfide ores, such as PbS (galena) and HgS (cinnabar). A series of related ores containing the disulfide ion, S_2^{2-} (analogous to the peroxide ion), are known as *pyrites*. Iron pyrite, FeS_2, occurs as golden-yellow cubic crystals (▶ **FIGURE 20.22**). Because it has been occasionally mistaken for gold by miners, it is often called fool's gold.

One of the most important sulfides is hydrogen sulfide (H_2S). This substance is not normally produced by direct union of the elements because it is unstable at elevated temperatures and decomposes into the elements. It is normally prepared by action of dilute acid on iron(II) sulfide.

$$FeS(s) + 2\,H^+(aq) \longrightarrow H_2S(aq) + Fe^{2+}(aq) \qquad [20.39]$$

▲ **FIGURE 20.22   Iron pyrite (FeS₂, on the right) with gold for comparison.**

One of hydrogen sulfide's most readily recognised properties is its smell; H_2S is largely responsible for the offensive odour of rotten eggs. Hydrogen sulfide is actually quite toxic. Fortunately, our noses are able to detect H_2S in extremely low, non-toxic concentrations.

Oxides, Oxyacids and Oxyanions of Sulfur

Sulfur dioxide is formed when sulfur is combusted in air; it has a choking odour and is poisonous. The gas is particularly toxic to lower organisms, such as fungi, so it is used to sterilise dried fruit and wine. At 1 bar pressure and room temperature, SO_2 dissolves in water to produce a solution of about 1.6 M concentration. The SO_2 solution is acidic, and we describe it as sulfurous acid (H_2SO_3).

Salts of SO_3^{2-} (sulfites) and HSO_3^- (hydrogen sulfites or bisulfites) are well known. Small quantities of Na_2SO_3 or $NaHSO_3$ are used as food additives to prevent bacterial spoilage. Because some people are extremely allergic to sulfites, all food products with sulfites must now carry a warning label disclosing their presence.

Although combustion of sulfur in air produces mainly SO_2, small amounts of SO_3 are also formed. The reaction produces chiefly SO_2 because the activation-energy barrier for further oxidation to SO_3 is very high unless the reaction is catalysed. Sulfur trioxide is of great commercial importance because it is the anhydride of sulfuric acid. In the manufacture of sulfuric acid, SO_2 is first obtained by burning sulfur. The SO_2 is then oxidised to SO_3, using a catalyst such as V_2O_5 or platinum. The SO_3 is dissolved in H_2SO_4 because it does not dissolve quickly in water, and then the $H_2S_2O_7$ formed in this reaction, called pyrosulfuric acid, is added to water to form H_2SO_4:

$$SO_3(g) + H_2SO_4(l) \longrightarrow H_2S_2O_7(l) \qquad [20.40]$$

$$H_2S_2O_7(l) + H_2O(l) \longrightarrow 2\,H_2SO_4(l) \qquad [20.41]$$

▲ CONCEPT CHECK 10
What is the net reaction of Equations 20.40 and 20.41?

Commercial sulfuric acid is 98% H_2SO_4. It is a dense, colourless, oily liquid that boils at 340 °C. Sulfuric acid has many useful properties. It is a strong acid, a good dehydrating agent and a moderately good oxidising agent. Its dehydrating ability is demonstrated in ▼ **FIGURE 20.23**.

Sulfuric acid is classified as a strong acid, but only the first hydrogen is completely ionised in aqueous solution. The second hydrogen ionises only partially.

$$H_2SO_4(aq) \longrightarrow H^+(aq) + HSO_4^-(aq)$$

$$HSO_4^-(aq) \rightleftharpoons H^+(aq) + SO_4^{2-}(aq) \qquad K_a = 1.1 \times 10^{-2}$$

Consequently, sulfuric acid forms two series of compounds: sulfates and bisulfates (or hydrogen sulfates). Bisulfate salts are common components of the 'dry

In this reaction, what has happened to the H and O atoms in the sucrose?

▲ **FIGURE 20.23** Sulfuric acid dehydrates table sugar to produce elemental carbon.

acids' used for adjusting the pH of swimming pools and hot tubs; they are also components of many toilet bowl cleaners.

The thiosulfate ion ($S_2O_3^{2-}$) is related to the sulfate ion and is formed by boiling an alkaline solution of SO_3^{2-} with elemental sulfur.

$$8\,SO_3^{2-}(aq) + S_8(s) \longrightarrow 8\,S_2O_3^{2-}(aq) \qquad [20.42]$$

The term *thio* indicates substitution of sulfur for oxygen. The structures of the sulfate and thiosulfate ions are compared in ▼ **FIGURE 20.24**. When acidified, the thiosulfate ion decomposes to form sulfur and H_2SO_3.

Thiosulfate ion is used in quantitative analysis as a reducing agent for iodine:

$$2\,S_2O_3^{2-}(aq) + I_2(s) \longrightarrow 2\,I^-(aq) + S_4O_6^{2-}(aq) \qquad [20.43]$$

What are the oxidation states of the sulfur atoms in the $S_2O_3^{2-}$ ion?

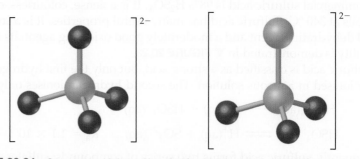

▲ **FIGURE 20.24** Structures of the sulfate (left) and thiosulfate (right) ions.

20.7 | NITROGEN

Nitrogen was discovered in 1772 by the Scottish botanist Daniel Rutherford (1749–1819). He found that when a mouse was enclosed in a sealed jar, the animal quickly consumed the life-sustaining component of air (oxygen) and died. When the 'fixed air' (CO_2) in the container was removed, a 'noxious air' remained that would not sustain combustion or life. That gas is known to us now as nitrogen.

Nitrogen constitutes 78% by volume of the Earth's atmosphere, where it occurs as N_2 molecules. Although nitrogen is a key element in living organisms, compounds of nitrogen are not abundant in the Earth's crust. The major natural deposits of nitrogen compounds are those of KNO_3 (saltpetre) in India and $NaNO_3$ (Chile saltpetre) in Chile and other desert regions of South America.

Properties of Nitrogen

Nitrogen is a colourless, odourless, tasteless gas composed of N_2 molecules. Its melting point is –210 °C and its normal boiling point is –196 °C.

The N_2 molecule is very unreactive because of the strong triple bond between nitrogen atoms (the N–N triple bond enthalpy is 941 kJ mol^{-1}, nearly twice that for the bond in O_2; see Table 8.4). When substances burn in air, they normally react with O_2 but not with N_2. When magnesium burns in air, however, it also reacts with N_2 to form magnesium nitride (Mg_3N_2). A similar reaction occurs with lithium, forming Li_3N.

$$3\,Mg(s) + N_2(g) \longrightarrow Mg_3N_2(s) \qquad [20.44]$$

The nitride ion is a strong Brønsted–Lowry base. It reacts with water to form ammonia (NH_3), as in the following reaction:

$$Mg_3N_2(s) + 6\,H_2O(l) \longrightarrow 2\,NH_3(aq) + 3\,Mg(OH)_2(s) \qquad [20.45]$$

The electron configuration of the nitrogen atom is $[He]2s^2 2p^3$. The element exhibits all formal oxidation states from +5 to −3, as shown in ▶ **TABLE 20.6**. The +5, 0 and −3 oxidation states are the most common and generally the most stable of these. Because nitrogen is more electronegative than all elements except fluorine, oxygen and chlorine, it exhibits positive oxidation states only in combination with these three elements.

Preparation and Uses of Nitrogen

Elemental nitrogen is obtained in commercial quantities by fractional distillation of liquid air.

Because of its low reactivity, large quantities of N_2 are used as an inert gaseous blanket to exclude O_2 during the processing and packaging of foods, the manufacture of chemicals, the fabrication of metals and the production of electronic devices. Liquid N_2 is employed as a coolant to freeze foods rapidly.

The largest use of N_2 is in the manufacture of nitrogen-containing fertilisers, which provide a source of *fixed* nitrogen. We discuss nitrogen fixation in the My World of Chemistry box later in Section 20.7 and in the My World of Chemistry box in Section 16.2. Our starting point in fixing nitrogen is the manufacture of ammonia via the Haber process (∞ Section 16.2, 'The Equilibrium Constant'). The ammonia can then be converted into a variety of useful, simple nitrogen-containing species, as shown in ▼ **FIGURE 20.25**. Many of the reactions along this chain of conversion are discussed in more detail later in this section.

Hydrogen Compounds of Nitrogen

Ammonia is one of the most important compounds of nitrogen. It is a colourless toxic gas that has a characteristic irritating odour. As we note in previous discussions, the NH_3 molecule is basic ($K_b = 1.8 \times 10^{-5}$) (∞ Section 17.7, 'Weak Bases').

TABLE 20.6 • Oxidation states of nitrogen	
Oxidation state	**Examples**
+5	N_2O_5, HNO_3, NO_3^-
+4	NO_2, N_2O_4
+3	HNO_2, NO_2^-, NF_3
+2	NO
+1	N_2O, $H_2N_2O_2$, $N_2O_2^{2-}$, HNF_2
0	N_2
−1	NH_2OH, NH_2F
−2	N_2H_4
−3	NH_3, NH_4^+, NH_2^-, CN^-

∞ Review this on page 618

∞ Review this on page 680

FIGURE IT OUT

In which of these species is the oxidation number of nitrogen +3?

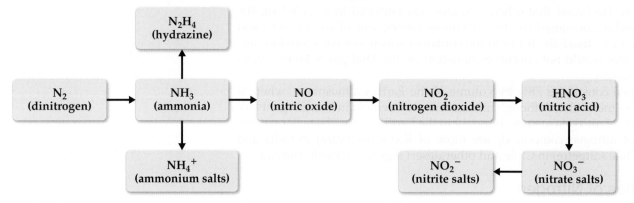

▲ **FIGURE 20.25** Sequence of conversion of N_2 into common nitrogen compounds.

FIGURE IT OUT

Is the N–N bond length in these molecules shorter or longer than the N–N bond length in N_2?

▲ **FIGURE 20.26** Hydrazine (top, N_2H_4) and methylhydrazine (bottom, CH_3NHNH_2).

In the laboratory NH_3 can be prepared by the action of NaOH on an ammonium salt. The NH_4^+ ion, which is the conjugate acid of NH_3, transfers a proton to OH^-. The resultant NH_3 is volatile and is driven from the solution by mild heating.

$$NH_4Cl(aq) + NaOH(aq) \longrightarrow NH_3(g) + H_2O(l) + NaCl(aq) \quad [20.46]$$

Commercial production of NH_3 is achieved by the Haber process.

$$N_2(g) + 3 H_2(g) \longrightarrow 2 NH_3(g) \quad [20.47]$$

About 8.0×10^5 tonnes of ammonia are produced annually in Australia. About 75% is used for fertiliser.

Hydrazine (N_2H_4) bears the same relationship to ammonia that hydrogen peroxide does to water. As shown in ◄ **FIGURE 20.26**, the hydrazine molecule contains an N–N single bond. Hydrazine is quite poisonous. It can be prepared by the reaction of ammonia with hypochlorite ion (OCl^-) in aqueous solution.

$$2 NH_3(aq) + OCl^-(aq) \longrightarrow N_2H_4(aq) + Cl^-(aq) + H_2O(l) \quad [20.48]$$

The reaction is complex, involving several intermediates, including chloramine (NH_2Cl). The poisonous NH_2Cl bubbles out of solution when household ammonia and chlorine bleach (which contains OCl^-) are mixed. This reaction is one reason for the frequently cited warning not to mix bleach and household ammonia.

Pure hydrazine is a colourless, oily liquid that explodes on heating in air. It can be handled safely in aqueous solution, where it behaves as a weak base ($K_b = 1.3 \times 10^{-6}$). The compound is a strong and versatile reducing agent. The major use of hydrazine and compounds related to it, such as methylhydrazine (Figure 20.26), is as rocket fuels.

SAMPLE EXERCISE 20.7 Writing a balanced equation

Hydroxylamine (NH_2OH) reduces copper(II) to the free metal in acid solutions. Write a balanced equation for the reaction, assuming that N_2 is the oxidation product.

SOLUTION

Analyse We are asked to write a balanced oxidation–reduction equation in which NH_2OH is converted to N_2 and Cu^{2+} is converted to Cu.

Plan Because this is a redox reaction, the equation can be balanced by the method of half-reactions discussed in Section 19.2. Thus we begin with two half-reactions, one involving NH_2OH and N_2 and the other involving Cu^{2+} and Cu.

Solve The unbalanced and incomplete half-reactions are

$$Cu^{2+}(aq) \longrightarrow Cu(s)$$

$$NH_2OH(aq) \longrightarrow N_2(g)$$

Balancing these equations as described in Section 19.2 gives

$$Cu^{2+}(aq) + 2\,e^- \longrightarrow Cu(s)$$

$$2\,NH_2OH(aq) \longrightarrow N_2(g) + 2\,H_2O(l) + 2\,H^+(aq) + 2\,e^-$$

Adding these half-reactions gives the balanced equation:

$$Cu^{2+}(aq) + 2\,NH_2OH(aq) \longrightarrow Cu(s) + N_2(g) + 2\,H_2O(l) + 2\,H^+(aq)$$

PRACTICE EXERCISE

(a) In power plants, hydrazine is used to prevent corrosion of the metal parts of steam boilers by the O_2 dissolved in the water. The hydrazine reacts with O_2 in water to give N_2 and H_2O. Write a balanced equation for this reaction. **(b)** Methylhydrazine, $N_2H_3CH_3(l)$, is used with the oxidiser dinitrogen tetroxide, $N_2O_4(l)$, to power the steering rockets of the space shuttle, *Orbiter*. The reaction of these two substances produces N_2, CO_2 and H_2O. Write a balanced equation for this reaction.

Answers: **(a)** $N_2H_4(aq) + O_2(aq) \longrightarrow N_2(g) + 2\,H_2O(l)$;
(b) $5\,N_2O_4(l) + 4\,N_2H_3CH_3(l) \longrightarrow 9\,N_2(g) + 4\,CO_2(g) + 12\,H_2O(g)$

(See also Exercise 20.46.)

MY WORLD OF CHEMISTRY

NITROGEN FIXATION AND NITROGENASE

Nitrogen is one of the essential elements in living organisms, found in many compounds vital to life, including proteins, nucleic acids, vitamins and hormones. Nitrogen is continually cycling through the biosphere in various forms, as shown in ▶ FIGURE 20.27. For example, certain microorganisms convert the nitrogen in animal waste and dead plants and animals into $N_2(g)$, which then returns to the atmosphere. For the food chain to be sustained, there must be a means of converting atmospheric $N_2(g)$ to a form plants can use. For this reason, if a chemist were asked to name the most important chemical reaction in the world, she might easily say *nitrogen fixation*, the process by which atmospheric $N_2(g)$ is converted into compounds suitable for plant use. Some fixed nitrogen results from the action of lightning on the atmosphere, and some is produced industrially using a process we discussed in Section 16.8. About 60% of fixed nitrogen, however, is a consequence of the action of the remarkable and complex enzyme *nitrogenase*. This enzyme is *not* present in humans or other animals; rather, it is found in bacteria that live in the root nodules of certain plants, such as the legumes clover and alfalfa.

Nitrogenase converts N_2 into NH_3, a process that, in the absence of a catalyst, has a very large activation energy. This process is a *reduction* reaction in which the oxidation state of N is reduced from 0 in N_2 to -3 in NH_3. The mechanism by which nitrogenase reduces N_2 is not fully understood. Like many other enzymes, including catalase, the active site of nitrogenase contains transition metal atoms; such enzymes are called *metalloenzymes*. Because transition metals can readily change

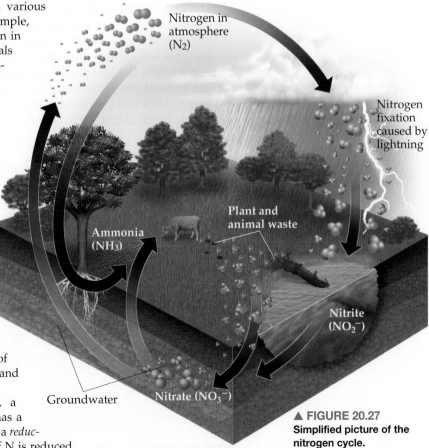

▲ FIGURE 20.27
Simplified picture of the nitrogen cycle.

oxidation state, metalloenzymes are especially useful for effecting transformations in which substrates are either oxidised or reduced.

It has been known for nearly 30 years that a portion of nitrogenase contains iron and molybdenum atoms. This portion, called the *FeMo-cofactor*, is thought to serve as the active site of the enzyme. The FeMo-cofactor of nitrogenase is a cluster of seven Fe atoms and one Mo atom, all linked by sulfur atoms (▶ FIGURE 20.28).

It is one of the wonders of life that simple bacteria can contain beautifully complex and vitally

important enzymes such as nitrogenase. Because of this enzyme, nitrogen is continually cycled between its comparatively inert role in the atmosphere and its critical role in living organisms. Without nitrogenase, life as we know it could not exist on Earth.

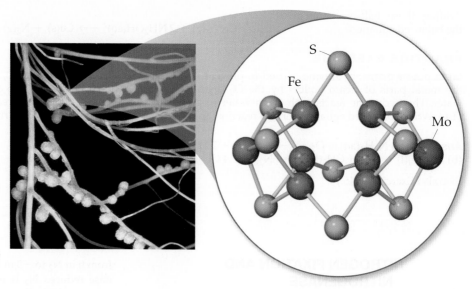

▶ FIGURE 20.28 **The FeMo-cofactor of nitrogenase.** Nitrogenase is found in nodules in the roots of certain plants, such as the white clover roots shown at the left. The cofactor, which is thought to be the active site of the enzyme, contains seven Fe atoms and one Mo atom, linked by sulfur atoms. The molecules on the outside of the cofactor connect it to the rest of the protein.

Oxides and Oxyacids of Nitrogen

Nitrogen forms three common oxides: N_2O (nitrous oxide), NO (nitric oxide) and NO_2 (nitrogen dioxide). It also forms two unstable oxides that we do not discuss, N_2O_3 (dinitrogen trioxide) and N_2O_5 (dinitrogen pentoxide).

Nitrous oxide (N_2O) is also known as laughing gas because a person becomes somewhat giddy after inhaling only a small amount of it. This colourless gas was the first substance used as a general anaesthetic. It is used as the compressed gas propellant in several aerosols and foams, such as in whipped cream. It can be prepared in the laboratory by carefully heating ammonium nitrate to about 200 °C.

$$NH_4NO_3(s) \xrightarrow{\Delta} N_2O(g) + 2\,H_2O(g) \qquad [20.49]$$

Nitric oxide (NO) is also a colourless gas but, unlike N_2O, it is slightly toxic. It can be prepared in the laboratory by reduction of dilute nitric acid, using copper or iron as a reducing agent, as shown in ▼ FIGURE 20.29.

$$3\,Cu(s) + 2\,NO_3^-(aq) + 8\,H^+(aq) \longrightarrow 3\,Cu^{2+}(aq) + 2\,NO(g) + 4\,H_2O(l)$$
$$[20.50]$$

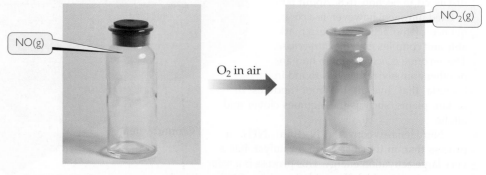

▲ FIGURE 20.29 **Formation of NO_2(g) as NO(g) combines with O_2(g) in the air.**

It is also produced by direct reaction of N_2 and O_2 at high temperatures. This reaction is a significant source of nitrogen oxide air pollutants (Section 13.2, 'Human Activities and Earth's Atmosphere'). The direct combination of N_2 and O_2 is not used for commercial production of NO, however, because the yield is low; the equilibrium constant K_P at 2400 K is only 0.05.

Review this on page 479

The commercial route to NO (and hence to other oxygen-containing compounds of nitrogen) is via the catalytic oxidation of NH_3.

$$4\,NH_3(g) + 5\,O_2(g) \xrightarrow[850\,°C]{Pt\ catalyst} 4\,NO(g) + 6\,H_2O(g) \qquad [20.51]$$

The catalytic conversion of NH_3 to NO is the first step in a three-step process known as the **Ostwald process**, by which NH_3 is converted commercially into nitric acid (HNO_3). Nitric oxide reacts readily with O_2, forming NO_2 when exposed to air (Figure 20.29).

$$2\,NO(g) + O_2(g) \longrightarrow 2\,NO_2(g) \qquad [20.52]$$

When dissolved in water, NO_2 forms nitric acid.

$$3\,NO_2(g) + H_2O(l) \longrightarrow 2\,H^+(aq) + 2\,NO_3^-(aq) + NO(g) \qquad [20.53]$$

Nitrogen is both oxidised and reduced in this reaction, so it disproportionates. The reduction product NO can be converted back into NO_2 by exposure to air (Equation 20.52) and thereafter dissolved in water to prepare more HNO_3.

Recently NO has been found to be an important neurotransmitter in the human body. It causes the muscles that line blood vessels to relax, thus allowing an increased passage of blood (see the My World of Chemistry box below).

Nitrogen dioxide (NO_2) is a yellow-brown gas (Figure 20.29). Like NO, it is a major constituent of smog (Section 13.2, 'Human Activities and Earth's Atmosphere'). It is poisonous and has a choking odour. As discussed in Section 16.1, NO_2 and N_2O_4 exist in equilibrium:

Review this on page 479

$$2\,NO_2(g) \rightleftharpoons N_2O_4(g) \qquad \Delta H° = -58\ kJ \qquad [20.54]$$

The two common oxyacids of nitrogen are nitric acid (HNO_3) and nitrous acid (HNO_2) (▶ FIGURE 20.30). *Nitric acid* is a colourless, corrosive liquid.

Nitric acid is a strong acid. It is also a powerful oxidising agent, as the following standard reduction potentials indicate:

$$NO_3^-(aq) + 2\,H^+(aq) + e^- \longrightarrow NO_2(g) + H_2O(l) \qquad E° = +0.79\ V \quad [20.55]$$

$$NO_3^-(aq) + 4\,H^+(aq) + 3\,e^- \longrightarrow NO(g) + 2\,H_2O(l) \qquad E° = +0.96\ V \quad [20.56]$$

Concentrated nitric acid will attack and oxidise most metals except Au, Pt, Rh and Ir.

FIGURE IT OUT

Which is the shortest NO bond in these two molecules?

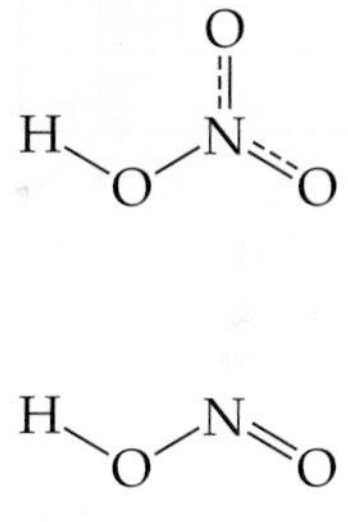

▲ FIGURE 20.30 **Structures of nitric acid (top) and nitrous acid (bottom).**

MY WORLD OF CHEMISTRY

NITROGLYCERIN AND HEART DISEASE

The 1870s an interesting observation was made in Alfred Nobel's dynamite factories. Workers who suffered from heart disease that caused chest pains when they exerted themselves found relief from the pains during the work-week. It quickly became apparent that nitroglycerin, present in the air of the factory, acted to enlarge blood vessels. Thus this powerfully explosive chemical became a standard treatment for angina pectoris, the chest pains accompanying heart failure. It took more than 100 years to discover that nitroglycerin was converted in the vascular smooth muscle to NO, which is the chemical agent actually causing dilation of the blood vessels. In 1998 the Nobel Prize in Physiology or Medicine was awarded to Robert F. Furchgott, Louis J. Ignarro and Ferid Murad for their discoveries of the detailed pathways by which NO acts in the cardiovascular system. It was a sensation that this simple, common air pollutant could exert important functions in the organism.

As useful as nitroglycerin is to this day in treating angina pectoris, it has a limitation in that prolonged administration results in development of tolerance, or desensitisation, of the vascular muscle to further vasorelaxation by nitroglycerin. The bioactivation of nitroglycerin is the subject of active research in the hope that a means of circumventing desensitisation can be found.

Its largest use is in the manufacture of NH_4NO_3 for fertilisers, which accounts for about 80% of that produced. HNO_3 is also used in the production of plastics, drugs and explosives.

Among the explosives made from nitric acid are nitroglycerin, trinitro-toluene (TNT) and nitrocellulose.

The following reaction occurs when nitroglycerin explodes:

$$4\,C_3H_5N_3O_9(l) \longrightarrow 6\,N_2(g) + 12\,CO_2(g) + 10\,H_2O(g) + O_2(g) \quad [20.57]$$

All the products of this reaction contain very strong bonds. As a result, the reaction is very exothermic. Furthermore, a tremendous amount of gaseous products forms from the liquid. The sudden formation of these gases, together with their expansion resulting from the heat generated by the reaction, produces the explosion.

Nitrous acid (HNO_2) (Figure 20.30) is considerably less stable than HNO_3 and tends to disproportionate into NO and HNO_3. It is normally made by action of a strong acid, such as H_2SO_4, on a cold solution of a nitrite salt, such as $NaNO_2$. Nitrous acid is a weak acid ($K_a = 4.5 \times 10^{-4}$).

 CONCEPT CHECK 11

What are the oxidation numbers of the nitrogen atoms in
a.   nitric acid
b.   nitrous acid?

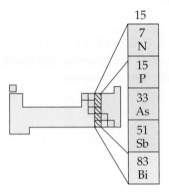

20.8 | THE OTHER GROUP 15 ELEMENTS: P, As, Sb AND Bi

Of the other group 15 elements—phosphorus, arsenic, antimony and bismuth—phosphorus has a central role in several aspects of biochemistry and environmental chemistry. In this section we explore the chemistry of these other group 15 elements, with an emphasis on the chemistry of phosphorus.

General Characteristics of the Group 15 Elements

The group 15 elements possess the outer-shell electron configuration ns^2np^3, where n has values ranging from 2 to 6. A noble gas configuration results from the addition of three electrons to form the −3 oxidation state. Ionic compounds containing X^{3-} ions are not common, however, except for salts of the more active metals, such as in Li_3N. More commonly, the group 15 element acquires an octet of electrons via covalent bonding. The oxidation number may range from −3 to +5, depending on the nature and number of the atoms to which the group 15 element is bonded.

Because of its lower electronegativity, phosphorus is found more frequently in positive oxidation states than is nitrogen. Some of the important properties of the group 15 elements are listed in ▼ TABLE 20.7.

The variation in properties among the elements of group 15 is more striking than that seen in groups 16 and 17. Nitrogen, at the one extreme, exists as a gaseous diatomic molecule; it is clearly non-metallic in character. At the other

TABLE 20.7 • Properties of the group 15 elements					
Property	**N**	**P**	**As**	**Sb**	**Bi**
Atomic radius (pm)	75	110	121	141	155
First ionisation energy (kJ mol^{-1})	1402	1012	947	834	703
Electron affinity (kJ mol^{-1})	>0	−72	−78	−103	−91
Electronegativity	3.0	2.1	2.0	1.9	1.9
X–X single-bond enthalpy (kJ mol^{-1})*	163	200	150	120	—
X–X triple-bond enthalpy (kJ mol^{-1})	941	490	380	295	192

* Approximate values only.

extreme, bismuth is a reddish-white, metallic-looking substance that has most of the characteristics of a metal.

Due to difficulties in obtaining X–X single-bond enthalpies from thermochemical experiments, the single bond data are not very reliable. However, one can see from the data in Table 20.7 that the N–N single bond has a low value for the bond enthalpy but this increases for the P–P single bond. However, the triple-bond enthalpy for nitrogen is much larger than that of phosphorus. This helps us to appreciate why nitrogen alone of the group 15 elements exists as a diatomic molecule in its stable state at 25 °C. All the other elements exist in structural forms with single bonds between the atoms.

Occurrence, Isolation and Properties of Phosphorus

Phosphorus occurs mainly in the form of phosphate minerals. The principal source of phosphorus is phosphate rock, which contains phosphate principally as $Ca_3(PO_4)_2$. The element is produced commercially by reduction of calcium phosphate with carbon in the presence of SiO_2:

$$2\,Ca_3(PO_4)_2(s) + 6\,SiO_2(s) + 10\,C(s) \xrightarrow{1500\,°C}$$
$$P_4(g) + 6\,CaSiO_3(l) + 10\,CO(g) \qquad [20.58]$$

The phosphorus produced in this fashion is the allotrope known as white phosphorus. This form distils from the reaction mixture as the reaction proceeds.

Phosphorus exists in several allotropic forms. White phosphorus consists of P_4 tetrahedra (▶ FIGURE 20.31). The 60° bond angles in P_4 are unusually small for molecules, so there is much strain in the bonding, which is consistent with the high reactivity of white phosphorus. This allotrope bursts spontaneously into flames if exposed to air. It is a white waxlike solid that melts at 44.2 °C and boils at 280 °C. When heated in the absence of air to about 400 °C, it is converted to a more stable allotrope known as red phosphorus. This form does not ignite on contact with air. It is also considerably less poisonous than the white form. We denote elemental phosphorus simply as P(s).

Phosphorus Halides

Phosphorus forms a wide range of compounds with the halogens, the most important of which are the trihalides and pentahalides.

Phosphorus chlorides, bromides and iodides can be made by the direct oxidation of elemental phosphorus with the elemental halogen. PCl_3, for example, which is a liquid at room temperature, is made by passing a stream of dry chlorine gas over white or red phosphorus.

$$2\,P(s) + 3\,Cl_2(g) \longrightarrow 2\,PCl_3(l) \qquad [20.59]$$

If excess chlorine gas is present, an equilibrium is established between PCl_3 and PCl_5.

$$PCl_3(l) + Cl_2(g) \rightleftharpoons PCl_5(s) \qquad [20.60]$$

The phosphorus halides hydrolyse on contact with water. The reactions occur readily, and most of the phosphorus halides fume in air as a result of reaction with water vapour. In the presence of excess water the products are the corresponding phosphorus oxyacid and hydrogen halide.

$$PBr_3(l) + 3\,H_2O(l) \longrightarrow H_3PO_3(aq) + 3\,HBr(aq) \qquad [20.61]$$

$$PCl_5(l) + 4\,H_2O(l) \longrightarrow H_3PO_4(aq) + 5\,HCl(aq) \qquad [20.62]$$

Oxy Compounds of Phosphorus

Probably the most significant compounds of phosphorus are those in which the element is combined with oxygen. Phosphorus(III) oxide (P_4O_6) is obtained by

White phosphorus   Red phosphorus

▲ **FIGURE 20.31   White and red phosphorus.** Despite the fact that both contain nothing but phosphorus atoms, these two forms of phosphorus differ greatly in reactivity. The white allotrope, which reacts violently with oxygen, must be stored under water so that it is not exposed to air. The much less reactive red form does not need to be stored this way.

How do the electron domains about P in P_4O_6 differ from those about P in P_4O_{10}?

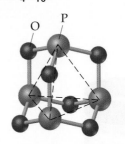

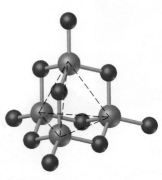

▲ **FIGURE 20.32** Structures of P_4O_6 (top) and P_4O_{10} (bottom).

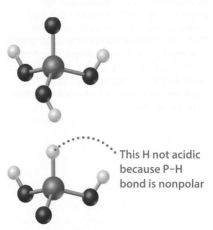

This H not acidic because P–H bond is nonpolar

▲ **FIGURE 20.33** Structures of H_3PO_4 (top) and H_3PO_3 (bottom).

---

* Note that the element phosphor*us* (FOS · for · us) has a -*us* suffix, whereas phosphor*ous* (fos · FOR · us) acid has an -*ous* suffix.

allowing white phosphorus to oxidise in a limited supply of oxygen. When oxidation takes place in the presence of excess oxygen, phosphorus(V) oxide (P_4O_{10}) forms. This compound is also readily formed by oxidation of P_4O_6. These two oxides represent the two most common oxidation states for phosphorus, +3 and +5. The structural relationship between P_4O_6 and P_4O_{10} is shown in ◀ **FIGURE 20.32**. Notice the resemblance these molecules have to the P_4 molecule (Figure 20.31); all three substances have a P_4 core.

SAMPLE EXERCISE 20.8 | **Calculating a standard enthalpy change**

The reactive chemicals on the tip of a 'strike anywhere' match are usually P_4S_3 and an oxidising agent such as $KClO_3$. When the match is struck on a rough surface, the heat generated by the friction ignites the P_4S_3 and the oxidising agent brings about rapid combustion. The products of the combustion of P_4S_3 are P_4O_{10} and SO_2. Calculate the standard enthalpy change for the combustion of P_4S_3 in air, given the following standard enthalpies of formation: P_4S_3 (-154.4 kJ mol^{-1}), P_4O_{10} (-2940 kJ mol^{-1}), SO_2 (-296.9 kJ mol^{-1}).

SOLUTION

Analyse We are given the reactants (P_4S_3 and O_2 from air) and the products (P_4O_{10} and SO_2) for a reaction, together with their standard enthalpies of formation, and asked to calculate the standard enthalpy change for the reaction.

Plan We first need a balanced chemical equation for the reaction. The enthalpy change for the reaction is then equal to the enthalpies of formation of products minus those of reactants (Equation 14.27, Section 14.7). We also need to recall that the standard enthalpy of formation of any element in its standard state is zero. Thus $\Delta_f H°(O_2) = 0$.

Solve The balanced chemical equation for the combustion is

$$P_4S_3(s) + 8\,O_2(g) \longrightarrow P_4O_{10}(s) + 3\,SO_2(g)$$

Thus we can write

$$\Delta H° = \Delta_f H°(P_4O_{10}) + 3\,\Delta_f H°(SO_2) - \Delta_f H°(P_4S_3) - 8\,\Delta_f H°(O_2)$$

$$= -2940\text{ kJ} + 3(-296.9)\text{ kJ} - (-154.4\text{ kJ}) - 8(0)$$

$$= -3676\text{ kJ}$$

Comment The reaction is strongly exothermic, making it evident why P_4S_3 is used on match tips.

PRACTICE EXERCISE

Write the balanced equation for the reaction of P_4O_{10} with water, and calculate $\Delta H°$ for this reaction using data from Appendix C.

Answer: $P_4O_{10}(s) + 6\,H_2O(l) \longrightarrow 4\,H_3PO_4(aq)$; -498.0 kJ

(See also Exercise 20.52.)

Phosphorus(V) oxide is the anhydride of phosphoric acid (H_3PO_4), a weak triprotic acid. In fact, P_4O_{10} has a very high affinity for water and is consequently used as a drying agent. Phosphorus(III) oxide is the anhydride of phosphorous acid (H_3PO_3), a weak diprotic acid.* The structures of H_3PO_4 and H_3PO_3 are shown in ◀ **FIGURE 20.33**. The hydrogen atom that is attached directly to phosphorus in H_3PO_3 is not acidic because the P–H bond is essentially non-polar.

One characteristic of phosphoric and phosphorous acids is their tendency to undergo condensation reactions when heated. A **condensation reaction** is one in which two or more molecules combine to form a larger molecule by eliminating a small molecule, such as H_2O. The reaction in which two H_3PO_4 molecules are joined by the elimination of one H_2O molecule to form $H_4P_2O_7$ is shown below.

These atoms are eliminated as H_2O

[22.63]

Phosphoric acid and its salts find their most important uses in detergents and fertilisers. The phosphates in detergents are often in the form of sodium tripolyphosphate ($Na_5P_3O_{10}$).

The phosphate ions form bonds with metal ions that contribute to the hardness of water. This keeps the metal ions from interfering with the action of the surfactant. The phosphates also keep the pH above 7 and thus prevent the surfactant molecules from becoming protonated.

Most mined phosphate rock is converted to fertilisers. The $Ca_3(PO_4)_2$ in phosphate rock is insoluble ($K_{sp} = 2.0 \times 10^{-29}$). It is converted to a soluble form for use in fertilisers by treating the phosphate rock with sulfuric or phosphoric acid.

$$Ca_3(PO_4)_2(s) + 3\,H_2SO_4(aq) \longrightarrow 3\,CaSO_4(s) + 2\,H_3PO_4(aq) \quad [20.64]$$

$$Ca_3(PO_4)_2(s) + 4\,H_3PO_4(aq) \longrightarrow 3\,Ca^{2+}(aq) + 6\,H_2PO_4^{-}(aq) \quad [20.65]$$

The mixture formed when ground phosphate rock is treated with sulfuric acid and then dried and pulverised is known as *superphosphate*. The $CaSO_4$ formed in this process is of little use in soil except when deficiencies in calcium or sulfur exist. It also dilutes the phosphorus, which is the nutrient of interest. If the phosphate rock is treated with phosphoric acid, the product contains no $CaSO_4$ and has a higher percentage of phosphorus. This product is known as *triple superphosphate*. Although the solubility of $Ca(H_2PO_4)_2$ allows it to be assimilated by plants, it also allows it to be washed from the soil and into bodies of water, thereby contributing to water pollution (∞ Section 13.4, 'Human Activities and Earth's Water').

Phosphorus compounds are important in biological systems. The element occurs in phosphate groups in RNA and DNA, the molecules responsible for the control of protein biosynthesis and transmission of genetic information (∞ Section 29.4, 'Nucleic Acids and DNA').

∞ Review this on page 485

∞ Find out more on page 1182

20.9 | CARBON

Carbon constitutes only 0.027% of the Earth's crust, so it is not an abundant element. Although some carbon occurs in elemental form as graphite and diamond, most is found in combined form. Over half occurs in carbonate compounds, such as $CaCO_3$. Carbon is also found in coal, petroleum and natural gas. The importance of the element stems in large part from its occurrence in all living organisms. Life as we know it is based on carbon compounds.

∞ Review this on page 413

Elemental Forms of Carbon

Carbon exists in five allotropic crystalline forms: graphite, diamond, graphene fullerenes and carbon nanotubes (∞ Section 11.9, 'Bonding in Solids'). *Graphite* is a soft, black, slippery solid that has a metallic lustre and conducts electricity. It consists of parallel sheets of carbon atoms; the sheets are held together by London forces (Figure 11.48(b)).

Diamond is a clear, hard solid in which the carbon atoms form a covalent network (Figure 11.48(a)). Diamond is denser than graphite ($\rho = 2.25$ g cm^{-3} for graphite; $\rho = 3.51$ g cm^{-3} for diamond). At very high pressures and temperatures (in the order of 100 000 bar at 3000 °C) graphite converts to diamond (▶ FIGURE 20.34). About 3×10^4 kg of industrial-grade diamonds are synthesised each year, mainly for use in cutting, grinding and polishing tools.

Graphite has a well-defined crystalline structure, but it also exists in two common amorphous forms: **carbon black** and **charcoal**.

Carbon black is formed when hydrocarbons such as methane are heated in a very limited supply of oxygen.

$$CH_4(g) + O_2(g) \longrightarrow C(s) + 2\,H_2O(g) \quad [20.66]$$

It is used as a pigment in black inks; large amounts are also used in making car tyres. Charcoal is formed when wood is heated strongly in the absence of air.

▲ **FIGURE 20.34 Synthetic diamonds.** Graphite and synthetic diamonds prepared from graphite. Most synthetic diamonds lack the size, colour and clarity of natural diamonds and are therefore not used in jewellery.

Charcoal has a very open structure, giving it an enormous surface area per unit mass. Activated charcoal, a pulverised form whose surface is cleaned by heating with steam, is widely used to adsorb molecules. It is used in filters to remove offensive odours from air and coloured or bad-tasting impurities from water. *Coke* is an impure form of carbon formed when coal is heated strongly in the absence of air. It is widely used as a reducing agent in metallurgical operations. Other less common allotropes of carbon include buckminsterfullerene, graphene and carbon nanotubes (⚭ Section 22.6, 'Cycloalkanes').

⚭ Find out more on page 908

Oxides of Carbon

Carbon forms two principal oxides: carbon monoxide (CO) and carbon dioxide (CO_2). *Carbon monoxide* is formed when carbon or hydrocarbons are burned in a limited supply of oxygen.

$$2\,C(s) + O_2(g) \longrightarrow 2\,CO(g) \qquad [20.67]$$

⚭ Find out more on page 864

It is a colourless, odourless, tasteless gas (mp = $-205\,°C$; bp = $-192\,°C$). It is toxic because it can bind to haemoglobin and thus interfere with oxygen transport (⚭ Section 21.3, 'Ligands with More than One Donor Atom'). Low-level poisoning results in headaches and drowsiness; high-level poisoning can cause death. Carbon monoxide is produced by car engines and is a major air pollutant.

Carbon monoxide is unusual in that it has a lone pair of electrons on carbon ($:C{\equiv}O:$). It is also isoelectronic with N_2, so you might imagine that CO would be equally unreactive. Moreover, both substances have high bond energies ($1072\ \text{kJ mol}^{-1}$ for C–O triple bond and $941\ \text{kJ mol}^{-1}$ for N–N triple bond). Because of the lower nuclear charge on carbon (compared with either N or O), however, the lone pair on carbon is not held as strongly as that on N or O.

MY WORLD OF CHEMISTRY

CARBON FIBRES AND COMPOSITES

The properties of graphite are anisotropic; that is, they differ in different directions through the solid. Along the carbon planes, graphite possesses great strength because of the number and strength of the carbon–carbon bonds in this direction. The bonds between planes are relatively weak, however, making graphite weak in that direction.

Fibres of graphite can be prepared in which the carbon planes are aligned to varying extents parallel to the fibre axis. These fibres are lightweight (density of about 2 g cm^{-3}) and chemically quite unreactive. The oriented fibres are made by first slowly pyrolysing (decomposing by action of heat) organic fibres at about 150 °C to 300 °C. These fibres are then heated to about 2500 °C to graphitise them (convert amorphous carbon to graphite). Stretching the fibre during pyrolysis helps orient the graphite planes parallel to the fibre axis. More amorphous carbon fibres are formed by pyrolysis of organic fibres at lower temperatures (1200 °C to 400 °C). These amorphous materials, commonly called carbon fibres, are the type most often used in commercial materials.

Composite materials that take advantage of the strength, stability and low density of carbon fibres are widely used. Composites are combinations of two or more materials. These materials are present as separate phases and are combined to form structures that take advantage of certain desirable properties of each component. In carbon composites the graphite fibres are often woven into a fabric that is embedded in a matrix that binds them into a solid structure. The fibres transmit loads evenly throughout the matrix. The finished composite is thus stronger than any one of its components.

Carbon composite materials are used widely in a number of applications, including high-performance graphite sports equipment such as tennis racquets, golf clubs and bicycle wheels (▼ FIGURE 20.35). Heat-resistant composites are required for many aerospace applications, where carbon composites now find wide use.

▲ **FIGURE 20.35** Carbon composites in commercial products.

Consequently CO is better able to function as an electron-pair donor (Lewis base) than is N_2. It forms a wide variety of covalent compounds, known as metal carbonyls, with transition metals. $Ni(CO)_4$, for example, is a volatile, toxic compound that is formed by simply warming metallic nickel in the presence of CO. The formation of metal carbonyls is the first step in the transition metal catalysis of a variety of reactions of CO.

Carbon monoxide has several commercial uses. Because it burns readily, forming CO_2, it is employed as a fuel.

$$2\,CO(g) + O_2(g) \longrightarrow 2\,CO_2(g) \qquad \Delta H° = -566\,\text{kJ} \qquad [20.68]$$

It is also an important reducing agent, widely used in metallurgical operations to reduce metal oxides, such as the iron oxides in blast furnaces.

$$Fe_3O_4(s) + 4\,CO(g) \longrightarrow 3\,Fe(s) + 4\,CO_2(g) \qquad [20.69]$$

Carbon dioxide is produced when carbon or carbon-containing substances are burned in excess oxygen.

$$C(s) + O_2(g) \longrightarrow CO_2(g) \qquad [20.70]$$

$$C_2H_5OH(l) + 3\,O_2(g) \longrightarrow 2\,CO_2(g) + 3\,H_2O(g) \qquad [20.71]$$

It is also produced when many carbonates are heated.

$$CaCO_3(s) \xrightarrow{\Delta} CaCO(s) + CO_2(g) \qquad [20.72]$$

Large quantities are also obtained as a by-product of the fermentation of sugar during the production of ethanol.

$$\underset{\text{Glucose}}{C_6H_{12}O_6(aq)} \xrightarrow{\text{yeast}} \underset{\text{Ethanol}}{2\,C_2H_5OH(aq)} + 2\,CO_2(g) \qquad [20.73]$$

In the laboratory, CO_2 can be produced by the action of acids on carbonates, as shown in ▶ **FIGURE 20.36**:

$$CO_3{}^{2-}(aq) + 2\,H^+(aq) \longrightarrow CO_2(g) + H_2O(l) \qquad [20.74]$$

Carbon dioxide is a colourless and odourless gas. It is a minor component of the Earth's atmosphere but a major contributor to the so-called greenhouse effect (∞ Section 13.2, 'Human Activities and Earth's Atmosphere'). Although it is not toxic, high concentrations increase respiration rate and can cause suffocation. It is readily liquefied by compression. When cooled at atmospheric pressure, however, it condenses as a solid rather than a liquid. The solid sublimes at atmospheric pressure at $-78\,°C$. This property makes solid CO_2 valuable as a refrigerant that is always free of the liquid form. Solid CO_2 is known as dry ice. About half the CO_2 consumed annually is used for refrigeration. The other major use is in the production of carbonated beverages. Large quantities are also used to manufacture *washing soda* ($Na_2CO_3 \cdot 10\,H_2O$) and *baking soda* ($NaHCO_3$). Baking soda is so named because the following reaction occurs in baking:

∞ Review this on page 480

$$NaHCO_3(s) + H^+(aq) \longrightarrow Na^+(aq) + CO_2(g) + H_2O(l) \qquad [20.75]$$

The $H^+(aq)$ is provided by vinegar, sour milk or the hydrolysis of certain salts. The bubbles of CO_2 that form are trapped in the dough, causing it to rise. Washing soda is used to precipitate metal ions that interfere with the cleansing action of soap.

CONCEPT CHECK 12

Yeast are living organisms that make bread rise in the absence of baking powder and acid. What must the yeast be producing to make the bread rise?

Carbonic Acid and Carbonates

Carbon dioxide is moderately soluble in H_2O at atmospheric pressure. The resultant solutions are moderately acidic because of the formation of carbonic acid (H_2CO_3).

$$CO_2(aq) + H_2O(l) \rightleftharpoons H_2CO_3(aq) \qquad [20.76]$$

▲ **FIGURE 20.36** CO_2 formation from the reaction between an acid and calcium carbonate in rock.

Carbonic acid is a weak diprotic acid. Its acidic character causes carbonated beverages to have a sharp, slightly acidic taste.

Although carbonic acid cannot be isolated as a pure compound, hydrogen carbonates (bicarbonates) and carbonates can be obtained by neutralising carbonic acid solutions. Partial neutralisation produces HCO_3^- and complete neutralisation gives CO_3^{2-}.

The HCO_3^- ion is a stronger base than acid ($K_b = 2.3 \times 10^{-8}$; $K_a = 5.6 \times 10^{-11}$). Consequently aqueous solutions of HCO_3^- are weakly alkaline.

$$HCO_3^-(aq) + H_2O(l) \rightleftharpoons H_2CO_3(aq) + OH^-(aq) \qquad [20.77]$$

The carbonate ion is much more strongly basic ($K_b = 1.8 \times 10^{-4}$).

$$CO_3^{2-}(aq) + H_2O(l) \rightleftharpoons HCO_3^-(aq) + OH^-(aq) \qquad [20.78]$$

Minerals containing the carbonate ion are plentiful. The principal carbonate minerals are calcite ($CaCO_3$), magnesite ($MgCO_3$), dolomite ($MgCa(CO_3)_2$) and siderite ($FeCO_3$). Calcite is the principal mineral in limestone rock, large deposits of which occur in many parts of the world. It is also the main constituent of marble, chalk, pearls, coral reefs and the shells of marine animals such as clams and oysters. Although $CaCO_3$ has low solubility in pure water, it dissolves readily in acidic solutions with evolution of CO_2.

$$CaCO_3(s) + 2H^+(aq) \rightleftharpoons Ca^{2+}(aq) + H_2O(l) + CO_2(g) \qquad [20.79]$$

Because water containing CO_2 is slightly acidic (Equation 20.76), $CaCO_3$ dissolves slowly in this medium:

$$CaCO_3(s) + H_2O(l) + CO_2(g) \longrightarrow Ca^{2+}(aq) + 2HCO_3^-(aq) \qquad [20.80]$$

∞ Review this on page 488

This reaction occurs when surface waters move underground through limestone deposits. It is the principal way that Ca^{2+} enters groundwater, producing 'hard water' (∞ Section 13.4, 'Human Activities and Earth's Water'). If the limestone deposit is deep enough underground, the dissolution of the limestone produces a cave.

One of the most important reactions of $CaCO_3$ is its decomposition into CaO and CO_2 at elevated temperatures, given earlier in Equation 20.72. Calcium oxide reacts with water to form $Ca(OH)_2$ and it is an important commercial base. It is also important in making mortar, which is a mixture of sand, water and CaO used in construction to bind bricks, blocks and rocks together. Calcium oxide reacts with water and CO_2 to form $CaCO_3$, which binds the sand in the mortar.

$$CaO(s) + H_2O(l) \rightleftharpoons Ca^{2+}(aq) + 2OH^-(aq) \qquad [20.81]$$

$$Ca^{2+}(aq) + 2OH^-(aq) + CO_2(aq) \longrightarrow CaCO_3(s) + H_2O(l) \qquad [20.82]$$

Carbides

The binary compounds of carbon with metals, metalloids and certain non-metals are called **carbides**. There are three types: ionic, interstitial and covalent. The ionic carbides are formed by the more active metals. The most common ionic carbides contain the *acetylide* ion (C_2^{2-}). This ion is isoelectronic with N_2 and its Lewis structure, $[:C\equiv C:]^{2-}$, has a carbon–carbon triple bond. The most important ionic carbide is calcium carbide (CaC_2), which is produced by the reduction of CaO with carbon at high temperature:

$$2CaO(s) + 5C(s) \longrightarrow 2CaC_2(s) + CO_2(g) \qquad [20.83]$$

The carbide ion is a very strong base that reacts with water to form acetylene ($H-C\equiv C-H$), as in the following reaction:

$$CaC_2(s) + 2H_2O(l) \longrightarrow Ca(OH)_2(aq) + C_2H_2(g) \qquad [20.84]$$

Calcium carbide is therefore a convenient solid source of acetylene, which is used in welding.

Interstitial carbides are formed by many transition metals. The carbon atoms occupy open spaces (interstices) between metal atoms in a manner analogous to the interstitial hydrides. Tungsten carbide, for example, is very hard and very heat-resistant, and so is used to make cutting tools.

Covalent carbides are formed by boron and silicon. Silicon carbide (SiC), known as Carborundum®, is used as an abrasive and in cutting tools. Almost as hard as diamond, SiC has a diamond-like structure with alternating Si and C atoms.

Other Inorganic Compounds of Carbon

Hydrogen cyanide, HCN (▶ **FIGURE 20.37**), is an extremely toxic gas that has the odour of bitter almonds. It is produced by the reaction of a cyanide salt, such as NaCN, with an acid (see Sample Exercise 20.2(c)).

Aqueous solutions of HCN are known as hydrocyanic acid. Neutralisation with a base, such as NaOH, produces cyanide salts, such as NaCN. Cyanides find use in the manufacture of several well-known plastics, including nylon and Orlon®. The CN^- ion forms very stable complexes with most transition metals.

Carbon disulfide, CS_2 (Figure 20.37), is an important industrial solvent for waxes, greases, celluloses and other non-polar substances. It is a colourless volatile liquid (bp 46.3 °C). The vapour is very poisonous and highly flammable. The compound is formed by direct reaction of carbon and sulfur at high temperature.

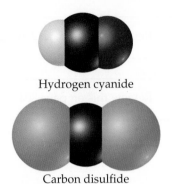

Hydrogen cyanide

Carbon disulfide

▲ **FIGURE 20.37**   **Structures of hydrogen cyanide and carbon disulfide.**

> ⚠ **CONCEPT CHECK 13**
>
> Based on what you know of their physical properties, does CS_2 have stronger intermolecular forces than CO_2? Explain.

20.10 | THE OTHER GROUP 14 ELEMENTS: Si, Ge, Sn AND Pb

The other elements of group 14, in addition to carbon, are silicon, germanium, tin and lead. The general trend from non-metallic to metallic character as we go down a family is strikingly evident in group 14. Carbon is a non-metal; silicon and germanium are metalloids; tin and lead are metals. In this section we consider a few general characteristics of group 14 and then look more thoroughly at silicon.

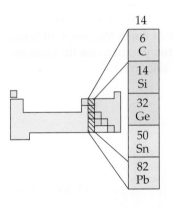

General Characteristics of the Group 14 Elements

Some properties of the group 14 elements are given in ▼ **TABLE 20.8**. The elements possess the outer-shell electron configuration ns^2np^2. The electronegativities of the elements are generally low; carbides that formally contain C^{4-} ions are observed only in the case of a few compounds of carbon with very active metals. Formation of 4+ ions by electron loss is only observed for Sn^{4+}; in general the ionisation energies are too high. The +2 oxidation state is found in the chemistry of germanium, tin and lead, however, and it is the principal oxidation state for lead. The vast majority of the compounds of the group 14 elements are covalently bonded. Carbon forms a maximum of four bonds. The other members of

TABLE 20.8 • Some properties of the group 14 elements

Property	C	Si	Ge	Sn	Pb
Atomic radius (pm)	77	117	122	140	146
First ionisation energy (kJ mol^{-1})	1086	786	762	709	716
Electronegativity	2.5	1.8	1.8	1.8	1.9
X–X single-bond enthalpy (kJ mol^{-1})	348	226	188	151	—

∞ Review this on page 276

∞ Review this on page 804

the family are able to form higher coordination numbers through valence-shell expansion (∞ Section 8.7, 'Exceptions to the Octet Rule').

Carbon differs from the other group 14 elements in its pronounced ability to form multiple bonds both with itself and with other non-metals, especially N, O and S. The origin of this behaviour was considered earlier (∞ Section 20.1, 'Periodic Trends and Chemical Reactions').

Table 20.8 shows that the strength of a bond between two atoms of a given element decreases as we go down group 14. Carbon–carbon bonds are quite strong. Carbon, therefore, has a striking ability to form compounds in which carbon atoms are bonded to one another in extended chains and rings, which accounts for the large number of organic compounds that exist. Other elements, especially those in the vicinity of carbon in the periodic table, can also form chains and rings, but these bonds are far less important in the chemistries of these other elements. The Si–Si bond strength (226 kJ mol^{-1}), for example, is much smaller than the Si–O bond strength (386 kJ mol^{-1}). As a result, the chemistry of silicon is dominated by the formation of Si–O bonds, and Si–Si bonds play a rather minor role.

Occurrence and Preparation of Silicon

Silicon is the second most abundant element, after oxygen, in the Earth's crust. It occurs in **silica** (SiO_2) and in an enormous variety of silicate minerals. The element is obtained by the reduction of molten silicon dioxide with carbon at high temperature.

$$SiO_2(l) + 2 C(s) \longrightarrow Si(l) + 2 CO(g) \qquad [20.85]$$

Elemental silicon has a diamond-type structure. Crystalline silicon is a grey metallic-looking solid that melts at 1410 °C. The element is a semiconductor and so is used in making transistors and solar cells. To be used as a semiconductor, it must be extremely pure, possessing less than $10^{-7}\%$ (1 ppb) impurities. One method of purification is to treat the element with Cl_2 to form $SiCl_4$. The $SiCl_4$ is a volatile liquid that is purified by fractional distillation and then converted back to elemental silicon by reduction with H_2:

$$SiCl_4(g) + 2 H_2(g) \longrightarrow Si(s) + 4 HCl(g) \qquad [20.86]$$

The element can be further purified by the process of zone refining. In the zone-refining process, a heated coil is passed slowly along a silicon rod, as shown in ◀ FIGURE 20.38. A narrow band of the element is thereby melted. As the molten area is swept slowly along the length of the rod, the impurities concentrate in the molten region, following it to the end of the rod. The purified top portion of the rod is retained for manufacture of electronic devices.

Silicates

Silicon dioxide and other compounds that contain silicon and oxygen comprise over 90% of the Earth's crust. **Silicates** are compounds in which a silicon atom in the 4+ oxidation state is surrounded in a tetrahedral fashion by four oxygen atoms, for example, the SiO_4^- ion as shown in ▶ FIGURE 20.39(a). It is known as the *orthosilicate* ion. This ion, which is found in very few silicate materials, can be regarded as the building block for more complex silicates. If two silicate tetrahedral are linked by sharing one oxygen atom as shown in Figure 20.39(b), we obtain the disilicate ion, $Si_2O_7^{2-}$. By combining two vertices of each tetrahedron to two other tetrahedra, we obtain an infinite chain with an ··· O–Si–O–Si ··· backbone as shown in ▶ FIGURE 20.40(a). It is called a single-strand silicate chain and consists of repeating units of the $Si_2O_6^{4-}$ ion. So by linking large numbers of silicate tetrahedra we can obtain chains, sheets (Figure 20.40(b)) and three-dimensional structures such as that of silicon dioxide. *Asbestos*, as shown in ▶ FIGURE 20.41, is a general term applied to a group of fibrous silicate minerals possessing either strand-like arrangements of the silicate tetrahedral or sheet structures in which the sheets are formed into rolls. Formerly widely used in the construction industry, it is now banned due to incontestable proof of its role in causing lung cancer.

▲ **FIGURE IT OUT**

What limits the range of temperatures you can use for zone refining of silicon?

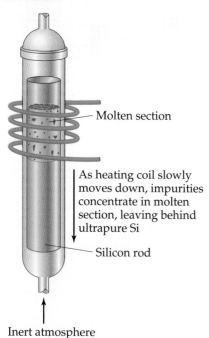

— Molten section

As heating coil slowly moves down, impurities concentrate in molten section, leaving behind ultrapure Si

— Silicon rod

Inert atmosphere

▲ FIGURE 20.38   **Zone-refining apparatus for production of ultrapure silicon.**

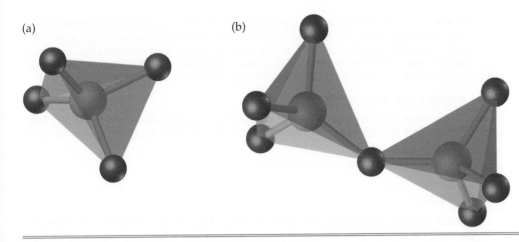

(a)

(b)

◀ **FIGURE 20.39   Silicate structures.** (a) Structure of the SiO_4 tetrahedron of the SiO_4^{4-} ion. This ion is found in several minerals, including zircon ($ZrSiO_4$). (b) Geometrical structure of the $Si_2O_7^{6-}$ ion, which is formed when two SiO_4 tetrahedra share a corner oxygen atom. This ion occurs in several minerals, including hardystonite ($Ca_2Zn(Si_2O_7)$).

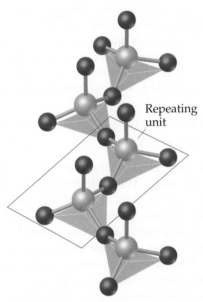

Repeating unit

Single-strand silicate chain, $Si_2O_6^{4-}$

(a)

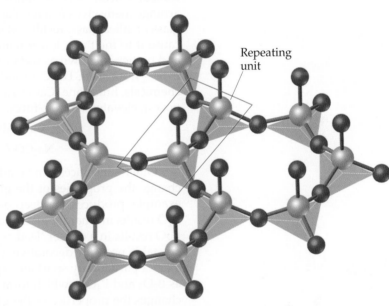

Repeating unit

Repeating unit

Two-dimensional silicate sheet, $Si_2O_5^{2-}$

(b)

▲ **FIGURE 20.40   Silicate chains and sheets.** Silicate structures consist of tetrahedra linked together through their vertices by a shared oxygen atom. (a) Representation of an infinite single-strand silicate chain. Each tetrahedron is linked to two others. The box shows the repeating unit of the chain, which is similar to the unit cell of solids (∞ Section 11.8, 'Structures of Solids'); the chain can be viewed as an infinite number of repeating units, laid side by side. The repeating unit has a formula of $Si_2O_6^{4-}$ or, the simplest formula, SiO_3^{2-}. (b) Representation of a two-dimensional sheet structure. Each tetrahedron is linked to three others. The repeating unit of the sheet has the formula $Si_2O_5^{2-}$.

SAMPLE EXERCISE 20.9   Determining an empirical formula

The mineral *chrysotile* is a non-carcinogenic asbestos mineral that is based on the sheet structure shown in Figure 20.40(b). In addition to silicate tetrahedra, the mineral contains Mg^{2+} and OH^- ions. Analysis of the mineral shows that there are 1.5 Mg atoms per Si atom. What is the empirical formula for chrysotile?

SOLUTION

Analyse A mineral is described that has a sheet silicate structure with Mg^{2+} and OH^- ions to balance charge and 1.5 Mg for each 1 Si. We are asked to write the empirical formula for the mineral.

Plan As shown in Figure 20.40(b), the silicate sheet structure is based on the $Si_2O_5^{2-}$ ion. We first add Mg^{2+} to give the proper Mg : Si ratio. We then add OH^- ions to obtain a neutral compound.

▲ **FIGURE 20.41   Serpentine asbestos.** Note the fibrous character of this silicate mineral.

Solve The observation that the Mg:Si ratio equals 1.5 is consistent with three Mg^{2+} ions per $Si_2O_5{}^{2-}$ ion. The addition of three Mg^{2+} ions would make $Mg_3(Si_2O_5)^{4+}$. In order to achieve charge balance in the mineral, there must be four OH^- ions per $Si_2O_5{}^{2-}$ ion. Thus the formula of chrysotile is $Mg_3(Si_2O_5)(OH)_4$. Since this is not reducible to a simpler formula, this is the empirical formula.

PRACTICE EXERCISE
The cyclosilicate ion consists of three silicate tetrahedra linked together in a ring. The ion contains three Si atoms and nine O atoms. What is the overall charge on the ion?

Answer: 6−

Glass

Quartz melts at approximately 1600 °C, forming a tacky liquid. In the course of melting, many silicon–oxygen bonds are broken. When the liquid is rapidly cooled, silicon–oxygen bonds are re-formed before the atoms are able to arrange themselves in a regular fashion. An amorphous solid, known as quartz glass or silica glass, results. Many different substances can be added to SiO_2 to cause it to melt at a lower temperature. The common **glass** used in windows and bottles is known as soda-lime glass. It contains CaO and Na_2O in addition to SiO_2 from sand. The CaO and Na_2O are produced by heating two inexpensive chemicals, limestone ($CaCO_3$) and soda ash (Na_2CO_3). These carbonates decompose at elevated temperatures:

$$CaCO_3(s) \longrightarrow CaO(s) + CO_2(g) \qquad [20.87]$$

$$Na_2CO_3(s) \longrightarrow Na_2O(s) + CO_2(g) \qquad [20.88]$$

Other substances can be added to soda-lime glass to produce colour or to change the properties of the glass in various ways. The addition of CoO, for example, produces the deep blue colour of 'cobalt glass'. Replacing Na_2O by K_2O results in a harder glass that has a higher melting point. Replacing CaO by PbO results in a denser 'lead crystal' glass with a higher refractive index. Lead crystal is used for decorative glassware; the higher refractive index gives this glass a particularly sparkling appearance. Addition of non-metal oxides, such as B_2O_3 and P_4O_{10}, which form network structures related to the silicates, also changes the properties of the glass. Adding B_2O_3 creates a glass with a higher melting point and a greater ability to withstand temperature changes. Such glasses, sold commercially under trade names such as Pyrex® and Kimax®, are used where resistance to thermal shock is important, such as in laboratory glassware or coffeemakers.

Silicones

Silicones consist of O–Si–O chains in which the remaining bonding positions on each silicon are occupied by organic groups such as CH_3. The chains are terminated by $-Si(CH_3)_3$ groups:

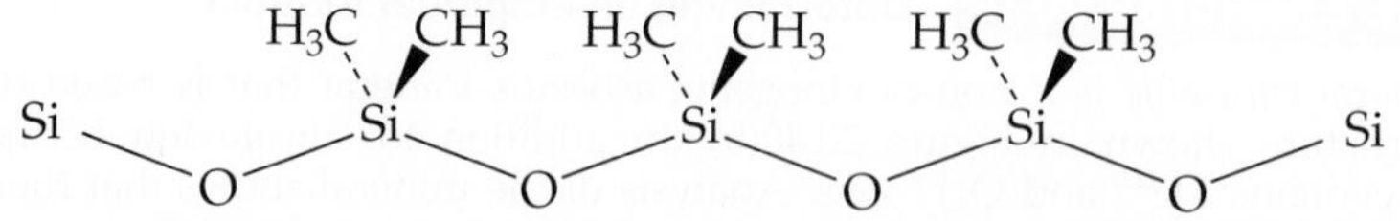

Depending on the length of the chain and the degree of cross-linking between chains, silicones can be either oils or rubber-like materials. Silicones are non-toxic and have good stability towards heat, light, oxygen and water. They are used commercially in a wide variety of products, including lubricants, car polishes, sealants and gaskets. They are also used for waterproofing fabrics. When applied to a fabric, the oxygen atoms form hydrogen bonds with the molecules on the surface of the fabric. The hydrophobic (water-repelling) organic groups of the silicone are then left pointing away from the surface as a barrier.

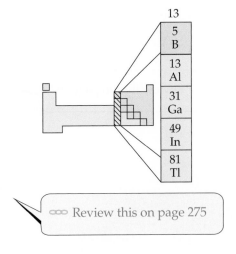

∞ Review this on page 275

20.11 | BORON

Boron is the only element of group 13 that is non-metallic. The element has an extended network structure. Its melting point (2300 °C) is intermediate between that of carbon (3550 °C) and that of silicon (1410 °C). The electron configuration of boron is $[He]2s^2 2p^1$.

In a family of compounds called **boranes**, the molecules contain only boron and hydrogen. The simplest borane is BH_3. This molecule contains only six valence electrons and is therefore an exception to the octet rule (∞ Section 8.7, 'Exceptions to the Octet Rule'). As a result, BH_3 reacts with itself to form *diborane* (B_2H_6). This reaction can be viewed as a Lewis acid–base reaction (Section 17.11), in which one B–H bonding pair of electrons in each BH_3 molecule is donated to the other. As a result, diborane is an unusual molecule in which hydrogen atoms appear to form two bonds (▶ **FIGURE 20.42**).

Sharing hydrogen atoms between the two boron atoms compensates somewhat for the deficiency in valence electrons around each boron. Nevertheless, diborane is an extremely reactive molecule, spontaneously flammable in air. The reaction of B_2H_6 with O_2 is highly exothermic.

$$B_2H_6(g) + 3\,O_2(g) \longrightarrow B_2O_3(s) + 3\,H_2O(g) \qquad \Delta H° = -2030\ \text{kJ} \qquad [20.89]$$

Other boranes, such as pentaborane (B_5H_9), are also very reactive. Decaborane ($B_{10}H_{14}$) is stable in air at room temperature, but it undergoes a very exothermic reaction with O_2 at higher temperatures. Boranes were at one time explored as solid fuels for rockets.

Boron and hydrogen also form a series of anions, called *borane anions*. Salts of the borohydride ion (BH_4^-) are widely used as reducing agents. Sodium borohydride ($NaBH_4$) is a commonly used reducing agent for certain organic compounds.

Diborane B_2H_6

▲ **FIGURE 20.42 The structure of diborane (B_2H_6).** Two of the H atoms bridge the two B atoms, giving a planar B_2H_2 core to the molecule. Two of the remaining H atoms lie on either side of the B_2H_2 core, giving a nearly tetrahedral bonding environment about the B atoms.

 CONCEPT CHECK 14
Distinguish between the substances silicon, silica and silicone.

 CONCEPT CHECK 15
Recall that the hydride ion is H^-. What is the oxidation state of boron in sodium borohydride?

The only important oxide of boron is boric oxide (B_2O_3). This substance is the anhydride of boric acid, and can be written H_3BO_3 or $B(OH)_3$. Boric acid is so weak an acid ($K_a = 5.8 \times 10^{-10}$) that solutions of H_3BO_3 are used as an eyewash. Upon heating, boric acid loses water by a condensation reaction similar to that described for phosphorus in Section 20.8:

$$4\,H_3BO_3(s) \longrightarrow H_2B_4O_7(s) + 5\,H_2O(g) \qquad [20.90]$$

The diprotic acid $H_2B_4O_7$ is called tetraboric acid. The hydrated sodium salt $Na_2B_4O_7 \cdot 10\,H_2O$, called borax, occurs in dry lake deposits in California and can also be readily prepared from other borate minerals. Solutions of borax are alkaline, and the substance is used in various laundry and cleaning products.

SAMPLE INTEGRATIVE EXERCISE   Putting concepts together

The interhalogen compound BrF_3 is a volatile, straw-coloured liquid. The compound exhibits appreciable electrical conductivity because of autoionisation.

$$2\,BrF_3(l) \rightleftharpoons BrF_2^+(solv) + BrF_4^-(solv)$$

(a) What are the molecular structures of the BrF_2^+ and BrF_4^- ions? **(b)** The electrical conductivity of BrF_3 decreases with increasing temperature. Is the autoionisation process exothermic or endothermic? **(c)** One chemical characteristic of BrF_3 is that it acts as a Lewis acid towards fluoride ions. What do you expect will happen when KBr is dissolved in BrF_3?

SOLUTION

(a) The BrF_2^+ ion has a total of $7 + 2(7) - 1 = 20$ valence-shell electrons. The Lewis structure for the ion is

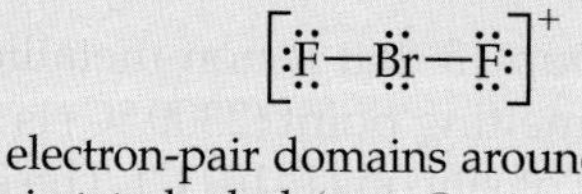

$$\left[\ddot{\textrm{:}}\ddot{\textrm{F}}\textrm{—}\ddot{\textrm{Br}}\textrm{—}\ddot{\textrm{F}}\textrm{:} \right]^+$$

Because there are four electron-pair domains around the central Br atom, the resulting electron-pair geometry is tetrahedral (Section 9.2, 'The VSEPR Model'). Because two of these domains are occupied by bonding pairs of electrons, the molecular geometry is non-linear.

Review this on page 297

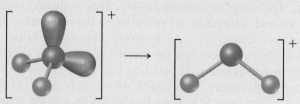

The BrF_4^- ion has a total of $7 + 4(7) + 1 = 36$ electrons, leading to the following Lewis structure.

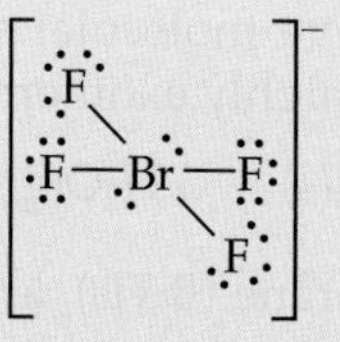

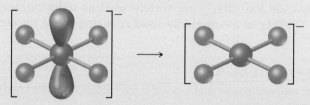

Because there are six electron-pair domains around the central Br atom in this ion, the electron-pair geometry is octahedral. The two non-bonding pairs of electrons are located opposite each other on the octahedron, leading to a square-planar molecular geometry.

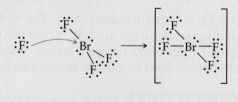

Review this on page 639

(b) The observation that conductivity decreases as temperature increases indicates that there are fewer ions present in the solution at the higher temperature. Thus increasing the temperature causes the equilibrium to shift to the left. According to Le Châtelier's principle, this shift indicates that the reaction is exothermic as it proceeds from left to right (Section 16.8, 'Le Chatelier's Principle').

Review this on page 694

(c) A Lewis acid is an electron-pair acceptor (Section 17.11, 'Lewis Acids and Bases'). The fluoride ion has four valence-shell electron pairs and can act as a Lewis base (an electron-pair donor). Thus we can imagine the following reaction occurring:

$$\textrm{F}^- \qquad \textrm{BrF}_3 \longrightarrow \textrm{BrF}_4^-$$

CHAPTER SUMMARY AND KEY TERMS

SECTION 20.1 The periodic table is useful for organising and remembering the descriptive chemistry of the elements. Among elements of a given group, size increases with increasing atomic number, and electronegativity and ionisation energy decrease. Non-metallic character parallels electronegativity, so the most non-metallic elements are found in the upper right portion of the periodic table.

Among the non-metallic elements, the first member of each group differs dramatically from the other members; it forms a maximum of four bonds to other atoms and exhibits a much greater tendency to form bonds than the heavier elements in its group.

Because O_2 and H_2O are abundant in our world, we focus on two important and general reaction types as we discuss the non-metals: oxidation by O_2 and proton-transfer reactions involving H_2O or aqueous solutions.

SECTION 20.2 Hydrogen has three isotopes: **protium** (^1_1H), **deuterium** (^2_1H) and **tritium** (^3_1H). Hydrogen is not a member of any particular periodic group, although it is usually placed above lithium. The hydrogen atom can either lose an electron, forming H^+, or gain one, forming H^- (the hydride ion). Because the H–H bond is relatively strong, H_2 is fairly unreactive unless activated by heat or a catalyst. Hydrogen forms a very strong bond to oxygen, so the reactions of H_2 with oxygen-containing compounds usually lead to the formation of H_2O. Because the bonds in CO and CO_2 are even stronger than the O–H bond, the reaction of H_2O with carbon or certain organic compounds leads to the formation of H_2. The H^+(aq) ion is able to oxidise many metals, forming H_2(g). The electrolysis of water also forms H_2(g).

The binary compounds of hydrogen are of three general types: **ionic hydrides** (formed by active metals), **metallic hydrides** (formed by transition metals) and **molecular hydrides** (formed by non-metals). The ionic hydrides contain the H^- ion; because this ion is extremely basic, ionic hydrides react with H_2O to form H_2 and OH^-.

SECTIONS 20.3 and 20.4 The noble gases (group 18) exhibit a very limited chemical reactivity because of the exceptional stability of their electron configurations. The xenon fluorides and oxides and KrF_2 are the best-established compounds of the noble gases.

The halogens (group 17) occur as diatomic molecules. All except fluorine exhibit oxidation states varying from −1 to +7. Fluorine is the most electronegative element, so it is restricted to the oxidation states 0 and −1. The oxidising power of the element (the tendency to form the −1 oxidation state) decreases as we proceed down the group. The hydrogen halides are among the most useful compounds of these elements; these gases dissolve in water to form the hydrohalic acids, such as HCl(aq). Hydrofluoric acid reacts with **silica**. The **interhalogens** are compounds formed between two different halogen elements. Chlorine, bromine and iodine form a series of oxyacids, in which the halogen atom is in a positive oxidation state. These compounds and their associated oxyanions are strong oxidising agents.

SECTIONS 20.5 and 20.6 Oxygen has two allotropes, O_2 and O_3 (ozone). Ozone is unstable compared to O_2, and it is a stronger oxidising agent than O_2. Most reactions of O_2 lead to oxides, compounds in which oxygen is in the −2 oxidation state.

The soluble oxides of non-metals generally produce acidic aqueous solutions; they are called **acidic anhydrides** or **acidic oxides**. In contrast, soluble metal oxides produce basic solutions and are called **basic anhydrides** or **basic oxides**. Many metal oxides that are insoluble in water dissolve in acid, accompanied by the formation of H_2O. Peroxides contain O–O bonds and oxygen in the −1 oxidation state. Peroxides are unstable, decomposing to O_2 and oxides. In such reactions peroxides are simultaneously oxidised and reduced, a process called **disproportionation**. Superoxides contain the O_2^- ion in which oxygen is in the $-\frac{1}{2}$ oxidation state.

Sulfur is the most important of the other group 16 elements. It has several allotropic forms; the most stable one at room temperature consists of S_8 rings. Sulfur forms two oxides, SO_2 and SO_3, and both are important atmospheric pollutants. Sulfur trioxide is the anhydride of sulfuric acid, the most important sulfur compound and the most-produced industrial chemical. Sulfuric acid is a strong acid and a good dehydrating agent. Sulfur forms several oxyanions as well, including the SO_3^{2-} (sulfite), SO_4^{2-} (sulfate) and $S_2O_3^{2-}$ (thiosulfate) ions. Sulfur is found combined with many metals as a sulfide, in which sulfur is in the −2 oxidation state. These compounds often react with acids to form hydrogen sulfide (H_2S), which smells like rotten eggs.

SECTIONS 20.7 and 20.8 Nitrogen is found in the atmosphere as N_2 molecules. Molecular nitrogen is chemically very stable because of the strong N–N triple bond. Molecular nitrogen can be converted into ammonia via the Haber process. Once the ammonia is made, it can be converted into a variety of different compounds that exhibit nitrogen oxidation states ranging from −3 to +5. The most important industrial conversion of ammonia is the **Ostwald process**, in which ammonia is oxidised to nitric acid (HNO_3). Nitrogen has three important oxides: nitrous oxide (N_2O), nitric oxide (NO) and nitrogen dioxide (NO_2). Nitrous acid (HNO_2) is a weak acid; its conjugate base is the nitrite ion (NO_2^-). Another important nitrogen compound is hydrazine (N_2H_4).

Phosphorus is the most important of the remaining group 15 elements. It occurs in nature as phosphate minerals. Phosphorus has several allotropes, including white phosphorus, which consists of P_4 tetrahedra. In reaction with the halogens, phosphorus forms trihalides PX_3 and pentahalides PX_5. These compounds undergo hydrolysis to produce an oxyacid of phosphorus and HX. Phosphorus forms two oxides, P_4O_6 and P_4O_{10}. Their corresponding acids, phosphorous acid and phosphoric acid, undergo **condensation reactions** when heated. Phosphorus compounds are important in biochemistry and as fertilisers.

SECTIONS 20.9 and 20.10 The allotropes of carbon include diamond, graphite, fullerenes, carbon nanotubes and graphene. Amorphous forms of graphite include **charcoal** and **carbon black**. Carbon forms two common oxides, CO and CO_2. Aqueous solutions of CO_2 produce the weak diprotic acid, carbonic acid (H_2CO_3), which is the parent acid of hydrogen carbonate and carbonate salts. Binary compounds of carbon are called **carbides**. Carbides may be ionic, interstitial or covalent. Calcium carbide (CaC_2) contains the strongly basic acetylide ion (C_2^{2-}), which reacts with water to form acetylene. Other important inorganic carbon compounds include hydrogen cyanide (HCN) and carbon disulfide (CS_2).

The other group 14 elements show great diversity in physical and chemical properties. Silicon, the second most abundant element, is a semiconductor. It reacts with Cl_2 to form $SiCl_4$, a liquid at room temperature, a reaction that is used to help purify silicon from its native minerals. Silicon forms strong Si–O bonds and therefore occurs in a variety of silicate minerals. Silica is SiO_2; **silicates** consist of SiO_4 tetrahedra, linked together at their vertices to form chains, sheets or three-dimensional structures. The most common three-dimensional silicate is quartz (SiO_2). **Glass** is an amorphous (non-crystalline) form of SiO_2. Silicones contain O–Si–O chains with organic groups bonded to the Si atoms. Like silicon, germanium is a metalloid; tin and lead are metallic.

SECTION 20.11 Boron is the only group 13 element that is a non-metal. It forms a variety of compounds with hydrogen called boron hydrides, or **boranes**. Diborane (B_2H_6) has an unusual structure with two hydrogen atoms that bridge between the two boron atoms. Boranes react with oxygen to form boric oxide (B_2O_3), in which boron is in the +3 oxidation state. Boric oxide is the anhydride of boric acid (H_3BO_3). Boric acid readily undergoes condensation reactions.

KEY SKILLS

- Be able to use periodic trends to explain the basic differences between the elements of a group or period. (Section 20.1)
- Explain why the first element in a group differs from the subsequent elements in the group. (Section 20.1)
- Be able to determine electron configurations and oxidation states of the elements. (Sections 20.2–20.11)
- Know the sources of the common non-metals, how they are obtained and their uses. (Sections 20.2–20.11)
- Understand how phosphoric and phosphorous acids undergo condensation reactions. (Section 20.8)
- Explain how the bonding and structure of silicates relate to their chemical structure and properties. (Section 20.10)

EXERCISES

VISUALISING CONCEPTS

20.1 One of these structures is a stable compound; the other is not. Identify the stable compound, and explain why it is stable. Explain why the other compound is not stable. [Section 20.1]

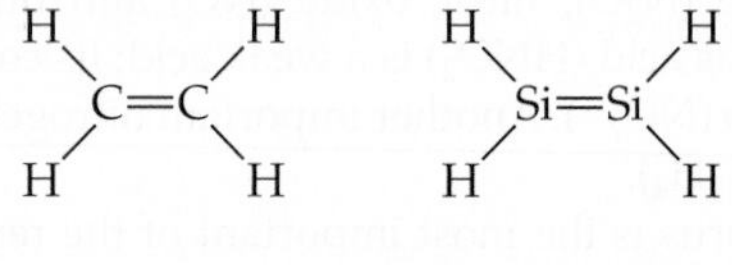

20.2 **(a)** Identify a *type* of chemical reaction represented by the diagram below. **(b)** Place appropriate charges on the species on both sides of the equation. **(c)** Write the chemical equation for the reaction. [Section 20.1]

 + ⟶ +

20.3 Which of the following species (there may be more than one) is/are likely to have the structure shown below: **(a)** XeF_4, **(b)** $BrF_4{}^+$, **(c)** SiF_4, **(d)** $TeCl_4$, **(e)** $HClO_4$? (The colours shown do not reflect the identity of any element.) [Sections 20.3, 20.4, 20.6 and 20.10]

20.4 Draw an energy profile for the reaction shown in Equation 20.30, assuming that the barrier to dissociation of $O_3(g)$ is 105 kJ. [Section 20.5]

20.5 Write the molecular formula and Lewis structure for each of the following oxides of nitrogen. [Section 20.7]

(a)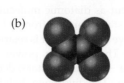

(b)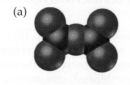

(c)

(d)

(e)

(f)

20.6 The atomic and ionic radii of the first three group 16 elements are

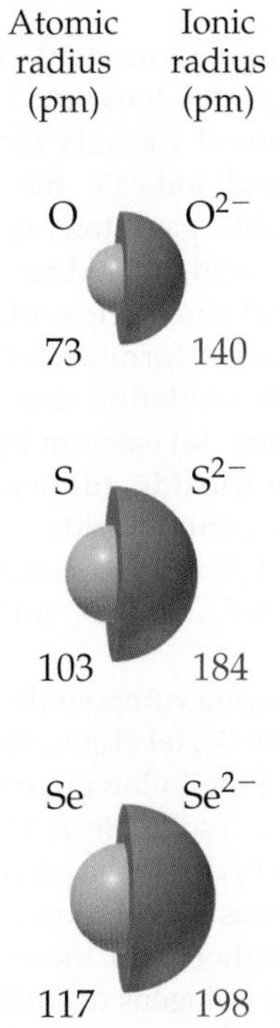

	Atomic radius (pm)	Ionic radius (pm)
O / O^{2-}	73	140
S / S^{2-}	103	184
Se / Se^{2-}	117	198

(a) Explain why the atomic radius increases in moving downward in the group. **(b)** Explain why the ionic radii are larger than the atomic radii. **(c)** Which of the three anions would you expect to be the strongest base in water? Explain. [Sections 20.7 and 20.6]

20.7 The structures of white and red phosphorus are shown here. Explain, using these structural models, why white phosphorus is much more reactive than red phosphorus. [Section 20.8]

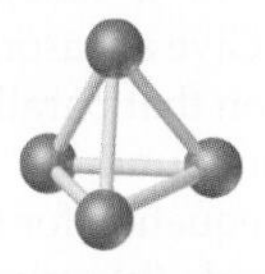

White phosphorus

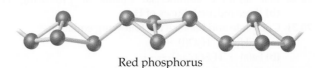

Red phosphorus

20.8 **(a)** Draw the Lewis structures for at least four species that have the general formula

$$\left[:X\equiv Y:\right]^{n}$$

where X and Y may be the same or different, and n may have a value from $+1$ to -2. **(b)** Which of the compounds is likely to be the strongest Brønsted–Lowry base? Explain. [Sections 20.1, 20.7 and 20.9]

PERIODIC TRENDS AND CHEMICAL REACTIONS (Section 20.1)

20.9 Identify each of the following elements as a metal, non-metal or metalloid: **(a)** phosphorus, **(b)** strontium, **(c)** manganese, **(d)** selenium, **(e)** rhodium, **(f)** krypton.

20.10 Identify each of the following elements as a metal, non-metal or metalloid: **(a)** rhenium, **(b)** arsenic, **(c)** argon, **(d)** zirconium, **(e)** tellurium, **(f)** gallium.

20.11 Consider the elements O, Ba, Co, Be, Br and Se. From this list select the element that **(a)** is most electronegative, **(b)** exhibits a maximum oxidation state of $+7$, **(c)** loses an electron most readily, **(d)** forms π-bonds most readily, **(e)** is a transition metal.

20.12 Consider the elements Li, K, Cl, C, Ne and Ar. From this list select the element that **(a)** is most electronegative, **(b)** has the greatest metallic character, **(c)** most readily forms a positive ion, **(d)** has the smallest atomic radius, **(e)** forms π-bonds most readily.

20.13 Explain the following observations: **(a)** The highest fluoride compound formed by nitrogen is NF_3, whereas phosphorus readily forms PF_5. **(b)** Although CO is a well-known compound, SiO doesn't exist under ordinary conditions. **(c)** AsH_3 is a stronger reducing agent than NH_3.

20.14 Explain the following observations: **(a)** HNO_3 is a stronger oxidising agent than H_3PO_4. **(b)** Silicon can form an ion with six fluorine atoms, SiF_6^{2-}, whereas carbon is able to bond to a maximum of four, CF_4. **(c)** There are three compounds formed by carbon and hydrogen that contain two carbon atoms each (C_2H_2, C_2H_4 and C_2H_6), whereas silicon forms only one analogous compound (Si_2H_6).

20.15 Complete and balance the following equations:
(a) $Mg_3N_2(s) + H_2O(l) \longrightarrow$
(b) $C_3H_7OH(l) + O_2(g) \longrightarrow$
(c) $MnO_2(s) + C(s) \xrightarrow{\Delta}$
(d) $AlP(s) + H_2O(l) \longrightarrow$
(e) $Na_2S(s) + HCl(aq) \longrightarrow$

20.16 Complete and balance the following equations:
(a) $NaOCH_3(s) + H_2O(l) \longrightarrow$
(b) $CuO(s) + HNO_3(aq) \longrightarrow$
(c) $WO_3(s) + H_2(g) \xrightarrow{\Delta}$
(d) $NH_2OH(l) + O_2(g) \longrightarrow$
(e) $Al_4C_3(s) + H_2O(l) \longrightarrow$

HYDROGEN, THE NOBLE GASES AND THE HALOGENS (Sections 20.1, 20.3 and 20.4)

20.17 Give a reason why hydrogen might be placed along with the group 1 elements of the periodic table.

20.18 In which of the following substances is hydrogen behaving analogously to a halogen: **(a)** H_2O, **(b)** H_2SO_4, **(c)** NaH, **(d)** H_2? Give a reason for your choice, and write a chemical reaction that parallels a reaction for an analogous halogen compound.

20.19 Write a balanced equation for the preparation of H_2 using **(a)** Mg and an acid, **(b)** carbon and steam, **(c)** methane and steam.

20.20 List **(a)** three commercial means of producing H_2, **(b)** three industrial uses of H_2.

20.21 Complete and balance the following equations:
(a) $NaH(s) + H_2O(l) \longrightarrow$
(b) $Fe(s) + H_2SO_4(aq) \longrightarrow$
(c) $H_2(g) + Br_2(g) \longrightarrow$
(d) $Na(l) + H_2(g) \longrightarrow$
(e) $PbO(s) + H_2(g) \longrightarrow$

20.22 Write balanced equations for each of the following reactions (some of these are analogous to reactions shown in the chapter). **(a)** Aluminium metal reacts with acids to form hydrogen gas. **(b)** Steam reacts with magnesium metal to give magnesium oxide and hydrogen. **(c)** Manganese(IV) oxide is reduced to manganese(II) oxide by hydrogen gas. **(d)** Calcium hydride reacts with water to generate hydrogen gas.

20.23 Identify the following hydrides as ionic, metallic or molecular: **(a)** BaH_2, **(b)** H_2Te, **(c)** $TiH_{1.7}$.

20.24 Identify the following hydrides as ionic, metallic or molecular: **(a)** B_2H_6, **(b)** RbH, **(c)** $Th_4H_{1.5}$.

20.25 Why does xenon form stable compounds with fluorine, whereas argon does not?

20.26 Write the chemical formula for each of the following compounds, and indicate the oxidation state of the halogen or noble gas atom in each: **(a)** chlorate ion, **(b)** hydroiodic acid, **(c)** iodine trichloride, **(d)** sodium hypochlorite, **(e)** perchloric acid, **(f)** xenon tetrafluoride.

20.27 Write the chemical formula for each of the following, and indicate the oxidation state of the halogen or noble gas atom in each: **(a)** calcium hypobromite, **(b)** bromic acid, **(c)** xenon trioxide, **(d)** perchlorate ion, **(e)** iodous acid, **(f)** iodine pentafluoride.

20.28 Name the following compounds: **(a)** $Fe(ClO_3)_3$, **(b)** $HClO_2$, **(c)** XeF_6, **(d)** BrF_5, **(e)** $XeOF_4$, **(f)** HIO_3 (named as an acid).

20.29 Name the following compounds: **(a)** $KClO_3$, **(b)** $Ca(IO_3)_2$, **(c)** $AlCl_3$, **(d)** $HBrO_3$, **(e)** H_5IO_6, **(f)** XeF_4.

20.30 Explain each of the following observations: **(a)** At room temperature I_2 is a solid, Br_2 is a liquid and Cl_2 and F_2 are both gases. **(b)** F_2 cannot be prepared by electrolytic oxidation of aqueous F^- solutions. **(c)** The boiling point of HF is much higher than those of the other hydrogen halides. **(d)** The halogens decrease in oxidising power in the order $F_2 > Cl_2 > Br_2 > I_2$.

20.31 Explain the following observations: **(a)** For a given oxidation state, the acid strength of the oxyacid in aqueous solution decreases in the order chlorine > bromine > iodine. **(b)** Hydrofluoric acid cannot be stored in glass bottles. **(c)** HI cannot be prepared by treating NaI with sulfuric acid. **(d)** The interhalogen ICl_3 is known, but $BrCl_3$ is not.

OXYGEN AND THE GROUP 16 ELEMENTS (Sections 20.5 and 20.6)

20.32 **(a)** List three industrial uses of O_2. **(b)** List two industrial uses of O_3.

20.33 Draw the Lewis structure of ozone. Explain why the O–O bond (128 pm) is longer in ozone than in O_2 (121 pm).

20.34 Write balanced equations for each of the following reactions. **(a)** When mercury(II) oxide is heated, it decomposes to form O_2 and mercury metal. **(b)** When copper(II) nitrate is heated strongly, it decomposes to form copper(II) oxide, nitrogen dioxide and oxygen. **(c)** Lead(II) sulfide, PbS(s), reacts with ozone to form $PbSO_4(s)$ and $O_2(g)$. **(d)** When heated in air, ZnS(s) is converted to ZnO. **(e)** Potassium peroxide reacts with $CO_2(g)$ to give potassium carbonate and O_2.

20.35 Complete and balance the following equations:
(a) $CaO(s) + H_2O(l) \longrightarrow$
(b) $Al_2O_3(s) + H^+(aq) \longrightarrow$
(c) $N_2O_2(g) + H_2O(l) \longrightarrow$
(d) $N_2O_3(g) + H_2O(l) \longrightarrow$
(e) $KO_2(s) + H_2O(l) \longrightarrow$
(d) $NO(g) + O_3(g) \longrightarrow$

20.36 Predict whether each of the following oxides is acidic, basic, amphoteric or neutral: **(a)** NO_2, **(b)** CO_2, **(c)** BaO, **(d)** Al_2O_3.

20.37 Select the more acidic member of each of the following pairs: **(a)** Mn_2O_7 and MnO_2, **(b)** SnO and SnO_2, **(c)** SO_2 and SO_3, **(d)** SiO_2 and SO_2, **(e)** Ga_2O_3 and In_2O_3, **(f)** SO_2 and SeO_2.

20.38 Write the chemical formula for each of the following compounds, and indicate the oxidation state of the group 16 element in each: **(a)** selenous acid, **(b)** potassium hydrogen sulfite, **(c)** hydrogen telluride, **(d)** carbon disulfide, **(e)** calcium sulfate.

20.39 In aqueous solution, hydrogen sulfide reduces **(a)** Fe^{3+} to Fe^{2+}, **(b)** Br_2 to Br^-, **(c)** MnO_4^- to Mn^{2+}, **(d)** HNO_3 to NO_2. In all cases, under appropriate conditions, the product is elemental sulfur. Write a balanced net ionic equation for each reaction.

20.40 An aqueous solution of SO_2 reduces **(a)** aqueous $KMnO_4$ to $MnSO_4(aq)$, **(b)** acidic aqueous $K_2Cr_2O_7$ to aqueous Cr^{3+}, **(c)** aqueous $Hg_2(NO_3)_2$ to mercury metal. Write balanced equations for these reactions.

20.41 Write the Lewis structure for each of the following species, and indicate the structure of each: **(a)** SeO_3^{2-}, **(b)** S_2Cl_2, **(c)** chlorosulfonic acid, HSO_3Cl (chlorine is bonded to sulfur).

20.42 Write a balanced equation for each of the following reactions. (You may have to guess at one or more of the reaction products, but you should be able to make a reasonable guess, based on your study of this chapter.) **(a)** Hydrogen selenide can be prepared by reaction of an aqueous acid solution on aluminium selenide.

(b) Sodium thiosulfate is used to remove excess Cl_2 from chlorine-bleached fabrics. The thiosulfate ion forms SO_4^{2-} and elemental sulfur, whereas Cl_2 is reduced to Cl^-.

NITROGEN AND THE GROUP 15 ELEMENTS (Sections 20.7 and 20.8)

20.43 Write the chemical formula for each of the following compounds, and indicate the oxidation state of nitrogen in each: **(a)** sodium nitrite, **(b)** ammonia, **(c)** nitrous oxide, **(d)** sodium cyanide, **(e)** nitric acid, **(f)** nitrogen dioxide.

20.44 Write the chemical formula for each of the following compounds, and indicate the oxidation state of nitrogen in each: **(a)** nitric acid, **(b)** hydrazine, **(c)** potassium cyanide, **(d)** sodium nitrate, **(e)** ammonium chloride, **(f)** lithium nitride.

20.45 Write the Lewis structure for each of the following species, and describe its geometry: **(a)** HNO_2, **(b)** N_3^-, **(c)** $N_2H_5^+$, **(d)** NO_3^-.

20.46 Complete and balance the following equations:
(a) $Ba_3N_2(s) + H_2O(l) \longrightarrow$
(b) $NO(g) + O_2(g) \longrightarrow$
(c) $N_2O_5(g) + H_2O(l) \longrightarrow$
(d) $NH_3(aq) + H^+(aq) \longrightarrow$
(e) $N_2H_4(l) + O_2(g) \longrightarrow$

20.47 Write complete balanced half-reactions for **(a)** oxidation of nitrous acid to nitrate ion in acidic solution, **(b)** oxidation of N_2 to N_2O in acidic solution.

20.48 Write complete balanced half-reactions for **(a)** reduction of nitrate ion to NO in acidic solution, **(b)** oxidation of HNO_2 to NO_2 in acidic solution.

20.49 Write a molecular formula for each compound, and indicate the oxidation state of the group 15 element in each formula: **(a)** orthophosphoric acid, **(b)** arsenous acid, **(c)** antimony(III) sulfide, **(d)** calcium dihydrogen phosphate, **(e)** potassium phosphide.

20.50 Account for the following observations: **(a)** Phosphorus forms a pentachloride, but nitrogen does not. **(b)** H_3PO_2 is a monoprotic acid. **(c)** Phosphonium salts, such as PH_4Cl, can be formed under anhydrous conditions, but they can't be made in aqueous solution. **(d)** White phosphorus is extremely reactive.

20.51 Account for the following observations: **(a)** H_3PO_3 is a diprotic acid. **(b)** Nitric acid is a strong acid, whereas phosphoric acid is weak. **(c)** Phosphate rock is ineffective as a phosphate fertiliser. **(d)** Phosphorus does not exist at room temperature as diatomic molecules, but nitrogen does. **(e)** Solutions of Na_3PO_4 are quite basic.

20.52 Write a balanced equation for each of the following reactions: **(a)** preparation of white phosphorus from calcium phosphate, **(b)** hydrolysis of PBr_3, **(c)** reduction of PBr_3 to P_4 in the gas phase, using H_2.

20.53 Write a balanced equation for each of the following reactions: **(a)** hydrolysis of PCl_5, **(b)** dehydration of orthophosphoric acid to form pyrophosphoric acid, **(c)** reaction of P_4O_{10} with water.

CARBON, THE OTHER GROUP 14 ELEMENTS AND BORON (Sections 20.9, 20.10 and 20.11)

20.54 Complete and balance the following equations:
(a) $ZnCO_3(s) \xrightarrow{\Delta}$
(b) $BaC_2(s) + H_2O(l) \longrightarrow$
(c) $C_2H_2(g) + O_2(g) \longrightarrow$
(d) $CS_2(g) + O_2(g) \longrightarrow$
(e) $Ca(CN)_2(s) + HBr(aq) \longrightarrow$

20.55 Write a balanced equation for each of the following reactions: **(a)** Hydrogen cyanide is formed commercially by passing a mixture of methane, ammonia and air over a catalyst at 800 °C. Water is a by-product of the reaction. **(b)** Baking soda reacts with acids to produce carbon dioxide gas. **(c)** When barium carbonate reacts in air with sulfur dioxide, barium sulfate and carbon dioxide form.

20.56 Write the formulae for the following compounds, and indicate the oxidation state of the group 14 element or of boron in each: **(a)** boric acid, **(b)** silicon tetrabromide, **(c)** lead(II) chloride, **(d)** sodium tetraborate decahydrate (borax), **(e)** boric oxide.

20.57 Write the formulae for the following compounds, and indicate the oxidation state of the group 14 element or of boron in each: **(a)** silicon dioxide, **(b)** germanium tetrachloride, **(c)** sodium borohydride, **(d)** stannous chloride, **(e)** diborane.

20.58 Select the member of group 14 that best fits each description: **(a)** has the lowest first ionisation energy, **(b)** is found in oxidation states ranging from −4 to +4, **(c)** is most abundant in the Earth's crust.

20.59 Select the member of group 14 that best fits each description: **(a)** forms chains to the greatest extent, **(b)** forms the most basic oxide, **(c)** is a metalloid that can form 2+ ions.

20.60 **(a)** What is the characteristic geometry about silicon in all silicate minerals? **(b)** Metasilicic acid has the empirical formula H_2SiO_3. Which of the structures shown in Figure 20.40 would you expect metasilicic acid to have?

20.61 Two silicate anions are known in which the linking of the tetrahedra forms a closed ring. One of these cyclic silicate anions contains three silicate tetrahedra, linked into a ring. The other contains six silicate tetrahedra. **(a)** Sketch these cyclic silicate anions. **(b)** Determine the formula and charge of each of the anions.

20.62 **(a)** How does the structure of diborane (B_2H_6) differ from that of ethane (C_2H_6)? **(b)** By using concepts discussed in Chapter 8, explain why diborane adopts the geometry that it does. **(c)** What is the significance of the statement that the hydrogen atoms in diborane are described as hydridic?

20.63 Write a balanced equation for each of the following reactions: **(a)** Diborane reacts with water to form boric acid and molecular hydrogen. **(b)** Upon heating, boric acid undergoes a condensation reaction to form tetraboric acid. **(c)** Boron oxide dissolves in water to give a solution of boric acid.

ADDITIONAL EXERCISES

20.64 In your own words, define the following terms: **(a)** reducing agent, **(b)** allotrope, **(c)** disproportionation, **(d)** interhalogen, **(e)** acidic anhydride, **(f)** condensation reaction.

20.65 Starting with D_2O, suggest preparations of **(a)** ND_3, **(b)** D_2SO_4, **(c)** NaOD, **(d)** DNO_3, **(e)** C_2D_2, **(f)** DCN.

20.66 Although the ClO_4^- and IO_4^- ions have been known for a long time, BrO_4^- was not synthesised until 1965. The ion was synthesised by oxidising the bromate ion with xenon difluoride, producing xenon, hydrofluoric acid and the perbromate ion. Write the balanced equation for this reaction.

20.67 Which of the following substances will burn in oxygen: SiH_4, SiO_2, CO, CO_2, Mg, CaO? Why won't some of these substances burn in oxygen?

20.68 Write a balanced equation for the reaction of each of the following compounds with water: **(a)** $SO_2(g)$, **(b)** $Cl_2O_7(g)$, **(c)** $Na_2O_2(s)$, **(d)** $BaC_2(s)$, **(e)** $RbO_2(s)$, **(f)** $Mg_3N_2(s)$, **(g)** NaH(s).

20.69 What is the anhydride for each of the following acids: **(a)** H_2SO_4, **(b)** $HClO_3$, **(c)** HNO_2, **(d)** H_2CO_3, **(e)** H_3PO_4?

20.70 Elemental sulfur is capable of reacting under suitable conditions with Fe, F_2, O_2 or H_2. Write balanced equations to describe these reactions. In which reactions is sulfur acting as a reducing agent and in which as an oxidising agent?

20.71 A sulfuric acid plant produces a considerable amount of heat. This heat is used to generate electricity, which helps reduce operating costs. The synthesis of H_2SO_4 consists of three main chemical processes: (1) oxidation of S to SO_2, (2) oxidation of SO_2 to SO_3, (3) the dissolving of SO_3 in H_2SO_4 and its reaction with water to form H_2SO_4. If the third process produces 130 kJ mol^{-1}, how much heat is produced in preparing a mole of H_2SO_4 from a mole of S? How much heat is produced in preparing a tonne of H_2SO_4?

20.72 **(a)** What is the oxidation state of P in PO_4^{3-} and of N in NO_3^-? **(b)** Why doesn't N form a stable NO_4^{3-} ion analogous to P?

20.73 **(a)** The P_4, P_4O_6 and P_4O_{10} molecules have a common structural feature of four P atoms arranged in a tetrahedron (Figures 20.31 and 20.32). Does this mean that the bonding between the P atoms is the same in all these cases? Explain. **(b)** Sodium trimetaphosphate ($Na_3P_3O_9$) and sodium tetrametaphosphate ($Na_4P_4O_{12}$) are used as water-softening agents. They contain cyclic $P_3O_9^{3-}$ and $P_4O_{12}^{4-}$ ions, respectively. Propose reasonable structures for these ions.

20.74 Silicon has a limited capacity to form linear, Si–Si bonded structures similar to those formed by carbon. **(a)** Predict the molecular formula of a hydride of silicon that contains a chain of three silicon atoms. **(b)** Write a balanced equation for the reaction between oxygen and the compound you predicted in part (a).

20.75 Ultrapure germanium, like silicon, is used in semiconductors. Germanium of 'ordinary' purity is prepared by the high-temperature reduction of GeO_2 with carbon. The Ge is converted to $GeCl_4$ by treatment with Cl_2 and then purified by distillation; $GeCl_4$ is then hydrolysed in water to GeO_2 and reduced to the elemental form with H_2. The element is then zone refined. Write a balanced chemical equation for each of the chemical transformations in the course of forming ultrapure Ge from GeO_2.

20.76 Complete and balance the following equations:

(a) $Li_3N(s) + H_2O(l) \longrightarrow$

(b) $NH_3(aq) + H_2O(l) \longrightarrow$

(c) $NO_2(g) + H_2O(l) \longrightarrow$

(d) $2 NO_2(g) \longrightarrow$

(e) $NH_3(g) + O_2(g) \xrightarrow{catalyst}$

(f) $CO(g) + O_2(g) \longrightarrow$

(g) $H_2CO_3(aq) \xrightarrow{\Delta}$

(h) $Ni(s) + CO(g) \longrightarrow$

(i) $CS_2(g) + O_2(g) \longrightarrow$

(j) $CaO(s) + SO_2(g) \longrightarrow$

(k) $Na(s) + H_2O(l) \longrightarrow$

(l) $CH_4(g) + H_2O(g) \xrightarrow{\Delta}$

(m) $LiH(s) + H_2O(l) \longrightarrow$

(n) $Fe_2O_3(s) + 3 H_2(g) \longrightarrow$

INTEGRATIVE EXERCISES

20.77 Using the thermochemical data in Table 20.1 and Appendix C, calculate the average Xe–F bond enthalpies in XeF_2, XeF_4 and XeF_6, respectively. What is the significance of the trend in these quantities?

20.78 Hydrogen gas has a higher fuel value than natural gas on a mass basis but not on a volume basis. Thus, hydrogen is not competitive with natural gas as a fuel transported long distances through pipelines. Calculate the heats of combustion of H_2 and CH_4 (the principal component of natural gas) **(a)** per mole of each, **(b)** per gram of each, **(c)** per cubic metre of each at STP. Assume $H_2O(l)$ as a product.

20.79 The solubility of Cl_2 in 100 g of water at STP is 310 cm^3. Assume that this quantity of Cl_2 is dissolved and equilibrated as follows:

$$Cl_2(aq) + H_2O \rightleftharpoons Cl^-(aq) + HClO(aq) + H^+(aq)$$

If the equilibrium constant for this reaction is 4.7×10^{-4}, calculate the equilibrium concentration of HClO formed.

20.80 When ammonium perchlorate decomposes thermally, the products of the reaction are $N_2(g)$, $O_2(g)$, $H_2O(g)$ and HCl(g). **(a)** Write a balanced equation for the reaction. [*Hint:* You might find it easier to use fractional coefficients for the products.] **(b)** Calculate the enthalpy change in the reaction per mole of NH_4ClO_4. The standard enthalpy of formation of $NH_4ClO_4(s)$ is -295.8 kJ. **(c)** When $NH_4ClO_4(s)$ is employed in solid-fuel booster rockets, it is packed with powdered aluminium. Given the high temperature needed for $NH_4ClO_4(s)$ decomposition and what the products of the reaction are, what role does the aluminium play?

20.81 Manganese silicide has the empirical formula MnSi and melts at 1280 °C. It is insoluble in water but does dissolve in aqueous HF. **(a)** What type of compound do you expect MnSi to be, in terms of Table 11.7? **(b)** Write a probable balanced chemical equation for the reaction of MnSi with concentrated aqueous HF.

20.82 Hydrazine has been employed as a reducing agent for metals. Using standard reduction potentials, predict whether the following metals can be reduced to the metallic state by hydrazine under standard conditions in acidic solution: **(a)** Fe^{2+}, **(b)** Sn^{2+}, **(c)** Cu^{2+}, **(d)** Ag^+, **(e)** Cr^{3+}.

20.83 Both dimethylhydrazine, $(CH_3)_2NNH_2$, and methylhydrazine, CH_3NHNH_2, have been used as rocket fuels. When dinitrogen tetroxide (N_2O_4) is used as the oxidiser, the products are H_2O, CO_2 and N_2. If the thrust of the rocket depends on the volume of the products produced, which of the substituted hydrazines produces a greater thrust per gram total mass of oxidiser plus fuel? [Assume that both fuels generate the same temperature and that $H_2O(g)$ is formed.]

20.84 Boron nitride has a graphite-like structure with B–N bond distances of 145 pm within sheets and a separation of 330 pm between sheets. At high temperatures the BN assumes a diamond-like form that is harder than diamond. Rationalise the similarity between BN and elemental carbon.

MasteringChemistry (www.pearson.com.au/masteringchemistry)

Make learning part of the grade. Access:

- tutorials with peronalised coaching
- study area
- Pearson eText

PHOTO/ART CREDITS

802 © Hemul/Dreamstime.com; **810, 815 820 (Figure 20.16), 823, 831** Richard Megna, Fundamental Photographs, NYC; **813, 828 (Figure 20.29)** Donald Clegg and Roxy Wilson Pearson Education/PH College Author series; **817** © Corbis/Bettman; **820 (Figure 20.17)** Maksym Gorpenyuk/Fotolia; **822 (Figure 20.19)** Jeffrey A. Scovil, **(Figure 20.20)** Lawrence Migdale/Science Source/Photo Researchers, Inc; **824** Kristen Brochmann, Fundamental Photographs, NYC; **828 (Figure 20.28)** Science Photo Library/Photo Researchers, Inc; **834 (aeroplane)** Chris H. Galbraith/Shutterstock.com, **(car)** Jordan Polizzi/Shutterstock.com, **(kayak)** marekuliasz/Shutterstock.com, **(artificial limbs)** John Kropewnicki/Shutterstock.com.

21

CHEMISTRY OF THE TRANSITION METALS

The colours in the panels of this stained-glass window were obtained by adding oxides of the transition metals to the molten glass before the glass sheets were formed.

KEY CONCEPTS

21.1 TRANSITION METALS
We examine the physical properties, electron configurations, oxidation states and magnetic properties of the *transition metals*.

21.2 TRANSITION METAL COMPLEXES
We introduce the concepts of *metal complexes* and *ligands* and provide a brief history of the development of *coordination chemistry*.

21.3 LIGANDS WITH MORE THAN ONE DONOR ATOM
We examine some *polydentate ligands (chelating agents)* and observe their presence in living systems.

21.4 NOMENCLATURE AND ISOMERISM IN COORDINATION CHEMISTRY
We introduce the nomenclature used for coordination compounds. We see that coordination compounds exhibit *isomerism*, in which two compounds have the same composition but different structures, and then look at two types: *structural isomers* and *stereoisomers*.

21.5 COLOUR AND MAGNETISM IN COORDINATION CHEMISTRY
We discuss colour and magnetism in coordination compounds, emphasising the visible portion of the electromagnetic spectrum and the notion of *complementary colours*. We then see that many transition metal complexes are paramagnetic because they contain unpaired electrons.

21.6 CRYSTAL-FIELD THEORY
We explore how *crystal-field theory* allows us to explain some of the interesting spectral and magnetic properties of coordination compounds.

The colours associated with chemistry are not only beautiful, they are also informative, providing insights into the structure and bonding of matter. Compounds of the transition metals constitute an important group of coloured substances. Some of them are used in paint pigments; others produce the colours in glass and precious gems. For example, the colours in the stained-glass art shown in the chapter-opening photograph are due mainly to compounds of the transition metals. Why do these compounds have colour, and why do these colours change as the ions or molecules bonded to the metal change? The chemistry we explore in this chapter will help to answer these questions.

In earlier chapters we have seen that metal ions can function as Lewis acids, forming covalent bonds with a variety of molecules and ions that function as Lewis bases. We have encountered many ions and compounds that result from such interactions, for example $[Fe(H_2O)_6]^{3+}$ and $[Ag(NH_3)_2]^+$ in Sections 17.11 and 18.5 and haemoglobin, which is an important iron compound responsible for the oxygen-carrying capacity of blood in Section 18.2. The extraction of gold from its ores depends on the formation of species such as $[Au(CN)_2]^-$. In this chapter we discuss the properties of transition metals and look at the chemistry associated with the complex assemblies of metals surrounded by molecules and ions. Metal compounds of this kind are called *coordination compounds*, and the branch of chemistry that focuses on them is called *coordination chemistry*.

21.1 | TRANSITION METALS

Many of the important metals of modern society are the transition metals (elements), which occupy the *d* block of the periodic table (▼ FIGURE 21.1). According to the IUPAC definition, a **transition element** is one for which an atom has an incomplete *d* subshell or which readily gives rise to cations having an incomplete *d* subshell. Consequently, the elements of group 12 (Zn, Cd, Hg) are not transition elements and in fact their properties are those of the representative group metals (*s*-block and *p*-block metals); however, they are included for comparison purposes. The period 4 elements Sc to Cu are also referred to as the first transition series (first series), the period 5 elements Y to Ag as the second transition series (second series) and period 6 containing La to Au is referred to as the third transition series (third series). It should be noted that after lanthanum the next 14 electrons enter the 4*f* energy level in Ce to Lu. It is only after this energy level is filled that the electrons then fill the 5*d* energy level at hafnium (Hf, *Z* = 72).

Physical Properties

Several physical properties of the elements of the first transition series are listed in ▼ TABLE 21.1. Some of these properties, such as ionisation energy and atomic radius, are characteristic of isolated atoms. Others, including density and melting point, are characteristic of the bulk solid metal.

The atomic properties vary in similar ways across each series. Notice, for example, that the bonding atomic radii (∞ Section 7.3, 'Sizes of Atoms and Ions') of the transition metals shown in ▶ FIGURE 21.2 exhibit the same pattern in all three series. The trend in atomic radii is complex because it is the product of several factors, some of which work in opposite directions. In general, we would expect the atomic radius to decrease steadily as we proceed from left to right across the transition series because of the increasing effective nuclear charge

 Review this on page 228

▼ **FIGURE 21.1** *d*-block elements in the periodic table. Transition metals are those elements that occupy groups 3–11 of the periodic table.

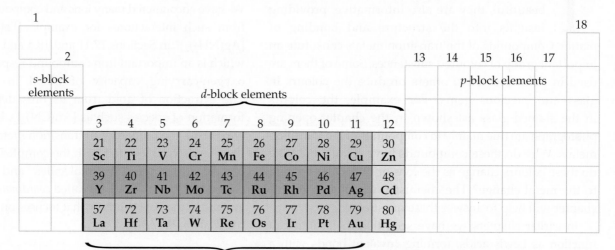

TABLE 21.1 • Properties of the first *d*-block series elements										
Group:	3	4	5	6	7	8	9	10	11	12
Element:	Sc	Ti	V	Cr	Mn	Fe	Co	Ni	Cu	Zn
Electron configuration	$3d^14s^2$	$3d^24s^2$	$3d^34s^2$	$3d^54s^1$	$3d^54s^2$	$3d^64s^2$	$3d^74s^2$	$3d^84s^2$	$3d^{10}4s^1$	$3d^{10}4s^2$
First ionisation energy (kJ mol^{-1})	631	658	650	653	717	759	758	737	745	906
Bonding atomic radius (pm)	144	136	125	127	139	125	126	121	138	131
Density (g cm^{-3})	3.0	4.5	6.1	7.9	7.2	7.9	8.7	8.9	8.9	7.1
Melting point (°C)	1541	1660	1917	1857	1244	1537	1494	1455	1084	420

experienced by the valence electrons. Indeed, for groups 3, 4 and 5 this is the trend observed. As the number of d electrons increases, however, not all of them are employed in bonding. Non-bonding electrons exert repulsive effects that cause increased bond distances, and we see their effects in the maximum that occurs at group 7 and in the increase seen as we move past the group 10 elements. The bonding atomic radius is an empirical quantity that is especially difficult to define for elements such as the transition metals, which can exist in various oxidation states. Nevertheless, comparisons of the variations from one series to another are valid.

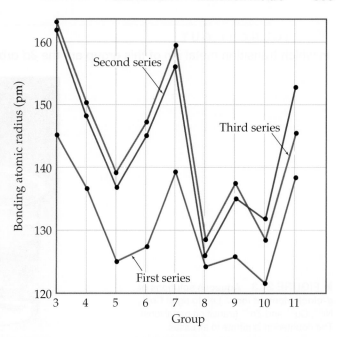

▲ **FIGURE 21.2   Bonding atomic radii.** Variation in the atomic radius of transition metals as a function of the periodic table group number. First series: Sc to Cu; second series: Y to Ag; third series: La to Au.

∞ Review this on page 227

> ### ⚠ CONCEPT CHECK 1
> Which element has the largest bonding atomic radius: Sc, Fe or Au?

The incomplete screening of the nuclear charge by added electrons produces an interesting and important effect in the third transition metal series. In general, the bonding atomic radius increases as we move down in a group, because of the increasing principal quantum number of the outer-shell electrons (∞ Section 7.3, 'Sizes of Atoms and Ions'). Once we move beyond the first series elements, however, the second and third transition series elements have virtually the same bonding atomic radii. In group 5, for example, tantalum has virtually the same radius as niobium. This effect has its origin in the lanthanide series, the elements with atomic numbers 57 to 70, in which the $4f$ subshell is being filled. The filling of $4f$ orbitals through the lanthanide elements causes a steady increase in the effective nuclear charge, producing a contraction in size, called the **lanthanide contraction**. This contraction just offsets the increase we would expect as we go from the second to the third series. Thus the second and third series transition metals in each group have about the same radii all the way across the series. As a consequence, the second and third series metals in a given group have great similarity in their chemical properties. For example, the chemical properties of zirconium and hafnium are remarkably similar. They always occur together in nature and they are very difficult to separate.

Electron Configurations and Oxidation States

Transition metals owe their location in the periodic table to the filling of the d subshells. When these metals are oxidised, however, they lose their outer s electrons before they lose electrons from the d subshell (∞ Section 7.4, 'Ionisation Energy'). The electron configuration of Fe is $[Ar]3d^64s^2$, for example, whereas that of Fe^{2+} is $[Ar]3d^6$. Formation of Fe^{3+} requires the loss of one additional $3d$ electron, giving $[Ar]3d^5$. Most transition metal ions contain partially occupied d subshells. The existence of these d electrons is partly responsible for several characteristics of transition metals.

∞ Review this on page 234

1. They often exhibit more than one stable oxidation state.
2. Many of their compounds are coloured, as shown in ▼ **FIGURE 21.3**.
3. Transition metals and their compounds often have unpaired electrons and exhibit interesting and important magnetic properties.

▼ **FIGURE 21.4** shows the common non-zero oxidation states for the period 4 transition metals. The +2 oxidation state, which is common for most transition metals, is due to the loss of the two outer $4s$ electrons. This oxidation state is found for all these elements except Sc, where the 3+ ion with an [Ar] configuration is particularly stable.

In which transition metal ion of this group are the 3d orbitals completely filled?

▶ **FIGURE 21.3** **Aqueous solutions of *d*-block element ions.** Left to right: Co^{2+}, Ni^{2+}, Cu^{2+} and Zn^{2+} (transition metal ions). The counterion is nitrate in all cases.

Why does the maximum oxidation state increase linearly from Sc to Mn?

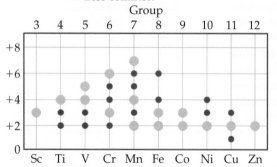

▲ **FIGURE 21.4** **Non-zero oxidation states of the period 4 transition metals.**

Oxidation states above +2 are due to successive losses of 3d electrons. From Sc to Mn the maximum oxidation state increases from +3 to +7 equalling in each case the total number of 4s plus 3d electrons in the atom. Thus manganese has a maximum oxidation state of $2 + 5 = +7$. As we move to the right beyond Mn in the first transition series, the maximum oxidation state decreases. This decrease is due in part to the attraction of d orbital electrons to the nucleus, which increases faster than the attraction of the s orbital electrons to the nucleus as we move from left to right across the periodic table. In other words, in each period the d electrons become more core-like as the atomic number increases. By the time we get to zinc, it is not possible to remove electrons from the 3d orbitals through chemical oxidation.

In the transition metals of periods 5 and 6, the increased size of the 4d and 5d orbitals makes it possible to attain maximum oxidation states as high as +8, which is achieved in RuO_4 and OsO_4. In general, the maximum oxidation states are found only when the metals are combined with the most electronegative elements, especially O, F and in some cases Cl.

Why doesn't Ti^{5+} exist?

Magnetism

The spin an electron possesses gives the electron a *magnetic moment*, a property that causes the electron to behave like a tiny magnet. In a *diamagnetic* solid, defined as one in which all the electrons in the solid are paired, the spin-up and spin-down electrons cancel one another (∞ Section 9.8, 'Second-Period Diatomic Molecules'). Diamagnetic substances are generally described as being non-magnetic, but when a diamagnetic substance is placed in a magnetic field, the motions of the electrons cause the substance to be very weakly repelled by the magnet. In other words, these supposedly non-magnetic substances do show some very faint magnetic character in the presence of a magnetic field.

A substance in which the atoms or ions have one or more unpaired electrons is *paramagnetic*. In a paramagnetic solid, the electrons on one atom or ion do not

∞ Review this on page 323

influence the unpaired electrons on neighbouring atoms or ions. As a result, the magnetic moments on the atoms or ions are randomly oriented, as shown in ▶ **FIGURE 21.5(a)**. When a paramagnetic substance is placed in a magnetic field, however, the magnetic moments tend to align parallel to one another, producing a net attractive interaction with the magnet. Thus unlike a diamagnetic substance, which is weakly repulsed by a magnetic field, a paramagnetic substance is attracted to a magnetic field.

When you think of a magnet, you probably picture a simple iron magnet. Iron exhibits **ferromagnetism**, a form of magnetism much stronger than paramagnetism. Ferromagnetism arises when the unpaired electrons of the atoms or ions in a solid are influenced by the orientations of the electrons in neighbouring atoms or ions. The most stable (lowest-energy) arrangement is when the spins of electrons on adjacent atoms or ions are aligned in the same direction, as in Figure 21.5(b). When a ferromagnetic solid is placed in a magnetic field, the electrons tend to align strongly in a direction parallel to the magnetic field. The attraction to the magnetic field that results may be as much as one million times stronger than that for a paramagnetic substance.

When a ferromagnet is removed from an external magnetic field, the interactions between the electrons cause the ferromagnetic substance to maintain a magnetic moment. We then refer to it as a *permanent magnet* (▶ **FIGURE 21.6**).

The only ferromagnetic transition metals are Fe, Co and Ni, but many alloys also exhibit ferromagnetism, which is in some cases stronger than the ferromagnetism of the pure metals. Particularly powerful ferromagnetism is found in alloys containing both transition metals and lanthanide metals. Two of the most important examples are $SmCo_5$ and $Nd_2Fe_{14}B$.

Two additional types of magnetism involving ordered arrangements of unpaired electrons are depicted in Figure 21.5. In materials that exhibit **antiferromagnetism** (Figure 21.5(c)), the unpaired electrons on a given atom or ion align so that their spins are oriented in the direction opposite the spin direction on neighbouring atoms. This means that the spin-up and spin-down electrons cancel each other. Examples of antiferromagnetic substances are chromium metal, FeMn alloys and such transition metal oxides as Fe_2O_3, $LaFeO_3$ and MnO.

A substance that exhibits **ferrimagnetism** (Figure 21.5(d)) has both ferromagnetic and antiferromagnetic properties. Like an antiferromagnet, the unpaired electrons align so that the spins in adjacent atoms or ions point in opposite directions. However, unlike an antiferromagnet, the net magnetic moments of the spin-up electrons are not fully cancelled by the spin-down electrons. This can happen because the magnetic centres have different numbers of unpaired electrons ($NiMnO_3$), because the number of magnetic sites aligned in one direction is larger than the number aligned in the other direction ($Y_3Fe_5O_{12}$) or because both these conditions apply (Fe_3O_4). Because the magnetic moments do not cancel, the properties of ferrimagnetic materials are similar to the properties of ferromagnetic materials.

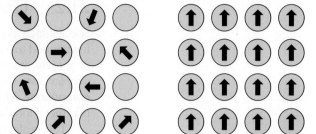

(a) Paramagnetic; spins random; spins do align if in magnetic field

(b) Ferromagnetic; spins aligned; spins become random at high temperature

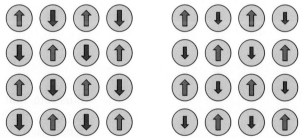

(c) Antiferromagnetic; spins opposed and cancel; spins become random at high temperature

(d) Ferrimagnetic; unequal spins opposed but do not cancel; spins become random at high temperature

▲ **FIGURE 21.5**   **The relative orientation of electron spins in various types of magnetic substances.**

▲ **FIGURE 21.6**   **A permanent magnet.** Permanent magnets are made from ferromagnetic and ferrimagnetic materials.

▲ **CONCEPT CHECK 3**

How do you think spin–spin interactions of unpaired electrons on adjacent atoms in a substance are affected by the interatomic distance?

Ferromagnets, ferrimagnets and antiferromagnets all become paramagnetic when heated above a critical temperature. This happens when the thermal energy is sufficient to overcome the forces determining the spin directions of the electrons. This temperature is called the *Curie temperature*, T_C, for ferromagnets and ferrimagnets and the *Néel temperature*, T_N, for antiferromagnets.

21.2 | TRANSITION METAL COMPLEXES

Species such as $[Ag(NH_3)_2]^+$ and $[Fe(OH_2)_6]^{3+}$ that are assemblies of a central metal ion bonded to a group of surrounding molecules or ions are called **metal complexes** or merely *complexes*. If the complex carries a net charge, it is generally called a *complex ion*. Compounds that contain complexes are known as **coordination compounds**. Most of the coordination compounds that we examine contain transition metal ions, although ions of other metals can form complexes as well.

The molecules or ions that bond to the metal ion in a complex are known as **ligands** (from the Latin word *ligare*, meaning 'to bind'). There are two NH_3 ligands bonded to Ag^+ in $[Ag(NH_3)_2]^+$. Each ligand functions as a Lewis base, donating a pair of electrons to the metal to form a bond with the metal. Thus every ligand has at least one unshared pair of valence electrons. Four most frequently encountered ligands,

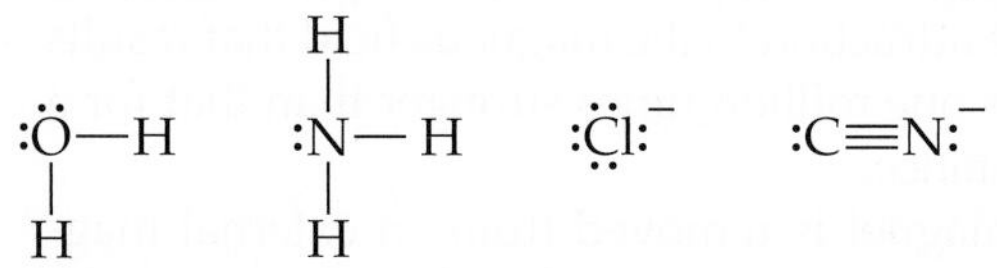

:Ö—H   :N—H   :C̈l:⁻   :C≡N:⁻

illustrate that ligands are usually polar molecules or anions. In forming a complex, the ligands are said to *coordinate* to the metal, and these sort of bonds, where two electrons are donated from one species to form the bond, are called coordinate bonds.

> ### ⚠ CONCEPT CHECK 4
> Is the interaction between an ammonia ligand and a metal cation a Lewis acid–base interaction? If so, which species acts as a Lewis acid?

The Development of Coordination Chemistry: Werner's Theory

Because compounds of the transition metals exhibit beautiful colours, the chemistry of these elements greatly fascinated chemists even before the periodic table was introduced. In the late 1700s to the 1800s, many coordination compounds were isolated and studied. These compounds showed properties that seemed puzzling in light of the bonding theories at the time. ▼ TABLE 21.2, for example, lists a series of compounds that result from the reaction of cobalt(III) chloride with ammonia. These compounds have strikingly different colours. Even the last two listed, which were both formulated as $CoCl_3 \cdot 4\,NH_3$, have different colours.

The modern formulations of the compounds in Table 21.2 are based on various lines of experimental evidence. For example, all four compounds are strong electrolytes (Section 4.1), but they yield different numbers of ions when

TABLE 21.2 • **Properties of some ammonia complexes of cobalt(III)**

Original formulation	Colour	Ions per formula unit	'Free' Cl⁻ ions per formula unit	Modern formulation
$CoCl_3 \cdot 6\,NH_3$	Orange	4	3	$[Co(NH_3)_6]Cl_3$
$CoCl_3 \cdot 5\,NH_3$	Purple	3	2	$[Co(NH_3)_5Cl]Cl_2$
$CoCl_3 \cdot 4\,NH_3$	Green	2	1	*trans*-$[Co(NH_3)_4Cl_2]Cl$
$CoCl_3 \cdot 4\,NH_3$	Violet	2	1	*cis*-$[Co(NH_3)_4Cl_2]Cl$

dissolved in water. For example, dissolving $CoCl_3 \cdot 6\,NH_3$ in water yields four ions per formula unit (that is, the $[Co(NH_3)_6Cl]^{3+}$ ion and three Cl^- ions), whereas $CoCl_3 \cdot 5\,NH_3$ yields only three ions per formula unit (that is, the $[Co(NH_3)_5Cl]^{2+}$ ion and two Cl^- ions). Furthermore, the reaction of the compounds with excess aqueous silver nitrate leads to the precipitation of variable amounts of $AgCl(s)$; the precipitation of $AgCl(s)$ in this way is often used to test for the number of 'free' Cl^- ions in an ionic compound. When $CoCl_3 \cdot 6\,NH_3$ is treated with excess $AgNO_3(aq)$, 3 mol of $AgCl(s)$ are produced per mole of complex, so all three Cl^- ions in the formula can react to form $AgCl(s)$. By contrast, when $CoCl_3 \cdot 5\,NH_3$ is treated with $AgNO_3(aq)$ in an analogous fashion, only 2 mol of $AgCl(s)$ precipitate per mole of complex; one of the Cl^- ions in the compound does not react to form $AgCl(s)$. These results are summarised in Table 21.2.

In 1893 the Swiss chemist Alfred Werner (1866–1919) proposed a theory that successfully explained the observations in Table 21.2, and it became the basis for understanding coordination chemistry. Werner proposed that metal ions exhibit both 'primary' and 'secondary' valences. The primary valence is the oxidation state of the metal, which for the complexes in Table 21.2 is +3. The secondary valence is the number of atoms directly bonded to the metal ion, which is also called the **coordination number**. For these cobalt complexes, Werner deduced a coordination number of 6 with the ligands in an octahedral arrangement around the Co^{3+} ion.

Werner's theory provided a beautiful explanation for the results in Table 21.2. The NH_3 molecules in the complexes are ligands that are bonded to the Co^{3+} ion; if there are fewer than six NH_3 molecules, the remaining ligands are Cl^- ions. The central metal and the ligands bound to it constitute the **coordination sphere** of the complex.

In writing the chemical formula for a coordination compound, Werner suggested using square brackets to set off the groups within the coordination sphere from other parts of the compound. He therefore proposed that $CoCl_3 \cdot 6\,NH_3$ and $CoCl_3 \cdot 5\,NH_3$ are better written as $[Co(NH_3)_6]Cl_3$ and $[Co(NH_3)_5Cl]Cl_2$, respectively. He further proposed that the chloride ions that are part of the coordination sphere are bound so tightly that they do not dissociate when the complex is dissolved in water. Thus dissolving $[Co(NH_3)_5Cl]Cl_2$ in water produces a $[Co(NH_3)_5Cl]^{2+}$ ion and two Cl^- ions; only the two 'free' Cl^- ions are able to react with $Ag^+(aq)$ to form $AgCl(s)$.

Werner's ideas also explained why there are two distinctly different forms of $CoCl_3 \cdot 4\,NH_3$. Using Werner's postulates, we formulate the compound as $[Co(NH_3)_4Cl_2]Cl$. As shown in ▶ **FIGURE 21.7**, there are two different ways to arrange the ligands in the $[Co(NH_3)_4Cl_2]^+$ complex, called the *cis* and *trans* forms. In *cis*-$[Co(NH_3)_4Cl_2]^+$ the two chloride ligands occupy adjacent vertices of the octahedral arrangement. In *trans*-$[Co(NH_3)_4Cl_2]^+$ the chlorides are opposite one another. As seen in Figure 21.7, the difference in these arrangements causes the complexes to have different colours.

The insight that Werner provided into the bonding in coordination comounds is even more remarkable when we realise that his theory

▲ **FIGURE 21.7**   **Isomers of $[Co(NH_3)_4Cl_2]^+$.** The *cis* isomer is violet, and the *trans* isomer is green.

predated Lewis's ideas of covalent bonding by more than 20 years. Because of his tremendous contributions to coordination chemistry, Werner was awarded the 1913 Nobel Prize in Chemistry.

SAMPLE EXERCISE 21.1 Identifying the coordination sphere of a complex

Palladium(II) tends to form complexes with a coordination number of 4. One such compound was originally formulated as $PdCl_2 \cdot 3 NH_3$. **(a)** Suggest the appropriate coordination compound formulation for this compound. **(b)** Suppose an aqueous solution of the compound is treated with excess $AgNO_3(aq)$. How many moles of $AgCl(s)$ are formed per mole of $PdCl_2 \cdot 3 NH_3$?

SOLUTION

Analyse We are given the coordination number of Pd(II) and a chemical formula indicating that the complex contains NH_3 and Cl^-. We are asked to determine **(a)** which ligands are attached to Pd(II) in the compound and **(b)** how the compound behaves towards $AgNO_3$ in aqueous solution.

Plan (a) Because of their charge, the Cl^- ions can be either in the coordination sphere, where they are bonded directly to the metal, or outside the coordination sphere, where they are bonded ionically to the complex. The electrically neutral NH_3 ligands must be in the coordination sphere, if we assume four ligands bonded to the Pd(II) ion. **(b)** Any chlorides in the coordination sphere do not precipitate as AgCl.

Solve
(a) By analogy to the ammonia complexes of cobalt(III), we predict that the three NH_3 groups of $PdCl_2 \cdot 3 NH_3$ serve as ligands attached to the Pd(II) ion. The fourth ligand around Pd(II) is one of the chloride ions. The second chloride ion is not a ligand; it serves only as an anion in this ionic compound. We conclude that the correct formulation is $[Pd(NH_3)_3Cl]Cl$.

(b) The chloride ion that serves as a ligand will not be precipitated as $AgCl(s)$ following the reaction with $AgNO_3(aq)$. Thus only the single 'free' Cl^- can react. We therefore expect to produce 1 mol of $AgCl(s)$ per mole of complex. The balanced equation is

$$[Pd(NH_3)_3Cl]Cl(aq) + AgNO_3(aq) \longrightarrow [Pd(NH_3)_3Cl]NO_3(aq) + AgCl(s)$$

PRACTICE EXERCISE

Predict the number of ions produced per formula unit in an aqueous solution of $CoCl_2 \cdot 6 H_2O$.

Answer: Three (the complex ion, $[Co(H_2O)_6]^{2+}$, and two chloride ions)

(See also Exercise 21.25.)

The Metal–Ligand Bond

The bond between a ligand and a metal ion is an example of an interaction between a Lewis base and a Lewis acid. Because the ligands have unshared pairs of electrons, they can function as Lewis bases (electron-pair donors). Metal ions (particularly transition metal ions) have empty valence orbitals, so they can act as Lewis acids (electron-pair acceptors). We can picture the bond between the metal ion and ligand as the result of their sharing a pair of electrons that was initially on the ligand:

$$Ag^+(aq) + 2 \underset{\underset{H}{|}}{\overset{\overset{H}{|}}{:N}} - H(aq) \longrightarrow \left[\underset{\underset{H}{|}}{\overset{\overset{H}{|}}{H-N}} : Ag : \underset{\underset{H}{|}}{\overset{\overset{H}{|}}{N}} - H \right]^+ (aq) \qquad [21.1]$$

The formation of metal–ligand bonds can markedly alter the properties we observe for the metal ion. A metal complex is a distinct chemical species that has physical and chemical properties different from the metal ion and the ligands from which it is formed. Complexes, for example, may have colours that differ dramatically from those of their component metal ions and ligands. ▶ FIGURE 21.8 shows the colour change that occurs when aqueous solutions of SCN^- and Fe^{3+} are mixed, forming $[Fe(OH_2)_5NCS]^{2+}$.

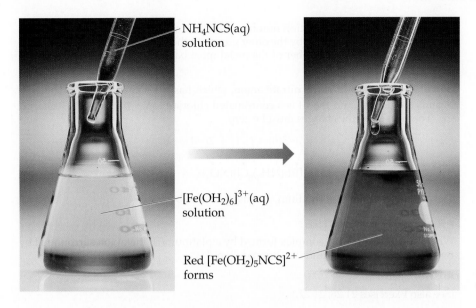

NH$_4$NCS(aq)
solution

[Fe(OH$_2$)$_6$]$^{3+}$(aq)
solution

Red [Fe(OH$_2$)$_5$NCS]$^{2+}$
forms

◀ **FIGURE 21.8**   **Reaction of Fe^{3+}(aq)
and NCS⁻(aq).**

Complex formation can also significantly change other properties of metal ions, such as their ease of oxidation or reduction. The silver ion, Ag$^+$, for example, is readily reduced in water:

$$Ag^+(aq) + e^- \longrightarrow Ag(s) \qquad E° = +0.799 \text{ V} \qquad [21.2]$$

In contrast, the [Ag(CN)$_2$]⁻ ion is not so easily reduced because complexation by CN⁻ ions stabilises silver in the +1 oxidation state:

$$[Ag(CN)_2]^-(aq) + e^- \longrightarrow Ag(s) + 2\,CN^-(aq) \qquad E° = -0.31 \text{ V} \qquad [21.3]$$

Hydrated metal ions are actually complex ions in which the ligand is water. Thus Fe^{3+}(aq) consists largely of [Fe(OH$_2$)$_6$]$^{3+}$. Complex ions form in aqueous solutions from reactions in which ligands such as NH$_3$, SCN⁻ and CN⁻ replace H$_2$O molecules in the coordination sphere of the metal ion.

CONCEPT CHECK 5

Write a balanced chemical equation for the reaction that causes the colour change in Figure 21.8.

Charges, Coordination Numbers and Geometries

The charge of a complex is the sum of the charges on the central metal and on its surrounding ligands. In [Cu(NH$_3$)$_4$]SO$_4$ we can deduce the charge on the complex ion if we first recognise that SO$_4$ represents the sulfate ion and therefore has a 2− charge. Because the compound is neutral, the complex ion must have a 2+ charge, [Cu(NH$_3$)$_4$]$^{2+}$. We can then use the charge of the complex ion to deduce the oxidation number of copper. Because the NH$_3$ ligands are neutral molecules, the oxidation number of copper must be +2.

$$\underset{[Cu(NH_3)_4]^{2+}}{+2 + 4(0) = +2}$$

SAMPLE EXERCISE 21.2   **Determining the oxidation number of a metal in a
complex**

What is the oxidation number of the central metal in [Rh(NH$_3$)$_5$Cl](NO$_3$)$_2$?

SOLUTION

Analyse We are given the chemical formula of a coordination compound and asked to determine the oxidation number of its metal atom.

Plan To determine the oxidation number of the Rh metal ion, we need to figure out what charges are contributed by the other groups in the substance. The overall charge is zero, so the oxidation number of the metal must balance the charge that is due to the rest of the compound.

Solve The NO_3 group is the nitrate anion, which has a $1-$ charge, NO_3^-. The NH_3 ligands are neutral and the Cl is a coordinated chloride ion, which has a $1-$ charge, Cl^-. The sum of all the charges must be zero.

$$x + 5(0) + (-1) + 2(-1) = 0$$

$$[Rh(NH_3)_5Cl](NO_3)_2$$

The oxidation number of rhodium, x, must therefore be $+3$.

PRACTICE EXERCISE

What is the charge of the complex formed by a platinum(II) metal ion surrounded by two ammonia molecules and two bromide ions?

Answer: zero

(See also Exercises 21.26, 21.27.)

SAMPLE EXERCISE 21.3 | **Determining the formula of a complex ion**

A complex ion contains a chromium(III) bound to four water molecules and two chloride ions. What is its formula?

SOLUTION

Analyse We are given a metal, its oxidation number and the number of ligands of each kind in a complex ion containing the metal and we are asked to write the chemical formula and charge of the ion.

Plan We write the metal first, then the ligands. We can use the charges of the metal ion and ligands to determine the charge of the complex ion. The oxidation state of the metal is $+3$, water is neutral and chloride has a -1 charge.

Solve

$$+3 + 4(0) + 2(-1) = +1$$

$$Cr(OH_2)_4Cl_2$$

The charge on the ion is $1+$, $[Cr(OH_2)_4Cl_2]^+$.

PRACTICE EXERCISE

Write the formula for the complex described in the Practice Exercise accompanying Sample Exercise 21.2.

Answer: $[Pt(NH_3)_2Br_2]$

Recall that the number of atoms directly bonded to the metal ion in a complex is called the *coordination number* of the complex. The atom of the ligand bound directly to the metal is called the **donor atom**. Nitrogen, for example, is the donor atom in the $[Ag(NH_3)_2]^+$ complex shown in Equation 21.1. The silver ion in $[Ag(NH_3)_2]^+$ has a coordination number of 2, whereas each cobalt ion in the Co(III) complexes in Table 21.2 has a coordination number of 6.

Some metal ions exhibit constant coordination numbers. The coordination number of chromium(III) and cobalt(III) is invariably 6, for example, and that of platinum(II) is always 4. The coordination numbers of most metal ions vary with the ligand, however. The most common coordination numbers are 4 and 6.

The coordination number of a metal ion is often influenced by the relative sizes of the metal ion and the surrounding ligands. As the ligand gets larger, fewer can coordinate to the metal ion. Thus iron(III) is able to coordinate to six fluorides in $[FeF_6]^{3-}$ but coordinates to only four chlorides in $[FeCl_4]^-$.

Ligands that transfer substantial negative charge to the metal also produce reduced coordination numbers. For example, six neutral ammonia molecules can coordinate to nickel(II), forming $[Ni(NH_3)_6]^{2+}$, but only four negatively charged cyanide ions can coordinate, forming $[Ni(CN)_4]^{2-}$.

Complexes in which the coordination number is 4 have two common geometries, tetrahedral and square planar, as shown in ▶ **FIGURE 21.9**. The tetrahedral geometry is the more common of the two, especially among non-transition metals. The square planar geometry is characteristic of transition metal ions with eight d electrons in the valence shell, such as platinum(II) and gold(III).

The vast majority of complexes in which the coordination number is 6 have an octahedral geometry. The octahedron, which has eight faces and six vertices, is often represented as a square with ligands above and below the plane, as in ▼ **FIGURE 21.10**. Recall, however, that all six vertices on an octahedron are geometrically equivalent (∞ Section 9.2, 'The VSEPR Model'). The octahedron can also be thought of as two square pyramids that share the same square base.

FIGURE IT OUT

What is the size of the NH_3–Zn–NH_3 bond angle? Of the NH_3–Pt–NH_3 bond angle?

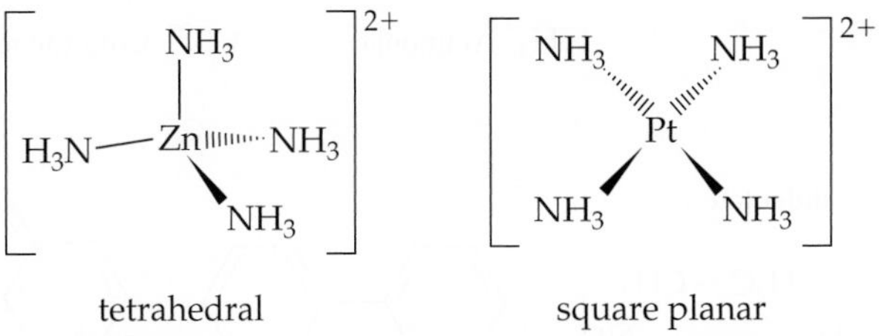

tetrahedral   square planar

▲ **FIGURE 21.9** In complexes having coordination number 4, the molecular geometry can be tetrahedral or square planar.

∞ Review this on page 303

CONCEPT CHECK 6

What are the geometries most commonly associated with

a. coordination number 4?

b. coordination number 6?

21.3 | LIGANDS WITH MORE THAN ONE DONOR ATOM

The ligands discussed so far, such as NH_3 and Cl^-, are called **monodentate ligands** (from the Latin, meaning 'one-toothed'). These ligands possess a single donor atom and are able to occupy only one site in a coordination sphere. Ligands having two donor atoms are **bidentate ligands** ('two-toothed') and those having three or more donor atoms are **polydentate ligands** ('many-toothed'). In both bidentate and polydentate species, the multiple donor atoms can simultaneously bond to the metal ion, thereby occupying two or more sites in a coordination sphere. ▼ **TABLE 21.3** gives examples of all three types of ligands. Because they appear to grasp the metal between two or more donor atoms, polydentate ligands are also known as **chelating agents** (from the Greek word *chele*, 'claw'). One such ligand is *ethylenediamine*, en:

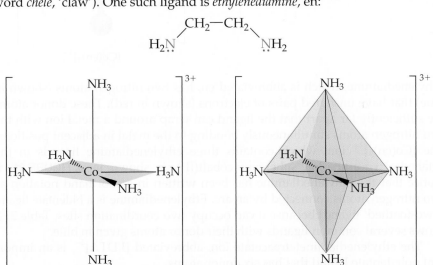

◀ **FIGURE 21.10** In complexes having coordination number 6, the molecular geometry is almost always octahedral. Two ways to draw octahedral geometry are shown.

TABLE 21.3 • Some common ligands

Ligand type   Examples

Monodentate

$H_2\ddot{O}$: Water   :$\ddot{\underset{..}{F}}$:⁻ Fluoride ion   [:C≡N:]⁻ Cyanide ion   [:$\ddot{O}$—H]⁻ Hydroxide io

:NH₃ Ammonia   :$\ddot{\underset{..}{Cl}}$:⁻ Chloride ion   [:$\ddot{S}$=C=$\ddot{N}$:]⁻ Thiocyanate ion   [:$\ddot{O}$—N=$\ddot{O}$:]⁻ Nitrite ion
　　　　　　　　　　　　　　　　　└──or──┘　　　　　　　　　　└─or─┘

Bidentate

H_2C—CH_2
$H_2\ddot{N}$　　$\ddot{N}H_2$
Ethylenediamine (en)

Bipyridine
(bipy)

Ortho-phenanthroline
(*o*-phen)

:$\ddot{O}$:⁻ Phenoxide ion

Oxalate ion

Carbonate ion

Polydentate

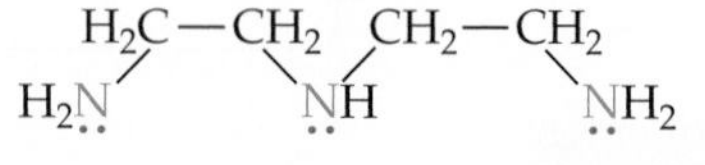

H_2C—CH_2　CH_2—CH_2
H_2N　　　$\overset{..}{N}H$　　　NH_2
Diethylenetriamine

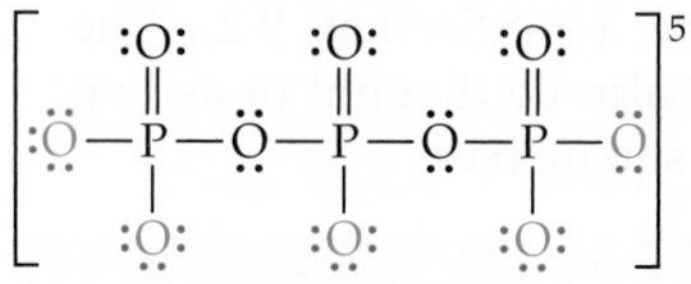

Triphosphate ion

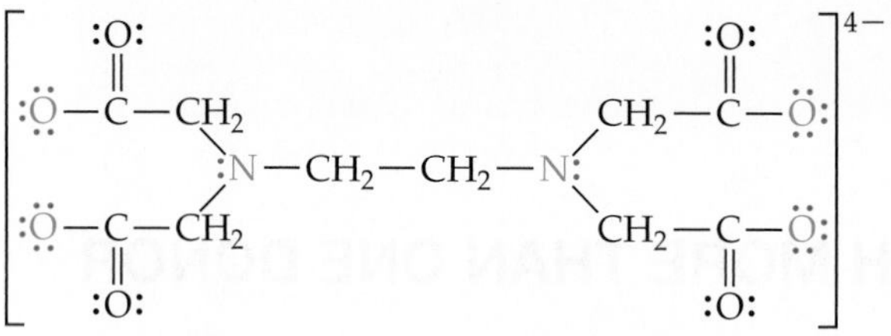

Ethylenediaminetetraacetate ion (EDTA⁴⁻)

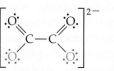

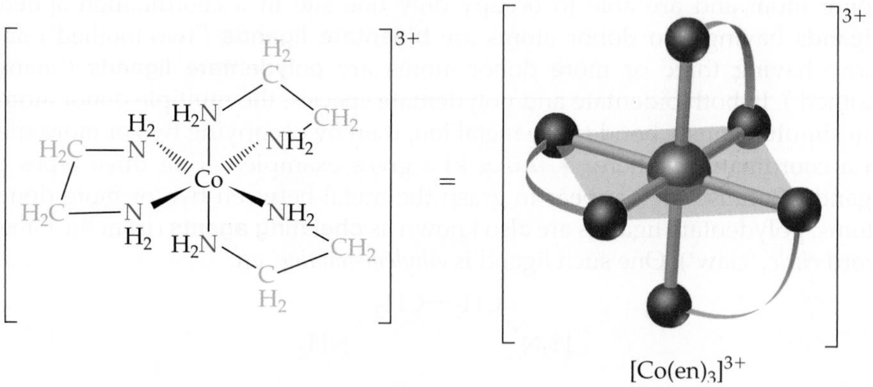

=

$[Co(en)_3]^{3+}$

► **FIGURE 21.11   The [Co(en)₃]³⁺ ion.**
The ligand is ethylenediamine.

Ethylenediamine, which is abbreviated en, has two nitrogen atoms (shown in blue) that have unshared pairs of electrons (shown in red). These donor atoms are sufficiently far apart that the ligand can wrap around a metal ion with the two nitrogen atoms simultaneously bonding to the metal in adjacent positions. The $[Co(en)_3]^{3+}$ ion, which contains three ethylenediamine ligands in the octahedral coordination sphere of cobalt(III), is shown in ▲ **FIGURE 21.11**. Notice that the ethylenediamine has been written in a shorthand notation as two nitrogen atoms connected by an arc. Ethylenediamine is a bidentate ligand ('two-toothed' ligand) because it can occupy two coordination sites. Table 21.3 shows several common ligands with their donor atoms given in blue.

The ethylenediaminetetraacetate ion, abbreviated $[EDTA]^{4-}$, is an important polydentate ligand that has six donor atoms:

[EDTA]$^{4-}$

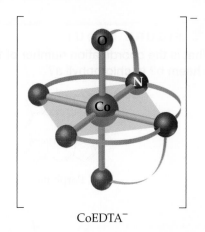

CoEDTA$^-$

▲ **FIGURE 21.12   The [CoEDTA]$^-$ ion.** Notice how the ethylenediaminetetraacetate ion, a polydentate ligand, is able to wrap around a metal ion, occupying six positions in the coordination sphere.

EDTA can wrap around a metal ion using all six of these donor atoms, as shown in ▶ FIGURE 21.12, although it sometimes binds to a metal using only five of its six donor atoms.

In general, chelating ligands form more stable complexes than do related monodentate ligands. The equilibrium formation constants for [Ni(NH$_3$)$_6$]$^{2+}$ and [Ni(en)$_3$]$^{2+}$, shown in Equations 21.4 and 21.5, illustrate this observation.

$$[Ni(H_2O)_6]^{2+}(aq) + 6\,NH_3(aq) \rightleftharpoons [Ni(NH_3)_6]^{2+}(aq) + 6\,H_2O(l)$$

$$K_f = 1.2 \times 10^9 \qquad [21.4]$$

$$[Ni(H_2O)_6]^{2+}(aq) + 3\,en(aq) \rightleftharpoons [Ni(en)_3]^{2+}(aq) + 6\,H_2O(l)$$

$$K_f = 6.8 \times 10^{17} \qquad [21.5]$$

Although the donor atom is nitrogen in both instances, [Ni(en)$_3$]$^{2+}$ has a formation constant that is more than 10^8 times larger than that of [Ni(NH$_3$)$_6$]$^{2+}$. The generally larger formation constants for polydentate ligands as compared with the corresponding monodentate ligands is known as the **chelate effect**.

Chelating agents are often used to prevent one or more of the customary reactions of a metal ion without actually removing it from solution. For example, a metal ion that interferes with a chemical analysis can often be complexed and its interference thereby removed. In a sense, the chelating agent hides the metal ion. For this reason, scientists sometimes refer to these ligands as *sequestering agents*. (The word *sequester* means to remove, set apart or separate.)

Phosphates such as sodium triphosphate, [Na$_5$OPO$_2$OPO$_2$OPO$_3$], shown here, are used to complex or sequester metal ions such as Ca^{2+} and Mg^{2+} in hard water so these ions cannot interfere with the action of soap or detergents:

$$Na_5 \left[\begin{array}{c} O \quad\quad O \quad\quad O \\ \parallel \quad\quad \parallel \quad\quad \parallel \\ {}^-O-P-O-P-O-P-O^- \\ \mid \quad\quad \mid \quad\quad \mid \\ {}_-O \quad\quad {}_-O \quad\quad {}_-O \end{array} \right]$$

Chelating agents such as EDTA are used in consumer products, including many prepared foods such as salad dressings and frozen desserts, to complex trace metal ions that catalyse decomposition reactions. Chelating agents are used in medicine to remove metal ions such as Hg^{2+}, Pb^{2+} and Cd^{2+}, which are detrimental to health. One method of treating lead poisoning is to administer Na$_2$[Ca(EDTA)]. The EDTA chelates the lead, allowing it to be removed from the body via urine.

 CONCEPT CHECK 7

Cobalt(III) has a coordination number of 6 in all its complexes. Is the carbonate ion acting as a monodentate or as a bidentate ligand in the [Co(NH$_3$)$_4$CO$_3$]$^+$ ion?

Metals and Chelates in Living Systems

Of the 29 elements known to be necessary for human life, nine are transition metals (V, Cr, Mn, Fe, Co, Ni, Cu, Mo, Cd) and one is a *d*-block element (Zn).

What is the coordination number of the metal ion in haem b? In chlorophyll *a*?

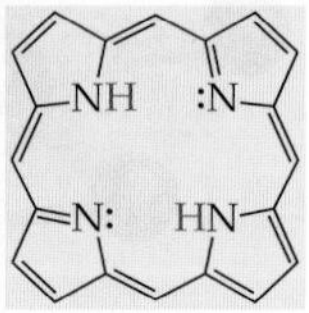

Porphine

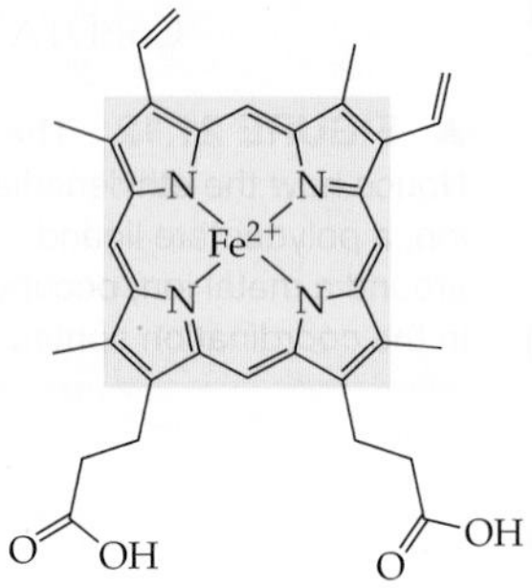

Haem b

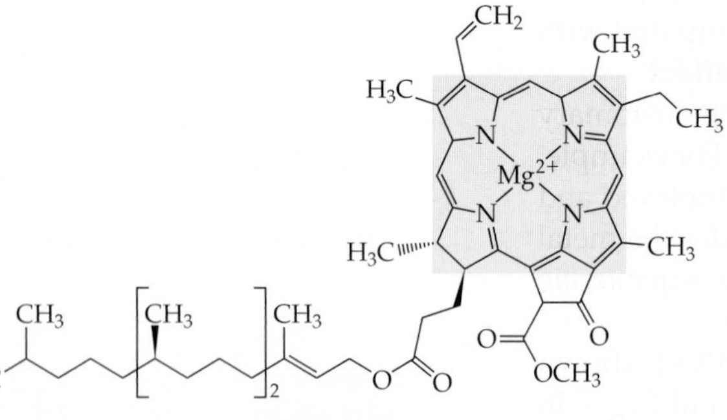

Chlorophyll *a*

▲ **FIGURE 21.13** **Porphine and two porphyrins, haem b and chlorophyll *a*.** Fe(II) and Mg(II) ions replace the two blue H atoms in porphine and bond with all four nitrogens in haem b and chlorophyll *a*, respectively.

They owe their roles in living systems mainly to their ability to form complexes with a variety of groups present in biological systems.

Although our bodies require only small quantities of metals, deficiencies can lead to serious illness. A deficiency of manganese, for example, can lead to convulsive disorders. Some epilepsy patients have been helped by the addition of manganese to their diets.

Among the most important chelating agents in nature are those derived from the *porphine* molecule, shown in ◀ FIGURE 21.13. This molecule can coordinate to a metal using the four nitrogen atoms as donors. Upon coordination to a metal, the two H^+ ions shown bonded to nitrogen are displaced. Complexes derived from porphine are called **porphyrins**. Different porphyrins contain different metal ions and have different substituent groups attached to the carbon atoms at the ligand's periphery. Two of the most important porphyrin or porphyrin-like compounds are *haem*, which contains Fe(II), and *chlorophyll*, which contains Mg(II).

▼ FIGURE 21.14 shows a schematic structure of myoglobin, a protein that contains one haem group. Myoglobin is a *globular protein*, one that folds into a compact, roughly spherical shape. Globular proteins are generally soluble in water and are mobile within cells. Myoglobin is found in the cells of skeletal muscle, particularly in seals, whales and porpoises. It stores oxygen in cells until it is needed for metabolic activities. Haemoglobin, the protein that transports oxygen in human blood, is made up of four haem-containing subunits, each of which is very similar to myoglobin. One haemoglobin can bind up to four O_2 molecules.

The coordination environment of the iron in myoglobin and haemoglobin is illustrated schematically in ▶ FIGURE 21.15. The iron is coordinated to the four nitrogen atoms of the porphyrin and to a nitrogen atom from the protein chain. The sixth position around the iron is occupied either by O_2 (in oxyhaemoglobin, the bright red form) or by water (in deoxyhaemoglobin, the purplish-red form). The oxy form is shown in Figure 21.15. Carbon monoxide is poisonous because the equilibrium binding constant of human haemoglobin for CO is about 210 times greater than that for O_2. As a

▶ **FIGURE 21.14** **Myoglobin.** Myoglobin is a protein that stores oxygen in cells. The molecule has a molecular weight of about 18 000 u and contains one haem unit, shown in orange. The haem unit is bound to the protein through a nitrogen-containing ligand, as shown on the left side of the haem. In the oxygenated form an O_2 molecule is coordinated to the haem group, as shown on the right side of the haem. The three-dimensional structure of the protein chain is represented by the continuous purple cylinder. The helical sections are denoted by the dashed lines. The protein wraps around to make a pocket for the haem group.

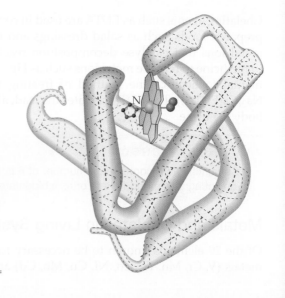

result, a relatively small quantity of CO can inactivate a substantial fraction of the haemoglobin in the blood. For example, a person breathing air that contains only 0.1% CO takes in enough CO after a few hours to convert up to 60% of the haemoglobin (Hb) into COHb, thereby reducing the blood's normal oxygen-carrying capacity by 60%.

Under normal conditions, a non-smoker breathing unpolluted air has about 0.3 to 0.5% COHb in her or his blood. This amount arises mainly from the production of small quantities of CO in the course of normal body chemistry and from the small amount of CO present in clean air. Exposure to higher concentrations of CO causes the COHb level to increase, which in turn leaves fewer Hb sites to which O_2 can bind. If the level of COHb becomes too high, oxygen transport is effectively shut down and death occurs. Because CO is colourless and odourless, CO poisoning occurs with very little warning. Improperly ventilated combustion devices, such as kerosene lanterns and stoves, thus pose a potential health hazard.

The **chlorophylls**, which are porphyrins that contain Mg(II), are the key components in the conversion of solar energy into forms that can be used by living organisms. This process, called **photosynthesis**, occurs in the leaves of green plants. In photosynthesis, carbon dioxide and water are converted to carbohydrate, with the release of oxygen:

$$6\,CO_2(g) + 6\,H_2O(l) \longrightarrow C_6H_{12}O_6(aq) + 6\,O_2(g) \qquad [21.6]$$

The product of this reaction is the sugar glucose, $C_6H_{12}O_6$, which serves as a fuel in biological systems. The formation of one mole of glucose requires the absorption of 48 moles of photons from sunlight or other sources of light. The photons are absorbed by chlorophyll-containing pigments in the leaves of plants. The structure of the most abundant chlorophyll, called chlorophyll *a*, is shown in Figure 21.13.

Chlorophylls contain a Mg^{2+} ion bound to four nitrogen atoms arranged around the metal in a planar array. The nitrogen atoms are part of a porphine-like ring (Figure 21.13). The series of alternating, or *conjugated*, double bonds in the ring surrounding the metal ion is similar to ones found in many organic dyes. This system of conjugated double bonds makes it possible for chlorophyll to absorb light strongly in the visible region of the spectrum, which corresponds to wavelengths ranging from 400 nm to 700 nm. Chlorophyll is green because it absorbs red light (maximum absorption at 655 nm) and blue light (maximum absorption at 430 nm) and transmits green light (▶ FIGURE 21.16).

The solar energy absorbed by chlorophyll is converted by a complex series of steps into chemical energy. This stored energy is then used to drive the reaction in Equation 21.6 to the right, a direction in which it is highly endothermic. Plant photosynthesis is nature's solar-energy-conversion machine; all living systems on the Earth depend on it for continued existence.

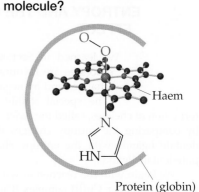

▲ **FIGURE IT OUT**

Where would CO bind in this molecule?

Haem

Protein (globin)

▲ **FIGURE 21.15** **Coordination sphere of the haems in oxymyoglobin and oxyhaemoglobin.**

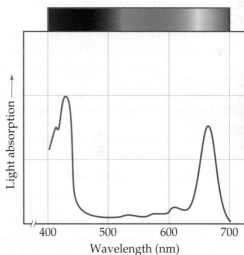

▲ **FIGURE 21.16** **The absorption of sunlight by chlorophyll.**

CONCEPT CHECK 8

What property of the porphine ligand makes it possible for chlorophyll to play a role in photosynthesis?

21.4 | NOMENCLATURE AND ISOMERISM IN COORDINATION CHEMISTRY

When complexes were first discovered and few were known, they were named after the chemist who originally prepared them. A few of these names persist; for example, the dark-red substance $NH_4[Cr(NH_3)_2(NCS)_4]$ is still known as

A CLOSER LOOK

ENTROPY AND THE CHELATE EFFECT

We learned in Section 14.12 that chemical processes are favoured by positive entropy changes and by negative enthalpy changes. The special stability associated with the formation of chelates, called the *chelate effect*, can be explained by comparing the entropy changes that occur with monodentate ligands with the entropy changes that occur with polydentate ligands.

We begin with the reaction in which two H_2O ligands of the square planar Cu(II) complex $[Cu(OH_2)_4]^{2+}$ are replaced by monodentate NH_3 ligands at 27 °C:

$$[Cu(OH_2)_4]^{2+}(aq) + 2\,NH_3(aq) \rightleftharpoons$$
$$[Cu(OH_2)_2(NH_3)_2]^{2+}(aq) + 2\,H_2O(l)$$

$$\Delta H° = -46 \text{ kJ}; \Delta S° = -8.4 \text{ J K}^{-1}; \Delta G° = -43 \text{ kJ}$$

The thermodynamic data tell us about the relative abilities of H_2O and NH_3 to serve as ligands in this reaction. In general, NH_3 binds more tightly to metal ions than does H_2O, so this substitution reaction is exothermic ($\Delta H < 0$). The stronger bonding of the NH_3 ligands also causes the $[Cu(OH_2)_2(NH_3)_2]^{2+}$ ion to be more rigid, which is probably the reason $\Delta S°$ is slightly negative.

We can use Equation 16.25, $\Delta G° = -RT \ln K$, to calculate the equilibrium constant of the reaction at 27 °C. The result, $K = 3.1 \times 10^7$, tells us that the equilibrium lies far to the right, favouring replacement of H_2O by NH_3. For this equilibrium, therefore, the enthalpy change, $\Delta H° = -46$ kJ, is large enough and negative enough to overcome the entropy change, $\Delta S° = -8.4$ J K^{-1}.

Now let's use a single bidentate ethylenediamine ligand in our substitution reaction:

$$[Cu(OH_2)_4]^{2+}(aq) + en(aq) \rightleftharpoons$$
$$[Cu(OH_2)_2(en)]^{2+}(aq) + 2\,H_2O(l)$$

$$\Delta H° = -54 \text{ kJ}; \Delta S° = +23 \text{ J K}^{-1}; \Delta G° = -61 \text{ kJ}$$

The en ligand binds slightly more strongly to the Cu^{2+} ion than two NH_3 ligands, so the enthalpy change here (–54 kJ) is

slightly more negative than for $[Cu(H_2O)_2(NH_3)_2]^{2+}$ (–46 kJ). There is a big difference in the entropy change, however: $\Delta S°$ is –8.4 J K^{-1} for the NH_3 reaction but +23 J K^{-1} for the en reaction. We can explain the positive $\Delta S°$ value by using concepts discussed in Section 14.10. Because a single en ligand occupies two coordination sites, two molecules of H_2O are released when one en ligand bonds. Thus there are three product molecules in the reaction but only two reactant molecules. The greater number of product molecules leads to the positive entropy change for the equilibrium.

The slightly more negative value of $\Delta H°$ for the en reaction (–54 kJ versus –46 kJ) coupled with the positive entropy change leads to a much more negative value of $\Delta G°$ (–61 kJ for en, –43 kJ for NH_3) and a correspondingly larger equilibrium constant: $K = 4.2 \times 10^{10}$.

We can combine our two equations using Hess's law (Section 14.6, 'Hess's Law') to calculate the enthalpy, entropy and free energy changes that occur for en to replace ammonia as ligands on Cu(II):

$$[Cu(NH_3)_2(OH_2)_2]^{2+}(aq) + en(aq) \rightleftharpoons$$
$$[Cu(H_2O)_2(en)]^{2+}(aq) + 2\,NH_3(aq)$$

$$\Delta H° = (-54 \text{ kJ}) - (-46 \text{ kJ}) = -8 \text{ kJ}$$

$$\Delta S° = (+23 \text{ J K}^{-1}) - (-8.4 \text{ J K}^{-1}) = +31 \text{ J K}^{-1}$$

$$\Delta G° = (-61 \text{ kJ}) - (-43 \text{ kJ}) = -18 \text{ kJ}$$

Notice that at 27 °C, the entropic contribution ($-T\Delta S°$) to the free energy change, $\Delta G° = \Delta H° - T\Delta S°$ (Equation 14.38), is negative and greater in magnitude than the enthalpic contribution ($\Delta H°$). The equilibrium constant for the NH_3–en reaction, 1.4×10^3, shows that the replacement of NH_3 by en is thermodynamically favourable.

The chelate effect is important in biochemistry and molecular biology. The additional thermodynamic stabilisation provided by entropy effects helps stabilise biological metal–chelate complexes, such as porphyrins, and can allow changes in the oxidation state of the metal ion while retaining the structural integrity of the complex.

RELATED EXERCISE: 21.33

Reinecke's salt. Once the structures of complexes were more fully understood, it became possible to name them in a more systematic manner. Let's consider two examples.

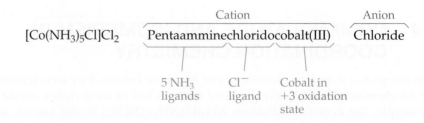

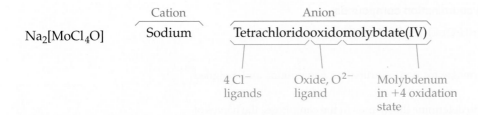

These examples illustrate how coordination compounds are named. The rules that govern the naming of this class of substances are as follows.

1. *In naming salts, the name of the cation is given before the name of the anion.* Thus in $[Co(NH_3)_5Cl]Cl_2$ we name the cation and then Cl^-.

2. *Within a complex ion or molecule, the ligands are named before the metal. Ligands are listed in alphabetical order, regardless of charge on the ligand. Prefixes that give the number of ligands are not considered part of the ligand name in determining alphabetical order.* Thus in the $[Co(NH_3)_5Cl]^{2+}$ ion we name the ammonia ligands first, then the chloride, then the metal: pentaamminechloro-cobalt(III). In writing the formula, however, the metal is listed first.

3. *The names of the anionic ligands end in the letter o, whereas neutral ones ordinar-ily bear the name of the molecules.* Some common ligands and their names are listed in ▼ **TABLE 21.4**. Special names are given to H_2O (aqua), NH_3 (ammine) and CO (carbonyl). For example, $[Fe(NH_3)_2(OH_2)_2(CN)_2]^+$ would be named as diamminediaquadicyanoiron(III) ion.

4. *Greek prefixes (di-, tri-, tetra-, penta-, hexa-) are used to indicate the number of each kind of ligand when more than one is present. If the ligand itself contains a prefix of this kind (for example, ethylenediamine), alternative prefixes are used (bis-, tris-, tetrakis-, pentakis-, hexakis-) and the ligand name is placed in parentheses.* For example, the name for $[Co(en)_3]Br_3$ is tris(ethylenediamine)cobalt(III) bromide.

5. *If the complex is an anion, its name ends in -ate.* The compound $K_4[Fe(CN)_6]$ is named potassium hexacyanidoferrate(II), for example, and the ion $[CoCl_4]^{2-}$ is named tetrachloridocobaltate(II) ion. The names of most metals change when they are part of an anionic complex. For example, iron = ferrate, copper = cuprate, gold = aurate, silver = argentate, molybdenum = molybdate, tungsten = tungstate.

6. *The oxidation number of the metal is given in Roman numerals in parentheses following the name of the metal.*

The following substances and their names demonstrate the application of these rules:

$[Ni(NH_3)_6]Br_2$ Hexaamminenickel(II) bromide

$[Co(H_2O)(CN)(en)_2]Cl_2$ Aquacyanidobis(ethylenediamine)cobalt(III) chloride

$Na_2[MoOCl_4]$ Sodium tetrachloridooxidomolybdate(IV)

TABLE 21.4 • Some common ligands			
Ligand	**Name in complexes**	**Ligand**	**Name in complexes**
Azide, N_3^-	Azido	Oxalate, $C_2O_4^{2-}$	Oxalato
Bromide, Br^-	Bromido	Oxide, O^{2-}	Oxido
Chloride, Cl^-	Chlorido	Ammonia, NH_3	Ammine
Cyanide, CN^-	Cyanido	Carbon monoxide, CO	Carbonyl
Fluoride, F^-	Fluorido	Ethylenediamine, en	Ethylenediamine
Hydroxide, OH^-	Hydroxido	Pyridine, C_5H_5N	Pyridine
Carbonate, CO_3^{2-}	Carbonato	Water, H_2O	Aqua

SAMPLE EXERCISE 21.4   Naming coordination compounds

Name the following compounds: **(a)** [Cr(H₂O)₄Cl₂]Cl, **(b)** K₄[Ni(CN)₄].

SOLUTION

Analyse We are given the chemical formulae for two coordination compounds and assigned the task of naming them.

Plan To name the complexes, we need to determine the ligands in the complexes, the names of the ligands and the oxidation state of the metal ion. We then put the information together following the rules listed above.

Solve

(a) As ligands, there are four water molecules, which are indicated as tetraaqua, and two chloride ions, indicated as dichlorido. The oxidation state of Cr is +3.

$$+3 + 4(0) + 2(-1) + (-1) = 0$$
$$[Cr(H_2O)_4Cl_2]Cl$$

Thus we have chromium(III). Finally, the anion is chloride. Putting these parts together, the name of the compound is

tetraaquadichloridochromium(III) chloride

(b) The complex has four cyanide ions, CN⁻, as ligands which we indicate as tetracyanido. The oxidation state of the nickel is zero.

$$4(+1) + 0 + 4(-1) = 0$$
$$K_4[Ni(CN)_4]$$

Because the complex is an anion, the metal is indicated as nickelate(0). Putting these parts together and naming the cation first, we have

potassium tetracyanidonickelate(0)

PRACTICE EXERCISE

Name the following compounds: **(a)** [Mo(NH₃)₃Br₃]NO₃, **(b)** (NH₄)₂[CuBr₄]. **(c)** Write the formula for sodium diaquadioxalatoruthenate(III).

Answers: **(a)** Triamminetribromomolybdenum(IV) nitrate, **(b)** ammonium tetrabromidocuprate(II), **(c)** Na[Ru(H₂O)₂(C₂O₄)₂]

(See also Exercises 21.36, 21.37.)

Isomerism

When two or more compounds have the same composition but a different arrangement of atoms, we call them **isomers**. Isomerism—the existence of isomers—is a characteristic feature of both organic and inorganic compounds. Although isomers are composed of the same collection of atoms, they usually differ in one or more physical or chemical properties such as colour, solubility or rate of reaction with some reagents. We consider two main kinds of isomers in coordination compounds: **structural isomers** (which have different bonds) and **stereoisomers** (which have the same bonds but different ways in which the ligands occupy the space around the metal centre). Each of these classes also has subclasses, as shown in ▶ **FIGURE 21.17**.

Structural Isomerism

Many different types of structural isomerism are known in coordination chemistry. Figure 21.17 gives two examples: linkage isomerism and coordination-sphere isomerism. **Linkage isomers** are relatively rare but arise when a particular ligand is capable of coordinating to a metal in two different ways. The nitrite ion, NO₂⁻, for example, can coordinate through either a nitrogen or an oxygen atom, as shown in ▶ **FIGURE 21.18**. When it coordinates through the nitrogen atom, the NO₂⁻ ligand is called *nitro*; when it coordinates through the oxygen atom, it is called *nitrito* and is generally written ONO⁻. The isomers

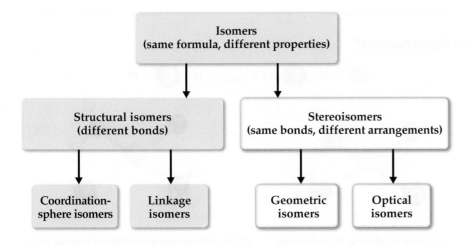

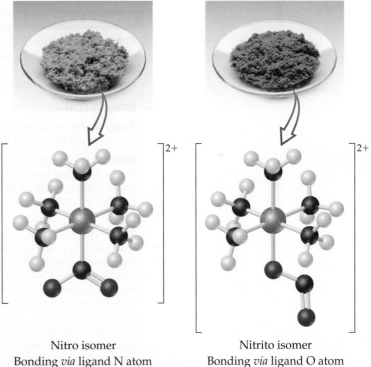

◀ **FIGURE 21.17**   **Forms of isomerism in coordination compounds.**

shown in Figure 21.18 have different properties. When the ligand is attached through the nitrogen, the isomer is yellow; when attached through the oxygen, it is red. Another ligand capable of coordinating through either of two donor atoms is thiocyanate, SCN^-, whose potential donor atoms are N and S.

Coordination-sphere isomers differ in the ligands that are directly bonded to the metal, as opposed to being outside the coordination sphere in the solid lattice. For example, there are three compounds whose molecular formula is $CrCl_3(H_2O)_6$: $[Cr(H_2O)_6]Cl_3$ (a violet compound), $[Cr(H_2O)_5Cl]Cl_2 \cdot H_2O$ (a green compound) and $[Cr(H_2O)_4Cl_2]Cl \cdot 2\,H_2O$ (also a green compound). In the two green compounds the water has been displaced from the coordination sphere by chloride ions and occupies a site in the crystal lattice.

> ⚠ **CONCEPT CHECK 9**
>
> Can the ammonia ligand engage in linkage isomerism? Explain.

Nitro isomer
Bonding *via* ligand N atom

Nitrito isomer
Bonding *via* ligand O atom

▲ **FIGURE 21.18**   **Linkage isomerism.**

Stereoisomerism

Stereoisomerism occurs when isomers have the same chemical bonds but different spatial arrangements. In the square planar complex $[Pt(NH_3)_2Cl_2]$, for example, the chloro ligands can be either adjacent to or opposite each other, as illustrated in ▼ **FIGURE 21.19**. This particular form of isomerism, in which the arrangement of the constituent atoms is different though the same bonds are present, is called geometric isomerism. The isomer on the left in Figure 21.19, with like ligands in adjacent positions, is called the *cis* isomer. The isomer on the right in Figure 21.19, with like ligands across from one another, is called the *trans* isomer. **Geometric isomers** generally have different properties, such as colours, solubilities, melting points and boiling points. They may also have markedly different chemical reactivities. For example, *cis*-$[Pt(NH_3)_2Cl_2]$, also called *cisplatin*, is effective in the treatment of testicular, ovarian and certain other cancers, whereas the *trans* isomer is ineffective.

Geometric isomerism is also possible in octahedral complexes when two or more different ligands are present. The *cis* and *trans* isomers of the tetra-amminedichloridocobalt(III) ion were shown in Figure 21.7. As noted in

◢

FIGURE IT OUT

Which of these isomers has a non-zero dipole moment?

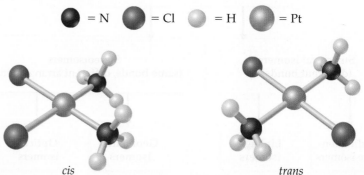

| ● = N | ● = Cl | ○ = H | ○ = Pt |

cis
Cl ligands adjacent to each other
NH₃ ligands adjacent to each other

trans
Cl ligands on opposite sides of central atom
NH₃ ligands on opposite sides of central atom

▶ **FIGURE 21.19 Geometric isomerism.**

Section 21.2 and Table 21.2, these two isomers have different colours. Their salts also possess different solubilities in water.

Because all the corners of a tetrahedron are adjacent to one another, *cis–trans* isomerism is not observed in tetrahedral complexes.

SAMPLE EXERCISE 21.5 | **Determining the number of geometric isomers**

The Lewis structure of the CO molecule indicates that the molecule has a lone pair of electrons on the C atom and a lone pair on the O atom ($:C \equiv O:$). When CO binds to a transition metal centre, it nearly always does so by using the lone pair on the C atom. How many geometric isomers are there for tetracarbonyldichloridoiron(II)?

SOLUTION

Analyse We are given the name of a complex containing only monodentate ligands, and we must determine the number of isomers the complex can form.

Plan We can count the number of ligands to determine the coordination number of the Fe and then use the coordination number to predict the geometry. We can then either make a series of drawings with ligands in different positions to determine the number of isomers or deduce the number of isomers by analogy to cases we have discussed.

Solve The name indicates that the complex has four carbonyl (CO) ligands and two chlorido (Cl⁻) ligands, so its formula is $Fe(CO)_4Cl_2$. The complex therefore has a coordination number of 6, and we can assume that it has an octahedral geometry. Like $[Co(NH_3)_4Cl_2]^+$ (Figure 21.7), it has four ligands of one type and two of another. Consequently it possesses two isomers: one with the Cl⁻ ligands across the metal from each other, *trans*-$[Fe(CO)_4Cl_2]$, and one with the Cl⁻ ligands adjacent, *cis*-$[Fe(CO)_4Cl_2]$.

In principle, the CO ligand could exhibit linkage isomerism by binding to a metal atom via the lone pair on the O atom. When bonded this way, a CO ligand is called an *isocarbonyl* ligand. Metal isocarbonyl complexes are extremely rare, and we do not normally have to consider the possibility that CO will bind in this way.

Comment It is easy to overestimate the number of geometric isomers. Sometimes different orientations of a single isomer are incorrectly thought to be different isomers. If two structures can be rotated so that they are equivalent, they are not isomers of each other. The problem of identifying isomers is compounded by the difficulty we often have in visualising three-dimensional molecules from their two-dimensional representations. It is sometimes easier to determine the number of isomers if we use three-dimensional models.

PRACTICE EXERCISE

How many isomers exist for square planar $[Pt(NH_3)_2ClBr]$?

Answer: Two
(See also Exercise 21.39.)

A second type of stereoisomerism is known as optical iso-merism. **Optical isomers**, called **enantiomers**, are mirror images that cannot be superimposed on each other. They bear the same resemblance to each other that your left hand bears to your right hand as discussed in Section 23.3. An example of a complex that exhibits this type of isomerism is the $[Co(en)_3]^{3+}$ ion. ▶ FIGURE 21.20 shows the two enantiomers of $[Co(en)_3]^{3+}$ and their mirror-image relationship to each other. Just as there is no way that we can twist or turn our right hand to make it look identical to our left, so there is no way to rotate one of these enantiomers to make it identical to the other. Molecules or ions that are not superimposable on their mirror image are said to be **chiral** (pronounced KY-rul).

Mirror

▲ FIGURE 21.20   **Optical isomerism.** Just as our hands are non-superimposable mirror images of each other, so too are optical isomers such as the two optical isomers of $[Co(en)_3]^{3+}$.

SAMPLE EXERCISE 21.6   **Predicting whether a complex has optical isomers**

Does either *cis*- or *trans*-$[CoCl_2(en)_2]^+$ exhibit optical isomerism?

SOLUTION

Analyse We are given the chemical formula for two geometric isomers and asked to determine whether either one has optical isomers. Because en is a bidentate ligand, we know that both complexes are octahedral and both have coordination number 6.

Plan We need to sketch the structures of the *cis* and *trans* isomers and their mirror images. We can draw the en ligand as two N atoms connected by a line, as is done in Figure 21.20. If the mirror image cannot be superimposed on the original structure, the complex and its mirror image are optical isomers.

Solve The *trans* isomer of $[Co(en)_2Cl_2]^+$ and its mirror image is

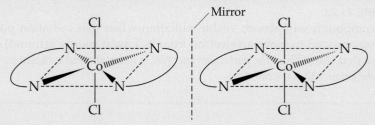

where the dashed vertical line represents a mirror. Notice that the mirror image of the *trans* isomer is identical to the original. (If you rotate the left drawing of the *trans* pair 180° about its vertical axis, the structure you end up with is identical to the right drawing of the pair.) Consequently, *trans*-$[Co(en)_2Cl_2]^+$ does not exhibit optical isomerism.

The mirror image of the *cis* isomer of $[Co(en)_2Cl_2]^+$, however, cannot be superimposed on the original:

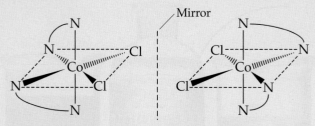

Thus the two *cis* structures are optical isomers (enantiomers): *cis*-$[Co(en)_2Cl_2]^+$ is a chiral complex.

PRACTICE EXERCISE

Does the square planar complex ion $[Pt(NH_3)(N_3)ClBr]^-$ have optical isomers?

Answer: No, because the complex is flat.

(See also Exercise 21.42.)

The properties of two optical isomers differ only if the isomers are in a chiral environment—that is, an environment in which there is a sense of right- and left-handedness. A chiral enzyme, for example, might catalyse the reaction of one optical isomer but not the other. Consequently, one optical isomer may produce a specific physiological effect in the body, with its mirror image producing either a different effect or none at all. Chiral reactions are also extremely important in the synthesis of pharmaceuticals and other industrially important chemicals.

21.5 | COLOUR AND MAGNETISM IN COORDINATION CHEMISTRY

Studies of the colours and magnetic properties of transition metal complexes have played an important role in the development of modern models for metal–ligand bonding.

Colour

We have seen that the salts of transition metal ions and their aqueous solutions exhibit a variety of colours. In general, the colour of a complex depends on the particular element, its oxidation state and the ligands bound to the metal. ▼ FIGURE 21.21 shows how the pale blue colour characteristic of $[Cu(H_2O)_4]^{2+}$ changes to a deep blue colour as NH_3 ligands replace H_2O ligands to form $[Cu(NH_3)_4]^{2+}$.

For a compound to have colour, it must absorb some portion of the spectrum of visible light. Visible light consists of electromagnetic radiation with wavelengths ranging from approximately 400 nm to 700 nm. White light contains all wavelengths in this visible region. It can be dispersed into a spectrum of colours, each of which has a characteristic range of wavelengths, as shown in ▶ FIGURE 21.22.

A compound will absorb visible radiation when that radiation possesses the energy needed to move an electron from its lowest energy (ground) state to some excited state. Thus *the particular energies of radiation that a substance absorbs dictate the colours that it exhibits.*

◢ **FIGURE IT OUT**

Is the equilibrium binding constant of ammonia for Cu(II) likely to be larger or smaller than that of water for Cu(II)?

▶ **FIGURE 21.21** **The colour of a coordination complex changes when the ligand changes.**

$[Cu(H_2O)_4]^{2+}$(aq) NH_3(aq) $[Cu(NH_3)_4]^{2+}$(aq)

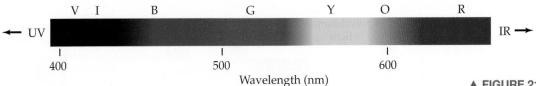

Wavelength (nm)

▲ **FIGURE 21.22   Visible spectrum.** The relationship between colour and wavelength for visible light.

When some portion of the spectrum of visible light is absorbed by an object, the colour we perceive is the sum of the remaining colours that are reflected or transmitted by the object and strike our eyes. An opaque object reflects light whereas a transparent one transmits it. If an object absorbs all wavelengths of visible light, none reaches our eyes from that object and consequently it appears black. If it absorbs no visible light it is white or colourless. If it absorbs all but the orange part of the spectrum the material appears orange. An interesting phenomenon of vision is that we also perceive an orange colour when visible light of all colours except blue strikes our eyes, as shown in ▼ **FIGURE 21.23**. Orange and blue are **complementary colours**: the removal of blue from white light makes the light look orange and *vice versa*. Thus an object has a particular colour for one of two reasons: (1) it reflects or transmits light of that colour; or (2) it absorbs light of the complementary colour. Complementary colours can be determined using an artist's colour wheel, shown in ▼ **FIGURE 21.24**. The wheel shows the colours of the visible spectrum, from red to violet. Complementary colours, such as orange and blue, appear as wedges opposite each other on the wheel.

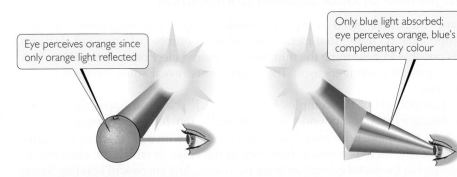

Eye perceives orange since only orange light reflected

Only blue light absorbed; eye perceives orange, blue's complementary colour

▼ **FIGURE 21.23   Two ways of perceiving the colour orange.** An object appears orange either when it reflects orange light to the eye (left), or when it transmits to the eye all colours except blue, the complement of orange (middle).

SAMPLE EXERCISE 21.7 Relating colour absorbed to colour observed

The complex ion *trans*-$[Co(NH_3)_4Cl_2]^+$ absorbs light primarily in the red region of the visible spectrum (the most intense absorption is at 680 nm). What is the colour of the complex?

SOLUTION

Analyse We need to relate the colour absorbed by a complex (red) to the colour observed for the complex.

Plan For an object that absorbs only one colour from the visible spectrum, the colour we see is complementary to the colour absorbed. We can use the colour wheel of Figure 21.24 to determine the complementary colour.

Solve From Figure 21.24, we see that green is complementary to red, so the complex appears green.

Comment As noted in Section 21.2, this green complex was one of those that helped Werner establish his theory of coordination (Table 21.2). The other geometric isomer of this complex, *cis*-$[Co(NH_3)_4Cl_2]^+$, absorbs yellow light and therefore appears violet.

PRACTICE EXERCISE

A certain transition metal complex ion absorbs at 630 nm. Which colour is this ion most likely to be—blue, yellow, green or orange?

Answer: Blue

(See also Exercises 21.44, 21.45.)

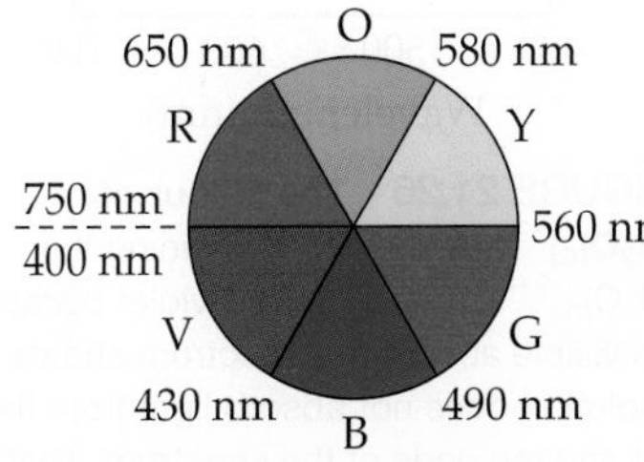

650 nm   O   580 nm

R   Y

750 nm   560 nm
400 nm

V   G

430 nm   B   490 nm

▲ **FIGURE 21.24   Artist's colour wheel.** Colours with approximate wavelength ranges are shown as wedges. The colours that are complementary to each other lie opposite each other.

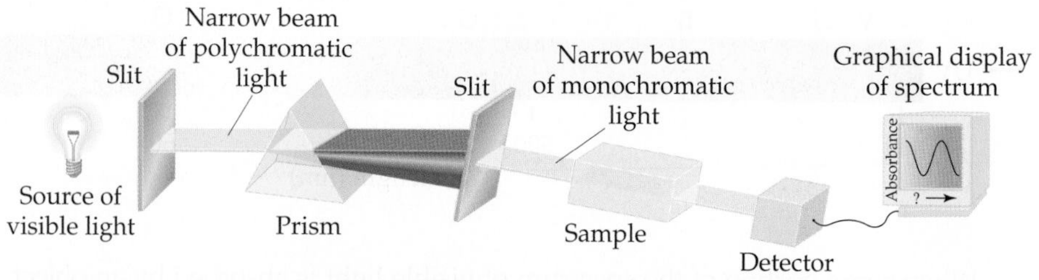

▲ **FIGURE 21.25** **Experimental determination of an absorption spectrum.** The prism is rotated so that different wavelengths of light pass through the sample. The detector measures the amount of light reaching it, and this information can be displayed as the absorption at each wavelength. The absorbance is a measure of the amount of light absorbed.

The amount of light absorbed by a sample as a function of wavelength is known as its **absorption spectrum**. The visible absorption spectrum of a transparent sample can be determined as shown in ▲ FIGURE 21.25. The spectrum of $[Ti(H_2O)_6]^{3+}$ which we discuss in Section 21.6, is shown in ◀ FIGURE 21.26. The absorption maximum of $[Ti(H_2O)_6]^{3+}$ is at about 500 nm. Because the sample absorbs most strongly in the green and yellow regions of the visible spectrum, what we see is the unabsorbed red and violet light, which we perceive as red violet.

Magnetism of Coordination Compounds

Many transition metal complexes exhibit simple paramagnetism. In such compounds the individual metal ions possess a number of unpaired electrons. It is possible to determine the number of unpaired electrons per metal ion from the degree of paramagnetism. The experiments reveal some interesting comparisons.

Compounds of the complex ion $[Co(CN)_6]^{3-}$ have no unpaired electrons, for example, but compounds of the $[CoF_6]^{3-}$ ion have four unpaired electrons per metal ion. Both complexes contain Co(III) with a $3d^6$ electron configuration. Clearly, there is a major difference in the ways in which the electrons are arranged in the metal orbitals in these two cases. Any successful bonding theory must explain this difference, and we present such a theory in Section 21.6.

> ◤ **CONCEPT CHECK 10**
>
> What is the electron configuration for
> **a.** the Co atom
> **b.** the Co^{3+} ion?
> How many unpaired electrons does each possess? (See Section 8.4 to review electron configurations of ions.)

FIGURE IT OUT

How would this absorbance spectrum change if you decreased the concentration of the $[Ti(H_2O)_6]^{3+}$ in solution?

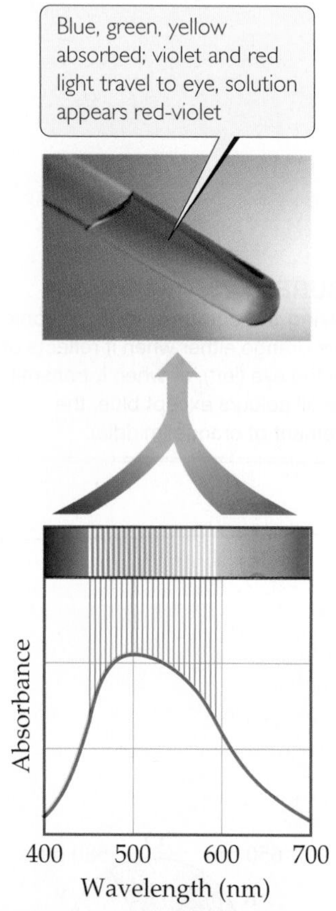

Blue, green, yellow absorbed; violet and red light travel to eye, solution appears red-violet

▲ **FIGURE 21.26** **The colour of $[Ti(H_2O)_6]^{3+}$.** A solution containing the $[Ti(H_2O)_6]^{3+}$ ion appears red-violet because, as its visible absorption spectrum shows, the solution does not absorb light from the violet and red ends of the spectrum. That unabsorbed light is what reaches our eyes.

21.6 | CRYSTAL-FIELD THEORY

Scientists have long recognised that many of the magnetic properties and colours of transition metal complexes are related to the presence of d electrons in metal orbitals. In this section we consider a model for bonding in transition metal complexes, called **crystal-field theory**, that accounts for many of the observed properties of these substances.*

The ability of a metal ion to attract ligands such as water around itself is a Lewis acid–base interaction. The base—that is, the ligand—donates a pair of electrons into a suitable empty orbital on the metal, as shown in

---

* The name *crystal field* arose because the theory was first developed to explain the properties of solid crystalline materials, such as ruby. The same theoretical model applies to complexes in solution.

MY WORLD OF CHEMISTRY

THE BATTLE FOR IRON IN LIVING SYSTEMS

Because living systems have difficulty assimilating enough iron to satisfy their nutritional needs, iron-deficiency anaemia is a common problem in humans. Chlorosis, an iron deficiency in plants that makes leaves turn yellow, is also commonplace.

Living systems have difficulty assimilating iron because most iron compounds found in nature are not very soluble in water. Microorganisms have adapted to this problem by secreting an iron-binding compound, called a *siderophore*, that forms an extremely stable water-soluble complex with iron(III). One such complex is *ferrichrome* (▼ FIGURE 21.27). The iron-binding strength of a siderophore is so great that it can extract iron from metal cooking pots and the iron in iron oxides.

When ferrichrome enters a living cell, the iron it carries is removed through an enzyme-catalysed reaction that reduces the strongly bonding iron(III) to iron(II), which is only weakly complexed by the siderophore (▶ FIGURE 21.28). Microorganisms thus acquire iron by excreting a siderophore into their immediate environment and then taking the resulting iron complex into the cell.

In humans, iron is assimilated from food in the intestine. A protein called *transferrin* binds iron and transports it across the intestinal wall to distribute it to other tissues in

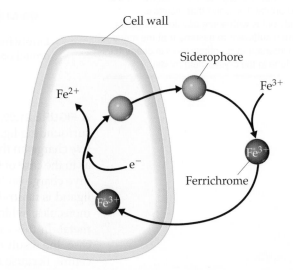

▲ **FIGURE 21.28**   **The iron-transport system of a bacterial cell.**

the body. The normal adult body contains about 4 g of iron. At any one time, about 3 g of this iron is in the blood, mostly in the form of haemoglobin. Most of the remainder is carried by transferrin.

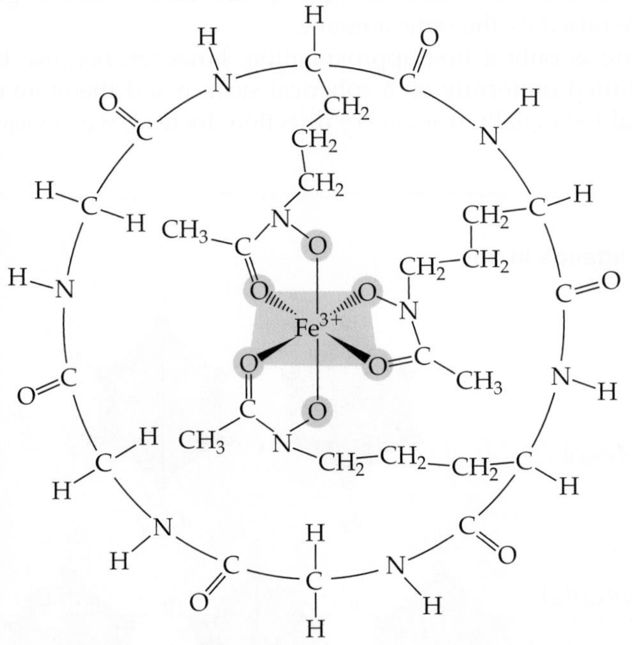

A bacterium that infects the blood requires a source of iron if it is to grow and reproduce. The bacterium excretes a siderophore into the blood to compete with transferrin for iron. The equilibrium constants for forming the iron complex are about the same for transferrin and siderophores. The more iron available to the bacterium, the more rapidly it can reproduce and thus the more harm it can do.

Several years ago, New Zealand clinics regularly gave iron supplements to infants soon after birth. However, the incidence of certain bacterial infections was eight times higher in treated than in untreated infants. Presumably, the presence of more iron in the blood than was absolutely necessary makes it easier for bacteria to obtain the iron needed for growth and reproduction.

In the United States it is common medical practice to supplement infant formula with iron sometime during the first year of life. However, iron supplements are not necessary for infants who breast-feed because breast milk contains two specialised proteins, lactoferrin and transferrin, which provide sufficient iron while denying its availability to bacteria. Even for infants fed with infant formulas, supplementing with iron during the first several months of life may be ill-advised.

For bacteria to continue to multiply in the blood, they must synthesise new supplies of siderophores. Synthesis of siderophores in bacteria slows, however, as the temperature is increased above the normal body temperature of 37 °C and stops completely at 40 °C. This suggests that fever in the presence of an invading microbe is a mechanism used by the body to deprive bacteria of iron.

RELATED EXERCISE: 21.63

▲ **FIGURE 21.27**   **Ferrichrome.**

▶ **FIGURE 21.29 Metal–ligand bond formation.** The ligand, which acts as a Lewis base, donates charge to the metal via a metal hybrid orbital. The bond that results is strongly polar, with some covalent character. It is often sufficient to assume that the metal–ligand interaction is entirely ionic in character, as is done in the crystal-field model.

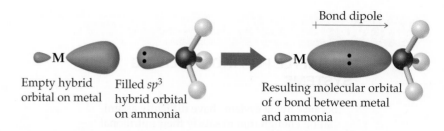

Empty hybrid orbital on metal   Filled sp^3 hybrid orbital on ammonia   Resulting molecular orbital of σ bond between metal and ammonia

▲ **FIGURE 21.29.** Much of the attractive interaction between the metal ion and the surrounding ligands is due, however, to the electrostatic forces between the positive charge on the metal and negative charges on the ligands. If the ligand is ionic, as in the case of Cl^- or SCN^-, the electrostatic interaction occurs between the positive charge on the metal centre and the negative charge on the ligand. When the ligand is neutral, as in the case of H_2O or NH_3, the negative ends of these polar molecules, which contain an unshared electron pair, are directed towards the metal. In this case, the attractive interaction is of the ion–dipole type. In either case, the result is the same: the ligands are attracted strongly towards the metal centre. Because of the electrostatic attraction between the positive metal ion and the electrons of the ligands, the complex formed by the metal ion and the ligands has a lower energy than the fully separated metal and ligands.

Although the positive metal ion is attracted to the electrons in the ligands, the d electrons on the metal ion feel a repulsion from the ligands (negative charges repel one another). Let's examine this effect more closely, and particularly for the case in which the ligands form an octahedral array around a metal ion which has a coordination number of 6.

The energy diagram in ▼ **FIGURE 21.30** shows how these ligand point charges affect the energies of the d orbitals. First we imagine the complex as having all the ligand point charges uniformly distributed on the surface of a sphere centred on the metal ion. The *average* energy of the metal ion's d orbitals is raised by the presence of this uniformly charged sphere. Hence, the energies of all five d orbitals are raised by the same amount.

This energy picture is only a first approximation, however, because the ligands are not distributed uniformly on a spherical surface and therefore do not approach the metal ion equally from every direction. Instead, we envisage

FIGURE IT OUT

Which d orbitals have lobes that point directly towards the ligands in an octahedral crystal field?

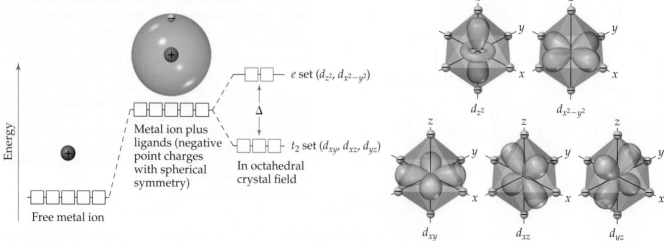

▲ **FIGURE 21.30 Energies of d orbitals in a free metal ion, a spherically symmetric crystal field and an octahedral crystal field.**

the six ligands approaching along x-, y- and z-axes, as shown on the right in Figure 21.30. This arrangement of ligands is called an *octahedral crystal field*. Because the metal ion's d orbitals have different orientations and shapes, they do not all experience the same repulsion from the ligands and therefore do not all have the same energy under the influence of the octahedral crystal field. To see why, we must consider the shapes of the d orbitals and how their lobes are oriented relative to the ligands.

Figure 21.30 shows that the lobes of the d_{z^2} and $d_{x^2-y^2}$ orbitals are directed *along* the x-, y- and z-axes and so point directly towards the ligand point charges. In the d_{xy}, d_{xz} and d_{yz} orbitals, however, the lobes are directed *between* the axes and so do not point directly towards the charges. The result of this difference in orientation—$d_{x^2-y^2}$ and d_{z^2} lobes point directly towards the ligand charges; d_{xy}, d_{xz} and d_{yz} lobes do not—is that the energy of the $d_{x^2-y^2}$ and d_{z^2} orbitals is higher than the energy of the d_{xy}, d_{xz} and d_{yz} orbitals. This difference in energy is represented by the red boxes in the energy diagram of Figure 21.30.

It might seem like the energy of the $d_{x^2-y^2}$ orbital should be different from that of the d_{z^2} orbital because the $d_{x^2-y^2}$ has four lobes pointing at ligands and the d_{z^2} has only two lobes pointing at ligands. However, the d_{z^2} orbital does have electron density in the xy plane, represented by the ring encircling the point where the two lobes meet. More advanced calculations show that two orbitals do indeed have the same energy in the presence of the octahedral crystal field.

Because their lobes point directly at the negative ligand charges, electrons in the metal ion's d_{z^2} and $d_{x^2-y^2}$ orbitals experience stronger repulsions than those in the d_{xz}, d_{xy} and d_{yz} orbitals. As a result, the energy splitting shown in Figure 21.30 occurs. The three lower-energy d orbitals are called the t_2 set of orbitals, and the two higher-energy ones are called the e set.* The energy gap, Δ, between the two sets is often called the *crystal-field splitting energy*.

The crystal-field model helps us account for the observed colours in transition metal complexes. The energy gap between the d orbitals, Δ, is of the same order of magnitude as the energy of a photon of visible light. It is therefore possible for a transition metal complex to absorb visible light, which excites an electron from the lower-energy d orbitals into the higher-energy ones. In the $[Ti(H_2O)_6]^{3+}$ ion, for example, the Ti(III) ion has an $[Ar]3d^1$ electron configuration (recall that, when determining the electron configurations of transition metal ions, we remove the s electrons first). Ti(III) is thus called a 'd^1 ion'. In the ground state of $[Ti(H_2O)_6]^{3+}$ the single $3d$ electron resides in one of the three lower-energy orbitals in the t_2 set. Absorption of light with a wavelength of 495 nm (242 kJ mol^{-1}) excites the $3d$ electron from the lower t_2 set to the upper e set of orbitals, as shown in ▶ **FIGURE 21.31**, generating the absorption spectrum shown in Figure 21.26. Because this transition involves exciting an electron from one set of d orbitals to the other, we call it a **d-d transition**. As noted earlier, the absorption of visible radiation that produces this d-d transition causes the $[Ti(H_2O)_6]^{3+}$ ion to appear purple.

▲ **FIGURE IT OUT**

How would you calculate the energy gap between the t_2 and e orbitals from this diagram?

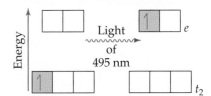

▲ **FIGURE 21.31** The *d-d* transition in $[Ti(H_2O)_6]^{3+}$ is produced by the absorption of 495 nm light.

 CONCEPT CHECK 11

Why are compounds of Ti(IV) colourless?

The magnitude of the energy gap, Δ, and consequently the colour of a complex depend on both the metal and the surrounding ligands. For example, $[Fe(H_2O)_6]^{3+}$ is light violet, $[Cr(H_2O)_6]^{3+}$ is violet and $[Cr(NH_3)_6]^{3+}$ is yellow. Ligands can be arranged in order of their abilities to increase the energy gap, Δ.

---

* The labels t_2 for the d_{xy}, d_{xz} and d_{yz} orbitals and e for the d_{z^2} and $d_{x^2-y^2}$ orbitals come from the application of a branch of mathematics called group theory to crystal-field theory. Group theory can be used to analyse the effects of symmetry on molecular properties.

The following is an abbreviated list of common ligands arranged in order of increasing Δ.

$$\xrightarrow{\text{Increasing } \Delta}$$

$$Cl^- < F^- < H_2O < NH_3 < en < NO_2^-(\text{N-bonded}) < CN^-$$

This list is known as the **spectrochemical series**. The magnitude of Δ increases by roughly a factor of 2 from the far left to the far right of the spectrochemical series.

Ligands that lie on the low-Δ end of the spectrochemical series are termed *weak-field ligands*; those on the high-Δ end are termed *strong-field ligands*. ▼ FIGURE 21.32 shows schematically what happens to the crystal-field splitting when the ligand is varied in a series of chromium(III) complexes. Because a Cr atom has an $[Ar]3d^54s^1$ electron configuration, Cr^{3+} has an $[Ar]3d^3$ electron configuration; Cr(III), therefore, is a d^3 ion. Consistent with Hund's rule, the three $3d$ electrons occupy the t_2 set of orbitals, with one electron in each orbital and all the spins the same. As the field exerted by the six surrounding ligands increases, the splitting of the metal d orbitals increases. Because the absorption spectrum is related to this energy separation, these complexes vary in colour.

FIGURE IT OUT

If you were told to add a colourless Cr(III) complex to this diagram, what would you draw and where would you place it?

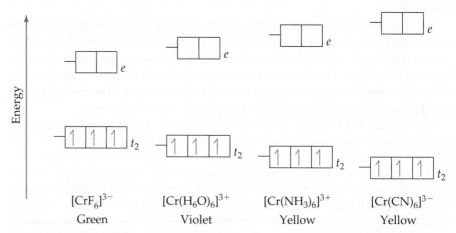

▲ **FIGURE 21.32 Effect of ligand on crystal-field splitting.** The greater the crystal-field strength of the ligand, the greater the energy gap, Δ, it causes between the t_2 and e sets of the metal ion's d orbitals.

SAMPLE EXERCISE 21.8   Using the spectrochemical series

Which of the following complexes of Ti^{3+} exhibits the shortest wavelength absorption in the visible spectrum: $[Ti(H_2O)_6]^{3+}$, $[Ti(en)_3]^{3+}$ or $[TiCl_6]^{3-}$?

SOLUTION

Analyse We are given three octahedral complexes, each containing Ti in the +3 oxidation state. We need to predict which complex absorbs the shortest wavelength of visible light.

Plan Ti(III) is a d^1 ion, so we anticipate that the absorption is due to a d-d transition in which the $3d$ electron is excited from the lower-energy t_2 set to the higher-energy e set. The wavelength of the light absorbed is determined by the magnitude of the energy difference, Δ. Thus we use the position of the ligands in the spectrochemical series to predict the relative values of Δ. The larger the energy, the smaller the wavelength.

Solve Of the three ligands involved, H_2O, en and Cl^-, we see that ethylenediamine (en) is highest in the spectrochemical series and will therefore cause the largest splitting, Δ, of the t_2 and e sets of orbitals. The larger the splitting, the shorter the wavelength of the light absorbed. Thus the complex that absorbs the shortest wavelength of light is $[Ti(en)_3]^{3+}$.

PRACTICE EXERCISE

The absorption spectrum of $[Ti(NCS)_6]^{3-}$ shows a band that lies intermediate in wavelength between those for $[TiCl_6]^{3-}$ and $[TiF_6]^{3-}$. What can we conclude about the place of NCS^- in the spectrochemical series?

Answer: It lies between Cl^- and F^-; that is, $Cl^- < NCS^- < F^-$.

(See also Exercise 21.58.)

Electron Configurations in Octahedral Complexes

The crystal-field model also helps us understand the magnetic properties and some important chemical properties of transition metal ions. From Hund's rule, we expect that electrons will always occupy the lowest-energy vacant orbitals first and that they will occupy a set of orbitals one at a time with their spins parallel (Section 6.8, 'Electron Configurations'). Thus if we have a d^1, d^2 or d^3 octahedral complex, the electrons will go into the lower-energy t_2 set of orbitals, with their spins parallel.

When a fourth electron must be added, we have the two choices shown in ▶ **FIGURE 21.33**. The electron can either go into an e orbital, where it will be the sole electron in the orbital, or become the second electron in a t_2 orbital. Because the energy difference between the t_2 and e sets is the splitting energy, Δ, the energy cost of going into an e orbital rather than a t_2 orbital is also Δ. Thus the goal of filling lowest-energy available orbitals first is met by putting the electron in a t_2 orbital.

There is a penalty for doing this, however, because the electron must now be paired with the electron already occupying the orbital. The difference between the energy required to pair an electron in an occupied orbital and the energy required to place that electron in an empty orbital is called the **spin-pairing energy**. The spin-pairing energy arises from the fact that the electrostatic repulsion between two electrons that share an orbital (and so must have opposite spins) is greater than the repulsion between two electrons that are in different orbitals and have parallel spins.

In coordination complexes the nature of the ligands and the charge on the metal ion often play major roles in determining which of the two electron arrangements shown in Figure 21.33 is used. In $[CoF_6]^{3-}$ and $[Co(CN)_6]^{3-}$, both ligands have a 1– charge. The F^- ion, however, is on the low end of the spectrochemical series, so it is a weak-field ligand. The CN^- ion is on the high end and so is a strong-field ligand, which means it produces a larger energy gap, Δ, than the F^- ion. The splittings of the d-orbital energies in these two complexes are compared in ▶ **FIGURE 21.34**.

Cobalt(III) has an $[Ar]3d^6$ electron configuration, so these are both d^6 complexes. Let's imagine that we add these six electrons one at a time to the d orbitals of the $[CoF_6]^{3-}$ ion. The first three will go into the lower-energy t_2 orbitals with their spins parallel. The fourth electron could go into one of the t_2 orbitals, pairing up with one of those already present. The F^- ion is a weak-field ligand, however, and so the energy gap, Δ, between the t_2 set and the e set is small. In this case, the more stable arrangement is the fourth electron in one of the e orbitals. Similarly, the fifth electron we add goes into the other e orbital. With all the orbitals containing at least one electron, the sixth must be paired up and it goes into a lower-energy t_2 orbital; we end up with four of the electrons in the t_2 set of orbitals and two electrons in the e set. Figure 21.34 shows that in the case of the $[Co(CN)_6]^{3-}$ complex, the crystal-field splitting is much larger. The spin-pairing energy is smaller than Δ, so all six electrons are paired in the t_2 orbitals.

◌◌◌ Review this on page 203

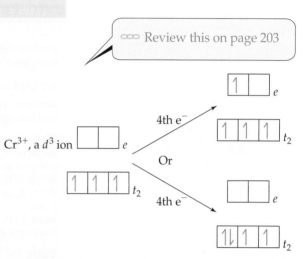

▲ FIGURE 21.33 Two possibilities for adding a fourth electron to a d^3 octahedral complex. Whether the fourth electron goes into a t_2 orbital or into an e orbital depends on the relative energies of the crystal-field splitting energy and the spin-pairing energy.

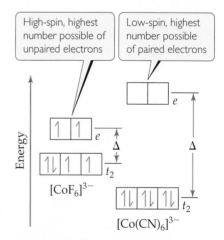

▲ FIGURE 21.34 High-spin and low-spin complexes. The high-spin $[CoF_6]^{3-}$ ion has a weak-field ligand and so a small Δ value. The spin-pairing energy required to pair electrons in the t_2 orbitals is greater than the e-t_2 gap energy, Δ. Therefore, filling e orbitals before any electrons are paired in t_2 orbitals is the lower-energy state. The low-spin $[Co(CN)_6]^{3-}$ ion has a strong-field ligand and so a large Δ value. Here the spin-pairing energy is less than Δ, making three sets of t_2-paired electrons the lower-energy state.

The $[CoF_6]^{3-}$ complex is a **high-spin complex**; that is, the electrons are arranged so that they remain unpaired as much as possible. The $[Co(CN)_6]^{3-}$ ion, is a **low-spin complex**; that is, the electrons are arranged so that they remain paired as much as possible. These two different electronic arrangements can be readily distinguished by measuring the magnetic properties of the complex. The absorption spectrum also shows characteristic features that indicate the electronic arrangement.

SAMPLE EXERCISE 21.9 Predicting the number of unpaired electrons in an octahedral complex

Predict the number of unpaired electrons in 6-coordinate high-spin and low-spin complexes of Fe^{3+}.

SOLUTION

Analyse We must determine how many unpaired electrons there are in the high-spin and low-spin complexes of Fe^{3+}.

Plan We need to consider how the electrons populate the d orbitals in Fe^{3+} when the metal is in an octahedral complex. There are two possibilities: one giving a high-spin complex and the other giving a low-spin complex. The electron configuration of Fe^{3+} gives us the number of d electrons. We then determine how these electrons populate the t_2 and e sets of d orbitals. In the high-spin case, the energy difference between the t_2 and e orbitals is small, and the complex has the maximum number of unpaired electrons. In the low-spin case, the energy difference between the t_2 and e orbitals is large, causing the t_2 orbitals to be filled before any electrons occupy the e orbitals.

Solve Fe^{3+} is a d^5 ion. In a high-spin complex, all five of these electrons are unpaired, with three in the t_2 orbitals and two in the e orbitals. In a low-spin complex, all five electrons reside in the t_2 set of d orbitals, so there is one unpaired electron:

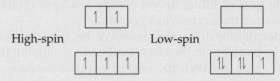

High-spin   Low-spin

PRACTICE EXERCISE

For which d electron configurations in octahedral complexes is it possible to distinguish between high-spin and low-spin arrangements?

Answer: d^4, d^5, d^6, d^7

(See also Exercise 21.52.)

Tetrahedral and Square Planar Complexes

So far we have considered the crystal-field model only for complexes with an octahedral geometry. When there are only four ligands about the metal, the geometry is generally tetrahedral, except for the special case of metal ions with a d^8 electron configuration, which we discuss in a moment. The crystal-field splitting of the metal d orbitals in tetrahedral complexes differs from that in octahedral complexes. Four equivalent ligands can interact with a central metal ion most effectively by approaching along the vertices of a tetrahedron. It turns out that the splitting of the metal d orbitals in a tetrahedral crystal is just the opposite of that for the octahedral case; those orbitals that were directed between the ligands are now pointed directly at them and *vice versa*. That is, the three metal d orbitals in the t_2 set are raised in energy, and the two orbitals in the e set are lowered, as illustrated in ◀ **FIGURE 21.35**. Because there are only four ligands instead of six, as in the octahedral case, the crystal-field splitting is much smaller for tetrahedral complexes. Calculations show that for the same metal ion and ligand set, the crystal-field splitting for a tetrahedral complex is only four-ninths as large as for the octahedral complex. For

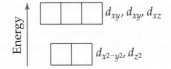

▲ **FIGURE 21.35** **Energies of the d orbitals in a tetrahedral crystal field.**

this reason, all tetrahedral complexes are high spin; the crystal field is never large enough to overcome the spin-pairing energies.

Square planar complexes, in which four ligands are arranged about the metal ion in a plane, can be envisaged as formed by removing two ligands from along the vertical z-axis of the octahedral complex. The changes that occur in the energy levels of the d orbitals are illustrated in ▶ FIGURE 21.36. Note in particular that the d_{z^2} orbital is now considerably lower in energy than the $d_{x^2-y^2}$ orbital. To understand why this is so, recall from Figure 21.30 that in an octahedral field the d_{z^2} orbital of the metal ion interacts with the ligands positioned above and below the xy plane. There are no ligands in these two positions in a square planar complex, which means that the d_{z^2} orbital experiences no repulsive force and so remains in a lower-energy, more stable state.

Square planar complexes are characteristic of metal ions with a d^8 electron configuration. They are nearly always low spin; that is, the eight d electrons are spin-paired to form a diamagnetic complex. Such an electronic arrangement is particularly common among the ions of heavier metals, such as Pt^{2+}, Ir^+ and Au^{3+}.

FIGURE IT OUT

Why is the $d_{x^2-y^2}$ orbital the highest-energy orbital in the square planar crystal field?

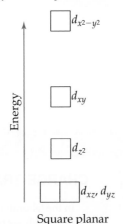

Square planar

▲ FIGURE 21.36   Energies of the d orbitals in a square planar crystal field.

CONCEPT CHECK 12

Why is the energy of the d_{xz} and d_{yz} orbitals in a square planar complex lower than that of the d_{xy} orbital?

SAMPLE EXERCISE 21.10 | **Populating d orbitals in tetrahedral and square planar complexes**

Nickel(II) complexes in which the metal coordination number is 4 can have either square planar or tetrahedral geometry. $[NiCl_4]^{2-}$ is paramagnetic, and $[Ni(CN)_4]^{2-}$ is diamagnetic. One of these complexes is square planar, and the other is tetrahedral. Use the relevant crystal-field splitting diagrams in the text to determine which complex has which geometry.

SOLUTION

Analyse We are given two complexes containing Ni^{2+} and their magnetic properties. We are given two molecular geometry choices and asked to use crystal-field splitting diagrams from the text to determine which complex has which geometry.

Plan We need to determine the number of d electrons in Ni^{2+} and then use Figure 21.35 for the tetrahedral complex and Figure 21.36 for the square planar complex.

Solve Nickel(II) has an electron configuration of $[Ar]3d^8$. The population of the d electrons in the two geometries is

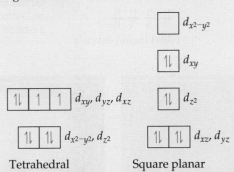

Comment Notice that the tetrahedral complex is paramagnetic with two unpaired electrons, whereas the square planar complex is diamagnetic.

PRACTICE EXERCISE

How many unpaired electrons do you predict for the tetrahedral $[CoCl_4]^{2-}$ ion?

Answer: Three

We have seen that the crystal-field model provides a basis for explaining many features of transition metal complexes. In fact, it can be used to explain many observations in addition to those we have discussed. Many lines of evidence show, however, that the bonding between transition metal ions and ligands must have some covalent character. The crystal-field model, although not entirely accurate in all details, provides an adequate and useful first description of the electronic structure of complexes.

A CLOSER LOOK

CHARGE-TRANSFER COLOUR

In the laboratory portion of your course, you have probably seen many colourful compounds of the transition metals. Many of these exhibit colour because of d-d transitions, in which visible light excites electrons from one d orbital to another. There are other colourful transition metal complexes, however, that derive their colour from a rather different type of excitation involving the d orbitals. Two such common substances are the deep violet permanganate ion (MnO_4^-) and the bright yellow chromate ion (CrO_4^{2-}), salts of which are shown in ▼ FIGURE 21.37. Both MnO_4^- and CrO_4^{2-} are tetrahedral complexes.

The permanganate ion strongly absorbs visible light with a maximum absorption at a wavelength of 565 nm. The strong absorption in the yellow portion of the visible spectrum is responsible for the violet appearance of salts and solutions of the ion (violet is the complementary colour to yellow). What is happening during this absorption? The MnO_4^- ion is a complex of Mn(VII), which has a d^0 electron configuration. As such, the absorption in the complex cannot be due to a d-d transition because there are no d electrons to excite. That does not mean, however, that the d orbitals are not involved in the transition. The excitation in the MnO_4^- ion is due to a *charge-transfer transition*, in which an electron on one of the oxygen ligands is excited into a vacant d orbital on the Mn atom (▶ FIGURE 21.38). In essence, an electron is transferred from a ligand to the metal, so this transition is called a *ligand-to-metal charge-transfer (LMCT) transition*. An LMCT transition is also

responsible for the colour of the CrO_4^{2-}, which is a d^0 Cr(VI) complex. Also shown in Figure 21.37 is a salt of the perchlorate ion (ClO_4^-). Like MnO_4^-, the ClO_4^- is tetrahedral and has its central atom in the +7 oxidation state. However, because the Cl atom doesn't have low-lying d orbitals, exciting an electron requires a more energetic photon than for MnO_4^-. The first absorption for ClO_4^- is in the ultraviolet portion of the spectrum, so all the visible light is transmitted and the salt appears white.

Other complexes exhibit charge-transfer excitations in which an electron from the metal atom is excited to an empty orbital on a ligand. Such an excitation is called a *metal-to-ligand charge-transfer (MLCT) transition*.

Charge-transfer transitions are generally more intense than d-d transitions. Many metal-containing pigments used for oil painting, such as cadmium yellow (CdS), chrome yellow ($PbCrO_4$) and red ochre (Fe_2O_3), have intense colours because of charge-transfer transitions.

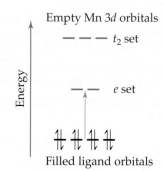

◀ FIGURE 21.38 **Ligand-to-metal charge-transfer (LMCT) transition in MnO_4^-.** As shown by the blue arrow, an electron is excited from a non-bonding pair on O into one of the empty d orbitals on Mn.

▲ FIGURE 21.37 **Effect of charge-transfer transitions.** The compounds $KMnO_4$, K_2CrO_4 and $KClO_4$ are shown from left to right. The $KMnO_4$ and K_2CrO_4 are intensely coloured because of ligand-to-metal charge transfer (LMCT) transitions in the MnO_4^- and CrO_4^{2-} anions. There are no valence d orbitals on Cl, so the charge-transfer transition for ClO_4^- requires ultraviolet light and $KClO_4$ is white.

SAMPLE INTEGRATIVE EXERCISE Putting concepts together

The oxalate ion has the Lewis structure shown in Table 21.3. **(a)** Show the geometrical structure of the complex formed by coordination of oxalate to cobalt(II), forming $[Co(C_2O_4)(H_2O)_4]$. **(b)** Write the formula for the salt formed upon coordination of three oxalate ions to Co(II), assuming that the charge-balancing cation is Na^+. **(c)** Sketch all the possible geometric isomers for the cobalt complex formed in part (b). Are any of these isomers chiral? Explain. **(d)** The equilibrium constant for the formation of the cobalt(II) complex produced by coordination of three oxalate anions, as in part (b), is 5.0×10^9. By comparison, the formation constant for formation of the cobalt(II) complex with three molecules of *ortho*-phenanthroline (Table 21.3) is 9×10^{19}. From these results, what conclusions can you draw regarding the relative Lewis base properties of the two ligands towards cobalt(II)?

Solve
(a) The complex formed by coordination of one oxalate ion is octahedral:

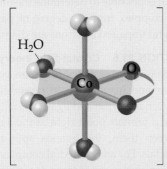

(b) Because the oxalate ion has a charge of 2− the net charge of a complex with three oxalate anions and one Co^{2+} ion is 4−. Therefore the coordination compound has the formula $Na_4[Co(C_2O_4)_3]$.

(c) There is only one geometric isomer. The complex is chiral, however, in the same way as the $[Co(en)_3]^{3+}$ complex, shown in Figure 21.20. These two mirror images are not superimposable, so there are two enantiomers:

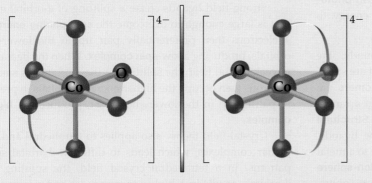

(d) The *ortho*-phenanthroline ligand is bidentate, like the oxalate ligand, so they both exhibit the chelate effect. Thus we can conclude that *ortho*-phenanthroline is a stronger Lewis base towards Co^{2+} than oxalate. This conclusion is consistent with what we learned about bases—namely, that nitrogen bases are generally stronger than oxygen bases. (Recall, for example, that NH_3 is a stronger base than H_2O.)

CHAPTER SUMMARY AND KEY TERMS

SECTION 21.1 **Transition metals** are characterised by incomplete filling of the *d* orbitals. The presence of *d* electrons in transition elements leads to multiple oxidation states. As we proceed through the transition metals in a given row of the periodic table, the attraction between the nucleus and the valence electrons increases more markedly for *d* electrons than for *s* electrons. As a result, the later transition elements in a period tend to have lower oxidation states.

The atomic and ionic radii of period 5 transition metals are larger than those of period 4 metals. The transition metals of periods 5 and 6 have comparable atomic and ionic radii and are also similar in other properties. This similarity is due to the **lanthanide contraction**.

The presence of unpaired electrons in valence orbitals leads to magnetic behaviour in transition metals and their compounds. In **ferromagnetic, ferrimagnetic** and **antiferromagnetic** substances the unpaired electron spins on atoms in a solid are affected by spins on neighbouring atoms. In a ferromagnetic substance the spins all point in the same direction. In an antiferromagnetic substance the spins point in opposite directions and cancel one another. In a ferrimagnetic substance the spins point in opposite directions but do not fully cancel. Ferromagnetic and ferrimagnetic substances are used to make permanent magnets.

SECTION 21.2   Coordination compounds are substances that contain **metal complexes**. Metal complexes contain metal ions bonded to several surrounding anions or molecules known as **ligands**. The metal ion and its ligands make up the **coordination sphere** of the complex. The number of atoms attached to the metal ion is the **coordination number** of the metal ion. The most common coordination numbers are 4 and 6; the most common coordination geometries are tetrahedral, square planar and octahedral. The atom of the ligand that bonds to the metal ion is the **donor atom**.

SECTION 21.3   Ligands that occupy only one site in a coordination sphere are called **monodentate ligands**. Ligands that have two donor atoms are **bidentate ligands**. **Polydentate ligands** have three or more donor atoms. Bidentate and polydendate ligands are also called **chelating agents**. In general, chelating agents form more stable complexes than do related monodentate ligands, an observation known as the **chelate effect**. Many biologically important molecules, such as the **porphyrins**, are complexes of chelating agents. A related group of plant pigments known as **chlorophylls** is important in **photosynthesis**, the process by which plants use solar energy to convert CO_2 and H_2O into carbohydrates.

SECTION 21.4   In naming coordination compounds, the number and type of ligands attached to the metal ion are specified, as is the oxidation state of the metal ion. **Isomers** are compounds with the same composition but different arrangements of atoms and therefore different properties. **Structural isomers** differ in the bonding arrangements of the ligands. **Linkage isomers** occur when a ligand can coordinate to a metal ion through either of two donor atoms. **Coordination-sphere isomers** contain different ligands in the coordination sphere. **Stereoisomers** are isomers with the same chemical bonding arrangements but different spatial arrangements of ligands. The most common forms of stereoisomers are **geometric isomers** and **optical isomers**. Geometric isomers differ from one another in the relative locations of donor atoms in the coordination

sphere; the most common are *cis–trans* isomers. Optical isomers are non-superimposable mirror images of each other. Geometric isomers differ from one another in their chemical and physical properties; optical isomers, or **enantiomers**, are **chiral**, however, meaning that they have a specific 'handedness' and differ only in the presence of a chiral environment.

SECTION 21.5   A substance has a particular colour because it either reflects or transmits light of that colour or absorbs light of the **complementary colour**. The amount of light absorbed by a sample as a function of wavelength is known as its **absorption spectrum**. The light absorbed provides the energy to excite electrons to higher-energy states.

It is possible to determine the number of unpaired electrons in a complex from its degree of paramagnetism. Compounds with no unpaired electrons are diamagnetic.

SECTION 21.6   Crystal-field theory successfully accounts for many properties of coordination compounds, including their colour and magnetism. In crystal-field theory, the interaction between metal ion and ligand is viewed as electrostatic. Because some *d* orbitals point right at the ligands whereas others point between them, the ligands split the energies of the metal *d* orbitals. For an octahedral complex, the *d* orbitals are split into a lower-energy set of three degenerate orbitals (the t_2 set) and a higher-energy set of two degenerate orbitals (the *e* set). Visible light can cause a ***d-d* transition**, in which an electron is excited from a lower-energy *d* orbital to a higher-energy *d* orbital. The **spectrochemical series** lists ligands in order of their ability to increase the split in *d*-orbital energies in octahedral complexes.

Strong-field ligands create a splitting of *d*-orbital energies that is large enough to overcome the **spin-pairing energy**. The *d* electrons then preferentially pair up in the lower-energy orbitals, producing a **low-spin complex**. When the ligands exert a weak crystal field, the splitting of the *d* orbitals is small. The electrons then occupy the higher-energy *d* orbitals in preference to pairing up in the lower-energy set, producing a **high-spin complex**.

Crystal-field theory also applies to tetrahedral and square planar complexes, which leads to different *d*-orbital splitting patterns. In a tetrahedral crystal field, the splitting of the *d* orbitals results in a higher-energy t_2 set and a lower-energy *e* set, the opposite of the octahedral case. The splitting by a tetrahedral crystal field is much smaller than that by an octahedral crystal field, so tetrahedral complexes are always high-spin complexes.

KEY SKILLS

- Describe the periodic trends in radii and oxidation states of the transition metal ions, including the origin and effect of the lanthanide contraction. (Section 21.1)

- Determine the oxidation number and the number of *d* electrons for metal ions in complexes. (Section 21.2)

- Distinguish between chelating and non-chelating ligands. (Section 21.3)

- Name coordination compounds given their formula and write their formula given their name. (Section 21.4)

- Recognise and draw the geometric isomers of a complex. (Section 21.4)

- Recognise and draw the optical isomers of a complex. (Section 21.4)

- Use crystal-field theory to explain the colours and to determine the number of unpaired electrons in a complex. (Section 21.5 and Section 21.6)

EXERCISES

VISUALISING CONCEPTS

21.1 **(a)** Draw the structure of [Pt(en)Cl$_2$]. **(b)** What is the coordination number for platinum in this complex, and what is the coordination geometry? **(c)** What is the oxidation state of the platinum? [Section 21.2]

21.2 Draw the Lewis structure for the ligand shown below. **(a)** Which atoms can serve as donor atoms? Classify this ligand as monodentate, bidentate or tridentate. **(b)** How many of these ligands are needed to fill the coordination sphere in an octahedral complex? [Section 21.3]

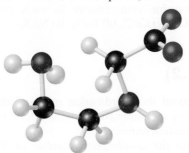

$$NH_2CH_2CH_2NHCH_2CO_2^-$$

21.3 The complex ion shown below has a 1− charge. Name the complex ion. [Section 21.4]

- = N
- = Cl
- = H
- = Pt

21.4 There are two geometric isomers of octahedral complexes of the type MA$_3$X$_3$, where M is a metal and A and X are monodentate ligands. Of the complexes shown here, which are identical to (1) and which are the geometric isomers of (1)? [Section 21.4]

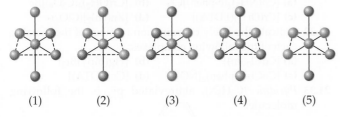

(1)   (2)   (3)   (4)   (5)

21.5 Which of the following complexes are chiral? Explain. [Section 21.4]

- = Cr   = NH$_2$CH$_2$CH$_2$NH$_2$ = Cl   = NH$_3$

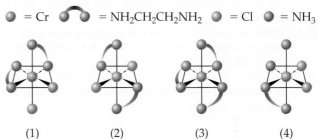

(1)   (2)   (3)   (4)

21.6 Consider the following crystal-field splitting diagrams. Select the one that fits each of the following descriptions: **(a)** a weak-field octahedral complex of Fe^{3+}, **(b)** a strong-field octahedral complex of Fe^{3+}, **(c)** a tetrahedral complex of Fe^{3+}, **(d)** a tetrahedral complex of Ni^{2+}? (The diagrams do not indicate the relative magnitudes of Δ.) [Section 21.6]

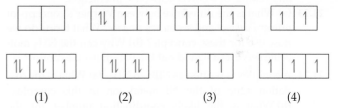

(1)   (2)   (3)   (4)

21.7 Consider the linear crystal field shown below, in which the negative charges are on the z-axis. Using Figure 21.30 as a guide, predict which d orbital has lobes closest to the charges. Which two have lobes furthest from the charges? Predict the crystal-field splitting of the d orbitals in linear complexes. [Section 21.6]

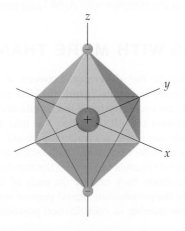

TRANSITION METALS (Section 21.1)

21.8 Which of the following properties are considered characteristic of the free isolated atoms, and which are characteristic of the bulk metal: **(a)** electrical conductivity, **(b)** first ionisation energy, **(c)** atomic radius, **(d)** melting point, **(e)** heat of vaporisation, **(f)** electron affinity?

21.9 Zirconium and hafnium are the group 4 elements in the second and third transition series. The atomic radii of these elements are virtually the same (Figure 21.2). Explain this similarity in atomic radius.

21.10 What is meant by the term *lanthanide contraction*? What properties of the transition elements are affected by the lanthanide contraction?

21.11 Write the formula for the fluoride corresponding to the highest expected oxidation state for **(a)** Sc, **(b)** Co, **(c)** Zn, **(d)** Br.

21.12 Write the formula for the oxide corresponding to the highest expected oxidation state for **(a)** Cd, **(b)** Ti, **(c)** Nb, **(d)** Ni.

21.13 Why does chromium exhibit several oxidation states in its compounds, whereas aluminium exhibits only the +3 oxidation state?

21.14 Write the expected electron configuration for **(a)** Cr^{3+}, **(b)** Au^{3+}, **(c)** Ru^{2+}, **(d)** Cu^+, **(e)** Mn^{4+}, **(f)** Ir^+.

21.15 What is the expected electron configuration for **(a)** Ti^{2+}, **(b)** Co^{3+}, **(c)** Pd^{2+}, **(d)** Mo^{3+}, **(e)** Ru^{3-}, **(f)** Ni^{4+}?

21.16 Which would you expect to be more easily oxidised, Ti^{2+} or Ni^{2+}?

21.17 Which would you expect to be the stronger reducing agent, Cr^{2+} or Fe^{2+}?

21.18 How does the presence of air affect the relative stabilities of Fe^{2+} and Fe^{3+} in solution?

21.19 On the atomic level, what distinguishes a paramagnetic material from a diamagnetic one? How does each behave in a magnetic field?

21.20 **(a)** What characteristics of a ferromagnetic material distinguish it from one that is paramagnetic? **(b)** What type of interaction must occur in the solid to bring about ferromagnetic behaviour? **(c)** Must a substance contain iron to be ferromagnetic? Explain.

21.21 Indicate whether each of the following compounds is expected to be diamagnetic or paramagnetic, and give a reason for your answer in each case: **(a)** $NbCl_5$, **(b)** $CrCl_2$, **(c)** $CuCl$, **(d)** RuO_4, **(e)** $NiCl_2$.

TRANSITION METAL COMPLEXES (Section 21.2)

21.22 **(a)** Define the italicised phrases in the following sentence: A *metal complex* is found in which the *coordination number* is 6, with four H_2O and two NH_3 *ligands*. **(b)** Explain why the formation of a metal–ligand bond is an example of a Lewis acid–base interaction.

21.23 **(a)** What is the difference between Werner's concepts of *primary valence* and *secondary valence*? What terms do we now use for these concepts? **(b)** Why can the NH_3 molecule serve as a ligand but the BH_3 molecule cannot?

21.24 A complex is written as $NiBr_2 \cdot 6\,NH_3$. **(a)** What is the oxidation state of the Ni metal ion in this complex? **(b)** What is the likely coordination number for the complex? **(c)** If the complex is treated with excess $AgNO_3$(aq), how many moles of AgBr will precipitate per mole of complex?

21.25 A certain complex of metal, M, is formulated as $MCl_3 \cdot 3\,H_2O$. The coordination number of the complex is not known, but is expected to be 4 or 6. **(a)** Would reaction of the complex with $AgNO_3$(aq) provide information about the coordination number? **(b)** Would conductivity measurements provide information about the coordination number?

21.26 Indicate the coordination number of the metal and the oxidation number of the metal in each of the following complexes:
(a) $Na_2[CdCl_4]$ **(b)** $K_2[Mo(OH_2)_2Cl_4]$
(c) $[Co(NH_3)_4Cl_2]Cl$ **(d)** $[Ni(OH_2)_2(CN)_5]^{3-}$
(e) $K_3[V(C_2O_4)_3]$ **(f)** $[Zn(en)_2]Br_2$

21.27 Indicate the coordination number of the metal and the oxidation number of the metal in each of the following complexes:
(a) $K_4[Fe(CN)_6]$ **(b)** $[Pd(NH_3)_4]Br_2$
(c) $[Mn(CO)_5Br]$ **(d)** $[Co(en)_2(C_2O_4)]^+$
(e) $NH_4[Cr(NH_3)_2(NCS)_4]$ **(f)** $[Cu(OH_2)(bipy)_2I]I$

21.28 Determine the number and type of each donor atom in each of the complexes in Exercise 21.26.

21.29 What are the number and types of donor atoms in each of the complexes in Exercise 21.27?

LIGANDS WITH MORE THAN ONE DONOR ATOM (Section 21.3)

21.30 **(a)** What is the difference between a monodentate ligand and a bidentate ligand? **(b)** How many bidentate ligands are necessary to fill the coordination sphere of a 6-coordinate complex? **(c)** You are told that a certain molecule can serve as a tridentate ligand. Based on this statement, what do you know about the molecule?

21.31 Polydentate ligands can vary in the number of coordination positions they occupy. In each of the following, identify the polydentate ligand present and indicate the probable number of coordination positions it occupies:

(a) $[Co(NH_3)_4(o\text{-phen})]Cl_3$ **(b)** $[Cr(OH_2)_4(C_2O_4)]Br$
(c) $[Cr(OH_2)(EDTA)]^-$ **(d)** $[Zn(en)_2](ClO_4)_2$

21.32 Indicate the likely coordination number of the metal in each of the following complexes:
(a) $[CdCl_2(en)]$ **(b)** $[Hg(bipy)Br_2]$
(c) $[CoCl_2(o\text{-phen})_2]NO_3$ **(d)** $[Ce(EDTA)]$

21.33 *Pyridine* (C_5H_5N), abbreviated py, is the following molecule:

(a) Is pyridine a monodentate or polydentate ligand?

(b) Consider the following equilibrium reaction:

$$[Ru(py)_4(bipy)]^{2+} + 2\,py \rightleftharpoons [Ru(py)_6]^{2+} + bipy$$

What would you predict for the magnitude of the equilibrium constant for this equilibrium? Explain the basis for your answer.

NOMENCLATURE AND ISOMERISM IN COORDINATION CHEMISTRY (Section 21.4)

21.34 Write the formula for each of the following compounds, being sure to use brackets to indicate the coordination sphere:
(a) hexaamminechromium(III) nitrate
(b) tetraamminecarbonatocobalt(III) sulfate
(c) dichloridobis(ethylenediamine)platinum(IV) bromide
(d) potassium diaquatetrabromidovanadate(III)
(e) bis(ethylenediamine)zinc(II) tetraiodidomercurate(II)

21.35 Write the formula for each of the following compounds, being sure to use brackets to indicate the coordination sphere:
(a) pentaaquaiodidomanganese(III) perchlorate
(b) tris(bipyridine)ruthenium(II) nitrate
(c) dichlorobis(*ortho*-phenanthroline)rhodium(III) sulfate
(d) sodium diamminetetrabromochromate(III)
(e) tris(ethylenediamine)cobalt(III) tris(oxalato)ferrate(II)

21.36 Write the names of the following compounds, using the standard nomenclature rules for coordination complexes:
(a) $[Rh(NH_3)_4Cl_2]Cl$ **(b)** $K_2[TiCl_6]$
(c) $MoOCl_4$ **(d)** $[Pt(OH_2)_4(C_2O_4)]Br_2$

21.37 Write names for the following coordination compounds:
(a) $[Cd(en)Cl_2]$ **(b)** $K_4[Mn(CN)_6]$
(c) $[Cr(NH_3)_5CO_3]Cl$ **(d)** $[Ir(NH_3)_4(OH_2)_2](NO_3)_3$

21.38 By writing formulae or drawing structures related to any one of these three complexes,

$$[Co(NH_3)_4Br_2]Cl$$

$$[Pd(NH_3)_2(ONO)_2]$$

$$cis\text{-}[VCl_2(en)_2]^+$$

illustrate **(a)** geometric isomerism, **(b)** linkage isomerism, **(c)** optical isomerism, **(d)** coordination-sphere isomerism.

21.39 **(a)** Draw the two linkage isomers of $[Co(NH_3)_5SCN]^{2+}$.
(b) Draw the two geometric isomers of $[Co(NH_3)_3Cl_3]^{2+}$.
(c) Two compounds with the formula $Co(NH_3)_5ClBr$ can be prepared. Use structural formulae to show how they differ. What kind of isomerism does this illustrate?

21.40 A 4-coordinate complex MA_2B_2 is prepared and found to have two different isomers. Is it possible to determine from this information whether the complex is square planar or tetrahedral? If so, which is it?

21.41 Consider an octahedral complex MA_3B_3. How many geometric isomers are expected for this compound? Will any of the isomers be optically active? If so, which ones?

21.42 Draw the *cis* and *trans* isomers of the $[Co(NH_3)(en)_2Cl]^{2+}$ ion. Is either or both of these isomers chiral? If so, draw the two enantiomers.

21.43 Sketch all the possible stereoisomers of **(a)** tetrahedral $[Cd(OH_2)_2Cl_2]$, **(b)** square planar $[IrCl_2(PH_3)_2]^-$, **(c)** octahedral $[Fe(o\text{-phen})_2Cl_2]^+$

COLOUR AND MAGNETISM IN COORDINATION CHEMISTRY; CRYSTAL-FIELD THEORY (Sections 21.5 and 21.6)

21.44 **(a)** To the closest 100 nm, what are the largest and smallest wavelengths of visible light? **(b)** What is meant by the term *complementary colour*? **(c)** What is the significance of complementary colours in understanding the colours of metal complexes? **(d)** If a complex absorbs light at 610 nm, what is the energy of this absorption in kJ mol^{-1}?

21.45 **(a)** A complex absorbs light with wavelength of 530 nm. Do you expect it to have colour? **(b)** A solution of a compound appears green. Does this observation necessarily mean that all colours of visible light other than green are absorbed by the solution? Explain. **(c)** What information is usually presented in a *visible absorption spectrum* of a compound? **(d)** What energy is associated with the absorption at 530 nm in kJ mol^{-1}?

21.46 In crystal-field theory, ligands are modelled with point negative charges. What is the basis of this assumption, and how does it relate to the nature of metal–ligand bonds?

21.47 Explain why the d_{xy}, d_{xz} and d_{yz} orbitals lie lower in energy than the d_{z^2} and $d_{x^2-y^2}$ orbitals in the presence of an octahedral arrangement of ligands about the central metal ion.

21.48 **(a)** Sketch a diagram that shows the definition of the *crystal-field splitting energy* (Δ) for an octahedral crystal field. **(b)** What is the relationship between the magnitude of Δ and the energy of the d-d transition for a d^1 complex? **(c)** Calculate Δ in kJ mol^{-1} if a d^1 complex has an absorption maximum at 590 nm.

21.49 The d-d transition of $[Ti(H_2O)_6]^{3+}$ produces an absorption maximum at a wavelength of 500 nm. **(a)** What is the magnitude of Δ for $[Ti(H_2O)_6]^{3+}$ in kJ mol^{-1}? **(b)** What is the *spectrochemical series*? How would the magnitude of Δ change if the H_2O ligands in $[Ti(H_2O)_6]^{3+}$ were replaced with NH_3 ligands?

21.50 Explain why many cyano complexes of divalent transition metal ions are yellow, whereas many aqua complexes of these ions are blue or green.

21.51 Give the number of d electrons associated with the central metal ion in each of the following complexes:
(a) $K_3[Fe(CN)_6]$ (b) $[Mn(OH_2)_6](NO_3)_2$
(c) $Na[Ag(CN)_2]$ (d) $[Cr(NH_3)_4Br_2]ClO_4$
(e) $[Sc(EDTA)]^{2-}$

21.52 Draw the crystal-field energy-level diagrams and show the placement of d electrons for each of the following:
(a) $[Cr(H_2O)_6]^{2+}$ (high spin), (b) $[Mn(OH_2)_6]^{2+}$ (high spin), (c) $[Ru(NH_3)_5OH_2]^{2+}$ (low spin), (d) $[IrCl_6]^{2-}$ (low spin), (e) $[Cr(en)_3]^{3+}$, (f) $[NiF_6]^{4-}$.

21.53 The complex $[Mn(NH_3)_6]^{2+}$ contains five unpaired electrons. Sketch the energy-level diagram for the d orbitals, and indicate the placement of electrons for this complex ion. Is the ion a high-spin or a low-spin complex?

21.54 The ion $[Fe(CN)_6]^{3-}$ has one unpaired electron, whereas $[Fe(NCS)_6]^{3-}$ has five unpaired electrons. From these results, what can you conclude about whether each complex is high spin or low spin? What can you say about the placement of NCS^- in the spectrochemical series?

21.55 One of the more famous species in coordination chemistry is the Creutz–Taube complex.

$$\left[(NH_3)_5RuN \bigcirc NRu(NH_3)_5 \right]^{5+}$$

It is named for the two scientists who discovered it and initially studied its properties. The central ligand is pyrazine, a planar six-membered ring with nitrogens at opposite sides. (a) How can you account for the fact that the complex, which has only neutral ligands, has an odd overall charge? (b) The metal is in a low-spin configuration in both cases. Assuming octahedral coordination, draw the d-orbital energy-level diagram for each metal. (c) In many experiments the two metal ions appear to be in exactly equivalent states. Can you think of a reason why this might appear to be so, recognising that electrons move very rapidly compared with nuclei?

ADDITIONAL EXERCISES

21.56 Based on the molar conductance values listed here for the series of platinum(IV) complexes, write the formula for each complex so as to show which ligands are in the coordination sphere of the metal. By way of example, the molar conductances of NaCl and $BaCl_2$ are 107 ohm^{-1} and 197 ohm^{-1}, respectively.

Complex	Molar conductance (ohm^{-1})* of 0.050 M solution
$Pt(NH_3)_6Cl_4$	523
$Pt(NH_3)_4Cl_4$	228
$Pt(NH_3)_3Cl_4$	97
$Pt(NH_3)_2Cl_4$	0
$KPt(NH_3)Cl_5$	108

*The ohm is a unit of resistance; conductance is the inverse of resistance.

21.57 (a) A compound with formula $RuCl_3 \cdot 5 H_2O$ is dissolved in water, forming a solution that is approximately the same colour as the solid. Immediately after forming the solution, the addition of excess $AgNO_3(aq)$ forms 2 mol of solid AgCl per mole of complex. Write the formula for the compound, showing which ligands are likely to be present in the coordination sphere. (b) After a solution of $RuCl_3 \cdot 5 H_2O$ has stood for about a year, addition of $AgNO_3(aq)$ precipitates 3 mol of AgCl per mole of complex. What has happened during that time?

21.58 Although the *cis* configuration is known for $[Pt(en)Cl_2]$, no *trans* form is known. (a) Explain why the *trans* compound is not possible. (b) Suggest what type of ligand would be required to form a *trans*-bidentate coordination to a metal atom.

21.59 The complexes $[V(H_2O)_6]^{3+}$ and $[VF_6]^{3-}$ are both known. (a) Draw the d-orbital energy-level diagram for V(III) octahedral complexes. (b) What gives rise to the colours of these complexes? (c) Which of the two complexes would you expect to absorb light of higher energy? Explain.

21.60 One of the more famous species in coordination chemistry is the Creutz–Taube complex,

21.61 Solutions of $[Co(NH_3)_6]^{2+}$, $[Co(H_2O)_6]^{2+}$ (both octahedral) and $[CoCl_4]^{2-}$ (tetrahedral) are coloured. One is pink, one is blue and one is yellow. Based on the spectrochemical series and remembering that the energy splitting in tetrahedral complexes is normally much less than that in octahedral ones, assign a colour to each complex.

21.62 In each of the following pairs of complexes, which would you expect to absorb at the longer wavelength:
(a) $[FeF_6]^{4-}$ or $[Fe(NH_3)_6]^{2+}$, (b) $[V(H_2O)_6]^{2+}$ or $[Cr(H_2O)_6]^{3+}$, (c) $[Co(NH_3)_6]^{2+}$ or $[CoCl_4]^{2-}$? Explain your reasoning in each case.

21.63 Suppose that a transition metal ion was in a lattice in which it was in contact with just two nearby anions, located on opposite sides of the metal. Diagram the splitting of the metal d orbitals that would result from such a crystal field. Assuming a strong field, how many unpaired electrons would you expect for a metal ion with six d electrons? [*Hint:* Consider the linear axis to be the z-axis.]

21.64 Give brief statements about the relevance of the following complexes in living systems: (a) haemoglobin, (b) chlorophylls, (c) siderophores.

INTEGRATIVE EXERCISES

21.65 Metallic elements are essential components of many important enzymes operating within our bodies. *Carbonic anhydrase*, which contains Zn^{2+}, is responsible for rapidly interconverting dissolved CO_2 and bicarbonate ion, HCO_3^-. The zinc in carbonic anhydrase is coordinated by three nitrogen-containing groups and a water molecule. The enzyme's action depends on the fact that the coordinated water molecule is more acidic than the bulk solvent molecules. Explain this fact in terms of Lewis acid–base theory.

21.66 Two different compounds have the formulation $CoBr(SO_4) \cdot 5\,NH_3$. Compound A is dark violet and compound B is red-violet. When compound A is treated with $AgNO_3(aq)$ no reaction occurs, whereas compound B reacts with $AgNO_3(aq)$ to form a white precipitate. When compound A is treated with $BaCl_2(aq)$ a white precipitate is formed, whereas compound B has no reaction with $BaCl_2(aq)$. **(a)** Is Co in the same oxidation state in these complexes? **(b)** Explain the reactivity of compounds A and B with $AgNO_3(aq)$ and $BaCl_2(aq)$. **(c)** Are compounds A and B isomers of one another? If so, which category from Figure 21.17 best describes the isomerism observed for these complexes? **(d)** Would compounds A and B be expected to be strong electrolytes, weak electrolytes or non-electrolytes?

21.67 A manganese complex formed from a solution containing potassium bromide and oxalate ion is purified and analysed. It contains 10.0% Mn, 28.6% potassium, 8.8% carbon and 29.2% bromine by mass. The remainder of the compound is oxygen. An aqueous solution of the complex has about the same electrical conductivity as an equimolar solution of $K_4[Fe(CN)_6]$. Write the formula of the compound, using brackets to denote the manganese and its coordination sphere.

21.68 The total concentration of Ca^{2+} and Mg^{2+} in a sample of hard water was determined by titrating a $0.100\,dm^3$ sample of the water with a solution of $EDTA^{4-}$. The $EDTA^{4-}$ chelates the two cations:

$$Mg^{2+} + [EDTA]^{4-} \longrightarrow [Mg(EDTA)]^{2-}$$

$$Ca^{2+} + [EDTA]^{4-} \longrightarrow [Ca(EDTA)]^{2-}$$

It requires $31.5\,cm^3$ of 0.0104 M $[EDTA]^{4-}$ solution to reach the end point in the titration. A second $0.100\,dm^3$ sample was then treated with sulfate ion to precipitate Ca^{2+} as calcium sulfate. The Mg^{2+} was then titrated with $21.7\,cm^3$ of 0.0104 M $[EDTA]^{4-}$. Calculate the concentrations of Mg^{2+} and Ca^{2+} in the hard water in mg dm^{-3}

21.69 The value of Δ for the $[CrF_6]^{3-}$ complex is $182\,kJ\,mol^{-1}$. Calculate the expected wavelength of the absorption corresponding to promotion of an electron from the lower-energy to the higher-energy *d*-orbital set in this complex. Should the complex absorb in the visible range? (Remember to divide by Avogadro's number.)

MasteringChemistry (www.pearson.com.au/masteringchemistry)

Make learning part of the grade. Access:

- tutorials with personalised coaching
- study area
- Pearson eText

PHOTO/ART CREDITS

850 Robert Garvey/Corbis/Bettmann; **854, 857, 859, 869, 872** Richard Megna, Fundamental Photographs, NYC; **855** Joel Arem/Photo Researchers, Inc; **874** Nigel Forrow, photographersdirect.com.

22

THE CHEMISTRY OF ORGANIC COMPOUNDS

Organic compounds are so named because they were originally derived entirely from living or once-living sources. Honey is a naturally occurring substance produced by bees using the nectar from flowers.

KEY CONCEPTS

22.1 GENERAL CHARACTERISTICS OF ORGANIC MOLECULES
We begin with a review of the main features of organic compounds.

22.2 AN INTRODUCTION TO HYDROCARBONS
We consider *hydrocarbons*, compounds containing only carbon atoms and hydrogen atoms. *Alkanes* are a general class of hydrocarbon that form the structural basis of organic chemistry.

22.3 APPLICATIONS AND PHYSICAL PROPERTIES OF ALKANES
We discuss common alkanes and their applications as fuels. Straight-chain *homologous series* are introduced.

22.4 STRUCTURES OF ALKANES
We explore the use of hybridisation and valence-bond theory to describe the bonding outcomes for carbon. We also look at isomers, compounds with identical compositions but different molecular structures.

22.5 ALKANE NOMENCLATURE
Using a set of systematic rules, we will learn how to name alkanes and their substituent alkyl branches unambiguously.

22.6 CYCLOALKANES
We explore the cyclic form of alkanes and discuss *axial* and *equatorial positions* using cycloalkane examples.

22.7 ORGANIC FUNCTIONAL GROUPS
We discuss the breadth of general *functional groups* attributed to organic compounds as well as their role as the central organising principle of organic chemistry.

22.8 REACTIONS OF ALKANES
We explore the two main chemical reactions of alkanes: combustion and free-radical reactions. Complete *combustion* is a process that requires oxygen to produce carbon dioxide and water. *Free-radical* chemistry is often used to introduce *functional groups* to the alkane skeleton. The *regioselectivity* of this reaction is described through the example of *halogenation.* Added to this is the concept of the two-electron bond, which we explain as the bond-breaking and bond-making processes of free-radical chemistry.

When a bee visits a flower, it obtains two things: nectar and pollen. Bees produce honey from the nectar of flowers as a source of high-energy food. Honey is above all a carbohydrate material, with over 95% of the solids being sugars. Glucose and fructose are the two simple sugars that make up most of the honey. There are small amounts of at least 22 other more complex sugars, including sucrose (table sugar). All sugars are composed of carbon, hydrogen and oxygen.

From honey to home heating, materials to medicine and recreation to rice, carbon-containing compounds form an indispensable part of our quality of life. Over 16 million carbon-containing compounds are known, and approximately 90% of the new compounds synthesised each year contain carbon. The study of compounds of carbon constitutes a branch of chemistry called **organic chemistry**.

The term organic chemistry arose from the eighteenth-century belief that organic compounds could only be formed by living systems. This idea was disproved in 1828 when the German chemist Friedrich Wöhler synthesised urea (H_2NCONH_2), an organic substance found in the urine of mammals, by heating the inorganic substance ammonium cyanate (NH_4NCO).

Over the course of the next nine chapters, we will explore the importance of organic chemistry as well as the diversity and complexity of its structures.

22.1 | GENERAL CHARACTERISTICS OF ORGANIC MOLECULES

What is it about carbon that leads to the tremendous diversity of its compounds and allows them to play such a diverse role in biology and industry? Let's consider some general features of organic molecules and, as we do, review principles we learned in earlier chapters.

The Structure of Organic Molecules

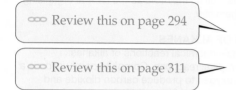
∞ Review this on page 294

∞ Review this on page 311

Because carbon has four valence electrons ($[He]2s^2 2p^2$), it forms four bonds in virtually all its compounds. When all four bonds are single bonds, the electron pairs are disposed in a tetrahedral arrangement (∞ Section 9.2, 'The VSEPR Model'). In the hybridisation model, the carbon 2s and 2p orbitals are then sp^3 hybridised. (∞ Section 9.5, 'Hybrid Orbitals').

Although it is only a model, this approach has been used widely to describe qualitatively both structure and reactivity of organic compounds. When there is one double bond, the arrangement is trigonal planar (and modelled as sp^2 hybridisation). With a triple bond, it is linear (and modelled as sp hybridisation). Examples are shown in ▼ FIGURE 22.1.

Almost every organic molecule contains C–H bonds. Because the valence shell of H can hold only two electrons, hydrogen forms only one covalent bond. As a result, hydrogen atoms are always located on the surface of organic molecules whereas the C–C bonds form the backbone, or skeleton, of the molecule, as in the propane molecule:

The Stabilities of Organic Molecules

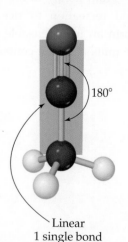
∞ Review this on page 279

∞ Review this on page 586

Carbon forms strong bonds with a variety of elements, especially H, O, N and the halogens (∞ Section 8.8, 'Strengths of Covalent Bonds'). Carbon also has an exceptional ability to bond to itself, forming a variety of molecules made up of chains or rings of carbon atoms. Most reactions with low or moderate activation energy (∞ Section 15.5, 'Temperature and Rate') begin when a region of high electron density on one molecule encounters a region of low electron density on another molecule. The regions of high electron density may be due

▶ **FIGURE 22.1 Carbon geometries.** The three common geometries around carbon are tetrahedral as in methane (CH_4), trigonal planar as in formaldehyde (CH_2O) and linear as in acetonitrile (CH_3CN). Notice that in all cases each carbon atom forms four bonds.

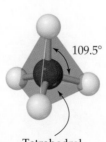

109.5°

Tetrahedral
4 single bonds
sp^3 hybridisation

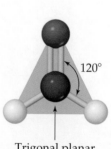

120°

Trigonal planar
2 single bonds
1 double bond
sp^2 hybridisation

180°

Linear
1 single bond
1 triple bond
sp hybridisation

to the presence of a multiple bond or to the more electronegative atom in a polar bond. Because of their strength and lack of polarity, both C–C single bonds and C–H bonds are relatively unreactive. To understand the implications of these facts better, consider ethanol:

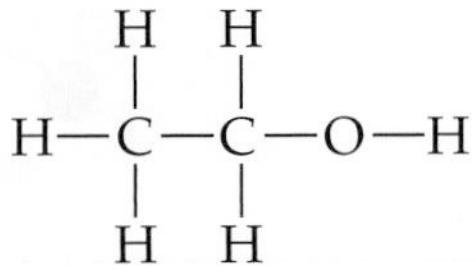

$$
\begin{array}{ccc}
 & \overset{\displaystyle H}{|} & \overset{\displaystyle H}{|} \\
H- & C- & C-O-H \\
 & \underset{\displaystyle H}{|} & \underset{\displaystyle H}{|}
\end{array}
$$

The differences in the electronegativity values of C (2.5) and O (3.5) and of O and H (2.1) indicate that the C–O and O–H bonds are quite polar (∞ Section 8.4, 'Bond Polarity and Electronegativity'). Thus many reactions of ethanol involve these bonds while the hydrocarbon portion of the molecule remains intact. A group of atoms such as the C–O–H group, which determines how an organic molecule reacts (in other words, how the molecule *functions*), is called a **functional group**. As we progress through the chapter we will see that the functional group is the centre of reactivity in an organic molecule.

∞ Review this on page 260

22.2 | AN INTRODUCTION TO HYDROCARBONS

Because carbon compounds are so numerous, it is convenient to organise them into families that have structural similarities. Compounds that contain *only* carbon and hydrogen are called **hydrocarbons**. The key structural feature of hydrocarbons (and of most organic substances) is the presence of stable carbon–carbon bonds. Carbon is the only element capable of forming stable, extended chains of atoms bonded through single, double or triple bonds.

Hydrocarbons can be divided into four types, depending on the kinds of carbon–carbon bonds in the molecules. ▼ TABLE 22.1 shows an example of each

TABLE 22.1 • The four hydrocarbon types

Type				Example		
Alkane	Saturated	Ethane	CH_3CH_3			
Alkene	Unsaturated	Ethene	$CH_2{=}CH_2$			
Alkyne	Unsaturated	Ethene	$CH{\equiv}CH$			
Aromatic	Unsaturated	Benzene	C_6H_6			

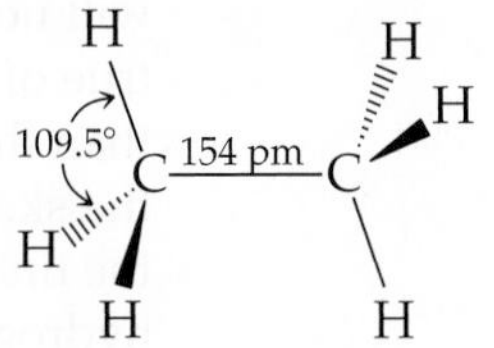

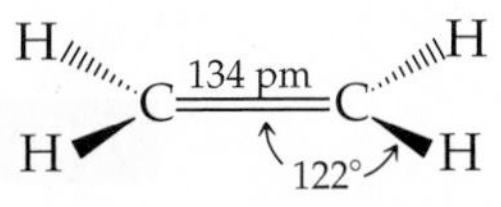

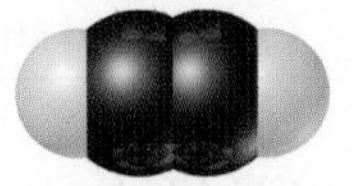

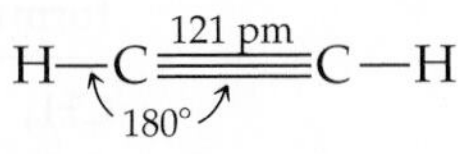

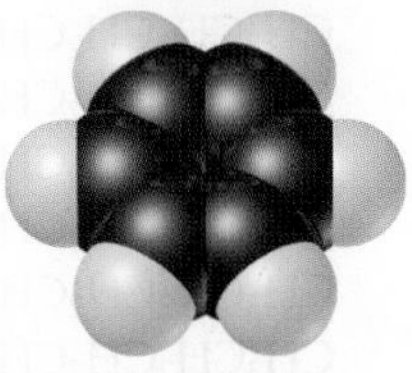

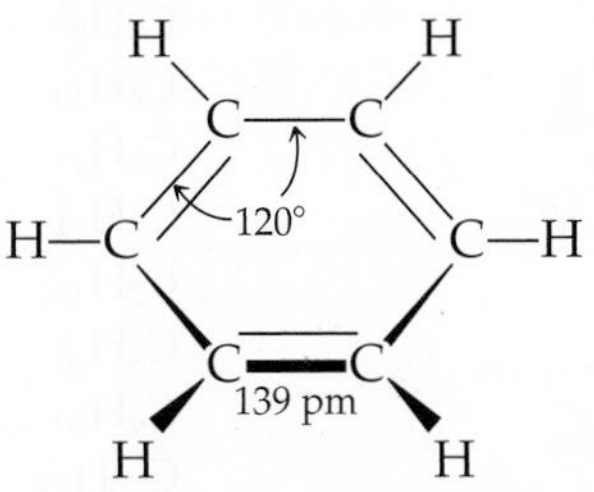

TABLE 22.2 • **The simplest alkanes**

Common name	Structural formula	Condensed structural formula	Ball-and-stick model	Space-filling representation
Methane	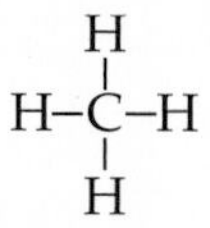 H–C–H with H above and H below	CH_4	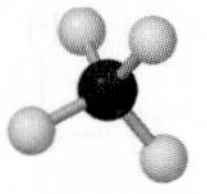	
Ethane	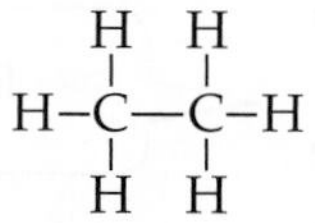 H–C–C–H with H above and below each C	CH_3CH_3	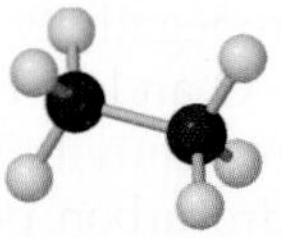	
Propane	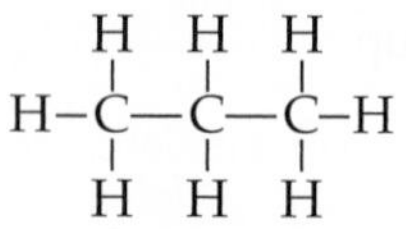 H–C–C–C–H with H above and below each C	$CH_3CH_2CH_3$	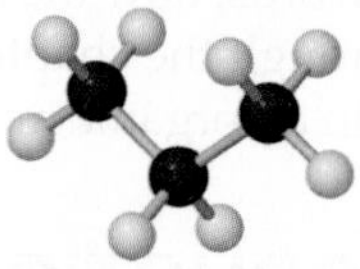	

∞ Review this on page 56

type. These hydrocarbons are either *aliphatic* (from the Greek work *aleiphar*, meaning oil) (∞ Section 2.9, 'Some Simple Organic Compounds') or *aromatic*, and are classified as either saturated or unsaturated. We will now explore aliphatic hydrocarbons that are **saturated** (that is, all carbons are bound to four other atoms, there are no double or triple bonds). Such hydrocarbons are called **alkanes**.

Alkanes

The three simplest alkanes, which contain one, two and three carbon atoms, are methane (CH_4), ethane (C_2H_6) and propane (C_3H_8), respectively. These are illustrated in ▲ TABLE 22.2.

The names of the first 10 *straight-chain* alkanes are given in ▼ TABLE 22.3. You will notice that each aliphatic alkane has a name that ends in *-ane*. This suffix is true of all alkanes as it denotes this class of organic compounds. Alkanes longer than those described in Table 22.3 can be made by adding further CH_2 groups to the skeleton of the molecule. Notice the trend within the molecular formulae in the first column of Table 22.3. For *n* carbon atoms ($n = 1, 2, 3 \ldots$) there are ($2n + 2$) hydrogen atoms, so alkanes have the general formula:

$$C_nH_{2n+2}$$

[22.1]

TABLE 22.3 • **First 10 members of the straight-chain alkane series**

Molecular formula	Condensed structural formula	Name	Boiling point (°C)
CH_4	CH_4	Methane	−161
C_2H_6	CH_3CH_3	Ethane	−89
C_3H_8	$CH_3CH_2CH_3$	Propane	−44
C_4H_{10}	$CH_3CH_2CH_2CH_3$	Butane	−0.5
C_5H_{12}	$CH_3CH_2CH_2CH_2CH_3$	Pentane	36
C_6H_{14}	$CH_3CH_2CH_2CH_2CH_2CH_3$	Hexane	68
C_7H_{16}	$CH_3CH_2CH_2CH_2CH_2CH_2CH_3$	Heptane	98
C_8H_{18}	$CH_3CH_2CH_2CH_2CH_2CH_2CH_2CH_3$	Octane	125
C_9H_{20}	$CH_3CH_2CH_2CH_2CH_2CH_2CH_2CH_2CH_3$	Nonane	151
$C_{10}H_{22}$	$CH_3CH_2CH_2CH_2CH_2CH_2CH_2CH_2CH_2CH_3$	Decane	174

The condensed structural formula reveals the way in which atoms are bonded to one another but does not require us to draw in all the bonds. It is one of the most useful ways of representing organic molecules.

 CONCEPT CHECK 1
How many C–H and C–C bonds are formed by the central carbon atom of propane?

22.3 | APPLICATIONS AND PHYSICAL PROPERTIES OF ALKANES

Alkanes are very important in our society, whether they are gaseous, liquid or solid. Many of their common applications, illustrated in ▼ FIGURE 22.2, will be familiar to you. Methane, a colourless and odourless gas, is the major component of natural gas and is used for home heating and in gas stoves and water heaters. Propane is the major component of the bottled gas used for home heating and cooking in areas where natural gas is not available. Butane is used in disposable lighters and in fuel canisters for gas camping stoves and lanterns. Alkanes with 5–12 carbon atoms per molecule (C_5–C_{12}) are found in petrol. This difference in application between different groups of alkanes is a result of their physical properties, such as boiling point (bp) and melting point (mp), which determine whether the alkane is a solid, liquid or gas at room temperature.

Hydrocarbon molecules such as alkanes are relatively non-polar because carbon and hydrogen do not differ greatly in their electronegativities. Thus alkanes are almost completely insoluble in water, but they dissolve readily in other non-polar solvents. Cholesterol is a good example of this (▼ FIGURE 22.3). For an organic molecule to be soluble in water or in very polar solvents it must itself be hugely polar (▼ FIGURE 22.4).

Homologous Series

A series of compounds, like the *n*-alkanes, that differ only by the number of methine (–CH_2–) units is called a **homologous series**. A homologous series of organic compounds contains a similar general formula, possessing similar chemical properties due to the presence of the same functional group (Section 22.7), and shows a gradation in physical properties as a result of increase in molecular size and mass (Table 22.3). The individual members of this series are called

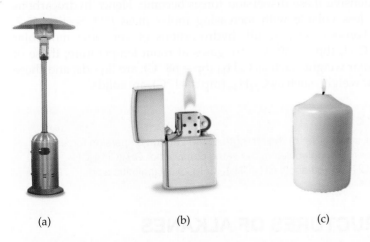

(a)   (b)   (c)

▲ **FIGURE 22.2 Applications of alkanes.** Alkanes are very important in our modern society, whether gaseous, liquid or solid. (a) Methane is used as a source of heat. (b) Butane is commonly used in lighter fluid. (c) Candles are traditionally made from waxes, which are long-chain alkanes.

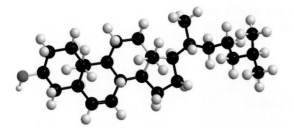
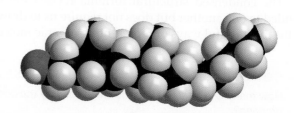

▲ **FIGURE 22.3** Cholesterol is soluble in non-polar solvents.

FIGURE IT OUT

How would replacing OH groups on ascorbic acid with CH_3 groups affect the substance's solubility in (a) polar solvents and (b) non-polar solvents?

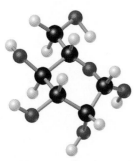
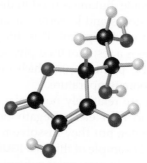

Glucose ($C_6H_{12}O_6$)   Ascorbic acid ($C_6H_8O_6$)   Stearate ($C_{17}H_{35}COO^-$)

▲ **FIGURE 22.4** Organic molecules soluble in polar solvents.

homologues. For example, pentane, decane and octane are homologues. Physical properties in a homologous series, such as melting and boiling points or density, tend to increase smoothly with an increasing number of carbon atoms.

The difference in melting points and boiling points of hydrocarbons, including alkanes, result from *dispersion forces*. The larger the alkane molecules are, the more extensive these dispersion forces become. Hence hydrocarbons tend to become less volatile with increasing molar mass (∞ Section 11.2, 'Intermolecular Forces'). As a result, hydrocarbons of very low molecular weight, such as C_2H_6 (bp = −89 °C), are gases at room temperature; those of moderate molecular weight, such as C_6H_{14} (bp = 68 °C), are liquids; and those of high molecular weight, such as $C_{22}H_{46}$ (mp = 44 °C), are solids.

∞ Review this on page 381

CONCEPT CHECK 2

Rank the following in order of the strength of the London dispersion forces between two or more molecules of the same compound: $CH_3(CH_2)_4CH_3$ (hexane); CH_3CH_2OH (ethanol); $CH_3(CH_2)_{13}CH_2COOH$ (palmitic acid).

22.4 | STRUCTURES OF ALKANES

∞ Review this on page 308

According to *valence-bond theory*, a covalent bond is formed when a portion of an atomic orbital of one atom overlaps with a portion of an atomic orbital of another atom (∞ Section 9.4, 'Covalent Bonding and Orbital Overlap').

MY WORLD OF CHEMISTRY

PETROLEUM PRODUCTS

Petroleum, also called crude oil, is a complex mixture of organic compounds, mainly hydrocarbons, with smaller quantities of other organic compounds containing nitrogen, oxygen or sulfur. The usual first step in the refining of petroleum is to separate it on the basis of boiling point as shown in ▼ TABLE 22.4. Because petrol is the most commercially important of these separated mixtures, called *fractions*, various processes are used to maximise its yield.

Petrol is a blend of volatile hydrocarbons containing alkanes. In a car engine a mixture of air and petrol vapour is compressed by a piston and then ignited by a spark plug. The burning of petrol vapour should create a strong, smooth expansion of gas, forcing the piston outward and imparting force along the drive shaft of the engine. If the petrol vapour burns too rapidly, the piston receives a single hard slam rather than a strong, smooth push. The result is a 'knocking' or 'pinging' sound and a reduction in the efficiency with which energy produced by the combustion is converted to work.

The octane number of a petrol blend (▼ FIGURE 22.5) is a measure of its resistance to knocking. Petrols with a high octane number burn more smoothly and are thus more effective fuels. The octane number of unleaded petrol is obtained by comparing its knocking characteristics with those of 2,2,4-trimethylpentane (also known as isooctane) and heptane. 2,2,4-Trimethylpentane is assigned an octane number of 100, and heptane is assigned zero. Petrol with the same knocking characteristics as a mixture of 90% isooctane and 10% heptane has an octane rating of 90.

The petrol obtained directly from fractionation of petroleum contains mainly straight-chain hydrocarbons and has an octane number of around 50. It is therefore subjected to processes such as *cracking*, which convert the straight-chain alkanes into more desirable branched-chain and aromatic hydrocarbons (▼ FIGURE 22.6). Cracking is also used to convert some of the less volatile kerosene and fuel-oil fractions into compounds with lower molecular weights that are suitable for use as motor fuel. The octane rating of petrol is also increased by adding aromatic compounds such as toluene ($C_6H_5CH_3$) and oxygenated hydrocarbons such as ethanol (CH_3CH_2OH).

RELATED EXERCISES: 22.19, 22.20

▲ **FIGURE 22.5 Octane rating.** The octane rating of petrol measures its resistance to knocking when burned in an engine. The octane rating of this petroleum product is 89, as shown on the face of the pump.

▲ **FIGURE 22.6 Fractional distillation.** Petroleum is separated into fractions by distillation and subjected to catalytic cracking in a refinery.

	Carbon-chain	Boiling-point	
TABLE 22.4 • Hydrocarbon fractions from petroleum			
Fraction	**lengths**	**range (°C)**	**Uses**
Gas	C_1 to C_5	−160 to 30	Gaseous fuel, production of H_2
Petrol	C_5 to C_{12}	30 to 200	Motor fuel
Kerosene, fuel oil	C_{12} to C_{18}	180 to 400	Diesel fuel, furnace fuel, cracking
Lubricants	C_{16} and up	350 and up	Lubricants
Paraffins	C_{20} and up	Low melting solids	Candles
Asphalt	C_{36} and up	Gummy residues	Surfacing roads

∞ Review this on pages 199 and 309

Energy

These two electrons occupy different atomic orbitals

✓

Energy

These two electrons pair up, which is not favoured

✗

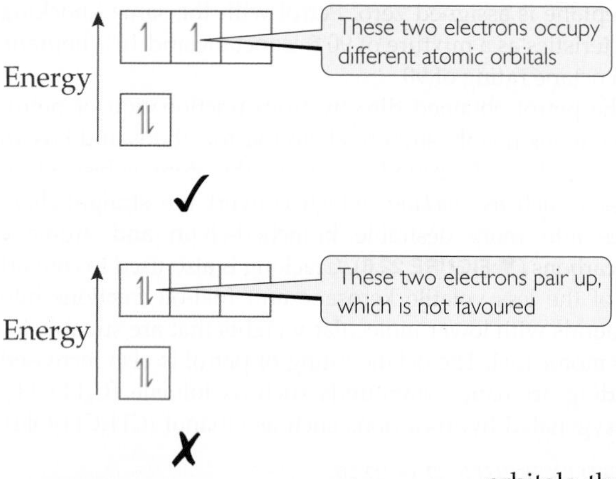

▲ **FIGURE 22.7** **Hund's rule** states that, when there is a choice of two or more degenerate atomic orbitals (of the same energy), electrons will occupy separate orbitals before pairing up in the same orbital.

Hence the bond angle is dictated in part by the atomic orbital of the atom in question. If, however, we were to predict the bond angles in a simple alkane based purely on the number and type of s and p orbitals and taking into account *Hund's rule* (◄ FIGURE 22.7), then angles of about 90° would be observed.

Figure 22.8 illustrates how p orbitals align at right angles to each other (∞ Section 6.6, 'Representations of Orbitals'). The spherical shape of the s orbital does not contribute to the direction the bond will take. But experimental evidence (for example, diffraction studies, such as those found in X-ray crystallography) indicates that the bond angles in simple straight-chain alkanes are approximately 109.5°, giving a tetrahedral geometry. Why is this so? To explain this, organic chemists use a model called **hybridisation** (∞ Section 9.5, 'Hybrid Orbitals').

Hybridisation

Recall that earlier in the text we described hybridisation as the process of mixing atomic orbitals in a mathematical operation. The newly formed orbitals, called *hybrid orbitals*, are based on the atomic orbitals that constitute them (▼ FIGURES 22.8 and ►22.9). Hybridisation and valence-bond theory are used predominantly by organic chemists to describe the bonding outcomes for carbon, irrespective of whether it forms one, two or three covalent bonds with another carbon atom. In essence, the hybrid orbitals are defined by the geometry of the bonds formed by carbon that influence the overall

🔺 **FIGURE IT OUT**

How many atomic orbitals contribute to form the four sp^3 hybrid orbitals?

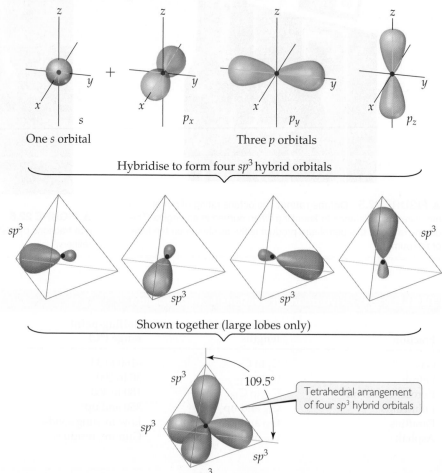

One s orbital

Three p orbitals

Hybridise to form four sp^3 hybrid orbitals

Shown together (large lobes only)

109.5°

Tetrahedral arrangement of four sp^3 hybrid orbitals

▶ **FIGURE 22.8** **Formation of sp^3 hybrid orbitals.** One s orbital and three p orbitals can hybridise to form four equivalent sp^3 hybrid orbitals. The large lobes of the hybrid orbitals point towards the corners of a tetrahedron.

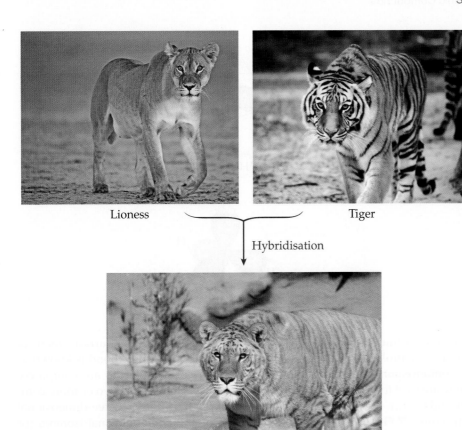

Lioness

Tiger

Hybridisation

Liger

◀ **FIGURE 22.9 Hybridisation.** Consider hybridisation in the following way. The mixing of a lion with a tiger leads to a hybrid, the liger. This hybrid has features of both the lion and the tiger but is neither one nor the other. The mixing process is called *hybridisation*.

shape and structure of the molecule. Any carbon bearing four single bonds must be tetrahedral and hence sp^3 hybridised.

In the case of sp^3 hybridisation of carbon, we envisage that the occupied 2s atomic orbitals 'mix' with all three 2p orbitals, two of which are occupied, to create four equivalent sp^3 *hybrid orbitals*. This hybridisation may occur as shown in ▶ **FIGURE 22.10**.

Each of the newly formed sp^3 hybrid orbitals contains one electron and forms large lobes that point towards a particular vertex of a *tetrahedron* (Figure 22.8), consistent with VSEPR theory (∞ Section 9.2, 'The VSEPR Model'). The tetrahedral geometry leads to bond angles of 109.5°, which is the largest possible separation for the four one-electron sp^3 hybrid orbitals in free space. By way of example, the three-dimensional structure of methane can be represented as shown in ▼ FIGURES 22.11 and 22.12.

Total energy of the four sp^3 hybrid orbitals is the same as the total energy of the 2s and 2p orbitals

Energy

$2p$

$2s$

sp^3

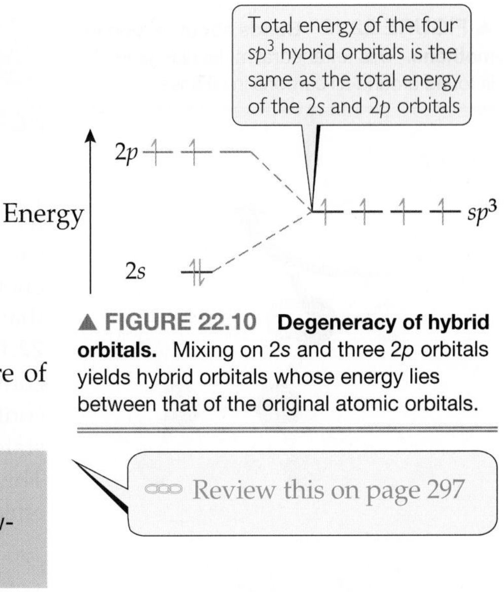

▲ **FIGURE 22.10 Degeneracy of hybrid orbitals.** Mixing on 2s and three 2p orbitals yields hybrid orbitals whose energy lies between that of the original atomic orbitals.

⚠ **CONCEPT CHECK 3**
Draw representations of the following orbitals: s, p_x, p_y, p_z. What do your drawings, in fact, represent? How do they relate to an sp^3 hybrid orbital?

A covalent bond that is formed by the end-to-end interaction of two one-electron orbitals (either atomic, hybridised or a mixture of both), as described in Figures 22.11 and 22.12, is called a sigma (σ) bond. As we will see in Chapter 24, carbon can also form π-bonds (∞ Section 24.1, 'The Structure of Unsaturated Hydrocarbons').

∞ Review this on page 297

∞ Find out more on page 954

Alkane Shape and Conformations

Although organic molecules are drawn in a static way, remember that they are always dynamic. One of the properties of alkanes is their ability to change

▶ **FIGURE 22.11 Hybridisation of atomic orbitals of carbon.** The interaction of the four sp^3 hybrid orbitals with four $1s$ orbitals of hydrogen yields methane. The hybrid-atomic orbital overlap causes two electrons to interact, leading to a single covalent bond, satisfying valence-bond theory.

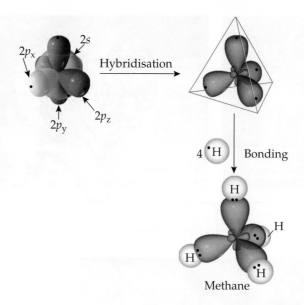

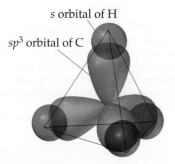

▲ **FIGURE 22.12 Bonds about carbon in methane.** This tetrahedral molecular geometry is found around all carbons in alkanes.

shape continuously (▼ **FIGURE 22.13**). The three-dimensional arrangement of atoms in a molecule that results from rotation about a single bond is known as a **conformation**. When different conformations of a molecule are compared, they are said to be **isomers** of the same structure. Isomers are two or more compounds with the same molecular formula but different three-dimensional structure. ▼ **FIGURE 22.14** illustrates some of the **conformational isomers** (or *conformers*) of pentane. The **Newman projection** shown in ▼ **FIGURE 22.15** is a useful way of viewing a molecule and for gaining an idea of its energy relative to other possible conformers. Newman projections are derived by looking along the carbon–carbon bond being rotated.

> ### CONCEPT CHECK 4
>
> Two major forms of carbon network exist, diamond and graphite. Which is the sp^3 hybridised form?

The number of conformational isomers within alkanes of higher complexity than methane is infinite because of free rotation about any of the carbon–carbon bonds. Each conformer, however, has a certain amount of energy associated with it. In this regard, there are two conformational isomers that are readily identifiable. They are called **staggered** and **eclipsed** (▶ **FIGURES 22.15** and **22.16**) and define the energetic limits of all conformations. Staggered conformations represent the lowest energy conformation, whereas eclipsed conformations are higher in energy. The difference in energy between the two states is due to added *non-bonded interaction* strain in the eclipsed conformation. This strain is derived from both **steric** (van der Waals interactions) and *torsional* effects, which are a resistance to twisting of the bonds. The magnitude depends

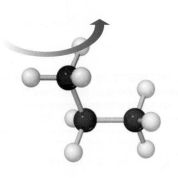

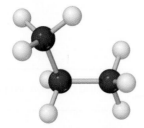

▲ **FIGURE 22.13 Rotation about a C–C bond occurs easily and rapidly in all alkanes.**

▲ **FIGURE 22.14 Conformers of pentane.** Rotation about carbon–carbon single bonds can result in noticeable changes in molecular shape.

on how close neighbouring atoms become and on the size of the two neighbouring groups. For example, hydrogen atoms on adjacent carbons, as in ethane, have minimal but still measurable strain (approximately 12.1 kJ mol^{-1}). With larger groups, such as bromo, iodo or methyl groups, the strain is larger.

The vast majority of bonds in acyclic aliphatic alkanes are found in a staggered conformation. The eclipsed conformation represents the activation barrier between two staggered conformations.

Let's now investigate the non-bonding strain effects for rotation about C2–C3 in butane (Figure 22.16). We will consider the staggered conformation of butane with the lowest energy as a useful starting point. This would be the case where the two large methyl (CH$_3$) groups that terminate the butane chain are as far apart as possible. Make sure at this point that you can identify that this is a Newman projection for butane. Rotation about C2–C3 by 60° yields the first eclipsed conformer. Non-bonding interactions between the methyl groups and adjacent hydrogen atoms in this eclipsed form strain the molecule and impart instability (not that the molecule is going to break apart or anything like that) giving a higher-energy conformer (15.8 kJ mol^{-1}). A further rotation of 60° relieves the strain somewhat but, in doing so, the two large methyl groups are able to interact weakly (worth approximately 3.7 kJ mol^{-1}), so that a *local* energy minimum is reached. Rotation of the C2–C3 bond by another 60° causes the two methyl groups to interact strongly, forming the conformer with the highest relative energy. Note that, although the hydrogen atoms also contribute to the overall energy, it is the methyl interactions that dominate in this case. Any further rotation from this point gives a structure that has already been discussed. Finally, the methyl groups are again opposed, resulting once more in the orientation that yields the minimum energy conformation (sometimes referred to as the *global* minimum).

CONCEPT CHECK 5

How many different staggered and eclipsed conformations exist for propane?

Staggered

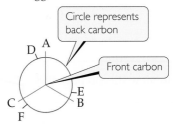

Circle represents back carbon

Front carbon

Eclipsed

▲ **FIGURE 22.15 Newman projection.** In a Newman projection representation, the two bonded carbon atoms about which rotation will occur are drawn as a vertex (front carbon) and a large circle (back carbon) from which all other bonds associated with those atoms radiate. The eclipsed conformation causes the groups to align one behind the other. By convention, eclipsed conformations are drawn slightly staggered so that the rear bonds can be seen separately from those in the front.

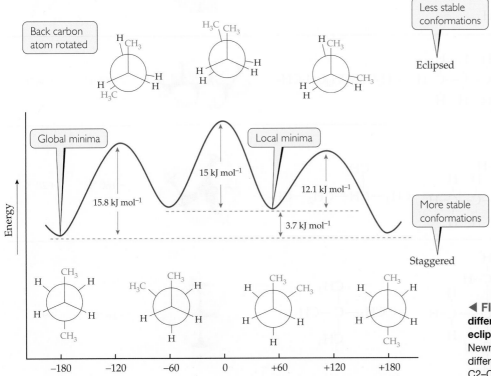

◀ **FIGURE 22.16 Relative energy differences between staggered and eclipsed conformations of butane.** Newman projections are used to illustrate the different conformers upon rotation about the C2–C3 bond.

Constitutional/Structural Isomers

The alkanes listed in Table 22.3 are *straight-chain aliphatic hydrocarbons* because all the carbon atoms are joined in a continuous chain. Alkanes consisting of four or more carbon atoms can also have branched chains and are called *branched-chain alkanes*. ▼ **TABLE 22.5** shows space-filling models, full structural formulae and condensed structural formulae for all the possible structures of alkanes containing four or five carbon atoms. Note that the molecular formula for a branched alkane *is the same* as for a straight-chain alkane with the same number of carbon atoms.

As we saw in Chapter 21, compounds with the same molecular formula but with different bonding arrangements (and hence different structures) are called **structural** or **constitutional isomers**) (∞ Section 21.4, 'Nomenclature and Isomerism in Coordination Chemistry').

∞ Review this on page 868

This is easily envisaged if you think of alkanes as a series of building blocks made up of C, CH, CH_2 and CH_3 groups. ▶ **FIGURE 22.17** demonstrates the principle.

TABLE 22.5 • Isomers of C_4H_{10} and C_5H_{12}

Systematic name (common name)	Structural formula	Condensed structural formula	Space-filling model	Melting point (°C)	Boiling point (°C)
Butane (*n*-butane)	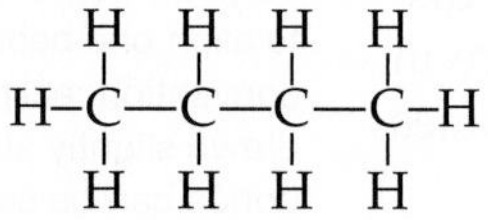	$CH_3CH_2CH_2CH_3$		−138 °C	−0.5 °C
2-Methylpropane (isobutane)	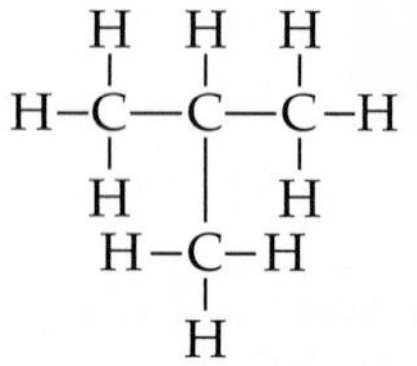	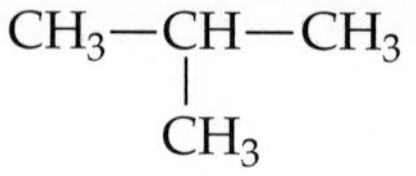 $CH_3-CH-CH_3$ CH_3		−159 °C	−12 °C
Pentane (*n*-pentane)	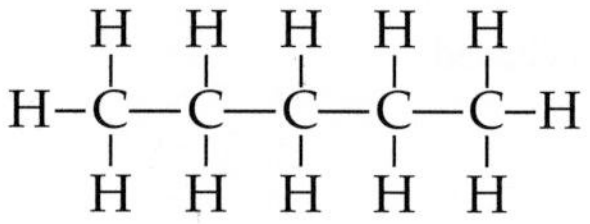	$CH_3CH_2CH_2CH_2CH_3$		−130 °C	+36 °C
2-Methylbutane (isopentane)	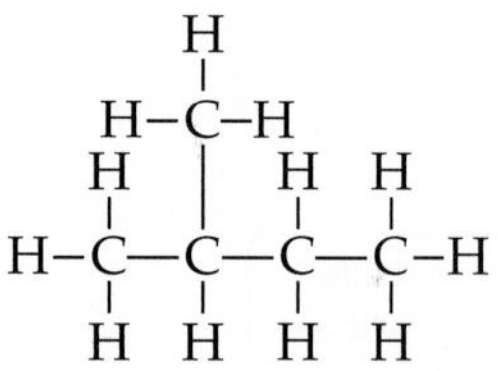	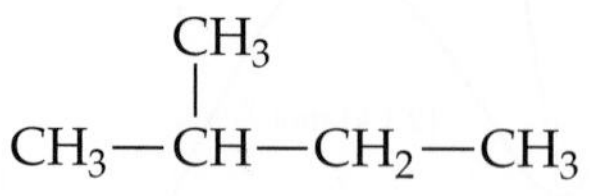 CH_3 $CH_3-CH-CH_2-CH_3$	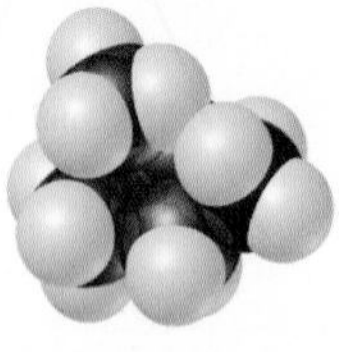	−160 °C	+28 °C
2,2-Dimethylpropane (neopentane)	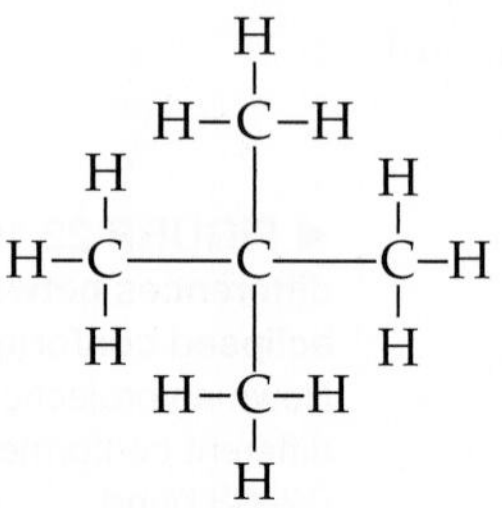	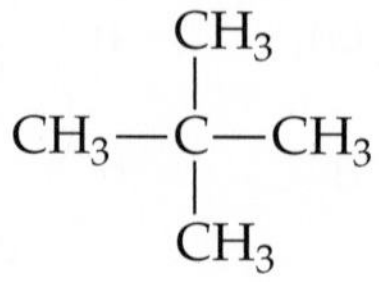 CH_3 CH_3-C-CH_3 CH_3	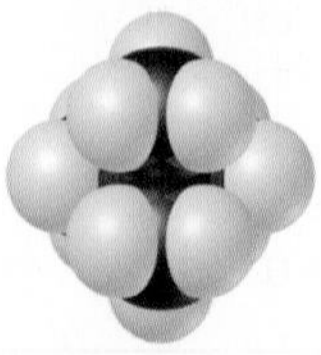	−16 °C	+9 °C

▲ **FIGURE 22.17** **Constitutional isomers of butane.** If we were to remove a terminal (end) carbon unit in butane and attach it to the middle remaining carbon, a structure different from the initial straight chain of butane would be obtained. The connectivity changes, but the molecular formula is exactly the same. There are still four carbons within the molecule and, if we were to complete the structure again, there would still be 10 hydrogens to include. The two structures differ not only in their shape but also in their physical and chemical properties.

Constitutional isomers of a given alkane differ slightly from one another in physical properties. Note the melting and boiling points of the constitutional isomers of butane and pentane given in Table 22.5. These melting and boiling points are a measure of the energy needed to overcome weak and favourable interactions that give rise to the solid or liquid properties, respectively. The variation in melting and boiling points for a set of constitutional isomers is a result of the subtleties of molecular interactions (∞ Section 11.2, 'Intermolecular Forces'). The number of possible structural isomers increases rapidly with the number of carbon atoms in the alkane. There are 18 possible constitutional isomers with the molecular formula C_8H_{18}, for example, and 75 possible isomers with the molecular formula $C_{10}H_{22}$.

∞ Review this on page 382

⚠ CONCEPT CHECK 6
How many constitutional isomers are possible for hexane? Draw them.

22.5 | ALKANE NOMENCLATURE

Constitutional isomerism complicates how we might describe organic molecules since a description based solely on molecular formulae is not nearly enough. Apart from ethane and propane, a number of different structures can be obtained from the same molecular formula. All of this means that we need a systematic means of naming organic compounds.

This need was recognised early in the history of organic chemistry. In 1892 the International Union of Chemistry met in Geneva, Switzerland, to formulate a set of rules for systematically naming organic substances. Since that time, the task of updating the rules for naming all compounds has fallen to the International Union of Pure and Applied Chemistry (IUPAC). Chemists everywhere, regardless of their nationality or political affiliation, subscribe to a common system for naming compounds. This naming process, called **nomenclature**, provides a systematic way to define unambiguously which atoms are bonded to one another within a molecule.

The IUPAC names for the isomers of butane and pentane are given in Table 22.5. These names, as well as those of other organic compounds, have three parts to them:

Prefix − Parent − Suffix

The following steps summarise the procedures used to arrive at the names of alkanes, all of which end with the suffix *-ane*. The prefix relates to the number, type and position of substituents (for example, the prefix 2,3-dibromo would indicate the compound contains two bromine atoms attached at positions 2 and 3). The parent name indicates the number of carbons within the longest chain (for example, *eth* = two carbons, *pent* = five carbons). We use a similar approach to writing the names of other organic compounds and we develop a further understanding of organic nomenclature in Chapters 23–29.

In order to understand the steps to derive an appropriate nomenclature, let's examine how we would name the following compound:

$$CH_2CH_2CH_2CHCH_3$$
$$| \qquad |$$
$$CH_2CH_3 \quad CH_3$$

1. *Find the longest continuous chain of carbon atoms and use the name of this chain (Table 22.3) as the base name of the compound.*

   Because this compound's longest chain is made up of seven C atoms (coloured), it is named a substituted heptane (see Table 22.3). Groups attached to the main chain are called *substituents* because they are in substitution of a hydrogen atom on the main chain.

2. *Number the carbon atoms in the longest chain, beginning with the end of the chain that is nearest to a substituent.*

   In our example, we number the C atoms from the right because that places the CH_3 substituent on the second C atom of the chain; if we number from the lower left, the CH_3 would be on the sixth C atom. The chain is numbered from the end that gives the lowest combined number for all substituent positions.

$$CH_2CH_2CH_2CHCH_3$$
$$|5 \quad 4 \quad 3 \quad |2 \quad 1$$
$$CH_2CH_3 \quad CH_3$$
$$6 \quad 7$$

3. *Name and give the location of each substituent group.*

   A substituent group that is formed by removing a hydrogen atom from an alkane is called an **alkyl group**. We have already been informally introduced to this nomenclature. Alkyl groups are named by replacing the *-ane* ending of the alkane name with *-yl*. ◀ TABLE 22.6 lists the three simplest alkyl groups.

   The name of our example is 2-methylheptane. Note the hyphen after the number. Note also that there is *no* space, comma or hyphen between methyl and heptane. The name should also allow you to draw the structure using these same three simple steps.

4. *Where two or more substituents are present, list them in alphabetical order, ignoring multiplying prefixes.*

   Where there are two or more of the same substituent, the number of substituents of that type is indicated by a prefix: *di-* (two), *tri-* (three), *tetra-* (four), *penta-* (five) and so on. The position of each substituent is separated by a comma (for example, 2,3-dibromopentane). In the case where substituents are different, each prefix is separated from the next by a hyphen. Substituents are also listed alphabetically (ethyl, methyl, propyl and so on). Notice how the following example is named (and numbered):

$$CH_2CH_3 \qquad\qquad\qquad CH_2CH_3$$
$$| \qquad\qquad\qquad\qquad\qquad |$$
$$CH_3{-}CHCHCH_2CHCH_3 \quad \text{is better than} \quad CH_3{-}CHCHCH_2CHCH_3$$
$$5| \quad 4 \quad 3 \quad |2 \; 1 \qquad\qquad\qquad 3| \quad 4 \quad 5 \quad |6 \; 7$$
$$CH_2CH_3 \quad CH_3 \qquad\qquad\qquad CH_2CH_3 \quad CH_3$$
$$6 \quad 7 \qquad\qquad\qquad\qquad\qquad 2 \quad 1$$

for numbering because *4-ethyl-2,5-dimethylheptane* gives a lower numbering than *4-ethyl-3,6-dimethylheptane* does.

Table 22.6 • Naming simple alkyl substituents

Substituent	Substituent name
CH_3	*methyl*
CH_3CH_2	*ethyl*
$CH_3CH_2CH_2$	*propyl*

SAMPLE EXERCISE 22.1 Naming alkanes

Give the systematic name for the following alkane

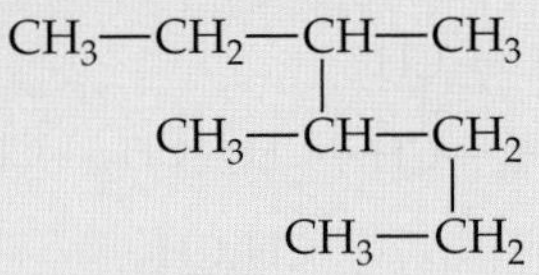

$$CH_3—CH_2—CH—CH_3$$
$$CH_3—CH—CH_2$$
$$CH_3—CH_2$$

SOLUTION

Analyse We are given the condensed structural formula of an alkane and asked to give its name.

Plan Because the hydrocarbon is an alkane, its name ends in *-ane*. The name of the parent hydrocarbon is based on the longest continuous chain of carbon atoms. Branches are alkyl groups, named after the number of C atoms in the branch and located by counting C atoms along the longest continuous chain.

Solve The longest continuous chain of C atoms extends from the upper left CH_3 group to the lower left CH_3 group and is seven C atoms long:

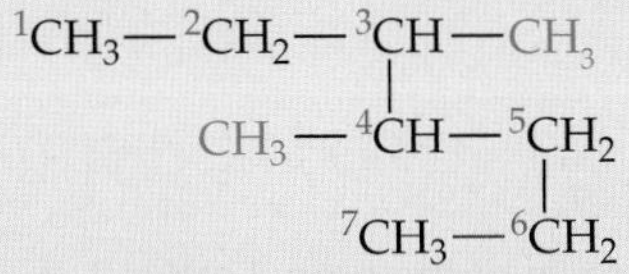

$$^1CH_3—^2CH_2—^3CH—CH_3$$
$$CH_3—^4CH—^5CH_2$$
$$^7CH_3—^6CH_2$$

The parent compound is thus heptane. There are two methyl groups branching off the main chain. Hence this compound is a dimethylheptane. To specify the location of the two methyl groups, we must number the C atoms from the end that gives the lower two numbers to the carbons bearing side chains. This means that we should start numbering at the upper left carbon. There is a methyl group on C3 and one on C4. The compound is thus 3,4-dimethylheptane.

PRACTICE EXERCISE

Name the following alkane:

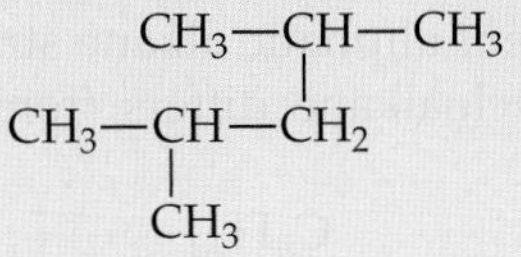

$$CH_3—CH—CH_3$$
$$CH_3—CH—CH_2$$
$$CH_3$$

Answer: 2,4-dimethylpentane
(See also Exercise 22.38.)

SAMPLE EXERCISE 22.2 Writing condensed structural formulae

Write the condensed structural formula for 3-ethyl-2-methylpentane.

SOLUTION

Analyse We are asked to draw the molecular structure of an organic compound from its systematic name.

Plan First we need to determine the classification of the organic compound, then work backwards from the parent compound, adding in functionality and side groups.

Solve Because the compound's name ends in *-ane*, it is an alkane, meaning that all the carbon–carbon bonds are single bonds. The parent hydrocarbon is pentane, indicating five C atoms (Table 22.3). There are two alkyl groups specified, an ethyl group (two carbon atoms, C_2H_5) and a methyl group (one carbon atom, CH_3). Counting from left to right along the five-carbon chain, the ethyl group will be attached to the third C atom and the methyl group will be attached to the second C atom.

We begin by writing a string of five C atoms attached to each other by single bonds. These represent the backbone of the parent pentane chain:

$$C—C—C—C—C$$

We next place a methyl group on the second C and an ethyl group on the third C atom of the chain. Hydrogens are then added to all the other C atoms to make the four bonds to each saturated carbon, giving the following condensed structure:

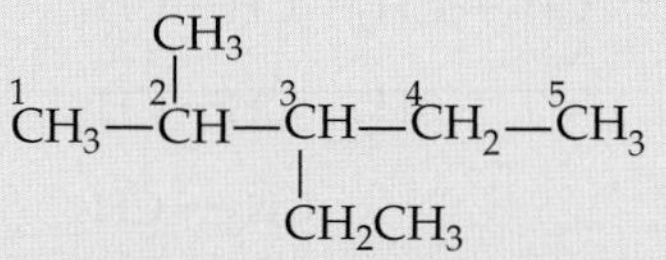

$$\overset{\displaystyle CH_3}{\underset{\displaystyle CH_2CH_3}{\overset{1}{CH_3}-\overset{2}{CH}-\overset{3}{CH}-\overset{4}{CH_2}-\overset{5}{CH_3}}}$$

The formula can be written even more concisely in the following style:

$$CH_3CH(CH_3)CH(CH_2CH_3)_2$$

In this formula the branching alkyl groups are indicated in parentheses.

PRACTICE EXERCISE

Write the condensed structural formula for 2,3-dimethylhexane.

Answer:

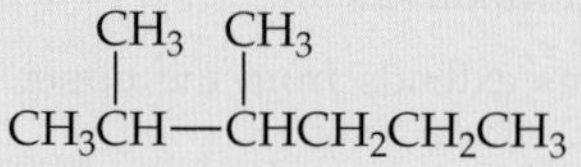

$$\overset{\displaystyle CH_3 \quad CH_3}{CH_3CH-CHCH_2CH_2CH_3} \quad \text{or} \quad CH_3CH(CH_3)CH(CH_3)CH_2CH_2CH_3$$

(See also Exercise 22.39.)

⚠ **CONCEPT CHECK 7**

What is the chemical formula of the propyl group?

22.6 | CYCLOALKANES

Hydrocarbons can form not only straight and branched chains but also rings, or cycles. Alkanes with this latter form of structure are called **cycloalkanes**. Cycloalkanes are formed by the loss of two C–H bonds and the formation of a C–C bond from the equivalent aliphatic alkane structure. As such, the general molecular formula for a cycloalkane differs from that of an alkane by two hydrogen atoms.

$$C_nH_{2n} \tag{22.2}$$

▶ **TABLE 22.7** illustrates a few simple cycloalkanes. Cycloalkanes containing from three to more than 30 carbon atoms are known. The most common forms contain five- (cyclopentane) and six- (cyclohexane) membered rings. Hydrocarbon rings containing fewer than five carbon atoms are strained because the C–C–C bond angle in these smaller rings must be much less than the 109.5° angle expected for an sp^3 hybridised atom. The amount of *ring strain* increases as the ring size gets smaller than that for cyclohexane. In cyclopropane, which has the shape of an equilateral triangle, the angle is only 60°; as a result, this molecule is much more reactive than propane, its straight-chain analogue.

Apart from model and structural formula representations, Table 22.7 also demonstrates how cycloalkane structures can be drawn as simple polygons in which each corner of the polygon represents a CH_2 group and each side represents a C–C single bond. The hydrogen atoms are still present in this *line drawing* representation—we just don't include them in the drawing. When using this representation for cycloalkanes, however, remember that you are drawing a three-dimensional structure in two dimensions and that, as a consequence, geometric subtleties are lost. In fact, cycloalkane rings with four or more carbons are not planar.

To illustrate this point, consider the structure of cyclohexane as shown in ▶ **FIGURE 22.18**. Looking at the structure from the top certainly gives the same hexagonal shape as represented by a line drawing. However, the view from any side tells a different and more complicated story. Two major conformations for

TABLE 22.7 • Cycloalkanes

Cycloalkane	Structural formula	Ball-and-stick representation	Line drawing
Cyclopropane	H_2C CH_2 H_2C		△
Cyclobutane	H_2C—CH_2 H_2C—CH_2		□
Cyclopentane	H_2C C H_2 H_2C CH_2 C H_2		⬠
Cyclohexane	H_2C C H_2 CH_2 H_2C CH_2 C H_2		⬡

▲ FIGURE IT OUT

In which conformation of cyclohexane are neighbouring hydrogens staggered and in which are they eclipsed?

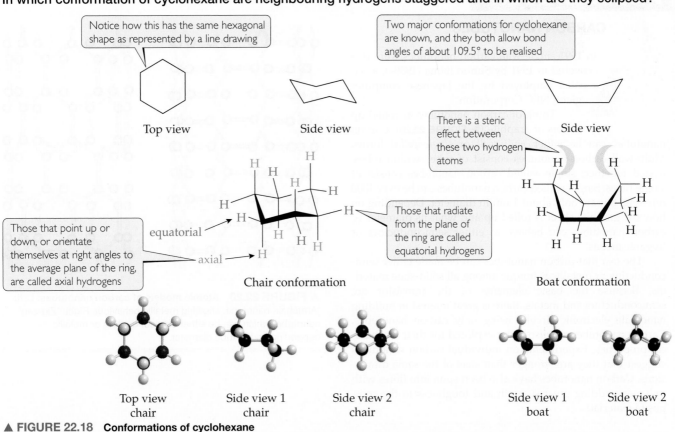

Notice how this has the same hexagonal shape as represented by a line drawing

Two major conformations for cyclohexane are known, and they both allow bond angles of about 109.5° to be realised

Top view

Side view

There is a steric effect between these two hydrogen atoms

Side view

Those that radiate from the plane of the ring are called equatorial hydrogens

equatorial

axial

Those that point up or down, or orientate themselves at right angles to the average plane of the ring, are called axial hydrogens

Chair conformation

Boat conformation

Top view chair

Side view 1 chair

Side view 2 chair

Side view 1 boat

Side view 2 boat

▲ FIGURE 22.18 **Conformations of cyclohexane**

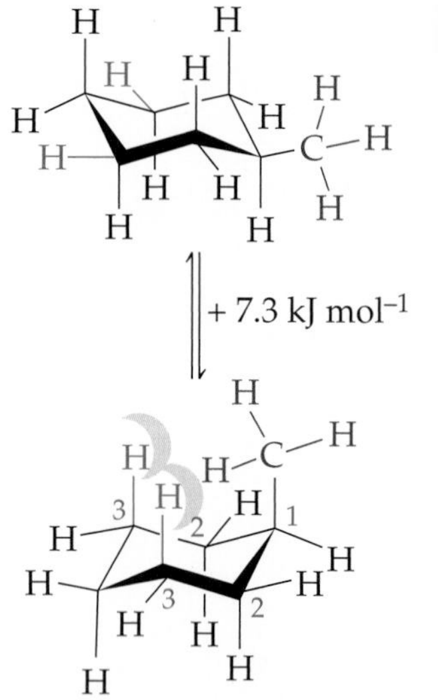

$+ 7.3 \text{ kJ mol}^{-1}$

▲ **FIGURE 22.19** **1,3-Diaxial repulsion.** Substituents placed in an axial conformation interact strongly with the two axial hydrogens (red, right side), destabilising this conformer relative to the equivalent equatorial conformer.

cyclohexane are known, and they both allow bond angles of about 109.5° to be realised. The **chair conformation** is the more stable of the two types of conformers. It is called a chair structure because of its resemblance to a chair. In this case, the hydrogen atoms adopt one of two orientations. Those that point up or down, or orientate themselves at right angles to the average plane of the ring, are called **axial hydrogens**; those that radiate from the plane of the ring are called **equatorial hydrogens**. Substituents at these positions are also described in the same way. Note that all bonds on adjacent carbons in cyclohexane are staggered, implying that in any given hemisphere the hydrogens alternate between axial and equatorial.

The less stable of the two forms (by about 30 kJ mol⁻¹) is described as the **boat conformation**. Its relative instability is caused by the non-bonding interactions generated by the close proximity of the two hydrogen atoms on C1 and C4 that overhang the ring system in this conformation (Figure 22.18), and by the eclipsing of hydrogens along the C2–C3 and C5–C6 bonds.

Now let's consider the effect of placing a single substituent on a cyclohexane ring. The substituent has two possible orientations, axial or equatorial. The two conformers do interconvert but the equatorial conformation is preferred because the methyl group experiences destabilising **1,3-diaxial interactions** with the other two axial hydrogens on that face (◄ **FIGURE 22.19**) if it is axial. These types of interactions are minimised in the equatorial conformation as the methyl group and, more importantly, the hydrogen atoms on that methyl group point away from the ring system.

We do not need to number the position of the substituent for a cycloalkane with only one substituent, nor do we need to define the conformation as either *ax* (for axial) or *eq* (for equatorial). For example, both structures in Figure 22.19 are called methylcyclohexane, as this name is suitable enough to avoid ambiguity.

A CLOSER LOOK

CARBON NANOTUBES

In 1991 single-wall carbon nanotubes were discovered in 1991 by Sumio Iijima (1939–), a scientist employed by the Japanese computer giant, NEC Corporation.

Think of carbon nanotubes as rolled up sheets of graphite (▶ **FIGURE 22.20**). Carbon nanotubes can be either *multi-wall* or *single-walled* forms. Multi-wall carbon nanotubes consist of tubes within tubes, nested together; single-walled carbon nanotubes consist of single tubes. Single-walled carbon nanotubes can be over 1000 nm long, but are only about 1 nm in diameter. Depending on how the graphite sheet is rolled up and what its diameter is, carbon nanotubes can behave as either semiconductors or 'organic metals'.

The fact that carbon nanotubes can be made either semiconducting or metallic is unique among all solid-state materials. Because the basic elements of the transistor are semiconductors and metals, there is great interest in building nanoscale electronic circuits using only carbon nanotubes. Carbon nanotubes are also being explored for their mechanical properties. Experiments on individual carbon nanotubes suggest that they are stronger than steel of the same dimensions. Carbon nanotubes have also been spun into fibres with polymers, adding great strength and toughness to the composite material.

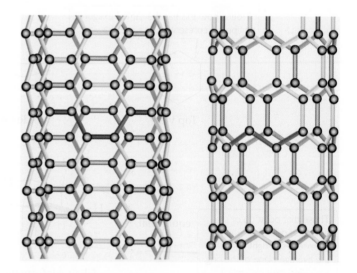

▲ **FIGURE 22.20** **Atomic models of carbon nanotubes.** Left: 'Armchair' nanotube, showing metallic behaviour. Right: 'Zig-zag' nanotube, which can be either semiconducting or metallic, depending on the tube diameter.

The concept of geometrical isomerism was introduced in Section 21.4 for coordination compounds. We will investigate the concept of *geometrical isomerism* in cycloalkanes in Chapter 23. Geometrical isomers occur in cycloalkanes when two or more substituents are found around the ring in axial, equatorial or both positions. Geometrical isomers differ in their physical properties and spatial orientation (∞∞ Section 23.2, '*Cis–Trans* Isomerism in Cycloalkanes').

∞∞ Find out more on page 929

> ⚠️ **CONCEPT CHECK 8**
>
> Why is it unnecessary to name the two conformers of methylcyclohexane as two distinct molecules?

22.7 | ORGANIC FUNCTIONAL GROUPS

The reactivity of organic compounds can be attributed to particular atoms or groups of atoms within the molecule. A site of reactivity in an organic molecule is called a **functional group** because it characteristically controls how the molecule behaves, or functions under a given set of conditions. Functional groups define both the reactivity and class descriptions of organic molecules. As we will see in (∞∞ Section 24.4, 'Electrophilic Addition and Substitution Reactions'), the presence of carbon–carbon double or triple bonds in a hydrocarbon will markedly increase that compound's reactivity. Furthermore, these two functional groups undergo similar yet characteristic reactions. The same is true of other functional groups. Each distinct kind of functional group often undergoes the same kinds of reactions, regardless of the size and complexity of the molecule. Thus the chemistry of an organic compound is largely determined by the functional groups it contains.

∞∞ Find out more on page 966

▼ **TABLE 22.8** lists the most common functional groups and gives examples of each. Notice that, in addition to C–C double bonds or C–C triple bonds, there are also many functional groups that contain elements other than just C and H. In fact, many of the functional groups found within organic molecules contain other non-metals such as O and N. In later chapters we examine the structure and chemical properties of important functional groups containing oxygen, that is, alcohols and ethers (Chapter 25), aldehydes and ketones (Chapter 26), carboxylic acids and esters (Chapter 27) and those containing nitrogen, such as amines and amides (Chapter 29).

Alkane groups constitute the less reactive portions of any compound and so are typically abbreviated to the symbol 'R'. In describing the general features of organic compounds, chemists often use the designation R to represent any structural group: methyl, ethyl, propyl and so on (▶ **FIGURE 22.21**); or, in fact, to represent a collective of atoms that do not participate in the reaction of interest. Alkanes, for example, which contain no functional group, are represented as R–H. **Alcohols**, which contain –OH, the alcohol functional group, are represented as R–OH. If two or more different structural groups are present in a molecule, we designate them R, R′, R″ and so forth.

In organic chemistry, a **heteroatom** is defined as an atom other than hydrogen or carbon. Of the different types of heteroatoms associated with a hydrocarbon skeleton, oxygen is the most prevalent. For oxygen to be added to a hydrocarbon, oxidation must occur. Because our atmosphere contains approximately 21% oxygen, oxidation reactions are always happening, and not only in the process of rusting. Carbon dioxide, water and most free radicals contain oxygen. Simple

$$CH_3CH_2CH_2OH \text{ equals } R\!-\!OH \text{ when } R = CH_3CH_2CH_2$$

$$CH_3OH \text{ equals } R'\!-\!OH \text{ when } R' = CH_3$$

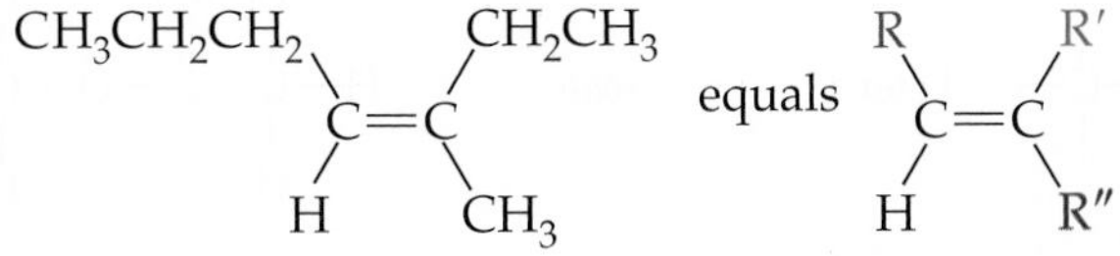

$$\underset{H}{\overset{CH_3CH_2CH_2}{\diagdown}}C\!=\!C\underset{CH_3}{\overset{CH_2CH_3}{\diagup}} \quad \text{equals} \quad \underset{H}{\overset{R}{\diagdown}}C\!=\!C\underset{R''}{\overset{R'}{\diagup}}$$

$$\text{where} \quad R = CH_3CH_2CH_3$$
$$R' = CH_2CH_3$$
$$R'' = CH_3$$

▲ **FIGURE 22.21** The generalisations of organic chemistry.

TABLE 22.8 • Common functional groups

Functional group	Compound type	Suffix or prefix	Example Structural formula	Example Ball-and-stick model	Systematic name (Common name)
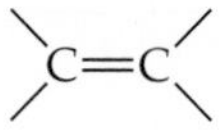 >C=C<	Alkene	-ene	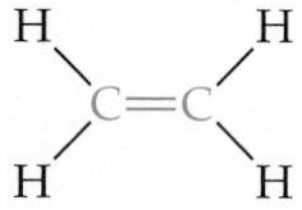 H₂C=CH₂ (H C=C H / H H)	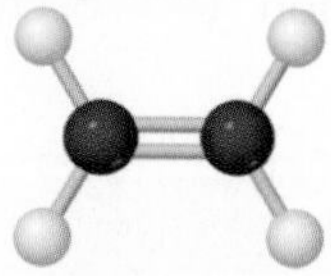	Ethene (Ethylene)
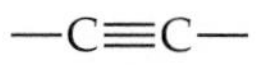 —C≡C—	Alkyne	-yne	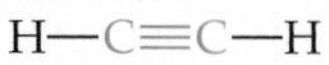 H—C≡C—H		Ethyne (Acetylene)
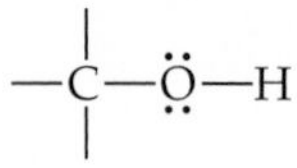 —C—Ö—H	Alcohol	-ol	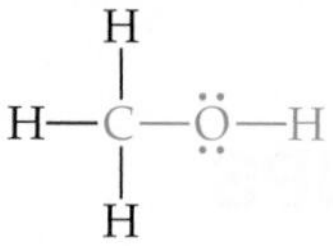 H—C—Ö—H (with H above and below)	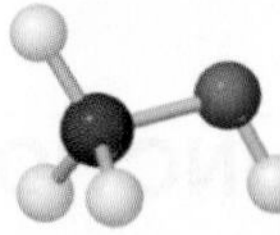	Methanol (Methyl alcohol)
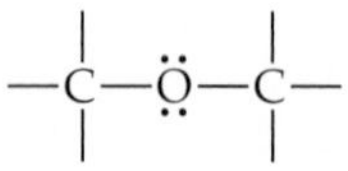 —C—Ö—C—	Ether	ether	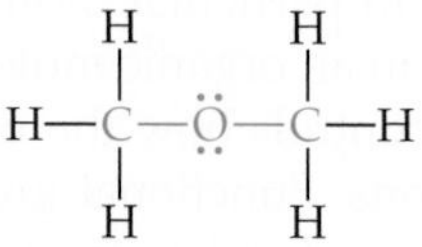 H—C—Ö—C—H	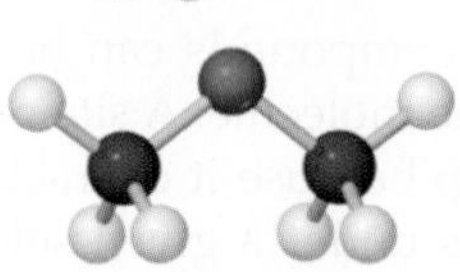	Dimethyl ether
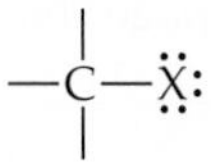 —C—Ẍ: (X = halogen)	Haloalkane	halo-	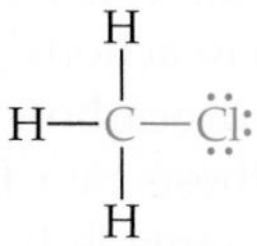 H—C—C̈l:	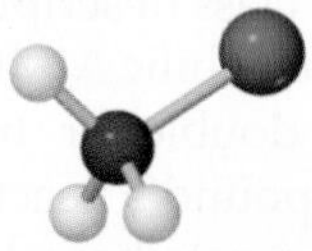	Chloromethane (Methyl chloride)
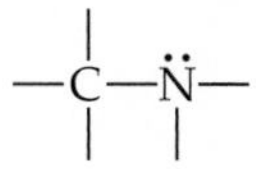 —C—N—	Amine	-amine	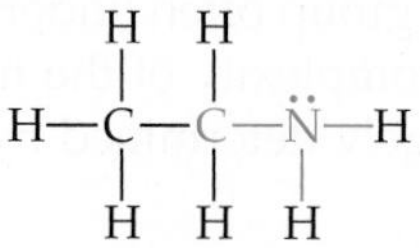 H—C—C—N̈—H	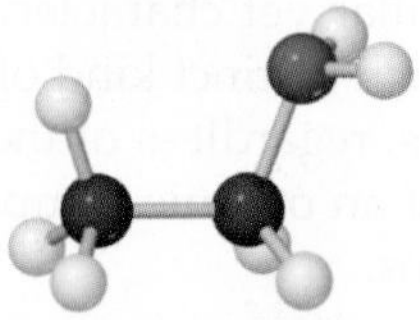	Ethylamine
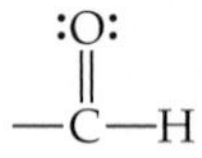 :O: —C—H	Aldehyde	-al	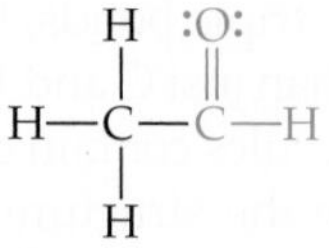 H—C—C(=O)—H	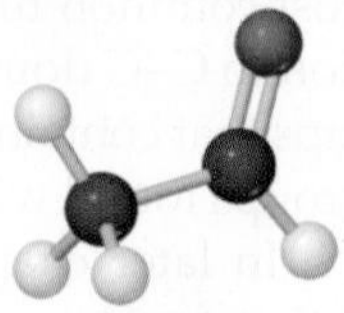	Ethanal (Acetaldehyde)
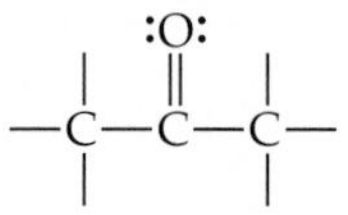 :O: —C—C—C—	Ketone	-one	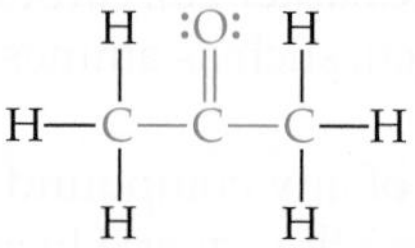 H—C—C(=O)—C—H	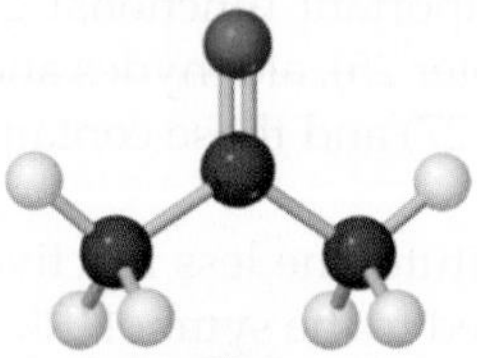	Propanone (Acetone)
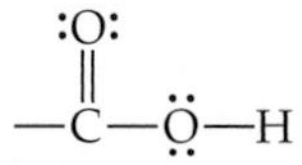 :O: —C—Ö—H	Carboxylic acid	-oic acid	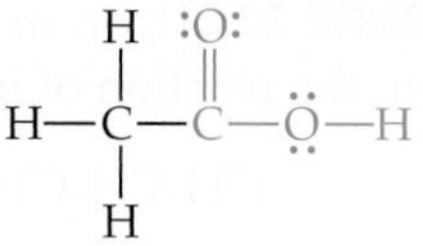 H—C—C(=O)—Ö—H	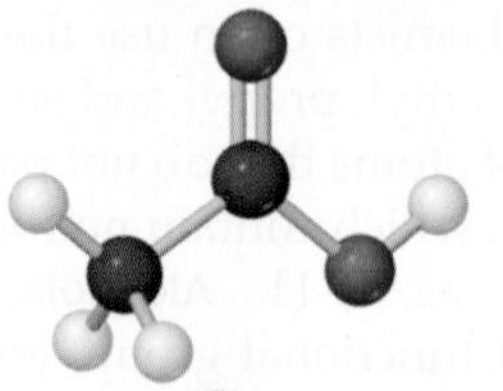	Ethanoic acid (Acetic acid)
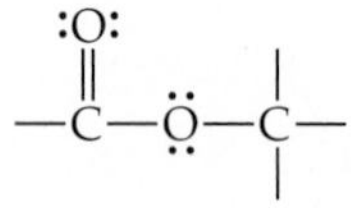 :O: —C—Ö—C—	Ester	-oate	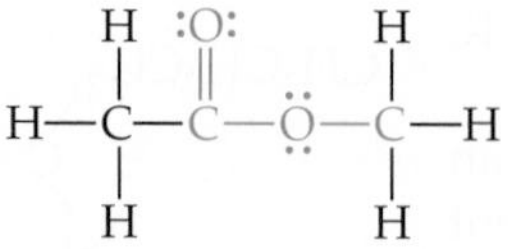 H—C—C(=O)—Ö—C—H	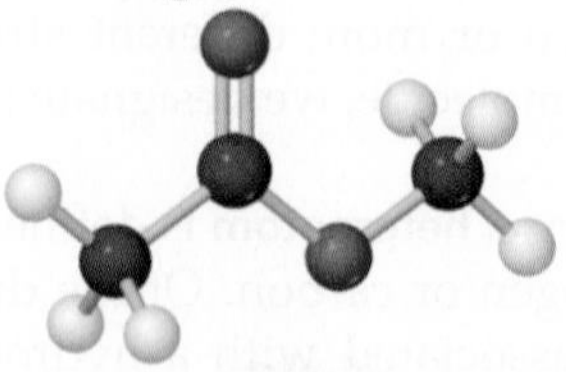	Methyl ethanoate (Methyl acetate)
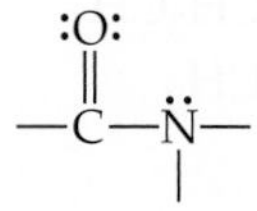 :O: —C—N̈—	Amide	-amide	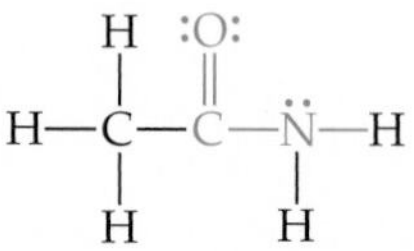 H—C—C(=O)—N̈—H	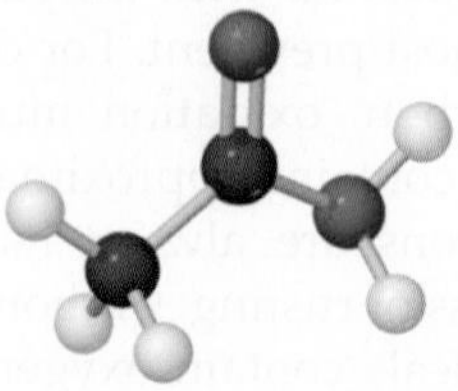	Ethanamide (Acetamide)

MY WORLD OF CHEMISTRY

STRUCTURE–ACTIVITY RELATIONSHIPS

Each of the following molecules has a similar structural element highlighted in bold. This suggests they have similar physiological responses since many enzymes and receptors recognise a molecule's shape—like a simple lock and key.

Morphine is an alkaloid derived from the opium poppy. It is effective at blocking the perception of pain by the brain while allowing the normal function of the nervous system. It is also able to produce a feeling of eupho-ria. This is also the case with heroin, which is devastatingly addictive. Heroin users become physically dependent on opioids, which means they have to continue to take them to avoid withdrawal symptoms such as chills, sweating, stiffness, cramps and anxiety. Methadone is used to treat heroin addiction. Because of its shape, methadone binds to opioid receptors and causes an opioid response. It is prescribed as a way of regulating, and ultimately reducing, heroin addiction. Naloxone is used to treat heroin overdose. It binds to the opioid receptors in the brain, displacing heroin and reversing the effects of the narcotic.

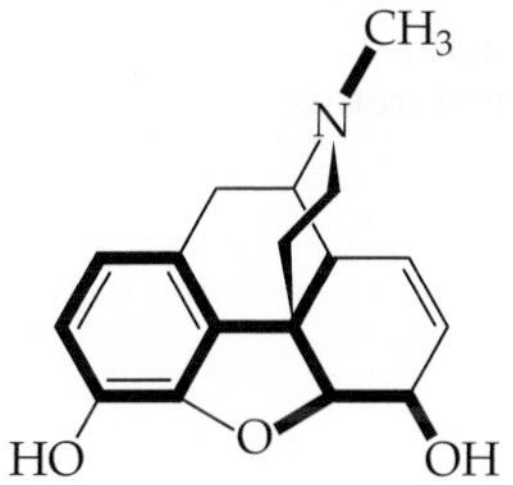

Morphine 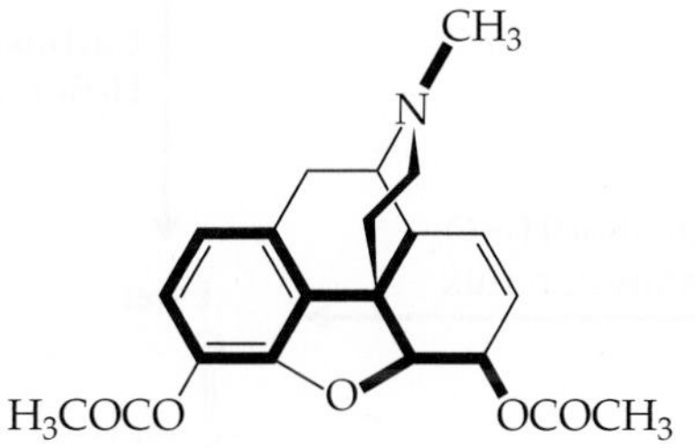   Heroin 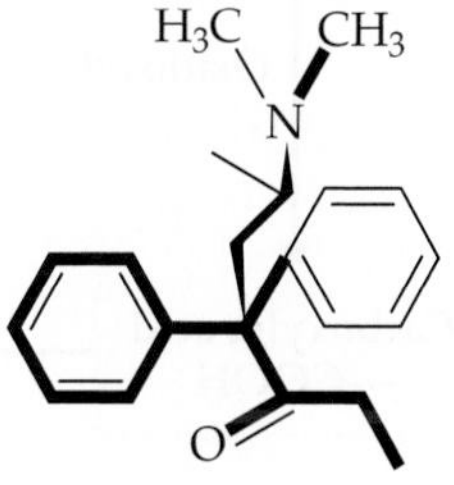   Methadone 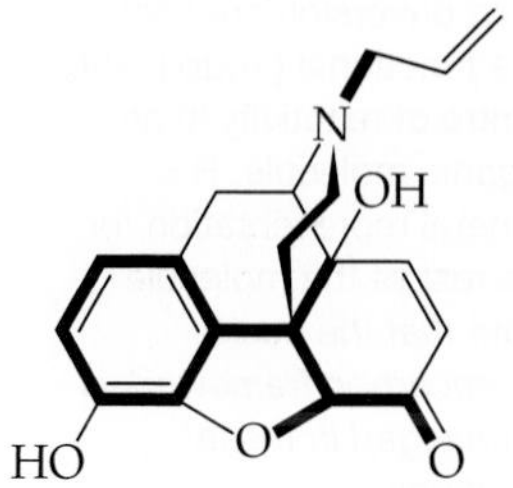   Naloxone

molecules containing oxygen, such as alcohols, are easily transformed into alkenes, haloalkanes, ethers, aldehydes and ketones and carboxylic acids and their derivatives. For the rest of this chapter we will investigate some of the common reactions used to functionalise alkanes.

22.8 | REACTIONS OF ALKANES

Alkanes contain only C–C and C–H bonds and as a result are relatively unreactive. At room temperature, for example, alkanes do not react with any common acids, bases or strong oxidising agents, and they are not even attacked by boiling nitric acid. Their low chemical reactivity is due primarily to the strength and lack of polarity of both C–C and C–H single bonds, and should be expected based on their important structural role. The whole basis of organic chemistry—that is, converting one functional group into another by making or breaking bonds, leading to the formation of new compounds (▼ FIGURE 22.22)—would fall apart if C–H and C–C bonds reacted under mild conditions. Functional groups define the reactivity and classes of organic molecules.

The following chapters show that hydrocarbons can be modified to impart greater reactivity by introducing unsaturation into the carbon–carbon framework and by attaching other reactive functional groups to the hydrocarbon backbone.

Combustion

Alkanes are not completely inert, however. Their relative high stability means that either harsh reaction conditions or the use of highly reactive reagents is needed to initiate a chemical reaction. One of the most commercially important reactions of alkanes is their **combustion** reaction in air, which is the basis of their use as fuels (∞ Section 3.2, 'Some Simple Patterns of Chemical Reactivity'). The complete combustion of an alkane in the presence of oxygen *always* generates carbon dioxide and water as the products of the reaction. For example, the balanced equation for the complete combustion of ethane is:

∞ Review this on page 74

$$2\,C_2H_6(g) + 7\,O_2(g) \rightarrow 4\,CO_2(g) + 6\,H_2O(l) \quad \Delta H = -2855\ \text{kJ} \quad [22.3]$$

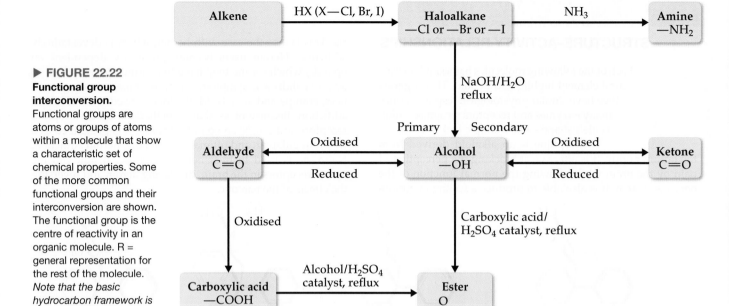

▶ **FIGURE 22.22**
Functional group interconversion.
Functional groups are atoms or groups of atoms within a molecule that show a characteristic set of chemical properties. Some of the more common functional groups and their interconversion are shown. The functional group is the centre of reactivity in an organic molecule. R = general representation for the rest of the molecule. *Note that the basic hydrocarbon framework is unchanged in these reactions.*

▲ **FIGURE 22.23** **Combustion.** Ignition of a mixture of oxygen with flammable gases such as hydrogen or methane can have explosive consequences.

∞ Review this on pages 118 and 514

The reaction of any organic molecule with oxygen is another example of an **oxidation** reaction (∞ Section 4.4, 'Oxidation–Reduction Reactions'). We investigate these types of reactions in the context of organic chemistry in greater detail in Chapter 26.

We know from calorimetric measurements that the reaction of a simple alkane such as methane (CH_4) with oxygen is highly exothermic (∞ Section 14.4, 'Enthalpies of Reaction'). Indeed, the combustion of methane (which is the major component of natural gas) keeps many of our homes warm during the winter months. Although the reactions of alkanes with oxygen are exothermic, most alkanes are stable indefinitely at room temperature in the presence of air because the activation energy required for combustion is large. This activation barrier is usually overcome by heat generated from a flame or spark, typically with explosive consequences (◀ **FIGURE 22.23**).

> ⚠ **CONCEPT CHECK 9**
> What products might you expect to get from the *incomplete* combustion of propane?

SAMPLE EXERCISE 22.3 **Oxidation of alkanes**

Write the equation for the complete combustion of 2-methylpropane.

SOLUTION

Analyse We are asked to react an alkane with oxygen under conditions that give the complete combustion products, namely CO_2 and H_2O.

Plan Write the molecular formula for 2-methylpropane and the reagent molecular oxygen on the reactants side of the equation and carbon dioxide and water on the products side. Then balance the equation for carbon, then hydrogen and finally oxygen.

Solve The molecular formula for 2-methylpropane follows the general formula for alkanes. 2-methylpropane has four carbons so its molecular formula is C_4H_{10}. So the unbalanced equation is

$$C_4H_{10}(g) + O_2(g) \rightarrow CO_2(g) + H_2O(l)$$

Balancing for carbon and hydrogen yields

$$C_4H_{10}(g) + O_2(g) \rightarrow 4\,CO_2(g) + 5\,H_2O(l)$$

Now, balancing for oxygen using the total of oxygen present on the product side yields the balanced equation

$$2\,C_4H_{10}(g) + 13\,O_2(g) \rightarrow 8\,CO_2(g) + 10\,H_2O(l)$$

PRACTICE EXERCISE

Write the equation for the complete combustion of 2,3-dimethylbutane.

Answer: $2\,C_6H_{14}(g) + 19\,O_2(g) \rightarrow 12\,CO_2(g) + 14\,H_2O(l)$

(See also Exercises 22.63, 22.64.)

Classification of C and H

An important skill to develop in organic chemistry is the ability to recognise which functional group within a molecule will react under a given set of conditions. This ability allows the organic chemist to synthesise very complex molecules. Sometimes, however, there is a choice of positions within a molecule at which reaction is likely to occur. This is as valid with simple organic molecules, like alkanes, as it is for highly complex molecular structures. Before going on to discuss one of the few important reactions of the alkanes, we need to have a way of distinguishing between the different kinds of carbons that appear. These classifications are important whenever we are trying to recognise which functional group or part of a molecule will react.

The classification used to distinguish between types of hydrogens and carbons and the appropriate symbolism is:

Primary (1°) carbon: a carbon atom bonded to one other carbon atom
1° H: a hydrogen bonded to a 1° carbon
Secondary (2°) carbon: a carbon atom bonded to two other carbon atoms
2° H: a hydrogen atom bonded to a 2° carbon
Tertiary (3°) carbon: a carbon bonded to three other carbon atoms
3° H: a hydrogen atom bonded to a 3° carbon
Quaternary (4°) carbon: a carbon bonded to four other carbon atoms

▶ **FIGURE 22.24** illustrates this classification by identifying the different carbon atoms of 2,2,3-trimethylpentane. This method of classification is useful, as we see in the following chapters, for determining the reactivity of a range of different functional groups.

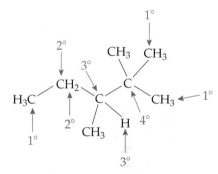

▲ **FIGURE 22.24   Classification of carbon and hydrogen atom types.** The classification of a selected set of hydrogen (green) and carbon (red) atoms is shown for 2,2,3-trimethylpentane.

 CONCEPT CHECK 10

How many primary, secondary and tertiary hydrogens are there in 2-methylbutane?

Free-Radical Reactions and Electron Movement

The majority of alkane reactions occur through the formation of **free radicals**, either by the scission of a C–C bond or by hydrogen abstraction. Free radicals gain their reactivity from having an unpaired electron. Both C–C bond scission and hydrogen abstraction are examples of a **homolytic bond cleavage** because one electron contained with the two-electron σ-bond is transferred to each atom, yielding two free radicals.

At this point, we will use one form of free-radical chemistry to emphasise the point that a single covalent bond contains two electrons and that reactions occur by making and breaking such bonds.

The most common of all the free-radical reactions of alkanes is the **halogenation**. This term simply means the incorporation of halogen (that is, F, Cl, Br, I), although chlorine and bromine atoms are the easiest and hence the most common to incorporate. This type of free-radical halogenation is generally called a **substitution reaction** because the reaction proceeds by abstracting a hydrogen atom and replacing or *substituting* it with a halogen. The chlorination of methane, for example, to yield the mono-, di-, tri- and tetra-chlorinated products (and HCl) occurs only in the presence of light (photolysis) with radiation in the range 200–500 nm most common. The reaction produces all permutations of substitution, as shown in Equation 22.5. The ratio of the four products formed is dictated by the amount of chlorine gas present, the photolysis time and the wavelength of light.

$$CH_4 + Cl_2 \xrightarrow{\text{dark}} \text{no reaction} \qquad [22.4]$$

$$CH_4 + Cl_2 \xrightarrow{\text{light}} CH_3Cl + CH_2Cl_2 + CHCl_3 + CCl_4 \qquad [22.5]$$

Electron movement in organic chemistry is a useful way of rationalising the conversion of starting materials to products. We call this the *mechanism* of a reaction. Remember that there are two electrons in any single covalent bond and how these electrons move gives rise to either a carbocation, carbanion or free radical, which are less stable (and hence more reactive) than the covalent bond from which they are formed. In Chapter 24, we will investigate a shorthand way of illustrating electron movement that simplifies the fundamental approach we are about to take to explain the free-radical halogenation of methane (∞ Section 24.3, 'Arrow Notation and Resonance Structures: Electron Counting').

∞ Find out more on page 963

Here, we will begin our study of mechanisms and the three steps to a radical chain reaction:

1. **Initiation**
2. **Propagation**
3. **Termination**

using Equation 22.5 as an example. We will employ Lewis structures to demonstrate the electron movement.

1. *Initiation step:* This step produces the first radicals. In this example, the absorption of light by Cl_2 causes the homolytic cleavage of the Cl–Cl bond, leading to the formation of two chlorine radicals:

$$:\!\ddot{C}l\!:\!\ddot{C}l\!: \xrightarrow{\text{light}} 2\,:\!\ddot{C}l\!\cdot \qquad [22.6]$$

The generation of chlorine or bromine radicals can also be initiated by heat.

2. *Propagation steps:* As these newly formed radicals collide with other molecules (which are not radicals), they generate or *propagate* new radicals. In the example, a newly formed chlorine radical (chlorine atom) can remove a hydrogen atom from a methane molecule. This step produces a stable product molecule, HCl, and an unstable methyl radical that is highly reactive because its carbon has only seven valence electrons.

Step 1

$$:\!\ddot{C}l\!\cdot\; +\; H\!:\!\underset{\underset{\textstyle H}{|}}{\overset{\overset{\textstyle H}{|}}{C}}\!:\!H \longrightarrow :\!\ddot{C}l\!:\!H \;+\; \cdot\underset{\underset{\textstyle H}{|}}{\overset{\overset{\textstyle H}{|}}{C}}\!:\!H \qquad [22.7]$$

When a molecule of Cl_2 encounters the methyl radical, one chlorine atom is transferred to the carbon to form a stable molecule of chloromethane. The second chlorine atom, now formed, can react with a new molecule of methane as in step 1.

Step 2

$$\cdot\underset{\underset{\textstyle H}{|}}{\overset{\overset{\textstyle H}{|}}{C}}\!:\!H \;+\; :\!\ddot{C}l\!:\!\ddot{C}l\!: \longrightarrow :\!\ddot{C}l\!:\!\underset{\underset{\textstyle H}{|}}{\overset{\overset{\textstyle H}{|}}{C}}\!:\!H \;+\; :\!\ddot{C}l\!\cdot \qquad [22.8]$$

The reaction of the methyl radical with chlorine (step 2) yields another chlorine radical which can propagate causing the *chain reaction*. Since each step produces a molecule of product and a newly formed radical, one radical gives rise to many product molecules. This means that only a small number of radicals need be generated during the initiation to allow the chain reaction to proceed.

Once the amount of CH_3Cl begins to build up in the mixture, it too can have a hydrogen atom abstracted by a chlorine radical, leading to the formation of CH_2Cl_2. Chloroform ($CHCl_3$) and CCl_4 are formed by the same process.

3. **Termination steps:** These occur when two radicals combine to form an unreactive product. For example:

$$H_3C\!\cdot + \cdot CH_3 \longrightarrow H_3C\!:\!CH_3 \qquad [22.9]$$

$$\text{or} \quad H_3C\!\cdot + :\!\ddot{C}l\cdot \longrightarrow :\!\ddot{C}l\!:\!CH_3 \qquad [22.10]$$

Termination steps such as those displayed in Equation 22.9 complicate the reaction by extending the carbon chain. Products such as those in Equations 22.9 and 22.10, which also contain C–H bonds, are also prone to further radical generation and substitution steps. The generation of a single radical can lead to the formation of many product molecules before the chain is terminated.

A CLOSER LOOK

REACTIVITY BY CARBON CLASSIFICATION

When propane is mono-halogenated two products are likely since both primary and secondary hydrogens exist (Equation 22.11).

$$
\begin{array}{l}
\quad\quad\quad \dot{C}H_2CH_2CH_3 \xrightarrow{\ Cl^{\bullet}\ } ClCH_2CH_2CH_3 \\
\quad\quad\quad\quad\quad\quad\quad\quad\quad\quad \text{1-chloropropane} \\[4pt]
CH_3CH_2CH_3 \\
\quad\quad\quad\quad\quad\quad\quad\quad\quad\quad\quad\quad\quad [22.11] \\[6pt]
\quad\quad\quad CH_3\dot{C}HCH_3 \xrightarrow{\ Cl^{\bullet}\ } CH_3\overset{Cl}{\underset{|}{C}}HCH_3 \\
\quad\quad\quad\quad\quad\quad\quad\quad\quad\quad \text{2-chloropropane}
\end{array}
$$

The direction of the reaction—that is, via abstraction of the primary or secondary hydrogen atom—depends on two factors: the C–H bond dissociation energy and the relative stability of the two radical intermediates formed (∞ Section 8.8, 'Strengths of Covalent Bonds'). The two factors are interrelated. There is a trend that suggests the more stable the radical, the lower the energy needed to cause C–H bond homolysis. There is approximately 15 kJ mol^{-1} difference in C–H bond dissociation energies between primary C–H bonds and secondary C–H bonds, with the latter being easier to cleave. However, the size of the energy difference between homolysis of a primary and secondary C–H bond cannot fully account for the approximate 1 : 1 product ratio.

There is also a valence bond argument that supports the bond dissociation energy arguments. Alkyl free radicals have a potentially filled *p* orbital and are stabilised by **electron-donating groups** such as alkyl groups. As a consequence, the order of alkyl radical stability is:

Methyl radical < Primary radical < Secondary radical < Tertiary radical

(least stable)   (most stable)

At this point, it appears there is little difference between the two pathways leading to the monochlorinated products if the product ratio is 1 : 1. Actually, there is a significant difference as it is statistically *three times* more likely to abstract a primary hydrogen (there are six of these in propane) as it is to abstract a secondary hydrogen (there are two of these in propane). So, in fact, there is a large discrimination for the formation of 2-chloropropane. Typically, the ratio of selectivity for monochlorination is of the order 1° : 2° : 3° = 1 : 4 : 5.

Bromine also reacts with alkanes via a free-radical mechanism and bromination is a more selective process. The ratio of **regioselectivity** for bromination is of the order 1° : 2° : 3° = 1 : 100 : 1500.

RELATED EXERCISES: 22.70, 22.71

Free-radical chain reactions are also found in biological systems, in which harmful radicals such as ˙OH and molecular O_2 attack functional and structural molecules within cells, leading to dysfunction or cell death. Natural **antioxidants** such as vitamin C (ascorbic acid) and vitamin E (α-tocopherol), as shown below, react readily with free radicals, terminating their propagation.

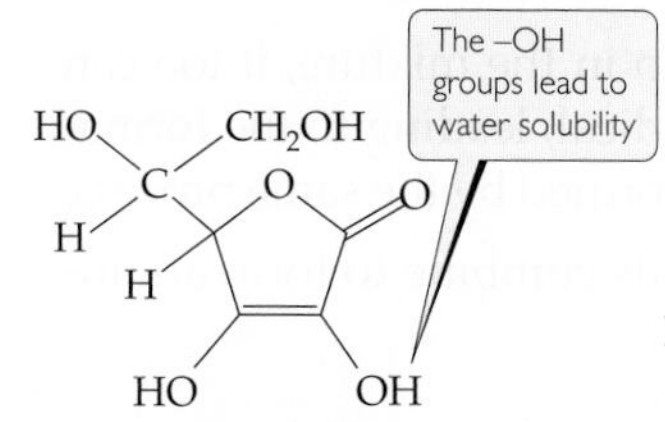

The –OH groups lead to water solubility

The long alkyl chain gives lipid solubility

Vitamin C
(*water soluble*)

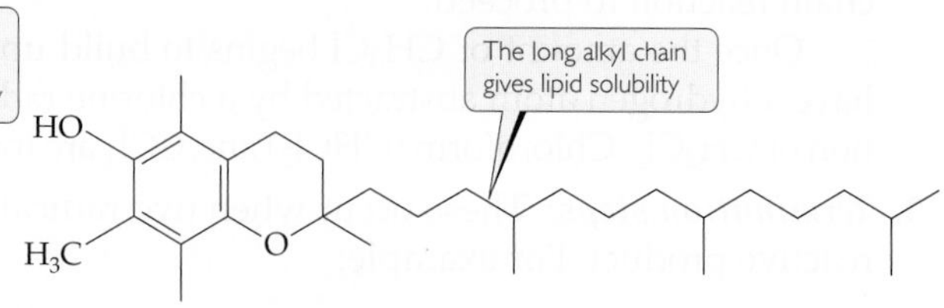

Vitamin E
(*lipid soluble*)

▲ CONCEPT CHECK 11

How is molecular O_2 represented as a free radical?

SAMPLE INTEGRATIVE EXERCISE   Putting concepts together

Combustion analysis of a hydrocarbon determined it contained 82.8% carbon and 17.2% hydrogen by mass. **(a)** What is the empirical formula of this compound? **(b)** Explain why your empirical formula cannot also be the molecular formula of this hydrocarbon. **(c)** If the molar mass of the compound is 58.1 g mol^{-1}, draw the structural formula of all possible constitutional isomers and give their names. **(d)** Does an equilibrium exist between the constitutional isomers at room temperature?

SOLUTION

Analyse In this problem we determine the formula of a compound and examine the relationship between its isomers.

Plan First determine the empirical formula of the compound then examine the bonding arrangement of the atoms involved to see if the empirical formula and molecular formula could be the same. Next determine the molecular formula and name all possible structures with this formula. Finally consider whether the constitutional isomers may be easily interconverted at room temperature.

Solve

(a) A 100 g sample of the hydrocarbon will contain 82.8 g of carbon and 17.2 g of hydrogen. This represents

$$\frac{82.8 \text{ g}}{12.01 \text{ g mol}^{-1}} = 6.89 \text{ mol carbon}$$

$$\frac{17.2 \text{ g}}{1.008 \text{ g mol}^{-1}} = 17.1 \text{ mol hydrogen}$$

The mole ratio of C : H is 6.89 : 17.1 = 1 : 2.48 or 2 : 5

The empirical formula of the hydrocarbon is therefore C_2H_5.

(b) In a stable bonding arrangement hydrogen forms one bond and carbon forms four bonds. The odd number of hydrogen atoms mean that it is not possible to construct a compound in which all atoms have a stable bonding arrangement; one carbon has just three bonds.

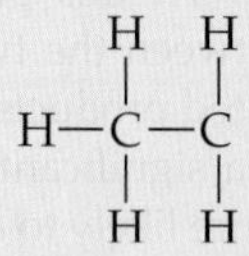

(c) An empirical formula of C_2H_5 would have a molar mass of 29.06 g mol^{-1}; the actual molar mass is double this and so the molecular formula is C_4H_{10}. The structural isomers with this formula are:

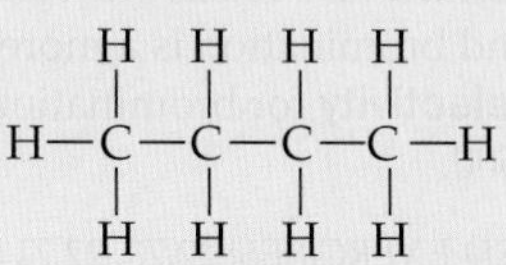

butane 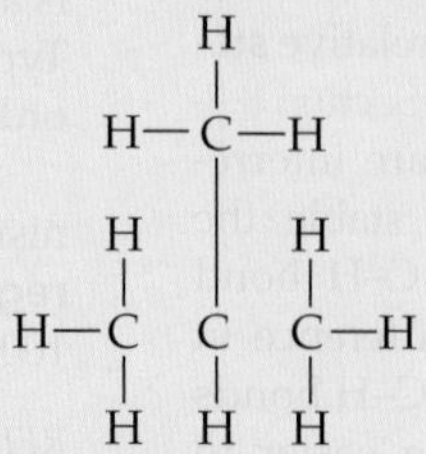   2-methylpropane

(d) To convert one constitutional isomer to another involves the breaking of chemical bonds. This requires quite a lot of energy and does not spontaneously occur at room temperature. So no equilibrium exists between the two isomers at room temperature.

(See also Exercise 22.83.)

CHAPTER SUMMARY AND KEY TERMS

INTRODUCTION and SECTION 22.1 **Organic chemistry** is the study of compounds that contain carbon. We have encountered many aspects of organic chemistry in earlier chapters. Carbon forms four bonds in its stable compounds.

SECTION 22.2 The simplest class of organic molecules consists of **hydrocarbons**, which contain only carbon and hydrogen. Alkanes are hydrocarbons that contain only single bonds. Alkanes have names that end in -*ane*, such as methane and ethane. **Alkanes** are **aliphatic**, forming straight-chain, branched-chain and cyclic arrangements.

SECTION 22.3 Alkanes are used primarily as fuels. Their properties, such as melting and boiling point, change in a regular fashion as the length of the chain increases. These properties are known as a **homologous series**.

SECTION 22.4 **Hybridisation** is used as a model to rationalise the bonding geometry of a saturated carbon atom (109.5°). The rotation about C–C single bonds gives rise to different **conformations**, ranging from **staggered** to **eclipsed** forms. The staggered forms dominate due to the added **steric** interactions involved in the eclipsed forms of a molecule. We can use **Newman projections** to describe the different conformers easily. **Structural** or **constitutional isomers** have the same molecular formula but a different connectivity of atoms.

SECTION 22.5 The **nomenclature** used for hydrocarbons is based on the longest continuous chain of carbon atoms in the structure and contains a prefix, parent and suffix. The locations of **alkyl groups**, which branch off the chain, are specified by numbering along the carbon chain.

SECTION 22.6 Alkanes with ring structures are called **cycloalkanes**. Five- and six-membered rings are most common. Cyclohexane has two conformations, **chair** and **boat** forms. The chair form is more stable as it reduces unfavourable steric interactions and has two types of hydrogen positions, **axial** and **equatorial**. When a substituent is added to the ring, it typically adopts an equatorial position to minimise **1,3-diaxial interactions**.

SECTION 22.7 The chemistry of organic compounds is dominated by the nature of their **functional groups**. The functional groups we have considered are:

R—O—H   alcohol

R—C(=O)—H   aldehyde

$\diagdown$C=C$\diagup$ alkene

—C≡C—   alkyne

R—C(=O)—N$\diagup$ amide

R—N$\diagup$(R' (or H)) R" (or H)   amine

R—C(=O)—O—H   carboxylic acid

R—C(=O)—O—R'   ester

R—O—R'   ether

R—C(=O)—R'   ketone

R, R' and R" represent hydrocarbon groups—for example, methyl (CH_3) or ethyl (CH_2CH_3). **Alcohols**, which contain –OH, the alcohol functional group, are represented as R–OH. Organic molecules need the reactivity imparted by **heteroatoms** in order to demonstrate their usefulness. Of the types of heteroatoms available, oxygen plays a major role.

SECTION 22.8 Alkanes and cycloalkanes are relatively unreactive. They do, however, undergo **combustion** in air (**oxidation**), and their main use is as sources of heat energy (exothermic reaction). Recognition of carbon and hydrogen atoms as primary, secondary or tertiary is important for determining the product distribution in organic reactions. **Free-radical** reactions are typically initiated by heat or light which leads to **homolytic bond cleavage** of halogens, leading to **halogenation**. The three phases of a free-radical reaction are **initiation**, **propagation** and **termination**. The preference of substitution is 3° H > 2° H > 1° H and is assisted by the presence of **electron-donating groups** around the carbon-centred radical. The preference for one product over its isomer is called **regioselectivity**. **Antioxidants** such as vitamins E and C eliminate free-radicals within the human body.

KEY SKILLS

- Draw alkane structures based on their names and name alkanes based on their structures. (Sections 22.2 and 22.4)
- Understand trends in the alkane homologous series. (Section 22.3)
- Be able to describe the concept of sp^3 hybridisation and σ-bonds. (Section 22.4)
- Be able to name systematically alkanes from structures and draw structures from systematic names. (Section 22.5)

- Be able to draw cycloalkanes containing 3–8 carbon atoms. (Section 22.6)
- Understand the difference between equatorial and axial positions in cyclohexane. (Section 22.6)
- Know the structure of the functional groups: alkene, alkyne, alcohol, ether, aldehyde, ketone, carboxylic acid, amine, amide. (Section 22.7)
- Be able to identify primary, secondary and tertiary carbon atoms. (Section 22.7)
- Be able to demonstrate the making and breaking of σ-bonds in a free-radical reaction. (Section 22.8)

KEY EQUATIONS

- General formula for alkanes C_nH_{2n+2} [22.1]
- General formula for cycloalkanes C_nH_{2n} [22.2]

EXERCISES

VISUALISING CONCEPTS

22.1 All the structures that follow have the same molecular formula, C_8H_{18}. Which structures are of the same molecule? [Section 22.2]

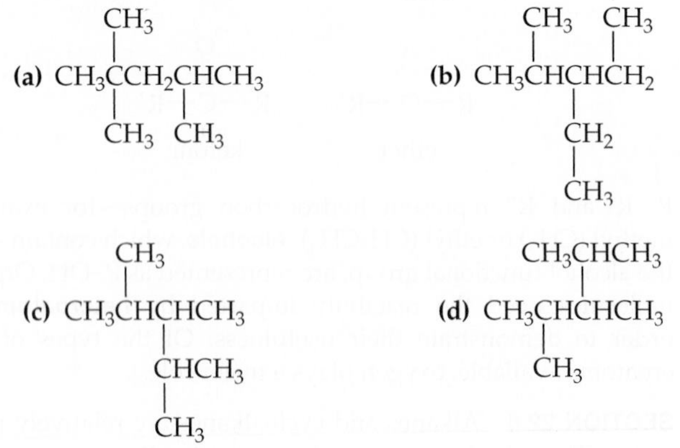

22.2 Which constitutional isomer is missing? [Section 22.4]

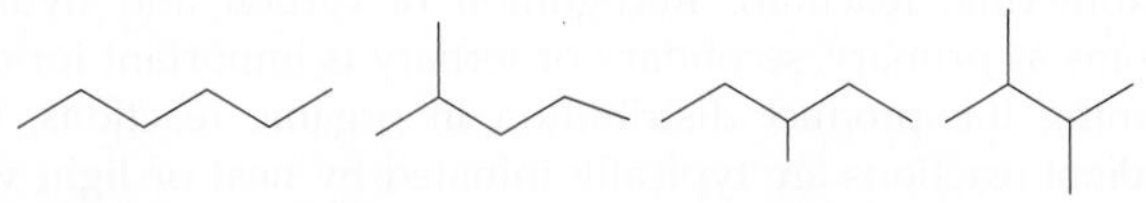

22.3 Define the following bond. [Section 22.4]

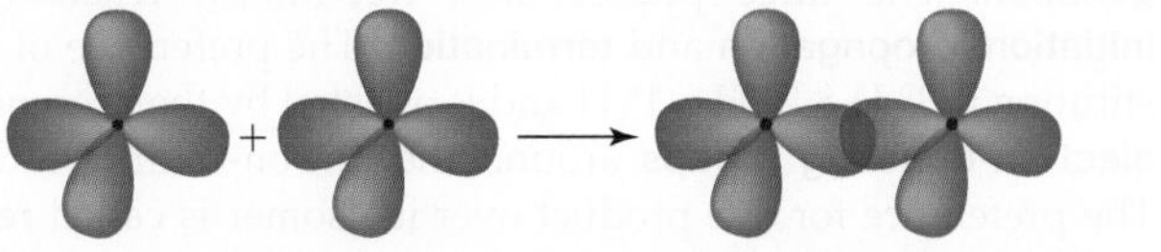

22.4 The structure of an alkane is drawn below as a line drawing representation. Is 2-propyl-3-methylbutane a suitable systematic name for this compound? Provide a better name using the guidelines in Section 22.5.

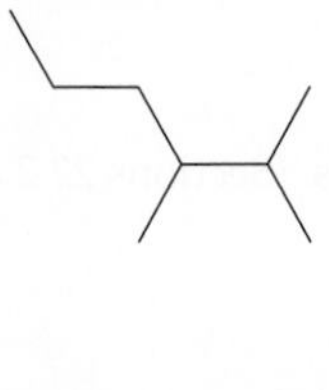

22.5 Draw the structural formula for the following molecular model. [Section 22.4]

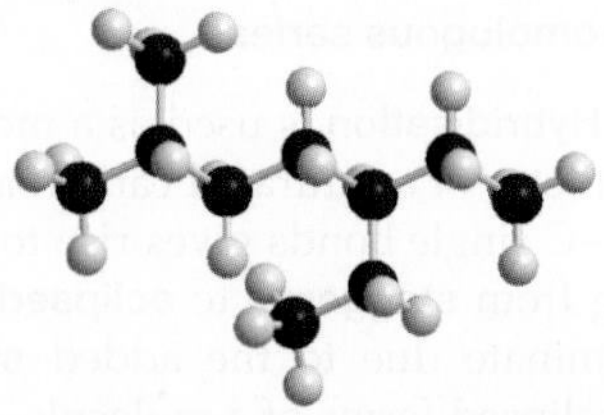

22.6 A selection of hydrogen atoms on the chair conformation of cyclohexane are coloured red, green, blue and yellow. Identify the position of each coloured ball (that is, axial or equatorial). What happens to these positions if an interconversion occurs to produce the other possible chair conformation. [Section 22.6]

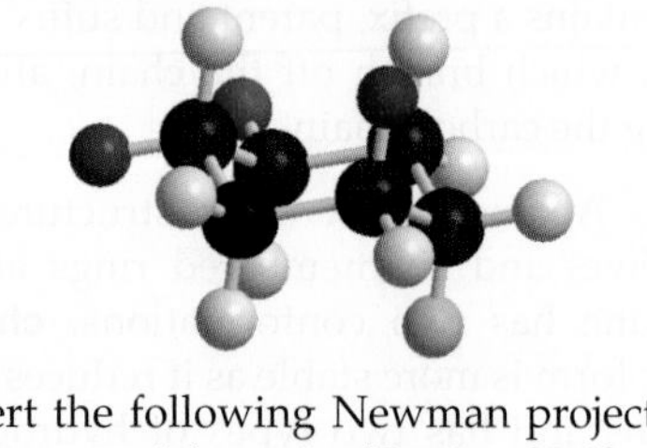

22.7 Convert the following Newman projection into a standard structural formula. [Section 22.4]

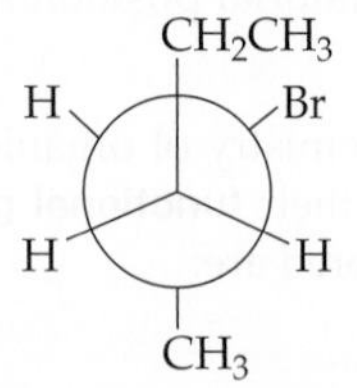

22.8 **(a)** Write the propagation steps leading to the formation of dichloromethane from chloromethane.

$$CH_3Cl + Cl^{\cdot} \rightarrow CH_2Cl_2$$

(b) Explain why free-radical halogenations usually give rise to mixtures of products. **(c)** Comment on the stoichiometry of chloromethane and chlorine needed to form carbon tetrachloride as the major product. [Section 22.8]

$$CH_3Cl + Cl^{\cdot} \rightarrow CCl_4$$

22.9 Identify each of the functional groups in the following molecules. [Section 22.7]

(a)

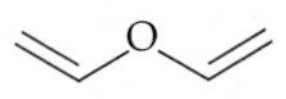

an anaesthetic

(b)

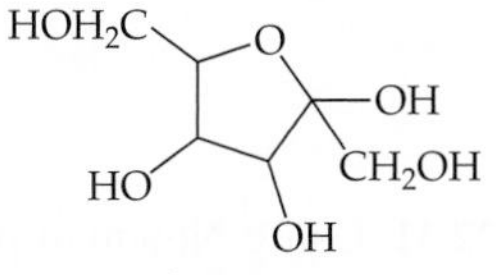

fructose

(c)

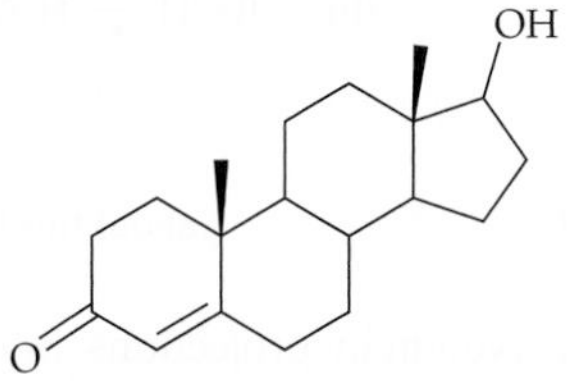

testosterone
(a male sex hormone)

(d)

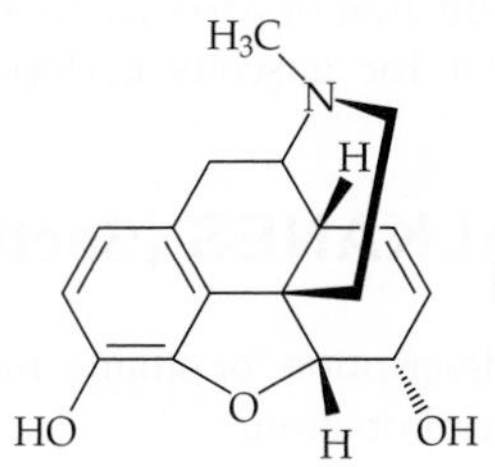

morphine
(an alkaloid extracted from poppies)

(e)

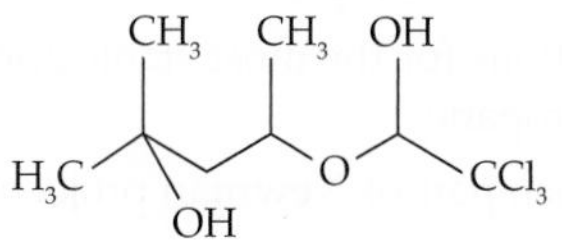

Chlorhexadol
(a depressant)

GENERAL CHARACTERISTICS OF ORGANIC MOLECULES (Section 22.1)

22.10 List five elements that are commonly found in organic compounds. Which ones are more electronegative than carbon?

22.11 Organic compounds containing C–O and C–Cl bonds are more reactive than simple alkane hydrocarbons. Considering the comparative values of C–H, C–C, C–O and C–Cl bond energies (Table 8.4), why is this so?

22.12 Are carbon monoxide or ammonia considered organic molecules? Why or why not?

22.13 Urea has been identified as an organic molecule. What features make it so?

22.14 (a) What are the approximate bond angles about carbon in an alkane? (b) What are the characteristic hybrid orbitals employed by carbon in an alkane?

AN INTRODUCTION TO HYDROCARBONS (Section 22.2)

22.15 What can we tell about a compound when we know the empirical formula? What additional information is conveyed by the molecular formula? By the structural formula? Explain in each case using CH_2 as an empirical formula.

22.16 Give the molecular formula of a hydrocarbon containing six carbon atoms that is (a) an alkane, (b) a cycloalkane.

22.17 (a) What is the difference between a straight-chain and a branched-chain alkane? (b) What is the difference

between an alkane and an alkyl group? (c) Why are alkanes said to be saturated?

22.18 What structural features help us identify a compound as (a) an alkane, (b) a cycloalkane, (c) a saturated hydrocarbon?

22.19 (a) What is a hydrocarbon? (b) Are all alkanes hydrocarbons? (c) Write the structural formula for propane (C_3H_8). (d) Butane is an alkane containing four carbon atoms in a line. Write a structural formula for this compound and determine its molecular and empirical formulae.

APPLICATIONS AND PHYSICAL PROPERTIES OF ALKANES (Section 22.3)

22.20 What is the octane number of a mixture of 35% heptane and 65% isooctane?

22.21 (a) Describe two ways in which the octane number of petrol consisting of alkanes can be increased. (b) Suggest a reason why 2,2,4-trimethylpentane is also known as isooctane.

22.22 (a) What ending is used for the names of alkanes? (b) *Undecane* is an alkane with all its 11 carbon atoms in one chain. It is used as an attractant to cockroaches. Write a structural formula for this compound and determine its molecular and empirical formulae.

22.23 Butane and 2-methylpropane are both non-polar molecules and have the same molecular formula, yet butane has the higher boiling point (–0.5 °C compared with –11.7 °C). Explain.

22.24 Rank the following compounds in order of increasing boiling points and explain your choice of order.

$CH_3(CH_2)_5CH_3$

heptane

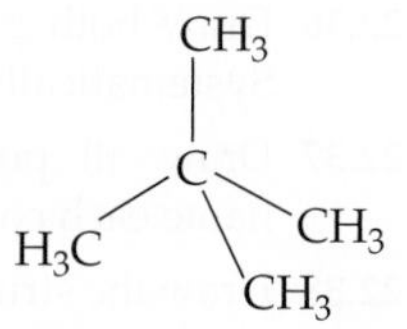

2,2-dimethylpropane
(neopentane)

$(CH_3)_2CHCH_2CH(CH_3)_2$

2,4-dimethylpentane

CH_3CH_3

ethane

22.25 The molar heat of combustion of gaseous cyclopropane is -2089 kJ mol^{-1}; that for gaseous cyclopentane is -3317 kJ mol^{-1}. Calculate the heat of combustion per CH_2 group in the two cases, and account for the difference.

STRUCTURE OF ALKANES (Section 22.4)

22.26 Draw a valence-bond description for ethane. Include the hybridisation for each carbon atom.

22.27 Distinguish between constitutional and conformational isomers using butane as an example.

22.28 Draw Newman projections for the most stable and least stable conformers of propane.

22.29 Which conformer in each pair of Newman projections is lower in energy?

(a)

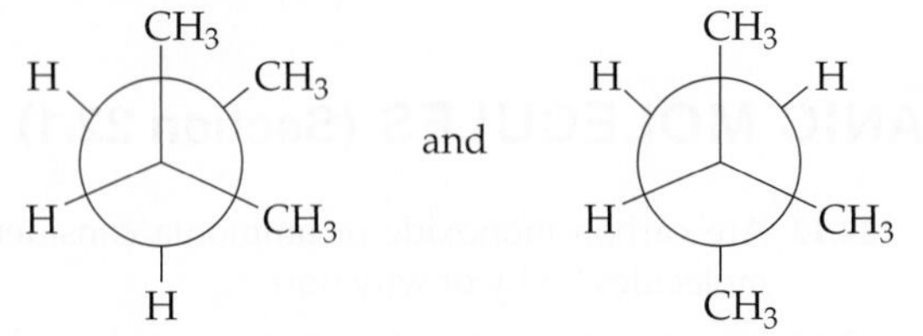

and

(b)

and

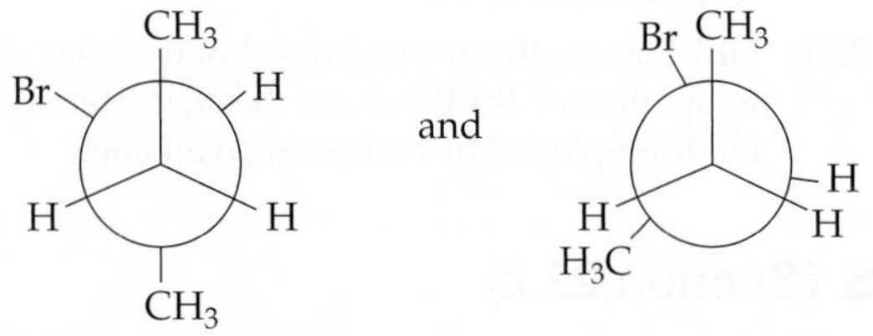

22.30 Ethylene glycol ($HOCH_2CH_2OH$) prefers to exist almost entirely in the following eclipsed form in the vapour phase. Give an explanation for this.

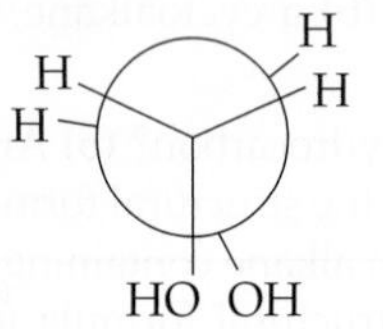

22.31 Using Newman projections, draw the most stable staggered form of the following compounds:

(a)

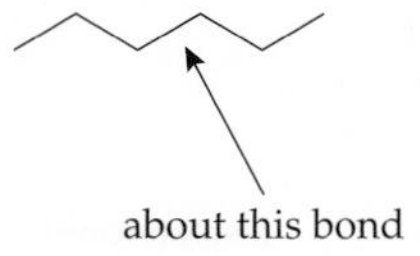

about this bond

(b) $BrCH_2\text{—}CH_2Br$

about this bond

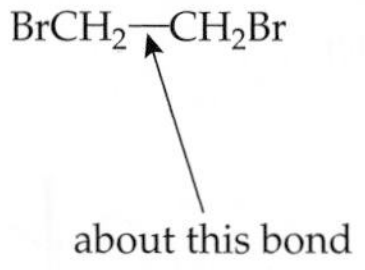

22.32 Convert the following Newman projections into their standard structural formula.

(a)   (b)   (c)

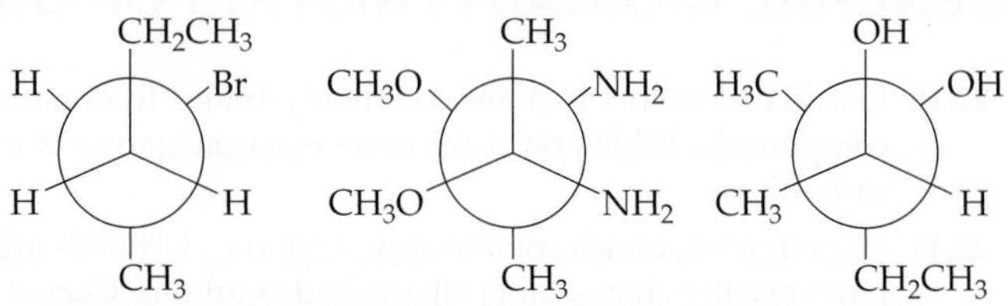

22.33 (a) Why are carbon nanotubes not classed as alkanes? (b) Would it be possible to construct the same cylindrical shape using fused cyclohexane rings?

22.34 (a) What are the structural differences between diamond and carbon nanotubes? (b) What are the properties of carbon nanotubes that make them so exciting for applications in the area of nanotechnology?

22.35 Polyethylene is a polymer used in many plastic applications, such as food wrap and soft-drink bottles. A general structure for polyethylene is shown below. Would such a polymer be more soluble in an organic solvent such as hexane or in water? Why?

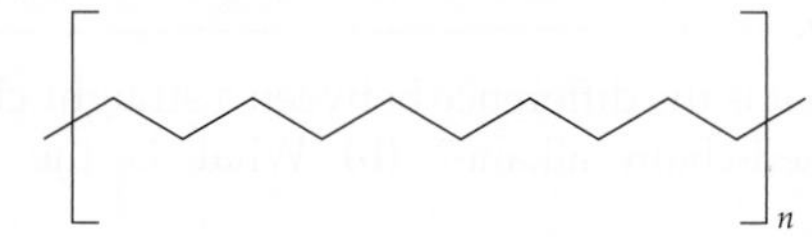

ALKANE NOMENCLATURE (Section 22.5)

22.36 Draw both possible cyclic structural isomers of C_4H_8. Systematically name each compound.

22.37 Draw all possible isomers of C_5H_{12}. Systematically name each compound.

22.38 Draw the structural formula or give the IUPAC name, as appropriate, for the following:

(a)

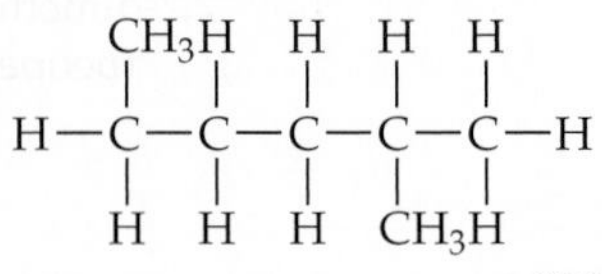

(b) $CH_3CH_2CH_2CH_2CH_2CH_2CCH_2CHCH_3$

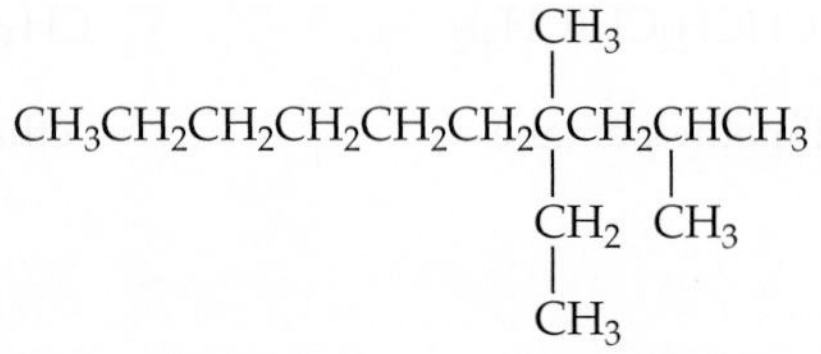

(c) 3-methylhexane

(d) 4-ethyl-2,2-dimethyloctane

(e) methylcyclohexane

22.39 Draw the structural formula or give the systematic name, as appropriate, for the following:

(a)

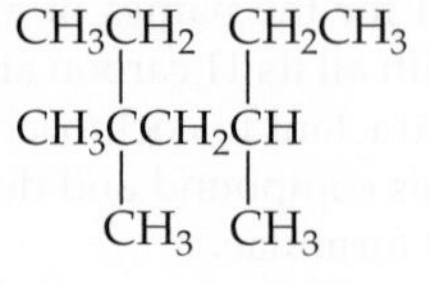

(b)

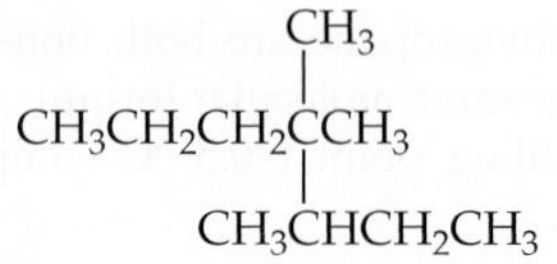

(c) 2,5-dimethylnonane

(d) 3-ethyl-4,4-dimethylheptane

(e) 1-ethyl-4-methylcyclohexane

22.40 Name the following compounds systematically:

(a) 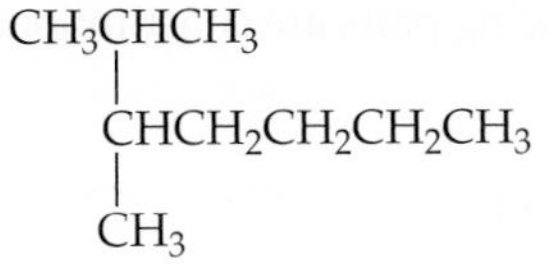
$$CH_3CHCH_3$$
$$\underset{\overset{|}{CH_3}}{CHCH_2CH_2CH_2CH_3}$$

(b) 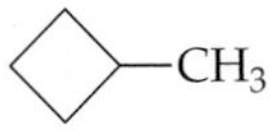 ◇—CH₃

(c) 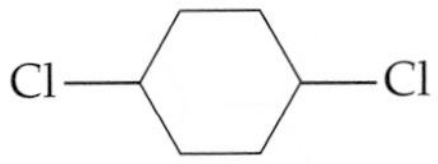
Cl—⬡—Cl

22.41 Give IUPAC names for the following alkanes:

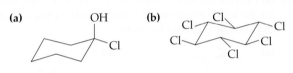

(a) 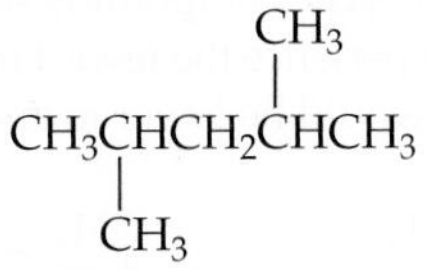 $CH_3CHCH_2CHCH_3$ with CH_3 groups

(b) 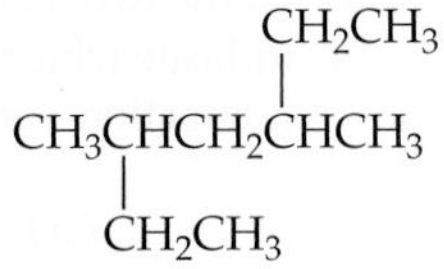 $CH_3CHCH_2CHCH_3$ with CH_2CH_3 groups

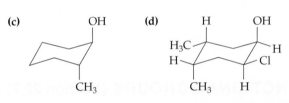

(c) 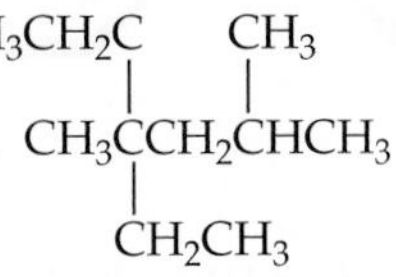 $CH_3CCH_2CHCH_3$ with H_3CH_2C, CH_3, and CH_2CH_3 groups

(d) 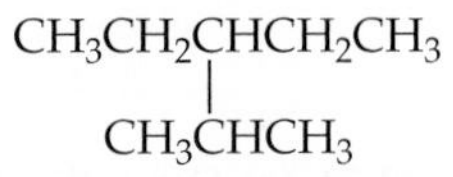 $CH_3CH_2CHCH_2CH_3$ with CH_3CHCH_3 group

22.42 What do ethane, ethanol and ethylene have in common with respect to their nomenclature? What can you infer about the number of carbon atoms in each compound?

22.43 Draw 3-ethyl-2,3-dimethylbutane. What would be the correct systematic name for this compound?

CYCLOALKANES (Section 22.6)

22.44 The compound cyclohexane is an alkane in which six carbon atoms form a ring. The partial structural formula of the compound is as follows:

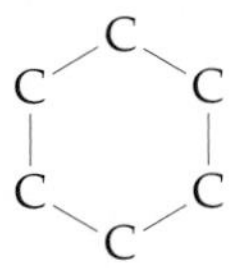

C—C—C—C—C—C (ring)

(a) Complete the structural formula for cyclohexane.
(b) Is the molecular formula for cyclohexane the same as that for hexane, in which the carbon atoms are in a straight line? If not, comment on the source of any differences. (c) Propose a structural formula for cyclohexene, which has one carbon–carbon double bond. Does it have the same molecular formula as cyclohexane?

22.45 (a) Draw line drawings for cyclopentane and cyclobutane. (b) Which of the two cycloalkanes has the higher ring strain? Why?

22.46 (a) 1-Bromo-4-*tert*-butylcyclohexane exists almost entirely in the conformation shown. Explain why based on the relative size of the two groups. (b) Label the Br and C(CH₃)₃ groups as axial or equatorial.

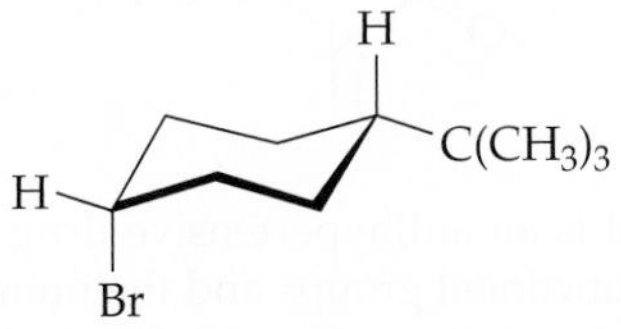

22.47 Label each CH₃, OH or Cl group in the following compounds as axial or equatorial.

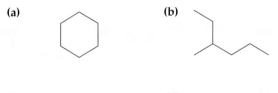

22.48 Place the following dimethylcycloalkanes in order of relative stability.

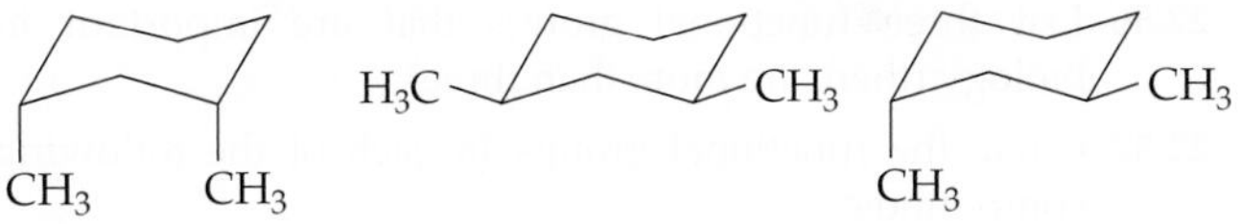

22.49 Why is geometric isomerism possible for cycloalkanes but not for alkanes?

22.50 Draw a cyclic, straight-chain or branched-chain constitutional isomer for each of the following:

(a) ⬡

(b) branched chain

(c)  ∧∨

(d) 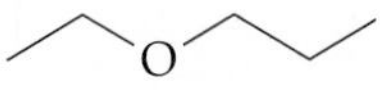 ∧O∧

(e) cyclohexane with tert-butyl group

22.51 Of the two disubstituted cyclic compounds shown below, indicate which would experience the fewest non-bonding interactions, and hence would be lower in energy.

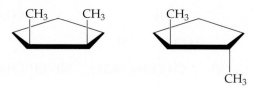

22.52 (a) Draw the structure of cyclohexane in the chair conformation including hydrogen atoms. Highlight those hydrogens that are axial and those that are equatorial. **(b)** Draw a 1,2-disubstituted cyclohexane in the chair conformation with your choice of alkyl substituents in axial positions. Now flip the structure to the other chair conformation. What do you notice about the position of the two substituents? Are both substituents axial, or equatorial, or is there one of each? Is this general? Explain. **(c)** Systematically name the structure you drew in part (b).

22.53 (a) Provide the molecular formula and systematic name for the following structures. **(b)** Draw the structure of a constitutional isomer for each.

(a)

(b)

(c) **(d)**
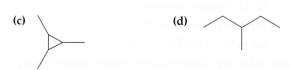

22.54 Which of the following pairs are constitutional isomers?

(a) and

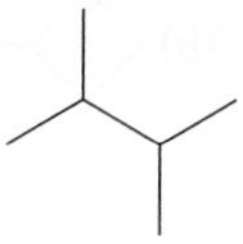

(b) 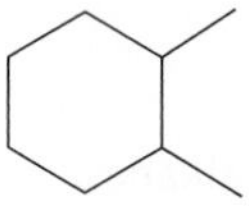 and

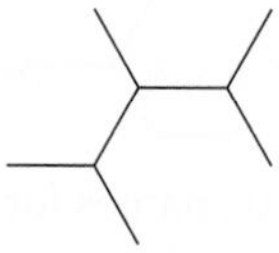

(c) 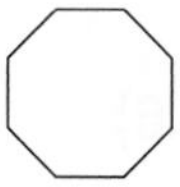 and

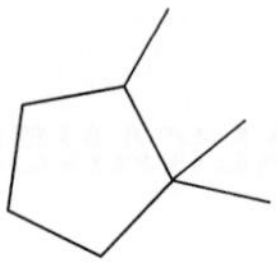

(d) 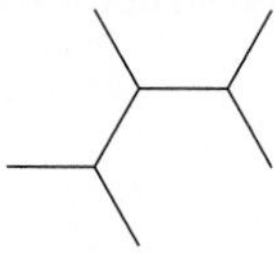 and
$$CH_2CH_2CH_3$$
$$CH_3CH_2CH_2CHCH_3$$

FUNCTIONAL GROUPS (Section 22.7)

22.55 (a) What is a functional group? **(b)** Why are functional groups important?

22.56 List three functional groups that are important in biology (there are more than three).

22.57 Circle the functional groups in each of the following compounds:

(a) NH_2 **(d)** O

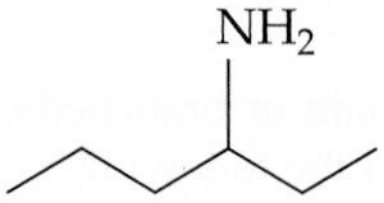

(b) CHO   **(e)** OH

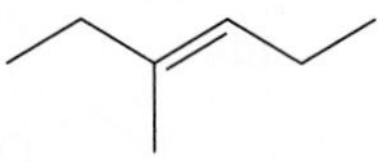

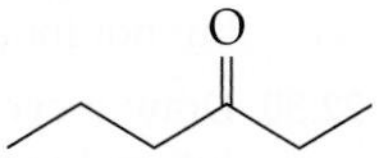

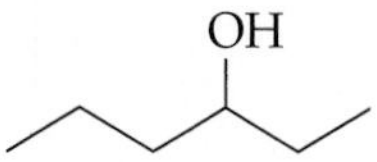

(c)   **(f)** O

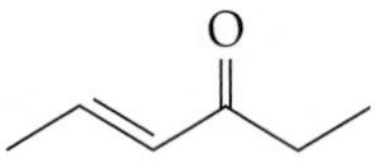

22.58 Indinavir is a HIV protease inhibitor. Identify the following functional groups and the number of each present in the structure of Indinavir: **(a)** alcohol; **(b)** amide; **(c)** amine; **(d)** aromatic hydrocarbon.

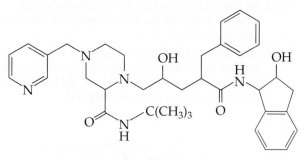

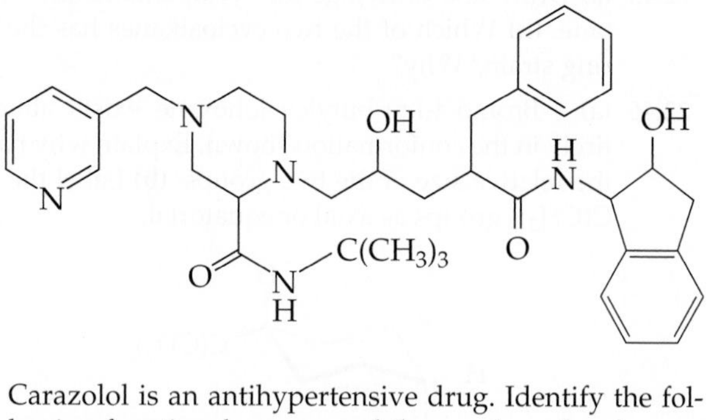

22.59 Carazolol is an antihypertensive drug. Identify the following functional groups and the number of each present in the structure Carazolol: **(a)** alcohol; **(b)** ether; **(c)** amine; **(d)** aromatic hydrocarbon.

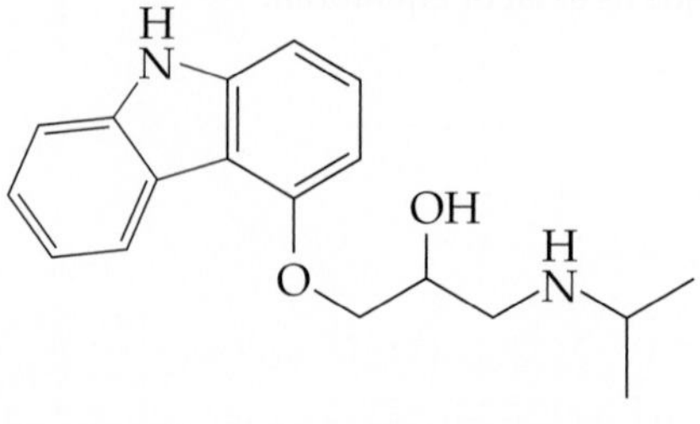

22.60 Taxol is a clinically important anticancer compound isolated from the bark of *Taxus*, the Pacific yew tree. Identify the following functional groups and the number of each present in the structure of Taxol: **(a)** alcohol; **(b)** ether; **(c)** amine; **(d)** amide; **(e)** aromatic hydrocarbon; **(f)** carboxylic ester; **(g)** alkene; **(h)** ketone.

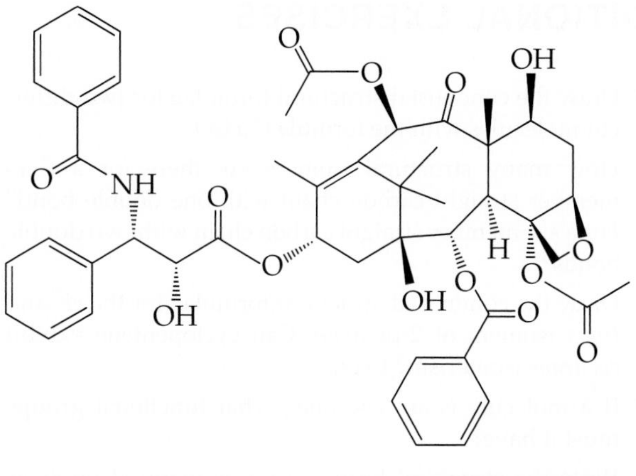

REACTIONS OF ALKANES (Section 22.8)

22.61 **(a)** What is the difference between a substitution reaction and an oxidation reaction? **(b)** Using condensed structural formulae, write the balanced equation for the free-radical substitution reaction of 2,2,4-trimethylpentane with Br_2 in the presence of light. Explain, using this reaction as an example, what is meant by a regioselective reaction. **(c)** Write a balanced chemical equation for the complete combustion of 2,2-dimethylpropane.

22.62 When cyclopropane is treated with HI, 1-iodopropane is formed. A similar type of reaction does not occur with cyclopentane or cyclohexane. What physical aspect of cyclopropane might account for its reactivity?

22.63 **(a)** Write a balanced equation for the complete combustion of propane. **(b)** Use Table 8.4 and your answer to part (a) to estimate ΔH for this reaction.

22.64 An unknown aliphatic alkane, A, yields 4.800 g CO_2 and 2.475 g H_2O when fully combusted. Suggest a plausible molecular formula for A that is less than or equal to C_6H_{14}.

22.65 **(a)** How much oxygen is required for the complete combustion of 70 g of cyclohexane? **(b)** How much carbon dioxide is produced in this reaction?

22.66 Propose structures for compounds with formulae less than C_6H_{14} containing:
(a) two tertiary carbons
(b) three primary carbons
(c) a quaternary carbon
(d) only two methyl groups.

22.67 The structure of lanostane, a biogenic precursor of steroids, is shown below. Classify each carbon atom as primary, secondary, tertiary or quaternary.

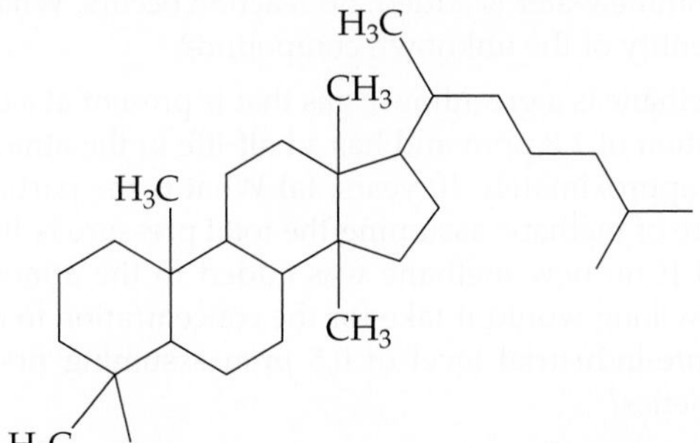

lanostane

22.68 Draw Lewis structures for the following radicals:
(a) methyl radical
(b) chlorine atom
(c) isopropyl radical.

22.69 **(a)** Write the propagation steps leading to the formation of dichloromethane, CH_2Cl_2, from methane. **(b)** Describe how termination of this chain reaction is possible.

22.70 Rank each radical in order of its relative stability.

(a)

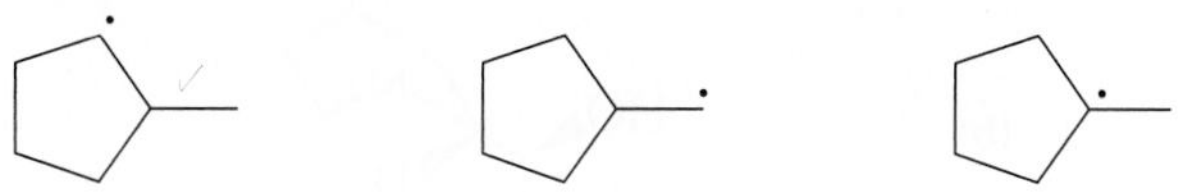

(b) $(CH_3)_2\overset{\bullet}{C}CH_2CH_3$ $(CH_3)_2CHCH_2\overset{\bullet}{C}H_2$ $\overset{\bullet}{C}H_3$

22.71 Which of the following alkyl halides can be prepared in good yield by radical halogenation?

(a) Cl   **(b)** Br   **(c)** Br   **(d)** Br

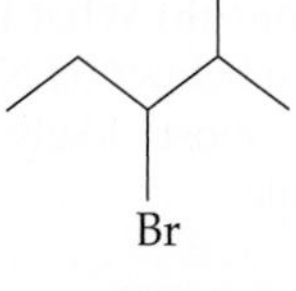

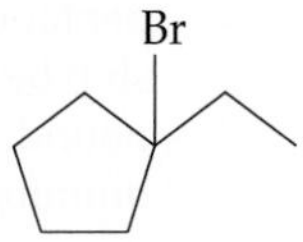

22.72 Draw all monobromo derivatives of 3-methylhexane. Which one of these configurational isomers would be produced in highest yield by the reaction of bromine with 3-methylhexane?

22.73 **(a)** What ratio of cyclohexane to chlorine would you suggest in order to achieve a monochlorination? **(b)** The free-radical chlorination of cyclohexane is much cleaner and occurs in far greater yield than that of hexane under the same conditions. Account for this difference.

ADDITIONAL EXERCISES

22.74 Draw the condensed structural formulae for two different molecules with the formula C_3H_4O.

22.75 How many structural isomers are there for a five-member straight carbon chain with one double bond? For a six-member straight carbon chain with two double bonds?

22.76 Draw the condensed structural formulae for the *cis* and *trans* isomers of 2-pentene. Can cyclopentene exhibit *cis–trans* isomerism? Explain.

22.77 If a molecule is an 'ene one', what functional groups must it have?

22.78 Write the structural formulae for as many alcohols as you can think of that have empirical formula C_3H_6O.

22.79 Identify each of the functional groups in these molecules:

(a)

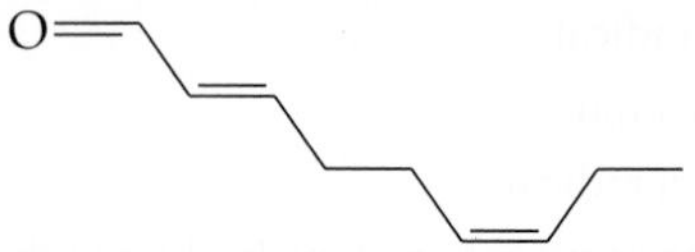

(Responsible for the odour of cucumbers)

(b)

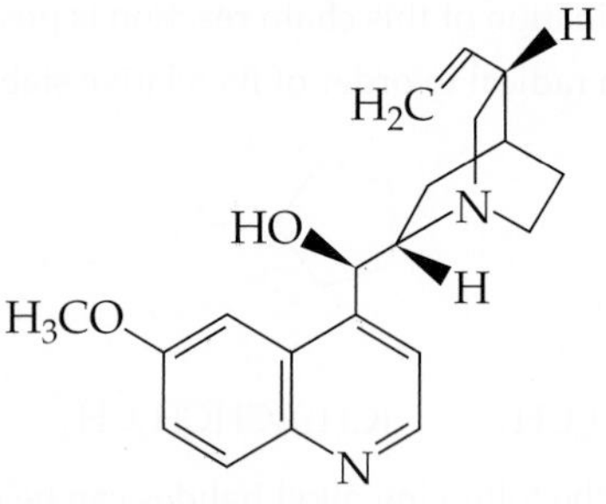

(Quinine—an antimalarial drug)

(c)

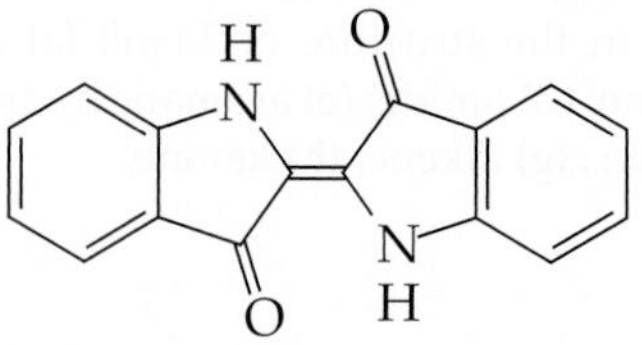

(Indigo—a blue dye)

(d)

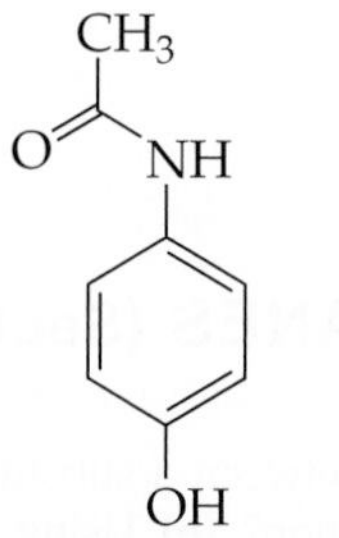

(Acetaminophen—aka Paracetamol)

22.80 Write a condensed structural formula for each of the following: **(a)** an acid with the formula $C_4H_8O_2$, **(b)** a cyclic ketone with the formula C_5H_8O, **(c)** a dihydroxy compound with the formula $C_3H_8O_2$, **(d)** a cyclic ester with the formula $C_5H_8O_2$.

INTEGRATIVE EXERCISES

22.81 **(a)** What type of isomeric relationship exists between pentane and 2-methylbutane? **(b)** What is the relationship between the two shown below? **(c)** Name this compound. **(d)** Identify the most likely point(s) for bromination in each molecule.

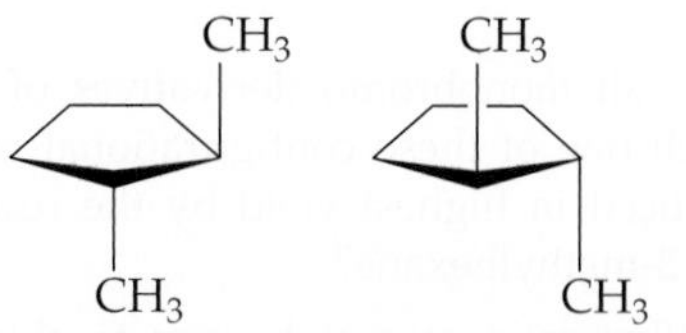

22.82 **(a)** Draw a relative energy profile similar to that shown in Figure 22.16 to illustrate the stability of the different conformers produced by rotation about one C–C single bond in propane. **(b)** What is the product of the complete combustion of propane? **(c)** How much energy is released upon the complete combustion of 44 dm^3 of propane? **(d)** Propane is converted into 2-bromopropane by a first-order process. The rate constant is $5.4 \times 10^2 \ h^{-1}$. If the initial concentration of propane is 0.150 M, what will its concentration be after 22.0 h?

22.83 An unknown substance is found to contain only carbon and hydrogen. It is a liquid that boils at 49 °C at 1 bar pressure. Upon analysis it is found to contain 85.7% carbon and 14.3% hydrogen by mass. At 100 °C and 735 torr, the vapour of this unknown has a density of 2.21 $g \ dm^{-3}$. When it is dissolved in hexane solution and bromine water is added, no reaction occurs. What is the identity of the unknown compound?

22.84 Methane is a greenhouse gas that is present at a concentration of 1.8 ppm and has a half-life in the atmosphere of approximately 10 years. **(a)** What is the partial pressure of methane assuming the total pressure is 102 kPa? **(b)** If no new methane was added to the atmosphere, how long would it take for the concentration to drop to a pre-industrial level of 0.5 ppm assuming first-order kinetics?

MasteringChemistry (www.pearson.com.au/masteringchemistry)

Make learning part of the grade. Access:

- tutorials with personalised coaching
- study area
- Pearson eText

PHOTO/ART CREDITS

890 AndreyKlepikov/Shutterstock; **895 (a)** Barbeques Galore (Aust) Pty Ltd, **(b)** Sheldon and Hammond; **897 (Figure 22.5)** © Natalia Bratslavsky/Shutterstock; **899** Bridgena Barnard/Shutterstock, © kristian sekulic/iStockphoto.com, © Tom Brakefield/Corbis.

23
STEREOCHEMISTRY OF ORGANIC COMPOUNDS

Smell is a very specific sense. Limonene is the compound responsible for the smell of oranges and lemons. The subtle difference between the three-dimensional shapes of the two isomers of limonene, related by their mirror image, can be differentiated by your nose.

KEY CONCEPTS

23.1 STEREOCHEMISTRY IN ORGANIC CHEMISTRY
Molecules occupy three-dimensional space and the arrangement of the atoms of a molecule within this space leads to isomerism. Compounds that have the same molecular formula are classed as *constitutional isomers* or *stereoisomers*.

23.2 *CIS–TRANS* ISOMERISM IN CYCLOALKANES
We discuss *geometric cis–trans isomers* using cycloalkane examples.

23.3 CHIRALITY IN ORGANIC COMPOUNDS
Chirality, or handedness, occurs when a molecule contains a *stereogenic centre*. Molecules with one stereogenic centre can exist as a pair of *enantiomers*.

23.4 MEASURING OPTICAL ACTIVITY
Chiral molecules exhibit optical activity by rotating plane polarised light. Optical activity is measured using a polarimeter.

23.5 ABSOLUTE STEREOCHEMISTRY
Defining the absolute configuration of a molecule about a chiral centre using the Cahn–Prelog–Ingold priority rules unambiguously defines molecular structure in three-dimensional space.

23.6 MOLECULES WITH MORE THAN ONE STEREOCENTRE
Compounds that contain two or more stereogenic centres can exist as both enantiomers and *diastereomers*. We discuss the physical and chemical relationship between enantiomers and diastereomers.

Stereochemistry is about the three-dimensional shape of a molecule. This may sometimes be of crucial importance in determining the physical properties of a compound. Recall from Section 21.4 that coordination compounds that exhibit stereochemistry give rise to different properties such as colour. Stereochemistry may also play a role in the chemical reactions of a compound. This is most clearly seen when an organic molecule interacts with a complex biological molecule such as an enzyme.

For example, would you believe that the compound limonene is responsible for the distinct smells of both oranges and lemons? Limonene is a colourless liquid hydrocarbon containing 10 carbon atoms yet it can adopt two different shapes. These shapes are mirror images of one another—rather like a right hand and a left hand—and this subtle difference in shape results in a different interaction with the smell receptors in our noses. One mirror image form smells of lemons and the other of oranges.

In this chapter we will explore the different shapes organic molecules adopt. We begin with an overview of isomers and examine each type. We use structural and bonding features within a molecule to determine the type of isomer that is formed. In that way we can predict the occurrence of isomers in all organic compounds. We will look at the physical characteristics of different isomers and how we can distinguish them in the nomenclature of the compound.

23.1 | STEREOCHEMISTRY IN ORGANIC CHEMISTRY

∞ Find out more on page 1161

We have already discussed the dimensionality of simple organic molecules and seen that structure is responsible for a compound's physical properties, such as melting point, boiling point and density. As we see in this chapter, and again in Chapter 29, the three-dimensional structure of a molecule is also important for its chemical properties, especially in a biological environment (∞ Section 29.2, 'Amino Acids'). Our objective in this chapter is to gain an appreciation of the importance of the three-dimensional shape of organic molecules. By the end of this chapter, you should be comfortable with visualising molecules in three dimensions from a two-dimensional representation and to understand common stereochemical terms and notation used to describe all compounds.

∞ Review this on page 868

The term *isomers* is used to describe molecules with the same molecular formula but some difference in structure (∞ Section 21.4, 'Nomenclature and Isomerism in Coordination Chemistry'). ▼ FIGURE 23.1 illustrates the relationship between the different classes of isomers.

The fact that two molecules can be quite different, even though they have the same molecular formula and the same connectivity of atoms, is a rather unusual concept but one that we concentrate on in this chapter. Stereoisomers can exist in several forms. Of these forms, there are three distinct classes: **geometric isomers**, **enantiomers** and **diastereomers** (also called diastereoisomers). Stereoisomers cannot be interconverted without the breaking and remaking of a covalent bond (∞ Section 23.2, '*Cis–Trans* Isomerism in Cycloalkanes'; Section 24.2, 'Isomerism and Nomenclature').

∞ Find out more on page 929

We can begin our understanding of stereochemistry by considering simple representations of atoms, such as four coloured balls—blue, green, red and yellow. When the balls are laid in a row, there are 12 possible permutations, taking into account the ability to read from right to left or left to right (▶ FIGURE 23.2). The relationship between any or all permutations can be described as *constitutional isomers* (∞ Section 22.4, 'Structures and Alkanes'). In essence, the balls act as building blocks, in the same way as we described

∞ Review this on page 902

◢ FIGURE IT OUT

Can propane demonstrate constitutional isomerism?

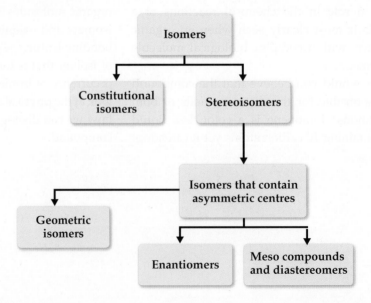

▶ FIGURE 23.1 **There are several different forms of isomerism.** Although constitutional, geometric and diastereomers vary in their physical properties, enantiomers vary only when in a chiral environment.

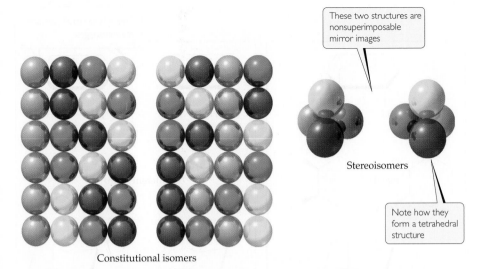

▲ **FIGURE 23.2   Understanding isomerism.** Orientation of coloured balls can be used as imagery to describe constitutional isomerism and stereoisomerism.

atoms or groups of atoms in Table 22.8. You can see from this example how easy it is to build up a library of compounds with the same molecular formula.

What happens when we change from a pseudo one-dimensional shape of the four balls, as occurs in a row, to one that is three-dimensional, such as a pyramid? The answer is a simplification in the number of permutations to just two. Notice that there is no way of superimposing the two pyramidal structures in Figure 23.2 onto one another—they are different. The two structures represented are, in fact, mirror images of each other. In chemical terms, they are stereoisomers and, more specifically, *enantiomers*.

23.2 | *CIS–TRANS* ISOMERISM IN CYCLOALKANES

The three isomers of 1,2-dimethylcyclopentane shown in ▼ **FIGURE 23.3** differ in the relative locations of their methyl groups. These three compounds are *geometric isomers* of each other, a form of *stereoisomerism*. That is, they are compounds that have the same molecular formula and the same connectivity between atoms, but differ in the spatial arrangement of these groups. Thus the **cis** isomer of 1,2-dimethylcyclohexane has the two methyl groups orientated on the same side of the plane of the ring, whereas the **trans** isomers have the methyl groups orientated on opposite sides.

The three-dimensional orientation of groups within a structural formula or line drawing can be described using bold and dashed wedges, as shown in Figure 23.3. The notation employs the two-dimensional plane of the paper to draw a three-dimensional object. A bold wedge (◀) indicates that this bond extends towards you out of the plane of the paper. A dashed wedge (⸱⸱⸱⸱ꞮꞮꞮꞮꞮ) indicates that this bond is directed into the paper. This symbolism will become important in Section 23.3, where we investigate three-dimensional structure in more detail.

Despite the ability of C–C bonds to rotate under normal conditions, their ability to do so in a cycloalkane is greatly restricted by ring strain. Hence the three stereoisomers shown in Figure 23.3 cannot be interconverted without breaking and remaking C–C bonds.

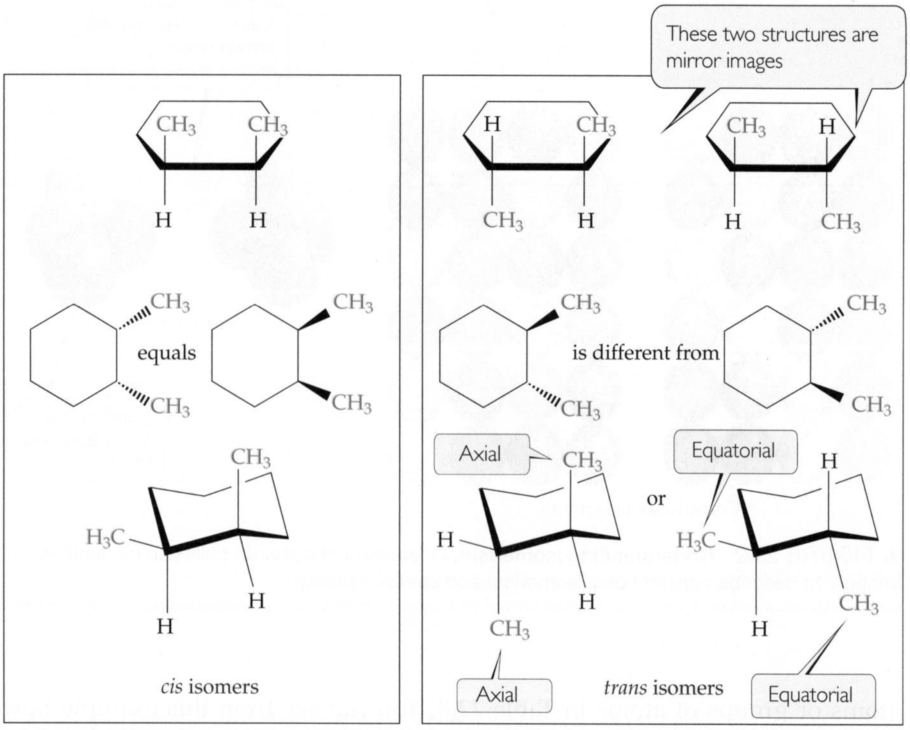

These two structures are mirror images

CH₃ CH₃

H H

equals

CH₃

CH₃

CH₃

CH₃

CH₃

H₃C

H

H

cis isomers

H CH₃

CH₃ H

CH₃ H

H CH₃

is different from

CH₃

CH₃

CH₃

CH₃

Axial CH₃

or

Equatorial H

H

H₃C

H

H

CH₃

H

CH₃

Axial

trans isomers

Equatorial

▶ **FIGURE 23.3 Geometric isomerism in disubstituted cycloalkanes.** *Cis* and *trans* isomerism is a feature of cycloalkanes. The easiest way to remember which isomer is which is to remember that *trans* (as in transpacific or transport) means 'across'.

⚠ **CONCEPT CHECK 1**

Which structure best represents the Newman projection?

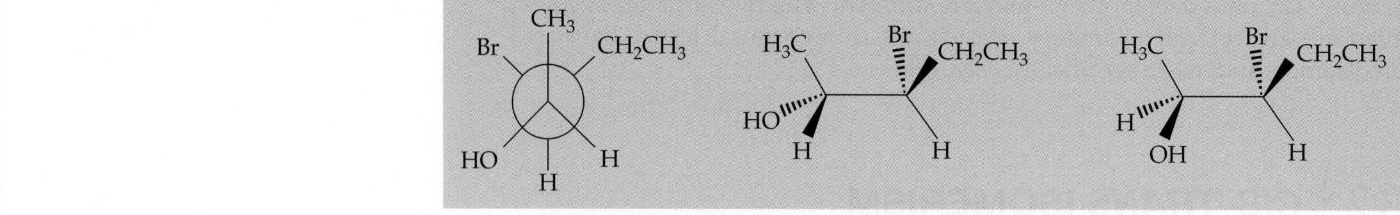

CH₃

Br CH₂CH₃

HO H

H

H₃C Br CH₂CH₃

HO''''

H H

H₃C Br CH₂CH₃

H''''

OH H

SAMPLE EXERCISE 23.1 **Naming *cis* and *trans* isomers**

Write the systematic name for the following cycloalkane:

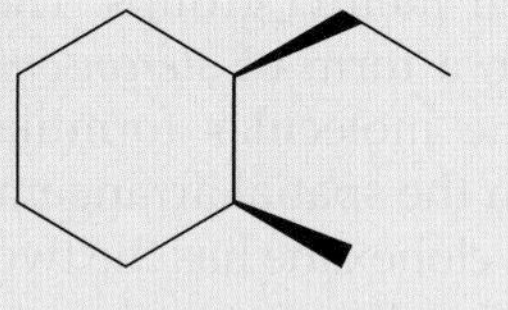

SOLUTION

Analyse We are given the structure of a cycloalkane bearing two substituents and asked to name it, taking into account stereochemistry.

Plan Understand the type of cycloalkane, the names of the substituents and their relative orientation to each other using the wedged nature of the bonds.

Solve The cycloalkane has six carbons associated with the ring structure, so the full name is derived from cyclohexane. The two substituent groups are an ethyl group at position 1 and a methyl group at position 2 based on nomenclature rules (∞ Section 22.5, 'Alkane Nomenclature').

The name for this alkane would be 1-ethyl-2-methylcyclohexane. However, this name tells us nothing about the relative orientations of the alkyl substituents. These two groups are *cis* to one another because the symbolism used in the line drawing has both substituents extending out of the page in the same direction. Hence, a full systematic name would be cis-1-ethyl-2-methylcyclohexane.

∞ Review this on page 904

PRACTICE EXERCISE

Write the systematic name for the following cycloalkane:

Answer: *trans*-1,3-dimethylcyclopentanes

(See also Exercises 23.16, 23.17.)

23.3 | CHIRALITY IN ORGANIC COMPOUNDS

Let us begin our discussion of **optical isomers** in organic chemistry by looking at ▶ FIGURE 23.4(a). Molecule **A** consists of a tetrahedral carbon atom with four different groups attached, represented by the white, blue, red and yellow balls. Is **A** identical to its mirror image, **B**? This question is best explored by making three-dimensional models and trying to overlay them; alternatively, try to imagine rotating **A** by 180° and placing it on top of **B**. What happens? The blue, black and white balls align, but the red and yellow are reversed. In fact, any way you try to overlay these two molecules, the result is the same—three of the balls are superimposable but the other two are not.

Molecules possessing non-superimposable mirror images, such as **A** and **B**, are termed **chiral** (Greek *cheir*, 'hand'; pronounced *ky-ral*)—that is, they show handedness (Figure 23.4(b)). In fact, *all molecules containing a single carbon atom bearing four different groups are chiral.* The fact that molecules **A** and **B** cannot be overlaid demonstrates that they are *different substances.* The relationship between two molecules that are non-superimposable mirror images is described by the term **enantiomer**. Hence, molecules **A** and **B** are enantiomers. For example, ▶ FIGURE 23.5 shows the two enantiomers of 2-bromopentane.

All enantiomers have a **chiral centre** (or **stereogenic centre**)—a central atom (usually carbon) to which four different groups are attached. This feature of the molecule, also known as the **stereocentre**, gives the molecule its chirality. Molecules that are superimposable are **achiral**, meaning *not chiral.* Achiral molecules can usually be identified because they possess a plane of symmetry (▼ FIGURE 23.6).

We can make the following general rules at this point:

1. *Stereocentres within any compound can be recognised because they have four different groups bonded to them.*

2. *The presence of one or more stereocentres always results in stereoisomers.*

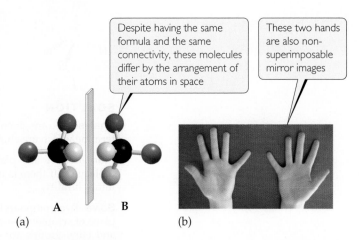

Despite having the same formula and the same connectivity, these molecules differ by the arrangement of their atoms in space

These two hands are also non-superimposable mirror images

A   **B**

(a)   (b)

▲ **FIGURE 23.4   Mirror images.**
(a) These two molecules are non-superimposable mirror images. Being able to move molecules around mentally in three-dimensional space to prove two molecules are non-superimposable is extremely difficult and will require practice before it is mastered. (b) Handedness is a term often used to describe chirality.

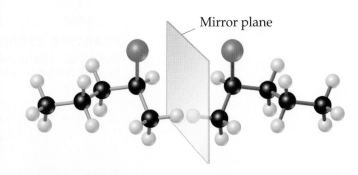

Mirror plane

▲ **FIGURE 23.5   The two enantiomeric forms of 2-bromopentane.** The mirror-image isomers are not superimposable with each other.

CONCEPT CHECK 2

What guiding principle determines whether something is chiral?

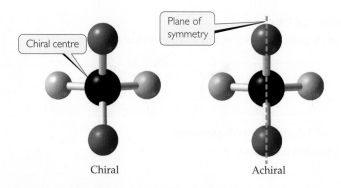

▶ **FIGURE 23.6 Chiral and achiral.**
A molecule that contains one chiral centre (stereocentre) is chiral (left). If a plane of symmetry exists, as in the molecule on the right, that molecule is said to be achiral.

SAMPLE EXERCISE 23.2 **Predicting chirality**

Determine which of the following structures is chiral:

I   II   III   IV

SOLUTION

Analyse We are given four structures containing wedged bonds and are asked to determine which are chiral.

Plan Identify any planes of symmetry among the structures to determine achiral molecules. If there is no plane of symmetry, identify the stereogenic centre to determine chirality.

Solve Two compounds are chiral: II and IV. Compounds I and III have at least one plane of symmetry—in fact, compound I has two planes of symmetry. Note that methyl and ethyl groups are classed as different groups, as are the halogens, chlorine and bromine. Remember, the alternative (and sometimes easier) way to determine whether a molecule is chiral or not is to check each carbon atom to see whether it is a stereocentre.

I
achiral

II
chiral

III
achiral

IV
chiral

PRACTICE EXERCISE

Identify the chiral centre in the following molecule and label it with an asterisk. Draw the molecule's mirror image.

H
|
C
H₃C   NH₂
COOCH₃

Answer:

mirror
plane

H
|
*C
H₃C   NH₂
COOCH₃

H
|
C
H₂N   CH₃
H₃COOC

(See also Exercises 23.7, 23.14.)

Most of the physical and chemical properties of enantiomers are identical because each enantiomer contains the same functional groups and the same non-bonding interactions. Enantiomers, such as the two forms of 2-bromopentane shown in Figure 23.5, have identical physical properties, such as melting and boiling points, and identical chemical properties when they react with achiral reagents. The properties of two enantiomers *differ only if they are in a chiral environment*. An easy way of demonstrating the differences between enantiomers in a chiral environment is to purchase caraway seeds and spearmint leaves from your local supermarket. Each product contains a different enantiomer of carvone (▶ FIGURE 23.7). Now smell them and note the stark difference.

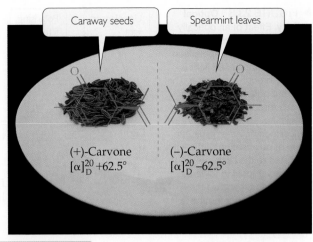

▲ **FIGURE 23.7 Enantiomers of carvone.** Caraway seeds smell very different from spearmint leaves or oil. The difference is a result of how each molecule interacts with the chiral receptors in the nose.

 CONCEPT CHECK 3

Find the chiral centre in carvone.

The difference in their smell is a result of the chirality of the receptors contained within your nose. Each enantiomer fits into the receptor's binding site in a similar way to how our right or left hands would fit into a left-hand glove. For one enantiomer, the fit is perfect, leading to a maximum interaction between the molecule and receptor. The other enantiomer fits awkwardly, leading to less than perfect binding and a different response, perhaps a different smell or taste.

23.4 | MEASURING OPTICAL ACTIVITY

Optical isomers are usually distinguished from each other by their interaction with **plane-polarised light**. If light is polarised—for example, by passage through a polarising filter—the light in only one particular plane is allowed to pass through, as shown in ▼ FIGURE 23.8. If the plane-polarised light is passed through a solution containing one enantiomer, this plane is rotated through an angle α, measured in degrees, to the right (clockwise, or +) or to the left (anticlockwise, or −) as viewed by an observer. The isomer that rotates plane-polarised light to the right is **dextrorotatory** (Latin *dexter*, 'right'); it is labelled the *d*-(+)-isomer. Its enantiomer rotates plane-polarised light to the left and is called **levorotatory** (Latin *laevus*, 'left'); it is labelled the *l*-(−)-isomer. Because of their effect on plane-polarised light, chiral molecules are said to be **optically active**.

 CONCEPT CHECK 4

Is *d*-(−) an appropriate descriptor?

Figure 23.7 provides information on the results of the plane-polarised light experiment for the enantiomers of carvone. Notice that one enantiomer rotates light in a positive direction while the other rotates light in a negative direction. The term $[\alpha]_D$ is known as the **specific rotation** (Equation 23.1) and takes into account the concentration of the sample c (g cm^{-3}) and the path length l of the sample (dm, 1 dm = 0.1 m) confined within a solution cell, as well as the observed rotation α (degrees). The specific rotation is usually expressed in degrees, even though the units formally should be cm^3 dm^{-1} g^{-1}.

$$[\alpha]_D^{20} = \frac{\alpha}{c \times l}$$ [23.1]

The subscript D refers to the source of monochromatic radiation used for the measurement (typically the sodium-D line, λ = 589 nm). The superscript 20 above the D refers to the temperature (in degrees Celsius) of the measurement,

▶ **FIGURE 23.8 Optical activity.** Effect of an optically active solution on the plane of polarisation of plane-polarised light. The unpolarised light is passed through a polariser. The resultant polarised light then passes through a solution containing a dextrorotatory optical isomer. As a result, the plane of polarisation of the light is rotated to the right relative to an observer looking towards the light source and so the optically active compound is dextrorotatory.

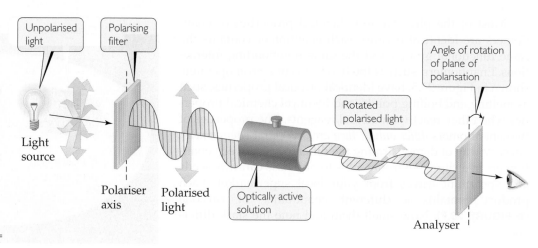

(a)

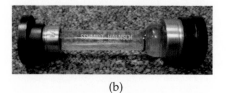

(b)

▲ **FIGURE 23.9 A polarimeter and solution cell.** A polarimeter (a) is used to measure the rotation of plane-polarised light as it passes through a sample contained within a 1 dm cell (b).

as optical activity is temperature dependent. Note that the magnitudes of the specific rotation for d-(+)- and l-(−)-carvone are identical—only the sign changes. An example of the instrument used to perform this optical experiment, called a **polarimeter**, as well as the type of solution cell used, is shown in ◀ **FIGURE 23.9**.

Optical rotation is a useful way of determining whether a compound is **enantiomerically pure**—that is, contains a single enantiomer. Failure to obtain the same specific rotation for a compound prepared in the laboratory as that reported in the literature most often happens because some of the other enantiomer is present. Often, no rotation can be observed for a substance whose molecules contain a chiral centre. This usually means a 50 : 50 mixture of the two enantiomers has formed under the reaction conditions. A mixture with an equal amount of both enantiomers is called a **racemic mixture**, or *racemate*. A racemic mixture of enantiomers does not rotate plane-polarised light because the two enantiomeric forms rotate the light to equal extents in opposite directions, cancelling out any optical effect. Since a racemic mixture contains equal amounts of the (+) and (−) enantiomer it is often designated (±) or (*dl*).

⚠ **CONCEPT CHECK 5**

What requirement is needed by the four substituents on carbon to call it a chiral centre?

In 1848 Louis Pasteur (1822–1895) achieved the first reported separation of a racemic mixture into optical isomers (▶ **FIGURE 23.11**). Working with the compound sodium ammonium tartrate, he separated 'right-handed' crystals from 'left-handed' crystals, using a microscope and a pair of tweezers. He achieved this separation (called a *resolution*) after noticing that the two optical isomers formed separate crystals whose shapes were mirror images of each other. Dissolving the two crystalline forms in water and subjecting them to plane-polarised light, Pasteur was able to demonstrate that the optical activity of the mirror images was opposite and different from the original racemic mixture. Another salt of tartaric acid, calcium tartrate, is occasionally formed on the corks of white wines (Figure 23.10(b)).

▶ **TABLE 23.1** lists the specific rotation for several well-known organic compounds. Note that *specific rotation is an experimental quantity* and neither the magnitude nor the sign is easily predicted from the structure. An example of the difficulties in prediction is shown in the acid hydrolysis of sucrose, a compound that contains nine stereogenic centres. Here we need not concentrate so much on the complex sugar structures as on the sign and magnitudes of the specific rotation of starting materials and products.

▲ **FIGURE 23.10** Louis Pasteur (1822–1895) was a French chemist and biologist. He was responsible for some of the most important concepts and applications of modern science, including the discovery of the science of microbiology, the process of pasteurisation and molecular asymmetry.

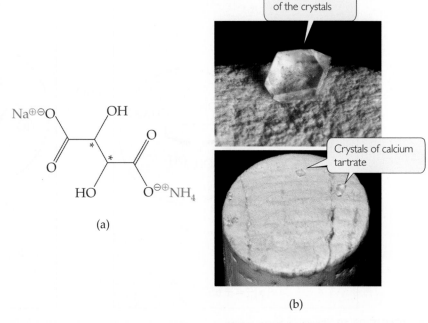

(a)

Notice the facets of the crystals

Crystals of calcium tartrate

(b)

▲ **FIGURE 23.11** **Resolution of a racemic mixture.** Crystals of sodium ammonium tartrate were separated by Pasteur into the two enantiomeric forms. (a) Pasteur was able to separate the crystals based on the two mirror-image forms of the crystal facets. (b) Because of the high concentration of tartaric acid in white wine, it sometimes crystallises as calcium tartrate on corks after long cellaring.

Sucrose, $[\alpha]_D^{25} = +65°$, is hydrolysed (broken up by the addition of water) under acidic conditions to give glucose, $[\alpha]_D^{25} = +53°$, and fructose, $[\alpha]_D^{25} = -89.5°$ (▼ **FIGURE 23.12**). The 1 : 1 mixture of glucose and fructose that is obtained upon the hydrolysis of sucrose is called *invert sugar* because the specific rotation of the mixture is dominated by the negative rotation of fructose. Since the total specific rotation is effectively the sum of individual specific rotations, hydrolysis changes the optical properties of the solution from rotating light clockwise (+) to anticlockwise (−), despite very little molecular difference between the two sides of the equation. Hence, specific rotation is extremely sensitive to molecular structure.

 CONCEPT CHECK 6

What are the similarities and the differences between *d* and *l* isomers of a compound?

TABLE 23.1 • **Specific rotations, $[\alpha]_D$, of some common organic compounds**

Compound	$[\alpha]_D$ (°)	Compound	$[\alpha]_D$ (°)
Cholesterol	−39	*d*-Camphor	+44
Penicillin-V	+233	*l*-lactic acid	−4
Sucrose	+66.5	Quinine	−178
Morphine	−132	*d*-Tartaric acid	+12
Strychnine	−139	Maltose	+141
Fructose	−92	Glucose	+53

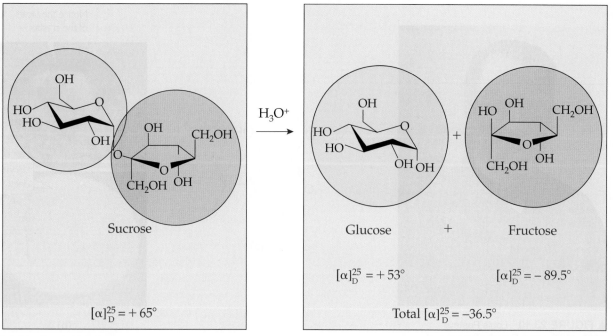

▲ **FIGURE 23.12 Forming invert sugar.** The complete hydrolysis of sucrose gives glucose and fructose as products. Complete hydrolysis of sucrose yields a specific rotation of −36.5°. Fructose is the sweetest common sugar, being about twice as sweet as sucrose; consequently, invert sugar is sweeter than sucrose. The enzyme *invertase*, which bees use in making honey, accomplishes the same chemical result as the acid-catalysed hydrolysis of sucrose.

SAMPLE EXERCISE 23.3 **Specific rotation**

The specific rotation of (−)-carvone is −62.5°, measured at 20 °C. Calculate the observed rotation at 20 °C for a solution prepared by dissolving 1000 mg of (−)-carvone in 30 cm^3 of ethanol and placing it in a solution cell 10 cm long.

SOLUTION

Analyse We are given the specific rotation, temperature, concentration and path length and asked to determine the observed rotation.

Plan These variables are all related by Equation 23.1, where $[\alpha]_D = -62.5$, $l = 10$ cm = 1 dm, $c = 1/30$ g cm^{-3}. We can solve for α.

Solve
Substituting into Equation 23.1 yields

$$-62.5 = \frac{30\alpha}{1 \times 1}$$
$$\alpha = -2.1°$$

PRACTICE EXERCISE

The observed rotation of a solution of camphor prepared by dissolving 1.6 g in 20 cm^3 of hexane was +2.4°. Determine the specific rotation of this sample, which was contained in a 1 dm-long cell, and use Table 23.1 to conclude whether the sample is that of a single enantiomer.

Answer: $[\alpha]_D = +30°$. This value is lower than that listed in Table 23.1; hence the sample contains a small amount of the other enantiomer.

(See also Exercises 23.22, 23.23.)

23.5 | ABSOLUTE STEREOCHEMISTRY

Using Priority Rules to Find a Stereocentre's Absolute Configuration

Although the (*d*)-(+)- and (*l*)-(−)-notation yields important experimental information about a compound, it tells us nothing about the **absolute configuration** (that is, the exact three-dimensional structure) of the molecule. To remove any ambiguity when describing the absolute configuration at a stereogenic centre, the **Cahn–Ingold–Prelog *R,S* notation** has been employed. The Cahn–Ingold–Prelog notation uses **sequence priority rules** to determine a stereocentre's absolute configuration.

The British chemists Robert Cahn (1899–1981) and Christopher Ingold (1893–1970), together with Vladimir Prelog (1906–1998), a Yugoslav chemist based in Switzerland, proposed these rules in 1956 to bring some order to the complex world of organic chemistry. Ingold was the first to propose many of the reaction mechanisms for organic reactions that you read about in textbooks such as this one, and Prelog was awarded the Nobel Prize in Chemistry in 1975 (jointly with Australian-born Sir John Cornforth) for his work on stereochemistry and complex naturally occurring chemical products.

Use the following steps to determine the absolute configuration of a stereogenic centre:

1. Locate the stereocentre.

2. Assign a priority to each substituent from 1 (highest) to 4 (lowest) based on the *atomic number* of the four atoms directly attached to the stereocentre. The higher the atomic number, the higher the priority. A representative set of groups in increasing priority order is

$$-H \quad -CH_3 \quad -NH_2 \quad -OH \quad -Cl \quad -Br$$

Increasing priority

If two atoms are identical in terms of priority, you need to move out along the respective chains until a difference in priority is encountered. Continue to reapply rule 2 to determine priority or until the end of the group is reached. If no difference is found, the molecule has symmetry and is therefore achiral. Examples of increasing priority fragments are

$$-CH_2-H \quad -CH_2-CH_3 \quad -CH_2-NH_2 \quad -CH_2-OH$$

Increasing priority

3. Orient the molecule so that the group with the lowest priority (usually hydrogen) is directed away from you, that is, it is connected to the stereocentre by a dashed wedge ($\cdots$ıιll).

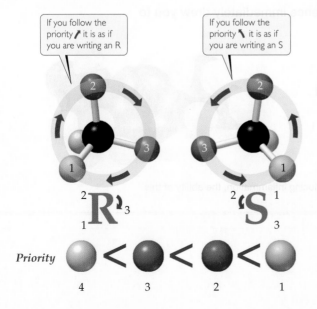

If you follow the priority ↗ it is as if you are writing an R

If you follow the priority ↖ it is as if you are writing an S

Priority   ○ < ● < ● < ○

4   3   2   1

MY WORLD OF CHEMISTRY

CHIRAL DRUGS

Many drugs of importance in medicine are chiral substances. When a drug is prepared, sold and administered as a racemic mixture, it often turns out that only one of the enantiomers has beneficial results. The other enantiomer is either inert or nearly so, or may even have a harmful effect. For the last reason, a great deal of attention has been given in recent years to methods for synthesising the desired enantiomer of chiral drugs. Chemical research directed at understanding how to control the stereochemistry of a reaction has led to new synthetic methods that make it not only possible, but relatively cost-effective, to produce chiral molecules in *enantiomerically pure* form. Worldwide sales of single-enantiomer drugs amounts to over AUS$150 billion annually.

Albuterol is a good example of the different physiological behaviour of enantiomers. The drug (*R*)-albuterol (▼ FIGURE 23.13) is an effective bronchodilator used to relieve the symptoms of asthma. Its enantiomer, (*S*)-albuterol, is not only ineffective as a bronchodilator but actually counters the effects of (*R*)-albuterol.

The non-steroidal analgesic ibuprofen (marketed under the trade names Advil™ and Nurofen™) is a chiral molecule usually sold as a racemic mixture. The more active enantiomer, (*S*)-ibuprofen (▼ FIGURE 23.14), relieves pain and reduces inflammation, whereas the weakly active *R* enantiomer, which isn't as easily utilised, is actually converted by enzymes within the body to the active *S* form.

In the 1950s and 1960s the drug thalidomide was administered to pregnant women as an anti-emetic to combat morning sickness. It was found to be teratogenic, being responsible for severe deformities in over 12 000 foetuses worldwide. The case against thalidomide has two sides, however. First, rigorous testing of all chiral drug candidates has now been made mandatory for both enantiomers after the thalidomide experience. The *R* enantiomer of thalidomide was active in preventing morning sickness during pregnancy, whereas the *S* enantiomer was responsible for the terrible side-effects in humans. Unfortunately, racemisation of (*R*)-thalidomide (to form the *S* enantiomer) occurs easily within the body, so even the useful *R* enantiomer has no therapeutic use for pregnant women. Australian researchers began a trial of thalidomide in 2002, involving cancer patients. The study found thalidomide sparked an increase in T-cells, which allowed their own immune response to fight the cancer. Thalidomide has also been approved by the US Food and Drug Administration for the treatment of leprosy. Trials are also under way for epilepsy and HIV.

RELATED EXERCISES: 23.27, 23.29

(*R*)-thalidomide

▲ FIGURE 23.13 (*R*)-Albuterol. This compound acts as a bronchodilator in patients with asthma. In contrast, (*S*)-albuterol counters this effect.

▲ FIGURE IT OUT

Locate the chiral centre in Ibuprofen. What evidence immediately drew you to that choice?

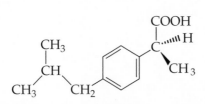

▲ FIGURE 23.14 (*S*)-Ibuprofen. For relieving pain and reducing inflammation, the ability of this enantiomer far outweighs that of the *R* enantiomer.

4. Read the three groups orientated towards you in order from highest (1) to lowest (3) priority.

5. If the numbers increase from 1–3 in a clockwise manner, the configuration is *R* (from the Latin *rectus*, 'straight, correct, right'). If the numbers increase in an anticlockwise manner, the configuration is *S* (from the Latin *sinister*, 'left').

6. If the compound containing the stereocentre is *R*, then its enantiomer is *S*, and *vice versa*.

SAMPLE EXERCISE 23.4   *R* and *S* notation

Assign the absolute configuration as *R* or *S* to the following compound:

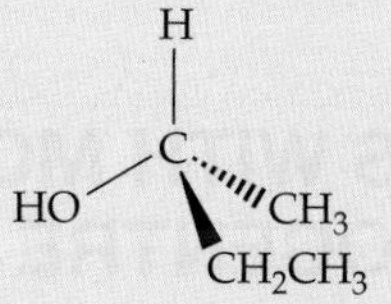

SOLUTION

Analyse We are asked to identify this molecule as *R* or *S*.

Plan Follow the sequence priority rules after identifying the stereocentre.

Solve
Step 1

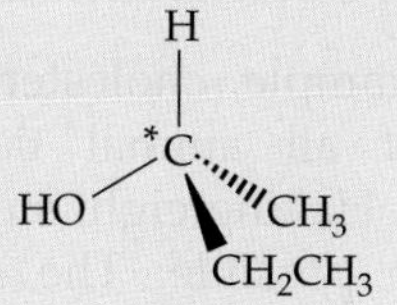

Step 2

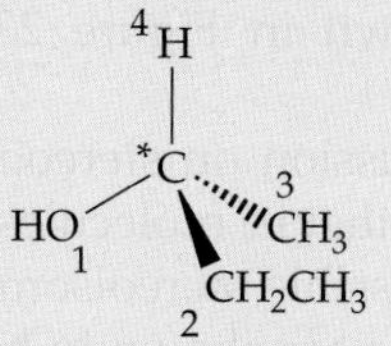

Step 3

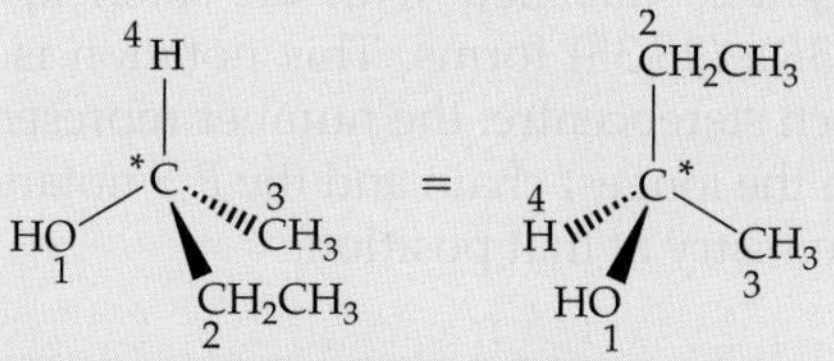

Step 4

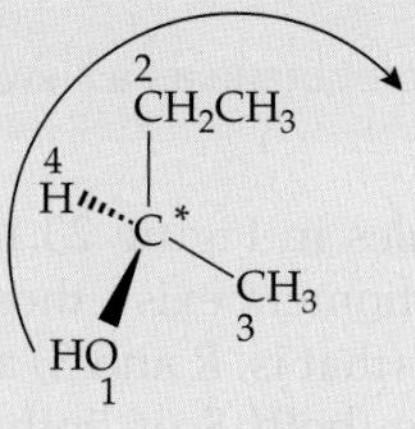

Step 5 *R* configuration

Name (*R*)-2-butanol

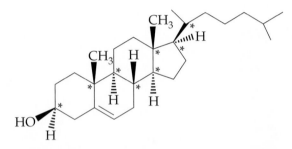
23.6 | MOLECULES WITH MORE THAN ONE STEREOCENTRE

A chiral molecule with one stereocentre can exist as one of two enantiomers, R or S. Generally, for a molecule with n stereocentres, the maximum number of stereoisomers possible is given by Equation 23.2

$$\text{Number of stereoisomers} = 2^n \qquad [23.2]$$

By way of example, cholesterol, a waxy solid found in the blood plasma of all animal tissue, has eight stereocentres (◄ FIGURE 23.15). Having eight stereocentres means that $2^8 = 256$ stereoisomers are possible. The specificity of natural systems towards one chiral form is illustrated dramatically in this case. The stereoisomer shown in Figure 23.15 is the *only* one produced naturally.

▲ FIGURE 23.15 **Cholesterol.** Although 256 possible stereoisomers exist, remarkably only a single stereoisomer exists in nature.

Let us complete our discussion on stereochemistry by considering molecules with two stereocentres—that is, molecules that can exist as a maximum of four stereoisomers. All the possible stereoisomeric permutations for the compound 2-bromo-3-chlorobutane are shown in ▶ FIGURE 23.16. This compound contains two stereocentres (labelled with an asterisk) and so can exist in (2R,3S), (2S,3R), (2R,3R), (2S,3S) forms. This notation is used to describe the stereochemistry at each stereocentre: the number represents the position of the chiral carbon atom on the longest chain and the R,S notation is used to describe the absolute stereochemistry at that position.

⚠ CONCEPT CHECK 7

Draw (2R,3S) 2-bromo-3-chlorobutane as a Newman projection.

Each of the chiral molecules in Figure 23.16 is related to the other three. Notice that two pairs of enantiomers exist: those containing opposing stereochemistry at each stereocentre (that is, R and S) and those with the same stereochemistry at each stereocentre (both S or both R). Thus the enantiomer of a molecule with (R,R) notation always has (S,S) notation. The two pairs of enantiomers are also related. They are neither non-superimposable nor mirror images and are called **diastereomers**. Diastereomers, by definition, are any set of stereoisomers that are not enantiomers. *Whereas enantiomers have identical physical and chemical properties in an achiral environment, diastereomers always have different chemical and physical properties.*

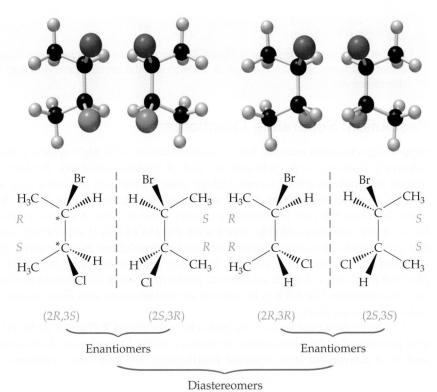

◄ FIGURE 23.16 Enantiomers and diastereomers. Compounds with two stereocentres have a maximum of four stereoisomers. These stereoisomers have a relationship to each other. The name describing the relationship differs if the two molecules in question are non-superimposable mirror images (enantiomers) or not (diastereomers).

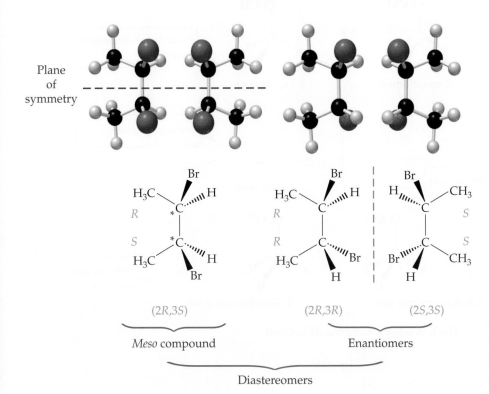

◄ FIGURE 23.17 *Meso* compounds. Despite possessing two stereocentres, 2,3-dibromobutane has only three stereoisomers. The *meso* form is achiral because it contains a plane of symmetry.

Let's now consider 2,3-dibromobutane as an example (▲ **FIGURE 23.17**). This compound has two stereocentres. However, a plane of symmetry exists through the C–C single bond between the two stereocentres in the (*R,S*) form, making it identical to the (*S,R*) form. From our earlier definition, a molecule possessing a plane of symmetry is not chiral. An achiral compound possessing two or more stereocentres is known as a **_meso_ compound**.

> ⚠️ **CONCEPT CHECK 8**
>
> Revisit the geometric isomers in Figure 23.3 for 1,2-dimethylcyclohexane. Can these structures be related using *R,S* notation and the terms enantiomer and diastereomer?

Resolution: Separating Enantiomers

Separating a racemic mixture into two enantiomers with high optical purity is extremely difficult to do without the aid of a chiral auxiliary. Pasteur was indeed fortunate that the racemic tartaric salts crystallised in such a way as to contain a single enantiomer within one crystalline form (Figure 23.10). More often than not, the two enantiomers co-crystallise to form a single crystalline form, or are indistinguishable. Even with the aid of chiral separating agents, this process of separating enantiomers (called **resolution**) is extremely expensive and time-consuming. We can, however, use the different chemical and physical properties of diastereomers to perform a clever experiment to resolve two enantiomers. The trick is to convert the enantiomers into diastereomers in order to separate them.

The resolution process is explained schematically in ▼ **FIGURE 23.18**. Whenever possible, the chemical reactions involved in the formation of diastereomers and their conversion into separate enantiomers are acid–base reactions. The

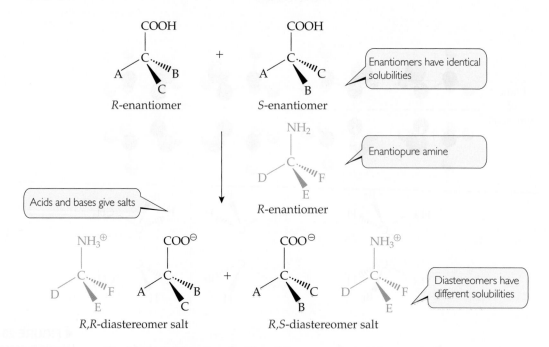

▶ **FIGURE 23.18 Resolution.**
Enantiomers, especially those containing a carboxylic acid functionality, can be separated by converting them into diastereomeric salts. Typically, crystallisation leads to the separation of one diastereomeric salt over the other. Conversion of the salt back to the carboxylic acid yields a purified enantiomer.

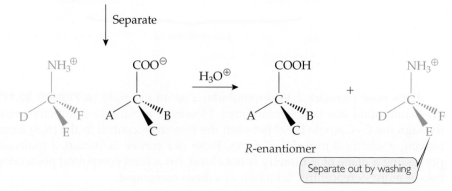

reason for this is simple: organic salts crystallise well in almost any solvent (including water) and the acid–base chemistry employed is uncomplicated. Here, an *enantiopure* amine such as (*R*)-α-methylbenzylamine (RAMBA) or (+)-chinchonine is added to a racemic mixture of a carboxylic acid. The two salts that ensue are diastereomeric because they each contain two chiral centres, yet are non-superimposable. Separation of one diastereomeric salt from the other is typically achieved by crystallisation methods. In order to maximise enantiomeric purity, the crystallisation is repeated several times. The pure enantiomer is subsequently liberated by treatment with aqueous acid.

The method described in Figure 23.18 is also widely used for the resolution of chiral amines, using enantiopure carboxylic acids such as the naturally occurring carboxylic acid, (*S*)-malic acid. Other methods, depending on the functionality present within the *racemate*, have also been developed. One of the more interesting methods involves an enzymatic resolution. Since biological systems such as enzymes are highly specific for a particular chirality, they can be conveniently used to perform simple derivatisations to a single enantiomer in a racemic mix. Derivatisation (for example, hydrolysis of an ester functionality) changes both the physical and chemical properties, allowing for easy separation and hence resolution.

SAMPLE INTEGRATIVE EXERCISE   **Putting concepts together**

Alanine is one of the 20 most common naturally occurring amino acids. A methyl ester derivative (RCOOCH$_3$), prepared in the laboratory, had the following stereochemistry:

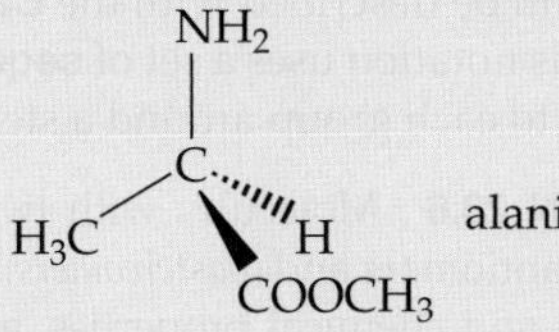

alanine methyl ester

(a) Identify the chiral centre in alanine methyl ester. How many possible stereoisomers exist for alanine? **(b)** List the groups attached to this chiral centre in *decreasing order* of sequence rule priority, then state whether the compound has the *R* or *S* configuration about this centre. **(c)** Draw the other enantiomer of alanine methyl ester. **(d)** Describe how a pure sample of this compound may be obtained from a racemic mixture.

SOLUTION

(a) The central carbon atom in alanine methyl ester is the only chiral centre since it has four different groups attached to it. Having only one stereocentre means that only two stereoisomers are possible.

(b) Using the sequence priority rules in Section 23.5, NH$_2$ > COOCH$_3$ > CH$_3$ > H. Since the lowest-priority group is already oriented back into the page, this compound has an *R* configuration.

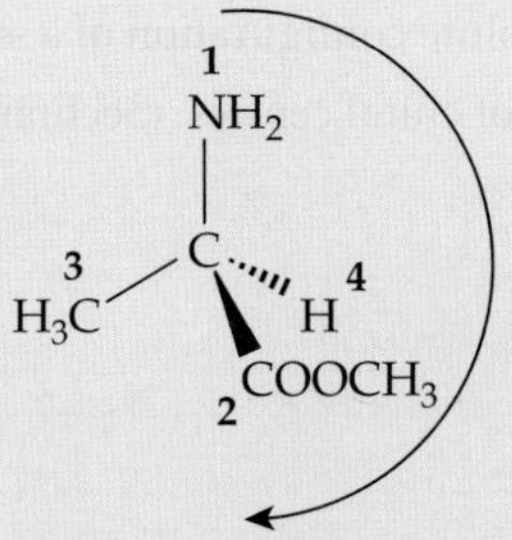

(c)

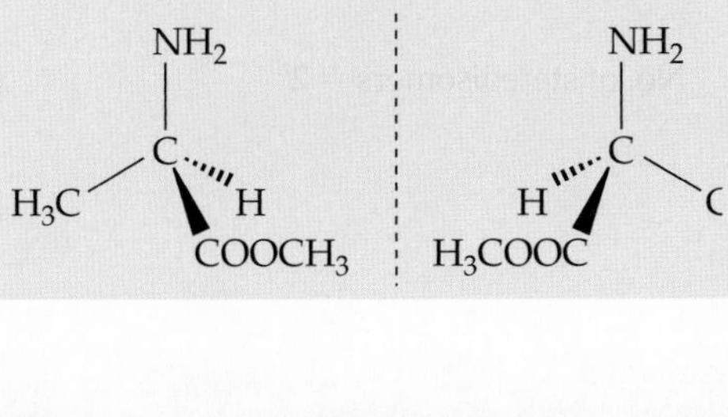

(d) It is very difficult to separate enantiomers directly, and in quantity. Since this compound contains an amine functional group, the best way to achieve resolution would be to react it with a single enantiomer of a chiral carboxylic acid such as (*S*)-malic acid. This would form the *R,S* and *S,S* diastereomeric salts, which can be separated by fractional crystallisation. Once isolated, addition of base would re-generate the amine, which could be separated by extraction from the water-soluble carboxylate salt and dried. Removing the solvent leaves the pure enantiomer.

(See also Exercise 23.8.)

CHAPTER SUMMARY AND KEY TERMS

SECTION 23.1 **Stereoisomers** are isomers with the same connectivity of atoms (bonding) but different arrangement of their atoms in space. The most common forms of stereoisomerism are **geometric isomers**, **enantiomers** and **diastereomers**.

SECTION 23.2 Two substituents on a cycloalkane can be *cis* or *trans* to each other. *Cis* and *trans* isomers are called geometric isomers. Geometric isomers differ from one another in their chemical and physical properties.

SECTION 23.3 **Optical isomers** are non-superimposable mirror images of one another. They are molecules that contain a **stereocentre**. Optical isomers, or enantiomers, are **chiral**, meaning that they have a specific 'handedness' and their physical and chemical properties differ only in the presence of a chiral environment. Molecules that are superimposable are **achiral**, meaning not chiral.

SECTION 23.4 Optical isomers can be distinguished from each other by their interactions with **plane-polarised light**; solutions of one enantiomer rotate the plane of polarisation to the right (**dextrorotatory**, +) and solutions of its mirror image rotate the plane to the left (**levorotatory**, −) with the same magnitude. Chiral molecules, therefore, are **optically active**. A **polarimeter** is used to determine the **specific rotation** of a chiral sample. A compound that is **enantiomerically pure** contains a single enantiomer. A mixture with an equal amount of two enantiomers is called a **racemic mixture**, or *racemate*.

SECTION 23.5 The **absolute configuration** around a stereocentre can be described with the **Cahn–Ingold–Prelog *R,S* notation**. This notation uses a set of **sequence priority rules** to assign priority to each group around a stereocentre.

SECTION 23.6 Molecules with two or more chiral centres can form enantiomers and *diastereomers*. Diastereomers have different physical and chemical properties, a trait that can be used in the **resolution** of enantiomers from a *racemic mixture* to yield an enantiomerically pure compound. An achiral compound possessing two or more stereocentres is known as a *meso* **compound**.

KEY SKILLS

- Be able to draw three-dimensional structures in two dimensions using bold and dashed wedges. (Sections 23.1 and 23.3)
- Understand the concept of chirality and be able to identify a chiral centre. (Section 23.3)
- Use the Cahn–Ingold–Prelog rules to describe unambiguously the absolute configuration of a stereocentre. (Section 23.5)
- Calculate the number of potential stereoisomers based on the number of chiral centres. (Section 23.6)

KEY EQUATIONS

- Specific rotation

$$[\alpha]_D^{20} = \frac{\alpha}{c \times l}$$ [23.1]

- Maximum number of stereoisomers for a compound with *n* stereocentres

$$\text{No. of stereoisomers} = 2^n$$ [23.2]

EXERCISES

VISUALISING CONCEPTS

23.1 List some of the many items used in everyday life that are chiral.

23.2 Which of the following compounds is capable of possessing isomerism? In each case where isomerism is possible, identify the type or types of isomerism.

(a)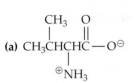

(b)

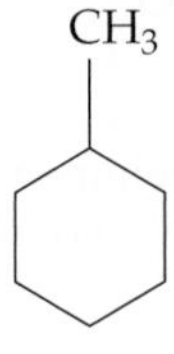

(c)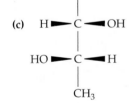

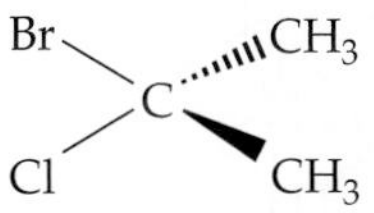

(d) $CH_3CH_2CH_2CH_3$

23.3 Which of the following have a non-superimposable mirror image?

(a)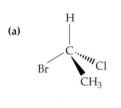

(b)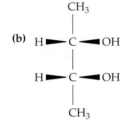

(c) (see figure)

(d) (see figure)

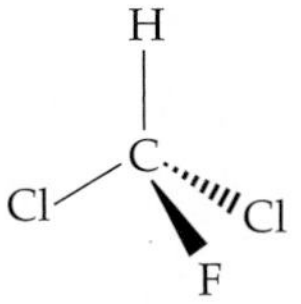

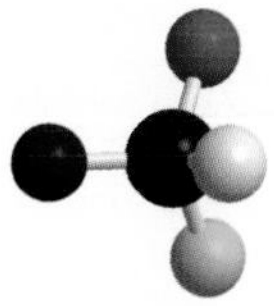

(e)

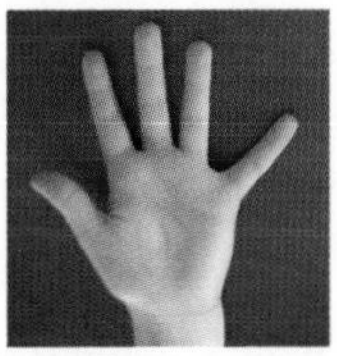

(f)

23.4 Decalin is a C_{10} compound composed of two fused six-membered rings. It is often used as a model to describe conformations in steroids, such as cholesterol. Two geometric isomers are possible, *cis*-decalin and *trans*-decalin. Which of these two isomers is the most stable? Explain your answer.

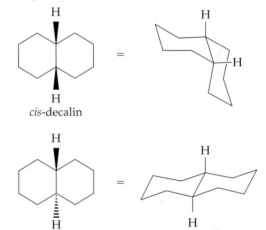

cis-decalin

trans-decalin

CHIRALITY IN ORGANIC COMPOUNDS AND *CIS–TRANS* ISOMERISM (Sections 23.2 and 23.3)

23.5 Locate the chiral carbon atoms, if any, in each of the following natural products:

(a)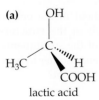
lactic acid

(b)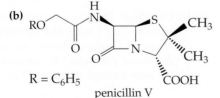
$R = C_6H_5$
penicillin V

(c)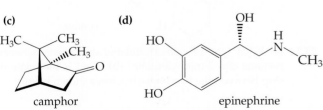
camphor

(d)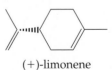
epinephrine

23.6 You are very familiar with limonene. (+)-Limonene gives oranges and lemons their odour.

(+)-limonene

(a) Is it possible to tell the absolute configuration of (+)-limonene from its name? **(b)** Identify the chiral centre in (+)-limonene. **(c)** Draw the mirror image of (+)-limonene. **(d)** Indicate the absolute stereochemistry of (+)-limonene.

23.7 Brevetoxin A, a cyclic polyether made up of 11 rings, is a potent neurotoxin. It is named from the *brevis* family of algae from which it was extracted. Locate all the stereocentres in Brevetoxin A.

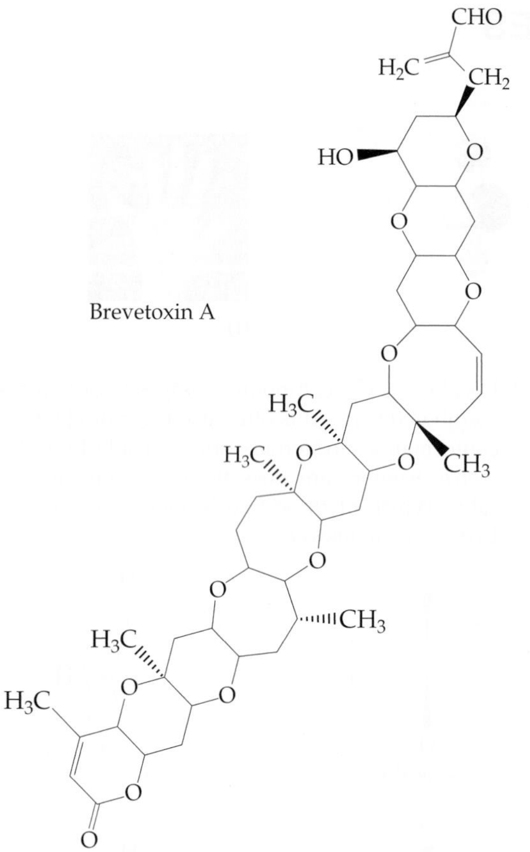

Brevetoxin A

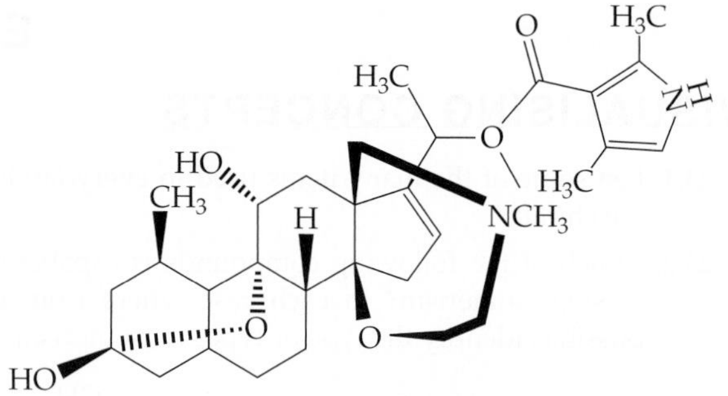

batrachotoxin

23.11 Quinine is a well-know antimalarial drug. How many chiral centres exist in quinine?

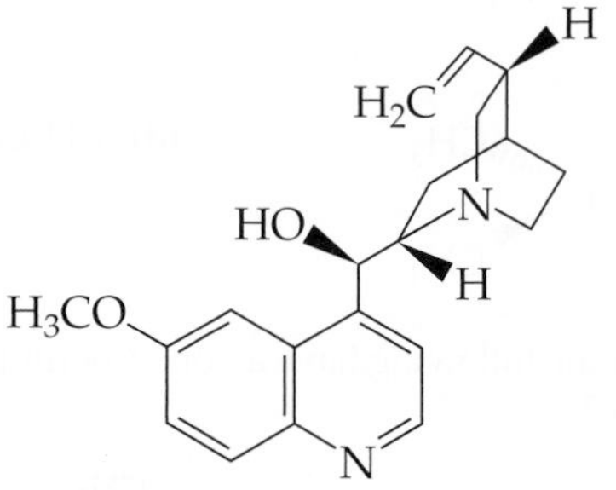

(quinine—an antimalarial drug)

23.8 (a) What conditions are necessary for an organic molecule to be classed as optically active? **(b)** What is a *racemic mixture*, and why doesn't it rotate plane-polarised light?

23.9 Which of the following molecules are chiral?

(a)

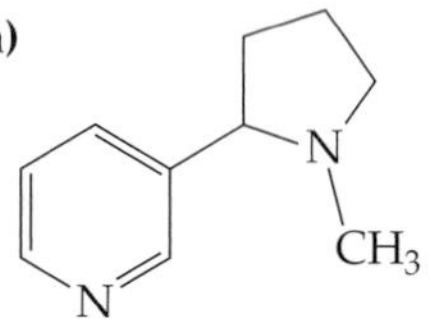

nicotine

(b)

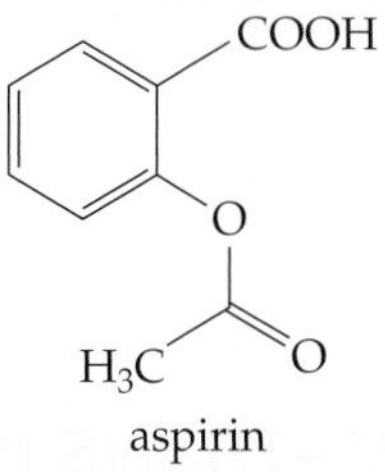

aspirin

(c)

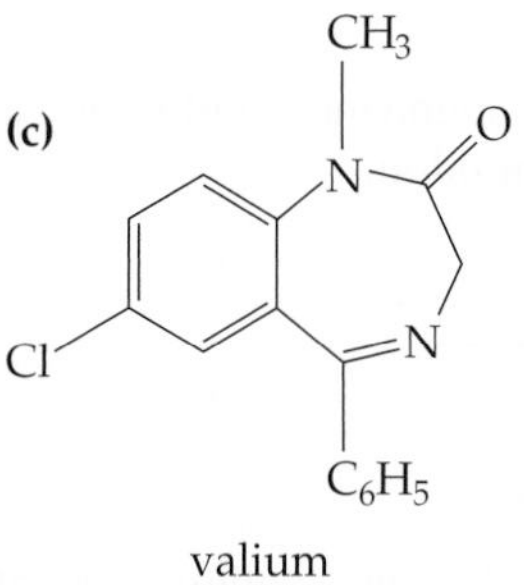

valium

(d)

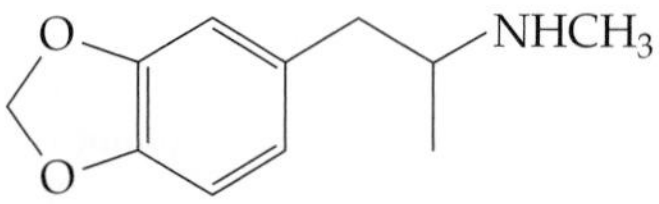

ecstasy

23.10 Batrachotoxin is one of the most potent toxins ever discovered. The poison is secreted from the skin of frogs and is used to coat the tip of blow darts. Only 200 μg is needed to kill a healthy adult. Locate all the stereocentres in batrachotoxin.

23.12 Which of the following molecules contain stereogenic centres?

(a)

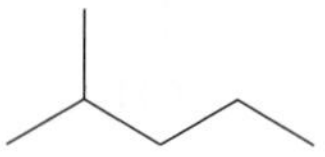

(b)

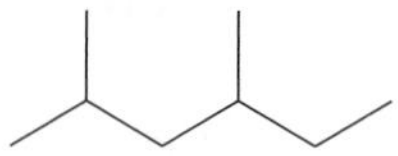

(c)

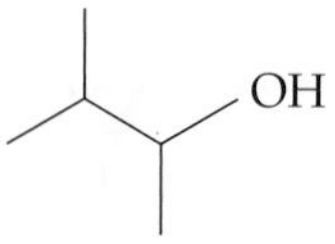

(d)

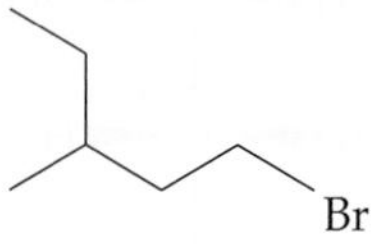

(e)

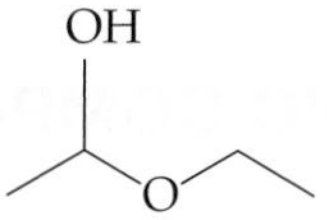

(f)

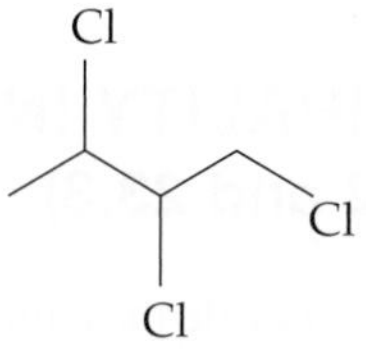

23.13 Coniine is the active natural ingredient in the poison hemlock. **(a)** Locate the stereocentre in the structure below and assign priorities to each of the substituents. **(b)** Would you expect the unnatural enantiomer to have the same biological activity? Why or why not?

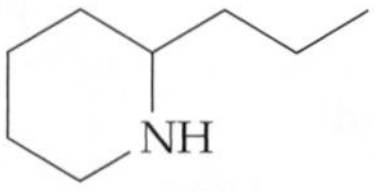

coniine

23.14 (a) What type of isomeric relationship exists between pentane and 2-methylbutane? **(b)** What is the relationship between the two *trans*-structures shown below?

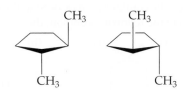

23.15 **(a)** Give the systematic or IUPAC name for each structure, assigning absolute stereochemistry to each centre. Are any of these compounds achiral?

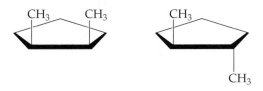

(b) Draw a *trans*-1,2-disubstituted cyclohexane in the chair conformation with your choice of alkyl substituent. What do you notice about the position of the two substituents? Are both substituents axial, or equatorial, or is there one of each? Is this general? Explain. **(c)** Systematically name the structure you drew in part (b), taking into account absolute stereochemistry.

23.16 Draw the structure for 2-bromo-2-chloro-3-methylpentane, and indicate any stereogenic centres in the molecule.

23.17 Does 3-chloro-3-methylhexane have optical isomers? Why or why not?

23.18 Are these two molecules mirror images?

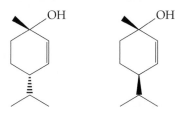

23.19 Is the following compound optically active?

OPTICAL ACTIVITY—MEASURING CHIRALITY (Section 23.4)

23.20 An 8 g sample of sucrose, $[\alpha]_D = +65°$, was dissolved in 25 cm^3 of water and placed in a solution cell of 5 cm length. What was the observed rotation? Is sucrose levorotatory or dextrorotatory?

23.21 The specific rotation, $[\alpha]_D$, of a pure compound obtained from plant material was found to be −20°. Explain the symbolism used to describe specific rotation and comment on what the result means.

23.22 The specific rotation $[\alpha]_D$ of enantiomerically pure *d*-lactic acid is +4°. Calculate the observed rotation of a 5 cm^3 solution of lactic acid containing 6 g of *d*-lactic acid and 2 g of *l*-lactic acid in a standard 10 cm solution cell.

ABSOLUTE STEREOCHEMISTRY (Section 23.5)

23.23 **(a)** What is meant by the term 'enantiomer'? **(b)** What relationship, if any, exists between *R,S* enantiomers and (*d*)-(+) or (*l*)-(−)?

23.24 Assign priorities to the following sets of substituents:
(a) −Br, −CH$_2$CH$_3$, −CH$_3$, −OH
(b) −COOH, −CH(CH$_3$)$_2$, −H, −OCH$_3$
(c) −CH$_2$CH$_2$CH$_2$COOH, −NH$_2$, −COOH, −H
(d) −COOH, −COOCH$_3$, −CH$_2$OH, −OH

23.25 The *S* enantiomer of naproxen is a potent anti-inflammatory agent whereas the *R* enantiomer is a harmful liver toxin. Draw the *S* enantiomer of naproxen and its mirror image.

23.26 Assign an *R,S* configuration to each stereocentre.

23.27 The fastest growing new treatment for asthma is the single-enantiomer drug Singulair (montelukast sodium) from Merck. Sales of the product in 2001 were worth $1.4 billion worldwide. Antidepressants such as GlaxoSmithKline's Paxil (paroxetine hydrochloride) and Pfizer's Zoloft (sertraline) had sales worth over $2 billion worldwide in 2005. Assign an *R,S* configuration to each stereocentre in each drug.

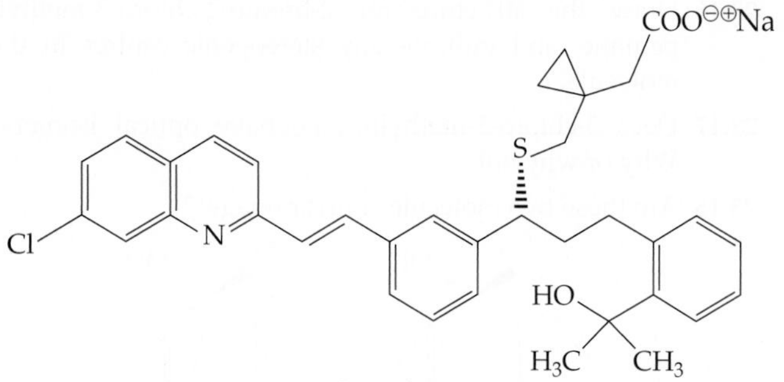

Singulair

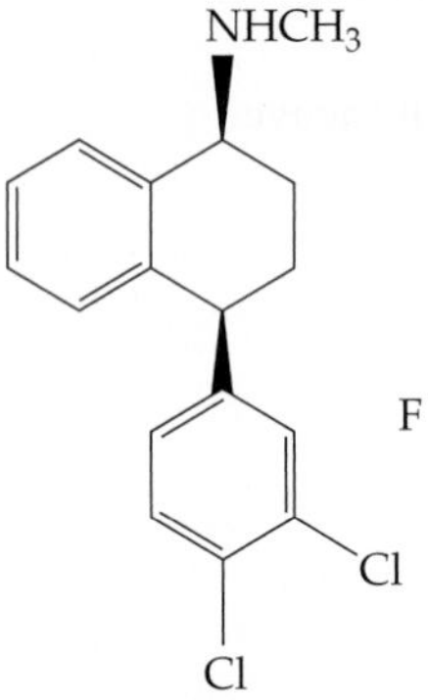

zoloft

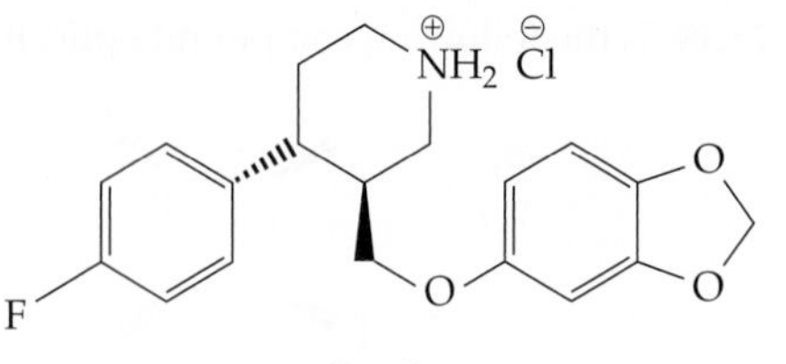

Paxil

23.28 Assign the configuration (*R* or *S*) to the stereogenic centre(s) in the following compounds.

(a)

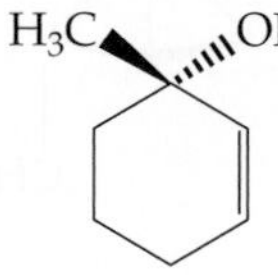

(b)

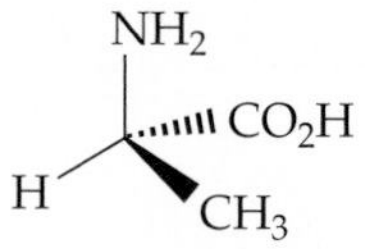

(c)

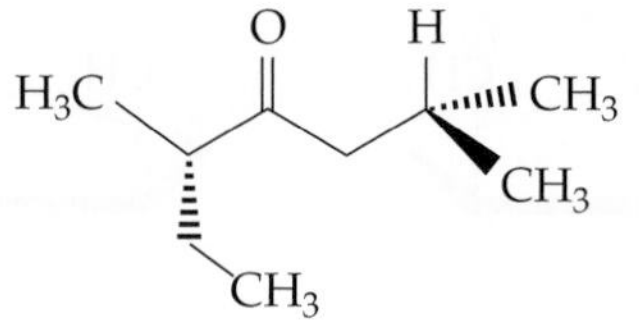

MOLECULES WITH MORE THAN ONE STEREOCENTRE (Section 23.6)

23.29 Draw all possible stereoisomers of tartaric acid and comment on their relationship. Identify any *meso* compounds.

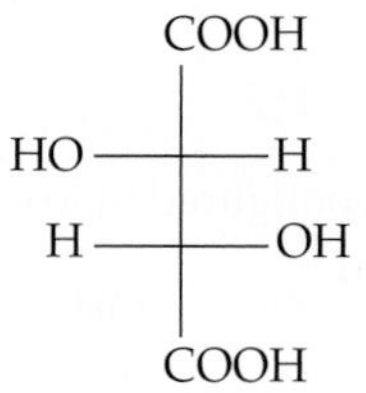

23.30 Assign *R* or *S* configurations to each stereocentre in the following molecules:

(a)

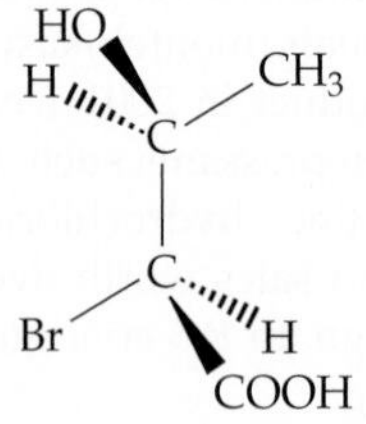

(b)

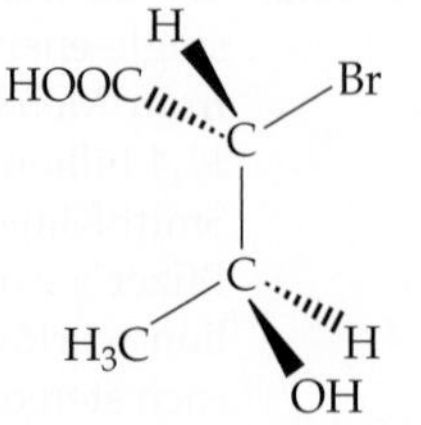

(c)

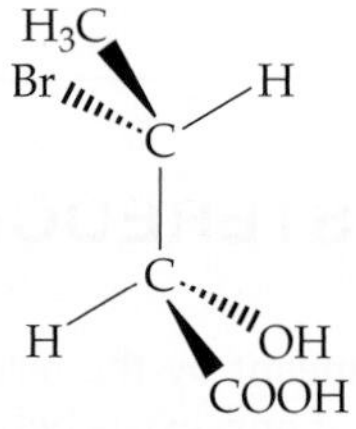

23.31 Draw the *meso* form of each of the following molecules:

(a)

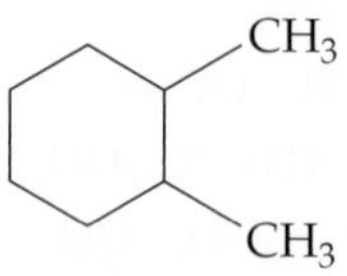

(b)

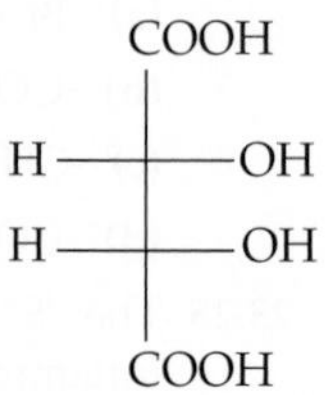

(c)

CH₃CHCH₂CHCH₃

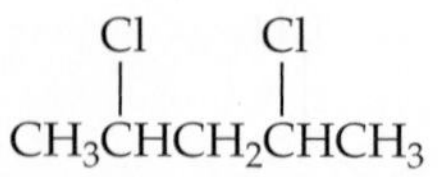

(d)

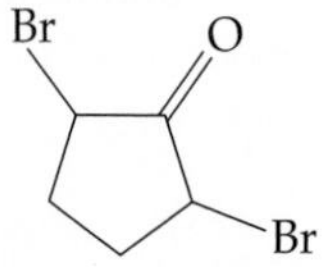

23.32 **(a)** Draw the structure of (2*R*,3*S*)-2-chloro-3-methyl pentane. **(b)** How many different 2,3-dichlorobutanes are there?

ADDITIONAL EXERCISES

23.33 What stereoisomers exist for pentane-2,4-diol?

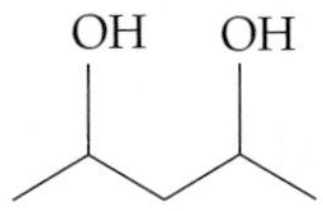

23.34 **(a)** Determine the stereochemistry at the two stereocentres within the Newman projection shown below. **(b)** Draw a Newman projection that is enantiomeric with the one shown below. **(c)** What relationship in specific rotation exists between these two structures?

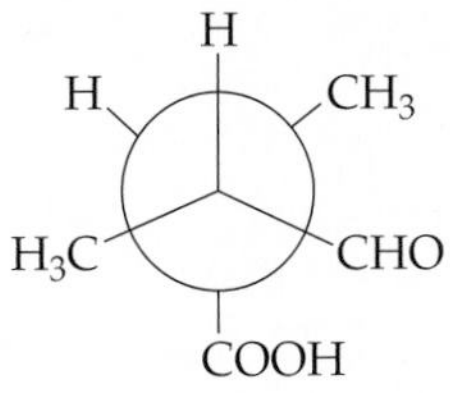

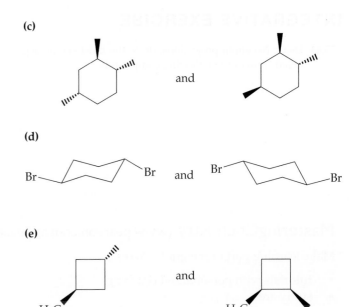

23.35 Convert the following Newman projections into their standard structural formula, retaining stereochemistry.

(a)

(b)

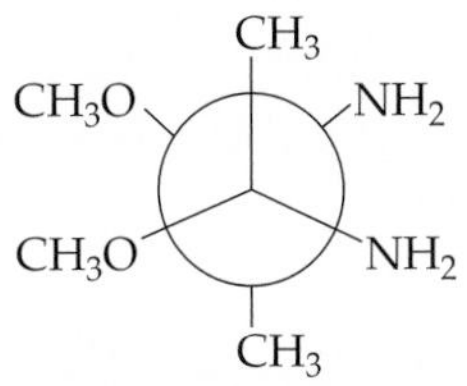

(c)

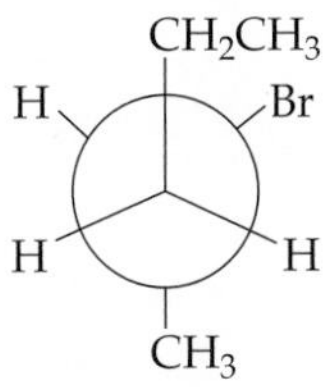

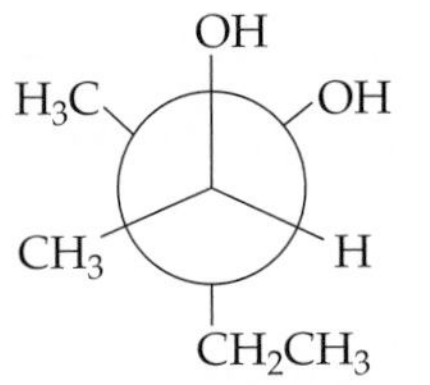

23.36 Draw the structure for 2-bromo-2-chloro-3-methylpentane, and indicate any stereogenic centres in the molecule.

23.37 Does 3-chloro-3-methylhexane have optical isomers? Why or why not?

23.38 Grandisol is a pheromone of the male cotton boll weevil. What is the maximum number of stereoisomers that could exist for this molecule?

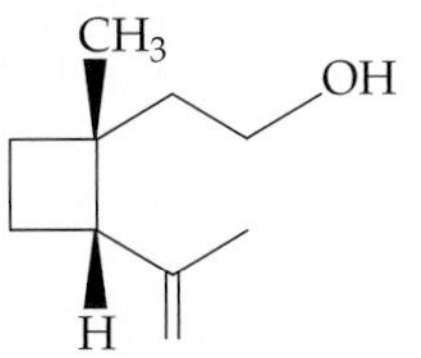

23.39 Indicate whether each of the following pairs of compounds are identical, enantiomers, diasteromers or constitutional isomers.

(a)

and

(b)

and

(c)

and

(d)

and

(e)

and

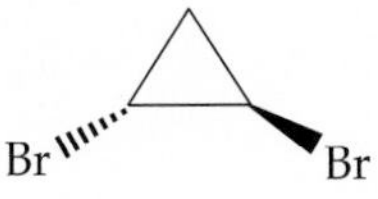

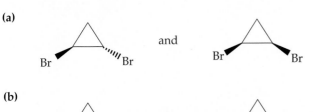

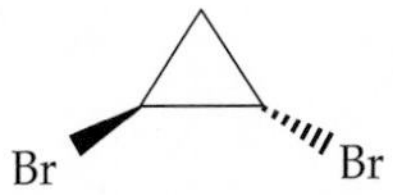

23.40 Indicate whether each of the following pairs of compounds are identical, enantiomers, diasteromers or constitutional isomers.

(a)

and

(b)

and

(c)

and

(d)

and

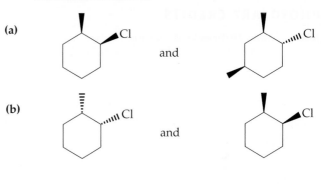

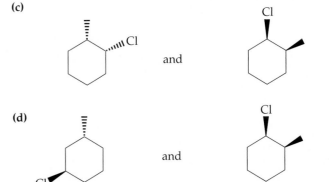

INTEGRATIVE EXERCISE

23.41 Using Newman projections, draw the most stable staggered form of the following compounds:

(a)

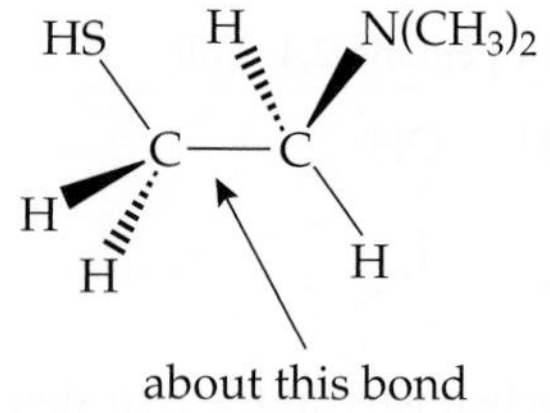

HS H N(CH₃)₂

H C — C H

H H

about this bond

(b)

Cl HO H

Cl C — C CBr₃

Cl

about this bond

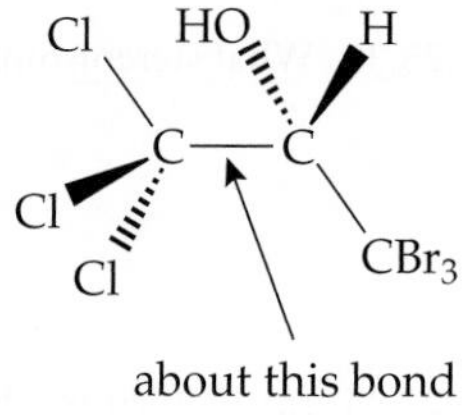

MasteringChemistry (www.pearson.com.au/masteringchemistry)

Make learning part of the grade. Access:

- tutorials with personalised coaching
- study area
- Pearson eText

PHOTO/ART CREDITS

926 Fancy/Alamy; **935 (Figure 23.10)** © Bettmann/Corbis.

24

CHEMISTRY OF ALKENES AND ALKYNES

Colour pigments for use as dyes are formed from unsaturated organic compounds such as those discussed in this chapter.

24.1 THE STRUCTURE OF UNSATURATED HYDROCARBONS

We will begin by investigating the different electronic and geometric features of unsaturated hydrocarbons which have unhybridised p orbitals that overlap in a paired arrangement to form π-bonds. There are two major classes of unsaturated hydrocarbons. *Alkenes,* also known as olefins, are hydrocarbons that contain a C–C double bond, as in ethene (C_2H_4). *Alkynes* contain a C–C triple bond, as in acetylene (C_2H_2).

24.2 ISOMERISM AND NOMENCLATURE

The nomenclature associated with alkenes can be complicated by *isomerism*. We describe the *E* and *Z* nomenclature for alkenes.

24.3 ARROW NOTATION AND RESONANCE STRUCTURES: ELECTRON COUNTING

The use of arrow notation is a means of schematically drawing electron redistribution during a reaction or to explain an explicit phenomenon.

24.4 ELECTROPHILIC ADDITION REACTIONS

The electron-rich nature of unsaturated hydrocarbons leads to their reactivity with *electrophiles*. These reactions are called *addition* reactions. We describe four classes of addition reactions: *halogenation*, *hydrohalogenation*, *hydration* and *halohydrination*. Markovnikov's rules are used to predict the product distribution of an addition reaction.

24.5 ALKANES FROM ALKENES: CATALYTIC HYDROGENATION

Catalytic hydrogenation can be used to convert alkenes and alkynes into alkanes. This reaction requires a metal catalyst such as Ni, Pd or Pt in the presence of H_2 gas. The use of a solid catalyst, a compound in solution and gaseous hydrogen means this reaction is *heterogeneous*.

24.6 ADDITION POLYMERISATION

Finally, we develop an insight into the greater relevance of chemistry to our society by investigating the *free-radical polymerisation* of alkenes to form common polymers such as polyethene. Addition polymerisations such as these are characterised by three steps: initiation, propagation and termination.

The chemistry of colour has fascinated people since ancient times. A large proportion of the brilliant colours around us—those in our clothes, the photographs in this book, the foods we eat—are due to the selective absorption of light by polyunsaturated hydrocarbons. Members of this subclass of hydrocarbon are often called *organic dyes*, organic molecules that strongly absorb selected wavelengths of visible light while allowing others to be transmitted. For example, a red T-shirt appears red because only red light is reflected from the fabric. The other wavelengths of visible light are absorbed by the dyes in the fabric.

Organic dyes are characterised by a distinct bonding pattern that features the extensive alternation of double and single bonds within a chain. The longer the alternating chain, the longer the wavelength at which absorption takes place. For example, butadiene, a four-carbon chain containing two double bonds separated by a single bond, absorbs light at 217 nm—well into the ultraviolet part of the spectrum. It is therefore colourless to the eye. However, β-carotene contains 11 pairs of alternating double and single bonds, and absorbs light at 500 nm, in the middle of the visible part of the spectrum. This is the substance chiefly responsible for the bright orange colour of carrots. The human body converts β-carotene into vitamin A, which in turn is converted into retinal, a component of *rhodopsin*, found in the retina of the eye. The absorption of visible light by rhodopsin is central to the mechanism of vision (see My World of Chemistry Section 24.3). Thus there seems to be a good basis for the maxim that eating carrots is good for your eyesight!

In this chapter, we concentrate on unsaturated hydrocarbons—hydrocarbons that contain multiple bonds—and investigate the effect this bonding has on the physical properties, structure and reactivity of these molecules.

butadiene

β-carotene

24.1 | THE STRUCTURE OF UNSATURATED HYDROCARBONS

In Chapter 22 we discussed the structure of both aliphatic alkanes and cycloalkanes. We described the carbon atoms that form the molecular skeleton as *saturated*—that is, they contain the largest possible number of hydrogen atoms. The tetrahedral bonding geometry of a saturated carbon atom was rationalised based on the complete hybridisation of the one 2*s* and three 2*p* orbitals available to carbon, forming a set of *sp*^{3} hybrid orbitals (Section 22.1 'General Characteristics of Organic Molecules'). But what happens if only the 2*s* and either one or two 2*p* orbitals hybridise? The answer is unsaturation. ▼ **FIGURE 24.1** gives examples of the two classes of **unsaturated hydrocarbons** we deal with in this chapter and compares the resulting bonding geometry about carbon with that of ethane.

Review this on page 892

Alkenes, also known as olefins, are unsaturated hydrocarbons that contain a C–C double bond. They have the general formula:

$$C_nH_{2n} \qquad [24.1]$$

The simplest alkene is $CH_2{=}CH_2$, called ethene (IUPAC name) or ethylene (common name). Ethene is a plant hormone that plays important roles in seed germination and fruit ripening. The C–C double bond makes ethene much more reactive than ethane.

Alkynes are hydrocarbons that contain a C–C triple bond, as in acetylene (C_2H_2), whose systematic name is ethyne. When acetylene is burned in a stream of oxygen, as in an oxy-acetylene torch, the flame reaches about 3200 K. The oxy-acetylene torch is used widely in welding, which requires high temperatures. Acetylene, and alkynes in general, are highly reactive molecules. Because of their higher reactivity, they are not as widely distributed in nature as alkenes; alkynes are, however, important intermediates in many industrial processes. Alkynes have the general molecular formula

$$C_nH_{2n-2} \qquad [24.2]$$

The π-bond

In the types of covalent bonds we have considered so far, the electron density is concentrated symmetrically about the line connecting the nuclei (the *internu-*

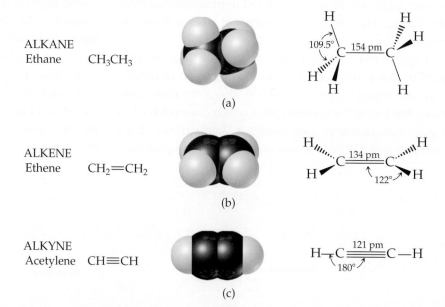

ALKANE
Ethane CH_3CH_3

(a)

ALKENE
Ethene $CH_2{=}CH_2$

(b)

ALKYNE
Acetylene $CH{\equiv}CH$

(c)

▶ **FIGURE 24.1 Three hydrocarbon types for a two-carbon system.**

clear axis). In other words, the line joining the two nuclei passes through the middle of the overlap region. These bonds are all called **sigma (σ)-bonds** (∞ Section 9.6, 'Multiple Bonds').

To describe the bonding in unsaturated hydrocarbons, we must consider a second kind of bond, one that results from the overlap between two *p* orbitals perpendicular to the internuclear axis (▶ FIGURE 24.2). This 'sideways' overlap of *p* orbitals produces a **pi (π)-bond**. A π-bond is a covalent bond in which the overlap regions lie above and below the internuclear axis. Unlike a σ-bond, in a π-bond there is no probability of finding an electron on the internuclear axis. Because the *p* orbitals in a π-bond overlap sideways rather than directly facing each other, the total overlap in a π-bond tends to be less than that in a σ-bond. As a consequence, π-bonds are generally weaker than σ-bonds.

We can summarise unsaturation in the following way:
C–C single bonds are σ-bonds.
C–C double bonds consist of one σ-bond and one π-bond.
C–C triple bonds consists of one σ-bond and two π-bonds.

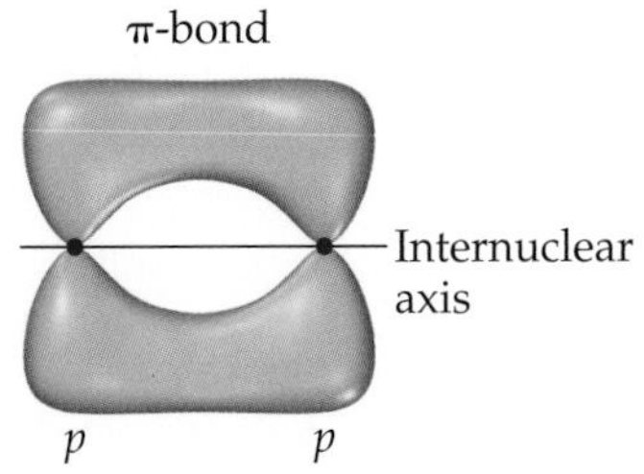

∞ Review this on page 314

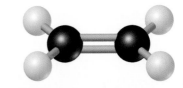

▲ **FIGURE 24.2 The π-bond.** When two *p* orbitals overlap in a sideways fashion, the result is a π-bond. Note that the two regions of overlap constitute a *single* π-bond.

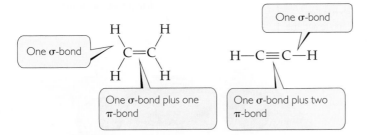

One σ-bond

One σ-bond plus one π-bond

One σ-bond

One σ-bond plus two π-bond

Bonding in Alkenes

In order to rationalise the bonding and geometry of an alkene (▶ FIGURE 24.3), let us consider ethene (C_2H_4). The bond angles about carbon in ethene are all approximately 120° (Figure 24.3), suggesting that each carbon atom uses sp^2 hybrid orbitals (Figure 9.18) to form σ-bonds with the other carbon and with two hydrogens. Because carbon has four valence electrons, after sp^2 hybridisation carbon still has one electron remaining in the *unhybridised* 2*p* orbital (▶ FIGURE 24.4).

The unhybridised 2*p* orbital is directed perpendicular to the plane that contains the three sp^2 hybrid orbitals. The hybrid orbitals separate within the same plane to give inter-orbital angles of 120°. This orbital pattern gives rise to a trigonal planar arrangement of σ-bonds.

Each sp^2 hybrid orbital on a carbon atom contains one electron. ▼ FIGURE 24.5 shows how the four C–H σ-bonds are formed by overlap of sp^2 hybrid orbitals on C with the 1*s* orbitals on each H atom. We use eight electrons to form these four electron-pair bonds. The C–C σ-bond is formed by the overlap of two sp^2 hybrid orbitals, one on each carbon atom, and requires two more electrons. Thus 10 of the 12 valence electrons in C_2H_4 are used to form five σ-bonds.

The remaining two valence electrons reside in the unhybridised 2*p* orbitals, one electron on each carbon atom. These two 2*p* orbitals can overlap with each other in a sideways fashion, as shown in Figure 24.2. The resultant electron density is concentrated above and below the C–C bond axis, forming a single π-bond. Thus the C–C double bond in ethene consists of one σ-bond and one π-bond. Note carefully that the unhybridised *p*-orbital overlap above and below the plane forms a single π-bond and not two π-bonds.

Although we cannot experimentally observe a π-bond directly (all we can observe are the positions of the atoms), the structure of ethene provides strong

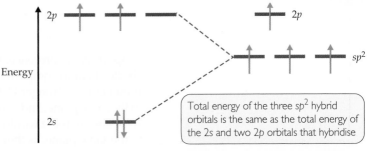

▲ **FIGURE 24.3 The molecular geometry of ethene.** Ethene, C_2H_4, has one C–C σ-bond and one C–C π-bond.

Energy

2*p*

2*s*

2*p*

sp^2

Total energy of the three sp^2 hybrid orbitals is the same as the total energy of the 2*s* and two 2*p* orbitals that hybridise

▲ **FIGURE 24.4 Degeneracy of hybrid orbitals.** Mixing 2*s* and two 2*p* orbitals yields hybrid orbitals whose energies lie between those of the original atomic orbitals.

FIGURE IT OUT

Why is it important that the *sp*^{2} hybrid orbitals of the two carbon atoms lie in the same plane?

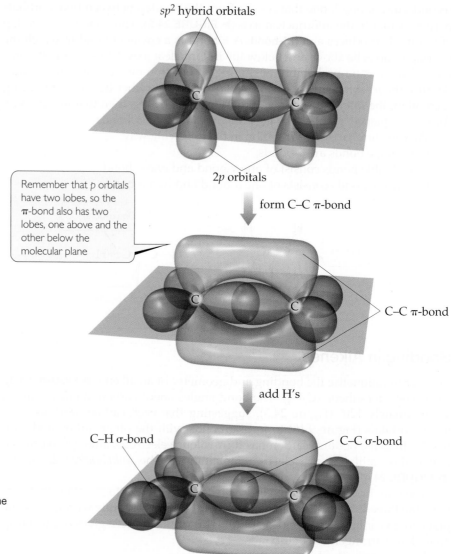

sp^2 hybrid orbitals

2*p* orbitals

form C–C π-bond

Remember that *p* orbitals have two lobes, so the π-bond also has two lobes, one above and the other below the molecular plane

C–C π-bond

add H's

C–H σ-bond

C–C σ-bond

▶ **FIGURE 24.5** **Hybridisation in ethene.** The σ-bonding framework is formed from *sp*^{2} hybrid orbitals on the carbon atoms. The unhybridised 2*p* orbitals on the C atoms are used to make a π-bond.

support for its presence. First, the C–C bond length in ethene (134 pm) is much shorter than in compounds with C–C single bonds (154 pm), consistent with the presence of a stronger C–C double bond (▼ TABLE 24.1). Second, all six atoms lie in the same plane, and stay in the same plane (Figures 24.3 and 24.5). If the π-bond were absent, there would be no reason for the two CH$_2$ fragments of ethene to lie in the same plane, as they could rotate with respect to each other in a similar way

TABLE 24.1 • **Carbon–carbon bond lengths, general molecular formulae and bond enthalpies for ethane, ethene and ethyne**

Molecule	General formula of class	C–C bond length (pm)	Approximate bond enthalpy (kJ mol^{-1})
Ethane	CH$_3$ — CH$_3$	154	350
Ethene	CH$_2$ = CH$_2$	134	620
Ethyne	CH ≡ CH	121	840

to alkanes. As a result, the introduction of one or more π-bonds introduces a degree of rigidity into molecules.

Note the difference in bond strength (bond enthalpy) between ethane and ethene in Table 24.1. The C–C double bond is not twice as strong (meaning twice as hard to break) as a single bond, nor is a triple bond three times as strong as a single bond. The fact that each π-bond is weaker than a σ-bond is part of the reason why alkenes and alkynes are far more reactive than alkanes.

CONCEPT CHECK 1

The molecule called diazine has the formula N_2H_2 and the Lewis structure

$$H—\ddot{N}=\ddot{N}—H$$

Would you expect diazine to be a linear molecule (all four atoms on the same line)? If not, would you expect the molecule to be planar (all four atoms in the same plane)?

Bonding in Alkynes

The bonding and geometry of C–C triple bonds can also be explained by hybridisation. Ethyne (more commonly known as acetylene, C_2H_2), for example, is a linear molecule containing a triple bond: H—C≡C—H. The linear geometry about the triple bond suggests that each carbon atom uses *sp* hybrid orbitals to form σ-bonds with the other carbon with one hydrogen. Each carbon atom thus has two remaining unhybridised 2p orbitals at right angles to each other and to the axis of the *sp* hybrid set (▶ **FIGURE 24.6**). These *p* orbitals overlap to form a pair of π-bonds. Thus the triple bond in acetylene consists of one σ-bond and two π-bonds.

| **SAMPLE EXERCISE 24.1** | Describing σ-bonds and π-bonds in a molecule |

Formaldehyde has the Lewis structure

H
\
C=Ö:
/
H

Describe how the bonds in formaldehyde are formed in terms of overlap of appropriate hybridised and unhybridised orbitals.

SOLUTION

Analyse We are given a structure containing several types of bonds and are asked to interpret these bonds in terms of hybrid orbitals.

Plan Single bonds are of the σ type, whereas double bonds consist of one σ-bond and one π-bond. The ways in which these bonds form can be deduced from the geometry of the molecule, which we predict using the VSEPR model.

Solve The C atom has three electron domains around it, which suggests a trigonal planar geometry with bond angles of about 120°. This geometry implies sp^2 hybrid orbitals on C. These hybrids are used to make the two C–H and one C–O σ-bonds to C. There remains an unhybridised 2p orbital on carbon, perpendicular to the plane of the three sp^2 hybrids.

The O atom also has three electron domains around it, so we assume that it has sp^2 hybridisation as well. One of these hybrids participates in the C–O σ-bond, while the other two hybrids hold the two non-bonding electron pairs of the O atom. Like the C atom, therefore, the O atom has an unhybridised 2p orbital that is perpendicular to the plane of the molecule. The unhybridised 2p orbitals on the C and O atoms overlap to form a C–O π-bond, as illustrated in ▼ **FIGURE 24.7**.

FIGURE IT OUT

Based on the models of bonding in ethene and acetylene, which molecule should have the higher carbon–carbon bond energy?

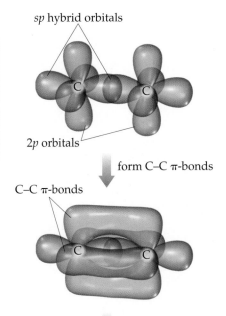

sp hybrid orbitals

2p orbitals

form C–C π-bonds

C–C π-bonds

add H's

C–H σ-bond   C–C σ-bond

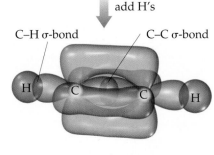

H   C   C   H

▲ **FIGURE 24.6   Formation of two π-bonds.** In acetylene, C_2H_2, the overlap of two sets of unhybridised carbon 2p orbitals leads to the formation of two π-bonds.

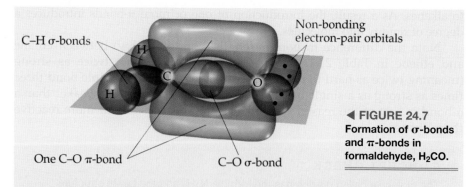

◀ FIGURE 24.7
Formation of σ-bonds and π-bonds in formaldehyde, H₂CO.

PRACTICE EXERCISE

Consider the acetonitrile molecule:

$$H{-}\underset{\displaystyle H}{\overset{\displaystyle H}{C}}{-}C{\equiv}N\text{:}$$

(a) Predict the bond angles around each carbon atom; **(b)** describe the hybridisation at each of the carbon atoms; **(c)** determine the total number of σ-bonds and π-bonds in the molecule.

Answers: **(a)** Approximately 109° around the left C and 180° on the right C; **(b)** sp^3, sp; **(c)** five σ-bonds and two π-bonds

(See also Exercises 24.13, 24.14.)

SAMPLE EXERCISE 24.2 | **Determining hybridisation**

Locate the sp^3, sp^2 and sp hybridised carbon atoms in the following compounds.

(a)

$$\underset{\displaystyle CH_3}{\overset{\displaystyle CH_3}{\underset{\displaystyle H}{\overset{\displaystyle H}{C{=}C}}}}$$

(b) $H_3C{-}C{\equiv}C{-}CH_3$

SOLUTION

Analyse We are given two compounds and are asked to relate the bonding types to a level of hybridisation.

Plan This type of question is made easier by remembering that all carbons are sp^3 hybridised, except for those containing a double bond (in which case they are sp^2 hybridised) and those containing a triple bond (which are sp hybridised).

Solve Using these observations:

(a)

sp^3
sp^2 CH₃
H C=C
 C H
sp^2 CH₃
sp^3

(b)

sp^3 sp sp^3
$H_3C{-}C{\equiv}C{-}CH_3$

PRACTICE EXERCISE

Locate the sp^3, sp^2 and sp hybridised carbon atoms in the following compounds.

(a)   **(b)**

Answers: (a) The molecule has $2 \times sp$, $2 \times sp^2$ and $1 \times sp^3$ hybridised carbon atoms.
(b) The molecule has $5 \times sp^3$ and $1 \times sp^2$ hybridised carbon atoms.

(See also Exercise 24.15.)

MY WORLD OF CHEMISTRY

TERPENES AND ISOPRENE

Terpenes are a class of naturally occurring compounds whose carbon skeleton can be divided into two or more isoprene units—a five-carbon unit whose systematic name is 2-methyl-1,3-butadiene (▼ FIGURE 24.8). Terpenes are among the most widely distributed and utilised of all compound classes. Compounds such as camphor, menthol, pinene, geraniol, carvone and limonene are terpenes and each is divisible into two isoprene units.

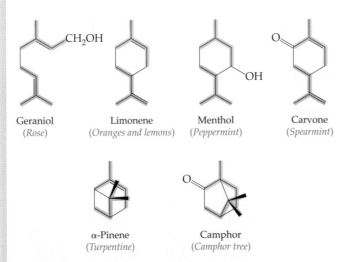

Geraniol
(Rose)

Limonene
(Oranges and lemons)

Menthol
(Peppermint)

Carvone
(Spearmint)

α-Pinene
(Turpentine)

Camphor
(Camphor tree)

▲ **FIGURE 24.8 Well-known terpenes.** Geraniol, limonene, menthol, pinene and camphor are common terpenes. The positions of the isoprene units are highlighted. Note that the isoprene units are 'linked' head-to-tail.

Isoprene is partly responsible for the bluish haze of the Blue Mountains of New South Wales (▼ FIGURE 24.9) when viewed from a distance, especially on hot days. This haze is rich in isoprene, oxidised derivatives of isoprene and other simple terpenes. The worldwide emission rate of these natural hydrocarbons by vegetation has been estimated to be 5×10^{11} kg year^{-1}. Gum trees are a particularly rich source of isoprene.

Isoprene is a colourless liquid with a distinct aromatic odour. Large petroleum-based companies such as Shell Chemicals manufacture large quantities of it as a by-product of ethene production. The polymer polyisoprene is used in a wide variety of rubber applications including medical equipment, baby bottle teats/nipples, toys, shoe soles, tyres and the elastic films used inside golf balls. It is also used in adhesives and in paints and coatings.

As shown in the My World of Chemistry box in Section 24.3, several biologically important terpenes, such as vitamin A, play a large role in vision. The terpenes demonstrate effectively how nature uses simple building blocks such as isoprene to construct larger and more complex molecules with even more complex functions.

▲ **FIGURE 24.9 The Blue Mountains in New South Wales.** The blue haze is caused in part by isoprene emitted by the plants of the region.

CONCEPT CHECK 2

Observe the structure of β-carotene on page 953. Is β-carotene a terpene? If so, highlight the individual isoprene units in β-carotene.

24.2 | ISOMERISM AND NOMENCLATURE

Alkene nomenclature is based on the longest continuous chain of carbon atoms that *contains* the double bond. The name given to this longest chain is obtained from the name of the corresponding alkane (Table 22.3), by noting the position of unsaturation, and by changing the ending from *-ane* to *-ene*. Examples of an alkene and cycloalkene and their systematic names are shown in ▼ FIGURE 24.10. Note that the longest carbon chain in 2-ethyl-3-methylpent-1-ene is not the one that contains the double bond. This chain is ignored because the longest chain *containing the double bond* takes precedence in naming. Note too that the double bond is always given the lowest position number possible.

The general formula for a cycloalkene is

$$C_nH_{2n-2} \qquad [24.3]$$

which is the same as that of an alkyne. Hence, cycloalkenes and alkynes with the same number of carbon atoms are constitutional isomers.

The location of the double bond along an alkene chain is indicated by a prefix number. The chain is always numbered from the end that brings us to the double bond sooner and hence gives the smallest-numbered prefix. In propene the only possible location for the double bond is between the first and second carbons; thus a prefix indicating its location is unnecessary. For butene (▼ FIGURE 24.11) there are two possible positions for the double bond: either after the first carbon (but-1-ene) or after the second carbon (but-2-ene).

If a substance contains two or more double bonds, each one is located by a numerical prefix. The ending of the name is altered to identify the number of double bonds: diene (two), triene (three) and so forth. For example, $CH_2=CH-CH_2-CH=CH_2$ is named penta-1,4-diene.

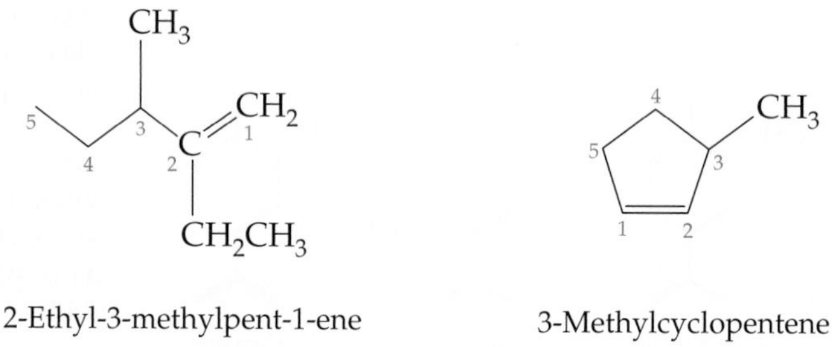

▶ FIGURE 24.10 **Nomenclature of alkenes.** Alkene nomenclature is derived from the longest carbon chain containing the double bond. The position of the double bond is always assigned the lowest position number possible. The position of the double bond in a cycloalkene is arbitrarily assigned as spanning positions 1 and 2.

2-Ethyl-3-methylpent-1-ene

3-Methylcyclopentene

FIGURE IT OUT

How many isomers are there for propene, C_3H_6?

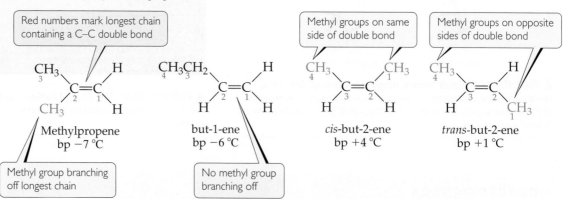

Red numbers mark longest chain containing a C–C double bond

Methyl groups on same side of double bond

Methyl groups on opposite sides of double bond

Methylpropene
bp −7 °C

Methyl group branching off longest chain

but-1-ene
bp −6 °C

No methyl group branching off

cis-but-2-ene
bp +4 °C

trans-but-2-ene
bp +1 °C

▲ FIGURE 24.11 **C_4H_8 structural isomers.** Full structural formulae, names and boiling points of alkenes with the molecular formula C_4H_8.

CONCEPT CHECK 3

How many distinct locations are there for one double bond in a five-carbon cyclic chain?

Isomerism in Alkenes—The *E, Z* System

Alkenes with two or more different substituents that lie on either side of the C–C double bond occur as geometrical isomers (Section 23.2, 'Cis–Trans Isomerism in Cycloalkanes'). For example, the two structures *cis*-but-2-ene and *trans*-but-2-ene (Figure 24.11) are geometric isomers. Their molecular formula and connectivity are the same; they differ only by the spatial arrangement of the two methyl groups. Geometric isomers, like diastereomers, possess distinct physical properties, such as boiling point, and often differ significantly in their chemical behaviour (Section 23.6, 'Molecules with More than One Stereocentre'). This isomerism is a result of the restricted rotation about the C–C double bond, which in turn is caused by the *p*-orbital overlap required for formation of the π-bond (▶ **FIGURE 24.12**). The carbon–carbon bond axis and the bonds to the hydrogen atoms and to the alkyl groups (designated R) are all in a plane. The *p* orbitals that overlap sideways to form the π-bond are perpendicular to the molecular plane. As Figure 24.12 shows, rotation around the carbon–carbon double bond requires the π-bond to be broken, a process that requires considerable energy (about 250 kJ mol⁻¹). Although rotation about a double bond doesn't occur easily, it is a key process in the chemistry of vision (see My World of Chemistry box later in Section 24.3).

Review this on page 929

Review this on page 941

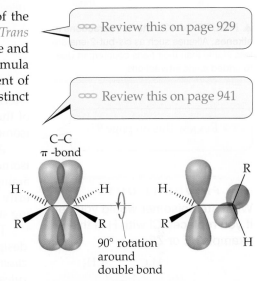

▲ **FIGURE 24.12 Rotation about a carbon–carbon double bond.** In an alkene, the overlap of the *p* orbitals that form the π-bond is lost in the rotation. For this reason, rotation about carbon–carbon double bonds does not occur readily, usually requiring the absorption of energy.

 CONCEPT CHECK 4

Why is it not necessary to name methylpropene in any other way?

SAMPLE EXERCISE 24.3 Drawing isomers

Draw all the structural and geometric isomers of pentene, C₅H₁₀, that have an unbranched hydrocarbon chain.

SOLUTION

Analyse We are given the name of an alkene and asked to derive all possible isomer permutations.

Plan Because the compound is named pentene and not pentadiene or pentatriene, we know that the five-carbon chain contains only one carbon–carbon double bond. Thus we can begin by first placing the double bond in various locations along the chain, remembering that the chain can be numbered from either end. After finding the different distinct locations for the double bond, we can consider whether the molecule can have *cis* and *trans* isomers.

Solve There can be a double bond after either the first carbon (pent-1-ene) or second carbon (pent-2-ene). These are the only two possibilities because the chain can be numbered from either end. (Thus what we might erroneously call pent-4-ene is actually pent-1-ene, as seen by numbering the carbon chain from the other end.) Because the first C atom in pent-1-ene is bonded to two H atoms, there are no *cis–trans* isomers. However, there are *cis* and *trans* isomers for pent-2-ene. Thus the three isomers for pentene are

CH₃—CH₂—CH₂—CH=CH₂
pent-1-ene

cis-pent-2-ene

trans-pent-2-ene

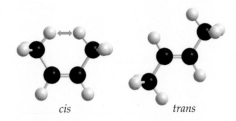

▲ **FIGURE 24.13** **Steric interactions in alkenes.** Alkenes such as *cis*-but-2-ene are less stable than their *trans* counterpart due to added steric interactions.

∞ Review this on page 937

⚠ **FIGURE IT OUT**

What stereoisomer would you get if you replaced Cl with H in this example? E or Z?

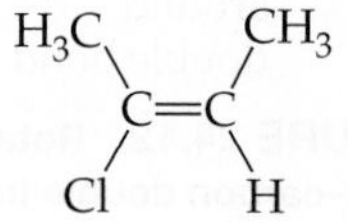

$$H_3C \quad CH_3$$
$$C=C$$
$$Cl \quad H$$

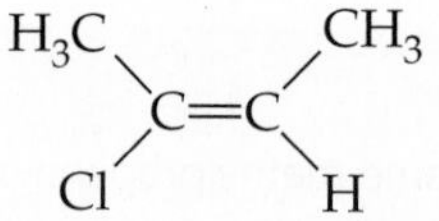

Step 1

Step 2

Step 3

Step 4

Therefore,

E

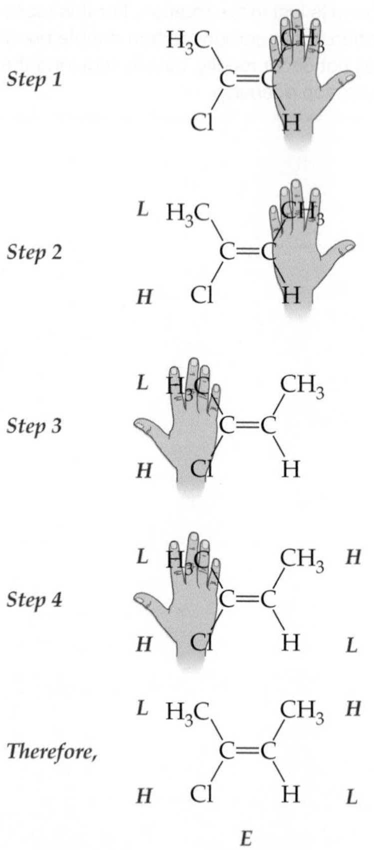

▲ **FIGURE 24.14** **Assignment of *E* and *Z* isomers.** The hand acts to remove one side of the molecular structure from consideration.

(You should convince yourself that *cis*- or *trans*-pent-3-ene is identical to *cis*- or *trans*-pent-2-ene, respectively.)

PRACTICE EXERCISE

How many straight-chain isomers are there of hexene, C_6H_{12}?

Answer: Five (hex-1-ene, *cis*-hex-2-ene, *trans*-hex-2-ene, *cis*-hex-3-ene, *trans*-hex-3-ene)

(See also Exercises 24.8, 24.11.)

Cis-alkenes tend to be less stable than *trans*-alkenes because of the extra steric interactions that occur when both alkyl substituents are on the same side of the double bond. ◄ **FIGURE 24.13** illustrates this point using the geometric isomers of but-2-ene.

Although *cis* and *trans* nomenclature is useful to describe simple geometric isomers, it fails to describe more complex isomers accurately. To remove any ambiguity in naming geometric isomers, the IUPAC has established a set of priority rules similar to the Cahn–Ingold–Prelog priority rules used for determining stereochemistry (∞ Section 23.5, 'Absolute Stereochemistry').

The system for naming isomers is known as the ***E, Z* system**. An alkene is designated *E* (from German *entgegen*, meaning 'opposite') or *Z* (from German *zusammen*, meaning 'together') based on specific priority rules. The priority rules are listed below and their application to the geometric isomer of 2-chloro-but-2-ene is shown in ◄ **FIGURE 24.14**.

$$H_3C \quad CH_3$$
$$C=C$$
$$Cl \quad H$$

2-chloro-but-2-ene

1. Split the molecule through the C–C double bond into two parts and concentrate on setting the priority for one side at a time.

2. Assign a priority to each substituent using H for highest priority and L for lowest priority, based on the *atomic number* of the two atoms directly attached to the sp^2 carbon atom. The higher the atomic number, the higher the priority. A representative set of groups in increasing priority order is:

$$—H, \ —CH_3, \ —NH_2, \ —OH, \ —Cl, \ —Br$$

Increasing priority

If two atoms are identical in terms of priority, you need to move out along the respective chains until you encounter a difference in priority. Reapply rule 2 to determine priority. If no difference is found, the two substituents must be identical and no *E*/*Z* isomerism is possible. An example of increasing priority fragments is:

$$—CH_2—H, \ —CH_2—CH_3, \ —CH_2—NH_2, \ —CH_2—OH$$

Increasing priority

3. Repeat step 2 for the other side of the C–C double bond.

4. If the two higher-priority groups are on the same side of the double bond, the alkene is the *Z* isomer. If they occur on opposite sides, as in the example shown in Figure 24.14, the alkene is the *E* isomer.

Alkynes

Alkynes are unsaturated hydrocarbons containing one or more C–C triple bonds. Alkynes are named by identifying the longest continuous chain in the molecule *containing the triple bond* and modifying the ending of the name, as listed in Table 22.3, from *-ane* to *-yne*, as shown in Sample Exercise 24.4. Alkynes are linear about the C–C triple bonds, so geometric isomerism is not possible about the triple bond. Alkynes are, however, constitutional isomers of cycloalkenes.

SAMPLE EXERCISE 24.4 **Naming unsaturated hydrocarbons**

Name the following compounds:

(a) 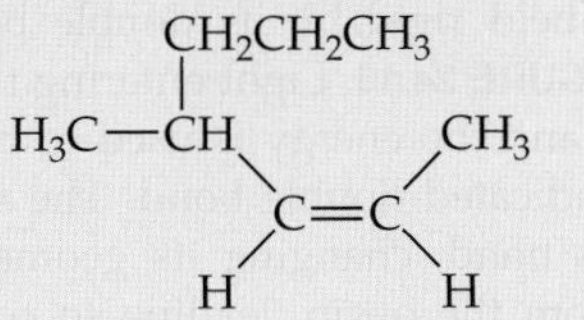 H_3C-CH with $CH_2CH_2CH_3$ branch, $C=C$, CH_3, H, H

(b) 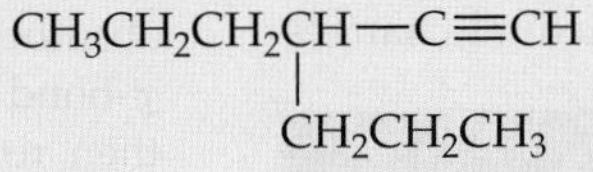 $CH_3CH_2CH_2CH-C\equiv CH$ with $CH_2CH_2CH_3$ branch

SOLUTION

Analyse We are given the structural formulae for two compounds—the first an alkene and the second an alkyne—and asked to name them.

Plan In each case, the name is based on the number of carbon atoms in the longest continuous carbon chain that contains the multiple bond. In the case of (a), care must be taken to indicate whether E, Z isomerism is possible and, if so, which isomer is given.

Solve **(a)** The longest continuous chain of carbons that contains the double bond is seven carbons in length. The parent compound is therefore heptene. Because the double bond begins at carbon 2 (numbering from the end closest to the double bond), the parent hydrocarbon chain is named hept-2-ene. A methyl group is located at carbon atom 4. Thus the compound is 4-methylhept-2-ene. The geometrical configuration at the double bond is Z (that is, the alkyl groups are bonded to the double bond on the same side). Thus the full name is Z-4-methylhept-2-ene.

(b) The longest continuous chain of carbon atoms containing the triple bond is six, so this compound is a derivative of hexyne. The triple bond comes after the first carbon (numbering from the right), making it a derivative of hex-1-yne. The branch from the hexyne chain contains three carbon atoms, making it a propyl group. Because it is located on the third carbon atom of the hexyne chain, the molecule is 3-propylhex-1-yne.

PRACTICE EXERCISE

Draw the condensed structural formula for 4-methylpent-2-yne.

Answer: $CH_3-C\equiv C-CH-CH_3$ with CH_3 branch

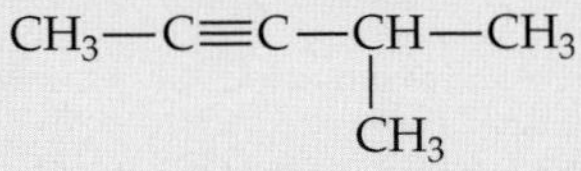

(See also Exercises 24.16, 24.17.)

In Chapter 30 we will discuss the techniques used to determine the molecular structure of alkenes and alkynes—namely nuclear magnetic resonance spectroscopy, infrared spectroscopy and mass spectrometry.

24.3 | ARROW NOTATION AND RESONANCE STRUCTURES: ELECTRON COUNTING

Organic chemists use **curved arrow notation** to describe the redistribution of valence electrons within a molecule. A useful analogy is to look at arrow notation as an electron accounting system, keeping track of electrons when describing resonance (∞ Section 8.6, 'Resonance Structures'), or in a chemical

∞ Review this on page 273

MY WORLD OF CHEMISTRY

THE CHEMISTRY OF VISION

Scientists have begun to understand the complex chemistry of vision. Vision begins when light is focused by the lens onto the retina, the layer of cells lining the interior of the eyeball. The retina contains *photoreceptor* cells known as rods and cones (▼ FIGURE 24.15). The human retina contains about 3 million cones and 100 million rods. The rods are sensitive to dim light and are used in night vision. The cones are sensitive to colours. The tops of the rods and cones contain a substance called *rhodopsin*. Rhodopsin consists of a protein, called *opsin*, bonded to a reddish-purple pigment called *retinal*, which is similar in structure to β-carotene (see chapter opening). Structural changes

around a double bond in the retinal portion of the molecule trigger a series of chemical reactions that result in vision. Our recent discussions now allow us to appreciate another aspect of double bonds: the rigidity that they introduce into molecules.

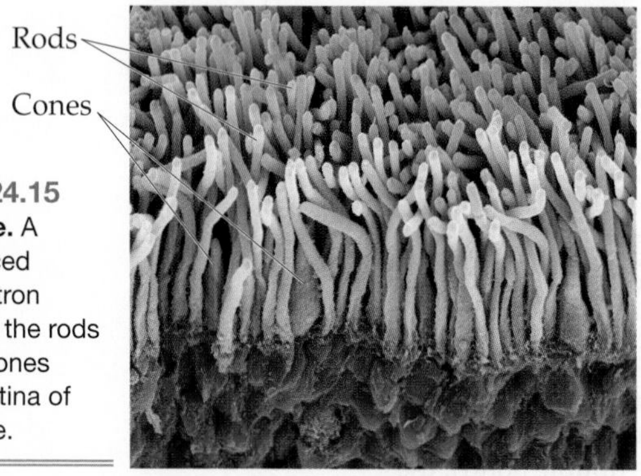

Imagine taking the –CH₂ group of the ethene molecule and rotating it relative to the other –CH₂ group, as shown in Figure 24.12. This rotation destroys the overlap of p orbitals, breaking the π-bond, a process that requires considerable energy. Thus the presence of a double bond restricts the rotation of the bonds in a molecule.

Our vision depends on this rigidity in retinal. In its normal form, retinal is held rigid by its double bonds, as shown on the left in ▼ FIGURE 24.16. Light entering the eye is absorbed by rhodopsin, and the energy is used to break the π-bond portion of the indicated double bond. The molecule then rotates around this bond, changing its geometry. The retinal then separates from the opsin, leading to conformational changes that trigger the reactions producing a nerve impulse. The brain interprets this impulse as the sensation of vision. It takes as few as five closely spaced molecules reacting in this fashion to produce the sensation of vision. Thus only five photons of light are necessary to stimulate the eye.

Enzymes in the retina revert the retinal to its original form and reattach it to the opsin. The slowness of this process helps to explain why intense bright light causes temporary blindness. The light causes all the retinal to separate from the opsin, leaving no further molecules to absorb light.

RELATED EXERCISES: 24.23, 24.63

▶ **FIGURE 24.15**
Inside the eye. A colour-enhanced scanning electron micrograph of the rods (yellow) and cones (blue) in the retina of the human eye.

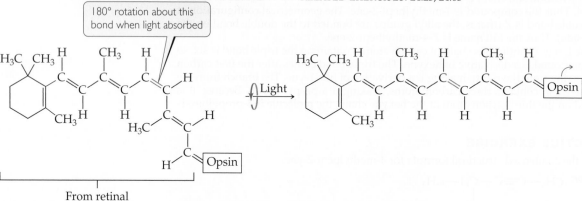

▲ **FIGURE 24.16** **The chemical basis of vision.** When rhodopsin absorbs visible light, the π component of the double bond (shown in red) breaks, allowing rotation that produces a change in molecular geometry.

reaction, when describing a reaction *mechanism*. Understanding a reaction mechanism using curved arrows is a logical and plausible way for an organic chemist to describe how a reaction proceeds—it not only gives the product but also defines potential intermediate structures through valence electron redistribution. *Essentially, the curved arrows start from where electrons are coming from and end where electrons are going to.*

Curved arrows are used for three common types of electron redistribution:

1. From a bond to an adjacent atom—this forms a lone pair on the receiving atom, or joins two molecules together.

2. From an atom to an adjacent bond—this uses a lone pair on the donor atom to form a π-bond.

3. From one bond to an adjacent bond—this redistributes the electrons although, overall, the same number of σ- and π-bonds is retained.

The use of curved arrows is a useful way to rationalise all your organic reactions or to demonstrate stability. Following this set of simple rules will help you master this technique. All structures generated using curved arrows must:

1. have the same number of valence electrons as the starting structure,

2. obey the rules of covalent bonding—for example, no more than four formal bonds to carbon,

3. differ only in the distribution of valence electrons,

4. have the same number of paired or unpaired electrons.

The concept of resonance was advanced in the 1930s by Linus Pauling (1901–1994) (▶ FIGURE 24.17). He rationalised that many molecules and ions are best described by writing two or more Lewis structures, and that individual Lewis structures are *contributing structures*. Let us consider the two contributing (resonance) structures of nitromethane using curved arrow notation to inter-convert contributor A and contributor B:

▲ **FIGURE 24.17** **Linus Pauling,** a scientist of the highest calibre who won a Nobel Prize in 1954 for his contributions to chemistry. He was also very socially aware, using his scientific knowledge and his influence to combat nuclear testing in the 1950s and 1960s. For this effort, he was awarded the Nobel Peace Prize in 1962 and is still the only person in history to have won two unshared Nobel Prizes.

$$\left[\; H_3C{-}\overset{\oplus}{N}\overset{\textstyle :\ddot{O}:^{\ominus}}{\underset{\ddot{O}:}{\big|}} \quad \longleftrightarrow \quad H_3C{-}\overset{\oplus}{N}\overset{\textstyle \ddot{O}:}{\underset{:\ddot{O}:^{\ominus}}{\big\|}} \; \right]$$

Contributor A   Contributor B

The two different contributing structures of nitromethane are linked by a double-headed (resonance) arrow as opposed to a set of equilibrium arrows (⇌) and are grouped together in square brackets since they both contribute to the same structure. This means that the actual molecule (or ion) is a hybrid of the various contributing structures. In other words the 'real' structure of nitromethane is neither contributor A nor contributor B, but something in between—that is, a structure in which the negative charge is spread across both oxygens. We say that the negative charge is *delocalised* across the two N–O bonds. We can observe this delocalisation experimentally by looking at the N–O bond lengths in *nitro* compounds. The two bond lengths are equal at 121 pm and lie between the size of a single N–O bond and a double N–O bond.

The presence, or absence, of contributing resonance structures plays a large part in the reactivity of molecules and ions. We will see a lot more of this, especially when discussing what constitutes a good leaving group (∞ Section 25.5, 'Nucleophilic Substitution Reactions of Haloalkanes') and the reactivity of phenols (∞ Section 28.4, 'Acidity of Phenols').

∞ Find out more on pages 1011 and 1124

A CLOSER LOOK

DESCRIBING CHARGE

⊕ or +? Organic chemists tend to circle positive and negative signs on molecules to distinguish charge from bonds and plus signs. It also helps when describing electron redistribution. From here on, we will use this form of notation in our mechanistic considerations.

Draw two equivalent resonance structures for the formate ion, HCO_2^- using curved arrows to indicate movement of electrons.

SOLUTION

Analyse Draw the two contributing forms and note the differences.

Plan Redistribute valence electrons on the oxygen bearing the negative charge and within the double bond.

Solve

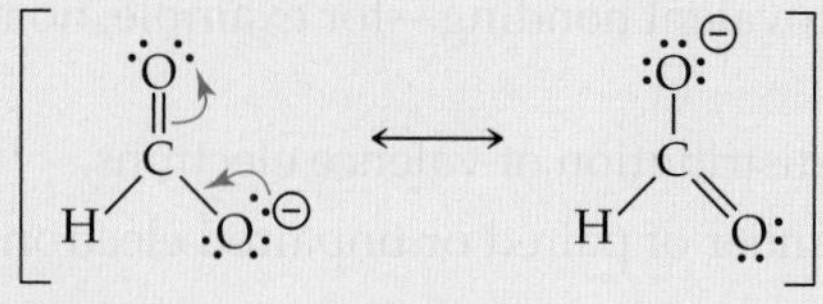

PRACTICE EXERCISE

Which of the following molecules or ions will exhibit delocalised bonding: SO_3^{2-}, H_2CO, O_3, NH_4^+?

Answer: SO_3^{2-} and O_3, as indicated by the presence of two or more resonance structures involving π-bonding for each of these molecules.

(See also Exercises 24.32, 24.33.)

Arrow notation can also be used to rationalise a reaction's pathway or *mechanism*. A simple and general reaction may look like this:

$$Nu^{\ominus}: + E^{\oplus} \longrightarrow Nu\!\!-\!\!E \qquad [24.4]$$

where 'Nu' is not the fictional element 'nubrium' but a nucleophile—that is, any molecule or ion with a lone pair of electrons available for reaction. We can use the arrow notation to describe pictorially how the lone pair on Nu is able to form a new covalent bond with an electrophile E (electro, meaning *negative*; phile, meaning *loving*), which is an electron-pair acceptor:

$$Nu^{\ominus}: \curvearrowright E^{\oplus} \longrightarrow Nu\!\!-\!\!E \qquad [24.5]$$

We will gain significant experience at using arrow notation in the next section and throughout the next five chapters as we discuss the reactivity of functional groups.

> **CONCEPT CHECK 5**
>
> Polyacetylene is an example of a semiconducting polymer. That is, if an electron is put into the system at one end, another electron will be accessible from the other end. What properties are responsible for its ability to conduct electricity?

(a) Halogenation

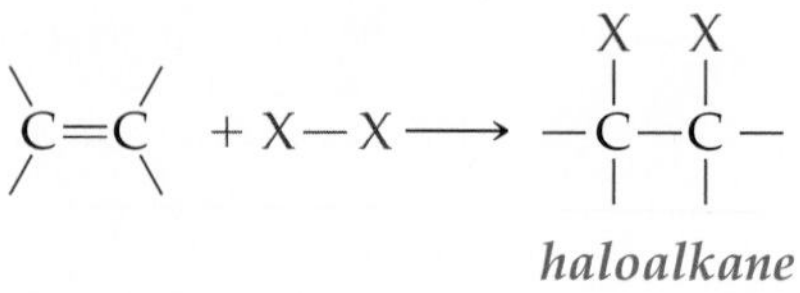

haloalkane

(b) Hydrohalogenation

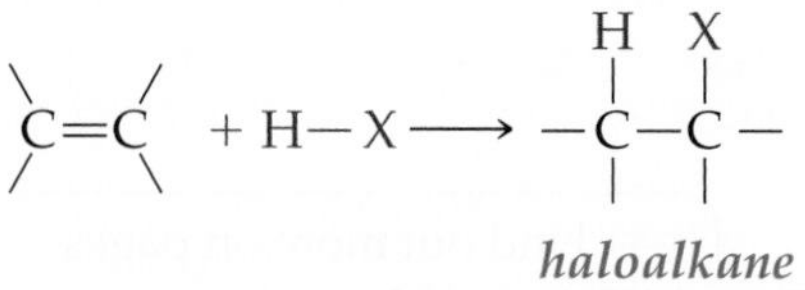

haloalkane

(c) Hydration

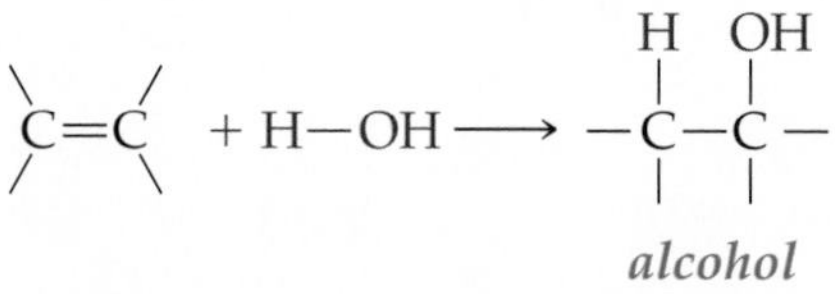

alcohol

▲ **FIGURE 24.18 Addition reactions of alkenes and alkynes (X = halogen).** Halogenation, hydrohalogenation and hydration of an alkene or alkyne are useful and efficient ways of preparing alkyl halides and alcohols. The addition reactions of alkynes follow those of the alkenes.

24.4 ELECTROPHILIC ADDITION REACTIONS

The presence of carbon–carbon double or triple bonds in a compound markedly increases its chemical reactivity. The most characteristic reactions of alkenes and alkynes are **addition reactions**, in which a reactant is added across the two carbon atoms that form the multiple bond. Three main types of addition reactions are typical of alkenes and alkynes, leading to increasing the degree of functionality within a hydrocarbon: **halogenation**, **hydrohalogenation** and **hydration**. These reactions involve the addition of X_2 (X = Cl, Br), HX (X = Cl, Br, I) and HOH (water), respectively. ◀ **FIGURE 24.18** shows a general representation of each reaction.

The reason why alkenes and alkynes are so reactive is twofold. In both cases, the unhybridised *p* orbitals that form the π-bonds are electron-rich, so would be expected to attract electron-deficient species. Also, we have seen that in carbon, π-bonds are weaker than σ-bonds, so would be expected to react to form σ-bonds wherever possible. Moreover, the shape of alkenes and alkynes make them easily accessible to reagents (▶ FIGURE 24.19). The electron-rich nature of a multiple bond leads to enhanced reactivity with electrophiles. Hence, **electrophiles** are molecules that accept a pair of electrons in a chemical reaction. Reactions of alkenes and alkynes with electrophiles are called electrophilic addition reactions.

Addition Reactions Involving HX (X = Cl, Br, I)

Let's consider the addition reaction between HBr and but-2-ene as a representative of this reaction type. The reaction proceeds in two steps. In the first step, which is rate-determining (∞ Section 15.6, 'Reaction Mechanism'), the electron-rich double bond attacks the hydrogen of HBr (the electrophile), transferring a proton to one of the two alkene carbons and producing a bromide ion in the process:

$$\text{[24.6]}$$

The pair of electrons that formed the π-bond between the carbon atoms in the alkene are used to form the new C–H bond. As a result, one carbon has fewer valence electrons associated with it. This carbon is now electron-deficient, as indicated by the positive charge. Organic molecules bearing a positive charge on carbon are called **carbocations**. Carbocations are classified as *primary* (1°), *secondary* (2°) or *tertiary* (3°), depending on the number of carbon units bonded to the positively charged carbon (∞ Section 22.8, 'Reactions of Alkanes'). For example, the carbocation formed on the right-hand side in Equation 24.6 is secondary, as the carbon bearing the positive charge has two other carbons bonded directly to it.

The second step, involving the addition of Br⁻ to the carbocation, is faster:

$$\text{[24.7]}$$

In this step the bromide ion donates a pair of electrons to the carbocation, forming the new C–Br bond. The bromide ion is behaving as a **nucleophile** (*nucleus-loving*). A nucleophile is a species that donates a pair of electrons in a chemical reaction. Nucleophiles are electron-rich species.

Since the first step in the reaction involves both the alkene and hydrogen halide, the rate law for the reaction is second order, first order in both the alkene and HBr:

$$\text{Rate} = -\frac{\Delta[CH_3CH=CHCH_3]}{\Delta t} = k[CH_3CH=CHCH_3][HBr] \quad \text{[24.8]}$$

The energy profile for the reaction is shown in ▼ FIGURE 24.20. The first energy maximum represents the transition state in the first step of the mechanism. The second maximum represents the transition state for the second step. Notice that there is an energy minimum between the first and second steps of the reaction. This energy minimum corresponds to the intermediate carbocation.

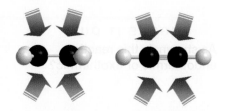

▲ FIGURE 24.19 **The accessibility of alkenes and alkynes.** The trigonal planar and linear geometries of alkenes and alkynes makes the addition of halogens easier through the exposed faces, as indicated by the arrows. Alkynes have two additional regions of accessibility, one directed out of the page and the other into the page.

∞ Review this on page 594

∞ Review this on page 913

▲ **FIGURE IT OUT**

**As shown, is this reaction
endothermic or exothermic?**

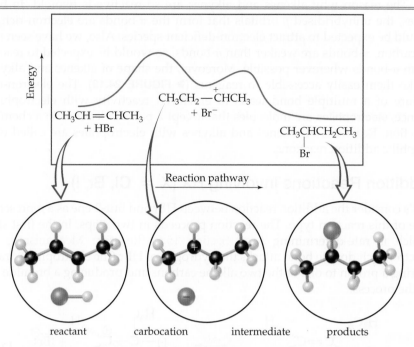

▶ **FIGURE 24.20** Energy profile for
addition of HBr to but-2-ene.

reactant   carbocation   intermediate   products

Let us now consider the reaction of 2-methylpropene with HBr. This
reaction is a *regioselective reaction* meaning that one product, 2-bromo-2-
methylpropane, is formed predominantly. The first step of the addition leads to
the possible formation of two intermediate carbocations:

[24.9]

From a large body of experiments, we know that 3° carbocations are more
stable and require a lower activation barrier to formation than either 1° or 2°
carbocations. Hence the product distribution can be rationalised based on
which intermediate is formed in preference.

The relative stability of primary, secondary and tertiary carbocations is:

less stable 1° < 2° < 3° *most stable*

and is based on how easily the alkyl groups around the positive charge can
spread the charge away. This is known as an *inductive effect*. The larger the area
the positive charge is spread over, the more stable the carbocation. In the case of

hydrocarbons, as the number of alkyl groups increases so too does the stability of the carbocation increase.

The first person to note the regioselectivity in hydrohalogenations (and hydrations) was Vladimir Markovnikov (1837–1904). The general rule for electrophilic additions to alkenes bears his name. **Markovnikov's rule** states:

In the addition of HX to an alkene, the hydrogen adds to the carbon atom of the double bond bearing the greater number of hydrogen atoms bonded directly to it.

This can also be written:

In the addition of HX to an alkene, the hydrogen adds to the least substituted carbon atom of the double bond.

This rule provides an easy way to predict the major product of an addition reaction involving HX.

Returning to Equation 24.9, C1 has two hydrogens directly attached whereas C2 has none (those on the methyl group hydrogens do not count because they are not directly attached to the sp^2 hybridised carbon of the double bond). The product predicted by using Markovnikov's rule is certainly the one predicted on the basis of carbocation stability.

SAMPLE EXERCISE 24.6 **Addition of HX to alkenes**

Name and draw the structural formula for the major product formed when 1-methylcyclohexene reacts with HCl. Justify your choice for the major product.

SOLUTION

Analyse We are asked for several things here, including knowing something about nomenclature and converting names into structures, as well as the reactivity of alkenes.

Plan Begin by drawing the structure of the reactants and identify the reaction as the addition of HCl to the double bond:

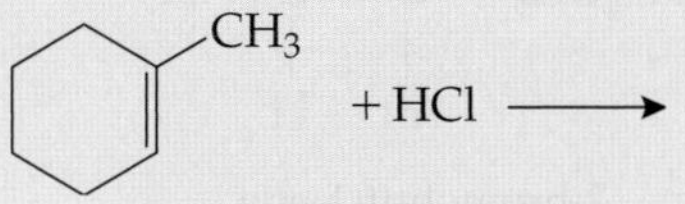

Solve Using Markovnikov's rule, the hydrogen atom of HX should add to the double bond on the side with the most number of hydrogens:

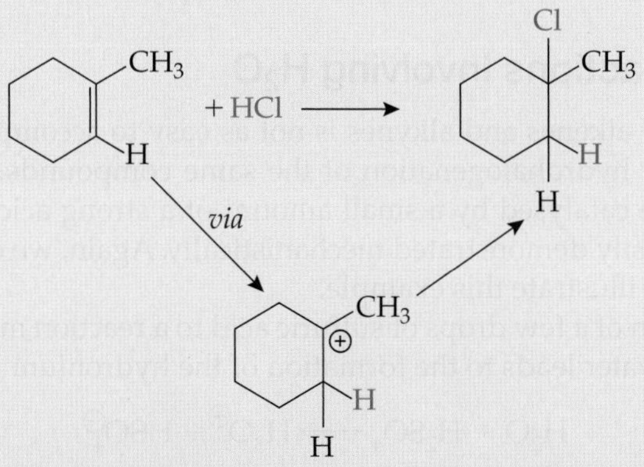

Hence our answer is *1-chloro-1-methylcyclohexane*. Our justification comes from the relative stabilities of the possible carbocation intermediates. Our chosen product forms a 3° carbocation intermediate over the alternative 2° carbocation intermediate because it is more stable due to enhanced inductive effects. To say *'because Markovnikov said so'* is not satisfactory.

PRACTICE EXERCISE

Addition of HCl to an alkene forms 2-chloro-3-methylbutane. What is the alkene?

Answer: 3-methylbut-1-ene

(See also Exercises 24.39, 24.40.)

The addition of HX across an alkyne forms a *geminal* dihalide. The term *geminal* refers to two groups occupying the same carbon. Equation 24.10 demonstrates reactions of this class.

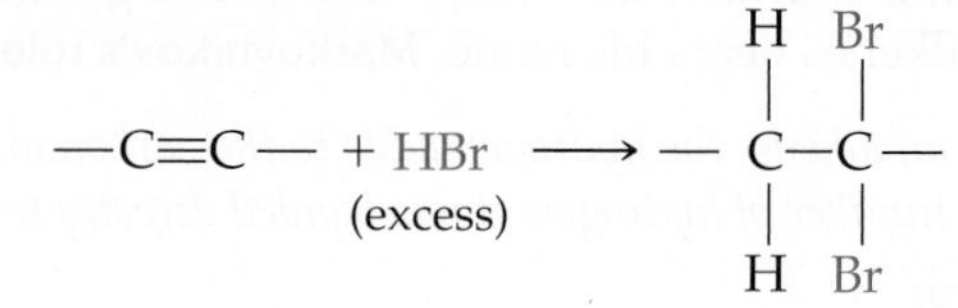

$$-C\equiv C- + HBr \xrightarrow{\text{(excess)}} \underset{\substack{| \\ H \;\; Br}}{\overset{\substack{H \;\; Br \\ |}}{-\underset{|}{C}-\underset{|}{C}-}}$$

geminal dihalide [24.10]

In the case of a terminal alkyne, for example but-1-yne, protonation occurs at the terminal carbon due to the formation of a stabilised secondary vinylic carbocation, which is relatively more stable than the alternative primary carbocation. Notice that the trend of carbocation stability is the same as that discussed for the addition of HBr to alkenes. Further reaction of this carbocation with the bromide nucleophile leads to 2-bromobut-1-ene as the sole product. Further addition of HBr to the alkene follows Markovnikov's rule, leading to the geminal dibromide 2,2-dibromobutane.

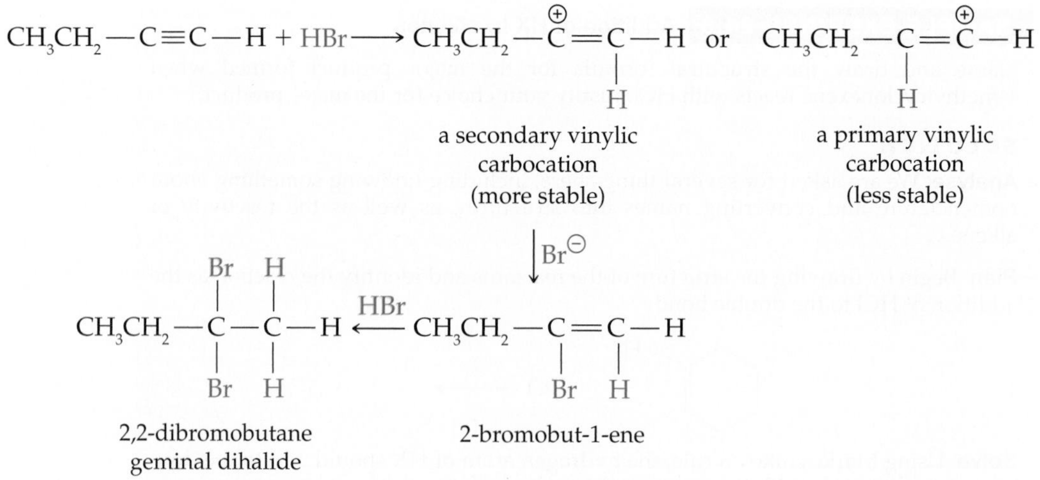

$$CH_3CH_2-C\equiv C-H + HBr \longrightarrow CH_3CH_2-\overset{\oplus}{C}=C-H \quad \text{or} \quad CH_3CH_2-C=\overset{\oplus}{C}-H$$

a secondary vinylic carbocation (more stable)

a primary vinylic carbocation (less stable)

$$\downarrow Br^{\ominus}$$

$$\underset{\substack{| \\ Br \;\; H}}{\overset{\substack{Br \;\; H \\ |}}{CH_3CH_2-\underset{|}{C}-\underset{|}{C}-H}} \xleftarrow{HBr} \underset{\substack{| \\ Br \;\; H}}{CH_3CH_2-C=C-H}$$

2,2-dibromobutane geminal dihalide

2-bromobut-1-ene

Addition Reactions Involving H₂O

The *hydration* of alkenes and alkynes is not as easy to accomplish as either the halogenation or hydrohalogenation of the same compounds. Hydration reactions need to be catalysed by a small amount of a strong acid, such as H_2SO_4. The reason is easily demonstrated mechanistically. Again, we can use but-2-ene as our alkene to illustrate this example.

The addition of a few drops of sulfuric acid to a reaction mixture containing but-2-ene and water leads to the formation of the hydronium ion, as shown:

$$H_2O + H_2SO_4 \longrightarrow H_3O^{\oplus} + HSO_4^{\ominus}$$

hydronium ion (an electrophile)

[24.11]

In terms of the addition of water, the hydronium species is the active electrophile. The reaction proceeds as shown in Equations 24.12 to 24.14. The first step involves the addition of a proton from the hydronium ion to one of the carbons that make up the double bond. This step is slow and rate-determining.

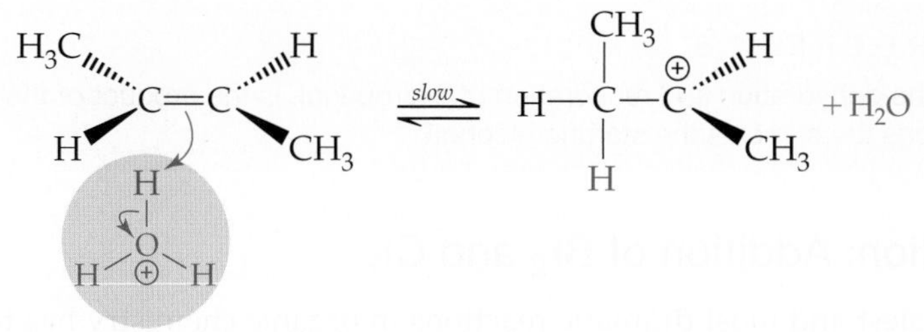

$$[24.12]$$

A molecule of water can then act as a nucleophile forming a C–O bond with the valence electrons of one of its two lone pairs. Note that a nucleophile need not be negatively charged; it just needs to be a species capable of donating a pair of electrons. In this case, an oxonium ion is formed.

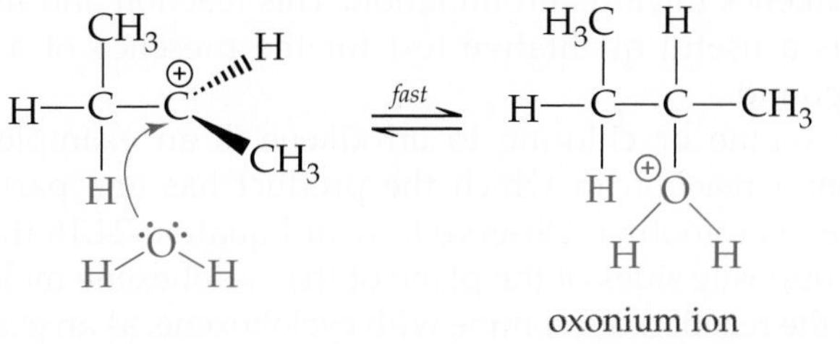

$$[24.13]$$

oxonium ion

Loss of a proton from the oxonium ion yields the final alcohol product. The other product formed is a new hydronium ion. This ion is capable of reacting further with another alkene in the same way described in Equations 24.12 to 24.14. Theoretically, an amount of sulfuric acid large enough to initiate the formation of a single hydronium ion is enough to allow the reaction to proceed. In practice, more is required to make sure the reaction is completed in a timely manner.

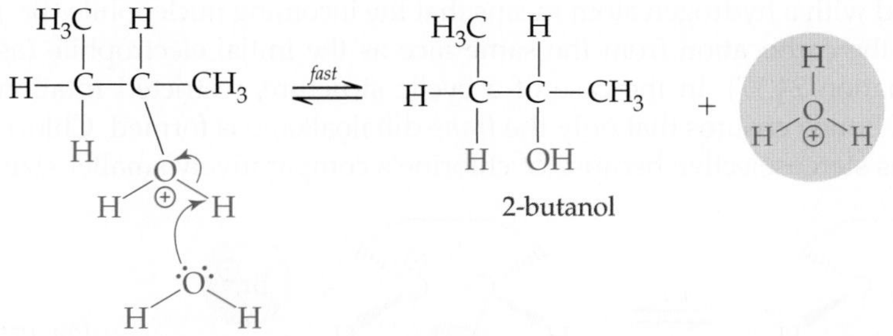

2-butanol

$$[24.14]$$

You may have noticed in Equations 24.12 to 24.14 that equilibrium arrows are used to describe each step of the reaction mechanism. This is because the reverse reaction—that is, the *dehydration* of alcohols—is also catalysed by H_2SO_4. In fact, this reverse method is a very useful way of preparing alkenes.

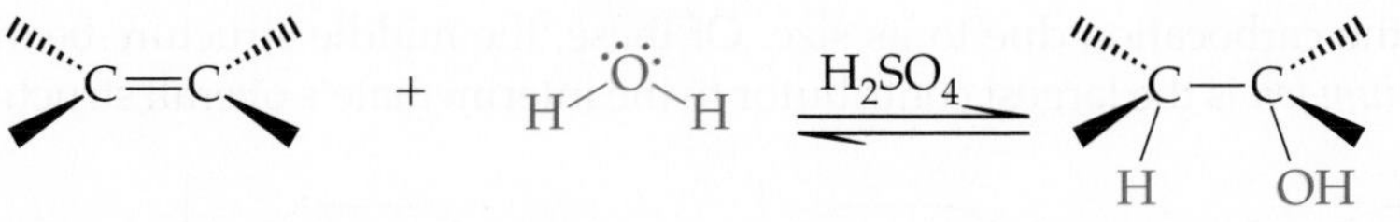

$$[24.15]$$

The direction of the equilibrium—that is, to maximise the yield of the alkene from the alcohol (right to left) or the alcohol from the alkene (left to right)—can be manipulated by adjusting the amount of water present in the reaction mixture. Large amounts of water (usually as aqueous acid) favour alcohol formation, whereas using concentrated acid and eliminating water favour alkene formation. By using these simple techniques, the alkene or alcohol can be prepared in high yield.

▲ **FIGURE 24.21 Bromine test for alkenes.** The bright orange colour of bromine solution is in contrast to the colourless solution produced by the addition of an alkene to the same solution.

> ### ▲ CONCEPT CHECK 6
>
> Consider the dehydration and rehydration of 1-propanol. Is the product of the two reactions the same as the starting alcohol?

Halogenation: Addition of Br_2 and Cl_2

One of the easiest and most dramatic reactions in organic chemistry has to be the reaction of Br_2 with an alkene. Bromine is a dense orange liquid, whose colour persists if made into a solution in hexane. The addition of a slight excess of an alkene to this solution (◄ FIGURE 24.21) at room temperature immediately turns the solution colourless. Undertaking this reaction in the presence of an alkane also demonstrates the large difference in reactivity between alkanes and alkenes towards bromination. This reaction and its dramatic colour change is a useful qualitative test for the presence of a C=C double bond in a compound.

The addition of bromine or chlorine to an alkene is an example of a **stereoselective reaction**: a reaction in which the product has one particular geometry in preference over another. Observe how in Equation 24.16 the two bromine atoms add to opposite sides of the plane of the cyclohexene molecule. Let us now investigate the reaction of bromine with cyclohexene, as an example of a halogenation, in more detail.

$$+ \ Br_2 \ \xrightarrow{\ CH_2Cl_2\ } \qquad\qquad\qquad [24.16]$$

The reasons for the stereoselectivity observed in the bromination of alkenes involve both steric and electronic components. The enormous size of a bromine atom compared with a hydrogen atom means that the incoming nucleophile (Br^-) cannot attack the carbocation from the same face as the initial electrophile (as shown in Equation 24.17). In the case of a cyclic structure, restricted rotation about the C–C bonds ensures that only the *trans*-dihaloalkane is formed. Chlorination is not as stereoselective because of chlorine's comparatively smaller size.

$$\xrightarrow{\ slow\ } \qquad\qquad\qquad [24.17]$$

Three contributors to the resonance structure of the intermediate carbocation are possible because of the electron-rich nature of the bromine atom and its accessibility to the carbocation due to its size. Of these, the middle structure bearing the *bromonium ion* is the largest contributor to the intermediate's overall structure.

$$[24.18]$$

The bromination and chlorination of alkynes follows a similar mechanism. The stepwise addition of bromine, by way of example, is shown in Equation 24.19. In each step, dichloromethane (CH_2Cl_2) is the solvent.

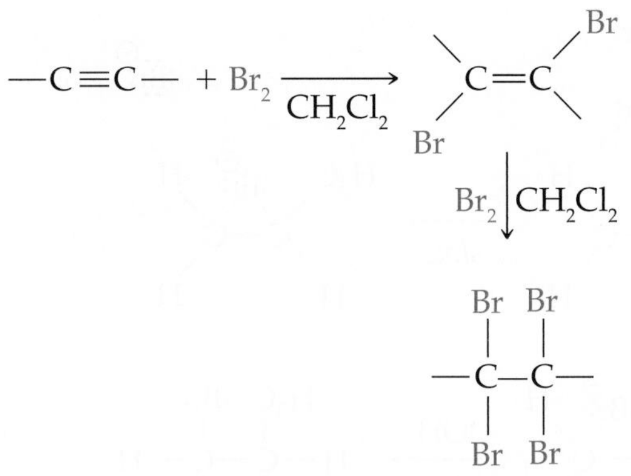

$$-C\equiv C- + Br_2 \xrightarrow{CH_2Cl_2} \overset{Br}{\underset{Br}{C=C}}$$

$$\downarrow Br_2 \Big| CH_2Cl_2$$

$$\overset{Br\ \ Br}{\underset{Br\ \ Br}{-C-C-}}$$

[24.19]

Addition of one equivalent of bromine leads to the formation of the dibromo-alkene in which the two bromo groups occupy a *trans* arrangement across the double bond as a result of the bromonium intermediate formed. A further equivalent of bromine leads to the tetrabrominated alkane in which all four positions around the C–C triple bond are occupied by halogen. Tetrahalogenation occurs when an excess of bromine or chlorine to the alkyne is used.

 CONCEPT CHECK 7

What would be the major organic product for the reaction of but-2-yne with excess bromine?

Halohydrin Formation

Let us now extend what we have learned about the mechanisms associated with the bromination and hydration of alkenes to more complicated additions. To do this, we will investigate the formation of the *halohydrins*—a class of compound containing halogen and OH functional groups. If the halogen and OH groups are on adjacent carbon atoms, the compound is called a *vicinal halohydrin*. Equations 24.20 and 24.21 illustrate the reaction type:

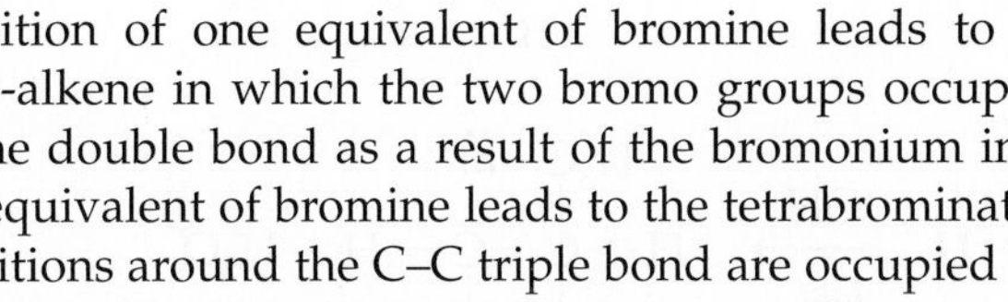

$$\overset{}{C=C} + Br_2 \xrightarrow{H_2O} \overset{}{\underset{Br\ \ \ OH}{-C-C-}} + HBr$$

[24.20]

a vicinal halohydrin

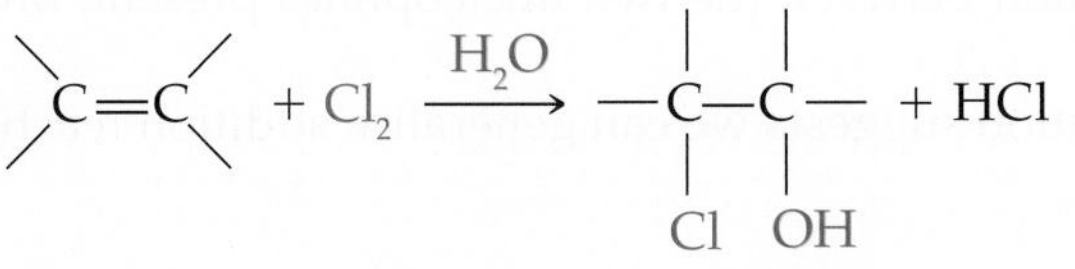

$$\overset{}{C=C} + Cl_2 \xrightarrow{H_2O} \overset{}{\underset{Cl\ \ OH}{-C-C-}} + HCl$$

[24.21]

The halogen of the halohydrin originates from the source you are familiar with (being Br_2 or Cl_2) and the OH group of the halohydrin occurs typically through the introduction of water to the reaction. This can be achieved either by adding water to the solvent of choice or as the solvent, depending on practicality.

The mechanism of bromohydrin formation has three steps. Here we look at the formation of the bromohydrin derived from propene in more detail:

Step 1

$$\text{slow} \qquad [24.22]$$

Step 2

$$\text{fast} \qquad [24.23]$$

Step 3

$$\text{fast} \qquad [24.24]$$

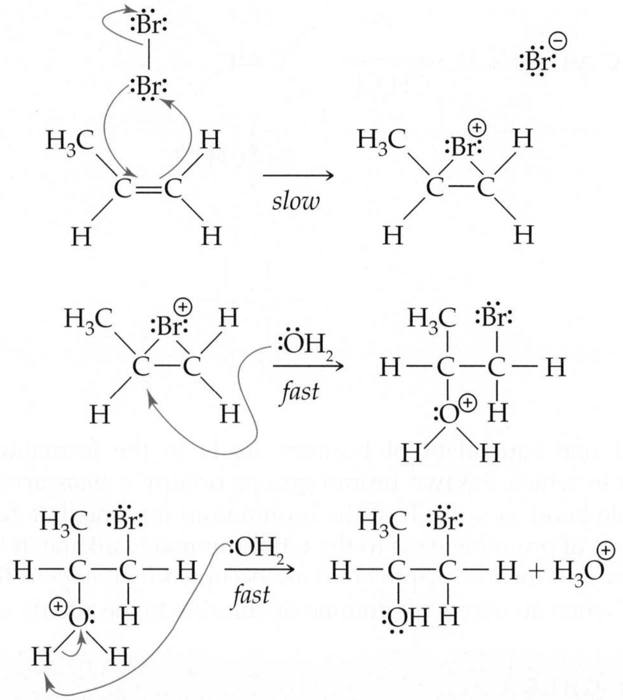

A cyclic bromonium ion is formed in the first step because Br^+ is the only electrophile in the reaction mixture (the corresponding cyclic chloronium ion would form if chlorine is used). This cyclic bromonium ion rapidly reacts with any nucleophile present. In the present case, two nucleophiles are present in solution. The two nucleophiles are Br^- and H_2O. But because H_2O is in the solvent, it is present in a higher concentration than Br^-. Consequently, reaction between the cyclic bromonium ion and H_2O is more likely. The protonated bromohydrin is a strong acid, and so reacts readily with water to form the halohydrin.

Notice throughout Equations 24.23–24.24 that the *regioselectivity* is well-defined with very little of the alternative bromohydrin formed. The product distribution follows *Markovnikov's rule,* as do all the additions we have studied so far. You may recall that for the addition of HCl across a double bond, the hydrogen adds to the side of the double bond with the most hydrogens. In the example of HCl addition, H^+ is considered to be the electrophile. In the case of halohydrin formation, the electrophile is Br^+, yet the outcome is still the same, that is, the electrophile adds to the side of the double bond with the most hydrogens.

While the major organic product of the reactions shown in Equations 24.20 and 24.21 is the vicinal halohydrin (called a bromohydrin or chlorohydrin, respectively), the corresponding dihalides are also formed to some extent. This is due to the competition between the two nucleophiles present, bromide ion and water.

Halohydrin formation suggests we can generalise addition reactions in the following way:

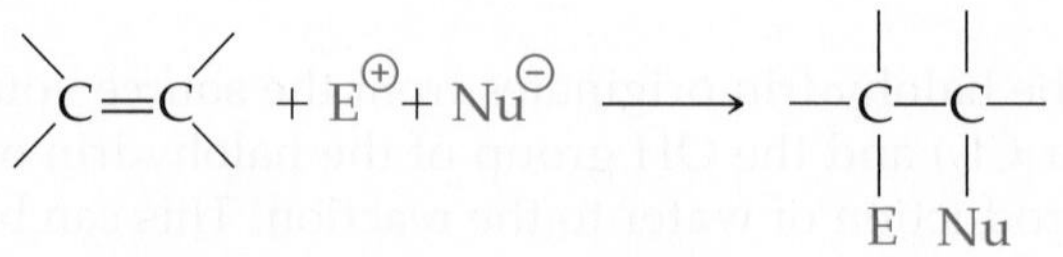

$$\diagup\text{C}=\text{C}\diagdown + \text{E}^\oplus + \text{Nu}^\ominus \longrightarrow -\overset{|}{\underset{\text{E}}{\text{C}}}-\overset{|}{\underset{\text{Nu}}{\text{C}}}- \qquad [24.25]$$

where the requisite nucleophile and electrophile can come from different starting materials. This means that provided you are able to identify a range of electrophiles (E^+) and nucleophiles (Nu^-) from a broad range of reagents, you will be able to convert alkenes into many highly functionalised compounds. ▶ TABLE 24.2 lists some common electrophile (E^+) and nucleophile (Nu^-) combinations and the addition products.

TABLE 24.2 • Summary of addition reactions to an alkene by electrophile (E^+) and nucleophile (Nu^-) from Equation 24.25.

Reagents	Electrophile	Nucleophile	Product
Br_2	Br_2	Br^-	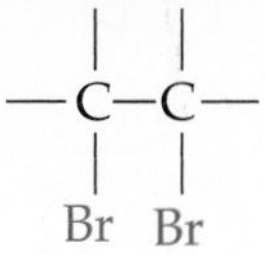 C–C with Br, Br
H_2O/H_2SO_4	H_3O^+	H_2O	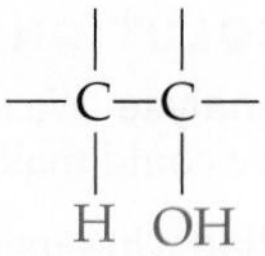 C–C with H, OH
Cl_2	Cl_2	Cl^-	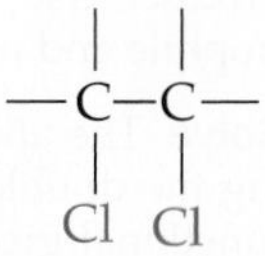 C–C with Cl, Cl
Br_2/H_2O	Br_2	H_2O	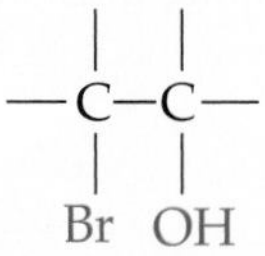 C–C with Br, OH
NBS/H_2O	Br_2	H_2O	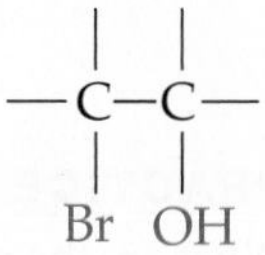 C–C with Br, OH
Cl_2/CH_3OH	Cl_2	CH_3OH	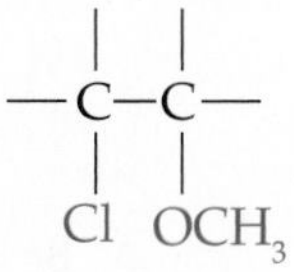 C–C with Cl, OCH_3
$Br_2/NaCl$	Br_2	Cl^-	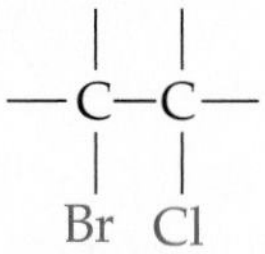 C–C with Br, Cl
$Cl_2/NaBr$	Cl_2	Br^-	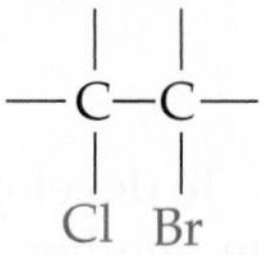 C–C with Cl, Br
NBS/NH_4Cl	Br_2	Cl^-	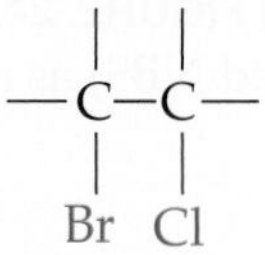 C–C with Br, Cl
NBS/NH_4F	Br_2	F^-	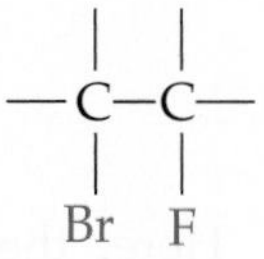 C–C with Br, F

SAMPLE EXERCISE 24.7 Addition reactions leading to halohydrins

Indicate how the following alcohol can be prepared from an alkene:

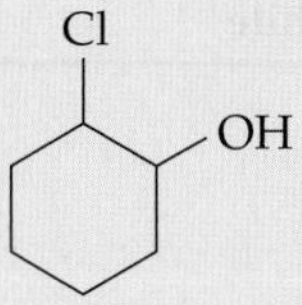

SOLUTION

Analyse We are asked to work backwards from a given structure to determine how we could make it from an alkene.

Plan Understand the differences between the starting alkene and the chlorohydrin product. Use our general equation [24.25] to determine from what reagents the electrophile and nucleophile are derived.

Solve The alkene required for this reaction is cyclohexene. This is deduced by placing the double bond between the positions occupied by the introduced Cl and OH functional groups. To get the addition of Cl and OH to the alkene, the reagents of choice must be Cl_2 and water. The water acts as both solvent and nucleophile.

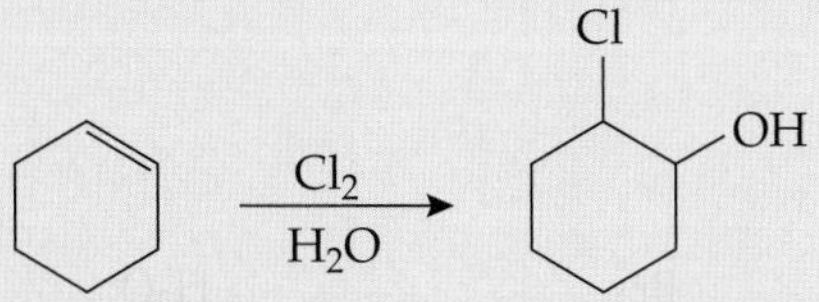

PRACTICE EXERCISE

What are the major and minor products of the following reaction?

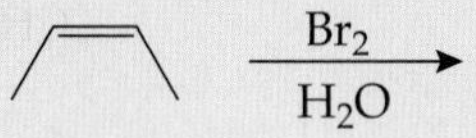

Answer:

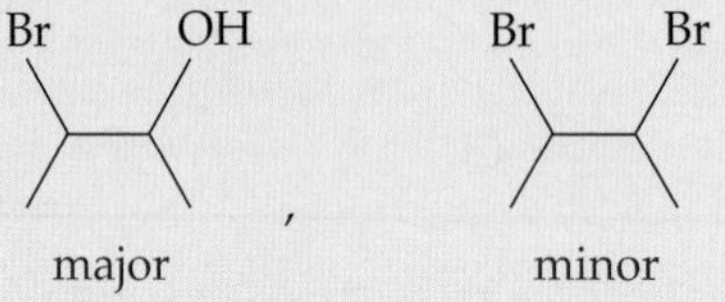

(See also Exercises 24.50, 24.51.)

To develop the concept shown in Equation 24.22 further, let us now investigate another source of electrophilic bromine, which is *N*-bromosuccinimide (◄ FIGURE 24.22). A general reaction utilising *N*-bromosuccinimide (abbreviated NBS) as an electrophilic source of bromine is shown in Equation 24.26.

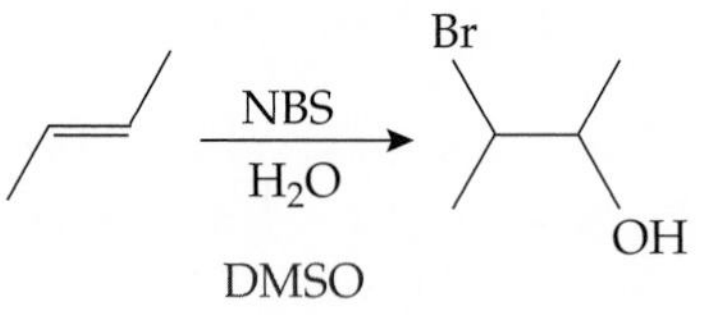

$$[24.26]$$

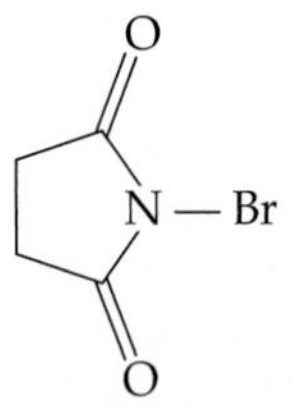

▲ FIGURE 24.22 *N*-Bromosuccinimide, or NBS, can act as an electrophilic source of bromine by decomposing slowly in the presence of water to form a tiny amount of Br_2.

Here, the NBS reacts with an alkene in the presence of water and dimethylsulfoxide (DMSO) $((CH_3)_2S{=}O)$, which is a polar organic solvent miscible with water. Under these conditions, the NBS decomposes slowly to form a small amount of Br_2, which then undergoes reaction to form the bromohydrin. Because only a small amount of Br^- is ever generated, the concentration of water is always in excess and so the bromohydrin is always formed in high yield.

A CLOSER LOOK

STEREOCHEMISTRY IN HALOHYDRIN FORMATION

Stereoselectivity plays a part in the formation of the bromohydrin. Recall from Equation 24.16 that the product of the bromination of cyclohexene is the *trans* dibromide.

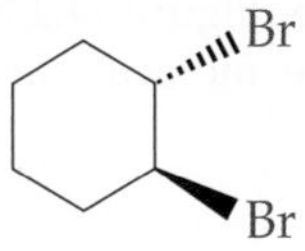

trans-1,2-dibromocyclohexane

Because the bridged halonium ion ring is opened by backside attack of H_2O in the halohydrin formation, addition of X and OH occurs in an *anti* (opposite) fashion and the *trans* product is most often formed.

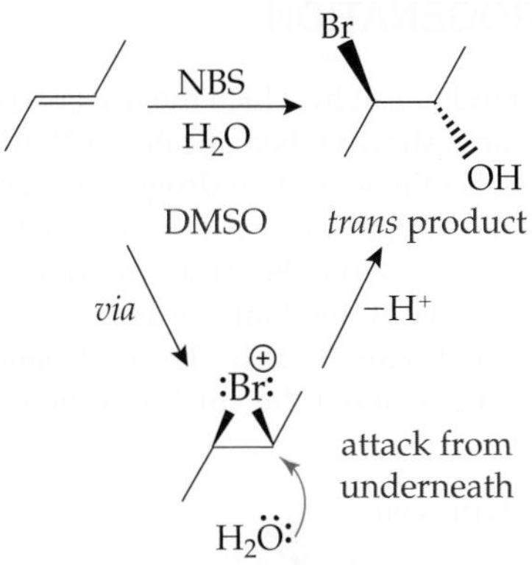

[24.27]

We can see the similarity more clearly in the following example where ammonium fluoride (NH_4F) provides a source of fluoride, which acts as the nucleophile, and DMSO acts as the solvent. Again, the *trans* product is formed.

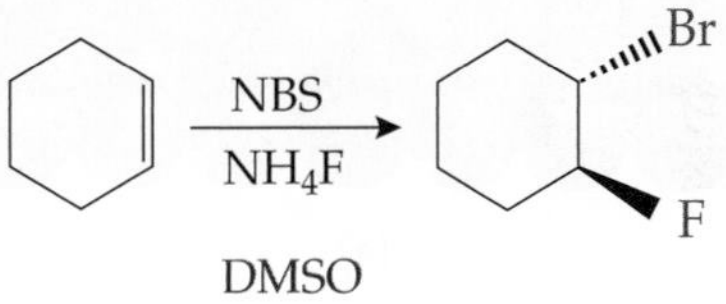

[24.28]

24.5 | ALKANES FROM ALKENES: CATALYTIC HYDROGENATION

The reaction between an alkene and molecular hydrogen (H_2), referred to as hydrogenation, does not occur readily under ordinary conditions of temperature and pressure. One reason for the lack of reactivity of H_2 towards alkenes is the high bond enthalpy of the H_2 molecule. To promote the reaction, it is necessary to use a catalyst that assists in rupturing the H–H bond. The most widely used catalysts are finely divided metals capable of adsorbing H_2, such as Pt, Pd and Ni. The hydrogenation reaction is often called a **catalytic hydrogenation**, since only a small amount of the solid metal is necessary to promote the reduction of the alkene to the corresponding alkane in solution. Because the alkene and catalyst exist in different phases, this type of reaction undergoes *heterogeneous catalysis*.

The reaction of hydrogen gas with ethene gas to form ethane gas provides an example of heterogeneous catalysis:

$$C_2H_4(g) + H_2(g) \xrightarrow{Pt} C_2H_6(g) \quad \Delta H° = -137 \text{ kJ mol}^{-1} \qquad [24.29]$$

ethene   ethane

We can understand the role of the catalyst in this process by considering the bond enthalpies involved (Table 8.4). In the course of the hydrogenation reaction, the H–H σ-bond and the C–C π-bond must be broken, and to do so requires the input of energy, which we can liken to the activation energy of the reaction. The formation of the new C–H σ-bonds *releases* an even greater amount of energy, making the reaction *exothermic*. When H_2 and C_2H_4 are bonded to the surface of the catalyst, less energy is required to break the bonds, lowering the activation energy of the reaction.

A CLOSER LOOK

HYDROGENATION

The mechanism by which the reaction is thought to occur is shown schematically in ▼ FIGURE 24.23. Both ethene and hydrogen are adsorbed on the metal surface (Figure 24.23(a)). Upon adsorption the H–H bond of H_2 breaks, leaving two H atoms that are bonded to the metal surface, as shown in Figure 24.23(b). The hydrogen atoms are relatively free to move about the surface. When a hydrogen

atom encounters an adsorbed ethene molecule, it can form a σ-bond to one of the carbon atoms, effectively destroying the C–C π-bond and leaving an *ethyl group* (C_2H_5) bonded to the surface *via* a metal-to-carbon bond (Figure 24.23(c)). This bond is relatively weak, and when the other carbon atom also encounters a hydrogen atom, a second C–H σ-bond is readily formed and the resulting ethane molecule is released from the metal surface (Figure 24.23(d)). The site is ready to adsorb another ethene molecule and thus begin the cycle again.

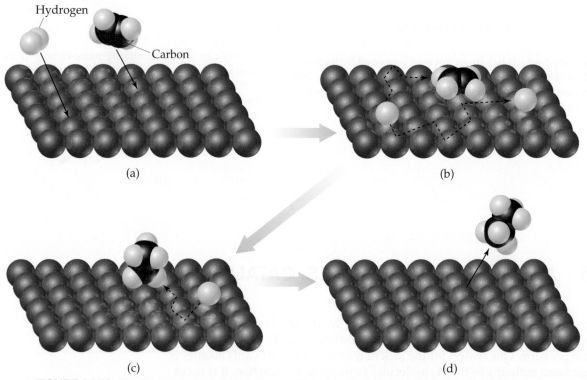

▲ FIGURE 24.23 **Mechanism for reaction of ethene with hydrogen on a catalytic surface.** (a) The hydrogen and ethene are adsorbed at the metal surface. (b) The H–H bond is broken to give two adsorbed hydrogen atoms. (c) These hydrogens migrate to the adsorbed ethene and bond to the carbon atoms. (d) As C–H bonds are formed, the adsorption of the molecule to the metal surface is decreased and ethane is released.

Catalytic hydrogenations favour a *cis* stereochemistry. For example, the reaction of 1,2-dimethylcyclohexene with hydrogen in the presence of 5% (w/w) palladium on carbon (Pd/C) yields the alkane in 100% yield, with >75% of the alkane having the *cis* stereochemistry. Palladium on carbon is a standard synthetic catalyst for hydrogenation.

$$\text{1,2-dimethylcyclohexene} + H_2\,(g) \xrightarrow[\text{5\% } (w/w)]{\text{Pd/C}} \text{cis-1,2-dimethylcyclohexane} \qquad [24.30]$$

> 75% yield

SAMPLE EXERCISE 24.8	Identifying the product of a hydrogenation reaction

Write the structural formula for the product of the hydrogenation of 3-methyl-1-pentene.

SOLUTION

Analyse We are asked to predict the compound formed when a particular alkene undergoes hydrogenation (reaction with H_2).

Plan To determine the structural formula of the reaction product, we must first write the structural formula or Lewis structure of the reactant. In the hydrogenation of the alkene, H_2 adds to the double bond, producing an alkane. (That is, each carbon atom of the double bond forms a bond to an H atom, and the double bond is converted to a single bond.)

Solve The name of the starting compound tells us that we have a chain of five C atoms with a double bond at one end (position 1) and a methyl group on the third C from that end (position 3):

$$CH_2{=}CH{-}\underset{\underset{\displaystyle CH_3}{|}}{CH}{-}CH_2{-}CH_3$$

Hydrogenation—the addition of two H atoms to the carbons of the double bond—leads to the following alkane:

$$CH_3{-}CH_2{-}\underset{\underset{\displaystyle CH_3}{|}}{CH}{-}CH_2{-}CH_3$$

Comment The longest chain in the product alkane has five carbon atoms; its name is therefore 3-methylpentane.

PRACTICE EXERCISE

Draw all the possible products (including stereochemistry) for the complete hydrogenation of the theoretical compound hexamethylcyclohexatriene, assuming 100% *cis* addition per double bond. Do not account for chair and boat conformations.

Answer: There are two products which, as line drawings, are:

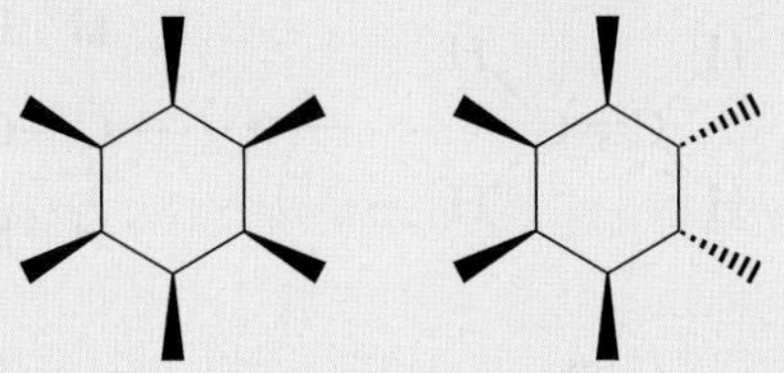

(See also Exercise 24.48.)

Hydrogenation is a useful commercial process for the preparation of fats from *polyunsaturated oils* (∞ Section 27.4, 'Fats, Oils and Waxes'). These oils, as the name suggests, contain many alkenes. Oils are generally **triglycerides** rich in linoleic ($C_{18}H_{32}O_2$), oleic ($C_{18}H_{34}O_2$), arachidonic ($C_{20}H_{32}O_2$) and other unsaturated fatty esters which, because of their structure, are liquids at room temperature (Section 26.2). Their conversion to stearic ($C_{18}H_{36}O_2$) and arachidic esters ($C_{20}H_{40}O_2$) by hydrogenation leads to **fats**, whose properties are more those of solids or semi-solids at room temperature. The reason for the change in physical properties has to do with the ease of packing the long alkyl chains over those chains that are 'kinked' by the introduction of double bonds (▼ FIGURE 24.24).

∞ Find out more on page 1090

The process of converting oils to fats is called **hardening**. This process is especially relevant to the manufacture of margarine and cooking fats. The final hardened product is typically white. In the past, β-carotene was added to colour the margarine so that it looks like butter.

▲ CONCEPT CHECK 8

What is the major product of the complete hydrogenation of ethyne?

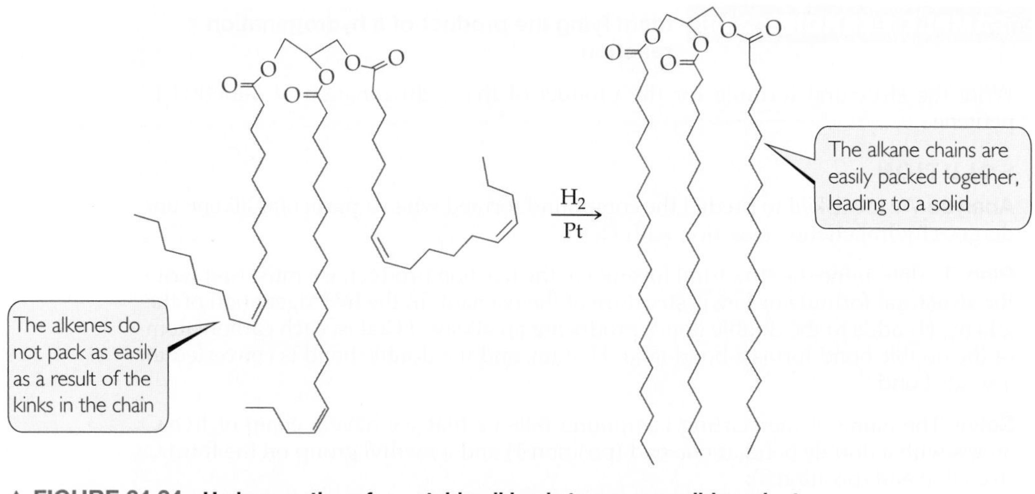

▲ **FIGURE 24.24** Hydrogenation of vegetable oil leads to a more solid product.

24.6 | ADDITION POLYMERISATION

In nature we find many substances of very high molecular mass that make up much of the structure of living organisms and tissues. Some examples include starch and cellulose (which abound in plants) and proteins (found in both plants and animals). In 1827 Jons Jakob Berzelius (1779–1848) coined the word **polymer** (from the Greek *poly*, 'many', and *meros*, 'parts') to denote molecular substances of high molecular mass formed by the *polymerisation* (joining together) of **monomers**, that is, molecules with low molecular mass. An example is the reaction of ethene monomers to form polyethene:

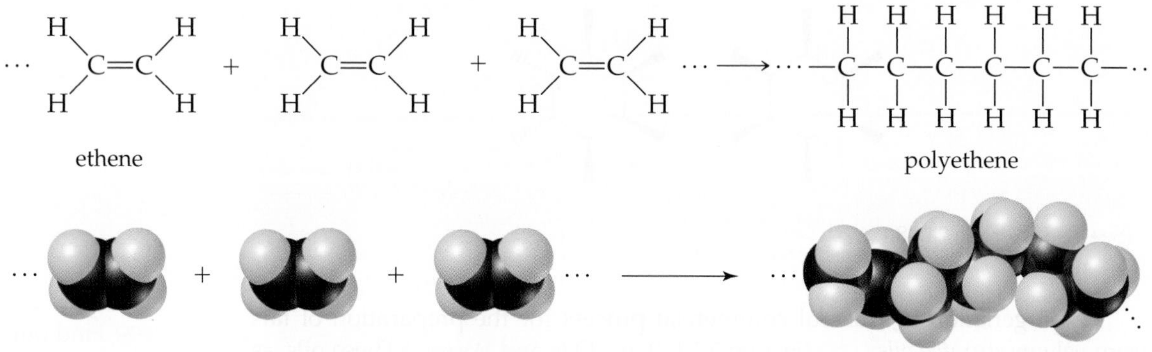

For a long time, humans have processed naturally occurring polymers, such as wool, leather, silk and natural rubber, into usable materials. Over the last century, however, chemists have learned to form synthetic polymers by polymerising compounds through controlled chemical reactions. A great many of these synthetic polymers have a backbone of carbon–carbon bonds because carbon atoms have an exceptional ability to form strong stable bonds with one another.

Plastics are materials that can be formed into various shapes, usually by the application of heat and pressure. **Thermoplastic** materials can be reshaped. For example, plastic milk containers are made from a polymer known as *polyethene*, which has a high molecular mass. These containers can be melted down and the polymer recycled for some other use. In contrast, a **thermosetting plastic** is shaped through irreversible chemical processes and thus cannot be readily reshaped.

Making Polymers

Polyethene (▶ **FIGURE 24.25**) is a solid substance that is used to make thousands of plastic products, such as plastic bags, food containers and laboratory equipment. In making polyethene, a reactive centre is formed that adds sequentially to the double bond in each ethene, extending the polymer chain and regenerating the reactive centre. Such reactions are classified as chain-growth or **addition polymerisations** and polyethene is an **addition polymer**. In ethene polymerisation, the reactive centre is frequently a free radical, which can be derived from a radical initiator such as lauroyl peroxide. Peroxides dissociate to form two radicals upon heating or irradiation with light because of the very weak O–O bond, forming two radicals:

$$\text{heat or light} \qquad 2 \qquad [24.31]$$

Note the use of *fishhook arrows* in Equation 24.31. They differ from the arrows we have studied to date, which were 'arrow'-headed, because they represent the movement of a single electron rather than an electron pair.

Once formed, the benzoyl radical can react with ethene to form a new carbon-centred radical. This step is called the **initiation** step (∞ Section 22.8, 'Reactions of Alkanes'). The relative stabilities of alkyl radicals follow a similar pattern to that of carbocations, namely 3° > 2° > 1° alkyl radicals.

$$\text{Carbon-centred radical} \qquad [24.32]$$

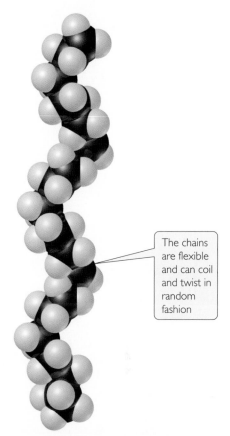

▲ **FIGURE 24.25   A segment of a polyethene chain.** The segment shown here consists of 28 carbon atoms. In commercial polyethene, the chain lengths range from about 10^3 to 10^5 CH_2 units.

> The chains are flexible and can coil and twist in random fashion

∞ Review this on page 914

The next stage of reaction is known as **propagation**. In this stage, the chain is extended by the consecutive addition of more ethene units through C–C bond formation:

Propagation

$$[24.33]$$

The propagation continues until two radical polymer chains react to form a completed polymer by a final C–C bond formation that terminates the radical. This is known as the **termination** step. In Equation 24.34, n and m are normally large integers, which may or may not have the same value.

Termination

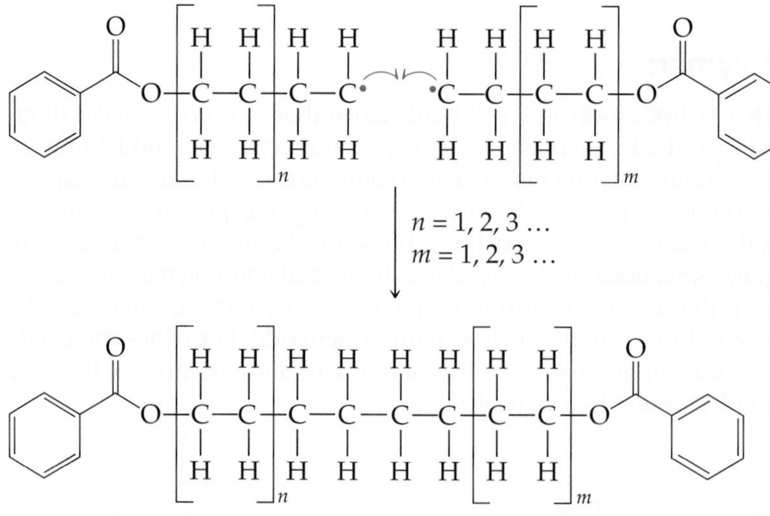

$$n = 1, 2, 3 \ldots$$
$$m = 1, 2, 3 \ldots$$

[24.34]

The reactive centre in a chain-growth polymerisation need not be a radical: depending on the monomer used, it may be an anion, cation or coordination complex. The most common process for polyethene production is *Ziegler–Natta* polymerisation, named after the two scientists who developed the reaction in the 1950s. This process typically involves titanium and aluminium complexes, with a coordination complex as the reactive centre, and is used in the formation of high-density polyethene (HDPE), a hard plastic. Radical polymerisation of polyethene gives a polymer with occasional long branches, rather than simple straight chains, which makes it harder to pack the molecules together—this is low-density polyethene (LDPE), used in stretchy plastic wrap.

We can write the overall equation for the polymerisation of ethene as follows:

$$n\,CH_2{=}CH_2 \longrightarrow \left[\begin{array}{c} H \;\; H \\ | \;\;\; | \\ C{-}C \\ | \;\;\; | \\ H \;\; H \end{array}\right]_n$$

[24.35]

Here, n represents the large number—ranging from hundreds to many thousands—of monomer molecules (ethene, in this case) that react to form one large polymer molecule. Within the polymer, a repeat unit (the unit shown in brackets in Equation 24.35) appears along the entire chain.

Polyethene is a very important commercial material; more than 10 billion kilograms are produced globally each year. Although its composition is simple, the polymer is not easy to make. Only after many years of research were the right conditions identified for manufacturing it. Today many different forms of polyethene, varying widely in physical properties, are known. Polymers of other chemical compositions provide still greater variety in physical and chemical properties. ▶ TABLE 24.3 lists several other common polymers obtained by addition polymerisation.

The names of polymers are simply derived from the monomers employed (Table 24.3). Poly(vinyl chloride) (PVC), for example, is made from vinyl chloride, poly(methyl methacrylate) is made from methyl methacrylate and polystyrene is made from styrene.

Structure and Physical Properties of Addition Polymers

The simple structural formulae given for polyethene and other polymers are deceptive. Since each carbon atom in polyethene is surrounded by four bonds, the atoms are arranged in a tetrahedral fashion, so that the chain is not straight

TABLE 24.3 • Addition polymers of commercial importance

Monomers	Addition Polymer	Structure	Uses
CH_2CH_2	Polyethene	$\{CH_2-CH_2\}_n$	Films, packaging, bottles
(propene, with CH_3)	Polypropylene	$\begin{bmatrix}CH_2-CH\\ \qquad \mid \\ \qquad CH_3\end{bmatrix}_n$	Kitchenware, fibres, appliances
(styrene)	Polystyrene	$\begin{bmatrix}CH_2-CH\\ \qquad \mid \\ \qquad C_6H_5\end{bmatrix}_n$	Packaging, disposable food containers, insulation
(vinyl chloride, Cl)	Poly(vinyl chloride) (PVC)	$\begin{bmatrix}CH_2-CH\\ \qquad \mid \\ \qquad Cl\end{bmatrix}_n$	Pipe fittings, clear film for meat packaging
H_3C ... $COOCH_3$	Poly(methyl methacrylate)	(structure with H_3C and $COOCH_3$) $_n$	Perspex, paints, moulded articles

as we sometimes depict it. Furthermore, the atoms are relatively free to rotate around the C–C single bonds. Thus rather than being straight and rigid, the chains of an addition polymer are very flexible, folding readily. The flexibility in the molecular chains can cause the polymer material itself to be very flexible

Both synthetic and naturally occurring polymers consist commonly of a collection of *macromolecules* (large molecules) of different molecular weights. Depending on the conditions of formation, the molecular weights may be distributed over a wide range or be closely clustered around an average value. In part, because of this distribution in molecular weights, polymers are largely amorphous (non-crystalline) materials. Rather than exhibiting a well-defined crystalline phase with a sharp melting point, they soften over a range of temperatures. They may, however, possess short-range order in some regions of the solid, with chains lined up in regular arrays, as shown in ▶ FIGURE 24.26. The extent of such ordering is indicated by the degree of **crystallinity** of the polymer. The crystallinity of a polymer can be enhanced with mechanical stretching, which aligns the polymer chains as the molten polymer is drawn out. Intermolecular forces between the polymer chains hold the chains together in the ordered crystalline regions, making the polymer denser, harder, less soluble and more resistant to heat. ▼ TABLE 24.4 shows how the properties of polyethene change as the degree of crystallinity increases.

The simple linear structure of polyethene is conducive to intermolecular interactions that lead to crystallinity. However, the degree of crystallinity in polyethene strongly depends on the average molecular mass. Polymerisation

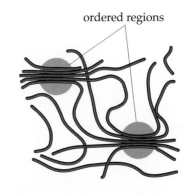

ordered regions

▲ FIGURE 24.26 **Interactions between polymer chains.** In the circled regions, the forces that operate between adjacent polymer-chain segments lead to ordering analogous to the ordering in crystals, though less regular.

TABLE 24.4 • Properties of polyethene as a function of crystallinity

	Crystallinity				
	55%	62%	70%	77%	85%
Melting point (°C)	109	116	125	130	133
Density (g cm^{-3})	0.92	0.93	0.94	0.95	0.96
Stiffness* (MPa)	12	17	13	29	35
Yield stress* (MPa)	170	320	520	830	1140

* These test results show that the mechanical strength of the polymer increases with increased crystallinity. Discussion of the exact meaning and significance of these tests is beyond the scope of this text.

MY WORLD OF CHEMISTRY

RECYCLING PLASTICS

If you look at the bottom of a plastic container, you are likely to see a recycle symbol (▶ FIGURE 24.27) containing a number, as shown in ▼ TABLE 24.5. The number and letter abbreviation below it indicate the kind of polymer from which the container is made, as summarised in Table 24.5. These symbols

TABLE 24.5 • Categories used for recycling polymeric materials

Recycling number	Abbreviation	Polymer
1	PET or PETE	Polyethene terephthalate
2	HDPE	High-density polyethene
3	V	Polyvinyl chloride (PVC)
4	LDPE	Low-density polyethene
5	PP	Polypropylene
6	PS	Polystyrene
7	—	Others

▲ FIGURE 24.27 Recycling symbols.

make it possible to sort containers by composition. In general, the lower the number, the greater the ease with which the material can be recycled.

results in a mixture of macromolecules with varying values of n and hence varying molecular masses. So-called low-density polyethene (LDPE), used in forming films and sheets, has an average molecular mass in the range of 10^4 u and has substantial chain branching. That is, side chains occur off the main chain of the polymer, much like spur lines that branch from a main railway line. These branches inhibit the formation of crystalline regions, reducing the density of the material. High-density polyethene (HDPE), used to form bottles, drums and pipes, has an average molecular mass in the range of 10^6 u. This form has less branching and thus a higher degree of crystallinity. Low-density and high-density polyethene are illustrated in ▼ FIGURE 24.28. Thus the properties of

▶ FIGURE 24.28 **Two types of polyethene.** (a) Schematic illustration of the structure of low-density polyethene (LDPE) and food-storage bags made of LDPE film. (b) Schematic illustration of the structure of high-density polyethene (HDPE) and containers formed from HDPE.

(a)   (b)

polyethene can be 'tuned' by varying the average length, crystallinity and branching of the chains, making it a very versatile material.

The physical properties of polymeric materials can be modified extensively by adding substances with lower molecular mass, called *plasticisers*, to reduce the extent of interactions between chains and thus to make the polymer more pliable. Poly(vinyl chloride) (PVC) (Table 24.3), for example, is a hard, rigid material of high molecular mass that is used to manufacture drainpipes. When blended with a suitable substance of lower molecular mass, however, it forms a flexible polymer that can be used to make rainboots and doll parts. In some applications, the plasticiser in a plastic object may be lost over time because of evaporation. As this happens, the plastic loses its flexibility and becomes subject to cracking.

Polymers can be made stiffer by introducing chemical bonds between the polymer chains, as illustrated in ▶ FIGURE 24.29. Forming bonds between chains is called **cross-linking**. The greater the number of cross-links in a polymer, the more rigid the material. Whereas thermoplastic materials consist of independent polymer chains, thermosetting materials become cross-linked when heated. The cross-links allow them to hold their shapes. An important example of cross-linking is the **vulcanisation** of natural rubber, a process discovered by Charles Goodyear in 1839 (see the My World of Chemistry box, page 986). Natural rubber is formed from a liquid resin derived from the inner bark of *Hevea brasiliensis,* or rubber tree. Chemically, it is a terpene, a polymer of isoprene, C_5H_8.

▲ **FIGURE 24.29 Cross-linking of polymer chains.** The cross-linking groups (red) constrain the relative motions of the polymer chains, making the material harder and less flexible than when the cross-links are not present.

$(n+2)$ [isoprene structure] $\longrightarrow$

Isoprene

[rubber polymer structure] [24.36]

rubber

[space-filling model of rubber]

MY WORLD OF CHEMISTRY

THE ACCIDENTAL DISCOVERY OF TEFLON®

In 1938 a scientist at Du Pont named Roy J. Plunkett (1910–1994) made a rather curious observation: a tank of gaseous tetrafluoroethene (CF_2=CF_2) that was supposed to be full seemed to have no gas in it. Rather than discard the tank, Plunkett decided to cut it open. He found that the inside of the tank was coated with a waxy white substance that was remarkably unreactive towards even the most corrosive reagents. That compound was formed by the addition polymerisation of tetrafluoroethene:

$$nCF_2\text{=}CF_2 \xrightarrow{\text{polymerisation}} \text{(}CF_2-CF_2\text{)}_n$$

As it turned out, the properties of this material, called Teflon®, were ideal for an immediate and important application in the development of the first atomic bomb. Uranium hexafluoride (UF_6), which was used to separate fissionable [235]U by gaseous diffusion, is an extremely corrosive material. Teflon® was used as a gasket material in the gaseous diffusion plant. It is now used in a variety of applications, from non-stick cookware to space suits.

Plunkett's desire to know more about something that just didn't seem right is a wonderful example of how natural scientific curiosity can lead to remarkable discoveries.

MY WORLD OF CHEMISTRY

VULCANISATION

Natural rubber is not a useful polymer because it is too soft and too chemically reactive. Charles Goodyear (1800–1860) accidentally discovered that adding sulfur to rubber and then heating the mixture makes the rubber harder and reduces its susceptibility to oxidation or other chemical attack. The sulfur changes rubber into a thermosetting polymer by cross-linking the polymer chains through reactions at some of the double bonds, as shown schematically in ▼ FIGURE 24.30. Cross-linking of about 5% of the double bonds creates a flexible, resilient rubber. When the rubber is stretched, the cross-links help to prevent the chains from slipping, so that the rubber retains its elasticity. Nature also uses a sulfide crosslink to rigidify and preserve certain protein structures, such as insulin (∞ Section 29.3, 'Proteins, Peptides and Enzymes').

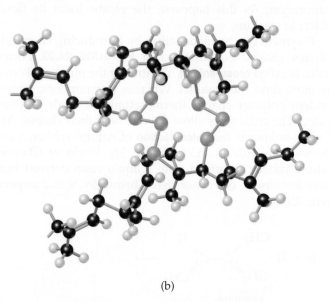

(a)   (b)

▲ FIGURE 24.30   **Vulcanisation of rubber.** (a) In natural rubber, carbon–carbon double bonds occur at regular intervals along the chain. (b) Chains of sulfur atoms have been added across two polymer chains, by the opening of a carbon–carbon double bond on each chain.

CONCEPT CHECK 9

How would you expect the properties of vulcanised rubber to vary as the percentage of sulfur increases?

SAMPLE INTEGRATIVE PROBLEM | Putting concepts together

In an example of an electrophilic addition reaction 1-butene reacts with hydrogen bromide to give 2-bromobutane. **(a)** Draw out the mechanism of this reaction including the carbocation intermediate. Show all the atoms attached to the double bond and use curved arrows to indicate electron-pair movements. **(b)** What are the hybridisation and molecular geometry of the two atoms C1 and C2 in the starting material, intermediate and product? **(c)** Indicate the stereogenic carbon centre in the product with *. **(d)** Explain why the product is formed as a racemic mixture.

SOLUTION

Analyse In this problem we examine the mechanism of an electrophilic addition reaction and how the shape of the intermediate relates to the chirality of the product.

Plan First identify the structure of the starting material and product from the names given. Then draw the mechanism of the reaction and identify how the molecular geometry of C1 and C2 changes as the reaction progresses. Finally identify the features of the geometry change that result in a racemic mixture being produced.

Solve

(a)

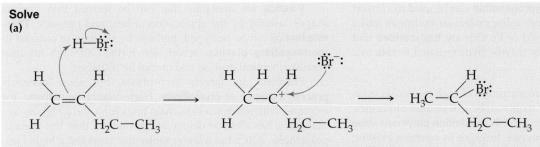

(b)

	Hybridisation—C1	Geometry—C1	Hybridisation—C2	Geometry—C2
Starting material	sp^2	Trigonal planar	sp^2	Trigonal planar
Intermediate	sp^3	Tetrahedral	sp^2	Trigonal planar
Product	sp^3	Tetrahedral	sp^3	Tetrahedral

(c)

$$H_3C-\overset{*}{C}\overset{H}{\underset{H_2C-CH_3}{\diagdown}}\ddot{B}r:$$

(d) The carbocation is sp^2 hybridised and trigonal planar in shape. The positive charge may be thought of as an empty p orbital pointing above and below the plane containing the carbocation. The final stage of this reaction involves attack by the bromide ion on the carbocation and there is equal probability of this occurring to the top or the bottom face of the intermediate.

$$:\ddot{B}r:^- \longrightarrow \quad \begin{array}{c} Br \\ H_3C \overset{\text{\tiny|||||}}{\diagdown} H \\ CH_2CH_3 \end{array} \quad \begin{array}{l} 50\% \\ (R)\text{-2-bromobutane} \end{array}$$

$$:\ddot{B}r:^- \longrightarrow \quad \begin{array}{c} H_3C \diagdown \overset{H}{\text{\tiny|||||}} CH_2CH_3 \\ Br \end{array} \quad \begin{array}{l} 50\% \\ (S)\text{-2-bromobutane} \end{array}$$

CHAPTER SUMMARY AND KEY TERMS

SECTION 24.1 Unsaturated hydrocarbons are those hydrocarbons that contain multiple bonds. Two main classes—**alkenes** and **alkynes**—are dealt with in this chapter. Alkenes contain one or more double bonds. Each double bond consists of one **σ-bond** and one **π-bond**. The π-bond is formed by the overlap of the unhybridised p orbitals on adjacent sp^2 hybridised carbon atoms. Alkynes contain one or more triple bonds. Each triple bond consists of one σ-bond and two π-bonds. The order of bond lengths is triple bond < double bond < single bond.

SECTION 24.2 The names of alkenes and alkynes are based on the longest continuous chain of carbon atoms that contains the multiple bond, and the location of the multiple bond is specified by a numerical prefix. The **E,Z system** is used to name the geometric isomers of alkenes and is based on the same priority system used in R,S nomenclature.

SECTION 24.3 Organic chemists use **curved arrow notation** to describe the redistribution of valence electrons. Arrow notation can also be used to describe resonance structures or to rationalise a reaction mechanism.

SECTION 24.4 Alkenes and alkynes readily undergo **addition reactions** to the carbon–carbon multiple bonds. There are three main classes of addition reactions: **halogenation, hydrohalogenation** and **hydration**. Additions of acids, such as HBr, proceed via a rate-determining step in which a proton is transferred to one of the alkene or alkyne carbon atoms, forming a **carbocation**. This addition follows **Markovnikov's rule**, which states that the hydrogen of HX (X = halogen, OH) adds to the side of the double bond with the most hydrogens. The carbocation is a form of **electrophile** that readily reacts with a **nucleophile**, in this case Br$^-$, to form the product. The addition of bromine to cyclohexene is an example of a **stereoselective reaction**.

SECTION 24.5 Catalytic hydrogenation can be used to convert alkenes or alkynes to alkanes using molecular hydrogen and a metal catalyst such as Ni, Pd or Pt. Oils are **triglycerides** that contain double bonds. They may be hydrogenated to **fats** in a process called **hardening**.

SECTION 24.6 Polymers are molecules of high molecular mass formed by joining together large numbers of smaller molecules, called **monomers**. In a chain growth or **addition polymerisation** reaction, the molecules form new linkages by opening existing bonds. Polyethene is an example of an **addition polymer**. Three stages exist in this type of polymerisation: **initiation**, **propagation** and **termination**.

Plastics are materials that can be formed into various shapes, usually by the application of heat and pressure. **Thermoplastics** can be reshaped, perhaps by heating, in contrast to **thermosetting plastics**, which are formed through an irreversible chemical process and cannot be reshaped.

Polymers are largely amorphous, but some materials possess a degree of **crystallinity**. High-density polyethene, for example, with not much side-chain branching and a high molecular mass, has a higher degree of crystallinity than low-density polyethene, which has a lower molecular mass and a higher relative degree of branching. Polymer properties are also strongly affected by **cross-linking**, in which short chains of atoms connect the long polymer chains. Rubber is cross-linked by short chains of sulfur atoms in the process called **vulcanisation**.

KEY SKILLS

- Understand how to name alkenes and alkynes. In the case of alkenes, be able to determine E, Z isomerism. (Sections 24.1 and 24.2)
- Describe and understand the rules for using curved arrow notation and their application to describe the mechanism of a chemical reaction. (Section 24.3)
- Be able to draw the products of addition reactions using a range of reactants (e.g. Br_2, HCl) and understand their product distribution. (Sections 24.4 and 24.5)
- Understand the importance of alkenes to the industrial manufacturing of plastics. (Section 24.6)
- Be able to draw polymers formed by the addition polymerisation of alkene monomers and understand some of their properties. (Section 24.6)

KEY EQUATIONS

- Addition reactions

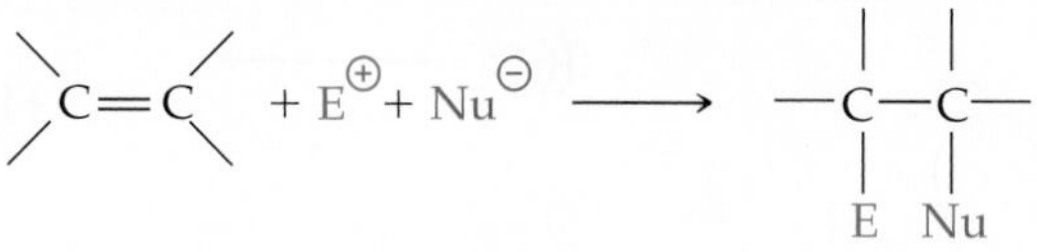

$$\text{C=C} + E^{\oplus} + Nu^{\ominus} \longrightarrow \begin{matrix} | & | \\ -C-C- \\ | & | \\ E & Nu \end{matrix} \qquad [24.25]$$

- Hydrogenation

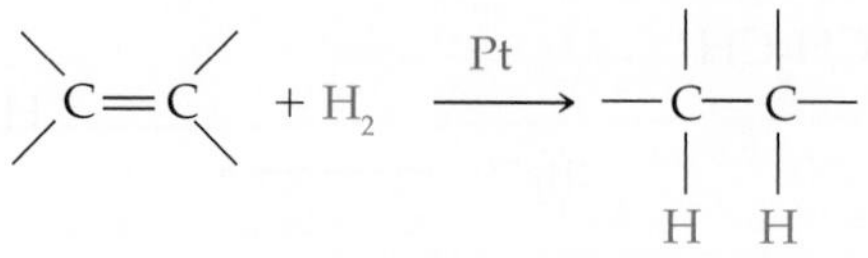

$$\text{C=C} + H_2 \xrightarrow{\text{Pt}} \begin{matrix} | & | \\ -C-C- \\ | & | \\ H & H \end{matrix} \qquad [24.29]$$

- Addition polymerisation

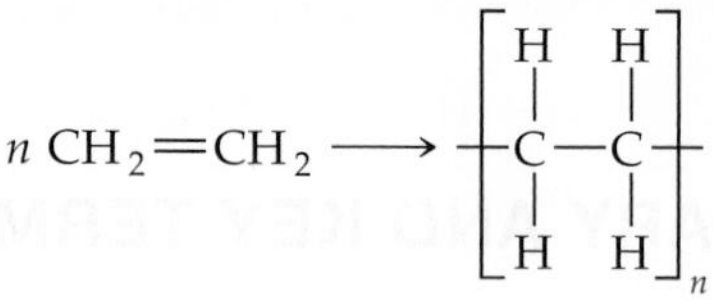

$$n\,CH_2{=}CH_2 \longrightarrow \begin{bmatrix} H & H \\ | & | \\ C-C \\ | & | \\ H & H \end{bmatrix}_n \qquad [24.35]$$

EXERCISES

VISUALISING CONCEPTS

24.1 Which of the following molecules is unsaturated? [Section 24.1]

$CH_3CH_2CH_2CH_3$
(a)

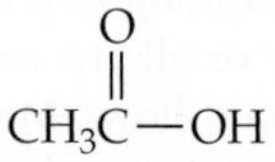

$CH_3\overset{\displaystyle O}{\overset{\|}{C}}{-}OH$
(b)

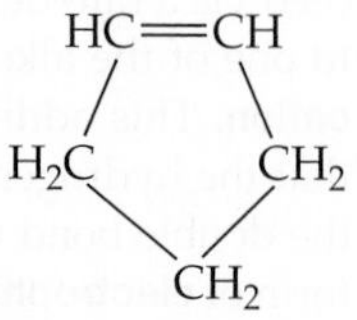

HC≡CH structure with H_2C and CH_2 and CH_2
(c)

$CH_3CH{≡}CHCH_3$
(d)

24.2 Consider the hydrocarbon drawn below. **(a)** What is the hybridisation at each carbon atom in the molecule? **(b)** How many σ-bonds are there in the molecule? **(c)** How many π-bonds? [Section 24.1]

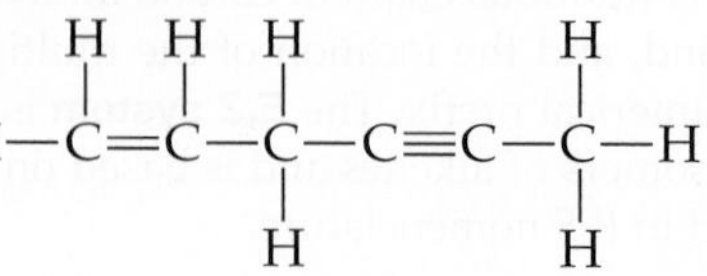

$$H{-}\underset{H}{\overset{H}{C}}{=}\underset{H}{\overset{H}{C}}{-}\underset{H}{\overset{H}{C}}{-}C{≡}C{-}\underset{H}{\overset{H}{C}}{-}H$$

24.3 Which of the following molecules will most readily undergo an addition reaction? [Section 24.4]

(a)

$$CH_3CH_2\overset{\displaystyle O}{\overset{\displaystyle \|}{C}}-OH$$

(b)

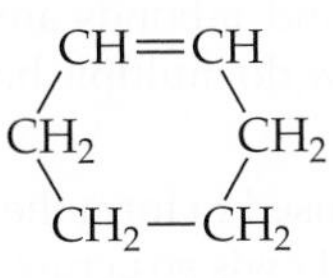

(c)

$$CH_3\overset{\displaystyle O}{\overset{\displaystyle \|}{C}}HC-OH$$
$$\underset{\displaystyle NH_2}{|}$$

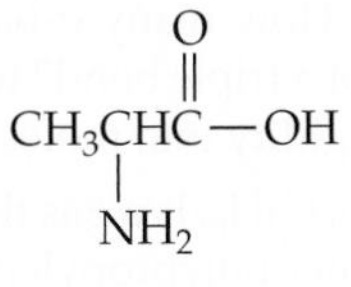

(d)

24.4 Consider the molecules shown below. Which of these would be most likely to form an addition polymer? [Section 24.4]

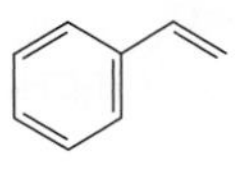

(i) 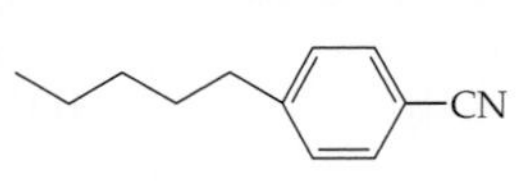   (ii) 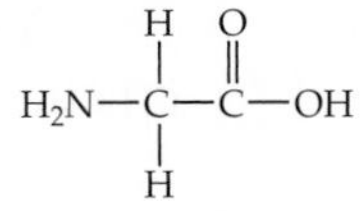   (iii)

24.5 Draw the product of the following valence electron movements. [Section 24.3]

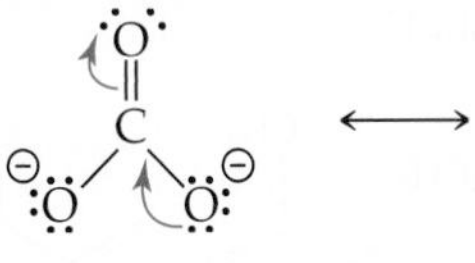

STRUCTURE, BONDING AND NAMING (Sections 24.1, 24.2 and 24.3)

24.6 What structural features help us identify a compound as **(a)** an alkene, **(b)** an alkyne?

24.7 Give the molecular formula of a hydrocarbon containing five carbon atoms that is **(a)** an alkane, **(b)** an alkene, **(c)** an alkyne. Which are saturated and which are unsaturated hydrocarbons?

24.8 Give the molecular formula of a cyclic alkane, a cyclic alkene and a linear alkyne that in each case contains six carbon atoms. Which are saturated and which are unsaturated hydrocarbons?

24.9 Give the general molecular formula for any cyclic alkene. What other general class of hydrocarbon has the same general formula?

24.10 Give the general molecular formula for any linear dialkene—that is, an aliphatic hydrocarbon with two double bonds along the chain.

24.11 How many straight-chain structural isomers are there for a five-member aliphatic hydrocarbon with one double bond? Name each structure.

24.12 Draw all the possible non-cyclic structural isomers of C_6H_{12}. Name each compound.

24.13 What are the approximate bond angles **(a)** about carbon in an alkane, **(b)** about a C–C double bond in an alkene, **(c)** about a C–C triple bond in an alkyne?

24.14 What are the characteristic hybrid orbitals employed by **(a)** carbon in an alkane, **(b)** carbon in a double bond in an alkene, **(c)** carbon in a triple bond in an alkyne?

24.15 Identify the carbon atom/s in the structure below that has/have each of the following hybridisations: **(a)** sp^3, **(b)** sp, **(c)** sp^2.

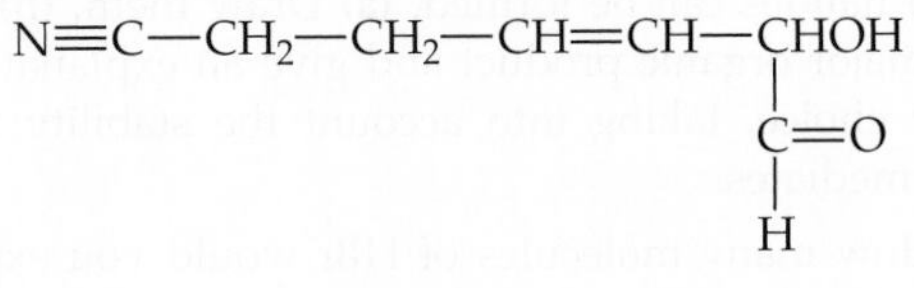

$$N{\equiv}C-CH_2-CH_2-CH{=}CH-CHOH$$
$$\underset{\displaystyle \underset{\displaystyle H}{|}}{\overset{\displaystyle |}{C}{=}O}$$

24.16 Name the following compounds systematically:

(a)

$$CH_3CH_2\underset{}{\diagdown}\!\!\!\!\!\!\!\!\underset{H}{\overset{}{C}}{=}\underset{H}{\overset{CH_2\overset{CH_3}{\overset{|}{C}}HCH_2CH_3}{C}}$$

(b) $HC{\equiv}CCH_2\overset{CH_2CH_3}{\underset{CH_3}{\overset{|}{\underset{|}{C}}}}CH_3$

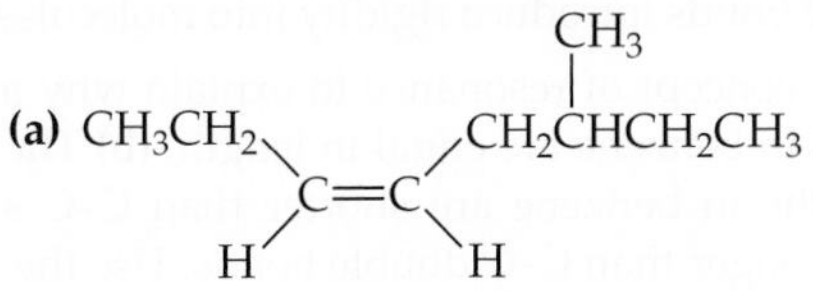

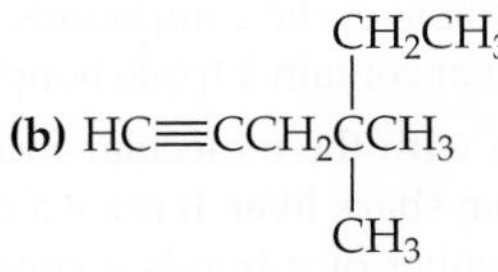

(c) $HC{\equiv}C-CH_2-Cl$

24.17 Name the following compounds systematically:

(a) *trans*-$CH_3CH{=}CHCH_2CH_2CH_3$

(b)

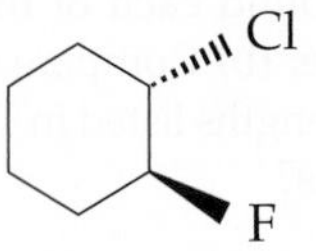

(c)

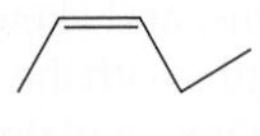

24.18 Why is geometric isomerism possible for alkenes, but not for aliphatic alkanes and alkynes?

24.19 **(a)** Using butene as an example, distinguish between structural (constitutional) and geometric isomers. **(b)** The compound 4-chloropent-2-ene exists as four stereoisomers. Explain how these four isomers arise by drawing the structures of all four stereoisomers and showing the relationship between them.

24.20 Draw the condensed structural formulae for the *E* and *Z* isomers of pent-2-ene. Can cyclopentene exhibit *E, Z* isomerism? Explain.

24.21 Explain why (*E*)-1,2-dichloroethene has no dipole moment, whereas (*Z*)-1,2-dichloroethene has a dipole moment.

24.22 Systematically name the following alkenes, assigning their geometry as *E* or *Z* where appropriate.

(a)
$$Cl\diagdown_{CH_3CH_2}C=C\diagup^{CH_2CH_2CH_3}_{CH_3}$$

(b)
$$Cl\diagdown_{H_3C}C=C\diagup^{H}_{CH_2CH_3}$$

(c)
$$Br\diagdown_{Cl}C=C\diagup^{CH_3}_{CH_2CH_3}$$

(d)
$$H\diagdown_{NH_2CH_2}C=C\diagup^{CH_2CH_2CH_3}_{CH_3}$$

24.23 How many *cis–trans* isomers are possible for this triterpene alcohol?

24.24 (a) If an atom is *sp* hybridised, how many unhybridised *p* orbitals remain in the valence shell? How many π-bonds can the atom form? (b) How many σ-bonds and π-bonds are generally part of a triple bond? (c) How do multiple bonds introduce rigidity into molecules?

24.25 (a) Use the concept of resonance to explain why all six C–C bonds in benzene are equal in length. (b) The C–C bond lengths in benzene are shorter than C–C single bonds but longer than C–C double bonds. Use the resonance model to explain this observation.

24.26 Why are there no known stable cyclic compounds with ring sizes of seven or less that contain a triple bond?

24.27 Squalene is a hydrocarbon with the molecular formula $C_{30}H_{50}$, and is obtained from shark liver. It is *not* a cyclic structure. Calculate the number of π-bonds it contains.

24.28 1,3-butadiene, C_4H_6, is a planar molecule that has the following carbon–carbon bond lengths:

$$H_2C\underset{134\,pm}{=\!\!=\!\!=}CH\underset{148\,pm}{-\!\!-\!\!-}CH\underset{134\,pm}{=\!\!=\!\!=}CH_2$$

(a) Predict the bond angles around each of the carbon atoms, and sketch the molecule. (b) Compare the bond lengths with the average bond lengths listed in Table 24.1. Can you explain any differences?

24.29 (a) Sketch a σ-bond that is constructed from *s* orbitals. (b) Sketch a σ-bond that is constructed from *p* orbitals. (c) Sketch a π-bond that is constructed from *p* orbitals. (d) Which is generally the stronger, a σ-bond or a π-bond? Explain.

24.30 (a) If the valence atomic orbitals of an atom are *sp* hybridised, how many unhybridised *p* orbitals remain in the valence shell? How many π-bonds can the atom form? (b) How many σ-bonds and π-bonds are generally part of a triple bond? (c) How do multiple bonds introduce rigidity into molecules?

24.31 Propylene, C_3H_6, is a gas that is used to form the important polymer polypropylene. Its Lewis structure is

(a) What is the total number of valence electrons in the propylene molecule? (b) How many valence electrons are used to make σ-bonds in the molecule? (c) How many valence electrons are used to make π-bonds in the molecule? (d) How many valence electrons remain in non-bonding pairs in the molecule? (e) What is the hybridisation at each carbon atom in the molecule?

24.32 (a) Write a single Lewis structure for acetate, $CH_3CO_2^-$. (b) Are there other equivalent Lewis structures for the molecule? (c) Would you expect $CH_3CO_2^-$ to exhibit delocalised π-bonding? Explain.

24.33 (a) What is the difference between a localised π-bond and a delocalised one? (b) How can you determine whether a molecule or ion will exhibit delocalised π-bonding? (c) Is the π-bond in NO_2^- localised or delocalised?

24.34 (a) Draw Lewis structures for methane (CH_4) and ethene (CH_2CH_2). (b) What is the hybridisation at the carbon atoms in CH_4 and CH_2CH_2? (c) The carbon atoms in CH_2CH_2 participate in multiple bonding, while the carbon atom in CH_4 does not. What does this imply about the hybridisation in these two molecules?

24.35 The nitrogen atoms in N_2 participate in multiple bonding, whereas those in hydrazine, N_2H_4, do not. How can you explain this observation in light of the hybridisation at the nitrogen atoms in the two molecules?

ADDITION REACTIONS OF UNSATURATED HYDROCARBONS (Sections 24.4 and 24.5)

24.36 (a) What is the difference between a substitution and an addition reaction? Which one is commonly observed with alkenes and which one with an alkane? (b) Using condensed structural formulae, write the balanced equation for the addition reaction of 2,4-dimethylpent-2-ene with Br_2.

24.37 Using condensed structural formulae, write a chemical equation for each of the following reactions: (a) hydrogenation of cyclohexene; (b) addition of H_2O to *trans*-pent-2-ene, using H_2SO_4 as a catalyst (draw both likely products); (c) reaction of propene with bromine in water solvent.

24.38 (a) Suggest a method of preparing *trans*-1,2-dichlorocyclohexane. (b) Suggest a method of preparing 1-chloro-2-fluorocyclohexane from cyclohexene.

24.39 In the hydrohalogenation of but-1-ene with HCl, two alkyl halides can be formed. (a) Draw them. (b) Circle the major organic product and give an explanation for your choice, taking into account the stability of any intermediates.

24.40 (a) How many molecules of HBr would you expect to react readily with each molecule of styrene? (b) What is the major product of the reaction? (c) State

Markovnikov's rule. **(d)** Does your answer to part (b) follow Markovnikov's rule?

24.41 Draw the intermediate that is thought to form in the addition of a hydrogen halide to an alkene, showing how it is formed and how it reacts, using cyclohexene as the alkene in your description.

24.42 The rate law for the addition of Br_2 to an alkene is first order in Br_2 and first order in alkene. Does this fact prove that the mechanism of addition of Br_2 to an alkene proceeds in the same manner as for addition of HBr? Explain.

24.43 Suggest suitable reagents for the following reactions:

(a)

(b)

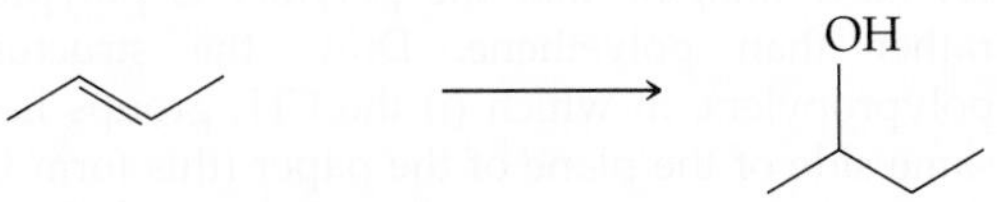

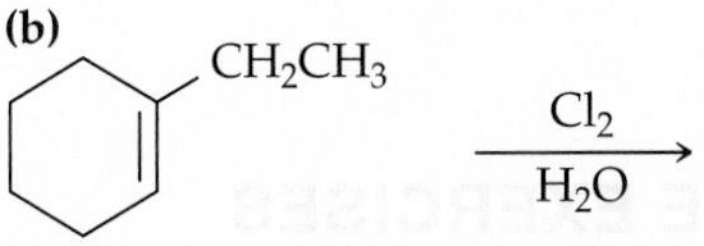

(c) $H_3C-C\equiv C-CH_3$ ⟶

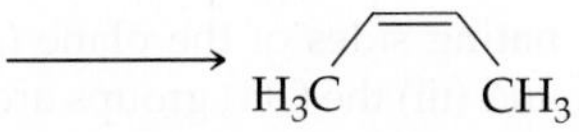

(d) $H_3C-C\equiv C-CH_3$ ⟶

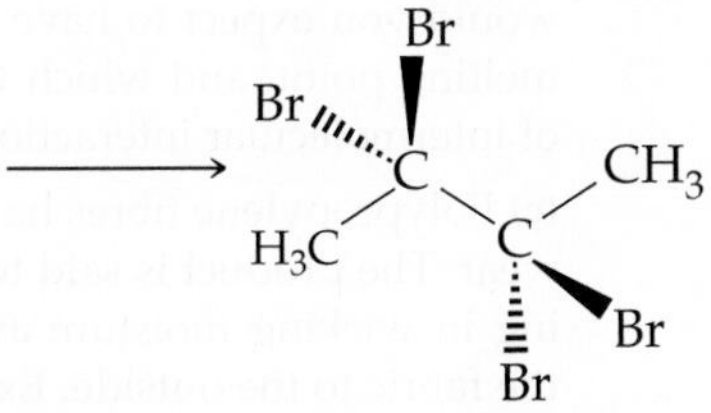

(e)

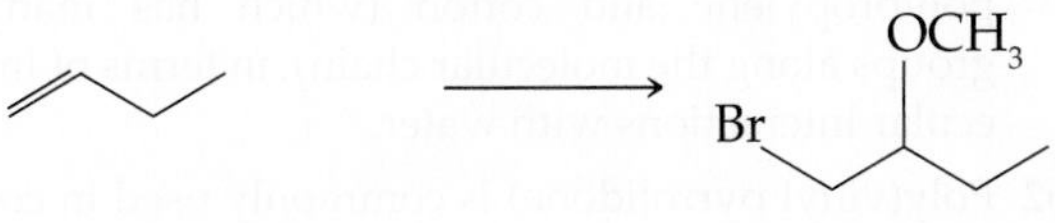

24.44 Draw another resonance form for the carbocations shown below.

(a)

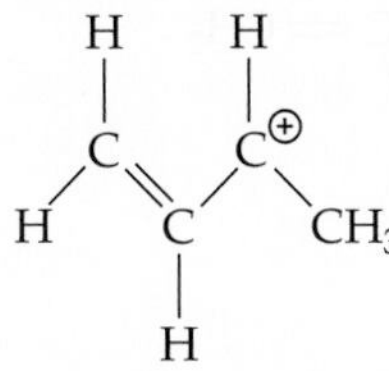

(b)

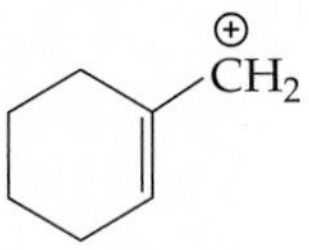

(c)

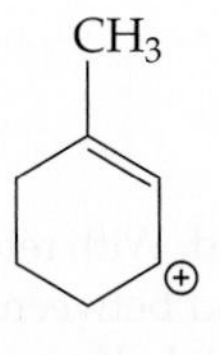

24.45 Place the following carbocations in an increasing order of relative stability:

(a) **(b)** 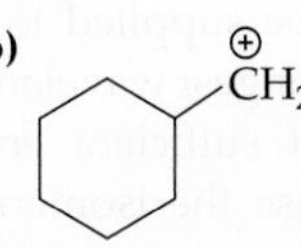 **(c)**

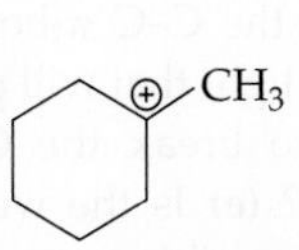

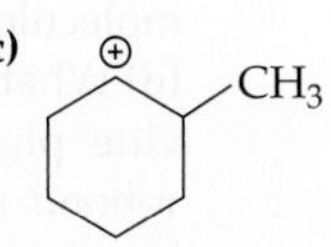

24.46 What alkenes would you hydrate to obtain the following alcohols?

(a) **(b)**

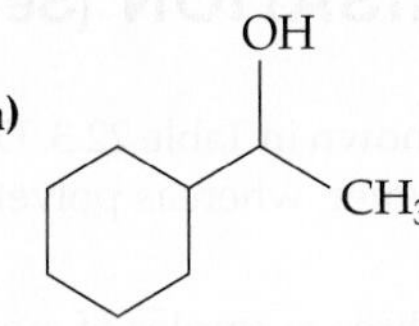

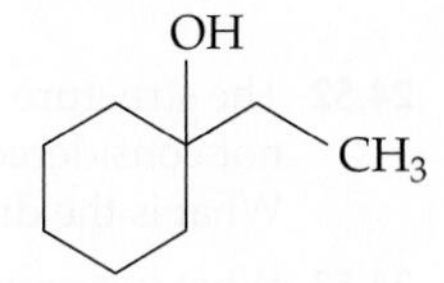

(c) 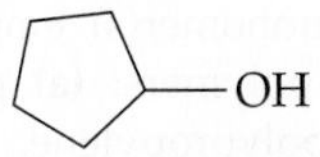 **(d)**

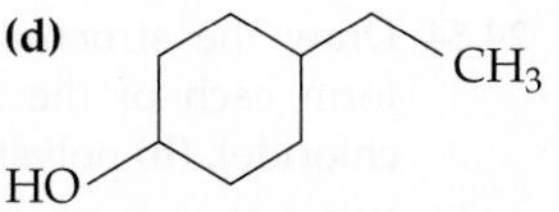

24.47 When 3-methyl but-1-ene reacts with H_2O in the presence of sulfuric acid, two alcohols are formed. **(a)** Draw the structures of these products. **(b)** Circle the major product and give an explanation for your choice, taking into account the stability of any intermediates.

24.48 **(a)** Write the structural formula for the product of the hydrogenation of 3-methyl pent-1-ene. What is the systematic name of the product? **(b)** When D_2 reacts with ethene (C_2H_4) in the presence of a finely divided catalyst, ethane with two deuteriums, CH_2D-CH_2D, is formed. (Deuterium, D, is an isotope of hydrogen of mass 2.) Very little ethane forms in which two deuteriums are bound to one carbon (for example, CH_3-CHD_2). Use the sequence of steps involved in the reaction to explain why this is so.

24.49 **(a)** When but-1-yne is hydrohalogenated with HCl a geminal dichloride is formed in preference. Draw its structure. **(b)** Give a reason for your choice.

24.50 **(a)** In the bromination of propene, a cyclic bromonium ion is formed in the first step. Draw it. **(b)** The addition of a nucleophile ($Nu^\ominus$) to this bromonium ion yields a single major product. Why is this the case and what product is formed? **(c)** Is this product chiral if Nu=OH? **(d)** What if Nu=Br?

24.51 What is the major product of the following reactions?

(a)
$$\xrightarrow[H_2O]{Br_2}$$

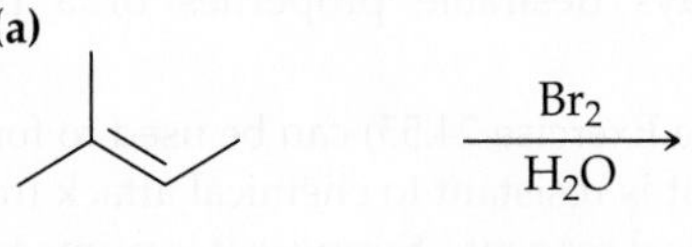

(b) CH_2CH_3
$$\xrightarrow[H_2O]{Cl_2}$$

(c)
$$\xrightarrow[H_2O]{Br_2}$$

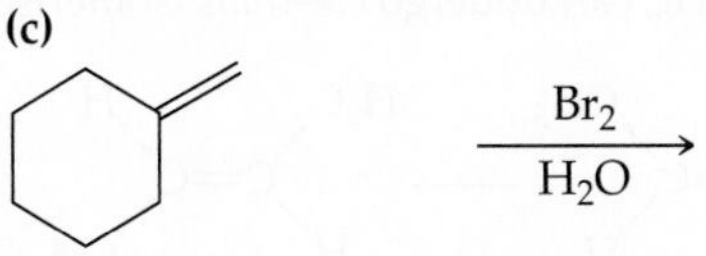

(d)
$$\xrightarrow[CH_3OH]{Br_2}$$

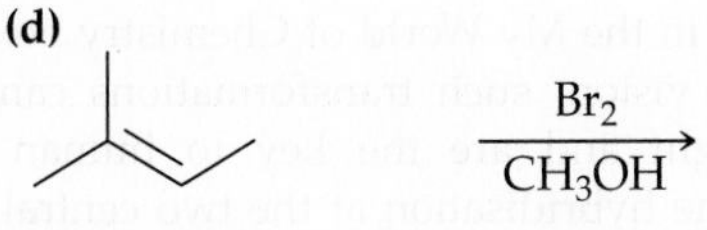

ADDITION POLYMERISATION (Section 24.6)

24.52 The structure of decane is shown in Table 22.3. Decane is not considered to be a polymer, whereas polyethene is. What is the distinction?

24.53 What is a monomer? Give three examples of monomers, taken from the examples given in this chapter.

24.54 Draw the structure of the monomer(s) employed to form each of the following polymers: **(a)** poly(vinyl chloride), **(b)** polystyrene, **(c)** polypropylene.

24.55 Write the chemical equation that represents the formation of **(a)** polychloroprene (neoprene) from chloroprene

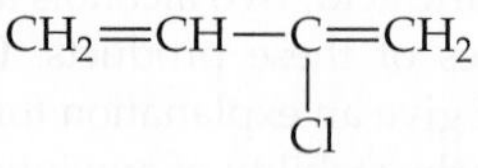

$$CH_2{=}CH{-}\underset{\underset{Cl}{|}}{C}{=}CH_2$$

(Neoprene is used in road pavement seals, expansion joints, conveyor belts and wire and cable jackets.);
(b) polyacrylonitrile

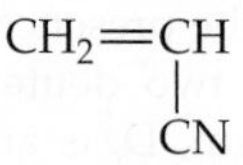

$$CH_2{=}\underset{\underset{CN}{|}}{CH}$$

(Polychloroprene is used in home furnishings, craft yarns, clothing and many other items.)

24.56 Teflon® is a polymer formed by the polymerisation of $F_2C{=}CF_2$. Draw the structure of a section of this polymer. What type of polymerisation reaction is required to form it?

24.57 What molecular features make a polymer flexible? Explain how cross-linking affects the chemical and physical properties of the polymer.

24.58 What molecular structural features cause high-density polyethene (HDPE) to be denser than low-density polyethene (LDPE)?

24.59 Are high molecular masses and a high degree of crystallinity always desirable properties of a polymer? Explain.

24.60 Neoprene (see Exercise 24.55) can be used to form flexible tubing that is resistant to chemical attack from a variety of chemical reagents. Suppose it is proposed to use neoprene tubing as a coating for the wires running to the heart from an implanted pacemaker. What questions would you ask to determine whether it might be suitable for such an application?

24.61 Although polyethene can twist and turn in random ways, the most stable form is a linear one with the carbon backbone oriented as shown in the figure below:

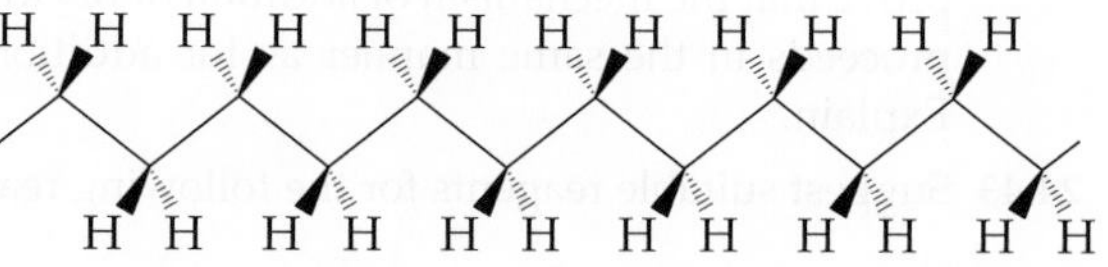

(a) What is the hybridisation of orbitals at each carbon atom? What angles do you expect between the bonds?

(b) Now imagine that the polymer is polypropylene rather than polyethene. Draw the structures for polypropylene in which (i) the CH_3 groups lie on the same side of the plane of the paper (this form is called isotactic polypropylene), (ii) the CH_3 groups lie on alternating sides of the plane (syndiotactic polypropylene), and (iii) the CH_3 groups are randomly distributed on either side (atactic polypropylene). Which of these forms would you expect to have the highest crystallinity and melting point, and which the lowest? Explain in terms of intermolecular interactions and molecular shapes.

(c) Polypropylene fibres have been employed in athletic wear. The product is said to be superior to cotton clothing in wicking moisture away from the body through the fabric to the outside. Explain the difference between polypropylene and cotton (which has many –OH groups along the molecular chain), in terms of intermolecular interactions with water.

24.62 Poly(vinyl pyrrolidone) is commonly used in cosmetics and in wine clarification. Draw a representative segment of the polymer prepared from *N*-vinyl pyrrolidone:

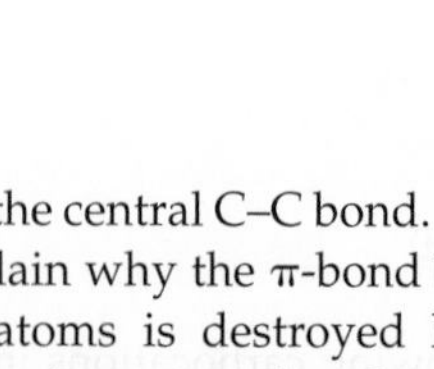

INTEGRATIVE EXERCISES

24.63 But-2-ene, C_4H_8, can undergo *cis–trans* isomerisation:

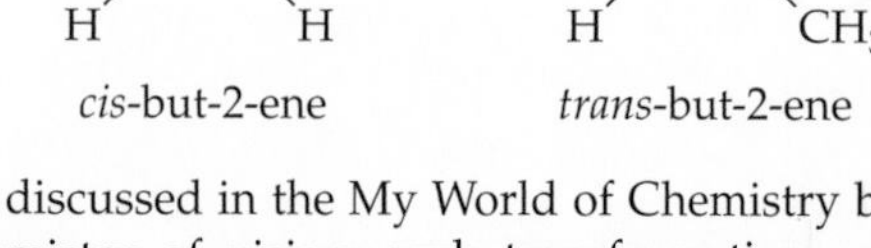

As discussed in the My World of Chemistry box on the chemistry of vision, such transformations can be induced by light and are the key to human vision. **(a)** What is the hybridisation at the two central carbon atoms of but-2-ene? **(b)** The isomerisation occurs by rotation about the central C–C bond. With reference to Figure 24.5, explain why the π-bond between the two central carbon atoms is destroyed halfway through the rotation from *cis*- to *trans*-but-2-ene. **(c)** Based on average bond enthalpies (Table 8.4), how much energy per molecule must be supplied to break the C–C π-bond? **(d)** What is the longest wavelength of light that will provide photons of sufficient energy to break the C–C π-bond and cause the isomerisation? **(e)** Is the wavelength in your answer to part (d) in the visible portion of the electromagnetic spectrum? Comment on the importance of this result for human vision.

24.64 Ethene (CH_2═CH_2) reacts with hydrogen gas in the presence of a catalyst to give ethane (CH_3CH_3). **(a)** Draw the structures of the reactants and the product for this reaction and indicate the number of σ- and π-bonds that have been broken and have been formed in this reaction. Use the average bond enthalpies in Table 8.4 to estimate **(b)** the enthalpy change for this reaction and **(c)** the average energy of a carbon–carbon σ-bond and π-bond. **(d)** Draw the electron distribution in a π-bond and explain why it is weaker than a σ-bond between the same atoms.

24.65 Consider the electrophilic addition of hydrogen chloride to propene. **(a)** Draw the reaction showing the primary and secondary carbocation intermediates and the products that arise from each intermediate. **(b)** Sketch an energy profile, similar to Figure 24.20, for the two possible pathways in this reaction showing the relative energies of the intermediates and indicating which carbocation is the more stable. **(c)** If the temperature of the reaction mixture was increased would the relative proportions of the two possible products change?

24.66 A sample of a polyunsaturated oil is hydrolysed to give either linoleic acid ($C_{18}H_{32}O_2$) or arachidonic acid ($C_{20}H_{32}O_2$) and you have to find out which one. You take a 1.0 g sample of the unknown acid and perform a hydrogenation reaction under catalytic conditions. This will hydrogenate the carbon–carbon double bonds present but not the carbon–oxygen double bond of the carboxylic acid group. **(a)** Calculate the double bond equivalents in linoleic acid and arachidonic acid. **(b)** Determine the number of carbon–carbon double bonds present in each of the acids. Your sample of unknown acid absorbs 299 cm^3 of hydrogen gas under STP conditions. **(c)** Calculate the number of moles of hydrogen gas used and identify which acid you have.

24.67 Acetylene (H—C≡C—H) can explosively form benzene (⬡). **(a)** Write a chemical equation for this reaction and **(b)** calculate the heat of reaction given $\Delta_f H°$ (benzene) = 83 kJ mol^{-1} and $\Delta_f H°$ (acetylene) = 227 kJ mol^{-1}.

24.68 Polyvinyl alcohol is a polymer with the following structure.

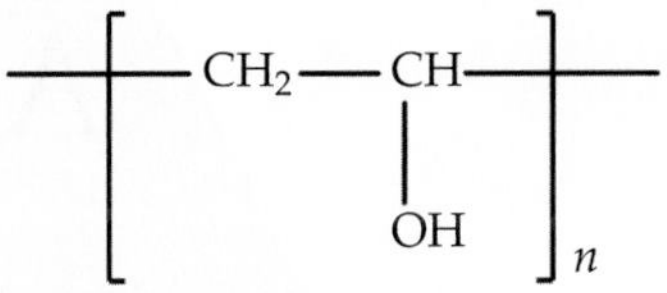

(a) What intermolecular forces exist between polymer chains in polyvinyl alcohol? **(b)** Do you expect polyvinyl alcohol to be water soluble?

MasteringChemistry (www.pearson.com.au/masteringchemistry)

Make learning part of the grade. Access:

- tutorials with peronalised coaching
- study area
- Pearson eText

PHOTO/ART CREDITS

952 turtix/Shutterstock; **959** Blue Mountains Tourism; **964** Science Photo Library RF/Photolibrary; **965** Linus Pauling, photographed by Ray Huff Studios, 1931; **984** © Sava Miokovic/iStockphoto.com, © Photographer Tyler Olson/Dreamstime.com, © Photographer Jack Schiffer/Dreamstime.com.

25

ALCOHOLS, HALOALKANES AND ETHERS

Australian wines are world renowned and represent a complex blend of organic molecules that produce the flavour of the wine with, of course, alcohol.

KEY CONCEPTS

25.1 ALCOHOLS: STRUCTURE, PROPERTIES AND NOMENCLATURE

We begin by examining the structure, classification and nomenclature of *alcohols*. Our discussions include non-covalent bonding interactions—the hydrogen bond, and its effects on the properties of alcohols such as boiling point and miscibility.

25.2 HALOALKANES

Haloalkanes are hydrocarbon derivatives where one or more hydrogens are replaced with a halogen: F, Cl, Br or I. Although some haloalkanes are useful in organic synthesis, others are harmful to the environment, for example CFCs.

25.3 ETHERS: STRUCTURE, PROPERTIES AND NOMENCLATURE

Ethers are a class of oxygen-containing organic compounds. We investigate their structure, classification and nomenclature.

25.4 REACTIONS OF ALCOHOLS

A useful characteristic of alcohols is their versatility in organic synthesis. Alcohols are shown to form ethers, aldehydes, ketones, carboxylic acids and esters. Here we concentrate on the formation of alkoxides, haloalkanes and alkenes from alcohols.

25.5 NUCLEOPHILIC SUBSTITUTION REACTIONS OF HALOALKANES

We discuss the reactivity of haloalkanes towards nucleophiles such as alcohols and alkoxide ions. We use the S_N nomenclature to describe nucleophilic substitution reactions.

25.6 HALOALKANES TO ALKENES: β-ELIMINATION

We consider the reaction of haloalkanes with bases to form alkenes by the process of elimination. The product distribution is dictated by Zaitsev's rule.

25.7 SUBSTITUTION VERSUS ELIMINATION

Finally, we discuss the competition between substitution and elimination. The nature of the haloalkane and nucleophile/base dictates which, if any, reaction is favoured.

Whether white, red or something in between, wine is a complex mixture of chemicals. One of the main constituents of wine is, of course, ethanol. Produced by the yeast fermentation of the sugars within the fruits from which it is made, much of the popularity of wine comes from its intoxicating effect. The science of wine and wine making is called *oenology* and today it helps sustain a multi-billion dollar industry worldwide.

Currently, ethanol is valued more than ever. Its uses are as diverse as a substitute for, or additive in, petrol; as a solvent in the manufacture of varnishes and perfumes; as a fluid in thermometers; in the manufacture of acetic acid; in the preparation of essences and flavourings and as a disinfectant. Its major drawback is its miscibility with water. Absolute ethanol contains >99% *v/v* ethanol and is purified from industrial ethanol by first applying fractional distillation, which produces a liquid that is 95% *v/v* ethanol, then by azeotropic distillation.

In this chapter we take an introductory look at the chemistry of alcohols and the related chemistry of haloalkanes and ethers. Each class has specific reactivity and properties that are intertwined, demonstrating important chemical principles.

25.1 ALCOHOLS: STRUCTURE, PROPERTIES AND NOMENCLATURE

Alcohols are hydrocarbon derivatives in which one or more hydrogens have been replaced by a *hydroxyl* group, OH. ▼ FIGURE 25.1 shows the structural formulae and names of the three simplest straight-chain alcohols together with five other important alcohols. The simple alcohols are named by changing the last letter in the name of the corresponding alkane to *-ol*; for example, ethan*e* becomes ethan*ol*. Where necessary, the location of the OH group is designated by an appropriate numeral prefix that indicates the position of the OH group (Figure 25.1).

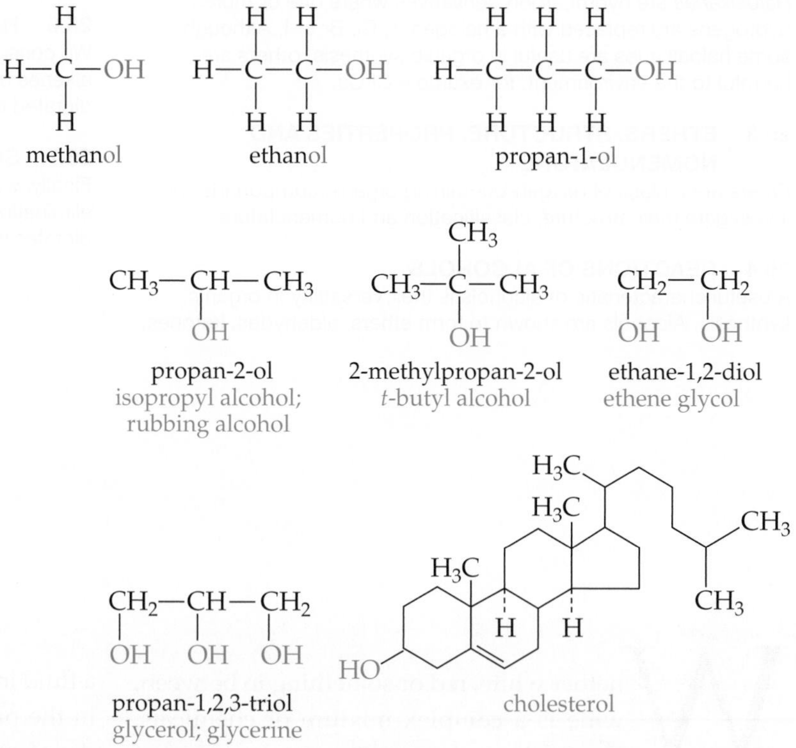

methanol

ethanol

propan-1-ol

propan-2-ol
isopropyl alcohol;
rubbing alcohol

2-methylpropan-2-ol
t-butyl alcohol

ethane-1,2-diol
ethene glycol

propan-1,2,3-triol
glycerol; glycerine

cholesterol

▶ **FIGURE 25.1** **Structural formulae of six important alcohols.** Common names are given in blue.

The oxygen atom of the hydroxyl group is sp^3 hybridised, having a tetrahedral arrangement of its bonds and lone pairs (▼ FIGURE 25.2). Two of the sp^3 hybrid orbitals of the oxygen atom form σ-bonds with carbon and hydrogen. Each of the other two orbitals contain a pair of unshared electrons, or **lone pair**. The bond angles about the oxygen atom of an alcohol are very similar to those around sp^3 carbon centres. Any variation from the ideal

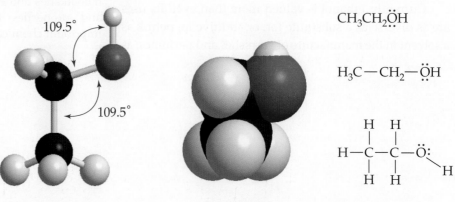

$CH_3CH_2\overset{..}{\underset{..}{O}}H$

$H_3C—CH_2—\overset{..}{\underset{..}{O}}H$

▶ **FIGURE 25.2** **The structure of ethanol.** A number of representations for ethanol are shown.

Ball-and-stick model

Space-filling representation

Structural formulae

▲ FIGURE IT OUT

What is the maximum number of hydrogen bonds a single ethanol molecule could possess?

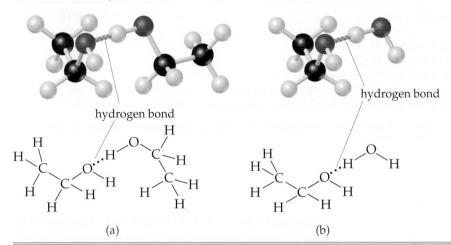

◀ FIGURE 25.3 Hydrogen-bonding interactions. (a) Between two ethanol molecules and (b) between an ethanol molecule and a water molecule.

109.5° bond angle is due to the comparatively large size of the lone pairs (∞ Section 9.2, 'The VSEPR Model').

∞ Review this on page 298

The properties of alcohols are very different from those of the alkanes from which they are derived. For example, methane, ethane and propane are all colourless gases under normal conditions, whereas methanol, ethanol and propanol are colourless liquids. The reason for the difference in physical properties lies in the fact that the O–H bond is polar, with the difference in electronegativity values between the oxygen and hydrogen atoms being 1.4 (∞ Section 8.4, 'Bond Polarity and Electronegativity'). As a result, the –OH functional group can participate in intermolecular hydrogen bonding, which increases the attractive forces between the molecules of the alcohol. The strength of a hydrogen bond is approximately 20 kJ mol^{-1}, which is considerably weaker than an O–H covalent bond (about 460 kJ mol^{-1}); however, it is strong enough to have a significant effect on the alcohol's physical properties. As a result, the boiling points of alcohols are much higher than those of their parent alkanes. The ability of alcohols to form hydrogen bonds is also why they are much more soluble in polar solvents such as water, than in the corresponding hydrocarbons (▲ FIGURE 25.3). For example, the solute–solute, solvent–solvent and solute–solvent forces are not greatly different within a mixture of CH_3CH_2OH and H_2O. As a result, there is no major change in the environments of the molecules as they are mixed.

∞ Review this on page 261

Liquids that mix completely in another liquid are said to be *miscible*, whereas solids that dissolve in a liquid are said to be *soluble* in that liquid. Pairs of liquids, such as ethanol and water, that mix in all proportions are miscible; those that do not mix are *immiscible*. For example, hexane (C_6H_{14}) is immiscible in water. The attraction between the polar water molecules and the non-polar hexane molecules is not sufficiently strong to allow the formation of a solution.

The number of carbon atoms in an alcohol affects its solubility in water. As the length of the carbon chain increases, the polar OH group becomes an ever-smaller part of the molecule, and the compound behaves more like a hydrocarbon. The solubility of the alcohol in water decreases correspondingly. In contrast, the solubility of the alcohol in a non-polar solvent like hexane increases as the non-polar hydrocarbon chain increases in length. The series of alcohols in ▼ TABLE 25.1 demonstrates that polar liquids tend to be miscible in other polar liquids and non-polar liquids in non-polar ones. This phenomenon is simply stated as '*like dissolves like*'.

TABLE 25.1 • Solubilities of some alcohols in water and in hexane*

Alcohol	Solubility in H_2O	Solubility in C_6H_{14}
CH_3OH (methanol)	∞	0.12
CH_3CH_2OH (ethanol)	∞	∞
$CH_3CH_2CH_2OH$ (propan-1-ol)	∞	∞
$CH_3CH_2CH_2CH_2OH$ (butan-1-ol)	0.11	∞
$CH_3CH_2CH_2CH_2CH_2OH$ (pentan-1-ol)	0.030	∞
$CH_3CH_2CH_2CH_2CH_2CH_2OH$ (hexan-1-ol)	0.0058	∞
$CH_3CH_2CH_2CH_2CH_2CH_2CH_2OH$ (heptan-1-ol)	0.0008	∞

* Expressed in mol alcohol/100 g solvent at 20 °C. The infinity symbol indicates that the alcohol is completely miscible with the solvent.

MY WORLD OF CHEMISTRY

VITAMIN D

Vitamin D, an alcohol, is a steroid hormone derived from cholesterol. The biologically active form of the hormone is vitamin D_3 whose primary function is to regulate calcium and phosphorus homeostasis and maintain bone density in the body. Food provides us with a precursor molecule which is converted to vitamin D_3 under the skin by the action of sunlight. Hence it is sometimes called the sunshine vitamin (▼ FIGURE 25.4). Milk is fortified with vitamin D as a dietary safeguard in countries where outdoor exposure is often limited. In places like Australia, the amount of vitamin D made in the skin on exposure to sunlight is sufficient for most people. Recent evidence suggests there is a need, for children especially, to spend more time outdoors. The 'slip-slop-slap' and 'stay out of the sun' campaigns, in conjunction with the great popularity of computer games, have been linked to an increased risk of vitamin D deficiency and the resulting underdevelopment of bone in children, called *rickets*. Rickets is characterised by improper mineralisation during the development of bones and results in soft bones, curved spine and knock-knees. Nevertheless, a sensible approach to sun exposure is also required.

vitamin D_3

Food sources:
Cheese
Margarine
Butter
Fortified milk
Healthy cereals
Fatty fish

▲ FIGURE 25.4   Vitamin D and its food sources.

⚠ CONCEPT CHECK 1

Suppose the hydrogens on the OH groups in glycerol (Figure 25.1) were replaced with methyl groups, CH_3. Would you expect the solubility in water of the resulting molecule to be higher than, lower than or about the same as the solubility of glycerol?

SAMPLE EXERCISE 25.1   **Predicting solubility patterns in organic compounds**

Predict whether the following compounds are more likely to dissolve in water or hexane (C_6H_{14}) solvent: cyclohexanol, acetic acid, chloroform ($CHCl_3$), heptane (C_7H_{16}), geraniol (Figure 24.8), and glycerol ($C_3H_8O_3$).

SOLUTION

Analyse We are asked to differentiate between various compounds by their solubility or miscibility in either hexane or water.

Plan *Like disolves in like.*

Solve Although cyclohexanol contains an OH group, the addition of the C_6 ring system would make its properties comparable with 1-hexanol. Given the choice of solvents, cyclohexanol dissolves better in hexane because the molecule is more 'alkane-like'.

Acetic acid is a very polar compound, containing little alkane character. This compound dissolves better in water as a solvent.

Chloroform is a moderately polar liquid because of its tetrahedral geometry. Chlorine atoms are poor hydrogen-bonding groups, however, and so this compound would be more miscible with hexane.

Heptane is an alkane, so it is non-polar. This compound is miscible with hexane.

Geraniol is a C_{10} hydrocarbon bearing a single OH group. This compound behaves like cyclohexanol and is more soluble in hexane.

Glycerol is a small organic compound containing three OH groups. It is very soluble in water.

PRACTICE EXERCISE

Arrange the following substances in order of increasing solubility in water:

$$H-\overset{\overset{H}{|}}{\underset{\underset{H}{|}}{C}}-\overset{\overset{H}{|}}{\underset{\underset{H}{|}}{C}}-\overset{\overset{H}{|}}{\underset{\underset{H}{|}}{C}}-\overset{\overset{H}{|}}{\underset{\underset{H}{|}}{C}}-\overset{\overset{H}{|}}{\underset{\underset{H}{|}}{C}}-H \qquad HO-\overset{\overset{H}{|}}{\underset{\underset{H}{|}}{C}}-\overset{\overset{H}{|}}{\underset{\underset{H}{|}}{C}}-\overset{\overset{H}{|}}{\underset{\underset{H}{|}}{C}}-\overset{\overset{H}{|}}{\underset{\underset{H}{|}}{C}}-\overset{\overset{H}{|}}{\underset{\underset{H}{|}}{C}}-OH$$

$$H-\overset{\overset{H}{|}}{\underset{\underset{H}{|}}{C}}-\overset{\overset{H}{|}}{\underset{\underset{H}{|}}{C}}-\overset{\overset{H}{|}}{\underset{\underset{H}{|}}{C}}-\overset{\overset{H}{|}}{\underset{\underset{H}{|}}{C}}-\overset{\overset{H}{|}}{\underset{\underset{H}{|}}{C}}-OH \qquad H-\overset{\overset{H}{|}}{\underset{\underset{H}{|}}{C}}-\overset{\overset{H}{|}}{\underset{\underset{H}{|}}{C}}-\overset{\overset{H}{|}}{\underset{\underset{H}{|}}{C}}-\overset{\overset{H}{|}}{\underset{\underset{H}{|}}{C}}-\overset{\overset{H}{|}}{\underset{\underset{H}{|}}{C}}-Cl$$

Answer: $C_5H_{12} < C_5H_{11}Cl < C_5H_{11}OH < C_5H_{10}(OH)_2$ (in order of increasing polarity and hydrogen-bonding ability)

(See also Exercises 25.2, 25.8.)

Common Alcohols

▶ **FIGURE 25.5** shows several commercial products that consist entirely or in part of an alcohol. Let's consider how some of the more important alcohols are formed and used. The simplest alcohol, methanol, has many important industrial uses and is produced on a large scale. Carbon monoxide and hydrogen are heated together under pressure in the presence of a metal oxide catalyst:

$$CO(g) + 2\,H_2(g) \xrightarrow[400\,°C]{200-300\,bar} CH_3OH(g) \qquad [25.1]$$

Because methanol has a very high octane rating as a car fuel, it is used as a petroleum additive and as a fuel in its own right.

Ethanol is a product of the fermentation of carbohydrates such as sugar and starch. In the absence of air, yeast cells convert carbohydrates into a mixture of ethanol and CO_2, as shown in Equation 25.2. In the process, yeast derives energy necessary for growth.

▲ **FIGURE 25.5   Everyday alcohols.** Many of the products we use every day—from rubbing alcohol and throat lozenges to hair spray and radiator coolant—are composed either entirely or mainly of alcohols.

MY WORLD OF CHEMISTRY

THE SOLUBILITY NEXUS

The solubility of organic molecules is an important issue, especially in the area of medicine. Not only must a drug be effective, but it must have sufficient solubility in water for aqueous administration in the blood. These two criteria are seldom satisfied collectively and are complicated further by the need for a drug candidate to pass through hydrophobic cell membranes and/or the blood–brain barrier.

One way to enhance the solubility of a substance in water is to increase the number of polar groups it contains. For example, increasing the number of OH groups along a carbon chain of a solute increases the extent of hydrogen bonding between that solute and water, thereby increasing solubility. Glucose ($C_6H_{12}O_6$) has five OH groups on a five-carbon cyclic framework, which makes the molecule very soluble in water (83 g dissolves in 100 cm^3 of water at 17.5 °C). Glucose is shown in ▼ **FIGURE 25.6** and is compared structurally with cyclohexane, its hydrocarbon analogue. The solubility of cyclohexane in water is extremely low due to its inability to hydrogen-bond to water. The second approach to increasing polarity is to add charge. As we will see in Chapters 27 and 29, negative charge can be added through *carboxylate groups* and positive charge through *ammonium salts*. Both are proven and common ways to make drugs soluble in water.

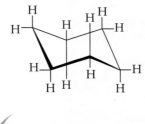

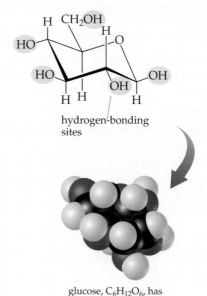

hydrogen-bonding sites

cyclohexane, C_6H_{12}, which has no polar OH groups, is essentially insoluble in water

glucose, $C_6H_{12}O_6$, has five OH groups and is highly soluble in water

▲ **FIGURE 25.6** **Hydrogen bonding and aqueous solubility.** The presence of OH groups capable of hydrogen bonding with water enhances the aqueous solubility of organic molecules.

$$C_6H_{12}O_6(aq) \xrightarrow{\text{yeast}} 2\, C_2H_5OH(aq) + 2\, CO_2(g) \qquad [25.2]$$

This reaction is conducted under carefully controlled conditions to produce beer, wine and other beverages in which ethanol is the active ingredient.

Many polyhydroxyl alcohols (those containing more than one OH group) are known. The simplest of these is ethane-1,2-diol (ethene glycol, $HOCH_2CH_2OH$). This substance is the major ingredient in car radiator coolant. Another common polyhydroxyl alcohol is propane-1,2,3-triol (glycerol, $HOCH_2CH(OH)CH_2OH$). This is a viscous liquid that dissolves readily in water and is widely used as a skin softener in cosmetic preparations. It is also used in foods and confectionary to keep them moist.

∞ Review this on page 897

Cholesterol is classed as a *steroid* (∞ Section 22.3, 'Applications and Physical Properties of Alkanes') but here we will look at it as a biochemically important alcohol. The OH group forms only a small component of this rather large molecule, so cholesterol is virtually insoluble in water (0.2 g per 100 cm^3

of H_2O) and has a waxy feel. Cholesterol is an essential component of our bodies; when present in excessive amounts, however, it may precipitate from solution. If it precipitates in the gall bladder, it may form crystalline lumps called *gallstones*. Cholesterol may also precipitate against the walls of veins and arteries and thus contribute to high blood pressure and other cardiovascular problems. The amount of cholesterol in our blood is determined not only by how much cholesterol we eat but also by total energy intake. There is evidence that excessive caloric intake leads the body to synthesise excessive amounts of cholesterol.

Naming Alcohols

Alcohol nomenclature is based on selecting the longest chain containing the OH group. This chain is numbered to give the OH group the lowest possible number. Alcohols that are *normal*—that is, they contain the OH functional group on the first carbon of a straight-chain alkyl group—are listed in Table 25.1. Sometimes you may see these compounds named as *n*-butanol, *n*-hexanol and so on. In these cases the *n* stands for normal. It is more appropriate, however, to use IUPAC nomenclature, which names these compounds using a positional prefix—for example, butan-1-ol, hexan-1-ol. Note that methanol and ethanol do not need a prefix.

The names of some simple alcohols are given below. Each molecule contains extra functionality or branching, yet all the names are derived from the name of the alcohol. Their names illustrate, to some extent, the hierarchy of nomenclature.

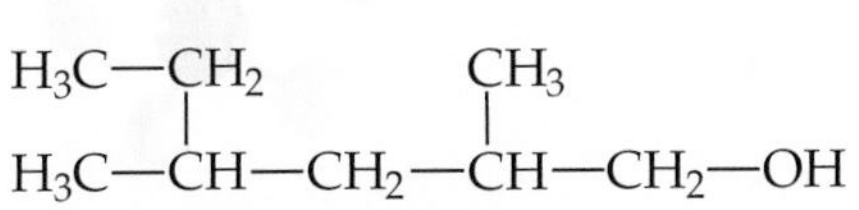

2,4-dimethylhexan-1-ol

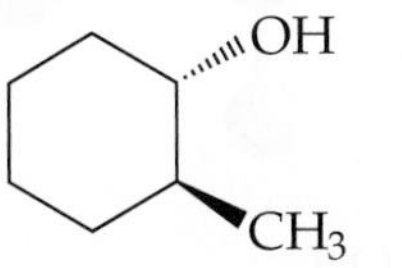

(1*S*,2*S*)-2-methylcyclohexanol
is better than
trans-2-methylcyclohexanol

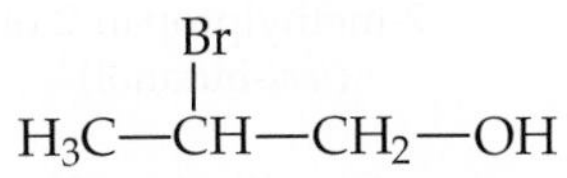

2-bromopropan-1-ol

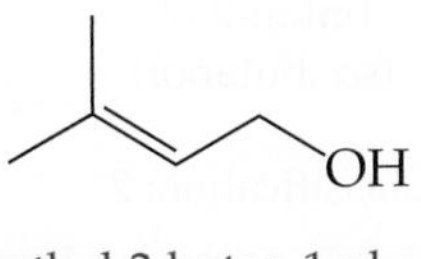

3-methyl-2-buten-1-ol

Systematic naming using the IUPAC conventions does not distinguish between, for example, 1-hexanol and hexan-1-ol as the most satisfactory way to name a functionalised alcohol. Both methods unambiguously describe the position of the OH group and could be used. The latter naming style is better suited to molecules that contain an alkene as well as the OH group. In such cases, '*an*' would be replaced by '*en*' to represent the incorporation of the alkene. Compounds of this type are called *unsaturated alcohols*.

The 2-methylcyclohexanol example illustrated above demonstrates two nomenclature principles. The alcohol functional group, when bonded to a cyclic structure, is assigned position 1, based on the arguments we used to discuss the nomenclature of cycloalkenes (Section 24.2). Although *trans* satisfactorily describes the orientation of the OH and CH_3 groups, it does not unambiguously describe the molecule (there is another enantiomer with a *trans* arrangement—can you draw it?). In this case, the absolute stereochemistry at each position should be described (Section 23.5, 'Absolute Stereochemistry').

Review this on page 936

FIGURE IT OUT

Which of the alcohols shown in this figure are chiral?

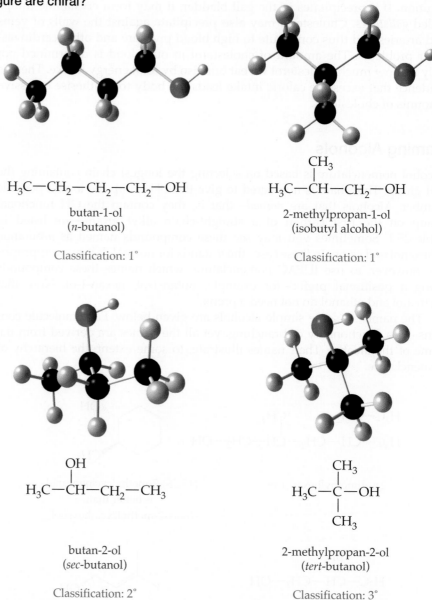

H₃C—CH₂—CH₂—CH₂—OH

butan-1-ol
(*n*-butanol)

Classification: 1°

CH_3
|
H₃C—CH—CH₂—OH

2-methylpropan-1-ol
(isobutyl alcohol)

Classification: 1°

OH
|
H₃C—CH—CH₂—CH₃

butan-2-ol
(*sec*-butanol)

Classification: 2°

CH_3
|
H₃C—C—OH
|
CH_3

2-methylpropan-2-ol
(*tert*-butanol)

Classification: 3°

▶ **FIGURE 25.7 Isomeric alcohols.** The structure, names and classifications of the four isomeric alcohols related to butanol.

The OH group is not always located at position 1. ▲ **FIGURE 25.7** illustrates the different $C_4H_{10}O$ alcohols that are possible, as well as their systematic and common names. Alcohols, like alkanes, are classified as primary, secondary or tertiary. We discuss this in more detail in the following section.

SAMPLE EXERCISE 25.2 | **Naming alcohols**

Systematically name the following alcohols, taking stereochemistry into account where appropriate.

(a)

(b)

CH_3

C......CH₃
HO   CH₂CH₃

SOLUTION

Analyse We are given two structures and asked to name them.

Plan Name each compound using the longest chain possessing the OH group. Assign stereochemistry based first on symmetry then on the Cahn–Ingold–Prelog rules in Chapter 23.

Solve
(a) This example has no stereochemical consequence because a plane of symmetry exists within the molecule. The compound is derived from cyclopentane. Its systematic name is 1-methylcyclopentanol.

(b) This example also has no stereochemical consequences because a plane of symmetry exists within the molecule. The compound is derived from butanol, as the longest chain bearing the OH group is four carbons long. Its systematic name is 2-methylbutan-2-ol.

PRACTICE EXERCISE
Draw the structures corresponding to the following names:
(a) (R)-4-methylpentan-2-ol **(b)** (Z)-3-penten-1-ol

Answers: **(a)**

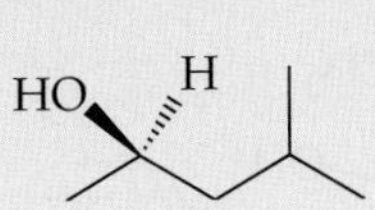

(b)

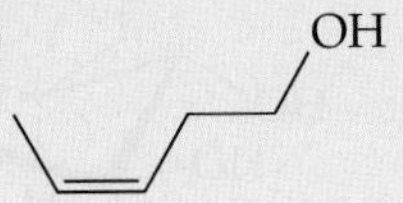

(See also Exercises 25.11, 25.12.)

Alcohols containing two or three OH groups are called **diols** and **triols**, respectively. Common examples are shown in Figure 25.1.

Classifying Alcohols

The classification and notation used to distinguish between types of alcohols are:

Primary (1°) alcohol: The carbon atom bearing the OH group is bonded to only one other carbon atom.

Secondary (2°) alcohol: The carbon atom bearing the OH group is bonded to two carbon atoms.

Tertiary (3°) alcohol: The carbon atom bearing the OH group is bonded to three carbon atoms.

Figure 25.7 illustrates this classification within a series of butanol isomers. There is a big difference in reactivity between 1°, 2° and 3° alcohols. An important part of your understanding of the chemistry of alcohols is the ability to recognise an alcohol's class and to use this knowledge to predict the product of a reaction. Each classification of alcohol—namely primary, secondary and tertiary—reacts differently to a range of reactants. As we see in the following chapters, this classification allows us to predict the type of product and/or the necessary reagents with which to undertake certain functional group manipulations.

SAMPLE EXERCISE 25.3 **Classifying alcohols**

Classify each of the following alcohols as primary, secondary or tertiary.

(a) **(b)** 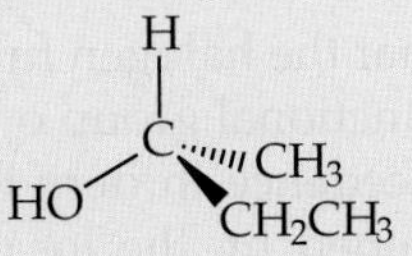 **(c)**

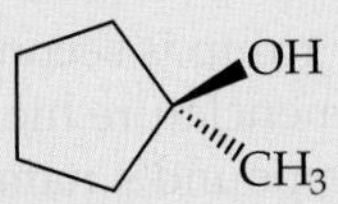

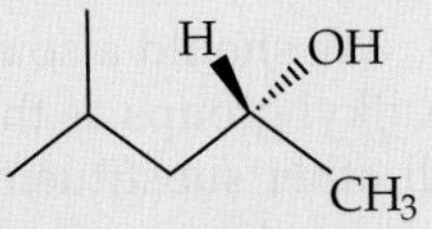

SOLUTION
Analyse We are given the structures of three alcohols and asked to classify them.

Plan Locate the carbon bearing the OH group and determine the number of carbons attached to this carbon.

Solve

(a) The carbon bearing the OH group has three carbons attached to it: two from the ring and one from the methyl group. Therefore this is a tertiary alcohol.

(b) The carbon bearing the OH group has two carbons attached to it, one from the methyl group and the other from the ethyl group. This compound can be classed as a secondary alcohol.

(c) The carbon bearing the OH group has two other carbons attached. Hence it too is secondary.

PRACTICE EXERCISE

Classify all alcohol functional groups in the following molecule as 1°, 2° or 3°.

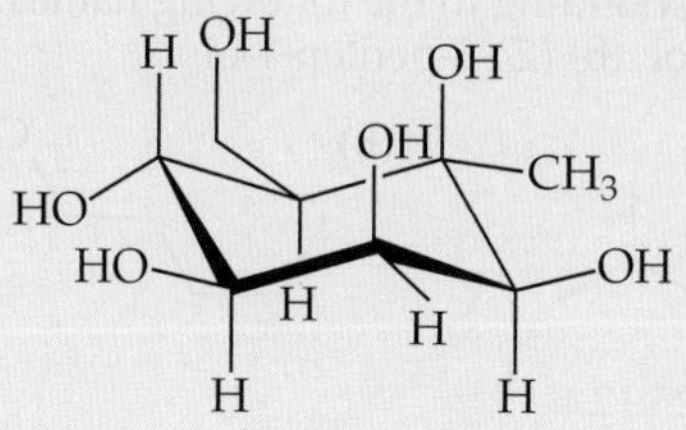

Answer: There are four 2° and one 3° OH groups (around the ring), and one 1° OH group.

(See also Exercises 25.13, 25.14.)

25.2 | HALOALKANES

Aliphatic or cyclic hydrocarbons containing a halogen directly bonded to an sp^3 hybridised carbon are called **haloalkanes**. They are generally written as:

$$R—X \qquad \text{where X = F, Cl, Br, I}$$

Haloalkanes are also commonly referred to as alkyl halides.

The haloalkane nomenclature is derived from the longest chain bearing the halogen. The position number is taken to be the lowest possible, bearing all other substituents in mind. Halogen substituents are indicated by the prefixes fluoro-, chloro-, bromo- and iodo-. The derived name places these substituents in alphabetical order:

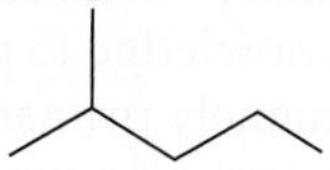

2-fluoropentane

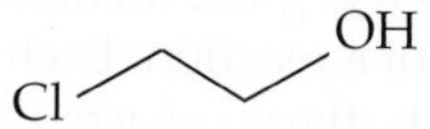

2-chloroethanol

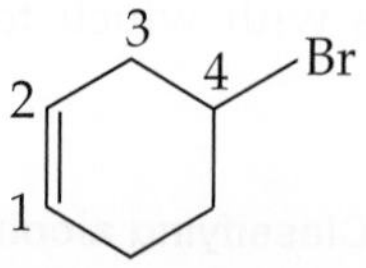

4-bromocyclohexene

We should also note here that the halogen functional group is second only to alkyl groups as the lowest functional group on the nomenclature hierarchy. All other substituents take precedence in deriving a compound's name, and the numbering used is determined by the location of the other group. For example, 2-chloroethanol is named as the alcohol and not the haloalkane. 4-Bromocyclohexene is named by taking into account the position of the double bond before the halogen.

Many haloalkanes have common names. Some examples are:

CH$_2$Cl$_2$

methylene chloride
(dichloromethane)

CHCl$_3$

chloroform
(trichloromethane)

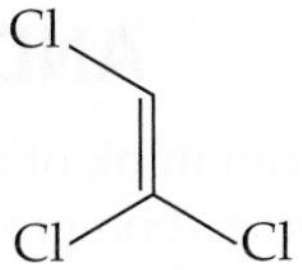

trichlor
(trichloroethene)

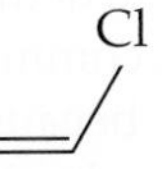

vinyl chloride
(chloroethene)

Compounds such as dichloromethane and chloroform are useful solvents in organic synthesis. Compounds of the form CHX$_3$ are often called **haloforms**, of which chloroform is an example. Perfluorinated compounds (in which all hydrogens are replaced by fluorine) are also very useful. One example we are all familiar with is Teflon™ (polytetrafluoroethene, (CF$_2$CF$_2$)$_n$) with its application in non-stick cookware among others (∞ Section 24.6, 'Addition Polymerisation').

∞ Review this on page 985

Having properties very different from both water and hydrocarbons, perfluorinated compounds have been used as solvents since they are easily partitioned from both water and most organic solvents. The polar structure also has application in medicine. For example, perfluorocarbons are able to dissolve oxygen at levels greater than blood. As a result they are undergoing trials in dramatic experiments as blood substitutes.

Another class of haloalkanes you may be familiar with are the **chlorofluoro-carbons**, or **CFCs**. This class of compound was typically used as refrigerants, as propellants in aerosols and as industrial solvents. They were usually marketed under the name *Freon*. CFCs are odourless, non-toxic, non-flammable and non-corrosive compounds and so found great industrial utility. During the mid-1970s, however, research into CFCs showed they had a harmful effect on the environment (∞ Section 13.2, 'Human Activities and the Earth's Atmosphere'). Being relatively chemically inert means that CFCs can persist in the atmosphere for a long time. Thus they are not broken down in the troposphere but rather find their way into the stratosphere where they are decomposed by the action of ultraviolet (UV) radiation. In doing so, a reaction sequence was initiated that depleted the ozone layer—a thin layer of the Earth's atmosphere that acts as a shield against excess UV radiation. The attributed result is not only a greater prevalence in skin cancer, but also a higher risk of global warming as chlorofluorocarbons are also much more effective as greenhouse gases than carbon dioxide. In the late 1980s the United Nations Environment Program put in place a number of agreements to phase out the use of CFCs by 1996. For their work in this area and forewarning of the implications, Sherwood Rowland, Mario Molina and Paul Crutzen were awarded the 1995 Nobel Prize in Chemistry.

∞ Review this on page 476

Haloalkanes have a wealth of chemistry associated with them. We have already seen how they can be formed from alkenes and alkanes (∞ Section 24.4, 'Electrophilic Addition Reactions' and ∞ Section 22.8, 'Reactions of Alkanes'). In this chapter we will explore two reaction types in detail—nucleophilic substitution (Section 25.5) and β-elimination (Section 25.6)—in which haloalkanes are reactants. Later, in Chapter 26, we will explore the use of haloalkanes for the preparation of Grignard reagents.

∞ Review this on pages 966 and 911

25.3 | ETHERS: STRUCTURE, PROPERTIES AND NOMENCLATURE

We can think of alcohols as compounds in which one hydrogen atom and one hydrocarbon group are bonded to an oxygen atom. Compounds in which two hydrocarbon groups are bonded to one oxygen atom are called **ethers**. As for alcohols, the oxygen atom is sp^3 hybridised with the two C–O bonds subtending an angle of approximately 109.5°. Two of the sp^3 hybrid orbitals of the oxygen atom form σ-bonds with two carbon atoms adjacent to them. Each of the other two hybridised orbitals contains a pair of unshared electrons. ▼ FIGURE 25.8 illustrates the Lewis structure and other molecular model representations of diethyl ether, the most commonly used of all the ethers.

Ethers are polar molecules, but because of the steric hindrance due to the alkyl groups attached, only weak dipole–dipole interactions exist between molecules. As a result of having alkyl groups bonded to the central oxygen atom, there is no opportunity for hydrogen bonding or other strong electrostatic interactions, so the physical properties of ethers are closer to those of alkanes than to those of alcohols. This moderate polarity is, however, advantageous as it allows ethers to dissolve a wider range of organic compounds than hexane would as a comparative solvent.

Ethers resemble hydrocarbons in their inertness to chemical reactions. As a consequence, ethers make excellent solvents in which to undertake chemical reactions. Two commonly used solvents are diethyl ether and tetrahydrofuran (THF). Diethyl ether, as we have already seen, is an acyclic ether (Figure 25.8), whereas THF (▼ FIGURE 25.9) is a cyclic ether derived, as the name suggests, from furan, that has a structure given on page 1059.

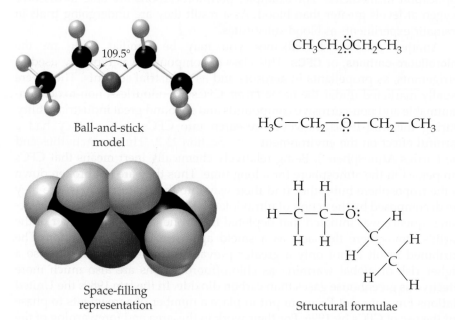

Ball-and-stick model

Space-filling representation

Structural formulae

▶ **FIGURE 25.8 The structure of diethyl ether.** A number of representations for diethyl ether are shown.

▶ **FIGURE 25.9 Cyclic ethers.** Simple cyclic ethers such as tetrahydrofuran (THF) and 1,4-dioxane are excellent solvents. Tetrahydropyrans (THP) are useful reagents in organic synthesis.

tetrahydrofuran (THF)

tetrahydropyran (THP)

1,4-dioxane

The structural difference between alcohols and ethers has a marked effect on the properties and chemical reactivity. Butanol and diethyl ether are representative molecules within their respective classes and are also constitutional isomers. ▼ TABLE 25.2 illustrates the differences in three of their principal properties.

CONCEPT CHECK 2

List two criteria that would make a liquid a good solvent in which to undertake chemical reactions.

Naming Ethers

Acyclic ethers are named systematically by selecting the longest chain and adding to it the name of the attached **alkoxy** (–OR) **group**. Common names are used more freely with this class of compounds, as the IUPAC method leads to lengthy names quite quickly. These common ether names are derived from the two alkyl groups that are attached to the ether oxygen. ▼ FIGURE 25.10 illustrates how to derive the systematic and common names of acyclic ethers.

Cyclic ethers, such as tetrahydrofuran, are **heterocyclic compounds** whose names are derived in some instances from the parent heterocycle. Heterocyclic compounds are derivatives of cyclic hydrocarbons in which one of the ring carbons is replaced with another atom, usually O or N, but many other hetero atoms have also been incorporated (for example, S, Se, P). Figure 25.9 showed the structure of the most common cyclic ethers.

Cyclic polyethers such as 18-crown-6 (▼ FIGURE 25.11) are useful in organic synthesis as phase transfer catalysts and metal ion complexation

TABLE 25.2 • Physical properties of the constitutional isomers butan-1-ol and diethyl ether

Compound	Structure	Melting point	Boiling point	Density g cm^{-3}
butan-1-ol	$CH_3CH_2CH_2CH_2OH$	–90 °C	118 °C	0.81
diethyl ether	$CH_3CH_2OCH_2CH_3$	–116 °C	35 °C	0.70

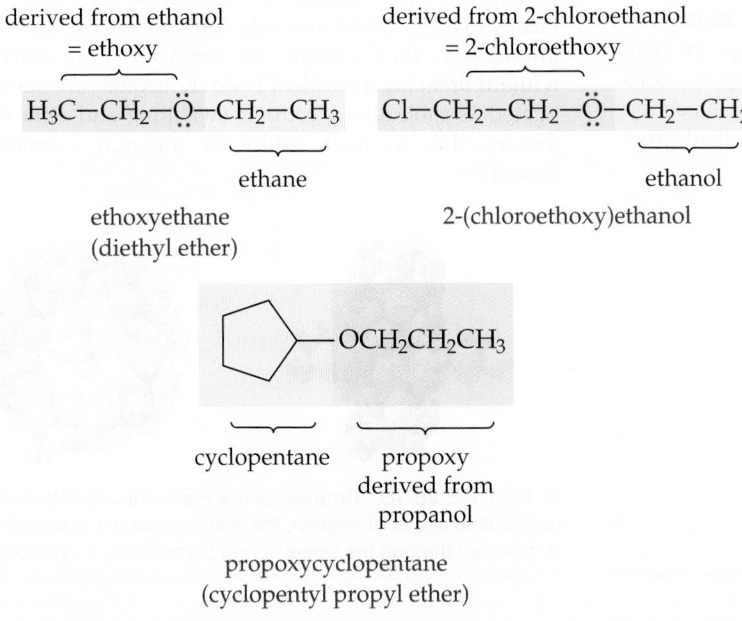

derived from ethanol
= ethoxy

$$H_3C-CH_2-\overset{..}{\underset{..}{O}}-CH_2-CH_3$$

ethane

ethoxyethane
(diethyl ether)

derived from 2-chloroethanol
= 2-chloroethoxy

$$Cl-CH_2-CH_2-\overset{..}{\underset{..}{O}}-CH_2-CH_2-OH$$

ethanol

2-(chloroethoxy)ethanol

$$-OCH_2CH_2CH_3$$

cyclopentane   propoxy
derived from
propanol

propoxycyclopentane
(cyclopentyl propyl ether)

◀ **FIGURE 25.10 Ether nomenclature.** Systematic names are derived from the longest alkyl chain present that is attached to the ether oxygen and the residual alkoxy group. Examples of common names are shown in brackets.

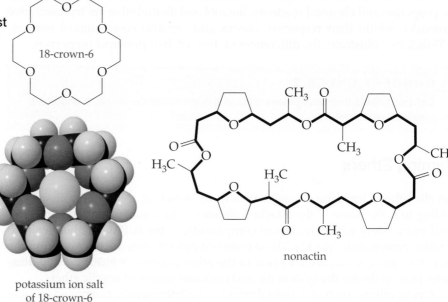

▲ **FIGURE IT OUT**

Which atoms in these molecules most closely interact with a metal cation?

18-crown-6

▶ **FIGURE 25.11 Macrocyclic ethers.**
Naturally occurring cyclic ethers like nonactin can transport ions across biological membranes. In doing so, they act as potent antibiotics. Crown ethers such as 18-crown-6 are synthetic analogues of nonactin that are also useful for the separation of heavy metal salts from waterways, as catalysts of reactions involving carbanions and as phase transfer catalysts.

potassium ion salt of 18-crown-6

nonactin

agents. Discovered and developed by Charles Pedersen in the 1960s, **crown ethers** are a synthetic analogue of naturally occurring antibiotics such as nonactin.

A CLOSER LOOK

CROWN ETHERS

Supramolecular chemistry, also known as 'chemistry beyond the molecule', is a multidisciplinary science that came to prominence in the late 1980s. This was in part due to the award of the 1987 Nobel Prize in Chemistry to Charles Pedersen (1904–1989), Donald Cram (1919–2001) and Jean-Marie Lehn (b 1939), three of the founders and early developers of supramolecular chemistry. This field began with Pedersen's pioneering work on crown ethers and their metal ion complexes (▼ FIGURE 25.12). The potential of supramolecular chemistry to bridge the gap between chemistry and biology, thereby finding applications in medicine, materials and environmental science, biotechnology and nanotechnology, has seen interest in this field grow rapidly over the past 10–15 years.

Supramolecular chemistry is seen by some as either a molecular art form or as *molecular engineering*. During its development throughout the 1980s and 1990s, supramolecular chemistry was used as a tool to create the most unlikely molecular topologies. For example, rotaxanes (*rota*, wheel; *axa*, axle) are a chemical curiosity containing two molecular components (▼ FIGURE 25.13). The molecular shapes of the two components differ, one resembling a wheel and the other an axle. The axle component is threaded through the centre of the wheel component, and is usually terminated by bulky groups to stop the wheel falling off. The use of the bulky stoppers means that the wheel and axle components are mechanically interlocked—they cannot be removed from each other without breaking a covalent bond. Rotaxanes are being investigated as molecular electronic switching and optical binary devices due to their ability to undergo conformational isomerism.

▲ **FIGURE 25.12 Crown ether complex of K⁺.**

▲ **FIGURE 25.13 Rotaxanes are mechanically interlocked molecules.** In this illustration, the axle component, a simple alkane, is threaded through the wheel or ring component, a cyclodextrin.

⚠ **CONCEPT CHECK 3**

What properties of 18-crown-6 make it possible to complex metal ions such as Na^+ or K^+?

25.4 | REACTIONS OF ALCOHOLS

Alcohols are central to organic chemistry. ▼ FIGURE 25.14 illustrates the relationships between alcohols and the other seven major classes of functional groups introduced in this text. Four of these functional groups, highlighted in red, can be arrived at through oxidation of a suitable alcohol (∞ Section 26.2, 'Preparation of Aldehydes and Ketones').

∞ Find out more on page 1039

In this section we explore further the interconversion of alcohols to alkenes and haloalkanes. Ethers are formed by taking advantage of the high reactivity of the conjugate base of an alcohol, called an **alkoxide**. Alkoxides are as strong a base as hydroxide and also make excellent nucleophiles.

Alkoxides

Alcohols react with alkali metals to form alkoxides in the same way that water reacts with alkali metals to form hydroxide. The reaction is irreversible, since the other product of the reaction, hydrogen gas, is evolved (▶ FIGURE 25.15).

$$2\,H_2O + 2\,Na \longrightarrow 2\,HO^-Na^+ + H_2(g) \qquad [25.3]$$

sodium hydroxide

$$2\,ROH + 2\,K \longrightarrow 2\,RO^-K^+ + H_2(g) \qquad [25.4]$$

potassium alkoxide

The name of the alkoxide is derived from the alcohol undergoing reaction. For example, the reaction of ethanol with sodium gives sodium ethoxide (Figure 25.15), the reaction of *tert*-butanol (2-methyl-2-propanol) with potassium metal yields potassium *tert*-butoxide. Reactions involving alkoxides invariably use the corresponding alcohol as the solvent for ease of preparation.

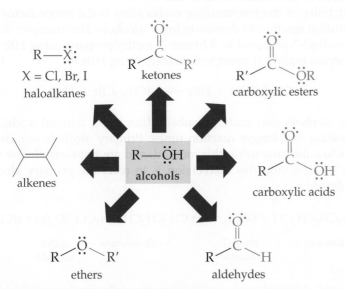

▲ FIGURE 25.14 **Functional group interconversion.**

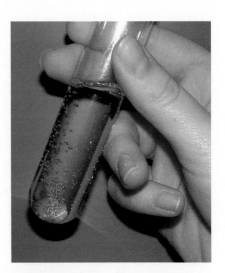

▲ FIGURE 25.15 **Alkoxide formation.** Sodium metal reacts with ethanol forming the alkoxide. The reaction is characterised by the evolution of hydrogen gas.

∞ Review this on page 966

> △ **CONCEPT CHECK 4**
>
> Why would you not prepare sodium methoxide by reacting sodium in a mixture of methanol and *tert*-butanol?

Basicity of Alcohols

The ability of an alcohol's OH group to accept a proton is an important property that is important in the context of dehydration (−H₂O). An alcohol can act as a weak base, forming an oxonium ion under acidic conditions. The oxonium ion ($pK_a \sim -2$) is a slightly better acid than the hydronium ion (H_3O^+, $pK_a \sim -1.7$), but of greater importance to organic synthesis (and to a large part of the rest of this chapter) is the ability of the oxonium ion to lose water, forming a carbocation intermediate:

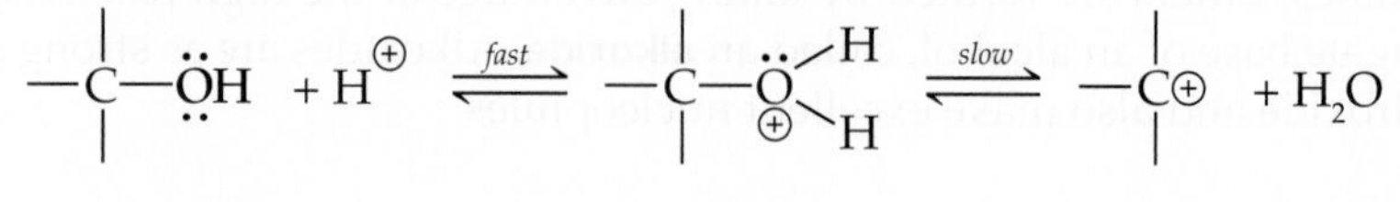

$$-\overset{|}{\underset{|}{C}}-\ddot{\overset{..}{O}}H \; + \; H^{\oplus} \quad \underset{}{\overset{fast}{\rightleftharpoons}} \quad -\overset{|}{\underset{|}{C}}-\underset{\oplus}{\ddot{O}}\overset{H}{\underset{H}{\diagdown}} \quad \overset{slow}{\rightleftharpoons} \quad -\overset{|}{\underset{|}{C}}\oplus \; + \; H_2O \qquad [25.5]$$

an oxonium ion   a carbocation

As we discussed in the context of Markovnikov's rule, the stability of the carbocation intermediate is an enormous driving force for the reactivity of a species, and even dictates the reaction pathway (∞ Section 24.4, 'Electrophilic Addition Reactions').

Alcohols to Haloalkanes

Alcohols can be converted into haloalkanes by treatment of a halogenating agent, usually under acidic conditions. In this reaction, the OH group of the alcohol is substituted by a halogen. The reaction is classed as a **nucleophilic substitution (S$_N$) reaction** and we will see much more of this in the next section.

A nucleophilic substitution reaction is any reaction in which one nucleophile is substituted for another.

In using mineral acids (e.g. HCl, HBr, HI) as the halogenating agent, the first stage of substitution is protonation of the alcohol to form an oxonium ion. This is followed by elimination of water to form the intermediate carbocation as indicated in Equation 25.5. This last step is rate-determining, being faster for formation of a 3° carbocation than 2° and 1°.

The stability of the intermediate carbocation is the major factor allowing the substitution reaction to dominate for 3° alcohols. For example, the conversion of 2-methyl-2-propanol to 2-bromo-2-methylpropane using HBr is a reaction that occurs readily at room temperature using HBr.

$$(CH_3)_3COH + HBr \longrightarrow (CH_3)_3CBr + H_2O \qquad [25.6]$$

Secondary alcohols also undergo substitution using mineral acids, but they require heating and longer reaction times. Primary alcohols require extreme conditions to substitute using this method, and are impractical. The most efficient way of achieving this conversion is to react the primary alcohol with thionyl chloride:

$$CH_3CH_2CH_2CH_2OH + SOCl_2 \longrightarrow CH_3CH_2CH_2CH_2Cl + SO_2(g) + HCl$$

butan-1-ol   thionyl chloride   1-chlorobutane   sulfur dioxide

$$[25.7]$$

The reason this reaction works well is that one of the products (SO₂) is gaseous, so once it is liberated the reaction goes to completion.

Dehydration of Alcohols

We have already indicated that alcohols can be dehydrated to form alkenes (see Equation 24.15) (◦◦◦ Section 24.4, 'Electrophilic Addition Reactions'). Equation 25.8 illustrates the accepted mechanism. This acid-catalysed dehydration process has, as one of its intermediates, an oxonium ion. The rate-determining step is the elimination of water to form the corresponding carbocation. Removal of a small group from a larger molecule, as in this case, is called an *elimination reaction*. Water is classed as a *good leaving group* because of its stability and relatively unreactive nature. Hydroxide ion, however, does not eliminate easily and so is classed as a *poor leaving group*.

◦◦◦ Review this on page 966

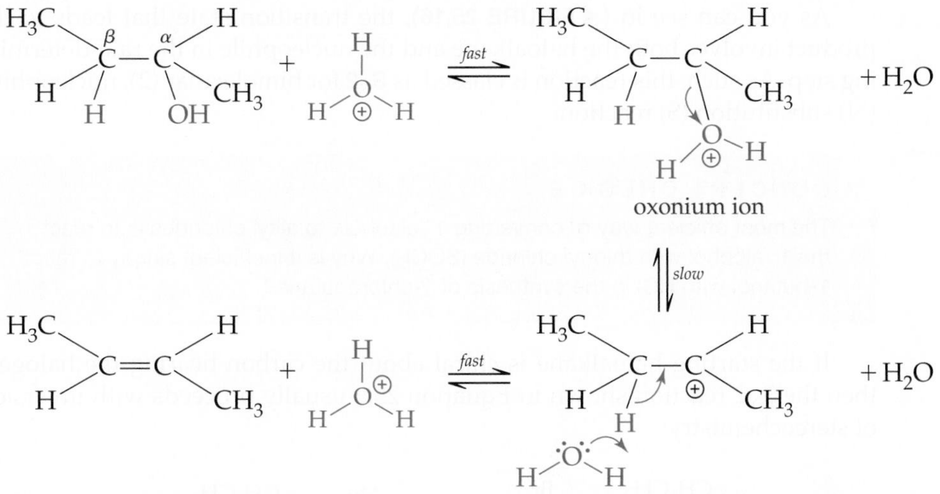

$$
\text{[25.8]}
$$

oxonium ion

slow

Leaving groups can be classified as good or poor based on their reactivity. In general, the better leaving groups are those that are conjugate bases of strong acids. A relative order of leaving group is:

$$I^- > H_2O > Br^- > Cl^- \gg F^- > CH_3COO^- > OH^- > NH_2^-$$

Good leaving   *Poor leaving*
group   *group*

Once water has been eliminated from the oxonium ion in Equation 25.8, the reaction proceeds rapidly to the formation of the alkene by the loss of a proton from a carbon atom located in a neighbouring position (called the β-position) to the original C–OH bond. The two electrons that once formed the C_β–H bond will now form the π-bond of the alkene. Without the β-hydrogen atom, the dehydration reaction does not proceed.

The nature of the equilibrium is such that removal of water or distillation of the alkene (which generally will have a lower boiling point than the corresponding alcohol) will maximise the formation of the alkene by driving the equilibrium in the forward direction.

 CONCEPT CHECK 5

Why would the boiling point of an alcohol be higher than that of the corresponding dehydrated product?

25.5 | NUCLEOPHILIC SUBSTITUTION REACTIONS OF HALOALKANES

The conversion of alcohols to haloalkanes by substitution opens up a large library of functional group manipulation, limited only by imagination. The reactivity of haloalkanes manifests itself from the difference in electronega-

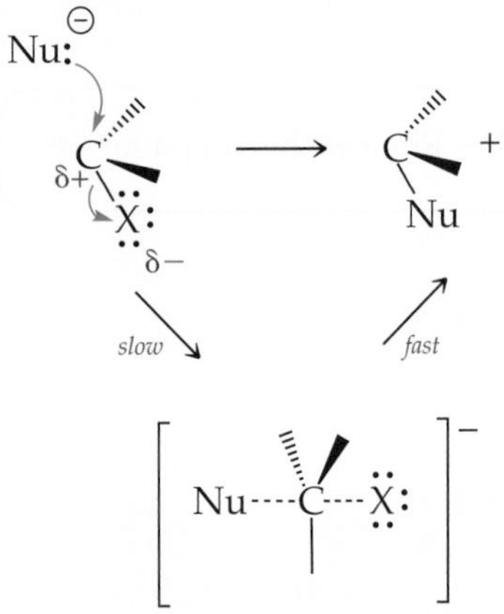

▲ FIGURE 25.16 Nucleophilic substitution reactions.

tivity between carbon (2.5) and the halogens (2.6–4.0), which means the carbon bearing the halogen is susceptible to nucleophilic attack.

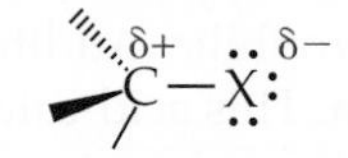

$$\overset{\delta+}{C}-\overset{\delta-}{X} \qquad \text{Where} \qquad X = F, Cl, Br, I$$

Nucleophilic substitution reactions of haloalkanes proceed by attack of the nucleophile (Nu:) at the carbon bearing the halogen. The trigonal-bipyramidal transition state formed in this reaction has long Nu–C and C–X bonds, signifying the strengthening bond formation of the incoming nucleophile and the weakening C–X bond of the outgoing halide. One of the driving forces for this reaction is the stability of the halide ion, which is an excellent leaving group.

As you can see in (◀ **FIGURE 25.16**), the transition state that leads to the product involves both the haloalkane and the nucleophile in the rate-determining step. As such, this reaction is classed as **S$_N$2** for bimolecular (2), nucleophilic (N) substitution (S) reaction.

> ### CONCEPT CHECK 6
> The most efficient way of converting 1° alcohols to alkyl chlorides is to react the 1° alcohol with thionyl chloride (SOCl$_2$). Why is it inefficient simply to react 1-butanol with HCl in the synthesis of 1-chlorobutane?

If the starting haloalkane is chiral about the carbon bearing the halogen, then the S$_N$2 reaction shown in Equation 25.9 usually proceeds with inversion of stereochemistry:

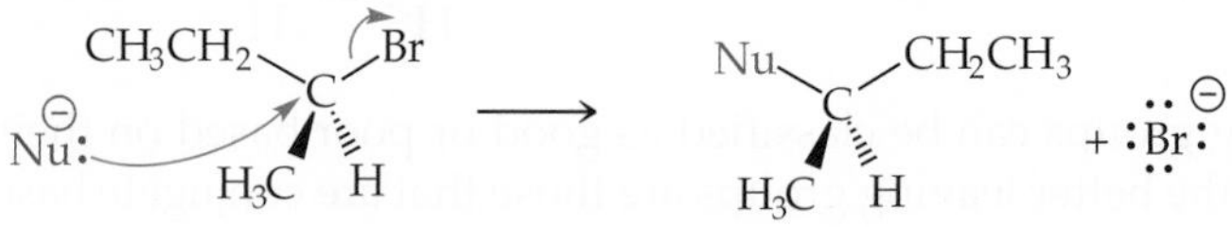

$$\text{(S) stereochemistry} \longrightarrow \text{(R) stereochemistry} \qquad [25.9]$$

Inversion of stereochemistry can be thought of as the inversion of an umbrella that happens on a windy day (▼ **FIGURE 25.17**). In this analogy, consider the umbrella canopy as three groups around a tetrahedral centre that do not participate in an S$_N$2 reaction. Upon nucleophilic attack, the canopy is inverted, with some strain, to yield a new conformation (▶ **FIGURE 25.18**).

However, the inversion of stereochemistry depends on the nature of the nucleophile and leaving group, the substituents about the stereocentre and the reaction conditions (for example, the solvent used).

▶ FIGURE 25.17 The mechanism of an S$_N$2 reaction is analogous to an umbrella on a windy day.

FIGURE IT OUT

If (*R*)-2-bromobutane was used in this reaction what would be the absolute stereochemistry of the alcohol produced?

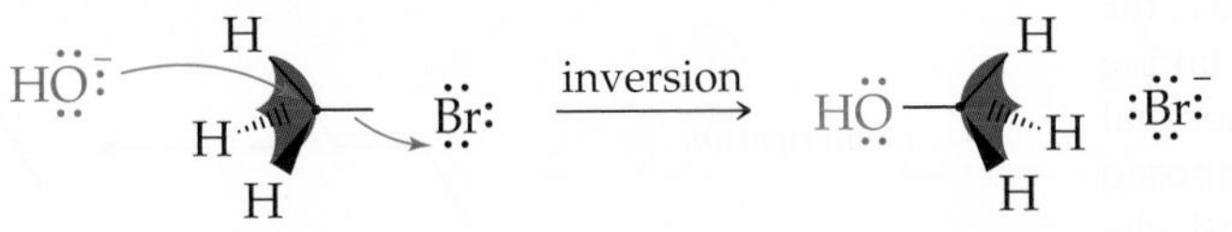

◀ **FIGURE 25.18 S$_N$2 reactions.**
Inversion of configuration occurs at the carbon atom undergoing nucleophilic attack.

▼ **TABLE 25.3** lists common nucleophiles employed in nucleophilic substitution reactions, as well as the classes of organic compound formed by the substitution. We have already investigated some of these in detail. Two nucleophile classes we have yet to discuss in any detail involve amines (which are neutral molecules) and the cyanide ion (CN$^-$). The addition of ammonia to a haloalkane, for example, leads initially to an **alkyl ammonium salt**. Such salts are characterised by four groups around nitrogen of which at least one is an alkyl group. Provided the ammonium ion is of the form R$_3$NH$^+$, R$_2$NH$_2^+$ or RNH$_3^+$, they are easily transformed into **amines** by the addition of base.

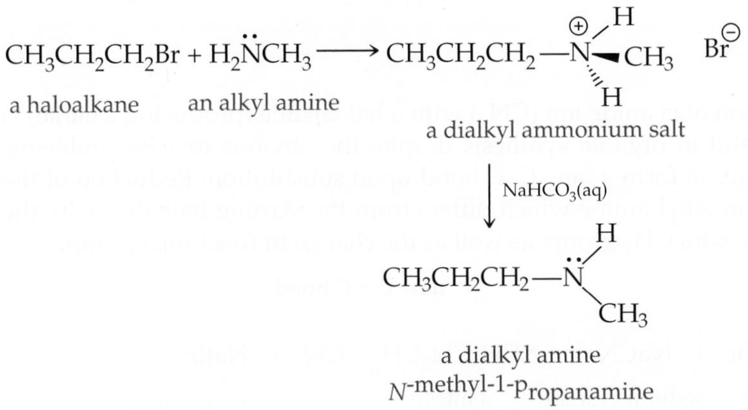

[25.10]

TABLE 25.3 • Common nucleophilic substitution reactions based on the reaction:

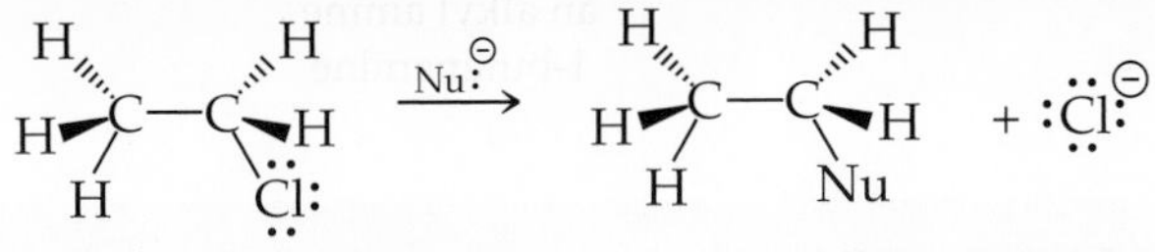

Nucleophile	Product	Class of compound
$^\ominus$:ÖH	CH$_3$CH$_2$OH	an alcohol
$^\ominus$:ÖR	CH$_3$CH$_2$OR	an ether
$^\ominus$:Ï:	CH$_3$CH$_2$I	an alkyl iodide
$^\ominus$:CN	CH$_3$CH$_2$CN	a nitrile
:NH$_3$	CH$_3$CH$_2$NH$_3^+$X$^-$	an alkyl ammonium salt
ÖH$_2$	CH$_3$CH$_2$OH	an alcohol

A CLOSER LOOK

MOLECULARITY

The molecularity of a reaction describes the number of atoms or molecules of reactant taking part in an elementary reaction. Remember that an overall reaction mechanism is composed of several elementary reactions. Most elementary reactions involve one (*unimolecular*) or two (*bimolecular*) species. Though most reactions are monitored by the loss of reactants or the formation of products, ultimately it is the molecularity of the transition state to the rate-determining step that defines the observed kinetics of the overall reaction and hence whether a reaction is unimolecular or bimolecular. Molecularity should not be confused with reaction order (∞ Section 15.6, 'Reaction Mechanisms').

Simple unimolecular reactions include decomposition of chlorine by homolytic bond cleavage or isomerisation reactions. In both cases, the rate of reaction depends directly on the concentration of the only reactant. In this case, it is easy to see why this is a unimolecular reaction.

homolytic bond cleavage $\quad Cl_2 \xrightarrow{h\nu} 2Cl\cdot$

isomerisation

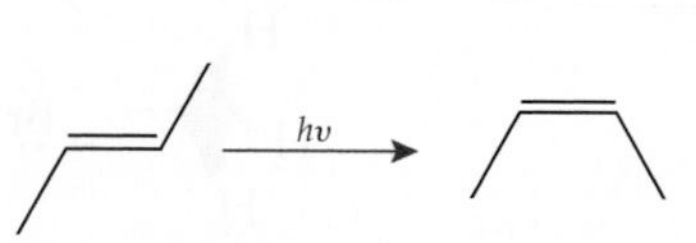

In a bimolecular reaction, two species are required to give rise to the product, for example the dimerisation of $NO_2\cdot$ or the addition of HCl to an alkene. In each case, the rate of reaction depends directly on the concentration of the two reactants, which need to collide before reaction can proceed.

$$2NO_2\cdot \longrightarrow N_2O_4$$

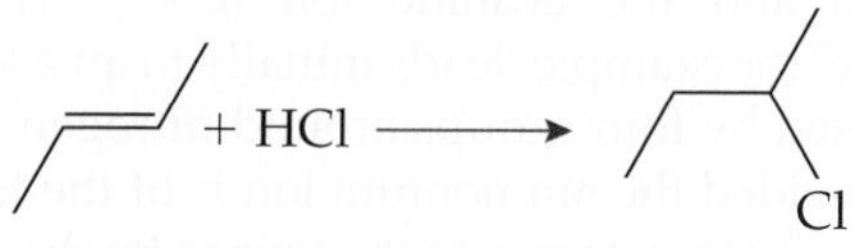

RELATED EXERCISES: 25.52, 25.56, 25.60

The reaction of cyanide ion (CN^-) with a haloalkane (producing a **nitrile**) is extremely useful in organic synthesis despite the obvious toxicity problems, due to its ability to form a new C–C bond upon substitution. Reduction of the nitrile yields an alkyl amine which differs from the starting haloalkane by the addition of an extra CH_2 group, as well as the change in functional group.

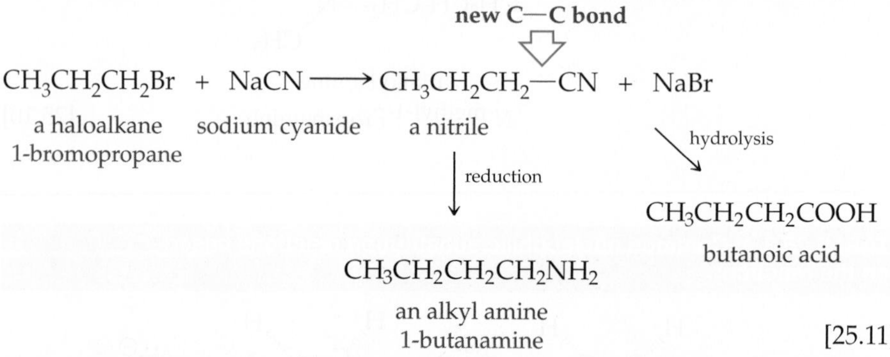

new C—C bond

$$CH_3CH_2CH_2Br \ + \ NaCN \longrightarrow CH_3CH_2CH_2{-}CN \ + \ NaBr$$

a haloalkane   sodium cyanide   a nitrile
1-bromopropane

hydrolysis

reduction

$CH_3CH_2CH_2COOH$

butanoic acid

$$CH_3CH_2CH_2CH_2NH_2$$

an alkyl amine
1-butanamine

[25.11]

⚠ CONCEPT CHECK 7

Iodide is a better leaving group than chloride. Why, then, if NaI is added to a solution of 1-chloroethane in acetone, does the nucleophilic substitution occur? To answer this, think about the possible equilibrium reactions and ways to influence these.

SAMPLE EXERCISE 25.4   Nucleophilic substitution reactions

Using 2-butanol as a starting material, design a synthesis of 2-methylbutyronitrile in two steps:

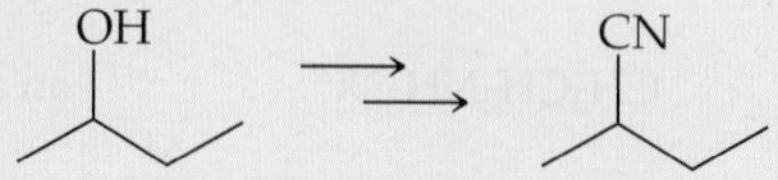

SOLUTION

Analyse We are asked to design a substitution reaction for replacing a secondary OH group by CN.

Plan For a synthesis problem like this, it is usually helpful to consider the sequence of steps starting from the product you want. You know from Table 25.3 that nitriles can be synthesised from the haloalkane by substitution.

Solve In the following example the chloride is chosen, although the bromide is just as efficient:

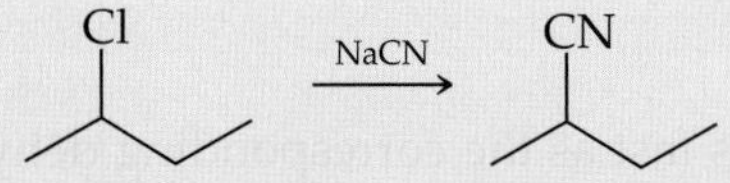

Formation of the alkyl chloride from the alcohol can occur in two ways. Since 2-butanol is a secondary alcohol, reaction with either HCl or thionyl chloride, $SOCl_2$, will yield 2-chlorobutane.

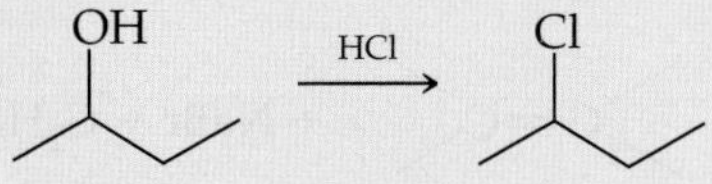

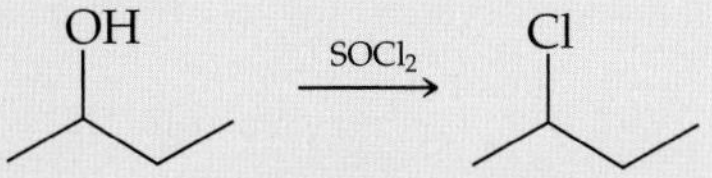

Thus the overall two-step reaction sequence is:

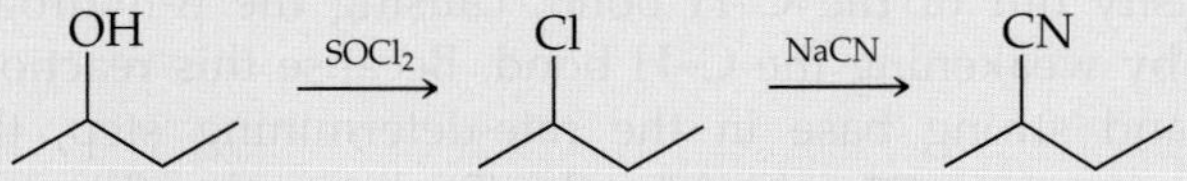

PRACTICE EXERCISE

Describe how you would prepare this ether from cyclohexanol.

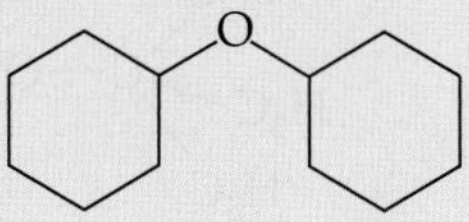

Answer: There are two parts to answering this question. Disconnection of the ether leads to the following reaction:

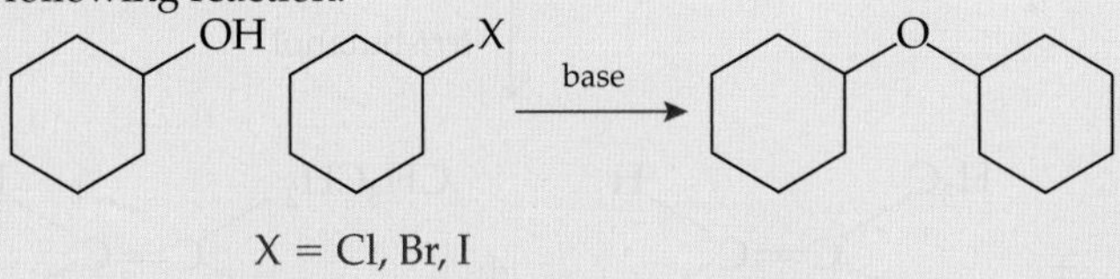

X = Cl, Br, I

The haloalkane can be synthesised from cyclohexanol by treatment with a haloacid (HCl, HBr, HI, respectively) or use of $SOCl_2$ in the case where X = Cl.

(See also Exercises 25.33, 25.34.)

25.6 | HALOALKANES TO ALKENES: β-ELIMINATION

The process of eliminating a good leaving group, as outlined for the dehydration of an alcohol, is quite general. For example, haloalkanes undergo elimination under basic conditions. This process, called a **β-elimination reaction**, involves the *dehydrohalogenation* of the haloalkane—that is, the loss of a hydrogen atom and a halogen atom from the same molecule. For the reaction to proceed, a hydrogen atom on the carbon atom adjacent to the carbon bearing the halide (that is, the β-carbon) must be present and be removed.

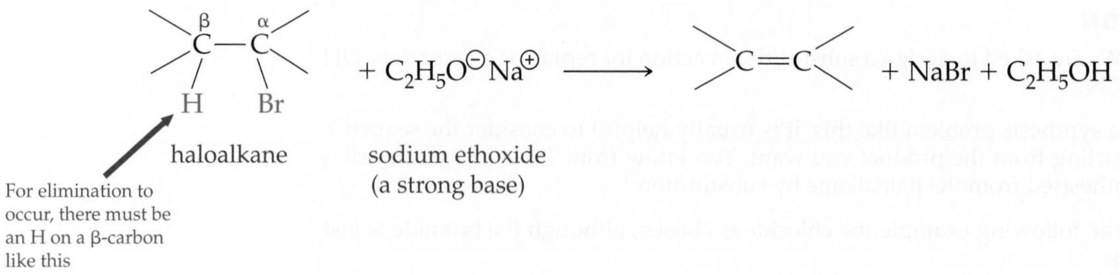

haloalkane   sodium ethoxide
(a strong base)

For elimination to occur, there must be an H on a β-carbon like this

$$[25.12]$$

The reaction is not as fast as the corresponding dehydration of an alcohol, requiring the use of a strong base (for example, sodium ethoxide) to proceed. The mechanism can be shown to occur in a single (and essentially irreversible) step:

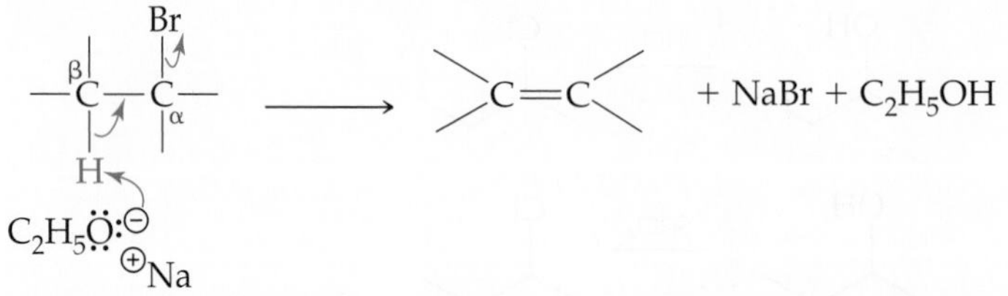

$$[25.13]$$

Typically, the hydrogen atoms of hydrocarbons do not react in this way. However, the inductive effect of the halogen atom on the α-carbon draws electron density out of the C–H bond, causing the β-hydrogen to become more acidic by weakening the C–H bond. Because this reaction involves the haloalkane and strong base in the rate-determining step, this **elimination reaction** is known as an **E2** or bimolecular (2) elimination (E) reaction.

Let us now consider the reaction of 2-chlorobutane with potassium *tert*-butoxide, another strong organic base:

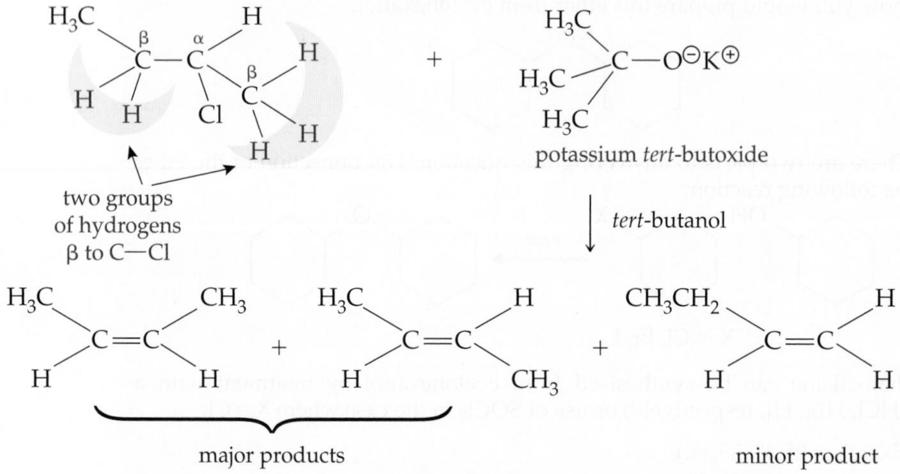

potassium *tert*-butoxide

two groups of hydrogens β to C—Cl

tert-butanol

$$[25.14]$$

major products   minor product

In this case, there are two sets of β-hydrogens as highlighted in Equation 25.14. Although β-elimination can lead to the formation of two alkenes, one elimination product is formed in preference to the other.

The major alkene formed by dehydrohalogenation is the one containing the more substituted double bond. This preference is known as **Zaitsev's rule**. The more substituted double bond is the one with the greater number of substituents (excluding hydrogen) attached to the two carbons of the double bond. For example, the but-2-ene formed in Equation 25.14 has two methyl groups connected to the carbons of the double bond, whereas but-1-ene has a single ethyl group connected. Because but-2-ene has two non-hydrogen substituents, it is the more substituted of the two alkenes formed. ▶ **FIGURE 25.19** illustrates what is meant by a substituted alkene.

▲ FIGURE IT OUT
Which of these alkenes are isomers?

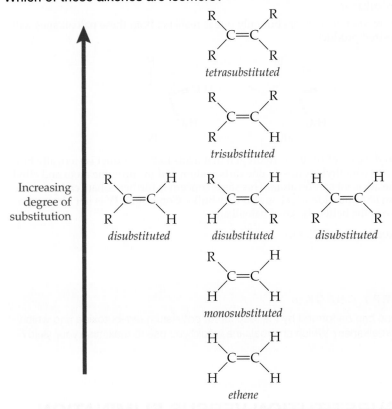

◀ **FIGURE 25.19   Substituted alkenes.**

SAMPLE EXERCISE 25.5   **β-Elimination in haloalkanes**

Draw the structural formula for the major product formed when 1-bromo-1- methyl-cyclohexane is reacted with sodium ethoxide in ethanol. Systematically name the major product.

SOLUTION

Analyse We are asked to determine the product of the reaction of a sterically hindered haloalkane with a strong base.

Plan Begin by structurally identifying the compounds involved. Look for β-hydrogens then determine the types of alkenes possible.

Solve

There are two different sets of β-hydrogens highlighted in yellow, those on the methyl group and two equivalent sets on the cyclohexane ring. Hence two different alkenes, I and II, are formed. The major product will be the one that satisfies the criterion of Zaitsev's rule. The more substituted double bond is found in II, and so it is formed in preference.

   Alkene II is named *1-methylcyclohexene.*

PRACTICE EXERCISE

2-Methylpent-2-ene can be formed by the reaction of potassium *tert*-butoxide and which bromoalkane?

Answer: There are two answers but only one is realistic. Both these haloalkanes will yield the desired product:

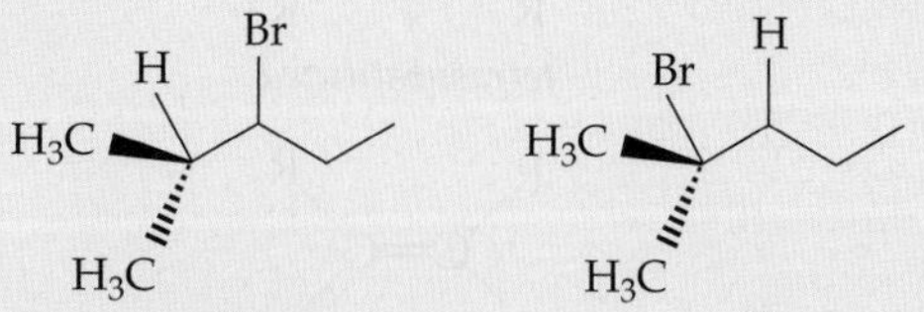

However, in the case of III, the β-hydrogen that leads to the product is sterically hindered by the two methyl groups on one carbon atom and the bromine atom and ethyl group on the adjacent carbon atom. This steric concern is further enhanced by the use of potassium *tert*-butoxide which is also very bulky. Compound IV is not hindered as much and so is the better choice of haloalkane.

(See also Exercises 25.43, 25.44.)

CONCEPT CHECK 8

Dec-2-ene can be formed by the reaction of potassium *tert*-butoxide and which two chloroalkanes? Which chloroalkane would you use to maximise your yield?

25.7 | SUBSTITUTION VERSUS ELIMINATION

We consider four types of reactions in this section, E1, E2, S_N1 and S_N2. We have already encountered two of these formally (E2 and S_N2). As we will see, each reaction type does not operate in isolation, but together they form a competing process. Which direction a reaction proceeds in depends on several factors, including starting material and reagents, type of solvent, and electronic and steric effects.

We have already seen in this chapter that alkoxides are a strong base, able to deprotonate haloalkanes in a β-elimination reaction.

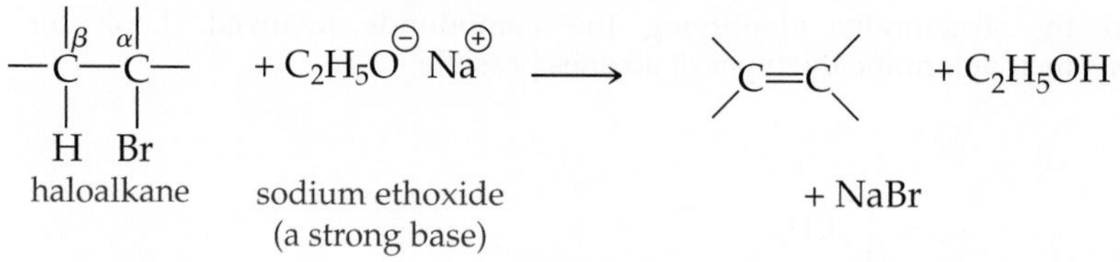

$$\overset{\beta\quad\alpha}{-\overset{|}{\underset{\underset{H}{|}}{C}}-\overset{|}{\underset{\underset{Br}{|}}{C}}-} \; + \; C_2H_5O^{\ominus}\,Na^{\oplus} \longrightarrow \quad \overset{}{C}=\overset{}{C} \; + \; C_2H_5OH$$

haloalkane   sodium ethoxide   + NaBr
                 (a strong base)

[25.15]

However, as we saw in Section 25.5, alkoxides are able to react with haloalkanes in another way. Instead of providing a means of elimination, alkoxides can also interact with haloalkanes to form ethers. The classic example is the *Williamson ether synthesis*, in which the halide is replaced by the alkoxide in a substitution reaction.

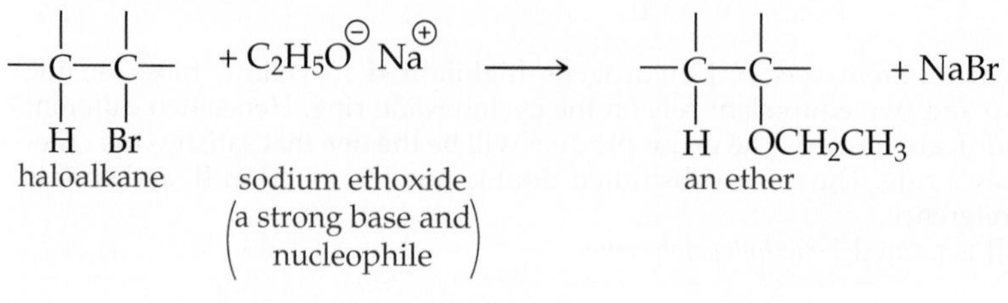

$$-\overset{|}{\underset{\underset{H}{|}}{C}}-\overset{|}{\underset{\underset{Br}{|}}{C}}- \; + \; C_2H_5O^{\ominus}\,Na^{\oplus} \longrightarrow \quad -\overset{|}{\underset{\underset{H}{|}}{C}}-\overset{|}{\underset{\underset{OCH_2CH_3}{|}}{C}}- \; + \; NaBr$$

haloalkane   sodium ethoxide   an ether
                 (a strong base and
                  nucleophile)

[25.16]

A CLOSER LOOK

NUCLEOPHILE OR LEWIS BASE?

Whether a reaction follows an elimination or substitution pathway is dependent on whether the added reagent acts as a Lewis base or nucleophile, respectively. Recall from Section 17.11 that a Lewis base is an electron-pair donor, the same as a nucleophile. In broader definitions, a Lewis base can donate its electron pair to something other than H⁺, though this is also acceptable. By this definition, both a nucleophile and Lewis base can form new covalent bonds. So what's the difference?

The answer is nothing. We usually invoke the term Lewis base when there is a proton abstraction process, as in an elimination reaction, and the term nucleophile in all other cases. So, a Lewis base can be a good nucleophile if there are no sterically bulky groups around the electron donor.

RELATED EXERCISE: 25.44

The mechanism for this substitution reaction involves attack of the nucleophilic alkoxide on the weakly electrophilic carbon atom bearing the halide. Note that strong bases are typically good nucleophiles. The carbon atom bearing the halogen is made more electrophilic by the inductive effects of the halogen. This reaction type is called a *nucleophilic substitution (S$_N$) reaction*. The reaction goes through a trigonal-planar carbon state with elongated bonds for the incoming alkoxide and the outgoing halide. Eventual loss of Br⁻ yields the ether. The reaction shown in Equation 25.17 can be further defined as an S$_N$2 reaction, since the rate-determining step requires the addition of the alkoxide in concert with the loss of halide.

$$[25.17]$$

The β-elimination reaction (Equation 25.12) is in fact in competition with the substitution reaction. Generally, *S$_N$2 reactions are in competition with E2 reactions.* Which mechanism predominates in the reaction is dictated by a variety of factors:

1. The size of the alkoxide: the bulkier the alkoxide, the more elimination is favoured. ▶ FIGURE 25.20 illustrates this change in size. Alkoxides such as that generated from 2-methyl-2-propanol (*tert*-butanol) favour E2 because of their large size, while those generated from methanol favour S$_N$2.

2. The accessibility of the carbon bearing the halogen: the more accessible it is, the more substitution is favoured.

3. The accessibility of the hydrogen atom in the β-position: the more accessible it is, the more elimination is favoured.

Any alkyl groups positioned off the α-carbon of the haloalkane tend to favour β-elimination. This is particularly true of the larger halogens, Br and I.

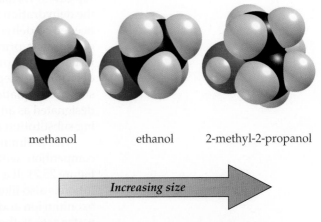

methanol   ethanol   2-methyl-2-propanol

Increasing size

▲ FIGURE 25.20 **Size increases with substitution.**

SAMPLE EXERCISE 25.6 Predicting substitution or elimination

What is the major organic product from the following reaction? Indicate whether this is an S_N2 or E2 reaction.

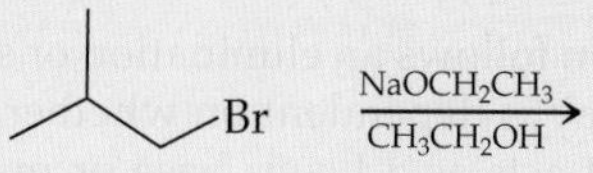

$$\xrightarrow[\text{CH}_3\text{CH}_2\text{OH}]{\text{NaOCH}_2\text{CH}_3}$$

SOLUTION

Analyse We are given the reaction of a haloalkane with a strong base also capable of acting as a good nucleophile. We are asked to determine the product and the mechanism.

Plan We need to consider both elimination and substitution products in our deliberation, as well as the nature of the base and whether a β-hydrogen is present and accessible.

Solve The elimination and substitution products are:

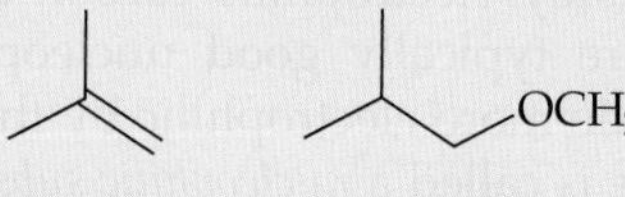

elimination
product

substitution
product

The α-carbon in the starting bromoalkane is essentially unhindered despite the bulk of the two methyl groups and the bromine atom. This implies that substitution would occur readily. In contrast, the β-hydrogen is more restricted by the same steric groups. This factor suggests that substitution predominates. The substitution follows an S_N2 mechanism.

The practice exercise below shows how the yield of alkene may be optimised.

PRACTICE EXERCISE

What is the major organic product from the following reaction?

$$\xrightarrow[\text{CH}_3\text{CH}_2\text{OH}]{\text{NaOCH}_2\text{CH}_3}$$

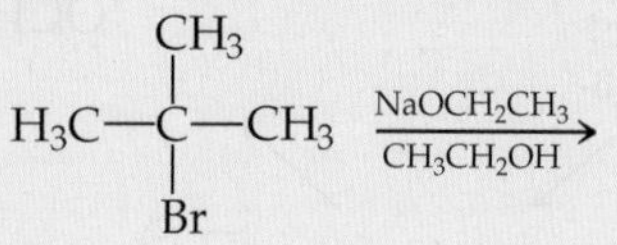

Answer: 2-methylpropene

(See also Exercises 25.50, 25.51.)

E1 and S_N1 Reactions

In Section 25.4 we discovered that generating a carbocation intermediate leads rapidly to formation of the alkene. The stability of the carbocation formed in the dehydration is also the driving force for that reaction. That is, the conditions needed to dehydrate a tertiary alcohol are much milder than those needed to dehydrate a primary alcohol, due to the relative stabilities of the intermediate tertiary and primary carbocations, respectively. In this case, the rate-determining step is the loss of water to form the carbocation intermediate as shown in Equation 25.5. As such, this elimination reaction is *unimolecular* and is designated as an **E1** mechanism. A summary of this mechanism and its competing substitution reaction is shown in ▶ **FIGURE 25.21**.

The β-elimination reaction, shown as route 2 in Figure 25.21, is also in competition with a nucleophilic substitution reaction, shown as route 1 in Figure 25.21. If a nucleophile is present within the reaction mixture, then substitution is also likely. The equilibrium arrows shown in Figure 25.21 between the oxonium ion and the carbocation in fact display this very principle. In this case, water acts as the leaving group in one reaction direction (from top to bottom) and as the nucleophile in the other direction (from bottom to top). The result of

FIGURE 25.21 Substitution and elimination reactions. The rate-determining step is the generation of the carbocation intermediate common to both reactions.

the latter reaction, however, leads only to formation of more of the starting alcohol. However, what if another nucleophile, such as Cl⁻, is also present? The result is the formation of a haloalkane. Since the rate-determining step is still the same (and unimolecular), this type of substitution is known as an **S$_N$1** reaction.

As opposed to the typical inversion of stereochemistry seen in S$_N$2 reactions, S$_N$1 reactions are indiscriminant, leading to a racemic mix—or at the very least a less optically pure compound than the starting material (∞ Section 23.4, 'Optical Activity: Measuring Chirality'). This is because in the S$_N$1 mechanism the carbocation intermediate is sp^2 hybridised and planar (▼ **FIGURE 25.22**). A nucleophile can attack the carbocation from either face leading to retention or inversion of stereochemistry.

▼ **TABLE 25.4** summarises the important parameters for substitution and elimination reactions discussed in this chapter.

∞ Review this on page 934

> ⚠ **CONCEPT CHECK 9**
> Is the reaction of 2-chloro-2-methylpropane with hydroxide S$_N$1, S$_N$2, E1 or E2?

▲ **FIGURE 25.22 S$_N$1 Racemisation.**

TABLE 25.4 • **General patterns of reactivity for substitution and elimination reactions at a sp^3 carbon**

Property	S_N2	S_N1
Rate of reaction	$1° > 2° \gg 3°$	$3° > 2° \gg 1°$
Reason:	Steric	Carbocation stability
Products	Substitution (major)	Substitution (major)
	E2 (minor)	E1 (major)
Stereochemistry	Stereospecific, inversion	Racemic product
Kinetics	2nd order (substrate, base)	1st order (substrate)
Leaving group	Important	Important
Nucleophile effect on rate	Important	No effect

	E2	E1
Rate of reaction	$3° > 2° \gg 1°$	$3° > 2° \gg 1°$
Cause:	Steric	Carbocation stability
Products	Alkene (major),	Most stable alkene
	S_N2 (minor)	
Kinetics	2nd order (substrate, base)	1st order (substrate)
Leaving group	Important	Important
Effect of base on rate	Strongest base fastest	No base necessary

SAMPLE INTEGRATIVE EXERCISE **Putting concepts together**

An unknown organic product was characterised by combustion analysis. You find the compound contains 59.96 % carbon and 13.42 % hydrogen and the remainder is oxygen.

(a) Determine the empirical formula of the compound. **(b)** If the molar mass is found to be 60.09 g mol^{-1}, draw and name all possible constitutional isomers of the compound and identify the functional groups present. **(c)** Your compound is not very soluble in water and has a boiling point close to room temperature; suggest the identity of the compound. **(d)** How would you synthesise this compound?

SOLUTION

Analyse This question provides some data which we use to obtain the formula of an unknown compound. Using Lewis structures we represent the structure of the possible compounds and, with the aid of some observations of the physical properties, we are asked to suggest the identity of the compound.

Plan Combustion analysis gives the empirical formula of a compound. This, together with a molar mass, enables us to determine the molecular formula from which we can construct all constitutional isomers. This allows us to identify the functional groups present and relate them to likely physical properties of the material.

Solve

(a) The percentage of oxygen in the compound is 100% − (59.96 + 13.42)% = 26.62%. The number of moles of each element in 100 g of the compound is;

$$\text{Moles of C} = \frac{59.96 \text{ g}}{12.01 \text{ g mol}^{-1}} = 4.992 \text{ mol C}$$

$$\text{Moles of H} = \frac{13.12 \text{ g}}{1.008 \text{ g mol}^{-1}} = 13.31 \text{ mol H}$$

$$\text{Moles of O} = \frac{26.62 \text{ g}}{10.00 \text{ g mol}^{-1}} = 1.664 \text{ mol O}$$

Dividing by the smallest number of moles present (1.664) gives the whole number ratio of the elements.

$$\text{Moles C : H : O} = \frac{4.992 \text{ mol}}{1.664 \text{ mol}} : \frac{13.31 \text{ mol}}{1.664 \text{ mol}} : \frac{1.664 \text{ mol}}{1.664 \text{ mol}} = 3 : 8 : 1$$

So the empirical formula is C_3H_8O.

(b) The molar mass of C_3H_8O is 60.09 g mol^{-1} which is the same as that determined for the unknown compound so we can conclude the molecular formula of the compound is also C_3H_8O. There are three possible constitutional isomers.

$$H_3C-CH_2-CH_2-OH$$

1-propanol

$$\underset{\underset{OH}{|}}{H_3C-CH-CH_3}$$

2-propanol

$$H_3C-CH_2-O-CH_3$$

methoxyethane

The first two structures are alcohols (a primary alcohol and secondary alcohol, respectively) and the final structure is an ether.

(c) Boiling point and solubility are directly related to the intermolecular forces present. As all the molecules have the same formula we assume the strength of the dispersion forces are similar in the three cases. Two compounds contain an alcohol functional group which can form hydrogen bonds and the third compound is an ether which, when pure, shows no hydrogen bonds between molecules. As a consequence of the hydrogen bonds the alcohols are likely to display an appreciably higher boiling point and greater solubility in water than the ether. The unknown compound is likely to be methoxyethane.

(d) A Williamson ether synthesis of iodomethane and ethanol in the presence of sodium metal would form methoxyethane. Note that forming the product from iodoethane and methanol will lead to the formation of ethene as a side product. This occurs through an elimination of HI using the sodium methoxide generated *in situ* as a base.

MY WORLD OF CHEMISTRY

POLYMERISATION VERSUS MACROCYCLISATION

The preparation of cyclic ethers (Section 25.3) has always involved two competing processes: the desired macrocyclisation and and unwanted polymerisation. ▶ FIGURE 25.23 illustrates these two processes using the crown ether, 18-crown-6, as the example. Intramolecular reaction of the OH group on the bromide in the presence of base yields the crown ether, 18-C-6, via substitution. Intermolecular reaction of the OH group on the bromide of a *second* molecule in the presence of base yields the polyether, also via substitution. Since both processes are competing, the result has always been a low (<20%) isolated yield of the macrocycle. Intramolecular processes are usually faster than intermolecular processes because of the proximity of reactive functional groups; however, this is true only of small molecules. In the example of crown ether synthesis shown in Figure 25.23 the longer the starting acyclic polyether, the lower the yield of the crown ether.

We can, however, take advantage of the affinity of polyethers for metal cations to bias the reaction outcome. In this case, the spherical cation acts as a *template* to promote the formation of the cyclic crown ether by allowing the polyether to wrap around the cation (▼ FIGURE 25.24). In this orientation, the carbon bearing the bromine atom and the OH group are brought into close proximity, enabling the intramolecular substitution reaction to occur in preference to the polymerisation upon the addition of base—for example, hydroxide ion. Removal of the cationic template yields the macrocycle in 60–80% yield. This very practical reaction is analogous to the

approach taken by nature in the formation of highly complex and functioning molecules such as DNA, in which a template strand is necessary to build the double helix using DNA polymerases I and II (⬯ Section 29.4, 'Nucleic Acids and DNA').

RELATED EXERCISE: 25.36

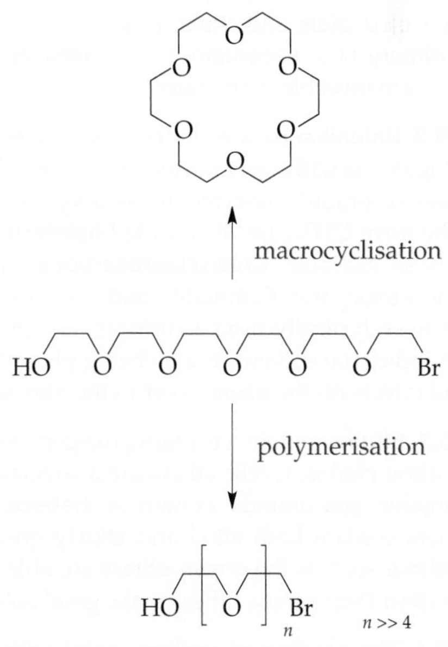

▲ FIGURE 25.23 **Competition between (formation of) a crown ether and a polymer.**

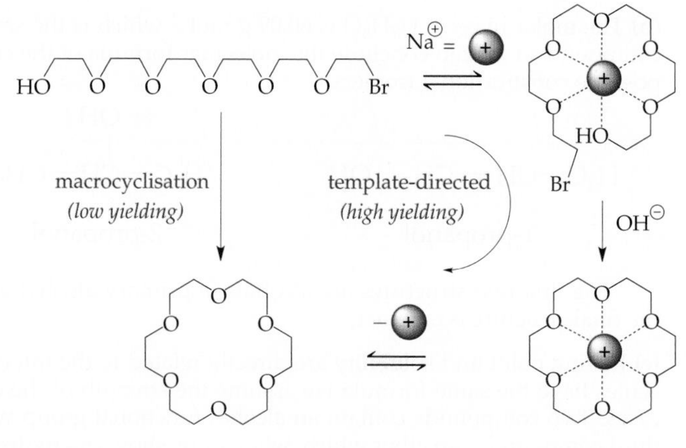

▶ **FIGURE 25.24 The template effect.**

CONCEPT CHECK 10

We have just looked at the competing reactions of polymerisation and macrocyclisation in My World of Chemistry. How would concentration—that is, concentrated versus dilute solutions of the acyclic polyether—favour either of these reactions?

CHAPTER SUMMARY AND KEY TERMS

SECTION 25.1 Alcohols are hydrocarbon derivatives containing one or more OH groups. Alcohols are more **miscible** with water than ethers of comparable size, due to their ability to hydrogen-bond as a result of the **lone pairs** on oxygen. Alcohols that have two or three OH groups, such as ethene glycol and glycerol, are called **diols** and **triols**, respectively. Alcohols are classed as **primary** (1°), **secondary** (2°) or **tertiary** (3°). Short-chain alcohols are miscible with water.

SECTION 25.2 Haloalkanes are hydrocarbons containing one or more halogens, usually generalised as RX (X = F, Cl, Br, I). They are also commonly referred to as alkyl halides. Compounds of the form CHX_3 are often called **haloforms**, of which chloroform is an example. **Chlorofluorocarbons**, or **CFCs**, are odourless, non-toxic, non-flammable and non-corrosive compounds that were typically used as refrigerants, propellants in aerosols and industrial solvents before being phased out due to their harmful effects on the ozone layer in the atmosphere.

SECTION 25.3 Ethers contain a bridging oxygen atom between two hydrocarbon chains. Cyclic ethers are a subclass of a large group of organic compounds known as **heterocyclic compounds**. Ethers contain both alkyl and **alkoxy** groups. Larger **cyclic polyethers** such as the **crown ethers** are able to complex metal ions within their cavity. Ethers make good solvents.

SECTION 25.4 The reaction of sodium metal with an alcohol leads to an **alkoxide**. Alkoxides are as strong a base as hydroxide. Potassium *tert*-butoxide and sodium ethoxide are examples cited frequently in this chapter. Alcohols act as a weak base, generating an oxonium ion under acidic conditions. Alcohols can be converted into haloalkanes by treatment of a halogenating agent, usually under acidic conditions. In this reaction, the OH group of the alcohol is substituted by a halogen in a **nucleophilic substitution (S_N) reaction**. Alcohols can also dehydrate to form alkenes.

SECTION 25.5 Nucleophilic substitution reactions of haloalkanes proceed by attack of the nucleophile at the carbon bearing the halogen. Because of the difference in electronegativity between carbon and any of the halides, this bond is polarised such that the carbon becomes mildly electrophilic. If the tranisition state that leads to the product involves both the haloalkane and the nucleophile in the rate-determining step, the reaction is classed as an **S_N2** or bimolecular (2), nucleophilic (N), substitution (S) reaction. The reaction of haloalkanes with **amines** leading to **alkyl ammonium salts,** is one such example. **Nitriles** are formed when haloalkanes are reacted with cyanide.

SECTION 25.6 Alkenes are readily formed from haloalkanes through a process called a **β-elimination reaction**. An **elimination reaction** involves the removal of a small group from a larger molecule. An **E2** reaction is a bimolecular (2) elimination (E) reaction. The alkene that forms in preference is the one that follows **Zaitsev's rule**, which states that the more substituted double bond predominates.

SECTION 25.7 Substitution and elimination reactions are competing reactions. Bimolecular substitution reactions (S_N2) compete against bimolecular elimination reactions (E2); which mechanism predominates in the reaction is dictated by a variety of steric factors. Tertiary alcohols undergo unimolecular reactions in which the rate-determining step is the elimination of water leading to a carbocation intermediate. In this case **E1** or **S_N1** reactions are likely depending on the reaction conditions.

KEY SKILLS

- Understand how to derive the names of simple alcohols, ethers and haloalkanes. Learn the common names for the simplest cases. Learn how to classify alcohols as 1°, 2° or 3°. (Sections 25.1, 25.2 and 25.3)
- Learn the basic functional group manipulation shown in Figure 25.14. (Section 25.4)
- Be able to explain the difference between S_N1, S_N2, E1 and E2 mechanisms and have a qualitative understanding of when one predominates over another to predict reaction outcomes. (Sections 25.5, 25.6 and 24.7)

KEY EQUATIONS

- Substitution

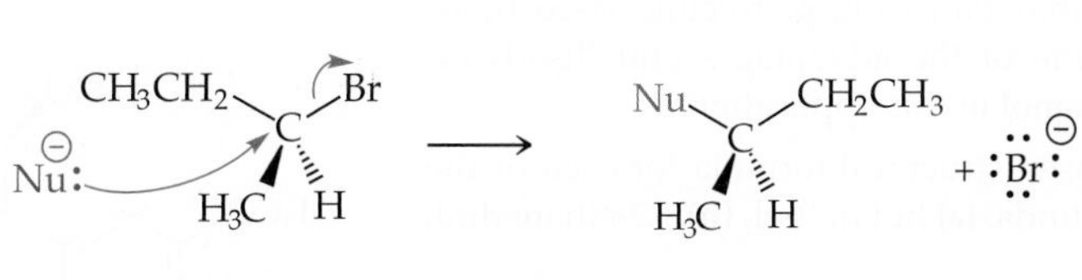

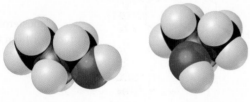

$$[25.9]$$

- β-Elimination

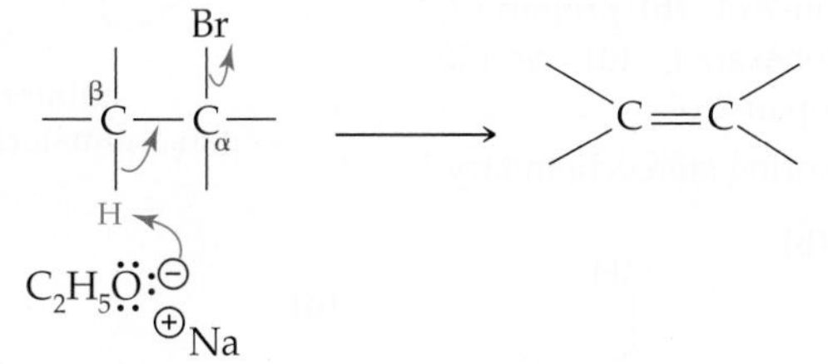

$$[25.13]$$

EXERCISES

VISUALISING CONCEPTS

25.1 Propan-1-ol ($CH_3CH_2CH_2OH$) and propan-2-ol (also known as isopropyl alcohol ($CH_3)_2CHOH$), whose space-filling models are shown, have boiling points of 97.2 °C and 82.5 °C, respectively. Explain why the boiling point of propan-1-ol is higher, even though both have the molecular formula of C_3H_8O. [Section 25.1]

propan-1-ol   propan-z-ol

25.2 What features must a molecule have to participate in hydrogen bonding with other molecules of the same kind? Which of the following compounds would you expect to **(a)** have the highest boiling point, **(b)** be least soluble in water? Explain. [Section 25.1]

$$CH_3OCH_3 \quad CH_3CH_2OH \quad CH_3C\equiv CH \quad \overset{\overset{\displaystyle O}{\|}}{H}COCH_3$$

(i)   (ii)   (iii)   (iv)

25.3 Piperonyl butoxide is used commercially as an additive to pyrethrum-based insecticides to enhance their effect. Copy the structure and highlight the atoms that constitute the cyclic ether and the acyclic polyether. [Section 25.3]

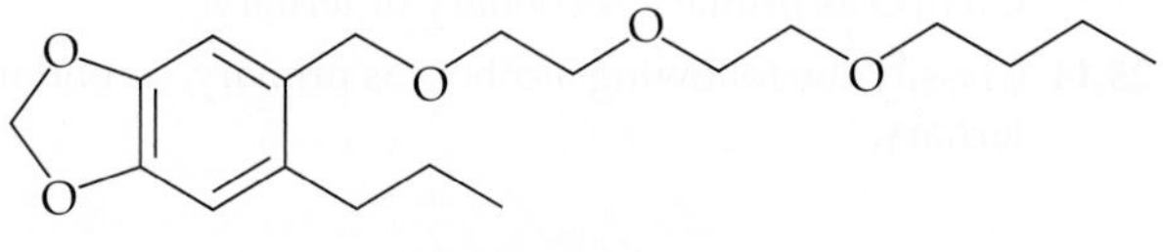

piperonyl butoxide

25.4 Draw the carbocation that triol 1 would form under the indicated conditions. Explain your answer. [Section 25.4]

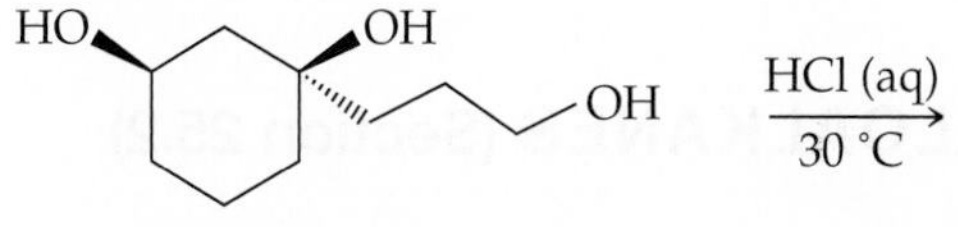

$$\xrightarrow[\text{30 °C}]{\text{HCl (aq)}}$$

1

25.5 How many distinct alkene products (including stereochemistry) are possible when the iodoalkane below undergoes E2 elimination? [Section 25.5]

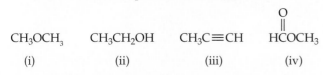

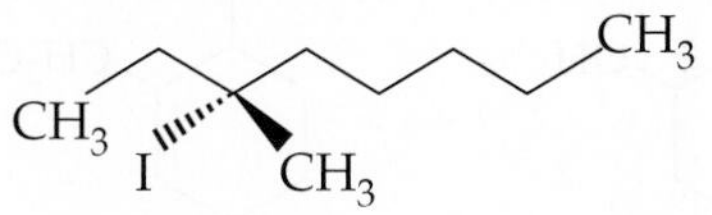

ALCOHOLS: STRUCTURE PROPERTIES AND NOMENCLATURE (Section 25.1)

25.6 Write the structural formulae for as many alcohols as you can think of that have the empirical formula C_3H_6O.

25.7 Two oxygen-containing compounds have the same empirical formula. One substance is a gas, the other is a liquid. How is it possible for two substances with the same empirical formula to have markedly different properties?

25.8 Explain why propan-1-ol is miscible with water yet hexan-1-ol is not.

25.9 Attention has focused on ethanol as either an alternative fuel to, or in combination with, petroleum-based fuels. Comment on some of the advantages and disadvantages of using ethanol in this application.

25.10 Write the condensed structural formula for each of the following compounds: **(a)** butan-2-ol, **(b)** 1,2-ethanediol, **(c)** diethyl ether.

25.11 Draw the structures of **(a)** hexan-2-ol, **(b)** propane-1, 2, 3-triol, **(c)** 4,4-dimethylcyclohexanol, **(d)** *cis*-1,2-cyclopentanediol, **(e)** 2-methylpropan-2-ol.

25.12 Name the following alcohols, ignoring stereochemistry:

(a)

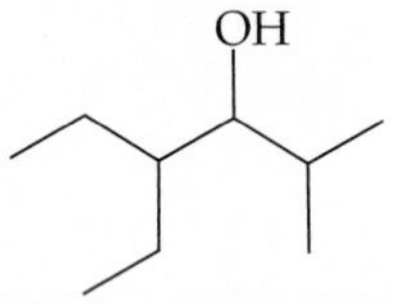

(b)

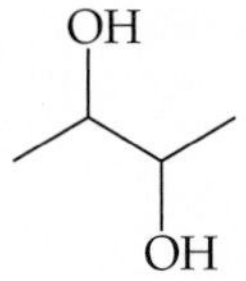

(c)

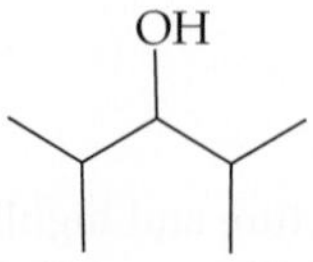

(d)

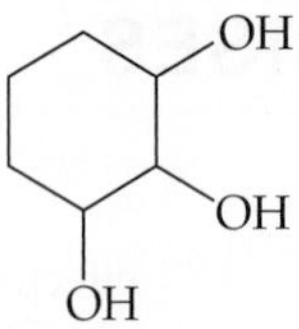

25.13 Classify the isomeric alcohols with molecular formula $C_4H_{10}O$ as primary, secondary or tertiary.

25.14 Classify the following alcohols as primary, secondary or tertiary.

(a)

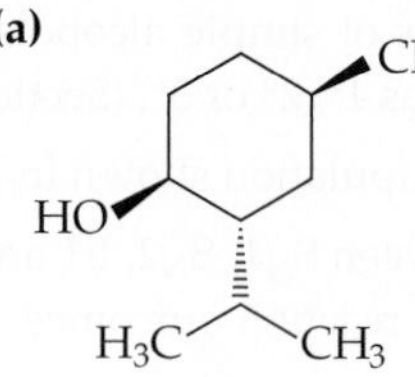

menthol

(b)

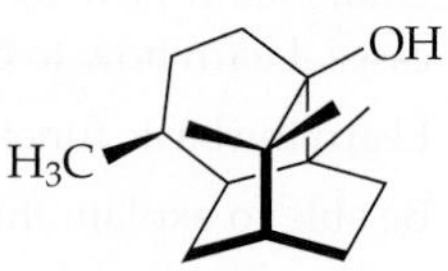

patchoulol

(c)

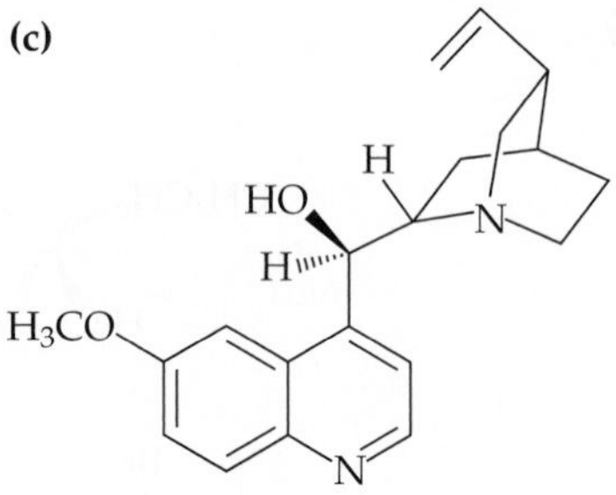

quinine
(an antimalarial drug)

(d)

chloramphenicol
(an antibiotic)

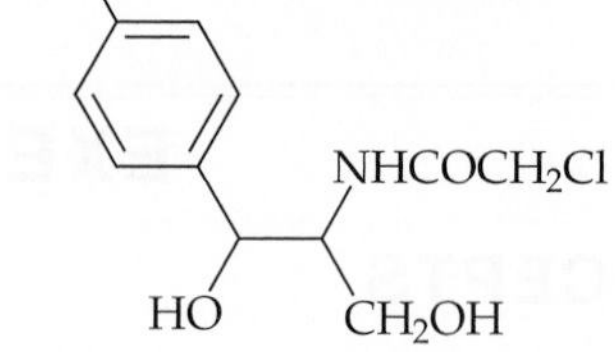

(e)

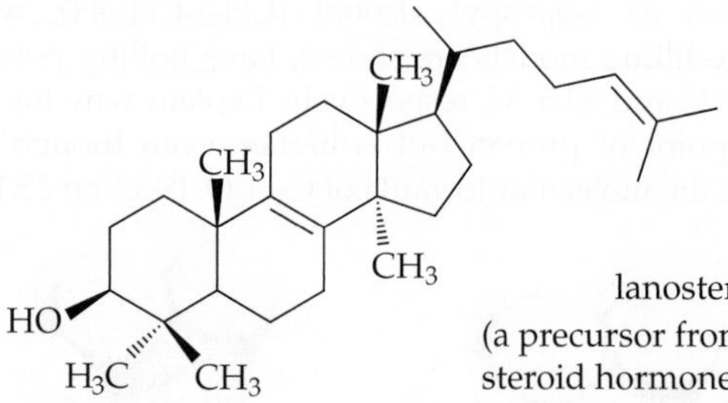

lanosterol
(a precursor from which all
steroid hormones are made)

HALOALKANES (Section 25.2)

25.15 Eight isomers have the molecular formula $C_5H_{11}Cl$. How many of these are tertiary chloroalkanes?

25.16 Give the IUPAC name for the following structures:

(a)

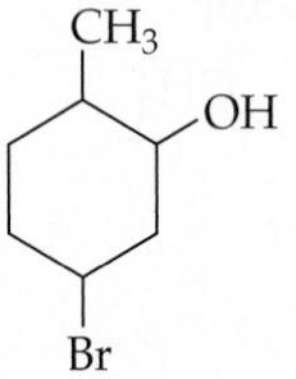

(b)

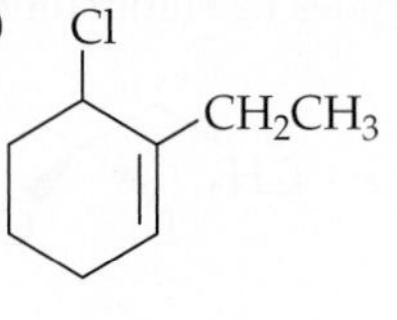

25.17 Describe the carbon–chlorine bond of chloroalkanes as polar or non-polar. Why?

ETHERS: STRUCTURE PROPERTIES AND NOMENCLATURE (Section 25.3)

25.18 Ethene glycol (HOCH$_2$CH$_2$OH), the major substance in antifreeze, has a normal boiling point of 198 °C. By comparison, ethanol (CH$_3$CH$_2$OH) boils at 78 °C at one atmospheric pressure. Ethene glycol dimethyl ether (CH$_3$OCH$_2$CH$_2$OCH$_3$) has a normal boiling point of 83 °C, and ethyl methyl ether (CH$_3$CH$_2$OCH$_3$) has a normal boiling point of 11 °C. **(a)** Explain why replacement of a hydrogen on the oxygen by CH$_3$ generally results in a lower boiling point. **(b)** What are the major factors responsible for the difference in boiling points of the two ethers?

25.19 **(a)** Give the empirical formula and structural formula for the cyclic ether containing four carbon atoms in the ring. **(b)** Write the structural formula for a straight-chain compound that is a structural isomer of your answer to part (a).

25.20 Why do ethers make good solvents, whereas alcohols usually make poor solvents in which to undertake a chemical reaction? When might an alcohol make an appropriate solvent?

25.21 Explain why the boiling point of ethanol (78 °C) is much higher than that of its isomer, dimethyl ether (–25 °C),

and why the boiling point of CH$_2$F$_2$ (–52 °C) is far above that of CF$_4$ (–128 °C).

25.22 Give names for the following ethers (ignore stereochemistry):

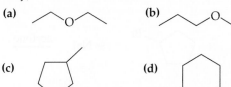

(a)   **(b)**

(c)   **(d)**

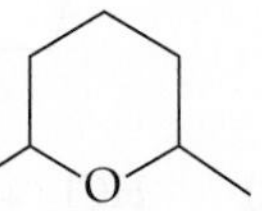

25.23 Write the structures for all the isomeric ethers with molecular formula C$_5$H$_{12}$O and give a name for each.

25.24 Which would you expect to be more soluble in water, cyclohexane or dioxane? Explain.

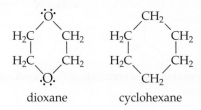

dioxane   cyclohexane

REACTIONS OF ALCOHOLS AND HALOALKANES (Sections 25.4 and 25.5)

25.25 Which of the following carbocations is most stable? Which is least stable?

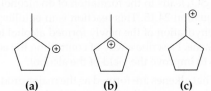

(a)   **(b)**   **(c)**

25.26 **(a)** Write a chemical equation for the substitution reaction of butan-2-ol with HBr. **(b)** Write a stepwise mechanism for this reaction. **(c)** Would butan-1-ol undergo the substitution reaction more quickly or more slowly than butan-2-ol?

25.27 Suggest a possible stable reaction product for each of the following reactions:

(a)

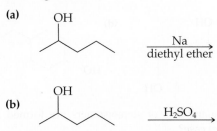

$$\xrightarrow[\text{diethyl ether}]{\text{Na}}$$

(b)

$$\xrightarrow{\text{H}_2\text{SO}_4}$$

25.28 Briefly explain why 2-methylpropan-2-ol makes an excellent solvent for oxidation reactions while propan-1-ol is a poor solvent choice.

25.29 Show how you could make each compound, starting with any alcohol of your choice.

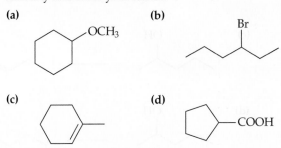

(a)   **(b)**

(c)   **(d)**

25.30 Suggest a plausible set of conditions that could be used to convert ethanol to diethyl ether.

25.31 What is wrong with the following synthesis?

$$\text{CH}_3\text{CH}_2\text{OH} + \text{H}^{\oplus} \longrightarrow \text{CH}_3\text{CH}_2\overset{\oplus}{\text{O}}\text{H}_2 \xrightarrow{\text{CH}_3\text{O}^{\ominus}} \text{CH}_3\text{CH}_2\text{OCH}_3$$

25.32 Sodium wire is normally used to 'dry' tetrahydrofuran before distillation and use in water-sensitive reactions. **(a)** Write an equation for the reaction of sodium metal with water. **(b)** This reaction is very exothermic and care should be taken as to how 'wet' the tetrahydrofuran is. What alternative to water might you use to destroy excess sodium, and what precautions should you take?

25.33 Write the structure of the major organic product expected from the following reactions:

(a)

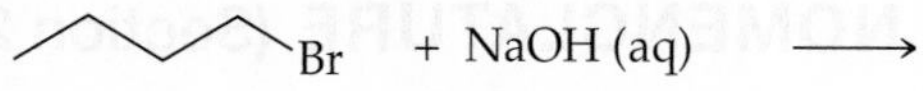

+ NaOH (aq) ⟶

(b)

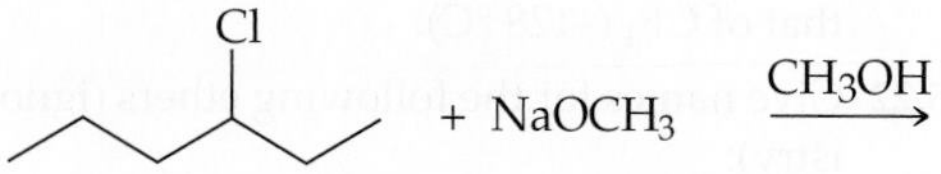

+ NaOCH₃ $\xrightarrow{CH_3OH}$

(c)

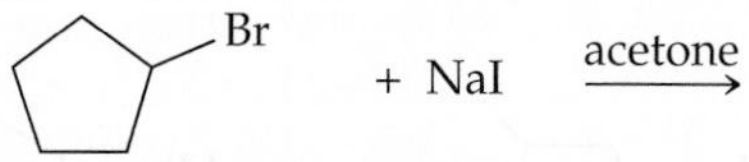

+ NaI $\xrightarrow{acetone}$

(d) HCl + CH₃CH₂NH₂ ⟶

(e)

+ NaCN ⟶

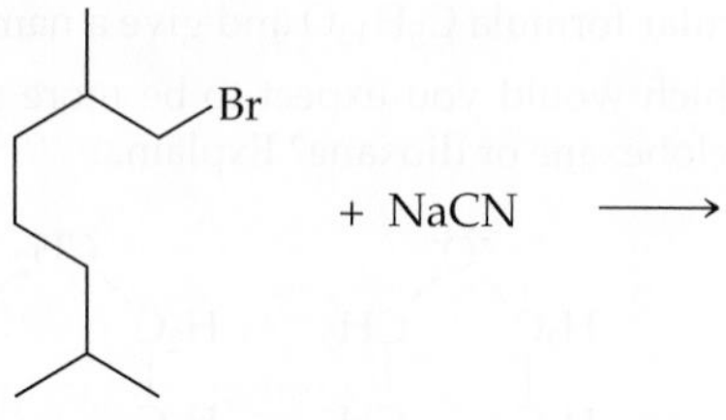

25.34 Suggest ways of conducting the transformations shown below.

(a)

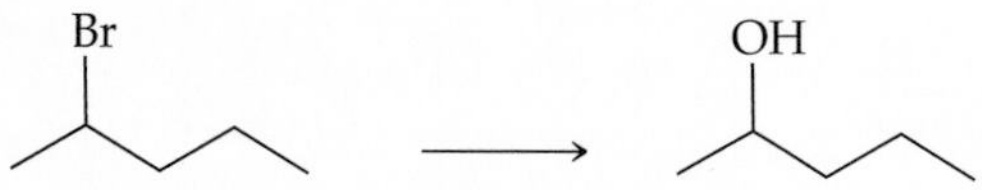

(b)

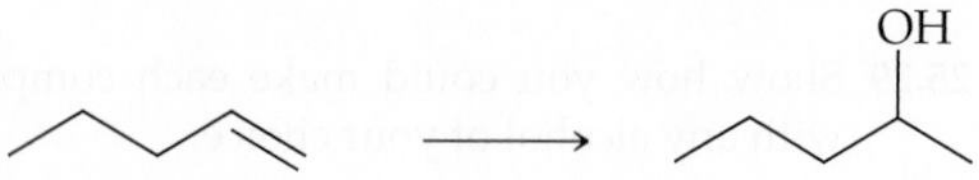

(c)

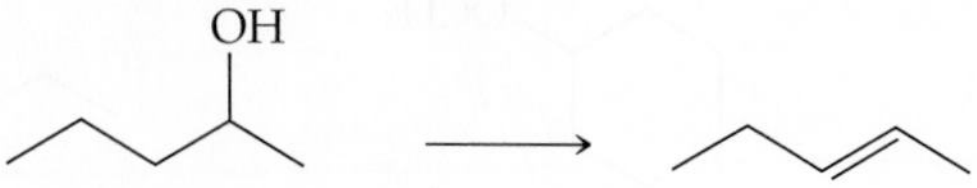

(d)

25.35 Provide the structure of the major organic product of the following reactions.

(a)

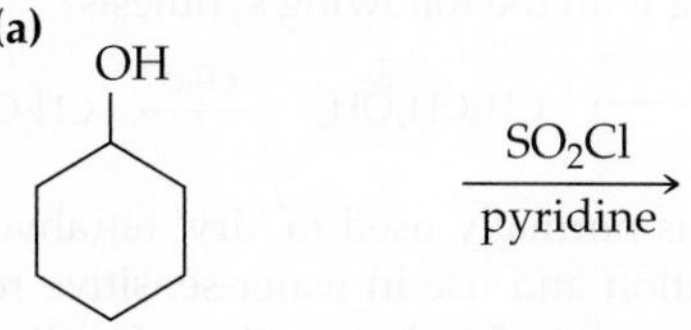

$\xrightarrow[\text{pyridine}]{SO_2Cl}$

(b)

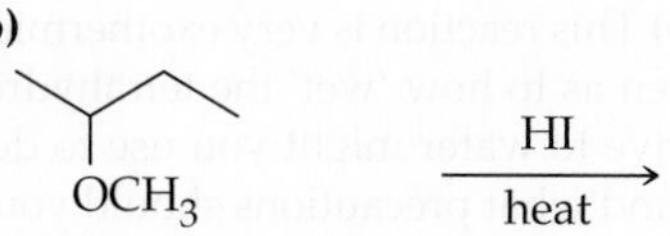

$\xrightarrow[\text{heat}]{HI}$

(c)

$\xrightarrow{Na \overset{\oplus}{} \overset{\ominus}{SCH_3}}$

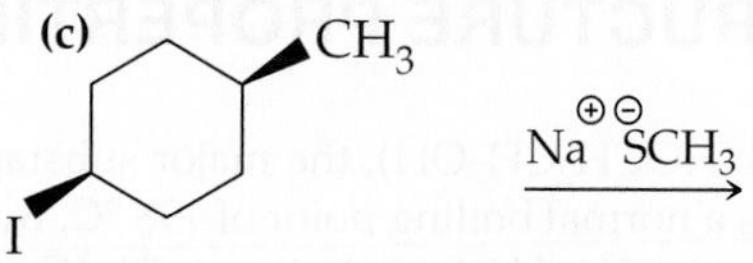

25.36 In the My World of Chemistry box on page 1023, a competition between macrocyclisation and polymerisation is discussed. **(a)** How might you optimise macrocyclisation without the addition of a template? **(b)** Would any metal ion be useful for serving as a template?

25.37 (a) Which of the following is *not* a nucleophile: (i) water (ii) cyanide ion (iii) methoxide ion (iv) $^{\oplus}CH_3$ (v) R≡C:$^{\oplus}$? **(b)** Rank the following species in order of increasing nucleophilicity in protic solvents: $CH_3CO_2^-$, HO^-, H_2O.

25.38 (a) Which of the following is the *best* leaving group: (i) fluoride, (ii) iodide, (iii) hydroxide, (iv) chloride or (v) bromide? **(b)** Why, in a polar protic solvent, is iodide a better nucleophile than fluoride?

25.39 Describe a synthesis of A starting from 3-methylcyclohexene-3-ol.

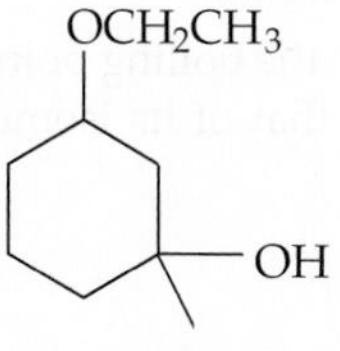

A

25.40 What is the major product from the acid-catalysed hydration of 2-methylpent-2-ene?

25.41 The addition of water to an alkene in the presence of H_2SO_4 leads to the formation of an alcohol, as shown in Equation 24.15. This reaction is in equilibrium with the dehydration of the newly formed alcohol to re-form the alkene. Describe a way of controlling the equilibrium so as to improve the yield of the alcohol.

25.42 What alkenes are formed as the major products from the dehydration of the following alcohols?

(a)

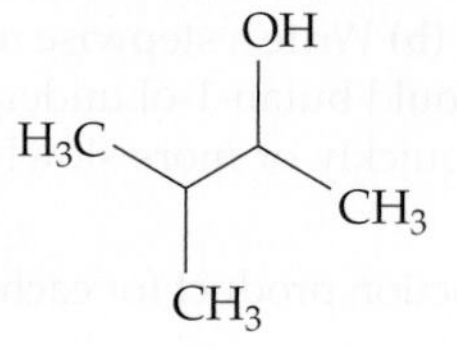

(b)

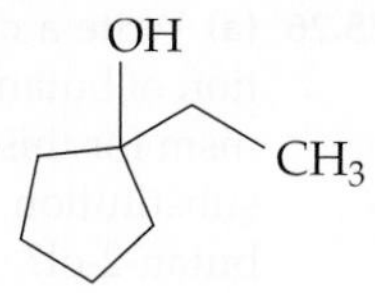

(c)

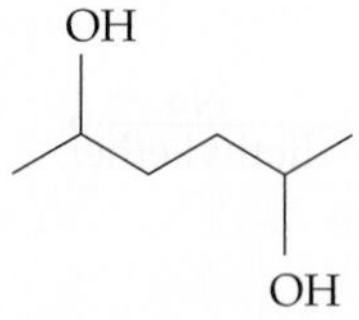

(d)

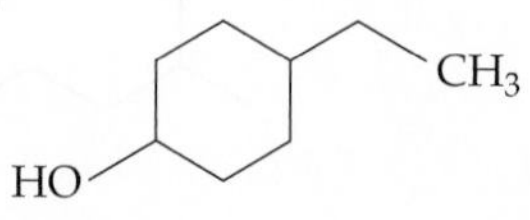

How many sets of stereoisomers are formed in each product case?

HALOALKANES TO ALKENES: β-ELIMINATION (Section 25.6)

25.43 When sodium ethoxide (NaOCH$_2$CH$_3$) reacts with 2-chlorobutane, a haloalkane and three unsaturated products are likely (including stereochemistry). **(a)** Draw them. **(b)** Which ones are likely to be formed in higher yield?

25.44 Which base, sodium ethoxide or potassium *tert*-butoxide, would you employ for the efficient dehydrohalogenation of 1-bromo-2-ethylbutane. Why?

25.45 The dehydration of norborneol yields none of the Zaitsev product. Why is this so? [*Hint:* Think about the geometry of the products.]

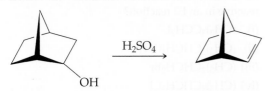

25.46 **(a)** Define bimolecular. **(b)** Using diagrams, give an example of a bimolecular reaction using diagrams.

25.47 What is the structure of the major alkene product formed when 2-methylpropane is reacted consecutively with Br$_2$, in the presence of light, and then potassium *tert*-butoxide? [*Hint:* You will need to recall the regioselectivity of free-radical bromination, covered in Chapter 22.]

25.48 Insert the missing reagents or products in the multi-step reactions shown below.

(a)

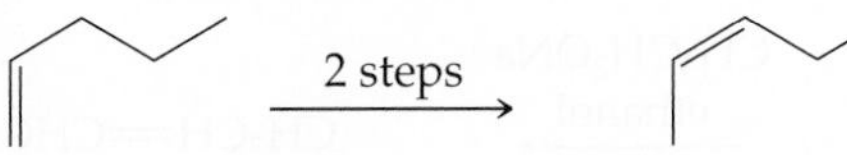

$$\xrightarrow{\text{2 steps}}$$

(b)

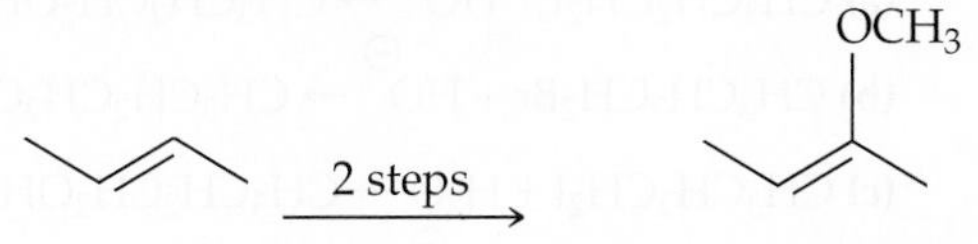

$$\xrightarrow{\text{2 steps}}$$

(c)

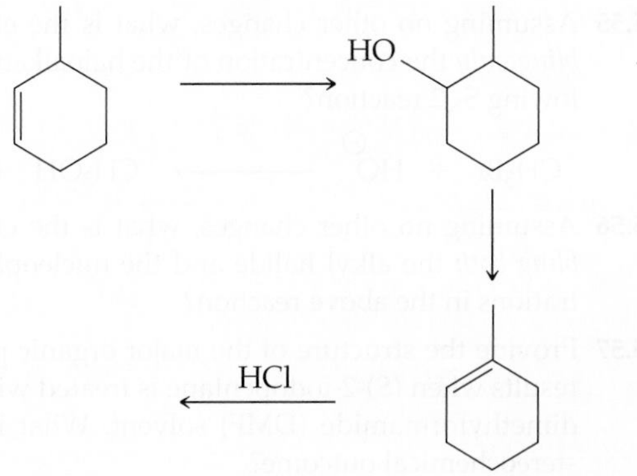

(d)

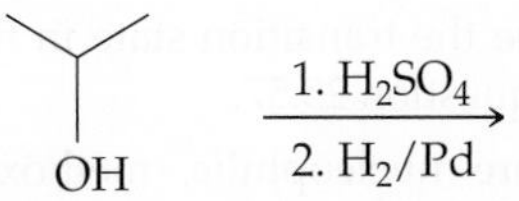

$$\xrightarrow[\text{2. H}_2/\text{Pd}]{\text{1. H}_2\text{SO}_4}$$

(e)

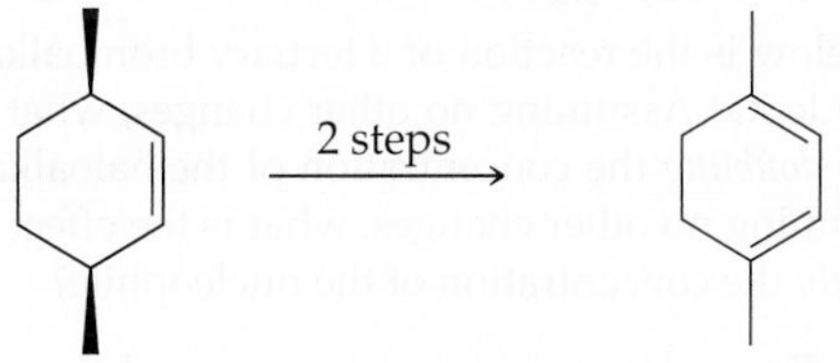

$$\xrightarrow{\text{2 steps}}$$

(f)

$$\xrightarrow[\text{2. KOBu}^+ \text{ (1 equiv.)}]{\text{1. excess HBr}}$$

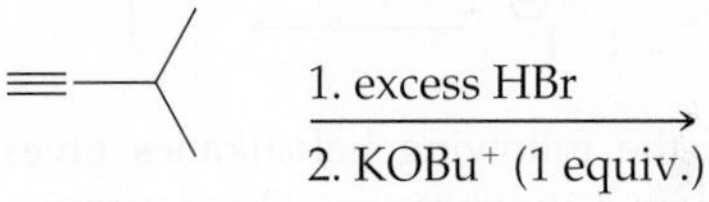

SUBSTITUTION VERSUS ELIMINATION (Section 25.7)

25.49 When haloalkanes react with alkoxides a competition exists between elimination and substitution reactions. Which product is formed can be biased by the judicious choice of haloalkane and alkoxide. The following ethers can only be prepared in high yield by the combination of one alkoxide ion and one haloalkane.

What is the correct combination in each case?

(a)

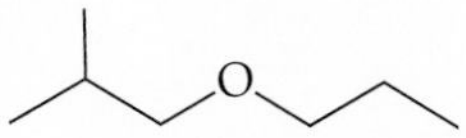

(b)  C(CH$_3$)$_3$

(c) OCH$_2$CH$_3$

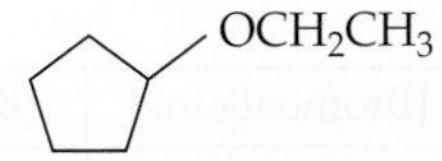

25.50 When sodium ethoxide (NaOCH$_2$CH$_3$) reacts with an haloalkane, two types of products are likely. Illustrate this point by drawing the *major* organic products when sodium ethoxide acts as a base or as a nucleophile in its reaction with 2-chlorobutane.

25.51 **(a)** What is the substitution product of the reaction shown below? **(b)** What is the elimination product of the reaction shown below?

$$\text{CH}_3\text{CH}_2\text{Br} + \overset{\ominus}{\text{HO}} \rightarrow$$

25.52 **(a)** Would all primary alkyl iodides undergo S$_N$2 reactions with sodium cyanide at identical rates? Explain. **(b)** A student attempted to prepare 1-chlorobutane by treating 1-butanol with NaCl in acetone. Was the student successful? Explain

25.53 Which of the following haloalkanes gives the *slowest* S$_N$2 reaction? Why?

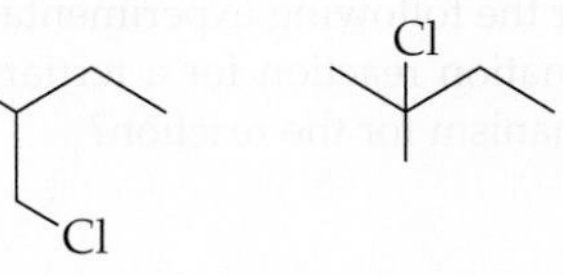

25.54 Which of the following S_N2 reactions is the *fastest*? Why?

(a) $CH_3CH_2CH_2I + HO^{\ominus} \rightarrow CH_3CH_2CH_2OH + I^{\ominus}$

(b) $CH_3CH_2CH_2Br + HO^{\ominus} \rightarrow CH_3CH_2CH_2OH + Br^{\ominus}$

(c) $CH_3CH_2CH_2I + H_2O \rightarrow CH_3CH_2CH_2OH + HI$

(d) $CH_3CHICH_3 + HO^{\ominus} \rightarrow CH_3CHOHCH_3 + I^{\ominus}$

25.55 Assuming no other changes, what is the effect of *doubling only* the concentration of the haloalkane in the following S_N2 reaction?

$$CH_3Br + HO^{\ominus} \longrightarrow CH_3OH + Br^{\ominus}$$

25.56 Assuming no other changes, what is the effect of *doubling both* the alkyl halide and the nucleophile concentrations in the above reaction?

25.57 Provide the structure of the major organic product that results when (S)-2-iodopentane is treated with NaCN in dimethyformamide (DMF) solvent. What is the likely stereochemical outcome?

25.58 Draw and describe the transition state in the S_N2 reaction described in question 25.57.

25.59 (a) Which is more nucleophilic, methoxide or *tert*-butoxide? Why? (b) Show the best way to prepare $CH_3OCH(CH_3)_2$ by an S_N2 reaction.

25.60 Below is the reaction of a tertiary bromoalkane with iodide. (a) Assuming no other changes, what is the effect of *doubling* the concentration of the haloalkane? (b) Assuming no other changes, what is the effect of doubling *only* the concentration of the nucleophile?

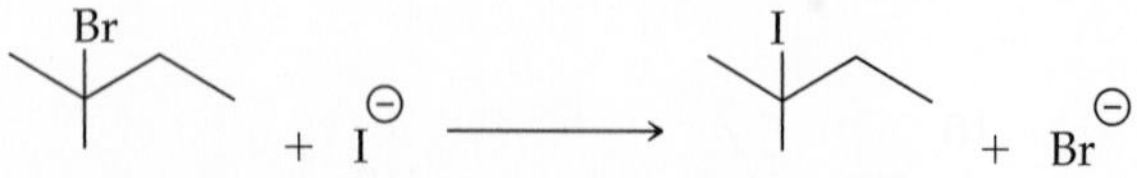

25.61 Which of the following haloalkanes gives the fastest S_N1 reaction: 1-iodopropane, 2-iodopropane, 1-bromopropane or 2-bromopropane?

25.62 (a) Does the concentration of the nucleophile have any effect on the rate of an S_N1 reaction? Explain. (b) Based on your answer, describe the rate law for an S_N1 reaction. (c) What type of solvents favour S_N1 reactions?

25.63 (a) Provide the structure of the major organic products that result in the reaction below (including stereochemistry). (b) What mechanism predominates?

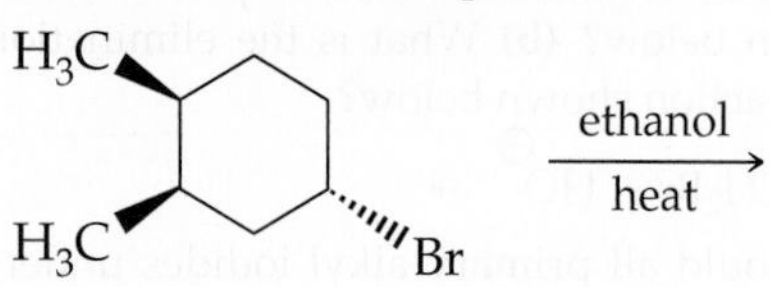

25.64 Explain why alkyl chloride undergoes S_N1 reactions even though it is a 1° halide? [*Hint*: Look at the intermediate carbocation.]

25.65 When 2-bromo-3-methyl-1-phenylbutane is treated with sodium methoxide, why is the major product 3-methyl-1-phenyl-1-butene?

25.66 When 1-iodo-1-methylcyclohexane is treated with sodium ethoxide, the more highly substituted alkene product predominates. When potassium *tert*-butoxide is used instead, the less substituted alkene predominates. Why?

25.67 (a) Which of the following halides is most reactive in an E2 reaction with sodium methoxide? (b) Which is least reactive in an E2 reaction?

(i) $(CH_3)_3CCH_2I$

(ii) $(CH_3)_2CHCHICH_3$

(iii) $(CH_3)_2CHCH_2Br$

(iv) $(CH_3)_2CHCH_2Cl$

25.68 Which of the following haloalkanes gives the fastest E2 reaction: 1-iodopropane, 2-iodopropane, 1-bromopropane or 2-bromopropane?

25.69 Supply the missing haloalkane reactant in the elimination reactions shown below.

(i) CH_3CH_2ONa $\xrightarrow[\text{heat}]{\text{ethanol}}$

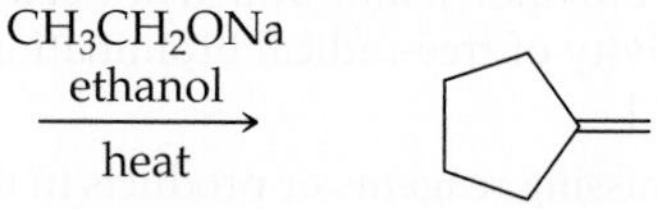

(ii) CH_3CH_2ONa $\xrightarrow[\text{heat}]{\text{ethanol}}$ $CH_3CH=CHCH_2CH_3$

(iii) CH_3CH_2ONa $\xrightarrow[\text{heat}]{\text{ethanol}}$

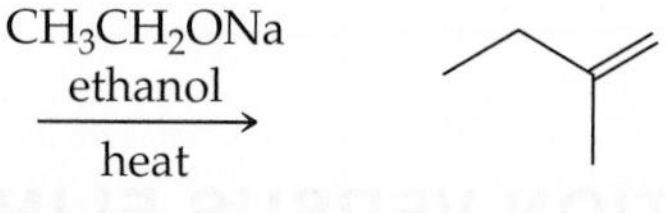

25.70 What is the major organic product of the following reaction? Why?

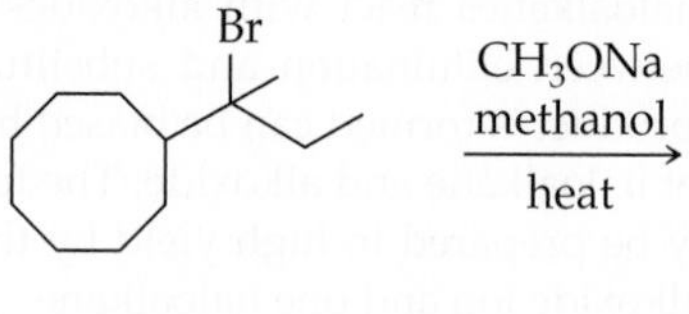

ADDITIONAL EXERCISES

25.71 Consider the following experimental data for the rate of an elimination reaction for a tertiary bromide. What is the mechanism for the reaction?

Experiment No.	[Bromoalkane]	[Base]	Rate
1	0.01	0.01	1
2	0.02	0.01	2
3	0.01	0.02	1

25.72 Which of the following bromides forms the most stable carbocation? Why?

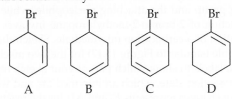

A   B   C   D

25.73 What is the major product formed in the following reactions and what mechanisms predominate?

(a)

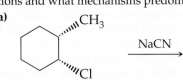

NaCN →

(b)

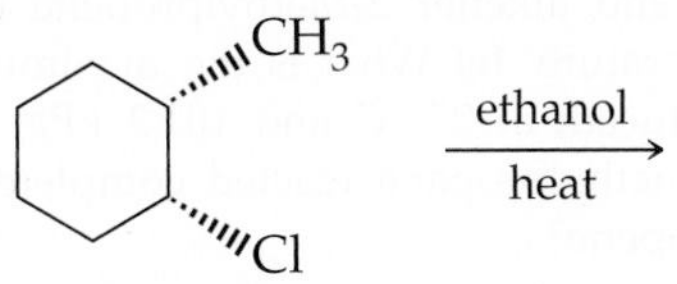

ethanol
heat →

(c)

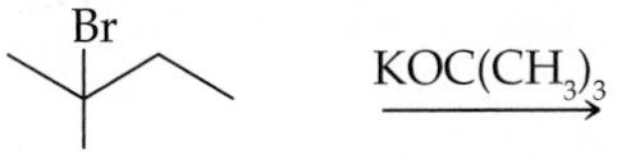

Br

KOC(CH₃)₃ →

(d)

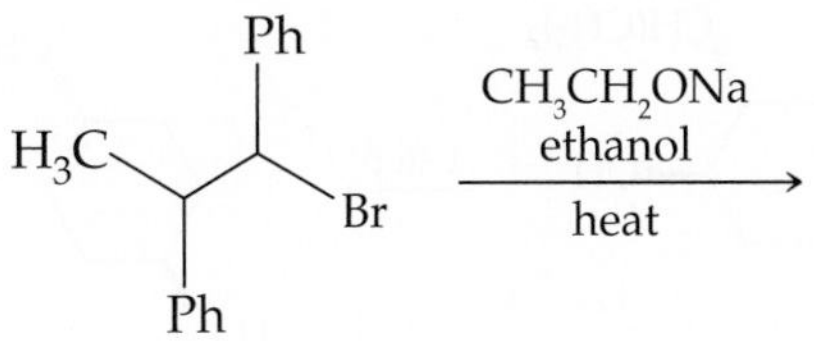

Ph
H₃C
Br
Ph

CH₃CH₂ONa
ethanol
heat →

(e)

OH

HCl →

25.74 **(a)** Dehydrohalogenation of 2-bromobutane in the presence of a strong base proceeds via which mechanistic pathways? **(b)** Why is the E1 not a likely mechanism when 1-chloropentane is heated in ethanol? **(c)** Predict the two most likely mechanisms (substitution and elimination) for the reaction of 2-iodohexane with sodium ethoxide and give the products of these reactions. **(d)** Can primary haloalkanes react by S_N2, S_N1, E2 and E1 mechanisms? Are any of these mechanisms prohibited? What conditions favour a particular mechanism?

INTEGRATIVE EXERCISES

25.75 This question relates to the hydration of 3,3-dimethylcyclohexene and the development of an appropriate mechanism to account for the formation of 1,2-dimethoxy-1-cyclohexanol.

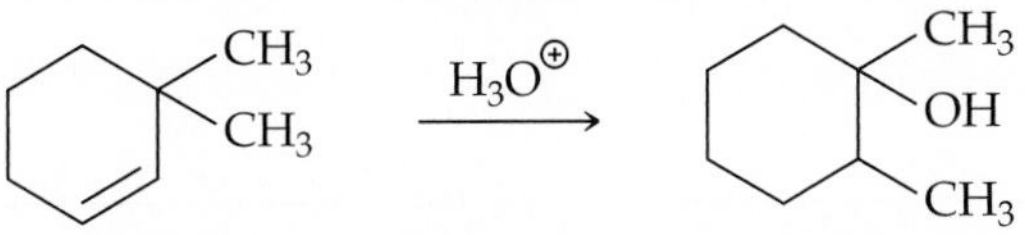

CH₃
CH₃

$H_3O^⊕$ →

CH₃
OH
CH₃

(a) Draw a mechanism leading to the first carbocation intermediate. **(b)** The next step involves a methyl migration from carbon 3 to carbon 2. Draw this intermediate and comment on why this is favoured. **(c)** Use water as a nucleophile and draw the next step of the mechanism. **(d)** Complete the mechanism to generate the product. What does this mechanism tell you about the requirement of acid?

25.76 Provide a detailed, step-by-step mechanism for the reaction shown below.

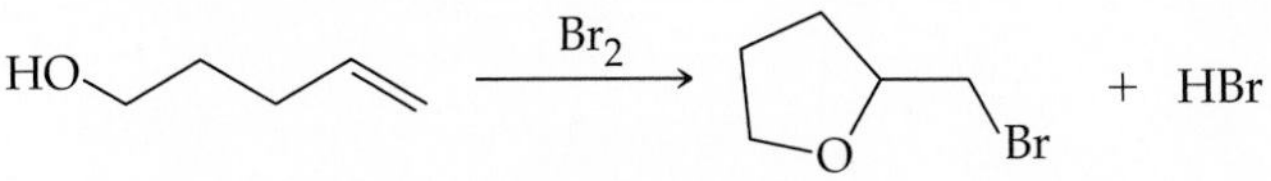

HO
Br_2 →

O
Br

+ HBr

25.77 Describe a way of converting but-2-ene into but-2-yne using a bromination as the first step.

25.78 Describe a way of converting hex-3-ene into hex-3-yne.

25.79 Propose a mechanism for the following reaction:

HO
OH

$H_3O^⊕$ →

O

+ H₂O

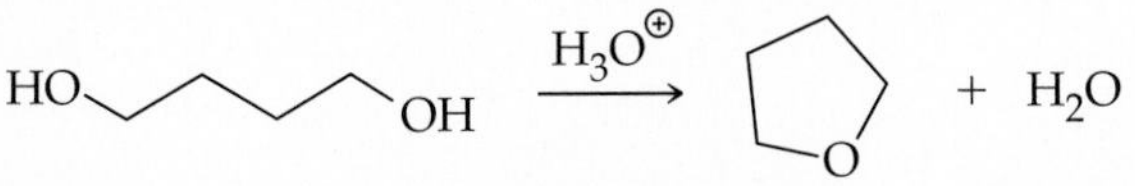

25.80 Propose a mechanism for the following reaction:

O

+ HBr →

Br
Br

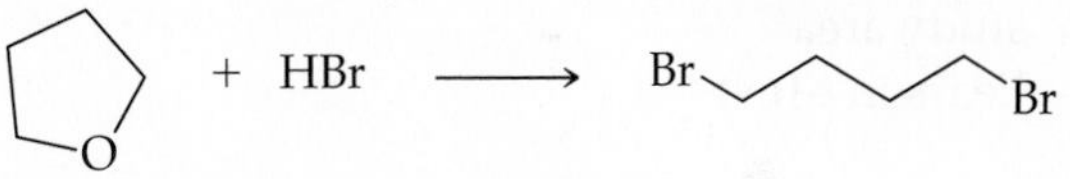

25.81 Starting with propan-1-ol as the only organic material, synthesise isopropyl propyl ether.

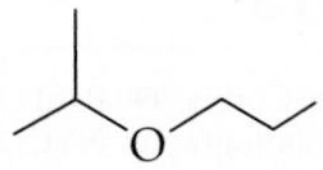

O

25.82 Describe how you would prepare this ether from cyclohexene.

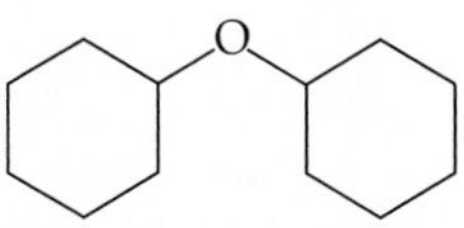

O

25.83 How would you accomplish the transformations shown below?

(a)

(CH₃)₃COH $\xrightarrow{\text{1 step}}$ (CH₃)₃CCl

(b)

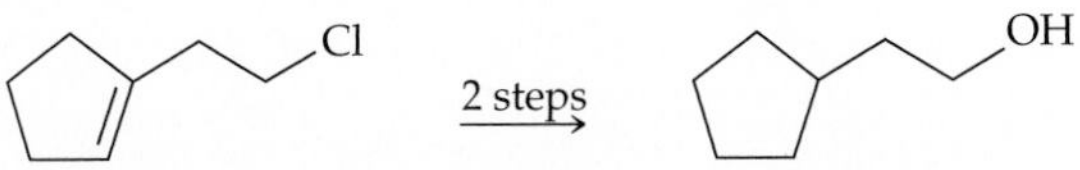

Cl

$\xrightarrow{\text{2 steps}}$

OH

(c)

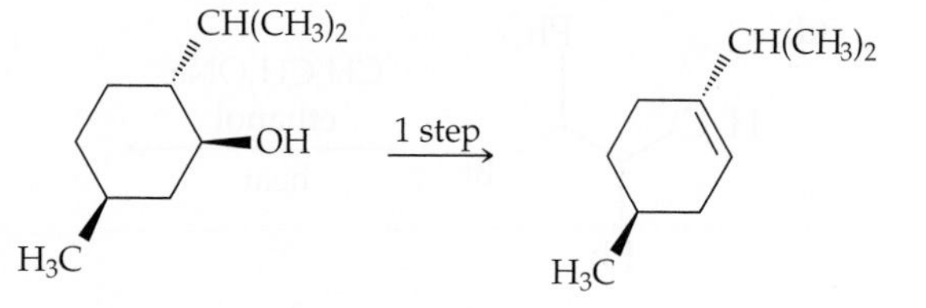

(d)

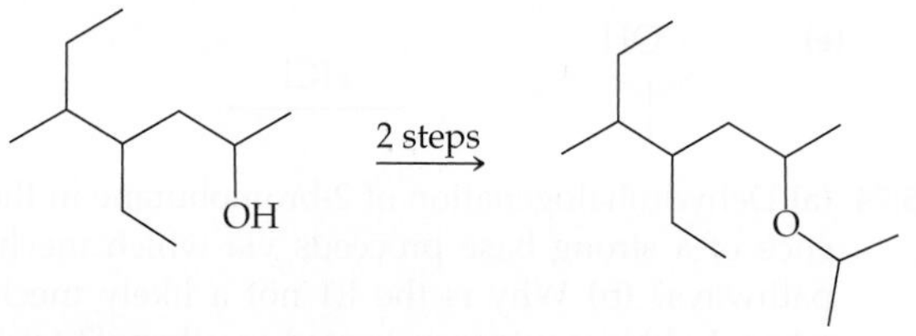

25.84 Consider the three alkenes: ethene, propene and 2-methylpropene. **(a)** For each alkene, draw the reaction showing the acid-catalysed hydration to form one major product. Include the carbocation intermediate in each case. **(b)** Write the systematic name of the alcohols produced. Classify the intermediate carbocations and the alcohol products as *primary, secondary* or *tertiary*. **(c)** What is the connection between the classification of the carbocations and the alcohols that arise from them?

25.85 In the S_N2 reaction of 1-bromobutane with hydroxide ions we imagine a smooth interchange between the leaving group and the attaching nucleophile, while the S_N1 reaction of 2-bromo-2-methylpropane involves the formation of a carbocation intermediate. Sketch the energy profile (similar to Figure 15.17) for the two mechanisms. Use labels to show which part of your sketch represents a transition state, which an intermediate and which energy changes represents E_a and ΔH for the reaction.

25.86 The conversion of 2-methyl-2-propanol to 2-bromo-2-methylpropane using concentrated hydrobromic acid occurs at room temperature. As the reaction proceeds, the reaction mixture separates into two layers. **(a)** Why does this occur? Treatment of the product, 2-bromo-2-methylpropane, with sodium ethoxide gives 2-methylpropene. **(b)** What is the geometry and hybridisation observed for the central carbon atom in both the haloalkane and alkene? 2-Methylpropene is a gas at room temperature. **(c)** What is the maximum volume of gas produced, at 25 °C and 101.3 kPa, if 5.0 g of 2-bromo-2-methylpropane reacted completely to form 2-methylpropene?

MasteringChemistry (www.pearson.com.au/masteringchemistry)

Make learning part of the grade. Access:

- tutorials with peronalised coaching
- study area
- Pearson eText

PHOTO/ART CREDITS

994 © Les and Dave Jacobs/cultura/Corbis; **998** Paul Orr/Shutterstock, Imageman/Shutterstock, OZaiachin/Shutterstock, coloursinmylife/Shutterstock; **999** Richard Megna, Fundamental Photographs, NYC; **1012** © Kentannenbaum/Dreamstime.com.

ALDEHYDES, KETONES AND CARBOHYDRATES

How sweet it is. Sucrose, a sugar comprising glucose and fructose, is manufactured from sugar cane.

KEY CONCEPTS

26.1 ALDEHYDES, KETONES AND THE CARBONYL GROUP
We begin by introducing the carbonyl group and two functional groups, aldehydes (RCHO) and ketones (RCOR'), that are formed around it.

26.2 PREPARATION OF ALDEHYDES AND KETONES
We investigate the oxidation of alcohols and alkenes as an efficient way of preparing aldehydes and ketones.

26.3 REACTIONS OF ALDEHYDES AND KETONES
We then explore the reactivity about the carbonyl group using nucleophiles, including organometallic reagents such as Grignard reagents. The formation of imines and acetals from aldehydes and ketones is discussed.

26.4 CARBOHYDRATES
We discuss the nomenclature and structure of carbohydrates, from simple monosaccharides to complex biopolymers such as starch and cellulose. The anomeric position is highlighted as important to saccharide properties. We tie up this chapter by investigating the formation of carbohydrates through an intermolecular hemiacetal formation.

Sugar is one of the oldest and best documented of all commodities. The crystallinity of the popular sugars—mainly sucrose, lactose, and fructose—as well as their sweet flavour has made them popular for over two millennia.

Sugar is classified as a member of the carbohydrate food group and its main natural sources are sugar cane (*Saccharum* spp.) and sugar beets (*Beta vulgaris*), in which sugar can account for 12% to 20% of the plant's dry weight. During the times of Alexander the Great, sugar cane arose as a popular alternative to honey, with people chewing on the raw cane to extract its sweetness.

Today most sugar cane comes from countries with warm climates, such as Brazil, India, China, Thailand, Mexico and Australia, the top sugar-producing countries in the world. From a regional perspective, Asia produces more sugar cane than any other region. Australia is the third largest raw sugar supplier in the world, with around 3.9 million tonnes of raw sugar being produced in 2011–2012.

Refined sugar can be made by dissolving raw sugar and purifying it using phosphoric acid. Alternative methods of purification include a carbonation process involving calcium hydroxide and carbon dioxide, or various filtration strategies. The sugar is then further purified by filtration through a bed of activated carbon or bone char, depending on where the processing takes place. White refined sugar is typically sold as granulated sugar, which has been dried to prevent clumping.

Understanding the chemistry of sugars and appreciating their structure requires a grounding in the chemistry of the carbonyl (C=O) group from which aldehydes and ketones are formed. In this chapter, we discuss themes in carbonyl chemistry, specifically nucleophilic addition, leading to a detailed examination of carbohydrates of which sugars are an important example.

26.1 | ALDEHYDES, KETONES AND THE CARBONYL GROUP

Many important functional groups contain a double bond between carbon and oxygen—a combination of atoms called the **carbonyl group**. In Table 22.8, the carbonyl group appears in the aldehyde, ketone, carboxylic acid, ester and amide functional groups. The chemistry of the latter three functional groups differs from that of aldehydes and ketones and so will be dealt with in the next chapter.

The geometry of the carbonyl group is similar to that of an alkene. The presence of the C–O double bond (▼ FIGURE 26.1) signifies that both the carbon and oxygen atoms are sp^2 hybridised. That is, a set of hybridised orbitals forms the σ-bond and a set of unhybridised p orbitals forms the π-bond of the C–O double bond by direct overlap. Figure 26.1(c) illustrates the bonding in formaldehyde (methanal), the simplest of all the carbonyl-containing compounds. Because the oxygen in the carbonyl group is sp^2 hybridised, its lone pairs have a trigonal planar arrangement. Notice the polarity of the C–O group, which is due to the difference in electronegativity between the carbon atom and the more electron-rich oxygen atom. This polarity, and the nature of the dipole, is very important for the reactivity of carbonyl compounds.

▲ **FIGURE IT OUT**

What does the structure of CO_2 look like?

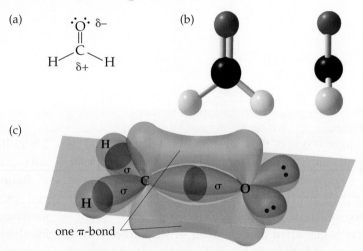

▲ **FIGURE 26.1** **The carbonyl group.** (a) The structural formula; (b) a molecular model representation, in front-on and side-on views, for formaldehyde; (c) a molecular orbital view of the bonding, indicating σ- and π-bonds.

The carbonyl double bond is shorter, stronger and more polarised than a C–C double bond.

	Length	Bond enthalpy
C=O	123 pm	745 kJ mol^{-1}
C=C	134 pm	611 kJ mol^{-1}

An **aldehyde** is characterised by the functional group CHO. The structures and names of five simple aldehydes are:

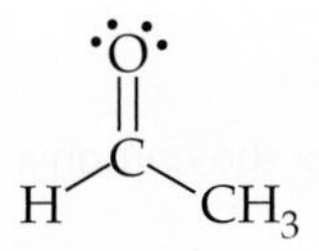

ethanal

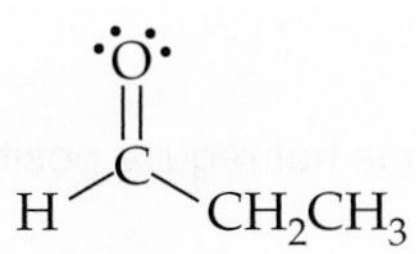

propanal

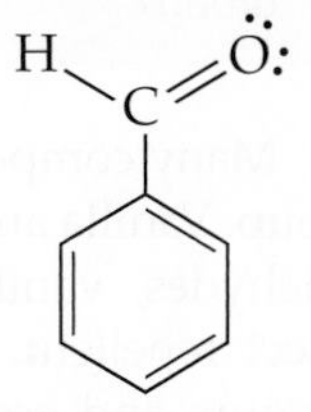

benzaldehyde

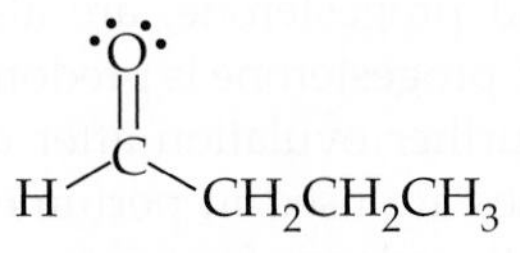

butanal

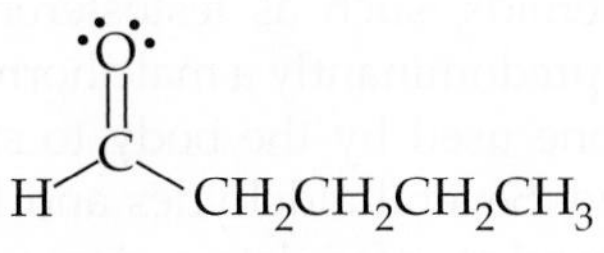

pentanal

Aldehyde nomenclature requires the selection of the parent alkane (longest alkyl chain) containing the carbonyl group. The carbon of the carbonyl group constitutes part of that parent alkane's name. The suffix *-ane* is partly retained, replacing the *e* with *al*. (Note the difference with benzaldehyde. Here the aldehyde group lies outside the ring system. Aromatic compounds such as this have their own unique nomenclature (∞ Section 28.2, 'Isomerism and Nomenclature in Aromatic Compounds').) As you can see from the examples above, an aldehyde group, by definition, must also occur at the beginning (or end) of a hydrocarbon chain. As a result, there is no need to number the position of the aldehyde—it is always assigned position 1.

∞ Find out more on page 1117

In terms of hierarchy, the aldehyde group takes precedence over alkene and alkyne functionality. Two aliphatic aldehyde examples illustrate this:

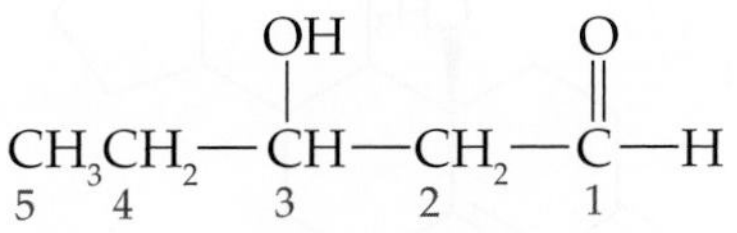

3-hydroxypentanal

$CH_3-CH_2-CH=CH-CHO$
pent-2-enal

In **ketones** the carbonyl group occurs within the interior of a hydrocarbon chain and is therefore flanked on either side by carbon atoms:

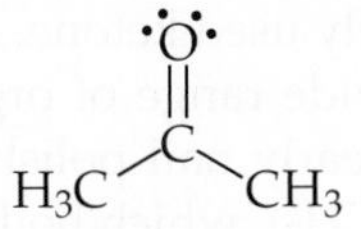

propanone
(acetone)

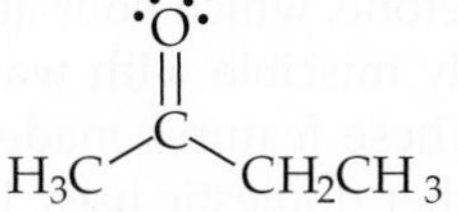

butanone

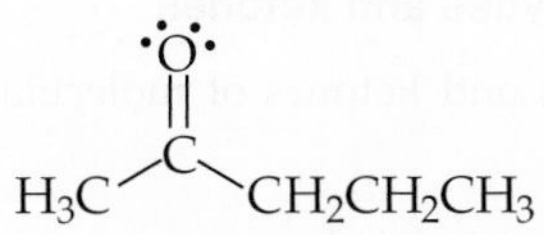

2-pentanone or pentan-2-one

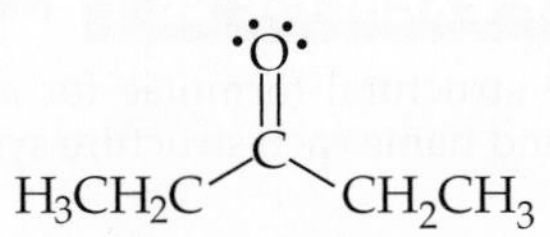

3-pentanone or pentan-3-one

In naming ketones, the parent alkane is derived from the longest chain containing the carbonyl group. The suffix *-ane* is partly retained, replacing the *e* with *one*. Ketones containing parent alkanes greater than C$_4$ can exist as constitutional isomers, so require numbering to define unambiguously the position of the carbonyl group. Two representations to describe the compound are shown and are commonly interconverted.

CONCEPT CHECK 1

Why do propanone and butanone not require position numbers for the carbonyl group?

Many compounds found in nature possess an aldehyde or ketone functional group. Vanilla and cinnamon flavourings are derived from the naturally occurring aldehydes, vanillin and cinnamaldehyde. Citronellal is a naturally occurring insect repellent. Carvone and camphor are ketones that impart characteristic flavours and scents. Steroids, such as testosterone and progesterone, are also ketones. Testosterone is predominantly a male hormone; progesterone is predominantly a female hormone used by the body to stop further ovulation after an ovum has been fertilised. Several aldehydes and ketones are used in perfumes. The ionones produce the odour of violets, α-damascone the odour of roses.

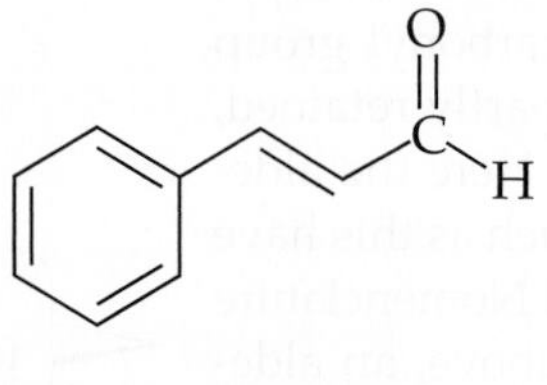

cinnamaldehyde (cinnamon)

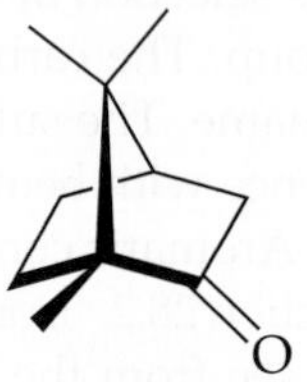

camphor

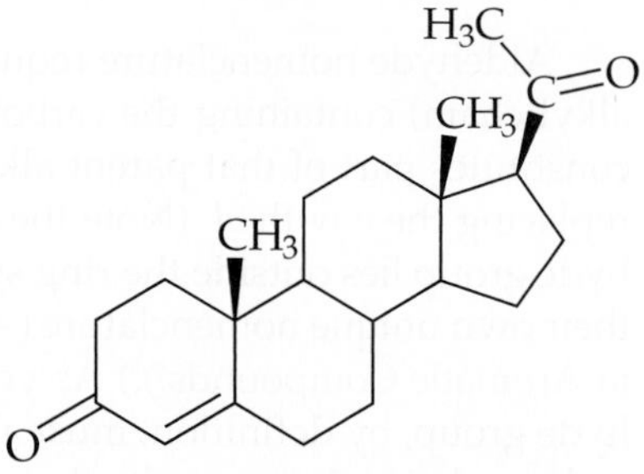

progesterone

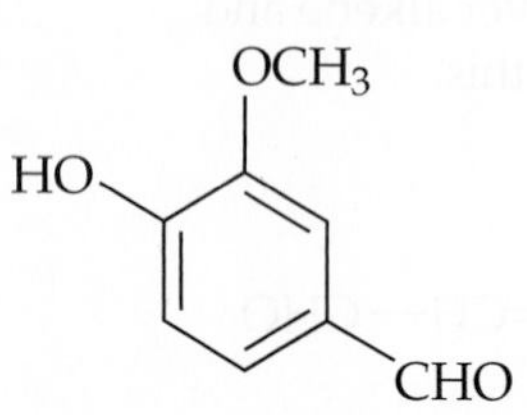

4-hydroxy-3-methoxybenzaldehyde (vanillin)

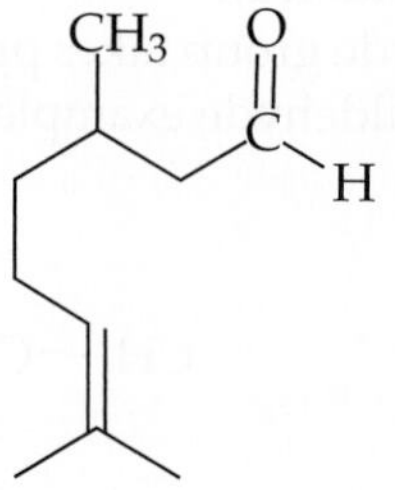

citronellal

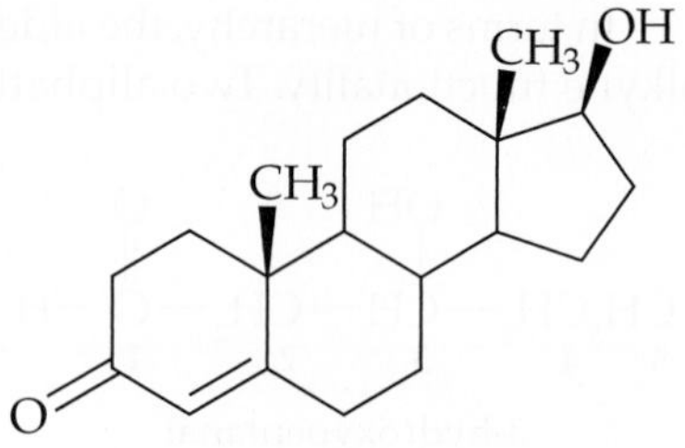

testosterone

Ketones are less reactive than aldehydes and are used extensively as solvents. Acetone, which boils at 56 °C, is the most widely used ketone. Acetone is completely miscible with water, yet it dissolves a wide range of organic substances. These features made it ideal for use as an early nail polish remover, among other domestic uses. Butanone ($CH_3COCH_2CH_3$), which boils at 80 °C, is used as an industrial solvent.

SAMPLE EXERCISE 26.1 Naming aldehydes and ketones

Write the structural formulae for all aldehydes and ketones of molecular formula $C_5H_{10}O$ and name each structure systematically.

SOLUTION

Analyse We are given a molecular formula and asked to generate as many different aldehydes and ketones as possible. We are given one oxygen atom in the molecular formula.

Plan Draw out the carbon skeleton, add the carbonyl group starting at one end (this will give an aldehyde) then progress this along the chain.

Solve

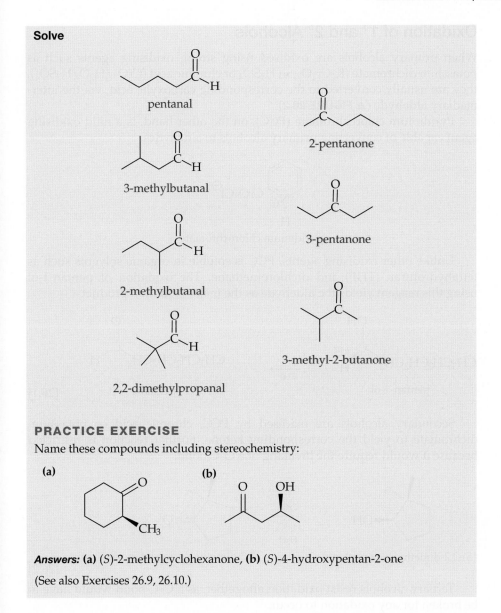

pentanal

2-pentanone

3-methylbutanal

3-pentanone

2-methylbutanal

2,2-dimethylpropanal

3-methyl-2-butanone

PRACTICE EXERCISE

Name these compounds including stereochemistry:

(a)

(b)

Answers: (a) (*S*)-2-methylcyclohexanone, **(b)** (*S*)-4-hydroxypentan-2-one

(See also Exercises 26.9, 26.10.)

26.2 | PREPARATION OF ALDEHYDES AND KETONES

Aldehydes and ketones are generally prepared by using similar reactions. Here, we concentrate on two sample types. The first involves the oxidation of alcohols, which is the standard technique for their preparation. The second involves the ozonolysis of alkenes.

FIGURE IT OUT

What is the major organic product formed by the oxidation of 1-propanol, 2-propanol?

$$\overset{:\ddot{O}H}{\underset{\underset{R}{\overset{|}{\underset{H}{\text{—}}}}{C}}{\text{—}}H} \xrightarrow{[O]} \overset{\cdot\ddot{O}\cdot}{\underset{R}{\overset{\|}{C}}}\text{—}H \xrightarrow{[O]} \overset{\cdot\ddot{O}\cdot}{\underset{R}{\overset{\|}{C}}}\text{—}OH$$

a 1° alcohol	an aldehyde	a carboxylic acid
Number of C–O equivalents: 1	2	3

◀ **FIGURE 26.2** **Oxidation of primary alcohols.** Primary alcohols are converted to carboxylic acids under standard experimental conditions. The symbol [O] represents an oxidation. To demonstrate the oxidation (that is, addition of oxygen) imagine C=O as being 2 × C—O.

Oxidation of 1° and 2° Alcohols

When primary alcohols are oxidised using strong oxidising agents such as potassium dichromate ($K_2Cr_2O_7$) in H_2SO_4 or chromic acid ($CrO_3/H_2O/H_2SO_4$), they are usually converted to the corresponding carboxylic acid, via the intermediary aldehyde (▲ FIGURE 26.2).

Pyridinium chlorochromate (PCC), on the other hand, is a mild oxidising agent capable of converting primary alcohols to aldehydes.

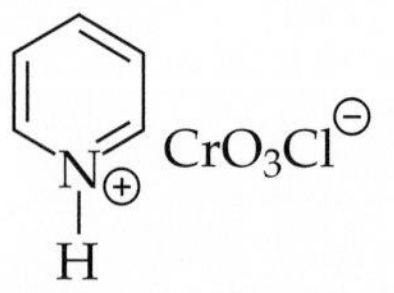

pyridinium chlorochromate

Unlike other oxidising agents, PCC is soluble in organic solvents such as tetrahydrofuran (THF) and dichloromethane. The oxidation of pentan-1-ol using this reagent yields the aldehyde as the major oxidation product.

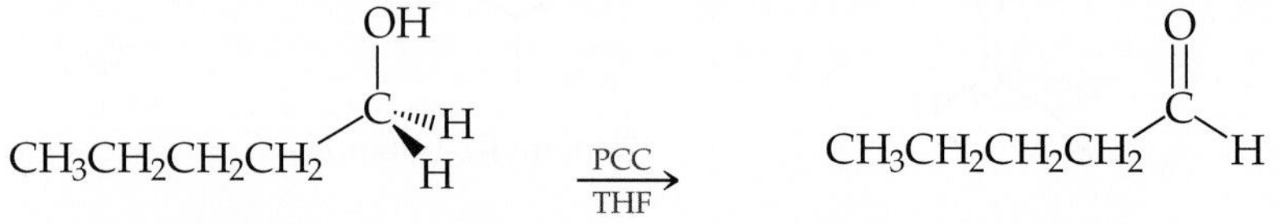

pentan-1-ol   pentanal   [26.1]

Secondary alcohols are oxidised by PCC, chromic acid or potassium dichromate to yield the corresponding ketone. Further reaction is restricted because it would require the breaking of a C–C bond.

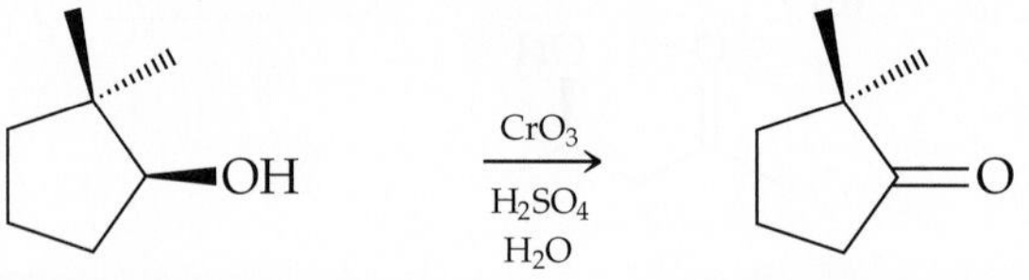

(S)-2,2-dimethylcyclopentanol   2,2-dimethylcyclopentanone
                                                                              [26.2]

Tertiary alcohols resist oxidation altogether, as a C–C bond would have to be broken for any oxidation to occur.

> ### ⚠️ CONCEPT CHECK 2
>
> A chemical oxidation requires the species being oxidised (the reductant) to lose electrons (Section 19.1). Write the half-equations for the oxidation of 2-propanol to propanone.

SAMPLE EXERCISE 26.2 Oxidation of alcohols

What reagent would you use to convert retinol (vitamin A) into retinal, the principal non-peptide component of rhodopsin, in the laboratory?

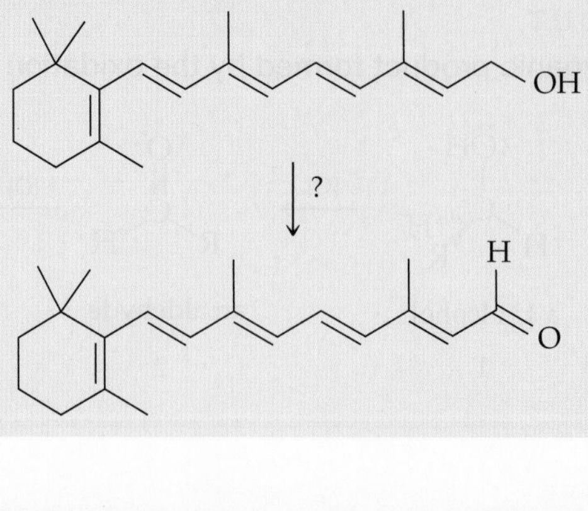

SOLUTION

Analyse We are asked how to convert an alcohol to an aldehyde.

Plan Determine the type of alcohol, where the conversion takes place and what is retained from starting material to product.

Solve This conversion is the oxidation of a primary alcohol to an aldehyde. To achieve this, pyridinium chlorochromate in acetone is the best reagent. The mild reactivity of this reagent does not affect the alkene groups and produces the aldehyde in excellent yield. Note that the reaction is conducted in acetone solvent. This solvent is useful for oxidations because it cannot be further oxidised under the experimental conditions employed.

PRACTICE EXERCISE

What starting alcohol would you use to prepare the following compounds?

(a)

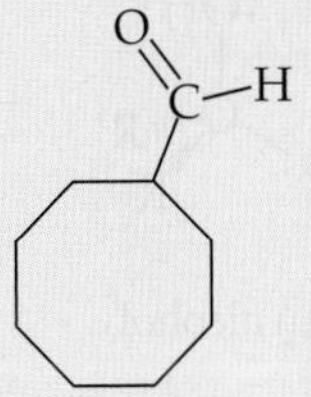

(b)

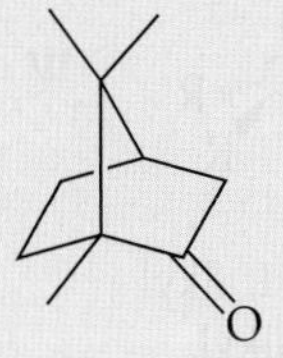

Answers:

(a)

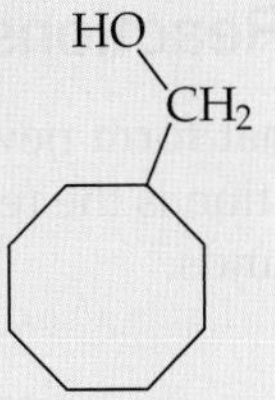

(b)

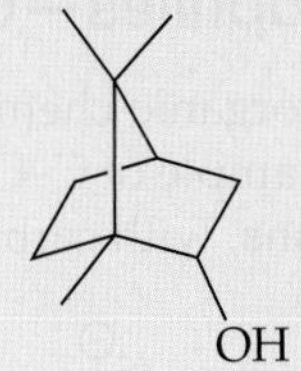

(See also Exercise 26.17.)

Ozonolysis

Aldehydes and ketones can be prepared in good yield by the reaction of alkenes with ozone, followed by a mild reduction using dimethylsulfide. This process, called **ozonolysis**, is a synthetically useful way of generating aldehydes and ketones.

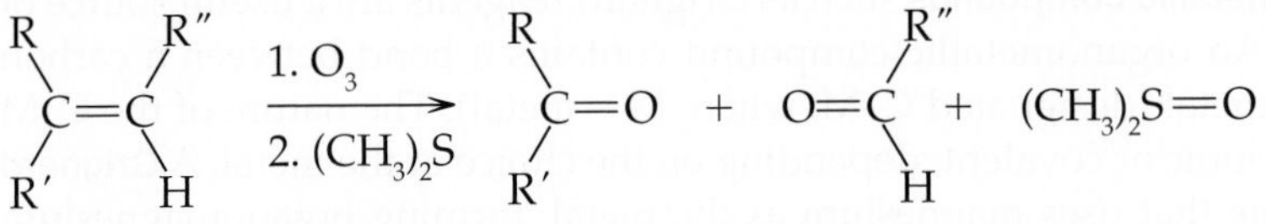

$$\underset{\text{an alkene}}{\overset{R}{\underset{R'}{>}}C=C\overset{R''}{\underset{H}{<}}} \xrightarrow[\text{2. (CH}_3)_2S]{\text{1. O}_3} \underset{\text{a ketone}}{\overset{R}{\underset{R'}{>}}C=O} + \underset{\text{an aldehyde}}{O=C\overset{R''}{\underset{H}{<}}} + \underset{\text{dimethylsulfoxide}}{(CH_3)_2S=O} \qquad [26.3]$$

Whether an aldehyde, ketone or a mixture of the two are formed depends solely on the substitution of the alkene. For symmetrical alkenes, only one product is formed.

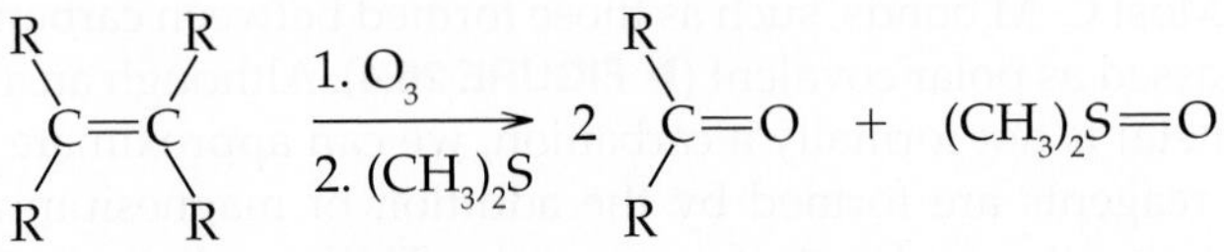

$$\underset{R}{\overset{R}{>}}C=C\underset{R}{\overset{R}{<}} \xrightarrow[\text{2. (CH}_3)_2S]{\text{1. O}_3} 2 \underset{R}{\overset{R}{>}}C=O + (CH_3)_2S=O \qquad [26.4]$$

In some instances, ozonolysis leads to the formation of one high molecular weight and one low molecular weight product, which can easily be separated by distillation.

26.3 | REACTIONS OF ALDEHYDES AND KETONES

Because the carbon in a carbonyl group has a partial positive charge, it is susceptible to nucleophilic attack. Many important synthetic reactions depend on this fact. In this section we concentrate on reactions where the tetrahedral intermediate formed from the initial attack on an aldehyde or ketone by a nucleophile is reacted with acid to produce an alcohol. Generally, the reaction can be described as:

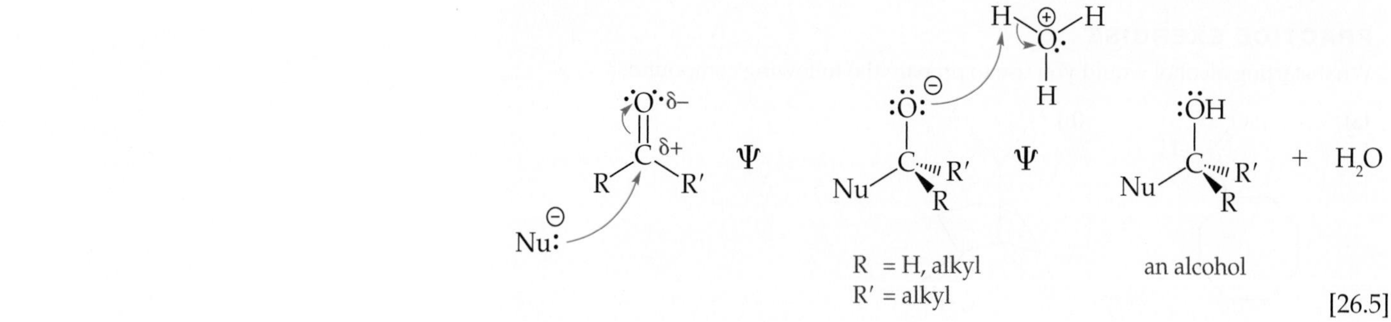

R = H, alkyl
R′ = alkyl

an alcohol

[26.5]

Addition of Carbon Nucleophiles—Grignard Reactions

Of the reaction types available to organic chemists, those that form new C–C bonds are especially useful. One example of C–C bond formation is the reaction of carbon nucleophiles, or **carbanions**, with carbonyl compounds:

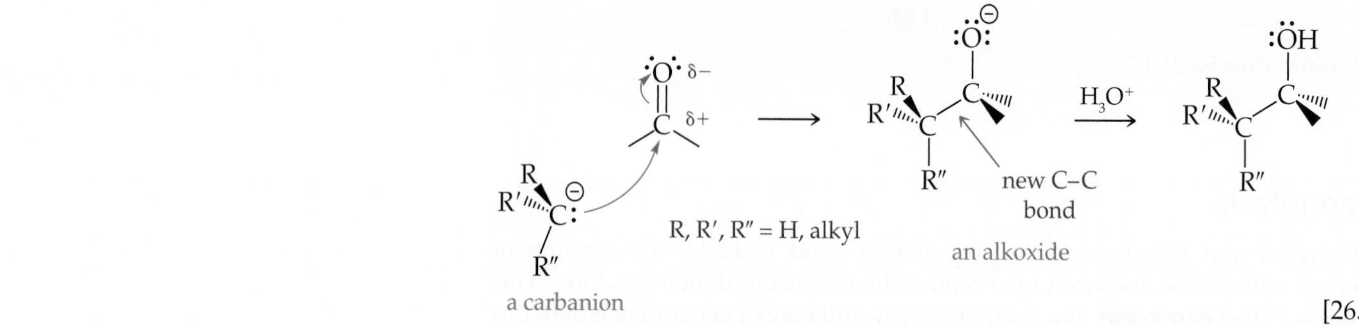

a carbanion

R, R′, R″ = H, alkyl

new C–C bond

an alkoxide

[26.6]

Organometallic compounds such as Grignard reagents are a useful source of carbanions. An organometallic compound contains a bond between a carbon atom and a metal (designated C–M, where M = metal). The nature of the C–M bond can be ionic or covalent, depending on the choice of the metal. A **Grignard reagent** is one that uses magnesium as the metal, forming organomagnesium compounds of the general type RMgX, where R = alkyl or aryl, and X = halide. For this discovery in 1902 and the general usefulness of these reagents in organic synthesis (▶ FIGURE 26.3), Victor Grignard (1871–1935) shared the 1912 Nobel Prize for Chemistry with Paul Sabatier (1854–1941).

Many C–M bonds are covalent, but in bonds between carbon and more electropositive metals, the electrons that make up the bond reside closer to the carbon atom. Most C–M bonds, such as those formed between carbon and magnesium, are classed as polar covalent (▶ FIGURE 26.4). Although an alkyl group bonded to a metal is not formally a carbanion, we can approximate it as such.

Grignard reagents are formed by the addition of magnesium metal to a haloalkane in an ether solvent (for example, THF or diethyl ether). This exothermic reaction is sometimes called an oxidative metallation because the magnesium metal is oxidised to Mg^{2+}. The Grignard reagents are named as the alkyl magnesium halide.

$$CH_3CH_2CH_2\overset{\delta+}{C}H_2\overset{\delta-}{Br} + Mg \xrightarrow{THF} CH_3CH_2CH_2\overset{\delta-}{C}H_2\overset{\delta+}{M}gBr$$

1- bromobutane   butyl magnesium bromide

[26.7]

The effect of magnesium insertion changes the polarity of the carbon that used to bear the halogen from electropositive to electronegative. This change has a dramatic effect on reactivity.

Grignard reagents are strong bases (as well as nucleophiles), so they are moisture-sensitive and tend to be prepared when needed rather than stored in a bottle. Magnesium metal is commercially available as turnings and is typically used in that form.

$$CH_3CH_2CH_2\overset{\delta-}{C}H_2\overset{\delta+}{M}gBr \xrightarrow[THF]{H_2O} CH_3CH_2CH_2CH_3$$

butyl magnesium bromide   butane

[26.8]

The type of alcohol formed (1°, 2° or 3°) in a Grignard reaction is dictated by the nature of the carbonyl compound used. For example, addition of the Grignard reagent propyl magnesium bromide to formaldehyde yields butan-1-ol, a 1° alcohol, upon aqueous workup.

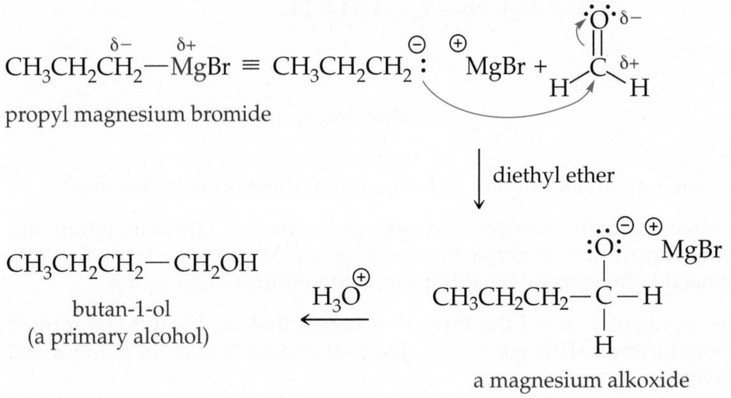

$$CH_3CH_2\overset{\delta-}{C}H_2-\overset{\delta+}{M}gBr \equiv CH_3CH_2CH_2\overset{\ominus}{:} \quad \overset{\oplus}{M}gBr + \underset{H}{\overset{\overset{\displaystyle \overset{..}{O}: \, \delta-}{\parallel}}{\underset{H}{C}}\overset{\delta+}{}$$

propyl magnesium bromide

↓ diethyl ether

$$:\overset{..}{O}:^{\ominus} \; \overset{\oplus}{M}gBr$$
$$CH_3CH_2CH_2-\underset{H}{\overset{|}{C}}-H$$

$$CH_3CH_2CH_2-CH_2OH \xleftarrow{\overset{\oplus}{H_3O}}$$

butan-1-ol
(a primary alcohol)

a magnesium alkoxide   [26.9]

The addition of propyl magnesium bromide to any other aldehyde—for example, propanal—yields a 2° alcohol upon aqueous workup.

$$CH_3CH_2\overset{\delta-}{C}H_2-\overset{\delta+}{M}gBr \equiv CH_3CH_2CH_2\overset{\ominus}{:} \quad \overset{\oplus}{M}gBr + \underset{H}{\overset{\overset{\displaystyle \overset{..}{O}: \, \delta-}{\parallel}}{\underset{}{C}}\overset{\delta+}{\underset{CH_2CH_3}{}}$$

propyl magnesium bromide

↓ diethyl ether

$$:\overset{..}{O}:^{\ominus} \; \overset{\oplus}{M}gBr$$
$$CH_3CH_2CH_2-\underset{H}{\overset{|}{C}}-CH_2CH_3$$

$$\underset{\underset{H}{\overset{|}{}}}{\overset{:\overset{..}{O}H}{CH_3CH_2CH_2-\overset{|}{C}-CH_2CH_3}} \xleftarrow{\overset{\oplus}{H_3O}}$$

hexan-3-ol
(a secondary alcohol)

a magnesium alkoxide

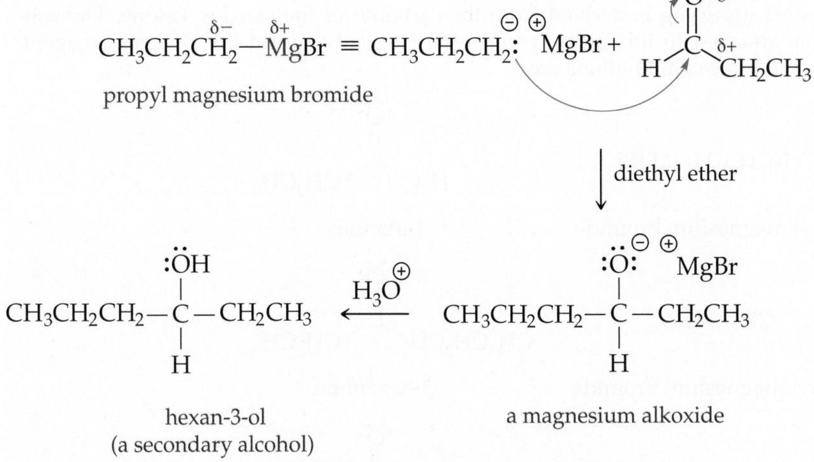

[26.10]

The addition of propyl magnesium bromide to any ketone, such as 2-butanone, yields a 3° alcohol upon aqueous workup.

$$R'-X \xrightarrow{RMgX} R'-R$$

a haloalkane   an alkane

$$R'-CN \xrightarrow{RMgX} \underset{R'}{\overset{\overset{\displaystyle O}{\parallel}}{C}}\underset{R}{}$$

a nitrile   a ketone

$$\underset{R'}{\overset{\overset{\displaystyle O}{\parallel}}{C}}\underset{Cl}{} \xrightarrow{RMgX} \underset{R'}{\overset{\overset{\displaystyle O}{\parallel}}{C}}\underset{R}{}$$

an acid chloride   a ketone

$$\underset{R'}{\overset{\overset{\displaystyle O}{\parallel}}{C}}\underset{OR''}{} \xrightarrow{RMgX} \underset{R'}{\overset{\overset{\displaystyle O}{\parallel}}{C}}\underset{R}{}$$

an ester   a ketone

$$\underset{R'}{\overset{\overset{\displaystyle O}{\parallel}}{C}}\underset{R}{} \xrightarrow{RMgX} \underset{R''}{\overset{\overset{\displaystyle OH}{|}}{\underset{|}{R'-C-R}}}$$

a ketone   an alcohol

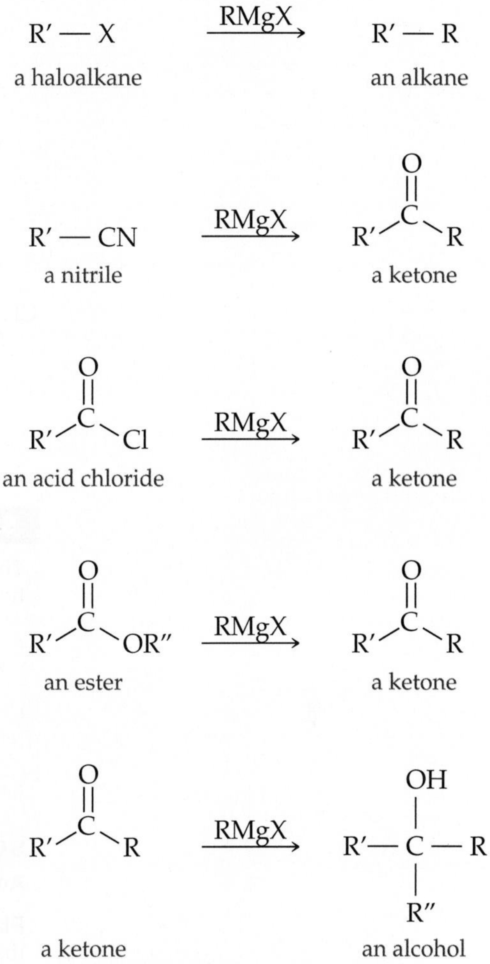

▲ FIGURE 26.3 **The versatility of Grignard reagents.**

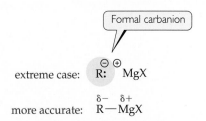

Formal carbanion

extreme case: $\overset{\ominus}{R}:\;\overset{\oplus}{M}gX$

more accurate: $\overset{\delta-}{R}-\overset{\delta+}{M}gX$

▲ FIGURE 26.4 **Grignard reagents as carbanion equivalents.** The bonding between carbon and magnesium is best described as polar covalent. It is still useful, particularly at the first-year level, to think of Grignard reagents as a form of carbanion.

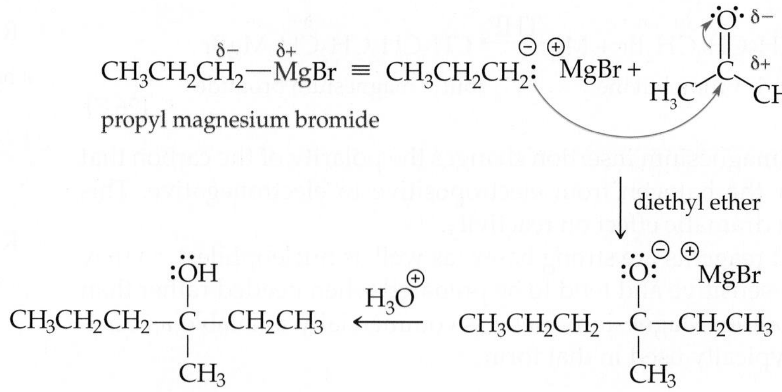

CH₃CH₂CH₂—MgBr ≡ CH₃CH₂CH₂:⁻ MgBr⁺ +

propyl magnesium bromide

$$\overset{\cdots}{O} \overset{\delta-}{} \\ \parallel \\ H_3C \overset{\;}{\underset{\delta+}{C}} CH_2CH_3$$

↓ diethyl ether

:ÖH
|
CH₃CH₂CH₂—C—CH₂CH₃   ⟵ H₃O⁺   :Ö:⁻ MgBr⁺
| |
CH₃   CH₃CH₂CH₂—C—CH₂CH₃
                                        |
                                        CH₃

3-methyl-3-hexanol   a magnesium alkoxide
(a tertiary alcohol)

[26.11]

SAMPLE EXERCISE 26.3 Addition of Grignard reagents

The compound 3-methylhexan-3-ol can be prepared from three different combinations of ketones and Grignard reagents. Show each combination.

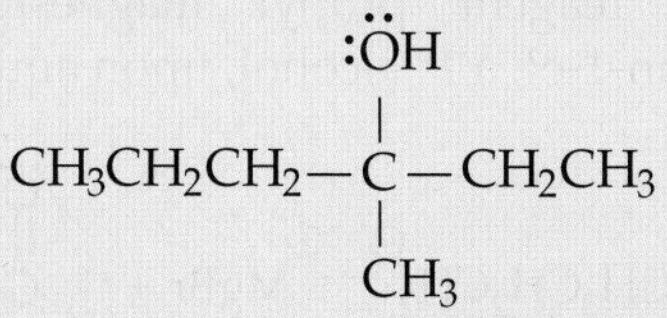

:ÖH
|
CH₃CH₂CH₂—C—CH₂CH₃
|
CH₃

3-methyl-3-hexanol

SOLUTION

Analyse How can 3-methylhexan-3-ol be formed from three different ketones?

Plan Tertiary alcohols can be formed from ketones by reaction with carbanions and their equivalents. Since reaction about the C=O group leads to the formation of a C–C bond, we need to investigate the groups tethered off the C–OH group.

Solve First, let us identify in red the three C–C bonds that are likely to be formed about the carbonyl carbon. This gives us a clue to the identification of ketones and Grignard reagents:

:ÖH
|
CH₃CH₂CH₂—C—CH₂CH₃
|
CH₃

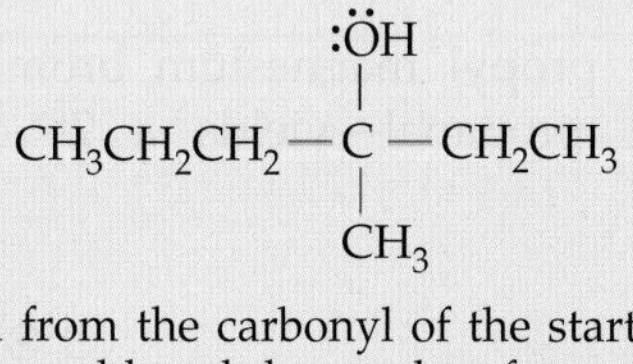

The C–OH grouping is derived from the carbonyl of the starting ketone. The substituents attached to this group could each be used to form the Grignard reagent. Hence, the three combinations are:

CH₃CH₂CH₂–MgBr   +
$$\overset{\cdots}{\underset{}{O}} \\ \parallel \\ H_3C \overset{\;}{} CH_2CH_3$$

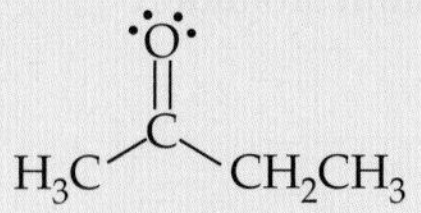

propyl magnesium bromide   butanone

CH₃–MgBr   +
$$\overset{\cdots}{\underset{}{O}} \\ \parallel \\ CH_3CH_2CH_2 \overset{\;}{} CH_2CH_3$$

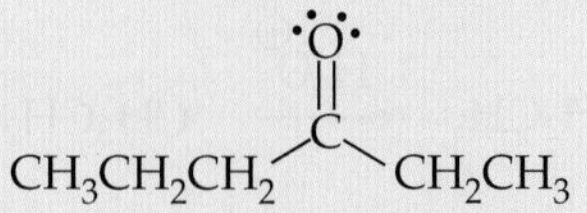

methyl magnesium bromide   3-hexanone

CH₂CH₃–MgBr   +
$$\overset{\cdots}{\underset{}{O}} \\ \parallel \\ H_3C \overset{\;}{} CH_2CH_2CH_3$$

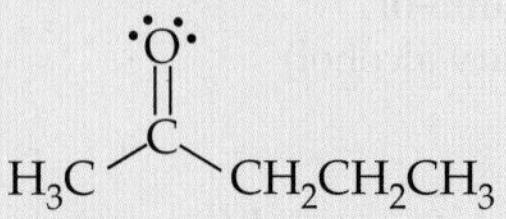

ethyl magnesium bromide   2-pentanone

PRACTICE EXERCISE

What reagents would you use to bring about the following transformations?

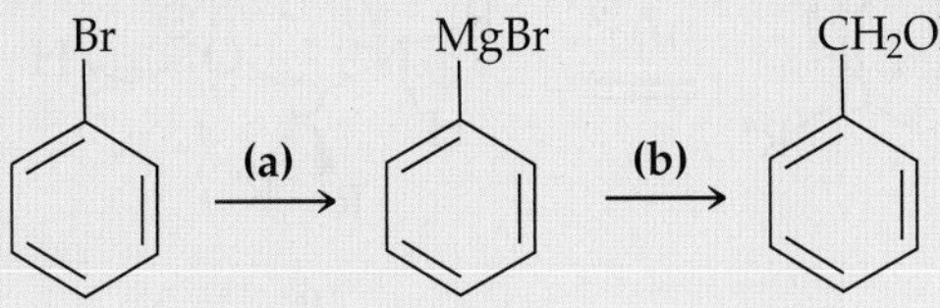

Br   MgBr   CH₂OH

(a)   **(b)**

Answers: **(a)** Mg in diethyl ether, **(b)** formaldehyde
(See also Exercises 26.24, 26.25.)

Addition of Nitrogen and Oxygen Nucleophiles: Formation of Imines and Acetals

The formation of new C–C bonds through the reaction of a carbonyl compound with a Grignard reagent is not only high yielding, but also occurs only in the forward direction. This means that the reaction is not under the same level of equilibrium control we have discussed for the hydration of an alkene or the dehydration of an alcohol. However, other nucleophiles are able to react with carbonyl compounds, under equilibrium conditions, to form stable compounds. The two most important reactions are those of carbonyl compounds with amines (nitrogen-containing weak organic bases) to form *imines*, and those of carbonyl compounds with alcohols to form *acetals* and *hemiacetals*. Both reactions have biological significance (∞ Section 26.4, 'Carbohydrates').

Imines, also known as Schiff bases, are formed by the acid-catalysed condensation of a *primary amine* (RNH₂, R = alkyl, aryl) (∞ Section 29.1, 'Amines and the Amide Bond') an aldehyde or ketone. For example, the reaction of propanamine, an aliphatic 1° amine, with propanone is shown below. Imines are characterised by a C–N double bond. They have geometries similar to those of alkenes, with the lone pair occupying one of the trigonal planar positions.

∞ Find out more on page 1054

∞ Find out more on page 1158

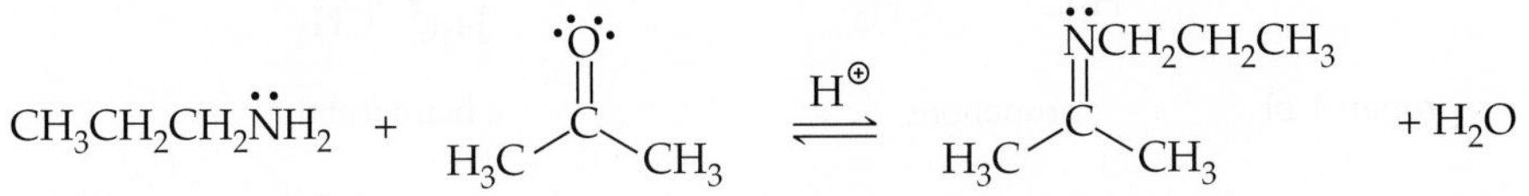

$$CH_3CH_2CH_2\overset{\cdot\cdot}{N}H_2 \;+\; \underset{\underset{\text{propanone}}{}}{\overset{\overset{\cdot\cdot}{O}\cdot}{\underset{H_3C}{\overset{\|}{C}}\underset{CH_3}{}}} \; \underset{}{\overset{H^{\oplus}}{\rightleftharpoons}} \; \underset{\underset{\text{an imine}}{}}{\overset{\overset{\cdot\cdot}{N}CH_2CH_2CH_3}{\underset{H_3C}{\overset{\|}{C}}\underset{CH_3}{}}} \;+\; H_2O$$

propanamine

[26.12]

The mechanism of imine formation is shown in Equation 26.13. Attack of the nucleophilic amine on the electrophilic carbonyl carbon atom leads to a tetrahedral intermediate amino alcohol. The acid-catalysed dehydration of this intermediate yields the imine. The reaction usually occurs at room temperature and in high yield. Removal of water as the reaction proceeds ensures near quantitative yields.

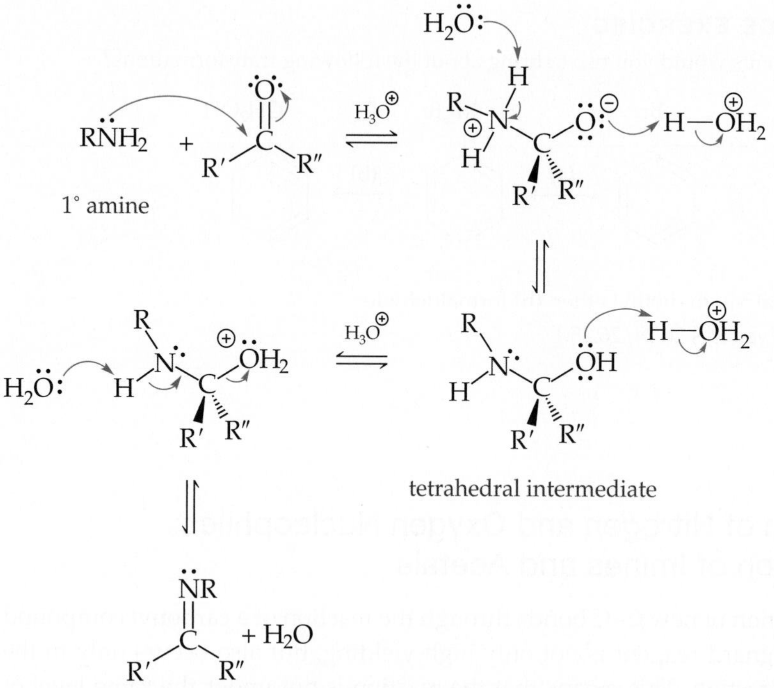

[26.13]

⚠ CONCEPT CHECK 3

Why is it only possible to form imines with primary amines?

Acetals and hemiacetals are formed by the reaction of alcohols with aldehydes and ketones in the presence of an acid catalyst. In such reactions, the hemiacetal is formed in minor yield because it is relatively unstable. Further reaction with alcohol usually yields the acetal and water. Acetals are stable under basic and nucleophilic conditions, so provide a common way to protect aldehydes and ketones during complex reaction sequences in organic synthesis. Because the reaction is in equilibrium, acetal formation can be reversed by the addition of H_3O^+.

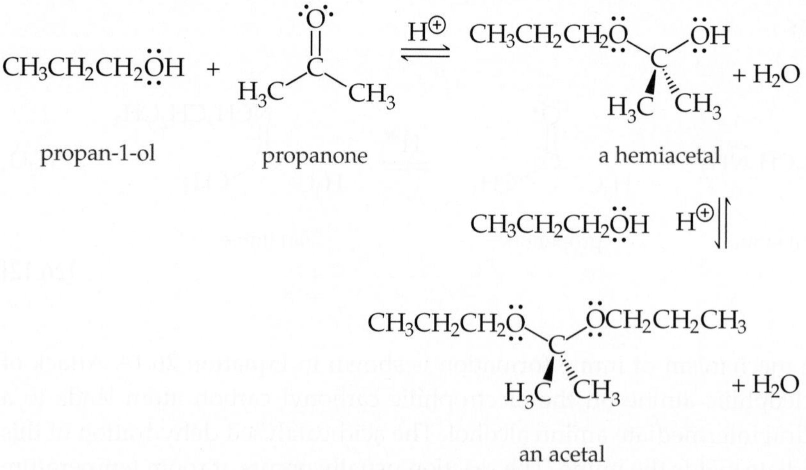

| propan-1-ol | propanone | | a hemiacetal |

[26.14]

an acetal

Hemiacetals are important in the function of *carbohydrates*, a class of biologically active molecules, of which sugars and cellulose are members (∞ Section 26.4, 'Carbohydrates'). Carbohydrates gain their stability by undergoing *intramolecular* hemiacetal formation. To do this, carbohydrates must have aldehyde and alcohol functionality within the same molecule and at a distance apart that allows for cyclisation. Typically, cyclic hemiacetals are five-membered

∞ Find out more on page 1054

(furanose) or six-membered (pyranose) systems. The equilibrium between cyclic and acyclic compounds is heavily favoured towards the cyclic hemiacetal. The cyclisation of 5-hydroxypentanal demonstrates the principle.

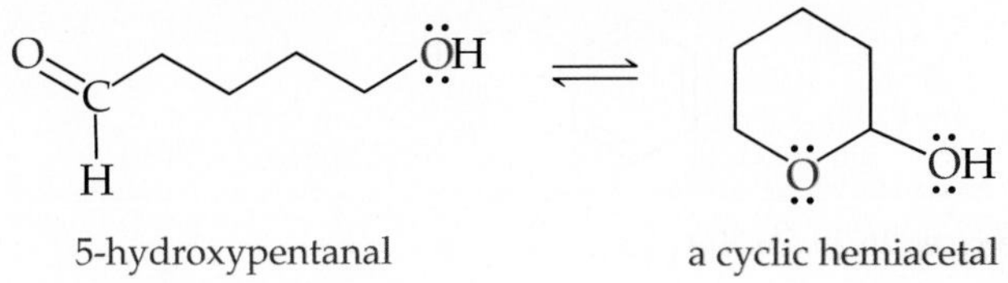

5-hydroxypentanal   a cyclic hemiacetal   [26.15]

The mechanism of acetal formation can be divided into two steps. The first, formation of the hemiacetal, follows a mechanism similar to that described for the reaction of amines with carbonyl compounds (without loss of water, Equation 26.13). Formation of the acetal from the hemiacetal is an acid-catalysed process involving the generation of an oxonium ion. Loss of water leads to a resonance-stabilised carbocation. This carbocation reacts with another molecule of alcohol, which, after proton transfer, yields the acetal.

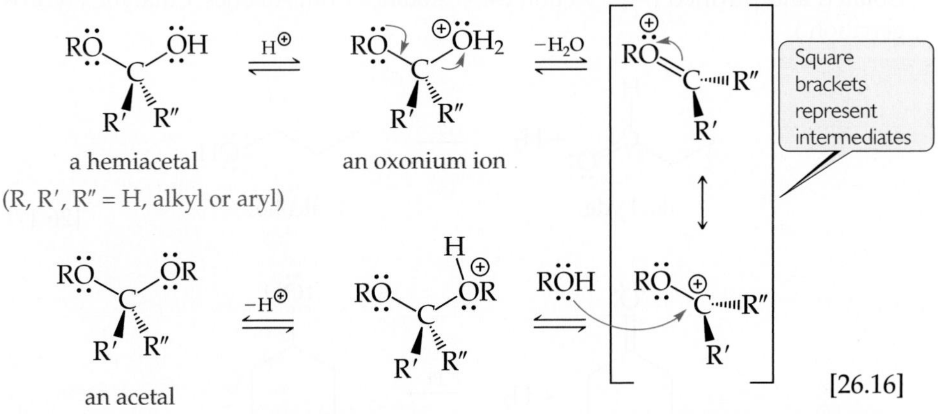

a hemiacetal   an oxonium ion

(R, R', R" = H, alkyl or aryl)

an acetal   [26.16]

Square brackets represent intermediates

SAMPLE EXERCISE 26.4 **Acetals**

Diols such as ethene glycol (1,2-ethanediol) are useful reagents with which to protect aldehydes and ketones in chemical syntheses. Give the product of the acid-catalysed reaction of ethene glycol and cyclohexanone.

SOLUTION

Analyse We are asked for the outcome of a reaction between an alcohol functional group and a ketone.

Plan Acetals require the acid-catalysed addition of two alcohols to a ketone:

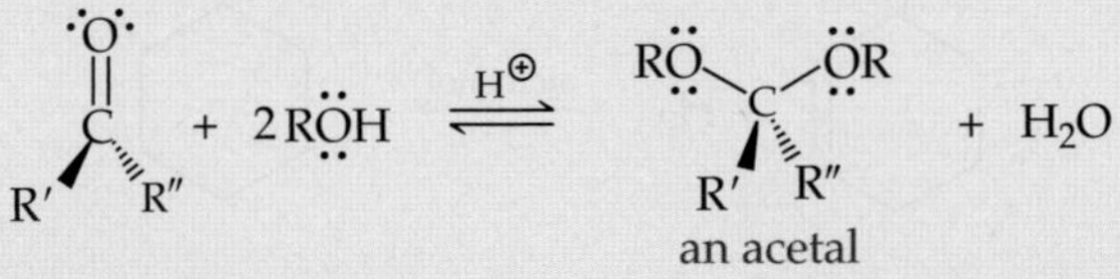

an acetal

Solve Ethene glycol acts to satisfy this need because it is a diol. Hence the product of this reaction is

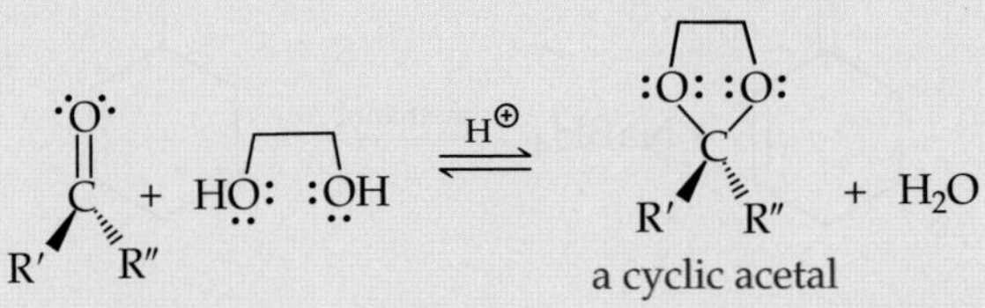

a cyclic acetal

PRACTICE EXERCISE

What product would you expect from the acid-catalysed reaction of benzaldehyde and ethanol?

Answer:

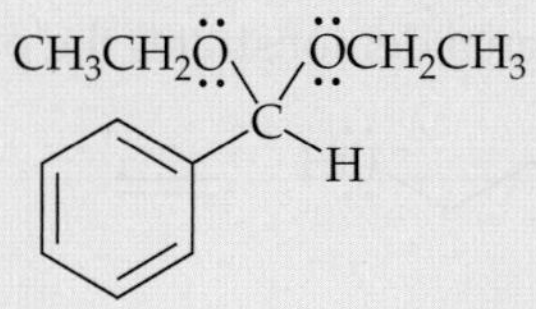

$$CH_3CH_2\overset{..}{\underset{..}{O}}\quad\overset{..}{\underset{..}{O}}CH_2CH_3$$

(See also Exercises 26.29, 26.30.)

Reduction Reactions

Aldehydes and ketones are easily reduced to the corresponding primary and secondary alcohols, respectively. Catalytic hydrogenation, using Pd, Pt, Ni or Rh metal catalysts, is typically a high-yielding reaction and the product is easily isolated and purified (∞ Section 24.5, 'Alkanes from Alkenes: Catalytic Hydrogenation').

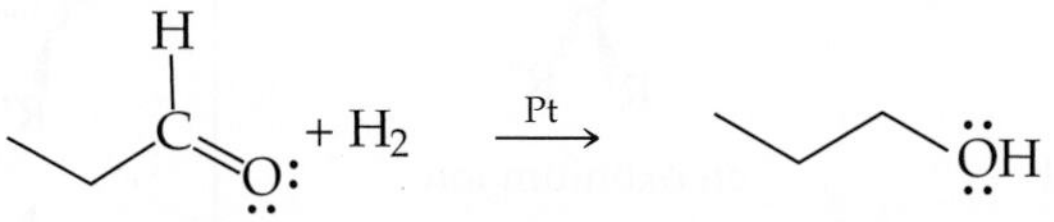

∞ Review this on page 977

[26.17]

aldehyde   1° alcohol

[26.18]

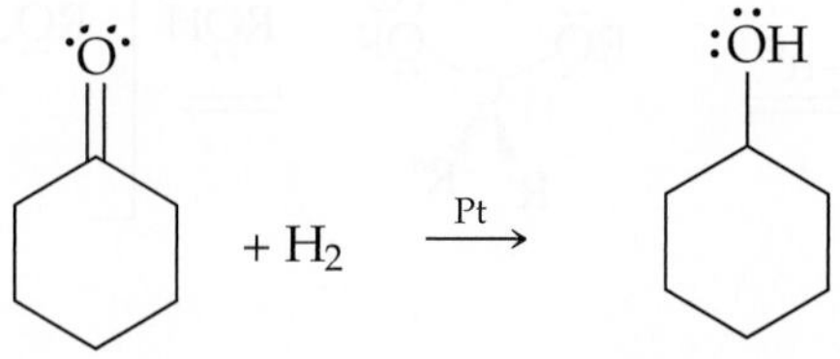

ketone   2° alcohol

In a laboratory situation, the reduction of aldehydes and ketones using **metal hydrides** is preferred over catalytic hydrogenation because it selectively reduces the carbonyl group rather than any alkene double bonds present and no gas-handling equipment is required.

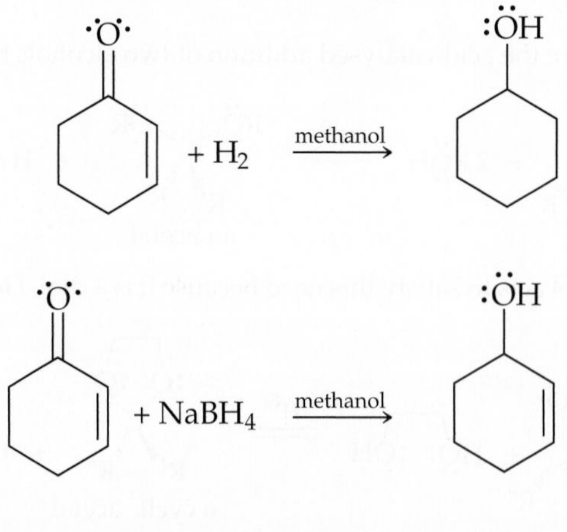

methanol

methanol

[26.19]

The two most common metal hydride reagents are sodium borohydride ($NaBH_4$) and lithium aluminium hydride ($LiAlH_4$). These compounds behave as a source of **hydride ion** (H^-), which is a powerful nucleophile. ▶ FIGURE 26.5 shows the structural formula of both metal hydrides and compares their chemical reductive properties.

FIGURE IT OUT

How soluble would sodium borohydride be in typical organic solvents? Why?

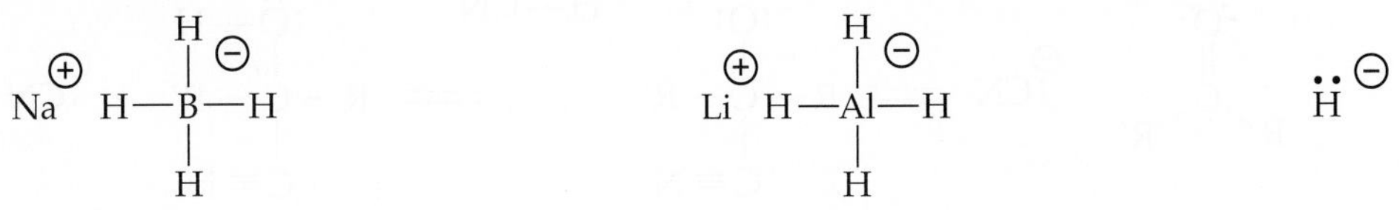

sodium borohydride	lithium aluminium hydride	hydride
Non-reactive in water or methanol	Highly reactive in water or methanol	
Mild reductant	Powerful reductant	
Selective for aldehydes and ketones	Reduces aldehydes, ketones, carboxylic acids and esters	

◀ **FIGURE 26.5 Metal hydrides.** The two most common metal hydrides for reduction are sodium borohydride and lithium aluminium hydride.

Metal hydrides act as a source of hydride ion (H^-) because they contain polar metal–hydrogen bonds, which place a partial negative charge on hydrogen in the same way as a negative charge is placed on carbon within a Grignard reagent:

$$\overset{\delta+}{M}-\overset{\delta-}{H} \equiv :H^-$$

In fact, the mechanism for the reduction of an aldehyde or ketone by a metal hydride, which involves nucleophilic attack on the carbonyl group by hydride, followed by protonation, follows a similar pathway to that demonstrated for the Grignard reaction in Equation 26.10. A typical mechanism, showing the reduction of butanal using hydride, is

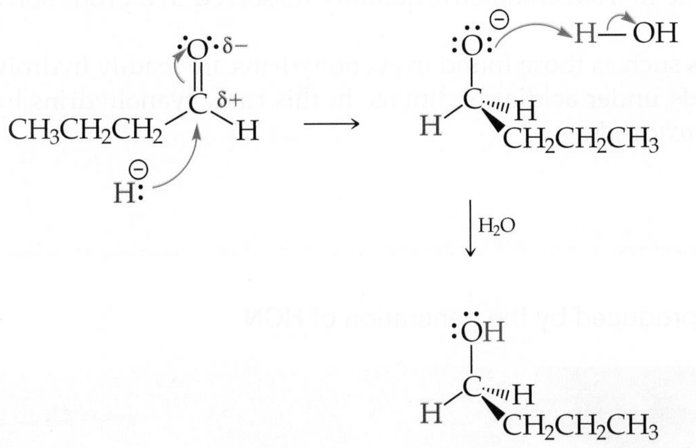

[26.20]

Cyanohydrins

Cyanohydrins are a class of compounds containing a hydroxy (OH) and nitrile (CN) group, both directly bonded to an α-carbon. Organic compounds containing the CN group are called **nitriles**. A cyanohydrin is therefore an *α-hydroxynitrile*.

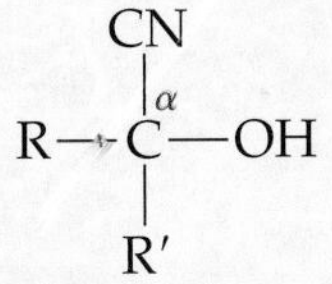

a cyanohydrin

Cyanohydrins are formed by the addition of the cyanide ion (^-CN) to an aldehyde or ketone. Cyanide ion is the conjugate base of hydrogen cyanide (HCN), which is itself a weak acid ($pK_a \sim 9.2$). The cyanide ion is small, and because it is both a base and a strong nucleophile, it can add to the carbonyl group by a **nucleophilic addition** mechanism.

$$R-\overset{\overset{\displaystyle :O:}{\|}}{\underset{\displaystyle R'}{C}} \quad \overset{\ominus}{:}CN \quad \rightleftharpoons \quad R-\overset{\overset{\displaystyle :O:^{\ominus}}{|}}{\underset{\displaystyle \underset{C\equiv N}{|}}{C}}-R' \quad H\!-\!CN \quad \rightleftharpoons \quad R-\overset{\overset{\displaystyle :O-H}{|}}{\underset{\displaystyle \underset{C\equiv N}{|}}{C}}-R' \quad + \quad \overset{\ominus}{:}CN$$

R, R′ = H, alkyl   a tetrahedral intermediate   a cyanohydrin   [26.21]

In the first step, the strong ⁻CN nucleophile adds to the electropositive carbonyl carbon to give a tetrahedral alkoxide intermediate. Protonation of the alkoxide yields the cyanohydrin. Note that in the example above, the protonation step regenerates cyanide ion. Thus the process is truly catalytic in the addition of cyanide ion and this reaction is classed as a *base-catalysed reaction*.

As shown by the mechanism in Equation 26.21, cyanohydrin formation is reversible. The formation of cyanohydrins is best with aldehydes, with formaldehyde being quantitative. Ketones are slower to react, though for simple ketones the yields are still greater than 90%. Bulky ketones do not react well, being slow and poor yielding. Note that the reversibility of cyanohydrin formation is put to use by the millipede *Apheloria corrugata*, which releases the cyanohydrin mandelonitrile from a storage gland to an outer chamber (▼ **FIGURE 26.6**). There it is enzymatically degraded to benzaldehyde and hydrogen cyanide before being sprayed at an enemy as a defence mechanism.

Cyanohydrins can be formed by using HCN with a catalytic amount of sodium or potassium cyanide to initiate reaction. Hydrogen cyanide, however, is highly toxic and volatile. A more practical method is to use sodium or potassium cyanide in a stoichiometric quantity dissolved in a protic solvent such as ethanol.

Nitriles such as those found in cyanohydrins are readily hydrolysed to carboxylic acids under acidic conditions. In this case, cyanohydrins hydrolyse to give *α-hydroxy acids*.

◢**FIGURE IT OUT**

Suggest the structure of the organic compound produced by the generation of HCN from mandelonitrile.

▶ **FIGURE 26.6 Millipede defence.** *Apheloria corrugata* releases mandelonitrile as part of its defence strategy. The nitrile is degraded to hydrogen cyanide which is deadly to predators.

mandelonitrile

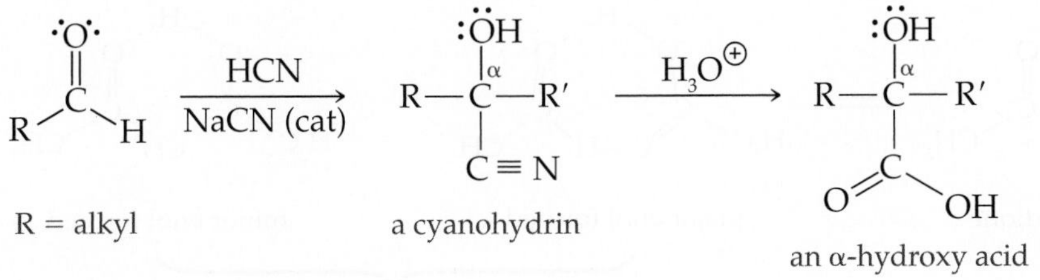

an α-hydroxy acid [26.22]

Tautomerism in Aldehydes and Ketones

Aldehydes and ketones bearing an α-hydrogen (that is, a hydrogen on a carbon atom next to the carbonyl carbon) are able to undergo tautomerism to form enols ('ene ols', or a compound bearing a double bond (alkene) and a hydroxyl group (alcohol)). Enols are constitutional isomers of their respective *keto* form (that is, an aldehyde or ketone). **Tautomerism** is the process by which two isomers, in this case the aldehyde or ketone and the enol, are interconverted by a formal movement of an atom or group, following normal valency rules. Hence an enol and its keto form are *tautomers* of the same structure.

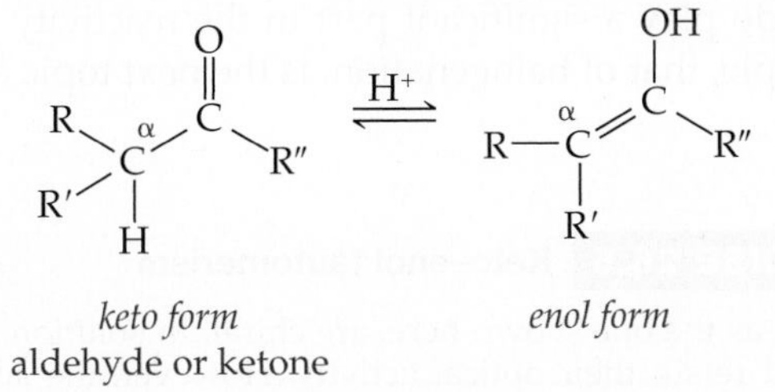

keto form
aldehyde or ketone

enol form

[26.23]

Aldehydes and ketones are in equilibrium with their enol forms and this equilibrium is acid catalysed. The process involves two separate proton transfers rather than a single proton jump, as might be expected just by looking at Equation 26.23. What tautomerism exploits is essentially the relative acidity of the α-hydrogen (pK_a ~9–13).

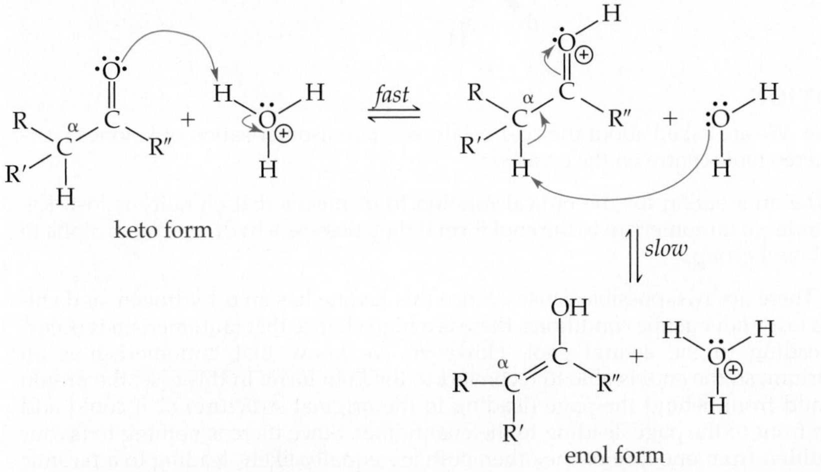

keto form

fast

slow

enol form [26.24]

The extent of tautomerisation depends very much on the aldehyde or ketone. For example, acetone exists >99.9% in the ketone form, whereas 2,4-pentanedione exists preferentially as the enol (~80%) by virtue of stabilising hydrogen bonding.

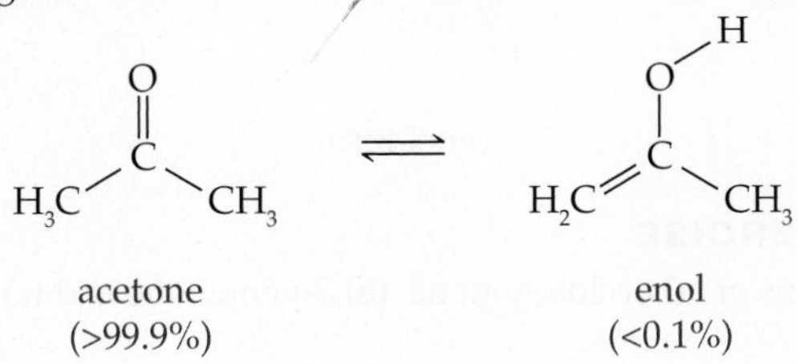

acetone
(>99.9%)

enol
(<0.1%)

[26.25]

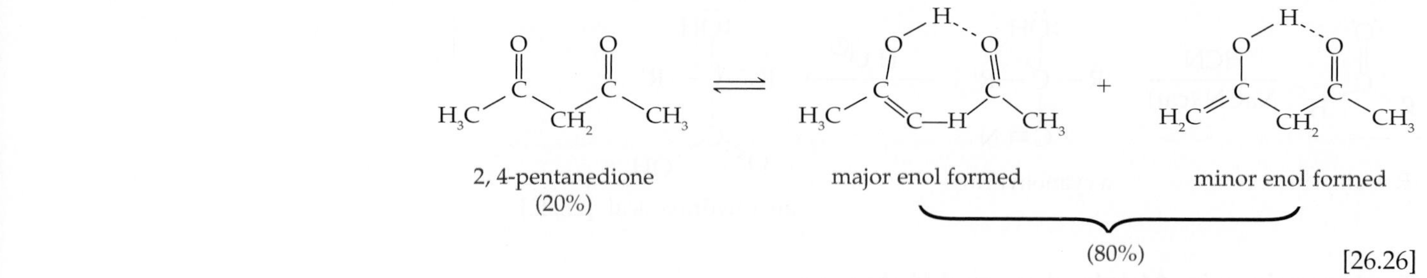

2, 4-pentanedione
(20%)

major enol formed

minor enol formed

(80%)

[26.26]

⚠ **CONCEPT CHECK 4**

Of the enols formed in Equation 26.26, can you explain why the one shown on the left is the major enol formed?

Though many aldehydes and ketones do not exist in the enol form to a large extent, enols do play a significant part in the reactivity of aldehydes and ketones. One example, that of halogenation, is the next topic of discussion.

SAMPLE EXERCISE 26.5 | **Keto–enol tautomerism**

Some ketones, such as the one shown here, are chiral. In solution, such compounds are quite stable and retain their optical activity. However, the addition of a small amount of acid causes the optical rotation to tend to 0° over time. Comment on why this is so.

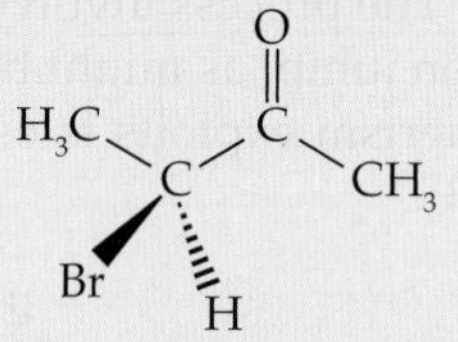

SOLUTION

Analyse We are asked about the acid-catalysed stereoisomerisation of ketones bearing a stereogenic centre on the α-carbon.

Plan The attenuation for the optical rotation to 0° means that chirality is lost. Ketones undergo tautomerism to the enol form if they possess a hydrogen atom alpha to the carbonyl group.

Solve There are two possible causes. Since this ketone has an α-hydrogen, and chirality is lost under acidic conditions, there is a high chance that tautomerism is occurring, leading to the achiral enol. However, we know that tautomerism is an equilibrium, so the enol is able to reconvert to the keto form. In this case, the proton could add from behind the page (leading to the original structure) or it could add from in front of the page, leading to the enantiomer. Since there is nothing to favour the addition from one face or the other, both are equally likely, leading to a racemic mix. This can be summarised as the following:

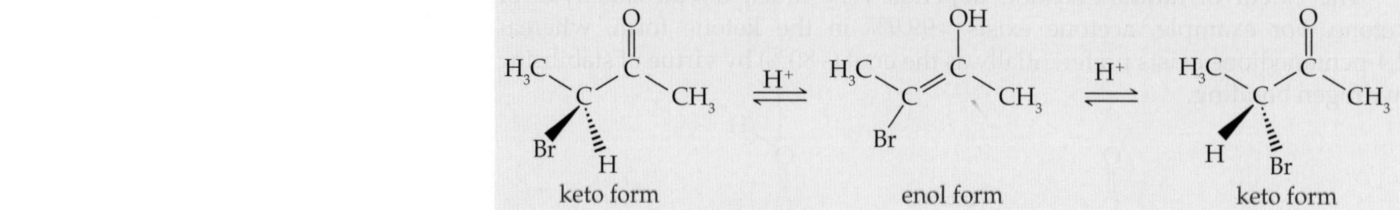

keto form

enol form

keto form

PRACTICE EXERCISE

Draw the enol forms of **(a)** cyclohexanone, **(b)** 3-pentanone and **(c)** propanal.

Answers:

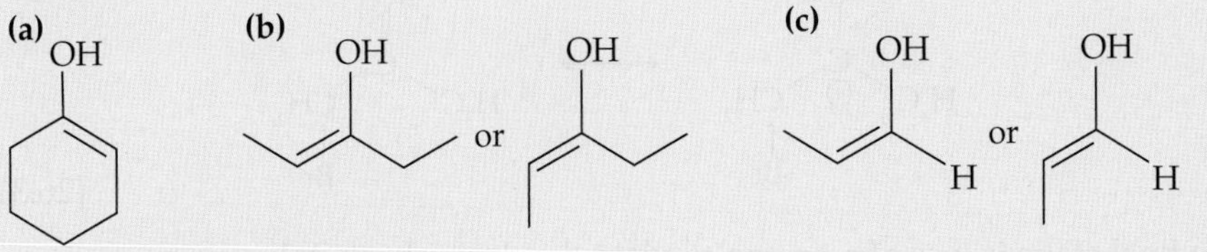

(a)

OH

(b)

OH

or

OH

(c)

OH

or

OH

(See also Exercises 26.43, 26.44.)

Halogenation of Aldehydes and Ketones

Aldehydes and ketones undergo a substitution reaction at the α-carbon in the presence of molecular chlorine or bromine in good yield.

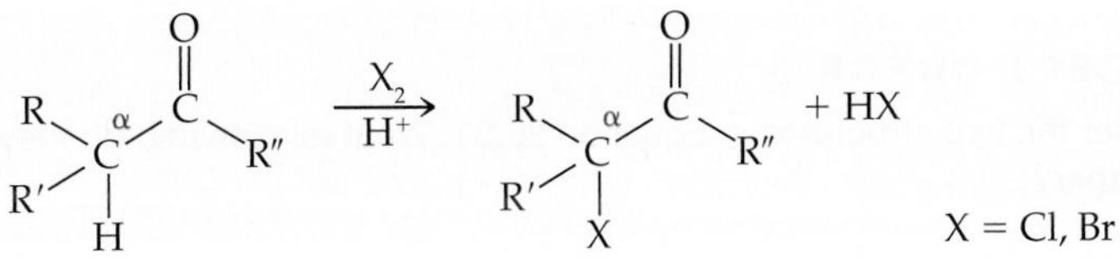

$$\text{aldehyde or ketone} \xrightarrow[\text{H}^+]{X_2} \text{α-halo aldehyde or ketone} + HX$$

X = Cl, Br

aldehyde or ketone   α-halo aldehyde or ketone   [26.27]

This reaction is *regiospecific* in that *only* the α-hydrogen is substituted under these conditions. Interestingly, when the reaction was first studied mechanistically by Arthur Lapworth (1872–1941) in 1904, he discovered that the rates of chlorination, bromination and iodination were all the same. This meant that halogenation occurred after the rate-determining step. We now know that the rate-determining step is the conversion of the ketone to its enol isomer. Let us look at this reaction in more detail, using the bromination of acetone as the example.

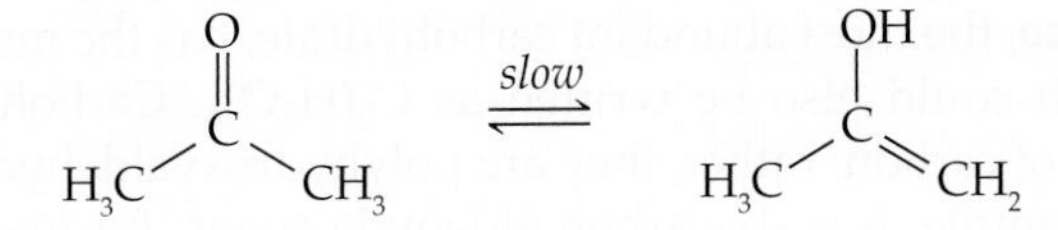

acetone   [26.28]

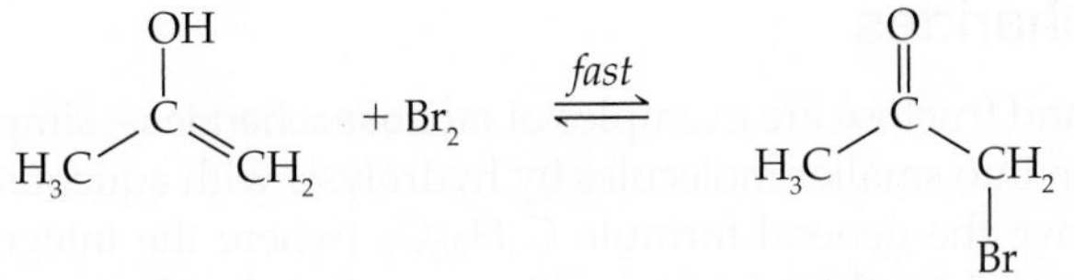

+ Br₂   *fast*   [26.29]

We can understand the enol halogenation by comparing it to the halogenation of an alkene (∞ Section 24.4, 'Electrophilic Addition and Substitution Reactions'). The first step involves attack of the double bond of the enol on bromine.

∞ Review this on page 972

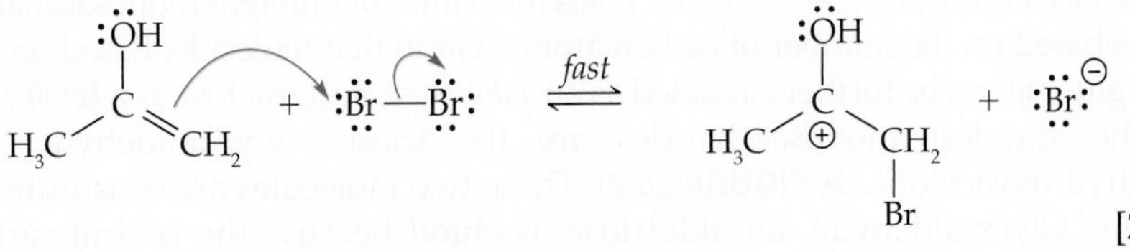

[26.30]

If the reaction followed the bromination of an alkene, then the next phase of the mechanism would be the formation of a new C–Br bond by attack of the bromide ion on the carbocation intermediate in an *intermolecular* reaction. However, in this case, the lone pair on the oxygen can stabilise the carbocation by resonance with the major resonance contributor being the oxonium ion.

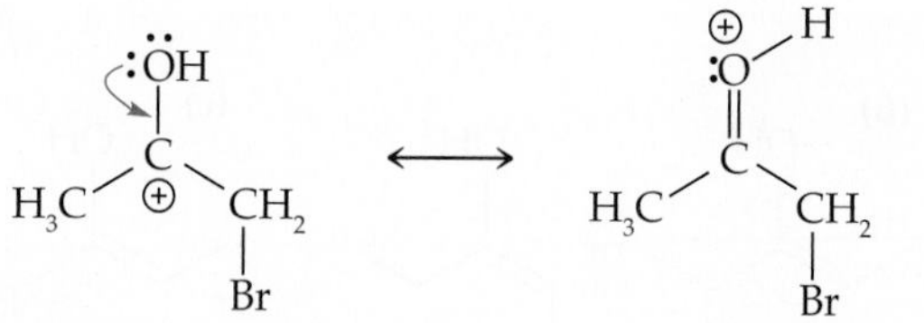

[26.31]

The final stage is loss of a proton to yield the desired α-haloketone.

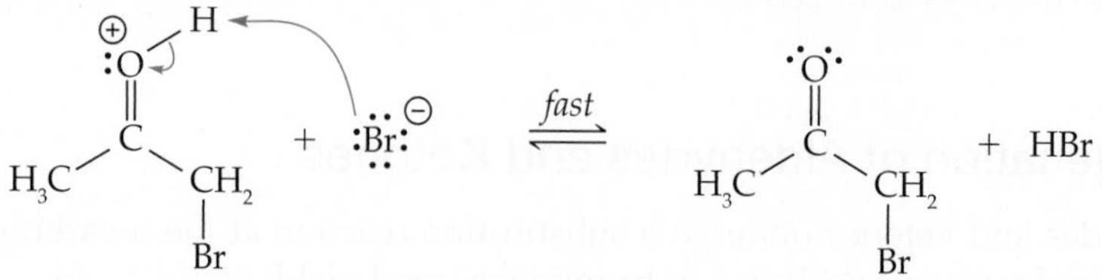

[26.32]

∞ Find out more on page 1183

> **CONCEPT CHECK 5**
> Consider the two structures in Equation 26.31. What relationship do they have to each other?

FIGURE IT OUT

How many chiral centres are there in the open-chain form of glucose?

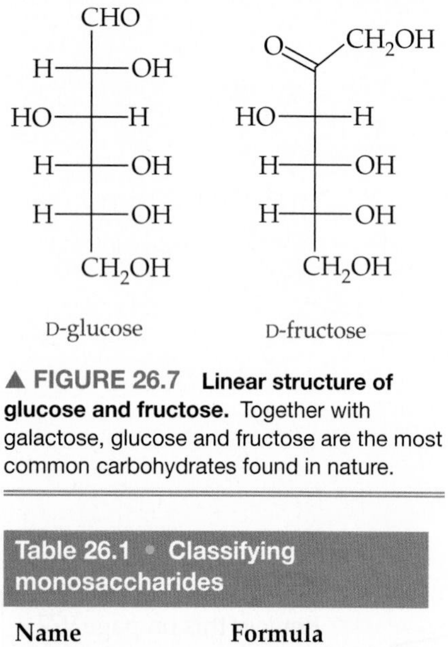

D-glucose   D-fructose

▲ **FIGURE 26.7 Linear structure of glucose and fructose.** Together with galactose, glucose and fructose are the most common carbohydrates found in nature.

Table 26.1 • Classifying monosaccharides

Name	Formula
Triose	$C_3H_6O_3$
Tetrose	$C_4H_8O_4$
Pentose	$C_5H_{10}O_5$
Hexose	$C_6H_{12}O_6$
Heptose	$C_7H_{14}O_7$
Octose	$C_8H_{16}O_8$

∞ Review this on page 931

26.4 | CARBOHYDRATES

Carbohydrates are an important class of naturally occurring substances, found in both plant and animal matter. They perform a variety of vital functions, such as energy storage; they act as structural components in plants; and they are an essential component of the nucleic acids, DNA and RNA (∞ Section 29.4, 'Nucleic Acids and DNA'), which store genetic information. The name **carbohydrate** (hydrate of carbon) comes from the empirical formulae for most substances in this class, which can be written as $C_x(H_2O)_y$ (x, y integers). For example, **glucose**, the most abundant carbohydrate, has the molecular formula $C_6H_{12}O_6$, which could also be written as $C_6(H_2O)_6$. Carbohydrates are not really hydrates of carbon; rather, they are polyhydroxyaldehydes and ketones. Glucose, for example, is a six-carbon aldehyde sugar; *fructose*, the sugar that occurs widely in fruit, is a six-carbon ketone sugar (◀ **FIGURE 26.7**).

Monosaccharides

Both glucose and fructose are examples of **monosaccharides**—simple sugars that can't be broken into smaller molecules by hydrolysis with aqueous acids. Monosaccharides have the general formula $C_nH_{2n}O_n$ (where the integer $n = 3$ to 8). They are classified by their functionality as well as by the number of carbon atoms they contain.

Polyhydroxyaldehydes are known as **aldoses** to signify the aldehyde functional group, while polyhydroxyketones are classified as **ketoses** to signify the ketone group within the molecule. For example, *glucose* is an *aldose* and *fructose* is a *ketose* (see Figure 26.7). ◀ **TABLE 26.1** lists the names of different monosaccharide classes based on the number of carbon atoms within that molecule. Based on this table, glucose can be further classified as an *aldohexose*, and fructose as a *ketohexose*.

The simplest monosaccharides are the trioses: glyceraldehyde and 1,3-dihydroxyacetone (▶ **FIGURE 26.8**). These two molecules are constitutional isomers. Glyceraldehyde, an aldotriose, is chiral because the central carbon atom is a stereocentre (∞ Section 23.3, 'Chirality in Organic Compounds'). The achiral dihydroxyacetone is classified as a ketotriose.

> **CONCEPT CHECK 6**
> Write the molecular and empirical formulae for D-glucose.

⚠ **FIGURE IT OUT**

Why is dihydroxyacetone a chiral?

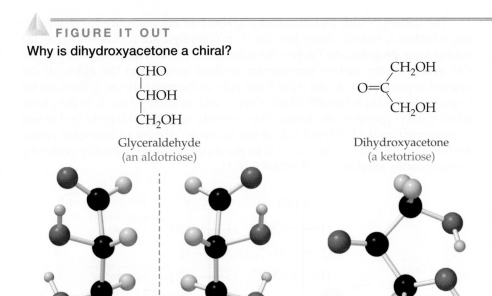

Glyceraldehyde
(an aldotriose)

Dihydroxyacetone
(a ketotriose)

| S | R | achiral |

◀ **FIGURE 26.8   Glyceraldehyde and dihydroxyacetone.** A structural formula and ball-and-stick representations of the two trioses are shown. Glyceraldehyde, which contains a stereocentre, is chiral and can exist as an *S*- or *R*-enantiomer.

Let's return to Figure 26.7. Notice that the structures have been named D-glucose and D-fructose, yet there is no indication of stereochemistry using the now customary bold and dashed wedges. In fact, these two molecules do infer stereochemistry and are being shown as **Fischer projections**. This representation, which is most common among carbohydrates, is a two-dimensional representation of chiral centres. The carbon stereocentres are not shown, which is an indication that Fischer projections are being used. ▼ **FIGURE 26.9** illustrates how the Fischer projections for glyceraldehyde relate to the types of structures you have become familiar with. Fischer projections are drawn so that the most oxidised carbon (usually the aldehyde or ketone) is positioned at the top of the drawing. The vertical lines represent bonds oriented back into the page, while the horizontal lines represent bonds coming out of the page. The point at which they cross indicates a stereocentre.

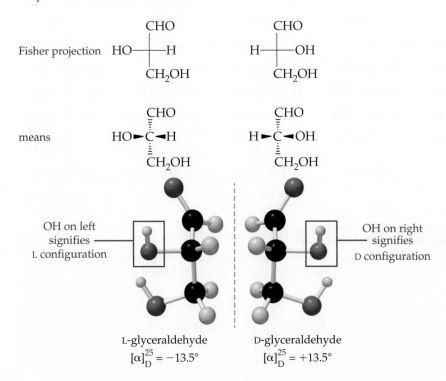

◀ **FIGURE 26.9   Fischer projections.** The D and L nomenclature for monosaccharides is referenced from D- and L-glyceraldehyde. The reference point is the position of the carbon stereocentre furthest away from the carbonyl group (aldehyde or ketone).

The trioses D- and L-glyceraldehyde serve as the reference for determining whether a carbohydrate has the D or L configuration. According to the conventions proposed by Fischer, the D form of a carbohydrate implies that the OH group on the carbon stereocentre furthest away from the aldehyde (or ketone) is positioned on the right-hand side of the molecule, as is the case for D-glyceraldehyde (▼ FIGURE 26.10). Conversely, a carbohydrate is in the L form when the OH group on the furthest stereocentre from the aldehyde (or ketone) is positioned on the left-hand side of the molecule. Almost all biological carbohydrates are of the D form. Several of the more common naturally occurring carbohydrates are shown in ▼ FIGURE 26.11.

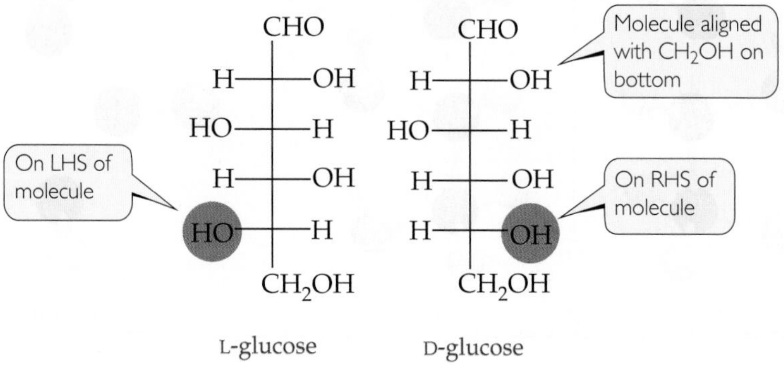

▶ FIGURE 26.10 Fischer projections of D- and L-glucose.

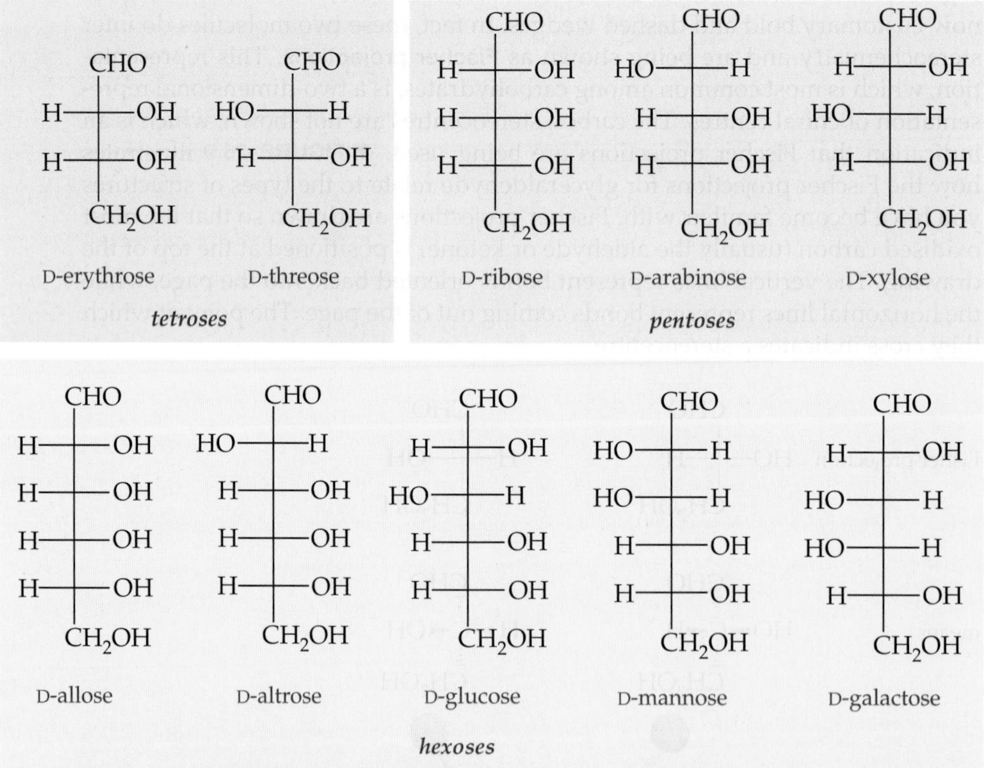

▶ FIGURE 26.11 Configurational relationships between selected naturally occurring aldoses.

SAMPLE EXERCISE 26.6 Fischer projections

Draw the Fischer projection of the following carbohydrate. By what name would you classify this molecule? Is this molecule in the D or L configuration?

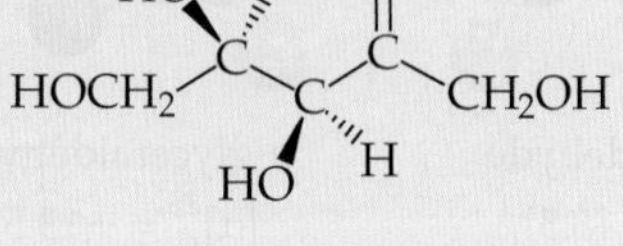

SOLUTION

Analyse We are asked to convert the open-chain form of a carbohydrate into its Fischer projection, class it as aldose or ketose, and determine its configuration using the D, L nomenclature.

Plan To begin, we need to arrange the molecule in such a way that the OH groups and H atoms point out of the page and convert this to a Fischer projection. The Fischer projection will also help determine D, L configuration based on glyceraldehyde. We can classify this carbohydrate based on the ring size formed and the carbonyl functional group contained—whether aldehyde or ketone.

Solve Arranging the structure in such a way that the OH groups and H atoms point out of the page is achieved by rotating about the C3–C4 bond:

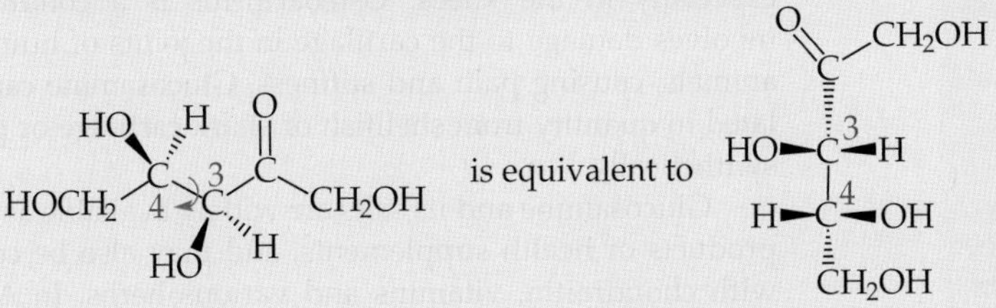

Position the structure in such a way that the ketone group is at the top of the molecule. From this point, the conversion from a stereoview to the Fischer projection is:

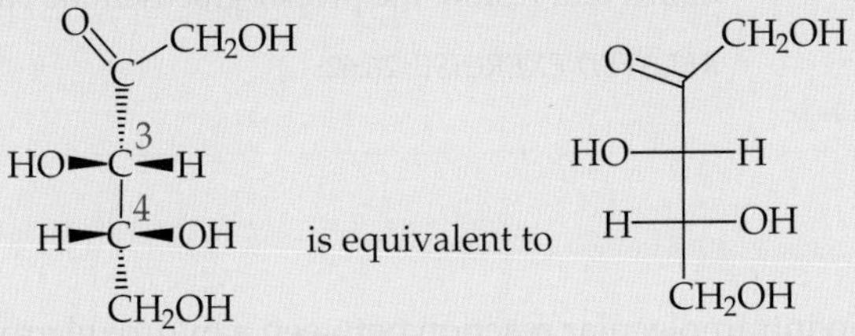

This molecule is classified as a *ketopentose*. It has the D configuration because the OH group on the stereocentre furthest from the ketone is positioned on the RHS, comparable with D-glyceraldehyde.

PRACTICE EXERCISE

Which molecules are D-monosaccharides and which are L-monosaccharides? What are their classifications?

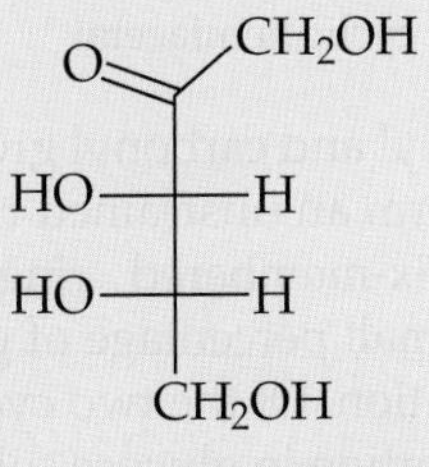

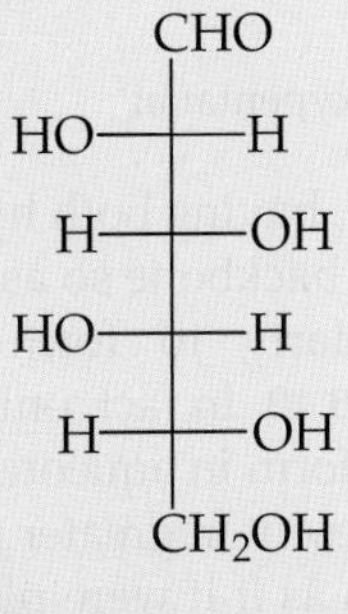

Answers: **(a)** L-aldopentose, **(b)** L-ketopentose, **(c)** D-aldohexose

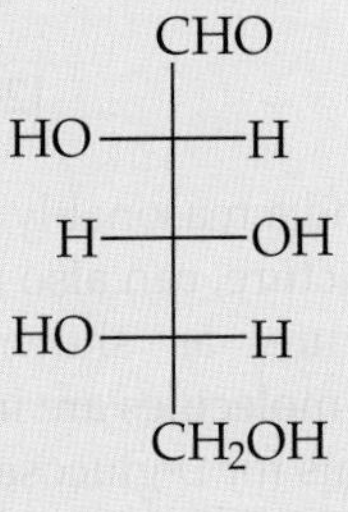

(See also Exercises 26.51, 26.64.)

△ **CONCEPT CHECK 7**

What is the relationship between D- and L-erythrose? What is the relationship between D- and L-ribose?

Cyclic versus Open-Chain Structures

In Section 26.3 we discussed the reaction of aldehydes and ketones with alcohols to form hemiacetals and acetals. We also discussed the formation of cyclic

MY WORLD OF CHEMISTRY

GLUCOSAMINE

Amino sugars such as glucosamine contain an NH_2 group in place of one of the OH groups (▼ FIGURE 26.12). Only three amino sugars are commonly found in nature: D-glucosamine, D-mannosamine and D-galactosamine.

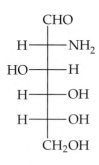

D-glucosamine

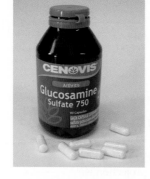

▲ FIGURE 26.12   Glucosamine is sold as a dietary supplement

D-Glucosamine is an amino derivative of glucose that is found especially in the polysaccharides contained within chitin, shellfish and cell membranes. Its use in the formation and repair of cartilage has more recently seen its popularity increase as a *nutraceutic*—a natural product with the capacity to prevent or heal disease.

Glucosamine and its salts, most commonly the hydrochloride or sulfate salts, have been included in recent guidelines as therapy options for people with symptomatic osteoarthritis, especially in the knees. Osteoarthritis is a condition that involves damage to the cartilage in the joints of humans and animals, causing pain and stiffness. Glucosamine can be isolated in quantity from shellfish or shark cartilage or prepared synthetically.

Glucosamine and its salts are widely available as licensed products or health supplements, and may also be combined with chondroitin, vitamins and various herbs. In Australia, glucosamine is classified as a complementary medicine and is available as over-the-counter supplements. It is thought that supplementing glucosamine levels in the body will replace the deficit and restore the proper glucosamine balance.

RELATED EXERCISE: 26.62

hemiacetals by an intramolecular reaction between a hydroxyl group and an aldehyde or ketone. For example, 5-hydroxypentanal forms a cyclic six-membered hemiacetal. Notice that this reaction converts an acyclic, achiral molecule into a cyclic, chiral molecule. The new stereocentre created in forming the cyclic structure is called the **anomeric carbon**.

$$\text{5-hydroxypentanal} \rightleftharpoons \text{a cyclic hemiacetal} \qquad [26.33]$$

Glucose, having both hydroxyl and carbonyl groups, and a reasonably long and flexible backbone so as to form an unstrained ring structure, can also react intramolecularly to form a six-membered ring structure, as shown in ▶ FIGURE 26.13. In fact, only a small percentage of glucose molecules are in the open-chain form in aqueous solution. Of the two cyclic forms for D-glucose, the β form is present in greater proportion in glucose solutions. Although the ring is often drawn as if it were planar, the molecules are actually non-planar because of the tetrahedral bond angles around the C and O atoms of the ring. This planar way of representing the cyclic structures of monosaccharides is known as a **Haworth projection**. Named after W. Norman Haworth (1883–1950), recipient of the 1937 Nobel Prize for Chemistry for research in carbohydrate chemistry, this representation is a simplified way to visualise the complexities of cyclic saccharides. Convention is to write the structure with the anomeric carbon on the right side and the hemiacetal ring oxygen in the back right position.

Figure 26.13 indicates that the ring structure of glucose can have two relative orientations at the anomeric carbon. In the α form, the OH group on the anomeric carbon (C1) and the CH_2OH group on C5 point in opposite directions. In the more stable β form, the OH group on the anomeric carbon and the CH_2OH group on C5 point in the same direction. The α and β forms are called

▲FIGURE IT OUT

How many hydroxyl groups are in an equatorial position in β-glucose?

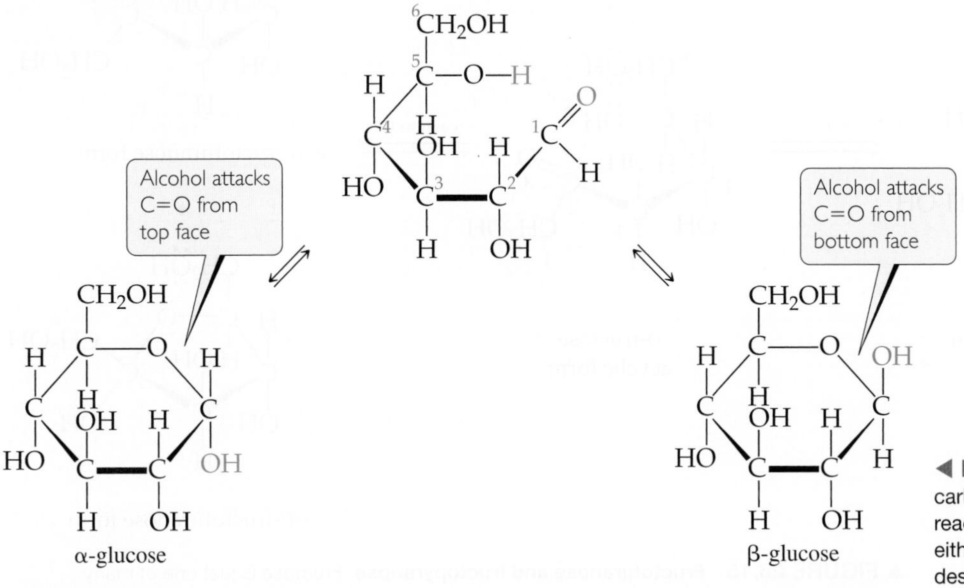

Alcohol attacks C=O from top face

Alcohol attacks C=O from bottom face

α-glucose

β-glucose

◀ **FIGURE 26.13 Cyclic glucose.** The carbonyl group of a glucose molecule can react with one of the hydroxyl groups to form either of two six-membered ring structures, designated α and β.

anomers (▶ **FIGURE 26.14**). Although the difference between the α and β forms might seem small, it has enormous biological consequences. As we will soon see, this one small change in structure accounts for the vast difference in properties between the biopolymers starch and cellulose.

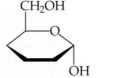

α anomer

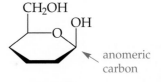

β anomer

▲ **FIGURE 26.14 Anomers.** The α and β anomers differ by their orientation to the CH_2OH group outside the ring.

▲ **CONCEPT CHECK 8**

Why would the β anomer of a cyclic monosaccharide be considered more stable than the α anomer?

Fructose can cyclise to form either five- or six-membered rings (▼ **FIGURE 26.15**). The five-membered ring forms when the OH group on C5 reacts with the carbonyl group on C2 (reaction B in Figure 26.15). The six-membered ring results from the reaction between the OH group on C6 and the carbonyl group on C2 (reaction A in Figure 26.15). Monosaccharides that exist as six-membered rings are named as derivatives of the oxygen-containing heterocycle, pyran. Hence, a cyclic monosaccharide that forms a six-membered ring is named a *pyranose*. Five-membered ringed monosaccharides are named as derivatives of the oxygen-containing heterocycle, furan. Hence, a cyclic monosaccharide that forms a five-membered ring is named a *furanose*.

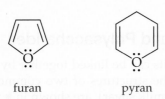

furan   pyran

We are now armed with enough information to name a cyclic monosaccharide unambiguously. For example, the top right-hand structure shown in Figure 26.15 has the name β-D-fructofuranose. This name indicates that the

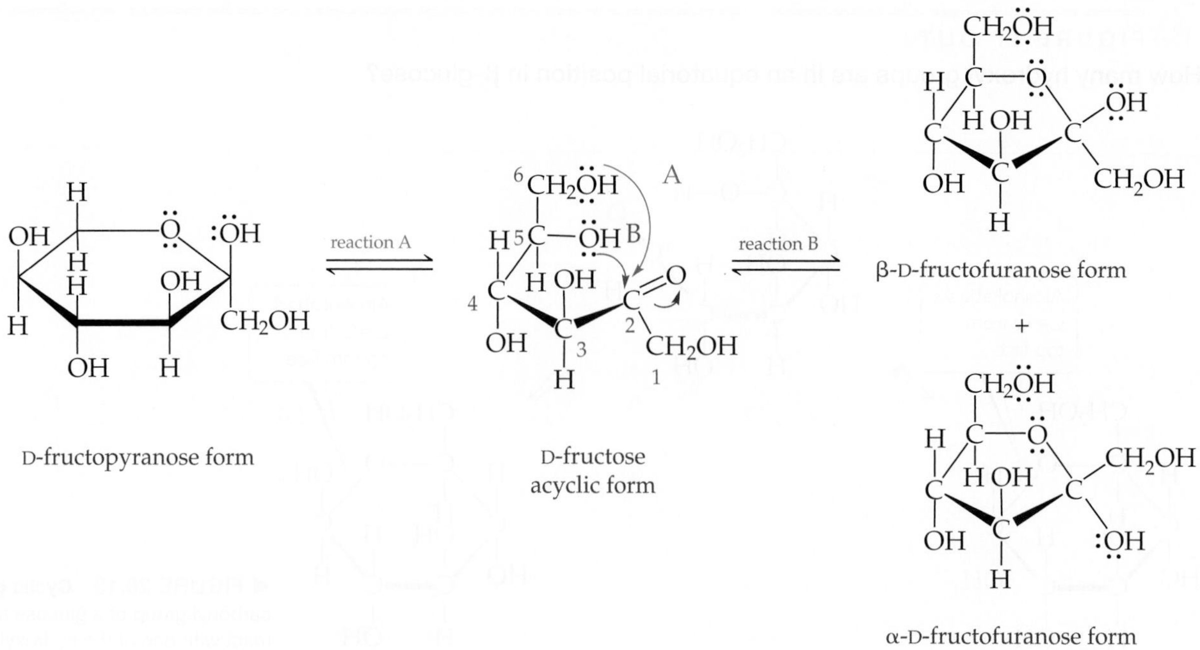

▲ **FIGURE 26.15**   **Fructofuranose and fructopyranose.** Fructose is just one of many monosaccharides that are able to exist in furanose and pyranose forms.

five-membered cyclic monosaccharide is derived from D-fructose. The β anomer has the OH and CH$_2$OH group found on C5 of the furanose form on the same side, which allows us to put in place the only unanswered variable from the open-chain form. As a point of completion, the two structures in Figure 26.13 are fully named α-D-glucopyranose and β-D-glucopyranose, although α-D-glucose and β-D-glucose are also acceptable.

The inference that the α and β anomers are interconvertible through an acyclic intermediate, as shown in Figure 26.15, is entirely true. This process, called **mutarotation**, is common to all carbohydrates that exist in a cyclic hemi-acetal form. Mutarotation is easily observable by monitoring the specific rotation of an aqueous solution of either anomer over time. For example, a solution prepared by dissolving crystalline β-D-galactose in water shows an initial rotation of +151°. Over time, this value decreases until, at equilibrium, a rotation value of +80° is observed (▶ **FIGURE 26.16**). The value of the rotation corresponds to a 72 : 28 ratio of α : β anomers in solution. A similar experiment can be performed using crystalline α-D-galactose. Initially, the rotation of a solution of α-D-galactose in water is +53°. Over time, this value rises as a result of the mutarotation, forming larger quantities of β-D-galactose. At equilibrium, a rotation value of +80° is also achieved (Figure 26.16). The value of the rotation still corresponds to a 72 : 28 ratio of α : β anomers in solution at equilibrium.

Oligosaccharides and Polysaccharides

Two monosaccharide units can be linked together by a condensation reaction to form a *disaccharide*. The structures of two common disaccharides, *sucrose* (table sugar) and *lactose* (milk sugar), are shown in ▶ **FIGURE 26.17**. The word *sugar* makes us think of sweetness. All sugars are sweet, but they differ in the

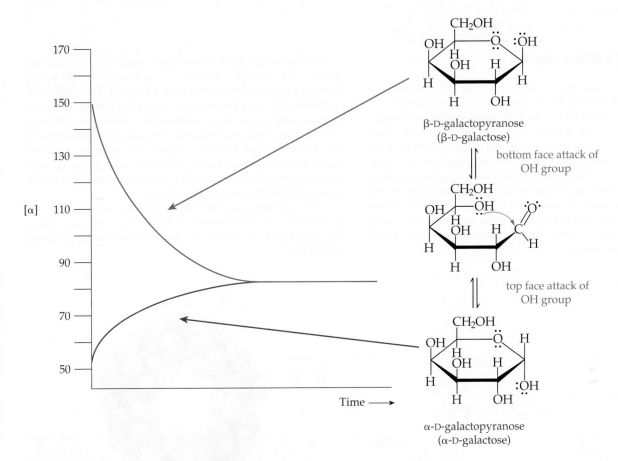

▲ **FIGURE 26.16** **Mutarotation of D-galactose.** At equilibrium, D-galactose exists as a 72 : 28 mixture of α : β anomers. The mutarotation process is easily followed by changes in the specific rotation of pure samples of both anomers.

degree of sweetness we perceive when we taste them. Sucrose is about six times sweeter than lactose, slightly sweeter than glucose, but only about half as sweet as fructose. Disaccharides can react with water (hydrolysis) in the presence of an acid catalyst to form two monosaccharides. When sucrose is hydrolysed, the mixture of glucose and fructose that forms, called *invert sugar*, is sweeter to the taste than the original sucrose. The sweet syrup present in canned fruits and lollies is largely invert sugar formed from the hydrolysis of added sucrose.

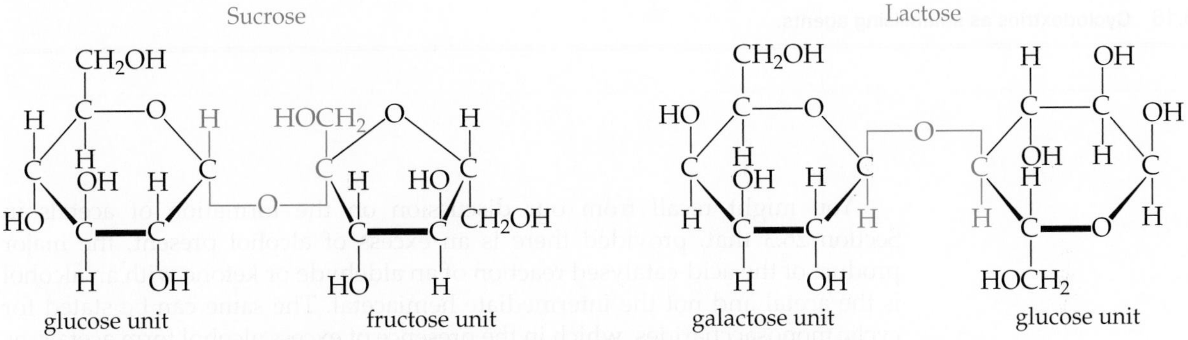

▲ **FIGURE 26.17** **Two disaccharides.** The structures of sucrose (left) and lactose (right).

MY WORLD OF CHEMISTRY

CYCLODEXTRINS

Cyclodextrins, such as the one shown in ▼ FIGURE 26.18, are *macrocyclic* compounds with the ability to include small organic molecules within their cavity. These cyclic molecules, which are composed of glucose units, have a *hydrophobic* (water-hating) interior and a *hydrophilic* (water-loving) exterior, due to the fact that the OH groups radiate out from the macrocycle rather than in towards the centre. This feature causes cyclodextrins to be water soluble. Molecules that are hydrophobic (that is, do not mix with water) are easily included within the cyclodextrin cavity, provided they are not too large.

Cyclodextrins have found uses recently as solubilising agents. For example, toluene is immiscible with water, forming a separate layer when added to water. The addition of cyclodextrin to the aqueous layer causes the two layers to mix by providing a way of optimising the interactions between cyclodextrin, water and toluene. The use of cyclodextrins in the pharmaceutical industry is ever-increasing as scientists realise their benefits as both a non-toxic solubilising agent at physiological pH and as a chiral 'reaction vessel' with which to carry out enantioselective reactions.

Cyclodextrins are formed by the enzymatic action of cyclodextrin glucanotransferase (CGTase) on glucose. Using this green process, tonnes of cyclodextrin can be prepared industrially.

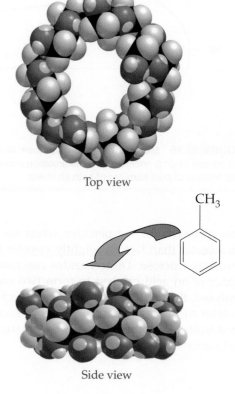

Top view

Side view

▲ FIGURE 26.18 Cyclodextrins as solubilising agents.

You might recall from our discussion on the formation of acetals in Section 26.3 that, provided there is an excess of alcohol present, the major product of the acid-catalysed reaction of an aldehyde or ketone with an alcohol is the acetal and not the intermediate hemiacetal. The same can be stated for cyclic monosaccharides, which in the presence of excess alcohol form acetals, as illustrated by the following reactions:

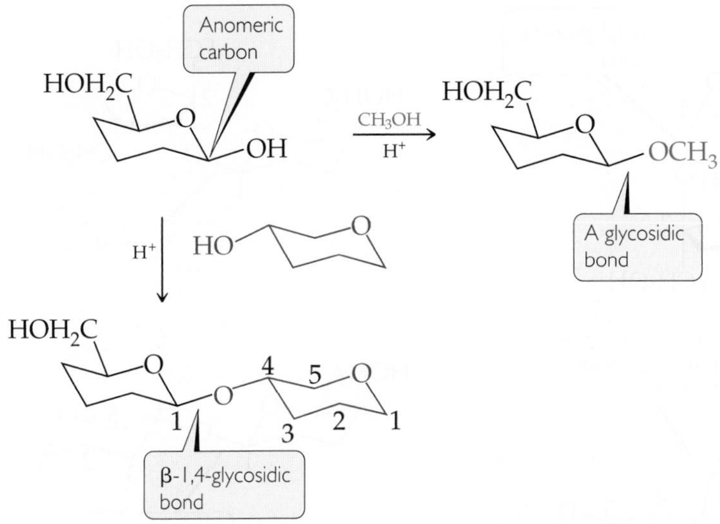

[26.34]

A cyclic acetal formed from a monosaccharide is called a **glycoside**. Both sucrose and lactose are *glycosides*. The bond between the anomeric carbon and the exocyclic alkoxy group of the acetal is called a **glycosidic bond**. Glycosides are stable in water, unlike simple acetals, and *do not* undergo mutarotation since, unlike the hemiacetal, the glycoside is not in equilibrium with its open-chain form in neutral solution. Hydrolysis of the glycoside is possible in aqueous acid, as is shown by the example of invert sugar.

The mechanism for the formation of the glycoside from the cyclic hemiacetal is illustrated in the set of equilibria given in Equation 26.35 for a simplified β-monosaccharide and the general alcohol, ROH. The acid-catalysed reaction occurs mainly by an elimination pathway. Elimination of H_2O by use of one of the lone pairs on the endocyclic oxygen yields the cationic intermediate (Equation 26.35). Attack at what used to be the anomeric carbon by ROH leads to an intermediate oxonium ion, which on proton loss yields the glycoside. This reaction pathway is unique to carbohydrate chemistry.

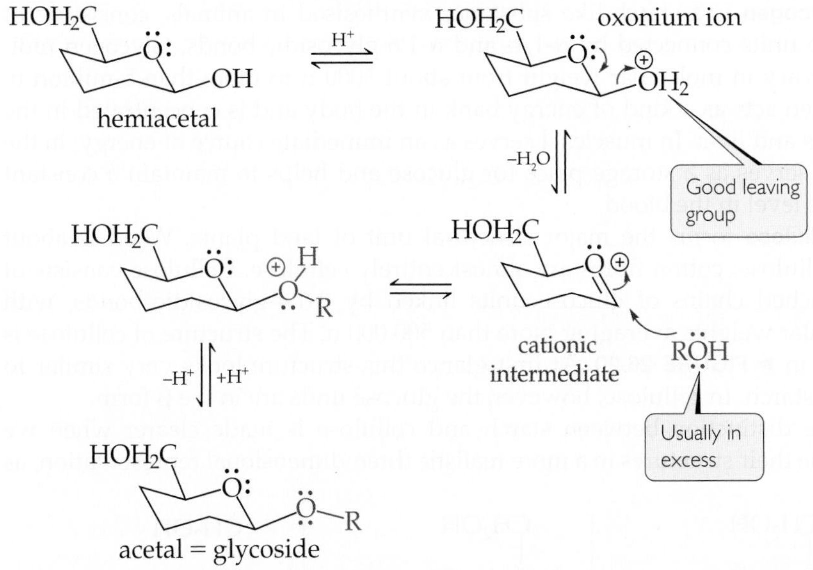

[26.35]

The position of the glycosidic bond in a carbohydrate is important for the properties it exhibits. Glycosidic bonds are classified by their orientation—that is, whether they are α or β bonds (equating to axial and equatorial bonds in the chair conformation, respectively), as well as on which carbons link the two monosaccharide subunits. Some examples of naming glycosidic bonds are:

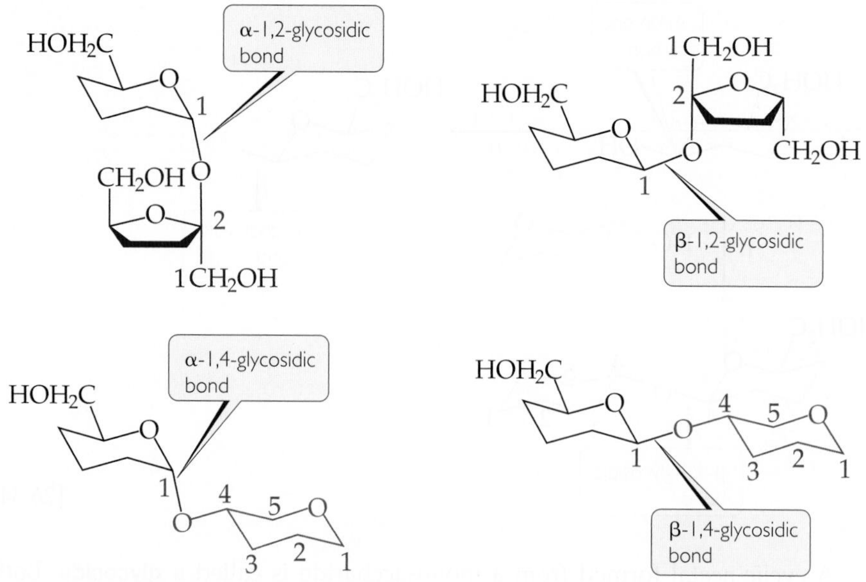

α-1,2-glycosidic bond

β-1,2-glycosidic bond

α-1,4-glycosidic bond

β-1,4-glycosidic bond

Polysaccharides are made up of many monosaccharide units, joined together by a bonding arrangement similar to those shown for the disaccharides in Figure 26.17. The most important polysaccharides are starch, glycogen and cellulose, which are formed from repeating glucose units.

Starch is not a pure substance. The term refers to a group of polysaccharides found in plants, composed of glucose units bound by α-1,4-glycosidic bonds. Starches serve as a major method of food storage in plant seeds and tubers. Corn, potatoes, wheat and rice all contain substantial amounts of starch. These plant products serve as major sources of food energy for humans. Enzymes within the digestive system called *amylases* catalyse the hydrolysis of starch to glucose. Starch molecules that form unbranched chains are called **amyloses**, whereas those that are branched are called **amylopectins**. ▼ FIGURE 26.19 illustrates an unbranched structure of starch. In this structure, the glucose units are in the α form.

Glycogen is a starch-like substance synthesised in animals, consisting of glucose units connected by α-1,4- and α-1,6-glycosidic bonds. Glycogen molecules vary in molecular weight from about 5000 u to more than 5 million u. Glycogen acts as a kind of energy bank in the body and is concentrated in the muscles and liver. In muscles it serves as an immediate source of energy; in the liver it serves as a storage place for glucose and helps to maintain a constant glucose level in the blood.

Cellulose forms the major structural unit of land plants. Wood is about 50% cellulose; cotton fibres are almost entirely cellulose. Cellulose consists of unbranched chains of glucose units linked by β-1,4-glycosidic bonds, with molecular weights averaging more than 500 000 u. The structure of cellulose is shown in ▶ FIGURE 26.20. At first glance this structure looks very similar to that of starch. In cellulose, however, the glucose units are in the β form.

The distinction between starch and cellulose is made clearer when we examine their structures in a more realistic three-dimensional representation, as

▶ **FIGURE 26.19 Structure of a starch molecule.** The molecule consists of many glucose units of the kind enclosed in brackets, joined by linkages of the α form. Note the glycosidic bond (blue) is often shown with a 90°-kink in it. This is for ease of representation and does *not* indicate an extra carbon atom.

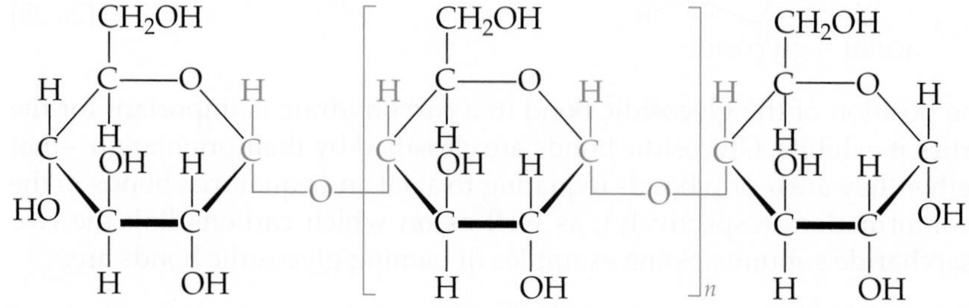

MY WORLD OF CHEMISTRY

VITAMIN C

L-Ascorbic acid, more commonly known as vitamin C, is derived from D-glucose via the uronic acid pathway. Although many animals synthesise vitamin C, primates (and guinea pigs) cannot. The structural formula of L-ascorbic acid resembles the furanose form of a cyclic monosaccharide.

The most important reaction requiring ascorbate as a cofactor is the hydroxylation of proline residues in collagen. This reaction is required for the maintenance of normal connective tissue, as well as for healing wounds. The best-known functions of vitamin C are its use as a reducing agent and antioxidant. Deficiency in vitamin C leads to the disease *scurvy*. Scurvy is characterised by easily bruised skin, muscle fatigue, soft swollen gums, decreased wound healing, haemorrhaging, osteoporosis and anaemia.

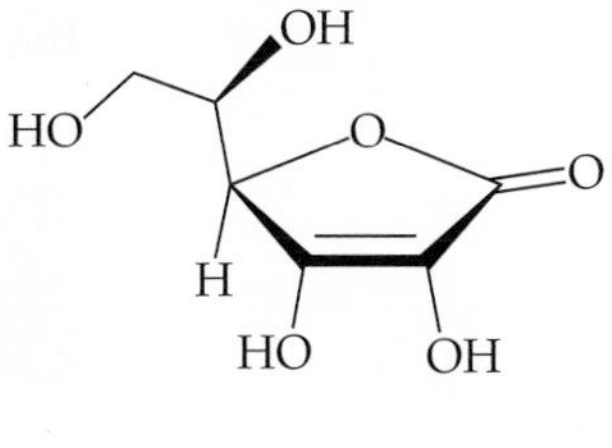

vitamin C

RELATED EXERCISE: 26.66

shown in ▼ **FIGURE 26.21** The individual glucose units have different relationships to one another in the two structures. Because of this fundamental difference, enzymes that readily hydrolyse starches do not hydrolyse cellulose.

You might eat a kilogram of cellulose and receive no caloric value from it whatsoever, even though the heat of combustion per unit weight is essentially the

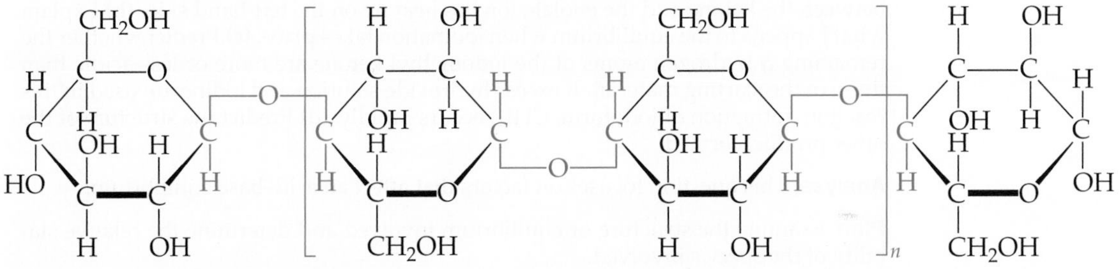

▲ **FIGURE 26.20 Structure of cellulose.** Like starch, cellulose is a polymer. The repeating unit is shown between brackets. The linkage in cellulose is of the β form, different from that in starch (see Figure 26.19).

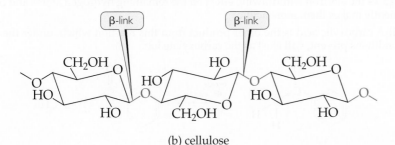

(a) starch

α-1,4-glycosidic linkages

β-link   β-link

(b) cellulose

◀ **FIGURE 26.21 Structural differences in starch and cellulose.** These representations show the geometrical arrangements of bonds about each carbon atom. The glucose rings are orientated differently with respect to one another in the two structures.

same for both cellulose and starch. A kilogram of starch, in contrast, would represent a substantial caloric intake. The difference is that the starch is hydrolysed to glucose, which is eventually oxidised with release of energy. Cellulose, however, is not readily hydrolysed by enzymes present in the body, so it passes through the digestive system relatively unchanged. Many bacteria contain enzymes, called *cellulases*, that hydrolyse cellulose. These bacteria are present in the digestive systems of grazing animals, such as cattle, which use cellulose for food.

CONCEPT CHECK 9

Which type of linkage would you expect to join the sugar molecules of a glycogen?

SAMPLE INTEGRATIVE PROBLEM   Putting concepts together

The halogenation of an aldehyde or ketone is promoted by the presence of a base, for example:

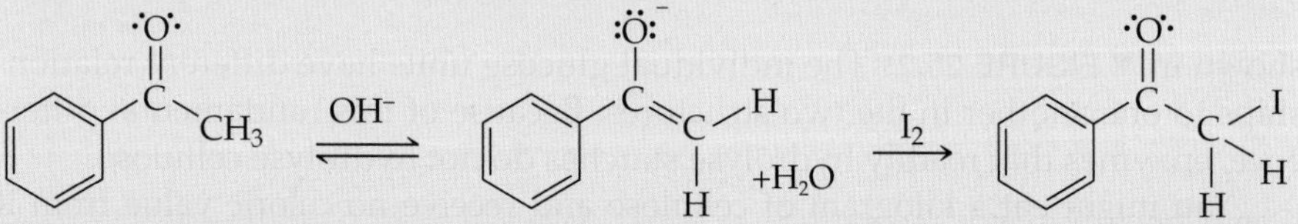

A hydrogen attached to a carbon atom is not normally acidic. **(a)** Suggest a reason why the hydrogen atom α to the carbonyl group may be lost as H^+. The equilibrium between the ketone and the enolate ion lies heavily on the left-hand side. **(b)** Explain what happens to the equilibrium when iodination takes place. **(c)** Predict whether the remaining α hydrogen atoms of the iodomethyl ketone are more or less acidic than those of the starting material. If excess hydroxide solution and iodine are used in this reaction, formation of iodoform, CHI_3, occurs rapidly. **(d)** Predict the structure of the other product formed.

Analyse This question focuses on factors that affect an acid–base equilibrium.

Plan Examine the structure or equilibrium involved and determine the relative stability of the species involved.

Solve
(a) There are two factors at work here. The carbonyl group has an electron withdrawing influence on the adjacent atoms which increases the ease with which H^+ is lost from the α carbon atom and the resulting anion is resonance stabilised.

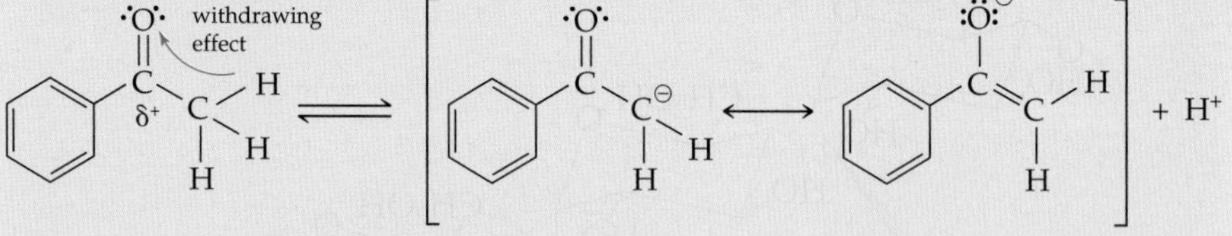

(b) Reaction of iodine with the enolate ion lowers the concentration of the enolate ion, which shifts the equilibrium to the right-hand side. Even though there is a very small amount of enolate present at any time, the reaction can go almost to completion as more enolate ion is generated as the iodination reaction proceeds.

(c) Substitution of a hydrogen atom with a more electronegative halogen atom increases the electron withdrawing effect on the remaining hydrogen atoms and consequently makes them more acidic.

(d) A carboxylic acid is the other product from this reaction which, under the basic conditions present, will exist as the carboxylate ion.

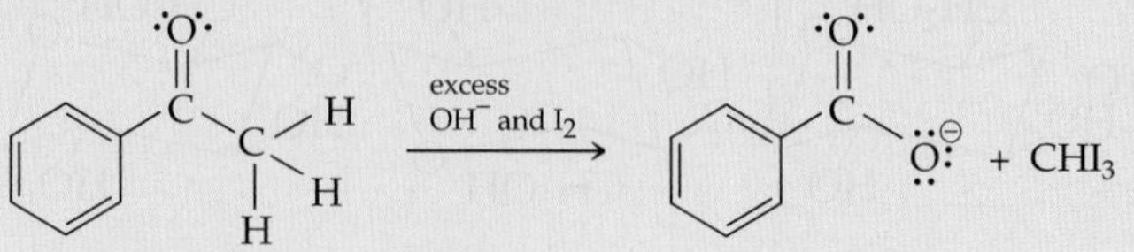

CHAPTER SUMMARY AND KEY TERMS

SECTIONS 26.1 and 26.2 **Aldehydes** and **ketones** are compounds containing a **carbonyl group** (C=O). Aldehydes contain the functional group CHO, whereas ketones contain a carbonyl group within a hydrocarbon chain. The polarity of the carbonyl group makes it susceptible to nucleophilic attack on carbon. Aldehydes and ketones can be prepared by the oxidation of 1° or 2° alcohols, respectively. Another approach to their preparation is through the **ozonolysis** of alkenes.

SECTION 26.3 The conversion of ketones and aldehydes to alcohols can occur by attack of an **organometallic compound** such as a **Grignard reagent** (RMgX) that acts as the nucleophile. Grignard reagents are equivalent to **carbanions** in reactivity. Reaction of aldehydes and ketones with amines yield **imines** and with alcohols yield **hemiacetals** and **acetals**. Imines can be catalytically hydrogenated to yield amines. **Metal hydrides** such as sodium borohydride act as a source of **hydride ion**, which can reduce ketones and aldehydes to alcohols. The **nucleophilic addition** reaction of aldehydes and ketones with cyanide ion (⁻CN) leads to the formation of **cyanohydrins**, which can be converted to α-hydroxy acids by hydrolysis of the nitrile group. Compounds containing the CN functional group are often called **nitriles**. **Tautomerism** occurs in carbonyl compounds bearing an **α-hydrogen**.

SECTION 26.4 **Carbohydrates**, which are polyhydroxyaldehydes and ketones, are the major structural constituents of plants and are a source of energy in both plants and animals.

Glucose is the most common **monosaccharide**, or simple sugar. Monosaccharides are classed as **aldoses** or **ketoses** depending on what functional group is present in its open-chain form. The open-chain form of a monosaccharide is often drawn as a **Fischer projection**, which is an easy way to determine whether the monosaccharide is of the D- or L-form.

Monosaccharides are usually drawn as a **Haworth projection** in their cyclic form. Two cyclic forms are common: the pyranose (six-membered ring) and furanose (five-membered ring) forms. Monosaccharides are classed as one of two **anomers** according to the stereochemistry of the groups on the **anomeric carbon**. **Mutarotation** occurs when the two anomeric forms come to equilibrium in solution. A **glycoside** is formed when some group other than OH is bonded to the anomeric carbon. Glycosides do not interconvert between anomeric forms and do not exist in open chain and cyclic forms. The link between two monosaccharide units via the anomeric carbon is termed the **glycosidic bond**.

Two monosaccharides can be linked together by means of a condensation reaction to form a disaccharide. **Polysaccharides** are complex carbohydrates made up of many monosaccharide units joined together. The three most important polysaccharides are **starch**, which is found in plants; **glycogen**, which is found in animals; and **cellulose**, which is also found in plants. Starch molecules that form largely unbranched chains are called **amylose**, whereas those that are highly branched are called **amylopectins**.

KEY SKILLS

- Understand how to identify aldehyde and ketone functional groups and use systematic approaches to naming. (Section 26.1)
- Be able to predict that the oxidation of primary alcohols gives aldehydes and oxidation of secondary alcohols give ketones and be able to quote the reagents necessary to undertake these interconversions. (Section 26.2)
- Understand the generality of nucleophilic addition to carbonyl compounds in the context of acetals, hemiacetals, imines and cyanohydrins. (Section 26.3)
- Be able to form and use Grignard reagents for nucleophilic addition. (Section 26.3)
- Be able to identify the major carbohydrates and identify whether a sugar is a mono-, di- or polysaccharide. (Section 26.4)
- Understand the stereochemistry of carbohydrates, the difference between α- and β-anomers, D and L and how it relates to carbohydrates. (Section 26.4)
- Be able to draw and interpret Fischer projections. (Section 26.4)

KEY EQUATIONS

- Oxidation of 1° alcohols

$$RCH_2OH \xrightarrow{\text{PCC}} RCHO$$
$$\xrightarrow{\text{chromic acid}} RCOOH$$

[26.1]

- Oxidation of 2° alcohols

[26.2]

- Ozonolysis

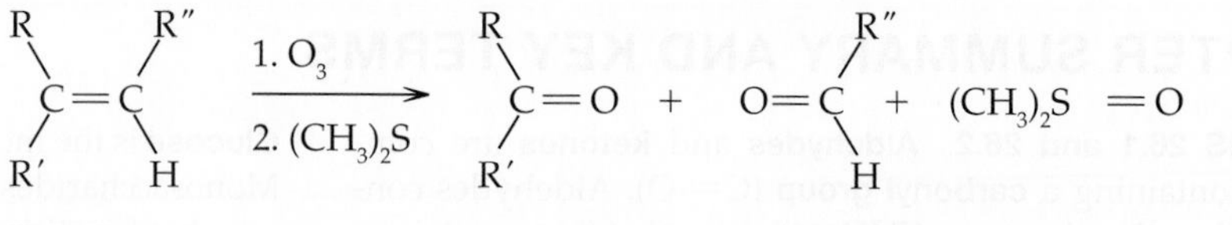

$$[26.3]$$

an alkene   a ketone   an aldehyde   dimethylsulfoxide

- Nucleophilic addition reaction

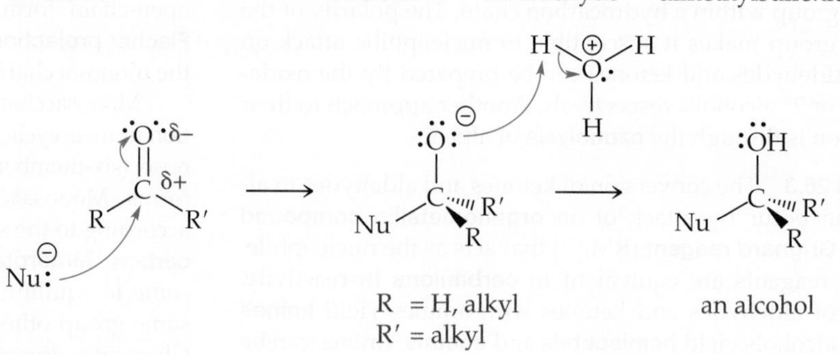

$$[26.5]$$

R = H, alkyl   an alcohol
R′ = alkyl

- Tautomerisation

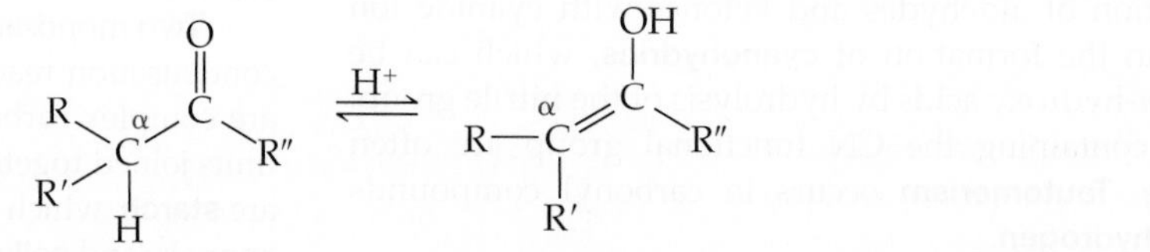

$$[26.23]$$

keto form   *enol form*
aldehyde or ketone

EXERCISES

VISUALISING CONCEPTS

26.1 Draw the structure of the alcohol precursor, its systematic name and the reaction conditions you might use to obtain the following compounds by oxidation. [Section 26.2]

(a)

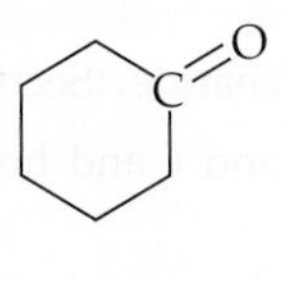

(b)

$CH_3CH_2CH_2CH_2CH_2CHO$

(c)

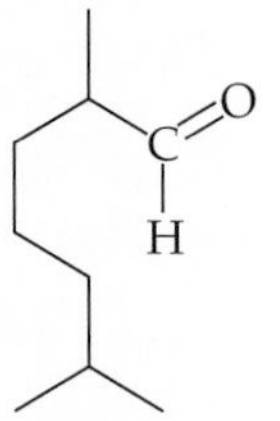

(d)

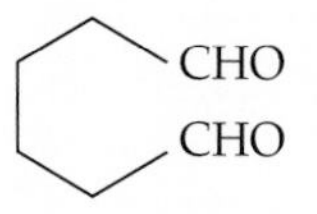

26.2 Piperonyl butoxide is used commercially as an additive to pyrethrum-based insecticides to enhance their effect. In this case, the cyclic ether is prepared by the action of formaldehyde on a catechol derivative. By what other functional group might this cyclic ether be named? [Section 26.3]

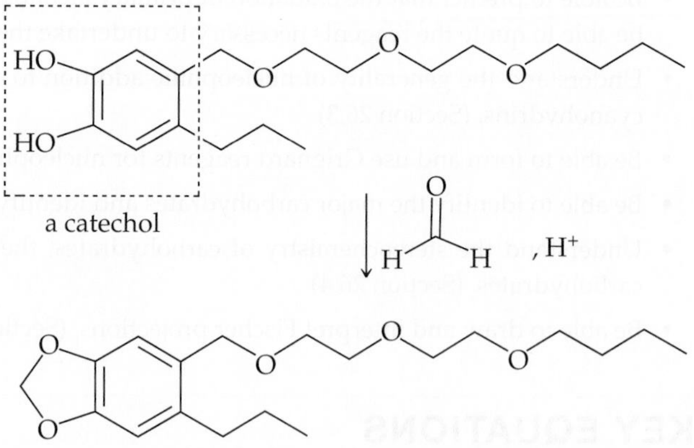

a catechol

piperonyl butoxide

26.3 The molecule *n*-octylglucoside, shown below, is widely used in biochemical research as a non-ionic detergent for 'solubilising' large hydrophobic protein molecules. What characteristics of this molecule are important for its use in this way? [Section 26.4]

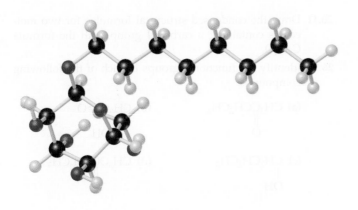

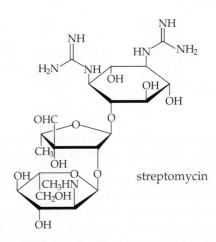

streptomycin

26.4 Streptomycin is a potent antibiotic produced by certain soil types. Its structure is shown here. Highlight the furanose and pyranose rings in streptomycin. [Section 26.4]

ALDEHYDES, KETONES AND THE CARBONYL GROUP (Section 26.1)

26.5 Classify the following compounds as acetals, alcohols, hemiacetals, esters or ethers:

(a)

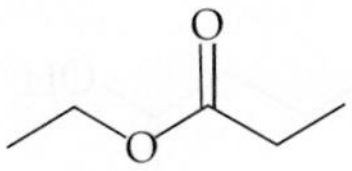

(b)

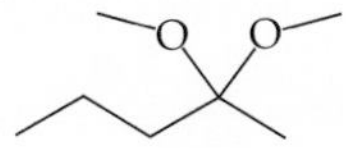

(c)

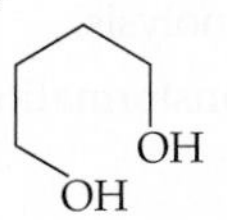

(d)

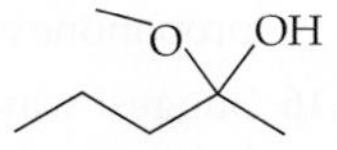

(e)

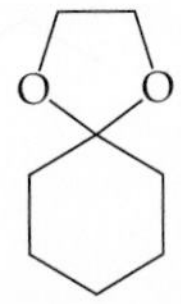

(f)

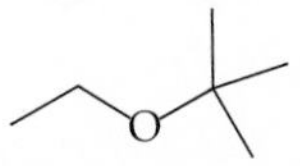

26.6 Compounds that contain both a carbonyl and alcohol functional groups usually exist as either the cyclic acetal or hemiacetal forms in preference to the open-chain form. Some examples are shown below. Deduce the structure of the open-chain form.

(a)

(b)

(c)

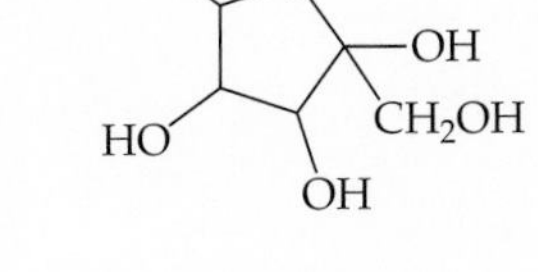

(d)

fructose

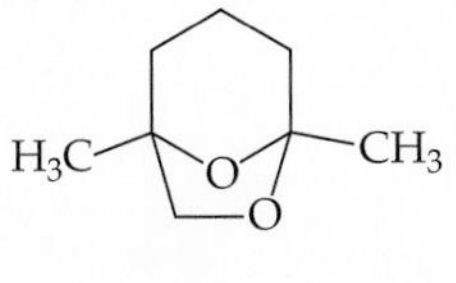

frontalin
(beetle pheremone)

26.7 Identify the functional groups in each of the following compounds:

(a) $HC\equiv C-CH_2-\overset{\overset{\displaystyle O}{\|}}{C}-H$

(b) $-CH_2-CH=CHCH_2COOH$

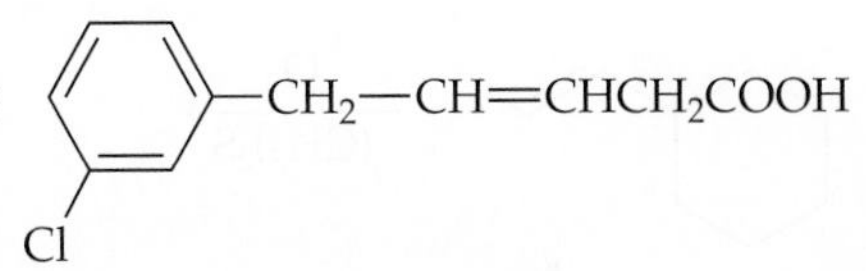

(c) H_2C ... CH_2 ... H_2C-CH ... Cl

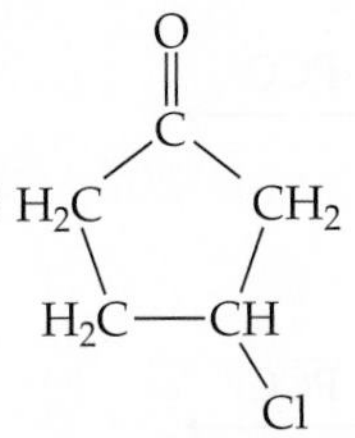

(d) $CH_3CH=CH\overset{\overset{\displaystyle O}{\|}}{C}-OCH_2CH_3$

(e) $CH_2-\overset{\overset{\displaystyle O}{\|}}{C}-N(CH_3)_2$

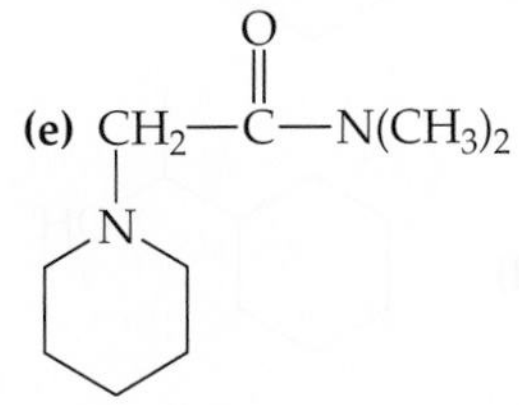

26.8 Give the structural formula for **(a)** an aldehyde that is an isomer of acetone, **(b)** an ether that is an isomer of propanone.

26.9 Name the following compounds:

(a)

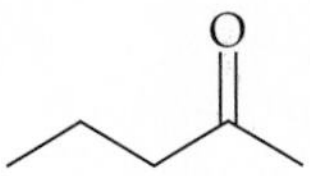

(b)

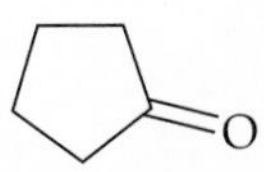

(c)

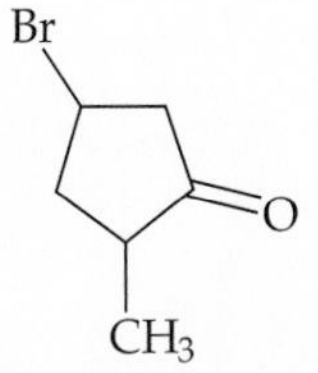

(d)

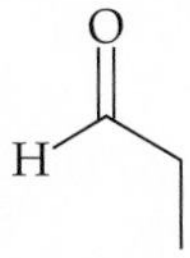

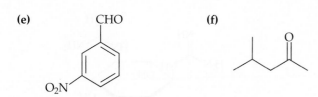

(e) CHO, O₂N substituent

(f)

26.10 Aldehydes and ketones can be named in a systematic way by counting the number of carbon atoms (including the carbonyl carbon) they contain. The name of the aldehyde or ketone is based on the hydrocarbon with the same number of carbon atoms. The ending *-al* for aldehyde or *-one* for ketone is added as appropriate. Draw the structural formulae for the following aldehydes or ketones: **(a)** propanal, **(b)** 2-pentanone, **(c)** 3-methylbutanone, **(d)** 2-methylbutanal.

26.11 Draw the condensed structural formulae for two molecules containing a carbonyl group with the formula C_3H_4O.

26.12 Identify the functional groups in each of the following compounds:

(a) $CH_3CCH_2CH_3$, with $\overset{\|}{O}$

(b) $CH_3C=O$, with OH

(c) $CH_2CH_2CH_3$, with OH

(d) $CH_3OCCH_2CH_3$, with $\overset{\|}{O}$

(e) H_2NCCH_3, with $\overset{\|}{O}$

(f) $CH_3CH_2NHCH_3$

PREPARATION OF ALDEHYDES AND KETONES (Section 26.2)

26.13 Provide the structure of the major organic products in the following reactions:

(a)

OH
$\xrightarrow[\text{H}_2\text{SO}_4]{\text{Na}_2\text{Cr}_2\text{O}_7}$

(b)

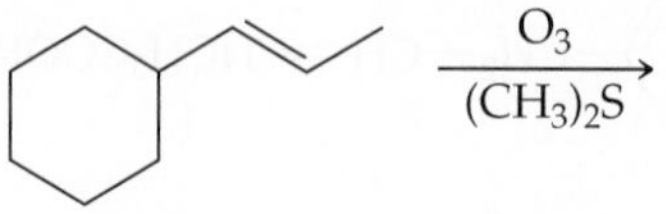

$\xrightarrow[\text{(CH}_3)_2\text{S}]{\text{O}_3}$

(c)

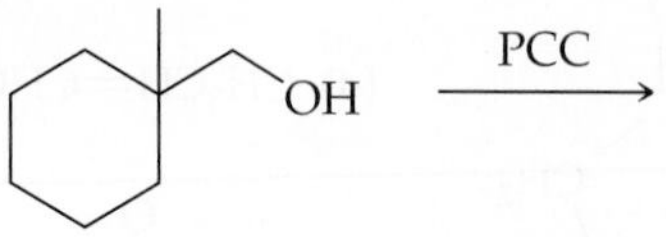

OH
$\xrightarrow{\text{PCC}}$

(d)

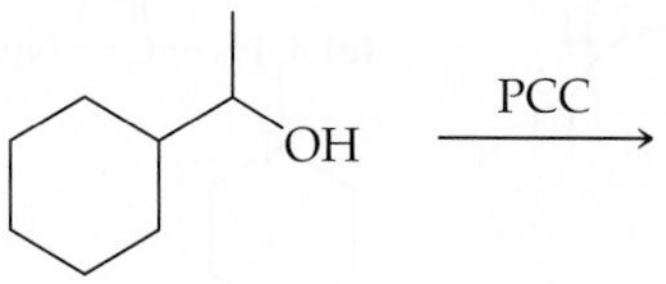

OH
$\xrightarrow{\text{PCC}}$

(e)

$\xrightarrow[\text{(CH}_3)_2\text{S}]{\text{O}_3}$

26.14 An unknown alkene was treated with ozone/dimethylsulfide and then with $K_2Cr_2O_7$, forming the following product. What was the alkene?

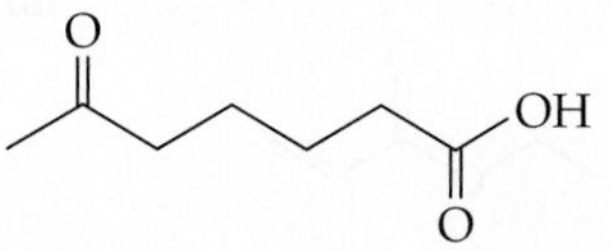

26.15 Give the structure of the alkene which would yield propanone and propanal upon ozonolysis.

26.16 Suggest ways of conducting the transformations shown below.

(a)

OH
$\longrightarrow$
O

(b)

O
$\longrightarrow$
OH

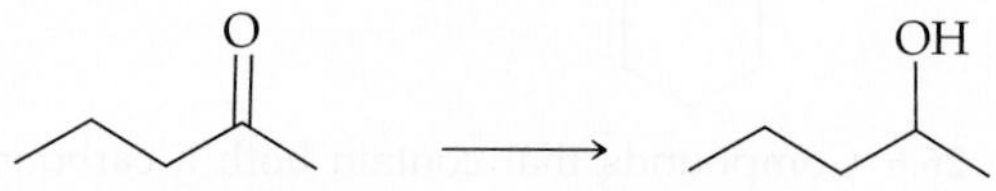

26.17 Suggest a possible stable reaction product for each of the following reactions:

(a)

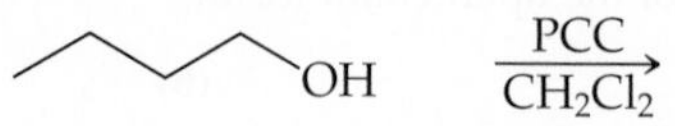

OH
$\xrightarrow{\text{PCC}}{\text{CH}_2\text{Cl}_2}$

(b)

OH
$\xrightarrow[\text{H}_2\text{SO}_4]{\text{CrO}_3}$

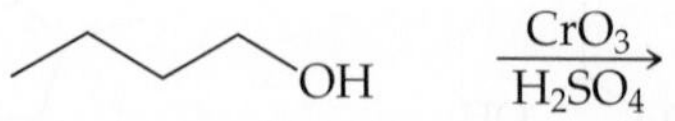

REACTIONS OF ALDEHYDES AND KETONES (Section 26.3)

26.18 Provide the major organic product(s) of the reactions below:

(a)

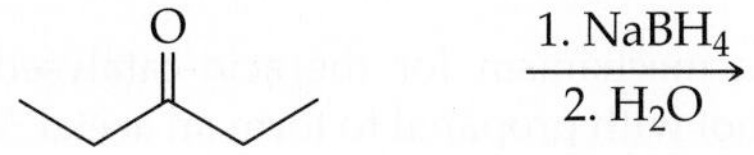

$$\xrightarrow{\text{1. NaBH}_4}_{\text{2. H}_2\text{O}}$$

(b)

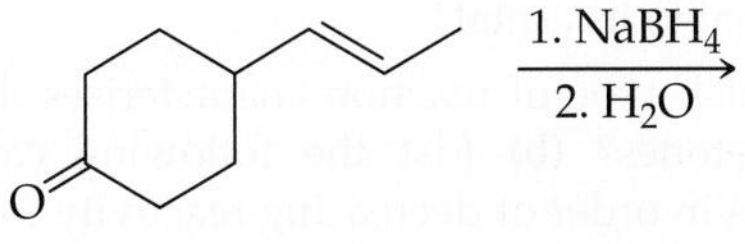

$$\xrightarrow{\text{1. NaBH}_4}_{\text{2. H}_2\text{O}}$$

(c)

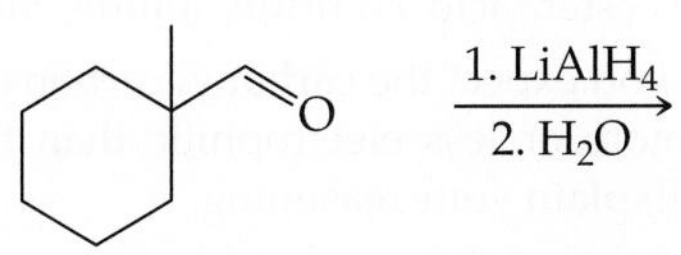

$$\xrightarrow{\text{1. LiAlH}_4}_{\text{2. H}_2\text{O}}$$

(d)

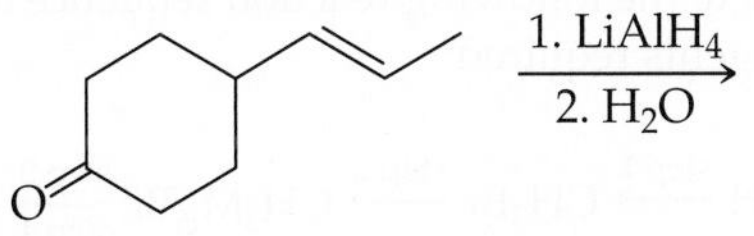

$$\xrightarrow{\text{1. LiAlH}_4}_{\text{2. H}_2\text{O}}$$

(e)

$$\xrightarrow{\text{1. NaBH}_4}_{\text{2. H}_2\text{O}}$$

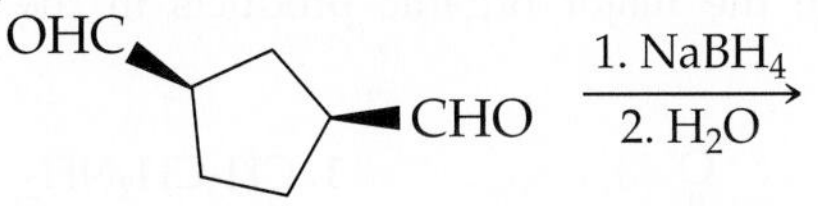

26.19 Which reducing agent is best used in the reaction shown below?

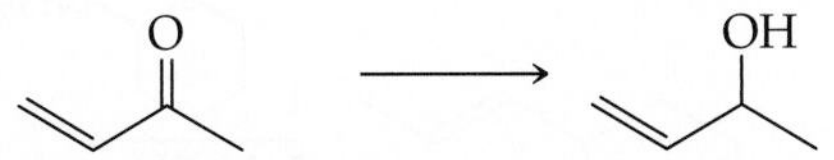

26.20 How would you accomplish the transformations shown below?

(a)

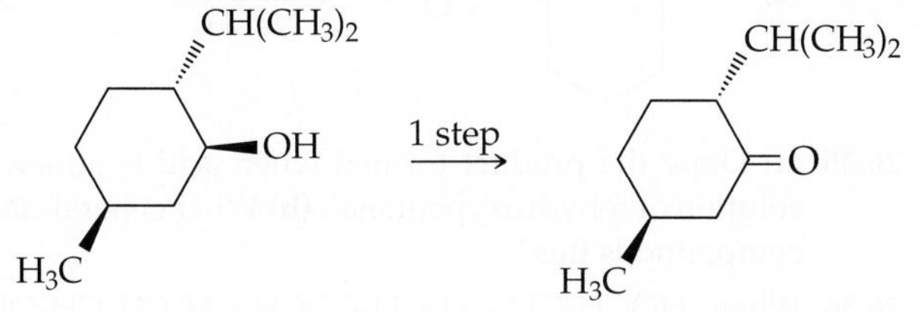

CH(CH₃)₂ ⋯ OH $\xrightarrow{\text{1 step}}$ CH(CH₃)₂ ⋯ =O

H₃C

(b)

CH(CH₃)₂ ⋯ OH $\xrightarrow{\text{1 step}}$ CH(CH₃)₂

H₃C

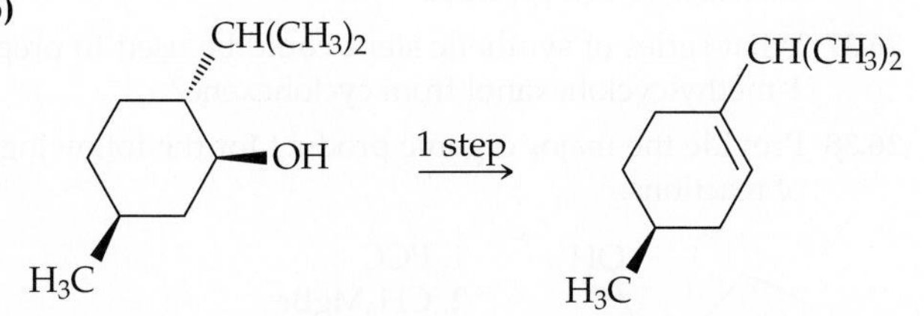

26.21 (a) What reagents would you use to undertake the following chemical transformation? **(b)** What solvent would you use to convert benzaldehyde into benzyl alcohol?

(c) Name two features of your solvent choice that make it an appropriate solvent.

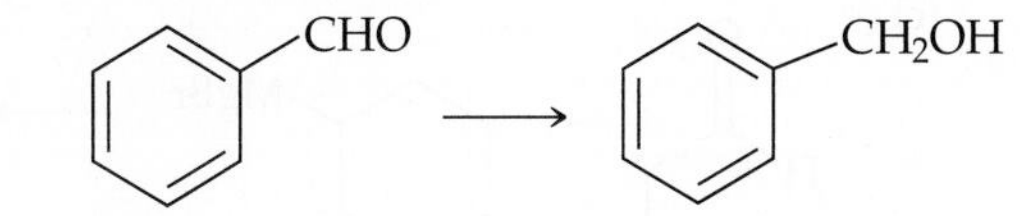

CHO → CH₂OH

benzaldehyde   benzyl alcohol

26.22 Draw the likely product for each transformation.

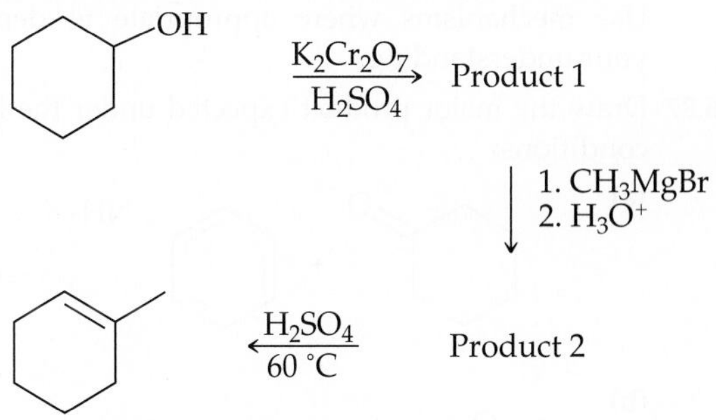

OH $\xrightarrow[\text{H}_2\text{SO}_4]{\text{K}_2\text{Cr}_2\text{O}_7}$ Product 1

$\downarrow$ 1. CH₃MgBr 2. H₃O⁺

$\xleftarrow[\text{60 °C}]{\text{H}_2\text{SO}_4}$ Product 2

26.23 Write down the reagents necessary to make the following conversions:

(a)

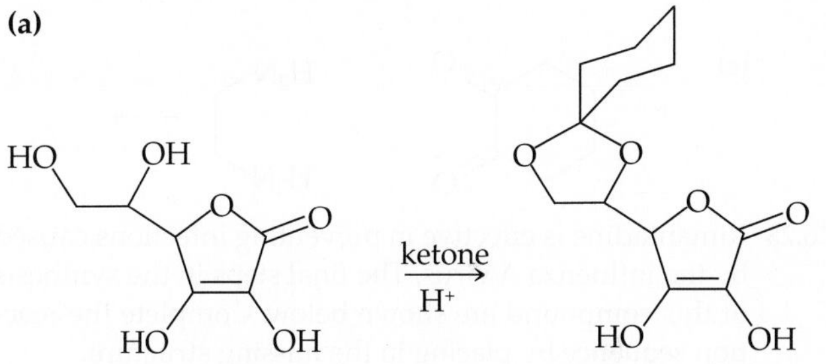

HO OH $\xrightarrow[\text{H}^+]{\text{ketone}}$

HO OH

ascorbic acid

(b)

H₃CO → H₃CO CH₂OH

Br   Br

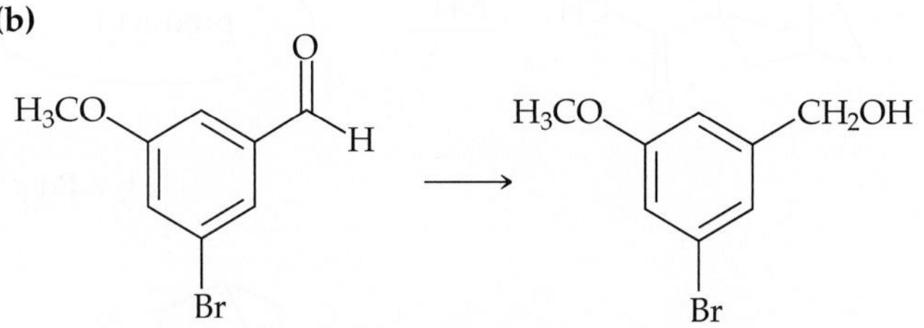

26.24 Write the structure of the Grignard reagent formed from each of the following alkyl halides upon reaction with magnesium in diethyl ether.

(a) 2-bromopropane

(b) 4-bromotoluene

(c) iodocyclobutane

(d) 3-chloro-2-methylpentane

26.25 Draw the major product expected under the following conditions (Ph = C₆H₅):

(a)

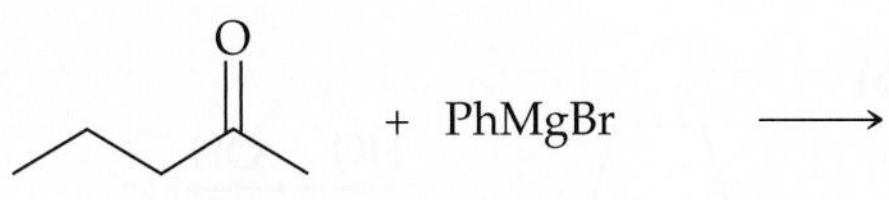

$$+ \text{PhMgBr} \longrightarrow$$

(b)

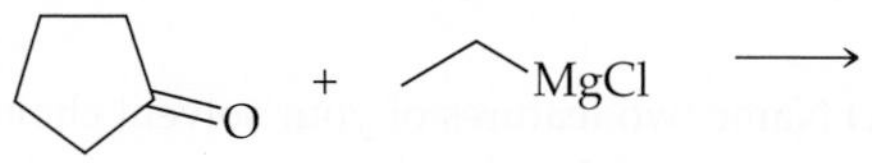

(c)

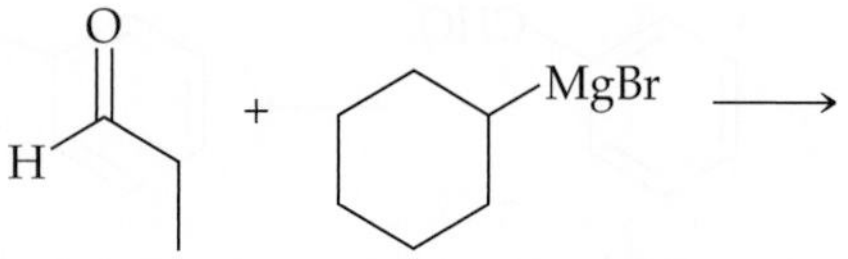

26.26 Give a plausible synthetic route to the formation of 4-methyl-2-pentanol from 1-bromo-2-methylpropane. Use mechanisms where appropriate to demonstrate your understanding.

26.27 Draw the major product expected under the following conditions:

(a)

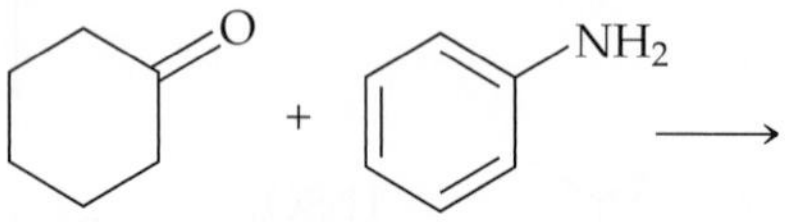

(b)

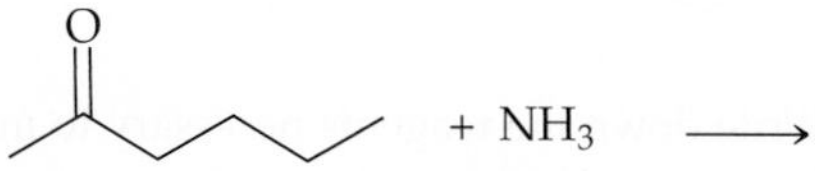

(c)

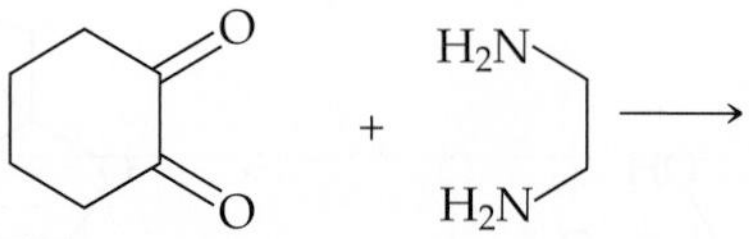

26.28 Rimantadine is effective in preventing infections caused by the influenza A virus. The final steps in the synthesis of this compound are shown below. Complete the reaction sequence by placing in the missing structure.

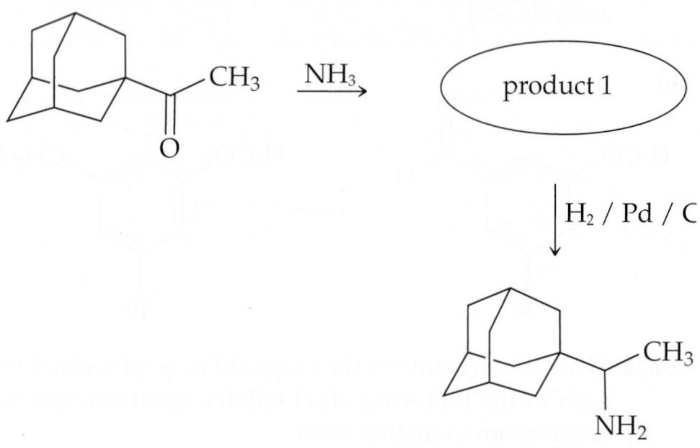

26.29 Draw the product of each reaction:

(a)

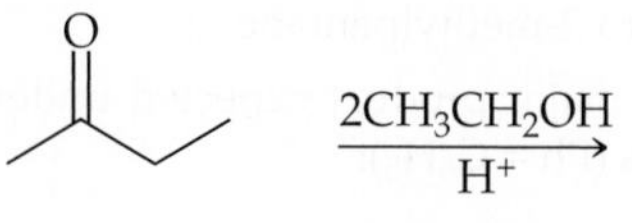

(b)

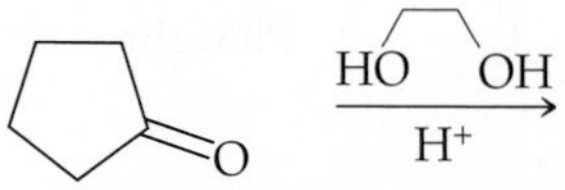

(c)

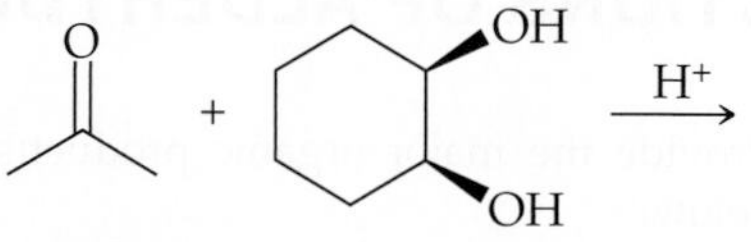

26.30 Write a mechanism for the acid-catalysed reaction of methanol with propanal to form an acetal. What solvent would you choose to maximise the conversion of propanal to its acetal?

26.31 (a) Which type of reaction characterises the aldehydes and ketones? **(b)** List the following carbonyl compounds in order of decreasing reactivity towards nucleophiles: ester, acid chloride, amide, aldehyde, ketone.

26.32 Would you expect the carbonyl carbon of benzaldehyde to be more or less electrophilic than that of acetaldehyde? Explain your reasoning.

26.33 Complete the following reaction sequence by supplying the reagents required:

$$CH_3OH \xrightarrow{\text{step 1}} CH_3Br \xrightarrow{\text{step 2}} CH_3MgBr \xrightarrow[\text{step 4}]{\text{step 3}} CH_3CH_2OH$$

26.34 Provide the major organic products in the reactions below:

(a)

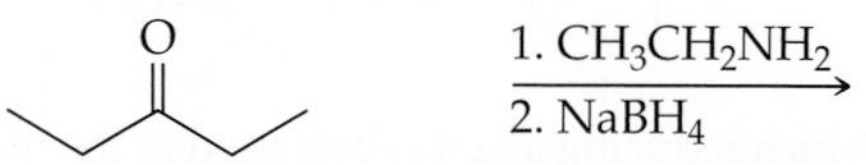

1. $CH_3CH_2NH_2$
2. $NaBH_4$

(b)

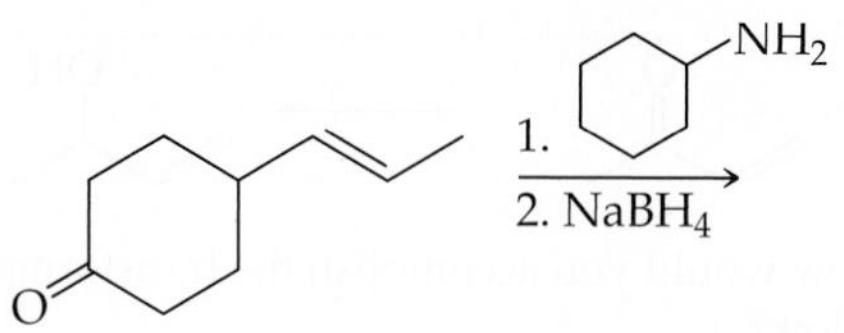

1.
2. $NaBH_4$

(c)

1. CH_3NH_2
2. $NaBH_4$

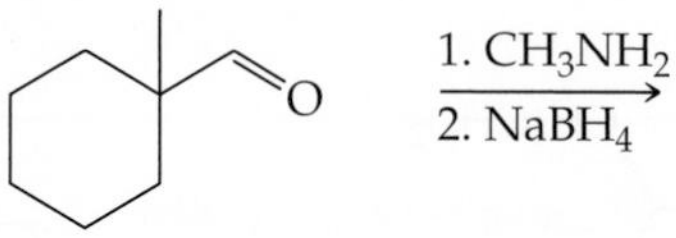

26.35 (a) Draw the product formed when acid is added to a solution of 5-hydroxypentanal. **(b)** What general class of compound is this?

26.36 When $HOCH_2CH_2CH_2CH_2COCH_2CH_2CH_2CH_2OH$ is heated in the presence of an acid catalyst, a reaction occurs. The product has the formula $C_9H_{16}O_2$. Provide the structure of this product.

26.37 What series of synthetic steps could be used to prepare 1-methylcyclohexanol from cyclohexene?

26.38 Provide the major organic product for the following set of reactions.

1. PCC
2. CH_3MgBr
3. H_3O^+
4. PCC

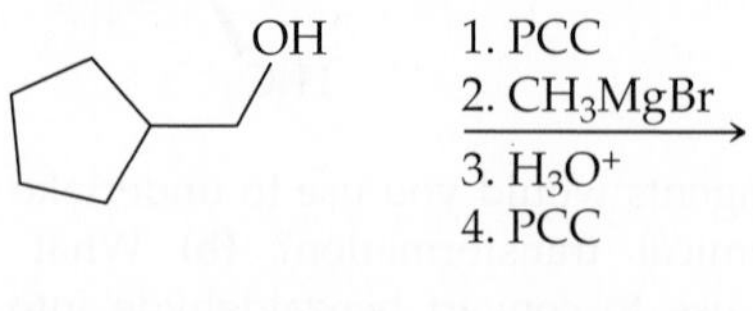

26.39 Provide the major organic product for the following set of reactions.

(a)

NaCN/HCN

(b)

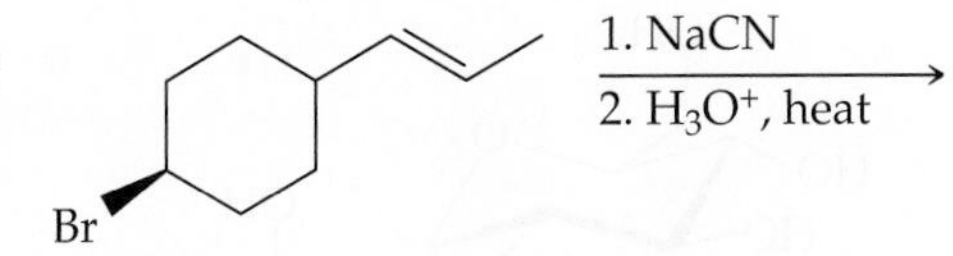

$$\xrightarrow[\text{2. H}_3\text{O}^+,\text{ heat}]{\text{1. NaCN}}$$

(c)

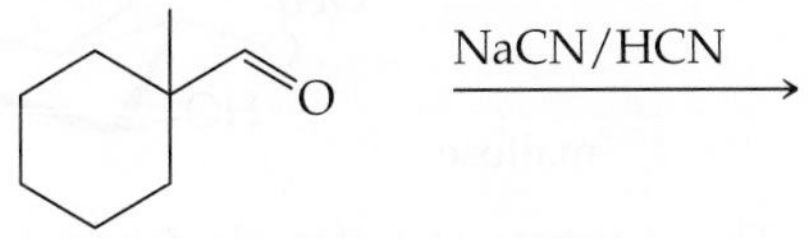

$$\xrightarrow{\text{NaCN/HCN}}$$

(d)

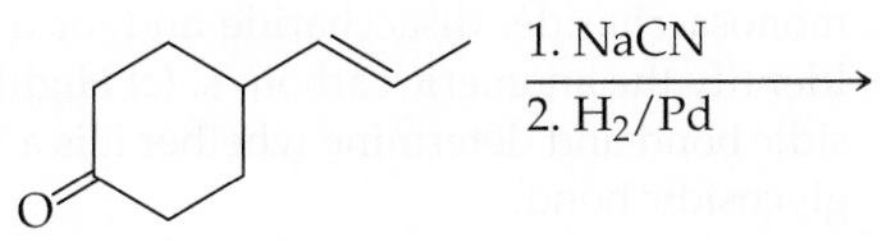

$$\xrightarrow[\text{2. H}_2/\text{Pd}]{\text{1. NaCN}}$$

(e)

$$\xrightarrow[\text{2. H}_3\text{O}^+]{\text{1. NaCN/HCN}}$$

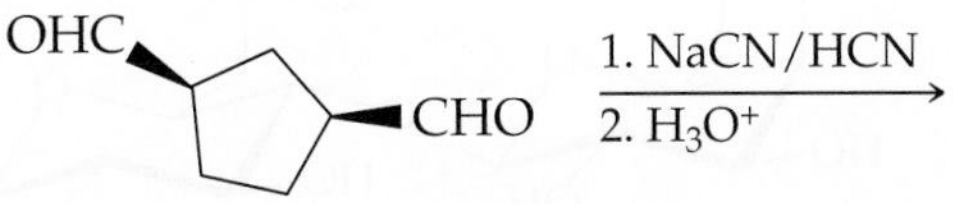

26.40 What would be the best way of converting 2-methylbutanal into the following?

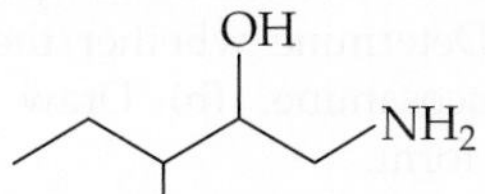

26.41 **(a)** Draw the resonance contributors for the structure you would form by deprotonating at the methyl group indicated with the arrow. **(b)** Identify the keto and enolate forms. **(c)** What might you say about the acidity of these methyl protons compared to pentane? **(d)** What other tautomer is possible?

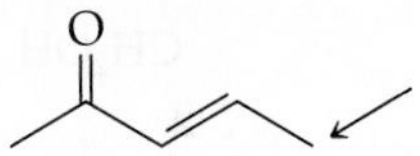

26.42 Draw all major resonance contributors of the following:

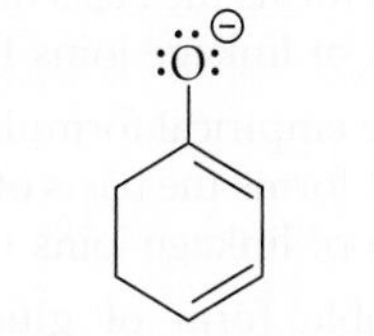

26.43 Draw the most stable enol tautomer of the ketone shown below. Explain your choice.

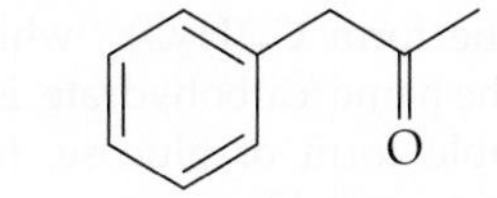

26.44 How many Hs in the compound below are replaced by Ds when it is shaken in D_2O containing trace acid?

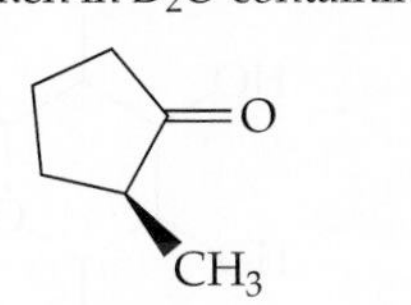

26.45 How would you prepare 2-cyanocyclohexanone from cyclohexanone?

26.46 What is the carbon nucleophile that attacks molecular bromine in the acid-catalysed α-bromination of a ketone?

26.47 Identify the major organic product of the following sequence.

$$\xrightarrow[\text{2. NaOH}]{\text{1. Br}_2,\text{ H}^+\text{(cat)}}$$

CARBOHYDRATES (Section 26.4)

26.48 In your own words, define the following terms: **(a)** carbohydrate, **(b)** monosaccharide, **(c)** disaccharide, **(d)** anomer, **(e)** glycoside.

26.49 **(a)** What is the difference between α-glucose and β-glucose? **(b)** Show the condensation of two glucose molecules to form a disaccharide with a 1,4-α-linkage; **(c)** with a 1,4-β-linkage.

26.50 The structural formula for the linear form of galactose is:

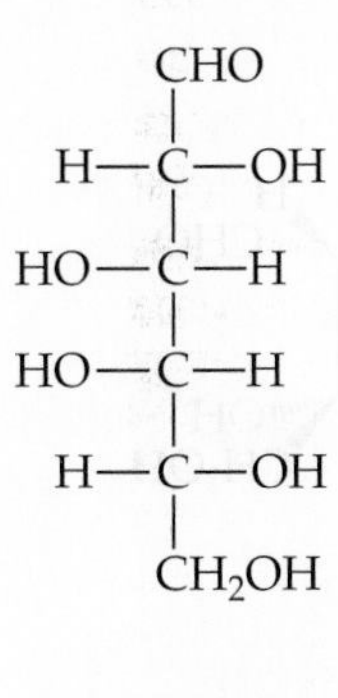

```
        CHO
         |
   H —  C — OH
         |
  HO —  C — H
         |
  HO —  C — H
         |
   H —  C — OH
         |
        CH₂OH
```

(a) How many chiral carbons are present in the molecule? **(b)** Draw the structure of the pyranose ring form of this sugar.

26.51 The structural formula for the linear form of D-mannose is as follows:

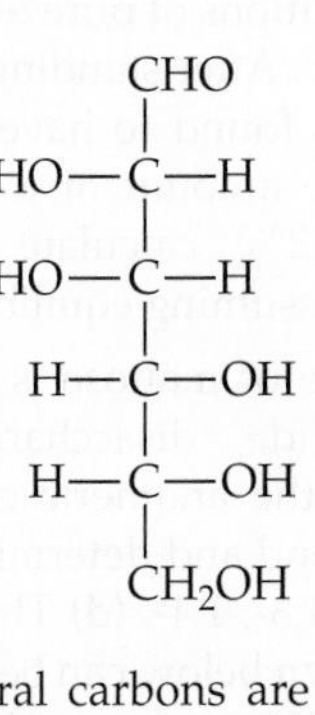

```
        CHO
         |
  HO —  C — H
         |
  HO —  C — H
         |
   H —  C — OH
         |
   H —  C — OH
         |
        CH₂OH
```

(a) How many chiral carbons are present in the molecule? **(b)** Draw the structure of the pyranose ring form of this sugar.

26.52 (a) What is the empirical formula of glycogen? (b) What is the unit that forms the basis of the glycogen polymer? (c) What form of linkage joins these monomeric units?

26.53 (a) What is the empirical formula of cellulose? (b) What is the unit that forms the basis of the cellulose polymer? (c) What form of linkage joins these monomeric units?

26.54 The most stable form of glucose contains the six-membered ring in a chair conformation with all substituents in equatorial positions. (a) What is the molecular formula for glucose? (b) Rewrite the molecular formula in the form $C_m(H_2O)_n$, which is the formula from which the name 'carbohydrate' is derived. (c) Draw the most stable form of glucose. (d) Identify which anomer is the most stable.

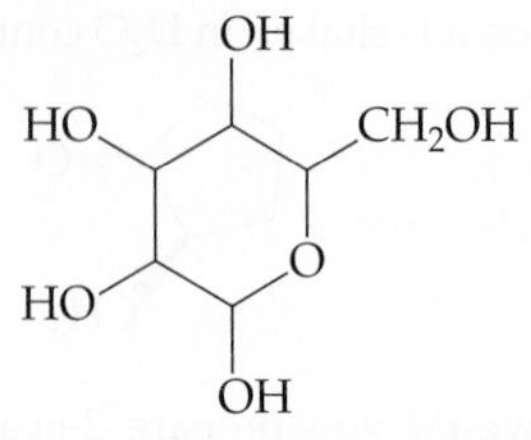

glucose

26.55 Starch, glycogen and cellulose are all polymers of glucose. (a) What are the structural differences between them? (b) Explain the structural differences that cause the difference in water solubility between these polymers.

26.56 Monosaccharides can be categorised in terms of the number of carbon atoms (pentoses have five carbons and hexoses have six carbons) and according to whether they contain an aldehyde (prefix *aldo-*, as in aldopentose) or ketone group (prefix *keto-*, as in ketopentose). Classify galactose and ribose in this way.

26.57 (a) How many stereogenic carbon atoms are there in the open-chain form of arabinose? (b) How many stereogenic carbon atoms are there in the open-chain form of xylose?

26.58 Aqueous solutions of pure α-D-galactose and β-D-galactose have specific rotations of +52.8° and +150.7°, respectively. After standing for several hours, a solution of α-D-galactose is found to have a new specific rotation of +80°. (a) Name the process observed. (b) Calculate the percentage of the α-anomer in the solution.

26.59 Aqueous solutions of pure α-D-xylose have a specific rotation of +92°. After standing for 10 hours, a solution of α-D-xylose is found to have a new specific rotation of +18.6°. If the amount of α-D-xylose in solution is reduced to 12%, calculate the specific rotation of β-D-xylose, assuming equilibrium has been reached.

26.60 The structure of maltose is drawn below. (a) Is this a monosaccharide, disaccharide and/or a glycoside? (b) Identify the anomeric carbon/s. (c) Highlight the glycosidic bond and determine its connectivity—for example, 1,2-, 1,3-, 1,4-. (d) The identity of the anomer of maltose drawn below can be determined by the orientation of the OH group on the non-glycosidic anomeric carbon. Which anomer is it?

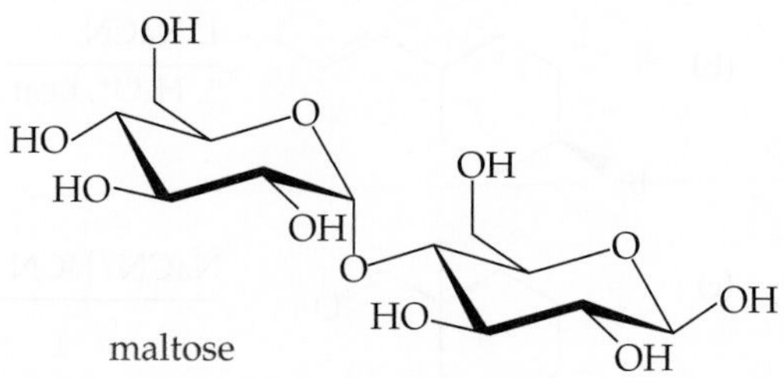

maltose

26.61 The structure of lactose is drawn below. (a) Is this a monosaccharide, disaccharide and/or a glycoside? (b) Identify the anomeric carbon/s. (c) Highlight the glycosidic bond and determine whether it is a 1,2-, 1,3- or 1,4-glycosidic bond.

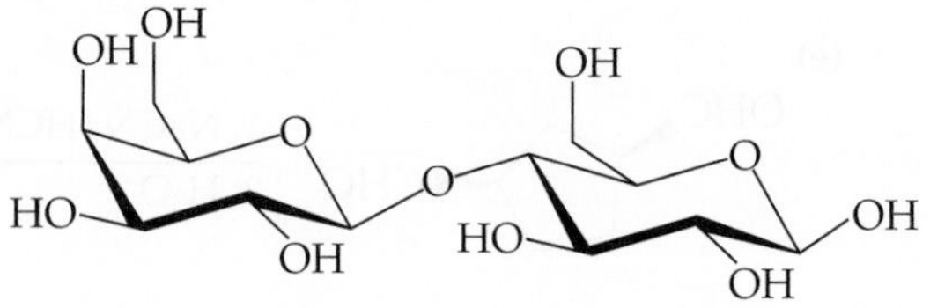

lactose

26.62 The open-chain structure of glucosamine is drawn below. (a) Determine whether the structure drawn is D- or L-glucosamine. (b) Draw glucosamine in its β-pyranose form.

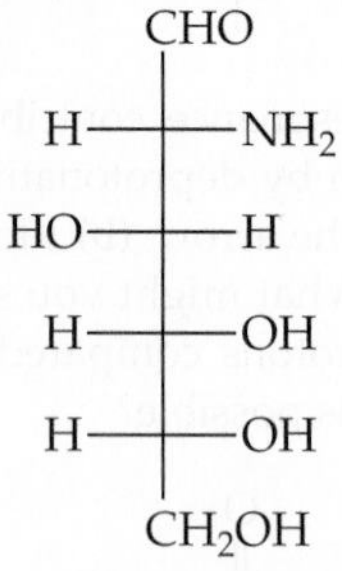

26.63 *Arbutin* is the principal antibacterial agent of the traditional herbal medicine called *uva-ursi*. Its systematic name is 4-hydroxyphenyl-β-D-glucopyranoside. Draw the structure of arbutin.

26.64 Convert each compound to a Fischer projection and hence class it as having D or L configuration.

(a)

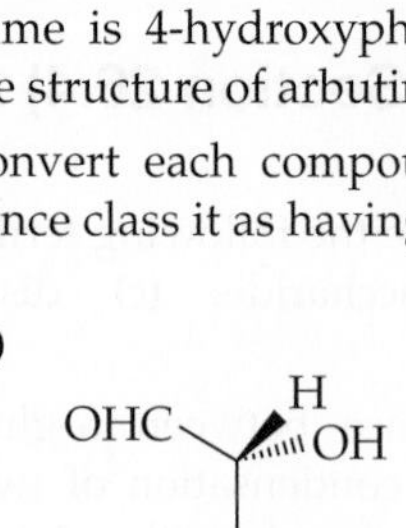

(b)

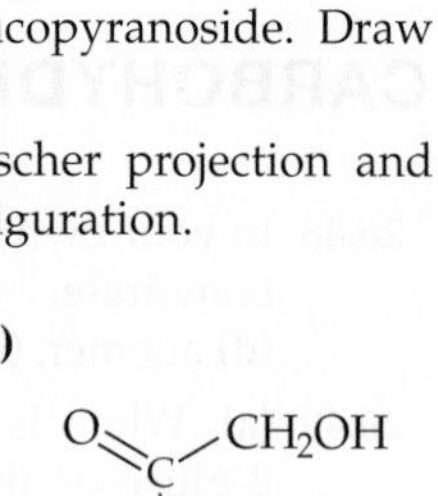

(c)

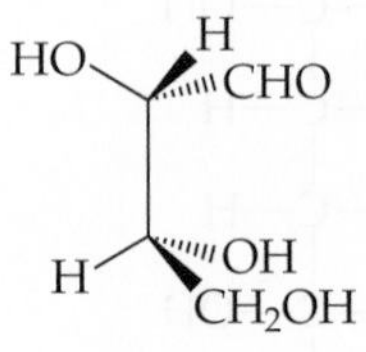

(d)

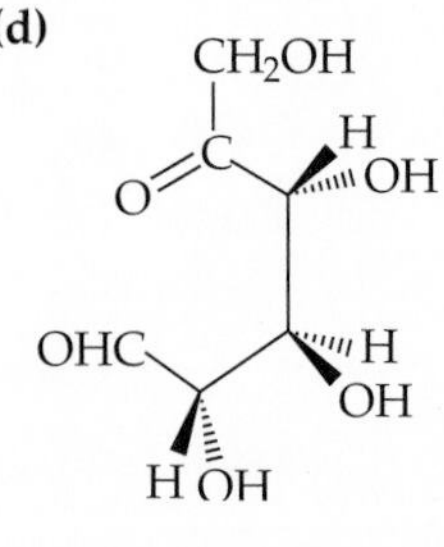

26.65 Convert each into a Haworth projection.

(a) Convert into an α-pyranose form   **(b)**

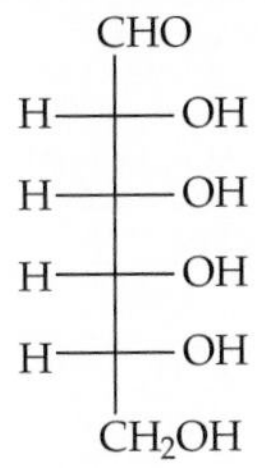

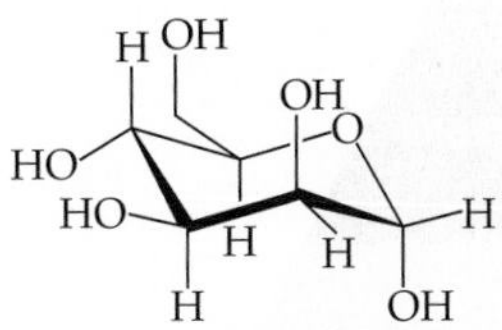

(c)

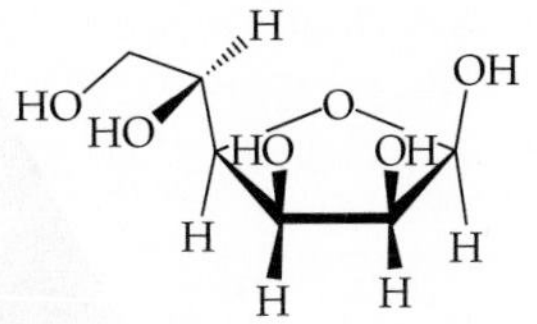

(d) Convert into a β-furanose form

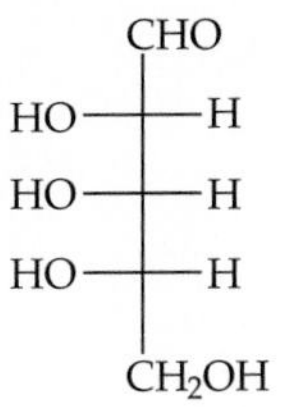

26.66 (a) How many stereogenic centres are there in L-ascorbic acid? **(b)** How many isomers are possible? **(c)** Would any of these other isomers be found in nature? Why?

INTEGRATIVE EXERCISES

26.67 2-Phenyl-2-butanol can be synthesised by three different combinations of a Grignard reagent and a ketone. Show each combination.

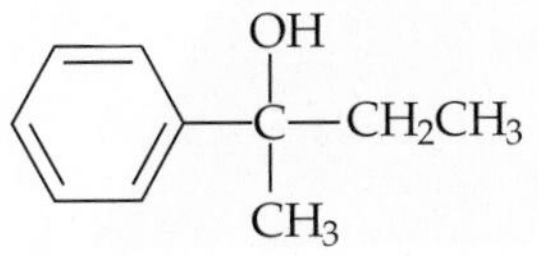

26.68 Acetone, $(CH_3)_2CO$, is widely used as an industrial solvent. **(a)** Draw the Lewis structure for the acetone molecule, and predict the geometry around each carbon atom. **(b)** Is the acetone molecule polar or non-polar? **(c)** What kinds of intermolecular attractive forces exist between acetone molecules? **(d)** 1-Propanol, $CH_3CH_2CH_2OH$, has a molecular weight that is very similar to that of acetone, yet acetone boils at 56.5 °C and 1-propanol boils at 97.2 °C. Explain the difference.

26.69 (a) How many stereocentres are present in 3-methylpentan-2-ol? List the configuration of the stereo-centres for each possible isomer and indicate the relationship (enantiomers or diastereoisomers) that exists between the isomers. **(b)** Write the equation for the reaction of 3-methylpentan-2-ol with acidified dichromate solution. **(c)** How many stereoisomers are there in the product of this reaction? **(d)** Draw two structures representing the enol forms of this product. **(e)** Describe the change in optical rotation that you might expect if you performed this reaction on a single stereoisomer of starting material.

26.70 Use Figure 8.5 on page 261 to determine the difference in electronegativity of the following pairs of atoms and classify the bond between them as ionic, polar covalent or non-polar covalent: Na–Cl; Mg–Cl; H–Cl; C–Cl; C–Mg; C–C.

26.71 Use the following data to determine whether glucose $(C_6H_{12}O_6)$ or sucrose $(C_{12}H_{22}O_{11})$ provides the greatest energy per gram. $\Delta_{combustion}H°$ (glucose) $= -2803$ kJ mol^{-1}; $\Delta_{combustion}H°$ (sucrose) $= -5640$ kJ mol^{-1}.

MasteringChemistry (www.pearson.com.au/masteringchemistry)

Make learning part of the grade. Access:

- tutorials with peronalised coaching
- study area
- Pearson eText

PHOTO/ART CREDITS

27

CARBOXYLIC ACIDS AND THEIR DERIVATIVES

A disc heart valve. The term *cardiovascular* pertains to the heart, blood and blood vessels. It often happens that only a part of the heart, such as the aortic valve, fails and needs replacement. About 250 000 valve-replacement procedures are performed annually worldwide, most using a St Jude valve. The valve is secured to the surrounding tissue via a biocompatible polyester sewing ring. The polyester used in commonly known as Dacron®.

KEY CONCEPTS

27.1 CARBOXYLIC ACIDS

We begin by discussing the properties and significance of carboxylic acids.

27.2 PREPARATION OF CARBOXYLIC ACIDS

We discuss the preparation of carboxylic acids by oxidation of primary alcohols and aldehydes. Hydrolysis of nitriles and Grignard reactions with carbon dioxide to form carboxylic acids are also discussed.

27.3 ESTERS AND ESTERIFICATION

We elaborate on the chemistry of carboxylic acids by exploring their reactivity with alcohols to form esters.

27.4 FATS, OILS AND WAXES

Fats, waxes and oils are important subclasses of esters that we discuss in detail. We recognise that this large class of molecules are used for energy storage.

27.5 ACID CHLORIDES, ANHYDRIDES AND NUCLEOPHILIC ACYL SUBSTITUTION

We investigate the susceptibility of the acyl group to nucleophilic attack, using acid chlorides, anhydrides and esters as examples.

27.6 CONDENSATION POLYMERISATION

Finally, we discuss the formation of polyesters and polyamides, such as nylon, which are industrially important polymers.

Polymers are an integral part of modern life. They are very large molecules constructed of just a few small molecules that are repeated many times. Their use in our lives is incredibly diverse and ranges from soft-drink bottles to components of life-saving heart valve replacements. A St Jude valve uses the polyester Dacron® to attach it to the surrounding heart tissue. Dacron® is Du Pont's trade name for the fibre formed from polyethene terephthalate (a polyester). The polar, oxygen-containing functional groups along the biocompatible polyester chain afford attractive interactions to facilitate tissue growth.

The starting materials of polymers of this type are derived from molecules containing the carboxylic acid functional group. One of the most common carboxylic acids is acetic acid, which is the primary ingredient of vinegar. Acetic acid is also an important industrial chemical used primarily for the formation of acetate esters. For example, aspirin (acetylsalicylic acid) is formed by the reaction of acetic acid with salicylic acid. Vinyl acetate is an important polymer precursor that is also derived from acetic acid. When vinyl acetate is polymerised, the product is poly(vinyl acetate), which is used in water-based latex paints and in glues for paper and wood.

In this chapter we study the chemistry of carboxylic acids and their common derivatives: esters, acid chlorides, acid anhydrides and, briefly, amides. We examine biological examples of esters and look at soaps and detergents. We conclude this chapter by discussing condensation polymerisation.

▲ **FIGURE 27.1 Everyday carboxylic acids and esters.** Spinach, rhubarb and some cleaners contain oxalic acid; vinegar contains acetic acid; vitamin C is ascorbic acid; citrus fruits contain citric acid; and aspirin is acetylsalicylic acid (which is both an acid and an ester).

∞ Find out more on page 1172

∞ Review this on page 383

(a)

(b)

▲ **FIGURE 27.2 Cellulose acetate fibre.** (a) The fabric woven from cellulose acetate fibre has a silky, vibrant texture compared with (b) the T-shirt fabric which is woven from simple cotton fibre.

27.1 | CARBOXYLIC ACIDS

Carboxylic acids contain the *carboxyl* functional group (a combination of *carb*onyl and hydr*oxyl* groups), which is often written as COOH or CO_2H. These weak acids are widely distributed in nature and are commonly used in consumer products (◄ FIGURE 27.1). They are also important in the manufacture of synthetic polymers (◄ FIGURE 27.2) used to make fibres, films and paints, and biopolymers such as proteins (∞ Section 29.3, 'Proteins, Peptides and Enzymes').

▶ FIGURE 27.3 shows the structural formulae of several carboxylic acids. Notice that oxalic acid and citric acid contain two and three carboxyl groups, respectively. Oxalic acid, which is found in high concentration in rhubarb leaves, is poisonous if ingested. Terephthalic acid is a key component of PET, a durable polymer used in packaging. Lactic acid is generated in muscle cells and tissue during the synthesis of adenosine triphosphate (ATP) from carbohydrates during sustained exercise. Ibuprofen is an analgesic, used to treat pain and inflammation. Benzoic acid is often used as a fungicide.

Structure, Properties and Nomenclature

The carboxyl functional group (COOH) that characterises carboxylic acids is constituted by an sp^2 hybridised carbon atom bonded to two oxygen atoms. One oxygen is sp^2 hybridised, leading to the formation of σ- and π-bonds with carbon (leading to a C–O double bond). The other oxygen is sp^3 hybridised, leading to a C–O single bond. The bonding within the carboxyl group yields two contributing resonance structures that are able to undergo proton exchange (▶ FIGURE 27.4).

In Chapter 11 we studied the phenomenon of hydrogen bonding (∞ Section 11.2, 'Intermolecular Forces'). We have already described the need for two components: a hydrogen bond donor (such as O–H, N–H) and a hydrogen bond acceptor, which is usually an atom containing a lone pair of electrons. In the case of carboxylic acids, both hydrogen bond donor and acceptor are present in the same functional group in a favourable geometry. As a result, carboxylic acids are able to dimerise (▶ FIGURE 27.5) in concentrated solutions or as neat (without solvent) liquids. Dimerisation is as a result of the intermolecular hydrogen bonding between the carbonyl oxygen and the OH group. As a result, carboxylic acids have relatively high boiling points (▼ TABLE 27.1).

Alkane-derived carboxylic acids have the general formula $C_nH_{2n}O_2$, and are named based on the longest carbon chain (parent alkane) bearing the carboxyl group, although sometimes common names are still employed (▼ FIGURE 27.6). Propanoic acid (▼ FIGURE 27.7) is derived from the three-carbon alkane, propane. The *e* of the alkane is replaced by *oic acid* to signify the carboxylic acid. Propanoic acid is also known by its common name, propionic acid. Table 27.1 gives the IUPAC names of the first 10 straight-chain carboxylic acids, their condensed formulae and their melting points.

In terms of the nomenclature hierarchy we have been developing for organic chemistry, the carboxylic acids sit at the pinnacle. All other functional groups we have come across are named as substituents if a carboxylic acid group is present. For example, 3-hydroxy-4-methoxybutanoic acid is *not* named as the alcohol, but rather the carboxylic acid.

$$CH_3O\!-\!\overset{\overset{\displaystyle OH}{|}}{\underset{4}{C}}\!-\!\underset{3}{C}\!-\!\underset{2}{C}\!-\!\overset{\overset{\displaystyle O}{\|}}{\underset{1}{C}}\!-\!OH$$

3-hydroxy-4-methoxybutanoic acid

▲ FIGURE IT OUT

Which of these substances have both a carboxylic acid functional group and an alcohol functional group?

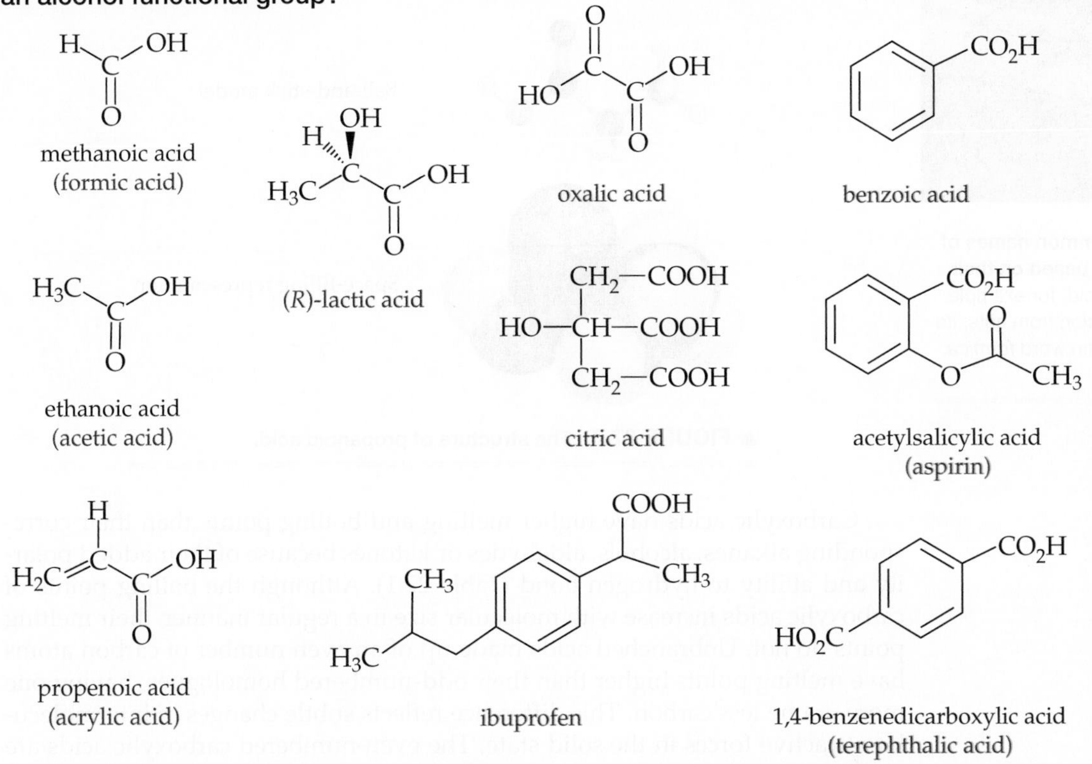

methanoic acid
(formic acid)

(*R*)-lactic acid

oxalic acid

benzoic acid

ethanoic acid
(acetic acid)

citric acid

acetylsalicylic acid
(aspirin)

propenoic acid
(acrylic acid)

ibuprofen

1,4-benzenedicarboxylic acid
(terephthalic acid)

◀ FIGURE 27.3
Structural formulae of some common carboxylic acids. The IUPAC names are given in black type for the monocarboxylic acids, but they are generally referred to by their common names. Note the three ways of writing this functional group.

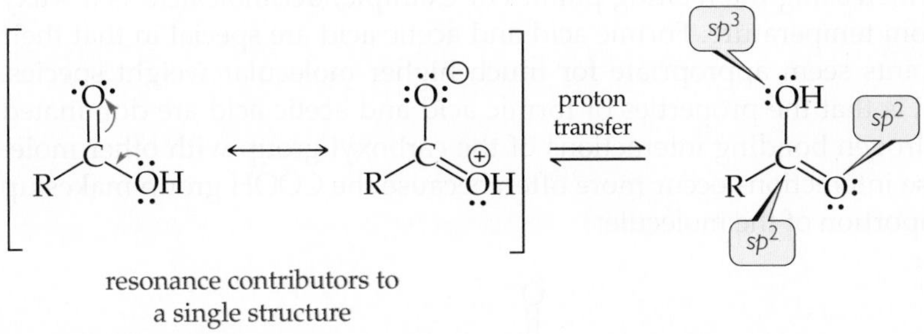

resonance contributors to
a single structure

proton
transfer

◀ FIGURE 27.4 The structure of the carboxyl group. The carboxyl group is constituted by a single sp^2 hybridised carbon bonded to an sp^2 hybridised oxygen atom and an sp^3 hybridised oxygen atom. The special arrangement of oxygen atoms around the central carbon atom allows the carboxyl group to exist as a hybrid of two resonance structures. The right-hand structure undergoes tautomerism (proton transfer) to also yield a carboxylic acid.

▲ FIGURE IT OUT

Would dimerisation be aided by polar or non-polar solvents?

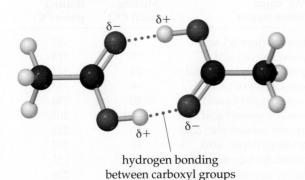

hydrogen bonding
between carboxyl groups

◀ FIGURE 27.5 Carboxylic acid dimerisation. In concentrated solution or as neat liquids, carboxylic acids are able to dimerise through intermolecular hydrogen bonding.

▲ FIGURE 27.6 **The common names of many carboxylic acids are based on their historical origins.** Formic acid, for example, was first prepared by extraction from ants; its name is derived from the Latin word *formica*, meaning 'ant'.

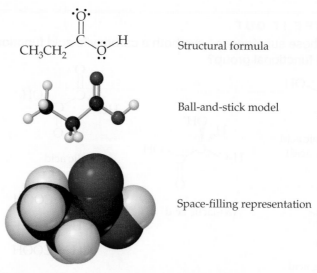

Structural formula

Ball-and-stick model

Space-filling representation

▲ FIGURE 27.7 **The structure of propanoic acid.**

Carboxylic acids have higher melting and boiling points than their corresponding alkanes, alcohols, aldehydes or ketones because of their added polarity and ability to hydrogen bond (Table 27.1). Although the boiling points of carboxylic acids increase with molecular size in a regular manner, their melting points do not. Unbranched acids made up of an even number of carbon atoms have melting points higher than their odd-numbered homologues, having one more or one less carbon. This difference reflects subtle changes in intermolecular attractive forces in the solid state. The even-numbered carboxylic acids are able to interact more strongly than their odd-numbered homologues, which increases the amount of energy needed to break the interactions in order to melt, thus increasing the melting point. For example, decanoic acid is a waxy solid at room temperature. Formic acid and acetic acid are special in that their melting points seem appropriate for much higher molecular weight species. The reason is that the properties of formic acid and acetic acid are dominated by the hydrogen bonding interactions of the carboxyl group with other molecules. These interactions occur more often because the COOH group makes up a large proportion of the molecule.

formic acid

Molecular formula	Condensed structural formula	IUPAC name (common name)	Melting point (°C)[*]	Boiling point (°C)[*]
CH_2O_2	HCOOH	Methanoic (formic) acid	8	101
$C_2H_4O_2$	CH_3COOH	Ethanoic (acetic) acid	17	118
$C_3H_6O_2$	CH_3CH_2COOH	Propanoic (propionic) acid	−24	141
$C_4H_8O_2$	$CH_3CH_2CH_2COOH$	Butanoic (butyric) acid	−6	163
$C_5H_{10}O_2$	$CH_3CH_2CH_2CH_2COOH$	Pentanoic (valeric) acid	−20	186
$C_6H_{12}O_2$	$CH_3CH_2CH_2CH_2CH_2COOH$	Hexanoic (caproic) acid	−4	205
$C_7H_{14}O_2$	$CH_3CH_2CH_2CH_2CH_2CH_2COOH$	Heptanoic (enanthylic) acid	−10	223
$C_8H_{16}O_2$	$CH_3CH_2CH_2CH_2CH_2CH_2CH_2COOH$	Octanoic (caprylic) acid	16	239
$C_9H_{18}O_2$	$CH_3CH_2CH_2CH_2CH_2CH_2CH_2CH_2COOH$	Nonanoic (pelargonic) acid	9	253
$C_{10}H_{20}O_2$	$CH_3CH_2CH_2CH_2CH_2CH_2CH_2CH_2CH_2COOH$	Decanoic (capric) acid	27	219

TABLE 27.1 • **First 10 members of the aliphatic carboxylic acid series**

[*] Rounded to the nearest degree.

The number of carbon atoms in the carboxylic acid chain not only affects its melting point, but also its solubility in water. Typically, a carboxylic acid is more water soluble than its alcohol analogue because of more extensive hydrogen bonding capacity. For example, formic, acetic, propanoic and butanoic acids are all infinitely soluble in water. As the length of the carbon chain increases, however, the polar COOH group becomes an ever decreasing proportion of the molecule, and the molecule behaves more like a hydrocarbon in terms of its solubility. The solubility of the carboxylic acids in water decreases correspondingly. In contrast, the solubility of carboxylic acids in non-polar solvents increases as the non-polar hydrocarbon chain length increases. Hence decanoic acid is more 'alkane-like' than acetic acid.

CONCEPT CHECK 1

Formic acid and acetic acid are common names given to carboxylic acids bearing one or two carbon atoms, respectively. What are the IUPAC names for these two acids?

Acidity

We have already mentioned that carboxylic acids are weak acids (∞ Section 17.10, 'Acid–Base Behaviour and Chemical Structure'). ▼ **TABLE 27.2** compares the pK_a values for several organic and inorganic acids. Being a weak acid, carboxylic acids dissociate by the following equation:

∞ Review this on page 693

$$RCOOH + H_2O \rightleftharpoons RCOO^- + H_3O^+ \qquad [27.1]$$

One of the main contributing factors that allows the pK_a of a carboxylic acid to be lower than that of the corresponding alcohol—for example, acetic acid (pK_a 4.8) compared with ethanol (pK_a 16)—is the relative stability of the two conjugate bases, that is, the **carboxylate anion** versus the alkoxide anion. The carboxylate anion is resonance stabilised (▼ **FIGURE 27.8**), having two contributors to the overall structure, whereas the alkoxide is not resonance stabilised. As a result, the carboxylate anion is not as reactive as the alkoxide.

The pK_a of a carboxylic acid can be influenced inductively by substituents (Table 27.2). The addition of an electron-withdrawing substituent such as a halide adjacent to the COOH group reduces the pK_a by weakening the carboxyl

TABLE 27.2 • Relative pK_a values for selected carboxylic acids*

Compound	Structural formula	pK_a	
Hydrogen chloride	HCl	−7	**Strong acid**
Sulfuric acid	H_2SO_4	−5	
Trichloroacetic acid	Cl_3CCOOH	0.7	
Dichloroacetic acid	$Cl_2CHCOOH$	1.5	
Chloroacetic acid	$ClCH_2COOH$	2.8	
Acetic acid	CH_3COOH	4.8	
Water	H_2O	15.7	**Weak acid**

* Error in measurement ± 0.2.

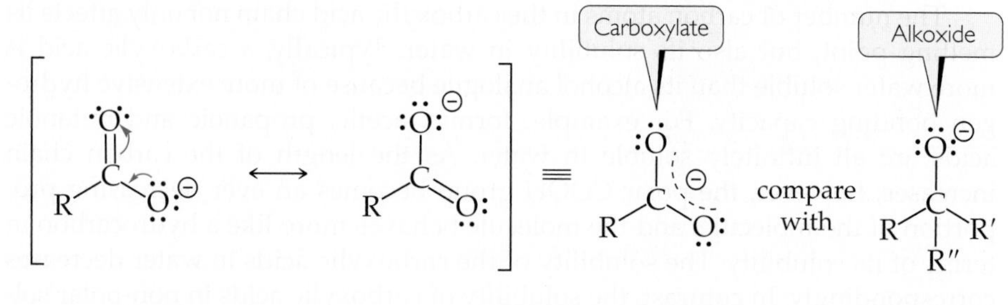

▶ **FIGURE 27.8** The carboxylate anion is resonance stablilised.

O–H bond (▼ FIGURE 27.9). For example, the pK_a of acetic acid is 4.8 whereas the pK_a of chloroacetic acid is 2.8. The addition of more chlorine atoms reduces the pK_a further (for example, pK_a of trichloroacetic acid is 0.7). The inductive effect is reduced dramatically as the number of intervening bonds between the electron-withdrawing substituent and the COOH group increases. Such an effect is seen in amino acids (∞ Section 29.2, 'Amino Acids').

Carboxylic acids react with bases such as NaOH, $NaHCO_3$ or Na_2CO_3 to form the corresponding sodium carboxylate salt. For example, when benzoic acid reacts with an aqueous solution of NaOH, sodium benzoate and water are formed. The conversion of the acid form to the carboxylate salt enhances aqueous solubility. Note that the salts of carboxylic acids are named in the same order as other salts: first cation, then anion. The name of the anion is derived from the carboxylic acid by replacing *-oic acid* with *-oate*.

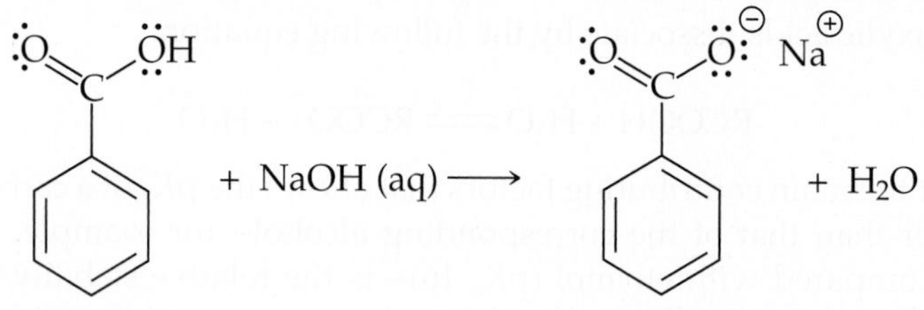

∞ Find out more on page 1166

benzoic acid + NaOH (aq) ⟶ sodium benzoate + H_2O

[27.2]

When propanoic acid is reacted with a solution of sodium hydrogen carbonate ($NaHCO_3$), the products are sodium propanoate, carbon dioxide and water. The reaction of carboxylic acids with carbonates such as $NaHCO_3$, Na_2CO_3, K_2CO_3 and $CaCO_3$ has special significance. Oral tablets such as Berocca™ and Aspro Clear™ combine carbonates with vitamin acids or citric acid to dissolve the tablet contents in water. The fizz associated with their dissolution is caused by the release of CO_2.

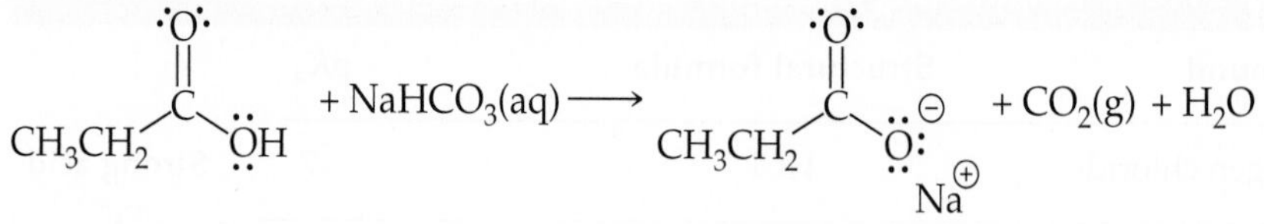

propanoic acid + $NaHCO_3$(aq) ⟶ sodium propanoate + CO_2(g) + H_2O

[27.3]

▶ **FIGURE 27.9** Inductive effect on pK_a. Electron-withdrawing substituents (for example, Cl, NO_2) adjacent to the carboxyl group are able to weaken the carboxyl O–H bond relative to acetic acid, making the compound a stronger acid. Any effect of a substituent that acts 'through bonds' rather than by providing an alternative resonance structure, is called an **inductive effect**.

O–H bond is lengthened as electron density is removed inductively

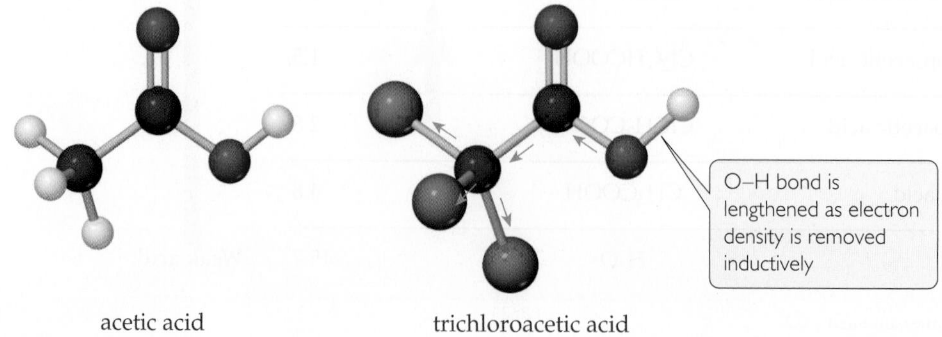

acetic acid   trichloroacetic acid

⚠ **CONCEPT CHECK 2**

What are the products of the reaction between butanoic acid and sodium carbonate?

Carboxylic acids react with amines to form ammonium salts. For example, the reaction of acetic acid with ethanamine yields the ammonium salt, ethyl ammonium acetate:

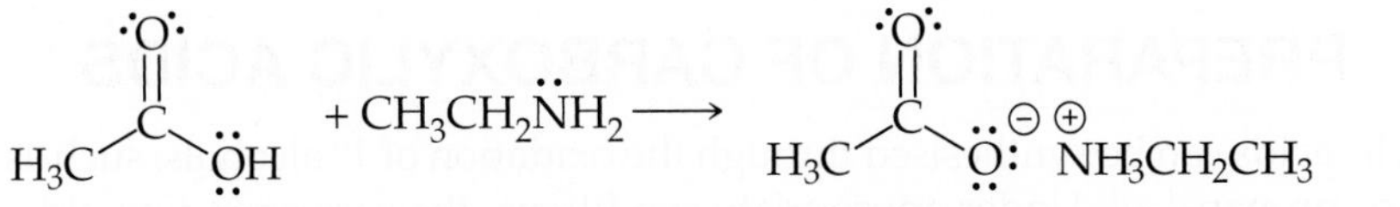

$$[27.4]$$

The reaction types demonstrated in Equations 27.2–27.4 are extremely useful in the preparation of water-soluble compounds, particularly for the pharmaceutical industry. The salt products are highly soluble in aqueous solutions and can be achieved under mild conditions or formed *in situ*, as in the case of soluble aspirin. Water solubility aids in absorption of the therapeutic agent. This reaction is also useful in the *resolution* of chiral carboxylic acids (∞ Section 23.6, 'Molecules with More than One Stereocentre').

∞ Review this on page 941

SAMPLE EXERCISE 27.1 **Properties of carboxylic acids**

Describe a way of separating benzoic acid from benzyl alcohol.

SOLUTION

Analyse We are asked how to separate an alcohol from a carboxylic acid containing a similar structure.

Plan The carboxylic acid and alcohol functional group have a different reactivity and so can be separated using chemical reactions.

Solve In this case, we can use the differences in acidity between the two compounds and the aqueous solubility of the benzoate anion to separate them.

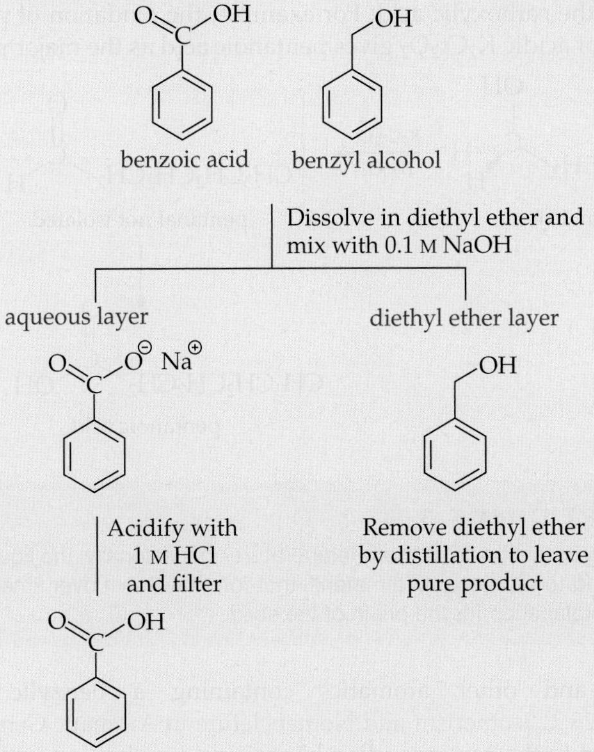

ᴏᴏᴏ Review this on page 1040

ᴏᴏᴏ Find out more on page 1118

PRACTICE EXERCISE

List the following set of carboxylic acids in order of acidity (lowest first):
(a) butanoic acid, **(b)** 2-iodobutanoic acid, **(c)** 2-fluorobutanoic acid, **(d)** trifluoroacetic acid, **(e)** 3-iodobutanoic acid.

Answer: a < e < b < c < d

(See also Exercises 27.11, 27.13.)

27.2 | PREPARATION OF CARBOXYLIC ACIDS

Carboxylic acids can be synthesised through the oxidation of 1° alcohols, such as ethanol or propan-1-ol. Under appropriate conditions, the corresponding aldehyde may also be isolated as the first product of oxidation (ᴏᴏᴏ Section 26.2, 'Preparation of Aldehydes and Ketones').

The oxidation of ethanol to acetic acid in air (with help from enzymes within wine) is responsible for causing wines to turn sour, producing vinegar.

Acetic acid is produced industrially by the reaction of methanol with carbon monoxide in the presence of a rhodium catalyst:

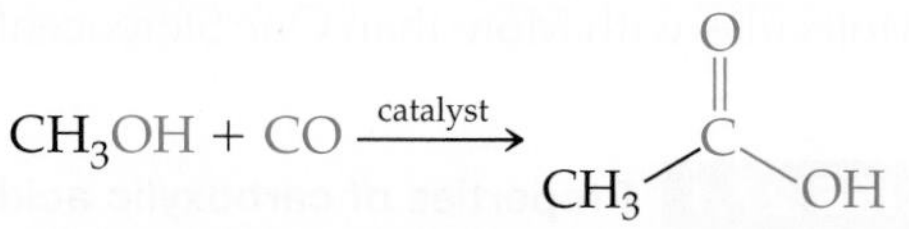

$$CH_3OH + CO \xrightarrow{\text{catalyst}} \underset{CH_3}{\overset{O}{\underset{\shortmid}{C}}} OH \qquad [27.5]$$

This reaction involves, in effect, the insertion of a carbon monoxide molecule between the CH_3 and OH groups. A reaction of this kind is called *carbonylation*.

When primary alcohols are oxidised using strong oxidising agents such as potassium dichromate ($K_2Cr_2O_7$) in H_2SO_4 or chromic acid ($CrO_3/H_2O/H_2SO_4$), they are converted to the corresponding carboxylic acid via the intermediary aldehyde. Aldehydes are more prone to oxidation than alcohols. In fact, left to stand, aldehydes are oxidised readily by the oxygen within air, also forming carboxylic acids. Hence, when the oxidation reaction of a primary alcohol proceeds, any aldehyde formed within the reaction is quickly converted to the carboxylic acid. For example, the oxidation of pentan-1-ol in the presence of acidic $K_2Cr_2O_7$ gives pentanoic acid as the major product:

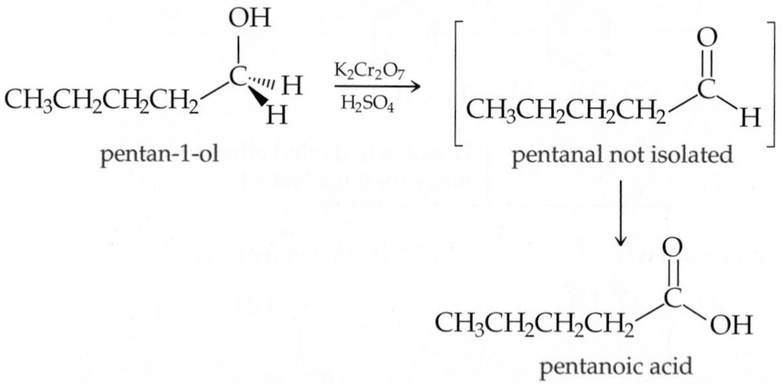

pentan-1-ol

pentanal not isolated

pentanoic acid [27.6]

> ⚠ **CONCEPT CHECK 3**
>
> Benzaldehyde is a liquid at room temperature. Occasionally, the liquid needs to be filtered to separate solid material that forms within it over time. Give a possible explanation for the origin of the solid.

Toluene and other aromatics containing a benzylic CH group (ᴏᴏᴏ Section 28.2, 'Isomerism and Nomenclature in Aromatic Compounds') are easily oxidised to the corresponding benzoic acid on boiling with solutions of potassium dichromate ($K_2Cr_2O_7$) or potassium permanganate ($KMnO_4$). This occurs even if the benzylic carbon contains one, two or three C–C bonds.

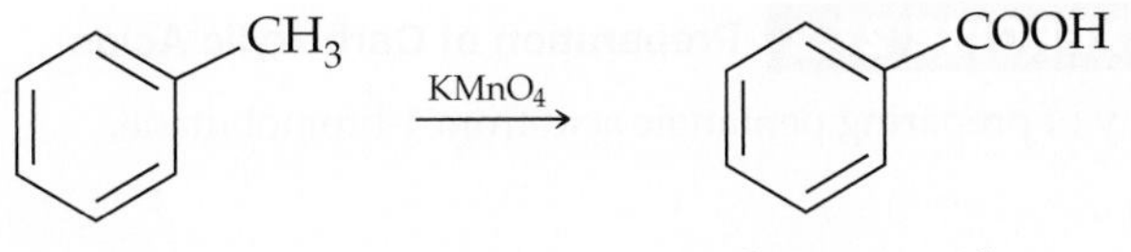

toluene → benzoic acid [27.7]

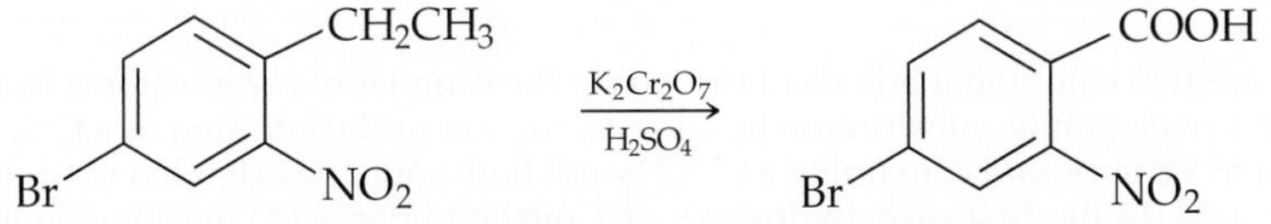

4-bromo-1-ethyl-2-nitrobenzene → 4-bromo-2-nitrobenzoic acid [27.8]

Benzoic acids can also be prepared by the reaction of phenylmagnesium bromide with carbon dioxide (in the form of dry ice), followed by protonation.

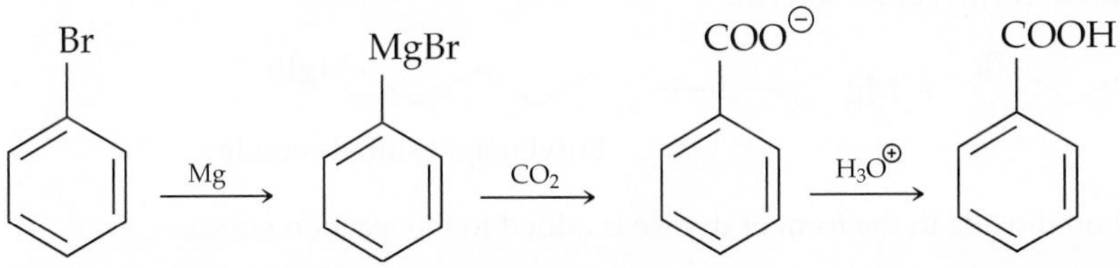

[27.9]

Recall from Chapter 26 that Grignard reagents make excellent nucleophiles, capable of reacting with carbonyl-containing compounds such as aldehydes and ketones (∞ Section 26.3, 'Reactions of Aldehydes and Ketones'). Generally speaking, the action of Grignard reagents (RMgX) on solid carbon dioxide to form carboxylic acids occurs by the following mechanism:

∞ Review this on page 1042

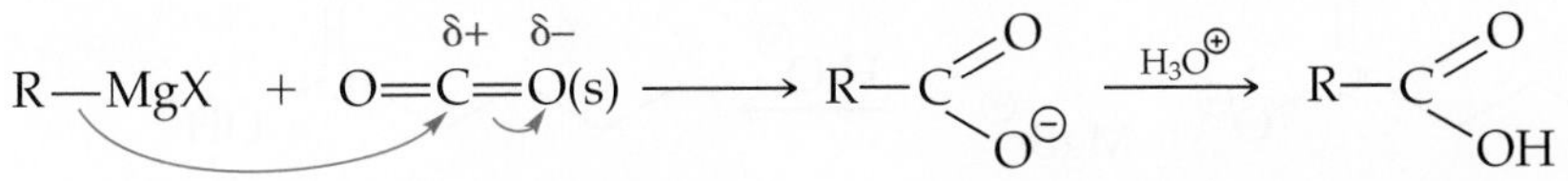

[27.10]

Carboxylic acids are also formed by the hydrolysis of certain functional groups. In the next section we discuss the reaction of carboxylic acids with alcohols to give an ester and with amines to give amides. We will also discuss the hydrolysis of esters to give carboxylic acids, using both acidic and basic conditions.

Carboxylic acids can be prepared by the hydrolysis of amides. It is fair to say that secondary and tertiary amides hydrolyse more slowly than primary amides under the conditions shown below (∞ Section 29.1, 'Amines and the Amide Bond'). Finally, carboxylic acids can be formed readily by the hydrolysis of nitriles (RCN) under acidic conditions.

∞ Find out more on page 1160

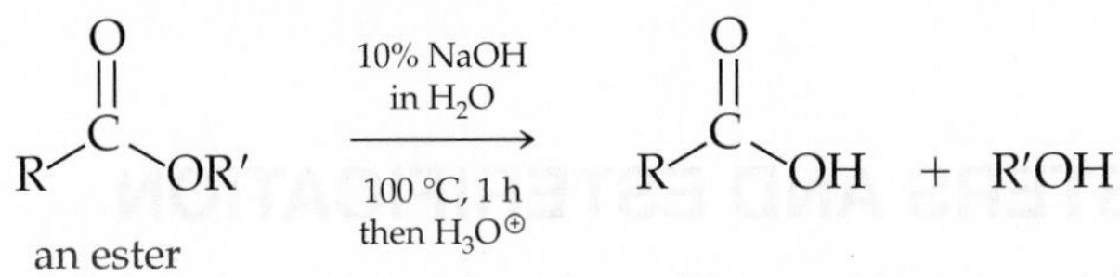

an ester [27.11]

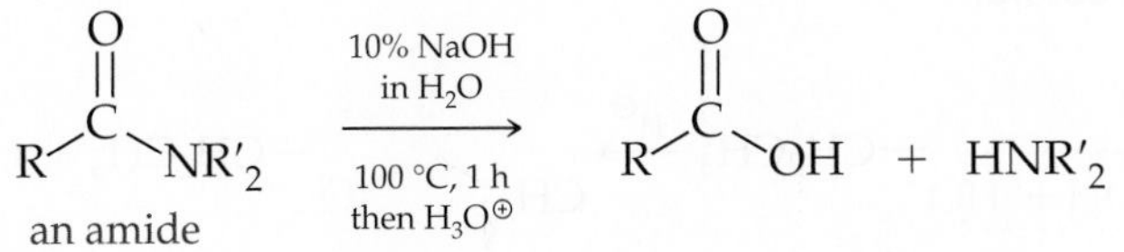

an amide [27.12]

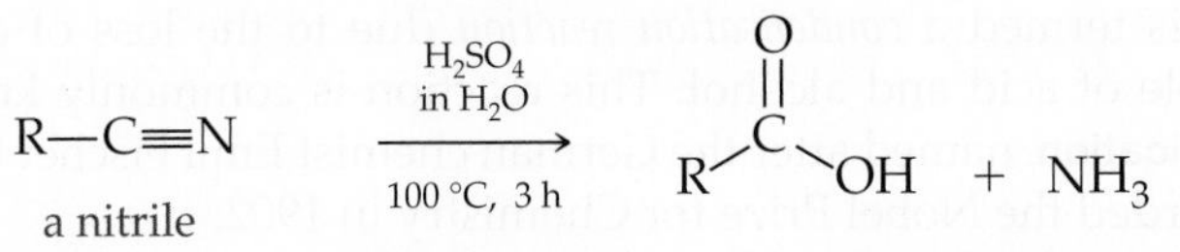

a nitrile [27.13]

SAMPLE EXERCISE 27.2 Preparation of Carboxylic Acids

Describe a way of preparing pentanoic acid from 1-bromobutane.

SOLUTION

Analyse We are asked to propose a synthesis of a carboxylic acid from a specific starting material and note that the target compound contains one more carbon atom than the reactant.

Plan There are two common methods of extending the number of carbon atoms in a chain: either a nucleophilic substitution by the cyanide ion or the addition of a Grignard reagent to a compound containing a C=O bond. Both could be extended to form a carboxylic acid. In the first case, hydrolysis of a nitrile under acid conditions will yield a carboxylic acid. In the second case, reaction of a Grignard reagent with carbon dioxide followed by dilute acid will furnish a carboxylic acid. We will choose the second route.

Solve Grignard reagents are very sensitive to moisture so are usually formed at the time they will be used in a 'one pot' reaction. First, 1-bromobutane is treated with magnesium in diethyl ether solvent.

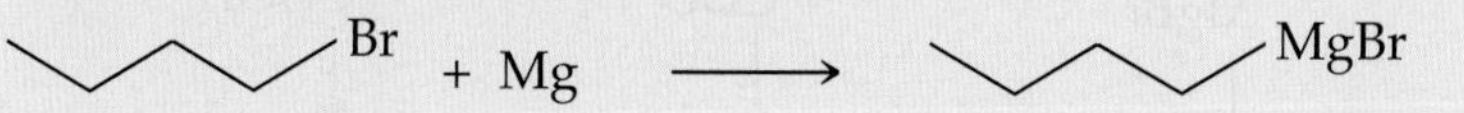

butyl magnesium bromide

Next, carbon dioxide in the form of dry ice is added to the reaction vessel.

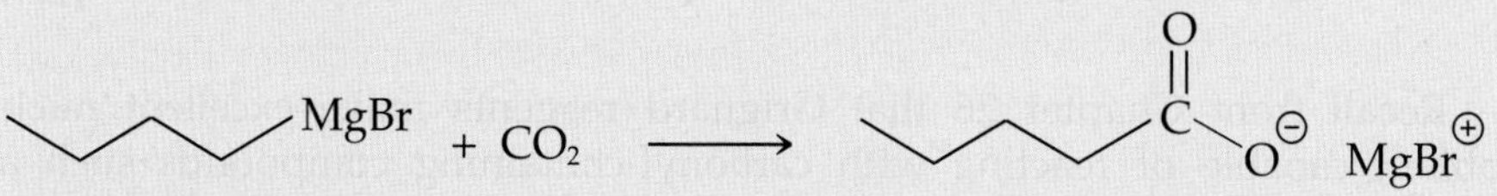

Finally, dilute acid is added to give the desired carboxylic acid.

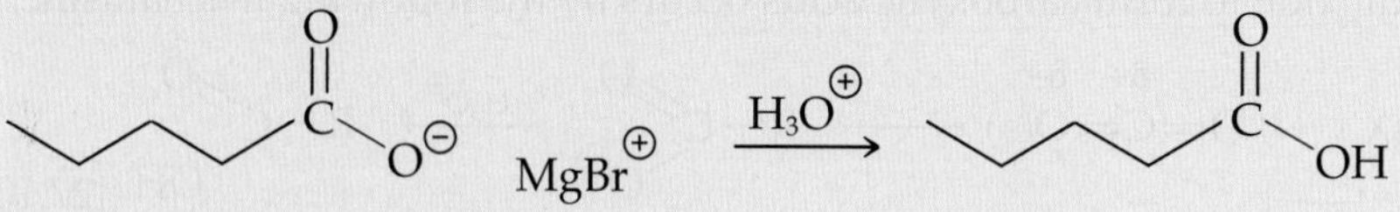

PRACTICE EXERCISE

Write the structural formula of the carboxylic acid formed by treatment of chlorobenzene with magnesium, followed by carbon dioxide and finally aqueous acid.

Answer:

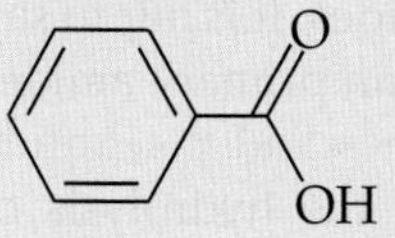

(See also Exercises 27.16, 27.17.)

27.3 | ESTERS AND ESTERIFICATION

Carboxylic acids can undergo acid-catalysed condensation reactions with alcohols to form **esters**:

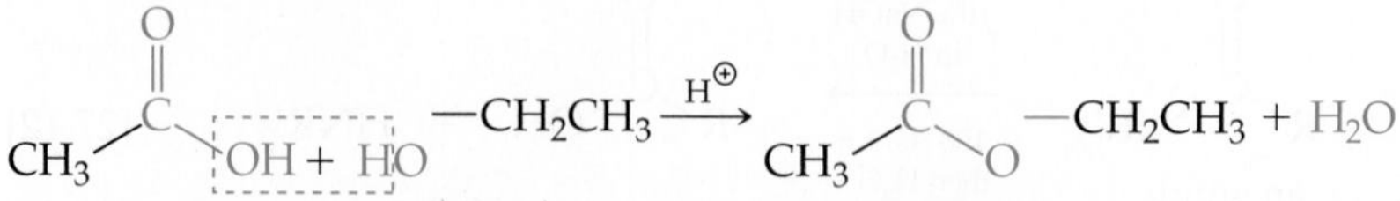

| acetic acid | ethanol | ethyl acetate | [27.14] |

The reaction is termed a *condensation reaction* due to the loss of one mole of water per mole of acid and alcohol. This reaction is commonly known as the **Fischer esterification**, named after the German chemist Emil Fischer (1852–1919), who was awarded the Nobel Prize for Chemistry in 1902.

Esters are compounds in which the OH group of a carboxylic acid is replaced by an OR group:

$$-\overset{\overset{\displaystyle O}{\|}}{C}-O-R, \quad -COOR, \quad or \quad -CO_2R$$

▶ **FIGURE 27.10** shows some common esters, which are named by using first the group from which the alcohol is derived and then the group from which the acid is derived. For example, ethyl butanoate is derived from ethanol and butanoic acid:

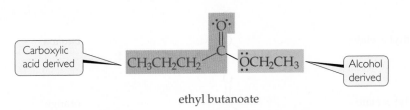

ethyl butanoate

▲ **FIGURE 27.10** **Everyday carboxylic acids and esters.** Many sunburn lotions contain the ester benzocaine; some nail polish remover is ethyl acetate; vegetable oils are also esters.

Esters generally have pleasant odours and are largely responsible for the pleasant aromas of fruit. ▼ **TABLE 27.4** lists some common esters and the structures, fragrances and flavours associated with them.

The mechanism of formation of an ester by the acid-catalysed condensation of an alcohol and a carboxylic acid can be shown as follows:

a carboxylic acid

tetrahedral carbonyl addition intermediates

proton transfer

an ester

[27.15]

As with aldehydes and ketones, the basis of reactivity of the carboxylic acid occurs at the carbonyl group. The electronegativity of two oxygen atoms on the carbonyl carbon makes this carbon more electropositive than the corresponding carbon of either an aldehyde or ketone. Note from Equation 27.15 that the esterification reaction is fully reversible at each step and that formation of the ester from the carboxylic acid requires an excess of alcohol to drive the reaction in the direction that favours ester formation.

From the mechanism shown in Equation 27.15, it should become apparent that carboxylic acids can be formed from esters by the addition of excess water in the presence of an acid catalyst. This reaction, called hydrolysis, is the reverse of that described and is an important reaction in organic synthesis, as esters are often prepared to protect carboxylic acids.

TABLE 27.4 • Simple esters as flavours and fragrances

Ester	Formula	Flavour/fragrance
Ethyl methanoate	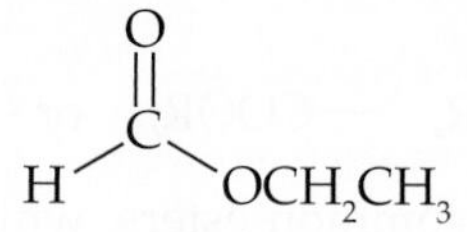	rum
Butyl acetate	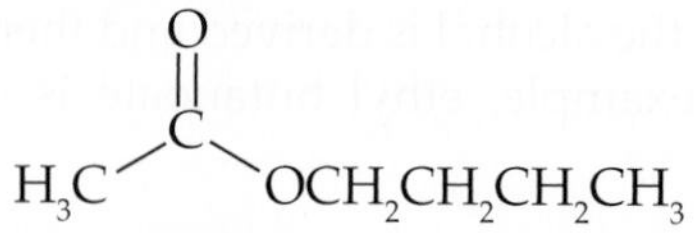	raspberry
Pentyl acetate	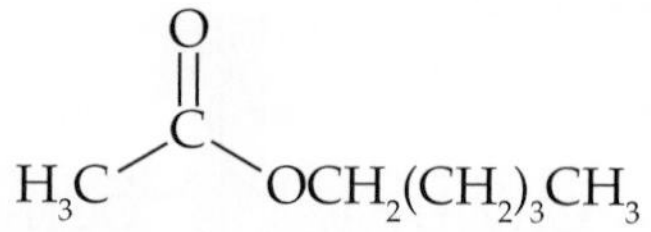	banana
Octyl acetate	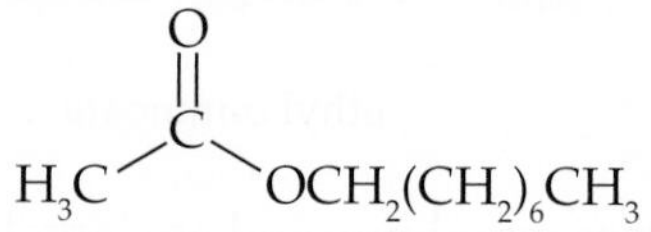	orange
Benzyl acetate	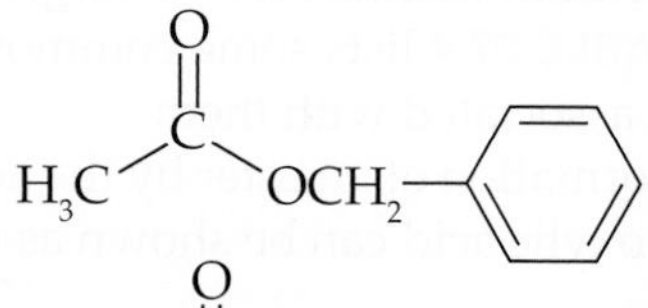	jasmine
Methyl butanoate	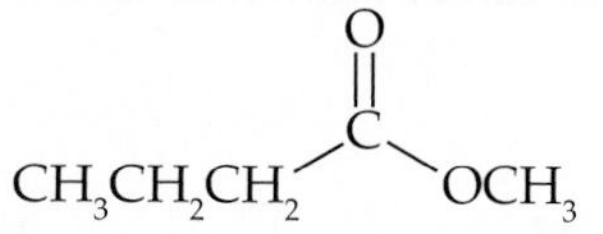	apple
Ethyl butanoate	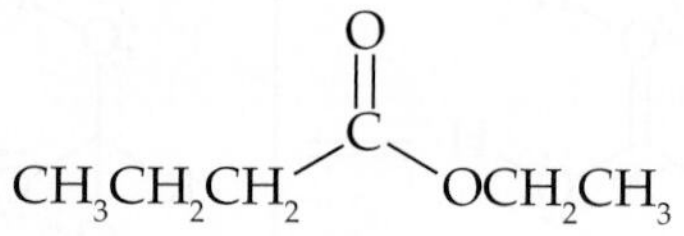	pineapple
Pentyl butanoate	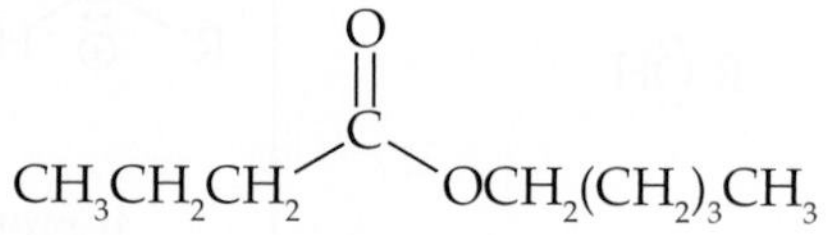	pear or apricot
Methyl benzoate	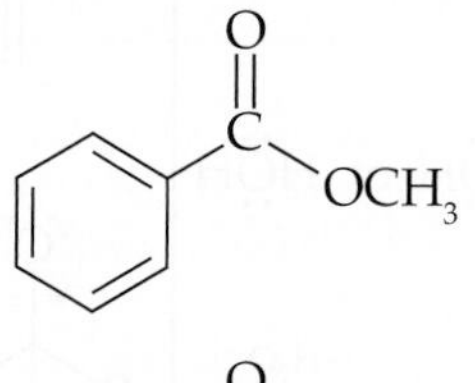	marzipan
Methyl salicylate	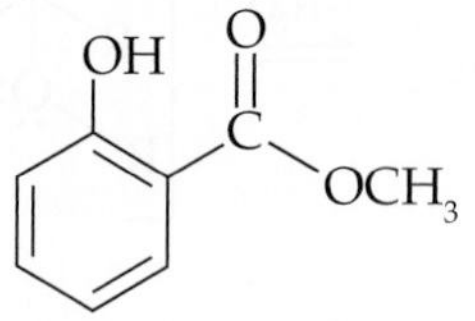	oil of wintergreen
Ethyl salicylate	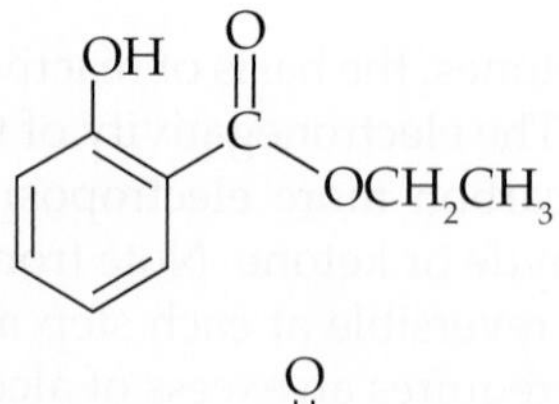	mint
Ethyl heptanoate	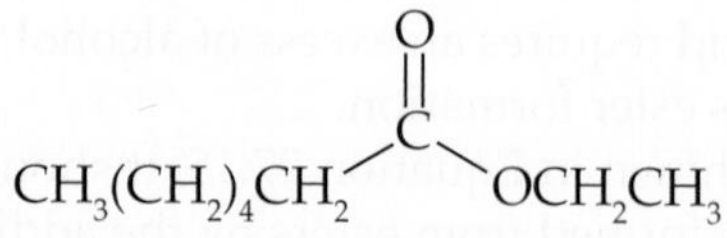	grape

The hydrolysis of esters is also possible in a basic aqueous medium, such as sodium hydroxide solution:

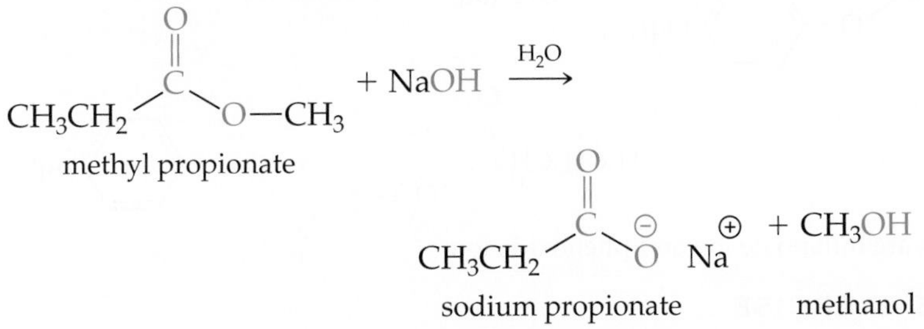

methyl propionate

sodium propionate   methanol   [27.16]

The products of the reaction are the sodium salt of the carboxylic acid and the alcohol. Distillation of the methanol and acidification of the aqueous solution of the sodium carboxylate yields the carboxylic acid. The hydrolysis of an ester in the presence of a base is called **saponification**, a term that comes from the Latin word for soap (*sapon*). The traditional synthesis of soap from animal fat is an example of this reaction.

SAMPLE EXERCISE 27.3   **Naming esters and predicting hydrolysis products**

Name each of the following esters and indicate the products of their reaction with aqueous base.

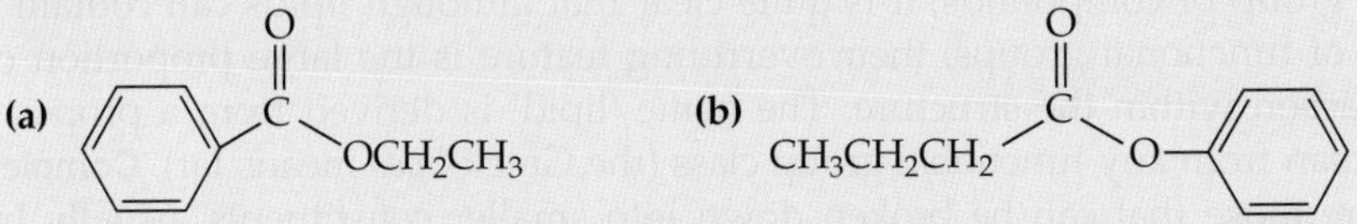

(a)

(b)

SOLUTION

Analyse We are asked to name each ester and to determine their saponification products.

Plan To name an ester, we must analyse its structure and determine the identities of the alcohol and acid from which it is formed. Identifying these two components will also allow us to determine the products of the saponification.

Solve Esters are formed by the condensation reaction between an alcohol and a carboxylic acid. We can identify the alcohol by adding an OH to the alkyl group attached to the O atom of the carboxyl (COO) group. We can identify the acid by adding an H group to the O atom of the carboxyl group. We've learned that the first part of an ester name indicates the alcohol portion and the second part indicates the acid portion. The name conforms to how the ester undergoes hydrolysis, reacting with base to form an alcohol and a carboxylate anion.

(a) This ester is derived from ethanol (CH_3CH_2OH) and benzoic acid (C_6H_5COOH). Its name is therefore ethyl benzoate. The net ionic equation for reaction of ethyl benzoate with hydroxide ion is:

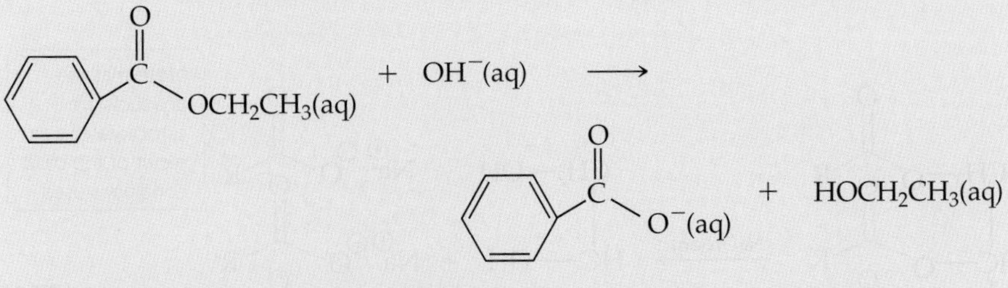

The products are benzoate ion and ethanol.

(b) This ester is derived from phenol (C_6H_5OH) and butanoic acid (commonly called butyric acid) ($CH_3CH_2CH_2COOH$). The residue from the phenol is called the phenyl group. The ester is therefore named phenyl butanoate. The net ionic equation for the reaction of phenyl butanoate with hydroxide ion is

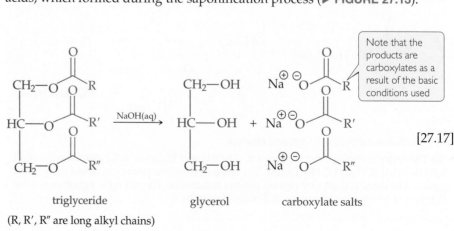

The products are butanoate ion and phenoxide ion.

PRACTICE EXERCISE

Write the structural formula for the ester formed from propanol and propionic acid.

Answer:

(See also Exercises 27.24, 27.25.)

27.4 | FATS, OILS AND WAXES

Lipids are naturally occurring substances that can be extracted from cells and tissues by non-polar organic solvents. Common examples of lipids include cholesterol (a steroid), turpentine (a terpene), beeswax and olive oil. Even from this very limited group of compounds, it is quite clear that although lipids can contain a variety of functional groups, their overriding feature is the large proportion of hydrocarbon within the structure. The name 'lipid' is derived from a property rather than from any functional group class (the Greek *lipos* means fat). **Complex lipids** are those that can be broken down into smaller constituents, usually by hydrolysis. **Simple lipids** are those that are not easily hydrolysed.

In this section we discuss a group of complex lipids that are esters of long-chain carboxylic acids called **fatty acids**. Two major groups of esters are derived from fatty acids: the waxes and triglycerides. **Waxes**, such as beeswax, are single esters made from fatty acids and long-chain alcohols (◀ FIGURE 27.11). **Triglycerides** (more correctly known as triacyl glycerols) are triesters of glycerol and include common fats and oils. ▶ FIGURE 27.12 shows two general examples of complex lipids. As you can see, even without investigating the type of functional group, both these molecules have alkyl chains of significant length. Their properties are derived from this feature. Waxes are typically greasy solids or semi-solids, while triglycerides can be solid or liquid at room temperature.

Historically, in the soap-making process a triglyceride (animal fat or a vegetable oil) was boiled with a strong base, usually NaOH. The resultant soap consisted of a mixture of sodium salts of long-chain carboxylic acids, the fatty acids, which formed during the saponification process (▶ FIGURE 27.13).

▲ **FIGURE 27.11 Beeswax.** The mouthpieces of these didgeridoos contain beeswax to help create a seal around the mouth. Beeswax is predominantly composed of the ester of palmitic acid, $C_{16}H_{31}CO_2H$, and the straight-chain alcohol, $C_{30}H_{61}OH$.

triglyceride
(R, R', R″ are long alkyl chains)

NaOH(aq) →

glycerol

carboxylate salts

Note that the products are carboxylates as a result of the basic conditions used

[27.17]

▲ **FIGURE IT OUT**

What structural features of a triglyceride molecule cause it to be insoluble in water?

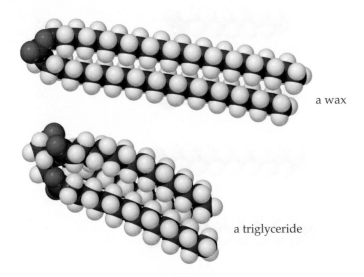

a wax

a triglyceride

▲ **FIGURE 27.13 Saponification.**
Saponification of fats and oils has long been used to make soap. This etching shows a step in the soap-making process during the mid-nineteenth century.

▲ **FIGURE 27.12 Complex lipids.** A general example of the structure of a wax ($C_{25}H_{51}CO_2C_{24}H_{49}$) and a triglyceride using the C_{16} fatty acid, palmitic acid.

The most common triglycerides are those in which each fatty acid component is unbranched and has an alkyl chain more than 12 carbon atoms long. ▼ TABLE 27.5 shows some of the more common saturated and unsaturated fatty acids found naturally. Note the difference in overall shape between saturated and unsaturated fatty acids. This shape difference is responsible for the different physical properties exhibited by the fatty acids and their corresponding triglycerides and waxes. For example, the melting point of stearic acid (mp = 70 °C), a saturated fatty acid, is 57 °C higher than oleic acid (mp = 13 °C) and 75 °C higher than linoleic acid (mp = −5 °C), despite the fact they all contain 18 carbon atoms. The reason is that 'kinked' fatty acids—that is, fatty acids that contain one or more double bonds—cannot pack as closely or efficiently as the saturated fatty acids.

Most saturated triglycerides are **fats** because they are solid at room temperature. Most triglycerides containing double bonds do not pack as well as their saturated analogues and so are classed as **oils**. Oils are triglycerides that are liquid at room temperature. The term *polyunsaturated* simply means there is more than one double bond contained within the triglyceride.

Consumers have come to realise that polyunsaturated vegetable oils are more easily digested than saturated ones. However, this has not always been the case. For many decades, the use of *lard*—a soft, white solid obtained by rendering pig fat—for cooking and baking meant that consumers wanted solid fats instead of liquid oils. Vegetable oils, cheaper to produce and polyunsaturated, are still converted to fats through a process called **hardening**. As chemists, we know this process as catalytic hydrogenation (⚬⚬ Section 24.5, 'Alkanes from Alkenes: Catalytic Hydrogenation'). The basic principle is not to reduce all the double bonds within an unsaturated triglyceride, but to reduce enough of them to yield the required properties and consistency. Needless to say, the process is carefully controlled. Margarine obtained from canola oil is a classic example.

⚬⚬ Review this on page 980

TABLE 27.5 • Structure and names of some common fatty acids

Name	Chain length	Structure	Melting point (°C)
		Saturated acids	
Lauric acid	C12	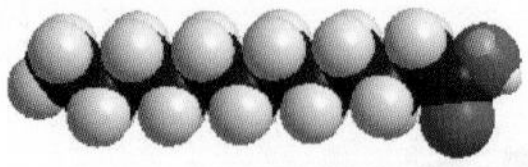	45
Myristic acid	C14		53
Palmitic acid	C16	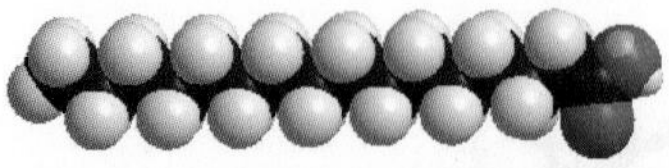	62
Stearic acid	C18		70
Arachidic acid	C20		75
		Unsaturated acids	
Oleic acid	C18 (1 × *cis* double bond)	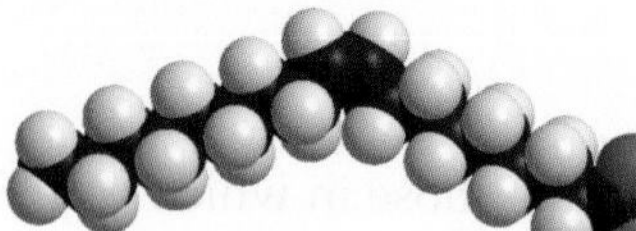	13
Linoleic acid	C18 (2 × *cis* double bonds)	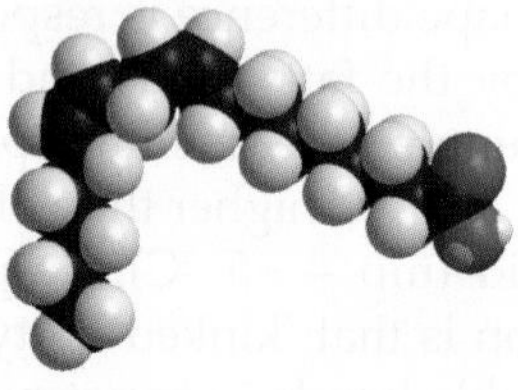	−5
Arachidonic acid	C20 (4 × *cis* double bonds)	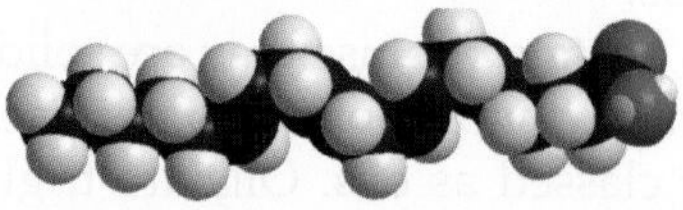	−50

SAMPLE EXERCISE 27.4   Lipids

Which of the following compounds is *not* classified as a lipid?

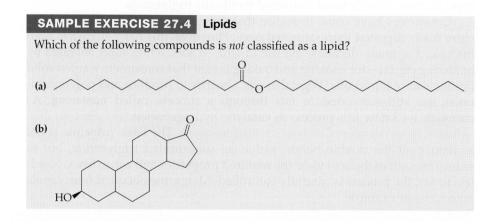

(a)

(b)

(c)

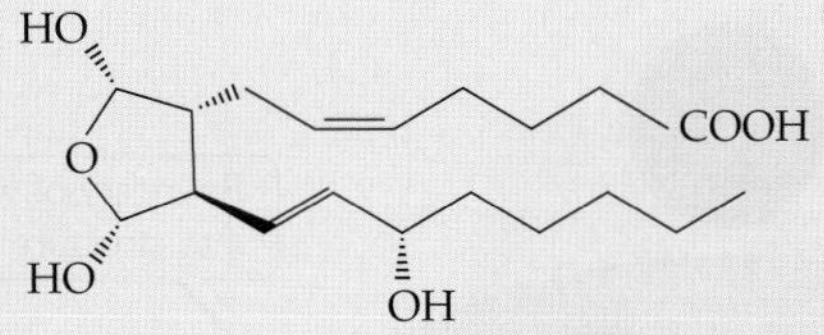

(d)

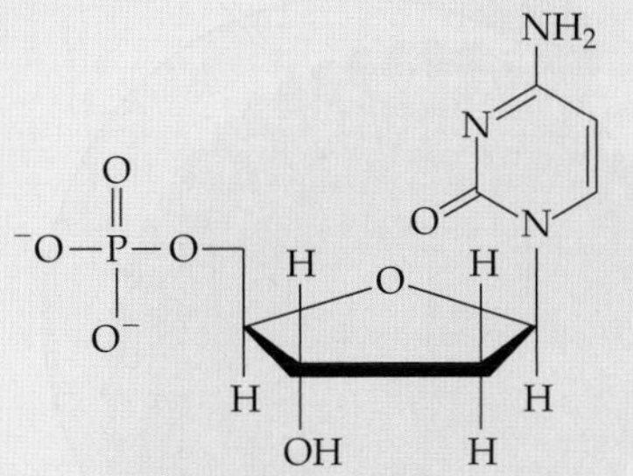

SOLUTION

Analyse We are asked to differentiate between compounds based on their functionality and to class them accordingly

Plan By definition, *lipids* are naturally occurring substances that can be extracted from cells and tissue by non-polar organic solvents such as hexane. So we will use this definition as a means of classification.

Solve Structure **(a)** is a wax, similar to beeswax. It is soluble in hexane and so is classed as a lipid. Structure **(b)** is a member of the steroid hormones. Its large carbon framework suggests this compound will be soluble in non-polar solvents. Hence this compound can be generally classed as a lipid. Structure **(c)** is a prostaglandin. Despite the presence of three OH and a carboxylic acid group, the two large alkyl chains would suggest this molecule is hexane soluble. This compound is also a lipid. Structure **(d)** is a nucleotide. Its collection of nitrogen- and oxygen-containing heterocycles and polar phosphate group would mean this compound is more aqueous than hexane soluble. Therefore, this class of compound is not a lipid.

PRACTICE EXERCISE

Draw the structure of a triglyceride containing lauric acid.

Answer:

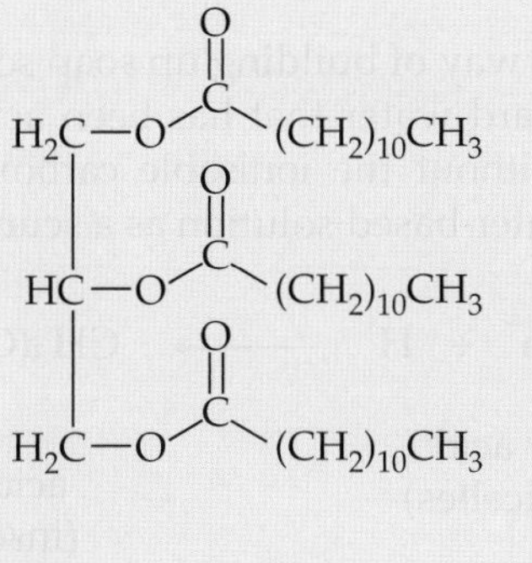

(See also Exercise 27.26.)

Soaps and Detergents

Chemically, a **soap** is the sodium or potassium salt of a fatty acid. These fatty acids are produced by the base-promoted hydrolysis of a triglyceride in a process known as *saponification*. The cleansing power of a soap is due to its ability to act as an emulsifying agent. That is, it is a substance that can sit at the interface between a hydrophobic liquid and a hydrophilic liquid, allowing the hydrophobic liquid to be dispersed in microscopic droplets. Many common forms of dirt (for example, grease, oil, fat) are hydrophobic and soaps allow them to be dispersed in waste water.

In water, soap forms a cloudy solution of **micelles,** as shown in ▼ **FIGURE 27.14.** Micelles form as a result of the hydrophobicity of the alkyl

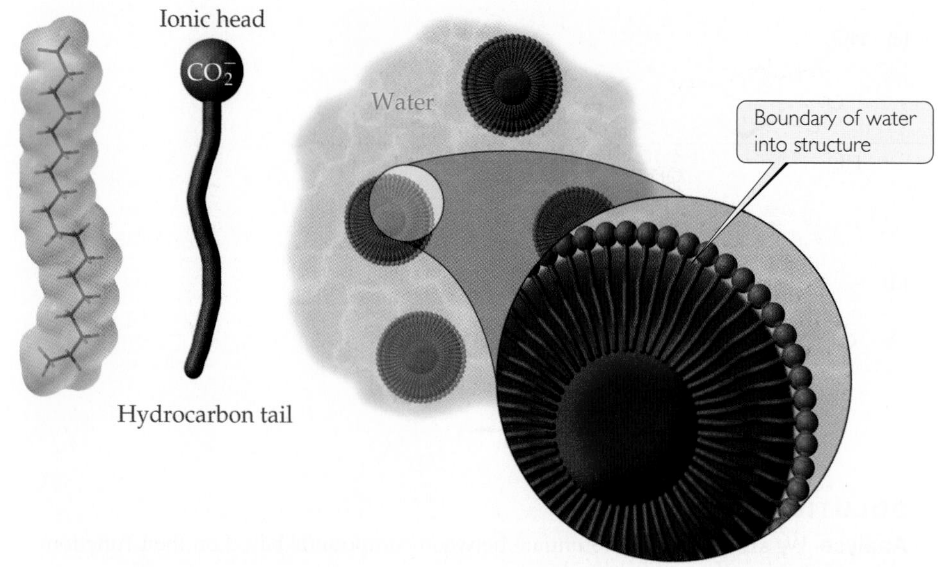

▶ **FIGURE 27.14 Soap micelle.** In water, soap forms a cloudy solution of micelles. The hydrophobic hydrocarbon chains cluster in the middle of the micelle and the polar (hydrophilic) carboxylate groups are on the surface. The Na^+ counterions are dissolved in the water around the micelle. Note: Micelles are not circular as drawn, but spherical (this is just a slice through a micelle).

chains and the hydrophilicity of the polar carboxylate group. Micelles are stabilised by dispersion forces between alkyl groups and by hydrogen bonding between carboxylate anions and water molecules.

A common disadvantage of soap is the presence of *soap scum*, an insoluble precipitate or film that is difficult to remove from household areas such as bathrooms. This scum can be produced in two common ways. Foremost is the reaction of a fatty acid with Ca^{2+}, Mg^{2+} or Fe^{2+} ions found in *hard water*. This reaction, called a *counterion exchange* reaction, converts the sodium salt to the calcium salt, which is insoluble in water.

$$2\ CH_3(CH_2)_{16}CO_2^-\ Na^+\ +\ Ca^{2+} \longrightarrow\ Ca[CH_3(CH_2)_{16}CO_2]_2 + 2\ Na^+$$

sodium salt of a fatty acid   calcium salt of a fatty acid
(soluble in water as micelles)   hard-water scum
       (insoluble in water)   [27.18]

The other common way of building up soap scum is through the precipitation of fatty acids in hard water that has been acidified by some means—for example, acid rain. Without the ionisable carboxylate group, the fatty acid floats to the top of a water-based solution as a scum.

$$2\ CH_3(CH_2)_{16}CO_2^-\ Na^-\ +\ H^+ \longrightarrow\ CH_3(CH_2)_{16}CO_2H(s) + Na^+$$

sodium salt of a fatty acid   fatty acid
(soluble in water as micelles)   acidic-water scum
       (insoluble in water)   [27.19]

> ### ▲ CONCEPT CHECK 4
>
> Many commercial laundry soaps contain sodium carbonate as a water-softening agent. How does it work in water that is hardened by a low pH or dissolved Ca^{2+}, Mg^{2+} or Fe^{2+} ions?

Synthetic detergents have been designed with the problem of soap scum in mind. Essentially, the same desirable characteristics of a fatty-acid soap must be maintained; that is, the detergent should contain a long hydrocarbon chain and a polar head group. It is desirable that this polar head group should not form insoluble salts with Ca^{2+}, Mg^{2+} or Fe^{2+} ions, in particular. Sodium salts of sulfonic acids, called sulfonates or sulfates, provide useful analogues to the

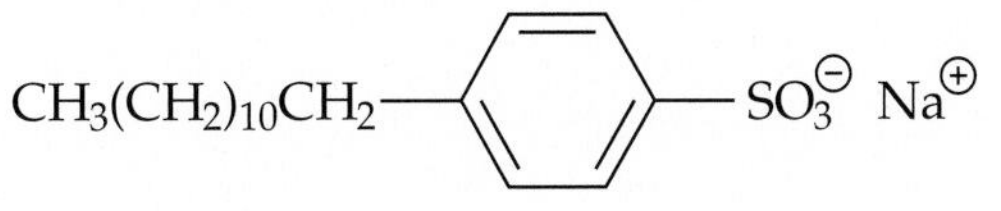

CH₃(CH₂)₁₀CH₂—⟨benzene⟩—SO₃⊖ Na⊕

sodium 4-dodecylbenzene sulfonate
(SDS)

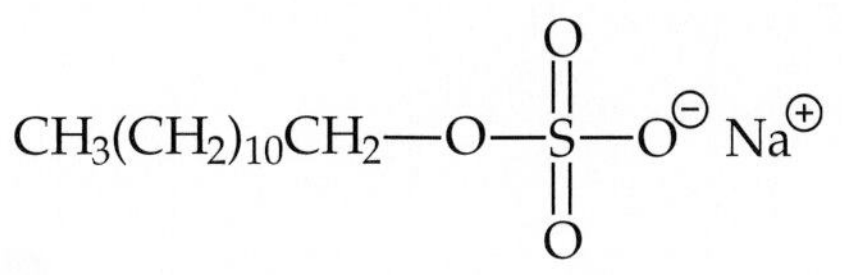

$$CH_3(CH_2)_{10}CH_2-O-\overset{\overset{\displaystyle O}{\|}}{\underset{\underset{\displaystyle O}{\|}}{S}}-O^{\ominus}\ Na^{\oplus}$$

sodium dodecyl sufate
(sodium lauryl sulfate)

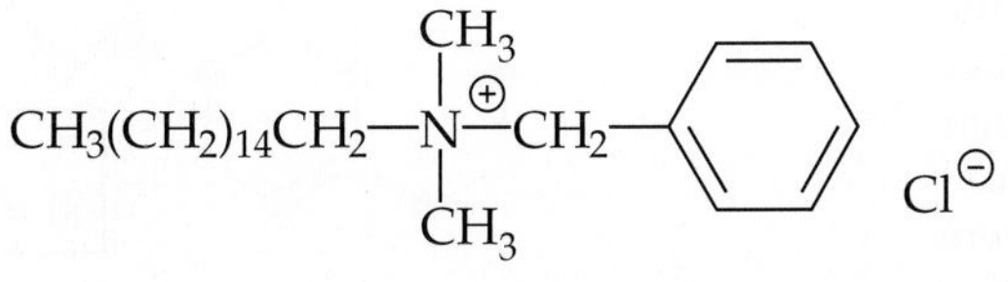

$$CH_3(CH_2)_{14}CH_2-\overset{\overset{\displaystyle CH_3}{|}}{\underset{\underset{\displaystyle CH_3}{|}}{\overset{\oplus}{N}}}-CH_2-⟨\text{benzene}⟩\qquad Cl^{\ominus}$$

benzylcetyldimethylammonium chloride
(benzalkonium chloride)

◀ **FIGURE 27.15 Synthetic detergents.** Detergents can be anionic or cationic, but still retain the features of a soap derived from a fatty acid.

carboxylate group. Being stronger acids than carboxylic acids means that sulfonic acids are not as easily protonated, even under the lowest pH conditions found domestically. Of even greater significance is the fact that calcium, magnesium and iron salts of sulfonic acids are soluble in water.

Not all detergents need to contain an anion. Benzalkonium chloride is one used in a variety of domestic liquid cleaning and disinfectant products. ▲ FIGURE 27.15 illustrates some common examples of synthetic detergents.

SAMPLE EXERCISE 27.5 | **Soap structure**

Draw the general structure of the main constituent of a soap micelle and indicate the hydrophilic and hydrophobic regions.

SOLUTION

Analyse We are asked to draw a schematic of the sodium or potassium salt of a fatty acid and indicate polar and non-polar regions.

Plan Since soaps are prepared from the carboxylate salt of a fatty acid, this question requires you to draw a long alkyl chain and a polar carboxylate head group.

Solve Using a simple circle for a head group and the line drawing of a long alkane, we get:

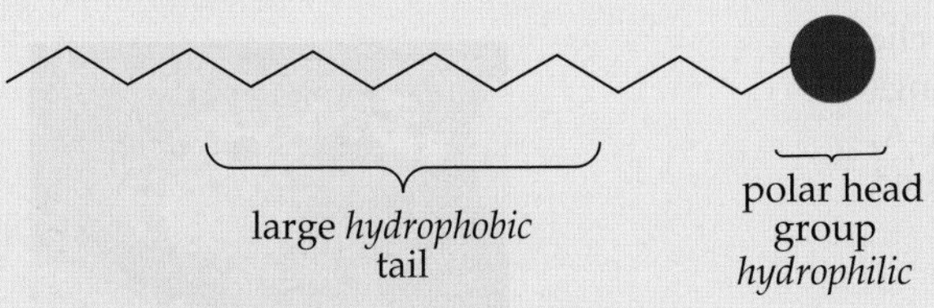

large *hydrophobic*
tail

polar head
group
hydrophilic

PRACTICE EXERCISE

Detail the common features of a soap and a detergent.

Answer: In both cases a hydrophobic tail and polar head group are present. The difference lies in both the nature of the head group (negatively or positively charged) and in its functionality.

(See also Exercise 27.28.)

27.5 | ACID CHLORIDES, ANHYDRIDES AND NUCLEOPHILIC ACYL SUBSTITUTION

Several functional group classes are derived from the carboxyl group. We have already introduced carboxylic acids and esters. Two others of note are the **acid anhydrides** and **acid halides**. The functional group known as an acid anhydride

MY WORLD OF CHEMISTRY

STEROIDS

Hormones are chemical messengers capable of stimulating or inhibiting particular biochemical processes. One major class of hormones is the steroids. Because steroids are non-polar compounds, they are considered lipids. Their non-polar character allows them to cross cell membranes so they can leave the cells in which they were synthesised and enter their target cells. There have been five Nobel prizes in the area of steroid chemistry giving it a rich and important history across chemistry and medicine.

Bile salts are steroids with detergent properties able to emulsify lipids in foodstuff passing through the intestine to enable fat digestion and absorption through the intestinal wall. *Cholesterol* is a steroid found in animal tissues and various foods that is normally synthesised by the liver and is important as a constituent of cell membranes and influences membrane fluidity; it's actually a precursor to all steroid hormones. *Mineralcorticoids* such as aldosterone are secreted by the adrenal cortex and regulate the balance of water and electrolytes or salts (Na^+, Cl^-, HCO_3^-) in the body. The addition of aldosterone causes increased Na^+ readsorption, for example. *Glucocorticoids* such as cortisol (also called hydrocortisone) is also produced by the adrenal cortex and are involved in carbohydrate, protein and fat metabolism. They also have excellent anti-inflammatory properties. The most noted of the steroids are the sex hormones such as the estrogens (for example, progesterone and estrone) and the androgens (for example, testosterone). The estrogens and androgens affect the growth and function of the reproductive organs, the development of secondary sex characteristics and the behavioural patterns of animals. The structure of progesterone is shown in ▼ **FIGURE 27.16**. Anabolic steroids such as stanolozol aid in the development of muscles and are man made.

All steroids contain the characteristic rigid tetracyclic backbone based on the androstane ring system: the four rings are designated A, B, C and D as shown in Figure 27.16. The A, B and C rings are six-membered rings; D is a five-membered ring. In steroids, the rings are all transfused.

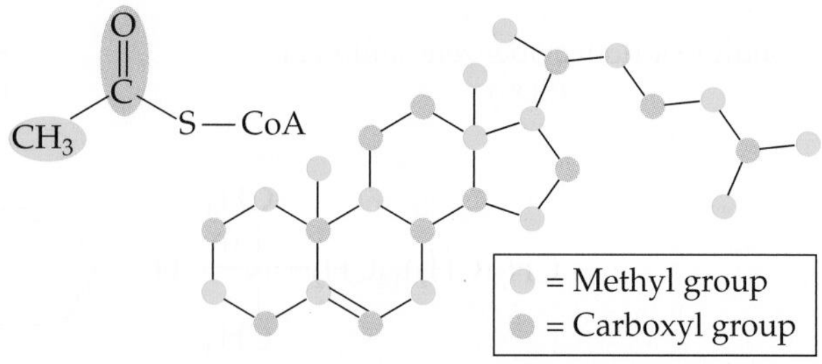

▲ **FIGURE 27.17** **The tetracyclic backbone of all steroids can be traced back to acetic acid.** The methyl carbons in acetic acid are indicated as yellow dots, and the carbonyl carbons that make up the backbone are represented in blue.

Steroids are made available from terpenes (○○ Section 24.1, 'The Structure of Unsaturated Hydrocarbons'), which are readily made in nature from acetic acid.

▲ **FIGURE 27.17** shows the origin of the acetic acid carbon atoms in a typical steroid structure. There are about 40 enzyme-catalysed steps to prepare cholesterol from acetic acid through biosynthesis.

Poison dart frogs are fascinating creatures in that they secrete four steroid toxins from glands located on the back and behind the ears—one of them being among the most toxic substances known to man, batrachotoxin (▼ **FIGURE 27.18**). Batrachotoxin contains the tetracyclic ring structure of the steroids. As a neurotoxin it affects the nervous system by increasing Na^+ permeability of cellular ion channels. It becomes lethal by permanently blocking nerve signal transmission to muscles, leading to cardiac arrest. The estimated lethal dose for this steroid poison in humans is somewhere between 2 and 200 µg.

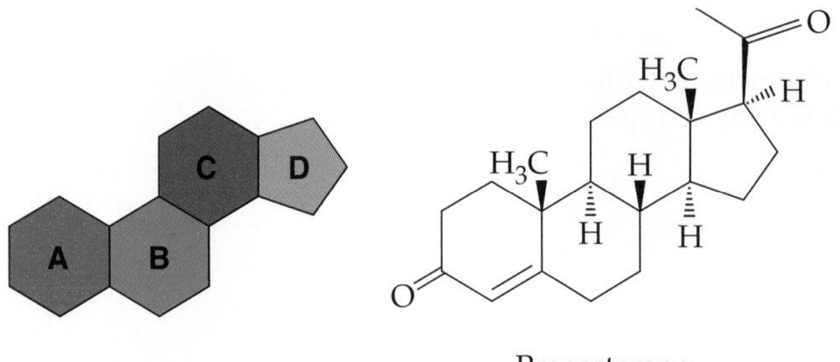

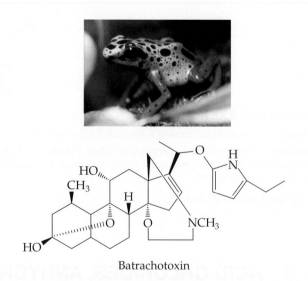

Progesterone

▲ **FIGURE 27.16** **The central tetracyclic structure to all steroids.** Progesterone is an estrone that plays a role in maintaining pregnancy.

Batrachotoxin

▲ **FIGURE 27.18** **Poison dart frogs excrete batrachotoxin, a steroid-based structure and one of the most toxic compounds known.**

has two **acyl groups** (RCO–) linked by an oxygen atom. An acid halide has a single acyl group bonded to a halogen atom:

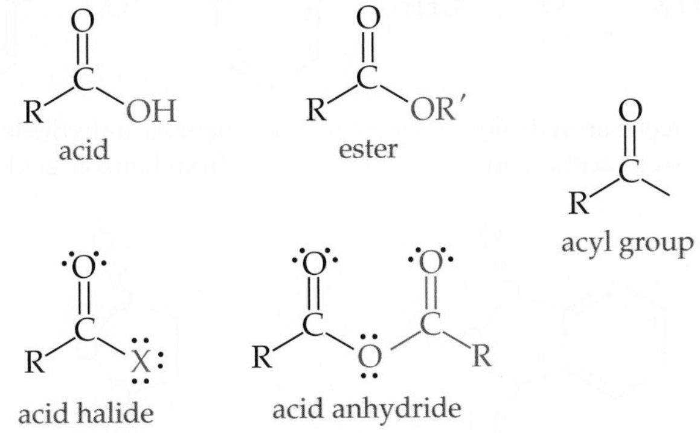

acid   ester   acyl group

acid halide   acid anhydride

Acid chlorides, which are the most common and practical acid halides, are usually prepared by treatment of a carboxylic acid with thionyl chloride ($SOCl_2$). You may recall this reagent as a useful way of converting alcohols to chloroalkanes (Section 24.5). Acid chlorides are named by replacing the suffix *-oic acid* with *-oyl halide*. For example, $CH_3CH_2CH_2COCl$ is named butanoyl chloride and is derived from butanoic acid.

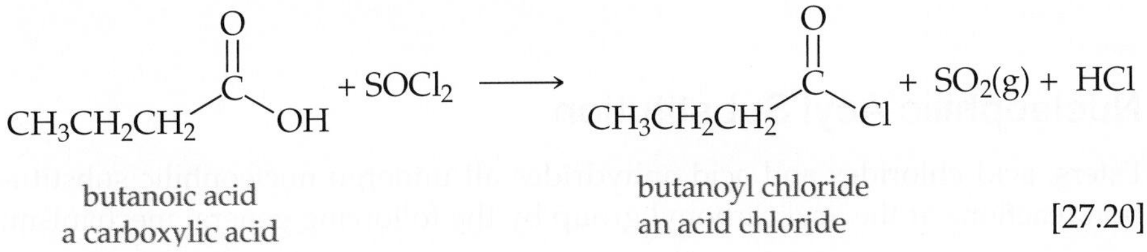

butanoic acid
a carboxylic acid

butanoyl chloride
an acid chloride

[27.20]

Acid anhydrides such as acetic anhydride, $(CH_3CO)_2O$, are usually prepared by reaction of a carboxylate anion with an acid chloride. In this way, both symmetrical anhydrides (same acid group) and unsymmetrical anhydrides (different acid groups) can be produced. Another general method is to dehydrate the carboxylic acid using a suitable dehydrating reagent, such as a more reactive anhydride. In some cases, especially 1,2-dicarboxylic acids such as phthalic acid or succinic acid, anhydrides can be formed by heating.

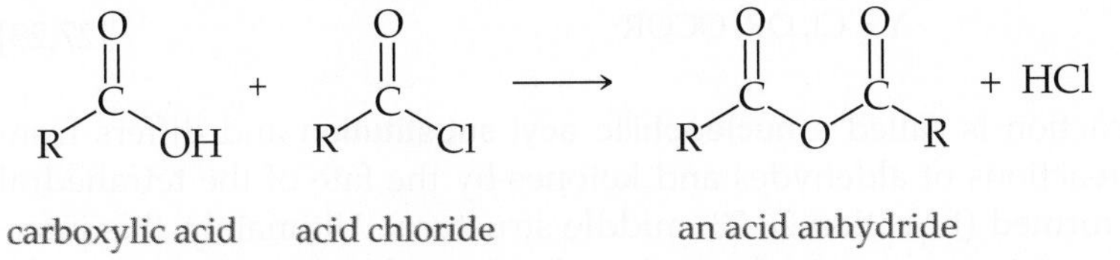

carboxylic acid   acid chloride   an acid anhydride

[27.21]

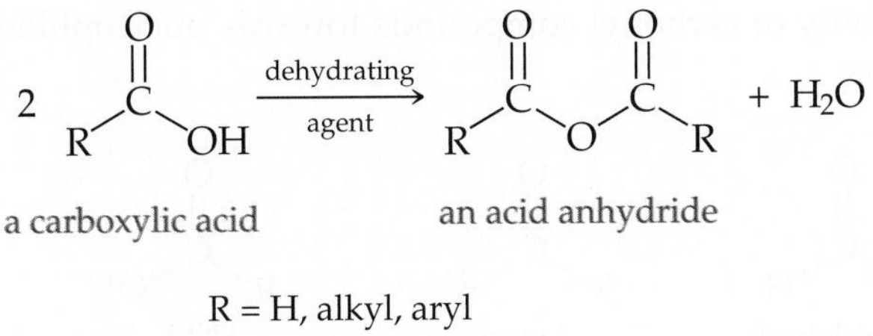

a carboxylic acid   an acid anhydride

R = H, alkyl, aryl

[27.22]

Anhydrides from simple carboxylic acids and cyclic anhydrides from dicarboxylic acids are named by replacing the word *acid* with *anhydride*.

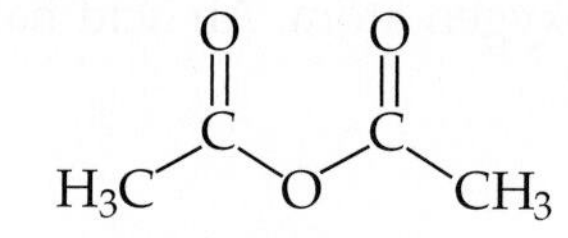

acetic anhydride
from acetic acid

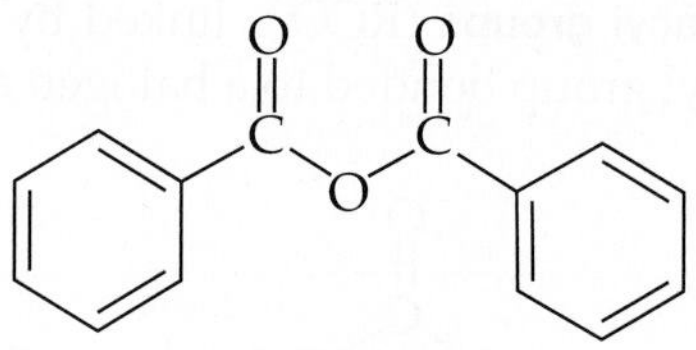

benzoic anhydride
from benzoic acid

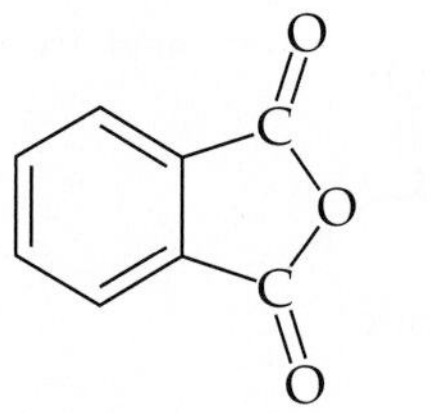

phthalic anhydride
from phthalic acid (below)

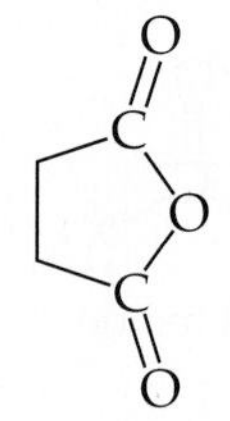

succinic anhydride
from succinic acid (below)

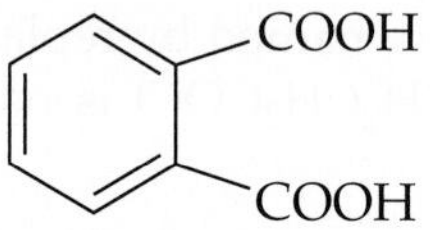

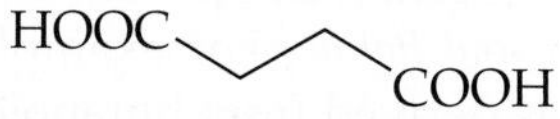

Nucleophilic Acyl Substitution

Esters, acid chlorides and acid anhydrides all undergo nucleophilic substitution reactions at the acyl carbonyl group by the following general mechanism:

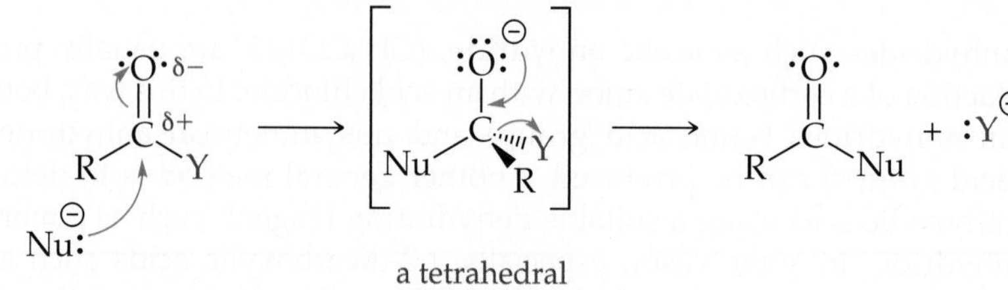

an acid chloride, anhydride or ester

a tetrahedral intermediate

R = H, alkyl, aryl
Y = Cl, OR, OCOR

[27.23]

Such a reaction is called a **nucleophilic acyl substitution** and differs from nucleophilic reactions of aldehydes and ketones by the fate of the tetrahedral intermediate formed (Equation 26.23, middle structure). Ultimately, this intermediate collapses to regenerate the carboxyl group, thereby eliminating the group (Y)—which is typically a *good leaving group*—as a stable anion.

The reactivity of carbonyl compounds towards nucleophilic attack can be summarised as:

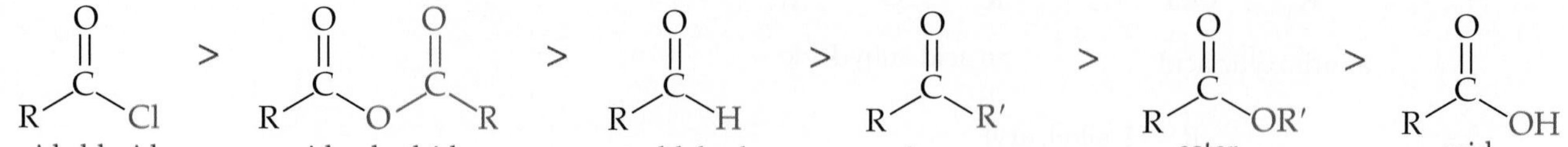

acid chloride > acid anhydride > aldehyde > ketone > ester > acid

This relative reactivity is governed by two factors: the electropositive nature of the carbonyl carbon atom by substituent inductive effects, and the stability of the leaving group (Y^-) displaced by substitution. Some useful interconversions using this reaction type are shown in ▶ **FIGURE 27.19** including the

FIGURE IT OUT

What role do water, alcohol and amine play in these interconversions?

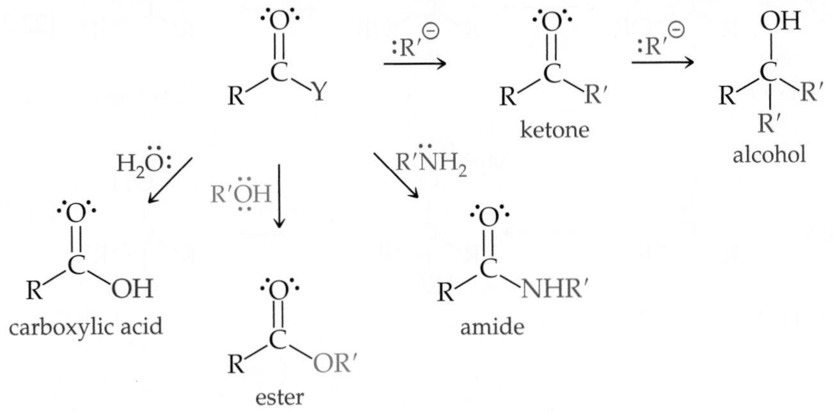

▲ **FIGURE 27.19 Functional group interconversions.** Acid halides, anhydrides and esters are versatile starting materials, leading to a range of different functional groups. Y = Cl, OR or OCOR.

reaction of amines with acid chlorides, anhydrides and esters to form **amides**. Amides have an acyl group bonded to a trivalent nitrogen atom, usually of the form RCONHR′. Amide bonds are the key structural features that join amino acids together to form polypeptides and proteins (∞ Section 29.1, 'Amines and the Amide Bond'). Amides are named by replacing the suffix *-oic acid* with *amide*. The *N* in the names shown below refers to functionalisation on nitrogen. So, for *N*-methylbutanamide, the methyl group is bonded to nitrogen.

∞ Find out more on page 1148

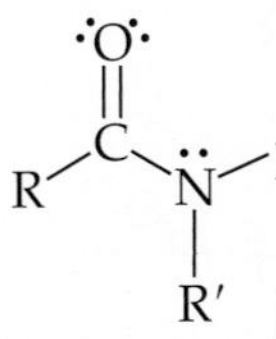

$$R = H, alkyl, aryl$$
$$R' = H, alkyl, aryl$$

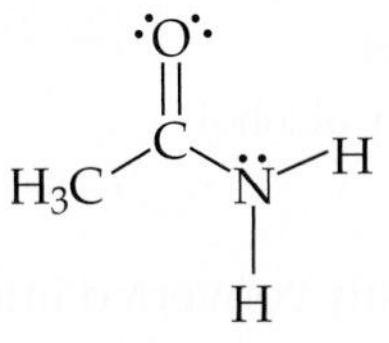

acetamide 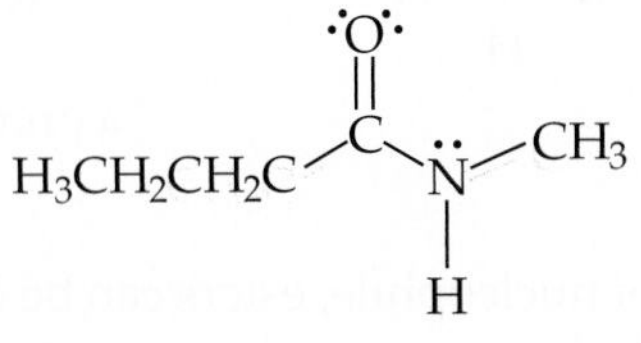   *N*-methylbutanamide 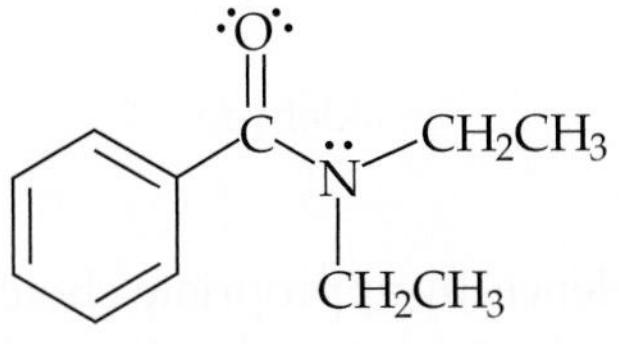   *N*,*N*-diethylbenzamide

A second interconversion from Figure 27.19 to highlight involves the reaction of acid chlorides, anhydrides and esters with carbanions, such as those generated through Grignard reagents, to form alcohols via a ketone intermediate (∞ Section 26.3, 'Reactions of Aldehydes and Ketones'). The addition of a strong nucleophile such as a Grignard reagent to an ester (or acid chloride or anhydride) usually yields the corresponding 3° alcohol as the final product instead of the expected ketone (based on a stoichiometric addition). For this reason, usually two equivalents of the Grignard reagent are added to the reaction mixture to drive the reaction to completion. Mechanistically, collapse of the tetrahedral intermediate from the ester, after attack of the first equivalent of RMgX, yields a ketone. The ketone is more electrophilic than the ester (and so more reactive towards a second equivalent of RMgX), so it undergoes further reaction to yield the stable alkoxide in preference. Upon the addition of acid, the tertiary alcohol is formed.

∞ Review this on page 1042

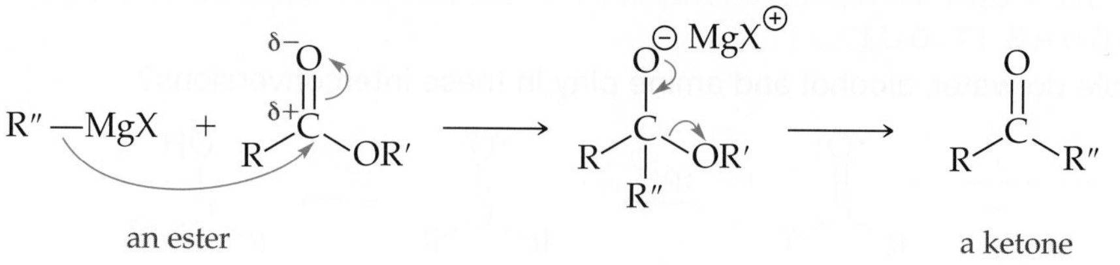

$$[27.24]$$

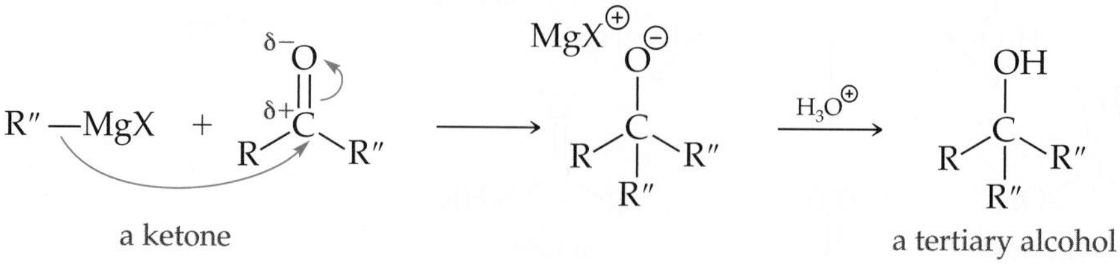

an ester / a ketone

$$[27.25]$$

a ketone / a tertiary alcohol

We can extrapolate this result to other nucleophiles such as hydride (H^-), which comes in the form of lithium aluminium hydride ($LiAlH_4$), a powerful reducing agent. As shown in Equation 27.26, collapse of the tetrahedral intermediate from the ester yields an aldehyde, which again is more reactive than the ester, and which undergoes a second addition of hydride to form the alkoxide. The addition of mild acid protonates the alkoxide, forming a primary alcohol.

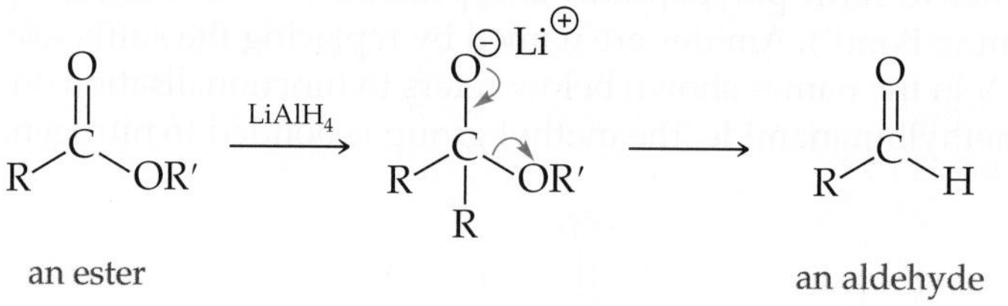

an ester / an aldehyde

$$[27.26]$$

an aldehyde / a primary alcohol

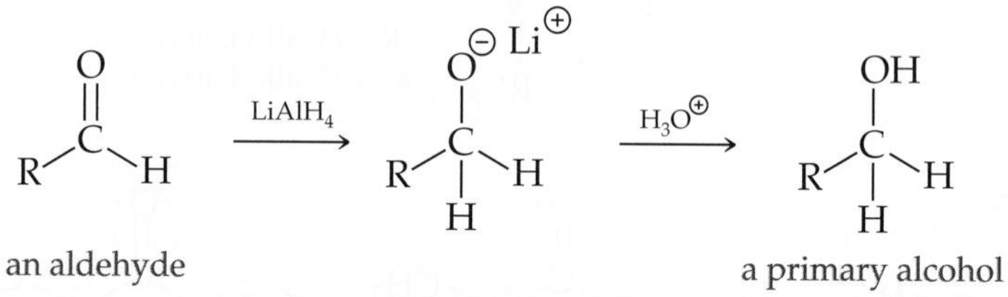

$$[27.27]$$

Hence, by appropriate choice of nucleophile, esters can be easily converted into both primary and tertiary alcohols.

27.6 | CONDENSATION POLYMERISATION

∞ Review this on page 980

In Chapter 24 we discussed one of the major classes of commercially important polymers and their formation by addition polymerisation (∞ Section 24.6, 'Addition Polymerisation'). A second general reaction used to synthesise commercially important polymers is **condensation polymerisation** or *step-growth polymerisation*. This type of polymerisation usually employs difunctional molecules as the monomers, with each new bond being created in a separate step. We discuss two types of condensation polymers in this section: **polyesters** and **polyamides**.

Polyesters are formed by the reaction of a diacid, diacid chloride or diester with a diol. Polymers formed from two different monomers are called **copolymers**. PET (polyethene terephthalate), or *Dacron®*, is the most common polyester copolymer used today.

MY WORLD OF CHEMISTRY

BINDEEZ

The 2007 'Australian Toy of the Year,' traded under the registered name *Bindeez* (*Aqua Dots* in North America), was urgently, and voluntarily, recalled in December 2007 as a result of several serious child illness cases.

Bindeez were small coloured beads that, when aligned into shapes and patterns and sprayed with water, bind together under a cross-linking action to create pictures. However, children who swallowed the beads suffered headaches, nausea, vomiting, confusion and reduced levels of consciousness.

The manufacturing process to produce the popular toy was supposed to use 1,5-pentanediol, a relatively non-toxic chemical found in many glues. However, the manufacturers used 1,4-butanediol, which is cheaper and widely used in cleaning fluids and plastics, but is a harmful substance (◀ FIGURE 27.20). The action of this substance on the beads was essentially to prevent the water-soluble glues from becoming sticky before they were needed.

An Australian biochemical geneticist, Kevin Carpenter, determined what effect the 1,4-butanediol was having. Using mass spectrometry (Section 30.4) on metabolised samples taken from urine, and correlating them to authentic samples of the substances, Carpenter found that the toxicity was due to gamma-hydroxybutanoic acid (GHB)*. 1,4-Butanediol is metabolised into GHB by the enzymes alcohol dehydrogenase and aldehyde dehydrogenase, which are the enzymes responsible for metabolising alcohol in the body.

▲ **FIGURE 27.20   Bindeez.** A recall of all Bindeez products in 2007 was made as a result of the body's action on one of the constituent chemicals 1,4-butanediol.

On 21 December 2007 the Australian Minister of Consumer Affairs issued a permanent order that prohibits the sale of all toys containing 1,4-butanediol.

RELATED EXERCISE: 27.17

* Gamma-hydroxybutanoic acid, or GHB, is a controlled substance and recreational drug, often known as the 'date rape drug'.

dimethyl   ethene glycol   PET
terephthalate   (Dacron®)

[27.28]

Polyamides are formed by the polycondensation of a diamine with a diacid, diacid chloride or diester. The best known of the polyamides is nylon 6,6, discovered by Wallace Carothers (1896–1937) and his associates at Du Pont in 1934. It is named with the notation 6,6 because it is made by the reaction of adipic acid (a diacid within a six-carbon chain) and 1,6-hexanediamine (a six-carbon chain diamine). Nylons are used in home furnishings, apparel, carpet, fishing line and toothbrush bristles, to name but a few.

$$n \, H_2N \, (CH_2)_6NH_2 + n \, HOC \, (CH_2)_4COH \longrightarrow \left[NH(CH_2)_6NH-C(CH_2)_4C \right]_n + 2n \, H_2O$$

diamine   adipic acid   nylon 6,6

[27.29]

Nylon 6, also called *Perlon*®, is made from ε-caprolactam, a cyclic amide. Its uses include ropes and ties. When ε-caprolactam is heated with a small amount of water, some of it is hydrolysed to the amino acid. Continued heating gives condensation and polymerisation to yield molten nylon 6. Note that a single-digit notation is used because nylon 6 is a polymer made up of one monomer containing six carbons. The monomer can be symbolised as A, which in polymer form creates the same simple repeating unit -A-A-A-A-. Nylon 6,6 is a polymer comprising two different monomers, symbolised as A and B, which forms copolymers of the form -A-B-A-B-.

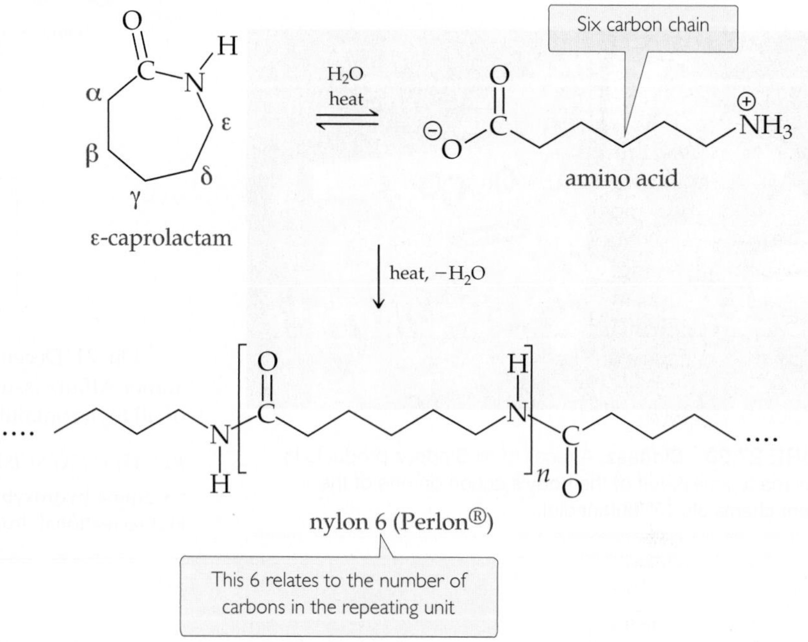

ε-caprolactam

Six carbon chain

amino acid

heat, −H₂O

nylon 6 (Perlon®)

[27.30]

This 6 relates to the number of carbons in the repeating unit

Polymers such as nylon and polyesters can also be formed into *fibres* that, like hair, are very long relative to their cross-sectional area and are not elastic. These fibres can be woven into fabrics or cords and fashioned into clothing and other useful objects.

CONCEPT CHECK 5

Would it be possible to make a condensation polymer out of this molecule alone?

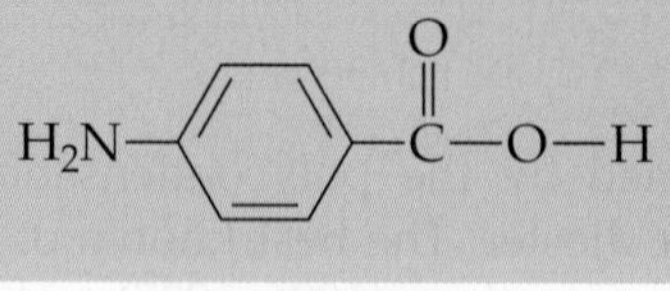

Fibres of nylon 6,6 are cold-drawn, meaning they are drawn out at room temperature. As the fibres are drawn, hydrogen bonds form between individual polymer strands, such that the carbonyl of one amide bond hydrogen bonds to an amide hydrogen atom on another strand (▶ **FIGURE 27.21**). The tensile strength and stiffness of nylon 6,6 are attributable to the hydrogen bonding that occurs in these fibres.

FIGURE IT OUT

What role do the hydrogen bonds play in a polymeric material such as nylon?

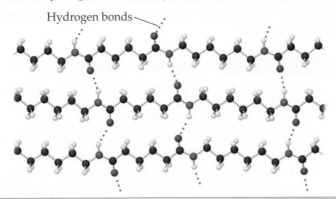

Hydrogen bonds

◀ **FIGURE 27.21 Hydrogen bonding in nylon 6,6.** The unusual and desirable physical properties of nylon 6,6 are a result of intermolecular hydrogen bonding.

MY WORLD OF CHEMISTRY

TOWARDS THE PLASTIC CAR

Many polymers can be formulated and processed to have sufficient structural strength, rigidity and heat stability to replace metals, glass and other materials in a variety of applications. The housings of electric motors and kitchen appliances such as coffee makers and can openers, for example, are now commonly made from specially formulated polymers. Engineering polymers are tailored to particular applications through choice of polymers, blending of polymers and modifications of processing steps. They generally have lower costs or superior performance over the materials they replace. In addition, shaping and colouring of the individual parts and their assembly to form the final product are often much easier.

The modern car provides many examples of the inroads engineering polymers have made in car design and construction. Car interiors have long been formed mainly of plastics. Through development of high-performance materials, significant progress has been made in introducing engineering polymers as engine components. ▼ FIGURE 27.22, for example, shows the manifold in a series of Ford V-8 utility and van engines. The manifold, which is made of nylon, must be stable at high temperatures. The use of an engineering polymer in this application eliminates machining and several assembly steps.

Car body parts can also be formed from engineering polymers. Components formed from engineering polymers usually weigh less than the components they replace, thus improving fuel economy. The bumper bars of Volkswagen's New Beetle (▼ FIGURE 27.23), for example, are made of nylon reinforced with a second polymer, poly(phenylene ether) (PPE), which has the following structure:

Because the poly(phenylene ether) polymer is linear and rather rigid, it confers rigidity and ensures shape retention.

A big advantage of most engineering polymers over metals is that they eliminate the need for costly corrosion protection steps in manufacture. In addition, some engineering polymer formulations permit manufacturing with the colour built in, thus eliminating painting steps.

RELATED EXERCISES: 27.51, 27.53

▲ **FIGURE 27.23 Plastic fenders.** The bumper bars of this Volkswagen Beetle are made of General Electric Noryl GTX, a composite of nylon and poly(phenylene ether).

▲ **FIGURE 27.22 Plastic engines.** The intake manifold of some Ford V-8 engines is formed from nylon.

Various substances may also be added to polymers to provide protection against the effects of sunlight or against degradation by oxidation. For example, manganese(II) salts and copper(I) salts, in concentrations as low as 5×10^{-4} % (w/w), are added to nylons to provide protection from light and oxidation and to help maintain whiteness.

SAMPLE EXERCISE 27.6 | **Condensation polymers**

What is the structure of the polymer formed from lactic acid, a naturally occurring compound obtained from carbohydrates?

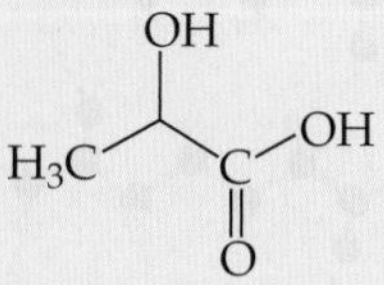

SOLUTION

Analyse We are given the structure of lactic acid and asked to derive the structure of the polymer made through self-condensation.

Plan Identify functional groups capable of forming a condensation polymer, then draw the polymer linking these functional groups up.

Solve Identification of the alcohol and carboxylic acid functional groups is important to determine the place of condensation. Condensation occurs by the formation of ester groups.

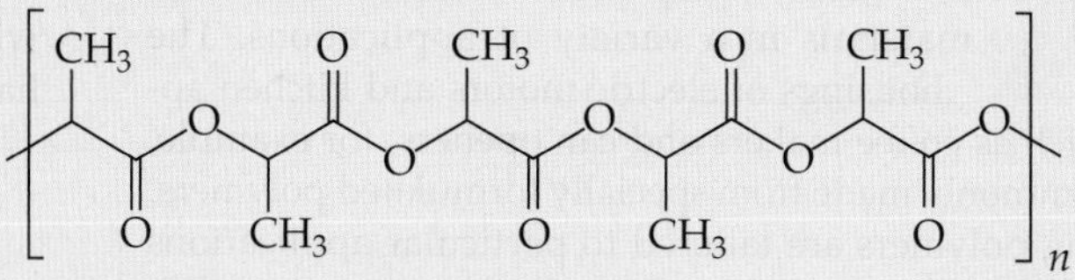

poly(lactic acid)

PRACTICE EXERCISE

Indicate whether the following polymer is a step (condensation) or chain (addition) growth polymer. Draw the structure of the monomer/s used in the preparation of this polymer.

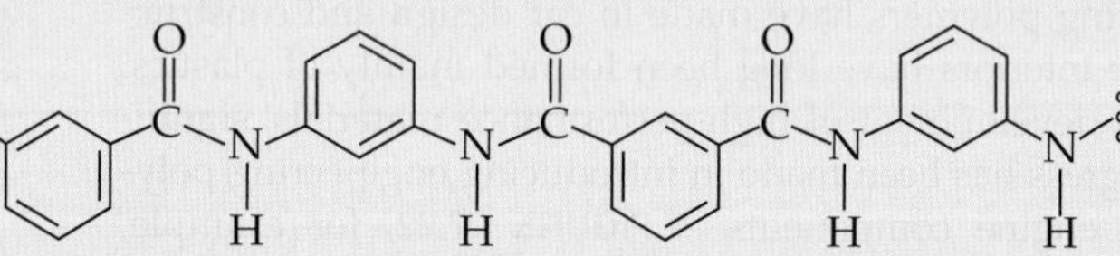

Answer: This is a polyamide, a step-growth polymer. The monomers that are used to prepare it are:

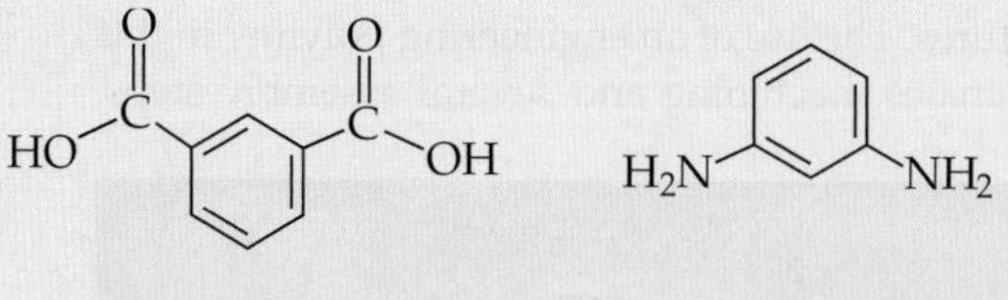

(See also Exercises 27.6, 27.42.)

Polymers for Medicine

Increasingly, modern synthetic materials such as polymers are being used in medical and biological applications. For our discussion here, a **biomaterial** is any material that has a biomedical application. The material might have a therapeutic use in a treatment, injury or disease or it might have a diagnostic use as part of a system for identifying a disease or for monitoring a quantity such as the glucose level in blood. Whether the use is therapeutic or diagnostic, the biomaterial is in contact with biological fluids, so it must have properties that meet the demands of that application. For example, a polymer used to make a disposable contact lens must be soft and have an easily wetted surface, whereas the polymer used to fill a dental cavity must be hard and wear resistant. The most important characteristics that influence the choice of a biomaterial are biocompatibility, physical requirements and chemical requirements, as illustrated in ▶ FIGURE 27.24.

Specialised polymers have been developed for a variety of biomedical applications. The degree to which the body accepts or rejects the foreign polymer is determined by the nature of the groups along the chain and the possibilities for interactions with the body's own molecules. Our bodies are composed largely of biopolymers such as proteins, polysaccharides (sugars) and polynucleotides (RNA, DNA). We will learn more about these molecules in Chapter 29. For now, we can simply note that the body's biopolymers have complex structures, with polar groups along the polymer chain. Proteins, for example, are long strings of amino acids (the monomers) that have formed a condensation polymer. The protein chain has the following structure:

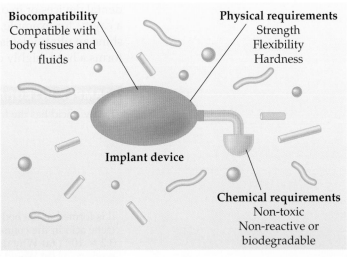

▲ **FIGURE 27.24 Schematic illustration of a human-made device implanted in a biological system.** To function successfully, the device must be biocompatible with its surroundings and meet the necessary physical and chemical requirements, some of which are listed for illustrative purposes.

$$\left[\begin{array}{c}\underset{\underset{R_1}{|}}{\overset{\overset{H}{|}}{C}}-\underset{}{\overset{\overset{O}{\|}}{C}}-\underset{\underset{H}{|}}{\overset{\overset{H}{|}}{N}}-\underset{\underset{R_2}{|}}{\overset{\overset{H}{|}}{C}}-\underset{}{\overset{\overset{O}{\|}}{C}}-\underset{\underset{H}{|}}{\overset{\overset{H}{|}}{N}}-\underset{\underset{R_3}{|}}{\overset{\overset{H}{|}}{C}}-\underset{}{\overset{\overset{O}{\|}}{C}}-\underset{\underset{H}{|}}{\overset{\overset{H}{|}}{N}}\end{array}\right]_n$$

where the R groups vary along the chain ($-CH_3$, $-CH(CH_3)_2$, and so on). Twenty different amino acids are used in different combinations to make up proteins in humans. By contrast, synthetic polymers are simpler, formed from a single repeating unit or perhaps two different repeating units. This difference in complexity is one of the reasons why synthetic polymers are identified by the body as foreign objects. Another reason is that there may be few or no polar groups along the chain that can interact with the body's aqueous medium.

Polymers can be characterised by their physical properties. Elastomers are used as biomaterials in flexible tubing, over leads for implanted heart pacemakers and as catheters (tubes implanted into the body to administer a drug or to drain fluids). Thermoplastics, such as polyethene or polyesters, are employed as membranes in blood dialysis machines and as replacements for blood vessels. Thermoset plastics find limited, but important, uses. Because they are hard, inflexible and somewhat brittle, they are most often used in

MY WORLD OF CHEMISTRY

BIODEGRADABLE SUTURES

Until the 1980s, if you had to have stitches in a wound, you had to go back later to get the stitches removed. Stitches that are used to hold living tissue together are called *sutures*. Since the

1980s, biodegradable sutures have been available. After the sutures have been applied to the tissue, they slowly dissolve away and do not release any harmful by-products. Today biodegradable sutures are made out of lactic acid—glycolic acid copolymers that slowly hydrolyse over time.

$$n\,\underset{\text{glycolic acid}}{HOCH_2\overset{\overset{O}{\|}}{C}-OH} + n\,\underset{\underset{CH_3}{|}\atop\text{lactic acid}}{HOCHC-OH} \longrightarrow \underset{\text{copolymer}}{-O-CH_2\overset{\overset{O}{\|}}{C}-O\left[\underset{\underset{CH_3}{|}}{CHC}-O-CH_2\overset{\overset{O}{\|}}{C}-O\right]_n\underset{\underset{CH_3}{|}}{CHC-}}$$

[27.31]

dental devices or in orthopaedic applications, such as joint replacements. To fill a cavity, for example, the dentist may pack some material into the cavity, then shine an ultraviolet lamp on it. The light initiates a photochemical reaction that forms a hard, highly cross-linked thermoset polymer.

SAMPLE INTEGRATIVE EXERCISE | Putting concepts together

Pyruvic acid has the following structure:

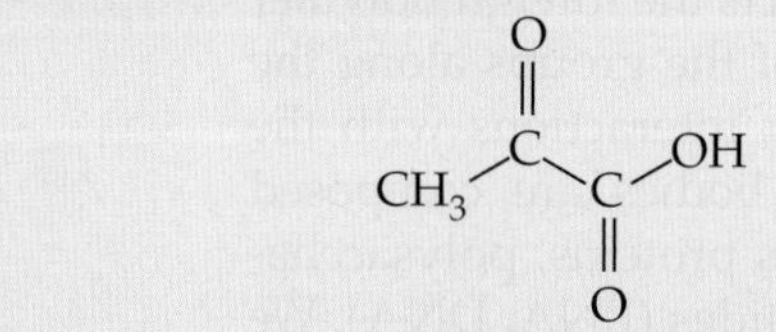

It is formed in the body from carbohydrate metabolism. In the muscle it is reduced to lactic acid in the course of exertion. The acid-dissociation constant for pyruvic acid is 3.2×10^{-3}. **(a)** Why does pyruvic acid have a higher acid-dissociation constant than acetic acid? **(b)** Would you expect pyruvic acid to exist primarily as the neutral acid or as dissociated ions in muscle tissue, assuming a pH of about 7.4 and an acid concentration of 2×10^{-4} M? **(c)** What would you predict for the solubility properties of pyruvic acid? Explain. **(d)** What is the hybridisation of each carbon atom in pyruvic acid? **(e)** Assuming H atoms as the reducing agent, write a balanced chemical equation for the reduction of pyruvic acid to lactic acid. (Although H atoms don't exist as such in biochemical systems, biochemical reducing agents deliver hydrogen for such reductions.)

SOLUTION

(a) The acid ionisation constant for pyruvic acid should be somewhat greater than that of acetic acid because the carbonyl function on the α-carbon atom exerts an electron-withdrawing effect on the carboxylic acid group. In the C–O–H bond system, the electrons are shifted from hydrogen, facilitating loss of the hydrogen as a proton (Section 17.10, 'Acid–Base Behaviour and Chemical Structure').

Review this on page 692

(b) To determine the extent of ionisation, we first set up the ionisation equilibrium and equilibrium constant expression. Using HPv as the symbol for the acid, we have

$$HPv \rightleftharpoons H^+ + Pv^-$$

$$K_a = \frac{[H^+][Pv^-]}{[HPv]} = 3.2 \times$$

Let $[Pv^-] = x$. Then the concentration of undissociated acid is $2 \times 10^{-4} - x$. The concentration of $[H^+]$ is fixed at 4.0×10^{-8} (the antilog of the pH value). Substituting, we obtain

$$3.2 \times 10^{-3} = \frac{[4.0 \times 10^{-8}]}{[2 \times 10^{-4}}$$

Solving for x, we obtain $x[3.2 \times 10^{-3} + 4.0 \times 10^{-8}] = 6.4 \times 10^{-7}$.

The second term in the brackets is negligible compared with the first, so $x = [Pv^-] = 6.4 \times 10^{-7}/3.2 \times 10^{-3} = 2 \times 10^{-4}$ M. This is the initial concentration of acid, which means that essentially all the acid has dissociated. We might have expected this result because the acid is quite dilute and the acid-dissociation constant is fairly high.

(c) Pyruvic acid should be quite soluble in water because it has polar functional groups and a small hydrocarbon component. It is miscible with water, ethanol and diethyl ether.

(d) The methyl group carbon has sp^2 hybridisation. The carbon carrying the carbonyl group has sp^2 hybridisation because of the double bond to oxygen. Similarly, the carboxylic acid carbon is sp^2 hybridised.

(e) The balanced chemical equation for this reaction is

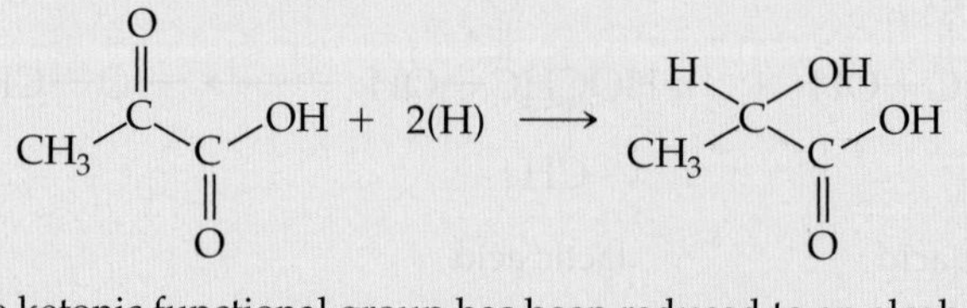

Essentially, the ketonic functional group has been reduced to an alcohol.

CHAPTER SUMMARY AND KEY TERMS

SECTION 27.1 Carboxylic acids are weak acids. They react with bases to form **carboxylate anions**, which are very stable due to resonance. Carboxylic acids readily form hydrogen bonded dimers in concentrated solution or neat.

SECTION 27.2 Carboxylic acids are commonly prepared by oxidation of primary alcohols or aldehydes or by the oxidation of benzylic carbon atoms, yielding benzoic acid derivatives. They can also be prepared by hydrolysis of nitriles, amides and esters.

SECTION 27.3 Esters are formed by the condensation reaction of carboxylic acids and alcohols. This reaction is called the **Fischer esterification**. Esters can form amides by a condensation reaction with amines. Esters undergo hydrolysis, called **saponification**, in the presence of strong bases.

SECTION 27.4 Esters such as **fats**, **oils**, **waxes** and **triglycerides** are abundant in biological systems. They form part of a general class called **lipids**. Lipids come in two forms; **complex** and **simple**, depending on the chemical properties of the lipid. Polyunsaturated triglycerides can be hardened by catalytic

hydrogenation, which changes the texture of the triglyceride from a liquid form to a solid/semi-solid form. This process is called **hardening**. The saponification of a triglyceride leads to a **soap**. In water, soaps form a **micelle**, encapsulating dirt particles. Hard waters that contain Ca^{2+} often leave an insoluble scum. **Synthetic detergents** are designer soaps that eliminate the unwanted properties associated with the formation of calcium salts.

SECTION 27.5 Acid anhydrides and **acid halides** also contain an **acyl group** and are more reactive than their ester or carboxylic acid analogues. Both the anhydrides and acid halides are useful starting materials with which to make esters and **amides** through **nucleophilic acyl substitution** reactions.

SECTION 27.6 A **condensation polymerisation** occurs when a small molecule is lost upon reaction. The two main types of condensation polymers are **polyesters** and **polyamides**. A polymer formed from two different monomers is called a **copolymer**. A **biomaterial** is any biocompatible material that has a biomedical application. Biomaterials are commonly polymers with special properties matched to the application.

KEY SKILLS

- Be able to prepare carboxylic acids from a range of starting materials. (Section 27.2)
- Know how to name carboxylic acids, esters, amides and acid chlorides and to discern these functional groups. (Sections 27.3–27.5)
- Be able to define terms such as lipid, oil and fat and reference these terms to classes of compounds. (Section 27.4)
- Understand the differences and similarities between a soap and a detergent. (Section 27.4)
- Be able to determine the products of acyl nucleophilic substitution of esters, acid chlorides and anhydrides with a range of nucleophiles. (Section 27.5)
- Be familiar with the chemistry of carboxylic acids—including their acidity, the stability of carboxylate salts, their reaction with alcohols to form esters and their reaction with amines to form amides. (Sections 27.2–27.5)
- Be able to draw and name polyesters from monomers and identify monomers within polyesters. (Section 27.6)

KEY EQUATIONS

- Acid–base reaction

benzoic acid + NaOH (aq) ⟶ sodium benzoate + H_2O [27.2]

- Esterification

$$CH_3-\overset{O}{\overset{\|}{C}}-OH + HO-CH_2CH_3 \xrightarrow{H^{\oplus}} CH_3-\overset{O}{\overset{\|}{C}}-O-CH_2CH_3 + H_2O$$ [27.14]

acetic acid   ethanol   ethyl acetate

• Nucleophilic acyl substitution

$$
\begin{array}{c}
\text{an acid chloride,}\\
\text{anhydride or ester}
\end{array}
\longrightarrow
\begin{array}{c}
\text{a tetrahedral}\\
\text{intermediate}\\
R = H, \text{alkyl, aryl}\\
Y = Cl, OR, OCOR
\end{array}
\longrightarrow
$$

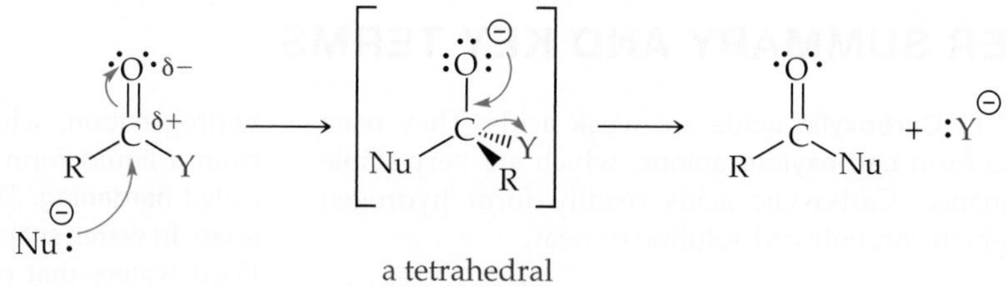

[27.23]

EXERCISES

VISUALISING CONCEPTS

27.1 Identify each of the functional groups in the following molecules. [Sections 27.1 and 27.2]

(a)

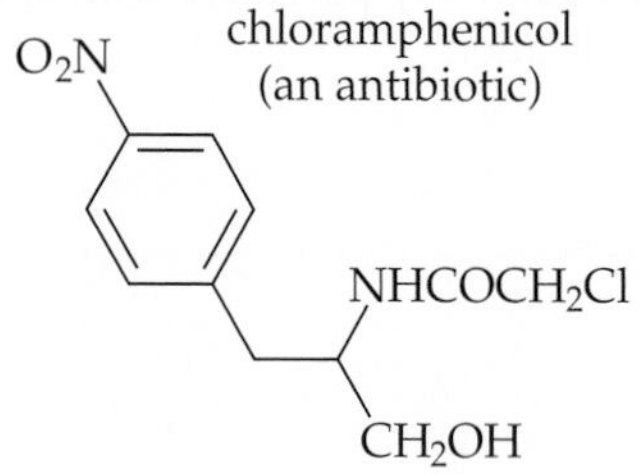

chloramphenicol
(an antibiotic)

O₂N

NHCOCH₂Cl

CH₂OH

(b)

COOH

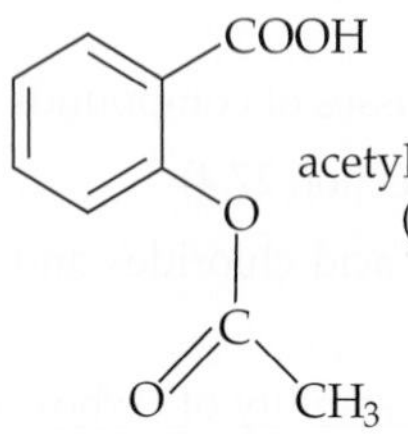

acetylsalicylic acid
(aspirin)

O
‖
C
CH₃

(c)

H
N
CH₃
‖
O

HO

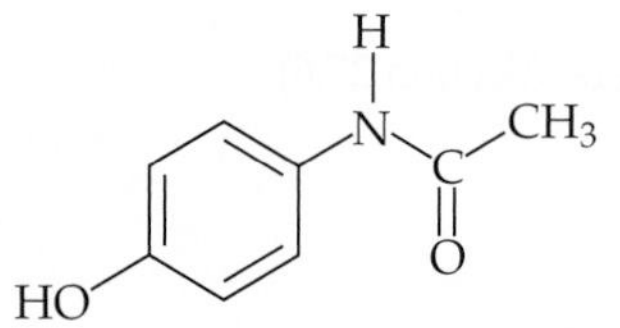

paracetamol

(d)

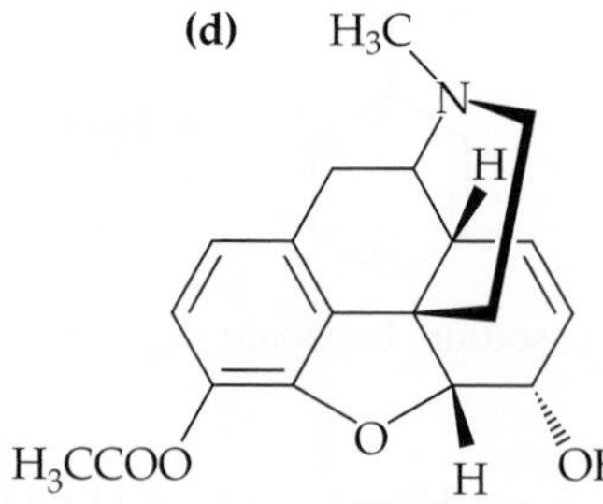

H₃C
N
H

heroin

H₃CCOO
O
H
OH

27.2 What features must a molecule have to participate in hydrogen bonding with other molecules of the same kind? Which of the following compounds would you expect to **(a)** have the highest melting point, **(b)** be most soluble in mild aqueous base? Explain. [Section 27.1]

$$
\begin{array}{cccc}
\overset{O}{\underset{\|}{\text{CH}_3\text{COH}}} & \text{CH}_3\text{CH}_2\text{OH} & \text{CH}_3\text{C}{\equiv}\text{CH} & \overset{O}{\underset{\|}{\text{HCOCH}_3}} \\
\text{(i)} & \text{(ii)} & \text{(iii)} & \text{(iv)}
\end{array}
$$

27.3 Draw the structure of the alcohol precursor, its systematic name and the reaction conditions you might use to obtain the following compounds by oxidation. [Section 27.1]

(a)

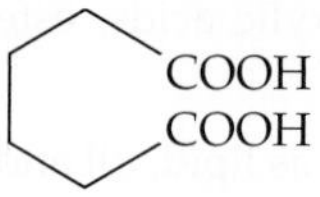

COOH
COOH

(b)

CH₃CH₂CH₂CH₂CH₂COOH

(c)

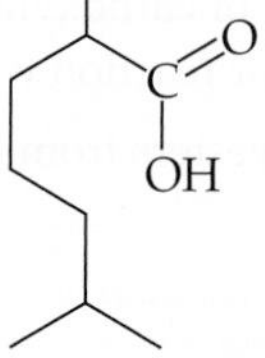

O
C
OH

(d)

COOH

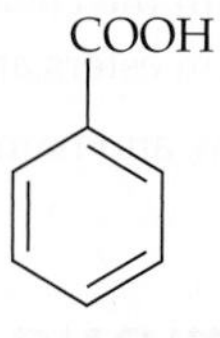

27.4 Which of the following waxes will have the lower melting point? [Section 27.4]

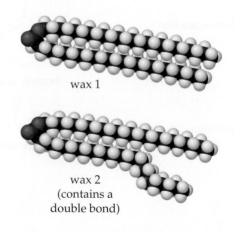

wax 1

wax 2
(contains a
double bond)

27.5 Which of the following fatty acids, when reacted with potassium hydroxide, would form better micelles and hence make the best soap? Explain. [Section 27.4]

arachidic acid

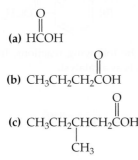

arachidonic acid

CARBOXYLIC ACIDS (Section 27.1)

27.7 Give the IUPAC name for each of the following acids:

(a) HCOH

(b) $CH_3CH_2CH_2COH$

(c) $CH_3CH_2CHCH_2COH$ with CH_3 branch

27.8 Draw the structure for 2-bromobutanoic acid and indicate any stereogenic centres within the molecule. How many stereoisomers are possible?

27.9 (a) Would you expect pure acetic acid to be a strongly hydrogen bonded substance? (b) How do the melting and boiling points of the substance support your answer? (See Table 27.1.)

27.10 Although carboxylic acids and alcohols both contain an –OH group, one is acidic in water and the other is not. Explain the difference.

27.11 Describe how you would distinguish between a carboxylic acid and its ester, using nothing more than an aqueous solution of sodium hydrogencarbonate ($NaHCO_3$, sodium bicarbonate).

27.12 'Bath bombs' contain calcium carbonate and citric acid. When placed into water, they fizz and dissolve rapidly. What are the products of this reaction?

27.13 Which of the following carboxylic acids is the most acidic? Why?

27.6 The polyamide Nomex is prepared from the condensation polymerisation of 1,3-benzenediamine with 1,3-benzenedicarboxylic acid. (a) Draw a structural formula representation of the polymer, being sure to indicate the repeat unit. (b) Would you expect such a polymer to have applications requiring high strength and high temperature, or to be more useful as a *glad-wrap* alternative? Explain your answer. [Section 27.6]

1,3-benzenediamine   1,3-benzenedicarboxylic acid

Acid A   Acid B   Acid C

27.14 (a) Sodium monofluoroacetate, known as '1080', is a commonly used rabbit and fox poison. What reagent could be used to convert fluoroacetic acid into 1080? (b) Explain why the pK_a for fluoroacetic acid ($pK_a = 2.59$) is lower than the pK_a of acetic acid (CH_3COOH, $pK_a = 4.76$).

fluoroacetic acid   1080

27.15 The structural formula for acetic acid is shown in the table below. Replacing hydrogen atoms on the carbon with chlorine atoms causes an increase in acidity, as follows:

Acid	Formula	K_a (25 °C)
Acetic	CH_3COOH	1.8×10^{-5}
Chloroacetic	$CH_2ClCOOH$	1.4×10^{-3}
Dichloroacetic	$CHCl_2COOH$	3.3×10^{-2}
Trichloroacetic	CCl_3COOH	2×10^{-1}

Using Lewis structures as the basis of your discussion, explain the observed trend in acidities in the series. Calculate the pH of a 0.010 M solution of each acid.

PREPARATION OF CARBOXYLIC ACIDS (Section 27.2)

27.16 Suggest plausible reagents for the transformations shown below.

(a)

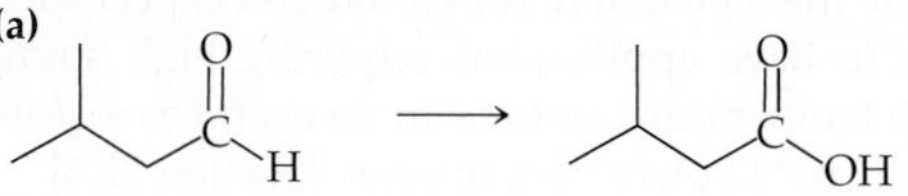

(b)

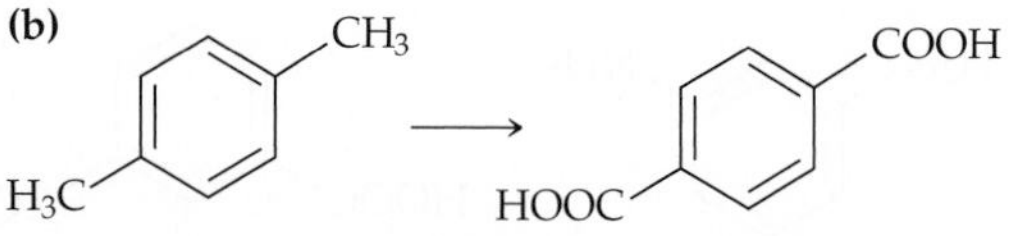

27.17 Describe methods for preparing butanoic acid from each of the following:

(a) butan-1-ol

(b) butanal

(c) methyl butanoate

(d) butanoyl chloride

ESTERS AND ESTERIFICATION (Section 27.3)

27.18 Classify the following compounds as acetals, alcohols, hemiacetals, carboxylic acids, esters or ethers:

(a)

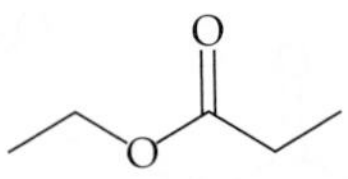

(b)

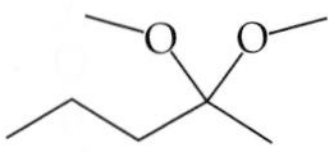

(c)

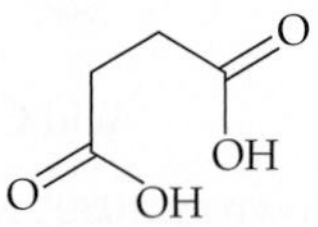

(d)

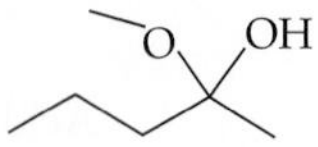

27.19 Identify the functional groups in each of the following compounds:

(a) $CH_3CCH_2CH_3$
         $\|$
         O

(b) $CH_3C=O$
         $|$
         OH

(c) $CH_2CH_2CH_3$
         $|$
         OH

(d) $CH_3OCCH_2CH_3$
         $\|$
         O

27.20 Write a condensed structural formula for each of the following: **(a)** an acid with the formula $C_4H_8O_2$, **(b)** a cyclic ketone with the formula C_5H_8O, **(c)** a cyclic ester with formula $C_5H_8O_2$.

27.21 Give the condensed formulae for the carboxylic acid and the alcohol from which each of the following esters is formed:

(a)

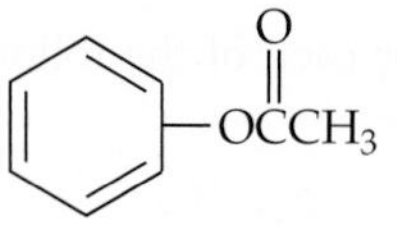

(b)

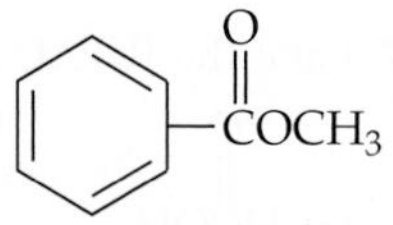

27.22 Draw the products of the following reactions. In each case, the sulfuric acid acts as a catalyst.

(a)

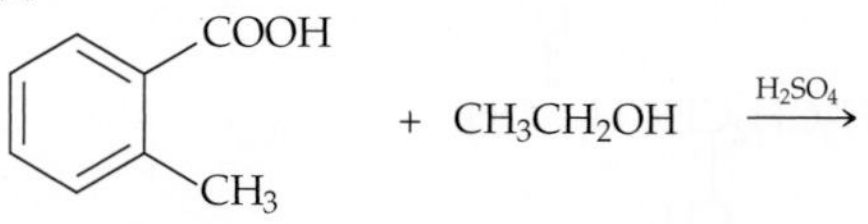

(b)

(c)

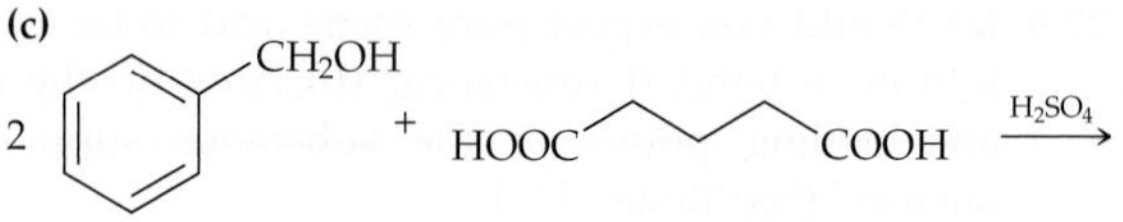

(d)

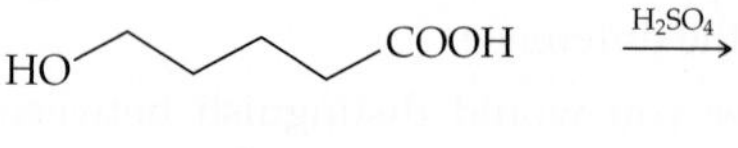

FATS, OILS AND WAXES (Section 27.4)

27.23 **(a)** What is the difference between a fat and an oil? **(b)** What is the difference between an fat and a wax? Draw an example of each to illustrate your point.

27.24 Write a balanced chemical equation using condensed structural formulae for the saponification (base hydrolysis) of **(a)** methyl propionate, **(b)** phenyl acetate.

27.25 Write a balanced chemical equation using condensed structural formulae for **(a)** the formation of butyl propanoate from the appropriate acid and alcohol, **(b)** the saponification (base hydrolysis) of methyl benzoate.

27.26 Triglycerides are formed from fatty acids and glycerol (1,2,3-propanetriol). **(a)** Draw a general formula for the structure of a triglyceride. **(b)** Describe the difference between a saturated and an unsaturated fatty acid. **(c)** Indicate the difference in melting point between a saturated triglyceride and a comparable polyunsaturated triglyceride.

27.27 Locate the stereogenic centre/s in a triglyceride bearing one palmitic acid, one stearic acid and one arachidic side chain (see Table 27.5).

27.28 The structure of a synthetic detergent is shown below. **(a)** What features make this compound act like a natural soap? **(b)** Draw a simple micelle using this molecule to demonstrate the detergent's application. Label the hydrophobic and hydrophilic regions of the micelle.

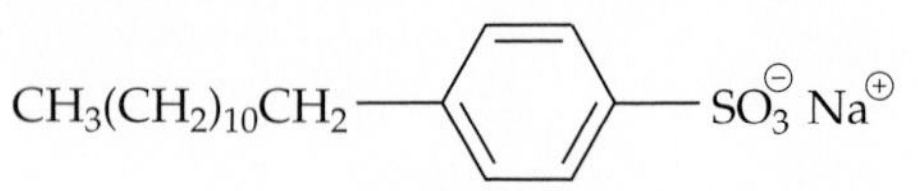

$$CH_3(CH_2)_{10}CH_2 - \text{[benzene ring]} - SO_3^{\ominus} \ Na^{\oplus}$$

sodium 4-dodecylbenzene sulfonate
(SDS)

NUCLEOPHILIC ACYL SUBSTITUTION (Section 27.5)

27.29 2-Phenyl-2-butanol can be synthesised by a combination of a Grignard reagent and an ester. Show the combination.

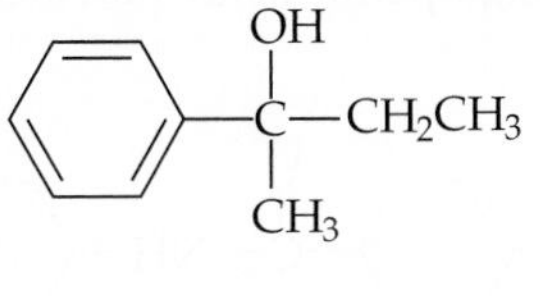

27.30 Draw the major product expected under the following conditions (Ph = C_6H_5):

(a)

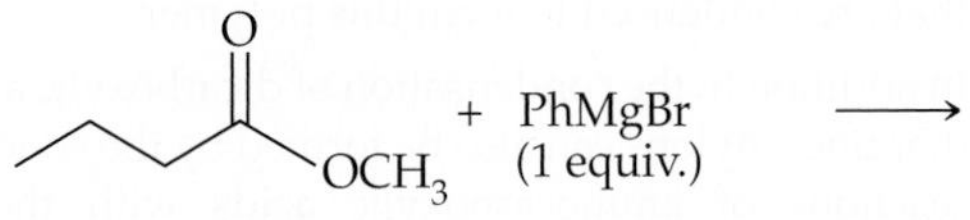

(b)

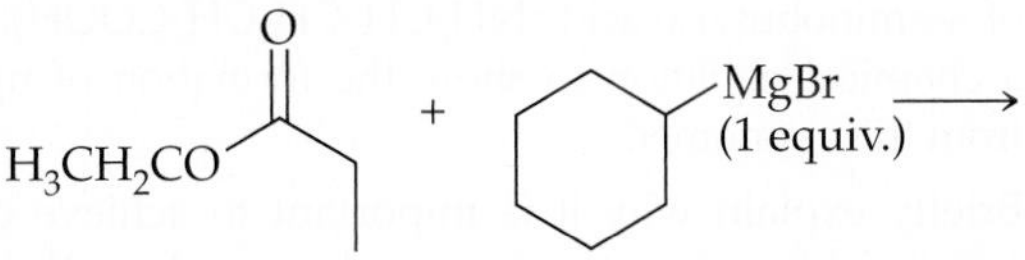

27.31 Draw the condensed structure of the compounds formed by reactions between **(a)** benzoic acid and ethanol, **(b)** ethanoyl chloride and methylamine, **(c)** acetic acid and phenol. Name the compound in each case.

27.32 Draw the condensed structures of the products formed from **(a)** butanoic acid and methanol, **(b)** benzoic acid and 2-propanol, **(c)** the reaction of propanoic acid with $SOCl_2$ and then dimethylamine. Name the compound in each case.

27.33 Acetic anhydride is formed from acetic acid in a condensation reaction that involves the removal of a mole of water from between two acetic acid molecules. Write the chemical equation for this process, and show the structure of acetic anhydride.

27.34 Suggest plausible reagents for the transformations below.

(a)

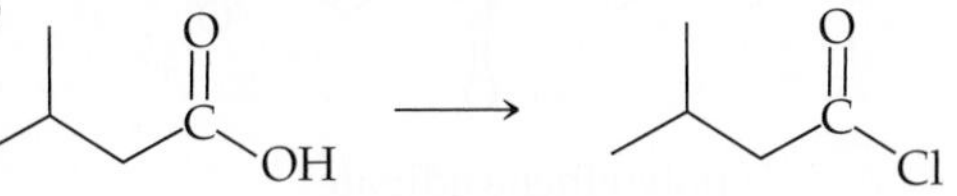

(b)

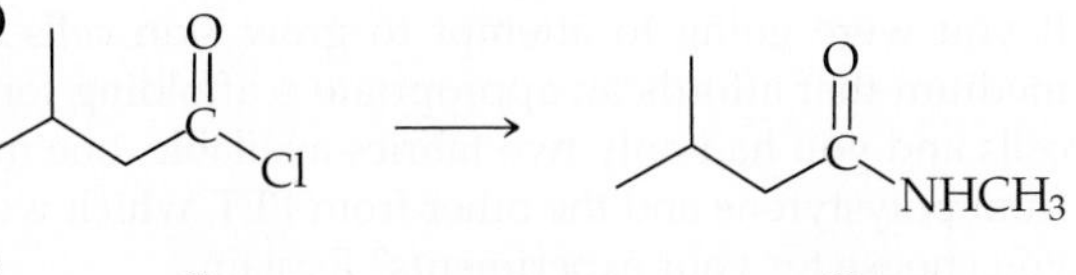

(c)

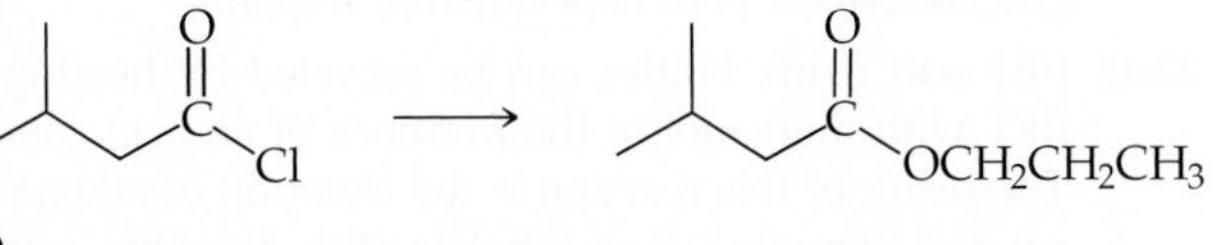

(d)

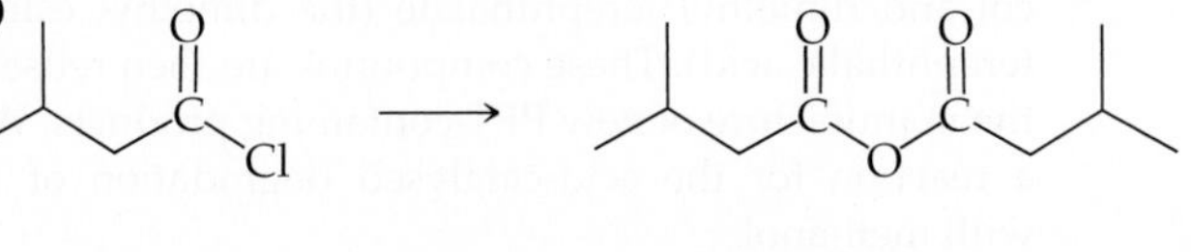

27.35 γ-Hydroxybutanoic acid (GHB) can be prepared enzymatically from 1,4-butanediol. **(a)** Devise a method of preparing GHB in the laboratory from 1,4-butanediol. What compromise did you have to make? **(b)** GHB can form a cyclic ester under acidic conditions. Draw the structure of this ester.

27.36 List the reagents necessary to undertake the following transformations:

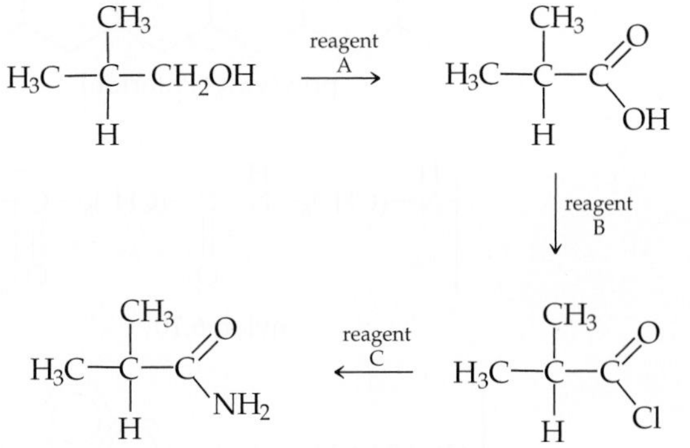

27.37 Several pharmaceuticals were originally derived from ergot, a fungus that affects grain crops which is a derivative of lysergic acid. Suggest reagents for transformations involving reagents A and B.

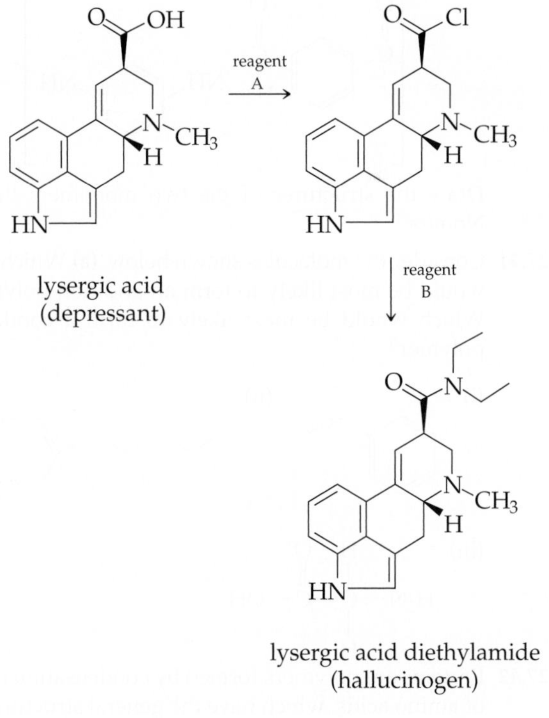

lysergic acid
(depressant)

lysergic acid diethylamide
(hallucinogen)

CONDENSATION POLYMERISATION (Section 27.6)

27.38 Write a chemical equation for the formation of a polymer via a condensation reaction from the monomers succinic acid ($HOOCCH_2CH_2COOH$) and ethenediamine ($H_2NCH_2CH_2NH_2$).

27.39 Draw the structure of the monomer/s employed to form each of the following polymers: **(a)** poly(vinyl chloride), **(b)** nylon 6,10 and **(c)** poly(ethene naphthalate).

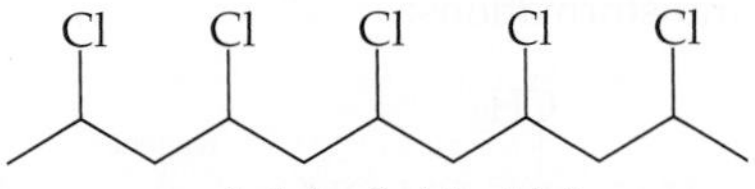

poly(vinyl chloride)

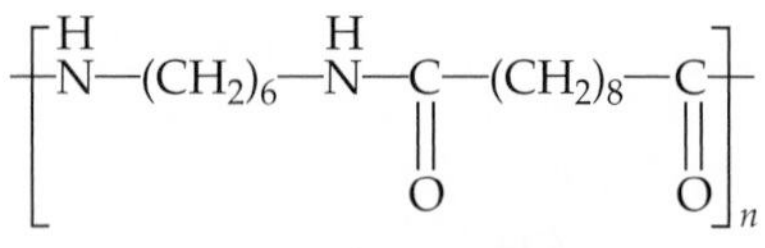

nylon 6,10

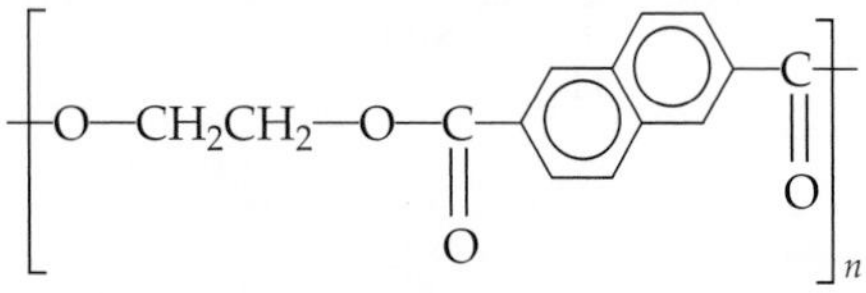

PEN

27.40 The nylon Nomex®, a condensation polymer, has the following structure:

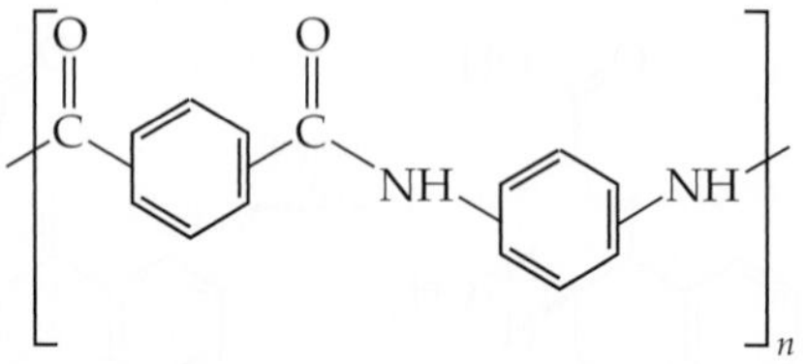

Draw the structures of the two monomers that yield Nomex®.

27.41 Consider the molecules shown below. **(a)** Which of these would be most likely to form an addition polymer? **(b)** Which would be most likely to form a condensation polymer?

(i)

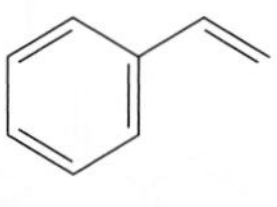

(ii)

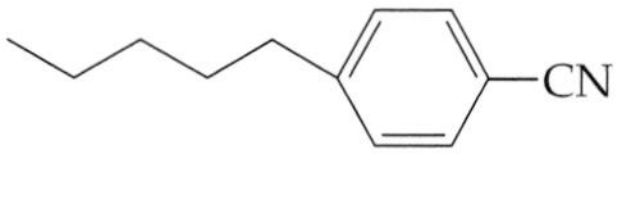

(iii)

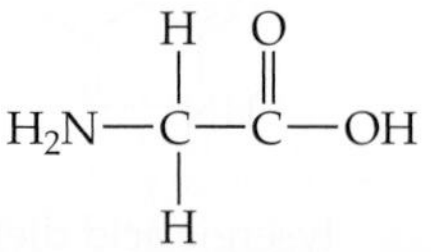

27.42 Proteins are polymers formed by condensation reactions of amino acids, which have the general structure

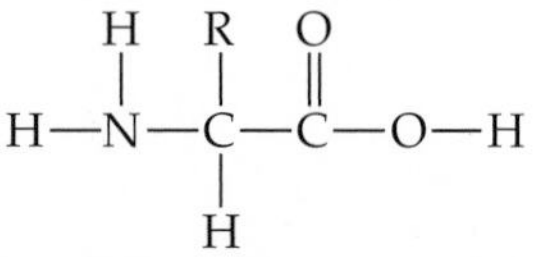

In this structure, R represents –H, –CH_3 or another group of atoms. Draw the general structure for a poly(amino acid) polymer formed by condensation polymerisation of the molecule shown here.

27.43 Kevlar®, a high-performance polymer, has the following structure:

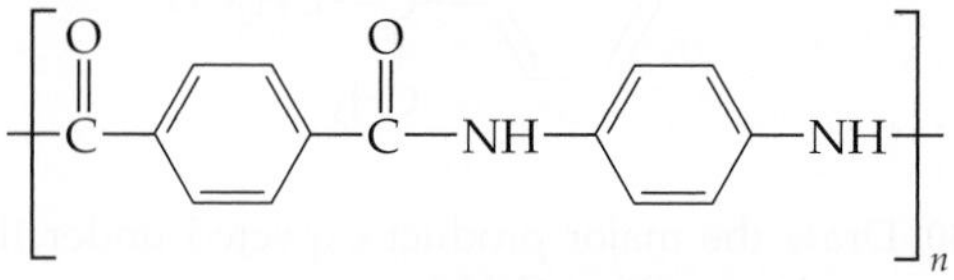

Write the structural formulae for the two compounds that are condensed to form this polymer.

27.44 In addition to the condensation of dicarboxylic acids with diamines, nylons can also be formed by the condensation reactions of aminocarboxylic acids with themselves. Nylon 4, for example, is formed by the polycondensation of 4-aminobutyric acid ($NH_2CH_2CH_2CH_2COOH$). Write a chemical equation to show the formation of nylon 4 from this monomer.

27.45 Briefly explain why it is important to achieve correct $1:1$ stoichiometry of monomers in a condensation polymerisation.

27.46 On the basis of the structure of polystyrene and poly(ethene adipate) shown below, which class of polymer would you expect to form the most effective interface with biological systems? Explain.

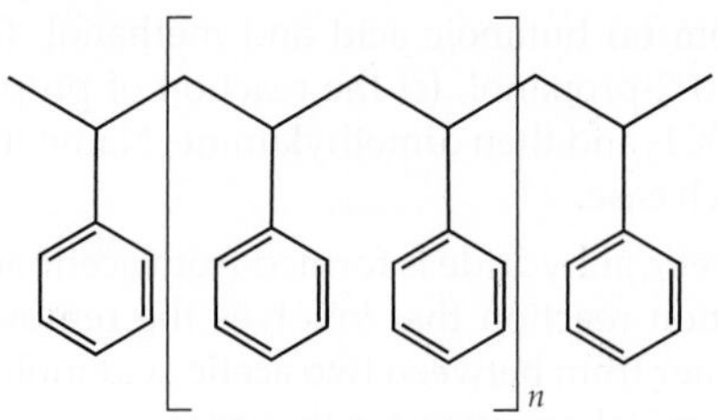

polystyrene

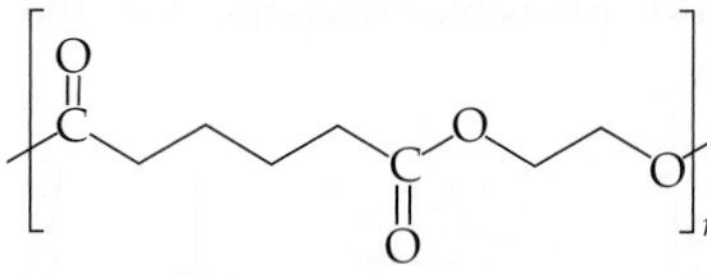

poly(ethene adipate)

27.47 If you were going to attempt to grow skin cells in a medium that affords an appropriate scaffolding for the cells and you had only two fabrics available, one made from polystyrene and the other from PET, which would you choose for your experiments? Explain.

27.48 PET soft drink bottles can be recycled by heating the PET with methanol in the presence of an acid catalyst. The result of this reaction is the isolation of ethene glycol and dimethyl terephthalate (the dimethyl ester of terephthalic acid). These compounds are then reused in the manufacture of new PET-containing products. Write a reaction for the acid-catalysed degradation of PET with methanol.

27.49 Poly(vinyl alcohol), PVA, is a useful water-soluble polymer. It cannot be prepared directly from vinyl alcohol, because vinyl alcohol tautomerises spontaneously to acetaldehyde. It can, however, be prepared from poly(vinyl acetate). Design an experiment for the conversion of poly(vinyl acetate) to poly(vinyl alcohol).

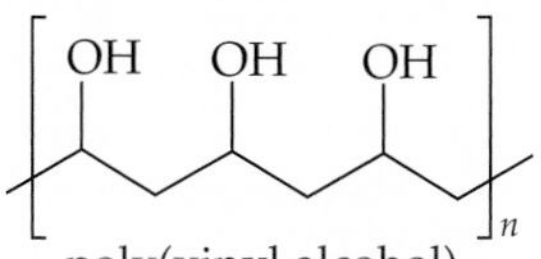

poly(vinyl alcohol)

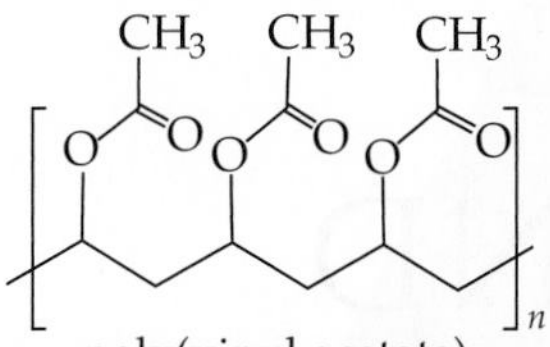

poly(vinyl acetate)

27.50 **(a)** Draw the structure of Qiana®, a polyamide synthesised by the reaction of hexanedioic acid with *trans*-1,4-cyclohexanediamine. **(b)** What effect, if any, would there be in using *cis*-1,4-cyclohexanediamine instead of *trans*-1,4-cyclohexanediamine?

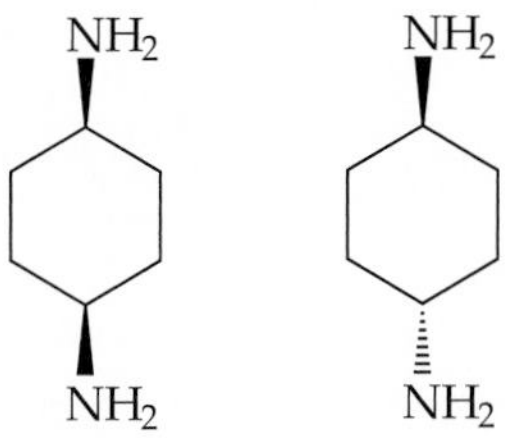

cis *trans*

1,4-cyclohexane diamine

27.51 Glyptal® resin makes a strong thermoset polymer with applications in the electronics and car industries. It is prepared by the condensation of terephthalic acid and glycerol. Draw the structure of Glyptal® and explain its remarkable strength and rigidity.

27.52 What would be the effect of adding poly(phenylene ether) to nylon?

INTEGRATIVE EXERCISES

27.53 Bromobenzene is reacted first with magnesium, then carbon dioxide followed by dilute acid in a Grignard reaction. **(a)** Draw the structure of the product of this reaction. A 0.0050 M solution of the product has a pH of 3.25. **(b)** What is the K_a of the product? **(c)** The conjugate base of the product is a resonance-stabilised anion. Draw all the major resonance structures of this anion.

27.54 Many esters are volatile compounds with distinct fragrances. The boiling point of an ester is often lower than you might expect from the boiling points of the carboxylic acid and alcohol from which it is derived. For example, ethyl methanoate has a boiling point of 54 °C and the boiling points of ethanol and methanoic acid are 78 °C and 100 °C, respectively. **(a)** Draw the structure of ethyl methanoate and indicate the approximate bond angle associated with the carbonyl carbon and ether oxygen atom. **(b)** What intermolecular force is responsible for an increase in boiling point as molecular mass increases in a series of similar compounds? **(c)** What additional intermolecular forces are present in carboxylic acids and alcohols that are not present in the related ester?

27.55 Ethene is an important product from the petrochemical industry. Devise a synthesis of *N*-ethyl acetamide using ethene as the only carbon source. [*Hint*: NH_2^- is a powerful nucleophile.]

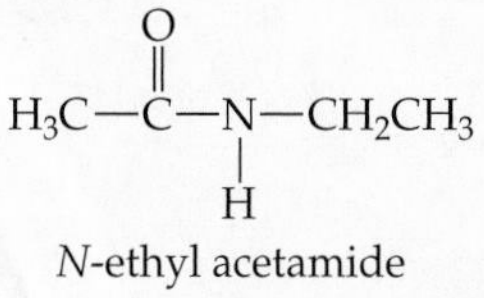

N-ethyl acetamide

MasteringChemistry (www.pearson.com.au/masteringchemistry)

Make learning part of the grade. Access:

- tutorials with peronalised coaching
- study area
- Pearson eText

PHOTO/ART CREDITS

1078 Christy Liem/Shutterstock, © Roman Ivaschenko/Fotolia 2004-2013, all rights reserved, n7atal7i/Shutterstock, James Steidl/Shutterstock, © Photographer Denisa Moorehouse/Dreamstime.com, © Photographer Jurgita Genyte/Dreamstime.com; **1080** Andrey Pavlov/Shutterstock; **1087** Plowes-ProteaPix; **1091** Culver Pictures; **1094** Wade, Leroy G. (2010) *Organic Chemistry*, (7th ed.), © 2010. Printed and Electronically reproduced by permission of Pearson Education, Inc., Upper Saddle River, New Jersey; **1096** Brandon Alms/Shutterstock; **1101** AAP Image; **1103** Reproduced with permission by Bob Roemer, Courtesy of Volkswagen Group Australia.

28

BENZENE AND ITS DERIVATIVES

Aspirin is the common name of acetylsalicylic acid, a benzene-containing compound used commonly as an analgesic for pain relief. Asprin was developed by Bayer in 1853 though salicylic acid present in willow bark had been known to relieve headaches since around 450 BC.

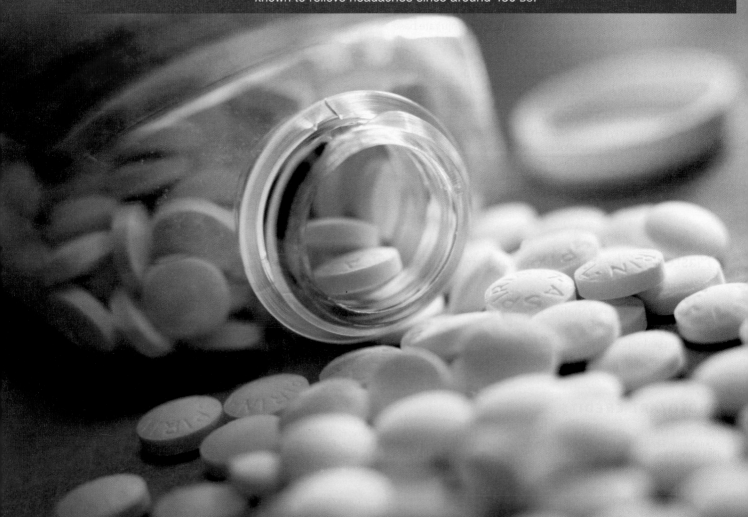

KEY CONCEPTS

28.1 THE STRUCTURE OF BENZENE
We begin by describing the molecular structure of benzene and use hybridisation to explain its unusual bonding characteristics.

28.2 ISOMERISM AND NOMENCLATURE IN AROMATIC COMPOUNDS
We develop an appreciation of the terminology used to describe isomerisation in benzene-containing compounds as well as the names of common aromatic compounds.

28.3 RESONANCE AND AROMATICITY
We introduce the concept of resonance to explain the stability of aromatic molecules and hence their reactivity.

28.4 ACIDITY OF PHENOLS
We study the physical and chemical properties of phenols as a class of aromatic compounds.

28.5 ELECTROPHILIC AROMATIC SUBSTITUTION (EAS) REACTIONS
Benzene derivatives react with electrophiles to form a very wide variety of compounds. We explore the reactivity of different substituted benzenes, the ability of different functional groups to direct the place of reaction in an activating or deactivating sense, and the mechanistic rationale leading to the final product.

Benzene and its derivatives play an important role in much of society. For example, when polymerised styrene forms polystyrene, a substance used for insulation and as a building material; tyrosine and phenylalanine are important naturally occurring amino acids that are benzene derivatives; phenols, the hydroxylated form of benzene, are the principal components of the tannins and are responsible for the antioxidant capacity of red wine; vanillin is the compound responsible for the very pleasant sweet, buttery smell associated with the extract of the vanilla plant. Finally, many drugs are composed around a benzene core, including aspirin and paracetamol.

The history of benzene and the discovery of its unusual structure is an interesting story. In 1825, Michael Faraday (see page 774) isolated a pure compound from the oily mixture that condensed from illuminating gas, the fuel burned in gaslights common at the time in England. The compound had a boiling point of 80 °C and a pungent odour. Elemental analysis of the compound showed it had a carbon-to-hydrogen ratio of 1:1, corresponding to an empirical formula of CH. About nine years later, German chemist Eilhard Mitscherlich (1794–1863) synthesised the same compound by heating benzoic acid in the presence of lime:

$$C_6H_5COOH \ + \ CaO \ \xrightarrow{\text{heat}} \ C_6H_6 \ + \ CaCO_3$$

gum of benzoin   calcium   benzene   calcium
(benzoic acid)   oxide   carbonate

He further used vapour-density measurements to determine the molecular weight of 78 u. He named the new compound *benzin*, after the gum benzoin from which he derived it. We now know this compound as benzene.

However, benzoic acid had been known for several hundred years before Mitscherlich's experiment. Benzoic acid is the white odourless compound isolated from benzoin resin, a substance extruded by several types of Southeast Asian trees of the genus *Styrax*. It is often used as a topical fungicide.

Although the experiments of Mitscherlich showed that benzene must have the formula C_6H_6, it was not clear what the structure of the molecule could be. Only after 1860, when the concept of unsaturation was introduced, did August Kekulé (1829–1896) propose a suitable model. Kekulé was instrumental in developing much of the structural theory of organic chemistry, but he is best known for his work on benzene (1865). How he came to propose the structure of benzene is the truly fascinating part of this story. Kekulé recalled falling asleep on a London bus, and while he was dozing, he conjured images of 'dancing atoms' that connected into snakes. As the snakes twisted and turned, eventually one caught its own tail, giving the image leading to the now famous structure of alternating double and single bonds. In his theory, however, two interconvertible structures were possible. As we know now, the structure is subtly different, but considering the level of chemical knowledge at this point in time, and more importantly, on the chemistry of benzene, Kekulé's proposed structure was revolutionary.

Many other compounds isolated during the 19th century were related to benzene. For example, toluene was extracted by distillation from the balsam of the tolu tree in the 1840s. Each of these compounds had a low C : H ratio and many, such as vanillin, had pleasant rather than pungent aromas. As a result, this group of compounds were called aromatic compounds. We will see in Section 28.3 that, as the unusual properties of aromatic compounds were investigated, the term 'aromatic' came to be applied to compounds with an unusual stability, regardless of their odour.

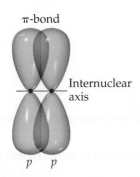

▲ **FIGURE 28.1** **The π-bond.** When two *p* orbitals overlap in a sideways fashion, the result is a π-bond. Note that the two regions of overlap constitute a *single* π-bond.

∞ Review this on page 955

28.1 | THE STRUCTURE OF BENZENE

In Chapter 24 we discussed the structure of both alkenes and alkynes and described the carbon atoms that form the molecular skeleton as *unsaturated*. The trigonal-planar bonding geometry of an alkene carbon was rationalised based on the hybridisation of one 2*s* and two 2*p* orbitals available to carbon. Recall that we also discussed the overlap of unhybridised *p* orbitals to form a π-bond (◀ FIGURE 28.1) (∞ Section 24.1, 'The Structure of Unsaturated Hydrocarbons'). One other class of hydrocarbon also takes advantage of π-bonding—the *aromatic hydrocarbons*.

Benzene is the simplest example of an **aromatic hydrocarbon** (also called an **arene**). This class of hydrocarbon contains a planar ring structure, usually drawn with either a circle in the middle (for reasons we discuss shortly), or alternating single and double bonds (▼ FIGURE 28.2). Although aromatic hydrocarbons involve sp^2-hybridised carbons, they are not alkenes, and undergo very different chemical reactions.

Bonding in Benzene

In many of the molecules we have discussed so far, the bonding electrons are *localised*. By this we mean that the σ- and π-electrons are associated totally with the two atoms that form the bond. In some molecules, however, we cannot adequately describe the bonding as being entirely localised. One molecule that cannot be described with localised π-bonds is benzene (C_6H_6).

To describe the bonding in benzene, we first choose a hybridisation scheme consistent with the geometry of the molecule (see Figure 28.2). As each carbon is surrounded by three atoms at 120° angles, the appropriate carbon hybrid orbital set is sp^2. Six localised C–C σ-bonds and six localised C–H σ-bonds are formed from the sp^2 hybrid orbitals, as shown in ▶ FIGURE 28.3(a). This leaves a 2*p* orbital on each carbon atom that is oriented perpendicularly to the plane of the hydrocarbon ring. The situation is very much like that in ethene, except we now have six carbon 2*p* orbitals arranged in a ring (Figure 28.3(b)). Each of the unhybridised 2*p* orbitals is occupied by one electron, meaning a total of six electrons have to be accounted for by π-bonding.

We could imagine using the unhybridised 2*p* orbitals of benzene to form three *localised* π-bonds—that is, π-bonds defined formally between neighbouring carbon atoms. As shown in ▶ FIGURE 28.4(a) and (b), there are two equivalent ways to make these localised bonds and each diagram corresponds to one of the resonance structures of the molecule. A representation that reflects *both* resonance structures has the six π-electrons distributed among all six carbon

▲ FIGURE IT OUT

How does the structure of benzene relate to cyclohexane?

aromatic benzene C_6H_6

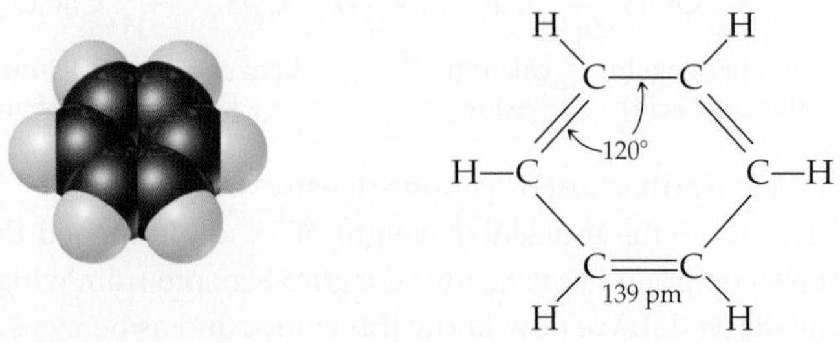

▲ **FIGURE 28.2** **Benzene.** Molecular formulae and geometrical representations for benzene.

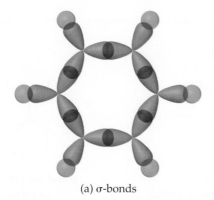

(a) σ-bonds

(b) 2p atomic orbitals

◄ **FIGURE 28.3**   **The σ- and π-bond networks in benzene, C_6H_6.** (a) The C–C and C–H σ-bonds all lie in the plane of the molecule and are formed by using carbon sp^2 hybrid orbitals. (b) Each carbon atom has an unhybridised 2p orbital that lies perpendicular to the molecular plane. These six 2p orbitals form the π-orbitals of benzene.

atoms, as shown in Figure 28.4(c). Notice how this figure corresponds to the 'circle-in-a-hexagon' drawing often used to represent benzene.

The model described in Figure 28.4(c) leads to the description of each carbon–carbon bond as having an identical bond length, which is between the length of a C–C single bond (154 pm) and the length of a C–C double bond (134 pm). This is consistent with the observed bond lengths in benzene (139 pm).

Since we cannot describe the π-bonds in benzene as individual electron-pair bonds between neighbouring atoms, we say that the π-bonds are *delocalised* among the six carbon atoms. In Section 28.3 we investigate the effects of delocalisation on reactivity and the concept of *aromaticity*. ▼ **FIGURE 28.5** shows the structures of some common aromatic compounds.

> ⚠ **CONCEPT CHECK 1**
>
> Pyridine is an analogue of benzene in which one of the sp^2 carbon atoms is replaced by nitrogen. Does this molecule have delocalised π-bonds? Draw a representation for pyridine that illustrates how this delocalisation comes about. How might the lone pair on nitrogen orientate itself within pyridine?
>
>

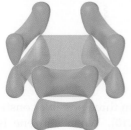

(a) Localised π-bonds

(b) Localised π-bonds

(c) Delocalised π-bonds

◄ **FIGURE 28.4**   **Delocalised π-bonds in benzene.** The six 2p orbitals of benzene, shown in Figure 28.3(b), can be used to make three C–C π-bonds. (a) and (b) show two equivalent ways to make localised π-bonds. These π-bonds correspond to the two resonance structures for benzene. (c) A representation of the delocalisation of the three C–C π-bonds among the six C atoms.

28.2 | ISOMERISM AND NOMENCLATURE IN AROMATIC COMPOUNDS

Aromatic compounds or *arenes* are a large and important class of organic compounds. The simplest member of the series is benzene (C_6H_6) and further

◢ **FIGURE IT OUT**

Identify the functional groups attached to benzene in these examples.

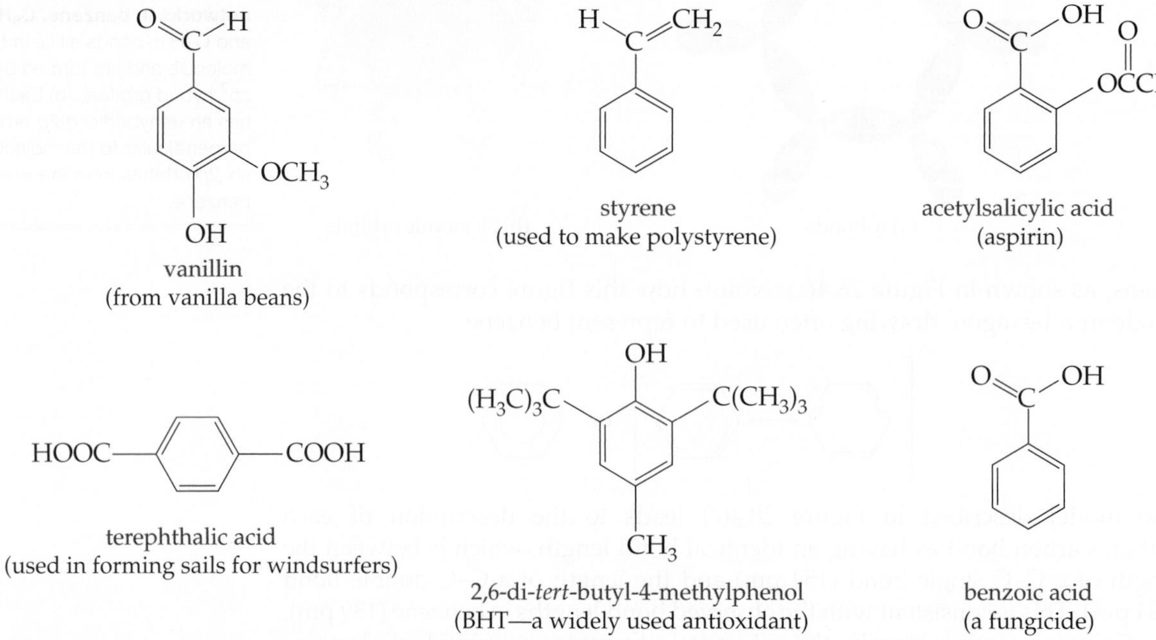

vanillin
(from vanilla beans)

styrene
(used to make polystyrene)

acetylsalicylic acid
(aspirin)

terephthalic acid
(used in forming sails for windsurfers)

2,6-di-*tert*-butyl-4-methylphenol
(BHT—a widely used antioxidant)

benzoic acid
(a fungicide)

▲ **FIGURE 28.5 Important aromatic compounds.** A small sample of common aromatic compounds containing benzene. These compounds are important in biochemistry, as pharmaceuticals, and in the production of modern materials.

arenes are formed essentially by fusing another C_6 skeleton to benzene. Each aromatic ring system is given a common name, as shown in ▼ **FIGURE 28.8**. The systematic names of monosubstituted benzenes are usually derived from benzene, although some common names are still used. For example, methylbenzene is known more commonly as toluene. The common names phenol, anisole, benzaldehyde, benzoic acid and aniline are also retained. ▼ **FIGURE 28.9** illustrates some systematic and common names.

The substituent group C_6H_5 formed by the loss of a hydrogen atom from benzene is called a **phenyl group**. The loss of a hydrogen atom from toluene leads to a **benzyl group**.

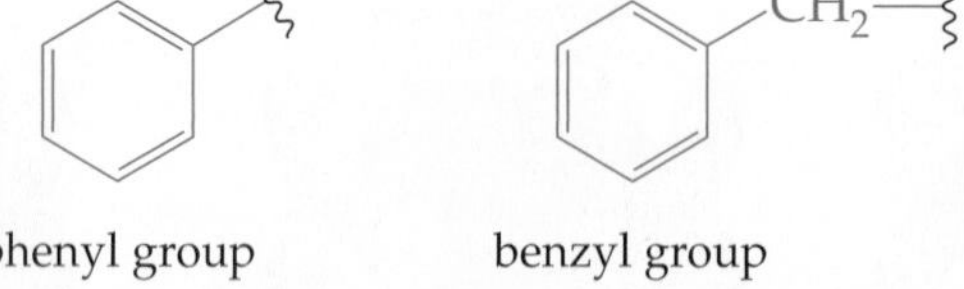

phenyl group

benzyl group

When a benzene ring has two substituents, they can form three possible constitutional isomers—*ortho*, *meta* and *para* (▼ **FIGURE 28.10**). When benzene is warmed in a mixture of nitric and sulfuric acids, for example, hydrogen is substituted by the nitro group, NO_2:

$$\text{C}_6\text{H}_6 + \text{HNO}_3 \xrightarrow{\text{H}_2\text{SO}_4} \text{C}_6\text{H}_5\text{NO}_2 + \text{H}_2\text{O} \qquad [28.1]$$

MY WORLD OF CHEMISTRY

THE DISCOVERY OF LIQUID CRYSTALS

In 1888 Frederick Reinitzer (1857–1927), an Austrian botanist, discovered that an organic compound he was studying, cholesteryl benzoate, had interesting and unusual properties. When heated, the substance melts at 145 °C to form a viscous milky liquid; at 179 °C the milky liquid suddenly becomes clear. When the substance is cooled, the reverse process occurs. Reinitzer's work represents the first systematic report of what we now call a *liquid crystal* (▼ FIGURE 28.6).

Instead of passing directly from a solid to the liquid phase when heated, some substances, such as cholesteryl benzoate, pass through an intermediate liquid crystalline phase that has some of the structure of solids and some of the freedom of motion possessed by liquids. Because of partial ordering, liquid crystals may be very viscous and possess properties between that of a solid and a liquid. The region in which they exhibit these properties is marked by sharp transition temperatures, as in Reinitzer's example of cholesteryl benzoate.

From the time of their discovery in 1888 until about 30 years ago, liquid crystals were largely a laboratory curiosity. However, researchers found that the weak intermolecular forces that hold the molecules together in a liquid crystal are easily affected by changes in temperature, pressure and electric fields. As a result, liquid crystals are now widely used as temperature and pressure sensors and in the displays of electrical devices such as digital watches, calculators, mobile phones and laptop and hand-held calculators.

Substances that form liquid crystals are often composed of long, rod-like molecules. They usually contain a long-chain alkane component (minimum chain length of five carbons) linked to an aromatic group such as benzene (▼ FIGURE 28.7). The liquid crystalline properties they exhibit are in fact due to the ordered interplay of London dispersion forces (∞ Section 11.2, 'Intermolecular Forces') and aromatic–aromatic interactions, also called π–π interactions.

RELATED EXERCISES: 28.3, 28.11, 28.12

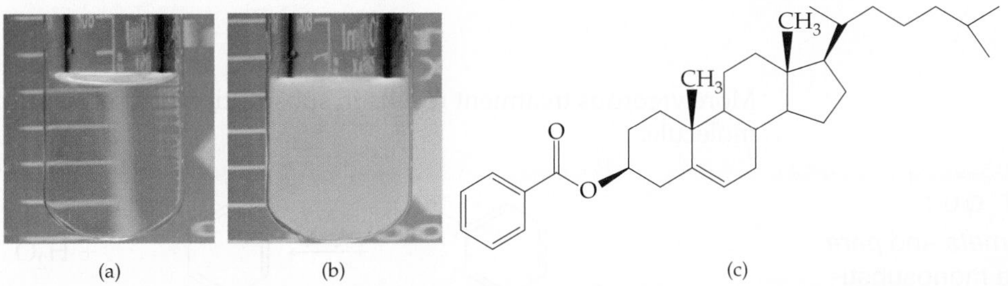

▲ **FIGURE 28.6   Liquid and liquid crystal.** (a) Molten cholesteryl benzoate at a temperature above 179 °C. In this temperature region the substance is a clear liquid. Note that the printing on the surface of the beaker behind the sample test tube is readable. (b) Cholesteryl benzoate at a temperature between 179 °C and its melting point, 145 °C. In this temperature interval, cholesteryl benzoate exists as a milky liquid crystalline phase. (c) The structure of cholesteryl benzoate.

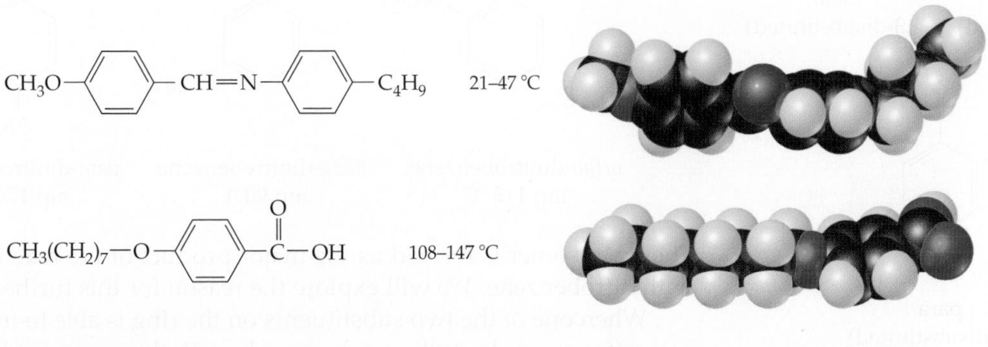

▲ **FIGURE 28.7   Structures and liquid crystal temperature intervals of two typical liquid crystalline materials.** The temperature interval indicates the temperature range in which the substance exhibits liquid crystalline behaviour.

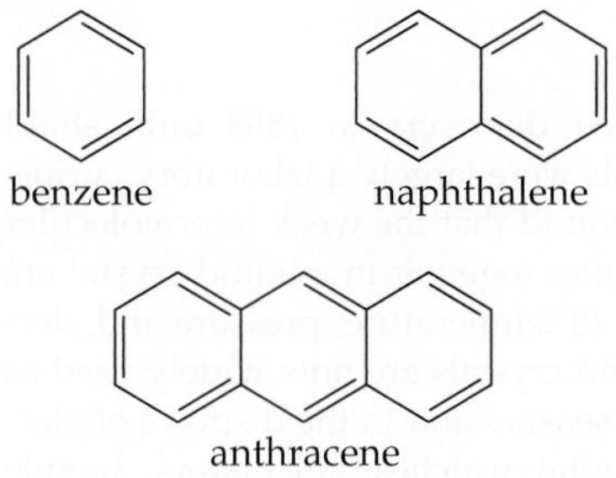
benzene
naphthalene
anthracene

▲ **FIGURE 28.8 Condensed structural formulae and common names of several aromatic compounds.** The aromatic rings are represented by hexagons with a circle inscribed inside to denote aromatic character. Each corner represents a carbon atom. Each carbon is bound to three other atoms—either three carbons or two carbons and a hydrogen. The hydrogen atoms are not shown.

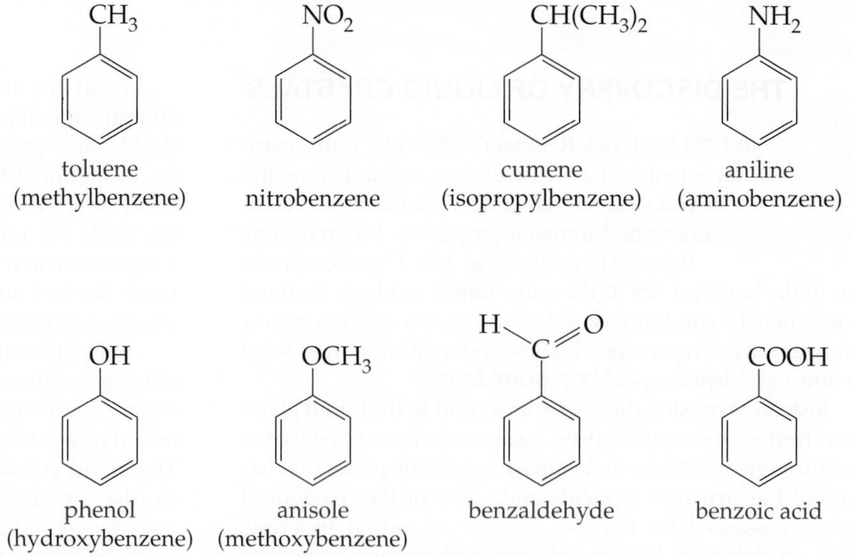
CH3
toluene
(methylbenzene)
NO2
nitrobenzene
CH(CH3)2
cumene
(isopropylbenzene)
NH2
aniline
(aminobenzene)
OH
phenol
(hydroxybenzene)
OCH3
anisole
(methoxybenzene)
H
C
O
benzaldehyde
COOH
benzoic acid

▲ **FIGURE 28.9 Benzene derivatives containing a single functional group.** The eight most common monosubstituted benzenes are listed together with their systematic and common names.

More vigorous treatment results in substitution of a second nitro group into the molecule:

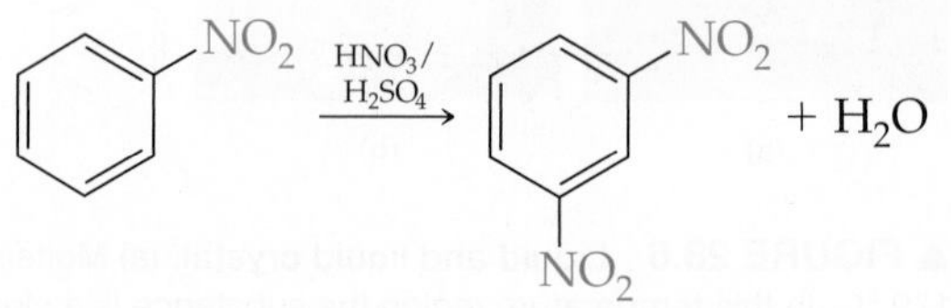
NO2
HNO3/
H2SO4
NO2
NO2
+ H2O

[28.2]

The three possible isomers of benzene with two nitro groups attached are *ortho-*, *meta-* and *para-*dinitrobenzene:

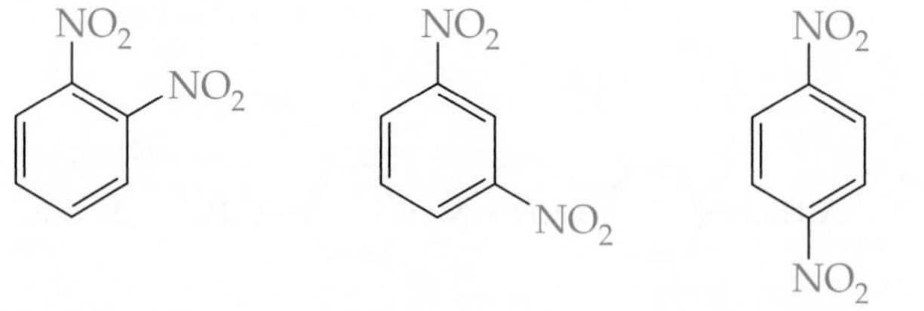
NO2
NO2
NO2
NO2
NO2
NO2

ortho-dinitrobenzene mp 118 °C | *meta*-dinitrobenzene mp 90 °C | *para*-dinitrobenzene mp 174 °C

The *meta* isomer is formed as the major product of the reaction of nitric acid with nitrobenzene. We will explore the reason for this further in Section 28.5.

When one of the two substituents on the ring is able to impart a common name (for example, aniline, toluene, phenol), the compound is named based on that parent molecule. The substituent that imparts the common name occupies position number 1. When neither group imparts a common name, the two substituents are listed alphabetically (with the lowest position numbers possible) before ending in *benzene*. The same rules apply to three or more substituents.

FIGURE IT OUT

How many *ortho*, *meta* and *para* positions exist in a monosubstituted benzene?

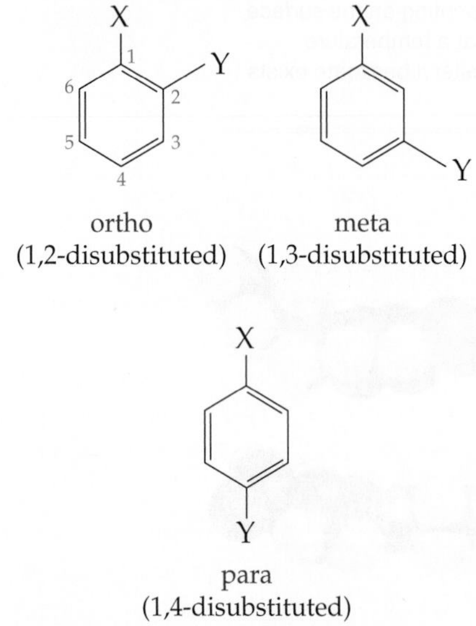
X
Y
1
2
3
4
5
6
ortho
(1,2-disubstituted)
X
Y
meta
(1,3-disubstituted)
X
Y
para
(1,4-disubstituted)

▲ **FIGURE 28.10 Substitution on benzene.** There are three ways of arranging two substituents on benzene—as *ortho*, *meta* and *para* substitution. Each isomer has different physical and chemical properties.

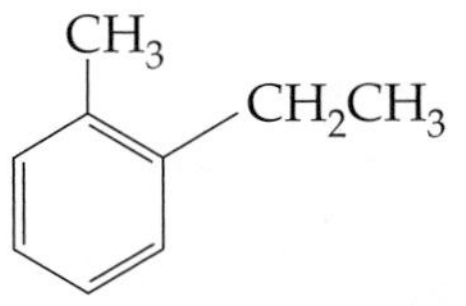

2-ethyltoluene or
o-ethyltoluene

4-chlorophenol or
p-chlorophenol

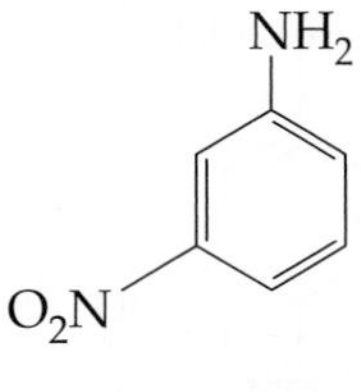

3-nitroaniline or
m-nitroaniline

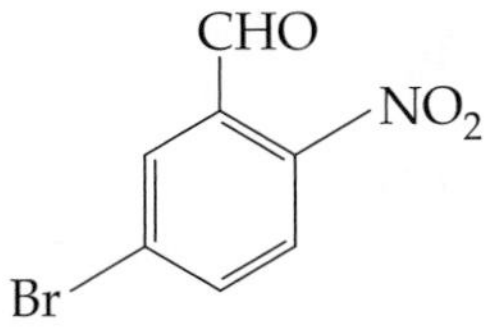

5-bromo-2-nitrobenzaldehyde

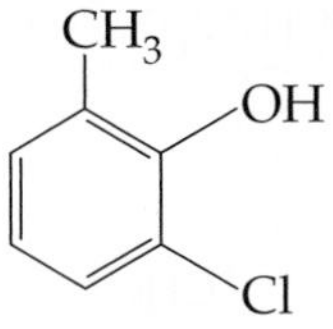

2-chloro-6-methylphenol

In naming disubstituted benzenes, either position numbers or the use of *o* for *ortho*, *m* for *meta* and *p* for *para* is acceptable (see above). For the more substituted benzenes, only position numbers are used. As with other nomenclature we have studied, these position numbers need to be made as small as possible while maintaining the parent molecule group at position 1. For example, 2-chloro-5-methylphenol is named as such and not as 3-chloro-2-hydroxy-toluene. This hierarchical method of naming is based on which functional group is of higher order. A useful, though not exhaustive, order is:

toluene < aniline < phenol < benzaldehyde < benzoic acid

 CONCEPT CHECK 2

When naphthalene is reacted with nitric and sulfuric acids, two compounds containing one nitro group are formed. Draw the structures of these two compounds.

Phenols

Phenols, or hydroxybenzenes, are named after the parent compound, phenol, which was first isolated from coal tar in 1834. The phenol structure is based on the direct attachment of the OH group to a benzene ring by an O–C bond. This connectivity confers unique properties on phenols. Some phenol derivatives are named as substituted phenols, others as hydroxy compounds, depending on which other functional groups are present. Many have trivial names.

Phenols are an important class of natural product, widely distributed throughout nature. Several examples of naturally occurring phenols are shown below and in ▼ **FIGURE 28.11**. Gallic acid and catechin are phenols found in wines. One of their properties is to act as an antioxidant. Compare their structure with another phenolic antioxidant, vitamin E. Vanillin is the compound responsible for the odour produced by vanilla beans. The cresols, of which *p*-cresol is shown, are isolated from coal tar. They found use in the 1960s–1980s as a wood preservative. There is evidence to suggest that due to their antioxidant properties, the polyphenols in fruit and vegetable juice, and in wine, may play an important role in delaying the onset of colds and in more specific conditions such as Alzheimer's disease.

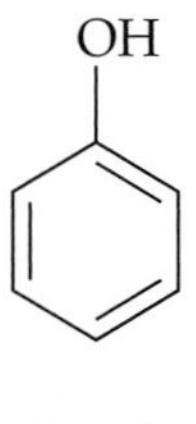

phenol

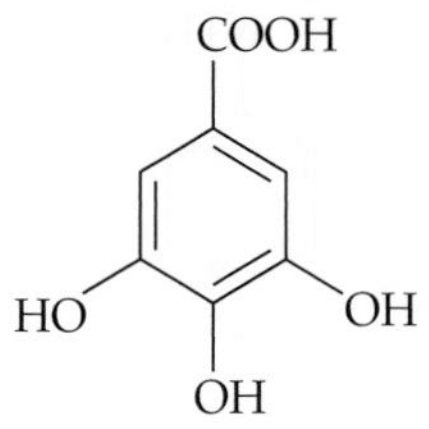

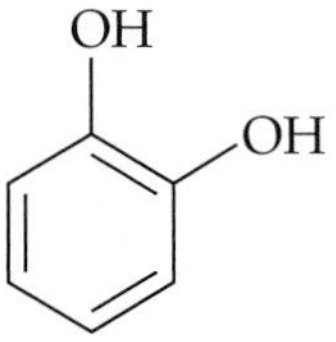

1,2-dihydroxybenzene
or
o-dihydroxybenzene
(catechol)

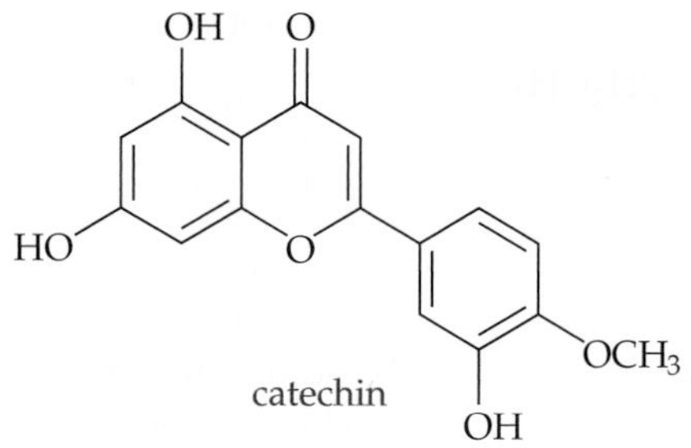

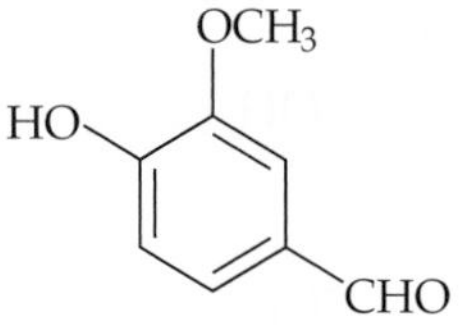

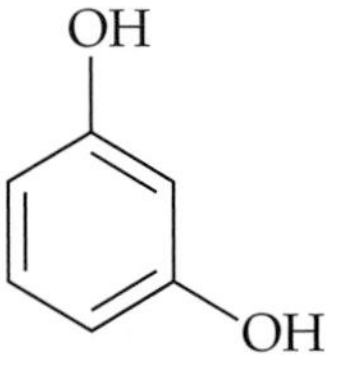

1,3-dihydroxybenzene
or
m-dihydroxybenzene
(resorcinol)

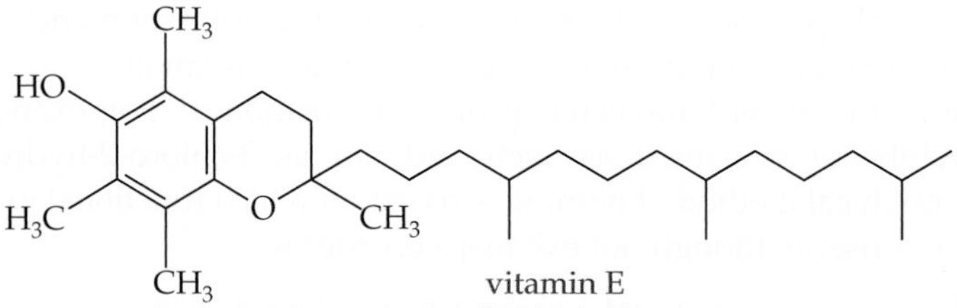

1,4-dihydroxybenzene
or
p-dihydroxybenzene
(hydroquinone)

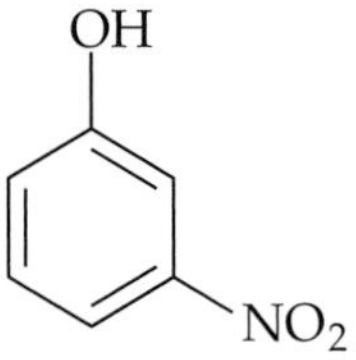

m-nitrophenol
or
3-nitrophenol

▲ **FIGURE 28.11 Common phenols.**
The systematic and common names are
listed.

gallic acid

catechin

4-hydroxy-
3-methoxybenzaldehyde
(vanillin)

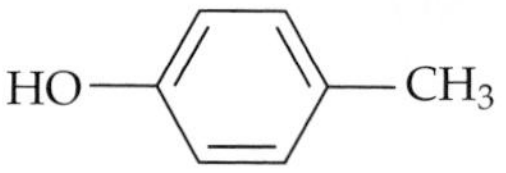

p-cresol or
4-methylphenol

vitamin E
(*lipid soluble*)

The largest single use of phenol is as an intermediate in the production of low-cost thermoset resins (called *phenolic resins*), used in the production of plywood adhesive as well as in the construction and car industries. Medicinally, phenol and its derivatives have been used in the past as a disinfectant and as a topical anaesthetic in preparations such as ointments, ear and nose drops, cold sore lotions, throat lozenges and antiseptic lotions.

28.3 | RESONANCE AND AROMATICITY

The planar, highly symmetrical structure of benzene, with its 120° bond angles, suggests a high degree of unsaturation. You might therefore expect benzene to resemble the unsaturated hydrocarbons and be highly reactive. The chemical behaviour of benzene, however, is unlike that of alkenes or alkynes. Benzene and the other aromatic hydrocarbons are much more stable than alkenes and alkynes because the π-electrons are delocalised in the unhybridised *p* orbitals (Section 28.1).

Let's consider two molecules, cyclohexene and benzene, and their reactions with bromine. Although cyclohexene reacts readily with bromine at room temperature, benzene (drawn as one contributing structure, that is, Kekulé's cyclohexatriene) does not react with bromine at all under these conditions.

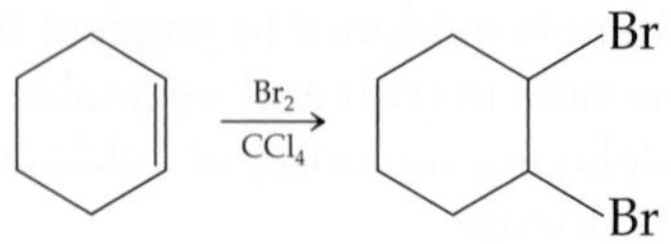

$$\xrightarrow[\text{CCl}_4]{\text{Br}_2}$$ [28.3]

$$\xrightarrow[\text{CCl}_4]{\text{Br}_2} \text{No reaction}$$ [28.4]

FIGURE IT OUT

What special feature diffentiates cyclohexatriene from benzene? How can you measure this?

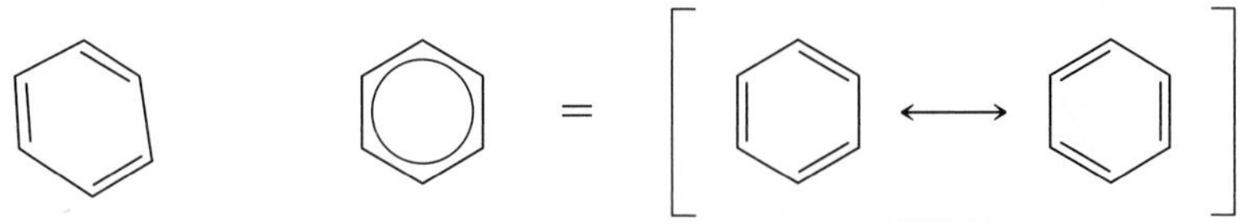

cyclohexatriene
(localised double bonds
one of two Kekulé structures)

benzene
(two canonical structures that relate to
the limits of the electronic structure)

▲ **FIGURE 28.12** **Resonance in benzene.** Benzene does not react chemically as if its double bonds are localised, as in cyclohexatriene. The delocalised nature of the π-electrons enhances the stability of benzene through resonance. The shorthand notation on the right reminds us that benzene is a blend of two resonance structures—it emphasises that the C–C double bonds cannot be assigned to specific edges of the hexagon. Chemists use both representations of benzene interchangeably.

▲ **FIGURE 28.12** illustrates the structure of cyclohexatriene and its difference from benzene. This difference is due to the property of **resonance** (∞ Section 8.6, 'Resonance Structures'). Resonance is an extremely important concept in describing the bonding in and stability of organic molecules, particularly aromatic and other highly conjugated molecules. For many molecules and ions, no single Lewis structure provides a truly accurate representation, so a set of **resonance structures** is often used to describe a molecule or ion. For example, there are two equivalent Lewis structures for benzene, each of which satisfies the octet rule. These two structures are in resonance, each resonance structure showing three C–C single bonds and three C–C double bonds, but the double bonds are in different places in the two structures. The actual structure is neither of these two but somewhere intermediate to both.

∞ Review this on page 272

In the case of benzene, curved arrows can also be used to interconvert the two canonical structures as shown in ▼ **FIGURE 28.13** (∞ Section 24.3, 'Arrow Notation and Resonance Structures: Electron Counting').

∞ Review this on page 963

Aromaticity

So far we have discussed benzene as an aromatic compound, but what is aromaticity? Originally, molecules derived from benzene were called aromatic purely because of their smell. Later, however, the German chemist Eric Hückel (1896–1980) defined what an aromatic compound is through a set of criteria. It must:

1. be planar;
2. have at least one unhybridised p orbital on each atom of the ring; and
3. have $[4n + 2]$ π-electrons, where n is an integer $n = 0, 1, 2, 3 \ldots$

Aromatic molecules resist any move to break that aromaticity. This resistance is quantified as aromatic stabilisation or **resonance energy**. We can estimate the stabilisation of the π-electrons in benzene by comparing the energy released when hydrogen gas reacts with benzene (a reaction known as *hydrogenation*) to form a saturated compound, with the energy required to hydrogenate certain alkenes. The hydrogenation of benzene to form cyclohexane can be represented as

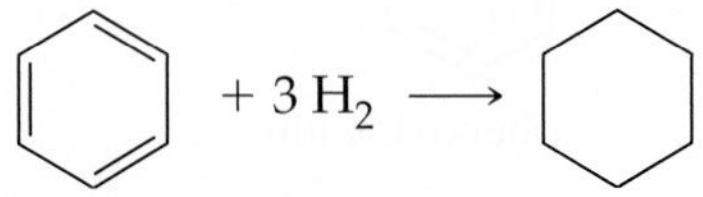

$$+ \; 3\,H_2 \longrightarrow \qquad \Delta H° = -208 \text{ kJ mol}^{-1} \qquad [28.5]$$

Convert contributor B into contributor A using curved arrows.

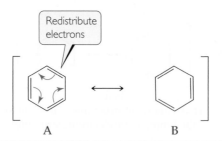

▲ FIGURE 28.13 **Resonance in benzene.** Two resonance structures contribute equally to the actual structure of benzene.

The enthalpy change in this reaction is -208 kJ mol^{-1}. The heat of hydrogenation of the cyclic alkene cyclohexene is -120 kJ mol^{-1}:

$$+ \ H_2 \longrightarrow \qquad \Delta H^\circ = -120 \text{ kJ mol}^{-1} \qquad [28.6]$$

Similarly, the heat released on hydrogenating 1,4-cyclohexadiene is -232 kJ mol^{-1}:

$$+ \ 2 \ H_2 \longrightarrow \qquad \Delta H^\circ = -232 \text{ kJ mol}^{-1} \qquad [28.7]$$

From these last two reactions, it appears that the heat of hydrogenating each double bond is roughly -116 kJ mol^{-1}. Since benzene has the equivalent of three double bonds, we might expect that the heat of hydrogenating benzene would be about three times -116 kJ mol^{-1} (or -348 kJ mol^{-1}). Instead, the heat released is 140 kJ less, indicating that benzene is more stable than would be expected for three double bonds. The difference of 140 kJ mol^{-1} between the 'expected' heat of hydrogenation, -348 kJ mol^{-1}, and the observed heat of hydrogenation, -208 kJ mol^{-1}, is due to stabilisation of the π-electrons through delocalisation.

Many other molecules apart from aromatic hydrocarbons show aromaticity. These molecules typically contain one or more **heteroatoms** (for example, O, N, S). Such compounds, called **heterocyclic compounds**, tend to be biologically active and are dealt with in more detail in Chapter 29.

28.4 | ACIDITY OF PHENOLS

Phenols are approximately 1 000 000 times stronger acids than alcohols. This may sound like a significant difference, but in essence all it means is that phenols are weak acids and alcohols are even weaker acids (comparable with water). Phenol was originally named carbolic acid, due to its unusually high acidity for an alcohol. To illustrate the difference in acidity of alcohols and phenols, let us consider two molecules: phenol and cyclohexanol. The chemical equations for their acid-dissociation constants are:

$$\text{OH} \qquad + \ H_2O \xrightleftharpoons{K_a} \qquad \text{O}^\ominus \qquad + \ H_3\overset{\oplus}{O}$$

Stabilised by delocalisation

phenoxide ion

$$[28.8]$$

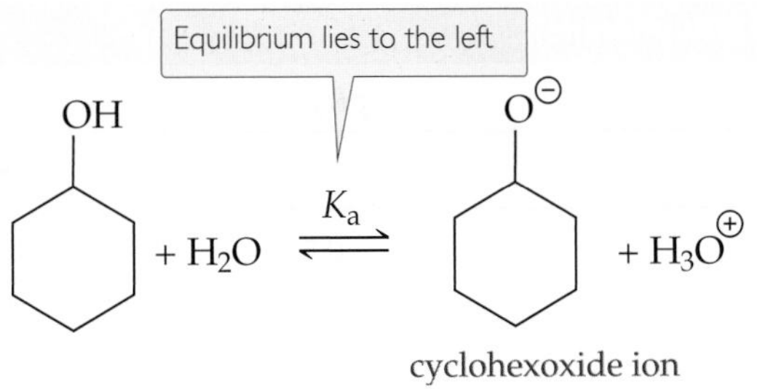

Equilibrium lies to the left

cyclohexoxide ion

[28.9]

The pK_a of phenol is about 10, while that for cyclohexanol is about 16 (recall that $pK_a = -\log_{10}K_a$) (∞ Section 17.8, 'Relationships Between K_a and K_b'). The difference in acidity between phenol and cyclohexanol is due to the stability of the corresponding anion products. Phenols are more acidic because of the **resonance-stabilised** nature of the anion formed upon deprotonation relative to the alkoxide. This delocalisation can be seen in the contributing structures illustrated in ▼ **FIGURE 28.14**.

∞ Review this on page 684

The stability of the phenoxide ion arises from resonance. The last three structures on the right in Figure 28.14 illustrate, using curved arrow notation, how the charge is distributed to the *ortho* and *para* positions of the aromatic ring. This charge distribution through resonance is important to the reactivity of aromatic compounds.

In contrast to phenoxide, the cyclohexoxide ion doesn't allow for charge delocalisation; the charge is formally on oxygen. Although arrow notation and the use of contributing structures allow us to rationalise why phenols are more acidic, they do not give us quantitative information on how strong an acid they might be. The pK_a value needs to be determined experimentally.

▼ **TABLE 28.1** compares the pK_a of several alcohols and phenols. Electron-withdrawing groups on the phenyl ring of phenol increase the acidity through **inductive effects**. **Electron-withdrawing groups** are those that are more electronegative than carbon. They can be atoms (for example, F, Cl, Br) or a collection of atoms such as the *nitro group*, NO_2, which, combined, are electron deficient. Each of these groups is able partially to withdraw electron density out of the O–H bond, making this bond weaker and hence the phenol more acidic. They can also contribute to the charge delocalisation of the resulting phenoxide ion.

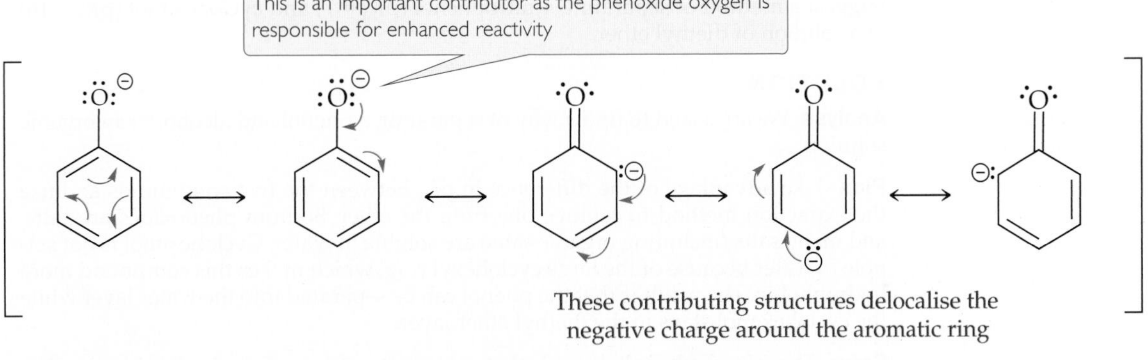

This is an important contributor as the phenoxide oxygen is responsible for enhanced reactivity

These contributing structures delocalise the negative charge around the aromatic ring

▲ **FIGURE 28.14 Resonance stabilisation.** Phenoxide ions are stabilised relative to alkoxide ions because phenoxide ions are able to resonance stabilise the negative charge around the aromatic ring.

TABLE 28.1 • Relative pK_a values for selected alcohols and phenols*

Compound	Structural formula	pK_a	
Hydrogen chloride	HCl	–7	**Stronger acid**
Picric acid	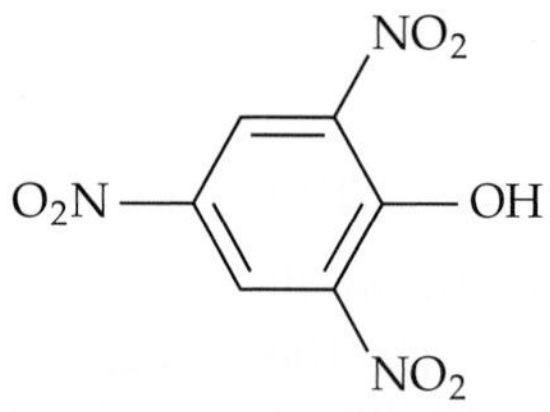	0.5	
2,4-Dinitrophenol	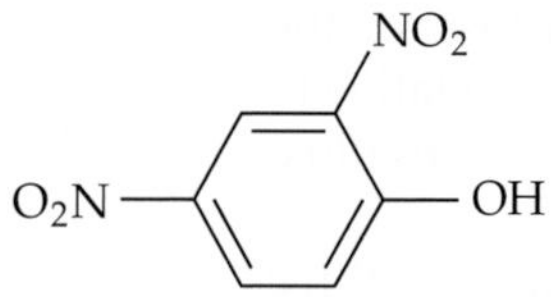	4	
4-Nitrophenol		7	
4-Chlorophenol		9	
p-Cresol		10	
Phenol	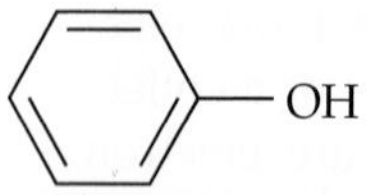	10	
Water	H$_2$O	15.7	
1-Propanol	CH$_3$CH$_2$CH$_2$OH	16	**Weaker acid**

* Error in measurement ±0.2

SAMPLE EXERCISE 28.1 **Utilising the acidity difference**

Suggest a method of separating 4-nitrophenol (pK_a = 7) and cyclohexanol (pK_a = 16) in a solution of diethyl ether.

SOLUTION

Analyse We are asked to find a way of separating a phenol and alcohol in an organic solution.

Plan Take advantage of the difference in pK_a between the two compounds and use the extraction method to remove one from the ether. Sodium phenoxides are salts, and many salts (including organic salts) are soluble in water. Cyclohexanol is not soluble in water because of the large cyclohexyl ring, which makes this compound more hydrophobic. The result is that the phenol can be separated into the water layer while the cyclohexanol stays in the diethyl ether layer.

Solve The pK_a of phenols is such that an aqueous 1 M sodium hydroxide (NaOH) solution is basic enough to deprotonate the phenol, but not the cyclohexanol, by the following chemical equation:

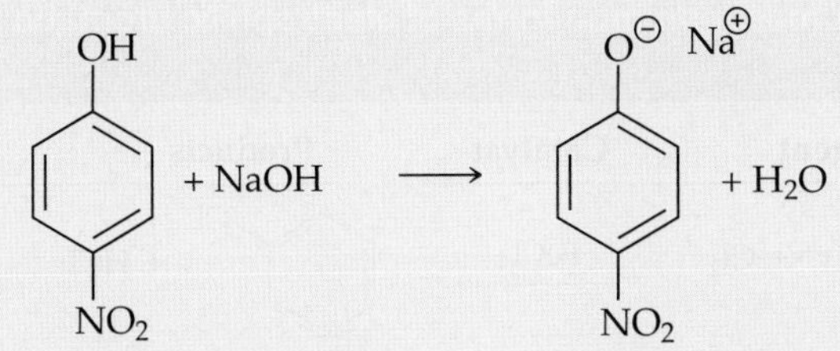

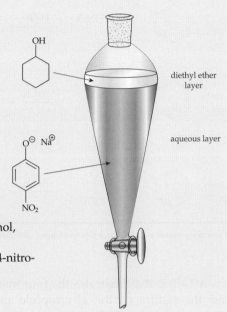

a sodium phenoxide

Once reaction has occurred, two layers are formed—an diethyl ether layer and an aqueous layer. The layers are then separated using a separating funnel.

Once the two solutions are separated, the diethyl ether can be distilled off (bp 35 °C) to leave the pure cyclohexanol. Acidification of the aqueous layer using 1 M HCl yields 4-nitrophenol as a solid, which can be filtered off and dried.

diethyl ether layer

aqueous layer

PRACTICE EXERCISE

Arrange these compounds in increasing order of acidity:

2,4-dichlorophenol, phenol, 2,6-dinitrophenol and 4-nitro-2-chlorophenol, using the trends in Table 28.1.

Answer: phenol < 2,4-dichlorophenol < 4-nitro-2-chlorophenol < 2,6-dinitrophenol

(See also Exercises 28.25, 28.27.)

 CONCEPT CHECK 3

So far we have seen that the phenolate ion gains stability through contributing resonance structures and that alkoxides do not have the same luxury. What does this tell you about the relative basicity of the alkoxide and phenolate ions?

28.5 | ELECTROPHILIC AROMATIC SUBSTITUTION (EAS) REACTIONS

Although aromatic hydrocarbons are unsaturated, they do not readily undergo addition reactions. The delocalised π-bonding causes aromatic compounds to behave quite differently from alkenes and alkynes (∞ Section 28.3, 'Resonance and Aromaticity').

∞ Review this on page 1122

Benzene, for example, does not react with Cl_2 or Br_2 under ordinary conditions. In contrast, aromatic hydrocarbons undergo substitution reactions relatively easily. Such a reaction is called an **electrophilic aromatic substitution (EAS)** reaction. In a substitution reaction, one atom of a molecule is replaced (substituted) by another atom or group of atoms. These reactions occur in the presence of an electrophile, usually given the symbol E^+. A catalyst is usually needed to generate the electrophile.

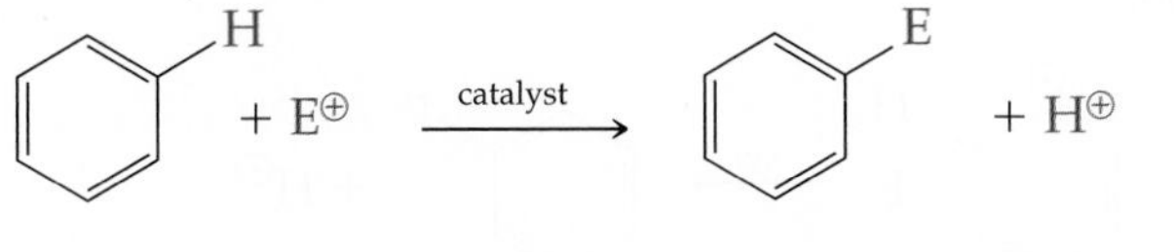

$$\text{(benzene)}\ H\ +\ E^{\oplus}\ \xrightarrow{\text{catalyst}}\ \text{(benzene)}\ E\ +\ H^{\oplus} \qquad [28.10]$$

TABLE 28.2 • The four most common forms of electrophilic aromatic substitution reaction

Reaction name	Reagent	Catalyst	Products	E⁺
Halogenation	+ Cl_2	$FeCl_3$	+ HCl	$Cl^{\oplus}$ chloronium
Nitration	+ HNO_3	H_2SO_4	+ H_2O	$NO_2^{\oplus}$ nitronium
Friedel–Crafts alkylation	+ RX	$AlCl_3$	+ HX	$R^{\oplus}$ alkyl carbocation
Friedel–Crafts acylation	+ RCX	$AlCl_3$	+ HX	$RC^{\oplus}$ acylium ion

Note: The R group represents a general group (for example, alkyl, aryl).

FIGURE IT OUT

Suggest an explanation for the dotted circle in the ring.

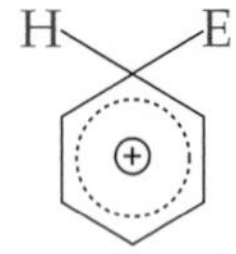

▲ **FIGURE 28.15 Wheland intermediate.** The delocalised intermediate carbocation is sometimes called a Wheland intermediate or sigma complex.

▲ **TABLE 28.2** illustrates the four most common EAS reactions, their names and the nature of the electrophile used. Halogenation, employing either the *bromonium ion* (Br^+) or *chloronium ion* (Cl^+), and nitration, employing the *nitronium ion* (NO_2^+), are useful ways of introducing new functionality to the benzene ring.

The Friedel-Crafts alkylation and acylation reactions generate new C–C bonds. Discovered in the late 1870s, they are two of the most important methods in forming new C–C bonds to the aromatic ring. The alkylation reaction uses carbocations to accomplish the substitution reaction. Tertiary carbocations react more cleanly and more efficiently than 2° or 1° carbocations. The *acylium ion* is formed from an activated carboxylic acid as the electrophile in a Friedel-Crafts acylation. This reaction generates aromatic ketones in high yields.

The mechanism of an electrophilic aromatic substitution reaction involves two steps. The initial step is rate-determining and leads to the formation of a resonance-stabilised carbocation intermediate, sometimes called a Wheland intermediate or sigma complex (◄ **FIGURE 28.15**).

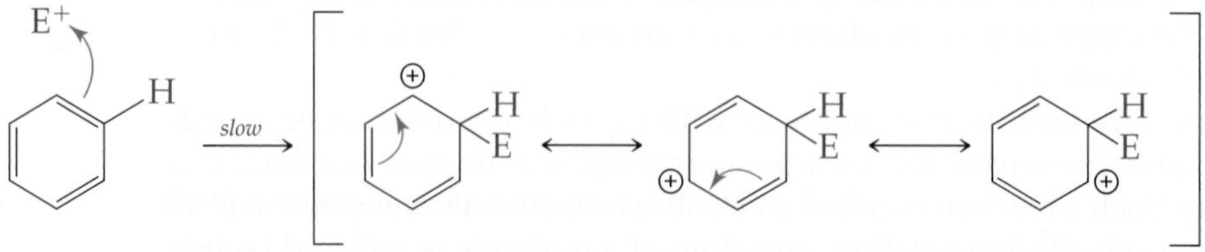

$$E^+ \quad \xrightarrow{slow} \quad [\text{ resonance structures }] \qquad [28.11]$$

This resonance stabilised intermediate is able to lose H⁺ to regain aromaticity—the driving force in making this reaction a substitution rather than an addition reaction.

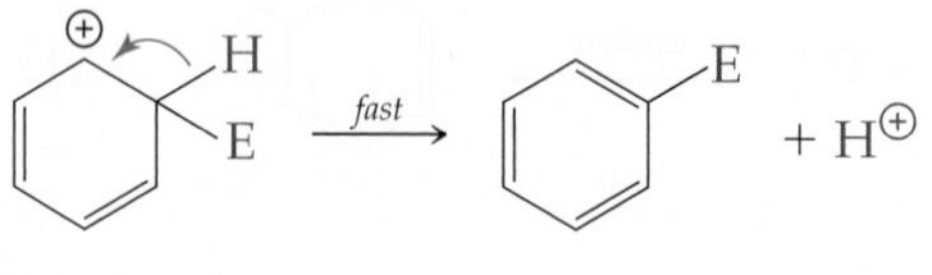

$$\xrightarrow{fast} \quad + \ H^{\oplus} \qquad [28.12]$$

CONCEPT CHECK 4

Compare and contrast the EAS bromination reaction with the electrophilic bromination of an alkene. What features are similar, and which are different?

SAMPLE EXERCISE 28.2 **Electrophilic aromatic substitution**

Draw the structural formula for the major product formed when *p*-xylene (1,4-dimethylbenzene) is reacted with HNO_3/H_2SO_4. Justify your choice for the major product.

SOLUTION

Analyse We are asked to determine the major product of the reaction of 1,4-dimethylbenzene with a mixture of nitric and sulfuric acid.

Plan The reaction of nitric acid with sulfuric acid yields the nitronium ion, NO_2^+, which is a useful electrophile. We will use this electrophile in a reaction with 1,4-dimethylbenzene.

Solve The major product is identified through the following reaction:

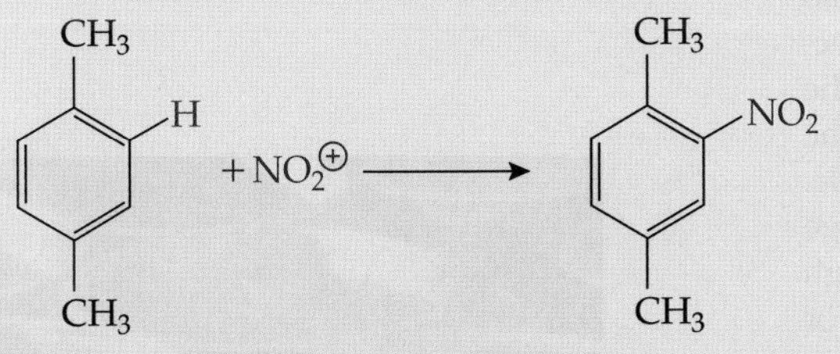

Because of the symmetry of *p*-xylene, substitution occurs in one of four equivalent positions.

PRACTICE EXERCISE

Addition of iodomethane to an aromatic compound, in the presence of $AlCl_3$, yields significant amounts of 4-chlorotoluene. What is the aromatic compound?

Answer: chlorobenzene

(See also Exercises 28.32, 28.33.)

Directing Groups and Substitution Effects

So far we have only considered benzene as the starting material for an electrophilic aromatic substitution reaction. Now we ask, what effects will other functional groups have on electrophilic aromatic substitution?

Let us begin to answer this question by looking at the nitration of toluene. In this case, three products, corresponding to the three possible isomers, are formed.

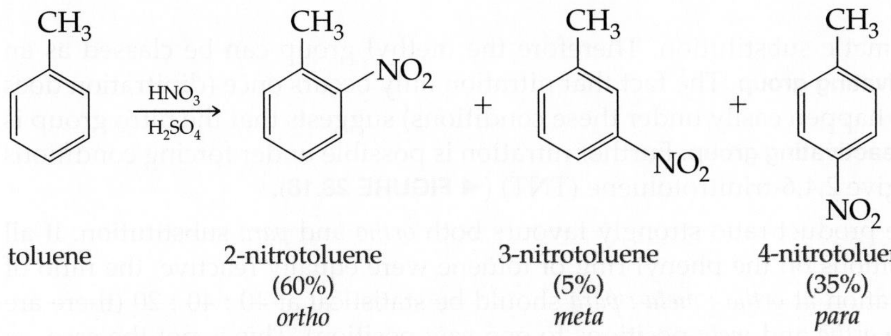

[28.13]

toluene — 2-nitrotoluene (60%) *ortho* — 3-nitrotoluene (5%) *meta* — 4-nitrotoluene (35%) *para*

Two things are to be noted here:

1. The nitration of toluene is approximately 25 times faster than that of benzene. In this case we say that toluene is **activated** towards electrophilic

A CLOSER LOOK

ORGANIC DYES

Light excites electrons in molecules. In a molecular orbital picture, we can envisage light exciting an electron from a filled molecular orbital (MO) to an empty one at higher energy. Because the MOs have definite energies, only light of the proper wavelength can excite electrons. The situation is analogous to that of atomic line spectra (∞ Section 6.3, 'Line Spectra and the Bohr Model'). If the appropriate wavelength for exciting electrons is in the visible portion of the electromagnetic spectrum, the substance will appear coloured as certain wavelengths of white light are absorbed, whereas others are not.

In using MO theory to discuss the absorption of light by molecules, we focus on two MOs in particular. The *highest occupied molecular orbital* (HOMO) is the MO of highest energy that has electrons in it. The *lowest unoccupied molecular orbital* (LUMO) is the MO of lowest energy that does not have electrons in it. The energy difference between the HOMO and the LUMO—known as the HOMO–LUMO gap—is related to the minimum energy needed to excite an electron in the molecule. Colourless or white substances usually have such a large HOMO–LUMO gap that visible light is not energetic enough to excite an electron to the higher level (▶ FIGURE 28.16). For example, the typical energy needed to excite an electron in naphthalene corresponds to light with a wavelength of 200–300 nm, which is far into the ultraviolet part of the spectrum. As a result, naphthalene cannot absorb any visible light and is therefore colourless.

Many of nature's rich colours, such as the green of plants and the bright hues of flowers, and the colours found in an inkjet printer, are due to *organic dyes* ▶ FIGURE 28.17, organic molecules that strongly absorb selected wavelengths of visible light. Organic dyes contain extensively delocalised π-electrons as a result of *conjugation*. The HOMO–LUMO gap in such molecules decreases as the number of conjugated double bonds increases, making it easier to excite the molecule with lower-energy (longer-wavelength) radiation (Figure 28.16). Generally, then, the more conjugated a compound is, the more coloured that compound will be.

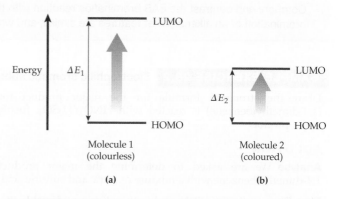

▲ **FIGURE 28.16** **Molecules absorbing light.** The lower the energy gap between the HOMO and LUMO within a molecule, the greater the likelihood that the molecule is coloured. This is because a light of longer wavelength (lower energy) can be absorbed by the molecule to allow the excitation.

▲ **FIGURE 28.17** **Organic dyes.** Many of the bright colours we see in our lives are a result of aromatic compounds.

FIGURE IT OUT

Why might TNT be explosive?

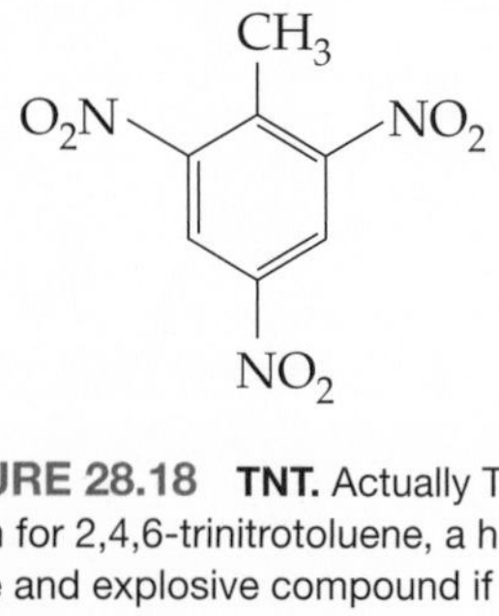

O_2N — CH_3 — NO_2

NO_2

▲ **FIGURE 28.18** **TNT.** Actually TNT is an acronym for 2,4,6-trinitrotoluene, a highly unstable and explosive compound if a force is applied to it.

aromatic substitution. Therefore the methyl group can be classed as an **activating group**. The fact that nitration only occurs once (dinitration does not happen easily under these conditions) suggests that the nitro group is a **deactivating group**. Further nitration is possible under forcing conditions to give 2,4,6-trinitrotoluene (TNT) (◀ FIGURE 28.18).

2. The product ratio strongly favours both *ortho* and *para* substitution. If all positions on the phenyl ring of toluene were equally reactive, the ratio of nitration at *ortho : meta : para* should be statistical at 40 : 40 : 20 (there are two *ortho* and *meta* positions to one *para* position). This is not the case, so this means that the methyl group is an **ortho, para-director**. In other words, the methyl group directs a second substituent to the *ortho* and *para* positions, but not to the *meta* position.

Both the activation and product distribution can be rationalised by looking at the intermediates formed after the addition of the electrophile and their relative stabilities, and then comparing this to the nitration of benzene (▼ FIGURE 28.19). We can look at these intermediates because the rate-determining step in this reaction is the addition of the electrophile rather than the re-aromatisation leading to the nitrobenzene product.

In both *ortho*- and *para*-attack, an intermediate resonance contributor can be drawn that form a 3° carbocation. As we saw in previous chapters, tertiary carbocations are more stable than either primary or secondary carbocations. In the case of benzene (shown) and for *meta*-attack, only 2° carbocation intermediates are formed. As a result, the overall stability of the intermediates for *ortho*- and *para*-attack are both favoured and lower in energy.

CONCEPT CHECK 5

Draw out the intermediates for *meta*-attack for the nitration of toluene and verify that all carbocations formed are secondary.

ortho attack

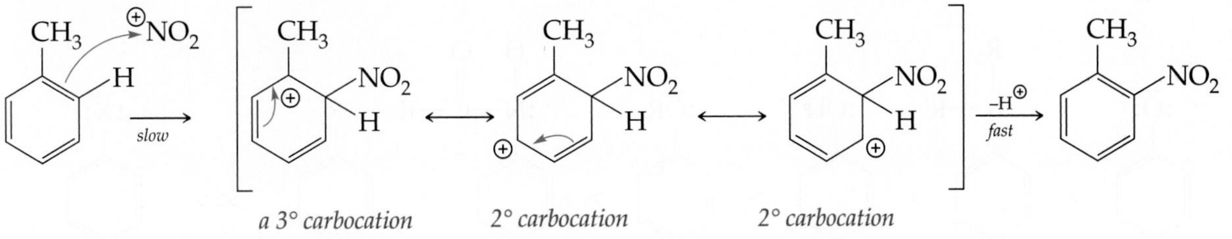

para attack

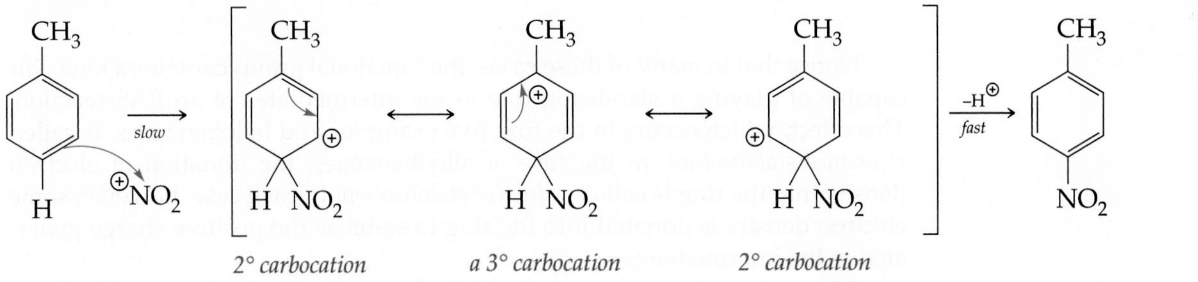

benzene

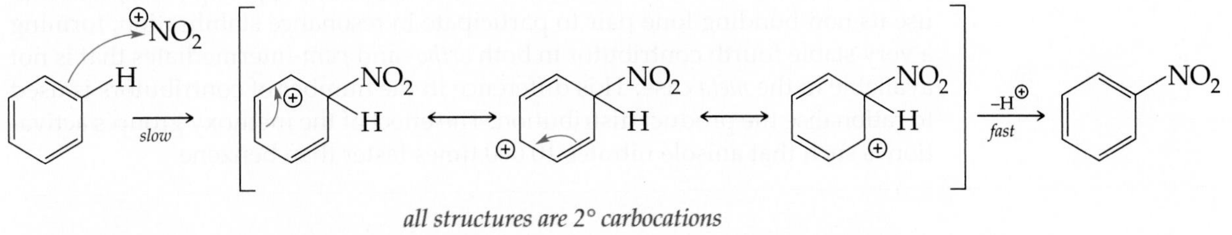

▲ FIGURE 28.19 **Intermediates of nitrating toluene and benzene.** The intermediates of electrophilic aromatic substitution reactions reveal information about which isomer dominates in a reaction.

Bromination of ethylbenzene using bromine and iron(III) tribromide (ferric bromide) also follows the same trend, with *ortho*- and *para*-substituted isomers favoured. Here, because of steric effects around the ethyl group, the *para*-isomer is favoured over the *ortho*-isomer.

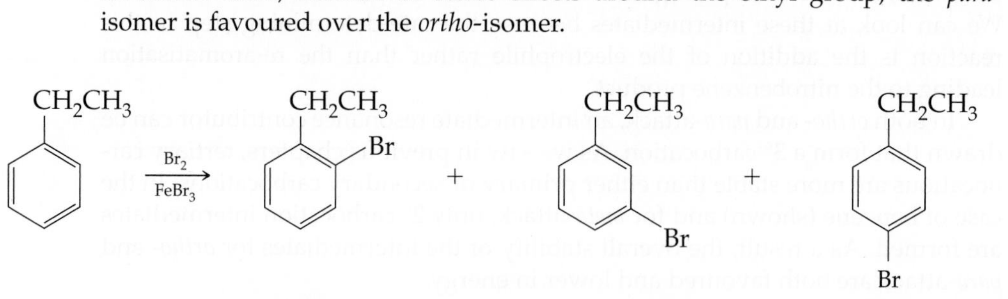

toluene | 2-bromoethylbenzene (38%) *ortho* | 3-bromoethylbenzene (<1%) *meta* | 4-bromoethylbenzene (62%) *para*

The methyl group in toluene or the ethyl group in ethylbenzene are slightly electron donating, stabilising the intermediates formed upon the first step of an EAS reaction. Generally speaking, *electron-donating groups are ortho, para-directors and activating towards electrophilic aromatic substitution.*

Ortho-, para-directing groups and their relative activation for EAS are:

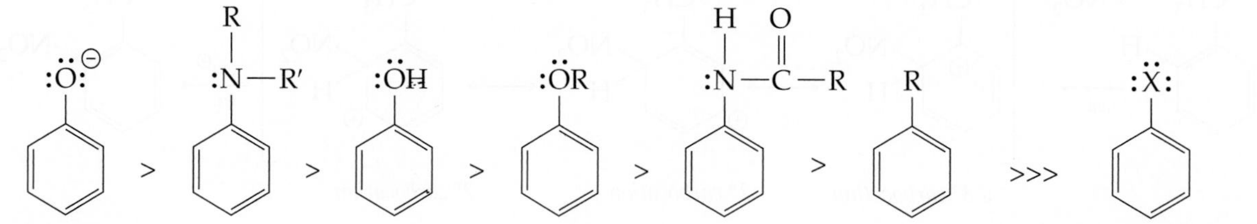

activating

phenoxides   anilines   phenols   phenyl ethers e.g. anisole   anilides   alkylbenzenes e.g. toluene   *deactivating* halobenzenes e.g. bromobenzene

Notice that in many of these cases, the functional group contains a lone pair capable of playing a stabilising role in the intermediates of an EAS reaction. This effect, which occurs in the first five examples and halobenzenes, is called *resonance stabilisation*. In the case of alkylbenzenes, the donation of electron density into the ring is called *inductive stabilisation*. In each case, however, some electron density is donated into the ring to stabilise the positive charge generated in the intermediates.

Although the halobenzenes (F, Cl, Br, I) are weakly deactivating by virtue of their inductive electron-withdrawing effects, they are still *o,p*-directing because of their ability to resonance stabilise the cationic intermediates.

To help with your understanding of resonance stabilisation, let's look at the intermediates of the nitration of anisole. Here, the alkoxy group (–OR) is able to use its non-bonding lone pair to participate in resonance stabilisation, forming a very stable fourth contributor in both *ortho*- and *para*-intermediates that is not available in the *meta* case. This difference in the number of contributors is used to rationalise the product distribution. The effect of the methoxy group's activation is such that anisole nitrates 10 000 times faster than benzene.

ortho-attack

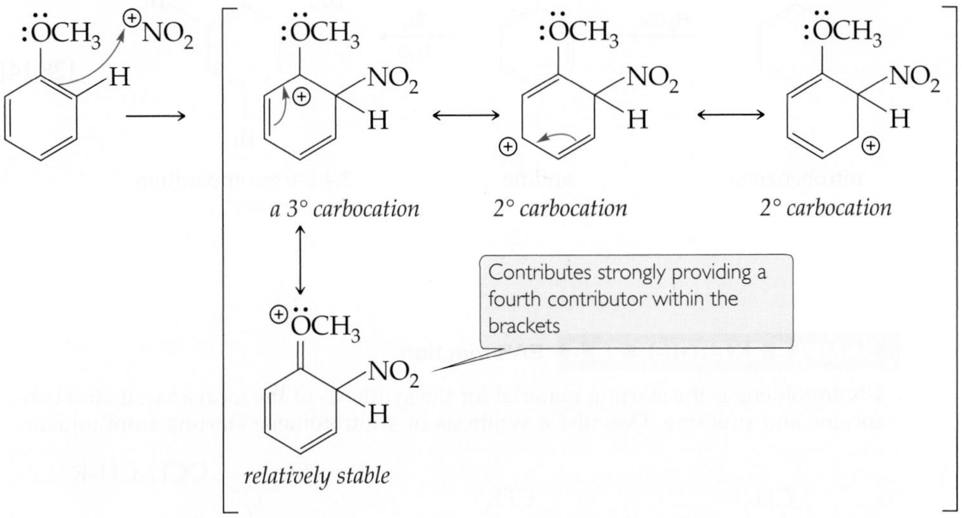

a 3° carbocation   2° carbocation   2° carbocation

Contributes strongly providing a
fourth contributor within the
brackets

relatively stable

para-attack

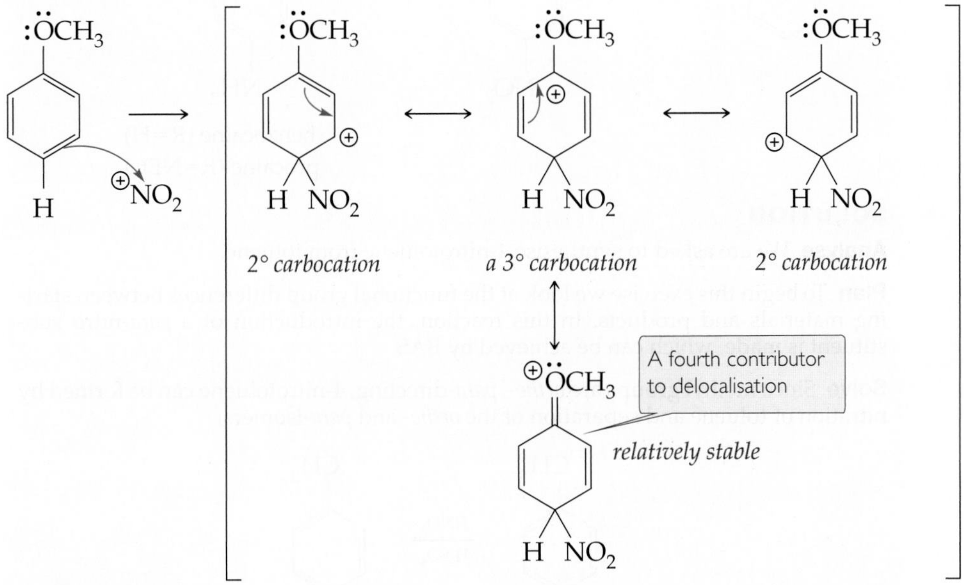

2° carbocation   a 3° carbocation   2° carbocation

A fourth contributor
to delocalisation

relatively stable

meta-attack

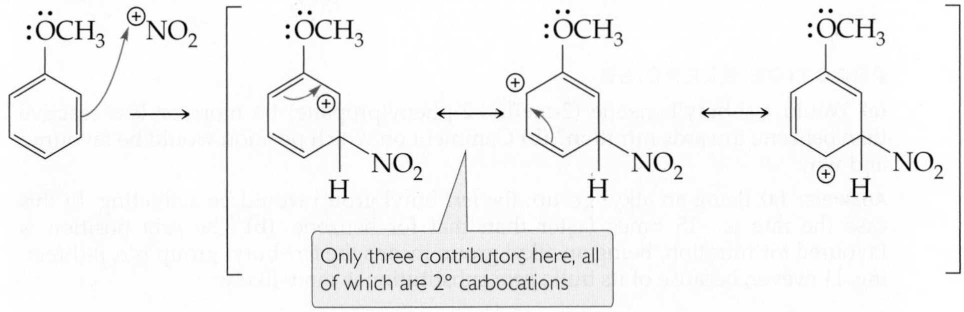

Only three contributors here, all
of which are 2° carbocations

Anilines are very reactive towards EAS and are highly activated *ortho*-, *para*-directors. For example, the bromination of aniline occurs rapidly and without the addition of a catalyst to give the tribromoaniline in excellent yield:

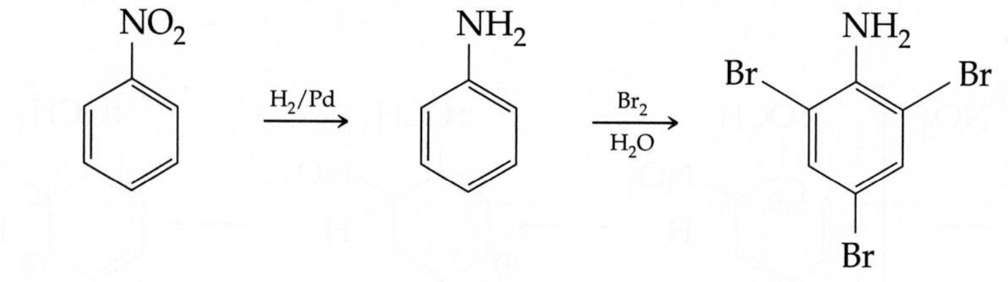

nitrobenzene   aniline   2,4,6-tribromoaniline
(>95%)

[28.14]

SAMPLE EXERCISE 28.3   **EAS reaction**

4-Nitrotoluene is the starting material for the synthesis of the local anaesthetics benzocaine and procaine. Describe a synthesis of 4-nitrotoluene starting from toluene.

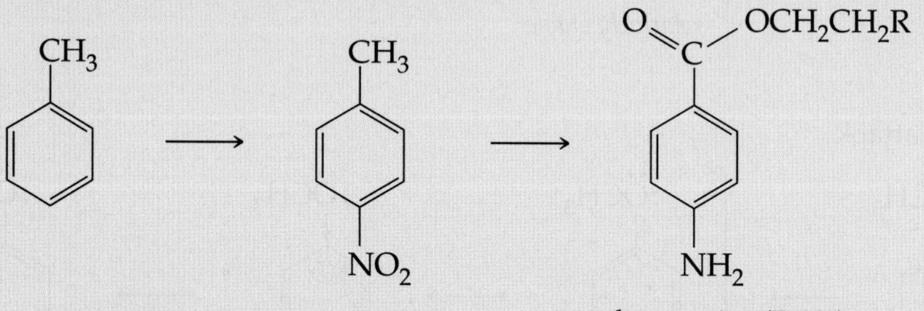

benzocaine (R=H)
procaine (R=NEt$_3$)

SOLUTION

Analyse We are asked to synthesise 4-nitrotoluene from toluene.

Plan To begin this exercise we look at the functional group differences between starting materials and products. In this reaction, the introduction of a *para*-nitro substituent is made, which can be achieved by EAS.

Solve Since methyl groups are *ortho-, para*-directing, 4-nitrotoluene can be formed by nitration of toluene and separation of the *ortho-* and *para*-isomers:

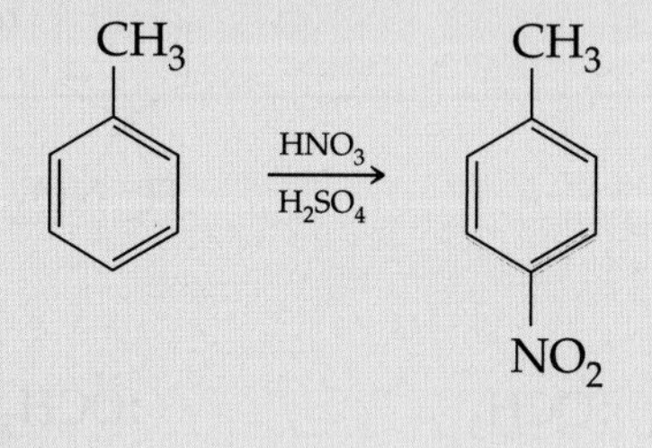

PRACTICE EXERCISE

(a) Would *tert*-butylbenzene (2-methyl-2-phenylpropane) be more or less reactive than benzene towards nitration? **(b)** Comment on which position would be favoured and why.

Answers: **(a)** Being an alkyl group, the *tert*-butyl group would be activating. In this case the rate is ~15 times faster than that for benzene. **(b)** The *para* position is favoured for nitration. Being an alkyl group means the *tert*-butyl group is *o, p*-directing. However, because of its bulk, *para* substitution is more likely.

(See also Exercises 28.41, 28.43.)

So far we have dealt with electron-donating substituents. Now let's see what happens with the nitration of (trifluoromethyl)benzene, an analogue of toluene where each of the methyl hydrogens is replaced with fluorine. In this case, the *meta* product 3-nitro(trifluoromethyl)benzene is almost exclusively formed (~90% yield).

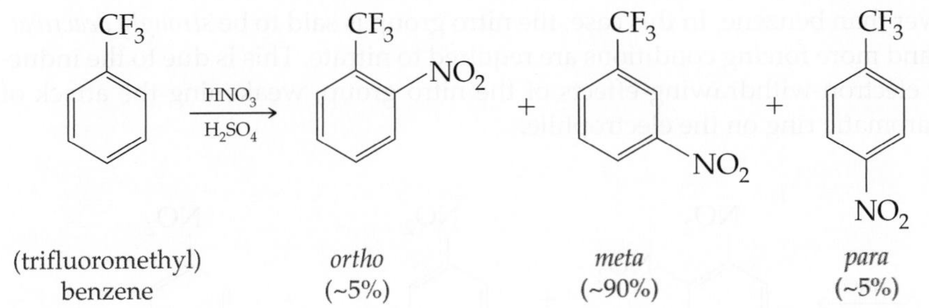

| (trifluoromethyl) benzene | ortho (~5%) | meta (~90%) | para (~5%) |

[28.15]

The near exclusive formation of the *meta* isomer means that the trifluoromethyl group is a ***meta*-director**. Though similar in size, there is a marked difference between the inductive properties of methyl and trifluoromethyl groups. The electronegativity of F compared to H means that inductively the trifluoromethyl group is electron withdrawing. This has two implications.

1. The electron-withdrawing nature means that attack on the electrophile by the electrons of the aromatic ring is impeded. This leads to deactivation.

2. The second is that in the case of *ortho*- and *para*-attack, the once favourable 3° carbocation of toluene is now significantly destabilised by the CF_3 group. This makes the *meta* substitution pathway more favourable. The instability relates to the neighbouring location of the carbocation with the electropositive carbon bearing the three fluoro groups. This situation is analogous to that for carbonyl compounds attached to an aromatic ring where the carbonyl carbon is also δ+.

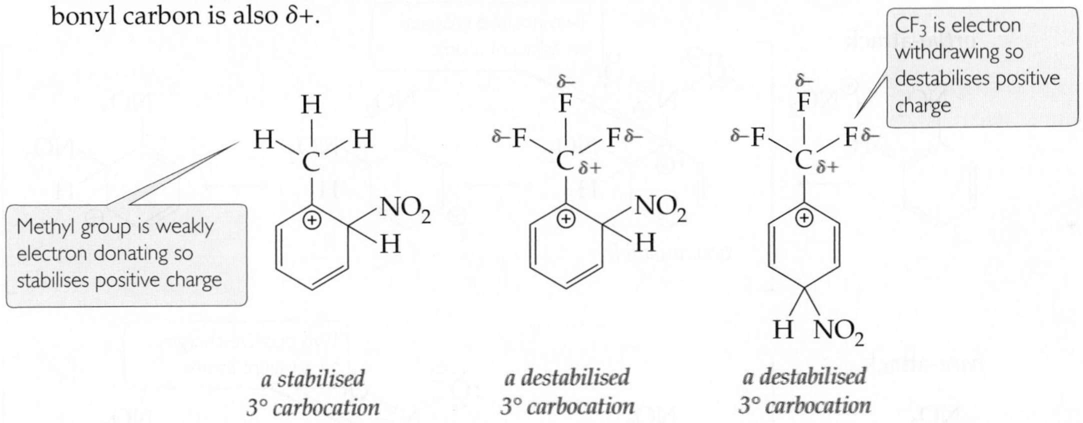

Methyl group is weakly electron donating so stabilises positive charge

CF_3 is electron withdrawing so destabilises positive charge

| a stabilised 3° carbocation | a destabilised 3° carbocation | a destabilised 3° carbocation |

Many groups are *meta*-directing and deactivating. A series of monosubstituted benzenes and their relative deactivation for EAS include:

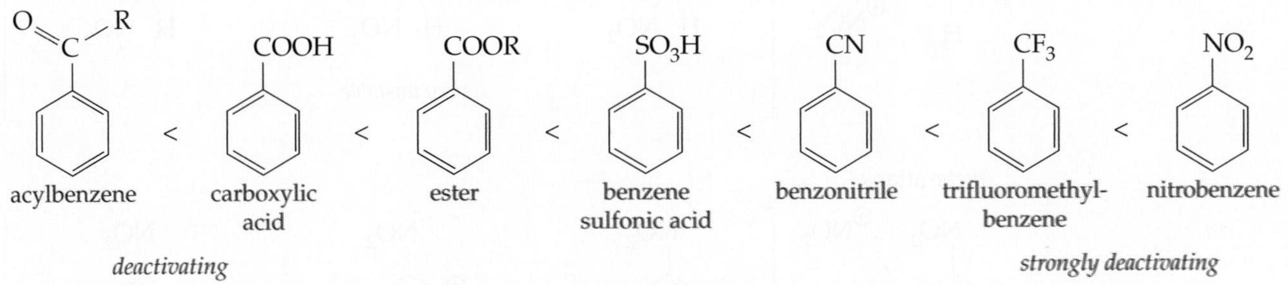

| acylbenzene | < | carboxylic acid | < | ester | < | benzene sulfonic acid | < | benzonitrile | < | trifluoromethyl-benzene | < | nitrobenzene |

deactivating   *strongly deactivating*

Generally speaking, electron-withdrawing groups are *meta*-directors and deactivating towards electrophilic aromatic substitution.

Let us complete our appreciation of directing groups by studying the nitration of nitrobenzene. The nitration of nitrobenzene is approximately 100 000 times slower than benzene. In this case, the nitro group is said to be *strongly deactivating* and more forcing conditions are required to nitrate. This is due to the inductive electron-withdrawing effects of the nitro group, weakening the attack of the aromatic ring on the electrophile.

[28.16]

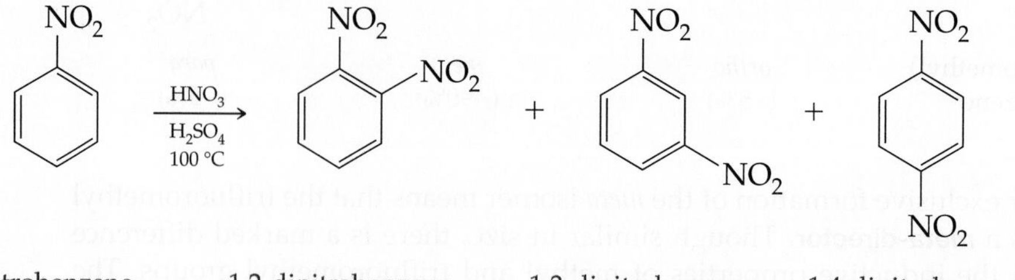

nitrobenzene 1,2-dinitrobenzene (6%) 1,3-dinitrobenzene (93%) 1,4-dinitrobenzene (>1%)

The reason why nitro groups are *meta*-directors becomes more obvious when investigating the resonance forms of the intermediates upon nitration. The proximity of the two positive charges in one resonance form of the *ortho* and *para* cases, in which they are on adjacent atoms, severely destabilises this group of intermediates, disfavouring *ortho-*, *para*-attack in favour of *meta*-attack.

ortho-attack

para-attack

meta-attack

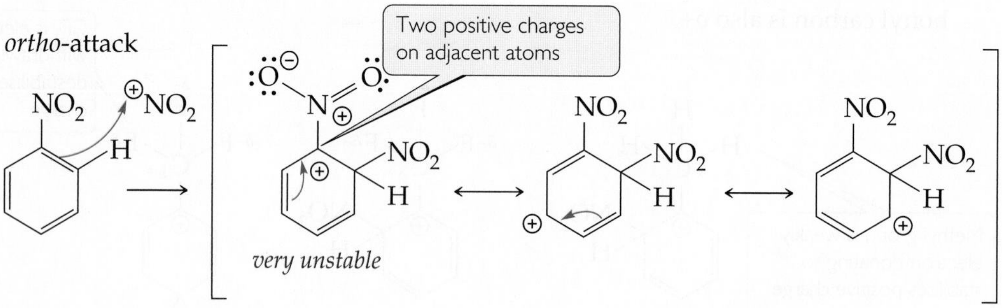

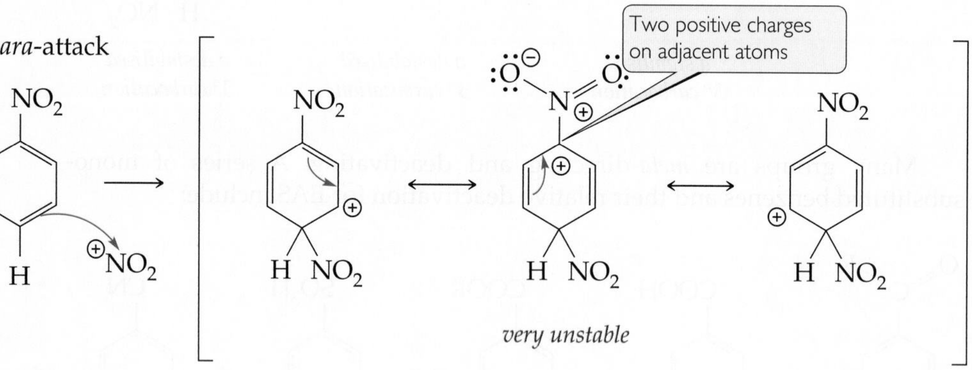

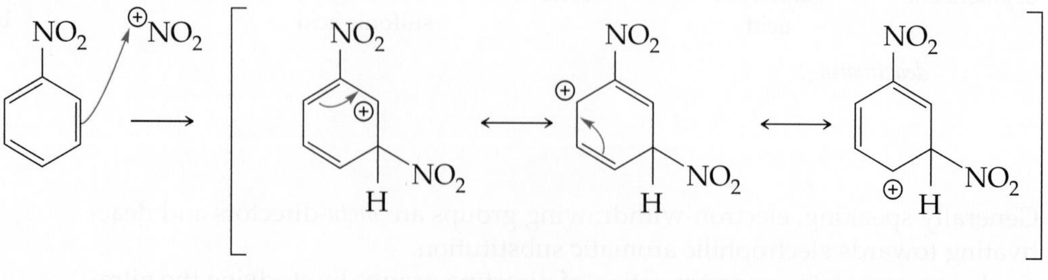

SAMPLE EXERCISE 28.4   Directing groups

Determine a suitable pathway for forming 2-nitrophenylacetamide from nitrobenzene.

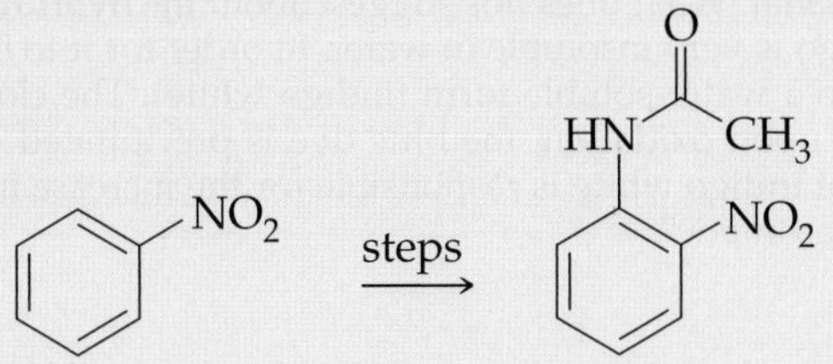

SOLUTION

Analyse We are asked to form 2-nitrophenylacetamide from nitrobenzene.

Plan Identify the differences in structure to determine what functionality needs to be added. Our knowledge of directing groups will also play a significant role in the pathway we follow. We also need to have some idea of synthetic transformations.

Solve The acetamide group can be formed in two steps from the nitro group by reduction, using catalytic hydrogenation or $SnCl_2$ forming the amine (aniline in this case). Acetylation using acetic anhydride or acetyl chloride leads to the acetamide:

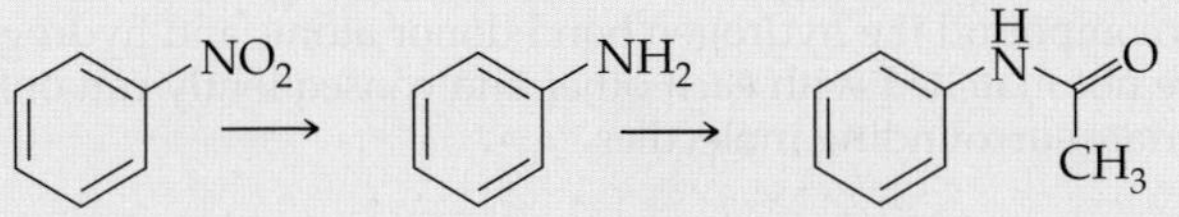

Acetamides and anilines are *o*-, *p*-directing and activating but nitro groups are *m*-directing and deactivating. Hence it makes sense to nitrate either the aniline or acetamide to give the desired product. You will need to separate the *o*- from *p*-isomer, however.

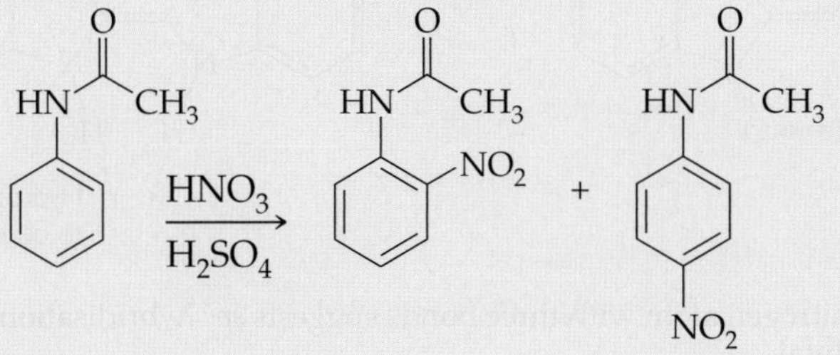

Trying to elaborate a synthesis from the nitrobenzene will undoubtedly lead to *meta* substitution, so this pathway should not be followed in this case.

PRACTICE EXERCISE

Which reacts more rapidly with HNO_3/H_2SO_4: anisole, toluene or nitrobenzene?

Answer: anisole

(See also Exercises 28.51, 28.52.)

SAMPLE INTEGRATIVE PROBLEM   Putting concepts together

Indigo is a dark blue dye with the following structure.

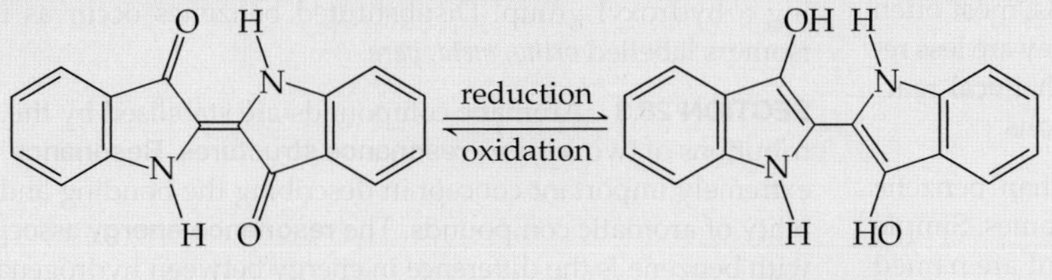

Indigo   Indigo white

(a) Classify the central double bond as either *E* or *Z*. **(b)** Intramolecular hydrogen bonding occurs in indigo but only intermolecular hydrogen bonding occurs in the related diastereoisomer. Explain why this is so. **(c)** The ring containing the nitrogen atom in indigo is planar. What does this suggest about the hybridisation of the nitrogen atom? **(d)** Indigo is very insoluble in water. In order for it to be used as a dye, it is first converted to a water-soluble form (indigo white). The cloth is impregnated with solution and, upon oxidation, the blue dye is precipitated in the cloth. What structural feature of indigo white is responsible for the increase in water solubility? **(e)** Is indigo white aromatic?

SOLUTION

Analyse This exercise focuses on the structure of indigo and the effect this has on hydrogen bonding and solubility.

Plan Determine the priority of the groups attached to the central double bond in order to classify it as *E* or *Z* then identify hydrogen bonding sites within the molecule and its isomer. Work out the hybridisation of the N atom and examine factors that may influence water solubility.

Solve

(a) In assigning priorities to the groups at either end of the central double bond, nitrogen has a higher priority than carbon. Consequently indigo has the two high priority groups in an *E* arrangement.

(b) The hydrogen bonding in indigo forms six-member rings which are stable. In the *Z* isomer of this compound the hydrogen bond donor atoms and hydrogen bond acceptor atoms are not matched with each other and consequently can only hydrogen bond to appropriate surrounding molecules.

(c) A planar nitrogen atom with three bonds suggests *sp*^{2} hybridisation with the lone pair in a *p* orbital.

(d) Reduction of a ketone to an alcohol group will be accompanied by an increased opportunity for hydrogen bonding and will increase the water solubility of the material.

(e) Yes, it follows the (4*n* + 2) rule for aromaticity. The lone pair on nitrogen forms part of the aromatic structure.

CHAPTER SUMMARY AND KEY TERMS

SECTION 28.1 Aromatic hydrocarbons, or **arenes**, are compounds containing multiple *sp*^{2}-hybridised carbons, most often in a six-membered ring, which we call benzene. They are less reactive than alkenes and undergo very different chemical reactions because of the delocalisation of their π-electrons.

SECTION 28.2 Aromatic hydrocarbons larger than benzene that incorporate multiple rings are given trivial names. Simple aromatic hydrocarbons bearing a single substituent are named based on benzene—for example, nitrobenzene. The substituent C_6H_5 is called a **phenyl group**. Loss of a hydrogen from toluene

leads to a **benzyl group**. **Phenols** are a benzene derivative bearing a hydroxyl group. Disubstituted benzenes occur as three isomers labelled *ortho*, *meta*, *para*.

SECTION 28.3 Aromatic compounds are stabilised by the contributions of two or more **resonance structures**. **Resonance** is an extremely important concept in describing the bonding and stability of aromatic compounds. The **resonance energy** associated with benzene is the difference in energy between hydrogenating benzene and the theoretical molecule 1,3,5-cyclohexatriene. Compounds are aromatic if they follow Hückel's criteria. The major criterion states that a compound is aromatic if it has [4*n* + 2] π-electrons and is cyclic. **Heterocyclic compounds** contain a **heteroatom**—that is, an atom other than carbon.

SECTION 28.4 Phenols are an unusual class of compound prevalent in nature. Their acidity differs greatly from another common and closely related class, the alcohols, because of the relative stability of the phenoxide ion. The anion is **resonance-stabilised** as a result of the phenolate's ability to distribute the electron density around the aromatic ring. **Electron-withdrawing groups** on the phenyl ring of phenol increase the acidity through **inductive effects**.

SECTION 28.5 Addition reactions are difficult to carry out with aromatic hydrocarbons, but **electrophilic aromatic substitution (EAS) reactions** are easily accomplished in the presence of a catalyst. Two general reactions are halogenation, which substitutes hydrogen for a single halogen (typically Cl or Br), and nitration, which substitutes a hydrogen for NO_2. Two new types of reactions discussed are the **Friedel–Crafts alkylation** and **acylation reactions**. Each of these reaction types is useful as they produce new C–C bonds. Substituted benzenes also undergo EAS, though the substituent can act as an **activating** or **deactivating** depending on whether it is electron donating or electron withdrawing. Generally, electron-donating groups are *ortho-*, *para-***directing** whereas electron-deficient groups are *meta-***directing**.

KEY SKILLS

- Know how to derive the name of simple mono- and disubstituted aromatic compounds by positional numbering, for example 1,2-dibromobenzene, as well as by *ortho*, *meta* and *para* nomenclature, for example *o*-dibromobenzene. (Section 28.2)

- Learn the common names such as toluene, phenol, aniline, etc. and the hierarchy of functional groups. (Section 28.2)

- Be able to use the curved arrow notation introduced in Chapter 24 to understand the stability and directive properties of different functional groups on the aromatic structure. (Section 28.5)

- Be able to draw Wheland intermediates to show the directive influences of substituents on benzene in EAS. (Section 28.5)

- Understand why phenols are acidic and be able to describe this through drawing resonance contributors to an intermediate structure. (Section 28.4)

- Be able to identify electrophiles from a given set of reagents, and in the case of substituted benzenes understand qualitatively their directive influences on the product outcome. (Section 28.5)

KEY EQUATIONS

- Aromaticity

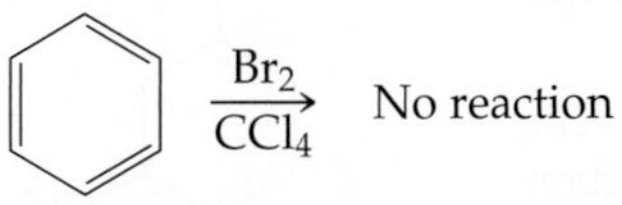

[28.4]

- Acidity

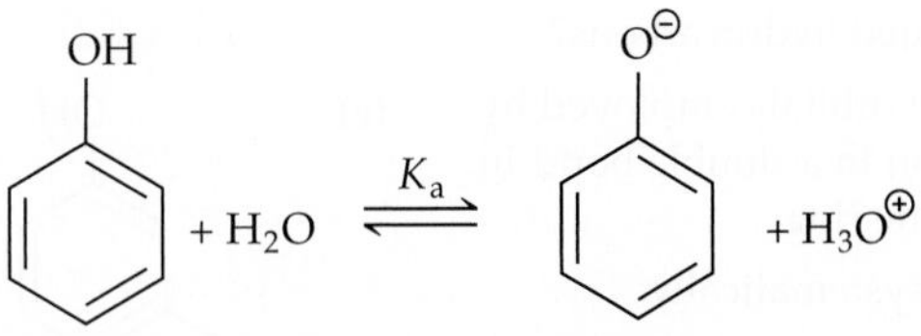

phenoxide ion

[28.8]

- Electrophilic aromatic substitution

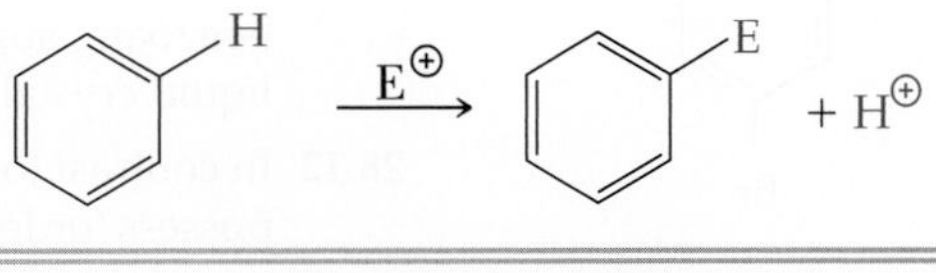

[28.10]

EXERCISES

VISUALISING CONCEPTS

28.1 **(a)** Draw a six-membered ring as a line drawing and demonstrate how sp^2 hybrid orbitals interact, as in benzene. **(b)** Illustrate how the *p* orbitals overlap to form a stable aromatic structure. [Section 28.1]

28.2 Which of the following molecules will readily undergo an electrophilic substitution reaction? [Section 28.5]

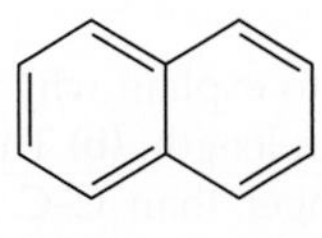

(a)

$$\underset{\text{(b)}}{CH_3CH_2\overset{\overset{\textstyle O}{\|}}{C}-OH}$$

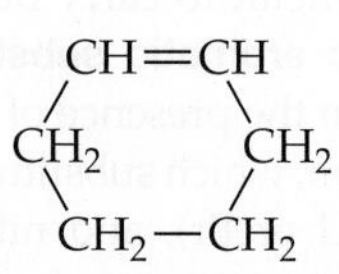

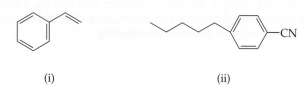

(c) 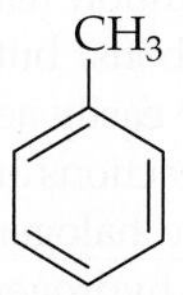   **(d)**

28.3 Consider the following molecules. Which of these would be most likely to exhibit liquid crystalline behaviour? [Section 28.3]

(i)   (ii)

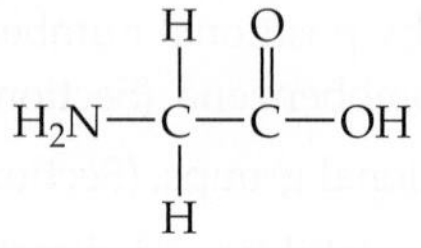

(iii)

28.4 Draw the product of the following valence electron movements. [Section 28.4]

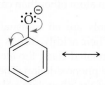

28.5 Which of the following structures is aromatic? [Section 28.3]

(a)   **(b)**

(c) 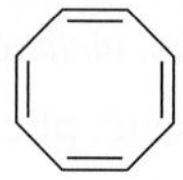   **(d)**

BENZENE STRUCTURE AND NOMENCLATURE (Sections 28.1 and 28.2)

28.6 What structural features help us identify a compound as an aromatic hydrocarbon?

28.7 Name another compound that has the same empirical formula as benzene.

28.8 Give the molecular formula of a cyclic alkane, a cyclic alkene, a linear alkyne and an aromatic hydrocarbon that in each case contains six carbon atoms. Which are saturated and which are unsaturated hydrocarbons?

28.9 What are the characteristic hybrid orbitals employed by **(a)** carbon in an alkane, **(b)** carbon in a double bond in an alkene, **(c)** carbon in the benzene ring.

28.10 Name the following compounds systematically:

(a)   **(b)**

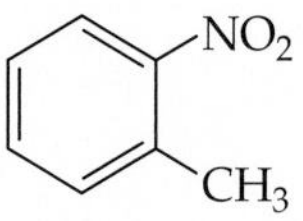

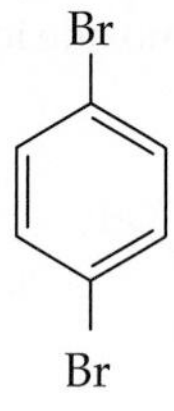

(c)

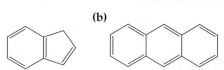

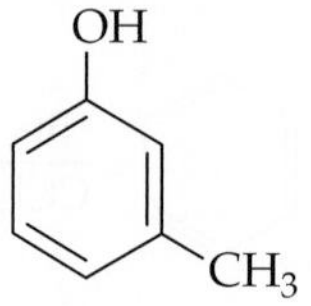

OH

CH₃

(d) $CH_3CH_2-\overset{\overset{\displaystyle CH_3}{|}}{\underset{}{C}}-CH_2Cl$

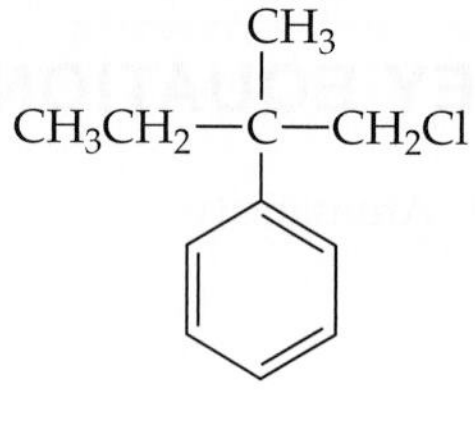

(e)

OH

OHC

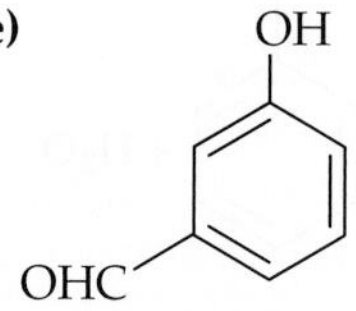

28.11 What observations made by Reinitzer on cholesteryl benzoate suggested that this substance possesses a liquid crystalline phase?

28.12 In contrast to ordinary liquids, liquid crystals are said to possess 'order'. What does this mean?

RESONANCE AND AROMATICITY (Section 28.3)

28.13 Which hydrogens are more acidic, those of cyclohexane or benzene? Why?

28.14 **(a)** Use the concept of resonance to explain why all six C–C bonds in benzene are equal in length. **(b)** The C–C bond lengths in benzene are longer than C–C single bonds but shorter than C–C double bonds. Use the resonance model to explain this observation.

28.15 Mothballs are composed of naphthalene, $C_{10}H_8$, the structure of which consists of two six-membered rings of carbon that are fused along an edge, as shown in the following incomplete Lewis structure:

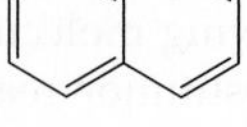

(a) Write two different complete Lewis structures for naphthalene. (b) The observed C–C bond lengths in the molecule are intermediate between C–C single and C–C double bonds. Explain. (c) Represent the resonance in naphthalene in a way analogous to that used to represent it in benzene.

28.16 (a) What is the difference between a localised π-bond and a delocalised one? (b) How can you determine whether a molecule or ion will exhibit delocalised π-bonding? (c) Demonstrate with the aid of a diagram the delocalisation in benzene. (d) Is the π-bond in $CH_3CO_2^-$ localised or delocalised?

28.17 (a) Briefly discuss some of the evidence that has been used to demonstrate the phenomenon of aromaticity. (b) Briefly discuss the valence-bond model of aromaticity.

28.18 (a) What *two* experimental results prove that cyclohexatriene is a poor representation of the real structure of benzene? (b) Benzene has six sp^2-hybridised carbons. How many electrons are left in each of the six p orbitals of benzene? How many electrons in the π-system of benzene? (c) With the aid of resonance structures, describe how the structure of benzene and the hypothetical cyclohexatriene differ.

28.19 (a) Draw a benzylic cation. (b) Use arrow notation to describe why this intermediate is more stable than that for the corresponding methylcyclohexane primary carbocation.

28.20 Which of the following compounds is more stable? Explain.

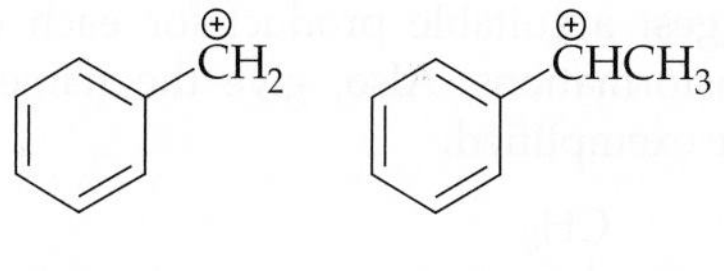

28.21 Which of the following alcohols undergoes dehydration more rapidly when heated with H_2SO_4? Why?

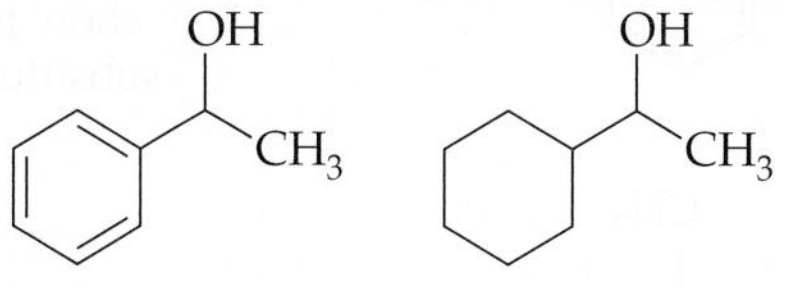

28.22 List the criteria that compounds must meet in order to be considered aromatic.

28.23 (a) How many π-electrons should an aromatic molecule contain? (b) Is the following molecule aromatic?

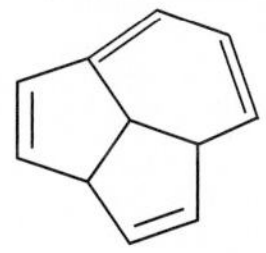

28.24 (a) One test for the presence of an alkene is to add a small amount of molecular bromine and look for the disappearance of the brown colour. This test does not work for detecting the presence of an aromatic hydrocarbon. Explain.

PROPERTIES OF PHENOLS (Section 28.4)

28.25 Place the following in decreasing order of acidity:

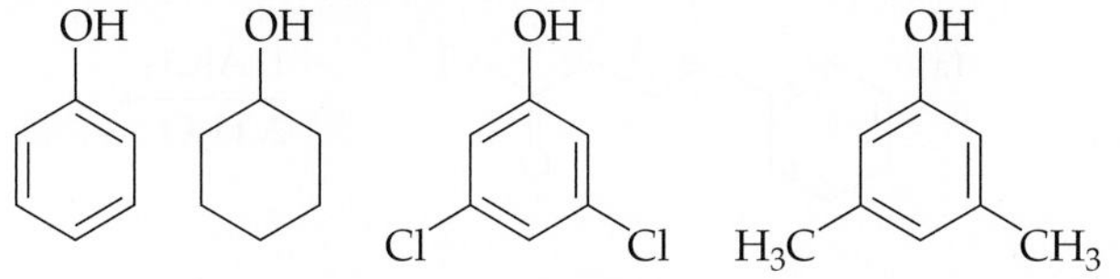

28.26 (a) Draw the structure of phenoxide. (b) Use arrow notation to describe why this intermediate is more stable than that for the corresponding cyclohexanol anion.

28.27 Butylated hydroxytoluene (BHT) has the following molecular structure:

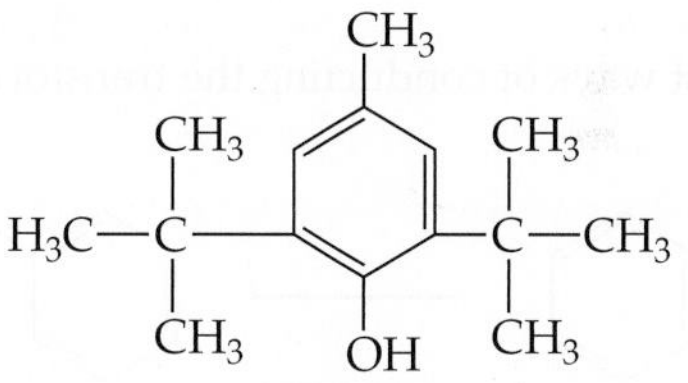

BHT

It is widely used as a preservative in a variety of foods, including dried cereals. (a) Based on its structure, would you expect BHT to be more soluble in water or in hexane (C_6H_{14})? Explain. (b) Identify phenol within this structure.

ELECTROPHILIC AROMATIC SUBSTITUTION REACTIONS (Section 28.5)

28.28 (a) What is the difference between a substitution and an addition reaction? Which one is commonly observed with alkenes and which one with aromatic hydrocarbons? (b) Using condensed structural formulae, write the balanced equation for the addition reaction of 2,4-dimethyl-2-pentene with Br_2. (c) Write a balanced equation for the substitution reaction of Cl_2 with *para*-dichlorobenzene in the presence of $FeCl_3$ as a catalyst.

28.29 Using condensed structural formulae, write a chemical equation for each of the following reactions: (a) hydrogenation of cyclohexatriene; (b) bromination of 1,4-dimethoxybenzene using $FeBr_3$ as a catalyst; (c) reaction of 2-chloropropane with benzene in the presence of $AlCl_3$.

28.30 Suggest a method of preparing ethylbenzene, starting with benzene and ethene as the only organic reagents.

28.31 Write a series of reactions leading to *para*-bromoethylbenzene, beginning with benzene and using other reagents as needed.

28.32 Suggest a suitable product for each of the following transformations. Also, give the name of the reaction type exemplified.

(a)

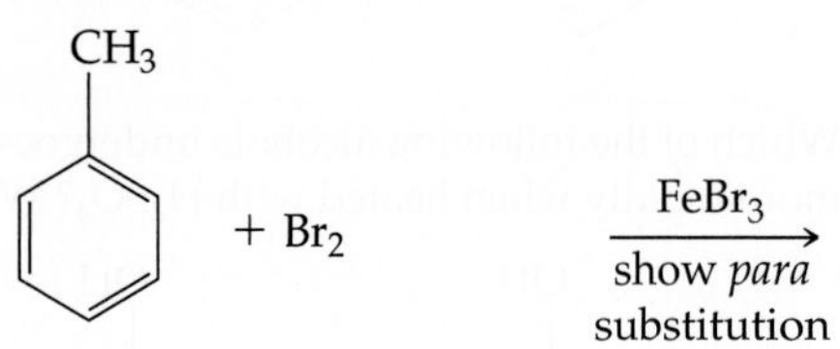

+ Br$_2$ $\xrightarrow[\text{show } para\text{ substitution}]{\text{FeBr}_3}$

(b)

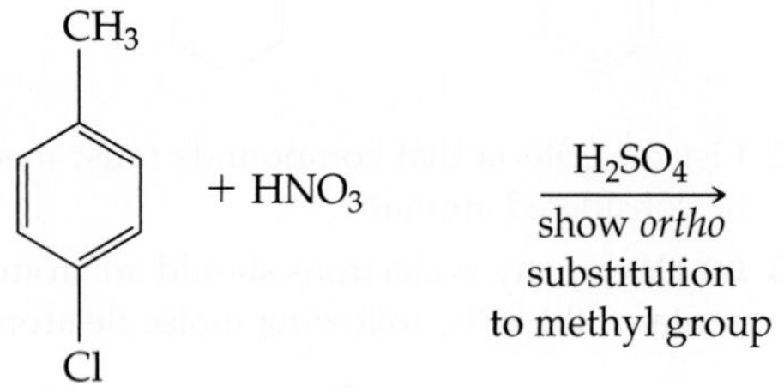

+ HNO$_3$ $\xrightarrow[\substack{\text{show } ortho \\ \text{substitution} \\ \text{to methyl group}}]{\text{H}_2\text{SO}_4}$

(c)

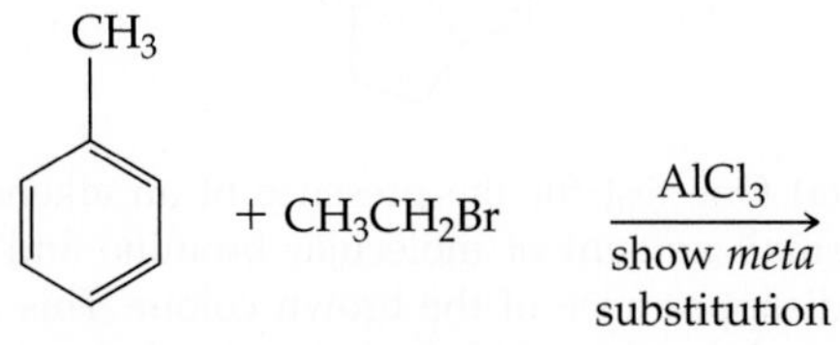

+ CH$_3$CH$_2$Br $\xrightarrow[\text{show } meta\text{ substitution}]{\text{AlCl}_3}$

(d)

+ $\xrightarrow{\text{AlCl}_3}$

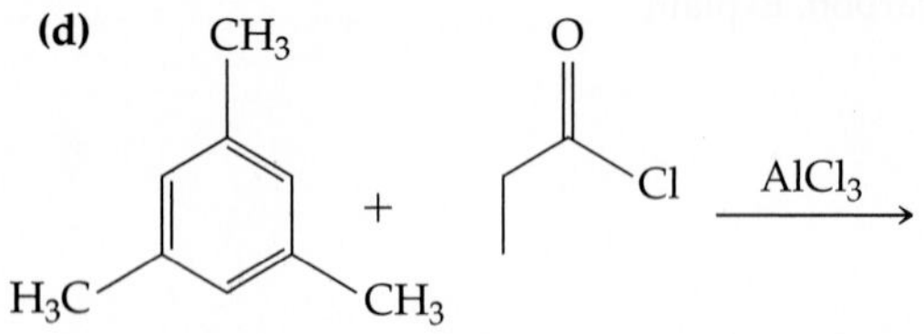

28.33 Suggest ways of conducting the transformations shown below.

(a)

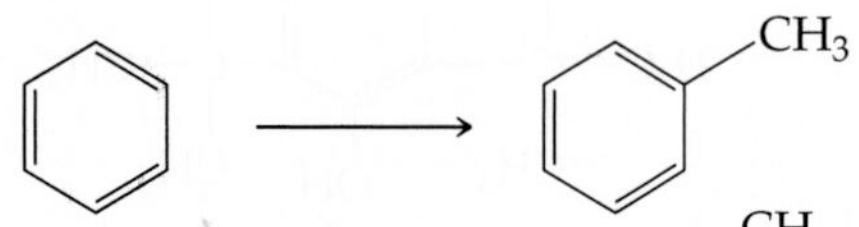

(b)

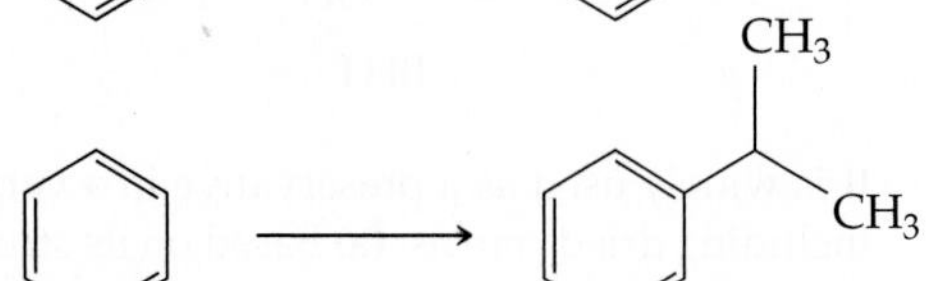

(c)

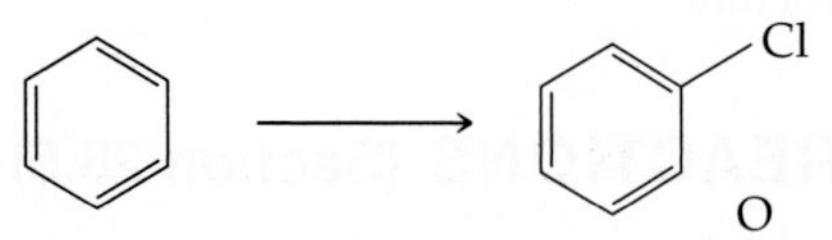

(d)

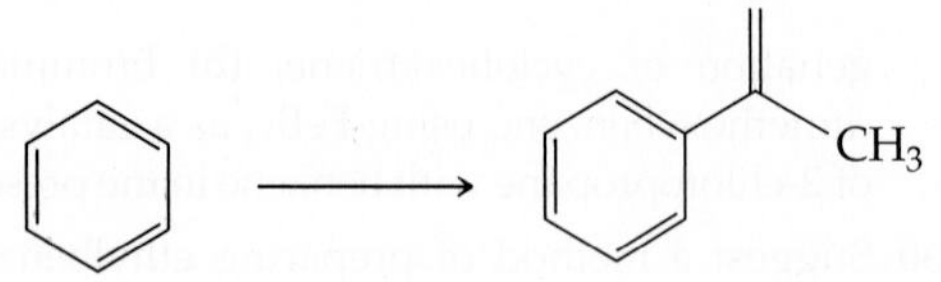

28.34 Identify and draw the electrophiles used in Exercise 28.33.

28.35 Draw another resonance form for these carbocations.

(a)

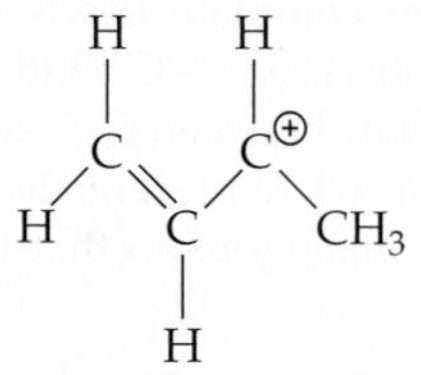

(b)

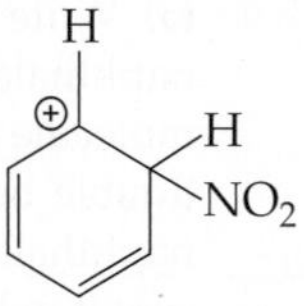

(c)

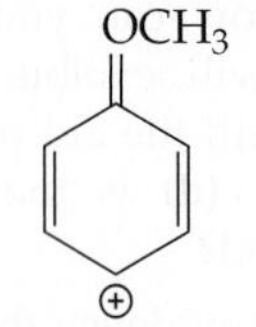

(d)

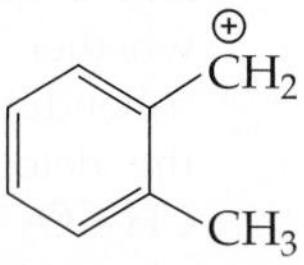

28.36 The structure of butylated hydroxytoluene (BHT) is shown in Exercise 28.27. How might you form this from 4-methylphenol?

28.37 Why does benzene undergo electrophilic substitution rather than electrophilic addition?

28.38 **(a)** Provide the major resonance structures of the Wheland intermediate in the reaction of benzene with the general electrophile E$^+$. **(b)** Is this intermediate carbocation aromatic?

28.39 Which step in the general mechanism for EAS is rate-determining? Explain.

28.40 Draw all the resonance structures of the species shown below.

28.41 Provide the major organic product of the reactions shown below. [*Hint*: Think intramolecular.]

(a) $\xrightarrow[\text{2. H}_2\text{O}]{\text{1. AlCl}_3}$

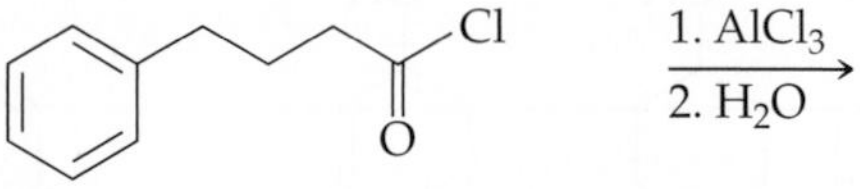

(b) $\xrightarrow[\text{2. H}_2\text{O}]{\text{1. AlBr}_3}$

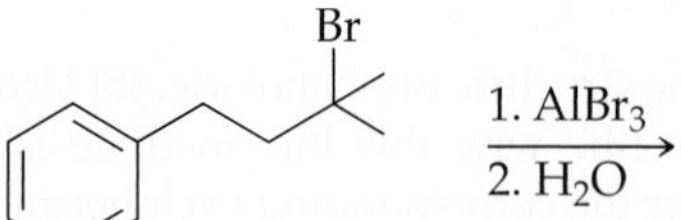

28.42 **(a)** What would be the electrophile in the following reaction? **(b)** What is the EAS product?

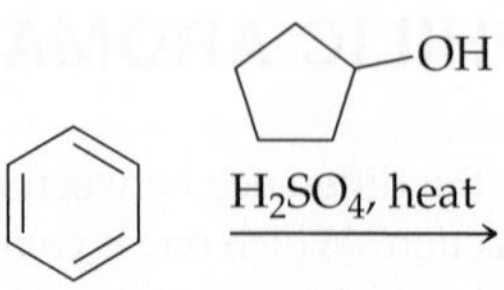

$\xrightarrow{\text{H}_2\text{SO}_4, \text{ heat}}$

28.43 Which of the following functional groups is the most deactivating for EAS? **(a)** methoxy, **(b)** phenol **(c)** ethyl, **(d)** acetamide, **(e)** benzene?

28.44 Rank the following groups in order of increasing activating power in electrophilic aromatic substitution reactions: $-$OCH$_3$, $-$CH$_2$CH$_3$, $-$Br.

28.45 Draw the most important contributor to the resonance hybrid formed when toluene undergoes *para* nitration.

28.46 What factors affect the *ortho*:*para* ratio in electrophilic aromatic substitution reactions?

28.47 (a) What is the effect of a chlorine substituent on EAS? (b) What is the effect of a nitro substituent on EAS?

28.48 What is the *major* product (including regioisomer) when ethylbenzene is reacted with Br_2/$FeBr_3$?

28.49 Give a reason why nitrobenzene can be used as a solvent for Friedel-Crafts alkylation.

28.50 Give an explanation why direct nitration of aniline using HNO_3 yields, among other products, *m*-nitroaniline.

28.51 Suggest a method for preparing 3-nitrobenzoic acid from benzene.

28.52 Devise a method for preparing *m*-chloroaniline from benzene.

28.53 Provide a series of synthetic steps by which 2-bromo-4-nitrobenzoic acid can be prepared from toluene.

28.54 What is the *major* product of the bromination of 4-methylphenol by Br_2 in the presence of $FeBr_3$?

28.55 What is the *major* product of the nitration of 4-bromoanisole?

28.56 What is the major mononitration product of 1-ethoxy-4-ethylbenzene?

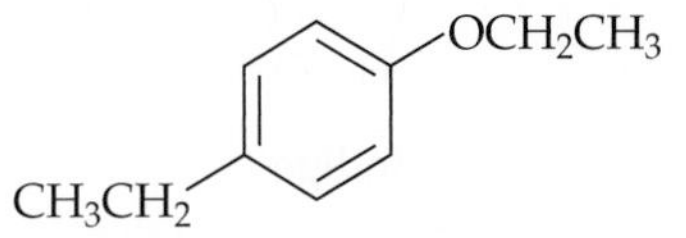

28.57 (a) What is the *major* product when 4-ethyltoluene is oxidised with potassium dichromate? (b) Name the product.

28.58 What is the *major* product of the following reaction?

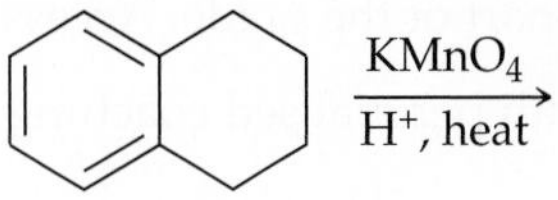

INTEGRATIVE EXERCISES

28.59 (a) Devise a synthesis of the analgesic acetaminophen (paracetamol) from phenol.

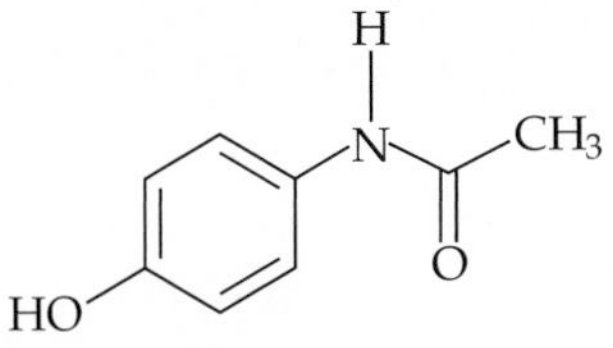

(b) Does the name *paracetamol* tell you anything about the structure of the molecule?

28.60 4-Nitrotoluene is the starting material for the synthesis of the local anaesthetics benzocaine and procaine. Describe a synthesis of benzocaine starting from 4-nitrotoluene.

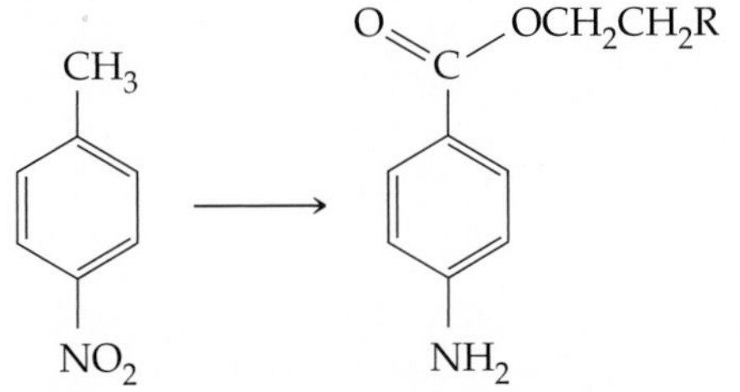

benzocaine (R=H)
procaine (R=NEt$_3$)

28.61 The heat of combustion of decahydronaphthalene ($C_{10}H_{18}$) is -6286 kJ mol^{-1}. The heat of combustion of naphthalene ($C_{10}H_8$) is -5157 kJ mol^{-1}. (In both cases, CO_2(g) and H_2O(l) are the products.) Using these data and data in Appendix C, calculate the heat of hydrogenation of naphthalene. Does this value provide any evidence for aromatic character in naphthalene?

28.62 The *cyclopentadienide ion* has the formula $C_5H_5^-$. The ion consists of a regular pentagon of C atoms, each bonded to two C neighbours, with a hydrogen atom bonded to

each C atom. All the atoms lie in the same plane. (a) Draw a Lewis structure for the ion. According to your structure, do all five C atoms have the same hybridisation? Explain. (b) Chemists generally view this ion as having sp^2 hybridisation at each C atom. Is that view consistent with your answer to part (a)? (c) Your Lewis structure should show one non-bonding pair of electrons. Under the assumption of part (b), in what type of orbital must this non-bonding pair reside? (d) Are there resonance structures equivalent to the Lewis structure you drew in part (a)? If so, how many? (e) The ion is often drawn as a pentagon enclosing a circle. Is this representation consistent with your answer to part (d)? Explain. (f) Both benzene and the cyclopentadienide ion are often described as systems containing six π-electrons. What do you think is meant by this description?

28.63 Both 3-aminobenzoic acid and 4-aminobenzoic acid are precursors to condensation polymers. (a) What type of isomers are these molecules? (b) Draw the structure of both of these molecules and the repeating unit of the polymer each forms. (c) Both compounds may be synthesised from toluene using the following reactions (not given in the correct order).

- Reduction of a nitro group using hydrogen gas and a palladium catalyst.
- Nitration using concentrated nitric acid and concentrated sulfuric acid.
- Oxidation of the aromatic alkyl group with acidified potassium permanganate.

Use your knowledge of the directing effect of a substituent to give the correct sequence of reactions for the synthesis of both compounds.

28.64 The pK_b of methylamine and analine is 3.36 and 9.40 respectively. Which molecule is the stronger base? Explain the reason for this difference.

28.65 Indigo is a blue dye that was originally extracted from plants but now is produced synthetically.

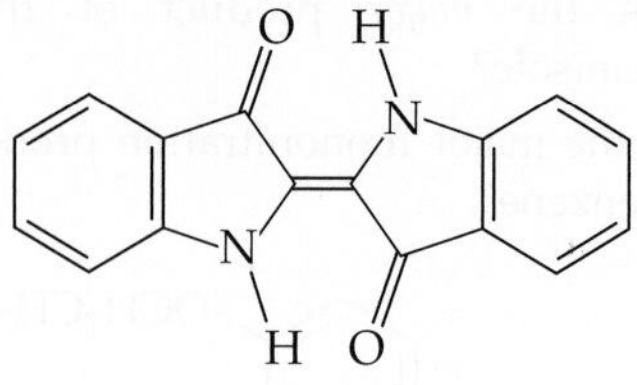

Indigo

(a) List the functional groups present in indigo.
(b) What is the hybridisation of the carbon atoms?
(c) Explain why the overall shape of indigo is planar.
(d) The molecule absorbs at 613 nm in the visible region of the spectrum. What energy does this correspond to?

MasteringChemistry (www.pearson.com.au/masteringchemistry)

Make learning part of the grade. Access:

- tutorials with peronalised coaching
- study area
- Pearson eText

PHOTO/ART CREDITS

NITROGEN-CONTAINING ORGANIC COMPOUNDS

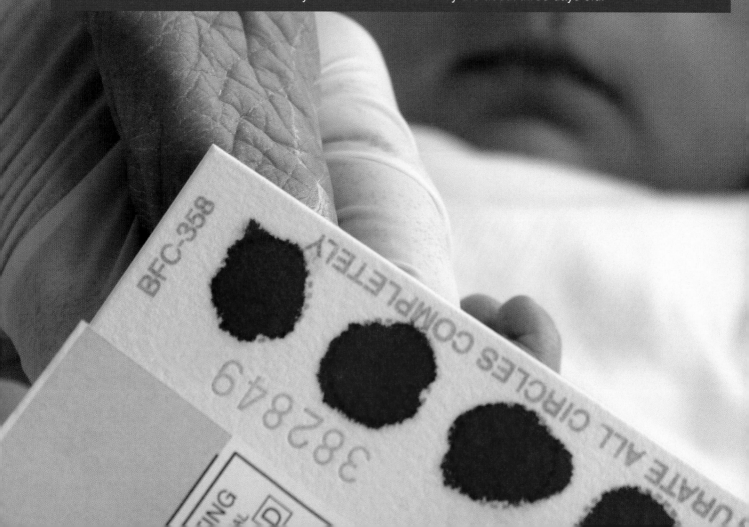

PKU testing. Phenylketoneuria (PKU) is a disease that causes significant mental retardation and seizures. It is caused by a genetic defect that leads to the inability of the body to convert the amino acid phenylalanine to tyrosine. Newborns are routinely tested for PKU when they are about three days old.

KEY CONCEPTS

29.1 AMINES AND THE AMIDE BOND
We begin by discussing nomenclature, properties and synthesis of amines and amides. A number of biologically important amines and amides are also introduced.

29.2 AMINO ACIDS
We introduce compounds that contain amine and carboxylic acid functional groups within the one molecule and study their complex acid–base properties. The synthesis of amino acids is briefly discussed.

29.3 PROTEINS, PEPTIDES AND ENZYMES
We combine our knowledge of carboxylic acids, amines and amides to describe primary, secondary and tertiary structure of proteins and peptides. Protein classes, including enzymes, and sequencing complete the picture.

29.4 NUCLEIC ACIDS AND DNA
We discuss the importance of DNA and RNA in the biosynthesis of proteins based on genetic information.

Phenylalanine is one of the two amino acids that make up the sweetener aspartame, an artificial sweetener sold under the trade names 'Equal®' and 'Nutrasweet®'. Phenylalanine also happens to be one of the essential amino acids, which means that we must have it in our diets to survive long term. However, in about one out of 10 000 to 20 000 Caucasian or Asian births, an enzyme that converts phenylalanine to another amino acid, tyrosine, is completely or nearly completely absent because of a genetic defect. The result is that phenylalanine accumulates in the blood and in body tissues. The disease that results is called phenylketonuria (PKU), which causes intellectual disability and seizures. This is why cans of diet soft drink sometimes carry the warning 'Phenylketonurics: contains phenylalanine'.

Although biological systems are unimaginably complex, they are nevertheless constructed of molecules of quite modest size. The example above illustrates that in order to understand biology we need to understand the chemical behaviour of molecules with low molar mass, as well as much larger molecules, which form the basis of *biological chemistry*.

This chapter serves as a brief introduction to other aspects of biological chemistry, building on important biochemical applications already introduced throughout the text.

Research at the interface of chemistry and biology, termed *biological chemistry*, is an exciting and dynamic area for scientists in both disciplines. Biological chemists apply the principles of chemistry to understand the molecular basis of biological processes relevant to medicine, biotechnology and other life sciences, such as structural biology, biochemistry, genetics, pharmacology and physiology. They then apply this knowledge to new chemical transformations, in materials science or in the area of medicinal chemistry. This differs from the area of science known as *biochemistry*, which aims to understand life in terms of its processes and reactions, but not necessarily on a molecular scale. In this chapter we present a brief overview of some of the elementary aspects of biological chemistry—proteins, vitamins and nucleic acids—from a molecular point of view and discuss the importance of nitrogen in each case.

Many biologically important molecules are very large (▼ FIGURE 29.1). The synthesis of these molecules is remarkable and one that places large demands on the chemical systems in living organisms. Organisms build complex biomolecules (for example, palytoxin) from much smaller and simpler substances that are readily available in the biosphere. The synthesis of large molecules requires energy because most of the reactions required are endothermic. The ultimate source of this energy is the sun. Mammals and other animals have no capacity for using solar energy directly; rather, they depend on plant photosynthesis to supply the bulk of their energy needs.

In addition to requiring large amounts of energy, living organisms are highly organised. The complexity of all the substances that make up even the simplest single-cell organisms and the relationships between all the many chemical processes that occur are truly amazing. In thermodynamic terms, this high degree of organisation means that the entropy of living systems is much lower than the entropy of the random mix of raw materials from which the systems are formed. Thus living systems must continuously work against the spontaneous tendency towards increased entropy that is characteristic of all organised systems. Maintaining a high degree of order places additional energetic requirements on organisms (see My World of Chemistry box in Section 14.13).

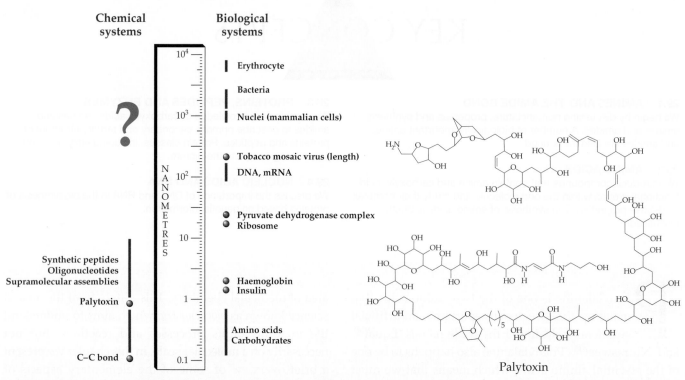

▲ FIGURE 29.1 **The scale of biomolecules.** A comparison of scale between biology and chemistry. Note that the scale is not linear, but logarithmic. Inset: Palytoxin, a natural product first isolated from soft marine corals.

We have introduced some important biochemical applications of fundamental chemical ideas in the My World of Chemistry boxes that appear throughout this text. This chapter serves as a brief introduction to other aspects of biological chemistry. Nevertheless, you will see some patterns emerging. Hydrogen bonding (∞ Section 11.2, 'Intermolecular Forces') for example, is critical to the function of many biological systems. The geometry of, and functional groups within, molecules can govern their biological activity. Many of the large molecules in living systems are polymers of much smaller molecules. These polymers, called *biopolymers*, can be classified into three broad categories: proteins, polysaccharides (carbohydrates) and nucleic acids. We have already discussed polysaccharides in Chapter 26, so in this chapter we complete our appreciation of biopolymers in Sections 29.3 and 29.4.

∞ Review this on page 383

29.1 | AMINES AND THE AMIDE BOND

Chapters 25–27 introduced the chemistry of oxygen-containing compounds; in this chapter we will examine the complementary role played by nitrogen-containing molecules. In fact, nitrogen-containing organic compounds form an incredibly large and important class in their own right.

Amines

Amines are organic bases with the general formula

where R, R′ and R″ may be H or a hydrocarbon group. Amines containing alkyl groups are called **aliphatic amines**. Aniline is the parent molecule of a class of amines known as **aromatic amines**. This class is characterised by a covalent bond between the nitrogen atom and at least one aromatic group (for example, phenyl or naphthyl). Amines that form part of a cyclic structure are called **cyclic amines** or *heterocycles*. A special class of heterocyclic compounds contains a nitrogen atom within an aromatic ring (for example, pyridine). We also class these amines as aromatic amines. In many cases the lone pair found on nitrogen in an aromatic amine can contribute to the aromaticity. ▼ **FIGURE 29.2** gives some examples of amine classes, including the five- and six-membered heterocycles pyrrolidine and piperidine, and the aromatic heterocycles pyrrole and pyridine.

Amines are further classified as primary, secondary or tertiary.

- **Primary (1°) amine**: contains *one* alkyl or aryl group bonded to nitrogen.
- **Secondary (2°) amine**: contains *two* alkyl or aryl groups bonded to nitrogen.
- **Tertiary (3°) amine**: contains *three* alkyl or aryl groups bonded to nitrogen.
- **Quaternary (4°) ammonium salts**: contain *four* alkyl or aryl groups directly attached to nitrogen.

▼ **FIGURE 29.3** shows the structure of some industrially and biologically important amines. Aniline, for example, is an aromatic amine used heavily in the manufacture of dye stuffs. A significant number of amines are derived from plant material, such as roots, leaves and bark. Amines that are derived from plant material and contain a basic nitrogen atom are commonly called **alkaloids**. Alkaloids form an enormous class of compounds whose physiological properties have been studied intensively for over 140 years. Coniine, for example, is a highly toxic alkaloid, found to be the active ingredient in 'poison hemlock', which was responsible for the death of Socrates. Nicotine is a highly addictive alkaloid found in tobacco leaves. Codeine, a pain-reliever, and morphine, an analgesic, are both alkaloids extracted from opium poppies. Adrenaline is a principal compound in a group of compounds known as the *phenylethylamines*. Adrenaline is not an alkaloid; it is a *hormone* that is released into the body in response to stress. It causes increased blood flow to the muscles and brain, accelerates respiration and heart rate, and stimulates the release of stored energy into the blood. When injury to the body occurs, it potentiates the healing process. At times, adrenaline also functions as a neurotransmitter. Pyridoxamine, or vitamin B_6, is a water soluble vitamin that exists in three major

FIGURE IT OUT

What makes an amine secondary?

$CH_3CH_2\ddot{N}H_2$	$(CH_3CH_2)_2\ddot{N}H$	$(CH_3CH_2)_3\ddot{N}$	$(CH_3CH_2)_4\overset{\oplus}{N} \overset{\ominus}{I}$
ethylamine	diethylamine	triethylamine	tetraethylammonium iodide
aliphatic amine	aliphatic amine	aliphatic amine	ammonium salt
1° amine	2° amine	3° amine	4° ammonium salt

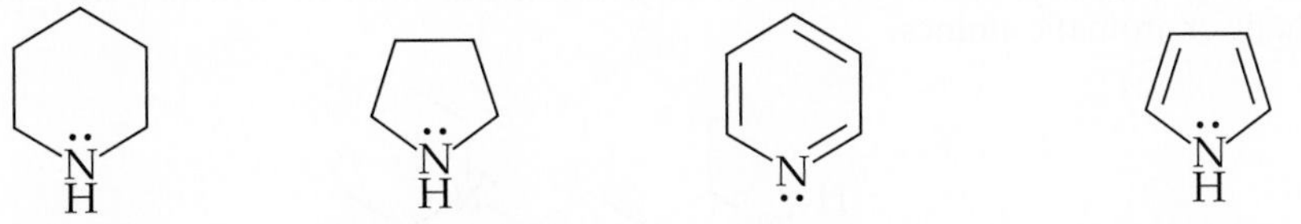

piperidine	pyrrolidine	pyridine	pyrrole
aliphatic amine	aliphatic amine	aromatic amine	aromatic amine
heterocyclic	heterocyclic	heterocyclic	heterocyclic
2° amine	2° amine	3° amine	2° amine

◀ **FIGURE 29.2** Amine classes.

▲ FIGURE IT OUT

Can you identify which amine is 1° and which is 3° in vitamin B₆?

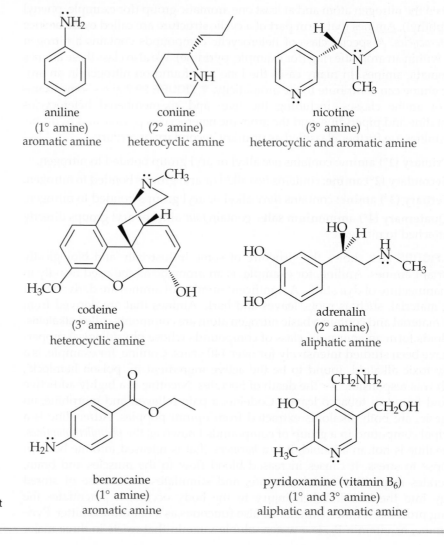

▶ **FIGURE 29.3 Industrially and biologically important amines.** The lone pair on nitrogen has been added to highlight its position in these compounds.

aniline
(1° amine)
aromatic amine

coniine
(2° amine)
heterocyclic amine

nicotine
(3° amine)
heterocyclic and aromatic amine

codeine
(3° amine)
heterocyclic amine

adrenalin
(2° amine)
aliphatic amine

benzocaine
(1° amine)
aromatic amine

pyridoxamine (vitamin B₆)
(1° and 3° amine)
aliphatic and aromatic amine

chemical forms: pyridoxine, pyridoxal and pyridoxamine. Its role in the body is varied but essential for good health. Benzocaine is an effective local and topical anaesthetic used in sting medication, sunburn cream and by dentists in mucosal tissue.

SAMPLE EXERCISE 29.1 Classifying amines

Classify each amino group of chloroquine, a compound used for the treatment of malaria, as primary, secondary or tertiary. Comment also on whether they are aliphatic or aromatic amines.

chloroquine

SOLUTION

Analyse We are asked to classify each amine in chloroquine.

Plan Look for each nitrogen atom and determine if aromatic or aliphatic as a result of the nature of the bonding around them. Classify each as primary, secondary or tertiary by determining how many N–C bonds exist.

Solve

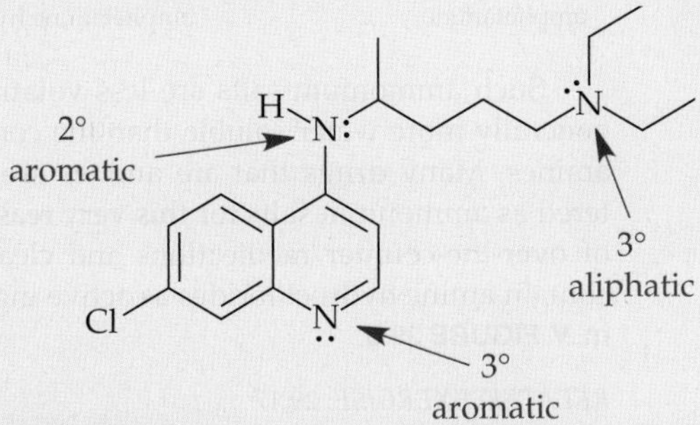

PRACTICE EXERCISE

Classify each amino group of spermine, a compound isolated from semen, as primary, secondary or tertiary.

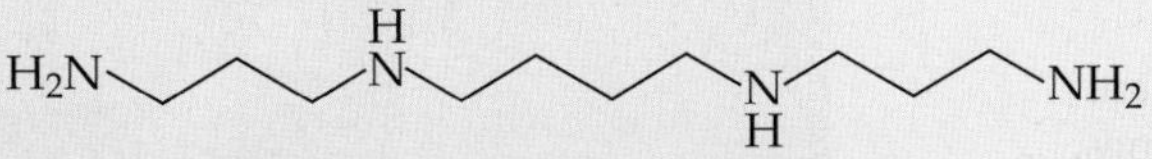

spermine

Answer: Spermine contains primary and secondary amines. The primary amines occupy positions at the end of the chain. The secondary amines appear within the chain.

(See also Exercises 29.12, 29.13.)

The systematic naming of simple monoaliphatic amines follows the same guidelines as for alcohols (Section 25.1, 'Alcohols: Structure, Properties and Nomenclature'). The *e* of the parent alkane is replaced by *amine* (as opposed to *ol* for alcohols). Otherwise, the name is derived from the alkyl group attached to nitrogen and named as the alkylamine. For example:

Review this on page 1001

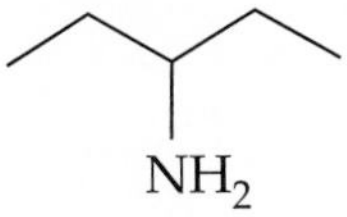

3-pentanamine    propane-1,3-diamine 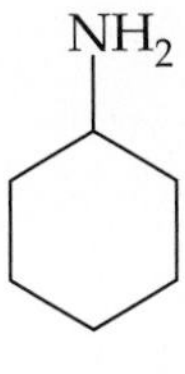   cyclohexylamine

For amines containing multiple alkyl groups bonded to carbon, the amine is named based on those alkyl groups (Figure 29.2). Amines have a low priority in the nomenclature hierarchy. In many instances, the **amino group** is named as a substituent rather than as the class of an organic compound:

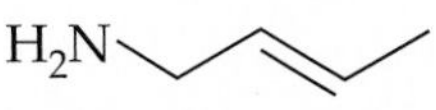

1-amino-but-2-ene    3-amino-propan-1-ol 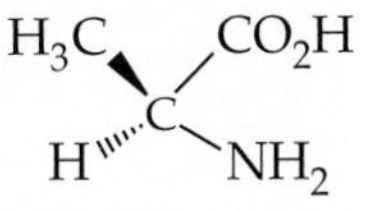   (*S*)-2-aminopropanoic acid

AMINES AND AMINE HYDROCHLORIDES

Many amines with low molecular weights have unpleasant 'fishy' odours. Amines and ammonia (NH_3) are produced by the anaerobic (absence of O_2) decomposition of dead animal or plant matter. Two such amines with very disagreeable odours are $H_2N(CH_2)_4NH_2$, known as *putrescine*, and $H_2N(CH_2)_5NH_2$, known as *cadaverine*. Despite their usually unpleasant odours, amines and the products of their reactions are very important compounds in biological systems.

The pigment of vision, rhodopsin, is prepared by the formation of an imine bond between the biologically active aldehyde 11-*cis*-retinal and the protein opsin (▼ FIGURE 29.4). When rhodopsin absorbs a photon of light, the *cis* double bond at C11 undergoes an isomerisation to the *trans* geometry, which triggers a change in the shape of rhodopsin, leading to nerve impulses that are detected by the brain as a visual image (∞ Section 24.2, 'Isomerism and Nomenclature').

Many drugs, including quinine, codeine, caffeine and amphetamine (benzedrine), are amines. Like other amines, these substances are weak bases; the amine nitrogen is readily protonated by treatment with an acid. The resulting products are called ammonium salts. For example, amphetamine hydrochloride is the ammonium salt formed by treating amphetamine with HCl.

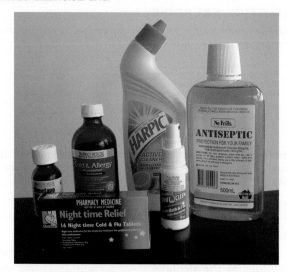

amphetamine   amphetamine hydrochloride

[29.1]

Such ammonium salts are less volatile, more stable and generally more water soluble than the corresponding neutral amines. Many drugs that are amines are sold and administered as ammonium salts for this very reason. Some examples of over-the-counter medications and cleaning products that contain amine hydrochlorides as active ingredients are shown in ▼ FIGURE 29.5.

RELATED EXERCISE: 29.17

▲ **FIGURE 29.5 Ammonium salts.** Many commercial products, such as over-the-counter medications and antiseptics, contain amine hydrochlorides as the major active ingredient.

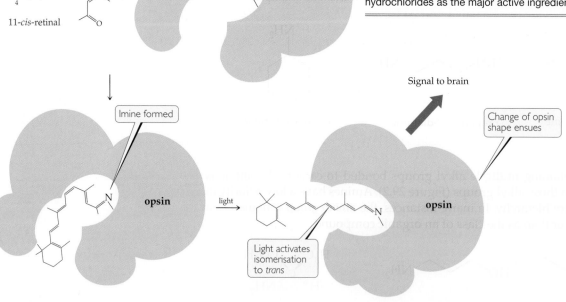

11-*cis*-retinal

H₂N **opsin**

Imine formed

opsin

light

Light activates isomerisation to *trans*

Signal to brain

opsin

Change of opsin shape ensues

▲ **FIGURE 29.4 Rhodopsin.** Imine formation between 11-*cis*-retinal and the protein opsin is important for vision.

The names of heterocyclic amines are derived from the parent aliphatic or aromatic heterocycle; for example:

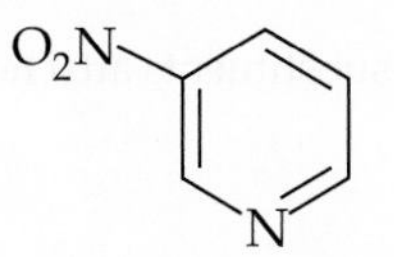

3-nitropyridine

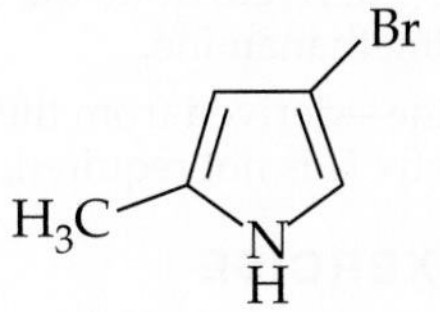

4-bromo-2-methylpyrrole

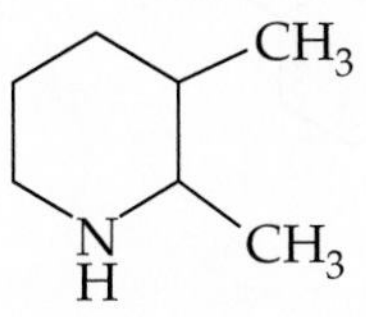

2,3-dimethylpiperidine

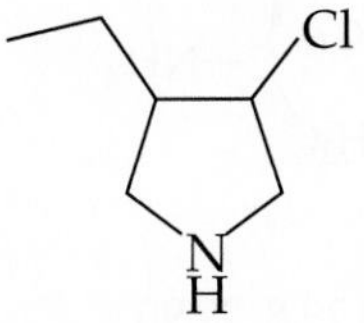

3-chloro-4-ethylpyrrolidine

When more than one type of alkyl group (for example, methyl and ethyl groups) are bonded to nitrogen in the same molecule, the name must also reflect the position of the alkyl groups on nitrogen by using the prefix *N*. For example, the name *N*-methylpiperidine indicates that the methyl substituent is located on the nitrogen of the piperidine ring and not located on the carbon framework (for example, 2-methylpiperidine). The name *N,N*-diethyl-1-butanamine is used to describe the fact that both ethyl groups are bonded to nitrogen. This compound differs from *N*-ethyl-2-ethyl-1-butanamine and 4-ethyl-3-hexanamine.

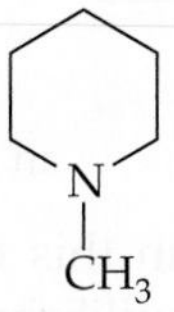

differs
from

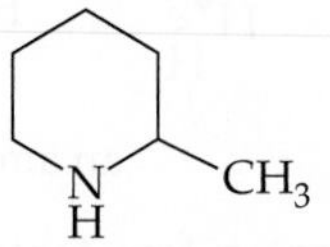

N-methylpiperidine

2-methylpiperidine

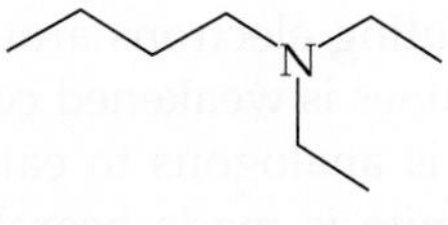

differs
from

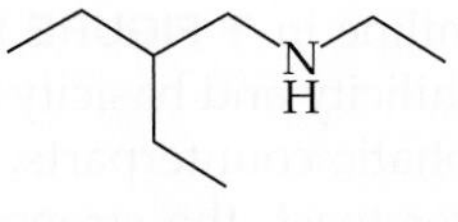

differs
from

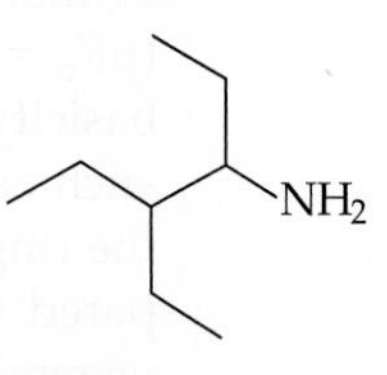

N,N-diethyl-1-butanamine

N-ethyl-2-ethyl-1-butanamine

4-ethyl-3-hexanamine

SAMPLE EXERCISE 29.2   Naming amines

Write the systematic or IUPAC name for the following amines:

(a) $CH_3CH_2CH_2CH_2CH_2NH_2$ (b)

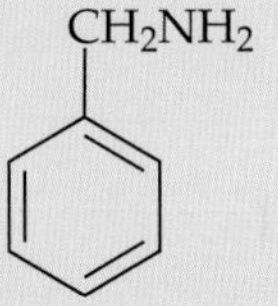

CH_2NH_2

(c) $(CH_3)_3N$

SOLUTION

Analyse We are asked to provide systematic names for three different amines

Plan Use the rules described above to structure the names.

Solve

(a) 1-pentanamine—derived from pentane.

(b) Benzylamine—derived from the benzyl group. This is better than naming the structure phenylmethanamine.

(c) Trimethylamine—derived from the three methyl substituents attached to nitrogen. Note that the prefix *N* is not required.

PRACTICE EXERCISE

Draw the structures for the following amines:
(a) *tert*-butylamine, **(b)** 4-nitroaniline, **(c)** *N*-methyl-2-hydroxypiperidine

Answers: (a)

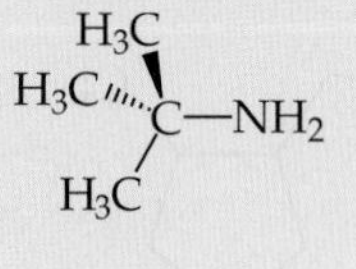

(b)

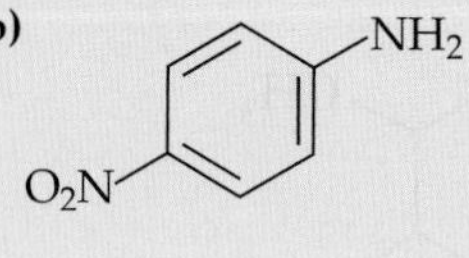

(c)

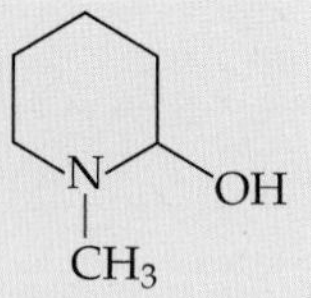

(See also Exercises 29.8, 29.9.)

Reactivity of Amines

We introduced the nucleophilic nature of amines in Chapter 26. Along with this nucleophilicity is the ability of primary, secondary and tertiary amines to act as a weak base. The presence of the amine functional group allows for extensive solvation in aqueous media. This means that most amines have some solubility in water and hence can be tested with litmus paper to determine their presence. The acid–base reaction for a primary amine in aqueous solution can be written generally as:

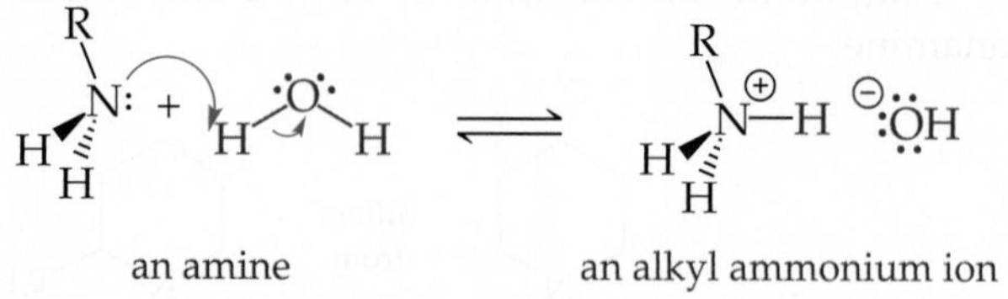

$$\text{an amine} \qquad \text{an alkyl ammonium ion}$$

[29.2]

There are a few features to note in this reaction. First, the reaction is an equilibrium reaction, although this equilibrium lies to the left for all amines. Primary, secondary and tertiary alkylamines are over 10^6 times more basic ($pK_b = 3$–4) than comparable aromatic amines ($pK_b = 9$–10). The difference in basicity can be attributed to the resonance effects available to aromatic amines, such as shown for aniline in ▼ **FIGURE 29.6**. By distributing electrons around the ring, the nucleophilicity and basicity of aromatic amines is weakened compared with their aliphatic counterparts. This situation is analogous to eating vegemite. Spread over toast, the strong taste of vegemite is made bearable, indeed enjoyable, for most people (▶ **FIGURE 29.7**). Compare this to eating the equivalent amount of vegemite from a spoon. In this case, the taste is too strong. Similarly, the strength of a nucleophile is reduced significantly when the electrons can be spread around an aromatic ring.

FIGURE IT OUT

Why is cyclohexylamine a stronger base than aniline?

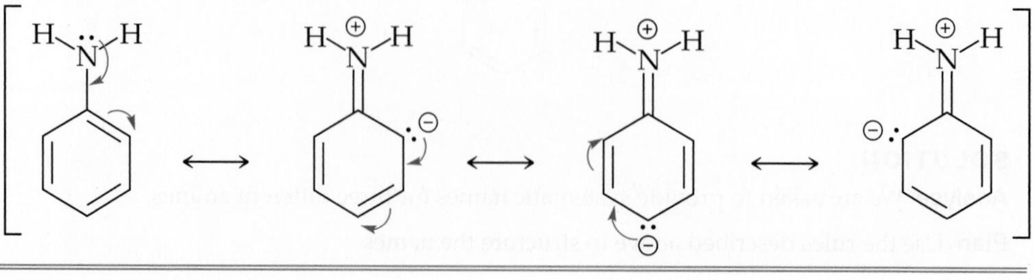

▶ **FIGURE 29.6 Resonance contributors for aniline.** The electron pair distribution responsible for the mild basicity and nucleophilicity in aromatic amines.

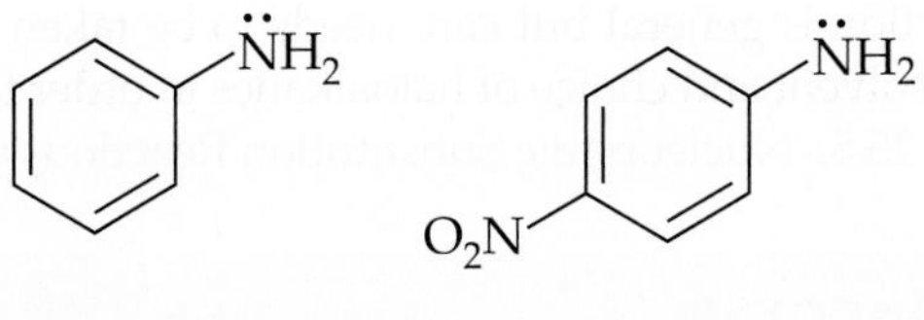

$pK_b = 9.4$ $pK_b = 13.0$

The basicity of aromatic amines is also influenced by the substituents on the aromatic ring. For example, 4-nitroaniline (▼ FIGURE 29.8) is a much weaker base than aniline, due to resonance and inductive effects leading to a redistribution of electron density onto the NO_2 group.

The second feature to note is that the acid–base reaction is written so that it involves the lone pair on nitrogen. The two electrons in this lone pair form a new N–H covalent bond. Finally, the product of the reaction is an **alkyl ammonium salt**, also known as a 4° ammonium salt. Such ammonium salts are easily formed in essentially quantitative yield by reaction of an amine with strong acid such as aqueous HCl:

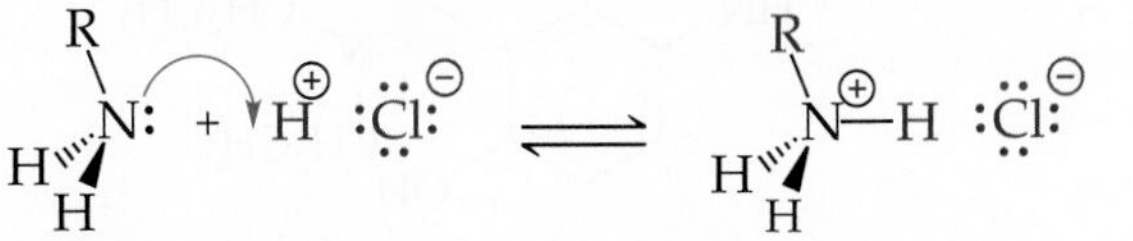

▲ FIGURE 29.7 **Understanding nucleophilic strength.** Nucleophiles in which the electron density associated with a negative charge can be spread around the molecule by resonance are typically weaker than those in which the negative charge resides solely on one atom. The situation is very similar to eating vegemite on toast. By spreading the vegemite over the toast, the taste is no where near as strong as eating the same amount from a spoon.

[29.3]

a primary amine   an alkyl ammonium salt

Such salts are usually highly soluble in water. Alkyl ammonium salts can also be formed by the reaction of amines with haloalkanes. In the following example, a secondary amine is reacted with a haloalkane to form an alkyl ammonium salt:

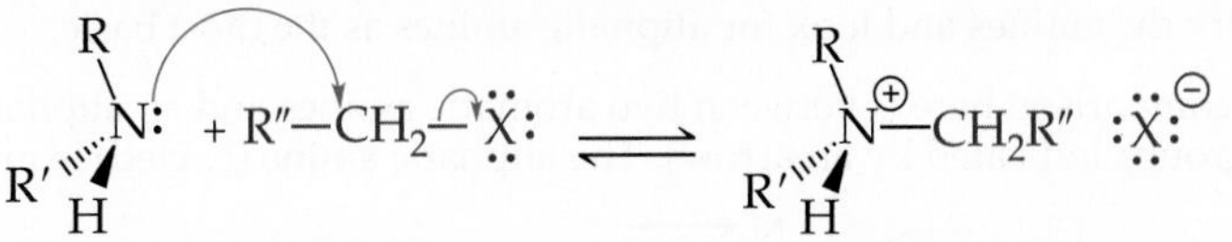

[29.4]

a secondary amine   a trialkyl ammonium salt

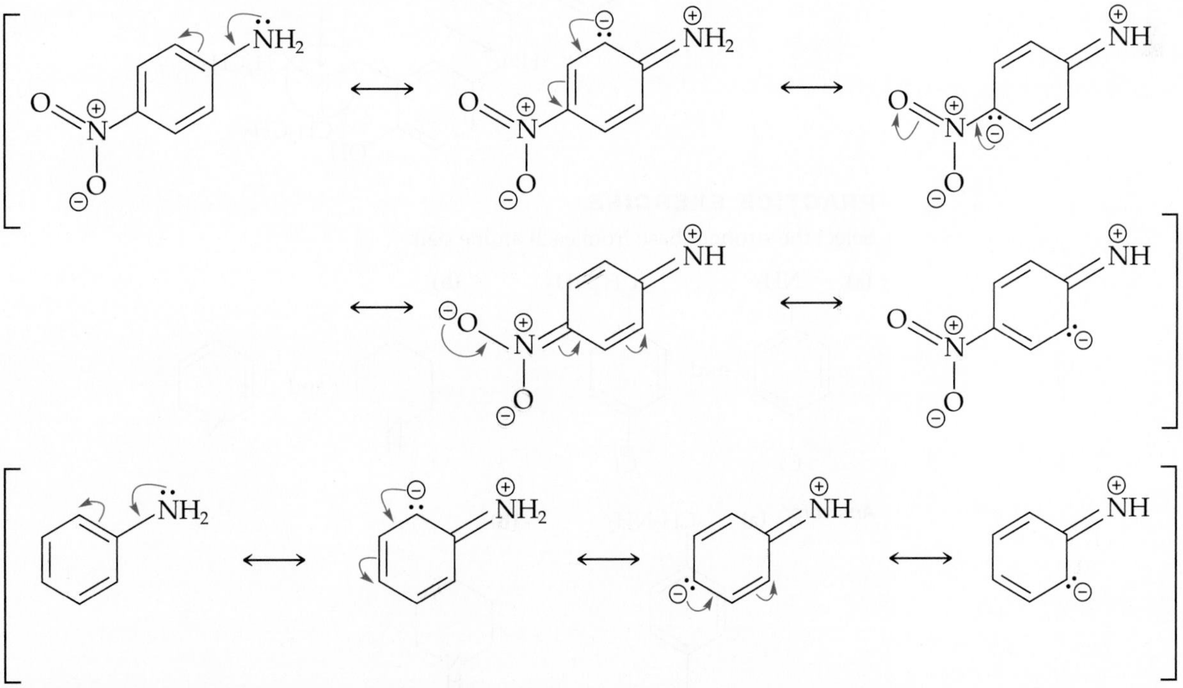

▲ FIGURE 29.8 **Rationalising the basicity of 4-nitroaniline.** There are more resonance contributors to the structure of 4-nitroaniline when compared to aniline, suggesting the basicity (and nucleophilicity) of the substituted aniline is less. This situation is accentuated by the electron-withdrawing effect of the nitro group, which is inductive throughout the molecule.

∞ Review this on page 1013

The alkylation reaction is general but care needs to be taken in stoichiometry of reagents, choice of solvent and choice of haloalkanes in order to limit further alkylation (∞ Section 25.5, 'Nucleophilic Substitution Reactions of Haloalkanes').

⚠ CONCEPT CHECK 1

Which is the better nucleophile: aniline or aminohexane?

SAMPLE EXERCISE 29.3 Amine basicity

Amodiaquin is commonly used to treat quinine-resistant malaria. Identify the amine groups within this compound and indicate which of the nitrogen atoms of amodiaquin is the most basic.

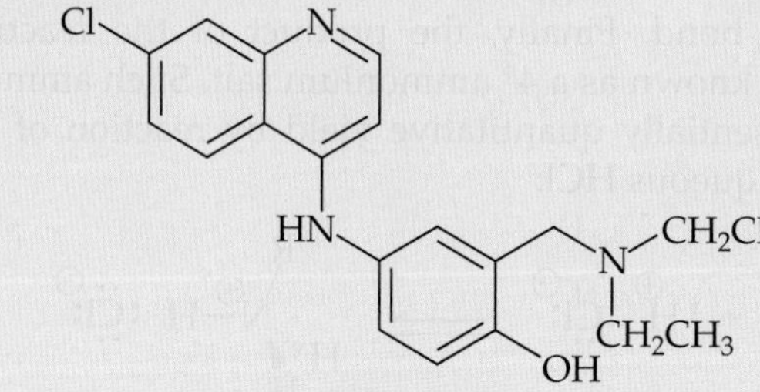

amodiaquin

SOLUTION

Analyse We are asked to classify the amines located within the structure of Amodiaquin and from there, which is the most basic.

Plan Identify the amines and look for aliphatic amines as the most basic.

Solve The comparison here is between two aromatic amines and an aliphatic amine (all amine groups indicated by an arrow). The aliphatic amine (circled) is most basic.

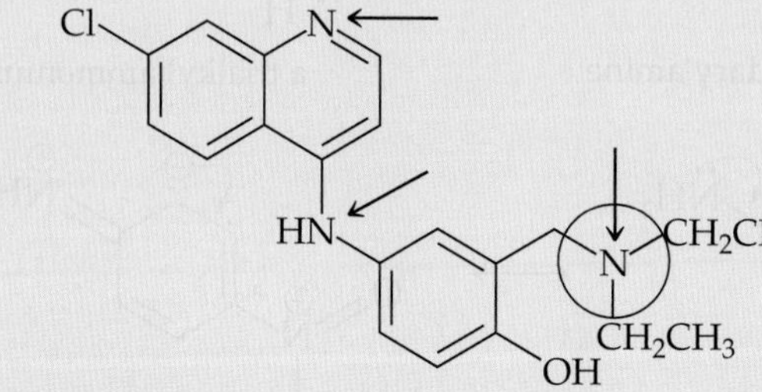

PRACTICE EXERCISE

Select the stronger base from each amine pair.

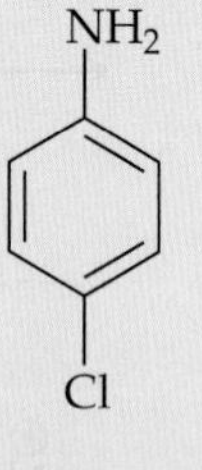

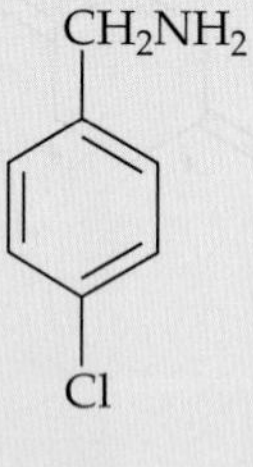

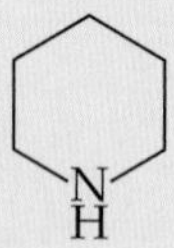

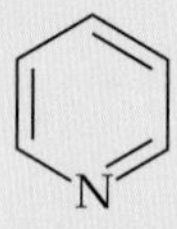

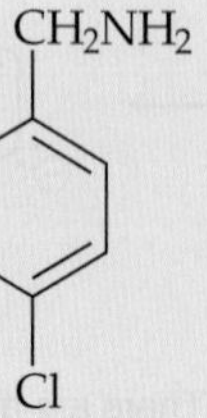

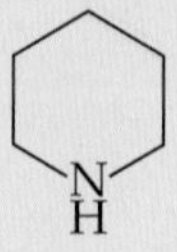

(See also Exercises 29.14, 29.15.)

Synthesis of amines

Amines can be formed in several ways, some simple and some complex. We look at three of the most common ways to form amines. Equation 29.4 illustrates the nucleophilicity of amines and their reactivity towards alkyl halides. The ammonium ion formed in the reaction is a weak acid ($pK_a = 9–11$), comparable in pK_a to phenols. Addition of an aqueous base during workup (see Equation 29.5) leads in this case to the tertiary amine. Hence, by a simple *alkylation* we have chemically transformed a secondary amine into a tertiary amine.

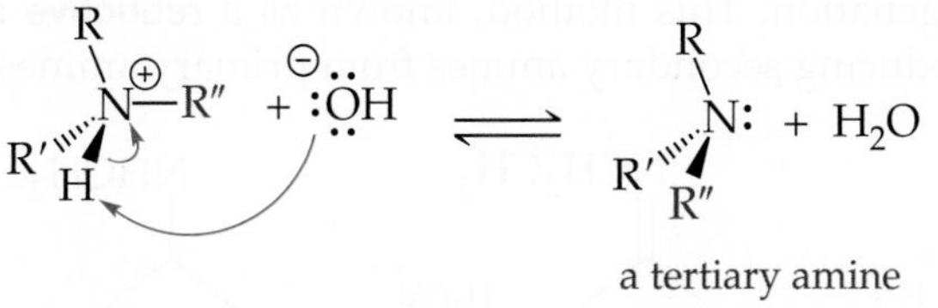

a tertiary amine

[29.5]

Aromatic amines such as 4-methoxyaniline are easily prepared by reduction of the corresponding nitro compound. This reduction is usually achieved by catalytic hydrogenation—that is, using H_2 in the presence of a metal catalyst such as Ni, Pd or Pt. Although this reaction is very general, easily accomplished and high yielding, care needs to be taken to ensure that other groups susceptible to hydrogenation, such as double bonds, are not also reduced (recall that benzene does not formally contain double bonds).

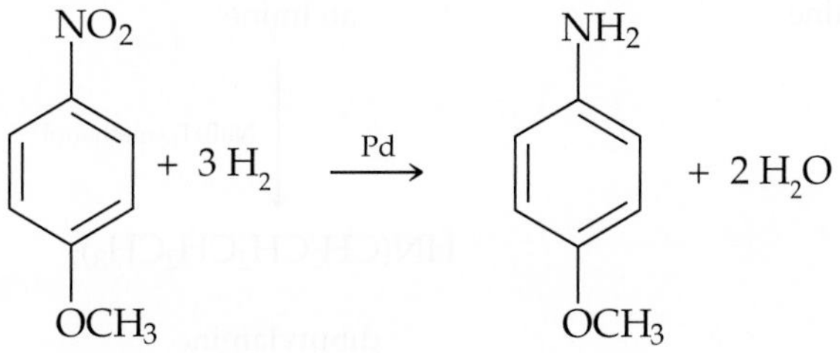

[29.6]

A metal reduction can also be used to effect the same transformation. Metals such as Fe and Zn, and metal salts such as $SnCl_2$, can be used. For example, reduction of the nitro group in Equation 29.6 by hydrogenation (H_2/Pd) would also reduce the alkene group. The use of stannous chloride as a reducing agent is more selective for the nitro group, yielding the desired product.

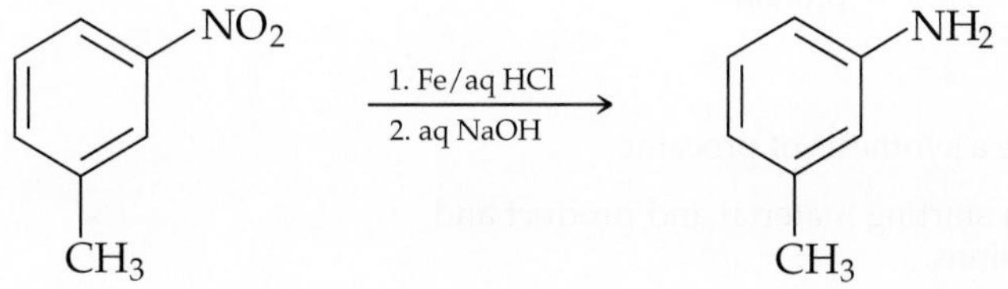

[29.7]

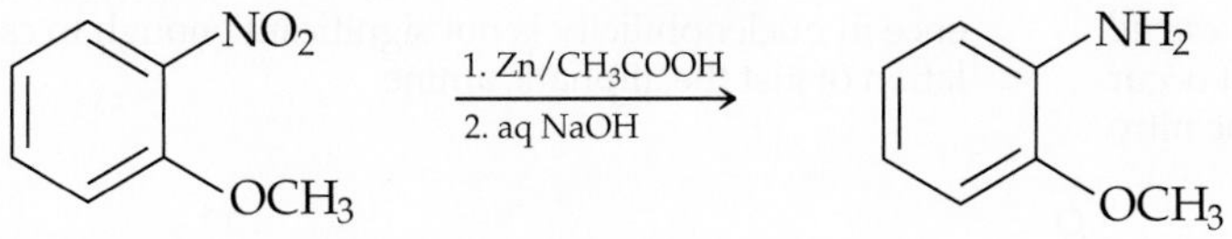

[29.8]

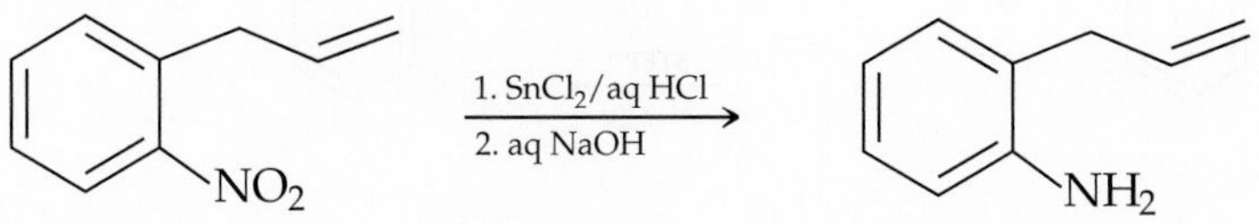

[29.9]

CONCEPT CHECK 2

Explain why it is necessary to add aqueous NaOH after reduction in Equations 29.7, 29.8 and 29.9 in order to isolate the amine.

Review this on page 1045

Recall that imines are formed by the reaction of aldehydes and ketones with primary amines (Section 26.3, 'Reactions of Aldehydes and Ketones'). The reactivity of the imine group resembles that of a carbonyl group, meaning that both functional groups are susceptible to hydride reduction and catalytic hydrogenation. This method, known as a **reductive amination**, is an easy way of producing secondary amines from primary amines.

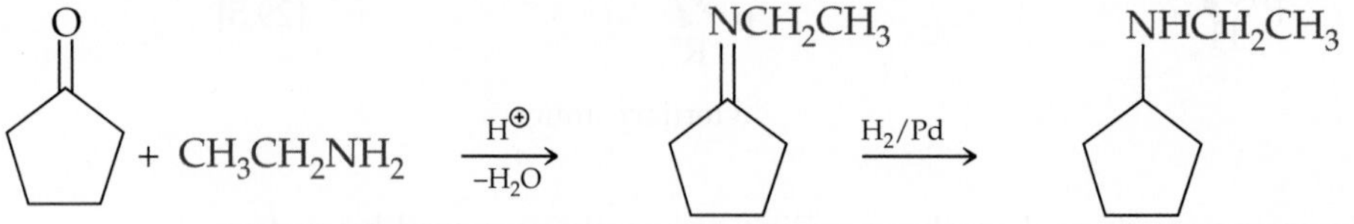

O
+ $CH_3CH_2NH_2$
$H^{\oplus}$
$-H_2O$
NCH_2CH_3
H_2/Pd
$NHCH_2CH_3$

[29.10]

cyclopentanone ethanamine *N*-ethyl-cyclopentanimine an imine *N*-ethyl-cyclopentanamine

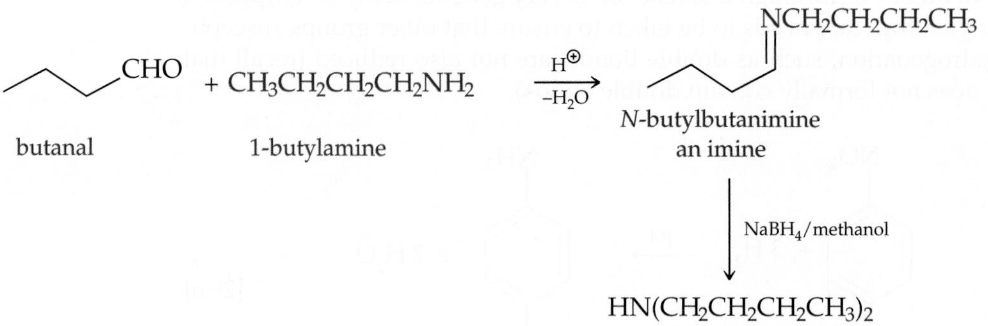

CHO
+ $CH_3CH_2CH_2CH_2NH_2$
$H^{\oplus}$
$-H_2O$
$NCH_2CH_2CH_2CH_3$
$NaBH_4$/methanol
$HN(CH_2CH_2CH_2CH_3)_2$

[29.11]

butanal 1-butylamine *N*-butylbutanimine an imine

dibutylamine

SAMPLE EXERCISE 29.4 Synthesising amines

Describe a synthesis for procaine, a local anaesthetic, from the suggested starting material. You may choose any reagents and conditions you like that will effect the transformations needed.

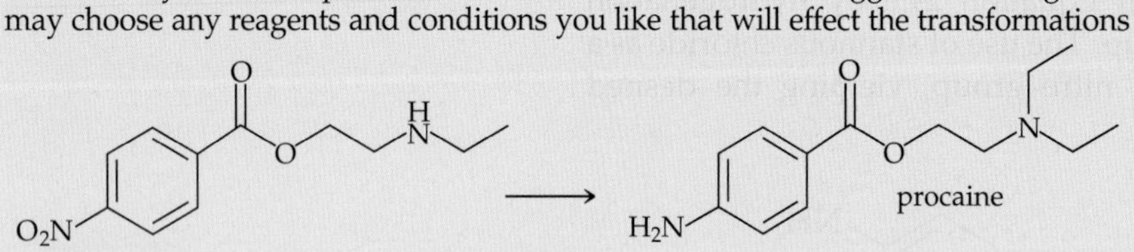

O
H
N
O
O_2N
O
O
N
H_2N
procaine

SOLUTION

Analyse We are given a compound and asked to devise a synthesis of procaine.

Plan Identify the functional group differences between starting material and product and then a plan of attack to make the necessary interconversions.

Solve In this case, the differences in functional groups can be achieved by an alkylation, introducing the ethyl group, and a reduction of the nitro group to give the amine. *Be very careful with the order of reaction.* In this example, alkylation must occur before reduction in order to yield procaine. Reducing the nitro group to the amino group, followed by alkylation, would be likely to cause a reaction to occur on both nitrogens. The difference in nucleophilicity is not significant enough to cause alkylation of just the aliphatic amine.

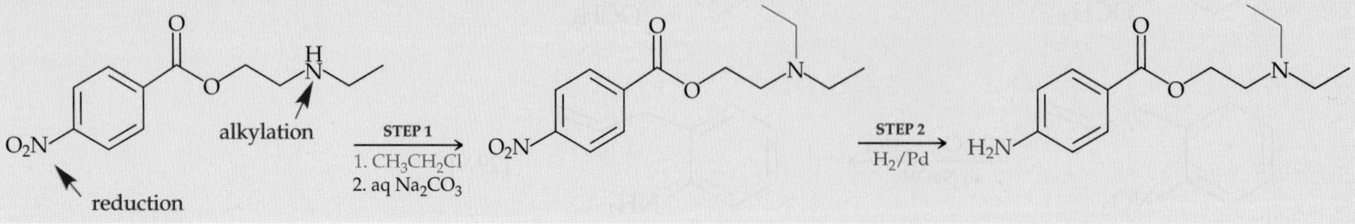

O_2N
alkylation
reduction
STEP 1
1. CH_3CH_2Cl
2. aq Na_2CO_3
O_2N
STEP 2
H_2/Pd
H_2N

PRACTICE EXERCISE

Describe a method for separating a mixture of nitrobenzene from aniline in diethyl ether solution.

Answer: The easiest way to achieve the separation is to take advantage of the basicity of aniline and the solubility of the resultant ammonium salt.

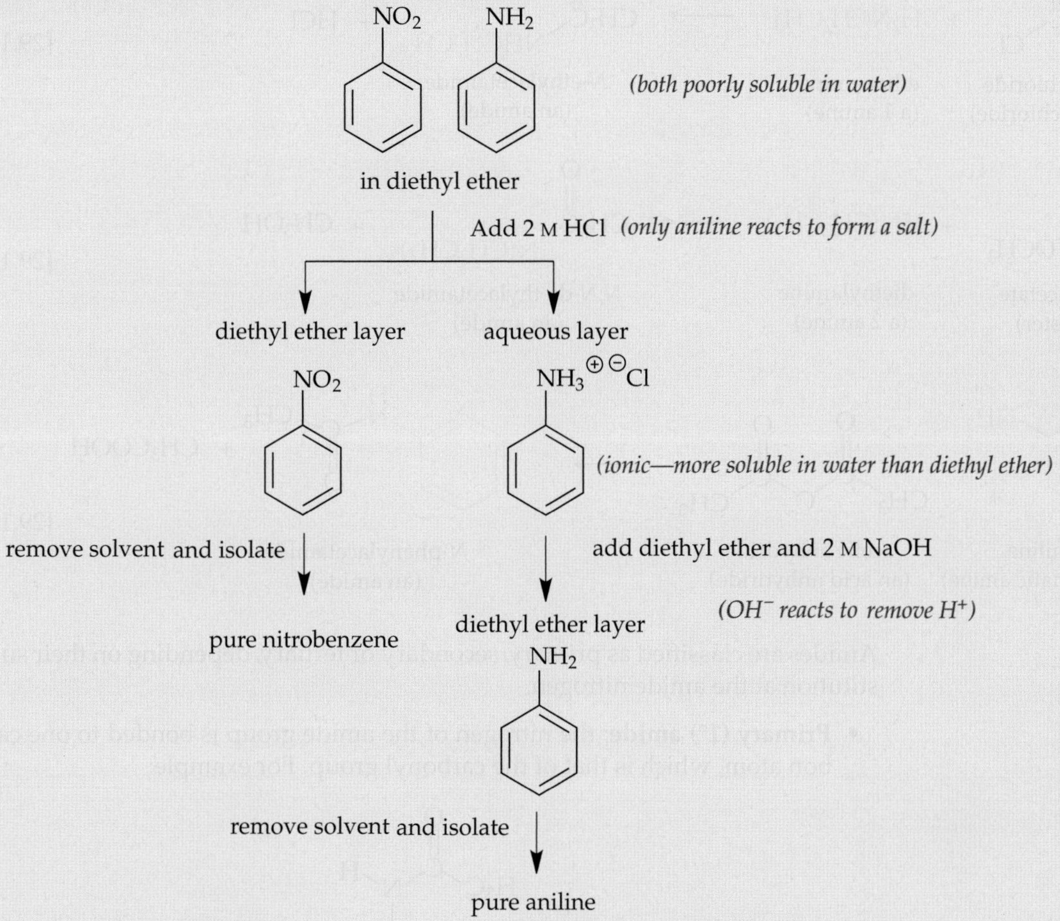

Comment This method is quite general. Any basic organic compound can be separated from a non-basic one in this way.

(See also Exercises 29.16, 29.17.)

Amides

Amides are derivatives of carboxylic acids in which the OH group of the carboxylic acid group has been replaced with a NRR′ group (R, R′ = H, alkyl, aryl), as in these examples:

ethanamide — acetamide
phenylmethanamide — benzamide

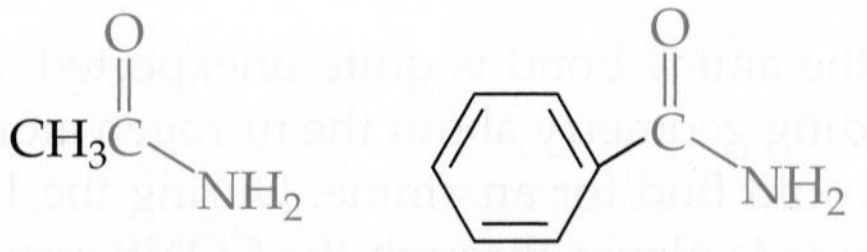

An amide can be considered as the condensation product of an amine and a carboxylic acid:

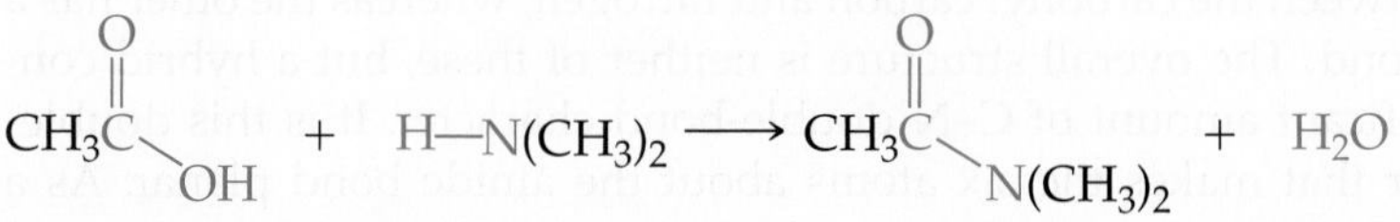

[29.12]

In Section 27.5 we saw that amides can also be formed by reaction of an acid anhydride, acid chloride or ester with ammonia, a primary or secondary amine. Examples of such reactions are:

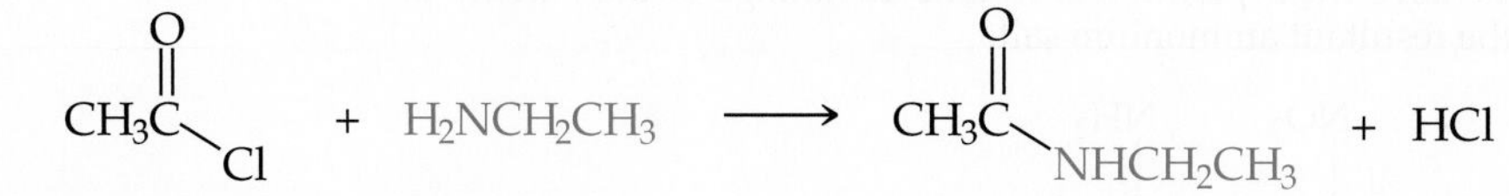

$$CH_3C\overset{O}{\underset{Cl}{\parallel}} + H_2NCH_2CH_3 \longrightarrow CH_3C\overset{O}{\underset{NHCH_2CH_3}{\parallel}} + HCl \qquad [29.13]$$

acetyl chloride (an acid chloride)	ethanamine (a 1 amine)	N-ethylacetamide (an amide)

$$CH_3C\overset{O}{\underset{OCH_3}{\parallel}} + HN(CH_2CH_3)_2 \longrightarrow CH_3C\overset{O}{\underset{N(CH_2CH_3)_2}{\parallel}} + CH_3OH \qquad [29.14]$$

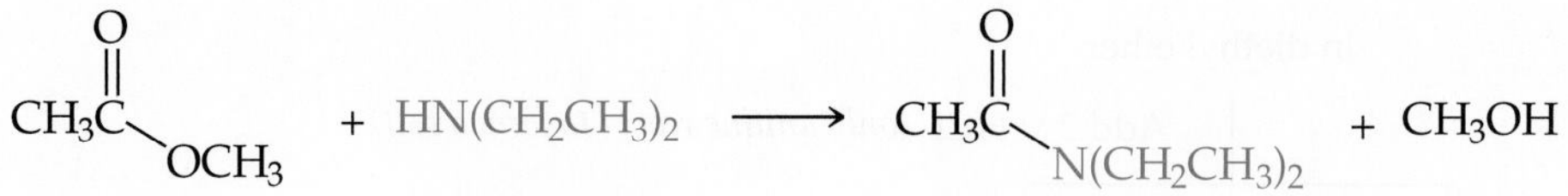

ethyl acetate (an ester)	diethylamine (a 2 amine)	N,N-diethylacetamide (an amide)

[29.15]

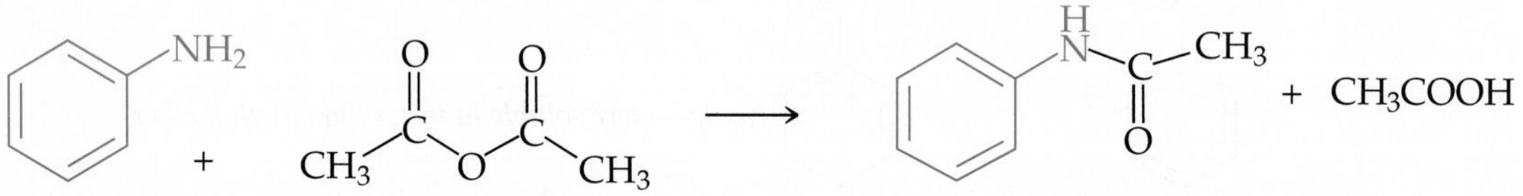

aniline (an aromatic amine)	acetic anhydride (an acid anhydride)	N-phenylacetamide (an amide)

Amides are classified as primary, secondary or tertiary, depending on their substitution at the amide nitrogen.

- **Primary (1°) amide**: the nitrogen of the amide group is bonded to one carbon atom, which is that of the carbonyl group. For example:

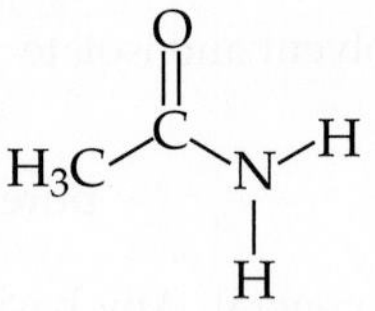

acetamide
1° amide

- **Secondary (2°) amide**: the nitrogen of the amide group is bonded to two carbon atoms, one of which is a carbonyl group; for example N-phenylacetamide.
- **Tertiary (3°) amide**: the nitrogen of the amide group is bonded to three carbon atoms, one of which is a carbonyl group. The other two carbon atoms usually belong to alkyl or aryl groups; for example N,N-diethylacetamide.

The geometry of the amide bond is quite unexpected. At first glance, you would expect the bonding geometry about the nitrogen atom to be tetrahedral, similar to what you would find for an amine. During the 1930s Linus Pauling discovered that an amide is planar through the CONR group (▶ FIGURE 29.9), commensurate with the observed bond angles of approximately 120° about the amide nitrogen.

Pauling rationalised the unusual geometry of the amide bond by describing two resonance hybrid structures (▶ FIGURE 29.10(a)). One structure has a single bond between the carbonyl carbon and nitrogen, whereas the other has a C–N double bond. The overall structure is neither of these, but a hybrid containing a significant amount of C–N double-bond character. It is this double-bond character that makes the six atoms about the amide bond planar. As a

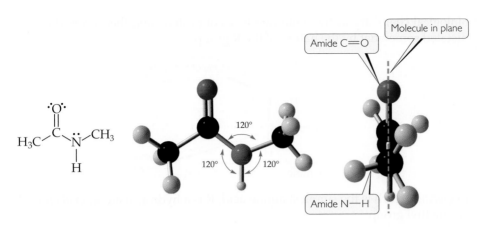

◀ **FIGURE 29.9 The geometry of the amide bond.** Bond angles of approximately 120° are seen about the amide nitrogen. This trigonal-planar geometry is coplanar with the amide carbonyl group.

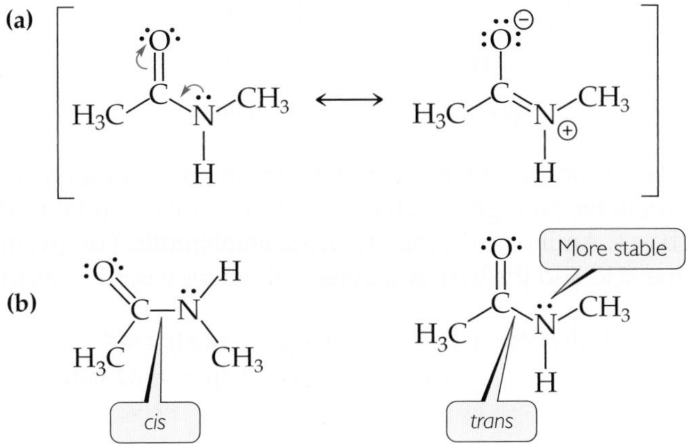

◀ **FIGURE 29.10 Conformation and rationalisation.** (a) The planar nature of the amide bond can be rationalised by resonance delocalisation. (b) Amides can exist as *cis* or *trans* conformers known as rotamers because there is limited rotation about the C–N bond. The *trans* conformer is favoured because it reduces any steric strain caused by substituents.

result of the restricted rotation about the C–N bond, two conformations are possible for the amide bond, *cis* and *trans* (Figure 29.10(b)). The *trans* isomer is favoured because it reduces any steric strain imposed by substituents on either side of the amide bond.

Cyclic amides are called **lactams**. We do not deal with these in any detail except to say that a very potent form of antibiotic, the **penicillins**, all contain a four-membered lactam ring, known as a β-lactam, fused to a five-membered ring containing sulfur (▼ **FIGURE 29.11**).

29.2 | AMINO ACIDS

Amino acids contain both an amino group and a carboxyl group. An **α-amino acid** (see below) is an amino acid in which the amino group is bonded to the carbon adjacent to the carboxyl group. α-Amino acids play a large role in

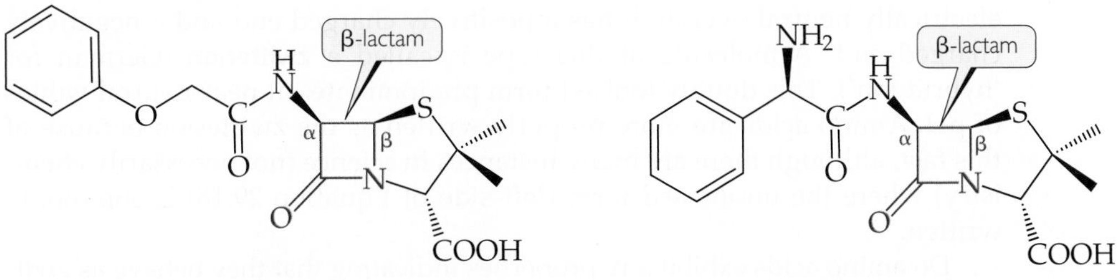

penicillin V

amoxicillin

◀ **FIGURE 29.11 Structure of penicillins.** Penicillin V and amoxicillin are two examples of lactams.

biological chemistry as the building blocks of protein, and their properties are dictated somewhat by the nature of the R group.

In *glycine*, which is the simplest amino acid, R is a hydrogen atom; in *alanine*, R is a methyl group.

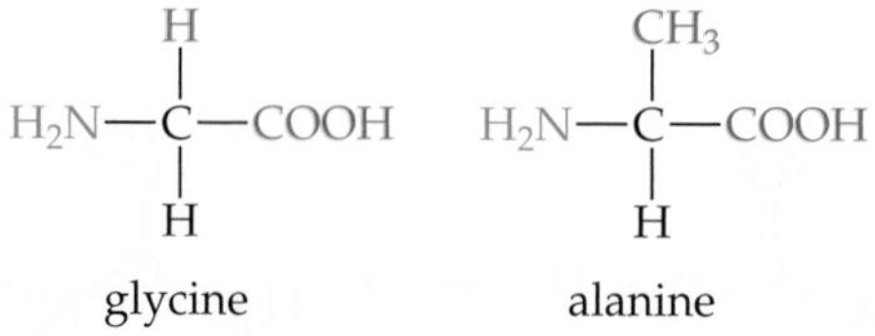

glycine   alanine

Because amino acids contain a carboxyl group, they can serve as weak acids. They also contain the NH_2 group, characteristic of amines, and thus they can also act as weak bases. Amino acids, therefore, are **amphiprotic**. For glycine, we might expect that the acid and the base reactions with water would be as follows:

$$Acid: H_2N-CH_2-COOH(aq) + H_2O(l) \rightleftharpoons$$
$$H_2N-CH_2-COO^-(aq) + H_3O^+(aq) \qquad [29.16]$$

$$Base: H_2N-CH_2-COOH(aq) + H_2O(l) \rightleftharpoons$$
$$^+H_3N-CH_2-COOH(aq) + OH^-(aq) \qquad [29.17]$$

The pH of a solution of glycine in water is about 6.0, indicating that it is a slightly stronger acid than a base.

The acid–base chemistry of amino acids is somewhat more complicated than shown in Equations 29.16 and 29.17, however. As a result of the fact that the COOH group can act as an acid and the NH_2 group can act as a base, amino acids undergo a 'self-contained' Brønsted–Lowry acid–base reaction in which the proton of the carboxyl group is transferred to the basic nitrogen atom:

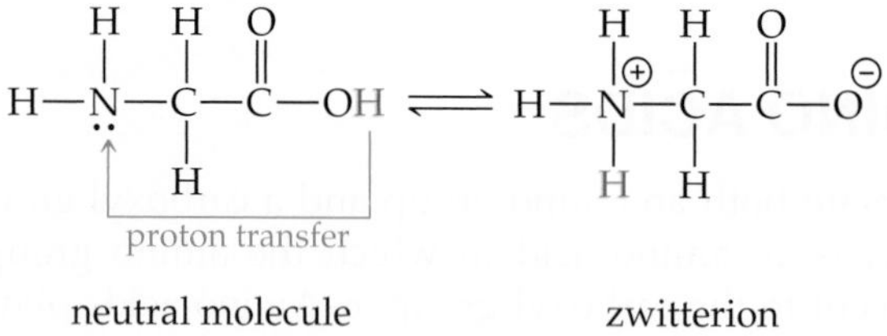

proton transfer

neutral molecule   zwitterion   [29.18]

Although the form of the amino acid on the right side of Equation 29.18 is electrically neutral overall, it has a positively charged end and a negatively charged end. A molecule of this type is called a **zwitterion** (German for 'hybrid ion'). This doubly ionised form predominates at near-neutral values of pH. Amino acids are more properly written as the zwitterion because of this fact, although there are many instances in science (not necessarily chemistry) where the un-ionised form (left side of Equation 29.18) is commonly written.

Do amino acids exhibit any properties indicating that they behave as zwitterions? If so, they should behave in a similar manner to ionic substances

(∞ Section 8.2, 'Ionic Bonding'). Crystalline amino acids (▶ FIGURE 29.12) have relatively high melting points, usually above 200 °C, which is characteristic of ionic solids. Amino acids are far more soluble in water than in non-polar solvents. In addition, the dipole moments of amino acids are large, consistent with a large separation of charge in the molecule. Thus the ability of amino acids to act simultaneously as acids and bases has important effects on their properties.

∞ Review this on page 253

With the exception of glycine, all naturally occurring amino acids have at least one stereocentre, the α-carbon. ▼ FIGURE 29.13 shows the two enantiomers of the amino acid alanine. For historical reasons, the two enantiomers are distinguished by the labels D and L (∞ Section 23.3, 'Chirality in Organic Compounds'). All amino acids normally found in proteins have the L configuration at the stereocentre (except for glycine, which is not chiral). In terms of absolute stereochemistry, all naturally occurring α-amino acids have the S configuration (Section 23.3). This means that the enantiomer or D-amino acid has R configuration.

∞ Review this on page 931

▼ FIGURE 29.14 shows the structural formulae of the 20 most common α-amino acids found in proteins. Our bodies can synthesise 10 of these α-amino acids in sufficient amounts for our needs. The other 10 must be ingested and are called *essential amino acids* because they are necessary components of our diet.

α-Amino acids, which make up the 20 most common naturally occurring amino acids, can be classified by the nature of their **side chain**, which is also bonded to the amino acid's α-carbon. These side chains are used to classify amino acids as polar and non-polar. The polar amino acids can be further classified as ionised (for example, containing acidic and basic side chains) and un-ionised (that is, neutral-polar side chains). Proline is a cyclic amino acid, which differs from the acyclic nature of the other 19 amino acids.

Although all amino acids are chiral, with the exception of glycine, isoleucine and threonine also contain a second stereocentre. Four stereoisomers are possible, but only *one* of the four stereoisomers is found in proteins.

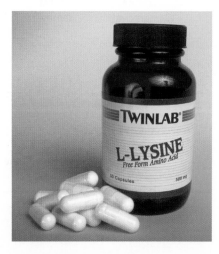

▲ **FIGURE 29.12   Lysine.** One of the amino acids found in proteins, lysine is available as a dietary supplement. The L on the label refers to a specific configuration of atoms that is found in naturally occurring amino acids.

CONCEPT CHECK 3

How many of the amino acids in Figure 29.14 contain
a. aromatic side groups,
b. sulfur, and
c. hydrocarbon side chains?

FIGURE IT OUT

Are all α-amino acids chiral?

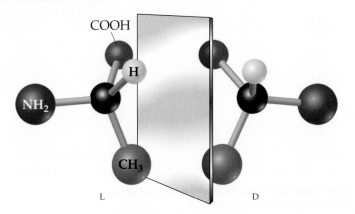

◀ **FIGURE 29.13   Alanine enantiomers.** The middle carbon of alanine is stereogenic, and therefore there are two enantiomers, which by definition are non-superimposable mirror images of each other.

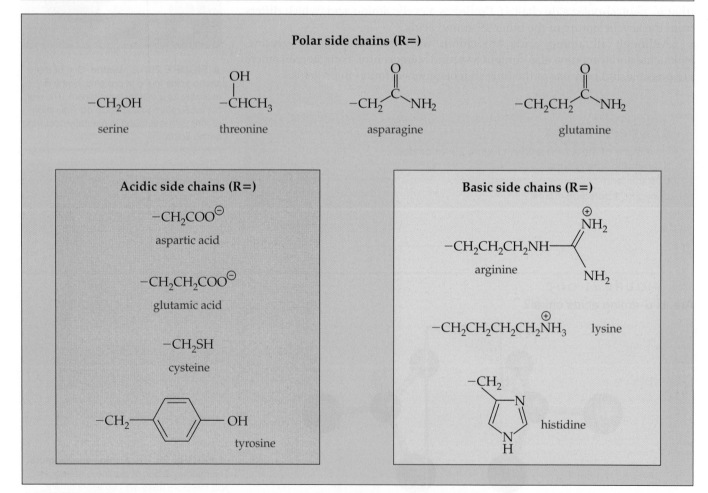

▲ **FIGURE 29.14** **The 20 common α-amino acids found in the human body.** Each ionisable group within the acidic and basic classes is shown in the form present in highest concentration at pH = 7.0.

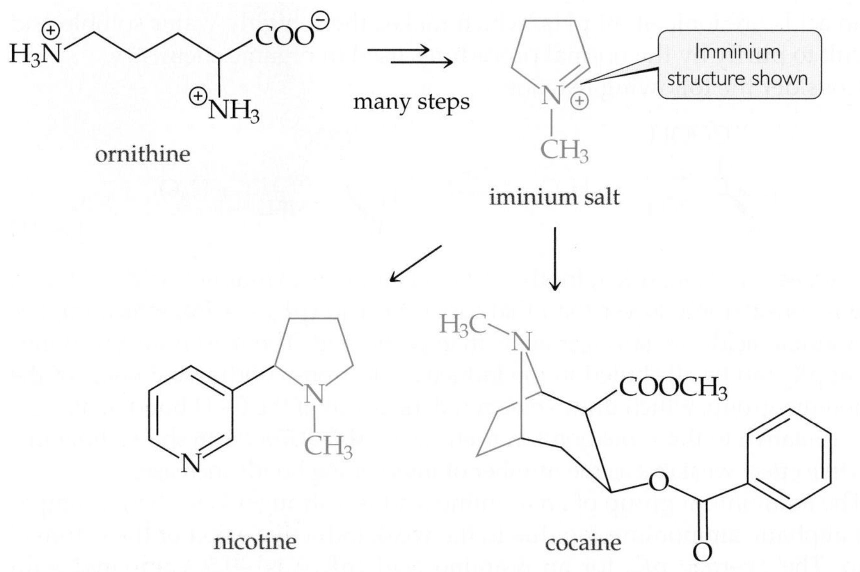

◀ **FIGURE 29.15   Ornithine.** This amino acid is a relative of lysine but with a shorter side chain.

Two amino acids not listed in Figure 29.14 deserve special mention because of their social and medicinal importance. Ornithine is an amino acid found in the liver, with importance in the urea cycle (▲ **FIGURE 29.15**). It is also a precursor for the biosynthesis of nicotine and tropane alkaloids such as cocaine. GABA (γ-aminobutyric acid) is a neurotransmitter (a substance that transmits nerve impulses across a synapse), found exclusively in the brain and central nervous system (CNS). It has been implicated in several neurodegenerative conditions, including memory loss. GABA is formed within the CNS by the enzyme-catalysed decarboxylation of glutamic acid. Its systematic name, 4-aminobutanoic acid, is less commonly used than its acronym, GABA.

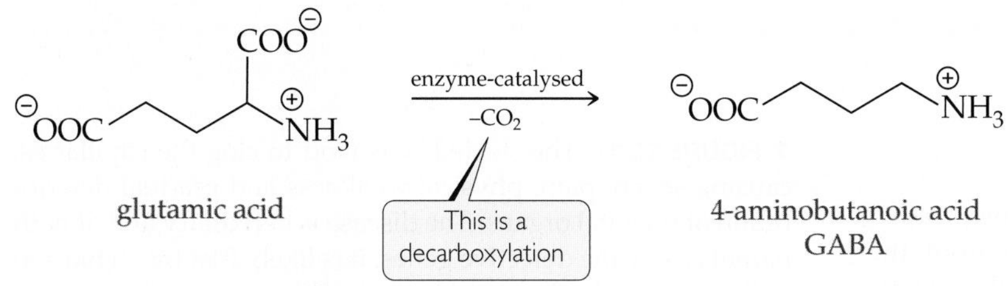

[29.19]

Acid–Base Properties

The amphiprotic nature of α-amino acids can be described by the following equation:

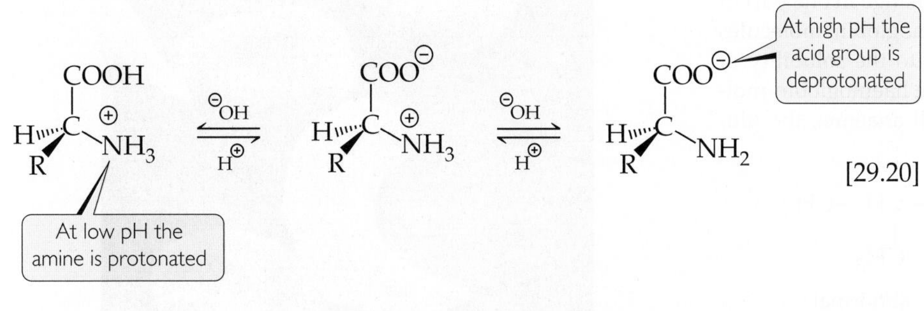

[29.20]

At low pH the basic amino group is protonated, forming the ammonium ion. In neutral solution the zwitterion is formed. The pH at which the zwitterion concentration is the highest is called the **isoelectric point**. In basic solution, the carboxylic acid group is ionised to form the carboxylate group. Notice that

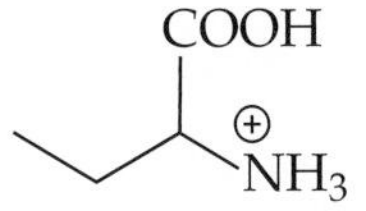

COOH

α-amino acid
pK_a 2.3

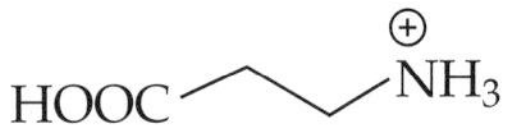

$HOOC$‒‒‒$\overset{\oplus}{N}H_3$

β-amino acid
pK_a ~ 3.6

$HOOC$‒‒‒‒$\overset{\oplus}{N}H_3$

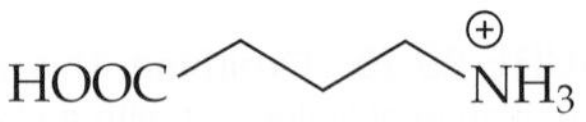

γ-amino acid
pK_a ~ 4.2

▲ **FIGURE 29.16 Carboxyl group acidity.** The inductive effect of the ammonium ion changes the pK_a of the COOH of amino acids. The effect of the ammonium group is connectivity dependent and approaches that of a normal carboxylic acid as the number of intervening bonds increase.

amino acids are ionic at all pHs, which makes them highly water soluble and difficult to purify by the normal procedures used in organic chemistry.

Consider the following reaction:

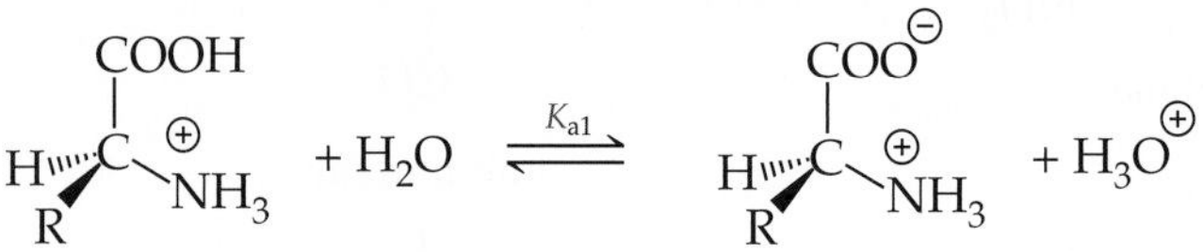

$$H\text{w}\overset{COOH}{\underset{R}{C}}\overset{\oplus}{NH_3} + H_2O \xrightarrow{K_{a1}} H\text{w}\overset{COO^\ominus}{\underset{R}{C}}\overset{\oplus}{NH_3} + H_3\overset{\oplus}{O} \qquad [29.21]$$

The average pK_a value (pK_{a1}) for the carboxyl group of an α-amino acid is ~2. This value is considerably lower than that for acetic acid ($pK_a = 4.76$), which implies that α-amino acids are stronger acids than acetic acid. The reason for this difference in pK_a can be attributed to the inductive electron-withdrawing effect of the ammonium group, which draws electron density out of the O–H bond, making it weaker relative to the same bond in acetic acid. ◀ **FIGURE 29.16** shows how this inductive effect weakens as the number of intervening bonds increases.

The ammonium group of an α-amino acid is a stronger acid than a comparable aliphatic ammonium ion due to the weak inductive effect of the carboxyl group. The average pK_a for an α-amino acid (pK_{a2}) is ~9.5, compared with pK_a ~10.5 for an aliphatic amine. This implies that the amine of an α-amino acid is generally a weaker base than a 1° aliphatic amine.

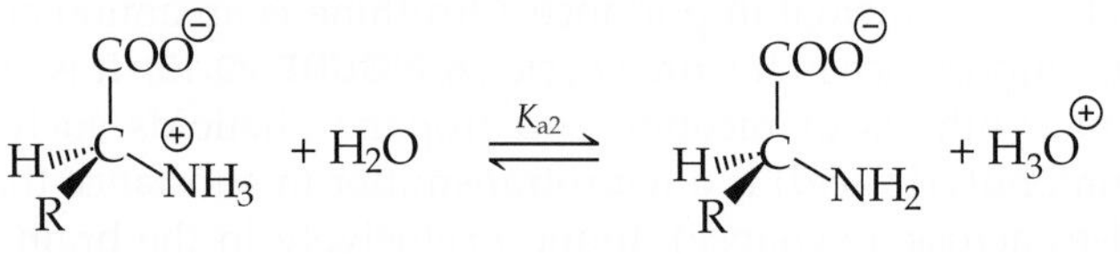

$$H\text{w}\overset{COO^\ominus}{\underset{R}{C}}\overset{\oplus}{NH_3} + H_2O \xrightarrow{K_{a2}} H\text{w}\overset{COO^\ominus}{\underset{R}{C}}NH_2 + H_3\overset{\oplus}{O} \qquad [29.22]$$

Let us now investigate what the pK_a value of an α-amino acid means and what information it provides. To do this, we need to consider the pH titration of

A CLOSER LOOK

SICKLE-CELL ANAEMIA

Our blood contains a complex protein called haemoglobin, which carries oxygen from the lungs to other parts of the body. In the genetic disease known as sickle-cell anaemia, haemoglobin molecules are abnormal and have a lower solubility, especially in their unoxygenated form. Consequently, as much as 85% of the haemoglobin in red blood cells crystallises from solution.

The reason for the insolubility of haemoglobin in sickle-cell anaemia can be traced to a structural change in one part of an amino acid side chain. Normal haemoglobin molecules contain glutamic acid, which contributes to the solubility of the haemoglobin molecule in water. In the haemoglobin molecules of people suffering from sickle-cell anaemia, the glutamic acid residue is replaced by valine:

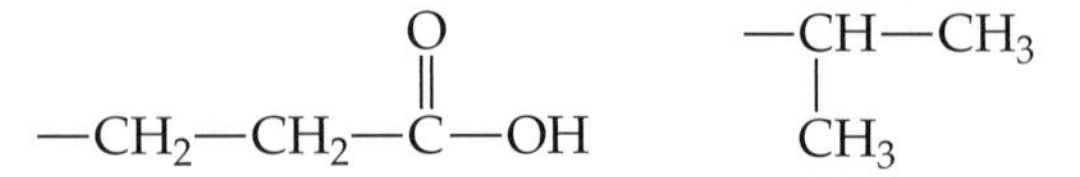

$$-CH_2-CH_2-\overset{\overset{O}{\|}}{C}-OH$$
normal

$$-CH-CH_3$$
$$\underset{CH_3}{|}$$
abnormal

Valine is non-polar (hydrophobic), and its presence leads to the aggregation of the defective form of haemoglobin into particles too large to remain suspended in biological fluids. It also causes the cells to distort into a sickle shape, as shown in

▼ FIGURE 29.17. The sickled cells tend to clog the capillaries, causing severe pain, physical weakness and gradual deterioration of the vital organs. The disease is hereditary and, if both parents carry the defective genes, it is likely that their children will possess only abnormal haemoglobin.

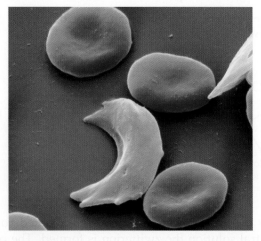

▲ **FIGURE 29.17 Normal and sickled red blood cells.** Normal red blood cells are about 1 μm in diameter.

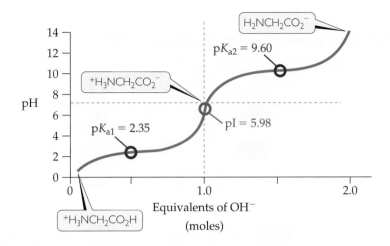

◀ **FIGURE 29.18** **Titration of glycine with aqueous NaOH solution.** pK_{a1} relates to the pK_a of the carboxylic acid, while pK_{a2} relates to the pK_a of the ammonium ion.

a simple amino acid, such as glycine, with aqueous sodium hydroxide solution (▲ **FIGURE 29.18**).

At low pH, glycine is in its fully protonated form, $^{+}H_3NCH_2CO_2H$. The addition of OH^- sees a gradual rise in pH until a plateau is reached (also described as a *point of inflexion* on the titration curve). At this point, which corresponds to the addition of 0.5 mol of OH^-, the pH = pK_{a1}. This is the point at which the concentration of the zwitterion equals that of the ammonium ion:

$$\text{when pH} = pK_{a1}, \quad [^{+}H_3NCH_2CO_2H] = [^{+}H_3NCH_2CO_2^{-}]$$

A pH value *lower* than pK_{a1} means that the ammonium ion predominates. A pH value *higher* than pK_{a1} means the zwitterion predominates in solution.

The endpoint of this first reaction is reached when 1 mole equivalent of OH^- is added. This point, known as the *isoelectric point*, is where the zwitterion $^{+}H_3NCH_2CO_2^{-}$ is at its maximum concentration. At the isoelectric point, pI, the amino acid has no net charge. The pI of an amino acid usually lies at the midpoint of pK_{a1} and pK_{a2}. We can represent this mathematically by:

$$pI = \tfrac{1}{2}(pK_{a1} + pK_{a2}) \qquad [29.23]$$

for glycine, $\qquad pI = \tfrac{1}{2}(2.35 + 9.60) = 5.98$

The addition of another 0.5 molar equivalents of OH^- causes the pH to rise, then plateau. This point of inflexion indicates pK_{a2}, the pK_a value associated with the ammonium group. This is the point at which the concentration of the zwitterion equals that of the carboxylate ion:

$$\text{when pH} = pK_{a2}, \quad [^{+}H_3NCH_2CO_2^{-}] = [H_2NCH_2CO_2^{-}]$$

A value of pH *lower* than pK_{a2} means that the zwitterion ion predominates. A value for the pH *higher* than pK_{a2} means that the amine and carboxylate groups predominate in solution.

The net charge of an amino acid is easily estimated using the pI value. As pH values fall below pI, the amino acid exists more and more as the ammonium ion, which is positively charged. At pH values higher than the pI, the amino acid exists as the carboxylate ion, which is negatively charged.

Knowledge about the pI of an amino acid is also useful for purification by **electrophoresis**, a process used to separate molecules based on charge (▶ **FIGURE 29.19**). This useful technique uses either paper or gel as a separating medium and electric charge to separate amino acids, proteins and other charged molecules (▼ **FIGURE 29.20**). You will learn more about this technique in courses devoted to biochemistry and cellular biology, where the separation of amino acids (and proteins) by this technique is a fundamental tool in the laboratory.

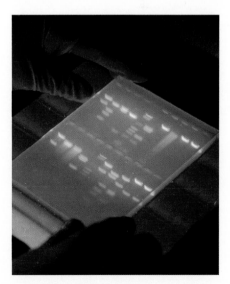

▲ **FIGURE 29.19** **A typical electrophoresis result.** The bands highlighted under UV light represent different amino acids, peptides or proteins that have been separated in a gel matrix. Each species can be isolated and then identified.

◢ **FIGURE IT OUT**

How would tyrosine, pI = 5.66, act at pH 5.66 in electrophoresis? Would this change at pH 9?

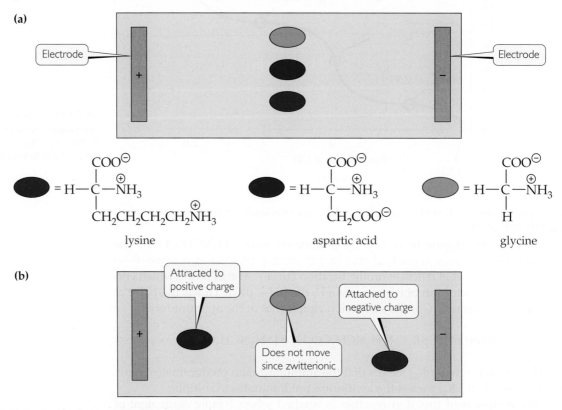

▲ **FIGURE 29.20 Separation of amino acids by electrophoresis.** By running an electrophoresis experiment at pH 5.98, we can separate three amino acids, based on their pI values. At pH 5.98, glycine has no net overall charge because of its zwitterionic state. Aspartic acid (pI 2.77) is deprotonated at pH 5.98 and is attracted to the positive electrode. Lysine (pI 9.74) is protonated at pH 5.98 and is attracted to the negative electrode. (a) At time = 0, (b) at time > 0.

SAMPLE EXERCISE 29.5 **Acid–base properties of amino acids**

Draw structural formulae for the predominant form of glycine (pK_a 2.35, 9.60) in aqueous solution at pH 1, 6 and 12.

SOLUTION

Analyse We are asked to draw the form of an amino acid at a given pH knowing the pK_a.

Plan When working through a problem like this, it is important to identify the relationship between the pH and the closest pK_a value.

Solve At pH 1, glycine is predominantly in the fully protonated form:

$$^+H_3NCH_2CO_2H$$

The reason is that pH < pK_{a1}.
   pH 6 is between the two given pK_a values, so the zwitterion is the major form:

$$^+H_3NCH_2CO_2{}^-$$

At pH 12, the carboxylate form predominates:

$$H_2NCH_2CO_2{}^-$$

The reason is that pH > pK_{a2}.
   To verify the answer to this problem, you could use Figure 29.18.

PRACTICE EXERCISE

Draw structural formulae for the predominant forms of serine (pK_a 2.21, 9.15) in aqueous solution at pH 2, 7 and 10.

Answer: At pH 2 $^+H_3NCHRCO_2H$

At pH 7 $^+H_3NCHRCO_2^-$

At pH 10 $H_2NCHRCO_2^-$

R = CH_2OH

(See also Exercises 29.35, 29.37.)

We can predict which species predominates at a certain pH in a much more quantitative way. For example, the equation for the acid-dissociation constant K_{a1}, shown in Equation 29.21, can be written as follows:

$$K_{a1} = \frac{[H_3O^{\oplus}][R'-CO_2^{\ominus}]}{[R'-COOH]}$$ [29.24]

where R' = $RCHNH_3^{\oplus}$

Rearrangement gives

$$\frac{K_{a1}}{[H_3O^{\oplus}]} = \frac{[R'-CO_2^{\ominus}]}{[R'-COOH]}$$ [29.25]

Knowing the value of pK_{a1} and the pH of the solution (which determines the value of $[H_3O^+]$), we can calculate the exact ratio of the fully protonated and zwitterion forms at values between pH = 0 and the pI, as well as the net charge on the α-carboxyl group at that pH.

In a similar way, the equation for the acid-dissociation constant K_{a2}, shown in Equation 29.22, can be written as follows:

$$K_{a2} = \frac{[H_3O^{\oplus}][R'-NH_2]}{[R'-NH_3^{\oplus}]}$$ [29.26]

where R' = $RCHCO_2^{\ominus}$

Rearrangement gives

$$\frac{K_{a2}}{[H_3O^{\oplus}]} = \frac{[R'-NH_2]}{[R'-NH_3^{\oplus}]}$$ [29.27]

Knowing the value of pK_{a2} and the pH of the solution, we can calculate the exact ratio of the amine and zwitterion forms at values between pI and pH = 14, as well as the net charge on the α-amino group at that pH.

These calculations are similar to those used for buffer solutions using the Henderson–Hasselbalch equation for weak acid/conjugate base pairs (∞ Section 18.2, 'Buffer Solutions'):

∞ Review this on page 713

$$pH = pK_a + \log\frac{[\text{conjugate base}]}{[\text{weak acid}]}$$ [29.28]

SAMPLE EXERCISE 29.6 **Estimating net charge**

Let's reconsider the question posed in Practice Exercise 29.5. Estimate the net charge on serine at pH = 2.

SOLUTION

Analyse We are asked to determine the percentage of each species in solution to determine the net charge at a given pH.

Plan Use the Henderson–Hasselbalch equation to determine the ratio of acid to conjugate base, hence the percentage of each in solution.

Solve Using Equation 29.28 and substituting in values:

$$pH = 2 \text{ and } pK_a = 2.21$$

gives

$$2 = 2.21 + \log \frac{[R'CO_2^{\ominus}]}{[R'CO_2H]}$$

$$\frac{[R'CO_2^{\ominus}]}{[R'CO_2H]} = 0.62$$

This ratio concludes that the amino acid exists as 38% $R'CO_2^-$ and can be calculated using:

$$\frac{x}{100-x} = 0.62$$

where x = the percentage of $R'CO_2^-$.

A similar calculation for the amino group gives

$$2 = 9.15 + \log \frac{[R'NH_2]}{[R'NH_3^{\oplus}]}$$

$$\frac{[R'NH_2]}{[R'NH_3^{\oplus}]} = 7.1 \times 10^{-8}$$

which indicates that the amino group is > 99.9% protonated at pH 3.

The average charge of the amino acid in solution can be calculated from the contributions of each charged species at pH 2. That is, at pH 2 the following two species exist:

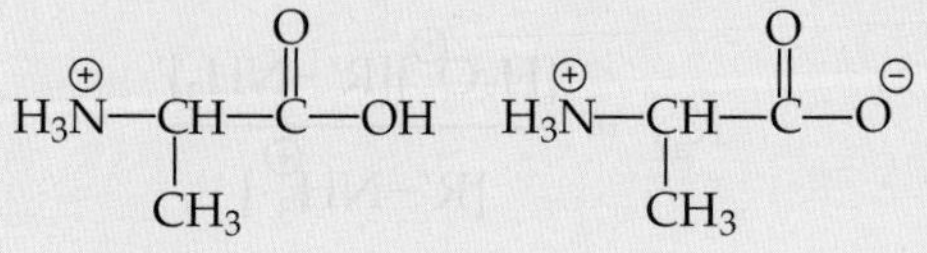

62%   38%

Their respective charges are +1 and 0.

Therefore, based on their relative proportions in solution,

$$\text{Average charge} = 0.62 \times 1.0 + 0.38 \times 0 = +0.62$$

PRACTICE EXERCISE

Estimate the net charge on serine at pH = 7 and pH = 10.

Answer: At pH 7 net charge = 0, at pH 10 net charge = −0.88

(See also Exercises 29.38, 29.39.)

Reactions Involving Amino Acids

There are several biological pathways for synthesising amino acids. For example, the bacterium *Bacillus subtilis* synthesises alanine from pyruvic acid in the following way:

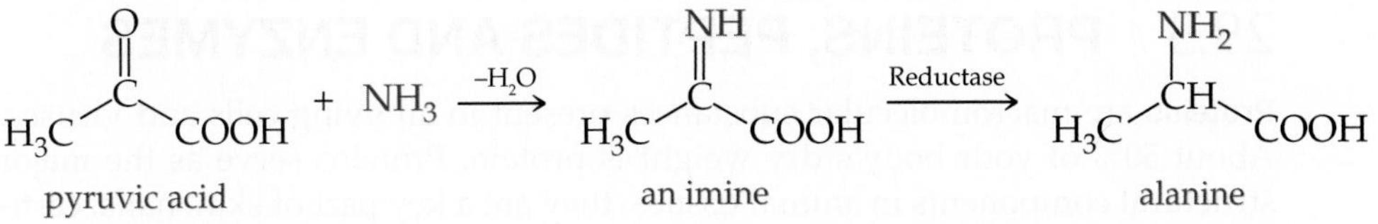

pyruvic acid   an imine   alanine   [29.29]

In a laboratory, there are two main ways of synthesising α-amino acids. The first method involves nucleophilic substitution of an α-halo carboxylic acid with ammonia:

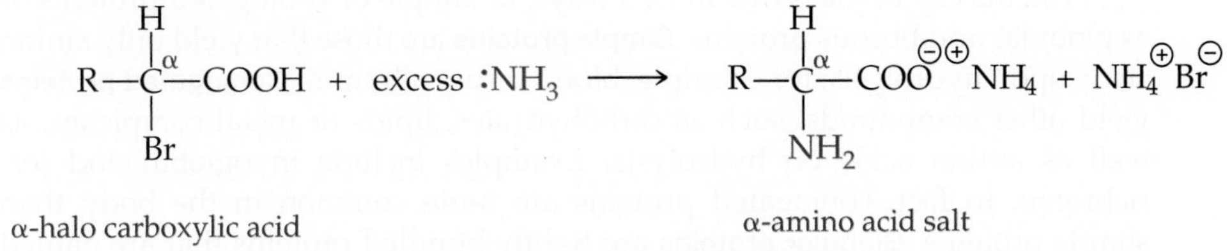

α-halo carboxylic acid   α-amino acid salt   [29.30]

The second method is called **Strecker synthesis** of α-amino acids. This three-step reaction converts aldehydes to α-amino acids in the presence of cyanide ion.

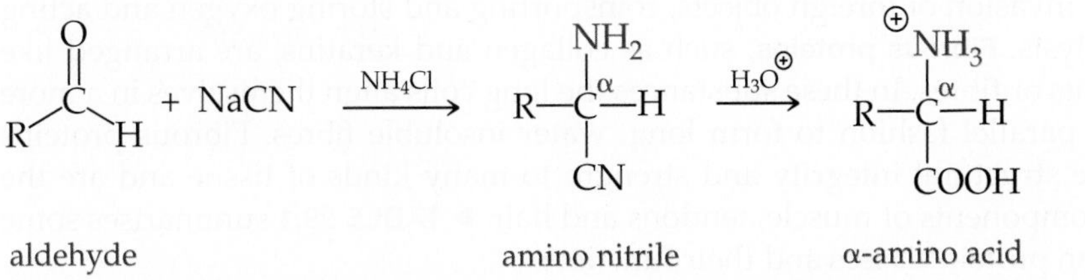

aldehyde   amino nitrile   α-amino acid   [29.31]

Treatment of the aldehyde with cyanide ion in the presence of ammonium chloride yields the intermediate amino nitrile. This reaction is twofold: reaction of the aldehyde with ammonium chloride yields an imine, which undergoes addition of the cyanide ion to form the amino nitrile; the nitrile group is then hydrolysed to the carboxylic acid in the presence of aqueous acid.

The methods described in Equations 29.30 and 29.31 are not ideal because the products are typically formed as racemates, which need to be separated to form the enantiomerically pure compounds. In some instances, separation is easily achieved by chiral resolution techniques. In other cases, more complicated and expensive techniques are required.

Since most amino acids are neither coloured nor fluorescent, their detection requires the use of a dye. The most common dye used, called ninhydrin, reacts specifically with α-amino acids to produce a purple-coloured anion, an aldehyde, CO_2 and water.

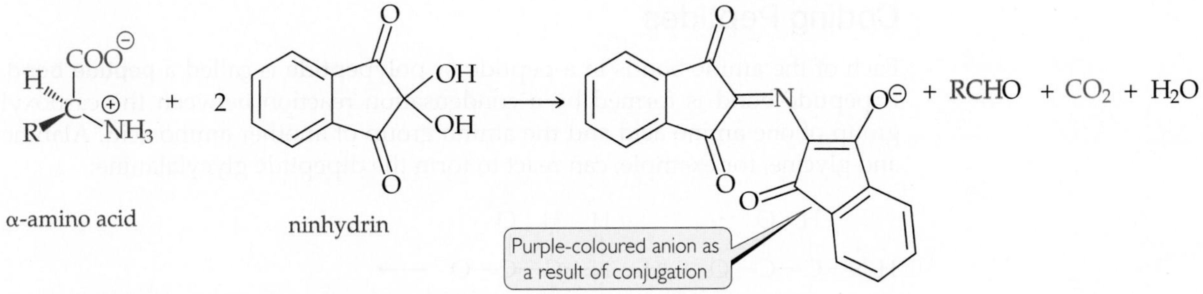

α-amino acid   ninhydrin   Purple-coloured anion as a result of conjugation   + RCHO + CO_2 + H_2O [29.32]

The fact that a purple-coloured solution is formed from colourless starting materials is important for the use of ninhydrin in quantifying the amount of α-amino acid present in a sample using absorption spectroscopy at 580 nm (∞ Section 30.1, 'The Electromagnetic Spectrum'). Note that only α-amino acids containing a primary amino group react to form the purple-coloured anion. Proline, which has a secondary amino group, does not react in the same way.

∞ Find out more on page 1200

29.3 | PROTEINS, PEPTIDES AND ENZYMES

Proteins are macromolecular substances present in all living cells and viruses. About 50% of your body's dry weight is protein. Proteins serve as the major structural components in animal tissues; they are a key part of skin, nails, cartilage and muscles. Other proteins catalyse reactions, transport oxygen, serve as hormones to regulate specific body processes and perform other tasks. Whatever their function, all proteins are chemically similar, being composed of the same basic building blocks—amino acids.

Proteins can be classified in two ways, as simple or conjugated proteins or as globular and fibrous proteins. **Simple proteins** are those that yield only amino acids upon hydrolysis; for example, blood serum albumin. **Conjugated proteins** yield other compounds, such as carbohydrates, lipids or metal complexes, as well as amino acids on hydrolysis. Examples include myoglobin and ferrichrome. In fact, conjugated proteins are more common in the body than simple proteins. **Globular proteins** are tightly bundled proteins that are named because of their globular or spherical shape. These proteins, which include albumins, enzymes, immunoglobulins and insulin, are water soluble and typically mobile within cells. They have non-structural functions, such as combating the invasion of foreign objects, transporting and storing oxygen and acting as catalysts. **Fibrous proteins**, such as collagen and keratins, are arranged like filaments or fibres. In these substances the long coils align themselves in a more or less parallel fashion to form long, water insoluble fibres. Fibrous proteins provide structural integrity and strength to many kinds of tissue and are the main components of muscle, tendons and hair. ▶ **TABLE 29.1** summarises some common protein classes and their functions.

Peptide is the name given to a short polymer (oligomer) of amino acids joined by *peptide amide bonds*. Peptides are classified by the number of amino acids in the chain. For example, a *dipeptide* is a molecule containing two amino acids joined by a peptide bond. A *tripeptide* is a molecule containing three amino acids joined by peptide bonds. **Polypeptides** are formed when many amino acids are linked together by peptide bonds. They are classed as *macromolecules* or *biopolymers*.

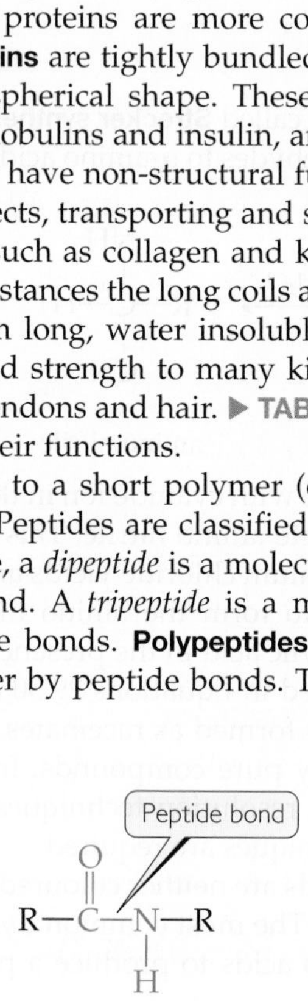

Coding Peptides

Each of the amide bonds in a peptide or polypeptide is called a **peptide bond**. A peptide bond is formed by a condensation reaction between the carboxyl group of one amino acid and the amino group of another amino acid. Alanine and glycine, for example, can react to form the dipeptide glycylalanine:

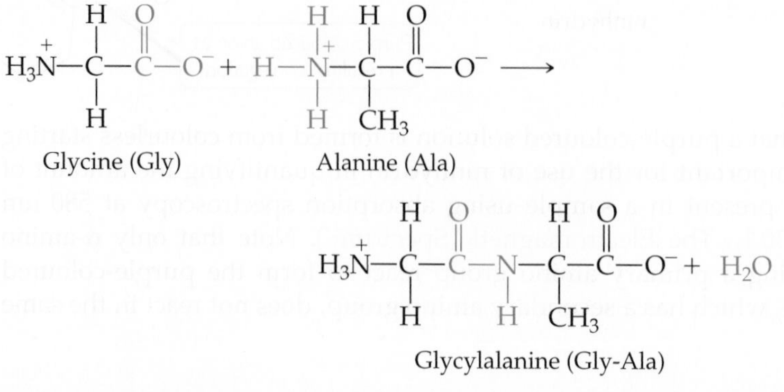

Glycine (Gly)   Alanine (Ala)

[29.33]

Glycylalanine (Gly-Ala)

TABLE 29.1 • Some common protein classes and their function		
Name of class	**Composition examples**	**Definition/functions**
Glycoproteins	Contain carbohydrates, e.g. ovalbumin (egg white)	A ubiquitous family of proteins containing oligosaccharides (carbohydrates), which impart properties of solubility, ligand affinity, cellular targeting and stability. Many proteins released by cells to the blood and other fluids are glycoproteins. For example, Antarctic fish survive near-freezing water temperatures as a result of freezing-point depression of their blood serum by a globular glycoprotein.
Lipoproteins	Contain fats, oils or steroids (lipids), e.g. low-density lipoproteins (LDLs) and high-density lipoproteins (HDLs)	Lipoproteins serve as carriers of hydrophobic ligands through solubility barriers such as aqueous body fluids. LDLs are the main transport for cholesterol through the body. HDLs carry excess cholesterol to the liver for processing.
Metalloproteins	Contain coordinated metal ions, e.g. haemoglobin, ferredoxin, NADH dehydrogenase, coenzyme Q, cytochrome c reductase	Metalloproteins are proteins that contain a metal cofactor. The metal may be an isolated ion or may be coordinated with a non-proteinaceous organic compound, such as the porphyrin found in haemoproteins. In some cases, the metal is coordinated with a side chain of the protein and an inorganic non-metallic ion.
Nucleoproteins	Contain nucleic acids, e.g. chromatin, RNA binding proteins, ribosomes	Nucleoproteins are any supramolecular complex of protein and nucleic acid; or any protein usually found closely associated with nucleic acid. They contain a diverse group of viral and genetic regulatory proteins which bind in a sequence-specific manner to DNA.
Phosphoproteins	Contain a phosphate group, e.g. synapsins, caesins (milk)	Phosphoproteins are any of a group of proteins containing chemically bound phosphoric acid. Protein phosphorylation is probably the most ubiquitous and diverse molecular mechanism of biological signalling and regulation.

The amino acid that furnishes the carboxyl group for peptide bond formation is named first, with a *-yl* ending; then the amino acid furnishing the amino group is named second. To make the naming of peptides easier, especially those that are more than 25 amino acids long, series of three-letter and single-letter codes have been devised (▼ TABLE 29.2). Based on the three-letter codes for the amino acids from Table 29.2, glycylalanine can be abbreviated Gly-Ala. In this notation it is understood that the unreacted amino group is on the left and the unreacted

TABLE 29.2 • Amino acid codes for proteins and peptides		
Amino acid	**Three-letter code**	**Single-letter code**
Glycine	Gly	G
Alanine	Ala	A
Valine	Val	V
Leucine	Leu	L
Isoleucine	Ile	I
Methionine	Met	M
Phenylalanine	Phe	F
Tryptophan	Trp	W
Proline	Pro	P
Asparagine	Asn	N
Glutamine	Gln	Q
Serine	Ser	S
Threonine	Thr	T
Aspartic acid	Asp	D
Glutamic acid	Glu	E
Cysteine	Cys	C
Tyrosine	Tyr	Y
Arginine	Arg	R
Histidine	His	H
Lysine	Lys	K

carboxyl group on the right. The artificial sweetener *aspartame* (◀ FIGURE 29.21) is the methyl ester of the dipeptide of aspartic acid and phenylalanine, Asp-Phe:

▲ **FIGURE 29.21** **Sweet stuff.** The artificial sweetener aspartame is the methyl ester of the dipeptide Asp-Phe.

SAMPLE EXERCISE 29.7 **Drawing the structural formula of a tripeptide**

Draw the full structural formula for alanylglycylserine.

SOLUTION

Analyse We are asked to draw a tripeptide.

Plan The name of this substance suggests that three amino acids—alanine, glycine and serine—have been linked together, forming a *tripeptide*. Note that the ending *-yl* has been added to each amino acid except for the last one, serine. By convention, the first-named amino acid (alanine, in this case) has a free amino group and the last-named one (serine) has a free carboxyl group. Thus we can construct the structural formula of the tripeptide from its amino acid building blocks.

Solve We first combine the carboxyl group of alanine with the amino group of glycine to form a peptide bond and then the carboxyl group of glycine with the amino group of serine to form another peptide bond. The resulting tripeptide consists of three 'building blocks' connected by peptide bonds:

Amino group ⟶ Carboxyl group

	Ala	Gly	Ser

We can abbreviate this tripeptide as Ala-Gly-Ser, or AGS.

PRACTICE EXERCISE

Name the dipeptide that has the following structure and give its abbreviation:

$$H_3\overset{+}{N}-\underset{\underset{HOCH_2}{|}}{\overset{\overset{H}{|}}{C}}-\overset{\overset{O}{||}}{C}-\underset{\underset{H}{|}}{N}-\underset{\underset{\underset{\underset{COOH}{|}}{CH_2}}{|}}{\overset{\overset{H}{|}}{C}}-\overset{\overset{O}{||}}{C}-O^-$$

Answer: Serylaspartic acid: Ser-Asp, SD

(See also Exercise 29.48.)

Proteins are linear (that is, unbranched) polypeptide molecules with molecular weights ranging from about 5000 to over 50 million u. Proteins differ from polypeptides in that proteins are capable of some function, whether chemical or structural, as a result of their three-dimensional structure. Because 20 different amino acids are commonly linked together in proteins and because proteins consist of hundreds of amino acids, the number of possible arrangements of amino acids within proteins is virtually limitless.

▲ **CONCEPT CHECK 4**

Does Gly-Ala have the same structure as Ala-Gly?

Protein Structure

The arrangement, or sequence, of amino acids along a protein chain is called its **primary structure**. The primary structure gives the protein its unique identity. A change in even one amino acid can alter the biochemical and physiological characteristics of the protein. For example, sickle-cell anaemia is a genetic disorder resulting from a single replacement in a protein chain in haemoglobin. The chain that is affected contains 146 amino acids. The substitution of a single amino acid with an acidic functional group for one that has a hydrocarbon side chain alters the solubility properties of the haemoglobin, and normal blood flow is impeded (see the A Closer Look box in Section 29.2).

Another example occurs within the free-radical scavenging Zn/Cu enzyme, superoxide dismutase (SOD1). A single mutation in the primary structure of SOD1—for example, by replacing Gly with Ala at the 93rd position of the peptide sequence—is enough to implicate this mutant enzyme as a contributor to motor neurone disease (MND).

Proteins in living organisms are not simply long, flexible chains with random shapes. Rather, the chains coil and stretch in particular ways. The **secondary structure** of a protein refers to how segments of the protein chain are orientated in a regular pattern as seen in ▼ FIGURE 29.22.

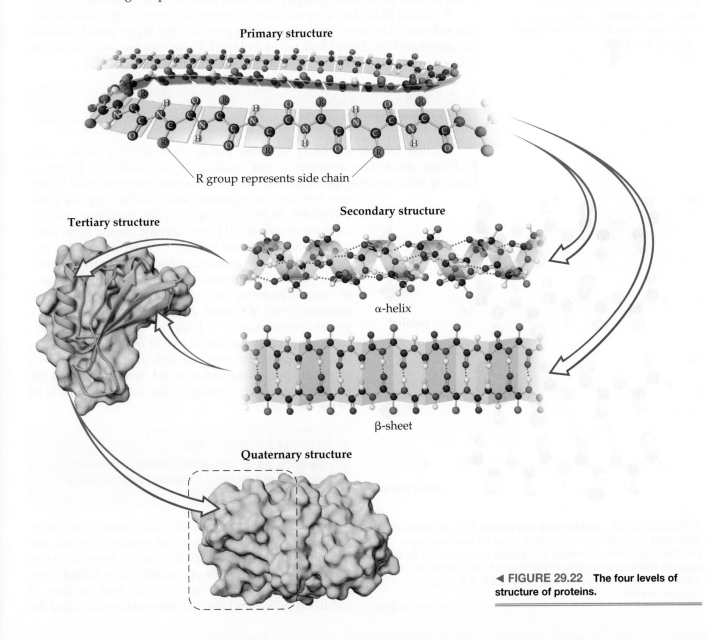

Primary structure

R group represents side chain

Secondary structure

α-helix

β-sheet

Tertiary structure

Quaternary structure

◀ FIGURE 29.22 **The four levels of structure of proteins.**

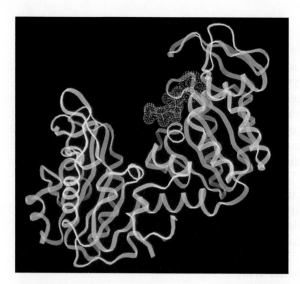

▲ **FIGURE 29.23 An enzyme containing a substrate.** A computer-generated model of an enzyme showing the amino acid backbone as a green ribbon. The substrate (violet) is shown to interact with a specific part of the protein called the active site.

One of the most important and common secondary-structure arrangements is the **α-helix** (Figure 29.22), first proposed by Linus Pauling and Robert B. Corey (1897–1971). Imagine winding a long protein chain in a helical fashion around a long cylinder. The helix is held in position by hydrogen bond interactions between N–H bonds and the oxygen atoms of nearby carbonyl groups. The pitch of the helix and the diameter of the cylinder must be such that (1) no bond angles are strained and (2) the N–H and C=O functional groups on adjacent turns are in a proper position for hydrogen bonding. In fact, each C=O group is hydrogen bonded to an N–H group four amino acid units away from it. An arrangement of this kind is possible for some amino acids along the chain, but not for others. Large protein molecules may contain segments of the chain that have the α-helical arrangement, interspersed with sections in which the chain is in a random coil. Typically, the α-helix has 3.6 amino acids per turn. The N–H groups of the amino acid point along the axis of the helix and all orientate themselves in the same direction. The C=O groups of the amino acids also point along the axis but in the opposite direction to the N–H groups, as a result of the *trans* arrangement of the peptide bond. All R groups point away from the helix structure.

◀ **FIGURE 29.23** shows the structure of an enzyme containing a small molecular substrate. The important aspect of this protein is the highly coiled nature of the amino acid backbone (depicted as a ribbon). These coils are α-helices, which help give rise to the overall three-dimensional structure of the protein.

CONCEPT CHECK 5

What bonding interaction is important in stabilising the α-helix form of a protein?

Another form of secondary structure that is often seen in proteins is the **β-pleated sheet**. This structure is not restricted to *intramolecular* hydrogen bonding structure, as is the α-helix, but may also form an intermolecular hydrogen bonding arrangement with another peptide chain (◀ **FIGURE 29.24**). Since the peptide backbone contains only the α-carbon, NH and carbonyl carbon in an alternating manner from the *N*- to *C*-terminus, the two peptide chains can hydrogen bond in one of two orientations: parallel or anti-parallel. Both orientations are seen in natural proteins, although anti-parallel is the most common form. ◀ **FIGURE 29.25** shows an example of a protein with extensive β-sheets. These β-sheets are formed by the interaction of two or more peptide chains in an anti-parallel manner (as indicated by the direction of the arrows). Note that it is not necessary for all the peptide chains contributing to the sheet structure to be of the same size.

CONCEPT CHECK 6

Describe the similarities and differences between the structure of a protein β-sheet and the structure of nylon 6,6.

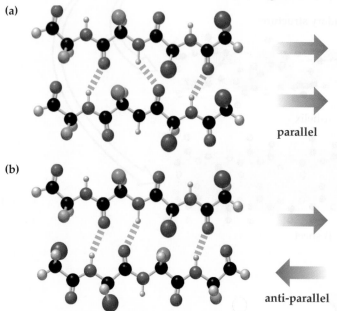

(a)

parallel

(b)

anti-parallel

▲ **FIGURE 29.24 β-sheet form of a protein.** F, Cl, Br and I are used along the peptide backbone to describe the two orientations possible for a β-sheet formation. (a) A parallel sheet, in which the two peptides orientate in the same direction from *N*- to *C*-terminus. (b) An anti-parallel sheet in which the two halide trends (F to I) occur in opposite directions.

Proteins are not active biologically unless they are in a particular form in solution. The process by which the protein adopts the biologically active form is called **folding**. The overall shape of a protein in its folded form, determined by all the bends, kinks and sections of rod-like α-helical and flat β-sheet structures, is called the

tertiary structure. Figure 21.14 shows the tertiary structure of myoglobin, a protein with a molecular weight of about 18 000 u, containing one haem group (∞ Section 21.3, 'Ligands with More than One Donor Atom').

∞ Review this on page 864

The tertiary structure of a protein is maintained by many different interactions. Certain folding of the protein chain leads to lower-energy (more stable) arrangements than other folding patterns. For example, a globular protein dissolved in aqueous solution folds in such a way that the non-polar hydrocarbon portions are tucked within the molecule, away from the polar water molecules. Most of the more polar acidic and basic side chains, however, project into the solution, where they can interact with water molecules through ion–dipole, dipole–dipole or hydrogen bonding interactions. The result of a mis-folding based on the change of a polar group was seen clearly in the example of sickle-cell anaemia.

Some proteins are more complicated and have a fourth level of structure known as the **quaternary structure**. This level of structure involves the aggregation of two or more protein substructures into a larger macromolecular assembly. For example, the protein shown in Figure 29.25 has a quaternary structure, made up of two identical protein substructures that interact non-covalently with each other. The function the protein exhibits is only possible due to the quaternary (dimeric) structure and not by either substructure separately.

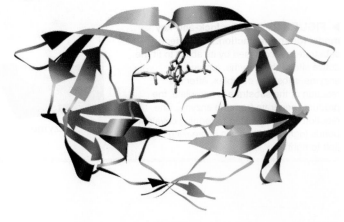

▲ **FIGURE 29.25** **A dimeric protein containing β-sheets.** The arrowed ribbons represent the directions of peptide chains that undergo intermolecular hydrogen bonding, forming β-sheet structures. The substrate in the middle fits nicely into the void made by the two halves of this protein dimer.

Enzymes

One of the most important classes of proteins are the **enzymes**, large protein molecules that serve as catalysts (∞ Section 15.7, 'Catalysis'). Enzymes usually catalyse only very specific reactions. Their tertiary and quaternary structures generally dictate that only certain substrate molecules can interact with the enzyme (Figures 29.23 and 29.25).

∞ Review this on page 598

Many of the most interesting and important examples of catalysis involve reactions within living systems. The human body is characterised by an extremely complex system of interrelated chemical reactions. All these reactions must occur at carefully controlled rates and at 37 °C in order to maintain life. Enzymes enable these reactions to take place in the proper order and at the correct rate. Most enzymes are proteins with molecular weights ranging from about 10 000 to about 1 million u. They are very selective in the reactions they catalyse, and some are absolutely specific, operating for only one substance in only one reaction. The decomposition of hydrogen peroxide, for example, is an important biological process. Because hydrogen peroxide is strongly oxidising, it can be physiologically harmful. For this reason, the blood and livers of mammals contain an enzyme, *catalase*, which catalyses the decomposition of hydrogen peroxide into water and oxygen (Equation 15.33). ▶ **FIGURE 29.26** shows the dramatic acceleration of this chemical reaction by the catalase in beef liver.

Although an enzyme is a large molecule, the reaction is catalysed at a very specific location in the enzyme, called the **active site**. The substances that undergo reaction at this site are called **substrates**. A simple explanation for the specificity demonstrated by enzymes is provided by the **lock-and-key model** illustrated in ▼ **FIGURE 29.27**. The substrate is pictured as fitting neatly into a special place on the enzyme (the active site), much like a specific key fitting into a lock. The active site is created by the coiling and folding of long protein chains

▲ **FIGURE 29.26** **Effect of an enzyme.** Ground-up beef liver causes hydrogen peroxide to decompose rapidly into water and oxygen. The decomposition is catalysed by the enzyme *catalase*. Grinding the liver breaks open the cells, so that the reaction takes place more rapidly. The frothing is due to escape of oxygen gas from the reaction mixture.

FIGURE IT OUT
Why might this be called a lock-and-key model?

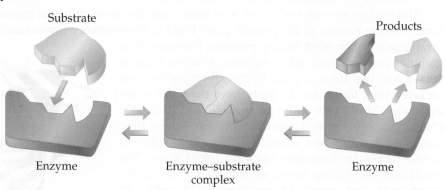

▶ **FIGURE 29.27 The lock-and-key model for enzyme action.** The correct substrate is recognised by its ability to fit the active site of the enzyme, forming the enzyme–substrate complex. After the reaction of the substrate is complete, the products separate from the enzyme. The reaction is just as suited to forming covalent bonds (right to left) as it is in breaking them (left to right).

Substrate

Products

Enzyme   Enzyme–substrate complex   Enzyme

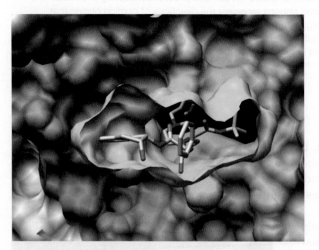

▲ **FIGURE 29.28 Molecular model of an active site.** The active site (shown in yellow) has been cut away to show how well the substrate fits into the different depressions and pockets, like a hand in a glove. Compare this representation to that in Figure 29.25 of the same protein.

to form a space, something like a pocket, into which the substrate molecule fits. ◀ FIGURE 29.28 shows a model of an enzyme's active site (cross-sectioned) with a bound substrate molecule.

The combination of the enzyme and the substrate is called the enzyme–substrate complex. Although Figure 29.29 shows both the active site and its complementary substrate as having rigid shapes, there is often a fair amount of flexibility in the active site. Thus the active site may change shape as it binds the substrate. The binding between the substrate and the active site involves intermolecular forces such as dipole–dipole attractions, hydrogen bonds and London dispersion forces.

As the substrate molecules enter the active site, they are activated so that they are capable of extremely rapid reaction at mild (for example, 37 °C) temperatures. This activation may result from the enzyme withdrawing or donating electron density at a particular bond. In addition, in the process of fitting into the active site, the substrate molecule may be distorted and thus made more reactive. Once the reaction occurs, the products then depart, allowing another substrate molecule to enter. This means that the products must bind more weakly to the active site than does the substrate. If the reverse were true, inhibition of the enzyme's activity would be observed.

The activity of an enzyme is also destroyed if some molecule in the solution is able to bind strongly to the active site and block the entry of the substrate. Such substances are called *enzyme inhibitors*. Nerve poisons and some toxic metal ions, such as lead and mercury, are believed to act in this way to inhibit enzyme activity. Other poisons act by attaching elsewhere on the enzyme, thereby distorting the active site so that the substrate no longer fits.

Enzymes are typically more efficient than human-made catalysts. The number of individual catalysed reaction events occurring at a particular active site, called the *turnover number*, is generally in the range of 10^3 to 10^7 per second. Such large turnover numbers correspond to very low activation energies. This is an area of chemistry of intense study because of its benefits associated with (a) mild reaction conditions, (b) high turnover numbers, (c) the sustainability of using microbes, bacteria and other organisms to promote reaction and (d) the removal of hazardous chemicals and solvents. Remember, most biological processes occur in water.

> ### CONCEPT CHECK 7
> What names are given to the following aspects of enzymes and enzyme catalysis:
> a. the place on the enzyme where catalysis occurs,
> b. the substances that undergo catalysis?

Sequencing of Peptides and Proteins

As a result of the almost endless permutations of amino acid sequences and the very specific nature of a protein's function, chemists and biochemists have come up with some clever ways of determining the precise structure of a protein. Let's consider the problem. There are three aspects of sequencing that need to be elucidated: (1) which amino acids are present; (2) what amounts of each amino acid are present; and (3) what order do these amino acids take in the primary structure of the protein.

The first two of these questions are easily answered provided the protein is pure. Amide bonds are hydrolysed in excess aqueous 6M HCl to yield the ammonium salt and carboxylic acid:

$$R\overset{\overset{\displaystyle O}{\|}}{C}\!-\!\underset{H}{N}\!-\!R' + \text{excess } H_3O^{\oplus} \longrightarrow RCOOH + R'NH_3^{\oplus} + H_2O \qquad [29.34]$$

amide   carboxylic   ammonium
                            acid   salt

For a simple dipeptide such as Gly-Ala, the products are the ammonium salts of glycine and alanine:

$$\text{Gly-Ala} + \text{excess } H_3O^{\oplus} \longrightarrow H_3N^{\oplus}\text{—}\overset{\overset{\displaystyle}{C}}{}\text{—OH (Gly)} + H_3N^{\oplus}\text{—COOH (Ala)} + 2\,H_2O \qquad [29.35]$$

Gly-Ala   Gly   Ala

Amino acids can be separated from each other by electrophoresis (Figure 29.20) and other separating methods such as chromatography. The amount of each amino acid is then quantified using ninhydrin by the reaction shown in Equation 29.32. ▼ FIGURE 29.29 summarises this entire process using a general protein.

Knowing which amino acids are present, as well as their relative amounts, is only the beginning. The specific sequence of amino acids within the protein, which gives rise to the protein's function, is critical. Remember, the simple dipeptide Gly-Ala is different from Ala-Gly. The processes required to sequence a

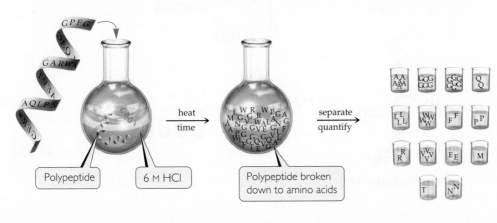

◀ FIGURE 29.29 **Amino acid analysis.** Note that the single-letter codes for the amino acids (Table 29.2) have been used in this example.

TABLE 29.3 • Useful reagents for protein cleavage	
Reagent	**Cleavage point(s) at the carboxyl group of**
Cyanogen bromide (BrCN)	Methionine
Trypsin	Arginine and lysine
Chymotrypsin	Phenylalanine, tryptophan and tyrosine
6 M HCl (partial hydrolysis)	Indiscriminant

protein need to be very specific and efficient. Two approaches are possible. Either the whole protein can be sequenced, or the protein can be cleaved at specific points to make smaller peptides, which are then sequenced. Both approaches have their advantages, although the latter is more practical. ▲ TABLE 29.3 lists three common methods of undertaking the cleavage of peptide bonds at specific amino acid sites.

 CONCEPT CHECK 8

An unknown hexapeptide contains six amino acid residues. The partial hydrolysis of the hexapeptide gave two fragments, identified as Val-Tyr-His-Glu and Ala-Ile-Val. What is the likely sequence of the hexapeptide?

The most widely used method of sequencing proteins was developed by Pehr Edman (1916–1977) in the 1950s. The process is a laborious one and has been automated to save time and sample amounts. Modern instruments can sequence as little as 0.1 μg of a protein. The general idea of his approach, called the **Edman degradation**, is to cleave and identify one amino acid at a time from the *N*-terminus of the peptide chain. The process is repeated until the entire sequence is known. The Edman degradation requires treatment of the peptide with phenyl isothiocyanate (Ph—N=C=S), followed by mild acid hydrolysis. The product of this reaction, a phenylthiohydantoin, is collected and compared with known phenylthiohydantoin derivatives of the 20 common amino acids to identify which amino acid is cleaved. The new peptide (which is now one residue shorter) is also subjected to the same conditions to characterise the second amino acid residue and so on.

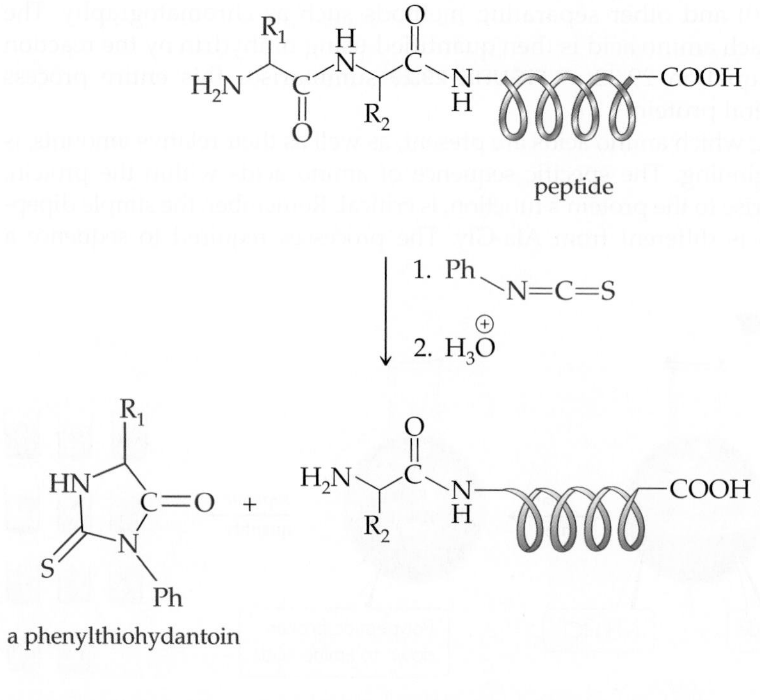

peptide

1. Ph—N=C=S

2. $H_3O^{\oplus}$

a phenylthiohydantoin

[29.36]

MY WORLD OF CHEMISTRY

B GROUP VITAMINS

Vitamins are small organic molecules that must be obtained through diet. They are usually required in trace amounts for proper growth. Many vitamins are used as coenzymes by the body; that is, they are required for the catalytic function of certain enzymes. All of the major B-group vitamins are nitrogen-containing compounds.

Vitamin B_1 is also known as thiamin, a water soluble heterocyclic structure made up of a substituted pyrimidine and a thiazole ring coupled by a methylene (CH_2) bridge. Thiamin is converted to its active form, thiamin pyrophosphate (TPP), in the brain and liver. In this form, thiamin acts as a cofactor for reductase enzymes responsible for, among other things, the reduction of pyruvate (Equation 29.29). A deficiency in thiamin may lead to *fatigue* and *depression*. Riboflavin, or vitamin B_2, is employed as a coenzyme for flavoproteins, which have a wide range of redox roles within the body. Riboflavin is found in useful quantities in eggs, milk, meat and cereal. A deficiency in vitamin B_2 leads, among other things, to *cracked lips* and *scaly skin*.

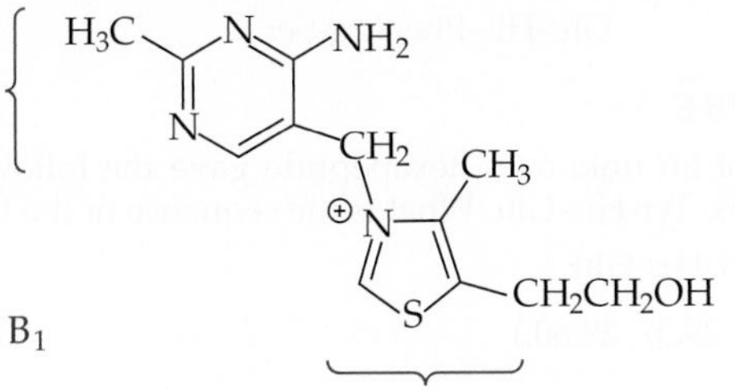

vitamin B_1

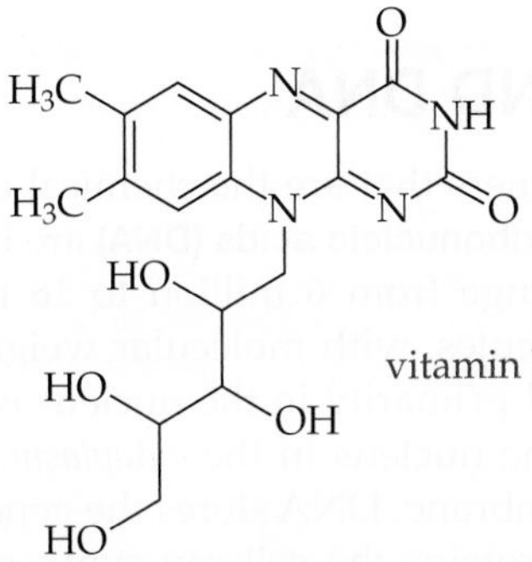

vitamin B_2

Nicotinic acid is the most common source of Vitamin B_3, also known as niacin. The human body uses vitamin B_3 in the form of NAD^+ (nicotinamide adenine dinucleotide) and $NADP^+$ (nicotinamide adenine dinucleotide phosphate) as a coenzyme in more than 50 enzyme-based reactions. Vitamin B_3 is instrumental in the release of energy from carbohydrates. It is necessary for proper brain function and is also involved in fat and cholesterol metabolism. A deficiency in vitamin B_3 may lead to *dermatitis, diarrhoea* and *dementia*. Vitamin B_5, also known as pantothenic acid, is found in abundance in whole-grain cereals, legumes and meat, and is usually formed from β-alanine and pantoic acid. Vitamin B_5 is required for the synthesis of coenzyme A (CoA), an acyltransferase, and for the metabolism of carbohydrates, fats and proteins. A deficiency in vitamin B_5 leads to *weight loss* and *irritability*.

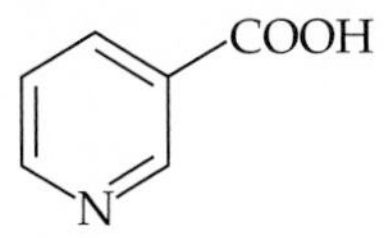

nicotinic acid

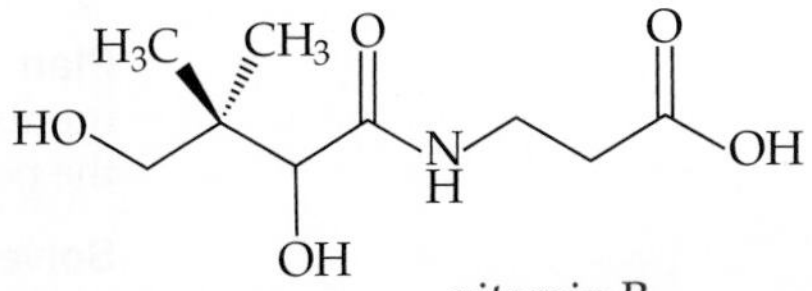

vitamin B_5

One of the most complicated and essential of the vitamins is vitamin B_{12}. Vitamin B_{12} is unusual as it is the only metal-ion-containing vitamin. The tetrapyrrole macrocycle (purple) of vitamin B_{12} is called a *corrin*, which is related to the porphyrin structure found in myoglobin, haemoglobin and chlorophyll. Centrally coordinated to the corrin macrocycle is a cobalt(II) ion with an octahedral geometry (red). The cyano (CN) and benzimidazole ligands (green) bind axially to the metal ion.

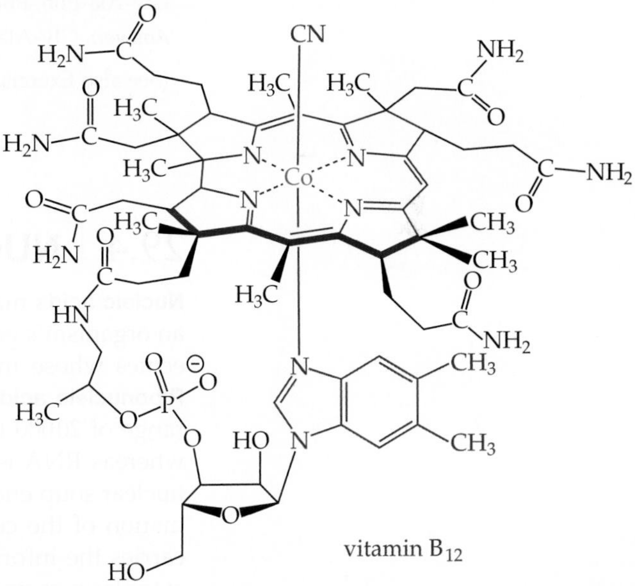

vitamin B_{12}

Vitamin B_{12} is synthesised exclusively by microorganisms and is found in the liver of animals. The liver can store up to six years' supply of vitamin B_{12}, so deficiencies in this vitamin are rare. Vitamin B_{12} helps to protect nerves and is involved in the formation of red blood cells in bone marrow which prevents anaemia. Vitamin B_{12} is necessary in the catabolism of fatty acids.

RELATED EXERCISES: 29.82, 29.83

| **SAMPLE EXERCISE 29.8** | **Sequencing peptides** |

An unknown pentapeptide was subjected to a range of different experiments to ascertain its structure. The following results were obtained:

Experiment	Fragment result
Total hydrolysis	Arg, Glu, His, Phe, Ser
Edman degradation	Glu
Chymotrypsin hydrolysis	Glu-His-Phe, Arg-Ser
Trypsin hydrolysis	Glu-His-Phe-Arg, Ser

What is the pentapeptide sequence?

SOLUTION

Analyse We are asked to sequence a small peptide based on the fragments obtained by different chemical reactions.

Plan Identify the reactions and what type of fragment each leads to. Determine the number of different amino acids listed through the hydrolysis products. Construct the peptide one fragment at a time.

Solve Total hydrolysis yields five amino acids, which is the total required for the pentapeptide. Hence, each amino acid occurs only once in the sequence. Peptides are always written from the *N*- to *C*-terminus. The Edman degradation yields the *N*-terminus residue, Glu. Chymotrypsin hydrolyses at the carboxyl end of Phe, which indicates the first three residues are Glu-His-Phe. At this point, the sequence can be only one of two possibilities: Glu-His-Phe-Arg-Ser or Glu-His-Phe-Ser-Arg. Since trypsin cleaves at the carboxyl group of Arg, leaving the single amino acid residue, Ser, the pentapeptide sequence was:

Glu-His-Phe-Arg-Ser

PRACTICE EXERCISE

The partial hydrolysis of an unknown hexapeptide gave the following fragments: Gly-Ala-Phe, Phe-Tyr-His, Tyr-His-Glu. What is the sequence of the hexapeptide?

Answer: Gly-Ala-Phe-Tyr-His-Glu

(See also Exercises 29.49, 29.57, 29.60.)

29.4 | NUCLEIC ACIDS AND DNA

Nucleic acids make up a class of biopolymers that are the chemical carriers of an organism's genetic information. **Deoxyribonucleic acids (DNA)** are huge molecules whose molecular weights may range from 6 million to 16 million u. **Ribonucleic acids (RNA)** are smaller molecules, with molecular weights in the range of 20000 to 40000 u. DNA is found primarily in the nucleus of the cell, whereas RNA is found mostly outside the nucleus in the *cytoplasm*, the nonnuclear soup enclosed within the cell membrane. DNA stores the genetic information of the cell and specifies which proteins the cell can synthesise. RNA carries the information stored by DNA out of the nucleus of the cell into the cytoplasm as part of the *transcription* process. Once in the cytoplasm, the information, transcribed into messenger RNA (mRNA), is used in protein synthesis. The process of protein synthesis from genetic material, known as *translation*, also involves two other forms of RNA, called ribosomal RNA (rRNA) and transfer RNA (tRNA), which bring together the programmed amino acids in the correct sequence for polypeptide synthesis.

The monomers of nucleic acids can come in two forms. **Nucleosides** are compounds containing a five-carbon sugar and a nitrogen-containing organic base. **Nucleotides** are nucleosides containing an additional phosphate group.

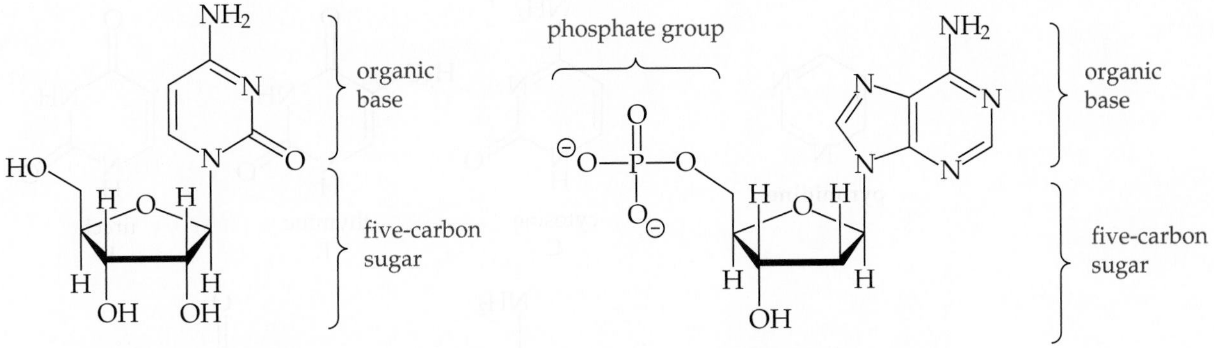

cytidine
a ribonucleoside

deoxyadenine monophosphate
a deoxyribonucleotide

The sugar component of RNA is *ribose*, whereas that in DNA is *deoxyribose*. *Deoxy*ribose differs from ribose only in that it lacks the OH group at carbon 2:

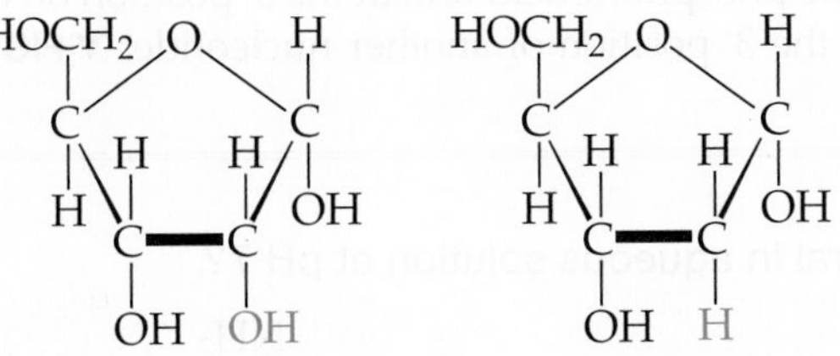

ribose   deoxyribose

Hence nucleosides and nucleotides can be either ribose or deoxyribose based. For example, a nucleoside derived from ribose is called a *ribonucleoside* and is found in RNA. A nucleotide derived from deoxyribose is called a *deoxyribonucleotide* and is found in DNA.

Nucleosides are also known generally as **N-glycosides**. An *N*-glycoside is formed when a monosaccharide reacts with an amine in the presence of acid (∞ Section 26.4, 'Carbohydrates'). A general reaction for the formation of an *N*-glycoside is

∞ Review this on page 1063

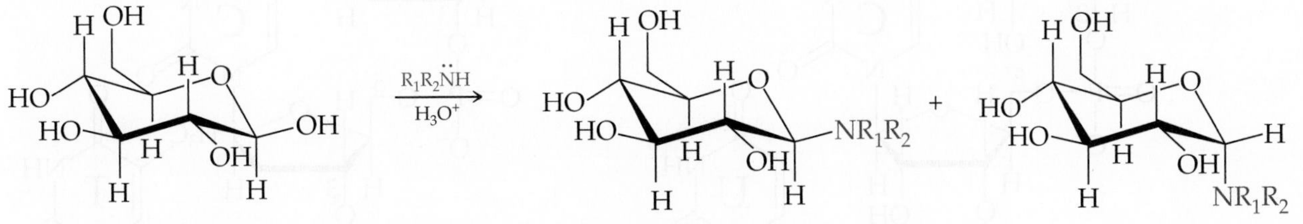

β-*N*-glycoside   α-*N*-glycoside   [29.37]

In the case of DNA and RNA, the amines used are derived from the aromatic heterocycles **pyrimidine** and **purine**. The heterocycles adenine (A), thymine (T), cytosine (C) and guanine (G) form the four bases of DNA used to store the genetic code.

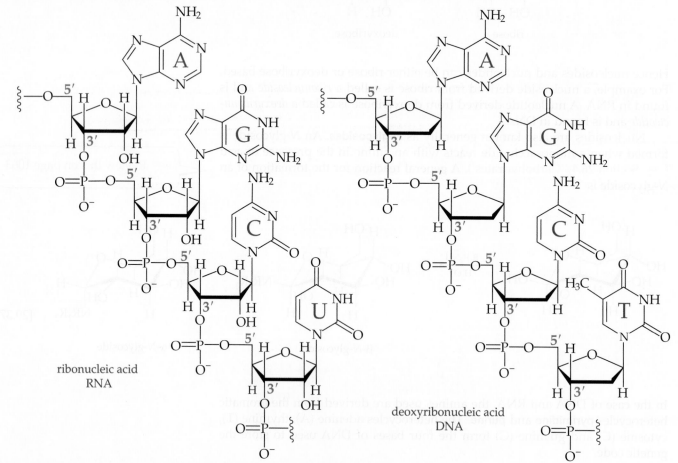

pyrimidine

cytosine
C

thymine
T

uracil
U

purine

adenine
A

guanine
G

RNA differs from DNA not only in the monosaccharide used, ribose or deoxyribose, but also in the replacement of thymine (T) by uracil (U) within the base code.

Nucleic acids are polynucleotides formed by condensation reactions between an OH group of the phosphoric acid unit at the 5' position on one nucleotide with the OH group at the 3' position of another nucleotide. ▼ FIGURE 29.30 shows a

▲ **FIGURE IT OUT**

Is DNA positively charged, negatively charged or neutral in aqueous solution at pH 7?

ribonucleic acid
RNA

deoxyribonucleic acid
DNA

▶ **FIGURE 29.30 Structure of a polynucleotide.** The general structure of DNA and RNA is shown, demonstrating the differences between the two types of nucleic acids.

portion of the polymeric chain of DNA and RNA, illustrating the 3'–5' linkage and the different bases employed by each nucleic acid.

DNA molecules consist of two deoxyribonucleic acid chains or strands that are wound together in the form of a **double helix**, as shown in ▼ **FIGURE 29.31**. The drawing in Figure 29.31(b) has been simplified to show the essential features of the structure. The sugar and phosphate groups form the backbone of each strand. The bases (represented by the letters A, T, C and G) are attached to the sugars. The two strands are held together by attractions between the bases in one strand and those in the other strand. These attractions involve both London dispersion interactions and hydrogen bonds. (∞ Section 11.2, 'Intermolecular Forces'). As shown in ▼ **FIGURE 29.32**, the structures of thymine (T) and adenine (A) make them perfect partners for hydrogen bonding. Likewise, cytosine (C) and guanine (G) form ideal hydrogen bonding partners. In the double-helix

∞ Review this on page 381

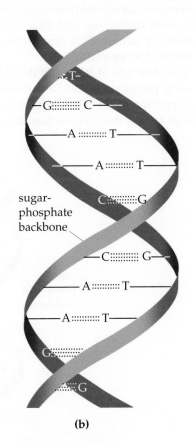

(a)   (b)

◀ **FIGURE 29.31** **Two views of DNA.** (a) A computer-generated model of a DNA double helix. The dark blue and light blue atoms represent the sugar-phosphate chains that wrap around the outside. Inside the chains are the bases, shown in red and yellow-green. (b) A schematic illustration of the double helix showing the hydrogen bond interactions between complementary base pairs.

▲ FIGURE IT OUT

Which pair of complementary bases, A–T or G–C, would you expect to bind more strongly?

Thymine   Adenine

T≡A

Cytosine   Guanine

C≡G

▲ **FIGURE 29.32** **Hydrogen bonding between complementary base pairs.** The hydrogen bonds are responsible for the formation and stability of the double-stranded helical structure of DNA, shown in Figure 29.31(b).

▲ **FIGURE 29.33** **James Watson (1928–) and Francis Crick (1916–2004).** Nobel laureates 1962, along with Maurice Wilkins (1916–2004), for their role in the determination of the structure of nucleic acids.

structure, therefore, each thymine on one strand is opposite an adenine on the other strand. Likewise, each cytosine is opposite a guanine. This *base-pairing* of C–G and A–T is called **Watson–Crick pairing**, after two of the scientists who were instrumental in solving the structure of DNA (◀ FIGURE 29.33). The double-helix structure with complementary bases on the two strands is the key to understanding how DNA functions.

▲ **CONCEPT CHECK 9**
Is the DNA double helix chiral?

The two strands of DNA unwind during cell division, and new complementary strands are constructed on the unravelling strands (▼ FIGURE 29.34). This enzyme-catalysed process results in two identical double-helix DNA structures, each containing one strand from the original structure and one newly synthesised strand. This replication process allows genetic information to be transmitted when cells divide.

The structure of RNA and its relationship to DNA is also the key to understanding protein synthesis, the means by which viruses infect cells, and many other problems of central importance to modern biology and medicine. The basis of taking the information stored in DNA and processing it to form pro-

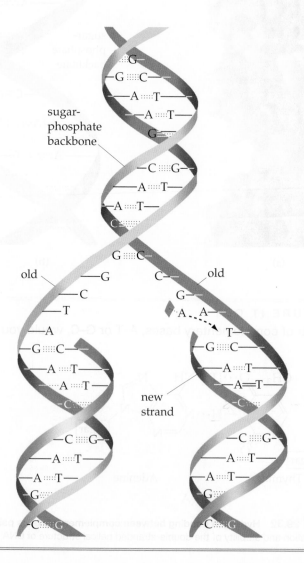

▶ **FIGURE 29.34** **DNA replication.** The original DNA double helix partially unwinds, and new nucleotides line up on each strand in complementary fashion. Hydrogen bonds help align the new nucleotides with the original DNA chain. When the new nucleotides are joined by condensation reactions, two identical double-helix DNA molecules result.

◄ FIGURE 29.35 **From DNA to protein.**

teins is shown in ▲ FIGURE 29.35. The mechanism for protein biosynthesis takes place within a *ribosome*, a complex piece of biological machinery composed of protein and rRNA, which reads the code contained within the mRNA to be translated. This code determines the order of the amino acid sequence within the polypeptide being synthesised.

So how is it possible for just four bases to code for all 20 naturally occurring amino acids? The answer is by using the ribonucleotides, A, C, G and U, in a triplet coding called a **codon**. Each codon is specific for a given amino acid (▼ TABLE 29.4). For example, the codon CAU on mRNA codes for histidine to be incorporated into the growing peptide chain, whereas the codon GCA on mRNA codes for alanine to be incorporated into the growing peptide chain.

The reason why translation uses a triplet code is very easy to understand. Using a triplet code based on the four ribonucleotides means that there are $4^3 = 64$ possible triplet codes, which is the smallest number of permutations possible greater than 20, which counts for the 20 naturally occurring amino acids. Of the possible 64 permutaions for the codons, 61 are used for specific amino acids. Most amino acids are specified by more than one codon, with the exception of methionine (Met) and tryptophan (Trp). The amino acids serine, leucine and arginine are all specified by any one of six codons. The three remaining codons are used to terminate the synthesis of the polypeptide and release it from the ribosome. These codons are called *stop codons*.

TABLE 29.4 ● Codon assignments

First base 5′ end	Second base	Third base 3′ end			
		U	C	A	G
U	U	UUU = Phe	UUC = Phe	UUA = Leu	UUG = Leu
	C	UCU = Ser	UCC = Ser	UCA = Ser	UCG = Ser
	A	UAU = Tyr	UAC = Tyr	UAA = Stop	UAG = Stop
	G	UGU = Cys	UGC = Cys	UGA = Stop	UGG = Trp
C	U	CUU = Leu	CUC = Leu	CUA = Leu	CUG = Leu
	C	CCU = Pro	CCC = Pro	CCA = Pro	CCG = Pro
	A	CAU = His	CAC = His	CAA = Gln	CAG = Gln
	G	CGU = Arg	CGC = Arg	CGA = Arg	CGG = Arg
A	U	AUU = Ile	AUC = Ile	AUA = Ile	AUG = Met
	C	ACU = Thr	ACC = Thr	ACA = Thr	ACG = Thr
	A	AAU = Asn	AAC = Asn	AAA = Lys	AAG = Lys
	G	AGU = Ser	AGC = Ser	AGA = Arg	AGG = Arg
G	U	GUU = Val	GUC = Val	GUA = Val	GUG = Val
	C	GCU = Ala	GCC = Ala	GCA = Ala	GCG = Ala
	A	GAU = Asp	GAC = Asp	GAA = Glu	GAG = Glu
	G	GGU = Gly	GGC = Gly	GGA = Gly	GGG = Gly

▲ **CONCEPT CHECK 10**
If the genetic code reads from left to right as 3′ to 5′ in one strand, how does it read from left to right in the complementary strand?

The construction of the polypeptide using the 61 codons sequenced on mRNA requires complementary sequence codons, called *anticodons*, on tRNA. For example, the codon series, 5′-UCA GAC, sequenced on mRNA requires a serine-bound tRNA with the complementary sequence, 3′-AGU-5′, as well as an aspartic acid-bound tRNA with the complementary sequence, 3′-CUG-5′. Once bound in the correct orientation, the two units are added to the growing peptide chain by an enzymatic process that cleaves the amino acid from the tRNA anticodon. ▼ **FIGURE 29.36** illustrates the process schematically.

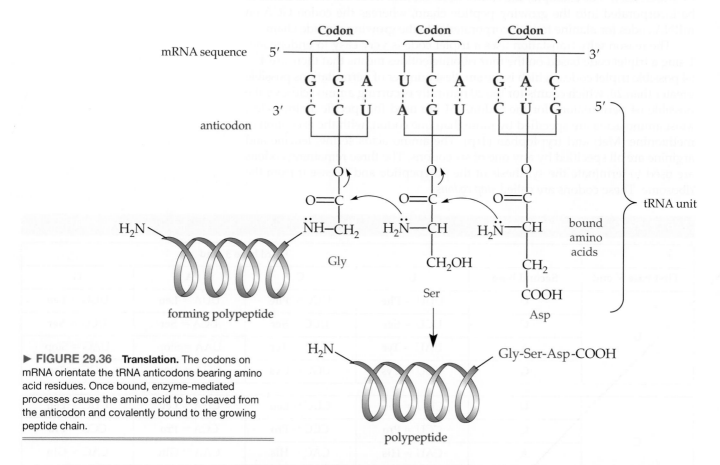

▶ **FIGURE 29.36 Translation.** The codons on mRNA orientate the tRNA anticodons bearing amino acid residues. Once bound, enzyme-mediated processes cause the amino acid to be cleaved from the anticodon and covalently bound to the growing peptide chain.

SAMPLE INTEGRATIVE EXERCISE | **Putting concepts together**

Glycine, NH_2CH_2COOH, the simplest of all naturally occurring amino acids, has a melting point of 292 °C. The pK_a of the acid group is 2.35 and the pK_a associated with the amino group is 9.78. **(a)** Draw a Lewis structure that indicates the charges on the molecule at the physiological pH of 7.4. **(b)** Use your structure to illustrate the concept of resonance. **(c)** Describe the hybridisation of the two carbon atoms and the nitrogen atom in glycine and the geometry of the atoms surrounding these three atoms. **(d)** Glycine has an unusually high melting point for a small molecule. Suggest a reason for this.

Analyse We are asked about the structure of a simple amino acid at a particular pH. This includes considering the hybridisation and molecular geometry of some of the atoms and commenting on a physical property.

Plan First draw the Lewis structure of glycine and determine the ionisation state of the amine and the carboxylic acid groups.

Solve

(a) A pH of 7.4 is several units on the base side of 2.35, consequently the carboxylic acid (pK_a = 2.35) will exist in the conjugate base form. Conversely physiological pH is more than two units on the acid side of 9.78 and so the amine will exist in the conjugate acid form (pK_a = 9.78). The Lewis structure will be:

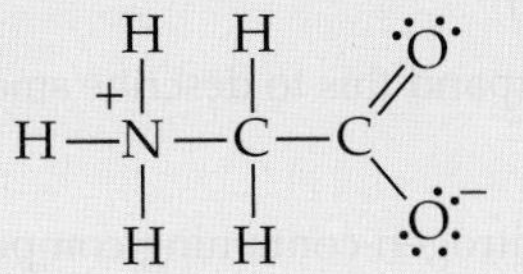

(b)

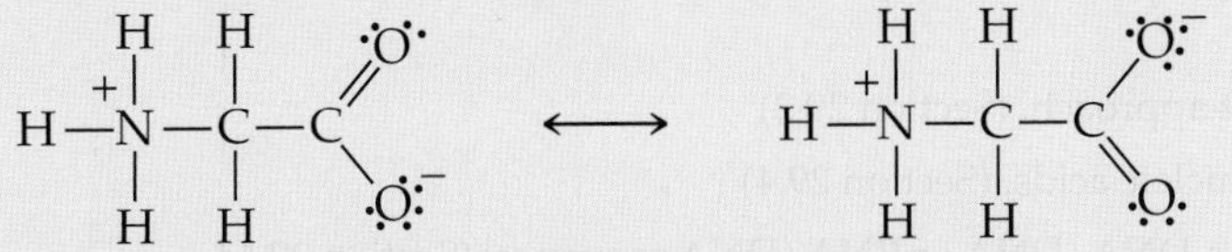

(c) N: sp^3 hybridised, tetrahedral geometry, CH$_2$: sp^3 hybridised, tetrahedral geometry, COO$^-$: sp^2 hybridised, trigonal-planar geometry

(d) Glycine exists as a zwitterion and strong electrostatic attraction between molecules results in some ionic-like characteristics such as a relatively high melting point.

CHAPTER SUMMARY AND KEY TERMS

SECTION 29.1 **Amines** are hydrocarbon derivatives containing one or more basic nitrogen atoms. **Aliphatic amines** contain alkyl chains. **Heterocycles** containing a basic nitrogen atom are classed as either **cyclic** or **aromatic amines**. **Alkaloids** are amines found in the bark, leaves and roots of plants, for example cocaine, nicotine and morphine. The **amino group** of a primary (RNH$_2$), secondary (RR'NH) or tertiary (RR'R''N) amine can be converted to an **alkyl ammonium salt** by the addition of acid or alkyl halide. Amines can be synthesised by the **reductive amination** of aldehydes and ketones. The condensation of an amine with a carboxylic acid yields an amide. Amides can be **primary**, **secondary** or **tertiary** based on the number of hydrogen atoms bonded to the amide nitrogen. The amide bond is highly planar due to resonance. **Lactams**, such as those found in the **penicillins**, are cyclic amides.

SECTION 29.2 Compounds containing both carboxylic acid and amine functional groups are called **amino acids**. α-**Amino acids** are chiral substances, with only the L enantiomers found naturally. Since both acidic and basic groups are found within the same molecule, amino acids are said to be **amphiprotic**. At neutral pH, amino acids exist as **zwitterions**. The **isoelectric point** of an amino acid is the pH where the neutral zwitterion predominates in solution. The 20 common naturally occurring amino acids are classed as α-amino acids. They differ only in the identity of their **side chains**. These side chains give the α-amino acids their different physical properties and allow their separation from each other by techniques such as **electrophoresis**. Amino acids can be formed by the **Strecker synthesis**, which converts aldehydes to α-amino acids in the presence of cyanide ion.

SECTION 29.3 **Proteins** are functional polymers of amino acids. The amino acids within proteins are linked by **peptide** (amide) **bonds**. A **polypeptide** is a polymer formed by linking many amino acids by peptide bonds. A **peptide** is the name given to a short sequence of amino acids. Proteins are classed as either **simple** or **conjugated** based on their constituents. They can be classed also as **fibrous** or **globular** based on their macromolecular shape. Protein structure is determined by the sequence of amino acids in the chain (its **primary structure**), the coiling or stretching of the chain (its **secondary structure**), the overall shape of the complete molecule (its **tertiary structure**) or its interaction with other proteins (its **quaternary structure**). The most important secondary structure arrangements are the **α-helix** and **β-pleated sheet**. The process by which a protein assumes its biologically active tertiary structure is called **folding**. The primary structure of a peptide or protein can be elucidated using the **Edman degradation**. Some proteins, called **enzymes**, function as catalysts. Enzymes contain an **active site**, which is the area of reactivity for a **substrate**. The active site and substrate form a very specific and functioning complex, much like a **lock and key**.

SECTION 29.4 **Nucleic acids** are biopolymers that carry the genetic information necessary for cell reproduction and development through control of protein synthesis. The building blocks of these biopolymers are **nucleosides** and phosphoric acid, which react to form **nucleotides**. Nucleosides are also generally known as **N-glycosides**. There are two types of nucleic acids: the **ribonucleic acids (RNA)** and the **deoxyribonucleic acids (DNA)**. These substances consist of a polymeric backbone of alternating phosphate and sugar groups, with **pyrimidine-** or **purine**-derived organic bases attached to the sugars. The DNA forms a **double helix** held together by hydrogen bonding between matching organic bases on the two strands. The specific nature of **Watson–Crick pairing** is the key to gene replication and protein synthesis. Each set of three nucleotides, or **codon**, within messenger RNA (mRNA) codes for a particular amino acid. This process, called translation, converts the genetic code into protein.

KEY SKILLS

- Be able to name and classify amines and amides. (Section 29.1)
- Recall that amines are both bases and nucleophiles and appreciate relative pK_a values. (Section 29.1)
- Prepare amines and amides from other functional groups. (Section 29.1)
- Know the 20 most common amino acids and their three-letter codes. Expand this to describe small peptides. (Section 29.2)
- Be able to use the Henderson–Hasselbalch equation. (Section 29.2)
- Have an understanding of the importance to biological processes of nitrogen-containing compounds. (Sections 29.3 and 29.4)
- Know the general classes of proteins and describe what is meant by primary, secondary, tertiary and quaternary structure of a protein. (Section 29.3)
- Be able to sequence a peptide through the fragment approach. (Section 29.3)
- Identify the purine and pyrimidine bases used in nucleic acids. (Section 29.4)
- Be able to show A/T and C/G complementarity in a DNA/DNA or RNA/DNA sequence. (Section 29.4)
- Have an understanding of the process of translation and transcription. (Section 29.4)

KEY EQUATIONS

- Isoelectric point

$$pI = \tfrac{1}{2}(pK_{a1} + pK_{a2})$$ [29.23]

- Henderson–Hasselbalch equation

$$pH = pK_a + \log \frac{[\text{conjugate base}]}{[\text{weak acid}]}$$ [29.28]

EXERCISES

VISUALISING CONCEPTS

29.1 From examination of the following ball-and-stick molecular models, choose the substance that **(a)** can be hydrolysed to form a solution containing glucose, **(b)** is capable of forming a zwitterion, **(c)** is one of the four bases present in DNA. [Sections 29.2 and 29.4]

(i)

(iii)

(iv)

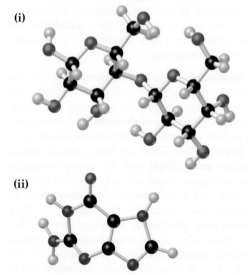

(ii)

29.2 Ceftibuten is an orally active cephalosporin antibiotic. Identify the β-lactam, secondary amide, primary amine and heterocyclic amine in this structure. [Section 29.1]

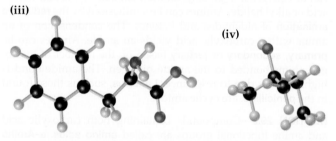

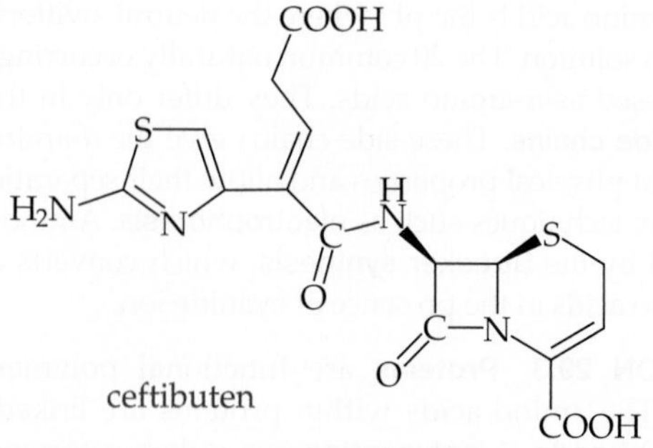

ceftibuten

29.3 What are the three-letter and one-letter codes for the following tetrapeptide? [Section 29.3]

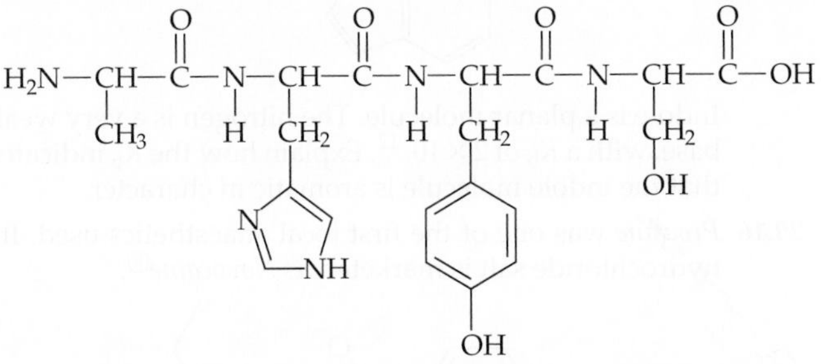

29.4 Cyclic GMP is a cellular regulatory agent found in animal and bacteria cells in concentrations of 10^{-6} moles kg^{-1}. Cyclic GMP is also implicated in the action and potency of Viagra. **(a)** Can you suggest what GMP might stand for? **(b)** What is the name given to the monosaccharide central to the structure of cyclic GMP? **(c)** Can cyclic GMP be classified as a glycoside? If so, what type of glycoside? [Section 29.4]

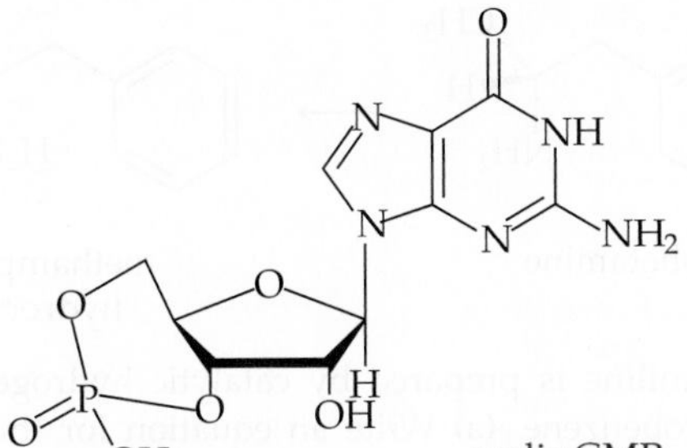

cyclic GMP

29.5 The configuration of an amino acid is drawn below as a Fischer projection. **(a)** Name the amino acid. **(b)** Assign its absolute stereochemistry as *R* or *S*. [Section 29.2]

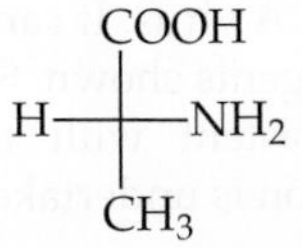

29.6 Classify the following compounds as an alkaloid, amino acid, nucleoside or carbohydrate. [Sections 29.1, 29.2 and 29.4]

(a)

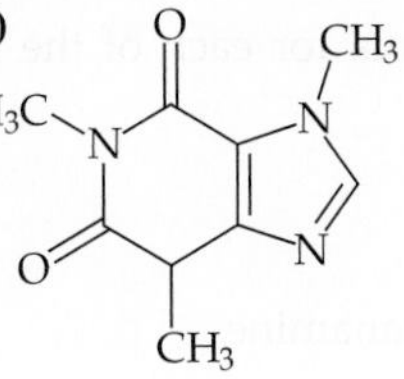

(b)

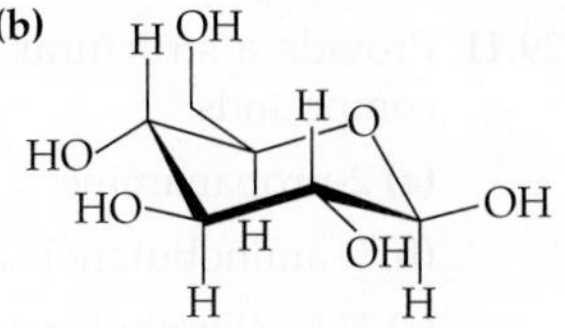

(c)

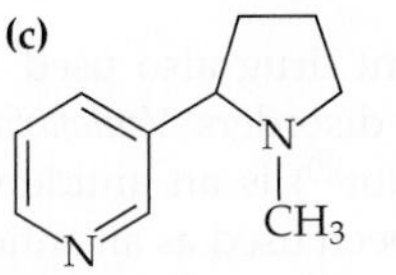

(d)

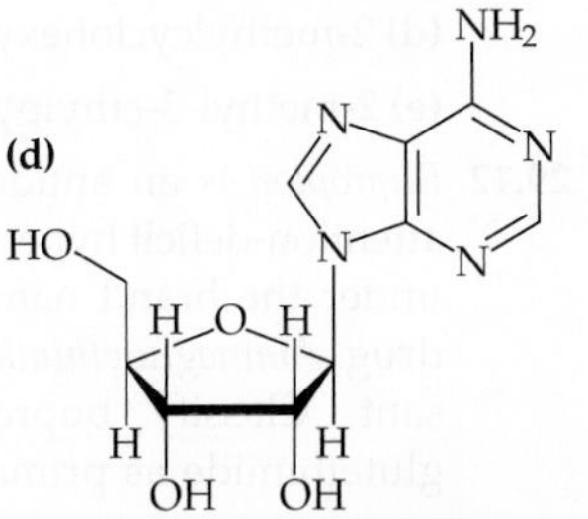

(e)

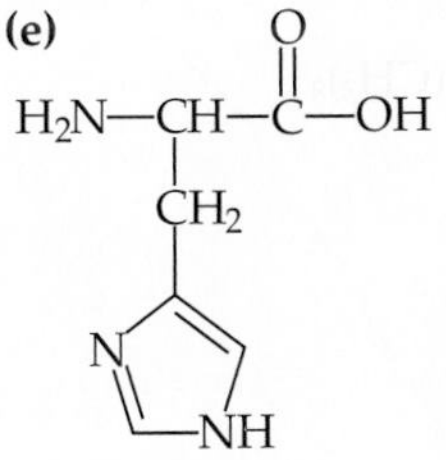

(f)

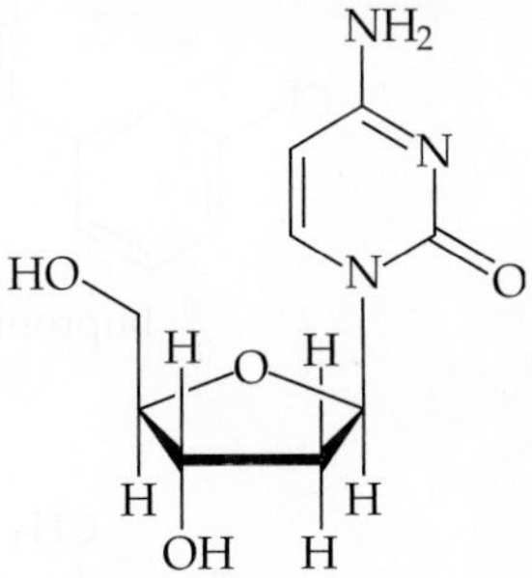

29.7 Identify the peptide formed by the mRNA sequence following translation. [Section 29.4]

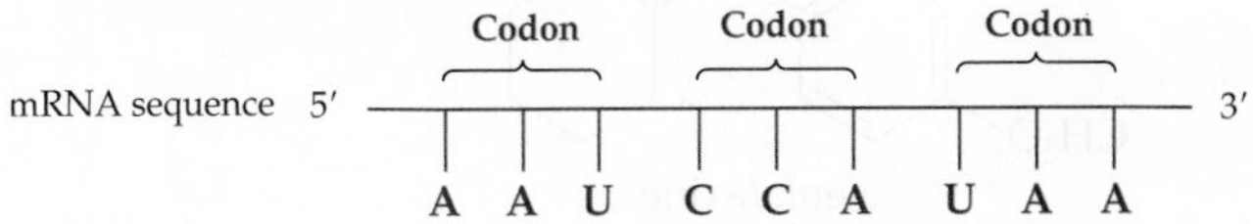

mRNA sequence 5′ —————————————— 3′

A A U   C C A   U A A

Codon   Codon   Codon

AMINES AND THE AMIDE BOND (Section 29.1)

29.8 Give names for the following amines:

(a) **(b)**

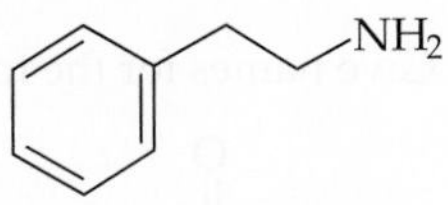

(c) CH₃CH₂NHCH₃ **(d)** CH₃CH₂CHCH₂CH₃ / NH₂

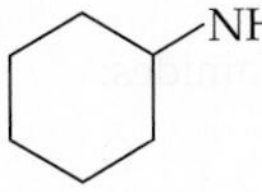

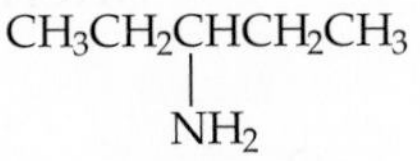

(e) **(f)**

29.9 Give names for the following amines:

(a) **(b)**

(c) **(d)** H₂NCH₂CH₂CH₂CH₂NH₂

29.10 Provide a structural formula for each of the following compounds:

(a) 4-nitropyridine

(b) benzylamine

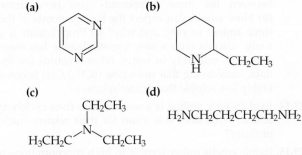

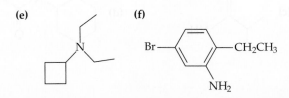

(c) *N,N*-dimethylpropylamine

(d) 2,4-diaminobenzoic acid

(e) 2-ethyl-1-butanamine

29.11 Provide a structural formula for each of the following compounds:

(a) 2-propanamine

(b) 2-aminobutanoic acid

(c) *N,N*-dimethylcyclopentanamine

(d) 2-methylcyclohexylamine

(e) 2-methyl-3-ethylpyrrole

29.12 *Bupropion* is an antidepressant drug also used to treat attention-deficit hyperactivity disorders. *Venlafaxine* (sold under the brand name Effexor®) is an antidepressant drug. *Aminoglutethimide* has been used as an anticonvulsant. Classify bupropion, venlafaxine and aminoglutethimide as primary, secondary or tertiary amines.

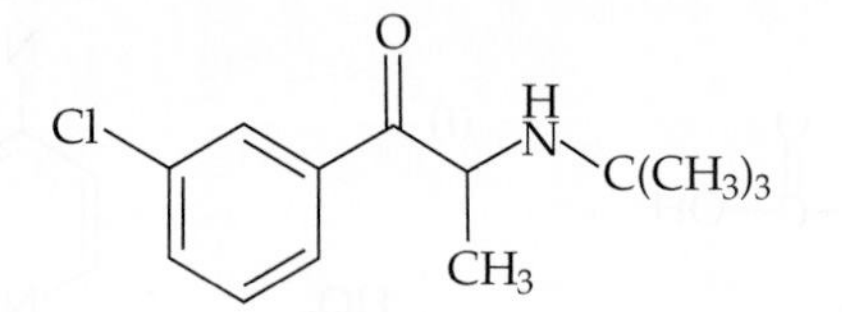

bupropion

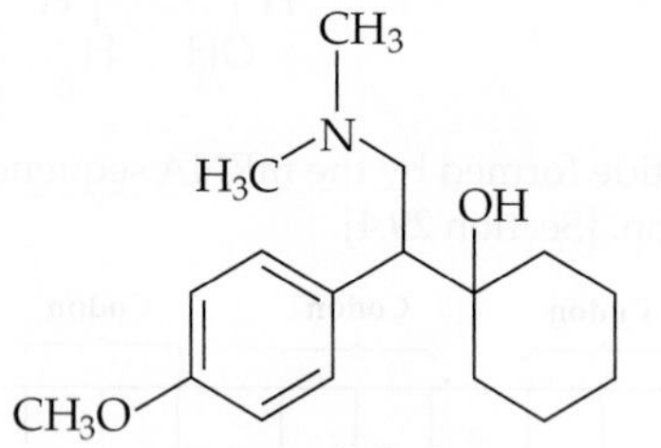

venlafaxine

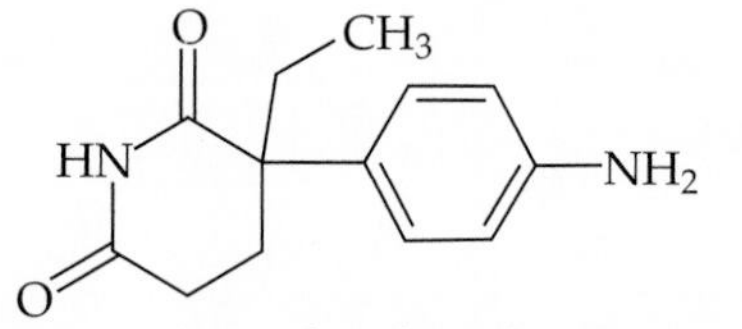

aminoglutethimide

29.13 (a) Draw a primary, secondary and tertiary amine with the molecular formula $C_6H_{15}N$. What relationship exists between the three compounds you have drawn? (b) How would you expect the boiling points of these three amines to vary, and why? (c) Propylamine is entirely miscible with water; trimethylamine has reasonably high solubility in water. What accounts for these data, considering that isobutane $((CH_3)_3CH)$ is considerably less soluble than trimethylamine?

29.14 Explain why aniline is a weaker base than cyclohexylamine. What does this mean for their relative nucleophilicity?

29.15 Indole smells rather terrible in high concentrations but has a pleasant floral odour when highly diluted. It has the following structure:

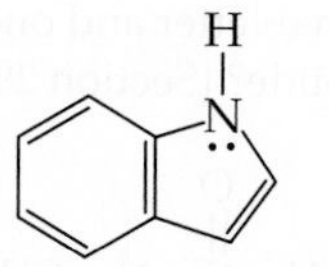

Indole is a planar molecule. The nitrogen is a very weak base, with a K_b of 2×10^{-12}. Explain how the K_b indicates that the indole molecule is aromatic in character.

29.16 *Procaine* was one of the first local anaesthetics used. Its hydrochloride salt is marketed as *Novocaine®*.

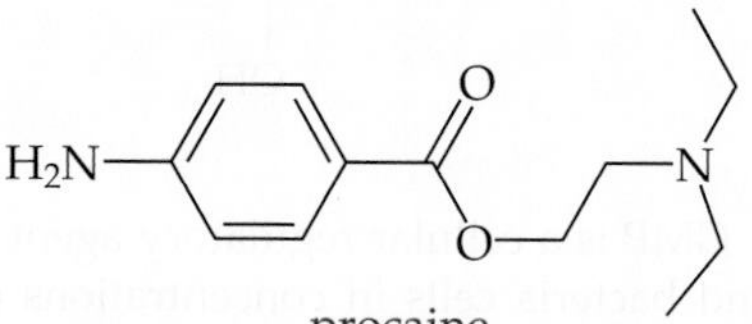

procaine

(a) Draw the product of the reaction of procaine with 1 mole equivalent of HCl. (b) Suggest a synthesis for procaine starting with 4-nitrobenzoic acid.

29.17 Suggest a single-step synthesis of methamphetamine hydrochloride (*speed*) from amphetamine.

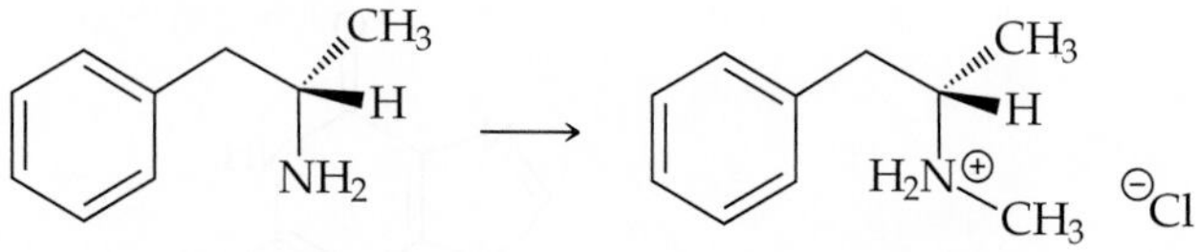

amphetamine   methamphetamine hydrochloride

29.18 Aniline is prepared by catalytic hydrogenation of nitrobenzene. (a) Write an equation for this reaction including reagents and catalyst. (b) Devise a chemical procedure based on the basicity of aniline to separate it from any unreacted nitrobenzene. [*Hint:* Both compounds are soluble in diethyl ether and insoluble in water.]

29.19 *Rimantadine* is effective in preventing infections caused by the influenza A virus. It can be formed from the ketone and the reagents shown. Suggest a structure for rimantadine consistent with the reaction conditions. [*Hint:* The reaction is undertaken in two steps.]

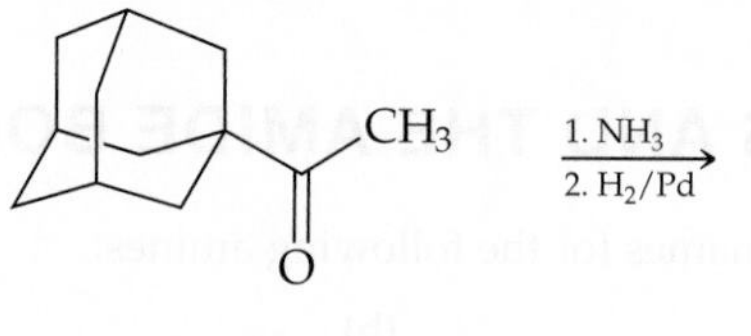

1. NH_3
2. H_2/Pd

29.20 Give names for the following amides:

(a)

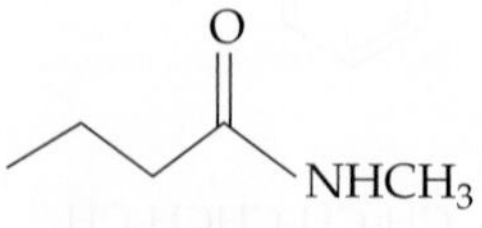

(b)

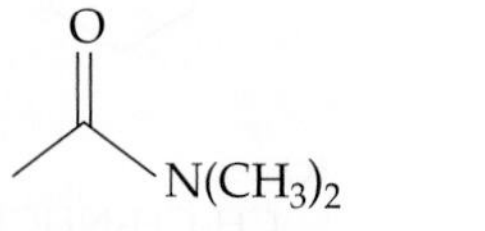

(c)

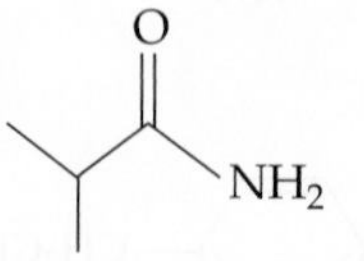

(d)

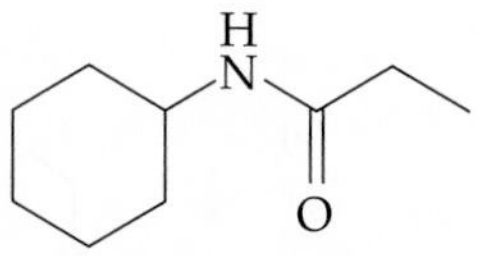

29.21 Provide a structural formula for each of the following compounds:

(a) 2-methylpropanamide

(b) acetamide

(c) N-phenylacetamide

(d) N,N-dimethyl-1-pentanamide

(e) N-ethylbenzamide

29.22 Define the following amides as primary, secondary or tertiary:

(a)

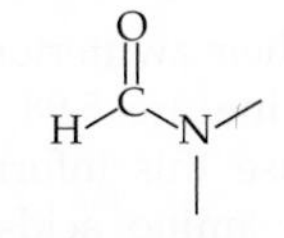

N,N-dimethylformamide
(solvent for organic synthesis)

(b)

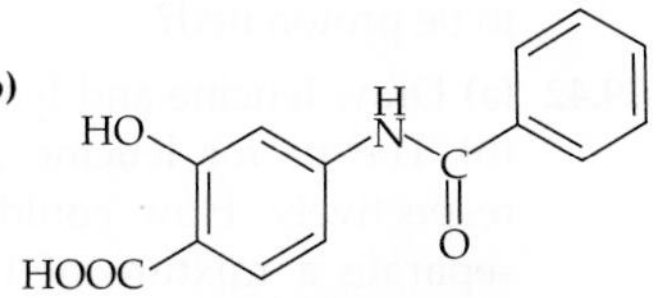

benzoylpas
(an antibacterial)

(c)

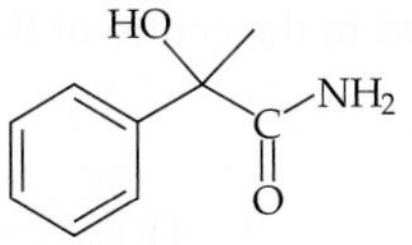

atrolactamide
(an anticonvulsant)

(d)

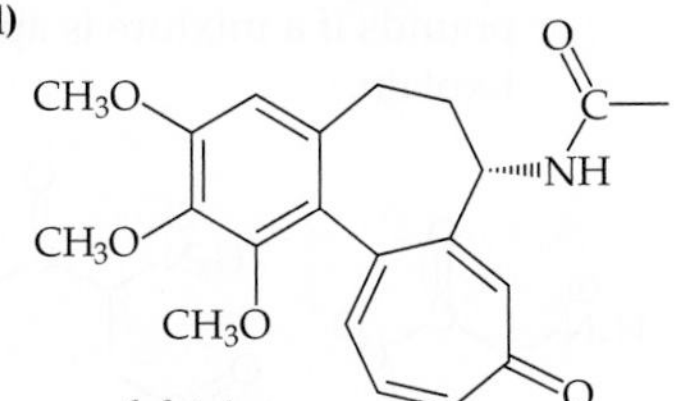

colchicine
(gout suppressant)

(e)

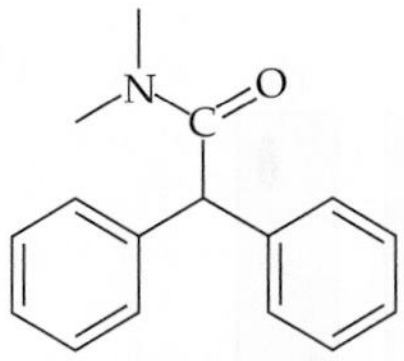

dipenamide
(a herbicide)

(f)

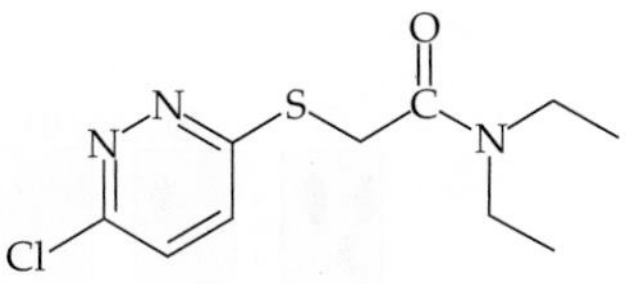

azintamide
(a choleretic)

29.23 Draw a primary, secondary and tertiary amide with the molecular formula $C_5H_{11}NO$. What isomeric relationship exists between the three compounds you have drawn?

29.24 *Capsaicin*, shown below, is the compound responsible for the sensation felt when chillis are eaten. It is often used as a tool in neurobiological research and has application as a topical analgesic. (a) Draw the structure of the carboxylic acid and amine that lead to capsaicin. (b) What is the general name given to this type of reaction?

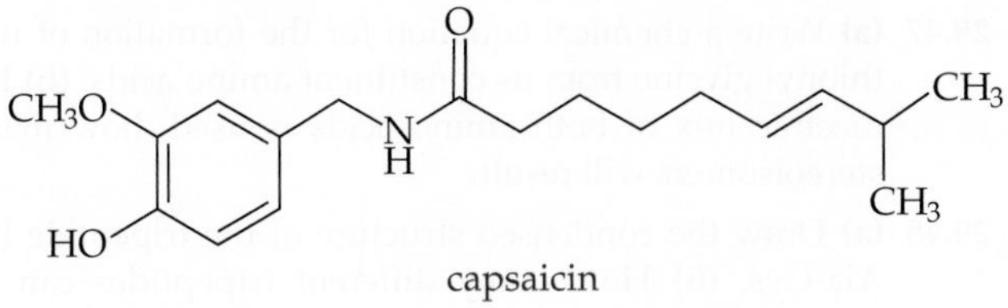

capsaicin

29.25 Devise a synthesis for *paracetamol* beginning with 4-nitrophenol.

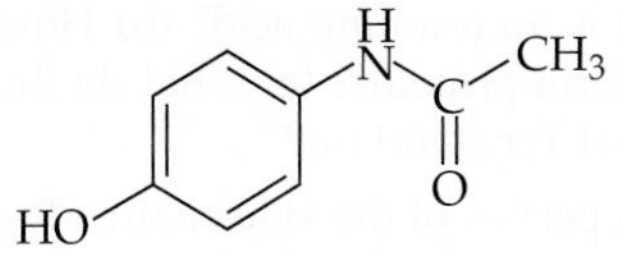

paracetamol

29.26 Propose a mechanism for the following reactions:

(a)

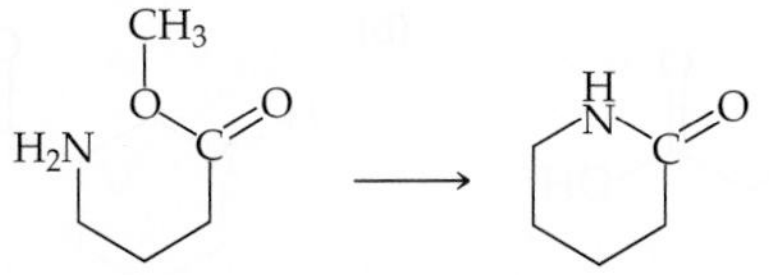

(b)

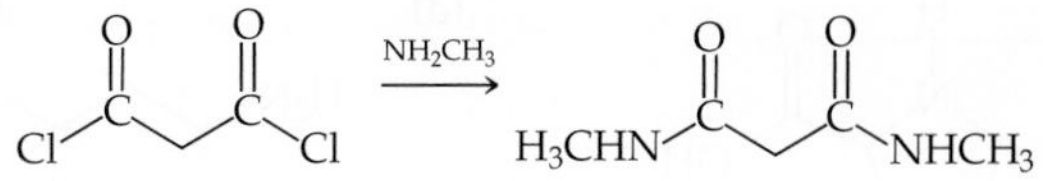

(c)

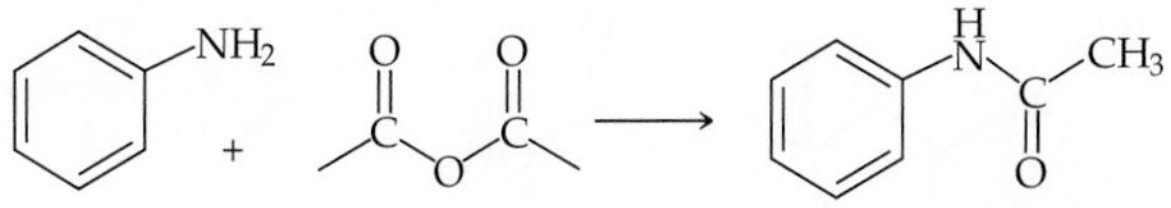

(d)

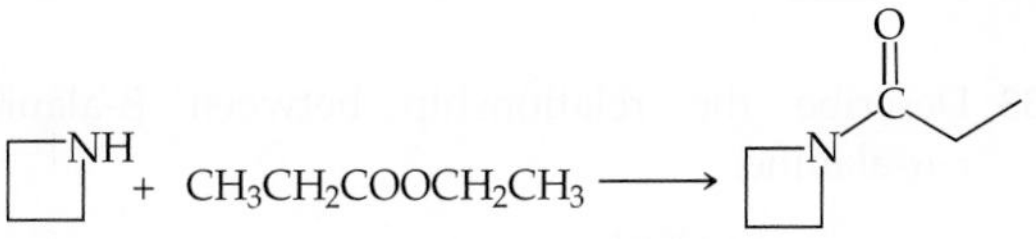

29.27 (a) Propose a stepwise synthesis of *lidocaine*, a local anaesthetic, from 2,6-dimethylaniline, diethylamine and chloroacetyl chloride ($ClCH_2COCl$). (b) To improve the solubility of lidocaine in aqueous solution, it is often administered as the hydrochloride salt. Draw the structure of lidocaine hydrochloride.

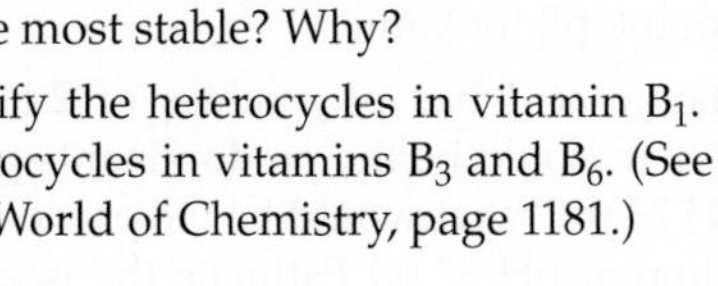

lidocaine

29.28 (a) Draw the product of benzaldehyde and aniline. (b) How many isomers are present? (c) Which one would be most stable? Why?

29.29 (a) Identify the heterocycles in vitamin B_1. (b) Identify the heterocycles in vitamins B_3 and B_6. (See Figure 29.3 and My World of Chemistry, page 1181.)

AMINO ACIDS, PROTEINS AND PEPTIDES (Sections 29.2 and 29.3)

29.30 (a) What is an α-amino acid? (b) How do amino acids react to form proteins? (c) What do the abbreviations F, Q, Pro and Tyr stand for?

29.31 What properties of the side chains (R groups) of amino acids are important in affecting the amino acid's overall biochemical behaviour? Give examples to illustrate your answer.

29.32 Classify the following five molecules as α-, β-, γ- or δ-amino acids:

(a)

(b)

(c)

(d)

(e)

29.33 Describe the relationship between β-alanine and L-α-alanine.

L-α-alanine   β-alanine

29.34 Alanine can exist in water in several ionic forms. (a) Suggest the form of alanine at low pH and at high pH. (b) Amino acids are reported to have two pK_a values, one in the range of 2 to 3 and the other in the range of 9 to 10. Alanine, for example, has pK_a values of about 2.3 and 9.6. Suggest the origin of the two pK_a values.

29.35 Aspartic acid is reported to have three pK_a values, one in the range of 2 to 3, another in the range 3 to 4, and the last in the range of 9 to 10. Suggest the origin of the three pK_a values.

29.36 (a) What is meant by the term *isoelectric point*? (b) Valine has pK_a values of 2.32 and 9.69. Determine the isoelectric point, pI, for valine.

29.37 Glutamic acid has pK_a values of 2.19, 4.25 and 9.67. (a) What would be the predominant species in solution at pH 2? (b) What would be the predominant species in solution at pH 8? (c) Estimate the isoelectric point, pI, for glutamic acid.

29.38 Using Equation 29.28, estimate the average charge on glutamine (a) at pH 3, (b) at pH 7. Glutamine has pK_a values of 2.17 and 9.13.

29.39 Using Equation 29.28, estimate the average charge on phenylalanine (a) at pH 1, (b) at pH 5.5, (c) at pH 8. Phenylalanine has pK_a values of 2.58 and 9.24.

29.40 Why is the average pK_a value (pK_{a1}) for the carboxyl group of an α-amino acid ~2 when that for acetic acid is about 4.8?

29.41 (a) What is meant by the term *zwitterion*? (b) Draw the zwitterion for lysine. (c) Why did you choose that amine to be protonated?

29.42 (a) Draw leucine and lysine in their zwitterionic forms. (b) The pI for leucine and lysine are 5.98 and 9.74, respectively. How could you use this information to separate a mixture of the two amino acids by electrophoresis?

29.43 Which of the following electrophoresis results could represent the separation of the following three compounds if a mixture is applied to the centre of the strip? Explain.

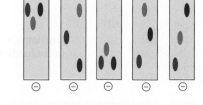

amino acid A   amino acid B   amino acid C

29.44 Outline a synthesis of phenylalanine (a) from 2-bromo-3-phenylpropanoic acid and (b) by the Strecker synthesis from 2-phenylethanal.

29.45 Outline a synthesis of isoleucine (a) from 2-bromo-3-methylpentanoic acid and (b) by the Strecker synthesis.

29.46 (a) Draw the two possible dipeptides formed by the condensation reaction between serine and lysine. (b) If a racemic mixture of both amino acids is used, how many stereoisomers will result?

29.47 (a) Write a chemical equation for the formation of methionyl glycine from its constituent amino acids. (b) If a racemic mix of both amino acids is used, how many stereoisomers will result?

29.48 (a) Draw the condensed structure of the tripeptide Ile-Ala-Cys. (b) How many different tripeptides can be made from the amino acids serine, threonine and phenylalanine? Give the abbreviations for each of these tripeptides, using the three-letter and one-letter codes for amino acids.

29.49 (a) What amino acids would be obtained by hydrolysis of the following tripeptide?

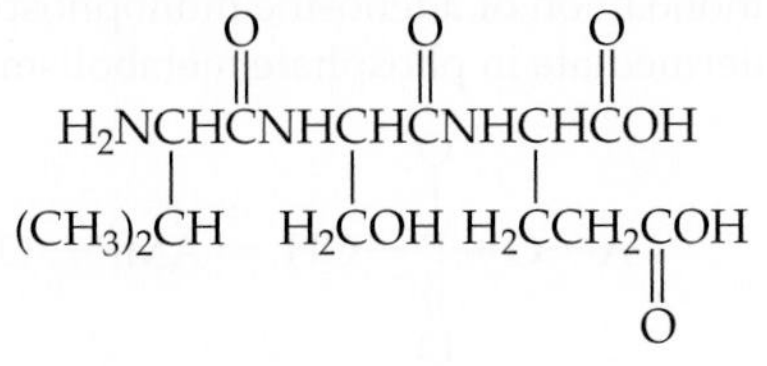

$$H_2NCHCNHCHCNHCHCOH$$

with groups: O (×3 carbonyls), $(CH_3)_2CH$, H_2COH, H_2CCH_2COH with O

(b) How many different tripeptides can be made from the amino acids glycine, serine and glutamic acid? Give the abbreviation for each of these tripeptides, using the three-letter codes for amino acids.

29.50 Describe what is meant by the primary, secondary and tertiary structures of proteins. Give examples to illustrate your answer.

29.51 (a) Describe the role of hydrogen bonding in determining the α-helical structure of a protein. **(b)** Suggest a reason why proline is never found within a protein α-helix.

29.52 (a) Draw the condensed structural formula of each of these tripeptides: Val-Gly-Asp, Phe-Ser-Ala. **(b)** What general properties would you expect each tripeptide to have based on their side chains—for example, are they hydrophobic or hydrophilic, acidic or basic?

29.53 Glutathione is a tripeptide found in most living cells. Partial hydrolysis yields Cys-Gly and Glu-Cys. What is the structure of glutathione?

29.54 Draw the full structural formula for alanylglycylserine.

29.55 Name the dipeptide that has the following structure, and give its one-letter and three-letter abbreviations:

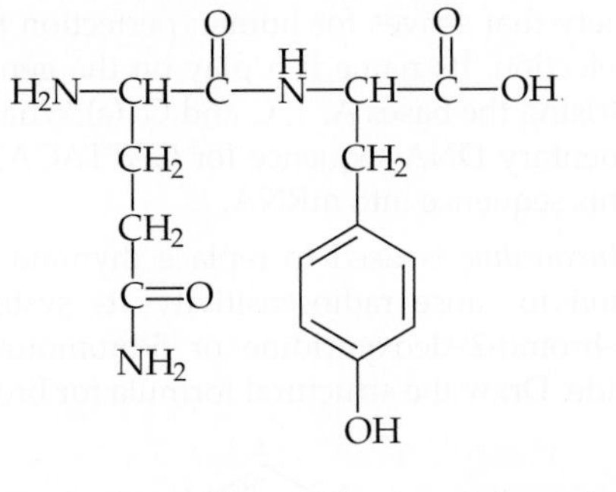

Structure showing: $H_2N—CH—C(=O)—N(H)—CH—C(=O)—OH$ with side chains $CH_2CH_2C(=O)NH_2$ and CH_2-(phenol ring with OH)

29.56 Bradykinin is a small peptide containing nine amino acids. Partial hydrolysis gave the following fragments:

Arg-Pro-Pro, Ser-Pro, Pro-Phe-Arg, Gly-Phe-Ser, Pro-Pro-Gly

What is the primary structure of bradykinin?

29.57 Draw the structure of the phenylthiohydantoin you would expect to obtain from the Edman degradation of the following peptides:

(a) Ala-Gly-Phe

(b) Gly-Phe-Ala

(c) Phe-Ala-Gly

29.58 A new cyclic pentapeptide has been isolated from a marine sponge. Total hydrolysis of the cyclic peptide yielded the following amino acid composition:

Lys　　Gly　　Phe　　Cys

(a) What experimental conditions and reagents could be used to perform the above hydrolysis? **(b)** What techniques could be used to cleave the cyclic peptide at specific sites to help identify the structure?

29.59 The abbreviation for a peptide is VAKG. **(a)** Draw the structural formula for this peptide. **(b)** Is this peptide neutral, acidic or basic? **(c)** What would be the products of a trypsin digestion? **(d)** Which amino acid would be identified in VAKG by the Edman degradation?

29.60 Using the information from the following experiments, write the sequence of the linear peptide.

Experimental procedure	Amino acid composition	
1. Edman degradation	Cys	
2. Hydrolysis using chymotrypsin	Fragment A	Cys, Phe, Gly
	Fragment B	Cys, Lys
3. Hydrolysis using trypsin	Fragment C	Cys, Gly, Lys, Phe
	Fragment D	Cys

29.61 Which of the following represent/s the best approach for sequencing a 50-amino acid polypeptide? Why?

(a) Edman degradation, followed by hydrolysis and then trypsin digestion.

(b) Use of cyanogen bromide, followed by Edman degradation and finally a chymotrypsin digestion.

(c) Edman degradation, trypsin digestion and reaction with cyanogen bromide.

29.62 The standard free energy of formation of solid glycine is -369 kJ mol^{-1}, whereas that of solid glycylglycine is -488 kJ mol^{-1}. What is $\Delta G°$ for the condensation of glycine to form glycylglycine?

29.63 The protein ribonuclease A in its native, or most stable, form is folded into a compact globular shape. **(a)** Does the native form have a lower or higher free energy than the denatured form, in which the protein is an extended chain? **(b)** What is the sign of the entropy change in going from the denatured to the folded form? **(c)** In the folded form the ribonuclease A has four –S–S– bonds that bridge parts of the chain, as shown in the diagram below. What effect do you predict that these four linkages have on the free energy and entropy of the folded form as compared with a hypothetical folded structure that does not possess the four –S–S– linkages? Explain. **(d)** A gentle reducing agent converts the four –S–S– linkages to eight –S–H bonds. What effect do you predict this would have on the tertiary structure of the protein?

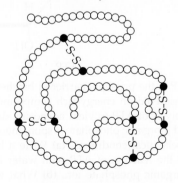

native ribonuclease A

NUCLEIC ACIDS AND DNA (Section 29.4)

29.64 Describe a nucleotide. Draw the structural formula for the nucleotide deoxycytidine monophosphate, which contains cytosine in the organic base.

29.65 A nucleoside consists of an organic base of the kind shown in Section 29.4, bound to ribose or deoxyribose. Draw the structure for deoxyguanosine, formed from guanine and deoxyribose.

29.66 Write a balanced chemical equation using condensed formulae for the condensation reaction between a mole of deoxyribose and a mole of doubly ionised phosphoric acid, HPO_4^{2-}.

29.67 A nucleotide undergoes hydrolysis under neutral conditions to yield 1 mol of $H_2PO_4^-$ and an organic product. The same starting material undergoes hydrolysis under acidic conditions to yield thymine and deoxyribose monophosphate. Draw the structure of the unknown substance.

29.68 When samples of double-stranded DNA are analysed, the molar quantity of adenine present equals that of thymine. Similarly, the molar quantity of guanine equals that of cytosine. Explain the significance of these observations.

29.69 Imagine a single DNA strand containing a section with the following base sequence: ACTCGA. What is the base sequence of the complementary strand?

29.70 **(a)** Write a nucleic acid sequence for the strand GGTACT, using the concept of complementary base pairing. **(b)** Is this strand DNA or RNA based? **(c)** An alternative notation is 5'-GGTACT-3'. What does this mean?

29.71 An mRNA strand has the following sequence: GUCAGGAAUCUU. What would be the sequence of the parent DNA strand that led to this transcription?

29.72 One of the most important molecules in biochemical systems is adenosine triphosphate (ATP), for which the structure is:

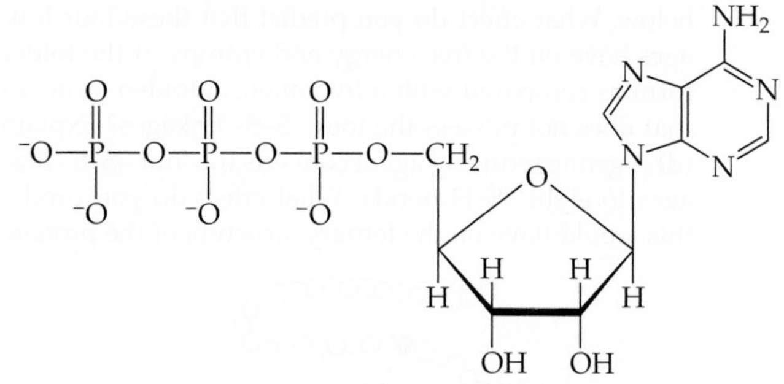

ATP

ATP is the principal carrier of biochemical energy. It is considered an energy-rich compound because the hydrolysis of ATP to yield adenosine diphosphate (ADP) and inorganic phosphate is spontaneous under aqueous biochemical conditions. **(a)** Write a balanced equation for the reaction of ATP with water to yield ADP and inorganic phosphate ion. **(b)** What would you expect for the sign of the free energy change for this reaction? **(c)** ADP can undergo further hydrolysis. What would you expect for the product of that reaction?

29.73 The monoanion of adenosine monophosphate (AMP) is an intermediate in phosphate metabolism:

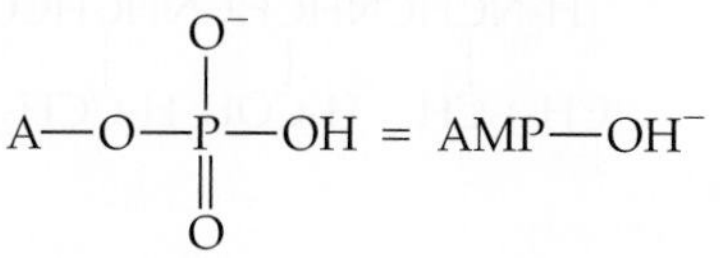

where A = adenosine. If the pK_a for this anion is 7.21, what is the ratio of [AMP—OH⁻] to [AMP—O²⁻] in blood at pH 7.4?

29.74 *Chargaff's rule* states that DNA contains equimolar amounts of guanine and cytosine and also equal amounts of adenine and thymine. **(a)** Does this imply that there are equimolar amounts of purine and pyrimidine bases in DNA? **(b)** Does Chargaff's rule apply only to double-stranded DNA or is it also applicable to single-stranded DNA? Explain.

29.75 **(a)** Draw the structure of adenine and thymine and indicate how they interact through Watson–Crick base pairing. **(b)** Draw the structure of cytosine and guanine and indicate how they interact through Watson–Crick base pairing. **(c)** Why doesn't guanine base-pair with thymine? Explain.

29.76 Poly U added to a cell-free system containing all the necessary materials for biosynthesis yields a polypeptide comprising a single amino acid. **(a)** What is that amino acid? **(b)** What synthetic addition polymer might have similar properties to the one formed by translating poly U?

29.77 The motion picture *GATTACA* (1997) is set around a society that strives for human perfection through genetic selection. Its name is a play on the genetic code, comprising the bases A, T, C and G. **(a)** What is the complementary DNA sequence for GATTACA? **(b)** Transcribe this sequence into mRNA.

29.78 *Broxuridine* is used to replace thymine during replication to cause radiosensitivity. Its systematic name is 5-bromo-2'-deoxyuridine or 5-bromouracil deoxyriboside. Draw the structural formula for broxuridine.

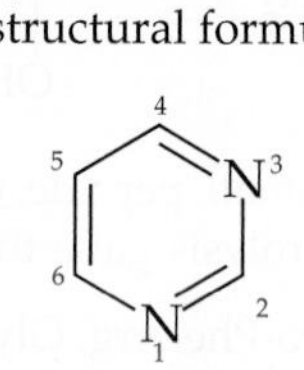

pyrimidine

29.79 *Cladribine* is a substituted purine nucleoside with antileukaemic activity. Its systematic name is 2-chloro-2'-deoxyadenosine. Draw the structural formula for cladribine.

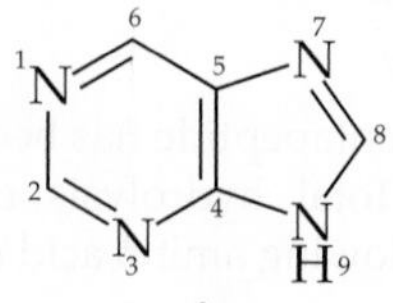

purine

29.80 What amino acid sequence is encoded by the following mRNA sequence?

UUU CCA GUG GAU CCG AUC UAA

29.81 **(a)** What amino acid sequence is encoded by the following mRNA sequence?

AGG ACU GCA UCG CAA

(b) What anticodon sequence of tRNAs is needed for the synthesis of this peptide?

INTEGRATIVE EXERCISES

29.82 Devise a synthesis for nicotinamide starting from nicotinic acid.

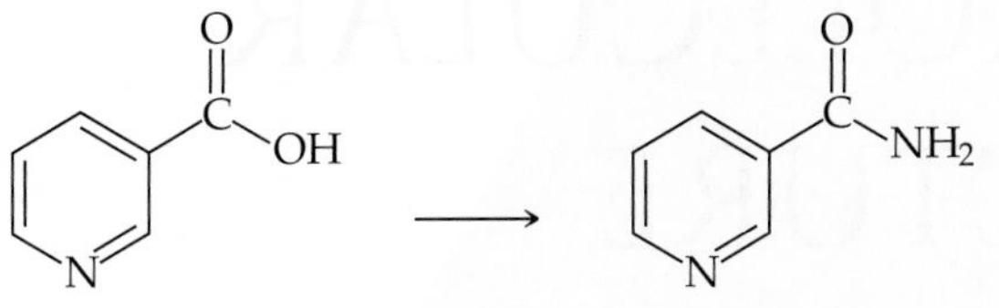

nicotinic acid   nicotinamide

29.83 **(a)** How many amide groups are there in vitamin B_{12}? See page 1181. **(b)** What type of glycoside is found in vitamin B_{12}? **(c)** Is the sugar component a furanose or pyranose form?

29.84 What general features of a vitamin allow you to predict whether it is a hydrophilic or a lipophilic vitamin?

29.85 Quinine is a natural product that has anti-malarial properties. It was originally extracted for therapeutic use from the bark of the cinchona tree, but is now synthesised by the pharmaceutical industry. Quinine is generally administered as the hydrochloride salt ($C_{20}H_{24}N_2O_2 \cdot HCl$). The p$K_a$ of this salt is 4.32. **(a)** What are the advantages of using the hydrochloride salt of quinine in medicine? **(b)** What is the pH of a 0.053 M solution of quinine hydrochloride?

29.86 Imidazole is a cyclic amine with the following structure.

(a) Draw a tautomer of imidazole. **(b)** Imidazole is a highly polar molecule; draw a resonance structure that demonstrates this polarity. **(c)** Imidazole is aromatic with six π-electrons; identify the π-electrons involved. **(d)** Protonation of imidazole occurs on only one of the nitrogen atoms. Which nitrogen atom is protonated and explain why this is so? Imidazole forms part of the side chain of the amino acid histidine and coordinates to iron in the oxygen transport protein, haemoglobin. **(e)** What part of the imidazole molecule acts as a donor in this coordination?

MasteringChemistry (www.pearson.com.au/masteringchemistry)

Make learning part of the grade. Access:

- tutorials with peronalised coaching
- study area
- Pearson eText

PHOTO/ART CREDITS

1146 Reproduced with permission of the March of Dimes Foundation; **1155** © Robyn Mackenzie/Dreamstime.com; **1163** Frank LaBua/Pearson Education/PH College; **1166** Meckes/Ottawa/Photo Researchers; **1167** © D. Vo Trung/Look at Sciences/Phototake; **1174** © photocuisine/Corbis; **1177** Richard Megna, Fundamental Photographs, NYC; **1186** © Bettmann/Corbis.

30

SOLVING MOLECULAR STRUCTURE

Rainbows give us insight into the different energies associated with light. Blue light is higher in energy while red is lower in energy. Chemists use a range of spectroscopic techniques that use this same energy distribution to distinguish between compounds at a molecular level and to confirm their structures.

KEY CONCEPTS

30.1 THE ELECTROMAGNETIC SPECTRUM
We begin by revisiting the electromagnetic spectrum and describe the regions of the spectrum in terms of electronic, molecular and nuclear transitions for organic compounds. The concept of an absorption spectrum is reinforced.

30.2 INFRARED (IR) SPECTROSCOPY
We then investigate IR spectroscopy as a means of identifying the most common functional groups within organic molecules. We discuss the concept of a wavenumber (cm^{-1}) and its relationship to wavelength.

30.3 NUCLEAR MAGNETIC RESONANCE (NMR) SPECTROSCOPY
The most valued tool for organic chemists is NMR spectroscopy. We discuss two types only, ^{1}H and ^{13}C NMR spectroscopy, but the principles hold for many more nuclei. We explore the concepts of *chemical shift* (δ), *integration* and *spin–spin coupling* (*J*), then we employ them to solve the structure of simple organic molecules.

30.4 MASS SPECTROMETRY
We then investigate electron impact mass spectrometry (EIMS) to describe the main principles of mass spectrometry techniques. *Fragmentation* and *molecular ion* (M$^+$) signal give us information on functionality and molecular weight, respectively. We identify isotope patterns for Cl and Br, providing useful information on molecular formula.

30.5 COMPOUND IDENTIFICATION USING SPECTRA
Finally, we combine the techniques in the above sections, together with simple chemical functional group tests and information within the molecular formula through the *index of hydrogen deficiency* (IHD), to piece together the molecular structure of organic unknowns.

As chemists we regularly need to prove, unambiguously, the presence and molecular structure of a compound, whether it is newly synthesised in the laboratory or a routine anti-doping test of Olympic competitors. Modern spectroscopic analysis is the backbone of the techniques used for this determination. Such analysis can be quick and accurate and, with skilful interpretation, may provide a detailed view of the exact shape of a molecule.

In Chapters 22 to 29 we discussed aspects of molecular structure and functionality within simple organic structures, and we have seen that the activity of a compound depends not only on the types of functional groups present but also on the shape and, specifically, the stereochemistry present. A compound is so much more than just a random collection of different atoms. Spectroscopy is capable of providing us with the information we need to construct accurately the three-dimensional structure of a molecule.

In the past, chemical reactions characteristic of particular functional groups were used to prove the presence of, for example, an aldehyde or amine in the compound under investigation. Unfortunately such analysis requires more material than may be available and are destructive in that the compound being tested is irreversibly changed. Analytical techniques that work on very small amounts and do not destroy or compromise the sample are far more suited to modern-day chemistry. Furthermore, the general complexity of molecules in terms of the functional groups present in natural products (for example, morphine or batrachotoxin), and the tendency of isolated natural products to be of one optically active form, mean that chemical reaction analysis techniques are often not sufficient to give the whole molecular structure.

morphine
(Gk *morphos*, god of dreams)

batrachotoxin
(toxin from the poison dart frog)

Modern spectroscopic techniques use only minute amounts of material, which in most cases may be recovered unchanged after the analysis. In this chapter we look at the core instrumental methods used to determine the molecular weight of a compound, the functional groups present and the three-dimensional shape that they adopt.

30.1 | THE ELECTROMAGNETIC SPECTRUM

∞ Review this on page 180

In Chapter 6 we introduced the electromagnetic spectrum and the concepts of wavelength (λ) and frequency (v) (∞ Section 6.1, 'The Wave Nature of Light').

Three regions of the electromagnetic spectrum are particularly important to the organic chemist (▼ FIGURE 30.1). The UV (ultraviolet) region of the spectrum is of an energy that equates to the transition of electrons between electronic energy levels within a molecule (▶ FIGURE 30.2). The technique, aptly named **UV-visible spectroscopy**, deals with this region and is particularly useful for aromatic and π-conjugated compounds. The infrared (IR) region of the spectrum corresponds to transitions between the vibrational energy levels within a molecule. This region of the electromagnetic spectrum is lower in energy than that required to excite an electron from the ground state using ultraviolet radiation, and provides unique information on the types of functional groups present in a molecule. The third region of importance is found at the frequency of radio waves. In this region, the energies relate to transitions between nuclear-spin energy levels. The energies required here are much lower than for either IR- or UV-visible spectroscopy.

When an organic molecule is exposed to a spectrum of electromagnetic radiation, it absorbs energy at certain wavelengths in the ultraviolet and visible region and allows other wavelengths to pass through. The absorption of energy by a molecule leads to an absorption spectrum and highlights the transition between different energy levels by an electron. Figure 30.2 illustrates the process. Two different energy levels (E_1 and E_2) are related by the absorption of energy with a particular wavelength (λ). This wavelength is related to the dif-

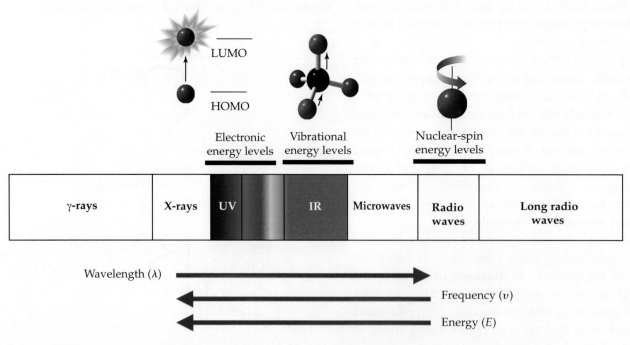

▲ **FIGURE 30.1 The electromagnetic spectrum.** Energy depends upon wavelength. The regions are described in terms of atomic transitions and the type of related spectroscopy.

FIGURE IT OUT

The absorption characteristics displayed by molecules have an analogy in everyday life. Why do dark-coloured cars get hotter inside than light-coloured cars?

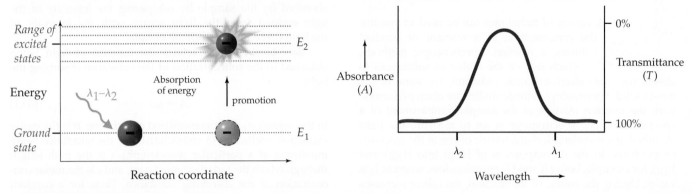

▲ **FIGURE 30.2   An absorption spectrum** is generated when a molecule absorbs electromagnetic radiation of a particular wavelength range ($\lambda_1 - \lambda_2$).

ference between E_1 and E_2 by Equations 6.1 and 6.2, so $\lambda = ch/(E_2 - E_1)$. A molecule initially in the lower-energy E_1 state, referred to as the *ground state*, can undergo a transition to E_2, the *excited state*, only by absorption of electromagnetic energy of wavelength λ. In gas-phase systems, absorption spectra can be sharp lines, but in practice a range of excited states of similar energies are possible. In liquids and solids, absorption invariably spreads over a range of wavelengths. Most spectroscopic techniques detect this absorption in order to glean information about the electronic nature of the molecule.

An absorption spectrum is a way of recording the wavelength maximum at which the absorption occurs. Such a spectrum also indicates how much radiation is absorbed.

In this chapter we deal with three powerful and complementary techniques that form the cornerstone of modern synthetic organic chemistry. Each technique is useful because it provides a vital piece of evidence, using very small amounts of sample. The chemical detective can draw a logical conclusion to the structural problem by combining two, but usually more, pieces of evidence. The techniques are:

- *infrared (IR) spectroscopy*, which observes the vibrations of covalent bonds and provides useful evidence for the presence of functional groups;
- *nuclear magnetic resonance (NMR) spectroscopy*, the technique of choice for the organic chemist. This method observes nuclear transitions within a strong magnetic field and yields information on the connectivity of nuclei (atoms);
- *mass spectrometry (MS)*, which typically bombards a very small amount of sample with electrons to ionise the molecules. Analysis of the mass of these ions gives the molecular mass. Mass spectrometry often provides clues to the structure and functional groups within a molecule, through either fragmentation or isotopic patterns. This technique is often used routinely because it needs only µg amounts of sample.

CONCEPT CHECK 1

Porphyrins, the major pigments of photosynthetic bacteria and all plants, have molar extinction coefficients $\varepsilon = 10^5 - 10^6$ dm^3 mol^{-1} cm^{-1} in the visible spectrum, whereas simple aromatic compounds like benzene or naphthalene have $\varepsilon = 10^2 - 10^3$ dm^3 mol^{-1} cm^{-1}. Explain why nature would choose to employ porphyrins over naphthalene for photosynthetic processes.

A CLOSER LOOK

USING SPECTROSCOPIC METHODS TO MEASURE REACTION RATES

A variety of techniques can be used to monitor the concentration of a reactant or product during a reaction. Spectroscopic methods, which rely on the ability of substances to absorb (or emit) electromagnetic radiation, are some of the most useful. Spectroscopic kinetic studies are often performed with the reaction mixture in the sample compartment of a spectrometer. The spectrometer is set to measure the light absorbed at a wavelength characteristic of one of the reactants or products. In the decomposition of HI(g) into H_2(g) and I_2(g), for example, both HI and H_2 are colourless, whereas I_2 is violet. During the course of the reaction, the colour increases in intensity as I_2 forms. As the concentration of I_2 increases and its colour becomes more intense, the amount of light absorbed by the reaction mixture increases, causing less light to reach the detector.

▼ FIGURE 30.3 shows the basic components of a spectrometer. The spectrometer measures the amount of light absorbed by the sample by comparing the intensity of the light emitted from the light source with the intensity of the light that emerges from the sample.

Beer–Lambert's law relates the amount of light being absorbed to the concentration of the substance absorbing the light:

$$A = \varepsilon cl \qquad [30.1]$$

In this equation, A is the measured absorbance, ε is the molar extinction coefficient (a characteristic of the substance being monitored at a particular wavelength), l is the path length through which the radiation must pass, and c is the molar concentration of the absorbing substance. Thus, for a constant path length, the concentration is directly proportional to absorbance.

RELATED EXERCISE: 30.1

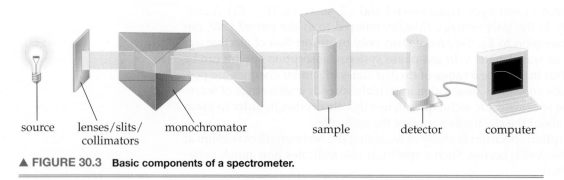

source | lenses/slits/collimators | monochromator | sample | detector | computer

▲ **FIGURE 30.3** **Basic components of a spectrometer.**

30.2 | INFRARED (IR) SPECTROSCOPY

Infrared photons (*infra*, Latin, meaning below or beneath) do not have enough energy to cause electronic transitions, but they can cause groups of atoms to vibrate about their respective bonds. These vibration transitions correspond to distinct energy levels. Molecules and, more importantly, the different functional groups contained within them, absorb IR radiation at specific wavelengths to induce the different vibrational modes within a covalent bond (that is, single, double, triple bonds). **Infrared spectroscopy** is the technique that records the absorption of infrared energy of a molecule as a function of wavelength. It identifies functional groups, but gives little structural information.

The position of an infrared band in an *IR spectrum* is defined by a **wavenumber** value (abscissa), which corresponds to the number of cycles the wave produces in a centimetre (and is therefore written in terms of cm^{-1}); and the band's percentage transmittance (%T) (ordinate), as shown in ▶ FIGURE 30.4. The wavenumber has traditionally been the most common method of specifying IR absorption, although sometimes the wavelength value (in microns) is also employed.

The absorbance of a sample is related to the **transmittance** by the equation:

$$\text{Absorbance} = -\log_{10}(\text{transmittance})$$

or

$$A = -\log_{10}T \qquad [30.2]$$

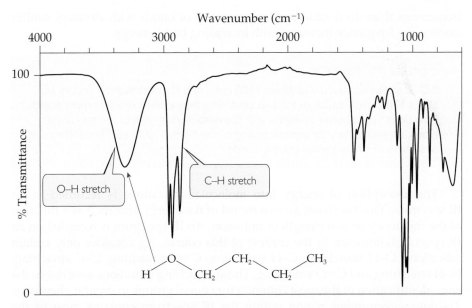

▲ **FIGURE 30.4 IR spectrum of butan-1-ol.** This IR spectrum is measured as a function of wavenumber (cm⁻¹). The ordinate describes the intensity, measured as percentage transmission (%*T*). In this spectrum of butan-1-ol, the broad, strong absorption at 3300 cm⁻¹ is due to the strong O–H stretching vibrations that occur at this frequency.

The value of transmittance (*T*) always lies between the values of 0 and 1 and the %*T* between 0 and 100.

The Spring Model

Before discussing the characteristics of IR absorptions, it is helpful to understand some theory on molecular vibrations. To do this, let's consider the covalent bond between two atoms as a spring (▶ FIGURE 30.5). This spring can be stretched (shown), compressed, bent or twisted, and then released from the input leading to the distortion. The output, which can be observed as a band within the IR spectrum, is in fact a vibration. The force needed to cause the vibration follows **Hooke's law**, which states that the elongation of a spring is proportional to the load applied. Mathematically:

$$F_s = -kx \tag{30.3}$$

where F_s is the applied force (which equals the force exerted by the spring, only opposite in sign) in units of newtons (N), k is the spring constant (N m⁻¹) and x is the deformation of the spring (for example, stretching, with units m) from an equilibrium position (▶ FIGURE 30.6). The strain in the spring is directly proportional to the load. In any system that can be described by Hooke's law, there exists a fundamental frequency of vibration, $v = \frac{1}{2}\pi\left(\frac{k}{m}\right)^{\frac{1}{2}}$. This is also true for molecules, so the frequency of vibration of any chemical bond depends on the mass (m) of atoms at the moving end of the bond and the stiffness (k) of the bond.

In a group of bonds with similar bond energies, the frequency decreases with an increase in atomic weight of the atoms. Of more interest to the organic chemist is the fact that stronger bonds (for example, double bonds) are generally 'stiffer' than single bonds and require more force to stretch or compress them. As a result, stronger bonds usually vibrate faster than weaker bonds. For example, O–H bonds (*D* O–H = 463 kJ mol⁻¹) are stronger than C–H bonds (*D* C–H = 413 kJ mol⁻¹), so O–H bonds vibrate at higher wavenumber (∞ Section 8.8, 'Strengths of Covalent Bonds'). Carbon–carbon triple bonds are stronger than C–C double bonds, so triple bonds vibrate at higher

◣ **FIGURE IT OUT**

Do loose springs oscillate with a higher or lower frequency than tight springs?

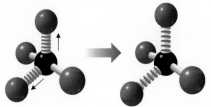

Symmetric stretching

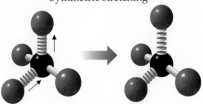

Asymmetric stretching

▲ **FIGURE 30.5 Spring model.** IR spectroscopy can be more easily understood by thinking of covalent bonds as springs. The stretching, bending or torsion of the spring leads to vibrations of a particular frequency. The vibrations can be modelled using Hooke's law.

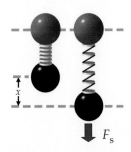

▲ **FIGURE 30.6 Hooke's law** of elasticity applied to a spring system states that the elongation of a spring is proportional to the load applied.

∞ Review this on page 279

frequencies than do double bonds. In a group of bonds with atoms of similar masses, the frequency increases with increasing bond energy.

> **CONCEPT CHECK 2**
>
> If C–O stretching occurs at about 1000 cm^{-1} and C=O stretching occurs at about 1700 cm^{-1}, which of the two stretching frequencies requires more energy? How does your answer correlate with the relative strengths of the two bond types, as well as to your understanding of the relative 'stiffness' in bending and stretching of single versus double bonds?

The absorption of energy into molecular vibrations is recorded in an IR spectrum. This spectrum gives a record of the energy absorbed as a function of the frequency or wavelength of radiation. An IR spectrum is recorded on an IR spectrophotometer. In the context of this course, we consider only certain vibrations: O–H stretching, C–H stretching, C=C stretching, C–C stretching, N–H stretching and C=O stretching. These stretching vibrations give rise to the easy identification of the most common functional groups in organic chemistry. The low-wavenumber region within the IR spectrum contains most of the complex vibrations, commonly called the **fingerprint region** of the spectrum (300–1400 cm^{-1}). Although it is useful to know of this region, its use in identifying organic molecules is limited because of the complexity of the spectrum in this region. Inorganic chemists find this region useful for the identification of metal–ligand (M–L) interactions. Organic chemists use the fingerprint region primarily to confirm the identity of a sample, by comparing it with an authentic spectrum of a known compound thought to be the same as the sample under investigation.

Stretching vibrations in the region 1550–3800 cm^{-1} are the most characteristic and predictable.

At this point we should also recognise that not all molecular vibrations absorb IR radiation. Generally, only bonds with dipole moments (for example, O–H, C=O, N–H bonds) are IR active and hence result in IR absorptions. A dipole moment between two covalently bonded atoms is achieved when the two atoms differ significantly in their electronegativity (∞∞ Section 8.4, 'Bond Polarity and Negativity').

∞∞ Review this on page 262

If the molecule is placed in an electromagnetic field, its bonds interact with the field, usually leading to a change in the vibration of the covalent bond, and a change in the dipole moment of the bond. If a bond is symmetrical or has zero dipole moment, the applied field does not interact with the bond as strongly, and this very weak vibration is said to be *IR inactive*.

Measuring IR Spectra

IR spectra are measured using liquid, solid or gaseous samples placed in a beam of infrared light. Typically, if the sample is a liquid, the compound can be thinly spread *neat* (without solvent) between two thick plates of sodium chloride, which are transparent to infrared light for most of the important wavelength regions. Solid samples are usually prepared by grinding a sample into a paste with an organic oil known as nujol (a type of paraffin oil) forming a mull, which is also sandwiched between two thick NaCl plates. A more modern approach is to place a solid on an ATR diamond or ZnSe window cell and a spectrum taken directly (▶ FIGURE 30.7). ATR stands for attenuated total reflectance. Gases typically require specialised cells.

▶ TABLE 30.1 illustrates the predicted regions where the most common functional groups absorb IR radiation. As mentioned above, strong bonds generally absorb at higher frequencies because of their greater 'stiffness'. Thus C–C single bonds absorb around 1200 cm^{-1}, C=C double bonds around 1600–1680 cm^{-1} and C≡C triple bonds around 2200 cm^{-1}. The absorption of

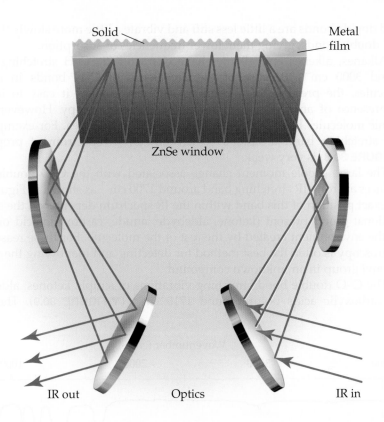

Solid   Metal film

ZnSe window

IR out   Optics   IR in

◀ **FIGURE 30.7 Attenuated total reflectance (ATR) IR spectroscopy.** Modern IR spectrometers use a direct approach by measuring the transmittance of electromagnetic radiation through a sample placed on a diamond or ZnSe surface.

TABLE 30.1 • Bond-stretching frequencies and their relationship to functional groups

Functional group	Representation	Stretching vibration frequency (cm^{-1})	Strength of IR band	Bond energy (kJ mol^{-1})*
O — H		3200–3500	Strong, broad	465
N — H		3100–3500	Medium	390
C — H		2700–3100	Strong to medium	415
C ≡ N		2200–2400	Strong to medium	890
C ≡ C		2000–2200	Weak	840
C = O		1630–1800	Strong	800
C = N		1630–1680	Strong	615
C = C		1600–1680	Weak	615

* Bond energies rounded to the nearest 5 kJ mol^{-1}.

C=C double bond vibrations in an IR spectrum are diagnostic and useful for structure determination. The position of C–C single bond absorptions in an IR spectrum are not very diagnostic because they occur in most organic compounds as part of the molecules' framework. Most double bonds produce observable absorptions in the region 1600–1680 cm^{-1}. Conjugation of the double bond has the effect of reducing the stretching frequencies from 1620–1640 cm^{-1} (in the case of an aliphatic C=C) to approximately 1600 cm^{-1} for an aromatic C=C. The lowering in frequency can be attributed to the loss of electron density in the double bonds due to the conjugation. As a result, conjugated double bonds are a little less stiff and vibrate a little more slowly than isolated double bonds, giving them lower frequencies of absorption.

Alkanes, alkenes and alkynes have characteristic C–H stretching bands around 3000 cm^{-1}. Because of the prevalence of C–H bonds in organic molecules, the presence of a band at 3000 cm^{-1} makes it easy to identify the presence of an organic compound by IR spectroscopy. However, some organic molecules have very weak absorbances in this area. For example, the C–H stretching band around 2900 cm^{-1} in an IR spectrum of propanone (▼ FIGURE 30.8) is very weak.

The large dipole moment change associated with the C–O double bond produces a strong IR stretching band around 1700 cm^{-1} as shown in Figure 30.8. The exact position of this band within the IR spectrum depends on the specific functional group present (ketone, aldehyde, amide, carboxylic acid or ester) and the environment created by the rest of the molecule. For these reasons, IR spectroscopy is often the best method for detecting and identifying the type of carbonyl group in an unknown compound.

The C–O double bond stretching vibrations of simple ketones, aldehydes and carboxylic acids occur around 1710 cm^{-1} (▼ FIGURE 30.9). These are

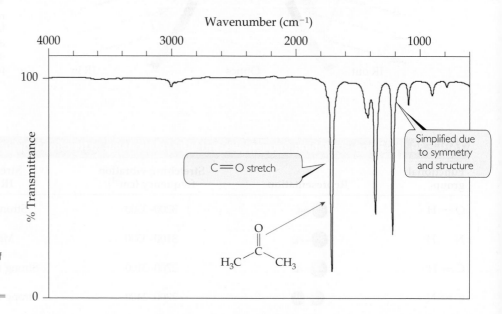

▶ **FIGURE 30.8   IR spectrum of propanone (acetone).** Note the simplicity of the spectrum and the presence of the dominant C═O stretch at 1710 cm^{-1}.

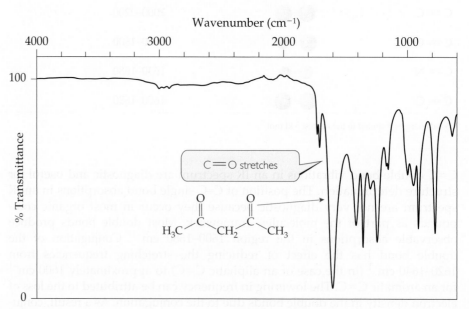

▶ **FIGURE 30.9   IR spectrum of 2,4-pentanedione.** Note the intense carbonyl stretching band at 1700 cm^{-1}.

slightly higher than those of a C–C double bond because of the relative bond strengths (C=C < C=O, see Table 30.1). Some carbonyl absorptions do occur above 1710 cm^{-1}. For example, simple carboxylic esters absorb around 1735 cm^{-1} and ring-strained carbonyl groups, such as those found in cyclobutanone, occur around 1785 cm^{-1}. In addition to the C=O stretching band, an aldehyde shows the characteristics of two C–H stretching bands around 2700 cm^{-1} and 2800 cm^{-1} corresponding to the aldehyde C–H stretch, whereas a ketone or an acid does not produce these absorptions. Carboxylic acids produce a characteristic O–H absorption around 3000 cm^{-1} in addition to the intense carbonyl stretching absorption. The position of this O–H stretch within the IR spectrum is usually complicated as a result of the carboxyl group participating in hydrogen bonding (∞ Section 27.1, 'Carboxylic Acids'), which also results in the broadening of the strong carbonyl absorption.

∞ Review this on page 1078

The carbonyl groups of amides absorb at particularly low IR frequencies (1640–1680 cm^{-1}). This very low frequency might be mistaken for an alkene C=C stretch if it weren't for two important features of its IR spectrum. The amide carbonyl absorption is much stronger than that of an alkene bond due to the difference in dipole moment between the two unsaturated bond types. Primary and secondary amides are also identified by the addition of a N–H stretching absorption around 3200 cm^{-1} (▼ **FIGURE 30.10**).

> ⚠ **CONCEPT CHECK 3**
>
> Sodium chloride and potassium bromide were often used as sample supports in IR spectroscopy. Describe one or two features that may have made them so popular in this role.

The only other multiple bond stretch that we need to consider is that of the nitrile bond (C≡N). Because of the polarity of this C≡N triple bond, nitriles usually exhibit very strong absorbances around 2200 cm^{-1} (▼ **FIGURE 30.11**) within their IR spectra. This value is similar to that for an alkyne (C≡C) stretch, although nitriles tend to absorb above 2200 cm^{-1} whereas alkynes absorb below 2200 cm^{-1}.

The vibrational frequencies of O–H bonds (alcohols) and N–H bonds (amines) occur at higher frequencies than for C–H bonds. Alcohol O–H bands are typically broad, absorbing over a wide range of frequencies centred around 3300 cm^{-1}. Their broadness is attributed to hydrogen-bonding arrangements between functional groups within the same molecule—that is, they are

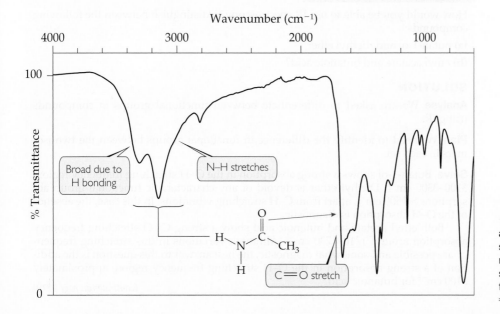

◄ **FIGURE 30.10** **IR spectrum of acetamide.** This spectrum was taken of a sample as a crystalline solid, hence there is more information contained within the spectrum than is usually needed for functional group identification.

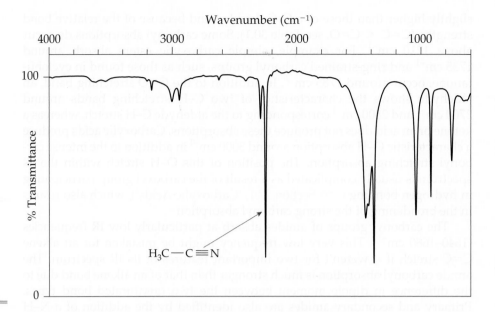

▶ **FIGURE 30.11   IR spectrum of acetonitrile.** This simple nitrile shows the characteristic and distinctive IR absorption band for C ≡ N at 2280 cm⁻¹.

intramolecular—or to interactions with other molecules, including solvent or absorbed water, interactions that are described as *intermolecular*. Therefore, the purity of a sample with respect to absorbed water needs to be taken into account. Any water that is present will also absorb at these frequencies, complicating the results by giving a false positive for an O–H group. This is usually overcome by extended drying of the sample before analysis by IR spectroscopy.

The amine N–H bond also has a stretching frequency around 3300 cm⁻¹ and, like alcohols, these amines participate in *intra*- and *inter*-molecular hydrogen bonding. The difference between O–H and N–H stretches lies in the shape of the band. Although O–H stretches are broad, N–H stretches are sharper. Primary amines typically have two peaks in their N–H absorption band, whereas secondary amines have only a single absorption band. The presence of water in an amine sample typically leads to one or two sharp spikes superimposed on the broad area of the water O–H stretching absorption. Again, drying the sample distinguishes a wet sample from one containing both an O–H and N–H functional group.

SAMPLE EXERCISE 30.1   **Distinguishing between constitutional isomers**

How would you be able to use IR spectroscopy to distinguish between the following compounds:

(a) butan-1-ol and diethyl ether,

(b) ethyl acetate and butanoic acid?

SOLUTION

Analyse We are asked to differentiate between functional groups in compounds using IR.

Plan We need to identify the difference in functional groups between the two isomers in each set.

Solve Butan-1-ol shows a strong absorption in the O–H stretching frequency region, 3100–3300 cm⁻¹. Diethyl ether is devoid of any characteristic functional group absorptions > 1500 cm⁻¹, apart from C–H stretching vibrations. In this case, the absence of the O–H absorption is diagnostic.

Both ethyl acetate and butanoic acid show a strong C=O stretching frequency absorption around 1710–1730 cm⁻¹. Although variations in this stretching frequency are possible and somewhat diagnostic, the best answer to this question is the addition of a strong absorption in the O–H stretching frequency region, approximately 3100 cm⁻¹ for butanoic acid.

(continues on page 1210)

MY WORLD OF CHEMISTRY

IR SPECTROSCOPY FOR BIOLOGICAL IMAGING

Imaging is a powerful technique in medicine and biology. It can be used to visualise the components of a cell, determine the concentration of a molecular species *in vitro* or monitor change in the progression of disease. The simplest way of generating contrast in an image is to add a chemical stain or dye that binds to the biomolecule (such as protein) of interest. High concentrations of the dye (and hence the biomolecule) are then easily visualised using visible microscopy. A more modern and sensitive imaging technique is to add a fluorescence marker. The fluorescent emission that occurs upon irradiation then becomes the contrast agent. ▼ FIGURE 30.12 demonstrates this principle using two differently coloured fluorescent markers, fluorescein and diamidenophenylindole (DAPI), with two different biological samples, a neuronal cell and a spinal cord section.

These methods are powerful and well used, but are indirect methods of identifying chemical changes and cellular structures. A method that directly monitors the chemical of interest adds a powerful technique to the repertoire of a medical researcher, pathologist or molecular biologist. IR spectroscopy, with its fingerprinting ability and its specific functional group monitoring, provides such a method.

In the late 1990s, imaging IR spectrometers became available after the development of two-dimensional infrared detector arrays, which were based on military technology developed for the Javelin anti-tank missile. These missiles use a heat-detecting array that acts as a camera and searches for a hot source emitting infrared radiation. By coupling these array detectors to an IR spectrometer, a full IR spectrum can be obtained at each point on a two-dimensional tissue section. An image can then be formed by plotting the values obtained for a functional group absorption over the tissue area, and colour-coding the absorbance value based on intensity. ▶ FIGURE 30.13(b) is such an image, where the red area indicates the highest absorbance at a specific wavenumber, 1020 cm^{-1}, and the blue area the lowest.

The spectra in Figure 30.13(c), taken from the red and blue areas, respectively, show a clear difference in absorbance due to changes in *proteoglycan** concentration. Staining using toluidine blue, Figure 30.13(a), is the traditional method used

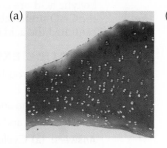
(a)

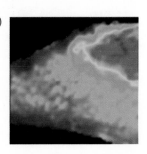
(b)

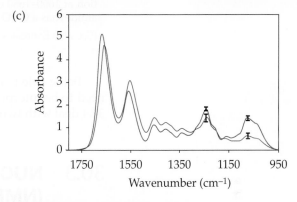
(c)

▲ **FIGURE 30.13** **Fluorescent dyes.**

for determining these types of changes. The chemical image generated using IR spectroscopy contains much more information and, importantly, is non-invasive.

Even more information can be derived by using the full IR spectrum, also known as the *IR fingerprint*, for imaging rather than a single wavenumber. ▼ FIGURE 30.14(b) shows an image obtained by using a clustering technique to colour-code IR fingerprints of liver cells and fibrous tissue, based on different functional groups. A traditional chemical stained section is shown for comparison. Note how the IR maps different frequencies and hence chemical composition in the region. These imaging methods offer both imaging contrast and spatially resolved molecular information, providing a new way of determining disease progression.

RELATED EXERCISE: 30.14

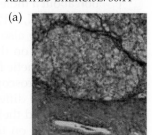
(a)

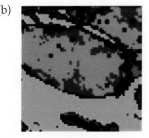
(b)

▲ **FIGURE 30.14** **Human fibrotic liver.** (a) Stained with Masson's trichrome to show the areas of fibrosis (green) and normal liver cells; (b) analysis of a serial section analysed by IR spectroscopy using cluster analysis, showing four clusters.

(a)

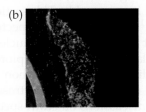
(b)

▲ **FIGURE 30.12** **Fluorescent dyes.** These two images are prepared using fluorescein (green) and DAPI (blue) fluorescent dyes. Notice how the two dyes localise differently in (a) neuronal cells and (b) spinal cord.

* Proteoglycans, found in connective tissue, are a class of glycoproteins with a very high proportion of carbohydrate.

Note that the positions of both the O–H and C=O stretch absorptions in the carboxylic acid are concentration dependent as a result of hydrogen bonding, whereas the C=O stretch absorption for the ester is concentration independent. The more concentrated the acid solution becomes, the greater the hydrogen bonding.

PRACTICE EXERCISE

How would you be able to use IR spectroscopy to distinguish between these compounds:

(a) cyclohexane and cyclohexene,

(b) acetamide and ethylamine?

Answers: **(a)** Cyclohexene shows a weak but characteristic C=C stretching absorption at 1600–1680 cm^{-1}. **(b)** Both show an N–H absorption, but acetamide (being an amide) has a C=O absorption at 1600–1680 cm^{-1}.

(See also Exercises 30.2, 30.9.)

Being able to interpret an IR spectrum is a useful skill. Typically, IR spectra need to be interpreted in conjunction with other techniques. Sections 30.3 and 30.4 deal with two other powerful spectroscopic techniques.

30.3 | NUCLEAR MAGNETIC RESONANCE (NMR) SPECTROSCOPY

As a technique, **nuclear magnetic resonance (NMR) spectroscopy** is probably the most powerful tool in the organic chemist's research arsenal. It is a non-destructive technique that requires little sample (typically 2–10 mg for ^{1}H NMR and 20–40 mg for ^{13}C NMR spectroscopy). It has become a routine technique, due mainly to the advent of Fourier Transform NMR (FT-NMR), which provides for very rapid data acquisitions (typically 10 minutes), although more demanding experiments take longer (overnight). NMR gives us information about structure, so it is extremely useful for identifying unknown substances or confirming known structures. In the latter case, NMR spectroscopy is also useful for indicating the purity of a sample (to approximately 97%) and for observing the stability of a substance in solution. Even if you are not going on with chemistry, the logical nature of the problem-solving skills developed by using this and the other techniques discussed in this chapter will be beneficial for your chosen science.

NMR spectroscopy is a technique that yields important structural information about a molecule by observing the interplay of the different nuclei, such as ^{1}H, ^{13}C, ^{15}N, ^{31}P, contained within it. The sensitivity of NMR spectroscopy allows us to also identify subtle differences in the same nuclei within a molecule based on their neighbouring functional groups, whose influence over a strong magnetic field is measurable. This section concentrates on two forms of NMR spectroscopy: ^{1}H and ^{13}C NMR spectroscopy. Information about the number and different types of hydrogen nuclei within a compound, their connectivity and their relationships will be explored using ^{1}H NMR spectroscopy. Information on the different types of carbon environment present in a compound will be gained using ^{13}C NMR spectroscopy.

Let us begin our discussions on NMR spectroscopy by looking at a simple example. The ^{13}C NMR spectrum shown in ▶ FIGURE 30.15 has three signals (or peaks), indicating that there are three different types of carbon environments within this compound, methyl acetate. The positions of these signals along the axis yields information about neighbouring functional groups that allows us to assign each carbon environment to a signal with confidence. However, to understand the reasons why we can interpret the spectrum shown in Figure 30.15 the way we do, we need to gain an understanding of the background theory to NMR spectroscopy.

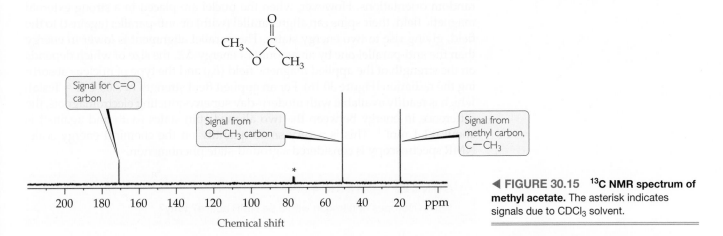

From Chapter 6 you should already be familiar with the concept that an electron has an associated magnetic field. Any atomic nucleus that has an odd number of protons or an odd number of neutrons will have a non-zero nuclear *spin* and resulting nuclear magnetic moment (μ). Like electron spin, nuclear spin is quantised (fixed). The allowed nuclear magnetic spin states are determined by the nuclear spin quantum number (I) of the nucleus. A nucleus with a spin quantum number of I ($I = 0, \frac{1}{2}, 1, \frac{3}{2}, 2$, etc.) has ($2I + 1$) spin states. If $I = \frac{1}{2}$, as for ^{1}H and ^{13}C, there are two allowed spin states with values of $+\frac{1}{2}$ and $-\frac{1}{2}$. This spinning charge creates an associated magnetic field that behaves as if it is a tiny bar magnet with a magnetic moment (μ).

In the absence of external effects, such as an applied magnetic field, the two spin states have the same energy, as shown in ▼ **FIGURE 30.16(a)**, and completely

---

△ **FIGURE IT OUT**

Why would an increase in applied magnetic field strength increase the population differences between two spin states?

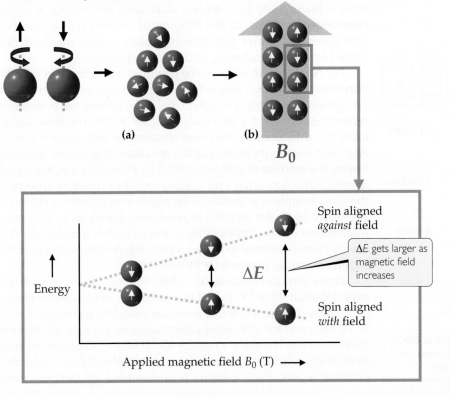

(a)

(b)

B_0

Energy

ΔE

Spin aligned *against* field

ΔE gets larger as magnetic field increases

Spin aligned *with* field

Applied magnetic field B_0 (T) ⟶

◀ **FIGURE 30.16** **Nuclear spin.** Like electron spin, nuclear spin generates a small magnetic field and has two allowed values for ^{1}H and ^{13}C. In the absence of an external magnetic field (a), the two spin states have the same energy and a random orientation. If an external magnetic field is applied (b), the parallel alignment of the nuclear spin to the magnetic field is lower in energy than the anti-parallel alignment. The energy difference, ΔE, corresponds to the radiofrequency portion of the electromagnetic spectrum. The weaker the applied magnetic field, the smaller this difference is.

random orientations. However, when the nuclei are placed in a strong external magnetic field, their spins can align parallel (with) or anti-parallel (against) to the field, giving rise to two energy states. The parallel alignment is lower in energy than the anti-parallel one by an amount of energy ΔE, the size of which depends on the strength of the applied magnetic field (B_0) and the type of nucleus absorbing the radiation (Figure 30.16). For an applied field strength of 7.05 T (T = Tesla), which is readily available with modern-day superconducting electromagnets, the difference in energy between the two aligned spin states (with and against) is about 0.1 J mol^{-1}. This is a very small quantity on the chemical energy scale. NMR spectroscopy is considered a ground-state phenomenon.

> ### CONCEPT CHECK 4
> Which isotopes of nitrogen would be NMR active? Why?

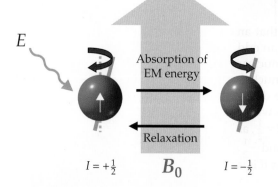

▲ **FIGURE 30.17   Nuclear spin flip.**
When Rf radiation of a frequency that corresponds to ΔE is applied, the parallel-spin nuclei absorb this energy and their spins flip to an anti-parallel orientation. Although nuclei are described as aligning in a parallel or anti-parallel fashion, the axis of rotation is tilted, causing the nucleus to precess about its own axis of spin in a similar manner to a spinning top.

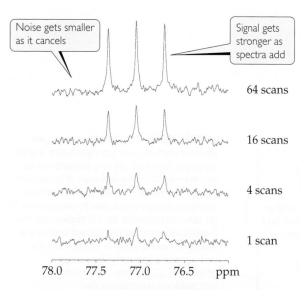

▲ **FIGURE 30.18   75 MHz ^{13}C NMR spectrum of CDCl$_3$** showing improvement of the signal-to-noise ratio from one scan (bottom spectrum) to 64 scans (top spectrum).

Nuclear Magnetic Resonance Frequencies

Based on your knowledge of thermodynamics, you would expect that the lower-energy state would have a higher population of nuclear spins aligned with the field compared with those aligned anti-parallel to the field (despite the small value of ΔE) because these aligned spins are lower in energy. If the nuclei being monitored by NMR spectroscopy are irradiated with electromagnetic radiation equal to the difference in energy between the two spin states in the applied magnetic field (ΔE), energy is absorbed and the spin of the nuclei can be 'flipped'—that is, excited from a parallel to an anti-parallel alignment, as shown in ◀ **FIGURE 30.17**. The radiation used in ^{1}H NMR spectroscopic experiments is in the radiofrequency (Rf) range, typically 200–900 MHz. This radiofrequency is known as the spectrometer's **operating frequency**. The frequency at which the nuclei absorb energy is called the **Larmor frequency**. The absorption of this electromagnetic radiation, and the flip of its nuclear spin from a lower state to a higher state (and *vice versa*), is called **resonance**. Detection of the flipping of nuclei between the two spin states leads, after detection and mathematical manipulation (called Fourier Transform), to the NMR spectrum.

For NMR spectroscopy to be useful, the difference in population between the two aligned spin states needs to be significant. The greater the applied magnetic field, the greater the population difference between the two states (Figure 30.16). The more significant this difference is, the more *sensitive* the technique becomes. A greater sensitivity improves the spectrum acquisition time and lowers the amount of sample needed by providing a better *signal-to-noise ratio*. Acquiring and adding together multiple spectra of the same sample also increases the *signal-to-noise ratio* as shown in ◀ **FIGURE 30.18**. An increase in resolution allows us to gain more information on the connectivity of nuclei within a molecule. This is why chemists desire to have spectrometers with more powerful superconducting magnets (▶ **FIGURE 30.19**).

From our discussions on NMR theory so far, you might expect that all ^{1}H or ^{13}C nuclei in a molecule would absorb energy at the same frequency. If this were the case, this technique would be of little use for structural determination. Luckily, the absorption frequency is not the same for all ^{1}H or all ^{13}C nuclei because, in organic molecules, these nuclei are not isolated from each other or other atoms. Each nucleus is surrounded by electrons from

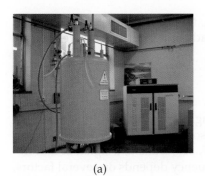
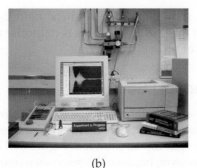

(a) (b)

◀ **FIGURE 30.19 A modern NMR spectrometer.** This 400 MHz NMR spectrometer shows (a) the superconducting magnet in the foreground with the return lines required to pump the liquid helium into the system to cool the magnet. In the background is the computer mainframe required for Fourier transformations. (b) The desktop computer, showing an initial free induction decay (FID) curve, is interfaced with the spectrometer for data manipulation.

MY WORLD OF CHEMISTRY

NUCLEAR SPIN AND MAGNETIC RESONANCE IMAGING

A major challenge facing medical diagnosis is seeing inside the human body from the outside. Until recently this was accomplished mainly by using X-rays to image human bones, muscles and organs. However, there are several drawbacks to using X-rays for medical imaging. First, X-rays do not give well-resolved images of overlapping physiological structures. Moreover, because damaged or diseased tissue often yields the same image as healthy tissue, X-rays frequently fail to detect illnesses or injuries. Finally, X-rays are high-energy radiation that can cause physiological harm, even in low doses.

In the 1980s a new technique called magnetic resonance imaging (MRI) moved to the forefront of medical imaging technology. The foundation of MRI is nuclear magnetic resonance (NMR) spectroscopy, which was discovered in the mid-1940s.

Because hydrogen is a major constituent of aqueous body fluids and fatty tissue, the hydrogen nucleus is the most convenient one for study by MRI. In MRI, a person's body is placed in a strong magnetic field. By irradiating the body with pulses of radiofrequency radiation and using sophisticated detection techniques, tissue can be imaged at specific depths within the body, giving pictures with spectacular detail (▶ **FIGURE 30.20**). The ability to sample at different depths allows medical technicians to construct a three-dimensional picture of the body.

MRI has none of the disadvantages of X-rays. Diseased tissue appears very different from healthy tissue, resolving

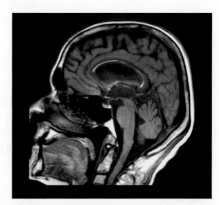

▲ **FIGURE 30.20 MRI image.** This image of a human head, obtained using MRI, shows the structures of a normal brain, airways and facial tissue.

overlapping structures at different depths in the body is much easier, and the radiofrequency radiation is not harmful to humans in the doses used. The technique has had such a profound influence on the modern practice of medicine that Paul Lauterbur (1929–2007), a chemist, and Peter Mansfield (1933–), a physicist, were awarded the 2003 Nobel Prize in Physiology or Medicine for their discoveries concerning MRI. The major drawback of this technique is expense: the current cost of a new MRI instrument for clinical applications is several million dollars.

RELATED EXERCISE: 30.3

other atoms, or groups of atoms, that can create local magnetic fields (B_{local}) that influence the applied field. This statement can be written in the following way:

$$B_{effective} = B_{applied} - B_{local} \qquad [30.4]$$

The result of the influence of these local magnetic fields is a **shielding** of the nuclei in question from the applied field by reducing its effect. In some instances, the local magnetic field of neighbouring electrons can add to the applied field, **deshielding** the nuclei in question. The difference in resonance frequencies for various ^{1}H or ^{13}C nuclei within a molecule due to shielding or deshielding is generally very small but measurable. Such a difference leads the two nuclei in question to be *magnetically inequivalent*.

In fact, for a particular magnetic field B_0, the resonance frequencies of ^{1}H and ^{13}C are quite different. This is naturally advantageous when investigating either type of nuclei in isolation.

The Chemical Shift

Let us now consider the simplest case of a single ^{1}H nucleus within an applied magnetic field. Absorption of energy at the resonance frequency, which leads to the nuclear spin flip as shown in Figure 30.17, also leads to a signal at that frequency. However, because the resonance frequency depends on several factors, including the magnetic field strength (B_0) and the operating frequency of the spectrometer, chemists have developed a dimensionless quantity called the **chemical shift (δ)**, expressed in parts per million, to quantify the result. The chemical shift scale takes into account the shift in frequency of a particular nucleus from a standard (accepted universally to be tetramethylsilane, TMS), as well as the operating frequency of the spectrometer:

$$\delta = \frac{\text{Shift in resonance frequency from TMS (Hz)}}{\text{Operating frequency of the spectrometer (MHz)}} \qquad [30.5]$$

The chemical shift of an NMR absorption signal is constant, regardless of the operating frequency of the instrument. This is because the chemical shift scale is relative. Using δ allows a comparison between NMR spectra obtained on different spectrometers in different laboratories across the world.

tetramethylsilane

The chemical shift of a ^{1}H nucleus that is strongly influenced by the electrons of an adjacent atom has a different chemical shift from a nucleus that is weakly influenced by neighbouring electrons. This change in chemical shift is important and is diagnostic. Any competent organic chemist can use the information of functional group effects on chemical shift to identify potential 'leads'. For example, the ^{1}H NMR spectrum of methyl acetate is shown in ▼ FIGURE 30.21. This compound has two 'different' types of methyl groups: one adjacent to the carbonyl group and the other directly bonded to oxygen. The difference lies in the fact that the two methyl groups are *not* related by symmetry, that is, they are non-identical. Their effective magnetic fields, and hence δ, differ because the effects of the electrons associated with the neighbouring carbonyl group and oxygen atom on the applied field are different—this is called *magnetic inequivalence*.

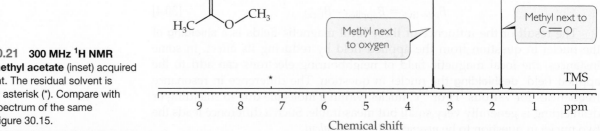

▶ **FIGURE 30.21** **300 MHz ^{1}H NMR spectrum of methyl acetate** (inset) acquired in CDCl$_3$ solvent. The residual solvent is marked with an asterisk (*). Compare with the ^{13}C NMR spectrum of the same compound in Figure 30.15.

FIGURE IT OUT

Is a methyl group resonance at 4 ppm shielded or deshielded compared with one at 1 ppm?

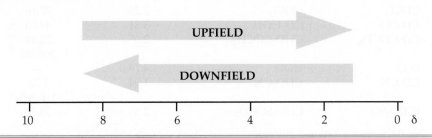

◀ **FIGURE 30.22**   **Chemical shift.** The term *downfield* refers to a higher chemical shift. Nuclei at high chemical shift are relatively *deshielded* by neighbouring electron motions. A hydrogen resonance that is *upfield* resonates at a lower chemical shift value. These hydrogens are relatively *shielded* and are the furthest from the resonance of a bare hydrogen atom.

If a hydrogen is partially shielded from the applied field by local magnetic fields, the signal occurs *upfield* towards tetramethylsilane (TMS) (▲ FIGURE 30.22). This is a shift to lower frequency. If the magnetic environment reinforces the applied field, the hydrogen is deshielded and the signal moves *downfield*, away from TMS. This is a shift to higher frequency. So, in the example of methyl acetate, the signal at δ3.39 is *downfield* of the signal at δ1.78.

Notice that the hydrogen atoms on each methyl group are equivalent to each other but different from the other methyl hydrogen atom set. Hence only two signals are seen, at 1.78 ppm and 3.39 ppm, in the ^{1}H NMR spectrum. The reason why the hydrogens of a methyl group form part of the same signal lies in the speed of rotation about the C–CH$_3$ or O–CH$_3$ bonds in methyl acetate. We mentioned in Chapter 22 that aliphatic alkanes are dynamic molecules, with free rotation about any single bond (∞ Section 22.4, 'Structures of Alkanes'). This rotation is faster than the timescale of the NMR spectrometer at 300 K (which is the standard operating temperature of a spectrometer), so each hydrogen on the methyl group appears to be equivalent.

∞ Review this on page 898

A spectrometer with an operating frequency of 400 MHz has a timescale of 400 000 000 times per second. Although several molecular dynamic events, such as rotation about a C–C bond, are faster than this, quite a few are slower—for example, hydrogen exchange rates and some conformational processes. These differences in timescale mean that NMR spectroscopy is a useful technique with which to study kinetic events.

Sample Preparation

Two other signals in the ^{1}H NMR spectrum of methyl acetate need to be accounted for. The first signal is arbitrarily assigned to 0.00 ppm and is a result of the 12 equivalent hydrogens of the reference compound, TMS, which is added to the NMR solvent. Historically, all other ^{1}H signals are reported by their shift from TMS. Modern NMR spectroscopy, however, typically uses the *residual solvent signal* as the spectrum reference (see ▼ TABLE 30.2). There are two reasons for this. The addition of TMS (1% *v/v*) to a sample compromises the purity of the sample, even though it is usually removed easily by evaporation. Second, the sample needs to be in solution for this technique to be effective and, as there is usually some source of ^{1}H present in the solvent, a marker independent of the sample exists. This leads us to the last signal to be discussed from Figure 30.21, that of solvent.

Recall from our initial discussion on the theory of NMR spectroscopy that any nuclei with a magnetic moment other than zero would produce an NMR spectrum. This means that ^{13}C produces an NMR signal while ^{12}C does not. Another way of saying this is that ^{12}C is *transparent* to the technique of NMR spectroscopy at the frequencies employed.

TABLE 30.2 • Common solvents used for sample preparation in ^{1}H and ^{13}C NMR spectroscopy			
Solvent	Residual solvent	δ_H (ppm)	δ_C (ppm)
$CDCl_3$	$CHCl_3$	7.26	77.16
CD_3OD	CD_2HOH	3.31	49.00
CD_3COCD_3	CHD_2COCD_3	2.05	29.84
			206.26[a]
D_2O	HOD	4.79	—
CD_3CN	CHD_2CN	1.94	1.32
			118.26[b]
$(CD_3)_2SO$	CHD_2SOCD_3	2.50	39.52

Data: H.E. Gottlieb, V. Kotlyar, A. Nudelman, *Journal of Organic Chemistry*, 1997, *62*, 7512.
[a] Value corresponds to carbonyl carbon.
[b] Value corresponds to nitrite carbon.

Typically, organic chemists use solvents to prepare an NMR sample, as a means of making the sample *homogeneous*. The type of solvent used is important because we need to distinguish between signals attributable to the sample and those from the solvent, especially when the solvent is in concentrations greater than 500 times that of the compound of interest in the preparation. This means that any solvent that contains hydrogens will also give rise to a ^{1}H NMR signal, which will be at least 500 times stronger than that of the sample. The effect is to 'lose' the important sample signals.

To counter this problem, organic chemists use a range of ^{1}H NMR transparent (known as *deuterated*) solvents. Solvents such as deuterated chloroform ($CDCl_3$) or d_4-methanol (CD_3OD) become useful and common solvents in ^{1}H NMR spectroscopy because ^{2}H absorbs at a completely different frequency from ^{1}H, so does not interfere. However, although $CHCl_3$ or CH_3OH become unattractive in ^{1}H NMR spectroscopy for reasons described above, they are acceptable solvents in ^{13}C NMR spectroscopy. It is almost impossible (and certainly very expensive) to produce large quantities (litres) of 100% (*v/v*) pure deuterated solvents. Typically, an NMR spectroscopy laboratory would stock, for example, 99.8% $CDCl_3$ solvent. The extra 0.2% would contain protium rather than deuterium. This amount of $CHCl_3$ is still enough to produce a significant signal (known as the *residual solvent signal*) in the ^{1}H NMR spectrum of any sample (see Figure 30.21).

 CONCEPT CHECK 5

What other solvents are practical for ^{1}H and ^{13}C NMR spectroscopy? What criteria are you using to make your judgment?

Interpreting NMR Spectra

Three components of an NMR spectrum provide useful structural information about a molecule. They are:

- *chemical shift* (δ)—gives evidence on neighbouring functional groups;
- *integration* of the NMR signal—relative areas of signals in a ^{1}H NMR spectrum are proportional to the number of hydrogens giving rise to each signal;
- *coupling constant* (J)—measured in hertz (Hz), the coupling constant is extremely useful for determining the connectivity of nuclei within a molecule.

As you might expect, the result of an NMR spectroscopic experiment could become very complicated even with simple molecular structures. Luckily,

FIGURE IT OUT

How many signals would you expect in the ^{13}C NMR spectrum of the molecules shown in this figure?

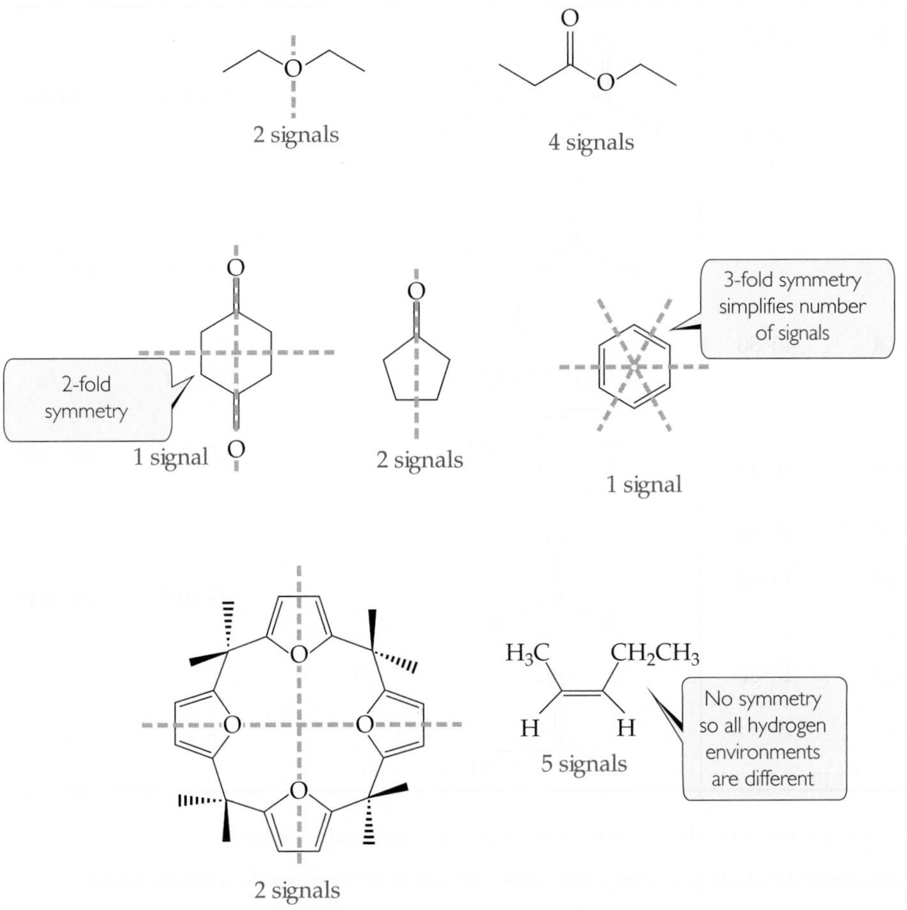

2 signals

4 signals

2-fold symmetry

1 signal

2 signals

3-fold symmetry simplifies number of signals

1 signal

2 signals

H$_3$C   CH$_2$CH$_3$

H   H

5 signals

No symmetry so all hydrogen environments are different

◀ **FIGURE 30.23   Symmetry.** The number of NMR signals present in a spectrum depends on the symmetry of the compound in question. The higher the symmetry, the smaller the number of signals in the spectrum. Molecules with a single set of equivalent hydrogens will give rise to one ^1H NMR signal. Two or more sets of equivalent hydrogens will give different ^1H NMR signals for each set. Here, the number of signals seen in the ^1H NMR spectrum of each compound is indicated.

molecular symmetry elements help to simplify matters. These symmetry elements occur in two ways. If the molecule has a plane of symmetry, then the two halves are identical and we need only concern ourselves with one-half of the molecule in order to work out its NMR spectrum. A few examples of the effects of symmetry on ^1H are shown in ▲ **FIGURE 30.23**.

The second factor that simplifies a NMR spectrum was introduced in our earlier discussion on rotation about C–CH$_3$ bonds. In fact, this dynamic behaviour of rotation about C–C bonds usually leads to the two hydrogens on a methylene (CH$_2$) group being chemically and magnetically equivalent (Figure 30.23). This simplification using symmetry is a powerful way to differentiate between two constitutional isomers.

The typical ^1H and ^{13}C NMR chemical shifts for hydrogens and carbons with neighbouring common functional groups are shown in ▼ **TABLE 30.3**. The ^1H signal associated with simple alkanes tends to resonate between 0.8 and 1.8 ppm, depending on their classification as primary, secondary or tertiary hydrogen atoms. The general trend for all other organic molecules is that δ depends on the electronegativity of nearby atoms and the hybridisation of adjacent atoms. Thus the greater the electronegativity of a neighbouring group (F > Cl > Br > I), the more deshielded the ^1H nucleus will be and the further downfield (higher δ) the ^1H NMR signal will lie. More than one electronegative group influences the signal even more, causing the ^1H signal to move further downfield (▼ **FIGURE 30.24**).

TABLE 30.3 • Approximate chemical shift regions for ¹H and ¹³C nuclei (coloured) neighbouring the most common functional groups

Type of neighbouring group[a]	δ_H[b,c]	δ_C[c]	Type of neighbouring group[a]	δ_H[c]	δ_C[c]
RCH_3, RCH_2R', R_3CH	0.8–1.6	10–55	(ester –CH group)	3.5–4.5	40–80
(alkene with adjacent CH)	1.5–2.5	40–80	(vinyl =CH)	4.5–6.0	100–150
(ketone α-CH)	2.0–2.8	40–80	ROH, RNH_2[d]	0.5–6.0	n/a
$ArCH_3$, $ArCH_2R$	2.0–3.0	20–40	(aromatic C—H)	6.5–8.5	110–160
$R—C{\equiv}C—H$	2.0–3.0	65–85	(aldehyde)	9.5–10.5	180–210
RCH_2X (X = Cl, Br, I)	3.0–4.0	40–80			
RCH_2OR'	3.0–4.0	40–80	(carboxylic acid)	10–13	175–185
RCN	n/a	110–130			

[a] R and R' refer to simple aliphatic groups.
[b] Note that, when the symbol δ is used, there is no need to add ppm. Remember, chemical shift is dimensionless and ppm is not a dimension.
[c] Chemical shifts referenced to TMS.
[d] Both OH and NH NMR absorptions are typically broad as a result of the hydrogen exchange they exhibit with solvent and traces of water within the solvent.

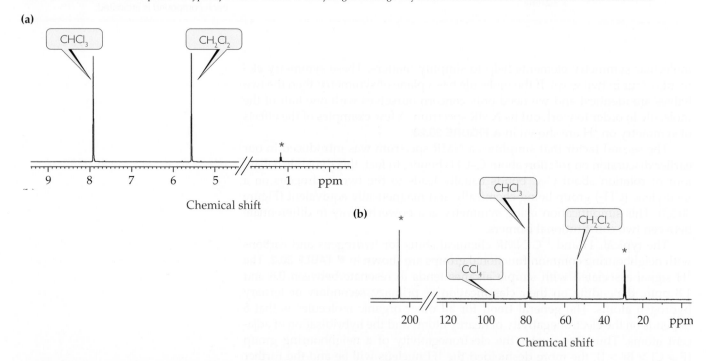

▲ **FIGURE 30.24** **Effect of electronegative atoms neighbouring ¹H and ¹³C nuclei on their chemical shift.** (a) 300 MHz ¹H NMR spectrum of $CHCl_3$ and CH_2Cl_2 in $(CD_3)_2CO$. (b) 75.5 MHz ¹³C NMR spectrum of CCl_4, $CHCl_3$ and CH_2Cl_2 in $(CD_3)_2CO$. Asterisks indicate solvent or residual solvent signals.

SAMPLE EXERCISE 30.2 Determining the number of NMR signals in a spectrum

How many absorptions would you expect in the ^{1}H NMR spectra of:
(a) 1,1-dimethylcyclohexane,
(b) ethyl methyl ether?

SOLUTION

Analyse We are asked to determine the number of 'different' hydrogen environments in two molecules.

Plan We need to draw out the structure of each compound and look for symmetry. If no symmetry elements such as mirror planes exist, we then count up the number of *different* hydrogen atoms, remembering that hydrogens on the same carbon are equivalent in an alkane.

Solve
(a) The answer is four absorptions.

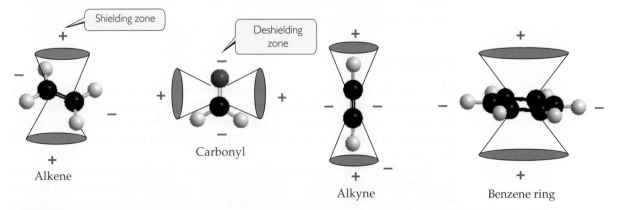

(b) The answer is three absorptions. In this case, there is no symmetry across the molecule.

PRACTICE EXERCISE

How many absorptions would you expect in the ^{13}C NMR spectra of:
(a) cyclohexanone
(b) 2-methyl-2-butene?

Answers: (a) 4, **(b)** 5

(See also Exercises 30.14, 30.15.)

The chemical shift is also influenced by the hybridisation of neighbouring atoms (▼ **FIGURE 30.25**). Generally, the more *p* character in a neighbouring atom ($sp > sp^2 > sp^3$), the higher the chemical shift will be for the observed nucleus. This change in δ is due to the deshielding effect from the applied magnetic field produced by these bonds, as indicated in Figure 30.25. Note that orientation is an important factor in this generalisation. The hydrogens on benzene are in a deshielding (−) zone, whereas those of ethyne, the carbons of which have more *p* character, are in a shielding (+) zone. In any case, the chemical shift of the signals attributable to either set of hydrogens is higher than that of a simple alkane.

▲ **FIGURE 30.25** Deshielding effects in alkenes, carbonyls, alkynes and benzene. The magnetic fields associated with the π-electrons of the *sp* and *sp*^{2} hybridised atoms cause regions of shielding (+) and deshielding (−) that influence the chemical shift of adjacent nuclei.

CONCEPT CHECK 6

Based on your understanding of Figure 30.25, would you expect the signal due to the central CH_2 hydrogens of the macrocyclic compound at the right to occur upfield or downfield from the central CH_2 group of heptane?

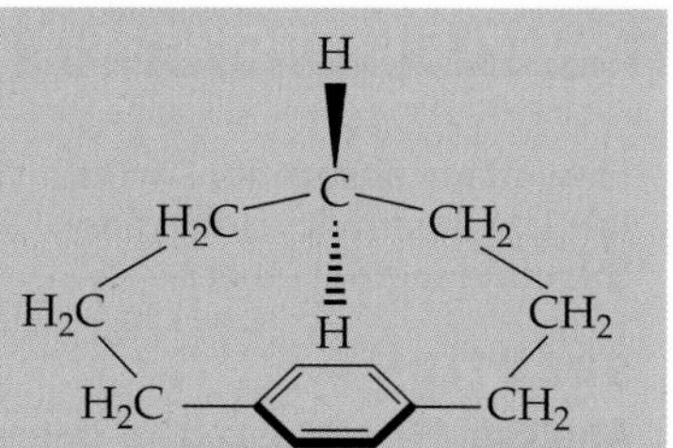

SAMPLE EXERCISE 30.3 Differentiating between products of a reaction

3,4-Dibromohexane can undergo base-induced double dehydrobromination to yield either 3-hexyne or 2,4-hexadiene. How could you use NMR spectroscopy to identify the product formed?

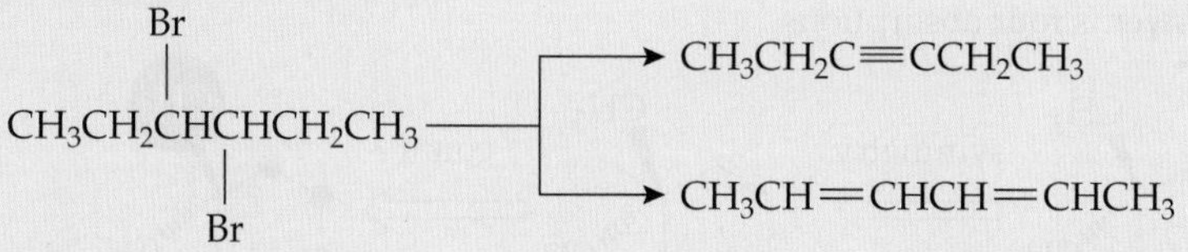

$$CH_3CH_2CHCHCH_2CH_3 \longrightarrow$$
(with Br above and Br below the two central carbons)

$$\longrightarrow CH_3CH_2C{\equiv}CCH_2CH_3$$

$$\longrightarrow CH_3CH{=}CHCH{=}CHCH_3$$

SOLUTION

Analyse We are asked to distinguish between the two possible products using NMR spectroscopy.

Plan As with all NMR questions of this kind, start with symmetry. Once identified, or not, then count up the number of different carbon or hydrogen environments. Look for differences.

Solve Each potential product has a mirror plane, as shown below. Bearing this in mind, both products will exhibit three absorptions in their ^{13}C NMR spectrum.

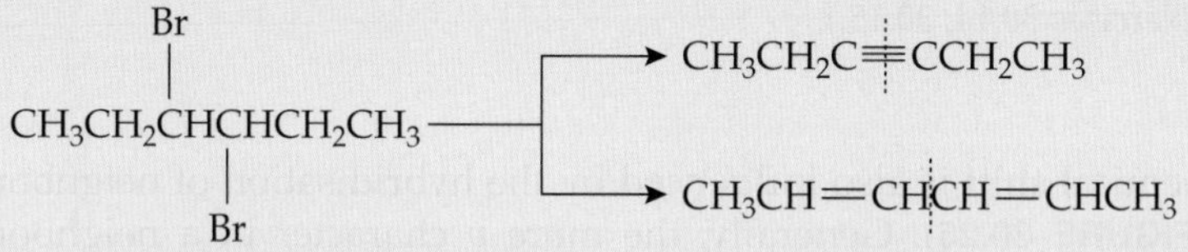

$$CH_3CH_2CHCHCH_2CH_3 \longrightarrow$$
(with Br above and Br below)

$$\longrightarrow CH_3CH_2C{\equiv}CCH_2CH_3$$

$$\longrightarrow CH_3CH{=}CHCH{=}CHCH_3$$

However, the numbers of absorptions in the two ^1H NMR spectra are different. The alkyne will exhibit two absorptions while the diene exhibits three. Purely on this basis, the two potential products are distinguishable.

PRACTICE EXERCISE

Propene reacts with HBr potentially to produce two alkylbromides. Only one is formed, the outcome following Markovnikov's rule. How could you use NMR to determine which alkylbromide is the Markovnikov product?

Answer: Draw the two possibilities. The Markovnikov product is symmetrical, which leads to only two absorptions in either the ^1H or ^{13}C NMR spectrum.

(See also Exercises 30.18, 30.19.)

Integration

Recall that in mathematical terms, **integration** refers to the area under a curve. The term *integration* in NMR is used to describe the area under each signal and is proportional to the number of equivalent nuclei that give rise to that signal. Integration is diagnostic in ^1H NMR experiments but used rarely in ^{13}C NMR experiments.

Two conventions describing the integration of a signal are used, as shown in ▶ **FIGURE 30.26**. The first method uses the height of the integral to give an area relative to other signals. For example, a ratio of areas that is 1 : 1 may in fact refer to a hydrogen ratio within the compound of 3 : 3, as for methyl acetate. An alternative method (and the one we use more often) does all the hard work for you. The numbers above the signals in Figure 30.26(b) equate to

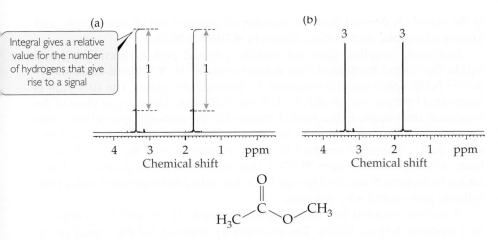

(a)

Integral gives a relative value for the number of hydrogens that give rise to a signal

1   1

4   3   2   1   ppm
Chemical shift

(b)

3   3

4   3   2   1   ppm
Chemical shift

$$H_3C - \overset{\overset{\textstyle O}{\|}}{C} - O - CH_3$$

◀ **FIGURE 30.26 Integration.** (a) Integral lines indicate the area under the signal. This technique doesn't necessarily indicate how many hydrogens each signal accounts for, only their relative areas. (b) This method indicates the number of hydrogens associated with each signal. Both (a) and (b) yield the same information.

the numbers of protons that give rise to those signals. This result is based on the integration shown in Figure 30.26(a) and its application to the compound in question—methyl acetate in this example. Integration can help in structural analysis, particularly when determining symmetry.

Spin–Spin Coupling

The last piece of NMR evidence we consider is probably the most powerful in determining structure. The concept is called **spin–spin coupling** or *J coupling*. So far we have looked at ^1H and ^{13}C nuclei in isolation. When two or more nuclei are close enough for their spins to interact or *couple*, the resulting signals split. The **splitting pattern** formed is characteristic of the coupling and yields more information about the nucleus's neighbour than it does about itself. For example, diethyl ether ($CH_3CH_2OCH_2CH_3$) has two sets of different hydrogens (as a result of symmetry), leading to two signals in the ^1H NMR spectrum (▼ **FIGURE 30.27**). The chemical shift of these signals is influenced by the ether oxygen, with the hydrogens of the CH_2 being more deshielded. Closer inspection of the signals shows them to be complicated by very uniform patterns. This complicated pattern is a result of the *through-bond* communication between ^1H nuclei across the CH_2 and CH_3 groups. For any set of coupled hydrogens, the distance between peaks in each splitting pattern, termed the **coupling constant** (J), is identical. The magnitude of the coupling depends on the nature and number of intervening bonds between the two (or more) nuclei. The further away the coupling nuclei are from each other, the smaller the value for J. Therefore, a ^4J coupling (in which there are four intervening bonds) is smaller than a ^3J coupling (in which there are three intervening bonds). The pattern, or **multiplicity**,

◢ **FIGURE IT OUT**

What does the very small signal at 7.3 ppm represent?

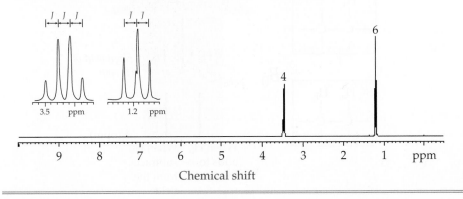

J J J

3.5   ppm

J J

1.2   ppm

6

4

9   8   7   6   5   4   3   2   1   ppm
Chemical shift

◀ **FIGURE 30.27 300 MHz ^1H NMR spectrum of diethyl ether in CDCl$_3$.** Inset: Expansions of the two signals showing the multiplicity of each signal.

singlet			1		
doublet			1 1		
triplet		1	2 1		
quartet		1 3	3 1		
quintet	1	4 6	4 1		
sextet	1 5	10 10	5 1		

▲ **FIGURE 30.28 Multiplicity.** *J* is a quantitative measure of the magnetic interaction of coupled nuclei. The splitting pattern for nuclei with $I = \frac{1}{2}$ follows a binomial distribution. For example, if a signal is split into two peaks of equal intensity, that signal is called a doublet.

of the signals is derived from the number of identical nuclei that couple, and follows a binomial distribution, shown in ◄ **FIGURE 30.28**.

Hydrogen coupling does not usually proceed past a functional group within the carbon framework that doesn't contain ¹H. ▼ **FIGURE 30.29** shows the ¹H NMR spectrum of 2-butanone to illustrate this point. The spectrum is dominated by three signals at 0.90, 2.00 and 2.30 ppm. An enlarged view of the resonance attributable to the methyl attached to the carbonyl group shows this signal to be a single sharp peak (or *singlet*), demonstrating that no coupling across the C=O group from the CH₂ group is seen. As with the ¹H NMR spectrum of diethyl ether (Figure 30.27), a characteristic splitting pattern is shown for the hydrogens of an ethyl group. Let's now take a more detailed look at this splitting pattern and why it occurs.

A nucleus coupled to *n* other equivalent nuclei (each with $I = \frac{1}{2}$) appears as a *multiplet* (Figure 30.28). The multiplicity attained for the signal of the coupled nuclei can be described as containing (*n* + 1) peaks. The origin of this phenomenon, known as the **(*n* + 1) rule**, can be ideally described by considering two non-equivalent hydrogens, H_a and H_b, found on adjacent carbons. ▼ **FIGURES 30.30** and ▶ **30.31** illustrate the principle as well as the underlying reasoning. The chemical shift of H_a is influenced by H_b. In fact, this influence (and hence the multiplicity) is a result of the two possible spin orientations of H_b (Figure 30.30). The parallel orientation of nuclear spins adds to the applied field, causing the signal attributable to H_a to appear at a lower applied field (higher δ), whereas the anti-parallel orientation causes H_a to appear at a higher field (lower δ). These two signals (combined they are termed a *doublet*) occur on either side of the position H_a would occupy in the absence of the influence of H_b.

If we turn our attention back to the expanded region of the ¹H NMR spectrum of diethyl ether (Figure 30.27), we see that the CH₂ group is *coupled* to the three equivalent hydrogens of the methyl group. This coupling causes the signal attributable to the CH₂ hydrogens to split into four (3 + 1) peaks with intensities in the ratio 1 : 3 : 3 : 1. This specific pattern is called a *quartet*

▶ **FIGURE 30.29 300 MHz ¹H NMR spectrum of butan-2-one in CDCl₃.** Despite the four intervening bonds linking ¹H nuclei, no ⁴*J* coupling is observed at this operating frequency due to the carbonyl group. Residual solvent is marked with an asterisk (*).

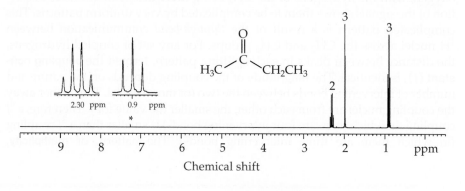

▶ **FIGURE 30.30 Origins of signal splitting.** The influence of the spin states of H_b on the applied field causes the signal for H_a to split.

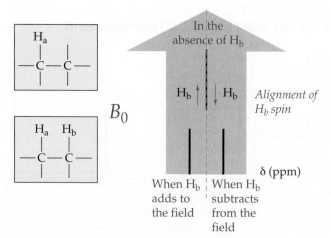

FIGURE IT OUT

Determine the possible connectivity between hydrogen nuclei that show as a doublet and septet multiplicity with integration in the ratio 6 : 1, respectively, within the ^{1}H NMR spectrum.

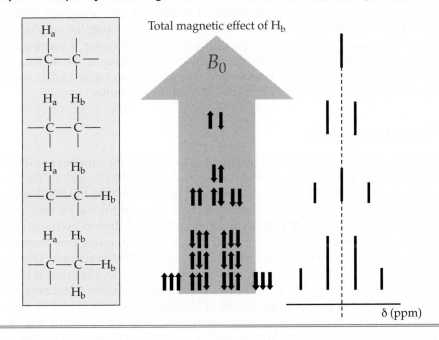

◀ **FIGURE 30.31 Origins of complex signal splitting.** The influence of the spin states of n neighbouring H_b on the applied field causes the signal for H_a to split into '$n+1$' peaks. The splitting pattern and the relative peak heights are attributable to the additive and cancelling effect of H_b spins.

(see Figure 30.28). Conversely, the methyl group couples to the CH_2 group, splitting into three (2 + 1) signals in the ratio 1 : 2 : 1, with the distance between each signal in the *triplet* (known as the *J* value) being equivalent to those in the quartet. The permutations of nuclear spin 'up' and 'down' leading to the triplet and quartet are shown in Figure 30.31.

As organic chemists, we can use the information derived through individual *J* values to determine the connectivity of a carbon framework. ▼ **FIGURE 30.32** illustrates how the connectivity of two nuclei can be expressed by the magnitude of *J*. This spectrum contains two sets of triplets and two sets of quartets for an unknown compound. The triplet/quartet pattern is diagnostic of an ethyl group. So, intuitively, we might expect two sets of ethyl groups within the

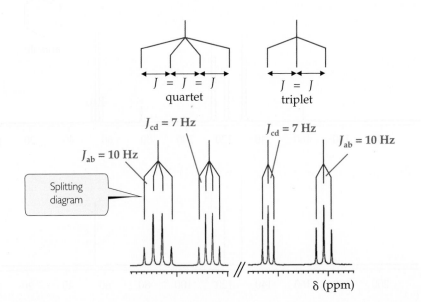

◀ **FIGURE 30.32 Coupling constants.** The spacing between each peak in a multiplet is called the coupling constant, *J*, expressed in hertz (Hz). Note that the *J* value between two mutually coupled nuclei is the same and different from that attained for other coupled pairs of nuclei. The coupling constant yields information on connectivity.

structure of the unknown. The difficulty lies in distinguishing between the two sets. Determination of the coupling constants, however, soon identifies which triplet is *coupled* to which quartet. We can use this information, together with the chemical shifts of the quartets, to determine the likely connectivity of the ethyl groups to other functional groups. Looking for sets of signals with the same J allows information to be extracted from very complex overlapping spectra. Because J values do not depend on the spectrometer, a spectrometer with a higher operating frequency will show the splitting of the signal compressed to a shorter distance on a δ scale, one more reason why higher operating frequencies are preferred for studying complex molecules.

^{13}C NMR Spectra

^{13}C NMR spectra are often used by chemists to indicate the number of different carbon atoms, as well as to provide some evidence on the nature of the functional groups contained within a molecule. In order to do this, ^{13}C NMR spectra often remove the multiplicity displayed by C–H coupling—a process called **decoupling**. As shown in ▼ FIGURE 30.33, coupling of carbon to hydrogen complicates the spectrum to an unnecessary degree. By decoupling the ^{1}H and ^{13}C nuclei, all multiplicity is lost and the signals are simplified to singlets. This type of experiment is called a *heteronuclear decoupling* experiment because it decouples the effect of ^{1}H nuclei on ^{13}C nuclei. Note that the reverse is not necessary, that is, the need to decouple the effects of ^{13}C coupling on the ^{1}H NMR experiment. The reason is that ^{1}H is approximately 99% naturally abundant and ^{13}C is only 1% naturally abundant. Having such a low natural abundance means that, for the most part, the effect of ^{13}C nuclei contained within the sample is below the level of sensitivity for ^{1}H NMR spectroscopy.

▲ **CONCEPT CHECK 7**

Describe the multiplicity seen in
a. the hydrogen-coupled and
b. the hydrogen-decoupled ^{13}C NMR spectrum of 1-propanol.

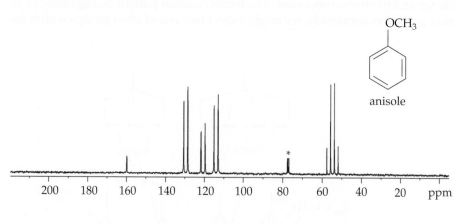

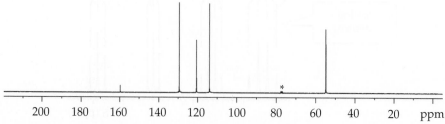

▶ **FIGURE 30.33** ^{13}C NMR spectra— **when too much information complicates interpretation.** Routinely, organic chemists undertake ^{13}C NMR experiments that involve decoupling the effect of ^{1}H. *Top spectrum:* Coupled ^{13}C NMR spectrum of anisole. *Bottom spectrum:* Decoupled ^{13}C NMR spectrum of anisole. Asterisks (*) denote CDCl$_3$ solvent.

SAMPLE EXERCISE 30.4   Using NMR to solve a chemical unknown

Compound A, with molecular formula $C_4H_8O_2$, has the following ^1H and ^{13}C NMR spectra. Deduce its structure using this information.

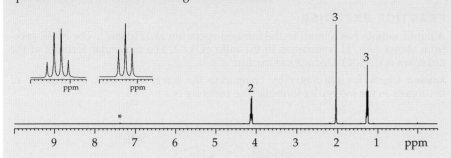

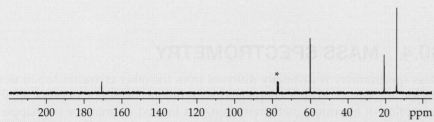

Note Numbers on top of signals represent integration values. Asterisk indicates signals due to solvent.

SOLUTION

Analyse We have been asked to use NMR spectra to deduce the molecular structure of a compound containing C, H and O.

Plan Based on chemical shift (δ) and splitting patterns in the ^1H NMR spectrum, we can deduce the structure of compound A. We will use the ^{13}C NMR spectrum to confirm the choice of structure.

Solve The ^1H NMR spectrum contains three signals at about 1.25, 2.00 and 4.10 ppm, with integration in the ratio 3 : 3 : 2. This integration totals the number of hydrogens in the molecular formula and so represents the molecule as a whole (no further symmetry).

Two of these hydrogen signals are split, forming a triplet and a quartet. This combined pattern is indicative of an ethyl group (CH_3CH_2) following the $n+1$ rule, a deduction helped by the integration ratio.

The third signal is a three-hydrogen singlet, which means that the hydrogens on this methyl group (because it integrates for three hydrogens) do not couple. The chemical shift of this resonance implies that this methyl group is bonded to a carbonyl group, consistent with the lack of coupling within the signal (that is, a singlet).

At this stage we have accounted for all atoms in the molecule except for an oxygen atom. This must be linked between the ethyl and carbonyl groups in order to give rise to the chemical shift seen by the CH_2 group at 4.10 ppm. The other ester

$$\underset{H_3CH_2C}{\overset{\overset{\displaystyle O}{\|}}{C}}\underset{OCH_3}{}$$

would have a chemical shift of 2.2 ppm for the CH_2 group, and a three-hydrogen singlet at 4.10 ppm. Placing all this information together, our unknown must be ethyl acetate.

$$\underset{H_3C}{\overset{\overset{\displaystyle O}{\|}}{C}}\underset{OCH_2CH_3}{}$$

Check Our deduction is confirmed by the four signals within the ^{13}C NMR spectrum. The signal at 171 ppm indicates the presence of a carbonyl group. The signal at 60 ppm is indicative of a carbon attached to oxygen (in this case, as part of the ester group), with the two other signals at 13 ppm and 21 ppm consistent with the two methyl carbon atoms. Contrast this result with the spectrum of butan-2-one (Figure 30.29).

PRACTICE EXERCISE

A liquid sample has a band in the infrared spectrum at 1710 cm^{-1}. The NMR spectrum shows two ^{1}H resonances in the ratio of 3 : 2. The molecular formula of the unknown is $C_5H_{10}O$. Deduce its structure.

Answer: Check for symmetry by comparing the integration with the number of hydrogens in the molecular formula. The structure is

$$CH_3CH_2 \underset{\overset{\|}{\underset{C}{}}}{\overset{O}{}} CH_2CH_3$$

30.4 | MASS SPECTROMETRY

Mass spectrometry is distinctly different from the other characterisation techniques introduced in this chapter. This technique does not depend on the absorption of electromagnetic radiation, but instead examines what happens when a molecule is bombarded with high-energy electrons. These electrons are used for two purposes. The molecule is *ionised* by the electrons and the electrons are used to break the molecule apart or **fragment** it, in a similar way to hitting a cricket ball through a pane of glass. The masses of the fragments are measured and brought together (in analysis) in order to reconstruct the molecule. Apart from determining the masses of the fragments, mass spectrometry can also indicate the mass of the *parent structure*, which in most instances, coincides with the signal of highest mass. This signal is called the **molecular ion**, or parent ion peak. Mass spectrometry is a technique used routinely to determine the molecular weight of a compound. It also provides information about fragmentation patterns and hence functional groups in molecules.

Mass spectrometry employs many different ionisation techniques, including electron impact ionisation (EI), chemical ionisation (CI), fast atom bombardment (FAB) and more advanced techniques such as matrix-assisted laser desorption ionisation time-of-flight (MALDI-TOF) and electrospray ionisation (ESI) methods. Each of these different techniques has certain advantages, and the modern organic chemist uses more than one mass spectrometry technique to solve the structure of an unknown.

Mass spectrometry is also routinely employed in analytical laboratories as part of the identification of compounds in complex reaction mixtures, in the separation and identification of minute quantities of biologically active compounds, or in the identification of drugs in hair samples or trace explosives on luggage. ◀ **FIGURE 30.34** shows an ESI mass spectrometer (right) linked to a high-performance liquid chromatography (LC) unit (left), which is an instrument routinely used for separating complex mixtures. The combination, known as LCMS, is a powerful analytical technique.

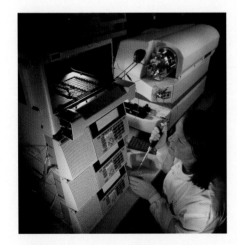

▲ **FIGURE 30.34   A modern LCMS.**
Mass spectrometry is often used in combination with other techniques, typically separation techniques such as liquid chromatography (LC) or gas-phase chromatography (GC). This LCMS is made up of three components: auto-sampler (left), liquid chromatography instrument (centre) and electrospray mass spectrometer (right). As different compounds are separated from a mixture by LC, they are analysed for their molecular mass.

Electron Impact Ionisation Mass Spectrometry

In this chapter we concentrate on **electron impact ionisation (EI) mass spectrometry (EIMS)** because it most clearly shows the principles of mass spectrometry, even though more complex methods are used more often today. Mass spectrometers using the EI technique operate by determining the *mass-to-charge ratio* (m/z) for ions in the gas phase. ▶ **FIGURE 30.35** illustrates how a result is obtained using an EI mass spectrometer. There are three principal components to EIMS: an ion source, which ionises molecules in a high vacuum and acceler-

FIGURE IT OUT

Why are neutral species not detected by this method?

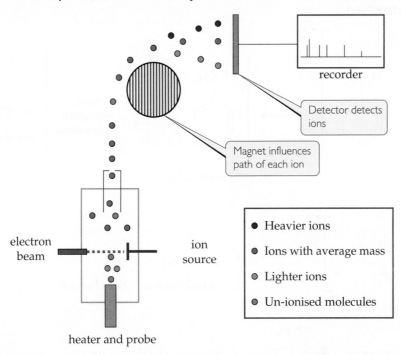

recorder

Detector detects ions

Magnet influences path of each ion

electron beam

ion source

- Heavier ions
- Ions with average mass
- Lighter ions
- Un-ionised molecules

heater and probe

◀ **FIGURE 30.35   The principle behind electron impact ionisation mass spectrometry.** The technique relies on generating charged species (ions) by bombarding a gas-phase sample with high-energy electrons.

ates them to form a beam; a strong magnet to deflect or separate the ions according to their masses; and a detector that records the masses of the ions generated. The intensity of each signal is then plotted as a function of the mass-to-charge ratio (m/z), see page 1229.

The sample to be analysed by EIMS is vaporised by heating to the gas phase within the probe. Ionisation of the sample occurs when a beam of high-energy electrons (10–70 eV, 1 eV = 96 kJ mol^{-1}) collides with the sample. Energy is transferred as a result of that collision, removing an electron from the molecule and yielding a positive ion. In this case, we say that the organic molecule has been *ionised* by electron impact and the signal that results is called the molecular ion signal (M^+). This molecular ion has the same mass as the neutral molecule less the mass of a single electron, which is of course negligible compared with the mass of the molecule.

CONCEPT CHECK 8

Which of these bonds will fragment first, C–I, C–H or C–C, under the same conditions?

If the electron collides with the molecule at higher energy (that is, around 70 eV), it not only brings about the ionisation of the molecule but also causes dissociation or *fragmentation* of the molecule, as shown in ▶ FIGURE 30.36. This fragmentation process gives a characteristic mixture of charged ions and radicals. The ions are detected easily, whereas the uncharged radicals are neither accelerated nor detected, but can be inferred by the difference in the mass between the molecular ion and the observed fragment. The type of cation (and radical) formed depends on the nature of the functional groups present, so fragmentation becomes diagnostic for functional group types. Those fragments able to stabilise the positive charge, such as allylic cations (derived from alkenes), benzylic cations ($C_6H_5CH_2^+$), acylium cations (derived from aldehydes and ketones), iminium cations (derived from amines) or oxonium cations (derived from alcohols and ethers), are always favoured (▼ FIGURE 30.37).

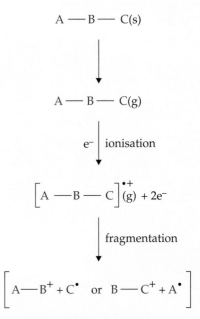

$$A - B - C(s)$$

$$\downarrow$$

$$A - B - C(g)$$

$$e^- \Big| \text{ionisation}$$

$$\left[A - B - C \right]^{\bullet +}(g) + 2e^-$$

$$\Big| \text{fragmentation}$$

$$\left[A - B^+ + C^\bullet \quad \text{or} \quad B - C^+ + A^\bullet \right]$$

▲ **FIGURE 30.36   Ionisation and fragmentation.**

◢ **FIGURE IT OUT**

Using resonance contributors, describe how a benzylic cation is able to be stabilised.

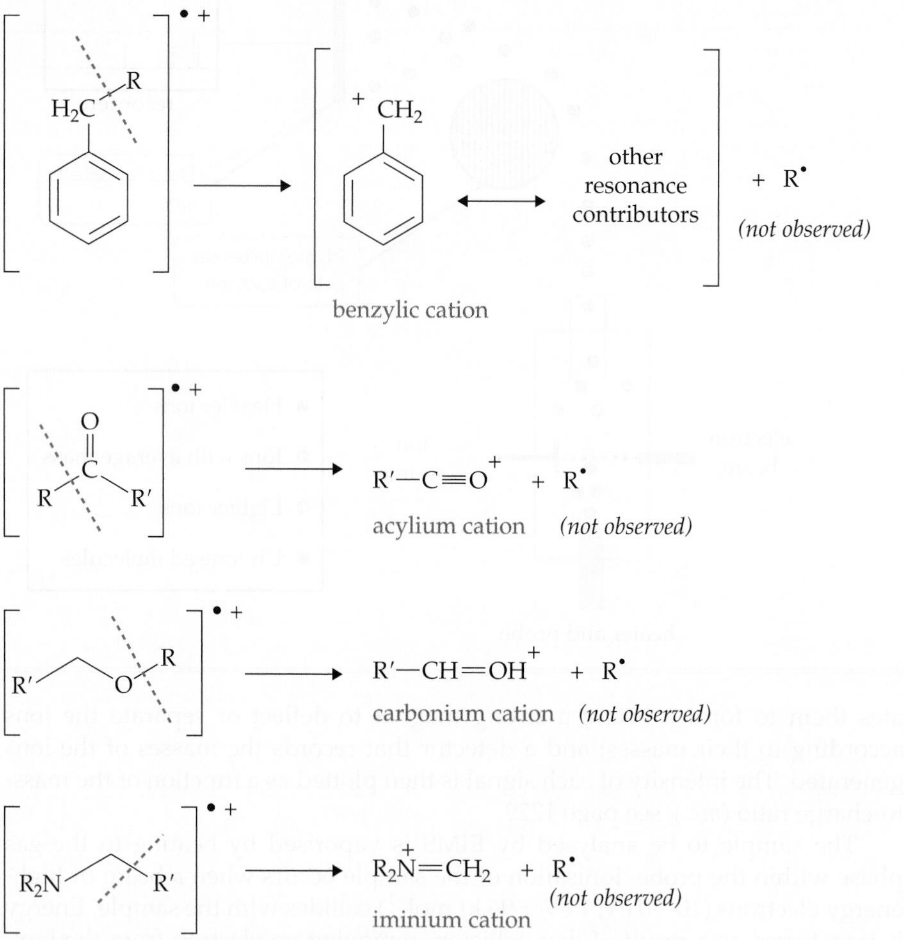

▲ **FIGURE 30.37** **Fragmentation of molecular ions.**

Once the ionisation and fragmentation processes have occurred, the ions formed are separated and detected. Initially, the positively charged ions are attracted to a negatively charged accelerator plate that forms a narrow slit, allowing only a stream of ions to pass through. Once past the slit, the positive ion enters an evacuated tube where its flight path is perturbed by the poles of a large magnet. As the positive ion passes through the magnetic field, its path is bent. The degree of bending depends on the mass of the fragments but also on their charge. The detector then converts the signal it receives into a mass spectrum.

Interpreting Mass Spectra

Two aspects of a mass spectrum lead the chemical detective to the identity of a compound. The first, called the **mass-to-charge ratio,** is symbolised by m/z (see ▶ **FIGURE 30.38**), where m equals the mass of the ion (in u) and z is its charge. Typically, organic chemists use the term m/z as a dimensionless quantity. As the charges on ionised or fragmented molecules almost always equate to $+1$, the mass-to-charge ratio for a particular signal in the mass spectrum is equal to the mass of that ion. These signals are assigned as a percentage abundance with respect to the most intense signal in the spectrum, called the **base peak**. The base signal does not necessarily correspond to the mass of the molecular ion—it is simply the most intense signal in the spectrum. For example, the base peak in Figure 30.38 occurs at $m/z = 31$ and is the result of fragmentation

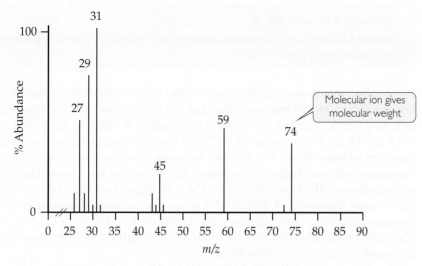

◀ **FIGURE 30.38** **EI mass spectrum of diethyl ether.** The fragmentation pattern seen within the EI spectrum is also deduced, showing the loss of different radicals from the initially ionised diethyl ether. Remember, uncharged fragments such as $CH_3^{\bullet}$ and $CH_3CH_2^{\bullet}$ are not observed.

leading to the cation CH_2OH^+. The molecular ion is found at $m/z = 74$. Figure 30.38 also illustrates the possible fragmentations leading to each of the signals within the mass spectrum. Note that only the positively charged fragments lead to the signals in the spectrum. All the neutral fragments are unaffected by the magnet and do not impact on the detector.

▼ **FIGURE 30.39** illustrates the principle of the mass-to-charge ratio by looking at species containing z values greater than 1. The signals at $m/z = 546$, 682, 909 and 1364 all equate to the same molecular weight, 2726 u. The multiplicity is due to this molecule's ability to react with H^+ at multiple sites, leading to multiple-charged species. This effect is common in subtle techniques such as ESI rather than in the chemically harsher techniques of EI and CI.

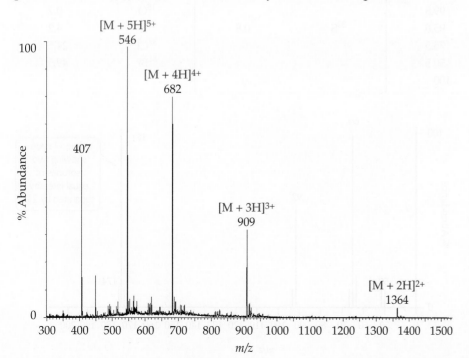

◀ **FIGURE 30.39** **Mass-to-charge ratio.** Multiple charged fragments yield signals within the mass spectrum corresponding to greater masses. For example, a molecule with molecular mass M will yield a signal at $M/2$ when doubly charged. This electrospray spectrum is of a new type of therapeutic agent designed to interfere with the translation of the mRNA code to protein. Notice that this type of ionisation leads to very little fragmentation, but does allow the sample molecules to accumulate protons, leading to multiple charged ions.

⚠️ **CONCEPT CHECK 9**

In what fundamental way is mass spectrometry related to Thomson's cathode-ray experiments (Figure 2.3)?

Routine mass spectrometry such as the type we have been describing is a low-resolution technique, meaning that the particle masses are rounded typically to a whole number. This means that carbon is taken as 12 u (^{12}C), oxygen as 16 u (^{16}O) and nitrogen as 14 u (^{14}N). However, most elements exist in more than one isotopic form. The heavier isotopes also give rise to small signals in the mass spectrum, higher than that of the molecular ion signal. These signals are sometimes called the $[M + 1]^+$ or $[M + 2]^+$ signals and can be very diagnostic in some instances. ▼ TABLE 30.4 gives the isotopic composition of some common elements, showing how they contribute to $[M + 1]^+$ or $[M + 2]^+$ signals. The isotopes for chlorine and bromine are highlighted as they show the most diagnostic abundance distributions.

For the most part, the $[M + 1]^+$ and $[M + 2]^+$ signals are significantly smaller (less than 1%) than the parent ion or molecular ion signal. However, chlorine exists as two isotopes, ^{35}Cl and ^{37}Cl, with respective isotope abundances of 75.5% and 24.5%. Bromine is the most easily recognised of the halogens. It has two signals, one for ^{79}Br and one for ^{81}Br, of near equal intensity (50.5% and 49.5%, respectively). What this means is that, if bromine is present in a molecule, the $[M + 2]^+$ ion has an intensity nearly equal to the molecular ion. This is illustrated in the EIMS spectrum of 4-bromoaniline, shown in ▼ FIGURE 30.40. Iodine is typically recognised by the presence of iodenium ion (I^+) at m/z 127. This clue is also combined with a characteristic 127 separation in the mass spectrum between the molecular ion signal and the first fragmentation. If chlorine is present, the $[M + 2]^+$ signal is about one-third as large as the M^+ signal.

TABLE 30.4 • Isotopic distributions for some common elements

Element	M	% Abundance	M + 1	% Abundance	M + 2	% Abundance
Hydrogen	^1H	100				
Carbon	^{12}C	98.9	^{13}C	1.1		
Nitrogen	^{14}N	99.6	^{15}N	0.4		
Oxygen	^{16}O	99.8			^{18}O	0.2
Sulfur	^{32}S	95.0	^{33}S	0.8	^{34}S	4.2
Chlorine	^{35}Cl	75.5			^{37}Cl	24.5
Bromine	^{79}Br	50.5			^{81}Br	49.5
Iodine	^{127}I	100				

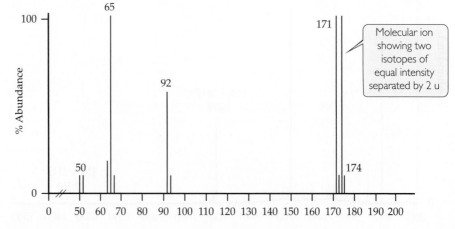

▶ **FIGURE 30.40 EI mass spectrum of 4-bromoaniline.** Note the diagnostic M^+ and $[M + 2]^+$ signals of equal intensity and the spacing of 79 u between the molecular ion and first fragmentation. These clues all indicate that the molecule contains bromine.

SAMPLE EXERCISE 30.5 | Interpreting mass spectra

An EI mass spectrum of compound B, a simple haloalkane, is shown below. Deduce its structure.

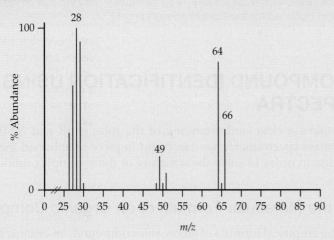

SOLUTION

Analyse We are asked to deduce the structure and hence molecular formula based on the mass spectrum observed.

Plan Use the M^+ and $[M+2]^+$ signals to deduce the halide present, as well as the molecular formula. Use the fragmentation patterns to confirm your deduction.

Solve The relative abundance of the M^+ and $[M+2]^+$ signals suggests that chlorine is the likely halide, although the observed ratio of these two signals is slightly different from the expected 3 : 1 ratio. The presence of Cl within the structure is confirmed by the difference between the M^+ and the second fragmentation of 29 u (difference = 35). The position of this second fragmentation also coincides with the cation $CH_3CH_2^+$. Hence, the structure of our unknown must correspond to *chloroethane*.

Check The fragmentation signal at 49 u corresponds to the loss of a methyl radical (15 u) from M^+.

PRACTICE EXERCISE

The mass spectrum of compound C contains major peaks at $m/z = 88, 73, 45, 43$ and 29. Confirm that this fragmentation pattern is consistent with compound C being ethyl acetate.

Answer: Fragmentation around the carbonyl group yielding acylium cations is extremely likely. For example:

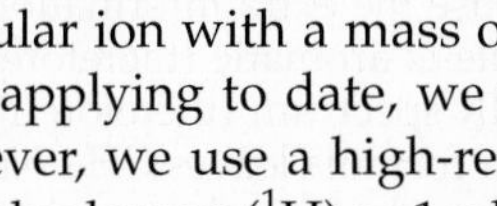

$$\underset{H_3C}{\overset{43}{}}\underset{}{\overset{O}{\underset{\parallel}{C}}}\underset{OCH_2CH_3}{\overset{45}{}} \quad \text{and} \quad \underset{H_3C}{\overset{}{}}\underset{15}{\overset{O}{\underset{\parallel}{C}}}\underset{OCH_2CH_3}{\overset{73}{}}$$

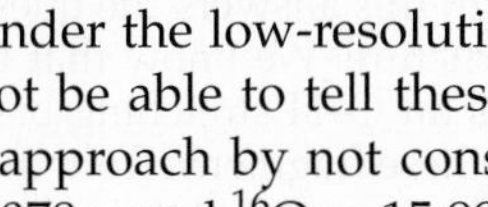

(See also Exercise 30.46.)

There is one last aspect of mass spectrometry that we should discuss. To do this, we consider the molecules propane, acetaldehyde and CO_2. Each one produces a molecular ion with a mass of 44 u. Under the low-resolution methods we have been applying to date, we would not be able to tell these molecules apart. If, however, we use a high-resolution approach by not considering the mass of atomic hydrogen (^1H) as 1 u but as 1.0078 u and ^{16}O as 15.9949, then the three molecules in question (with different molecular formulae) will correspond to different exact masses—C_3H_8, C_2H_4O and CO_2 have exact masses of 44.0626, 44.0262 and 43.9898 u. For this high-resolution technique to be useful, accurate masses are always quoted to four decimal places. Analysis requires a highly accurate measurement (± 1 part in 10^6). The result can then be compared with exact masses calculated for various element combinations. Note that, by definition, ^{12}C has an atomic weight of 12.0000 u.

> **△ CONCEPT CHECK 10**
>
> How could mass spectrometry be used to determine the identity of a white powder sample found in a clandestine drug laboratory? Is the value of the molecular ion obtained for the sample enough proof?

30.5 | COMPOUND IDENTIFICATION USING SPECTRA

Now that we have a clear understanding of the roles of IR and NMR spectroscopy and mass spectrometry, we can begin to piece together all the necessary information in order to solve the structure of more complex unidentified compounds.

Deducing the Molecular Formula of an Organic Compound

To work out the empirical formula of an organic compound, an organic chemist subjects a small amount of sample to *microanalysis*. This destructive technique combusts the sample and analyses the amount and type of gas released. From these data, a percentage C, H, N is obtained based on weight. The remaining percentage is inferred as the content of O in the sample. Manipulation of these percentages, by taking into account the atomic masses of C, H, N and O leads to an empirical formula, from which a molecular formula can be determined, normally in conjunction with the value for the molecular ion signal in the mass spectrum.

> **SAMPLE EXERCISE 30.6** Microanalysis to molecular formula
>
> The IR spectrum of a liquid aromatic compound has a broad band in the range 3000–3500 cm^{-1}. EIMS indicates a molecular ion signal at $m/z = 108$ and a sample of the unknown does not dissolve in 1 M NaOH. Microanalytical results for this unknown are: C, 78%; H, 7.4%. Deduce the molecular formula of this unknown.
>
> **SOLUTION**
>
> **Analyse** We are asked to use a set of analytical and chemical data for an unknown and asked to calculate its molecular formula.
>
> **Plan** Work out the empirical formula and use the molecular ion result from EIMS to yield the molecular formula.
>
> **Solve** The elements present are C = 78%, H = 7.4% and O = 14.6%. Taking into account their atomic weights:
>
> $$C : H : O = \frac{78}{12} : \frac{7.4}{1} : \frac{14.6}{16}$$
>
> $$= 6.5 : 7.4 : 0.91 \text{ (divide by smallest number)}$$
> $$= 7 : 8 : 1$$

Therefore the empirical formula is C_7H_8O. The empirical mass is equivalent to the molecular ion signal, so the molecular formula is also C_7H_8O.

Although this answers the question, let's use the extra information given to us to deduce a structure. We know that the molecule is aromatic (therefore it contains C_6) and contains an O–H stretching band in the IR spectrum (therefore it contains OH). Eliminating these fragments from the molecular formula leaves CH_7. Not being soluble in 1 M NaOH rules out that the unknown is a phenol. Using the remaining carbon atom to bridge the aromatic and OH groups yields benzyl alcohol as a likely suspect.

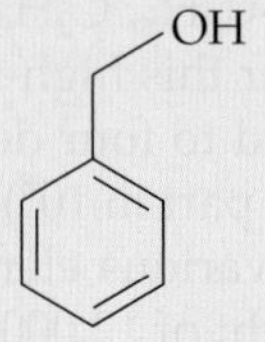

PRACTICE EXERCISE

(a) An unknown liquid has an IR spectrum with a band at 3330 cm^{-1}. EIMS indicates a molecular ion signal at $m/z = 60$. Microanalytical results are: C, 60%; H, 13.4%. Deduce the molecular formula of the unknown.

(b) An unknown compound has a band at 1710 cm^{-1} in the IR spectrum. EIMS indicates a molecular ion signal at $m/z = 88$ and a base signal at $m/z = 43$. Microanalysis gives: C, 54.5%; H, 9.15%. Deduce the molecular formula of the unknown.

Answers: **(a)** C_3H_8O, **(b)** $C_4H_8O_2$

(See also Exercises 30.51, 30.52.)

In Chapters 22 and 24 we identified the general molecular formula of an alkane as C_nH_{2n+2}, a cycloalkane is C_nH_{2n} and the corresponding alkene also has the same general molecular formula, C_nH_{2n}. This difference in molecular formula between, for example, alkanes and alkenes actually yields important structural information. While the example is trivial, knowing something about the **index of hydrogen deficiency (IHD)**, which is essentially the summation of all π-bonds and rings within a molecule, is useful when determining possible structures of compounds, especially alkenes, ketones and aldehydes, amines, alkyl halides and so on. The IHD quantity is determined by comparing the number of hydrogens in the molecular formula of an unknown with that of an *alkane* reference containing the same number of carbon atoms as the unknown. This last statement can be summarised in the following way:

IHD = Number double bond or ring equivalents

$$= \frac{\text{maximum number H possible per C} - \text{actual number H per C}}{2} \quad [30.6]$$

Let's investigate the molecular formula, $C_4H_8O_2$, used in Practice Exercise 30.6(b). The index of hydrogen deficiency is

$$\text{IHD} = \frac{(2n + 2) - 8}{2}$$

where $n = 4$

$$\text{IHD} = \frac{(2 \times 4 + 2) - 8}{2}$$

$$= 1$$

which means that one double bond (or a ring system) must exist within the molecule. This unknown is ethyl acetate. The presence of a C=O double bond in ethyl acetate supports our IHD calculations. While this calculation works for compounds containing C, H and O, other elements can complicate matters.

To help us, several guidelines have been devised to determine IHD.

1. Work out the molecular formula of the *reference alkane* based on the number of carbon atoms in the unknown's molecular formula.

2. For each halogen in the unknown's molecular formula, *subtract* one hydrogen off the reference's molecular formula.

3. For each nitrogen in the unknown's molecular formula, *add* one hydrogen to the reference's molecular formula.

4. Make no modifications for oxygen or sulfur in the formula.

Chemical Wet Testing: Tests for Functional Groups

As long as enough of an unknown compound is present, chemical wet tests are useful for providing clues to the types of functional groups present. These tests may be as simple as solubility in aqueous solution or as complex as two- or three-step processes in a micro-test tube. ▼ FIGURE 30.41 illustrates the process of using functional group tests to offer clues to an unknown's structure. This flow diagram has a significant amount of chemistry, so we will briefly point out the requirements for a positive test for each functional group. Much of this chemistry has been dealt with in the preceding chapters.

Alkenes: The decolourisation of a bromine solution when applied to an organic molecule usually indicates the presence of an alkene.

Alcohols: Alcohols of low molar mass, such as CH_3OH and CH_3CH_2OH, mix completely with water, but those with larger non-polar alkyl groups have low solubility. The addition of sodium metal to alcohols will evolve hydrogen gas (bubbles) generating the alkoxide.

$$2\,ROH + 2\,Na \longrightarrow 2\,RO^- + 2\,Na^+ + H_2(g) \qquad [30.7]$$

Alcohols do not react with aqueous hydroxide.

Phenols: Phenols behave as weak acids with pK_a about 10. As a result, they are soluble in 1 M NaOH but not soluble in saturated aqueous sodium bicarbonate solution.

Carboxylic acids: Carboxylic acids react with saturated aqueous sodium bicarbonate solution, with the release of CO_2 from solution.

$$RCOOH + HCO_3^-\,(aq) \longrightarrow RCOO^-\,(aq) + H_2O + CO_2\,(g) \qquad [30.8]$$

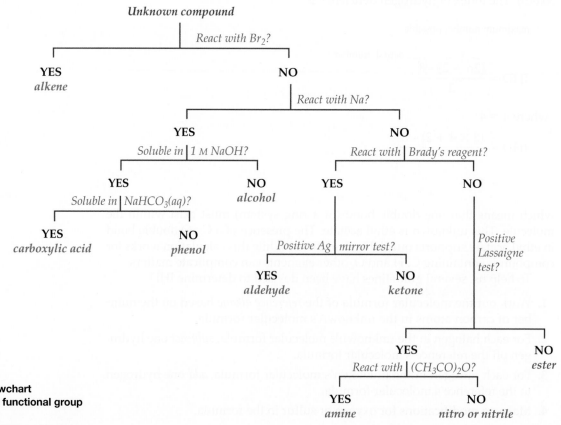

▶ **FIGURE 30.41 Flowchart illustrating some simple functional group tests.**

Aldehydes and ketones: Both aldehydes and ketones react with 2,4-dinitrophenylhydrazine (DNP), also known as Brady's reagent, to form highly coloured precipitates. A generalised equation for this reaction is:

$$R_2C{=}O + H_2NR' \longrightarrow R_2C{=}NR' + H_2O \qquad [30.9]$$

To distinguish aldehydes from ketones, a silver mirror test is usually employed.

$$RCHO + 2\,Ag^+\,(aq) + 3\,OH^-\,(aq) \longrightarrow RCOO^-\,(aq) + 2\,Ag(s) + 2H_2O \quad [30.10]$$

Nitrogen: The Lassaigne sodium fusion test is a dramatic test that requires the use of molten sodium metal and ferrous sulfate. The bright blue colour of the product, $NaFe[Fe(CN)_6]$ (also known as Prussian blue), indicates a positive test. The nitrogen can be present in amines, nitro compounds, nitriles and amides.

Amines: Amines are soluble in 1 M HCl and give alkaline solutions when mixed with water. Primary aromatic amines yield bright red/orange-coloured azo-dyes when mixed with nitrous acid and 2-naphthol. Amines form amides by reaction with acetic anhydride.

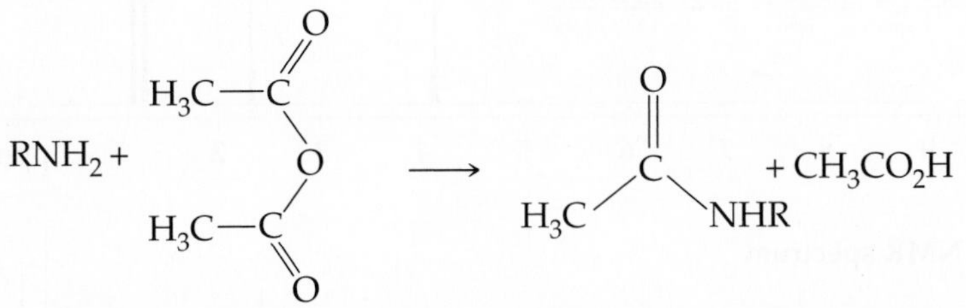

$$[30.11]$$

We are now ready to tackle an advanced problem as a chemical detective. To help you to become familiar with the process, here are some guidelines.

1. *Determine the molecular formula and index of hydrogen deficiency.* Calculate the empirical formula, determine the mass of the molecular ion (from the mass spectrum) and hence calculate the molecular formula and IHD.

2. *Check the IR spectrum.* Look for stretching bands indicative of functional groups O–H, N–H, C=O and so on.

3. *^1H NMR spectrum: number of signals and their positioning.* Determine the number of non-equivalent hydrogens by counting the number of signals present. Determine their chemical shift and relate this to potential neighbouring functional groups—for example, CH_2–O at about 4 ppm, $CH_2C{=}O$ at about 2.5 ppm. Remember, this is only a guide.

4. *^1H NMR spectrum: integration.* Determine the relative number of equivalent hydrogens in each signal by measuring the area under each signal. Of course, this may already be done for you (Figure 30.26). The integral value could be exact in terms of the number of hydrogens or, based on extra symmetry within the molecule, may be a fraction of it.

5. *^1H NMR spectrum: splitting pattern.* Use your knowledge of splitting patterns and the $(n + 1)$ rule to determine the connectivity between groups of atoms. Learn to recognise ethyl groups by their characteristic splitting pattern.

6. *Deduce the structure.* Using all this information (including chemical tests), try to deduce the structure. Remember, it might not be the first one you draw.

7. *Check.* Use the ^{13}C NMR spectrum and any distinguishable fragmentation patterns in the mass spectrum to confirm your choice of structure. Use this information to distinguish isomers.

SAMPLE INTEGRATIVE PROBLEM **Putting concepts together**

Unknown D is a sweet-smelling liquid. Microanalytical results for this unknown give C, 58.80%; H, 9.87%, and the characterisation spectra acquired for this unknown are shown below. Use the information to deduce the structure of unknown D.

EI mass spectrum

IR spectrum

^{1}H NMR spectrum

^{13}C NMR spectrum

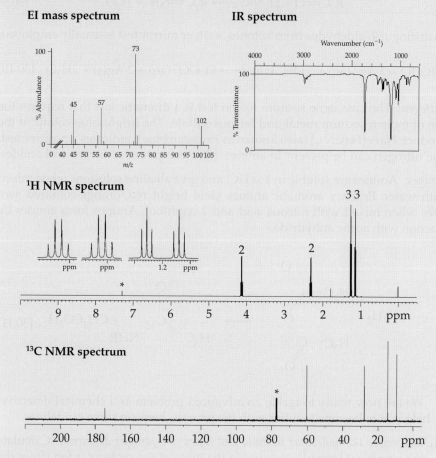

Asterisks denote solvent signals.

Analyse We are asked to use a set of analytical and chemical data for an unknown and asked to determine its structural formula.

Plan Use all the information given in the form of spectra and microanalytical data to deduce the structure of the unknown.

Solve The microanalytical data give an empirical formula of $C_5H_{10}O_2$, which has a mass equal to the molecular ion signal. Hence the molecular formula is also $C_5H_{10}O_2$. The IHD is 1, suggesting the presence of either a ring or double bond within the structure, but not both.

The IR spectrum shows a strong band at about 1750 cm^{-1}, attributable to a C=O stretching frequency. There is no O–H stretch, which means the extra oxygen atom in the molecular formula must be of ester or ether origin.

There are four signals in the ^{1}H NMR spectrum with integration in the ratio 2 : 2 : 3 : 3, which accounts for all the hydrogens in the molecular formula.

The chemical shifts of the two sets of quartets indicates that one set (approximately 4 ppm) neighbours oxygen and the other neighbours a carbonyl C=O group. Using the $(n + 1)$ rule, the hydrogens responsible for the quartet splitting patterns must be coupled to methyl groups. Hence, two different ethyl groups exist in this structure.

Putting together the clues to date, we have the following fragments:

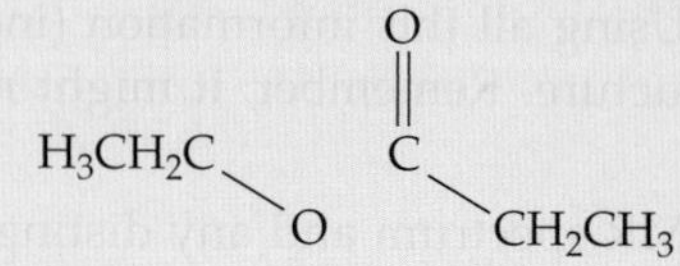

These fragments make up the molecular formula of the unknown. Knowing that the unknown shows a positive test for esters (and is sweet-smelling) would confirm the choice in structure. Hence, our unknown is *ethyl propanoate*.

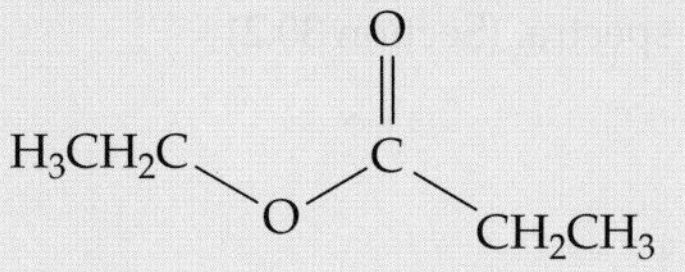

Check Use the ^{13}C NMR data to confirm the structure. The structure of ethyl propanoate has five different carbon atoms, the same number of signals as found in the ^{13}C NMR spectrum. In terms of chemical shift, the signals would be distributed as follows:

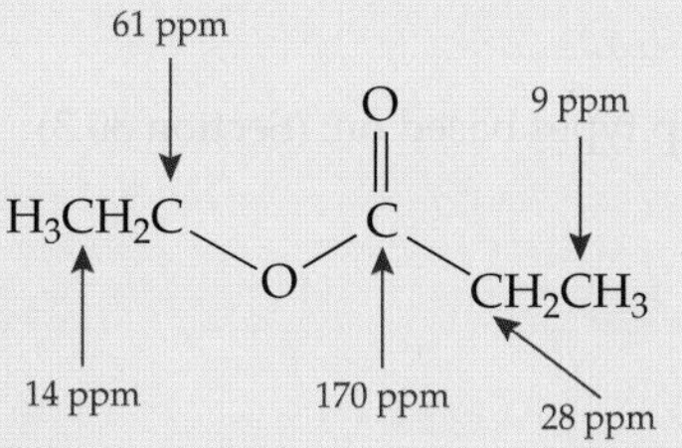

PRACTICE EXERCISE

A number of these problems can be found in the end-of-chapter Exercises.

CHAPTER SUMMARY AND KEY TERMS

SECTION 30.1 All molecules absorb electromagnetic radiation to some degree and hence give rise to absorption spectra. The amount and type of absorption depends on the energy of radiation and the nature of the molecule, its bond types and its electronic structure. **Beer–Lambert's law** can be used to relate the amount of energy being absorbed to the concentration of the substance absorbing the energy in **UV-visible spectroscopy**.

SECTION 30.2 Infrared spectroscopy observes the vibrations of covalent bonds. The difference in vibration energy is a result of the stiffness of the covalent bond. This principle follows **Hooke's law** and provides useful evidence for the presence of functional groups in molecules. Compounds are measured either as solids, liquids or gases. An **infrared spectrum** is usually plotted as percentage **transmittance** (%T) versus **wavenumber** (cm^{-1}). The region below 1400 cm^{-1} is known as the **fingerprint region**.

SECTION 30.3 Nuclear magnetic resonance (NMR) spectroscopy is a method for observing nuclear spin transitions, giving information on the connectivity of nuclei (atoms). The technique relies on the difference in energy levels between nuclear spins of $I > 0$ ($I = \frac{1}{2}$ for ^{1}H and ^{13}C), within a strong magnetic field. The absorption of radiation at the **Larmor frequency**, which leads to an NMR spectrum, is termed **resonance**. The frequency of an NMR spectrometer is called its **operating frequency**.

The electrons within these functional groups also have a spin that creates local magnetic fields that **shield** or **deshield** the neighbouring nuclei from the applied magnetic field. Three aspects of an NMR spectrum are important: the **chemical shift (δ)** of a signal, the area under the signal (available by integration) and the **splitting pattern** of the signal all yield diagnostic information about the molecular structure. Equivalent hydrogens or carbons within a molecule by symmetry give rise to identical chemical shifts in ^{1}H or ^{13}C NMR spectra, respectively.

An NMR signal is split by the influence of adjacent atoms on the local magnetic field of the nucleus in question. This splitting is called **spin–spin coupling**. The splitting pattern that results follows the **(n + 1) rule**. Splitting patterns are commonly described in terms of their **multiplicity**—for example, singlets, doublets, triplets, quartets, multiplets. Two (or more) coupled nuclei share the same **coupling constant** (J), which allows for easy tracing of the molecule's connectivity. ^{13}C NMR spectra are commonly devoid of coupling because they are recorded in a hydrogen-decoupled mode. **Decoupling** is used either to simplify a spectrum or to indicate coupled nuclei.

SECTION 30.4 Electron impact ionisation mass spectrometry (EIMS) uses high-energy electrons to bombard a sample compound, causing it to *ionise* or **fragment**. The **molecular ion** (M^{+}) formed upon ionisation has a **mass-to-charge ratio** (m/z) equal in quantity to the molecular weight of the sample being tested. Signals with a mass spectrum are assigned a percentage abundance with respect to the **base peak**. Mass spectrometry often provides clues to the structure and functional groups contained within a molecule through either fragmentation or isotopic patterns.

SECTION 30.5 Using the **index of hydrogen deficiency (IHD)**, the chemist can determine the number of double bond or ring equivalents within a molecular formula.

KEY SKILLS

- Understand how to use Beer–Lambert's law. (Section 30.1)
- Be able to identify O–H, C–H, C≡C, C=C, C=O stretches in IR spectra. (Section 30.2)

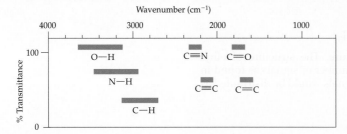

- Apply NMR chemical shifts to determine possible functional group types based on: (Section 30.3)

(a) ^{1}H NMR spectroscopy

- Apply multiplicity and J in ^{1}H NMR to determine connectivity within organic molecules. (Section 30.3)
- Be able to identify the molecular ion signal and apply fragmentation patterns to determine functional groups and molecular weight. (Section 30.4)
- Be able to determine molecular formula from percentage composition and mass spectra. (Section 30.5)
- Calculate IHD based on a molecular formula. (Section 30.5)

KEY EQUATIONS

- Beer–Lambert's law $\qquad A = \varepsilon c l$ [30.1]

- Transmittance
$$\text{Absorbance} = -\log_{10}(\text{transmittance})$$
or
$$A = -\log_{10} T$$
[30.2]

- Effective magnetic field $\qquad B_{\text{effective}} = B_{\text{applied}} - B_{\text{local}}$ [30.4]

- Chemical shift $\qquad \delta = \dfrac{\text{Shift in resonance frequency from TMS (Hz)}}{\text{Operating frequency of the spectrometer (MHz)}}$ [30.5]

- IHD $\qquad$ IHD = Number double bond or ring equivalents
$$= \frac{\text{maximum number H possible per C} - \text{actual number H per C}}{}$$
[30.6]

2

EXERCISES

VISUALISING CONCEPTS

30.1 (a) An aqueous solution of 4-aminobenzoic acid gives a maximum absorbance of 0.21. What is the percentage transmittance? **(b)** The absorption spectrum of a solution of naphthalene ($\varepsilon = 300\ dm^3\ mol^{-1}\ cm^{-1}$) in hexane obtained through a path length of 1.0 cm is given below. What is the concentration of the sample? [*Hint:* You will need to estimate the absorbance value by reading off the graph.] [Section 30.1]

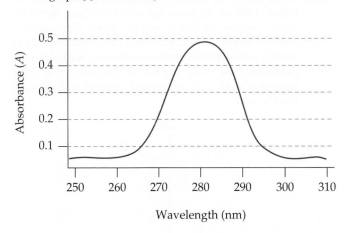

Wavelength (nm)

30.2 Determine which of the following compounds would be likely to give this IR spectrum. [Section 30.2]

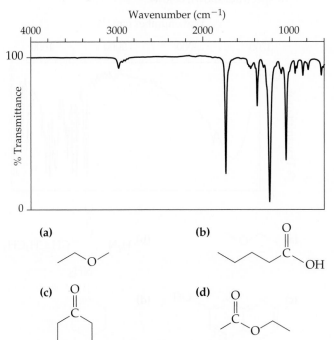

30.3 The My World of Chemistry box in Section 30.3 describes the technique called MRI, a medical equivalent to NMR spectroscopy. **(a)** Instruments for obtaining MRI data are typically labelled with a frequency, such as 600 MHz. Why do you suppose this label is relevant to the experiment? **(b)** In general, the stronger the magnetic field, the greater the information obtained from an NMR or MRI experiment. Why do you suppose this is the case? **(c)** What safety precautions do you think are relevant around superconducting magnets? [Section 30.3]

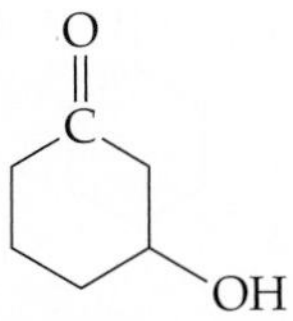

30.4 How could you use both ^1H and ^{13}C NMR to distinguish between these isomers? [*Hint:* Look for symmetry elements.] [Section 30.3]

(a) $CH_3CH_2OCH_2CH_3$ and $CH_3CH_2CH_2CH_2OH$

(b)

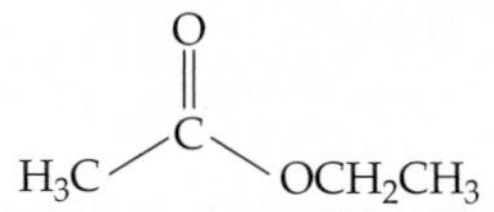

$$H_3C-\overset{\overset{\displaystyle O}{\|}}{C}-OCH_2CH_3 \quad \text{and} \quad H_3CH_2C-\overset{\overset{\displaystyle O}{\|}}{C}-OCH_3$$

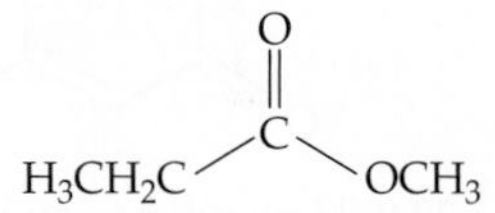

(c) $CH_3CH_2CH_2Br$ and $CH_3CHBrCH_3$

30.5 The two compounds shown below have a molecular ion signal at $m/z = 90$. Describe two ways, using mass spectrometry, of distinguishing between them. [Section 30.4]

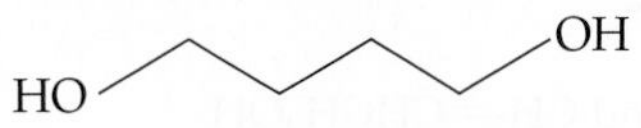

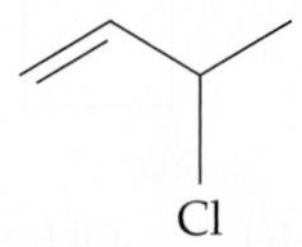

30.6 Confirm that the fragmentation pattern seen in the following EI mass spectrum is consistent with the structure shown below. What chemical wet test(s) could you perform to support the structure shown? [Sections 30.4 and 30.5]

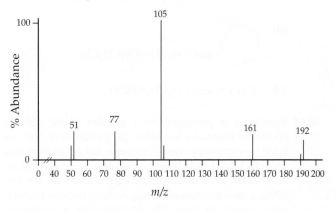

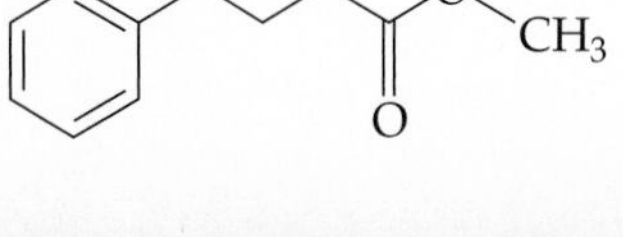

m/z

INFRARED SPECTROSCOPY (Section 30.2)

30.7 Identify a possible functional group that might be responsible for the following IR absorptions:

(a) 3500 cm^{-1}

(b) 1705 cm^{-1}

(c) 1650 cm^{-1}

(d) 1710 and 3400 cm^{-1}

(e) 1680 and 3200 cm^{-1}

30.8 Norethidrone is a steroid used primarily as an oral contraceptive. Its synthesis was first achieved in 1951. Indicate which parts of the structure of norethidrone have characteristic absorptions identifiable by IR spectroscopy. Indicate their approximate positions in wavenumbers (cm^{-1}).

30.9 How might you use IR spectroscopy to help distinguish between the following pairs of compounds?

(a) $CH_3CH_2OCH_2CH_3$ and $CH_3CH_2CH_2CH_2OH$

(b) and $CH_2{=}CHCH_2OH$

(c) and

(d) and $CH_3CH{=}CHCH_2CH_3$

(e) $(CH_3)_3N$ and $CH_3CH_2NHCH_3$

30.10 Ephedrine is produced by a number of the *Ephedra* species of plants. Its medicinal properties have been related to respiratory conditions since 200 AD. Its structure was elucidated in 1923 and ephedrine began clinical trials in 1926 as a bronchodilator for asthma. In the late 1990s a new recreational drug, ecstacy, became popular with people in their 20s. Its structure is similar to ephedrine, which also acts as a heart stimulant.

ephedrine   ecstacy

Could police and customs officials use IR spectroscopy to distinguish between the two compounds? Explain.

30.11 Picric acid is an aromatic compound used extensively before 1960 as an agent for co-crystallising liquid organic samples so that a melting point determination could be made. Picric acid is not a carboxylic acid but instead a phenol bearing nitro substituents at the 2,4 and 6 positions. How could you use IR spectroscopy to determine that picric acid was indeed a phenol and not a carboxylic acid?

30.12 Which of the following structures is consistent with the IR spectrum shown below?

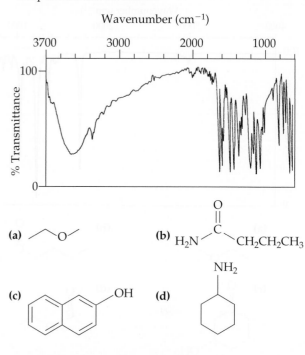

(a)   (b)

(c)   (d)

30.13 (a) Of the functional groups O–H, C=O, C≡N, which would be most diagnostic for bioimaging? Why? (b) Can you name another functional group that could be used for diagnostic purposes using the IR technique? What problems might arise?

NMR SPECTROSCOPY (Section 30.3)

30.14 Indicate the number of different absorptions and the relative intensities expected in the ^{1}H NMR spectra of:

(a) 2,2-dimethylbutane   (b) ethanol

(c) acetone   (d) 2-chloropropane

(e) cyclohexane   (f) cyclohexanone

(g) ethyl methyl ether

30.15 Indicate the number of different absorptions expected in the ^{13}C NMR spectra of:

(a) 2,2-dimethylbutane   (b) ethanol

(c) acetone   (d) 2-chloropropane

(e) cyclohexane   (f) cyclohexanone

(g) ethyl methyl ether

30.16 The ^{1}H NMR spectrum of acetone recorded on a spectrometer with an operating frequency of 200 MHz has a single resonance signal at 2.1 ppm.

(a) What is the difference in Hz between this signal and that of tetramethylsilane (TMS)?

(b) At what δ value does this signal lie if recorded using a 400 MHz spectrometer?

(c) How many Hz away from TMS does the absorption in the 400 MHz spectrum correspond to?

30.17 The ^{13}C NMR spectrum of acetone recorded on a spectrometer with an operating frequency of 400 MHz has resonance signals at 30 and 180 ppm.

(a) Assign these two signals.

(b) What is the difference in Hz between the two acetone signals?

(c) What structural or electronic features of the molecule cause this large difference in chemical shift?

30.18 When the antioxidant BHT was first prepared, chemists were not sure of its structure. The chemical formula was determined by elemental analysis, but NMR, which would have instantaneously revealed the structure, had not yet been discovered. The problem arose because BHT failed the usual tests for phenols; for example, it was not soluble in dilute basic solution. Because of this, chemists thought the second 2-methylpropyl (*t*-butyl) group was linked to oxygen to form an ether. Of course, the reason BHT does not behave like other phenols is because the OH group is hindered by the two neighbouring 2-methylpropyl groups, a protective quality that also allows this molecule to be used as a commercial antioxidant.

In terms of the number of equivalent hydrogens present, how could ^{1}H NMR spectroscopy be used to verify the correct structure for BHT?

30.19 Two bottles labelled propanol were found in a laboratory. Because of their differing boiling points, the bottles clearly contained the two propanol constitutional isomers. To determine which propanol was in which bottle, a ^{1}H NMR spectrum was obtained on each sample. Draw the two possible isomers and describe how NMR might be used to distinguish between them.

30.20 Predict the splitting pattern observed in the ^{1}H NMR spectrum for each of these compounds:

(a) CH_3CHBr_2 (b) $(CH_3)_2CHCl$ (c) $(CH_3)_3CBr$

30.21 Predict the splitting pattern observed in the hydrogen-coupled ^{13}C NMR spectrum for each of these compounds:

(a) CH_3CHBr_2 (b) $(CH_3)_2CHCl$ (c) $(CH_3)_3CBr$

30.22 The effective magnetic field around a nucleus has contributions from both the applied magnetic field and the influence of neighbouring electrons, which create their own magnetic field. The result of these influences is a difference in chemical shift (δ) between nuclei in a molecule. Demonstrate this principle by illustrating graphically the effect the three H_b nuclei have on the splitting pattern of H_a in $Cl_2CH_aC(H_b)_3$.

30.23 The spin of an electron generates a magnetic field, with 'spin-up' and 'spin-down' electrons having opposite fields. In the absence of a magnetic field, a 'spin-up' and a 'spin-down' electron have the same energy. **(a)** Why do you think that the use of a magnet was important in the discovery of electron spin? **(b)** A phenomenon called electron spin resonance (ESR) is closely related to nuclear magnetic resonance. In ESR a compound with an unpaired electron is placed in a magnetic field, causing the unpaired electron to have two different energy states similar to our discussion of nuclear spin (see Figure 30.16). ESR uses microwave radiation instead of radiofrequency radiation to excite the unpaired electron from one state to the other. Does an ESR experiment require photons of greater or lesser energy than an NMR experiment?

30.24 2-Bromobutane undergoes a β-elimination reaction in the presence of potassium *tert*-butoxide to produce one of two isomeric alkenes.

How could you use NMR spectroscopy to determine which alkene is formed as the major product?

30.25 2-methylbut-2-ene contains signals at δ20.3, 22.5, 22.9, 110.3 and 111.5 in the ^{13}C NMR hydrogen-decoupled spectrum, whereas 2-methylbutane contains only four signals. Explain why this is so.

30.26 The energy, E, of electromagnetic radiation can be determined by the formula

$$E = hv$$

where h = Planck's constant.

Using this formula, calculate the amount of energy required to spin-flip a proton (expressed in kJ mol^{-1}) in a spectrometer with operating frequency of 300 MHz, which is the typical operating frequency in the modern NMR laboratory.

30.27 What is meant by these terms?

(a) chemical shift   (b) operating frequency

(c) tesla   (d) spin–spin coupling

(e) local magnetic field   (f) integration

30.28 Predict the number of ^{1}H NMR signals and also the splitting pattern of each of these compounds in the ^{1}H NMR spectrum.

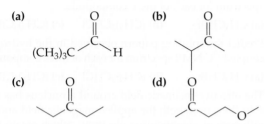

30.29 Predict the number of ^{13}C NMR signals in the hydrogen-decoupled spectrum of the following:

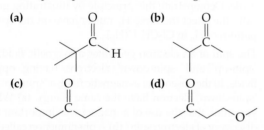

30.30 The reaction of an unknown with ethanol in the presence of a small amount of H_2SO_4 leads to a compound that smells distinctly like rum. The hydrogen-decoupled ^{13}C NMR spectrum of the product is shown below (* = solvent).

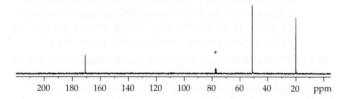

(a) What is the structure of the unknown?

(b) What other structure(s) would be consistent with this ^{13}C NMR spectrum?

30.31 Two bottles labelled *cis*-cyclohexandiol were found in a laboratory. Because of their differing melting points, the bottles clearly contained two different *cis*-cyclohexandiol isomers. To determine which isomer was in which bottle, the chemist decided to run ^{13}C NMR spectra on a sample of each. The three possible isomers were:

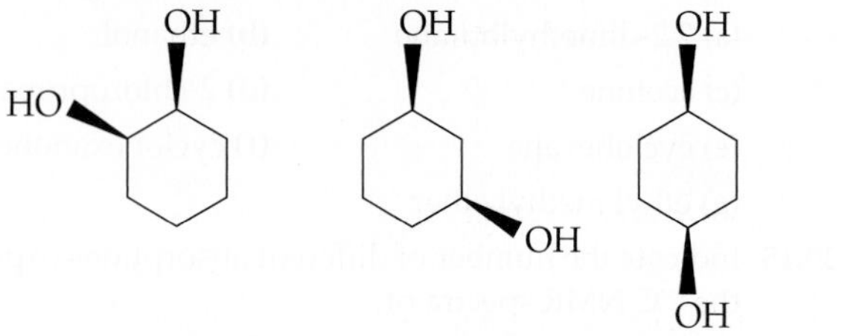

The two ^{13}C NMR hydrogen-decoupled spectra contained the following data:

- spectrum of sample in bottle A: 20.20, 62.55
- spectrum of sample in bottle B: 16.09, 29.52, 36.15, 63.08

Identify which isomer was present in each bottle.

30.32 (a) Why is ^{13}C NMR spectroscopy less sensitive than ^{1}H NMR spectroscopy? (b) In Figure 30.21, there exist small signals either side of the main singlets. If these are not from an impurity, what are they from? (c) Figure 30.33 shows hydrogen-coupled and hydrogen-decoupled ^{13}C NMR spectra. Why doesn't the splitting pattern of the solvent change from a hydrogen-coupled to a hydrogen-decoupled spectrum?

30.33 ^{31}P is NMR active ($I = \frac{1}{2}$) and is also the only naturally occurring phosphorus isotope. Describe the splitting you would see in both the ^{1}H NMR and ^{31}P NMR spectra of PH_2Cl.

30.34 The 200 MHz ^{1}H NMR spectrum of propanal (shown) has three distinctive resonances: δ9.5 (1H, singlet), 2.2 (2H, quartet), 1.0 (3H, triplet). Assign each resonance to the structure.

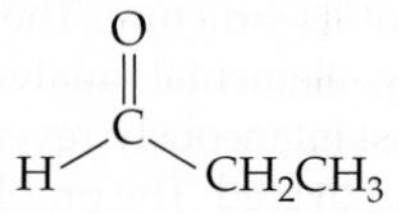

30.35 The 200 MHz ^{1}H NMR spectrum of 2-cyanopent-2-ene has four distinctive resonances: δ5.3 (1H, singlet), 2.1 (2H, quartet), 1.9 (3H, singlet), 1.0 (3H, triplet). Assign each resonance to the structure.

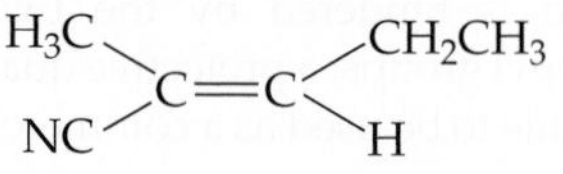

MASS SPECTROMETRY (Section 30.4)

30.36 Carbon tetrabromide shows nine signals in the region $m/z = 328$–337 of the EI mass spectrum. Account for each of these signals.

30.37 Triethylborane has two signals at m/z 97 and 98 in its EI mass spectrum. Account for the two signals.

30.38 An aliphatic compound gives the following result from microanalysis: C, 66.6%; H, 11.2%. The mass spectrum reveals a molecular ion signal at $m/z = 72$. Determine the compound's molecular formula and IHD.

30.39 An aliphatic compound contains C, 39.76% and H, 7.34% but doesn't contain oxygen. The EI mass spectrum shows a molecular ion at $m/z = 150$ and a second signal of equal intensity at $m/z = 152$. Determine this compound's molecular formula.

30.40 (a) The mass spectrometer in Figure 30.35 has a magnet as one of its components. What is the purpose of the magnet? (b) The atomic weight of Cl is 35.5 u. However, the mass spectrum of Cl, shown below, does not have a signal at this mass. Explain.

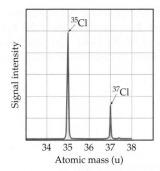

34 35 36 37 38
Atomic mass (u)

30.41 What is meant by these terms?

   (a) M^+ **(b)** $[M + 2]^+$ **(c)** base peak

   (d) m/z **(e)** EIMS

30.42 **(a)** What are the labels on the axes of a mass spectrum?

   (b) In order to measure a mass spectrum of an atom or molecule, the atom or molecule must first lose (or gain) one or more electrons. Why is this so?

30.43 Illustrate the stability of the following cations by drawing any contributing resonance structures.

   (a) acylium cation   **(b)** carbonium cation

   (c) iminium cation

30.44 Calculate the IHD for the following compounds:

   (a) cholesterol, $C_{27}H_{46}O$ **(b)** acetic acid, $C_2H_4O_2$

   (c) aspirin, $C_9H_8O_4$ **(d)** naphthalene, $C_{10}H_8$

   (e) ascorbic acid, $C_6H_8O_6$ **(f)** cocaine, $C_{17}H_{21}NO_4$

30.45 The guidelines for determining the IHD of a compound are listed on page 1233. Using examples as appropriate, discuss why, for each nitrogen in the unknown's molecular formula, one hydrogen needs to be added to the reference molecular formula, but for oxygen nothing is added.

INTEGRATED EXERCISES (Section 30.5)

30.48 After a stocktake in a chemical store, two partially labelled bottles containing liquids with different odours remained. The label on both bottles was C_4H_xO where the value of x was unreadable. The bottles were marked A and B. The store's card index showed that there were five possibilities for A and B, numbered 1–5.

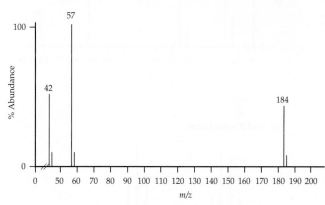

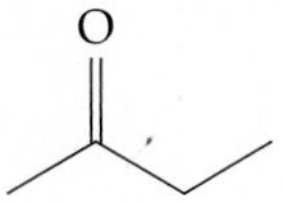

The ^1H NMR and IR spectra of A and B were recorded and are described below. For each bottle and its contents, suggest a possible conclusion for the information presented—for example, IR band at 2950 cm^{-1} indicates

30.46 Following a literature procedure, a student added butyl bromide (three equivalents) to 2,4-pentanedione in the presence of excess potassium carbonate and potassium iodide, as shown in the following reaction scheme:

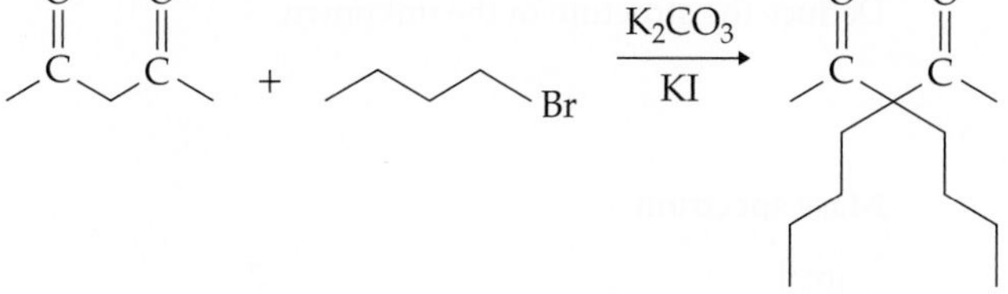

As well as isolating the product, the student also isolated a clear oil which she promptly submitted for EIMS. The results are shown below. What is the structure of the by-product? [*Hint:* Look at the difference in mass between the molecular ion and the base peak.]

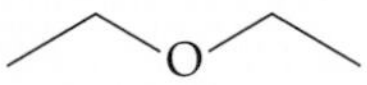

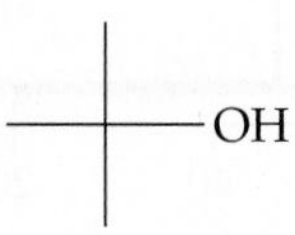

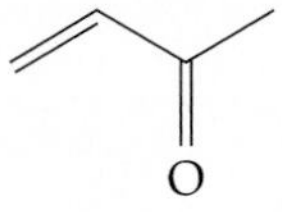

30.47 With the aid of a diagram, describe what the EI mass spectrum of 2-chloropropane might look like.

C–H present—and deduce the structures of A and B.

- *Bottle A:* Strong IR band near 1700 cm^{-1}, ^1H NMR spectrum: three signals in the ratio 3 : 3 : 2.
- *Bottle B:* Strong IR band near 3500 cm^{-1}, ^1H NMR spectrum: two signals in the ratio 9 : 1.

30.49 The product formed upon oxidation of an unknown organic compound with chromic acid has the molecular formula $C_5H_{10}O$. The ^1H NMR spectrum of the product is shown below and its IR spectrum has a strong band at 1700 cm^{-1}. Using this information, deduce the structure of the product and hence the structure of the unknown compound.

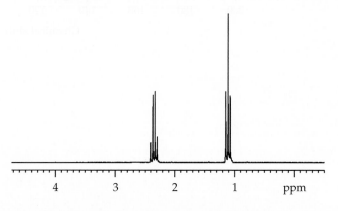

30.50 An acidic liquid has a broad band at 3300 cm^{-1} and a sharp band at 1710 cm^{-1} in the infrared spectrum. One mole of this unknown reacts with one mole of NaOH. The ^{1}H NMR spectrum shows three signals in the ratio 2 : 2 : 1. Microanalysis gave C, 33.2%; H, 4.6%; Cl, 32.7%. Deduce the structure of the unknown.

30.51 This basic compound gives a positive result to the Lassaigne test. Microanalysis gave the following result: C, 71.22; H, 14.94; N, 13.84. Use this information and the spectra below to deduce the structure of the unknown.

Mass spectrum

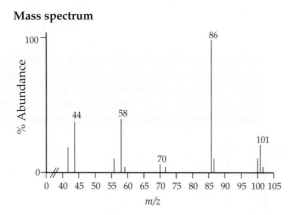

IR spectrum

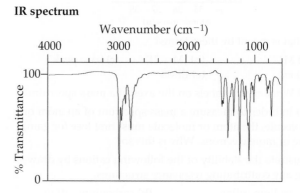

^{1}H NMR spectrum

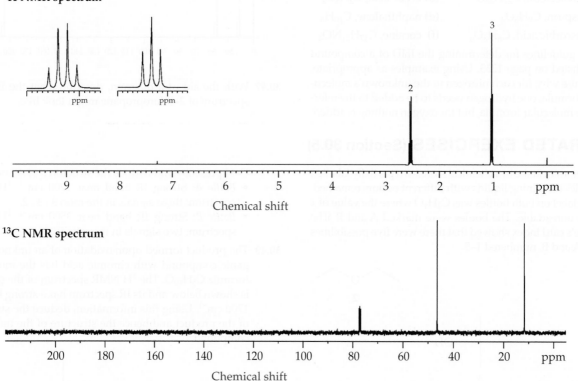

^{13}C NMR spectrum

30.52 This unknown evolves gas when a small amount of sodium metal is added. Microanalysis gave the following result: C, 50.69; H, 7.09; N, 19.71. Use all the information available to deduce the structure of the unknown.

Mass spectrum

IR spectrum

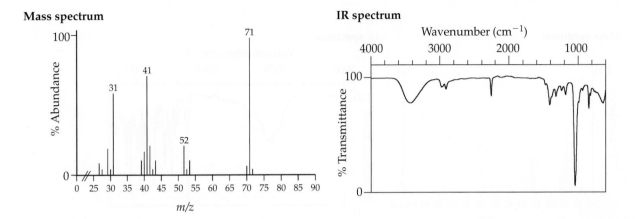

¹H NMR spectrum

¹³C NMR spectrum

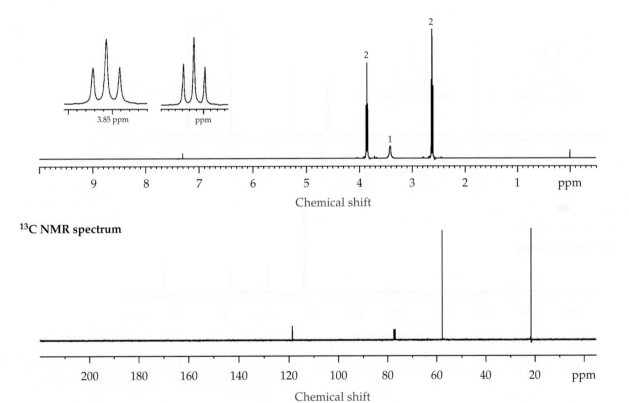

30.53 This unknown is a derivative of an amino acid. Micro-analysis gave the following result: C, 46.59; H, 8.80; N, 13.58. Solvent signals within the ^{1}H NMR spectrum are labelled with an asterisk. Use all the information available to deduce the structure of the unknown.

Mass spectrum

IR spectrum

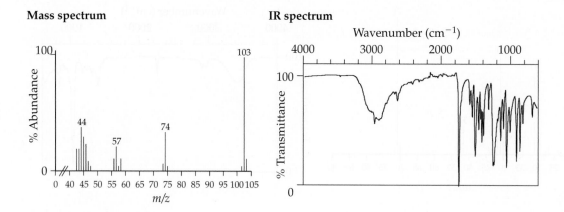

^{1}H NMR spectrum

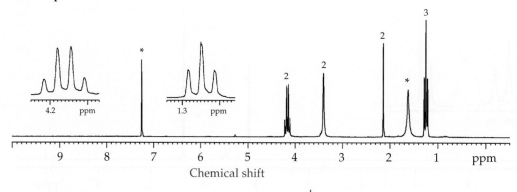

^{13}C NMR spectrum

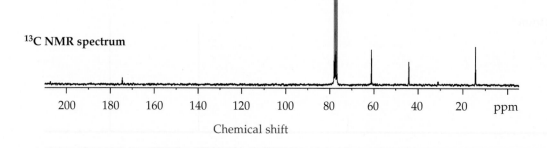

30.54 This unknown liquid is miscible with water, giving a mildly basic solution. Microanalysis gave the following result: C, 47.97; H, 12.08; N, 18.65. The broad, three-hydrogen signal within the ^{1}H NMR spectrum (indi-cated) disappears on the addition of D_2O. Use all the information available to deduce the structure of the unknown.

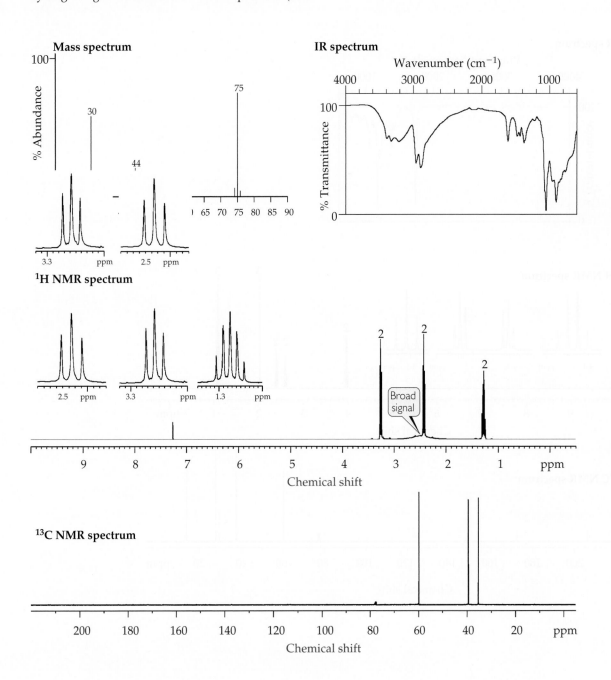

30.55 This unknown liquid reacts with Brady's reagent to form a highly coloured precipitate. The electrospray mass spectrum (not shown) has a signal at $m/z = 145$ corresponding to $[M + H]^+$. Microanalysis gave the following result: C, 58.32; H, 8.39. Use all the information available to deduce the structure of the unknown.

IR spectrum

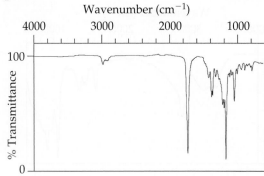

¹H NMR spectrum

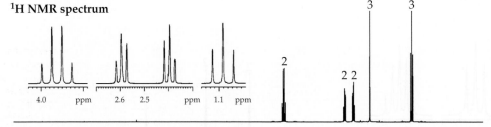
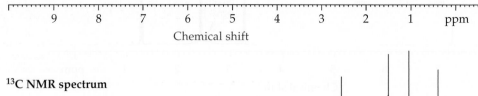

¹³C NMR spectrum

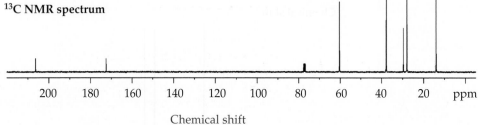

30.56 When this unknown is saponified, it leads to the isolation of benzyl alcohol as one of the products. Microanalysis gave the following result: C, 68.74; H, 6.29. Use all the information available to deduce the structure of the unknown.

Mass spectrum

IR spectrum

¹H NMR spectrum

¹³C NMR spectrum

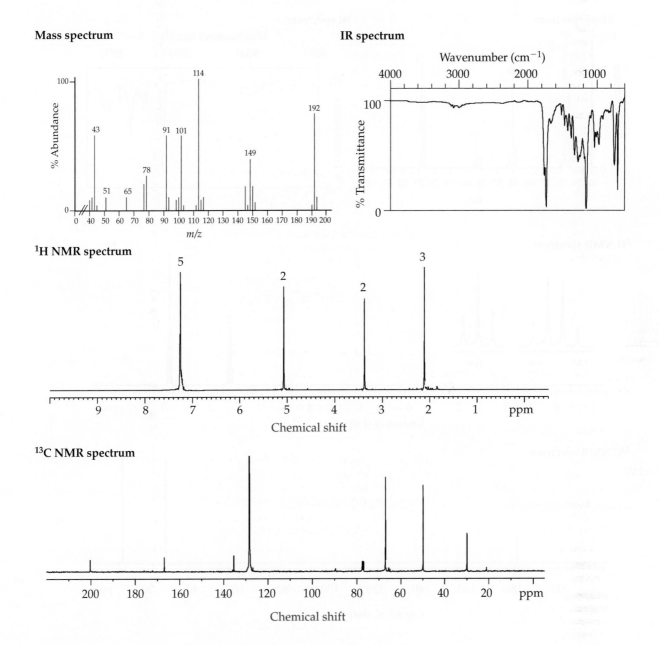

30.57 This unknown liquid evolves a gas when dissolved in aqueous sodium bicarbonate. Microanalysis gave the following result: C, 48.64; H, 8.16. Use all the information available to deduce the structure of the unknown.

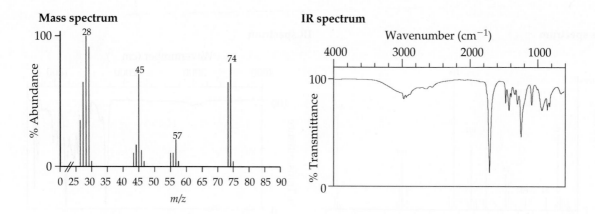

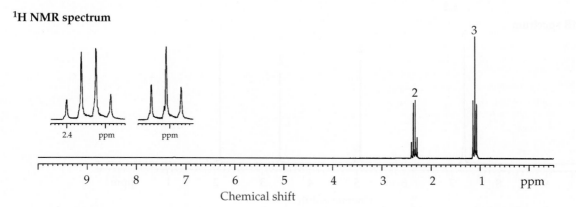

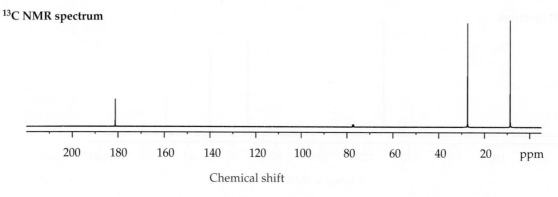

MasteringChemistry (www.pearson.com.au/masteringchemistry)

Make learning part of the grade. Access:

- tutorials with peronalised coaching
- study area
- Pearson eText

PHOTO/ART CREDITS

1198 oriontrail/Shutterstock; **1209** Professor Donald McNaughton, Monash University; **1213 (Figure 30.20)** Photo Researchers, Inc; **1126** © Colin Cuthbert/Science Photo Library.

MATHEMATICAL OPERATIONS

A.1 | EXPONENTIAL NOTATION

The numbers used in chemistry are often either extremely large or extremely small. Such numbers are conveniently expressed in the form

$$N \times 10^n$$

where N is a number between 1 and 10, and n is the exponent. Some examples of this *exponential notation*, which is also called *scientific notation*, follow.

1 200 000 is 1.2×10^6 (read 'one point two times ten to the sixth power')

0.000604 is 6.04×10^{-4} (read 'six point zero four times ten to the negative fourth power')

A positive exponent, as in the first example, tells us how many times a number must be multiplied by 10 to give the long form of the number:

$$1.2 \times 10^6 = 1.2 \times 10 \times 10 \times 10 \times 10 \times 10 \times 10 \text{ (six tens)}$$
$$= 1\ 200\ 000$$

It is also convenient to think of the *positive exponent* as the number of places the decimal point must be moved to the *left* to obtain a number greater than 1 and less than 10. If we begin with 3450 and move the decimal point three places to the left, we end up with 3.45×10^3.

In a related fashion, a negative exponent tells us how many times we must divide a number by 10 to give the long form of the number.

$$6.04 \times 10^{-4} = \frac{6.04}{10 \times 10 \times 10 \times 10} = 0.000604$$

It is convenient to think of the *negative exponent* as the number of places the decimal point must be moved to the *right* to obtain a number greater than 1 but less than 10. For example, if we begin with 0.0048 and move the decimal point three places to the right, we end up with 4.8×10^{-3}.

In the system of exponential notation, with each shift of the decimal point one place to the right, the exponent *decreases* by 1:

$$4.8 \times 10^{-3} = 48 \times 10^{-4}$$

Similarly, with each shift of the decimal point one place to the left, the exponent *increases* by 1:

$$4.8 \times 10^{-3} = 0.48 \times 10^{-2}$$

Many scientific calculators have a key labelled EXP or EE, which is used to enter numbers in exponential notation. To enter the number 5.8×10^3 on such a calculator, the key sequence is

$$\boxed{5}\ \boxed{\cdot}\ \boxed{8}\ \boxed{\text{EXP}}\ \text{(or}\ \boxed{\text{EE}}\text{)}\ \boxed{3}$$

On some calculators the display will show 5.8, then a space, followed by 03, the exponent. On other calculators, a small 10 is shown with an exponent 3.

To enter a negative exponent, use the key labelled $+/-$. For example, to enter the number 8.6×10^{-5}, the key sequence is

$$\boxed{8}\;\boxed{\cdot}\;\boxed{6}\;\boxed{\text{EXP}}\;\boxed{+/-}\;\boxed{5}$$

When entering a number in exponential notation, do not key in the 10 if you use the EXP or EE button.

In working with exponents, it is important to recall that $10^0 = 1$. The following rules are useful for carrying exponents through calculations.

1. **Addition and Subtraction:** In order to add or subtract numbers expressed in exponential notation, the powers of 10 must be the same.

$$(5.22 \times 10^4) + (3.21 \times 10^2) = (522 \times 10^2) + (3.21 \times 10^2)$$
$$= 525 \times 10^2 \text{ (3 significant figures)}$$
$$= 5.25 \times 10^4$$

$$(6.25 \times 10^{-2}) - (5.77 \times 10^{-3}) = (6.25 \times 10^{-2}) - (0.577 \times 10^{-2})$$
$$= 5.67 \times 10^{-2} \text{ (3 significant figures)}$$

When you use a calculator to add or subtract, you need not be concerned with having numbers with the same exponents because the calculator automatically takes care of this matter.

2. **Multiplication and Division:** When numbers expressed in exponential notation are multiplied, the exponents are added; when numbers expressed in exponential notation are divided, the exponent of the denominator is subtracted from the exponent of the numerator.

$$(5.4 \times 10^2)(2.1 \times 10^3) = (5.4)(2.1) \times 10^{2+3}$$
$$= 11 \times 10^5$$
$$= 1.1 \times 10^6$$

$$(1.2 \times 10^5)(3.22 \times 10^{-3}) = (1.2)(3.22) \times 10^{5-3} = 3.9 \times 10^2$$

$$\frac{3.2 \times 10^5}{6.5 \times 10^2} = \frac{3.2}{6.5} \times 10^{5-2} = 0.49 \times 10^3 = 4.9 \times 10^2$$

$$\frac{5.7 \times 10^7}{8.5 \times 10^{-2}} = \frac{5.7}{8.5} \times 10^{7-(-2)} = 0.67 \times 10^9 = 6.7 \times 10^8$$

3. **Powers and Roots:** When numbers expressed in exponential notation are raised to a power, the exponents are multiplied by the power. When the roots of numbers expressed in exponential notation are taken, the exponents are divided by the root.

$$(1.2 \times 10^5)^3 = (1.2)^3 \times 10^{5 \times 3}$$
$$= 1.7 \times 10^{15}$$

$$\sqrt[3]{2.5 \times 10^6} = \sqrt[3]{2.5} \times 10^{6/3}$$
$$= 1.3 \times 10^2$$

Scientific calculators usually have keys labelled x^2 and $\sqrt{x}$ for squaring and taking the square root of a number, respectively. To take higher powers or roots, many calculators have y^x and $\sqrt[x]{y}$ (or INV y^x) keys. For example, to perform the operation $\sqrt[3]{7.5 \times 10^{-4}}$ on such a calculator, you would key in 7.5×10^{-4}, press the $\sqrt[x]{y}$ key (or the INV and then the y^x keys), enter the root, 3, and finally press $=$. The result is 9.1×10^{-2}.

SAMPLE EXERCISE 1 | Using exponential notation

Perform each of the following operations, using your calculator where possible:

(a) Write the number 0.0054 in standard exponential notation.
(b) $(5.0 \times 10^{-2}) + (4.7 \times 10^{-3})$
(c) $(5.98 \times 10^{12})(2.77 \times 10^{-5})$
(d) $\sqrt[4]{1.75 \times 10^{-12}}$

SOLUTION

(a) Because we move the decimal point three places to the right to convert 0.0054 to 5.4, the exponent is -3:

$$5.4 \times 10^{-3}$$

Scientific calculators are generally able to convert numbers to exponential notation using one or two keystrokes; frequently 'SCI' for 'scientific notation' will convert a number into exponential notation. Consult your instruction manual to see how this operation is accomplished on your calculator.
(b) To add these numbers longhand, we must convert them to the same exponent.

$$(5.0 \times 10^{-2}) + (0.47 \times 10^{-2}) = (5.0 + 0.47) \times 10^{-2} = 5.5 \times 10^{-2}$$

(Note that the result has only two significant figures.) To perform this operation on a calculator, we enter the first number, strike the $+$ key, then enter the second number and strike the $=$ key.
(c) Performing this operation longhand, we have

$$(5.98 \times 2.77) \times 10^{12-5} = 16.6 \times 10^{7} = 1.66 \times 10^{8}$$

On a scientific calculator, we enter 5.98×10^{12}, press the $\times$ key, enter 2.77×10^{-5}, and press the $=$ key.
(d) To perform this operation on a calculator, we enter the number, press the $\sqrt[x]{y}$ key (or the INV and y^x keys), enter 4, and press the $=$ key. The result is 1.15×10^{-3}.

PRACTICE EXERCISE

Perform the following operations:

(a) Write 67000 in exponential notation, showing two significant figures.
(b) $(3.378 \times 10^{-3}) - (4.97 \times 10^{-5})$
(c) $(1.84 \times 10^{15})(7.45 \times 10^{-2})$
(d) $(6.67 \times 10^{-8})^3$

Answers: **(a)** 6.7×10^4, **(b)** 3.328×10^{-3}, **(c)** 2.47×10^{16}, **(d)** 2.97×10^{-22}

A.2 | LOGARITHMS

Common Logarithms

The common, or base-10, logarithm (abbreviated log) of any number is the power to which 10 must be raised to equal the number. For example, the common logarithm of 1000 (written log 1000) is 3 because raising 10 to the third power gives 1000.

$$10^3 = 1000, \text{ therefore, } \log 1000 = 3$$

Further examples are

$$\log 10^5 = 5$$
$$\log 1 = 0 \quad \text{Remember that } 10^0 = 1$$
$$\log 10^{-2} = -2$$

In these examples the common logarithm can be obtained by inspection. However, it is not possible to obtain the logarithm of a number such as 31.25 by inspection. The logarithm of 31.25 is the number x that satisfies the following relationship:

$$10^x = 31.25$$

Most electronic calculators have a key labelled LOG that can be used to obtain logarithms. For example, on many calculators we obtain the value of log 31.25 by entering 31.25 and pressing the LOG key. We obtain the following result:

$$\log 31.25 = 1.4949$$

Notice that 31.25 is greater than 10 (10^1) and less than 100 (10^2). The value for log 31.25 is accordingly between log 10 and log 100, that is, between 1 and 2.

Significant Figures and Common Logarithms

For the common logarithm of a measured quantity, the number of digits after the decimal point equals the number of significant figures in the original number. For example, if 23.5 is a measured quantity (three significant figures), then log 23.5 = 1.371 (three significant figures after the decimal point).

Antilogarithms

The process of determining the number that corresponds to a certain logarithm is known as obtaining an *antilogarithm*. It is the reverse of taking a logarithm. For example, we saw previously that log 23.5 = 1.371. This means that the antilogarithm of 1.371 equals 23.5.

$$\log 23.5 = 1.371$$
$$\text{antilog } 1.371 = 23.5$$

The process of taking the antilog of a number is the same as raising 10 to a power equal to that number.

$$\text{antilog } 1.371 = 10^{1.371} = 23.5$$

Many calculators have a key labelled 10^x that allows you to obtain antilogs directly. On others, it will be necessary to press a key labelled INV (for *inverse*), followed by the LOG key.

Natural Logarithms

Logarithms based on the number e are called natural, or base e, logarithms (abbreviated ln). The natural log of a number is the power to which e (which has the value 2.71828 ...) must be raised to equal the number. For example, the natural log of 10 equals 2.303.

$$e^{2.303} = 10, \text{ therefore } \ln 10 = 2.303$$

Your calculator probably has a key labelled LN that allows you to obtain natural logarithms. For example, to obtain the natural log of 46.8, you enter 46.8 and press the LN key.

$$\ln 46.8 = 3.846$$

The natural antilog of a number is e raised to a power equal to that number. If your calculator can calculate natural logs, it will also be able to calculate natural antilogs. On some calculators there is a key labelled e^x that allows you to calculate natural antilogs directly; on others, it will be necessary to first press the INV key followed by the LN key. For example, the natural antilog of 1.679 is given by

$$\text{Natural antilog } 1.679 = e^{1.679} = 5.36$$

The relation between common and natural logarithms is as follows:

$$\ln a = 2.303 \log a$$

Notice that the factor relating the two, 2.303, is the natural log of 10, which we calculated above.

Mathematical Operations Using Logarithms

Because logarithms are exponents, mathematical operations involving logarithms follow the rules for the use of exponents. For example, the product of z^a and z^b (where z is any number) is given by

$$z^a \cdot z^b = z^{(a+b)}$$

Similarly, the logarithm (either common or natural) of a product equals the *sum* of the logs of the individual numbers.

$$\log ab = \log a + \log b \quad \ln ab = \ln a + \ln b$$

For the log of a quotient,

$$\log (a/b) = \log a - \log b \quad \ln(a/b) = \ln a - \ln b$$

Using the properties of exponents, we can also derive the rules for the logarithm of a number raised to a certain power.

$$\log a^n = n \log a \qquad \ln a^n = n \ln a$$
$$\log a^{1/n} = (1/n) \log a \quad \ln a^{1/n} = (1/n) \ln a$$

pH Problems

∞ Review this on page 664

One of the most frequent uses for common logarithms in general chemistry is in working pH problems. The pH is defined as $-\log[H^+]$, where $[H^+]$ is the hydrogen ion concentration of a solution (∞ Section 17.4, 'The pH Scale'). The following sample exercise illustrates this application.

SAMPLE EXERCISE 2 **Using logarithms**

(a) What is the pH of a solution whose hydrogen ion concentration is 0.015 M?

(b) If the pH of a solution is 3.80, what is its hydrogen ion concentration?

SOLUTION

1. We are given the value of $[H^+]$. We use the LOG key of our calculator to calculate the value of $\log[H^+]$. The pH is obtained by changing the sign of the value obtained. (Be sure to change the sign *after* taking the logarithm.)

$$[H^+] = 0.015$$
$$\log[H^+] = -1.82 \qquad \text{(2 significant figures)}$$
$$pH = -(-1.82) = 1.82$$

2. To obtain the hydrogen ion concentration when given the pH, we must take the antilog of $-pH$.

$$pH = -\log[H^+] = 3.80$$
$$\log[H^+] = -3.80$$
$$[H^+] = \text{antilog}(-3.80) = 10^{-3.80} = 1.6 \times 10^{-4} \text{ M}$$

PRACTICE EXERCISE

Perform the following operations: **(a)** $\log (2.5 \times 10^{-5})$, **(b)** $\ln 32.7$, **(c)** antilog -3.47, **(d)** $e^{-1.89}$.

Answers: **(a)** -4.60, **(b)** 3.487, **(c)** 3.4×10^{-4}, **(d)** 1.5×10^{-1}

A.3 | QUADRATIC EQUATIONS

An algebraic equation of the form $ax^2 + bx + c = 0$ is called a *quadratic equation*. The two solutions to such an equation are given by the quadratic formula:

$$x = \frac{-b \pm \sqrt{b^2 - 4ac}}{2a}$$

SAMPLE EXERCISE 3   **Using the quadratic formula**

Find the values of x that satisfy the equation $2x^2 + 4x = 1$.

SOLUTION

To solve the given equation for x, we must first put it in the form

$$ax^2 + bx + c = 0$$

and then use the quadratic formula. If

$$2x^2 + 4x = 1$$

then

$$2x^2 + 4x - 1 = 0$$

Using the quadratic formula, where $a = 2$, $b = 4$, and $c = -1$, we have

$$x = \frac{-4 \pm \sqrt{(4)(4) - 4(2)(-1)}}{2(2)}$$

$$= \frac{-4 \pm \sqrt{16 + 8}}{4} = \frac{-4 \pm \sqrt{24}}{4} = \frac{-4 \pm 4.899}{4}$$

The two solutions are

$$x = \frac{0.899}{4} = 0.225 \quad \text{and} \quad x = \frac{-8.899}{4} = -2.225$$

Often in chemical problems the negative solution has no physical meaning and only the positive answer is used.

A.4 | GRAPHS

Often the clearest way to represent the interrelationship between two variables is to graph them. Usually, the variable that is being experimentally varied, called the *independent variable*, is shown along the horizontal axis (x-axis). The variable that responds to the change in the independent variable, called the *dependent variable*, is then shown along the vertical axis (y-axis). For example, consider an experiment in which we vary the temperature of an enclosed gas and measure its pressure. The independent variable is temperature, and the dependent variable is pressure. The data shown in ▶ TABLE 1 can be obtained by means of this experiment. These data are shown graphically in ▶ FIGURE 1. The relationship between temperature and pressure is linear. The equation for any straight-line graph has the form

$$y = mx + b$$

where m is the slope of the line and b is the intercept with the y-axis. In the case of Figure 1, we could say that the relationship between temperature and pressure takes the form

$$P = mT + b$$

where P is pressure in kPa and T is temperature in °C. As shown in Figure 1, the slope is 4.0×10^{-2} kPa °C^{-1}, and the intercept—the point where the line crosses the y-axis—is 11.2 kPa. Therefore, the equation for the line is

$$P = \left(4.0 \times 10^{-2} \text{ kPa °C}^{-1}\right)T + 11.2 \text{ kPa}$$

TABLE 1 • **Interrelation between pressure and temperature**

Temperature (°C)	Pressure (kPa)
20.0	12.0
30.0	12.4
40.0	12.8
50.0	13.2

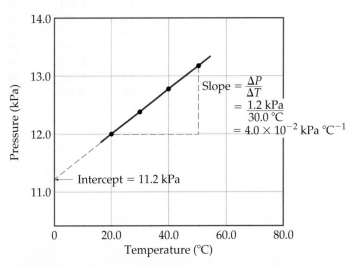

▲ **FIGURE 1**   A graph of pressure against temperature yields a straight line for the data.

PROPERTIES OF WATER

Density:	0.99987 g cm^{-3} at 0 °C
	1.00000 g cm^{-3} at 4 °C
	0.99707 g cm^{-3} at 25 °C
	0.95838 g cm^{-3} at 100 °C
Heat of fusion:	6.008 kJ mol^{-1} at 0 °C
Heat of vaporisation:	44.94 kJ mol^{-1} at 0 °C
	44.02 kJ mol^{-1} at 25 °C
	40.67 kJ mol^{-1} at 100 °C
Ion-product constant, K_w:	1.14×10^{-15} at 0 °C
	1.01×10^{-14} at 25 °C
	5.47×10^{-14} at 50 °C
Specific heat:	Ice (at −3 °C) 2.092 J g^{-1} K^{-1}
	Water (at 14.5 °C) 4.184 J g^{-1} K^{-11}
	Steam (at 100 °C) 1.841 J g^{-1} K^{-1}

Vapour Pressure (kPa)

T(°C)	P	T(°C)	P	T(°C)	P	T(°C)	P
0	0.611	21	2.486	35	5.626	92	75.59
5	0.872	22	2.644	40	7.373	94	81.45
10	1.228	23	2.809	45	9.586	96	87.67
12	1.403	24	2.984	50	12.33	98	94.30
14	1.599	25	3.168	55	15.73	100	101.32
16	1.817	26	3.361	60	19.92	102	108.78
17	1.937	27	3.565	65	25.00	104	116.67
18	2.064	28	3.780	70	31.16	106	125.04
19	2.197	29	4.005	80	47.34	108	133.91
20	2.338	30	4.242	90	70.10	110	143.27

THERMODYNAMIC QUANTITIES FOR SELECTED SUBSTANCES AT 298.15 K (25 °C)

Substance	$\Delta_f H°$ (kJ mol^{-1})	$\Delta_f G°$ (kJ mol^{-1})	$S°$ (J mol^{-1} K)	Substance	$\Delta_f H°$ (kJ mol^{-1})	$\Delta_f G°$ (kJ mol^{-1})	$S°$ (J mol^{-1} K)
Aluminum				$CCl_4(g)$	−106.7	−64.0	309.4
Al(s)	0	0	28.32	$CCl_4(l)$	−139.3	−68.6	214.4
$AlCl_3(s)$	−705.6	−630.0	109.3	$CF_4(g)$	−679.9	−635.1	262.3
$Al_2O_3(s)$	−1669.8	−1576.5	51.00	$CH_4(g)$	−74.8	−50.8	186.3
				$C_2H_2(g)$	226.77	209.2	200.8
Barium				$C_2H_4(g)$	52.30	68.11	219.4
Ba(s)	0	0	63.2	$C_2H_6(g)$	−84.68	−32.89	229.5
$BaCO_3(s)$	−1216.3	−1137.6	112.1	$C_3H_8(g)$	−103.85	−23.47	269.9
BaO(s)	−553.5	−525.1	70.42	$C_4H_{10}(g)$	−124.73	−15.71	310.0
				$C_4H_{10}(l)$	−147.6	−15.0	231.0
Beryllium				$C_6H_6(g)$	82.9	129.7	269.2
Be(s)	0	0	9.44	$C_6H_6(l)$	49.0	124.5	172.8
BeO(s)	−608.4	−579.1	13.77	$CH_3OH(g)$	−201.2	−161.9	237.6
$Be(OH)_2(s)$	−905.8	−817.9	50.21	$CH_3OH(l)$	−238.6	−166.23	126.8
				$C_2H_5OH(g)$	−235.1	−168.5	282.7
Bromine				$C_2H_5OH(l)$	−277.7	−174.76	160.7
Br(g)	111.8	82.38	174.9	$C_6H_{12}O_6(s)$	−1273.02	−910.4	212.1
Br$^-$(aq)	−120.9	−102.8	80.71	CO(g)	−110.5	−137.2	197.9
$Br_2(g)$	30.71	3.14	245.3	$CO_2(g)$	−393.5	−394.4	213.6
$Br_2(l)$	0	0	152.3	$HC_2H_3O_2$	−487.0	−392.4	159.8
HBr(g)	−36.23	−53.22	198.49				
				Cesium			
Calcium				Cs(g)	76.50	49.53	175.6
Ca(g)	179.3	145.5	154.8	Cs(l)	2.09	0.03	92.07
Ca(s)	0	0	41.4	Cs(s)	0	0	85.15
$CaCO_3$(s, calcite)	−1207.1	−1128.76	92.88	CsCl(s)	−442.8	−414.4	101.2
$CaCl_2(s)$	−795.8	−748.1	104.6				
$CaF_2(s)$	−1219.6	−1167.3	68.87	Chlorine			
CaO(s)	−635.5	−604.17	39.75	Cl(g)	121.7	105.7	165.2
$Ca(OH)_2(s)$	−986.2	−898.5	83.4	Cl$^-$(aq)	−167.2	−131.2	56.5
$CaSO_4(s)$	−1434.0	−1321.8	106.7	$Cl_2(g)$	0	0	222.96
				HCl(aq)	−167.2	−131.2	56.5
Carbon				HCl(g)	−92.30	−95.27	186.69
C(g)	718.4	672.9	158.0				
C(s, diamond)	1.88	2.84	2.43	Chromium			
C(s, graphite)	0	0	5.69	Cr(g)	397.5	352.6	174.2

Substance	$\Delta_f H°$ (kJ mol^{-1})	$\Delta_f G°$ (kJ mol^{-1})	$S°$ (J mol^{-1} K)	Substance	$\Delta_f H°$ (kJ mol^{-1})	$\Delta_f G°$ (kJ mol^{-1})	$S°$ (J mol^{-1} K)
Cr(s)	0	0	23.6	Li$^+$(aq)	−278.5	−273.4	12.2
Cr$_2$O$_3$(s)	−1139.7	−1058.1	81.2	Li$^+$(g)	685.7	648.5	133.0
Cobalt				LiCl(s)	−408.3	−384.0	59.30
Co(g)	439	393	179	**Magnesium**			
Co(s)	0	0	28.4	Mg(g)	147.1	112.5	148.6
Copper				Mg(s)	0	0	32.51
Cu(g)	338.4	298.6	166.3	MgCl$_2$(s)	−641.6	−592.1	89.6
Cu(s)	0	0	33.30	MgO(s)	−601.8	−569.6	26.8
CuCl$_2$(s)	−205.9	−161.7	108.1	Mg(OH)$_2$(s)	−924.7	−833.7	63.24
CuO(s)	−156.1	−128.3	42.59	**Manganese**			
Cu$_2$O(s)	−170.7	−147.9	92.36	Mn(g)	280.7	238.5	173.6
Fluorine				Mn(s)	0	0	32.0
F(g)	80.0	61.9	158.7	MnO(s)	−385.2	−362.9	59.7
F$^-$(aq)	−332.6	−278.8	−13.8	MnO$_2$(s)	−519.6	−464.8	53.14
F$_2$(g)	0	0	202.7	MnO$_4^-$(aq)	−541.4	−447.2	191.2
HF(g)	−268.61	−270.70	173.51	**Mercury**			
Hydrogen				Hg(g)	60.83	31.76	174.89
H(g)	217.94	203.26	114.60	Hg(l)	0	0	77.40
H$^+$(aq)	0	0	0	HgCl$_2$(s)	−230.1	−184.0	144.5
H$^+$(g)	1536.2	1517.0	108.9	Hg$_2$Cl$_2$(s)	−264.9	−210.5	192.5
H$_2$(g)	0	0	130.58	**Nickel**			
Iodine				Ni(g)	429.7	384.5	182.1
I(g)	106.60	70.16	180.66	Ni(s)	0	0	29.9
I$^-$(aq)	−55.19	−51.57	111.3	NiCl$_2$(s)	−305.3	−259.0	97.65
I$_2$(g)	62.25	19.37	260.57	NiO(s)	−239.7	−211.7	37.99
I$_2$(s)	0	0	116.73	**Nitrogen**			
HI(g)	25.94	1.30	206.3	N(g)	472.7	455.5	153.3
Iron				N$_2$(g)	0	0	191.50
Fe(g)	415.5	369.8	180.5	NH$_3$(aq)	−80.29	−26.50	111.3
Fe(s)	0	0	27.15	NH$_3$(g)	−46.19	−16.66	192.5
Fe^{2+}(aq)	−87.86	−84.93	113.4	NH$_4^+$(aq)	−132.5	−79.31	113.4
Fe^{3+}(aq)	−47.69	−10.54	293.3	N$_2$H$_4$(g)	95.40	159.4	238.5
FeCl$_2$(s)	−341.8	−302.3	117.9	NH$_4$CN(s)	0.0	—	—
FeCl$_3$(s)	−400	−334	142.3	NH$_4$Cl(s)	−314.4	−203.0	94.6
FeO(s)	−271.9	−255.2	60.75	NH$_4$NO$_3$(s)	−365.6	−184.0	151
Fe$_2$O$_3$(s)	−822.16	−740.98	89.96	NO(g)	90.37	86.71	210.62
Fe$_3$O$_4$(s)	−1117.1	−1014.2	146.4	NO$_2$(g)	33.84	51.84	240.45
FeS$_2$(s)	−171.5	−160.1	52.92	N$_2$O(g)	81.6	103.59	220.0
Lead				N$_2$O$_4$(g)	9.66	98.28	304.3
Pb(s)	0	0	68.85	NOCl(g)	52.6	66.3	264
PbBr$_2$(s)	−277.4	−260.7	161	HNO$_3$(aq)	−206.6	−110.5	146
PbCO$_3$(s)	−699.1	−625.5	131.0	HNO$_3$(g)	−134.3	−73.94	266.4
Pb(NO$_3$)$_2$(aq)	−421.3	−246.9	303.3	**Oxygen**			
Pb(NO$_3$)$_2$(s)	−451.9	—	—	O(g)	247.5	230.1	161.0
PbO(s)	−217.3	−187.9	68.70	O$_2$(g)	0	0	205.0
Lithium				O$_3$(g)	142.3	163.4	237.6
Li(g)	159.3	126.6	138.8	OH$^-$(aq)	−230.0	−157.3	−10.7
Li(s)	0	0	29.09	H$_2$O(g)	−241.82	−228.57	188.83

Substance	$\Delta_f H°$ (kJ mol^{-1})	$\Delta_f G°$ (kJ mol^{-1})	$S°$ (J mol^{-1} K)	Substance	$\Delta_f H°$ (kJ mol^{-1})	$\Delta_f G°$ (kJ mol^{-1})	$S°$ (J mol^{-1} K)
$H_2O(l)$	−285.83	−237.13	69.91	$Ag^+(aq)$	105.90	77.11	73.93
$H_2O_2(g)$	−136.10	−105.48	232.9	$AgCl(s)$	−127.0	−109.70	96.11
$H_2O_2(l)$	−187.8	−120.4	109.6	$Ag_2O(s)$	−31.05	−11.20	121.3
Phosphorus				$AgNO_3(s)$	−124.4	−33.41	140.9
$P(g)$	316.4	280.0	163.2	**Sodium**			
$P_2(g)$	144.3	103.7	218.1	$Na(g)$	107.7	77.3	153.7
$P_4(g)$	58.9	24.4	280	$Na(s)$	0	0	51.45
$P_4(s, red)$	−17.46	−12.03	22.85	$Na^+(aq)$	−240.1	−261.9	59.0
$P_4(s, white)$	0	0	41.08	$Na^+(g)$	609.3	574.3	148.0
$PCl_3(g)$	−288.07	−269.6	311.7	$NaBr(aq)$	−360.6	−364.7	141.00
$PCl_3(l)$	−319.6	−272.4	217	$NaBr(s)$	−361.4	−349.3	86.82
$PF_5(g)$	−1594.4	−1520.7	300.8	$Na_2CO_3(s)$	−1130.9	−1047.7	136.0
$PH_3(g)$	5.4	13.4	210.2	$NaCl(aq)$	−407.1	−393.0	115.5
$P_4O_6(s)$	−1640.1	—	—	$NaCl(g)$	−181.4	−201.3	229.8
$P_4O_{10}(s)$	−2984.0	−2697.8	228.9	$NaCl(s)$	−410.9	−384.0	72.33
$POCl_3(g)$	−542.2	−502.5	325	$NaHCO_3(s)$	−947.7	−851.8	102.1
$POCl_3(l)$	−597.0	−520.9	222	$NaNO_3(aq)$	−446.2	−372.4	207
$H_3PO_4(aq)$	−1288.3	−1142.6	158.2	$NaNO_3(s)$	−467.9	−367.0	116.5
Potassium				$NaOH(aq)$	−469.6	−419.2	49.8
$K(g)$	89.99	61.17	160.2	$NaOH(s)$	−425.6	−379.5	64.46
$K(s)$	0	0	64.67	**Strontium**			
$KCl(s)$	−435.9	−408.3	82.7	$SrO(s)$	−592.0	−561.9	54.9
$KClO_3(s)$	−391.2	−289.9	143.0	$Sr(g)$	164.4	110.0	164.6
$KClO_3(aq)$	−349.5	−284.9	265.7	**Sulfur**			
$K_2CO_3(s)$	−1150.18	−1064.58	155.44	$S(s, rhombic)$	0	0	31.88
$KNO_3(s)$	−492.70	−393.13	132.9	$S_8(g)$	102.3	49.7	430.9
$K_2O(s)$	−363.2	−322.1	94.14	$SO_2(g)$	−296.9	−300.4	248.5
$KO_2(s)$	−284.5	−240.6	122.5	$SO_3(g)$	−395.2	−370.4	256.2
$K_2O_2(s)$	−495.8	−429.8	113.0	$SO_4^{2-}(aq)$	−909.3	−744.5	20.1
$KOH(s)$	−424.7	−378.9	78.91	$SOCl_2(l)$	−245.6	—	—
$KOH(aq)$	−482.4	−440.5	91.6	$H_2S(g)$	−20.17	−33.01	205.6
Rubidium				$H_2SO_4(aq)$	−909.3	−744.5	20.1
$Rb(g)$	85.8	55.8	170.0	$H_2SO_4(l)$	−814.0	−689.9	156.1
$Rb(s)$	0	0	76.78	**Titanium**			
$RbCl(s)$	−430.5	−412.0	92	$Ti(g)$	468	422	180.3
$RbClO_3(s)$	−392.4	−292.0	152	$Ti(s)$	0	0	30.76
Scandium				$TiCl_4(g)$	−763.2	−726.8	354.9
$Sc(g)$	377.8	336.1	174.7	$TiCl_4(l)$	−804.2	−728.1	221.9
$Sc(s)$	0	0	34.6	$TiO_2(s)$	−944.7	−889.4	50.29
Selenium				**Vanadium**			
$H_2Se(g)$	29.7	15.9	219.0	$V(g)$	514.2	453.1	182.2
Silicon				$V(s)$	0	0	28.9
$Si(g)$	368.2	323.9	167.8	**Zinc**			
$Si(s)$	0	0	18.7	$Zn(g)$	130.7	95.2	160.9
$SiC(s)$	−73.22	−70.85	16.61	$Zn(s)$	0	0	41.63
$SiCl_4(l)$	−640.1	−572.8	239.3	$ZnCl_2(s)$	−415.1	−369.4	111.5
$SiO_2(s, quartz)$	−910.9	−856.5	41.84	$ZnO(s)$	−348.0	−318.2	43.9
Silver							
$Ag(s)$	0	0	42.55				

AQUEOUS EQUILIBRIUM CONSTANTS

TABLE 1 • Dissociation Constants for Acids at 25 °C

Name	Formula	K_{a1}	K_{a2}	K_{a3}
Acetic	CH_3COOH	1.8×10^{-5}		
Arsenic	H_3AsO_4	5.6×10^{-3}	1.0×10^{-7}	3.0×10^{-12}
Arsenous	H_3AsO_3	5.1×10^{-10}		
Ascorbic	$H_2C_6H_6O_6$	8.0×10^{-5}	1.6×10^{-12}	
Benzoic	C_6H_5COOH	6.3×10^{-5}		
Boric	H_3BO_3	5.8×10^{-10}		
Butanoic	C_3H_7COOH	1.5×10^{-5}		
Carbonic	H_2CO_3	4.3×10^{-7}	5.6×10^{-11}	
Chloroacetic	$CH_2ClCOOH$	1.4×10^{-3}		
Chlorous	$HClO_2$	1.1×10^{-2}		
Citric	$HOOCC(OH)(CH_2COOH)_2$	7.4×10^{-4}	1.7×10^{-5}	4.0×10^{-7}
Cyanic	$HCNO$	3.5×10^{-4}		
Formic	$HCOOH$	1.8×10^{-4}		
Hydroazoic	HN_3	1.9×10^{-5}		
Hydrocyanic	HCN	4.9×10^{-10}		
Hydrofluoric	HF	6.8×10^{-4}		
Hydrogen chromate ion	$HCrO_4^-$	3.0×10^{-7}		
Hydrogen peroxide	H_2O_2	2.4×10^{-12}		
Hydrogen selenate ion	$HSeO_4$	2.2×10^{-2}		
Hydrosulfuric	H_2S	9.5×10^{-8}	1×10^{-19}	
Hypobromous	$HBrO$	2.5×10^{-9}		
Hypochlorous	$HClO$	3.0×10^{-8}		
Hypoiodous	HIO	2.3×10^{-11}		
Iodic	HIO_3	1.7×10^{-1}		
Lactic	$CH_3CH(OH)COOH$	1.4×10^{-4}		
Malonic	$CH_2(COOH)_2$	1.5×10^{-3}	2.0×10^{-6}	
Nitrous	HNO_2	4.5×10^{-4}		
Oxalic	$(COOH)_2$	5.9×10^{-2}	6.4×10^{-5}	
Paraperiodic	H_5IO_6	2.8×10^{-2}	5.3×10^{-9}	
Phenol	C_6H_5OH	1.3×10^{-10}		
Phosphoric	H_3PO_4	7.5×10^{-3}	6.2×10^{-8}	4.2×10^{-13}
Propionic	C_2H_5COOH	1.3×10^{-5}		
Pyrophosphoric	$H_4P_2O_7$	3.0×10^{-2}	4.4×10^{-3}	
Selenous	H_2SeO_3	2.3×10^{-3}	5.3×10^{-9}	
Sulfuric	H_2SO_4	Strong acid	1.2×10^{-2}	
Sulfurous	H_2SO_3	1.7×10^{-2}	6.4×10^{-8}	
Tartaric	$HOOC(CHOH)_2COOH$	1.0×10^{-3}	4.6×10^{-5}	

TABLE 2 • Dissociation Constants for Bases at 25 °C

Name	Formula	K_b
Ammonia	NH_3	1.8×10^{-5}
Aniline	$C_6H_5NH_2$	4.3×10^{-10}
Dimethylamine	$(CH_3)_2NH$	5.4×10^{-4}
Ethylamine	$C_2H_5NH_2$	6.4×10^{-4}
Hydrazine	H_2NNH_2	1.3×10^{-6}
Hydroxylamine	$HONH_2$	1.1×10^{-8}
Methylamine	CH_3NH_2	4.4×10^{-4}
Pyridine	C_5H_5N	1.7×10^{-9}
Trimethylamine	$(CH_3)_3N$	6.4×10^{-5}

TABLE 3 • Solubility-Product Constants for Compounds at 25 °C

Name	Formula	K_{sp}	Name	Formula	K_{sp}
Barium carbonate	$BaCO_3$	5.0×10^{-9}	Lead(II) fluoride	PbF_2	3.6×10^{-8}
Barium chromate	$BaCrO_4$	2.1×10^{-10}	Lead(II) sulfate	$PbSO_4$	6.3×10^{-7}
Barium fluoride	BaF_2	1.7×10^{-6}	Lead(II) sulfide*	PbS	3×10^{-28}
Barium oxalate	$Ba(COO^-)_2$	1.6×10^{-6}	Magnesium hydroxide	$Mg(OH)_2$	1.6×10^{-12}
Barium sulfate	$BaSO_4$	1.1×10^{-10}	Magnesium carbonate	$MgCO_3$	3.5×10^{-8}
Cadmium carbonate	$CdCO_3$	1.8×10^{-14}	Magnesium oxalate	MgC_2O_4	8.6×10^{-5}
Cadmium hydroxide	$Cd(OH)_2$	2.5×10^{-14}	Manganese(II) carbonate	$MnCO_3$	5.0×10^{-10}
Cadmium sulfide*	CdS	8×10^{-28}	Manganese(II) hydroxide	$Mn(OH)_2$	1.6×10^{-13}
Calcium carbonate (calcite)	$CaCO_3$	4.5×10^{-9}	Manganese(II) sulfide*	MnS	2×10^{-53}
Calcium chromate	$CaCrO_4$	7.1×10^{-4}	Mercury(I) chloride	Hg_2Cl_2	1.2×10^{-18}
Calcium fluoride	CaF_2	3.9×10^{-11}	Mercury(I) iodide	Hg_2I_2	1.1×10^{-28}
Calcium hydroxide	$Ca(OH)_2$	6.5×10^{-6}	Mercury(II) sulfide*	HgS	2×10^{-53}
Calcium phosphate	$Ca_3(PO_4)_2$	2.0×10^{-29}	Nickel(II) carbonate	$NiCO_3$	1.3×10^{-7}
Calcium sulfate	$CaSO_4$	2.4×10^{-5}	Nickel(II) hydroxide	$Ni(OH)_2$	6.0×10^{-16}
Chromium(III) hydroxide	$Cr(OH)_3$	1.6×10^{-30}	Nickel(II) sulfide*	NiS	3×10^{-20}
Cobalt(II) carbonate	$CoCO_3$	1.0×10^{-10}	Silver bromate	$AgBrO_3$	5.5×10^{-5}
Cobalt(II) hydroxide	$Co(OH)_2$	1.3×10^{-15}	Silver bromide	$AgBr$	5.0×10^{-13}
Cobalt(II) sulfide*	CoS	5×10^{-22}	Silver carbonate	Ag_2CO_3	8.1×10^{-12}
Copper(I) bromide	$CuBr$	5.3×10^{-9}	Silver chloride	$AgCl$	1.8×10^{-10}
Copper(II) carbonate	$CuCO_3$	2.3×10^{-10}	Silver chromate	Ag_2CrO_4	1.2×10^{-12}
Copper(II) hydroxide	$Cu(OH)_2$	4.8×10^{-20}	Silver iodide	AgI	8.3×10^{-17}
Copper(II) sulfide*	CuS	6×10^{-37}	Silver sulfate	Ag_2SO_4	1.5×10^{-5}
Iron(II) carbonate	$FeCO_3$	2.1×10^{-11}	Silver sulfide*	Ag_2S	6×10^{-51}
Iron(II) hydroxide	$Fe(OH)_2$	7.9×10^{-16}	Strontium carbonate	$SrCO_3$	9.3×10^{-10}
Lanthanum fluoride	LaF_3	2×10^{-19}	Tin(II) sulfide*	SnS	1×10^{-26}
Lanthanum iodate	$La(IO_3)_3$	6.1×10^{-12}	Zinc carbonate	$ZnCO_3$	1.0×10^{-10}
Lead(II) carbonate	$PbCO_3$	7.4×10^{-14}	Zinc hydroxide	$Zn(OH)_2$	3.0×10^{-16}
Lead(II) chloride	$PbCl_2$	1.7×10^{-5}	Zinc oxalate	$Zn(COO^-)_2$	2.7×10^{-8}
Lead(II) chromate	$PbCrO_4$	2.8×10^{-13}	Zinc sulfide*	ZnS	2×10^{-25}

* For a solubility equilibrium of the type $MS(s) + H_2O(l) \rightleftharpoons M^{2+}(aq) + HS^-(aq) + OH^-(aq)$

APPENDIX E

STANDARD REDUCTION POTENTIALS AT 25 °C

Half-reaction	E°(V)
$Ag^+(aq) + e^- \longrightarrow Ag(s)$	+0.799
$AgBr(s) + e^- \longrightarrow Ag(s) + Br^-(aq)$	+0.095
$AgCl(s) + e^- \longrightarrow Ag(s) + Cl^-(aq)$	+0.222
$Ag(CN)_2^-(aq) + e^- \longrightarrow Ag(s) + 2\,CN^-(aq)$	−0.31
$Ag_2CrO_4(s) + 2\,e^- \longrightarrow 2\,Ag(s) + CrO_4^{2-}(aq)$	+0.446
$AgI(s) + e^- \longrightarrow Ag(s) + I^-(aq)$	−0.151
$Ag(S_2O_3)_2^{3-}(aq) + e^- \longrightarrow Ag(s) + 2\,S_2O_3^{2-}(aq)$	+0.01
$Al^{3+}(aq) + 3\,e^- \longrightarrow Al(s)$	−1.66
$H_3AsO_4(aq) + 2\,H^+(aq) + 2\,e^- \longrightarrow$ $H_3AsO_3(aq) + H_2O(l)$	+0.559
$Ba^{2+}(aq) + 2\,e^- \longrightarrow Ba(s)$	−2.90
$BiO^+(aq) + 2\,H^+(aq) + 3\,e^- \longrightarrow Bi(s) + H_2O(l)$	+0.32
$Br_2(l) + 2\,e^- \longrightarrow 2\,Br^-(aq)$	+1.065
$2\,BrO_3^-(aq) + 12\,H^+(aq) + 10\,e^- \longrightarrow$ $Br_2(l) + 6\,H_2O(l)$	+1.52
$2\,CO_2(g) + 2\,H^+(aq) + 2\,e^- \longrightarrow H_2C_2O_4(aq)$	−0.49
$Ca^{2+}(aq) + 2\,e^- \longrightarrow Ca(s)$	−2.87
$Cd^{2+}(aq) + 2\,e^- \longrightarrow Cd(s)$	−0.403
$Ce^{4+}(aq) + e^- \longrightarrow Ce^{3+}(aq)$	+1.61
$Cl_2(g) + 2\,e^- \longrightarrow 2\,Cl^-(aq)$	+1.359
$2\,HClO(aq) + 2\,H^+(aq) + 2\,e^- \longrightarrow$ $Cl_2(g) + 2\,H_2O(l)$	+1.63
$ClO^-(aq) + H_2O(l) + 2\,e^- \longrightarrow$ $Cl^-(aq) + 2\,OH^-(aq)$	+0.89
$2\,ClO_3^-(aq) + 12\,H^+(aq) + 10\,e^- \longrightarrow$ $Cl_2(g) + 6\,H_2O(l)$	+1.47
$Co^{2+}(aq) + 2\,e^- \longrightarrow Co(s)$	−0.277
$Co^{3+}(aq) + e^- \longrightarrow Co^{2+}(aq)$	+1.842
$Cr^{3+}(aq) + 3\,e^- \longrightarrow Cr(s)$	−0.74
$Cr^{3+}(aq) + e^- \longrightarrow Cr^{2+}(aq)$	−0.41
$CrO_7^{2-}(aq) + 14\,H^+(aq) + 6\,e^- \longrightarrow$ $2\,Cr^{3+}(aq) + 7\,H_2O(l)$	+1.33
$CrO_4^{2-}(aq) + 4\,H_2O(l) + 3\,e^- \longrightarrow$ $Cr(OH)_3(s) + 5\,OH^-(aq)$	−0.13
$Cu^{2+}(aq) + 2\,e^- \longrightarrow Cu(s)$	+0.337
$Cu^{2+}(aq) + e^- \longrightarrow Cu^+(aq)$	+0.153
$Cu^+(aq) + e^- \longrightarrow Cu(s)$	+0.521
$CuI(s) + e^- \longrightarrow Cu(s) + I^-(aq)$	−0.185
$F_2(g) + 2\,e^- \longrightarrow 2\,F^-(aq)$	+2.87
$Fe^{2+}(aq) + 2\,e^- \longrightarrow Fe(s)$	−0.440
$Fe^{3+}(aq) + e^- \longrightarrow Fe^{2+}(aq)$	+0.771
$Fe(CN)_6^{3-}(aq) + e^- \longrightarrow Fe(CN)_6^{4-}(aq)$	+0.36
$2\,H^+(aq) + 2\,e^- \longrightarrow H_2(g)$	0.000

Half-reaction	E°(V)
$2\,H_2O(l) + 2\,e^- \longrightarrow H_2(g) + 2\,OH^-(aq)$	−0.83
$HO_2^-(aq) + H_2O(l) + 2\,e^- \longrightarrow 3\,OH^-(aq)$	+0.88
$H_2O_2(aq) + 2\,H^+(aq) + 2\,e^- \longrightarrow 2\,H_2O(l)$	+1.776
$Hg_2^{2+}(aq) + 2\,e^- \longrightarrow 2\,Hg(l)$	+0.789
$2\,Hg^{2+}(aq) + 2\,e^- \longrightarrow Hg_2^{2+}(aq)$	+0.920
$Hg^{2+}(aq) + 2\,e^- \longrightarrow Hg(l)$	+0.854
$I_2(s) + 2\,e^- \longrightarrow 2\,I^-(aq)$	+0.536
$2\,IO_3^-(aq) + 12\,H^+(aq) + 10\,e^- \longrightarrow$ $I_2(s) + 6\,H_2O(l)$	+1.195
$K^+(aq) + e^- \longrightarrow K(s)$	−2.925
$Li^+(aq) + e^- \longrightarrow Li(s)$	−3.05
$Mg^{2+}(aq) + 2\,e^- \longrightarrow Mg(s)$	−2.37
$Mn^{2+}(aq) + 2\,e^- \longrightarrow Mn(s)$	−1.18
$MnO_2(s) + 4\,H^+(aq) + 2\,e^- \longrightarrow$ $Mn^{2+}(aq) + 2\,H_2O(l)$	+1.23
$MnO_4^-(aq) + 8\,H^+(aq) + 5\,e^- \longrightarrow$ $Mn^{2+}(aq) + 4\,H_2O(l)$	+1.51
$MnO_4^-(aq) + 2\,H_2O(l) + 3\,e^- \longrightarrow$ $MnO_2(s) + 4\,OH^-(aq)$	+0.59
$HNO_2(aq) + H^+(aq) + e^- \longrightarrow NO(g) + H_2O(l)$	+1.00
$N_2(g) + 4\,H_2O(l) + 4\,e^- \longrightarrow 4\,OH^-(aq) + N_2H_4(aq)$	−1.16
$N_2(g) + 5\,H^+(aq) + 4\,e^- \longrightarrow N_2H_5^+(aq)$	−0.23
$NO_3^-(aq) + 4\,H^+(aq) + 3\,e^- \longrightarrow NO(g) + 2\,H_2O(l)$	+0.96
$Na^+(aq) + e^- \longrightarrow Na(s)$	−2.71
$Ni^{2+}(aq) + 2\,e^- \longrightarrow Ni(s)$	−0.28
$O_2(g) + 4\,H^+(aq) + 4\,e^- \longrightarrow 2\,H_2O(l)$	+1.23
$O_2(g) + 2\,H_2O(l) + 4\,e^- \longrightarrow 4\,OH^-(aq)$	+0.40
$O_2(g) + 2\,H^+(aq) + 2\,e^- \longrightarrow H_2O_2(aq)$	+0.68
$O_3(g) + 2\,H^+(aq) + 2\,e^- \longrightarrow O_2(g) + H_2O(l)$	+2.07
$Pb^{2+}(aq) + 2\,e^- \longrightarrow Pb(s)$	−0.126
$PbO_2(s) + HSO_4^-(aq) + 3\,H^+(aq) + 2\,e^- \longrightarrow$ $PbSO_4(s) + 2\,H_2O(l)$	+1.685
$PbSO_4(s) + H^+(aq) + 2\,e^- \longrightarrow Pb(s) + HSO_4^-(aq)$	−0.356
$PtCl_4^{2-}(aq) + 2\,e^- \longrightarrow Pt(s) + 4\,Cl^-(aq)$	+0.73
$S(s) + 2\,H^+(aq) + 2\,e^- \longrightarrow H_2S(g)$	+0.141
$H_2SO_3(aq) + 4\,H^+(aq) + 4\,e^- \longrightarrow S(s) + 3\,H_2O(l)$	+0.45
$HSO_4^-(aq) + 3\,H^+(aq) + 2\,e^- \longrightarrow$ $H_2SO_3(aq) + H_2O(l)$	+0.17
$Sn^{2+}(aq) + 2\,e^- \longrightarrow Sn(s)$	−0.136
$Sn^{4+}(aq) + 2\,e^- \longrightarrow Sn^{2+}(aq)$	+0.154
$VO_2^+(aq) + 2\,H^+(aq) + e^- \longrightarrow VO^{2+}(aq) + H_2O(l)$	+1.00
$Zn^{2+}(aq) + 2\,e^- \longrightarrow Zn(s)$	−0.763

CHAPTER 1

1 (a) About 100 elements. (b) Atoms and molecules.

2 The water molecule contains the atoms of two different elements in definite proportions.

3 (a) This is a chemical change because a new substance is being formed. (b) This is a physical change because the water merely changes its physical state and not its composition.

4 1 pg (picogram), which equals 10^{-12} g.

5 (b) It is inexact because it is a measured quantity. Both (a) and (c) are exact; (a) involves counting and (c) is a defined value.

CHAPTER 2

1 (a) The law of multiple proportions. (b) The second compound must contain two oxygen atoms for each carbon atom (that is, twice as many oxygen atoms as the first compound).

2 Most α particles pass through the foil without being deflected because most of the volume of the atoms that comprise the foil is empty space.

3 (a) The atom has 15 electrons because atoms have equal numbers of electrons and protons. (b) The protons reside in the nucleus of the atom.

4 Any single atom of chromium must be one of the isotopes of that element. The isotope mentioned has a mass of 52.94 u and is probably ^{53}Cr. The average atomic mass differs from the mass of any particular atom because it is the average mass of the naturally occurring isotopes of the element.

5 (a) Cl, (b) third period and group 17, (c) 17, (d) non-metal.

6 (a) C_2H_6, (b) CH_3, (c) probably the ball-and-stick model because the angles between the sticks indicate the angles between the atoms.

7 We write the empirical formulae for ionic compounds as the smallest whole number ratio of ions. Thus the formula is CaO.

8 The transition metals can form more than one type of cation, and the charges of these ions are therefore indicated explicitly with Roman numerals: chromium(II) ion is Cr^{2+}. Calcium, on the other hand, *always* forms the Ca^{2+} ion, so there is no need to distinguish it from other calcium ions with different charges.

9 An *-ide* ending usually means a monatomic ion; an *-ate* ending indicates an oxyanion. The most common oxyanions for each element have the *-ate* ending. An *-ite* ending also indicates an oxyanion, but one having less O than the anion whose name ends in *-ate*.

10 BO_3^{3-} and SiO_4^{4-}. The borate has three O atoms, like the other oxyanions of the second period in Figure 2.21, and its charge is 3–, following the trend of increasing negative charge as you move to the left in the period. The silicate has four O atoms, as do the other oxyanions in the third period in Figure 2.21, and its charge is 4–, also following the trend of increasing charge moving to the left.

11

CHAPTER 3

1 Each $Mg(OH)_2$ has 1 Mg, 2 O and 2 H; thus 3 $Mg(OH)_2$ represents 3 Mg, 6 O and 6 H.

2 The product is an ionic compound involving Na^+ and S^{2-}, and its chemical formula is therefore Na_2S.

3 There are experimental uncertainties in the measurements.

4 3.14 mol because 2 mol H_2 reacts with 1 mol O_2 based on the coefficients in the balanced equation.

5 The number of grams of product formed is the sum of the masses of the two reactants, 50.0 g. When two substances react in a combination reaction, only one substance is formed as a product. According to the law of combination of mass, the mass of the product must equal the masses of the two reactants.

CHAPTER 4

1 (a) K^+(aq) and CN^-(aq), (b) Na^+(aq) and ClO_4^-(aq).

2 NaOH because it is the only solute which is a strong electrolyte.

3 Na^+(aq) and NO_3^-(aq).

4 HBr is a strong acid (Table 4.2 lists strong acids).

5 SO_2.

6 (a) Neon, (b) zero.

7 (a) Yes, nickel is below zinc in the activity series so Ni^{2+}(aq) will oxidise Zn(s) to form Ni(s) and Zn^{2+}(aq). (b) No reaction will occur because the Zn^{2+}(aq) ions cannot be oxidised further.

8 The molarity is halved to 0.25 M by doubling its volume.

9 12.5 cm^3.

CHAPTER 5

1 The mass number decreases by 4.

2 Only the neutron as it is the only neutral particle listed.

3 From their atomic numbers we see that they each have an odd number of protons. Given the rarity of stable isotopes with odd numbers of neutrons and protons, we expect that each isotope will possess an even number of neutrons. From their atomic numbers we see that this is the case. F (10 neutrons), Na (12 neutrons), Al (14 neutrons) and P (16 neutrons).

4 From Figure 5.1 we see that 30 neutrons would mean that the nucleus was outside the belt of stability. Since stable isotopes with an odd number of neutrons are rare, the number of neutrons would be 24.

5 (a) Yes, doubling the mass would double the amount of radioactivity of the sample as shown in Equation 5.18(b). No, changing the mass would not change the half-life as shown in Equation 5.20.

6 No. Alpha particles are more readily absorbed by matter than beta or gamma rays. Geiger counters must be calibrated for the radiation they are being used to detect.

7 No. Stable nuclei having mass numbers around 100 are the most stable nuclei. They could not form a still more stable nucleus with an accompanying release of energy.

CHAPTER 6

1 Visible light is one form of electromagnetic radiation; all visible light is electromagnetic radiation, but not all electromagnetic radiation is visible.

2 They both travel at the speed of light, *c*. The differing ability to penetrate skin is due to the different energies of visible light and X-rays (discussed in the next section).

3 As temperature increases the average energy of the emitted radiation increases. Blue-white light is at the short end of the visible spectrum (at about 400 nm) whereas red light is closer to the other end of the visible spectrum (about 700 nm). Thus the blue-white light has a higher frequency, is more energetic and is consistent with higher temperatures.

4 Electrons would be ejected, and the ejected electrons would have greater kinetic energy than they did when ejected by yellow light.

5 According to the third postulate, photons of only certain allowed frequencies can be absorbed or emitted as the electron changes energy state. The lines in the spectrum correspond to the allowed frequencies.

6 Absorb, because it is moving from a lower-energy state ($n = 3$) to a higher-energy state ($n = 7$).

7 Yes, all moving objects produce matter waves, but the wavelengths associated with macroscopic objects, such as the cricket ball, are too small to allow for any way of observing them.

8 The small size and mass of a subatomic particle. The term $h/4\pi$ in the uncertainty principle is a very small number that becomes important only when considering extremely small objects, such as electrons.

9 Yes, there is a difference. The first statement says that the electron's position is known exactly, which violates the uncertainty principle. The second statement says that there is a high probability of where the electron is, but there is still uncertainty in its position.

10 Bohr proposed that the electron in the hydrogen atom moves in a well-defined circular orbit about the nucleus, which violates the uncertainty principle. An orbital is a wave function that gives the probability of finding the electron at any point in space, in accord with the uncertainty principle.

11 The energy of an orbital is proportional to $1/n^2$. The difference between $1/2^2$ and $1/1^2$ is much greater than the difference between $1/3^2$ and $1/2^2$.

12 The radial probability function for the $3s$ orbital will have three maxima and two nodes.

13 A node, where the electron density is zero.

14 No. We know that the $4s$ orbital is higher in energy than the $3s$ orbital. Likewise, we know that the $3d$ orbitals are higher in energy than the $3s$ orbital. But, without more information, we do not know whether the $4s$ orbital is higher or lower in energy than the $3d$ orbitals.

15 The $6s$ orbital, which starts to hold electrons at element 55, Cs.

16 Each of the three elements has a different valence electron configuration for its nd and $(n + 1)s$ subshells: for Ni, $3d^8 4s^2$; for Pd, $4d^{10}$; and for Pt, $5d^9 6s^1$. This strongly suggests that the energies of the nd and $(n + 1)s$ subshells for the three elements are very similar.

CHAPTER 7

1 The atomic number of an element depends on the number of protons in the nucleus, whereas the atomic mass depends (mainly) on the number of protons *and* the number of neutrons in the nucleus, and for neighbouring elements the atomic mass of the most common isotopes is not necessarily related in a simple way.

2 Z_{eff} increases across a period so the $2p$ electron of an Ne atom has a greater Z_{eff}.

3 No. Atomic mass is determined almost entirely by the composition of the nucleus. Atomic radius is determined by the distribution of the electrons.

4 The photoelectric effect, first explained by Einstein, involves the ejection of electrons using the energy of light. If the energy of a photon, $h\nu$, is greater than the ionisation energy of an atom, then the light can, in principle, be used to ionise the atom.

5 I_2 for a carbon atom. In each process an electron is being removed from an atom or ion with five electrons, either B(g) or C$^+$(g). The higher nuclear charge of C$^+$ makes I_2 for carbon higher than I_1 for boron.

6 Since electrons in the *s*-subshell are lost first when transition metals are ionised, these two ions will have the same electron configuration: [Ar]$3d^3$.

7 The first ionisation energy for Cl$^-$(g) is the energy needed to remove an electron from Cl$^-$, forming Cl(g) + e$^-$. That is the reverse process of Equation 7.5, so the first ionisation energy of Cl$^-$(g) is +349 kJ mol^{-1}.

8 In general, increasing ionisation energy correlates with decreasing metallic character.

9 Molecular, since ionic compounds have high melting points; P, since it is a non-metal and hence more likely to form molecular compounds.

CHAPTER 8

1 Cl has seven valence electrons. The first and second Lewis symbols are both correct—they both show seven valence electrons, and it doesn't matter which of the four sides has the single electron. The third symbol shows only five electrons and is incorrect.

2 Magnesium fluoride is an ionic substance, MgF$_2$, which consists of Mg^{2+} and F$^-$ ions (Section 2.7). Each magnesium atom loses two electrons, and each fluorine atom gains one electron. Thus we can say that each Mg atom transfers one electron to each of two F atoms.

3 Rhodium, Rh.

4 Attractive forces are between each electron and either nucleus. Repulsive forces are those between the two nuclei and those between any two electrons. Because He$_2$ doesn't exist, it is likely that the repulsions are greater than the attractions.

5 CO$_2$ has C–O double bonds. Because the C–O bond in carbon monoxide is shorter, it is likely to be a triple bond.

6 Electron affinity measures the energy released when an isolated atom gains an electron to form an ion. The electronegativity measures the ability of the atom to hold on to its own electrons and attract electrons from other atoms in compounds.

7 Polar covalent. Based on the examples of F$_2$, HF and LiF, the difference in electronegativity is great enough to introduce some polarity to the bond, but not great enough to cause a complete electron transfer from one atom to the other.

8 IF. Because the difference in electronegativity between I and F is greater than that between Cl and F, the magnitude of q should be greater for IF. In addition, because I has a larger atomic radius than Cl, the bond length in IF is longer than that in ClF. Thus both q and r are larger for IF, and therefore $\mu = qr$ will be larger for IF.

9 Red indicates very high electron density on one end of the molecule, which means a large separation of positive and negative charges. In HBr and HI, the electronegativity differences are too small to lead to large charge separations in the molecules.

10 OsO$_4$. The data suggest that the yellow substance is a molecular species, with its low melting and boiling points.

Os in OsO_4 has an oxidation number of +8 and Cr in CrO_3 has an oxidation number of +3. In Section 8.4 we saw that a compound with a metal in a high oxidation state should show a high degree of covalence and OsO_4 fits this situation.

11 There is probably a better choice of Lewis structure than the one chosen. Because the formal charges must add up to 0 and the formal charge on the F atom is +1, there must be an atom that has a formal charge of –1. Because F is the most electronegative element, we don't expect it to carry a positive formal charge.

12 Yes. There are two resonance structures for ozone that both contribute equally to the overall description of the molecule. Each O–O bond is therefore an average of a single bond and a double bond, which is a 'one-and-a-half' bond.

13 As 'one-and-a-third' bonds. There are three resonance structures, and each of the three N–O bonds is single in two of those structures and double in the third. Each bond in the actual ion is an average of these: $(1 + 1 + 2)/3 = 1\,1/3$.

CHAPTER 9

1 Octahedral. Removing two atoms that are opposite each other leads to a square planar geometry.

2 The molecule does not follow the octet rule because it has 10 electrons around the central A atom. There are four electron domains around A: two single bonds, one double bond and one non-bonding pair.

3 Yes. If there were only one resonance structure, we would expect the electron domain that is due to the double bond to 'push' the domains that are due to the single bonds, leading to angles slightly different from 120°. However, we must remember that there are two other equivalent resonance structures—each of the three O atoms has a double bond to N in one of the three resonance structures (Section 8.6). Because of resonance, all three O atoms are equivalent, and they will experience the same amount of repulsion, which leads to bond angles equal to 120°.

4 No. The directions of the C–O and C–S bond dipoles are directly opposite. The bond dipoles will not cancel each other and the OCS molecule has a non-zero dipole moment.

5 Yes. In both F_2 and Cl_2 the bond is formed by the overlap of singly occupied p orbitals, as in Figure 9.14(c). In F_2, however, the overlapping orbitals are $2p$ orbitals, whereas in Cl_2 they are $3p$ orbitals. Because $3p$ orbitals extend further in space, the optimum overlap in Cl_2 is achieved at a longer distance than in F_2. Thus the bond length in Cl_2 is longer than that in F_2.

6 The three $2p$ orbitals are equivalent to one another; they differ only in their orientation. Thus the two Be–F bonds would be equivalent to each other. Because p orbitals are perpendicular to one another, we would expect an F–Be–F bond angle of 90°. Experimentally, the bond angle is 180°.

7 The unhybridised p orbital is oriented perpendicular to the plane defined by the three sp^2 hybrids, with one lobe on each side of the plane.

8 The molecule would fall apart. With one electron in the bonding MO and one in the antibonding MO, there is no net stabilisation of the electrons relative to two separate H atoms.

9 Zero, because there is one electron in the bonding MO and one electron in the antibonding MO.

10 Yes. In Be_2^+ there would be two electrons in the σ_{2s} MO but only one electron in the σ_{2s}^* MO, and so the ion is predicted to have a bond order of $\frac{1}{2}$. It should (and does) exist.

11 No. If the σ_{2p} MO were lower in energy than the π_{2p} MOs, we would expect the σ_{2p} MO to hold two electrons and the π_{2p} MOs to hold one electron each, with the same spin. The molecule would therefore be paramagnetic.

CHAPTER 10

1 The major reason gases have similar properties is the relatively large distances between molecules, so they are all mostly empty space.

2 As the pressure increases the volume decreases. Doubling the pressure causes the volume to decrease to half its original value.

3 The volume decreases, but it doesn't decrease to half because the volume is proportional to the temperature on the Kelvin scale but not on the Celsius scale. The volume will decrease by a factor of approximately $323/373$.

4 Because 22.71 dm^3 is the volume of one mole of the gas at STP, it contains Avogadro's number of molecules, 6.022×10^{23}.

5 Because water has a lower molar mass (18.0 g mol^{-1}) than N_2 (28.0 g mol^{-1}) the water vapour is less dense.

6 According to Dalton's law of partial pressures, the pressure that is due to N_2 (its partial pressure) doesn't change. However, the total pressure that is due to the partial pressures of both N_2 and O_2 increases.

7 Higher speeds are correlated with lower masses, so HCl < O_2 < H_2.

8 The average kinetic energies depend only on temperature and not on the identity of the gas. Thus the trend in average kinetic energies is HCl (298 K) ≈ H_2 (298 K) < O_2 (350 K).

9 (a) Mean free path decreases because the molecules are crowded closer together. (b) There is no effect. Although the molecules are moving faster at the higher temperature, they are not crowded any closer together.

10 (b) Gases deviate from ideality most at low temperatures and high pressures. Thus the helium gas would deviate most from ideal behaviour at 100 K (the lowest temperature listed) and 5 bar (the highest pressure listed).

11 The fact that real gases deviate from ideal behaviour can be attributed to the finite sizes of the molecules and to the attractions that exist between molecules.

CHAPTER 11

1 (a) In a gas the energy of attraction is less than the average kinetic energy. (b) In a solid, the energy of attraction is greater than the average kinetic energy.

2 CH_4 < CCl_4 < CBr_4. Because all three molecules are non-polar, the strength of dispersion forces determines the relative boiling points. Polarisibility increases in order of increasing molecular size and molecular mass, CH_4 < CCl_4 < CBr_4. Hence the dispersion forces and the boiling points increase in the same order.

3 $Ca(NO_3)_2$ in water, because calcium nitrate is a strong electrolyte that forms ions and water is a polar molecule with a dipole moment. Ion–dipole forces cannot be present in a CH_3OH/H_2O mixture because CH_3OH does not form ions.

4 (a) Both viscosity and surface tension decrease with increasing temperature because of the increased molecular motion. (b) Both properties increase as the strength of intermolecular forces increases.

5 Cohesive. The properties reflect the attraction of the molecules of a substance for one another.

6 Melting (or fusion), endothermic. Ice absorbs heat from the surroundings as it melts.

7 Both compounds are non-polar. Consequently, only dispersion forces exist between the molecules. (a) Because dispersion forces are stronger for the larger, heavier CBr_4 than for CCl_4, CBr_4 has a lower vapour pressure than CCl_4. (b) The substance with the lower vapour pressure is less volatile. Thus CBr_4 is less volatile than CCl_4.

8 Crystalline solids melt at a specific temperature, whereas amorphous ones tend to melt over a temperature range.

9 As summarised in Table 11.6 and shown in Figure 11.41, one-eighth of an atom is at each of the eight corners, and a whole atom is at the centre. Thus the total number of atoms is $(\frac{1}{8} \times 8) + 1 = 2$.

10 The higher the coordination number of the particles in a crystal, the greater the packing efficiency.

11 Molecular solids are composed of molecules or non-metal atoms. Because Co is a metal and K_2O is an ionic substance, they do not form molecular solids. C_6H_6, however, is a molecular substance and forms a molecular solid.

CHAPTER 12

1 (a) Separating solvent molecules from each other requires energy and is therefore endothermic. (b) Forming the solvent–solute interactions is exothermic.

2 No, because the number of Ag^+ and Cl^- entering into the solution would be very small.

3 The added solute provides a template for the solid to begin to crystallise from solution, and the excess dissolved solute comes out of solution, leaving a saturated solution.

4 The solubility in water would be considerably lower because there would no longer be hydrogen bonding with water, which promotes solubility.

5 Dissolved gases become less soluble as temperature increases and they come out of solution, forming bubbles below the boiling point of water.

6 230 ppm (1 ppm is 1 part in 10^6); 2.30×10^5 ppb (1 ppb is 1 part in 10^9).

7 X_A denotes mole fraction of A and has no units.

8 Not necessarily; if the solute was a strong or a weak electrolyte, it could have a lower molality and still cause an increase of 0.51 °C. The total molality of all the particles would have to be 1 M.

9 The 0.20 M solution is hypotonic with respect to the 0.5 M solution. (A hypotonic solution will have a lower concentration and hence a lower osmotic pressure.)

10 They would have the same osmotic pressure because they would have the same concentration of particles. (Both are strong electrolytes that are 0.20 M in total ions.)

11 The smaller droplets carry negative charges because of the embedded stearate ions and thus repel each other.

CHAPTER 13

1 Photoionisation is a process in which a molecule breaks into ions upon irradiation with light; photodissociation is a process in which molecules break up upon irradiation with light but the products bear no charge.

2 Because these molecules do not absorb light at those wavelengths.

3 SO_2 in the atmosphere reacts with oxygen to form SO_3. SO_3 in the atmosphere reacts with water to form H_2SO_4, sulfuric acid. The sulfuric acid dissolves in water droplets that fall to Earth, causing acid rain that has a pH of around 4.

4 NO_2 photodissociates to NO and O; the O atoms react with O_2 in the atmosphere to form ozone, which is a key ingredient in photochemical smog.

5 Higher humidity means there is more water in the air. Water absorbs infrared light which we feel as heat. After sunset the ground, which has been warmed earlier in the day, reradiates heat out. In locations with higher humidity, this energy is somewhat absorbed by the water and in turn is to some extent reradiated back to the Earth, resulting in warmer temperatures as compared to a location with low humidity.

6 The pollutants are capable of being oxidised, either directly by reaction with dissolved oxygen or indirectly by the action of organisms such as bacteria.

7 The overall number and types of atoms are the same at the end of the reaction as at the start. The greater the proportion of these atoms that are focused into the desired product, the fewer that are left over as 'waste'.

CHAPTER 14

1 (a) No, the potential energy is lower at the bottom of the hill. (b) Once the bicycle stops, its kinetic energy is zero.

2 The balance (current state), does not depend on the ways the money may have been transferred into the account or on the particular expenditure made in withdrawing money from the account. It depends only on the net total of all the transactions.

3 Because U, P and V are state functions which do not depend on the path, $H = U + PV$ is also a state function.

4 No, because work is defined as the pressure multiplied by the change in volume. So if ΔV is zero, then $w = -P\,\Delta V$ is zero.

5 A thermometer to measure temperature change.

6 No, because only half as much matter is involved, the value of ΔH would be $\frac{1}{2}(-483.6\text{ kJ}) = -241.8$ kJ.

7 The heat gained or lost by the system is equal in magnitude but opposite in sign to the heat gained or lost by its surroundings.

8 (a) The sign of ΔH changes. (b) The magnitude of ΔH doubles.

9 $2\,C(\text{graphite}) + H_2(g) \longrightarrow C_2H_2(g)$
$\Delta_f H° = 226.8$ kJ mol^{-1}.

10 No, non-spontaneous processes can occur so long as they receive some continuous outside assistance. An example is the building of a brick wall or the electrolysis of water to form hydrogen and oxygen gas.

11 No. Just because the system is restored to its original condition does not mean that the surroundings have also been restored to their original condition, so it is not necessarily reversible.

12 ΔS depends not merely on q but on q_{rev}. Although there are many possible paths that could take a system from its initial to its final state, there is only *one* reversible, isothermal path between two states. Thus ΔS has only one particular value regardless of the path taken between states.

13 Because rusting is a spontaneous process, $\Delta_{univ}S$ must be positive. Therefore, the entropy of the surroundings must increase and that increase must be larger than the entropy decrease of the system.

14 $S = 0$, based on Equation 14.31 and the fact that $\ln 1 = 0$.

15 A molecule can vibrate (atoms moving relative to one another) and rotate (tumble), whereas a single atom cannot undergo these motions.

16 It must be a perfect crystal at 0 K (third law of thermodynamics), which means it has only one accessible microstate.

17 $\Delta_{surr}S$ always increases. For simplicity, assume the process is isothermal. The change in entropy of the surroundings in an isothermal process is $\Delta_{surr}S = -q_{rev}/T$. For an exothermic process $-q_{rev}$ is a positive number, so $\Delta_{surr}S$ is positive, that is, the entropy of the surroundings increases.

18 (a) In any spontaneous process the entropy of the universe increases. (b) In any spontaneous process operating at constant pressure, the free energy of the system decreases.

19 It indicates that the process to which the thermodynamic quantity refers has taken place under standard conditions as summarised in Table 14.7.

20 Above the boiling point vaporisation is spontaneous, and $\Delta G < 0$, therefore $\Delta H - T\Delta S < 0$, and $\Delta H < T\Delta S$.

CHAPTER 15

1 The rates will increase.

2 Average rate is for a large time interval; instantaneous rate is for an 'instant' in time. Yes, they can have the same numerical value, especially if a plot of concentration versus time is linear.

3 Reaction rate is what we measure as a reaction proceeds—change in concentration with time for one or more of the components in the mixture. Reaction rate always has units of concentration per time, usually $M\ s^{-1}$. A rate constant is what we calculate from reaction rate data and its magnitude is proportional to the reaction rate, but its units depend on the reaction order. The rate law of a reaction is an equation that relates reaction rate to the rate constant: Rate = $k[A]^m[B]^n$, for components A and B in the reaction.

4 No. Rate is always change in concentration per time; the rate constant has units that depend on the form of the rate law.

5 (a) The reaction is second order in NO, first order in H_2 and third order overall. (b) No. Doubling NO concentration will quadruple the rate, but doubling H_2 concentration will merely double the rate.

6 The pressure (initial pressure) of CH_3NC at the start of the reaction ($t = 0$).

7 The half-life will increase.

8 Molecules must collide to react, so the greater the number of collisions per second the greater the reaction rate.

9 The collision may not have occurred with enough energy for reaction to occur, and/or the collision may not have occurred with the proper orientation of reactant molecules to favour product formation.

10 Bimolecular.

11 Most reactions occur in elementary steps; the rate law is governed by the elementary steps, not by their sum (which is the overall balanced equation).

12 The odds of three molecules colliding with each other properly to react is very low.

13 By lowering the activation energy for the reaction or by increasing the frequency factor.

CHAPTER 16

1 (a) The rates of the forward and reverse reactions. (b) Greater than 1.

2 When the concentrations of reactants and products are no longer changing.

3 It does not depend on starting concentrations.

4 Units of moles, dm^{-3}, are used to calculate K_c; units of partial pressure are used to calculate K_P.

5 It is cubed.

6 $K_P = P_{H_2O}$

7 (a) It shifts to the right. (b) It shifts to the left.

8 It will shift to the left, the side with a larger number of moles of gas.

9 No.

10 As the temperature increases, a larger fraction of molecules in the liquid phase have enough energy to overcome their intermolecular attractions and go into the vapour; the evaporation process is endothermic.

CHAPTER 17

1 H^+ ion and OH^- ion.

2 $NH_3(aq)$: it accepts a proton to form $NH_4^+(aq)$.

3 HNO_3 is a strong acid which means that NO_3^- has negligible basicity.

4 pH is defined as $-\log[H^+]$. This quantity will become negative if the H^+ ion concentration exceeds 1 M, which is possible. Such a solution would be highly acidic.

5 pH = 14.00 − 3.00 = 11.00. The solution is basic because pH > 7.

6 Both NaOH and $Ba(OH)_2$ are soluble hydroxides. Therefore the hydroxide concentrations will be 0.001 M for NaOH and 0.002 M for $Ba(OH)_2$. Because the $Ba(OH)_2$ solution has a higher hydroxide ion concentration it will have a higher pH.

7 Because it is the conjugate base of a substance that has negligible acidity, CH_3^- must be a strong base. Bases stronger than OH^- abstract H^+ from water molecules: $CH_3^- + H_2O \longrightarrow CH_4 + OH^-$.

8 Because weak acids typically undergo very little ionisation, often less than 1%.

9 This is the acid-dissociation constant for the loss of the third and final proton from H_3PO_4: $HPO_4^{2-} \rightleftharpoons H^+ + PO_4^{3-}$.

10 Because the NO_3^- ion is the conjugate base of a strong acid, it will have no effect on pH. (NO_3^- ion has negligible basicity.) Because the CO_3^{2-} ion is the conjugate base of a weak acid, it will have an effect on pH, increasing the pH.

11 The increasing acidity down a group is due mainly to the decrease in H–X bond strength. The trend across a period is due mainly to the increasing electronegativity.

12 $HBrO_3$. For an oxyacid, acidity increases as the electromagnetivity of the central atom increases, which would make $HBrO_2$ more acidic than HIO_2. Acidity also increases as the number of oxygens bound to the central atom increases, which would make $HBrO_3$ more acidic than $HBrO_2$. Combining these two relationships we can order these acids in terms of increasing acid-dissociation constants: $HIO_2 < HBrO_2 < HBrO_3$.

13 The carboxyl group, –COOH.

14 It must have an unshared pair of electrons.

CHAPTER 18

1 (a) $[NH_4^+] = 0.10$ M; $[Cl^-] = 0.10$ M; $[NH_3] = 0.12$ M, (b) Cl^-, (c) Equation 18.2.

2 HNO_3 and NO_3^-. This is a strong acid and its conjugate base. Buffers are composed of weak acids and their conjugate bases. The NO_3^- ion is merely a spectator in any acid–base chemistry and is therefore ineffective in helping control the pH of a solution.

3 (a) The NaOH (a strong base) reacts with the acid member of the buffer (CH_3COOH), abstracting a proton. Thus

[CH$_3$COOH] decreases and [CH$_3$COO$^-$] increases. (b) The HCl (a strong acid) reacts with the base member of the buffer (CH$_3$COO$^-$). Thus [CH$_3$COO$^-$] decreases and [CH$_3$COOH] increases.

4   The solution will most effectively resist a change in either direction when its pH = pK_a. Thus the optimal pH is pH = pK_a = −log(1.8 × 10^{-5}) = 4.74, and the pH range of the buffer is about 4.7±1.

5   The pH will increase on addition of the base.

6   The following titration curve shows the titration of 25 cm^3 of Na$_2$CO$_3$ with HCl, both with 0.1 M concentrations. The overall reaction between the two is

Na$_2$CO$_3$(aq) + HCl(aq) ⟶ 2 NaCl(aq) + CO$_2$(g) + H$_2$O(l)

The initial pH (sodium carbonate in water only) is near 11 because CO$_3^{2-}$ is a weak base in water. The graph shows two equivalence points, A and B. The first point, A, is reached at a pH of about 9:

Na$_2$CO$_3$(aq) + HCl(aq) ⟶ NaCl(aq) + NaHCO$_3$(aq)

HCO$_3^-$ is weakly basic in water and is a weaker base than the carbonate ion. The second point, B, is reached at a pH of about 4:

NaHCO$_3$(aq) + HCl(aq) ⟶ NaCl(aq) + CO$_2$(g) + H$_2$O(l)

H$_2$CO$_3$, a weak acid, forms and decomposes to carbon dioxide and water.

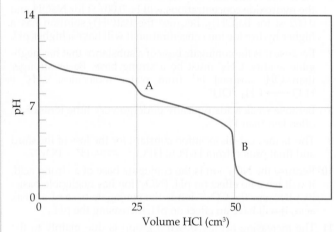

7   The nearly vertical portion of the titration curve at the equivalence point is smaller for a weak acid–strong base titration; as a result fewer indicators undergo their colour change within this narrow range.

8   AgCl. Because all three compounds produce the same number of ions, their relative solubilities correspond directly to their K_{sp} values, with the compound with the largest K_{sp} value being most soluble.

9   Amphoteric substances are insoluble in water but dissolve in sufficient acid or base. Amphiprotic substances can both donate and accept protons.

10 A high concentration of H$_2$S and a low concentration of H$^+$ (that is, high pH) will shift the equilibrium to the right, reducing [Cu^{2+}].

11 The solution must contain one or more of the cations in group 1 of the qualitative analysis scheme shown in Figure 18.21: Ag$^+$, Pb^{2+} or Hg$_2^{2+}$.

CHAPTER 19

1   Oxygen is first assigned an oxidation number of −2. Nitrogen must then have a +3 oxidation number for the sum of oxidation numbers to equal −1, the charge on the ion.

2   No. Electrons should appear in the two half-reactions but cancel when they are added properly.

3   To preserve electrical neutrality of the solution as the Zn^{2+} are released into solution.

4   Yes. A redox reaction with a positive cell potential is spontaneous under standard conditions.

5   1 bar pressure of Cl$_2$(g) and 1 M concentration of Cl$^-$(aq).

6   True.

7   Using data from Appendix E, we have $E°_{red}$ = −0.126V for Pb^{2+}(aq) ⟶ Pb(s) and $E°_{red}$ = 0.854V for Hg^{2+}(aq) ⟶ Hg(l). Because Pb(s) has the most negative value for $E°_{red}$, it is the stronger reducing agent (see Figure 19.12). The comparison can also be made by reference to the activity series where Pb lies also above Hg, indicating that Pb is oxidised more readily than Hg. The more readily a substance is oxidised, the stronger it is as a reducing agent.

8   Al, Zn. Both are easier to oxidise than Fe.

CHAPTER 20

1   No.

2   No. N can form triple bonds but P cannot, as it would have to form P$_2$.

3   H$^-$, hydride.

4   +1 for everything except H$_2$, for which the oxidation state of H is 0.

5   No—it is the volume of Pd that can increase to accommodate hydrogen, not its mass.

6   0 for Cl$_2$; −1 for Cl$^-$; +1 for ClO$^-$.

7   They should both be strong, since the central halogen is in the oxidation state for both of them. We need to look up the redox potentials to see which ion, BrO$_3^-$ or ClO$_3^-$, has the larger reduction potential. The ion with the larger reduction potential is the stronger oxidising agent. BrO$_3^-$ is the stronger oxidising agent on this basis (+1.52 V standard reduction potential in acid compared to +1.4 V for ClO$_3^-$).

8   The standard energy to dissociate one mole of oxygen atoms from one mole of ozone was given as 105 kJ. If we assume, as usual, that one photon will dissociate one molecule, that means the energy of the photons should be 105 kJ per mole (of photons). Using Avogadro's number, we can calculate that one photon would then have 1.744 × 10^{-19} J of energy. Using the equations $c = \lambda \nu$ and $E = h\nu$, we can find that a photon with 1.744 × 10^{-19} J of energy will have a wavelength, λ, of 1140 nm, or 1.14 × 10^6 m, which is in the infrared part of the spectrum.

9   HIO$_3$.

10 SO$_3$(g) + H$_2$O(l) ⟶ H$_2$SO$_4$(l).

11 (a) +5 (b) +3.

12 CO$_2$(g).

13 Yes, it must, since CS$_2$ is a liquid at room temperature and pressure, and CO$_2$ is a gas.

14 Silicon is the element, Si. Silica is SiO$_2$. Silicones are polymers that have an O–Si–O backbone and hydrocarbon groups on the Si.

15 +3.

CHAPTER 21

1   Sc is the biggest.

2   To form Ti^{5+} you would have to remove core electrons.

3   The larger the distance, the weaker the spin–spin interactions.

4   Yes, it is a Lewis acid–base interaction; the metal ion is the Lewis acid (electron pair acceptor).

5   $[Fe(H_2O)_6]^{3+}(aq) + SCN^-(aq) \longrightarrow$ $[Fe(H_2O)_5SCN]^{2+}(aq) + H_2O(l)$.

6   (a) Tetrahedral (b) octahedral.

7   Bidentate.

8   Its conjugation (alternating single and double C–C bonds) allows it to absorb visible light.

9   No, ammonia cannot engage in linkage isomerism—the only atom that can coordinate to a metal is the nitrogen.

10 (a) Co is $1s^2\ 2s^2\ 2p^6\ 3s^2\ 3p^6\ 4s^2\ 3d^7$. (b) Co^{3+} is $1s^2\ 2s^2\ 2p^6\ 3s^2$ $3p^6\ 3d^6$. Co has three unpaired electrons; Co^{3+} has four unpaired electrons, assuming all five d orbitals have the same energy.

11 Ti(IV) has lost all of the Ti valence electrons; only core electrons remain, and the energy gap between filled and empty orbitals is large, corresponding to light in the ultraviolet which we cannot perceive as coloured.

12 The ligands are in the xy plane. The d_{xy} orbital has its lobes mostly in that plane, so its energy is higher than d_{xz} and d_{yz}.

CHAPTER 22

1   Two C–H bonds and two C–C bonds.

2   Palmitic acid > hexane > ethanol.

3

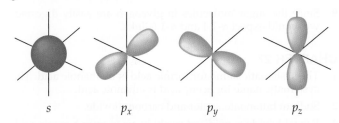

s p_x p_y p_z

The drawings represent the probability of an electron occupying a given space. The blending (hybridisation) of all these orbitals leads to one sp^3 hybrid orbital.

4   Diamond, as a result of its tetrahedral structure.

5   Only two conformations exist, one staggered and one eclipsed.

6   Five: $CH_3CH_2CH_2CH_2CH_2CH_3$; $CH_3CH(CH_3)CH_2CH_2CH_3$; $CH_3CH_2CH(CH_3)CH_2CH_3$; $CH_3CH(CH_3)CH(CH_3)CH_3$; $CH_3C(CH_3)_2CH_2CH_3$.

7   $CH_3CH_2CH_2$.

8   Because they interconvert between having the methyl group axial and equatorial.

9   CO and CO_2, among other products.

10 Nine primary, two secondary hydrogens and one tertiary.

11 Three ways of representing molecular oxygen as a free radical are shown below.

$:\ddot{O}:::\ddot{O}:$ $:\ddot{O}::\ddot{O}:$ $:\ddot{O}::\dot{O}: \leftrightarrow :\dot{O}::\ddot{O}:$
       (1)   (2)   (3)

We can distinguish between them using molecular orbital theory, which tells us that O_2 has a double bond (eliminating (1)) and only two unpaired electrons (eliminating (2)); thus the resonance hybrid (3) is the best Lewis structure for molecular oxygen.

CHAPTER 23

1   The structure on the left is the one only shown by the Newman projection.

2   There must exist a chiral centre. This means there will be non-superparable mirror images.

3

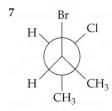

* marks the chiral centre

4   No. An optical isomer can only be d-(+) or l-(–).

5   They must all be different.

6   The magnitudes of their specific rotations will be identical. They only differ in their sign. d-Isomers rotate light in a clockwise (positive) direction, whereas l-isomers rotate light in an anticlockwise (or negative) direction.

7

<pre>
        Br
  H   Cl

  H   CH_3
       CH_3
</pre>

8   Yes. While the *cis* isomer is achiral, the two *trans* isomers are enantiomers which can be described as (R,R) and (S,S).

CHAPTER 24

1   The molecule should not be linear. Because there are three electron domains around each N atom, we expect sp^2 hybridisation and H–N–N angles of approximately 120°. The molecule is expected to be planar; the unhybridised $2p$ orbitals on the N atoms can form a π-bond only if all four atoms lie in the same plane. You might notice that there are two ways in which the H atoms can be arranged: they can be both on the same side of the N–N bond or on opposite sides of the N–N bond.

2   Yes, β-carotene is a terpene, comprised of eight isoprene units.

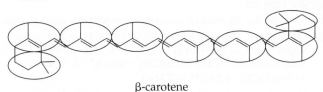

β-carotene

3   Only one.

4   The methyl group in methyl propene can only be at the 2 position; if it was in the 1 or 3 position, the compound would be butene, not methyl propene.

5   The presence of many double bonds separated only by single bonds allows polyacetylene to conduct electricity. All the π orbitals that form the double bonds in the structure lie in the same plane and can overlap with each other, allowing electrons to flow from one end of the molecule to the other. Such a molecule is said to be conjugated.

6   No. The dehydration of 1-propanol leads to the formation of propene. The hydration of this alkene leads, by Markovnikov's rule, to 2-propanol.

7   2,2,3,3-Tetrabromobutane.

8 Ethane.

9 The amount of cross-linking would increase, leading to an inflexible, hard and less soluble product.

CHAPTER 25

1 Considerably lower because there would no longer be hydrogen bonding with water.

2 (1) Chemical inertness. (2) Ability to solubilise a wide range of compounds and reagents.

3 The cyclic nature of 18-crown-6 means that the electronegative oxygen atoms are arranged to coordinate with the positive metal ion.

4 The *tert*-butanol would also react, leading to a complex mixture.

5 An alcohol group is capable of hydrogen bonding, given alcohol's higher boiling points than the corresponding alkenes formed on dehydration.

6 Using HCl will also lead to some formation of the elimination product.

7 NaI is more soluble than NaCl in organic solvents. Whenever Cl^- ions are released into a solution, they will precipitate to form NaCl(s), driving the equilibrium towards the 1-iodoethane product by Le Châtelier's principle.

8 2-Chlorodecane and 3-chlorodecane could both give dec-2-ene on reaction with potassium *tert*-butoxide. 2-Chlorodecane would be a better substrate because the formation of dec-2-ene over dec-1-ene is favoured by Zaitsev's rule, while 3-chlorodecane would be expected to give a mixed product with similar amounts of the equally substituted alkenes dec-2-ene and dec-3-ene.

9 The hydroxide ion will displace chloride from 2-chloro-2-methylpropane in a nucleophilic substitution reaction. Because the carbon-bearing chlorine is not very accessible, this substitution will proceed by an S_N1 mechanism.

10 Concentrated solutions favour polymerisation as there are plenty of reactive species per volume solution. Dilution, and in fact high dilution, favours macrocyclisation as the number of reacting species per volume solvent is low, so all one end has to react with is the other.

CHAPTER 26

1 Because the alternative isomers are aldehydes not ketones.

2 Oxidation: $CH_3CHOHCH_3 \longrightarrow CH_3COCH_3 + 2\,H^+ + 2\,e^-$
Reduction: $Cr_2O_7^{2-} + 14\,H^+ + 6\,e^- \longrightarrow 2\,Cr^{3+} + 7\,H_2O$
Total equation: $3\,CH_3CHOHCH_3 + Cr_2O_7^{e-} + 8\,H^+$
$3\,CH_3COCH_3 + 2\,Cr^{3+} + 7\,H_2O$

3 From the reactions shown in Equation 26.13, the reactions that form an imine require the elimination of two sequential hydrogens from nitrogen as H_3O^+. Formation and elimination of the analogous RH_2O^+ species will be vastly more unlikely, so a secondary or tertiary amine will only be able to form a single bond with a carbonyl carbon.

4 A hydrogen on a carbon adjacent to carbonyl, an α-hydrogen, will have some acidic character due to the possibility of resonance stabilisation of the product anion. It makes sense that a hydrogen that is adjacent to two carbonyl groups would be more acidic than a hydrogen adjacent to just one, hence the major enol is formed by loss of one of the CH_2 hydrogens.

5 The two structures are resonance structures. The true structure will be a resonance hybrid with unequal contributions from the two.

6 $C_6H_{12}O_6$, CH_2O.

7 D- and L-erythrose are diastereomers. So are D- and L-ribose.

8 In the β anomer, it is possible for both the exocyclic CH_2OH group and the hydroxyl group at the anomeric carbon to adopt an equatorial conformation, minimising steric hindrance. In the α anomer, if the exocyclic group is axial, the hydroxy group at the anomeric carbon is forced into the less favourable axial conformation.

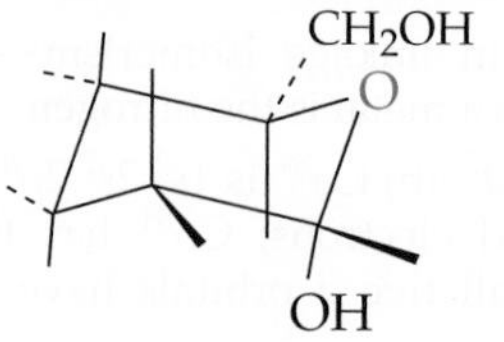

$$CH_2OH$$

α anomer

$$CH_2OH$$

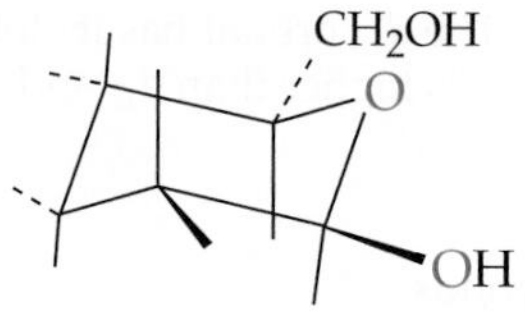

β anomer

9 Since the sugar molecules in glycogen are easily digested, we would expect α-1,4 and α-1,6 links.

CHAPTER 27

1 The systematic name for formic acid is methanoic acid. The systematic name for acetic acid is ethanoic acid.

2 Sodium butanoate, water and carbon dioxide.

3 Benzaldehyde is oxidised easily in air to form benzoic acid. The solid formed is, in fact, benzoic acid.

4 Sodium carbonate can buffer aqueous solutions to a moderate pH and provide another anion (carbonate) that can compete with carboxylate anion for the 2+ metal ions. Water softeners typically pass hard water through sodium carbonate so that the Ca^{2+}, Mg^{2+}, and Fe^{2+} ions will be precipitated out as the carbonate.

5 Yes, the molecule has both the $-NH_2$ and $-COOH$ groups in it, which can react as in nylon to make a polymer.

CHAPTER 28

1 Nitrogen in pyridine is also sp^2 hybridised, so the lone pair will be oriented as shown in the figure—in the plane of the ring, pointing directly away from it.

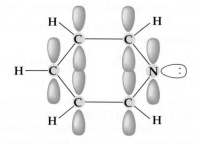

2 All other possible structures can be converted to one of these by rotating the molecule.

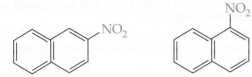

3 Since the phenolate ion can gain stability through resonance while alkoxide ions cannot, alkoxide ions will be less stable and will react more readily with hydrogen-donating compounds, that is, be more basic.

4 In both reactions, the initial step is the formation of a single bond by reaction of an electrophile with a π-bond, and the final result is a compound with additional carbon–bromine bonds. In electrophilic bromination of an alkene, the intermediate carbocation reacts with a Br^- ion to form a second C–Br bond, whereas in EAS the intermediate carbocation loses H^+ to regenerate the π-bond.

5 All carbocations are secondary.

$$CH_3 \quad\quad CH_3 \quad\quad CH_3$$

CHAPTER 29

1 Aminohexane is more basic than the aromatic amine aniline, so it will also be a stronger nucleophile.

2 Under acid conditions, amines will form ammonium ions, NR_3H^+, which will generally be water soluble. In order to isolate the products of the reactions in Equations 28.7, 28.8 and 28.9 from the aqueous reaction mixture, it is necessary to ensure that the pH is such that they will be present as the amines, NR_3, which will be insoluble in water.

3 (a) Four. Phenylalanine, tryptophan, tyrosine and histidine. Although histidine does not have a benzene ring, it meets the condition outlined in Section 27.3 of having $4n + 2$ π-electrons (the two double bonds and the lone pair on nitrogen) (b) Two. Cysteine and methionine. (c) Four. Alanine, valine, isoleucine, leucine.

4 No. In Gly-Ala, glycine has a free amino group and alanine has a free carboxylate group, while in Ala-Gly, alanine has a free amine group and glycine has a free carboxylate group.

5 Intramolecular hydrogen bonding.

6 The individual strands contain amide bonds that undergo intermolecular hydrogen bonding to form a sheet-like structure. Nylon-6,6 is a polymer made up of two monomer types, a diacid and a diamine. A β-sheet typically comprises a permutation of the 20 naturally occurring amino acids.

7 (a) Active site, (b) substrates.

8 Ala-Ile-Val-Tyr-His-Glu.

9 Yes, it can occur as a right-handed or left-handed helix. These two forms are non-superimposable mirror images, that is, enantiomers.

10 5' to 3'.

CHAPTER 30

1 High molar extinction coefficients mean that the material can absorb more energy per mole. The difference in molar mass between porphyrins and benzene is far less than a factor of a thousand, so porphyrins can also absorb much more energy per gram of compound. An organism that gets the highest energy absorption for the amount of material synthesised will be making the most efficient use of the resources available to it and be at a competitive advantage to other organisms.

2 C=O requires more energy. This is consistent with the greater strength of the C=O bond and our experience on a much larger scale that bending or stretching two springs at once is more difficult than bending or stretching one.

3 The main answer is that both KBr and NaCl are transparent in the region 4000–550 cm^{-1}. As solids that are abundant, KBr and NaCl are easily pressed into pellets and polished plates, respectively.

4 ^{15}N (I = $\frac{1}{2}$) is NMR active. This is because it has a non-zero magnetic moment (μ).

5 For ^1H NMR spectroscopy, any deuterated compound that is a liquid at room temperature would be suitable. Any solvent that does not contain hydrogens would also be suitable. For ^{13}C NMR spectroscopy, any solvent that doesn't contain carbon or only a small number of inequivalent carbons would be suitable. Solvents such as CS_2 would be useful for both ^1H and ^{13}C NMR spectroscopy as it contains no hydrogens and only one type of carbon.

6 Upfield (lower ppm).

7 Coupled: 10 ppm, quartet; 20 ppm, triplet; 60 ppm, triplet. Decoupled: singlets at 10 ppm, 20 ppm and 60 ppm.

8 C–I is the weakest and most easily fragmented bond.

9 In Thomson's cathode-ray experiments and in mass spectrometry, a stream of charged particles is passed through the poles of a magnet. The charged particles are deflected by the magnetic field according to their mass and charge.

10 Analysis of the white powder by high-resolution mass spectrometry will be evidence enough, since the value obtained for the molecular ion can only correspond to a single molecular formula. A low-resolution mass spectrum will provide proof, but not sufficient to determine the identity of the compound unambiguously. The same fragmentation patterns will also yield evidence.

ANSWERS TO FIGURE IT OUT

CHAPTER 1

Figure 1.1 There are nine carbon atoms in aspirin.

Figure 1.3 Molecules of a compound are composed of more than one type of atom whereas molecules of an element are composed of only one type of atom.

Figure 1.10 True.

Figure 1.11 1000.

Figure 1.16 The darts would be scattered (poor precision) but their average position would be at the centre (good accuracy).

CHAPTER 2

Figure 2.3 We know the rays travel from the cathode because of the way the beam is deflected by the magnet as shown in Figure 2.3(c).

Figure 2.6 The beta rays, whose path is diverted away from the negative plate towards the positive plate, consist of electrons. Because the electrons are much less massive than the alpha particles, their motion is affected more by the electric field.

Figure 2.7 The beam consists of alpha particles that carry a +2 charge.

Figure 2.10 Based on the periodic trend, we would expect that elements that precede a non-reactive gas, as F does, will also be reactive non-metals. The elements fitting this pattern are H and Cl.

Figure 2.15 The ball-and-stick model shows more clearly the connections between atoms, so we can see the angles at which the atoms are attached in the molecule.

Figure 2.20 Removing one O atom from the perbromate ion gives the bromate ion BrO_3^-.

CHAPTER 3

Figure 3.3 The formula CO_2 represents one molecule containing one C and 2 O atoms, whereas 2 CO represents two molecules, each containing one C and one O atom for a total of two C and two O atoms.

Figure 3.8 Both figures show combustion reactions in which the fuel is a hydrocarbon (CH_4 in Figure 3.4 and C_3H_8 in Figure 3.8). In both cases the reactants are the hydrocarbon and O_2, and the products are CO_2 and H_2O.

Figure 3.9 As shown, 18 g H_2O = 1 mol H_2O = 6.02×10^{23} molecules H_2O. Thus 9.00 g H_2O = 0.500 mol H_2O = 3.01×10^{23} molecules H_2O.

Figure 3.11 (a) The molar mass of CH_4, 16.0 g mol^{-1}. (b) Avogadro's Number, since 6.02×10^{23} molecules of CH_4 form 1 mol of CH_4.

Figure 3.12 The mole ratio is obtained by dividing the larger number of moles of any element in the compound by the smallest number of moles of an element.

Figure 3.16 There are 7 mol O_2 and each mol O_2 yields 2 mol H_2O. Thus 14 mol H_2O would have formed.

CHAPTER 4

Figure 4.3 NaCl (aq), because it dissociates into Na^+(aq) ions and Cl^-(aq) ions in solution. CH_3OH is a molecular compound and does not dissociate in water.

Figure 4.4 K^+ and NO_3^-.

Figure 4.9 Two moles of hydrochloric acid are required to react with each mole of $Mg(OH)_2$.

Figure 4.20 The volume needed to reach the end point if $Ba(OH)_2$(aq) were used would be half that if NaOH(aq) were used.

CHAPTER 5

Figure 5.1 From Figure 5.1 we see that the belt of stability for a nucleus containing 70 protons lies at approximately 102 neutrons.

Figure 5.5 6.25 g. After one half-life, the amount of the radioactive material will have dropped to 25.0 g. After two half-lives, it will have dropped to 12.5 g. After three half lives, it will have dropped to 6.25 g.

Figure 5.16 Because large quantities of water are needed to condense the secondary coolant once it passes through the turbine.

Figure 5.19 Alpha rays are less dangerous when they are outside the body because they cannot penetrate the skin. However, once inside the body they can do great harm to any cells they come in contact with.

CHAPTER 6

Figure 6.3 The wavelength of (a) is twice that of (b) and the frequency of (a) is consequently half that of (b). Thus the wavelength of (b) is 0.50 m and its frequency is 6.0×10^8 cycles s^{-1}. The wavelength and frequency of (c) are the same as that of (b) but its amplitude is lower.

Figure 6.4 The X-ray has a shorter frequency and consequently a longer wavelength than red light.

Figure 6.11 The $n = 2$ to $n = 1$ transition involves a larger energy change than the $n = 3$ to $n = 2$ transition. (Compare the space differences between the states in the figure). If the $n = 2$ to $n = 1$ transition produces visible light, the $n = 3$ to $n = 2$ transition produces visible light. The infrared radiation has lower frequency and hence lower energy than visible light, whereas the ultraviolet has greater frequency and higher energy. Thus the $n = 2$ to $n = 1$ transition is most likely to produce ultraviolet radiation.

Figure 6.15 The region of the highest electron density is where the density of dots is highest, which is near the nucleus.

Figure 6.16 The fourth shell ($n = 4$) would contain four subshells labelled $4s$, $4p$, $4d$ and $4f$.

Figure 6.17 There would be four maxima and three nodes.

Figure 6.19 (a) The intensity of the colour indicates that the probability of finding the electron is greater at the interior of the lobes than on the edges (b) $2p_x$.

Figure 6.23 The $4d$ and $4f$ subshells are not shown.

CHAPTER 7

Figure 7.6 Bottom and left.

Figure 7.7 They get larger just like the atoms do.

Figure 7.9 Ar, as it has a larger Z_{eff}.

Figure 7.10 The halogens (Group 17). It does make sense because we know that they are very stable as ions.

Figure 7.11 Ionisation energy—lower ionisation energy is correlated with increasing metallic character.

Figure 7.14 Anions are above the line, cations are below the line.

CHAPTER 8

Figure 8.2 Yes, the same sort of reaction should occur between any of the alkali metals and any of the elemental halogens.

Figure 8.3 Cations have a smaller radius than their neutral atoms and anions have a larger radius. Because Na and Cl are in the same period of the periodic table, we would expect Na^+ to have a smaller radius than Cl^-, so we would guess that the larger green spheres represent Cl^-.

Figure 8.4 The repulsion between the nuclei would decrease, the attractions between the nuclei and the electrons would decrease, and the repulsions between the electrons would be unaffected.

Figure 8.8 The change in electronegativity difference is the most important in this series.

Figure 8.10 The lengths of the bonds of the outer O–O atoms to the inner O atom are the same.

Figure 8.11 Yes. The electron densities on the left and right parts of the molecule are the same, indicating that resonance has made the two O–O bonds equivalent to one another.

Figure 8.12 Exothermic.

CHAPTER 9

Figure 9.1 The atomic radii.

Figure 9.3 Octahedral.

Figure 9.7 The electron pair in the bonding domain is attracted to two nuclear centres, whereas the non-bonding pair is attracted towards just one.

Figure 9.10 The non-bonding electron pairs exert a greater repulsive force than the bonding electron pairs.

Figure 9.11 The heads of the arrows point toward regions of highest electron density, as indicated by the red colour.

Figure 9.15 As the internuclear distance decreases, nucleus–nucleus repulsion becomes a dominant component of the potential energy.

Figure 9.17 The small lobes of the sp hybrid orbitals are very much smaller in spatial extent and therefore provide very little overlap with the F orbitals.

Figure 9.23 The σ_{1s}^*.

Figure 9.24 The two electrons in the σ_{1s} MO.

Figure 9.25 The $1s$ orbitals of Li are small in spatial extent because they experience a strong nuclear attraction. In addition both the bonding and antibonding MOs formed from them are occupied so there is no significant net bonding.

Figure 9.26 Nodal planes between the atoms are found in anti-bonding MOs.

Figure 9.32 The σ_{2p} and the π_{2p} orbitals. Because the σ_{2p} orbital mixes with the σ_{2s} it is pushed to higher energy and the σ_{2s} is moved to lower energy. The σ_{2p} orbital thus rises above the π_{2p} in energy.

Figure 9.35 Because N_2 has no unpaired electrons it is diamagnetic. Therefore it would simply flow down with no tendency to remain in the magnetic field.

Figure 9.36 All the electrons in the $n = 2$ level are valence-shell electrons.

CHAPTER 10

Figure 10.2 The height of the mercury column will increase.

Figure 10.6 It would be linear.

Figure 10.10 One mole.

Figure 10.12 Helium is small and inert.

Figure 10.16 About a third.

Figure 10.17 The smallest molecules are the fastest at a particular temperature.

Figure 10.19 n, Moles of a gas.

Figure 10.21 Not really. CO_2 is least ideal and does have the largest molar mass, but H_2 the lightest gas, deviates from the ideal line more than the heavier N_2.

Figure 10.22 True.

Figure 10.24 The pressure would increase.

CHAPTER 11

Figure 11.1 The density in a liquid is much closer to a solid than it is to a gas.

Figure 11.8 Both compounds are non-polar and incapable of forming hydrogen bonds. Therefore the boiling point is determined by the dispersion forces, which are stronger for the larger heavier SnH_4.

Figure 11.9 The non-hydrogen atom must possess a non-bonding electron pair.

Figure 11.10 There are four electron pairs surrounding oxygen in a water molecule. Two of the electron pairs are used to make covalent bonds to hydrogen within the H_2O molecule, whereas the other two are available to make hydrogen bonds to neighbouring molecules. Because the electron-pair geometry is tetrahedral (four electron domains around the central atom), the H–O–H bond angle is approximately 109°.

Figure 11.18 Wax is a hydrocarbon that cannot form hydrogen bonds. Therefore coating the inside of tube with wax will dramatically decrease the adhesive forces between water and the tube and change the shape of the water meniscus to an inverted U-shape. Neither wax nor glass can form metallic bonds with mercury so the shape of the mercury meniscus will be qualitatively the same, an inverted U-shape.

Figure 11.20 Because we are dealing with a state function, the energy of going straight from a solid to a gas must be the same as going from a solid to a gas through an intermediate liquid state. Therefore, the heat of sublimation must be equal to the sum of the heat of fusion and the heat of vaporisation: $\Delta_{sub}H = \Delta_{fus}H + \Delta_{vap}H$.

Figure 11.23 Increases, because the molecules have more kinetic energy as the temperature increases and can escape more easily.

Figure 11.24 All liquids including ethylene glycol reach their normal boiling points when their vapour pressure is equal to atmospheric pressure, that is, 101.3 kPa.

Figure 11.26 Freezing, because for most substances the solid phase is denser than the liquid phase and increasing the pressure will eventually drive a phase transition from the liquid to the solid state (provided the temperature is below the critical temperature).

Figure 11.47 The intermolecular forces are stronger in toluene as shown by its higher boiling point. The molecules pack more efficiently in benzene which explains its higher melting point, even though the intermolecular forces are weaker.

CHAPTER 12

Figure 12.1 The negative end of the water dipole (the oxygen is attracted to the positive Na^+ ions, whereas the positive end of the dipole (the hydrogen) is attracted to the negative Cl^- ions.

Figure 12.7 The dissolving of the crystal and the crystallisation by which ions become reattached to the solid. At equilibrium, these rates are equal.

Figure 12.8 When a crystal is added to it, it does not dissolve but seeds crystallisation of the excess sodium acetate.

Figure 12.13 If the partial pressure of a gas over a solution is doubled, the concentration of gas in the solution would double.

Figure 12.14 The slopes increase as the molecular weight increases. The larger the molecular mass, the larger the polarisability of the molecules, leading to greater intermolecular attractive forces between gas molecules and water molecules.

Figure 12.17 From the graphs we see that the solubility of KCl is about 51 g per 100 g H_2O, whereas NaCl has a solubility of about 39 g per 100 g H_2O. So KCl is more soluble than NaCl at this temperature.

Figure 12.18 N_2 has the same molecular weight as CO but is non-polar, so we can predict that its curve will be just below that of CO.

Figure 12.24 The water will move through the semi-permeable membrane towards the more concentrated solution. Thus the liquid level in the left arm will increase.

Figure 12.25 Water will move towards the more concentrated solute solution that is inside the red blood cells, causing them to undergo haemolysis.

Figure 12.28 The two negatively charged groups both have the composition $-COO^-$.

CHAPTER 13

Figure 13.1 About 85 km.

Figure 13.2 The atmosphere absorbs a significant fraction of the solar radiation.

Figure 13.3 The peak value is about 5×10^{12} molecules cm^{-3}. If we use Avogadro's number to convert molecules to moles, and the conversion factor 1000 cm^3 = 1 dm^3, we find that the concentration of ozone at the peak is 8×10^{-9} mol dm^{-3}.

Figure 13.13 This is ambiguous; both temperature and salinity vary with density in similar ways, but temperature seems to parallel density better. Temperature decreases down to 1000 m then remains relatively constant; density increases down to 1000 m and then remains relatively constant.

Figure 13.14 The depth of the aquifer and the nature of the intervening layers (how porous or dense they are).

Figure 13.16 Water is crossing the membrane from a more concentrated solution to a more dilute solution.

CHAPTER 14

Figure 14.3 Yes, the system is still closed since matter cannot escape the system to the surroundings unless the piston is pulled completely out of the cylinder.

Figure 14.4 If U_{final} equals $U_{initial}$ then $\Delta U = 0$.

Figure 14.5

Figure 14.6 No, the sign on w is positive and the sign on q is negative. We need to know the magnitudes of q and w to determine whether $\Delta U = q + w$ is positive or negative.

Figure 14.9 The battery is doing work on the surroundings, so $W = < 0$.

Figure 14.10 We need to know whether Zn(s) or HCl(aq) is the limiting reagent of the reaction. If it is Zn(s), then the addition of more Zn will lead to the generation of more $H_2(g)$ and more work will be done.

Figure 14.15 Two cups provide more thermal insulation so less heat will escape the system.

Figure 14.16 The stirrer ensures that all the water in the bomb is at the same temperature.

Figure 14.17 The condensation of 2 H_2O(g) to 2 H_2O(l).

Figure 14.19 Because the final volume would be less than twice the volume of flask A, the final pressure would be greater than 0.5 bar.

Figure 14.21 Freezing of water into ice is exothermic.

Figure 14.22 δT must be infinitesimally small.

Figure 14.27 Ice, because it is the phase in which the molecules are held most rigidly.

Figure 14.29 The decrease in the number of molecules due to the formation of bonds.

Figure 14.30 During a phase change, the temperature remains constant but the entropy change can be large.

Figure 14.31 Based on the three molecules shown, the addition of each carbon increases $S°$ by 40–45 J mol^{-1} K^{-1}. We would therefore predict that $S°$ for C_4H_{10} would be 310–315 J mol^{-1} K^{-1}. Appendix C gives $S°$ for C_4H_{10} as 310 J mol^{-1} K^{-1}.

Figure 14.33 If we plot progress of the reaction versus free energy, equilibrium is at a minimum point in free energy, as shown in the figure. In that sense, the reaction runs 'downhill' until it reaches the minimum point.

CHAPTER 15

Figure 15.3 B.

Figure 15.4 It decreases.

Figure 15.7 At early times in the reaction; both graphs look linear close to $t = 0$.

Figure 15.9 The reaction is first order in CH_3NC.

Figure 15.17 The energy needed to overcome the energy barrier (the activation energy) looks about twice as large as the overall energy change for the reaction.

Figure 15.23 For the blue curve: The transition states are at the top of the peaks (2) and the intermediate is in the 'valley' between the two peaks. For the red curve: The top of the peak is the transition state; no intermediates are shown.

CHAPTER 16

Figure 16.1 The colour in the tube stops changing.

Figure 16.2 No, as can be seen from Figure 16.2.

Figure 16.6 The boxes would be approximately the same size.

Figure 16.7 It will be lower; some CO_2 has to react with CaO to make some $CaCO_3$.

Figure 16.9 500 bar and 400 °C.

Figure 16.10 Nitrogen (and some of the added hydrogen) is converted into ammonia.

Figure 16.14 About two to three times faster, based on the graph.

Figure 16.15 About 5×10^{-4}.

CHAPTER 17

Figure 17.1 Hydrogen bonds.

Figure 17.2 $O^{2-}(aq) + H_2O(l) \longrightarrow 2\ OH^-(aq)$.

Figure 17.6 Phenolphthalein change from colourless for pH values < 8, to pink for pH values > 10. A pink colour indicates a pH > 10.

Figure 17.7 Bromothymol blue would be most suitable because it changes pH over a range that brackets pH = 7. Methyl red is not sensitive to pH changes when pH > 6, while phenolphthalein is not sensitive to pH changes when pH < 8, so neither changes colour at pH = 7.

Figure 17.10 Yes. The equilibrium in question is $CH_3COOH \rightleftharpoons H^+ + CH_3COO^-$.

If the percent dissociation remained constant as the acid concentration increased, the concentration of all three species would increase at the same rate. However, because there are two products and only one reactant, the total concentration of products would increase faster than the concentration of reactants. To offset this effect the percent dissociation decreases as the acid concentration increases.

Figure 17.11 The nitrogen atom in hydroxylamine accepts a proton to form NH_3OH^+. As a general rule, non-bonding electron pairs on nitrogen atoms are more basic than non-bonding electron pairs on oxygen atoms.

CHAPTER 18

Figure 18.5 $25.00\ cm^3$. The number of moles of base needed to reach the equivalence point remains the same. Therefore by doubling the concentration of added base, the volume needed to reach the equivalence point is halved.

Figure 18.7 The volume of base needed to reach the equivalence point would not change because this quantity does not depend on the strength of the acid. However, the pH at the equivalence point, which is greater than 7 for a weak acid–strong base titration would decrease to 7 because hydrochloric acid is a strong acid.

Figure 18.9 The pH at the equivalence point increases (becomes more basic) as the acid becomes weaker. The volume of added base needed to reach the equivalence point remains unchanged.

Figure 18.11 Yes, any indicator that changes colour between pH = 3 and pH = 11 could be used for a strong acid–strong base titration. Methyl red changes colour between pH values of approximately 4 and 6.

Figure 18.20 ZnS and CuS would both precipitate on addition of H_2S, preventing the separation of the two ions.

Figure 18.21 Yes, CuS would precipitate in Step 2 on addition of H_2S to an acidic solution while Zn^{2+} remained in solution.

CHAPTER 19

Figure 19.1 (a) The bubbling is caused by the hydrogen gas formed in the reaction. (b) The reaction is exothermic and the heat liberated causes the formation of steam.

Figure 19.2 The permanganate, MnO_4^- is reduced. The oxalate ion is the reducing agent.

Figure 19.3 The blue colour is due to $Cu^{2+}(aq)$. As this ion is reduced, forming $Cu(s)$, the blue colour fades.

Figure 19.4 The Zn is oxidised and thus serves as the anode of the cell.

Figure 19.5 The electrical balance is maintained in two ways. Anions migrate into the half-cell and cations move out.

Figure 19.9 As the cell operates, H^+ is reduced to H_2 in the cathode half-cell. As H^+ is depleted, the positive Na^+ ions are drawn into the half-cell to maintain electrical neutrality.

Figure 19.10 The reduction occurs at the cathode. The substance that is reduced most easily is the one with the larger standard reduction potential, $E°_{red}$.

Figure 19.12 Oxidation is the loss of electrons. An oxidising agent causes another substance to lose electrons by gaining them itself. A strong oxidising agent readily gains electrons, meaning that it is easily reduced.

Figure 19.13 The variable n is the number of moles of electrons transferred in the process.

Figure 19.14 The $Ni^{2+}(aq)$ and the cations in the salt bridge migrate towards the cathode. The $NO_3^-(aq)$ and the anions in the salt bridge migrate towards the anode.

Figure 19.18 The cathode consists of $PbO_2(s)$. Because each oxygen has an oxidation state of –2, lead must have an oxidation state of +4 in this compound.

Figure 19.21 The oxidising agent is $O_2(g)$ from the air.

CHAPTER 20

Figure 20.6 The beaker on the right is warmer.

Figure 20.10 More soluble in CCl_4—the colours are deeper.

Figure 20.11 CF_2.

Figure 20.12 Redox reactions: the halides are being oxidised.

Figure 20.16 No.

Figure 20.18 Based on this structure—yes, it would have a dipole moment. In fact, if you look it up, hydrogen peroxide's dipole moment is larger than that of water.

Figure 20.23 They have been converted into water.

Figure 20.24 Formally they could both be +2. If we consider that the central sulfur is like SO_4^{2-}, however, then the central sulfur would be +6, like SO_4^{2-}, and then the terminal sulfur would be –2.

Figure 20.25 Nitrite.

Figure 20.26 Longer.

Figure 20.30 The N=O double bond in the nitrous acid structure.

Figure 20.32 In P_4O_6 the electron domains about the P atoms are trigonal pyramidal; in P_4O_{10} the electron domains about the P atoms are tetrahedral.

Figure 20.38 The minimum temperature should be the melting point of silicon; the temperature of the heating coil should not be so high that the silicon rod starts to melt outside the zone of the heating coil.

CHAPTER 21

Figure 21.3 Zn (it is colourless).

Figure 21.4 The increase parallels the linear increase in valence electron count.

Figure 21.5 All the electrons would align with the direction of the magnetic field.

Figure 21.9 109.5 degrees for the tetrahedral Zn complex; 90 degrees for the square planar Pt complex.

Figure 21.13 4 for both haem b and chlorophyll a, assuming no other ligands come in to bind.

Figure 21.15 In the same place as oxygen, by displacing it.

Figure 21.19 The cis one.

Figure 21.21 Larger since ammonia displaces water from the Cu^{2+}.

Figure 21.26 The peak would stay in the same position in terms of wavelength, but the absorbance would decrease.

Figure 21.30 $d_{x^2-y^2}$ and d_{z^2}.

Figure 21.31 Convert the wavelength of light, 495 nm, into energy in joules using $E = hc/\lambda$.

Figure 21.32 It would be to the right of the 'yellow' member of the series, but the energy gap between filled and empty d orbitals would be even larger than that of the 'yellow' one.

Figure 21.36 That orbital has lobes that point directly at the ligands.

CHAPTER 22

Figure 22.5 (a) Decrease solubility as H-bonding is eliminated. (b) Increase solubility—like dissolves like.

Figure 22.8 Four.

Figure 22.19 Staggered = chair, eclipsed = boat.

CHAPTER 23

Figure 23.1 No, there are no other ways of positioning three C–C bonds.

Figure 23.14 There is one that appears at the sharp end of the wedge.

CHAPTER 24

Figure 24.5 To achieve maximum overlap of the unhybridised orbitals thereby creating a π-bond.

Figure 24.6 Acetylene, since it has a triple bond.

Figure 24.11 None.

Figure 24.14 Z.

Figure 24.20 Exothermic.

CHAPTER 25

Figure 25.3 Three, one from the hydrogen attached to the oxygen and two involving the lone pairs on the oxygen.

Figure 25.7 Butan-2-ol is chiral. It has one stereocentre and so can exist as R or S stereoisomers.

Figure 25.11 The oxygen atom.

Figure 25.18 S.

Figure 25.19 The distributed alkenes; the 1,1-disubstituted alkene is a constitutional isomer of the 1,2-disubstituted alkenes and the two 1,2-disubstituted alkenes are stereoisomers (geometric isomers) of one another.

CHAPTER 26

Figure 26.1 O=C=O.

Figure 26.2 Propanoic acid, propanone.

Figure 26.5 Being a salt, $NaBH_4$ would not be very soluble in common organic solvents. It is slightly soluble in polar solvents like methanol where ion pairing can be disrupted.

Figure 26.6 Benzaldehyde.

Figure 26.7 Three.

Figure 26.8 Because of a plane of symmetry.

Figure 26.13 Four. There are four hydroxyl groups attached directly to the ring, all of which are in an equatorial position.

CHAPTER 27

Figure 27.3 Lactic acid and citric acid.

Figure 27.5 Non-polar solvents as polar solvents compete for hydrogen bonding.

Figure 27.12 Long-chain fatty acids.

Figure 27.19 They are nucleophiles.

Figure 27.21 It acts to cross-link the polymers.

CHAPTER 28

Figure 28.2 Apart from each having six carbons and both being hydrocarbons, there are no similarities. Their shape, electronic characteristics and reactivity are very different.

Figure 28.5 Vanillin: aldehyde, phenol, ether; styrene: alkene; asprin: carboxylic acid, ester; terephthalic acid: carboxylic acid; BHT: phenol; benzoic acid: carboxylic acid.

Figure 28.10 Two *ortho*, two *meta* and one *para*.

Figure 28.12 Aromaticity brought about by resonance. Heats of hydrogenation can be used to measure this.

Figure 28.13

Figure 28.15 The dotted circle is a representation of the total bonding distribution as a result of resonance.

Figure 28.18 The nitro groups are converted to the very stable N_2 upon reaching the activation energy barrier, thereby releasing heat due to the exothermic nature of the reaction.

CHAPTER 29

Figure 29.2 The number of C–N bonds.

Figure 29.3 The pyridine N is 3°, the aliphatic amine is 1°.

Figure 29.6 The lone pair in cyclohexylamine cannot be redistributed and so is concentrated on the nitrogen.

Figure 29.13 All but glycine.

Figure 29.20 At pH 5.66, tyrosine has no net overall charge and will not move in an electrophoresis experiment. At pH 9, tyrosine will be deprotonated and attracted to the positive electrode.

Figure 29.27 A molecule of specific shape is required to unlock the activity the enzyme, similar to a specific key needed to unlock a lock.

Figure 29.30 Negatively charged.

Figure 29.32 G–C. There are three hydrogen bonds holding this base pair together and there are two hydrogen bonds binding adenine to thymine.

CHAPTER 30

Figure 30.2 Dark-coloured cars absorb electromagnetic radiation within the visible region while light-coloured cars reflect it.

Figure 30.5 Lower.

Figure 30.16 An applied magnetic field increases the energy difference between spin-aligned and spin-not-aligned nuclei. Thermodynamically, most nuclei would like to minimise their energy and so would populate the lower energy state.

Figure 30.22 Deshielded.

Figure 30.23 Top to bottom, left to right: 2, 5, 2, 3, 1, 4, 5 signals.

Figure 30.27 The solvent. There is a small amount of $CHCl_3$ in the $CDCl_3$ solvent.

Figure 30.31 $(CH_3)_2CH–$.

Figure 30.35 The magnetic field only influences charged particles and so it is only the charged particles that are bent towards the detector.

Figure 30.37

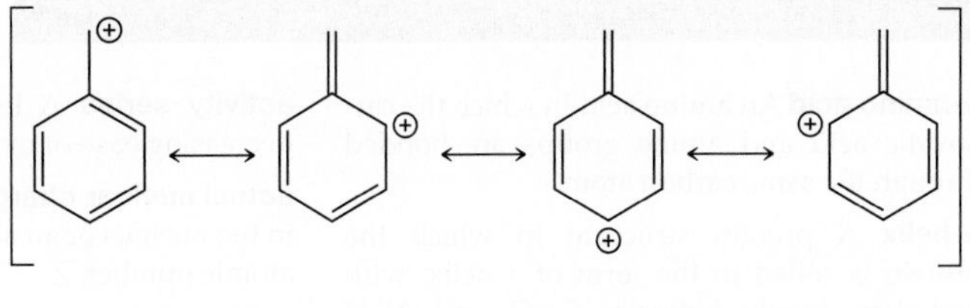

GLOSSARY

α-amino acid An amino acid in which the carboxylic acid and amine groups are bonded through the same carbon atom.

α-helix A protein structure in which the protein is coiled in the form of a helix, with hydrogen bonds between C=O and N–H groups on adjacent turns.

absolute configuration The configuration of a molecule defined unambiguously using the Cahn–Ingold–Prelog rules.

absorption spectrum The amount of light absorbed by a sample as a function of wavelength.

accuracy A measure of how closely individual measurements agree with the correct value.

acetal The product formed by the reaction of an aldehyde or ketone with two moles of an alcohol.

achiral A term used to describe molecules with a plane of symmetry that is superimposable on its mirror image.

acid A substance that is able to donate a H^+ ion (a proton) and hence increase the concentration of $H^+(aq)$ when it dissolves in water.

acid anhydride An organic compound that has two carbonyl groups linked by a single oxygen.

acid-dissociation constant (K_a) An equilibrium constant that expresses the extent to which an acid transfers a proton to solvent water.

acid halide A compound with molecular formula RCOX, where R = alkyl, aryl and X = halogen substituent.

acid rain Rainwater that has become excessively acidic because of absorption of pollutant oxides, notably SO_3 produced by human activities.

acidic anhydride (acidic oxide) An oxide that forms an acid when added to water; soluble non-metal oxides are acidic anhydrides.

acidic oxide See **acidic anhydride**.

actinide element Element in which the $5f$ orbitals are only partially occupied.

activated complex (transition state) The particular arrangement of atoms found at the top of the potential energy barrier as a reaction proceeds from reactants to products.

activating group A substituent that increases the reactivity of an aromatic ring.

activation energy (E_a) The minimum energy needed for reaction; the height of the energy barrier to formation of products.

activator group A substituent that makes the functional group to which it is attached more reactive. Often applied to substituents on aromatic rings undergoing an electrophilic aromatic substitution reaction.

active site Specific site on a heterogeneous catalyst or an enzyme where catalysis occurs.

activity The decay rate of a radioactive material, generally expressed as the number of disintegrations per unit time.

activity series A list of metals in order of decreasing ease of oxidation.

actual nuclear charge The number of protons in the nucleus of an atom. It corresponds to the atomic number, Z.

acyl group The group R—C=O, where R = alkyl, aryl.

acylation reaction A reaction that introduces an acyl group (RCO) onto a molecule.

addition polymer A polymer formed from alkenes by a free-radical process.

addition polymerisation Polymerisation that occurs through coupling of monomers with one another, with no other products formed in the reaction.

addition reaction A reaction in which a reagent adds to the two carbon atoms of a carbon–carbon multiple bond.

adsorption The binding of molecules to a surface.

alcohol An organic compound obtained by substituting a hydroxyl group (−OH) for a hydrogen on a hydrocarbon.

aldehyde An organic compound that contains a carbonyl group to which at least one hydrogen atom is attached.

aldose A carbohydrate that contains an aldehyde group when in its open-chain form.

aliphatic Meaning 'fat-like', refers to a non-aromatic, non-cyclic hydrocarbon such as an alkane.

aliphatic amine An amine in which alkyl chains are bonded to nitrogen.

alkali metals Members of group 1 in the periodic table.

alkaline earth metals Members of group 2 in the periodic table.

alkaloid A large class of naturally occurring compounds with biological activity. Alkaloids are cyclic amines derived from plant material and contain a basic nitrogen atom.

alkanes Compounds of carbon and hydrogen containing only carbon–carbon single bonds.

alkenes Hydrocarbons containing one or more carbon–carbon double bonds.

alkoxide The conjugate base of an alcohol.

alkoxy group A group of the form −OR, where R is generally alkyl or aryl.

alkyl ammonium salt An ionic class of molecule containing four groups around nitrogen, of which one is an alkyl group; for example, R_3NH^+, $R_2NH_2^+$, RNH_3^+, where R = alkyl group.

alkyl group A group that is formed by removing a hydrogen atom from an alkane.

alkyl halide A compound of general formula RX where R = alkyl group and X = halogen substituent.

alkynes Hydrocarbons containing one or more carbon–carbon triple bonds.

alloy A substance that has the characteristic properties of a metal and contains more than one element. Often there is one principal metallic component, with other elements present in smaller amounts. Alloys may be homogeneous or heterogeneous in nature.

alpha (α) decay A radioactive process in which a particle identical to a helium nucleus ($^4_2$He) is emitted from an unstable nucleus.

alpha (α) particles Particles that are identical to helium-4 nuclei, consisting of two protons and two neutrons, symbol $^4_2$He or $^4_2\alpha$.

alpha (α) radiation A stream of helium-4 nuclei represented as $^4_2$He or $^4_2\alpha$.

amide An organic compound that has an NR_2 group attached to a carbonyl.

amine A compound that has the general formula R_3N, where R may be H or a hydrocarbon group.

amino acid A carboxylic acid that contains an amino (–NH_2) group attached to the carbon atom adjacent to the carboxylic acid (−COOH) functional group.

amino group A nitrogen-containing group of the type –NH_2, –NHR or –NR_2.

amorphous solid A solid whose molecular arrangement lacks a regular, long-range pattern.

amphiprotic Refers to the capacity of a substance to either lose or gain a proton; that is, to act as an acid or a base.

amphoteric Capable of behaving as either an acid or a base.

amphoteric oxides and hydroxides Substances which dissolve in both acidic and basic solutions.

amplitude The maximum extent of the oscillation of the wave.

amylopectin A branched derivative of starch, a polysaccharide.

amylose An unbranched derivative of starch, a polysaccharide.

angstrom A common non-SI unit of length, denoted Å, that is used to measure atomic dimensions: $1\,\text{Å} = 10^{-10}\,\text{m}$.

angular momentum quantum number The number (l) that defines the shape of an orbital.

anion A negatively charged ion.

anode An electrode at which oxidation occurs.

anomer Diastereoisomers that are formed about the anomeric carbon upon hemiacetal formation. The two forms are α and β anomers.

anomeric carbon The new stereocentre formed in a monosaccharide as a result of intramolecular hemiacetal formation.

antibonding molecular orbital A molecular orbital in which electron density is concentrated outside the region between the two nuclei of bonded atoms. Such orbitals, designated as σ* or π*, are less stable (of higher energy) than bonding molecular orbitals.

antiferromagnetism A form of magnetism in which unpaired electrons spins in adjacent sites point in opposite directions and cancel each other's effects.

antioxidants Compounds able to terminate the harmful effects of naturally occurring free radicals.

aqueous solution A solution in which water is the solvent.

arenes See **aromatic hydrocarbons**.

aromatic amine An amine in which at least one aryl group is bonded to nitrogen, or the nitrogen forms part of the aromatic unit, as in pyridine.

aromatic hydrocarbons Hydrocarbon compounds that contain a planar, cyclic arrangement of carbon atoms linked by both σ-bonds and delocalised π-bonds.

Arrhenius equation An equation that relates the rate constant for a reaction to the frequency factor, A, the activation energy, E_a and the temperature, T : $k = Ae^{-E_a/RT}$. In its logarithmic form it is written $\ln k = -E_a/RT + \ln A$.

ascorbic acid Vitamin C. Required as a cofactor in the hydroxylation of proline residues in collagen and as an antioxidant.

atmosphere (atm) A non-SI unit of pressure equal to 760 torr; 1 atm = 101.325 kPa.

atom The smallest representative particle of an element.

atomic mass The mass of an atom relative to the isotope of carbon, ^{12}C, which is assigned a mass of 12 exactly.

atomic number The number of protons in the nucleus of an atom of an element.

atomic radius An estimate of the size of an atom. See **bonding atomic radius**.

autoionisation The process whereby water spontaneously forms low concentrations of $H^+(aq)$ and $OH^-(aq)$ ions by proton transfer from one water molecule to another.

average atomic mass The average mass of the isotopes of an element in unified atomic mass units (u).

Avogadro's hypothesis A statement that equal volumes of gases at the same temperature and pressure contain equal numbers of molecules.

Avogadro's law A statement that the volume of a gas maintained at constant temperature and pressure is directly proportional to the number of moles of the gas.

Avogadro's number The number of ^{12}C atoms in exactly 12 g of ^{12}C; it equals 6.022×10^{23}.

axial hydrogens Those hydrogens on a chair conformation that extend perpendicular to the rough plane of the ring.

β-elimination reaction A reaction that produces an unsaturated hydrocarbon by the loss of atoms from adjacent carbons in a molecule.

β-pleated sheet A form of protein secondary structure characterised by an intermolecular

hydrogen bonding arrangement with another peptide chain.

band An array of closely spaced molecular orbitals occupying a discrete range of energy.

band gap The energy gap between an occupied valence band and a vacant band called the conduction band.

bar A unit of pressure equal to 10^5 Pa.

base A substance that is an H^+ acceptor; a base produces an excess of OH^-(aq) ions when it dissolves in water.

base-dissociation constant (K_b) An equilibrium constant that expresses the extent to which a base reacts with solvent water, accepting a proton and forming OH^-(aq).

base peak The peak within the mass spectrum that has 100% relative intensity. This does not necessarily correspond to the molecular ion.

basic anhydride (basic oxide) An oxide that forms a base when added to water; soluble metal oxides are basic anhydrides.

basic oxide See **basic anhydride**.

battery A self-contained electrochemical power source that contains one or more voltaic cells.

Bayer process A hydrometallurgical procedure for purifying bauxite in the recovery of aluminium from bauxite-containing ores.

becquerel (Bq) The SI unit of radioactivity. It corresponds to one nuclear disintegration per second.

Beer–Lambert law The light absorbed by a substance (A) equals the product of its molar absorption coefficient (ε), the path length through which the light passes (l) and the molar concentration of the substance (c): $A = \varepsilon \times c \times l$.

benzyl group A hydrocarbon group with formula $C_6H_5CH_2$.

beta (β) particles Energetic electrons emitted from the nucleus, symbol $^0_{-1}e$ or $^0_{-1}\beta$.

beta (β) radiation A stream of high-energy electrons emitted by an unstable nucleus, represented by $^0_{-1}e$ or $^0_{-1}\beta$. When we consider the decay of a specific nucleus we call it **beta emission**.

bidentate ligand A ligand in which two linked coordinating atoms are bound to a metal.

bimolecular reaction An elementary reaction that involves two molecules.

biochemistry The study of the chemistry of living systems.

biocompatible Describes any substance or material that can be compatibly placed within living systems.

biodegradable Refers to organic material that bacteria are able to oxidise.

biological chemistry A branch of chemistry that applies the principles of chemistry to understand the molecular basis of biological processes.

biomaterial Any material that has a biomedical application.

biopolymer A polymeric molecule of high molecular weight found in living systems. The

three major classes of biopolymers are proteins, carbohydrates and nucleic acids.

biotin Vitamin H, the cofactor required for carboxylase enzymes. Biotin is a water-soluble vitamin found in a variety of foods and is also synthesised by intestinal bacteria.

boat conformation The least stable conformation of a cyclohexane ring.

body-centred cubic cell A cubic unit cell in which the lattice points occur at the corners and the centre.

bomb calorimeter A device for measuring the heat evolved in the combustion of a substance under constant-volume conditions.

bond angles The angles made by the lines joining the nuclei of the atoms in a molecule.

bond dipole The dipole moment that is due to unequal electron sharing between two atoms in a covalent bond.

bond enthalpy The enthalpy change, ΔH, required to break a particular bond when the substance is in the gas phase.

bond length The distance between the centres of two bonded atoms.

bond order The number of bonding electron pairs shared between two atoms; in MO theory, it is (number of bonding electrons − number of antibonding electrons)/2.

bond polarity A measure of the degree to which the electrons are shared unequally between two atoms in a chemical bond.

bonding atomic radius Half the distance between the nuclei of two bonded atoms.

bonding molecular orbital A molecular orbital in which the electron density is concentrated in the internuclear region. The energy of a bonding molecular orbital is lower than the energy of the separate atomic orbitals from which it forms.

bonding pair In a Lewis structure, a pair of electrons that is shared by two atoms.

boranes Covalent hydrides of boron.

Born–Haber cycle A thermodynamic cycle based on Hess's law that relates the lattice energy of an ionic substance to its enthalpy of formation and to other measurable quantities.

Boyle's law A law stating that at constant temperature, the product of the volume and pressure of a given amount of gas is a constant.

branching chain reactions A series of reactions in which one reaction initiates the next.

Brønsted–Lowry acid A substance (molecule or ion) that acts as a proton donor.

Brønsted–Lowry base A substance (molecule or ion) that acts as a proton acceptor.

buffer capacity The amount of acid or base a buffer can neutralise before the pH begins to change appreciably.

buffer solution (buffer) A solution that undergoes a limited change in pH upon addition of a small amount of acid or base.

Cahn–Ingold–Prelog *R,S* notation Use of *R,S* notation to describe the configuration of a stereocentre as defined by a set of sequence priority rules.

calcination The heating of an ore to bring about its decomposition and the elimination of a volatile product. For example, a carbonate ore might be calcined to drive off CO_2.

calorie A non-SI unit of energy, it is the amount of energy needed to raise the temperature of 1 g of water by 1 °C. A related unit is the joule: 1 cal = 4.184 J.

calorimeter An apparatus that measures the evolution of heat.

calorimetry The experimental measurement of heat produced in chemical and physical processes.

capillary action The process by which a liquid rises in a tube because of a combination of adhesion to the walls of the tube and cohesion between liquid particles.

carbanion A species containing a carbon atom that bears a negative charge.

carbide A binary compound of carbon with a metal or metalloid.

carbocation A species containing a carbon atom that bears a positive charge.

carbohydrates A class of substances formed from polyhydroxy aldehydes or ketones, typically with a molecular formula in the form $Cr(H_2O)_m$.

carbon black A microcrystalline form of carbon.

carbonyl group The C–C double bond, a characteristic feature of several organic functional groups, such as ketones and aldehydes.

carboxylate anion The conjugate base of a carboxylic acid.

carboxylic acid A compound that contains the –COOH functional group.

catalyst A substance that changes the speed of a chemical reaction without itself undergoing a permanent chemical change in the process.

catalytic hydrogenation A reaction used to add molecular hydrogen to a molecule in the presence of a catalyst. Most commonly used for the conversion of unsaturated hydrocarbons to saturated hydrocarbons.

cathode An electrode at which reduction occurs.

cathode rays Streams of electrons that are produced when a high voltage is applied to electrodes in an evacuated tube.

cathodic protection A means of protecting a metal against corrosion by making it the cathode in a voltaic cell. This can be achieved by attaching a more easily oxidised metal, which serves as an anode, to the metal to be protected.

cation A positively charged ion.

cell potential A measure of the driving force, or 'electrical pressure', for an electrochemical reaction; it is measured in volts: $1 \text{ V} = 1 \text{ JC}^{-1}$. Also called electromotive force.

cellulose A polysaccharide of glucose; it is the major structural element in plant matter.

Celsius scale A temperature scale on which water freezes at 0 °C and boils at 100 °C at sea level.

ceramic A solid inorganic material, either crystalline (oxides, carbides, silicates) or amorphous (glasses). Most ceramics melt at high temperatures.

chair conformation The most stable conformation of a cyclohexane ring.

changes of state Transformations of matter from one state to a different one, for example from a gas to a liquid.

charcoal A form of carbon produced when wood is heated strongly in a deficiency of air.

Charles's law A law stating that, at constant pressure, the volume of a given quantity of gas is proportional to absolute temperature.

chelate effect The generally larger formation constants for polydentate ligands as compared with the corresponding *monodentate* ligands.

chelating agent A polydentate ligand that is capable of occupying two or more sites in the coordination sphere.

chemical bond A strong attractive force that exists between atoms in a molecule.

chemical changes Processes in which one or more substances are converted into other substances; also called **chemical reactions**.

chemical equation A representation of a chemical reaction using the chemical formulae of the reactants and products; a balanced chemical equation contains equal numbers of atoms of each element on both sides of the equation.

chemical equilibrium A state of dynamic balance in which the rate of formation of the products of a reaction from the reactants equals the rate of formation of the reactants from the products; at equilibrium the concentrations of the reactants and products remain constant.

chemical formula A notation that uses chemical symbols with numerical subscripts to convey the relative proportions of atoms of the different elements in a substance.

chemical kinetics The area of chemistry concerned with the speeds, or rates, at which chemical reactions occur.

chemical nomenclature The rules used in naming substances.

chemical properties Properties that describe a substance's composition and its reactivity; how the substance reacts or changes into other substances.

chemical reactions Processes in which one or more substances are converted into other substances; also called **chemical changes**.

chemical shift (δ) The frequency of a signal, expressed in parts per million (ppm), from tetramethylsilane divided by the operating frequency of the spectrometer.

chemical symbol Chemical abbreviation of the name of an element.

chemistry The scientific discipline that treats the composition, properties and transformations of matter.

chiral A term describing a molecule or an ion that cannot be superimposed on its mirror image.

chiral (stereogenic) centre (stereocentre) A tetrahedral atom, usually carbon, bearing four different substituents.

chlorofluorocarbons (CFCs) Compounds composed entirely of chlorine, fluorine and carbon.

chlorophyll A plant pigment that plays a major role in conversion of solar energy to chemical energy in photosynthesis.

cholesteric liquid crystal A liquid crystal formed from flat, disc-shaped molecules that align through a stacking of the molecular discs.

cis A geometric isomer in which two substituents attached to a cyclic compound are orientated on different sides of the plane of the ring.

coal A naturally occurring solid containing hydrocarbons of high molecular weight, as well as compounds containing sulfur, oxygen and nitrogen.

codon The sets of three successive nucleotides in mRNA that are encoded for amino acids. Used in translation.

coenzyme A compound necessary for the action of an enzyme.

coke An impure form of carbon, formed when coal is heated strongly in the absence of air.

colligative properties Those properties of a solvent (vapour-pressure lowering, freezing-point lowering, boiling-point elevation, osmotic pressure) that depend on the total concentration of solute particles present.

collision model A model of reaction rates based on the idea that molecules must collide to react; it explains the factors influencing reaction rates in terms of the frequency of collisions, the number of collisions with energies exceeding the activation energy, and the probability that the collisions occur with suitable orientations.

colloids (colloidal dispersions) Mixtures containing particles larger than normal solutes but small enough to remain suspended in the dispersing medium.

combination reaction A chemical reaction in which two or more substances combine to form a single product.

combustion A chemical reaction that proceeds with evolution of heat and usually also a flame; most combustion involves reaction with oxygen, as in the burning of a match.

common-ion effect A shift of an equilibrium induced by an ion common to the equilibrium. For example, added Na_2SO_4 decreases the solubility of the slightly soluble salt $BaSO_4$, or added CH_3COONa decreases the percent ionisation of CH_3COOH.

complementary colours Colours that, when mixed in proper proportions, appear white or colourless.

complete ionic equation A chemical equation in which dissolved strong electrolytes (such as dissolved ionic compounds) are written as separate ions.

complex ion (complex) An assembly of a metal ion and the Lewis bases (ligands) bonded to it.

complex lipids Lipids that can be broken down into smaller constituents, usually by hydrolysis.

compound A substance composed of two or more elements united chemically in definite proportions.

concentration The quantity of solute present in a given quantity of solvent or solution.

concentration cell A voltaic cell containing the same electrolyte and the same electrode materials in both the anode and cathode compartments. The emf of the cell is derived from a difference in the concentrations of the same electrolyte solutions in the compartments.

condensation polymerisation Polymerisation in which molecules are joined together through condensation reactions.

condensation reaction A chemical reaction in which a small molecule (such as a molecule of water) is split out from between two reacting molecules.

conduction band A band of molecular orbitals lying higher in energy than the occupied valence band, and distinctly separated from it.

conformational isomers Isomers which are differentiated by their three-dimensional shape at any instance by rotation about a C–C bond.

conjugate acid A substance formed by addition of a proton to a Brønsted–Lowry base.

conjugate acid–base pair An acid and a base, such as H_2O and OH^-, that differ only in the presence or absence of a proton.

conjugate base A substance formed by the loss of a proton from a Brønsted–Lowry acid.

conjugated protein A protein that yields compounds, such as carbohydrates, lipids, metal complexes and so on, as well as amino acids on hydrolysis.

constitutional isomers See **structural isomers**.

continuous spectrum A spectrum that contains radiation distributed over all wavelengths.

conversion factor A ratio relating the same quantity in two systems of units that is used to convert the units of measurement.

coordination compound (complex) A compound containing a metal ion bonded to a group of surrounding molecules or ions that act as ligands.

coordination number The number of adjacent atoms to which an atom is directly bonded. In a complex, the coordination number of the metal ion is the number of donor atoms to which it is bonded.

coordination sphere The metal ion and its surrounding ligands.

coordination-sphere isomers Structural isomers of coordination compounds in which the ligands within the coordination sphere differ but whose molecular formula is the same.

copolymer A complex polymer resulting from the polymerisation of two or more chemically different monomers.

core electrons The electrons that are not in the outermost shell of an atom.

corrosion The process by which a metal is oxidised by substances in its environment.

counterion exchange The process by which one ion within a salt is replaced with another.

coupling constant A measure of the spin–spin interactions between two coupled nuclei.

covalent bond A bond formed between two or more atoms by a sharing of electrons.

covalent-network solids Solids in which the units that make up the three-dimensional network are joined by covalent bonds.

critical mass The amount of fissionable material necessary to maintain a chain reaction.

critical pressure The pressure at which a gas at its critical temperature is converted to a liquid state.

critical temperature The highest temperature at which it is possible to convert the gaseous form of a substance to a liquid. The critical temperature increases with an increase in the magnitude of intermolecular forces.

cross-linking The formation of bonds between polymer chains.

crown ethers Cyclic compounds containing three or more ether linkages.

crystal-field theory A theory that accounts for the colours and the magnetic and other properties of transition metal complexes in terms of the splitting of the energies of metal ion d orbitals by the electrostatic interaction with the ligands.

crystal lattice An imaginary network of points on which the repeating unit of the structure of a solid (the contents of the unit cell) may be imagined to be laid down so that the structure of the crystal is obtained. Each point represents an identical environment in the crystal.

crystalline solid (crystal) A solid whose internal arrangement of atoms, molecules or ions shows a regular repetition in any direction through the solid.

crystallinity A measure of the extent of crystalline character (order) in a polymer.

crystallisation The process in which a dissolved solute comes out of solution and forms a crystalline solid.

cubic close packing A close-packing arrangement in which the atoms of the third layer of a solid are not directly over those in the first layer.

curie A non-SI measure of radioactivity: 1 curie = 3.7×10^{10} nuclear disintegrations per second.

curved arrow notation A way of accounting for the movement of electrons within a molecule or group of molecules. Double-headed arrows signify the movement of two electrons, while fishhook arrows indicate the movement of one electron.

cyanocobalamine Vitamin B_{12}. This vitamin is unusual as it is the only metal-ion-containing vitamin. Vitamin B_{12} helps protect nerves and is involved in the formation of red blood cells in bone marrow.

cyanohydrins Compounds containing both a hydroxy (OH) and nitrile (CN) group typically on the same carbon atom.

cyclic amine A cycloalkane in which at least one of the carbons within the ring is replaced by nitrogen. A heterocyclic compound.

cyclic polyether A cyclic compound containing more than one ether group within the ring.

cycloalkanes Saturated hydrocarbons of general formula C_nH_{2n} in which the carbon atoms form a closed ring.

d-d transition The transition of an electron from a lower-energy d orbital to a higher-energy d orbital.

Dalton's law of partial pressures A law stating that the total pressure of a mixture of gases is the sum of the pressures that each gas would exert if it were present alone.

deactivating group A substituent that makes the functional group to which it is attached less reactive. Often applied to substituents on aromatic rings undergoing an electrophilic aromatic substitution reaction.

decomposition reaction A chemical reaction in which a single compound reacts to give two or more products.

decoupling An experiment where information on coupling between two nuclei is lost by saturation.

degenerate Describes a situation in which two or more orbitals have the same energy.

dehydration A reaction in which water is lost from a molecule. Most commonly used to describe the formation of alkenes from alcohols in which H and OH are removed from adjacent atoms.

delocalised electrons Electrons that are spread over a number of atoms in a molecule rather than localised between a pair of atoms.

density The ratio of an object's mass to its volume.

deoxyribonucleic acid (DNA) A polynucleotide in which the sugar component is deoxyribose.

desalination The removal of salts from seawater, brine or brackish water to make it fit for human consumption.

deshield See **shielding/deshielding**.

deuterium The isotope of hydrogen whose nucleus contains a proton and a neutron: ^2_1H.

dextrorotatory, or **dextro** or **d**. A term used to label a chiral molecule that rotates the plane of polarisation of plane-polarised light to the right (clockwise).

diamagnetism A type of magnetism that causes a substance with no unpaired electrons to be weakly repelled from a magnetic field.

diastereoisomerism The existence of configurational stereoisomers that are not enantiomers.

diastereomers Stereoisomers that have a non-mirror-image relationship.

diatomic molecule A molecule composed of only two atoms.

1,3-diaxial interaction Interactions between two groups in an axial position on the same side of a chair conformation of a cyclohexane ring separated by three carbon atoms.

diffusion The spreading of one substance through a space occupied by one or more other substances.

dilution The process of preparing a less concentrated solution from a more concentrated one by adding solvent.

diol An organic molecule characterised by two OH functional groups.

dipole A molecule with one end having a slight negative charge and the other end having a slight positive charge; a polar molecule.

dipole–dipole force The force that exists because of the interactions of dipoles on polar molecules in close contact.

dipole moment A measure of the separation and magnitude of the positive and negative charges in polar molecules.

dispersion forces Attractive interaction between atoms.

displacement reaction A reaction in which an element reacts with a compound, displacing an element from it.

disproportionation A reaction in which a species undergoes simultaneous oxidation and reduction (as in $N_2O_3(g) \rightarrow NO(g) + NO_2(g)$).

donor atom The atom of a ligand that bonds to the metal.

doping Incorporation of a heteroatom into a solid to change its electrical properties; for example, incorporation of P into Si.

double bond A covalent bond involving two electron pairs.

double helix The structure for DNA that involves the winding of two DNA polynucleotide chains together in a helical arrangement. The two strands of the double helix are complementary in that the organic bases on the two strands are paired for optimal hydrogen bond interaction.

Downs cell A cell used to obtain sodium metal by electrolysis of molten NaCl.

dynamic equilibrium A state of balance in which opposing processes occur at the same rate.

E1 An elimination reaction displaying first-order kinetics.

E2 An elimination reaction displaying second-order kinetics.

eclipsed A conformation in which all groups around a C–C single bond are as close as possible. This arrangement results in the bonds about a C–C single bond being behind each other when viewed as a Newman projection.

Edman degradation A systematic process designed to cleave and identify one amino acid at a time from the N-terminus of the peptide chain.

effective nuclear charge The net positive charge experienced by an electron in a many-electron atom; this charge is not the full nuclear charge because there is some shielding of the nucleus by the other electrons in the atom.

effusion The escape of a gas through a tiny hole.

elastomer A material that can undergo a substantial change in shape via stretching, bending or compression and return to its original shape upon release of the distorting force.

electrochemistry The branch of chemistry that deals with the relationships between electricity and chemical reactions.

electrolysis reaction A reaction in which a non-spontaneous redox reaction is brought about by the passage of current under a sufficient external electrical potential. The devices in which electrolysis reactions occur are called electrolytic cells.

electrolyte A solute that produces ions in solution; an electrolytic solution conducts an electric current.

electrolytic cell A device in which a non-spontaneous oxidation–reduction reaction is caused to occur by passage of current under a sufficient external electrical potential.

electromagnetic radiation (radiant energy) A form of energy that has wave characteristics and propagates through a vacuum at the characteristic speed of 3.00×10^8 m s^{-1}.

electrometallurgy The use of electrolysis to reduce or refine metals.

electromotive force (emf) A measure of the driving force, or *electrical pressure*, for the completion of an electrochemical reaction. Electromotive force is measured in volts: $1\,V = 1\,JC^{-1}$. Also called the **cell potential**.

electron A negatively charged subatomic particle found outside the atomic nucleus; it is a part of all atoms. An electron has a mass 1/1836 times that of a proton.

electron affinity The energy change that occurs when an electron is added to a gaseous atom or ion.

electron capture A mode of radioactive decay in which an inner-shell orbital electron is captured by the nucleus.

electron configuration A particular arrangement of electrons in the orbitals of an atom.

electron density The probability of finding an electron at any particular point in an atom; this probability is equal to ψ^2, the square of the wave function.

electron domain In the VSEPR model, regions about a central atom in which electrons are concentrated.

electron-domain geometry The three-dimensional arrangement of the electron domains around an atom, according to the VSEPR model.

electron-donating group A functional group that has a positive inductive effect.

electron impact ionisation A mass spectrometry method used to produce positively charged ions by bombarding a sample with high-energy electrons.

electron impact ionisation mass spectrometry (EIMS) A popular technique used to determine the molecular weight of a compound through bombardment with high-energy electrons.

electron-sea model A model for the behaviour of electrons in metals.

electron shell A collection of orbitals that have the same value of n. For example, the orbitals with $n = 3$ (the 3s, 3p and 3d orbitals) comprise the third shell.

electron spin A property of the electron that makes it behave as though it were a tiny magnet. The electron behaves as if it were spinning on its axis; electron spin is quantised.

electron subshell Set of one or more orbitals with the same n and l values.

electron-withdrawing group A group that has an attraction for electrons that it shares in a chemical bond with other atoms.

electronegativity A measure of the ability of an atom that is bonded to another atom to attract electrons to itself.

electronic charge The negative charge carried by an electron; it has a magnitude of 1.602×10^{-19} C.

electronic structure The arrangement of electrons of an atom or molecule.

electrophile Any atom, molecule or ion that accepts a pair of electrons from a nucleophile to form a new covalent bond.

electrophilic aromatic substitution (EAS) The mechanism by which benzene and its derivatives react to form more highly substituted compounds.

electrophoresis A method for separating compounds based on their electric charge at a given pH.

element A substance that cannot be separated into simpler substances by chemical means.

elementary reaction A process in a chemical reaction that occurs in a single event or step. An overall chemical reaction consists of one or more elementary reactions or steps.

elimination reaction A reaction that proceeds by the loss of atoms or groups of atoms, resulting in the formation of multiple bonds.

empirical formula (simplest formula) A chemical formula that shows the kinds of atoms and their relative numbers in a substance in the smallest possible whole-number ratios.

enantiomerically pure A term used to describe a sample containing 100% of a single enantiomer.

enantiomers Two mirror-image molecules of a chiral substance. The enantiomers are non-superimposable.

endothermic process A process in which a system absorbs heat from its surroundings.

energy The capacity to do work or to transfer heat.

energy-level diagram A diagram that shows the energies of molecular orbitals relative to the atomic orbitals from which they are derived. Also called a **molecular orbital diagram**.

enthalpy A quantity defined by the relationship $H = E + PV$; the enthalpy change, ΔH, for a reaction that occurs at constant pressure is the heat evolved or absorbed in the reaction: $\Delta H = q_P$.

enthalpy (heat) of formation The enthalpy change that accompanies the formation of a substance from the most stable forms of its component elements.

enthalpy of reaction The enthalpy change associated with a chemical reaction.

entropy A thermodynamic function associated with the number of different equivalent energy states or spatial arrangements in which a system may be found. It is a thermodynamic state function, which means that once we specify the conditions for a system—that is, the temperature, pressure and so on—the entropy is defined.

enzyme A protein molecule that acts to catalyse specific biochemical reactions.

equatorial hydrogens Those hydrogens on a chair conformation that radiate out in the rough plane of the ring.

equilibrium constant The numerical value of the equilibrium constant expression for a system at equilibrium. The equilibrium constant is most usually denoted by K_P for gas-phase systems or K_c for solution-phase systems.

equilibrium constant expression The expression that describes the relationship between the concentrations (or partial pressures) of the substances present in a system at equilibrium. The numerator is obtained by multiplying the concentrations of the substances on the product side of the equation, each raised to a power equal to its coefficient in the chemical equation. The denominator similarly contains the concentrations of the substances on the reactant side of the equation.

equivalence point The point in a titration at which the added solute reacts completely with the solute present in the solution.

ester An organic compound that has an –OR group attached to a carbonyl; it is the product of a reaction between a carboxylic acid and an alcohol.

ether A compound in which two hydrocarbon groups are bonded to one oxygen.

exchange (metathesis) reaction A reaction between compounds that, when written as a molecular equation, appears to involve the exchange of ions between the two reactants.

excited state A higher energy state than the ground state.

exothermic process A process in which a system releases heat to its surroundings.

extensive property A property that depends on the amount of material considered, for example mass or volume.

E, Z system A system used to distinguish between geometrical isomers in alkenes. The term E for *entgegen* means opposite. The term Z for *zusammen* means together (or same side).

f-block metals Lanthanide and actinide elements in which the $4f$ or $5f$ orbitals are partially occupied.

face-centred cubic cell A cubic unit cell that has lattice points at each corner as well as at the centre of each face.

faraday A unit of charge that equals the total charge of one mole of electrons: $1\,F = 96\,500\,C$.

Faraday's constant Quantity of electrical charge on one mole of electrons.

fats Triglycerides that are solids or semi-solids at room temperature. They are usually rich in saturated fatty acids.

fatty acids Carboxylic acids with > C12 alkyl chains.

ferrimagnetism A form of magnetism in which unpaired electron spins on different-type ions point in opposite directions but do not fully cancel out.

ferromagnetism The ability of some substances to become permanently magnetised through alignment of unpaired electron spins.

fibrous proteins Proteins that are arranged like filaments or fibres, for example collagen.

fingerprint region Usually defined as the most complex region of the IR spectrum $(300–1400\ cm^{-1})$.

first law of thermodynamics A statement of our experience that energy is conserved in any process. We can express the law in many ways. One of the more useful expressions is that the change in internal energy, ΔU, of a system in any process is equal to the heat, q, added to the system, plus the work, w, done on the system by its surroundings: $\Delta U = q + w$.

first-order reaction A reaction in which the reaction rate is proportional to the concentration of a single reactant, raised to the first power.

Fischer esterification The acid-catalysed condensation of an alcohol and carboxylic acid.

Fischer projection A convention for representing the configuration of a stereocentre in two dimensions.

fission The splitting of a large nucleus into two smaller ones.

folding The process by which a protein adopts the biologically active form.

folic acid Vitamin M. A conjugated molecule consisting of a pteridine ring structure linked through *p*-aminobenzoic acid to glutamic acid. Folic acid is obtained primarily from leafy vegetables, such as spinach and silverbeet, and acts as a cofactor for methyltransferases, enzymes required in the biosynthesis of serine, methionine, glycine and the purine nucleotides.

force A push or a pull.

formal charge The number of valence electrons in an isolated atom minus the number of electrons assigned to the atom in the Lewis structure.

formation constant For a metal ion complex, the equilibrium constant for formation of the complex from the metal ion and base species present in solution. It is a measure of the tendency of the complex to form.

formula mass Of a substance, it is the sum of the atomic masses of each atom in its chemical formula.

formula weight The mass of the collection of atoms represented by a chemical formula. For example, the formula weight of NO_2 (46.0 u) is the sum of the masses of one nitrogen atom and two oxygen atoms.

fossil fuels Coal, oil and natural gas, which are currently our major sources of energy.

fragment A chemical species produced by the cleavage of a covalent bond. Usually associated with mass spectrometry.

fragmentation The process in mass spectrometry whereby the molecular ion dissociates into smaller ions.

free energy See **Gibbs free energy (G)**.

free radical A substance with one or more unpaired electrons.

frequency The number of times per second that one complete wavelength passes a given point.

frequency factor (A) A term in the Arrhenius equation that is related to the frequency of collision and the probability that the collisions are favourably oriented for reaction.

Friedel–Crafts alkylation A reaction that introduces a hydrocarbon substituent onto an aromatic ring.

fuel cell A voltaic cell that utilises the oxidation of a conventional fuel, such as H_2 or CH_4, in the cell reaction.

fuel value The energy released when 1 g of a substance is combusted.

functional group An atom or group of atoms that imparts characteristic chemical properties to an organic compound.

furanose Five-membered ring form of a carbohydrate.

fusion The joining of two light nuclei to form a more massive one. Also known as a *thermonuclear reaction*.

galvanic cell See **voltaic (galvanic) cell**.

gamma (γ) radiation Energetic electromagnetic radiation emanating from the nucleus of a radioactive atom.

gas Matter that has no fixed volume or shape; it conforms to the volume and shape of its container.

gas constant (R) The constant of proportionality in the ideal-gas equation.

Geiger counter A device that can detect and measure radioactivity.

geometric isomers Compounds with the same type and number of atoms and the same chemical bonds but different spatial arrangements of these atoms and bonds.

Gibbs free energy (G) A thermodynamic state function that combines enthalpy and entropy, in the form $G = H - TS$. For a change occurring at constant temperature and pressure, the change in free energy is $\Delta G = \Delta H - T\Delta S$.

glass An amorphous solid formed by fusion of SiO_2, CaO and Na_2O. Other oxides may also be used to form other types of glass with differing characteristics.

globular protein Tightly bundled proteins, so named because of their globular or spherical shape.

glucose A polyhydroxy aldehyde whose formula is $CH_2OH(CHOH)_4CHO$; it is the most important of the monosaccharides.

glycogen The general name given to a group of polysaccharides of glucose that are synthesised in mammals and used to store energy from carbohydrates.

glycoside A carbohydrate derivative in which the hydroxy group (–OH) located at the anomeric carbon is replaced by another group, typically an alkoxy group (–OR) or amine.

glycosidic bond The bond from the anomeric carbon to the OR group in a glycoside.

Graham's law A law stating that the rate of effusion of a gas is inversely proportional to the square root of its molecular weight.

gray (Gy) The SI unit for radiation dose corresponding to the absorption of 1 J of energy per kilogram of tissue: 1 Gy = 100 rads.

green chemistry initiative Chemistry that promotes the design and application of chemical products and processes that are compatible with human health and preserve the environment.

greenhouse gases Gases such as H_2O and CO_2 which trap infrared radiation in the atmosphere.

Grignard reagent A class of reagent with general formula RMgX, where R = alkyl, aryl and X = halide.

ground state The lowest-energy, or most stable, state.

group Elements that are in the same column of the periodic table; elements within the same group or family exhibit similarities in their chemical behaviour.

Haber process The catalyst system and conditions of temperature and pressure developed by Fritz Haber and co-workers for the formation of NH_3 from H_2 and N_2.

haemoglobin An iron-containing protein responsible for oxygen transport in the blood.

half-life The time required for the concentration of a reactant substance to decrease to half its initial value.

half-reaction An equation for either an oxidation or a reduction that explicitly shows the electrons involved; for example, Zn^{2+}(aq) + $2e^- \rightarrow Zn$(s).

Hall process A process used to obtain aluminium by electrolysis of Al_2O_3 dissolved in molten cryolite, Na_3AlF_6.

haloalkanes Of general formula R–X, where X = F, Cl, Br, I. A haloalkane is any hydrocarbon containing one or more halo groups.

haloforms Compounds of the form CHX_3, for example chloroform.

halogenation A reaction in which at least one halide is added. In the case of addition reactions to alkenes, two halogen atoms are added across the double bond.

halogens Members of group 17 in the periodic table.

hard water Water that contains appreciable concentrations of Ca^{2+} and Mg^{2+}; these ions react with soaps to form an insoluble material.

hardening A term used to describe the conversion of oils to fats by hydrogenation of unsaturated triglycerides.

Haworth projection A convention for representing the configuration of functional groups around a furanose or pyranose ring.

heat The flow of energy from a body at higher temperature to one at lower temperature when they are placed in thermal contact.

heat capacity The quantity of heat required to raise the temperature of a sample of matter by 1 °C (or 1 K).

heat of fusion The enthalpy change, ΔH, for melting a solid.

heat of sublimation The enthalpy change, ΔH, for vaporisation of a solid.

heat of vaporisation The enthalpy change, ΔH, for vaporisation of a liquid.

Heisenberg's uncertainty principle States that the dual nature of matter places a fundamental limitation on the precision with which we can know the location and momentum of any object.

hemiacetal The product formed by the reaction of an aldehyde or ketone with one mole of alcohol.

Henderson–Hasselbalch equation The relationship between the pH, pK_a and the concentrations of acid and conjugate base in an aqueous solution:

$$pH = pK_a + \log\frac{[base]}{[acid]}$$

Henry's law A law stating that the concentration of a gas in a solution, S_g, is proportional to the pressure of gas over the solution: $S_g = kP_g$.

Hess's law The heat evolved in a given process can be expressed as the sum of the heats of several processes that, when added, yield the process of interest.

heteroatom An atom other than carbon or hydrogen used to form functional groups in organic compounds.

heterocyclic compound A cyclic compound in which one or more carbon atoms within the ring are replaced by heteroatoms.

heterogeneous alloy An alloy in which the components are not distributed uniformly; instead, two or more distinct phases with characteristic compositions are present.

heterogeneous catalyst A catalyst that is in a different phase from that of the reactant substances.

heterogeneous equilibrium The equilibrium established between substances in two or more different phases; for example, between a gas and a solid or between a solid and a liquid.

hexagonal close packing A close-packing arrangement in which the atoms of the third layer of a solid lie directly over those in the first layer.

high-spin complex A complex whose electrons populate the d orbitals to give the maximum number of unpaired electrons.

high-temperature superconductivity The 'frictionless' flow of electrical current (superconductivity) at temperatures above 30 K in certain complex metal oxides.

homogeneous catalyst A catalyst that is in the same phase as the reactant substances.

homogeneous equilibrium The equilibrium established between reactant and product substances that are all in the same phase.

homologous series A series of compounds that differ only by the number of methene (–CH_2–) units.

homolytic bond cleavage Symmetrical breaking of a covalent bond to yield two radicals.

Hooke's law States that the elongation of a spring is proportional to the load applied. Mathematically: $F_s = -kx$

Hund's rule A rule stating that electrons occupy degenerate orbitals in such a way as to maximise the number of electrons with the same spin. In other words, each orbital has one electron placed in it before pairing of electrons in orbitals occurs. Note that this rule applies only to orbitals that are degenerate, which means that they have the same energy.

hybrid orbital An orbital that results from the mixing of different kinds of atomic orbitals on the same atom. For example, an sp^3 hybrid results from the mixing, or hybridising, of one s orbital and three p orbitals.

hybridisation The mixing of different types of atomic orbitals to produce a set of equivalent hybrid orbitals.

hydration Solvation when the solvent is water.

hydride ion An ion formed by the addition of an electron to a hydrogen atom: H^-.

hydrocarbons Compounds composed of only carbon and hydrogen.

hydrogen bonding Bonding that results from intermolecular attractions between molecules containing hydrogen bonded to an electronegative element. The most important examples involve OH, NH and HF.

hydrohalogenation A reaction that adds HX across a C=C double bond (X is typically Cl, Br, I).

hydrolysis A reaction with water. When a cation or anion reacts with water, it changes the pH.

hydrometallurgy Aqueous chemical processes for recovery of a metal from an ore.

hydronium ion (H₃O⁺) The predominant form of the proton in aqueous solution.

hydrophilic Water-attracting. The term is often used to describe a type of colloid.

hydrophobic Water-repelling. The term is often used to describe a type of colloid.

hypothesis A tentative explanation of a series of observations or of a natural law.

ideal gas A hypothetical gas whose pressure, volume and temperature behaviour is completely described by the ideal-gas equation.

ideal-gas equation An equation of state for gases that embodies Boyle's law, Charles's law and Avogadro's hypothesis in the form $PV = nRT$.

ideal solution A solution that obeys Raoult's law.

imine The product formed by the condensation of an aldehyde or ketone with a primary amine.

immiscible liquids Liquids that do not dissolve in one another to any significant extent.

index of hydrogen deficiency (IHD) A summation of all π-bonds and rings within a molecule. The index of hydrogen deficiency quantity is determined by comparing the number of hydrogens in the molecular formula of an unknown with that of an *alkane* reference that contains the same number of carbon atoms as the unknown.

indicator A substance added to a solution to indicate by a colour change the point at which the added solute has reacted with all the solute present in solution.

inductive effect An electronic effect transmitted through bonds by a more electronegative neighbouring atom.

inductive stabilisation The stabilisation obtained when a neighbouring electron donating (or withdrawing) group can add (or withdraw) electron density from a positive (or negative) charge through σ-bonds.

infrared (IR) spectroscopy An analytical technique used to characterise functional groups that is based on the absorption of energy by a molecule as it vibrates by stretching and bending.

infrared (IR) spectrum The result of undertaking an IR spectroscopic experiment; usually expressed as a plot of percentage transmittance (%T) as a function of wavenumber.

initiation The first step in a chain reaction.

instantaneous rate The reaction rate at a particular time as opposed to the average rate over an interval of time.

insulator A solid with extremely low electrical conductivity.

integration The area under a signal within a ¹H NMR spectrum. The area is proportional to the number of equivalent protons that gives rise to that signal.

intensive property A property that is independent of the amount of material considered, for example density.

interhalogens Compounds formed between two different halogen elements. Examples include IBr and BrF₃.

intermediate A substance formed in one elementary step of a multistep mechanism and consumed in another; it is neither a reactant nor an ultimate product of the overall reaction.

intermetallic compound A homogeneous alloy with definite properties and composition. Intermetallic compounds are stoichiometric compounds, but their compositions are not readily explained in terms of ordinary chemical bonding theory.

intermolecular forces The short-range attractive forces operating between the particles that make up the units of a liquid or solid substance. These same forces also cause gases to liquefy or solidify at low temperatures and high pressures.

internal energy The total energy possessed by a system. When a system undergoes a change, the change in internal energy, ΔU, is defined as the heat, q, added to the system, plus the work, w, done on the system by its surroundings: $\Delta U = q + w$.

ion Electrically charged atom or group of atoms (polyatomic ion); ions can be positively or negatively charged, depending on whether electrons are lost (positive) or gained (negative) by the atoms.

ion–dipole force The force that exists between an ion and a neutral polar molecule that possesses a permanent dipole moment.

ion exchange A process in which ions in solution are exchanged for other ions held on the surface of an ion-exchange resin; the exchange of a hard-water cation such as Ca^{2+} for a soft-water cation such as Na^+ is used to soften water.

ion-product constant For water, K_w is the product of the aquated hydrogen ion and hydroxide ion concentrations: $[H^+][OH^-] = K_w = 1.0 \times 10^{-14}$ at 25 °C.

ionic bond A bond between oppositely charged ions. The ions are formed from atoms by transfer of one or more electrons.

ionic compound A compound composed of cations and anions.

ionic hydrides Compounds formed when hydrogen reacts with alkali metals and also the heavier alkaline earths (Ca, Sr and Ba); these compounds contain the hydride ion, H^-.

ionic solids Solids that are composed of ions.

ionisation energy The energy required to remove an electron from a gaseous atom when the atom is in its ground state.

ionising radiation Radiation that has sufficient energy to remove an electron from a molecule, thereby ionising it.

irreversible process A process that is not reversible; as a result, some of its potential for accomplishing work is dissipated as heat. Any spontaneous process is irreversible in practice.

isoelectric point The pH at which the maximum number of molecules of a compound in aqueous solution contain both positively and negatively charged regions. That is, the pH at which the zwitterion form is maximised.

isoelectronic series A series of atoms, ions or molecules having the same number of electrons.

isomers Compounds whose molecules have the same overall composition but different structures.

isothermal (isothermic) process A process that occurs at constant temperature.

isotopes Atoms of the same element containing different numbers of neutrons and therefore having different masses.

joule (J) The SI unit of energy, 1 kg m⁻² s⁻².

Kelvin scale The absolute temperature scale; the SI unit for temperature is the Kelvin. Zero on the Kelvin scale corresponds to −273.15 °C; therefore, K = °C + 273.15.

ketone A compound in which the carbonyl group occurs at the interior of a carbon chain and is therefore flanked by carbon atoms.

ketose A carbohydrate that contains a ketone group when in its open-chain form.

kilogram (kg) Base unit of mass in the SI system.

kilopascal (kPa) The commonly used unit of pressure in chemical calculations, which is equivalent to 1000 Pa.

kinetic energy The energy that an object possesses by virtue of its motion.

kinetic-molecular theory A set of assumptions about the nature of gases. These assumptions, when translated into mathematical form, yield the ideal-gas equation.

L-ascorbic acid See **ascorbic acid**.

lactam A cyclic amide in which the amide bond forms part of the ring.

lanthanide contraction The gradual decrease in atomic and ionic radii with increasing atomic number among the lanthanide elements, atomic numbers 57 to 70. The decrease arises because of a gradual increase in effective nuclear charge through the lanthanide series.

lanthanide (rare earth) element Element in which the 4*f* subshell is only partially occupied.

Larmor frequency The frequency at which a nucleus absorbs energy. This frequency is a function of the strength of the applied field B_0.

lattice energy The energy required to separate completely the ions in an ionic solid.

law of conservation of mass A scientific law stating that the total mass of the products of a chemical reaction is the same as the total mass of the reactants, so that mass remains constant during the reaction.

law of constant composition A law stating that the elemental composition of a pure substance is always the same, regardless of its source; also called the **law of definite proportions**.

law of definite proportions see **law of constant composition**.

law of mass action The rules by which the equilibrium constant is expressed in terms of the concentrations of reactants and products, in accordance with the balanced chemical equation for the reaction.

Le Châtelier's principle A principle stating that, when we disturb a system at chemical equilibrium, the relative concentrations of reactants and products shift so as to partially undo the effects of the disturbance.

leaching The selective dissolution of a desired mineral by passing an aqueous reagent solution through an ore.

levorotatory, or **levo** or *l*. A term used to label a chiral molecule that rotates the plane of polarisation of plane-polarised light to the left (anticlockwise).

Lewis acid An electron-pair acceptor.

Lewis base An electron-pair donor.

Lewis structure A representation of covalent bonding in a molecule that is drawn using Lewis symbols. Shared electron pairs are shown as lines, and unshared electron pairs are shown as pairs of dots. Only the valence-shell electrons are shown.

Lewis symbol (electron-dot symbol) The chemical symbol for an element, with a dot for each valence electron.

ligand An ion or molecule that coordinates to a metal atom or to a metal ion to form a complex.

lime-soda process A method used in large-scale water treatment to reduce water hardness by removing Mg^{2+} and Ca^{2+}. The substances added to the water are lime, CaO (or slaked lime, $Ca(OH)_2$) and soda ash, Na_2CO_3, in amounts determined by the concentrations of the undesired ions.

limiting reactant (limiting reagent) The reactant present in the smallest stoichiometric quantity in a mixture of reactants; the amount of product that can form is limited by the complete consumption of the limiting reactant.

line spectrum A spectrum that contains radiation at only certain specific wavelengths.

linkage isomers Structural isomers of coordination compounds in which a ligand differs in its mode of attachment to a metal ion.

lipids A heterogeneous class of naturally occurring organic compounds that are more soluble in aprotic organic solvents but not soluble in water.

liquid Matter that has a distinct volume but no specific shape.

liquid crystal A substance that exhibits one or more partially ordered liquid phases above the melting point of the solid form. By contrast, in non-liquid crystalline substances the liquid phase that forms upon melting is completely unordered.

lithosphere That portion of our environment consisting of the solid Earth.

lock-and-key model A model of enzyme action in which the substrate molecule is pictured as fitting rather specifically into the active site on the enzyme. It is assumed that, in being bound to the active site, the substrate is somehow activated for reaction.

London dispersion forces Intermolecular forces resulting from attractions between induced dipoles.

lone pair A pair of electrons that do not participate in covalent bonding in neutral molecules.

low-spin complex A metal complex in which the electrons are paired in lower-energy orbitals.

magic numbers Numbers of protons and neutrons that result in very stable nuclei.

magnetic quantum number A number (m_l) that describes the orientation in space of an orbital.

main-group elements Elements in the *s* and *p* blocks of the periodic table.

Markovnikov's rule A rule stating that, for the addition of HX to an alkene, the hydrogen of HX adds to the carbon of the double bond with the most hydrogens already attached to it. In essence, the most stable carbocation is formed in preference.

mass A measure of the amount of material in an object. It measures the resistance of an object to being moved. In SI units, mass is measured in kilograms.

mass defect The difference between the mass of a nucleus and the total masses of the individual nucleons that it contains.

mass number The sum of the number of protons and neutrons in the nucleus of a particular atom.

mass percentage The number of grams of solute in each 100 g of solution.

mass spectrometer An instrument used to measure the precise masses and relative amounts of atomic and molecular ions.

mass-to-charge ratio Defined as m/z, where m = mass of the ionised species and z = charge on that species. Organic chemists use the term m/z as a dimensionless quantity.

matter Anything that occupies space and has mass; the physical material of the universe.

matter waves The wave characteristics of a particle.

mean free path The average distance travelled by a gas molecule between collisions.

meso compound An achiral compound containing two or more stereocentres.

mesosphere The region of the atmosphere stretching from an altitude of about 50 km to about 85 km.

meta-director A substituent on an aromatic compound that preferentially directs an incoming substituent in an electrophillic aromatic substitution to the *meta* position.

metal complex (complex ion or **complex)** An assembly of a metal ion and the Lewis bases bonded to it.

metal hydride A compound containing a metal-to-hydrogen bond. Typically used as a source of hydride H^- in organic chemistry.

metallic bond Bonding, usually in solid metals, in which the bonding electrons are relatively free to move throughout the three-dimensional structure.

metallic character The extent to which an element exhibits the physical and chemical properties characteristic of metals, such as lustre, malleability, ductility and good thermal and electrical conductivity.

metallic elements (metals) Elements that are usually solids at room temperature, exhibit high electrical and heat conductivity, and appear lustrous. Most of the elements in the periodic table are metals.

metallic hydrides Compounds formed when hydrogen reacts with transition metals; these compounds contain the hydride ion, H^-.

metallic solids Solids that are composed of metal atoms.

metalloids Elements that lie along the diagonal line separating the metals from the non-metals in the periodic table; the properties of metalloids are intermediate between those of metals and non-metals.

metallurgy The science of extracting metals from their natural sources by a combination of chemical and physical processes. It is also concerned with the properties and structures of metals and alloys.

metathesis (exchange) reaction A reaction in which two substances react through an exchange of their component ions: AX + BY → AY + BX. Precipitation and acid–base neutralisation reactions are examples of metathesis reactions.

metre The SI unit of length. It is the distance light travels in a vacuum in 1/299792458 seconds.

metric system A system of measurement used in science and in most countries. The metre and the gram are examples of metric units.

micelle A spherical aggregate of molecules such as the sodium or potassium salts of fatty acids, oriented in such a way that they have a hydrophilic (charged) surface and a hydrophobic core.

microanalysis A method used to determine the elemental composition of a compound or sample.

microstate The state of a system at a particular instant; one of many possible states of the system.

mineral A solid, inorganic substance occurring in nature, such as calcium carbonate, which occurs as calcite.

miscible Describes liquids that mix in all proportions.

mixture A combination of two or more substances in which each substance retains its own chemical identity.

molal boiling-point-elevation constant (K_b) A constant characteristic of a particular solvent that gives the increase in boiling point as a function of solution molality: $\Delta T_b = K_b m$.

molal freezing-point-depression constant (K_f) A constant characteristic of a particular solvent that gives the decrease in freezing point as a function of solution molality: $\Delta T_f = K_f m$.

molality The concentration of a solution expressed as moles of solute per kilogram of solvent; abbreviated m.

molar heat capacity The heat required to raise the temperature of one mole of a substance by 1 °C.

molar mass The mass of one mole of a substance in grams.

molar volume The volume (22.71 dm^3) occupied by 1 mole of a gas at 273.15 K and 1 bar pressure.

molarity The concentration of a solution expressed as moles of solute per litre of solution; abbreviated M.

mole A collection of Avogadro's number (6.022×10^{23}) of objects; for example, a mole of H_2O is 6.022×10^{23} H_2O molecules.

mole fraction The ratio of the number of moles of one component of a mixture to the total moles of all components; abbreviated X, with a subscript to identify the component.

molecular compound A compound that consists of molecules.

molecular equation A chemical equation in which the formula for each substance is written without regard for whether it is an electrolyte or a non-electrolyte.

molecular formula A chemical formula that indicates the actual number of atoms of each element in one molecule of a substance.

molecular geometry The arrangement in space of the atoms of a molecule.

molecular hydrides Compounds formed when hydrogen reacts with non-metals and metalloids.

molecular ion Typically, the species formed by the loss of an electron from a molecule.

molecular mass The average mass of a molecule of a substance relative to that of a carbon-12 atom; it is the masses of the atoms in the molecules.

molecular orbital (MO) An allowed state for an electron in a molecule. According to molecular orbital theory, a molecular orbital is entirely analogous to an atomic orbital, which is an allowed state for an electron in an atom. A molecular orbital may be classified as σ or π depending on the disposition of electron density with respect to the internuclear axis.

molecular orbital diagram A diagram that shows the energies of molecular orbitals relative to the atomic orbitals from which they are derived; also called an energy-level diagram.

molecular orbital theory A theory that accounts for the allowed states for electrons in molecules.

molecular solids Solids that are composed of molecules.

molecularity The number of molecules participating as reactants in an elementary reaction.

molecule A chemical combination of two or more atoms.

momentum The product of the mass (m) and velocity (u) of a particle.

monodentate ligand A ligand that binds to the metal ion via a single donor atom. It occupies one position in the coordination sphere.

monomers Molecules with low molecular weights that can be joined together (polymerised) to form a polymer.

monosaccharide A simple sugar, most commonly containing six carbon atoms. The joining together of monosaccharide units by condensation reactions results in formation of polysaccharides.

multidentate ligand A ligand containing two or more donor atoms capable of coordinating to the same metal ion.

multiple bonding Bonding involving two or more electron pairs.

multiplicity The number of peaks an NMR signal is split into due to spin–spin coupling.

mutarotation The change in optical rotation that occurs upon the equilibration of α and β anomers from a single anomer.

(n + 1) rule States that the signal for a nucleus coupled to n inequivalent nuclei will be split into n + 1 peaks.

N-glycoside Formed particularly when a monosaccharide is reacted with an amine in the presence of acid. Nucleosides are also N-glycosides.

nanomaterial A material whose useful characteristics are the result of features in the range from 1 to 100 nm.

nanotechnology Technology that relies on the properties of matter at the nanoscale, that is, in the range from 1 nm to 100 nm.

natural gas A naturally occurring mixture of gaseous hydrocarbon compounds composed of hydrogen and carbon.

nematic liquid crystal A liquid crystal in which the molecules are aligned in the same general direction, along their long axes, but in which the ends of the molecules are not aligned.

Nernst equation An equation that relates the cell emf, E, to the standard emf, $E°$, and the reaction quotient, Q: $E = E° - (RT/nF) \ln Q$.

net ionic equation A chemical equation for a solution reaction in which soluble strong electrolytes are written as ions and spectator ions are omitted.

neutralisation reaction A reaction in which an acid and a base react in stoichiometrically equivalent amounts; the neutralisation reaction between an acid and a metal hydroxide produces water and a salt.

neutron An electrically neutral particle found in the nucleus of an atom; it has approximately the same mass as a proton.

Newman projection A representation of a molecule's spatial arrangement looking along a C–C single bond.

niacin Vitamin B$_3$. Used in the form of NAD$^+$ (nicotinamide adenine dinucleotide) and NADP$^+$ (nicotinamide adenine dinucleotide phosphate) in more than 50 enzyme-based reactions. Vitamin B$_3$ is instrumental in the release of energy from carbohydrates, is necessary for proper brain function and is also involved in fat and cholesterol metabolism.

nitriles Any hydrocarbon containing the C≡N group.

nitro group A functional group of formula –NO$_2$.

noble gases Members of group 18 in the periodic table.

node A locus of points in an atom at which the electron density is zero.

nomenclature The systematic naming of a compound.

non-electrolyte A substance that does not ionise in water and consequently gives a non-conducting solution.

non-ionising radiation Radiation that does not have sufficient energy to remove an electron from a molecule.

non-bonding pair (lone pair) In a Lewis structure a pair of electrons assigned completely to one atom.

non-metallic elements (non-metals) Elements in the upper right corner of the periodic table; non-metals differ from metals in their physical and chemical properties.

non-polar covalent bond A covalent bond in which the electrons are shared equally between two atoms.

normal boiling point The boiling point at 1 bar pressure.

normal melting point The melting point at 1 bar pressure.

nuclear binding energy The energy required to decompose an atomic nucleus into its component protons and neutrons.

nuclear disintegration series A series of nuclear reactions that begins with an unstable nucleus and terminates with a stable one; also called a **radioactive series**.

nuclear magnetic resonance (NMR) spectroscopy A non-destructive technique able to provide information on the types and relative number of atoms, atom connectivity and spatial relationships. This technique observes nuclear spin transitions within a strong magnetic field.

nuclear reactions Reactions that involve changes in atomic nuclei.

nuclear transmutation A conversion of one kind of nucleus to another.

nucleic acids Polymers of high molecular weight that carry genetic information and control protein synthesis.

nucleon A particle found in the nucleus of an atom.

nucleophile Any atom, molecule or ion that donates a pair of electrons to an electrophile to form a new covalent bond.

nucleophilic acyl substitution The mechanism by which one carboxylic acid derivative is converted into a less reactive one.

nucleophilic addition A reaction that involves the addition of a nucleophile to a reagent, usually a carbonyl compound.

nucleophilic substitution (S_N) reaction Any reaction in which one nucleophile is substituted by another.

nucleoside Compounds containing a five-carbon sugar and a nitrogen-containing organic base.

nucleotides Compounds formed from a molecule of phosphoric acid, a sugar molecule and an organic nitrogen base. Nucleotides form linear polymers called DNA and RNA, which are involved in protein synthesis and cell reproduction.

nucleus The very small, very dense, positively charged portion of an atom; it is composed of protons and neutrons.

nuclide A nucleus of a specific isotope of an element.

nujol A paraffin-like substance used for preparing mull samples for IR spectroscopy.

octet rule A rule stating that bonded atoms tend to possess or share a total of eight valence-shell electrons.

oils Triglycerides that are liquids at room temperature. They are usually rich in unsaturated fatty acids.

operating frequency The frequency at which a NMR spectrometer operates to observe a given nucleus, usually ^1H.

optical isomers Stereoisomers in which the two forms of the compound are non-superimposable mirror images.

optically active Possessing the ability to rotate the plane of polarised light.

orbital An allowed energy state of an electron in the quantum mechanical model of the atom; the term *orbital* is also used to describe the spatial distribution of the electron. An orbital is defined by the values of three quantum numbers: n, l and m_l.

ore A source of a desired element or mineral, usually accompanied by large quantities of other materials such as sand and clay.

organic chemistry The study of carbon-containing compounds, typically containing carbon–carbon bonds.

organometallic compound A compound containing a carbon-to-metal bond.

***ortho*, *para*-director** A substituent on an aromatic compound that directs an incoming substituent in an electrophilic aromatic substitution, to the *ortho* and *para* positions in preference.

osmosis The net movement of solvent through a semi-permeable membrane towards the solution with greater solute concentration.

osmotic pressure The pressure that must be applied to a solution to stop osmosis from pure solvent into the solution.

Ostwald process An industrial process used to make nitric acid from ammonia. The NH_3 is catalytically oxidised by O_2 to form NO; NO in air is oxidised to NO_2; HNO_3 is formed in a disproportionation reaction when NO_2 dissolves in water.

overall reaction order The sum of the reaction orders of all the reactants appearing in the rate expression when the rate can be expressed as $rate = k[A]^a[B]^b \cdots$.

overlap The extent to which atomic orbitals on different atoms share the same region of space. When the overlap between two orbitals is large, a strong bond may be formed.

oxidation A process in which a substance loses one or more electrons.

oxidation number (oxidation state) A positive or negative whole number assigned to an element in a molecule or ion on the basis of a set of formal rules; to some degree it reflects the positive or negative character of that atom.

oxidation–reduction (redox) reaction A chemical reaction in which the oxidation states of certain atoms change.

oxidising agent (oxidant) The substance that is reduced and thereby causes the oxidation of some other substance in an oxidation–reduction reaction.

oxyacid A compound in which one or more OH groups, and possibly additional oxygen atoms, are bonded to a central atom.

oxyanion A polyatomic ion that contains one or more oxygen atoms.

ozone The name given to O_3, an allotrope of oxygen.

ozonolysis A synthetically useful way of generating aldehydes and ketones.

π-bond See **pi (π)-bond**.

pantothenic acid Vitamin B_5. A water-soluble vitamin found in abundance in whole grain cereals, legumes and meat. Vitamin B_5 is required for the synthesis of coenzyme A (CoA).

paramagnetism A property that a substance possesses if it contains one or more unpaired electrons. A paramagnetic substance is drawn into a magnetic field.

partial pressure The pressure exerted by a particular gas in a mixture.

particle accelerator A device that uses strong magnetic and electrostatic fields to accelerate charged particles.

parts per billion (ppb) The concentration of a solution in grams of solute per 10^9 (billion) grams of solution; equals micrograms of solute per litre of solution for aqueous solutions.

parts per million (ppm) The concentration of a solution in grams of solute per 10^6 (million) grams of solution; equals milligrams of solute per litre of solution for aqueous solutions.

pascal (Pa) The SI unit of pressure: $1\ Pa = 1\ N\ m^{-2}$.

Pauli exclusion principle A rule stating that no two electrons in an atom may have the same four quantum numbers (n, l, m_l and m_s). As a reflection of this principle, there can be no more than two electrons in any one atomic orbital.

penicillin A lactam with antibiotic activity.

peptide The name given to a short polymer (oligomer) of amino acids joined by amide bonds, usually without significant tertiary structure.

peptide bond A bond formed between two amino acids.

percent ionisation Measure of acid strength.

percent yield The ratio of the actual (experimental) yield of a product to its theoretical (calculated) yield, multiplied by 100.

period The row of elements that lie in a horizontal row in the periodic table.

periodic table The arrangement of elements in order of increasing atomic number, with elements having similar properties placed in vertical columns.

petroleum A naturally occurring combustible liquid composed of hundreds of hydrocarbons and other organic compounds.

pH The negative log in base 10 of the aquated hydrogen ion concentration: $pH = -\log[H^+]$.

pH range The range of pH at which the buffer is able to keep the acidity of the solution at an almost constant level.

pH titration curve See **titration curve**.

phase change The conversion of a substance from one state of matter to another. The phase changes we consider are melting and freezing (solid ↔ liquid), sublimation and deposition (solid ↔ gas) and vaporisation and condensation (liquid ↔ gas).

phase diagram A graphic representation of the equilibria among the solid, liquid and gaseous phases of a substance as a function of temperature and pressure.

phenol A compound with molecular formula C_6H_5OH.

phenyl group A hydrocarbon group with formula C_6H_5.

photochemical smog A complex mixture of undesirable substances produced by the action of sunlight on an urban atmosphere polluted by car emissions. The major starting ingredients are nitrogen oxides and organic substances, notably olefins and aldehydes.

photodissociation The breaking of a molecule into two or more neutral fragments as a result of absorption of light.

photoelectric effect The emission of electrons from a metal surface, induced by light.

photoionisation The removal of an electron from an atom or molecule by absorption of light.

photon The smallest increment (a quantum) of radiant energy; a photon of light with frequency has an energy equal to $h\nu$.

photosynthesis The process that occurs in plant leaves by which light energy is used to convert carbon dioxide and water to carbohydrates and oxygen.

phylloquinone Vitamin K, responsible for the regulation of prothrombin (factor X) and other proteins responsible for blood clotting.

physical changes Changes (such as a phase change) that occur with no change in chemical composition.

physical properties Properties that can be measured without changing the composition of a substance, for example colour and freezing point.

pi (π)-bond A covalent bond in which electron density is concentrated above and below the line joining the bonded atoms.

pi (π) molecular orbital A molecular orbital that concentrates the electron density on opposite sides of a line that passes through the nucleus.

picometre (pm) SI unit used to express atomic dimensions.

Planck's constant (h) The constant that relates the energy and frequency of a photon, $E = h\nu$. Its value is 6.626×10^{-34} J s.

plane-polarised light Light that has its vectors occur in only parallel planes.

plastic A material that can be formed into particular shapes by application of heat and pressure.

polar covalent bond A covalent bond in which the electrons are not shared equally.

polar molecule A molecule that possesses a non-zero dipole moment.

polarimeter An instrument used to measure the rotation of plane-polarised light.

polarisability The ease with which the electron cloud of an atom or a molecule is distorted by an outside influence, thereby inducing a dipole moment.

polyamides Polymers in which the repeating structural units are joined together by amide bonds. Usually formed by the condensation of dicarboxylic acids and their derivatives with diamines.

polyatomic ion An electrically charged group of two or more atoms.

polyatomic molecule Molecule that contains more than two atoms.

polydentate ligand See **multidentate ligand**.

polyesters Polymers in which the repeating structural units are joined together by ester bonds. Usually formed by the condensation of dicarboxylic acids and their derivatives with diols.

polymer A large molecule of high molecular mass, formed by the joining together, or polymerisation, of a large number of molecules of low molecular mass. The individual molecules forming the polymer are called monomers.

polypeptide A polymer of amino acids that has a molecular weight of less than 10000.

polyprotic acid A substance capable of dissociating more than one proton in water; H_2SO_4 is an example.

polysaccharide A substance made up of several monosaccharide units joined together.

porphyrin A complex derived from the porphine molecule.

positron A particle with the same mass as an electron but with a positive charge, symbol $_1^0e$ or $_1^0\beta$.

positron emission The emission of short-lived particles which have the same mass as an electron but are opposite in charge. They are represented as $_1^0e$ or $_1^0\beta$.

potential energy The energy that an object possesses as a result of its composition or its position with respect to another object.

precipitate An insoluble substance that forms in, and separates from, a solution.

precipitation reaction A reaction that occurs between substances in solution in which one of the products is insoluble.

precision The closeness of agreement between several measurements of the same quantity; the reproducibility of a measurement.

pressure A measure of the force exerted on a unit area. It is expressed in units of Pa or kPa.

pressure-volume (P-V) work Work performed by expansion of a gas against a resisting pressure.

primary (1°) alcohol A compound containing the hydroxy (OH) group of an alcohol on a primary carbon atom. Oxidation leads to the formation of aldehydes and carboxylic acids.

primary amide A compound of the type RCONR′R″ in which R′ and R″ are H.

primary amine An amine of the form R–NH₂, where R = alkyl, aryl.

primary structure The sequence of amino acids along a protein chain.

primitive cubic cell A cubic unit cell in which the lattice points are at the corners only.

principal quantum number A number (n) that indicates the energy level of an orbital.

probability density (ψ^2) A value that represents the probability that an electron will be found at a given point in space.

product A substance produced in a chemical reaction; it appears to the right of the arrow in a chemical equation.

propagation Fundamental steps in a free-radical polymerisation that are repeated over and over again in a chain reaction.

property A characteristic that gives a sample of matter its unique identity.

protein A biopolymer formed from amino acids.

protium The most common isotope of hydrogen.

proton A positively charged subatomic particle found in the nucleus of an atom.

pure substance Matter that has a fixed composition and distinct properties.

purine A nitrogen-containing heterocyclic compound comprising fused five- and six-membered rings. Used as an organic base for genetic information storage within DNA.

pyranose Six-membered ring form of a carbohydrate.

pyridoxine Vitamin B₆. This water-soluble vitamin is required as a cofactor in enzymes involved in transamination reactions which are required for the synthesis of amino acids.

pyrimidine A nitrogen-containing six-membered heterocyclic compound used as an organic base for genetic information storage within DNA.

pyrometallurgy A process in which heat converts a mineral in an ore from one chemical form to another and eventually to the free metal.

qualitative analysis The determination of the presence or absence of a particular substance in a mixture.

quantised Energy absorbed or emitted by an atom can only have certain allowed values which are integral multiples of $h\nu$.

quantitative analysis The determination of the amount of a given substance that is present in a sample.

quantum The smallest increment of radiant energy that may be absorbed or emitted; the magnitude of radiant energy is $h\nu$.

quantum mechanics Describes the arrangement of electrons in an atom.

quantum numbers Describe the shell, subshell, orbital and spin of an electron.

quaternary ammonium salts A nitrogen-containing compound in which the nitrogen atom holds a positive charge and an anion.

quaternary structure The aggregation of two or more proteins into a larger macromolecular assembly.

racemic mixture A mixture of equal amounts of the dextrorotatory and levorotatory forms of a chiral molecule. A racemic mixture will not rotate polarised light. Also called a racemate.

rad A non-SI unit of measurement of the energy absorbed from radiation by tissue or other biological material; 1 rad = transfer of 1×10^{-2} J of energy per kilogram of material.

radial probability function The probability that the electron will be found at a certain distance from the nucleus.

radioactive Possessing radioactivity, the spontaneous disintegration of an unstable atomic nucleus with accompanying emission of radiation.

radioactive series A series of nuclear reactions that begins with an unstable nucleus and terminates with a stable one. Also called **nuclear disintegration series**.

radioactivity Spontaneous emission of radiation.

radiocarbon dating Process whereby the isotope carbon-14 is used to date objects.

radioisotope An isotope that is radioactive; that is, it is undergoing nuclear changes with emission of radiation.

radiometric dating Process whereby isotopes are used to date objects.

radionuclide A radioactive nuclide.

radiotracer A radioisotope that can be used to trace the path of an element in a chemical system.

Raoult's law A law stating that the partial pressure of a solvent over a solution, P_A, is given by the vapour pressure of the pure solvent, P_A^5, times the mole fraction of a solvent in the solution, X_A: $P_A = X_A P_A^5$.

rare earth element See **lanthanide element**.

rate constant A constant of proportionality between the reaction rate and the concentrations of reactants that appear in the rate law.

rate-determining step The slowest elementary step in a reaction mechanism.

rate law An equation that relates the reaction rate to the concentrations of reactants (and sometimes of products also).

RBE Relative biological effectiveness of radiation dosage. A multiplication factor to convert Gy to Sv. Sv = Gy × RBE.

reactant A starting substance in a chemical reaction; it appears to the left of the arrow in a chemical equation.

reaction mechanism A detailed picture, or model, of how the reaction occurs; that is, the order in which bonds are broken and formed and the changes in relative positions of the atoms as the reaction proceeds.

reaction order The power to which the concentration of a reactant is raised in a rate law.

reaction quotient (Q) The value that is obtained when concentrations of reactants and products are inserted into the equilibrium expression. If the concentrations are equilibrium concentrations, $Q = K$ otherwise, $Q \neq K$.

reaction rate The decrease in concentration of a reactant or the increase in concentration of a product with time.

redox (oxidation–reduction) reaction A reaction in which certain atoms undergo changes in oxidation states. The substance increasing in oxidation state is oxidised; the substance decreasing in oxidation state is reduced.

reducing agent (reductant) The substance that is oxidised and thereby causes the reduction of some other substance in an oxidation–reduction reaction.

reduction A process in which a substance gains one or more electrons.

reductive amination A reaction between an aldehyde or ketone and primary amine, or ammonia, that occurs in the presence of a reducing agent, leading to the formation of a new amine.

refining The process of converting an impure form of a metal into a more usable substance of well-defined composition. For example, crude pig iron from the blast furnace is refined in a converter to produce steels of desired compositions.

regioselective reaction A reaction in which one direction of bond forming and bond breaking occurs in preference to all others.

regioselectivity The preferential formation of one constitutional isomer above others in a reaction.

rem A measure of the biological damage caused by radiation: rems = rads × RBE.

renewable energy Energy such as solar energy, wind energy and hydroelectric energy, derived from essentially inexhaustible sources.

representative (main-group) element An element from within the s and p blocks of the periodic table.

resolution The act of separating a racemic mixture into its constituent enantiomers.

resonance 1. The term given to the absorption of electromagnetic radiation at the Larmor frequency, which leads to the flip of its nuclear spin from a lower state to a higher state (and *vice versa*). 2. The way electron delocalisation is shown using Lewis structures.

resonance energy The difference in energy between a resonance hybrid and the most stable contributing structure in which the bonds are localised.

resonance stabilisation The stabilisation given to any molecule in which its electrons can be delocalised over three or more atoms.

resonance structures (resonance forms) Individual Lewis structures in cases where two or more Lewis structures are equally good descriptions of a single molecule. The resonance structures in such an instance are 'averaged' to give a more accurate description of the real molecule.

retinol Vitamin A. A lipophilic vitamin found in foods such as milk, cheese, butter, eggs and cod liver oil needed for that part of the vision process that responds to dim light.

reverse osmosis The process by which water molecules move under high pressure through a semi-permeable membrane from the more concentrated to the less concentrated solution.

reversible process A process that can go back and forth between states along exactly the same path; a system at equilibrium is reversible if equilibrium can be shifted by an infinitesimal modification of a variable such as temperature.

riboflavin Vitamin B_2. An aqueous soluble vitamin employed as a coenzyme for flavoproteins. Riboflavin is found in useful quantities in eggs, milk, meat and cereal.

ribonucleic acid (RNA) A polynucleotide in which ribose is the sugar component.

roasting Thermal treatment of an ore to bring about chemical reactions involving the furnace atmosphere. For example, a sulfide ore might be roasted in air to form a metal oxide and SO_2.

root-mean-square (rms) speed (u_{rms}) The square root of the average of the squared speeds of the gas molecules in a gas sample.

rotational motion Movement of a molecule as though it is spinning like a top.

σ-bond See **sigma (σ)-bond**.

salinity A measure of the salt content of seawater, brine or brackish water. It is equal to the mass in grams of dissolved salts present in 1 kg of seawater.

salt An ionic compound formed by replacing one or more H^+ of an acid by other cations.

saponification Hydrolysis of an ester in the presence of a base.

saturated solution A solution in which undissolved solute and dissolved solute are in equilibrium.

scientific law A concise verbal statement or a mathematical equation that summarises a wide range of observations and experiences.

scientific method The general process of advancing scientific knowledge by making experimental observations and by formulating hypotheses, theories and laws.

scintillation counter An instrument that is used to detect and measure radiation by the fluorescence it produces in a fluorescing medium.

second law of thermodynamics A statement of our experience that there is a direction to the way events occur in nature. When a process occurs spontaneously in one direction, it is non-spontaneous in the reverse direction. It is possible to state the second law in many different forms, but they all relate back to the same idea about spontaneity. One of the most common statements found in chemical contexts is that in any spontaneous process the entropy of the universe increases.

second-order reaction A reaction in which the overall reaction order (the sum of the concentration-term exponents) in the rate law is 2.

secondary (2°) alcohol A compound containing the hydroxy (–OH) group of an alcohol on a secondary carbon atom. Oxidation leads to the formation of ketones.

secondary amide An amide in which two N–C bonds exist about the amide N.

secondary amine An amine that has one hydrogen atom bonded to nitrogen.

secondary structure The manner in which a protein is coiled or stretched.

semiconductor A solid with limited electrical conductivity.

sequence priority rules A process of determining priority based on the atomic weights of atoms connected to a stereocentre.

shielding/deshielding Any effect that causes a nucleus to come into resonance at a higher/lower field, respectively.

SI units The preferred metric units for use in science.

side chain That part of an α-amino acid that distinguishes it from other α-amino acids.

sievert (Sv) The SI unit for effective dosage of radiation. It is the product of the radiation dose in grays and the RBE.

sigma (σ)-bond A covalent bond in which electron density is concentrated along the internuclear axis.

sigma (σ) molecular orbital A molecular orbital that centres the electron density about an imaginary line passing through two nuclei.

significant figures The digits that indicate the precision with which a measurement is made; all digits of a measured quantity are significant, including the last digit, which is uncertain.

silica Any of the various forms of SiO_2 in nearly all of which the lattice structures consist of interconnected tetrahedral building blocks.

silicates Compounds containing silicon and oxygen, structurally based on SiO_4 tetrahedra.

simple lipids Lipids that cannot be easily broken down into smaller components.

simple protein A protein that yields only amino acids upon hydrolysis.

single bond A covalent bond involving one electron pair.

slag A mixture of molten silicate minerals. Slags may be acidic or basic, according to the acidity or basicity of the oxide added to silica.

smectic liquid crystal A liquid crystal in which the molecules are aligned along their long axes and arranged in sheets, with the ends of the molecules aligned. There are several different kinds of smectic phases.

smelting A melting process in which the materials formed in the course of the chemical reactions that occur separate into two or more layers. For example, the layers might be slag and molten metal.

S_N1 A nucleophilic substitution reaction displaying first-order kinetics.

S_N2 A nucleophilic substitution reaction displaying second-order kinetics.

soap A mixture obtained by the hydrolysis of fats in aqueous base, used primarily for cleaning; it usually aggregates into micelles.

sol A colloid consisting of a solid material dispersed in a liquid medium.

sol-gel process A process in which extremely small particles (0.003 µm to 0.1 µm in diameter) of uniform size are produced in a series of chemical steps, followed by controlled heating.

solid State of matter that has both a definite shape and a definite volume.

solubility The amount of a substance that dissolves in a given quantity of solvent at a given temperature to form a saturated solution.

solubility-product constant (solubility product) (K_{sp}) An equilibrium constant related to the equilibrium between a solid salt and its ions in solution. It provides a quantitative measure of the solubility of a slightly soluble salt.

solute A substance dissolved in a solvent to form a solution; it is normally the component of a solution present in the smaller amount.

solution A mixture of substances that has a uniform composition; a homogeneous mixture.

solution alloy A homogeneous alloy, with the components distributed uniformly throughout.

solvation The clustering of solvent molecules around a solute particle.

solvent The dissolving medium of a solution; it is normally the component of a solution present in the greater amount.

specific heat capacity The heat capacity of 1 g of a substance; the heat required to raise the temperature of 1 g of a substance by 1 °C.

specific rotation The amount an optically active compound of concentration 1.0 g per 100 cm^{-3} rotates plane-polarised light through a 1-dm-long cell at 25 °C.

spectator ions Ions that go through a reaction unchanged and appear on both sides of the complete ionic equation.

spectrochemical series A list of ligands arranged in order of their ability to split the d-orbital energies (using the terminology of the crystal-field model).

spectrum The distribution among various wavelengths of the radiant energy emitted or absorbed by an object.

spin magnetic quantum number (m_s) A quantum number associated with the electron spin; it may have values of or $+\frac{1}{2}$ or $-\frac{1}{2}$.

spin-pairing energy The energy required to pair an electron with another electron occupying an orbital.

spin–spin coupling The communication of nuclear spin information between nuclei expressed in an NMR spectrum by multiplicity.

splitting pattern The output of spin–spin coupling within a spectrum.

spontaneous process A process that is capable of proceeding in a given direction, as written or described, without needing to be driven by an outside source of energy. A process may be spontaneous even though it is very slow.

staggered A conformation in which all groups around a C–C single bond are separated as far as possible.

standard atmospheric pressure Defined as 1 bar = 1×10^5 Pa = 1×10^2 kPa.

standard cell potential See **standard emf**.

standard electrode potential See **standard reduction potential**.

standard emf The emf of a cell when all reagents are at standard conditions. (Also called the standard cell potential.)

standard enthalpy change (Δ*H*°) The change in enthalpy in a process when all reactants and products are in their stable forms at 1 bar pressure and a specified temperature, commonly 25 °C.

standard enthalpy of formation (Δ$_f$*H*°) The change in enthalpy that accompanies the formation of one mole of a substance from its elements, with all substances in their standard states.

standard free energy of formation (Δ$_f$*G*°) The change in free energy associated with the formation of a substance from its elements under standard conditions.

standard hydrogen electrode (SHE) An electrode based on the half-reaction $2H^+$ (1 M) + $2e^- \rightarrow H_2$ (1 bar). The standard electrode potential of the standard hydrogen electrode is defined as 0 V.

standard molar entropy (*S*°) The entropy value for a mole of a substance in its standard state.

standard reduction potential (*E*$°_{red}$) The potential of a reduction half-reaction under standard conditions, measured relative to the standard hydrogen electrode. A standard reduction potential is also called a **standard electrode potential**.

standard solution A solution of known concentration.

standard atmospheric pressure Defined as 1 bar = 100 kPa.

standard temperature and pressure (STP) Defined as 0 °C and 1 bar pressure; frequently used as reference conditions for a gas.

starch The general name given to a group of polysaccharides that act as energy storage substances in plants.

state function A property of a system that is determined by the state or condition of the system and not by how it got to that state; its value is fixed when temperature, pressure, composition and physical form are specified; *P*, *V*, *T*, *E* and *H* are state functions.

states of matter The three forms that matter can assume: solid, liquid and gas.

stereocentre See **chiral centre**.

stereogenic centre See **chiral centre**.

stereoisomerism A property of molecules that have the same molecular formula and same connectivity but differ in the arrangement of their atoms in space.

stereoisomers Compounds possessing the same formula and bonding arrangement but differing in the spatial arrangements of the atoms.

stereoselective reaction A reaction in which one stereoisomer is formed (or destroyed) in preference to all other possible stereoisomers.

steric An effect due to the size of an atom, or groups of atoms.

stoichiometry The relationships between the quantities of reactants and products involved in chemical reactions.

stratosphere The region of the atmosphere directly above the troposphere.

Strecker synthesis A method of synthesising α-amino acids from aldehydes.

strong acid An acid that ionises completely in water.

strong base A base that ionises completely in water.

strong electrolyte A substance (strong acids, strong bases and most salts) that is completely ionised in solution.

structural formula A formula that shows not only the number and kinds of atoms in the molecule but also the arrangement (connections) of the atoms.

structural isomers Compounds possessing the same formula but differing in the bonding arrangements of the atoms. Also often called **constitutional isomers**.

subatomic particles Particles such as protons, neutrons and electrons that are smaller than an atom.

subshell One or more orbitals with the same set of quantum numbers n and l. For example, we speak of the $2p$ subshell ($n = 2, l = 1$). which is composed of three orbitals ($2p_x$, $2p_y$ and $2p_z$).

substitution reactions Reactions in which one atom (or group of atoms) replaces another atom (or group) within a molecule; substitution reactions are typical for alkanes and aromatic hydrocarbons.

substrate A substance that undergoes a reaction at the active site in an enzyme.

superconducting ceramic A complex metal oxide that undergoes a transition to a superconducting state at a low temperature.

superconducting transition temperature (T_c) The temperature below which a substance exhibits superconductivity.

superconductivity The 'frictionless' flow of electrons that occurs when a substance loses all resistance to the flow of electrical current.

supercritical fluid State of a substance where the temperature exceeds the critical temperature and the pressure exceeds the critical pressure and the liquid and gas are indistinguishable from one another.

supercritical mass An amount of fissionable material larger than the critical mass.

supersaturated solutions Solutions containing more solute than an equivalent saturated solution.

surface tension The intermolecular, cohesive attraction that causes a liquid to minimise its surface area.

surroundings In thermodynamics, everything that lies outside the system that we study.

synchrotron A device used to produce very intense radiation, in any range, from the infrared to short wavelength X-rays.

synthetic detergent A substance that cleans by forming micelles but is not made by the hydrolysis of fats. A synthetic analogue of a soap.

system In thermodynamics, the portion of the universe that we single out for study. We must be careful to state exactly what the system contains and what transfers of energy it may have with its surroundings.

tautomerism The interconversion of equilibrating isomers that differ in the location of their bonding electrons, usually accompanying a proton movement.

termination The final step in a chain reaction that stops further reaction.

termolecular reaction An elementary reaction that involves three molecules. Termolecular reactions are rare.

tertiary (3°) alcohol A compound containing the hydroxy (OH) group of an alcohol on a tertiary carbon.

tertiary amide An amide in which three N–C bonds exist about the amide N.

tertiary amine An amine that has no hydrogen atoms bonded to nitrogen.

tertiary structure The overall shape of a large protein, specifically the manner in which sections of the protein fold back upon themselves or intertwine.

theoretical yield The quantity of product that is calculated to form when all of the limiting reagent reacts.

theory A tested model or explanation that satisfactorily accounts for a certain set of phenomena.

thermochemistry The relationship between chemical reactions and energy changes.

thermodynamics The study of energy and its transformation.

thermonuclear reaction Another name for fusion reactions, reactions in which two light nuclei are joined to form a more massive one.

thermoplastic A polymeric material that can be readily reshaped by application of heat and pressure.

thermosetting plastic A plastic that, once formed in a particular mould, is not readily reshaped by application of heat and pressure.

thermosphere The region of the Earth's atmosphere stretching from an altitude of about 85 km to about 110 km.

thiamin Vitamin B_1. A water-soluble heterocyclic structure made up of a substituted pyrimidine and thiazole rings. Thiamin acts as a cofactor for reductase enzymes, responsible for, among other things, the reduction of pyruvate.

third law of thermodynamics A law stating that the entropy of a pure, crystalline solid at absolute zero temperature is zero: $S (0 \text{ K}) = 0$.

titration The process of reacting a solution of unknown concentration with one of known concentration (a standard solution).

titration curve A graph of pH as a function of added titrant.

tocopherols Vitamin E. Of this group, the α-tocopherol molecule is the most potent. Due to its lipophilic nature, vitamin E accumulates in cellular membranes, fat deposits and other circulating lipoproteins. The major function of vitamin E is to act as a natural antioxidant, thereby preventing peroxidation of polyunsaturated membrane fatty acids.

torr An outdated unit of pressure (1 torr = 1 mm Hg).

trans A geometric isomer in which two substituents attached to a cyclic compound are oriented on different sides of the plane of the ring.

transistor An electrical device that forms the heart of an integrated circuit.

transition elements (transition metals) Elements in which the d orbitals are partially occupied.

transition state See **activated complex**.

translational motion Movement in which an entire molecule moves in a definite direction.

transmittance Expressed by the ratio I/I_0, where I = intensity of light through the sample and I_0 = intensity of light used for the measurement. The value of transmittance always lies between 0 and 1.

transuranium elements Elements that follow uranium in the periodic table.

triglycerides Triesters of glycerol, including common fats and oils.

triol An organic molecule characterised by three OH functional groups.

triple bond A covalent bond involving three electron pairs.

triple point The temperature at which solid, liquid and gas phases coexist in equilibrium.

tritium The isotope of hydrogen whose nucleus contains a proton and two neutrons.

troposphere The region of the Earth's atmosphere extending from the surface to about 12 km altitude.

Tyndall effect The scattering of a beam of visible light by the particles in a colloidal dispersion.

uncertainty principle A principle stating there is an inherent uncertainty in the precision with which we can simultaneously specify the position and momentum of a particle. This uncertainty is significant only for extremely small particles, such as electrons.

unified atomic mass unit (u) One-twelfth the mass of one carbon-12 atom.

unimolecular reaction An elementary reaction that involves a single molecule.

unit cell The smallest portion of a crystal that reproduces the structure of the entire crystal when repeated in different directions in space. It is the repeating unit or building block of the crystal lattice.

unsaturated hydrocarbon A hydrocarbon containing one or more double or triple bonds.

unsaturated solutions Solutions containing less solute than a saturated solution.

UV-visible spectroscopy An analytical technique based on the transitions between electronic energy states within a molecule.

valence band A band of closely spaced molecular orbitals that is essentially fully occupied by electrons.

valence-bond theory A model of chemical bonding in which an electron-pair bond is formed between two atoms by the overlap of orbitals on the two atoms.

valence electrons The outermost electrons of an atom; those that occupy orbitals not occupied in the nearest noble gas element of lower atomic number. The valence electrons are the ones the atom uses in bonding.

valence orbitals Orbitals that contain the outer-shell electrons of an atom.

valence-shell electron-pair repulsion (VSEPR) model A model that accounts for the geometric arrangements of shared and unshared electron pairs around a central atom in terms of the repulsions between electron pairs.

van der Waals equation An equation of state for non-ideal gases that is based on adding corrections to the ideal-gas equation. The correction terms account for intermolecular forces of attraction and for the volumes occupied by the gas molecules themselves.

vapour Gaseous state of any substance that normally exists as a liquid or solid.

vapour pressure The pressure exerted by a vapour in equilibrium with its liquid or solid phase.

vibrational motion Movement of the atoms within a molecule in which they move periodically towards and away from one another.

viscosity A measure of the resistance of fluids to flow.

vitamin D_3 A steroid hormone derived from cholesterol whose primary function is to regulate calcium and phosphorus homeostasis and maintain bone in the body.

vitamins Organic substances other than proteins, carbohydrates and fats that must be obtained through diet; required in trace amounts for proper growth.

volatile Tending to evaporate readily.

voltaic (galvanic) cell A device in which a spontaneous oxidation–reduction reaction occurs with the passage of electrons through an external circuit.

volume (V) SI unit for volume is the cubic metre (m^3).

vulcanisation The process of cross-linking polymer chains in rubber.

Watson–Crick pairing The selective pairing by hydrogen bonding of purine and pyrimidine bases that gives rise to the DNA double helix.

watt A unit of power: $1\,W = 1\,J\,s^{-1}$.

wave function A mathematical description of an allowed energy state (an orbital) for an electron in the quantum mechanical model of the atom; it is usually symbolised by the Greek letter ψ.

wavelength The distance between identical points on successive waves.

wavenumber A conventional unit in IR spectroscopy expressed in reciprocal centimetres (cm^{-1}).

waxes Single esters made from fatty acids and long-chain alcohols.

weak acid An acid that only partly ionises in water.

weak base A base that only partly ionises in water.

weak electrolyte A substance that only partly ionises in solution.

work The movement of an object against some force.

Zaitsev's rule A rule stating that when a mixture of isomeric alkenes is formed through an elimination reaction, the more substituted alkene will generally predominate.

zero-order reaction A reaction for which the rate is independent of the concentration of the reacting species.

Ziegler–Natta polymerisation A specific olefin polymerisation process using organo-metallic catalysts. Used extensively for the polymerisation of ethene and propylene.

zwitterion A neutral molecule containing both positive and negative charges by virtue of internal proton exchange.

COMMON IONS

Positive Ions (Cations)

1+

Ammonium (NH_4^+)
Cesium (Cs^+)
Copper(I) or cuprous (Cu^+)
Hydrogen (H^+)
Lithium (Li^+)
Potassium (K^+)
Silver (Ag^+)
Sodium (Na^+)

2+

Barium (Ba^{2+})
Cadmium (Cd^{2+})
Calcium (Ca^{2+})
Chromium(II) or chromous (Cr^{2+})
Cobalt(II) or cobaltous (Co^{2+})
Copper(II) or cupric (Cu^{2+})
Iron(II) or ferrous (Fe^{2+})
Lead(II) or plumbous (Pb^{2+})
Magnesium (Mg^{2+})
Manganese(II) or manganous (Mn^{2+})
Mercury(I) or mercurous (Hg_2^{2+})

Mercury(II) or mercuric (Hg^{2+})
Strontium (Sr^{2+})
Nickel(II) (Ni^{2+})
Tin(II) or stannous (Sn^{2+})
Zinc (Zn^{2+})

3+

Aluminium (Al^{3+})
Chromium(III) or chromic (Cr^{3+})
Iron(III) or ferric (Fe^{3+})

Negative Ions (Anions)

1−

Acetate (CH_3COO^-)
Bromide (Br^-)
Chlorate (ClO_3^-)
Chloride (Cl^-)
Cyanide (CN^-)
Dihydrogen phosphate ($H_2PO_4^-$)
Fluoride (F^-)
Hydride (H^-)
Hydrogen carbonate or bicarbonate (HCO_3^-)

Hydrogen sulfite or bisulfite (HSO_3^-)
Hydroxide (OH^-)
Iodide (I^-)
Nitrate (NO_3^-)
Nitrite (NO_2^-)
Perchlorate (ClO_4^-)
Permanganate (MnO_4^-)
Thiocyanate (SCN^-)

2−

Carbonate (CO_3^{2-})
Chromate (CrO_4^{2-})
Dichromate ($Cr_2O_7^{2-}$)
Hydrogen phosphate (HPO_4^{2-})
Oxide (O^{2-})
Peroxide (O_2^{2-})
Sulfate (SO_4^{2-})
Sulfide (S^{2-})
Sulfite (SO_3^{2-})

3−

Arsenate (AsO_4^{3-})
Phosphate (PO_4^{3-})

FUNDAMENTAL CONSTANTS*

Unified atomic mass unit	1 u	$= 1.660\,538\,86 \times 10^{-24}$ g
	1 g	$= 6.022\,141\,5 \times 10^{23}$ u
Avogadro's number	N_A	$= 6.022\,141\,5 \times 10^{23}$ mol^{-1}
Boltzmann constant	k	$= 1.380\,650\,53 \times 10^{-23}$ J K^{-1}
Electron charge	e	$= 1.602\,176\,53 \times 10^{-19}$ C
Faraday constant	F	$= 9.648\,533\,383 \times 10^{4}$ C mol^{-1}
Gas constant	R	$= 8.314\,472$ J mol^{-1} K^{-1}
		$= 8.314\,472$ kPa dm^3 mol^{-1} K^{-1}
Mass of electron	m_e	$= 9.109\,382\,6 \times 10^{-28}$ g
		$= 5.485\,799\,1 \times 10^{-4}$ u
Mass of neutron	m_n	$= 1.674\,927\,16 \times 10^{-24}$ g
		$= 1.008\,664\,91$ u
Mass of proton	m_p	$= 1.672\,621\,71 \times 10^{-24}$ g
		$= 1.007\,276\,46$ u
Planck's constant	h	$= 6.626\,068\,76 \times 10^{-34}$ J s
Speed of light	c	$= 2.997\,924\,58 \times 10^{8}$ m/s^{-1}

USEFUL CONVERSION FACTORS AND RELATIONSHIPS

Length

SI unit: metre (m)

$$1 \text{ km} = 1000 \text{ m}$$
$$1 \text{ m} = 100 \text{ cm}$$
$$1 \text{ cm} = 10 \text{ mm}$$
$$1 \text{ pm} = 10^{-12} \text{ m}$$
$$1 \text{ Å} = 10^{-10} \text{ m}$$
$$1 \text{ pm} = 10^{-2} \text{ Å}$$

Mass

SI unit: kilogram (kg)

$$1 \text{ kg} = 10^3 \text{ g}$$
$$1 \text{ tonne} = 10^3 \text{ kg}$$
$$1 \text{ u} = 1.660\,538\,86 \times 10^{-24} \text{ g}$$

Temperature

SI unit: kelvin (K)

$$0 \text{ K} = -273.15 \text{ °C}$$
$$\text{K} = \text{°C} + 273.15$$

Energy (derived)

SI unit: joule (J)

$$1 \text{ J} = 1 \text{ kg m}^2 \text{ s}^{-1}$$
$$1 \text{ J} = 1 \text{ C V}$$
$$1 \text{ cal} = 4.184 \text{ J}$$
$$1 \text{ eV} = 1.602 \times 10^{-19} \text{ J}$$

Pressure (derived)

SI unit: pascal (Pa)

$$1 \text{ Pa} = 1 \text{ N m}^{-2}$$
$$= 1 \text{ kg m}^{-1} \text{ s}^{-2}$$
$$1 \text{ atm} = 101\,325 \text{ Pa}$$
$$1 \text{ bar} = 10^5 \text{ Pa}$$
$$= 100 \text{ kPa}$$

Volume (derived)

SI unit: cubic metre (m³)

$$1 \text{ m}^3 = 1000 \text{ dm}^3$$
$$1 \text{ dm}^3 = 1000 \text{ cm}^3$$

INDEX OF USEFUL TABLES AND FIGURES